Lectures on Convex Sets

Second Edition

Lectures on Convex Sets

Second Edition

Valeriu Soltan
George Mason University, USA

World Scientific

NEW JERSEY • LONDON • SINGAPORE • BEIJING • SHANGHAI • HONG KONG • TAIPEI • CHENNAI • TOKYO

Published by

World Scientific Publishing Co. Pte. Ltd.
5 Toh Tuck Link, Singapore 596224
USA office: 27 Warren Street, Suite 401-402, Hackensack, NJ 07601
UK office: 57 Shelton Street, Covent Garden, London WC2H 9HE

Library of Congress Control Number: 2019045356

British Library Cataloguing-in-Publication Data
A catalogue record for this book is available from the British Library.

LECTURES ON CONVEX SETS
Second Edition

ISBN 978-981-120-211-7
ISBN 978-981-120-351-0 (pbk)

For any available supplementary material, please visit
https://www.worldscientific.com/worldscibooks/10.1142/11328#t=suppl

Printed in Singapore

To Valentina, my wife

Preface

This book provides a systematic treatment of basic algebraic, set-theoretic, and topological properties of convex sets (possibly non-closed or unbounded) in the n-dimensional Euclidean space $\mathbb{R}^n$. Its purpose is to give the reader a self-contained and detailed presentation of the topic, with unified terminology, notation, and complete proofs.

As any well-established mathematical discipline, convex geometry is divided into various subfields related to other parts of mathematics. The present text contains the topics of convex geometry that have close ties with convex analysis, optimization, operations research, and their applications. In this regard, the present book can be also viewed as a geometric addition to various texts on convex analysis.

The first systematic study of convex sets within convex analysis was given in the lecture notes of Fenchel [105] and in the seminal monograph of Rockafellar [241]. Afterward, the list of topics on convex sets became almost canonical: it includes general properties of convex sets and convex hulls, cones and conic hulls, support and separation properties of convex sets, the extreme and exposed structures, polyhedral sets.

The contents of this book grew out of graduate courses on Euclidean geometry and convexity, taught by the author for almost two decades to students with various backgrounds in mathematics, optimization, and operations research. The style is similar to that of lecture notes: it is rather detailed, with plenty of cross-references, which allow reading of this book practically from any page. Every chapter is concluded with a selection of problems and comments on related research topics and results.

The text can be used for a one-semester advanced undergraduate or an entry-level graduate course on convex sets, or as a supplementary material for a course on convex geometry or convex analysis. Also, it may be viewed as a source for independent study of the subject, suitable to non-geometers.

Prerequisites include proof-based undergraduate courses on linear algebra and analysis. All the necessary facts from these disciplines are given in the introductory Chapter 1.

This book is partitioned into chapters and sections as shown in the table of contents. Chapter 2, which describes the affine structure of the vector space $\mathbb{R}^n$, can be viewed as a set of additional topics to linear algebra. This chapter defines the language and methods of proof for the whole book.

The main body of the book, divided into Chapters 3–13, deals with general properties of convex sets and convex hulls, cones and conic hulls, recession cones and lineality spaces, normal directions, cone polarity, support and separation properties, the extreme and exposed structures of convex sets, polytopes and polyhedra.

The concluding Chapter 14 contains solutions of all problems listed in the book. As a whole, these are research-type problems; they complement the main text with additional, more specific results.

The second edition gives an essentially revised and extended version of the original text and corrects inevitable errors. Every chapter is rewritten, with many new theorems, examples, problems, and bibliographical references included. The second edition contains three new chapters and 100 additional problems, all provided with solutions.

For further comments, information, and corrections, please contact the author at vsoltan@gmu.edu.

Valeriu Soltan

Notation

$f^{-1}(Y)$	inverse f-image of a set Y	3
$\mathbb{R}^n$	n-dimensional coordinate space	3
$X + Y$	(Minkowski) sum of sets X and Y	4
$\operatorname{span} X$	(linear) span of a set X	5
$\dim S$	dimension of a subspace S	5
$\operatorname{null} f$	null space of a linear transformation f	7
$\boldsymbol{x}\cdot\boldsymbol{y}$	dot product of vectors $\boldsymbol{x}$ and $\boldsymbol{y}$	7
$X^{\perp}$	orthogonal complement of a set X	7
$\|\boldsymbol{x}\|$	norm of a vector $\boldsymbol{x}$	8
$\|\boldsymbol{x} - \boldsymbol{y}\|$	distance between points $\boldsymbol{x}$ and $\boldsymbol{y}$	8
f_S	orthogonal projection on a subspace S	8
$B_\rho(\boldsymbol{a})$	closed ball with center $\boldsymbol{a}$ and radius ρ	9
$U_\rho(\boldsymbol{a})$	open ball with center $\boldsymbol{a}$ and radius ρ	9
$S_\rho(\boldsymbol{a})$	sphere with center $\boldsymbol{a}$ and radius ρ	9
$\mathbb{B}$	closed unit ball with center $\boldsymbol{o}$	9
$\mathbb{U}$	open unit ball with center $\boldsymbol{o}$	9
$\mathbb{S}$	unit sphere with center $\boldsymbol{o}$	9
$B_\rho(X)$	closed ρ-neighborhood of a set X	9
$U_\rho(X)$	open ρ-neighborhood of a set X	9
$\operatorname{int} X$	interior of a set X	10
$\operatorname{cl} X$	closure of a set X	10
$\operatorname{bd} X$	boundary of a set X	10
$\operatorname{diam} X$	diameter of a bounded set X	11
$\delta(X, Y)$	*inf*-distance between sets X and Y	11
$\operatorname{dir} L$	direction space of a plane L	17
$\operatorname{ort} L$	orthogonal space of a plane L	17
$\dim L$	dimension of a plane L	18
$\langle \boldsymbol{x}, \boldsymbol{y} \rangle$	line through points $\boldsymbol{x}$ and $\boldsymbol{y}$	29
$[\boldsymbol{x}, \boldsymbol{y}\rangle$	closed halfline through $\boldsymbol{y}$ with endpoint $\boldsymbol{x}$	30
$(\boldsymbol{x}, \boldsymbol{y}\rangle$	open halfline through $\boldsymbol{y}$ with endpoint $\boldsymbol{x}$	30
$[\boldsymbol{x}, \boldsymbol{y}]$	closed segment with endpoints $\boldsymbol{x}$ and $\boldsymbol{y}$	30
$[\boldsymbol{x}, \boldsymbol{y})$	semi-open segment with endpoints $\boldsymbol{x}$ and $\boldsymbol{y}$	30
$(\boldsymbol{x}, \boldsymbol{y})$	open segment with endpoints $\boldsymbol{x}$ and $\boldsymbol{y}$	30
$\operatorname{aff} X$	affine span of a set X	30
$\operatorname{nor} V$	normal halfline of a halfspace V	32
$\operatorname{dir} X$	direction space of a set X	42
$\operatorname{ort} X$	orthogonal space of a set X	42

$\dim X$	dimension of a set X	52
f_L	orthogonal projection on a plane L	69
$\Delta(\boldsymbol{x}_1, \ldots, \boldsymbol{x}_{r+1})$	r-simplex with vertices $\boldsymbol{x}_1, \ldots, \boldsymbol{x}_{r+1}$	78
$\operatorname{rint} X$	relative interior of a set X	84
$\operatorname{rbd} X$	relative boundary of a set X	113
$\operatorname{conv} X$	convex hull of a set X	123
$\operatorname{conv}_r X$	union of convex hulls of r-pointed subsets of X	129
$\Gamma_{\boldsymbol{a}}(\boldsymbol{x}_1, \ldots, \boldsymbol{x}_r)$	r-simplicial cone generated by $\boldsymbol{x}_1, \ldots, \boldsymbol{x}_r$	154
$\operatorname{ap} C$	apex set of a cone C	160
$\operatorname{cone}_{\boldsymbol{a}} X$	(convex) conic hull with apex $\boldsymbol{a}$ of a set X	175
$C_{\boldsymbol{a}}(X)$	cone with apex $\boldsymbol{a}$ generated by a set X	175
$\operatorname{rec} X$	recession cone of a set X	207
$\operatorname{rec}_{\boldsymbol{a}} X$	recession cone of a set X at $\boldsymbol{a} \in X$	211
$\operatorname{lin} X$	lineality space of a set X	224
$E_{\boldsymbol{a}}(X)$	normalized set of X with respect to $\boldsymbol{a}$	242
X°	polar cone of a set X	269
p_K	metric projection on a convex set K	289
$D_{\boldsymbol{z}}(K)$	support cone of K at point $\boldsymbol{z} \in \operatorname{cl} K$	291
$T_{\boldsymbol{z}}(K)$	tangent cone of K at point $\boldsymbol{z} \in \operatorname{cl} K$	291
$Q_{\boldsymbol{z}}(K)$	normal set of K at point $\boldsymbol{z} \in \operatorname{cl} K$	291
$N_{\boldsymbol{z}}(K)$	normal cone of K at point $\boldsymbol{z} \in \operatorname{cl} K$	291
$\operatorname{nor} K$	normal cone of a convex set K	291
$\operatorname{bar} X$	barrier cone of a set X	298
$F_X(K)$	extreme face of K generated by a set $X \subset K$	380
$\operatorname{ext} K$	set of extreme points of a convex set K	388
$\operatorname{extr} K$	union of extreme halflines of a convex set K	395
$\operatorname{ext}_1 K$	set of 1-extreme points of a convex set K	403
$G_X(K)$	exposed face of K generated by a set $X \subset K$	418
$\exp K$	set of exposed points of a convex set K	426
$\operatorname{expr} K$	union of exposed halflines of a convex set K	431
$\exp_1 K$	set of 1-exposed points of a convex set K	440

Contents

Chapter 1

Prerequisites

1.1 Basic Set Theory and Notation

This chapter contains a brief account of necessary prerequisites, written as an informal list of definitions, assertions, and formulas. Topics include some standard facts from the proof-based undergraduate courses on linear algebra and analysis: basic set theory, the vector space $\mathbb{R}^n$ and its Euclidean structure, elementary topology and analysis in $\mathbb{R}^n$. Some necessary results, usually treated as additional topics in these courses, are given in the form of problems at the end of the chapter. It is assumed that the reader starts the book from Chapter 2 and uses the introductory chapter as a reference material.

In what follows, $\mathbb{R}$ stands for the set of real numbers, called *scalars*, and for the number line as well. Given a scalar α, its absolute, ceiling, floor, and sign values are denoted $|\alpha|$, $\lceil \alpha \rceil$, $\lfloor \alpha \rfloor$, and $\operatorname{sgn}(\alpha)$, respectively. The symbol $\Box$ marks the end of a proof or an assertion without proof.

For a nonempty set $A \subset \mathbb{R}$, the symbols $\inf A$ and $\sup A$ mean the infimum and supremum values of scalars from A; they may be $-\infty$ and ∞, respectively. If A and B are nonempty sets in $\mathbb{R}$, then

$$\inf (A + B) = \inf A + \inf B \quad \text{and} \quad \sup (A + B) = \sup A + \sup B.$$

The minimum and maximum values of scalars from A, denoted, respectively, $\min A$ and $\max A$, are always assumed to be finite.

Scalars α and β, with $\alpha < \beta$, determine the closed interval $[\alpha, \beta]$, two semi-open intervals $[\alpha, \beta)$ and $(\alpha, \beta]$, and open interval (α, β) of the number line $\mathbb{R}$. Similarly, $[\alpha, \infty), (\alpha, \infty), (-\infty, \alpha]$, and $(-\infty, \alpha)$ are the closed halflines and open halflines of $\mathbb{R}$ determined by α.

Given a universal set U, the union, intersection, and set difference of sets X and Y from U are denoted $X \cup Y$, $X \cap Y$, and $X \setminus Y$, respectively.

The symbol $\varnothing$ stands for the empty set.

In a standard way, $x \in X$ means that an element x of U belongs to X, and $X \subset Y$ shows that X is a subset of Y (possibly, $X = Y$). A subset X of Y is called *proper* if $X \neq Y$. Sets X and Y are called *disjoint* provided $X \cap Y = \varnothing$.

A set $X \subset U$ is often described by a suitable condition $P(x)$ on U, like $X = \{x \in U : P(x)\}$, or by its indexed elements, like $X = \{x_\alpha : \alpha \in I\}$ (briefly, $X = \{x_\alpha\}$). We distinguish finite, denumerable, countable (finite or denumerable), and uncountable sets. Nonempty finite sets will be usually expressed as $X = \{x_1, \ldots, x_r\}$. The maximum number of pairwise distinct elements in a finite set X is denoted $\operatorname{card} X$. A set consisting of a single element is called singleton.

The operations of union and intersection of two sets can be extended to the case of a nonempty indexed family $\mathcal{F} = \{X_\alpha : \alpha \in I\}$ (briefly, $\mathcal{F} = \{X_\alpha\}$) of sets in U; they will be denoted $\underset{\alpha}{\cup} X_\alpha$ and $\underset{\alpha}{\cap} X_\alpha$, respectively (if $I = \varnothing$, then $\underset{\alpha}{\cup} X_\alpha = \varnothing$ and $\underset{\alpha}{\cap} X_\alpha = U$). Alternative notations for $\underset{\alpha}{\cup} X_\alpha$ and $\underset{\alpha}{\cap} X_\alpha$ are

$$\cup\,(X_\alpha : X_\alpha \in \mathcal{F}) \quad \text{and} \quad \cap\,(X_\alpha : X_\alpha \in \mathcal{F}).$$

These operations assume the *axiom of choice*, which states that one can choose an element x_α in every set $X_\alpha \in \mathcal{F}$.

If $\mathcal{F} = \{X_\alpha\}$ is an indexed family of sets in the universal set U, then *De Morgan's Laws* state that

$$U \setminus (\underset{\alpha}{\cup} X_\alpha) = \underset{\alpha}{\cap} (U \setminus X_\alpha) \quad \text{and} \quad U \setminus (\underset{\alpha}{\cap} X_\alpha) = \underset{\alpha}{\cup} (U \setminus X_\alpha).$$

Furthermore, for any set $Y \subset U$, one has

$$(\underset{\alpha}{\cup} X_\alpha) \cap Y = \underset{\alpha}{\cup} (X_\alpha \cap Y) \quad \text{and} \quad (\underset{\alpha}{\cap} X_\alpha) \cup Y = \underset{\alpha}{\cap} (X_\alpha \cup Y).$$

Given a family $\mathcal{F} = \{X_\alpha\}$ of sets in U, a set $X_\beta \in \mathcal{F}$ is called *maximal* in $\mathcal{F}$ provided $X_\beta = X_\gamma$ for any set $X_\gamma \in \mathcal{F}$ satisfying the condition $X_\beta \subset X_\gamma$. If $X_\gamma \subset X_\beta$ for all $X_\gamma \in \mathcal{F}$, then X_β is called the *largest* in $\mathcal{F}$. We say that $\mathcal{F}$ is *nested* if $X_\beta \subset X_\gamma$ or $X_\gamma \subset X_\beta$ whenever $X_\beta, X_\gamma \in \mathcal{F}$. An alternative form of the axiom of choice is the *Kuratowski-Zorn Lemma*, which states that the family $\mathcal{F}$ has a maximal element provided the union of every nested subfamily of sets from $\mathcal{F}$ belongs to $\mathcal{F}$.

The family $\mathcal{F} = \{X_\alpha\}$ is called *disjoint* if $\underset{\alpha}{\cap} X_\alpha = \varnothing$, and is called *pairwise disjoint* provided $X_\beta \cap X_\gamma = \varnothing$ for any distinct sets $X_\beta, X_\gamma \in \mathcal{F}$.

A *mapping* f from a set A into a set B (notation $f : A \to B$) assigns to every element $x \in A$ a unique element $f(x) \in B$. The set A, denoted dom f, is called the *domain* of f, and the set $\{f(x) : x \in A\}$, denoted rng f, is called the *range* of f. The mapping $f : A \to B$ is called *onto* if every element $y \in B$ is an f-image of a suitable element $x \in A$; it is *one-to-one* if $f(x) \neq f(z)$ whenever $x \neq z$; finally, f is *invertible* if it is both onto and one-to-one. An invertible mapping $f : A \to B$ has the *inverse mapping* $f^{-1} : B \to A$ defined by the conditions

$$f^{-1}(f(x)) = x \quad \text{and} \quad f(f^{-1}(y)) = y \quad \text{for all} \quad x \in A \quad \text{and} \quad y \in B.$$

For a mapping $f : A \to B$ (not necessarily invertible), the *image* of a set $X \subset A$ and the *inverse image* of a set $Y \subset B$ are defined by

$$f(X) = \{f(x) : x \in X\} \quad \text{and} \quad f^{-1}(Y) = \{x \in A : f(x) \in Y\}.$$

We let $f(\varnothing) = \varnothing$ and $f^{-1}(\varnothing) = \varnothing$. It is well known that

$$X \subset f^{-1}(f(X)),\ f^{-1}(Y) = f^{-1}(Y \cap \operatorname{rng} f),\ f(f^{-1}(Y)) = Y \cap \operatorname{rng} f.$$

For a mapping $f : A \to B$ and nonempty families of subsets $\mathcal{F} = \{X_\alpha\}$ and $\mathcal{G} = \{Y_\alpha\}$ in A and B, respectively, the following assertions hold.

1. $f(\bigcup_\alpha X_\alpha) = \bigcup_\alpha f(X_\alpha)$ and $f(\bigcap_\alpha X_\alpha) \subset \bigcap_\alpha f(X_\alpha)$, with equality provided f is one-to-one.
2. $f^{-1}(\bigcup_\alpha Y_\alpha) = \bigcup_\alpha f^{-1}(Y_\alpha)$ and $f^{-1}(\bigcap_\alpha Y_\alpha) = \bigcap_\alpha f^{-1}(Y_\alpha)$.
3. If $X_1, X_2 \subset A$, then $f(X_1) \backslash f(X_2) \subset f(X_1 \backslash X_2)$, with equality provided f is one-to-one.
4. If $Y_1, Y_2 \subset B$, then $f^{-1}(Y_1) \setminus f^{-1}(Y_2) = f^{-1}(Y_1 \setminus Y_2)$.

The *composition* of mappings $f : A \to B$ and $g : B \to C$ is defined by

$$(g \circ f)(x) = g(f(x)) \quad \text{for all} \quad x \in A.$$

The following properties of compositions are well known.

1. If both f and g are one-to-one (respectively, onto), then $g \circ f$ is one-to-one (respectively, onto).
2. If $g \circ f$ is one-to-one (respectively, onto), then f is one-to-one (respectively, g is onto).

1.2 The Vector Space $\mathbb{R}^n$

Throughout this book we deal with elements and subsets of the n-dimensional vector space $\mathbb{R}^n$, $n \geqslant 1$, consisting of all n-tuples $\boldsymbol{x} = (x_1, \ldots, x_n)$, where $x_1, \ldots, x_n$ are scalars, named the *coordinates* of $\boldsymbol{x}$. The elements of $\mathbb{R}^n$ are called *vectors*, or *points*. We make no difference between these two notions, and their use should support the reader's intuition. To distinguish similarly looking elements, 0 will stand for the number zero, and $\boldsymbol{o}$ for the *origin* (*zero vector*) of $\mathbb{R}^n$.

Given vectors $\boldsymbol{x} = (x_1, \dots, x_n)$ and $\boldsymbol{y} = (y_1, \dots, y_n)$ and a scalar λ, the *sum* $\boldsymbol{x} + \boldsymbol{y}$ and the *product* $\lambda\boldsymbol{x}$ are defined, respectively, as

$$\boldsymbol{x} + \boldsymbol{y} = (x_1 + y_1, \dots, x_n + y_n) \quad \text{and} \quad \lambda\boldsymbol{x} = (\lambda x_1, \dots, \lambda x_n).$$

These operations can be extended to the case of nonempty sets X and Y in $\mathbb{R}^n$:

$$X + Y = \{\boldsymbol{x} + \boldsymbol{y} : \boldsymbol{x} \in X, \boldsymbol{y} \in Y\} \quad \text{and} \quad \lambda X = \{\lambda\boldsymbol{x} : \boldsymbol{x} \in X\}.$$

The set $X + Y$ is often called the *Minkowski sum* of sets X and Y, and λX is the *scalar multiple* of X. We let $X + Y = \varnothing$ if at least one of the sets X and Y is empty; also, $\lambda\varnothing = \varnothing$ for all scalars λ. If X consists of a single vector $\boldsymbol{x}$, then $\boldsymbol{x} + Y$ will stand for $\{\boldsymbol{x}\} + Y$ and called a translate of Y. We write $-\boldsymbol{x}$ and $-X$ for $(-1)\boldsymbol{x}$ and $(-1)X$, respectively.

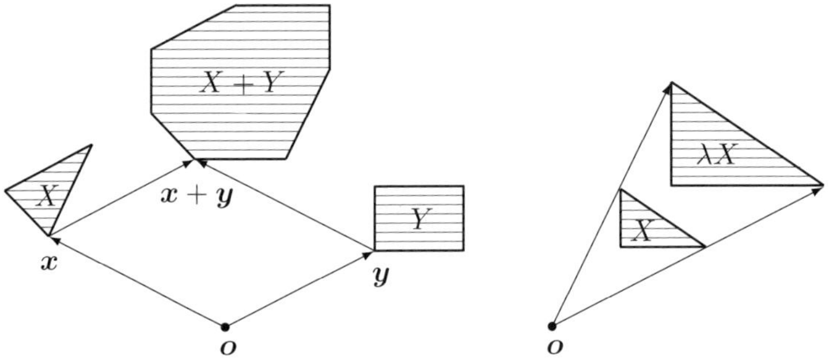

Fig. 1.1 The sum $X + Y$ and a scalar multiple λX.

For sets $X, Y, Z, V \subset \mathbb{R}^n$ and scalars λ, μ, one has:

1. $X + Y = Y + X$,
2. $(X + Y) + Z = X + (Y + Z)$,
3. $(X \cup Y) + Z = (X + Z) \cup (Y + Z)$,
4. $(X \cap Y) + Z \subset (X + Z) \cap (Y + Z)$,
5. $(X \cap Y) + (Z \cap V) \subset (X + Z) \cap (Y + V)$,
6. $\lambda(X + Y) = \lambda X + \lambda Y$,
7. $(\lambda + \mu)X \subset \lambda X + \mu X$.

A nonempty set $S \subset \mathbb{R}^n$ is called a *subspace* if it is closed under vector addition and scalar multiplication: $\boldsymbol{x} + \boldsymbol{y} \in S$ and $\lambda\boldsymbol{x} \in S$ whenever $\boldsymbol{x}, \boldsymbol{y} \in$

S and $\lambda \in \mathbb{R}$. We say that a subspace $S \subset \mathbb{R}^n$ is *nontrivial* provided $\{\boldsymbol{o}\} \neq S \neq \mathbb{R}^n$. It is easy to see that the sum of finitely many subspaces and the intersection of any family of subspaces also are subspaces.

Subspaces S and T of $\mathbb{R}^n$ are called *independent* if $S \cap T = \{\boldsymbol{o}\}$ (possibly, $S = \{\boldsymbol{o}\}$ or $T = \{\boldsymbol{o}\}$); equivalently, S and T are independent if every vector $\boldsymbol{x}$ in $S + T$ is uniquely expressible as $\boldsymbol{x} = \boldsymbol{y} + \boldsymbol{z}$, where $\boldsymbol{y} \in S$ and $\boldsymbol{z} \in T$. The sum of independent subspaces S and T (and sum of nonempty subsets $X \subset S$ and $Y \subset T$ as well) is named *direct sum*. Independent subspaces S and T are called *complementary* provided $S + T = \mathbb{R}^n$.

The intersection of all subspaces containing a given set $X \subset \mathbb{R}^n$ is called the (linear) *span* of X and denoted $\operatorname{span} X$; it is the smallest subspace of $\mathbb{R}^n$ containing X (we let $\operatorname{span} \varnothing = \{\boldsymbol{o}\}$). If $X \neq \varnothing$, then $\operatorname{span} X$ is the set of all (finite) linear combinations of vectors from X:

$$\operatorname{span} X = \{\lambda_1 \boldsymbol{x}_1 + \cdots + \lambda_k \boldsymbol{x}_k : k \geqslant 1,\ \lambda_1, \ldots, \lambda_k \in \mathbb{R},\ \boldsymbol{x}_1, \ldots, \boldsymbol{x}_k \in X\}.$$

For nonempty sets X and Y in $\mathbb{R}^n$,

1. $\operatorname{span}(X + Y) \subset \operatorname{span}(X \cup Y) = \operatorname{span} X + \operatorname{span} Y$.
2. If $\boldsymbol{o} \in X \cap Y$, then $\operatorname{span}(X + Y) = \operatorname{span} X + \operatorname{span} Y$.

A set $\{\boldsymbol{c}_1, \ldots, \boldsymbol{c}_r\}$ of vectors in $\mathbb{R}^n$ is called *linearly dependent* if there are scalars $\lambda_1, \ldots, \lambda_r$, not all zero, such that $\lambda_1 \boldsymbol{c}_1 + \cdots + \lambda_r \boldsymbol{c}_r = \boldsymbol{o}$; otherwise $\{\boldsymbol{c}_1, \ldots, \boldsymbol{c}_r\}$ is *linearly independent*. The empty set $\varnothing$ is assumed to be linearly independent.

1. Any subset of a linearly independent set is linearly independent, and any set containing a linearly dependent set is linearly dependent.
2. Any linearly independent subset of $\mathbb{R}^n$ consists of n or fewer pairwise distinct vectors.

A *basis* for a subspace $S \subset \mathbb{R}^n$ is a linearly independent set whose span is S (the empty set $\varnothing$ forms a basis for $\{\boldsymbol{o}\}$).

1. Every subspace $S \subset \mathbb{R}^n$ has a basis.
2. Any two bases for S have the same cardinality.
3. If a set X spans S and Y is a linearly independent subset of X, then there is a basis Z for S such that $Y \subset Z \subset X$.

The *standard basis* for $\mathbb{R}^n$ consists of the vectors

$$\boldsymbol{e}_1 = (1, 0, \ldots, 0),\ \boldsymbol{e}_2 = (0, 1, \ldots, 0),\ \ldots,\ \boldsymbol{e}_n = (0, 0, \ldots, 1).$$

The (unique) number of vectors in a basis for a subspace $S \subset \mathbb{R}^n$ is called the *dimension* of S and denoted $\dim S$. For subspaces S and T of $\mathbb{R}^n$, one has:

1. If $S \subset T$, then $\dim S \leqslant \dim T$, with $\dim S = \dim T$ if and only if $S = T$.
2. $\dim(S+T) = \dim S + \dim T - \dim(S \cap T)$.
3. If $m = \dim S$, then any linearly independent subset of m vectors from S and any set of m vectors spanning S are bases for S.
4. If a basis $B = \{\boldsymbol{b}_1, \dots, \boldsymbol{b}_r\}$ for $S \cap T$ is completed into bases

   $$B \cup C = \{\boldsymbol{b}_1, \dots, \boldsymbol{b}_r, \boldsymbol{c}_1, \dots, \boldsymbol{c}_p\} \quad \text{and} \quad B \cup E = \{\boldsymbol{b}_1, \dots, \boldsymbol{b}_r, \boldsymbol{e}_1, \dots, \boldsymbol{e}_q\}$$

   for S and T, respectively, then the combined list

   $$B \cup C \cup E = \{\boldsymbol{b}_1, \dots, \boldsymbol{b}_r, \boldsymbol{c}_1, \dots, \boldsymbol{c}_p, \boldsymbol{e}_1, \dots, \boldsymbol{e}_q\}$$

   is a basis for $S + T$ (any of the sets $B, B \cup C, B \cup E$ is empty if the corresponding subspace is 0-dimensional).

It is known that any subspace $S \subset \mathbb{R}^n$ of dimension m, where $m \leqslant n-1$, is the intersection of $n-m$ suitable subspaces of dimension $n-1$. Furthermore, if S is the intersection of a family $\mathcal{H}$ of $(n-1)$-dimensional subspaces, then $\mathcal{H}$ contains a subfamily of $n-m$ members whose intersection is S.

A mapping $f : \mathbb{R}^n \to \mathbb{R}^m$ is called a *linear transformation* provided

$$f(\boldsymbol{x} + \boldsymbol{y}) = f(\boldsymbol{x}) + f(\boldsymbol{y}) \quad \text{and} \quad f(\lambda \boldsymbol{x}) = \lambda f(\boldsymbol{x})$$

whenever $\boldsymbol{x}, \boldsymbol{y} \in \mathbb{R}^n$ and $\lambda \in \mathbb{R}$. A linear transformation $f : \mathbb{R}^n \to \mathbb{R}$ is called a *linear functional*.

Any linear transformation $f : \mathbb{R}^n \to \mathbb{R}^m$ preserves linear combinations of finitely many vectors:

$$f(\lambda_1 \boldsymbol{x}_1 + \cdots + \lambda_r \boldsymbol{x}_r) = \lambda_1 f(\boldsymbol{x}_1) + \cdots + \lambda_r f(\boldsymbol{x}_r)$$

whenever $\boldsymbol{x}_1, \dots, \boldsymbol{x}_r \in \mathbb{R}^n$ and $\lambda_1, \dots, \lambda_r \in \mathbb{R}$. Furthermore, for sets $X, Y \subset \mathbb{R}^n$ and $\lambda \in \mathbb{R}$,

$$f(X + Y) = f(X) + f(Y) \quad \text{and} \quad f(\lambda X) = \lambda f(X).$$

For a linear transformation $f : \mathbb{R}^n \to \mathbb{R}^m$, the following assertions hold.

1. If f is one-to-one, then $n \leqslant m$.
2. If f is onto, then $n \geqslant m$.
3. If f is invertible, then $n = m$ and the inverse mapping f^{-1} is a linear transformation.
4. If $\{\boldsymbol{c}_1, \dots, \boldsymbol{c}_r\}$ is a linearly independent set in $\mathbb{R}^n$ and f is one-to-one, then the set $\{f(\boldsymbol{c}_1), \dots, f(\boldsymbol{c}_r)\}$ is linearly independent.

5. If $\{\boldsymbol{e}_1, \ldots, \boldsymbol{e}_r\}$ is a linearly independent set in $\mathbb{R}^m$ and vectors $\boldsymbol{c}_1, \ldots, \boldsymbol{c}_r \in \mathbb{R}^n$ satisfy the conditions $f(\boldsymbol{c}_i) = \boldsymbol{e}_i$, $1 \leqslant i \leqslant r$, then the set $\{\boldsymbol{c}_1, \ldots, \boldsymbol{c}_r\}$ is linearly independent.
6. The composition $g \circ f$ of linear transformations $f : \mathbb{R}^n \to \mathbb{R}^m$ and $g : \mathbb{R}^m \to \mathbb{R}^k$ is a linear transformation.

For a linear transformation $f : \mathbb{R}^n \to \mathbb{R}^m$, the subspaces

$$\operatorname{null} f = \{\boldsymbol{x} \in \mathbb{R}^n : f(\boldsymbol{x}) = \boldsymbol{o}\} \quad \text{and} \quad \operatorname{rng} f = \{f(\boldsymbol{x}) : \boldsymbol{x} \in \mathbb{R}^n\}$$

are called, respectively, the *null space* and the *range* of f. One has

$$\dim(\operatorname{null} f) + \dim(\operatorname{rng} f) = n.$$

Given complementary subspaces S and T of $\mathbb{R}^n$, any vector $\boldsymbol{x} \in \mathbb{R}^n$ is uniquely expressible as the sum $\boldsymbol{x} = \boldsymbol{y} + \boldsymbol{z}$ of vectors $\boldsymbol{y} \in S$ and $\boldsymbol{z} \in T$, and the linear transformation $f : \mathbb{R}^n \to \mathbb{R}^n$ defined by $f(\boldsymbol{x}) = \boldsymbol{y}$ is called the *linear projection* on S along T. Clearly, $f(\boldsymbol{x}) = S \cap (\boldsymbol{x} + T)$ for all $\boldsymbol{x} \in \mathbb{R}^n$.

1.3 The Euclidean Structure of $\mathbb{R}^n$

The *dot product* $\boldsymbol{x} \cdot \boldsymbol{y}$ of vectors $\boldsymbol{x} = (x_1, \ldots, x_n)$ and $\boldsymbol{y} = (y_1, \ldots, y_n)$ is defined by $\boldsymbol{x} \cdot \boldsymbol{y} = x_1 y_1 + \cdots + x_n y_n$. One has

$$\boldsymbol{x} \cdot \boldsymbol{y} = \boldsymbol{y} \cdot \boldsymbol{x}, \quad (\boldsymbol{x} + \boldsymbol{z}) \cdot \boldsymbol{y} = \boldsymbol{x} \cdot \boldsymbol{y} + \boldsymbol{z} \cdot \boldsymbol{y}, \quad (\lambda \boldsymbol{x}) \cdot \boldsymbol{y} = \lambda(\boldsymbol{x} \cdot \boldsymbol{y}),$$
$$\boldsymbol{x} \cdot \boldsymbol{x} \geqslant 0, \text{ with } \boldsymbol{x} \cdot \boldsymbol{x} = 0 \text{ if and only if } \boldsymbol{x} = \boldsymbol{o}.$$

Any real-valued linear functional $\varphi(\boldsymbol{x})$ on $\mathbb{R}^n$ is expressible in the form

$$\varphi(\boldsymbol{x}) = \boldsymbol{x} \cdot \boldsymbol{e} = x_1 e_1 + \cdots + x_n e_n,$$

where $\boldsymbol{e} = (e_1, \ldots, e_n) \in \mathbb{R}^n$ is a suitable vector.

Vectors $\boldsymbol{x}$ and $\boldsymbol{y}$ are called *orthogonal* provided $\boldsymbol{x} \cdot \boldsymbol{y} = 0$ (possibly, $\boldsymbol{x} = \boldsymbol{o}$ or $\boldsymbol{y} = \boldsymbol{o}$). Nonempty sets X and Y in $\mathbb{R}^n$ are said to be (mutually) *orthogonal* if $\boldsymbol{x} \cdot \boldsymbol{y} = 0$ whenever $\boldsymbol{x} \in X$ and $\boldsymbol{y} \in Y$. It is easy to see that orthogonal subspaces S and T of $\mathbb{R}^n$ are independent.

The *orthogonal complement* of a nonempty set $X \subset \mathbb{R}^n$ is defined by

$$X^{\perp} = \{\boldsymbol{y} \in \mathbb{R}^n : \boldsymbol{x} \cdot \boldsymbol{y} = 0 \text{ for all } \boldsymbol{x} \in X\}.$$

We let $\varnothing^{\perp} = \mathbb{R}^n$. The following properties of orthogonal complements are well known.

1. The set $X^{\perp}$ is a subspace of $\mathbb{R}^n$ and $X^{\perp} = (\operatorname{span} X)^{\perp}$.
2. Any subspaces $S \subset \mathbb{R}^n$ and its orthogonal complement $S^{\perp}$ are complementary subspaces.

3. If S and T are subspaces of $\mathbb{R}^n$, then

$$(S+T)^\perp = S^\perp \cap T^\perp \quad \text{and} \quad (S\cap T)^\perp = S^\perp + T^\perp.$$

If subspaces S and T of $\mathbb{R}^n$ satisfy the condition $S \subset T$, then the subspace $S^\perp \cap T$ is called the *orthogonal complement* of S within T.

The *norm* of a vector $\boldsymbol{x} = (x_1, \ldots, x_n)$ is defined by

$$\|\boldsymbol{x}\| = \sqrt{\boldsymbol{x}\cdot\boldsymbol{x}} = \sqrt{x_1^2 + \cdots + x_n^2}\,.$$

Basic properties of norms:

1. $\|\boldsymbol{x}\| \geqslant 0$, with $\|\boldsymbol{x}\| = 0$ if and only if $\boldsymbol{x} = \boldsymbol{o}$.
2. $\|\lambda\boldsymbol{x}\| = |\lambda|\,\|\boldsymbol{x}\|$.
3. $|\boldsymbol{x}\cdot\boldsymbol{y}| \leqslant \|\boldsymbol{x}\|\,\|\boldsymbol{y}\|$.
4. $\|\boldsymbol{x}+\boldsymbol{y}\| \leqslant \|\boldsymbol{x}\| + \|\boldsymbol{y}\|$.
5. $\|\boldsymbol{x}+\boldsymbol{y}\|^2 = \|\boldsymbol{x}\|^2 + \|\boldsymbol{y}\|^2$ if $\boldsymbol{x}\cdot\boldsymbol{y} = 0$ (Pythagorean Theorem).

A nonempty set of pairwise orthogonal nonzero vectors is called *orthogonal set.* An orthogonal set consisting of unit vectors (i. e., vectors of unit norm) is called *orthonormal.* Any orthogonal set in $\mathbb{R}^n$ is linearly independent. Furthermore, every subspace S of $\mathbb{R}^n$ has an orthogonal basis.

For a pair of points $\boldsymbol{x} = (x_1, \ldots, x_n)$ and $\boldsymbol{y} = (y_1, \ldots, y_n)$, the norm

$$\|\boldsymbol{x}-\boldsymbol{y}\| = \sqrt{(x_1-y_1)^2 + \cdots + (x_n-y_n)^2}$$

is called the *distance* between $\boldsymbol{x}$ and $\boldsymbol{y}$. Clearly,

1. $\|\boldsymbol{x}-\boldsymbol{y}\| \geqslant 0$, with $\|\boldsymbol{x}-\boldsymbol{y}\| = 0$ if and only if $\boldsymbol{x} = \boldsymbol{y}$.
2. $\|\lambda\boldsymbol{x} - \lambda\boldsymbol{y}\| = |\lambda|\,\|\boldsymbol{x}-\boldsymbol{y}\|$.
3. $\|\boldsymbol{x}-\boldsymbol{y}\| \leqslant \|\boldsymbol{x}-\boldsymbol{z}\| + \|\boldsymbol{z}-\boldsymbol{y}\|$ whenever $\boldsymbol{x},\boldsymbol{y},\boldsymbol{z} \in \mathbb{R}^n$.
4. $\|\boldsymbol{x}+\boldsymbol{y}\|^2 + \|\boldsymbol{x}-\boldsymbol{y}\|^2 = 2\|\boldsymbol{x}\|^2 + 2\|\boldsymbol{y}\|^2$ for any $\boldsymbol{x},\boldsymbol{y} \in \mathbb{R}^n$.

The linear projection of $\mathbb{R}^n$ on a subspace S along its orthogonal complement $S^\perp$, denoted f_S, is called the *orthogonal projection* on S. For any vectors $\boldsymbol{x} \in \mathbb{R}^n$ and $\boldsymbol{y} \in S$, one has

$$\|\boldsymbol{x}-\boldsymbol{y}\|^2 = \|\boldsymbol{x} - f_S(\boldsymbol{x})\|^2 + \|f_S(\boldsymbol{x}) - \boldsymbol{y}\|^2.$$

The orthogonal projection f_S may be defined as the metric projection on S: $f_S(\boldsymbol{x}) = \boldsymbol{z}$, where $\boldsymbol{z}$ is the nearest to $\boldsymbol{x}$ vector in S. If, additionally, $\dim S = n-1$, then $S = \{\boldsymbol{x} \in \mathbb{R}^n : \boldsymbol{x}\cdot\boldsymbol{e} = 0\}$, where $\boldsymbol{e}$ is a suitable nonzero vector, and the orthogonal projection on S can be expressed as

$$f_S(\boldsymbol{x}) = \boldsymbol{x} - \frac{\boldsymbol{x}\cdot\boldsymbol{e}}{\|\boldsymbol{e}\|^2}\boldsymbol{e} \quad \text{for all} \quad \boldsymbol{x} \in \mathbb{R}^n.$$

1.4 Elementary Topology and Analysis in $\mathbb{R}^n$

Given a point $\boldsymbol{a} \in \mathbb{R}^n$ and a scalar $\rho > 0$, the sets

$$B_\rho(\boldsymbol{a}) = \{\boldsymbol{x} \in \mathbb{R}^n : \|\boldsymbol{x} - \boldsymbol{a}\| \leqslant \rho\},$$
$$U_\rho(\boldsymbol{a}) = \{\boldsymbol{x} \in \mathbb{R}^n : \|\boldsymbol{x} - \boldsymbol{a}\| < \rho\},$$
$$S_\rho(\boldsymbol{a}) = \{\boldsymbol{x} \in \mathbb{R}^n : \|\boldsymbol{x} - \boldsymbol{a}\| = \rho\}$$

are called, respectively, the *closed ball*, *open ball*, and *sphere* with center $\boldsymbol{a}$ and radius ρ. The sets $\mathbb{B} = B_1(\boldsymbol{o})$, $\mathbb{U} = U_1(\boldsymbol{o})$, and $\mathbb{S} = S_1(\boldsymbol{o})$ are called, respectively, the *closed unit ball*, *open unit ball*, and *unit sphere*. Obviously,

$$B_{\rho+\gamma}(\mu\boldsymbol{a} + \boldsymbol{b} + \boldsymbol{c}) = \mu B_{\rho/|\mu|}(\boldsymbol{a}) + B_\gamma(\boldsymbol{b}) + \boldsymbol{c},$$
$$U_{\rho+\gamma}(\mu\boldsymbol{a} + \boldsymbol{b} + \boldsymbol{c}) = \mu U_{\rho/|\mu|}(\boldsymbol{a}) + U_\gamma(\boldsymbol{b}) + \boldsymbol{c}$$

whenever $\boldsymbol{b}, \boldsymbol{c} \in \mathbb{R}^n$ and $\mu \neq 0$.

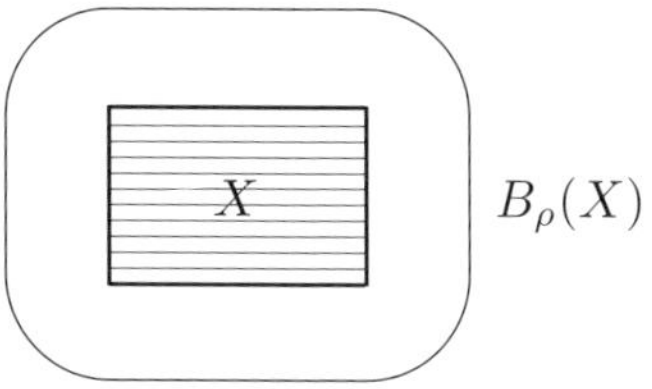

Fig. 1.2 A closed ρ-neighborhood of the set X.

For a scalar $\rho > 0$, the closed and open ρ-*neighborhoods* of a nonempty set $X \subset \mathbb{R}^n$, denoted $B_\rho(X)$ and $U_\rho(X)$, respectively, are defined as

$$B_\rho(X) = \cup\,(B_\rho(\boldsymbol{x}) : \boldsymbol{x} \in X) = B_\rho(\boldsymbol{o}) + X,$$
$$U_\rho(X) = \cup\,(U_\rho(\boldsymbol{x}) : \boldsymbol{x} \in X) = U_\rho(\boldsymbol{o}) + X.$$

We let $B_\rho(\varnothing) = U_\rho(\varnothing) = \varnothing$. Clearly,

$$U_\delta(X) \subset B_\delta(X) \subset U_\rho(X) \quad \text{whenever} \quad 0 < \delta < \rho.$$

We say that an infinite sequence of points $\boldsymbol{x}_1, \boldsymbol{x}_2, \ldots$ *converges* to a point $\boldsymbol{x}$ (and write $\lim\limits_{i\to\infty} \boldsymbol{x}_i = \boldsymbol{x}$, or $\boldsymbol{x}_i \to \boldsymbol{x}$) if $\lim\limits_{i\to\infty} \|\boldsymbol{x}_i - \boldsymbol{x}\| = 0$. With

$$\boldsymbol{x} = (x_1, \ldots, x_n) \quad \text{and} \quad \boldsymbol{x}_i = (x_1^{(i)}, \ldots, x_n^{(i)}), \ i \geqslant 1,$$

one has $\lim_{i\to\infty} \boldsymbol{x}_i = \boldsymbol{x}$ if and only if $\lim_{i\to\infty} x_j^{(i)} = x_j$ for all $1 \leqslant j \leqslant n$.

A point $\boldsymbol{x} \in \mathbb{R}^n$ is called *interior* for a set $X \subset \mathbb{R}^n$ provided there is a scalar $\rho > 0$ such that $U_\rho(\boldsymbol{x}) \subset X$. A set $Y \subset \mathbb{R}^n$ is said to be *open* if every point $\boldsymbol{y} \in Y$ is interior for Y (the empty set $\varnothing$ is defined to be open).

The union of interior points of a given set X, denoted $\operatorname{int} X$, is called the *interior* of X (we let $\operatorname{int} \varnothing = \varnothing$); it is the largest open set contained in a set $X \subset \mathbb{R}^n$. Furthermore,

$$\operatorname{int} X \subset X, \quad \operatorname{int}(\operatorname{int} X) = \operatorname{int} X, \quad \operatorname{int} X \subset \operatorname{int} Y \text{ if } X \subset Y.$$

If $\{X_\alpha\}$ is a family of sets in $\mathbb{R}^n$, then

$$\operatorname{int}\big(\bigcap_\alpha X_\alpha\big) \subset \bigcap_\alpha \operatorname{int} X_\alpha \quad \text{and} \quad \bigcup_\alpha \operatorname{int} X_\alpha \subset \operatorname{int}\big(\bigcup_\alpha X_\alpha\big).$$

For finitely many sets $X_1, \dots, X_r$ in $\mathbb{R}^n$, one has

$$\operatorname{int}(X_1 \cap \dots \cap X_r) = \operatorname{int} X_1 \cap \dots \cap \operatorname{int} X_r.$$

The union of a family of open sets and the intersection of finitely many open sets are open sets. Moreover, $\operatorname{int} X + Y \subset \operatorname{int}(X + Y)$ for sets X and Y in $\mathbb{R}^n$.

A point $\boldsymbol{x} \in \mathbb{R}^n$ is called a *closure point* of a set $X \subset \mathbb{R}^n$ if every open ball $U_\rho(\boldsymbol{x}) \subset \mathbb{R}^n$, $\rho > 0$, meets X; similarly, $\boldsymbol{x}$ is an *accumulation point* of X if every open ball $U_\rho(\boldsymbol{x}) \subset \mathbb{R}^n$, $\rho > 0$, meets $X \setminus \{\boldsymbol{x}\}$. All closure points of X form the *closure* of X, denoted $\operatorname{cl} X$, which can be expressed as

$$\operatorname{cl} X = \cap\,(B_\rho(X) : \rho > 0) = \cap\,(U_\rho(X) : \rho > 0).$$

We let $\operatorname{cl} \varnothing = \varnothing$. Given sets X and Y in $\mathbb{R}^n$, one has

$$X \subset \operatorname{cl} X, \quad \operatorname{cl}(\operatorname{cl} X) = \operatorname{cl} X, \quad \operatorname{cl} X \subset \operatorname{cl} Y \text{ if } X \subset Y.$$

If $\{X_\alpha\}$ is a family of sets in $\mathbb{R}^n$, then

$$\operatorname{cl}\big(\bigcap_\alpha X_\alpha\big) \subset \bigcap_\alpha \operatorname{cl} X_\alpha \quad \text{and} \quad \bigcup_\alpha \operatorname{cl} X_\alpha \subset \operatorname{cl}\big(\bigcup_\alpha X_\alpha\big) = \operatorname{cl}\big(\bigcup_\alpha \operatorname{cl} X_\alpha\big).$$

For finitely many sets $X_1, \dots, X_r$ in $\mathbb{R}^n$, we have

$$\operatorname{cl}(X_1 \cup \dots \cup X_r) = \operatorname{cl} X_1 \cup \dots \cup \operatorname{cl} X_r.$$

A set $X \subset \mathbb{R}^n$ is called *closed* if $\operatorname{cl} X = X$ (the empty set $\varnothing$ is closed). Equivalently, X is closed if and only if its complement $\mathbb{R}^n \setminus X$ is open. The intersection of a family of closed sets and the union of finitely many closed sets are closed sets.

The *boundary* of a set $X \subset \mathbb{R}^n$, denoted $\operatorname{bd} X$, is defined by $\operatorname{bd} X = \operatorname{cl} X \setminus \operatorname{int} X$. Clearly, $\operatorname{bd} X$ is the set of all points $\boldsymbol{x} \in \mathbb{R}^n$ such that every open ball $U_\rho(\boldsymbol{x})$, $\rho > 0$, meets both X and $\mathbb{R}^n \setminus X$ (we let $\operatorname{bd} \varnothing = \varnothing$). The set X is closed if and only if $\operatorname{bd} X \subset X$. Furthermore, $\operatorname{bd} X \neq \varnothing$ if and only if $\varnothing \neq X \neq \mathbb{R}^n$.

A subset Y of a set X is called *dense* in X provided $X \subset \operatorname{cl} Y$. Every set X in $\mathbb{R}^n$ has a countable dense subset. If X is a dense subset of Y and Y is a dense subset of Z, then X is a dense subset of Z.

A nonempty set $X \subset \mathbb{R}^n$ is called *connected* if it cannot be partitioned into two nonempty subsets such that each of them does not meet the closure of the other. A countable union of closed sets is called an F_σ-set. A countable intersection of open sets is called a G_δ-set. If $Z \subset \mathbb{R}^n$ is a closed set containing an F_σ-set $X \subset \mathbb{R}^n$, then $Z \setminus X$ is a G_δ-set.

A set $X \subset \mathbb{R}^n$ is called *bounded* if it lies in a closed ball of positive radius. Properties of bounded sets:

1. A set $X \subset \mathbb{R}^n$ is bounded if and only if for any given point $\boldsymbol{a} \in \mathbb{R}^n$, there is an open ball $U_\rho(\boldsymbol{a})$ containing X.
2. A nonempty set $X \subset \mathbb{R}^n$ is bounded if and only if every infinite sequence of points from X contains a convergent subsequence.
3. A set $X \subset \mathbb{R}^n$ is bounded if and only if $\operatorname{cl} X$ is bounded.
4. Given sets X and Y in $\mathbb{R}^n$ and nonzero scalars λ, μ, the sum $\lambda X + \mu Y$ is bounded if and only if both sets X and Y are bounded.
5. If $X \subset \mathbb{R}^n$ is a nonempty bounded set and $\rho > 0$, then there is a finite set $\{\boldsymbol{x}_1, \dots, \boldsymbol{x}_r\} \subset X$ satisfying the condition
$$X \subset U_\rho(\boldsymbol{x}_1) \cup \cdots \cup U_\rho(\boldsymbol{x}_r).$$

The *diameter* of a bounded set X is defined by

$$\operatorname{diam} X = \sup\{\|\boldsymbol{x} - \boldsymbol{y}\| : \boldsymbol{x}, \boldsymbol{y} \in X\}, \quad \operatorname{diam} \varnothing = 0.$$

If $X, Y \subset \mathbb{R}^n$ and $\mu \in \mathbb{R}$, then $\operatorname{diam}(\mu X + Y) \leqslant |\mu| \operatorname{diam} X + \operatorname{diam} Y$.

A set $X \subset \mathbb{R}^n$ is called *compact* if every family of open sets in $\mathbb{R}^n$ whose union covers X contains a finite subfamily that also covers X. Criteria for the compactness of a nonempty set X in $\mathbb{R}^n$:

1. X is bounded and closed.
2. Every infinite sequence of points from X contains a subsequence converging to a point in X (Heine-Borel Theorem).

The *inf*-distance between nonempty sets X and Y in $\mathbb{R}^n$ is defined by

$$\delta(X, Y) = \inf\{\|\boldsymbol{x} - \boldsymbol{y}\| : \boldsymbol{x} \in X,\ \boldsymbol{y} \in Y\}.$$

Nonempty sets X and Y in $\mathbb{R}^n$ are called *strongly disjoint* provided $\delta(X, Y) > 0$.

A mapping $f : \mathbb{R}^n \to \mathbb{R}^m$ is called *continuous* at a point $\boldsymbol{x}_0 \in \mathbb{R}^n$ if for any $\varepsilon > 0$ there is a $\delta = \delta(\varepsilon) > 0$ such that $\|f(\boldsymbol{x}) - f(\boldsymbol{x}_0)\| < \varepsilon$ whenever $\|\boldsymbol{x} - \boldsymbol{x}_0\| < \delta$. The mapping f is said to be *continuous* on $\mathbb{R}^n$ if it is continuous at every point of $\mathbb{R}^n$. A continuous real-valued function f on $\mathbb{R}^n$ attains its maximum and minimum values on any nonempty compact subset of $\mathbb{R}^n$.

Problems for Chapter 1

Problem 1.1. Let $f : \mathbb{R}^n \to \mathbb{R}^m$ be a linear transformation and S be a subspace of $\mathbb{R}^n$ complementary to $\operatorname{null} f$. Prove that the mapping $h : S \to \operatorname{rng} f$ defined by $h(\boldsymbol{x}) = f(\boldsymbol{x})$ is an invertible linear transformation and the inverse image $f^{-1}(\boldsymbol{x})$ of every vector $\boldsymbol{x} \in \operatorname{rng} f$ equals $h^{-1}(\boldsymbol{x}) + \operatorname{null} f$.

Problem 1.2. Prove that a linear transformation $f : \mathbb{R}^n \to \mathbb{R}^n$ is a linear projection if and only if $f \circ f = f$.

Problem 1.3. (Cauchy-Schwarz Inequality) Prove that $|\boldsymbol{x} \cdot \boldsymbol{y}| \leqslant \|\boldsymbol{x}\| \, \|\boldsymbol{y}\|$, with $|\boldsymbol{x} \cdot \boldsymbol{y}| = \|\boldsymbol{x}\| \, \|\boldsymbol{y}\|$ if and only if one of the vectors $\boldsymbol{x}$ and $\boldsymbol{y}$ is a scalar multiple of the other. Consequently, $\boldsymbol{x} \cdot \boldsymbol{y} = \|\boldsymbol{x}\| \, \|\boldsymbol{y}\|$ if and only if one of the vectors $\boldsymbol{x}$ and $\boldsymbol{y}$ is a nonnegative scalar multiple of the other.

Problem 1.4. (Triangle Inequality) Prove that $\|\boldsymbol{x} + \boldsymbol{y}\| \leqslant \|\boldsymbol{x}\| + \|\boldsymbol{y}\|$, with $\|\boldsymbol{x} + \boldsymbol{y}\| = \|\boldsymbol{x}\| + \|\boldsymbol{y}\|$ if and only if one of the vectors $\boldsymbol{x}$ and $\boldsymbol{y}$ is a nonnegative scalar multiple of the other.

Problem 1.5. Let $\boldsymbol{x}, \boldsymbol{y}$ and $\boldsymbol{z}$ be points in $\mathbb{R}^n$, where $\boldsymbol{x} \neq \boldsymbol{y}$. Prove that

$$\|\boldsymbol{x} - \boldsymbol{y}\| \leqslant \|\boldsymbol{x} - \boldsymbol{z}\| + \|\boldsymbol{z} - \boldsymbol{y}\|,$$

with $\|\boldsymbol{x} - \boldsymbol{y}\| = \|\boldsymbol{x} - \boldsymbol{z}\| + \|\boldsymbol{z} - \boldsymbol{y}\|$ if and only if $\boldsymbol{x} - \boldsymbol{z} = \lambda(\boldsymbol{x} - \boldsymbol{y})$ for a suitable scalar $0 \leqslant \lambda \leqslant 1$.

Problem 1.6. Let $\boldsymbol{x}, \boldsymbol{y}$ and $\boldsymbol{z}$ be distinct points in $\mathbb{R}^n$. Prove that

$$\big| \|\boldsymbol{x} - \boldsymbol{y}\| - \|\boldsymbol{z} - \boldsymbol{y}\| \big| \leqslant \|\boldsymbol{x} - \boldsymbol{z}\|,$$

with $\big| \|\boldsymbol{x} - \boldsymbol{y}\| - \|\boldsymbol{z} - \boldsymbol{y}\| \big| = \|\boldsymbol{x} - \boldsymbol{z}\|$ if and only if one of the vectors $\boldsymbol{x} - \boldsymbol{y}$ and $\boldsymbol{z} - \boldsymbol{y}$ is a nonnegative scalar multiple of the other.

Problem 1.7. Given vectors $\boldsymbol{x}, \boldsymbol{z} \in \mathbb{R}^n$, with $\boldsymbol{x} \neq \boldsymbol{o}$, prove that

$$\lim_{t \to 0} \frac{\|\boldsymbol{x} + t\boldsymbol{z}\| - \|\boldsymbol{x}\|}{t} = \frac{\boldsymbol{x} \cdot \boldsymbol{z}}{\|\boldsymbol{x}\|}.$$

Problem 1.8. Prove the following properties of balls in $\mathbb{R}^n$.

(*a*) Any of the inclusions $B_\rho(\boldsymbol{c}) \subset B_\mu(\boldsymbol{e})$ and $U_\rho(\boldsymbol{c}) \subset U_\mu(\boldsymbol{e})$ holds if and only if $\|\boldsymbol{c} - \boldsymbol{e}\| \leqslant \mu - \rho$.

(*b*) $B_\rho(\boldsymbol{c}) = B_\mu(\boldsymbol{e})$ (respectively, $U_\rho(\boldsymbol{c}) = U_\mu(\boldsymbol{e})$) if and only if $\boldsymbol{c} = \boldsymbol{e}$ and $\rho = \mu$.

Problem 1.9. Prove that the distance function $\delta_{\boldsymbol{c}}(\boldsymbol{x}) = \|\boldsymbol{x} - \boldsymbol{c}\|$ is continuous on $\mathbb{R}^n$ for any choice of the point $\boldsymbol{c} \in \mathbb{R}^n$.

Problem 1.10. Let $\{\boldsymbol{b}_1, \ldots, \boldsymbol{b}_r\} \subset \mathbb{R}^n$ be a linearly independent set. Prove that an infinite sequence $\boldsymbol{x}_i = \lambda_1^{(i)}\boldsymbol{b}_1 + \cdots + \lambda_r^{(i)}\boldsymbol{b}_r$, $i \geqslant 1$, converges to a vector $\boldsymbol{x} = \lambda_1\boldsymbol{b}_1 + \cdots + \lambda_r\boldsymbol{b}_r$ if and only if $\lim_{i\to\infty} \lambda_j^{(i)} = \lambda_j$ for all $1 \leqslant j \leqslant r$.

Problem 1.11. Let $\{\boldsymbol{b}_1, \ldots, \boldsymbol{b}_r\} \subset \mathbb{R}^n$ be a linearly independent set. Prove the existence of a scalar $\rho > 0$ such that for any vectors $\boldsymbol{b}'_i \in U_\rho(\boldsymbol{b}_i)$, $1 \leqslant i \leqslant r$, the set $\{\boldsymbol{b}'_1, \ldots, \boldsymbol{b}'_r\}$ is linearly independent.

Problem 1.12. (Lindelöf's Theorem) Let a set $X \subset \mathbb{R}^n$ be covered with a family $\mathcal{F}$ of open sets. Prove that $\mathcal{F}$ contains a countable subfamily also covering X.

Problem 1.13. For sets X, Y in $\mathbb{R}^n$ and scalars λ, μ, prove that

$$\lambda \operatorname{cl} X + \mu \operatorname{cl} Y \subset \operatorname{cl}(\lambda X + \mu Y) = \operatorname{cl}(\lambda \operatorname{cl} X + \mu \operatorname{cl} Y),$$

where equality holds provided any of the following conditions is satisfied.

(*a*) At most one of the sets X and Y is unbounded.

(*b*) The planes $\operatorname{span} X$ and $\operatorname{span} Y$ are independent.

Problem 1.14. Let X and Y be nonempty set in $\mathbb{R}^n$ such that at least one of them is bounded. Prove the existence of points $\boldsymbol{x} \in \operatorname{cl} X$ and $\boldsymbol{y} \in \operatorname{cl} Y$ with the property $\|\boldsymbol{x} - \boldsymbol{y}\| = \delta(X, Y)$. Consecutively, $\delta(X, Y) > 0$ if and only if $\operatorname{cl} X \cap \operatorname{cl} Y = \varnothing$.

Problem 1.15. Given nonempty sets X and Y in $\mathbb{R}^n$ and a scalar $\rho > 0$, prove the following assertions.

(*a*) $\delta(X, Y) > 0$ if and only if $\boldsymbol{o} \notin \operatorname{cl}(X - Y)$.

(*b*) $\delta(X, Y) \geqslant \rho$ if and only if $U_{\rho/2}(X) \cap U_{\rho/2}(Y) = \varnothing$.

(*c*) $\delta(X, Y) > \rho$ if and only if $B_{\rho/2}(X) \cap B_{\rho/2}(Y) = \varnothing$.

Problem 1.16. Let $f : \mathbb{R}^n \to \mathbb{R}^m$ be a continuous mapping and $X \subset \mathbb{R}^n$ and $Y \subset \mathbb{R}^m$ be sets. Prove the following assertions.

(*a*) $f(\operatorname{cl} X) \subset \operatorname{cl} f(X) = \operatorname{cl} f(\operatorname{cl} X)$.

(*b*) $f(\operatorname{cl} X) = \operatorname{cl} f(X)$ provided X is bounded (consequently, $f(X)$ is compact if X is compact).

(*c*) $\operatorname{cl} f^{-1}(Y) = \operatorname{cl} f^{-1}(Y \cap \operatorname{rng} f) \subset f^{-1}(\operatorname{cl}(Y \cap \operatorname{rng} f)) \subset f^{-1}(\operatorname{cl} Y)$.

Problem 1.17. Given a continuous mapping $f : \mathbb{R}^n \to \mathbb{R}^m$, prove the following assertions.

(*a*) The inverse image $f^{-1}(Y)$ of every closed set $Y \subset \mathbb{R}^m$ is a closed set.

(*b*) The inverse image $f^{-1}(Y)$ of every open set $Y \subset \mathbb{R}^m$ is an open set.

Problem 1.18. Given a linear transformation $f : \mathbb{R}^n \to \mathbb{R}^m$ and the standard basis $\{\boldsymbol{e}_1, \dots, \boldsymbol{e}_n\}$ for $\mathbb{R}^n$, let $f(\boldsymbol{e}_i) = (a_{i1}, \dots, a_{im})$, $1 \leqslant i \leqslant n$. Prove that f satisfies the Lipschitz condition

$$\|f(\boldsymbol{x})\| \leqslant L\|\boldsymbol{x}\|, \quad \boldsymbol{x} \in \mathbb{R}^n, \quad \text{where } L = \Big(\sum_{i=1}^{m}\sum_{j=1}^{n} a_{ji}^2\Big)^{1/2}.$$

Deduce from here that every linear transformation $f : \mathbb{R}^n \to \mathbb{R}^m$ is continuous on $\mathbb{R}^n$. In particular, any real-valued linear functional $\varphi(\boldsymbol{x}) = \boldsymbol{x}\cdot\boldsymbol{e}$, where $\boldsymbol{e} \in \mathbb{R}^n$, is continuous on $\mathbb{R}^n$.

Problem 1.19. Let $f : \mathbb{R}^n \to \mathbb{R}^m$ be a one-to-one linear transformation. Prove that $f(\operatorname{cl} X) = \operatorname{cl} f(X)$ for any set $X \subset \mathbb{R}^n$.

Problem 1.20. Let $f : \mathbb{R}^n \to \mathbb{R}^m$ be a linear transformation and $Y \subset \mathbb{R}^m$ be a set. Prove that $\operatorname{cl} f^{-1}(Y \cap \operatorname{rng} f) = f^{-1}(\operatorname{cl}(Y \cap \operatorname{rng} f))$.

Chapter 2

The Affine Structure of $\mathbb{R}^n$

2.1 Planes and Hyperplanes

2.1.1 *Basic Properties of Planes*

Definition 2.1. A *plane* $L \subset \mathbb{R}^n$ is a set which is either empty or a translate of a subspace:

$$L = \boldsymbol{a} + S\,(= \{\boldsymbol{a} + \boldsymbol{x} : \boldsymbol{x} \in S\}),$$

where $\boldsymbol{a}$ is a point in $\mathbb{R}^n$ and S is a subspace of $\mathbb{R}^n$.

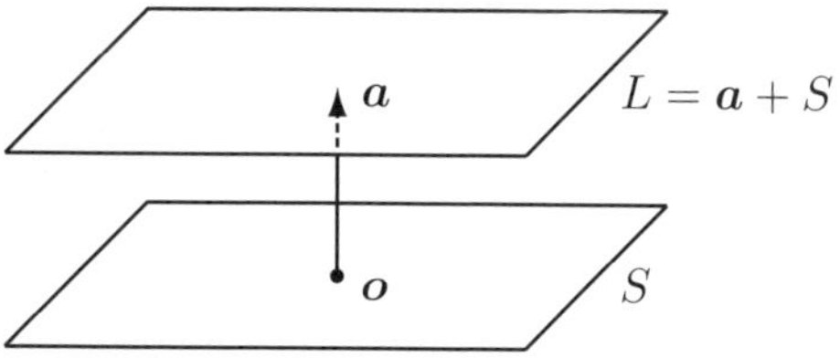

Fig. 2.1 A nonempty plane L as a translate of a subspace S.

For instance, planes in $\mathbb{R}^3$ are the empty set, singletons (as translates of the trivial subspace $\{\boldsymbol{o}\}$), lines, usual planes, and $\mathbb{R}^3$. Synonyms for *plane*, used by various authors, are affine set, affine subspace, affine variety, and flat. We prefer the term plane for its accordance with the well established in the literature term *hyperplane* (see Definition 2.16). Despite the fact that all subspaces in $\mathbb{R}^n$ are nonempty (they contain the zero vector $\boldsymbol{o}$), we allow planes be empty to simplify various arguments.

Theorem 2.2. *A nonempty plane $L \subset \mathbb{R}^n$ is a translate of a unique subspace S of $\mathbb{R}^n$, given by*

$$S = L - L \,(= \{\boldsymbol{x} - \boldsymbol{y} : \boldsymbol{x}, \boldsymbol{y} \in L\}). \tag{2.1}$$

Furthermore, the following assertions take place.

(*a*) *For a point $\boldsymbol{c} \in \mathbb{R}^n$, the equality $L = \boldsymbol{c} + S$ holds if and only if $\boldsymbol{c} \in L$.*
(*b*) *$L = X + S$ and $S = L - X$ for any nonempty subset X of L.*
(*c*) *L is a subspace if and only if $\boldsymbol{o} \in L$.*

Proof. According to Definition 2.1, L can be expressed as $L = \boldsymbol{a} + S$, where $\boldsymbol{a} \in \mathbb{R}^n$ and S is a subspace of $\mathbb{R}^n$. To prove (2.1), choose points $\boldsymbol{x}$ and $\boldsymbol{y}$ in L. Then $\boldsymbol{x} = \boldsymbol{a} + \boldsymbol{x}'$ and $\boldsymbol{y} = \boldsymbol{a} + \boldsymbol{y}'$ for suitable vectors $\boldsymbol{x}', \boldsymbol{y}' \in S$. This argument gives

$$\boldsymbol{x} - \boldsymbol{y} = (\boldsymbol{a} + \boldsymbol{x}') - (\boldsymbol{a} + \boldsymbol{y}') = \boldsymbol{x}' - \boldsymbol{y}' \in S.$$

Hence

$$L - L = \{\boldsymbol{x} - \boldsymbol{y} : \boldsymbol{x}, \boldsymbol{y} \in L\} \subset S.$$

On the other hand, any vector $\boldsymbol{x} \in S$ can be written as

$$\boldsymbol{x} = (\boldsymbol{a} + \boldsymbol{x}) - (\boldsymbol{a} + \boldsymbol{o}) \in (\boldsymbol{a} + S) - (\boldsymbol{a} + S) = L - L,$$

which implies the opposite inclusion $S \subset L - L$. The obtained equality $S = L - L$ shows the desired uniqueness of S.

(*a*) If $L = \boldsymbol{c} + S$ for a point $\boldsymbol{c} \in \mathbb{R}^n$, then $\boldsymbol{c} = \boldsymbol{c} + \boldsymbol{o} \in \boldsymbol{c} + S = L$. Conversely, if $\boldsymbol{c} \in L$, then $\boldsymbol{c} = \boldsymbol{a} + \boldsymbol{c}'$, where $\boldsymbol{c}' \in S$. Since $\boldsymbol{c}' + S = S$, we have

$$\boldsymbol{c} + S = (\boldsymbol{a} + \boldsymbol{c}') + S = \boldsymbol{a} + (\boldsymbol{c}' + S) = \boldsymbol{a} + S = L.$$

(*b*) Let X be a nonempty subset of L. By assertion (*a*),

$$X + S = \cup\,(\boldsymbol{x} + S : \boldsymbol{x} \in X) = \cup\,(L : \boldsymbol{x} \in X) = L.$$

Similarly,

$$L - X = \cup\,(L - \boldsymbol{x} : \boldsymbol{x} \in X) = \cup\,(S : \boldsymbol{x} \in X) = S.$$

(*c*) If L is a subspace, then the inclusion $\boldsymbol{o} \in L$ is obvious. Conversely, let $\boldsymbol{o} \in L$. Then $\boldsymbol{o} = \boldsymbol{a} + \boldsymbol{x}$ for a suitable vector $\boldsymbol{x} \in S$. Since $\boldsymbol{o} \in S$, one has $\boldsymbol{a} = \boldsymbol{o} - \boldsymbol{x} \in S$. Consequently, $L = \boldsymbol{a} + S = S$. □

In view of Theorem 2.2, we may assign to a nonempty plane $L \subset \mathbb{R}^n$ a pair of subspaces, $\operatorname{dir} L$ and $\operatorname{ort} L$, defined below. These subspaces are mutually orthogonal and may be viewed as local coordinate subspaces associated with L.

Definition 2.3. For a nonempty plane $L \subset \mathbb{R}^n$, the subspace $S = L - L$ is called the *direction space* of L and is denoted $\operatorname{dir} L$. The orthogonal complement of $\operatorname{dir} L$, denoted $\operatorname{ort} L$, is called the *orthospace* of L. We let $\operatorname{dir} L = \varnothing$ and $\operatorname{ort} L = \mathbb{R}^n$ if $L = \varnothing$.

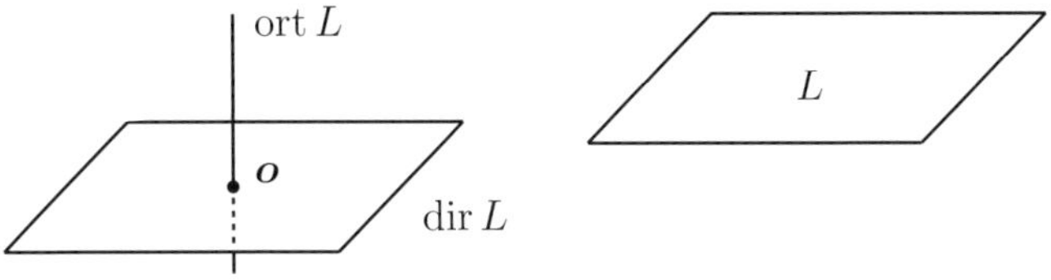

Fig. 2.2 Direction space and orthospace of a plane L.

Theorem 2.2 implies that a nonempty plane $L \subset \mathbb{R}^n$ can be expressed as $L = \boldsymbol{c} + \operatorname{dir} L$ whenever $\boldsymbol{c} \in L$. We observe that $\operatorname{ort} L$ may be distinct from the orthogonal complement $L^{\perp}$ of L (see page 7). Clearly, $\operatorname{ort} L = L^{\perp}$ if and only if L is a subspace. The next theorem deals with various operations on planes (see also Corollary 2.84 and Problems 2.1 and 2.2).

Theorem 2.4. *The following assertions hold.*

(a) *For any family* $\mathcal{F} = \{L_\alpha\}$ *of planes in* $\mathbb{R}^n$, *their intersection* $\bigcap_\alpha L_\alpha$ *is a plane. Furthermore,* $\operatorname{dir}(\bigcap_\alpha L_\alpha) = \bigcap_\alpha \operatorname{dir} L_\alpha$ *provided* $\bigcap_\alpha L_\alpha \neq \varnothing$.

(b) *If* $L_1, \dots, L_r \subset \mathbb{R}^n$ *are planes and* $\mu_1, \dots, \mu_r$ *are scalars, then the sum* $\mu_1 L_1 + \cdots + \mu_r L_r$ *is a plane. Furthermore,*

$$\operatorname{dir}(\mu_1 L_1 + \cdots + \mu_r L_r) = \mu_1 \operatorname{dir} L_1 + \cdots + \mu_r \operatorname{dir} L_r. \tag{2.2}$$

Proof. (a) Since the case $\bigcap_\alpha L_\alpha = \varnothing$ is obvious (the empty set is a plane), we may suppose that $\bigcap_\alpha L_\alpha \neq \varnothing$. Choose a point $\boldsymbol{c} \in \bigcap_\alpha L_\alpha$. According to Theorem 2.2, every plane $L_\alpha \in \mathcal{F}$ can be written as $L_\alpha = \boldsymbol{c} + \operatorname{dir} L_\alpha$. Because $\bigcap_\alpha \operatorname{dir} L_\alpha$ is a subspace, the equality

$$\bigcap_\alpha L_\alpha = \bigcap_\alpha (\boldsymbol{c} + \operatorname{dir} L_\alpha) = \boldsymbol{c} + \bigcap_\alpha \operatorname{dir} L_\alpha$$

shows that $\bigcap_\alpha L_\alpha$ is a plane as a translate of $\bigcap_\alpha \operatorname{dir} L_\alpha$. By the same theorem,

$$\operatorname{dir}(\bigcap_\alpha L_\alpha) = \bigcap_\alpha L_\alpha - \boldsymbol{c} = \bigcap_\alpha (L_\alpha - \boldsymbol{c}) = \bigcap_\alpha \operatorname{dir} L_\alpha.$$

(*b*) If at least one of the planes $L_1, \ldots, L_r$ is empty, then both sides of (2.2) are empty sets. Suppose that all $L_1, \ldots, L_r$ are nonempty. Let $L_i = \boldsymbol{c}_i + \operatorname{dir} L_i$ for suitable points $\boldsymbol{c}_i \in L_i$, $1 \leqslant i \leqslant r$. Then

$$\mu_1 L_1 + \cdots + \mu_r L_r = (\mu_1 \boldsymbol{c}_1 + \cdots + \mu_r \boldsymbol{c}_r) + (\mu_1 \operatorname{dir} L_1 + \cdots + \mu_r \operatorname{dir} L_r).$$

Hence the set $\mu_1 L_1 + \cdots + \mu_r L_r$ is a plane as a translate of the subspace $\mu_1 \operatorname{dir} L_1 + \cdots + \mu_r \operatorname{dir} L_r$. By Theorem 2.2,

$$\begin{aligned} \operatorname{dir}(\mu_1 L_1 + \cdots + \mu_r L_r) &= (\mu_1 L_1 + \cdots + \mu_r L_r) - (\mu_1 \boldsymbol{c}_1 + \cdots + \mu_r \boldsymbol{c}_r) \\ &= \mu_1 (L_1 - \boldsymbol{c}_1) + \cdots + \mu_r (L_r - \boldsymbol{c}_r) = \mu_1 \operatorname{dir} L_1 + \cdots + \mu_r \operatorname{dir} L_r. \qquad \square \end{aligned}$$

2.1.2 *Dimension of Planes*

Theorem 2.2 gives a base to the definition below.

Definition 2.5. If $L \subset \mathbb{R}^n$ is a nonempty plane and S is the direction space of L, then the *dimension* of L, denoted $\dim L$, is defined by $\dim L = \dim S$. We let $\dim L = -1$ if $L = \varnothing$.

Planes of dimension 0 and 1 are usually called *points* and *lines*, respectively (compare with Definition 2.22).

Theorem 2.6. *The following assertions take place.*

(*a*) *If $L \subset \mathbb{R}^n$ is a plane, then for a point $\boldsymbol{a} \in \mathbb{R}^n$ and a scalar $\mu \neq 0$, the plane $\boldsymbol{a} + \mu L$ is a translate of L and has the same dimension as L.*
(*b*) *If planes L_1 and L_2 in $\mathbb{R}^n$ satisfy the inclusion $L_1 \subset L_2$, then $\dim L_1 \leqslant \dim L_2$, with $\dim L_1 = \dim L_2$ if and only if $L_1 = L_2$.*

Proof. Since both assertions are obvious for the cases $L = \varnothing$ and $L_1 = \varnothing$, we may assume that all planes involved are nonempty.

(*a*) Let $L = \boldsymbol{c} + S$, where $S = \operatorname{dir} L$. Because $\mu S = S$, we have

$$\begin{aligned} \boldsymbol{a} + \mu L &= \boldsymbol{a} + \mu(\boldsymbol{c} + S) = \boldsymbol{a} + \mu \boldsymbol{c} + S \\ &= \boldsymbol{a} + (\mu - 1)\boldsymbol{c} + (\boldsymbol{c} + S) = \boldsymbol{a} + (\mu - 1)\boldsymbol{c} + L. \end{aligned}$$

Hence $\boldsymbol{a} + \mu L$ is a translate of L, and of S as well. Therefore,

$$\dim(\boldsymbol{a} + \mu L) = \dim S = \dim L.$$

(*b*) Choose a point $\boldsymbol{b} \in L_1$. Then $\boldsymbol{b} \in L_2$. By Theorem 2.2, $L_i = \boldsymbol{b} + S_i$, where $S_i = \operatorname{dir} L_i$, $i = 1, 2$. Consequently,

$$S_1 = L_1 - \boldsymbol{b} \subset L_2 - \boldsymbol{b} = S_2.$$

Therefore, $\dim S_1 \leqslant \dim S_2$, with $\dim S_1 = \dim S_2$ if and only if $S_1 = S_2$ (see page 5). Hence

$$\dim L_1 = \dim S_1 \leqslant \dim S_2 = \dim L_2,$$

and $\dim L_1 = \dim L_2$ if and only if $L_1 = \boldsymbol{b} + S_1 = \boldsymbol{b} + S_2 = L_2$. □

Theorem 2.7. *If* L_1 *and* L_2 *are nonempty planes in* $\mathbb{R}^n$, *then*

$$\dim (L_1 + L_2) = \dim L_1 + \dim L_2 - \dim (\operatorname{dir} L_1 \cap \operatorname{dir} L_2).$$

If, additionally, $L_1 \cap L_2 \neq \varnothing$, *then*

$$\dim (L_1 + L_2) = \dim L_1 + \dim L_2 - \dim (L_1 \cap L_2).$$

Proof. By Theorem 2.2, we can write $L_i = \boldsymbol{a}_i + S_i$, where $\boldsymbol{a}_i \in L_i$ and $S_i = \operatorname{dir} L_i$, $i = 1, 2$. Then

$$L_1 + L_2 = \boldsymbol{a}_1 + \boldsymbol{a}_2 + (S_1 + S_2),$$

and Definitions 2.5 gives

$$\begin{aligned} \dim (L_1 + L_2) &= \dim (S_1 + S_2) \\ &= \dim S_1 + \dim S_2 - \dim (S_1 \cap S_2) \qquad (2.3) \\ &= \dim L_1 + \dim L_2 - \dim (S_1 \cap S_2). \end{aligned}$$

Suppose, additionally, that $L_1 \cap L_2 \neq \varnothing$, and choose a point $\boldsymbol{c} \in L_1 \cap L_2$. By Theorem 2.2, $L_i = \boldsymbol{c} + S_i$, $i = 1, 2$. Consequently,

$$L_1 \cap L_2 = (\boldsymbol{c} + L_1) \cap (\boldsymbol{c} + L_2) = \boldsymbol{c} + S_1 \cap S_2.$$

Therefore, $\dim (L_1 \cap L_2) = \dim (S_1 \cap S_2)$. Combined with (2.3), this implies

$$\dim (L_1 + L_2) = \dim L_1 + \dim L_2 - \dim (L_1 \cap L_2). \qquad \square$$

We conclude this subsection with a useful condition for the nonemptiness of the intersection of two planes (compare with Problem 2.3).

Theorem 2.8. *For nonempty planes* L_1 *and* L_2 *in* $\mathbb{R}^n$, *the following conditions are equivalent.*

(*a*) $L_1 \cap L_2 \neq \varnothing$.
(*b*) *The union* $L_1 \cup L_2$ *lies in a translate of* $L_1 + L_2$.

Proof. (*a*) ⇒ (*b*) Choose a point $\boldsymbol{c} \in L_1 \cap L_2$. By Theorem 2.2, $L_i = \boldsymbol{c} + S_i$, where $S_i = \operatorname{dir} L_i$, $i = 1, 2$. Then

$$\begin{aligned} L_1 \cup L_2 &= (\boldsymbol{c} + S_1) \cup (\boldsymbol{c} + S_2) = \boldsymbol{c} + (S_1 \cup S_2) \subset \boldsymbol{c} + (S_1 + S_2) \\ &= (\boldsymbol{c} + S_1) + (\boldsymbol{c} + S_2) - \boldsymbol{c} = (L_1 + L_2) - \boldsymbol{c}. \end{aligned}$$

$(b) \Rightarrow (a)$ By the assumption, $L_1 \cup L_2 \subset \boldsymbol{b} + (L_1 + L_2)$ for a suitable point $\boldsymbol{b} \in \mathbb{R}^n$. Choose a point $\boldsymbol{c} \in L_1 + L_2$ and let

$$S = (L_1 + L_2) - \boldsymbol{c} \quad \text{and} \quad L_i' = L_i - (\boldsymbol{b} + \boldsymbol{c}), \ i = 1, 2.$$

Then

$$\begin{aligned} L_i' = L_i - (\boldsymbol{b} + \boldsymbol{c}) &\subset (L_1 \cup L_2) - (\boldsymbol{b} + \boldsymbol{c}) \\ &\subset (L_1 + L_2) - \boldsymbol{c} = S, \quad i = 1, 2. \end{aligned}$$

Furthermore, $\boldsymbol{o} \in S$ due to the inclusion $\boldsymbol{c} \in L_1 + L_2$. Hence S is a subspace (see Theorem 2.2), implying that $S = \operatorname{dir}(L_1 + L_2)$. Since L_i' is a translate of L_i, we can write $L_i' = \boldsymbol{a}_i + S_i$, where $\boldsymbol{a}_i \in \mathbb{R}^n$ and $S_i = \operatorname{dir} L_i$, $i = 1, 2$. By Theorem 2.4,

$$S = \operatorname{dir}(L_1 + L_2) = \operatorname{dir} L_1 + \operatorname{dir} L_2 = S_1 + S_2.$$

Furthermore, $\boldsymbol{a}_i \in L_i' \subset S$, $i = 1, 2$.

Choose a basis B for $S_1 \cap S_2$ and complete it into bases $B \cup C$ and $B \cup E$ for S_1 and S_2, respectively (see page 6). The combined list $B \cup C \cup E$ is a basis for $S_1 + S_2 = S$, where any of the sets $B, B \cup C$, and $B \cup E$ is empty if the corresponding subspace is 0-dimensional. Therefore, $\boldsymbol{a}_1$ and $\boldsymbol{a}_2$ (which belong to S) can be expressed as $\boldsymbol{a}_i = \boldsymbol{u}_i + \boldsymbol{v}_i + \boldsymbol{w}_i$, where

$$\boldsymbol{u}_i \in \operatorname{span} B, \quad \boldsymbol{v}_i \in \operatorname{span} C, \quad \boldsymbol{w}_i \in \operatorname{span} E, \quad i = 1, 2.$$

Let

$$\boldsymbol{z}_0 = \boldsymbol{v}_2 + \boldsymbol{w}_1, \quad \boldsymbol{z}_1 = -\boldsymbol{u}_1 + (\boldsymbol{v}_2 - \boldsymbol{v}_1), \quad \boldsymbol{z}_2 = -\boldsymbol{u}_2 + (\boldsymbol{w}_1 - \boldsymbol{w}_2).$$

Obviously, $\boldsymbol{z}_0 = \boldsymbol{a}_1 + \boldsymbol{z}_1 = \boldsymbol{a}_2 + \boldsymbol{z}_2$. Moreover $\boldsymbol{z}_i + S_i = S_i$ due to the obvious inclusions

$$\boldsymbol{z}_1 \in \operatorname{span}(B \cup C) = S_1, \quad \boldsymbol{z}_2 \in \operatorname{span}(B \cup E) = S_2.$$

Hence

$$L_i' = \boldsymbol{a}_i + S_i = \boldsymbol{a}_i + (\boldsymbol{z}_i + S_i) = (\boldsymbol{a}_i + \boldsymbol{z}_i) + S_i = \boldsymbol{z}_0 + S_i, \quad i = 1, 2.$$

Thus $\boldsymbol{z}_0 \in L_1' \cap L_2'$ according to Theorem 2.2. Finally,

$$\boldsymbol{b} + \boldsymbol{c} + \boldsymbol{z}_0 \in \boldsymbol{b} + \boldsymbol{c} + L_1' \cap L_2' = (\boldsymbol{b} + \boldsymbol{c} + L_1') \cap (\boldsymbol{b} + \boldsymbol{c} + L_2') = L_1 \cap L_2,$$

which implies the desired conclusion $L_1 \cap L_2 \neq \varnothing$. □

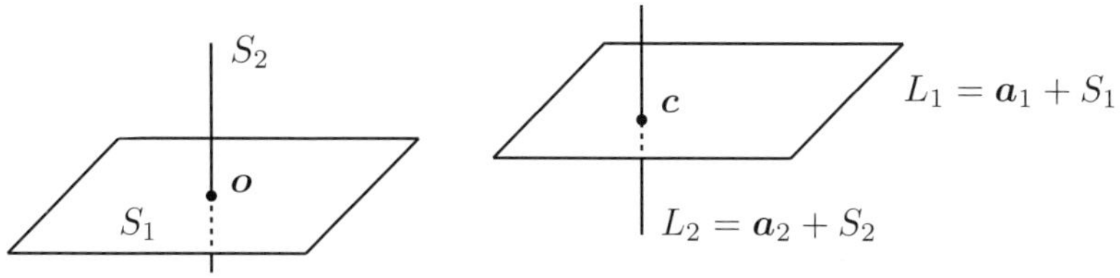

Fig. 2.3 Complementary planes L_1 and L_2.

2.1.3 *Independent and Complementary Planes*

We recall that subspaces S_1 and S_2 are *independent* provided $S_1 \cap S_2 = \{\boldsymbol{o}\}$ (possibly, $S_1 = \{\boldsymbol{o}\}$ or $S_2 = \{\boldsymbol{o}\}$). If, additionally, $S_1 + S_2 = \mathbb{R}^n$, then S_1 and S_2 are called *complementary* subspaces.

Definition 2.9. Nonempty planes L_1 and L_2 in $\mathbb{R}^n$ are called *independent* (respectively, *complementary*) if their direction spaces, $\operatorname{dir} L_1$ and $\operatorname{dir} L_2$, are independent (respectively, complementary) subspaces. The sum of independent planes L_1 and L_2 in $\mathbb{R}^n$, is called their *direct sum.*

Theorem 2.10. *Nonempty planes L_1 and L_2 in $\mathbb{R}^n$ are independent if and only if*

$$\dim (L_1 + L_2) = \dim L_1 + \dim L_2.$$

Furthermore, L_1 and L_2 are complementary if and only if

$$\dim (L_1 + L_2) = \dim L_1 + \dim L_2 = n.$$

Proof. Let $L_i = \boldsymbol{a}_i + S_i$, where $S_i = \operatorname{dir} L_i$, $i = 1, 2$. Then

$$L_1 + L_2 = (\boldsymbol{a}_1 + \boldsymbol{a}_2) + (S_1 + S_2),$$

which yields the equality $\dim (L_1 + L_2) = \dim (S_1 + S_2)$.

The subspaces S_1 and S_2 are independent if and only if

$$\dim (S_1 + S_2) = \dim S_1 + \dim S_2.$$

Furthermore, S_1 and S_2 are complementary if and only if

$$\dim S_1 + \dim S_2 = n.$$

Now, both assertions of the theorem follows from Definition 2.9 and the equalities $\dim L_i = \dim S_i$, $i = 1, 2$. $\square$

Theorem 2.11. *For nonempty planes L_1 and L_2 in $\mathbb{R}^n$, the following conditions are equivalent.*

(*a*) *L_1 and L_2 are complementary.*
(*b*) *$L_1 + L_2 = \mathbb{R}^n$ and $L_1 \cap L_2$ is a singleton.*
(*c*) *There are complementary subspaces S_1 and S_2 and a point $\boldsymbol{c} \in \mathbb{R}^n$ such that $L_1 = \boldsymbol{c} + S_1$ and $L_2 = \boldsymbol{c} + S_2$.*

Proof. Let $L_i = \boldsymbol{a}_i + S_i$, where $S_i = \operatorname{dir} L_i$, $i = 1, 2$.

(*a*) $\Rightarrow$ (*b*) If L_1 and L_2 are complementary, then the subspaces S_1 and S_2 are complementary. Therefore,

$$L_1 + L_2 = (\boldsymbol{a}_1 + \boldsymbol{a}_2) + (S_1 + S_2) = (\boldsymbol{a}_1 + \boldsymbol{a}_2) + \mathbb{R}^n = \mathbb{R}^n.$$

Hence $L_1 \cup L_2 \subset \mathbb{R}^n = L_1 + L_2$, and Theorem 2.8 gives $L_1 \cap L_2 \neq \varnothing$. By Theorem 2.7,

$$\dim(L_1 \cap L_2) = \dim L_1 + \dim L_2 - \dim \mathbb{R}^n = \dim S_1 + \dim S_2 - n = 0.$$

These equalities shows that $L_1 \cap L_2$ is a singleton.

(*b*) $\Rightarrow$ (*c*) Denote by $\boldsymbol{c}$ the unique point of $L_1 \cap L_2$. Then $L_i = \boldsymbol{c} + S_i$, $i = 1, 2$ (see Theorem 2.2). Furthermore,

$$S_1 + S_2 = (L_1 + L_2) - 2\boldsymbol{c} = \mathbb{R}^n - 2\boldsymbol{c} = \mathbb{R}^n,$$
$$\{\boldsymbol{c}\} = L_1 \cap L_2 = (\boldsymbol{c} + S_1) \cap (\boldsymbol{c} + S_2) = \boldsymbol{c} + S_1 \cap S_2.$$

Hence $S_1 \cap S_2 = \{\boldsymbol{o}\}$, and the subspaces S_1 and S_2 are complementary.

(*c*) $\Rightarrow$ (*a*) This part trivially follows from Definition 2.9. $\square$

See Problem 2.5 for another characterization of complementary planes.

2.1.4 *Parallel Planes*

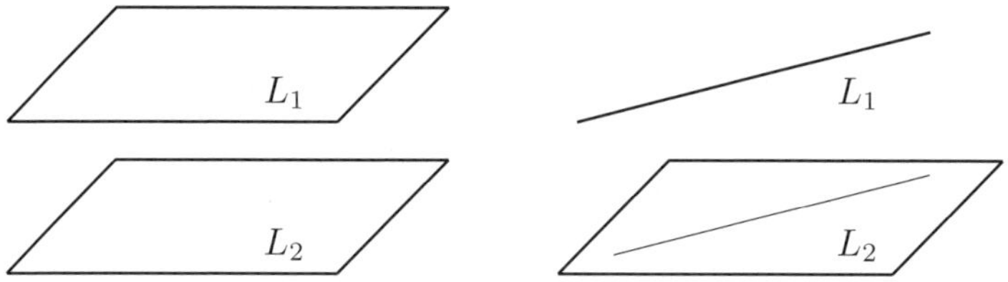

Fig. 2.4 Parallel planes L_1 and L_2.

Definition 2.12. Nonempty planes L_1 and L_2 in $\mathbb{R}^n$ are called *parallel* if one of them contains a translate of the other.

We observe that Definition 2.12 allows parallel planes to have common points or even to coincide. Furthermore, this type of parallelism is not an equivalence relation on the family of all planes in $\mathbb{R}^n$. For instance, the coordinate x and y axes of $\mathbb{R}^3$ are not parallel, while both are parallel to the coordinate xy plane. Obviously, any singleton is parallel to every nonempty plane.

Theorem 2.13. *If* L_1 *and* L_2 *are nonempty planes in* $\mathbb{R}^n$ *satisfying the inequality* $\dim L_1 \leqslant \dim L_2$*, then* L_1 *and* L_2 *are parallel if and only if* $\operatorname{dir} L_1 \subset \operatorname{dir} L_2$.

Proof. Let $L_i = \boldsymbol{a}_i + \operatorname{dir} L_i$, $i = 1, 2$. Suppose first that L_1 and L_2 are parallel. Theorem 2.6 and the assumption $\dim L_1 \leqslant \dim L_2$ imply that namely L_2 contains a translate of L_1. Choose a vector $\boldsymbol{c} \in \mathbb{R}^n$ satisfying the condition $\boldsymbol{c} + L_1 \subset L_2$. Put $\boldsymbol{b} = \boldsymbol{c} + \boldsymbol{a}_1$. Then

$$\boldsymbol{b} = \boldsymbol{c} + \boldsymbol{a}_1 \in \boldsymbol{c} + \boldsymbol{a}_1 + \operatorname{dir} L_1 = \boldsymbol{c} + L_1 \subset L_2.$$

Thus $L_2 = \boldsymbol{b} + \operatorname{dir} L_2$ according to Theorem 2.2. Therefore,

$$\operatorname{dir} L_1 = L_1 - \boldsymbol{a}_1 = (\boldsymbol{c} + L_1) - (\boldsymbol{c} + \boldsymbol{a}_1) \subset L_2 - \boldsymbol{b} = \operatorname{dir} L_2.$$

Conversely, if $\operatorname{dir} L_1 \subset \operatorname{dir} L_2$, then

$$\begin{aligned}(\boldsymbol{a}_2 - \boldsymbol{a}_1) + L_1 &= (\boldsymbol{a}_2 - \boldsymbol{a}_1) + (\boldsymbol{a}_1 + \operatorname{dir} L_1) \\ &= \boldsymbol{a}_2 + \operatorname{dir} L_1 \subset \boldsymbol{a}_2 + \operatorname{dir} L_2 = L_2,\end{aligned}$$

implying that L_1 and L_2 are parallel. □

Corollary 2.14. *For nonempty planes* L_1 *and* L_2 *in* $\mathbb{R}^n$ *of the same dimension, the following conditions are equivalent.*

(*a*) L_1 *and* L_2 *are parallel.*
(*b*) $\operatorname{dir} L_1 = \operatorname{dir} L_2$.
(*c*) *Each of the planes* L_1 *and* L_2 *is a translate of the other.* □

Theorem 2.15. *Given nonempty planes* L_1 *and* L_2 *in* $\mathbb{R}^n$*, let*

$$L_1' = L_1 + \operatorname{dir} L_2 \quad \textit{and} \quad L_2' = \operatorname{dir} L_1 + L_2.$$

Then L_1' *and* L_2' *form a unique pair of parallel planes containing* L_1 *and* L_2*, respectively, and both having the smallest possible dimension,* $\dim (L_1 + L_2)$. *Furthermore,* L_1' *and* L_2' *are disjoint if and only if* L_1 *and* L_2 *are disjoint.*

Proof. Let $L_i = \boldsymbol{a}_i + S_i$, where $S_i = \operatorname{dir} L_i$, $i = 1, 2$. Corollary 2.14 shows that the planes

$$\begin{aligned} L_1' &= L_1 + S_2 = \boldsymbol{a}_1 + (S_1 + S_2), \\ L_2' &= S_1 + L_2 = \boldsymbol{a}_2 + (S_1 + S_2) \end{aligned} \tag{2.4}$$

are parallel as translates of the same subspace $S_1 + S_2$. Clearly,

$$L_1 \subset L_1 + S_2 = L_1' \quad \text{and} \quad L_2 \subset S_1 + L_2 = L_2',$$
$$\dim L_1' = \dim L_2' = \dim (S_1 + S_2) = \dim (L_1 + L_2).$$

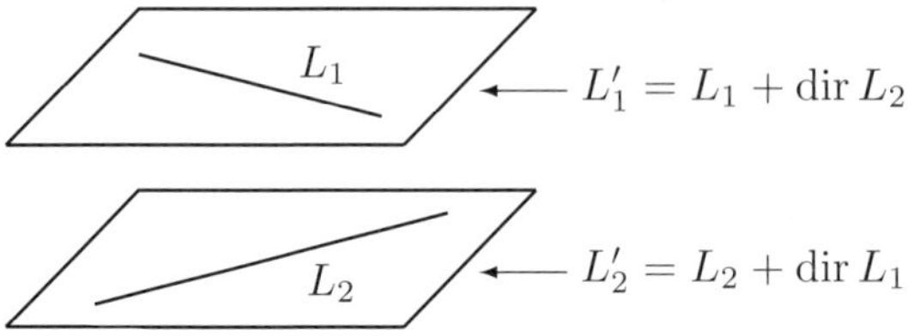

Fig. 2.5 Parallel planes L_1' and L_2' through planes L_1 and L_2.

If M_1 and M_2 is any pair of parallel planes of the same dimension, containing L_1 and L_2, respectively, then $\boldsymbol{a}_i \in L_i \subset M_i$, $i = 1, 2$, and a combination of Theorem 2.13 and Corollary 2.14 gives

$$S_1 \subset \operatorname{dir} M_1, \quad S_2 \subset \operatorname{dir} M_2, \quad \operatorname{dir} M_1 = \operatorname{dir} M_2.$$

Therefore, $S_1 + S_2 \subset \operatorname{dir} M_i$, with equality if and only if $\dim (S_1 + S_2) = \dim M_i$, $i = 1, 2$. Consequently, if the dimension of both planes M_1 and M_2 is $\dim (S_1 + S_2)$, then

$$M_i = \boldsymbol{a}_i + (S_1 + S_2) = L_i', \quad i = 1, 2.$$

If the planes L_1' and L_2' are disjoint, then so are L_1 and L_2. Suppose that $L_1' \cap L_2' \neq \varnothing$ and choose a point $\boldsymbol{c} \in L_1' \cap L_2'$. Theorem 2.2 and the above argument imply that $L_1' = \boldsymbol{c} + S_1 + S_2 = L_2'$. Comparing these equalities with (2.4), we conclude from the same theorem that

$$\boldsymbol{a}_i - \boldsymbol{c} \in L_i' - L_i' = S_1 + S_2, \quad i = 1, 2.$$

Furthermore,

$$L_i - \boldsymbol{c} \subset L_i + S_j - \boldsymbol{c} = L_i' - \boldsymbol{c} = S_1 + S_2, \quad i, j \in \{1, 2\},\ i \neq j.$$

This argument gives the inclusions

$$\begin{aligned} (L_1 - \boldsymbol{c}) \cup (L_2 - \boldsymbol{c}) \subset S_1 + S_2 &= (\boldsymbol{a}_1 - \boldsymbol{c}) + (\boldsymbol{a}_2 - \boldsymbol{c}) + (S_1 + S_2) \\ &= (\boldsymbol{a}_1 + S_1 - \boldsymbol{c}) + (\boldsymbol{a}_2 + S_2 - \boldsymbol{c}) = (L_1 - \boldsymbol{c}) + (L_2 - \boldsymbol{c}), \end{aligned}$$

and Theorem 2.8 shows that $(L_1 - \boldsymbol{c}) \cap (L_2 - \boldsymbol{c}) \neq \varnothing$. Finally,

$$L_1 \cap L_2 = \boldsymbol{c} + (L_1 - \boldsymbol{c}) \cap (L_2 - \boldsymbol{c}) \neq \varnothing. \qquad \square$$

2.1.5 *Hyperplanes*

Definition 2.16. A *hyperplane* in $\mathbb{R}^n$ is a plane of dimension $n-1$.

Theorem 2.17. *A set $H \subset \mathbb{R}^n$ is a hyperplane if and only if there is a nonzero vector $\boldsymbol{e} \in \mathbb{R}^n$ and a scalar γ such that*

$$H = \{\boldsymbol{x} \in \mathbb{R}^n : \boldsymbol{x} \cdot \boldsymbol{e} = \gamma\}. \tag{2.5}$$

Furthermore, this representation of H is unique up to a common nonzero scalar multiple of $\boldsymbol{e}$ and γ.

Proof. A combination of Definitions 2.1 and 2.16 implies that a set $H \subset \mathbb{R}^n$ is a hyperplane if and only if $H = \boldsymbol{a} + S$, where $\boldsymbol{a} \in \mathbb{R}^n$ and $S \subset \mathbb{R}^n$ is a subspace of dimension $n-1$. We know (see page 8) that S can be expressed as

$$S = \{\boldsymbol{u} \in \mathbb{R}^n : \boldsymbol{u} \cdot \boldsymbol{e} = 0\}, \tag{2.6}$$

where $\boldsymbol{e}$ is a suitable nonzero vector. Furthermore, this expression for S is unique up to a nonzero scalar multiple of $\boldsymbol{e}$. Put $\gamma = \boldsymbol{a} \cdot \boldsymbol{e}$. Then H is a hyperplane if and only if

$$\begin{aligned} H &= \boldsymbol{a} + S = \{\boldsymbol{a} + \boldsymbol{u} \in \mathbb{R}^n : \boldsymbol{u} \cdot \boldsymbol{e} = 0\} = \{\boldsymbol{x} \in \mathbb{R}^n : (\boldsymbol{x} - \boldsymbol{a}) \cdot \boldsymbol{e} = 0\} \\ &= \{\boldsymbol{x} \in \mathbb{R}^n : \boldsymbol{x} \cdot \boldsymbol{e} = \gamma\}. \end{aligned}$$

Assume now that H has another representation

$$H = \{\boldsymbol{x} \in \mathbb{R}^n : \boldsymbol{x} \cdot \boldsymbol{e}' = \gamma'\}$$

for a nonzero vector $\boldsymbol{e}' \in \mathbb{R}^n$ and a scalar γ'. Then $\boldsymbol{a} \cdot \boldsymbol{e}' = \gamma'$ because $\boldsymbol{a} \in H$. Furthermore,

$$\begin{aligned} S = H - \boldsymbol{a} &= \{\boldsymbol{x} - \boldsymbol{a} \in \mathbb{R}^n : \boldsymbol{x} \cdot \boldsymbol{e}' = \gamma'\} \\ &= \{\boldsymbol{u} \in \mathbb{R}^n : (\boldsymbol{u} + \boldsymbol{a}) \cdot \boldsymbol{e}' = \gamma'\} \\ &= \{\boldsymbol{u} \in \mathbb{R}^n : \boldsymbol{u} \cdot \boldsymbol{e}' = 0\}. \end{aligned}$$

A comparison of these equalities with (2.6) yields that $\boldsymbol{e}' = \lambda \boldsymbol{e}$ for a suitable scalar $\lambda \neq 0$. Consequently,

$$\gamma' = \boldsymbol{a} \cdot \boldsymbol{e}' = \boldsymbol{a} \cdot (\lambda \boldsymbol{e}) = \lambda(\boldsymbol{a} \cdot \boldsymbol{e}) = \lambda\gamma. \qquad \square$$

Theorem 2.18. *If a hyperplane $H \subset \mathbb{R}^n$ is given by (2.5), then the following assertions hold.*

(*a*) *The direction space and the orthospace of H are described as*

$$\operatorname{dir} H = \{\boldsymbol{x} \in \mathbb{R}^n : \boldsymbol{x} \cdot \boldsymbol{e} = 0\}, \quad \operatorname{ort} H = \operatorname{span}\{\boldsymbol{e}\} = \{\lambda \boldsymbol{e} : \lambda \in \mathbb{R}\}.$$

(*b*) *A set* $H' \subset \mathbb{R}^n$ *is a translate of* H *if and only if* H' *can be expressed as*

$$H' = \{\boldsymbol{x} \in \mathbb{R}^n : \boldsymbol{x}\cdot\boldsymbol{e} = \gamma'\}, \quad \gamma' \in \mathbb{R}.$$

(*c*) *If hyperplanes* $H_1, \ldots, H_r \subset \mathbb{R}^n$ *are given by*

$$H_i = \{\boldsymbol{x} \in \mathbb{R}^n : \boldsymbol{x}\cdot\boldsymbol{e} = \gamma_i\}, \quad 1 \leqslant i \leqslant r,$$

and if $\mu_1, \ldots, \mu_r$ *are scalars, not all nonzero, then*

$$\mu_1 H_1 + \cdots + \mu_r H_r = \{\boldsymbol{x} \in \mathbb{R}^n : \boldsymbol{x}\cdot\boldsymbol{e} = \mu_1\gamma_1 + \cdots + \mu_r\gamma_r\}.$$

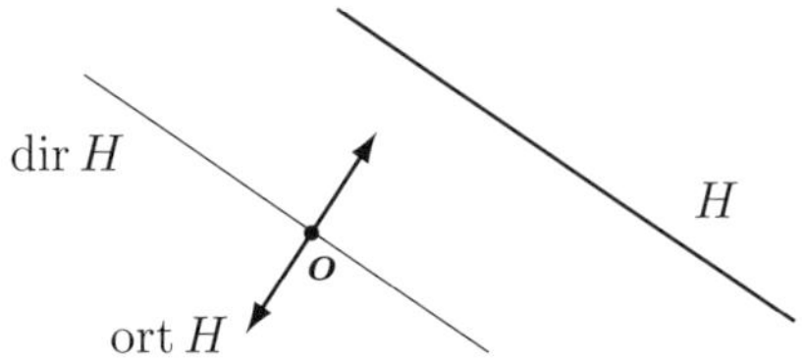

Fig. 2.6 Direction space and orthospace of a hyperplane H.

Proof. (*a*) This assertion follows from the proof of Theorem 2.17.

(*b*) Corollary 2.14 shows that H' is a translate of H if and only if $\operatorname{dir} H' = \operatorname{dir} H$. Given a vector $\boldsymbol{a}' \in H'$, Theorem 2.2 shows that

$$H' = \boldsymbol{a}' + \operatorname{dir} H' = \boldsymbol{a}' + \operatorname{dir} H.$$

Let $\gamma' = \boldsymbol{a}'\cdot\boldsymbol{e}$. Then H' is a translate of H if and only if

$$H' = \boldsymbol{a}' + \operatorname{dir} H = \boldsymbol{a}' + \{\boldsymbol{x} \in \mathbb{R}^n : \boldsymbol{x}\cdot\boldsymbol{e} = 0\} = \{\boldsymbol{x} \in \mathbb{R}^n : \boldsymbol{x}\cdot\boldsymbol{e} = \gamma'\}.$$

(*c*) Let $\boldsymbol{a}_i = (\gamma_i/\|\boldsymbol{e}\|^2)\boldsymbol{e}$, $1 \leqslant i \leqslant r$. Then

$$\boldsymbol{a}_i\cdot\boldsymbol{e} = \frac{\gamma_i}{\|\boldsymbol{e}\|^2}\boldsymbol{e}\cdot\boldsymbol{e} = \gamma_i, \quad 1 \leqslant i \leqslant r.$$

Consequently, $\boldsymbol{a}_i \in H_i$, $1 \leqslant i \leqslant r$. A combination of Corollary 2.14 and above assertion (*b*) shows that $H_i = \boldsymbol{a}_i + \operatorname{dir} H$ for all $1 \leqslant i \leqslant r$. Then

$$\mu_1 H_1 + \cdots + \mu_r H_r = (\mu_1\boldsymbol{a}_1 + \cdots + \mu_r\boldsymbol{a}_r) + (\mu_1 \operatorname{dir} H + \cdots + \mu_r \operatorname{dir} H).$$

We have $\mu_1 \operatorname{dir} H + \cdots + \mu_r \operatorname{dir} H = \operatorname{dir} H$ due to

$$\mu_i \operatorname{dir} H = \begin{cases} \{\boldsymbol{o}\} & \text{if } \mu_i = 0, \\ \operatorname{dir} H & \text{if } \mu_i \neq 0, \end{cases}$$

and the assumption that not all scalars $\mu_1, \dots, \mu_r$ are zero. Since

$$(\mu_1 \boldsymbol{a}_1 + \cdots + \mu_r \boldsymbol{a}_r) \cdot \boldsymbol{e} = \mu_1 \gamma_1 + \cdots + \mu_r \gamma_r,$$

we conclude that

$$\begin{aligned}\mu_1 H_1 + \cdots + \mu_r H_r &= (\mu_1 \boldsymbol{a}_1 + \cdots + \mu_r \boldsymbol{a}_r) + \operatorname{dir} H \\ &= (\mu_1 \boldsymbol{a}_1 + \cdots + \mu_r \boldsymbol{a}_r) + \{\boldsymbol{x} \in \mathbb{R}^n : \boldsymbol{x} \cdot \boldsymbol{e} = 0\} \\ &= \{\boldsymbol{x} \in \mathbb{R}^n : \boldsymbol{x} \cdot \boldsymbol{e} = \mu_1 \gamma_1 + \cdots + \mu_r \gamma_r\}. \qquad \square\end{aligned}$$

The next theorem provides useful criteria for a plane to be (or not to be) parallel to a given hyperplane. We recall that $X^{\perp}$ stands for the orthogonal complement of a set $X \subset \mathbb{R}^n$ (see page 7).

Theorem 2.19. *For a hyperplane $H \subset \mathbb{R}^n$ and a nonempty proper plane $L \subset \mathbb{R}^n$, the following assertions hold.*

(*a*) *H and L are parallel if and only if either $H \cap L = \varnothing$ or $L \subset H$.*
(*b*) *If H is given by* (2.5), *then H and L are parallel if and only if $\boldsymbol{e} \in \operatorname{ort} L$.*
(*c*) *H and L are not parallel if and only if $\varnothing \neq H \cap L \neq L$.*
(*d*) *H and L are not parallel if and only if $H + L = \mathbb{R}^n$.*
(*e*) *H and L are not parallel if and only if $H \cap L$ is a plane of dimension* $\dim L - 1$.
(*f*) *If H is given by* (2.5), *then $H \cap L$ is a plane of dimension* $\dim L - 1$ *if and only if $\boldsymbol{e} \notin \operatorname{ort} L$.*

Proof. In what follows, let $H = \boldsymbol{a} + S$ and $L = \boldsymbol{b} + T$, where $\boldsymbol{a}$ and $\boldsymbol{b}$ are points and S and T are the respective direction spaces (see Definition 2.3).

(*a*) Assume first that H and L are parallel. Then $T \subset S$ according to Theorem 2.13. If $H \cap L \neq \varnothing$ and $\boldsymbol{c}$ is a point in $H \cap L$, then, due to Theorem 2.2, $L = \boldsymbol{c} + T \subset \boldsymbol{c} + S = H$.

Conversely, suppose that either $H \cap L = \varnothing$ or $L \subset H$. Since the case $L \subset H$ obviously implies the parallelism of L and H, we may suppose that $H \cap L = \varnothing$. Assume, for contradiction, that H and L are not parallel. Then $T \not\subset S$ by Theorem 2.13. Choose a vector $\boldsymbol{b}_1 \in T \setminus S$ and a basis $\{\boldsymbol{b}_2, \dots, \boldsymbol{b}_n\}$ for S. Then $\{\boldsymbol{b}_1, \dots, \boldsymbol{b}_n\}$ is a basis for $\mathbb{R}^n$. Therefore, $\boldsymbol{b} - \boldsymbol{a}$ can be expressed as a linear combination $\boldsymbol{b} - \boldsymbol{a} = \lambda_1 \boldsymbol{b}_1 + \cdots + \lambda_n \boldsymbol{b}_n$. Put $\boldsymbol{c} = \lambda_2 \boldsymbol{b}_2 + \cdots + \lambda_n \boldsymbol{b}_n$. Clearly, $\boldsymbol{c} \in S$ and $\boldsymbol{b} - \lambda_1 \boldsymbol{b}_1 = \boldsymbol{a} + \boldsymbol{c}$. Furthermore,

$$\boldsymbol{b} - \lambda_1 \boldsymbol{b}_1 \in \boldsymbol{b} + T = L \quad \text{and} \quad \boldsymbol{a} + \boldsymbol{c} \in \boldsymbol{a} + S = H,$$

contrary to the assumption $H \cap L = \varnothing$. Hence H and L are parallel.

(*b*) Suppose that H is given by (2.5). By Corollary 2.14, H and L are parallel if and only if $T \subset S$, which is equivalent to $\operatorname{ort} H = S^{\perp} \subset T^{\perp} =$

$\operatorname{ort} L$. Since $\operatorname{ort} H = \operatorname{span}\{\boldsymbol{e}\}$, the inclusion $\operatorname{ort} H \subset \operatorname{ort} L$ is equivalent to $\boldsymbol{e} \in \operatorname{ort} L$.

(c) This assertion is the contrapositive of (a).

(d) According to Theorem 2.13, H and L are not parallel if and only if $T \not\subset S$. Because $\dim S = n - 1$, the condition $T \not\subset S$ is equivalent to $S + T = \mathbb{R}^n$, or, to

$$H + L = (\boldsymbol{a} + \boldsymbol{b}) + (S + T) = \mathbb{R}^n.$$

(e) Assume first that H and L are not parallel. Then $H + L = \mathbb{R}^n$ by statement (d) above, and Theorem 2.8 gives $H \cap L \neq \varnothing$. Therefore, from Theorem 2.7 we obtain

$$\begin{aligned}\dim (H \cap L) &= \dim H + \dim L - \dim \mathbb{R}^n \\ &= (n-1) + \dim L - n = \dim L - 1.\end{aligned}$$

Conversely, if $H \cap L$ is a plane of dimension $\dim L - 1$, then $\varnothing \neq H \cap L \neq L$, and assertion ($c$) shows that H and L are not parallel.

(f) This assertion follows from assertions (b) and (e). □

The theorem below describes in geometric terms the fact that a nonempty plane in $\mathbb{R}^n$ is the solution set of a system of linear equations.

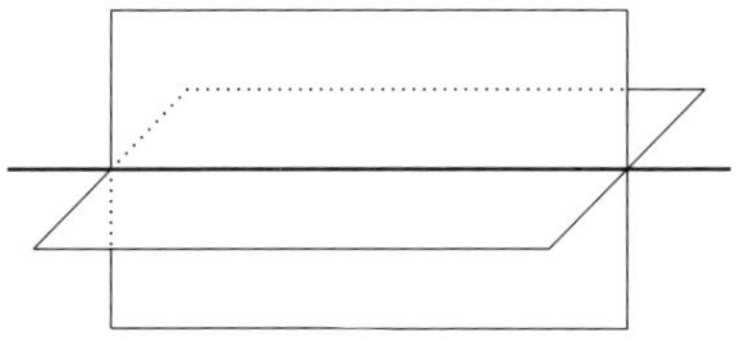

Fig. 2.7 A line as the intersection of two planes in $\mathbb{R}^3$.

Theorem 2.20. *Any plane $L \subset \mathbb{R}^n$ of dimension m, where $0 \leqslant m \leqslant n-1$, can be expressed as the intersection of $n - m$ suitable hyperplanes. Furthermore, if L is the intersection of a family $\mathcal{H}$ of hyperplanes, then $\mathcal{H}$ contains a subfamily of $n - m$ members whose intersection is L.*

Proof. Let $L = \boldsymbol{a} + S$, where $S = \operatorname{dir} L$. We know (see page 6) that S can be expressed as the intersection of suitable $(n-1)$-dimensional subspaces $S_1, \dots, S_{n-m} \subset \mathbb{R}^n$. The sets $G_i = \boldsymbol{a} + S_i$, $1 \leqslant i \leqslant n - m$, are hyperplanes, and

$$\begin{aligned} L &= \boldsymbol{a} + S = \boldsymbol{a} + S_1 \cap \cdots \cap S_{n-m} \\ &= (\boldsymbol{a} + S_1) \cap \cdots \cap (\boldsymbol{a} + S_{n-m}) \\ &= G_1 \cap \cdots \cap G_{n-m}. \end{aligned}$$

Suppose that L is the intersection of a family $\mathcal{H} = \{H_\alpha\}$ of hyperplanes. Choose a point $\boldsymbol{c} \in L$. According to Theorem 2.2, all sets $L - \boldsymbol{c}$ and $H_\alpha - \boldsymbol{c}$, where $H_\alpha \in \mathcal{H}$, are subspaces. Clearly,

$$L - \boldsymbol{c} = \bigcap_\alpha H_\alpha - \boldsymbol{c} = \bigcap_\alpha (H_\alpha - \boldsymbol{c}),$$

$$\dim (L - \boldsymbol{c}) = m, \quad \dim (H_\alpha - \boldsymbol{c}) = n - 1 \quad \text{for all} \quad H_\alpha \in \mathcal{H}.$$

Consequently, the family $\{H_\alpha - \boldsymbol{c} : H_\alpha \in \mathcal{H}\}$ contains $n - m$ members, say $H_1 - \boldsymbol{c}, \ldots, H_{n-m} - \boldsymbol{c}$, whose intersection is $L - \boldsymbol{c}$. Finally,

$$\begin{aligned} L &= \boldsymbol{c} + (L - \boldsymbol{c}) = \boldsymbol{c} + (H_1 - \boldsymbol{c}) \cap \cdots \cap (H_{n-m} - \boldsymbol{c}) \\ &= H_1 \cap \cdots \cap H_{n-m}. \end{aligned}$$ □

Corollary 2.21. *Every plane* L *in* $\mathbb{R}^n$ *is a closed set.*

Proof. The assertion is obvious if $L = \varnothing$ or $L = \mathbb{R}^n$. Assume that $\varnothing \neq L \neq \mathbb{R}^n$. Suppose first that L is a hyperplane. By Theorem 2.17, L can be expressed as $L = \{\boldsymbol{x} \in \mathbb{R}^n : \boldsymbol{x} \cdot \boldsymbol{e} = \gamma\}$. Because the linear functional $\varphi(\boldsymbol{x}) = \boldsymbol{x} \cdot \boldsymbol{e}$ is continuous on $\mathbb{R}^n$, the hyperplane L is closed as the inverse image $\varphi^{-1}(\gamma)$ of the closed set $\{\gamma\}$ (see Problems 1.17 and 1.18).

If $\dim L \leqslant n - 2$, then, by Theorem 2.20, L is the intersection of $n - m$ hyperplanes, say $H_1, \ldots, H_{n-m}$. Since all hyperplanes $H_1, \ldots, H_{n-m}$ are closed sets, their intersection also is closed. □

2.2 Lines and Halfspaces

2.2.1 *Lines, Halflines, and Segments*

Definition 2.22. The *line* through distinct points $\boldsymbol{x}$ and $\boldsymbol{y}$ in $\mathbb{R}^n$, denoted $\langle \boldsymbol{x}, \boldsymbol{y} \rangle$, is defined by

$$\langle \boldsymbol{x}, \boldsymbol{y} \rangle = \{(1 - \lambda)\boldsymbol{x} + \lambda \boldsymbol{y} : \lambda \in \mathbb{R}\}. \tag{2.7}$$

Remarks. 1. For distinct points $\boldsymbol{x}, \boldsymbol{y} \in \mathbb{R}^n$, the mapping $\lambda \mapsto (1-\lambda)\boldsymbol{x} + \lambda \boldsymbol{y}$ is an invertible mapping from $\mathbb{R}$ onto the line $\langle \boldsymbol{x}, \boldsymbol{y} \rangle$. Consequently, any point $\boldsymbol{u} \in \langle \boldsymbol{x}, \boldsymbol{y} \rangle$ is uniquely expressible in the form $\boldsymbol{u} = (1 - \lambda)\boldsymbol{x} + \lambda \boldsymbol{y}$, where $\lambda \in \mathbb{R}$.

2. Rewriting (2.7) as $\langle \boldsymbol{x}, \boldsymbol{y} \rangle = \boldsymbol{x} + \{\lambda(\boldsymbol{y} - \boldsymbol{x}) : \lambda \in \mathbb{R}\}$, we see that $\langle \boldsymbol{x}, \boldsymbol{y} \rangle$ is a translate of the 1-dimensional subspace span $\{\boldsymbol{y} - \boldsymbol{x}\}$ (compare with Definition 2.5).

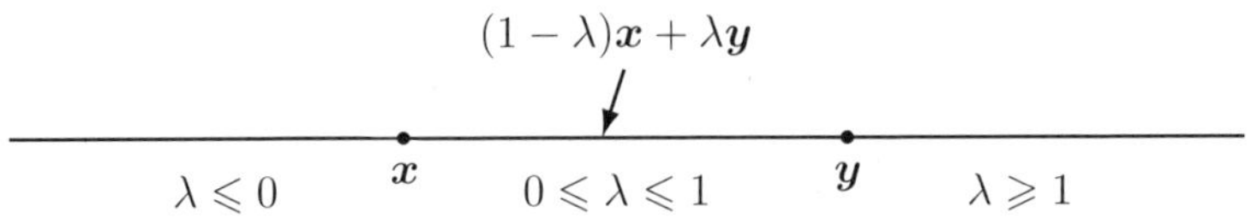

Fig. 2.8 The line through distinct points $\boldsymbol{x}$ and $\boldsymbol{y}$.

Definition 2.23. For distinct points $\boldsymbol{x}$ and $\boldsymbol{y}$ in $\mathbb{R}^n$, the *closed halflines* $[\boldsymbol{x}, \boldsymbol{y}\rangle$ and $\langle \boldsymbol{x}, \boldsymbol{y}]$, and the *open halflines* $(\boldsymbol{x}, \boldsymbol{y}\rangle$ and $\langle \boldsymbol{x}, \boldsymbol{y})$ are defined, respectively, as

$$[\boldsymbol{x}, \boldsymbol{y}\rangle = \{(1-\lambda)\boldsymbol{x} + \lambda\boldsymbol{y} : \lambda \geqslant 0\}, \quad (\boldsymbol{x}, \boldsymbol{y}\rangle = \{(1-\lambda)\boldsymbol{x} + \lambda\boldsymbol{y} : \lambda > 0\},$$
$$\langle \boldsymbol{x}, \boldsymbol{y}] = \{(1-\lambda)\boldsymbol{x} + \lambda\boldsymbol{y} : \lambda \leqslant 1\}, \quad \langle \boldsymbol{x}, \boldsymbol{y}) = \{(1-\lambda)\boldsymbol{x} + \lambda\boldsymbol{y} : \lambda < 1\}.$$

Definition 2.24. For distinct points $\boldsymbol{x}$ and $\boldsymbol{y}$ in $\mathbb{R}^n$, the *closed segment* $[\boldsymbol{x}, \boldsymbol{y}]$, the *semi-open segments* $[\boldsymbol{x}, \boldsymbol{y})$ and $(\boldsymbol{x}, \boldsymbol{y}]$, and the *open segment* $(\boldsymbol{x}, \boldsymbol{y})$ are defined, respectively, as

$$\begin{aligned}
[\boldsymbol{x}, \boldsymbol{y}] &= \{(1-\lambda)\boldsymbol{x} + \lambda\boldsymbol{y} : 0 \leqslant \lambda \leqslant 1\}, \\
[\boldsymbol{x}, \boldsymbol{y}) &= \{(1-\lambda)\boldsymbol{x} + \lambda\boldsymbol{y} : 0 \leqslant \lambda < 1\}, \\
(\boldsymbol{x}, \boldsymbol{y}] &= \{(1-\lambda)\boldsymbol{x} + \lambda\boldsymbol{y} : 0 < \lambda \leqslant 1\}, \\
(\boldsymbol{x}, \boldsymbol{y}) &= \{(1-\lambda)\boldsymbol{x} + \lambda\boldsymbol{y} : 0 < \lambda < 1\}.
\end{aligned}$$

For any point $\boldsymbol{x} \in \mathbb{R}^n$, we let

$$[\boldsymbol{x}, \boldsymbol{x}] = \{\boldsymbol{x}\} \quad \text{and} \quad [\boldsymbol{x}, \boldsymbol{x}) = (\boldsymbol{x}, \boldsymbol{x}] = (\boldsymbol{x}, \boldsymbol{x}) = \varnothing.$$

As expected, $[\boldsymbol{x}, \boldsymbol{y}\rangle = \langle \boldsymbol{y}, \boldsymbol{x}]$. Indeed,

$$\begin{aligned}
[\boldsymbol{x}, \boldsymbol{y}\rangle &= \{(1-\lambda)\boldsymbol{x} + \lambda\boldsymbol{y} : \lambda \geqslant 0\} = \{\mu\boldsymbol{x} + (1-\mu)\boldsymbol{y} : \mu = 1 - \lambda \leqslant 1\} \\
&= \{(1-\mu)\boldsymbol{y} + \mu\boldsymbol{x} : \mu \leqslant 1\} = \langle \boldsymbol{y}, \boldsymbol{x}].
\end{aligned}$$

In a similar way,

$$(\boldsymbol{x}, \boldsymbol{y}\rangle = \langle \boldsymbol{y}, \boldsymbol{x}), \quad [\boldsymbol{x}, \boldsymbol{y}] = [\boldsymbol{y}, \boldsymbol{x}], \quad [\boldsymbol{x}, \boldsymbol{y}) = (\boldsymbol{y}, \boldsymbol{x}].$$

Furthermore,

$$\boldsymbol{a} + \mu[\boldsymbol{x}, \boldsymbol{y}] = [\boldsymbol{a} + \mu\boldsymbol{x}, \boldsymbol{a} + \mu\boldsymbol{y}], \quad \boldsymbol{a} + \mu[\boldsymbol{x}, \boldsymbol{y}\rangle = [\boldsymbol{a} + \mu\boldsymbol{x}, \boldsymbol{a} + \mu\boldsymbol{y}\rangle$$

whenever $\boldsymbol{a} \in \mathbb{R}^n$ and $\mu \in \mathbb{R}$. Similar assertions hold for all other types of segments and halflines.

Lemmas 2.25–2.27 below describe some standard properties of halflines and segments. Although these properties are intuitively clear and can be justified by suitable pictures, we provide their formal proofs in Problem 2.9.

Lemma 2.25. *For distinct points $\boldsymbol{x}$ and $\boldsymbol{y}$ in $\mathbb{R}^n$, the following assertions hold.*

(a) *If $\boldsymbol{u}$ and $\boldsymbol{v}$ are distinct points in $\langle \boldsymbol{x}, \boldsymbol{y} \rangle$, then $\langle \boldsymbol{u}, \boldsymbol{v} \rangle = \langle \boldsymbol{x}, \boldsymbol{y} \rangle$.*
(b) *If $\boldsymbol{u} \in (\boldsymbol{x}, \boldsymbol{y}\rangle$, then $[\boldsymbol{x}, \boldsymbol{u}\rangle = [\boldsymbol{x}, \boldsymbol{y}\rangle$ and $(\boldsymbol{x}, \boldsymbol{u}\rangle = (\boldsymbol{x}, \boldsymbol{y}\rangle$.*
(c) *If $\boldsymbol{w} \in \langle \boldsymbol{x}, \boldsymbol{y} \rangle$, then $[\boldsymbol{x}, \boldsymbol{y}] \subset [\boldsymbol{x}, \boldsymbol{w}] \cup [\boldsymbol{w}, \boldsymbol{y}]$.*
(d) *If distinct points $\boldsymbol{u}, \boldsymbol{v} \in \langle \boldsymbol{x}, \boldsymbol{y} \rangle$ are expressed as*

$$\boldsymbol{u} = (1 - \lambda)\boldsymbol{x} + \lambda \boldsymbol{y} \quad \text{and} \quad \boldsymbol{v} = (1 - \mu)\boldsymbol{x} + \mu\, \boldsymbol{y}, \tag{2.8}$$

then $\boldsymbol{u} \in (\boldsymbol{x}, \boldsymbol{v})$ if and only if one of the conditions $0 < \lambda < \mu$ and $\mu < \lambda < 0$ is satisfied. □

Lemma 2.26. *For points $\boldsymbol{x}, \boldsymbol{y}$, and $\boldsymbol{z}$ in $\mathbb{R}^n$, the following assertions hold.*

(a) *If $\boldsymbol{u} \in [\boldsymbol{x}, \boldsymbol{y}]$ and $\boldsymbol{v} \in [\boldsymbol{x}, \boldsymbol{z}]$, then the closed segments $[\boldsymbol{u}, \boldsymbol{z}]$ and $[\boldsymbol{v}, \boldsymbol{y}]$ meet.*
(b) *If $\boldsymbol{q} \in [\boldsymbol{u}, \boldsymbol{z}]$, then $\boldsymbol{q} \in [\boldsymbol{w}, \boldsymbol{y}]$ for a suitable point $\boldsymbol{w} \in [\boldsymbol{x}, \boldsymbol{z}]$.* □

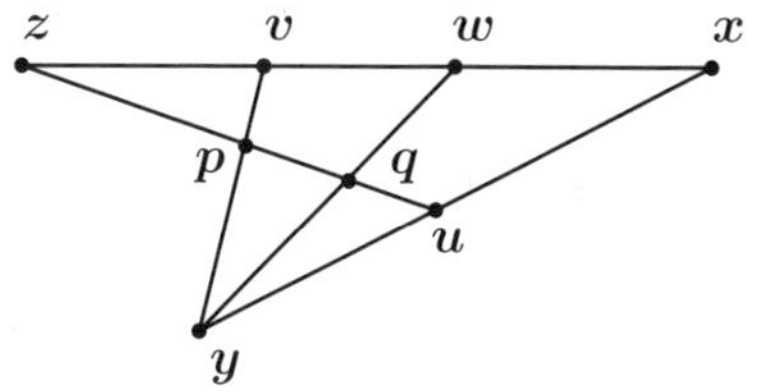

Lemma 2.27. *If closed segments $[\boldsymbol{u}_1, \boldsymbol{v}_1]$ and $[\boldsymbol{u}_2, \boldsymbol{v}_2]$ in $\mathbb{R}^n$ meet at a point $\boldsymbol{z}$, then, for any point $\boldsymbol{u} \in [\boldsymbol{u}_1, \boldsymbol{u}_2]$, there is a point $\boldsymbol{v} \in [\boldsymbol{v}_1, \boldsymbol{v}_2]$ such that $\boldsymbol{z} \in [\boldsymbol{u}, \boldsymbol{v}]$.* □

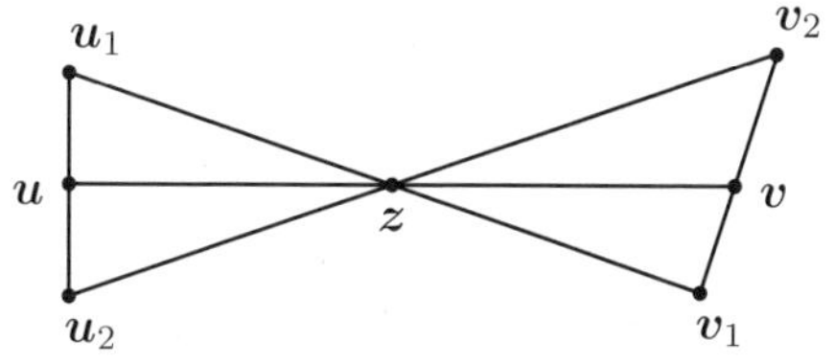

Lemma 2.28. *Let $\boldsymbol{x}, \boldsymbol{y}$, and $\boldsymbol{z}$ be pairwise distinct points in $\mathbb{R}^n$ such that $\boldsymbol{x} \notin (\boldsymbol{y}, \boldsymbol{z})$. Given points $\boldsymbol{u} \in (\boldsymbol{y}, \boldsymbol{z})$, $\boldsymbol{y}' \in (\boldsymbol{x}, \boldsymbol{y}\rangle$, and $\boldsymbol{z}' \in (\boldsymbol{x}, \boldsymbol{z}'\rangle$, the open segment $(\boldsymbol{y}', \boldsymbol{z}')$ meets the open halfline $(\boldsymbol{x}, \boldsymbol{u}\rangle$.* □

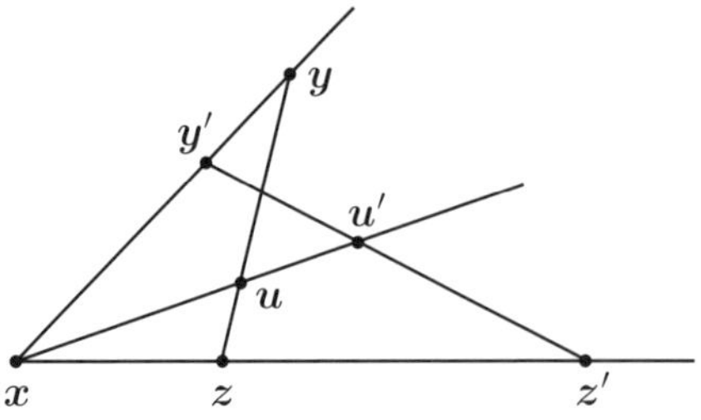

2.2.2 *Halfspaces*

Definition 2.29. If a hyperplane $H \subset \mathbb{R}^n$ is given by (2.5), then the sets

$$V_1 = \{\boldsymbol{x} \in \mathbb{R}^n : \boldsymbol{x}\cdot\boldsymbol{e} \leqslant \gamma\} \quad \text{and} \quad V_2 = \{\boldsymbol{x} \in \mathbb{R}^n : \boldsymbol{x}\cdot\boldsymbol{e} \geqslant \gamma\} \tag{2.9}$$

are called the opposite *closed halfspaces* determined by H, and the sets

$$W_1 = \{\boldsymbol{x} \in \mathbb{R}^n : \boldsymbol{x}\cdot\boldsymbol{e} < \gamma\} \quad \text{and} \quad W_2 = \{\boldsymbol{x} \in \mathbb{R}^n : \boldsymbol{x}\cdot\boldsymbol{e} > \gamma\} \tag{2.10}$$

are called the opposite *open halfspaces* determined by H.

Theorems 2.17 and 2.18 imply the following assertions.

Corollary 2.30. *Let closed halfspaces V_1 and V_2 of $\mathbb{R}^n$ be given by (2.9) and open halfspaces W_1 and W_2 of $\mathbb{R}^n$ be given by (2.10). The following assertions hold.*

(*a*) *The expressions (2.9) and (2.10) are unique up to a common positive multiple of $\boldsymbol{e}$ and γ.*

(*b*) *Sets V_1' and V_2' are, respectively, translates of V_1 and V_2 if and only if they can be expressed as*

$$V_1' = \{\boldsymbol{x} \in \mathbb{R}^n : \boldsymbol{x}\cdot\boldsymbol{e} \leqslant \gamma'\} \quad \textit{and} \quad V_2' = \{\boldsymbol{x} \in \mathbb{R}^n : \boldsymbol{x}\cdot\boldsymbol{e} \geqslant \gamma'\}.$$

(*c*) *Sets W_1' and W_2' are, respectively, translates of W_1 and W_2 if and only if they can be expressed as*

$$W_1' = \{\boldsymbol{x} \in \mathbb{R}^n : \boldsymbol{x}\cdot\boldsymbol{e} < \gamma'\} \quad \textit{and} \quad W_2' = \{\boldsymbol{x} \in \mathbb{R}^n : \boldsymbol{x}\cdot\boldsymbol{e} > \gamma'\}.$$ □

Corollary 2.30 justifies the following definition.

Definition 2.31. In terms of descriptions (2.9) and (2.10), the closed halfline $\{\lambda\boldsymbol{e} : \lambda \geqslant 0\}$, denoted $\operatorname{nor} V_1$, is called the *normal halfline* of

V_1, and all nonzero vectors from $\operatorname{nor} V_1$ are called the *normal vectors* of V_1. We let $\operatorname{nor} W_1 = \operatorname{nor} V_1$.

Similarly, the closed halfline $\{\lambda \boldsymbol{e} : \lambda \leqslant 0\}$, denoted $\operatorname{nor} V_2$, is called the *normal halfline* of V_2, and all nonzero vectors from $\operatorname{nor} V_2$ are called the *normal vectors* of V_2. We let $\operatorname{nor} W_2 = \operatorname{nor} V_2$.

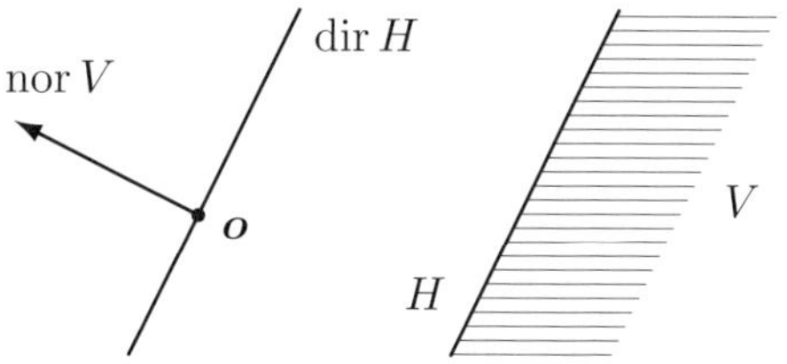

Fig. 2.9 Normal halfline of a closed halfspace V determined by the hyperplane H.

Remark. If a closed halfspace $V \subset \mathbb{R}^n$ is determined by a hyperplane $H \subset \mathbb{R}^n$, then $\operatorname{nor} V$ is a closed halfline of the line $\operatorname{ort} H$ (see Theorem 2.18).

The following result is useful in the study of support properties of convex sets.

Theorem 2.32. *If closed halfspaces $V_1, \ldots, V_r \subset \mathbb{R}^n$ are given by*

$$V_i = \{\boldsymbol{x} \in \mathbb{R}^n : \boldsymbol{x}\cdot\boldsymbol{e} \leqslant \gamma_i\}, \quad \boldsymbol{e} \neq \boldsymbol{o}, \quad 1 \leqslant i \leqslant r,$$

and if $\mu_1, \ldots, \mu_r$ are nonnegative scalars, not all nonzero, then

$$\mu_1 V_1 + \cdots + \mu_r V_r = \{\boldsymbol{x} \in \mathbb{R}^n : \boldsymbol{x}\cdot\boldsymbol{e} \leqslant \mu_1\gamma_1 + \cdots + \mu_r\gamma_r\}.$$

Proof. Consider the closed halfspace $V = \{\boldsymbol{x} \in \mathbb{R}^n : \boldsymbol{x}\cdot\boldsymbol{e} \leqslant 0\}$, and let $\boldsymbol{a}_i = (\gamma_i/\|\boldsymbol{e}\|^2)\boldsymbol{e}$, $1 \leqslant i \leqslant r$. Then

$$\boldsymbol{a}_i\cdot\boldsymbol{e} = \frac{\gamma_i}{\|\boldsymbol{e}\|^2}\boldsymbol{e}\cdot\boldsymbol{e} = \gamma_i, \quad 1 \leqslant i \leqslant r.$$

Consequently, $V_i = \boldsymbol{a}_i + V$, $1 \leqslant i \leqslant r$, which gives

$$\mu_1 V_1 + \cdots + \mu_r V_r = (\mu_1\boldsymbol{a}_1 + \cdots + \mu_r\boldsymbol{a}_r) + (\mu_1 V + \cdots + \mu_r V).$$

We have $\mu_1 V + \cdots + \mu_r V = V$ due to

$$\mu_i V = \begin{cases} \{\boldsymbol{o}\} & \text{if } \mu_i = 0, \\ V & \text{if } \mu_i > 0, \end{cases}$$

and the assumption that not all scalars $\mu_1, \dots, \mu_r$ are zero. Since

$$(\mu_1 \boldsymbol{a}_1 + \dots + \mu_r \boldsymbol{a}_r)\cdot \boldsymbol{e} = \mu_1\gamma_1 + \dots + \mu_r\gamma_r,$$

we conclude that

$$\begin{aligned}\mu_1 V_1 + \dots + \mu_r V_r &= (\mu_1 \boldsymbol{a}_1 + \dots + \mu_r \boldsymbol{a}_r) + V \\ &= (\mu_1 \boldsymbol{a}_1 + \dots + \mu_r \boldsymbol{a}_r) + \{\boldsymbol{x} \in \mathbb{R}^n : \boldsymbol{x}\cdot\boldsymbol{e} \leqslant 0\} \\ &= \{\boldsymbol{x} \in \mathbb{R}^n : \boldsymbol{x}\cdot\boldsymbol{e} \leqslant \mu_1\gamma_1 + \dots + \mu_r\gamma_r\}. \qquad \square\end{aligned}$$

Topological properties of halfspaces are described in the next theorem.

Theorem 2.33. *Let a hyperplane $H \subset \mathbb{R}^n$ be described by* (2.5). *The halfspaces V_1 and V_2 given by* (2.9) *are closed sets, and the halfspaces W_1 and W_2 given by* (2.10) *are open sets. Furthermore,*

$$V_i = \operatorname{cl} W_i, \quad W_i = \operatorname{int} V_i, \quad H = \operatorname{bd} V_i = \operatorname{bd} W_i, \quad i = 1, 2. \tag{2.11}$$

Proof. According to Problem 1.18, the linear functional $\varphi(\boldsymbol{x}) = \boldsymbol{x}\cdot\boldsymbol{e}$ is continuous on $\mathbb{R}^n$. Therefore, the halfspaces (2.9) are closed sets as inverse φ^{-1}-images of the closed halflines $(-\infty, \gamma]$ and $[\gamma, \infty)$ of $\mathbb{R}$ (see Problem 1.17). Similarly, the halfspaces (2.10) are open sets as φ^{-1}-images of the respective open halflines of $\mathbb{R}$.

For assertions (2.11), it suffices to consider the case $i = 1$. The inclusion $W_1 \subset V_1$ shows that $W_1 = \operatorname{int} W_1 \subset \operatorname{int} V_1$. Hence

$$\operatorname{bd} V_1 = V_1 \setminus \operatorname{int} V_1 \subset V_1 \setminus W_1 = H.$$

To prove the opposite inclusion, choose a point $\boldsymbol{u} \in H$ and a scalar $\rho > 0$. Let

$$\boldsymbol{v}_1 = \boldsymbol{u} - \frac{\rho}{2\|\boldsymbol{e}\|}\boldsymbol{e} \quad \text{and} \quad \boldsymbol{v}_2 = \boldsymbol{u} + \frac{\rho}{2\|\boldsymbol{e}\|}\boldsymbol{e}.$$

Since $\|\boldsymbol{v}_1 - \boldsymbol{u}\| = \|\boldsymbol{v}_2 - \boldsymbol{u}\| = \rho/2$, both points $\boldsymbol{v}_1$ and $\boldsymbol{v}_2$ belong to the open ball $U_\rho(\boldsymbol{u})$. Furthermore, the inequalities

$$\boldsymbol{v}_1\cdot\boldsymbol{e} = (\boldsymbol{u} - \frac{\rho}{2\|\boldsymbol{e}\|}\boldsymbol{e})\cdot\boldsymbol{e} = \gamma - \frac{\rho\|\boldsymbol{e}\|}{2} < \gamma$$

and

$$\boldsymbol{v}_2\cdot\boldsymbol{e} = (\boldsymbol{u} + \frac{\rho}{\|\boldsymbol{e}\|}\boldsymbol{e})\cdot\boldsymbol{e} = \gamma + \frac{\rho\|\boldsymbol{e}\|}{2} > \gamma$$

show that $\boldsymbol{v}_1 \in W_1 \subset V_1$ and $\boldsymbol{v}_2 \in W_2 = \mathbb{R}^n \setminus V_1$. Hence $U_\rho(\boldsymbol{u})$ meets both sets V_1 and $\mathbb{R}^n \setminus V_1$, implying the inclusion $\boldsymbol{u} \in \operatorname{bd} V_1$. Thus $H \subset \operatorname{bd} V_1$. Consequently,

$$\operatorname{bd} V_1 = H \quad \text{and} \quad \operatorname{int} V_1 = V_1 \setminus \operatorname{bd} V_1 = V_1 \setminus H = W_1.$$

Similarly, the inclusion $W_1 \subset V_1$ gives $\operatorname{cl} W_1 \subset V_1$. Therefore,

$$\operatorname{bd} W_1 = \operatorname{cl} W_1 \setminus W_1 \subset V_1 \setminus W_1 = H.$$

Repeating the argument involving points $\boldsymbol{v}_1$ and $\boldsymbol{v}_2$ defined above, we conclude that every point $\boldsymbol{u} \in H$ belongs to $\operatorname{bd} W_1$. Thus $H \subset \operatorname{bd} W_1$. Summing up,

$$\operatorname{bd} W_1 = H \quad \text{and} \quad V_1 = W_1 \cup H = W_1 \cup \operatorname{bd} W_1 = \operatorname{cl} W_1. \qquad \square$$

The next result complements Theorem 2.19.

Theorem 2.34. *For a hyperplane $H \subset \mathbb{R}^n$ and a nonempty proper plane $L \subset \mathbb{R}^n$, the following conditions are equivalent.*

(*a*) *H and L are parallel.*
(*b*) *Either $L \subset H$ or L lies in an open halfspace determined by H.*
(*c*) *L lies in a closed halfspace determined by H.*

Proof. Let $H = \boldsymbol{a}+S$ and $L = \boldsymbol{b}+T$, where S and T are the corresponding direction spaces. If H is given by (2.5), then $S = \{\boldsymbol{x} \in \mathbb{R}^n : \boldsymbol{x}\cdot\boldsymbol{e} = 0\}$ due to Theorem 2.18.

(*a*) $\Rightarrow$ (*b*) Let H and L be parallel. According to Theorem 2.13, $T \subset S$, and Theorem 2.19 shows that either $H \cap L = \varnothing$ or $L \subset H$. Since the case $L \subset H$ is obvious, we may assume that $H \cap L = \varnothing$. Then $\boldsymbol{b} \notin H$, because otherwise $L = \boldsymbol{b}+T \subset \boldsymbol{b}+S = H$ (see Theorem 2.2). Put $\beta = \boldsymbol{b}\cdot\boldsymbol{e}$. Clearly, $\beta \neq \gamma$ due to $\boldsymbol{b} \notin H$. Suppose that $\beta < \gamma$ (the case $\beta > \gamma$ is similar). We are going to prove that L lies in the open halfspace $W = \{\boldsymbol{x} \in \mathbb{R}^n : \boldsymbol{x}\cdot\boldsymbol{e} < \gamma\}$. Indeed, any point $\boldsymbol{x} \in L$ can be written as $\boldsymbol{x} = \boldsymbol{b}+\boldsymbol{x}'$, where $\boldsymbol{x}' \in T$, which gives

$$\boldsymbol{x}\cdot\boldsymbol{e} = \boldsymbol{b}\cdot\boldsymbol{e} + \boldsymbol{x}'\cdot\boldsymbol{e} = \beta + 0 < \gamma.$$

(*b*) $\Rightarrow$ (*a*) Since the case $L \subset H$ obviously implies that L is parallel to H, we may assume that L lies in an open halfspace determined by H. Then $H \cap L = \varnothing$, and Theorem 2.19 shows that H and L are parallel.

(*b*) $\Rightarrow$ (*c*) Denote by W and open halfspace determined by H and satisfying one of the conditions $L \subset H$ and $L \subset W$. Then L lies in the closed halfspace $V = H \cup W$, as desired.

(*c*) $\Rightarrow$ (*b*) Let V be a closed halfspace determined by H and containing L. According to Definition 2.31, we can write $V = \{\boldsymbol{x} \in \mathbb{R}^n : \boldsymbol{x}\cdot\boldsymbol{e} \leqslant \gamma\}$ (the case of opposite inequality is similar). Denote by W the open halfspace $V \setminus H$. Since the case $L \subset W$ is obvious, we may assume that L meets

H. Choose points $\boldsymbol{u} \in L \cap H$ and $\boldsymbol{v} \in L$, and let $\boldsymbol{w} = 2\boldsymbol{u} - \boldsymbol{v}$. Because $L = \boldsymbol{b} + T$, we can write

$$\boldsymbol{u} = \boldsymbol{b} + \boldsymbol{u}', \quad \boldsymbol{v} = \boldsymbol{b} + \boldsymbol{v}', \quad \text{where} \quad \boldsymbol{u}', \boldsymbol{v}' \in T.$$

Consequently,

$$\boldsymbol{w} = 2\boldsymbol{u} - \boldsymbol{v} = \boldsymbol{b} + (2\boldsymbol{u}' - \boldsymbol{v}') \in \boldsymbol{b} + T = L \subset V.$$

So, $\boldsymbol{u} \cdot \boldsymbol{e} = \gamma$, $\boldsymbol{v} \cdot \boldsymbol{e} \leqslant \gamma$, and $\boldsymbol{w} \cdot \boldsymbol{e} \leqslant \gamma$. Therefore,

$$\gamma \leqslant 2\gamma - \boldsymbol{v} \cdot \boldsymbol{e} = 2\boldsymbol{u} \cdot \boldsymbol{e} - \boldsymbol{v} \cdot \boldsymbol{e} = (2\boldsymbol{u} - \boldsymbol{v}) \cdot \boldsymbol{e} = \boldsymbol{w} \cdot \boldsymbol{e} \leqslant \gamma.$$

This argument gives $2\gamma - \boldsymbol{v} \cdot \boldsymbol{e} = \gamma$, or $\boldsymbol{v} \cdot \boldsymbol{e} = \gamma$. Hence $\boldsymbol{v} \in H$. Since $\boldsymbol{v}$ was chosen arbitrarily in L, we conclude that $L \subset H$. □

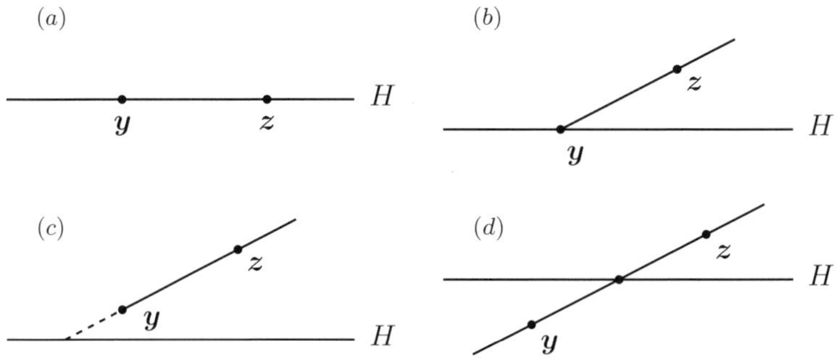

Fig. 2.10 Illustration to Theorem 2.35.

The theorem below plays an important role in dealing with various properties of convex sets.

Theorem 2.35. *For a hyperplane $H \subset \mathbb{R}^n$ and distinct points $\boldsymbol{y}, \boldsymbol{z} \in \mathbb{R}^n$, the following assertions hold.*

(a) *If both $\boldsymbol{y}$ and $\boldsymbol{z}$ belong to H, then the line $\langle \boldsymbol{y}, \boldsymbol{z} \rangle$ lies in H.*
(b) *If $\boldsymbol{y} \in H$ and $\boldsymbol{z}$ belongs to an open halfspace determined by H, then the open halfline $(\boldsymbol{y}, \boldsymbol{z}\rangle$ lies in this halfspace.*
(c) *If both $\boldsymbol{y}$ and $\boldsymbol{z}$ belong to an open halfspace determined by H, then at least one of the closed halflines $[\boldsymbol{y}, \boldsymbol{z}\rangle$, $\langle \boldsymbol{y}, \boldsymbol{z}]$ lies in this halfspace.*
(d) *If $\boldsymbol{y}$ and $\boldsymbol{z}$ belong, respectively, to the opposite open halfspaces determined by H, then $\langle \boldsymbol{y}, \boldsymbol{z} \rangle$ meets H at a single point, which belongs to the open segment $(\boldsymbol{y}, \boldsymbol{z})$.*

Proof. Suppose that H is given by (2.5). By Definition 2.22, any point $\boldsymbol{u} \in \langle \boldsymbol{y}, \boldsymbol{z} \rangle$ can be written as $\boldsymbol{u} = (1-\lambda)\boldsymbol{y} + \lambda \boldsymbol{z}$, where $\lambda \in \mathbb{R}$. Consequently,

$$\boldsymbol{u}\cdot\boldsymbol{e} = ((1-\lambda)\boldsymbol{y} + \lambda\boldsymbol{z})\cdot\boldsymbol{e} = (1-\lambda)\boldsymbol{y}\cdot\boldsymbol{e} + \lambda\,\boldsymbol{z}\cdot\boldsymbol{e}. \tag{2.12}$$

(*a*) If $\boldsymbol{y}, \boldsymbol{z} \in H$ and $\boldsymbol{u} \in \langle \boldsymbol{y}, \boldsymbol{z} \rangle$, then $\boldsymbol{y}\cdot\boldsymbol{e} = \boldsymbol{z}\cdot\boldsymbol{e} = \gamma$, and (2.12) gives the equality $\boldsymbol{u}\cdot\boldsymbol{e} = \gamma$. Hence $\langle \boldsymbol{y}, \boldsymbol{z} \rangle \subset H$.

(*b*) Assume that $\boldsymbol{y} \in H$ and $\boldsymbol{z}$ belongs to the open halfspace $W = \{\boldsymbol{x} \in \mathbb{R}^n : \boldsymbol{x}\cdot\boldsymbol{e} < \gamma\}$ (the case of the opposite open halfspace is similar). Thus $\boldsymbol{y}\cdot\boldsymbol{e} = \gamma$ and $\boldsymbol{z}\cdot\boldsymbol{e} < \gamma$. If $\boldsymbol{u} \in (\boldsymbol{y}, \boldsymbol{z}\rangle$, then $\lambda > 0$, and (2.12) gives

$$\boldsymbol{u}\cdot\boldsymbol{e} = \boldsymbol{y}\cdot\boldsymbol{e} + \lambda(\boldsymbol{z}\cdot\boldsymbol{e} - \boldsymbol{y}\cdot\boldsymbol{e}) < \boldsymbol{y}\cdot\boldsymbol{e} = \gamma.$$

Hence $(\boldsymbol{y}, \boldsymbol{z}\rangle \subset W$.

(*c*) Suppose that $\boldsymbol{y}$ and $\boldsymbol{z}$ belong to the same open halfspace W determined by H. As above, we may assume that $W = \{\boldsymbol{x} \in \mathbb{R}^n : \boldsymbol{x}\cdot\boldsymbol{e} < \gamma\}$. If $\boldsymbol{z}\cdot\boldsymbol{e} \leqslant \boldsymbol{y}\cdot\boldsymbol{e}$ and $\boldsymbol{u} \in [\boldsymbol{y}, \boldsymbol{z}\rangle$, then $\lambda \geqslant 0$, and (2.12) gives

$$\boldsymbol{u}\cdot\boldsymbol{e} = \boldsymbol{y}\cdot\boldsymbol{e} + \lambda(\boldsymbol{z}\cdot\boldsymbol{e} - \boldsymbol{y}\cdot\boldsymbol{e}) \leqslant \boldsymbol{y}\cdot\boldsymbol{e} < \gamma.$$

Hence $[\boldsymbol{y}, \boldsymbol{z}\rangle \subset W$. Similarly, if $\boldsymbol{y}\cdot\boldsymbol{e} \leqslant \boldsymbol{z}\cdot\boldsymbol{e}$ and $\boldsymbol{u} \in \langle \boldsymbol{y}, \boldsymbol{z}]$, then $\lambda \leqslant 1$, and (2.12) gives

$$\boldsymbol{u}\cdot\boldsymbol{e} = \boldsymbol{y}\cdot\boldsymbol{e} + \lambda(\boldsymbol{z}\cdot\boldsymbol{e} - \boldsymbol{y}\cdot\boldsymbol{e}) \leqslant \boldsymbol{y}\cdot\boldsymbol{e} + (\boldsymbol{z}\cdot\boldsymbol{e} - \boldsymbol{y}\cdot\boldsymbol{e}) = \boldsymbol{z}\cdot\boldsymbol{e} < \gamma.$$

Hence $\langle \boldsymbol{y}, \boldsymbol{z}] \subset W$.

(*d*) Let $\boldsymbol{y}$ and $\boldsymbol{z}$ belong, respectively, to the opposite open halfspaces determined by H. Without loss of generality, we assume that $\boldsymbol{y}\cdot\boldsymbol{e} < \gamma < \boldsymbol{z}\cdot\boldsymbol{e}$. Put

$$\mu = \frac{\gamma - \boldsymbol{y}\cdot\boldsymbol{e}}{\boldsymbol{z}\cdot\boldsymbol{e} - \boldsymbol{y}\cdot\boldsymbol{e}}. \tag{2.13}$$

Since $0 < \mu < 1$, Definition 2.24 shows that the point $\boldsymbol{u} = (1-\mu)\boldsymbol{y} + \mu\boldsymbol{z}$ belongs to $(\boldsymbol{y}, \boldsymbol{z})$. Combining (2.12) and (2.13), we obtain

$$\begin{aligned}\boldsymbol{u}\cdot\boldsymbol{e} &= (1-\mu)\boldsymbol{y}\cdot\boldsymbol{e} + \mu\boldsymbol{z}\cdot\boldsymbol{e} \\ &= \Big(1 - \frac{\gamma - \boldsymbol{y}\cdot\boldsymbol{e}}{\boldsymbol{z}\cdot\boldsymbol{e} - \boldsymbol{y}\cdot\boldsymbol{e}}\Big)\boldsymbol{y}\cdot\boldsymbol{e} + \frac{\gamma - \boldsymbol{y}\cdot\boldsymbol{e}}{\boldsymbol{z}\cdot\boldsymbol{e} - \boldsymbol{y}\cdot\boldsymbol{e}}\,\boldsymbol{z}\cdot\boldsymbol{e} = \gamma.\end{aligned}$$

Hence $\boldsymbol{u} \in H$. By assertion (*b*), the open halflines $(\boldsymbol{u}, \boldsymbol{y}\rangle$ and $(\boldsymbol{u}, \boldsymbol{z}\rangle$ lie, respectively, in distinct open halfspaces determined by H. So, $\boldsymbol{u}$ is the unique point from $\langle \boldsymbol{y}, \boldsymbol{z} \rangle$ which belongs to H. □

Corollary 2.36. *Let $\boldsymbol{y}$ and $\boldsymbol{z}$ be distinct points in a closed halfspace V determined by a hyperplane $H \subset \mathbb{R}^n$. If $(\boldsymbol{y}, \boldsymbol{z}) \cap H \neq \varnothing$, then $\langle \boldsymbol{y}, \boldsymbol{z} \rangle \subset H$.*

Proof. Choose a point $\boldsymbol{u} \in (\boldsymbol{y}, \boldsymbol{z}) \cap H$. By Theorem 2.35, the open halfline $h_1 = (\boldsymbol{u}, \boldsymbol{y}\rangle$ lies either in H (if $\boldsymbol{y} \in H$) or in the open halfspace $W = V \setminus H$ (if $\boldsymbol{y} \in W$). Therefore, $[\boldsymbol{u}, \boldsymbol{y}\rangle \subset V$. Similarly, $[\boldsymbol{u}, \boldsymbol{z}\rangle \subset V$. Consequently, the line $\langle \boldsymbol{y}, \boldsymbol{z} \rangle = h_1 \cup h_2$ lies in V. Because $\boldsymbol{u} \in \langle \boldsymbol{y}, \boldsymbol{z} \rangle \cap H$, Theorem 2.34 shows that $\langle \boldsymbol{y}, \boldsymbol{z} \rangle \subset H$. □

2.3 Affine Spans

2.3.1 *Planes and Affine Combinations of Points*

Definition 2.37. An *affine combination* of point $\boldsymbol{a}_1, \ldots, \boldsymbol{a}_r \in \mathbb{R}^n$ (or an affine combination of the set $\{\boldsymbol{a}_1, \ldots, \boldsymbol{a}_r\}$), $r \geqslant 1$, is a linear combination

$$\lambda_1 \boldsymbol{a}_1 + \cdots + \lambda_r \boldsymbol{a}_r, \quad \text{where} \quad \lambda_1 + \cdots + \lambda_r = 1.$$

Example. Given distinct points $\boldsymbol{x}$ and $\boldsymbol{y}$ in $\mathbb{R}^n$, the affine combinations of the set $\{\boldsymbol{x}, \boldsymbol{y}\}$ fulfill the line $\langle \boldsymbol{x}, \boldsymbol{y} \rangle$ (see Definition 2.22).

The next two theorems show that affine combinations provide suitable characteristic properties of planes.

Theorem 2.38. *A nonempty set $X \subset \mathbb{R}^n$ is a plane if and only if it contains all affine combinations $(1-\lambda)\boldsymbol{x} + \lambda \boldsymbol{y}$ whenever $\boldsymbol{x}, \boldsymbol{y} \in X$ and $\lambda \in \mathbb{R}$.*

Proof. Suppose first that X is a plane and express it as $X = \boldsymbol{a} + S$ for a suitable point $\boldsymbol{a} \in \mathbb{R}^n$ and a subspace $S \subset \mathbb{R}^n$. Choose points $\boldsymbol{x}, \boldsymbol{y} \in X$ and a scalar $\lambda \in \mathbb{R}$. Then $\boldsymbol{x} = \boldsymbol{a} + \boldsymbol{x}'$ and $\boldsymbol{y} = \boldsymbol{a} + \boldsymbol{y}'$, where $\boldsymbol{x}', \boldsymbol{y}' \in S$. Consequently,

$$(1-\lambda)\boldsymbol{x} + \lambda \boldsymbol{y} = \boldsymbol{a} + ((1-\lambda)\boldsymbol{x}' + \lambda \boldsymbol{y}') \in \boldsymbol{a} + S = X.$$

Conversely, assume that $(1-\lambda)\boldsymbol{x} + \lambda \boldsymbol{y} \in X$ whenever $\boldsymbol{x}, \boldsymbol{y} \in X$ and $\lambda \in \mathbb{R}$. Choose a point $\boldsymbol{c} \in X$ and let $T = X - \boldsymbol{c}$. Clearly, $\boldsymbol{o} = \boldsymbol{c} - \boldsymbol{c} \in T$. We are going to show that $(1-\lambda)\boldsymbol{x} + \lambda \boldsymbol{y} \in T$ for all $\boldsymbol{x}, \boldsymbol{y} \in T$ and $\lambda \in \mathbb{R}$. Indeed, expressing $\boldsymbol{x}$ and $\boldsymbol{y}$ as $\boldsymbol{x} = \boldsymbol{x}' - \boldsymbol{c}$ and $\boldsymbol{y} = \boldsymbol{y}' - \boldsymbol{c}$, with $\boldsymbol{x}', \boldsymbol{y}' \in X$, we obtain

$$(1-\lambda)\boldsymbol{x} + \lambda \boldsymbol{y} = ((1-\lambda)\boldsymbol{x}' + \lambda \boldsymbol{y}') - \boldsymbol{c} \in X - \boldsymbol{c} = T.$$

Based on this argument, for points $\boldsymbol{x}, \boldsymbol{y} \in T$ and a scalar λ, one has

$$\lambda \boldsymbol{x} = (1-\lambda)\boldsymbol{o} + \lambda \boldsymbol{x} \in T, \quad \boldsymbol{x} + \boldsymbol{y} = 2((1-\tfrac{1}{2})\boldsymbol{x} + \tfrac{1}{2}\boldsymbol{y}) \in T.$$

Hence T is a subspace, which shows that $X = \boldsymbol{c} + T$ is a plane. □

Remark. We can reformulate Theorem 2.38, saying that a set $X \subset \mathbb{R}^n$, with $\operatorname{card} X \geqslant 2$, is a plane if and only if with every pair of distinct points $\boldsymbol{x}$ and $\boldsymbol{y}$ it contains the line $\langle \boldsymbol{x}, \boldsymbol{y} \rangle$. See Problems 3.2 and 3.3 for refinements of this assertion.

Theorem 2.39. *Let $X \subset \mathbb{R}^n$ be a nonempty set and $r \geqslant 2$ be an integer. The set X is a plane if and only if it contains all affine combinations of every set of r points from X.*

Proof. Suppose that X is a plane. By induction on r, we are going to prove that X contains all affine combinations of every set of r points from X. The case $r = 2$ is confirmed in Theorem 2.38. Assuming that the induction hypothesis holds for all positive integers $r \leqslant k - 1$, where $k \geqslant 3$, choose an affine combination $\boldsymbol{y} = \lambda_1 \boldsymbol{x}_1 + \cdots + \lambda_k \boldsymbol{x}_k$ of points $\boldsymbol{x}_1, \ldots, \boldsymbol{x}_k \in X$. At least one of the scalars $\lambda_1, \ldots, \lambda_k$ is not 1, since otherwise $\lambda_1 + \cdots + \lambda_k = k \geqslant 3$, contrary to the hypothesis $\lambda_1 + \cdots + \lambda_k = 1$. Let, for instance, $\lambda_1 \neq 1$. By the induction hypothesis, the affine combination

$$\boldsymbol{z} = \frac{\lambda_2}{1 - \lambda_1} \boldsymbol{x}_2 + \cdots + \frac{\lambda_k}{1 - \lambda_1} \boldsymbol{x}_k$$

belongs to X. Finally, the equality $\boldsymbol{y} = \lambda_1 \boldsymbol{x}_1 + (1 - \lambda_1)\boldsymbol{z}$ and Theorem 2.38 imply the inclusion $\boldsymbol{y} \in X$.

Conversely, suppose that X contains all affine combinations of every set of r points from X. If $r = 2$, then Theorem 2.38 shows that X is a plane. Let $r \geqslant 3$. Choose points $\boldsymbol{x}, \boldsymbol{y} \subset X$ and append to them any points $\boldsymbol{x}_3, \ldots, \boldsymbol{x}_r$ from X. By the assumption, all affine combinations

$$(1 - \lambda)\boldsymbol{x} + \lambda \boldsymbol{y} = (1 - \lambda)\boldsymbol{x} + \lambda \boldsymbol{y} + 0\boldsymbol{x}_3 + \cdots + 0\boldsymbol{x}_r, \quad \lambda \in \mathbb{R},$$

belong to X. Now, the same Theorem 2.38 implies that X is a plane. □

Theorem 2.40. *Let $X \subset \mathbb{R}^n$ be a nonempty set and $r \geqslant 2$ be an integer. Then X is a plane if and only*

$$\mu_1 X + \cdots + \mu_r X = (\mu_1 + \cdots + \mu_r) X \tag{2.14}$$

for every choice of scalars $\mu_1, \ldots, \mu_r$, with $\mu_1 + \cdots + \mu_r \neq 0$.

Proof. Suppose that X is a plane. Because the inclusion

$$(\mu_1 + \cdots + \mu_r) X \subset \mu_1 X + \cdots + \mu_r X$$

holds for any set $X \subset \mathbb{R}^n$ and any choice of scalars $\mu_1, \ldots, \mu_r$, it suffices to prove the opposite inclusion. For this, put

$$\mu = \mu_1 + \cdots + \mu_r \quad \text{and} \quad \lambda_i = \mu_i / \mu, \quad 1 \leqslant i \leqslant r.$$

Then $\mu \neq 0$ and $\lambda_1 + \cdots + \lambda_r = 1$. According to Theorem 2.39,

$$\lambda_1 X + \cdots + \lambda_r X = \{\lambda_1 \boldsymbol{x}_1 + \cdots + \lambda_r \boldsymbol{x}_r : \boldsymbol{x}_1, \ldots, \boldsymbol{x}_r \in X\} \subset X.$$

Multiplying both sides of this inclusion by μ, we obtain that

$$\mu_1 X + \cdots + \mu_r X \subset (\mu_1 + \cdots + \mu_r) X.$$

Hence the equality (2.14) holds.

Conversely, let X satisfy (2.14). Choose an affine combination

$$\mu_1 \boldsymbol{x}_1 + \cdots + \mu_r \boldsymbol{x}_r, \quad \mu_1 + \cdots + \mu_r = 1,$$

of points $\boldsymbol{x}_1, \ldots, \boldsymbol{x}_r \in X$. Then

$$\mu_1 \boldsymbol{x}_1 + \cdots + \mu_r \boldsymbol{x}_r \in \mu_1 X + \cdots + \mu_r X = (\mu_1 + \cdots + \mu_r) X = X,$$

and Theorem 2.39 implies that X is a plane. □

2.3.2 *Basic Properties of Affine Spans*

Definition 2.41. For a given set $X \subset \mathbb{R}^n$, the intersection of all planes containing X is called the *affine span* of X and denoted $\operatorname{aff} X$.

For instance, the affine span of a point $\boldsymbol{x} \in \mathbb{R}^n$ is the singleton $\{\boldsymbol{x}\}$, the affine span of two distinct points $\boldsymbol{x}, \boldsymbol{y} \in \mathbb{R}^n$ is the line $\langle \boldsymbol{x}, \boldsymbol{y} \rangle$, and the affine span of three non-collinear points is the 2-dimensional plane through these points.

Theorem 2.4 implies that the affine span of any set $X \subset \mathbb{R}^n$ exists and is the smallest plane containing X. Furthermore, $\operatorname{aff} \varnothing = \varnothing$. Some elementary properties of affine spans, whose proofs do not involve affine combinations of points, are described below.

Theorem 2.42. *For sets X and Y in $\mathbb{R}^n$, the following assertions hold.*

(a) $X \subset \operatorname{aff} X$, *with* $X = \operatorname{aff} X$ *if and only if* X *is a plane.*
(b) $\operatorname{aff} (\operatorname{aff} X) = \operatorname{aff} X$.
(c) $\operatorname{aff} X \subset \operatorname{span} X = \operatorname{aff} (\{\boldsymbol{o}\} \cup X)$.
(d) $\operatorname{aff} X = \operatorname{span} X$ *if and only if* $\boldsymbol{o} \in \operatorname{aff} X$.
(e) $\operatorname{aff} X \subset \operatorname{aff} Y$ *if* $X \subset Y$.
(f) $\operatorname{aff} X = \operatorname{aff} Y$ *if* $X \subset Y \subset \operatorname{aff} X$.
(g) $\operatorname{aff} X = \operatorname{aff} (X \cup Z)$ *if and only if* $Z \subset \operatorname{aff} X$.

Furthermore, if $\{X_\alpha\}$ is a family of sets in $\mathbb{R}^n$, then the assertions below are true.

(h) $\operatorname{aff} (\bigcap_\alpha X_\alpha) \subset \bigcap_\alpha \operatorname{aff} X_\alpha$.
(i) $\bigcup_\alpha \operatorname{aff} X_\alpha \subset \operatorname{aff} (\bigcup_\alpha X_\alpha)$.
(j) $\bigcup_\alpha \operatorname{aff} X_\alpha = \operatorname{aff} (\bigcup_\alpha X_\alpha)$ *if the family* $\{X_\alpha\}$ *is nested.*
(k) $\operatorname{aff} (\bigcup_\alpha \operatorname{aff} X_\alpha) = \operatorname{aff} (\bigcup_\alpha X_\alpha)$.

Proof. Let $\mathcal{P}(X)$ denote the family of all planes containing a given set $X \subset \mathbb{R}^n$. The proofs of assertions (a)–(k) derive from the following simple arguments.

1. $\operatorname{aff} X$ is the smallest element in $\mathcal{P}(X)$.
2. $\mathcal{P}(X)$ is exactly the family of all planes containing $\operatorname{aff} X$.
3. $\mathcal{P}(\{\boldsymbol{o}\} \cup X)$ is exactly the family of all subspaces containing X.
4. If $X \subset Y$, then $\mathcal{P}(Y) \subset \mathcal{P}(X)$. □

Corollary 2.43. *If X is a set in $\mathbb{R}^n$, then* $\operatorname{aff} X = \operatorname{aff} (\operatorname{cl} X)$. *Consequently,* $\operatorname{aff} X = \operatorname{aff} Y$ *for any dense subset Y of X.*

Proof. Since the plane aff X is a closed set (see Corollary 2.21), the inclusion $X \subset \operatorname{aff} X$ gives $X \subset \operatorname{cl} X \subset \operatorname{aff} X$. By Theorem 2.42,

$$\operatorname{aff} X \subset \operatorname{aff}(\operatorname{cl} X) \subset \operatorname{aff}(\operatorname{aff} X) = \operatorname{aff} X.$$

Hence $\operatorname{aff} X = \operatorname{aff}(\operatorname{cl} X)$. If Y is a dense subset of X, then $\operatorname{cl} X = \operatorname{cl} Y$. By the above argument,

$$\operatorname{aff} X = \operatorname{aff}(\operatorname{cl} X) = \operatorname{aff}(\operatorname{cl} Y) = \operatorname{aff} Y. \qquad \square$$

The next theorem gives an important description of affine spans in terms of affine combinations of points.

Theorem 2.44. *The affine span of a nonempty set $X \subset \mathbb{R}^n$ is the collection of all (finite) affine combinations of points from X:*

$$\operatorname{aff} X = \{\lambda_1 \boldsymbol{x}_1 + \cdots + \lambda_k \boldsymbol{x}_k : k \geqslant 1,\ \boldsymbol{x}_1, \ldots, \boldsymbol{x}_k \in X,\ \lambda_1 + \cdots + \lambda_k = 1\}.$$

Proof. Because aff X is a plane containing X, Theorem 2.39 shows that the set

$$L = \{\lambda_1 \boldsymbol{x}_1 + \cdots + \lambda_k \boldsymbol{x}_k : k \geqslant 1,\ \boldsymbol{x}_1, \ldots, \boldsymbol{x}_k \in X,\ \lambda_1 + \cdots + \lambda_k = 1\}$$

lies in aff X. We are going to prove that L is a plane. For this, choosing points $\boldsymbol{x}, \boldsymbol{y} \in L$ and expressing them as affine combinations

$$\boldsymbol{x} = \gamma_1 \boldsymbol{x}_1 + \cdots + \gamma_p \boldsymbol{x}_p \quad \text{and} \quad \boldsymbol{y} = \mu_1 \boldsymbol{y}_1 + \cdots + \mu_q \boldsymbol{y}_q$$

of suitable points $\boldsymbol{x}_1, \ldots, \boldsymbol{x}_p, \boldsymbol{y}_1, \ldots, \boldsymbol{y}_q$ from X, we see that $(1-\lambda)\boldsymbol{x} + \lambda\boldsymbol{y}$ also is an affine combination of these points for every choice of $\lambda \in \mathbb{R}$:

$$\begin{aligned}(1-\lambda)\boldsymbol{x} + \lambda\boldsymbol{y} &= (1-\lambda)\gamma_1 \boldsymbol{x}_1 + \cdots + (1-\lambda)\gamma_p \boldsymbol{x}_p \\ &\quad + \lambda\mu_1 \boldsymbol{y}_1 + \cdots + \lambda\mu_q \boldsymbol{y}_q, \\ (1-\lambda)\gamma_1 + \cdots &+ (1-\lambda)\gamma_p + \lambda\mu_1 + \cdots + \lambda\mu_q = 1.\end{aligned}$$

Hence $(1-\lambda)\boldsymbol{x} + \lambda\boldsymbol{y} \in L$, and Theorem 2.38 shows that L is a plane.

Finally, because $X \subset L$ (every point $\boldsymbol{x} \in X$ can be written as $1\boldsymbol{x} \in L$), the inclusions $X \subset L \subset \operatorname{aff} X$ and Theorem 2.42 give $\operatorname{aff} X = L$. $\square$

Corollary 2.45. *If $\boldsymbol{x}_1, \ldots, \boldsymbol{x}_r$ are points in $\mathbb{R}^n$, $r \geqslant 1$, then*

$$\operatorname{aff}\{\boldsymbol{x}_1, \ldots, \boldsymbol{x}_r\} = \{\lambda_1 \boldsymbol{x}_1 + \cdots + \lambda_r \boldsymbol{x}_r : \lambda_1 + \cdots + \lambda_r = 1\}.$$

Proof. By Theorem 2.44, a point $\boldsymbol{x} \in \mathbb{R}^n$ belongs to $\operatorname{aff}\{\boldsymbol{x}_1, \ldots, \boldsymbol{x}_r\}$ if and only if it can be written as an affine combination $\boldsymbol{x} = \lambda_{i_1} \boldsymbol{x}_{i_1} + \cdots + \lambda_{i_s} \boldsymbol{x}_{i_s}$ of suitable points $\boldsymbol{x}_{i_1}, \ldots, \boldsymbol{x}_{i_s}$ from $\{\boldsymbol{x}_1, \ldots, \boldsymbol{x}_r\}$. Renumbering the indices $i_1, \ldots, i_s$, we may suppose that $1 \leqslant i_1 \leqslant \cdots \leqslant i_s \leqslant r$. For every index $i \in \{1, \ldots, r\} \setminus \{i_1, \ldots, i_s\}$, if any, we add $0\boldsymbol{x}_i$ to the right-hand side of $\boldsymbol{x} = \lambda_{i_1} \boldsymbol{x}_{i_1} + \cdots + \lambda_{i_s} \boldsymbol{x}_{i_s}$ to express $\boldsymbol{x}$ as an affine combination $\boldsymbol{x} = \lambda_1 \boldsymbol{x}_1 + \cdots + \lambda_r \boldsymbol{x}_r$. $\square$

Theorem 2.44 allows us to establish the following important algebraic property of affine spans.

Theorem 2.46. *If $X_1, \dots, X_r$ are sets in $\mathbb{R}^n$ and $\mu_1, \dots, \mu_r$ are scalars, then*

$$\operatorname{aff}(\mu_1 X_1 + \cdots + \mu_r X_r) = \mu_1 \operatorname{aff} X_1 + \cdots + \mu_r \operatorname{aff} X_r. \tag{2.15}$$

Proof. Excluding the obvious case when at least one of the sets $X_1, \dots, X_r$ is empty (then both sides of (2.15) are empty sets), we may assume that all these sets are nonempty.

An induction argument shows that the proof can be reduced to the case $r = 2$ (let $\mu_1 X_1 = \mu_1 X_1 + \mu_2\{\boldsymbol{o}\}$ if $r = 1$). Because the set $\mu_1 \operatorname{aff} X_1 + \mu_2 \operatorname{aff} X_2$ is a plane (see Theorem 2.4), the obvious inclusion

$$\mu_1 X_1 + \mu_2 X_2 \subset \mu_1 \operatorname{aff} X_1 + \mu_2 \operatorname{aff} X_2$$

and Theorem 2.42 give

$$\operatorname{aff}(\mu_1 X_1 + \mu_2 X_2) \subset \mu_1 \operatorname{aff} X_1 + \mu_2 \operatorname{aff} X_2.$$

For the opposite inclusion, choose any point $\boldsymbol{x} \in \mu_1 \operatorname{aff} X_1 + \mu_2 \operatorname{aff} X_2$. Then $\boldsymbol{x} = \mu_1 \boldsymbol{x}_1 + \mu_1 \boldsymbol{x}_2$ for suitable points $\boldsymbol{x}_1 \in \operatorname{aff} X_1$ and $\boldsymbol{x}_2 \in \operatorname{aff} X_2$. By Theorem 2.44, $\boldsymbol{x}_1$ and $\boldsymbol{x}_2$ can be written as affine combinations

$$\boldsymbol{x}_1 = \lambda_1 \boldsymbol{u}_1 + \cdots + \lambda_p \boldsymbol{u}_p \quad \text{and} \quad \boldsymbol{x}_2 = \gamma_1 \boldsymbol{v}_1 + \cdots + \gamma_q \boldsymbol{v}_q,$$

where $\boldsymbol{u}_1, \dots, \boldsymbol{u}_p \in X_1$ and $\boldsymbol{v}_1, \dots, \boldsymbol{v}_q \in X_2$. Because

$$\mu_1 \boldsymbol{u}_i + \mu_2 \boldsymbol{v}_j \in \mu_1 X_1 + \mu_2 X_2 \quad \text{for all} \quad 1 \leqslant i \leqslant p \quad \text{and} \quad 1 \leqslant j \leqslant q,$$

the equalities

$$\begin{aligned}\boldsymbol{x} = \mu_1 \boldsymbol{x}_1 + \mu_2 \boldsymbol{x}_2 &= \mu_1 \Big(\sum_{j=1}^{q} \gamma_j\Big)\Big(\sum_{i=1}^{p} \lambda_i \boldsymbol{u}_i\Big) + \mu_2 \Big(\sum_{i=1}^{p} \lambda_i\Big)\Big(\sum_{j=1}^{q} \gamma_j \boldsymbol{v}_j\Big) \\ &= \sum_{i=1}^{p} \sum_{j=1}^{q} \lambda_i \gamma_j (\mu_1 \boldsymbol{u}_i + \mu_2 \boldsymbol{v}_j), \quad \sum_{i=1}^{p} \sum_{j=1}^{q} \lambda_i \gamma_j = 1,\end{aligned}$$

show that $\boldsymbol{x}$ is an affine combination of points from $\mu_1 X_1 + \mu_2 X_2$. Therefore, Theorem 2.44 gives $\boldsymbol{x} \in \operatorname{aff}(\mu_1 X_1 + \mu_2 X_2)$. Summing up,

$$\mu_1 \operatorname{aff} X_1 + \mu_2 \operatorname{aff} X_2 \subset \operatorname{aff}(\mu_1 X_1 + \mu_2 X_2). \qquad \square$$

A sharper version of Theorem 2.46 for the case of large sums is considered in Problem 2.14.

Definition 2.47. For a nonempty set $X \subset \mathbb{R}^n$, the subspaces

$$\operatorname{dir} X = \operatorname{aff} X - \operatorname{aff} X \quad \text{and} \quad \operatorname{ort} X = (\operatorname{dir} X)^{\perp}$$

are called, respectively, the *direction space* and the *orthospace* of X. We let $\operatorname{dir} \varnothing = \varnothing$ and $\operatorname{ort} \varnothing = \mathbb{R}^n$.

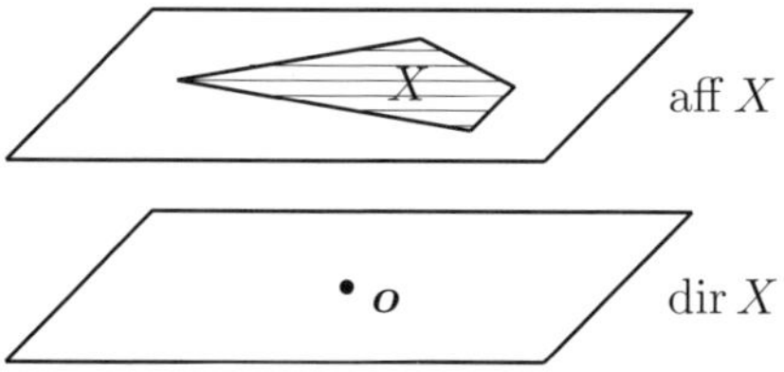

Fig. 2.11 The direction space dir X of the plane aff X.

Theorem 2.48. *For a nonempty set $X \subset \mathbb{R}^n$ and a point $\boldsymbol{c} \in \operatorname{aff} X$, the direction space of X can be described in the following ways.*

(*a*) $\operatorname{dir} X = \operatorname{aff} X - \operatorname{aff} X = \operatorname{aff}(X - X) = \operatorname{span}(X - X)$.
(*b*) $\operatorname{dir} X = \operatorname{aff} X - \boldsymbol{c} = \operatorname{aff}(X - \boldsymbol{c}) = \operatorname{span}(X - \boldsymbol{c})$.
(*c*) $\operatorname{dir} X = \operatorname{dir}(\operatorname{cl} X)$. *Consequently,* $\operatorname{dir} X = \operatorname{dir} Y$ *for any dense subset Y of X.*
(*d*) $\operatorname{dir} X = \{\lambda_1\boldsymbol{x}_1 + \cdots + \lambda_k\boldsymbol{x}_k : k \geqslant 1,\ \boldsymbol{x}_1, \dots, \boldsymbol{x}_k \in X,\ \lambda_1 + \cdots + \lambda_k = 0\}$.

Proof. Assertions (*a*) and (*b*) immediately follow from a combination of Theorems 2.42 and 2.46.

(*c*) This part follows from Corollary 2.43.

(*d*) Choose a point $\boldsymbol{a} \in X$. If $\boldsymbol{x} \in \operatorname{dir} X$, then $\boldsymbol{x} \in \operatorname{aff} X - \boldsymbol{a}$, and Theorem 2.44 shows that $\boldsymbol{x}$ can be expressed as

$$\boldsymbol{x} = \lambda_1\boldsymbol{x}_1 + \cdots + \lambda_p\boldsymbol{x}_p + (-1)\boldsymbol{a}, \quad \boldsymbol{x}_1, \dots, \boldsymbol{x}_p \in X,\ \lambda_1 + \cdots + \lambda_p = 1.$$

Therefore, $\boldsymbol{x}$ belongs to the set

$$S(X) = \{\lambda_1\boldsymbol{x}_1 + \cdots + \lambda_k\boldsymbol{x}_k : k \geqslant 1,\ \boldsymbol{x}_1, \dots, \boldsymbol{x}_k \in X,\ \lambda_1 + \cdots + \lambda_k = 0\}.$$

Conversely, if $\boldsymbol{x} \in S(X)$, then $\boldsymbol{x} = \mu_1\boldsymbol{x}_1 + \cdots + \mu_q\boldsymbol{x}_q$ for suitable points $\boldsymbol{x}_1, \dots, \boldsymbol{x}_q \in X$ and scalars $\mu_1, \dots, \mu_q$ whose sum is 0. Because $\boldsymbol{x} + \boldsymbol{a}$ can be written an affine combination

$$\boldsymbol{x} + \boldsymbol{c} = \mu_1\boldsymbol{x}_1 + \cdots + \mu_q\boldsymbol{x}_q + 1\boldsymbol{a}$$

of points from X, Theorem 2.39 shows that $\boldsymbol{x} + \boldsymbol{a} \in \operatorname{aff} X$. Consequently,

$$\boldsymbol{x} = (\boldsymbol{x} + \boldsymbol{a}) - \boldsymbol{a} \in \operatorname{aff} X - \boldsymbol{a} = \operatorname{dir} X. \qquad \square$$

Theorem 2.49. *For nonempty sets X and Y in $\mathbb{R}^n$, the following assertions hold.*

(*a*) $\operatorname{aff}(X \cup Y) = (\operatorname{aff} X + \operatorname{dir} Y) \cup (\operatorname{dir} X + \operatorname{aff} Y) \cup L(X, Y)$, *where*

$$L(X, Y) = \{\lambda\boldsymbol{x} + \mu\boldsymbol{y} : \boldsymbol{x} \in \operatorname{aff} X,\ \boldsymbol{y} \in \operatorname{aff} Y,\ \lambda + \mu = 1\}.$$

(*b*) *If* $\text{aff}\, X \cap \text{aff}\, Y \neq \varnothing$ *and* $\boldsymbol{c} \in \text{aff}\, X \cap \text{aff}\, Y$, *then*

$$\text{aff}\,(X \cup Y) = \text{aff}\,(X + Y) - \boldsymbol{c}. \tag{2.16}$$

(*c*) *If* $\text{aff}\, X \cap \text{aff}\, Y = \varnothing$, $\boldsymbol{c} \in \text{aff}\, X$, $\boldsymbol{e} \in \text{aff}\, Y$, *and* l *is the line through* $\boldsymbol{c}$ *and* $\boldsymbol{e}$, *then*

$$\text{aff}\,(X \cup Y) = \text{aff}\,(X + Y) + l - \boldsymbol{c} - \boldsymbol{e}. \tag{2.17}$$

Proof. (*a*) By Theorem 2.44, a point $\boldsymbol{z} \in \mathbb{R}^n$ belongs to $\text{aff}\,(X \cup Y)$ if and only if it can be written as an affine combination

$$\boldsymbol{z} = \lambda_1 \boldsymbol{x}_1 + \cdots + \lambda_p \boldsymbol{x}_p + \mu_1 \boldsymbol{y}_1 + \cdots + \mu_q \boldsymbol{y}_q,$$

where $\boldsymbol{x}_1 \ldots, \boldsymbol{x}_p \in X$ and $\boldsymbol{y}_1, \ldots, \boldsymbol{y}_q \in Y$. If one of the sums

$$\lambda = \lambda_1 + \cdots + \lambda_p \quad \text{and} \quad \mu = \mu_1 + \cdots + \mu_q$$

equals 0, then $\boldsymbol{x}$ belongs to the respective set $\text{aff}\, X + \text{dir}\, Y$ or $\text{dir}\, X + \text{aff}\, Y$, as follows from Theorems 2.44 and 2.48. Suppose that both scalars λ and μ are not 0. Then, by Theorem 2.44,

$$\boldsymbol{x}' = \lambda^{-1}(\lambda_1 \boldsymbol{x}_1 + \cdots + \lambda_p \boldsymbol{x}_p) \in \text{aff}\, X,$$
$$\boldsymbol{y}' = \mu^{-1}(\mu_1 \boldsymbol{y}_1 + \cdots + \mu_q \boldsymbol{y}_q) \in \text{aff}\, Y,$$

implying the inclusion $\boldsymbol{z} = \lambda \boldsymbol{x}' + \mu \boldsymbol{y}' \in L(X, Y)$.

(*b*) Let $X' = X \cup \{\boldsymbol{c}\}$ and $Y' = Y \cup \{\boldsymbol{c}\}$. Theorem 2.42 and the inclusions

$$X \subset X' \subset \text{aff}\, X, \quad Y \subset Y' \subset \text{aff}\, Y, \quad X \cup Y \subset X' \cup Y' \subset \text{aff}\,(X \cup Y)$$

give

$$\text{aff}\, X' = \text{aff}\, X, \quad \text{aff}\, Y' = \text{aff}\, Y, \quad \text{aff}\,(X' \cup Y') = \text{aff}\,(X \cup Y).$$

Since $\boldsymbol{o} \in (X' - \boldsymbol{c}) \cap (Y' - \boldsymbol{c})$, the same theorem and the properties of spans (see page 5) imply

$$\begin{aligned} &\text{aff}\,((X' - \boldsymbol{c}) \cup (Y' - \boldsymbol{c})) = \text{span}\,((X' - \boldsymbol{c}) \cup (Y' - \boldsymbol{c})) \\ &= \text{span}\,((X' - \boldsymbol{c}) + (Y' - \boldsymbol{c})) = \text{aff}\,((X' - \boldsymbol{c}) + (Y' - \boldsymbol{c})). \end{aligned}$$

Combining Theorems 2.42 and 2.46, we obtain

$$\begin{aligned} \text{aff}\,(X \cup Y) &= \text{aff}\,(X' \cup Y') = \text{aff}\,((X' - \boldsymbol{c}) \cup (Y' - \boldsymbol{c}) + \boldsymbol{c}) \\ &= \text{aff}\,((X' - \boldsymbol{c}) \cup (Y' - \boldsymbol{c})) + \boldsymbol{c} \\ &= \text{aff}\,((X' - \boldsymbol{c}) + (Y' - \boldsymbol{c})) + \boldsymbol{c} \\ &= \text{aff}\,(X' + Y') - \boldsymbol{c} = \text{aff}\, X' + \text{aff}\, Y' - \boldsymbol{c} \\ &= \text{aff}\, X + \text{aff}\, Y - \boldsymbol{c} = \text{aff}\,(X + Y) - \boldsymbol{c}. \end{aligned}$$

(*c*) By Definition 2.22, any point $\boldsymbol{x} \in l$ can be written as $\boldsymbol{x} = (1-\lambda)\boldsymbol{c} + \lambda\boldsymbol{e}$, where $\lambda \in \mathbb{R}$. So, Theorem 2.44 shows that

$$l \subset \operatorname{aff}(\operatorname{aff} X \cup \operatorname{aff} Y) = \operatorname{aff}(X \cup Y).$$

Hence

$$X \cup Y \subset X \cup Y \cup l \subset \operatorname{aff}(X \cup Y),$$

and Theorem 2.42 implies the equality

$$\operatorname{aff}(X \cup Y) = \operatorname{aff}(X \cup Y \cup l).$$

Since $\boldsymbol{e} \in \operatorname{aff} Y \cap l$ and $\boldsymbol{c} \in \operatorname{aff} X \cap \operatorname{aff}(Y \cup l)$, a combination of Theorem 2.42 and assertion (*b*) gives

$$\begin{aligned}\operatorname{aff}(X \cup Y) &= \operatorname{aff}(X \cup Y \cup l) = \operatorname{aff}(X \cup \operatorname{aff}(Y \cup l)) \\ &= \operatorname{aff} X + \operatorname{aff}(Y \cup l) - \boldsymbol{c} = \operatorname{aff} X + \operatorname{aff} Y + l - \boldsymbol{c} - \boldsymbol{e} \\ &= \operatorname{aff}(X + Y) + l - \boldsymbol{c} - \boldsymbol{e}.\end{aligned}$$ □

2.4 Affinely Independent Sets and Affine Bases

2.4.1 *Affinely Independent Sets*

Definition 2.50. A set $\{\boldsymbol{a}_1, \dots, \boldsymbol{a}_r\} \subset \mathbb{R}^n$, $r \geqslant 2$, is called *affinely dependent* if there are scalars $\nu_1, \dots, \nu_r$, not all zero, such that

$$\nu_1 \boldsymbol{a}_1 + \cdots + \nu_r \boldsymbol{a}_r = \boldsymbol{o} \quad \text{and} \quad \nu_1 + \cdots + \nu_r = 0. \tag{2.18}$$

The set $\{\boldsymbol{a}_1, \dots, \boldsymbol{a}_r\}$ is called *affinely independent* if it is not affinely dependent. The empty set $\varnothing$ and every singleton $\{\boldsymbol{a}\}$ in $\mathbb{R}^n$ are assumed to be affinely independent.

Remarks. 1. Every subset of an affinely independent set in $\mathbb{R}^n$ is affinely independent, and every set in $\mathbb{R}^n$ containing an affinely dependent set is affinely dependent.

2. Every affinely independent set $\{\boldsymbol{a}_1, \dots, \boldsymbol{a}_r\} \subset \mathbb{R}^n$ consists of pairwise distinct points. Indeed, assuming $\boldsymbol{a}_i = \boldsymbol{a}_j$ for some indices $1 \leqslant i < j \leqslant r$, we could write $1\boldsymbol{a}_i + (-1)\boldsymbol{a}_j = \boldsymbol{o}$; this argument and the above remark would imply the affine dependence of $\{\boldsymbol{a}_1, \dots, \boldsymbol{a}_r\}$.

Theorem 2.51. *The affine span of a nonempty set* $X \subset \mathbb{R}^n$ *is the collection of all (finite) affine combinations of affinely independent subsets of* X.

Proof. By Theorem 2.44, a point $\boldsymbol{x} \in \mathbb{R}^n$ belongs to aff X if and only if it is expressible as an affine combination $\boldsymbol{x} = \lambda_1\boldsymbol{x}_1 + \cdots + \lambda_p\boldsymbol{x}_p$ of suitable points $\boldsymbol{x}_1, \ldots, \boldsymbol{x}_p \in X$. We are going to show that $\boldsymbol{x}$ can be written as an affine combination of $p-1$ of fewer points from $\{\boldsymbol{x}_1, \ldots, \boldsymbol{x}_p\}$ provided the set $\{\boldsymbol{x}_1, \ldots, \boldsymbol{x}_p\}$ is affinely dependent. So, let $\{\boldsymbol{x}_1, \ldots, \boldsymbol{x}_p\}$ be affinely dependent. Then there are scalars $\nu_1, \ldots, \nu_p$, not all zero, such that

$$\nu_1\boldsymbol{x}_1 + \cdots + \nu_p\boldsymbol{x}_p = \boldsymbol{o} \quad \text{and} \quad \nu_1 + \cdots + \nu_p = 0.$$

Without loss of generality, we may assume that $\nu_1 \neq 0$. Then

$$\boldsymbol{x}_1 = -\frac{\nu_2}{\nu_1}\boldsymbol{x}_2 - \cdots - \frac{\nu_p}{\nu_1}\boldsymbol{x}_p, \quad \frac{\nu_2}{\nu_1} + \cdots + \frac{\nu_p}{\nu_1} = -1,$$

and $\boldsymbol{x}$ can be written as an affine combination

$$\boldsymbol{x} = \lambda_1\boldsymbol{x}_1 + \cdots + \lambda_p\boldsymbol{x}_p = \Big(\lambda_2 - \lambda_1\frac{\nu_2}{\nu_1}\Big)\boldsymbol{x}_2 + \cdots + \Big(\lambda_p - \lambda_1\frac{\nu_p}{\nu_1}\Big)\boldsymbol{x}_p,$$

which involves $p-1$ or fewer points from $\{\boldsymbol{x}_1, \ldots, \boldsymbol{x}_p\}$.

Consecutively repeating this argument, we obtain a desired expression for $\boldsymbol{x}$. □

Remark. Unlike Theorem 2.44, the length of affine combinations in Theorem 2.51 is bounded above by $m+1$, where m is the dimension of X (see Corollary 2.66).

Theorem 2.52. *For points $\boldsymbol{a}_1, \ldots, \boldsymbol{a}_r \in \mathbb{R}^n$, $r \geqslant 1$, the following conditions are equivalent.*

(*a*) *The set $\{\boldsymbol{a}_1, \ldots, \boldsymbol{a}_r\}$ is affinely independent.*
(*b*) *None of $\boldsymbol{a}_1, \ldots, \boldsymbol{a}_r$ is an affine combination of the others.*
(*c*) *For every index $i = 1, \ldots, r$, the set*

$$\{\boldsymbol{a}_1 - \boldsymbol{a}_i, \ldots, \boldsymbol{a}_{i-1} - \boldsymbol{a}_i, \boldsymbol{a}_{i+1} - \boldsymbol{a}_i, \ldots, \boldsymbol{a}_r - \boldsymbol{a}_i\} \tag{2.19}$$

is linearly independent (this set is empty if $r = 1$).
(*d*) *There is an index $i \in \{1, \ldots, r\}$ such that the set (2.19) is linearly independent.*
(*e*) *The plane* aff $\{\boldsymbol{a}_1, \ldots, \boldsymbol{a}_r\}$ *has dimension $r-1$.*
(*f*) *Every point $\boldsymbol{x} \in$* aff $\{\boldsymbol{a}_1, \ldots, \boldsymbol{a}_r\}$ *is uniquely expressible as an affine combination of $\boldsymbol{a}_1, \ldots, \boldsymbol{a}_r$.*
(*g*) *Every point $\boldsymbol{x} \in$* dir $\{\boldsymbol{a}_1, \ldots, \boldsymbol{a}_r\}$ *is uniquely expressible as a linear combination*

$$\boldsymbol{x} = \lambda_1\boldsymbol{a}_1 + \cdots + \lambda_r\boldsymbol{a}_r, \quad \textit{where } \lambda_1 + \cdots + \lambda_r = 0.$$

Proof. Since the case $r = 1$ is obvious, we assume that $r \geqslant 2$.

$(a) \Rightarrow (b)$ Assume for a moment the existence of a point $\boldsymbol{a}_i$, $1 \leqslant i \leqslant r$, which is an affine combination of $\boldsymbol{a}_1, \dots, \boldsymbol{a}_{i-1}, \boldsymbol{a}_{i+1}, \dots, \boldsymbol{a}_r$:

$$\boldsymbol{a}_i = \lambda_1 \boldsymbol{a}_1 + \cdots + \lambda_{i-1}\boldsymbol{a}_{i-1} + \lambda_{i+1}\boldsymbol{a}_{i+1} + \cdots + \lambda_r \boldsymbol{a}_r,$$
$$\lambda_1 + \cdots + \lambda_{i-1} + \lambda_{i+1} + \cdots + \lambda_r = 1.$$

Rewriting these equalities as

$$\lambda_1 \boldsymbol{a}_1 + \cdots + \lambda_{i-1}\boldsymbol{a}_{i-1} + (-1)\boldsymbol{a}_i + \lambda_{i+1}\boldsymbol{a}_{i+1} + \cdots + \lambda_r \boldsymbol{a}_r = 0,$$
$$\lambda_1 + \cdots + \lambda_{i-1} + (-1) + \lambda_{i+1} + \cdots + \lambda_r = 0,$$

we conclude that the set $\{\boldsymbol{a}_1, \dots, \boldsymbol{a}_r\}$ is affinely dependent, in contradiction with condition (a).

$(b) \Rightarrow (c)$ Suppose the existence of an index $i \in \{1, \dots, r\}$ such that the set (2.19) is linearly dependent. Then at least one of these vectors, say $\boldsymbol{a}_1 - \boldsymbol{a}_i$, $i \neq 1$, is a linear combination of the others:

$$\begin{aligned}\boldsymbol{a}_1 - \boldsymbol{a}_i = \lambda_2(\boldsymbol{a}_2 - \boldsymbol{a}_i) &+ \cdots + \lambda_{i-1}(\boldsymbol{a}_{i-1} - \boldsymbol{a}_i) \\ &+ \lambda_{i+1}(\boldsymbol{a}_{i+1} - \boldsymbol{a}_i) + \cdots + \lambda_r(\boldsymbol{a}_r - \boldsymbol{a}_i).\end{aligned}$$

With

$$\lambda_i = 1 - \lambda_2 - \cdots - \lambda_{i-1} - \lambda_{i+1} - \cdots - \lambda_r,$$

the equality $\boldsymbol{a}_1 = \lambda_2 \boldsymbol{a}_2 + \cdots + \lambda_r \boldsymbol{a}_r$ shows that $\boldsymbol{a}_1$ is an affine combination of $\boldsymbol{a}_2, \dots, \boldsymbol{a}_r$, contradicting condition (b).

Since (c) obviously implies (d), we show next that $(d) \Rightarrow (e)$. For this, consider the $(r-1)$-dimensional subspace

$$S_i = \operatorname{span}\{\boldsymbol{a}_1 - \boldsymbol{a}_i, \dots, \boldsymbol{a}_{i-1} - \boldsymbol{a}_i, \boldsymbol{a}_{i+1} - \boldsymbol{a}_i, \dots, \boldsymbol{a}_r - \boldsymbol{a}_i\}. \tag{2.20}$$

A combination of Theorems 2.42 and 2.46 gives

$$\begin{aligned}S_i &= \operatorname{aff}\{\boldsymbol{a}_1 - \boldsymbol{a}_i, \dots, \boldsymbol{a}_{i-1} - \boldsymbol{a}_i, \boldsymbol{o}, \boldsymbol{a}_{i+1} - \boldsymbol{a}_i, \dots, \boldsymbol{a}_r - \boldsymbol{a}_i\} \\ &= \operatorname{aff}\{\boldsymbol{a}_1, \dots, \boldsymbol{a}_r\} - \boldsymbol{a}_i.\end{aligned}$$

According to Definition 2.5, the plane $\operatorname{aff}\{\boldsymbol{a}_1, \dots, \boldsymbol{a}_r\} = \boldsymbol{a}_i + S_i$ has dimension $r - 1$.

$(e) \Rightarrow (f)$ Choose a point $\boldsymbol{x} \in \operatorname{aff}\{\boldsymbol{a}_1, \dots, \boldsymbol{a}_r\}$. By Corollary 2.45, $\boldsymbol{x}$ is an affine combination of the form $\boldsymbol{x} = \lambda_1 \boldsymbol{a}_1 + \cdots + \lambda_r \boldsymbol{a}_r$. To show its uniqueness, consider an arbitrary expression for $\boldsymbol{x}$ as an affine combination $\boldsymbol{x} = \mu_1 \boldsymbol{a}_1 + \cdots + \mu_r \boldsymbol{a}_r$. Then the vector $\boldsymbol{y} = \boldsymbol{x} - \boldsymbol{a}_1$ can be simultaneously written as

$$\begin{aligned}\boldsymbol{y} &= \lambda_2(\boldsymbol{a}_2 - \boldsymbol{a}_1) + \cdots + \lambda_r(\boldsymbol{a}_r - \boldsymbol{a}_1) \\ &= \mu_2(\boldsymbol{a}_2 - \boldsymbol{a}_1) + \cdots + \mu_r(\boldsymbol{a}_r - \boldsymbol{a}_1).\end{aligned}$$

The subspace $S = \operatorname{span}\{\boldsymbol{a}_2 - \boldsymbol{a}_1, \dots, \boldsymbol{a}_r - \boldsymbol{a}_1\}$ is $(r-1)$-dimensional as a translate of the plane $\operatorname{aff}\{\boldsymbol{a}_1, \dots, \boldsymbol{a}_r\}$. Hence the set $\{\boldsymbol{a}_2 - \boldsymbol{a}_1, \dots, \boldsymbol{a}_r - \boldsymbol{a}_1\}$ is linearly independent, which implies that the above expressions for $\boldsymbol{y}$ are identical: $\lambda_i = \mu_i$ for all $2 \leqslant i \leqslant r$. Finally,

$$\lambda_1 = 1 - \lambda_2 - \dots - \lambda_r = 1 - \mu_2 - \dots - \mu_r = \mu_1.$$

$(f) \Rightarrow (a)$ Assume, for contradiction, that the set $\{\boldsymbol{a}_1, \dots, \boldsymbol{a}_r\}$ is affinely dependent. Then there are scalars $\nu_1, \dots, \nu_r$, not all zero, such that (2.18) holds. Choose a point $\boldsymbol{x} \in \operatorname{aff}\{\boldsymbol{a}_1, \dots, \boldsymbol{a}_r\}$. By Corollary 2.45, $\boldsymbol{x}$ can be written as an affine combination $\boldsymbol{x} = \lambda_1\boldsymbol{a}_1 + \dots + \lambda_r\boldsymbol{a}_r$. Consequently,

$$\boldsymbol{x} = (\lambda_1 + \nu_1)\boldsymbol{a}_1 + \dots + (\lambda_r + \nu_r)\boldsymbol{a}_r$$

is another expression for $\boldsymbol{x}$ as an affine combination of $\boldsymbol{a}_1, \dots, \boldsymbol{a}_r$, contrary to condition (f).

$(f) \Rightarrow (g)$ Choose any point $\boldsymbol{x} \in \operatorname{dir}\{\boldsymbol{a}_1, \dots, \boldsymbol{a}_r\}$. By Theorem 2.48, $\boldsymbol{a}_1 + \boldsymbol{x} \in \operatorname{aff}\{\boldsymbol{a}_1, \dots, \boldsymbol{a}_r\}$. By condition (f), $\boldsymbol{a}_1 + \boldsymbol{x}$ is uniquely expressible as an affine combination $\boldsymbol{a}_1 + \boldsymbol{x} = \lambda_1\boldsymbol{a}_1 + \dots + \lambda_r\boldsymbol{a}_r$. With $\lambda_1' = 1 - \lambda_1$, the point $\boldsymbol{x}$ is uniquely expressible as

$$\boldsymbol{x} = \lambda_1'\boldsymbol{a}_1 + \lambda_2\boldsymbol{a}_2 + \dots + \lambda_r\boldsymbol{a}_r, \quad \text{where} \quad \lambda_1' + \lambda_2 + \dots + \lambda_r = 0.$$

$(g) \Rightarrow (f)$ Choose any point $\boldsymbol{x} \in \operatorname{aff}\{\boldsymbol{a}_1, \dots, \boldsymbol{a}_r\}$. By Theorem 2.48, $\boldsymbol{x} - \boldsymbol{a}_1 \in \operatorname{dir}\{\boldsymbol{a}_1, \dots, \boldsymbol{a}_r\}$. Condition (g) shows that $\boldsymbol{x} - \boldsymbol{a}_1$ is uniquely expressible as a linear combination

$$\boldsymbol{x} - \boldsymbol{a}_1 = \lambda_1\boldsymbol{a}_1 + \dots + \lambda_r\boldsymbol{a}_r, \quad \text{where} \quad \lambda_1 + \dots + \lambda_r = 0.$$

Then $\boldsymbol{x}$ is uniquely expressible as an affine combination

$$\boldsymbol{x} = \lambda_1'\boldsymbol{a}_1 + \lambda_2\boldsymbol{a}_2 + \dots + \lambda_r\boldsymbol{a}_r, \quad \text{where} \quad \lambda_1' = \lambda_1 + 1. \qquad \square$$

Corollary 2.53. *If $L \subset \mathbb{R}^n$ is a nonempty plane of dimension m, then any affinely independent subset of L consists of $m+1$ or fewer points. In particular, any affinely independent set in $\mathbb{R}^n$ has $n+1$ of fewer points.*

Proof. Let $X = \{\boldsymbol{a}_1, \dots, \boldsymbol{a}_r\}$ be an affinely independent subset of L. By Theorem 2.52, the plane $M = \operatorname{aff} X$ has dimension $r-1$. Since $\operatorname{aff} X \subset \operatorname{aff} L = L$ (see Theorem 2.42), we obtain from Theorem 2.6 that

$$r - 1 = \dim(\operatorname{aff} X) \leqslant \dim L = m.$$

Consequently, $r \leqslant m + 1$. $\square$

Theorem 2.54. *Let a set $\{\boldsymbol{a}_1, \dots, \boldsymbol{a}_r\} \subset \mathbb{R}^n$, $r \geqslant 1$, be affinely independent and $\boldsymbol{a}_{r+1}$ be a point in $\mathbb{R}^n$. Then the set $\{\boldsymbol{a}_1, \dots, \boldsymbol{a}_{r+1}\}$ is affinely independent if and only if $\boldsymbol{a}_{r+1} \notin \operatorname{aff}\{\boldsymbol{a}_1, \dots, \boldsymbol{a}_r\}$.*

Proof. Assume first that the set $\{\boldsymbol{a}_1, \ldots, \boldsymbol{a}_{r+1}\}$ is affinely independent. By Theorem 2.52, $\boldsymbol{a}_{r+1}$ is not an affine combination of $\boldsymbol{a}_1, \ldots, \boldsymbol{a}_r$. Consequently, Theorem 2.44 implies that $\boldsymbol{a}_{r+1} \notin \operatorname{aff}\{\boldsymbol{a}_1, \ldots, \boldsymbol{a}_r\}$.

Conversely, let $\boldsymbol{a}_{r+1} \notin \operatorname{aff}\{\boldsymbol{a}_1, \ldots, \boldsymbol{a}_r\}$. Assume, for contradiction, that the set $\{\boldsymbol{a}_1, \ldots, \boldsymbol{a}_{r+1}\}$ is affinely dependent. Then there are scalars $\nu_1, \ldots, \nu_{r+1}$, not all zero, such that

$$\nu_1 \boldsymbol{a}_1 + \cdots + \nu_{r+1} \boldsymbol{a}_{r+1} = \boldsymbol{o} \quad \text{and} \quad \nu_1 + \cdots + \nu_{r+1} = 0. \tag{2.21}$$

We observe that $\nu_{r+1} = 0$. Indeed, suppose that $\nu_{r+1} \neq 0$. Then $\boldsymbol{a}_{r+1}$ can be written as an affine combination

$$\boldsymbol{a}_{r+1} = -\frac{\nu_1}{\nu_{r+1}} \boldsymbol{a}_1 + \cdots - \frac{\nu_r}{\nu_{r+1}} \boldsymbol{a}_r,$$

and Theorem 2.44 shows that $\boldsymbol{a}_{r+1} \in \operatorname{aff}\{\boldsymbol{a}_1, \ldots, \boldsymbol{a}_r\}$, contrary to the assumption. Hence $\nu_{r+1} = 0$, implying that (2.21) has the form

$$\nu_1 \boldsymbol{a}_1 + \cdots + \nu_r \boldsymbol{a}_r = \boldsymbol{o} \quad \text{and} \quad \nu_1 + \cdots + \nu_r = 0,$$

where not all $\nu_1, \ldots, \nu_r$ are zero. Consequently, the set $\{\boldsymbol{a}_1, \ldots, \boldsymbol{a}_r\}$ is affinely dependent, a contradiction. □

The following assertion is useful in the study of topological properties of convex sets.

Theorem 2.55. *Let* $\{\boldsymbol{c}_1, \ldots, \boldsymbol{c}_r\}$ *be an affinely independent set in* $\mathbb{R}^n$ *and* $\boldsymbol{x}, \boldsymbol{x}_1, \boldsymbol{x}_2, \ldots$ *be an infinite sequence of points in* $\operatorname{aff}\{\boldsymbol{c}_1, \ldots, \boldsymbol{c}_r\}$:

$$\begin{aligned} \boldsymbol{x} &= \lambda_1 \boldsymbol{c}_1 + \cdots + \lambda_r \boldsymbol{c}_r, \quad \lambda_1 + \cdots + \lambda_r = 1, \\ \boldsymbol{x}_i &= \lambda_1^{(i)} \boldsymbol{c}_1 + \cdots + \lambda_r^{(i)} \boldsymbol{c}_r, \quad \lambda_1^{(i)} + \cdots + \lambda_r^{(i)} = 1, \quad i \geqslant 1. \end{aligned}$$

Then $\lim_{i \to \infty} \boldsymbol{x}_i = \boldsymbol{x}$ *if and only if* $\lim_{i \to \infty} \lambda_j^{(i)} = \lambda_j$ *for all* $1 \leqslant j \leqslant r$.

Proof. Since the case $r = 1$ is obvious, we assume that $r \geqslant 2$. By Theorem 2.52, the vector set $\{\boldsymbol{c}_2 - \boldsymbol{c}_1, \ldots, \boldsymbol{c}_r - \boldsymbol{c}_1\}$ is linearly independent. Put $\boldsymbol{y} = \boldsymbol{x} - \boldsymbol{c}_1$ and $\boldsymbol{y}_i = \boldsymbol{x}_i - \boldsymbol{c}_1$, $i \geqslant 2$. Then

$$\begin{aligned} \boldsymbol{y} &= \lambda_2(\boldsymbol{c}_2 - \boldsymbol{c}_1) + \cdots + \lambda_r(\boldsymbol{c}_r - \boldsymbol{c}_1) \\ \boldsymbol{y}_i &= \lambda_2^{(i)}(\boldsymbol{c}_2 - \boldsymbol{c}_1) + \cdots + \lambda_r^{(i)}(\boldsymbol{c}_r - \boldsymbol{c}_1), \ i \geqslant 2. \end{aligned}$$

Since $\|\boldsymbol{x} - \boldsymbol{x}_i\| = \|\boldsymbol{y} - \boldsymbol{y}_i\|$, one has $\boldsymbol{x}_i \to \boldsymbol{x}$ if and only if $\boldsymbol{y}_i \to \boldsymbol{y}$. Furthermore, $\boldsymbol{y}_i \to \boldsymbol{y}$ if and only if $\lambda_j^{(i)} \to \lambda_j$ for all $2 \leqslant j \leqslant r$ (see Problem 1.10). In this case,

$$\lim_{i \to \infty} \lambda_1^{(i)} = \lim_{i \to \infty} (1 - \lambda_2^{(i)} - \cdots - \lambda_r^{(i)}) = 1 - \lambda_2 - \cdots - \lambda_r = \lambda_1. \quad \square$$

2.4.2 Affine Bases

Definition 2.56. Given a nonempty plane $L \subset \mathbb{R}^n$, we say that a set $X = \{\boldsymbol{a}_1, \ldots, \boldsymbol{a}_r\} \subset \mathbb{R}^n$, $r \geqslant 1$, is an *affine basis* for L if X is affinely independent and $L = \operatorname{aff} X$. The empty set $\varnothing$ is an affine basis for the empty plane.

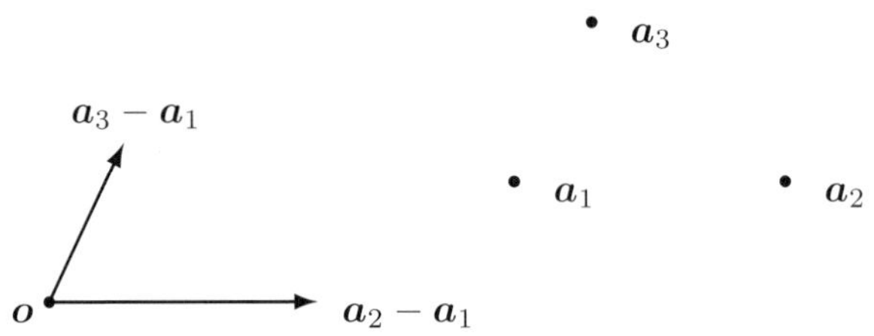

Fig. 2.12 Affine basis $\{\boldsymbol{a}_1, \boldsymbol{a}_2, \boldsymbol{a}_3\}$ and basis $\{\boldsymbol{a}_2 - \boldsymbol{a}_1, \boldsymbol{a}_3 - \boldsymbol{a}_1\}$ for $\mathbb{R}^2$.

Theorem 2.57. *Every plane $L \subset \mathbb{R}^n$ has an affine basis. Furthermore, if $m = \dim L$, then every affine basis for L consists of $m+1$ points.*

Proof. Since the case $-1 \leqslant m \leqslant 0$ is obvious, we may assume that $m \geqslant 1$. Choose a point $\boldsymbol{a} \in L$ and consider the m-dimensional subspace $S = L - \boldsymbol{a}$. If $\{\boldsymbol{b}_1, \ldots, \boldsymbol{b}_m\}$ is a basis for S, then, according to Theorem 2.42,

$$S = \operatorname{span}\{\boldsymbol{b}_1, \ldots, \boldsymbol{b}_m\} = \operatorname{aff}\{\boldsymbol{o}, \boldsymbol{b}_1, \ldots, \boldsymbol{b}_m\}.$$

Therefore,

$$L = \boldsymbol{a} + S = \boldsymbol{a} + \operatorname{aff}\{\boldsymbol{o}, \boldsymbol{b}_1, \ldots, \boldsymbol{b}_m\} = \operatorname{aff}\{\boldsymbol{a}, \boldsymbol{a} + \boldsymbol{b}_1, \ldots, \boldsymbol{a} + \boldsymbol{b}_m\}$$

(see Theorem 2.46). Finally, Theorem 2.52 shows that the set $\{\boldsymbol{a}, \boldsymbol{a} + \boldsymbol{b}_1, \ldots, \boldsymbol{a} + \boldsymbol{b}_m\}$ is an affine basis for L. The same theorem shows that every affine basis for L consists of $m+1$ points. $\square$

Theorem 2.52 implies the corollary below.

Corollary 2.58. *For a nonempty plane $L \subset \mathbb{R}^n$ of dimension m and points $\boldsymbol{a}_1, \ldots, \boldsymbol{a}_{m+1} \in L$, the following conditions are equivalent.*

(*a*) *The set $\{\boldsymbol{a}_1, \ldots, \boldsymbol{a}_{m+1}\}$ is an affine basis for L.*
(*b*) *The set $\{\boldsymbol{a}_1, \ldots, \boldsymbol{a}_{m+1}\}$ is affinely independent.*
(*c*) $L = \operatorname{aff}\{\boldsymbol{a}_1, \ldots, \boldsymbol{a}_{m+1}\}$.

(*d*) *For every index* $i = 1, \ldots, m+1$, *the set*

$$\{\boldsymbol{a}_1 - \boldsymbol{a}_i, \ldots, \boldsymbol{a}_{i-1} - \boldsymbol{a}_i, \boldsymbol{a}_{i+1} - \boldsymbol{a}_i, \ldots, \boldsymbol{a}_{m+1} - \boldsymbol{a}_i\} \tag{2.22}$$

is a basis for the subspace $\operatorname{dir} L$ (*this set is empty if* $m = 0$).

(*e*) *There is an index* $i \in \{1, \ldots, m+1\}$ *such that the set* (2.22) *is a basis for the subspace* $\operatorname{dir} L$.

(*f*) *Every point* $\boldsymbol{x} \in L$ *is uniquely expressible as an affine combination of* $\boldsymbol{a}_1, \ldots, \boldsymbol{a}_{m+1}$.

(*g*) *Every point* $\boldsymbol{x} \in \operatorname{dir} L$ *is uniquely expressible as a linear combination*

$$\boldsymbol{x} = \lambda_1 \boldsymbol{a}_1 + \cdots + \lambda_{m+1} \boldsymbol{a}_{m+1}, \quad \text{where } \lambda_1 + \cdots + \lambda_{m+1} = 0. \qquad \square$$

Corollary 2.58 gives a base to the following definition.

Definition 2.59. Let $L \subset \mathbb{R}^n$ be an m-dimensional plane with an affine basis $\{\boldsymbol{a}_1, \ldots, \boldsymbol{a}_{m+1}\}$. If a point $\boldsymbol{x} \in L$ is (uniquely) expressed as an affine combination $\boldsymbol{x} = \lambda_1 \boldsymbol{a}_1 + \cdots + \lambda_r \boldsymbol{a}_r$, then $\lambda_1, \ldots, \lambda_r$ are called the *affine coordinates* of $\boldsymbol{x}$ relative to $\{\boldsymbol{a}_1, \ldots, \boldsymbol{a}_r\}$.

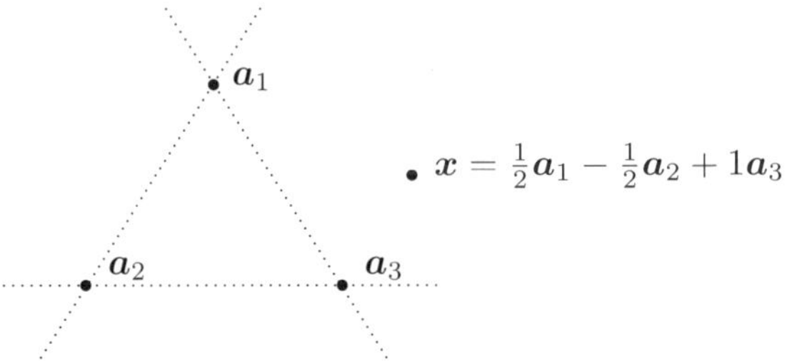

Fig. 2.13 Affine coordinates of $\boldsymbol{x}$ relative to $\{\boldsymbol{a}_1, \boldsymbol{a}_2, \boldsymbol{a}_3\}$.

Corollary 2.60. *If* $L \subset \mathbb{R}^n$ *is a proper plane of dimension* m, *and* $\boldsymbol{a}$ *is a point in* $\mathbb{R}^n \setminus L$, *then* $\operatorname{aff}(\{\boldsymbol{a}\} \cup L)$ *is a plane of dimension* $m+1$. *Furthermore, if* X *is an affine basis for* L, *then* $\{\boldsymbol{a}\} \cup X$ *is an affine basis for* $\operatorname{aff}(\{\boldsymbol{a}\} \cup L)$.

Proof. Since the case $L = \varnothing$ is obvious, we let $L \neq \varnothing$. By Theorem 2.57, L has an affine basis $\{\boldsymbol{a}_1, \ldots, \boldsymbol{a}_{m+1}\}$, and Theorem 2.54 shows that the set $\{\boldsymbol{a}, \boldsymbol{a}_1, \ldots, \boldsymbol{a}_{m+1}\}$ is affinely independent. Theorem 2.42 gives

$$\operatorname{aff}(\{\boldsymbol{a}\} \cup L) = \operatorname{aff}(\{\boldsymbol{a}\} \cup \operatorname{aff}\{\boldsymbol{a}_1, \ldots, \boldsymbol{a}_{m+1}\}) = \operatorname{aff}\{\boldsymbol{a}, \boldsymbol{a}_1, \ldots, \boldsymbol{a}_{m+1}\}.$$

Consequently, Corollary 2.58 shows that $\{\boldsymbol{a}, \boldsymbol{a}_1, \ldots, \boldsymbol{a}_{m+1}\}$ is an affine basis for aff $(\{\boldsymbol{a}\} \cup L)$. Finally, Theorem 2.52 implies the equality

$$\dim(\operatorname{aff}(\{\boldsymbol{a}\} \cup L)) = m + 1. \qquad \square$$

The next result refines Theorem 2.57 by putting a restriction on the choice of affine bases.

Theorem 2.61. *If Y is an affinely independent subset of a set $X \subset \mathbb{R}^n$, then there is an affine basis Z for* aff X *such that $Y \subset Z \subset X$.*

Proof. The case $X = \varnothing$ is obvious. So, we may assume that $X \neq \varnothing$. Choose a point $\boldsymbol{a} \in Y$ if $Y \neq \varnothing$ or any point $\boldsymbol{a} \in X$ if $Y = \varnothing$. Consider the set $Y' = (Y - \boldsymbol{a}) \setminus \{\boldsymbol{o}\}$. Clearly, $Y' \subset X - \boldsymbol{a}$. By Theorem 2.52, the set Y' is linearly independent (if card $Y \leqslant 1$, then $Y' = \varnothing$). We know (see page 5) that Y' can be extended to a basis U for the subspace span $(X - \boldsymbol{a})$. Furthermore, span $(X - \boldsymbol{a}) =$ aff $(X - \boldsymbol{a})$ because $\boldsymbol{o} \in X - \boldsymbol{a}$ (see Theorem 2.42).

Finally, let $Z = \{\boldsymbol{a}\} \cup (\boldsymbol{a} + U)$. Then $Y \subset Z \subset X$ and, according to Corollary 2.58, Z is an affine basis for the plane

$$\boldsymbol{a} + \operatorname{span}(X - \boldsymbol{a}) = \boldsymbol{a} + \operatorname{aff}(X - \boldsymbol{a}) = \operatorname{aff} X. \qquad \square$$

Corollary 2.62. *Any affinely independent subset of a plane $L \subset \mathbb{R}^n$ can be extended to an affine basis for L.* $\square$

2.4.3 *Dimension of Sets*

Definition 2.63. The *dimension* of a set $X \subset \mathbb{R}^n$, denoted $\dim X$, is defined by $\dim X = \dim(\operatorname{aff} X)$. In particular, $\dim X = -1$ if $X = \varnothing$.

Theorem 2.64. *For nonempty sets X and Y in $\mathbb{R}^n$, the following assertions hold.*

(*a*) $0 \leqslant \dim X \leqslant n$, *with* $\dim X = 0$ *if and only if X is a singleton.*
(*b*) *If $X \subset Y$, then* $\dim X \leqslant \dim Y$, *with* $\dim X = \dim Y$ *if and only if* $\operatorname{aff} X = \operatorname{aff} Y$.
(*c*) $\dim X = \dim Y$ *if* $X \subset Y \subset \operatorname{aff} X$.
(*d*) $\dim X = \dim(\operatorname{cl} X)$.
(*e*) $\dim(X + Y) \leqslant \dim X + \dim Y$, *and* $\dim(X + Y) = \dim X + \dim Y$ *if and only if the planes* aff X *and* aff Y *are independent.*
(*f*) $\dim(X' + Y') \leqslant \dim(X + Y)$ *for any subsets $X' \subset X$ and $Y' \subset Y$. Furthermore,* $\dim(X' + Y') = \dim(X + Y)$ *if* $\dim X' = \dim X$ *and* $\dim Y' = \dim Y$.

(*g*) *If* $\operatorname{aff} X \cap \operatorname{aff} Y \neq \varnothing$, *then* $\dim(X \cup Y) = \dim(X + Y)$.
(*h*) *If* $\operatorname{aff} X \cap \operatorname{aff} Y = \varnothing$, *then* $\dim(X \cup Y) = \dim(X + Y) + 1$.

Proof. (*a*) This assertion obviously follows from the definitions.

(*b*) According to Theorem 2.42, $\operatorname{aff} X \subset \operatorname{aff} Y$. This argument and Theorem 2.6 give

$$\dim X = \dim(\operatorname{aff} X) \leqslant \dim(\operatorname{aff} Y) = \dim Y.$$

Again by Theorems 2.6, the equality $\dim X = \dim Y$ holds if and only if $\operatorname{aff} X = \operatorname{aff} Y$.

(*c*) As above, the inclusions $X \subset Y \subset \operatorname{aff} X$ give $\operatorname{aff} X = \operatorname{aff} Y$; so, $\dim X = \dim Y$.

(*d*) Since $\operatorname{aff} X = \operatorname{aff}(\operatorname{cl} X)$ (see Corollary 2.43), one has $\dim X = \dim(\operatorname{cl} X)$.

(*e*) From Theorems 2.7 and 2.46, we conclude:

$$\begin{aligned}\dim(X + Y) &= \dim(\operatorname{aff}(X + Y)) = \dim(\operatorname{aff} X + \operatorname{aff} Y)\\ &\leqslant \dim(\operatorname{aff} X) + \dim(\operatorname{aff} Y) = \dim X + \dim Y.\end{aligned}$$

According to Theorem 2.10, the equality

$$\dim(X + Y) = \dim X + \dim Y$$

holds if and only if the planes $\operatorname{aff} X$ and $\operatorname{aff} Y$ are independent.

(*f*) Since $X' + Y' \subset X + Y$, the inequality

$$\dim(X' + Y') \leqslant \dim(X + Y)$$

follows from assertion (*b*). Suppose that $\dim X' = \dim X$ and $\dim Y' = \dim Y$. Then $\operatorname{aff} X' = \operatorname{aff} X$ and $\operatorname{aff} Y' = \operatorname{aff} Y$, and Theorem 2.44 gives

$$\operatorname{aff}(X' + Y') = \operatorname{aff} X' + \operatorname{aff} Y' = \operatorname{aff} X + \operatorname{aff} Y = \operatorname{aff}(X + Y).$$

Consecutively,

$$\dim(X' + Y') = \dim(\operatorname{aff}(X' + Y')) = \dim(\operatorname{aff}(X + Y)) = \dim(X + Y).$$

(*g*) Choose a point $\boldsymbol{c} \in \operatorname{aff} X \cap \operatorname{aff} Y$. A combination of Theorems 2.46 and 2.49 gives

$$\begin{aligned}\dim(X \cup Y) &= \dim(\operatorname{aff}(X \cup Y)) = \dim(\operatorname{aff}(X + Y) - \boldsymbol{c})\\ &= \dim(\operatorname{aff}(X + Y)) = \dim(X + Y).\end{aligned}$$

(*h*) Choose points $\boldsymbol{c} \in \operatorname{aff} X$ and $\boldsymbol{e} \in \operatorname{aff} Y$ and denote by l the line through $\boldsymbol{c}$ and $\boldsymbol{e}$. By Theorem 2.49,

$$\begin{aligned}\dim(\operatorname{aff}(X \cup Y)) &= \dim(\operatorname{aff}(X + Y) + l - \boldsymbol{c} - \boldsymbol{e})\\ &= \dim(\operatorname{aff}(X + Y) + l).\end{aligned}$$

Since $\operatorname{aff} X = \boldsymbol{c} + \operatorname{dir} X$ and $\operatorname{aff} Y = \boldsymbol{e} + \operatorname{dir} Y$ (see Theorem 2.48), the one-dimensional subspace $l - \boldsymbol{c}$ does not lie in $\operatorname{dir}(X+Y)$. Therefore,

$$\begin{aligned}\dim(X \cup Y) &= \dim(\operatorname{aff}(X \cup Y)) = \dim(\operatorname{aff}(X+Y)+l)\\ &= \dim(\operatorname{dir}(X+Y) + (l - \boldsymbol{c}))\\ &= \dim(\operatorname{dir}(X+Y)) + 1\\ &= \dim(\operatorname{aff}(X+Y)) + 1\\ &= \dim(X+Y) + 1.\end{aligned}$$ □

Corollary 2.65. *For a set X and a plane L in $\mathbb{R}^n$, the following assertion hold.*

(*a*) *If $X \subset L$, then* $\dim X \leqslant \dim L$, *with* $\dim X = \dim L$ *if and only if* $\operatorname{aff} X = L$.
(*b*) *If X meets L such that $X \not\subset L$, then* $\dim(X \cap L) \leqslant \dim X - 1$. □

A combination of Theorems 2.51 and 2.61 and Corollaries 2.53 and 2.58 implies one more corollary.

Corollary 2.66. *For a set $X \subset \mathbb{R}^n$ of dimension m, the following assertions hold.*

(*a*) *Any affinely independent subset of X contains $m+1$ or fewer points.*
(*b*) *If Y is and affinely independent subset of X, then X contains an affinely independent subset Z of $m+1$ points such that $Y \subset Z$ and* $\operatorname{aff} X = \operatorname{aff} Y$.
(*c*) $\operatorname{card} X \geqslant m+1$, *with* $\operatorname{card} X = m+1$ *if and only if the set X is affinely independent.*
(*d*) *Every affinely independent set of $m+1$ points from X is an affine basis for* $\operatorname{aff} X$.
(*e*) $\operatorname{aff} X$ *is the collection of all affine combinations of affinely independent subsets of X each consisting of $m+1$ or fewer points.* □

Corollary 2.67. *If $X \subset \mathbb{R}^n$ is a set of dimension m, where $m \leqslant n-1$, and $\boldsymbol{a}$ is a point in $\mathbb{R}^n \setminus \operatorname{aff} X$, then* $\dim(\{\boldsymbol{a}\} \cup X) = m+1$.

Proof. Let $L = \operatorname{aff} X$. Then $\dim L = m$, and, according to Theorem 2.42,

$$\operatorname{aff}(\{\boldsymbol{a}\} \cup X) = \operatorname{aff}(\{\boldsymbol{a}\} \cup \operatorname{aff} X) = \operatorname{aff}(\{\boldsymbol{a}\} \cup L).$$

Because $\dim(\operatorname{aff}(\{\boldsymbol{a}\} \cup L)) = m+1$ (see Corollary 2.60), we conclude that $\dim(\{\boldsymbol{a}\} \cup X) = m+1$. □

Theorem 2.68. *If* $L \subset \mathbb{R}^n$ *is a nonempty plane of dimension* m *and* $U_\rho(\boldsymbol{a}) \subset \mathbb{R}^n$ *is an open ball centered at a point* $\boldsymbol{a} \in L$*, then*

$$\operatorname{aff}(L \cap U_\rho(\boldsymbol{a})) = L \quad \textit{and} \quad \dim(L \cap U_\rho(\boldsymbol{a})) = m.$$

Consequently, $\operatorname{aff}(L \cap B_\rho(\boldsymbol{a})) = L$ *and* $\dim(L \cap B_\rho(\boldsymbol{a})) = m$*, where* $B_\rho(\boldsymbol{a})$ *is the respective closed ball.*

Proof. The case $m = 0$ is obvious. So, we may assume that $m \geqslant 1$. Consider the subspace $S = L - \boldsymbol{a}$ and the open ball $U_\rho(\boldsymbol{o}) = U_\rho(\boldsymbol{a}) - \boldsymbol{a}$. Because $\dim S = \dim L = m$, one can choose in $U_\rho(\boldsymbol{o}) \cap S$ a basis $\{\boldsymbol{b}_1, \ldots, \boldsymbol{b}_m\}$ for S. Put $\boldsymbol{a}_i = \boldsymbol{a} + \boldsymbol{b}_i$, $1 \leqslant i \leqslant m$. By Corollary 2.58, the set $\{\boldsymbol{a}, \boldsymbol{a}_1, \ldots, \boldsymbol{a}_m\}$ is an affine basis for L. Furthermore, $\boldsymbol{a}_i \in U_\rho(\boldsymbol{a})$ for all $1 \leqslant i \leqslant m$ due to $\|\boldsymbol{a}_i - \boldsymbol{a}\| = \|\boldsymbol{b}_i\| < \rho$. From the inclusions

$$\{\boldsymbol{a}, \boldsymbol{a}_1, \ldots, \boldsymbol{a}_m\} \subset L \cap U_\rho(\boldsymbol{a}) \subset L \cap B_\rho(\boldsymbol{a}) \subset L = \operatorname{aff}\{\boldsymbol{a}, \boldsymbol{a}_1, \ldots, \boldsymbol{a}_m\}$$

and Theorem 2.42 we conclude that

$$\operatorname{aff}(L \cap U_\rho(\boldsymbol{a})) = \operatorname{aff}(L \cap B_\rho(\boldsymbol{a})) = L.$$

Finally, Definition 2.63 shows that

$$\dim(L \cap U_\rho(\boldsymbol{a})) = \dim(L \cap B_\rho(\boldsymbol{a})) = \dim L = m. \qquad \square$$

2.5 Halfplanes

2.5.1 *Basic Properties of Halfplanes*

Definition 2.69. Let $L \subset \mathbb{R}^n$ be a plane of positive dimension (possibly, $L = \mathbb{R}^n$). A *closed halfplane* of L is a set of the form $D = V \cap L$, where V is a closed halfspace of $\mathbb{R}^n$ satisfying the condition $\varnothing \neq V \cap L \neq L$. The set $\operatorname{int} V \cap L$ is called the *relative interior* of D.

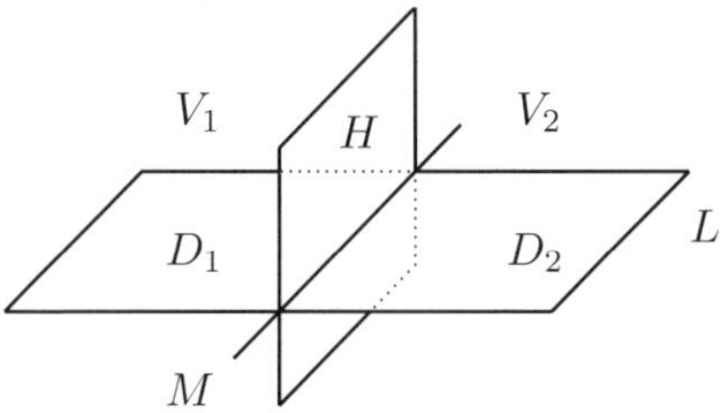

Fig. 2.14 Closed halfplanes D_1 and D_2 of the plane L.

Similarly, an *open halfplane* of L is a set of the form $E = W \cap L$, where W is an open halfspace of $\mathbb{R}^n$ such that $\varnothing \neq W \cap L \neq L$. Furthermore, if $H \subset \mathbb{R}^n$ is the boundary hyperplane of V (respectively, of W), then the subplane $M = H \cap L$ is called the *boundary plane* of D (respectively, of E).

Theorem 2.70. *For a plane $L \subset \mathbb{R}^n$ of positive dimension, the following assertions hold.*

(*a*) *A set $D \subset \mathbb{R}^n$ is a closed halfplane of L if and only if there is a vector $\boldsymbol{e} \in \mathbb{R}^n \setminus \operatorname{ort} L$ and a scalar γ such that D is expressible as*

$$D = \{\boldsymbol{x} \in L : \boldsymbol{x}\cdot\boldsymbol{e} \leqslant \gamma\} \quad \textit{or} \quad D = \{\boldsymbol{x} \in L : \boldsymbol{x}\cdot\boldsymbol{e} \geqslant \gamma\}. \tag{2.23}$$

(*b*) *A set $E \subset \mathbb{R}^n$ is an open halfplane of L if and only if there is a vector $\boldsymbol{e} \in \mathbb{R}^n \setminus \operatorname{ort} L$ and a scalar γ such that E is expressible as*

$$E = \{\boldsymbol{x} \in L : \boldsymbol{x}\cdot\boldsymbol{e} < \gamma\} \quad \textit{or} \quad E = \{\boldsymbol{x} \in L : \boldsymbol{x}\cdot\boldsymbol{e} > \gamma\}. \tag{2.24}$$

Proof. (*a*) Let D be a closed halfplane of L. According to Definition 2.69, $D = V \cap L$ for a suitable closed halfspace $V \subset \mathbb{R}^n$ satisfying the condition $\varnothing \neq V \cap L \neq L$. Without loss of generality, we may write $V = \{\boldsymbol{x} \in \mathbb{R}^n : \boldsymbol{x}\cdot\boldsymbol{e} \leqslant \gamma\}$, where $\boldsymbol{e} \neq \boldsymbol{o}$ and γ is a suitable scalar (the case of the opposite inequality $\boldsymbol{e} \geqslant \gamma$ is similar). If H is the boundary hyperplane of V, then $H = \{\boldsymbol{x} \in \mathbb{R}^n : \boldsymbol{x}\cdot\boldsymbol{e} = \gamma\}$. Theorem 2.34 shows that H and L are not parallel. Consequently, Theorem 2.19 implies that $\boldsymbol{e} \notin \operatorname{ort} L$. Finally,

$$D = V \cap L = \{\boldsymbol{x} \in \mathbb{R}^n : \boldsymbol{x}\cdot\boldsymbol{e} \leqslant \gamma\} \cap L = \{\boldsymbol{x} \in L : \boldsymbol{x}\cdot\boldsymbol{e} \leqslant \gamma\}.$$

Conversely, if a set D is given by one of conditions (2.23), then $D = V \cap L$, where V is a closed halfspace of one of the forms

$$V = \{\boldsymbol{x} \in \mathbb{R}^n : \boldsymbol{x}\cdot\boldsymbol{e} \leqslant \gamma\} \quad \text{or} \quad V = \{\boldsymbol{x} \in \mathbb{R}^n : \boldsymbol{x}\cdot\boldsymbol{e} \geqslant \gamma\}.$$

Repeating the above argument, we conclude that $\varnothing \neq V \cap L \neq L$. Hence D is a closed halfplane of L.

The proof of assertion (*b*) is similar. □

Corollary 2.71. *If $L \subset \mathbb{R}^n$ is a plane of positive dimension m, then the boundary plane of any halfplane of L has dimension $m - 1$.*

Proof. Let D be a closed halfplane of L (the case of open halfplanes is similar). By Theorem 2.70, there is a vector $\boldsymbol{e} \in \mathbb{R}^n \setminus \operatorname{ort} L$ and a scalar γ such that D is expressed by one of the forms (2.23). According to Definition 2.69, the boundary plane M of D equals $H \cap L$, where H is the hyperplane given by $H = \{\boldsymbol{x} \in \mathbb{R}^n : \boldsymbol{x}\cdot\boldsymbol{e} = \gamma\}$. The condition $\boldsymbol{e} \notin \operatorname{ort} L$ and Theorem 2.19 imply that $\dim M = \dim L - 1 = m - 1$. □

Theorem 2.73 below describes halfplanes (closed or open) which have a given boundary plane (see Definition 2.69). We will need the following lemma.

Lemma 2.72. *Let* $L \subset \mathbb{R}^n$ *be a plane of positive dimension* m*,* M *be a nonempty proper subplane* M *of* L*, and* $\boldsymbol{a}$ *be a point in* M*. Choose a vector* $\boldsymbol{e} \in \operatorname{ort} M \setminus \operatorname{ort} L$ *and let* $H = \{\boldsymbol{x} \in \mathbb{R}^n : \boldsymbol{x}\cdot\boldsymbol{e} = \gamma\}$*, where* $\gamma = \boldsymbol{a}\cdot\boldsymbol{e}$*. Then*

$$M \subset H \cap L = \{\boldsymbol{x} \in L : \boldsymbol{x}\cdot\boldsymbol{e} = \gamma\},$$

with $M = H \cap L$ *if and only if* $\dim M = m - 1$.

Proof. According to Definition 2.3 and Theorem 2.18,

$$\operatorname{dir} M = (\operatorname{ort} M)^{\perp} \subset \{\boldsymbol{e}\}^{\perp} = \{\boldsymbol{x} \in \mathbb{R}^n : \boldsymbol{x}\cdot\boldsymbol{e} = 0\} = \operatorname{dir} H.$$

This argument and Theorem 2.2 give the inclusion

$$M = \boldsymbol{a} + \operatorname{dir} M \subset \boldsymbol{a} + \operatorname{dir} H = H.$$

So, $M \subset H \cap L$. By Theorem 2.19, $\dim(H \cap L) = m - 1$. Consequently, $M = H \cap L$ is and only if $\dim M = m - 1$ (see Theorem 2.6). □

Theorem 2.73. *Let* $L \subset \mathbb{R}^n$ *be a plane of positive dimension* m*,* M *be a subplane of* L *of dimension* $m-1$*, and* $\boldsymbol{a}$ *be a point in* M*. Choose a vector* $\boldsymbol{e} \in \operatorname{ort} M \setminus \operatorname{ort} L$ *and let* $\gamma = \boldsymbol{a}\cdot\boldsymbol{e}$*. Then the following assertions hold.*

(*a*) *There are precisely two closed halfplanes* D_1 *and* D_2 *(respectively, precisely two open halfplanes* E_1 *and* E_2*) of* L *whose boundary plane is* M*.*
(*b*) *These halfplanes are described, respectively, by* (2.23) *and* (2.24).

Proof. (*a*) This part immediately follows from assertion (*b*).

(*b*) Let D be a closed halfplane of L with boundary plane M. According to Theorem 2.70, there is a vector $\boldsymbol{c} \in \mathbb{R}^n \setminus \operatorname{ort} L$ and the scalar $\beta = \boldsymbol{a}\cdot\boldsymbol{c}$ such that D can be described by one of the conditions

$$D = \{\boldsymbol{x} \in L : \boldsymbol{x}\cdot\boldsymbol{c} \leqslant \beta\} \quad \text{or} \quad D = \{\boldsymbol{x} \in L : \boldsymbol{x}\cdot\boldsymbol{c} \geqslant \beta\}.$$

Assume that $D = \{\boldsymbol{x} \in L : \boldsymbol{x}\cdot\boldsymbol{c} \leqslant \beta\}$ (the second case is similar). Since $M = H \cap L$, where $H = \{\boldsymbol{x} \in \mathbb{R}^n : \boldsymbol{x}\cdot\boldsymbol{c} = \beta\}$ (see Lemma 2.72), one has $\boldsymbol{c} \in \operatorname{ort} H \subset \operatorname{ort} M$. Furthermore, the assumptions $M \subset L$ and $\dim M = \dim L - 1 = m - 1$ give $\operatorname{ort} L \subset \operatorname{ort} M$ and

$$\dim(\operatorname{ort} M) = n - m + 1 = \dim(\operatorname{ort} L) + 1.$$

Because $\boldsymbol{c} \in \operatorname{ort} M \setminus \operatorname{ort} L$, we can write $\operatorname{ort} M = \operatorname{ort} L + l$, where $l = \operatorname{span}\{\boldsymbol{e}\}$. Consequently, $\boldsymbol{c} = \boldsymbol{b} + \lambda \boldsymbol{e}$, where $\boldsymbol{b} \in \operatorname{ort} L$. Furthermore, $\lambda \neq 0$ due to $\boldsymbol{c} \notin \operatorname{ort} L$. If $\lambda > 0$, then

$$\begin{aligned} D &= \{\boldsymbol{x} \in L : \boldsymbol{x} \cdot \boldsymbol{c} \leqslant \beta\} = \{\boldsymbol{x} \in \boldsymbol{a} + \operatorname{ort} L : \boldsymbol{x} \cdot \boldsymbol{c} \leqslant \boldsymbol{a} \cdot \boldsymbol{c}\} \\ &= \boldsymbol{a} + \{\boldsymbol{x} \in \operatorname{ort} L : \boldsymbol{x} \cdot \boldsymbol{c} \leqslant 0\} = \boldsymbol{a} + \{\boldsymbol{x} \in \operatorname{ort} L : \boldsymbol{x} \cdot (\boldsymbol{b} + \lambda \boldsymbol{e}) \leqslant 0\} \\ &= \boldsymbol{a} + \{\boldsymbol{x} \in \operatorname{ort} L : \boldsymbol{x} \cdot \boldsymbol{e} \leqslant 0\} = \{\boldsymbol{x} \in \boldsymbol{a} + \operatorname{ort} L : \boldsymbol{x} \cdot \boldsymbol{e} \leqslant \boldsymbol{a} \cdot \boldsymbol{e}\} \\ &= \{\boldsymbol{x} \in L : \boldsymbol{x} \cdot \boldsymbol{e} \leqslant \gamma\} = D_1. \end{aligned}$$

Similarly, $D = D_2$ if $\lambda < 0$.

The case of open halfplanes of L is similar. □

2.5.2 *Algebraic Descriptions of Halfplanes*

Theorem 2.74. *Given a plane $L \subset \mathbb{R}^n$ of positive dimension m, the following assertions hold.*

(*a*) *A set $D \subset \mathbb{R}^n$ is a closed halfplane of L if and only if D can be expressed as $D = h + M$, where M is an $(m-1)$-dimensional subplane of L and h is a closed halfline with endpoint $\boldsymbol{o}$, which lies in* $\operatorname{dir} L$ *but not in* $\operatorname{dir} M$.

(*b*) *A set $E \subset \mathbb{R}^n$ is an open halfplane of L if and only if E can be expressed as $E = h + M$, where M is an $(m-1)$-dimensional subplane of L and h is an open halfline with endpoint $\boldsymbol{o}$, which lies in* $\operatorname{dir} L$ *but not in* $\operatorname{dir} M$.

Proof. (*a*) Let $D \subset \mathbb{R}^n$ be a closed halfplane of L. By Theorem 2.70, $D = \{\boldsymbol{x} \in L : \boldsymbol{x} \cdot \boldsymbol{e} \leqslant \gamma\}$ for a suitable vector $\boldsymbol{e} \in \mathbb{R}^n \setminus \operatorname{ort} L$ and a scalar γ. The boundary plane M of D has dimension $m - 1$ and is given by $M = \{\boldsymbol{x} \in L : \boldsymbol{x} \cdot \boldsymbol{e} = \gamma\}$ (see Corollary 2.71). Choose a point $\boldsymbol{a} \in M$ and let $S = L - \boldsymbol{a}$, $T = M - \boldsymbol{a}$, and $F = D - \boldsymbol{a}$. By Theorem 2.2, $L = \boldsymbol{a} + S$ and $M = \boldsymbol{a} + T$. Clearly, $F = \{\boldsymbol{x} \in S : \boldsymbol{x} \cdot \boldsymbol{e} \leqslant 0\}$ is a closed halfplane of S with the boundary plane T.

Choose a basis $\{\boldsymbol{b}_1, \ldots, \boldsymbol{b}_{m-1}\}$ for T and a vector $\boldsymbol{b}_m \in F \setminus T$. Then $\{\boldsymbol{b}_1, \ldots, \boldsymbol{b}_m\}$ is a basis for S such that $\boldsymbol{b}_i \cdot \boldsymbol{e} = 0$ for all $1 \leqslant i \leqslant m-1$, and $\boldsymbol{b}_m \cdot \boldsymbol{e} < 0$. Since every vector $\boldsymbol{x} \in S$ can be written as $\boldsymbol{x} = \xi_1 \boldsymbol{b}_1 + \cdots + \xi_m \boldsymbol{b}_m$, one has

$$\begin{aligned} F = \{\boldsymbol{x} \in S : \boldsymbol{x} \cdot \boldsymbol{e} \leqslant 0\} &= \{\xi_1 \boldsymbol{b}_1 + \cdots + \xi_m \boldsymbol{b}_m : (\xi_m \boldsymbol{b}_m) \cdot \boldsymbol{e} \leqslant 0\} \\ &= \{\xi_1 \boldsymbol{b}_1 + \cdots + \xi_m \boldsymbol{b}_m : \xi_m \geqslant 0\}. \end{aligned}$$

Consider the closed halfline $h = \{\lambda \boldsymbol{b}_m : \lambda \geqslant 0\}$. Obviously, h lies in S but not in T. The above argument shows that a point $\boldsymbol{x} \in S$ belongs to F if and only if $\boldsymbol{x} = \boldsymbol{y} + \boldsymbol{z}$, where $\boldsymbol{y} \in h$ and $\boldsymbol{z} \in T$. Thus $F = h + T$, which gives

$$D = \boldsymbol{a} + F = \boldsymbol{a} + h + T = h + M.$$

Repeating the above argument in the inverse order, we conclude that any set of the form $D = h + M$ is expressible as $D = \{\boldsymbol{x} \in L : \boldsymbol{x} \cdot \boldsymbol{e} \leqslant \gamma\}$; so, it is a closed halfplane of L.

The proof of assertion (b) is similar. □

A modification in the proof of Theorem 2.74 allows us to describe half-planes in terms of affine coordinates.

$$D = \{\lambda_1 \boldsymbol{a}_1 + \lambda_2 \boldsymbol{a}_2 + \lambda_3 \boldsymbol{a}_3 :$$
$$\lambda_1 + \lambda_2 + \lambda_3 = 1, \lambda_3 \geqslant 0\}$$

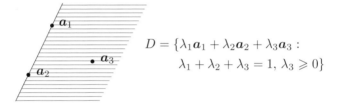

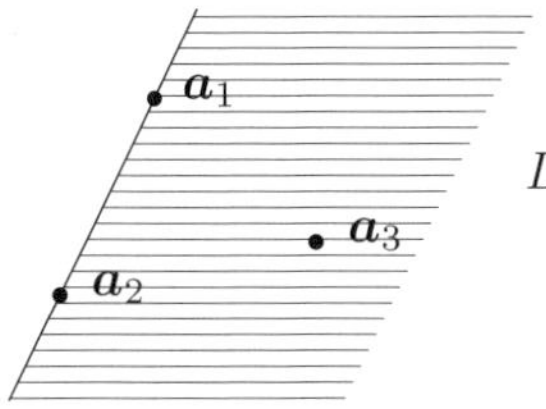

Fig. 2.15   A closed 2-dimensional halfplane in affine coordinates.

Theorem 2.75. *Let $L \subset \mathbb{R}^n$ be a plane of positive dimension m. The following assertions hold.*

(a) *A set $D \subset \mathbb{R}^n$ is a closed halfplane of L if and only if there is an affine basis $\{\boldsymbol{a}_1, \ldots, \boldsymbol{a}_{m+1}\}$ for L such that*

$$D = \{\lambda_1 \boldsymbol{a}_1 + \cdots + \lambda_{m+1} \boldsymbol{a}_{m+1} : \lambda_1 + \cdots + \lambda_{m+1} = 1, \ \lambda_{m+1} \geqslant 0\}. \quad (2.25)$$

(b) *A set $E \subset \mathbb{R}^n$ is an open halfplane of L if and only if there is an affine basis $\{\boldsymbol{a}_1, \ldots, \boldsymbol{a}_{m+1}\}$ for L such that*

$$E = \{\lambda_1 \boldsymbol{a}_1 + \cdots + \lambda_{m+1} \boldsymbol{a}_{m+1} : \lambda_1 + \cdots + \lambda_{m+1} = 1, \ \lambda_{m+1} > 0\}. \quad (2.26)$$

Proof. We will consider the case of closed halfplanes (the case of open halfplanes is similar). Repeating the proof of Theorem 2.74, we obtain that a set $D \subset \mathbb{R}^n$ is a closed halfplane of L if and only if there is a translate $F = D - \boldsymbol{a}$ expressible in the form

$$F = \{\xi_1 \boldsymbol{b}_1 + \cdots + \xi_m \boldsymbol{b}_m : \xi_m \geqslant 0\},$$

where $\{\boldsymbol{b}_1, \ldots, \boldsymbol{b}_m\}$ is a suitable basis for the direction space dir L.

Let $\boldsymbol{a}_1 = \boldsymbol{a}$ and $\boldsymbol{a}_{i+1} = \boldsymbol{a} + \boldsymbol{b}_i$, $1 \leqslant i \leqslant m$. By Theorem 2.52, the set $\{\boldsymbol{a}_1, \ldots, \boldsymbol{a}_{m+1}\}$ is affinely independent. Furthermore, with

$$\lambda_1 = 1 - \xi_1 - \cdots - \xi_m, \quad \lambda_{i+1} = \xi_i, \quad 1 \leqslant i \leqslant m,$$

the halfplane D can be described as

$$\begin{aligned} D &= \boldsymbol{a} + F = \boldsymbol{a} + \{\xi_1 \boldsymbol{b}_1 + \cdots + \xi_m \boldsymbol{b}_m : \xi_m \geqslant 0\} \\ &= \{\lambda_1 \boldsymbol{a} + \xi_1(\boldsymbol{a} + \boldsymbol{b}_1) + \cdots + \xi_m(\boldsymbol{a} + \boldsymbol{b}_m) : \xi_m \geqslant 0\} \\ &= \{\lambda_1 \boldsymbol{a}_1 + \cdots + \lambda_{m+1} \boldsymbol{a}_{m+1} : \lambda_1 + \cdots + \lambda_{m+1} = 1, \ \lambda_{m+1} \geqslant 0\}. \end{aligned}$$

In a similar way, we obtain that every set of the form (2.25) is a closed halfplane of L. □

2.6 Affine Transformations

2.6.1 *Basic Properties of Affine Transformations*

Definition 2.76. A mapping $f : \mathbb{R}^n \to \mathbb{R}^m$ is called an *affine transformation* provided it can be expressed as $f(\boldsymbol{x}) = \boldsymbol{a} + g(\boldsymbol{x})$, where $\boldsymbol{a} \in \mathbb{R}^m$ and $g : \mathbb{R}^n \to \mathbb{R}^m$ is a linear transformation. An *affine functional* on $\mathbb{R}^n$ is an affine transformation $f : \mathbb{R}^n \to \mathbb{R}$.

Theorem 2.77. *Let an affine transformation $f : \mathbb{R}^n \to \mathbb{R}^m$ be expressed as $f(\boldsymbol{x}) = \boldsymbol{a} + g(\boldsymbol{x})$, where $\boldsymbol{a} \in \mathbb{R}^m$ and $g : \mathbb{R}^n \to \mathbb{R}^m$ is a linear transformation. The following assertions hold.*

(*a*) *f is uniquely expressible in the form $f(\boldsymbol{x}) = \boldsymbol{a} + g(\boldsymbol{x})$.*
(*b*) *f is one-to-one (respectively, onto) if and only if so is g.*
(*c*) *f is a continuous mapping.*
(*d*) *If $X \subset \mathbb{R}^n$, $Y \subset \mathbb{R}^m$, and $\boldsymbol{c} \in g^{-1}(\boldsymbol{a})$, then*

$$f(X) = \boldsymbol{a} + g(X) \quad \text{and} \quad f^{-1}(Y) = -\boldsymbol{c} + g^{-1}(Y).$$

(*e*) *If f is invertible, then its inverse, given by $f^{-1}(\boldsymbol{x}) = -g^{-1}(\boldsymbol{a}) + g^{-1}(\boldsymbol{x})$, also is an affine transformation.*

Proof. (*a*) Let $f(\boldsymbol{x}) = \boldsymbol{a}_1 + g_1(\boldsymbol{x}) = \boldsymbol{a}_2 + g_2(\boldsymbol{x})$, where $\boldsymbol{a}_1, \boldsymbol{a}_2 \in \mathbb{R}^m$ and both g_1 and g_2 are linear transformations. Then $\boldsymbol{a}_1 = f(\boldsymbol{o}) = \boldsymbol{a}_2$, and

$$g_1(\boldsymbol{x}) = f(\boldsymbol{x}) - \boldsymbol{a}_1 = f(\boldsymbol{x}) - \boldsymbol{a}_2 = g_2(\boldsymbol{x}) \quad \text{for all} \quad \boldsymbol{x} \in \mathbb{R}^n.$$

(*b*) This assertion is obvious.

(*c*) The transformation f is continuous as the composition of continuous mappings $g : \mathbb{R}^n \to \mathbb{R}^m$ and $t : \mathbb{R}^m \to \mathbb{R}^m$, where $t(\boldsymbol{y}) = \boldsymbol{a} + \boldsymbol{y}$.

(*d*) The equality $f(X) = \boldsymbol{a} + g(X)$ is obvious. Next, given a point $\boldsymbol{x} \in \mathbb{R}^n$, the equalities

$$f(-\boldsymbol{c} + \boldsymbol{x}) = \boldsymbol{a} + g(-\boldsymbol{c} + \boldsymbol{x}) = \boldsymbol{a} - \boldsymbol{a} + g(\boldsymbol{x}) = g(\boldsymbol{x}),$$

show that $\boldsymbol{x} \in g^{-1}(Y)$ if and only if $-\boldsymbol{c} + \boldsymbol{x} \in f^{-1}(Y)$.

(*e*) Suppose that f is invertible. Then g is invertible and g^{-1} is a linear transformation (see page 7). Consider the affine transformation $h(\boldsymbol{x}) = -g^{-1}(\boldsymbol{a}) + g^{-1}(\boldsymbol{x})$. The identities

$$f(h(\boldsymbol{x})) = \boldsymbol{a} + g(-g^{-1}(\boldsymbol{a}) + g^{-1}(\boldsymbol{x})) = \boldsymbol{a} - \boldsymbol{a} + \boldsymbol{x} = \boldsymbol{x},$$
$$h(f(\boldsymbol{x})) = -g^{-1}(\boldsymbol{a}) + g^{-1}(\boldsymbol{a} + g(\boldsymbol{x})) = -g^{-1}(\boldsymbol{a}) + g^{-1}(\boldsymbol{a}) + \boldsymbol{x} = \boldsymbol{x}$$

show that $h = f^{-1}$. □

Theorem 2.78. *For an affine transformation $f : \mathbb{R}^n \to \mathbb{R}^m$, the following assertions hold.*

(*a*) *If $\{\boldsymbol{a}_1, \dots, \boldsymbol{a}_r\}$ is an affinely independent set in $\mathbb{R}^n$ and f is one-to-one, then the set $\{f(\boldsymbol{a}_1), \dots, f(\boldsymbol{a}_r)\}$ is affinely independent.*

(*b*) *If $\{\boldsymbol{c}_1, \dots, \boldsymbol{c}_r\}$ is an affinely independent set in $\mathbb{R}^m$ and points $\boldsymbol{a}_1, \dots, \boldsymbol{a}_r \in \mathbb{R}^n$ satisfy the conditions $f(\boldsymbol{a}_i) = \boldsymbol{c}_i$, $1 \leqslant i \leqslant r$, then the set $\{\boldsymbol{a}_1, \dots, \boldsymbol{a}_r\}$ is affinely independent.*

Proof. Let $f(\boldsymbol{x}) = \boldsymbol{a} + g(\boldsymbol{x})$, where $\boldsymbol{a} \in \mathbb{R}^m$ and $g : \mathbb{R}^n \to \mathbb{R}^m$ is a linear transformation. Since both assertions (*a*) and (*b*) are obvious when $r = 1$, we may assume that $r \geqslant 2$.

(*a*) By Theorem 2.52, the set $\{\boldsymbol{a}_2 - \boldsymbol{a}_1, \dots, \boldsymbol{a}_r - \boldsymbol{a}_1\}$ is linearly independent. Because g is one-to-one, the set $\{g(\boldsymbol{a}_2 - \boldsymbol{a}_1), \dots, g(\boldsymbol{a}_r - \boldsymbol{a}_1)\}$ also is linearly independent (see page 6). The equalities

$$f(\boldsymbol{a}_i) - f(\boldsymbol{a}_1) = g(\boldsymbol{a}_i) - g(\boldsymbol{a}_1) = g(\boldsymbol{a}_i - \boldsymbol{a}_1), \quad 2 \leqslant i \leqslant r,$$

and Theorem 2.52 show that $\{f(\boldsymbol{a}_1), \dots, f(\boldsymbol{a}_r)\}$ is affinely independent.

(*b*) As above, the set $\{\boldsymbol{c}_2 - \boldsymbol{c}_1, \dots, \boldsymbol{c}_r - \boldsymbol{c}_1\}$ is linearly independent, and the equalities

$$\boldsymbol{a}_i - \boldsymbol{a}_1 = f(\boldsymbol{a}_i) - f(\boldsymbol{a}_1) = g(\boldsymbol{a}_i) - g(\boldsymbol{a}_1) = g(\boldsymbol{c}_i - \boldsymbol{c}_1), \quad 2 \leqslant i \leqslant r,$$

show that the set $\{\boldsymbol{a}_2 - \boldsymbol{a}_1, \dots, \boldsymbol{a}_r - \boldsymbol{a}_1\}$ also is linearly independent (see page 6). Finally, Theorem 2.52 implies that the set $\{\boldsymbol{a}_1, \dots, \boldsymbol{a}_r\}$ is affinely independent. □

Theorem 2.79. *If $f : \mathbb{R}^n \to \mathbb{R}^m$ is an affine transformation and $X \subset \mathbb{R}^n$ is a set, then $\dim X \geqslant \dim f(X)$, with $\dim X = \dim f(X)$ provided f is one-to-one.*

Proof. Let $m = \dim f(X)$. Since the case $X = \varnothing$ is obvious, we assume that $m \geqslant 0$. Choose in $f(X)$ an affinely independent set $\{\boldsymbol{c}_1, \ldots, \boldsymbol{c}_{m+1}\}$ (see Corollary 2.66). If $\boldsymbol{a}_1, \ldots, \boldsymbol{a}_{m+1}$ are points in X satisfying the conditions $f(\boldsymbol{a}_i) = \boldsymbol{c}_i$, $1 \leqslant i \leqslant m+1$, then Theorem 2.78 shows that the set $\{\boldsymbol{a}_1, \ldots, \boldsymbol{a}_{m+1}\}$ is affinely independent. By the same corollary, $\dim X \geqslant m = \dim f(X)$.

Suppose that f is one-to-one. As above, X contains an affinely independent set $\{\boldsymbol{a}_1, \ldots, \boldsymbol{a}_{m+1}\}$. By Theorem 2.78, the set $\{f(\boldsymbol{a}_1), \ldots, f(\boldsymbol{a}_{m+1})\}$ is affinely independent. Corollary 2.66 implies that $\dim f(X) \geqslant m = \dim X$, and the above argument gives the equality $\dim f(X) = \dim X$. □

Because any affine transformation $f : \mathbb{R}^n \to \mathbb{R}^m$ is continuous (see Theorem 2.77), the following result is an immediate consequence of Problems 1.16, 1.19, and 1.20.

Corollary 2.80. *Let $f : \mathbb{R}^n \to \mathbb{R}^m$ be an affine transformation and $X \subset \mathbb{R}^n$ and $Y \subset \mathbb{R}^m$ be sets. Then*

$$f(\operatorname{cl} X) \subset \operatorname{cl} f(X) = \operatorname{cl} f(\operatorname{cl} X), \tag{2.27}$$

with $f(\operatorname{cl} X) = \operatorname{cl} f(X)$ if f is one-to-one or X is bounded. Furthermore,

$$\operatorname{cl} f^{-1}(Y) = \operatorname{cl} f^{-1}(Y \cap \operatorname{rng} f) = f^{-1}(\operatorname{cl}(Y \cap \operatorname{rng} f)) \subset f^{-1}(\operatorname{cl} Y). \qquad \square$$

Remark. The inclusion $f(\operatorname{cl} X) \subset \operatorname{cl} f(X)$ in Corollary 2.80 may be proper if X is unbounded and f is not one-to-one. For instance, let f be the orthogonal projection of $\mathbb{R}^2$ on its x-axis, and let $X = \{(x, y) : x > 0, xy \geqslant 1\}$. Then X is a closed set, while $f(X)$ is the open halfline $\{(x, 0) : x > 0\}$.

2.6.2 *Affine Transformations and Affine Combinations*

Theorem 2.81. *For a mapping $f : \mathbb{R}^n \to \mathbb{R}^m$, the following conditions are equivalent.*

(a) f is an affine transformation.
(b) $f(\lambda \boldsymbol{x} + \mu \boldsymbol{y}) = (1 - \lambda - \mu) f(\boldsymbol{o}) + \lambda f(\boldsymbol{x}) + \mu f(\boldsymbol{y})$ whenever $\boldsymbol{x}, \boldsymbol{y} \in \mathbb{R}^n$ and $\lambda, \mu \in \mathbb{R}$.
(c) $f((1-\lambda)\boldsymbol{x} + \lambda \boldsymbol{y}) = (1-\lambda) f(\boldsymbol{x}) + \lambda f(\boldsymbol{y})$ whenever $\boldsymbol{x}, \boldsymbol{y} \in \mathbb{R}^n$ and $\lambda \in \mathbb{R}$.

Proof. $(a) \Rightarrow (b)$ Let f be an affine transformation, expressed as $f(\boldsymbol{x}) = \boldsymbol{a} + g(\boldsymbol{x})$, where $\boldsymbol{a} \in \mathbb{R}^m$ and $g : \mathbb{R}^n \to \mathbb{R}^m$ is a linear transformation. Then $f(\boldsymbol{o}) = \boldsymbol{a}$. For points $\boldsymbol{x}, \boldsymbol{y} \in \mathbb{R}^n$ and scalars λ, μ, one has

$$\begin{aligned} f(\lambda\boldsymbol{x} + \mu\boldsymbol{y}) &= \boldsymbol{a} + g(\lambda\boldsymbol{x} + \mu\boldsymbol{y}) = \boldsymbol{a} + \lambda g(\boldsymbol{x}) + \mu g(\boldsymbol{y}) \\ &= (1 - \lambda - \mu)\boldsymbol{a} + \lambda(\boldsymbol{a} + g(\boldsymbol{x})) + \mu(\boldsymbol{a} + g(\boldsymbol{y})) \\ &= (1 - \lambda - \mu)f(\boldsymbol{o}) + \lambda f(\boldsymbol{x}) + \mu f(\boldsymbol{y}). \end{aligned}$$

The assertion $(b) \Rightarrow (c)$ is obvious; so, it suffices to show that (c) implies (a). Assuming (c), we are going to prove that the mapping $g(\boldsymbol{x}) = f(\boldsymbol{x}) - f(\boldsymbol{o})$ is a linear transformation. Indeed, for every $\boldsymbol{x} \in \mathbb{R}^n$ and $\lambda \in \mathbb{R}$,

$$\begin{aligned} g(\lambda\boldsymbol{x}) &= f((1 - \lambda)\boldsymbol{o} + \lambda\boldsymbol{x}) - f(\boldsymbol{o}) \\ &= (1 - \lambda)f(\boldsymbol{o}) + \lambda f(\boldsymbol{x}) - f(\boldsymbol{o}) \\ &= \lambda(f(\boldsymbol{x}) - f(\boldsymbol{o})) = \lambda g(\boldsymbol{x}). \end{aligned}$$

With $\lambda = \frac{1}{2}$, the last equality gives

$$\begin{aligned} g(\boldsymbol{x} + \boldsymbol{y}) &= 2(\tfrac{1}{2}g(\boldsymbol{x} + \boldsymbol{y})) = 2g(\tfrac{1}{2}\boldsymbol{x} + \tfrac{1}{2}\boldsymbol{y}) \\ &= 2g((1 - \tfrac{1}{2})\boldsymbol{x} + \tfrac{1}{2}\boldsymbol{y}) \\ &= 2f((1 - \tfrac{1}{2})\boldsymbol{x} + \tfrac{1}{2}\boldsymbol{y}) - 2f(\boldsymbol{o}) \\ &= 2((1 - \tfrac{1}{2})f(\boldsymbol{x}) + \tfrac{1}{2}f(\boldsymbol{y})) - 2f(\boldsymbol{o}) \\ &= (f(\boldsymbol{x}) - f(\boldsymbol{o})) + (f(\boldsymbol{y}) - f(\boldsymbol{o})) \\ &= g(\boldsymbol{x}) + g(\boldsymbol{y}). \end{aligned}$$

Hence $g(\boldsymbol{x})$ is a linear transformation, and $f(\boldsymbol{x}) = f(\boldsymbol{o}) + g(\boldsymbol{x})$ is an affine one. □

Theorem 2.82. *A mapping $f : \mathbb{R}^n \to \mathbb{R}^m$ is an affine transformation if and only if for a given integer $r \geqslant 2$ it preserves all affine combinations of r points from $\mathbb{R}^n$:*

$$f(\lambda_1\boldsymbol{x}_1 + \cdots + \lambda_r\boldsymbol{x}_r) = \lambda_1 f(\boldsymbol{x}_1) + \cdots + \lambda_r f(\boldsymbol{x}_r)$$

whenever $\boldsymbol{x}_1, \ldots, \boldsymbol{x}_r \in \mathbb{R}^n$ and $\lambda_1 + \cdots + \lambda_r = 1$.

Proof. Suppose that f is an affine transformation. By induction on r, we are going to prove that f preserves affine combinations of any r points of $\mathbb{R}^n$. The case $r = 2$ is proved in Theorem 2.81. Assume that the assertion holds for all positive integers $r \leqslant k - 1$, where $k \geqslant 3$, and choose an affine combination

$$\lambda_1\boldsymbol{x}_1 + \cdots + \lambda_k\boldsymbol{x}_k, \quad \text{where} \quad \boldsymbol{x}_1, \ldots, \boldsymbol{x}_k \in \mathbb{R}^n.$$

At least one of $\lambda_1, \ldots, \lambda_k$ is not 1, since otherwise $\lambda_1 + \cdots + \lambda_k = k \geqslant 3$. Let, for instance, $\lambda_1 \neq 1$. Consider the affine combination

$$\boldsymbol{z} = \frac{\lambda_2}{1-\lambda_1}\boldsymbol{x}_2 + \cdots + \frac{\lambda_k}{1-\lambda_1}\boldsymbol{x}_k.$$

By the induction hypothesis,

$$f(\boldsymbol{z}) = \frac{\lambda_2}{1-\lambda_1} f(\boldsymbol{x}_2) + \cdots + \frac{\lambda_k}{1-\lambda_1} f(\boldsymbol{x}_k).$$

Using the case $r = 2$, one has

$$\begin{aligned} f(\lambda_1\boldsymbol{x}_1 + \cdots + \lambda_k\boldsymbol{x}_k) &= f(\lambda_1\boldsymbol{x}_1 + (1-\lambda_1)\boldsymbol{z}) = \lambda_1 f(\boldsymbol{x}_1) + (1-\lambda_1) f(\boldsymbol{z}) \\ &= \lambda_1 f(\boldsymbol{x}_1) + (1-\lambda_1)\Big(\frac{\lambda_2}{1-\lambda_1} f(\boldsymbol{x}_2) + \cdots + \frac{\lambda_k}{1-\lambda_1} f(\boldsymbol{x}_k)\Big) \\ &= \lambda_1 f(\boldsymbol{x}_1) + \cdots + \lambda_k f(\boldsymbol{x}_k). \end{aligned}$$

Conversely, suppose that f preserves affine combinations of any r points from $\mathbb{R}^n$. If $r = 2$, then Theorem 2.81 shows that f is an affine transformation. Let $r \geqslant 3$. Choose points $\boldsymbol{x}, \boldsymbol{y} \in \mathbb{R}^n$ and append to them some points $\boldsymbol{x}_3, \ldots, \boldsymbol{x}_r \in \mathbb{R}^n$. Given a scalar $\lambda \in \mathbb{R}$, one has

$$\begin{aligned} f((1-\lambda)\boldsymbol{x} + \lambda\boldsymbol{y}) &= f((1-\lambda)\boldsymbol{x} + \lambda\boldsymbol{y} + 0\boldsymbol{x}_3 + \cdots + 0\boldsymbol{x}_r) \\ &= (1-\lambda) f(\boldsymbol{x}) + \lambda f(\boldsymbol{y}) + 0 f(\boldsymbol{x}_3) + \cdots + 0 f(\boldsymbol{x}_r) \\ &= (1-\lambda) f(\boldsymbol{x}) + \lambda f(\boldsymbol{y}). \end{aligned}$$

Consecutively, Theorem 2.81 implies that f is an affine transformation. □

Theorem 2.83. *For an affine transformation $f : \mathbb{R}^n \to \mathbb{R}^m$ and sets $X \subset \mathbb{R}^n$ and $Y \subset \mathbb{R}^m$, one has*

$$\operatorname{aff} f(X) = f(\operatorname{aff} X), \tag{2.28}$$

$$\operatorname{aff} f^{-1}(Y) = f^{-1}(\operatorname{aff}(Y \cap \operatorname{rng} f)) \subset f^{-1}(\operatorname{aff} Y). \tag{2.29}$$

Proof. Excluding the obvious cases $X = \varnothing$ and $Y \cap \operatorname{rng} f = \varnothing$, we assume that both sets X and $Y \cap \operatorname{rng} f$ are nonempty. Consequently, $f^{-1}(Y) \neq \varnothing$.

For (2.28), choose a point $\boldsymbol{x} \in \operatorname{aff} f(X)$. By Theorem 2.44, $\boldsymbol{x}$ can be written as an affine combination

$$\boldsymbol{x} = \lambda_1\boldsymbol{x}_1 + \cdots + \lambda_p\boldsymbol{x}_p, \quad \text{where} \quad \boldsymbol{x}_1, \ldots, \boldsymbol{x}_p \in f(X).$$

Let $\boldsymbol{z}_1, \ldots, \boldsymbol{z}_p$ be points in X such that $f(\boldsymbol{z}_i) = \boldsymbol{x}_i$ for all $1 \leqslant i \leqslant p$. Put $\boldsymbol{z} = \lambda_1\boldsymbol{z}_1 + \cdots + \lambda_p\boldsymbol{z}_p$. Then $\boldsymbol{z} \in \operatorname{aff} X$ by the same theorem, and

$$\boldsymbol{x} = \lambda_1\boldsymbol{x}_1 + \cdots + \lambda_p\boldsymbol{x}_p = \lambda_1 f(\boldsymbol{z}_1) + \cdots + \lambda_p f(\boldsymbol{z}_p) = f(\boldsymbol{z})$$

according to Theorem 2.82. Hence $\boldsymbol{x} = f(\boldsymbol{z}) \in f(\operatorname{aff} X)$, which proves the inclusion $\operatorname{aff} f(X) \subset f(\operatorname{aff} X)$.

Conversely, let $\boldsymbol{x} \in \operatorname{aff} X$. Similarly to the above, $\boldsymbol{x}$ can be expressed as an affine combination

$$\boldsymbol{x} = \mu_1 \boldsymbol{x}_1 + \cdots + \mu_q \boldsymbol{x}_q, \quad \text{where} \quad \boldsymbol{x}_1, \ldots, \boldsymbol{x}_q \in X.$$

Then $f(\boldsymbol{x}) = \mu_1 f(\boldsymbol{x}_1) + \cdots + \mu_q f(\boldsymbol{x}_q)$ according to Theorem 2.82, and Theorem 2.44 gives $f(\boldsymbol{x}) \in \operatorname{aff} f(X)$. Hence $f(\operatorname{aff} X) \subset \operatorname{aff} f(X)$.

It remains to prove (2.29). Letting $X = f^{-1}(Y)$ in (2.28), one has

$$f(\operatorname{aff} f^{-1}(Y)) = \operatorname{aff} f(f^{-1}(Y)) = \operatorname{aff}(Y \cap \operatorname{rng} f).$$

Hence

$$\operatorname{aff} f^{-1}(Y) \subset f^{-1}(f(\operatorname{aff} f^{-1}(Y))) = f^{-1}(\operatorname{aff}(Y \cap \operatorname{rng} f)).$$

For the opposite inclusion, let $\boldsymbol{x} \in f^{-1}(\operatorname{aff}(Y \cap \operatorname{rng} f))$. Then $f(\boldsymbol{x})$ belongs to $\operatorname{aff}(Y \cap \operatorname{rng} f)$, and Theorem 2.44 implies that $f(\boldsymbol{x})$ can be written as an affine combination

$$f(\boldsymbol{x}) = \gamma_1 \boldsymbol{y}_1 + \cdots + \gamma_r \boldsymbol{y}_r, \quad \text{where} \quad \boldsymbol{y}_1, \ldots, \boldsymbol{y}_r \in Y \cap \operatorname{rng} f.$$

Choose points $\boldsymbol{z}_1, \ldots, \boldsymbol{z}_r \in f^{-1}(Y)$ such that $f(\boldsymbol{z}_i) = \boldsymbol{y}_i$, $1 \leqslant i \leqslant r$, and put $\boldsymbol{z} = \gamma_1 \boldsymbol{z}_1 + \cdots + \gamma_r \boldsymbol{z}_r$. Then $\boldsymbol{z} \in \operatorname{aff} f^{-1}(Y)$ and

$$f(\boldsymbol{z}) = \gamma_1 f(\boldsymbol{z}_1) + \cdots + \gamma_r f(\boldsymbol{z}_r) = \gamma_1 \boldsymbol{y}_1 + \cdots + \gamma_r \boldsymbol{y}_r = f(\boldsymbol{x}).$$

Let $\boldsymbol{u}_i = \boldsymbol{z}_i + (\boldsymbol{x} - \boldsymbol{z})$, $1 \leqslant i \leqslant r$. Since $\boldsymbol{u}_i$ is an affine combination, one has

$$f(\boldsymbol{u}_i) = f(\boldsymbol{z}_i) + f(\boldsymbol{x}) - f(\boldsymbol{z}) = f(\boldsymbol{z}_i) = \boldsymbol{y}_i.$$

Hence $\boldsymbol{u}_i \in f^{-1}(\boldsymbol{y}_i) \subset f^{-1}(Y)$, $1 \leqslant i \leqslant r$. The equalities

$$\boldsymbol{x} = \boldsymbol{u}_i - \boldsymbol{z}_i + \boldsymbol{z}, \quad 1 \leqslant i \leqslant r,$$

show that $\boldsymbol{x}$ can be written as an affine combination of $\boldsymbol{u}_1, \ldots, \boldsymbol{u}_r$:

$$\begin{aligned}
\boldsymbol{x} &= \gamma_1 \boldsymbol{x} + \cdots + \gamma_r \boldsymbol{x} = \gamma_1(\boldsymbol{u}_1 - \boldsymbol{z}_1 + \boldsymbol{z}) + \cdots + \gamma_r(\boldsymbol{u}_r - \boldsymbol{z}_r + \boldsymbol{z}) \\
&= (\gamma_1 \boldsymbol{u}_1 + \cdots + \gamma_r \boldsymbol{u}_r) - (\gamma_1 \boldsymbol{z}_1 + \cdots + \gamma_r \boldsymbol{z}_r) + (\gamma_1 + \cdots + \gamma_r)\boldsymbol{z} \\
&= \gamma_1 \boldsymbol{u}_1 + \cdots + \gamma_r \boldsymbol{u}_r - \boldsymbol{z} + \boldsymbol{z} = \gamma_1 \boldsymbol{u}_1 + \cdots + \gamma_r \boldsymbol{u}_r.
\end{aligned}$$

Therefore, $\boldsymbol{x} \in \operatorname{aff} f^{-1}(Y)$. Summing up,

$$f^{-1}(\operatorname{aff}(Y \cap \operatorname{rng} f)) \subset \operatorname{aff} f^{-1}(Y). \qquad \square$$

Remark. The inclusion $f^{-1}(\text{aff}\,(Y \cap \text{rng}\, f)) \subset f^{-1}(\text{aff}\, Y)$ in (2.29) may be proper. Indeed, let f be the orthogonal projection of $\mathbb{R}^2$ on its x-axis, and let $Y = \{(0,1),(0,-1)\}$. Then $f^{-1}(\text{aff}\,(Y \cap \text{rng}\, f)) = \varnothing$, while $f^{-1}(\text{aff}\, Y)$ is the y-axis.

Corollary 2.84. *If $f : \mathbb{R}^n \to \mathbb{R}^m$ is an affine transformation, then the set* $\text{rng}\, f$ *is a plane in $\mathbb{R}^m$. Furthermore, for any planes $L \subset \mathbb{R}^n$ and $M \subset \mathbb{R}^m$, both sets $f(L)$ and $f^{-1}(M)$ also are planes.*

Proof. Theorems 2.42 and 2.83, combined with the equalities

$$f(L) = f(\text{aff}\, L) = \text{aff}\, f(L),$$

show that $f(L)$ is a plane. In particular, the set $\text{rng}\, f = f(\mathbb{R}^n)$ is a plane.

The set $N = M \cap \text{rng}\, f$ is a plane as the intersection of the planes M and $\text{rng}\, f$ (see Theorem 2.4), and the same Theorem 2.83 gives

$$f^{-1}(M) = f^{-1}(N) = f^{-1}(\text{aff}\, N) = \text{aff}\, f^{-1}(N) = \text{aff}\, f^{-1}(M).$$

Hence $f^{-1}(M)$ is a plane. □

2.6.3 *Homotheties and Affine Projections*

Homotheties and projections are important classes of affine transformations, widely used in the study of convex sets.

Definition 2.85. Given a point $\boldsymbol{a} \in \mathbb{R}^n$ and a scalar $\mu \neq 0$, the affine transformation

$$f(\boldsymbol{x}) = \boldsymbol{a} + \mu \boldsymbol{x}, \quad \boldsymbol{x} \in \mathbb{R}^n,$$

is called a *homothety* with ratio μ. Furthermore, f is a *direct homothety* if $\mu > 0$ (respectively, an *inverse homothety* if $\mu < 0$). The homothety f is called a *contraction* if $0 < |\mu| < 1$ (respectively, an *expansion* if $|\mu| > 1$).

Among all homotheties, those with ratio $\mu = 1$ are the simplest ones: a transformation $f(\boldsymbol{x}) = \boldsymbol{a} + \boldsymbol{x}$ is called the *translate* on a vector $\boldsymbol{a}$. If $\mu \neq 1$, then $f(\boldsymbol{x}) = \boldsymbol{a} + \mu \boldsymbol{x}$ can be written as $f(\boldsymbol{x}) = \boldsymbol{c} + \mu(\boldsymbol{x} - \boldsymbol{c})$, where the point $\boldsymbol{c} = \boldsymbol{a}/(1-\mu)$ is called the *center* of f.

Theorem 2.86. *The following assertions hold.*

(*a*) *Every homothety $f(\boldsymbol{x}) = \boldsymbol{a} + \mu \boldsymbol{x}$ is invertible, and its inverse is a homothety, given by*

$$f^{-1}(\boldsymbol{x}) = -\mu^{-1}\boldsymbol{a} + \mu^{-1}\boldsymbol{x}.$$

(*b*) *The composition of homotheties* $f(\boldsymbol{x}) = \boldsymbol{a} + \mu\boldsymbol{x}$ *and* $h(\boldsymbol{x}) = \boldsymbol{c} + \gamma\boldsymbol{x}$ *also is a homothety, given by*

$$(f \circ h)(\boldsymbol{x}) = (\boldsymbol{a} + \mu\boldsymbol{c}) + \mu\gamma\,\boldsymbol{x}.$$

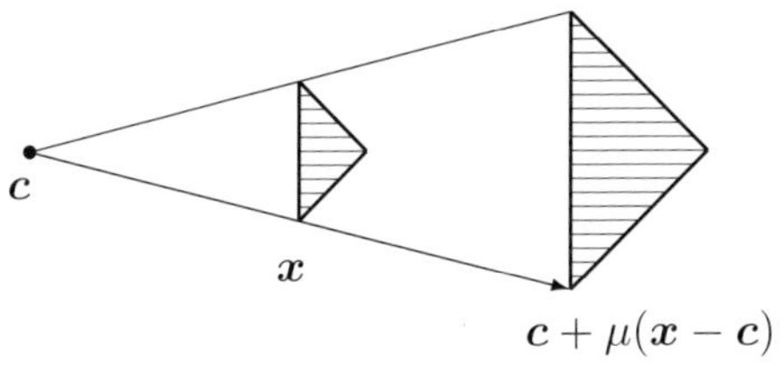

Fig. 2.16 A direct homothety with center $\boldsymbol{c}$.

Proof. (*a*) With $g(\boldsymbol{x}) = -\mu^{-1}\boldsymbol{a} + \mu^{-1}\boldsymbol{x}$, we have

$$(f \circ g)(\boldsymbol{x}) = f(-\mu^{-1}\boldsymbol{a} + \mu^{-1}\boldsymbol{x}) = \boldsymbol{a} + \mu(-\mu^{-1}\boldsymbol{a} + \mu^{-1}\boldsymbol{x}) = \boldsymbol{x},$$
$$(g \circ f)(\boldsymbol{x}) = g(\boldsymbol{a} + \mu\boldsymbol{x}) = -\mu^{1}\boldsymbol{a} + \mu^{-1}(\boldsymbol{a} + \mu\boldsymbol{x}) = \boldsymbol{x}.$$

Hence f is invertible, and $f^{-1}(\boldsymbol{x}) = g(\boldsymbol{x})$.

(*b*) Since $\mu\gamma \neq 0$ and

$$(f \circ h)(\boldsymbol{x}) = f(\boldsymbol{c} + \gamma\boldsymbol{x}) = \boldsymbol{a} + \mu(\boldsymbol{c} + \gamma\boldsymbol{x}) = (\boldsymbol{a} + \mu\boldsymbol{c}) + \mu\gamma\boldsymbol{x},$$

we obtain that $(f \circ h)$ is a homothety. □

Definition 2.87. A mapping $f : \mathbb{R}^n \to \mathbb{R}^n$ is called an *affine projection* if it can be expressed as $f(\boldsymbol{x}) = \boldsymbol{c} + g(\boldsymbol{x} - \boldsymbol{c})$, where $\boldsymbol{c}$ is a point in $\mathbb{R}^n$ and $g : \mathbb{R}^n \to \mathbb{R}^n$ is a linear projection (see page 7).

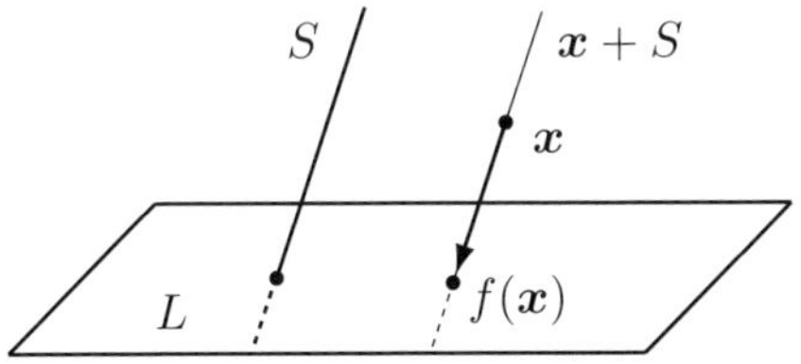

Fig. 2.17 The affine projection on a plane L along a complementary subspace S.

Remark. An affine projection of the form $f(\boldsymbol{x}) = \boldsymbol{c} + g(\boldsymbol{x} - \boldsymbol{c})$ can also be expressed as $f(\boldsymbol{x}) = \boldsymbol{a} + g(\boldsymbol{x})$, where $\boldsymbol{a} = \boldsymbol{c} - g(\boldsymbol{c})$, pointing to an analogy with Definition 2.1.

Theorem 2.88. *A mapping $f : \mathbb{R}^n \to \mathbb{R}^n$ is an affine projection if and only if there is a plane $L \subset \mathbb{R}^n$ and a complementary to L subspace $S \subset \mathbb{R}^n$ such that $f(\boldsymbol{x}) = L \cap (\boldsymbol{x} + S)$ for all $\boldsymbol{x} \in \mathbb{R}^n$.*

Proof. Assume first that f is an affine projection. By the definition, f can be written as $f(\boldsymbol{x}) = \boldsymbol{c} + g(\boldsymbol{x} - \boldsymbol{c})$, where $\boldsymbol{c} \in \mathbb{R}^n$ and $g : \mathbb{R}^n \to \mathbb{R}^n$ is a linear projection. Since the subspaces $T = \operatorname{rng} g$ and $S = \operatorname{null} g$ are complementary (see page 7), Definition 2.9 shows that the planes $L = \boldsymbol{c} + T$ and S also are complementary. Furthermore, $g(\boldsymbol{x}) = T \cap (\boldsymbol{x} + S)$ for all $\boldsymbol{x} \in \mathbb{R}^n$, which gives

$$\begin{aligned} f(\boldsymbol{x}) &= \boldsymbol{c} + g(\boldsymbol{x} - \boldsymbol{c}) = \boldsymbol{c} + T \cap (\boldsymbol{x} - \boldsymbol{c} + S) \\ &= (\boldsymbol{c} + T) \cap (\boldsymbol{x} + S) = L \cap (\boldsymbol{x} + S). \end{aligned}$$

Conversely, suppose the existence of a plane $L \subset \mathbb{R}^n$ and a complementary to L subspace $S \subset \mathbb{R}^n$ such that $f(\boldsymbol{x}) = L \cap (\boldsymbol{x} + S)$ for all $\boldsymbol{x} \in \mathbb{R}^n$. Denote by $\boldsymbol{c}$ the only point of $L \cap S$ (see Theorem 2.11) and let $T = \operatorname{dir} L$. By Theorem 2.2, $L = \boldsymbol{c} + T$. As mentioned above, the mapping $g(\boldsymbol{x}) = T \cap (\boldsymbol{x} + S)$ is the linear projection on T along S. Therefore,

$$\begin{aligned} f(\boldsymbol{x}) &= L \cap (\boldsymbol{x} + S) = (\boldsymbol{c} + T) \cap (\boldsymbol{c} + \boldsymbol{x} - \boldsymbol{c} + S) \\ &= \boldsymbol{c} + T \cap (\boldsymbol{x} - \boldsymbol{c} + S) = \boldsymbol{c} + g(\boldsymbol{x} - \boldsymbol{c}), \end{aligned}$$

which shows that $f(\boldsymbol{x})$ is an affine projection. □

Remarks. 1. In terms of Theorem 2.88, we will say that the affine transformation $f(\boldsymbol{x}) = \boldsymbol{c} + g(\boldsymbol{x} - \boldsymbol{c})$ projects $\mathbb{R}^n$ onto the plane $L = \boldsymbol{c} + \operatorname{rng} g$ *along the subspace* $S = \operatorname{null} g$.

2. In Definition 2.87, one can write $f(\boldsymbol{x}) = \boldsymbol{c}' + g(\boldsymbol{x} - \boldsymbol{c}')$ for any point $\boldsymbol{c}' \in \operatorname{rng} f$. Indeed, $\boldsymbol{c}' - \boldsymbol{c} \in \operatorname{dir}(\operatorname{rng} f)$ (see Theorem 2.2), which gives the equality $g(\boldsymbol{c}' - \boldsymbol{c}) = \boldsymbol{c}' - \boldsymbol{c}$ because g is a linear projection on $\operatorname{rng} g = \operatorname{dir}(\operatorname{rng} f)$. Consequently, $\boldsymbol{c}' - g(\boldsymbol{c}') = \boldsymbol{c} - g(\boldsymbol{c})$ due to the linearity of g. Finally

$$\begin{aligned} f(\boldsymbol{x}) &= \boldsymbol{c} + g(\boldsymbol{x} - \boldsymbol{c}) = (\boldsymbol{c} - g(\boldsymbol{c})) + g(\boldsymbol{x}) \\ &= (\boldsymbol{c}' - g(\boldsymbol{c}')) + g(\boldsymbol{x}) = \boldsymbol{c}' + g(\boldsymbol{x} - \boldsymbol{c}'). \end{aligned}$$

We recall (see page 8) that the linear projection of $\mathbb{R}^n$ on a subspace $S \subset \mathbb{R}^n$ along its orthogonal complement $S^\perp$ is denoted f_S and called the orthogonal projection on S.

Definition 2.89. In terms of Definition 2.87, f will be called the *orthogonal projection* on the plane $L = \boldsymbol{c} + \operatorname{rng} g$, and denoted f_L, provided g is an orthogonal linear projection.

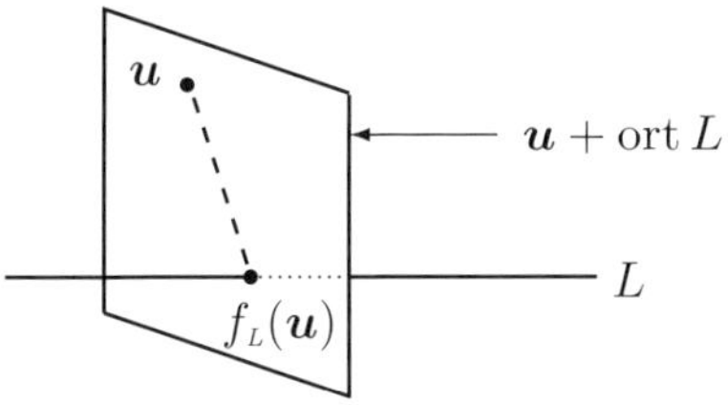

Fig. 2.18 Illustration to Theorem 2.90.

Theorem 2.90. *Let $L \subset \mathbb{R}^n$ be a nonempty plane and f_L be the orthogonal projection on L. For points $\boldsymbol{u} \in \mathbb{R}^n$ and $\boldsymbol{v} \in L$, one has*

$$\|\boldsymbol{u} - \boldsymbol{v}\|^2 = \|\boldsymbol{u} - f_L(\boldsymbol{u})\|^2 + \|f_L(\boldsymbol{u}) - \boldsymbol{v}\|^2.$$

Furthermore, $f_L(\boldsymbol{u})$ is the unique nearest to $\boldsymbol{u}$ point in L, and this property characterizes $f_L(\boldsymbol{u})$. Consequently, $\delta(\boldsymbol{u}, L) = \|\boldsymbol{u} - f_L(\boldsymbol{u})\|$.

Proof. Let T be the direction space of L (see Definition 2.3). Then f_L is the affine projection on L along the subspace $S = T^\perp$. By Theorem 2.11, the set $L \cap S$ is a singleton, say $\{\boldsymbol{c}\}$. Theorem 2.2 shows that $L = \boldsymbol{c} + T$. If g_T is the orthogonal projection on T, then $g_T(\boldsymbol{c}) = \boldsymbol{o}$ and

$$f_L(\boldsymbol{x}) = \boldsymbol{c} + g_T(\boldsymbol{x} - \boldsymbol{c}) = \boldsymbol{c} + g_T(\boldsymbol{x}) \quad \text{for all} \quad \boldsymbol{x} \in \mathbb{R}^n,$$

as shown in the above remark. Let $\boldsymbol{w} = g_T(\boldsymbol{v})$. Then $\boldsymbol{v} = \boldsymbol{c} + \boldsymbol{w}$. We know (see page 8) that

$$\|(\boldsymbol{u} - \boldsymbol{c}) - \boldsymbol{w}\|^2 = \|(\boldsymbol{u} - \boldsymbol{c}) - g_T(\boldsymbol{u} - \boldsymbol{c})\|^2 + \|g_T(\boldsymbol{u} - \boldsymbol{c}) - \boldsymbol{w}\|^2.$$

Combining this equality with $g_T(\boldsymbol{u} - \boldsymbol{c}) = g_T(\boldsymbol{u})$, we obtain

$$\begin{aligned}
\|\boldsymbol{u} - \boldsymbol{v}\|^2 &= \|(\boldsymbol{u} - \boldsymbol{c}) - \boldsymbol{w}\|^2 \\
&= \|(\boldsymbol{u} - \boldsymbol{c}) - g_T(\boldsymbol{u} - \boldsymbol{c})\|^2 + \|g_T(\boldsymbol{u} - \boldsymbol{c}) - \boldsymbol{w}\|^2 \\
&= \|(\boldsymbol{u} - \boldsymbol{c}) - g_T(\boldsymbol{u})\|^2 + \|g_T(\boldsymbol{u}) - \boldsymbol{w}\|^2 \\
&= \|(\boldsymbol{u} - (\boldsymbol{c} + g_T(\boldsymbol{u})))\|^2 + \|(\boldsymbol{c} + g_T(\boldsymbol{u})) - (\boldsymbol{c} + \boldsymbol{w})\|^2 \\
&= \|\boldsymbol{u} - f_L(\boldsymbol{u})\|^2 + \|f_L(\boldsymbol{u}) - \boldsymbol{v}\|^2.
\end{aligned}$$

Since $g_T(\boldsymbol{u}-\boldsymbol{c})$ is the unique nearest to $\boldsymbol{u}-\boldsymbol{c}$ point in T, and this property characterizes $g_T(\boldsymbol{u}-\boldsymbol{c})$ (see page 8), we obtain that $f_L(\boldsymbol{u}) = \boldsymbol{c}+g_T(\boldsymbol{u}-\boldsymbol{c})$ is the unique nearest to $\boldsymbol{u}$ $(= \boldsymbol{c}+\boldsymbol{u}-\boldsymbol{c})$ point in $L = \boldsymbol{c}+T$, and this property characterizes $f_L(\boldsymbol{u})$. □

Theorem 2.91. *For a hyperplane $H = \{\boldsymbol{x} \in \mathbb{R}^n : \boldsymbol{x}\cdot\boldsymbol{e} = \gamma\} \subset \mathbb{R}^n$ and a point $\boldsymbol{u} \in \mathbb{R}^n$, the following assertions hold.*

(*a*) *The orthogonal projection $f_H(\boldsymbol{u})$ is given by*

$$f_H(\boldsymbol{u}) = \boldsymbol{u} + \frac{\gamma - \boldsymbol{u}\cdot\boldsymbol{e}}{\|\mathbf{e}\|^2}.$$

(*b*) *$f_H(\boldsymbol{u})$ is the nearest to $\boldsymbol{u}$ point in H and*

$$\delta(\boldsymbol{u}, H) = \|\boldsymbol{u} - f_H(\boldsymbol{u})\| = \frac{|\gamma - \boldsymbol{u}\cdot\boldsymbol{e}|}{\|\mathbf{e}\|}.$$

Proof. (*a*) Let $T = \{\boldsymbol{x} \in \mathbb{R}^n : \boldsymbol{x}\cdot\boldsymbol{e} = 0\}$ be the direction space of H (see Theorem 2.18). We know (see page 8) that the orthogonal projection on T is given by

$$g_T(\boldsymbol{x}) = \boldsymbol{x} - \frac{\boldsymbol{x}\cdot\boldsymbol{e}}{\|\mathbf{e}\|^2}\boldsymbol{e} \quad \text{for all} \quad \boldsymbol{x} \in \mathbb{R}^n. \tag{2.30}$$

Since the point $\boldsymbol{a} = (\gamma/\|\mathbf{e}\|^2)\boldsymbol{e}$ belongs to H, the equalities (2.30) and $f_H(\boldsymbol{u}) = \boldsymbol{a} + g_T(\boldsymbol{u}-\boldsymbol{a})$ give

$$\begin{aligned} f_H(\boldsymbol{u}) &= \boldsymbol{a} + g_T(\boldsymbol{u}-\boldsymbol{a}) = \frac{\gamma}{\|\mathbf{e}\|^2}\boldsymbol{e} + g_T(\boldsymbol{u} - \frac{\gamma}{\|\mathbf{e}\|^2}\boldsymbol{e}) \\ &= \frac{\gamma}{\|\mathbf{e}\|^2}\boldsymbol{e} + (\boldsymbol{u} - \frac{\gamma}{\|\mathbf{e}\|^2}\boldsymbol{e}) - \Big((\boldsymbol{u} - \frac{\gamma}{\|\mathbf{e}\|^2}\boldsymbol{e})\cdot\frac{\boldsymbol{e}}{\|\mathbf{e}\|^2}\Big)\boldsymbol{e} \\ &= \boldsymbol{u} + \frac{\gamma - \boldsymbol{u}\cdot\boldsymbol{e}}{\|\mathbf{e}\|^2}\boldsymbol{e}. \end{aligned}$$

(*b*) Theorem 2.90 shows that $f_H(\boldsymbol{u})$ is the nearest to $\boldsymbol{u}$ point in H. Therefore,

$$\delta(\boldsymbol{u}, H) = \|\boldsymbol{u} - f_H(\boldsymbol{u})\| = \left|\frac{\gamma - \boldsymbol{u}\cdot\boldsymbol{e}}{\|\mathbf{e}\|^2}\right| \|\mathbf{e}\| = \frac{|\gamma - \boldsymbol{u}\cdot\boldsymbol{e}|}{\|\mathbf{e}\|}. \qquad \square$$

Corollary 2.92. *For a hyperplane $H = \{\boldsymbol{x} \in \mathbb{R}^n : \boldsymbol{x}\cdot\boldsymbol{e} = \gamma\} \subset \mathbb{R}^n$ and a nonempty set $X \subset \mathbb{R}^n$, the following assertions hold.*

(*a*) $\delta(H, X) = \inf\{|\gamma - \boldsymbol{x}\cdot\boldsymbol{e}| : \boldsymbol{x} \in X\}/\|\mathbf{e}\|$.
(*b*) *If X lies in the closed halfspace $V = \{\boldsymbol{x} \in \mathbb{R}^n : \boldsymbol{x}\cdot\boldsymbol{e} \leqslant \gamma\}$, then*

$$\delta(H, X) = (\gamma - \sup\{\boldsymbol{x}\cdot\boldsymbol{e} : \boldsymbol{x} \in X\})/\|\mathbf{e}\|.$$

(*c*) *If* X *lies in the closed halfspace* $V = \{\boldsymbol{x} \in \mathbb{R}^n : \boldsymbol{x}\cdot\boldsymbol{e} \geqslant \gamma\}$, *then*

$$\delta(H, X) = (\inf\{\boldsymbol{x}\cdot\boldsymbol{e} : \boldsymbol{x} \in X\} - \gamma)/\|\boldsymbol{e}\|.$$

(*d*) *If* X *lies in a hyperplane* $H' = \{\boldsymbol{x} \in \mathbb{R}^n : \boldsymbol{x}\cdot\boldsymbol{e} = \gamma'\}$, *then*

$$\delta(H, X) = |\gamma - \gamma'|/\|\boldsymbol{e}\|. \qquad \square$$

Theorem 2.93. *For a closed halfspace* $V = \{\boldsymbol{x} \in \mathbb{R}^n : \boldsymbol{x}\cdot\boldsymbol{e} \leqslant \gamma\}$ *and a point* $\boldsymbol{u} \in \mathbb{R}^n \setminus V$, *the following assertions hold.*

(*a*) *The orthogonal projection* $f_H(\boldsymbol{u})$ *of* $\boldsymbol{u}$ *on the boundary hyperplane* H *of* V *is the nearest to* $\boldsymbol{u}$ *point in* V.
(*b*) *For a point* $\boldsymbol{v} \in H$, *one has* $f_H(\boldsymbol{u}) = \boldsymbol{v}$ *if and only if* $\boldsymbol{u} \in \boldsymbol{v} + \operatorname{nor} V$.

Proof. (*a*) Choose any point $\boldsymbol{v} \in V$. Theorem 2.35 implies that the closed segment $[\boldsymbol{u}, \boldsymbol{v}]$ meets the hyperplane H at a unique point $\boldsymbol{z} \in (\boldsymbol{u}, \boldsymbol{v}]$. Clearly, $\|\boldsymbol{u} - \boldsymbol{z}\| \leqslant \|\boldsymbol{u} - \boldsymbol{v}\|$, where the equality holds if and only if $\boldsymbol{z} = \boldsymbol{v}$. By Theorem 2.91, $\|\boldsymbol{u} - f_H(\boldsymbol{u})\| \leqslant \|\boldsymbol{u} - \boldsymbol{z}\|$, where the equality holds if and only if $f_H(\boldsymbol{u}) = \boldsymbol{z}$. So, $f_H(\boldsymbol{u})$ is the unique nearest to $\boldsymbol{u}$ point in V.

(*b*) By Theorem 2.91, $f_H(\boldsymbol{u}) = \boldsymbol{v}$ if and only if $\boldsymbol{u} \in \boldsymbol{v} + \operatorname{ort} H$. Now, the desired assertion follows from the obvious equality

$$(\mathbb{R}^n \setminus V) \cap (\boldsymbol{v} + \operatorname{ort} H) = (\boldsymbol{v} + \operatorname{nor} V) \setminus \{\boldsymbol{v}\}. \qquad \square$$

Problems and Notes for Chapter 2

Problems for Chapter 2

Problem 2.1. Let $\mathcal{F} = \{L_\alpha\}$ be a family of planes in $\mathbb{R}^n$, with $\bigcap_\alpha L_\alpha \neq \varnothing$, and let $\mathcal{F}' = \{L'_\alpha\}$ be another family of planes, with $\bigcap_\alpha L'_\alpha \neq \varnothing$, such that every $L'_\alpha \in \mathcal{F}'$ is a translate of the corresponding plane L_α. Prove that $\bigcap_\alpha L'_\alpha$ is a translate of $\bigcap_\alpha L_\alpha$.

Problem 2.2. Let $\mathcal{F} = \{L_\alpha\}$ be a countable family of planes in $\mathbb{R}^n$. Prove that the set $M = \bigcup_\alpha L_\alpha$ is a plane if and only if one of the planes from $\mathcal{F}$ contains all the others.

Problem 2.3. Prove that the union of nonempty planes L_1 and L_2 of $\mathbb{R}^n$ lies in a plane $L \subset \mathbb{R}^n$ of dimension $\dim(L_1 + L_2) + 1$ or less.

Problem 2.4. Let L_1 and L_2 be disjoint nonempty planes in $\mathbb{R}^n$. Prove that their sum $L_1 + L_2$ is a proper plane.

Problem 2.5. Let L_1 and L_2 be nonempty planes in $\mathbb{R}^n$. Prove that L_1 and L_2 are complementary planes if and only if every vector $\boldsymbol{u} \in \mathbb{R}^n$ is uniquely expressible as $\boldsymbol{u} = \boldsymbol{u}_1 + \boldsymbol{u}_2$, where $\boldsymbol{u}_1 \in L_1$ and $\boldsymbol{u}_2 \in L_2$.

Problem 2.6. Let L_1 and L_2 be nonempty planes in $\mathbb{R}^n$, with $\dim L_1 \leqslant \dim L_2$. Prove that L_1 and L_2 are parallel if and only if the following two conditions are satisfied:

(a) either $L_1 \cap L_2 = \varnothing$ or $L_1 \subset L_2$,

(b) the union $L_1 \cup L_2$ lies within a plane $L \subset \mathbb{R}^n$ of dimension $\dim L_2 + 1$ or less.

Problem 2.7. Prove that the intersection of r hyperplanes in $\mathbb{R}^n$, where $1 \leqslant r \leqslant n$, of the form $H_i = \{\boldsymbol{x} \in \mathbb{R}^n : \boldsymbol{x} \cdot \boldsymbol{e}_i = \gamma_i\}$, $1 \leqslant i \leqslant r$, is an $(n - r)$-dimensional plane if and only if the set $\{\boldsymbol{e}_1, \dots, \boldsymbol{e}_r\}$ is linearly independent.

Problem 2.8. Show that a nonempty set $X \subset \mathbb{R}^n$ of dimension m is a plane if and only if there is a scalar $\rho > 0$ satisfying the condition: for every point $\boldsymbol{c} \in X$, the set $B_\rho(\boldsymbol{c}) \cap X$ is an m-dimensional ball. If, additionally, X is closed, then the value of ρ may depend on $\boldsymbol{c} \in X$.

Problem 2.9. Prove Lemmas 2.25–2.28.

Problem 2.10. Let $H \subset \mathbb{R}^n$ be a hyperplane, $\boldsymbol{a}$ and $\boldsymbol{b}$ be points in $\mathbb{R}^n \setminus H$, and h be an open halfline with endpoint $\boldsymbol{a}$ which meets H. Prove that the open halfline $h' = (\boldsymbol{b} - \boldsymbol{a}) + h$ meets H if and only if $\boldsymbol{a}$ and $\boldsymbol{b}$ belong to the same open halfspace determined by H.

Problem 2.11. Let $\mathcal{F} = \{F_\alpha\}$ be a family of halfplanes of a nonempty plane $L \subset \mathbb{R}^n$ of positive dimension (each F_α may be closed or open). Show that the following statements hold.

(a) If the union $U = \underset{\alpha}{\cup} F_\alpha$ is a proper subset of L, then U is a halfplane of L if and only if $\mathcal{F}$ is nested.

(b) If the intersection $Z = \underset{\alpha}{\cap} F_\alpha$ is nonempty, then Z is a halfplane of L if and only if $\mathcal{F}$ is nested.

Definition. Given a nonzero vector $\boldsymbol{e} \in \mathbb{R}^n$ and scalars $\gamma < \gamma'$, the sets

$$\{\boldsymbol{x} \in \mathbb{R}^n : \gamma \leqslant \boldsymbol{x} \cdot \boldsymbol{e} \leqslant \gamma'\} \quad \text{and} \quad \{\boldsymbol{x} \in \mathbb{R}^n : \gamma < \boldsymbol{x} \cdot \boldsymbol{e} < \gamma'\}$$

are called, respectively, the *closed slab* and the *open slab* of $\mathbb{R}^n$.

Problem 2.12. Let $P \subset \mathbb{R}^n$ be a slab (open or closed), and let points $\boldsymbol{y}, \boldsymbol{z} \in P$ be such that the closed halfline $[\boldsymbol{y}, \boldsymbol{z}\rangle$ lies in P. Prove that the whole line $\langle \boldsymbol{y}, \boldsymbol{z}\rangle$ lies in P.

Problem 2.13. Let X and Y be nonempty sets in $\mathbb{R}^n$ such that a translate of Y lies in $\operatorname{aff} X$. Prove that for any point $\boldsymbol{y} \in Y$, one has

$$\operatorname{aff}(X + Y) = \operatorname{aff} X + Y = \operatorname{aff} X + \boldsymbol{y}.$$

Problem 2.14. Let $X_1, \ldots, X_r$ be nonempty sets in $\mathbb{R}^n$ and $\mu_1, \ldots, \mu_r$ be scalars such that $r > m$, where $m = \dim(\mu_1 X_1 + \cdots + \mu_r X_r)$. Given points $\boldsymbol{a}_i \in X_i$, $1 \leqslant i \leqslant r$, prove the existence of an index set $I \subset \{1, \ldots, r\}$ satisfying the conditions $\operatorname{card} I \leqslant m$ and

$$\operatorname{aff}(\mu_1 X_1 + \cdots + \mu_r X_r) = \sum_{i \notin I} \mu_i \boldsymbol{a}_i + \sum_{i \in I} \mu_i \operatorname{aff} X_i.$$

Problem 2.15. Given points $\boldsymbol{a}_i = (a_{i1}, \ldots, a_{in}) \in \mathbb{R}^n$, $1 \leqslant i \leqslant r$, consider the vectors $\boldsymbol{a}_i' = (a_{i1}, \ldots, a_{in}, 1) \in \mathbb{R}^{n+1}$, $1 \leqslant i \leqslant r$. Prove the equivalence of the following conditions:

(a) the set $\{\boldsymbol{a}_1, \ldots, \boldsymbol{a}_r\}$ is affinely independent,

(b) the set $\{\boldsymbol{a}_1', \ldots, \boldsymbol{a}_r'\}$ is linearly independent.

Problem 2.16. Let a set $\{\boldsymbol{a}_1, \ldots, \boldsymbol{a}_r\} \subset \mathbb{R}^n$ be affinely independent. Prove the existence of a scalar $\rho > 0$ such that for any choice of points $\boldsymbol{a}_i' \in U_\rho(\boldsymbol{a}_i)$, $1 \leqslant i \leqslant r$, the set $\{\boldsymbol{a}_1', \ldots, \boldsymbol{a}_r'\}$ is affinely independent.

Problem 2.17. Prove that an affine transformation $f : \mathbb{R}^n \to \mathbb{R}^n$ is an affine projection if and only if $f \circ f = f$.

Notes for Chapter 2

The affine structure of $\mathbb{R}^n$. A substantial part of Chapter 2 is scattered through the literature on linear algebra and high-dimensional geometry. A historiographic reference here is the paper of Steinitz [265, § 4], which contains some assertions from Theorems 2.2, 2.20, 2.38, 2.48, 2.52, and Corollary 2.45.

Iterative construction of affine spans. Given a nonempty set $X \subset \mathbb{R}^n$, let $\operatorname{aff}_r X$ denote the collection of all affine combinations of subsets $Y \subset X$ whose cardinality is r or less. In particular, $\operatorname{aff}_2 X$ is the union of points and lines through distinct points of X. Corollary 2.66 implies that $\operatorname{aff}_{m+1} X = \operatorname{aff} X$ for every m-dimensional set X in $\mathbb{R}^n$. It is easy to see that $\operatorname{aff}_r(\operatorname{aff}_s X) \subset \operatorname{aff}_{rs} X$, and this inclusion may be proper (compare with Theorem 4.10). For instance, if $X = \{\boldsymbol{x}, \boldsymbol{y}, \boldsymbol{z}, \boldsymbol{u}\} \subset \mathbb{R}^3$ is an affinely independent set, then $\operatorname{aff}_4 X \setminus \operatorname{aff}_2(\operatorname{aff}_2 X)$ is the four-point set

$$\tfrac{1}{2}\{-\boldsymbol{x} + \boldsymbol{y} + \boldsymbol{z} + \boldsymbol{u}, \boldsymbol{x} - \boldsymbol{y} + \boldsymbol{z} + \boldsymbol{u}, \boldsymbol{x} + \boldsymbol{y} - \boldsymbol{z} + \boldsymbol{u}, \boldsymbol{x} + \boldsymbol{y} + \boldsymbol{z} - \boldsymbol{u}\}.$$

For a set $X \subset \mathbb{R}^n$ and integers $r_1, \ldots, r_k \geqslant 2$, Klee [178] studied properties of sets $\operatorname{aff}_{r_1}(\cdots(\operatorname{aff}_{r_k} X)\cdots)$. In particular, he proved the following results:

1. $\operatorname{aff}_{r_1 + \cdots + r_k - 1} X \subset \operatorname{aff}_{r_1}(\cdots(\operatorname{aff}_{r_k} X)\cdots) \subset \operatorname{aff}_{r_1 + \cdots + r_k} X$,
2. the set $\operatorname{aff}_{r_1 + \cdots + r_k} X \setminus \operatorname{aff}_{r_1}(\cdots(\operatorname{aff}_{r_k} X)\cdots)$ is finite provided X is finite,
3. $\operatorname{aff}_{r_1 + \cdots + r_k} X = \operatorname{aff}_{r_1}(\cdots(\operatorname{aff}_{r_k} X)\cdots)$ if not all $r_1, \ldots, r_k$ are the same.

4. $\operatorname{aff} X = \operatorname{aff}_{r_1}(\cdots(\operatorname{aff}_{r_k} X)\cdots)$ for any integers $r_1, \dots, r_k \geqslant 2$ such that either (*i*) $r_1 + \cdots + r_k \geqslant n + 2$, or (*ii*) $r_1 + \cdots + r_k \geqslant n + 1$ and not all $r_1, \dots, r_k$ are the same.

Collineations. A mapping $f : \mathbb{R}^n \to \mathbb{R}^m$ is called *collineation* provided it takes any three collinear points $\boldsymbol{x}, \boldsymbol{y}, \boldsymbol{z}$ in $\mathbb{R}^n$ onto collinear (not necessarily distinct) points $f(\boldsymbol{x}), f(\boldsymbol{y}), f(\boldsymbol{z})$ in $\mathbb{R}^m$. Originated as a theorem of projective geometry (see, for instance, Darboux [84] and a sharper version by Swift [275]), the important theorem of affine geometry states that an invertible collineation $f : \mathbb{R}^n \to \mathbb{R}^n$, $n \geqslant 2$, is an affine transformation (see, e. g., Veblen and Whitehead [286, p. 12], and many others). Lenz [197] sharpened the assertion of Veblen and Whitehead by replacing the "invertible" condition with that of "one-to-one." Frenkel [111, p. 91] proved that a mapping $f : \mathbb{R}^n \to \mathbb{R}^m$, where $n \geqslant 2$ and $m \geqslant 2$, is an affine transformation provided its range is at least 2-dimensional and $f(\operatorname{aff}\{\boldsymbol{x}, \boldsymbol{z}\}) = \operatorname{aff}\{f(\boldsymbol{x}), f(\boldsymbol{z})\}$ whenever $\boldsymbol{x}, \boldsymbol{z} \in \mathbb{R}^n$. Further related results can be found in the papers of Chubarev and Pinelis [73], Jung [162], Li and Wang [200–202], and Artstein-Avidan and Slomka [5].

Chapter 3

Convex Sets

3.1 General Properties of Convex Sets

3.1.1 *Basic Properties of Convex Sets*

Definition 3.1. A nonempty set $X \subset \mathbb{R}^n$ is called *convex* provided it contains all points of the form $(1-\lambda)\boldsymbol{x} + \lambda\boldsymbol{y}$ whenever $\boldsymbol{x}, \boldsymbol{y} \in X$ and $\lambda \in [0,1]$. The empty set $\varnothing$ is assumed to be convex.

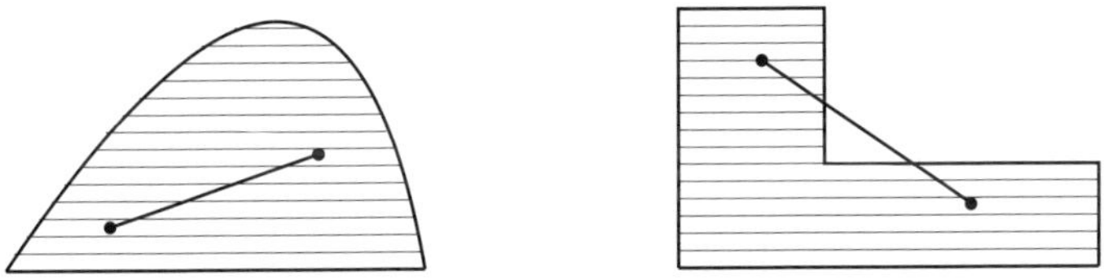

Fig. 3.1 Convex and nonconvex sets.

We can reformulate Definition 3.1, saying that a nonempty set X in $\mathbb{R}^n$ is convex if and only if it contains all closed segments $[\boldsymbol{x}, \boldsymbol{y}]$ with endpoints $\boldsymbol{x}, \boldsymbol{y} \in X$. Equivalently, the set X is convex if and only if $(\boldsymbol{x}, \boldsymbol{y}) \subset X$ whenever $\boldsymbol{x}, \boldsymbol{y} \in X$. The condition $\lambda \in [0,1]$ in Definition 3.1 can be weakened provided the set X is closed or relative open (see Problems 3.2 and 3.3).

Examples. 1. Convex subsets of a line $l \subset \mathbb{R}^n$ are the empty set, all singletons, segments and halflines of all kinds (see Definitions 2.23 and 2.24), and the whole line l.

2. Every plane L in $\mathbb{R}^n$ is a convex set. Indeed, the case $L = \varnothing$ is obvious. If $L \neq \varnothing$ and $\boldsymbol{x}$ and $\boldsymbol{y}$ are distinct points in L, then $[\boldsymbol{x}, \boldsymbol{y}] \subset \langle \boldsymbol{x}, \boldsymbol{y} \rangle \subset L$

according to Theorem 2.38.

3. Every halfplane of a plane $L \subset \mathbb{R}^n$ of positive dimension is a convex set. Indeed, if D is a closed halfplane of L, then, by Theorem 2.70, it can be expressed as $D = \{\boldsymbol{x} \in L : \boldsymbol{x} \cdot \boldsymbol{e} \leqslant \gamma\}$ for a suitable vector $\boldsymbol{e} \in \mathbb{R}^n \setminus \operatorname{ort} L$ and a scalar γ. Choose points $\boldsymbol{x}, \boldsymbol{y} \in D$ and a scalar $\lambda \in [0, 1]$. Then $(1-\lambda)\boldsymbol{x} + \lambda\boldsymbol{y} \in L$ because L is convex, and the inequality

$$((1-\lambda)\boldsymbol{x} + \lambda\,\boldsymbol{y})\cdot\boldsymbol{e} = (1-\lambda)\boldsymbol{x}\cdot\boldsymbol{e} + \lambda\boldsymbol{y}\cdot\boldsymbol{e} \leqslant (1-\lambda)\gamma + \lambda\gamma = \gamma$$

shows that $(1-\lambda)\boldsymbol{x} + \lambda\boldsymbol{y} \in D$. Similarly, every open halfplane of L is a convex set.

4. Every closed ball $B_\rho(\boldsymbol{a}) \subset \mathbb{R}^n$ is a convex set. Indeed, choose points $\boldsymbol{x}, \boldsymbol{y} \in B_\rho(\boldsymbol{a})$ and a scalar $\lambda \in [0, 1]$. Let $\boldsymbol{u} = (1-\lambda)\boldsymbol{x} + \lambda\boldsymbol{y}$. Then

$$\begin{aligned}\|\boldsymbol{a} - \boldsymbol{u}\| &= \|(1-\lambda)(\boldsymbol{a} - \boldsymbol{x}) + \lambda(\boldsymbol{a} - \boldsymbol{y})\| \\ &\leqslant (1-\lambda)\|\boldsymbol{a} - \boldsymbol{x}\| + \lambda\|\boldsymbol{a} - \boldsymbol{y}\| \leqslant (1-\lambda)\rho + \lambda\rho = \rho.\end{aligned}$$

So, $\boldsymbol{u} \in B_\rho(\boldsymbol{a})$, implying the inclusion $[\boldsymbol{x}, \boldsymbol{y}] \subset B_\rho(\boldsymbol{a})$. Similarly, every open ball $U_\rho(\boldsymbol{a}) \subset \mathbb{R}^n$ is a convex set.

Definition 3.2. *A convex combination* of points $\boldsymbol{a}_1, \dots, \boldsymbol{a}_r \in \mathbb{R}^n$ (or a convex combination of a set $\{\boldsymbol{a}_1, \dots, \boldsymbol{a}_r\}$), $r \geqslant 1$, is a linear combination

$$\lambda_1\boldsymbol{a}_1 + \cdots + \lambda_r\boldsymbol{a}_r, \quad \text{where} \quad \lambda_1, \dots, \lambda_r \geqslant 0 \quad \text{and} \quad \lambda_1 + \cdots + \lambda_r = 1.$$

If, additionally, all scalars $\lambda_1, \dots, \lambda_r$ are positive, then $\lambda_1\boldsymbol{a}_1 + \cdots + \lambda_r\boldsymbol{a}_r$ is called a *positive convex combination.*

For instance, all convex combinations of distinct points $\boldsymbol{x}$ and $\boldsymbol{y}$ in $\mathbb{R}^n$ fulfil the closed segment $[\boldsymbol{x}, \boldsymbol{y}]$ (see Definition 2.24), while all positive convex combinations of $\boldsymbol{x}$ and $\boldsymbol{y}$ fulfill the open segment $(\boldsymbol{x}, \boldsymbol{y})$. We observe that all convex (all positive convex) combinations of identical points $\boldsymbol{x}$ and $\boldsymbol{x}$ fulfill the singleton $[\boldsymbol{x}, \boldsymbol{x}] = \{\boldsymbol{x}\}$, while $(\boldsymbol{x}, \boldsymbol{x}) = \varnothing$.

The next result extends Definition 3.1 by considering convex combinations of any given number of points. See Theorem 3.53 for the case of convergent convex series.

Theorem 3.3. *Let $X \subset \mathbb{R}^n$ be a nonempty set and $r \geqslant 2$ be an integer. Then X is convex if and only if it contains all convex combinations (equivalently, all positive convex combinations) of r points from X.*

Proof. Let the set X be convex. By induction on r, we are going to prove that X contains all convex combinations of r points from X. The case $r = 2$

coincides with Definition 3.1. Assuming that the inductive hypothesis holds for all positive integers $r \leqslant k-1$, where $k \geqslant 3$, choose a convex combination $\boldsymbol{y} = \lambda_1 \boldsymbol{x}_1 + \cdots + \lambda_k \boldsymbol{x}_k$ of points $\boldsymbol{x}_1, \ldots, \boldsymbol{x}_k \in X$. At least one of the scalars $\lambda_1, \ldots, \lambda_k$ is not 1, since otherwise $\lambda_1 + \cdots + \lambda_k = k \geqslant 3$, contrary to the assumption $\lambda_1 + \cdots + \lambda_k = 1$. Let, for instance, $\lambda_1 \neq 1$. By the induction hypothesis, the convex combination

$$\boldsymbol{z} = \frac{\lambda_2}{1-\lambda_1}\boldsymbol{x}_2 + \cdots + \frac{\lambda_k}{1-\lambda_1}\boldsymbol{x}_k$$

belongs to X. Finally, the equality $\boldsymbol{y} = \lambda_1 \boldsymbol{x}_1 + (1-\lambda_1)\boldsymbol{z}$ and Definition 3.1 imply the inclusion $\boldsymbol{y} \in X$.

Conversely, suppose that X contains all positive convex combinations of r points from X. Assume first that $r = 2$ and choose points $\boldsymbol{x}, \boldsymbol{y} \in X$. If $\boldsymbol{x} = \boldsymbol{y}$, then $(\boldsymbol{x}, \boldsymbol{y}) = \varnothing \subset X$; otherwise,

$$(\boldsymbol{x}, \boldsymbol{y}) = \{(1-\lambda)\boldsymbol{x} + \lambda \boldsymbol{y} : 0 < \lambda < 1\} \subset X.$$

Because $[\boldsymbol{x}, \boldsymbol{y}] = \{\boldsymbol{x}\} \cup (\boldsymbol{x}, \boldsymbol{y}) \cup \{\boldsymbol{y}\}$, the closed segment $[\boldsymbol{x}, \boldsymbol{y}]$ lies in X, which gives the convexity of X. Let $r \geqslant 3$. Choose points $\boldsymbol{x}, \boldsymbol{y} \in X$ and let $\boldsymbol{x}_2 = \cdots = \boldsymbol{x}_r = \boldsymbol{y}$. By the assumption, every positive convex combination

$$(1-\lambda)\boldsymbol{x} + \lambda \boldsymbol{y} = (1-\lambda)\boldsymbol{x} + \frac{\lambda}{r-1}(\boldsymbol{x}_2 + \boldsymbol{x}_3 + \cdots + \boldsymbol{x}_r), \quad 0 < \lambda < 1,$$

belongs to X. Consequently, the above argument shows that X is a convex set by the above argument. □

Theorem 3.4. *Let $X \subset \mathbb{R}^n$ be a set and $r \geqslant 2$ be an integer. Then X is convex if and only if*

$$\mu_1 X + \cdots + \mu_r X = (\mu_1 + \cdots + \mu_r)X \tag{3.1}$$

for every choice of scalars $\mu_1, \ldots, \mu_r \geqslant 0$.

Proof. Because the assertion is obvious when $X = \varnothing$, we may assume that X is nonempty. Let X be convex. Due to the inclusion

$$(\mu_1 + \cdots + \mu_r)X \subset \mu_1 X + \cdots + \mu_r X,$$

which holds for any set $X \subset \mathbb{R}^n$ and any choice of scalars $\mu_1, \ldots, \mu_r$, it suffices to prove the opposite inclusion

$$\mu_1 X + \cdots + \mu_r X \subset (\mu_1 + \cdots + \mu_r)X$$

provided X is convex and $\mu_1, \ldots, \mu_r \geqslant 0$. Excluding one more obvious case, $\mu_1 = \cdots = \mu_r = 0$, we suppose that $\mu = \mu_1 + \cdots + \mu_r > 0$. Put $\lambda_i = \mu_i/\mu$,

$1 \leqslant i \leqslant r$. Then $\lambda_1, \ldots, \lambda_r \geqslant 0$ and $\lambda_1 + \cdots + \lambda_r = 1$. According to Theorem 3.3,

$$\lambda_1 X + \cdots + \lambda_r X = \{\lambda_1 \boldsymbol{x}_1 + \cdots + \lambda_r \boldsymbol{x}_r : \boldsymbol{x}_1, \ldots, \boldsymbol{x}_r \in X\} \subset X.$$

Multiplying both sides of this inclusion by μ, we obtain

$$\mu_1 X + \cdots + \mu_r X \subset (\mu_1 + \cdots + \mu_r) X.$$

Conversely, let X satisfy condition (3.1) for any choice of scalars $\mu_1, \ldots, \mu_r \geqslant 0$. Choose an affine combination $\mu_1 \boldsymbol{x}_1 + \cdots + \mu_r \boldsymbol{x}_r$ of points $\boldsymbol{x}_1, \ldots, \boldsymbol{x}_r \in X$. By the assumption,

$$\mu_1 \boldsymbol{x}_1 + \cdots + \mu_r \boldsymbol{x}_r \in \mu_1 X + \cdots + \mu_r X = X,$$

and Theorem 3.3 implies that X is a convex set. □

3.1.2 *Simplices*

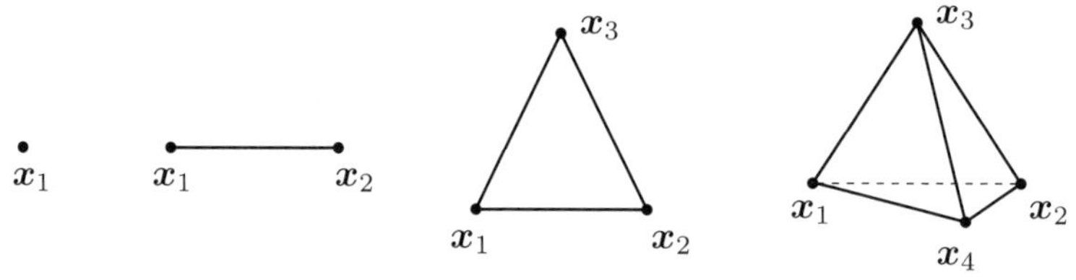

Fig. 3.2 Simplices in $\mathbb{R}^3$.

Definition 3.5. Let $\{\boldsymbol{x}_1, \ldots, \boldsymbol{x}_{r+1}\}$, $r \geqslant 0$, be an affinely independent set in $\mathbb{R}^n$ (see Definition 2.50). The *r-simplex* $\Delta = \Delta(\boldsymbol{x}_1, \ldots, \boldsymbol{x}_{r+1})$ with vertices $\boldsymbol{x}_1, \ldots, \boldsymbol{x}_{r+1}$ is defined as

$$\Delta = \{\lambda_1 \boldsymbol{x}_1 + \cdots + \lambda_{r+1} \boldsymbol{x}_{r+1} : \lambda_1, \ldots, \lambda_{r+1} \geqslant 0,\ \lambda_1 + \cdots + \lambda_{r+1} = 1\}.$$

Equivalently, Δ is the set of all convex combinations of $\boldsymbol{x}_1, \ldots, \boldsymbol{x}_{r+1}$.

Remark. For a point $\boldsymbol{x} \in \Delta(\boldsymbol{x}_1, \ldots, \boldsymbol{x}_{r+1})$, the scalars $\lambda_1, \ldots, \lambda_{r+1}$ in the expression $\boldsymbol{x} = \lambda_1 \boldsymbol{x}_1 + \cdots + \lambda_{r+1} \boldsymbol{x}_{r+1}$ are uniquely determined by $\boldsymbol{x}$. Indeed, since $\boldsymbol{x}$ can be written as an affine combination

$$\boldsymbol{x} = \lambda_1 \boldsymbol{x}_1 + \cdots + \lambda_r \boldsymbol{x}_r + (1 - \lambda_1 - \cdots - \lambda_r) \boldsymbol{x}_{r+1}$$

of an affinely independent set $\{\boldsymbol{x}_1, \ldots, \boldsymbol{x}_{r+1}\}$, the uniqueness of $\lambda_1, \ldots, \lambda_{r+1}$ follows from Theorem 2.52. Consequently, a point $\boldsymbol{x} \in \operatorname{aff} \{\boldsymbol{x}_1, \ldots, \boldsymbol{x}_{r+1}\}$ belongs to $\Delta(\boldsymbol{x}_1, \ldots, \boldsymbol{x}_{r+1})$ if and only if $\boldsymbol{x}$ is a convex combination of $\boldsymbol{x}_1, \ldots, \boldsymbol{x}_{r+1}$.

Theorem 3.6. *Every* r*-simplex* $\Delta = \Delta(\boldsymbol{x}_1, \dots, \boldsymbol{x}_{r+1}) \subset \mathbb{R}^n$ *is a compact convex set of dimension* r*. Moreover,* $\operatorname{aff} \Delta = \operatorname{aff}\{\boldsymbol{x}_1, \dots, \boldsymbol{x}_{r+1}\}$*.*

Proof. Choose points $\boldsymbol{x}, \boldsymbol{y} \in \Delta$, and express them as convex combinations

$$\boldsymbol{x} = \gamma_1 \boldsymbol{x}_1 + \cdots + \gamma_{r+1}\boldsymbol{x}_{r+1} \quad \text{and} \quad \boldsymbol{y} = \mu_1 \boldsymbol{x}_1 + \cdots + \mu_{r+1}\boldsymbol{x}_{r+1}.$$

Given a scalar $\lambda \in [0, 1]$, let

$$\alpha_i = (1 - \lambda)\gamma_i + \lambda\mu_i, \quad 1 \leqslant i \leqslant r + 1.$$

Clearly, $\alpha_1, \dots, \alpha_{r+1} \geqslant 0$, and the equalities

$$(1 - \lambda)\boldsymbol{x} + \lambda\boldsymbol{y} = \alpha_1\boldsymbol{x}_1 + \cdots + \alpha_{r+1}\boldsymbol{x}_{r+1}, \quad \alpha_1 + \cdots + \alpha_{r+1} = 1$$

show that $(1 - \lambda)\boldsymbol{x} + \lambda\boldsymbol{y} \in \Delta$. Hence Δ is a convex set.

Because convex combinations are particular cases of affine combinations, Corollary 2.45 implies the inclusions

$$\{\boldsymbol{x}_1, \dots, \boldsymbol{x}_{r+1}\} \subset \Delta \subset \operatorname{aff}\{\boldsymbol{x}_1, \dots, \boldsymbol{x}_{r+1}\},$$

and Theorem 2.42 gives $\operatorname{aff} \Delta = \operatorname{aff}\{\boldsymbol{x}_1, \dots, \boldsymbol{x}_{r+1}\}$. Furthermore, from Definition 2.63 and Theorem 2.52 one has

$$\dim \Delta = \dim(\operatorname{aff} \Delta) = \dim(\operatorname{aff}\{\boldsymbol{x}_1, \dots, \boldsymbol{x}_{r+1}\}) = r.$$

To prove the compactness of Δ, choose an infinite sequence of points $\boldsymbol{u}_1, \boldsymbol{u}_2, \dots \in \Delta$. By Definition 3.5, every $\boldsymbol{u}_i$ can be written as a convex combination $\boldsymbol{u}_i = \lambda_1^{(i)}\boldsymbol{x}_1 + \cdots + \lambda_{r+1}^{(i)}\boldsymbol{x}_{r+1}$, $i \geqslant 1$. Consider $r + 1$ scalar sequences $\lambda_j^{(1)}, \lambda_j^{(2)}, \dots$, $1 \leqslant j \leqslant r+1$. Since all scalars $\lambda_j^{(i)}$ belong to $[0, 1]$, Cantor's diagonal argument gives the existence of an infinite sequence of indices $i_1, i_2, \dots$ such that all subsequences $\lambda_j^{(i_1)}, \lambda_j^{(i_2)}, \dots$, $1 \leqslant j \leqslant r + 1$, converge. Put

$$\lambda_j = \lim_{s\to\infty} \lambda_j^{(i_s)}, \quad 1 \leqslant j \leqslant r + 1.$$

Obviously, $\lambda_1, \dots, \lambda_{r+1} \geqslant 0$ and

$$\begin{aligned}\lambda_1 + \cdots + \lambda_{r+1} &= \lim_{s\to\infty} \lambda_1^{(i_s)} + \cdots + \lim_{s\to\infty} \lambda_{r+1}^{(i_s)} \\ &= \lim_{s\to\infty} (\lambda_1^{(i_s)} + \cdots + \lambda_{r+1}^{(i_s)}) = 1.\end{aligned}$$

Similarly,

$$\begin{aligned}\lim_{s\to\infty} \boldsymbol{u}_{i_s} &= \lim_{s\to\infty} (\lambda_1^{(i_s)}\boldsymbol{x}_1 + \cdots + \lambda_{r+1}^{(i_s)}\boldsymbol{x}_{r+1}) \\ &= \lambda_1\boldsymbol{x}_1 + \cdots + \lambda_{r+1}\boldsymbol{x}_{r+1} \in \Delta.\end{aligned}$$

The Heine-Borel theorem (see page 11) implies that Δ is a compact set. $\square$

Combining Theorems 3.3 and 3.6 and Corollary 2.66, we obtain the corollary below.

Corollary 3.7. *For a nonempty convex set $K \subset \mathbb{R}^n$, the following assertions hold.*

(*a*) *If $\{\boldsymbol{x}_1, \ldots, \boldsymbol{x}_{r+1}\}$ is an affinely independent subset of K, then the r-simplex $\Delta(\boldsymbol{x}_1, \ldots, \boldsymbol{x}_{r+1})$ lies in K.*
(*b*) *The dimension of K equals the maximum dimension of a simplex contained in K.*
(*c*) $\operatorname{aff} K = \operatorname{aff} \Delta$ *for every simplex $\Delta \subset K$ of dimension* $\dim K$. □

3.1.3 *Algebra of Convex Sets*

Theorem 3.8. *If $\mathcal{F} = \{K_\alpha\}$ is a family of convex sets in $\mathbb{R}^n$, then the following assertions hold.*

(*a*) *The intersection $M = \bigcap_\alpha K_\alpha$ is a convex set.*
(*b*) *If the family $\mathcal{F}$ is nested, then the union $N = \bigcup_\alpha K_\alpha$ is a convex set.*

Proof. (*a*) Since the case $M = \varnothing$ is obvious, we may suppose that $M \neq \varnothing$. If $\boldsymbol{x}, \boldsymbol{y} \in M$, then $\boldsymbol{x}, \boldsymbol{y} \in K_\alpha$ for every $K_\alpha \in \mathcal{F}$, which gives $[\boldsymbol{x}, \boldsymbol{y}] \subset K_\alpha$ by the convexity of K_α. Hence $[\boldsymbol{x}, \boldsymbol{y}] \subset \bigcap_\alpha K_\alpha = M$, and M is convex.

(*b*) Excluding the obvious case $N = \varnothing$, we assume that N is nonempty. Let $\boldsymbol{x}, \boldsymbol{y} \in N$. Then $\boldsymbol{x} \in K_\gamma$ and $\boldsymbol{y} \in K_\mu$ for suitable sets $K_\gamma, K_\mu \in \mathcal{F}$. By the hypothesis, one of the sets K_γ, K_μ contains the other. Let, for instance, $K_\gamma \subset K_\mu$. In this case, $\boldsymbol{x}, \boldsymbol{y} \in K_\mu$, and $[\boldsymbol{x}, \boldsymbol{y}] \subset K_\mu$ by the convexity of K_μ. Thus $[\boldsymbol{x}, \boldsymbol{y}] \subset K_\mu \subset N$, and N is a convex set. □

Corollary 3.9. *Given a set $X \subset \mathbb{R}^n$ and a convex subset K of X, there is a maximal (under inclusion) convex set $M \subset \mathbb{R}^n$ satisfying the condition $K \subset M \subset X$.*

Proof. Denote by $\mathcal{F}$ the family of all convex sets $M \subset \mathbb{R}^n$ satisfying the inclusions $K \subset M \subset X$. The family $\mathcal{F}$ is nonempty since $K \in \mathcal{F}$. According to Theorem 3.8, the union of every nested subfamily of $\mathcal{F}$ belongs to $\mathcal{F}$. Consequently, the Kuratowski-Zorn Lemma (see page 2) shows the existence of a maximal element in $\mathcal{F}$. □

Theorem 3.10. *If $K_1, \ldots, K_r$ are convex sets in $\mathbb{R}^n$ and $\mu_1, \ldots, \mu_r$ are scalars, then the sum $\mu_1 K_1 + \cdots + \mu_r K_r$ is a convex set.*

Proof. If at least one of the sets $K_1, \ldots, K_r$ is empty, then the whole sum $\mu_1 K_1 + \cdots + \mu_r K_r$ is empty, and thus is a convex set. Suppose that all sets $K_1, \ldots, K_r$ are nonempty. An induction argument shows that the proof can be reduced to the case $r = 2$ (let $\mu_1 K_1 = \mu_1 K_1 + \mu_2\{\boldsymbol{o}\}$ if $r = 1$). Choose points $\boldsymbol{x}$ and $\boldsymbol{y}$ in $\mu_1 K_1 + \mu_2 K_2$ and a scalar $\lambda \in [0,1]$. Then $\boldsymbol{x} = \mu_1 \boldsymbol{x}_1 + \mu_2 \boldsymbol{x}_2$ and $\boldsymbol{y} = \mu_1 \boldsymbol{y}_1 + \mu_2 \boldsymbol{y}_2$ for suitable points $\boldsymbol{x}_1, \boldsymbol{y}_1 \in K_1$ and $\boldsymbol{x}_2, \boldsymbol{y}_2 \in K_2$. By a convexity argument, $(1-\lambda)\boldsymbol{x}_i + \lambda \boldsymbol{y}_i \in K_i$, $i = 1, 2$. Finally, the inclusion

$$\begin{aligned}(1-\lambda)\boldsymbol{x} + \lambda\boldsymbol{y} &= \mu_1((1-\lambda)\boldsymbol{x}_1 + \lambda\boldsymbol{y}_1) + \mu_2((1-\lambda)\boldsymbol{x}_2 + \lambda\boldsymbol{y}_2)\\ &\in \mu_1 K_1 + \mu_2 K_2\end{aligned}$$

implies that $\mu_1 K_1 + \mu_2 K_2$ is a convex set. □

Corollary 3.11. *If $K \subset \mathbb{R}^n$ is a nonempty convex set and ρ is a positive scalar, then both ρ-neighborhoods $B_\rho(K)$ and $U_\rho(K)$ are convex sets.*

Proof. The neighborhoods $B_\rho(K)$ and $U_\rho(K)$ are given by

$$B_\rho(K) = B_\rho(\boldsymbol{o}) + K \quad \text{and} \quad U_\rho(K) = U_\rho(\boldsymbol{o}) + K$$

(see page 9). Since any closed or open ball in $\mathbb{R}^n$ is convex (see Example 4 on page 76), Theorem 3.10 implies the convexity of both neighborhoods. □

The following lemma is useful in various arguments on convex sets.

Lemma 3.12. *Let $K \subset \mathbb{R}^n$ be a nonempty convex set and $\boldsymbol{a}$ be a point in K. Given scalars $0 < \lambda \leqslant \gamma$, one has*

$$\boldsymbol{a} + \lambda(K - \boldsymbol{a}) \subset \boldsymbol{a} + \gamma(K - \boldsymbol{a}).$$

Proof. Clearly, both sets $\boldsymbol{a} + \lambda(K - \boldsymbol{a})$ and $\boldsymbol{a} + \gamma(K - \boldsymbol{a})$ contain $\boldsymbol{a}$. If $\boldsymbol{u} \in \boldsymbol{a} + \lambda(K - \boldsymbol{a})$, then $\boldsymbol{u} = \boldsymbol{a} + \lambda(\boldsymbol{v} - \boldsymbol{a})$ for a suitable point $\boldsymbol{v} \in K$. Using Lemma 2.25 and the convexity of $\boldsymbol{a} + \gamma(K - \boldsymbol{a})$ (see Theorem 3.10), we obtain

$$\boldsymbol{u} = \boldsymbol{a} + \lambda(\boldsymbol{v} - \boldsymbol{a}) \in [\boldsymbol{a}, \boldsymbol{a} + \gamma(\boldsymbol{v} - \boldsymbol{a})] \subset \boldsymbol{a} + \gamma(K - \boldsymbol{a}).$$

Consecutively, $\boldsymbol{a} + \lambda(K - \boldsymbol{a}) \subset \boldsymbol{a} + \gamma(K - \boldsymbol{a})$. □

The next theorem shows an interesting relation between set-theoretic and algebraic operations on convex sets.

Theorem 3.13. *Let K be a convex set in $\mathbb{R}^n$. For any homothetic copy $\boldsymbol{a} + \mu K$ of K, $\mu \neq 0$, there are sets X and Y in $\mathbb{R}^n$ satisfying the conditions*

$$\boldsymbol{a} + \mu K = \begin{cases} \cap\,(\boldsymbol{z} + \operatorname{sgn}(\mu)K : \boldsymbol{z} \in X) & \text{if} \quad 0 < |\mu| < 1,\\ \cup\,(\boldsymbol{z} + \operatorname{sgn}(\mu)K : \boldsymbol{z} \in Y) & \text{if} \quad |\mu| > 1.\end{cases}$$

Proof. Because the case $K = \varnothing$ is obvious (choose any X and Y), we assume that K is nonempty. Also, we may suppose that $\mu > 0$. Indeed, if $\mu < 0$, then

$$\boldsymbol{a} + \mu K = \boldsymbol{a} + (-\mu)(-K) \quad \text{and} \quad \boldsymbol{z} + \operatorname{sgn}(\mu)K = \boldsymbol{z} + \operatorname{sgn}(-\mu)(-K),$$

reducing the assertion to the case of $-K$. Put $K' = \boldsymbol{a} + \mu K$.

1. Let $0 < \mu < 1$. For the existence of a desired set X, it suffices to show that for any $\boldsymbol{u} \in \mathbb{R}^n \setminus K'$ there is a point $\boldsymbol{z} \in \mathbb{R}^n$ satisfying the conditions $K' \subset \boldsymbol{z} + K$ and $\boldsymbol{u} \notin \boldsymbol{z} + K$. Then X is the collection of such points $\boldsymbol{z}$.

Fix a point $\boldsymbol{u} \in \mathbb{R}^n \setminus K'$. We consider separately the following cases.

1*a*. Suppose that $\boldsymbol{u} + \mu(\boldsymbol{x} - \boldsymbol{u}) \in K'$ for all $\boldsymbol{x} \in K'$. Put $\boldsymbol{z} = \boldsymbol{u} + \mu^{-1}(\boldsymbol{a} - \boldsymbol{u})$. Then the implications

$$\begin{aligned}
\boldsymbol{x} \in K' \ &\Rightarrow\ \boldsymbol{u} + \mu(\boldsymbol{x} - \boldsymbol{u}) \in K' = \boldsymbol{a} + \mu K \ \Rightarrow\ \mu\boldsymbol{x} \in \mu\boldsymbol{u} + (\boldsymbol{a} - \boldsymbol{u}) + \mu K \\
&\Rightarrow\ \boldsymbol{x} \in \boldsymbol{u} + \mu^{-1}(\boldsymbol{a} - \boldsymbol{u}) + K = \boldsymbol{z} + K, \\
\boldsymbol{u} \notin K' \ &\Rightarrow\ \boldsymbol{o} \notin (\boldsymbol{a} + \mu K) - \boldsymbol{u} \ \Rightarrow\ \mu\boldsymbol{u} \notin \mu\boldsymbol{u} + (\boldsymbol{a} - \boldsymbol{u}) + \mu K \\
&\Rightarrow\ \boldsymbol{u} \notin \boldsymbol{u} + \mu^{-1}(\boldsymbol{a} - \boldsymbol{u}) + K = \boldsymbol{z} + K
\end{aligned}$$

prove that $\boldsymbol{z}$ is a desired point.

1*b*. Assume the existence of a point $\boldsymbol{v} \in K'$ such that the point $\boldsymbol{y} = \boldsymbol{u} + \mu(\boldsymbol{v} - \boldsymbol{u})$ does not belong to K'. Clearly, $\boldsymbol{y} \in (\boldsymbol{u}, \boldsymbol{v})$ due to the assumption $0 < \mu < 1$. We observe that $(\boldsymbol{u}, \boldsymbol{y}) \cap K' = \varnothing$. Indeed, if $(\boldsymbol{u}, \boldsymbol{y})$ contained a point $\boldsymbol{s} \in K'$, then $\boldsymbol{y} \in (\boldsymbol{s}, \boldsymbol{v}) \subset K'$ due to the convexity of K' (see Theorem 3.10). Because $(1 - \mu)^i \to 0$ as $i \to \infty$, the point $\boldsymbol{y}_i = \boldsymbol{u} + (1 - \mu)^i(\boldsymbol{v} - \boldsymbol{u})$ tends to $\boldsymbol{u}$ at $i \to \infty$. Hence there is a smallest positive integer r such that $\boldsymbol{y}_r \notin K'$. Consequently, the point $\boldsymbol{w} = \boldsymbol{u} + (1 - \mu)^{r-1}(\boldsymbol{v} - \boldsymbol{u})$ belongs to K'.

Let $\boldsymbol{z} = \boldsymbol{w} + \mu^{-1}(\boldsymbol{a} - \boldsymbol{w})$. If $\boldsymbol{x} \in K'$, then $(1 - \mu)\boldsymbol{w} + \mu\boldsymbol{x} \in K'$ by the convexity of K', implying the inclusion

$$\begin{aligned}
\boldsymbol{x} &\in \mu^{-1}(K' - (1 - \mu)\boldsymbol{w}) = \mu^{-1}(\mu\boldsymbol{w} + (\boldsymbol{a} - \boldsymbol{w}) + \mu K) \\
&= \boldsymbol{w} + \mu^{-1}(\boldsymbol{a} - \boldsymbol{w}) + K = \boldsymbol{z} + K.
\end{aligned}$$

On the other hand, the relations

$$\mu\boldsymbol{u} + (1 - \mu)\boldsymbol{w} = \boldsymbol{u} + (1 - \mu)(\boldsymbol{w} - \boldsymbol{u}) = \boldsymbol{u} + (1 - \mu)^r(\boldsymbol{v} - \boldsymbol{u}) \notin K'$$

show that

$$\begin{aligned}
\boldsymbol{u} &\notin \mu^{-1}(K' - (1 - \mu)\boldsymbol{w}) = \mu^{-1}(\mu\boldsymbol{w} + (\boldsymbol{a} - \boldsymbol{w}) + \mu K) \\
&= \boldsymbol{w} + \mu^{-1}(\boldsymbol{a} - \boldsymbol{w}) + K = \boldsymbol{z} + K.
\end{aligned}$$

2. Let $\mu > 1$. Similarly to the above argument, it suffices to show that for any $\boldsymbol{u} \in K'$ there is a point $\boldsymbol{z} \in \mathbb{R}^n$ satisfying the inclusion $\boldsymbol{u} \in \boldsymbol{z} + K \subset K'$. Then Y is the collection of such points $\boldsymbol{z}$. Fix a point $\boldsymbol{u} \in K'$. We can write $\boldsymbol{u} = \boldsymbol{a} + \mu\boldsymbol{v}$ for a suitable point $\boldsymbol{v} \in K$. Let $\boldsymbol{z} = \boldsymbol{u} - \boldsymbol{v}$. Since $\boldsymbol{o} \in K - \boldsymbol{v}$, one has

$$\boldsymbol{u} = \boldsymbol{u} + \boldsymbol{o} \in \boldsymbol{u} + (K - \boldsymbol{v}) = \boldsymbol{z} + K.$$

By Lemma 3.12, $K - \boldsymbol{v} \subset \mu(K - \boldsymbol{v})$. Therefore,

$$\boldsymbol{z} + K = \boldsymbol{u} + (K - \boldsymbol{v}) \subset \boldsymbol{u} + \mu(K - \boldsymbol{v}) = \boldsymbol{a} + \mu\boldsymbol{v} + \mu K - \mu\boldsymbol{v} = K'. \quad \square$$

Problems 3.5 and 3.6 give additional results on algebraic properties of convex sets. The next theorem shows that affine spans of convex sets can be describes in terms of affine combinations of two points (compare with Theorem 2.44 and Corollary 2.66).

Theorem 3.14. *The affine span of a nonempty convex set $K \subset \mathbb{R}^n$ can be expressed as*

$$\operatorname{aff} K = \{(1-\lambda)\boldsymbol{x} + \lambda\boldsymbol{y} : \boldsymbol{x}, \boldsymbol{y} \in K,\ \lambda \in \mathbb{R}\}.$$

Proof. Because $\operatorname{aff} K$ is a plane containing K, Theorem 2.38 shows that the set

$$L = \{(1-\lambda)\boldsymbol{x} + \lambda\boldsymbol{y} : \boldsymbol{x}, \boldsymbol{y} \in K,\ \lambda \in \mathbb{R}\}$$

lies in $\operatorname{aff} K$. For the opposite inclusion, choose a point $\boldsymbol{x} \in \operatorname{aff} K$. By Theorem 2.44, $\boldsymbol{x}$ can be written as an affine combination $\boldsymbol{x} = \lambda_1\boldsymbol{x}_1 + \cdots + \lambda_r\boldsymbol{x}_r$ of suitable points $\boldsymbol{x}_1, \ldots, \boldsymbol{x}_r \in K$. Excluding all zero terms of the form $0\boldsymbol{x}_i$, we may assume that none of the scalars $\lambda_1, \ldots, \lambda_r$ is zero. Since $\lambda_1 + \cdots + \lambda_r = 1$, at least one of these scalars is positive. We suppose that all positive scalars among $\lambda_1, \ldots, \lambda_r$ are given by $\lambda_1, \ldots, \lambda_p$, where $1 \leqslant p \leqslant r$. If $p = r$, then $\boldsymbol{x}$ is a convex combination of $\boldsymbol{x}_1, \ldots, \boldsymbol{x}_r$, and $\boldsymbol{x} \in K \subset \operatorname{aff} K$ according to Theorem 3.3. Assume that $p \leqslant r - 1$ and let $\mu = \lambda_1 + \cdots + \lambda_p$. Clearly, $\mu > 1$. Put

$$\mu_i = \begin{cases} \lambda_i/\mu & \text{if } \ 1 \leqslant i \leqslant p, \\ \lambda_i/(1-\mu) & \text{if } \ p+1 \leqslant i \leqslant r. \end{cases}$$

Obviously,

$$\boldsymbol{u} = \mu_1\boldsymbol{x}_1 + \cdots + \mu_p\boldsymbol{x}_p \quad \text{and} \quad \boldsymbol{v} = \mu_{p+1}\boldsymbol{x}_{p+1} + \cdots + \mu_r\boldsymbol{x}_r$$

are convex combinations of points from K. Therefore, $\boldsymbol{u}, \boldsymbol{v} \in K$ according to Theorem 3.3. Because $\boldsymbol{x}$ can be written as $\boldsymbol{x} = (1-\mu)\boldsymbol{v} + \mu\boldsymbol{u}$, we have $\boldsymbol{x} \in L$. Summing up, $\operatorname{aff} K \subset L$. $\square$

Theorem 3.15. *If $f : \mathbb{R}^n \to \mathbb{R}^m$ is an affine transformation and $K \subset \mathbb{R}^n$ and $M \subset \mathbb{R}^m$ are convex sets, then both sets $f(K)$ and $f^{-1}(M)$ are convex.*

Proof. Excluding the obvious cases $K = \varnothing$ and $M \cap \operatorname{rng} f = \varnothing$, we assume that both sets K and $M \cap \operatorname{rng} f$ are nonempty. Consequently, both sets $f(K)$ and $f^{-1}(M)$ are nonempty.

Choose points $\boldsymbol{x}, \boldsymbol{y} \in f(K)$ and a scalar $\lambda \in [0, 1]$. Let $\boldsymbol{x}_0$ and $\boldsymbol{y}_0$ be points in K satisfying the conditions $f(\boldsymbol{x}_0) = \boldsymbol{x}$ and $f(\boldsymbol{y}_0) = \boldsymbol{y}$. Then $(1-\lambda)\boldsymbol{x}_0 + \lambda\boldsymbol{y}_0 \in K$ because K is convex. By Theorem 2.81,

$$\begin{aligned}(1-\lambda)\boldsymbol{x} + \lambda\boldsymbol{y} &= (1-\lambda)f(\boldsymbol{x}_0) + \lambda f(\boldsymbol{y}_0)\\ &= f((1-\lambda)\boldsymbol{x}_0 + \lambda\boldsymbol{y}_0) \in f(K),\end{aligned}$$

which shows the convexity of $f(K)$.

Similarly, if $\boldsymbol{x}, \boldsymbol{y} \in f^{-1}(M)$ and $\lambda \in [0, 1]$, then $f(\boldsymbol{x}), f(\boldsymbol{y}) \in M$, and $(1-\lambda)f(\boldsymbol{x}) + \lambda f(\boldsymbol{y}) \in M$ due to the convexity of M. Theorem 2.81 gives

$$f((1-\lambda)\boldsymbol{x} + \lambda\boldsymbol{y}) = (1-\lambda)f(\boldsymbol{x}) + \lambda f(\boldsymbol{y}) \in M \cap \operatorname{rng} f.$$

Hence

$$(1-\lambda)\boldsymbol{x} + \lambda\boldsymbol{y} \in f^{-1}(M \cap \operatorname{rng} f) = f^{-1}(M),$$

and $f^{-1}(M)$ is convex. □

See Problem 3.19 and various results on page 122 about mappings $\varphi : \mathbb{R}^n \to \mathbb{R}^m$ which preserve convexity of sets.

3.2 Relative Interior of Convex Sets

3.2.1 *Relative Interior of Arbitrary Sets*

Definition 3.16. Given a nonempty set $X \subset \mathbb{R}^n$, a point $\boldsymbol{a} \in X$ is a *relative interior* point of X provided there is a scalar $\rho > 0$ such that $\operatorname{aff} X \cap U_\rho(\boldsymbol{a}) \subset X$. The set of relative interior points of X is called the *relative interior* of X and denoted $\operatorname{rint} X$. The set X is called *relative open* if $\operatorname{rint} X = X$. The empty set $\varnothing$ is relative open.

Examples. 1. The relative interior of a plane $L \subset \mathbb{R}^n$ equals L; in particular, $\operatorname{rint}\{\boldsymbol{c}\} = \{\boldsymbol{c}\}$ for every point $\boldsymbol{c} \in \mathbb{R}^n$.

2. If $L \subset \mathbb{R}^n$ is a plane of positive dimension and D is a closed halfplane of L given by $D = \{\boldsymbol{x} \in L : \boldsymbol{x} \cdot \boldsymbol{e} \leqslant \gamma\}$, where $\boldsymbol{e} \in \mathbb{R}^n \setminus \operatorname{ort} L$ and γ is a scalar (see Theorem 2.70), then $\operatorname{rint} D$ coincides with the open halfplane $E = \{\boldsymbol{x} \in L : \boldsymbol{x} \cdot \boldsymbol{e} < \gamma\}$. Indeed, if D is given by $D = V \cap L$, where $V = \{\boldsymbol{x} \in \mathbb{R}^n : \boldsymbol{x} \cdot \boldsymbol{e} \leqslant \gamma\}$ is the closed halfspace, Corollary 3.31 below shows that $\operatorname{rint} D = \operatorname{int} V \cap L = E$.

The difference between the concepts of relative interior and topological interior in $\mathbb{R}^n$ can be illustrated by the following example (see Figure 3.3). The relative interior of a closed segment $[\boldsymbol{x}, \boldsymbol{y}]$ in $\mathbb{R}^2$, where $\boldsymbol{x} \neq \boldsymbol{y}$, is the open segment $(\boldsymbol{x}, \boldsymbol{y})$, while the topological interior of $[\boldsymbol{x}, \boldsymbol{y}]$ is empty (indeed, no open ball $U_\rho(\boldsymbol{z}) \subset \mathbb{R}^2$ centered at a point $\boldsymbol{z} \in [\boldsymbol{x}, \boldsymbol{y}]$ lies in $[\boldsymbol{x}, \boldsymbol{y}]$).

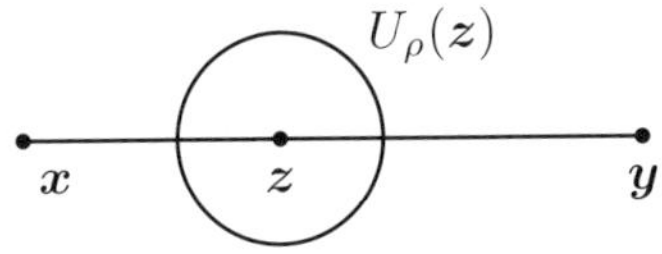

Fig. 3.3 The relative interior of a closed segment $[\boldsymbol{x}, \boldsymbol{y}]$ is the open segment $(\boldsymbol{x}, \boldsymbol{y})$.

The next result slightly sharpens Definition 3.16.

Theorem 3.17. *Let* $X \subset \mathbb{R}^n$ *be a set with nonempty relative interior. For any point* $\boldsymbol{a} \in \operatorname{rint} X$*, there is a scalar* $\delta > 0$ *such that* $\operatorname{aff} X \cap U_\delta(\boldsymbol{a}) \subset \operatorname{rint} X$*. Furthermore, if* $0 < \varepsilon \leqslant \delta$*, then*

$$\operatorname{aff}(\operatorname{rint} X \cap U_\varepsilon(\boldsymbol{a})) = \operatorname{aff}(X \cap U_\varepsilon(\boldsymbol{a})) = \operatorname{aff}(\operatorname{rint} X) = \operatorname{aff} X. \tag{3.2}$$

Consecutively, $\dim(\operatorname{rint} X) = \dim X$.

Proof. Choose a scalar $\rho > 0$ satisfying the condition $U_\rho(\boldsymbol{a}) \cap \operatorname{aff} X \subset X$ and put $\delta = \rho/2$. Let $\boldsymbol{x}$ be a point in $U_\delta(\boldsymbol{a}) \cap \operatorname{aff} X$. We first observe that $U_\delta(\boldsymbol{x}) \subset U_\rho(\boldsymbol{a})$, as follows from Problem 1.8. Consequently,

$$\operatorname{aff} X \cap U_\delta(\boldsymbol{x}) \subset \operatorname{aff} X \cap U_\rho(\boldsymbol{a}) \subset X.$$

Hence $\boldsymbol{x} \in \operatorname{rint} X$, implying the inclusion $\operatorname{aff} X \cap U_\delta(\boldsymbol{a}) \subset \operatorname{rint} X$.

For (3.2), choose a scalar $\varepsilon \in (0, \delta]$. Then

$$\operatorname{aff} X \cap U_\varepsilon(\boldsymbol{a}) \subset \operatorname{aff} X \cap U_\delta(\boldsymbol{a}) \subset \operatorname{rint} X.$$

Consequently,

$$\operatorname{aff} X \cap U_\varepsilon(\boldsymbol{a}) = (\operatorname{aff} X \cap U_\varepsilon(\boldsymbol{a})) \cap U_\varepsilon(\boldsymbol{a}) \subset \operatorname{rint} X \cap U_\varepsilon(\boldsymbol{a}).$$

Now, a combination of Theorems 2.42 and 2.68 gives

$$\begin{aligned} \operatorname{aff} X &= \operatorname{aff}(\operatorname{aff} X \cap U_\varepsilon(\boldsymbol{a})) \subset \operatorname{aff}(\operatorname{rint} X \cap U_\varepsilon(\boldsymbol{a})) \\ &\subset \operatorname{aff}(\operatorname{rint} X) \subset \operatorname{aff} X. \end{aligned}$$

Therefore, $\operatorname{aff}(\operatorname{rint} X \cap U_\varepsilon(\boldsymbol{a})) = \operatorname{aff}(\operatorname{rint} X) = \operatorname{aff} X$.

In a similar way, the inclusions

$$\operatorname{aff} X \cap U_\varepsilon(\boldsymbol{a}) \subset \operatorname{rint} X \cap U_\varepsilon(\boldsymbol{a}) \subset X \cap U_\varepsilon(\boldsymbol{a}) \subset X$$

imply that $\operatorname{aff}(X \cap U_\varepsilon(\boldsymbol{a})) = \operatorname{aff} X$. Finally, the above argument and Definition 2.63 give

$$\dim(\operatorname{rint} X) = \dim(\operatorname{aff}(\operatorname{rint} X)) = \dim(\operatorname{aff} X) = \dim X. \qquad \square$$

Theorem 3.18. *For sets X and Y in $\mathbb{R}^n$, the following assertions hold.*

(*a*) *If $X \subset Y$ and $\operatorname{aff} X = \operatorname{aff} Y$, then $\operatorname{rint} X \subset \operatorname{rint} Y$. Moreover, for any point $\boldsymbol{a} \in \operatorname{rint} X$, there is a scalar $\delta > 0$ such that*

$$U_\delta(\boldsymbol{a}) \cap \operatorname{rint} X = U_\delta(\boldsymbol{a}) \cap \operatorname{rint} Y.$$

(*b*) $\operatorname{rint}(\operatorname{rint} X) = \operatorname{rint} X$.
(*c*) *If $\operatorname{rint} X \subset Y \subset X$, then $\operatorname{rint} X = \operatorname{rint} Y$.*
(*d*) *If a translate of Y lies in $\operatorname{aff} X$, then $\operatorname{rint} X + Y \subset \operatorname{rint}(X + Y)$.*
(*e*) $\operatorname{rint}(\boldsymbol{c} + X) = \boldsymbol{c} + \operatorname{rint} X$ *for any point $\boldsymbol{c} \in \mathbb{R}^n$.*

Proof. Since all assertions of the theorem are obvious when $\operatorname{rint} X = \varnothing$, we may assume, in what follows, that $\operatorname{rint} X \neq \varnothing$.

(*a*) Choose a point $\boldsymbol{x} \in \operatorname{rint} X$ and a scalar $\rho > 0$ satisfying the condition $\operatorname{aff} X \cap U_\rho(\boldsymbol{x}) \subset X$. Since $\operatorname{aff} X = \operatorname{aff} Y$, we have

$$\operatorname{aff} Y \cap U_\rho(\boldsymbol{x}) = \operatorname{aff} X \cap U_\rho(\boldsymbol{x}) \subset X \subset Y.$$

Hence $\boldsymbol{x} \in \operatorname{rint} Y$, and $\operatorname{rint} X \subset \operatorname{rint} Y$. Furthermore, Theorem 3.17 implies the existence of a scalar $\delta > 0$ such that $\operatorname{aff} X \cap U_\delta(\boldsymbol{a}) \subset \operatorname{rint} X$. By the above argument, $\operatorname{rint} X \cap U_\delta(\boldsymbol{a}) \subset \operatorname{rint} Y \cap U_\delta(\boldsymbol{a})$.

Hence it remains to prove the opposite inclusion. Indeed,

$$\begin{aligned}\operatorname{rint} Y \cap U_\delta(\boldsymbol{a}) &\subset \operatorname{aff} Y \cap U_\delta(\boldsymbol{a}) = \operatorname{aff} X \cap U_\delta(\boldsymbol{a}) \\ &= (\operatorname{aff} X \cap U_\delta(\boldsymbol{a})) \cap U_\delta(\boldsymbol{a}) \subset \operatorname{rint} X \cap U_\delta(\boldsymbol{a}).\end{aligned}$$

(*b*) Because $\operatorname{rint} X \subset X$ and $\operatorname{aff}(\operatorname{rint} X) = \operatorname{aff} X$ (see Theorem 3.17), assertion (*a*) implies that $\operatorname{rint}(\operatorname{rint} X) \subset \operatorname{rint} X$. For the opposite inclusion, choose a point $\boldsymbol{x} \in \operatorname{rint} X$. By the same theorem, there is a scalar $\delta > 0$ such that $U_\delta(\boldsymbol{x}) \cap \operatorname{aff} X \subset \operatorname{rint} X$. Thus

$$\operatorname{aff}(\operatorname{rint} X) \cap U_\delta(\boldsymbol{x}) = \operatorname{aff} X \cap U_\delta(\boldsymbol{x}) \subset \operatorname{rint} X.$$

Therefore, $\boldsymbol{x} \in \operatorname{rint}(\operatorname{rint} X)$, and $\operatorname{rint} X \subset \operatorname{rint}(\operatorname{rint} X)$.

(*c*) We first observe that $\operatorname{aff} X = \operatorname{aff}(\operatorname{rint} X) = \operatorname{aff} Y$. Indeed, a combination of Theorems 2.42 and 3.17 gives

$$\operatorname{aff} X = \operatorname{aff}(\operatorname{rint} X) \subset \operatorname{aff} Y \subset \operatorname{aff} X.$$

Now, assertions (a) and (b) imply that

$$\operatorname{rint} X = \operatorname{rint}(\operatorname{rint} X) \subset \operatorname{rint} Y \subset \operatorname{rint} X.$$

Summing up, $\operatorname{rint} X = \operatorname{rint} Y$.

(d) Let $\boldsymbol{z} \in \operatorname{rint} X + Y$. Then $\boldsymbol{z} = \boldsymbol{x} + \boldsymbol{y}$, where $\boldsymbol{x} \in \operatorname{rint} X$ and $\boldsymbol{y} \in Y$. Choose a scalar $\rho > 0$ satisfying the condition $U_\rho(\boldsymbol{x}) \cap \operatorname{aff} X \subset X$. Because

$$\operatorname{aff}(X + Y) = \operatorname{aff} X + Y = \operatorname{aff} X + \boldsymbol{y}$$

(see Problem 2.13) and $U_\rho(\boldsymbol{z}) = U_\rho(\boldsymbol{x}) + \boldsymbol{y}$, one has

$$\begin{aligned}\operatorname{aff}(X + Y) \cap U_\rho(\boldsymbol{z}) &= (\operatorname{aff} X + \boldsymbol{y}) \cap (U_\rho(\boldsymbol{x}) + \boldsymbol{y}) \\ &= \operatorname{aff} X \cap U_\rho(\boldsymbol{x}) + \boldsymbol{y} \subset X + \boldsymbol{y} \subset X + Y.\end{aligned}$$

Therefore, $\boldsymbol{z} \in \operatorname{rint}(X + Y)$, and $\operatorname{rint} X + Y \subset \operatorname{rint}(X + Y)$.

(e) Since a translate of $\boldsymbol{c}$ lies in $\operatorname{aff} X$, assertion (d) gives the inclusion $\boldsymbol{c} + \operatorname{rint} X \subset \operatorname{rint}(\boldsymbol{c} + X)$. Hence it remains to show that

$$\operatorname{rint}(\boldsymbol{c} + X) \subset \boldsymbol{c} + \operatorname{rint} X.$$

Let $\boldsymbol{u} \in \operatorname{rint}(\boldsymbol{c} + X)$. Then $\boldsymbol{u} \in \boldsymbol{c} + X$, and we can write $\boldsymbol{u} = \boldsymbol{c} + \boldsymbol{x}$ for a suitable point $\boldsymbol{x} \in X$. Choose a scalar $\delta > 0$ such that

$$U_\delta(\boldsymbol{u}) \cap \operatorname{aff}(\boldsymbol{c} + X) \subset \boldsymbol{c} + X.$$

Using Theorem 2.46, we obtain

$$\begin{aligned}\operatorname{aff} X \cap U_\delta(\boldsymbol{x}) &= \operatorname{aff}(\boldsymbol{c} + X - \boldsymbol{c}) \cap U_\delta(\boldsymbol{u} - \boldsymbol{c}) \\ &= (\operatorname{aff}(\boldsymbol{c} + X) - \boldsymbol{c}) \cap (U_\delta(\boldsymbol{u}) - \boldsymbol{c}) \\ &= \operatorname{aff}(\boldsymbol{c} + X) \cap U_\delta(\boldsymbol{u}) - \boldsymbol{c} \\ &\subset (\boldsymbol{c} + X) - \boldsymbol{c} = X.\end{aligned}$$

Hence $\boldsymbol{x} \in \operatorname{rint} X$ and $\boldsymbol{u} = \boldsymbol{c} + \boldsymbol{x} \in \boldsymbol{c} + \operatorname{rint} X$. □

Remarks. 1. The condition $\operatorname{aff} X = \operatorname{aff} Y$ is essential in assertion (a). For instance, if X is a triangle in $\mathbb{R}^2$ and Y one of its sides, then $Y \subset X$, while $\operatorname{rint} X$ and $\operatorname{rint} Y$ are disjoint.

2. The inclusion $\operatorname{rint} X + Y \subset \operatorname{rint}(X + Y)$ in assertion (d) may be proper. Indeed, let $X = [0, 1] \cup \mathbb{Q}$ and $Y = [0, 1]$, where $\mathbb{Q}$ denotes the set of all rational numbers in $\mathbb{R}$. Then $\operatorname{rint} X = (0, 1)$ and

$$\operatorname{rint} X + Y = (0, 2) \neq \mathbb{R} = X + Y = \operatorname{rint}(X + Y).$$

The next theorem gives an important algebraic description of relative interior points (see also Theorem 3.26).

Theorem 3.19. *Let a set $X \subset \mathbb{R}^n$ have nonempty relative interior. Given a point $\boldsymbol{x} \in \operatorname{rint} X$, the following assertions hold.*

(a) *For any point* $\boldsymbol{y} \in \operatorname{aff} X$, *there is a scalar* $0 < \lambda < 1$ *satisfying the condition* $(1-\lambda)\boldsymbol{x} + \lambda \boldsymbol{y} \in X$.

(b) *For any point* $\boldsymbol{y} \in \operatorname{aff} X$, *there is a scalar* $\gamma > 1$ *satisfying the condition* $\gamma \boldsymbol{x} + (1-\gamma)\boldsymbol{y} \in X$.

Proof. (a) Choose a point $\boldsymbol{y} \in \operatorname{aff} X$ and a scalar $\rho > 0$ such that $\operatorname{aff} X \cap U_\rho(\boldsymbol{x}) \subset X$. Excluding the obvious case $\boldsymbol{x} = \boldsymbol{y}$, we suppose that $\boldsymbol{x} \neq \boldsymbol{y}$. Let a scalar δ satisfy the inequalities $0 < \delta < \min\{\rho, \|\boldsymbol{x} - \boldsymbol{y}\|\}$. The line $l = \langle \boldsymbol{x}, \boldsymbol{y} \rangle$ lies in $\operatorname{aff} X$ (see Theorem 2.38) and meets the sphere $S_\delta(\boldsymbol{x})$ at two distinct points, $\boldsymbol{u}$ and $\boldsymbol{v}$, as shown in Figure 3.4. Since $S_\delta(\boldsymbol{x}) \subset U_\rho(\boldsymbol{x})$, one has

$$\boldsymbol{u}, \boldsymbol{v} \in \operatorname{aff} X \cap S_\delta(\boldsymbol{x}) \subset \operatorname{aff} X \cap U_\rho(\boldsymbol{x}) \subset X.$$

By Lemma 2.25, one of the points $\boldsymbol{u}, \boldsymbol{v}$, say $\boldsymbol{v}$, belongs to the open halfline $(\boldsymbol{x}, \boldsymbol{y}\rangle$. Then $\boldsymbol{v} = (1-\lambda)\boldsymbol{x} + \lambda \boldsymbol{y}$ for a suitable scalar $\lambda > 0$. Consequently, $\lambda(\boldsymbol{x} - \boldsymbol{y}) = \boldsymbol{x} - \boldsymbol{v}$, which gives

$$\lambda \|\boldsymbol{x} - \boldsymbol{y}\| = \|\boldsymbol{x} - \boldsymbol{v}\| = \delta < \|\boldsymbol{x} - \boldsymbol{y}\|.$$

Hence $0 < \lambda < 1$.

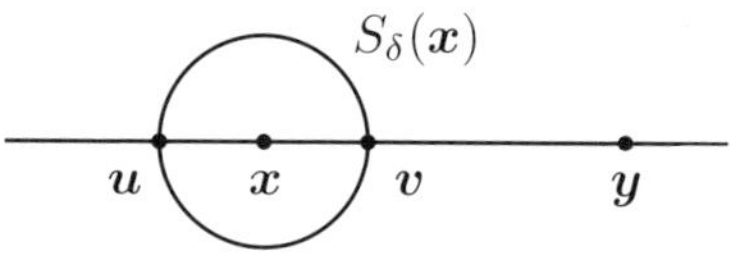

Fig. 3.4 Illustration to Theorem 3.19.

(b) Choose a point $\boldsymbol{y} \in \operatorname{aff} X$ and let $\boldsymbol{y}' = 2\boldsymbol{x} - \boldsymbol{y}$. Then $\boldsymbol{y}'$ belongs to $\operatorname{aff} X$ as an affine combination of $\boldsymbol{x}$ and $\boldsymbol{y}$ from $\operatorname{aff} X$ (see Theorem 2.38). By condition (b), there is a scalar $0 < \lambda < 1$ such that $(1-\lambda)\boldsymbol{x} + \lambda \boldsymbol{y}' \in X$. Put $\gamma = 1 + \lambda$. Then $\gamma > 1$ and

$$\gamma \boldsymbol{x} + (1-\gamma)\boldsymbol{y} = (1+\lambda)\boldsymbol{x} + (-\lambda)(2\boldsymbol{x} - \boldsymbol{y}') = (1-\lambda)\boldsymbol{x} + \lambda \boldsymbol{y}' \in X. \quad \square$$

3.2.2 *Relative Interior and Simplices*

Theorem 3.20. *The relative interior of an* r*-simplex* $\Delta = \Delta(\boldsymbol{x}_1, \dots, \boldsymbol{x}_{r+1})$ *in* $\mathbb{R}^n$ *is the set of all positive convex combinations of* $\boldsymbol{x}_1, \dots, \boldsymbol{x}_{r+1}$:

$$\operatorname{rint} \Delta = \{\lambda_1 \boldsymbol{x}_1 + \cdots + \lambda_{r+1}\boldsymbol{x}_{r+1} : \lambda_1, \dots, \lambda_{r+1} > 0, \\ \lambda_1 + \cdots + \lambda_{r+1} = 1\}.$$

Proof. First, we will show that the set

$$M = \{\lambda_1 \boldsymbol{x}_1 + \cdots + \lambda_{r+1}\boldsymbol{x}_{r+1} : \lambda_1, \dots, \lambda_{r+1} > 0,$$
$$\lambda_1 + \cdots + \lambda_{r+1} = 1\}$$

lies in $\operatorname{rint} \Delta$. For this, choose a point $\boldsymbol{x} = \lambda_1\boldsymbol{x}_1 + \cdots + \lambda_{r+1}\boldsymbol{x}_{r+1} \in M$ and put $\varepsilon = \min\{\lambda_1, \dots, \lambda_{r+1}\}$. Theorem 2.55 gives the existence of a scalar $\rho > 0$ with the following property: for every point $\boldsymbol{u} \in \operatorname{aff} \Delta \cap U_\rho(\boldsymbol{x})$, its affine coordinates $\mu_1, \dots, \mu_{r+1}$ in the expression

$$\boldsymbol{u} = \mu_1\boldsymbol{x}_1 + \cdots + \mu_{r+1}\boldsymbol{x}_{r+1}, \quad \mu_1 + \cdots + \mu_{r+1} = 1,$$

satisfy the inequalities $|\lambda_i - \mu_i| < \varepsilon$, $1 \leqslant i \leqslant r+1$. Consequently,

$$\mu_i > \lambda_i - \varepsilon \geqslant 0 \ \text{ for all } \ 1 \leqslant i \leqslant r+1,$$

which gives the inclusion $\boldsymbol{u} \in \Delta$ (see Definition 3.5). Hence $\operatorname{aff} \Delta \cap U_\rho(\boldsymbol{x}) \subset \Delta$, and $\boldsymbol{x} \in \operatorname{rint} \Delta$ according to Definition 3.16.

Conversely, let a point $\boldsymbol{x}$ belong to $\operatorname{rint} \Delta$. Then $\boldsymbol{x} \in \Delta$ and can be written as a convex combination $\boldsymbol{x} = \lambda_1\boldsymbol{x}_1 + \cdots + \lambda_{r+1}\boldsymbol{x}_{r+1}$. Assume for a moment that at least one of the scalars $\lambda_1, \dots, \lambda_{r+1}$, say λ_1, equals zero. Then $\lambda_2 + \cdots + \lambda_{r+1} = 1$. Choose a scalar $\rho > 0$, and put

$$\gamma = \frac{\rho}{2}\Big(\|\boldsymbol{x}_1\| + \frac{1}{r}(\|\boldsymbol{x}_2\| + \cdots + \|\boldsymbol{x}_{r+1}\|)\Big)^{-1},$$
$$\boldsymbol{v} = -\gamma\boldsymbol{x}_1 + \big(\lambda_2 + \frac{\gamma}{r}\big)\boldsymbol{x}_2 + \cdots + \big(\lambda_{r+1} + \frac{\gamma}{r}\big)\boldsymbol{x}_{r+1}.$$

Clearly, $\boldsymbol{v}$ is an affine combination of $\boldsymbol{x}_1, \dots, \boldsymbol{x}_{r+1}$. Therefore,

$$\boldsymbol{v} \in \operatorname{aff}\{\boldsymbol{x}_1, \dots, \boldsymbol{x}_{r+1}\} = \operatorname{aff} \Delta$$

according to Theorem 3.6. Furthermore, $\boldsymbol{v} \in U_\rho(\boldsymbol{x})$ because

$$\|\boldsymbol{v} - \boldsymbol{x}\| = \| -\gamma\boldsymbol{x}_1 + \frac{\gamma}{r}\boldsymbol{x}_2 + \cdots + \frac{\gamma}{r}\boldsymbol{x}_{r+1}\|$$
$$\leqslant \gamma\|\boldsymbol{x}_1\| + \frac{\gamma}{r}\|\boldsymbol{x}_2\| + \cdots + \frac{\gamma}{r}\|\boldsymbol{x}_{r+1}\| = \frac{\rho}{2} < \rho.$$

On the other hand, $\boldsymbol{v} \notin \Delta$ since its first affine coordinate, $-\gamma$, is negative (see a remark on page 78). So, $\operatorname{aff} \Delta \cap U_\rho(\boldsymbol{x}) \not\subset \Delta$ for any choice of $\rho > 0$, contrary to the assumption $\boldsymbol{x} \in \operatorname{rint} \Delta$. Thus all scalars $\lambda_1, \dots, \lambda_{r+1}$ should be positive, which gives the inclusion $\boldsymbol{x} \in M$. □

Corollary 3.21. *If $K \subset \mathbb{R}^n$ is a nonempty convex set, then* $\operatorname{rint} K$ *is a nonempty set of the same dimension as K. Furthermore,* $\operatorname{rint} \Delta \subset \operatorname{rint} K$ *for every simplex $\Delta \subset K$ whose dimension equals* $\dim K$*.*

Proof. Let $m = \dim K$. By Corollary 3.7, K contains an m-simplex Δ' such that $\operatorname{aff} \Delta' = \operatorname{aff} K$. A combination of Theorems 3.18 and 3.20 implies that $\varnothing \neq \operatorname{rint} \Delta' \subset \operatorname{rint} K$. The equality $\dim(\operatorname{rint} K) = \dim K$ follows from Theorem 3.17.

Similarly, if Δ is an m-simplex in K, then $\operatorname{aff} \Delta = \operatorname{aff} K$ according to Corollary 3.7, and Theorem 3.18 gives the inclusion $\operatorname{rint} \Delta \subset \operatorname{rint} K$. □

Theorem 3.22. *Let $X \subset \mathbb{R}^n$ be a nonempty set of dimension m. A point $\boldsymbol{x} \in \mathbb{R}^n$ belongs to $\operatorname{rint} X$ if and only if there is an m-simplex $\Delta \subset X$ with the property $\boldsymbol{x} \in \operatorname{rint} \Delta$.*

Proof. Suppose first that $\boldsymbol{x}$ belongs to the relative interior of an m-simplex $\Delta = \Delta(\boldsymbol{x}_1, \dots, \boldsymbol{x}_{m+1}) \subset X$. Since the set $\{\boldsymbol{x}_1, \dots, \boldsymbol{x}_{m+1}\}$ is affinely independent, it is an affine basis for $\operatorname{aff} X$ (see Corollary 2.66). This argument and Theorem 3.6 give $\operatorname{aff} \Delta = \operatorname{aff} \{\boldsymbol{x}_1, \dots, \boldsymbol{x}_{m+1}\} = \operatorname{aff} X$. Consequently, $\boldsymbol{x} \in \operatorname{rint} \Delta \subset \operatorname{rint} X$ according to Theorem 3.18.

Conversely, let $\boldsymbol{x} \in \operatorname{rint} X$. The case $m = 0$ is obvious (then X is a singleton, say $\{\boldsymbol{u}\}$, and $\Delta = \Delta(\boldsymbol{u})$ is the desired 0-simplex). Therefore, we may assume that $m \geqslant 1$. Choose a scalar $\rho > 0$ satisfying the inclusion $\operatorname{aff} X \cap U_\rho(\boldsymbol{x}) \subset X$. Because $\dim(\operatorname{aff} X) = \dim X = m$, the set $S = \operatorname{aff} X - \boldsymbol{x}$ is an m-dimensional subspace (see Theorem 2.2). Let $\{\boldsymbol{b}_1, \dots, \boldsymbol{b}_m\}$ be a basis for S such that $\|\boldsymbol{b}_i\| < \rho/m$, $1 \leqslant i \leqslant m$. Put $\boldsymbol{b}_{m+1} = -(\boldsymbol{b}_1 + \dots + \boldsymbol{b}_m)$. Clearly, $\boldsymbol{b}_{m+1} \in S$. Let $\boldsymbol{x}_i = \boldsymbol{x} + \boldsymbol{b}_i$, $1 \leqslant i \leqslant m+1$. Then

$$\{\boldsymbol{x}_1, \dots, \boldsymbol{x}_{m+1}\} = \boldsymbol{x} + \{\boldsymbol{b}_1, \dots, \boldsymbol{b}_{m+1}\} \subset \boldsymbol{x} + S = \operatorname{aff} X.$$

This inclusion and the inequalities

$$\|\boldsymbol{x}_i - \boldsymbol{x}\| = \|\boldsymbol{b}_i\| < \rho/m, \quad 1 \leqslant i \leqslant m,$$
$$\|\boldsymbol{x}_{m+1} - \boldsymbol{x}\| \leqslant \|\boldsymbol{b}_1\| + \dots + \|\boldsymbol{b}_m\| < m\,(\rho/m) = \rho$$

show that $\{\boldsymbol{x}_1, \dots, \boldsymbol{x}_{m+1}\} \subset \operatorname{aff} X \cap U_\rho(\boldsymbol{x}) \subset X$.

By Theorem 2.54, the set $\{\boldsymbol{b}_1, \dots, \boldsymbol{b}_{m+1}\}$ is affinely independent. Hence, its translate $\{\boldsymbol{x}_1, \dots, \boldsymbol{x}_{m+1}\}$ also is affinely independent, as follows from Theorem 2.78. Since the set $\operatorname{aff} X \cap U_\rho(\boldsymbol{x})$ is convex (as the intersection of convex sets $\operatorname{aff} X$ and $U_\rho(\boldsymbol{x})$), Corollary 3.7 shows that the m-simplex $\Delta(\boldsymbol{x}_1, \dots, \boldsymbol{x}_{m+1})$ lies in $\operatorname{aff} X \cap U_\rho(\boldsymbol{x}) \subset X$. Finally, the equalities

$$\begin{aligned}\boldsymbol{x} &= \frac{1}{m+1}((m+1)\boldsymbol{x} + \boldsymbol{o}) = \frac{1}{m+1}((m+1)\boldsymbol{x} + \boldsymbol{b}_1 + \dots + \boldsymbol{b}_{m+1}) \\ &= \frac{1}{m+1}((\boldsymbol{x} + \boldsymbol{b}_1) + \dots + (\boldsymbol{x} + \boldsymbol{b}_{m+1})) = \frac{1}{m+1}(\boldsymbol{x}_1 + \dots + \boldsymbol{x}_{m+1}),\end{aligned}$$

and Theorem 3.20 give the inclusion $\boldsymbol{x} \in \operatorname{rint} \Delta(\boldsymbol{x}_1, \dots, \boldsymbol{x}_{m+1})$. □

The next result refines Theorem 3.22 (see also Problem 3.11).

Theorem 3.23. *Let $K \subset \mathbb{R}^n$ be a convex set of positive dimension m and $\boldsymbol{u}$ be a point in K. A point $\boldsymbol{x} \in K \setminus \{\boldsymbol{u}\}$ belongs to* $\operatorname{rint} K$ *if and only if K contains an m-simplex Δ with vertex $\boldsymbol{u}$ such that $\boldsymbol{x} \in \operatorname{rint} \Delta$.*

Proof. Since the basis $\{\boldsymbol{b}_1, \dots, \boldsymbol{b}_m\}$ for the subspace S in the proof of Theorem 3.22 was chosen arbitrarily, we can modify the argument of that proof by letting

$$\boldsymbol{b}_1 = \delta(\boldsymbol{u} - \boldsymbol{x}), \quad \text{where} \quad 0 < \delta < \frac{\rho}{m\|\boldsymbol{u} - \boldsymbol{x}\|}.$$

As above, $\|\boldsymbol{b}_1\| < \rho/m$ and

$$\boldsymbol{x} = \frac{1}{m+1}(\boldsymbol{x}_1 + \cdots + \boldsymbol{x}_{m+1}), \tag{3.3}$$

where $\boldsymbol{x}_1 = \boldsymbol{x} + \boldsymbol{b}_1 = (1-\delta)\boldsymbol{x} + \delta\boldsymbol{u}$. Using this expression for $\boldsymbol{x}_1$ in (3.3), we obtain

$$\boldsymbol{x} = \frac{1}{m+1}((1-\delta)\boldsymbol{x} + \delta\boldsymbol{u} + \boldsymbol{x}_2 + \cdots + \boldsymbol{x}_{m+1}).$$

Solving this equation for $\boldsymbol{x}$, we express it as a positive convex combination

$$\boldsymbol{x} = \frac{1}{m+\delta}(\delta\boldsymbol{u} + \boldsymbol{x}_2 + \cdots + \boldsymbol{x}_{m+1}).$$

Finally, a combination of Corollary 3.7 and Theorem 3.20 gives

$$\Delta(\boldsymbol{u}, \boldsymbol{x}_2, \dots, \boldsymbol{x}_{m+1}) \subset K \quad \text{and} \quad \boldsymbol{x} \in \operatorname{rint} \Delta(\boldsymbol{u}, \boldsymbol{x}_2, \dots, \boldsymbol{x}_{m+1}). \qquad \square$$

3.2.3 *Geometric Properties of Relative Interior*

Theorem 3.24. *If $K \subset \mathbb{R}^n$ is a nonempty convex set, $\boldsymbol{x} \in \operatorname{rint} K$, and $\boldsymbol{y} \in K$, then $(1-\lambda)\boldsymbol{x} + \lambda\boldsymbol{y} \in \operatorname{rint} K$ for all $0 \leqslant \lambda < 1$. Consequently, the semi-open segment $[\boldsymbol{x}, \boldsymbol{y})$ lies in* $\operatorname{rint} K$.

Proof. Choose a point $\boldsymbol{z} = (1-\lambda)\boldsymbol{x} + \lambda\boldsymbol{y}$, where $0 \leqslant \lambda < 1$. Since the case $\lambda = 0$ is obvious, we may assume that $\lambda > 0$. Let a scalar $\rho > 0$ satisfy the condition $\operatorname{aff} K \cap U_\rho(\boldsymbol{x}) \subset K$. We are going to show that $\operatorname{aff} K \cap U_{(1-\lambda)\rho}(\boldsymbol{z}) \subset K$ (see Figure 3.5).

Indeed, let $\boldsymbol{u}$ be a point in $\operatorname{aff} K \cap U_{(1-\lambda)\rho}(\boldsymbol{z})$. Put $\boldsymbol{v} = \frac{1}{1-\lambda}\boldsymbol{u} - \frac{\lambda}{1-\lambda}\boldsymbol{y}$ (so that $\boldsymbol{u} = (1-\lambda)\boldsymbol{v} + \lambda\boldsymbol{y}$). Then $\boldsymbol{v} \in \operatorname{aff} K$ as an affine combination of $\boldsymbol{u}$ and $\boldsymbol{y}$ from $\operatorname{aff} K$ (see Theorem 2.38). Since $\boldsymbol{x} = \frac{1}{1-\lambda}\boldsymbol{z} - \frac{\lambda}{1-\lambda}\boldsymbol{y}$, one has

$$\|\boldsymbol{x} - \boldsymbol{v}\| = \frac{1}{1-\lambda}\|\boldsymbol{u} - \boldsymbol{z}\| < \frac{1}{1-\lambda}(1-\lambda)\rho = \rho.$$

Hence $\boldsymbol{v} \in \operatorname{aff} K \cap U_\rho(\boldsymbol{x}) \subset K$, and $\boldsymbol{u} = (1-\lambda)\boldsymbol{v} + \lambda\boldsymbol{y} \in K$ by the convexity of K. Thus $\operatorname{aff} K \cap U_{(1-\lambda)\rho}(\boldsymbol{z}) \subset K$, which gives the inclusion $\boldsymbol{z} \in \operatorname{rint} K$. Consequently, $(\boldsymbol{x}, \boldsymbol{y}) \subset \operatorname{rint} K$. $\square$

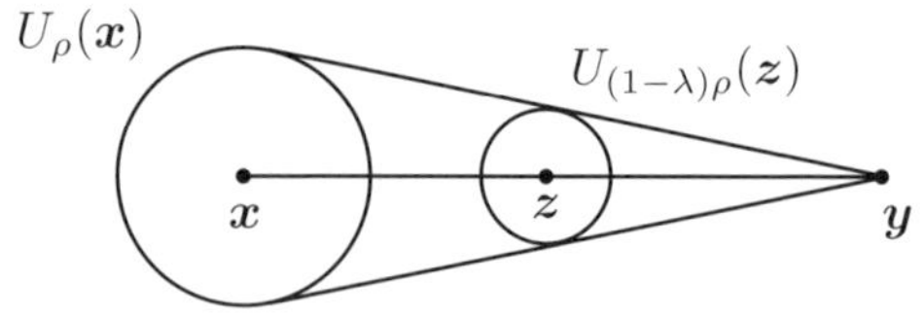

Fig. 3.5 Illustration to Theorem 3.24.

Corollary 3.25. *If $K \subset \mathbb{R}^n$ is a convex set, then its relative interior is a convex set whose dimension equals* $\dim K$. *Furthermore,*

$$\operatorname{aff}(\operatorname{rint} K) = \operatorname{aff} K.$$

Proof. The assertion is obvious when $K = \varnothing$. If $K \neq \varnothing$, then the nonemptiness of $\operatorname{rint} K$ is proved in Corollary 3.21, and the convexity of $\operatorname{rint} K$ follows from Theorem 3.24. By Theorem 2.42, the inclusion $\operatorname{rint} K \subset K$ implies that $\operatorname{aff}(\operatorname{rint} K) \subset \operatorname{aff} K$. Since

$$\dim(\operatorname{aff}(\operatorname{rint} K)) = \dim(\operatorname{rint} K) = \dim K = \dim(\operatorname{aff} K),$$

Theorem 2.6 gives the equality $\operatorname{aff}(\operatorname{rint} K) = \operatorname{aff} K$. □

The next theorem gives an important algebraic description of relative interior points of a convex set.

Theorem 3.26. *For a nonempty convex set $K \subset \mathbb{R}^n$ and a point $\boldsymbol{x} \in \mathbb{R}^n$, the following conditions are equivalent.*

(*a*) $\boldsymbol{x} \in \operatorname{rint} K$.
(*b*) *For any point $\boldsymbol{y} \in \operatorname{aff} K$, there is a scalar $0 < \lambda < 1$ such that $(1-\lambda)\boldsymbol{x} + \lambda\boldsymbol{y} \in K$.*
(*c*) *For any point $\boldsymbol{y} \in \operatorname{aff} K$, there is a scalar $\gamma > 1$ satisfying the inclusion $\gamma\boldsymbol{x} + (1-\gamma)\boldsymbol{y} \in K$.*
(*d*) *For any point $\boldsymbol{y} \in K$, there is a scalar $\gamma > 1$ with the property $\gamma\boldsymbol{x} + (1-\gamma)\boldsymbol{y} \in K$.*
(*e*) *There is a point $\boldsymbol{y} \in \operatorname{rint} K$ and a scalar $\gamma > 1$ satisfying the inclusion $\gamma\boldsymbol{x} + (1-\gamma)\boldsymbol{y} \in K$.*

Proof. The implications $(a) \Rightarrow (b)$ and $(a) \Rightarrow (c)$ follow from Theorem 3.19. Because $(c) \Rightarrow (d) \Rightarrow (e)$, it suffices to show that $(e) \Rightarrow (a)$.

Let a point $\boldsymbol{y} \in \operatorname{rint} K$ and a scalar $\gamma > 1$ satisfy condition (e). Since the case $\boldsymbol{x} = \boldsymbol{y}$ is obvious, we suppose that $\boldsymbol{x} \neq \boldsymbol{y}$. Put $\boldsymbol{z} = \gamma\boldsymbol{x} + (1-\gamma)\boldsymbol{y}$. By the assumption, $\boldsymbol{z} \in K$. Because $\boldsymbol{x}$ is expressible as

$$\boldsymbol{x} = (1-\gamma^{-1})\boldsymbol{y} + \gamma^{-1}\boldsymbol{z}, \quad \text{where} \quad 0 < \gamma^{-1} < 1,$$

one has $\boldsymbol{x} \in (\boldsymbol{y}, \boldsymbol{z}) \subset \operatorname{rint} K$ according to Theorem 3.24. □

Theorem 3.26 implies the corollary below.

Corollary 3.27. *For a nonempty convex set $K \subset \mathbb{R}^n$ and a point $\boldsymbol{x} \in \mathbb{R}^n$, the following conditions are equivalent.*

(*a*) $\boldsymbol{x} \in \operatorname{rint} K$.
(*b*) *For any $\boldsymbol{y} \in \operatorname{aff} K \setminus \{\boldsymbol{x}\}$, there is a point $\boldsymbol{u} \in (\boldsymbol{x}, \boldsymbol{y}) \cap K$.*
(*c*) *For any $\boldsymbol{y} \in \operatorname{aff} K \setminus \{\boldsymbol{x}\}$, there is a point $\boldsymbol{u} \in K \setminus \{\boldsymbol{x}\}$ such that $\boldsymbol{x} \in (\boldsymbol{u}, \boldsymbol{y})$.*
(*d*) *For any $\boldsymbol{y} \in K \setminus \{\boldsymbol{x}\}$, there is a point $\boldsymbol{u} \in K \setminus \{\boldsymbol{x}\}$ such that $\boldsymbol{x} \in (\boldsymbol{u}, \boldsymbol{y})$.*
(*e*) *There are points $\boldsymbol{y} \in \operatorname{rint} K \setminus \{\boldsymbol{x}\}$ and $\boldsymbol{u} \in K \setminus \{\boldsymbol{x}\}$ such that $\boldsymbol{x} \in (\boldsymbol{u}, \boldsymbol{y})$.* □

Theorem 3.28. *Let $K \subset \mathbb{R}^n$ be a nonempty convex set, and $\boldsymbol{a}$ be a point in $\operatorname{rint} K$. Then $\operatorname{rint} K = M_1 \cup M_2 \cup \cdots$, where every set M_i is the homothetic copy of K given by*

$$M_i = \boldsymbol{a} + (1 - \tfrac{1}{i+1})(K - \boldsymbol{a}), \quad i \geqslant 1.$$

Proof. Choose a point $\boldsymbol{y} \in \operatorname{rint} K$. By Theorem 3.26, there is a scalar $\gamma > 1$ such that the point $\boldsymbol{z} = (1-\gamma)\boldsymbol{a} + \gamma\boldsymbol{y}$ belongs to K. Obviously,

$$\boldsymbol{y} = (1-\gamma^{-1})\boldsymbol{a} + \gamma^{-1}\boldsymbol{z} = \boldsymbol{a} + \gamma^{-1}(\boldsymbol{z} - \boldsymbol{a}), \quad \text{where} \quad 0 < \gamma^{-1} < 1.$$

Choose an integer $i \geqslant 1$ such that $\gamma^{-1} \leqslant 1 - \frac{1}{i+1}$ and put

$$\boldsymbol{u} = \tfrac{1}{i+1}\boldsymbol{a} + (1 - \tfrac{1}{i+1})\boldsymbol{z} = \boldsymbol{a} + (1 - \tfrac{1}{i+1})(\boldsymbol{z} - \boldsymbol{a}).$$

Then $\boldsymbol{u} \in M_i$. This argument proves the inclusion $\operatorname{rint} K \subset M_1 \cup M_2 \cup \cdots$.

Conversely, choose a point $\boldsymbol{y} \in M_1 \cup M_2 \cup \cdots$. Then $\boldsymbol{y} \in M_i$ for a suitable index $i \geqslant 1$. Equivalently, there is a point $\boldsymbol{z} \in K$ such that

$$\boldsymbol{y} = \boldsymbol{a} + (1 - \tfrac{1}{i+1})(\boldsymbol{z} - \boldsymbol{a}) = \tfrac{1}{i+1}\boldsymbol{a} + (1 - \tfrac{1}{i+1})\boldsymbol{z}.$$

Therefore, $\boldsymbol{y} \in \operatorname{rint} K$ according to Theorem 3.24. Summing up, the set $M_1 \cup M_2 \cup \cdots$ lies in $\operatorname{rint} K$. □

3.2.4 *Relative Interior and Algebra of Convex Sets*

Theorem 3.29. *Let $\mathcal{F} = \{K_\alpha\}$ be a family of convex sets in $\mathbb{R}^n$ whose relative interiors have a point in common. Then*

$$\operatorname{rint}\,(\bigcap_\alpha K_\alpha) \subset \bigcap_\alpha \operatorname{rint} K_\alpha.$$

If, additionally, the family $\mathcal{F}$ is finite, then

$$\operatorname{rint}\,(\bigcap_\alpha K_\alpha) = \bigcap_\alpha \operatorname{rint} K_\alpha.$$

Proof. Since $\varnothing \neq \bigcap_\alpha \operatorname{rint} K_\alpha \subset \bigcap_\alpha K_\alpha$, Corollary 3.21 shows that the set $\operatorname{rint}(\bigcap_\alpha K_\alpha)$ is nonempty. Choose points $\boldsymbol{x} \in \operatorname{rint}(\bigcap_\alpha K_\alpha)$ and $\boldsymbol{y} \in \bigcap_\alpha \operatorname{rint} K_\alpha$. Then $\boldsymbol{y} \in \bigcap_\alpha K_\alpha$, and, by Theorem 3.26, there is a scalar $\gamma > 1$ such that the point $\boldsymbol{z} = \gamma\boldsymbol{x} + (1-\gamma)\boldsymbol{y}$ belongs to $\bigcap_\alpha K_\alpha$. Since $0 < \gamma^{-1} < 1$, Theorem 3.24 gives the inclusions

$$\boldsymbol{x} = (1-\gamma^{-1})\boldsymbol{y} + \gamma^{-1}\boldsymbol{z} \in \operatorname{rint} K_\alpha \quad \text{for all} \quad K_\alpha \in \mathcal{F}.$$

Therefore, $\boldsymbol{x} \in \bigcap_\alpha \operatorname{rint} K_\alpha$.

Suppose now that the family $\mathcal{F}$ is finite and let $\mathcal{F} = \{K_1, \dots, K_r\}$. By the above proved, it suffices to verify the inclusion

$$\operatorname{rint} K_1 \cap \cdots \cap \operatorname{rint} K_r \subset \operatorname{rint}(K_1 \cap \cdots \cap K_r).$$

For this, choose points

$$\boldsymbol{x} \in \operatorname{rint} K_1 \cap \cdots \cap \operatorname{rint} K_r \quad \text{and} \quad \boldsymbol{y} \in K_1 \cap \cdots \cap K_r.$$

According to Theorem 3.26, there are scalars $\gamma_i > 1$ such that

$$\boldsymbol{z}_i = \gamma_i\boldsymbol{x} + (1-\gamma_i)\boldsymbol{y} \in K_i, \quad 1 \leqslant i \leqslant r.$$

Put $\gamma = \min\{\gamma_1, \dots, \gamma_r\}$. Clearly, $\gamma > 1$. Furthermore, Lemma 2.25 and the convexity of K_i give $\gamma\boldsymbol{x} + (1-\gamma)\boldsymbol{y} \in [\boldsymbol{y}, \boldsymbol{z}_i] \subset K_i$, $1 \leqslant i \leqslant r$.

Thus $\gamma\boldsymbol{x} + (1-\gamma)\boldsymbol{y} \in K_1 \cap \cdots \cap K_r$, and Theorem 3.26 shows that $\boldsymbol{x}$ belongs to $\operatorname{rint}(K_1 \cap \cdots \cap K_r)$. □

Remark. The inclusion $\operatorname{rint}(\bigcap_\alpha K_\alpha) \subset \bigcap_\alpha \operatorname{rint} K_\alpha$ in Theorem 3.29 may be proper if the family $\{K_\alpha\}$ is infinite. Indeed, on the real line $\mathbb{R}$ consider the intervals $K_i = [0, 1 + \frac{1}{i}]$, $i \geqslant 1$. Then

$$\operatorname{rint}(\textstyle\bigcap_i K_i) = \operatorname{rint}[0,1] = (0,1) \neq (0,1] = \bigcap_i (0, 1 + \frac{1}{i}) = \bigcap_i \operatorname{rint} K_i.$$

Corollary 3.30. *If convex sets $K_1, \dots, K_r \subset \mathbb{R}^n$ satisfy the condition*

$$\operatorname{aff}(K_1 \cap \cdots \cap K_r) = \operatorname{aff} K_i \quad \textit{for all} \quad 1 \leqslant i \leqslant r,$$

then $\operatorname{rint}(K_1 \cap \cdots \cap K_r) = \operatorname{rint} K_1 \cap \cdots \cap \operatorname{rint} K_r$.

Proof. Since the assertion is obvious when all sets $K_1, \dots, K_r$ are empty, we may assume that at least one of them is nonempty. Then the hypothesis implies that all these sets, including $K_1 \cap \cdots \cap K_r$, are nonempty. A combination of Theorem 3.18 and Corollary 3.21 gives

$$\varnothing \neq \operatorname{rint}(K_1 \cap \cdots \cap K_r) \subset \operatorname{rint} K_i \quad \text{for all} \quad 1 \leqslant i \leqslant r.$$

Therefore, $\operatorname{rint} K_1 \cap \cdots \cap \operatorname{rint} K_r \neq \varnothing$, and the assertion follows from Theorem 3.29. □

Since $\operatorname{rint} L = L$ for every plane $L \subset \mathbb{R}^n$, Theorem 3.29 implies one more corollary.

Corollary 3.31. *If a plane $L \subset \mathbb{R}^n$ meets the relative interior of a convex set $K \subset \mathbb{R}^n$, then* $\operatorname{rint}(K \cap L) = \operatorname{rint} K \cap L$. □

Remark. Corollary 3.31 may fail if $L \cap \operatorname{rint} K = \varnothing$. For instance, given the square $K = \{(x, y) : 0 \leqslant x, y \leqslant 1\}$ and the line $L = \{(x, 1) : x \in \mathbb{R}\}$, one has $\operatorname{rint} K = \{(x, y) : 0 < x, y < 1\}$ and

$$\operatorname{rint}(K \cap L) = \{(x, 1) : 0 < x < 1\} \neq \varnothing = \operatorname{rint} K \cap L.$$

Theorem 3.32. *If $\mathcal{F} = \{K_\alpha\}$ is a nested family of convex sets in $\mathbb{R}^n$, then the set $\bigcup_\alpha \operatorname{rint} K_\alpha$ is convex and*

$$\operatorname{rint}\big(\bigcup_\alpha K_\alpha\big) \subset \bigcup_\alpha \operatorname{rint} K_\alpha. \tag{3.4}$$

If, additionally, $\operatorname{aff} K_\gamma = \operatorname{aff}\big(\bigcup_\alpha K_\alpha\big)$ *for every set $K_\gamma \in \mathcal{F}$, then*

$$\operatorname{rint}\big(\bigcup_\alpha K_\alpha\big) = \bigcup_\alpha \operatorname{rint} K_\alpha. \tag{3.5}$$

Proof. First, we are going to show that the set $\bigcup_\alpha \operatorname{rint} K_\alpha$ is convex. Excluding the obvious case $\bigcup_\alpha K_\alpha = \varnothing$, we suppose that $\bigcup_\alpha K_\alpha$ is nonempty. Then $\bigcup_\alpha \operatorname{rint} K_\alpha \neq \varnothing$ according to Corollary 3.21. Let $\boldsymbol{x}, \boldsymbol{y}$ be points in $\bigcup_\alpha \operatorname{rint} K_\alpha$. Then $\boldsymbol{x} \in \operatorname{rint} K_\beta$ and $\boldsymbol{y} \in \operatorname{rint} K_\delta$ for some sets K_β and K_δ from $\mathcal{F}$. Assuming, for instance, that $K_\beta \subset K_\delta$, we obtain $\boldsymbol{x} \in K_\delta$. Consequently, Theorem 3.24 gives $(\boldsymbol{x}, \boldsymbol{y}] \subset \operatorname{rint} K_\delta$. Thus

$$[\boldsymbol{x}, \boldsymbol{y}] = \{\boldsymbol{x}\} \cup (\boldsymbol{x}, \boldsymbol{y}] \subset \operatorname{rint} K_\beta \cup \operatorname{rint} K_\delta \subset \bigcup_\alpha \operatorname{rint} K_\alpha,$$

which shows the convexity of $\bigcup_\alpha \operatorname{rint} K_\alpha$.

To prove inclusion (3.4), let $\boldsymbol{x} \in \operatorname{rint}\big(\bigcup_\alpha K_\alpha\big)$. Since the family $\mathcal{F}$ is nested, the family $\{\operatorname{aff} K_\alpha\}$ of planes also is nested (see Theorem 2.42). By a dimension argument, the family $\{\operatorname{aff} K_\alpha\}$ contains at most finitely many distinct planes. Hence there is a set $K_\delta \in \mathcal{F}$ such that $\operatorname{aff} K_\delta = \operatorname{aff}\big(\bigcup_\alpha K_\alpha\big)$. Choose a point $\boldsymbol{y} \in \operatorname{rint} K_\delta$. Because the set $\bigcup_\alpha K_\alpha$ is convex (see Theorem 3.8) and $\boldsymbol{y} \in \bigcup_\alpha K_\alpha$, Theorem 3.26 shows the existence of a scalar $\gamma > 1$ such that the point $\boldsymbol{z} = \gamma\boldsymbol{x} + (1 - \gamma)\boldsymbol{y}$ belongs to $\bigcup_\alpha K_\alpha$. Therefore, $\boldsymbol{z} \in K_\beta$ for a suitable set $K_\beta \in \mathcal{F}$.

If $K_\beta \subset K_\delta$, then $\boldsymbol{z} \in K_\delta$ and the equality

$$\boldsymbol{x} = (1 - \gamma^{-1})\boldsymbol{y} + \gamma^{-1}\boldsymbol{z}, \quad 0 < \gamma^{-1} < 1, \tag{3.6}$$

together with Theorem 3.24, implies that $\boldsymbol{x} \in \operatorname{rint} K_\delta \subset \bigcup_\alpha \operatorname{rint} K_\alpha$.

Assume that $K_\delta \subset K_\beta$. By the choice of K_δ, Theorem 2.42 shows that $\operatorname{aff} K_\delta = \operatorname{aff} K_\beta = \operatorname{aff}(\bigcup_\alpha K_\alpha)$. Hence $\boldsymbol{y} \in \operatorname{rint} K_\delta \subset \operatorname{rint} K_\beta$ due to Theorem 3.18. As above, (3.6) and Theorem 3.24 give

$$\boldsymbol{x} \in \operatorname{rint} K_\beta \subset \bigcup_\alpha \operatorname{rint} K_\alpha.$$

Summing up, inclusion (3.4) holds.

Finally, if $\operatorname{aff} K_\gamma = \operatorname{aff}(\bigcup_\alpha K_\alpha)$ for every set $K_\gamma \in \mathcal{F}$, then Theorem 3.18 implies that $\operatorname{rint} K_\gamma \subset \operatorname{rint}(\bigcup_\alpha K_\alpha)$ for all $K_\gamma \in \mathcal{F}$. Hence the opposite to (3.4) inclusion holds, which results in (3.5). □

Remark. Inclusion (3.4) may be proper. Indeed, let

$$K_1 = \{(0, y) : 0 \leqslant y \leqslant 1\} \quad \text{and} \quad K_2 = \{(x, y) : 0 \leqslant x, y \leqslant 1\}.$$

Clearly, $K_1 \subset K_2$ and

$$\operatorname{rint} K_1 = \{(0, y) : 0 < y < 1\} \quad \text{and} \quad K_2 = \{(x, y) : 0 < x, y < 1\}.$$

Consequently, $\operatorname{rint} K_1 \cup \operatorname{rint} K_2 \not\subset \operatorname{rint} K_2 = \operatorname{rint}(K_1 \cup K_2)$.

Theorem 3.33. *If $\mathcal{F} = \{K_\alpha\}$ is a nested family of relative open convex sets in $\mathbb{R}^n$, then the set $K = \bigcup_\alpha K_\alpha$ is convex and relative open.*

Proof. The convexity of K follows from Theorem 3.8. So, it suffices to show that K is relative open. Since the family $\mathcal{F}$ is nested, the family $\{\operatorname{aff} K_\alpha\}$ of planes also is nested (see Theorem 2.42). By a dimension argument, the family $\{\operatorname{aff} K_\alpha\}$ contains at most finitely many distinct planes. Hence there is a set $K_\delta \in \mathcal{F}$ such that $\operatorname{aff} K_\delta = \operatorname{aff} K$. Clearly, $K_\beta \subset K_\delta$ for any set $K_\beta \in \mathcal{F}$ whose affine span is a proper subplane of $\operatorname{aff} K_\delta$. Let

$$\mathcal{G} = \{K_\alpha \in \mathcal{F} : \operatorname{aff} K_\alpha = \operatorname{aff} K\}.$$

The family $\mathcal{G}$ is nested, $K_\delta \in \mathcal{G}$, and $K = \cup(K_\alpha : K_\alpha \in \mathcal{G})$. By Theorem 3.32,

$$\begin{aligned}\operatorname{rint} K &= \operatorname{rint}(\cup(K_\alpha : K_\alpha \in \mathcal{G})) = \cup(\operatorname{rint} K_\alpha : K_\alpha \in \mathcal{G}) \\ &= \cup(K_\alpha : K_\alpha \in \mathcal{G}) = K.\end{aligned}$$

Hence K is relative open. □

Theorem 3.34. *If $K_1, \ldots, K_r$ are convex sets in $\mathbb{R}^n$ and $\mu_1, \ldots, \mu_r$ are scalars, then*

$$\operatorname{rint}(\mu_1 K_1 + \cdots + \mu_r K_r) = \mu_1 \operatorname{rint} K_1 + \cdots + \mu_r \operatorname{rint} K_r.$$

Proof. Because the assertion is obvious when at least one of the sets $K_1, \dots, K_r$ is empty (then both sides of the equality are empty sets), we may assume that all sets $K_1, \dots, K_r$ are nonempty. An induction argument shows that the proof can be reduced to the case $r = 2$ (let $\mu_1 K_1 = \mu_1 K_1 + \mu_2\{\boldsymbol{o}\}$ if $r = 1$). By Corollary 3.21, all three sets $\operatorname{rint} K_1$, $\operatorname{rint} K_2$, and $\operatorname{rint}(\mu_1 K_1 + \mu_2 K_2)$ are nonempty.

Let $\boldsymbol{x} \in \operatorname{rint}(\mu_1 K_1 + \mu_2 K_2)$. Choose points $\boldsymbol{y}_i \in \operatorname{rint} K_i$, $i = 1, 2$, and put $\boldsymbol{y} = \mu_1 \boldsymbol{y}_1 + \mu_2 \boldsymbol{y}_2$. Since $\boldsymbol{y} \in \mu_1 K_1 + \mu_2 K_2$, Theorem 3.26 shows the existence of a scalar $\gamma > 1$ such that

$$\boldsymbol{z} = \gamma \boldsymbol{x} + (1 - \gamma)\boldsymbol{y} \in \mu_1 K_1 + \mu_2 K_2.$$

Hence $\boldsymbol{z}$ can be written as $\boldsymbol{z} = \mu_1 \boldsymbol{z}_1 + \mu_2 \boldsymbol{z}_2$, where $\boldsymbol{z}_i \in K_i$, $i = 1, 2$. Put

$$\boldsymbol{x}_i = (1 - \gamma^{-1})\boldsymbol{y}_i + \gamma^{-1}\boldsymbol{z}_i, \quad i = 1, 2.$$

Due to $0 < \gamma^{-1} < 1$, Theorem 3.24 implies that $\boldsymbol{x}_i \in \operatorname{rint} K_i$, $i = 1, 2$. Therefore,

$$\begin{aligned} \boldsymbol{x} &= (1 - \gamma^{-1})\boldsymbol{y} + \gamma^{-1}\boldsymbol{z} = (1 - \gamma^{-1})(\mu_1 \boldsymbol{y}_1 + \mu_2 \boldsymbol{y}_2) + \gamma^{-1}(\mu_1 \boldsymbol{z}_1 + \mu_2 \boldsymbol{z}_2) \\ &= \mu_1 \boldsymbol{x}_1 + \mu_2 \boldsymbol{x}_2 \in \mu_1 \operatorname{rint} K_1 + \mu_2 \operatorname{rint} K_2. \end{aligned}$$

Summing up,

$$\operatorname{rint}(\mu_1 K_1 + \mu_2 K_2) \subset \mu_1 \operatorname{rint} K_1 + \mu_2 \operatorname{rint} K_2.$$

For the opposite inclusion, choose any point $\boldsymbol{x} \in \mu_1 \operatorname{rint} K_1 + \mu_2 \operatorname{rint} K_2$. Also, let $\boldsymbol{y} \in \mu_1 K_1 + \mu_2 K_2$. Then

$$\boldsymbol{x} = \mu_1 \boldsymbol{x}_1 + \mu_2 \boldsymbol{x}_2 \quad \text{and} \quad \boldsymbol{y} = \mu_1 \boldsymbol{y}_1 + \mu_2 \boldsymbol{y}_2$$

for suitable points $\boldsymbol{x}_i \in \operatorname{rint} K_i$ and $\boldsymbol{y}_i \in K_i$, $i = 1, 2$. By Theorem 3.26, there are scalars $\gamma_1, \gamma_2 > 1$ such that

$$\boldsymbol{u}_i = \gamma_i \boldsymbol{x}_i + (1 - \gamma_i)\boldsymbol{y}_i \in K_i, \quad i = 1, 2.$$

Let $\gamma = \min\{\gamma_1, \gamma_2\}$. Then $\gamma > 1$ and $0 < \gamma/\gamma_i \leqslant 1$, $i = 1, 2$. Put

$$\boldsymbol{v}_i = \gamma \boldsymbol{x}_i + (1 - \gamma)\boldsymbol{y}_i \in K_i, \quad i = 1, 2.$$

Clearly,

$$\begin{aligned} \boldsymbol{v}_i &= \gamma \boldsymbol{x}_i + (1 - \gamma)\boldsymbol{y}_i = \gamma\big(\frac{1}{\gamma_i}\boldsymbol{u}_i + (1 - \frac{1}{\gamma_i})\boldsymbol{y}_i\big) + (1 - \gamma)\boldsymbol{y}_i \\ &= \frac{\gamma}{\gamma_i}\boldsymbol{u}_i + \big(1 - \frac{\gamma}{\gamma_i}\big)\boldsymbol{y}_i \in [\boldsymbol{u}_i, \boldsymbol{y}_i] \subset K_i, \quad i = 1, 2. \end{aligned}$$

Thus

$$\begin{aligned}\gamma\boldsymbol{x} + (1-\gamma)\boldsymbol{y} &= \gamma(\mu_1\boldsymbol{x}_1 + \mu_2\boldsymbol{x}_2) + (1-\gamma)(\mu_1\boldsymbol{y}_1 + \mu_2\boldsymbol{y}_2) \\ &= \mu_1(\gamma\boldsymbol{x}_1 + (1-\gamma)\boldsymbol{y}_1) + \mu_2(\gamma\boldsymbol{x}_2 + (1-\gamma)\boldsymbol{y}_2) \qquad (3.7) \\ &= \mu_1\boldsymbol{v}_1 + \mu_2\boldsymbol{v}_2 \in \mu_1 K_1 + \mu_2 K_2.\end{aligned}$$

Finally, Theorem 3.26 and (3.7) give the inclusion $\boldsymbol{x} \in \operatorname{rint}(\mu_1 K_1 + \mu_2 K_2)$. Hence

$$\mu_1 \operatorname{rint} K_1 + \mu_2 \operatorname{rint} K_2 \subset \operatorname{rint}(\mu_1 K_1 + \mu_2 K_2). \qquad \square$$

A sharper version of Theorem 3.34 for the case of large sums is considered in Problem 3.14. The following corollary complements assertion (*d*) of Theorem 3.18.

Corollary 3.35. *If K_1 and K_2 be convex sets in $\mathbb{R}^n$ such that a translate of K_2 lies in* aff K_1, *then*

$$\operatorname{rint}(K_1 + K_2) = \operatorname{rint} K_1 + K_2. \qquad (3.8)$$

In particular, $\operatorname{rint}(2K) = \operatorname{rint} K + K$ *for any convex set $K \subset \mathbb{R}^n$.*

Proof. One has $\operatorname{rint} K_1 + K_2 \subset \operatorname{rint}(K_1 + K_2)$ according to Theorem 3.18. The opposite inclusion,

$$\operatorname{rint}(K_1 + K_2) = \operatorname{rint} K_1 + \operatorname{rint} K_2 \subset \operatorname{rint} K_1 + K_2,$$

follows from Theorem 3.34. The second assertion follows from (3.8) and the equality $2K = K + K$ (see Theorem 3.4). $\square$

3.2.5 *Relative Interior and Affine Structure*

The following result is a refinement of Theorem 3.14.

Theorem 3.36. *If $K \subset \mathbb{R}^n$ is a nonempty convex set and $\boldsymbol{a} \in \operatorname{rint} K$, then*

$$\operatorname{aff} K = \{(1-\lambda)\boldsymbol{a} + \lambda\boldsymbol{x} : \boldsymbol{x} \in \operatorname{rint} K,\ \lambda \geqslant 0\}.$$

Proof. Since aff K is a plane containing K, Theorem 2.38 implies that the set

$$L = \{(1-\lambda)\boldsymbol{a} + \lambda\boldsymbol{x} : \boldsymbol{x} \in \operatorname{rint} K,\ \lambda \geqslant 0\}$$

lies in aff K. For the opposite inclusion, choose any point $\boldsymbol{u} \in \operatorname{aff} K$. By Theorem 3.26, there is a scalar $0 < \mu < 1$ for which the point $\boldsymbol{v} = (1 -$

$\mu)\boldsymbol{a} + \mu\boldsymbol{u}$ belongs to K. Theorem 3.24 shows that the point $\boldsymbol{z} = \frac{1}{2}(\boldsymbol{a} + \boldsymbol{v})$ is in $\operatorname{rint} K$. Hence

$$\begin{aligned}\boldsymbol{u} &= (1 - \mu^{-1})\boldsymbol{a} + \mu^{-1}\boldsymbol{v} = (1 - \mu^{-1})\boldsymbol{a} + \mu^{-1}(2\boldsymbol{z} - \boldsymbol{a}) \\ &= (1 - 2\mu^{-1})\boldsymbol{a} + 2\mu^{-1}\boldsymbol{z} \in L. \end{aligned}$$ □

Theorem 3.37. *If $K \subset \mathbb{R}^n$ is a nonempty convex set and $\boldsymbol{a} \in \operatorname{rint} K$, then $\operatorname{aff} K = M_1 \cup M_2 \cup \cdots$, where every set $M_i = \boldsymbol{a} + i(K - \boldsymbol{a})$, $i \geqslant 1$, is a homothetic copy of K.*

Proof. Let $\boldsymbol{u}$ be any point in $\operatorname{aff} K$. By Theorem 3.36, there is a point $\boldsymbol{x} \in \operatorname{rint} K$ and a scalar $\lambda \geqslant 0$ such that $\boldsymbol{u} = (1 - \lambda)\boldsymbol{a} + \lambda\boldsymbol{x}$. Since the case $\lambda = 0$ is obvious (then $\boldsymbol{u} = \boldsymbol{a}$), we may assume that $\lambda > 0$. Choose an integer $r > \lambda$ and put $\boldsymbol{z} = (1 - r^{-1})\boldsymbol{a} + r^{-1}\boldsymbol{u}$. Then $0 < r^{-1}\lambda < 1$ and

$$\begin{aligned}\boldsymbol{z} &= (1 - r^{-1})\boldsymbol{a} + r^{-1}\boldsymbol{u} = (1 - r^{-1})\boldsymbol{a} + r^{-1}((1 - \lambda)\boldsymbol{a} + \lambda\boldsymbol{x}) \\ &= (1 - r^{-1}\lambda)\boldsymbol{a} + r^{-1}\lambda\boldsymbol{x}.\end{aligned}$$

Hence $\boldsymbol{z} \in (\boldsymbol{a}, \boldsymbol{x}) \subset \operatorname{rint} K \subset K$ (see Lemma 2.25 and Corollary 3.25). Therefore,

$$\boldsymbol{u} = \boldsymbol{a} + r(\boldsymbol{z} - \boldsymbol{a}) \in \boldsymbol{a} + r(K - \boldsymbol{a}) = M_r.$$

So, $\operatorname{aff} K \subset \operatorname{rint} M_1 \cup \operatorname{rint} M_2 \cup \cdots$.

For the opposite inclusion, choose any point $\boldsymbol{u} \in M_1 \cup M_2 \cup \cdots$. Then $\boldsymbol{u} \in M_j$ for a suitable index $j \geqslant 1$. We can write $\boldsymbol{u} = \boldsymbol{a} + j(\boldsymbol{x} - \boldsymbol{a})$, where $\boldsymbol{x} \in K$. Since $\boldsymbol{u}$ is an affine combination of the points $\boldsymbol{a}$ and $\boldsymbol{x}$ from K, Theorem 2.38 implies that $\boldsymbol{u} \in \operatorname{aff} K$. Thus $M_1 \cup M_2 \cup \cdots \subset \operatorname{aff} K$. □

Theorem 3.38. *If $K_1, \ldots, K_r$ are convex sets in $\mathbb{R}^n$ whose relative interiors have a point in common, then*

$$\operatorname{aff}(K_1 \cap \cdots \cap K_r) = \operatorname{aff} K_1 \cap \cdots \cap \operatorname{aff} K_r. \tag{3.9}$$

Proof. Theorem 2.42 gives

$$\operatorname{aff}(K_1 \cap \cdots \cap K_r) \subset \operatorname{aff} K_1 \cap \cdots \cap \operatorname{aff} K_r.$$

For the opposite inclusion, choose points

$$\boldsymbol{x} \in \operatorname{aff} K_1 \cap \cdots \cap \operatorname{aff} K_r \quad \text{and} \quad \boldsymbol{y} \in \operatorname{rint} K_1 \cap \cdots \cap \operatorname{rint} K_r.$$

By Theorem 3.26, there are scalars $\gamma_1, \ldots, \gamma_r > 1$ with the property

$$\boldsymbol{z}_i = (1 - \gamma_i)\boldsymbol{x} + \gamma_i\boldsymbol{y} \in K_i, \quad 1 \leqslant i \leqslant r.$$

Put $\gamma = \min\{\gamma_1, \ldots, \gamma_r\}$ and $\boldsymbol{u} = (1-\gamma)\boldsymbol{x} + \gamma\boldsymbol{y}$. Obviously, $\gamma > 1$ and $0 < \gamma\gamma_i^{-1} < 1$ for all $1 \leqslant i \leqslant r$. Consequently,

$$\begin{aligned}\boldsymbol{u} &= (1-\gamma)\boldsymbol{x} + \gamma\boldsymbol{y} = (1-\gamma)\boldsymbol{x} + \gamma((1-\gamma_i^{-1})\boldsymbol{x} + \gamma_i^{-1}\boldsymbol{z}_i) \\ &= (1-\gamma\gamma_i^{-1})\boldsymbol{x} + \gamma\gamma_i^{-1}\boldsymbol{z}_i \in [\boldsymbol{y}, \boldsymbol{z}_i], \quad 1 \leqslant i \leqslant r.\end{aligned}$$

Because $[\boldsymbol{y}, \boldsymbol{z}_i] \subset K_i$ by the convexity of K_i, we obtain

$$\boldsymbol{u} \in [\boldsymbol{y}, \boldsymbol{z}_1] \cap \cdots \cap [\boldsymbol{y}, \boldsymbol{z}_r] \subset K_1 \cap \cdots \cap K_r.$$

Since $\boldsymbol{x}$ can be expressed as an affine combination $\boldsymbol{x} = \frac{\gamma}{\gamma-1}\boldsymbol{y} - \frac{1}{\gamma-1}\boldsymbol{u}$ of the points $\boldsymbol{y}, \boldsymbol{u} \in K_1 \cap \cdots \cap K_r$, Theorem 2.44 implies the inclusion $\boldsymbol{x} \in \operatorname{aff}(K_1 \cap \cdots \cap K_r)$. Summing up,

$$\operatorname{aff} K_1 \cap \cdots \cap \operatorname{aff} K_r \subset \operatorname{aff}(K_1 \cap \cdots \cap K_r). \qquad \square$$

Remark. Equality (3.9) may fail if $\operatorname{rint} K_1 \cap \cdots \cap \operatorname{rint} K_r = \varnothing$. Indeed, let

$$K_1 = \{(x, y, 0) : x^2 + y^2 \leqslant 1\} \text{ and } K_2 = \{(0, y, z) : y^2 + (z-1)^2 \leqslant 1\}.$$

Then $\operatorname{aff}(K_1 \cap K_2) = \{\boldsymbol{o}\}$, while $\operatorname{aff} K_1 \cap \operatorname{aff} K_2$ is the y-axis.

Theorem 3.39. *If K_1 and K_2 are convex sets in $\mathbb{R}^n$ whose relative interiors have a point in common, then*

$$\dim(K_1 \cup K_2) = \dim(K_1 + K_2) = \dim K_1 + \dim K_2 - \dim(K_1 \cap K_2).$$

Proof. Because $\operatorname{aff}(K_1 \cup K_2)$ is a translate of $\operatorname{aff}(K_1 + K_2)$ (see Theorem 2.49), one has

$$\begin{aligned}\dim(K_1 \cup K_2) &= \dim(\operatorname{aff}(K_1 \cup K_2)) \\ &= \dim(\operatorname{aff}(K_1 + K_2)) = \dim(K_1 + K_2).\end{aligned}$$

By Theorems 2.46 and 3.38, one has

$$\operatorname{aff}(K_1 + K_2) = \operatorname{aff} K_1 + \operatorname{aff} K_2, \quad \operatorname{aff}(K_1 \cap K_2) = \operatorname{aff} K_1 \cap \operatorname{aff} K_2.$$

Finally, Theorem 2.7 gives

$$\begin{aligned}\dim(K_1 + K_2) &= \dim(\operatorname{aff}(K_1 + K_2)) = \dim(\operatorname{aff} K_1 + \operatorname{aff} K_2) \\ &= \dim(\operatorname{aff} K_1) + \dim(\operatorname{aff} K_2) - \dim(\operatorname{aff} K_1 \cap \operatorname{aff} K_2) \\ &= \dim(\operatorname{aff} K_1) + \dim(\operatorname{aff} K_2) - \dim(\operatorname{aff}(K_1 \cap K_2)) \\ &= \dim K_1 + \dim K_2 - \dim(K_1 \cap K_2). \qquad \square\end{aligned}$$

Definition 3.40. We will say that a hyperplane $H \subset \mathbb{R}^n$ *cuts* a convex set $K \subset \mathbb{R}^n$ if K meets both open halfspaces of $\mathbb{R}^n$ determined by H.

Theorem 3.41. *A hyperplane $H \subset \mathbb{R}^n$ cuts a convex sets $K \subset \mathbb{R}^n$ if and only if H meets* $\operatorname{rint} K$ *and does not contain K.*

Proof. Denote by W_1 and W_2 the open halfspaces determined by H.

Assume first that H cuts K and choose points $\boldsymbol{v}_1 \in K \cap W_1$ and $\boldsymbol{v}_2 \in K \cap W_2$. Since K does not contain K, it remains to show that $H \cap \operatorname{rint} K \neq \varnothing$. Choose any point $\boldsymbol{u} \in \operatorname{rint} K$. If $\boldsymbol{u} \in H$, then $\boldsymbol{u} \in H \cap \operatorname{rint} K$. Suppose that $\boldsymbol{u} \notin H$. Then $\boldsymbol{u}$ belongs to one of the halfspaces W_1 and W_2. Let, for instance $\boldsymbol{u} \in W_1$. By Theorem 3.24, the open segment $(\boldsymbol{u}, \boldsymbol{v}_2)$ is included in $\operatorname{rint} K$. On the other hand, Theorem 2.35 shows that $(\boldsymbol{u}, \boldsymbol{v}_2)$ meets H. So, $H \cap \operatorname{rint} K \neq \varnothing$.

Conversely, let H meet $\operatorname{rint} K$ and do not contain K. Choose points $\boldsymbol{u} \in H \cap \operatorname{rint} K$ and $\boldsymbol{v} \in K \setminus H$. Then $\boldsymbol{v}$ belongs to one of the halfspaces W_1 and W_2. Let $\boldsymbol{v} \in W_1$ (the case $\boldsymbol{v} \in W_2$ is similar). By Theorem 3.26, there is a scalar $\gamma > 1$ such that the point $\boldsymbol{w} = \gamma \boldsymbol{u} + (1-\gamma)\boldsymbol{v}$ belongs to K. Since $\boldsymbol{u} = (1-\gamma^{-1})\boldsymbol{v} + \gamma^{-1}\boldsymbol{w} \in (\boldsymbol{v}, \boldsymbol{w})$, Corollary 2.36 implies the inclusion $\boldsymbol{w} \in W_2$ (indeed, if $\boldsymbol{w} \in H \cup W_2$, then $[\boldsymbol{v}, \boldsymbol{w}]$ would lie in H, contrary to the assumption $\boldsymbol{v} \in W_1$). Hence H cuts K. □

The following corollary provides a useful property of convex sets (compare with Corollary 2.65).

Corollary 3.42. *If a hyperplane $H \subset \mathbb{R}^n$ cuts a convex set K, then*

$$\dim K = \dim (K \cap H) + 1.$$

Proof. A combination of Corollary 2.60 and Theorem 3.41 shows that $\dim (K \cup H) = n$. Also Theorem 3.41 states that $H \cap \operatorname{rint} K \neq \varnothing$. Since $\operatorname{rint} H = H$, Theorem 3.39 gives

$$\begin{aligned} \dim K &= \dim (K \cup H) - \dim H + \dim (K \cap H) \\ &= n - (n-1) + \dim (K \cap H) = \dim (K \cap H) + 1. \end{aligned}$$ □

Remark. The condition that H cuts K is essential in Corollary 3.42. Indeed, if K is a disk in $\mathbb{R}^2$ and H is a tangent line of K, then

$$\dim K = 2 \neq 1 = \dim (K \cap H) + 1.$$

Theorem 3.43. *Let $f : \mathbb{R}^n \to \mathbb{R}^m$ be an affine transformation. For convex sets $K \subset \mathbb{R}^n$ and $M \subset \mathbb{R}^m$, one has*

$$\operatorname{rint} f(K) = f(\operatorname{rint} K), \tag{3.10}$$

$$f^{-1}(\operatorname{rint} M) \subset \operatorname{rint} f^{-1}(M) = f^{-1}(\operatorname{rint}(M \cap \operatorname{rng} f)). \tag{3.11}$$

If, additionally, $\operatorname{rint} f^{-1}(M) \cap \operatorname{rng} f \neq \varnothing$, *then* $f^{-1}(\operatorname{rint} M) = \operatorname{rint} f^{-1}(M)$.

Proof. Excluding the obvious cases $K = \varnothing$ and $M \cap \operatorname{rng} f = \varnothing$, we suppose that both sets K and $M \cap \operatorname{rng} f$ are nonempty. Consecutively, both sets $f(K)$ and $f^{-1}(M)$ are nonempty. According to Theorem 3.15, these sets are convex.

For (3.10), choose points $\boldsymbol{x} \in \operatorname{rint} f(K)$ and $\boldsymbol{y}_0 \in \operatorname{rint} K$. Put $\boldsymbol{y} = f(\boldsymbol{y}_0)$. Then $\boldsymbol{y} \in f(\operatorname{rint} K) \subset f(K)$. By Theorem 3.26, there is a scalar $\gamma > 1$ such that the point $\boldsymbol{u} = \gamma\boldsymbol{x} + (1-\gamma)\boldsymbol{y}$ belongs to $f(K)$. Let a point $\boldsymbol{u}_0 \in K$ satisfy the condition $f(\boldsymbol{u}_0) = \boldsymbol{u}$. Put $\boldsymbol{x}_0 = (1-\gamma^{-1})\boldsymbol{y}_0 + \gamma^{-1}\boldsymbol{u}_0$. Since $0 < \gamma^{-1} < 1$, Theorem 3.24 yields the inclusion $\boldsymbol{x}_0 \in \operatorname{rint} K$. Theorem 2.81 gives

$$\begin{aligned}\boldsymbol{x} &= (1-\gamma^{-1})\boldsymbol{y} + \gamma^{-1}\boldsymbol{u} = (1-\gamma^{-1})f(\boldsymbol{y}_0) + \gamma^{-1}f(\boldsymbol{u}_0)\\ &= f((1-\gamma^{-1})\boldsymbol{y}_0 + \gamma^{-1}\boldsymbol{u}_0) = f(\boldsymbol{x}_0) \in f(\operatorname{rint} K).\end{aligned}$$

Summing up, $\operatorname{rint} f(K) \subset f(\operatorname{rint} K)$.

For the opposite inclusion, let $\boldsymbol{x} \in f(\operatorname{rint} K)$ and $\boldsymbol{y} \in \operatorname{rint} f(K)$. Then $\boldsymbol{x} = f(\boldsymbol{x}_0)$ and $\boldsymbol{y} = f(\boldsymbol{y}_0)$ for suitable points $\boldsymbol{x}_0 \in \operatorname{rint} K$ and $\boldsymbol{y}_0 \in K$. By Theorem 3.26, there is a scalar $\gamma > 1$ such that $\boldsymbol{u}_0 = \gamma\boldsymbol{x}_0 + (1-\gamma)\boldsymbol{y}_0 \in K$. Consequently,

$$\begin{aligned}\gamma\boldsymbol{x} + (1-\gamma)\boldsymbol{y} &= \gamma f(\boldsymbol{x}_0) + (1-\gamma)f(\boldsymbol{y}_0)\\ &= f(\gamma\boldsymbol{x}_0 + (1-\gamma)\boldsymbol{y}_0) = f(\boldsymbol{u}_0) \in f(K).\end{aligned}$$

Hence $\boldsymbol{x} \in \operatorname{rint} f(K)$ by the same theorem. Thus $f(\operatorname{rint} K) \subset \operatorname{rint} f(K)$.

To prove the equality $\operatorname{rint} f^{-1}(M) = f^{-1}(\operatorname{rint}(M \cap \operatorname{rng} f))$, choose a point $\boldsymbol{x} \in \operatorname{rint} f^{-1}(M) = \operatorname{rint} f^{-1}(M \cap \operatorname{rng} f)$.

Let $\boldsymbol{z}_0 \in \operatorname{rint}(M \cap \operatorname{rng} f)$ and $\boldsymbol{z} \in f^{-1}(\operatorname{rint}(M \cap \operatorname{rng} f))$, where $f(\boldsymbol{z}) = \boldsymbol{z}_0$. As above (since $\boldsymbol{z} \in f^{-1}(M \cap \operatorname{rng} f)$), there is a scalar $\gamma > 1$ such that

$$\gamma\boldsymbol{x} + (1-\gamma)\boldsymbol{z} \in f^{-1}(M \cap \operatorname{rng} f).$$

Therefore, $\gamma f(\boldsymbol{x}) + (1-\gamma)\boldsymbol{z}_0 \in M \cap \operatorname{rng} f$, and Theorem 3.26 implies the inclusion $f(\boldsymbol{x}) \in \operatorname{rint}(M \cap \operatorname{rng} f)$. Hence $\boldsymbol{x} \in f^{-1}(\operatorname{rint}(M \cap \operatorname{rng} f))$.

Conversely, let $\boldsymbol{x} \in f^{-1}(\operatorname{rint}(M \cap \operatorname{rng} f))$ and $\boldsymbol{z} \in \operatorname{rint} f^{-1}(M \cap \operatorname{rng} f)$. Then

$$f(\boldsymbol{x}) \in \operatorname{rint}(M \cap \operatorname{rng} f) \quad \text{and} \quad f(\boldsymbol{z}) \in M \cap \operatorname{rng} f.$$

Theorem 3.26 gives the existence of a scalar $\gamma > 1$ with the property

$$f(\gamma\boldsymbol{x} + (1-\gamma)\boldsymbol{z}) = \gamma f(\boldsymbol{x}) + (1-\gamma)f(\boldsymbol{z}) \in M \cap \operatorname{rng} f.$$

Hence $\gamma\boldsymbol{x} + (1-\gamma)\boldsymbol{z} \in f^{-1}(M \cap \operatorname{rng} f)$, which gives

$$\boldsymbol{x} \in \operatorname{rint} f^{-1}(M \cap \operatorname{rng} f) = \operatorname{rint} f^{-1}(M).$$

It remains to show that $f^{-1}(\operatorname{rint} M) \subset \operatorname{rint} f^{-1}(M)$. Since this inclusion is obvious if $\operatorname{rint} M \cap \operatorname{rng} f = \varnothing$ (then $f^{-1}(\operatorname{rint} M) = \varnothing$), we may assume that $\operatorname{rint} M \cap \operatorname{rng} f \neq \varnothing$. Because the set $\operatorname{rng} f$ is a plane (see Corollary 2.84), Corollary 3.31 implies

$$\operatorname{rint}(M \cap \operatorname{rng} f) = \operatorname{rint} M \cap \operatorname{rng} f.$$

By the above proved,

$$\begin{aligned}\operatorname{rint} f^{-1}(M) &= f^{-1}(\operatorname{rint}(M \cap \operatorname{rng} f))\\ &= f^{-1}(\operatorname{rint} M \cap \operatorname{rng} f) = f^{-1}(\operatorname{rint} M).\end{aligned}$$ □

3.2.6 *Homothetic Copies and Metric Neighborhoods*

We recall (see page 9) that $B_\rho(X)$ and $U_\rho(X)$ denote, respectively, the closed and open ρ-neighborhoods of a set $X \subset \mathbb{R}^n$.

Theorem 3.44. *Given a nonempty convex set $K \subset \mathbb{R}^n$, a point $\boldsymbol{a} \in \operatorname{rint} K$, and scalars $\rho > 0$ and $\mu > 1$, the following assertions hold.*

(*a*) *If* $\operatorname{aff} K \cap B_\rho(\boldsymbol{a}) \subset K$, *then* $\operatorname{aff} K \cap B_{(\mu-1)\rho}(K) \subset \boldsymbol{a} + \mu(K - \boldsymbol{a})$.
(*b*) *If* $\operatorname{aff} K \cap U_\rho(\boldsymbol{a}) \subset K$, *then* $\operatorname{aff} K \cap U_{(\mu-1)\rho}(K) \subset \boldsymbol{a} + \mu(K - \boldsymbol{a})$.

Proof. (*a*) Let $\boldsymbol{y} \in \operatorname{aff} K \cap B_{(\mu-1)\rho}(K)$. The case $\boldsymbol{y} \in K$ is obvious because $K \subset \boldsymbol{a} + \mu(K - \boldsymbol{a})$ (see Lemma 3.12). Suppose that $\boldsymbol{y} \in \operatorname{aff} K \setminus K$. Then

$$\boldsymbol{y} \in \operatorname{aff} K \setminus K \subset \operatorname{aff} K \setminus (\operatorname{aff} K \cap B_\rho(\boldsymbol{a})) = \operatorname{aff} K \setminus B_\rho(\boldsymbol{a}).$$

Hence $\boldsymbol{y} \notin B_\rho(\boldsymbol{a})$, which gives $\|\boldsymbol{a} - \boldsymbol{y}\| > \rho$. By the choice of $\boldsymbol{y}$, there is a point $\boldsymbol{u} \in K$ such that $\|\boldsymbol{u} - \boldsymbol{y}\| \leqslant (\mu - 1)\rho$. We may assume that $\|\boldsymbol{a} - \boldsymbol{u}\| \geqslant \rho$. Indeed, otherwise we can replace $\boldsymbol{u}$ with a point $\boldsymbol{u}' \in (\boldsymbol{u}, \boldsymbol{y})$ which satisfies the conditions $\|\boldsymbol{u}' - \boldsymbol{a}\| = \rho$ and $\|\boldsymbol{u}' - \boldsymbol{y}\| \leqslant (\mu - 1)\rho$.

1. Suppose first that $\boldsymbol{u}$ belongs to the line $\langle \boldsymbol{a}, \boldsymbol{y}\rangle$. Clearly, $\boldsymbol{u} \in (\boldsymbol{y}, \boldsymbol{a}\rangle$, since otherwise $\boldsymbol{y} \in [\boldsymbol{u}, \boldsymbol{a}] \subset K$, contrary to the assumption $\boldsymbol{y} \notin K$. Next, we may assume that $\boldsymbol{u} \in (\boldsymbol{a}, \boldsymbol{y})$, since otherwise we can replace $\boldsymbol{u}$ with a point $\boldsymbol{u}' \in (\boldsymbol{a}, \boldsymbol{y})$ which satisfies the condition $\|\boldsymbol{u}' - \boldsymbol{a}\| = \rho$.

Now, the inclusion $\boldsymbol{u} \in (\boldsymbol{a}, \boldsymbol{y})$ implies that $\boldsymbol{u} = (1 - \lambda)\boldsymbol{a} + \lambda \boldsymbol{y}$ for a suitable scalar $0 < \lambda < 1$. Consequently, $\boldsymbol{y} = \boldsymbol{a} + \lambda^{-1}(\boldsymbol{u} - \boldsymbol{a})$, where $\lambda^{-1} > 1$. Then

$$\begin{aligned}\lambda^{-1} &= \frac{\|\boldsymbol{a} - \boldsymbol{y}\|}{\|\boldsymbol{a} - \boldsymbol{u}\|} = \frac{\|\boldsymbol{a} - \boldsymbol{u}\| + \|\boldsymbol{u} - \boldsymbol{y}\|}{\|\boldsymbol{a} - \boldsymbol{u}\|} = 1 + \frac{\|\boldsymbol{u} - \boldsymbol{y}\|}{\|\boldsymbol{a} - \boldsymbol{u}\|}\\ &\leqslant 1 + \frac{(\mu - 1)\rho}{\rho} = \mu,\end{aligned}$$

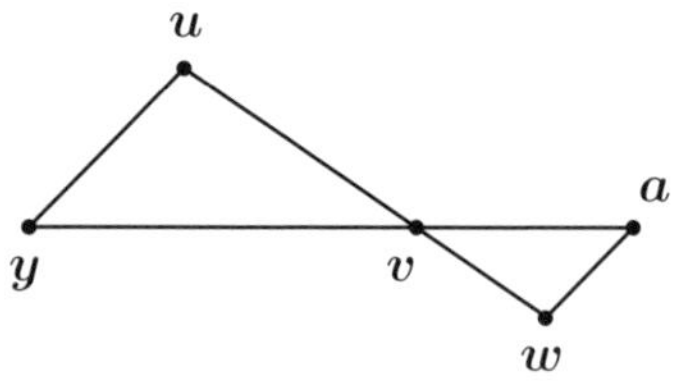

Fig. 3.6 Illustration to Theorem 3.44.

and Lemma 3.12 implies the inclusion

$$\boldsymbol{y} \in \boldsymbol{a} + \lambda^{-1}(K - \boldsymbol{a}) \subset \boldsymbol{a} + \mu(K - \boldsymbol{a}).$$

2. Next, suppose that $\boldsymbol{u} \notin \langle \boldsymbol{a}, \boldsymbol{y} \rangle$ and denote by L the 2-dimensional plane generated by the affinely independent set $\{\boldsymbol{a}, \boldsymbol{u}, \boldsymbol{y}\}$ from aff K. Then $L = \operatorname{aff}\{\boldsymbol{a}, \boldsymbol{u}, \boldsymbol{y}\} \subset \operatorname{aff} K$. Choose on the circle $L \cap S_\rho(\boldsymbol{a})$ the point $\boldsymbol{w}$ such that the vector $\boldsymbol{w} - \boldsymbol{a}$ is a positive multiple of $\boldsymbol{y} - \boldsymbol{u}$. Denote by $\boldsymbol{v}$ the point of intersection of the closed segments $[\boldsymbol{a}, \boldsymbol{y}]$ and $[\boldsymbol{u}, \boldsymbol{w}]$ (see Figure 3.6). Clearly, $\boldsymbol{v} = (1-\gamma)\boldsymbol{a} + \gamma \boldsymbol{y}$, where $0 < \gamma < 1$. Equivalently, $\boldsymbol{y} = \boldsymbol{a} + \gamma^{-1}(\boldsymbol{v} - \boldsymbol{a})$, with $\gamma^{-1} > 1$. Consequently, $\boldsymbol{y} - \boldsymbol{v} = (\gamma^{-1} - 1)(\boldsymbol{v} - \boldsymbol{a})$.

Since the triangles $\Delta(\boldsymbol{a}, \boldsymbol{v}, \boldsymbol{w})$ and $\Delta(\boldsymbol{y}, \boldsymbol{v}, \boldsymbol{u})$ are similar, we have

$$\gamma^{-1} - 1 = \frac{\|\boldsymbol{y} - \boldsymbol{v}\|}{\|\boldsymbol{v} - \boldsymbol{a}\|} = \frac{\|\boldsymbol{y} - \boldsymbol{u}\|}{\|\boldsymbol{w} - \boldsymbol{a}\|} \leqslant \frac{(\mu - 1)\rho}{\rho} = \mu - 1.$$

Hence $\gamma^{-1} \leqslant \mu$, and Lemma 3.12 implies the inclusion

$$\boldsymbol{y} \in \boldsymbol{a} + \gamma^{-1}(K - \boldsymbol{a}) \subset \boldsymbol{a} + \mu(K - \boldsymbol{a}).$$

(*b*) Choose a sequence of scalars $\rho_1 < \rho_2 < \dots$ which converges to ρ. Clearly, $B_{\rho_i}(\boldsymbol{a}) \subset U_\rho(\boldsymbol{a})$ for all $i \geqslant 1$. Furthermore,

$$U_\rho(\boldsymbol{a}) = \bigcup_{i \geqslant 1} B_{\rho_i}(\boldsymbol{a}) \quad \text{and} \quad U_{(\mu-1)\rho}(K) = \bigcup_{i \geqslant 1} B_{(\mu-1)\rho_i}(K).$$

If $\operatorname{aff} K \cap U_\rho(\boldsymbol{a}) \subset K$, then $\operatorname{aff} K \cap B_{\rho_i}(\boldsymbol{a}) \subset K$ for all $i \geqslant 1$, and assertion (*a*) gives

$$\operatorname{aff} K \cap U_{(\mu-1)\rho}(K) = \bigcup_{i \geqslant 1} (\operatorname{aff} K \cap B_{(\mu-1)\rho_i}(K)) \subset \boldsymbol{a} + \mu(K - \boldsymbol{a}). \qquad \square$$

Remark. The converse to Theorem 3.44 assertion does not hold. For instance, let $K = \{(x, y) : y \geqslant x^2\}$ and $\boldsymbol{a} = (0, 1)$. We are going to show that no ρ-neighborhood $B_\rho(K)$ of K contains its homothetic copy

$$K' = \boldsymbol{a} + 2(K - \boldsymbol{a}) = \{(x, y) : y \geqslant \tfrac{1}{2}x^2 - 1\}.$$

First, we observe that for any point $\boldsymbol{x} = (x, x^2) \in \operatorname{bd} K$, one has

$$\boldsymbol{a} + 2(\boldsymbol{x} - \boldsymbol{a}) = (2x, 2x^2 - 1) = (2x, \tfrac{1}{2}(2x)^2 - 1) \in \operatorname{bd} K'.$$

Next, choose a point $\boldsymbol{u} = (u, u^2) \in \operatorname{bd} K$. Then the line l tangent to K at $\boldsymbol{u}$ is given by $y - u^2 = 2u(x - u)$. Therefore, the orthogonal to l line l_1 is described by the equation $y - u^2 = -\frac{1}{2u}(x - u)$. The line l_1 meets $\operatorname{bd} K'$, given by $y = \frac{1}{2}x^2 - 1$, at two distinct points. An easy computation shows that one of these points, with both positive coordinates, is $\boldsymbol{v} = (v, v^2)$, where

$$v = \frac{\sqrt{8u^4 + 12u^2 + 1} - 1}{2u}.$$

Clearly, $\boldsymbol{u}$ is the nearest to $\boldsymbol{v}$ point in K; so, the distance $\delta(\boldsymbol{v}, K)$ equals $\|\boldsymbol{v} - \boldsymbol{u}\|$. It is easy to see that

$$\begin{aligned}\|\boldsymbol{v} - \boldsymbol{u}\| &= \sqrt{(v - u)^2 + (v^2 - u^2)^2} \geqslant v - u \\ &= \frac{\sqrt{8u^4 + 12u^2 + 1} - 2u^2 - 1}{2u} \geqslant \frac{u}{3}.\end{aligned}$$

Therefore, $\delta(\boldsymbol{v}, K) \to \infty$ as $u \to \infty$. Because $\boldsymbol{v} \in \operatorname{bd} K'$, no ρ-neighborhood $U_\rho(K)$ of K contains K'.

Theorem 3.45. *Given a bounded convex set $K \subset \mathbb{R}^n$ of diameter $\delta > 0$ and a scalar $\rho > 0$, the following assertions hold.*

(*a*) *For any point $\boldsymbol{a} \in K$, the set $\operatorname{aff} K \cap B_\rho(K)$ contains the homothetic copy $\boldsymbol{a} + \frac{\delta+\rho}{\delta}(K - \boldsymbol{a})$ of K.*

(*b*) *If, additionally, $\boldsymbol{a} \in \operatorname{rint} K$, then the set $\operatorname{aff} K \cap U_\rho(K)$ contains the homothetic copy $\boldsymbol{a} + \frac{\delta+\rho}{\delta}(K - \boldsymbol{a})$ of K.*

Proof. (*a*) Choose a point $\boldsymbol{y} \in \boldsymbol{a} + \frac{\delta+\rho}{\delta}(K - \boldsymbol{a})$. Then $\boldsymbol{y} = \boldsymbol{a} + \frac{\delta+\rho}{\delta}(\boldsymbol{u} - \boldsymbol{a})$ for a suitable point $\boldsymbol{u} \in K$. Since $\boldsymbol{y}$ is an affine combination of $\boldsymbol{a}$ and $\boldsymbol{u}$ from K, one has $\boldsymbol{y} \in \operatorname{aff} K$ (see Theorem 2.51). Furthermore, $\boldsymbol{y} - \boldsymbol{u} = \frac{\rho}{\delta}(\boldsymbol{u} - \boldsymbol{a})$, which gives

$$\|\boldsymbol{y} - \boldsymbol{u}\| = \frac{\rho}{\delta}\,\|\boldsymbol{u} - \boldsymbol{a}\| \leqslant \frac{\rho}{\delta}\operatorname{diam} K = \frac{\rho}{\delta}\delta = \rho.$$

Consequently, $\boldsymbol{y} \in \operatorname{aff} K \cap B_\rho(K)$.

(*b*) Suppose that $\boldsymbol{a} \in \operatorname{rint} K$. With above chosen points $\boldsymbol{y}$ and $\boldsymbol{u}$, it suffices to show that $\|\boldsymbol{u} - \boldsymbol{a}\| < \delta$. This inequality is obvious if $\boldsymbol{u} = \boldsymbol{a}$. Suppose that $\boldsymbol{u} \neq \boldsymbol{a}$. By Corollary 3.27, there is a point $\boldsymbol{v} \in K$ with the property $\boldsymbol{a} \in (\boldsymbol{u}, \boldsymbol{v})$. Therefore, $\|\boldsymbol{u} - \boldsymbol{a}\| < \|\boldsymbol{u} - \boldsymbol{v}\| \leqslant \operatorname{diam} K = \delta$, and

$$\|\boldsymbol{y} - \boldsymbol{u}\| = \frac{\rho}{\delta}\,\|\boldsymbol{u} - \boldsymbol{a}\| < \frac{\rho}{\delta}\delta = \rho.$$

So, $\boldsymbol{y} \in \operatorname{aff} K \cap U_\rho(K)$. □

3.3 Closure of Convex Sets

3.3.1 *Closure and Relative Interior*

Theorem 3.46. *For a convex set $K \subset \mathbb{R}^n$, its closure $\operatorname{cl} K$ is a convex set with the same affine span and whence the same dimension as K:*

$$\operatorname{aff}(\operatorname{cl} K) = \operatorname{aff} K \quad \textit{and} \quad \dim(\operatorname{cl} K) = \dim K.$$

Proof. Both parts of the assertion are obvious if $K = \varnothing$. So, we may assume that $K \neq \varnothing$. Because any open ball $U_\rho(\boldsymbol{o}) \subset \mathbb{R}^n$ is convex (see an example on page 75), Theorem 3.10 implies that every open neighborhood $U_\rho(K) = U_\rho(\boldsymbol{o}) + K$, $\rho > 0$, of K is convex. Consequently, Theorem 3.8 shows that $\operatorname{cl} K$ is convex as the intersection of the convex sets $U_\rho(K)$, $\rho > 0$. The second part of the assertion follows from Corollary 2.43. □

The next result is a sharper version of Theorem 3.24.

Theorem 3.47. *If $K \subset \mathbb{R}^n$ is a nonempty convex set, $\boldsymbol{x} \in \operatorname{rint} K$, and $\boldsymbol{y} \in \operatorname{cl} K$, then $(1-\lambda)\boldsymbol{x} + \lambda\boldsymbol{y} \in \operatorname{rint} K$ for all $0 \leqslant \lambda < 1$. Consequently, the semi-open segment $[\boldsymbol{x}, \boldsymbol{y})$ lies in $\operatorname{rint} K$.*

Proof. Choose a point $\boldsymbol{z} = (1-\lambda)\boldsymbol{x} + \lambda\boldsymbol{y}$, where $0 \leqslant \lambda < 1$. Since the case $\lambda = 0$ is obvious, we may assume that $\lambda > 0$. By Theorem 3.17, there is a scalar $\delta > 0$ such that $\operatorname{aff} K \cap U_\delta(\boldsymbol{x}) \subset \operatorname{rint} K$. Because $\boldsymbol{y} \in \operatorname{cl} K \subset \operatorname{aff} K$, the closed segment $[\boldsymbol{x}, \boldsymbol{y}]$ lies in $\operatorname{aff} K$. Let a point $\boldsymbol{u} \in K$ be such that $\|\boldsymbol{u} - \boldsymbol{y}\| < \frac{1-\lambda}{\lambda}\delta$. Put $\boldsymbol{v} = \frac{1}{1-\lambda}\boldsymbol{z} - \frac{\lambda}{1-\lambda}\boldsymbol{u}$ (whence $\boldsymbol{z} = (1-\lambda)\boldsymbol{v} + \lambda\boldsymbol{u}$, as depicted in Figure 3.7). The equalities

$$\boldsymbol{x} = \frac{1}{1-\lambda}\boldsymbol{z} - \frac{\lambda}{1-\lambda}\boldsymbol{y} \quad \text{and} \quad \|\boldsymbol{x} - \boldsymbol{v}\| = \frac{\lambda}{1-\lambda}\|\boldsymbol{u} - \boldsymbol{y}\| < \delta$$

show that $\boldsymbol{v} \in U_\delta(\boldsymbol{x})$. Furthermore, $\boldsymbol{v} \in \operatorname{aff} K$ because $\boldsymbol{v}$ is an affine combination of $\boldsymbol{u}$ and $\boldsymbol{z}$ from $\operatorname{aff} K$ (see Theorem 2.38). Summing up, $\boldsymbol{v} \in \operatorname{aff} K \cap U_\rho(\boldsymbol{x}) \subset \operatorname{rint} K$. By Theorem 3.24, $\boldsymbol{z} = (1-\lambda)\boldsymbol{v} + \lambda\boldsymbol{u} \in \operatorname{rint} K$. □

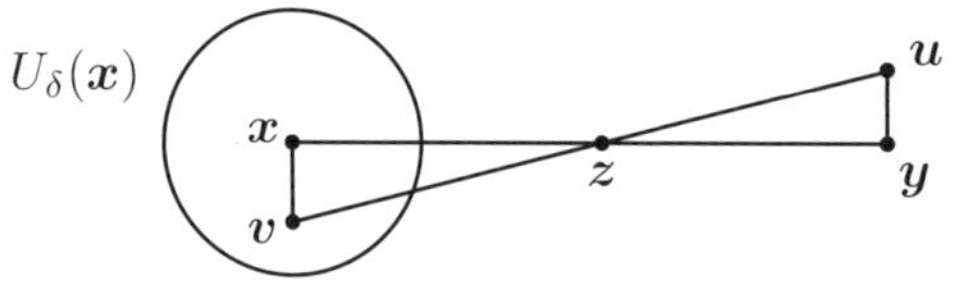

Fig. 3.7 Illustration to Theorem 3.47.

Remark. In Theorem 3.47, one cannot replace $\operatorname{rint} K$ with K. Indeed, let $K = \{\boldsymbol{o}\} \cup \{(x, y) : 0 < x, y < 1\}$. Clearly, K is a convex set and the point $\boldsymbol{u} = (0, 1)$ belongs to $\operatorname{cl} K$. On the other hand, $(\boldsymbol{o}, \boldsymbol{u}) \cap K = \varnothing$. A description of convex sets $K \subset \mathbb{R}^n$ satisfying the condition $(\boldsymbol{x}, \boldsymbol{y}) \subset K$ whenever $\boldsymbol{x} \in K$ and $\boldsymbol{y} \in \operatorname{cl} K$ is given in Problem 11.15.

Theorem 3.48. *For a convex set* $K \subset \mathbb{R}^n$, *one has*

$$\operatorname{cl}(\operatorname{rint} K) = \operatorname{cl} K \quad \textit{and} \quad \operatorname{rint}(\operatorname{cl} K) = \operatorname{rint} K.$$

Proof. Without loss of generality we may assume that $K \neq \varnothing$.

For the first equality, we observe that $\operatorname{cl}(\operatorname{rint} K) \subset \operatorname{cl} K$ due to the obvious inclusion $\operatorname{rint} K \subset K$ and monotonicity of closure. For the opposite inclusion, choose points $\boldsymbol{x} \in \operatorname{cl} K$ and $\boldsymbol{y} \in \operatorname{rint} K$. By Theorem 3.47, all points $\boldsymbol{z}_\lambda = (1 - \lambda)\boldsymbol{y} + \lambda \boldsymbol{x}$, $0 \leqslant \lambda < 1$, belong to $\operatorname{rint} K$. Since $\boldsymbol{z}_\lambda \to \boldsymbol{x}$ as $\lambda \to 1$, one has $\boldsymbol{x} \in \operatorname{cl}(\operatorname{rint} K)$.

Because $\operatorname{aff} K = \operatorname{aff}(\operatorname{cl} K)$ (see Theorem 3.46), the inclusion $K \subset \operatorname{cl} K$ and Theorem 3.18 give $\operatorname{rint} K \subset \operatorname{rint}(\operatorname{cl} K)$. Conversely, choose points $\boldsymbol{x} \in \operatorname{rint}(\operatorname{cl} K)$ and $\boldsymbol{z} \in \operatorname{rint} K$. Then $\boldsymbol{u} = \gamma \boldsymbol{x} + (1 - \gamma)\boldsymbol{z} \in \operatorname{cl} K$ for a suitable scalar $\gamma > 1$ (see Theorem 3.26). Therefore,

$$\boldsymbol{x} = (1 - \gamma^{-1})\boldsymbol{z} + \gamma^{-1}\boldsymbol{u} \in \operatorname{rint} K$$

according to Theorem 3.47. Hence $\operatorname{rint}(\operatorname{cl} K) \subset \operatorname{rint} K$. □

Corollary 3.49. *If* K_1 *and* K_2 *are convex sets in* $\mathbb{R}^n$, *then* $\operatorname{cl} K_1 = \operatorname{cl} K_2$ *if and only if* $\operatorname{rint} K_1 = \operatorname{rint} K_2$. □

Theorem 3.50. *A convex set* $K \subset \mathbb{R}^n$ *is a plane if and only if any of the following two conditions holds:* (*a*) $\operatorname{rint} K$ *is a plane,* (*b*) $\operatorname{cl} K$ *is a plane.*

In particular, $K = \mathbb{R}^n$ *if and only if any of the following two conditions is satisfied:* (*c*) $\operatorname{rint} K = \mathbb{R}^n$, (*d*) $\operatorname{cl} K = \mathbb{R}^n$.

Proof. Excluding the obvious case $K = \varnothing$, we assume that K is nonempty.

(*a*) If K is a plane, then $\operatorname{rint} K = K$, which shows that $\operatorname{rint} K$ also is a plane. Conversely, if $\operatorname{rint} K$ is a plane, then

$$\operatorname{rint} K \subset K \subset \operatorname{aff} K = \operatorname{aff}(\operatorname{rint} K) = \operatorname{rint} K$$

according to Theorem 3.17. Hence $K = \operatorname{rint} K$, and K is a plane.

(*b*) If K is a plane, then $\operatorname{cl} K = K$ (see Corollary 2.21), which shows that $\operatorname{cl} K$ is a plane. Conversely, assume that $\operatorname{cl} K$ is a plane. Then Theorem 3.48 gives $\operatorname{rint} K = \operatorname{rint}(\operatorname{cl} K) = \operatorname{cl} K$, implying that $\operatorname{rint} K$ is a plane. By the above proved, K is a plane. □

The next theorem describes a local property of the relative interior of a convex sets (compare with Theorem 3.17 and Problem 3.8).

Theorem 3.51. *If* $K \subset \mathbb{R}^n$ *is a nonempty convex set and* $\boldsymbol{a}$ *is a point in* $\operatorname{cl} K$*, then* $\operatorname{rint} K \cap U_\rho(\boldsymbol{a}) \neq \varnothing$ *for any open ball* $U_\rho(\boldsymbol{a}) \subset \mathbb{R}^n$, $\rho > 0$. *Furthermore,*

$$\begin{aligned} \operatorname{aff}(\operatorname{rint} K \cap U_\rho(\boldsymbol{a})) &= \operatorname{aff}(K \cap U_\rho(\boldsymbol{a})) \\ &= \operatorname{aff}(\operatorname{cl} K \cap U_\rho(\boldsymbol{a})) = \operatorname{aff} K. \end{aligned} \tag{3.12}$$

Proof. First, we contend that $\operatorname{rint} K \cap U_\rho(\boldsymbol{a}) \neq \varnothing$. Since the case $K = \{\boldsymbol{a}\}$ is obvious ($\operatorname{rint}\{\boldsymbol{a}\} = \{\boldsymbol{a}\}$), we may assume that K has positive dimension. Then $\dim(\operatorname{rint} K) > 0$ (see Theorem 3.17), implying that $\operatorname{rint} K \neq \{\boldsymbol{a}\}$. Choose points $\boldsymbol{u} \in \operatorname{rint} K \setminus \{\boldsymbol{a}\}$ and $\boldsymbol{z} \in (\boldsymbol{u}, \boldsymbol{a})$ such that $\|\boldsymbol{z} - \boldsymbol{a}\| < \rho$. Then $\boldsymbol{z} \in \operatorname{rint} K$ according to Theorem 3.47. By Theorem 3.17, there is a $\delta > 0$ such that $\operatorname{aff} K \cap U_\delta(\boldsymbol{z}) \subset \operatorname{rint} K$. Let $0 < \varepsilon < \min\{\delta, \rho - \|\boldsymbol{z} - \boldsymbol{a}\|\}$. For every point $\boldsymbol{x} \in U_\varepsilon(\boldsymbol{z})$, one has

$$\|\boldsymbol{x} - \boldsymbol{a}\| \leqslant \|\boldsymbol{x} - \boldsymbol{z}\| + \|\boldsymbol{z} - \boldsymbol{a}\| < \varepsilon + \|\boldsymbol{z} - \boldsymbol{a}\| < \rho,$$

implying the inclusion $\boldsymbol{x} \in U_\rho(\boldsymbol{a})$. Consequently,

$$U_\varepsilon(\boldsymbol{z}) \cap \operatorname{rint} K \subset U_\rho(\boldsymbol{a}) \cap \operatorname{rint} K.$$

A combination of Theorems 2.42, 3.17 (with $\boldsymbol{z}$ instead of $\boldsymbol{a}$), and 3.46 gives

$$\begin{aligned} \operatorname{aff} K &= \operatorname{aff}(\operatorname{rint} K \cap U_\varepsilon(\boldsymbol{z})) \subset \operatorname{aff}(\operatorname{rint} K \cap U_\rho(\boldsymbol{a})) \\ &\subset \operatorname{aff}(K \cap U_\rho(\boldsymbol{a})) \subset \operatorname{aff}(\operatorname{cl} K \cap U_\rho(\boldsymbol{a})) \subset \operatorname{aff}(\operatorname{cl} K) = \operatorname{aff} K. \end{aligned}$$

Summing up, the equalities (3.12) hold. □

Corollary 3.52. *If the closure of a convex set* $K \subset \mathbb{R}^n$ *meets an open halfspace* $W \subset \mathbb{R}^n$*, then* $\operatorname{rint}(K \cap W) = \operatorname{rint} K \cap W$.

Proof. Choose a point $\boldsymbol{u} \in \operatorname{cl} K \cap W$. Since W is an open set (see Theorem 2.33), there is an open ball $U_\rho(\boldsymbol{u}) \subset \mathbb{R}^n$ which lies in W. By Theorem 3.51, one can find a point $\boldsymbol{v} \in \operatorname{rint} K \cap W$. Because $\operatorname{rint} W = W$, Theorem 3.29 gives the equality $\operatorname{rint}(K \cap W) = \operatorname{rint} K \cap W$. □

The next result generalizes Theorem 3.3 for the case of convex convergent series.

Theorem 3.53. *A nonempty convex set* $K \subset \mathbb{R}^n$ *contains every convergent series*

$$\boldsymbol{z} = \sum_{i=1}^{\infty} \lambda_i \boldsymbol{x}_i, \ \text{where } \boldsymbol{x}_i \in K, \ \sum_{i=1}^{\infty} \lambda_i = 1, \ \text{and } \lambda_i \geqslant 0 \text{ for all } i \geqslant 1. \tag{3.13}$$

Proof. Choose a point $\boldsymbol{z}$ of the form (3.13). Excluding zero terms, we may assume that all scalars λ_i are positive. Put $X = \{\boldsymbol{x}_1, \boldsymbol{x}_2, \dots\}$ and $L = \operatorname{aff} X$. Let $r = \dim L$. Then X contains an affinely independent set Y of $r+1$ points with the property $\operatorname{aff} Y = L$ (see Corollary 2.66). Without loss of generality, we suppose that $Y = \{\boldsymbol{x}_1, \dots, \boldsymbol{x}_{r+1}\}$. So, $L = \operatorname{aff}\{\boldsymbol{x}_1, \dots, \boldsymbol{x}_{r+1}\}$. Consider the convex set $M = K \cap L$. The inclusions $Y \subset M \subset L$ and Theorem 2.42 imply that $\operatorname{aff} M = L$.

We first assert that $\boldsymbol{z} \in \operatorname{cl} M$. For this, we express $\boldsymbol{z}$ as

$$\boldsymbol{z} = \sum_{i=1}^{\infty} \lambda_i \boldsymbol{x}_i = \lim_{k\to\infty} \sum_{i=1}^{k} \lambda_i \boldsymbol{x}_i = \lim_{k\to\infty} \gamma_k \Big(\sum_{i=1}^{k} \eta_i^{(k)} \boldsymbol{x}_i\Big)$$

where $\gamma_k = \sum_{j=1}^{k} \lambda_j$ and $\eta_i^{(k)} = \dfrac{\lambda_i}{\gamma_k}$ for all $1 \leqslant i \leqslant k$.

Clearly, every $\boldsymbol{z}_k = \eta_1^{(k)} \boldsymbol{x}_1 + \cdots + \eta_k^{(k)} \boldsymbol{x}_k$, $k \geqslant 1$, is a positive convex combination of the points $\boldsymbol{x}_1, \dots, \boldsymbol{x}_k \in M$. Hence all $\boldsymbol{z}_1, \boldsymbol{z}_2, \dots$ belong to M (see Theorem 3.3). Since $\lim_{k\to\infty} \gamma_k = 1$, one has

$$\boldsymbol{z} = \lim_{k\to\infty} \gamma_k \boldsymbol{z}_k = \frac{\lim\limits_{k\to\infty} \gamma_k \boldsymbol{z}_k}{\lim\limits_{k\to\infty} \gamma_k} = \lim_{k\to\infty} \frac{\gamma_k \boldsymbol{z}_k}{\gamma_k} = \lim_{k\to\infty} \boldsymbol{z}_k \in \operatorname{cl} M.$$

To prove the inclusion $\boldsymbol{z} \in M$, we write

$$\begin{aligned} \boldsymbol{z} = \sum_{i=1}^{\infty} \lambda_i \boldsymbol{x}_i &= \sum_{i=1}^{r+1} \lambda_i \boldsymbol{x}_i + \sum_{i=r+2}^{\infty} \lambda_i \boldsymbol{x}_i \\ &= \gamma_{r+1} \Big(\sum_{i=1}^{r+1} \eta_i^{(r+1)} \boldsymbol{x}_i\Big) + (1 - \gamma_{r+1}) \Big(\sum_{i=r+2}^{\infty} \delta_i \boldsymbol{x}_i\Big), \end{aligned}$$

where γ_{r+1} and $\eta_i^{(r+1)}$ are defined as above, and

$$\delta_i = \frac{\lambda_i}{1 - \gamma_{r+1}} > 0 \quad \text{for all} \quad i \geqslant r+2.$$

Clearly, $0 < \gamma_{r+1} < 1$ and $\boldsymbol{z} = \gamma_{r+1} \boldsymbol{u} + (1 - \gamma_{r+1}) \boldsymbol{v}$, where

$$\boldsymbol{u} = \sum_{i=1}^{r+1} \eta_i^{(r+1)} \boldsymbol{x}_i, \quad \boldsymbol{v} = \sum_{i=r+2}^{\infty} \delta_i \boldsymbol{x}_i, \quad \sum_{i=r+2}^{\infty} \delta_i = 1.$$

Similarly to the above argument, one has $\boldsymbol{v} \in \operatorname{cl} M$. We are going to prove that $\boldsymbol{u} \in \operatorname{rint} M$. Indeed, since all scalars $\eta_i^{(r+1)}$ are positive, Theorem 3.20 shows that $\boldsymbol{u}$ belongs to the relative interior of the simplex $\Delta = \Delta(\boldsymbol{x}_1, \dots, \boldsymbol{x}_{r+1})$, which lies in M according to Corollary 3.7. By Theorem 3.6, $\operatorname{aff} \Delta = L = \operatorname{aff} M$. Now, Theorem 3.18 gives $\boldsymbol{u} \in \operatorname{rint} \Delta \subset \operatorname{rint} M$. Finally, the inclusions $\boldsymbol{u} \in \operatorname{rint} M$, $\boldsymbol{v} \in \operatorname{cl} M$, and Theorem 3.47 imply that the point $\boldsymbol{z} \in (\boldsymbol{u}, \boldsymbol{v})$ belongs to $\operatorname{rint} M$. Hence $\boldsymbol{z} \in M \subset K$. □

3.3.2 *Closure and Algebra of Convex Sets*

Theorem 3.54. *If* $\mathcal{F} = \{K_\alpha\}$ *is a family of convex sets in* $\mathbb{R}^n$*, then*

$$\mathrm{cl}\,(\bigcap_\alpha K_\alpha) \subset \bigcap_\alpha \mathrm{cl}\,K_\alpha. \tag{3.14}$$

Furthermore, $\mathrm{cl}\,(\bigcap_\alpha K_\alpha) = \bigcap_\alpha \mathrm{cl}\,K_\alpha$ *provided* $\bigcap_\alpha \mathrm{rint}\,K_\alpha \neq \varnothing$.

Proof. We observe that inclusion (3.14) holds for any family $\{K_\alpha\}$ of sets (see page 10), not necessarily convex. For the second assertion, it suffices to show that $\bigcap_\alpha \mathrm{cl}\,K_\alpha \subset \mathrm{cl}\,(\bigcap_\alpha K_\alpha)$ provided $\bigcap_\alpha \mathrm{rint}\,K_\alpha \neq \varnothing$. Indeed, choose points $\boldsymbol{x} \in \bigcap_\alpha \mathrm{cl}\,K_\alpha$ and $\boldsymbol{y} \in \bigcap_\alpha \mathrm{rint}\,K_\alpha$. By Theorem 3.47,

$$(1-\lambda)\boldsymbol{y} + \lambda\boldsymbol{x} \in \bigcap_\alpha \mathrm{rint}\,K_\alpha \quad \text{for all} \quad 0 \leqslant \lambda < 1.$$

Therefore,

$$\boldsymbol{x} = \lim_{\lambda\to 1}((1-\lambda)\boldsymbol{y} + \lambda\boldsymbol{x}) \in \mathrm{cl}\,(\bigcap_\alpha \mathrm{rint}\,K_\alpha) \subset \mathrm{cl}\,(\bigcap_\alpha K_\alpha). \qquad \square$$

Remark. The second assertion of Theorem 3.54 fails if $\bigcap_\alpha \mathrm{rint}\,K_\alpha = \varnothing$. Indeed, let H be a hyperplane in $\mathbb{R}^n$ and W be an open halfspace determined by H. Then $\mathrm{cl}\,(H \cap W) = H \cap W = \varnothing \neq H = \mathrm{cl}\,H \cap \mathrm{cl}\,W$. A related to Theorem 3.54 result is given in Problem 3.15.

Since $\mathrm{rint}\,L = \mathrm{cl}\,L = L$ for every plane $L \subset \mathbb{R}^n$, Theorem 3.54 implies the following result.

Corollary 3.55. *If a plane* $L \subset \mathbb{R}^n$ *meets the relative interior of a convex set* $K \subset \mathbb{R}^n$*, then* $\mathrm{cl}\,(K \cap L) = \mathrm{cl}\,K \cap L$. $\square$

Corollary 3.56. *If* $\mathcal{F} = \{K_\alpha\}$ *is a nested family of convex sets in* $\mathbb{R}^n$*, then* $\bigcup_\alpha \mathrm{cl}\,K_\alpha$ *is a convex set satisfying the conditions*

$$\bigcup_\alpha \mathrm{cl}\,K_\alpha \subset \mathrm{cl}\,(\bigcup_\alpha K_\alpha) = \mathrm{cl}\,(\bigcup_\alpha \mathrm{cl}\,K_\alpha). \tag{3.15}$$

Proof. The convexity of $\bigcup_\alpha \mathrm{cl}\,K_\alpha$ follows from Theorems 3.8 and 3.46, and (3.15) holds for any family $\{K_\alpha\}$ of sets, not necessarily convex (see page 10). $\square$

We will be using the following corollaries, which are particular cases of more general statements (see Problem 1.13 and Corollary 2.80).

Corollary 3.57. *For convex sets* $K_1, K_2 \subset \mathbb{R}^n$ *and scalars* μ_1, μ_2*, the sum* $\mu_1\,\mathrm{cl}\,K_1 + \mu_2\,\mathrm{cl}\,K_2$ *is a convex set satisfying the inclusion*

$$\mu_1\,\mathrm{cl}\,K_1 + \mu_2\,\mathrm{cl}\,K_2 \subset \mathrm{cl}\,(\mu_1 K_1 + \mu_2 K_2). \tag{3.16}$$

If at least one of the sets K_1, K_2 is bounded or the planes $\operatorname{aff} K_1$ *and* $\operatorname{aff} K_2$ *are independent, then the sum $\mu_1 \operatorname{cl} K_1 + \mu_2 \operatorname{cl} K_2$ is a closed set, and*

$$\mu_1 \operatorname{cl} K_1 + \mu_2 \operatorname{cl} K_2 = \operatorname{cl}(\mu_1 K_1 + \mu_2 K_2). \qquad \square$$

Remark. Inclusion (3.16) in Corollary 3.57 may be proper if both convex sets K_1 and K_2 are unbounded. Indeed, let

$$K_1 = \{(x, y) : x > 0,\ xy \geqslant 1\} \quad \text{and} \quad K_2 = \{(x, 0) : x \leqslant 0\}.$$

Then both K_1 and K_2 are closed, while their sum $K_1 + K_2 = \{(x, y) : y > 0\}$ is an open halfplane.

Corollary 3.58. *Let $f : \mathbb{R}^n \to \mathbb{R}^m$ be an affine transformation and $K \subset \mathbb{R}^n$ and $M \subset \mathbb{R}^m$ be convex sets. Then $f(\operatorname{cl} K)$ is a convex set and*

$$f(\operatorname{cl} K) \subset \operatorname{cl} f(K),$$
$$\operatorname{cl} f^{-1}(M) = f^{-1}(\operatorname{cl}(M \cap \operatorname{rng} f)).$$

Furthermore, the following assertions hold.

(a) $f(\operatorname{cl} K) = \operatorname{cl} f(K)$ *if f is one-to-one or K is bounded,*
(b) $\operatorname{cl} f^{-1}(M) = f^{-1}(\operatorname{cl} M)$ *provided* $\operatorname{rint} M \cap \operatorname{rng} f \neq \varnothing$. $\square$

Remark. The inclusion $f(\operatorname{cl} K) \subset \operatorname{cl} f(K)$ in Corollary 3.58 can be proper if f is not one-to-one and K is unbounded. Indeed, let f be the orthogonal projection of $\mathbb{R}^2$ on the x-axis, and $K = \{(x, y) : x > 0,\ xy \geqslant 1\}$. Then K is a closed convex set, while $f(K)$ is the open halfline $\{(x, 0) : x > 0\}$.

The next result complements Theorem 3.28.

Theorem 3.59. *Let $K \subset \mathbb{R}^n$ be a nonempty convex set, and $\boldsymbol{a}$ be a point in* $\operatorname{rint} K$. *Then* $\operatorname{cl} K = M_1 \cap M_2 \cap \cdots$, *where M_i is the homothetic copy of K with center $\boldsymbol{a}$ given by*

$$M_i = \boldsymbol{a} + (1 + \tfrac{1}{i})(K - \boldsymbol{a}), \quad i \geqslant 1.$$

Proof. Since the mappings $f_i(\boldsymbol{x}) = \boldsymbol{a} + (1 + \frac{1}{i})(\boldsymbol{x} - \boldsymbol{a})$, $i \geqslant 1$, are affine transformations, Theorem 3.43 gives

$$\operatorname{rint} M_i = \operatorname{rint} f_i(K) = f_i(\operatorname{rint} K) = \boldsymbol{a} + (1 + \tfrac{1}{i})(\operatorname{rint} K - \boldsymbol{a}), \quad i \geqslant 1.$$

Similarly, Problem 1.13 shows that

$$\operatorname{cl} M_i = \boldsymbol{a} + (1 + \tfrac{1}{i})(\operatorname{cl} K - \boldsymbol{a}) = f_i(\operatorname{cl} K), \quad i \geqslant 1.$$

For the inclusion $\operatorname{cl} K \subset M_1 \cap M_2 \cap \cdots$, choose any point $\boldsymbol{y} \in \operatorname{cl} K$ and put $\boldsymbol{z}_i = \boldsymbol{a} + (1 + \frac{1}{i})(\boldsymbol{y} - \boldsymbol{a})$, $i \geqslant 1$. Then

$$\boldsymbol{a} = f_i(\boldsymbol{a}) \in f_i(\operatorname{rint} K) = \operatorname{rint} M_i, \ \ \boldsymbol{z}_i = f_i(\boldsymbol{y}) \in f_i(\operatorname{cl} K) = \operatorname{cl} M_i, \ \ i \geqslant 1.$$

Consequently, Theorem 3.47 gives

$$\boldsymbol{y} = \tfrac{1}{i+1}\boldsymbol{a} + \tfrac{i}{i+1}\boldsymbol{z}_i \in \operatorname{rint} M_i \subset M_i, \quad i \geqslant 1.$$

Therefore, $\boldsymbol{y} \in M_1 \cap M_2 \cap \cdots$.

Conversely, let $\boldsymbol{y} \in M_1 \cap M_2 \cap \cdots$. The inclusion $\boldsymbol{y} \in M_i$ shows the existence of a point $\boldsymbol{u}_i \in K$ such that $\boldsymbol{y} = \boldsymbol{a} + (1 + \frac{1}{i})(\boldsymbol{u}_i - \boldsymbol{a})$, $i \geqslant 1$. Then

$$\boldsymbol{u}_i = \boldsymbol{a} + \tfrac{i}{i+1}(\boldsymbol{y} - \boldsymbol{a}), \quad i \geqslant 1,$$

which gives $\lim_{i \to \infty} \boldsymbol{u}_i = \lim_{i \to \infty} \big(\boldsymbol{a} + \frac{i}{i+1}(\boldsymbol{y} - \boldsymbol{a})\big) = \boldsymbol{a} + (\boldsymbol{y} - \boldsymbol{a}) = \boldsymbol{y}$.

Consequently, $\boldsymbol{y} \in \operatorname{cl} K$, as desired. □

The theorem below often is called *the cancellation law* of addition of closed convex sets in $\mathbb{R}^n$. A similar assertion (for the case of nonclosed convex sets K and M and a compact set X) is given in Theorem 12.15.

Theorem 3.60. *If K_1 and K_2 are convex sets and X is a nonempty bounded set in $\mathbb{R}^n$ such that $K_1 + X \subset K_2 + X$, then $\operatorname{cl} K_1 \subset \operatorname{cl} K_2$. Consequently, $\operatorname{cl} K_1 = \operatorname{cl} K_2$ provided $K_1 + X = K_2 + X$.*

Proof. Omitting the trivial case $K_1 = \varnothing$, we may assume that K_1 is nonempty. Choose points $\boldsymbol{u} \in K_1$ and $\boldsymbol{x}_0 \in X$. Due to

$$\boldsymbol{u} + \boldsymbol{x}_0 \in K_1 + X \subset K_2 + X,$$

there are points $\boldsymbol{v}_1 \in K_2$ and $\boldsymbol{x}_1 \in X$ such that $\boldsymbol{u} + \boldsymbol{x}_0 = \boldsymbol{v}_1 + \boldsymbol{x}_1$. Similarly, we select recursively points $\boldsymbol{v}_i \in K_2$ and $\boldsymbol{x}_i \in X$ such that

$$\boldsymbol{u} + \boldsymbol{x}_{i-1} = \boldsymbol{v}_i + \boldsymbol{x}_i \quad \text{for all} \quad i \geqslant 1.$$

Adding the first r of these equalities, one has

$$r\boldsymbol{u} + \boldsymbol{x}_0 = \boldsymbol{v}_1 + \cdots + \boldsymbol{v}_r + \boldsymbol{x}_r, \quad r \geqslant 1.$$

Equivalently, $\boldsymbol{u} - \frac{1}{r}(\boldsymbol{v}_1 + \cdots + \boldsymbol{v}_r) = \frac{1}{r}(\boldsymbol{x}_r - \boldsymbol{x}_0)$, $r \geqslant 1$.

By Theorem 3.3, every convex combination $\boldsymbol{w}_r = \frac{1}{r}(\boldsymbol{v}_1 + \cdots + \boldsymbol{v}_r)$, $r \geqslant 1$, lies in K_2. If $\delta = \operatorname{diam} X$, then $\|\boldsymbol{x}_0 - \boldsymbol{x}_r\| \leqslant \delta$, and

$$\|\boldsymbol{u} - \boldsymbol{w}_r\| = \tfrac{1}{r}\|\boldsymbol{x}_0 - \boldsymbol{x}_r\| \leqslant \tfrac{\delta}{r} \to 0 \quad \text{when} \quad r \to \infty.$$

Hence $\boldsymbol{u} = \lim_{r \to \infty} \boldsymbol{w}_r \in \operatorname{cl} K_2$, which gives the inclusion $K_1 \subset \operatorname{cl} K_2$. Consequently, $\operatorname{cl} K_1 \subset \operatorname{cl} K_2$, as desired. □

Remarks. 1. The sets K_1 and K_2 in Theorem 3.60 may be distinct. Indeed, suppose that $K_1 = X = \mathbb{U}$ and $K_2 = \mathbb{B}$, where $\mathbb{U}$ and $\mathbb{B}$ denote, respectively, the open and closed unit balls of $\mathbb{R}^n$. Then $K_1 \neq K_2$, while $K_1 + X = K_2 + X = \mathbb{U}$.

2. The assumption on boundedness of X in Theorem 3.60 is essential. For instance, consider the unit square $K_1 = \{(x, y) : 0 \leqslant x, y \leqslant 1\}$, the closed segment $K_2 = \{(0, y) : 0 \leqslant y \leqslant 1\}$, and the closed halfline $X = \{(x, 0) : x \geqslant 0\}$. Clearly, $K_1 \neq K_2$, while

$$K_1 + X = K_2 + X = \{(x, y) : 0 \leqslant x,\ 0 \leqslant y \leqslant 1\}.$$

3.3.3 *Relative Boundary of Convex Sets*

Similarly to the boundary of a set $X \subset \mathbb{R}^n$, defined by $\operatorname{bd} X = \operatorname{cl} X \setminus \operatorname{int} X$, we introduce the following concept.

Definition 3.61. The *relative boundary* of a set $X \subset \mathbb{R}^n$, denoted $\operatorname{rbd} X$, is defined by

$$\operatorname{rbd} X = \operatorname{cl} X \setminus \operatorname{rint} X.$$

Examples. 1. The relative boundary of the open unit disk $U = \{(x, y, 0) : x^2 + y^2 \leqslant 1\}$ of the coordinate xy-plane in $\mathbb{R}^3$ is the circle $C = \{(x, y, 0) : x^2 + y^2 = 1\}$.

2. If $L \subset \mathbb{R}^n$ is a plane, then $\operatorname{rbd} L = \operatorname{cl} L \setminus \operatorname{rint} L = L \setminus L = \varnothing$ (see Corollary 2.21 and an example on page 84).

3. If $L \subset \mathbb{R}^n$ is a plane of positive dimension and D is a closed halfplane of L given by $D = \{\boldsymbol{x} \in L : \boldsymbol{x} \cdot \boldsymbol{e} \leqslant \gamma\}$ for a suitable vector $\boldsymbol{e} \in \mathbb{R}^n \setminus \operatorname{ort} L$ and a scalar γ (see Theorem 2.70), then $\operatorname{rbd} D$ is the $(m-1)$-dimensional plane $M = \{\boldsymbol{x} \in L : \boldsymbol{x} \cdot \boldsymbol{e} = \gamma\}$, as follows from Corollary 3.64 below.

Theorem 3.62. *For a nonempty set $X \subset \mathbb{R}^n$, the following assertions hold.*

(*a*) $\operatorname{rbd} X \subset \operatorname{bd} X$*; furthermore,* $\operatorname{rbd} X = \operatorname{bd} X$ *if* $\operatorname{int} X \neq \varnothing$.
(*b*) $\operatorname{rbd} X \neq \varnothing$ *if and only if X is not a plane.*
(*c*) *The set* $\operatorname{rbd} X$ *is closed.*

Proof. (*a*) Due to the obvious inclusion $\operatorname{int} X \subset \operatorname{rint} X$, one has

$$\operatorname{rbd} X = \operatorname{cl} X \setminus \operatorname{rint} X \subset \operatorname{cl} X \setminus \operatorname{int} X = \operatorname{bd} X.$$

If, additionally, $\operatorname{int} X \neq \varnothing$, then $\operatorname{int} X = \operatorname{rint} X$ by the same theorem, which gives the equality $\operatorname{rbd} X = \operatorname{bd} X$.

(*b*) If X is a plane, then $\operatorname{rbd} X = \varnothing$ (see an example on page 113). Suppose that X is not a plane. Then $\varnothing \neq X \neq \operatorname{aff} X$. Choose points $\boldsymbol{x} \in X$ and $\boldsymbol{z} \in \operatorname{aff} X \setminus X$. Denote by λ' the supremum of all values $\lambda \in [0,1]$ such $(1-\lambda)\boldsymbol{x} + \lambda\boldsymbol{z} \in X$. Put $\boldsymbol{u} = (1-\lambda')\boldsymbol{x} + \lambda'\boldsymbol{z}$. Obviously, $\boldsymbol{u} \in \operatorname{cl} X \cap [\boldsymbol{x}, \boldsymbol{z}]$.

We assert that $\boldsymbol{u} \notin \operatorname{rint} X$. Because this fact is obvious if $\boldsymbol{u} = \boldsymbol{z}$, one may assume that $\boldsymbol{u} \neq \boldsymbol{z}$. Then $\lambda' < 1$. Given a scalar $\rho > 0$, there is a scalar μ satisfying the inequalities $\lambda' < \mu < \min\{1, \lambda' + \rho/\|\boldsymbol{x} - \boldsymbol{z}\|\}$. Put $\boldsymbol{v} = (1-\mu)\boldsymbol{x} + \mu\boldsymbol{z}$. Obviously, $\boldsymbol{v} \in [\boldsymbol{x}, \boldsymbol{z}]$, but $\boldsymbol{v} \notin X$ due to the choice of λ'. Since

$$\|\boldsymbol{u} - \boldsymbol{v}\| = (\mu - \lambda')\|\boldsymbol{x} - \boldsymbol{z}\| < \rho,$$

one has $[\boldsymbol{x}, \boldsymbol{z}] \cap U_\rho(\boldsymbol{u}) \not\subset X$. Consequently, $\operatorname{aff} X \cap U_\rho(\boldsymbol{u}) \not\subset X$, implying that $\boldsymbol{u} \notin \operatorname{rint} X$. Summing up, $\boldsymbol{u} \in \operatorname{rbd} X$.

(*c*) Since the case $\operatorname{rbd} X = \varnothing$ is obvious, we assume that $\operatorname{rbd} X$ is nonempty. Let $\boldsymbol{x}_1, \boldsymbol{x}_2, \ldots$ be a sequence of points in $\operatorname{rbd} X$ convergent to a point $\boldsymbol{x} \in \mathbb{R}^n$. Then $\boldsymbol{x} \in \operatorname{cl} X$ due to $\operatorname{rbd} X \subset \operatorname{cl} X$. We are going to prove that no set $\operatorname{aff} X \cap U_\rho(\boldsymbol{x})$, $\rho > 0$, lies in X. Indeed, choose an index i large enough to satisfy the inequality $\|\boldsymbol{x} - \boldsymbol{x}_i\| < \rho/2$. Since $\boldsymbol{x}_i \in \operatorname{rbd} X$, there is a point $\boldsymbol{y}_i \in (\operatorname{aff} X \setminus X) \cap U_{\rho/2}(\boldsymbol{x}_i)$. The inequalities

$$\|\boldsymbol{x} - \boldsymbol{y}_i\| \leqslant \|\boldsymbol{x} - \boldsymbol{x}_i\| + \|\boldsymbol{x}_i - \boldsymbol{y}_i\| < \rho$$

give $\boldsymbol{y}_i \in U_\rho(\boldsymbol{x})$. Therefore, $\operatorname{aff} X \cap U_\rho(\boldsymbol{x}) \not\subset X$, and $\boldsymbol{x} \notin \operatorname{rint} X$. Summing up, $\boldsymbol{x} \in \operatorname{cl} X \setminus \operatorname{rint} X = \operatorname{rbd} X$, which confirms the closedness of $\operatorname{rbd} X$. □

Corollary 3.63. *For any convex set $K \subset \mathbb{R}^n$, one has*

$$\operatorname{rbd} K = \operatorname{rbd}(\operatorname{cl} K) = \operatorname{rbd}(\operatorname{rint} K).$$

Proof. A combination of Theorems 3.18 and 3.48 yields

$$\operatorname{rbd}(\operatorname{cl} K) = \operatorname{cl}(\operatorname{cl} K) \setminus \operatorname{rint}(\operatorname{cl} K) = \operatorname{cl} K \setminus \operatorname{rint} K = \operatorname{rbd} K,$$
$$\operatorname{rbd}(\operatorname{rint} K) = \operatorname{cl}(\operatorname{rint} K) \setminus \operatorname{rint}(\operatorname{rint} K) = \operatorname{cl} K \setminus \operatorname{rint} K = \operatorname{rbd} K. \quad \square$$

Corollary 3.64. *If a plane $L \subset \mathbb{R}^n$ meets the relative interior of a convex set $K \subset \mathbb{R}^n$, then* $\operatorname{rbd}(K \cap L) = \operatorname{rbd} K \cap L$.

Proof. Since $\operatorname{rint} L = \operatorname{cl} L = L$, Corollaries 3.31 and 3.55 give

$$\begin{aligned}\operatorname{rbd}(K \cap L) &= \operatorname{cl}(K \cap L) \setminus \operatorname{rint}(K \cap L)\\ &= (\operatorname{cl} K \cap L) \setminus (\operatorname{rint} K \cap L)\\ &= (\operatorname{cl} K \setminus \operatorname{rint} K) \cap L = \operatorname{rbd} K \cap L.\end{aligned} \quad \square$$

The next result describes an important property of the relative boundary of a convex set (see Theorem 6.10 for the inclusion $[\boldsymbol{x}, \boldsymbol{z}\rangle \subset \operatorname{rint} K$).

Theorem 3.65. *Let $K \subset \mathbb{R}^n$ be a convex set of positive dimension, $\boldsymbol{x} \in \operatorname{rint} K$, and $\boldsymbol{z} \in \operatorname{aff} K \setminus \{\boldsymbol{x}\}$. Then either $[\boldsymbol{x}, \boldsymbol{z}\rangle \subset \operatorname{rint} K$, or the closed halfline $[\boldsymbol{x}, \boldsymbol{z}\rangle$ contains a unique point $\boldsymbol{u} \in \operatorname{rbd} K$ such that $[\boldsymbol{x}, \boldsymbol{u}) \subset \operatorname{rint} K$ and $[\boldsymbol{x}, \boldsymbol{z}\rangle \setminus [\boldsymbol{x}, \boldsymbol{u}] \subset \operatorname{aff} K \setminus \operatorname{cl} K$.*

Proof. Suppose that $[\boldsymbol{x}, \boldsymbol{z}\rangle \not\subset \operatorname{rint} K$ and choose a point $\boldsymbol{v} \in [\boldsymbol{x}, \boldsymbol{z}\rangle \setminus \operatorname{rint} K$. We observe first that no point from $[\boldsymbol{x}, \boldsymbol{z}\rangle \setminus [\boldsymbol{x}, \boldsymbol{v}]$ belongs to $\operatorname{cl} K$. Indeed, assuming the existence of a point $\boldsymbol{y} \in \operatorname{cl} K \cap ([\boldsymbol{x}, \boldsymbol{z}\rangle \setminus [\boldsymbol{x}, \boldsymbol{v}])$, we would obtain the inclusion $\boldsymbol{v} \in [\boldsymbol{x}, \boldsymbol{y}) \subset \operatorname{rint} K$ (see Theorem 3.47). The inclusion $\operatorname{cl} K \cap [\boldsymbol{x}, \boldsymbol{z}\rangle \subset [\boldsymbol{x}, \boldsymbol{v}]$ shows that the set $\operatorname{cl} K \cap [\boldsymbol{x}, \boldsymbol{z}\rangle$ is bounded. By the convexity of $\operatorname{cl} K$, this set should be a closed segment: $\operatorname{cl} K \cap [\boldsymbol{x}, \boldsymbol{z}\rangle = [\boldsymbol{x}, \boldsymbol{u}]$, as shown on Figure 3.8.

We assert that $\boldsymbol{u} \in \operatorname{rbd} K$. For if $\boldsymbol{u}$ belonged to $\operatorname{rint} K$, then, according to Corollary 3.27, there would be a point $\boldsymbol{y} \in K \cap [\boldsymbol{x}, \boldsymbol{z}\rangle$ satisfying the inclusion $\boldsymbol{u} \in [\boldsymbol{x}, \boldsymbol{y})$, in contradiction with $\operatorname{cl} K \cap [\boldsymbol{x}, \boldsymbol{z}\rangle = [\boldsymbol{x}, \boldsymbol{u}]$.

Finally, assume the existence of another point $\boldsymbol{u}' \in [\boldsymbol{x}, \boldsymbol{z}\rangle$ which belongs to $\operatorname{rbd} K$. Then either $\boldsymbol{u}' \in [\boldsymbol{x}, \boldsymbol{u})$ or $\boldsymbol{u} \in [\boldsymbol{x}, \boldsymbol{u}')$, and Theorem 3.47 shows that one of the points $\boldsymbol{u}, \boldsymbol{u}'$ should be in $\operatorname{rint} K$, which is impossible. □

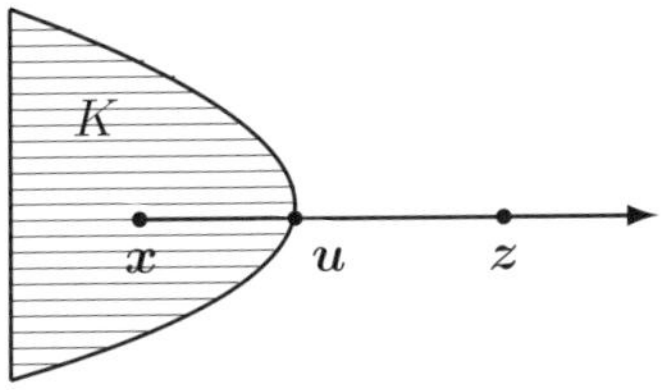

Fig. 3.8 Illustration to Theorem 3.65.

Corollary 3.66. *Let $K \subset \mathbb{R}^n$ be a convex set of positive dimension and $\boldsymbol{x}$ be a point in $\operatorname{rint} K$. Any closed halfline $h \subset \operatorname{aff} K$ with endpoint $\boldsymbol{x}$ either meets $\operatorname{rbd} K$ at a single point or entirely lies in $\operatorname{rint} K$.* □

Theorem 3.67. *If F and K are convex sets in $\mathbb{R}^n$, with $F \subset \operatorname{cl} K$, then the following assertions hold.*

(*a*) *Either $F \subset \operatorname{rbd} K$ or $\operatorname{rint} F \subset \operatorname{rint} K$.*

(*b*) *If* $F \subset \operatorname{rbd} K$, *then* $\operatorname{aff} F \cap \operatorname{cl} K \subset \operatorname{rbd} K$ *and* $\dim F \leqslant \dim K - 1$.

Proof. (*a*) Suppose that $F \not\subset \operatorname{rbd} K$. We observe first that $\operatorname{rint} F \cap \operatorname{rint} K \neq \varnothing$, since otherwise $\operatorname{rint} F \subset \operatorname{rbd} K$, and a combination of Theorems 3.48 and 3.62 would give $F \subset \operatorname{cl}(\operatorname{rint} F) \subset \operatorname{rbd} K$. Let $\boldsymbol{z} \in \operatorname{rint} F \cap \operatorname{rint} K$. Choose any point $\boldsymbol{x} \in \operatorname{rint} F$. Then $\boldsymbol{y} = \gamma\boldsymbol{x} + (1-\gamma)\boldsymbol{z} \in F$ for a suitable scalar $\gamma > 1$ (see Theorem 3.26). Because $\boldsymbol{y} \in F \subset \operatorname{cl} K$ and $0 < \gamma^{-1} < 1$, Theorem 3.47 shows that the point $\boldsymbol{x} = (1-\gamma^{-1})\boldsymbol{z} + \gamma^{-1}\boldsymbol{y}$ belongs to $\operatorname{rint} K$. Summing up, $\operatorname{rint} F \subset \operatorname{rint} K$.

(*b*) Assume for a moment that $\operatorname{aff} F \cap \operatorname{cl} K \not\subset \operatorname{rbd} K$. Then the set $\operatorname{aff} F \cap \operatorname{cl} K$ meets $\operatorname{rint} K$. Let $\boldsymbol{y}$ be a point in $\operatorname{aff} F \cap \operatorname{rint} K$. Choose any point $\boldsymbol{x} \in \operatorname{rint} F$. By Theorem 3.26, there is a scalar $0 < \lambda < 1$ such that the point $\boldsymbol{z} = (1-\lambda)\boldsymbol{x} + \lambda\boldsymbol{y}$ belongs to F. On the other hand, $\boldsymbol{z} \in \operatorname{rint} K$ due to the inclusion $\boldsymbol{x} \in \operatorname{rint} F \subset F \subset \operatorname{cl} K$ (see Theorem 3.47). So, $\boldsymbol{z} \in F \cap \operatorname{rint} K$, in contradiction with the assumption $F \subset \operatorname{rbd} K$. Summing up, $\operatorname{aff} F \cap \operatorname{cl} K \subset \operatorname{rbd} K$.

Finally, suppose that $\dim F = \dim K$. Then $\dim(\operatorname{aff} F) = \dim(\operatorname{aff} K)$, which, combined with the inclusion $\operatorname{aff} F \subset \operatorname{aff} K$, gives $\operatorname{aff} F = \operatorname{aff} K$ (see Theorem 2.6). Therefore, $\operatorname{rint} F \subset \operatorname{rint} K$ according to Theorem 3.18, in contradiction with $\operatorname{rint} F \subset F \subset \operatorname{rbd} K$. Hence $\dim F \leqslant \dim K - 1$. □

Theorem 3.68. *Let* $K \subset \mathbb{R}^n$ *be a convex set of positive dimension* m, *which is not a plane. Then* $m - 1 \leqslant \dim(\operatorname{rbd} K) \leqslant m$. *Furthermore, the following conditions are equivalent.*

(*a*) $\dim(\operatorname{rbd} K) = m - 1$.
(*b*) $\operatorname{cl} K$ *is a closed halfplane.*
(*c*) $\operatorname{rbd} K$ *is a plane.*
(*d*) $\operatorname{rbd} K$ *is a convex set.*

Proof. The inclusion $\operatorname{rbd} K \subset \operatorname{cl} K$ and Theorem 2.64 give

$$\dim(\operatorname{rbd} K) \leqslant \dim(\operatorname{cl} K) = \dim K = m.$$

Put $L = \operatorname{aff} K$. Then $\operatorname{cl} K \neq L$ by Theorem 3.50. Choose a point $\boldsymbol{z} \in L \setminus \operatorname{cl} K$ and let $N = \operatorname{aff}(\{\boldsymbol{z}\} \cup \operatorname{rbd} K)$. Obviously, $N \subset L$.

We will prove that $\operatorname{cl} K \subset N$. Indeed, suppose that $\operatorname{cl} K \not\subset N$. Then $\operatorname{rint} K \not\subset N$ by Theorem 3.48. Choose a point $\boldsymbol{y} \in \operatorname{rint} K \setminus N$ and consider the line $l = \langle \boldsymbol{y}, \boldsymbol{z} \rangle$. We observe that $l \cap N = \{\boldsymbol{z}\}$ (otherwise, $l \cap N$ would contain another point $\boldsymbol{w}$, and $\boldsymbol{y} \in l = \langle \boldsymbol{w}, \boldsymbol{z} \rangle \subset N$ by Lemma 2.25 and Theorem 2.38). On the other hand, the open segment $(\boldsymbol{y}, \boldsymbol{z})$ should contain a point of $\operatorname{rbd} K$ (see Theorem 3.65). The obtained contradiction proves the inclusion $\operatorname{cl} K \subset N$.

Now, the obvious inclusions $L = \operatorname{aff} K = \operatorname{aff}(\operatorname{cl} K) \subset N \subset L$ imply that $N = L$. Hence $\operatorname{aff}(\{\boldsymbol{z}\} \cup \operatorname{rbd} K) = L$. If $\boldsymbol{z} \in \operatorname{aff}(\operatorname{rbd} K)$, then $\operatorname{aff}(\operatorname{rbd} K) = L$, which gives the equality $\dim(\operatorname{aff} K) = m$. If $\boldsymbol{z} \notin \operatorname{aff}(\operatorname{rbd} K)$, then $\dim(\operatorname{rbd} K) = m - 1$, as follows from Corollary 2.67.

$(a) \Rightarrow (b)$ If $\dim(\operatorname{rbd} K) = m - 1$, then the above argument shows that the plane $M = \operatorname{aff}(\operatorname{rbd} K)$ is $(m-1)$-dimensional and the point $\boldsymbol{z} \in L \setminus \operatorname{cl} K$ belongs to $L \setminus M$. The set $L \setminus M$ is the union of two open halfplanes of L, say E_1 and E_2 (see Theorem 2.73). Without loss of generality, we may assume that $\boldsymbol{z} \in E_1$. We are going to show that $\operatorname{rint} K \subset E_2$. Indeed, assume for a moment that a point $\boldsymbol{u} \in \operatorname{rint} K$ belongs to E_1. Since E_1 is the intersection of L and a suitable open halfspace (see Definition 2.69), a combination of Theorems 2.38 and 2.35 implies the inclusions $(\boldsymbol{u}, \boldsymbol{z}) \subset (\boldsymbol{u}, \boldsymbol{z}\rangle \subset E_1$. On the other hand, $(\boldsymbol{u}, \boldsymbol{z})$ should contain a point from $\operatorname{rbd} K$ (see Theorem 3.65), contrary to the assumption. Hence $\operatorname{rint} K \subset E_2$. In a similar way, no point of E_2 lies in $L \setminus \operatorname{rint} K$. Hence $\operatorname{rint} K = E_2$, which gives the equality $\operatorname{cl} K = \operatorname{rbd} K \cup E_2 = \operatorname{cl} E_2$.

$(b) \Rightarrow (c)$ If $\operatorname{cl} K$ is a closed halfplane, then $\operatorname{rbd} K$ is a plane of dimension $m - 1$ (see Corollary 2.71).

$(c) \Rightarrow (d)$ This part is obvious because every plane is a convex set.

$(d) \Rightarrow (a)$ If $\operatorname{rbd} K$ is a convex set, then $\dim(\operatorname{rbd} K) \leqslant m - 1$ by Theorem 3.67. The opposite inequality is proved above. □

Problems and Notes for Chapter 3

Problems for Chapter 3

Problem 3.1. Prove that a nonempty convex set $K \subset \mathbb{R}^n$ is closed (respectively, relative open) if and only if for every line $l \subset \operatorname{aff} K$, the set $K \cap l$ is closed (respectively, relative open).

Problem 3.2. Given a nonempty relative open set $X \subset \mathbb{R}^n$, prove the following assertions.

(*a*) X is convex if and only if there is a scalar $0 < \lambda < 1$ such that $(1-\lambda)\boldsymbol{x} + \lambda \boldsymbol{y} \in X$ whenever $\boldsymbol{x}, \boldsymbol{y} \in X$.

(*b*) X is a plane if and only if there is a scalar $1 < \lambda < 2$ such that $(1-\lambda)\boldsymbol{x} + \lambda \boldsymbol{y} \in X$ whenever $\boldsymbol{x}, \boldsymbol{y} \in X$.

Problem 3.3. Given a nonempty closed set $X \subset \mathbb{R}^n$, prove the following assertions.

(*a*) X is convex if and only if for any pair of points $\boldsymbol{x}, \boldsymbol{y} \in X$ there is a scalar $0 < \lambda < 1$, possibly depending on $\boldsymbol{x}$ and $\boldsymbol{y}$, such that $(1-\lambda)\boldsymbol{x} + \lambda\boldsymbol{y} \in X$.

(*b*) X is a plane if and only if for any pair of points $\boldsymbol{x}, \boldsymbol{y} \in X$ there is a scalar $1 < \lambda < 2$, possibly depending on $\boldsymbol{x}$ and $\boldsymbol{y}$, such that $(1-\lambda)\boldsymbol{x} + \lambda\boldsymbol{y} \in X$.

Problem 3.4. Let a convex set K lie in the union of a countable family $\mathcal{F} = \{L_\alpha\}$ of planes in $\mathbb{R}^n$. Prove that one of the planes contains K.

Problem 3.5. (Grzybowski and Urbański [137]) Given a convex set $K \subset \mathbb{R}^n$ and scalars λ, μ, prove that

$$\lambda K + \mu K = (\lambda + \mu)K + \frac{|\lambda| + |\mu| - |\lambda + \mu|}{2}(K - K).$$

Problem 3.6. (Grzybowski and Urbański [137]) Let $K, M \subset \mathbb{R}^n$ be compact convex sets and $\lambda, \mu, \gamma, \delta$ be scalars, with $\lambda K + \mu M = \gamma K + \delta M$. Prove that

$$K = \frac{\delta - \mu}{\lambda - \gamma} M \text{ if } \lambda \neq \gamma, \text{ or } M = \frac{\lambda - \gamma}{\delta - \mu} K \text{ if } \delta \neq \mu.$$

Problem 3.7. Given nonempty convex sets K_1 and K_2 in $\mathbb{R}^n$, prove the convexity of the following sets:

$$A = \{\boldsymbol{x} \in \mathbb{R}^n : \boldsymbol{x} + K_1 \subset K_2\}, \quad B = \{\boldsymbol{x} \in \mathbb{R}^n : (\boldsymbol{x} + K_1) \cap K_2 \neq \varnothing\}.$$

Problem 3.8. Let $K \subset \mathbb{R}^n$ be a convex set and $B_\rho(\boldsymbol{a}) \subset \mathbb{R}^n$ be a closed ball which meets $\operatorname{cl} K$ at more than one point. Prove that $\operatorname{rint} K \cap U_\rho(\boldsymbol{a}) \neq \varnothing$ and deduce from here the following assertions.

(*a*) $\operatorname{cl} K \cap B_\rho(\boldsymbol{a}) = \operatorname{cl}(K \cap B_\rho(\boldsymbol{a}))$.

(*b*) $\operatorname{rint}(K \cap B_\rho(\boldsymbol{a})) = \operatorname{rint} K \cap U_\rho(\boldsymbol{a})$.

(*c*) $\operatorname{rbd} K \cap B_\rho(\boldsymbol{a}) \subset \operatorname{rbd}(K \cap B_\rho(\boldsymbol{a}))$.

Problem 3.9. Given a convex set $K \subset \mathbb{R}^n$ of positive dimension and a point $\boldsymbol{u} \in K$, prove that $\operatorname{rint} K \subset \cup([\boldsymbol{u}, \boldsymbol{v}) : \boldsymbol{v} \in K)$, with equality if and only if $\boldsymbol{u} \in \operatorname{rint} K$.

Problem 3.10. Let X and Y be sets in $\mathbb{R}^n$, both with nonempty relative interior. Prove that $\operatorname{rint} X + \operatorname{rint} Y \subset \operatorname{rint}(X+Y)$, with equality provided the planes $\operatorname{aff} X$ and $\operatorname{aff} Y$ are independent.

Problem 3.11. Let $K \subset \mathbb{R}^n$ be a nonempty convex set of dimension m and $\Delta(\boldsymbol{u}_1, \dots, \boldsymbol{u}_{m+1})$ be an m-simplex in $\operatorname{aff} K$. Prove that a point $\boldsymbol{z} \in \mathbb{R}^n$ belongs to $\operatorname{rint} K$ is and only if there is an m-simplex $\Delta(\boldsymbol{z}_1, \dots, \boldsymbol{z}_{m+1}) \subset K$ directly homothetic to $\Delta(\boldsymbol{u}_1, \dots, \boldsymbol{u}_{m+1})$, with $\boldsymbol{z} = \frac{1}{m+1}(\boldsymbol{z}_1 + \dots + \boldsymbol{z}_{m+1})$.

Problem 3.12. Let $X \subset \mathbb{R}^n$ be a nonempty set of dimension m and $\boldsymbol{a}$ be a point in $\operatorname{aff} X$. Prove that the set X is bounded if and only if there is an m-simplex $\Delta = \Delta(\boldsymbol{z}_1, \dots, \boldsymbol{z}_{m+1}) \subset \operatorname{aff} X$ with the properties $X \subset \Delta$ and $\boldsymbol{a} =$

$\frac{1}{m+1}(\boldsymbol{z}_1 + \cdots + \boldsymbol{z}_{m+1})$. Furthermore, Δ can be chosen directly homothetic to any given m-simplex in aff X.

Problem 3.13. Let $K \subset \mathbb{R}^n$ be a convex set and F be a convex subset of $\operatorname{cl} K$. Prove that the set $M = F \cup \operatorname{rint} K$ is convex and satisfies the conditions $\operatorname{cl} M = \operatorname{cl} K$ and $\operatorname{rint} M = \operatorname{rint} K$.

Problem 3.14. Let $K_1, \ldots, K_r$ be nonempty convex sets in $\mathbb{R}^n$ and $\mu_1, \ldots, \mu_r$ be scalars. Suppose that $r > m$, where $m = \dim(\mu_1 K_1 + \cdots + \mu_r K_r)$. Prove the existence of an index set $I \subset \{1, \ldots, r\}$ satisfying the conditions $\operatorname{card} I \leqslant m$ and

$$\operatorname{rint}(\mu_1 K_1 + \cdots + \mu_r K_r) = \sum_{i \in I} \mu_i \operatorname{rint} K_i + \sum_{i \notin I} \mu_i K_i.$$

Problem 3.15. Let K_1 and K_2 be convex sets in $\mathbb{R}^n$ such that K_1 is closed and $K_1 \cap \operatorname{rint} K_2 \neq \varnothing$. Prove that $\operatorname{cl}(K_1 \cap K_2) = K_1 \cap \operatorname{cl} K_2$.

Problem 3.16. Let K and M be nonempty convex sets in $\mathbb{R}^n$ such that $K \subset M$ and X be a dense subset of M. Prove the following assertions.

(*a*) If $\dim K = \dim M$, then the set $X \cap \operatorname{rint} K$ is dense in K. In particular, the set $X \cap \operatorname{rint} M$ is dense in M.

(*b*) If $\dim K < \dim M$, then the set $(X \cap \operatorname{rint} M) \setminus \operatorname{cl} K$ is dense in M.

Problem 3.17. Let $\mathcal{F} = \{K_\alpha\}$ be a finite family of convex sets whose relative interiors have a point in common, and let $K = \bigcap_\alpha K_\alpha$. Prove that

$$\bigcup_\alpha (\operatorname{rbd} K \cap \operatorname{rbd} K_\alpha) = \bigcup_\alpha (\operatorname{cl} K \cap \operatorname{rbd} K_\alpha) = \operatorname{rbd} K.$$

Problem 3.18. (Tietze [278]) A set $X \subset \mathbb{R}^n$ is called *locally convex* if for any point $\boldsymbol{x} \in X$ there is an open ball $U_\rho(\boldsymbol{x}) \subset \mathbb{R}^n$, $\rho = \rho(\boldsymbol{x})$, such that $U_\rho(\boldsymbol{x}) \cap X$ is convex. Prove that a closed connected set $X \subset \mathbb{R}^n$ is convex if and only if it is locally convex.

Problem 3.19. A *projective transformation* $h : \mathbb{R}^n \setminus H \to \mathbb{R}^m$ is defined by $h(\boldsymbol{x}) = (\boldsymbol{a} + g(\boldsymbol{x}))/(\alpha + \boldsymbol{x}\cdot\boldsymbol{e})$, where $\boldsymbol{a} \in \mathbb{R}^m$, $g : \mathbb{R}^n \to \mathbb{R}^m$ is a linear transformation, α is a scalar, $\boldsymbol{e}$ is a nonzero vector in $\mathbb{R}^n$, and $H = \{\boldsymbol{x} \in \mathbb{R}^n : \alpha + \boldsymbol{x}\cdot\boldsymbol{e} = 0\}$ is a hyperplane. Prove that the image $h(K)$ of a convex set $K \subset \mathbb{R}^n \setminus H$ is a convex set.

Problem 3.20. (Wolfe [294]) Let $\boldsymbol{a}_1, \ldots, \boldsymbol{a}_r$ be points in $\mathbb{R}^n$. Prove that a point $\boldsymbol{a} \in \mathbb{R}^n$ is a convex combination of the form $\boldsymbol{a} = \lambda \boldsymbol{a}_1 + \cdots + \lambda_r \boldsymbol{a}_r$ if and only if

$$\|\boldsymbol{x} - \boldsymbol{a}\| \leqslant \lambda_1 \|\boldsymbol{x} - \boldsymbol{a}_1\| + \cdots + \lambda_r \|\boldsymbol{x} - \boldsymbol{a}_r\| \quad \text{for all} \quad \boldsymbol{x} \in \mathbb{R}^n.$$

Problem 3.21. The solid quadrics in $\mathbb{R}^n$ which contain no line are the *solid ellipsoid*, *solid elliptic paraboloid*, *solid elliptic cone*, and *solid elliptic hyperboloid*, given, respectively, by

$$\begin{aligned} A &= \{(x_1, \dots, x_n) : a_1 x_1^2 + \dots + a_n x_n^2 \leqslant 1\}, \\ B &= \{(x_1, \dots, x_n) : a_1 x_1^2 + \dots + a_{n-1} x_{n-1}^2 \leqslant a_n x_n\}, \\ C &= \{(x_1, \dots, x_n) : a_1 x_1^2 + \dots + a_{n-1} x_{n-1}^2 \leqslant a_n x_n^2,\ x_n > 0\}, \\ D &= \{(x_1, \dots, x_n) : a_1 x_1^2 + \dots + a_{n-1} x_{n-1}^2 + 1 \leqslant a_n x_n^2,\ x_n > 0\}, \end{aligned}$$

where $a_1, \dots, a_n$ are positive scalars. Prove that these solid quadrics are convex sets.

Notes for Chapter 3

Historic sketch of convex geometry. The concept of convexity dates back to ancient Greece. For instance, Archimedes (in his book *On the sphere and cylinder*, see [1, p. 404]) defines a line *concave in the same direction* by the following property: "... if every two points on it are taken either, all straight lines connecting the points fall on the same side of the line, or some fall on one and the same side while others fall on itself, but none on the other side." Based on works of Euler, Cauchy, Schaläfly, and Steiner, convex geometry became an independent branch of mathematics around 1900, with notable contributions of Brunn [58, 59] and Minkowski [214, 215] (see Hancock [145] for an account of the work of Minkowski on the geometry of numbers). Interestingly, Minkowski [214, p. 38] used initially the terms "concave and everywhere convex" for convex and strictly convex sets, respectively.

Convex geometry has a modest historiography. Sketches on convex geometry history can be found in the articles of Fenchel [106] and Gruber [131, 132]. The book of Bonnesen and Fenchel [39] reflects various achievements in convex geometry prior to 1934. A selection of biographies can be found in the book of Gardner [117]. More recent publications are due to Kjeldsen [166–168].

Addition and subtraction of convex sets. Theorem 3.10 can be traced in the lectures of Fenchel [105, p. 35], and Theorem 3.13 is mainly due to Jamison [159]. Theorem 3.60 is due to Rådström [236]

The concept of addition of convex sets X and Y in $\mathbb{R}^n$ was introduced by Minkowski [215, p. 177] (afterwards called the Minkowski addition), while their difference, called the *Minkowski difference* and defined by

$$X \div Y = \{\boldsymbol{z} \in \mathbb{R}^n : \boldsymbol{z} + Y \subset X\} = \cap (X - \boldsymbol{z} : \boldsymbol{z} \in Y),$$

is due to Hadwiger [139] (see also [140, Chapter IV]). These operations are reciprocally quasi-inverse: $X \subset (X + Y) \div Y$ and $(X \div Y) + Y \subset X$.

Set-theoretic operations on convex sets. Borovikov [41] showed that the intersection of a nested sequence of simplices in $\mathbb{R}^n$ also is a simplex (possibly, of smaller dimension). In this regard, Gruber [129] posed the problem to describe the classes of closed convex sets in $\mathbb{R}^n$ which are closed under the intersections of nested sequences of their elements; he also proved in [129, 130] that the following two families have the above property: the family consisting of direct sums of simplices, simplicial cones and subspaces, and the family consisting of direct sums of parallelotopes, simplicial cones and subspaces. See Lawrence [193] for a description of convex bodies $K \subset \mathbb{R}^n$ for which the intersection of every decreasing sequences of affine copies of K is an affine copy of K.

Klee [184] posed the problem to describe, in intrinsic geometric terms, those n-dimensional convex sets in $\mathbb{R}^n$ which are not the intersection of any infinite strictly decreasing sequences of convex sets. This problem is solved in [184] for dimensions two and three. In particular, a convex set K in the plane is *not* the intersection of an infinite decreasing sequence of pairwise distinct convex sets if and only if $K = S \cup P \cup Q$, where S is an open convex m-gonal region (possibly, unbounded), P is the union of m open segments or halflines properly contained in the respective open sides of S, and every point of Q is a vertex of S which is an end point of two closed segments or halflines forming P.

A similar problem on the union of convex sets is much easier: the convex sets in $\mathbb{R}^n$ which are not unions of infinite increasing sequences of pairwise distinct convex sets are exactly the convex polytopes (see Problem 13.5).

Lattices of convex sets. A family of convex sets in $\mathbb{R}^n$ (or in a given convex set $S \subset \mathbb{R}^n$), closed with respect to the operations $K \wedge M = K \cap M$ and $K \vee M = \text{conv}\,(K \cup M)$, may be viewed as a *lattice* (a partially ordered set in which every two elements have a unique supremum and a unique infimum). Various properties and classifications of such lattices are given by Bennett [31], Bergman [33], Huhn [155], Wehrung and Semenova [292].

Topological properties of convex sets. Theorems 3.24 and 3.47 (for the case of convex bodies) are due to Brunn [61]. The notion of relative interior and some assertions from Theorems 3.22, 3.24, 3.48, 3.62, 3.65, and Corollary 3.21, and Theorem 3.43 for the case of affine projections can be found in Steinitz [267, § 26]. Theorem 3.53 is due to Blackwell and Girshick [37, pp. 48–49] (see also Rubin and Wesler [246], Cook and Webster [79]). Theorem 3.68 is a finite-dimensional version of an assertion of Klee [170], which characterizes hyperplanes in linear topological spaces.

Local convexity. The following result of Tamura [277] is related to Problem 3.18: If a connected set $X \subset \mathbb{R}^n$ is locally convex, then its relative interior is convex. A variety of similar assertions is summarized in Bonnesen and Fenchel [39, pp. 6–7], Burago and Zalgaller [64], Mani-Levitska [206], and Valentine [285, Part IV].

Universal convex sets. Mazur (see Mauldin [209, p. 111–112]) posed the problem on the existence of a convex body $K \subset \mathbb{R}^3$ symmetric about the origin of $\mathbb{R}^3$ such that every planar centrally symmetric convex disc is affinely equivalent to the intersection of K with a certain 2-dimensional subspace. This problem was solved in the negative by Bessaga [34] and Grünbaum [133] (see Klee [177] for further references). Grząślewicz [136] showed the existence of a compact convex set $K \subset \mathbb{R}^{n+2}$ such that every compact convex subset of the closed unit ball $\mathbb{B}$ of

$\mathbb{R}^n$ can be obtained (up to a rigid motion) as the intersection of K with a certain n-dimensional plane of $\mathbb{R}^{n+2}$.

Convexity-preserving mappings. A mapping $f : \mathbb{R}^n \to \mathbb{R}^m$ is called *convexity-preserving* if the f-image of every convex set in $\mathbb{R}^n$ is a convex set in $\mathbb{R}^m$. As shown in Theorem 3.15, affine transformations are convexity-preserving. The converse assertion is not true. For instance, the mapping $f : \mathbb{R}^n \to \mathbb{R}^n$, $n \geqslant 2$, defined by $f(x_1, x_2, \ldots, x_n) = (x_1^3, 0, \ldots, 0)$, is convexity-preserving but not affine.

Walsh [288] proved that a one-to-one convexity-preserving mapping of $\mathbb{R}^2$ or $\mathbb{R}^3$ into itself is an affine transformation. This result was extended by Meyer and Kay [213], who showed that a one-to-one and convexity-preserving mapping f of a real vector space V, $\dim V \geqslant 2$, into another real vector space W is an affine transformation. Kuz'minyh [190] observed that the one-to-one condition here cannot be dropped, by constructing a convexity-preserving mapping $\mathbb{R}^2 \to \mathbb{R}^2$ which is onto and discontinuous at every point of $\mathbb{R}^2$.

Positively answering Alexandrov's question, Kuz'minyh [190] proved that a one-to-one mapping of a real vector space L, $\dim L \geqslant 2$, into itself is an affine transformation if it maps every segment onto a convex set. Shaidenko-Künzi [250] (correcting a proposition from [190]) established the following result: If $K \subset \mathbb{R}^n$ is a convex body which has at least n tangent hyperplanes in general position, and if $\mathcal{H}$ is the family of all (positive or negative) homothetic copies of K, then a bijection $f : \mathbb{R}^n \to \mathbb{R}^n$ which maps every convex body $M \in \mathcal{H}$ onto a convex set is an affine transformation.

Ariyawansa, Davidon, and McKennon [3] showed that if D is an open connected subset of $\mathbb{R}^n$, $n \geqslant 2$, and $f : D \to \mathbb{R}^m$ is a continuous one-to-one convexity-preserving mapping, then there is a projective transformation $h : \mathbb{R}^n \to \mathbb{R}^m$ (see Problem 3.19), expressed as $h(\boldsymbol{x}) = g(\boldsymbol{x})/\varphi(\boldsymbol{x})$, where $g : \mathbb{R}^n \to \mathbb{R}^m$ is an affine transformation and φ is a nonzero affine functional on $\mathbb{R}^n$, such that $h|_D = f$.

Starshaped sets. A subset X of a vector space $\mathbb{E}$ is said to be *starshaped* with respect to a point $\boldsymbol{c} \in X$ if the segment $[\boldsymbol{x}, \boldsymbol{c}]$ belongs to X whenever $\boldsymbol{x} \in X$. The *kernel* of X consists of all those points with respect to which X is starshaped. Brunn [62] proved that the kernel of a compact starshaped set $X \subset \mathbb{R}^n$ is a compact convex set. Corollary 3.9 is due to Toranzos [279] (see also Smith [251]). Obviously, a set $X \subset \mathbb{E}$ is convex if and only if X is identical with its kernel.

The kernel of a star-shaped set $X \subset \mathbb{R}^n$ is called *proper* if its closure is different from the closure of X. Post [233] proved that a closed convex set $K \subset \mathbb{R}^2$ is a proper kernel if and only if it is neither a halfplane nor a slab between two parallel lines. Klee [180] established a much stronger result, by proving that a closed convex set K in a separable Banach space is a proper kernel of a starshaped set if and only if K contains no hyperplane. A convex set $K \subset \mathbb{R}^n$ containing a hyperplane is a proper kernel of a starshaped set if and only if the relative boundary of K contains a line which does not meet K (see Breen [47]).

Chapter 4

Convex Hulls

4.1 General Properties of Convex Hulls

4.1.1 *Basic Properties of Convex Hulls*

Definition 4.1. For a given set $X \subset \mathbb{R}^n$, the intersection of all convex sets containing X is called the *convex hull* of X and is denoted $\operatorname{conv} X$.

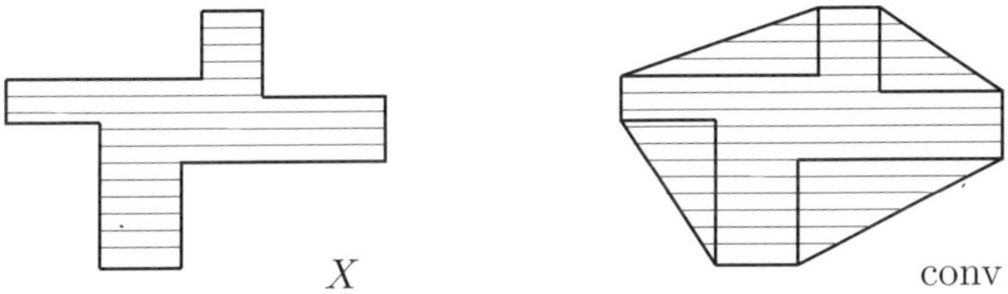

Fig. 4.1 The convex hull of a set X.

Theorem 3.8 implies that the convex hull of any set $X \subset \mathbb{R}^n$ exists and is the smallest convex set containing X. Furthermore, $\operatorname{conv} \varnothing = \varnothing$.

Example. Every closed ball $B_\rho(\boldsymbol{a}) \subset \mathbb{R}^n$ is the convex hull of its boundary sphere $S_\rho(\boldsymbol{a})$ (see Problem 4.1).

The first result of this chapter describes elementary properties of convex hulls.

Theorem 4.2. *For sets X and Y in $\mathbb{R}^n$, the following assertions hold.*

(*a*) $X \subset \operatorname{conv} X$, *with* $X = \operatorname{conv} X$ *if and only if* X *is convex.*
(*b*) $\operatorname{conv}(\operatorname{conv} X) = \operatorname{conv} X$.

(*c*) $\operatorname{conv} X \subset \operatorname{aff} X$ *and* $\operatorname{aff}(\operatorname{conv} X) = \operatorname{aff} X$.
(*d*) $\dim(\operatorname{conv} X) = \dim X$.
(*e*) $\operatorname{conv} X \subset \operatorname{conv} Y$ *if* $X \subset Y$.
(*f*) $\operatorname{conv} X = \operatorname{conv} Y$ *if* $X \subset Y \subset \operatorname{conv} X$.
(*g*) $\operatorname{conv} X = \operatorname{conv}(X \cup Z)$ *if and only if* $Z \subset \operatorname{conv} X$.

Furthermore, if $\{X_\alpha\}$ *is a family of sets in* $\mathbb{R}^n$*, then the assertions below are true.*

(*h*) $\operatorname{conv}(\bigcap_\alpha X_\alpha) \subset \bigcap_\alpha \operatorname{conv} X_\alpha$.
(*i*) $\bigcup_\alpha \operatorname{conv} X_\alpha \subset \operatorname{conv}(\bigcup_\alpha X_\alpha)$.
(*j*) $\bigcup_\alpha \operatorname{conv} X_\alpha = \operatorname{conv}(\bigcup_\alpha X_\alpha)$ *if the family* $\{X_\alpha\}$ *is nested.*
(*k*) $\operatorname{conv}(\bigcup_\alpha \operatorname{conv} X_\alpha) = \operatorname{conv}(\bigcup_\alpha X_\alpha)$.

Proof. Let $\mathcal{K}(X)$ denote the family of all convex sets containing a given set $X \subset \mathbb{R}^n$. The proofs of assertions (*a*)–(*k*) derive from the following simple arguments, combined with Theorems 2.42 and 3.8.

1. $\operatorname{conv} X$ is the smallest element in $\mathcal{K}(X)$.
2. $X \subset \operatorname{conv} X \subset \operatorname{aff} X$.
3. $\mathcal{K}(X)$ is exactly the family of all convex sets containing $\operatorname{conv} X$.
4. If $X \subset Y$, then $\mathcal{K}(Y) \subset \mathcal{K}(X)$. □

The next theorem gives an important description of convex hulls in terms of convex combinations of points (see Definition 3.2).

Theorem 4.3. *The convex hull of a nonempty set* $X \subset \mathbb{R}^n$ *is the collection of all (finite) convex combinations of points from* X*:*

$$\operatorname{conv} X = \{\lambda_1 \boldsymbol{x}_1 + \cdots + \lambda_k \boldsymbol{x}_k : k \geqslant 1,\ \boldsymbol{x}_1, \ldots, \boldsymbol{x}_k \in X,\ \lambda_1, \ldots, \lambda_k \geqslant 0,\ \lambda_1 + \cdots + \lambda_k = 1\}.$$

Equivalently, $\operatorname{conv} X$ *is the collection of all positive convex combinations of points from* X.

Proof. According to Theorem 3.8, $\operatorname{conv} X$ is a convex set containing X, and Theorem 3.3 implies that the set

$$M = \{\lambda_1 \boldsymbol{x}_1 + \cdots + \lambda_k \boldsymbol{x}_k : k \geqslant 1,\ \boldsymbol{x}_1, \ldots, \boldsymbol{x}_k \in X,\ \lambda_1, \ldots, \lambda_k \geqslant 0,\ \lambda_1 + \cdots + \lambda_k = 1\}$$

lies in $\operatorname{conv} X$. We assert that M is a convex set. Indeed, choosing points $\boldsymbol{x}, \boldsymbol{y} \in M$ and expressing them as convex combinations

$$\boldsymbol{x} = \gamma_1 \boldsymbol{x}_1 + \cdots + \gamma_p \boldsymbol{x}_p \quad \text{and} \quad \boldsymbol{y} = \mu_1 \boldsymbol{y}_1 + \cdots + \mu_q \boldsymbol{y}_q$$

of suitable points $\boldsymbol{x}_1, \dots, \boldsymbol{x}_p, \boldsymbol{y}_1, \dots, \boldsymbol{y}_q$ from X, we see that $(1-\lambda)\boldsymbol{x} + \lambda\boldsymbol{y}$ is a convex combination of these points for every choice of $\lambda \in [0, 1]$:

$$\begin{aligned}(1-\lambda)\boldsymbol{x} + \lambda\boldsymbol{y} = (1-\lambda)\gamma_1\boldsymbol{x}_1 + \cdots + (1-\lambda)\gamma_p\boldsymbol{x}_p \\ + \lambda\mu_1\boldsymbol{y}_1 + \cdots + \lambda\mu_q\boldsymbol{y}_q.\end{aligned}$$

Hence $(1-\lambda)\boldsymbol{x} + \lambda\boldsymbol{y} \in M$.

Since $X \subset M$ (every point $\boldsymbol{x} \in X$ can be written as $1\boldsymbol{x}$), the inclusions $X \subset M \subset \operatorname{conv} X$ and Theorem 4.2 give $\operatorname{conv} X = M$.

The second assertion follows from the first one. Indeed, if a point $\boldsymbol{x} \in \operatorname{conv} X$ is written as a convex combination $\boldsymbol{x} = \lambda_1\boldsymbol{x}_1 + \cdots + \lambda_k\boldsymbol{x}_k$ of points $\boldsymbol{x}_1, \dots, \boldsymbol{x}_k \in X$, then, excluding all terms of the form $0\boldsymbol{x}_i$, we obtain a positive convex combination. □

Corollary 4.4. *For a finite set $X = \{\boldsymbol{x}_1, \dots, \boldsymbol{x}_r\} \subset \mathbb{R}^n$, one has*

$$\operatorname{conv} X = \{\lambda_1\boldsymbol{x}_1 + \cdots + \lambda_r\boldsymbol{x}_r : \lambda_1, \dots, \lambda_r \geqslant 0, \lambda_1 + \cdots + \lambda_r = 1\}.$$

Proof. By Theorem 4.3, a point $\boldsymbol{x} \in \mathbb{R}^n$ belongs to $\operatorname{conv} X$ if and only if it can be written as a convex combination $\boldsymbol{x} = \lambda_{i_1}\boldsymbol{x}_{i_1} + \cdots + \lambda_{i_s}\boldsymbol{x}_{i_s}$ of suitable points $\boldsymbol{x}_{i_1}, \dots, \boldsymbol{x}_{i_s}$ from $\{\boldsymbol{x}_1, \dots, \boldsymbol{x}_r\}$. Renumbering the indices $i_1, \dots, i_s$, we may suppose that $1 \leqslant i_1 \leqslant \cdots \leqslant i_s \leqslant r$. For every index $i \in \{1, \dots, r\} \setminus \{i_1, \dots, i_s\}$, if any, we add $0\boldsymbol{x}_i$ to the right-hand side of $\boldsymbol{x} = \lambda_{i_1}\boldsymbol{x}_{i_1} + \cdots + \lambda_{i_s}\boldsymbol{x}_{i_s}$ to express $\boldsymbol{x}$ as a convex combination $\boldsymbol{x} = \lambda_1\boldsymbol{x}_1 + \cdots + \lambda_r\boldsymbol{x}_r$. □

A combination of Definition 3.5 and Corollary 4.4 gives the following corollary.

Corollary 4.5. *Any r-simplex $\Delta(\boldsymbol{x}_1, \dots, \boldsymbol{x}_{r+1}) \subset \mathbb{R}^n$, $0 \leqslant r \leqslant n$, is the convex hull of its vertices:*

$$\Delta(\boldsymbol{x}_1, \dots, \boldsymbol{x}_{r+1}) = \operatorname{conv}\{\boldsymbol{x}_1, \dots, \boldsymbol{x}_{r+1}\}.$$ □

Theorem 4.6. *The convex hull of a nonempty set $X \subset \mathbb{R}^n$ is the collection of all convex combinations (equivalently, of all positive convex combinations) of affinely independent subsets of X.*

Proof. By Theorem 4.3, a point $\boldsymbol{x} \in \mathbb{R}^n$ belongs to $\operatorname{conv} X$ if and only if it can be written as a convex combination $\boldsymbol{x} = \lambda_1\boldsymbol{x}_1 + \cdots + \lambda_p\boldsymbol{x}_p$ of suitable points $\boldsymbol{x}_1, \dots, \boldsymbol{x}_p$ from X. Excluding all terms of the form $0\boldsymbol{x}_i$, we may assume that the scalars $\lambda_1, \dots, \lambda_p$ are positive.

We are going to show that $\boldsymbol{x}$ can be written as a convex combination of $p-1$ of fewer points from the set $\{\boldsymbol{x}_1, \dots, \boldsymbol{x}_p\}$ provided $\{\boldsymbol{x}_1, \dots, \boldsymbol{x}_p\}$

is affinely dependent. So, let the set $\{\boldsymbol{x}_1, \ldots, \boldsymbol{x}_p\}$ be affinely dependent. According to Definition 2.50, there are scalars $\nu_1, \ldots, \nu_p$, not all zero, such that

$$\nu_1 \boldsymbol{x}_1 + \cdots + \nu_p \boldsymbol{x}_p = \boldsymbol{o} \quad \text{and} \quad \nu_1 + \cdots + \nu_p = 0.$$

Clearly, at least one of the scalar $\nu_1, \ldots, \nu_p$ is positive. Put

$$t = \min \{\lambda_i / \nu_i : \nu_i > 0, \ 1 \leqslant i \leqslant p\}.$$

Then $\lambda_i - t\nu_i \geqslant 0$ for all $1 \leqslant i \leqslant p$ (regardless of whether or not ν_i is positive), and $\lambda_i - t\nu_i = 0$ for at least one index $i \in \{1, \ldots, p\}$. Furthermore, $(\lambda_1 - t\nu_1) + \cdots + (\lambda_p - t\nu_p) = 1$. Hence

$$\boldsymbol{x} = (\lambda_1 - t\nu_1)\boldsymbol{x}_1 + \cdots + (\lambda_p - t\nu_p)\boldsymbol{x}_p$$

is a convex combination of $p - 1$ or fewer points from X. Consecutively repeating this argument, we obtain a required expression for $\boldsymbol{x}$. □

Remark. Although Theorem 4.6 looks similar to Theorems 2.51, convex hulls do not allow, in general, the existence of finite *convex bases* (compare with Theorem 2.61). For instance, every point $\boldsymbol{x}$ in the convex hull of the unit circle $C = \{(x, y) : x^2 + y^2 = 1\}$ of $\mathbb{R}^2$ is a convex combination of two or fewer points from C, while no finite subset X of C satisfies the condition $\operatorname{conv} X = \operatorname{conv} C$.

Theorem 4.7. *For a nonempty set $X \subset \mathbb{R}^n$ of dimension m, the following assertions hold.*

(*a*) $\dim(\operatorname{conv} X) = \dim X$.
(*b*) $\operatorname{conv} X$ *is the collection of all positive convex combinations of affinely independent subsets of X each of cardinality $m + 1$ or less.*
(*c*) $\operatorname{conv} X$ *is the collection of all convex combinations of affinely independent subsets of X each of cardinality $m + 1$.*
(*d*) $\operatorname{conv} X$ *is the union of all m-simplices with vertices in X.*

Proof. (*a*) This part follows from Definition 2.63 and the equality $\operatorname{aff}(\operatorname{conv} X) = \operatorname{aff} X$ (see Theorem 4.2).

(*b*) Since $\dim(\operatorname{aff} X) = \dim X = m$, Theorem 2.52 implies that every affinely independent subset of X contains $m + 1$ or fewer points. Thus assertion (*b*) follows from Theorem 4.6.

(*c*) By assertion (*b*), a point $\boldsymbol{x} \in \mathbb{R}^n$ belongs to $\operatorname{conv} X$ if and only $\boldsymbol{x}$ is a convex combination $\boldsymbol{x} = \lambda_1 \boldsymbol{x}_1 + \cdots + \lambda_p \boldsymbol{x}_p$ of an affinely independent set $\{\boldsymbol{x}_1, \ldots, \boldsymbol{x}_p\} \subset X$, where $p \leqslant m + 1$. Theorem 2.61 shows that the set $\{\boldsymbol{x}_1, \dots, \boldsymbol{x}_p\}$ can be extended to an affinely independent subset $\{\boldsymbol{x}_1, \dots, \boldsymbol{x}_{m+1}\}$ of X. Then

$$\boldsymbol{x} = \lambda_1 \boldsymbol{x}_1 + \cdots + \lambda_p \boldsymbol{x}_p + 0\boldsymbol{x}_{k+1} + \cdots + 0\boldsymbol{x}_{m+1}$$

is a desired expression for $\boldsymbol{x}$.

Corollary 4.5 shows that (d) is an equivalent form of (c). □

Theorem 4.7 implies that any point $\boldsymbol{x} \in \operatorname{conv} X$ belongs to the convex hull of a set which consists of $m+1$ or fewer points from X, where $m = \dim X$. A refinement of this assertion is given in the theorem below.

Theorem 4.8. *Given a nonempty set $X \subset \mathbb{R}^n$ and a point $\boldsymbol{x}_0 \in \mathbb{R}^n$, any point $\boldsymbol{x} \in \operatorname{conv} X$ can be written as a convex combination of points from an affinely independent set of the form $\{\boldsymbol{x}_0\} \cup Y$, where $Y \subset X$.*

Proof. By Theorem 4.6, a point $\boldsymbol{x} \in \mathbb{R}^n$ belongs to $\operatorname{conv} X$ if and only if can be written as a positive convex combination $\boldsymbol{x} = \lambda_1 \boldsymbol{x}_1 + \cdots + \lambda_p \boldsymbol{x}_p$ of an affinely independent set $\{\boldsymbol{x}_1, \dots, \boldsymbol{x}_p\} \subset X$. Without loss of generality, we may suppose that the number p is this expression is minimum possible. If $\boldsymbol{x}_0 \in \mathbb{R}^n \setminus \operatorname{aff}\{\boldsymbol{x}_1, \dots, \boldsymbol{x}_p\}$, then $\{\boldsymbol{x}_0, \boldsymbol{x}_1, \dots, \boldsymbol{x}_p\}$ is affinely independent (see Theorem 2.54), and $\boldsymbol{x} = 0\boldsymbol{x}_0 + \lambda_1 \boldsymbol{x}_1 + \cdots + \lambda_p \boldsymbol{x}_p$ is a desired expression. Let $\boldsymbol{x}_0 \in \operatorname{aff}\{\boldsymbol{x}_1, \dots, \boldsymbol{x}_p\}$. By Theorem 2.52, $\boldsymbol{x}_0$ can be uniquely written as an affine combination $\boldsymbol{x}_0 = \mu_1 \boldsymbol{x}_1 + \cdots + \mu_p \boldsymbol{x}_p$. Leaving aside the obvious case $\boldsymbol{x} = \boldsymbol{x}_0$ (then $\boldsymbol{x} = 1\boldsymbol{x}_0$ is the convex combination of $\boldsymbol{x}_0$), we will assume that $\boldsymbol{x} \neq \boldsymbol{x}_0$. Then at least one of the scalars $\lambda_1 - \mu_1, \dots, \lambda_p - \mu_p$ is not zero, and the equality $(\lambda_1 - \mu_1) + \cdots + (\lambda_p - \mu_p) = 0$ shows that at least one of them is negative. Since all scalars $\mu_i + t(\lambda_i - \mu_i)$, $1 \leqslant i \leqslant p$, are positive for $t = 1$, a continuity argument shows the existence of a $t_0 > 1$ such that

$$\mu_i + t_0(\lambda_i - \mu_i) \geqslant 0 \quad \text{for all} \quad 1 \leqslant i \leqslant p$$

and $\mu_i + t_0(\lambda_i - \mu_i) = 0$ for at least one index $i \in \{1, \dots, p\}$. Therefore, the expression

$$\begin{aligned} \boldsymbol{z} &= (1 - t_0)\boldsymbol{x}_0 + t_0 \boldsymbol{x} \\ &= (1 - t_0)(\mu_1 \boldsymbol{x}_1 + \cdots + \mu_p \boldsymbol{x}_p) + t_0(\lambda_1 \boldsymbol{x}_1 + \cdots + \lambda_p \boldsymbol{x}_p) \\ &= (\mu_1 + t_0(\lambda_1 - \mu_1))\boldsymbol{x}_1 + \cdots + (\mu_p + t_0(\lambda_p - \mu_p))\boldsymbol{x}_p \end{aligned}$$

is a convex combination of $p-1$ or fewer points from $\{\boldsymbol{x}_1, \dots, \boldsymbol{x}_p\}$. Let, for instance,

$$\boldsymbol{z} = \eta_1 \boldsymbol{x}_1 + \cdots + \eta_{p-1} \boldsymbol{x}_{p-1}, \ \eta_1, \dots, \eta_{p-1} \geqslant 0, \ \eta_1 + \cdots + \eta_{p-1} = 1.$$

Then

$$x = (1 - t_0^{-1})x_0 + t_0^{-1}z = (1 - t_0^{-1})x_0 + t_0^{-1}\eta_1 x_1 + \cdots + t_0^{-1}\eta_{p-1}x_{p-1}$$

is a convex combination of $x_0, x_1, \ldots, x_{p-1}$ (as depicted in Figure 4.2 for the case $p = 3$).

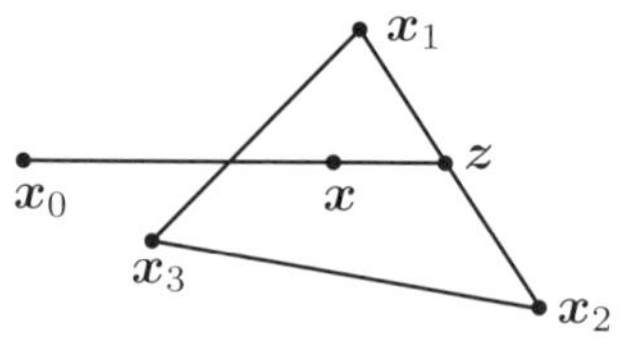

Fig. 4.2 Illustration to Theorem 4.8.

Suppose that the set $\{x_0, x_1, \ldots, x_{p-1}\}$ is affinely dependent. Then, according to Theorem 4.6, x is a convex combination of $p - 1$ or fewer points from $\{x_0, x_1, \ldots, x_{p-1}\}$. By the minimality of p, x is not a convex combination of $x_1, \ldots, x_{p-1}$. Hence x is a convex combination of x_0 and some $p - 2$ points from $\{x_1, \ldots, x_{p-1}\}$. Consecutively repeating this argument, we obtain a desired set $Y \subset \{x_1, \ldots, x_{p-1}\} \subset X$. □

The following assertion refines part (*d*) of Theorem 4.7.

Corollary 4.9. *Let $X \subset \mathbb{R}^n$ be a nonempty set of positive dimension m, and x_0 be a point in* aff X. *If $S(x_0)$ denotes the union of all m-simplices $\Delta(x_0, x_1, \ldots, x_m)$, where $x_1, \ldots, x_m \in X$, then* $\operatorname{conv} X \subset S(x_0)$. *Furthermore,* $\operatorname{conv} X = S(x_0)$ *if and only if* $x_0 \in \operatorname{conv} X$.

Proof. By Theorem 4.8, any point $x \in \operatorname{conv} X$ can be written as a convex combination $x = \lambda_0 x_0 + \lambda_1 x_1 + \cdots + \lambda_p x_p$ of an affinely independent set $\{x_0, x_1, \ldots, x_p\}$, where $x_1, \ldots, x_p \in X$. Theorem 2.61 shows that $\{x_0, x_1, \ldots, x_p\}$ can be extended to an affinely independent subset $\{x_0, x_1, \ldots, x_m\}$ of X. Rewriting x as

$$x = \lambda_0 x_0 + \lambda_1 x_1 + \cdots + \lambda_p x_p + 0x_{p+1} + \cdots + 0x_m,$$

we obtain the inclusion $x \in \Delta(x_0, x_1, \ldots, x_m)$. Thus $\operatorname{conv} X \subset S(x_0)$.

If $\operatorname{conv} X = S(x_0)$, then the inclusion $x_0 \in \operatorname{conv} X$ is obvious. Conversely, let $x_0 \in \operatorname{conv} X$. A combination of Corollary 3.7 and Theorem 4.2 implies that $\operatorname{conv} X$ contains every m-simplex $\Delta(x_0, x_1, \ldots, x_m)$, where $x_1, \ldots, x_m \in X$. Therefore, $S(x_0) \subset \operatorname{conv} X$, as desired. □

4.1.2 *Iterative Construction of Convex Hulls*

For a nonempty set $X \subset \mathbb{R}^n$ and a positive integer r, denote by $\text{conv}_r X$ the collection of all convex combination of r or fewer points from X:

$$\text{conv}_r X = \cup\, (\text{conv}\,\{\boldsymbol{x}_1, \dots, \boldsymbol{x}_k\} : k \leqslant r,\, \boldsymbol{x}_1, \dots, \boldsymbol{x}_k \in X).$$

We let $\text{conv}_0 X = \varnothing$ and $\text{conv}_r \varnothing = \varnothing$ for all $r \geqslant 0$.

Example. If $X = \{\boldsymbol{x}_1, \boldsymbol{x}_2, \boldsymbol{x}_3, \boldsymbol{x}_4\}$ is an affinely independent set in $\mathbb{R}^3$, then $\text{conv}_1 X = X$, $\text{conv}_2 X$ is the union of all edges of the simplex $\Delta = \Delta(\boldsymbol{x}_1, \boldsymbol{x}_2, \boldsymbol{x}_3, \boldsymbol{x}_4)$, $\text{conv}_3 X$ is the union of all faces of Δ, and $\text{conv}_4 X = \Delta$. Furthermore, $\text{conv}_2(\text{conv}_2 X) = \Delta$. Indeed, every point $\boldsymbol{x} \in \Delta$ belongs to a closed segment with endpoints on $[\boldsymbol{x}_1, \boldsymbol{x}_2]$ and $[\boldsymbol{x}_3, \boldsymbol{x}_4]$, respectively, as depicted in Figure 4.3.

A combination of Theorems 4.2 and 4.7 shows that

$$\varnothing = \text{conv}_0 X \subset X = \text{conv}_1 X \subset \cdots \subset \text{conv}_{m+1} X = \text{conv}\, X \qquad (4.1)$$

for any set $X \subset \mathbb{R}^n$ of dimension m, where $-1 \leqslant m \leqslant n$.

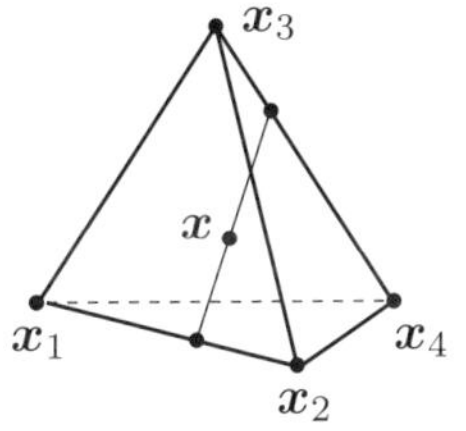

Fig. 4.3 $\Delta(\boldsymbol{x}_1, \boldsymbol{x}_2, \boldsymbol{x}_3, \boldsymbol{x}_4) = \text{conv}_2(\text{conv}_2\{\boldsymbol{x}_1, \boldsymbol{x}_2, \boldsymbol{x}_3, \boldsymbol{x}_4\})$.

Theorem 4.10. *For any set $X \subset \mathbb{R}^n$ and nonnegative integers r and s, one has*

$$\text{conv}_r(\text{conv}_s X) = \text{conv}_{rs} X.$$

Proof. Since the cases $X = \varnothing$ and $rs = 0$ are obvious, we may assume that $X \neq \varnothing$ and $rs > 0$.

Let $\boldsymbol{z} \in \text{conv}_r(\text{conv}_s X)$. Then $\boldsymbol{z}$ is a convex combination,

$$\boldsymbol{z} = \lambda_1 \boldsymbol{y}_1 + \cdots + \lambda_p \boldsymbol{y}_p, \quad \text{where} \quad \boldsymbol{y}_1, \dots, \boldsymbol{y}_p \in \text{conv}_s X,\ p \leqslant r.$$

Similarly, every $\boldsymbol{y}_i$, $1 \leqslant i \leqslant p$, can be written as a convex combination

$$\boldsymbol{y}_i = \mu_1^{(i)} \boldsymbol{x}_1^{(i)} + \cdots + \mu_{m_i}^{(i)} \boldsymbol{x}_{m_i}^{(i)}, \quad \text{where} \quad \boldsymbol{x}_1^{(i)}, \dots, \boldsymbol{x}_{m_i}^{(i)} \in X,\ m_i \leqslant s.$$

So, $\boldsymbol{z}$ is a convex combination of points from a subset $Y \subset X$ of cardinality rs or less:

$$\boldsymbol{z} = \lambda_1 \mu_1^{(1)} \boldsymbol{x}_1^{(1)} + \cdots + \lambda_1 \mu_{m_1}^{(1)} \boldsymbol{x}_{m_1}^{(1)} + \cdots + \lambda_p \mu_1^{(p)} \boldsymbol{x}_1^{(p)} + \cdots + \lambda_p \mu_{m_p}^{(p)} \boldsymbol{x}_{m_p}^{(p)},$$

implying that $\boldsymbol{z} \in \operatorname{conv}_{rs} X$. Hence $\operatorname{conv}_r(\operatorname{conv}_s X) \subset \operatorname{conv}_{rs} X$.

Conversely, let $\boldsymbol{z} \in \operatorname{conv}_{rs} X$. Then $\boldsymbol{z}$ can be expressed as a convex combination $\boldsymbol{z} = \eta_1 \boldsymbol{x}_1 + \cdots + \eta_q \boldsymbol{x}_q$ of suitable points $\boldsymbol{x}_1, \ldots, \boldsymbol{x}_q \in X$, $q \leqslant rs$. We may assume that $q = rs$ (otherwise choose points $\boldsymbol{x}_{q+1}, \ldots, \boldsymbol{x}_{rs}$ in X and append $0\boldsymbol{x}_{q+1} + \cdots + 0\boldsymbol{x}_{rs}$ to $\eta_1 \boldsymbol{x}_1 + \cdots + \eta_q \boldsymbol{x}_q$). Let

$$\lambda_i = \eta_{(i-1)s+1} + \eta_{(i-1)s+2} + \cdots + \eta_{is}, \quad 1 \leqslant i \leqslant r.$$

For every $\lambda_i \neq 0$, $1 \leqslant i \leqslant r$, consider the convex combination

$$\boldsymbol{y}_i = \lambda_i^{-1}(\eta_{(i-1)s+1}\boldsymbol{x}_{(i-1)s+1} + \eta_{(i-1)s+2}\boldsymbol{x}_{(i-1)s+2} + \cdots + \eta_{is}\boldsymbol{x}_{is}).$$

Then $\boldsymbol{y}_i \in \operatorname{conv}_s X$. Also, choose any $\boldsymbol{y}_i \in \operatorname{conv}_s X$ if $\lambda_i = 0$, $1 \leqslant i \leqslant r$.

Since $\boldsymbol{z} = \lambda_1 \boldsymbol{y}_1 + \cdots + \lambda_r \boldsymbol{y}_r$ is a convex combination of $\boldsymbol{y}_1, \ldots, \boldsymbol{y}_r$, one has $\boldsymbol{z} \in \operatorname{conv}_r(\operatorname{conv}_s X)$. Hence $\operatorname{conv}_{rs} X \subset \operatorname{conv}_r(\operatorname{conv}_s X)$. □

A combination of (4.1) and Theorem 4.10 gives the following assertion (here $\lceil a \rceil$ means the ceiling value of a scalar a).

Corollary 4.11. *If $X \subset \mathbb{R}^n$ is a set of dimension m and $r_1, \ldots, r_k$ are nonnegative integers satisfying the inequality $r_1 \cdot \ldots \cdot r_k \geqslant m + 1$, then*

$$\operatorname{conv}_{r_1}(\operatorname{conv}_{r_2}(\cdots(\operatorname{conv}_{r_k} X)\cdots)) = \operatorname{conv} X.$$

In particular, $\operatorname{conv}_2(\operatorname{conv}_2(\cdots(\operatorname{conv}_2 X)\cdots)) = \operatorname{conv} X$, *where the operation* conv_2 *is consequently applied* $\lceil \log_2(m+1) \rceil$ *times.* □

4.1.3 *Algebra of Convex Hulls*

Theorem 4.12. *For sets $X_1, \ldots, X_r \subset \mathbb{R}^n$ and scalars $\mu_1, \ldots, \mu_r$, one has*

$$\operatorname{conv}(\mu_1 X_1 + \cdots + \mu_r X_r) = \mu_1 \operatorname{conv} X_1 + \cdots + \mu_r \operatorname{conv} X_r.$$

Proof. Excluding the obvious case when at least one of the sets $X_1, \ldots, X_r$ is empty, we assume that all these sets are nonempty.

An induction argument shows that the proof can be reduced to the case $r = 2$ (let $\mu_1 X_1 = \mu_1 X_1 + \mu_2\{\boldsymbol{o}\}$ if $r = 1$). Because the set $\mu_1 \operatorname{conv} X_1 + \mu_2 \operatorname{conv} X_2$ is convex (see Theorem 3.10), the obvious inclusion

$$\mu_1 X_1 + \mu_2 X_2 \subset \mu_1 \operatorname{conv} X_1 + \mu_2 \operatorname{conv} X_2,$$

and Theorem 4.2 give

$$\operatorname{conv}(\mu_1 X_1 + \mu_2 X_2) \subset \mu_1 \operatorname{conv} X_1 + \mu_2 \operatorname{conv} X_2.$$

For the opposite inclusion, choose a point $\boldsymbol{x} \in \mu_1 \operatorname{conv} X_1 + \mu_2 \operatorname{conv} X_2$. Then $\boldsymbol{x} = \mu_1 \boldsymbol{x}_1 + \mu_1 \boldsymbol{x}_2$ for suitable points $\boldsymbol{x}_1 \in \operatorname{conv} X_1$ and $\boldsymbol{x}_2 \in \operatorname{conv} X_2$. By Theorem 4.3, $\boldsymbol{x}_1$ and $\boldsymbol{x}_2$ can be written as convex combinations

$$\boldsymbol{x}_1 = \lambda_1 \boldsymbol{u}_1 + \cdots + \lambda_p \boldsymbol{u}_p \quad \text{and} \quad \boldsymbol{x}_2 = \gamma_1 \boldsymbol{v}_1 + \cdots + \gamma_q \boldsymbol{v}_q$$

of points $\boldsymbol{u}_1, \dots, \boldsymbol{u}_p \in X_1$ and $\boldsymbol{v}_1, \dots, \boldsymbol{v}_q \in X_2$. Because

$$\mu_1 \boldsymbol{u}_i + \mu_2 \boldsymbol{v}_j \in \mu_1 X_1 + \mu_2 X_2 \quad \text{for all} \quad 1 \leqslant i \leqslant p \ \text{ and } \ 1 \leqslant j \leqslant q,$$

the equalities

$$\boldsymbol{x} = \mu_1 \boldsymbol{x}_1 + \mu_2 \boldsymbol{x}_2 = \sum_{i=1}^{p} \sum_{j=1}^{q} \lambda_i \gamma_j (\mu_1 \boldsymbol{u}_i + \mu_2 \boldsymbol{v}_j), \qquad \sum_{i=1}^{p} \sum_{j=1}^{q} \lambda_i \gamma_j = 1,$$

show that $\boldsymbol{x}$ is a convex combination of points from $\mu_1 X_1 + \mu_2 X_2$. Theorem 4.3 gives the inclusion $\boldsymbol{x} \in \operatorname{conv}(\mu_1 X_1 + \mu_2 X_2)$. □

A sharper version of Theorem 4.12 for the case of large sums is considered in Problem 4.14. The next theorem describes a decomposition of the convex hull of the union of sets (compare with Theorem 4.10).

Theorem 4.13. *For nonempty sets $X_1, \dots, X_r$ in $\mathbb{R}^n$, one has*

$$\begin{aligned} \operatorname{conv}(X_1 \cup \cdots \cup X_r) = \{ & \lambda_1 \boldsymbol{x}_1 + \cdots + \lambda_r \boldsymbol{x}_r : \boldsymbol{x}_1 \in \operatorname{conv} X_1, \dots, \\ & \boldsymbol{x}_r \in \operatorname{conv} X_r,\ \lambda_1, \dots, \lambda_r \geqslant 0,\ \lambda_1 + \cdots + \lambda_r = 1 \}. \end{aligned}$$

Proof. Theorem 4.2 implies the inclusion

$$\operatorname{conv} X_1 \cup \cdots \cup \operatorname{conv} X_r \subset \operatorname{conv}(X_1 \cup \cdots \cup X_r).$$

This argument and Theorem 4.3 show that the set

$$\begin{aligned} M = \{ & \lambda_1 \boldsymbol{x}_1 + \cdots + \lambda_r \boldsymbol{x}_r : \boldsymbol{x}_1 \in \operatorname{conv} X_1, \dots, \boldsymbol{x}_r \in \operatorname{conv} X_r, \\ & \lambda_1, \dots, \lambda_r \geqslant 0,\ \lambda_1 + \cdots + \lambda_r = 1 \} \end{aligned}$$

lies in $\operatorname{conv}(X_1 \cup \cdots \cup X_r)$. Conversely, let $\boldsymbol{x} \in \operatorname{conv}(X_1 \cup \cdots \cup X_r)$. By the same Theorem 4.3, $\boldsymbol{x}$ is expressible as a convex combination

$$\boldsymbol{x} = \mu_1 \boldsymbol{z}_1 + \cdots + \mu_k \boldsymbol{z}_k, \quad \text{where} \quad \boldsymbol{z}_1, \dots, \boldsymbol{z}_k \in X_1 \cup \cdots \cup X_r.$$

Renumbering the points $\boldsymbol{z}_1, \dots, \boldsymbol{z}_k$, we may assume that $\{\boldsymbol{z}_1, \dots, \boldsymbol{z}_k\}$ is partitioned into subsets $Y_1, \dots, Y_r$ (some of them may be empty) such that

$$Y_i = \{\boldsymbol{z}_{p_{i-1}+1}, \dots, \boldsymbol{z}_{p_i}\} \subset X_i, \quad 1 \leqslant i \leqslant r, \quad \text{where} \quad p_0 = 0,\ p_r = k.$$

Let $\lambda_i = \mu_{p_{i-1}+1} + \cdots + \mu_{p_i}$, $1 \leqslant i \leqslant r$. Clearly,

$$\lambda_1 + \cdots + \lambda_r = 1 \quad \text{and} \quad \lambda_i \geqslant 0 \quad \text{for all} \quad 1 \leqslant i \leqslant r.$$

Denote by I the set of all indices $i \in \{1, \ldots, r\}$ satisfying the condition: either Y_i is empty or $\lambda_i = 0$. Let $J = \{1, \ldots, r\} \setminus I$. For every $i \in I$, choose a point $\boldsymbol{x}_i \in X_i$, and for every $i \in J$, let

$$\boldsymbol{x}_i = \lambda_i^{-1}(\mu_{p_{i-1}+1}\boldsymbol{z}_{p_{i-1}+1} + \cdots + \mu_{p_i}\boldsymbol{z}_{p_i}).$$

Then every $\boldsymbol{x}_i$ is a convex combination of points from X_i, $i \in J$, and Theorem 4.3 shows that $\boldsymbol{x}_i \in \operatorname{conv} X_i$ for all $1 \leqslant i \leqslant r$. Finally, $\boldsymbol{x} = \lambda_1\boldsymbol{x}_1 + \cdots + \lambda_r\boldsymbol{x}_r$ is a desired convex combination. □

Corollary 4.14. *For nonempty sets $X_1, \ldots, X_r$ in $\mathbb{R}^n$, one has:*

$$\begin{aligned}\operatorname{conv}(X_1 \cup \cdots \cup X_r) &= \cup\,(\lambda_1 \operatorname{conv} X_1 + \cdots + \lambda_r \operatorname{conv} X_r : \\ &\qquad \lambda_1, \ldots, \lambda_r \geqslant 0, \lambda_1 + \cdots + \lambda_r = 1), \\ \operatorname{conv}(X_1 \cup \cdots \cup X_r) &= \cup\,(\operatorname{conv}\{\boldsymbol{x}_1, \ldots, \boldsymbol{x}_r\} : \boldsymbol{x}_i \in \operatorname{conv} X_i, \\ &\qquad 1 \leqslant i \leqslant r).\end{aligned}$$

□

Theorem 4.15. *For an affine transformation $f : \mathbb{R}^n \to \mathbb{R}^m$ and sets $X \subset \mathbb{R}^n$ and $Y \subset \mathbb{R}^m$, one has*

$$\operatorname{conv} f(X) = f(\operatorname{conv} X), \tag{4.2}$$

$$\operatorname{conv} f^{-1}(Y) = f^{-1}(\operatorname{conv}(Y \cap \operatorname{rng} f)) \subset f^{-1}(\operatorname{conv} Y). \tag{4.3}$$

Proof. Excluding the obvious cases $X = \varnothing$ and $Y \cap \operatorname{rng} f = \varnothing$, we may assume that both sets X and $Y \cap \operatorname{rng} f$ are nonempty.

For (4.2), choose a point $\boldsymbol{x} \in \operatorname{conv} f(X)$. By Theorem 4.3, $\boldsymbol{x}$ can be written as a convex combination

$$\boldsymbol{x} = \lambda_1\boldsymbol{x}_1 + \cdots + \lambda_p\boldsymbol{x}_p, \quad \text{where} \quad \boldsymbol{x}_1, \ldots, \boldsymbol{x}_p \in f(X).$$

Let $\boldsymbol{z}_1, \ldots, \boldsymbol{z}_p$ be points in X such that $f(\boldsymbol{z}_i) = \boldsymbol{x}_i$ for all $1 \leqslant i \leqslant p$. Put $\boldsymbol{z} = \lambda_1\boldsymbol{z}_1 + \cdots + \lambda_p\boldsymbol{z}_p$. Then $\boldsymbol{z} \in \operatorname{conv} X$ by same theorem, and

$$\boldsymbol{x} = \lambda_1\boldsymbol{x}_1 + \cdots + \lambda_p\boldsymbol{x}_p = \lambda_1 f(\boldsymbol{z}_1) + \cdots + \lambda_p f(\boldsymbol{z}_p) = f(\boldsymbol{z})$$

according to Theorem 2.82. Hence $\boldsymbol{x} = f(\boldsymbol{z}) \in f(\operatorname{conv} X)$, which proves the inclusion $\operatorname{conv} f(X) \subset f(\operatorname{conv} X)$.

Conversely, let $\boldsymbol{x} \in \operatorname{conv} X$. Similarly to the above argument, $\boldsymbol{x}$ can be written as a convex combination

$$\boldsymbol{x} = \mu_1\boldsymbol{x}_1 + \cdots + \mu_q\boldsymbol{x}_q, \quad \text{where} \quad \boldsymbol{x}_1, \ldots, \boldsymbol{x}_q \in X.$$

Then $f(\boldsymbol{x}) = \mu_1 f(\boldsymbol{x}_1) + \cdots + \mu_q f(\boldsymbol{x}_q)$ according to Theorem 2.82, and Theorem 4.3 gives $f(\boldsymbol{x}) \in \operatorname{conv} f(X)$. Hence $f(\operatorname{conv} X) \subset \operatorname{conv} f(X)$.

Next, we are going to prove (4.3). With $X = f^{-1}(Y)$, (4.2) gives

$$f(\operatorname{conv} f^{-1}(Y)) = \operatorname{conv} f(f^{-1}(Y)) = \operatorname{conv}(Y \cap \operatorname{rng} f).$$

Hence

$$\operatorname{conv} f^{-1}(Y) \subset f^{-1}(f(\operatorname{conv} f^{-1}(Y))) = f^{-1}(\operatorname{conv}(Y \cap \operatorname{rng} f)).$$

For the opposite inclusion, choose a point $\boldsymbol{x} \in f^{-1}(\operatorname{conv}(Y \cap \operatorname{rng} f))$. Then $f(\boldsymbol{x})$ belongs to $\operatorname{conv}(Y \cap \operatorname{rng} f)$, and Theorem 4.3 implies that $f(\boldsymbol{x})$ can be written as a convex combination

$$f(\boldsymbol{x}) = \gamma_1 \boldsymbol{x}_1 + \cdots + \gamma_r \boldsymbol{x}_r, \quad \text{where} \quad \boldsymbol{x}_1, \ldots, \boldsymbol{x}_r \in Y \cap \operatorname{rng} f.$$

Choose points $\boldsymbol{z}_1, \ldots, \boldsymbol{z}_r \in f^{-1}(Y)$ such that $f(\boldsymbol{z}_i) = \boldsymbol{x}_i$, $1 \leqslant i \leqslant r$, and put $\boldsymbol{z} = \gamma_1 \boldsymbol{z}_1 + \cdots + \gamma_r \boldsymbol{z}_r$. Then $\boldsymbol{z} \in \operatorname{conv} f^{-1}(Y)$ and

$$f(\boldsymbol{z}) = \gamma_1 f(\boldsymbol{z}_1) + \cdots + \gamma_r f(\boldsymbol{z}_r) = \gamma_1 \boldsymbol{x}_1 + \cdots + \gamma_r \boldsymbol{x}_r = f(\boldsymbol{x}).$$

Let $\boldsymbol{u}_i = \boldsymbol{z}_i + (\boldsymbol{x} - \boldsymbol{z})$, $1 \leqslant i \leqslant r$. Since $\boldsymbol{u}_i$ is an affine combination, we have

$$f(\boldsymbol{u}_i) = f(\boldsymbol{z}_i) + f(\boldsymbol{x}) - f(\boldsymbol{z}) = f(\boldsymbol{z}_i) = \boldsymbol{x}_i.$$

Hence $\boldsymbol{u}_i \in f^{-1}(\boldsymbol{x}_i) \subset f^{-1}(Y)$, $1 \leqslant i \leqslant r$. The equalities

$$\boldsymbol{x} = \boldsymbol{u}_i - \boldsymbol{z}_i + \boldsymbol{z}, \quad 1 \leqslant i \leqslant r,$$

show that $\boldsymbol{x}$ can be written as a convex combination of $\boldsymbol{u}_1, \ldots, \boldsymbol{u}_r$:

$$\begin{aligned}\boldsymbol{x} &= (\gamma_1 + \cdots + \gamma_r)\boldsymbol{x} = \gamma_1(\boldsymbol{u}_1 - \boldsymbol{z}_1 + \boldsymbol{z}) + \cdots + \gamma_r(\boldsymbol{u}_r - \boldsymbol{z}_r + \boldsymbol{z}) \\ &= (\gamma_1 \boldsymbol{u}_1 + \cdots + \gamma_r \boldsymbol{u}_r) - (\gamma_1 \boldsymbol{z}_1 + \cdots + \gamma_r \boldsymbol{z}_r) + (\gamma_1 + \cdots + \gamma_r)\boldsymbol{z} \\ &= \gamma_1 \boldsymbol{u}_1 + \cdots + \gamma_r \boldsymbol{u}_r.\end{aligned}$$

Therefore, $\boldsymbol{x} \in \operatorname{conv} f^{-1}(Y)$. Summing up,

$$f^{-1}(\operatorname{conv}(Y \cap \operatorname{rng} f)) \subset \operatorname{conv} f^{-1}(Y).$$

Finally Theorem 4.2 gives the inclusion $\operatorname{conv}(Y \cap \operatorname{rng} f) \subset \operatorname{conv} Y \cap \operatorname{rng} f$, which results in

$$f^{-1}(\operatorname{conv}(Y \cap \operatorname{rng} f)) \subset f^{-1}(\operatorname{conv} Y \cap \operatorname{rng} f) = f^{-1}(\operatorname{conv} Y). \qquad \square$$

Remark. The inclusion $f^{-1}(\operatorname{conv}(Y \cap \operatorname{rng} f)) \subset f^{-1}(\operatorname{conv} Y)$ in (4.3) may be proper. Indeed, let $f : \mathbb{R}^2 \to \mathbb{R}^2$ be the orthogonal projection of $\mathbb{R}^2$ on its x-axis and $Y = \{(0,1), (0,-1)\}$. Then $f^{-1}(\operatorname{conv}(Y \cap \operatorname{rng} f)) = \varnothing$, while $f^{-1}(\operatorname{conv} Y)$ is the y-axis.

4.2 Topological Properties of Convex Hulls

4.2.1 *Closure and Convex Hulls*

Theorem 4.16. *For a set* $X \subset \mathbb{R}^n$*, one has*

$$\operatorname{conv} X \subset \operatorname{conv}(\operatorname{cl} X) \subset \operatorname{cl}(\operatorname{conv} X) = \operatorname{cl}(\operatorname{conv}(\operatorname{cl} X)). \tag{4.4}$$

Furthermore, if X *bounded, then* $\operatorname{conv} X$ *also is bounded and*

$$\operatorname{conv}(\operatorname{cl} X) = \operatorname{cl}(\operatorname{conv} X).$$

Consequently, $\operatorname{conv} X$ *is compact provided* X *is compact.*

Proof. Excluding the obvious case $X = \varnothing$, we may assume that X is nonempty. Because $X \subset \operatorname{conv} X$, one has $\operatorname{cl} X \subset \operatorname{cl}(\operatorname{conv} X)$, as follows from Theorem 4.2. Since the set $\operatorname{cl}(\operatorname{conv} X)$ is convex (see Theorem 3.46), we obtain that $\operatorname{conv}(\operatorname{cl} X) \subset \operatorname{cl}(\operatorname{conv} X)$ according to Theorem 4.2. This inclusion gives $\operatorname{cl}(\operatorname{conv}(\operatorname{cl} X)) \subset \operatorname{cl}(\operatorname{conv} X)$.

Similarly, $\operatorname{conv} X \subset \operatorname{conv}(\operatorname{cl} X)$ due to the inclusion $X \subset \operatorname{cl} X$. Therefore, $\operatorname{cl}(\operatorname{conv} X) \subset \operatorname{cl}(\operatorname{conv}(\operatorname{cl} X))$. Summing up, all parts of (4.4) hold.

Let X be bounded and $B_\rho(\boldsymbol{a}) \subset \mathbb{R}^n$ be a closed ball containing X. Since $B_\rho(\boldsymbol{a})$ is a convex set (see an example on page 75), Theorem 4.2 implies the inclusion $\operatorname{conv} X \subset B_\rho(\boldsymbol{a})$. Hence $\operatorname{conv} X$ is a bounded set.

For the equality $\operatorname{conv}(\operatorname{cl} X) = \operatorname{cl}(\operatorname{conv} X)$, it remains to show, by the above proved, that $\operatorname{cl}(\operatorname{conv} X) \subset \operatorname{conv}(\operatorname{cl} X)$. Let $\boldsymbol{x} \in \operatorname{cl}(\operatorname{conv} X)$. Then $\boldsymbol{x} = \lim_{i\to\infty} \boldsymbol{x}_i$ for a sequence of points $\boldsymbol{x}_1, \boldsymbol{x}_2, \ldots$ from $\operatorname{conv} X$. Put $m = \dim X$. According to Theorem 4.7, every $\boldsymbol{x}_i$ can be written as a convex combination

$$\boldsymbol{x}_i = \lambda_1^{(i)} \boldsymbol{y}_1^{(i)} + \cdots + \lambda_{m+1}^{(i)} \boldsymbol{y}_{m+1}^{(i)}, \quad \boldsymbol{y}_1^{(i)}, \ldots, \boldsymbol{y}_{m+1}^{(i)} \in X, \quad i \geqslant 1.$$

Consider the $m+1$ sequences of points $\boldsymbol{y}_j^{(1)}, \boldsymbol{y}_j^{(2)}, \ldots$ from X and $m+1$ sequences of scalars $\lambda_j^{(1)}, \lambda_j^{(2)}, \ldots$ from $[0, 1]$, $1 \leqslant j \leqslant m+1$. Cantor's diagonal argument and boundedness of X imply the existence of an infinite sequence of indices $i_1, i_2, \ldots$ such that all $2m+2$ subsequences

$$\boldsymbol{y}_j^{(i_1)}, \boldsymbol{y}_j^{(i_2)}, \ldots, \quad \lambda_j^{(i_1)}, \lambda_j^{(i_2)}, \ldots, \quad 1 \leqslant j \leqslant m+1,$$

converge (see page 11). Put

$$\boldsymbol{y}_j = \lim_{r\to\infty} \boldsymbol{y}_j^{(i_r)} \quad \text{and} \quad \lambda_j = \lim_{r\to\infty} \lambda_j^{(i_r)}, \quad 1 \leqslant j \leqslant m+1.$$

Then

$$\boldsymbol{y}_1, \ldots, \boldsymbol{y}_{m+1} \in \operatorname{cl} X, \quad \lambda_1, \ldots, \lambda_{m+1} \geqslant 0, \quad \lambda_1 + \cdots + \lambda_{m+1} = 1.$$

Furthermore,

$$\begin{aligned} \boldsymbol{x} &= \lim_{r\to\infty} \boldsymbol{x}_{i_r} = \lim_{r\to\infty} (\lambda_1^{(i_r)} \boldsymbol{y}_1^{(i_r)} + \cdots + \lambda_{m+1}^{(i_r)} \boldsymbol{y}_{m+1}^{(i_r)}) \\ &= \lambda_1 \boldsymbol{y}_1 + \cdots + \lambda_{m+1} \boldsymbol{y}_{m+1}, \end{aligned}$$

which shows that $\boldsymbol{x}$ is a convex combination of $\boldsymbol{y}_1, \ldots, \boldsymbol{y}_{m+1}$. By Theorem 4.3, $\boldsymbol{x} \in \operatorname{conv}(\operatorname{cl} X)$, and the inclusion $\operatorname{cl}(\operatorname{conv} X) \subset \operatorname{conv}(\operatorname{cl} X)$ is proved. Finally, if X is compact, then the equality

$$\operatorname{conv} X = \operatorname{conv}(\operatorname{cl} X) = \operatorname{cl}(\operatorname{conv} X)$$

implies the closedness of $\operatorname{conv} X$. Because $\operatorname{conv} X$ is bounded, it also is compact. □

Remark. If a closed set $X \subset \mathbb{R}^n$ is unbounded, then $\operatorname{conv} X$ may be nonclosed. For instance, the set $X = \{(x, y) : y = 1/|x|\} \subset \mathbb{R}^2$ is closed, while its convex hull is the open halfplane $\{(x, y) : y > 0\}$.

4.2.2 *Relative Interior and Convex Hulls*

Theorem 4.17. *For a set X in $\mathbb{R}^n$, one has*

$$\operatorname{conv}(\operatorname{rint} X) \subset \operatorname{rint}(\operatorname{conv} X).$$

Furthermore, the set $\operatorname{conv} X$ is relative open provided X is relative open.

Proof. By Theorem 4.2, $\operatorname{aff} X = \operatorname{aff}(\operatorname{conv} X)$. The inclusion $X \subset \operatorname{conv} X$ and Theorem 3.18 imply that $\operatorname{rint} X \subset \operatorname{rint}(\operatorname{conv} X)$. Since $\operatorname{rint}(\operatorname{conv} X)$ is convex (see Corollary 3.25), one has $\operatorname{conv}(\operatorname{rint} X) \subset \operatorname{rint}(\operatorname{conv} X)$.

If X is relative open (that is, $X = \operatorname{rint} X$), then, by the above proved,

$$\operatorname{conv} X = \operatorname{conv}(\operatorname{rint} X) \subset \operatorname{rint}(\operatorname{conv} X) \subset \operatorname{conv} X.$$

Hence $\operatorname{rint}(\operatorname{conv} X) = \operatorname{conv} X$, and $\operatorname{conv} X$ is relative open. □

See Problem 4.12 for a related to Theorem 4.17 assertion.

Theorem 4.18. *Let $X \subset \mathbb{R}^n$ be a nonempty set. A point $\boldsymbol{x} \in \mathbb{R}^n$ belongs to $\operatorname{rint}(\operatorname{conv} X)$ if and only if $\boldsymbol{x}$ is a positive convex combination of finitely many points $\boldsymbol{x}_1, \ldots, \boldsymbol{x}_p \in X$ such that $\operatorname{aff}\{\boldsymbol{x}_1, \ldots, \boldsymbol{x}_p\} = \operatorname{aff} X$.*

Proof. Put $m = \dim X$. According to Corollary 2.66, X contains an affinely independent subset $\{\boldsymbol{y}_1, \ldots, \boldsymbol{y}_{m+1}\}$ with the property $\operatorname{aff} X = \operatorname{aff}\{\boldsymbol{y}_1, \ldots, \boldsymbol{y}_{m+1}\}$. Let

$$\Delta = \Delta(\boldsymbol{y}_1, \ldots, \boldsymbol{y}_{m+1}) \quad \text{and} \quad \boldsymbol{y} = \frac{1}{m+1}(\boldsymbol{y}_1 + \cdots + \boldsymbol{y}_{m+1}).$$

A combination of Theorems 3.18 and 3.20 gives $\boldsymbol{y} \in \operatorname{rint} \Delta \subset \operatorname{rint}(\operatorname{conv} X)$.

Assume first that $\boldsymbol{x} \in \operatorname{rint}(\operatorname{conv} X)$. Since the case $\boldsymbol{x} = \boldsymbol{y}$ is obvious, we assume that $\boldsymbol{x}$ and $\boldsymbol{y}$ are distinct. Theorem 3.26 shows the existence of a scalar $\gamma > 1$ such that the point $\boldsymbol{z} = \gamma\boldsymbol{x} + (1-\gamma)\boldsymbol{y}$ belongs to $\operatorname{conv} X$. By Theorem 4.3, $\boldsymbol{z}$ can be written as a positive convex combination $\boldsymbol{z} = \lambda_1\boldsymbol{z}_1 + \cdots + \lambda_r\boldsymbol{z}_r$ of suitable points $\boldsymbol{z}_1, \ldots, \boldsymbol{z}_r \in X$. Then

$$\begin{aligned}\boldsymbol{x} &= (1-\gamma^{-1})\boldsymbol{y} + \gamma^{-1}\boldsymbol{z} \\ &= \frac{1-\gamma^{-1}}{m+1}\boldsymbol{y}_1 + \cdots + \frac{1-\gamma^{-1}}{m+1}\boldsymbol{y}_{m+1} + \gamma^{-1}\lambda_1\boldsymbol{z}_1 + \cdots + \gamma^{-1}\lambda_r\boldsymbol{z}_r\end{aligned} \tag{4.5}$$

is a positive convex combination of $\boldsymbol{y}_1, \ldots, \boldsymbol{y}_{m+1}, \boldsymbol{z}_1, \ldots, \boldsymbol{z}_r$. Theorem 2.42 and the inclusions

$$\begin{aligned}\{\boldsymbol{y}_1, \ldots, \boldsymbol{y}_{m+1}\} &\subset \{\boldsymbol{y}_1, \ldots, \boldsymbol{y}_{m+1}, \boldsymbol{z}_1, \ldots, \boldsymbol{z}_r\} \\ &\subset \operatorname{aff} X = \operatorname{aff}\{\boldsymbol{y}_1, \ldots, \boldsymbol{y}_{m+1}\}\end{aligned}$$

show that $\operatorname{aff}\{\boldsymbol{y}_1, \ldots, \boldsymbol{y}_{m+1}, \boldsymbol{z}_1, \ldots, \boldsymbol{z}_r\} = \operatorname{aff} X$. Hence (4.5) gives a desired expression for $\boldsymbol{x}$.

Conversely, let $\boldsymbol{x} = \lambda_1\boldsymbol{x}_1 + \cdots + \lambda_p\boldsymbol{x}_p$ be a positive convex combination of suitable points $\boldsymbol{x}_1, \ldots, \boldsymbol{x}_p \in X$ satisfying the condition $\operatorname{aff}\{\boldsymbol{x}_1, \ldots, \boldsymbol{x}_p\} = \operatorname{aff} X$. According to Theorem 2.61, the set $\{\boldsymbol{x}_1, \ldots, \boldsymbol{x}_p\}$ contains an affinely independent subset Y of cardinality $m+1$ whose affine span equals $\operatorname{aff} X$. Without loss of generality, we may assume that $Y = \{\boldsymbol{x}_1, \ldots, \boldsymbol{x}_{m+1}\}$. If $p = m+1$, then $\operatorname{conv}\{\boldsymbol{x}_1, \ldots, \boldsymbol{x}_{m+1}\}$ is the m-simplex $\Delta = \Delta(\boldsymbol{x}_1, \ldots, \boldsymbol{x}_{m+1})$, and a combination of Theorems 3.18 and 3.20 gives

$$\boldsymbol{x} \in \operatorname{rint} \Delta \subset \operatorname{rint}(\operatorname{conv} X).$$

Suppose that $p \geqslant m+2$, and put $\lambda = \lambda_1 + \cdots + \lambda_{m+1}$. Clearly, $0 < \lambda < 1$, and

$$\boldsymbol{y} = \frac{\lambda_1}{\lambda}\boldsymbol{x}_1 + \cdots + \frac{\lambda_{m+1}}{\lambda}\boldsymbol{x}_{m+1}, \quad \boldsymbol{z} = \frac{\lambda_{m+2}}{1-\lambda}\boldsymbol{x}_{m+2} + \cdots + \frac{\lambda_p}{1-\lambda}\boldsymbol{x}_p$$

are convex combinations of points from X. Hence $\boldsymbol{y}, \boldsymbol{z} \in \operatorname{conv} X$ according to Theorem 4.3. As above, $\boldsymbol{y} \in \operatorname{rint} \Delta \subset \operatorname{rint}(\operatorname{conv} X)$. Finally, the obvious equality $\boldsymbol{x} = \lambda\boldsymbol{y} + (1-\lambda)\boldsymbol{z}$ and Theorem 3.24 imply the inclusion $\boldsymbol{x} \in \operatorname{rint}(\operatorname{conv} X)$. □

Remark. Unlike Theorem 4.6, the set $\{\boldsymbol{x}_1, \ldots, \boldsymbol{x}_p\}$ from Theorem 4.18 cannot, in general, be chosen affinely independent. Indeed, denote by X the set $\{\boldsymbol{x}_1, \boldsymbol{x}_2, \boldsymbol{x}_3, \boldsymbol{x}_4\}$ of points in $\mathbb{R}^2$ depicted in Figure 4.4. Clearly, $\boldsymbol{o} \in \operatorname{int}(\operatorname{conv} X)$ and X is affinely dependent. At the same time, no proper subset Y of X satisfies the condition $\boldsymbol{o} \in \operatorname{int}(\operatorname{conv} Y)$.

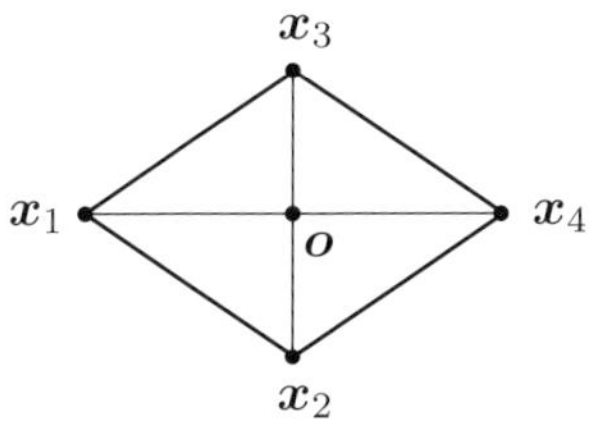

Fig. 4.4 Illustration to Theorem 4.18.

The next corollary can be viewed as a generalization of Theorem 3.20.

Corollary 4.19. *For a finite set* $X = \{\boldsymbol{x}_1, \ldots, \boldsymbol{x}_r\}$ *in* $\mathbb{R}^n$, *one has*

$$\text{rint}\,(\text{conv}\, X) = \{\lambda_1 \boldsymbol{x}_1 + \cdots + \lambda_r \boldsymbol{x}_r : \lambda_1, \ldots, \lambda_r > 0,\ \lambda_1 + \cdots + \lambda_r = 1\}.$$

Proof. Theorem 4.18 shows that every positive convex combination of the points $\boldsymbol{x}_1, \ldots, \boldsymbol{x}_r$ belongs to rint (conv X). Conversely, choose a point $\boldsymbol{x}$ in rint (conv X). Since the point $\boldsymbol{y} = \frac{1}{r}(\boldsymbol{x}_1 + \cdots + \boldsymbol{x}_r)$ is a convex combination of $\boldsymbol{x}_1, \ldots, \boldsymbol{x}_r$ and whence belongs to conv X, there is a scalar $\gamma > 1$ such that $\boldsymbol{z} = \gamma \boldsymbol{x} + (1 - \gamma)\boldsymbol{y} \in \text{conv}\, X$ (see Theorem 3.26). By Corollary 4.4, $\boldsymbol{z}$ can be written as a convex combination $\boldsymbol{z} = \mu_1 \boldsymbol{x}_1 + \cdots + \mu_r \boldsymbol{x}_r$. Consequently,

$$\begin{aligned} \boldsymbol{x} &= (1 - \gamma^{-1})\boldsymbol{y} + \gamma^{-1}\boldsymbol{z} \\ &= \Big(\frac{1 - \gamma^{-1}}{r} + \gamma^{-1}\mu_1\Big)\boldsymbol{x}_1 + \cdots + \Big(\frac{1 - \gamma^{-1}}{r} + \gamma^{-1}\mu_r\Big)\boldsymbol{x}_r \end{aligned}$$

is a positive convex combination of $\boldsymbol{x}_1, \ldots, \boldsymbol{x}_r$. □

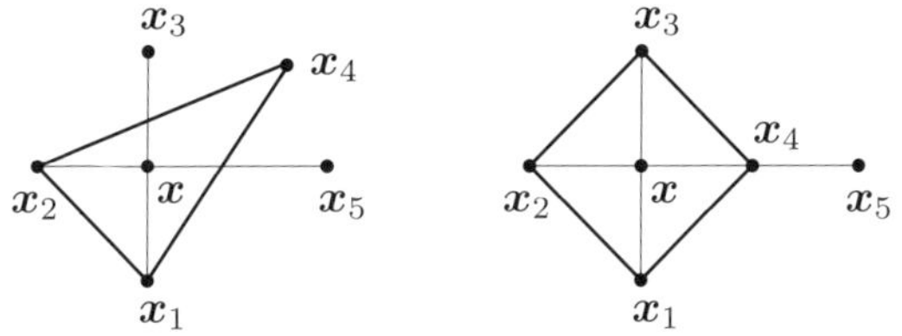

Fig. 4.5 Illustration to Theorem 4.20.

The next result sharpens Theorem 4.18 by establishing bounds on the cardinality of Y. Figure 4.5 illustrates two possible cases described in Theorem 4.20 for the particular case $m = n = 2$: here $X = \{\boldsymbol{x}_1, \boldsymbol{x}_2, \boldsymbol{x}_3, \boldsymbol{x}_4, \boldsymbol{x}_5\}$ and $\boldsymbol{x} \in \text{int}\,(\text{conv}\, X)$. Furthermore, $Y = \{\boldsymbol{x}_1, \boldsymbol{x}_2, \boldsymbol{x}_4\}$ corresponds to the first case, and $Y = \{\boldsymbol{x}_1, \boldsymbol{x}_2, \boldsymbol{x}_3, \boldsymbol{x}_4\}$ corresponds to the second case described in the theorem.

Theorem 4.20. *Let $X \subset \mathbb{R}^n$ be a set of positive dimension m. A point $\boldsymbol{x} \in \mathbb{R}^n$ belongs to* $\operatorname{rint}(\operatorname{conv} X)$ *if and only if there is a subset Y of X such that*

$$\boldsymbol{x} \in \operatorname{rint}(\operatorname{conv} Y), \quad m+1 \leqslant \operatorname{card} Y \leqslant 2m, \quad \operatorname{aff} Y = \operatorname{aff} X. \tag{4.6}$$

Furthermore, if Y is a minimal subset of X satisfying (4.6), *then either* $\operatorname{card} Y \leqslant 2m-1$ *or Y consists of $2m$ points collinear in pairs with $\boldsymbol{x}$.*

Proof. If a subset Y of X satisfies conditions (4.6), then

$$\operatorname{aff}(\operatorname{conv} Y) = \operatorname{aff} Y = \operatorname{aff} X = \operatorname{aff}(\operatorname{conv} X)$$

according to Theorem 4.2. Since $\operatorname{conv} Y \subset \operatorname{conv} X$, Theorem 3.18 implies the inclusion $\boldsymbol{x} \in \operatorname{rint}(\operatorname{conv} Y) \subset \operatorname{rint}(\operatorname{conv} X)$.

Conversely, let $\boldsymbol{x} \in \operatorname{rint}(\operatorname{conv} X)$. A combination of Theorem 4.18 and Corollary 4.19 shows the existence of a finite set $Z \subset X$ with the properties $\boldsymbol{x} \in \operatorname{rint}(\operatorname{conv} Z)$ and $\operatorname{aff} Z = \operatorname{aff} X$. We are going to choose in Z a subset Y satisfying conditions (4.6). Because the cases $m \leqslant 1$ are obvious, we may suppose that $m \geqslant 2$.

Denote by $\mathcal{F}$ the family of planes in $\mathbb{R}^n$, each of dimension $m-1$ or less, such that every plane $L \in \mathcal{F}$ contains $\boldsymbol{x}$ and is the affine span of points from $\{\boldsymbol{x}\} \cup Z$. Because the set Z is finite, the family $\mathcal{F}$ also is finite. Choose in $\operatorname{aff} X$ a line l through $\boldsymbol{x}$ which does not lie in any plane $L \in \mathcal{F}$ (see Problem 2.2). According to Theorem 4.16, $\operatorname{conv} Z$ is a compact set, and Theorem 3.65 shows that $l \cap \operatorname{conv} Z$ is a closed segment $[\boldsymbol{y}, \boldsymbol{y}']$, where $\boldsymbol{y}, \boldsymbol{y}' \in \operatorname{rbd}(\operatorname{conv} Z)$ and $\boldsymbol{x} \in (\boldsymbol{y}, \boldsymbol{y}')$. So, $\boldsymbol{x} = (1-\eta)\boldsymbol{y} + \eta\boldsymbol{y}'$ for a scalar $\eta \in (0,1)$.

By Theorem 4.7, $\boldsymbol{y}$ is a positive convex combination, $\boldsymbol{y} = \lambda_1\boldsymbol{x}_1 + \cdots + \lambda_k\boldsymbol{x}_k$, of an affinely independent set $\{\boldsymbol{x}_1, \ldots, \boldsymbol{x}_k\} \subset Z$, where $k \leqslant m+1$. Then $\boldsymbol{y} \in \operatorname{rint}(\operatorname{conv}\{\boldsymbol{x}_1, \ldots, \boldsymbol{x}_k\})$ due to Corollary 4.19.

We assert that $k = m$ and $\boldsymbol{x} \notin \operatorname{aff}\{\boldsymbol{x}_1, \ldots, \boldsymbol{x}_m\}$. Indeed, assume for a moment that $k = m+1$. Then $\dim(\operatorname{aff}\{\boldsymbol{x}_1, \ldots, \boldsymbol{x}_{m+1}\}) = m$ according to Theorem 2.52. Since

$$\operatorname{aff}\{\boldsymbol{x}_1, \ldots, \boldsymbol{x}_{m+1}\} \subset \operatorname{aff} Z \quad \text{and} \quad \dim(\operatorname{aff} Z) = m,$$

Theorem 2.6 implies that $\operatorname{aff}\{\boldsymbol{x}_1, \ldots, \boldsymbol{x}_{m+1}\} = \operatorname{aff} Z$. Consequently, Theorem 3.18 gives the inclusion

$$\boldsymbol{y} \in \operatorname{rint}(\operatorname{conv}\{\boldsymbol{x}_1, \ldots, \boldsymbol{x}_{m+1}\}) \subset \operatorname{rint}(\operatorname{conv} Z),$$

in contradiction with $\boldsymbol{y} \in \operatorname{rbd}(\operatorname{conv} Z)$. Hence $k \leqslant m$.

If $k \leqslant m-1$ or $\boldsymbol{x}$ belonged to $\operatorname{aff}\{\boldsymbol{x}_1, \ldots, \boldsymbol{x}_k\}$, then the plane $L = \operatorname{aff}\{\boldsymbol{x}, \boldsymbol{x}_1, \ldots, \boldsymbol{x}_k\}$ would be a member of $\mathcal{F}$, and $l = \langle \boldsymbol{x}, \boldsymbol{y} \rangle \subset L$, contrary to the choice of l. Summing up, $\boldsymbol{y} = \lambda_1 \boldsymbol{x}_1 + \cdots + \lambda_m \boldsymbol{x}_m$, where the set $\{\boldsymbol{x}, \boldsymbol{x}_1, \ldots, \boldsymbol{x}_m\}$ is affinely independent (see Theorem 2.54).

Similarly, $\boldsymbol{y}'$ is a positive convex combination, $\boldsymbol{y}' = \mu_1 \boldsymbol{x}'_1 + \cdots + \mu_m \boldsymbol{x}'_m$, of points $\boldsymbol{x}'_1, \ldots, \boldsymbol{x}'_m \in Z$ such that the set $\{\boldsymbol{x}, \boldsymbol{x}'_1, \ldots, \boldsymbol{x}'_m\}$ is affinely independent. Put $Y = \{\boldsymbol{x}_1, \ldots, \boldsymbol{x}_m, \boldsymbol{x}'_1, \ldots, \boldsymbol{x}'_m\}$. Clearly, $\operatorname{card} Y \leqslant 2m$. We assert that $\operatorname{aff} Y = \operatorname{aff} X$. Indeed, $\operatorname{aff} Y \subset \operatorname{aff} Z = \operatorname{aff} X$ because $Y \subset Z$. Assuming that $\operatorname{aff} Y \neq \operatorname{aff} X$, we would obtain

$$\boldsymbol{x} \in \langle \boldsymbol{y}_1, \boldsymbol{y}_2 \rangle \subset \operatorname{aff}\{\boldsymbol{x}_1, \ldots, \boldsymbol{x}_{m+1}, \boldsymbol{x}'_1, \ldots, \boldsymbol{x}'_{m+1}\} = \operatorname{aff} Y,$$

contrary to the above argument. Because $\boldsymbol{x}$ is expressible as a positive convex combination

$$\begin{aligned} \boldsymbol{x} &= (1-\eta)\boldsymbol{y} + \eta \boldsymbol{y}' \\ &= (1-\eta)\lambda_1 \boldsymbol{x}_1 + \cdots + (1-\eta)\lambda_m \boldsymbol{x}_m + \eta\mu_1 \boldsymbol{x}'_1 + \cdots + \eta\mu_m \boldsymbol{x}'_m, \end{aligned}$$

Theorem 4.18 implies that $\boldsymbol{x} \in \operatorname{rint}(\operatorname{conv} Y)$. Finally, $\operatorname{card} Y \geqslant m+1$ by Corollary 2.66.

Now, suppose that Y is a minimal subset of X satisfying (4.6), and $\operatorname{card} Y = 2m$. With the notation above, $\boldsymbol{y}$ and $\boldsymbol{y}'$ belong, respectively, to the relative interiors of the simplices

$$\Delta = \Delta(\boldsymbol{x}_1, \ldots, \boldsymbol{x}_m) \quad \text{and} \quad \Delta' = \Delta(\boldsymbol{x}'_1, \ldots, \boldsymbol{x}'_m).$$

We observe that both Δ and Δ' lie in $\operatorname{rbd}(\operatorname{conv} Z)$. Indeed, assume for a moment that Δ contains a point $\boldsymbol{u} \in \operatorname{rint}(\operatorname{conv} Z)$. According to Theorem 3.26, there is a scalar $\gamma > 1$ satisfying the condition

$$\boldsymbol{v} = \gamma \boldsymbol{y} + (1-\gamma)\boldsymbol{u} \in \Delta \subset \operatorname{conv} Z,$$

and Theorem 3.24 gives

$$\boldsymbol{y} = (1-\gamma^{-1})\boldsymbol{u} + \gamma^{-1}\boldsymbol{v} \in \operatorname{rint}(\operatorname{conv} Z),$$

in contradiction with $\boldsymbol{y} \in \operatorname{rbd}(\operatorname{conv} Z)$.

The above argument holds for every choice of $\boldsymbol{y}$ in $\operatorname{rint} \Delta$ and the corresponding point $\boldsymbol{y}' = \langle \boldsymbol{y}, \boldsymbol{x} \rangle \cap \operatorname{rint} \Delta'$, and vice versa. Therefore, $\boldsymbol{y} \to \boldsymbol{y}'$ is an invertible mapping from $\operatorname{rint} \Delta$ to $\operatorname{rint} \Delta'$. Allowing some scalars λ_i, μ_j from the representations

$$\boldsymbol{y} = \lambda_1 \boldsymbol{x}_1 + \cdots + \lambda_m \boldsymbol{x}_m \quad \text{and} \quad \boldsymbol{y}' = \mu_1 \boldsymbol{x}'_1 + \cdots + \mu_m \boldsymbol{x}'_m$$

take zero values, we extend this correspondence to an invertible mapping $f : \Delta \to \Delta'$ (see Theorem 2.55). We assert that

$$f(\{\boldsymbol{x}_1, \dots, \boldsymbol{x}_m\}) \subset \{\boldsymbol{x}'_1, \dots, \boldsymbol{x}'_m\}. \tag{4.7}$$

Indeed, assume for a moment that $f(\boldsymbol{x}_1) \notin \{\boldsymbol{x}'_1, \dots, \boldsymbol{x}'_m\}$, and express $f(\boldsymbol{x}_1)$ as a convex combination $f(\boldsymbol{x}_1) = \alpha_1 \boldsymbol{x}'_1 + \dots + \alpha_m \boldsymbol{x}'_m$, where at least two scalars, say α_1 and α_2, are not 0. Put

$$\begin{aligned} \boldsymbol{z}_1 &= (\alpha_1 + \alpha_2)\boldsymbol{x}'_1 + \alpha_3 \boldsymbol{x}'_3 + \dots + \alpha_m \boldsymbol{x}'_m, \\ \boldsymbol{z}_2 &= (\alpha_1 + \alpha_2)\boldsymbol{x}'_2 + \alpha_3 \boldsymbol{x}'_3 + \dots + \alpha_m \boldsymbol{x}'_m. \end{aligned}$$

Then $\boldsymbol{z}_1$ and $\boldsymbol{z}_2$ are distinct points in Δ'. Hence their pre-images $\boldsymbol{u}_1 = f^{-1}(\boldsymbol{z}_1)$ and $\boldsymbol{u}_2 = f^{-1}(\boldsymbol{z}_2)$ are distinct points in Δ. Because

$$\begin{aligned} f(\boldsymbol{x}_1) &= \frac{\alpha_1}{\alpha_1 + \alpha_2}\big((\alpha_1 + \alpha_2)\boldsymbol{x}'_1 + \alpha_3 \boldsymbol{x}'_3 + \dots + \alpha_m \boldsymbol{x}'_m\big) \\ &\quad + \frac{\alpha_2}{\alpha_1 + \alpha_2}\big((\alpha_1 + \alpha_2)\boldsymbol{x}'_2 + \alpha_3 \boldsymbol{x}'_3 + \dots + \alpha_m \boldsymbol{x}'_m\big) \\ &= \frac{\alpha_1}{\alpha_1 + \alpha_2}\boldsymbol{z}_1 + \frac{\alpha_2}{\alpha_1 + \alpha_2}\boldsymbol{z}_2 \in (\boldsymbol{z}_1, \boldsymbol{z}_2), \end{aligned}$$

the point $\boldsymbol{x}_1$ should belong to $(\boldsymbol{u}_1, \boldsymbol{u}_2)$, as shown on Figure 4.6.

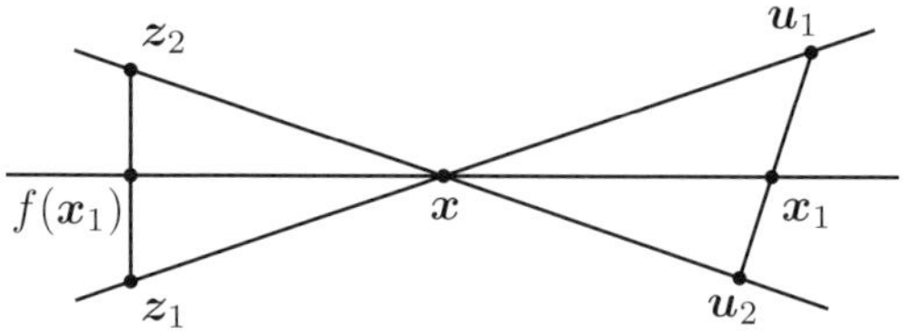

Fig. 4.6 Illustration to Theorem 4.20.

Express $\boldsymbol{x}_1$ as $\boldsymbol{x}_1 = (1 - \lambda)\boldsymbol{u}_1 + \lambda \boldsymbol{u}_2$, where $0 < \lambda < 1$. Since both $\boldsymbol{u}_1$ and $\boldsymbol{u}_2$ belong to Δ, they can be written as convex combinations

$$\boldsymbol{u}_1 = \beta_1 \boldsymbol{x}_1 + \dots + \beta_m \boldsymbol{x}_m \quad \text{and} \quad \boldsymbol{u}_2 = \gamma_1 \boldsymbol{x}_1 + \dots + \gamma_m \boldsymbol{x}_m.$$

Consequently,

$$\begin{aligned} \boldsymbol{x}_1 &= (1 - \lambda)\boldsymbol{u}_1 + \lambda \boldsymbol{u}_2 \\ &= ((1 - \lambda)\beta_1 + \lambda\gamma_1)\boldsymbol{x}_1 + \dots + ((1 - \lambda)\beta_m + \lambda\gamma_m)\boldsymbol{x}_m. \end{aligned}$$

Because $\boldsymbol{x}_1 = 1\boldsymbol{x}_1 + 0\boldsymbol{x}_2 + \dots + 0\boldsymbol{x}_m$ is the only way to express $\boldsymbol{x}_1$ as an affine combination of $\boldsymbol{x}_1, \dots, \boldsymbol{x}_m$ (see Theorem 2.52), one has

$$(1 - \lambda)\beta_1 + \lambda\gamma_1 = 1, \quad (1 - \lambda)\beta_i + \lambda\gamma_i = 0, \quad 2 \leqslant i \leqslant m. \tag{4.8}$$

Solving (4.8) for β_i and γ_i, we obtain $\beta_1 = \gamma_1 = 1$ and $\beta_i = \gamma_i = 0$ for all $2 \leqslant i \leqslant m$. Thus $\boldsymbol{u}_1 = \boldsymbol{u}_2$, contrary to the assumption.

So, inclusion (4.7) holds. Since f is invertible, one has

$$f(\{\boldsymbol{x}_1, \ldots, \boldsymbol{x}_m\}) = \{\boldsymbol{x}'_1, \ldots, \boldsymbol{x}'_m\}.$$

This argument implies that the set Y can be partitioned into pairs $\{\boldsymbol{x}_i, \boldsymbol{x}'_j\}$, each of them collinear with $\boldsymbol{x}$. □

We will need the following lemma.

Lemma 4.21. *Let* $\{\boldsymbol{x}_1, \ldots, \boldsymbol{x}_m\} \subset \mathbb{R}^n$ *be a linearly independent set, and let* $\boldsymbol{x}_0 = \lambda_1\boldsymbol{x}_1 + \cdots + \lambda_m\boldsymbol{x}_m$, *with exactly* r *nonzero scalars among* $\lambda_1, \ldots, \lambda_m$, $1 \leqslant r \leqslant m$. *For negative scalars* $\alpha_1, \ldots, \alpha_m$, *there is a subset* Y *of the set* $Z = \{\boldsymbol{x}_1, \alpha_1\boldsymbol{x}_1, \ldots, \boldsymbol{x}_m, \alpha_m\boldsymbol{x}_m\}$ *satisfying the conditions*

$$\boldsymbol{o} \in \operatorname{rint}(\operatorname{conv}(\{\boldsymbol{x}_0\} \cup Y)), \quad \operatorname{card} Y = 2m - r, \quad \dim(\{\boldsymbol{x}_0\} \cup Y) = m.$$

Proof. Renumbering the vectors $\boldsymbol{x}_1, \ldots, \boldsymbol{x}_m$, we assume that namely the scalars $\lambda_1, \ldots, \lambda_r$ are not zero. Furthermore, all $\lambda_1, \ldots, \lambda_r$ can be taken negative. Indeed, if there is a positive λ_i, $1 \leqslant i \leqslant r$, then we replace $\lambda_i\boldsymbol{x}_i$ and $\boldsymbol{x}_i$ with $(-\lambda_i)(-\boldsymbol{x}_i)$ and $-\boldsymbol{x}_i$, respectively.

Suppose first that $r = m$. Dividing both parts of the equality $\boldsymbol{o} = \boldsymbol{x}_0 - \lambda_1\boldsymbol{x}_1 - \cdots - \lambda_m\boldsymbol{x}_m$ by the positive scalar $\gamma = 1 - \lambda_1 - \cdots - \lambda_m$, we express $\boldsymbol{o}$ as a positive convex combination of $\boldsymbol{x}_0, \boldsymbol{x}_1, \ldots, \boldsymbol{x}_m$. Let $Y = \{\boldsymbol{x}_1, \ldots, \boldsymbol{x}_m\}$. By Corollary 4.19, $\boldsymbol{o} \in \operatorname{rint}(\operatorname{conv}(\{\boldsymbol{x}_0\} \cup Y))$, while the properties $\operatorname{card} Y = m$ and $\dim(\{\boldsymbol{x}_0\} \cup Y) = m$ are obvious.

Assume now that $r \leqslant m - 1$ and let $\boldsymbol{x}'_i = \alpha_i\boldsymbol{x}_i$, $r + 1 \leqslant i \leqslant m$. Adding the equalities

$$\boldsymbol{o} = \boldsymbol{x}_0 - \lambda_1\boldsymbol{x}_1 - \cdots - \lambda_r\boldsymbol{x}_r, \quad \boldsymbol{o} = -\alpha_i\boldsymbol{x}_i + \boldsymbol{x}'_i, \quad r + 1 \leqslant i \leqslant m,$$

we obtain

$$\begin{aligned}\boldsymbol{o} = \boldsymbol{x}_0 &- (\lambda_1\boldsymbol{x}_1 + \cdots + \lambda_r\boldsymbol{x}_r + \alpha_{r+1}\boldsymbol{x}_{r+1} + \cdots + \alpha_m\boldsymbol{x}_m) \\ &+ \boldsymbol{x}'_{r+1} + \cdots + \boldsymbol{x}'_m.\end{aligned}$$

Dividing both parts of this equality by the positive scalar

$$\gamma = m - r + 1 - (\lambda_1 + \cdots + \lambda_r + \alpha_{r+1} + \cdots + \alpha_m),$$

we express $\boldsymbol{o}$ as a positive convex combination of the points $\boldsymbol{x}_0, \boldsymbol{x}_1, \ldots, \boldsymbol{x}_m$, $\boldsymbol{x}'_{r+1}, \ldots, \boldsymbol{x}'_m$. Put $Y = \{\boldsymbol{x}_1, \ldots, \boldsymbol{x}_m, \boldsymbol{x}'_{r+1}, \ldots, \boldsymbol{x}'_m\}$. Theorem 2.52 shows that the set $\{\boldsymbol{o}, \boldsymbol{x}_1, \ldots, \boldsymbol{x}_m\}$ is affinely independent. Therefore, Theorem 4.18 and Corollary 4.19 give

$$\boldsymbol{o} \in \operatorname{rint}(\operatorname{conv}(\{\boldsymbol{x}_0\} \cup Y)) \quad \text{and} \quad \dim(\{\boldsymbol{x}_0\} \cup Y) = m. \qquad \square$$

The following result is a variation of Theorem 4.20.

Theorem 4.22. *Let $X \subset \mathbb{R}^n$ be a set of positive dimension m, and let $\boldsymbol{x}_0 \in X$. A point $\boldsymbol{x} \in \mathbb{R}^n$ belongs to* rint (conv X) *if and only if there is a subset $Y \subset X$ such that*

$$\begin{aligned} \boldsymbol{x} \in \operatorname{rint}(\operatorname{conv}(\{\boldsymbol{x}_0\} \cup Y)), \quad m+1 \leqslant \operatorname{card} Y \leqslant 2m-1, \\ \operatorname{aff}(\{\boldsymbol{x}_0\} \cup Y) = \operatorname{aff} X. \end{aligned} \tag{4.9}$$

Proof. If a subset Y of X satisfies conditions (4.9), then

$$\operatorname{aff}(\operatorname{conv}(\{\boldsymbol{x}_0\} \cup Y)) = \operatorname{aff}(\{\boldsymbol{x}_0\} \cup Y) = \operatorname{aff} X = \operatorname{aff}(\operatorname{conv} X)$$

according to Theorem 4.2. Since $\operatorname{conv}(\{\boldsymbol{x}_0\} \cup Y) \subset \operatorname{conv} X$, Theorem 3.18 implies the inclusion $\boldsymbol{x} \in \operatorname{rint}(\operatorname{conv}(\{\boldsymbol{x}_0\} \cup Y)) \subset \operatorname{rint}(\operatorname{conv} X)$.

Conversely, let $\boldsymbol{x} \in \operatorname{rint}(\operatorname{conv} X)$. By Theorem 4.20, X contains a subset Z such that

$$\boldsymbol{x} \in \operatorname{rint}(\operatorname{conv} Z), \quad m+1 \leqslant \operatorname{card} Z \leqslant 2m, \quad \operatorname{aff} Z = \operatorname{aff} X.$$

Without loss of generality, we assume that Z is minimal.

1. If $\operatorname{card} Z \leqslant 2m-1$, then we put $Y = Z \setminus \{\boldsymbol{x}_0\}$ (observe that $Y = Z$ if $\boldsymbol{x}_0 \notin Z$). Consequently,

$$m+1 \leqslant \operatorname{card} Y \leqslant 2m-1 \quad \text{and} \quad \operatorname{aff}(\{\boldsymbol{x}_0\} \cup Y) = \operatorname{aff} Z = \operatorname{aff} X$$

according to Theorem 2.42. Furthermore, Theorem 3.18 gives

$$\boldsymbol{x} \in \operatorname{rint}(\operatorname{conv} Z) \subset \operatorname{rint}(\operatorname{conv}(\{\boldsymbol{x}_0\} \cup Y)).$$

2. Suppose that $\operatorname{card} Z = 2m$. Translating X on $-\boldsymbol{x}$, we assume that $\boldsymbol{x} = \boldsymbol{o}$. By Theorem 4.20, Z consists of m disjoint pairs $\{\boldsymbol{x}_i, \boldsymbol{x}'_i\}$ such that $\boldsymbol{o} \in (\boldsymbol{x}_i, \boldsymbol{x}'_i)$. The equality $\dim Z = m$ implies that the set $\{\boldsymbol{x}_1, \dots, \boldsymbol{x}_m\}$ is linearly independent. Excluding the obvious case $\boldsymbol{x}_0 = \boldsymbol{o}$, we may suppose that $\boldsymbol{x}_0 \neq \boldsymbol{o}$. If $\boldsymbol{x}_0 \in Z$, then we put $Y = Z \setminus \{\boldsymbol{x}_0\}$. Suppose that $\boldsymbol{x}_0 \notin Z$. Since $\{\boldsymbol{x}_1, \dots, \boldsymbol{x}_m\}$ is a basis for span X, the vector $\boldsymbol{x}_0$ is uniquely expressed as a linear combination $\boldsymbol{x}_0 = \lambda_1\boldsymbol{x}_1 + \cdots + \lambda_m\boldsymbol{x}_m$. Let r be the number of nonzero scalars among $\lambda_1, \dots, \lambda_m$. Clearly, $r \geqslant 1$ due to $\boldsymbol{x}_0 \neq \boldsymbol{o}$. By Lemma 4.21, there is a subset Y of Z such that

$$\boldsymbol{o} \in \operatorname{rint}(\operatorname{conv}(\{\boldsymbol{x}_0\} \cup Y)), \quad \operatorname{card} Y = 2m - r, \quad \operatorname{aff}(\{\boldsymbol{x}_0\} \cup Y) = \operatorname{aff} X.$$

Thus $\operatorname{card} Y = 2m - r \leqslant 2m - 1$. By Corollary 2.66, one has $m + 1 \leqslant \operatorname{card}(\{\boldsymbol{x}_0\} \cup Y)$. Therefore, $m \leqslant \operatorname{card} Y$ due to $\boldsymbol{x}_0 \notin Y$. □

Remark. The inequality $m + 1 \leqslant \operatorname{card} Y$ in Theorem 4.22 cannot be replaced with $m \leqslant \operatorname{card} Y$. For instance, let $X = \{\boldsymbol{x}_0, \boldsymbol{x}_1, \boldsymbol{x}_2, \boldsymbol{x}_3\}$, where $\boldsymbol{x}_0 = (-1, 0)$, $\boldsymbol{x}_1 = (0, -1)$, $\boldsymbol{x}_2 = (0, 1)$, $\boldsymbol{x}_3 = (2, 0)$, as depicted in Figure 4.7. Clearly, the point $\boldsymbol{x} = (1, 0)$ belongs to int (conv X), and $Y = \{\boldsymbol{x}_1, \boldsymbol{x}_2, \boldsymbol{x}_3\}$ is the only subset of X satisfying (4.9).

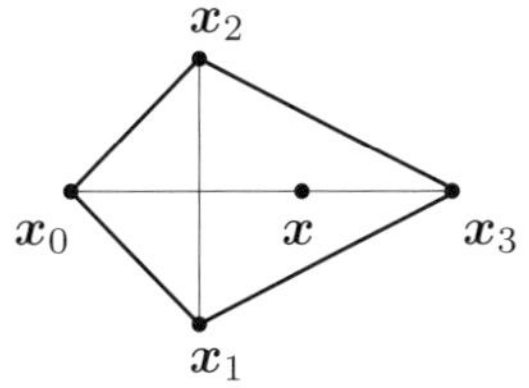

Fig. 4.7 Illustration to Theorem 4.22.

Problems and Notes for Chapter 4

Problems for Chapter 4

Problem 4.1. Prove that any closed ball $B_\rho(\boldsymbol{a}) \subset \mathbb{R}^n$ is the convex hull of its boundary sphere $S_\rho(\boldsymbol{a})$. Furthermore, $B_\rho(\boldsymbol{a}) = \text{conv}_2 S_\rho(\boldsymbol{a})$.

Problem 4.2. Let $\Delta = \Delta(\boldsymbol{x}_1, \ldots, \boldsymbol{x}_{r+1})$ be an r-simplex in $\mathbb{R}^n$. For a nonempty proper subset X of $\{\boldsymbol{x}_1, \ldots, \boldsymbol{x}_{r+1}\}$, denote by Δ_X and Δ_Y the simplices with the vertex sets X and $Y = \{\boldsymbol{x}_1, \ldots, \boldsymbol{x}_{r+1}\} \setminus X$, respectively. Prove that $\Delta = \text{conv}_2(\Delta_X \cup \Delta_Y)$.

Problem 4.3. Let $Q = \{(x_1, \ldots, x_n) : 0 \leqslant x_i \leqslant 1,\ i = 1, \ldots, n\}$ be the unit cube of $\mathbb{R}^n$ and X be the set of points $(x_1, \ldots, x_n)$ in Q with all irrational coordinates $x_1, \ldots, x_n$. Prove that $\text{conv}\, X = \text{conv}_2 X = \text{int}\, Q$.

Problem 4.4. Let $L \subset \mathbb{R}^n$ be a plane of positive dimension m, $M \subset L$ be a plane of dimension $m - 1$, and D be a closed halfplane of L determined by M. Prove that

$$D = \text{conv}\,(h \cup M) = \text{conv}_2(h \cup M),$$

where h is any given closed halfline with endpoint $\boldsymbol{a} \in M$, which meets $D \setminus M$.

Problem 4.5. Let L_1 and L_2 be nonempty planes in $\mathbb{R}^n$. Prove the following assertions.

(*a*) If $L_1 \cap L_2 \neq \varnothing$, then $\text{conv}\,(L_1 \cup L_2) = \text{conv}_2(L_1 \cup L_2) = \text{aff}\,(L_1 \cup L_2)$.

(*b*) If $L_1 \cap L_2 = \varnothing$ and G is the open slab of $\text{aff}\,(L_1 \cup L_2)$ between by the parallel planes $L_1' = L_1 + \text{dir}\, L_2$ and $L_2' = L_2 + \text{dir}\, L_1$ (see Theorem 2.15), then

$$\text{conv}\,(L_1 \cup L_2) = \text{conv}_2(L_1 \cup L_2) = L_1 \cup G \cup L_2.$$

Problem 4.6. Let L_1 and L_2 be nonempty planes in $\mathbb{R}^n$. Prove that the set $\text{conv}\,(L_1 \cup L_2)$ is closed if and only if any of the following two conditions holds: (*a*) $L_1 \cap L_2 \neq \varnothing$, (*b*) L_1 is a translate of L_2.

Problem 4.7. Prove that $\text{diam}\,(\text{conv}\, X) = \text{diam}\, X$ for a bounded set $X \subset \mathbb{R}^n$.

Problem 4.8. Given a set $X \subset \mathbb{R}^n$ and a scalar $\rho > 0$, prove the equalities

$$\operatorname{conv} B_\rho(X) = B_\rho(\operatorname{conv} X) \quad \text{and} \quad \operatorname{conv} U_\rho(X) = U_\rho(\operatorname{conv} X).$$

Problem 4.9. Given a convex set $K \subset \mathbb{R}^n$ and sets $X, Y \subset \mathbb{R}^n$, prove that

$$\operatorname{conv}((K + X) \cup (K + Y)) = K + \operatorname{conv}(X \cup Y).$$

Problem 4.10. Let a convex set $K \subset \mathbb{R}^n$ and points $\boldsymbol{x}, \boldsymbol{y} \in \mathbb{R}^n \setminus K$ satisfy the conditions $\boldsymbol{x} \in \operatorname{conv}(\{\boldsymbol{y}\} \cup K)$ and $\boldsymbol{y} \in \operatorname{conv}(\{\boldsymbol{x}\} \cup K)$. Prove that $\boldsymbol{x} = \boldsymbol{y}$.

Problem 4.11. (Radon [235]) Given a nonempty set $X \subset \mathbb{R}^n$, prove the equivalence of the following conditions:

(a) X is affinely independent,

(b) X does not contain a pair of disjoint subsets whose affine spans meet,

(c) X does not contain a pair of disjoint subsets whose convex hulls meet.

Problem 4.12. Let a set $X \subset \mathbb{R}^n$ have the property $X \subset \operatorname{cl}(\operatorname{rint} X)$. Prove that $\operatorname{conv}(\operatorname{rint} X) = \operatorname{rint}(\operatorname{conv} X)$.

Problem 4.13. Let X be a dense subset of a set $Y \subset \mathbb{R}^n$. Prove that

$$\operatorname{cl}(\operatorname{conv} X) = \operatorname{cl}(\operatorname{conv} Y) \quad \text{and} \quad \operatorname{rint}(\operatorname{conv} X) = \operatorname{rint}(\operatorname{conv} Y).$$

Problem 4.14. Let $X_1, \dots, X_r$ be nonempty sets in $\mathbb{R}^n$ and $\mu_1, \dots, \mu_r$ be scalars such that $r > m = \dim(\mu_1 X_1 + \cdots + \mu_r X_r)$. Prove that for any point $\boldsymbol{x}$ in $\operatorname{conv}(\mu_1 X_1 + \cdots + \mu_r X_r)$ there is an index set $I = I(\boldsymbol{x}) \subset \{1, \dots, r\}$, with $\operatorname{card} I \leqslant m$, and points $\boldsymbol{z}_i \in X_i$, $i \in J = \{1, \dots, r\} \setminus I$, satisfying the condition

$$\boldsymbol{x} \in \sum_{i \in J} \mu_i \boldsymbol{z}_i + \sum_{i \in I} \mu_i \operatorname{conv} X_i.$$

Problem 4.15. Given nonempty sets $X_1, \dots, X_r \subset \mathbb{R}^n$, prove that the relative interior of $\operatorname{conv}(X_1 \cup \cdots \cup X_r)$ is the collection of all positive convex combinations of the form $\lambda_1 \boldsymbol{x}_1 + \cdots + \lambda_r \boldsymbol{x}_r$, where $\boldsymbol{x}_i \in \operatorname{rint}(\operatorname{conv} X_i)$ for all $1 \leqslant i \leqslant r$.

Problem 4.16. Given convex sets $K_1, \dots, K_r \subset \mathbb{R}^n$, prove that

$$\operatorname{rint}(\operatorname{conv}(K_1 \cup \cdots \cup K_r)) \subset \operatorname{conv}(\operatorname{rint} K_1 \cup \cdots \cup \operatorname{rint} K_r).$$

Furthermore, if the relative interiors of $K_1, \dots, K_r$ have a point in common, then

$$\operatorname{rint}(\operatorname{conv}(K_1 \cup \cdots \cup K_r)) = \operatorname{conv}(\operatorname{rint} K_1 \cup \cdots \cup \operatorname{rint} K_r).$$

Problem 4.17. (Reay [237, Lemma 4.1]) Let $X \subset \mathbb{R}^n$ be a set of positive dimension m and Y be a subset of $\operatorname{conv} X$ of cardinality $r \geqslant 2$. Prove the existence of a set $Z \subset X$ of cardinality rm such that $Y \subset \operatorname{conv} Z$.

Problem 4.18. (Reay [237, Lemma 4.6]) Let $X \subset \mathbb{R}^n$ be a set of positive dimension m and Y be a subset of $\operatorname{rint}(\operatorname{conv} X)$ of cardinality $r \geqslant 2$. Prove the existence of a set $Z \subset X$ of cardinality rm such that $Y \subset \operatorname{rint}(\operatorname{conv} Z)$ and $\operatorname{aff} Z = \operatorname{aff} X$.

Problem 4.19. Let $X \subset \mathbb{R}^n$ and Y be a compact subset of $\operatorname{rint}(\operatorname{conv} X)$. Prove the existence of a finite subset Z of X such that $Y \subset \operatorname{rint}(\operatorname{conv} Z)$ and $\operatorname{aff} Z = \operatorname{aff} X$.

Notes for Chapter 4

The concept of convex hull. For a compact set $X \subset \mathbb{R}^n$, Carathéodory [66] considered the intersection of all closed halfspaces containing X such that their boundary hyperplanes support X. He observed that this intersection is the smallest convex set containing X. Clearly, this intersection determines the *closed* convex hull of X, which coincides with $\operatorname{conv} X$ due to the compactness of X (see Theorem 4.16). Steinitz [265, § 9] defines the convex hull of any set $X \subset \mathbb{R}^n$ (again calling it the smallest convex set containing X) as the intersection of all convex sets containing X. Bonnesen and Fenchel [39, p. 5] already use the term "convex hull."

Carathéodory-type results. Part (*c*) of Theorem 4.7, without involving affine independence of points, is considered by Carathéodory [66, p. 200] for the case of compact sets in $\mathbb{R}^n$ (although stated there for the case of closed sets). His method uses the support properties of compact convex sets and induction on n. Multiple priority references on the earlier paper [65] is a common mistake in the literature.

For the case of any set $X \subset \mathbb{R}^n$, Theorem 4.3, Corollary 4.4, Theorems 4.6 and 4.7 can be found in Steinitz [265, § 10]. A particular case of Theorem 4.8 (when the set X is finite and $\boldsymbol{x}_0 \in X$) can be found in the paper of Watson [290].

An earlier article of Brunn [60] gives an algebraic proof of the following assertion: if a point $\boldsymbol{x}$ is a convex combination of a set of $n+1+p$ points in $\mathbb{R}^n$, then $\boldsymbol{x}$ can be written as a convex combination of at most $n+1$ points of the set. Cook [78] established the following analogue of this result for the case of convex combinations with linear constraints. Let

$$\Omega = \{\overline{\lambda} = (\lambda_1, \dots, \lambda_m) \in \mathbb{R}^m : \lambda_1 + \cdots + \lambda_m = 1, \lambda_i \geqslant 0, D\overline{\lambda} \leqslant \boldsymbol{d}\},$$

where D is a $k \times m$ matrix and $\boldsymbol{d} \in \mathbb{R}^k$. If $\boldsymbol{x}_0 = \lambda_1 \boldsymbol{x}_1 + \cdots + \lambda_m \boldsymbol{x}_m$, where $\overline{\lambda} = (\lambda_1, \dots, \lambda_m) \in \Omega$, then $\boldsymbol{x}_0 = \mu_1 \boldsymbol{x}_1 + \cdots + \mu_m \boldsymbol{x}_m$, where $\overline{\mu} = (\mu_1, \dots, \mu_m) \in \Omega$ such that $\overline{\mu}$ has at most $n+k+1$ nonzero components.

The number $n+1$ in Theorem 4.7 can be lowered to n provided the set X has additional connectedness properties. Fenchel [103] (for $n = 3$) and Stoelinga [269, Theorem 25] (for all $n \geqslant 3$) proved that every point in the convex hull of an n-dimensional connected set can be expressed as a convex combination of n points from X. Bunt [63, p. 23] (see also Kramer [188]) further observed that connectedness of X can be replaced by the weaker condition "has n or fewer components."

Hanner and Rådström [146] defined a set $M \subset \mathbb{R}^n$ as *convexly connected* if there is no hyperplane $H \subset \mathbb{R}^n$ such that $H \cap M = \varnothing$ and M has points in both open halfspaces determined by H. They proved that if a set X in $\mathbb{R}^n$ is compact and has n or fewer convexly connected components, then every point $\boldsymbol{x} \in X$ can be expressed as a convex combination of n points from X. Bárány and Karasev [26] proved that $\operatorname{conv} X = \operatorname{conv}_{k+1} X$ for any set $X \subset \mathbb{R}^n$ satisfying the condition: every linear image of X to $\mathbb{R}^{n-k}$ is convex. One more result of this type belongs to Tverberg [282]: if a set $X \subset \mathbb{R}^n$ is a union of convex sets, all meeting a fixed hyperplane $H \subset \mathbb{R}^n$, then $\operatorname{conv} X = \operatorname{conv}_n X$.

The above number n can be further lowered in the case when X is a *simple curve* in $\mathbb{R}^n$ (defined as a 1–1 continuous image of a closed number interval such that no hyperplane meets X at more than n points): Egerváry [98], for $n = 3$, and Derry [91], for all $n \geqslant 3$, proved that $\operatorname{conv}_r X = \operatorname{conv} X$, where $n = 2r$ or $n = 2r+1$ (see an earlier paper of Karlin and Shapley [164] for the case of moment curves).

Iterative construction of convex hulls. The property of conv_r described in Theorem 4.10 is due to Lepin [198] and, independently, to Klee (see [40]). The first part of Corollary 4.11 is given in Bonnice and Klee [40], and its second part, with $r_1 = \cdots = r_k = 2$ and $m = n$, which implies the inequality $k \leqslant \lceil \log_2(n+1) \rceil$, is known due to Brunn [60], Hjelmslev [150], Straszewicz [272, pp. 29–36], and some others (see Danzer, Grünbaum, and Klee [83, p. 116] for additional references).

Topological properties of convex hulls. The second part of Theorem 4.16, which states that the convex hull of a compact set in $\mathbb{R}^n$ is compact, can be found in Steinitz [266, § 21]. Fischer [108] observed that the assertion of Theorem 4.18 can be formulated and proved similarly to Theorem 5.51. Theorem 4.20 is attributed to Steinitz [266, § 20], who proved an equivalent assertion in terms of positive bases for $\mathbb{R}^n$ (see page 205). Probably, Dines [94] was the first to formulate this theorem in terms of convex hulls as a geometric interpretation of an assertion on linear inequalities by Dines and McCoy [95]. Other proofs of Theorem 4.20 are due to Robinson [240], Gustin [138], and Valentine [285, Theorem 3.13].

Triangulations and barycentric coordinates. If $X = \{\boldsymbol{x}_1, \ldots, \boldsymbol{x}_r\}$ is a finite set of dimension m in $\mathbb{R}^n$, then a *triangulation* of $\operatorname{conv} X$ is any collection $\mathcal{F} = \{\Delta_1, \ldots, \Delta_p\}$ of m-simplices satisfying the following conditions:

(a) the union of these simplices is $\operatorname{conv} X$,

(b) any two simplices from $\mathcal{F}$ meet in a common face, possibly empty,

(c) if a point $\boldsymbol{x}_i \in X$ belongs to a simplex $\Delta_k \in \mathcal{F}$, then $\boldsymbol{x}_i$ is a vertex of Δ_k.

See the book of De Loera, Rambau, and Santos [88] for existence, methods of construction, and combinatorial properties of various types of triangulation.

According to Corollary 4.4, any point $\boldsymbol{u} \in \Delta_k(\boldsymbol{x}_{i_1}, \ldots, \boldsymbol{x}_{i_{m+1}}) \in \mathcal{F}$ can be written as a convex combination $\boldsymbol{u} = \lambda_{i_1}^{(k)}(\boldsymbol{u})\, \boldsymbol{x}_{i_1} + \cdots + \lambda_{i_{m+1}}^{(k)}(\boldsymbol{u})\, \boldsymbol{x}_{i_{m+1}}$. Since the affine coordinates $\lambda_{i_j}^{(k)}(\boldsymbol{u})$ continuously depend on $\boldsymbol{u}$ (see Theorem 2.55), the above conditions (a)–(c) allow suitable selections of $\lambda_{i_j}^{(s)}(\boldsymbol{u})$ to be "glued" into continuous functions $\lambda_i(\boldsymbol{u})$ on $\operatorname{conv} X$, $1 \leqslant i \leqslant r$, such that the following conditions are satisfied:

$$\lambda_1(\boldsymbol{u}), \dots, \lambda_r(\boldsymbol{u}) \geqslant 0, \quad \lambda_1(\boldsymbol{u}) + \dots + \lambda_r(\boldsymbol{u}) = 1,$$
$$\boldsymbol{u} = \lambda_1(\boldsymbol{u})\,\boldsymbol{x}_1 + \dots + \lambda_r(\boldsymbol{u})\,\boldsymbol{x}_r.$$

Any collection of functions $\lambda_1(\boldsymbol{u}), \dots, \lambda_r(\boldsymbol{u})$ satisfying these conditions are often called *barycentric coordinates* on $\operatorname{conv} X$ (see, e. g., Kalman [163], Fuglede [112], Brøndsted [56], Warren [289], and many others).

Convex hull and large sums. Problem 4.14 is a refinement of a lemma of Shapley and Folkman (see Starr [264]), which states that for every point $\boldsymbol{x}$ in $\operatorname{conv}(X_1 + \dots + X_r)$, where $X_1, \dots, X_r$ are compact subset of $\mathbb{R}^n$ and $r \geqslant n+1$, there is an index set $I \subset \{1, \dots, r\}$ such that $\operatorname{card} I \leqslant n$ and

$$\boldsymbol{x} \in \sum_{i \in I} \operatorname{conv} X_i + \sum_{i \notin I} X_i.$$

Their proof was gradually simplified and extended to the case of arbitrary sets by Artstein [4], Green and Heller [126], Howe [153], and Zhou [298]. Lawrence and Soltan [195] gave the following refinement of this assertion: If $X_1, \dots, X_r$ are nonempty sets in $\mathbb{R}^n$, then for every point $\boldsymbol{x} \in \operatorname{conv}(X_1 + \dots + X_r)$, there is an index set $I \subset \{1, \dots, r\}$ with $\operatorname{card} I \leqslant n$ and nonempty subsets $Y_i \subset X_i$, $1 \leqslant i \leqslant r$, such that

$$\boldsymbol{x} \in \sum_{i \in I} \operatorname{conv} Y_i + \sum_{i \notin I} Y_i, \quad \sum_{i \in I} \operatorname{card} Y_i \leqslant n + \operatorname{card} I, \quad \operatorname{card} Y_i = 1 \text{ if } i \notin I.$$

Convex closure. The correspondence $h : X \mapsto \operatorname{conv} X$ can be considered as a closure mapping on the family of all subsets of $\mathbb{R}^n$, since it has the following obvious properties: $X \subset hX$, $hhX = hX$, $hX \subset hY$ if $X \subset Y$.

Koenen [186] proved that $hchchX = chchX$, where cZ means the complement $\mathbb{R}^n \setminus Z$ of a set $Z \subset \mathbb{R}^n$. Therefore, consecutively applying to a given set $X \subset \mathbb{R}^n$ the operations h and c (and using the identities $ccX = X$ and $hhX = hX$), one can obtain 10 or fewer distinct sets:

$$X,\ cX,\ hX,\ chX,\ hcX,\ hchX,\ chcX,\ chchX,\ hchcX,\ chchcX.$$

Similar problems on semigroups generated by the mappings h, c, $i = chc$, and $f = h \cap hc$ are studied by Soltan [252, § 7]: if $m(\varphi, \eta, \dots)$ denotes the number of elements in the semigroup generated by mappings $\varphi, \eta, \dots$ on the family of all subsets of $\mathbb{R}^n$, then

$$m(c, f) = 6, \quad m(f, i) = 7, \quad m(h, i) = 5,$$
$$m(f, h) = 5, \quad m(f, h, i) = 12, \quad m(c, f, h) = 22.$$

Chapter 5

Cones and Conic Hulls

5.1 Cones

5.1.1 *Basic Properties of Cones*

Definition 5.1. A nonempty set $C \subset \mathbb{R}^n$ is called a *cone* with apex $\boldsymbol{a} \in \mathbb{R}^n$ provided $\boldsymbol{a} + \lambda(\boldsymbol{x} - \boldsymbol{a}) \in C$ whenever $\boldsymbol{x} \in C$ and $\lambda > 0$. A *convex cone* is a cone which is a convex set. An apex $\boldsymbol{a}$ of C is called *proper* if $\boldsymbol{a} \in C$; otherwise it is called *improper*. Given a point $\boldsymbol{a} \in \mathbb{R}^n$, we let the empty set $\varnothing$ be a cone with apex $\boldsymbol{a}$.

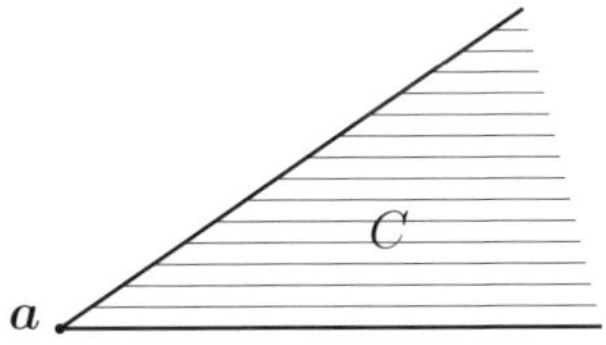

Fig. 5.1 A convex cone with apex $\boldsymbol{a}$.

Remark. Since an open halfline $(\boldsymbol{a}, \boldsymbol{x}\rangle$ consists of all affine combinations of the form $\boldsymbol{a} + \lambda(\boldsymbol{x} - \boldsymbol{a})$, $\lambda > 0$, we can reformulate Definition 5.1, saying that a nonempty set $C \subset \mathbb{R}^n$ is a cone with apex $\boldsymbol{a}$ if and only if the open halfline $(\boldsymbol{a}, \boldsymbol{x}\rangle$ lies in C whenever $\boldsymbol{x} \in C \setminus \{\boldsymbol{a}\}$. Hence a cone with apex $\boldsymbol{a}$ may be the empty set $\varnothing$, the singleton $\{\boldsymbol{a}\}$, or a union of halflines (open or closed) with common endpoint $\boldsymbol{a}$.

Definition 5.1 can be modified by imposing the weaker condition $\lambda \geqslant 0$, which is equivalent to the combination of conditions $\boldsymbol{a} \in C$ and $\lambda > 0$. Under this modification, every cone is nonempty and contains all its apices.

We observe that the assumption $\lambda > 0$ simplifies the study of the relative interior of convex cones and conic hulls, while the assumption $\lambda \geqslant 0$ is suitable for dealing with their closures.

Examples. 1. A nonempty plane $L \subset \mathbb{R}^n$ is a convex cone, and every point $\boldsymbol{a} \in L$ is its apex. Indeed, this fact is obvious if $L = \{\boldsymbol{a}\}$. If $\dim L \geqslant 1$ and $\boldsymbol{x}$ is a point in $L \setminus \{\boldsymbol{a}\}$, then $(\boldsymbol{a}, \boldsymbol{x}\rangle \subset \langle \boldsymbol{a}, \boldsymbol{x}\rangle \subset L$ (see a remark on page 38).

2. Any halfplane F in $\mathbb{R}^n$ (either closed or open, see Definition 2.69) is a convex cone, and every point $\boldsymbol{a} \in \operatorname{rbd} F$ is its apex. Indeed, let $\boldsymbol{x}$ be a point in $F \setminus \{\boldsymbol{a}\}$. If $\boldsymbol{x} \in \operatorname{rbd} F$, then $(\boldsymbol{a}, \boldsymbol{x}\rangle \subset \langle \boldsymbol{a}, \boldsymbol{x}\rangle \subset \operatorname{rbd} F$ because $\operatorname{rbd} F$ is a plane (see Theorem 3.68). If $\boldsymbol{x} \in F \setminus \operatorname{rbd} F$, then $(\boldsymbol{a}, \boldsymbol{x}\rangle \subset F$ (indeed, since F is the intersection of a plane and a halfspace, a combination of Theorems 2.38 and 2.35 implies the inclusion $(\boldsymbol{a}, \boldsymbol{x}\rangle \subset F$).

3. The positive octant of $\mathbb{R}^n$,

$$O^+ = \{(x_1, \ldots, x_n) \in \mathbb{R}^n : x_1 > 0, \ldots, x_n > 0\},$$

and the nonnegative octant of $\mathbb{R}^n$,

$$O = \{(x_1, \ldots, x_n) \in \mathbb{R}^n : x_1 \geqslant 0, \ldots, x_n \geqslant 0\},$$

are convex cones, and the origin $\boldsymbol{o}$ is the only apex of each of them (see Theorem 5.7 for a more general assertion).

Theorem 5.2. *For a set $X \subset \mathbb{R}^n$ and a point $\boldsymbol{a} \in \mathbb{R}^n$, the following conditions are equivalent.*

(*a*) *X is a cone with apex $\boldsymbol{a}$.*
(*b*) *$\mathbb{R}^n \setminus X$ is a cone with apex $\boldsymbol{a}$.*
(*c*) *$X - \boldsymbol{a}$ is a cone with apex $\boldsymbol{o}$.*
(*d*) *$X = \boldsymbol{a} + \lambda(X - \boldsymbol{a})$ for all $\lambda > 0$.*

Proof. Since the cases $X = \varnothing$ and $X = \mathbb{R}^n$ are obvious, we may assume that $\varnothing \neq X \neq \mathbb{R}^n$.

(*a*) $\Leftrightarrow$ (*b*) Let X be a cone with apex $\boldsymbol{a}$. Choose any point $\boldsymbol{x}$ in $\mathbb{R}^n \setminus X$ and a scalar $\lambda > 0$. Assume, for contradiction, that the point $\boldsymbol{z} = \boldsymbol{a} + \lambda(\boldsymbol{x} - \boldsymbol{a})$ does not belong to $\mathbb{R}^n \setminus X$. Then $\boldsymbol{z} \in X$. Since X is a cone with apex $\boldsymbol{a}$, one has $\boldsymbol{x} = \boldsymbol{a} + \lambda^{-1}(\boldsymbol{z} - \boldsymbol{a}) \in X$, contrary to the choice of $\boldsymbol{x}$. Hence $\boldsymbol{z} \in \mathbb{R}^n \setminus X$, and $\mathbb{R}^n \setminus X$ is a cone with apex $\boldsymbol{a}$.

Conversely, if $\mathbb{R}^n \setminus X$ is a cone with apex $\boldsymbol{a}$, then, by the above argument, the set $X = \mathbb{R}^n \setminus (\mathbb{R}^n \setminus X)$ is a cone with apex $\boldsymbol{a}$.

$(a) \Leftrightarrow (c)$ This assertion immediately follows from Definition 5.1 and the equivalence of inclusions

$$\boldsymbol{a} + \lambda(\boldsymbol{x} - \boldsymbol{a}) \in X \quad \text{and} \quad \boldsymbol{o} + \lambda\boldsymbol{z} \in X - \boldsymbol{a}, \quad \lambda > 0,$$

where $\boldsymbol{x} \in X$ and $\boldsymbol{z} = \boldsymbol{x} - \boldsymbol{a} \in X - \boldsymbol{a}$.

$(a) \Leftrightarrow (d)$ Let X be a cone with apex $\boldsymbol{a}$. If $\lambda > 0$, then

$$\boldsymbol{a} + \lambda(X - \boldsymbol{a}) = \{\boldsymbol{a} + \lambda(\boldsymbol{x} - \boldsymbol{a}) : \boldsymbol{x} \in X\} \subset X.$$

Similarly, for any $\boldsymbol{x} \in X$, the point $\boldsymbol{z} = \boldsymbol{a} + \lambda^{-1}(\boldsymbol{x} - \boldsymbol{a})$ belongs to X. Hence

$$\boldsymbol{x} = \boldsymbol{a} + \lambda(\boldsymbol{a} + \lambda^{-1}(\boldsymbol{x} - \boldsymbol{a}) - \boldsymbol{a}) = \boldsymbol{a} + \lambda(\boldsymbol{z} - \boldsymbol{a}) \in \boldsymbol{a} + \lambda(X - \boldsymbol{a}),$$

which gives the opposite inclusion $X \subset \boldsymbol{a} + \lambda(X - \boldsymbol{a})$.

Conversely, let X satisfy the condition $X = \boldsymbol{a} + \lambda(X - \boldsymbol{a})$ for all $\lambda > 0$. Choose a point $\boldsymbol{x} \in X$ and a scalar $\lambda > 0$. Then

$$\boldsymbol{a} + \lambda(\boldsymbol{x} - \boldsymbol{a}) \in \boldsymbol{a} + \lambda(X - \boldsymbol{a}) = X,$$

implying that X is a cone with apex $\boldsymbol{a}$. □

Theorem 5.3. *A convex set $K \subset \mathbb{R}^n$ is a cone with apex $\boldsymbol{a}$ if and only if there is a positive scalar $\lambda \neq 1$ such that $K = \boldsymbol{a} + \lambda(K - \boldsymbol{a})$.*

Proof. Since the case $K = \varnothing$ is obvious, we may assume that $K \neq \varnothing$. Due to Theorem 5.2, it suffices to verify the "if" part. Rewriting the equality $K = \boldsymbol{a} + \lambda(K - \boldsymbol{a})$ as $K - \boldsymbol{a} = \lambda(K - \boldsymbol{a})$, we easily obtain by induction on r that

$$\lambda^{-r}(K - \boldsymbol{a}) = K - \boldsymbol{a} = \lambda^{r}(K - \boldsymbol{a}) \quad \text{for all} \quad r \geqslant 1. \tag{5.1}$$

Let $\boldsymbol{x} \in K$ and $\mu > 0$. We assert that $\boldsymbol{a} + \mu(\boldsymbol{x} - \boldsymbol{a}) \in K$. For this, choose integers p and q such that $\lambda^p < \mu < \lambda^q$ (p is negative and q is positive if $\lambda > 1$, and p is positive and q is negative if $0 < \lambda < 1$). Due to (5.1), both points $\lambda^q(\boldsymbol{x} - \boldsymbol{a})$ and $\lambda^p(\boldsymbol{x} - \boldsymbol{a})$ belong to $K - \boldsymbol{a}$. By Lemma 2.25,

$$(\boldsymbol{o}, \lambda^p(\boldsymbol{x} - \boldsymbol{a})) \subset (\boldsymbol{o}, \mu(\boldsymbol{x} - \boldsymbol{a})) \subset (\boldsymbol{o}, \lambda^q(\boldsymbol{x} - \boldsymbol{a})).$$

This argument and the convexity of $K - \boldsymbol{a}$ give

$$\begin{aligned}\mu(\boldsymbol{x} - \boldsymbol{a}) &\in (\boldsymbol{o}, \lambda^p(\boldsymbol{x} - \boldsymbol{a})) \setminus (\boldsymbol{o}, \lambda^q(\boldsymbol{x} - \boldsymbol{a})) \\ &= [\lambda^q(\boldsymbol{x} - \boldsymbol{a}), \lambda^p(\boldsymbol{x} - \boldsymbol{a})) \subset K - \boldsymbol{a}.\end{aligned}$$

So, $\boldsymbol{a} + \mu(\boldsymbol{x} - \boldsymbol{a}) \in K$, which shows that K is a cone with apex $\boldsymbol{a}$. □

Remark. The convexity assumption in Theorem 5.3 is essential. Indeed, if X is the set of all points in $\mathbb{R}^2$ with rational coordinates, then $X = \frac{1}{2}X = \boldsymbol{o} + \frac{1}{2}(X - \boldsymbol{o})$, while X is not a cone with apex $\boldsymbol{o}$.

Theorem 5.4. *For a cone $C \subset \mathbb{R}^n$ with apex $\boldsymbol{a}$, the following assertions hold.*

(a) $\operatorname{conv} C$ *is a convex cone with apex* $\boldsymbol{a}$.
(b) $\{\boldsymbol{a}\} \cup C$ *is a cone with proper apex* $\boldsymbol{a}$, *and* $C \setminus \{\boldsymbol{a}\}$ *is a cone with improper apex* $\boldsymbol{a}$.
(c) *If* $C \neq \varnothing$, *then* $\operatorname{aff} C = \operatorname{aff}(\{\boldsymbol{a}\} \cup C)$.
(d) *If C is convex, then the cone $\{\boldsymbol{a}\} \cup C$ is convex.*
(e) *If $\boldsymbol{a}$ is a proper apex of C and the set $C \setminus \{\boldsymbol{a}\}$ is convex, then C is convex and $\boldsymbol{a}$ is the only apex of C.*

Proof. Because assertions (a), (b) and (d) are obvious when $C = \varnothing$, we may assume that C is nonempty.

(a) Let $\boldsymbol{x} \in \operatorname{conv} C$ and $\lambda > 0$. By Theorem 4.3, $\boldsymbol{x}$ can be written as a convex combination

$$\boldsymbol{x} = \mu_1 \boldsymbol{x}_1 + \cdots + \mu_p \boldsymbol{x}_p, \quad \text{where} \quad \boldsymbol{x}_1, \ldots, \boldsymbol{x}_p \in C.$$

Since $\boldsymbol{a} + \lambda(\boldsymbol{x}_i - \boldsymbol{a}) \in C$ for all $1 \leqslant i \leqslant p$, and since $\boldsymbol{a} + \lambda(\boldsymbol{x} - \boldsymbol{a})$ can be written as the convex combination

$$\boldsymbol{a} + \lambda(\boldsymbol{x} - \boldsymbol{a}) = \mu_1(\boldsymbol{a} + \lambda(\boldsymbol{x}_1 - \boldsymbol{a})) + \cdots + \mu_p(\boldsymbol{a} + \lambda(\boldsymbol{x}_p - \boldsymbol{a})),$$

the same theorem gives the inclusion $\boldsymbol{a} + \lambda(\boldsymbol{x} - \boldsymbol{a}) \in \operatorname{conv} C$. Hence $\operatorname{conv} C$ is a convex cone with apex $\boldsymbol{a}$.

(b) This part follows from Definition 5.1.

(c) The assertion is obvious when $C = \{\boldsymbol{a}\}$. Suppose the existence of a point $\boldsymbol{x} \in C \setminus \{\boldsymbol{a}\}$. Let $\boldsymbol{z} = \frac{1}{2}(\boldsymbol{a} + \boldsymbol{x})$. Then $\boldsymbol{z} \in (\boldsymbol{a}, \boldsymbol{x}\rangle \subset C$, and Theorem 2.39 implies the inclusion

$$\boldsymbol{a} = 2\boldsymbol{z} - \boldsymbol{x} \in \langle \boldsymbol{x}, \boldsymbol{z} \rangle \subset \operatorname{aff} C.$$

Summing up, $C \subset \{\boldsymbol{a}\} \cup C \subset \operatorname{aff} C$, and Theorem 2.42 gives the equality $\operatorname{aff} C = \operatorname{aff}(\{\boldsymbol{a}\} \cup C)$.

(d) Let C be convex. By assertion (b), $\{\boldsymbol{a}\} \cup C$ is a cone. It remains to show that this cone is convex. Indeed, let $\boldsymbol{x}$ and $\boldsymbol{y}$ be points in $\{\boldsymbol{a}\} \cup C$. If both $\boldsymbol{x}$ and $\boldsymbol{y}$ belong to C, then $[\boldsymbol{x}, \boldsymbol{y}] \subset C \subset \{\boldsymbol{a}\} \cup C$ by the convexity of C. Assume that at least one of the points $\boldsymbol{x}, \boldsymbol{y}$ equals $\boldsymbol{a}$. Because the case $\boldsymbol{x} = \boldsymbol{y} = \boldsymbol{a}$ is obvious, we assume that at least one of $\boldsymbol{x}, \boldsymbol{y}$ is not $\boldsymbol{a}$. Let, for instance, $\boldsymbol{x} = \boldsymbol{a}$ and $\boldsymbol{y} \neq \boldsymbol{a}$. Then

$$[\boldsymbol{a}, \boldsymbol{y}] = \{\boldsymbol{a}\} \cup (\boldsymbol{a}, \boldsymbol{y}] \subset \{\boldsymbol{a}\} \cup (\boldsymbol{a}, \boldsymbol{y}\rangle \subset \{\boldsymbol{a}\} \cup C.$$

(*e*) The convexity of C follows from the equality $C = \{\boldsymbol{a}\} \cup (C \setminus \{\boldsymbol{a}\})$ and assertion (*d*) above. Assume, for contradiction, the existence of another apex $\boldsymbol{a}'$ of C. Since $\boldsymbol{a} \in C$, the point $\boldsymbol{u} = \boldsymbol{a}' + 2(\boldsymbol{a} - \boldsymbol{a}')$ belongs to C. Consequently, $\boldsymbol{a} = \frac{1}{2}(\boldsymbol{u}+\boldsymbol{a}') \in (\boldsymbol{u}, \boldsymbol{a}')$. On the other hand, $(\boldsymbol{u}, \boldsymbol{a}') \subset C\setminus\{\boldsymbol{a}\}$ due to the convexity of $C \setminus \{\boldsymbol{a}\}$. The obtained contradiction shows that $\boldsymbol{a}$ is the unique apex of C. □

Remark. If $C \subset \mathbb{R}^n$ is a convex cone with proper apex $\boldsymbol{a}$, then the cone $C \setminus \{\boldsymbol{a}\}$ may be nonconvex (compare with assertion (*b*) of Theorem 5.4). For instance, the closed halfplane $D = \{(x, y) : y \geqslant 0\}$ of $\mathbb{R}^2$ is a convex cone with apex $\boldsymbol{o}$, while the cone $D \setminus \{\boldsymbol{o}\}$ is not convex.

The result below is analogous to Theorems 2.39 and 3.3.

Theorem 5.5. *For a nonempty set $X \subset \mathbb{R}^n$ and a point $\boldsymbol{a} \in \mathbb{R}^n$, the following conditions are equivalent.*

(*a*) *X is a convex cone with apex $\boldsymbol{a}$.*
(*b*) *$\boldsymbol{a} + \lambda(\boldsymbol{x} - \boldsymbol{a}) + \mu(\boldsymbol{y} - \boldsymbol{a}) \in X$ whenever $\lambda, \mu > 0$ and $\boldsymbol{x}, \boldsymbol{y} \in X$.*
(*c*) *Given an integer $r \geqslant 2$, one has*

$$\boldsymbol{a} + \mu_1(\boldsymbol{x}_1 - \boldsymbol{a}) + \cdots + \mu_r(\boldsymbol{x}_r - \boldsymbol{a}) \in X$$

whenever $\mu_1, \ldots, \mu_r > 0$ and $\boldsymbol{x}_1, \ldots, \boldsymbol{x}_r \in X$.
(*d*) *Given an integer $r \geqslant 2$, one has*

$$\mu_1 X + \cdots + \mu_r X = (\mu_1 + \cdots + \mu_r - 1)\boldsymbol{a} + X \tag{5.2}$$

whenever $\mu_1, \ldots, \mu_r > 0$.

Proof. We are going to show that (*a*) ⇔ (*b*) ⇔ (*c*) and (*a*) ⇔ (*d*).

(*a*) ⇒ (*c*) Let $\mu_1, \ldots, \mu_r > 0$ and $\boldsymbol{x}_1, \ldots, \boldsymbol{x}_r \in X$. Put $\mu = \mu_1 + \cdots + \mu_r$. By Theorem 3.3, X contains the convex combination

$$\boldsymbol{z} = \frac{\mu_1}{\mu}\boldsymbol{x}_1 + \cdots + \frac{\mu_r}{\mu}\boldsymbol{x}_r.$$

Because X is a cone with apex $\boldsymbol{a}$, we have

$$\begin{aligned} &\boldsymbol{a} + \mu_1(\boldsymbol{x}_1 - \boldsymbol{a}) + \cdots + \mu_r(\boldsymbol{x}_r - \boldsymbol{a}) \\ &= \boldsymbol{a} + \mu\Big(\frac{\mu_1}{\mu}\boldsymbol{x}_1 + \cdots + \frac{\mu_r}{\mu}\boldsymbol{x}_r - \frac{\mu_1 + \cdots + \mu_r}{\mu}\boldsymbol{a}\Big) \\ &= \boldsymbol{a} + \mu(\boldsymbol{z} - \boldsymbol{a}) \in X. \end{aligned}$$

$(c) \Rightarrow (b)$ Since the case $r = 2$ is obvious, we assume that $r > 2$. Choose any scalars $\lambda, \mu > 0$ and points $\boldsymbol{x}, \boldsymbol{y} \in X$. Let $\boldsymbol{x}_2 = \cdots = \boldsymbol{x}_r = \boldsymbol{y}$. By the hypothesis, the linear combination

$$\begin{aligned} &\boldsymbol{a} + \lambda(\boldsymbol{x} - \boldsymbol{a}) + \mu(\boldsymbol{y} - \boldsymbol{a}) \\ &= \boldsymbol{a} + \lambda(\boldsymbol{x} - \boldsymbol{a}) + \sum_{i=2}^{r} \frac{\mu}{r-1}(\boldsymbol{x}_i - \boldsymbol{a}) \end{aligned}$$

belongs to X.

$(b) \Rightarrow (a)$ Choosing $\boldsymbol{x} = \boldsymbol{y} \in X$ and $\lambda = \mu = \eta/2 > 0$, one has $\boldsymbol{a} + \eta(\boldsymbol{x} - \boldsymbol{a}) \in X$ whenever $\boldsymbol{x} \in X$ and $\eta > 0$. Hence X is a cone with apex $\boldsymbol{a}$. To prove that X is a convex set, let $\boldsymbol{x}, \boldsymbol{y} \in X$ and $0 < \lambda < 1$. By the assumption,

$$(1-\lambda)\boldsymbol{x} + \lambda\boldsymbol{y} = \boldsymbol{a} + (1-\lambda)(\boldsymbol{x} - \boldsymbol{a}) + \lambda(\boldsymbol{y} - \boldsymbol{a}) \in X.$$

Thus $(\boldsymbol{x}, \boldsymbol{y}) \subset X$, implying the convexity of X.

$(a) \Leftrightarrow (d)$ Let X be a convex cone with apex $\boldsymbol{a}$ and $\mu_1, \ldots, \mu_r$ be positive scalars. A combination of Theorems 3.4 and 5.2 gives

$$\mu_1(X - \boldsymbol{a}) + \cdots + \mu_r(X - \boldsymbol{a}) = (\mu_1 + \cdots + \mu_r)(X - \boldsymbol{a}) = X - \boldsymbol{a}.$$

Consequently,

$$(\mu_1 X - \mu_1 \boldsymbol{a}) + \cdots + (\mu_r X - \mu_r \boldsymbol{a}) = X - \boldsymbol{a},$$

which is equivalent to (5.2).

Conversely, if X satisfies condition (d), then (5.2) can be rewritten as

$$\boldsymbol{a} + \mu_1(X - \boldsymbol{a}) + \cdots + \mu_r(X - \boldsymbol{a}) = X$$

whenever $\mu_1, \ldots, \mu_r > 0$. Hence X contains all positive combinations

$$\boldsymbol{a} + \mu_1(\boldsymbol{x}_1 - \boldsymbol{a}) + \cdots + \mu_r(\boldsymbol{x}_r - \boldsymbol{a}), \quad \text{where} \quad \boldsymbol{x}_1, \ldots, \boldsymbol{x}_r \in X,$$

and the equivalence $(a) \Leftrightarrow (c)$ shows that X is a convex cone with apex $\boldsymbol{a}$. □

5.1.2 *Simplicial Cones*

Definition 5.6. Let $\{\boldsymbol{a}, \boldsymbol{x}_1, \ldots, \boldsymbol{x}_r\}$ be an affinely independent set in $\mathbb{R}^n$. The *r-simplicial cone* $\Gamma_{\boldsymbol{a}}(\boldsymbol{x}_1, \ldots, \boldsymbol{x}_r)$ with apex $\boldsymbol{a}$ generated by $\boldsymbol{x}_1, \ldots, \boldsymbol{x}_r$ is defined by

$$\Gamma_{\boldsymbol{a}}(\boldsymbol{x}_1, \ldots, \boldsymbol{x}_r) = \{\boldsymbol{a} + \lambda_1(\boldsymbol{x}_1 - \boldsymbol{a}) + \cdots + \lambda_r(\boldsymbol{x}_r - \boldsymbol{a}) : \lambda_1, \ldots, \lambda_r \geqslant 0\}.$$

Example. The nonnegative octant O of $\mathbb{R}^n$, given by

$$O = \{(x_1, \dots, x_n) \in \mathbb{R}^n : x_1 \geqslant 0, \dots, x_n \geqslant 0\},$$

is a simplicial cone with apex $\boldsymbol{o}$; indeed, it can be expressed as $O = \Gamma_{\boldsymbol{o}}(\boldsymbol{e}_1, \dots, \boldsymbol{e}_n)$, where $\{\boldsymbol{e}_1, \dots, \boldsymbol{e}_n\}$ is the standard basis for $\mathbb{R}^n$.

Remark. For a point $\boldsymbol{x} \in \Gamma_{\boldsymbol{a}}(\boldsymbol{x}_1, \dots, \boldsymbol{x}_r)$, the scalars $\lambda_1, \dots, \lambda_r$ in the expression

$$\boldsymbol{x} = \boldsymbol{a} + \lambda_1(\boldsymbol{x}_1 - \boldsymbol{a}) + \cdots + \lambda_r(\boldsymbol{x}_r - \boldsymbol{a}) \tag{5.3}$$

are uniquely determined by $\boldsymbol{x}$. Indeed, since $\boldsymbol{x}$ can be written as an affine combination

$$\boldsymbol{x} = (1 - \lambda_1 - \cdots - \lambda_r)\boldsymbol{a} + \lambda_1\boldsymbol{x}_1 + \cdots + \lambda_r\boldsymbol{x}_r$$

of the affinely independent set $\{\boldsymbol{a}, \boldsymbol{x}_1, \dots, \boldsymbol{x}_r\}$, the uniqueness of $\lambda_1, \dots, \lambda_r$ follows from Theorem 2.52. Consequently, a point $\boldsymbol{x} \in \operatorname{aff}\{\boldsymbol{a}, \boldsymbol{x}_1, \dots, \boldsymbol{x}_r\}$ belongs to $\Gamma_{\boldsymbol{a}}(\boldsymbol{x}_1, \dots, \boldsymbol{x}_r)$ if and only if all scalars $\lambda_1, \dots, \lambda_r$ in the expression (5.3) are nonnegative.

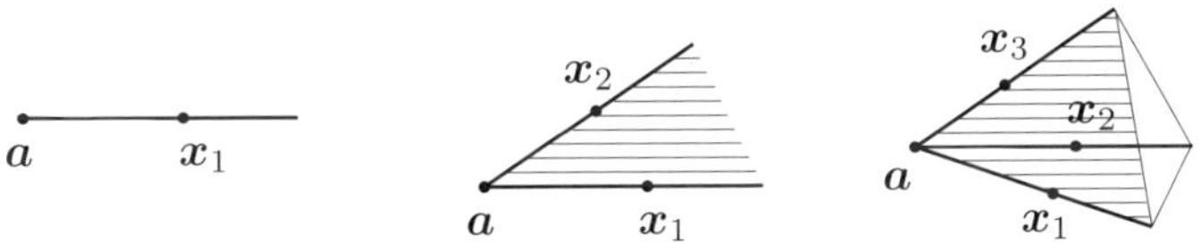

Fig. 5.2 Simplicial cones in $\mathbb{R}^3$.

Theorem 5.7. *Any r-simplicial cone $\Gamma = \Gamma_{\boldsymbol{a}}(\boldsymbol{x}_1, \dots, \boldsymbol{x}_r) \subset \mathbb{R}^n$ is a convex cone of dimension r, and $\boldsymbol{a}$ is its only apex. Furthermore,*

$$\operatorname{aff} \Gamma = \operatorname{aff}\{\boldsymbol{a}, \boldsymbol{x}_1, \dots, \boldsymbol{x}_r\}. \tag{5.4}$$

Proof. Choose a point $\boldsymbol{x} \in \Gamma$ and a scalar $\eta > 0$. By Definition 5.6,

$$\boldsymbol{x} = \boldsymbol{a} + \lambda_1(\boldsymbol{x}_1 - \boldsymbol{a}) + \cdots + \lambda_r(\boldsymbol{x}_r - \boldsymbol{a})$$

for suitable scalars $\lambda_1, \dots, \lambda_r \geqslant 0$. So,

$$\boldsymbol{a} + \eta(\boldsymbol{x} - \boldsymbol{a}) = \boldsymbol{a} + \lambda_1\eta(\boldsymbol{x}_1 - \boldsymbol{a}) + \cdots + \lambda_r\eta(\boldsymbol{x}_r - \boldsymbol{a}) \in \Gamma.$$

Hence Γ is a cone with apex $\boldsymbol{a}$.

For the convexity of Γ, choose points $\boldsymbol{y}, \boldsymbol{z} \in \Gamma$ and a scalar $\mu \in [0,1]$. Then $\boldsymbol{y}$ and $\boldsymbol{z}$ can be written as

$$\boldsymbol{y} = \boldsymbol{a} + \lambda_1(\boldsymbol{x}_1 - \boldsymbol{a}) + \cdots + \lambda_r(\boldsymbol{x}_r - \boldsymbol{a}), \ \lambda_1, \ldots, \lambda_r \geqslant 0,$$
$$\boldsymbol{z} = \boldsymbol{a} + \gamma_1(\boldsymbol{x}_1 - \boldsymbol{a}) + \cdots + \gamma_r(\boldsymbol{x}_r - \boldsymbol{a}), \ \gamma_1, \ldots, \gamma_r \geqslant 0.$$

Let $\mu_i = (1-\mu)\lambda_i + \mu\gamma_i$, $1 \leqslant i \leqslant r$. Clearly, $\mu_1, \ldots, \mu_r \geqslant 0$ and

$$(1-\mu)\boldsymbol{y} + \mu\boldsymbol{z} = \boldsymbol{a} + \mu_1(\boldsymbol{x}_1 - \boldsymbol{a}) + \cdots + \mu_r(\boldsymbol{x}_r - \boldsymbol{a}) \in \Gamma,$$

which shows that Γ is a convex set.

The obvious inclusions

$$\{\boldsymbol{a}, \boldsymbol{x}_1, \ldots, \boldsymbol{x}_r\} \subset \Gamma \subset \operatorname{aff}\{\boldsymbol{a}, \boldsymbol{x}_1, \ldots, \boldsymbol{x}_r\}$$

and Theorem 2.42 give (5.4). Also, Theorem 2.52 shows that

$$\dim \Gamma = \dim(\operatorname{aff}\{\boldsymbol{a}, \boldsymbol{x}_1, \ldots, \boldsymbol{x}_r\}) = r.$$

It remains to prove that $\boldsymbol{a}$ is the unique apex of Γ. Indeed, assume the existence of another apex $\boldsymbol{a}'$ of Γ. Let $\boldsymbol{c} = \frac{1}{2}(\boldsymbol{a} + \boldsymbol{a}')$ and $\boldsymbol{b} = 2\boldsymbol{a} - \boldsymbol{a}'$, as depicted in Figure 5.3.

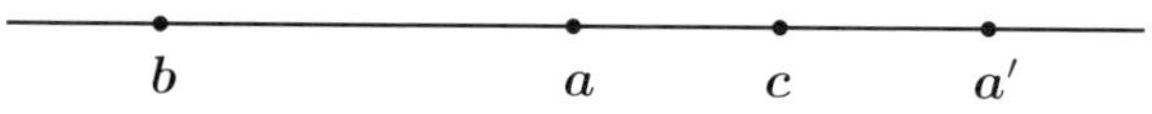

Fig. 5.3 Illustration to Theorem 5.7.

Then $\boldsymbol{b}, \boldsymbol{c} \in (\boldsymbol{a}', \boldsymbol{a}\rangle \subset \Gamma$. Similarly, $\boldsymbol{a}' \in (\boldsymbol{a}, \boldsymbol{c}\rangle \subset \Gamma$. This inclusion allows us to express $\boldsymbol{a}'$ as

$$\boldsymbol{a}' = \boldsymbol{a} + \alpha_1(\boldsymbol{x}_1 - \boldsymbol{a}) + \cdots + \alpha_r(\boldsymbol{x}_r - \boldsymbol{a}), \quad \alpha_1, \ldots, \alpha_r \geqslant 0.$$

Consequently,

$$\boldsymbol{b} = 2\boldsymbol{a} - \boldsymbol{a}' = \boldsymbol{a} + (-\alpha_1)(\boldsymbol{x}_1 - \boldsymbol{a}) + \cdots + (-\alpha_r)(\boldsymbol{x}_r - \boldsymbol{a}).$$

Since $\boldsymbol{b} \in \Gamma$, we should have $-\alpha_1, \ldots, -\alpha_r \geqslant 0$ (see a remark on page 155). Thus $\alpha_1 = \cdots = \alpha_r = 0$, and $\boldsymbol{a}' = \boldsymbol{a}$, contrary to the assumption $\boldsymbol{a}' \neq \boldsymbol{a}$. The obtained contradiction proves the uniqueness of the apex of Γ. $\square$

A combination Corollary 2.66 and Theorems 5.5 and 5.7 yields a frequently used assertion, given below.

Corollary 5.8. *For a convex cone $C \subset \mathbb{R}^n$ of positive dimension with apex $\boldsymbol{a}$, the following assertions hold.*

(*a*) *If* $\boldsymbol{x}_1, \ldots, \boldsymbol{x}_r$ *are points in* $C \setminus \{\boldsymbol{a}\}$ *such that the set* $\{\boldsymbol{a}, \boldsymbol{x}_1, \ldots, \boldsymbol{x}_r\}$ *is affinely independent, then the* r*-simplicial cone* $\Gamma_{\boldsymbol{a}}(\boldsymbol{x}_1, \ldots, \boldsymbol{x}_r)$ *lies in* $\{\boldsymbol{a}\} \cup C$.
(*b*) *The dimension of* C *equals the maximum dimension of a simplicial cone with apex* $\boldsymbol{a}$ *contained in* $\{\boldsymbol{a}\} \cup C$.
(*c*) *If* $m = \dim C$, *then* $\operatorname{aff} C = \operatorname{aff} \Gamma_{\boldsymbol{a}}(\boldsymbol{x}_1, \ldots, \boldsymbol{x}_m)$ *for every* m*-simplicial cone* $\Gamma_{\boldsymbol{a}}(\boldsymbol{x}_1, \ldots, \boldsymbol{x}_m)$ *contained in* $\{\boldsymbol{a}\} \cup C$. □

5.1.3 *Algebra of Cones*

The first result of this subsection deals with set-theoretic properties of cones.

Corollary 5.9. *For a family* $\{C_\alpha\}$ *of cones in* $\mathbb{R}^n$ *with a common apex* $\boldsymbol{a}$, *the following assertions hold.*

(*a*) *Both sets* $\bigcap_\alpha C_\alpha$ *and* $\bigcup_\alpha C_\alpha$ *are cones with apex* $\boldsymbol{a}$.
(*b*) *The set* $\operatorname{conv}(\bigcup_\alpha C_\alpha)$ *is the smallest convex cone with apex* $\boldsymbol{a}$ *containing* $\bigcup_\alpha C_\alpha$.
(*c*) *If the family* $\{C_\alpha\}$ *is nested and all cones* C_α *are convex, then the union* $\bigcup_\alpha C_\alpha$ *is a convex cone with apex* $\boldsymbol{a}$.

Proof. Assertion (*a*) is obvious, while the remaining two follow from Theorems 3.8 and 5.4. □

Remark. If in Corollary 5.9, $\boldsymbol{a}$ is an improper apex of at least one cone C_α, then $\boldsymbol{a}$ is an improper apex of the cone $\bigcap_\alpha C_\alpha$. Similarly, if $\boldsymbol{a}$ is a proper apex of at least one cone C_α, then $\boldsymbol{a}$ is a proper apex of the cone $\bigcup_\alpha C_\alpha$. See page 205 for a more general assertion on unions and intersections of cones.

Theorem 5.10. *Let* $C_1, \ldots, C_r \subset \mathbb{R}^n$ *be cones with apices* $\boldsymbol{a}_1, \ldots, \boldsymbol{a}_r$, *respectively, and* $\mu_1, \ldots, \mu_r$ *be scalars. The following assertions hold.*

(*a*) *The sum* $\mu_1 C_1 + \cdots + \mu_r C_r$ *is a cone with apex* $\mu_1 \boldsymbol{a}_1 + \cdots + \mu_r \boldsymbol{a}_r$.
(*b*) *The cone* $\mu_1 C_1 + \cdots + \mu_r C_r$ *is convex if all* $C_1, \ldots, C_r$ *are convex.*
(*c*) *If all* $C_1, \ldots, C_r$ *are convex cones with proper apices* $\boldsymbol{a}_1, \ldots, \boldsymbol{a}_r$, *then*

$$\mu_1 C_1 + \cdots + \mu_r C_r = \mu_1 \boldsymbol{a}_1 + \cdots + \mu_r \boldsymbol{a}_r + \operatorname{conv}(\mu_1(C_1 - \boldsymbol{a}_1) \cup \cdots \cup \mu_r(C_r - \boldsymbol{a}_r)).$$

Proof. An induction argument shows that the proof of the theorem can be reduced to the case $r = 2$ (let $\mu_1 C_1 = \mu_1 C_1 + \mu_2\{\boldsymbol{o}\}$ if $r = 1$).

(*a*) Choose a point $\boldsymbol{x} \in \mu_1 C_1 + \mu_2 C_2$. Then $\boldsymbol{x} = \mu_1 \boldsymbol{x}_1 + \mu_2 \boldsymbol{x}_2$, where $\boldsymbol{x}_1 \in C_1$ and $\boldsymbol{x}_2 \in C_2$. For a scalar $\lambda > 0$, one has

$$\begin{aligned}&(\mu_1 \boldsymbol{a}_1 + \mu_2 \boldsymbol{a}_2) + \lambda(\boldsymbol{x} - (\mu_1 \boldsymbol{a}_1 + \mu_2 \boldsymbol{a}_2))\\ &\qquad = \mu_1(\boldsymbol{a}_1 + \lambda(\boldsymbol{x}_1 - \boldsymbol{a}_1)) + \mu_2(\boldsymbol{a}_2 + \lambda(\boldsymbol{x}_2 - \boldsymbol{a}_2)) \in \mu_1 C_1 + \mu_2 C_2,\end{aligned}$$

which shows that $\mu_1 C_1 + \mu_2 C_2$ is a cone with apex $\mu_1 \boldsymbol{a}_1 + \mu_2 \boldsymbol{a}_2$.

(*b*) If the cones C_1 and C_2 are convex, then the convexity of $\mu_1 C_1 + \mu_2 C_2$ follows from Theorem 3.10.

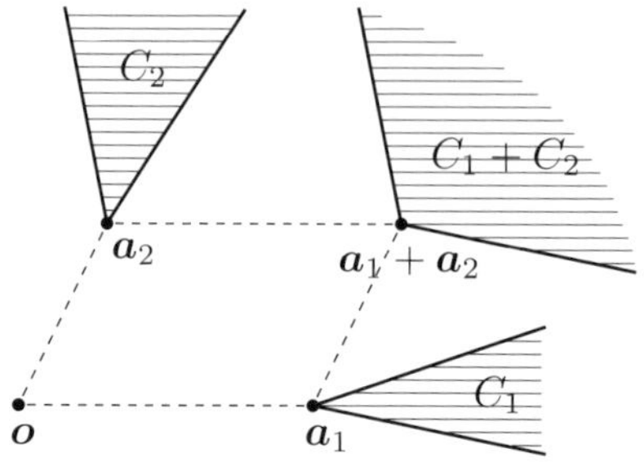

Fig. 5.4 The sum of cones C_1 and C_2.

(*c*) Let C_1 and C_2 be convex cones with proper apices $\boldsymbol{a}_1$ and $\boldsymbol{a}_2$, respectively. For the equality

$$\mu_1 C_1 + \mu_2 C_2 = \mu_1 \boldsymbol{a}_1 + \mu_2 \boldsymbol{a}_2 + \operatorname{conv}(\mu_1(C_1 - \boldsymbol{a}_1) \cup \mu_2(C_2 - \boldsymbol{a}_2)), \quad (5.5)$$

put $C_1' = C_1 - \boldsymbol{a}_1$ and $C_2' = C_2 - \boldsymbol{a}_2$. By Theorem 5.2, C_1' and C_2' are convex cones with common proper apex $\boldsymbol{o}$. Clearly, (5.5) can be rewritten as

$$\mu_1 C_1' + \mu_2 C_2' = \operatorname{conv}(\mu_1 C_1' \cup \mu_2 C_2'). \quad (5.6)$$

If $\boldsymbol{x} \in \mu_1 C_1' + \mu_2 C_2'$, then $\boldsymbol{x} = \mu_1 \boldsymbol{x}_1 + \mu_2 \boldsymbol{x}_2$, where $\boldsymbol{x}_1 \in C_1'$ and $\boldsymbol{x}_2 \in C_2'$. Since $2\boldsymbol{x}_1 \in C_1'$ and $2\boldsymbol{x}_2 \in C_2'$, Theorem 4.13 gives

$$\boldsymbol{x} = \tfrac{1}{2}(\mu_1(2\boldsymbol{x}_1) + \mu_2(2\boldsymbol{x}_2)) \in \operatorname{conv}(\mu_1 C_1' \cup \mu_2 C_2').$$

Conversely, let $\boldsymbol{x} \in \operatorname{conv}(\mu_1 C_1' \cup \mu_2 C_2')$. By the same theorem, $\boldsymbol{x}$ can be expressed as a convex combination

$$\boldsymbol{x} = \lambda_1 \boldsymbol{x}_1 + \lambda_2 \boldsymbol{x}_2, \quad \text{where} \quad \boldsymbol{x}_1 \in \mu_1 C_1' \quad \text{and} \quad \boldsymbol{x}_2 \in \mu_2 C_2'.$$

Hence

$$\boldsymbol{x} = \lambda_1 \boldsymbol{x}_1 + \lambda_2 \boldsymbol{x}_2 \in \mu_1(\lambda_1 C_1') + \mu_2(\lambda_2 C_2').$$

If both λ_1 and λ_2 are positive, then $\lambda_1 C_1' = C_1'$ and $\lambda_2 C_2' = C_2'$ (see Theorem 5.2), implying the inclusion $\boldsymbol{x} \in \mu_1 C_1' + \mu_2 C_2'$. If, for instance, $\lambda_1 = 1$ and $\lambda_2 = 0$, then

$$\boldsymbol{x} = \boldsymbol{x}_1 + \boldsymbol{o} \in \mu_1 C_1' + \mu_2 \boldsymbol{o} \subset \mu_1 C_1' + \mu_2 C_2'.$$

Summing up, (5.6) holds. □

Remark. In the last assertion of Theorem 5.10, the assumption that $\boldsymbol{a}_1, \ldots, \boldsymbol{a}_r$ are proper apices is essential. Indeed, consider the convex cones

$$C_1 = \{(x, 0) : x > 0\} \quad \text{and} \quad C_2 = \{(0, y) : y > 0\}$$

with a common improper apex $\boldsymbol{o}$. Clearly,

$$C_1 + C_2 = \{(x, y) : x, y > 0\} \neq \{(x, y) : x, y \geqslant 0\} \setminus \{\boldsymbol{o}\} = \operatorname{conv}(C_1 \cup C_2).$$

Corollary 5.11. *If $C \subset \mathbb{R}^n$ is a nonempty cone with apex $\boldsymbol{a}$, then for any point $\mathbf{e} \in \mathbb{R}^n$ and a scalar μ, the set $\mathbf{e} + \mu C$ is a cone with apex $\mathbf{e} + \mu\boldsymbol{a}$. If, additionally, C is convex or $\boldsymbol{a}$ is a proper apex of C, then so are $\mathbf{e} + \mu C$ and $\mathbf{e} + \mu\boldsymbol{a}$. Furthermore,*

$$\mathbf{e} + \mu C = \begin{cases} \mathbf{e} + (\mu - 1)\boldsymbol{a} + C & \textit{if} \quad \mu > 0, \\ \{\mathbf{e}\} & \textit{if} \quad \mu = 0, \\ \mathbf{e} + (\mu + 1)\boldsymbol{a} - C & \textit{if} \quad \mu < 0. \end{cases}$$

Proof. By Theorem 5.10, the set $\mathbf{e} + \mu C$ is a cone with apex $\mathbf{e} + \mu\boldsymbol{a}$. Clearly, $\mathbf{e} + \mu\boldsymbol{a}$ is a proper apex of $\mathbf{e} + \mu C$ if and only if $\boldsymbol{a}$ is a proper apex of C. If C is convex, the convexity of $\mathbf{e} + \mu C$ follows from Theorem 5.10. The last assertion is due to the equalities

$$\mathbf{e} + \mu C = \mathbf{e} + \mu\boldsymbol{a} + \mu(C - \boldsymbol{a}) \quad \text{and} \quad \mu(C - \boldsymbol{a}) = \operatorname{sgn}(\mu)(C - \boldsymbol{a}),$$

where the second one derives from Theorem 5.2. □

Theorem 5.12. *Let $f : \mathbb{R}^n \to \mathbb{R}^m$ be an affine transformation, and $C \subset \mathbb{R}^n$ and $D \subset \mathbb{R}^m$ be cones with apices $\boldsymbol{a}$ and $\mathbf{c}$, respectively, where $\mathbf{c} \in \operatorname{rng} f$. Then the sets $f(C)$ and $f^{-1}(D)$ are cones with apices $f(\boldsymbol{a})$ and $\boldsymbol{u} \in f^{-1}(\boldsymbol{c})$, respectively. If, additionally, both C and D are convex cones or $\boldsymbol{a}$ and $\boldsymbol{u}$ are proper apices, then so are, respectively, $f(C)$ and $f^{-1}(D)$, and $f(\boldsymbol{a})$ and $\boldsymbol{u}$.*

Proof. Choose a point $\boldsymbol{x} \in f(C)$ and a scalar $\lambda > 0$. Then $\boldsymbol{x} = f(\boldsymbol{x}_0)$ for a suitable point $\boldsymbol{x}_0 \in C$. Since $\boldsymbol{a} + \lambda(\boldsymbol{x}_0 - \boldsymbol{a})$ is an affine combination which lies in C, Theorem 2.81 gives

$$f(\boldsymbol{a}) + \lambda(\boldsymbol{x} - f(\boldsymbol{a})) = f(\boldsymbol{a} + \lambda(\boldsymbol{x}_0 - \boldsymbol{a})) \in f(C).$$

Therefore, $f(C)$ is a cone with apex $f(\boldsymbol{a})$.

If $f^{-1}(D)$ is empty, then it is a cone with any apex $\boldsymbol{u} \in f^{-1}(\boldsymbol{c})$. Suppose that $f^{-1}(D) \neq \varnothing$. Let $\boldsymbol{x} \in f^{-1}(D)$ and $\lambda > 0$. For any point $\boldsymbol{u} \in f^{-1}(\boldsymbol{c})$, one has

$$f(\boldsymbol{u} + \lambda(\boldsymbol{x} - \boldsymbol{u})) = \boldsymbol{c} + \lambda(f(\boldsymbol{x}) - \boldsymbol{c}) \in D.$$

Hence $\boldsymbol{u} + \lambda(\boldsymbol{x} - \boldsymbol{u}) \in f^{-1}(D)$, and $f^{-1}(D)$ is a cone with apex $\boldsymbol{u}$.

If, additionally, both C and D are convex cones, then Theorem 3.15 shows the convexity of $f(C)$ and $f^{-1}(D)$. □

5.1.4 *Apex Sets of Cones*

Definition 5.13. The *apex set* of a nonempty cone $C \subset \mathbb{R}^n$, denoted $\operatorname{ap} C$, is the set of all apices of C. We let $\operatorname{ap} C = \mathbb{R}^n$ if $C = \varnothing$.

Examples. 1. For the convex cones

$$C_1 = \{(x, y, z) : x > 0,\ y > 0\}, \quad C_2 = \operatorname{cl} C_1, \quad C_3 = C_1 \cup \{\boldsymbol{o}\},$$

the apex sets of C_1 and C_2 coincide with the z-axis of $\mathbb{R}^3$, while $\operatorname{ap} C_3 = \{\boldsymbol{o}\}$, as depicted in Figure 5.5.

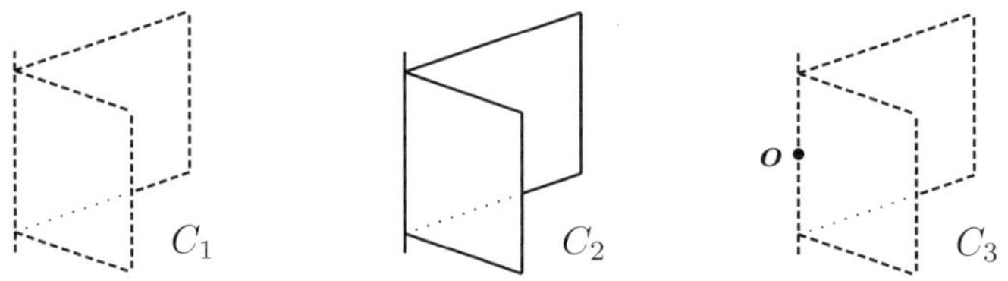

Fig. 5.5 Apex sets of convex cones.

2. The apex set of a nonempty plane $L \subset \mathbb{R}^n$ coincides with L (see an example on page 150).

3. If F is a halfplane in $\mathbb{R}^n$ (either closed or open), then $\operatorname{ap} F = \operatorname{rbd} F$ (see an example on page 150).

4. The apex set of a simplicial cone $\Gamma_{\boldsymbol{a}}(\boldsymbol{x}_1, \ldots, \boldsymbol{x}_r) \subset \mathbb{R}^n$ is the singleton $\{\boldsymbol{a}\}$, as follows from Theorem 5.7.

Theorem 5.14. *The apex set of any cone $C \subset \mathbb{R}^n$ is a plane. Furthermore, either $\operatorname{ap} C \subset C$ or $C \cap \operatorname{ap} C = \varnothing$.*

Proof. The assertion is obvious if $C = \varnothing$ or if $\operatorname{ap} C$ is a singleton. So, we may assume that $C \neq \varnothing$ and $\operatorname{ap} C$ contain more than one point. According to Theorem 2.38, it suffices to show that for any choice of distinct points $\boldsymbol{a}_1, \boldsymbol{a}_2 \in \operatorname{ap} C$, the line $l = \langle \boldsymbol{a}_1, \boldsymbol{a}_2 \rangle$ lies in $\operatorname{ap} C$. Choose any point $\boldsymbol{a} \in l$. Then $\boldsymbol{a} = (1-\gamma)\boldsymbol{a}_1 + \gamma \boldsymbol{a}_2$ for a suitable scalar γ. We express γ as a difference of positive scalars α and β: $\gamma = \alpha - \beta$.

Let $C_1 = C - \boldsymbol{a}_1$ and $C_2 = C - \boldsymbol{a}_2$. Clearly, $C_2 = (\boldsymbol{a}_1 - \boldsymbol{a}_2) + C_1$. By Theorem 5.2, both C_1 and C_2 are cones with apex $\boldsymbol{o}$. By the same theorem,

$$\begin{aligned}(\boldsymbol{a}_1 - \boldsymbol{a}_2) + C_1 = C_2 = \alpha C_2 &= \alpha((\boldsymbol{a}_1 - \boldsymbol{a}_2) + C_1) \\ &= \alpha(\boldsymbol{a}_1 - \boldsymbol{a}_2) + \alpha C_1 = \alpha(\boldsymbol{a}_1 - \boldsymbol{a}_2) + C_1.\end{aligned}$$

A similar argument gives

$$(\boldsymbol{a}_1 - \boldsymbol{a}_2) + C_1 = \beta(\boldsymbol{a}_1 - \boldsymbol{a}_2) + C_1.$$

Consequently,

$$\begin{aligned}&\gamma(\boldsymbol{a}_1 - \boldsymbol{a}_2) + C_1 = (\alpha(\boldsymbol{a}_1 - \boldsymbol{a}_2) + C_1) - \beta(\boldsymbol{a}_1 - \boldsymbol{a}_2) \\ &\quad = ((\boldsymbol{a}_1 - \boldsymbol{a}_2) + C_1) - \beta(\boldsymbol{a}_1 - \boldsymbol{a}_2) = (\boldsymbol{a}_1 - \boldsymbol{a}_2) + (C_1 - \beta(\boldsymbol{a}_1 - \boldsymbol{a}_2)) \\ &\quad = (\boldsymbol{a}_1 - \boldsymbol{a}_2) + (C_1 - (\boldsymbol{a}_1 - \boldsymbol{a}_2)) = C_1.\end{aligned}$$

From here we obtain

$$C - \boldsymbol{a} = C - (\boldsymbol{a}_1 + \gamma(\boldsymbol{a}_2 - \boldsymbol{a}_1)) = C_1 + \gamma(\boldsymbol{a}_1 - \boldsymbol{a}_2) = C_1.$$

Since C_1 is a cone with apex $\boldsymbol{o}$, Theorem 5.2 implies that C is a cone with apex $\boldsymbol{a}$. Summing up, $l \subset \operatorname{ap} C$, which shows that $\operatorname{ap} C$ is a plane.

For the second part, suppose the existence of an apex $\boldsymbol{a} \in C \cap \operatorname{ap} C$. If $\operatorname{ap} C = \{\boldsymbol{a}\}$, then the inclusion $\operatorname{ap} C \subset C$ is obvious. Assume the existence of another apex $\boldsymbol{a}'$ of C. Then the open halfline $(\boldsymbol{a}', \boldsymbol{a}\rangle$ lies in C. If $\boldsymbol{a}''$ is the midpoint of the open segment $(\boldsymbol{a}, \boldsymbol{a}')$, then $\boldsymbol{a}'' \in C$, implying that $\boldsymbol{a}' \in (\boldsymbol{a}, \boldsymbol{a}''\rangle \subset C$ (see Lemma 2.25). Hence $\operatorname{ap} C \subset C$. □

As mentioned in a remark on page 149, a nonempty cone $C \subset \mathbb{R}^n$ with apex $\boldsymbol{a}$ is either a singleton $\{\boldsymbol{a}\}$ or a union of halflines (closed or open) with common endpoint $\boldsymbol{a}$. In view of Theorem 5.14, this assertion can be generalized to the case of cones with arbitrary apex sets.

Theorem 5.15. *Let $C \subset \mathbb{R}^n$ be a cone whose apex set is a plane of dimension m, where $0 \leqslant m \leqslant n-1$. Then C has one of the following shapes.*

(*a*) *C is a plane.*
(*b*) *C is a union of open $(m+1)$-dimensional halfplanes bounded by* $\operatorname{ap} C$.

(c) *C is a union of closed $(m+1)$-dimensional halfplanes bounded by* $\operatorname{ap} C$.

Proof. (a) If C is a plane, then $C = \operatorname{ap} C$ (see an example on page 160). So, we may assume that C is not a plane. Then $C \setminus \operatorname{ap} C \neq \varnothing$ because C is nonempty.

Choose any point $\boldsymbol{u}$ in $C \setminus \operatorname{ap} C$ and put $L_{\boldsymbol{u}} = \operatorname{aff}(\operatorname{ap} C \cup \{\boldsymbol{u}\})$. Corollary 2.60 shows that $L_{\boldsymbol{u}}$ is a plane of dimension $m+1$. Let $E_{\boldsymbol{u}}$ be the open halfplane of $L_{\boldsymbol{u}}$ which is determined by $\operatorname{ap} C$ and contains $\boldsymbol{u}$ (see Theorem 2.73).

We assert that $E_{\boldsymbol{u}} \subset C$. For this, choose any point $\boldsymbol{x} \in E_{\boldsymbol{u}}$. Further, let $\{\boldsymbol{a}_1, \dots, \boldsymbol{a}_{m+1}\}$ be an affine basis for $\operatorname{ap} C$ (see Theorem 2.57). Corollary 2.60 implies that $\{\boldsymbol{a}_1, \dots, \boldsymbol{a}_{m+1}, \boldsymbol{u}\}$ is an affine basis for $L_{\boldsymbol{u}}$. By Theorem 2.75, $\boldsymbol{x}$ can be written as an affine combination

$$\boldsymbol{x} = \lambda_1 \boldsymbol{a}_1 + \dots + \lambda_{m+1} \boldsymbol{a}_{m+1} + \lambda_{m+2} \boldsymbol{u}, \quad \text{where } \lambda_{m+2} > 0. \tag{5.7}$$

We consider the cases $\lambda_{m+2} \neq 1$ and $\lambda_{m+2} = 1$ separately.

1. Assume first that $\lambda_{m+2} \neq 1$. Because the point

$$\boldsymbol{a} = \frac{\lambda_1}{1 - \lambda_{m+2}} \boldsymbol{a}_1 + \dots + \frac{\lambda_{m+1}}{1 - \lambda_{m+2}} \boldsymbol{a}_{m+1},$$

is an affine combination of $\boldsymbol{a}_1, \dots, \boldsymbol{a}_{m+1} \in \operatorname{ap} C$, one has $\boldsymbol{a} \in \operatorname{ap} C$ (see Theorem 2.39). Consequently,

$$\boldsymbol{x} = (1 - \lambda_{m+2})\boldsymbol{a} + \lambda_{m+2}\boldsymbol{u} = \boldsymbol{a} + \lambda_{m+2}(\boldsymbol{u} - \boldsymbol{a}) \in (\boldsymbol{a}, \boldsymbol{u}\rangle \subset C.$$

2. Let $\lambda_{m+2} = 1$. Then $\lambda_1 + \dots + \lambda_{m+1} = 1 - \lambda_{m+2} = 0$. Put

$$\boldsymbol{a} = (1 - \lambda_1)\boldsymbol{a}_1 - \lambda_2 \boldsymbol{a}_2 - \dots - \lambda_{m+1}\boldsymbol{a}_{m+1}, \quad \boldsymbol{z} = \boldsymbol{a}_1 + \tfrac{1}{2}(\boldsymbol{u} - \boldsymbol{a}_1).$$

Similarly to the above, $\boldsymbol{a} \in \operatorname{ap} C$ is an affine combination of $\boldsymbol{a}_1, \dots, \boldsymbol{a}_{m+1}$. Because $\boldsymbol{a}_1 \in \operatorname{ap} C$ and $\boldsymbol{u} \in C$, the point $\boldsymbol{z}$ also belongs to C. Therefore, (5.7) and the equalities

$$\begin{aligned}
\boldsymbol{x} &= \lambda_1 \boldsymbol{a}_1 + \dots + \lambda_{m+1}\boldsymbol{a}_{m+1} + 1\boldsymbol{u} \\
&= (\boldsymbol{a}_1 + \boldsymbol{u}) - ((1 - \lambda)\boldsymbol{a}_1 - \lambda_2 \boldsymbol{a}_2 - \dots - \lambda_{m+1}\boldsymbol{a}_{m+1}) \\
&= 2\boldsymbol{z} - \boldsymbol{a} = \boldsymbol{a} + 2(\boldsymbol{z} - \boldsymbol{a})
\end{aligned}$$

imply that $\boldsymbol{x} \in C$. Summing up, $E_{\boldsymbol{u}} \subset C$.

The above argument shows that the set $C \setminus \operatorname{ap} C$ is a union of $(m+1)$-dimensional open halfplanes of the form $E_{\boldsymbol{u}}$, where $\boldsymbol{u} \in C \setminus \operatorname{ap} C$.

(b) If $C \cap \operatorname{ap} C = \varnothing$, then

$$C = \cup\,(E_{\boldsymbol{u}} : \boldsymbol{u} \in C \setminus \operatorname{ap} C).$$

(*c*) If $C \cap \mathrm{ap}\, C = \varnothing$, then $\mathrm{ap}\, C \subset C$ according to Theorem 5.14. Consequently, every set $D_{\boldsymbol{u}} = \mathrm{ap}\, C \cup E_{\boldsymbol{u}}$, $\boldsymbol{u} \in C \setminus \mathrm{ap}\, C$, is a closed $(m+1)$-dimensional halfplane determined by $\mathrm{ap}\, C$ (see Theorem 2.73), and

$$\begin{aligned} C &= \mathrm{ap}\, C \cup (C \setminus \mathrm{ap}\, C) = \cup (\mathrm{ap}\, C \cup E_{\boldsymbol{u}} : \boldsymbol{u} \in C \setminus \mathrm{ap}\, C) \\ &= \cup (D_{\boldsymbol{u}} : \boldsymbol{u} \in C \setminus \mathrm{ap}\, C). \end{aligned}$$

□

Theorem 5.16. *If $C \subset \mathbb{R}^n$ is a nonempty cone with apex $\boldsymbol{a}$, then*

$$\boldsymbol{u} - \boldsymbol{a} + \mathrm{ap}\, C \subset C \quad \text{for any point} \quad \boldsymbol{u} \in C.$$

Proof. Let $\boldsymbol{u}$ be a point in C. Suppose first that $\boldsymbol{u} \in \mathrm{ap}\, C$. Then $\mathrm{ap}\, C \subset C$ according to Theorem 5.14. On the other hand, Theorem 2.2 and Definition 2.3 give

$$\boldsymbol{u} - \boldsymbol{a} \in \mathrm{ap}\, C - \mathrm{ap}\, C = \mathrm{dir}\, (\mathrm{ap}\, C),$$

implying the equality $\boldsymbol{u} - \boldsymbol{a} + \mathrm{ap}\, C = \mathrm{ap}\, C$ according to the same theorem. Summing up, $\boldsymbol{u} - \boldsymbol{a} + \mathrm{ap}\, C \subset C$.

Suppose now that $\boldsymbol{u} \in C \setminus \mathrm{ap}\, C$. Put $L = \mathrm{aff}\, (\{\boldsymbol{u}\} \cup \mathrm{ap}\, C)$. As follows from the proof of Theorem 5.15, the open halfplane $E_{\boldsymbol{u}}$ of L determined by $\mathrm{ap}\, C$ and containing $\boldsymbol{u}$, lies in C. Since $E_{\boldsymbol{u}}$ is the intersection of L and a suitable open halfspace (see Definition 2.69), Theorem 2.34 implies that the plane $\boldsymbol{u} - \boldsymbol{a} + \mathrm{ap}\, C$ lies in $E_{\boldsymbol{u}}$. So, $\boldsymbol{u} - \boldsymbol{a} + \mathrm{ap}\, C \subset E_{\boldsymbol{u}} \subset C$, as desired. □

Theorem 5.17. *For a convex cone $C \subset \mathbb{R}^n$ with proper apex $\boldsymbol{a}$, the following assertions hold.*

(*a*) $\mathrm{ap}\, C$ *is the largest plane among all planes containing $\boldsymbol{a}$ and lying in C.*
(*b*) $\mathrm{ap}\, C = C \cap (2\boldsymbol{a} - C)$.

Proof. (*a*) Because $\mathrm{ap}\, C$ is a plane contained in C (see Theorem 5.14), it suffices to show that every line l through $\boldsymbol{a}$ which lies in C also lies in $\mathrm{ap}\, C$. This assertion is obvious provided $\mathrm{ap}\, C = C$. So, we may assume that $\mathrm{ap}\, C \neq C$. Our goal is to prove that any point $\boldsymbol{a}_1 \in l \setminus \{\boldsymbol{a}\}$ belongs to $\mathrm{ap}\, C$. For this, choose a point $\boldsymbol{z} \in C \setminus \{\boldsymbol{a}_1\}$.

The inclusion $(\boldsymbol{a}_1, \boldsymbol{z}\rangle \subset C$ is obvious when $\boldsymbol{z} \in l$; so, we will suppose that $\boldsymbol{z} \in C \setminus l$. Put $\boldsymbol{a}_2 = 2\boldsymbol{a} - \boldsymbol{a}_1$. Clearly, $\boldsymbol{a}_2 \in l$ and $\boldsymbol{a} = (\boldsymbol{a}_1 + \boldsymbol{a}_2)/2$ (see Figure 5.6). Denote by $E_{\boldsymbol{z}}$ the open halfplane of $\mathrm{aff}\, \{\boldsymbol{a}_1, \boldsymbol{a}_2, \boldsymbol{z}\}$ which is determined by l and contains $\boldsymbol{z}$. According to Theorem 2.75,

$$E_{\boldsymbol{z}} = \{\lambda_1 \boldsymbol{a}_1 + \lambda_2 \boldsymbol{a}_2 + \lambda_3 \boldsymbol{z} : \lambda_1 + \lambda_2 + \lambda_3 = 1,\ \lambda_3 > 0\}.$$

We assert that $E_{\boldsymbol{z}} \subset C$. Indeed, let $\boldsymbol{x}$ be a point in $E_{\boldsymbol{z}}$. Then

$$\boldsymbol{x} = \mu_1 \boldsymbol{a}_1 + \mu_2 \boldsymbol{a}_2 + \mu_3 \boldsymbol{z}, \quad \mu_1 + \mu_2 + \mu_3 = 1,\ \mu_3 > 0.$$

Assume that $\mu_1 \leqslant \mu_2$ (the case $\mu_2 \leqslant \mu_1$ is similar). Then

$$\mu_1 \leqslant \tfrac{1}{2}(\mu_1 + \mu_2) = \tfrac{1}{2}(1 - \mu_3) < \tfrac{1}{2}.$$

Let $\gamma = \mu_3/(1 - 2\mu_1)$. Clearly, $\gamma \in [0, 1]$, which implies that the point $\boldsymbol{u} = (1-\gamma)\boldsymbol{a}_2 + \gamma\boldsymbol{z}$ belongs to C due to the convexity of C. Consequently,

$$\begin{aligned}\boldsymbol{x} &= \mu_1(\boldsymbol{a}_1 + \boldsymbol{a}_2) + (\mu_2 - \mu_1)\boldsymbol{a}_2 + \mu_3\boldsymbol{z} \\ &= 2\mu_1\boldsymbol{a} + (1 - 2\mu_1 - \mu_3)\,\boldsymbol{a}_2 + \mu_3\boldsymbol{z} \\ &= 2\mu_1\boldsymbol{a} + (1 - 2\mu_1)((1-\gamma)\,\boldsymbol{a}_2 + \gamma\boldsymbol{z}) \\ &= 2\mu_1\boldsymbol{a} + (1 - 2\mu_1)\,\boldsymbol{u} \in (\boldsymbol{a}, \boldsymbol{u}\rangle \subset C\end{aligned}$$

because $\boldsymbol{a} \in \operatorname{ap} C$.

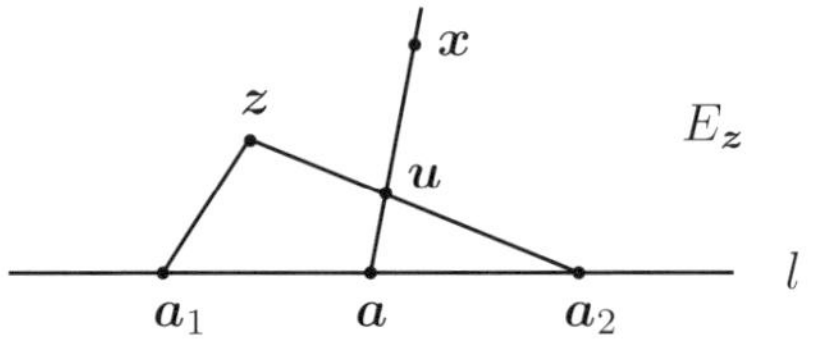

Fig. 5.6 Illustration to Theorem 5.17.

Summing up, $E_{\boldsymbol{z}} \subset C$. Since $E_{\boldsymbol{z}}$ is the intersection of $\operatorname{aff}\{\boldsymbol{a}_1, \boldsymbol{a}_2, \boldsymbol{z}\}$ and a suitable open halfspace (see Definition 2.69), Theorem 2.38 implies the inclusions $(\boldsymbol{a}_1, \boldsymbol{x}\rangle \subset E_{\boldsymbol{z}} \subset C$ for any point $\boldsymbol{x} \in E_{\boldsymbol{z}}$. Thus $\boldsymbol{a}_1 \in \operatorname{ap} C$, and $l \subset \operatorname{ap} C$.

(b) Choose any point $\boldsymbol{x} \in \operatorname{ap} C$. Then $\boldsymbol{x} \in C$ and $\boldsymbol{x}' = \boldsymbol{x} + 2(\boldsymbol{a} - \boldsymbol{x}) \in C$, which gives $\boldsymbol{x} = 2\boldsymbol{a} - \boldsymbol{x}' \in 2\boldsymbol{a} - C$. Hence $\boldsymbol{x} \in C \cap (2\boldsymbol{a} - C)$. Conversely, let $\boldsymbol{x} \in C \cap (2\boldsymbol{a} - C)$. Since the case $\boldsymbol{x} = \boldsymbol{a}$ is obvious, we assume that $\boldsymbol{x} \neq \boldsymbol{a}$. Then $\boldsymbol{x} \in C \setminus \{\boldsymbol{a}\}$ and $\boldsymbol{x} = 2\boldsymbol{a} - \boldsymbol{x}'$, where $\boldsymbol{x}' \in C \setminus \{\boldsymbol{a}\}$. Both open halflines $(\boldsymbol{a}, \boldsymbol{x}\rangle$ and $(\boldsymbol{a}, \boldsymbol{x}'\rangle$ lie in C due to $\boldsymbol{a} \in \operatorname{ap} C$. Consequently, the line $l = \langle \boldsymbol{x}, \boldsymbol{x}'\rangle$ lies in C, and $\boldsymbol{x} \in l \subset \operatorname{ap} C$ by assertion (a) above. □

Remark. Theorem 5.17 does not hold for the case of nonconvex cones. For instance, the nonconvex cone $C = \{(x, y) : xy = 0\} \subset \mathbb{R}^2$ has a unique apex, which is $\boldsymbol{o}$. However, $\{\boldsymbol{o}\}$ is not the largest subspace contained in C; furthermore, $C \cap (2\boldsymbol{o} - C) = C$.

Theorem 5.18. *If $C \subset \mathbb{R}^n$ is a cone and $C' = C \setminus \operatorname{ap} C$, then the following assertions hold.*

(a) *C' is a cone such that $\operatorname{ap} C \subset \operatorname{ap} C'$ and $C' \cap \operatorname{ap} C' = \varnothing$. If, additionally, C is convex, then C' also is convex.*

(*b*) *If L is a nonempty plane in* $\operatorname{ap} C$, *then the set* $C_L = L \cup C'$ *is a cone with* $L \subset \operatorname{ap} C_L$. *If, additionally,* C *is convex, then* C_L *also is convex and* $L = \operatorname{ap} C_L$.

Proof. Both assertions are obvious when $C = \varnothing$ or $\operatorname{ap} C = C$ (then $C' = \varnothing$ and $\operatorname{ap} C' = \mathbb{R}^n$). So, we may suppose that $\varnothing \neq C \neq \operatorname{ap} C$, which gives $C' \neq \varnothing$.

(*a*) First, we assert that C' is a cone, with $\operatorname{ap} C \subset \operatorname{ap} C'$. This fact is obvious when $C' = C$; so, we may assume that $C' \neq C$. Then $C \cap \operatorname{ap} C \neq \varnothing$, and Theorem 5.14 shows the inclusion $\operatorname{ap} C \subset C$. Choose points $\boldsymbol{a} \in \operatorname{ap} C$ and $\boldsymbol{x} \in C'$ and a scalar $\lambda > 0$. Assume, for contradiction, that the point $\boldsymbol{z} = \boldsymbol{a} + \lambda(\boldsymbol{x} - \boldsymbol{a})$ does not belong to C'. Clearly, $\boldsymbol{z} \in C \setminus C' = \operatorname{ap} C$ due to $\boldsymbol{z} \in C$. Then (see Theorem 5.14 again), $\boldsymbol{x} \in \langle \boldsymbol{a}, \boldsymbol{z} \rangle \subset \operatorname{ap} C$, contrary to the choice of $\boldsymbol{x}$. Hence $\boldsymbol{z} \in C'$, and C' is a cone with apex $\boldsymbol{a}$. Hence $\operatorname{ap} C \subset \operatorname{ap} C'$.

Next, assume for a moment that $C' \cap \operatorname{ap} C' \neq \varnothing$. Then $\operatorname{ap} C' \subset C'$ according to Theorem 5.14. In this case, $\operatorname{ap} C \subset \operatorname{ap} C' \subset C'$, contrary to $C' \cap \operatorname{ap} C = \varnothing$.

Finally, suppose that C is convex. To show the convexity of C', choose points $\boldsymbol{x}, \boldsymbol{y} \in C'$. Then $\boldsymbol{x}, \boldsymbol{y} \in C$ and $[\boldsymbol{x}, \boldsymbol{y}] \subset C$ by the convexity of C. Assume for a moment that $[\boldsymbol{x}, \boldsymbol{y}] \not\subset C'$. Then the open segment $(\boldsymbol{x}, \boldsymbol{y})$ contains a point $\boldsymbol{a} \in C \cap \operatorname{ap} C$. Consequently, $\langle \boldsymbol{x}, \boldsymbol{y} \rangle = \langle \boldsymbol{x}, \boldsymbol{a}] \cup [\boldsymbol{a}, \boldsymbol{y} \rangle \subset C$. By Theorem 5.17, the line $\langle \boldsymbol{x}, \boldsymbol{y} \rangle$ lies in $\operatorname{ap} C$, in contradiction with the choice of $\boldsymbol{x}$ and $\boldsymbol{y}$. Hence $[\boldsymbol{x}, \boldsymbol{y}] \subset C'$, implying the convexity of C'.

(*b*) First, we will show that C_L is a cone, with $L \subset \operatorname{ap} C_L$. Indeed, choose points $\boldsymbol{a} \in L$ and $\boldsymbol{x} \in C_L$. If $\boldsymbol{x} \in L \setminus \{\boldsymbol{a}\}$, then $(\boldsymbol{a}, \boldsymbol{x} \rangle \subset L \subset C_L$. If $\boldsymbol{x} \in C'$, then assertion (*a*) shows that $(\boldsymbol{a}, \boldsymbol{x} \rangle \subset C' \subset C_L$. Summing up, $L \subset \operatorname{ap} C_L$. Consequently, C_L is a cone.

Next, assume that C is convex. For the convexity of C_L, choose points $\boldsymbol{x}, \boldsymbol{y} \in C_L$. If both $\boldsymbol{x}$ and $\boldsymbol{y}$ belong to C', then $[\boldsymbol{x}, \boldsymbol{y}] \subset C' \subset C_L$ by the proved above. Similarly, if both $\boldsymbol{x}$ and $\boldsymbol{y}$ belong to L, then $[\boldsymbol{x}, \boldsymbol{y}] \subset L \subset C_L$. Suppose that $\boldsymbol{x} \in L$ and $\boldsymbol{y} \in C'$. Then $\boldsymbol{x} \in \operatorname{ap} C \subset \operatorname{ap} C'$ and $(\boldsymbol{x}, \boldsymbol{y} \rangle \subset C'$ according to assertion (*a*). Consequently,

$$[\boldsymbol{x}, \boldsymbol{y}] = \{\boldsymbol{x}\} \cup (\boldsymbol{x}, \boldsymbol{y}] \subset L \cup C' = C_L.$$

Summing up, the set C_L is convex. It remains to prove that $L = \operatorname{ap} C_L$. By the above argument, it suffices to show that $\operatorname{ap} C_L \subset L$. We will prove first the inclusion $\operatorname{ap} C_L \subset \operatorname{ap} C'$. Indeed, assume for a moment that $\operatorname{ap} C_L \not\subset \operatorname{ap} C'$. Then $\operatorname{ap} C_L \cap C' \neq \varnothing$. Choose a point $\boldsymbol{x} \in \operatorname{ap} C_L \cap C'$ and a point $\boldsymbol{a} \in L$. By Theorem 5.17, the line $l = \langle \boldsymbol{a}, \boldsymbol{x} \rangle$ lies in $\operatorname{ap} C_L$, implying that

$\boldsymbol{x} \in l \subset \operatorname{ap} C_L \subset \operatorname{ap} C'$, contrary to the choice of $\boldsymbol{x}$ in C'. Hence $\operatorname{ap} C_L \subset L$. Finally, since $\varnothing \neq L \subset C_L \cap \operatorname{ap} C_L$, Theorem 5.14 shows that $\operatorname{ap} C_L \subset C_L$. Because $\operatorname{ap} C_L \subset \operatorname{ap} C'$ and $C' \cap \operatorname{ap} C' = \varnothing$, one has $\operatorname{ap} C_L \subset C_L \backslash C' = L$. $\square$

A combination of Theorems 5.17 and 5.18 implies the following corollary.

Corollary 5.19. *Given a nonempty cone $C \subset \mathbb{R}^n$ with apex $\boldsymbol{a}$, let $C'' = \operatorname{ap} C \cup C$. Then C'' is a cone such that $\operatorname{ap} C \subset \operatorname{ap} C''$. Furthermore, if C is convex, then the following assertions hold.*

(*a*) *The cone C'' is convex and $\operatorname{ap} C = \operatorname{ap} C''$.*
(*b*) *$\operatorname{ap} C$ is the largest plane among all planes containing $\boldsymbol{a}$ and lying in C''.*
(*b*) *$\operatorname{ap} C = C'' \cap (2\boldsymbol{a} - C'')$.* $\square$

Remark. The inclusion $\operatorname{ap} C \subset \operatorname{ap} C''$ in Corollary 5.19 may be proper if the cone C is not convex. For instance, let L denote the x-axis of the plane $\mathbb{R}^2$. Then L is the improper apex set of the nonconvex cone $C = \{(x, y) : y \neq 0\}$, while the apex set of the cone $\operatorname{ap} C \cup C$ is the whole plane.

5.1.5 *Topological Properties of Cones*

Theorem 5.20. *For a nonempty cone $C \subset \mathbb{R}^n$ with apex $\boldsymbol{a}$ and a point $\boldsymbol{x} \in \mathbb{R}^n$, $\boldsymbol{x} \neq \boldsymbol{a}$, the following assertions hold.*

(*a*) *If $\boldsymbol{x} \in \operatorname{rint} C$, then the open halfline $(\boldsymbol{a}, \boldsymbol{x}\rangle$ lies in $\operatorname{rint} C$.*
(*b*) *If $\boldsymbol{x} \in \operatorname{cl} C$, then the closed halfline $[\boldsymbol{a}, \boldsymbol{x}\rangle$ lies in $\operatorname{cl} C$.*
(*c*) *If C is not a plane and $\boldsymbol{x} \in \operatorname{rbd} C$, then the closed halfline $[\boldsymbol{a}, \boldsymbol{x}\rangle$ lies in $\operatorname{rbd} C$; furthermore, if C is convex, then the line $\langle \boldsymbol{a}, \boldsymbol{x}\rangle$ is disjoint from $\operatorname{rint} C$.*

Proof. (*a*) Choose a point $\boldsymbol{z} \in (\boldsymbol{a}, \boldsymbol{x}\rangle$. Then $\boldsymbol{z} = \boldsymbol{a} + \lambda(\boldsymbol{x} - \boldsymbol{a})$ for a suitable scalar $\lambda > 0$. According to Definition 3.16, there is a scalar $\rho > 0$ such that $\operatorname{aff} C \cap U_\rho(\boldsymbol{x}) \subset C$. We assert that $\operatorname{aff} C \cap U_{\lambda\rho}(\boldsymbol{z}) \subset C$ (see Figure 5.7).

Indeed, choose any point $\boldsymbol{z}' \in \operatorname{aff} C \cap U_{\lambda\rho}(\boldsymbol{z})$ and let $\boldsymbol{x}' = (1 - \lambda^{-1})\boldsymbol{a} + \lambda^{-1}\boldsymbol{z}'$. Since $\boldsymbol{x} = (1 - \lambda^{-1})\boldsymbol{a} + \lambda^{-1}\boldsymbol{z}$, one has

$$\|\boldsymbol{x} - \boldsymbol{x}'\| = \|\lambda^{-1}(\boldsymbol{z} - \boldsymbol{z}')\| = \lambda^{-1}\|\boldsymbol{z} - \boldsymbol{z}'\| < \lambda^{-1}(\lambda\rho) = \rho,$$

which gives the inclusion $\boldsymbol{x}' \in U_\rho(\boldsymbol{x})$. Furthermore, $\boldsymbol{x}' \in \operatorname{aff} C$ because $\boldsymbol{x}'$ is an affine combination of $\boldsymbol{a}$ and $\boldsymbol{z}'$ from $\operatorname{aff} C$ (see Theorem 2.38). Hence $\boldsymbol{x}' \in \operatorname{aff} C \cap U_\rho(\boldsymbol{x}) \subset C$ by the choice of ρ. Therefore,

$$\boldsymbol{z}' = (1 - \lambda)\boldsymbol{a} + \lambda\boldsymbol{x}' \in (\boldsymbol{a}, \boldsymbol{x}\rangle \subset C$$

because C is a cone with apex $\boldsymbol{a}$. Summing up, $\operatorname{aff} C \cap U_{\lambda\rho}(\boldsymbol{z}) \subset C$, which implies the inclusion $\boldsymbol{z} \in \operatorname{rint} C$.

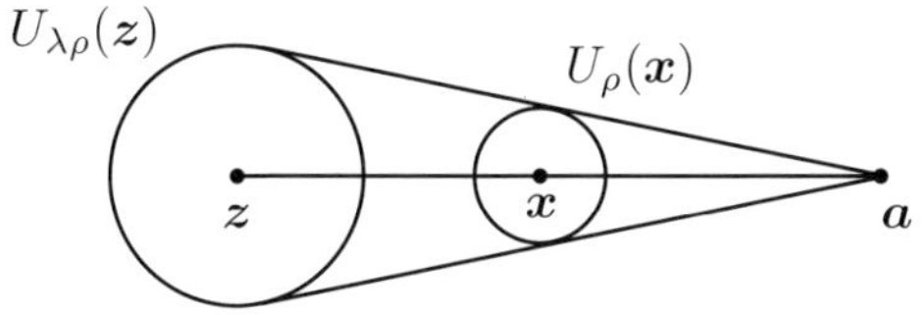

Fig. 5.7 Illustration to Theorem 5.20.

(b) Let $\boldsymbol{u} \in [\boldsymbol{a}, \boldsymbol{x}\rangle$. Because the case $\boldsymbol{u} = \boldsymbol{a}$ is obvious, we assume that $\boldsymbol{u} \neq \boldsymbol{a}$. Then $\boldsymbol{u} = \boldsymbol{a} + \lambda(\boldsymbol{x} - \boldsymbol{a})$ for a suitable scalar $\lambda > 0$. Choose a positive scalar ρ satisfying the condition $\rho < \lambda\|\boldsymbol{x} - \boldsymbol{a}\|$. The inclusion $\boldsymbol{x} \in \operatorname{cl} C$ shows the existence of a point $\boldsymbol{x}' \in C \cap U_{\rho/\lambda}(\boldsymbol{x})$. Clearly, $\boldsymbol{x}' \neq \boldsymbol{a}$ due to $\boldsymbol{a} \notin U_{\rho/\lambda}(\boldsymbol{x})$. Put $\boldsymbol{u}' = \boldsymbol{a} + \lambda(\boldsymbol{x}' - \boldsymbol{a})$. Then

$$\boldsymbol{u}' \in (\boldsymbol{a}, \boldsymbol{x}'\rangle \subset C \quad \text{and} \quad \|\boldsymbol{u} - \boldsymbol{u}'\| = \lambda\|\boldsymbol{x} - \boldsymbol{x}'\| < \lambda(\rho/\lambda) = \rho.$$

Thus the open ball $U_\rho(\boldsymbol{u})$, meets C, which implies the inclusion $\boldsymbol{u} \in \operatorname{cl} C$. Hence $[\boldsymbol{a}, \boldsymbol{x}\rangle = \{\boldsymbol{a}\} \cup (\boldsymbol{a}, \boldsymbol{x}\rangle \subset \operatorname{cl} C$.

(c) Since C is not a plane, one has $\operatorname{rbd} C \neq \varnothing$ (see Theorem 3.62). Choose a point $\boldsymbol{x} \in \operatorname{rbd} C$. Then $\boldsymbol{x} \in \operatorname{cl} C \setminus \operatorname{rint} C$, and a combination of assertions (a) and (b) gives

$$[\boldsymbol{a}, \boldsymbol{x}\rangle \subset \operatorname{cl} C \setminus \operatorname{rint} C = \operatorname{rbd} C.$$

It remains to show that $\langle \boldsymbol{a}, \boldsymbol{x}\rangle \cap \operatorname{rint} C = \varnothing$. Due to the above inclusion, it suffices to prove that the open halfline $h = \langle \boldsymbol{a}, \boldsymbol{x}\rangle \setminus [\boldsymbol{a}, \boldsymbol{x}\rangle$ is disjoint from $\operatorname{rint} C$. Indeed, if a point $\boldsymbol{u} \in h$ belonged to $\operatorname{rint} C$, then, by Theorem 3.47, $(\boldsymbol{a}, \boldsymbol{x}) \subset (\boldsymbol{u}, \boldsymbol{x}) \subset \operatorname{rint} C$, in contradiction with the inclusion $[\boldsymbol{a}, \boldsymbol{x}] \subset [\boldsymbol{a}, \boldsymbol{x}\rangle \subset \operatorname{rbd} C$. □

Corollary 5.21. *For a cone $C \subset \mathbb{R}^n$ with apex $\boldsymbol{a}$, all three sets $\operatorname{cl} C$, $\operatorname{rint} C$, and $\operatorname{rbd} C$ also are cones with apex $\boldsymbol{a}$. If, additionally, C is convex, then both cones $\operatorname{cl} C$ and $\operatorname{rint} C$ are convex and*

$$\operatorname{rint} C = C + \operatorname{rint} C = \operatorname{cl} C + \operatorname{rint} C. \tag{5.8}$$

Proof. Because the case $C = \varnothing$ is obvious, we may assume that $C \neq \varnothing$. Then the first assertion immediately follows from Theorem 5.20. If C is convex, then Theorem 3.46 and Corollary 3.25 imply that both cones $\operatorname{cl} C$ and $\operatorname{rint} C$ are convex. Because $2C = C$ and $\operatorname{rint}(\operatorname{cl} C) = \operatorname{rint} C$ (see Theorem 3.46), the equalities (5.8) follow from Corollary 3.35. □

Dealing with topological properties of cones, it is sometimes convenient to consider the intersections of cones with suitable closed balls or spheres. We recall that $B_\rho(\boldsymbol{a})$ and $S_\rho(\boldsymbol{a})$ denote, respectively, the closed ball and sphere of radius $\rho > 0$ centered at $\boldsymbol{a} \in \mathbb{R}^n$.

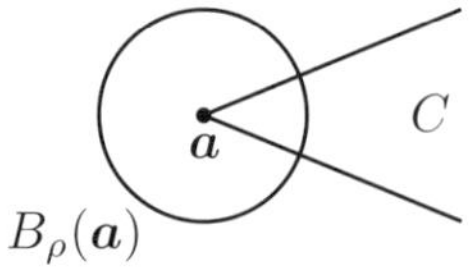

Fig. 5.8 Illustration to Theorem 5.22.

Theorem 5.22. *For a cone $C \subset \mathbb{R}^n$ with apex $\boldsymbol{a}$ and a scalar $\rho > 0$, the following assertions hold.*

(*a*) *The cone C is convex if and only if any of the sets $C \cap B_\rho(\boldsymbol{a})$ and $C \cap U_\rho(\boldsymbol{a})$ is convex.*
(*b*) $\operatorname{cl} C \cap B_\rho(\boldsymbol{a}) = \operatorname{cl}(C \cap B_\rho(\boldsymbol{a})) = \operatorname{cl}(C \cap U_\rho(\boldsymbol{a}))$.
(*c*) *The cone C is closed if and only if the set $C \cap B_\rho(\boldsymbol{a})$ is closed.*
(*d*) $\operatorname{cl} C \cap S_\rho(\boldsymbol{a}) = \operatorname{cl}(C \cap S_\rho(\boldsymbol{a}))$.
(*e*) *The cone $\{\boldsymbol{a}\} \cup C$ is closed if and only if the set $C \cap S_\rho(\boldsymbol{a})$ is closed.*

Proof. (*a*) If the cone C is convex, then the set $C \cap B_\rho(\boldsymbol{a})$ is convex as the intersection of convex sets C and $B_\rho(\boldsymbol{a})$ (see Theorem 3.8 and an example on page 76). Conversely, suppose that the set $C \cap B_\rho(\boldsymbol{a})$ is convex and choose points $\boldsymbol{x}, \boldsymbol{y} \in C$. If $\boldsymbol{x} = \boldsymbol{y} = \boldsymbol{a}$, then $\boldsymbol{a}$ is a proper apex of C and $[\boldsymbol{x}, \boldsymbol{y}] = \{\boldsymbol{a}\} \subset C$. Similarly, if exactly one of the points $\boldsymbol{x}$ and $\boldsymbol{y}$, say $\boldsymbol{x}$, coincides with $\boldsymbol{a}$, then $[\boldsymbol{x}, \boldsymbol{y}] = [\boldsymbol{a}, \boldsymbol{y}] \subset [\boldsymbol{a}, \boldsymbol{y}\rangle \subset C$. Finally, suppose that none of the points $\boldsymbol{x}$ and $\boldsymbol{y}$ is $\boldsymbol{a}$. To show the inclusion $[\boldsymbol{x}, \boldsymbol{y}] \subset C$, choose any point $\boldsymbol{u} \in [\boldsymbol{x}, \boldsymbol{y}]$ (see Figure 5.9).

Then $\boldsymbol{u} = (1-\lambda)\boldsymbol{x} + \lambda\boldsymbol{y}$ for a suitable scalar $0 \leqslant \lambda \leqslant 1$. Put

$$\boldsymbol{x}' = \boldsymbol{a} + \frac{\rho}{\|\boldsymbol{x}-\boldsymbol{a}\|}(\boldsymbol{x}-\boldsymbol{a}), \quad \boldsymbol{y}' = \boldsymbol{a} + \frac{\rho}{\|\boldsymbol{y}-\boldsymbol{a}\|}(\boldsymbol{y}-\boldsymbol{a}),$$

$$\mu = \frac{\lambda\|\boldsymbol{a}-\boldsymbol{y}\|}{(1-\lambda)\|\boldsymbol{a}-\boldsymbol{x}\| + \lambda\|\boldsymbol{a}-\boldsymbol{y}\|}, \quad \gamma = \frac{(1-\lambda)\|\boldsymbol{a}-\boldsymbol{x}\| + \lambda\|\boldsymbol{a}-\boldsymbol{y}\|}{\rho}.$$

Clearly, $\|\boldsymbol{x}'-\boldsymbol{a}\| = \|\boldsymbol{y}'-\boldsymbol{a}\| = \rho$, $0 \leqslant \mu \leqslant 1$, and $\gamma > 0$. Furthermore, $\boldsymbol{x}' \in [\boldsymbol{a}, \boldsymbol{x}\rangle \subset C$ and $\boldsymbol{y}' \in [\boldsymbol{a}, \boldsymbol{y}\rangle \subset C$. Thus $\boldsymbol{x}', \boldsymbol{y}' \in C \cap B_\rho(\boldsymbol{a})$. Now, let $\boldsymbol{u}' = (1-\mu)\boldsymbol{x}' + \mu\boldsymbol{y}'$. Then $\boldsymbol{u}' \in [\boldsymbol{x}', \boldsymbol{y}'] \subset C \cap B_\rho(\boldsymbol{a}) \subset C$ due to the convexity of $C \cap B_\rho(\boldsymbol{a})$. It is easy to verify that $\boldsymbol{u} = \boldsymbol{a} + \gamma(\boldsymbol{u}' - \boldsymbol{a})$. Hence $\boldsymbol{u} \in [\boldsymbol{a}, \boldsymbol{u}'\rangle \subset C$, implying the convexity of C. The case of the set $C \cap U_\rho(\boldsymbol{a})$ is similar.

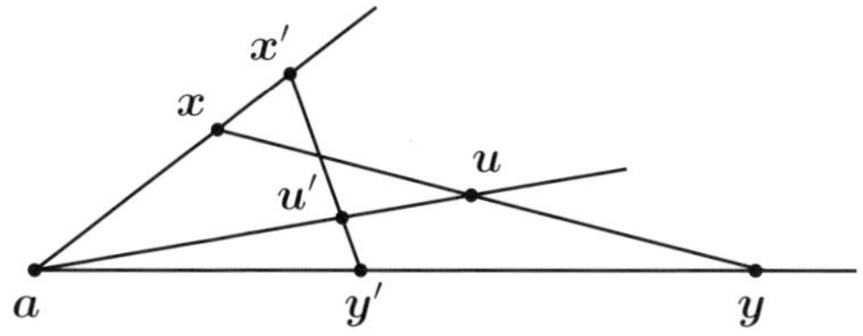

Fig. 5.9 Illustration to Theorem 5.22.

(*b*) The inclusions

$$\mathrm{cl}\,(C \cap U_\rho(\boldsymbol{a})) \subset \mathrm{cl}\,C \cap \mathrm{cl}\,U_\rho(\boldsymbol{a}) = \mathrm{cl}\,C \cap B_\rho(\boldsymbol{a})$$

follow from the properties of closure (see page 10). Hence it remains to show that $\mathrm{cl}\,C \cap B_\rho(\boldsymbol{a}) \subset \mathrm{cl}\,(C \cap U_\rho(\boldsymbol{a}))$. Choose a point $\boldsymbol{x} \in \mathrm{cl}\,C \cap B_\rho(\boldsymbol{a})$. Because the case $\boldsymbol{x} = \boldsymbol{a}$ is obvious, we may suppose that $\boldsymbol{x} \neq \boldsymbol{a}$. Let $\boldsymbol{x}_1, \boldsymbol{x}_2, \ldots$ be a sequence of points in C which converges to $\boldsymbol{x}$. By a continuity argument, we may assume that none of the points $\boldsymbol{x}_1, \boldsymbol{x}_2, \ldots$ is $\boldsymbol{a}$. Let

$$\boldsymbol{u}_i = \boldsymbol{a} + \frac{i\|\boldsymbol{x} - \boldsymbol{a}\|}{(i+1)\|\boldsymbol{x}_i - \boldsymbol{a}\|}(\boldsymbol{x}_i - \boldsymbol{a}), \quad i \geqslant 1.$$

Then $\boldsymbol{u}_i \in [\boldsymbol{a}, \boldsymbol{x}_i\rangle \subset C$ and $\boldsymbol{u}_i \in U_\rho(\boldsymbol{a})$ due to $\|\boldsymbol{u}_i - \boldsymbol{a}\| = \frac{i}{i+1}\|\boldsymbol{x} - \boldsymbol{a}\| < \rho$, $i \geqslant 1$. Thus $\boldsymbol{u}_i \in C \cap U_\rho(\boldsymbol{a})$, $i \geqslant 1$. Furthermore,

$$\lim_{i\to\infty} \boldsymbol{u}_i = \boldsymbol{a} + \lim_{i\to\infty} \frac{i\|\boldsymbol{x} - \boldsymbol{a}\|}{(i+1)\|\boldsymbol{x}_i - \boldsymbol{a}\|}(\boldsymbol{x}_i - \boldsymbol{a}) = \boldsymbol{a} + (\boldsymbol{x} - \boldsymbol{a}) = \boldsymbol{x}.$$

Therefore, $\boldsymbol{x} \in \mathrm{cl}\,(C \cap U_\rho(\boldsymbol{a}))$, implying the desired inclusion.

(*c*) If the cone C is closed, then the set $C \cap B_\rho(\boldsymbol{a})$ is closed as the intersections of closed sets. Conversely, assume that the set $C \cap B_\rho(\boldsymbol{a})$ is closed and choose a point $\boldsymbol{x} \in \mathrm{cl}\,C$. Because the case $\boldsymbol{x} = \boldsymbol{a}$ is obvious ($\boldsymbol{a} \in C \cap B_\rho(\boldsymbol{a}) \subset C$ due to the closedness of $C \cap B_\rho(\boldsymbol{a})$), we may suppose that $\boldsymbol{x} \neq \boldsymbol{a}$. Let $\boldsymbol{x}_1, \boldsymbol{x}_2, \ldots$ be a sequence of points in C which converges to $\boldsymbol{x}$. By a continuity argument, we may assume that none of the points $\boldsymbol{x}_1, \boldsymbol{x}_2, \ldots$ is $\boldsymbol{a}$. Let

$$\boldsymbol{u} = \boldsymbol{a} + \frac{\rho}{\|\boldsymbol{x} - \boldsymbol{a}\|}(\boldsymbol{x} - \boldsymbol{a}), \quad \boldsymbol{u}_i = \boldsymbol{a} + \frac{\rho}{\|\boldsymbol{x}_i - \boldsymbol{a}\|}(\boldsymbol{x}_i - \boldsymbol{a}), \quad i \geqslant 1.$$

Then $\boldsymbol{u}_i \in [\boldsymbol{a}, \boldsymbol{x}_i\rangle \subset C$ and $\boldsymbol{u}_i \in B_\rho(\boldsymbol{a})$ due to $\|\boldsymbol{a} - \boldsymbol{u}_i\| = \rho$ for all $i \geqslant 1$. Furthermore,

$$\lim_{i\to\infty} \boldsymbol{u}_i = \boldsymbol{a} + \lim_{i\to\infty} \frac{\rho}{\|\boldsymbol{x}_i - \boldsymbol{a}\|}(\boldsymbol{x}_i - \boldsymbol{a}) = \boldsymbol{a} + \frac{\rho}{\|\boldsymbol{x} - \boldsymbol{a}\|}(\boldsymbol{x} - \boldsymbol{a}) = \boldsymbol{u}.$$

By the closedness of $C \cap B_\rho(\boldsymbol{a})$,

$$\boldsymbol{u} = \lim_{i\to\infty} \boldsymbol{u}_i \in C \cap B_\rho(\boldsymbol{a}) \subset C \quad \text{and} \quad \boldsymbol{x} \in [\boldsymbol{a}, \boldsymbol{u}\rangle \subset C.$$

The proof of assertions (d) and (e) is similar to those of (b) and (c). □

Remark. The cone $\{\boldsymbol{a}\} \cup C$ in assertion (e) of Theorem 5.22 cannot be replaced with C. Indeed, the convex cone $C = \{(x, y) : x \geqslant 0, y \geqslant 0\} \setminus \{\boldsymbol{o}\}$ is nonclosed, while all sections of C by circles $S_\rho(\boldsymbol{o}) \subset \mathbb{R}^2$, $\rho > 0$, are closed sets.

Theorem 5.23. *Any r-simplicial cone $\Gamma = \Gamma_{\boldsymbol{a}}(\boldsymbol{x}_1, \ldots, \boldsymbol{x}_r) \subset \mathbb{R}^n$ is a closed set, and*

$$\operatorname{rint} \Gamma = \{\boldsymbol{a} + \lambda_1(\boldsymbol{x}_1 - \boldsymbol{a}) + \cdots + \lambda_r(\boldsymbol{x}_r - \boldsymbol{a}) : \lambda_1, \ldots, \lambda_r > 0\}.$$

Proof. For the first assertion, we observe that Γ is the intersection of closed halfplanes $D_1, \ldots, D_r$ of the plane $\operatorname{aff}\{\boldsymbol{a}, \boldsymbol{x}_1, \ldots, \boldsymbol{x}_r\}$, given by

$$D_i = \{\lambda_0\boldsymbol{a} + \lambda_1\boldsymbol{x}_1 + \cdots + \lambda_r\boldsymbol{x}_r : \lambda_0 + \cdots + \lambda_r = 1,\ \lambda_i \geqslant 0\},\ 1 \leqslant i \leqslant r$$

(see Theorem 2.75). Consequently, Γ is closed as the intersection of closed sets $D_1, \ldots, D_r$.

For the second part, we start by showing that any point of the form

$$\boldsymbol{x} = \boldsymbol{a} + \lambda_1(\boldsymbol{x}_1 - \boldsymbol{a}) + \cdots + \lambda_r(\boldsymbol{x}_r - \boldsymbol{a}),\ \lambda_1, \ldots, \lambda_{r+1} > 0,$$

belongs to $\operatorname{rint} \Gamma$. Indeed, rewriting $\boldsymbol{x}$ as an affine combination

$$\boldsymbol{x} = (1 - \lambda_1 - \cdots - \lambda_r)\,\boldsymbol{a} + \lambda_1\boldsymbol{x}_1 + \cdots + \lambda_r\boldsymbol{x}_r$$

and using Theorem 5.7, we see that $\boldsymbol{x} \in \operatorname{aff}\{\boldsymbol{a}, \boldsymbol{x}_1, \ldots, \boldsymbol{x}_r\} = \operatorname{aff} \Gamma$.

Put $\varepsilon = \min\{\lambda_1, \ldots, \lambda_{r+1}\}$. Theorem 2.55 implies the existence of a scalar $\rho > 0$ such that for every point

$$\boldsymbol{u} = \boldsymbol{a} + \mu_1(\boldsymbol{x}_1 - \boldsymbol{a}) + \cdots + \mu_r(\boldsymbol{x}_r - \boldsymbol{a}) \in \operatorname{aff} \Gamma \cap U_\rho(\boldsymbol{x}),$$

its affine coordinates $1 - \mu_1 - \cdots - \mu_r, \mu_1, \ldots, \mu_r$ relative to $\boldsymbol{a}, \boldsymbol{x}_1, \ldots, \boldsymbol{x}_r$ satisfy the inequalities

$$|(1 - \lambda_1 - \cdots - \lambda_r) - (1 - \mu_1 - \cdots - \mu_r)| < \varepsilon,$$
$$|\lambda_i - \mu_i| < \varepsilon,\quad 1 \leqslant i \leqslant r.$$

Hence $\mu_i > \lambda_i - \varepsilon > 0$ for all $1 \leqslant i \leqslant r$, which gives $\boldsymbol{u} \in \Gamma$. Definition 3.16 shows that $\boldsymbol{x} \in \operatorname{rint} \Gamma$.

Conversely, let $\boldsymbol{x} \in \operatorname{rint} \Gamma$. By the above argument, the point

$$\boldsymbol{y} = \boldsymbol{a} + 1(\boldsymbol{x}_1 - \boldsymbol{a}) + \cdots + 1(\boldsymbol{x}_r - \boldsymbol{a})$$

belongs to $\operatorname{rint} \Gamma$. Theorem 3.26 shows the existence of a scalar $\gamma > 1$ such that the point $\boldsymbol{z} = \gamma \boldsymbol{x} + (1 - \gamma)\boldsymbol{y}$ belongs to Γ. Expressing $\boldsymbol{z}$ as

$$\boldsymbol{z} = \boldsymbol{a} + \mu_1(\boldsymbol{x}_1 - \boldsymbol{a}) + \cdots + \mu_r(\boldsymbol{x}_r - \boldsymbol{a}), \ \mu_1, \ldots, \mu_r \geqslant 0,$$

we obtain

$$\begin{aligned} \boldsymbol{x} &= (1 - \gamma^{-1})\boldsymbol{y} + \gamma^{-1}\boldsymbol{z} \\ &= \boldsymbol{a} + (1 - \gamma^{-1} + \mu_1)(\boldsymbol{x}_1 - \boldsymbol{a}) + \cdots + (1 - \gamma^{-1} + \mu_r)(\boldsymbol{x}_r - \boldsymbol{a}). \end{aligned}$$

Since $0 < \gamma^{-1} < 1$, one has $1 - \gamma^{-1} + \mu_i > 0$ for all $1 \leqslant i \leqslant r$. □

The next result is an analog of Theorem 3.22.

Theorem 5.24. *For a cone $C \subset \mathbb{R}^n$ of positive dimension m and an apex $\boldsymbol{a}$ of C, the following assertions hold.*

(*a*) *If $\Gamma_{\boldsymbol{a}}(\boldsymbol{x}_1, \ldots, \boldsymbol{x}_m)$ is an m-simplicial cone lying in $\{\boldsymbol{a}\} \cup C$, then*

$$\operatorname{rint} \Gamma_{\boldsymbol{a}}(\boldsymbol{x}_1, \ldots, \boldsymbol{x}_m) \subset \operatorname{rint} C.$$

(*b*) *A point $\boldsymbol{x} \in \mathbb{R}^n \setminus \{\boldsymbol{a}\}$ belongs to $\operatorname{rint} C$ if and only if there is an m-simplicial cone $\Gamma_{\boldsymbol{a}}(\boldsymbol{x}_1, \ldots, \boldsymbol{x}_m) \subset \{\boldsymbol{a}\} \cup C$ containing $\boldsymbol{x}$ in its relative interior.*

Proof. (*a*) This part follows from Theorem 3.18 and Corollary 5.8.

(*b*) Assume first that $\boldsymbol{x} \in \operatorname{rint} C$. Then $\operatorname{aff} C \cap U_\rho(\boldsymbol{x}) \subset C$ for a suitable open ball $U_\rho(\boldsymbol{x}) \subset \mathbb{R}^n$. Let M be an $(m-1)$-dimensional subplane of $\operatorname{aff} C$ which contains $\boldsymbol{x}$ and misses $\boldsymbol{a}$. According to Theorem 2.68, $M \cap U_\rho(\boldsymbol{x})$ is an $(m-1)$-dimensional open ball. Similarly to the proof of Theorem 3.22, one can choose in $M \cap U_\rho(\boldsymbol{x})$ an $(m-1)$-simplex $\Delta(\boldsymbol{x}_1, \ldots, \boldsymbol{x}_m)$ satisfying the condition $\boldsymbol{x} = \frac{1}{m}(\boldsymbol{x}_1 + \cdots + \boldsymbol{x}_m)$. Since $\operatorname{aff}\{\boldsymbol{x}_1, \ldots, \boldsymbol{x}_m\} \subset M$ and $\boldsymbol{a} \notin M$, the set $\{\boldsymbol{a}, \boldsymbol{x}_1, \ldots, \boldsymbol{x}_m\}$ is affinely independent (see Theorem 2.54). Finally, the equality

$$\boldsymbol{x} = \frac{1}{m}(\boldsymbol{x}_1 + \cdots + \boldsymbol{x}_m) = \boldsymbol{a} + \frac{1}{m}(\boldsymbol{x}_1 - \boldsymbol{a}) + \cdots + \frac{1}{m}(\boldsymbol{x}_m - \boldsymbol{a})$$

and Theorem 5.23 give the inclusion $\boldsymbol{x} \in \operatorname{rint} \Gamma_{\boldsymbol{a}}(\boldsymbol{x}_1, \ldots, \boldsymbol{x}_m)$.

Conversely, let $\boldsymbol{x}$ belong to the relative interior of an m-simplicial cone $\Gamma_{\boldsymbol{a}}(\boldsymbol{x}_1, \ldots, \boldsymbol{x}_m) \subset \{\boldsymbol{a}\} \cup C$. Because $\operatorname{aff} \Gamma_{\boldsymbol{a}}(\boldsymbol{x}_1, \ldots, \boldsymbol{x}_m) = \operatorname{aff} C$ (see Corollary 5.8), the inclusion $\boldsymbol{x} \in \operatorname{rint} C$ follows from Theorem 3.18. □

Theorem 5.25. *If $C \subset \mathbb{R}^n$ is a nonempty cone, then* $\operatorname{ap} C \subset \operatorname{cl} C$. *Furthermore, if* $\operatorname{ap} C \neq C$, *then* $\operatorname{ap} C \subset \operatorname{cl}(C \setminus \operatorname{ap} C)$.

Proof. For the inclusion $\operatorname{ap} C \subset \operatorname{cl} C$, choose points $\boldsymbol{a} \in \operatorname{ap} C$ and $\boldsymbol{x} \in C$. Then the points $\boldsymbol{x}_\lambda = (1-\lambda)\boldsymbol{a}+\lambda\boldsymbol{x}$ belong to C for all $\lambda > 0$. Consequently, $\boldsymbol{a} = \lim_{\lambda\to 0} \boldsymbol{x}_\lambda \in \operatorname{cl} C$.

Suppose that $\operatorname{ap} C \neq C$. By Theorem 5.14, the set $C \setminus \operatorname{ap} C$ is nonempty. Choose points $\boldsymbol{a} \in \operatorname{ap} C$ and $\boldsymbol{c} \in C \setminus \operatorname{ap} C$. As above, $\boldsymbol{x}_\lambda = (1-\lambda)\boldsymbol{a}+\lambda\boldsymbol{x} \in C$ for all $\lambda > 0$. Furthermore, $\boldsymbol{x}_\lambda \notin \operatorname{ap} C$ since otherwise $\boldsymbol{x} \in \langle \boldsymbol{a}, \boldsymbol{x}_\lambda\rangle \subset \operatorname{ap} C$ (see Theorem 2.38). Hence $\boldsymbol{x}_\lambda \in C \setminus \operatorname{ap} C$, implying that

$$\boldsymbol{a} = \lim_{\lambda\to 0} \boldsymbol{x}_\lambda \in \operatorname{cl}(C \setminus \operatorname{ap} C). \qquad \square$$

Theorem 5.26. *For a nonempty convex cone $C \subset \mathbb{R}^n$, the following conditions are equivalent.*

(*a*) *C is a plane.*
(*b*) $\operatorname{ap} C = C$.
(*c*) $\operatorname{ap} C \subset \operatorname{rint} C$.
(*d*) $\operatorname{ap} C \cap \operatorname{rint} C \neq \varnothing$.

Proof. (*a*) $\Leftrightarrow$ (*b*) This part of the proof follows from Theorem 5.14 and an example on page 150.

(*b*) $\Rightarrow$ (*c*) By the above argument, C is a plane. Therefore, $\operatorname{ap} C = C = \operatorname{rint} C$ (see an example on page 84). Obviously, condition (*c*) implies (*d*).

(*d*) $\Rightarrow$ (*b*) Choose a point $\boldsymbol{a} \in \operatorname{ap} C \cap \operatorname{rint} C$. Leaving aside the obvious case $C = \{\boldsymbol{a}\}$, assume the existence of a point $\boldsymbol{x} \in C \setminus \{\boldsymbol{a}\}$. By Corollary 3.27, there is a point $\boldsymbol{u} \in C$ such that $\boldsymbol{a} \in (\boldsymbol{u}, \boldsymbol{x})$. Then

$$\langle \boldsymbol{a}, \boldsymbol{x}\rangle = \langle \boldsymbol{u}, \boldsymbol{a}) \cup \{\boldsymbol{a}\} \cup (\boldsymbol{a}, \boldsymbol{x}\rangle \subset C,$$

and Theorem 5.17 shows that $\boldsymbol{x} \in \operatorname{ap} C$. Consequently, $C \subset \operatorname{ap} C$, which gives $C = \operatorname{ap} C$. $\square$

Corollary 5.27. *For a nonempty convex cone $C \subset \mathbb{R}^n$, conditions* (*a*)–(*d*) *below are equivalent.*

(*a*) *C is not a plane.*
(*b*) $\operatorname{ap} C \neq C$.
(*c*) $\operatorname{ap} C \subset \operatorname{rbd} C$.
(*d*) $\operatorname{ap} C \cap \operatorname{rbd} C \neq \varnothing$.
(*e*) *Furthermore,* $\operatorname{ap} C = \operatorname{rbd} C$ *if and only if C is a halfplane of* $\operatorname{aff} C$.

Proof. Because the equivalence of conditions (a)–(d) follows from Theorems 5.25 and 5.26, it suffices to prove assertion (e). If C is a halfplane (either closed or open), then $\operatorname{ap} C = \operatorname{rbd} C$ (see an example on page 150). Conversely, suppose that $\operatorname{ap} C = \operatorname{rbd} C$. By Theorem 5.14, $\operatorname{ap} C$ is a plane, and Theorem 3.68 shows that $\operatorname{cl} C$ is a closed halfplane. According to Theorem 5.14, either $\operatorname{ap} C \subset C$ or $C \cap \operatorname{ap} C = \varnothing$. Hence C is a closed halfplane if $\operatorname{rbd} C = \operatorname{ap} C \subset C$, or is an open halfplane if $C \cap \operatorname{rbd} C = C \cap \operatorname{ap} C = \varnothing$. $\square$

The next result complements Corollary 5.21.

Theorem 5.28. *If $C \subset \mathbb{R}^n$ is a convex cone, then all three sets,*

$$\operatorname{cl} C, \quad C_1 = \operatorname{ap} C \cup \operatorname{rint} C, \quad C_2 = \operatorname{ap}(\operatorname{cl} C) \cup \operatorname{rint} C,$$

are convex cones such that $C_1 \subset C$ and $C_2 \subset \operatorname{cl} C$. Furthermore,

$$\operatorname{ap} C = \operatorname{ap} C_1 \subset \operatorname{ap}(\operatorname{cl} C) = \operatorname{ap} C_2. \tag{5.9}$$

Proof. Both assertions of the theorem are obvious if C is a plane. So, we may assume that C is not a plane. Then $\operatorname{ap} C \subset \operatorname{rbd} C$ according to Corollary 5.27. We divide our argument into several steps.

1. First, we assert that all three sets $\operatorname{cl} C$, C_1, and C_2 are convex. Indeed, the convexity of $\operatorname{cl} C$ follows from Theorem 3.46. For the convexity of C_1, choose points $\boldsymbol{x}, \boldsymbol{y} \in C_1$. If both $\boldsymbol{x}$ and $\boldsymbol{y}$ belong either to $\operatorname{ap} C$ or to $\operatorname{rint} C$, then, respectively, either $[\boldsymbol{x}, \boldsymbol{y}] \subset \operatorname{ap} C$ (see Theorem 5.14) or $[\boldsymbol{x}, \boldsymbol{y}] \subset \operatorname{rint} C$ (see Corollary 3.25). If $\boldsymbol{x} \in \operatorname{ap} C$ and $\boldsymbol{y} \in \operatorname{rint} C$, then $(\boldsymbol{x}, \boldsymbol{y}) \subset \operatorname{rint} C$ due to $\operatorname{ap} C \subset \operatorname{rbd} C$ and Theorem 3.24, implying the inclusion

$$[\boldsymbol{x}, \boldsymbol{y}] \subset \operatorname{ap} C \cup \operatorname{rint} C = C_1.$$

Similarly, let $\boldsymbol{x}, \boldsymbol{y} \in C_2$. If $\boldsymbol{x}, \boldsymbol{y} \in \operatorname{ap}(\operatorname{cl} C)$ or $\boldsymbol{x}, \boldsymbol{y} \in \operatorname{rint} C$, then, respectively, either $[\boldsymbol{x}, \boldsymbol{y}] \subset \operatorname{ap}(\operatorname{cl} C)$ by Theorem 5.14 or $[\boldsymbol{x}, \boldsymbol{y}] \subset \operatorname{rint} C$ by Corollary 3.25. Let $\boldsymbol{x} \in \operatorname{ap}(\operatorname{cl} C)$ and $\boldsymbol{y} \in \operatorname{rint} C$. Since $\operatorname{cl} C$ is not a plane, one has $(\boldsymbol{x}, \boldsymbol{y}) \subset \operatorname{rint} C$ due to $\operatorname{ap}(\operatorname{cl} C) \subset \operatorname{rbd}(\operatorname{cl} C) = \operatorname{rbd} C$ and Theorem 3.47; whence

$$[\boldsymbol{x}, \boldsymbol{y}] \subset \operatorname{ap}(\operatorname{cl} C) \cup \operatorname{rint} C = C_2.$$

The inclusions $C_1 \subset C$ and $C_2 \subset \operatorname{cl} C$ are obvious.

2. Next, we are going to show that $\operatorname{ap} C = \operatorname{ap} C_1$. Indeed, let $\boldsymbol{a} \in \operatorname{ap} C$ and $\boldsymbol{x} \in C_1 \setminus \{\boldsymbol{a}\}$. If $\boldsymbol{x} \in \operatorname{ap} C$, then $(\boldsymbol{a}, \boldsymbol{x}\rangle \subset \langle \boldsymbol{a}, \boldsymbol{x}\rangle \subset \operatorname{ap} C$ because $\operatorname{ap} C$ is a plane (see Theorem 5.14). If $\boldsymbol{x} \in \operatorname{rint} C$, then $(\boldsymbol{a}, \boldsymbol{x}\rangle \subset \operatorname{rint} C \subset C_1$, as follows from Theorem 5.20. Hence $\operatorname{ap} C \subset \operatorname{ap} C_1$.

For the opposite inclusion, we observe that $\operatorname{rint} C \subset C_1 \subset \operatorname{cl} C$, which gives $\operatorname{rint} C_1 = \operatorname{rint} C$ (see Corollary 3.49). Since C is not a plane, C_1 also is not a plane, and Corollary 5.27 implies

$$\operatorname{ap} C_1 \subset C_1 \setminus \operatorname{rint} C_1 = (\operatorname{ap} C \cup \operatorname{rint} C) \setminus \operatorname{rint} C = \operatorname{ap} C.$$

3. The inclusion $\operatorname{ap} C \subset \operatorname{ap}(\operatorname{cl} C)$ follows from Theorem 5.20.

4. Finally, we assert that $\operatorname{ap}(\operatorname{cl} C) = \operatorname{ap} C_2$. As above, let $\boldsymbol{a} \in \operatorname{ap}(\operatorname{cl} C)$ and $\boldsymbol{x} \in C_2 \setminus \{\boldsymbol{a}\}$. If $\boldsymbol{x} \in \operatorname{ap}(\operatorname{cl} C)$, then $(\boldsymbol{a}, \boldsymbol{x}\rangle \subset \langle \boldsymbol{a}, \boldsymbol{x}\rangle \subset \operatorname{ap}(\operatorname{cl} C)$ according to Theorem 5.14. Let $\boldsymbol{x} \in \operatorname{rint} C$. Then $(\boldsymbol{a}, \boldsymbol{x}\rangle \subset \operatorname{cl} C$ because $\boldsymbol{a} \in \operatorname{ap}(\operatorname{cl} C)$. For any $\boldsymbol{z} \in (\boldsymbol{a}, \boldsymbol{x}\rangle$, there is a point $\boldsymbol{u} \in (\boldsymbol{a}, \boldsymbol{x}\rangle$ such that $\boldsymbol{z} \in (\boldsymbol{x}, \boldsymbol{u})$, implying the inclusion $\boldsymbol{z} \in \operatorname{rint} C$ (see Theorem 3.47). Hence

$$(\boldsymbol{a}, \boldsymbol{x}\rangle \subset \operatorname{ap}(\operatorname{cl} C) \cup \operatorname{rint} C = C_2,$$

which shows that $\boldsymbol{a} \in \operatorname{ap} C_2$. Thus $\operatorname{ap}(\operatorname{cl} C) \subset \operatorname{ap} C_2$.

To prove the opposite inclusion, we observe that $\operatorname{rint} C \subset C_2 \subset \operatorname{cl} C$, which gives $\operatorname{rint} C_2 = \operatorname{rint} C$ according to Corollary 3.49. Since C is not a plane, C_2 also is not a plane, and Corollary 5.27 gives

$$\operatorname{ap} C_2 \subset C_2 \setminus \operatorname{rint} C_2 = (\operatorname{ap}(\operatorname{cl} C) \cup \operatorname{rint} C) \setminus \operatorname{rint} C = \operatorname{ap}(\operatorname{cl} C).$$

The fact that all three sets $\operatorname{cl} C$, C_1, and C_2 are cones follows from (5.9) and the assumption $\operatorname{ap} C \neq \varnothing$. □

Remarks. 1. All four convex cones from Theorem 5.28 may be pairwise distinct. Indeed, let

$$C = \{(x, y, z) : y > |x|\} \cup \{(x, x, 0) : x \geqslant 0\}.$$

Then $\operatorname{cl} C = \{(x, y, z) : y \geqslant |x|\}$ and $\operatorname{rint} C = \{(x, y, z) : y > |x|\}$. Furthermore, $\operatorname{ap} C = \{\boldsymbol{o}\}$ and $\operatorname{ap}(\operatorname{cl} C) = \{(0, 0, z) : z \in \mathbb{R}\}$, which shows that C, C_1, C_2, and $\operatorname{cl} C$ are pairwise distinct. It is easy to see that all these cones are convex, while the set $\operatorname{ap}(\operatorname{cl} C) \cup C$ is not a convex cone.

2. The inclusion $\operatorname{ap} C \subset C \cap \operatorname{ap}(\operatorname{cl} C)$ may be proper. For instance, if

$$C = \{(x, y) : y < 0\} \cup \{(x, 0) : x \geqslant 0\},$$

then $\operatorname{ap} C = \{\boldsymbol{o}\}$ and $\operatorname{ap}(\operatorname{cl} C)$ is the x-axis of $\mathbb{R}^2$, implying that $\operatorname{ap} C \neq C \cap \operatorname{ap}(\operatorname{cl} C)$.

5.2 Conic Hulls

5.2.1 *Basic Properties of Conic Hulls*

Definition 5.29. For a given set $X \subset \mathbb{R}^n$, the intersection of all convex cones with *proper* apex $\boldsymbol{a}$ containing X is called the *conic hull* of X with apex $\boldsymbol{a}$ and is denoted $\operatorname{cone}_{\boldsymbol{a}} X$. Similarly, the intersection of all (not necessarily convex) cones with *proper* apex $\boldsymbol{a}$ containing the set X is called the *generated cone* of X with apex $\boldsymbol{a}$ and is denoted $C_{\boldsymbol{a}}(X)$.

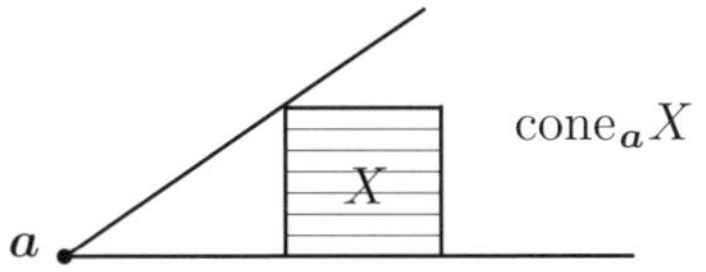

Fig. 5.10 The conic hull of a set X with apex $\boldsymbol{a}$.

Remarks. 1. In contrast with general cones (see Definition 5.1), conic hulls and generated cones involve only cones with *proper* apices. This restriction simplifies many arguments related to conic hulls, especially those related to closedness conditions on convex sets. The conic hull $\operatorname{cone}_{\boldsymbol{o}} X$ of a set $X \subset \mathbb{R}^n$ with apex $\boldsymbol{o}$ is often called the *positive hull* of X.

2. Corollary 5.9 shows for any set $X \subset \mathbb{R}^n$ and a point $\boldsymbol{a} \in \mathbb{R}^n$, the conic hull $\operatorname{cone}_{\boldsymbol{a}} X$ exists and is the smallest convex cone with *proper* apex $\boldsymbol{a}$ containing X. Similarly, the generated cone $C_{\boldsymbol{a}}(X)$ exists and is the smallest cone with *proper* apex $\boldsymbol{a}$ containing X. Clearly,

$$C_{\boldsymbol{a}}(X) \subset \operatorname{cone}_{\boldsymbol{a}} X \quad \text{and} \quad C_{\boldsymbol{a}}(\varnothing) = \operatorname{cone}_{\boldsymbol{a}} \varnothing = \{\boldsymbol{a}\}. \qquad (5.10)$$

Furthermore, if $X \not\subset \{\boldsymbol{a}\}$, then $C_{\boldsymbol{a}}(X)$ is the union of all closed halflines of the form $[\boldsymbol{a}, \boldsymbol{x}\rangle$, where $\boldsymbol{x} \in X \setminus \{\boldsymbol{a}\}$.

3. In view of Theorem 5.14, Definition 5.29 can be extended by considering the conic hull $\operatorname{cone}_L X$ (respectively, the generated cone $C_L(X)$) of a set $X \subset \mathbb{R}^n$ with respect to a nonempty plane $L \subset \mathbb{R}^n$ as the intersection of all convex cones (respectively, of all cones) $C \subset \mathbb{R}^n$ containing X such that $L \subset \operatorname{ap} C$ (see Problem 5.11 for a more general construction). Consequently, a considerable part of this section can be rewritten in terms of $\operatorname{cone}_L X$ and $C_L(X)$.

The next result describes elementary properties of conic hulls.

Theorem 5.30. *For sets X and Y in $\mathbb{R}^n$ and a point $\boldsymbol{a} \in \mathbb{R}^n$, the following assertions hold.*

(*a*) $X \subset \operatorname{cone}_{\boldsymbol{a}} X$, *and* $X = \operatorname{cone}_{\boldsymbol{a}} X$ *if and only if* X *is a convex cone with proper apex* $\boldsymbol{a}$.
(*b*) $\operatorname{cone}_{\boldsymbol{a}} X = \operatorname{cone}_{\boldsymbol{a}}(\{\boldsymbol{a}\} \cup X)$.
(*c*) $(\boldsymbol{c} - \boldsymbol{a}) + \operatorname{cone}_{\boldsymbol{a}} X = \operatorname{cone}_{\boldsymbol{c}}(\boldsymbol{c} - \boldsymbol{a} + X)$ *for any point* $\boldsymbol{c} \in \mathbb{R}^n$.
(*d*) $\operatorname{cone}_{\boldsymbol{a}}(\operatorname{cone}_{\boldsymbol{a}} X) = \operatorname{cone}_{\boldsymbol{a}} X$.
(*e*) $\operatorname{cone}_{\boldsymbol{a}} X \subset \operatorname{cone}_{\boldsymbol{a}} Y$ *if* $X \subset Y$.
(*f*) $\operatorname{cone}_{\boldsymbol{a}} X = \operatorname{cone}_{\boldsymbol{a}} Y$ *if* $X \subset Y \subset \operatorname{cone}_{\boldsymbol{a}} X$.
(*g*) $\operatorname{cone}_{\boldsymbol{a}} X = \operatorname{cone}_{\boldsymbol{a}}(X \cup Z)$ *if and only if* $Z \subset \operatorname{cone}_{\boldsymbol{a}} X$.

Furthermore, if $\{X_\lambda\}$ *is a family of sets in* $\mathbb{R}^n$ *and* $\boldsymbol{a} \in \mathbb{R}^n$, *then the statements below are true.*

(*h*) $\operatorname{cone}_{\boldsymbol{a}}(\bigcap_\lambda X_\lambda) \subset \bigcap_\lambda \operatorname{cone}_{\boldsymbol{a}} X_\lambda$.
(*i*) $\bigcup_\lambda \operatorname{cone}_{\boldsymbol{a}} X_\lambda \subset \operatorname{cone}_{\boldsymbol{a}}(\bigcup_\lambda X_\lambda)$.
(*j*) $\bigcup_\lambda \operatorname{cone}_{\boldsymbol{a}} X_\lambda = \operatorname{cone}_{\boldsymbol{a}}(\bigcup_\lambda X_\lambda)$ *if the family* $\{X_\lambda\}$ *is nested.*
(*k*) $\operatorname{cone}_{\boldsymbol{a}}(\bigcup_\lambda X_\lambda) = \operatorname{cone}_{\boldsymbol{a}}(\bigcup_\lambda \operatorname{conv} X_\lambda) = \operatorname{cone}_{\boldsymbol{a}}(\bigcup_\lambda \operatorname{cone}_{\boldsymbol{a}} X_\lambda)$.

Proof. Let $\mathcal{C}(X)$ denote the family of all convex cones with proper apex $\boldsymbol{a}$ containing a given set $X \subset \mathbb{R}^n$. The proofs of assertions (*a*)–(*k*) derive from the following arguments.

1. $\operatorname{cone}_{\boldsymbol{a}} X$ is the smallest element in $\mathcal{C}(X)$.
2. $\mathcal{C}(X)$ is exactly the family of convex cones with proper apex $\boldsymbol{a}$ containing $\operatorname{cone}_{\boldsymbol{a}} X$.
3. If $X \subset Y$, then $\mathcal{C}(Y) \subset \mathcal{C}(X)$.
4. The union of a nested family of convex cones with proper apex $\boldsymbol{a}$ is a convex cone with proper apex $\boldsymbol{a}$ (see Corollary 5.9). □

A similar argument results in a list of properties of generated cones, as shown in the next obvious theorem.

Theorem 5.31. *For sets X and Y in $\mathbb{R}^n$ and a point $\boldsymbol{a} \in \mathbb{R}^n$, the following assertions hold.*

(*a*) $X \subset C_{\boldsymbol{a}}(X)$, *and* $X = C_{\boldsymbol{a}}(X)$ *if and only if* X *is a cone with proper apex* $\boldsymbol{a}$.
(*b*) *If* $X \neq \varnothing$, *then* $C_{\boldsymbol{a}}(X) = \{\boldsymbol{a} + \lambda(\boldsymbol{x} - \boldsymbol{a}) : \boldsymbol{x} \in X,\ \lambda \geqslant 0\}$.

(*c*) $C_{\boldsymbol{a}}(X) = C_{\boldsymbol{a}}(\{\boldsymbol{a}\} \cup X)$.
(*d*) $(\boldsymbol{c} - \boldsymbol{a}) + C_{\boldsymbol{a}}(X) = C_{\boldsymbol{c}}(\boldsymbol{c} - \boldsymbol{a} + X)$ *for any point* $\boldsymbol{c} \in \mathbb{R}^n$.
(*e*) $C_{\boldsymbol{a}}(C_{\boldsymbol{a}}(X)) = C_{\boldsymbol{a}}(X)$.
(*f*) $C_{\boldsymbol{a}}(X) \subset C_{\boldsymbol{a}}(Y)$ *if* $X \subset Y$.
(*g*) $C_{\boldsymbol{a}}(X) = C_{\boldsymbol{a}}(Y)$ *if* $X \subset Y \subset C_{\boldsymbol{a}}(X)$.
(*h*) $C_{\boldsymbol{a}}(X) = C_{\boldsymbol{a}}(X \cup Z)$ *if and only if* $Z \subset C_{\boldsymbol{a}}(X)$.

Furthermore, if $\{X_\lambda\}$ *is a family of sets in* $\mathbb{R}^n$ *and* $\boldsymbol{a} \in \mathbb{R}^n$, *then the statements below are true.*

(*i*) $C_{\boldsymbol{a}}(\underset{\lambda}{\cap} X_\lambda) \subset \underset{\lambda}{\cap} C_{\boldsymbol{a}}(X_\lambda)$.
(*j*) $\underset{\lambda}{\cup} C_{\boldsymbol{a}}(X_\lambda) = C_{\boldsymbol{a}}(\underset{\lambda}{\cup} X_\lambda) = C_{\boldsymbol{a}}(\underset{\lambda}{\cup} C_{\boldsymbol{a}}(X_\lambda))$. □

The theorem below gives an important description of conic hulls in terms of nonnegative combinations of points.

Theorem 5.32. *For a nonempty set* $X \subset \mathbb{R}^n$ *and a point* $\boldsymbol{a} \in \mathbb{R}^n$, *one has*

$$\begin{aligned}\text{cone}_{\boldsymbol{a}} X = \{&\boldsymbol{a} + \lambda_1(\boldsymbol{x}_1 - \boldsymbol{a}) + \cdots + \lambda_k(\boldsymbol{x}_k - \boldsymbol{a}) : k \geqslant 1,\ \boldsymbol{x}_1, \ldots, \boldsymbol{x}_k \in X,\\ &\lambda_1, \ldots, \lambda_k \geqslant 0\}.\end{aligned}$$

Proof. Since $\text{cone}_{\boldsymbol{a}} X$ is a convex cone with proper apex $\boldsymbol{a}$ containing X, Theorem 5.5 implies that the set

$$\begin{aligned}C = \{&\boldsymbol{a} + \lambda_1(\boldsymbol{x}_1 - \boldsymbol{a}) + \cdots + \lambda_k(\boldsymbol{x}_k - \boldsymbol{a}) : k \geqslant 1,\ \boldsymbol{x}_1, \ldots, \boldsymbol{x}_k \in X,\\ &\lambda_1, \ldots, \lambda_k \geqslant 0\}\end{aligned}$$

lies in $\text{cone}_{\boldsymbol{a}} X$. We assert that C is a convex cone with proper apex $\boldsymbol{a}$. Indeed, the inclusion $\boldsymbol{a} \in C$ is obvious. If $\boldsymbol{x} \in C$, then

$$\boldsymbol{x} = \boldsymbol{a} + \lambda_1(\boldsymbol{x}_1 - \boldsymbol{a}) + \cdots + \lambda_k(\boldsymbol{x}_k - \boldsymbol{a})$$

for suitable points $\boldsymbol{x}_1, \ldots, \boldsymbol{x}_k \in X$ and scalars $\lambda_1, \ldots, \lambda_k \geqslant 0$. Given a scalar $\mu \geqslant 0$, one has

$$\boldsymbol{a} + \mu(\boldsymbol{x} - \boldsymbol{a}) = \boldsymbol{a} + \mu\lambda_1(\boldsymbol{x}_1 - \boldsymbol{a}) + \cdots + \mu\lambda_k(\boldsymbol{x}_k - \boldsymbol{a}) \in C,$$

implying that C is a cone with proper apex $\boldsymbol{a}$. For the convexity of C, choose points $\boldsymbol{u}, \boldsymbol{v} \in C$. Then $\boldsymbol{u}$ and $\boldsymbol{v}$ can be expressed as nonnegative combinations

$$\begin{aligned}\boldsymbol{u} &= \boldsymbol{a} + \gamma_1(\boldsymbol{u}_1 - \boldsymbol{a}) + \cdots + \gamma_p(\boldsymbol{u}_p - \boldsymbol{a}),\ \boldsymbol{u}_1, \ldots, \boldsymbol{u}_p \in X,\\ \boldsymbol{v} &= \boldsymbol{a} + \mu_1(\boldsymbol{v}_1 - \boldsymbol{a}) + \cdots + \mu_q(\boldsymbol{v}_q - \boldsymbol{a}),\ \boldsymbol{v}_1, \ldots, \boldsymbol{v}_q \in X.\end{aligned}$$

For a scalar $\lambda \in [0,1]$, one has

$$(1-\lambda)\boldsymbol{u} + \lambda\boldsymbol{v} = \boldsymbol{a} + (1-\lambda)\gamma_1(\boldsymbol{u}_1 - \boldsymbol{a}) + \cdots + (1-\lambda)\gamma_p(\boldsymbol{u}_p - \boldsymbol{a})$$
$$+ \lambda\mu_1(\boldsymbol{v}_1 - \boldsymbol{a}) + \cdots + \lambda\mu_q(\boldsymbol{v}_q - \boldsymbol{a}) \in C.$$

Hence C is a convex set.

Since $X \subset C$ (every point $\boldsymbol{x} \in X$ can be written as $\boldsymbol{x} = \boldsymbol{a} + 1(\boldsymbol{x} - \boldsymbol{a})$), the inclusions $X \subset C \subset \text{cone}_{\boldsymbol{a}} X$ and Theorem 5.30 give $\text{cone}_{\boldsymbol{a}} X = C$. □

Corollary 5.33. *For a finite set $X = \{\boldsymbol{x}_1, \ldots, \boldsymbol{x}_r\} \subset \mathbb{R}^n$ and a point $\boldsymbol{a} \in \mathbb{R}^n$, one has*

$$\text{cone}_{\boldsymbol{a}} X = \{\boldsymbol{a} + \lambda_1(\boldsymbol{x}_1 - \boldsymbol{a}) + \cdots + \lambda_r(\boldsymbol{x}_r - \boldsymbol{a}) : \lambda_1, \ldots, \lambda_r \geqslant 0\}.$$

Proof. By Theorem 5.32, a point $\boldsymbol{x} \in \mathbb{R}^n$ belongs to $\text{cone}_{\boldsymbol{a}} X$ if and only if it can be expressed as

$$\boldsymbol{x} = \boldsymbol{a} + \lambda_{i_1}(\boldsymbol{x}_{i_1} - \boldsymbol{a}) + \cdots + \lambda_{i_s}(\boldsymbol{x}_{i_s} - \boldsymbol{a}), \tag{5.11}$$

where $\lambda_{i_1}, \ldots, \lambda_{i_s} \geqslant 0$ and $\boldsymbol{x}_{i_1}, \ldots, \boldsymbol{x}_{i_s} \in \{\boldsymbol{x}_1, \ldots, \boldsymbol{x}_r\}$. Renumbering the indices $i_1, \ldots, i_s$, we may suppose that $1 \leqslant i_1 \leqslant \cdots \leqslant i_s \leqslant r$. For every index $i \in \{1, \ldots, r\} \setminus \{i_1, \ldots, i_s\}$, if any, we add $0(\boldsymbol{x}_i - \boldsymbol{a})$ to the right-hand side of (5.11) to express $\boldsymbol{x}$ in a desired form. □

A combination of Definition 5.6 and Corollary 5.33 gives the following assertion (compare with Corollary 5.56).

Corollary 5.34. *Any r-simplicial cone $\Gamma_{\boldsymbol{a}}(\boldsymbol{x}_1, \ldots, \boldsymbol{x}_r) \subset \mathbb{R}^n$, $1 \leqslant r \leqslant n$, can be expressed as*

$$\Gamma_{\boldsymbol{a}}(\boldsymbol{x}_1, \ldots, \boldsymbol{x}_r) = \text{cone}_{\boldsymbol{a}}\{\boldsymbol{x}_1, \ldots, \boldsymbol{x}_r\}.$$ □

The next result is analogous to Theorem 4.6.

Theorem 5.35. *Let $\boldsymbol{a} \in \mathbb{R}^n$ and $X \subset \mathbb{R}^n$ be a set such that $X \setminus \{\boldsymbol{a}\} \neq \varnothing$. Then $\text{cone}_{\boldsymbol{a}} X \setminus \{\boldsymbol{a}\}$ is the collection of all positive combinations*

$$\boldsymbol{x} = \boldsymbol{a} + \lambda_1(\boldsymbol{x}_1 - \boldsymbol{a}) + \cdots + \lambda_k(\boldsymbol{x}_k - \boldsymbol{a}), \quad \boldsymbol{x}_1, \ldots, \boldsymbol{x}_k \in X, \quad k \geqslant 1,$$

where $\{\boldsymbol{a}, \boldsymbol{x}_1, \ldots, \boldsymbol{x}_k\}$ is affinely independent.

Proof. Assume first that a point $\boldsymbol{x} \in \mathbb{R}^n$ belongs to $\text{cone}_{\boldsymbol{a}} X \setminus \{\boldsymbol{a}\}$. By Theorem 5.32, $\boldsymbol{x}$ can be written as

$$\boldsymbol{x} = \boldsymbol{a} + \lambda_1(\boldsymbol{x}_1 - \boldsymbol{a}) + \cdots + \lambda_k(\boldsymbol{x}_k - \boldsymbol{a}), \tag{5.12}$$

where $\boldsymbol{x}_1, \ldots, \boldsymbol{x}_k \in X$ and $\lambda_1, \ldots, \lambda_k \geqslant 0$. Since $\boldsymbol{x} \neq \boldsymbol{a}$, at least one of the scalars $\lambda_1, \ldots, \lambda_k$ is positive. By a finiteness argument, we may assume that (5.12) involves a minimum possible number of points $\boldsymbol{x}_1, \dots, \boldsymbol{x}_k$. Furthermore, excluding in (5.12) all terms $0(\boldsymbol{x}_i - \boldsymbol{a})$, we may suppose that the scalars $\lambda_1, \dots, \lambda_k$ are positive.

Under these assumptions, we are going to show first that the set $\{\boldsymbol{x}_1, \dots, \boldsymbol{x}_k\}$ is affinely independent. Indeed, suppose that it is affinely dependent. Then there are scalars $\nu_1, \dots, \nu_k$, not all zero, such that

$$\nu_1 \boldsymbol{x}_1 + \cdots + \nu_k \boldsymbol{x}_k = \boldsymbol{o} \quad \text{and} \quad \nu_1 + \cdots + \nu_k = 0. \tag{5.13}$$

Clearly, at least one of the scalar $\nu_1, \dots, \nu_k$ is positive. Put

$$\alpha = \min \{\lambda_i / \nu_i : \nu_i > 0, \ 1 \leqslant i \leqslant k\}.$$

Then $\lambda_i - \alpha\nu_i \geqslant 0$ for all $1 \leqslant i \leqslant k$, and $\lambda_i - \alpha\nu_i = 0$ for at least one $i \in \{1, \dots, k\}$. Based on (5.13), we obtain

$$\boldsymbol{x} = \boldsymbol{a} + (\lambda_1 - \alpha\nu_1)(\boldsymbol{x}_1 - \boldsymbol{a}) + \cdots + (\lambda_k - \alpha\nu_k)(\boldsymbol{x}_k - \boldsymbol{a}),$$

in contradiction with the minimality of k.

Next, we assert that the set $\{\boldsymbol{a}, \boldsymbol{x}_1, \dots, \boldsymbol{x}_k\}$ is affinely independent. By the above proved, it suffices to show that $\boldsymbol{a} \notin \operatorname{aff} \{\boldsymbol{x}_1, \dots, \boldsymbol{x}_k\}$, as follows from Theorem 2.54. Assume for a moment that $\boldsymbol{a} \in \operatorname{aff} \{\boldsymbol{x}_1, \dots, \boldsymbol{x}_k\}$. Then $\boldsymbol{a}$ can be written as an affine combination $\boldsymbol{a} = \mu_1 \boldsymbol{x}_1 + \cdots + \mu_k \boldsymbol{x}_k$ (see Corollary 2.45). Consequently,

$$\boldsymbol{o} = \mu_1(\boldsymbol{x}_1 - \boldsymbol{a}) + \cdots + \mu_k(\boldsymbol{x}_k - \boldsymbol{a}).$$

Clearly, at least one of the scalars $\mu_1, \dots, \mu_k$ is positive. Put

$$\beta = \min \{\lambda_i / \mu_i : \mu_i > 0, \ 1 \leqslant i \leqslant k\}.$$

Then $\lambda_i - \beta\mu_i \geqslant 0$ for all $1 \leqslant i \leqslant k$, and $\lambda_i - \beta\mu_i = 0$ for at least one $i \in \{1, \dots, k\}$. Since

$$\boldsymbol{x} = \boldsymbol{a} + (\lambda_1 - \beta\mu_1)(\boldsymbol{x}_1 - \boldsymbol{a}) + \cdots + (\lambda_k - \beta\mu_k)(\boldsymbol{x}_k - \boldsymbol{a}),$$

we obtain a contradiction with the minimality of k. Summing up, $\boldsymbol{x}$ has a desired form.

Conversely, assume that a point $\boldsymbol{x} \in \mathbb{R}^n$ is expressed in the form (5.12) such that all scalars $\lambda_1, \dots, \lambda_k$ are positive and the set $\{\boldsymbol{a}, \boldsymbol{x}_1, \dots, \boldsymbol{x}_k\}$ is affinely independent. By Theorem 5.32, $\boldsymbol{x} \in \operatorname{cone}_{\boldsymbol{a}} X$. It remains to prove that $\boldsymbol{x} \neq \boldsymbol{a}$. Assume, for contradiction, that $\boldsymbol{x} = \boldsymbol{a}$. From (5.12) we obtain

$$\lambda_1(\boldsymbol{x}_1 - \boldsymbol{a}) + \cdots + \lambda_k(\boldsymbol{x}_k - \boldsymbol{a}) = \boldsymbol{o}, \quad \lambda_1, \dots, \lambda_k > 0. \tag{5.14}$$

On the other hand, Theorem 2.52 implies that the vector set $\{\boldsymbol{x}_1 - \boldsymbol{a}, \dots, \boldsymbol{x}_k - \boldsymbol{a}\}$ is linearly independent, in contradiction with (5.14). Hence $\boldsymbol{x} \neq \boldsymbol{a}$. $\square$

Problem 5.4 describes an assertion which is similar to Theorem 5.35 but deals with the case of cones. By analogy with Theorem 4.8, we prove one more variation of Theorem 5.35.

Theorem 5.36. *Let $X \subset \mathbb{R}^n$ be a nonempty set, $\boldsymbol{a} \in \mathbb{R}^n$, and $\boldsymbol{x}_1$ be a given point in X. Then* $\operatorname{cone}_{\boldsymbol{a}} X$ *is the collection of all nonnegative combinations*

$$\boldsymbol{x} = \boldsymbol{a} + \lambda_1(\boldsymbol{x}_1 - \boldsymbol{a}) + \cdots + \lambda_k(\boldsymbol{x}_k - \boldsymbol{a}), \quad k \geqslant 1,$$

where $\{\boldsymbol{x}_1, \dots, \boldsymbol{x}_k\}$ is an affinely independent subset of X.

Proof. Since the case $\boldsymbol{x}_1 = \boldsymbol{a}$ immediately follows from Theorem 5.35, we may assume that $\boldsymbol{x}_1 \neq \boldsymbol{a}$. By Theorem 5.32, a point $\boldsymbol{x}$ belongs to $\operatorname{cone}_{\boldsymbol{a}} X$ if and only if it can be written as

$$\boldsymbol{x} = \boldsymbol{a} + \lambda_1(\boldsymbol{y}_1 - \boldsymbol{a}) + \cdots + \lambda_p(\boldsymbol{y}_p - \boldsymbol{a}),$$

where $\boldsymbol{y}_1, \dots, \boldsymbol{y}_p \in X$ and $\lambda_1, \dots, \lambda_p \geqslant 0$. Put $\lambda = \lambda_1 + \cdots + \lambda_p$. Because the case $\boldsymbol{x} = \boldsymbol{a}$ is obvious (let $\boldsymbol{x} = \boldsymbol{a} + 0\boldsymbol{x}_1$), we may suppose that $\boldsymbol{x} \neq \boldsymbol{a}$. Then $\lambda > 0$. Put

$$\boldsymbol{y} = \lambda^{-1}(\boldsymbol{x} - \boldsymbol{a}) = \lambda^{-1}\lambda_1(\boldsymbol{y}_1 - \boldsymbol{a}) + \cdots + \lambda^{-1}\lambda_p(\boldsymbol{y}_p - \boldsymbol{a}).$$

Clearly, $\boldsymbol{y}$ is a convex combination of $\boldsymbol{y}_1 - \boldsymbol{a}, \dots, \boldsymbol{y}_p - \boldsymbol{a}$. By Theorem 4.3, $\boldsymbol{y} \in \operatorname{conv}(X - \boldsymbol{a}) = \operatorname{conv} X - \boldsymbol{a}$, or $\boldsymbol{a} + \boldsymbol{y} \in \operatorname{conv} X$. Theorem 4.8 shows the existence of an affinely independent subset $\{\boldsymbol{x}_1, \dots, \boldsymbol{x}_k\}$ of X such that $\boldsymbol{a} + \boldsymbol{y}$ can be written as a convex combination

$$\boldsymbol{a} + \boldsymbol{y} = \mu_1\boldsymbol{x}_1 + \cdots + \mu_k\boldsymbol{x}_k.$$

Equivalently, $\boldsymbol{y} = \mu_1(\boldsymbol{x}_1 - \boldsymbol{a}) + \cdots + \mu_k(\boldsymbol{x}_k - \boldsymbol{a})$. Finally,

$$\boldsymbol{x} = \boldsymbol{a} + \lambda\boldsymbol{y} = \boldsymbol{a} + \lambda\mu_1(\boldsymbol{x}_1 - \boldsymbol{a}) + \cdots + \lambda\mu_k(\boldsymbol{x}_k - \boldsymbol{a})$$

is a desired expression for $\boldsymbol{x}$. □

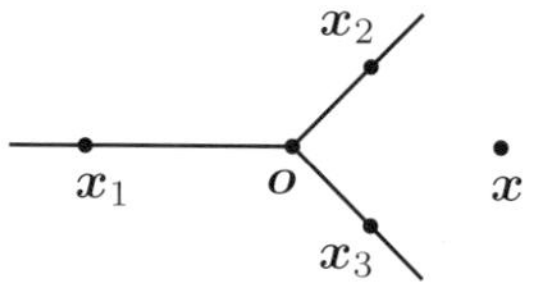

Fig. 5.11 Illustration to Theorem 5.36.

Remark. For a given point $\boldsymbol{x}$ of the form (5.12), the set $\{\boldsymbol{a}, \boldsymbol{x}_1, \dots, \boldsymbol{x}_k\}$ in Theorem 5.36 may be affinely dependent. Indeed, let $\boldsymbol{a} = \boldsymbol{o}$ and $X = \{\boldsymbol{x}_1, \boldsymbol{x}_2, \boldsymbol{x}_3\} \subset \mathbb{R}^2$, where $\boldsymbol{x}_1 = (-2, 0)$, $\boldsymbol{x}_2 = (1, 1)$, and $\boldsymbol{x}_3 = (1, -1)$, as depicted in Figure 5.11. Clearly, $\text{cone}_{\boldsymbol{o}} X = \mathbb{R}^2$, and the set $\{\boldsymbol{o}, \boldsymbol{x}_1, \boldsymbol{x}_2, \boldsymbol{x}_3\}$ is affinely dependent. On the other hand, given the point $\boldsymbol{x} = (2, 0)$, the equalities $\boldsymbol{x}_1 + \boldsymbol{x}_2 + \boldsymbol{x}_3 = \boldsymbol{o}$ and $\boldsymbol{x} = \boldsymbol{x}_2 + \boldsymbol{x}_3$ easily imply that the expression

$$\boldsymbol{x} = \boldsymbol{o} + \lambda(\boldsymbol{x}_1 - \boldsymbol{o}) + (1 + \lambda)(\boldsymbol{x}_2 - \boldsymbol{o}) + (1 + \lambda)(\boldsymbol{x}_3 - \boldsymbol{o}), \quad \lambda > 0,$$

is the only way to write $\boldsymbol{x}$ in the form (5.12).

5.2.2 *Affine Spans, Convex Hulls, and Conic Hulls*

The following result can be viewed as an analog of Theorem 4.7.

Theorem 5.37. *For a set $X \subset \mathbb{R}^n$ of positive dimension m and a point $\boldsymbol{a} \in \mathbb{R}^n$, the following assertions hold.*

(*a*) *If $\boldsymbol{a} \notin \text{aff}\, X$, then $\text{cone}_{\boldsymbol{a}} X \setminus \{\boldsymbol{a}\}$ is the collection of all positive combinations*

$$\boldsymbol{x} = \boldsymbol{a} + \lambda_1(\boldsymbol{x}_1 - \boldsymbol{a}) + \cdots + \lambda_k(\boldsymbol{x}_k - \boldsymbol{a}), \quad 1 \leqslant k \leqslant m + 1,$$

where $\boldsymbol{x}_1, \dots, \boldsymbol{x}_k \in X$ and $\{\boldsymbol{a}, \boldsymbol{x}_1, \dots, \boldsymbol{x}_k\}$ is affinely independent.

(*b*) *If $\boldsymbol{a} \in \text{aff}\, X$, then $\text{cone}_{\boldsymbol{a}} X \setminus \{\boldsymbol{a}\}$ is the collection of all positive combinations*

$$\boldsymbol{x} = \boldsymbol{a} + \lambda_1(\boldsymbol{x}_1 - \boldsymbol{a}) + \cdots + \lambda_k(\boldsymbol{x}_k - \boldsymbol{a}), \quad 1 \leqslant k \leqslant m,$$

where $\boldsymbol{x}_1, \dots, \boldsymbol{x}_k \in X$ and $\{\boldsymbol{a}, \boldsymbol{x}_1, \dots, \boldsymbol{x}_k\}$ is affinely independent.

(*c*) *If $\boldsymbol{a} \notin \text{aff}\, X$, then $\text{cone}_{\boldsymbol{a}} X$ is the union of $(m + 1)$-simplicial cones of the form $\Gamma_{\boldsymbol{a}}(\boldsymbol{x}_1, \dots, \boldsymbol{x}_{m+1})$, where $\boldsymbol{x}_1, \dots, \boldsymbol{x}_{m+1} \in X$.*

(*d*) *If $\boldsymbol{a} \in \text{aff}\, X$, then $\text{cone}_{\boldsymbol{a}} X$ is the union of m-simplicial cones of the form $\Gamma_{\boldsymbol{a}}(\boldsymbol{x}_1, \dots, \boldsymbol{x}_m)$, where $\boldsymbol{x}_1, \dots, \boldsymbol{x}_m \in X$.*

Proof. Corollary 2.67 shows that $\dim(\{\boldsymbol{a}\} \cup X) \leqslant m + 1$, with equality if and only if $\boldsymbol{a} \notin \text{aff}\, X$. Because $k + 1$ is the maximum cardinality of an affinely independent subset of a k-dimensional set $Z \subset \mathbb{R}^n$ (see Corollary 2.66), assertions (*a*) and (*b*) follows from Theorem 5.35.

(*c*) Let $\boldsymbol{a} \notin \text{aff}\, X$. If $\{\boldsymbol{x}_1, \dots, \boldsymbol{x}_{m+1}\}$ is an affinely independent subset of X, then $\{\boldsymbol{a}, \boldsymbol{x}_1, \dots, \boldsymbol{x}_{m+1}\}$ is also affinely independent (see Theorem 2.54). By Corollary 5.8, the simplicial cone $\Gamma_{\boldsymbol{a}}(\boldsymbol{x}_1, \dots, \boldsymbol{x}_{m+1})$ lies in $\text{cone}_{\boldsymbol{a}} X$. Conversely, let $\boldsymbol{x} \in \text{cone}_{\boldsymbol{a}} X$. Since the case $\boldsymbol{x} = \boldsymbol{a}$ is obvious, we assume that $\boldsymbol{x} \neq \boldsymbol{a}$. Theorem 5.35 implies that $\boldsymbol{x}$ can be expressed as a positive combination

$$\boldsymbol{x} = \boldsymbol{a} + \lambda_1(\boldsymbol{x}_1 - \boldsymbol{a}) + \cdots + \lambda_k(\boldsymbol{x}_k - \boldsymbol{a}), \quad \boldsymbol{x}_1, \ldots, \boldsymbol{x}_k \in X,$$

where the set $\{\boldsymbol{a}, \boldsymbol{x}_1, \ldots, \boldsymbol{x}_k\}$ is affinely independent. Theorem 2.61 shows that $\{\boldsymbol{x}_1, \ldots, \boldsymbol{x}_k\}$ is extendable to an affinely independent subset $\{\boldsymbol{x}_1, \ldots, \boldsymbol{x}_{m+1}\}$ of X. Because $\{\boldsymbol{a}, \boldsymbol{x}_1, \ldots, \boldsymbol{x}_{m+1}\}$ is affinely independent, the equality

$$\begin{aligned}\boldsymbol{x} = \boldsymbol{a} &+ \lambda_1(\boldsymbol{x}_1 - \boldsymbol{a}) + \cdots + \lambda_k(\boldsymbol{x}_k - \boldsymbol{a}) \\ &+ 0(\boldsymbol{x}_{k+1} - \boldsymbol{a}) + \cdots + 0(\boldsymbol{x}_{m+1} - \boldsymbol{a}),\end{aligned}$$

implies the inclusion $\boldsymbol{x} \in \Gamma_{\boldsymbol{a}}(\boldsymbol{x}_1, \ldots, \boldsymbol{x}_{m+1})$.

(*d*) The case $\boldsymbol{a} \in \operatorname{aff} X$ is similar. □

Various relations between affine spans, convex hulls, and conic hulls are given in the next theorem (compare with Theorem 5.30). We recall that $C_{\boldsymbol{a}}(X)$ stands for the cone with apex $\boldsymbol{a}$ generated by a set $X \subset \mathbb{R}^n$ (see Definition 5.29).

Theorem 5.38. *For a set $X \subset \mathbb{R}^n$ and a point $\boldsymbol{a} \in \mathbb{R}^n$, the following assertions hold.*

(*a*) $\operatorname{conv} X \subset \operatorname{cone}_{\boldsymbol{a}} X$.
(*b*) $\operatorname{conv}(\operatorname{cone}_{\boldsymbol{a}} X) = \operatorname{cone}_{\boldsymbol{a}}(\operatorname{conv} X) = \operatorname{cone}_{\boldsymbol{a}} X$.
(*c*) $\operatorname{cone}_{\boldsymbol{a}} X = \operatorname{conv} C_{\boldsymbol{a}}(X) = C_{\boldsymbol{a}}(\operatorname{conv} X)$.
(*d*) $C_{\boldsymbol{a}}(X) \subset \operatorname{cone}_{\boldsymbol{a}} X \subset \operatorname{aff}(\{\boldsymbol{a}\} \cup X)$ *and*

$$\operatorname{aff} C_{\boldsymbol{a}}(X) = \operatorname{aff}(\operatorname{cone}_{\boldsymbol{a}} X) = \operatorname{aff}(\{\boldsymbol{a}\} \cup X). \tag{5.15}$$

Consequently, $\dim C_{\boldsymbol{a}}(X) = \dim(\operatorname{cone}_{\boldsymbol{a}} X) = \dim(\{\boldsymbol{a}\} \cup X)$.
(*e*) $C_{\boldsymbol{a}}(X) \subset \operatorname{aff} X \Leftrightarrow \operatorname{cone}_{\boldsymbol{a}} X \subset \operatorname{aff} X \Leftrightarrow \boldsymbol{a} \in \operatorname{aff} X$. *So, any of the equalities* $\dim C_{\boldsymbol{a}}(X) = \dim X$ *and* $\dim(\operatorname{cone}_{\boldsymbol{a}} X) = \dim X$ *holds if and only if* $\boldsymbol{a} \in \operatorname{aff} X$.
(*f*) $\operatorname{cone}_{\boldsymbol{a}} X$ *is a plane if and only if* $\operatorname{cone}_{\boldsymbol{a}} X = \operatorname{aff}(\{\boldsymbol{a}\} \cup X)$.
(*g*) *If* $X \neq \varnothing$, *then* $\operatorname{cone}_{\boldsymbol{a}} X$ *is a plane if and only if* $\operatorname{cone}_{\boldsymbol{a}} X = \operatorname{aff} X$.
(*h*) *If* $\boldsymbol{a} \notin \operatorname{conv} X$, *then* $\operatorname{cone}_{\boldsymbol{a}} X \setminus \{\boldsymbol{a}\}$ *is the smallest convex cone with improper apex* $\boldsymbol{a}$ *containing* X. *Consequently,* $\operatorname{ap}(\operatorname{cone}_{\boldsymbol{a}} X) = \{\boldsymbol{a}\}$, *which shows that* $\operatorname{cone}_{\boldsymbol{a}} X$ *is not a plane provided* $X \neq \varnothing$ *and* $\boldsymbol{a} \notin \operatorname{conv} X$.
(*i*) *If* $C \subset \mathbb{R}^n$ *is a cone with proper apex* $\boldsymbol{a}$ *such that* $X \cap C \subset \{\boldsymbol{a}\}$, *then* $C_{\boldsymbol{a}}(X) \cap C = \{\boldsymbol{a}\}$.

(j) *If* $C \subset \mathbb{R}^n$ *is a cone with proper apex* $\boldsymbol{a}$ *such that* $\operatorname{conv} X \cap C \subset \{\boldsymbol{a}\}$, *then* $\operatorname{cone}_{\boldsymbol{a}} X \cap C = \{\boldsymbol{a}\}$.

Proof. Since all assertions of the theorem are obvious when $X = \varnothing$, we may assume that X is nonempty.

(a) Definition 5.29 shows $\operatorname{cone}_{\boldsymbol{a}} X$ is a convex cone containing X. Therefore, $\operatorname{conv} X \subset \operatorname{cone}_{\boldsymbol{a}} X$ by the definition of $\operatorname{conv} X$.

(b) The equality $\operatorname{conv}(\operatorname{cone}_{\boldsymbol{a}} X) = \operatorname{cone}_{\boldsymbol{a}} X$ immediately follows from the convexity of $\operatorname{cone}_{\boldsymbol{a}} X$. Similarly, the equality $\operatorname{cone}_{\boldsymbol{a}}(\operatorname{conv} X) = \operatorname{cone}_{\boldsymbol{a}} X$ follows from the inclusions $X \subset \operatorname{conv} X \subset \operatorname{cone}_{\boldsymbol{a}} X$ and Theorem 5.30.

(c) Because $C_{\boldsymbol{a}}(X) \subset \operatorname{cone}_{\boldsymbol{a}} X$ (see (5.10)) and because the set $\operatorname{cone}_{\boldsymbol{a}} X$ is convex, Theorem 4.2 gives $\operatorname{conv} C_{\boldsymbol{a}}(X) \subset \operatorname{cone}_{\boldsymbol{a}} X$. On the other hand, since $\operatorname{conv} C_{\boldsymbol{a}}(X)$ is a convex cone with proper apex $\boldsymbol{a}$ (see Theorem 5.4), the definition of $\operatorname{cone}_{\boldsymbol{a}} X$ gives $\operatorname{cone}_{\boldsymbol{a}} X \subset \operatorname{conv} C_{\boldsymbol{a}}(X)$. Thus $\operatorname{cone}_{\boldsymbol{a}} X = \operatorname{conv} C_{\boldsymbol{a}}(X)$.

Similarly, the inclusion $\operatorname{conv} X \subset \operatorname{cone}_{\boldsymbol{a}} X$ implies that $C_{\boldsymbol{a}}(\operatorname{conv} X) \subset \operatorname{cone}_{\boldsymbol{a}} X$. For the opposite inclusion, choose any point $\boldsymbol{x} \in \operatorname{cone}_{\boldsymbol{a}} X$. By Theorem 5.32, $\boldsymbol{x}$ can be written as

$$\boldsymbol{x} = \boldsymbol{a} + \lambda_1(\boldsymbol{x}_1 - \boldsymbol{a}) + \cdots + \lambda_k(\boldsymbol{x}_k - \boldsymbol{a})$$

for suitable $\boldsymbol{x}_1, \ldots, \boldsymbol{x}_k \in X$ and $\lambda_1, \ldots, \lambda_k \geqslant 0$. Let $\lambda = \lambda_1 + \cdots + \lambda_k$. Because the case $\lambda = 0$ is obvious, we assume that $\lambda > 0$ and consider the convex combination

$$\boldsymbol{z} = \boldsymbol{a} + \frac{\lambda_1}{\lambda}(\boldsymbol{x}_1 - \boldsymbol{a}) + \cdots + \frac{\lambda_k}{\lambda}(\boldsymbol{x}_k - \boldsymbol{a}) = \frac{\lambda_1}{\lambda}\boldsymbol{x}_1 + \cdots + \frac{\lambda_k}{\lambda}\boldsymbol{x}_k.$$

Then $\boldsymbol{z} \in \operatorname{conv} X$ by Theorem 4.3, and $\boldsymbol{x} = \boldsymbol{a} + \lambda(\boldsymbol{z} - \boldsymbol{a}) \in C_{\boldsymbol{a}}(\operatorname{conv} X)$.

(d) Because $C_{\boldsymbol{a}}(X) \subset \operatorname{cone}_{\boldsymbol{a}} X$ (see (5.10)), it suffices to show that $\operatorname{cone}_{\boldsymbol{a}} X \subset \operatorname{aff}(\{\boldsymbol{a}\} \cup X)$. For this, choose any point $\boldsymbol{x} \in \operatorname{cone}_{\boldsymbol{a}} X$. By Theorem 5.32, $\boldsymbol{x}$ can be written as

$$\boldsymbol{x} = \boldsymbol{a} + \lambda_1(\boldsymbol{x}_1 - \boldsymbol{a}) + \cdots + \lambda_k(\boldsymbol{x}_k - \boldsymbol{a})$$

for suitable $\boldsymbol{x}_1, \ldots, \boldsymbol{x}_k \in X$ and $\lambda_1, \ldots, \lambda_k \geqslant 0$. Since this expression for $\boldsymbol{x}$ is an affine combination of the points $\boldsymbol{a}, \boldsymbol{x}_1, \ldots, \boldsymbol{x}_k$, Theorem 2.44 gives the inclusion $\boldsymbol{x} \in \operatorname{aff}(\{\boldsymbol{a}\} \cup X)$. Hence $\operatorname{cone}_{\boldsymbol{a}} X \subset \operatorname{aff}(\{\boldsymbol{a}\} \cup X)$. Now, the inclusions

$$\{\boldsymbol{a}\} \cup X \subset C_{\boldsymbol{a}}(X) \subset \operatorname{cone}_{\boldsymbol{a}} X \subset \operatorname{aff}(\{\boldsymbol{a}\} \cup X)$$

and Theorem 2.42 imply (5.15). This argument and Definition 2.63 result in the second assertion of (d).

(e) This part follows from (d) and the fact that $\operatorname{aff} X = \operatorname{aff}(\{\boldsymbol{a}\} \cup X)$ if and only if $\boldsymbol{a} \in \operatorname{aff} X$ (see Theorem 2.42).

(f) If $\operatorname{cone}_{\boldsymbol{a}} X$ is a plane, then $\operatorname{aff}(\{\boldsymbol{a}\} \cup X) \subset \operatorname{cone}_{\boldsymbol{a}} X$ because $\operatorname{aff}(\{\boldsymbol{a}\} \cup X)$ is the smallest plane in $\mathbb{R}^n$ containing $\{\boldsymbol{a}\} \cup X$. The opposite inclusion, $\operatorname{cone}_{\boldsymbol{a}} X \subset \operatorname{aff}(\{\boldsymbol{a}\} \cup X)$, follows from assertion (d). Conversely, if $\operatorname{cone}_{\boldsymbol{a}} X = \operatorname{aff}(\{\boldsymbol{a}\} \cup X)$, then $\operatorname{cone}_{\boldsymbol{a}} X$ is a plane.

(g) If $\operatorname{cone}_{\boldsymbol{a}} X$ is a plane, then $\operatorname{aff} X \subset \operatorname{cone}_{\boldsymbol{a}} X$ because $\operatorname{aff} X$ is the smallest plane in $\mathbb{R}^n$ containing X. Since the opposite inclusion is obvious when $X = \{\boldsymbol{a}\}$, we may suppose that $X \setminus \{\boldsymbol{a}\} \neq \varnothing$. Choose a point $\boldsymbol{x} \in X \setminus \{\boldsymbol{a}\}$ and let $h = [\boldsymbol{a}, \boldsymbol{x}\rangle$. Consider the closed halfline $h' = 2\boldsymbol{a} - h$. Since $h \subset \operatorname{cone}_{\boldsymbol{a}} X$ and since $\operatorname{cone}_{\boldsymbol{a}} X$ is a plane, the halfline h' lies in $\operatorname{cone}_{\boldsymbol{a}} X$. Due to assertion ($c$), there is a point $\boldsymbol{z} \in \operatorname{conv} X$ such that $h' = [\boldsymbol{a}, \boldsymbol{z}\rangle$. Consequently, $\boldsymbol{a} \in \langle \boldsymbol{x}, \boldsymbol{z}\rangle \subset \operatorname{conv} X \subset \operatorname{aff} X$. By assertion ($e$), $\operatorname{cone}_{\boldsymbol{a}} X \subset \operatorname{aff} X$. Summing up, $\operatorname{cone}_{\boldsymbol{a}} X = \operatorname{aff} X$.

Conversely, if $\operatorname{cone}_{\boldsymbol{a}} X = \operatorname{aff} X$, then $\operatorname{cone}_{\boldsymbol{a}} X$ is a plane.

(h) By assertion (c),

$$\operatorname{cone}_{\boldsymbol{a}} X \setminus \{\boldsymbol{a}\} = \cup\,((\boldsymbol{a}, \boldsymbol{x}\rangle : \boldsymbol{x} \in \operatorname{conv} X).$$

Therefore, the set $C = \operatorname{cone}_{\boldsymbol{a}} X \setminus \{\boldsymbol{a}\}$ is the smallest cone with improper apex $\boldsymbol{a}$ containing X. Assume, for contradiction, that C is not convex. Then there points $\boldsymbol{x}, \boldsymbol{y} \in C$ such the open segment $(\boldsymbol{x}, \boldsymbol{y})$ does not lie in C. On the other hand, $(\boldsymbol{x}, \boldsymbol{y}) \subset \operatorname{cone}_{\boldsymbol{a}} X$ due to the convexity of $\operatorname{cone}_{\boldsymbol{a}} X$. Hence $\boldsymbol{a} \in (\boldsymbol{x}, \boldsymbol{y})$. By assertion ($b$), there are points $\boldsymbol{u}, \boldsymbol{v} \in \operatorname{cone} X$ such that $\boldsymbol{x} \in (\boldsymbol{a}, \boldsymbol{u}\rangle$ and $\boldsymbol{y} \in (\boldsymbol{a}, \boldsymbol{v}\rangle$. This argument obviously implies the inclusion $\boldsymbol{a} \in (\boldsymbol{u}, \boldsymbol{v}) \subset \operatorname{conv} X$, contrary to the assumption $\boldsymbol{a} \notin \operatorname{conv} X$. Summing up, the set C is convex. Since $\{\boldsymbol{a}\}$ is a 0-dimensional plane which lies in $\operatorname{ap} C$, Theorem 5.18 shows that $\boldsymbol{a}$ is the only apex of $\operatorname{cone}_{\boldsymbol{a}} X$.

(i) Assume for a moment that the set $C_{\boldsymbol{a}}(X) \cap C$ contains a point $\boldsymbol{u} \neq \boldsymbol{a}$. Then $\boldsymbol{u} = \boldsymbol{a} + \lambda(\boldsymbol{x} - \boldsymbol{a})$ for a suitable scalar $\lambda > 0$ and a point $\boldsymbol{x} \in X \setminus \{\boldsymbol{a}\}$. Consequently, $\boldsymbol{x} = \boldsymbol{a} + \lambda^{-1}(\boldsymbol{u} - \boldsymbol{a}) \in X \cap C$, contrary to the assumption $X \cap C \subset \{\boldsymbol{a}\}$.

(j) This part immediately follows from a combination of assertions (c) and (i). □

The next result gives an important corollary of Theorem 5.38 (see also Theorems 5.47 and 5.50).

Theorem 5.39. *Let a set $X \subset \mathbb{R}^n$ be contained in a nonempty proper plane L of $\mathbb{R}^n$. For any point $\boldsymbol{a} \in \mathbb{R}^n \setminus L$, one has*

$$L \cap \operatorname{cone}_{\boldsymbol{a}} X = \operatorname{conv} X \quad \textit{and} \quad (\boldsymbol{a} + \operatorname{dir} L) \cap \operatorname{cone}_{\boldsymbol{a}} X = \{\boldsymbol{a}\}.$$

Proof. Since the case $X = \varnothing$ is obvious, we may assume that $X \neq \varnothing$.

Because $\operatorname{conv} X \subset \operatorname{aff} X \subset L$ (see Theorem 4.2) and $\operatorname{conv} X \subset \operatorname{cone}_{\boldsymbol{a}} X$ (see Theorem 5.38), we have $\operatorname{conv} X \subset L \cap \operatorname{cone}_{\boldsymbol{a}} X$. For the opposite inclusion, choose a point $\boldsymbol{x} \in L \cap \operatorname{cone}_{\boldsymbol{a}} X$. By Theorem 5.32, we can write

$$\boldsymbol{x} = \boldsymbol{a} + \lambda_1(\boldsymbol{x}_1 - \boldsymbol{a}) + \cdots + \lambda_k(\boldsymbol{x}_k - \boldsymbol{a}),$$

where $\boldsymbol{x}_1, \ldots, \boldsymbol{x}_k \in X$ and $\lambda_1, \ldots, \lambda_k \geqslant 0$. Since $\boldsymbol{x} \neq \boldsymbol{a}$ (due to $\boldsymbol{a} \notin L$), at least one of the scalars $\lambda_1, \ldots, \lambda_k$ is positive. Put

$$\lambda = \lambda_1 + \cdots + \lambda_k \quad \text{and} \quad \boldsymbol{u} = \frac{\lambda_1}{\lambda}\boldsymbol{x}_1 + \cdots + \frac{\lambda_k}{\lambda}\boldsymbol{x}_k.$$

Clearly, $\boldsymbol{u}$ is a convex combination of $\boldsymbol{x}_1, \ldots, \boldsymbol{x}_k$, which gives $\boldsymbol{u} \in \operatorname{conv} X$. Moreover, $\boldsymbol{x} = \boldsymbol{a} + \lambda(\boldsymbol{u} - \boldsymbol{a})$. Assuming that $\lambda \neq 1$, we would obtain

$$\boldsymbol{a} = \frac{1}{1-\lambda}\boldsymbol{x} - \frac{\lambda}{1-\lambda}\boldsymbol{u} \in \langle \boldsymbol{x}, \boldsymbol{u} \rangle \subset \operatorname{aff} X \subset L,$$

contradicting the assumption $\boldsymbol{a} \notin L$. Hence $\lambda = 1$, and $\boldsymbol{x} = \boldsymbol{u} \in \operatorname{conv} X$.

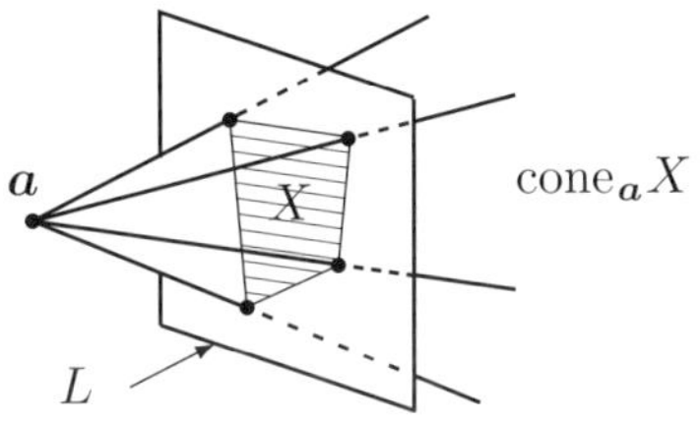

Fig. 5.12 Illustration to Theorem 5.39.

The inclusion $\{\boldsymbol{a}\} \subset (\boldsymbol{a} + \operatorname{dir} L) \cap \operatorname{cone}_{\boldsymbol{a}} X$ is obvious; so, it remains to prove the opposite one. Assume, for contradiction, the existence of a point $\boldsymbol{x} \in (\boldsymbol{a} + \operatorname{dir} L) \cap \operatorname{cone}_{\boldsymbol{a}} X$ which is not $\boldsymbol{a}$. Because $\boldsymbol{x} \in \operatorname{cone}_{\boldsymbol{a}} X \setminus \{\boldsymbol{a}\}$, Theorem 5.35 implies that $\boldsymbol{x}$ can be written as a positive combination

$$\boldsymbol{x} = \boldsymbol{a} + \lambda_1(\boldsymbol{x}_1 - \boldsymbol{a}) + \cdots + \lambda_r(\boldsymbol{x}_r - \boldsymbol{a}),$$

where $\{\boldsymbol{a}, \boldsymbol{x}_1, \ldots, \boldsymbol{x}_r\}$ is affinely independent subset of $\{\boldsymbol{a}\} \cup X$. Corollary 2.62 shows that the set $\{\boldsymbol{x}_1, \ldots, \boldsymbol{x}_r\}$ can be extended to an affine basis $Y = \{\boldsymbol{x}_1, \ldots, \boldsymbol{x}_{m+1}\}$ for L, where $m = \dim X$, and Corollary 2.60 implies that the set $Z = \{\boldsymbol{a}, \boldsymbol{x}_1, \ldots, \boldsymbol{x}_{m+1}\}$ is an affine basis for $\operatorname{aff}(\{\boldsymbol{a}\} \cup L)$. Clearly,

$$\begin{aligned} \boldsymbol{x} = \boldsymbol{a} &+ \lambda_1(\boldsymbol{x}_1 - \boldsymbol{a}) + \cdots + \lambda_r(\boldsymbol{x}_r - \boldsymbol{a}) \\ &+ 0(\boldsymbol{x}_{r+1} - \boldsymbol{a}) + \cdots + 0(\boldsymbol{x}_{m+1} - \boldsymbol{a}), \end{aligned} \tag{5.16}$$

is an affine combination of points $\boldsymbol{a}, \boldsymbol{x}_1, \ldots, \boldsymbol{x}_{m+1}$, and Theorem 2.52 shows that this representation is unique.

Similarly, the inclusion $\boldsymbol{x} \in \boldsymbol{a} + \operatorname{dir} L$ and Theorem 2.52 imply that $\boldsymbol{x}$ is uniquely expressible as an affine combination

$$\boldsymbol{x} = \boldsymbol{a} + \mu_1 \boldsymbol{x}_1 + \cdots + \mu_{m+1} \boldsymbol{x}_{m+1}, \quad \mu_1 + \cdots + \mu_{m+1} = 0. \tag{5.17}$$

Comparing (5.16) and (5.17), we obtain

$$\lambda_1 = \cdots = \lambda_r = \mu_1 = \cdots = \mu_{m+1} = 0.$$

Consequently, $\boldsymbol{x} = \boldsymbol{a}$, in contradiction with the assumption on $\boldsymbol{x}$. Summing up, $(\boldsymbol{a} + \operatorname{dir} L) \cap \operatorname{cone}_{\boldsymbol{a}} X \subset \{\boldsymbol{a}\}$. □

5.2.3 *Algebra of Conic Hulls*

Theorem 5.40. *If $X \subset \mathbb{R}^n$ is a set, $\boldsymbol{a}$ and $\boldsymbol{c}$ are points in $\mathbb{R}^n$, and μ is a scalar, then*

$$\operatorname{cone}_{\boldsymbol{c}+\mu\boldsymbol{a}}(\boldsymbol{c} + \mu X) = \boldsymbol{c} + \mu \operatorname{cone}_{\boldsymbol{a}} X.$$

Furthermore,

$$\operatorname{cone}_{\boldsymbol{c}+\mu\boldsymbol{a}}(\boldsymbol{c} + \mu X) = \begin{cases} \boldsymbol{c} + (\mu - 1)\boldsymbol{a} + \operatorname{cone}_{\boldsymbol{a}} X & \text{if} \quad \mu > 0, \\ \{\boldsymbol{c}\} & \text{if} \quad \mu = 0, \\ \boldsymbol{c} + (\mu + 1)\boldsymbol{a} - \operatorname{cone}_{\boldsymbol{a}} X & \text{if} \quad \mu < 0. \end{cases} \tag{5.18}$$

Proof. Since the case $X = \varnothing$ is obvious, we may assume that X is nonempty. Clearly, $\boldsymbol{c}+\mu X \subset \boldsymbol{c}+\mu \operatorname{cone}_{\boldsymbol{a}} X$. By Theorem 5.10, $\boldsymbol{c}+\mu \operatorname{cone}_{\boldsymbol{a}} X$ is a convex cone with apex $\boldsymbol{c} + \mu\boldsymbol{a}$. Therefore, Theorem 5.30 gives

$$\operatorname{cone}_{\boldsymbol{c}+\mu\boldsymbol{a}}(\boldsymbol{c} + \mu X) \subset \boldsymbol{c} + \mu \operatorname{cone}_{\boldsymbol{a}} X.$$

Conversely, let $\boldsymbol{u} \in \boldsymbol{c} + \mu \operatorname{cone}_{\boldsymbol{a}} X$. Then $\boldsymbol{u} = \boldsymbol{c} + \mu \boldsymbol{y}$, where $\boldsymbol{y} \in \operatorname{cone}_{\boldsymbol{a}} X$. By Theorem 5.32, $\boldsymbol{y}$ can be written as

$$\boldsymbol{y} = \boldsymbol{a} + \lambda_1(\boldsymbol{x}_1 - \boldsymbol{a}) + \cdots + \lambda_k(\boldsymbol{x}_k - \boldsymbol{a})$$

for suitable points $\boldsymbol{x}_1, \ldots, \boldsymbol{x}_k \in X$ and scalars $\lambda_1, \ldots, \lambda_k \geqslant 0$. By the same theorem, the equality

$$\begin{aligned} \boldsymbol{u} = (\boldsymbol{c} + \mu\boldsymbol{a}) &+ \lambda_1((\boldsymbol{c} + \mu\boldsymbol{x}_1) - (\boldsymbol{c} + \mu\boldsymbol{a})) + \cdots \\ &+ \lambda_k((\boldsymbol{c} + \mu\boldsymbol{x}_k) - (\boldsymbol{c} + \mu\boldsymbol{a})) \end{aligned}$$

implies the inclusion $\boldsymbol{u} \in \operatorname{cone}_{\boldsymbol{c}+\mu\boldsymbol{a}}(\boldsymbol{c} + \mu X)$. Hence

$$\boldsymbol{c} + \mu \operatorname{cone}_{\boldsymbol{a}} X \subset \operatorname{cone}_{\boldsymbol{c}+\mu\boldsymbol{a}}(\boldsymbol{c} + \mu X).$$

The equalities (5.18) follow from Corollary 5.11. □

Theorem 5.41. *For sets $X_1, \dots, X_r \subset \mathbb{R}^n$, points $\boldsymbol{a}_1, \dots, \boldsymbol{a}_r \in \mathbb{R}^n$, and scalars $\mu_1, \dots, \mu_r$, $r \geqslant 2$, one has*

$$\begin{aligned}&\mathrm{cone}_{\mu_1\boldsymbol{a}_1+\cdots+\mu_r\boldsymbol{a}_r}(\mu_1 X_1 + \cdots + \mu_r X_r)\\ &\quad\subset \mu_1\mathrm{cone}_{\boldsymbol{a}_1} X_1 + \cdots + \mu_r\mathrm{cone}_{\boldsymbol{a}_r} X_r\\ &\quad= \mathrm{cone}_{\mu_1\boldsymbol{a}_1+\cdots+\mu_r\boldsymbol{a}_r}(\mu_1(\{\boldsymbol{a}_1\}\cup X_1) + \cdots + \mu_r(\{\boldsymbol{a}_r\}\cup X_r)).\end{aligned}$$

In particular, if $\boldsymbol{a}_i \in X_i$ for all $1 \leqslant i \leqslant r$, then

$$\begin{aligned}&\mathrm{cone}_{\mu_1\boldsymbol{a}_1+\cdots+\mu_r\boldsymbol{a}_r}(\mu_1 X_1 + \cdots + \mu_r X_r)\\ &\quad= \mu_1\mathrm{cone}_{\boldsymbol{a}_1} X_1 + \cdots + \mu_r\mathrm{cone}_{\boldsymbol{a}_r} X_r.\end{aligned}$$

Proof. An induction argument shows that the proof can be reduced to the case $r = 2$ (for $r = 1$, we can write $\mu_1 X_1 = \mu_1 X_1 + \mu_2\boldsymbol{o}$). Assume that both sets X_1 and X_2 are nonempty (otherwise the assertion is obvious). Since

$$\mu_1 X_1 + \mu_2 X_2 \subset \mu_1\mathrm{cone}_{\boldsymbol{a}_1} X_1 + \mu_2\mathrm{cone}_{\boldsymbol{a}_2} X_2,$$

and since $\mu_1\mathrm{cone}_{\boldsymbol{a}_1} X_1 + \mu_2\mathrm{cone}_{\boldsymbol{a}_2} X_2$ is a convex cone with apex $\mu_1\boldsymbol{a}_1 + \mu_2\boldsymbol{a}_2$ (see Theorem 5.10), we obtain from Theorem 5.30 that

$$\mathrm{cone}_{\mu_1\boldsymbol{a}_1+\mu_2\boldsymbol{a}_2}(\mu_1 X_1 + \mu_2 X_2) \subset \mu_1\mathrm{cone}_{\boldsymbol{a}_1} X_1 + \mu_2\mathrm{cone}_{\boldsymbol{a}_2} X_2.$$

By Theorem 5.30, $\mathrm{cone}_{\boldsymbol{a}_i} X_i = \mathrm{cone}_{\boldsymbol{a}_i}(\{\boldsymbol{a}_i\}\cup X_i)$, $i = 1, 2$. Thus

$$\begin{aligned}&\mu_1\mathrm{cone}_{\boldsymbol{a}_1} X_1 + \mu_2\mathrm{cone}_{\boldsymbol{a}_2} X_2\\ &\quad= \mu_1\mathrm{cone}_{\boldsymbol{a}_1}(\{\boldsymbol{a}_1\}\cup X_1) + \mu_2\mathrm{cone}_{\boldsymbol{a}_2}(\{\boldsymbol{a}_2\}\cup X_2).\end{aligned}$$

Hence it remains to prove the equality

$$\begin{aligned}&\mu_1\mathrm{cone}_{\boldsymbol{a}_1}(\{\boldsymbol{a}_1\}\cup X_1) + \mu_2\mathrm{cone}_{\boldsymbol{a}_2}(\{\boldsymbol{a}_2\}\cup X_2)\\ &\quad= \mathrm{cone}_{\mu_1\boldsymbol{a}_1+\mu_2\boldsymbol{a}_2}(\mu_1(\{\boldsymbol{a}_1\}\cup X_1) + \mu_2(\{\boldsymbol{a}_2\}\cup X_2)).\end{aligned}$$

By the above argument (with $\{\boldsymbol{a}_i\}\cup X_i$ instead of X_i),

$$\begin{aligned}&\mathrm{cone}_{\mu_1\boldsymbol{a}_1+\mu_2\boldsymbol{a}_2}(\mu_1(\{\boldsymbol{a}_1\}\cup X_1) + \mu_2(\{\boldsymbol{a}_2\}\cup X_2))\\ &\quad\subset \mu_1\mathrm{cone}_{\boldsymbol{a}_1}(\{\boldsymbol{a}_1\}\cup X_1) + \mu_2\mathrm{cone}_{\boldsymbol{a}_2}(\{\boldsymbol{a}_2\}\cup X_2).\end{aligned}$$

For the opposite inclusion, choose a point

$$\boldsymbol{x} \in \mu_1\mathrm{cone}_{\boldsymbol{a}_1}(\{\boldsymbol{a}_1\}\cup X_1) + \mu_2\mathrm{cone}_{\boldsymbol{a}_2}(\{\boldsymbol{a}_2\}\cup X_2).$$

Then $\boldsymbol{x} = \mu_1\boldsymbol{x}_1 + \mu_2\boldsymbol{x}_2$, where

$$\boldsymbol{x}_1 \in \mathrm{cone}_{\boldsymbol{a}_1}(\{\boldsymbol{a}_1\}\cup X_1) \quad\text{and}\quad \boldsymbol{x}_2 \in \mathrm{cone}_{\boldsymbol{a}_2}(\{\boldsymbol{a}_2\}\cup X_2).$$

By Theorem 5.32, $\boldsymbol{x}_1$ and $\boldsymbol{x}_2$ can be expressed as nonnegative combinations

$$\begin{aligned}\boldsymbol{x}_1 &= \boldsymbol{a}_1 + \gamma_1(\boldsymbol{u}_1 - \boldsymbol{a}_1) + \cdots + \gamma_p(\boldsymbol{u}_p - \boldsymbol{a}_1), \quad \boldsymbol{u}_1, \dots, \boldsymbol{u}_p \in \{\boldsymbol{a}_1\}\cup X_1,\\ \boldsymbol{x}_2 &= \boldsymbol{a}_2 + \beta_1(\boldsymbol{v}_1 - \boldsymbol{a}_2) + \cdots + \beta_q(\boldsymbol{v}_q - \boldsymbol{a}_2), \quad \boldsymbol{v}_1, \dots, \boldsymbol{v}_q \in \{\boldsymbol{a}_1\}\cup X_2.\end{aligned}$$

For all $1 \leqslant i \leqslant p$ and $1 \leqslant j \leqslant q$, one has

$$\mu_1 \boldsymbol{u}_i + \mu_2 \boldsymbol{a}_2,\ \mu_1 \boldsymbol{a}_1 + \mu_2 \boldsymbol{v}_j \in \mu_1(\{\boldsymbol{a}_1\} \cup X_1) + \mu_2(\{\boldsymbol{a}_2\} \cup X_2).$$

Therefore, the equality

$$\begin{aligned}\boldsymbol{x} = (\mu_1 \boldsymbol{a}_1 + \mu_2 \boldsymbol{a}_2) &+ \gamma_1((\mu_1 \boldsymbol{u}_1 + \mu_2 \boldsymbol{a}_2) - (\mu_1 \boldsymbol{a}_1 + \mu_2 \boldsymbol{a}_2)) + \cdots \\ &+ \gamma_p((\mu_1 \boldsymbol{u}_p + \mu_2 \boldsymbol{a}_2) - (\mu_1 \boldsymbol{a}_1 + \mu_2 \boldsymbol{a}_2)) \\ &+ \beta_1((\mu_1 \boldsymbol{a}_1 + \mu_2 \boldsymbol{v}_1) - (\mu_1 \boldsymbol{a}_1 + \mu_2 \boldsymbol{a}_2)) + \cdots \\ &+ \beta_q((\mu_1 \boldsymbol{a}_1 + \mu_2 \boldsymbol{v}_q) - (\mu_1 \boldsymbol{a}_1 + \mu_2 \boldsymbol{a}_2))\end{aligned}$$

and Theorem 5.32 give the inclusion

$$\boldsymbol{x} \in \operatorname{cone}_{\mu_1 \boldsymbol{a}_1 + \mu_2 \boldsymbol{a}_2}(\mu_1(\{\boldsymbol{a}_1\} \cup X_1) + \mu_2(\{\boldsymbol{a}_2\} \cup X_2)). \qquad \square$$

Remark. The inclusion in Theorem 5.41 may be proper. Indeed, let $X_1 = (1, 0)$ and $X_2 = (0, 1)$. Then

$$\operatorname{cone}_{\boldsymbol{o}} X_1 = \{(x, 0) : x \geqslant 0\} \quad \text{and} \quad \operatorname{cone}_{\boldsymbol{o}} X_2 = \{(0, y) : y \geqslant 0\}.$$

Therefore,

$$\begin{aligned}\operatorname{cone}_{\boldsymbol{o}}(X_1 + X_2) &= \{(x, x) : x \geqslant 0\} \\ &\neq \{(x, y) : x, y \geqslant 0\} = \operatorname{cone}_{\boldsymbol{o}} X_1 + \operatorname{cone}_{\boldsymbol{o}} X_2.\end{aligned}$$

Theorem 5.42. *Let $X_1, \ldots, X_r$ be sets in $\mathbb{R}^n$ with the property* $\operatorname{conv} X_1 \cap \cdots \cap \operatorname{conv} X_r \neq \varnothing$. *For any point $\boldsymbol{a} \in \operatorname{conv} X_1 \cap \cdots \cap \operatorname{conv} X_r$, one has*

$$\operatorname{cone}_{\boldsymbol{a}}(\operatorname{conv} X_1 \cap \cdots \cap \operatorname{conv} X_r) = \operatorname{cone}_{\boldsymbol{a}} X_1 \cap \cdots \cap \operatorname{cone}_{\boldsymbol{a}} X_r.$$

Proof. Due to Theorem 5.30, it suffices to prove the inclusion

$$\operatorname{cone}_{\boldsymbol{a}} X_1 \cap \cdots \cap \operatorname{cone}_{\boldsymbol{a}} X_r \subset \operatorname{cone}_{\boldsymbol{a}}(\operatorname{conv} X_1 \cap \cdots \cap \operatorname{conv} X_r).$$

So, let $\boldsymbol{x} \in \operatorname{cone}_{\boldsymbol{a}} X_1 \cap \cdots \cap \operatorname{cone}_{\boldsymbol{a}} X_r$. Since the case $\boldsymbol{x} = \boldsymbol{a}$ is obvious, we may assume that $\boldsymbol{x} \neq \boldsymbol{a}$. Then $\boldsymbol{x} = \boldsymbol{a} + \lambda_i(\boldsymbol{z}_i - \boldsymbol{a})$ for suitable scalars $\lambda_i > 0$ and points $\boldsymbol{z}_i \in \operatorname{conv} X_i$, $1 \leqslant i \leqslant r$. We can write

$$\boldsymbol{z}_i = (1 - \lambda_i^{-1})\boldsymbol{a} + \lambda_i^{-1}\boldsymbol{x}, \quad 1 \leqslant i \leqslant r.$$

Without loss of generality, we may assume that λ_1 is the largest among $\lambda_1, \ldots, \lambda_r$. Then

$$\begin{aligned}\boldsymbol{z}_1 = (1 - \lambda_1^{-1})\boldsymbol{a} + \lambda_1^{-1}\boldsymbol{x} &= \big(1 - \frac{\lambda_i}{\lambda_1}\big)\boldsymbol{a} + \frac{\lambda_i}{\lambda_1}((1 - \lambda_i^{-1})\boldsymbol{a} + \lambda_i^{-1}\boldsymbol{x}) \\ &= \big(1 - \frac{\lambda_i}{\lambda_1}\big)\boldsymbol{a} + \frac{\lambda_i}{\lambda_1}\boldsymbol{z}_i \in [\boldsymbol{a}, \boldsymbol{z}_i] \subset \operatorname{conv} X_i, \quad 1 \leqslant i \leqslant r.\end{aligned}$$

Hence $z_1 \in \operatorname{conv} X_1 \cap \cdots \cap \operatorname{conv} X_r$ and

$$x = a + \lambda_1(z_1 - a) \in \operatorname{cone}_a(\operatorname{conv} X_1 \cap \cdots \cap \operatorname{conv} X_r). \qquad \square$$

Theorem 5.43. *For nonempty sets $X_1, \dots, X_r$ in $\mathbb{R}^n$ and a point $a \in \mathbb{R}^n$, one has*

$$\operatorname{cone}_a(X_1 \cup \cdots \cup X_r) = \{a + \lambda_1(x_1 - a) + \cdots + \lambda_r(x_r - a) : \\ x_1 \in \operatorname{conv} X_1, \dots, x_r \in \operatorname{conv} X_r,\ \lambda_1, \dots, \lambda_r \geqslant 0\}.$$

Equivalently,

$$\operatorname{cone}_a(X_1 \cup \cdots \cup X_r) = (1 - r)a + \operatorname{cone}_a X_1 + \cdots + \operatorname{cone}_a X_r. \tag{5.19}$$

If, additionally, $a \in \operatorname{conv} X_1 \cap \cdots \cap \operatorname{conv} X_r$, then

$$\operatorname{cone}_a(X_1 \cup \cdots \cup X_r) = \operatorname{cone}_a(X_1 + \cdots + X_r - (r-1)a). \tag{5.20}$$

Proof. Theorem 5.32 show that the set

$$C = \{a + \lambda_1(x_1 - a) + \cdots + \lambda_r(x_r - a) : \\ x_1 \in \operatorname{conv} X_1, \dots, x_r \in \operatorname{conv} X_r,\ \lambda_1, \dots, \lambda_r \geqslant 0\}$$

lies in $\operatorname{cone}_a(\operatorname{conv} X_1 \cup \cdots \cup \operatorname{conv} X_r)$, which equals $\operatorname{cone}_a(X_1 \cup \cdots \cup X_r)$ according to Theorem 5.30. Conversely, let $x \in \operatorname{cone}_a(X_1 \cup \cdots \cup X_r)$. By the same Theorem 5.32, x is expressible as a nonnegative combination

$$x = a + \mu_1(z_1 - a) + \cdots + \mu_k(z_k - a), \quad z_1, \dots, z_k \in X_1 \cup \cdots \cup X_r.$$

Renumbering the terms $\mu_1(z_1 - a), \dots, \mu_k(z_k - a)$, we may assume that the set $\{z_1, \dots, z_k\}$ is partitioned into subsets $Y_1, \dots, Y_r$ (some of them may be empty) such that

$$Y_i = \{z_{p_{i-1}+1}, \dots, z_{p_i}\} \subset X_i, \quad 1 \leqslant i \leqslant r, \quad \text{where } p_0 = 0,\ p_r = k.$$

Let $\lambda_i = \mu_{p_{i-1}+1} + \cdots + \mu_{p_i}$, $1 \leqslant i \leqslant r$. Denote by I the set of all indices $i \in \{1, \dots, r\}$ satisfying the property: either Y_i is empty or $\lambda_i = 0$. Put $J = \{1, \dots, r\} \setminus I$. For every $i \in I$, choose a point $x_i \in X_i$, and for every $i \in J$, let

$$x_i = \frac{\mu_{p_{i-1}+1}}{\lambda_i} z_{p_{i-1}+1} + \cdots + \frac{\mu_{p_i}}{\lambda_i} z_{p_i}.$$

Clearly, every point x_i is a convex combination of points from X_i, and Theorem 4.3 shows that $x_i \in \operatorname{conv} X_i$ for all $1 \leqslant i \leqslant r$. Finally,

$$x = a + \lambda_1(x_1 - a) + \cdots + \lambda_r(x_r - a)$$

is a desired representation.

The second assertion of the theorem follows from a combination of Theorems 5.31 and 5.38:

$$\begin{aligned} C &= (1-r)\boldsymbol{a} + \sum_{i=1}^{r} \{\boldsymbol{a} + \lambda_i(\boldsymbol{x}_i - \boldsymbol{a}) : \boldsymbol{x}_i \in \operatorname{conv} X_i, \lambda_i \geqslant 0\} \\ &= (1-r)\boldsymbol{a} + \sum_{i=1}^{r} C_{\boldsymbol{a}}(\operatorname{conv} X_i) = (1-r)\boldsymbol{a} + \sum_{i=1}^{r} \operatorname{cone}_{\boldsymbol{a}} X_i. \end{aligned}$$

Finally, assume that $\boldsymbol{a} \in \operatorname{conv} X_1 \cap \cdots \cap \operatorname{conv} X_r$. Combining Theorems 5.30 and 5.41, one has:

$$\begin{aligned} &\operatorname{cone}_{\boldsymbol{a}} X_1 + \cdots + \operatorname{cone}_{\boldsymbol{a}} X_r \\ &\quad = \operatorname{cone}_{\boldsymbol{o}}(X_1 - \boldsymbol{a}) + \cdots + \operatorname{cone}_{\boldsymbol{o}}(X_r - \boldsymbol{a}) + r\boldsymbol{a} \\ &\quad = \operatorname{cone}_{\boldsymbol{o}}((X_1 - \boldsymbol{a}) + \cdots + (X_r - \boldsymbol{a})) + r\boldsymbol{a} \\ &\quad = \operatorname{cone}_{\boldsymbol{o}}(X_1 + \cdots + X_r - r\boldsymbol{a}) + r\boldsymbol{a} \\ &\quad = \operatorname{cone}_{\boldsymbol{a}}(X_1 + \cdots + X_r - (r-1)\boldsymbol{a}) + (r-1)\boldsymbol{a}. \end{aligned}$$

This argument and (5.19) immediately imply (5.20). □

Theorem 5.44. *If $f : \mathbb{R}^n \to \mathbb{R}^m$ is an affine transformation, $\boldsymbol{a} \in \mathbb{R}^n$, $\boldsymbol{c} \in \operatorname{rng} f$, and $\boldsymbol{u} \in f^{-1}(\boldsymbol{c})$, then for sets $X \subset \mathbb{R}^n$ and $Y \subset \mathbb{R}^m$, one has*

$$\operatorname{cone}_{f(\boldsymbol{a})} f(X) = f(\operatorname{cone}_{\boldsymbol{a}} X), \tag{5.21}$$

$$\operatorname{cone}_{\boldsymbol{u}} f^{-1}(Y) = f^{-1}(\operatorname{cone}_{\boldsymbol{c}}(Y \cap \operatorname{rng} f)) \subset f^{-1}(\operatorname{cone}_{\boldsymbol{c}} Y). \tag{5.22}$$

Proof. Excluding the obvious cases $X = \varnothing$ and $Y = \varnothing$, we assume that both sets X and Y are nonempty.

For (5.21), choose a point $\boldsymbol{x} \in \operatorname{cone}_{f(\boldsymbol{a})} f(X)$. By Theorem 5.32, there are points $\boldsymbol{x}_1, \ldots, \boldsymbol{x}_p \in f(X)$ and scalars $\lambda_1, \ldots, \lambda_p \geqslant 0$ such that

$$\boldsymbol{x} = f(\boldsymbol{a}) + \lambda_1(\boldsymbol{x}_1 - f(\boldsymbol{a})) + \cdots + \lambda_p(\boldsymbol{x}_p - f(\boldsymbol{a})).$$

Choose points $\boldsymbol{z}_i \in X$ such that $f(\boldsymbol{z}_i) = \boldsymbol{x}_i$ for all $1 \leqslant i \leqslant p$. Put

$$\boldsymbol{z} = \boldsymbol{a} + \lambda_1(\boldsymbol{z}_1 - \boldsymbol{a}) + \cdots + \lambda_p(\boldsymbol{z}_p - \boldsymbol{a}).$$

Then $\boldsymbol{z} \in \operatorname{cone}_{\boldsymbol{a}} X$ by the same theorem. Since $\boldsymbol{z}$ is an affine combination of $\boldsymbol{a}, \boldsymbol{z}_1, \ldots, \boldsymbol{z}_p$, Theorem 2.82 gives

$$\begin{aligned} \boldsymbol{x} &= f(\boldsymbol{a}) + \lambda_1(\boldsymbol{x}_1 - f(\boldsymbol{a})) + \cdots + \lambda_p(\boldsymbol{x}_p - f(\boldsymbol{a})) \\ &= f(\boldsymbol{a}) + \lambda_p(f(\boldsymbol{z}_1) - f(\boldsymbol{a})) + \cdots + \lambda_p(f(\boldsymbol{z}_p) - f(\boldsymbol{a})) \\ &= f(\boldsymbol{a} + \lambda_1(\boldsymbol{z}_1 - \boldsymbol{a}) + \cdots + \lambda_p(\boldsymbol{z}_p - \boldsymbol{a})) = f(\boldsymbol{z}). \end{aligned}$$

Hence $\boldsymbol{x} = f(\boldsymbol{z}) \in f(\operatorname{cone}_{\boldsymbol{a}} X)$, which gives $\operatorname{cone}_{f(\boldsymbol{a})} f(X) \subset f(\operatorname{cone}_{\boldsymbol{a}} X)$.

Conversely, let $\boldsymbol{x} \in \text{cone}_{\boldsymbol{a}} X$. According to Theorem 5.32,

$$\boldsymbol{x} = \boldsymbol{a} + \mu_1(\boldsymbol{x}_1 - \boldsymbol{a}) + \cdots + \mu_q(\boldsymbol{x}_q - \boldsymbol{a}),$$

where $\boldsymbol{x}_1, \ldots, \boldsymbol{x}_q \in X$ and $\mu_1, \ldots, \mu_q \geqslant 0$. As above, $\boldsymbol{x}$ is an affine combination of $\boldsymbol{a}, \boldsymbol{x}_1, \ldots, \boldsymbol{x}_q$, which gives

$$f(\boldsymbol{x}) = f(\boldsymbol{a}) + \mu_1(f(\boldsymbol{x}_1) - f(\boldsymbol{a})) + \cdots + \mu_q(f(\boldsymbol{x}_q) - f(\boldsymbol{a})).$$

Therefore, $f(\boldsymbol{x}) \in \text{cone}_{f(\boldsymbol{a})} f(X)$ by the same theorem. Hence $f(\text{cone}_{\boldsymbol{a}} X) \subset \text{cone}_{f(\boldsymbol{a})} f(X)$.

It remains to prove (5.22). Letting $X = f^{-1}(Y)$ in (5.21), we obtain

$$f(\text{cone}_{\boldsymbol{u}} f^{-1}(Y)) = \text{cone}_{f(\boldsymbol{u})} f(f^{-1}(Y)) = \text{cone}_{\boldsymbol{c}}(Y \cap \text{rng}\, f).$$

Hence

$$\text{cone}_{\boldsymbol{u}} f^{-1}(Y) \subset f^{-1}(f(\text{cone}_{\boldsymbol{u}} f^{-1}(Y))) = f^{-1}(\text{cone}_{\boldsymbol{c}}(Y \cap \text{rng}\, f)).$$

Conversely, let $\boldsymbol{x} \in f^{-1}(\text{cone}_{\boldsymbol{c}}(Y \cap \text{rng}\, f))$. Then $f(\boldsymbol{x}) \in \text{cone}_{\boldsymbol{c}}(Y \cap \text{rng}\, f)$, and Theorem 5.32 implies that $f(\boldsymbol{x})$ can be written as

$$f(\boldsymbol{x}) = \boldsymbol{c} + \gamma_1(\boldsymbol{x}_1 - \boldsymbol{c}) + \cdots + \gamma_k(\boldsymbol{x}_k - \boldsymbol{c})$$

for suitable points $\boldsymbol{x}_1, \ldots, \boldsymbol{x}_k \in Y \cap \text{rng}\, f$ and scalars $\gamma_1, \ldots, \gamma_k \geqslant 0$. Choose points $\boldsymbol{z}_i \in f^{-1}(Y)$ such that $f(\boldsymbol{z}_i) = \boldsymbol{x}_i$, $1 \leqslant i \leqslant k$. Put

$$\boldsymbol{z} = \boldsymbol{u} + \gamma_1(\boldsymbol{z}_1 - \boldsymbol{u}) + \cdots + \gamma_k(\boldsymbol{z}_k - \boldsymbol{u}).$$

Then $\boldsymbol{z} \in \text{cone}_{\boldsymbol{u}} f^{-1}(Y)$ and

$$\begin{aligned} f(\boldsymbol{z}) &= f(\boldsymbol{u}) + \gamma_1(f(\boldsymbol{z}_1) - f(\boldsymbol{u})) + \cdots + \gamma_k(f(\boldsymbol{z}_k) - f(\boldsymbol{u})) \\ &= \boldsymbol{c} + \gamma_1(\boldsymbol{x}_1 - \boldsymbol{c}) + \cdots + \gamma_k(\boldsymbol{x}_k - \boldsymbol{c}) = f(\boldsymbol{x}). \end{aligned}$$

Let $\gamma = \gamma_1 + \cdots + \gamma_k$. The case $\gamma = 0$ is obvious, because $f(\boldsymbol{x}) = \boldsymbol{c}$ and $\boldsymbol{x} \in f^{-1}(Y) \subset \text{cone}_{\boldsymbol{u}} f^{-1}(Y)$. Assume that $\gamma > 0$, and let

$$\boldsymbol{v}_i = \boldsymbol{z}_i + \gamma^{-1}(\boldsymbol{x} - \boldsymbol{z}), \ 1 \leqslant i \leqslant k.$$

Since $\boldsymbol{v}_i$ is an affine combination of $\boldsymbol{x}_i, \boldsymbol{x}, \boldsymbol{z}$, Theorem 2.82 gives

$$f(\boldsymbol{v}_i) = f(\boldsymbol{z}_i) + \gamma^{-1} f(\boldsymbol{x}) - \gamma^{-1} f(\boldsymbol{z}) = f(\boldsymbol{z}_i) = \boldsymbol{x}_i.$$

Hence $\boldsymbol{v}_i \in f^{-1}(\boldsymbol{x}_i) \subset f^{-1}(Y)$, $1 \leqslant i \leqslant k$. Finally, the equalities

$$\begin{aligned} \boldsymbol{x} &= \boldsymbol{z} + (\boldsymbol{x} - \boldsymbol{z}) \\ &= \boldsymbol{u} + \gamma_1(\boldsymbol{z}_1 - \boldsymbol{u}) + \cdots + \gamma_k(\boldsymbol{z}_k - \boldsymbol{u}) + (\gamma_1 + \cdots + \gamma_k)\gamma^{-1}(\boldsymbol{x} - \boldsymbol{z}) \\ &= \boldsymbol{u} + \gamma_1(\boldsymbol{z}_1 + \gamma^{-1}(\boldsymbol{x} - \boldsymbol{z}) - \boldsymbol{u}) + \cdots + \gamma_k(\boldsymbol{z}_k + \gamma^{-1}(\boldsymbol{x} - \boldsymbol{z}) - \boldsymbol{u}) \\ &= \boldsymbol{u} + \gamma_1(\boldsymbol{v}_1 - \boldsymbol{u}) + \cdots + \gamma_k(\boldsymbol{v}_k - \boldsymbol{u}) \end{aligned}$$

and Theorem 5.32 show that $\boldsymbol{x} \in \text{cone}_{\boldsymbol{u}} f^{-1}(Y)$.

Finally Theorem 5.30 gives the inclusion

$$\text{cone}_{\boldsymbol{c}}(Y \cap \text{rng}\, f) \subset \text{cone}_{\boldsymbol{c}} Y \cap \text{rng}\, f,$$

which results in

$$f^{-1}(\text{cone}_{\boldsymbol{c}}(Y \cap \text{rng}\, f)) \subset f^{-1}(\text{cone}_{\boldsymbol{c}} Y \cap \text{rng}\, f) = f^{-1}(\text{cone}_{\boldsymbol{c}} Y). \qquad \square$$

Remark. The inclusion in (5.22) may be proper. Indeed, let $f : \mathbb{R}^2 \to \mathbb{R}^2$ be the orthogonal projection of $\mathbb{R}^2$ on its x-axis and $Y = \{(0,1), (0,-1)\}$. Then $\mathrm{cone}_{\boldsymbol{o}} f^{-1}(Y) = f^{-1}(Y) = \varnothing$, while

$$f^{-1}(\mathrm{cone}_{\boldsymbol{o}} Y) = \mathrm{cone}_{\boldsymbol{o}} Y = \{(0, y) : y \in \mathbb{R}\}.$$

5.2.4 *Closure and Conic Hulls*

We recall that $C_{\boldsymbol{a}}(X)$ denotes the cone with apex $\boldsymbol{a}$ generated by a set $X \subset \mathbb{R}^n$ (see Definition 5.29).

Theorem 5.45. *For any set $X \subset \mathbb{R}^n$ and a point $\boldsymbol{a} \in \mathbb{R}^n$, the following assertions hold.*

(*a*) $C_{\boldsymbol{a}}(\mathrm{cl}\, X) \subset \mathrm{cl}\, C_{\boldsymbol{a}}(X) = \mathrm{cl}\, C_{\boldsymbol{a}}(\mathrm{cl}\, X)$.
(*b*) *If X is bounded and $\boldsymbol{a} \notin \mathrm{cl}\, X$, then $C_{\boldsymbol{a}}(\mathrm{cl}\, X) = \mathrm{cl}\, C_{\boldsymbol{a}}(X)$.*
(*c*) $\mathrm{cone}_{\boldsymbol{a}}(\mathrm{cl}\, X) \subset \mathrm{cl}\,(\mathrm{cone}_{\boldsymbol{a}} X) = \mathrm{cl}\,(\mathrm{cone}_{\boldsymbol{a}}(\mathrm{cl}\, X))$.
(*d*) *If X is bounded and $\boldsymbol{a} \notin \mathrm{cl}\,(\mathrm{conv}\, X)$, then* $\mathrm{cone}_{\boldsymbol{a}}(\mathrm{cl}\, X) = \mathrm{cl}\,(\mathrm{cone}_{\boldsymbol{a}} X)$.
(*e*) *If Y is a dense subset of X, then*

$$\mathrm{cl}\, C_{\boldsymbol{a}}(X) = \mathrm{cl}\, C_{\boldsymbol{a}}(Y) \quad \text{and} \quad \mathrm{cl}\,(\mathrm{cone}_{\boldsymbol{a}} X) = \mathrm{cl}\,(\mathrm{cone}_{\boldsymbol{a}} Y). \tag{5.23}$$

Proof. Since the case $X = \varnothing$ is obvious ($C_{\boldsymbol{a}}(\varnothing) = \mathrm{cone}_{\boldsymbol{a}} \varnothing = \{\boldsymbol{a}\}$), we may assume that $X \neq \varnothing$.

(*a*) Clearly, $\mathrm{cl}\, X \subset \mathrm{cl}\, C_{\boldsymbol{a}}(X)$ due to $X \subset C_{\boldsymbol{a}}(X)$. Because $\mathrm{cl}\, C_{\boldsymbol{a}}(X)$ is a cone with apex $\boldsymbol{a}$ (see Theorem 5.28), one has $C_{\boldsymbol{a}}(\mathrm{cl}\, X) \subset \mathrm{cl}\, C_{\boldsymbol{a}}(X)$. Similarly, the inclusion $C_{\boldsymbol{a}}(X) \subset C_{\boldsymbol{a}}(\mathrm{cl}\, X)$ and the above argument give

$$C_{\boldsymbol{a}}(\mathrm{cl}\, X) \subset \mathrm{cl}\, C_{\boldsymbol{a}}(X) \subset \mathrm{cl}\, C_{\boldsymbol{a}}(\mathrm{cl}\, X).$$

Consequently, $\mathrm{cl}\, C_{\boldsymbol{a}}(X) = \mathrm{cl}\, C_{\boldsymbol{a}}(\mathrm{cl}\, X)$.

(*b*) Let X be bounded and $\boldsymbol{a} \notin \mathrm{cl}\, X$. To prove the equality $C_{\boldsymbol{a}}(\mathrm{cl}\, X) = \mathrm{cl}\, C_{\boldsymbol{a}}(X)$, it suffices to show (by the above argument) that $\mathrm{cl}\, C_{\boldsymbol{a}}(X) \subset C_{\boldsymbol{a}}(\mathrm{cl}\, X)$. Choose a point $\boldsymbol{x} \in \mathrm{cl}\, C_{\boldsymbol{a}}(X)$. Then $\boldsymbol{x} = \lim_{i \to \infty} \boldsymbol{x}_i$ for a suitable sequence of points $\boldsymbol{x}_1, \boldsymbol{x}_2, \ldots$ from $C_{\boldsymbol{a}}(X)$. We can write $\boldsymbol{x}_i = \boldsymbol{a} + \gamma_i(\boldsymbol{z}_i - \boldsymbol{a})$ for appropriate points $\boldsymbol{z}_i \in X$ and scalars $\gamma_i \geqslant 0$, $i \geqslant 1$. Since X is bounded, there is a subsequence $\boldsymbol{z}'_1, \boldsymbol{z}'_2, \ldots$ of $\boldsymbol{z}_1, \boldsymbol{z}_2, \ldots$ which converges to a point $\boldsymbol{z} \in \mathrm{cl}\, X$. Clearly, $\boldsymbol{z} \neq \boldsymbol{a}$. Denote by $\boldsymbol{x}'_1, \boldsymbol{x}'_2, \ldots$ and $\gamma'_1, \gamma'_2, \ldots$ the respective subsequences of $\boldsymbol{x}_1, \boldsymbol{x}_2, \ldots$ and $\gamma_1, \gamma_2, \ldots$. The sequence $\gamma'_1, \gamma'_2, \ldots$ converges to a scalar $\gamma \geqslant 0$ due to

$$\lim_{i \to \infty} \gamma'_i = \lim_{i \to \infty} \frac{\|\boldsymbol{x}'_i - \boldsymbol{a}\|}{\|\boldsymbol{z}'_i - \boldsymbol{a}\|} = \frac{\|\boldsymbol{x} - \boldsymbol{a}\|}{\|\boldsymbol{z} - \boldsymbol{a}\|}.$$

Thus

$$x = \lim_{i\to\infty} x'_i = \lim_{i\to\infty} (a + \gamma'_i(z'_i - a)) = a + \gamma(z - a) \in C_a(\mathrm{cl}\, X).$$

(c) Using Theorems 4.16 and 5.38 and assertion (a) above, we obtain

$$\begin{aligned}\mathrm{cone}_a(\mathrm{cl}\, X) &= C_a(\mathrm{conv}\,(\mathrm{cl}\, X)) \subset C_a(\mathrm{cl}\,(\mathrm{conv}\, X)) \\ &\subset \mathrm{cl}\,(C_a(\mathrm{conv}\, X)) = \mathrm{cl}\,(\mathrm{cone}_a X).\end{aligned}$$

Similarly, the inclusion $\mathrm{cone}_a X \subset \mathrm{cone}_a(\mathrm{cl}\, X)$ and the above argument give

$$\mathrm{cone}_a(\mathrm{cl}\, X) \subset \mathrm{cl}\,(\mathrm{cone}_a X) \subset \mathrm{cl}\,(\mathrm{cone}_a(\mathrm{cl}\, X)).$$

Consequently, $\mathrm{cl}\,(\mathrm{cone}_a X) = \mathrm{cl}\,(\mathrm{cone}_a(\mathrm{cl}\, X))$.

(d) If X is bounded, then $\mathrm{conv}\, X$ is bounded, as follows from Theorem 4.16. By Theorems 4.16 and 5.38 and assertion (b),

$$\begin{aligned}\mathrm{cone}_a(\mathrm{cl}\, X) &= C_a(\mathrm{conv}\,(\mathrm{cl}\, X)) = C_a(\mathrm{cl}\,(\mathrm{conv}\, X)) \\ &= \mathrm{cl}\,(C_a(\mathrm{conv}\, X)) = \mathrm{cl}\,(\mathrm{cone}_a X).\end{aligned}$$

(e) Since $Y \subset X \subset \mathrm{cl}\, Y$, a combination of Theorem 5.30 and assertion (a) gives

$$C_a(Y) \subset C_a(X) \subset C_a(\mathrm{cl}\, Y) \subset \mathrm{cl}\, C_a(Y),$$
$$\mathrm{cone}_a Y \subset \mathrm{cone}_a X \subset \mathrm{cone}_a(\mathrm{cl}\, Y) \subset \mathrm{cl}\,(\mathrm{cone}_a Y).$$

Consequently, (5.23) holds. □

Remarks. 1. The inclusion in assertions (a) and (c) of Theorem 5.45 may be proper. Indeed, if $X \subset \mathbb{R}^2$ is the unit disk $B_1(a)$, where $a = (0, 1)$, then $o \in \mathrm{bd}\, X$ and $C_a(X) = \mathrm{cone}_o X = \{o\} \cup \{(x, y) : y > 0\}$, which is a non-closed set.

2. The assumption on the boundedness of X in assertions (b) and (d) is essential. Indeed, let $X \subset \mathbb{R}^2$ be a closed convex set bounded by a branch of hyperbola: $X = \{(x, y) : x > 0,\ xy \geqslant 1\}$. Clearly, each of the sets $C_o(X)$ and $\mathrm{cone}_o X$ is the union of $\{o\}$ and the open quadrant $\{(x, y) : x, y > 0\}$, which is not a closed set.

Theorem 7.4 provides closedness criteria for generated cones and conic hulls of arbitrary sets in $\mathbb{R}^n$. The following result deals with an important particular case related to closedness conditions for conic hulls.

Theorem 5.46. *For a finite set $X \subset \mathbb{R}^n$ and a point $a \in \mathbb{R}^n$, both cones $C_a(X)$ and $\mathrm{cone}_a X$ are closed. Consequently, if $C \subset \mathbb{R}^n$ is a union of finitely many closed halflines with common endpoint $a \in \mathbb{R}^n$, then the convex cone $\mathrm{conv}\, C$ is closed.*

Proof. All assertions are obvious if $X \subset \{\boldsymbol{a}\}$. Thus we may assume that $X \not\subset \{\boldsymbol{a}\}$. Since the set X is finite, the generated cone $C_{\boldsymbol{a}}(X)$ is the union of finitely many closed halflines of the form $[\boldsymbol{a}, \boldsymbol{x}\rangle$, where $\boldsymbol{x} \in X \setminus \{\boldsymbol{a}\}$. Thus $C_{\boldsymbol{a}}(X)$ is a closed set.

For the closedness of $\mathrm{cone}_{\boldsymbol{a}} X$, let $m = \dim X$. By Theorem 5.37, $\mathrm{cone}_{\boldsymbol{a}} X$ is the union of simplicial cones of the form $\Gamma_{\boldsymbol{a}}(\boldsymbol{x}_1, \dots, \boldsymbol{x}_k)$, where $k \leqslant m+1$ and $\boldsymbol{x}_1, \dots, \boldsymbol{x}_k \in X$. The family of such cones if finite due to the finiteness of X. Because simplicial cones are closed (see Theorem 5.23), the conic hull $\mathrm{cone}_{\boldsymbol{a}} X$ is closed as a finite union of closed sets (see page 10).

For the second assertion, suppose that C is the union of closed halflines $h_1, \dots, h_r$ with common endpoint $\boldsymbol{a}$. Choose points $\boldsymbol{z}_i \in h_i \setminus \{\boldsymbol{a}\}$, $1 \leqslant i \leqslant r$, and let $Z = \{\boldsymbol{z}_1, \dots, \boldsymbol{z}_r\}$. Then $C = C_{\boldsymbol{a}}(Z)$, and Theorem 5.38 gives

$$\mathrm{conv}\, C = \mathrm{conv}\, C_{\boldsymbol{a}}(Z) = \mathrm{cone}_{\boldsymbol{a}} Z.$$

Finally, the above argument shows that the set $\mathrm{conv}\, C$ is closed. □

The next theorem plays an important role in many applications (compare with Theorems 5.39 and 5.50).

Theorem 5.47. *Let a set $X \subset \mathbb{R}^n$ be contained in a nonempty proper plane $L \subset \mathbb{R}^n$. For any point $\boldsymbol{a} \in \mathbb{R}^n \setminus L$, one has*

$$L \cap \mathrm{cl}\,(\mathrm{cone}_{\boldsymbol{a}} X) = \mathrm{cl}\,(\mathrm{conv}\, X).$$

Proof. Because the case $X = \varnothing$ is obvious, we may suppose that $X \neq \varnothing$. By Theorem 5.39, $L \cap \mathrm{cone}_{\boldsymbol{a}} X = \mathrm{conv}\, X$. Since the plane L is closed (see Corollary 2.21), we have

$$\mathrm{cl}\,(\mathrm{conv}\, X) = \mathrm{cl}\,(L \cap \mathrm{cone}_{\boldsymbol{a}} X) \subset L \cap \mathrm{cl}\,(\mathrm{cone}_{\boldsymbol{a}} X).$$

For the opposite inclusion, put $m = \dim L$ and choose an affine basis $\{\boldsymbol{c}_1, \dots, \boldsymbol{c}_{m+1}\}$ for L (see Theorem 2.57). A combination of Theorem 2.54 and Corollary 2.58 shows that $\{\boldsymbol{a}, \boldsymbol{c}_1, \dots, \boldsymbol{c}_{m+1}\}$ is an affine basis for the $(m+1)$-dimensional plane $\mathrm{aff}\,(\{\boldsymbol{a}\} \cup L)$.

Now, choose a point $\boldsymbol{x} \in L \cap \mathrm{cl}\,(\mathrm{cone}_{\boldsymbol{a}} X)$. Then $\boldsymbol{x}$ can be written as an affine combination $\boldsymbol{x} = \gamma_1 \boldsymbol{c}_1 + \cdots + \gamma_{m+1} \boldsymbol{c}_{m+1}$. At the same time, $\boldsymbol{x} = \lim_{i\to\infty} \boldsymbol{x}_i$ for a suitable sequence $\boldsymbol{x}_1, \boldsymbol{x}_2, \dots$ of points from $\mathrm{cone}_{\boldsymbol{a}} X$. Because $\mathrm{cone}_{\boldsymbol{a}} X \subset \mathrm{aff}\,(\{\boldsymbol{a}\} \cup L)$, every $\boldsymbol{x}_i$ can be written as an affine combination

$$\boldsymbol{x}_i = \mu_0^{(i)} \boldsymbol{a} + \mu_1^{(i)} \boldsymbol{c}_1 + \cdots + \mu_{m+1}^{(i)} \boldsymbol{c}_{m+1}, \quad i \geqslant 1.$$

Let $\lambda^{(i)} = \mu_1^{(i)} + \cdots + \mu_{m+1}^{(i)}$, $i \geqslant 1$. Theorem 2.55 shows that

$$\lim_{i\to\infty} \mu_0^{(i)} = 0 \quad \text{and} \quad \lim_{i\to\infty} \mu_i^{(i)} = \gamma_i \quad \text{for all} \quad i \geqslant 1.$$

Hence $\lambda^{(i)} \to 1$ as $i \to \infty$. In particular, there is an index j such that $\lambda^{(i)} > 0$ for all $i \geqslant j$. Furthermore,

$$\begin{aligned} \boldsymbol{x}_i &= \boldsymbol{a} + \mu_1^{(i)}(\boldsymbol{c}_1 - \boldsymbol{a}) + \cdots + \mu_{m+1}^{(i)}(\boldsymbol{c}_{m+1} - \boldsymbol{a}) \\ &= \boldsymbol{a} + \lambda^{(i)}\Big(\frac{\mu_1^{(i)}}{\lambda^{(i)}}\boldsymbol{c}_1 + \cdots + \frac{\mu_{m+1}^{(i)}}{\lambda^{(i)}}\boldsymbol{c}_{m+1} - \boldsymbol{a}\Big), \quad i \geqslant 1. \end{aligned}$$

Clearly, every point

$$\boldsymbol{u}_i = \frac{\mu_1^{(i)}}{\lambda^{(i)}}\boldsymbol{c}_1 + \cdots + \frac{\mu_{m+1}^{(i)}}{\lambda^{(i)}}\boldsymbol{c}_{m+1}, \quad i \geqslant 1,$$

is an affine combination of $\boldsymbol{c}_1, \ldots, \boldsymbol{c}_{m+1}$. Therefore, Corollary 2.58 shows that

$$\boldsymbol{u}_i \in \operatorname{aff}\{\boldsymbol{c}_1, \ldots, \boldsymbol{c}_{m+1}\} = L, \quad i \geqslant 1.$$

Because $\boldsymbol{x}_i = \boldsymbol{a} + \lambda^{(i)}(\boldsymbol{u}_i - \boldsymbol{a})$, Lemma 2.25 implies that the closed halflines $[\boldsymbol{a}, \boldsymbol{x}_i\rangle$ and $[\boldsymbol{a}, \boldsymbol{u}_i\rangle$ coincide for all $i \geqslant 1$. Hence the closed halfline $h_i = [\boldsymbol{a}, \boldsymbol{x}_i\rangle$ meets L at $\boldsymbol{u}_i$, $i \geqslant 1$. Therefore

$$\boldsymbol{u}_i \in L \cap h_i \subset L \cap \operatorname{cone}_{\boldsymbol{a}} X = \operatorname{conv} X.$$

Finally,

$$\begin{aligned} \boldsymbol{x} &= \gamma_1 \boldsymbol{c}_1 + \cdots + \gamma_{m+1}\boldsymbol{c}_{m+1} \\ &= \lim_{i\to\infty} \frac{\mu_1^{(i)}}{\lambda^{(i)}}\boldsymbol{c}_1 + \cdots + \lim_{i\to\infty} \frac{\mu_{m+1}^{(i)}}{\lambda^{(i)}}\boldsymbol{c}_{m+1} = \lim_{i\to\infty} \boldsymbol{u}_i \in \operatorname{cl}(\operatorname{conv} X). \qquad \square \end{aligned}$$

Remark. Unlike Theorems 5.39, we cannot state that $(\boldsymbol{a} + \operatorname{dir} L) \cap \operatorname{cl}(\operatorname{cone}_{\boldsymbol{a}} X) = \{\boldsymbol{a}\}$. Indeed, let L be the line in $\mathbb{R}^2$ given by the equation $y = 1$, and let $X = \{(x, 1) : x \geqslant 0\}$. Then $\operatorname{dir} L$ is the x-axis and $\operatorname{cl}(\operatorname{cone}_{\boldsymbol{o}} X) = \{(x, y) : x, y \geqslant 0\}$. Consequently, $(\boldsymbol{a} + \operatorname{dir} L) \cap \operatorname{cl}(\operatorname{cone}_{\boldsymbol{a}} X)$ is the closed halfline $\{(x, 0) : x \geqslant 0\}$.

5.2.5 *Relative Interior and Conic Hulls*

Theorem 5.48. *For a nonempty set $X \subset \mathbb{R}^n$ and a point $\boldsymbol{a} \in \mathbb{R}^n$, the following conditions are equivalent.*

(*a*) $\boldsymbol{a} \in \operatorname{rint}(\operatorname{conv} X)$.
(*b*) $\boldsymbol{a} \in \operatorname{rint}(\operatorname{cone}_{\boldsymbol{a}} X)$.
(*c*) $\operatorname{cone}_{\boldsymbol{a}}(\operatorname{rint}(\operatorname{conv} X)) = \operatorname{aff} X$.
(*d*) $\operatorname{cone}_{\boldsymbol{a}} X = \operatorname{aff} X$.
(*e*) $\operatorname{cone}_{\boldsymbol{a}} X$ *is a plane.*

Proof. Excluding the obvious case $X = \{\boldsymbol{a}\}$, we assume that $X \setminus \{\boldsymbol{a}\} \neq \varnothing$.

$(a) \Rightarrow (b)$ Let $\boldsymbol{a} \in \operatorname{rint}(\operatorname{conv} X)$. Then $\boldsymbol{a} \in \operatorname{aff} X$ and $\operatorname{cone}_{\boldsymbol{a}} X \subset \operatorname{aff} X$ according to Theorem 5.38. The inclusions

$$\operatorname{conv} X \subset \operatorname{cone}_{\boldsymbol{a}} X \subset \operatorname{aff} X = \operatorname{aff}(\operatorname{conv} X)$$

and Theorem 2.42 give $\operatorname{aff}(\operatorname{conv} X) = \operatorname{aff}(\operatorname{cone}_{\boldsymbol{a}} X)$. Theorem 3.18 implies that

$$\boldsymbol{a} \in \operatorname{rint}(\operatorname{conv} X) \subset \operatorname{rint}(\operatorname{cone}_{\boldsymbol{a}} X).$$

$(a) \Rightarrow (c)$ If $\boldsymbol{a} \in \operatorname{rint}(\operatorname{conv} X)$, then $\boldsymbol{a} \in \operatorname{aff} X$, and Theorem 5.38 gives the inclusion $\operatorname{cone}_{\boldsymbol{a}}(\operatorname{rint}(\operatorname{conv} X)) \subset \operatorname{aff} X$. Conversely, let $\boldsymbol{x} \in \operatorname{aff} X$. Since the case $\boldsymbol{x} = \boldsymbol{a}$ is obvious, we may assume that $\boldsymbol{x} \neq \boldsymbol{a}$. By Corollary 3.27, there is a point $\boldsymbol{z} \in \operatorname{rint}(\operatorname{conv} X)$ such that $\boldsymbol{z} \in (\boldsymbol{a}, \boldsymbol{x})$. Hence

$$\boldsymbol{x} \in (\boldsymbol{a}, \boldsymbol{z}\rangle \subset \operatorname{cone}_{\boldsymbol{a}}(\operatorname{rint}(\operatorname{conv} X)).$$

$(b) \Rightarrow (d)$ If $\boldsymbol{a} \in \operatorname{rint}(\operatorname{cone}_{\boldsymbol{a}} X)$, then Corollary 5.26 shows that $\operatorname{cone}_{\boldsymbol{a}} X$ is a plane, and Theorem 5.38 implies that $\operatorname{cone}_{\boldsymbol{a}} X = \operatorname{aff} X$.

$(c) \Rightarrow (d)$ If $\operatorname{cone}_{\boldsymbol{a}}(\operatorname{rint}(\operatorname{conv} X)) = \operatorname{aff} X$, then $\boldsymbol{a} \in \operatorname{aff} X$, and Theorem 5.38 gives

$$\operatorname{aff} X = \operatorname{cone}_{\boldsymbol{a}}(\operatorname{rint}(\operatorname{conv} X)) \subset \operatorname{cone}_{\boldsymbol{a}}(\operatorname{conv} X) = \operatorname{cone}_{\boldsymbol{a}} X \subset \operatorname{aff} X.$$

Hence $\operatorname{cone}_{\boldsymbol{a}} X = \operatorname{aff} X$.

$(d) \Leftrightarrow (e)$ This part is proved in Theorem 5.38.

$(d) \Rightarrow (a)$ Choose a point $\boldsymbol{x} \in \operatorname{conv} X \setminus \{\boldsymbol{a}\}$. Then $[\boldsymbol{a}, \boldsymbol{x}\rangle \subset \operatorname{cone}_{\boldsymbol{a}} X = \operatorname{aff} X$, which implies the inclusion $\langle \boldsymbol{a}, \boldsymbol{x}\rangle \subset \operatorname{cone}_{\boldsymbol{a}} X$. Let $\boldsymbol{y}$ be a point in $\langle \boldsymbol{a}, \boldsymbol{x}\rangle$ such that $\boldsymbol{a} \in (\boldsymbol{x}, \boldsymbol{y})$. Because $\operatorname{cone}_{\boldsymbol{a}} X = \operatorname{cone}_{\boldsymbol{a}}(\operatorname{conv} X)$, there is a point $\boldsymbol{z} \in \operatorname{conv} X$ such that $\boldsymbol{y} \in [\boldsymbol{a}, \boldsymbol{z}\rangle$. Clearly, $\boldsymbol{a} \in (\boldsymbol{x}, \boldsymbol{z})$. Finally, $\boldsymbol{a} \in \operatorname{rint}(\operatorname{conv} X)$ according to Corollary 3.27. □

Theorem 5.49. *For a set* $X \subset \mathbb{R}^n$ *and a point* $\boldsymbol{a} \in \mathbb{R}^n$, *one has*

$$\operatorname{cone}_{\boldsymbol{a}}(\operatorname{rint}(\operatorname{conv} X)) = \{\boldsymbol{a}\} \cup \operatorname{rint}(\operatorname{cone}_{\boldsymbol{a}} X). \tag{5.24}$$

Furthermore,

$$\operatorname{cone}_{\boldsymbol{a}}(\operatorname{rint}(\operatorname{conv} X)) = \operatorname{rint}(\operatorname{cone}_{\boldsymbol{a}} X) \tag{5.25}$$

if and only if $\operatorname{cone}_{\boldsymbol{a}} X = \operatorname{aff}(\{\boldsymbol{a}\} \cup X)$. *If, additionally,* $X \neq \varnothing$, *then* (5.25) *holds if and only if* $\operatorname{cone}_{\boldsymbol{a}} X = \operatorname{aff} X$.

Proof. Since the case $X \subset \{\boldsymbol{a}\}$ is obvious, we assume that $X \setminus \{\boldsymbol{a}\} \neq \varnothing$.

We will start with the second statement. If (5.25) holds, then

$$\boldsymbol{a} \in \operatorname{cone}_{\boldsymbol{a}}(\operatorname{rint}(\operatorname{conv} X)) = \operatorname{rint}(\operatorname{cone}_{\boldsymbol{a}} X),$$

and a combination of Theorems 5.30 and 5.48 gives

$$\operatorname{cone}_{\boldsymbol{a}} X = \operatorname{aff} X = \operatorname{aff}(\{\boldsymbol{a}\} \cup X).$$

Conversely, if any of the equalities

$$\operatorname{cone}_{\boldsymbol{a}} X = \operatorname{aff} X \quad \text{and} \quad \operatorname{cone}_{\boldsymbol{a}} X = \operatorname{aff}(\{\boldsymbol{a}\} \cup X)$$

holds then, by the same theorems, both sides of (5.25) are equal to aff X.

Next, we will prove (5.24). By the above argument, we may assume that $\boldsymbol{a} \notin \operatorname{rint}(\operatorname{cone}_{\boldsymbol{a}} X)$. Under this assumption, we assert that

$$\operatorname{rint}(\operatorname{conv} X) \subset \operatorname{rint}(\operatorname{cone}_{\boldsymbol{a}} X). \tag{5.26}$$

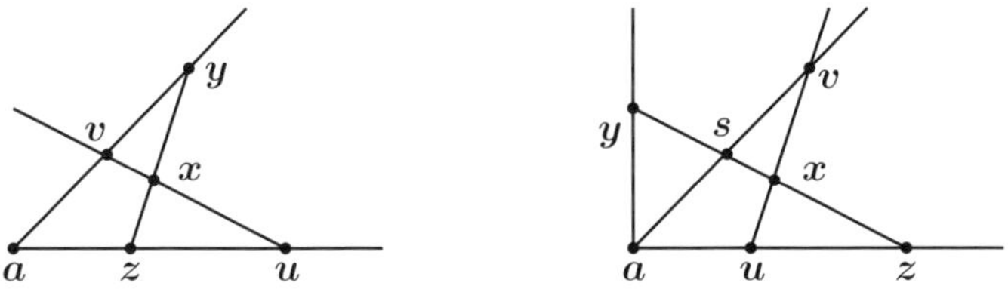

Fig. 5.13 Illustration to Theorem 5.49.

Indeed, let $\boldsymbol{x} \in \operatorname{rint}(\operatorname{conv} X)$. To prove the inclusion $\boldsymbol{x} \in \operatorname{rint}(\operatorname{cone}_{\boldsymbol{a}} X)$, it suffices to show that for every $\boldsymbol{u} \in \operatorname{cone}_{\boldsymbol{a}} X \setminus \{\boldsymbol{x}\}$ there is a point $\boldsymbol{v} \in \operatorname{cone}_{\boldsymbol{a}} X$ such that $\boldsymbol{x} \in (\boldsymbol{u}, \boldsymbol{v})$, as follows from Corollary 3.27. Because the case $\boldsymbol{u} \in [\boldsymbol{a}, \boldsymbol{x}\rangle$ is obvious, we assume that $\boldsymbol{u} \notin [\boldsymbol{a}, \boldsymbol{x}\rangle$. By Theorem 5.38, there is a point $\boldsymbol{z} \in \operatorname{conv} X$ such that $\boldsymbol{u} \in [\boldsymbol{a}, \boldsymbol{z}\rangle$. Due to $\boldsymbol{x} \in \operatorname{rint}(\operatorname{conv} X)$, one can find a point $\boldsymbol{y} \in \operatorname{conv} X$ with the property $\boldsymbol{x} \in (\boldsymbol{y}, \boldsymbol{z})$.

If $\boldsymbol{z} \in [\boldsymbol{a}, \boldsymbol{u}]$, then the closed halflines $[\boldsymbol{u}, \boldsymbol{x}\rangle$ and $[\boldsymbol{a}, \boldsymbol{y}\rangle$ have a common point, $\boldsymbol{v}$ (see Lemma 2.26 and Figure 5.13). Consequently, $\boldsymbol{v} \in [\boldsymbol{a}, \boldsymbol{y}\rangle \subset \operatorname{cone}_{\boldsymbol{a}}(\operatorname{conv} X)$ and $\boldsymbol{x} \in (\boldsymbol{u}, \boldsymbol{v})$. If $\boldsymbol{u} \in [\boldsymbol{a}, \boldsymbol{z}]$, then we can choose a point $\boldsymbol{s} \in [\boldsymbol{x}, \boldsymbol{y}]$ so close to $\boldsymbol{x}$ that the closed halflines $[\boldsymbol{a}, \boldsymbol{s}\rangle$ and $[\boldsymbol{u}, \boldsymbol{x}\rangle$ meet at a point, say $\boldsymbol{v}$. As above, $\boldsymbol{v} \in [\boldsymbol{a}, \boldsymbol{s}\rangle \subset \operatorname{cone}_{\boldsymbol{a}}(\operatorname{conv} X)$ and $\boldsymbol{x} \in (\boldsymbol{u}, \boldsymbol{v})$. Summing up, $\operatorname{rint}(\operatorname{conv} X) \subset \operatorname{rint}(\operatorname{cone}_{\boldsymbol{a}} X)$.

We return to the proof of (5.24). A combination of Corollary 5.21 and Theorem 5.18 (with $C = \operatorname{rint}(\operatorname{cone}_{\boldsymbol{a}} X)$ and $L = \{\boldsymbol{a}\}$) shows that the set $\{\boldsymbol{a}\} \cup \operatorname{rint}(\operatorname{cone}_{\boldsymbol{a}} X)$ is a convex cone with apex $\boldsymbol{a}$. Hence (5.26) gives

$$\operatorname{cone}_{\boldsymbol{a}}(\operatorname{rint}(\operatorname{conv} X)) \subset \{\boldsymbol{a}\} \cup \operatorname{rint}(\operatorname{cone}_{\boldsymbol{a}} X).$$

For the opposite inclusion, choose a point $\boldsymbol{x} \in \operatorname{rint}(\operatorname{cone}_{\boldsymbol{a}} X)$. Then $\boldsymbol{x} \in \operatorname{cone}_{\boldsymbol{a}} X \setminus \{\boldsymbol{a}\}$ due to the assumption $\boldsymbol{a} \notin \operatorname{rint}(\operatorname{cone}_{\boldsymbol{a}} X)$. Theorem 5.38 shows that the convex set $K = [\boldsymbol{a}, \boldsymbol{x}\rangle \cap \operatorname{conv} X$ is one-dimensional. Choose a point $\boldsymbol{y} \in \operatorname{rint} K$.

We intend to show that $\boldsymbol{y} \in \operatorname{rint}(\operatorname{conv} X)$. For this, it suffices to prove that for any $\boldsymbol{u} \in \operatorname{conv} X \setminus \{\boldsymbol{y}\}$ there is a point $\boldsymbol{v} \in \operatorname{conv} X$ such that $\boldsymbol{y} \in (\boldsymbol{u}, \boldsymbol{v})$, as described in Corollary 3.27. If $\boldsymbol{u} \in K$, then the inclusion $\boldsymbol{y} \in \operatorname{rint} K$ implies the existence of points $\boldsymbol{r}, \boldsymbol{t} \in K \setminus \{\boldsymbol{y}\}$ such that $\boldsymbol{y} \in (\boldsymbol{s}, \boldsymbol{t})$. By Lemma 2.25, either $\boldsymbol{y} \in (\boldsymbol{u}, \boldsymbol{t})$ or $\boldsymbol{y} \in (\boldsymbol{s}, \boldsymbol{u})$; so $\boldsymbol{v}$ may be chosen in $\{\boldsymbol{s}, \boldsymbol{t}\}$.

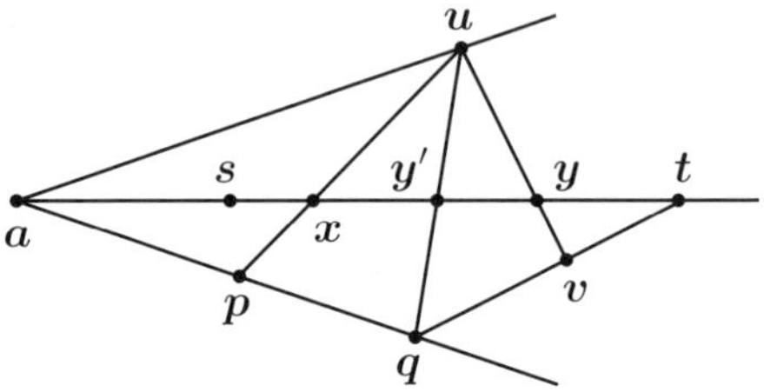

Fig. 5.14 Illustration to Theorem 5.49.

Assume that $\boldsymbol{u} \notin K$. Then $\boldsymbol{u} \notin [\boldsymbol{a}, \boldsymbol{x}\rangle$, as depicted in Figure 5.14. Because $\boldsymbol{u} \in \operatorname{conv} X \subset \operatorname{cone}_{\boldsymbol{a}} X$ and $\boldsymbol{x} \in \operatorname{rint}(\operatorname{cone}_{\boldsymbol{a}} X)$, Corollary 3.27 shows the existence of a point $\boldsymbol{p} \in \operatorname{cone}_{\boldsymbol{a}} X$ such that $\boldsymbol{x} \in (\boldsymbol{u}, \boldsymbol{p})$. Since $\operatorname{cone}_{\boldsymbol{a}} X = \operatorname{cone}_{\boldsymbol{a}}(\operatorname{conv} X)$, there is a point $\boldsymbol{q} \in \operatorname{conv} X$ such that $\boldsymbol{p} \in [\boldsymbol{a}, \boldsymbol{q}\rangle$, as follows from Theorem 5.38. Obviously, $[\boldsymbol{u}, \boldsymbol{q}] \subset \operatorname{conv} X$, and the point $\boldsymbol{y}' = [\boldsymbol{a}, \boldsymbol{x}\rangle \cap [\boldsymbol{u}, \boldsymbol{q}]$ belongs to K. If $\boldsymbol{y}' = \boldsymbol{y}$, then put $\boldsymbol{v} = \boldsymbol{q}$. Let $\boldsymbol{y}' \neq \boldsymbol{y}$. As above, there are points $\boldsymbol{s}, \boldsymbol{t} \in K \setminus \{\boldsymbol{y}\}$ such that $\boldsymbol{y} \in (\boldsymbol{s}, \boldsymbol{t})$. Then the closed halfline $[\boldsymbol{u}, \boldsymbol{y}\rangle$ meets one of the closed segments $[\boldsymbol{s}, \boldsymbol{q}]$ and $[\boldsymbol{t}, \boldsymbol{q}]$. Assume, for instance, that $[\boldsymbol{u}, \boldsymbol{y}\rangle$ meets $[\boldsymbol{t}, \boldsymbol{q}]$, and denote by $\boldsymbol{v}$ their common point. Clearly, $\boldsymbol{y} \in (\boldsymbol{u}, \boldsymbol{v})$ and $\boldsymbol{v} \in [\boldsymbol{t}, \boldsymbol{q}] \subset \operatorname{conv} X$. Summing up,

$$\boldsymbol{y} \in \operatorname{rint}(\operatorname{conv} X) \quad \text{and} \quad \boldsymbol{x} \in [\boldsymbol{a}, \boldsymbol{y}\rangle \subset \operatorname{cone}_{\boldsymbol{a}}(\operatorname{rint}(\operatorname{conv} X)).$$

Hence $\{\boldsymbol{a}\} \cup \operatorname{rint}(\operatorname{cone}_{\boldsymbol{a}} X) \subset \operatorname{cone}_{\boldsymbol{a}}(\operatorname{rint}(\operatorname{conv} X))$. □

The following result complements Theorem 5.47.

Theorem 5.50. *Let a set $X \subset \mathbb{R}^n$ be contained in a nonempty proper plane $L \subset \mathbb{R}^n$. For any point $\boldsymbol{a} \in \mathbb{R}^n \setminus L$, one has*

$$L \cap \operatorname{rint}(\operatorname{cone}_{\boldsymbol{a}} X) = \operatorname{rint}(\operatorname{conv} X).$$

Proof. Since the case $X = \varnothing$ is obvious, we may assume that $X \neq \varnothing$. Choose any point $\boldsymbol{x} \in \operatorname{rint}(\operatorname{conv} X)$. Theorem 5.49 shows that

$$\boldsymbol{x} \in \operatorname{rint}(\operatorname{conv} X) \setminus \{\boldsymbol{a}\} \subset \operatorname{cone}_{\boldsymbol{a}}(\operatorname{rint}(\operatorname{conv} X)) \setminus \{\boldsymbol{a}\} \subset \operatorname{rint}(\operatorname{cone}_{\boldsymbol{a}} X).$$

Since $\boldsymbol{x} \in L$, the plane L meets $\operatorname{rint}(\operatorname{cone}_{\boldsymbol{a}} X)$. Therefore, a combination of Corollary 3.31 and Theorem 5.39 gives

$$L \cap \operatorname{rint}(\operatorname{cone}_{\boldsymbol{a}} X) = \operatorname{rint}(L \cap \operatorname{cone}_{\boldsymbol{a}} X) = \operatorname{rint}(\operatorname{conv} X). \qquad \square$$

The following result is analogous to Theorem 4.18.

Theorem 5.51. *Let $\boldsymbol{a} \in \mathbb{R}^n$ and $X \subset \mathbb{R}^n$ be a nonempty set. A point $\boldsymbol{x} \in \mathbb{R}^n$ belongs to $\operatorname{rint}(\operatorname{cone}_{\boldsymbol{a}} X)$ if and only if there are finitely many points $\boldsymbol{x}_1, \dots, \boldsymbol{x}_k \in X$ and scalars $\lambda_1, \dots, \lambda_k > 0$ satisfying the conditions*

$$\begin{gathered} \boldsymbol{x} = \boldsymbol{a} + \lambda_1(\boldsymbol{x}_1 - \boldsymbol{a}) + \cdots + \lambda_k(\boldsymbol{x}_k - \boldsymbol{a}), \\ \operatorname{aff}\{\boldsymbol{x}_1, \dots, \boldsymbol{x}_k\} = \operatorname{aff} X. \end{gathered} \tag{5.27}$$

Proof. Since the case $X = \{\boldsymbol{a}\}$ is obvious, we may assume that $X \neq \{\boldsymbol{a}\}$. Let $\boldsymbol{x} \in \operatorname{rint}(\operatorname{cone}_{\boldsymbol{a}} X)$. If $\boldsymbol{x} = \boldsymbol{a}$, then Theorem 5.48 implies the inclusion $\boldsymbol{a} \in \operatorname{rint}(\operatorname{conv} X)$. In this case, according to Theorem 4.18, there are finitely many points $\boldsymbol{x}_1, \dots, \boldsymbol{x}_k \in X$ such that $\operatorname{aff}\{\boldsymbol{x}_1, \dots, \boldsymbol{x}_k\} = \operatorname{aff} X$ and $\boldsymbol{a}$ is their positive convex combination: $\boldsymbol{a} = \lambda_1 \boldsymbol{x}_1 + \cdots + \lambda_k \boldsymbol{x}_k$. Clearly,

$$\boldsymbol{a} = \boldsymbol{a} + \lambda_1(\boldsymbol{x}_1 - \boldsymbol{a}) + \cdots + \lambda_k(\boldsymbol{x}_k - \boldsymbol{a}).$$

Let $\boldsymbol{x} \neq \boldsymbol{a}$. Theorem 5.49 shows that $\boldsymbol{x} \in \operatorname{cone}_{\boldsymbol{a}}(\operatorname{rint}(\operatorname{conv} X))$, and Theorem 5.38 implies the existence of a point $\boldsymbol{z} \in \operatorname{rint}(\operatorname{conv} X)$ such that $\boldsymbol{x} = \boldsymbol{a} + \lambda(\boldsymbol{z} - \boldsymbol{a})$ for a suitable scalar $\lambda > 0$. As above, Theorem 4.18 shows the existence of finitely many points $\boldsymbol{x}_1, \dots, \boldsymbol{x}_k \in X$ such that $\operatorname{aff}\{\boldsymbol{x}_1, \dots, \boldsymbol{x}_k\} = \operatorname{aff} X$ and $\boldsymbol{z}$ is their positive convex combination: $\boldsymbol{z} = \mu_1 \boldsymbol{x}_1 + \cdots + \mu_k \boldsymbol{x}_k$. Put $\lambda_i = \lambda \mu_i$, $1 \leqslant i \leqslant k$. Then $\lambda_1, \dots, \lambda_k > 0$ and

$$\boldsymbol{x} = \boldsymbol{a} + \lambda(\boldsymbol{z} - \boldsymbol{a}) = \boldsymbol{a} + \lambda_1(\boldsymbol{x}_1 - \boldsymbol{a}) + \cdots + \lambda_k(\boldsymbol{x}_k - \boldsymbol{a}).$$

Conversely, suppose that $\boldsymbol{x}$ satisfies (5.27). Let

$$\lambda = \lambda_1 + \cdots + \lambda_k \quad \text{and} \quad \boldsymbol{u} = \lambda^{-1}(\lambda_1 \boldsymbol{x}_1 + \cdots + \lambda_k \boldsymbol{x}_k).$$

Clearly, $\boldsymbol{u}$ is a positive convex combination of $\boldsymbol{x}_1, \dots, \boldsymbol{x}_k$, and Theorem 4.18 shows that $\boldsymbol{u} \in \operatorname{rint}(\operatorname{conv} X)$. If $\boldsymbol{a} = \boldsymbol{u}$, Theorem 5.48 implies that $\operatorname{cone}_{\boldsymbol{a}} X$ is a plane; hence $\boldsymbol{x} \in \operatorname{cone}_{\boldsymbol{a}} X = \operatorname{rint}(\operatorname{cone}_{\boldsymbol{a}} X)$. If $\boldsymbol{a} \neq \boldsymbol{u}$, then Theorem 5.49 gives

$$\boldsymbol{x} = \boldsymbol{a} + \lambda(\boldsymbol{u} - \boldsymbol{a}) \in \operatorname{cone}_{\boldsymbol{a}}(\operatorname{rint}(\operatorname{conv} X)) \setminus \{\boldsymbol{a}\} = \operatorname{rint}(\operatorname{cone}_{\boldsymbol{a}} X). \qquad \square$$

An important particular case of Theorem 5.51, combined with Corollary 4.19, is given in the assertion below.

Corollary 5.52. *For a finite set* $X = \{\boldsymbol{x}_1, \dots, \boldsymbol{x}_r\} \subset \mathbb{R}^n$ *and a point* $\boldsymbol{a} \in \mathbb{R}^n$, *one has*

$$\mathrm{rint}\,(\mathrm{cone}_{\boldsymbol{a}} X) = \{\boldsymbol{a} + \lambda_1(\boldsymbol{x}_1 - \boldsymbol{a}) + \cdots + \lambda_r(\boldsymbol{x}_r - \boldsymbol{a}) : \lambda_1, \dots, \lambda_r > 0\}. \quad \square$$

The next result is an analog of Theorem 4.20 for the case of conic hulls.

Theorem 5.53. *Let* $X \subset \mathbb{R}^n$ *be a set of positive dimension* m *and* $\boldsymbol{a} \in \mathbb{R}^n$. *A point* $\boldsymbol{x} \in \mathbb{R}^n$ *belongs to* $\mathrm{rint}\,(\mathrm{cone}_{\boldsymbol{a}} X)$ *if and only if there is a subset* Y *of* X *such that*

$$\boldsymbol{x} \in \mathrm{rint}\,(\mathrm{cone}_{\boldsymbol{a}} Y), \quad m + 1 \leqslant \mathrm{card}\, Y \leqslant 2m, \quad \mathrm{aff}\, Y = \mathrm{aff}\, X.$$

Proof. Let $\boldsymbol{x} \in \mathrm{rint}\,(\mathrm{cone}_{\boldsymbol{a}} X)$. Then $\boldsymbol{x} \in \mathrm{cone}_{\boldsymbol{a}}(\mathrm{rint}\,(\mathrm{conv}\, X))$ according to Theorem 5.49. Assume first that $\boldsymbol{x} = \boldsymbol{a}$. Then Theorem 5.48 implies the equality $\mathrm{cone}_{\boldsymbol{a}} X = \mathrm{aff}\, X$. By Theorem 2.61, X contains an affine basis Y for $\mathrm{aff}\, X$. Clearly, $\mathrm{card}\, Y = m + 1$, and

$$\boldsymbol{x} = \boldsymbol{a} \in \mathrm{aff}\, X = \mathrm{aff}\, Y = \mathrm{rint}\,(\mathrm{aff}\, Y) = \mathrm{rint}\,(\mathrm{cone}_{\boldsymbol{a}} Y).$$

Suppose that $\boldsymbol{x} \neq \boldsymbol{a}$. Theorem 5.38 shows that $\boldsymbol{x} = \boldsymbol{a} + \lambda(\boldsymbol{z} - \boldsymbol{a})$, where $\lambda > 0$ and $\boldsymbol{z} \in \mathrm{rint}\,(\mathrm{conv}\, X)$. Theorem 4.20 implies the existence of a set $Y \subset X$ such that

$$\boldsymbol{z} \in \mathrm{rint}\,(\mathrm{conv}\, Y), \quad m + 1 \leqslant \mathrm{card}\, Y \leqslant 2m, \quad \text{and} \quad \mathrm{aff}\, Y = \mathrm{aff}\, X.$$

Finally,

$$\boldsymbol{x} \in \mathrm{cone}_{\boldsymbol{a}}(\mathrm{rint}\,(\mathrm{conv}\, Y)) \setminus \{\boldsymbol{a}\} = \mathrm{rint}\,(\mathrm{cone}_{\boldsymbol{a}} Y)$$

according to Theorem 5.49. The converse assertion follows from Theorem 5.51. $\square$

Remark. If $2m$ is the minimum cardinality of a set $Y \subset X$ satisfying the hypothesis of Theorem 5.53, then, unlike Theorem 4.20, Y cannot be partitioned into m pairs of points collinear with $\boldsymbol{x}$. For instance, let $\boldsymbol{x} = (3, 0, 0)$ and $X = \{\boldsymbol{x}_1, \boldsymbol{x}_2, \boldsymbol{x}_3, \boldsymbol{x}_4\}$, where

$$\boldsymbol{x}_1 = (1, 1, 0), \quad \boldsymbol{x}_2 = (1, -1, 0), \quad \boldsymbol{x}_3 = (2, 0, 2), \quad \boldsymbol{x}_4 = (2, 0, -2).$$

Then

$$\mathrm{cone}_{\boldsymbol{o}} X = \{(x, y, z) : x \geqslant |y| + |z|\} \quad \text{and} \quad \boldsymbol{x} \in \mathrm{int}\,(\mathrm{cone}_{\boldsymbol{o}} X).$$

Clearly, no proper subset Y of X satisfies the condition $\boldsymbol{x} \in \mathrm{int}\,(\mathrm{cone}_{\boldsymbol{o}} Y)$ and X cannot be partitioned into two pairs of points collinear with $\boldsymbol{x}$.

Corollary 5.54. *Let $X \subset \mathbb{R}^n$ be a set and $\boldsymbol{a}$ be a point in $\mathbb{R}^n$ such that* $\text{cone}_{\boldsymbol{a}} X$ *is a plane of positive dimension m. Then $\boldsymbol{a} \in \text{aff}\, X$ and X contains a subset Y which satisfies the following conditions:*

$$\text{cone}_{\boldsymbol{a}} Y = \text{aff}\, Y = \text{aff}\, X \quad \textit{and} \quad m + 1 \leqslant \text{card}\, Y \leqslant 2m. \tag{5.28}$$

Furthermore, if Y is a minimal subset of X satisfying (5.28), then either $\text{card}\, Y \leqslant 2m - 1$ *or Y consists of $2m$ points collinear in pairs with $\boldsymbol{a}$.*

Proof. Theorem 5.48 implies that $\text{cone}_{\boldsymbol{a}} X = \text{aff}\, X$ and $\boldsymbol{a} \in \text{rint}\,(\text{cone}_{\boldsymbol{a}} X)$. Theorem 5.53 shows the existence of a set $Y \subset X$ satisfying the conditions

$$\boldsymbol{x} \in \text{rint}\,(\text{cone}_{\boldsymbol{a}} Y), \quad m + 1 \leqslant \text{card}\, Y \leqslant 2m, \quad \text{aff}\, Y = \text{aff}\, X.$$

Again by Theorem 5.48, $\text{cone}_{\boldsymbol{a}} Y$ is a plane. Hence $\text{cone}_{\boldsymbol{a}} Y = \text{aff}\, Y$. □

5.2.6 *Generated Cones of Convex Sets*

Theorem 5.38 implies the following corollaries.

Corollary 5.55. *For any convex set $K \subset \mathbb{R}^n$ and a point $\boldsymbol{a} \in \mathbb{R}^n$, one has* $C_{\boldsymbol{a}}(K) = \text{cone}_{\boldsymbol{a}} K$. *Consequently, the generated cone $C_{\boldsymbol{a}}(K)$ is convex.* □

Corollary 5.56. *Any r-simplicial cone $\Gamma_{\boldsymbol{a}}(\boldsymbol{x}_1, \dots, \boldsymbol{x}_r) \subset \mathbb{R}^n$, $1 \leqslant r \leqslant n$, can be expressed as*

$$\Gamma_{\boldsymbol{a}}(\boldsymbol{x}_1, \dots, \boldsymbol{x}_r) = C_{\boldsymbol{a}}(\Delta),$$

where $\Delta = \Delta(\boldsymbol{x}_1, \dots, \boldsymbol{x}_r)$ is the r-simplex with vertices $\boldsymbol{x}_1, \dots, \boldsymbol{x}_r$.

Proof. A combination of Corollaries 4.5 and 5.34 and Theorem 5.38 gives

$$\begin{aligned} \Gamma_{\boldsymbol{a}}(\boldsymbol{x}_1, \dots, \boldsymbol{x}_m) &= \text{cone}_{\boldsymbol{a}}\{\boldsymbol{x}_1, \dots, \boldsymbol{x}_m\} \\ &= C_{\boldsymbol{a}}(\text{conv}\,\{\boldsymbol{x}_1, \dots, \boldsymbol{x}_m\}) = C_{\boldsymbol{a}}(\Delta). \end{aligned}$$ □

Theorem 5.57. *Let $K \subset \mathbb{R}^n$ be a nonempty convex set, $\boldsymbol{a}$ be a point in $\mathbb{R}^n$, and ρ be a positive scalar. The following assertions hold.*

(*a*) *If $\boldsymbol{a} \in K$, then*

$$C_{\boldsymbol{a}}(K) = C_{\boldsymbol{a}}(K \cap B_\rho(\boldsymbol{a})) = C_{\boldsymbol{a}}(K \cap U_\rho(\boldsymbol{a})).$$

(*b*) *If $\boldsymbol{a} \in \text{cl}\, K$, then*

$$C_{\boldsymbol{a}}(\text{rint}\, K) = C_{\boldsymbol{a}}(\text{rint}\, K \cap B_\rho(\boldsymbol{a})) = C_{\boldsymbol{a}}(\text{rint}\, K \cap U_\rho(\boldsymbol{a})).$$

Proof. (a) By Theorem 5.30,

$$C_{\boldsymbol{a}}(K \cap U_\rho(\boldsymbol{a})) \subset C_{\boldsymbol{a}}(K \cap B_\rho(\boldsymbol{a})) \subset C_{\boldsymbol{a}}(K).$$

For the opposite inclusions, choose a point $\boldsymbol{y} \in C_{\boldsymbol{a}}(K)$ Since the case $\boldsymbol{y} = \boldsymbol{a}$ is obvious, we may assume that $\boldsymbol{y} \neq \boldsymbol{a}$. Then $\boldsymbol{y} = \boldsymbol{a} + \lambda(\boldsymbol{x} - \boldsymbol{a})$ for a suitable point $\boldsymbol{x} \in K \setminus \{\boldsymbol{a}\}$ and a scalar $\lambda > 0$. One has $[\boldsymbol{a}, \boldsymbol{x}] \subset K$ by the convexity of K. Choose a point $\boldsymbol{z} \in (\boldsymbol{a}, \boldsymbol{x})$ such that $\|\boldsymbol{z} - \boldsymbol{a}\| < \rho$. Then $\boldsymbol{z} \in K \cap U_\rho(\boldsymbol{a})$. Expressing $\boldsymbol{z}$ as $\boldsymbol{z} = (1 - \mu)\boldsymbol{a} + \mu\boldsymbol{x}$ for a suitable scalar $0 < \mu < 1$, we can write $\boldsymbol{x} = (1 - \mu^{-1})\boldsymbol{a} + \mu^{-1}\boldsymbol{z}$, which gives

$$\boldsymbol{y} = \boldsymbol{a} + \lambda(\boldsymbol{x} - \boldsymbol{a}) = \boldsymbol{a} + \lambda((1 - \mu^{-1})\boldsymbol{a} + \mu^{-1}\boldsymbol{z} - \boldsymbol{a}) = \boldsymbol{a} + \lambda\mu^{-1}(\boldsymbol{z} - \boldsymbol{a}).$$

Thus $\boldsymbol{y} \in C_{\boldsymbol{a}}(K \cap U_\rho(\boldsymbol{a}))$.

(b) The proof of this assertion is similar; we provide it for the sake of completeness. As above, Theorem 5.30 gives the inclusions

$$C_{\boldsymbol{a}}(\operatorname{rint} K \cap U_\rho(\boldsymbol{a})) \subset C_{\boldsymbol{a}}(\operatorname{rint} K \cap B_\rho(\boldsymbol{a})) \subset C_{\boldsymbol{a}}(\operatorname{rint} K).$$

For the opposite inclusions, choose a point $\boldsymbol{y} \in C_{\boldsymbol{a}}(\operatorname{rint} K)$. Since the case $\boldsymbol{y} = \boldsymbol{a}$ is obvious, we may assume that $\boldsymbol{y} \neq \boldsymbol{a}$. Then $\boldsymbol{y} = \boldsymbol{a} + \lambda(\boldsymbol{x} - \boldsymbol{a})$ for a suitable point $\boldsymbol{x} \in \operatorname{rint} K \setminus \{\boldsymbol{a}\}$ and a scalar $\lambda > 0$. One has $(\boldsymbol{a}, \boldsymbol{x}) \subset \operatorname{rint} K$ according to Theorem 3.47. Choose a point $\boldsymbol{z} \in (\boldsymbol{a}, \boldsymbol{x})$ such that $\|\boldsymbol{z} - \boldsymbol{a}\| < \rho$. Then $\boldsymbol{z} \in \operatorname{rint} K \cap U_\rho(\boldsymbol{a})$. Expressing $\boldsymbol{z}$ as $\boldsymbol{z} = (1 - \mu)\boldsymbol{a} + \mu\boldsymbol{x}$ for a suitable scalar $0 < \mu < 1$, we can write $\boldsymbol{x} = (1 - \mu^{-1})\boldsymbol{a} + \mu^{-1}\boldsymbol{z}$, which gives

$$\boldsymbol{y} = \boldsymbol{a} + \lambda(\boldsymbol{x} - \boldsymbol{a}) = \boldsymbol{a} + \lambda((1 - \mu^{-1})\boldsymbol{a} + \mu^{-1}\boldsymbol{z} - \boldsymbol{a}) = \boldsymbol{a} + \lambda\mu^{-1}(\boldsymbol{z} - \boldsymbol{a}).$$

Thus $\boldsymbol{y} \in C_{\boldsymbol{a}}(\operatorname{rint} K \cap U_\rho(\boldsymbol{a}))$. □

A combination of Theorem 5.48 and Corollary 5.55 imply the following result.

Corollary 5.58. *For a nonempty convex set $K \subset \mathbb{R}^n$ and a point $\boldsymbol{a} \in \mathbb{R}^n$, the following conditions are equivalent.*

(a) $\boldsymbol{a} \in \operatorname{rint} K$.
(b) $\boldsymbol{a} \in \operatorname{rint}(C_{\boldsymbol{a}}(K))$.
(c) $C_{\boldsymbol{a}}(\operatorname{rint} K) = \operatorname{aff} K$.
(d) $C_{\boldsymbol{a}}(K) = \operatorname{aff} K$.
(e) $C_{\boldsymbol{a}}(K)$ *is a plane.* □

Theorem 5.59. *Given a convex set $K \subset \mathbb{R}^n$ which is not a plane, the following inclusions hold.*

$$K \subset \cap(C_{\boldsymbol{x}}(K) : \boldsymbol{x} \in \mathbb{R}^n) \subset \cap(C_{\boldsymbol{x}}(K) : \boldsymbol{x} \in \operatorname{aff} K)$$
$$\subset \cap(C_{\boldsymbol{x}}(K) : \boldsymbol{x} \in \operatorname{cl} K) \subset \cap(C_{\boldsymbol{x}}(K) : \boldsymbol{x} \in \operatorname{rbd} K) \subset \operatorname{cl} K.$$

Proof. By Theorem 3.62, $\operatorname{rbd} K \neq \varnothing$. A combination of Theorem 2.42 and Corollary 2.43 shows that

$$\operatorname{rbd} K \subset \operatorname{cl} K \subset \operatorname{aff}(\operatorname{cl} K) = \operatorname{aff} K.$$

Consequently, Theorem 5.31 gives the inclusions

$$\begin{aligned} K &\subset \cap\,(C_{\boldsymbol{x}}(K) : \boldsymbol{x} \in \mathbb{R}^n) \subset \cap\,(C_{\boldsymbol{x}}(K) : \boldsymbol{x} \in \operatorname{aff} K) \\ &\subset \cap\,(C_{\boldsymbol{x}}(K) : \boldsymbol{x} \in \operatorname{cl} K) \subset \cap\,(C_{\boldsymbol{x}}(K) : \boldsymbol{x} \in \operatorname{rbd} K). \end{aligned}$$

Hence it remains to show that $\cap\,(C_{\boldsymbol{x}}(K) : \boldsymbol{x} \in \operatorname{rbd} K) \subset \operatorname{cl} K$. It suffices to prove that for any point $\boldsymbol{u} \in \mathbb{R}^n \setminus \operatorname{cl} K$ there is a point $\boldsymbol{x} \in \operatorname{rbd} K$ satisfying the condition $\boldsymbol{u} \notin C_{\boldsymbol{x}}(K)$.

If $\boldsymbol{u} \in \mathbb{R}^n \setminus \operatorname{aff} K$, then choose any point $\boldsymbol{x} \in \operatorname{rbd} K$. Indeed, since $\operatorname{rbd} K \subset \operatorname{aff} K$, Theorem 5.38 gives $C_{\boldsymbol{x}}(K) \subset \operatorname{aff} K$, and thus $\boldsymbol{u} \notin C_{\boldsymbol{x}}(K)$.

Suppose that $\boldsymbol{u} \in \operatorname{aff} K \setminus \operatorname{cl} K$. Choose a point $\boldsymbol{z} \in \operatorname{rint} K$. By Theorem 3.65, the open segment $(\boldsymbol{u}, \boldsymbol{z})$ meets $\operatorname{rbd} K$ at a point $\boldsymbol{x}$. We assert that $\boldsymbol{u} \notin C_{\boldsymbol{x}}(K)$. Theorem 5.48 gives the inequality $C_{\boldsymbol{x}}(K) \neq \operatorname{aff} K$. This argument and Corollary 5.27 show that $\boldsymbol{x} \in \operatorname{rbd} C_{\boldsymbol{x}}(K)$. Assume, for contradiction, that $\boldsymbol{u} \in C_{\boldsymbol{x}}(K)$. Since $\boldsymbol{z} \in \operatorname{rint} K \subset \operatorname{rint} C_{\boldsymbol{x}}(K)$ due to Theorem 5.49, we obtain, by Theorem 3.47, that $\boldsymbol{x} \in (\boldsymbol{u}, \boldsymbol{z}) \subset \operatorname{rint} C_{\boldsymbol{x}}(K)$, contrary to the inclusion $\boldsymbol{x} \in \operatorname{rbd} C_{\boldsymbol{x}}(K)$. Summing up, $\boldsymbol{u} \notin C_{\boldsymbol{x}}(K)$. □

Theorems 5.41 and 5.42 imply two more useful corollaries.

Corollary 5.60. *For convex sets $K_1, \dots, K_r \subset \mathbb{R}^n$, points $\boldsymbol{a}_1 \in K_1, \dots, \boldsymbol{a}_r \in K_r$, and scalars $\mu_1, \dots, \mu_r$, $r \geqslant 2$, one has*

$$\begin{aligned} &C_{\mu_1 \boldsymbol{a}_1 + \cdots + \mu_r \boldsymbol{a}_r}(\mu_1 K_1 + \cdots + \mu_r K_r) \\ &\quad = \mu_1 C_{\boldsymbol{a}_1}(K_1) + \cdots + \mu_r C_{\boldsymbol{a}_r}(K_r). \end{aligned}$$ □

Corollary 5.61. *Let $K_1, \dots, K_r$ be convex sets in $\mathbb{R}^n$ with the property $K_1 \cap \cdots \cap K_r \neq \varnothing$. For any point $\boldsymbol{a} \in K_1 \cap \cdots \cap K_r$, one has*

$$C_{\boldsymbol{a}}(K_1 \cap \cdots \cap K_r) = C_{\boldsymbol{a}}(K_1) \cap \cdots \cap C_{\boldsymbol{a}}(K_r).$$ □

Problems and Notes for Chapter 5

Problems for Chapter 5

Problem 5.1. Let $\Gamma = \Gamma_{\boldsymbol{a}}(\boldsymbol{x}_1, \dots, \boldsymbol{x}_r)$ be an r-simplicial cone with apex $\boldsymbol{a}$. For a nonempty proper subset X of $\{\boldsymbol{x}_1, \dots, \boldsymbol{x}_r\}$, denote by Γ_X and Γ_Y the simplicial cones with apex $\boldsymbol{a}$ generated by X and $Y = \{\boldsymbol{x}_1, \dots, \boldsymbol{x}_r\} \setminus X$, respectively. Prove that $\Gamma_X \cap \Gamma_Y = \{\boldsymbol{a}\}$ and $\Gamma = \operatorname{conv}_2(\Gamma_X \cup \Gamma_Y)$.

Problem 5.2. Given an r-simplex $\Delta = \Delta(\boldsymbol{x}_1, \dots, \boldsymbol{x}_{r+1}) \subset \mathbb{R}^n$ and a point $\boldsymbol{a}$ in $\operatorname{rint} \Delta$, let $\Gamma_i = C_{\boldsymbol{a}}(\Delta_i)$, where

$$\Delta_i = \Delta(\boldsymbol{x}_1, \dots, \boldsymbol{x}_{i-1}, \boldsymbol{x}_{i+1}, \dots, \boldsymbol{x}_{r+1}), \quad 1 \leqslant i \leqslant r+1.$$

Prove that $\operatorname{aff} \Delta = \Gamma_1 \cup \dots \cup \Gamma_{r+1}$ and $\Gamma_i \cap \operatorname{rint} \Gamma_j = \varnothing$ whenever $i \neq j$.

Problem 5.3. Prove that simplicial cones $\Gamma_{\boldsymbol{a}}(\boldsymbol{x}_1, \dots, \boldsymbol{x}_r)$ and $\Gamma_{\boldsymbol{c}}(\boldsymbol{z}_1, \dots, \boldsymbol{z}_s)$ in $\mathbb{R}^n$ coincide if and only if the following conditions are satisfied: $\boldsymbol{a} = \boldsymbol{c}$, $r = s$, and the points $\boldsymbol{z}_1, \dots, \boldsymbol{z}_r$ can be renumbered such that $\boldsymbol{z}_i$ belongs to the open halfline $(\boldsymbol{a}, \boldsymbol{x}_i\rangle$ for all $1 \leqslant i \leqslant r$.

Problem 5.4. Let $C \subset \mathbb{R}^n$ be a cone with apex $\boldsymbol{a}$ such that $C \setminus \{\boldsymbol{a}\} \neq \varnothing$. Prove that $\operatorname{conv} C \setminus \{\boldsymbol{a}\}$ is the collection of all combinations

$$\boldsymbol{x} = (1-k)\boldsymbol{a} + \boldsymbol{x}_1 + \dots + \boldsymbol{x}_k, \quad \boldsymbol{x}_1, \dots, \boldsymbol{x}_k \in C, \quad k \geqslant 1,$$

where the set $\{\boldsymbol{a}, \boldsymbol{x}_1, \dots, \boldsymbol{x}_k\}$ is affinely independent.

Problem 5.5. Let $C \subset \mathbb{R}^n$ be a nonempty cone with apex $\boldsymbol{a}$. Prove that

$$\operatorname{aff} C = \boldsymbol{a} + \operatorname{conv}(C - C) = \operatorname{conv}(C \cup (2\boldsymbol{a} - C)).$$

Problem 5.6. (Bair [11]) Prove that a nonempty convex set $K \subset \mathbb{R}^n$ satisfies the equality $K = \boldsymbol{a} + \mu K$, where $\boldsymbol{a} \neq \boldsymbol{o}$ and $\mu > 0$, if and only if K is either a cone or contains a translate of the line $\langle \boldsymbol{o}, \boldsymbol{a} \rangle$.

Problem 5.7. (Klee [176]) Let $\boldsymbol{u}$ and $\boldsymbol{v}$ be distinct points of a convex set $K \subset \mathbb{R}^n$, and let $\boldsymbol{w} \in (\boldsymbol{u}, \boldsymbol{v})$. Prove that $C_{\boldsymbol{w}}(K) = \operatorname{conv}(C_{\boldsymbol{u}}(K) \cup C_{\boldsymbol{v}}(K))$.

Problem 5.8. Let C and D be convex cones of positive dimension in $\mathbb{R}^n$ with a common apex $\boldsymbol{a}$ such that $C \subset D$; let $S_\rho(\boldsymbol{a}) \subset \mathbb{R}^n$ be a sphere and X be a dense subset of the set $D \cap S_\rho(\boldsymbol{a})$. Prove the following assertions.

(a) If $\dim C = \dim D$, then the set $\operatorname{rint} C \cap X$ is dense in $C \cap S_\rho(\boldsymbol{a})$.

(b) If $\dim C < \dim D$, then the set $(\operatorname{rint} D \setminus \operatorname{cl} C) \cap X$ is dense in $D \cap S_\rho(\boldsymbol{a})$.

Problem 5.9. Let $L \subset \mathbb{R}^n$ be a nonempty proper plane, $\boldsymbol{a}$ be a point in $\mathbb{R}^n \setminus L$, and L' be a translate of L containing $\boldsymbol{a}$. Prove that the cone $C_{\boldsymbol{a}}(L)$ is the union of $\{\boldsymbol{a}\}$ and the open halfplane of the plane $M = \operatorname{aff}(\{\boldsymbol{a}\} \cup L)$ which is determined by L' and contains L.

Problem 5.10. Let $K \subset \mathbb{R}^n$ be a convex set and Λ be a convex subset of the closed halfline $[0, \infty)$ of $\mathbb{R}$. Prove that the set $\Lambda K = \{\lambda \boldsymbol{x} : \lambda \in \Lambda, \boldsymbol{x} \in K\}$ is convex.

Problem 5.11. (Rockafellar [241, p. 22]) For a pair of nonempty disjoint convex sets K_1 and K_2 in $\mathbb{R}^n$, the *umbra* of K_1 with respect to K_2 is defined by

$$U(K_1, K_2) = \cap((1-\gamma)\boldsymbol{x} + \gamma K_1 : \gamma \geqslant 1 \text{ and } \boldsymbol{x} \in K_2),$$

and the *penumbra* of K_1 with respect to K_2 is given by

$$P(K_1, K_2) = \cup\,((1-\gamma)\boldsymbol{x} + \gamma K_1 : \gamma \geqslant 1 \text{ and } \boldsymbol{x} \in K_2).$$

Prove that both sets $U(K_1, K_2)$ and $P(K_1, K_2)$ are convex.

Notes for Chapter 5

A big part of results from this chapter can be found in the literature for the case of cones and conic hulls with apex $\boldsymbol{o}$.

Convex cones. The basic paper on the subject is that of Steinitz [266]. For instance, part (a) of Theorem 5.4, Theorem 5.14 (for the case of proper apex), part (c) of Theorem 5.15, part (a) of Theorem 5.17, and various properties of generated cones of the form $C_L(X)$ can be found there. In present form, Theorems 5.14 and 5.15 are given by Lawrence and Soltan [196]. The same authors proved that the union and the intersection of a nested family of cones in $\mathbb{R}^n$ (not necessarily convex or with the same apex) also are cones.

Conic hulls. Theorem 5.32 is due to Steinitz [266, § 19], and Theorem 5.35 (for $\boldsymbol{a} = \boldsymbol{o}$) can be found in Fenchel [105, Theorem 6]. The first assertion of Theorem 5.37 and Theorem 5.46 are given in Fenchel [105, Chapter 1, § 5]. Assertion (d) of Theorem 5.45 is proved by Steinitz [266, § 17] for the case when X is a compact convex set. Further generalization of this assertion is due to Hu and Wang [154], who gave a sufficient condition for the closedness of $\operatorname{cone}_{\boldsymbol{o}} X$ (with a possibility $\boldsymbol{o} \in X$) in terms of the distance from $\boldsymbol{o}$ to the set

$$E(X) = \{\boldsymbol{x} \in \operatorname{cl} X : t\boldsymbol{x} \notin X \text{ for all } t > 1\}.$$

Theorem 5.51 is due to Fischer [108], and Corollary 5.52 is a generalization of an assertion of Gerstenhaber [118, Theorem 1]. Corollary 5.54 can be found in Steinitz [267, § 20] for the case $\operatorname{cone}_{\boldsymbol{a}} X = \mathbb{R}^n$.

Positive bases. Corollary 5.54 gives an affine version of the concept of positive basis, described below. A set X of vectors in $\mathbb{R}^n$ is called *positively independent* provided no vector $\boldsymbol{x} \in X$ belongs to $\operatorname{cone}_{\boldsymbol{o}}(X \setminus \{\boldsymbol{x}\})$. A set $X \subset \mathbb{R}^n$ is a *positive basis* for $\mathbb{R}^n$ provided X is positively independent and $\operatorname{cone}_{\boldsymbol{o}} X = \mathbb{R}^n$. As proved by Steinitz [266, § 20] (also, by Gale [115] and Davis [85]), every positive basis X for $\mathbb{R}^n$ satisfies the inequalities $n+1 \leqslant \operatorname{card} X \leqslant 2n$. Here the equality $\operatorname{card} X = n+1$ holds if and only if X consists of the vectors of a basis for $\mathbb{R}^n$ together with a sum of negative multiples of these vectors, and the equality $\operatorname{card} X = 2n$ holds if and only if X consists of the vectors of a basis for $\mathbb{R}^n$ together with their negative multiples (see also Audet [7] and Hare, Song [148]). Reay [237] proved that a positive basis X for $\mathbb{R}^n$ admits a partition $X = X_1 \cup \cdots \cup X_r$ such that:

(a) $\operatorname{card} X_1 \geqslant \cdots \geqslant \operatorname{card} X_r \geqslant 2$, (*b*) $\text{pos}(X_1 \cup \cdots \cup X_i)$ is a subspace of dimension $\text{card}\, X_1 + \cdots + \text{card}\, X_i - i$ for all $1 \leqslant i \leqslant r$.

See McKinney [211] for an algebraic description of positive bases for $\mathbb{R}^n$, who also studied the notion of *strong positive independence* of a set $X \subset \mathbb{R}^n$, given by the condition $\text{pos}\, U \cap \text{pos}\, V = \{\boldsymbol{o}\}$ for all pairs of disjoint subsets U and V of X. Hansen and Klee [147], generalizing results from [211], showed that a set $X \subset \mathbb{R}^n$ is strongly positively independent if and only if $X = X_0 \cup X_1 \cup \cdots \cup X_r$ such that:

(*a*) X_0 is linearly independent,

(*b*) the subspaces $\text{span}\, X_1, \ldots, \text{span}\, X_r$ are independent,

(*c*) $\text{card}\, X_i = \dim(\text{span}\, X_i) + 1$ for all $1 \leqslant i \leqslant r$.

A set X is called a *strongly positive basis* for $\mathbb{R}^n$ if X is strongly positively independent and $\text{pos}\, X = \mathbb{R}^n$. If X is a strongly positive basis for $\mathbb{R}^n$ then $X = X_1 \cup \cdots \cup X_k$ such that conditions (*b*) and (*c*) above are satisfied (see McKinney [211]). Given a positive basis $\{\boldsymbol{c}_1, \ldots, \boldsymbol{c}_r\}$ for $\mathbb{R}^n$, a representation $\boldsymbol{x} = \alpha_1 \boldsymbol{c}_1 + \cdots + \alpha_r \boldsymbol{c}_r$ of a vector $\boldsymbol{x} \in \mathbb{R}^n$ as a positive combination of $\boldsymbol{c}_1, \ldots, \boldsymbol{c}_r$ is called *minimal* provided it involves the minimum possible number of nonzero coefficients $\alpha_1, \ldots, \alpha_r$. Reay [238], expanding results of Bonnice and Klee [40] and McKinney [211], showed that a positive basis $\{\boldsymbol{c}_1, \ldots, \boldsymbol{c}_r\}$ for $\mathbb{R}^n$ is strongly positive if and only if, for every vector $\boldsymbol{x} \in \mathbb{R}^n$ its minimal representation as a positive combination of $\{\boldsymbol{c}_1, \ldots, \boldsymbol{c}_r\}$ is unique.

Chapter 6

Recession Directions

6.1 Recession Cones

6.1.1 *Basic Properties of Recession Cones*

Definition 6.1. The *recession cone* of a nonempty set $X \subset \mathbb{R}^n$, denoted $\operatorname{rec} X$, is defined by

$$\operatorname{rec} X = \{\boldsymbol{e} \in \mathbb{R}^n : \lambda \boldsymbol{e} + \boldsymbol{x} \in X \text{ whenever } \lambda \geqslant 0 \text{ and } \boldsymbol{x} \in X\}.$$

We let $\operatorname{rec} X = \varnothing$ if $X = \varnothing$.

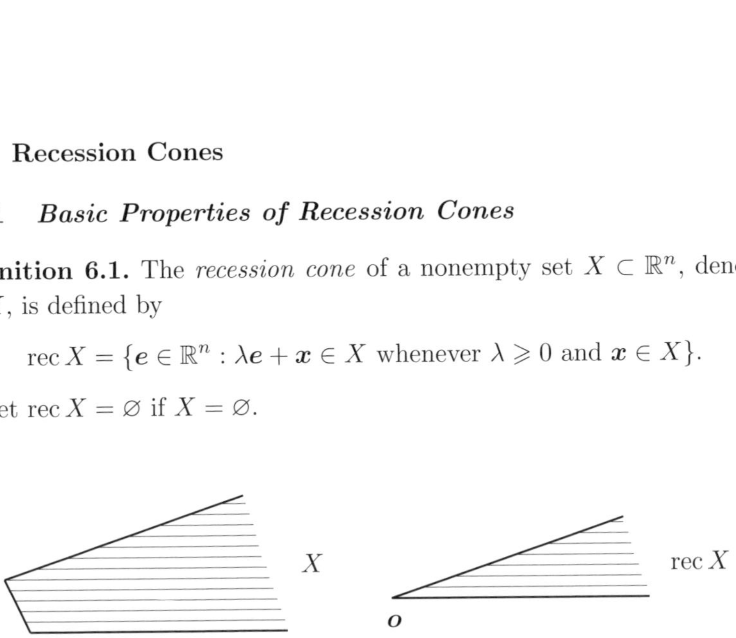

Fig. 6.1 The recession cone of a set X.

Given a vector $\boldsymbol{e} \in \mathbb{R}^n$, the scalar multiples $\lambda \boldsymbol{e}$, $\lambda \geqslant 0$, fulfil the closed halfline $[\boldsymbol{o}, \boldsymbol{e}\rangle$ if $\boldsymbol{e} \neq \boldsymbol{o}$, or the singleton $\{\boldsymbol{o}\}$ if $\boldsymbol{e} = \boldsymbol{o}$. This argument shows that the recession cone of a nonempty set $X \subset \mathbb{R}^n$ is either $\{\boldsymbol{o}\}$ or a union of closed halflines with endpoint $\boldsymbol{o}$.

Examples. 1. If $L \subset \mathbb{R}^n$ is a nonempty plane, then $\operatorname{rec} L$ is its direction space $\operatorname{dir} L$ (see Definition 2.3). Indeed, if $\boldsymbol{e} \in \operatorname{dir} L$ and $\lambda \geqslant 0$, then, for any point $\boldsymbol{x} \in L$, one has $\lambda \boldsymbol{e} + \boldsymbol{x} \in \operatorname{dir} L + \boldsymbol{x} = L$ according to Theorem 2.2. Thus, $\operatorname{dir} L \subset \operatorname{rec} L$. Conversely, if $\boldsymbol{e} \in \operatorname{rec} L$ and $\boldsymbol{x} \in L$, then $\boldsymbol{e} + \boldsymbol{x} \in L$, which gives $\boldsymbol{e} \in L - \boldsymbol{x} = \operatorname{dir} L$ by the same theorem. Hence $\operatorname{rec} L \subset \operatorname{dir} L$.

2. If $L \subset \mathbb{R}^n$ is a plane of positive dimension and D is a closed halfplane of L, expressed as $D = \{\boldsymbol{x} \in L : \boldsymbol{x}\cdot\boldsymbol{e} \leqslant \gamma\}$ for a suitable vector $\boldsymbol{e} \in \mathbb{R}^n \setminus \operatorname{ort} L$ and a scalar γ (see Theorem 2.70), then $\operatorname{rec} D$ is a translate of D. More exactly, $\operatorname{rec} D = G$, where G is the closed halfplane of the direction space $\operatorname{dir} L$ given by $G = \{\boldsymbol{x} \in \operatorname{dir} L : \boldsymbol{x}\cdot\boldsymbol{e} \leqslant 0\}$. Indeed, similarly to the example 1 above, if $\boldsymbol{c} \in G$ and $\lambda \geqslant 0$, then

$$\lambda\boldsymbol{c} + \boldsymbol{x} \in D + \boldsymbol{x} \subset \operatorname{dir} L + \boldsymbol{x} = L$$

for any point $\boldsymbol{x} \in D$. Furthermore, $(\lambda\boldsymbol{c}+\boldsymbol{x})\cdot\boldsymbol{e} \leqslant \boldsymbol{x}\cdot\boldsymbol{e} \leqslant \gamma$. Thus, $G \subset \operatorname{rec} D$. Conversely, if $\boldsymbol{c} \in \operatorname{rec} D$ and $\boldsymbol{x} \in D$, then $\boldsymbol{c} + \boldsymbol{x} \in D \subset L$, which gives $\boldsymbol{c} \in L - \boldsymbol{x} = \operatorname{dir} L$. Furthermore, the inequality $(\lambda\boldsymbol{c}+\boldsymbol{x})\cdot\boldsymbol{e} \leqslant \gamma$ holds for all $\lambda \geqslant 0$ only if $\boldsymbol{c}\cdot\boldsymbol{e} \leqslant 0$. Consequently, $\boldsymbol{c} \in G$.

Similarly, G is the recession cone of the open halfplane

$$E = \{\boldsymbol{x} \in L : \boldsymbol{x}\cdot\boldsymbol{e} < \gamma\}, \quad \boldsymbol{e} \in \mathbb{R}^n \setminus \operatorname{ort} L, \quad \gamma \in \mathbb{R}.$$

Theorem 6.2. *For a nonempty set $X \subset \mathbb{R}^n$, the following assertions hold.*

(*a*) $\operatorname{rec} X$ *is the largest set of vectors* $\mathbf{e} \in \mathbb{R}^n$ *satisfy the condition*

$$\lambda\mathbf{e} + X \subset X \quad \textit{for all} \quad \lambda \geqslant 0,$$

or, equivalently, the equality

$$\lambda\mathbf{e} + X = X \quad \textit{for all} \quad \lambda \geqslant 0.$$

(*b*) $\operatorname{rec} X$ *is the largest among all convex cones* $C \subset \mathbb{R}^n$ *with proper apex* $\boldsymbol{o}$ *which satisfy the inclusion* $C + X \subset X$*, or, equivalently, the equality* $C + X = X$.

(*c*) *If* X *is bounded, then* $\operatorname{rec} X = \{\boldsymbol{o}\}$.

(*d*) $\operatorname{rec} X \subset \cap\,(\lambda(X - \boldsymbol{u}) : \lambda > 0)$ *for any given point* $\boldsymbol{u} \in X$.

(*e*) $\operatorname{rec} X \subset \operatorname{dir} X$*, and* $\operatorname{rec} X = \operatorname{dir} X$ *if and only if* X *is a plane.*

Proof. (*a*) The inclusion $\lambda\boldsymbol{e}+X \subset X$ whenever $\lambda \geqslant 0$ immediately follows from Definition 6.19. Since $\boldsymbol{x} = \boldsymbol{o} + \boldsymbol{x}$ for every point $\boldsymbol{x} \in X$, and since $\boldsymbol{o} \in \operatorname{rec} X$, we obtain the opposite inclusion $X \subset \lambda\boldsymbol{e} + X$.

(*b*) First, we assert that $\operatorname{rec} X$ is a convex cone with apex $\boldsymbol{o}$. Indeed, choose any vectors $\boldsymbol{c}, \boldsymbol{e} \in \operatorname{rec} X$ and scalars $\lambda, \mu > 0$. Given a point $\boldsymbol{x} \in X$, we have $\boldsymbol{z} = \mu\boldsymbol{c} + \boldsymbol{x} \in X$. Consecutively,

$$(\lambda\boldsymbol{e} + \mu\boldsymbol{c}) + \boldsymbol{x} = \lambda\boldsymbol{e} + (\mu\boldsymbol{c} + \boldsymbol{x}) = \lambda\boldsymbol{e} + \boldsymbol{z} \in X,$$

which gives the inclusion $\lambda\boldsymbol{e}+\mu\boldsymbol{c} \in \operatorname{rec} X$. This argument and Theorem 5.5 prove that $\operatorname{rec} X$ is a convex cone with apex $\boldsymbol{o}$. Obviously, $\boldsymbol{o} \in \operatorname{rec} X$, implying that $\boldsymbol{o}$ is a proper apex of $\operatorname{rec} X$.

The inclusion $\operatorname{rec} X + X \subset X$ follows from Definition 6.1. If a cone $C \subset \mathbb{R}^n$ with proper apex $\boldsymbol{o}$ satisfies the condition $C + X \subset X$, then, for any vector $\boldsymbol{e} \in C$, a scalar $\lambda \geqslant 0$, and a point $\boldsymbol{x} \in X$, one has

$$\lambda \boldsymbol{e} + \boldsymbol{x} \in C + X \subset X,$$

implying that $\boldsymbol{e} \in \operatorname{rec} X$. Hence $C \subset \operatorname{rec} X$. Since $\boldsymbol{x} = \boldsymbol{o} + \boldsymbol{x}$ for every point $\boldsymbol{x} \in X$, and since $\boldsymbol{o} \in \operatorname{rec} X$, we obtain the opposite inclusion $X \subset \operatorname{rec} X + X$.

(*c*) Assume for a moment that $\operatorname{rec} X \neq \{\boldsymbol{o}\}$ and choose a nonzero vector $\boldsymbol{e} \in \operatorname{rec} X$. Given a point $\boldsymbol{u} \in X$, the closed halfline $h = \boldsymbol{u} + \{\lambda \boldsymbol{e} : \lambda \geqslant 0\}$ lies in X. Because h is an unbounded set, the set X also should be unbounded, in contradiction with the assumption. Summing up, $\operatorname{rec} X = \{\boldsymbol{o}\}$.

(*d*) If $\boldsymbol{u}$ is a given point in X, then, according to assertion (*b*) above, $\operatorname{rec} X$ is a convex cone with proper apex $\boldsymbol{o}$ such that $\boldsymbol{u} + \operatorname{rec} X \subset X$. Theorem 5.2 shows that $\lambda \operatorname{rec} X = \operatorname{rec} X$ for all $\lambda > 0$. Therefore, the inclusion $\operatorname{rec} X \subset X - \boldsymbol{u}$ gives

$$\operatorname{rec} X = \lambda \operatorname{rec} X \subset \lambda(X - \boldsymbol{u}), \quad \lambda > 0.$$

Hence $\operatorname{rec} X \subset \cap\,(\lambda(X - \boldsymbol{u}) : \lambda > 0)$.

(*e*) Let $\boldsymbol{e} \in \operatorname{rec} X$. Then $\boldsymbol{e} + \boldsymbol{x} \in X$ for any choice of $\boldsymbol{x} \in X$. According to Theorem 2.48, one has

$$\boldsymbol{e} \in X - \boldsymbol{x} \subset \operatorname{aff} X - \boldsymbol{x} = \operatorname{dir} X.$$

Consecutively, $\operatorname{rec} X \subset \operatorname{dir} X$. If X is a plane, then $\operatorname{rec} X = \operatorname{dir} X$ (see an example on page 207). Conversely, let $\operatorname{rec} X = \operatorname{dir} X$. Then $\operatorname{rec} X$ is a translate of $\operatorname{aff} X$ (see Theorem 2.48), and a combination of Theorem 2.2 and assertion (*b*) gives $X = \operatorname{rec} X + X = \operatorname{aff} X$, implying that X is a plane. □

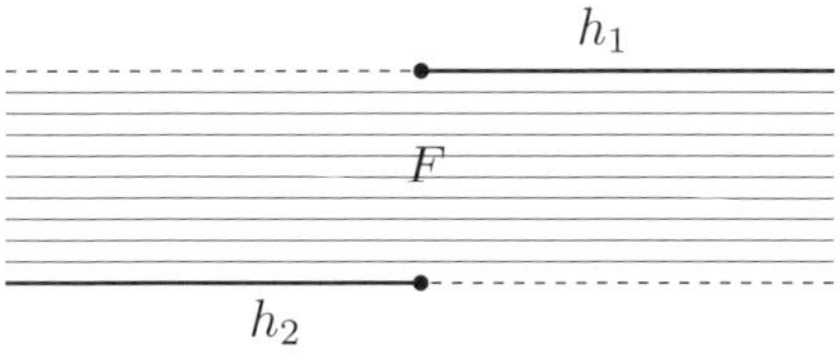

Fig. 6.2 Illustration to Theorem 6.2.

Remark. It might happen that $\operatorname{rec} X = \{\boldsymbol{o}\}$ for an unbounded set $X \subset \mathbb{R}^n$, even if the set X is convex. Furthermore, the inclusion $\operatorname{rec} X \subset \cap (\lambda(X - \boldsymbol{u}) : \lambda > 0)$ may be proper for any choice of $\boldsymbol{u} \in X$ (compare with Theorem 6.14). Indeed, let $X = F \cup h_1 \cup h_2$, where

$$F = \{(x, y) : 0 < y < 1\}, \quad h_1 = \{(x, 1) : x \geqslant 0\}, \quad h_2 = \{(x, 0) : x \leqslant 0\},$$

as depicted in Figure 6.2. Then $\operatorname{rec} X = \{\boldsymbol{o}\}$, while

$$\cap (\lambda(X - \boldsymbol{u}) : \lambda > 0) = \begin{cases} \{(x, 0) : x \geqslant 0\} & \text{if } \boldsymbol{u} \in h_1, \\ \{(x, 0) : x \in \mathbb{R}\} & \text{if } \boldsymbol{u} \in F, \\ \{(x, 0) : x \leqslant 0\} & \text{if } \boldsymbol{u} \in h_2. \end{cases}$$

Theorem 6.3. *If $C \subset \mathbb{R}^n$ is a nonempty cone with apex $\boldsymbol{a}$, then*

$$\operatorname{ap} C - \boldsymbol{a} \subset \operatorname{rec} C \subset (\operatorname{ap} C \cup C) - \boldsymbol{a}.$$

Furthermore, $\operatorname{rec} C = (\operatorname{ap} C \cup C) - \boldsymbol{a}$ *if and only if the cone C is convex.*

Proof. By Theorem 5.2, $C - \boldsymbol{a}$ is a cone with apex $\boldsymbol{o}$. By the same reason, $\operatorname{ap} C - \boldsymbol{a}$ is the apex set of $C - \boldsymbol{a}$. Theorem 5.16 shows that $\boldsymbol{u} + (\operatorname{ap} C - \boldsymbol{a}) \subset C$ for any point $\boldsymbol{u} \in C$. According to Theorem 5.14, $\operatorname{ap} C - \boldsymbol{a}$ is a subspace, which is a convex cone with proper apex $\boldsymbol{o}$. This argument and Theorem 6.2 imply the inclusion $\operatorname{ap} C - \boldsymbol{a} \subset \operatorname{rec} C$.

Since $\boldsymbol{o} = \boldsymbol{a} - \boldsymbol{a} \in (\operatorname{ap} C \cup C) - \boldsymbol{a}$, a combination of Corollary 5.19 and Theorem 6.2 gives

$$\begin{aligned} \operatorname{rec} C &= \operatorname{rec} C + \boldsymbol{o} \subset \operatorname{rec} C + ((\operatorname{ap} C \cup C) - \boldsymbol{a}) \\ &= (\operatorname{rec} (\operatorname{ap} C \cup C) + (\operatorname{ap} C \cup C)) - \boldsymbol{a} \qquad (6.1) \\ &\subset (\operatorname{ap} C \cup C) - \boldsymbol{a}. \end{aligned}$$

If, additionally, C is convex, then Corollary 5.19 implies that the cone $\operatorname{ap} C \cup C$ is convex, and Theorem 5.5 gives

$$\begin{aligned} ((\operatorname{ap} C \cup C) - \boldsymbol{a}) + (\operatorname{ap} C \cup C) &= (\operatorname{ap} C \cup C) + (\operatorname{ap} C \cup C) - \boldsymbol{a} \\ &= (\operatorname{ap} C \cup C) - \boldsymbol{a}. \end{aligned}$$

This argument, inclusion (6.1), and Theorem 6.2 how that $\operatorname{rec} C = (\operatorname{ap} C \cup C) - \boldsymbol{a}$. Conversely, if $\operatorname{rec} C = (\operatorname{ap} C \cup C) - \boldsymbol{a}$, then the cone $(\operatorname{ap} C \cup C)$ is convex as a translate of the convex cone $\operatorname{rec} C$ (see Theorem 6.2). Finally, the convexity of C follows from Theorem 5.18. □

Sometimes it is convenient to consider the *recession cone* of a nonempty set $X \subset \mathbb{R}^n$ at *a given point* $\boldsymbol{a} \in X$, denoted $\operatorname{rec}_{\boldsymbol{a}} X$ and defined by

$$\operatorname{rec}_{\boldsymbol{a}} X = \{\boldsymbol{e} \in \mathbb{R}^n : \lambda \boldsymbol{e} + \boldsymbol{a} \in X \text{ whenever } \lambda \geqslant 0\}.$$

Clearly, $\operatorname{rec} X = \cap\,(\operatorname{rec}_{\boldsymbol{a}} X : \boldsymbol{a} \in X)$. Obvious modifications in the proof of Theorem 6.2 show the following properties of $\operatorname{rec}_{\boldsymbol{a}} X$.

1. $\operatorname{rec}_{\boldsymbol{a}} X$ is the largest among all convex cones $C \subset \mathbb{R}^n$ with proper apex $\boldsymbol{o}$ which satisfy the inclusion $C + \boldsymbol{a} \subset X$.
2. $\operatorname{rec}_{\boldsymbol{a}} X = \cap\,(\lambda(X - \boldsymbol{a}) : \lambda > 0)$.

6.1.2 *Algebra of Sets and Recession Cones*

Theorem 6.4. *If a family $\mathcal{F} = \{X_\alpha\}$ of sets in $\mathbb{R}^n$ satisfies the condition $\underset{\alpha}{\cap} X_\alpha \neq \varnothing$, then*

$$\underset{\alpha}{\cap} \operatorname{rec} X_\alpha \subset \operatorname{rec}\,(\underset{\alpha}{\cap} X_\alpha) \quad \textit{and} \quad \underset{\alpha}{\cap} \operatorname{rec} X_\alpha \subset \operatorname{rec}\,(\underset{\alpha}{\cup} X_\alpha). \tag{6.2}$$

Proof. Choose a vector $\boldsymbol{e} \in \underset{\alpha}{\cap} \operatorname{rec} X_\alpha$. Then $\lambda \boldsymbol{e} + \boldsymbol{x} \in X_\alpha$ whenever $\lambda \geqslant 0$ and $\boldsymbol{x} \in X_\alpha$. Hence $\lambda \boldsymbol{e} + \boldsymbol{x} \in \underset{\alpha}{\cap} X_\alpha$ for all $\lambda \geqslant 0$ and $\boldsymbol{x} \in \underset{\alpha}{\cap} X_\alpha$, implying that $\boldsymbol{e} \in \operatorname{rec}\,(\underset{\alpha}{\cap} X_\alpha)$. Thus $\underset{\alpha}{\cap} \operatorname{rec} X_\alpha \subset \operatorname{rec}\,(\underset{\alpha}{\cap} X_\alpha)$.

Similarly, let $\boldsymbol{e} \in \underset{\alpha}{\cap} \operatorname{rec} X_\alpha$ and $\boldsymbol{x} \in \underset{\alpha}{\cup} X_\alpha$. Then $\boldsymbol{x} \in X_\beta$ for a suitable set $X_\beta \in \mathcal{F}$. Because $\boldsymbol{e} \in \operatorname{rec} X_\beta$, one has $\lambda \boldsymbol{e} + \boldsymbol{x} \in X_\beta \subset \underset{\alpha}{\cup} X_\alpha$ for all $\lambda \geqslant 0$. Consequently, $\boldsymbol{e} \in \operatorname{rec}\,(\underset{\alpha}{\cup} X_\alpha)$, and $\underset{\alpha}{\cap} \operatorname{rec} X_\alpha \subset \operatorname{rec}\,(\underset{\alpha}{\cup} X_\alpha)$. $\square$

Remark. The inclusions in Theorem 6.4 may be proper, even if the sets X_α are convex. Indeed, let

$$X_1 = \{(0,1)\} \cup \{(x,y) : -1 < y < 1\} \quad \text{and} \quad X_2 = \{(x,0) : x \in \mathbb{R}\}.$$

Then $\operatorname{rec} X_1 = \{\boldsymbol{o}\}$ and $\operatorname{rec} X_2 = X_2$. Furthermore,

$$\operatorname{rec} X_1 \cap \operatorname{rec} X_2 = \{\boldsymbol{o}\} \neq \operatorname{rec}\,(X_1 \cap X_2) = \operatorname{rec} X_2 = X_2.$$

Theorem 6.5. *Given a set $X \subset \mathbb{R}^n$, a point $\boldsymbol{a} \in \mathbb{R}^n$, and a scalar μ, one has*

$$\operatorname{rec}\,(\boldsymbol{a} + \mu X) = \mu \operatorname{rec} X = \operatorname{sgn}(\mu) \operatorname{rec} X. \tag{6.3}$$

Proof. Since both cases $X = \varnothing$ and $\mu = 0$ are obvious, we may assume that X is nonempty and $\mu \neq 0$. Clearly, $\boldsymbol{e} \in \operatorname{rec}\,(\boldsymbol{a} + \mu X)$ if and only if

$$\lambda \boldsymbol{e} + (\boldsymbol{a} + \mu \boldsymbol{x}) \in \boldsymbol{a} + \mu X \quad \text{whenever} \quad \lambda \geqslant 0 \quad \text{and} \quad \boldsymbol{x} \in X.$$

Equivalently, $\lambda(\boldsymbol{e}/\mu) + \boldsymbol{x} \in X$ for all $\lambda \geqslant 0$ and $\boldsymbol{x} \in X$. In other words, $\boldsymbol{e}/\mu \in \operatorname{rec} X$, or $\boldsymbol{e} \in \mu \operatorname{rec} X$. Summing up,

$$\operatorname{rec}(\boldsymbol{a} + \mu X) = \mu \operatorname{rec} X.$$

By Theorem 6.2, $\operatorname{rec} X$ is a convex cone with apex $\boldsymbol{o}$. Therefore, Theorem 5.2 gives

$$\mu \operatorname{rec} X = \left\{ \begin{array}{cl} \operatorname{rec} X & \text{if } \mu > 0 \\ \{\boldsymbol{o}\} & \text{if } \mu = 0 \\ -\operatorname{rec} X & \text{if } \mu < 0 \end{array} \right\} = \operatorname{sgn}(\mu) \operatorname{rec} X. \qquad \square$$

Theorem 6.6. *For sets* $X_1, \ldots, X_r \subset \mathbb{R}^n$ *and scalars* $\mu_1, \ldots, \mu_r$, *one has*

$$\mu_1 \operatorname{rec} X_1 + \cdots + \mu_r \operatorname{rec} X_r \subset \operatorname{rec}(\mu_1 X_1 + \cdots + \mu_r X_r). \tag{6.4}$$

If, additionally, all sets $X_1, \ldots, X_r$ *are nonempty and the planes* $\operatorname{aff} X_1, \ldots, \operatorname{aff} X_r$ *are independent, then*

$$\mu_1 \operatorname{rec} X_1 + \cdots + \mu_r \operatorname{rec} X_r = \operatorname{rec}(\mu_1 X_1 + \cdots + \mu_r X_r). \tag{6.5}$$

Proof. Since both assertion of the theorem are obvious when at least one of the sets $X_1, \ldots, X_r$ is empty, we may suppose that all of them are nonempty. Furthermore, the case $r = 1$ follows from Theorem 6.5, and an induction argument shows that the proof can be reduced to the case $r = 2$ and $\mu_1 \mu_2 \neq 0$.

We start with (6.4). Choose points $\boldsymbol{e}_1 \in \mu_1 \operatorname{rec} X_1$ and $\boldsymbol{e}_2 \in \mu_2 \operatorname{rec} X_2$. Let $\boldsymbol{x}$ be a point in $\mu_1 X_1 + \mu_2 X_2$. Then $\boldsymbol{x}$ can be expressed as $\boldsymbol{x} = \mu_1 \boldsymbol{x}_1 + \mu_2 \boldsymbol{x}_2$ for suitable points $\boldsymbol{x}_1 \in X_1$ and $\boldsymbol{x}_2 \in X_2$. For a scalar $\lambda \geqslant 0$, one has

$$\lambda(\mu_1 \boldsymbol{e}_1 + \mu_2 \boldsymbol{e}_2) + \boldsymbol{x} = \mu_1(\lambda \boldsymbol{e}_1 + \boldsymbol{x}_1) + \mu_2(\lambda \boldsymbol{e}_2 + \boldsymbol{x}_2) \in \mu_1 X_1 + \mu_2 X_2,$$

which implies the inclusion $\mu_1 \boldsymbol{e}_1 + \mu_2 \boldsymbol{e}_2 \in \operatorname{rec}(\mu_1 X_1 + \mu_2 X_2)$. Hence

$$\mu_1 \operatorname{rec} X_1 + \mu_2 \operatorname{rec} X_2 \subset \operatorname{rec}(\mu_1 X_1 + \mu_2 X_2).$$

Assume now that the planes $\operatorname{aff} X_1$ and $\operatorname{aff} X_2$ are independent. For (6.5), it suffices to prove, due to the above argument, that

$$\operatorname{rec}(\mu_1 X_1 + \mu_2 X_2) \subset \mu_1 \operatorname{rec} X_1 + \mu_2 \operatorname{rec} X_2. \tag{6.6}$$

Choose points $\boldsymbol{u}_1 \in X_1$ and $\boldsymbol{u}_2 \in X_2$, and put

$$X_1' = X_1 - \boldsymbol{u}_1 \quad \text{and} \quad X_2' = X_2 - \boldsymbol{u}_2.$$

Then $\boldsymbol{o}_1 \in X_1' \cap X_2'$, which implies that

$$S_1 = \operatorname{span} X_1' = \operatorname{aff} X_1 - \boldsymbol{u}_1 \quad \text{and} \quad S_2 = \operatorname{span} X_2' = \operatorname{aff} X_2 - \boldsymbol{u}_2$$

are independent subspaces. Furthermore, Theorem 6.5 shows that

$$\operatorname{rec} X_1' = \operatorname{rec} X_1 \subset S_1 \quad \text{and} \quad \operatorname{rec} X_2' = \operatorname{rec} X_2 \subset S_2.$$

By the same theorem,

$$\begin{aligned}\operatorname{rec}(\mu_1 X_1' + \mu_2 X_2') &= \operatorname{rec}(\mu_1 X_1 + \mu_2 X_2 - (\mu_1 \boldsymbol{u}_1 + \mu_2 \boldsymbol{u}_2)) \\ &= \operatorname{rec}(\mu_1 X_1 + \mu_2 X_2).\end{aligned}$$

We assert that

$$\operatorname{rec}(\mu_1 X_1' + \mu_2 X_2') \subset \mu_1 \operatorname{rec} X_1' + \mu_2 \operatorname{rec} X_2'.$$

Indeed, let $\boldsymbol{e} \in \operatorname{rec}(\mu_1 X_1' + \mu_2 X_2')$. By Theorem 6.2,

$$\begin{aligned}\boldsymbol{e} &\in \operatorname{dir}(\mu_1 X_1' + \mu_2 X_2') = \operatorname{span}(\mu_1 X_1' + \mu_2 X_2') \\ &= \operatorname{span}(\mu_1 X_1') + \operatorname{span}(\mu_2 X_2') = \operatorname{span} X_1' + \operatorname{span} X_2' = S_1 + S_2.\end{aligned}$$

Hence $\boldsymbol{e} = \boldsymbol{e}_1 + \boldsymbol{e}_2$ for suitable vectors $\boldsymbol{e}_1 \in S_1$ and $\boldsymbol{e}_2 \in S_2$.

Choose points $\boldsymbol{x}_1 \in X_1'$ and $\boldsymbol{x}_2 \in X_2'$. For all $\lambda \geqslant 0$,

$$\begin{aligned}&(\lambda \boldsymbol{e}_1 + \mu_1 \boldsymbol{x}_1) + (\lambda \boldsymbol{e}_2 + \mu_2 \boldsymbol{x}_2) \\ &\qquad = \lambda \boldsymbol{e} + (\mu_1 \boldsymbol{x}_1 + \mu_2 \boldsymbol{x}_2) \in \mu_1 X_1' + \mu_2 X_2'.\end{aligned}$$

Since the sum $\mu_1 X_1' + \mu_2 X_2'$ is direct (see page 5), the inclusions

$$\lambda \boldsymbol{e}_1 + \mu_1 \boldsymbol{x}_1 \in S_1 \quad \text{and} \quad \lambda \boldsymbol{e}_2 + \mu_2 \boldsymbol{x}_2 \in S_2$$

show that

$$\lambda \boldsymbol{e}_1 + \mu_1 \boldsymbol{x}_1 \in \mu_1 X_1' \quad \text{and} \quad \lambda \boldsymbol{e}_2 + \mu_2 \boldsymbol{x}_2 \in \mu_2 X_2'$$

for all $\lambda \geqslant 0$. Therefore (see Theorem 6.5),

$$\boldsymbol{e}_1 \in \operatorname{rec}(\mu_1 X_1') = \mu_1 \operatorname{rec} X_1' \quad \text{and} \quad \boldsymbol{e}_2 \in \operatorname{rec}(\mu_2 X_2') = \mu_2 \operatorname{rec} X_2',$$

implying the inclusion

$$\boldsymbol{e} = \boldsymbol{e}_1 + \boldsymbol{e}_2 \in \mu_1 \operatorname{rec} X_1' + \mu_2 \operatorname{rec} X_2'.$$

By the same theorem,

$$\begin{aligned}\operatorname{rec}(\mu_1 X_1 + \mu_2 X_2) &= \operatorname{rec}(\mu_1 X_1' + \mu_2 X_2') \\ &\subset \mu_1 \operatorname{rec} X_1' + \mu_2 \operatorname{rec} X_2' = \mu_1 \operatorname{rec} X_1 + \mu_2 \operatorname{rec} X_2.\end{aligned}$$

□

Remark. Inclusion (6.4) may be proper. Indeed, let

$$X_1 = \{(x, y) : y \geqslant x^2\} \quad \text{and} \quad X_2 = \{(x, y) : y \leqslant -x^2\}.$$

Then $\operatorname{rec} X_1$ and $\operatorname{rec} X_2$ are opposite closed halflines, given by

$$\operatorname{rec} X_1 = \{(0, y) : y \geqslant 0\} \quad \text{and} \quad \operatorname{rec} X_2 = \{(0, y) : y \leqslant 0\}.$$

Therefore, $\operatorname{rec} X_1 + \operatorname{rec} X_2 = \{(0, y) : y \in \mathbb{R}\}$. On the other hand, $X_1 + X_2 = \mathbb{R}^2$, implying that $\operatorname{rec}(X_1 + X_2) = \mathbb{R}^2$.

Theorem 6.7. *For a linear transformation $g : \mathbb{R}^n \to \mathbb{R}^m$ and sets $X \subset \mathbb{R}^n$ and $Y \subset \mathbb{R}^m$, one has*

$$g(\operatorname{rec} X) \subset \operatorname{rec} g(X) \quad \text{and} \quad \operatorname{rec} g^{-1}(Y) = g^{-1}(\operatorname{rec}(Y \cap \operatorname{rng} f)).$$

Furthermore, if g is one-to-one, then $g(\operatorname{rec} X) = \operatorname{rec} g(X)$.

Proof. Excluding the obvious cases $X = \varnothing$ and $Y \cap \operatorname{rng} g = \varnothing$, we assume that both sets X and $Y \cap \operatorname{rng} g$ are nonempty. Consequently, $g^{-1}(Y) \neq \varnothing$.

First, we will prove the inclusion $g(\operatorname{rec} X) \subset \operatorname{rec} g(X)$. Let $\boldsymbol{e} \in g(\operatorname{rec} X)$, $\boldsymbol{x} \in g(X)$, and $\lambda \geqslant 0$. Choose points $\boldsymbol{e}' \in \operatorname{rec} X$ and $\boldsymbol{x}' \in X$ such that $g(\boldsymbol{e}') = \boldsymbol{e}$ and $g(\boldsymbol{x}') = \boldsymbol{x}$. Then $\lambda\boldsymbol{e}' + \boldsymbol{x}' \in X$, which gives the inclusion

$$\lambda\boldsymbol{e} + \boldsymbol{x} = g(\lambda\boldsymbol{e}' + \boldsymbol{x}') \in g(X) \quad \text{for all} \quad \lambda \geqslant 0.$$

Hence $\boldsymbol{e} \in \operatorname{rec} g(X)$, and $g(\operatorname{rec} X) \subset \operatorname{rec} g(X)$.

With $X = g^{-1}(Y)$, we obtain, by the above proved,

$$g(\operatorname{rec} g^{-1}(Y)) \subset \operatorname{rec} g(g^{-1}(Y)) = \operatorname{rec}(Y \cap \operatorname{rng} g).$$

Thus

$$\operatorname{rec} g^{-1}(Y) \subset g^{-1}(\operatorname{rec}(Y \cap \operatorname{rng} g)).$$

Conversely, let $\boldsymbol{e} \in g^{-1}(\operatorname{rec}(Y \cap \operatorname{rng} g))$. Choose a point $\boldsymbol{x} \in g^{-1}(Y)$ and a scalar $\lambda \geqslant 0$. Let $\boldsymbol{e}' = g(\boldsymbol{e})$ and $\boldsymbol{x}' = g(\boldsymbol{x})$. Then $\boldsymbol{e}' \in \operatorname{rec}(Y \cap \operatorname{rng} g)$ and $\boldsymbol{x}' \in Y \cap \operatorname{rng} g$, which gives

$$\lambda\boldsymbol{e}' + \boldsymbol{x}' \in Y \cap \operatorname{rng} g \quad \text{for all} \quad \lambda \geqslant 0.$$

Therefore,

$$\lambda\boldsymbol{e} + \boldsymbol{x} \in g^{-1}(\lambda\boldsymbol{e}' + \boldsymbol{x}') \in g^{-1}(Y) \quad \text{for all} \quad \lambda \geqslant 0.$$

So, $\boldsymbol{e} \in \operatorname{rec} g^{-1}(Y)$, and

$$g^{-1}(\operatorname{rec}(Y \cap \operatorname{rng} g)) \subset \operatorname{rec} g^{-1}(Y).$$

Finally, we assert that $g(\operatorname{rec} X) = \operatorname{rec} g(X)$ provided g is one-to-one. By the above proved, it suffices to show the inclusion $\operatorname{rec} g(X) \subset g(\operatorname{rec} X)$. Choose points $\boldsymbol{e} \in \operatorname{rec} g(X)$ and $\boldsymbol{x} \in g(X)$. Then $\boldsymbol{e} + \boldsymbol{x} \in g(X)$. Let $\boldsymbol{u}, \boldsymbol{v} \in X$ be such that $g(\boldsymbol{u}) = \boldsymbol{x}$ and $g(\boldsymbol{v}) = \boldsymbol{e} + \boldsymbol{x}$. Then $\boldsymbol{e} = g(\boldsymbol{v}) - \boldsymbol{x} = g(\boldsymbol{v} - \boldsymbol{u})$. We assert that the vector $\boldsymbol{c} = \boldsymbol{v} - \boldsymbol{u}$ belongs to $\operatorname{rec} X$. Indeed, for $\boldsymbol{y} \in X$ and $\lambda \geqslant 0$, we have

$$g(\lambda\boldsymbol{c} + \boldsymbol{y}) = \lambda g(\boldsymbol{c}) + g(\boldsymbol{y}) = \lambda\boldsymbol{e} + g(\boldsymbol{y}) \in g(X).$$

Thus $\lambda\boldsymbol{c} + \boldsymbol{y} \in X$ because g is one-to-one. This argument gives $\boldsymbol{c} \in \operatorname{rec} X$, implying that $\boldsymbol{e} = g(\boldsymbol{c}) \in g(\operatorname{rec} X)$. □

Remark. Inclusion $g(\operatorname{rec} X) \subset \operatorname{rec} g(X)$ in Theorem 6.7 may be proper if g is not one-to-one. Indeed, let $X = \{(x, y) : y \geqslant x^2\}$ and g be the orthogonal projection of $\mathbb{R}^2$ on the x-axis. Clearly,

$$\operatorname{rec} X = \{(0, y) : y \geqslant 0\} \quad \text{and} \quad g(\operatorname{rec} X) = \{\boldsymbol{o}\}.$$

On the other hand, $g(X) = \{(x, 0) : x \in \mathbb{R}\}$ and $\operatorname{rec} g(X) = g(X)$.

6.1.3 *Topology and Recession Cones*

Theorem 6.8. *For a nonempty set $X \subset \mathbb{R}^n$, the following assertions hold.*

(*a*) *If* $\boldsymbol{x} \in \operatorname{cl} X$, *then* $\boldsymbol{x} + \operatorname{rec} X \subset \operatorname{cl} X$.
(*b*) *If* $\operatorname{rint} X \neq \varnothing$ *and* $\boldsymbol{x} \in \operatorname{rint} X$, *then* $\boldsymbol{x} + \operatorname{rec} X \subset \operatorname{rint} X$.

Proof. (*a*) Let $\boldsymbol{x} \in \operatorname{cl} X$ and $\boldsymbol{e} \in \operatorname{rec} X$. Express $\boldsymbol{x}$ as the limit of a sequence $\boldsymbol{x}_1, \boldsymbol{x}_2, \ldots$ of points from X. Then $\boldsymbol{x}_i + \boldsymbol{e} \in X$, $i \geqslant 1$, by the definition of $\operatorname{rec} X$, which results in

$$\boldsymbol{x} + \boldsymbol{e} = \lim_{i \to \infty} (\boldsymbol{x}_i + \boldsymbol{e}) \in \operatorname{cl} X.$$

Summing up, $\boldsymbol{x} + \operatorname{rec} X \subset \operatorname{cl} X$.

(*b*) Suppose that $\operatorname{rint} X \neq \varnothing$. Let $\boldsymbol{x} \in \operatorname{rint} X$ and $\boldsymbol{e} \in \operatorname{rec} X$. Put $\boldsymbol{z} = \boldsymbol{x} + \boldsymbol{e}$. Clearly, $\boldsymbol{z} \in X$. By Definition 3.16, there is a scalar $\rho > 0$ such that $\operatorname{aff} X \cap U_\rho(\boldsymbol{x}) \subset X$. Choose any point $\boldsymbol{z}' \in \operatorname{aff} X \cap U_\rho(\boldsymbol{z})$ and let $\boldsymbol{x}' = \boldsymbol{z}' - \boldsymbol{e}$. Then

$$\|\boldsymbol{x} - \boldsymbol{x}'\| = \|(\boldsymbol{x} + \boldsymbol{e}) - (\boldsymbol{e} + \boldsymbol{x}')\| = \|\boldsymbol{z} - \boldsymbol{z}'\| < \rho,$$

which gives the inclusion $\boldsymbol{x}' \in U_\rho(\boldsymbol{x})$. By Theorem 6.2, $\boldsymbol{e} \in \operatorname{dir} X$. Consequently,

$$\boldsymbol{x}' = \boldsymbol{z}' - \boldsymbol{e} \in \operatorname{aff} X - \boldsymbol{e} = \operatorname{aff} X,$$

according to Theorem 2.2. Hence $\boldsymbol{x}' \in \operatorname{aff} X \cap U_\rho(\boldsymbol{x}) \subset X$. Consequently, $\boldsymbol{z}' = \boldsymbol{x}' + \boldsymbol{e} \in X$. Thus $\operatorname{aff} X \cap U_\rho(\boldsymbol{z}) \subset X$, implying the inclusion $\boldsymbol{z} \in \operatorname{rint} X$. Summing up, $\boldsymbol{x} + \operatorname{rec} X \subset \operatorname{rint} X$. □

Theorem 6.9. *For any set $X \subset \mathbb{R}^n$, both cones* $\operatorname{rec}(\operatorname{cl} X)$ *and* $\operatorname{rec}(\operatorname{rint} X)$ *are closed. Furthermore,*

$$\operatorname{cl}(\operatorname{rec} X) \subset \operatorname{rec}(\operatorname{cl} X), \tag{6.7}$$

$$\operatorname{cl}(\operatorname{rec} X) \subset \operatorname{rec}(\operatorname{rint} X) \quad \textit{if} \quad \operatorname{rint} X \neq \varnothing. \tag{6.8}$$

Proof. Since both assertions are obvious when $X = \varnothing$ or $\operatorname{rint} X = \varnothing$, we may assume that both sets X and $\operatorname{rint} X$ are nonempty.

First, we are going to prove that the cone $\operatorname{rec}(\operatorname{cl} X)$ is closed. Indeed, suppose that a vector $\boldsymbol{e} \in \mathbb{R}^n$ is the limit of a sequence of vectors $\boldsymbol{e}_1, \boldsymbol{e}_2, \ldots$ from $\operatorname{rec}(\operatorname{cl} X)$. Since $\lambda \boldsymbol{e}_i + \boldsymbol{x} \in \operatorname{cl} X$ whenever $\boldsymbol{x} \in \operatorname{cl} X$, $\lambda \geqslant 0$, and $i \geqslant 1$, we obtain

$$\lambda \boldsymbol{e} + \boldsymbol{x} = \lim_{i \to \infty} (\lambda \boldsymbol{e}_i + \boldsymbol{x}) \in \operatorname{cl} X \quad \text{for all} \quad \lambda \geqslant 0.$$

Consequently, $\boldsymbol{e} \in \operatorname{rec}(\operatorname{cl} X)$, which shows the closedness of $\operatorname{rec}(\operatorname{cl} X)$.

Next, we will prove inclusion (6.7). Indeed, choose a vector $\boldsymbol{e} \in \operatorname{rec} X$, a point $\boldsymbol{x} \in \operatorname{cl} X$, and a scalar $\lambda \geqslant 0$. Expressing $\boldsymbol{x}$ as the limit of a sequence of points $\boldsymbol{x}_1, \boldsymbol{x}_2, \ldots$ from X, we have $\lambda \boldsymbol{e} + \boldsymbol{x}_i \in X$, $i \geqslant 1$. Therefore,

$$\lambda \boldsymbol{e} + \boldsymbol{x} = \lim_{i \to \infty} (\lambda \boldsymbol{e} + \boldsymbol{x}_i) \in \operatorname{cl} X.$$

Hence $\boldsymbol{e} \in \operatorname{rec}(\operatorname{cl} X)$, and $\operatorname{rec} X \subset \operatorname{rec}(\operatorname{cl} X)$. Finally, (6.7) follows from the closedness of $\operatorname{rec}(\operatorname{cl} X)$.

Now, we will show that the cone $\operatorname{rec}(\operatorname{rint} X)$ is closed. Indeed, suppose that a vector $\boldsymbol{e} \in \mathbb{R}^n$ is the limit of a sequence of vectors $\boldsymbol{e}_1, \boldsymbol{e}_2, \ldots$ from $\operatorname{rec}(\operatorname{rint} X)$. Choose a point $\boldsymbol{x} \in \operatorname{rint} X$, and a scalar $\rho > 0$ satisfying the condition $\operatorname{aff} X \cap U_\rho(\boldsymbol{x}) \subset X$. Given a scalar $\lambda \geqslant 0$, let

$$\boldsymbol{z} = \lambda \boldsymbol{e} + \boldsymbol{x} \quad \text{and} \quad \boldsymbol{z}_i = \lambda \boldsymbol{e}_i + \boldsymbol{x}, \quad i \geqslant 1.$$

Denote by k a positive integer such that

$$\|\boldsymbol{z}_k - \boldsymbol{z}\| = \|\lambda \boldsymbol{e}_k - \lambda \boldsymbol{e}\| < \rho/2.$$

Problem 1.8 gives the inclusion $U_{\rho/2}(\boldsymbol{z}) \subset U_\rho(\boldsymbol{z}_k)$. Furthermore, $\boldsymbol{e} = \lim_{i \to \infty} \boldsymbol{e}_i \in \operatorname{dir} X$ due to Theorem 6.2. Consecutively (see Theorem 2.48),

$$\boldsymbol{z} = \lambda \boldsymbol{e} + \boldsymbol{x} \in \operatorname{dir} X + \operatorname{aff} X = \operatorname{aff} X.$$

Since $\operatorname{aff} X \cap U_\rho(\boldsymbol{z}_k) \subset X$ (see the proof of Theorem 6.8), we obtain

$$\operatorname{aff} X \cap U_{\rho/2}(\boldsymbol{z}) \subset \operatorname{aff} X \cap U_\rho(\boldsymbol{z}_k) \subset X,$$

implying the inclusion $\boldsymbol{z} \in \operatorname{rint} X$. Summing up, $\boldsymbol{e} \in \operatorname{rec}(\operatorname{rint} X)$, which shows the closedness of $\operatorname{rec}(\operatorname{rint} X)$. Now, Theorem 6.8 gives

$$\operatorname{rec} X \subset \operatorname{rec}(\operatorname{cl} X) \quad \text{and} \quad \operatorname{rec} X \subset \operatorname{rec}(\operatorname{rint} X).$$

Finally, (6.8) immediately follow from the above argument. □

Remark. Inclusions (6.7) and (6.8) may be proper. Furthermore, the cones $\operatorname{rec}(\operatorname{cl} X)$ and $\operatorname{rec}(\operatorname{rint} X)$ may be incomparable by inclusion (according to Theorem 6.14, these cones coincide if the set X is convex). Indeed, let the set $X \subset \mathbb{R}^2$ be given by

$$X = \{(x, y) : 0 < y < 1\} \cup \{(x, y) : \text{both } x \text{ and } y \text{ are rational}\}.$$

Then $\operatorname{rec} X = \{\boldsymbol{o}\}$, $\operatorname{rec}(\operatorname{rint} X)$ is the x-axis of $\mathbb{R}^2$, and $\operatorname{rec}(\operatorname{cl} X) = \mathbb{R}^2$. Similarly, if

$$X = \{(x, y) : 0 < y < 1\} \cup \{(0, y) : y > 0\},$$

then $\operatorname{rec} X = \operatorname{rec}(\operatorname{cl} X) = \{\boldsymbol{o}\}$, while $\operatorname{rec}(\operatorname{rint} X)$ is the x-axis of $\mathbb{R}^2$.

6.1.4 *Unbounded Convex Sets and Halflines*

Theorem 6.10. *Let the closure of a convex set $K \subset \mathbb{R}^n$ contain a closed halfline h with endpoint $\boldsymbol{x}$. Then the following assertions hold.*

(*a*) *If $\boldsymbol{x} \in \operatorname{rint} K$, then $h \subset \operatorname{rint} K$.*
(*b*) *If $\boldsymbol{x} \in \operatorname{rbd} K$, then either $h \subset \operatorname{rbd} K$ or $h \setminus \{\boldsymbol{x}\} \subset \operatorname{rint} K$.*
(*c*) *For points $\boldsymbol{u} \in \operatorname{cl} K$ and $\boldsymbol{v} \in \operatorname{rint} K$, the closed halflines*

$$h' = (\boldsymbol{u} - \boldsymbol{x}) + h \quad \text{and} \quad h'' = (\boldsymbol{v} - \boldsymbol{x}) + h$$

lie in $\operatorname{cl} K$ *and* $\operatorname{rint} K$, *respectively.*

Proof. (*a*) Let $\boldsymbol{x} \in \operatorname{rint} K$. Assume for a moment that $h \not\subset \operatorname{rint} K$. By Theorem 3.65, there is a unique point $\boldsymbol{u} \in h \cap \operatorname{rbd} K$ with the property $h \setminus [\boldsymbol{x}, \boldsymbol{u}] \subset \mathbb{R}^n \setminus \operatorname{cl} K$, in contradiction with the assumption $h \subset \operatorname{cl} K$.

(*b*) Let $\boldsymbol{x} \in \operatorname{rbd} K$. Suppose that $h \not\subset \operatorname{rbd} K$, and choose a point $\boldsymbol{y} \in h \cap \operatorname{rint} K$. Clearly, the closed halfline $g = (\boldsymbol{y} - \boldsymbol{x}) + h$ lies in h and has $\boldsymbol{y}$ as endpoint. Furthermore, $g \subset \operatorname{rint} K$ by the above proved. Since $(\boldsymbol{x}, \boldsymbol{y}) \subset \operatorname{rint} K$ (see Theorem 3.47), one has $h \setminus \{\boldsymbol{x}\} = (\boldsymbol{x}, \boldsymbol{y}) \cup g \subset \operatorname{rint} K$.

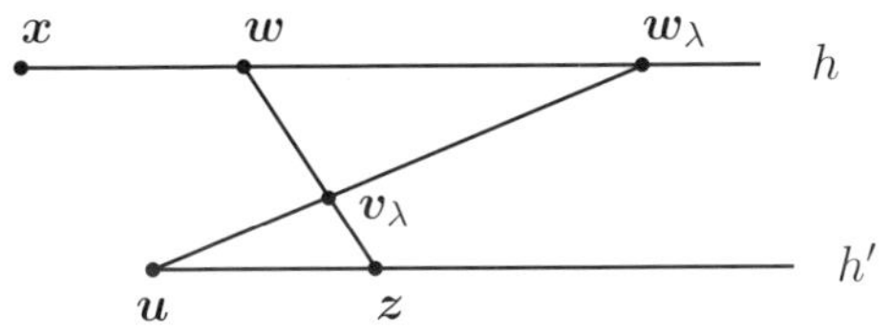

Fig. 6.3 Illustration to Theorem 6.10.

(*c*) We first observe that $\boldsymbol{u}$ is the endpoint of h'. To prove the inclusion $h' \subset \operatorname{cl} K$, choose a point $\boldsymbol{z} \in h'$. Excluding the obvious case $\boldsymbol{z} = \boldsymbol{u}$, we assume that $\boldsymbol{z} \neq \boldsymbol{u}$. Put $\boldsymbol{w} = \boldsymbol{x} - \boldsymbol{u} + \boldsymbol{z}$. Then $\boldsymbol{w} \in (\boldsymbol{x} - \boldsymbol{u}) + h' = h \subset \operatorname{cl} K$ and $\boldsymbol{x} \neq \boldsymbol{w}$. Let

$$\boldsymbol{v}_\lambda = (1 - \lambda)\boldsymbol{w} + \lambda \boldsymbol{z}, \quad \boldsymbol{w}_\lambda = \frac{1}{1-\lambda}\boldsymbol{w} - \frac{\lambda}{1-\lambda}\boldsymbol{x}, \quad 0 < \lambda < 1.$$

Clearly, $\boldsymbol{w} \in (\boldsymbol{x}, \boldsymbol{w}_\lambda)$ and $\boldsymbol{w}_\lambda \in [\boldsymbol{x}, \boldsymbol{w}\rangle = h$ (see Lemma 2.25). The equalities

$$\begin{aligned} \boldsymbol{v}_\lambda &= \boldsymbol{w} + \lambda(\boldsymbol{z} - \boldsymbol{w}) = \boldsymbol{w} + \lambda(\boldsymbol{u} - \boldsymbol{x}) \\ &= (1 - \lambda)\Big(\frac{1}{1-\lambda}\boldsymbol{w} - \frac{\lambda}{1-\lambda}\boldsymbol{x}\Big) + \lambda \boldsymbol{u} \\ &= (1 - \lambda)\boldsymbol{w}_\lambda + \lambda \boldsymbol{u} \end{aligned}$$

and the convexity of $\operatorname{cl} K$ (see Theorem 3.46) show that $\boldsymbol{v}_\lambda \in [\boldsymbol{u}, \boldsymbol{w}_\lambda] \subset \operatorname{cl} K$, as depicted in Figure 6.3. Finally,

$$\boldsymbol{z} = \lim_{\lambda \to 1} ((1-\lambda)\boldsymbol{w} + \lambda \boldsymbol{z}) = \lim_{\lambda \to 1} \boldsymbol{v}_\lambda \in \operatorname{cl} K,$$

Summing up, $h' \subset \operatorname{cl} K$. Since $\boldsymbol{v} \in \operatorname{rint} K \subset \operatorname{cl} K$, the closed halfline h'' lies in $\operatorname{cl} K$, as proved above. Finally, assertion (a) yields the inclusion $h'' \subset \operatorname{rint} K$. □

Corollary 6.11. *Let $K \subset \mathbb{R}^n$ be a convex set and C be a cone with apex $\boldsymbol{a}$ such that $C \subset \operatorname{cl} K$. For points $\boldsymbol{u} \in \operatorname{cl} K$ and $\boldsymbol{v} \in \operatorname{rint} K$, the cones*

$$C' = (\boldsymbol{u} - \boldsymbol{a}) + C \quad \text{and} \quad C'' = (\boldsymbol{v} - \boldsymbol{a}) + C$$

lie in $\operatorname{cl} K$ and $\operatorname{rint} K$, respectively.

Proof. Without loss of generality, we may assume that $\boldsymbol{a} \in C$. Since the corollary is obvious when $C = \{\boldsymbol{a}\}$, we may suppose that $C \neq \{\boldsymbol{a}\}$. Consequently, we can express C as the union of closed halflines $[\boldsymbol{a}, \boldsymbol{x}\rangle$, $\boldsymbol{x} \in C \setminus \{\boldsymbol{a}\}$. Then

$$C' = (\boldsymbol{u} - \boldsymbol{a}) + \cup\,([\boldsymbol{a}, \boldsymbol{x}\rangle : \boldsymbol{x} \in C \setminus \{\boldsymbol{a}\}),$$
$$C'' = (\boldsymbol{v} - \boldsymbol{a}) + \cup\,([\boldsymbol{a}, \boldsymbol{x}\rangle : \boldsymbol{x} \in C \setminus \{\boldsymbol{a}\}).$$

The inclusions $C' \subset \operatorname{cl} K$ and $C'' \subset \operatorname{rint} K$ follow from Theorem 6.10. □

Theorem 6.12. *For a nonempty convex set $K \subset \mathbb{R}^n$ and points $\boldsymbol{u} \in \operatorname{cl} K$ and $\boldsymbol{v} \in \operatorname{rint} K$, the following conditions are equivalent.*

(a) *K is unbounded.*
(b) *There is a closed halfline with endpoint $\boldsymbol{u}$ which lies in $\operatorname{cl} K$.*
(c) *There is a closed halfline with endpoint $\boldsymbol{v}$ which lies in $\operatorname{rint} K$.*

Proof. $(a) \Rightarrow (b)$ Choose a sequence of points $\boldsymbol{x}_1, \boldsymbol{x}_2, \dots$ in $K \setminus \{\boldsymbol{u}\}$ such that $\lim_{i\to\infty} \|\boldsymbol{x}_i - \boldsymbol{u}\| = \infty$. Put

$$\boldsymbol{e}_i = \frac{\boldsymbol{x}_i - \boldsymbol{u}}{\|\boldsymbol{x}_i - \boldsymbol{u}\|}, \quad i \geqslant 1. \tag{6.9}$$

Clearly, $\|\boldsymbol{e}_i\| = 1$ for all $i \geqslant 1$, which shows that the set $\{\boldsymbol{e}_1, \boldsymbol{e}_2, \dots\}$ lies on the unit sphere $\mathbb{S}$. Since the set $\mathbb{S}$ is compact, the sequence $\boldsymbol{e}_1, \boldsymbol{e}_2, \dots$ contains a subsequence which converges to a unit vector $\boldsymbol{e}$. Without loss of generality, we may assume that the sequence $\boldsymbol{e}_1, \boldsymbol{e}_2, \dots$ converges to $\boldsymbol{e}$. If an integer i_0 is such that $\|\boldsymbol{x}_i - \boldsymbol{u}\| > 1$ for all $i \geqslant i_0$, then, by (6.9), $\boldsymbol{u} + \boldsymbol{e}_i$ can be written as the convex combination

$$\boldsymbol{u} + \boldsymbol{e}_i = \Big(1 - \frac{1}{\|\boldsymbol{x}_i - \boldsymbol{u}\|}\Big)\boldsymbol{u} + \frac{1}{\|\boldsymbol{x}_i - \boldsymbol{u}\|}\boldsymbol{x}_i, \quad i \geqslant i_0.$$

Hence $\boldsymbol{u} + \boldsymbol{e}_i \in [\boldsymbol{u}, \boldsymbol{x}_i] \subset \operatorname{cl} K$ for all $i \geqslant i_0$ by the convexity of $\operatorname{cl} K$ (see Theorem 3.46). Consequently, $\boldsymbol{u} + \boldsymbol{e} = \lim_{i\to\infty}(\boldsymbol{u} + \boldsymbol{e}_i) \in \operatorname{cl} K$.

We assert that the closed halfline $[\boldsymbol{u}, \boldsymbol{u} + \boldsymbol{e}\rangle$ lies in $\operatorname{cl} K$. Indeed, choose a point $\boldsymbol{y}$ in $[\boldsymbol{u}, \boldsymbol{u} + \boldsymbol{e}\rangle$. Excluding the obvious case $\boldsymbol{y} = \boldsymbol{u}$, we may assume that

$$\boldsymbol{y} = (1 - \lambda)\boldsymbol{u} + \lambda(\boldsymbol{u} + \boldsymbol{e}) = \boldsymbol{u} + \lambda \boldsymbol{e}, \quad \lambda > 0.$$

Clearly, $\|\boldsymbol{y} - \boldsymbol{u}\| = \lambda$. Choose an integer $i_1(\geqslant i_0)$ such that $\|\boldsymbol{x}_i - \boldsymbol{u}\| > \lambda$ for all $i \geqslant i_1$, and put $\boldsymbol{y}_i = \boldsymbol{u} + \lambda \boldsymbol{e}_i$. The inclusion

$$\boldsymbol{y}_i = (1 - \lambda)\boldsymbol{u} + \lambda(\boldsymbol{u} + \boldsymbol{e}_i) \in [\boldsymbol{u}, \boldsymbol{u} + \boldsymbol{e}_i\rangle = [\boldsymbol{u}, \boldsymbol{x}_i\rangle$$

(see Lemma 2.25) and the inequality

$$\|\boldsymbol{y}_i - \boldsymbol{u}\| = \lambda < \|\boldsymbol{x}_i - \boldsymbol{u}\|, \quad i \geqslant i_1,$$

give $\boldsymbol{y}_i \in [\boldsymbol{u}, \boldsymbol{x}_i] \subset \operatorname{cl} K$ for all $i \geqslant i_1$. Because

$$\lim_{i\to\infty} \boldsymbol{y}_i = \lim_{i\to\infty} (\boldsymbol{u} + \lambda \boldsymbol{e}_i) = \boldsymbol{u} + \lambda \boldsymbol{e} = \boldsymbol{y},$$

the point $\boldsymbol{y}$ is in $\operatorname{cl} K$. Summing up, $[\boldsymbol{u}, \boldsymbol{u} + \boldsymbol{e}\rangle \subset \operatorname{cl} K$.

$(b) \Rightarrow (c)$ This part of the proof follows directly from Theorem 6.10.

$(c) \Rightarrow (a)$ Let a closed halfline h with endpoint $\boldsymbol{v}$ lie in $\operatorname{rint} K$. Select a point $\boldsymbol{z} \in h \setminus \{\boldsymbol{v}\}$. Then $h = [\boldsymbol{v}, \boldsymbol{z}\rangle$ according to Lemma 2.25. Choose a scalar $\rho > 0$ and a point

$$\boldsymbol{y} = (1 - \lambda)\boldsymbol{v} + \lambda \boldsymbol{z}, \quad \text{where} \quad \lambda > \frac{\rho}{\|\boldsymbol{v} - \boldsymbol{z}\|}.$$

Then $\boldsymbol{y} \in [\boldsymbol{v}, \boldsymbol{z}\rangle \subset \operatorname{rint} K \subset K$ and

$$\|\boldsymbol{v} - \boldsymbol{y}\| = \|\boldsymbol{v} - (1 - \lambda)\boldsymbol{v} - \lambda \boldsymbol{z}\| = \lambda\|\boldsymbol{v} - \boldsymbol{z}\| > \rho.$$

Hence K is unbounded. □

Remark. It might happen that for an unbounded convex set $K \subset \mathbb{R}^n$ and a point $\boldsymbol{u} \in K$, no closed halfline with endpoint $\boldsymbol{u}$ lies in K. Indeed, if $K \subset \mathbb{R}^2$ is given by $K = \{\boldsymbol{o}\} \cup \{(x, y) : 0 < y < 1\}$, then no closed halfline $h \subset \mathbb{R}^2$ with endpoint $\boldsymbol{o}$ lies in K.

The next result gives a slight generalization of Theorem 6.12.

Theorem 6.13. *Let $K \subset \mathbb{R}^n$ be a unbounded convex set; $\boldsymbol{u}$ and $\boldsymbol{v}$ be points in $\operatorname{cl} K$ and $\mathbb{R}^n$, respectively. The following assertions hold.*

(*a*) *If an unbounded sequence* $\boldsymbol{x}_1, \boldsymbol{x}_2, \ldots$ *of points in* $K \setminus \{\boldsymbol{u}, \boldsymbol{v}\}$ *is such that the unit vectors*

$$\boldsymbol{c}_i = \frac{\boldsymbol{x}_i - \boldsymbol{v}}{\|\boldsymbol{x}_i - \boldsymbol{v}\|}, \quad i \geqslant 1, \tag{6.10}$$

converge to a vector $\boldsymbol{c}$, *then the closed halfline* $[\boldsymbol{u}, \boldsymbol{u} + \boldsymbol{c}\rangle$ *lies in* $\operatorname{cl} K$.

(*b*) *Every unbounded sequence of points in* K *contains a subsequence, say* $\boldsymbol{x}_1, \boldsymbol{x}_2, \ldots$, *such that the respective sequence* (6.10) *converges.*

Proof. (*a*) The assumption on the unboundedness of $\boldsymbol{x}_1, \boldsymbol{x}_2, \ldots$ is equivalent to the condition $\lim_{i\to\infty} \|\boldsymbol{x}_i\| = \infty$. The inequalities

$$\begin{aligned} \|\boldsymbol{x}_i\| - \|\boldsymbol{u}\| \leqslant \|\boldsymbol{x}_i - \boldsymbol{u}\| \leqslant \|\boldsymbol{x}_i\| + \|\boldsymbol{u}\|, \\ \|\boldsymbol{x}_i\| - \|\boldsymbol{v}\| \leqslant \|\boldsymbol{x}_i - \boldsymbol{v}\| \leqslant \|\boldsymbol{x}_i\| + \|\boldsymbol{v}\| \end{aligned} \tag{6.11}$$

give

$$\lim_{i\to\infty} \|\boldsymbol{x}_i - \boldsymbol{u}\| = \lim_{i\to\infty} \|\boldsymbol{x}_i - \boldsymbol{v}\| = \infty \quad \text{and} \quad \lim_{i\to\infty} \frac{\|\boldsymbol{x}_i - \boldsymbol{v}\|}{\|\boldsymbol{x}_i - \boldsymbol{u}\|} = 1.$$

Therefore,

$$\begin{aligned} \lim_{i\to\infty} \frac{\boldsymbol{x}_i - \boldsymbol{u}}{\|\boldsymbol{x}_i - \boldsymbol{u}\|} &= \lim_{i\to\infty} \Big(\frac{\boldsymbol{x}_i - \boldsymbol{v}}{\|\boldsymbol{x}_i - \boldsymbol{u}\|} + \frac{\boldsymbol{v} - \boldsymbol{u}}{\|\boldsymbol{x}_i - \boldsymbol{u}\|}\Big) \\ &= \lim_{i\to\infty} \Big(\frac{\|\boldsymbol{x}_i - \boldsymbol{v}\|}{\|\boldsymbol{x}_i - \boldsymbol{u}\|} \frac{\boldsymbol{x}_i - \boldsymbol{v}}{\|\boldsymbol{x}_i - \boldsymbol{v}\|} + \frac{\boldsymbol{v} - \boldsymbol{u}}{\|\boldsymbol{x}_i - \boldsymbol{u}\|}\Big) \\ &= 1\boldsymbol{c} + \boldsymbol{o} = \boldsymbol{c}. \end{aligned}$$

Hence the vectors $\boldsymbol{e}_1, \boldsymbol{e}_2, \ldots$ from (6.9) converge to $\boldsymbol{c}$. As shown in the proof of Theorem 6.12, the closed halfline $[\boldsymbol{x}, \boldsymbol{x} + \boldsymbol{c}\rangle$ lies in $\operatorname{cl} K$.

(*b*) Let $\boldsymbol{y}_1, \boldsymbol{y}_2, \ldots$ be an unbounded sequence of points in K. By an argument from the proof of Theorem 6.12, $\boldsymbol{y}_1, \boldsymbol{y}_2, \ldots$ contains a subsequence $\boldsymbol{x}_1, \boldsymbol{x}_2, \ldots$ such that the respective sequence (6.9) converges to a unit vector $\boldsymbol{e}$. As above, (6.11) implies

$$\lim_{i\to\infty} \|\boldsymbol{x}_i - \boldsymbol{u}\| = \lim_{i\to\infty} \|\boldsymbol{x}_i - \boldsymbol{v}\| = \infty \quad \text{and} \quad \lim_{i\to\infty} \frac{\|\boldsymbol{x}_i - \boldsymbol{u}\|}{\|\boldsymbol{x}_i - \boldsymbol{v}\|} = 1.$$

Finally,

$$\begin{aligned} \lim_{i\to\infty} \frac{\boldsymbol{x}_i - \boldsymbol{v}}{\|\boldsymbol{x}_i - \boldsymbol{v}\|} &= \lim_{i\to\infty} \Big(\frac{\boldsymbol{x}_i - \boldsymbol{u}}{\|\boldsymbol{x}_i - \boldsymbol{v}\|} + \frac{\boldsymbol{u} - \boldsymbol{v}}{\|\boldsymbol{x}_i - \boldsymbol{v}\|}\Big) \\ &= \lim_{i\to\infty} \Big(\frac{\|\boldsymbol{x}_i - \boldsymbol{u}\|}{\|\boldsymbol{x}_i - \boldsymbol{v}\|} \frac{\boldsymbol{x}_i - \boldsymbol{u}}{\|\boldsymbol{x}_i - \boldsymbol{u}\|} + \frac{\boldsymbol{u} - \boldsymbol{v}}{\|\boldsymbol{x}_i - \boldsymbol{v}\|}\Big) \\ &= 1\boldsymbol{e} + \boldsymbol{o} = \boldsymbol{e}. \end{aligned}$$

□

6.1.5 *Recession Cones of Convex Sets*

Theorem 6.14. *For a convex set $K \subset \mathbb{R}^n$, the assertions below take place.*

(a) $\mathrm{rec}\,(\mathrm{rint}\, K) = \mathrm{rec}\,(\mathrm{cl}\, K)$.

(b) *The following conditions are equivalent:*

$$(b_1)\ K \text{ is unbounded}, \quad (b_2)\ \mathrm{rec}\,(\mathrm{cl}\, K) \not\subset \{\boldsymbol{o}\}, \quad (b_3)\ \mathrm{rec}\,(\mathrm{rint}\, K) \not\subset \{\boldsymbol{o}\}.$$

(c) *If $K \neq \varnothing$ and $\boldsymbol{u} \in \mathrm{cl}\, K$, then*

$$\mathrm{rec}\,(\mathrm{cl}\, K) = \cap\,(\lambda(\mathrm{cl}\, K - \boldsymbol{u}) : \lambda > 0).$$

(d) *If $K \neq \varnothing$ and $\boldsymbol{u} \in \mathrm{rint}\, K$, then*

$$\mathrm{rec}\,(\mathrm{rint}\, K) = \cap\,(\lambda(\mathrm{rint}\, K - \boldsymbol{u}) : \lambda > 0) = \cap\,(\lambda(K - \boldsymbol{u}) : \lambda > 0).$$

(e) *If $M \subset \mathbb{R}^n$ is a convex set containing K, then* $\mathrm{rec}\,(\mathrm{cl}\, K) \subset \mathrm{rec}\,(\mathrm{cl}\, M)$.

Proof. Since the case $K = \varnothing$ is obvious, we assume that K is nonempty.

(a) First, we assert that $\mathrm{rec}\,(\mathrm{rint}\, K) \subset \mathrm{rec}\,(\mathrm{cl}\, K)$. For this, choose a vector $\boldsymbol{e} \in \mathrm{rec}\,(\mathrm{rint}\, K)$. The case $\boldsymbol{e} = \boldsymbol{o}$ is obvious due to the inclusion $\boldsymbol{o} \in \mathrm{rec}\,(\mathrm{cl}\, K)$. Let $\boldsymbol{e} \neq \boldsymbol{o}$ and $\boldsymbol{u}$ be a point in $\mathrm{rint}\, K$. The definition of $\mathrm{rec}\,(\mathrm{rint}\, K)$ shows that the closed halfline $h = \{\boldsymbol{u} + \lambda \boldsymbol{e} : \lambda \geqslant 0\}$ lies in $\mathrm{rint}\, K$. If $\boldsymbol{v}$ is any point in $\mathrm{cl}\, K$, then Theorem 6.10 implies the inclusion $\boldsymbol{v} + (h - \boldsymbol{u}) \subset \mathrm{cl}\, K$. Thus $\boldsymbol{e} \in h - \boldsymbol{u} \subset \mathrm{rec}\,(\mathrm{cl}\, K)$. Summing up, $\mathrm{rec}\,(\mathrm{rint}\, K) \subset \mathrm{rec}\,(\mathrm{cl}\, K)$.

The opposite inclusion holds by a similar argument.

(b) First, we assert that $(b_1) \Leftrightarrow (b_2)$. Let K be unbounded. Choose a point $\boldsymbol{u} \in \mathrm{cl}\, K$. By Theorem 6.12, there is a closed halfline h with endpoint $\boldsymbol{u}$ which lies in $\mathrm{cl}\, K$. Theorem 6.10 shows that the closed halfline $(\boldsymbol{v} - \boldsymbol{u}) + h$ lies in $\mathrm{cl}\, K$ whenever $\boldsymbol{v} \in \mathrm{cl}\, K$. Therefore, the closed halfline $h - \boldsymbol{u}$ (whose endpoint is $\boldsymbol{o}$) lies in $\mathrm{rec}\,(\mathrm{cl}\, K)$. Thus $\mathrm{rec}\,(\mathrm{cl}\, K) \not\subset \{\boldsymbol{o}\}$. Conversely, if $\mathrm{rec}\,(\mathrm{cl}\, K) \not\subset \{\boldsymbol{o}\}$, then $\mathrm{rec}\,(\mathrm{cl}\, K)$ contains a closed halfline m with endpoint $\boldsymbol{o}$. Then $\boldsymbol{v} + m \subset \mathrm{cl}\, K$ for any given point $\boldsymbol{v} \in \mathrm{cl}\, K$, which shows that $\mathrm{cl}\, K$ (and whence K) is unbounded.

The equivalence of conditions (b_2) and (b_3) follows from assertion (a).

(c) Let $C = \cap\,(\lambda(\mathrm{cl}\, K - \boldsymbol{u}) : \lambda > 0)$. Theorem 6.2 implies the inclusions $\mathrm{rec}\,(\mathrm{cl}\, K) \subset C$. Hence it suffices to show that $C \subset \mathrm{rec}\,(\mathrm{cl}\, K)$. This inclusion is obvious when $K = \mathbb{R}^n$. So, we may assume that $K \neq \mathbb{R}^n$. Choose a point $\boldsymbol{e} \in \mathbb{R}^n \setminus \mathrm{rec}\,(\mathrm{cl}\, K)$ and let $h = \{\lambda \boldsymbol{e} : \lambda \geqslant 0\}$. Then there is a point $\boldsymbol{v} \in \mathrm{cl}\, K$ such that the closed halfline $\boldsymbol{v} + h$ does not lie in $\mathrm{cl}\, K$. Theorem 6.10 implies that the closed halfline $\boldsymbol{u} + h$ also does not lie in $\mathrm{cl}\, K$. Hence there is a scalar $\mu > 0$ such that $\boldsymbol{u} + \mu\boldsymbol{e} \notin \operatorname{cl} K$. Therefore, $\boldsymbol{e} \notin \mu^{-1}(\operatorname{cl} K - \boldsymbol{u})$. Thus $\boldsymbol{e} \notin C$, and the inclusion $C \subset \operatorname{rec}(\operatorname{cl} K)$ holds.

(d) Theorem 6.2 implies that

$$\operatorname{rec}(\operatorname{rint} K) \subset \cap\,(\lambda(\operatorname{rint} K - \boldsymbol{u}) : \lambda > 0) \subset \cap\,(\lambda(K - \boldsymbol{u}) : \lambda > 0).$$

Hence it suffices to show that the set $D = \cap\,(\lambda(K - \boldsymbol{u}) : \lambda > 0)$ lies in $\operatorname{rec}(\operatorname{rint} K)$. This assertion is obvious when $K = \mathbb{R}^n$. So, we may assume that $K \neq \mathbb{R}^n$. Choose a point $\boldsymbol{e} \in \mathbb{R}^n \backslash \operatorname{rec}(\operatorname{rint} K)$ and let $h = \{\lambda\boldsymbol{e} : \lambda \geqslant 0\}$. Then there is a point $\boldsymbol{x} \in \operatorname{rint} K$ such that the closed halfline $\boldsymbol{x} + h$ does not lie in $\operatorname{rint} K$. Consequently, $\boldsymbol{x} + h \not\subset K$. Hence there is a scalar $\mu > 0$ such that $\boldsymbol{u} + \mu\boldsymbol{e} \notin K$. Therefore, $\boldsymbol{e} \notin \mu^{-1}(K - \boldsymbol{u})$. Thus $\boldsymbol{e} \notin D$, and the inclusion $D \subset \operatorname{rec}(\operatorname{rint} K)$ holds.

(e) Let $\boldsymbol{e} \in \operatorname{rec}(\operatorname{cl} K)$. The case $\boldsymbol{e} = \boldsymbol{o}$ is obvious due to the inclusion $\boldsymbol{o} \in \operatorname{rec}(\operatorname{cl} M)$. Let $\boldsymbol{e} \neq \boldsymbol{o}$ and $h = \{\lambda\boldsymbol{e} : \lambda \geqslant 0\}$. Choose points $\boldsymbol{x} \in \operatorname{cl} K$ and $\boldsymbol{y} \in \operatorname{cl} M$. Then $\boldsymbol{x} + h \subset \operatorname{cl} K$ by the definition of $\operatorname{rec}(\operatorname{cl} K)$. Therefore, $\boldsymbol{x} + h \subset \operatorname{cl} K \subset \operatorname{cl} M$. Theorem 6.12 implies the inclusion $\boldsymbol{y} + h \subset \operatorname{cl} M$. Therefore, $h \subset \operatorname{rec}(\operatorname{cl} M)$. Summing up, $\operatorname{rec}(\operatorname{cl} K) \subset \operatorname{rec}(\operatorname{cl} M)$. □

A combination of Theorems 6.2 and 6.10 implies the corollary below.

Corollary 6.15. *For a nonempty convex set $K \subset \mathbb{R}^n$, the following assertions hold.*

(a) *If $\boldsymbol{u}$ is a given point in* $\operatorname{cl} K$, *then* $\operatorname{rec}(\operatorname{cl} K)$ *is the largest among all convex cones $C \subset \mathbb{R}^n$ with proper apex $\boldsymbol{o}$ which satisfy the inclusion $\boldsymbol{u} + C \subset \operatorname{cl} K$.*

(b) *If $\boldsymbol{u}$ is a given point in* $\operatorname{rint} K$, *then* $\operatorname{rec}(\operatorname{cl} K)$ *is the largest among all convex cones $C \subset \mathbb{R}^n$ with proper apex $\boldsymbol{o}$ which satisfy the inclusion $\boldsymbol{u} + C \subset \operatorname{rint} K$.* □

Remark. Corollary 6.15 does not hold if we replace $\operatorname{cl} K$ or $\operatorname{rint} K$ with K. Indeed, let X be a planar convex set depicted in Figure 6.2. Clearly, $\operatorname{rec} X = \{\boldsymbol{o}\}$. On the other hand, for every point $\boldsymbol{u} \in X$, there is a horizontal closed halfline h with endpoint $\boldsymbol{u}$ (the direction of h depends on $\boldsymbol{u}$) such that $h \subset X$.

Corollary 6.15 immediately implies the following obvious assertion.

Corollary 6.16. *If $K \subset \mathbb{R}^n$ is a nonempty convex set, then*

$$\operatorname{rec}(\operatorname{cl} K) = \{\boldsymbol{e} \in \mathbb{R}^n : \exists\, \boldsymbol{x} \in K \text{ such that } \lambda\boldsymbol{e} + \boldsymbol{x} \in K \text{ whenever } \lambda \geqslant 0\}.$$

Theorem 6.17. *If $K \subset \mathbb{R}^n$ is a nonempty convex set, then* $\operatorname{rec}(\operatorname{cl} K)$ *is the set of limits of all convergent sequences $\lambda_1 \boldsymbol{x}_1, \lambda_2 \boldsymbol{x}_2, \ldots$, where $\boldsymbol{x}_1, \boldsymbol{x}_2, \ldots$ belong to K and $\lambda_1, \lambda_2, \ldots$ are positive scalars tending to* 0.

Proof. We assert first that every point $\boldsymbol{e} \in \operatorname{rec}(\operatorname{cl} K)$ can be expressed in the described above form. This fact is obvious if $\boldsymbol{e} = \boldsymbol{o}$ (choose a point $\boldsymbol{x} \in K$ and put $\boldsymbol{x}_i = \boldsymbol{x}$ and $\lambda_i = \frac{1}{i}$ for all $i \geqslant 1$). So, we may suppose that $\boldsymbol{e} \neq \boldsymbol{o}$. Then the closed halfline $[\boldsymbol{o}, \boldsymbol{e}\rangle$ lies in $\operatorname{rec}(\operatorname{cl} K)$ and whence in $\operatorname{rec}(\operatorname{rint} K)$, as follows from Theorem 6.14. Choose a point $\boldsymbol{x} \in \operatorname{rint} K$. Then $[\boldsymbol{x}, \boldsymbol{x} + \boldsymbol{e}\rangle \subset \operatorname{rint} K$ according to Theorem 6.8. Therefore $\boldsymbol{x} + i\boldsymbol{e} \in \operatorname{rint} K \subset K$ for all $i \geqslant 1$. With $\boldsymbol{x}_i = \boldsymbol{x} + i\boldsymbol{e}$ and $\lambda_i = 1/i$ for all $i \geqslant 1$, we have

$$\lim_{i \to \infty} \lambda_i \boldsymbol{x}_i = \tfrac{1}{i}(\boldsymbol{x} + i\boldsymbol{e}) = \boldsymbol{e}.$$

Conversely, suppose that $\boldsymbol{e} = \lim_{i \to \infty} \lambda_i \boldsymbol{x}_i$, where all $\boldsymbol{x}_1, \boldsymbol{x}_2, \ldots$ are in K and positive scalars $\lambda_1, \lambda_2, \ldots$ tend to 0. Because the case $\boldsymbol{e} = \boldsymbol{o}$ is obvious, we may assume that $\boldsymbol{e} \neq \boldsymbol{o}$. Then the sequence $\boldsymbol{x}_1, \boldsymbol{x}_2, \ldots$ is unbounded, since otherwise $\lambda_i \boldsymbol{x}_i \to \boldsymbol{o}$ as $i \to \infty$. Excluding from $\{\boldsymbol{x}_1, \boldsymbol{x}_2, \ldots\}$ all zero vectors, let $\boldsymbol{c} = \boldsymbol{e}/\|\boldsymbol{e}\|$ and $\boldsymbol{c}_i = \boldsymbol{x}_i/\|\boldsymbol{x}_i\|$, $i \geqslant 1$. Then

$$\boldsymbol{c} = \frac{\boldsymbol{e}}{\|\boldsymbol{e}\|} = \frac{\lim\limits_{i \to \infty} \lambda_i \boldsymbol{x}_i}{\lim\limits_{i \to \infty} \lambda_i \|\boldsymbol{x}_i\|} = \lim_{i \to \infty} \frac{\lambda_i \boldsymbol{x}_i}{\lambda_i \|\boldsymbol{x}_i\|} = \lim_{i \to \infty} \frac{\boldsymbol{x}_i}{\|\boldsymbol{x}_i\|} = \lim_{i \to \infty} \boldsymbol{c}_i,$$

and Theorem 6.13 (with $\boldsymbol{v} = \boldsymbol{o}$) implies that the closed halfline $[\boldsymbol{u}, \boldsymbol{u}+\boldsymbol{c}\rangle$ lies in $\operatorname{cl} K$ for every choice of $\boldsymbol{u}$ in $\operatorname{cl} K$. Therefore, $\boldsymbol{e} \in [\boldsymbol{o}, \boldsymbol{c}\rangle \subset \operatorname{rec}(\operatorname{cl} K)$. $\square$

See Problem 6.4 for a variation of Theorem 6.17. We conclude this section with an important relation between recession cones and generated cones of a convex set (see Problems 11.14 and 12.8 for further refinements).

Theorem 6.18. *Let $K \subset \mathbb{R}^n$ be a convex set which is not a plane, and let X be a dense subset of* $\operatorname{rbd} K$. *Then*

$$\operatorname{cl} K = \cap\, (C_{\boldsymbol{a}}(\operatorname{cl} K) : \boldsymbol{a} \in X), \tag{6.12}$$

$$\operatorname{rec}(\operatorname{cl} K) = \cap\, (C_{\boldsymbol{a}}(\operatorname{cl} K) - \boldsymbol{a} : \boldsymbol{a} \in X). \tag{6.13}$$

Proof. First, we are going to prove (6.12). Because the inclusion $\operatorname{cl} K \subset \cap\, (C_{\boldsymbol{a}}(\operatorname{cl} K) : \boldsymbol{a} \in X)$ is obvious, it remains to show that $\cap\, (C_{\boldsymbol{a}}(\operatorname{cl} K) : \boldsymbol{a} \in X) \subset \operatorname{cl} K$. For this, it suffices to prove that for any point $\boldsymbol{x} \in \mathbb{R}^n \setminus \operatorname{cl} K$ there is a point $\boldsymbol{s} \in X$ satisfying the condition $\boldsymbol{x} \notin C_{\boldsymbol{s}}(\operatorname{cl} K)$. If $\boldsymbol{x} \in \mathbb{R}^n \setminus \operatorname{aff} K$, then, choosing any point $\boldsymbol{s} \in X$, we have $C_{\boldsymbol{s}}(\operatorname{cl} K) \subset \operatorname{aff} K$ according to Theorem 5.38. Consequently, $\boldsymbol{x} \notin C_{\boldsymbol{s}}(\operatorname{cl} K)$.

Suppose that $\boldsymbol{x} \in \operatorname{aff} K \setminus \operatorname{cl} K$. Choose a point $\boldsymbol{z} \in \operatorname{rint} K$ and a scalar $\rho > 0$ such that $\operatorname{aff} K \cap U_\rho(\boldsymbol{z}) \subset \operatorname{rint} K$ (see Theorem 3.17). By Theorem 3.65, the open segment $(\boldsymbol{x}, \boldsymbol{z})$ meets $\operatorname{rbd} K$ at a point $\boldsymbol{u}$. Let $\boldsymbol{u} = (1-\lambda)\boldsymbol{x} + \lambda \boldsymbol{z}$, where $0 < \lambda < 1$. Clearly, $\boldsymbol{z} = (1-\lambda^{-1})\boldsymbol{x} + \lambda^{-1}\boldsymbol{u}$.

Choose a positive scalar δ satisfying the condition $\delta < \lambda\rho$. Since X is dense in $\operatorname{rbd} K$, there is a point $\boldsymbol{s} \in X$ such that $\|\boldsymbol{s} - \boldsymbol{u}\| < \delta$. Put $\boldsymbol{v} = (1-\lambda^{-1})\boldsymbol{x} + \lambda^{-1}\boldsymbol{s}$ (so that $\boldsymbol{s} = (1-\lambda)\boldsymbol{x} + \lambda\boldsymbol{v}$). Clearly, $\boldsymbol{v} \in \operatorname{aff} K$ as an affine combination of points $\boldsymbol{s}$ and $\boldsymbol{x}$ from $\operatorname{aff} K$ (see Theorem 2.38). The inequality

$$\|\boldsymbol{z} - \boldsymbol{v}\| = \lambda^{-1}\|\boldsymbol{u} - \boldsymbol{s}\| < \lambda^{-1}\delta < \rho$$

shows that $\boldsymbol{v} \in \operatorname{aff} K \cap U_\rho(\boldsymbol{z}) \subset \operatorname{rint} K$.

We assert that $\boldsymbol{x} \notin C_{\boldsymbol{s}}(\operatorname{cl} K)$. Indeed, Theorem 5.48 gives $C_{\boldsymbol{s}}(\operatorname{cl} K) \neq \operatorname{aff} K$. Then Corollary 5.27 show the inclusion $\boldsymbol{s} \in \operatorname{rbd} C_{\boldsymbol{s}}(K)$. Assume, for contradiction, that $\boldsymbol{x} \in C_{\boldsymbol{s}}(\operatorname{cl} K)$. Then, due to Theorem 5.49, $\boldsymbol{z} \in \operatorname{rint} K \subset \operatorname{rint} C_{\boldsymbol{s}}(\operatorname{cl} K)$. By Theorem 3.47, $\boldsymbol{s} \in (\boldsymbol{x}, \boldsymbol{z}) \subset \operatorname{rint} C_{\boldsymbol{s}}(\operatorname{cl} K)$, contrary to the inclusion $\boldsymbol{s} \in \operatorname{rbd} C_{\boldsymbol{s}}(\operatorname{cl} K)$. Summing up, $\boldsymbol{x} \notin C_{\boldsymbol{s}}(\operatorname{cl} K)$.

For (6.13), we observe first that $\boldsymbol{a} + \operatorname{rec}(\operatorname{cl} K) \subset \operatorname{cl} K \subset C_{\boldsymbol{a}}(\operatorname{cl} K)$ for every point $\boldsymbol{a} \in X$ (see Theorem 6.2). Hence

$$\operatorname{rec}(\operatorname{cl} K) \subset \cap\,(C_{\boldsymbol{a}}(\operatorname{cl} K) - \boldsymbol{a} : \boldsymbol{a} \in X).$$

Conversely, suppose that an open halfline h with endpoint $\boldsymbol{o}$ does not lie in $\operatorname{rec}(\operatorname{cl} K)$. Choose a point $\boldsymbol{z} \in \operatorname{rint} K$ and a scalar $\rho > 0$ such that $\operatorname{aff} K \cap U_\rho(\boldsymbol{z}) \subset \operatorname{rint} K$ (see Theorem 3.17). Corollary 6.15 shows that the open halfline $h_{\boldsymbol{z}} = \boldsymbol{z} + h$ does not lie in $\operatorname{cl} K$. By Corollary 3.66, $h_{\boldsymbol{z}}$ meets $\operatorname{rbd} K$ at a point $\boldsymbol{u} \in \operatorname{rbd} K$ such that the open halfline $h_{\boldsymbol{u}} = \boldsymbol{u} + h$ is disjoint from $\operatorname{cl} K$. Since X is dense in $\operatorname{rbd} K$, there is a point $\boldsymbol{s} \in X$ such that $\|\boldsymbol{s} - \boldsymbol{u}\| < \rho$. Put $\boldsymbol{v} = \boldsymbol{z} + (\boldsymbol{s} - \boldsymbol{u})$ and $h_{\boldsymbol{v}} = \boldsymbol{v} + h$. By the above argument, $h_{\boldsymbol{v}}$ meets $\operatorname{rbd} K$ at $\boldsymbol{s}$ such that the open halfline $h_{\boldsymbol{s}} = \boldsymbol{s} + h$ is disjoint from $\operatorname{cl} K$. Furthermore, as shown above, $C_{\boldsymbol{s}}(\operatorname{cl} K)$ does not contain $h_{\boldsymbol{s}}$. Consequently, $h = h_{\boldsymbol{s}} - \boldsymbol{s} \not\subset C_{\boldsymbol{s}}(\operatorname{cl} K) - \boldsymbol{s}$. Summing up,

$$\cap\,(C_{\boldsymbol{a}}(\operatorname{cl} K) - \boldsymbol{a} : \boldsymbol{a} \in X) \subset \operatorname{rec}(\operatorname{cl} K). \qquad \square$$

6.2 Lineality Spaces

6.2.1 *Basic Properties of Lineality Spaces*

Definition 6.19. The *lineality space* of a nonempty set X in $\mathbb{R}^n$, denoted $\operatorname{lin} X$, is defined by

$$\operatorname{lin} X = \{\boldsymbol{e} \in \mathbb{R}^n : \lambda\boldsymbol{e} + \boldsymbol{x} \in X \text{ whenever } \lambda \in \mathbb{R} \text{ and } \boldsymbol{x} \in X\}.$$

We let $\operatorname{lin} X = \varnothing$ if $X = \varnothing$.

For instance, if $L \subset \mathbb{R}^n$ is a plane, then $\operatorname{lin} L$ is its direction space $\operatorname{dir} L$ (see an example on page 207).

Theorem 6.20. *For a nonempty set $X \subset \mathbb{R}^n$, the following assertions hold.*

(a) $\operatorname{lin} X = \operatorname{rec} X \cap (-\operatorname{rec} X) = \operatorname{lin}(\operatorname{rec} X) = \operatorname{ap}(\operatorname{rec} X)$.
(b) $\operatorname{lin} X$ *is the largest set of vectors* $\boldsymbol{e} \in \mathbb{R}^n$ *satisfy the condition*

$$\lambda \boldsymbol{e} + X \subset X \quad \text{for all} \quad \lambda \in \mathbb{R},$$

or, equivalently, the equality

$$\lambda \boldsymbol{e} + X = X \quad \text{for all} \quad \lambda \in \mathbb{R}.$$

(c) $\operatorname{lin} X$ *is the largest among all subspaces* $S \subset \mathbb{R}^n$ *satisfying the condition* $S + X \subset X$, *or, equivalently, the equality* $S + X = X$.
(d) $\operatorname{lin} X \subset \cap (\lambda(X - \boldsymbol{u}) : \lambda \neq 0)$ *for any given point* $\boldsymbol{u} \in X$.
(e) $\operatorname{lin} X \subset \operatorname{dir} X$, *with* $\operatorname{lin} X = \operatorname{dir} X$ *if and only if* X *is a plane.*

Proof. (a) The equality $\operatorname{lin} X = \operatorname{rec} X \cap (-\operatorname{rec} X)$ follows from the definitions. By Theorem 6.2, $\operatorname{rec} X$ is a convex cone with proper apex $\boldsymbol{o}$. Consequently, Theorem 5.17 gives $\operatorname{ap}(\operatorname{rec} X) = \operatorname{rec} X \cap (-\operatorname{rec} X)$. Similarly, Theorem 6.3 implies

$$\operatorname{lin}(\operatorname{rec} X) = \operatorname{rec}(\operatorname{rec} X) \cap (-\operatorname{rec}(\operatorname{rec} X)) = \operatorname{rec} X \cap (-\operatorname{rec} X) = \operatorname{lin} X.$$

(b) The inclusion $\lambda \boldsymbol{e} + X \subset X$ whenever $\lambda \in \mathbb{R}$ immediately follows from Definition 6.19. Since $\boldsymbol{x} = \boldsymbol{o} + \boldsymbol{x}$ for every point $\boldsymbol{x} \in X$, and since $\boldsymbol{o} \in \operatorname{lin} X$, we obtain the opposite inclusion $X \subset \lambda \boldsymbol{e} + X$.

(c) The assertion that $\operatorname{lin} X$ is a subspace follows from the equality $\operatorname{lin} X = \operatorname{ap}(\operatorname{rec} X)$ and the fact that the apex set of a cone is a plane (see Theorem 5.14).

The inclusion $\operatorname{lin} X + X \subset X$ follows from Definition 6.19. If a subspace $S \subset \mathbb{R}^n$ satisfies the inclusion $S + X \subset X$, then, for any $\boldsymbol{e} \in S$, $\lambda \in \mathbb{R}$, and $\boldsymbol{x} \in X$, one has $\lambda \boldsymbol{e} + \boldsymbol{x} \in S + X \subset X$, implying that $\boldsymbol{e} \in \operatorname{lin} X$. Hence $S \subset \operatorname{lin} X$, confirming that $\operatorname{lin} X$ is the largest among all subspaces satisfying the condition $S + X \subset X$. Since $\boldsymbol{x} = \boldsymbol{o} + \boldsymbol{x}$ for every point $\boldsymbol{x} \in X$, and since $\boldsymbol{o} \in \operatorname{lin} X$, we obtain the opposite inclusion $X \subset \operatorname{lin} X + X$.

(d) According to assertion (c) above, $\operatorname{lin} X$ is a subspace and $\boldsymbol{u} + \operatorname{lin} X \subset X$. Then $\lambda \operatorname{lin} X = \operatorname{lin} X$ for all $\lambda \neq 0$, and the inclusion $\operatorname{lin} X \subset X - \boldsymbol{u}$ gives

$$\operatorname{lin} X = \lambda \operatorname{lin} X \subset \lambda(X - \boldsymbol{u}), \quad \lambda \neq 0.$$

Hence $\operatorname{lin} X \subset \cap(\lambda(X - \boldsymbol{u}) : \lambda \neq 0)$.

(e) Let $\boldsymbol{e} \in \operatorname{lin} X$. Then $\boldsymbol{e} + \boldsymbol{x} \in X$ whenever $\boldsymbol{x} \in X$. According to Theorem 2.48, one has

$$\boldsymbol{e} \in X - \boldsymbol{x} \subset \operatorname{aff} X - \boldsymbol{x} = \operatorname{span}(X - \boldsymbol{x}) = \operatorname{dir} X.$$

Consecutively, $\operatorname{lin} X \subset \operatorname{dir} X$. If X is a plane, then $\operatorname{lin} X = \operatorname{dir} X$. Conversely, let $\operatorname{lin} X = \operatorname{dir} X$. Then $\operatorname{lin} X$ is a translate of $\operatorname{aff} X$ (see Theorem 2.48), and a combination of Theorem 2.2 and assertion (c) gives $X = \operatorname{lin} X + X = \operatorname{aff} X$, implying that X is a plane. □

Theorem 6.21. *If $C \subset \mathbb{R}^n$ is a nonempty cone with apex $\boldsymbol{a}$, then*

$$\operatorname{lin} C = \operatorname{ap} C - \boldsymbol{a}.$$

Proof. By Theorem 5.14, $\operatorname{ap} C$ is a plane through $\boldsymbol{a}$. Hence the set $\operatorname{ap} C - \boldsymbol{a}$ is a subspace (see Theorem 2.2). For a given point $\boldsymbol{u} \in C$, Theorem 5.16 shows that $\boldsymbol{u} + (\operatorname{ap} C - \boldsymbol{a}) \subset C$. This argument and Theorem 6.20 give the inclusion $\operatorname{ap} C - \boldsymbol{a} \subset \operatorname{lin} C$.

Hence it remains to prove the opposite inclusion. The case $\operatorname{lin} C = \{\boldsymbol{o}\}$ is obvious:

$$\operatorname{lin} C = \{\boldsymbol{o}\} = \{\boldsymbol{a}\} - \{\boldsymbol{a}\} \subset \operatorname{ap} C - \boldsymbol{a}.$$

Hence we may assume that $\operatorname{lin} C \neq \{\boldsymbol{o}\}$. Then $C \neq \{\boldsymbol{a}\}$. Let $L = \boldsymbol{a} + \operatorname{lin} C$. We assert that every point $\boldsymbol{c} \in L$ lies in $\operatorname{ap} C$. For this, choose any point $\boldsymbol{x} \in C \setminus \{\boldsymbol{c}\}$. If $\boldsymbol{x} \in L$, then $L = \boldsymbol{x} + \operatorname{lin} C \subset C$ (see Theorem 2.2), and Theorem 2.38 gives $(\boldsymbol{c}, \boldsymbol{x}\rangle \subset \langle \boldsymbol{c}, \boldsymbol{x}\rangle \subset L \subset C$.

Suppose that $\boldsymbol{x} \in C \setminus L$. Since $\boldsymbol{a} \in \operatorname{ap} C$, the open halfline $h = (\boldsymbol{a}, \boldsymbol{x}\rangle$ lies in C. For any point $\boldsymbol{u} \in h$, the plane $\boldsymbol{u} + \operatorname{lin} C$ lies in C (see Theorem 6.20). Thus the set

$$E = (h - \boldsymbol{a}) + L = h + \operatorname{lin} C = \cup(\boldsymbol{u} + \operatorname{lin} C : \boldsymbol{u} \in h)$$

lies in C. According to Theorem 2.74, E is an open halfplane of the plane $M = \operatorname{aff}(h \cup L)$. Since E is the intersection of M and a suitable open halfspace (see Definition 2.69), a combination of Theorems 2.38 and 2.35 implies the inclusions $(\boldsymbol{c}, \boldsymbol{x}\rangle \subset E \subset C$. Hence $\boldsymbol{c} \in \operatorname{ap} C$, and $L \subset \operatorname{ap} C$. Finally, $\operatorname{ap} C = L - \boldsymbol{a} \subset \operatorname{ap} C - \boldsymbol{a}$. □

6.2.2 *Algebra of Sets and Lineality Spaces*

Corollary 6.22. *Given a set $X \subset \mathbb{R}^n$, a point $\boldsymbol{a} \in \mathbb{R}^n$, and a scalar μ, one has*

$$\operatorname{lin}(\boldsymbol{a} + \mu X) = \mu \operatorname{lin} X = \operatorname{sgn}(\mu) \operatorname{lin} X.$$

Proof. By Theorems 6.5 and 6.20,

$$\begin{aligned}\operatorname{lin}(\boldsymbol{a}+\mu X) &= \operatorname{rec}(\boldsymbol{a}+\mu X)\cap(-\operatorname{rec}(\boldsymbol{a}+\mu X))\\ &= \mu\operatorname{rec}X\cap(-\mu\operatorname{rec}X) = \mu\operatorname{lin}X = \operatorname{sgn}(\mu)\operatorname{lin}X. \qquad \square\end{aligned}$$

Theorem 6.23. *For sets* $X_1,\dots,X_r\subset\mathbb{R}^n$ *and scalars* $\mu_1,\dots,\mu_r$, *one has*

$$\mu_1\operatorname{lin}X_1+\cdots+\mu_r\operatorname{lin}X_r\subset\operatorname{lin}(\mu_1X_1+\cdots+\mu_rX_r).$$

If, additionally, all sets $X_1,\dots,X_r$ *are nonempty and the planes* aff $X_1,\dots,$ aff X_r *are independent, then*

$$\mu_1\operatorname{lin}X_1+\cdots+\mu_r\operatorname{lin}X_r=\operatorname{lin}(\mu_1X_1+\cdots+\mu_rX_r).$$

Proof. Similarly to Theorems 6.6, the proof can be reduced to the case of two nonempty sets X_1 and X_2 and $\mu_1\mu_2\neq 0$. By the same theorem,

$$\begin{aligned}\mu_1\operatorname{lin}X_1+\mu_2\operatorname{lin}X_2 &= \mu_1\operatorname{rec}X_1\cap(-\operatorname{rec}X_1)+\mu_2\operatorname{rec}X_2\cap(-\operatorname{rec}X_2)\\ &\subset(\mu_1\operatorname{rec}X_1+\mu_2\operatorname{rec}X_2)\cap(-\mu_1\operatorname{rec}X_1-\mu_2\operatorname{rec}X_2)\\ &\subset\operatorname{rec}(\mu_1X_1+\mu_2X_2)\cap(-\operatorname{rec}(\mu_1X_1+\mu_2X_2))\\ &=\operatorname{lin}(\mu_1X_1+\mu_2X_2).\end{aligned}$$

Assume now that the planes aff X_1 and aff X_2 are independent. By the above proved, it suffices o show that

$$\operatorname{lin}(\mu_1X_1+\mu_2X_2)\subset\mu_1\operatorname{lin}X_1+\mu_2\operatorname{lin}X_2. \tag{6.14}$$

Choose points $\boldsymbol{u}_1\in X_1$ and $\boldsymbol{u}_2\in X_2$, and put

$$X_1'=X_1-\boldsymbol{u}_1 \quad\text{and}\quad X_2'=X_2-\boldsymbol{u}_2.$$

Then $\boldsymbol{o}_1\in X_1'\cap X_2'$, which implies that

$$S_1=\operatorname{span}X_1'=\operatorname{aff}X_1-\boldsymbol{u}_1 \quad\text{and}\quad S_2=\operatorname{span}X_2'=\operatorname{aff}X_2-\boldsymbol{u}_2$$

are independent subspaces. Furthermore, Corollary 6.22 shows that

$$\operatorname{lin}X_1'=\operatorname{lin}X_1\subset S_1 \quad\text{and}\quad \operatorname{lin}X_2'=\operatorname{lin}X_2\subset S_2.$$

By the same corollary,

$$\begin{aligned}\operatorname{lin}(\mu_1X_1'+\mu_2X_2') &= \operatorname{lin}(\mu_1X_1+\mu_2X_2-(\mu_1\boldsymbol{u}_1+\mu_2\boldsymbol{u}_2))\\ &=\operatorname{lin}(\mu_1X_1+\mu_2X_2).\end{aligned}$$

We assert that

$$\operatorname{lin}(\mu_1X_1'+\mu_2X_2')\subset\mu_1\operatorname{lin}X_1'+\mu_2\operatorname{lin}X_2'.$$

Indeed, let $\boldsymbol{e} \in \operatorname{lin}(\mu_1 X_1' + \mu_2 X_2')$. By Theorem 6.20,

$$\begin{aligned}\boldsymbol{e} &\in \operatorname{dir}(\mu_1 X_1' + \mu_2 X_2') = \operatorname{span}(\mu_1 X_1' + \mu_2 X_2')\\ &= \operatorname{span}(\mu_1 X_1') + \operatorname{span}(\mu_2 X_2') = \operatorname{span} X_1' + \operatorname{span} X_2' = S_1 + S_2.\end{aligned}$$

Hence $\boldsymbol{e} = \boldsymbol{e}_1 + \boldsymbol{e}_2$ for suitable vectors $\boldsymbol{e}_1 \in S_1$ and $\boldsymbol{e}_2 \in S_2$.

Choose points $\boldsymbol{x}_1 \in X_1'$ and $\boldsymbol{x}_2 \in X_2'$. For all $\lambda \in \mathbb{R}$,

$$\begin{aligned}&(\lambda \boldsymbol{e}_1 + \mu_1 \boldsymbol{x}_1) + (\lambda \boldsymbol{e}_2 + \mu_2 \boldsymbol{x}_2)\\ &\qquad = \lambda \boldsymbol{e} + (\mu_1 \boldsymbol{x}_1 + \mu_2 \boldsymbol{x}_2) \in \mu_1 X_1' + \mu_2 X_2'.\end{aligned}$$

Since the sum $\mu_1 X_1' + \mu_2 X_2'$ is direct (see page 5), the inclusions

$$\lambda \boldsymbol{e}_1 + \mu_1 \boldsymbol{x}_1 \in S_1 \quad \text{and} \quad \lambda \boldsymbol{e}_2 + \mu_2 \boldsymbol{x}_2 \in S_2$$

show that

$$\lambda \boldsymbol{e}_1 + \mu_1 \boldsymbol{x}_1 \in \mu_1 X_1' \quad \text{and} \quad \lambda \boldsymbol{e}_2 + \mu_2 \boldsymbol{x}_2 \in \mu_2 X_2'$$

for all $\lambda \in \mathbb{R}$. Therefore (see Corollary 6.22),

$$\boldsymbol{e}_1 \in \operatorname{lin}(\mu_1 X_1') = \mu_1 \operatorname{lin} X_1' \quad \text{and} \quad \boldsymbol{e}_2 \in \operatorname{lin}(\mu_2 X_2') = \mu_2 \operatorname{lin} X_2',$$

implying the inclusion

$$\boldsymbol{e} = \boldsymbol{e}_1 + \boldsymbol{e}_2 \in \mu_1 \operatorname{lin} X_1' + \mu_2 \operatorname{lin} X_2'.$$

By the same corollary

$$\begin{aligned}\operatorname{lin}(\mu_1 X_1 + \mu_2 X_2) &= \operatorname{lin}(\mu_1 X_1' + \mu_2 X_2')\\ &\subset \mu_1 \operatorname{lin} X_1' + \mu_2 \operatorname{lin} X_2' = \mu_1 \operatorname{lin} X_1 + \mu_2 \operatorname{lin} X_2.\end{aligned}$$ □

Remark. The inclusion in Theorem 6.23 may be proper. Indeed, if

$$X_1 = \{(x, y) : x \geqslant 0, y \geqslant 0\} \quad \text{and} \quad X_2 = \{(x, y) : x \leqslant 0, y \leqslant 0\},$$

then $\operatorname{lin} X_1 = \operatorname{lin} X_2 = \{\boldsymbol{o}\}$. On the other hand, $X_1 + X_2 = \mathbb{R}^2$, implying that $\operatorname{lin}(X_1 + X_2) = \mathbb{R}^2$.

Theorem 6.24. *For a linear transformation $g : \mathbb{R}^n \to \mathbb{R}^m$ and sets $X \subset \mathbb{R}^n$ and $Y \subset \mathbb{R}^m$, one has*

$$g(\operatorname{lin} X) \subset \operatorname{lin} g(X) \quad \textit{and} \quad \operatorname{lin} g^{-1}(Y) = g^{-1}(\operatorname{lin}(Y \cap \operatorname{rng} g)).$$

Furthermore, if g is one-to-one, then $g(\operatorname{lin} X) = \operatorname{lin} g(X)$.

Proof. Excluding the obvious cases $X = \varnothing$ and $Y \cap \operatorname{rng} g = \varnothing$, we assume that both sets X and $Y \cap \operatorname{rng} g$ are nonempty. Consequently, $g^{-1}(Y) \neq \varnothing$.

By Theorems 6.7 and 6.20,

$$\begin{aligned} g(\operatorname{lin} X) &= g(\operatorname{rec} X \cap (-\operatorname{rec} X)) \subset g(\operatorname{rec} X) \cap g(-\operatorname{rec} X) \\ &\subset \operatorname{rec} g(X) \cap (-\operatorname{rec} g(X)) = \operatorname{lin} g(X). \end{aligned}$$

The same theorems imply that

$$\begin{aligned} \operatorname{lin} g^{-1}(Y) &= \operatorname{rec} g^{-1}(Y) \cap (-\operatorname{rec} g^{-1}(Y)) \\ &= g^{-1}(\operatorname{rec}(Y \cap \operatorname{rng} g)) \cap (-g^{-1}(\operatorname{rec}(Y \cap \operatorname{rng} g))) \\ &= g^{-1}(\operatorname{rec}(Y \cap \operatorname{rng} g) \cap (-\operatorname{rec}(Y \cap \operatorname{rng} g))) \\ &= g^{-1}(\operatorname{lin}(Y \cap \operatorname{rng} g)). \end{aligned}$$

If g is one-to-one, then, according to Theorem 6.7, we have

$$\begin{aligned} g(\operatorname{lin} X) &= g(\operatorname{rec} X \cap (-\operatorname{rec} X)) = g(\operatorname{rec} X) \cap g(-\operatorname{rec} X) \\ &= \operatorname{rec} g(X) \cap (-\operatorname{rec} g(X)) = \operatorname{lin} g(X). \end{aligned}$$ □

Remark. The inclusion $g(\operatorname{lin} X) \subset \operatorname{lin} g(X)$ in Theorem 6.24 may be proper even if $g(\operatorname{rec} X) = \operatorname{rec} g(X)$. Indeed, let $X = \{(x, y) : y \geqslant |x|\}$ and g be the orthogonal projection of $\mathbb{R}^2$ on the x-axis. Then $\operatorname{rec} X = X$ and $g(X)$ is the x-axis, which gives $g(\operatorname{rec} X) = g(X) = \operatorname{rec} g(X)$. On the other hand, $\operatorname{lin} X = \{\boldsymbol{o}\}$, while $\operatorname{lin} g(X)$ is the x-axis, showing that $g(\operatorname{lin} X) \neq \operatorname{lin} g(X)$.

6.2.3 *Lineality Decompositions*

The next theorem gives a way to reduce the study of arbitrary sets to those with zero lineality. The method is based on a decomposition of a set into a direct sum of its lineality space and a set of smaller dimension. For instance, the set $P = \{(x, y) : 1 \leqslant x < 2\}$ is the direct sum of its lineality space (which is the y-axis) and the semi-open segment $m = \{(x, 0) : 1 \leqslant x < 2\}$, as depicted in Figure 6.4.

Theorem 6.25. *If $X \subset \mathbb{R}^n$ is a nonempty set and $L \subset \mathbb{R}^n$ is a plane complementary to $\operatorname{lin} X$, then X can be expressed as the direct sum*

$$X = (X \cap L) + \operatorname{lin} X, \quad \textit{where} \quad \operatorname{lin}(X \cap L) = \{\boldsymbol{o}\}. \tag{6.15}$$

Furthermore,

$$\operatorname{rec} X = (\operatorname{rec} X \cap \operatorname{dir} L) + \operatorname{lin} X = \operatorname{rec}(X \cap L) + \operatorname{lin} X. \tag{6.16}$$

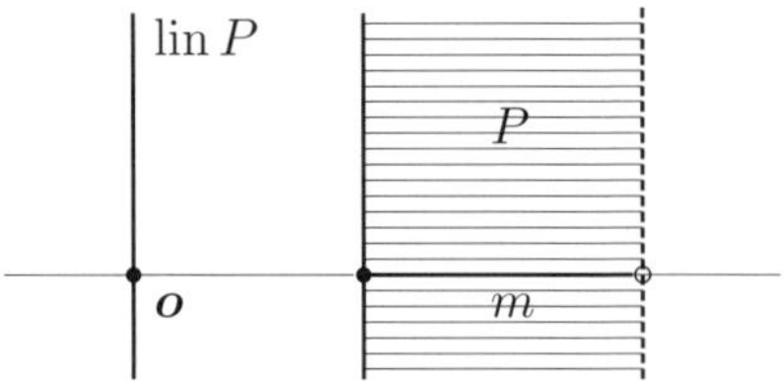

Fig. 6.4 Illustration to Theorem 6.25.

Proof. Choose a point $\boldsymbol{x} \in X$ and consider the plane $M = \boldsymbol{x} + \operatorname{lin} X$. Since L and M are complementary planes, Theorem 2.11 shows that $L \cap M$ is a singleton, say $\{\boldsymbol{z}\}$. By Theorem 6.20, $M \subset X$. Hence $\boldsymbol{z} \in M \cap L \subset X \cap L$. Because $\boldsymbol{x} + \operatorname{lin} X = \boldsymbol{z} + \operatorname{lin} X$ (see Theorem 2.2), we have

$$\boldsymbol{x} \in \boldsymbol{x} + \operatorname{lin} X = \boldsymbol{z} + \operatorname{lin} X \subset (X \cap L) + \operatorname{lin} X.$$

Thus $X \subset (X \cap L) + \operatorname{lin} X$.

Conversely, let $\boldsymbol{x} \in (X \cap L) + \operatorname{lin} X$. Then $\boldsymbol{x} = \boldsymbol{u} + \boldsymbol{e}$ such that $\boldsymbol{u} \in X \cap L$ and $\boldsymbol{e} \in \operatorname{lin} X$. By Theorem 6.20, $\boldsymbol{x} = \boldsymbol{u} + \boldsymbol{e} \in \boldsymbol{u} + \operatorname{lin} X \subset X$. Hence $(X \cap L) + \operatorname{lin} X \subset X$.

By Theorem 6.23,

$$\operatorname{lin} X = \operatorname{lin}(X \cap L) + \operatorname{lin}(\operatorname{lin} X) = \operatorname{lin}(X \cap L) + \operatorname{lin} X.$$

Therefore, $\operatorname{lin}(X \cap L) = \{\boldsymbol{o}\}$. Next, because $\operatorname{lin}(\operatorname{rec} X) = \operatorname{lin} X$ (see Theorem 6.20), the above argument, applied to $\operatorname{rec} X$ and $\operatorname{dir} L$ instead of X and L, respectively, gives the equality

$$\operatorname{rec} X = (\operatorname{rec} X \cap \operatorname{dir} L) + \operatorname{lin} X.$$

Finally, Theorem 6.6, used for the equality $X = (X \cap L) + \operatorname{lin} X$, gives

$$\operatorname{rec} X = \operatorname{rec}(X \cap L) + \operatorname{rec}(\operatorname{lin} X) = \operatorname{rec}(X \cap L) + \operatorname{lin} X. \qquad \square$$

Remark. If $X \subset \mathbb{R}^n$ is a nonempty set and $L \subset \mathbb{R}^n$ is a plane complementary to $\operatorname{lin} X$, then $X \cap L$ is a projection of X on L along $\operatorname{lin} X$ (compare with Theorem 2.88).

The next result gives a useful generalization of Theorem 6.25.

Theorem 6.26. *For a nonempty set $X \subset \mathbb{R}^n$, a subspace T of $\operatorname{lin} X$, and a complementary to T plane $L \subset \mathbb{R}^n$, one has*

$$X = (X \cap L) + T, \quad \textit{where} \quad \operatorname{lin}(X \cap L) = \operatorname{lin} X \cap \operatorname{dir} L.$$

Furthermore,

$$\operatorname{rec} X = (\operatorname{rec} X \cap \operatorname{dir} L) + T = \operatorname{rec}(X \cap L) + T.$$

Proof. Put $M = L \cap \text{lin}\, X$. Then

$$\text{dir}\, M = \text{dir}\, L \cap \text{lin}\, X \quad \text{and} \quad \text{lin}\, X = \text{dir}\, M + T.$$

Furthermore, $X + \text{dir}\, M = X$ according to Theorem 6.20. Denote by N a subplane of L complementary to M. Clearly,

$$L = N + \text{dir}\, M \quad \text{and} \quad \mathbb{R}^n = N + \text{lin}\, X.$$

Now Theorem 6.25 gives

$$\begin{aligned} X &= (X \cap N) + \text{lin}\, X = (X \cap N) + (\text{dir}\, M + T) \\ &= ((X + \text{dir}\, M) \cap (N + \text{dir}\, M)) + T = (X \cap L) + T. \end{aligned}$$

Because $\text{lin}\,(\text{rec}\, X) = \text{lin}\, X$ (see Theorem 6.20), the above argument, applied to $\text{rec}\, X$ and $\text{dir}\, L$ instead of X and L, respectively, gives the equality

$$\text{rec}\, X = (\text{rec}\, X \cap \text{dir}\, L) + T.$$

Finally, Theorem 6.6, used for the equality $X = (X \cap L) + T$, gives

$$\text{rec}\, X = \text{rec}\,(X \cap L) + \text{rec}\, T = \text{rec}\,(X \cap L) + T. \qquad \square$$

Corollary 6.27. *Let $X \subset \mathbb{R}^n$ be a nonempty set, T be a subspace of $\text{lin}\, X$, and $L \subset \mathbb{R}^n$ be a plane complementary to T. If a plane $L' \subset \mathbb{R}^n$ is a translate of L, then $X \cap L' = \boldsymbol{e} + X \cap L$ for a suitable vector $\boldsymbol{e} \in T$.*

Proof. Suppose that $L = \boldsymbol{a} + \text{dir}\, L$ and $L' = \boldsymbol{b} + L$ for suitable vectors $\boldsymbol{a}$ and $\boldsymbol{b}$. Since the subspaces $\text{dir}\, L$ and T are complementary, $\boldsymbol{b}$ can be written as $\boldsymbol{b} = \boldsymbol{c} + \boldsymbol{e}$, where $\boldsymbol{c} \in \text{dir}\, L$ and $\boldsymbol{e} \in T$. Therefore,

$$\begin{aligned} L' &= \boldsymbol{b} + L = (\boldsymbol{c} + \boldsymbol{e}) + \boldsymbol{a} + \text{dir}\, L = \boldsymbol{e} + \boldsymbol{a} + (\boldsymbol{c} + \text{dir}\, L) \\ &= \boldsymbol{e} + \boldsymbol{a} + \text{dir}\, L = \boldsymbol{e} + L. \end{aligned}$$

By Theorem 6.20, $X = \boldsymbol{e} + X$. Thus

$$X \cap L' = (\boldsymbol{e} + X) \cap (\boldsymbol{e} + L) = \boldsymbol{e} + X \cap L. \qquad \square$$

Sometimes it is desirable that the plane L in Theorem 6.25 satisfies the condition $\dim\,(X \cap L) = \dim L$. While, in general, this is not true, the result below shows that L can be replaced with a smaller plane still satisfying (6.15). For this, we will need an auxiliary lemma.

Lemma 6.28. *Let S and T be complementary subspaces of $\mathbb{R}^n$ and $L \subset \mathbb{R}^n$ be a plane whose translate contains S. Then $L = S + (L \cap T)$.*

Proof. Choose a vector $\boldsymbol{a} \in \mathbb{R}^n$ such that $S \subset \boldsymbol{a} + L$. The vector $\boldsymbol{a}$ can be uniquely written as $\boldsymbol{a} = \boldsymbol{b} + \boldsymbol{c}$, where $\boldsymbol{b} \in S$ and $\boldsymbol{c} \in T$. Therefore,

$$S = S - \boldsymbol{b} \subset \boldsymbol{a} + L - \boldsymbol{b} = \boldsymbol{c} + L.$$

The plane $M = \boldsymbol{c} + L$ is a subspace (see Theorem 2.2). From linear algebra we know that M can be expressed as the direct sum $M = S + (M \cap T)$. Finally,

$$\begin{aligned} L = M - \boldsymbol{c} &= S + (M \cap T) - \boldsymbol{c} \\ &= S + ((M - \boldsymbol{c}) \cap (T - \boldsymbol{c})) = S + (L \cap T). \end{aligned}$$ □

Theorem 6.29. *Let $X \subset \mathbb{R}^n$ be a nonempty set, and $L \subset \mathbb{R}^n$ be a plane complementary to $\operatorname{lin} X$. With $M = L \cap \operatorname{aff} X$, the set X can be expressed as the direct sum*

$$X = (X \cap M) + \operatorname{lin} X, \quad \textit{where} \quad \operatorname{lin}(X \cap M) = \{\boldsymbol{o}\}. \tag{6.17}$$

Furthermore, $\operatorname{aff}(X \cap M) = M$ *and*

$$\operatorname{rec} X = (\operatorname{rec} X \cap \operatorname{dir} M) + \operatorname{lin} X = \operatorname{rec}(X \cap M) + \operatorname{lin} X. \tag{6.18}$$

Proof. Clearly,

$$X \cap L = (X \cap \operatorname{aff} X) \cap L = X \cap (L \cap \operatorname{aff} X) = X \cap M.$$

This argument and Theorem 6.25 give

$$X = (X \cap L) + \operatorname{lin} X = (X \cap M) + \operatorname{lin} X.$$

By Theorem 6.23,

$$\operatorname{lin} X = \operatorname{lin}(X \cap M) + \operatorname{lin}(\operatorname{lin} X) = \operatorname{lin}(X \cap M) + \operatorname{lin} X.$$

Therefore, $\operatorname{lin}(X \cap M) = \{\boldsymbol{o}\}$.

For the equality $\operatorname{aff}(X \cap M) = M$, it suffices to show that $\dim(X \cap M) = \dim M$ (see Corollary 2.65). First, (6.17) gives

$$\dim(X \cap M) = \dim X - \dim(\operatorname{lin} X).$$

Since $\operatorname{lin} X \subset \operatorname{dir} X$ (see Theorem 6.20), and since $\operatorname{dir} X$ is a translate of $\operatorname{aff} X$ (see Theorem 2.48), Lemma 6.28 shows that

$$\operatorname{aff} X = \operatorname{lin} X + (S \cap \operatorname{aff} X),$$

where S is the direction space of L. According to Corollary 6.27, $S \cap \operatorname{aff} X$ is a translate of $L \cap \operatorname{aff} X$. Consequently,

$$\begin{aligned} \dim M &= \dim(L \cap \operatorname{aff} X) = \dim(S \cap \operatorname{aff} X) \\ &= \dim(\operatorname{aff} X) - \dim(\operatorname{lin} X) = \dim X - \dim(\operatorname{lin} X). \end{aligned}$$

Summing up, $\dim(X \cap M) = \dim M$, which gives $\operatorname{aff}(X \cap M) = M$.

Finally, (6.18) follows from (6.16). □

6.2.4 *Topology and Lineality Spaces*

A combination of Theorems 6.8, 6.9, and 6.20, implies the following corollary.

Corollary 6.30. *For any set* $X \subset \mathbb{R}^n$, *one has*

$$\mathrm{lin}\, X \subset \mathrm{lin}\,(\mathrm{cl}\, X), \quad \mathrm{lin}\, X \subset \mathrm{lin}\,(\mathrm{rint}\, X) \ \textit{if} \ \mathrm{rint}\, X \neq \varnothing. \qquad \square$$

Remark. Both inclusions in Corollary 6.30 may be proper. Furthermore, the subspaces $\mathrm{lin}\,(\mathrm{cl}\, X)$ and $\mathrm{lin}\,(\mathrm{rint}\, X)$ may be incomparable by inclusion. Indeed, let the set $X \subset \mathbb{R}^2$ be given by

$$X = \{(x, y) : 0 < y < 1\} \cup \{(x, y) : \text{both } x \text{ and } y \text{ are rational}\}.$$

Then $\mathrm{lin}\, X = \{\boldsymbol{o}\}$, $\mathrm{lin}\,(\mathrm{rint}\, X)$ is the x-axis of $\mathbb{R}^2$, and $\mathrm{lin}\,(\mathrm{cl}\, X) = \mathbb{R}^2$. Similarly, if

$$X = \{(x, y) : 0 < y < 1\} \cup \{(0, y) : y \in \mathbb{R}\},$$

then $\mathrm{lin}\, X = \mathrm{lin}\,(\mathrm{cl}\, X) = \{\boldsymbol{o}\}$, while $\mathrm{lin}\,(\mathrm{rint}\, X)$ is the x-axis of $\mathbb{R}^2$.

Theorem 6.31. *For a nonempty set* $X \subset \mathbb{R}^n$, *a subspace* T *of* $\mathrm{lin}\, X$, *and a complementary to* T *plane* $L \subset \mathbb{R}^n$, *one has*

$$\mathrm{cl}\, X = (\mathrm{cl}\, X \cap L) + T = \mathrm{cl}\,(X \cap L) + T, \tag{6.19}$$

$$\mathrm{rint}\, X = (\mathrm{rint}\, X \cap L) + T = \mathrm{rint}\,(X \cap L) + T, \tag{6.20}$$

$$\mathrm{rbd}\, X = (\mathrm{rbd}\, X \cap L) + T = \mathrm{rbd}\,(X \cap L) + T. \tag{6.21}$$

Proof. First, we will deal with (6.19). By Corollary 6.30, one has $T \subset \mathrm{lin}\, X \subset \mathrm{lin}\,(\mathrm{cl}\, X)$. This argument and Theorem 6.26 (applied to both X and $\mathrm{cl}\, X$) give

$$X = (X \cap L) + T \quad \text{and} \quad \mathrm{cl}\, X = (\mathrm{cl}\, X \cap L) + T.$$

Since the planes L and T are independent, Problem 1.13 gives

$$\mathrm{cl}\, X = \mathrm{cl}\,((X \cap L) + T) = \mathrm{cl}\,(X \cap L) + T.$$

Next, we are going to prove (6.20). Both inequalities are obvious when $\mathrm{rint}\, X = \varnothing$. So, we may assume that $\mathrm{rint}\, X \neq \varnothing$. Similarly to the above argument, $T \subset \mathrm{lin}\, X \subset \mathrm{lin}\,(\mathrm{rint}\, X)$, which gives

$$\mathrm{rint}\, X = (\mathrm{rint}\, X \cap L) + T.$$

Since the planes L and T are independent and since $\mathrm{rint}\, T = T$, Problem 3.10 gives

$$\mathrm{rint}\, X = \mathrm{rint}\,((X \cap L) + T) = \mathrm{rint}\,(X \cap L) + T.$$

Finally, consider (6.21). One has

$$\begin{aligned}\operatorname{rbd} X &= \operatorname{cl} X \setminus \operatorname{rint} X = (\operatorname{cl}(X \cap L) + T) \setminus (\operatorname{rint}(X \cap L) + T)\\ &= (\operatorname{cl}(X \cap L) \setminus (\operatorname{rint}(X \cap L)) + T = \operatorname{rbd}(X \cap L) + T.\end{aligned}$$

The second equality in (6.21) obviously follows from the identities

$$\begin{aligned}\operatorname{rbd} X \cap L &= (\operatorname{cl} X \setminus \operatorname{rint} X) \cap L = (\operatorname{cl} X \cap L) \setminus (\operatorname{rint} X \cap L)\\ &= \operatorname{cl}(X \cap L) \setminus \operatorname{rint}(X \cap L) = \operatorname{rbd}(X \cap L). \qquad \square\end{aligned}$$

In a similar way, Theorem 6.29 implies the corollary below.

Corollary 6.32. *Let $X \subset \mathbb{R}^n$ be a nonempty set, $L \subset \mathbb{R}^n$ be a plane complementary to* $\operatorname{lin} X$, *and* $M = L \cap \operatorname{aff} X$. *Then*

$$\begin{aligned}\operatorname{cl} X &= \operatorname{cl}(X \cap M) + \operatorname{lin} X = (\operatorname{cl} X \cap M) + \operatorname{lin} X,\\ \operatorname{rint} X &= \operatorname{rint}(X \cap M) + \operatorname{lin} X = (\operatorname{rint} X \cap M) + \operatorname{lin} X,\\ \operatorname{rbd} X &= \operatorname{rbd}(X \cap M) + \operatorname{lin} X = (\operatorname{rbd} X \cap M) + \operatorname{lin} X. \qquad \square\end{aligned}$$

Theorem 6.14 implies the corollary below.

Corollary 6.33. *For a convex set $K \subset \mathbb{R}^n$, the following assertions hold.*

(*a*) $\operatorname{lin}(\operatorname{cl} K) = \operatorname{lin}(\operatorname{rint} K)$.

(*b*) *If $K \neq \varnothing$ and $\boldsymbol{u} \in \operatorname{cl} K$, then*

$$\operatorname{lin}(\operatorname{cl} K) = \cap(\lambda(\operatorname{cl} K - \boldsymbol{u}) : \lambda \neq 0).$$

(*c*) *If $K \neq \varnothing$ and $\boldsymbol{u} \in \operatorname{rint} K$, then*

$$\operatorname{lin}(\operatorname{rint} K) = \cap(\lambda(\operatorname{rint} K - \boldsymbol{u}) : \lambda \neq 0).$$

(*d*) *If $M \subset \mathbb{R}^n$ is convex set containing K, then* $\operatorname{lin}(\operatorname{cl} K) \subset \operatorname{lin}(\operatorname{cl} M)$. $\square$

Corollary 6.15 implies one more result.

Corollary 6.34. *For a nonempty convex set $K \subset \mathbb{R}^n$, the following assertions hold.*

(*a*) *If $\boldsymbol{u}$ is a given point in* $\operatorname{cl} K$, *then* $\operatorname{lin}(\operatorname{cl} K)$ *is the largest among all subspaces $S \subset \mathbb{R}^n$ satisfying the inclusion $\boldsymbol{u} + S \subset \operatorname{cl} K$.*

(*b*) *If $\boldsymbol{u}$ is a given point in* $\operatorname{rint} K$, *then* $\operatorname{lin}(\operatorname{rint} K)$ *is the largest among all subspaces $S \subset \mathbb{R}^n$ satisfying the inclusion $\boldsymbol{u} + S \subset \operatorname{rint} K$.* $\square$

Corollary 6.34 immediately implies the following obvious assertion.

Corollary 6.35. *If $K \subset \mathbb{R}^n$ is a nonempty convex set, then*

$$\operatorname{lin}(\operatorname{cl} K) = \{\boldsymbol{e} \in \mathbb{R}^n : \exists \boldsymbol{x} \in K \text{ such that } \lambda\boldsymbol{e} + \boldsymbol{x} \in K \text{ whenever } \lambda \in \mathbb{R}\}.$$

6.2.5 *Line-Free Convex Sets and Bounded Plane Sections*

Definition 6.36. A convex set $K \subset \mathbb{R}^n$ is called *line-free* if it contains no line (in particular, the empty set $\varnothing$ is line-free).

Theorem 6.37. *For a convex set $K \subset \mathbb{R}^n$, the following conditions are equivalent.*

(*a*) *K is line-free (equivalently, $\mathrm{cl}\, K$ is line-free, or $\mathrm{rint}\, K$ is line-free).*
(*b*) *$\mathrm{rec}\,(\mathrm{cl}\, K)$ is line-free (equivalently, $\mathrm{rec}\,(\mathrm{rint}\, K)$ is line-free).*
(*c*) *$\mathrm{lin}\,(\mathrm{cl}\, K) \subset \{\boldsymbol{o}\}$ (equivalently, $\mathrm{lin}\,(\mathrm{rint}\, K) \subset \{\boldsymbol{o}\}$).*

Proof. Since all assertions of the theorem hold if $K = \varnothing$, we may assume that K is nonempty. First, we will show that the assertions in condition (*a*) are equivalent. Indeed, since $\mathrm{rint}\, K \subset K \subset \mathrm{cl}\, K$, we have

$$\mathrm{cl}\, K \text{ is line-free } \Rightarrow K \text{ is line-free } \Rightarrow \mathrm{rint}\, K \text{ is line-free.}$$

Conversely, suppose that $\mathrm{cl}\, K$ contains a line l. Choose points $\boldsymbol{u} \in l$ and $\boldsymbol{v} \in \mathrm{rint}\, K$. Then, by Corollary 6.11, the line $(\boldsymbol{v} - \boldsymbol{u}) + l$, considered as a cone with apex $\boldsymbol{v} - \boldsymbol{u}$, lies in $\mathrm{rint}\, K$.

(*a*) $\Leftrightarrow$ (*b*) According to Theorem 6.14, $\mathrm{rec}\,(\mathrm{cl}\, K) = \mathrm{rec}\,(\mathrm{rint}\, K)$. So, by the above argument, it suffices to prove that $\mathrm{cl}\, K$ is line-free if and only if $\mathrm{rec}\,(\mathrm{cl}\, K)$ is line-free.

Suppose that $\mathrm{cl}\, K$ is line-free and choose a point $\boldsymbol{u} \in \mathrm{cl}\, K$. Then $\boldsymbol{u} + \mathrm{rec}\,(\mathrm{cl}\, K) \subset \mathrm{cl}\, K$ according to Corollary 6.15. Hence the cone $\boldsymbol{u} + \mathrm{rec}\,(\mathrm{cl}\, K)$ is line-free, which implies that $\mathrm{rec}\,(\mathrm{cl}\, K)$ is line-free. Conversely, assume that $\mathrm{cl}\, K$ contains a line l. Choose a point $\boldsymbol{u} \in l$. If $\boldsymbol{v}$ is any point in $\mathrm{cl}\, K$, then the line $(\boldsymbol{v} - \boldsymbol{u}) + l$ lies in $\mathrm{cl}\, K$ (see Corollary 6.11). Hence the 1-dimensional subspace $l - \boldsymbol{u}$ lies in $\mathrm{lin}\,(\mathrm{cl}\, K)$, and whence in $\mathrm{rec}\,(\mathrm{cl}\, K)$.

(*b*) $\Leftrightarrow$ (*c*) It suffices to show that the cone $\mathrm{rec}\,(\mathrm{cl}\, K)$ contains a line l if and only if $\mathrm{dir}\, l \subset \mathrm{lin}\,(\mathrm{cl}\, K)$. Indeed, let $l \subset \mathrm{rec}\,(\mathrm{cl}\, K)$. By Theorem 5.17, the subspace $\mathrm{dir}\, l$ lies in $\mathrm{ap}\,(\mathrm{rec}\,(\mathrm{cl}\, K))$. Since $\mathrm{lin}\,(\mathrm{cl}\, K) = \mathrm{ap}\,(\mathrm{rec}\,(\mathrm{cl}\, K))$ according to Theorem 6.20, we obtain that $\mathrm{lin}\,(\mathrm{cl}\, K)$ contains $\mathrm{dir}\, l$. Conversely, if $\mathrm{lin}\,(\mathrm{cl}\, K)$ contains a 1-dimensional subspace m, then the equality $\mathrm{lin}\,(\mathrm{cl}\, K) = \mathrm{rec}\,(\mathrm{cl}\, K) \cap (-\mathrm{rec}\,(\mathrm{cl}\, K))$ shows that $\mathrm{rec}\,(\mathrm{cl}\, K)$ contains m. $\square$

Remark. If $K \subset \mathbb{R}^n$ is a line-free convex set, then the inclusions

$$\mathrm{rec}\, K \subset \mathrm{rec}\,(\mathrm{cl}\, K) \quad \text{and} \quad \mathrm{lin}\, K \subset \mathrm{lin}\,(\mathrm{cl}\, K),$$

proved in Theorem 6.9 and Corollary 6.30, respectively, combined with Theorem 6.37, show that the cone $\mathrm{rec}\, K$ is line-free and $\mathrm{lin}\, K = \{\boldsymbol{o}\}$. On the other hand, neither of these conditions is sufficient for K to be line-free. Indeed, the convex set $K = \{\boldsymbol{o}\} \cup \{(x, y) : 0 < y < 1\}$ in $\mathbb{R}^2$ contains a line, while $\operatorname{rec} K = \operatorname{lin} K = \{\boldsymbol{o}\}$.

A combination of Theorems 6.25, 6.29, and 6.37 implies the following result.

Corollary 6.38. *Let $K \subset \mathbb{R}^n$ be a nonempty convex set which is either closed or relative open, and let L be a plane (either in $\mathbb{R}^n$ or in $\operatorname{aff} K$) complementary to $\operatorname{lin} K$. Then the set $K \cap L$ is line-free and $K = (K \cap L) + \operatorname{lin} K$.* □

The following assertions about line-free convex sets are obvious:

1. The intersection of any family of line-free convex sets is a line-free convex set.
2. For a line-free convex set $K \subset \mathbb{R}^n$, a point $\boldsymbol{a} \in \mathbb{R}^n$, and a scalar $\mu \in \mathbb{R}$, the set $\boldsymbol{a} + \mu K$ is line-free.

The next two theorems describe conditions for the existence of bounded plane sections of a convex set in $\mathbb{R}^n$.

Theorem 6.39. *Let $K \subset \mathbb{R}^n$ be a nonempty convex set and $L \subset \mathbb{R}^n$ be a nonempty plane. If a translate of L meets K along a bounded set, then*

$$\operatorname{rec} K \cap \operatorname{dir} L = \{\boldsymbol{o}\}.$$

Consequently, $\dim L + \dim (\operatorname{lin} K) \leqslant n$.

Proof. Suppose that a translate L' of L meets K along a bounded set. Assume, for contradiction, that $\operatorname{rec} K \cap \operatorname{dir} L \neq \{\boldsymbol{o}\}$. Choose a nonzero vector $\boldsymbol{e} \in \operatorname{rec} K \cap \operatorname{dir} L$ and put $h = \{\lambda \boldsymbol{e} : \lambda \geqslant 0\}$. Let $\boldsymbol{u}$ be any point in $K \cap L'$. Then the closed halfline $\boldsymbol{u} + h$ lies in K by the definition of $\operatorname{rec} K$. We have $\operatorname{dir} L' = \operatorname{dir} L$ due to Theorem 2.4. Now, Theorem 2.2 gives

$$\boldsymbol{u} + h \subset \boldsymbol{u} + \operatorname{dir} L = \boldsymbol{u} + \operatorname{dir} L' = L'.$$

Thus $\boldsymbol{u} + h \subset K \cap L'$, contrary to the assumption on boundedness of $K \cap L'$. Hence $\operatorname{rec} K \cap \operatorname{dir} L = \{\boldsymbol{o}\}$.

Since $\operatorname{lin} K \subset \operatorname{rec} K$, the last equality results in $\operatorname{lin} K \cap \operatorname{dir} L = \{\boldsymbol{o}\}$, which shows that the subspaces $\operatorname{lin} K$ and $\operatorname{dir} L$ are independent. Consequently, the planes $\operatorname{lin} K$ and L are independent (see Definition 2.9), and Theorem 2.10 gives

$$\dim L + \dim (\operatorname{lin} K) = \dim (\operatorname{lin} K + L) \leqslant n. \qquad \square$$

Remark. We observe that boundedness of the set $K \cap L$ in Theorem 6.39 is not equivalent to the condition $\operatorname{rec} K \cap \operatorname{dir} L = \{\boldsymbol{o}\}$. Indeed, if X is a planar convex set depicted in Figure 6.2, then $\operatorname{rec} X = \{\boldsymbol{o}\}$, while every nonempty section of X by a horizontal line is unbounded.

Theorem 6.40. *Given a nonempty convex set $K \subset \mathbb{R}^n$ and a nonempty plane $L \subset \mathbb{R}^n$, the following conditions are equivalent.*

(*a*) *There is a translate of L which meets* $\operatorname{cl} K$ (*equivalently, meets* $\operatorname{rint} K$) *along a bounded set.*
(*b*) *All sections of K* (*equivalently, all sections of* $\operatorname{cl} K$, *or all sections of* $\operatorname{rint} K$) *by translates of L are bounded.*
(*c*) $\operatorname{rec}(\operatorname{cl} K) \cap \operatorname{dir} L = \operatorname{rec}(\operatorname{rint} K) \cap \operatorname{dir} L = \{\boldsymbol{o}\}$.

Proof. $(a) \Rightarrow (c)$ If a translate L' of L meets $\operatorname{cl} K$ (equivalently, $\operatorname{rint} K$) along a bounded set, then a combination of Theorems 6.14 and 6.39 show that condition (c) is satisfied.

$(c) \Rightarrow (b)$ Obviously, it suffices to show that all sections of $\operatorname{cl} K$ by translates of L are unbounded. Indeed, assume for a moment the existence of a translate L' of L which meets $\operatorname{cl} K$ along an unbounded set. According to Theorem 6.12, the set $\operatorname{cl} K \cap L'$ contains a closed halfline h with endpoint $\boldsymbol{u}$, say. By Corollary 6.15, the closed halfline $h' = h - \boldsymbol{u}$ lies in $\operatorname{rec}(\operatorname{cl} K)$. Since h' also lies in $\operatorname{dir} L'$ and $\operatorname{dir} L' = \operatorname{dir} L$ (see Theorem 2.4), we obtain a contradiction with condition (c).

$(b) \Rightarrow (a)$ The only nontrivial part of the proof of this assertion is to show that if all translates of L meet $\operatorname{rint} K$ along bounded sets, then so do the translates of L to $\operatorname{cl} K$. Assume for a moment the existence of a translate L' of L which meets $\operatorname{cl} K$ along an unbounded set. By Theorem 6.12, there is a closed halfline $h \subset \operatorname{cl} K \cap L'$ with endpoint $\boldsymbol{u}$, say. Choose a point $\boldsymbol{v} \in \operatorname{rint} K$. By Theorem 6.10, the closed halfline $h' = \boldsymbol{v} - \boldsymbol{u} + h$ lies in $\operatorname{rint} K$. Consequently, the plane $L' = (\boldsymbol{v} - \boldsymbol{u}) + L$ meets $\operatorname{rint} K$ along an unbounded set, a contradiction with condition (b). $\square$

Remark. If a nonempty convex set $K \subset \mathbb{R}^n$ is line-free, then the technique of polar cones allows us to prove the existence of a hyperplane $H \subset \mathbb{R}^n$ satisfying conditions (a)–(c) of Theorem 6.40 (see Corollary 8.9).

Theorem 6.41. *Let $K \subset \mathbb{R}^n$ be a nonempty convex set and $H \subset \mathbb{R}^n$ be a hyperplane which meets* $\operatorname{cl} K$ *along a bounded set. Then either K is line-free, or the subspace* $\operatorname{lin}(\operatorname{cl} K)$ *is one-dimensional and* $\operatorname{cl} K$ *is given as*

$$\operatorname{cl} K = (H \cap \operatorname{cl} K) + \operatorname{lin}(\operatorname{cl} K). \tag{6.22}$$

Proof. Suppose that K is not line-free. Then $\operatorname{lin}(\operatorname{cl} K) \neq \{\boldsymbol{o}\}$ according to Theorem 6.37. We observe that $\operatorname{dir} H \cap \operatorname{lin}(\operatorname{cl} K) = \{\boldsymbol{o}\}$. Otherwise, choosing a point $\boldsymbol{x} \in H \cap \operatorname{cl} K$ and a 1-dimensional subspace $l \subset \operatorname{dir} H \cap \operatorname{lin}(\operatorname{cl} K)$, one would have

$$\boldsymbol{x} + l \subset (\boldsymbol{x} + \operatorname{dir} H) \cap \operatorname{cl} K = H \cap \operatorname{cl} K$$

(see Corollary 6.15), contrary to the assumption that $H \cap \operatorname{cl} K$ is bounded. Hence $\operatorname{dir} H \cap \operatorname{lin}(\operatorname{cl} K) = \{\boldsymbol{o}\}$. This argument gives

$$\dim(\operatorname{lin}(\operatorname{cl} K)) \leqslant n - \dim(\operatorname{dir} H) = n - (n-1) = 1.$$

Consequently, $\dim(\operatorname{lin}(\operatorname{cl} K)) = 1$ due to the assumption $\operatorname{lin}(\operatorname{cl} K) \neq \{\boldsymbol{o}\}$. Therefore, the subspaces $\operatorname{dir} H$ and $\operatorname{lin}(\operatorname{cl} K)$ are complementary, and Theorem 6.25 gives (6.22). □

Problems and Notes for Chapter 6

Problems for Chapter 6

Problem 6.1. Let $X \subset \mathbb{R}^n$ be a nonempty closed set. Prove that $\operatorname{rec} X$ consists of all vectors $\boldsymbol{e} \in \mathbb{R}^n$ satisfying the condition: for every point $\boldsymbol{x} \in X$ there is a positive scalar $\varepsilon(\boldsymbol{e}, \boldsymbol{x})$ such that $\lambda\boldsymbol{e} + \boldsymbol{x} \in X$ whenever $0 < \lambda < \varepsilon(\boldsymbol{e}, \boldsymbol{x})$.

Problem 6.2. Given a nonempty convex set $K \subset \mathbb{R}^n$, prove that

$$\operatorname{rec} K = \{\boldsymbol{e} \in \mathbb{R}^n : \boldsymbol{e} + K \subset K\} = \cap(K - \boldsymbol{x} : \boldsymbol{x} \in K).$$

Problem 6.3. Given a nonempty convex set $K \subset \mathbb{R}^n$, prove that

$$\operatorname{lin} K = \{\boldsymbol{e} \in \mathbb{R}^n : \pm\boldsymbol{e} + K \subset K\} = \{\boldsymbol{e} \in \mathbb{R}^n : \boldsymbol{e} + K = K\}.$$

Problem 6.4. (Auslender and Teboulle [9, Proposition 2.1.3]) Let $K \subset \mathbb{R}^n$ be a nonempty convex set. Prove that $\operatorname{rec}(\operatorname{cl} K)$ consists of all vectors $\boldsymbol{e} \in \mathbb{R}^n$ satisfying the condition: for any sequence of positive scalars $\lambda_1, \lambda_2, \ldots$ tending to 0, there is a sequence of points $\boldsymbol{x}_1, \boldsymbol{x}_2, \ldots$ from K such that $\boldsymbol{e} = \lim_{i\to\infty} \lambda_i \boldsymbol{x}_i$.

Problem 6.5. Given closed convex sets K_1 and K_2 in $\mathbb{R}^n$, prove that

$$\operatorname{rec}(K_1 \cap K_2) = \operatorname{rec} K_1 \cap \operatorname{rec} K_2 \quad \text{and} \quad \operatorname{lin}(K_1 \cap K_2) = \operatorname{lin} K_1 \cap \operatorname{rec} K_2.$$

Definition. Let $L \subset \mathbb{R}^n$ be a plane of positive dimension. A *slab* of L is the intersection of L and a slab P of $\mathbb{R}^n$ such that $\varnothing \neq L \cap P \neq L$ (see page 72).

Problem 6.6. Let $K \subset \mathbb{R}^n$ be a closed convex set of positive dimension m. Prove the following assertions.

(a) $\dim(\operatorname{lin} K) = m$ if and only if K is a plane.

(*b*) $\dim(\operatorname{lin} K) = m - 1$ if and only if K is either a halfplane or a plane slab.

(*c*) K contains an $(m-1)$-dimensional plane if and only if K is a plane, a halfplane, or a plane slab.

Problem 6.7. (Steinitz [267, § 26], Stoker [270], Goberna, Jornet, Rodríguez [123]) Given a convex set $K \subset \mathbb{R}^n$, prove the equivalence of the following conditions.

(*a*) $\operatorname{rbd} K$ is disconnected.

(*b*) $\operatorname{rbd} K$ lies in the union of two distinct parallel planes of the same dimension.

(*c*) $\operatorname{cl} K$ is a closed plane slab.

Problem 6.8. Prove that the relative boundary of a closed convex set $K \subset \mathbb{R}^n$ is locally convex (see Problem 3.18) if and only if K is either a plane, a halfplane, or a plane slab.

Problem 6.9. Let $C \subset \mathbb{R}^n$ be a nonempty closed cone with apex $\boldsymbol{o}$. Prove the equivalence of the following conditions.

(*a*) $\operatorname{ap}(\operatorname{conv} C) = \{\boldsymbol{o}\}$.

(*b*) There is a scalar $\gamma > 0$ such that for any choice of nonnegative scalars $\lambda_1, \ldots, \lambda_r$ and any linearly independent set $\{\boldsymbol{x}_1, \ldots, \boldsymbol{x}_r\}$ of unit vectors in C, one has

$$\|\lambda_1 \boldsymbol{x}_1 + \cdots + \lambda_r \boldsymbol{x}_r\| \geqslant \gamma(\lambda_1 + \cdots + \lambda_r).$$

Problem 6.10. (Fenchel [105, Theorem 11]) Let a closed cone $C \subset \mathbb{R}^n$ with apex $\boldsymbol{a}$ have the property $\operatorname{ap}(\operatorname{conv} C) = \{\boldsymbol{a}\}$. Prove that $\operatorname{conv} C$ is a closed convex cone with apex $\boldsymbol{a}$.

Problem 6.11. Let $K \subset \mathbb{R}^n$ be a nonempty convex set and $P \subset \mathbb{R}^n$ be a closed slab whose boundary hyperplanes are translates of an $(n-1)$-dimensional subspace $S \subset \mathbb{R}^n$ (see page 72). Prove the equivalence of the following conditions.

(*a*) There is a translate of P which meets $\operatorname{cl} K$ (equivalently, meets $\operatorname{rint} K$) along a bounded set.

(*b*) $\operatorname{rec}(\operatorname{cl} K) \cap S = \operatorname{rec}(\operatorname{rint} K) \cap S = \{\boldsymbol{o}\}$.

Notes for Chapter 6

Recession cones. Given a nonempty closed convex set $K \subset \mathbb{R}^n$ and a point $\boldsymbol{a} \in K$, Steinitz [267, pp. 6–7] showed that K is bounded if and only if it contains no closed halfline with endpoint $\boldsymbol{a}$ (compare with Theorem 6.12). For the case when K is unbounded, Stoker [270] proved that the union $C_{\boldsymbol{a}}$ of all closed halflines enclosed in K and having $\boldsymbol{a}$ as an endpoint, is a convex cone. Since $C_{\boldsymbol{b}} = (\boldsymbol{b} - \boldsymbol{a}) + C_{\boldsymbol{a}}$ for any other point $\boldsymbol{b} \in K$ (see Theorem 6.10), Stoker called $C_{\boldsymbol{a}}$ the *characteristic cone* of K. Part (*b*) of Theorem 6.2 and Theorem 6.12 and can be found in Stoker [270]. Independently, Choquet [71] considered the closed convex cone $c(K) = \cap(\lambda(K - \boldsymbol{a}) : \lambda > 0)$, which he called the *asymptotic cone* of K. Choquet also mentioned that the set K is compact if and only if $c(K) = \{\boldsymbol{o}\}$ (compare with Theorem 6.14).

The term *recession cone* of K, denoted 0^+K, was coined by Rockafellar [241, p. 61], who argued that the term *asymptotic cone* does not agree with other uses of "asymptote" and "asymptotic" and may be misleading. Bair [10] defines the *asymptotic cone* C_X of an arbitrary set $X \subset \mathbb{R}^n$ as the union of $\boldsymbol{o}$ and all closed halflines starting at $\boldsymbol{o}$ whose translate may be included in $\operatorname{cl} X$; equivalently, $C_X = \cup(\cap(\lambda(\operatorname{cl} X - \boldsymbol{x}) : \lambda > 0) : \boldsymbol{x} \in \operatorname{cl} X)$. Recession properties of nonconvex sets are considered by Beer [30].

The second part of Theorem 6.18 is proved by Bair [16] for the case $X = \operatorname{rbd} K$.

Asymptotic cones. Theorem 6.17, which is due to Rockafellar [241, Theorem 8.2], is based on the approach of Fenchel [105, p. 42], who introduced the following definition: given a nonempty set $M \subset \mathbb{R}^n$ and a point $\boldsymbol{s} \in \mathbb{R}^n$, the union of closed halflines $[\boldsymbol{s}, \boldsymbol{x}\rangle$ which are the limits of sequences of closed halflines $[\boldsymbol{s}, \boldsymbol{x}_i\rangle$, where $\boldsymbol{x}_i \in M$ and $\|\boldsymbol{x}_i\| \to \infty$, is called the *asymptotic cone* $A_{\boldsymbol{s}}(M)$ of M with apex $\boldsymbol{s}$. Clearly, $A_{\boldsymbol{s}}(M)$ can be characterized as the set of limits of the form $\lambda_1\boldsymbol{x}_1, \lambda_2\boldsymbol{x}_2, \ldots$, where $\boldsymbol{x}_1, \boldsymbol{x}_2, \ldots$ belong to M and $\lambda_1, \lambda_2, \ldots$ are positive scalars tending to 0. Fenchel [105, pp. 42–44] observed that $A_{\boldsymbol{s}}(M)$ is a closed cone and that $A_{\boldsymbol{r}}(M) = A_{\boldsymbol{s}}(M) + (\boldsymbol{r} - \boldsymbol{s})$ for any choice of points $\boldsymbol{r}, \boldsymbol{s} \in \mathbb{R}^n$. He also proved that given a nonempty s-convex set $K \subset \mathbb{R}^n$ (the family of s-convex sets includes all closed and all relative open convex sets) and a point $\boldsymbol{r} \in K$, the cone $A_{\boldsymbol{r}}(K)$ coincides with the union of all closed halflines with apex $\boldsymbol{r}$ which are contained in K (that is, $A_{\boldsymbol{r}}(K) = \boldsymbol{r} + \operatorname{rec} K$).

The concept of asymptotic cone, denoted X_∞, for the case of any set X in a linear topological space was independently considered by Dedieu [90], who defined X_∞ as $\cap(\operatorname{cl}((0, \varepsilon]X) : \varepsilon > 0)$, or, equivalently, as the set of all vectors $\boldsymbol{e}$ which are limits of sequences $t_1^{-1}\boldsymbol{x}_1, t_2^{-1}\boldsymbol{x}_2, \ldots$, where all $\boldsymbol{x}_1, \boldsymbol{x}_2, \ldots$ belong to X and positive scalars $t_1, t_2, \ldots$ tend to ∞ (see Auslender and Teboulle [9, Chapter 2] for the properties and further references on asymptotic cones). Penot [228] introduced the concept of firm asymptotic cone of a set X.

Lineality spaces. In terms of lineality space, Theorem 6.25 was formulated by Fenchel [105, p. 44] for the case of s-convex sets.

Line-free convex sets and bounded plane sections. The term "line-free" can be found in Motzkin's PhD thesis [220, § 4]. The last assertion of Theorem 6.37 is contributed to Dragomirescu [96]. The equivalence of conditions (b) and (c) in Theorem 6.40 is due to Rockafellar [241, p. 64].

Chapter 7

Closedness Conditions

7.1 Closedness of Conic Hulls

7.1.1 *Conic Hulls and Normalized Sets*

According to Theorem 5.45, for any set $X \subset \mathbb{R}^n$ and a point $\boldsymbol{a} \in \mathbb{R}^n$, the following relations hold:

$$C_{\boldsymbol{a}}(\mathrm{cl}\, X) \subset \mathrm{cl}\, C_{\boldsymbol{a}}(X) \quad \text{and} \quad \mathrm{cone}_{\boldsymbol{a}}(\mathrm{cl}\, X) \subset \mathrm{cl}\,(\mathrm{cone}_{\boldsymbol{a}} X). \tag{7.1}$$

Furthermore, inclusions (7.1) become equalities provided X is bounded and $\boldsymbol{a} \notin \mathrm{cl}\, X$ and $\boldsymbol{a} \notin \mathrm{cl}\,(\mathrm{conv}\, X)$, respectively.

Theorem 7.4 below gives criteria for equalities in (7.1) regardless of the size of X and position of $\boldsymbol{a}$ with respect to X. These criteria are often called *closedness conditions* for generated cones and conic hulls. Our argument here is based on considering the intersection of a cone with apex $\boldsymbol{a}$ and a suitable sphere $S_\rho(\boldsymbol{a}) \subset \mathbb{R}^n$ (see Theorem 5.22). We will need the following auxiliary result.

Theorem 7.1. *Given a sphere $S_\rho(\boldsymbol{a}) \subset \mathbb{R}^n$, the following assertions hold.*

(*a*) *If $C \subset \mathbb{R}^n$ is a cone with apex $\boldsymbol{a}$, then $C_{\boldsymbol{a}}(C \cap S_\rho(\boldsymbol{a})) = \{\boldsymbol{a}\} \cup C$ and*

$$\mathrm{cl}\, C = C_{\boldsymbol{a}}(\mathrm{cl}\, C \cap S_\rho(\boldsymbol{a})) = C_{\boldsymbol{a}}(\mathrm{cl}\,(C \cap S_\rho(\boldsymbol{a}))). \tag{7.2}$$

(*b*) *If X is a subset of $S_\rho(\boldsymbol{a})$, then $C_{\boldsymbol{a}}(\mathrm{cl}\, X) = \mathrm{cl}\, C_{\boldsymbol{a}}(X)$.*

Proof. Since the case $C \subset \{\boldsymbol{a}\}$ is obvious, we may assume that $C \not\subset \{\boldsymbol{a}\}$.

(*a*) Combining the inclusion $C \cap S_\rho(\boldsymbol{a}) \subset \{\boldsymbol{a}\} \cup C$ and Theorem 5.30, we obtain that $C_{\boldsymbol{a}}(C \cap S_\rho(\boldsymbol{a})) \subset \{\boldsymbol{a}\} \cup C$. For the opposite inclusion, choose a point $\boldsymbol{x} \in \{\boldsymbol{a}\} \cup C$. Because the case $\boldsymbol{x} = \boldsymbol{a}$ is obvious, we may suppose that $\boldsymbol{x} \neq \boldsymbol{a}$. Let $\boldsymbol{u} = \boldsymbol{a} + \frac{\rho}{\|\boldsymbol{x}-\boldsymbol{a}\|}(\boldsymbol{x} - \boldsymbol{a})$. Clearly, $\boldsymbol{u} \in (\boldsymbol{a}, \boldsymbol{x}\rangle \subset C$ and $\boldsymbol{u} \in S_\rho(\boldsymbol{a})$ due to $\|\boldsymbol{a} - \boldsymbol{u}\| = \rho$. Thus $\boldsymbol{u} \in C \cap S_\rho(\boldsymbol{a})$. By Lemma 2.25,

$$\boldsymbol{x} \in (\boldsymbol{a}, \boldsymbol{u}\rangle \subset C_{\boldsymbol{a}}(C \cap S_\rho(\boldsymbol{a})).$$

Summing up, $\{\boldsymbol{a}\} \cup C \subset C_{\boldsymbol{a}}(C \cap S_\rho(\boldsymbol{a}))$.

Equalities (7.2) immediately follow from Theorems 5.22 and 5.45.

(*b*) This part follows from assertion (*b*) of Theorem 5.45. □

Definition 7.2. For a point $\boldsymbol{a} \in \mathbb{R}^n$ and a set $X \subset \mathbb{R}^n$, with $X \not\subset \{\boldsymbol{a}\}$, the *normalized set* of X with respect to $\boldsymbol{a}$ is defined by

$$E_{\boldsymbol{a}}(X) = \{\boldsymbol{e_x} = \boldsymbol{a} + \frac{\boldsymbol{x} - \boldsymbol{a}}{\|\boldsymbol{x} - \boldsymbol{a}\|} : \boldsymbol{x} \in X \setminus \{\boldsymbol{a}\}\}. \tag{7.3}$$

We let $E_{\boldsymbol{a}}(X) = \varnothing$ provided $X \subset \{\boldsymbol{a}\}$.

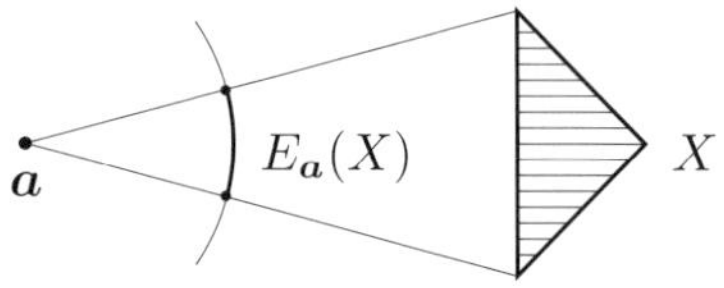

Fig. 7.1 Illustration to Definition 7.2.

Geometrically, the normalized set $E_{\boldsymbol{a}}(X)$ consists of all points at which the closed halflines $[\boldsymbol{a}, \boldsymbol{x}\rangle$, $\boldsymbol{x} \in X \setminus \{\boldsymbol{a}\}$, meet the unit sphere $S_1(\boldsymbol{a})$.

Theorem 7.3. *For sets X and Y in $\mathbb{R}^n$ and a point $\boldsymbol{a} \in \mathbb{R}^n$, the following assertions hold.*

(*a*) $E_{\boldsymbol{a}}(X) = C_{\boldsymbol{a}}(X) \cap S_1(\boldsymbol{a})$.
(*b*) $C_{\boldsymbol{a}}(X) = C_{\boldsymbol{a}}(E_{\boldsymbol{a}}(X))$.
(*c*) $C_{\boldsymbol{a}}(X) \subset C_{\boldsymbol{a}}(Y)$ *if and only if* $E_{\boldsymbol{a}}(X) \subset E_{\boldsymbol{a}}(Y)$.
(*d*) $C_{\boldsymbol{a}}(X) = C_{\boldsymbol{a}}(Y)$ *if and only if* $E_{\boldsymbol{a}}(X) = E_{\boldsymbol{a}}(Y)$.
(*e*) $\operatorname{cone}_{\boldsymbol{a}} X = C_{\boldsymbol{a}}(E_{\boldsymbol{a}}(\operatorname{conv} X))$.

Proof. Since the case $X \subset \{\boldsymbol{a}\}$ is obvious, we may suppose that $X \not\subset \{\boldsymbol{a}\}$. Then $E_{\boldsymbol{a}}(X) \neq \varnothing$.

(*a*) If a vector $\boldsymbol{e_x}$ of the form (7.3) belongs to $E_{\boldsymbol{a}}(X)$, then $\boldsymbol{e_x} \in [\boldsymbol{a}, \boldsymbol{x}\rangle \subset C_{\boldsymbol{a}}(X)$. Because $\|\boldsymbol{a} - \boldsymbol{e_x}\| = 1$, one has $\boldsymbol{e_x} \in S_1(\boldsymbol{a})$. Thus $\boldsymbol{e_x} \in C_{\boldsymbol{a}}(X) \cap S_1(\boldsymbol{a})$. Conversely, let $\boldsymbol{z} \in C_{\boldsymbol{a}}(X) \cap S_1(\boldsymbol{a})$. Then $\boldsymbol{z} = \boldsymbol{a} + \lambda(\boldsymbol{u} - \boldsymbol{a})$ for a suitable point $\boldsymbol{u} \in X \setminus \{\boldsymbol{a}\}$. Because $\boldsymbol{z} \in S_1(\boldsymbol{a})$, we have

$$\lambda = \|\boldsymbol{z} - \boldsymbol{a}\|/\|\boldsymbol{u} - \boldsymbol{a}\| = 1/\|\boldsymbol{u} - \boldsymbol{a}\|,$$

implying that $\boldsymbol{z} = \boldsymbol{e_u} \in E_{\boldsymbol{a}}(X)$.

(*b*) Let $\boldsymbol{u} \in C_{\boldsymbol{a}}(X)$. Because the case $\boldsymbol{u} = \boldsymbol{a}$ is obvious, we may suppose that $\boldsymbol{u} \neq \boldsymbol{a}$. Then $\boldsymbol{u} = \boldsymbol{a} + \lambda(\boldsymbol{x} - \boldsymbol{a})$ for a suitable point $\boldsymbol{x} \in X \setminus \{\boldsymbol{a}\}$ and a scalar $\lambda > 0$. With $\mu = \lambda\|\boldsymbol{x} - \boldsymbol{a}\|$, one has

$$\begin{aligned}\boldsymbol{u} &= \boldsymbol{a} + \lambda(\boldsymbol{x} - \boldsymbol{a}) = \boldsymbol{a} + \mu\big(\boldsymbol{a} + \frac{\boldsymbol{x} - \boldsymbol{a}}{\|\boldsymbol{x} - \boldsymbol{a}\|} - \boldsymbol{a}\big) \\ &= \boldsymbol{a} + \mu(\boldsymbol{e}_{\boldsymbol{x}} - \boldsymbol{a}) \in C_{\boldsymbol{a}}(E_{\boldsymbol{a}}(X)).\end{aligned}$$

Conversely, let $\boldsymbol{u} \in C_{\boldsymbol{a}}(E_{\boldsymbol{a}}(X))$. As above, we may assume that $\boldsymbol{u} \neq \boldsymbol{a}$. Then $\boldsymbol{u} = \boldsymbol{a} + \gamma(\boldsymbol{e}_{\boldsymbol{x}} - \boldsymbol{a})$ for a suitable vector $\boldsymbol{e}_{\boldsymbol{x}}$ of the form (7.3) and a scalar $\gamma > 0$. Then

$$\begin{aligned}\boldsymbol{u} &= \boldsymbol{a} + \gamma(\boldsymbol{e}_{\boldsymbol{x}} - \boldsymbol{a}) = \boldsymbol{a} + \gamma\big(\boldsymbol{a} + \frac{\boldsymbol{x} - \boldsymbol{a}}{\|\boldsymbol{x} - \boldsymbol{a}\|} - \boldsymbol{a}\big) \\ &= \boldsymbol{a} + \frac{\gamma}{\|\boldsymbol{x} - \boldsymbol{a}\|}(\boldsymbol{x} - \boldsymbol{a}) \in C_{\boldsymbol{a}}(X).\end{aligned}$$

(*c*) If $C_{\boldsymbol{a}}(X) \subset C_{\boldsymbol{a}}(Y)$, then the above argument gives

$$E_{\boldsymbol{a}}(X) = C_{\boldsymbol{a}}(X) \cap S_1(\boldsymbol{a}) \subset C_{\boldsymbol{a}}(Y) \cap S_1(\boldsymbol{a}) = E_{\boldsymbol{a}}(Y).$$

Similarly, if $E_{\boldsymbol{a}}(X) \subset E_{\boldsymbol{a}}(Y)$, then Theorem 5.31 implies that

$$C_{\boldsymbol{a}}(X) = C_{\boldsymbol{a}}(E_{\boldsymbol{a}}(X)) \subset C_{\boldsymbol{a}}(E_{\boldsymbol{a}}(Y)) = C_{\boldsymbol{a}}(Y).$$

Assertion (*d*) immediately follows from (*c*).

(*e*) Theorem 5.38 and assertion (*b*) give

$$\text{cone}_{\boldsymbol{a}} X = C_{\boldsymbol{a}}(\text{conv}\, X) = C_{\boldsymbol{a}}(E_{\boldsymbol{a}}(\text{conv}\, X)). \qquad \square$$

The next result follows assertion (*c*) of Theorem 5.22 and provides closedness criteria for generated cones and conic hulls.

Theorem 7.4. *For a set $X \subset \mathbb{R}^n$ and a point $\boldsymbol{a} \in \mathbb{R}^n$, the following assertions hold.*

(*a*) $E_{\boldsymbol{a}}(\text{cl}\, X) \subset \text{cl}\, E_{\boldsymbol{a}}(X)$.
(*b*) $C_{\boldsymbol{a}}(\text{cl}\, E_{\boldsymbol{a}}(X)) = \text{cl}\, C_{\boldsymbol{a}}(E_{\boldsymbol{a}}(X)) = \text{cl}\, C_{\boldsymbol{a}}(X)$.
(*c*) $C_{\boldsymbol{a}}(\text{cl}\, X) = \text{cl}\, C_{\boldsymbol{a}}(X)$ *if and only if* $E_{\boldsymbol{a}}(\text{cl}\, X) = \text{cl}\, E_{\boldsymbol{a}}(X)$.
(*d*) *The equalities below are equivalent:*

- (d_1) $\text{cone}_{\boldsymbol{a}}(\text{cl}\, X) = \text{cl}\,(\text{cone}_{\boldsymbol{a}} X)$,
- (d_2) $E_{\boldsymbol{a}}(\text{conv}\,(\text{cl}\, X)) = \text{cl}\, E_{\boldsymbol{a}}(\text{conv}\, X)$.

Proof. The case $X \subset \{\boldsymbol{a}\}$ is obvious; so, we may assume that $X \not\subset \{\boldsymbol{a}\}$.

(*a*) Combining Theorems 7.3, 5.45, and 5.22, we obtain

$$\begin{aligned}E_{\boldsymbol{a}}(\text{cl}\, X) &= C_{\boldsymbol{a}}(\text{cl}\, X) \cap S_1(\boldsymbol{a}) \subset \text{cl}\, C_{\boldsymbol{a}}(X) \cap S_1(\boldsymbol{a}) \\ &= \text{cl}\,(C_{\boldsymbol{a}}(X) \cap S_1(\boldsymbol{a})) = \text{cl}\, E_{\boldsymbol{a}}(X).\end{aligned}$$

(*b*) This assertion follows from Theorems 7.1 and 7.3.

(*c*) By the above argument,

$$C_{\boldsymbol{a}}(\mathrm{cl}\, X) = C_{\boldsymbol{a}}(E_{\boldsymbol{a}}(\mathrm{cl}\, X)) \subset C_{\boldsymbol{a}}(\mathrm{cl}\, E_{\boldsymbol{a}}(X)) = \mathrm{cl}\, C_{\boldsymbol{a}}(X).$$

So, $C_{\boldsymbol{a}}(\mathrm{cl}\, X) = \mathrm{cl}\, C_{\boldsymbol{a}}(X)$ if and only if $C_{\boldsymbol{a}}(E_{\boldsymbol{a}}(\mathrm{cl}\, X)) = C_{\boldsymbol{a}}(\mathrm{cl}\, E_{\boldsymbol{a}}(X))$, and Theorem 7.3 shows that the latter equality holds if and only if $E_{\boldsymbol{a}}(\mathrm{cl}\, X) = \mathrm{cl}\, E_{\boldsymbol{a}}(X)$.

(*d*) Based on Theorems 5.38 and 4.16, we obtain:

$$\begin{aligned}\mathrm{cone}_{\boldsymbol{a}}(\mathrm{cl}\, X) &= C_{\boldsymbol{a}}(\mathrm{conv}\,(\mathrm{cl}\, X)) = C_{\boldsymbol{a}}(E_{\boldsymbol{a}}(\mathrm{conv}\,(\mathrm{cl}\, X)))\\ &\subset C_{\boldsymbol{a}}(E_{\boldsymbol{a}}(\mathrm{cl}\,(\mathrm{conv}\, X))) \subset C_{\boldsymbol{a}}(\mathrm{cl}\, E_{\boldsymbol{a}}(\mathrm{conv}\, X))\\ &= \mathrm{cl}\, C_{\boldsymbol{a}}(E_{\boldsymbol{a}}(\mathrm{conv}\, X)) = \mathrm{cl}\, C_{\boldsymbol{a}}(\mathrm{conv}\, X) = \mathrm{cl}\,(\mathrm{cone}_{\boldsymbol{a}} X).\end{aligned}$$

So, equality (d_1) holds if and only if

$$C_{\boldsymbol{a}}(E_{\boldsymbol{a}}(\mathrm{conv}\,(\mathrm{cl}\, X))) = C_{\boldsymbol{a}}(\mathrm{cl}\, E_{\boldsymbol{a}}(\mathrm{conv}\, X)),$$

and Theorem 7.3 shows that the latter equality holds if and only if (d_2) is true. □

Remark. Condition (d_2) of Theorem 7.4 cannot be replaced with $E_{\boldsymbol{a}}(\mathrm{cl}\, X) = \mathrm{cl}\, E_{\boldsymbol{a}}(X)$, as shown in the example below. Let X be a set in $\mathbb{R}^3$, expressed as the union $l \cup h \cup Y$, where l is the y-axis, h is the nonnegative part of the z-axis, and

$$Y = \{(1, y, z) : y \neq 0,\ z = \tfrac{1}{|y|}\}.$$

It is easy to see that both sets X and $C = C_{\boldsymbol{o}}(X)$ are closed. Consequently, the set $E_{\boldsymbol{a}}(X)$ is closed. Thus $E_{\boldsymbol{a}}(\mathrm{cl}\, X) = \mathrm{cl}\, E_{\boldsymbol{a}}(X)$. On the other hand, the cone

$$\mathrm{conv}\, C = \mathrm{cone}_{\boldsymbol{o}} X = l \cup \{(x, y, z) : x \geqslant 0, z > 0\}$$

is not closed, implying that the set $E_{\boldsymbol{a}}(\mathrm{conv}\, X)$ also is not closed. (See in this regard Problem 6.10.)

7.1.2 *Generated Cones and Recession Directions*

While establishing the closedness of a generated cone $C_{\boldsymbol{a}}(K)$, where $K \subset \mathbb{R}^n$ is a convex set and $\boldsymbol{a}$ is a point in $\mathbb{R}^n$, it is often useful to distinguish the cases $\boldsymbol{a} \in \mathrm{cl}\, K$ and $\boldsymbol{a} \in \mathbb{R}^n \setminus \mathrm{cl}\, K$ (see Theorems 7.6 and Theorem 7.7).

We start with the case $\boldsymbol{a} \in \mathrm{cl}\, K$. The closedness of $C_{\boldsymbol{a}}(K)$ is obvious in the case $\boldsymbol{a} \in \mathrm{rint}\, K$; indeed, Theorem 5.48 shows that the cone $C_{\boldsymbol{a}}(K)$ is the plane $\mathrm{aff}\, K$, which is a closed set (see Corollary 2.21). So, the only nontrivial case here is that of $\boldsymbol{a} \in \operatorname{rbd} K$. Theorem 7.6 below deals with an important particular shape of K which guarantees the equality

$$C_{\boldsymbol{a}}(\operatorname{cl} K) = \operatorname{cl} C_{\boldsymbol{a}}(K).$$

Definition 7.5. A nonempty set $X \subset \mathbb{R}^n$ is called *locally conic* at a point $\boldsymbol{a} \in \operatorname{cl} X$ provided there is a cone $C \subset \mathbb{R}^n$ with apex $\boldsymbol{a}$ and a closed ball $B_\rho(\boldsymbol{a}) \subset \mathbb{R}^n$ such that $X \cap B_\rho(\boldsymbol{a}) = C \cap B_\rho(\boldsymbol{a})$.

We observe that the apex $\boldsymbol{a}$ of C in Definition 7.5 is proper if and only if $\boldsymbol{a} \in X$.

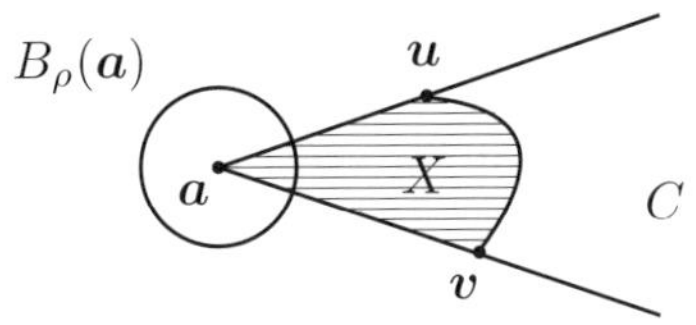

Fig. 7.2 A set X is locally conic at $\boldsymbol{a}$.

It is easy to see (as depicted in Figure 7.2) that a 2-dimensional convex set X in the plane $\mathbb{R}^2$ is locally conic at a point $\boldsymbol{a} \in \operatorname{bd} X$ if and only if $\operatorname{bd} X$ contains a pair of nontrivial closed segments $[\boldsymbol{a}, \boldsymbol{u}]$ and $[\boldsymbol{a}, \boldsymbol{v}]$ (possibly, $\boldsymbol{a} \in (\boldsymbol{u}, \boldsymbol{v})$). See Problem 11.10 in this regard.

Theorem 7.6. *Let a convex set $K \subset \mathbb{R}^n$ be locally conic at a point $\boldsymbol{a} \in \operatorname{cl} K$ and $C \subset \mathbb{R}^n$ be a cone with apex $\boldsymbol{a}$ such that $K \cap B_\rho(\boldsymbol{a}) = C \cap B_\rho(\boldsymbol{a})$ for a suitable closed ball $B_\rho(\boldsymbol{a}) \subset \mathbb{R}^n$. The following assertions hold.*

(*a*) *The cone C is convex and*
$$C \subset C_{\boldsymbol{a}}(K \cap B_\rho(\boldsymbol{a})) \subset C_{\boldsymbol{a}}(K).$$
(*b*) *The set $\operatorname{cl} K$ is locally conic at $\boldsymbol{a}$ and*
$$\operatorname{cl} K \cap B_\rho(\boldsymbol{a}) = \operatorname{cl} C \cap B_\rho(\boldsymbol{a}).$$
(*c*) $C_{\boldsymbol{a}}(\operatorname{cl} K) = \operatorname{cl} C = \operatorname{cl} C_{\boldsymbol{a}}(K)$.

Proof. Because the case $K \subset \{\boldsymbol{a}\}$ is obvious, we assume that $K \not\subset \{\boldsymbol{a}\}$.

(*a*) The set $C \cap B_\rho(\boldsymbol{a}) = K \cap B_\rho(\boldsymbol{a})$ is convex as the intersection of convex sets K and $B_\rho(\boldsymbol{a})$. Therefore, the convexity of C follows from Theorem 5.22. By Theorem 7.1,

$$\begin{aligned} C \subset \{\boldsymbol{a}\} \cup C &= C_{\boldsymbol{a}}(C \cap S_\rho(\boldsymbol{a})) \subset C_{\boldsymbol{a}}(C \cap B_\rho(\boldsymbol{a})) \\ &= C_{\boldsymbol{a}}(K \cap B_\rho(\boldsymbol{a})) \subset C_{\boldsymbol{a}}(K). \end{aligned}$$

(*b*) We observe that $\boldsymbol{a} \in \operatorname{cl} C$, as follows from Theorem 5.25. This argument and Theorem 3.51 imply that $B_\rho(\boldsymbol{a})$ meets both sets $\operatorname{rint} K$ and $\operatorname{rint} C$. Consequently, Theorem 3.54 gives

$$\operatorname{cl} K \cap B_\rho(\boldsymbol{a}) = \operatorname{cl}(K \cap B_\rho(\boldsymbol{a})) = \operatorname{cl}(C \cap B_\rho(\boldsymbol{a})) = \operatorname{cl} C \cap B_\rho(\boldsymbol{a}).$$

By Theorem 5.28, $\operatorname{cl} C$ is a convex cone with apex $\boldsymbol{a}$. Hence, according to Definition 7.5, $\operatorname{cl} K$ is locally conic at $\boldsymbol{a}$.

(*c*) Since $\operatorname{cl} C$ is a convex cone with apex $\boldsymbol{a}$, one has

$$C_{\boldsymbol{a}}(\operatorname{cl} C) = \operatorname{cl} C = \operatorname{cl} C_{\boldsymbol{a}}(\operatorname{cl} C).$$

Now, assertion (*b*), and Theorems 5.45 and 5.57 give

$$\begin{aligned} C_{\boldsymbol{a}}(\operatorname{cl} K) &= C_{\boldsymbol{a}}(\operatorname{cl} K \cap B_\rho(\boldsymbol{a})) = C_{\boldsymbol{a}}(\operatorname{cl} C \cap B_\rho(\boldsymbol{a})) \\ &= C_{\boldsymbol{a}}(\operatorname{cl} C) = \operatorname{cl} C = \operatorname{cl} C_{\boldsymbol{a}}(\operatorname{cl} C) = \operatorname{cl} C_{\boldsymbol{a}}(\operatorname{cl} C \cap B_\rho(\boldsymbol{a})) \\ &= \operatorname{cl} C_{\boldsymbol{a}}(\operatorname{cl} K \cap B_\rho(\boldsymbol{a})) = \operatorname{cl} C_{\boldsymbol{a}}(\operatorname{cl} K) = \operatorname{cl} C_{\boldsymbol{a}}(K). \end{aligned} \qquad \square$$

Remarks. 1. Both inclusions $C \subset C_{\boldsymbol{a}}(K \cap B_\rho(\boldsymbol{a})) \subset C_{\boldsymbol{a}}(K)$ in Theorem 7.6 may be proper if $\boldsymbol{a} \in \operatorname{cl} K \setminus K$ (compare with Theorem 5.57). Indeed, let $K \subset \mathbb{R}^2$ be a convex set, given by

$$K = \{(x, y) : x > 0, y > 0\} \cup h_1 \cup h_2,$$

where $h_1 = \{(x, 0) : x \geqslant 2\}$ and $h_2 = \{(0, y) : y \geqslant 2\}$. Clearly, K is locally conic at $\boldsymbol{o}$, and the convex cone $C = \{(x, y) : x > 0, y > 0\}$ satisfies the condition $K \cap \mathbb{B} = C \cap \mathbb{B}$, where $\mathbb{B}$ means the closed unit ball of $\mathbb{R}^n$. Therefore,

$$C \neq C_{\boldsymbol{o}}(K \cap \mathbb{B}) = \{\boldsymbol{o}\} \cup C \neq \operatorname{cl} C = C_{\boldsymbol{o}}(K).$$

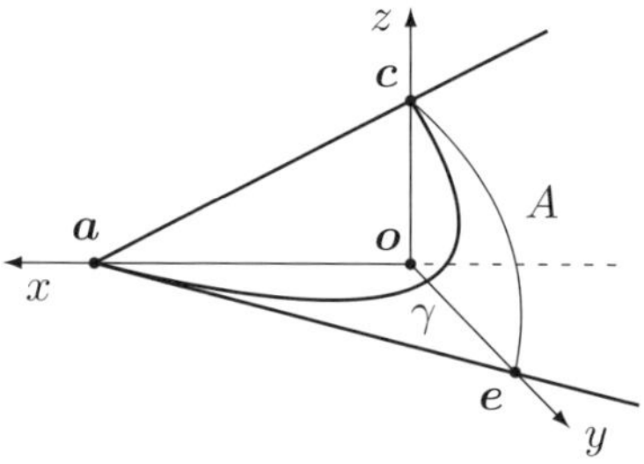

Fig. 7.3 Illustration to Theorem 7.6.

2. A convex set $K \subset \mathbb{R}^n$ satisfying the condition $C_{\boldsymbol{a}}(\operatorname{cl} K) = \operatorname{cl} C_{\boldsymbol{a}}(K)$, where $\boldsymbol{a} \in \operatorname{rbd} K$, may be locally non-conic at $\boldsymbol{a}$. Indeed, denote by A the quarter arc of the unit circle in the yz-coordinate plane of $\mathbb{R}^3$, as depicted in Figure 7.3. Let $\boldsymbol{c}$ and $\boldsymbol{e}$ be the endpoints of A. Choose a nonzero point $\boldsymbol{a}$ on the x-axis, and consider the closed convex cone $C = \mathrm{cone}_{\boldsymbol{a}}(A \cup \{\boldsymbol{o}\})$. Denote by γ a smooth curve with endpoints $\boldsymbol{a}$ and $\boldsymbol{c}$ on the boundary of C such that the closed segment $[\boldsymbol{a}, \boldsymbol{e}]$ is tangent to γ at $\boldsymbol{a}$ only. It is easy to see that the set $K = \mathrm{conv}\,([\boldsymbol{e}, \boldsymbol{o}] \cup \gamma)$ is compact and convex and $C = C_{\boldsymbol{a}}(K)$. At the same time, K is not locally conic at $\boldsymbol{a}$ because $K \cap B_\rho(\boldsymbol{a}) \neq C \cap B_\rho(\boldsymbol{a})$ for any $\rho > 0$.

The next result considers the generated cone $C_{\boldsymbol{a}}(K)$ of a convex set $K \subset \mathbb{R}^n$ under the condition $\boldsymbol{a} \in \mathbb{R}^n \setminus \mathrm{cl}\, K$.

Theorem 7.7. *If $K \subset \mathbb{R}^n$ is a convex set and $\boldsymbol{a} \in \mathbb{R}^n \setminus \mathrm{cl}\, K$, then*

$$\mathrm{cl}\, C_{\boldsymbol{a}}(K) = C_{\boldsymbol{a}}(\mathrm{cl}\, K) \cup (\boldsymbol{a} + \mathrm{rec}\,(\mathrm{cl}\, K)). \tag{7.4}$$

Proof. Excluding the obvious case $K = \varnothing$ (then both sides of (7.4) are equal to $\{\boldsymbol{a}\}$), we may suppose that K is nonempty.

If K is bounded, then $\mathrm{rec}\,(\mathrm{cl}\, K) = \{\boldsymbol{o}\}$ (see Theorem 6.14), and Theorem 5.45 confirms (7.4). Hence one can assume that K is unbounded. We first assert that

$$C_{\boldsymbol{a}}(\mathrm{cl}\, K) \cup (\boldsymbol{a} + \mathrm{rec}\,(\mathrm{cl}\, K)) \subset \mathrm{cl}\, C_{\boldsymbol{a}}(K). \tag{7.5}$$

Since $C_{\boldsymbol{a}}(\mathrm{cl}\, K) \subset \mathrm{cl}\, C_{\boldsymbol{a}}(K)$ according to Theorem 5.45, it suffices to prove the inclusion

$$\boldsymbol{a} + \mathrm{rec}\,(\mathrm{cl}\, K) \subset \mathrm{cl}\, C_{\boldsymbol{a}}(K). \tag{7.6}$$

For this, choose a point $\boldsymbol{z} \in \boldsymbol{a} + \mathrm{rec}\,(\mathrm{cl}\, K)$. Then $\boldsymbol{z} = \boldsymbol{a} + \boldsymbol{x}$, where $\boldsymbol{x} \in \mathrm{rec}\,(\mathrm{cl}\, K)$. By Theorem 6.17, $\boldsymbol{x} = \lim_{i\to\infty} \lambda_i \boldsymbol{x}_i$, where $\lambda_i > 0$ and $\boldsymbol{x}_i \in K$ for all $i \geqslant 1$ such that $\lim_{i\to\infty} \lambda_i = 0$. Theorem 5.38 gives

$$\boldsymbol{a} + \lambda_i(\boldsymbol{x}_i - \boldsymbol{a}) \in C_{\boldsymbol{a}}(K) \quad \text{for all} \quad i \geqslant 1.$$

Therefore,

$$\boldsymbol{z} = \boldsymbol{a} + \boldsymbol{x} = \boldsymbol{a} + \lim_{i\to\infty} \lambda_i \boldsymbol{x}_i = \lim_{i\to\infty} (\boldsymbol{a} + \lambda_i(\boldsymbol{x}_i - \boldsymbol{a})) \in \mathrm{cl}\, C_{\boldsymbol{a}}(K).$$

Consequently, (7.6) holds.

For the opposite to (7.5) inclusion, it suffices to show that

$$\mathrm{cl}\, C_{\boldsymbol{a}}(K) \setminus C_{\boldsymbol{a}}(\mathrm{cl}\, K) \subset \boldsymbol{a} + \mathrm{rec}\,(\mathrm{cl}\, K). \tag{7.7}$$

Because (7.7) is obvious when $\mathrm{cl}\, C_{\boldsymbol{a}}(K) = C_{\boldsymbol{a}}(\mathrm{cl}\, K)$, we may assume that $C_{\boldsymbol{a}}(\mathrm{cl}\, K)$ is a proper subset of $\mathrm{cl}\, C_{\boldsymbol{a}}(K)$ (see Theorem 5.45). Choose a point

$\boldsymbol{x} \in \operatorname{cl} C_{\boldsymbol{a}}(K) \setminus C_{\boldsymbol{a}}(\operatorname{cl} K)$ and a sequence of points $\boldsymbol{x}_1, \boldsymbol{x}_2, \ldots$ from $C_{\boldsymbol{a}}(K)$ converging to $\boldsymbol{x}$. Since $\boldsymbol{x} \neq \boldsymbol{a}$, we may suppose that none of the points $\boldsymbol{x}_1, \boldsymbol{x}_2, \ldots$ is $\boldsymbol{a}$. Then $\boldsymbol{x}_i = \boldsymbol{a} + \mu_i(\boldsymbol{z}_i - \boldsymbol{a})$ for suitable points $\boldsymbol{z}_i \in K$ and scalars $\mu_i > 0$, $i \geqslant 1$.

We assert that the sequence $\boldsymbol{z}_1, \boldsymbol{z}_2, \ldots$ is unbounded. Indeed, assume for a moment that $\boldsymbol{z}_1, \boldsymbol{z}_2, \ldots$ is bounded. Then it contains a subsequence $\boldsymbol{z}_{i_1}, \boldsymbol{z}_{i_2}, \ldots$ converging to a point $\boldsymbol{z} \in \operatorname{cl} K$. Clearly, $\boldsymbol{z} \neq \boldsymbol{a}$ due to the assumption $\boldsymbol{a} \notin \operatorname{cl} K$. By a continuity argument,

$$\lim_{j\to\infty} \mu_{i_j} = \lim_{j\to\infty} \frac{\|\boldsymbol{x}_{i_j} - \boldsymbol{a}\|}{\|\boldsymbol{z}_{i_j} - \boldsymbol{a}\|} = \frac{\|\boldsymbol{x} - \boldsymbol{a}\|}{\|\boldsymbol{z} - \boldsymbol{a}\|}.$$

With $\mu = \frac{\|\boldsymbol{x}-\boldsymbol{a}\|}{\|\boldsymbol{z}-\boldsymbol{a}\|}$, one has

$$\boldsymbol{x} = \lim_{j\to\infty} \boldsymbol{x}_{i_j} = \lim_{j\to\infty} (\boldsymbol{a} + \mu_{i_j}(\boldsymbol{z}_{i_j} - \boldsymbol{a})) = \boldsymbol{a} + \mu(\boldsymbol{z} - \boldsymbol{a}) \in C_{\boldsymbol{a}}(\operatorname{cl} K),$$

contradicting the choice of $\boldsymbol{x}$. Thus $\boldsymbol{z}_1, \boldsymbol{z}_2, \ldots$ must be unbounded.

Since the sequence $\boldsymbol{x}_1 - \boldsymbol{a}, \boldsymbol{x}_2 - \boldsymbol{a}, \ldots$ is bounded, one has

$$\lim_{i\to\infty} \mu_i = \lim_{i\to\infty} \frac{\|\boldsymbol{x}_i - \boldsymbol{a}\|}{\|\boldsymbol{z}_i - \boldsymbol{a}\|} = 0.$$

Therefore,

$$\boldsymbol{x} - \boldsymbol{a} = \lim_{i\to\infty} (\boldsymbol{x}_i - \boldsymbol{a}) = \lim_{i\to\infty} \mu_i(\boldsymbol{z}_i - \boldsymbol{a}) = \lim_{i\to\infty} \mu_i \boldsymbol{z}_i,$$

and Theorem 6.17 gives $\boldsymbol{x} - \boldsymbol{a} \in \operatorname{rec}(\operatorname{cl} K)$, or $\boldsymbol{x} \in \boldsymbol{a} + \operatorname{rec}(\operatorname{cl} K)$. Summing up, (7.7) holds, and (7.4) is proved. $\square$

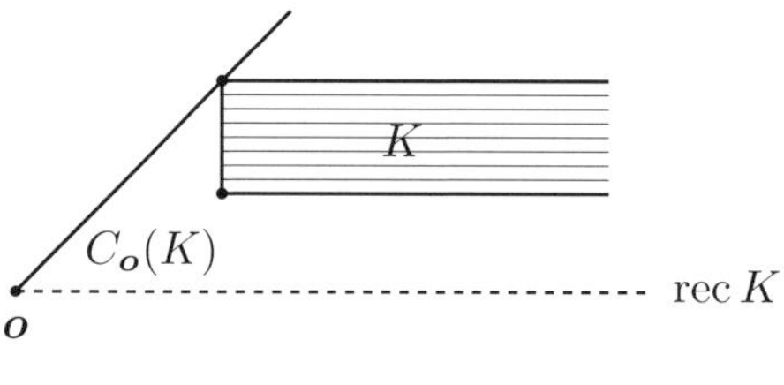

Fig. 7.4 Illustration to Theorem 7.7.

The following examples illustrate Theorem 7.7.

Examples. 1. For the closed convex set $K = \{(x, y) : 2 \leqslant x,\ 1 \leqslant y \leqslant 2\}$ in $\mathbb{R}^2$, the recession cone $\operatorname{rec} K$ is the closed halfline $\{(x, 0) : x \geqslant 0\}$. The generated cone $C_{\boldsymbol{o}}(K)$ is given by $C_{\boldsymbol{o}}(K) = \{\boldsymbol{o}\} \cup \{(x, y) : x > 0,\ 0 < y \leqslant x\}$, as depicted in Figure 7.4. Clearly, $\operatorname{rec} K \not\subset C_{\boldsymbol{o}}(K)$.

2. Theorem 7.7 does not hold if $\boldsymbol{a} \in \operatorname{rbd} K$. Indeed, consider the closed convex set $M = \{(x,y) : y^2 \geqslant x\}$ in $\mathbb{R}^2$. The cone $\operatorname{rec} M$ is the closed halfline $\{(0,y) : y \geqslant 0\}$. If $\boldsymbol{a}$ is a boundary point of M, say $\boldsymbol{a} = (1,1)$, then the closed halfline $\boldsymbol{a} + \operatorname{rec} M = \{(1,y) : y \geqslant 1\}$ lies in M. At the same time, the generated cone $C_{\boldsymbol{a}}(M)$ is the union of $\{\boldsymbol{a}\}$ and the open halfline $\{(x,y) : y > 2x - 1\}$. We observe that every cone $C_{\boldsymbol{s}}(M)$ is closed whenever $\boldsymbol{s} \notin M$.

Corollary 7.8. *If $K \subset \mathbb{R}^n$ is a proper convex set and $\boldsymbol{a}$ is a point in $\mathbb{R}^n \setminus \operatorname{cl} K$, then the equality*

$$\operatorname{cl} C_{\boldsymbol{a}}(K) = C_{\boldsymbol{a}}(\operatorname{cl} K) \tag{7.8}$$

holds if and only if $\boldsymbol{a} + \operatorname{rec}(\operatorname{cl} K) \subset C_{\boldsymbol{a}}(\operatorname{cl} K)$. In particular, if $\operatorname{aff} K \neq \mathbb{R}^n$ and $\boldsymbol{a} \in \mathbb{R}^n \setminus \operatorname{aff} K$, then (7.8) takes place if and only if the set K is bounded.

Proof. The first assertion immediately follows from Theorem 7.7.

Assume, additionally, that $\operatorname{aff} K \neq \mathbb{R}^n$ and $\boldsymbol{a} \in \mathbb{R}^n \setminus \operatorname{aff} K$. If the set K is bounded, then $\operatorname{rec}(\operatorname{cl} K) = \{\boldsymbol{o}\}$ (see Theorem 6.14), implying the inclusion $\boldsymbol{a} + \operatorname{rec}(\operatorname{cl} K) = \{\boldsymbol{a}\} \subset C_{\boldsymbol{a}}(\operatorname{cl} K)$. Then (7.8) holds by the above argument.

Conversely, suppose that (7.8) holds. Then $\boldsymbol{a} + \operatorname{rec}(\operatorname{cl} K) \subset C_{\boldsymbol{a}}(\operatorname{cl} K)$. By Theorem 6.2, $\operatorname{rec}(\operatorname{cl} K) \subset \operatorname{dir}(\operatorname{cl} K)$, and Theorem 5.39 gives

$$\boldsymbol{a} + \operatorname{rec}(\operatorname{cl} K) \subset (\boldsymbol{a} + \operatorname{dir}(\operatorname{cl} K)) \cap C_{\boldsymbol{a}}(\operatorname{cl} K) = \{\boldsymbol{a}\}.$$

Thus $\operatorname{rec}(\operatorname{cl} K) = \{\boldsymbol{o}\}$, and K is bounded (see Theorem 6.14). □

The next theorem describes an important property of linearity spaces.

Theorem 7.9. *If $K \subset \mathbb{R}^n$ is a nonempty convex set and $\boldsymbol{a}$ is a point in $\mathbb{R}^n \setminus \operatorname{cl} K$, then*

$$\operatorname{lin}(\operatorname{cl} C_{\boldsymbol{a}}(K)) = \operatorname{lin}(\operatorname{cl} K). \tag{7.9}$$

Proof. Because $\operatorname{cl} K \subset \operatorname{cl} C_{\boldsymbol{a}}(K)$, Corollary 6.33 shows that

$$\operatorname{lin}(\operatorname{cl} K) \subset \operatorname{lin}(\operatorname{cl} C_{\boldsymbol{a}}(K)).$$

Hence it suffices to prove the opposite inclusion

$$\operatorname{lin}(\operatorname{cl} C_{\boldsymbol{a}}(K)) \subset \operatorname{lin}(\operatorname{cl} K).$$

For this, choose a vector $\boldsymbol{e} \in \operatorname{lin}(\operatorname{cl} C_{\boldsymbol{a}}(K))$. The case $\boldsymbol{e} = \boldsymbol{o}$ is obvious. So, we may assume that $\boldsymbol{e} \neq \boldsymbol{o}$. By the definition,

$$\operatorname{lin}(\operatorname{cl} C_{\boldsymbol{a}}(K)) = \operatorname{rec}(\operatorname{cl} C_{\boldsymbol{a}}(K)) \cap (-\operatorname{rec}(\operatorname{cl} C_{\boldsymbol{a}}(K))).$$

Hence both vectors $\boldsymbol{e}$ and $-\boldsymbol{e}$ belong to $\operatorname{rec}(\operatorname{cl} C_{\boldsymbol{a}}(K))$. Theorem 6.2 shows that both points $\boldsymbol{a}+\boldsymbol{e}$ and $\boldsymbol{a}-\boldsymbol{e}$ belong to $\operatorname{cl} C_{\boldsymbol{a}}(K)$. Let $\boldsymbol{y}_1, \boldsymbol{y}_2, \ldots$ and $\boldsymbol{z}_1, \boldsymbol{z}_2, \ldots$ be sequences of points in $C_{\boldsymbol{a}}(K)$ such that

$$\lim_{i\to\infty} \boldsymbol{y}_i = \boldsymbol{a}+\boldsymbol{e} \quad \text{and} \quad \lim_{i\to\infty} \boldsymbol{z}_i = \boldsymbol{a}-\boldsymbol{e}.$$

Because $\boldsymbol{e} \neq \boldsymbol{o}$, we may suppose that none of the points $\boldsymbol{y}_i$ and $\boldsymbol{z}_i$, $i \geqslant 1$, is $\boldsymbol{a}$. Therefore, we can write

$$\boldsymbol{y}_i = \boldsymbol{a} + \lambda_i(\boldsymbol{u}_i - \boldsymbol{a}) \quad \text{and} \quad \boldsymbol{z}_i = \boldsymbol{a} + \mu_i(\boldsymbol{v}_i - \boldsymbol{a}), \quad i \geqslant 1, \tag{7.10}$$

where $\boldsymbol{u}_i, \boldsymbol{v}_i \in K$ and $\lambda_i, \mu_i > 0$ for all $i \geqslant 1$. Choosing suitable subsequences of $\boldsymbol{y}_1, \boldsymbol{y}_2, \ldots$ and $\boldsymbol{z}_1, \boldsymbol{z}_2, \ldots$, we may assume that the respective sequences $\lambda_1, \lambda_2, \ldots$ and $\mu_1, \mu_2, \ldots$ converge (possibly, to ∞).

We are going to show that, in fact,

$$\lim_{i\to\infty} \lambda_i = \lim_{i\to\infty} \mu_i = 0.$$

Indeed, assume, for instance, that $\lim_{i\to\infty} \lambda_i \geqslant \gamma > 0$. Then

$$0 < \frac{1}{\lambda_i + \mu_i} < \gamma^{-1} \quad \text{for all} \quad i \geqslant 1.$$

Consider the sequence of convex combinations

$$\boldsymbol{c}_i = \frac{\lambda_i}{\lambda_i+\mu_i}\boldsymbol{u}_i + \frac{\mu_i}{\lambda_i+\mu_i}\boldsymbol{v}_i, \quad i \geqslant 1.$$

Then all $\boldsymbol{c}_1, \boldsymbol{c}_2, \ldots$ belong to K by the convexity of K. Rewriting (7.10) as

$$\boldsymbol{u}_i = \boldsymbol{a} + \frac{\boldsymbol{y}_i - \boldsymbol{a}}{\lambda_i} \quad \text{and} \quad \boldsymbol{v}_i = \boldsymbol{a} + \frac{\boldsymbol{z}_i - \boldsymbol{a}}{\mu_i}, \quad i \geqslant 1,$$

we can express $\boldsymbol{c}_i$ as

$$\begin{aligned} \boldsymbol{c}_i &= \frac{\lambda_i}{\lambda_i+\mu_i}\Big(\boldsymbol{a} + \frac{\boldsymbol{y}_i - \boldsymbol{a}}{\lambda_i}\Big) + \frac{\mu_i}{\lambda_i+\mu_i}\Big(\boldsymbol{a} + \frac{\boldsymbol{z}_i - \boldsymbol{a}}{\mu_i}\Big) \\ &= \boldsymbol{a} + \frac{(\boldsymbol{y}_i - \boldsymbol{a}) + (\boldsymbol{z}_i - \boldsymbol{a})}{\lambda_i + \mu_i}, \quad i \geqslant 1. \end{aligned}$$

Since $\lim_{i\to\infty}((\boldsymbol{y}_i - \boldsymbol{a}) + (\boldsymbol{z}_i - \boldsymbol{a})) = \boldsymbol{e} - \boldsymbol{e} = \boldsymbol{o}$, one has

$$\lim_{i\to\infty} \boldsymbol{c}_i = \boldsymbol{a} + \lim_{i\to\infty} \frac{(\boldsymbol{y}_i - \boldsymbol{a}) + (\boldsymbol{z}_i - \boldsymbol{a})}{\lambda_i + \mu_i} = \boldsymbol{a}.$$

Thus $\boldsymbol{a} \in \operatorname{cl} K$, contrary to the assumption.

Hence $\lim_{i\to\infty} \lambda_i = \lim_{i\to\infty} \mu_i = 0$. Finally, the equalities

$$\begin{aligned} \boldsymbol{e} &= \lim_{i\to\infty} (\boldsymbol{y}_i - \boldsymbol{a}) = \lim_{i\to\infty} \lambda_i(\boldsymbol{u}_i - \boldsymbol{a}) = \lim_{i\to\infty} \lambda_i \boldsymbol{u}_i, \\ -\boldsymbol{e} &= \lim_{i\to\infty} (\boldsymbol{z}_i - \boldsymbol{a}) = \lim_{i\to\infty} \mu_i(\boldsymbol{v}_i - \boldsymbol{a}) = \lim_{i\to\infty} \mu_i \boldsymbol{v}_i \end{aligned}$$

and Theorem 6.17 show that both $\boldsymbol{e}$ and $-\boldsymbol{e}$ belong to $\operatorname{rec}(\operatorname{cl} K)$. Thus $\boldsymbol{e} \in \operatorname{lin}(\operatorname{cl} K)$, which proves the inclusion $\operatorname{lin}(\operatorname{cl} C_{\boldsymbol{a}}(K)) \subset \operatorname{lin}(\operatorname{cl} K)$. $\square$

Remark. Theorem 7.9 does not hold if $\boldsymbol{a} \in \operatorname{rbd} K$. Indeed, let K be the unit disk of $\mathbb{R}^2$ and $\boldsymbol{a} = (1, 0)$. Clearly, $\operatorname{lin} K = \{\boldsymbol{o}\}$. On the other hand, $\operatorname{cl} C_{\boldsymbol{a}}(K)$ is the closed halfplane $\{(x, y) : x \leqslant 1\}$ and $\operatorname{lin}(\operatorname{cl} C_{\boldsymbol{a}}(K))$ is the y-axis.

7.2 Closedness of Affine Images

7.2.1 *Affine Images of Arbitrary Sets*

As stated in Corollary 2.80, for a set $X \subset \mathbb{R}^n$ and an affine transformation $f : \mathbb{R}^n \to \mathbb{R}^m$, one has $f(\mathrm{cl}\, X) \subset \mathrm{cl}\, f(X)$, with the equality if X is bounded or f is one-to-one. Theorem 7.14 below shows that asymptotic planes provide a useful criterion for the *closedness condition* $f(\mathrm{cl}\, X) = \mathrm{cl}\, f(X)$ without any restrictions on X and f.

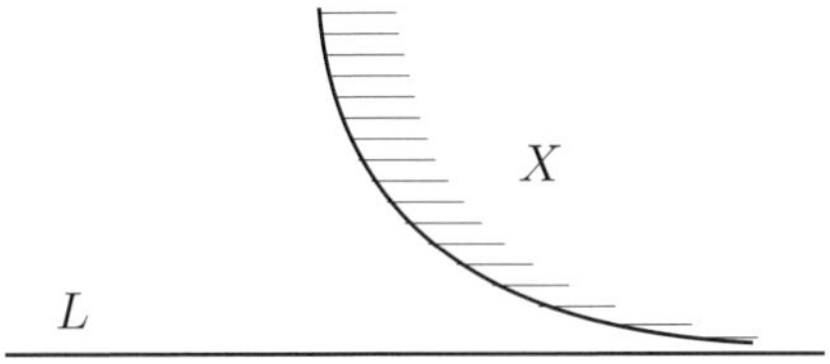

Fig. 7.5 An asymptotic plane L to a set X.

We recall (see page 11) that the *inf*-distance between nonempty sets X and Y in $\mathbb{R}^n$ is defined by

$$\delta(X, Y) = \inf \{\|\boldsymbol{x} - \boldsymbol{y}\| : \boldsymbol{x} \in X,\ \boldsymbol{y} \in Y\}.$$

Definition 7.10. A nonempty plane $L \subset \mathbb{R}^n$ is called *asymptotic* to a nonempty set $X \subset \mathbb{R}^n$ (or L is an *asymptote* of X) provided

$$L \cap \mathrm{cl}\, X = \varnothing \quad \text{and} \quad \delta(L, X) = 0.$$

The results below illustrate the role of asymptotic planes in closedness of vector sums.

Lemma 7.11. *For a nonempty set $X \subset \mathbb{R}^n$ and a subspace $S \subset \mathbb{R}^n$, the following assertions hold.*

(a) $X + S \subset \mathrm{cl}\, X + S \subset \mathrm{cl}\,(X + S) = \mathrm{cl}\,(\mathrm{cl}\, X + S)$.
(b) *The set* $\mathrm{cl}\, X + S$ *is closed if and only if* $\mathrm{cl}\, X + S = \mathrm{cl}\,(X + S)$.

Proof. Assertion (a) follows from standard properties of closure (see Problem 1.13), while assertion (b) is an immediate consequence of (a). □

Lemma 7.12. *Let $X \subset \mathbb{R}^n$ be a nonempty set and $S \subset \mathbb{R}^n$ be a subspace. A translate L of S is an asymptote of X if and only if*

$$L \subset \mathrm{cl}\,(X + S) \setminus (\mathrm{cl}\, X + S).$$

Proof. Let a translate $L = \boldsymbol{c} + S$ be an asymptote of X. We first assert that $L \cap (\mathrm{cl}\, X + S) = \varnothing$. Indeed, assume for a moment the existence of a point $\boldsymbol{z} \in L \cap (\mathrm{cl}\, X + S)$. Then $\boldsymbol{z} = \boldsymbol{c} + \boldsymbol{u} = \boldsymbol{x} + \boldsymbol{v}$, where $\boldsymbol{x} \in \mathrm{cl}\, X$ and $\boldsymbol{u}, \boldsymbol{v} \in S$. Consequently,

$$\boldsymbol{x} = \boldsymbol{c} + (\boldsymbol{u} - \boldsymbol{v}) \in \boldsymbol{c} + S = L,$$

contrary to the assumption $\mathrm{cl}\, X \cap L = \varnothing$.

For the inclusion $L \subset \mathrm{cl}\,(X + S)$, choose a point $\boldsymbol{z} \in L$. Then $\boldsymbol{z} = \boldsymbol{c} + \boldsymbol{u}$, where $\boldsymbol{u} \in S$. The condition $\delta(X, L) = 0$ means the existence of a sequence $\boldsymbol{x}_1, \boldsymbol{x}_2, \dots$ of points from X such that $\lim_{i\to\infty} \delta(\boldsymbol{x}_i, L) = 0$. Denote by $\boldsymbol{z}_i$ the orthogonal projection of $\boldsymbol{x}_i$ on L. Clearly, $\delta(\boldsymbol{x}_i, L) = \|\boldsymbol{x}_i - \boldsymbol{z}_i\|$ (see Theorem 2.90). Because $\boldsymbol{z}_i \in L = \boldsymbol{c} + S$, we can write $\boldsymbol{z}_i = \boldsymbol{c} + \boldsymbol{u}_i$ for a suitable point $\boldsymbol{u}_i \in S$, $i \geqslant 1$. Therefore,

$$\boldsymbol{x}_i + (\boldsymbol{z} - \boldsymbol{z}_i) = \boldsymbol{x}_i + (\boldsymbol{u} - \boldsymbol{u}_i) \in X + S.$$

Furthermore, from

$$\|(\boldsymbol{x}_i + \boldsymbol{z} - \boldsymbol{z}_i) - \boldsymbol{z}\| = \|\boldsymbol{x}_i - \boldsymbol{z}_i\| = \delta(\boldsymbol{x}_i, L) \to 0 \quad \text{as} \quad i \to \infty,$$

we obtain the inclusion $\boldsymbol{z} \in \mathrm{cl}\,(X + S)$. Hence $L \subset \mathrm{cl}\,(X + S)$.

Conversely, suppose that $L \subset \mathrm{cl}\,(X + S) \setminus (\mathrm{cl}\, X + S)$. From $L + S = L$ (see Theorem 2.2), one has

$$(\mathrm{cl}\, X \cap L) + S \subset (\mathrm{cl}\, X + S) \cap (L + S) = (\mathrm{cl}\, X + S) \cap L = \varnothing.$$

Therefore, $\mathrm{cl}\, X \cap L = \varnothing$. Now, choose a point $\boldsymbol{z} \in L$ and express it as $\boldsymbol{z} = \boldsymbol{c} + \boldsymbol{u}$, where $\boldsymbol{u} \in S$. The inclusion $\boldsymbol{z} \in L \subset \mathrm{cl}\,(X + S)$ implies the existence of an infinite sequence $\boldsymbol{z}_1, \boldsymbol{z}_2, \dots$ of points from $X + S$ such that $\lim_{i\to\infty} \|\boldsymbol{z}_i - \boldsymbol{z}\| = 0$. Let $\boldsymbol{z}_i = \boldsymbol{x} + \boldsymbol{u}_i$, where $\boldsymbol{x}_i \in X$ and $\boldsymbol{u}_i \in S$, $i \geqslant 1$. We have $\boldsymbol{c} + \boldsymbol{u} - \boldsymbol{u}_i \in \boldsymbol{c} + S = L$. Therefore,

$$\delta(\boldsymbol{x}_i, L) \leqslant \|\boldsymbol{x}_i - (\boldsymbol{c} + \boldsymbol{u} - \boldsymbol{u}_i)\| = \|\boldsymbol{z}_i - \boldsymbol{z}\| \to 0 \quad \text{as} \quad i \to \infty.$$

Hence $\delta(X, L) = 0$, and L is an asymptote of X. □

Theorem 7.13. *If $X \subset \mathbb{R}^n$ is a nonempty set and $S \subset \mathbb{R}^n$ is a subspace, then the sum $\mathrm{cl}\, X + S$ is not closed if and only if a suitable translate of S is an asymptote of X.*

Proof. If a suitable translate $L = \boldsymbol{c} + S$ of S is an asymptote of X, then a combination of Lemmas 7.11 and 7.12 gives

$$L \subset \mathrm{cl}\,(X + S) \setminus (\mathrm{cl}\, X + S) = \mathrm{cl}\,(\mathrm{cl}\, X + S) \setminus (\mathrm{cl}\, X + S),$$

implying that the sum $\mathrm{cl}\, X + S$ is not closed.

Conversely, suppose that the sum $\mathrm{cl}\, X + S$ is not closed. Choose a point

$$\boldsymbol{a} \in \mathrm{cl}\,(\mathrm{cl}\, X + S) \setminus (\mathrm{cl}\, X + S) = \mathrm{cl}\,(X + S) \setminus (\mathrm{cl}\, X + S)$$

and consider the plane $M = \boldsymbol{a} + S$. We assert that $(\mathrm{cl}\, X + S) \cap M = \varnothing$. Indeed, assume the existence of a point $\boldsymbol{z} \in (\mathrm{cl}\, X + S) \cap M$. Then $\boldsymbol{z} = \boldsymbol{x} + \boldsymbol{u} = \boldsymbol{a} + \boldsymbol{v}$, where $\boldsymbol{x} \in \mathrm{cl}\, X$ and $\boldsymbol{u}, \boldsymbol{v} \in S$. Consequently,

$$\boldsymbol{a} = \boldsymbol{x} + (\boldsymbol{u} - \boldsymbol{v}) \in \mathrm{cl}\, X + S,$$

which is impossible by the choice of $\boldsymbol{a}$.

Next, we assert that $M \subset \mathrm{cl}\,(X+S)$. Indeed, let $\boldsymbol{e} \in M$. Then $\boldsymbol{e} = \boldsymbol{a}+\boldsymbol{u}$ for a suitable point $\boldsymbol{u} \in S$. Because $\boldsymbol{a} \in \mathrm{cl}\,(X + S)$, there is a sequence of points $\boldsymbol{a}_1, \boldsymbol{a}_2, \ldots$ in $X+S$ such that $\lim_{i\to\infty} \|\boldsymbol{a}_i - \boldsymbol{a}\| = 0$. Let $\boldsymbol{e}_i = \boldsymbol{a}_i + \boldsymbol{u}$, $i \geqslant 1$. Then

$$\boldsymbol{e}_i \in (X + S) + S = X + S,$$
$$\lim_{i\to\infty} \|\boldsymbol{e}_i - \boldsymbol{e}\| = \lim_{i\to\infty} \|\boldsymbol{a}_i - \boldsymbol{a}\| = 0.$$

Hence $\boldsymbol{e} \in \mathrm{cl}\,(X + S)$, and $M \subset \mathrm{cl}\,(X + S)$. By Lemma 7.12, M is an asymptote of X. $\square$

The closure of the sum $X + S$ in terms of asymptotic translates of S is described in Problem 7.4.

Theorem 7.14. *Let $f : \mathbb{R}^n \to \mathbb{R}^m$ be an affine transformation, expressed as $f(\boldsymbol{x}) = \boldsymbol{a} + g(\boldsymbol{x})$, where $\boldsymbol{a} \in \mathbb{R}^m$ and $g : \mathbb{R}^n \to \mathbb{R}^m$ is a linear transformation, and let $X \subset \mathbb{R}^n$ be a nonempty set. The following conditions are equivalent.*

(*a*) $f(\mathrm{cl}\, X) = \mathrm{cl}\, f(X)$.
(*b*) *The sum* $\mathrm{cl}\, X + \mathrm{null}\, g$ *is closed.*
(*c*) *No translate of the subspace* $\mathrm{null}\, g$ *is an asymptote of* X.

Proof. Because the case $X = \varnothing$ is obvious, we may suppose that $X \neq \varnothing$. Furthermore, due to the obvious relations

$$f(\mathrm{cl}\, X) = \boldsymbol{a} + g(\mathrm{cl}\, X) \quad \text{and} \quad \mathrm{cl}\, f(X) = \boldsymbol{a} + \mathrm{cl}\, g(X),$$

we can replace condition (*a*) with $g(\mathrm{cl}\, X) = \mathrm{cl}\, g(X)$.

(*a*) $\Rightarrow$ (*b*) The equality $g(\mathrm{cl}\, X) = \mathrm{cl}\, g(X)$ shows that the set $g(\mathrm{cl}\, X)$ is closed. Choose a subspace $S \subset \mathbb{R}^n$ complementary to $\mathrm{null}\, g$ and denote by Y the projection of $\mathrm{cl}\, X$ on S along $\mathrm{null}\, g$. Then $\mathrm{cl}\, X + \mathrm{null}\, g = Y + \mathrm{null}\, g$, which gives

$$g(\mathrm{cl}\, X) = g(\mathrm{cl}\, X + \mathrm{null}\, g) = g(Y + \mathrm{null}\, g) = g(Y).$$

Consequently, the set $g(Y)$ is closed.

The mapping $h : S \to \operatorname{rng} f$ defined by $h(\boldsymbol{x}) = g(\boldsymbol{x})$ is an invertible linear transformation (see Problem 1.1). Clearly, $h(Y) = g(Y)$. Furthermore, the set $Y = h^{-1}(h(Y))$ is closed by the continuity of $h^{-1} : \operatorname{rng} g \to S$ and closedness of $h(Y)$, as follows from Problem 1.17. Because the subspaces S and $\operatorname{null} g$ are complementary, the sum $Y + \operatorname{null} g$ is closed (see Problem 1.13). Consequently, the sum $\operatorname{cl} X + \operatorname{null} g$ is closed.

$(b) \Rightarrow (a)$ In a similar way, we obtain that the set $g(\operatorname{cl} X)$ is closed. Therefore (see Theorem 2.79),

$$\operatorname{cl} g(X) \subset \operatorname{cl} g(\operatorname{cl} X) = g(\operatorname{cl} X) \subset \operatorname{cl} g(X),$$

implying that $g(\operatorname{cl} X) = \operatorname{cl} g(X)$.

The equivalence of conditions (b) and (c) follows from Theorem 7.13. □

The closure of the set $f(X)$ from Theorem 7.14 in terms of asymptotic translates of $\operatorname{null} g$ is described in Problem 7.5.

7.2.2 *Affine Images of Convex Sets*

This subsection describes a sufficient condition (in terms of recession cones and lineality spaces) for the equality $f(\operatorname{cl} K) = \operatorname{cl} f(K)$, where $f : \mathbb{R}^n \to \mathbb{R}^m$ is an affine transformation and $K \subset \mathbb{R}^n$ is a convex set (see Theorem 7.16).

Theorem 7.15. *If a translate of a subspace $S \subset \mathbb{R}^n$ is asymptotic to a nonempty convex set $K \subset \mathbb{R}^n$, then*

$$\{\boldsymbol{o}\} \neq S \cap \operatorname{rec}(\operatorname{cl} K) \not\subset \operatorname{lin}(\operatorname{cl} K).$$

Proof. Let a translate $L = \boldsymbol{a} + S$ of S be an asymptotic plane of K. Then $L \cap \operatorname{cl} K = \varnothing$ and a suitable sequence $\boldsymbol{x}_1, \boldsymbol{x}_2, \dots$ of points from K satisfies the condition $\lim_{i\to\infty} \delta(\boldsymbol{x}_i, L) = 0$. We observe that this sequence is unbounded. Indeed, assume for a moment that $\boldsymbol{x}_1, \boldsymbol{x}_2, \dots$ is bounded. Then it contains a subsequence $\boldsymbol{y}_1, \boldsymbol{y}_2, \dots$ which converges to a point $\boldsymbol{y} \in \operatorname{cl} K$. In this case,

$$\delta(\boldsymbol{y}, L) = \lim_{i\to\infty} \delta(\boldsymbol{y}_i, L) = 0.$$

Hence $\boldsymbol{y} \in L$, contrary to the assumption $L \cap \operatorname{cl} K = \varnothing$.

By Theorem 6.13, $\boldsymbol{x}_1, \boldsymbol{x}_2, \dots$ contains a subsequence $\boldsymbol{z}_1, \boldsymbol{z}_2, \dots$ such that $\lim_{i\to\infty} \|\boldsymbol{z}_i\| = \infty$ and the unit vectors $\boldsymbol{e}_i = \boldsymbol{z}_i/\|\boldsymbol{z}_i\|$, $i \geqslant 1$, converge to a unit vector $\boldsymbol{e}$. Theorem 6.17 implies that $\boldsymbol{e} \in \operatorname{rec}(\operatorname{cl} K)$. Since

$$\begin{aligned}\delta(\boldsymbol{z}_i, S) &= \delta(\boldsymbol{z}_i, L - \boldsymbol{a}) = \inf\{\|\boldsymbol{z}_i - \boldsymbol{x} - \boldsymbol{a}\| : \boldsymbol{x} \in L\}\\ &\leqslant \inf\{\|\boldsymbol{z}_i - \boldsymbol{x}\| + \|\boldsymbol{a}\| : \boldsymbol{x} \in L\} = \delta(\boldsymbol{z}_i, L) + \|\boldsymbol{a}\|,\end{aligned}$$

we obtain

$$\delta(\boldsymbol{e}, S) = \lim_{i\to\infty} \delta(\boldsymbol{e}_i, S) = \lim_{i\to\infty} \delta(\boldsymbol{z}_i, S)/\|\boldsymbol{z}_i\|$$
$$\leqslant \lim_{i\to\infty} (\delta(\boldsymbol{z}_i, L) + \|\boldsymbol{a}\|)/\|\boldsymbol{z}_i\| = 0.$$

Thus $\boldsymbol{e} \in S$. Summing up, $\boldsymbol{e} \in \operatorname{rec}(\operatorname{cl} K) \cap S$. A combination of Theorems 6.2 and 6.9 shows that $\operatorname{rec}(\operatorname{cl} K)$ is a closed convex cone with apex $\boldsymbol{o}$, and so is the set $C = \operatorname{rec}(\operatorname{cl} K) \cap S$.

Assume, for contradiction, that $C \subset \operatorname{lin}(\operatorname{cl} K)$. First, we observe that C is a subspace. Indeed, if $\boldsymbol{x} \in C$, then

$$\boldsymbol{x} \in \operatorname{lin}(\operatorname{cl} K) = \operatorname{rec}(\operatorname{cl} K) \cap (-\operatorname{rec}(\operatorname{cl} K)),$$

implying that $-\boldsymbol{x} \in \operatorname{rec}(\operatorname{cl} K) \cap S = C$. Therefore, C is a subspace as a convex cone symmetric about $\boldsymbol{o}$. Denote by T a subspace of $\mathbb{R}^n$ complementary to C. The inclusion $C \subset \operatorname{lin}(\operatorname{cl} K)$ implies that every plane $C_i = \boldsymbol{z}_i + C$, $i \geqslant 1$, lies in $\operatorname{cl} K$. Since the planes C_i and T are translates of complementary subspaces C and T, the set $C_i \cap T$ is a singleton, say $\boldsymbol{u}_i$ (see Theorem 2.11). Clearly, $\delta(\boldsymbol{u}_i, L) = \delta(\boldsymbol{z}_i, L)$.

We observe that the sequence $\boldsymbol{u}_1, \boldsymbol{u}_2, \dots$ is bounded. Indeed, otherwise, repeating the above argument, we would obtain a recession unit vector of $\operatorname{cl} K$ which belongs to $C \cap T$; the latter is impossible because $C \cap T = \{\boldsymbol{o}\}$. Hence the sequence $\boldsymbol{u}_1, \boldsymbol{u}_2, \dots$ contains a subsequence $\boldsymbol{u}'_1, \boldsymbol{u}'_2, \dots$ which converges to a point $\boldsymbol{u} \in \operatorname{cl} K \cap T$. Finally,

$$\delta(\boldsymbol{u}, L) = \lim_{i\to\infty} \delta(\boldsymbol{u}_i, L) = \lim_{i\to\infty} \delta(\boldsymbol{z}_i, L) = 0,$$

implying the inclusion $\boldsymbol{u} \in L$, in contradiction with $\operatorname{cl} K \cap L = \varnothing$. $\square$

Theorem 7.16. *Let $K \subset \mathbb{R}^n$ be a convex set and $f : \mathbb{R}^n \to \mathbb{R}^m$ be an affine transformation expressed as $f(\boldsymbol{x}) = \boldsymbol{a} + g(\boldsymbol{x})$, where $\boldsymbol{a} \in \mathbb{R}^m$ and $g : \mathbb{R}^n \to \mathbb{R}^m$ is a linear transformation. If*

$$\operatorname{null} g \cap \operatorname{rec}(\operatorname{cl} K) \subset \operatorname{lin}(\operatorname{cl} K), \tag{7.11}$$

then $f(\operatorname{cl} K) = \operatorname{cl}(f(K))$. Furthermore, under condition (7.11),

$$g(\operatorname{rec}(\operatorname{cl} K)) = \operatorname{rec}(g(\operatorname{cl} K)) \quad \text{and} \quad g(\operatorname{lin}(\operatorname{cl} K)) = \operatorname{lin}(g(\operatorname{cl} K)).$$

Proof. Since the case $K = \varnothing$ is obvious, we may assume that $K \neq \varnothing$. Theorem 7.15 shows that no translate of the subspace $\operatorname{null} g$ is asymptotic to K. By Theorem 7.14, $f(\operatorname{cl} K) = \operatorname{cl}(f(K))$.

Next, we assert that

$$g(\operatorname{rec}(\operatorname{cl} K)) = \operatorname{rec}(g(\operatorname{cl} K)). \tag{7.12}$$

For this, we embed $\mathbb{R}^n$ into $\mathbb{R}^{n+1}$ so that every point $\boldsymbol{x} \in \mathbb{R}^n$ is identified with the point $(\boldsymbol{x}, 0)$ from the n-dimensional subspace $H_0 \subset \mathbb{R}^{n+1}$ given by $x_{n+1} = 0$. Clearly, $\hat{\boldsymbol{o}} = (\boldsymbol{o}, 0)$ is the origin of $\mathbb{R}^{n+1}$, where $\boldsymbol{o}$ is the origin of $\mathbb{R}^n$. Put $\hat{K} = \{(\boldsymbol{x}, 1) : \boldsymbol{x} \in K\}$. Then

$$\hat{K} = (\boldsymbol{o}, 1) + K \quad \text{and} \quad \operatorname{cl} \hat{K} = (\boldsymbol{o}, 1) + \operatorname{cl} K.$$

Theorem 6.5 and Corollary 6.22 give, respectively, the equalities

$$\operatorname{rec}(\operatorname{cl} \hat{K}) = \operatorname{rec}(\operatorname{cl} K) \quad \text{and} \quad \operatorname{lin}(\operatorname{cl} \hat{K}) = \operatorname{lin}(\operatorname{cl} K).$$

Let

$$C = C_{\hat{o}}(\operatorname{cl} \hat{K}) = \{\lambda\,(\boldsymbol{x}, 1) : \boldsymbol{x} \in \operatorname{cl} K,\ \lambda \geqslant 0\}.$$

By Theorem 7.7,

$$\operatorname{cl} C = C \cup \operatorname{rec}(\operatorname{cl} \hat{K}) = C \cup \operatorname{rec}(\operatorname{cl} K).$$

Consider the linear transformation $\hat{g} : \mathbb{R}^{n+1} \to \mathbb{R}^{n+1}$ defined by

$$\hat{g}(\boldsymbol{x}, x_{n+1}) = (g(\boldsymbol{x}), x_{n+1}).$$

Since

$$\operatorname{null} \hat{g} = \{(\boldsymbol{x}, 0) : \boldsymbol{x} \in \operatorname{null} g\} \subset H_0,$$

the hypothesis and Theorem 7.9 give

$$\begin{aligned}
\operatorname{rec}(\operatorname{cl} C) \cap \operatorname{null} \hat{g} &= (\operatorname{rec}(\operatorname{cl} C) \cap H_0) \cap (\operatorname{null} \hat{g} \cap H_0) \\
&= \operatorname{rec}(\operatorname{cl} K) \cap \operatorname{null} g \\
&\subset \operatorname{lin}(\operatorname{cl} K) = \operatorname{lin}(\operatorname{cl} \hat{K}) = \operatorname{lin}(\operatorname{cl} C).
\end{aligned}$$

By the proved above, $\hat{g}(\operatorname{cl} C) = \operatorname{cl}(\hat{g}(C))$. Since the set $g(\operatorname{cl} K)$ is closed, Theorem 7.7 gives

$$\begin{aligned}
\hat{g}(\operatorname{cl} C) &= \hat{g}(C \cup \operatorname{rec}(\operatorname{cl} K)) = \hat{g}(C) \cup \hat{g}(\operatorname{rec}(\operatorname{cl} K)) \\
&= \{\lambda(\boldsymbol{x}, 1) : \boldsymbol{x} \in g(\operatorname{cl} K),\ \lambda \geqslant 0\} \cup \{(\boldsymbol{x}, 0) : \boldsymbol{x} \in g(\operatorname{rec}(\operatorname{cl} K))\}.
\end{aligned}$$

On the other hand,

$$\begin{aligned}
\operatorname{cl}(\hat{g}(C)) &= \operatorname{cl}\{\lambda(\boldsymbol{x}, 1) : \boldsymbol{x} \in g(\operatorname{cl} K),\ \lambda \geqslant 0\} \\
&= \{\lambda(\boldsymbol{x}, 1) : \boldsymbol{x} \in g(\operatorname{cl} K),\ \lambda \geqslant 0\} \cup \{(\boldsymbol{x}, 0) : \boldsymbol{x} \in \operatorname{rec}(g(\operatorname{cl} K))\}.
\end{aligned}$$

Finally, (7.12) follows from

$$\begin{aligned}
\{(\boldsymbol{x}, 0) : \boldsymbol{x} \in g(\operatorname{rec}(\operatorname{cl} K))\} &= \hat{g}(\operatorname{cl} C) \cap H_0 \\
&= \operatorname{cl}(\hat{g}(C)) \cap H_0 = \{(\boldsymbol{x}, 0) : \boldsymbol{x} \in \operatorname{rec}(g(\operatorname{cl} K))\}.
\end{aligned}$$

It remains to show that

$$g(\mathrm{lin}\,(\mathrm{cl}\,K)) = \mathrm{lin}\,(g(\mathrm{cl}\,K)).$$

Due to Theorem 6.24, it suffices to prove the inclusion

$$\mathrm{lin}\,(g(\mathrm{cl}\,K)) \subset g(\mathrm{lin}\,(\mathrm{cl}\,K)). \tag{7.13}$$

Since the case $\mathrm{lin}\,(g(\mathrm{cl}\,K)) = \{\boldsymbol{o}\}$ is obvious, we assume that $\mathrm{lin}\,(g(\mathrm{cl}\,K)) \neq \{\boldsymbol{o}\}$. Choose any 1-dimensional subspace $l \subset \mathrm{lin}\,(g(\mathrm{cl}\,K))$ and denote by h_1 and h_2 the oppose closed halflines of l with common endpoint $\boldsymbol{o}$. Since both h_1 and h_2 lie in $\mathrm{lin}\,(g(\mathrm{cl}\,K))$, the equality

$$g(\mathrm{rec}\,(\mathrm{cl}\,K)) = \mathrm{rec}\,(g(\mathrm{cl}\,K))$$

implies the existence of closed halflines h_1' and h_2' with common endpoint $\boldsymbol{o}$ both lying in $\mathrm{rec}\,(\mathrm{cl}\,K)$ such that $g(h_1') = h_1$ and $g(h_2') = h_2$. If the set $l' = h_1' \cup h_2'$ is a subspace, then $l' \subset \mathrm{lin}\,(\mathrm{cl}\,K)$, and $l = g(l') \subset g(\mathrm{lin}\,(\mathrm{cl}\,K))$. So, assume that $h_1' \cup h_2'$ is not a subspace. Choose any nonzero vectors $\boldsymbol{x}_1 \in h_1'$ and $\boldsymbol{x}_2 \in h_2'$. Since $g(\boldsymbol{x}_1) \in h_1$ and $g(\boldsymbol{x}_2) \in h_1$, there are positive scalars α_1 and α_2 such that $\alpha_1 g(\boldsymbol{x}_1) + \alpha_2 g(\boldsymbol{x}_2) = \boldsymbol{o}$. Hence $g(\alpha_1\boldsymbol{x}_1 + \alpha_2\boldsymbol{x}_2) = \boldsymbol{o}$, implying that $\boldsymbol{x} = \alpha_1\boldsymbol{x}_1 + \alpha_2\boldsymbol{x}_2 \in \mathrm{null}\,g$. Denote by h the closed halfline with endpoint $\boldsymbol{o}$ which contains $\boldsymbol{x}$. Clearly, $h \subset \mathrm{null}\,g$. Because $\mathrm{rec}\,(\mathrm{cl}\,K)$ is a convex cone (see Theorem 6.2), we have

$$h \subset \mathrm{conv}\,(h_1' \cup h_2') \subset \mathrm{rec}\,(\mathrm{cl}\,K).$$

Therefore, $h \subset \mathrm{lin}\,(\mathrm{cl}\,K)$ due to the assumption (7.11). Hence the line $l = h \cup (-h)$ lies in $\mathrm{lin}\,(\mathrm{cl}\,K)$, and the subspace $S = \mathrm{span}\,(h_1 \cup h_2)$ lies in $\mathrm{lin}\,(\mathrm{cl}\,K)$. As a result,

$$l = g(h_1 \cup (-h_1)) \subset g(S) \subset g(\mathrm{lin}\,(\mathrm{cl}\,K)).$$

Summing up, (7.13) holds. □

Remark. In general, $g(\mathrm{rec}\,K) \neq \mathrm{rec}\,(g(K))$ even if (7.11) is satisfied. Indeed, let

$$K = \{\boldsymbol{o}\} \cup \{(x, y) : 0 < x,\ 0 < y < 1\}$$

and g be the orthogonal projection of $\mathbb{R}^2$ on the x-axis. Then

$$\mathrm{rec}\,K = \{\boldsymbol{o}\} \quad \text{and} \quad \mathrm{rec}\,(\mathrm{cl}\,K) = \{(x, 0) : x \geqslant 0\}.$$

Clearly, $\mathrm{null}\,g \cap \mathrm{rec}\,(\mathrm{cl}\,K) = \{\boldsymbol{o}\} = \mathrm{lin}\,(\mathrm{cl}\,K)$. On the other hand,

$$g(\mathrm{rec}\,K) = \{\boldsymbol{o}\} \neq \mathrm{rec}\,(g(K)) = g(K) = \{(x, 0) : x \geqslant 0\}.$$

A combination of Theorems 6.3, 6.21, and 7.16 implies the following corollary.

Corollary 7.17. *Let $C \subset \mathbb{R}^n$ be a convex cone with apex $\mathbf{c}$ and $f : \mathbb{R}^n \to \mathbb{R}^m$ be an affine transformation expressed as $f(\boldsymbol{x}) = \boldsymbol{a} + g(\boldsymbol{x})$, where $\boldsymbol{a} \in \mathbb{R}^m$ and $g : \mathbb{R}^n \to \mathbb{R}^m$ is a linear transformation. Then*

$$f(\operatorname{cl} C) = \operatorname{cl}(f(C)) \quad \textit{and} \quad g(\operatorname{ap}(\operatorname{cl} C)) = \operatorname{ap}(g(\operatorname{cl} C)).$$

provided $(\boldsymbol{c} + \operatorname{null} g) \cap \operatorname{cl} C \subset \operatorname{ap}(\operatorname{cl} C)$.

Proof. A combination of Theorems 6.3 and 6.21 gives

$$\operatorname{rec}(\operatorname{cl} C) = \operatorname{cl} C - \boldsymbol{c} \quad \text{and} \quad \operatorname{lin}(\operatorname{cl} C) = \operatorname{ap}(\operatorname{cl} C) - \boldsymbol{c}.$$

Due to the assumption,

$$\begin{aligned} \operatorname{null} g \cap \operatorname{rec}(\operatorname{cl} C) &= (\boldsymbol{c} + \operatorname{null} g) \cap (\boldsymbol{c} + \operatorname{rec}(\operatorname{cl} C)) - \boldsymbol{c} \\ &= (\boldsymbol{c} + \operatorname{null} g) \cap \operatorname{cl} C - \boldsymbol{c} \subset \operatorname{ap}(\operatorname{cl} C) - \boldsymbol{c} = \operatorname{lin}(\operatorname{cl} C). \end{aligned}$$

Now, Theorem 7.16 gives $f(\operatorname{cl} C) = \operatorname{cl} f(C)$. Because $g(\operatorname{cl} C)$ is a closed convex cone with apex $g(\boldsymbol{c})$ (see Theorem 5.12 and 5.21), we obtain

$$\begin{aligned} g(\operatorname{ap}(\operatorname{cl} C)) &= g(\boldsymbol{c} + \operatorname{lin}(\operatorname{cl} C)) = g(\boldsymbol{c}) + g(\operatorname{lin}(\operatorname{cl} C)) \\ &= g(\boldsymbol{c}) + \operatorname{lin}(g(\operatorname{cl} C)) = \operatorname{ap}(g(\operatorname{cl} C)). \end{aligned}$$ □

7.3 Closedness of Sums and Convex Hulls

7.3.1 *Sums of Convex Sets*

The next theorem, providing a sufficient closedness condition for a sum of convex sets, describes the algebra of their recession directions.

Theorem 7.18. *If K_1 and K_2 are convex sets in $\mathbb{R}^n$, satisfying the condition*

$$\operatorname{rec}(\operatorname{cl} K_1) \cap (-\operatorname{rec}(\operatorname{cl} K_2)) \subset \operatorname{lin}(\operatorname{cl} K_1) \cap \operatorname{lin}(\operatorname{cl} K_2), \tag{7.14}$$

then

$$\begin{aligned} \operatorname{cl}(K_1 + K_2) &= \operatorname{cl} K_1 + \operatorname{cl} K_2, \\ \operatorname{rec}(\operatorname{cl}(K_1 + K_2)) &= \operatorname{rec}(\operatorname{cl} K_1) + \operatorname{rec}(\operatorname{cl} K_2), \\ \operatorname{lin}(\operatorname{cl}(K_1 + K_2)) &= \operatorname{lin}(\operatorname{cl} K_1) + \operatorname{lin}(\operatorname{cl} K_2). \end{aligned}$$

Proof. Since the case when at least one of the sets K_1 and K_2 is empty obviously holds, we may assume that both sets K_1 and K_2 are nonempty. In the space $\mathbb{R}^{2n} = \mathbb{R}^n \times \mathbb{R}^n$, consider the sets $\hat{K}_1 = X_1 \times \{\boldsymbol{o}\}$ and $\hat{K}_2 = \{\boldsymbol{o}\} \times X_2$. Because the sets $S_1 = \mathbb{R}^n \times \{\boldsymbol{o}\}$ and $S_2 = \{\boldsymbol{o}\} \times \mathbb{R}^n$ can be considered as complementary subspaces of $\mathbb{R}^{2n} = \mathbb{R}^n \times \mathbb{R}^n$, the inclusions $\hat{K}_1 \subset S_1$ and $\hat{K}_2 \subset S_2$, together with Problem 1.13, give

$$\mathrm{cl}\,(K_1 \times K_2) = \mathrm{cl}\,(\hat{K}_1 + \hat{K}_2) = \mathrm{cl}\,\hat{K}_1 + \mathrm{cl}\,\hat{K}_2.$$

A combination of Theorems 6.6 and 6.23 implies

$$\mathrm{rec}\,(\mathrm{cl}\,(\hat{K}_1 + \hat{K}_2)) = \mathrm{rec}\,(\mathrm{cl}\,\hat{K}_1 + \mathrm{cl}\,\hat{K}_2) = \mathrm{rec}\,(\mathrm{cl}\,\hat{K}_1) + \mathrm{rec}\,(\mathrm{cl}\,\hat{K}_2), \quad (7.15)$$

$$\mathrm{lin}\,(\mathrm{cl}\,(\hat{K}_1 + \hat{K}_2)) = \mathrm{lin}\,(\mathrm{cl}\,\hat{K}_1 + \mathrm{cl}\,\hat{K}_2) = \mathrm{lin}\,(\mathrm{cl}\,\hat{K}_1) + \mathrm{lin}\,(\mathrm{cl}\,\hat{K}_2). \quad (7.16)$$

Obviously,

$$\mathrm{rec}\,(\mathrm{cl}\,\hat{K}_1) = \mathrm{rec}\,(\mathrm{cl}\,K_1) \times \{\boldsymbol{o}\}, \quad \mathrm{rec}\,(\mathrm{cl}\,\hat{K}_2) = \{\boldsymbol{o}\} \times \mathrm{rec}\,(\mathrm{cl}\,K_2),$$
$$\mathrm{lin}\,(\mathrm{cl}\,\hat{K}_1) = \mathrm{lin}\,(\mathrm{cl}\,K_1) \times \{\boldsymbol{o}\}, \quad \mathrm{lin}\,(\mathrm{cl}\,\hat{K}_2) = \{\boldsymbol{o}\} \times \mathrm{lin}\,(\mathrm{cl}\,K_2).$$

Let $\varphi : \mathbb{R}^n \times \mathbb{R}^n \to \mathbb{R}^n$ be the linear transformation defined by

$$\varphi(\boldsymbol{x}_1, \boldsymbol{x}_2) = \boldsymbol{x}_1 + \boldsymbol{x}_2 \quad \text{for all} \quad \boldsymbol{x}_1, \boldsymbol{x}_2 \in \mathbb{R}^n. \quad (7.17)$$

Clearly, $\mathrm{null}\,\varphi$ is the n-dimensional subspace of $\mathbb{R}^{2n} = \mathbb{R}^n \times \mathbb{R}^n$ given as

$$\mathrm{null}\,\varphi = \{(\boldsymbol{x}, -\boldsymbol{x}) : \boldsymbol{x} \in \mathbb{R}^n\}. \quad (7.18)$$

Based on (7.15) and (7.16), inclusion (7.14) gives

$$\begin{aligned}
&\mathrm{rec}\,(\mathrm{cl}\,(\hat{K}_1 + \hat{K}_2)) \cap \mathrm{null}\,\varphi \\
&\quad = (\mathrm{rec}\,(\mathrm{cl}\,\hat{K}_1) + \mathrm{rec}\,(\mathrm{cl}\,\hat{K}_2)) \cap \{(\boldsymbol{x}, -\boldsymbol{x}) : \boldsymbol{x} \in \mathbb{R}^n\} \\
&\quad = \mathrm{rec}\,(\mathrm{cl}\,K_1) \cap (-\mathrm{rec}\,(\mathrm{cl}\,K_2)) \subset \mathrm{lin}\,(\mathrm{cl}\,K_1) \cap \mathrm{lin}\,(\mathrm{cl}\,K_2) \\
&\quad = (\mathrm{lin}\,(\mathrm{cl}\,\hat{K}_1) + \mathrm{lin}\,(\mathrm{cl}\,\hat{K}_2)) \cap \{(\boldsymbol{x}, -\boldsymbol{x}) : \boldsymbol{x} \in \mathbb{R}^n\} \\
&\quad \subset \mathrm{lin}\,(\mathrm{cl}\,\hat{K}_1) + \mathrm{lin}\,(\mathrm{cl}\,\hat{K}_2) = \mathrm{lin}\,(\mathrm{cl}\,(\hat{K}_1 + \hat{K}_2)).
\end{aligned}$$

Finally, from Theorem 7.16 we obtain

$$\begin{aligned}
\mathrm{cl}\,(K_1 + K_2) &= \mathrm{cl}\,(\varphi(\hat{K}_1 + \hat{K}_2)) = \varphi(\mathrm{cl}\,(\hat{K}_1 + \hat{K}_2)) \\
&= \varphi(\mathrm{cl}\,\hat{K}_1 + \mathrm{cl}\,\hat{K}_2) = \mathrm{cl}\,K_1 + \mathrm{cl}\,K_2.
\end{aligned}$$

Therefore,

$$\begin{aligned}
&\mathrm{rec}\,(\mathrm{cl}\,(K_1 + K_2)) = \mathrm{rec}\,(\varphi(\mathrm{cl}\,(\hat{K}_1 + \hat{K}_2))) = \varphi(\mathrm{rec}\,(\mathrm{cl}\,(\hat{K}_1 + \hat{K}_2))) \\
&\quad = \varphi(\mathrm{rec}\,(\mathrm{cl}\,\hat{K}_1 + \mathrm{cl}\,\hat{K}_2)) = \varphi(\mathrm{rec}\,(\mathrm{cl}\,\hat{K}_1) + \mathrm{rec}\,(\mathrm{cl}\,\hat{K}_2)) \\
&\quad = \mathrm{rec}\,(\mathrm{cl}\,K_1) + \mathrm{rec}\,(\mathrm{cl}\,K_2).
\end{aligned}$$

Similarly,

$$\begin{aligned}\text{lin}\,(\text{cl}\,(K_1+K_2)) &= \text{lin}\,(\varphi(\text{cl}\,(\hat{K}_1+\hat{K}_2))) = \varphi(\text{lin}\,(\text{cl}\,(\hat{K}_1+\hat{K}_2)))\\ &= \varphi(\text{lin}\,(\text{cl}\,\hat{K}_1)+\text{lin}\,(\text{cl}\,\hat{K}_2)) = \text{lin}\,(\text{cl}\,K_1)+\text{lin}\,(\text{cl}\,K_2). \qquad \square\end{aligned}$$

Corollary 7.19. *If convex cones C_1 and C_2 in $\mathbb{R}^n$ satisfy the condition $\varnothing \neq \text{cl}\,C_1 \cap (-\text{cl}\,C_2) \subset \text{ap}\,(\text{cl}\,C_1) \cap \text{ap}\,(\text{cl}\,C_2)$, then*

$$\begin{aligned}\text{cl}\,(C_1+C_2) &= \text{cl}\,C_1+\text{cl}\,C_2,\\ \text{ap}\,(\text{cl}\,(C_1+C_2)) &= \text{ap}\,(\text{cl}\,C_1)+\text{ap}\,(\text{cl}\,C_2).\end{aligned}$$

Proof. Choose a point $\boldsymbol{a} \in \text{cl}\,C_1 \cap (-\text{cl}\,C_2)$. A combination of Theorems 6.3 and 6.21 gives

$$\text{rec}\,(\text{cl}\,C_i) = \text{cl}\,C_i - \boldsymbol{a} \quad \text{and} \quad \text{lin}\,(\text{cl}\,C_i) = \text{ap}\,(\text{cl}\,C_i) - \boldsymbol{a}, \quad i = 1, 2.$$

Due to the assumption,

$$\begin{aligned}\text{rec}\,(\text{cl}\,C_1) \cap (-\text{rec}\,(\text{cl}\,C_2)) &= \text{cl}\,C_1 \cap (-\text{cl}\,C_2) - \boldsymbol{a}\\ &\subset \text{ap}\,(\text{cl}\,C_1) \cap \text{ap}\,(\text{cl}\,C_2) - \boldsymbol{a} = \text{lin}\,(\text{cl}\,C_1) \cap \text{lin}\,(\text{cl}\,C_2).\end{aligned}$$

Now, Theorem 7.18 gives

$$\text{cl}\,(C_1+C_2) = \text{cl}\,C_1+\text{cl}\,C_2, \quad \text{lin}\,(\text{cl}\,(C_1+C_2)) = \text{lin}\,(\text{cl}\,C_1)+\text{lin}\,(\text{cl}\,C_2).$$

Because $\text{cl}\,(C_1+C_2)$ is a closed convex cone with apex $2\boldsymbol{a}$ (see Theorem 5.10 and Corollary 5.21), we have

$$\begin{aligned}\text{ap}\,(\text{cl}\,(C_1+C_2)) &= 2\boldsymbol{a} + \text{lin}\,(\text{cl}\,(C_1+C_2))\\ &= (\boldsymbol{a} + \text{lin}\,(\text{cl}\,C_1)) + (\boldsymbol{a} + \text{lin}\,(\text{cl}\,C_2))\\ &= \text{ap}\,(\text{cl}\,C_1)+\text{ap}\,(\text{cl}\,C_2). \qquad \square\end{aligned}$$

7.3.2 *Convex Hulls of Unions of Convex Sets*

Theorem 7.20. *If K_1 and K_2 are nonempty convex sets in $\mathbb{R}^n$ satisfying (7.14), then*

$$\begin{aligned}\text{cl}\,(\text{conv}\,(K_1 \cup K_2)) &= \text{conv}\,(\text{cl}\,K_1 \cup \text{cl}\,K_2)\\ &\quad \cup (\text{cl}\,K_1 + \text{rec}\,(\text{cl}\,K_2)) \cup (\text{cl}\,K_2 + \text{rec}\,(\text{cl}\,K_1)).\end{aligned} \tag{7.19}$$

Furthermore,

$$\text{rec}\,(\text{cl}\,(\text{conv}\,(K_1 \cup K_2))) = \text{rec}\,(\text{cl}\,K_1) + \text{rec}\,(\text{cl}\,K_2), \tag{7.20}$$

$$\text{lin}\,(\text{cl}\,(\text{conv}\,(K_1 \cup K_2))) = \text{lin}\,(\text{cl}\,K_1) + \text{lin}\,(\text{cl}\,K_2). \tag{7.21}$$

Proof. We embed $\mathbb{R}^n$ into $\mathbb{R}^{n+1}$ so that every point $\boldsymbol{x} \in \mathbb{R}^n$ is identified with the point $(\boldsymbol{x}, 0)$ from the n-dimensional subspace S of $\mathbb{R}^{n+1}$ given by $x_{n+1} = 0$. Clearly, $\hat{\boldsymbol{o}} = (\boldsymbol{o}, 0)$ is the origin of $\mathbb{R}^{n+1}$, where $\boldsymbol{o}$ is the origin of $\mathbb{R}^n$. Consequently, K_1 and K_2 will be identified with the respective subsets of S. Let

$$\hat{K}_i = \{(\boldsymbol{x}, 1) : \boldsymbol{x} \in K_i\}, \quad i = 1, 2.$$

Then

$$\hat{K}_i = (\boldsymbol{o}, 1) + K_i \quad \text{and} \quad \operatorname{cl} \hat{K}_i = (\boldsymbol{o}, 1) + \operatorname{cl} K_i, \quad i = 1, 2.$$

Theorem 6.5 and Corollary 6.22 give, respectively, the equalities

$$\operatorname{rec}(\operatorname{cl} \hat{K}_i) = \operatorname{rec}(\operatorname{cl} K_i) \quad \text{and} \quad \operatorname{lin}(\operatorname{cl} \hat{K}_i) = \operatorname{lin}(\operatorname{cl} K_i), \quad i = 1, 2.$$

Put

$$C_i = C_{\hat{\boldsymbol{o}}}(\operatorname{cl} \hat{K}_i) = \{\lambda\, (\boldsymbol{x}, 1) : \boldsymbol{x} \in \operatorname{cl} K_i,\ \lambda \geqslant 0\}, \quad i = 1, 2.$$

Obviously, $\operatorname{rec}(\operatorname{cl} C_i) = \operatorname{cl} C_i$ (see Theorem 6.3), and Theorem 7.7 implies

$$\operatorname{cl} C_i = C_i \cup \operatorname{rec}(\operatorname{cl} \hat{K}_i) = C_i \cup \operatorname{rec}(\operatorname{cl} K_i), \quad i = 1, 2.$$

By Theorem 7.9,

$$\operatorname{lin}(\operatorname{cl} C_i) = \operatorname{lin}(\operatorname{cl} \hat{K}_i) = \operatorname{lin}(\operatorname{cl} K_i), \quad i = 1, 2. \tag{7.22}$$

We have $C_i \cap (-C_j) = \{\hat{\boldsymbol{o}}\}$, $i, j = 1, 2$, because $C_i \setminus \{\hat{\boldsymbol{o}}\}$ and $-C_j \setminus \{\hat{\boldsymbol{o}}\}$ lie in the opposite open halfspaces of $\mathbb{R}^{n+1}$ determined by S and described, respectively, by the inequalities $x_{n+1} > 0$ and $x_{n+1} < 0$. This argument and (7.14) give

$$\begin{aligned}
&\operatorname{rec}(\operatorname{cl} C_1) \cap (-\operatorname{rec}(\operatorname{cl} C_2)) = \operatorname{cl} C_1 \cap (-\operatorname{cl} C_2)\\
&\quad = (C_1 \cup \operatorname{rec}(\operatorname{cl} K_1)) \cap (-(C_2 \cup \operatorname{rec}(\operatorname{cl} K_2)))\\
&\quad = \operatorname{rec}(\operatorname{cl} K_1) \cap (-\operatorname{rec}(\operatorname{cl} K_2))\\
&\quad \subset \operatorname{lin}(\operatorname{cl} K_1) \cap \operatorname{lin}(\operatorname{cl} K_2) = \operatorname{lin}(\operatorname{cl} C_1) \cap \operatorname{lin}(\operatorname{cl} C_2).
\end{aligned}$$

According to Theorem 7.18,

$$\operatorname{cl}(C_1 + C_2) = \operatorname{cl} C_1 + \operatorname{cl} C_2.$$

Consequently,

$$\operatorname{cl}(C_1 + C_2) \cap H = (\operatorname{cl} C_1 + \operatorname{cl} C_2) \cap H, \tag{7.23}$$

$$\operatorname{cl}(C_1 + C_2) \cap S = (\operatorname{cl} C_1 + \operatorname{cl} C_2) \cap S, \tag{7.24}$$

where $H \subset \mathbb{R}^{n+1}$ is a hyperplane, given by $H = \{(\boldsymbol{x}, 1) : \boldsymbol{x} \in \mathbb{R}^n\}$.

We are going to show that the first two assertion of the theorem derive, respectively, from (7.23) and (7.24). By (7.19),

$$\operatorname{cl} C_1 + \operatorname{cl} C_2 = (C_1 \cup \operatorname{rec}(\operatorname{cl} C_1)) + (C_2 \cup \operatorname{rec}(\operatorname{cl} C_2)) = (C_1 + C_2)$$
$$\cup (C_1 + \operatorname{rec}(\operatorname{cl} C_2)) \cup (C_2 + \operatorname{rec}(\operatorname{cl} C_1)) \cup (\operatorname{rec}(\operatorname{cl} C_1) + \operatorname{rec}(\operatorname{cl} C_2)).$$

Next, a combination of Theorems 5.10, 5.31, and 5.38 gives

$$C_1 + C_2 = \operatorname{conv}(C_1 \cup C_2) = \operatorname{conv}(C_{\hat{\boldsymbol{o}}}(C_1 \cup C_2))$$
$$= \operatorname{conv}(C_{\hat{\boldsymbol{o}}}(C_{\hat{\boldsymbol{o}}}(\hat{K}_1) \cup C_{\hat{\boldsymbol{o}}}(\hat{K}_2)))$$
$$= \operatorname{conv}(C_{\hat{\boldsymbol{o}}}(\hat{K}_1 \cup \hat{K}_2)) = C_{\hat{\boldsymbol{o}}}(\operatorname{conv}(\hat{K}_1 \cup \hat{K}_2)).$$

Because H contains $\operatorname{conv}(\hat{K}_1 \cup \hat{K}_2)$, Theorems 5.39 and 5.47 imply

$$(C_1 + C_2) \cap H = \operatorname{conv}(\hat{K}_1 \cup \hat{K}_2),$$
$$\operatorname{cl}(C_1 + C_2) \cap H = \operatorname{cl}(C_{\hat{\boldsymbol{o}}}(\operatorname{conv}(\hat{K}_1 \cup \hat{K}_2))) \cap H$$
$$= \operatorname{cl}(\operatorname{conv}(\hat{K}_1 \cup \hat{K}_2)) = (\boldsymbol{o}, 1) + \operatorname{cl}(\operatorname{conv}(K_1 \cup K_2)).$$

On the other hand,

$$(\operatorname{cl} C_1 + \operatorname{cl} C_2) \cap H = ((C_1 + C_2) \cap H) \cup ((C_1 + \operatorname{rec}(\operatorname{cl} C_2))) \cap H$$
$$\cup ((C_1 + \operatorname{rec}(\operatorname{cl} C_2))) \cap H$$
$$= \operatorname{conv}(\hat{K}_1 \cup \hat{K}_2) \cup (\hat{K}_1 + \operatorname{rec}(\operatorname{cl} \hat{K}_2)) \cup (\hat{K}_2 + \operatorname{rec}(\operatorname{cl} \hat{K}_1))$$
$$= \operatorname{conv}(\hat{K}_1 \cup \hat{K}_2) \cup (\hat{K}_1 + \operatorname{rec}(\operatorname{cl} K_2)) \cup (\hat{K}_2 + \operatorname{rec}(\operatorname{cl} K_1))$$
$$= (\boldsymbol{o}, 1) + \operatorname{conv}(K_1 \cup K_2) \cup (K_1 + \operatorname{rec}(\operatorname{cl} K_2)) \cup (K_2 + \operatorname{rec}(\operatorname{cl} K_1)).$$

Comparing the obtained equalities, we conclude that (7.19) holds. Similarly,

$$\operatorname{rec}(\operatorname{cl}(\operatorname{conv}(K_1 \cup K_2))) = \operatorname{cl}(C_1 + C_2) \cap S$$
$$= (\operatorname{cl} C_1 + \operatorname{cl} C_2) \cap S = \operatorname{rec}(\operatorname{cl} K_1) + \operatorname{rec}(\operatorname{cl} K_2).$$

Finally, a combination of Theorems 7.9 and 7.18 gives

$$\operatorname{lin}(\operatorname{cl}(\operatorname{conv}(K_1 \cup K_2))) = \operatorname{lin}(\operatorname{cl}(\operatorname{conv}(\hat{K}_1 \cup \hat{K}_2)))$$
$$= \operatorname{lin}(\operatorname{cl}(\operatorname{conv}(C_1 \cup C_2))) = \operatorname{lin}(\operatorname{cl}(C_1 + C_2))$$
$$= \operatorname{lin}(\operatorname{cl} C_1 + \operatorname{cl} C_2) = \operatorname{lin}(\operatorname{cl} C_1) + \operatorname{lin}(\operatorname{cl} C_2)$$
$$= \operatorname{lin}(\operatorname{cl} \hat{K}_1) + \operatorname{lin}(\operatorname{cl} \hat{K}_2) = \operatorname{lin}(\operatorname{cl} K_1) + \operatorname{lin}(\operatorname{cl} K_2).$$ □

Remark. Generally, (7.19) does not hold if (7.14) is not satisfied. For instance, let

$$K_1 = \{(x, y) : x < 0, xy \leqslant -1\} \quad \text{and} \quad K_2 = \{(x, y) : x > 0, xy \geqslant 1\}.$$

Clearly, $\operatorname{conv}(K_1 \cup K_2) = \{(x, y) : y > 0\}$, and

$$\operatorname{rec} K_1 = \{(x, y) : x \leqslant 0, y \geqslant 0\}, \quad \operatorname{rec} K_2 = \{(x, y) : x \geqslant 0, y \geqslant 0\}.$$

Therefore, $\operatorname{conv}(K_1 \cup K_2) = K_1 + \operatorname{rec} K_2 = K_2 + \operatorname{rec} K_1$, while

$$\operatorname{cl}(\operatorname{conv}(K_1 \cup K_2)) = \{(x, y) : y \geqslant 0\}.$$

7.3.3 *Algebra of Line-Free Convex Sets*

A combination of Theorems 6.37 and 7.9 implies the corollary below.

Corollary 7.21. *A convex set $K \subset \mathbb{R}^n$ is line-free if and only if for any given point $\boldsymbol{a} \in \mathbb{R}^n \backslash \mathrm{cl}\, K$ the generated cone $C_{\boldsymbol{a}}(K)$ is line-free (equivalently, both cones $C_{\boldsymbol{a}}(\mathrm{cl}\, K)$ and $C_{\boldsymbol{a}}(\mathrm{rint}\, K)$ are line-free).* □

Theorem 7.22. *For line-free convex sets K_1 and K_2 in $\mathbb{R}^n$, the following conditions are equivalent.*

(*a*) *The sum $K_1 + K_2$ is line-free.*
(*b*) *The sum $\mathrm{cl}\, K_1 + \mathrm{cl}\, K_2$ is line-free.*
(*c*) *The set $\mathrm{conv}\,(K_1 \cup K_2)$ is line-free.*
(*d*) *The set $\mathrm{conv}\,(\mathrm{cl}\, K_1 \cup \mathrm{cl}\, K_2)$ is line-free.*
(*e*) *The sum $\mathrm{rec}\,(\mathrm{cl}\, K_1) + \mathrm{rec}\,(\mathrm{cl}\, K_2)$ is line-free.*
(*f*) $\mathrm{rec}\,(\mathrm{cl}\, K_1) \cap (-\mathrm{rec}\,(\mathrm{cl}\, K_2)) \subset \{\boldsymbol{o}\}$.

Proof. Since the case when at least one of the sets K_1 and K_2 is empty obviously holds, we may assume that both sets K_1 and K_2 are nonempty. Then condition (f) may be rewritten as follows:

(*f*) $\mathrm{rec}\,(\mathrm{cl}\, K_1) \cap (-\mathrm{rec}\,(\mathrm{cl}\, K_2)) = \{\boldsymbol{o}\}$.

The equivalence of conditions (a) and (b) follows from Theorem 6.37 and the inclusions

$$K_1 + K_2 \subset \mathrm{cl}\, K_1 + \mathrm{cl}\, K_2 \subset \mathrm{cl}\,(K_1 + K_2).$$

Similarly, $(c) \Leftrightarrow (d)$ due to

$$\mathrm{conv}\,(K_1 \cup K_2) \subset \mathrm{conv}\,(\mathrm{cl}\, K_1 \cup \mathrm{cl}\, K_2) \subset \mathrm{cl}\,(\mathrm{conv}\,(K_1 \cup K_2)).$$

$(b) \Leftrightarrow (f)$ Assume for a moment that $(b) \not\Rightarrow (f)$ and choose a nonzero vector $\boldsymbol{e} \in \mathrm{rec}\,(\mathrm{cl}\, K_1) \cap (-\mathrm{rec}\,(\mathrm{cl}\, K_2))$. Let $\boldsymbol{x}_1$ and $\boldsymbol{x}_2$ be points in $\mathrm{cl}\, K_1$ and $\mathrm{cl}\, K_2$, respectively. Then

$$h_1 = \boldsymbol{x}_1 + \{\lambda \boldsymbol{e} : \lambda \geqslant 0\} \subset \mathrm{cl}\, K_1 \quad \text{and} \quad h_2 = \boldsymbol{x}_2 + \{\lambda \boldsymbol{e} : \lambda \leqslant 0\} \subset \mathrm{cl}\, K_2.$$

Clearly, the sum $h_1 + h_2$ is the line $(\boldsymbol{x}_1 + \boldsymbol{x}_2) + \{\lambda \boldsymbol{e} : \lambda \in \mathbb{R}\}$ which lies in $\mathrm{cl}\, K_1 + \mathrm{cl}\, K_2$, contrary to condition (f).

Conversely, suppose that condition (f) is satisfied. Since

$$\mathrm{lin}\,(\mathrm{cl}\, K_1) = \mathrm{lin}\,(\mathrm{cl}\, K_2) = \{\boldsymbol{o}\},$$

Theorem 7.18 gives

$$\mathrm{lin}\,(\mathrm{cl}\,(K_1 + K_2)) = \mathrm{lin}\,(\mathrm{cl}\, K_1) + \mathrm{lin}\,(\mathrm{cl}\, K_2) = \{\boldsymbol{o}\}.$$

These equalities and Theorem 6.37 show that the set $\mathrm{cl}\,(K_1 + K_2)$ is line-free. Consequently, the sum $\mathrm{cl}\,K_1 + \mathrm{cl}\,K_2$ is line-free.

$(e) \Leftrightarrow (f)$ By the proved above, the sum $\mathrm{rec}\,(\mathrm{cl}\,K_1) + \mathrm{rec}\,(\mathrm{cl}\,K_2)$ is line-free if and only if

$$\mathrm{rec}\,(\mathrm{rec}\,(\mathrm{cl}\,K_1)) \cap (-\mathrm{rec}\,(\mathrm{rec}\,(\mathrm{cl}\,K_2))) = \{\boldsymbol{o}\}.$$

Since $\mathrm{rec}\,(\mathrm{cl}\,K_i) = \mathrm{rec}\,(\mathrm{rec}\,(\mathrm{cl}\,K_i))$, $i = 1, 2$, we obtain condition (d).

$(c) \Leftrightarrow (f)$ Assume first that the set $\mathrm{conv}\,(K_1 \cup K_2)$ is line-free. By Theorem 6.37, the set $\mathrm{rec}\,(\mathrm{cl}\,(\mathrm{conv}\,(K_1 \cup K_2)))$ is line-free. Theorem 6.14 gives

$$\mathrm{rec}\,(\mathrm{cl}\,K_1) \cup \mathrm{rec}\,(\mathrm{cl}\,K_2) \subset \mathrm{rec}\,(\mathrm{cl}\,(\mathrm{conv}\,(K_1 \cup K_2))).$$

This argument and Theorem 5.10 imply

$$\begin{aligned}\mathrm{rec}\,(\mathrm{cl}\,K_1) + \mathrm{rec}\,(\mathrm{cl}\,K_2) &= \mathrm{conv}\,(\mathrm{rec}\,(\mathrm{cl}\,K_1) \cup \mathrm{rec}\,(\mathrm{cl}\,K_2)) \\ &\subset \mathrm{rec}\,(\mathrm{cl}\,(\mathrm{conv}\,(K_1 \cup K_2))).\end{aligned}$$

Hence the sum $\mathrm{rec}\,(\mathrm{cl}\,K_1) + \mathrm{rec}\,(\mathrm{cl}\,K_2)$ is line-free. By the above proved, condition (f) is satisfied.

Conversely, the implication $(f) \Rightarrow (c)$ follows from Theorem 7.20. $\square$

Corollary 7.23. *Let $K \subset \mathbb{R}^n$ be a line-free convex set and $f : \mathbb{R}^n \to \mathbb{R}^m$ be an affine transformation expressed as $f(\boldsymbol{x}) = \boldsymbol{a} + g(\boldsymbol{x})$, where $\boldsymbol{a} \in \mathbb{R}^n$ and $g : \mathbb{R}^n \to \mathbb{R}^m$ is a linear transformation. If $\mathrm{rec}\,(\mathrm{cl}\,K) \cap \mathrm{null}\,g = \{\boldsymbol{o}\}$, then both sets $f(K)$ and $f(\mathrm{cl}\,K)$ are line-free and $g(\mathrm{rec}\,(\mathrm{cl}\,K)) = \mathrm{rec}\,(g(\mathrm{cl}\,K))$.*

Proof. Since the case $K = \varnothing$ is obvious, we may assume that $K \neq \varnothing$. By Theorem 6.37, the set $\mathrm{cl}\,K$ is line-free and $\mathrm{lin}\,(\mathrm{cl}\,K) = \{\boldsymbol{o}\}$. Theorem 7.16 gives

$$\mathrm{lin}\,(g(\mathrm{cl}\,K)) = g(\mathrm{lin}\,(\mathrm{cl}\,K)) = g(\{\boldsymbol{o}\}) = \{\boldsymbol{o}\}.$$

Hence $g(\mathrm{cl}\,K)$ is line-free. Consequently, $g(K)$ is line-free. Thus both sets $f(K) = \boldsymbol{a} + g(K)$ and $f(\mathrm{cl}\,K) = \boldsymbol{a} + g(\mathrm{cl}\,K)$ are line-free. Theorem 7.16 gives $g(\mathrm{rec}\,(\mathrm{cl}\,K)) = \mathrm{rec}\,(g(\mathrm{cl}\,K))$. $\square$

A combination of Theorems 7.18, 7.20, and 7.22 implies the corollary below.

Corollary 7.24. *For nonempty line-free convex sets K_1 and K_2 are in $\mathbb{R}^n$, the following assertions hold.*

(*a*) *If the sum* $K_1 + K_1$ *is line-free, then*

$$\operatorname{cl}(K_1 + K_2) = \operatorname{cl} K_1 + \operatorname{cl} K_2,$$
$$\operatorname{rec}(\operatorname{cl}(K_1 + K_2)) = \operatorname{rec}(\operatorname{cl} K_1) + \operatorname{rec}(\operatorname{cl} K_2).$$

(*b*) *If the set* $\operatorname{conv}(K_1 \cup K_1)$ *is line-free, then*

$$\operatorname{cl}(\operatorname{conv}(K_1 \cup K_2)) = \operatorname{conv}(\operatorname{cl} K_1 \cup \operatorname{cl} K_2) \cup \operatorname{rec}(\operatorname{cl} K_1) \cup \operatorname{rec}(\operatorname{cl} K_2),$$
$$\operatorname{rec}(\operatorname{cl}(\operatorname{conv}(K_1 \cup K_2))) = \operatorname{rec}(\operatorname{cl} K_1) + \operatorname{rec}(\operatorname{cl} K_2). \qquad \square$$

Problems and Notes for Chapter 7

Problems for Chapter 7

Problem 7.1. Let $K \subset \mathbb{R}^n$ be a nonempty convex set and $\boldsymbol{a}$ be a point in $\mathbb{R}^n \setminus \operatorname{cl} K$. Prove the following equivalence:

$$\operatorname{cl} C_{\boldsymbol{a}}(K) = \boldsymbol{a} + \operatorname{rec}(\operatorname{cl} K) \quad \Leftrightarrow \quad K \subset \boldsymbol{a} + \operatorname{rec}(\operatorname{cl} K).$$

Problem 7.2. Given nonempty sets $X_1, \dots, X_r \subset \mathbb{R}^n$ and scalars $\mu_1, \dots, \mu_r$, prove the following assertions.

(*a*) If a set X_i is locally conic at a point $\boldsymbol{a}_i \in \mathbb{R}^n$, $1 \leqslant i \leqslant r$, then the sum $\mu_1 X_1 + \cdots + \mu_r X_r$ is locally conic at the point $\mu_1 \boldsymbol{a}_1 + \cdots + \mu_r \boldsymbol{a}_r$.

(*b*) If $X_1 \cap \cdots \cap X_r \neq \varnothing$ and every set X_i is locally conic at a point $\boldsymbol{a} \in \mathbb{R}^n$, $1 \leqslant i \leqslant r$, then $X_1 \cap \cdots \cap X_r$ is locally conic at $\boldsymbol{a}$.

Problem 7.3. Let a plane $L \subset \mathbb{R}^n$ be expressed as $L = \boldsymbol{c} + S$, where $S \subset \mathbb{R}^n$ is a subspace and $\boldsymbol{c} \in \mathbb{R}^n$ is a nonzero vector orthogonal to S. Let $T \subset \mathbb{R}^n$ be the orthogonal complement of S and p_T be the orthogonal projection on T. Prove that L is asymptotic to a nonempty set X if and only if $\boldsymbol{c} \in \operatorname{cl} p_T(X) \setminus p_T(\operatorname{cl} X)$.

Problem 7.4. (Soltan [262]) Let a nonempty set $X \subset \mathbb{R}^n$ and a subspace $S \subset \mathbb{R}^n$ be such that the sum $\operatorname{cl} X + S$ is not closed. Prove that $\operatorname{cl}(X + S)$ is a disjoint union of nonempty sets $\operatorname{cl} X + S$ and S_X, where S_X denotes the union of all translates of S which are asymptotic planes of X.

Problem 7.5. (Soltan [262]) Let $f : \mathbb{R}^n \to \mathbb{R}^m$ be an affine transformation expressed as $f(\boldsymbol{x}) = \boldsymbol{a} + g(\boldsymbol{x})$, where $\boldsymbol{a} \in \mathbb{R}^m$ and $g : \mathbb{R}^n \to \mathbb{R}^m$ is a linear transformation, and let $X \subset \mathbb{R}^n$ be a nonempty set such that the image $f(\operatorname{cl} X)$ is not closed. Prove that $\operatorname{cl} f(X)$ is a disjoint union of nonempty sets $f(\operatorname{cl} X)$ and $f(N_X)$, where N_X denotes the union of all translates of the null space $\operatorname{null} g$ which are asymptotic planes of X.

Problem 7.6. Let a plane $L \subset \mathbb{R}^n$ be asymptotic to a convex set $K \subset \mathbb{R}^n$. Prove the existence of a closed halfline $h \subset \mathbb{R}^n$ parallel to L such that either $h \subset \operatorname{rbd} K$ or h is asymptotic to K.

Problem 7.7. Let $K \subset \mathbb{R}^n$ be a nonempty convex set and $C \subset \mathbb{R}^n$ be a convex cone with apex $\boldsymbol{o}$. Given a point $\boldsymbol{u} \in K$, prove that

$$\mathrm{cl}\,(K + C) = \mathrm{cl}\,(\mathrm{conv}\,(K \cup (\boldsymbol{u} + C))).$$

Problem 7.8. Prove the following assertions.

(a) Every F_σ-set (respectively, convex F_σ-set) is a countable union of an increasing sequence of compact sets (respectively, compact convex sets).

(b) An affine image of an F_σ-set (respectively, convex F_σ-set) is an F_σ-set (respectively, convex F_σ-set).

(c) If $X \subset \mathbb{R}^n$ is an F_σ-set, then both sets $\mathrm{conv}\, X$ and $\mathrm{cone}_{\boldsymbol{a}} X$, $\boldsymbol{a} \in \mathbb{R}^n$, are F_σ-sets.

(d) If X and Y are F_σ sets (respectively, convex F_σ-sets) in $\mathbb{R}^n$, then $X + Y$ is an F_σ-set (respectively, convex F_σ-set).

Notes for Chapter 7

Conic hulls. The last assertion of Theorem 7.4 is a simplified version of Theorem 2 from Komiya [187]. Closedness of $\mathrm{cone}_{\boldsymbol{o}} K$, where $K \subset \mathbb{R}^n$ is a closed convex set containing $\boldsymbol{o}$, is formulated by Meng, Roshchina, and Yang [212] in two different ways: (a) in terms of the inclusion $\mathrm{pos}^\infty K \subset \mathrm{pos}\, K$, where $\mathrm{pos}^\infty K$ stands for the union of relative boundary halflines of the cone $\mathrm{pos}\, K$, which are disjoint from K, and (b) in terms of exact tangent approximation (compare with Theorem 7.6). The second part of Corollary 7.8 and some closedness conditions for conic hulls are due to Bair and Jongmans [24, 25].

The notion *locally conic set* can be traced in the paper of Waksman and Epelman [287] (an equivalent notion *exactness of tangent approximation* is used by Hu and Wang [154] and Meng, Roshchina, and Yang [212]). Theorem 7.7 is obtained by Choquet [71] for the case when K is a line-free closed convex set in a separable topological space and $\boldsymbol{a} = \boldsymbol{o}$. For the case of any closed convex sets in $\mathbb{R}^n$ it is proved by Rockafellar [241, Theorem 9.6] (see also Bair and Jongmans [24, 25]). Theorem 7.9 is given in Soltan [260, Theorem 5.29].

Affine images. The equivalence of conditions (a) and (b) in Theorem 7.14 is due to Han and Mangasarian [144] (see also Borwein and Moors [42]), while the equivalence (a) $\Leftrightarrow$ (c) is proved by Soltan [260, Theorem 2.52]. Theorem 7.15 is due to Rockafellar [241, Theorem 9.1] (see Auslender [8], Auslender and Teboulle [9, Chapter 2] for various related results). Closedness conditions for linear images of convex cones are considered by Borwein and Moors [42, 43] and Pataki [226].

Sums and convex hulls. Theorem 7.18 is due to Rockafellar [241, Corollary 9.1.1]. A weaker version of this result (but for the case of any topological vector space) was previously obtained by Dieudonné [93]. Theorem 7.20 is proved by Rockafellar [241, Theorem 9.8]. Some closedness conditions for convex hulls of convex sets are obtained by Bair [17]. In particular, he proved that given nonempty planes L_1 and L_2 in $\mathbb{R}^n$, the set $\operatorname{conv}(L_1 \cup L_2)$ is closed if and only if any of the following two conditions are satisfied: (*a*) $L_1 \cap L_2 \neq \varnothing$, (*b*) L_1 is a translate of L_2.

The equivalence of conditions (*a*) and (*d*) of Theorem 7.22 is shown by Gritzmann and Klee [128].

Gale and Klee [116] defined (in an equivalent form) *continuous convex sets* as n-dimensional closed convex sets in $\mathbb{R}^n$ without boundary rays or asymptotic halflines. They showed that for a continuous convex set $K \subset \mathbb{R}^n$, the following assertions hold.

1. For any point $\boldsymbol{a} \in \mathbb{R}^n \setminus K$, the convex hull $\operatorname{conv}(\{\boldsymbol{a}\} \cup K)$ is closed.
2. For any point $\boldsymbol{a} \in \mathbb{R}^n \setminus K$, the generated cone $C_{\boldsymbol{a}}(K)$ is closed.
3. For any closed convex set $L \subset \mathbb{R}^n \setminus K$, the sum $K + L$ is closed.

Algebra of recession directions and closedness conditions. Theorems 7.16, 7.18, and 7.20, except the equalities

$$g(\operatorname{lin}(\operatorname{cl} K)) = \operatorname{lin}(g(\operatorname{cl} K)),$$
$$\operatorname{lin}(\operatorname{cl}(K_1 + K_2)) = \operatorname{lin}(\operatorname{cl} K_1) + \operatorname{lin}(\operatorname{cl} K_2),$$
$$\operatorname{lin}(\operatorname{cl}(\operatorname{conv}(K_1 \cup K_2))) = \operatorname{lin}(\operatorname{cl} K_1) + \operatorname{lin}(\operatorname{cl} K_2),$$

are due to Rockafellar [241, § 9]. An infinite-dimension version of Theorems 7.16, with a stronger condition, $\operatorname{rec} K_1 \cap (-\operatorname{rec} K_2) = \{\boldsymbol{o}\}$, was previously obtained by Dieudonné [93]. One more closedness criterion for linear images of closed sets is obtained by Auslender [8]. Some closedness conditions for convex hulls of unions of convex sets are considered by Bair [17].

Various closedness conditions for the sums and linear images of convex cones are considered by Auslender and Teboulle [9, Chapter 2], Borwein and Moors [42, 43], Gritzmann and Klee [127], Pataki [226], and Ujvári [283, 284].

Corollary 7.21 can be found in the paper of Dragomirescu [96].

F_σ-sets. Problem 7.8 immediately implies that basic operations on (convex) closed sets in $\mathbb{R}^n$ (like unions, sums, convex and conic hulls, affine images, etc.) make them F_σ-sets. It is natural to ask whether the resulting sets cover the whole variety of convex F_σ-set in $\mathbb{R}^n$. The following results show that is the case in many instances.

Klee [176, Theorem 6.1] proved that a convex set $K \subset \mathbb{R}^n$ is an F_σ-set if and only if in any space $\mathbb{R}^m$, $m \geqslant n+1$, there is a closed convex set $M \subset \mathbb{R}^m$ such that K is an orthogonal projection of M. This result was later obtained by Yaksubaev [296] for the case when $m = n+1$.

Bromek and Kaniewski [48] proved that for a cone $C \subset \mathbb{R}^n$ with apex $\boldsymbol{o}$, the following conditions are equivalent:

(*a*) C is a (convex) F_σ-set,

(*b*) C is the generated cone, $C_{\boldsymbol{o}}(X)$, of a suitable compact (convex) set $X \subset \mathbb{R}^n$,

(*c*) C is a linear projection of some (convex) line-free closed cone $C' \subset \mathbb{R}^{n+1}$,

(d) if C is convex, then it is the sum of a closed convex cone and a closed halfline,

(e) if C is convex, then it is the sum of two closed convex cones.

The equivalence of conditions (a) and (b) for the case of convex cones was independently established by Eifler [100] and Sung and Tam [274]. The last two authors also showed that (a) is equivalent to any of the conditions below:

(f) C is the union of an increasing sequence of closed cones with apex $\boldsymbol{o}$,

(g) if C is convex, then it is the barrier cone of some convex sets in $\mathbb{R}^n$ (see page 298 for the definition of barrier cones).

Larman [191] added one more equivalent property to the above list: (see page 308 for the definition of inner aperture).

(h) if C is convex, then it is an inner aperture of a suitable convex set $K \subset \mathbb{R}^n$.

Chapter 8

Cone Polarity and Normal Directions

8.1 Cone Polarity

8.1.1 *Polar Cones*

Definition 8.1. For a nonempty set $X \subset \mathbb{R}^n$, the set

$$X^\circ = \{\boldsymbol{u} \in \mathbb{R}^n : \boldsymbol{u}\cdot\boldsymbol{x} \leqslant 0 \text{ for all } \boldsymbol{x} \in X\}$$

is called the *polar cone* of X. We let $X^\circ = \mathbb{R}^n$ if $X = \varnothing$.

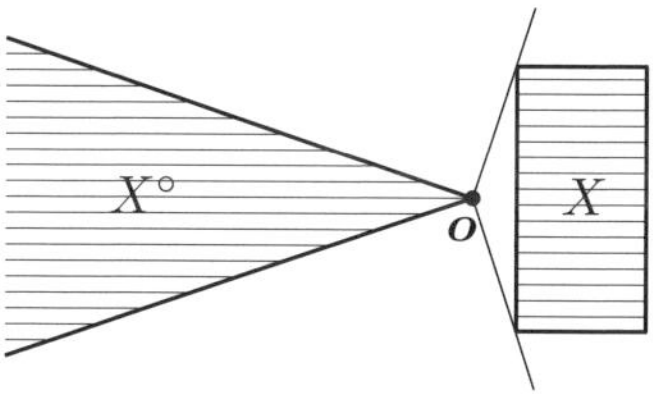

Fig. 8.1 The polar cone of a set X.

Remark. Clearly, $\boldsymbol{o} \in X^\circ$ for any set $X \subset \mathbb{R}^n$. If $X \subset \{\boldsymbol{o}\}$, then $X^\circ = \mathbb{R}^n$, and if $X \not\subset \{\boldsymbol{o}\}$, then the polar cone X° can be expressed as the intersection of closed halfspaces:

$$X^\circ = \cap\,(V_{\boldsymbol{x}} : \boldsymbol{x} \in X \setminus \{\boldsymbol{o}\}), \quad \text{where} \quad V_{\boldsymbol{x}} = \{\boldsymbol{u} \in \mathbb{R}^n : \boldsymbol{u}\cdot\boldsymbol{x} \leqslant 0\}. \tag{8.1}$$

Examples. 1. Given a nonzero vector $\boldsymbol{e} \in \mathbb{R}^n$, the polar cone $\{\boldsymbol{e}\}^\circ$ is the closed halfspace $V = \{\boldsymbol{u} \in \mathbb{R}^n : \boldsymbol{u}\cdot\boldsymbol{e} \leqslant 0\}$. Furthermore, V° is the closed halfline $[\boldsymbol{o}, \boldsymbol{e}\rangle = \{\lambda\boldsymbol{e} : \lambda \geqslant 0\}$.

2. If $S \subset \mathbb{R}^n$ is a subspace, then $S^\circ = S^\perp$. Indeed, the inclusion $S^\perp \subset S^\circ$ is obvious. Conversely, choose a point $\boldsymbol{u} \in S^\circ$ and let $\boldsymbol{x} \in S$. Then $-\boldsymbol{x} \in S$, and the inequalities $\boldsymbol{u}\cdot\boldsymbol{x} \leqslant 0$ and $\boldsymbol{u}\cdot(-\boldsymbol{x}) \leqslant 0$ give $\boldsymbol{u}\cdot\boldsymbol{x} = 0$, implying the inclusion $\boldsymbol{u} \in S^{\perp}$. In particular, $\{\boldsymbol{o}\}^{\circ} = \mathbb{R}^n$ and $(\mathbb{R}^n)^{\circ} = \{\boldsymbol{o}\}$.

The theorem below describes some general properties of polar set. We recall that $C_{\boldsymbol{o}}(X)$ and $\text{cone}_{\boldsymbol{o}}X$ denote, respectively, the generated cone and conic hull of a set $X \subset \mathbb{R}^n$ with apex $\boldsymbol{o}$ (see Definition 5.29).

Theorem 8.2. *For sets X and Y in $\mathbb{R}^n$, the following assertions hold.*

(*a*) $Y^{\circ} \subset X^{\circ}$ *provided* $X \subset Y$.
(*b*) $(X \cup Y)^{\circ} = X^{\circ} \cap Y^{\circ}$.
(*c*) $X^{\circ} = (\text{conv}\, X)^{\circ} = (C_{\boldsymbol{o}}(X))^{\circ} = (\text{cone}_{\boldsymbol{o}}X)^{\circ}$.
(*d*) $X^{\perp} \subset X^{\circ} = (\text{cl}\, X)^{\circ}$ *and* $(\mu X)^{\circ} = \mu X^{\circ}$ *whenever* $\mu \neq 0$.
(*e*) $(X + Y)^{\circ} = (\text{cl}\, X + \text{cl}\, Y)^{\circ}$. *If, additionally,* $\boldsymbol{o} \in \text{cl}\, X \cap \text{cl}\, Y$, *then* $(X + Y)^{\circ} = X^{\circ} \cap Y^{\circ}$.
(*f*) X° *is a closed convex cone with proper apex* $\boldsymbol{o}$, *and* $\text{cl}\, X \cap X^{\circ} \subset \{\boldsymbol{o}\}$.
(*g*) $(X^{\circ})^{\circ} = \text{cl}\,(\text{cone}_{\boldsymbol{o}}X)$.
(*h*) *If X is a convex cone with apex* $\boldsymbol{o}$, *then* $(X^{\circ})^{\circ} = \text{cl}\, X$.
(*i*) $X^{\circ} \neq \mathbb{R}^n$ *if and only if* $X \not\subset \{\boldsymbol{o}\}$.
(*j*) $X^{\circ} \not\subset \{\boldsymbol{o}\}$ *if and only if* $\text{cone}_{\boldsymbol{o}}X \neq \mathbb{R}^n$.

Proof. If at least one of the sets X and Y is empty, then all assertions are obvious. So, we may suppose that both sets X and Y are nonempty.

Assertions (*a*) and (*b*) immediately follow from Definition 8.1.

(*c*) We observe that

$$(\text{cone}_{\boldsymbol{o}}X)^{\circ} \subset (\text{conv}\, X)^{\circ} \subset X^{\circ} \quad \text{and} \quad (\text{cone}_{\boldsymbol{o}}X)^{\circ} \subset (C_{\boldsymbol{o}}(X))^{\circ} \subset X^{\circ}$$

due to assertion (*a*) and the inclusions

$$X \subset \text{conv}\, X \subset \text{cone}_{\boldsymbol{o}}X \quad \text{and} \quad X \subset C_{\boldsymbol{o}}(X) \subset \text{cone}_{\boldsymbol{o}}X.$$

So, it suffices to prove that $X^{\circ} \subset (\text{cone}_{\boldsymbol{o}}X)^{\circ}$. For this, choose a point $\boldsymbol{u} \in X^{\circ}$ and any point $\boldsymbol{x} \in \text{cone}_{\boldsymbol{o}}X$. By Theorem 5.32, $\boldsymbol{x}$ can be expressed as a nonnegative combination $\boldsymbol{x} = \lambda_1\boldsymbol{x}_1 + \cdots + \lambda_k\boldsymbol{x}_k$ of suitable points $\boldsymbol{x}_1, \ldots, \boldsymbol{x}_k \in X$. The equality

$$\boldsymbol{u}\cdot\boldsymbol{x} = \lambda_1\boldsymbol{u}\cdot\boldsymbol{x}_1 + \cdots + \lambda_k\boldsymbol{u}\cdot\boldsymbol{x}_k \leqslant 0,$$

shows that $\boldsymbol{u}$ belongs to $(\text{cone}_{\boldsymbol{o}}X)^{\circ}$. Thus $X^{\circ} \subset (\text{cone}_{\boldsymbol{o}}X)^{\circ}$.

(*d*) The inclusion $X^{\perp} \subset X^{\circ}$ immediately follows from the definitions (see page 7 for $X^{\perp}$). For the equality $X^{\circ} = (\text{cl}\, X)^{\circ}$, we first observe that $(\text{cl}\, X)^{\circ} \subset X^{\circ}$ due to $X \subset \text{cl}\, X$ and assertion (*a*) above.

Conversely, let $\boldsymbol{u} \in X^\circ$. If $\boldsymbol{u} = \boldsymbol{o}$, then the inclusion $\boldsymbol{u} \in (\operatorname{cl} X)^\circ$ is obvious. Assume that $\boldsymbol{u} \neq \boldsymbol{o}$. Definition 8.1 shows that X lies in the closed halfspace $V_{\boldsymbol{u}} = \{\boldsymbol{x} \in \mathbb{R}^n : \boldsymbol{x}\cdot\boldsymbol{u} \leqslant 0\}$. Since $V_{\boldsymbol{u}}$ is a closed set (see Theorem 2.33), one has $\operatorname{cl} X \subset V_{\boldsymbol{u}}$. Consequently, $\boldsymbol{u}\cdot\boldsymbol{x} = \boldsymbol{x}\cdot\boldsymbol{u} \leqslant 0$ for all $\boldsymbol{x} \in \operatorname{cl} X$, which gives $\boldsymbol{u} \in (\operatorname{cl} X)^\circ$. Hence $X^\circ \subset (\operatorname{cl} X)^\circ$.

The second part of assertion (d) follows from Definition 8.1 and the relation

$$\boldsymbol{u}\cdot(\mu\boldsymbol{x}) \leqslant 0 \quad \text{if and only if} \quad (\mu\boldsymbol{u})\cdot\boldsymbol{x} \leqslant 0$$

whenever $\boldsymbol{x} \in X$, $\boldsymbol{u} \in X^\circ$, and $\mu \neq 0$.

(e) Using twice assertion (d) and Problem 1.13, we obtain:

$$(X+Y)^\circ = (\operatorname{cl}(X+Y))^\circ = (\operatorname{cl}(\operatorname{cl} X + \operatorname{cl} Y))^\circ (\operatorname{cl} X + \operatorname{cl} Y)^\circ.$$

If, additionally, $\boldsymbol{o} \in \operatorname{cl} X \cap \operatorname{cl} Y$, then the above arguments and Theorem 5.43 imply that

$$\begin{aligned}(X+Y)^\circ &= (\operatorname{cl} X + \operatorname{cl} Y)^\circ = (\operatorname{cone}_{\boldsymbol{o}}(\operatorname{cl} X + \operatorname{cl} Y))^\circ \\ &= (\operatorname{cone}_{\boldsymbol{o}}(\operatorname{cl} X \cup \operatorname{cl} Y))^\circ = (\operatorname{cl} X \cup \operatorname{cl} Y)^\circ \\ &= (\operatorname{cl} X)^\circ \cap (\operatorname{cl} Y)^\circ = X^\circ \cap Y^\circ.\end{aligned}$$

(f) If $\boldsymbol{u} \in X^\circ$ and $\lambda \geqslant 0$, then $(\lambda\boldsymbol{u})\cdot\boldsymbol{x} \leqslant 0$ whenever $\boldsymbol{x} \in X$. Hence $\lambda\boldsymbol{u} \in X^\circ$, implying that X° is a cone with proper apex $\boldsymbol{o}$. For the convexity of X°, choose vectors $\boldsymbol{u}_1, \boldsymbol{u}_2 \in X^\circ$ and a scalar $\lambda \in [0,1]$. Then

$$((1-\lambda)\boldsymbol{u}_1 + \lambda\boldsymbol{u}_2)\cdot\boldsymbol{x} = (1-\lambda)\boldsymbol{u}_1\cdot\boldsymbol{x} + \lambda\boldsymbol{u}_2\cdot\boldsymbol{x} \leqslant 0 \quad \text{for all} \quad \boldsymbol{x} \in X,$$

implying that $(1-\lambda)\boldsymbol{u}_1 + \lambda\boldsymbol{u}_2 \in X^\circ$. Hence X° is a convex set.

If a sequence of points $\boldsymbol{u}_1, \boldsymbol{u}_2, \dots$ from X° converges to a point $\boldsymbol{u} \in \mathbb{R}^n$, then

$$\boldsymbol{u}\cdot\boldsymbol{x} = \lim_{i\to\infty} \boldsymbol{u}_i\cdot\boldsymbol{x} \leqslant 0 \quad \text{for every} \quad \boldsymbol{x} \in X,$$

giving the inclusion $\boldsymbol{u} \in X^\circ$. Thus X° is closed.

The inclusion $\operatorname{cl} X \cap X^\circ \subset \{\boldsymbol{o}\}$ is obvious if $\operatorname{cl} X \cap X^\circ = \varnothing$. Suppose that $\operatorname{cl} X \cap X^\circ \neq \varnothing$ and choose any vector $\boldsymbol{x} \in \operatorname{cl} X \cap X^\circ$. Then $\boldsymbol{x}\cdot\boldsymbol{x} \leqslant 0$ due to assertion (d) above, which gives $\boldsymbol{x} = \boldsymbol{o}$.

(g) We first assert that $\operatorname{cl}(\operatorname{cone}_{\boldsymbol{o}} X) \subset (X^\circ)^\circ$. Indeed, as in the proof of assertion (c), we have $\boldsymbol{u}\cdot\boldsymbol{x} \leqslant 0$ for all $\boldsymbol{u} \in X^\circ$ and $\boldsymbol{x} \in \operatorname{cone}_{\boldsymbol{o}} X$. This argument gives $\operatorname{cone}_{\boldsymbol{o}} X \subset (X^\circ)^\circ$. According to assertion (f), the set $(X^\circ)^\circ$ is closed. Consequently, $\operatorname{cl}(\operatorname{cone}_{\boldsymbol{o}} X) \subset (X^\circ)^\circ$.

So, it remains to prove the opposite inclusion $(X^\circ)^\circ \subset \operatorname{cl}(\operatorname{cone}_{\boldsymbol{o}} X)$. Assume, for contradiction, the existence of a point $\boldsymbol{v} \in (X^\circ)^\circ \setminus \operatorname{cl}(\operatorname{cone}_{\boldsymbol{o}} X)$.

By Theorem 8.22, there is a point $\boldsymbol{z} \in \text{cl}\,(\text{cone}_{\boldsymbol{o}}X)$ such that the closed halfspace

$$V = \{\boldsymbol{x} \in \mathbb{R}^n : (\boldsymbol{x} - \boldsymbol{z})\cdot(\boldsymbol{v} - \boldsymbol{z}) \leqslant 0\}$$

contains $\text{cl}\,(\text{cone}_{\boldsymbol{o}}X)$. We assert that $\boldsymbol{o}$ belongs to the boundary hyperplane

$$H = \{\boldsymbol{x} \in \mathbb{R}^n : (\boldsymbol{x} - \boldsymbol{z})\cdot(\boldsymbol{v} - \boldsymbol{z}) = 0\} \tag{8.2}$$

of V. Indeed, this fact is obvious if $\boldsymbol{z} = \boldsymbol{o}$ due to $\boldsymbol{z} \in H$. Let $\boldsymbol{z} \neq \boldsymbol{o}$. Then $2\boldsymbol{z} \in \text{cl}\,(\text{cone}_{\boldsymbol{o}}X)$ since $\text{cl}\,(\text{cone}_{\boldsymbol{o}}X)$ is a convex cone with apex $\boldsymbol{o}$ (see Theorem 5.28). The obvious inclusion $\boldsymbol{z} \in (\boldsymbol{o}, 2\boldsymbol{z}) \subset V$ shows that $[\boldsymbol{o}, 2\boldsymbol{z}] \subset \langle \boldsymbol{o}, 2\boldsymbol{z}\rangle \subset H$, as follows from Corollary 2.36. Hence $\boldsymbol{o} \in H$.

The inclusion $\boldsymbol{o} \in H$ and (8.2) give $\boldsymbol{z}\cdot(\boldsymbol{v} - \boldsymbol{z}) = 0$. Hence V can be expressed as

$$V = \{\boldsymbol{x} \in \mathbb{R}^n : \boldsymbol{x}\cdot(\boldsymbol{v} - \boldsymbol{z}) \leqslant 0\}.$$

Combining this argument with the inclusion $\text{cone}_{\boldsymbol{o}}X \subset V$ and assertions (a) and (c) above, we obtain

$$\boldsymbol{v} - \boldsymbol{z} \in V^\circ \subset (\text{cone}_{\boldsymbol{o}}X)^\circ = X^\circ.$$

Consequently, $\boldsymbol{v}\cdot(\boldsymbol{v} - \boldsymbol{z}) \leqslant 0$ due to the choice $\boldsymbol{v} \in (X^\circ)^\circ$. On the other hand,

$$\begin{aligned}\boldsymbol{v}\cdot(\boldsymbol{v} - \boldsymbol{z}) &= (\boldsymbol{v} - \boldsymbol{z} + \boldsymbol{z})\cdot(\boldsymbol{v} - \boldsymbol{z}) \\ &= \|\boldsymbol{v} - \boldsymbol{z}\|^2 + \boldsymbol{z}\cdot(\boldsymbol{v} - \boldsymbol{z}) = \|\boldsymbol{v} - \boldsymbol{z}\|^2 > 0.\end{aligned}$$

The obtained contradiction proves the inclusion $(X^\circ)^\circ \subset \text{cl}\,(\text{cone}_{\boldsymbol{o}}X)$.

(h) This statement follows from assertion (g) above.

(i) Suppose that $X \not\subset \{\boldsymbol{o}\}$ and choose a nonzero point $\boldsymbol{x}_0 \in X$. Then X° lies in the closed halfspace $V = \{\boldsymbol{u} \in \mathbb{R}^n : \boldsymbol{u}\cdot\boldsymbol{x}_0 \leqslant 0\}$, implying that $X^\circ \neq \mathbb{R}^n$. If $X \subset \{\boldsymbol{o}\}$, then $X^\circ = \mathbb{R}^n$.

(j) Suppose that $X^\circ \not\subset \{\boldsymbol{o}\}$ and choose a nonzero point $\boldsymbol{u}_0 \in X$. Then X lies in the closed halfspace $V = \{\boldsymbol{x} \in \mathbb{R}^n : \boldsymbol{x}\cdot\boldsymbol{u}_0 \leqslant 0\}$, implying that $\text{cone}_{\boldsymbol{o}}X \subset V$. Therefore, $\text{cone}_{\boldsymbol{o}}X \neq \mathbb{R}^n$. If $X^\circ = \{\boldsymbol{o}\}$, then assertion (g) gives

$$\text{cl}\,(\text{cone}_{\boldsymbol{o}}X) = (X^\circ)^\circ = (\{\boldsymbol{o}\})^\circ = \mathbb{R}^n.$$

According to Theorem 3.50, $\text{cone}_{\boldsymbol{o}}X = \mathbb{R}^n$. $\square$

Remark. It is easy to see that assertion (b) from Theorem 8.2 holds for any family $\{X_\alpha\}$ of sets in $\mathbb{R}^n$: $(\bigcup_\alpha X_\alpha)^\circ = \bigcap_\alpha X_\alpha^\circ$.

Theorem 8.3. *The following assertions hold.*

(*a*) *If* $\{C_\alpha\}$ *is a family of convex cones in* $\mathbb{R}^n$ *with common apex* $\boldsymbol{o}$, *then*

$$(\bigcap_\alpha \operatorname{cl} C_\alpha)^\circ = \operatorname{cl}(\operatorname{cone}_{\boldsymbol{o}}(\bigcup_\alpha C_\alpha^\circ)) = \operatorname{cl}(\operatorname{conv}(\bigcup_\alpha C_\alpha^\circ)).$$

(*b*) *If* $C_1, \ldots, C_r \subset \mathbb{R}^n$ *are nonempty convex cones with common apex* $\boldsymbol{o}$, *then*

$$\begin{aligned}(C_1 + \cdots + C_r)^\circ &= (C_1 \cup \cdots \cup C_r)^\circ = C_1^\circ \cap \cdots \cap C_r^\circ,\\ (\operatorname{cl} C_1 \cap \cdots \cap \operatorname{cl} C_r)^\circ &= \operatorname{cl}(\operatorname{conv}(C_1^\circ \cup \cdots \cup C_r^\circ)) \\ &= \operatorname{cl}(C_1^\circ + \cdots + C_r^\circ).\end{aligned} \tag{8.3}$$

Proof. (*a*) Since both sets $\bigcap_\alpha \operatorname{cl} C_\alpha$ and $\operatorname{cl}(\operatorname{conv}(\bigcup_\alpha C_\alpha^\circ))$ are closed convex cones with apex $\boldsymbol{o}$ (see Corollary 5.9), a combination of Theorem 8.2, Corollary 5.9, and the above remark gives

$$\begin{aligned}(\bigcap_\alpha \operatorname{cl} C_\alpha)^\circ &= (\bigcap_\alpha (C_\alpha^\circ)^\circ)^\circ = ((\bigcup_\alpha C_\alpha^\circ)^\circ)^\circ \\ &= \operatorname{cl}(\operatorname{cone}_{\boldsymbol{o}}(\bigcup_\alpha C_\alpha^\circ)) = \operatorname{cl}(\operatorname{conv}(\bigcup_\alpha C_\alpha^\circ)).\end{aligned}$$

(*b*) By Theorem 5.10,

$$C_1 + \cdots + C_r = \operatorname{conv}(C_1 \cup \cdots \cup C_r).$$

Consequently, Theorem 8.2 gives

$$\begin{aligned}(C_1 + \cdots + C_r)^\circ &= (\operatorname{conv}(C_1 \cup \cdots \cup C_r))^\circ \\ &= (C_1 \cup \cdots \cup C_r)^\circ = C_1^\circ \cap \cdots \cap C_r^\circ.\end{aligned}$$

Similarly, the equality

$$C_1^\circ + \cdots + C_r^\circ = \operatorname{conv}(C_1^\circ \cup \cdots \cup C_r^\circ)$$

and assertion (*a*) give

$$\begin{aligned}(\operatorname{cl} C_1 \cap \cdots \cap \operatorname{cl} C_r)^\circ &= \operatorname{cl}(\operatorname{conv}(C_1^\circ \cup \cdots \cup C_r^\circ)) \\ &= \operatorname{cl}(C_1^\circ + \cdots + C_r^\circ).\end{aligned}$$ □

8.1.2 *Polar Decomposition of Cones*

Theorem 8.4. *For a nonempty convex cone* $C \subset \mathbb{R}^n$ *with apex* $\boldsymbol{o}$, *the following assertions hold.*

(*a*) $C^\perp = (\operatorname{cl} C)^\perp = \operatorname{ap} C^\circ$ *and* $\operatorname{ap} C \subset \operatorname{ap}(\operatorname{cl} C) = (C^\circ)^\perp$.
(*b*) $\operatorname{span} C = (\operatorname{ap} C^\circ)^\perp$ *and* $\operatorname{span} C^\circ = (\operatorname{ap}(\operatorname{cl} C))^\perp$. *Consequently,* $\operatorname{ap} C^\circ$ *and* $\operatorname{ap}(\operatorname{cl} C)$ *are orthogonal subspaces.*

(*c*) $\dim(\operatorname{ap} C^\circ) = n - \dim C$ *and* $\dim C^\circ = n - \dim(\operatorname{ap}(\operatorname{cl} C))$.
(*d*) $\dim C = n$ *if and only if* $\operatorname{ap} C^\circ = \{\boldsymbol{o}\}$ (*equivalently,* $\operatorname{int} C \neq \varnothing$ *if and only if* C° *is line-free*).
(*e*) $\dim C^\circ = n$ *if and only if* $\operatorname{ap}(\operatorname{cl} C) = \{\boldsymbol{o}\}$ (*equivalently,* $\operatorname{int} C^\circ \neq \varnothing$ *if and only if* C *is line-free*).
(*f*) C *is a subspace if and only if* C° *is a subspace.*
(*g*) $C^\circ = C^\perp$ *if and only if* C *is a subspace.*
(*h*) $C \neq \operatorname{ap} C$ *if and only if* $C^\circ \neq \operatorname{ap} C^\circ$.

Proof. (*a*) For the first equality, it suffices to show that $C^\perp = \operatorname{ap} C^\circ$ (because $X^\perp = (\operatorname{cl} X)^\perp$ for any set $X \subset \mathbb{R}^n$). The inclusion $C^\perp \subset \operatorname{ap} C^\circ$ holds since $C^\perp$ is a subspace which lies in the convex cone C° (see Theorem 8.2) and $\operatorname{ap} C^\circ$ is the largest subspace lying in C° (see Theorem 5.17). Conversely, let $\boldsymbol{u} \in \operatorname{ap} C^\circ$. Then $-\boldsymbol{u} \in \operatorname{ap} C^\circ$, which gives $\boldsymbol{u}\cdot\boldsymbol{x} \leqslant 0$ and $(-\boldsymbol{u})\cdot\boldsymbol{x} \leqslant 0$ for every $\boldsymbol{u} \in C$. Thus $\boldsymbol{u}\cdot\boldsymbol{x} = 0$ whenever $\boldsymbol{u} \in C$, implying that $\boldsymbol{x} \in C^\perp$. Summing up, $C^\perp = \operatorname{ap} C^\circ$.

For the second part, we first observe that $\operatorname{ap} C \subset \operatorname{ap}(\operatorname{cl} C)$ according to Theorem 5.28. Next, the equality $(C^\circ)^\circ = \operatorname{cl} C$ (see Theorem 8.2) and the above argument give

$$\operatorname{ap}(\operatorname{cl} C) = \operatorname{ap}(C^\circ)^\circ = (C^\circ)^\perp.$$

(*b*) Using the equality $\operatorname{span} X = (X^\perp)^\perp$ (see page 7), we obtain from (*a*):

$$\operatorname{span} C = (C^\perp)^\perp = (\operatorname{ap} C^\circ)^\perp, \quad \operatorname{span} C^\circ = ((C^\circ)^\perp)^\perp = (\operatorname{ap}(\operatorname{cl} C))^\perp.$$

Consequently, the relations

$$\operatorname{ap} C^\circ = (\operatorname{span} C)^\perp \quad \text{and} \quad \operatorname{ap}(\operatorname{cl} C) \subset \operatorname{cl} C \subset \operatorname{span} C$$

show that the subspaces $\operatorname{ap} C^\circ$ and $\operatorname{ap}(\operatorname{cl} C)$ are orthogonal.

(*c*) Based on assertion (*b*), we have

$$\dim(\operatorname{ap} C^\circ) = n - \dim(\operatorname{span} C) = n - \dim C,$$
$$\dim C^\circ = \dim(\operatorname{span} C^\circ) = \dim(\operatorname{ap}(\operatorname{cl} C))^\perp = n - \dim(\operatorname{ap}(\operatorname{cl} C)).$$

Assertions (*d*) and (*e*) immediately follow from (*c*). Equivalent versions of assertions (*d*) and (*e*) are due to Theorem 6.37.

(*f*) If C is a subspace, then $C^\circ = C^\perp$ is a subspace (see an example on page 269). Conversely, if C° is a subspace, then $\operatorname{cl} C = (C^\circ)^\circ = (C^\circ)^\perp$ is a subspace, and Theorem 3.50 shows that C is a subspace.

(*g*) If $C^\circ = C^\perp$, then C° is a subspace because $C^\perp$ is subspace (see page 7). Since $C = \text{cone}_{\boldsymbol{o}} C$ and $(C^\circ)^\circ = (C^\circ)^\perp$ (see an example on page 269), Theorem 8.2 gives

$$\text{cl}\, C = \text{cl}\,(\text{cone}_{\boldsymbol{o}} C) = (C^\circ)^\circ = (C^\perp)^\circ = (C^\perp)^\perp = \text{span}\, C.$$

Hence $\text{cl}\, C$ is a subspace, and Theorem 3.50 implies that C is a subspace. Conversely, if C is subspace, then $C^\circ = C^\perp$, as follows from an example on page 269.

(*h*) This assertion is contrapositive to (*f*), as follows from Corollary 5.27. □

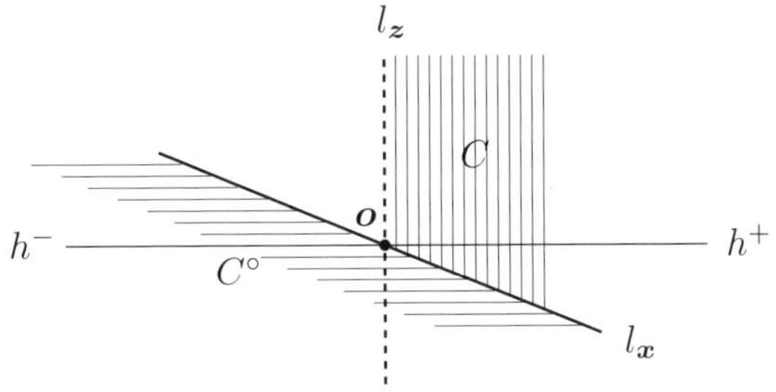

Fig. 8.2 Illustration to Theorem 8.5.

The next theorem describes a simultaneous decomposition of a convex cone and its polar. We illustrate this decomposition with the following example.

Example. For the convex cone $C = \{\boldsymbol{o}\} \cup \{(0, y, z) : y > 0\}$ in $\mathbb{R}^3$, its polar cone is $C^\circ = \{(x, y, 0) : y \leqslant 0\}$. Denote by $l_{\boldsymbol{x}}$ and $l_{\boldsymbol{z}}$ the x- and z-axes, respectively, and let h^+ and h^- stand for the nonnegative and nonpositive halflines of the y-axis. Then

$$\text{cl}\, C = h^+ + l_{\boldsymbol{z}} \quad \text{and} \quad C^\circ = h^- + l_{\boldsymbol{x}},$$

as depicted in Figure 8.2. Clearly,

$$\text{ap}\, C = \{\boldsymbol{o}\}, \quad \text{ap}\,(\text{cl}\, C) = l_{\boldsymbol{z}}, \quad \text{ap}\, C^\circ = l_{\boldsymbol{x}}.$$

In terms of Theorem 8.5, $\text{span}\, C$ is the yz-plane, and S is the y-axis. Then

$$\text{cl}\, C \cap S = h^+ \quad \text{and} \quad C^\circ \cap S = h^-.$$

Theorem 8.5. *Let* $C \subset \mathbb{R}^n$ *be a nonempty convex cone with apex* $\mathbf{o}$ *and* E *be the orthogonal complement of* $\operatorname{ap}(\operatorname{cl} C)$ *within* $\operatorname{span} C$ *(possibly,* $E = \{\mathbf{o}\}$*). Then* E *is the orthogonal complement of* $\operatorname{ap} C^\circ$ *within* $\operatorname{span} C^\circ$ *and* $\mathbb{R}^n$ *is the direct sums of pairwise orthogonal subspaces:*

$$\mathbb{R}^n = \operatorname{ap}(\operatorname{cl} C) + \operatorname{ap} C^\circ + E. \tag{8.4}$$

Consequently,

$$\operatorname{cl} C = (\operatorname{cl} C \cap E) + \operatorname{ap}(\operatorname{cl} C) \quad \text{and} \quad C^\circ = (C^\circ \cap E) + \operatorname{ap} C^\circ. \tag{8.5}$$

Furthermore, both sets $\operatorname{cl} C \cap E$ *and* $C^\circ \cap E$ *are line-free and*

$$\begin{aligned} \operatorname{span}(\operatorname{cl} C \cap E) &= \operatorname{span}(C^\circ \cap E) = E, \\ \operatorname{rint} C &= (\operatorname{rint} C \cap E) + \operatorname{ap}(\operatorname{cl} C) = \operatorname{rint}(C \cap E) + \operatorname{ap}(\operatorname{cl} C), \\ \operatorname{rint} C^\circ &= (\operatorname{rint} C^\circ \cap E) + \operatorname{ap} C^\circ = \operatorname{rint}(C^\circ \cap E) + \operatorname{ap} C^\circ. \end{aligned}$$

Proof. Equality (8.4) follows from Theorem 8.4 and the choice of E:

$$\mathbb{R}^n = \operatorname{ap} C^\circ + \operatorname{span} C = \operatorname{ap} C^\circ + \operatorname{ap}(\operatorname{cl} C) + E.$$

Combining Theorem 8.4 and (8.4), we obtain

$$\operatorname{span} C^\circ = (\operatorname{ap}(\operatorname{cl} C))^\perp = \operatorname{ap} C^\circ + E.$$

Hence E is the orthogonal complement of $\operatorname{ap} C^\circ$ within $\operatorname{span} C^\circ$.

By Theorem 6.21, $\operatorname{ap}(\operatorname{cl} C) = \operatorname{lin}(\operatorname{cl} C)$. Since E can be expressed as

$$E = (\operatorname{ap}(\operatorname{cl} C))^\perp \cap \operatorname{span}(\operatorname{cl} C) = \operatorname{span} C^\circ \cap (\operatorname{ap} C^\circ)^\perp,$$

Theorem 6.29 gives

$$\begin{aligned} \operatorname{cl} C &= (\operatorname{cl} C \cap E) + \operatorname{ap}(\operatorname{cl} C) \quad \text{and} \quad \operatorname{span}(\operatorname{cl} C \cap E) = E, \\ C^\circ &= (C^\circ \cap E) + \operatorname{ap} C^\circ \quad \text{and} \quad \operatorname{span}(C^\circ \cap E) = E. \end{aligned}$$

Corollary 6.38 shows that the set $\operatorname{cl} C \cap E$ is line-free, and a combination of Theorems 3.48 and 6.31 gives

$$\begin{aligned} \operatorname{rint} C = \operatorname{rint}(\operatorname{cl} C) &= (\operatorname{rint}(\operatorname{cl} C) \cap E) + \operatorname{ap}(\operatorname{cl} C) \\ &= (\operatorname{rint} C \cap E) + \operatorname{ap}(\operatorname{cl} C). \end{aligned}$$

Similarly,

$$\operatorname{rint} C^\circ = (\operatorname{rint} C^\circ \cap E) + \operatorname{ap} C^\circ = \operatorname{rint}(C^\circ \cap E) + \operatorname{ap} C^\circ. \qquad \square$$

Theorem 8.6. *Let* $C \subset \mathbb{R}^n$ *be a nonempty convex cone with apex* $\mathbf{o}$*, which is not a subspace. The following assertions hold.*

(*a*) *A vector* $\boldsymbol{e} \in \mathbb{R}^n$ *belongs to* $\operatorname{rint} C^\circ$ *if and only if* $\boldsymbol{x}\cdot\boldsymbol{e} < 0$ *for all vectors* $\boldsymbol{x} \in \operatorname{cl} C \setminus \operatorname{ap}(\operatorname{cl} C)$.

(*b*) *A vector* $\boldsymbol{e} \in \mathbb{R}^n$ *belongs to* $\operatorname{rint} C$ *if and only if* $\boldsymbol{x}\cdot\boldsymbol{e} < 0$ *for all vectors* $\boldsymbol{x} \in C^\circ \setminus \operatorname{ap} C^\circ$.

Proof. Since $\operatorname{ap}(\operatorname{cl} C)$ is a subspace (see Theorem 8.4) and C is not, we have $\operatorname{ap}(\operatorname{cl} C) \neq \operatorname{cl} C$. Consequently, $\operatorname{ap}(\operatorname{cl} C)$ is a proper subspace of $\operatorname{span}(\operatorname{cl} C) = \operatorname{span} C$. Denote by E the orthogonal complement of $\operatorname{ap}(\operatorname{cl} C)$ within $\operatorname{span} C$. By the above argument, $E \neq \{\boldsymbol{o}\}$. Theorem 8.5 shows that E is the orthogonal complement of $\operatorname{ap} C^\circ$ within $\operatorname{span} C^\circ$.

(*a*) Assume first that a vector $\boldsymbol{e} \in \mathbb{R}^n$ belongs to $\operatorname{rint} C^\circ$. By Theorem 8.5, $\boldsymbol{e}$ can be expressed as $\boldsymbol{e} = \boldsymbol{e}_1 + \boldsymbol{e}_2$, where $\boldsymbol{e}_1 \in \operatorname{rint}(C^\circ \cap E)$ and $\boldsymbol{e}_2 \in \operatorname{ap} C^\circ$. Let $\boldsymbol{x} \in \operatorname{cl} C \setminus \operatorname{ap}(\operatorname{cl} C)$. Due to the same theorem, we can write $\boldsymbol{x} = \boldsymbol{x}_1 + \boldsymbol{x}_2$, where $\boldsymbol{x}_1$ is a nonzero vector in $\operatorname{cl} C \cap E$ and $\boldsymbol{x}_2 \in \operatorname{ap}(\operatorname{cl} C)$. Because the subspaces $\operatorname{ap}(\operatorname{cl} C)$, $\operatorname{ap} C^\circ$, and E are pairwise orthogonal, we have

$$\boldsymbol{x}\cdot\boldsymbol{e} = (\boldsymbol{x}_1 + \boldsymbol{x}_2)\cdot(\boldsymbol{e}_1 + \boldsymbol{e}_2) = \boldsymbol{x}_1\cdot\boldsymbol{e}_1.$$

Since $\operatorname{span}(C^\circ \cap E) = E$ (see Theorem 8.5), and since $\boldsymbol{e}_1 + \boldsymbol{x}_1 \in E$, Theorem 3.26 shows the existence of a scalar $\varepsilon > 0$ such that

$$\boldsymbol{e}_1 + \varepsilon\boldsymbol{x}_1 = (1-\varepsilon)\boldsymbol{e}_1 + \varepsilon(\boldsymbol{e}_1 + \boldsymbol{x}_1) \in C^\circ \cap E \subset C^\circ.$$

By a polarity argument, $\boldsymbol{x}_1\cdot(\boldsymbol{e}_1 + \varepsilon\boldsymbol{x}_1) \leqslant 0$, implying that

$$\boldsymbol{x}\cdot\boldsymbol{e} = \boldsymbol{x}_1\cdot\boldsymbol{e}_1 \leqslant -\varepsilon\|\boldsymbol{x}_1\|^2 < 0.$$

Conversely, suppose that a vector $\boldsymbol{e} \in \mathbb{R}^n$ satisfies the condition $\boldsymbol{x}\cdot\boldsymbol{e} < 0$ for all vectors $\boldsymbol{x} \in \operatorname{cl} C \setminus \operatorname{ap}(\operatorname{cl} C)$. By Theorem 5.25, $\operatorname{ap}(\operatorname{cl} C) \subset \operatorname{cl}(C \setminus \operatorname{ap}(\operatorname{cl} C))$. Therefore, a continuity argument implies that $\boldsymbol{x}\cdot\boldsymbol{e} \leqslant 0$ for all vectors $\boldsymbol{x} \in \operatorname{cl} C$. Thus $\boldsymbol{e} \in (\operatorname{cl} C)^\circ = C^\circ$.

Assume for a moment that $\boldsymbol{e} \notin \operatorname{rint} C^\circ$. Then $\boldsymbol{e} \in \operatorname{rbd} C^\circ$. By Theorem 3.62, there is a sequence of vectors $\boldsymbol{e}_1, \boldsymbol{e}_2, \ldots$ in $\operatorname{span} C^\circ \setminus C^\circ$ converging to $\boldsymbol{e}$. Since $\boldsymbol{e} \neq \boldsymbol{o}$ (otherwise $\boldsymbol{x}\cdot\boldsymbol{e} = 0$ whenever $\boldsymbol{x} \in \operatorname{cl} C$), we may assume that none of $\boldsymbol{e}_1, \boldsymbol{e}_2, \ldots$ is $\boldsymbol{o}$. From the definition of C° we deduce the existence of nonzero vectors $\boldsymbol{x}_1, \boldsymbol{x}_2, \ldots$ in $\operatorname{cl} C$ such that $\boldsymbol{x}_i\cdot\boldsymbol{e}_i > 0$ for all $i \geqslant 1$. Due to (8.5), we can write

$$\boldsymbol{x}_i = \boldsymbol{v}_i + \boldsymbol{w}_i, \text{ where } \boldsymbol{v}_i \in \operatorname{cl} C \cap S \text{ and } \boldsymbol{w}_i \in \operatorname{ap}(\operatorname{cl} C).$$

According to Theorem 8.4, $\boldsymbol{w}_i\cdot\boldsymbol{e}_i = 0$, which gives

$$\boldsymbol{v}_i\cdot\boldsymbol{e}_i = (\boldsymbol{x}_i - \boldsymbol{w}_i)\cdot\boldsymbol{e}_i = \boldsymbol{x}_i\cdot\boldsymbol{e}_i > 0, \quad i \geqslant 1.$$

Let $\boldsymbol{v}_i' = \boldsymbol{v}_i/\|\boldsymbol{v}_i\|$, $i \geqslant 1$. Then $\boldsymbol{v}_i'\boldsymbol{e}_i > 0$ and $\boldsymbol{v}_i' \in \mathrm{cl}\, C \cap E$ because $\mathrm{cl}\, C \cap E$ is a cone with apex $\boldsymbol{o}$. Replacing the sequence $\boldsymbol{v}_1', \boldsymbol{v}_2', \dots$ with a converging subsequence, we may assume that $\boldsymbol{v}_1', \boldsymbol{v}_2', \dots$ converges to a unit vector $\boldsymbol{v}' \in \mathrm{cl}\, C \cap E$. Since $\mathrm{ap}\,(\mathrm{cl}\, C) \cap E = \{\boldsymbol{o}\}$, we have $\boldsymbol{v}' \in (\mathrm{cl}\, C \cap E) \setminus \mathrm{ap}\,(\mathrm{cl}\, C)$. On the other hand,

$$\boldsymbol{v}' \cdot \boldsymbol{e} = \lim_{i \to \infty} \boldsymbol{v}_i' \cdot \boldsymbol{e}_i \geqslant 0,$$

in contradiction with the choice of $\boldsymbol{e}$. Hence $\boldsymbol{e} \in \mathrm{rint}\, C^\circ$.

(b) By Theorem 8.4, $C^\circ \neq \mathrm{ap}\, C^\circ$. Furthermore, $\mathrm{rint}\, C = \mathrm{rint}\,(\mathrm{cl}\, C)$, as follows from Theorem 3.48. This argument and the equality $(C^\circ)^\circ = \mathrm{cl}\, C$ show that assertion (b) derives from (a). □

Remark. In assertion (a) of Theorem 8.6, the cone $\mathrm{cl}\, C$ cannot be replaced with C. Indeed, consider the convex cone

$$C = h \cup \{(x, y, z) : x > 0, y > 0\}, \quad \text{where} \quad h = \{(0, 0, z) : z \geqslant 0\},$$

and the vector $\boldsymbol{e} = (-1, -1, 0)$. Then $\boldsymbol{e} \in \mathrm{rint}\, C^\circ$, while $\boldsymbol{x} \cdot \boldsymbol{e} = 0$ for all $\boldsymbol{x} \in h$. Similarly, if $C = \{\boldsymbol{o}\} \cup \{(x, y, z) : x > 0, y > 0\}$ and $\boldsymbol{e} = (-1, 0, 0)$, then $\boldsymbol{x} \cdot \boldsymbol{e} < 0$ for all vectors $\boldsymbol{x} \in C \setminus \mathrm{ap}\, C$, while $\boldsymbol{e} \in \mathrm{rbd}\, C^\circ$.

The next theorem plays an important role in the study of support and separation properties of convex cones.

Theorem 8.7. *Let $C \subset \mathbb{R}^n$ be a nonempty proper convex cone with apex $\boldsymbol{o}$. If $S \subset \mathbb{R}^n$ is an $(n-1)$-dimensional subspaces, then the following assertions are equivalent.*

(a) $S \cap \mathrm{cl}\, C = \mathrm{ap}\,(\mathrm{cl}\, C)$.
(b) $(S^\perp \setminus \{\boldsymbol{o}\}) \cap \mathrm{rint}\, C^\circ \neq \varnothing$.

Furthermore, $S \cap \mathrm{rint}\, C = \varnothing$ if C is not a subspace and any of the conditions (a) and (b) holds.

Proof. First, we consider the case when C is a subspace. Then $C = \mathrm{cl}\, C$ (see Corollary 2.21), $\mathrm{ap}\, C = C$ (see an example on page 150), C° is a subspace, and $\mathrm{rint}\, C^\circ = C^\circ = C^\perp$ (see an example on page 84 and Theorem 8.4). Consequently, conditions (a) and (b) can be written, respectively, as (a') $C \subset S$ and (b') $(S^\perp \setminus \{\boldsymbol{o}\}) \subset C^\perp$, whose equivalence is obvious.

In what follows we suppose that C is not a subspace.

(a) ⇒ (b) Since C is not a subspace, one has $\mathrm{cl}\, C \neq \mathrm{ap}\,(\mathrm{cl}\, C)$, and $\mathrm{ap}\,(\mathrm{cl}\, C) \subset \mathrm{rbd}\, C$ according to Corollary 5.27.

First, we assert that $S \cap \operatorname{rint} C = \varnothing$. Indeed, the above argument gives

$$\begin{aligned} S \cap \operatorname{rint} C &= S \cap (\operatorname{cl} C \setminus \operatorname{rbd} C) \subset S \cap (\operatorname{cl} C \setminus \operatorname{ap}(\operatorname{cl} C)) \\ &= (S \cap \operatorname{cl} C) \setminus (S \cap \operatorname{ap}(\operatorname{cl} C)) = \operatorname{ap}(\operatorname{cl} C) \setminus \operatorname{ap}(\operatorname{cl} C) = \varnothing. \end{aligned}$$

Next, we state that $\operatorname{rint} C$ lies in one of the open opposite halfspaces determined by S. Indeed, assume for a moment the existence of distinct points $\boldsymbol{u}, \boldsymbol{v} \in \operatorname{rint} C$ which lie in the opposite open halfspaces determined by S. Theorem 2.35 shows that the open segment $(\boldsymbol{u}, \boldsymbol{v})$ meets S at a point $\boldsymbol{w}$. Then $\boldsymbol{w} \in \operatorname{rint} C$ due to the convexity of $\operatorname{rint} C$ (see Corollary 3.25). Hence, $\boldsymbol{w} \in S \cap \operatorname{rint} C$, contrary to the above argument.

Let $S = \{\boldsymbol{x} \in \mathbb{R}^n : \boldsymbol{x} \cdot \boldsymbol{e} = 0\}$, where $\boldsymbol{e} \neq \boldsymbol{o}$ (see page 8). The open halfspaces determined by S can be described as

$$W_1 = \{\boldsymbol{x} \in \mathbb{R}^n : \boldsymbol{x} \cdot \boldsymbol{e} < 0\} \quad \text{and} \quad W_2 = \{\boldsymbol{x} \in \mathbb{R}^n : \boldsymbol{x} \cdot (-\boldsymbol{e}) < 0\}.$$

Suppose that $\operatorname{rint} C \subset V_1$. Then $\operatorname{cl} C = \operatorname{cl}(\operatorname{rint} C) \subset \operatorname{cl} W_1 = V_1$, where V_1 is the closed halfspace given by $V_1 = \{\boldsymbol{x} \in \mathbb{R}^n : \boldsymbol{x} \cdot \boldsymbol{e} \leqslant 0\}$. Assumption (a) results in the inclusion

$$\operatorname{cl} C \setminus \operatorname{ap}(\operatorname{cl} C) = \operatorname{cl} C \setminus (S \cap \operatorname{cl} C) = \operatorname{cl} C \setminus S \subset V_1 \setminus S = W_1,$$

and Theorem 8.6 shows that $\boldsymbol{e} \in \operatorname{rint} C^\circ$. Similarly, $-\boldsymbol{e} \in \operatorname{rint} C^\circ$ if $\operatorname{rint} C \subset W_2$. Summing up,

$$\pm \boldsymbol{e} \in (\operatorname{span}\{\boldsymbol{e}\} \setminus \{\boldsymbol{o}\}) \cap \operatorname{rint} C^\circ = (S^\perp \setminus \{\boldsymbol{o}\}) \cap \operatorname{rint} C^\circ,$$

confirming condition (b).

$(b) \Rightarrow (a)$ Choose a vector $\boldsymbol{e} \in (S^\perp \setminus \{\boldsymbol{o}\}) \cap \operatorname{rint} C^\circ$. By Theorem 8.6,

$$\operatorname{cl} C \setminus \operatorname{ap}(\operatorname{cl} C) \subset W_1 = \{\boldsymbol{x} \in \mathbb{R}^n : \boldsymbol{x} \cdot \boldsymbol{e} < 0\}.$$

Since

$$\operatorname{ap}(\operatorname{cl} C) \subset \operatorname{cl}(\operatorname{cl} C \setminus \operatorname{ap}(\operatorname{cl} C)) \subset \operatorname{cl} W_1 = V_1$$

(see Theorem 5.25), one has $\operatorname{ap}(\operatorname{cl} C) \subset V_1 \setminus W_1 = H$. Consequently,

$$\begin{aligned} S \cap \operatorname{cl} C &= S \cap (\operatorname{ap}(\operatorname{cl} C)) \cup (\operatorname{cl} C \setminus \operatorname{ap}(\operatorname{cl} C)) \\ &= S \cap \operatorname{ap}(\operatorname{cl} C) = \operatorname{ap}(\operatorname{cl} C). \end{aligned}$$ □

Remarks. 1. The equality $S \cap \operatorname{rint} C = \varnothing$ in Theorem 8.7 does not imply any of conditions (a) and (b). Indeed, let S be the x-axis of $\mathbb{R}^2$ and $C = \{(x, y) : x, y \geqslant 0\}$. Then $S \cap \operatorname{rint} C = \varnothing$, while

$$S \cap C = \{(x, 0) : x \geqslant 0\} \neq \{\boldsymbol{o}\} = \operatorname{ap} C.$$

2. A generalization of Theorem 8.7 to the case of any nontrivial subspace $S \subset \mathbb{R}^n$ is given in Problem 9.7.

Corollary 8.8. *Given a nonempty line-free convex cone $C \subset \mathbb{R}^n$ with apex $\boldsymbol{o}$, the following assertions hold.*

(a) *There is an $(n-1)$-dimensional subspace $S \subset \mathbb{R}^n$ satisfying the condition $S \cap \operatorname{cl} C = \{\boldsymbol{o}\}$.*
(b) *An $(n-1)$-dimensional subspace $S \subset \mathbb{R}^n$ satisfies the condition $S \cap \operatorname{cl} C = \{\boldsymbol{o}\}$ if and only if $(S^{\perp} \setminus \{\boldsymbol{o}\}) \cap \operatorname{int} C^{\circ} \neq \varnothing$.*

Proof. By Theorem 6.37, C is line-free if and only if $\operatorname{ap}(\operatorname{cl} C) = \{\boldsymbol{o}\}$. Also, $\operatorname{int} C^{\circ} = \operatorname{rint} C^{\circ}$ provided C is line-free (see Theorem 8.4). Consequently, Corollary 8.8 follows from Theorem 8.7. □

Corollary 8.8 implies the following result, which complements Theorem 6.40.

Corollary 8.9. *Given a line-free convex set $K \subset \mathbb{R}^n$ of positive dimension, the following assertions hold.*

(a) *There is a hyperplane which meets $\operatorname{cl} K$ along a bounded set.*
(b) *Given a hyperplane $H \subset \mathbb{R}^n$, every translate of H meets $\operatorname{cl} K$ along a bounded set if and only if $(\operatorname{ort} H \setminus \{\boldsymbol{o}\}) \cap \operatorname{rint}(\operatorname{rec} K)^{\circ} \neq \varnothing$.* □

The next result (see Theorem 8.11) describes the intersection $\operatorname{cl} C \cap (-C^{\circ})$, where $C \subset \mathbb{R}^n$ is a convex cone with apex $\boldsymbol{o}$. We will need the following lemma.

Lemma 8.10. *Let $C \subset \mathbb{R}^n$ be a nonempty convex cone with apex $\boldsymbol{o}$, and E be the orthogonal complement of $\operatorname{ap}(\operatorname{cl} C)$ within $\operatorname{span} C$ (possibly, $E = \{\boldsymbol{o}\}$). Then*

$$(\operatorname{cl} C \cap E)^{\circ} = C^{\circ} + \operatorname{ap}(\operatorname{cl} C) \quad \textit{and} \quad (C^{\circ} \cap E)^{\circ} = \operatorname{cl} C + \operatorname{ap} C^{\circ}.$$

Proof. Since $S^{\circ} = S^{\perp}$ (see an example on page 269) and $(\operatorname{cl} C)^{\circ} = C^{\circ}$ (see Theorem 8.4), Theorems 8.3 and 8.5 give

$$\begin{aligned}(\operatorname{cl} C \cap E)^{\circ} &= \operatorname{cl}((\operatorname{cl} C)^{\circ} + E^{\circ}) = \operatorname{cl}(C^{\circ} + E^{\perp}) \\ &= \operatorname{cl}(C^{\circ} + \operatorname{ap}(\operatorname{cl} C) + \operatorname{ap} C^{\circ}).\end{aligned}$$

By Theorem 5.16, $C^{\circ} + \operatorname{ap} C^{\circ} = C^{\circ}$. Because C° and $\operatorname{ap}(\operatorname{cl} C)$ lie in orthogonal subspaces (see Theorem 8.4), we obtain from Problem 1.13 and closedness of C°:

$$\operatorname{cl}(C^{\circ} + \operatorname{ap}(\operatorname{cl} C) + \operatorname{ap} C^{\circ}) = \operatorname{cl}(C^{\circ} + \operatorname{ap}(\operatorname{cl} C)) = C^{\circ} + \operatorname{ap}(\operatorname{cl} C).$$

Similarly,

$$\begin{aligned}(C^\circ \cap S)^\circ &= \mathrm{cl}\,((C^\circ)^\circ + E^\circ) = \mathrm{cl}\,(\mathrm{cl}\,C + E^\perp) \\ &= \mathrm{cl}\,(\mathrm{cl}\,C + \mathrm{ap}\,(\mathrm{cl}\,C) + \mathrm{ap}\,C^\circ) \\ &= \mathrm{cl}\,(\mathrm{cl}\,C + \mathrm{ap}\,C^\circ) = \mathrm{cl}\,C + \mathrm{ap}\,C^\circ. \qquad \square\end{aligned}$$

Theorem 8.11. *Let $C \subset \mathbb{R}^n$ be a nonempty convex cone with apex $\boldsymbol{o}$. Also, let $D = \mathrm{cl}\,C \cap (-C^\circ)$ and E be the orthogonal complement of* $\mathrm{ap}\,(\mathrm{cl}\,C)$ *within* $\mathrm{span}\,C$ *(possibly, $E = \{\boldsymbol{o}\}$). Then the following assertions take place.*

(*a*) *D is a line-free closed convex cone with apex $\boldsymbol{o}$.*
(*b*) $\mathrm{span}\,D = E$, $\dim D = \dim E$, *and* $D^\circ = C^\circ - \mathrm{cl}\,C$.
(*c*) $\mathrm{rint}\,D = \mathrm{rint}\,C \cap (-\mathrm{rint}\,C^\circ) = \mathrm{rint}\,(C \cap E) \cap (-\mathrm{rint}\,(C^\circ \cap E))$.

Proof. First, we consider case when C is a subspace. Then $C^\circ = C^\perp$, implying that $D = E = \{\boldsymbol{o}\}$ and making assertions (*a*)–(*c*) obvious. So, in what follows, we may assume that C is not a subspace.

(*a*) The set D is a closed convex cone with apex $\boldsymbol{o}$ as the intersection of closed convex cones $\mathrm{cl}\,C$ and C° with the same apex (see Corollaries 5.9 and 5.21). Since the subspaces $\mathrm{ap}\,(\mathrm{cl}\,C)$, $\mathrm{ap}\,C^\circ$, and E are pairwise orthogonal (see Theorem 8.4), equalities (8.5) give

$$\begin{aligned}D &= \mathrm{cl}\,C \cap (-C^\circ) \\ &= ((\mathrm{cl}\,C \cap E) + \mathrm{ap}\,(\mathrm{cl}\,C)) \cap ((-C^\circ \cap E) + \mathrm{ap}\,C^\circ) \qquad (8.6) \\ &= (\mathrm{cl}\,C \cap E) \cap (-C^\circ \cap E) = D \cap S \subset E.\end{aligned}$$

This argument shows that D is a line-free cone as the intersection of line-free cones $\mathrm{cl}\,C \cap S$ and $-C^\circ \cap S$, as follows from Theorem 8.5.

(*b*) The equalities

$$(\mathrm{cl}\,C \cap E) \cap (C^\circ \cap E) = (\mathrm{cl}\,C \cap C^\circ) \cap E = \{\boldsymbol{o}\},$$

and a combination of Theorem 7.22 and Corollary 7.24 implies that the sum $F = (\mathrm{cl}\,C \cap E) + (-C^\circ \cap E)$ of line-free closed convex cones $\mathrm{cl}\,C \cap E$ and $-C^\circ \cap E$ also is a line-free closed convex cone. Thus $\dim F^\circ = n$, as follows from Theorem 8.4. On the other hand, a combination of Theorems 8.3, 8.5 and Lemma 8.10 gives

$$\begin{aligned}F^\circ &= ((\mathrm{cl}\,C \cap E) + (-C^\circ \cap E))^\circ = (\mathrm{cl}\,C \cap E)^\circ \cap (-C^\circ \cap E)^\circ \\ &= (C^\circ + \mathrm{ap}\,(\mathrm{cl}\,C)) \cap (-\mathrm{cl}\,C + \mathrm{ap}\,C^\circ) \\ &= ((C^\circ \cap E) + \mathrm{ap}\,(\mathrm{cl}\,C) + \mathrm{ap}\,C^\circ) \cap (-(\mathrm{cl}\,C \cap E) + \mathrm{ap}\,(\mathrm{cl}\,C) + \mathrm{ap}\,C^\circ) \\ &= ((C^\circ \cap E) + S^\perp) \cap (-(\mathrm{cl}\,C \cap E) + E^\perp) \\ &= (C^\circ \cap E) \cap (-(\mathrm{cl}\,C \cap E)) + E^\perp = (D \cap E) + E^\perp.\end{aligned}$$

Consequently,

$$\dim D = \dim (D \cap E) = \dim F^\circ - \dim E^\perp = n - \dim E^\perp = \dim E.$$

Finally, the inclusion $D \subset E$ and Corollary 2.65 give $\operatorname{span} D = E$.

A combination of Theorems 8.2 and 8.3 gives

$$\begin{aligned} D^\circ &= (\operatorname{cl} C \cap (-C^\circ))^\circ = \operatorname{cl}((\operatorname{cl} C)^\circ + (-C^\circ)^\circ) \\ &= \operatorname{cl}(C^\circ + (-\operatorname{cl} C)) = C^\circ - \operatorname{cl} C, \end{aligned}$$

where the last equality follows from Theorem 7.18 and

$$\operatorname{rec}(\operatorname{cl} C) \cap \operatorname{rec} C^\circ = \operatorname{cl} C \cap C^\circ = \{\boldsymbol{o}\}.$$

(c) The inclusion $\boldsymbol{o} \in D$ implies that the affine spans of C, C°, and D coincide, respectively, with their spans (see Theorem 2.42). Therefore, Corollary 3.30 and Theorem 3.46, combined with (8.6) and the equalities

$$\operatorname{span} D = \operatorname{span}(\operatorname{cl} C \cap E) = \operatorname{span}(C^\circ \cap E) = E,$$

show that

$$\begin{aligned} \operatorname{rint} D &= \operatorname{rint}(\operatorname{cl} C \cap E) \cap (-\operatorname{rint}(C^\circ \cap E)) \\ &= \operatorname{rint}(C \cap E) \cap (-\operatorname{rint}(C^\circ \cap E)). \end{aligned}$$

Finally, Theorem 8.5 gives

$$\begin{aligned} \operatorname{rint} C \cap (-\operatorname{rint} C^\circ) &= (\operatorname{rint}(C \cap E) + \operatorname{ap}(\operatorname{cl} C)) \\ &\qquad \cap (-\operatorname{rint}(C^\circ \cap E) + \operatorname{ap} C^\circ) \\ &= \operatorname{rint}(C \cap E) \cap (-\operatorname{rint}(C^\circ \cap E)). \end{aligned}$$ □

Corollary 8.12. *If $C \subset \mathbb{R}^n$ is a nonempty convex cone $C \subset \mathbb{R}^n$ with apex $\boldsymbol{o}$, then the following conditions are equivalent.*

(a) *C is a subspace.*
(b) $\operatorname{cl} C \cap (-C^\circ) = \{\boldsymbol{o}\}$.
(c) $\operatorname{rint} C \cap (-\operatorname{rint} C^\circ) = \{\boldsymbol{o}\}$.

Proof. If C is a subspace, say S, then

$$\operatorname{cl} C \cap (-C^\circ) = \operatorname{rint} C \cap (-\operatorname{rint} C^\circ) = S \cap S^\perp = \{\boldsymbol{o}\}.$$

If C is not a subspace, then the orthogonal complement E of $\operatorname{ap}(\operatorname{cl} C)$ within $\operatorname{span} C$ has positive dimension. Theorem 8.11 shows that both sets $\operatorname{cl} C \cap (-C^\circ)$ and $\operatorname{rint} C \cap (-\operatorname{rint} C^\circ)$ have positive dimension. □

8.1.3 *Cones with Base*

Definition 8.13. Let $C \subset \mathbb{R}^n$ be a cone with apex $\boldsymbol{a}$ such that $C \not\subset \{\boldsymbol{a}\}$. We say that a set $B \subset C$ is a *base* of C provided every open halfline $(\boldsymbol{a}, \boldsymbol{x}\rangle \subset C$, $\boldsymbol{x} \in C \setminus \{\boldsymbol{a}\}$, meets B at a single point. If B is a convex set, then it is called a *convex base* of C.

Examples. 1. If $C \subset \mathbb{R}^n$ a cone with apex $\boldsymbol{a}$ such that $C \not\subset \{\boldsymbol{a}\}$ and $S_\rho(\boldsymbol{a})$ is a sphere of radius $\rho > 0$ centered at $\boldsymbol{a}$, then $C \cap S_\rho(\boldsymbol{a})$ is a base of C. This base has a set of useful properties. For instance, Theorem 5.20 implies that C is closed (relative open) if and only if the set $C \cap S_\rho(\boldsymbol{a})$ is closed (relative open) in $S_\rho(\boldsymbol{a})$. Furthermore, if $\mathcal{H}$ stands for the family of all closed halflines with common apex $\boldsymbol{a}$ which lie in C, and if closed halflines $h, h' \in \mathcal{H}$ meet $S_\rho(\boldsymbol{a})$ at points $\boldsymbol{v}, \boldsymbol{v}'$, respectively, then the function $\theta(h, h') = \|\boldsymbol{v} - \boldsymbol{v}'\|$ defines a metric on $\mathcal{H}$. Hence $\mathcal{H}$ can be viewed as a bounded set with respect to the metric θ.

2. The set $B = \{(x, y, 1) : -1 < x < 1\}$ is a convex base of both cones

$$C_1 = \{(x, y, z) : |x| < z\} \quad \text{and} \quad C_2 = \{\boldsymbol{o}\} \cup C_1,$$

while the cone $C_3 = \{(x, y, z) : |x| \leqslant z\}$ has no convex base. Indeed, we observe first that $\operatorname{ap} C_3$ is the y-axis of $\mathbb{R}^3$. Now, if $\boldsymbol{a} = (0, \alpha, 0) \in \operatorname{ap} C_2$ and the open halflines $h_1 = \{(0, y, z) : y < \alpha\}$ and $h_2 = \{(0, y, z) : y > \alpha\}$ meet a convex subset B of C_2 at some points $\boldsymbol{u}_1$ and $\boldsymbol{u}_2$, respectively, then $(\boldsymbol{u}_1, \boldsymbol{u}_2) \subset B$ due to the convexity of B. Consequently, $(\boldsymbol{o}, \boldsymbol{u}_1\rangle \subset B \cap C_3$, implying that B cannot be a base of C_3.

The following theorem is useful in many arguments.

Theorem 8.14. *Let $C \subset \mathbb{R}^n$ be a convex cone with apex $\boldsymbol{a}$ such that $C \not\subset \{\boldsymbol{a}\}$. If C has a convex base B, then C is not a plane, $\boldsymbol{a} \notin B$, and the following assertions hold.*

(*a*) *Both sets $C' = C \setminus \{\boldsymbol{a}\}$ and $C'' = \{\boldsymbol{a}\} \cup C$ are convex cones, and B is a convex base of each of them.*
(*b*) *$C' = C_{\boldsymbol{a}}(B) \setminus \{\boldsymbol{a}\}$ and $C'' = C_{\boldsymbol{a}}(B)$.*
(*c*) *The plane $L = \operatorname{aff} B$ has dimension $\dim C - 1$ and $B = C \cap L$.*

Proof. Since $C \not\subset \{\boldsymbol{a}\}$, there is at least one open halfline $h = (\boldsymbol{a}, \boldsymbol{x}\rangle$ which meets B at a point $\boldsymbol{u}$, say. Assume for a moment that $\boldsymbol{a} \in B$. Then $(\boldsymbol{a}, \boldsymbol{u}] \subset (\boldsymbol{a}, \boldsymbol{u}\rangle = h$, and $(\boldsymbol{a}, \boldsymbol{u}] \subset h \cap B$, contrary to the assumption that h meets B at a single point. So, $\boldsymbol{a} \notin B$.

Next, assume for a moment that C is a plane and choose a line l through $\boldsymbol{a}$ which lies in C. Denote by h and h' the opposite open halflines of l, having $\boldsymbol{a}$ as a common endpoint. By the definition of B, the halflines h and h' meet B at some points $\boldsymbol{u}$ and $\boldsymbol{u}'$, respectively. Then $[\boldsymbol{u}, \boldsymbol{u}'] \subset l \cap B$, contrary to the assumption that each of h and h' meets B at a single point. Hence C cannot be a plane.

(a) Suppose first that $\boldsymbol{a}$ is a proper apex of C. Then $C'' = C$, implying that B is a convex base of C''. Furthermore,

$$C = \{\boldsymbol{a}\} \cup ((\boldsymbol{a}, \boldsymbol{x}\rangle : \boldsymbol{x} \in B) = \cup ([\boldsymbol{a}, \boldsymbol{x}\rangle : \boldsymbol{x} \in B) = C_{\boldsymbol{a}}(B),$$

and Theorem 5.38 implies that C' is a convex cone with apex $\boldsymbol{a}$. If $\boldsymbol{x} \in C'$, then, by the assumption on B, the open halfline $(\boldsymbol{a}, \boldsymbol{x}\rangle$ meets B at a unique point. Hence B is a convex base of C'.

Let $\boldsymbol{a}$ be an improper apex of C. Then $C' = C$, implying that B is a convex base of C'. By Theorem 5.14, $C \cap \operatorname{ap} C = \varnothing$. Therefore, Theorem 5.18 shows (with $L = \{\boldsymbol{a}\}$), that C'' is a convex cone and $\operatorname{ap} C'' = \{\boldsymbol{a}\}$. If $\boldsymbol{x} \in C'' \setminus \{\boldsymbol{a}\}$, then $(\boldsymbol{a}, \boldsymbol{x}\rangle \subset C$ and, by the assumption on B, the open halfline $(\boldsymbol{a}, \boldsymbol{x}\rangle$ meets B at a unique point. Hence B is a convex base of C''.

(b) This part follows from Theorem 5.31 and the above argument.

(c) First, we assert that $\boldsymbol{a} \notin L$. Indeed, assume, for contradiction, that $\boldsymbol{a} \in L$. By Theorem 3.14, there are points $\boldsymbol{x}, \boldsymbol{y} \in B$ such that $\boldsymbol{a} \in \langle \boldsymbol{x}, \boldsymbol{y}\rangle$. Clearly, $\boldsymbol{x} \neq \boldsymbol{y}$ (otherwise $\boldsymbol{a} = \boldsymbol{x} = \boldsymbol{y} \in B$, which is impossible by the proved above). Since $\boldsymbol{a} \notin [\boldsymbol{x}, \boldsymbol{y}]$ (otherwise $\boldsymbol{a} \in [\boldsymbol{x}, \boldsymbol{y}] \subset B$ due to the convexity of B), both $\boldsymbol{x}$ and $\boldsymbol{y}$ belong to the same open halfline h with endpoint $\boldsymbol{a}$. In this case, $[\boldsymbol{x}, \boldsymbol{y}] \subset h \cap B$, contrary to the assumption that $h \cap B$ is a singleton.

Next, we assert that $\dim L = m-1$, where $m = \dim C$. Indeed, $\dim L \leqslant m-1$ because L is a proper subplane of $\operatorname{aff} C$ (due to $\boldsymbol{a} \in \operatorname{aff} C \setminus L$). The obvious equality $C = \operatorname{cone}_{\boldsymbol{a}} B$ and Theorem 5.37 imply the existence of an affinely independent set $\{\boldsymbol{x}_1, \ldots, \boldsymbol{x}_m\} \subset B$. Then $\{\boldsymbol{x}_1, \ldots, \boldsymbol{x}_m\} \subset L$, and Corollary 2.53 gives $\dim L \geqslant m - 1$.

Finally, it remains to show that $B = C \cap L$. Because the inclusion $B \subset C \cap L$ is obvious, we need to prove that $C \cap L \subset B$. For this, let $\boldsymbol{x} \in C \cap L$. Then $\boldsymbol{x} \neq \boldsymbol{a}$ due to $\boldsymbol{a} \notin L$, and the open halfline $h = (\boldsymbol{a}, \boldsymbol{x}\rangle$ lies in C. By the definition of B, h meets B at a point $\boldsymbol{u}$. Consequently, $\boldsymbol{u} \in B \subset L$. Since $h \cap L = \{\boldsymbol{x}\}$ (otherwise $\boldsymbol{a} \in \operatorname{cl} h \subset L$), we conclude that $\boldsymbol{x} = \boldsymbol{u} \in B$. Summing up, $C \cap L \subset B$. $\square$

Theorem 8.15. *Let $C \subset \mathbb{R}^n$ be a convex cone with apex $\boldsymbol{a}$, which has a* *convex base. A convex set* $B \subset \mathbb{R}^n$ *is a base of* C *if and only if* $B = C \cap H$*, where* H *is a hyperplane of the form*

$$H = \{\boldsymbol{x} \in \mathbb{R}^n : \boldsymbol{x}\cdot\boldsymbol{e} = \gamma\}, \quad \boldsymbol{e} \in \operatorname{rint}(C - \boldsymbol{a})^\circ, \quad \gamma < \boldsymbol{a}\cdot\boldsymbol{e}. \tag{8.7}$$

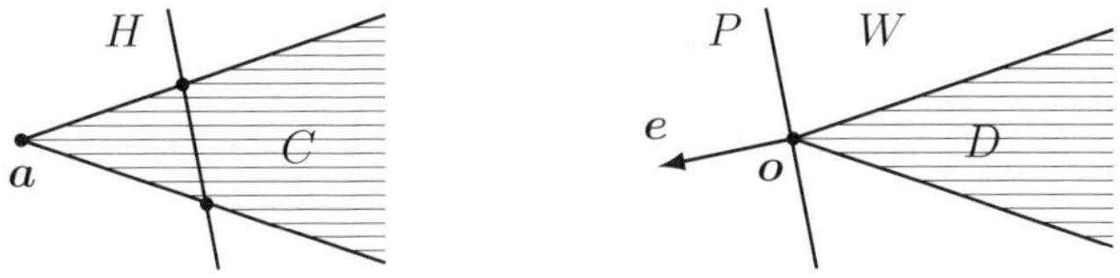

Fig. 8.3 Illustration to Theorem 8.15.

Proof. Let B be a convex base of C. Put $A = B - \boldsymbol{a}$ and $D = C - \boldsymbol{a}$. Then A is a convex base of the convex cone D with apex $\boldsymbol{o}$. By Theorem 8.14, the plane $L = \operatorname{aff} A$ misses $\boldsymbol{o}$ and has dimension $\dim D - 1 = \dim(\operatorname{aff} D) - 1$. Lemma 2.72 shows the existence of a hyperplane $G \subset \mathbb{R}^n$ of the form

$$G = \{\boldsymbol{x} \in \mathbb{R}^n : \boldsymbol{x}\cdot\boldsymbol{e} = \beta\}, \quad \boldsymbol{e} \neq \boldsymbol{o}, \quad \beta \in \mathbb{R},$$

such that $L = G \cap \operatorname{aff} D$. By Theorem 8.14,

$$A = D \cap L = D \cap G \cap \operatorname{aff} D = D \cap G.$$

Next, we observe that $\boldsymbol{o} \notin G$. Indeed, if $\boldsymbol{o}$ belonged to G, then, by Theorem 5.38,

$$\operatorname{aff} D = \operatorname{aff} C_{\boldsymbol{o}}(A) = \operatorname{aff}(\{\boldsymbol{o}\} \cup A) \subset G,$$

which is impossible because $L \neq \operatorname{aff} D$. Consider the hyperplane

$$P = \{\boldsymbol{x} \in \mathbb{R}^n : \boldsymbol{x}\cdot\boldsymbol{e} = 0\}.$$

Then $A \cap P \subset G \cap P = \varnothing$, and Theorem 2.34 shows that G (and whence A) lies in an open halfspace W determined by P. Accordingly, the whole set $D \setminus \{\boldsymbol{o}\}$, as the union of open halflines $(\boldsymbol{o}, \boldsymbol{x}\rangle$, $\boldsymbol{x} \in A$, lies in W (see Theorem 2.35). Thus $\operatorname{cl} D$ lies in the closed halfspace $V = \operatorname{cl} W$. Because D is not a plane, $\operatorname{ap}(\operatorname{cl} D)$ lies in $\operatorname{cl} D$ (see Theorem 5.25). Hence $\operatorname{ap}(\operatorname{cl} D) \subset V$. Furthermore, the inclusion $\boldsymbol{o} \in \operatorname{ap}(\operatorname{cl} D)$ gives $\operatorname{ap}(\operatorname{cl} D) \subset P$, as follows from Theorem 2.34. Theorem 8.6 implies that one of the vectors $\pm\boldsymbol{e}$, say $\boldsymbol{e}$, belongs to $\operatorname{rint}(D)^\circ = \operatorname{rint}(C - \boldsymbol{a})^\circ$. Clearly, $W = \{\boldsymbol{x} \in \mathbb{R}^n : \boldsymbol{x}\cdot\boldsymbol{e} < 0\}$. This argument and the inclusion $G \subset W$ give $\beta < 0$.

Finally, put $\gamma = \boldsymbol{a}\cdot\boldsymbol{e} + \beta$ and $H = \boldsymbol{a} + G$. Clearly,

$$\begin{aligned} H &= \boldsymbol{a} + G = \{(\boldsymbol{a}+\boldsymbol{x}) \in \mathbb{R}^n : (\boldsymbol{a}+\boldsymbol{x})\cdot\boldsymbol{e} = \boldsymbol{a}\cdot\boldsymbol{e} + \beta\} \\ &= \{\boldsymbol{x} \in \mathbb{R}^n : \boldsymbol{x}\cdot\boldsymbol{e} = \gamma\}, \quad \text{where} \quad \gamma < \boldsymbol{a}\cdot\boldsymbol{e}, \\ B &= \boldsymbol{a} + A = \boldsymbol{a} + D \cap G = (\boldsymbol{a}+D) \cap (\boldsymbol{a}+G) = C \cap H. \end{aligned}$$

Conversely, let $B = C \cap H$, where H is a hyperplane of the form (8.7). Then $\boldsymbol{a} \notin H$ and B is a convex set as the intersection of convex sets C and H. Choose any point $\boldsymbol{x} \in C \setminus \{\boldsymbol{a}\}$ and let $h = (\boldsymbol{a}, \boldsymbol{x}\rangle$. Then $h \subset C$. Put

$$\lambda = \frac{\gamma - \boldsymbol{a}\cdot\boldsymbol{e}}{(\boldsymbol{x}-\boldsymbol{a})\cdot\boldsymbol{e}}.$$

By the above argument, $(\boldsymbol{x}-\boldsymbol{a})\cdot\boldsymbol{e} < 0$, which gives $\lambda > 0$. Furthermore,

$$(\boldsymbol{a} + \lambda(\boldsymbol{x}-\boldsymbol{a}))\cdot\boldsymbol{e} = \boldsymbol{a}\cdot\boldsymbol{e} + \frac{\gamma - \boldsymbol{a}\cdot\boldsymbol{e}}{(\boldsymbol{x}-\boldsymbol{a})\cdot\boldsymbol{e}}(\boldsymbol{x}-\boldsymbol{a})\cdot\boldsymbol{e} = \gamma.$$

Therefore, $\boldsymbol{a} + \lambda(\boldsymbol{x}-\boldsymbol{a}) \in H$, implying that h meets H at a single point (see Theorem 2.35). Hence $C \cap H$ is a convex base of C. □

Theorem 8.16. *Let $C \subset \mathbb{R}^n$ be a convex cone with apex $\boldsymbol{a}$ such that $C \not\subset \{\boldsymbol{a}\}$. Then C has a convex base if and only if $C \cap \operatorname{ap}(\operatorname{cl} C) \subset \{\boldsymbol{a}\}$.*

Proof. Let C have a convex base B. By Theorem 8.15, $B = C \cap H$, where H is a hyperplane given by (8.7). Put $\mu = \boldsymbol{a}\cdot\boldsymbol{e}$ and consider the hyperplane Q and the open halfspace Y, given, respectively, by

$$Q = \{\boldsymbol{x} \in \mathbb{R}^n : \boldsymbol{x}\cdot\boldsymbol{e} = \mu\} \quad \text{and} \quad Y = \{\boldsymbol{x} \in \mathbb{R}^n : \boldsymbol{x}\cdot\boldsymbol{e} < \mu\}. \tag{8.8}$$

As shown in the proof of Theorem 8.15, H and Q are disjoint parallel hyperplanes such that $H \subset Y$. Furthermore, $C \setminus \{\boldsymbol{a}\} \subset Y$, which, combined with Theorem 8.7, gives $\operatorname{cl} C \cap Q = \operatorname{ap}(\operatorname{cl} C)$.

Assume, for contradiction, that $C \cap \operatorname{ap}(\operatorname{cl} C) \not\subset \{\boldsymbol{a}\}$ and choose a point $\boldsymbol{u} \in C \cap \operatorname{ap}(\operatorname{cl} C) \setminus \{\boldsymbol{a}\}$. Then the open halfline $h = (\boldsymbol{a}, \boldsymbol{u}\rangle$ lies in the cone $C \cap \operatorname{ap}(\operatorname{cl} C)$. Since B is a convex base of C, the halfline h meets B, and whence meets H. On the other hand, $h \subset \operatorname{ap}(\operatorname{cl} C) \subset Q$, and thus h is disjoint from H. The obtained contradiction gives $C \cap \operatorname{ap}(\operatorname{cl} C) \subset \{\boldsymbol{a}\}$.

Conversely, suppose that $C \cap \operatorname{ap}(\operatorname{cl} C) \subset \{\boldsymbol{a}\}$. Then C is not a plane. Choose any vector $\boldsymbol{e} \in \operatorname{rint}(C - \boldsymbol{a})^\circ$ and consider the hyperplane Q and the halfspace Y defined by (8.8). Let γ be a scalar satisfying the condition $\gamma < \beta$, and H be the hyperplane given by (8.7). As shown in the proof of Theorem 8.15, the set $B = C \cap H$ is a convex base of C. □

Theorem 8.17. *Let $C \subset \mathbb{R}^n$ be a convex cone with apex $\boldsymbol{a}$ such that $C \not\subset \{\boldsymbol{a}\}$. The following assertions hold.*

(*a*) *C is line-free if and only if it has a bounded convex base.*
(*b*) *C is line-free if and only if it has a line-free convex base.*
(*c*) *C is line-free if and only if every convex base of C is line-free.*
(*d*) *If C is closed, then C is line-free if and only if it has a compact convex base.*
(*e*) *If C is closed, then C is line-free if and only if every convex base of C is compact.*

Proof. Due to Theorem 8.14, we may assume that $\boldsymbol{a}$ is a proper apex of C. Translating C on $-\boldsymbol{a}$ we also may suppose that $\boldsymbol{a} = \boldsymbol{o}$. Then $\operatorname{rec} C = C$, as follows from Theorem 6.3.

(*a*) Suppose that C is line-free. Then $\operatorname{cl} C$ is a line-free convex cone (see Corollary 5.21 and Theorem 6.37). Therefore, $\operatorname{ap}(\operatorname{cl} C) = \{\boldsymbol{o}\}$. Choose a nonzero vector $\boldsymbol{e} \in \operatorname{rint} C^\circ$. According to Theorem 8.7, the $(n-1)$-dimensional subspace $T = \{\boldsymbol{x} \in \mathbb{R}^n : \boldsymbol{x} \cdot \boldsymbol{e} = 0\}$ satisfies the condition $T \cap \operatorname{cl} C = \{\boldsymbol{o}\}$. By Theorem 6.40, any hyperplane $H = \{\boldsymbol{x} \in \mathbb{R}^n : \boldsymbol{x} \cdot \boldsymbol{e} = \gamma\}$, with $\gamma < 0$, as a translate of T, meets C along a bounded set. Theorem 8.16 shows $C \cap H$ is a convex base of C.

Conversely, if C has a bounded convex base B, then the equality $C = C_{\boldsymbol{o}}(B)$ and Theorem 7.9 show that $\operatorname{lin}(\operatorname{cl} C) = \{\boldsymbol{o}\}$. Consequently, Theorem 6.37 implies that C is line-free.

(*b*) If C is line-free, then any convex base of C, as a subset of C is line-free. Conversely, if every convex base of C is line-free, then, a combination of Theorems 6.37 and 7.9 shows that C is line-free.

(*c*) If C is line-free, then any convex base of C, as a subset of C, must be line-free. The converse statement follows from assertion (*b*) above.

(*d*) Suppose that C is closed. By assertion (*a*) above, C has a bounded base B. Theorem 8.15 shows that $B = C \cap H$, where H is a hyperplane given by (8.7). Consequently, B is closed as the intersection of closed sets C and H. Thus, B is compact.

Conversely, suppose that C has a compact base B. Since $\boldsymbol{a} \notin B$, Theorem 5.45 implies that the cone $C = C_{\boldsymbol{a}}(B)$ is closed.

(*e*) Due to assertion (*d*) above, it suffices to prove that every convex base B of C is compact provided C is closed. If C is line-free and closed, then $\operatorname{ap} C = \{\boldsymbol{a}\}$. According to Theorem 8.16, any convex base B of C has the form $B = C \cap H$, where H is a hyperplane given by (8.7). Thus B is closed as the intersection of closed sets C and H. Obviously, the hyperplane $H - \boldsymbol{a}$ meets the cone $\operatorname{rec} C = C - \boldsymbol{a}$ at $\boldsymbol{o}$ only. By Theorem 6.40, $C \cap H$ is bounded. Summing up, B is compact. □

Remark. Assertion (e) of Theorem 8.17 does not hold if C is not closed. Indeed, the closed halfline $h = \{(x, 1) : x \geqslant 0\}$ is a convex base of the line-free convex cone $C = \{\boldsymbol{o}\} \cup \{(x, y) : x \geqslant 0, y > 0\}$.

8.2 Normal and Barrier Cones

8.2.1 *Metric Projections*

Definition 8.18. Given a nonempty set $X \subset \mathbb{R}^n$ and a point $\boldsymbol{a} \in \mathbb{R}^n$, we will say that a point $\boldsymbol{c} \in X$ is a *nearest* to $\boldsymbol{a}$ point in X provided

$$\|\boldsymbol{a} - \boldsymbol{c}\| = \delta(\boldsymbol{a}, X) = \inf\{\|\boldsymbol{a} - \boldsymbol{x}\| : \boldsymbol{x} \in X\}.$$

Remark. If the set X in Definition 8.18 is closed, then any point $\boldsymbol{a} \in \mathbb{R}^n$ has at least one nearest point in X. Indeed, the continuous function $\delta_{\boldsymbol{a}}(\boldsymbol{x}) = \|\boldsymbol{x} - \boldsymbol{a}\|$ (see Problem 1.9) attains a minimum value on X.

Examples. 1. If L is a nonempty plane in $\mathbb{R}^n$, then every point $\boldsymbol{a} \in \mathbb{R}^n$ has a unique nearest point in L, which is the orthogonal projection of $\boldsymbol{a}$ on L (see Theorem 2.90).

2. If $B_\rho(\boldsymbol{u}) \subset \mathbb{R}^n$ is a closed ball, then every point $\boldsymbol{a} \in \mathbb{R}^n \setminus B_\rho(\boldsymbol{u})$ has a unique nearest point $\boldsymbol{c}$ in $B_\rho(\boldsymbol{u})$, which is the point of intersection of the segment $[\boldsymbol{a}, \boldsymbol{u}]$ and the sphere $S_\rho(\boldsymbol{u})$.

The following theorem reveals an important property of convex sets (see also page 307 for the respective characteristic property).

Theorem 8.19. *Let $K \subset \mathbb{R}^n$ be a nonempty convex set and $\boldsymbol{a}$ be a point in $\mathbb{R}^n$. Then* $\operatorname{cl} K$ *contains a unique nearest to $\boldsymbol{a}$ point $\boldsymbol{c}$.*

Proof. Since the case $\boldsymbol{a} \in \operatorname{cl} K$ is obvious (put $\boldsymbol{c} = \boldsymbol{a}$), we assume that $\boldsymbol{a} \in \mathbb{R}^n \setminus \operatorname{cl} K$. The continuous function $\delta_{\boldsymbol{a}}(\boldsymbol{x}) = \|\boldsymbol{x} - \boldsymbol{a}\|$ achieves its minimum value on $\operatorname{cl} K$ at a suitable point $\boldsymbol{c} \in \operatorname{cl} K$ (see page 12). Assume for a moment that $\operatorname{cl} K$ contains another point $\boldsymbol{c}'$ which is nearest to $\boldsymbol{a}$, as depicted in Figure 8.4. Let $\delta = \|\boldsymbol{a}-\boldsymbol{c}\| = \|\boldsymbol{a}-\boldsymbol{c}'\|$. By the convexity of $\operatorname{cl} K$ (see Theorem 3.46), the closed segment $[\boldsymbol{c}, \boldsymbol{c}']$ lies in $\operatorname{cl} K$. In particular, the midpoint $\boldsymbol{e} = (\boldsymbol{c} + \boldsymbol{c})/2$ belongs to $\operatorname{cl} K$.

By the parallelogram law (see page 8),

$$\begin{aligned}\|\boldsymbol{a} - \boldsymbol{e}\|^2 &= \|\boldsymbol{a} - \tfrac{1}{2}(\boldsymbol{c} + \boldsymbol{c}')\|^2 = \tfrac{1}{4}\|(\boldsymbol{a} - \boldsymbol{c}) + (\boldsymbol{a} - \boldsymbol{c}')\|^2 \\ &= \tfrac{1}{4}(2\|\boldsymbol{a} - \boldsymbol{c}\|^2 + 2\|\boldsymbol{a} - \boldsymbol{c}'\|^2 - \|\boldsymbol{c} - \boldsymbol{c}'\|^2) \\ &= \tfrac{1}{4}(4\delta^2 - \|\boldsymbol{c} - \boldsymbol{c}'\|^2) < \delta^2.\end{aligned}$$

Hence $\|\boldsymbol{a}-\boldsymbol{e}\| < \delta$, contrary to the choice of $\boldsymbol{c}$. The obtained contradiction shows the uniqueness of a nearest to $\boldsymbol{a}$ point in $\operatorname{cl} K$. □

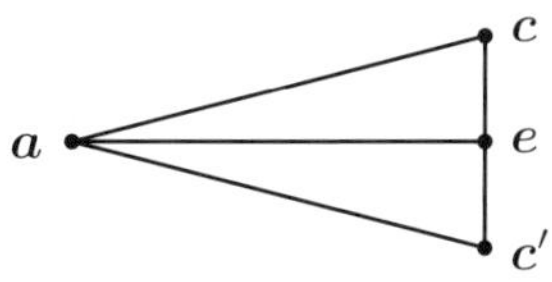

Fig. 8.4 Illustration to Theorem 8.19.

Theorem 8.19 gives a base to the following definition.

Definition 8.20. Let $K \subset \mathbb{R}^n$ be a nonempty convex set. The mapping $p_K : \mathbb{R}^n \to \operatorname{cl} K$ defined for every $\boldsymbol{x} \in \mathbb{R}^n$ by $p_K(\boldsymbol{x}) = \boldsymbol{z}$, where $\boldsymbol{z}$ is the (unique) nearest to $\boldsymbol{x}$ point of $\operatorname{cl} K$, is called the *metric projection* on K.

Example. The metric projection on a nonempty plane $L \subset \mathbb{R}^n$ is precisely the orthogonal projection on L (see Definition 2.89 and Theorem 2.90).

Theorem 8.21. *For a nonempty convex set $K \subset \mathbb{R}^n$, the metric projection p_K satisfies the Lipschitz condition*

$$\|p_K(\boldsymbol{x}) - p_K(\boldsymbol{y})\| \leqslant \|\boldsymbol{x} - \boldsymbol{y}\|, \quad \boldsymbol{x}, \boldsymbol{y} \in \mathbb{R}^n. \tag{8.9}$$

Consequently, p_K is a continuous mapping on $\mathbb{R}^n$.

Proof. Choose any points $\boldsymbol{x}, \boldsymbol{y} \in \mathbb{R}^n$. By Theorem 8.22 $\operatorname{cl} K$ is included in the set

$$V = \{\boldsymbol{u} \in \mathbb{R}^n : (\boldsymbol{u} - p_K(\boldsymbol{x}))\cdot(\boldsymbol{x} - p_K(\boldsymbol{x})) \leqslant 0\}.$$

Therefore, $p_K(\boldsymbol{y}) \in \operatorname{cl} K \subset V$, which results in

$$(p_K(\boldsymbol{y}) - p_K(\boldsymbol{x}))\cdot(\boldsymbol{x} - p_K(\boldsymbol{x})) \leqslant 0. \tag{8.10}$$

Similarly,

$$(p_K(\boldsymbol{x}) - p_K(\boldsymbol{y}))\cdot(\boldsymbol{y} - p_K(\boldsymbol{y})) \leqslant 0. \tag{8.11}$$

With $\boldsymbol{u} = \boldsymbol{x} - p_K(\boldsymbol{x})$ and $\boldsymbol{v} = \boldsymbol{y} - p_K(\boldsymbol{y})$, the inequalities (8.10) and (8.11) can be rewritten as

$$(p_K(\boldsymbol{x}) - p_K(\boldsymbol{y}))\cdot\boldsymbol{u} \geqslant 0, \quad (p_K(\boldsymbol{x}) - p_K(\boldsymbol{y}))\cdot\boldsymbol{v} \leqslant 0.$$

Therefore, $(p_K(\boldsymbol{x}) - p_K(\boldsymbol{y}))\cdot(\boldsymbol{u} - \boldsymbol{v}) \geqslant 0$, which implies

$$\begin{aligned}\|\boldsymbol{x} - \boldsymbol{y}\|^2 &= \|(p_K(\boldsymbol{x}) - p_K(\boldsymbol{y})) + (\boldsymbol{u} - \boldsymbol{v})\|^2 \\ &= \|p_K(\boldsymbol{x}) - p_K(\boldsymbol{y})\|^2 + 2(p_K(\boldsymbol{x}) - p_K(\boldsymbol{y}))\cdot(\boldsymbol{u} - \boldsymbol{v}) + \|\boldsymbol{u} - \boldsymbol{v}\|^2 \\ &\geqslant \|p_K(\boldsymbol{x}) - p_K(\boldsymbol{y})\|^2.\end{aligned}$$

Hence $\|p_K(\boldsymbol{x}) - p_K(\boldsymbol{y})\| \leqslant \|\boldsymbol{x} - \boldsymbol{y}\|$. □

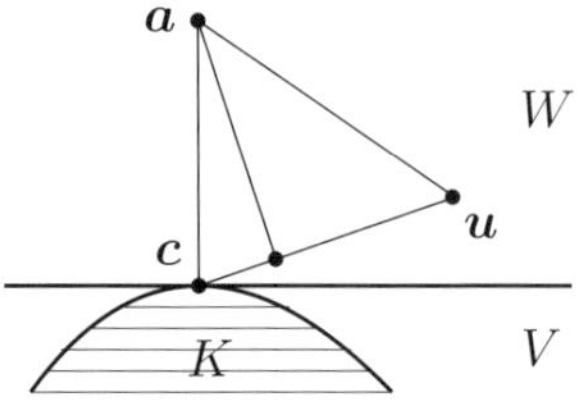

Fig. 8.5 Illustration to Theorem 8.22.

Theorem 8.22. *Let $K \subset \mathbb{R}^n$ be a nonempty convex set. A point $\boldsymbol{c} \in \operatorname{cl} K$ is the metric projection of a point $\boldsymbol{a} \in \mathbb{R}^n \setminus \operatorname{cl} K$ on K if and only if K is included in the closed halfspace*

$$V = \{\boldsymbol{x} \in \mathbb{R}^n : (\boldsymbol{x} - \boldsymbol{c})\cdot(\boldsymbol{a} - \boldsymbol{c}) \leqslant 0\}. \tag{8.12}$$

Proof. Assume first that $p_K(\boldsymbol{a}) = \boldsymbol{c}$. Put $\boldsymbol{e} = \boldsymbol{a} - \boldsymbol{c}$ and $\gamma = \boldsymbol{c}\cdot\boldsymbol{e}$. Then the set V from (8.12) can be expressed as $V = \{\boldsymbol{x} \in \mathbb{R}^n : \boldsymbol{x}\cdot\boldsymbol{e} \leqslant \gamma\}$, implying that V is a closed halfspace determined by the hyperplane $H = \{\boldsymbol{x} \in \mathbb{R}^n : \boldsymbol{x}\cdot\boldsymbol{e} = \gamma\}$ (see Definition 2.29 and Figure 8.5). Clearly, $\boldsymbol{c} \in H$.

Indeed, assume for a moment that $K \not\subset V$ and choose a point $\boldsymbol{u} \in K\setminus W$, where $W = \{\boldsymbol{x} \in \mathbb{R}^n : \boldsymbol{x}\cdot\boldsymbol{e} > \gamma\}$. Then $\boldsymbol{u} \in \operatorname{cl} K \setminus W$ and $\boldsymbol{u}\cdot\boldsymbol{e} > \gamma$. Clearly, $(\boldsymbol{u} - \boldsymbol{c})\cdot\boldsymbol{e} > 0$. By the convexity of $\operatorname{cl} K$ (see Theorem 3.46), all points

$$\boldsymbol{x}_\lambda = (1 - \lambda)\boldsymbol{c} + \lambda\boldsymbol{u}, \quad 0 \leqslant \lambda \leqslant 1,$$

belong to $\operatorname{cl} K$. The equalities

$$\begin{aligned}\|\boldsymbol{a} - \boldsymbol{x}_\lambda\|^2 &= \|\boldsymbol{e} - \lambda(\boldsymbol{u} - \boldsymbol{c})\|^2 = (\boldsymbol{e} - \lambda(\boldsymbol{u} - \boldsymbol{c}))\cdot(\boldsymbol{e} - \lambda(\boldsymbol{u} - \boldsymbol{c})) \\ &= \|\boldsymbol{e}\|^2 + \lambda(\lambda\|\boldsymbol{u} - \boldsymbol{c}\|^2 - 2\,\boldsymbol{e}\cdot(\boldsymbol{u} - \boldsymbol{c}))\end{aligned}$$

imply

$$\|\boldsymbol{a} - \boldsymbol{x}_\lambda\| < \|\boldsymbol{e}\| = \|\boldsymbol{a} - \boldsymbol{c}\| \quad \text{if} \quad 0 < \lambda < 2\,\boldsymbol{e}\cdot(\boldsymbol{u} - \boldsymbol{c})/\|\boldsymbol{u} - \boldsymbol{c}\|^2.$$

Since the inequality $\|\boldsymbol{a} - \boldsymbol{x}_\lambda\| < \|\boldsymbol{a} - \boldsymbol{c}\|$ contradicts the choice of $\boldsymbol{c}$, no point of K can be in W. Hence $K \subset V$.

Conversely, suppose that $K \subset V$. Since $\boldsymbol{a} - \boldsymbol{c} \in \operatorname{nor} V$ (see Definition 2.31), Theorem 2.93 shows that $\boldsymbol{c}$ is the nearest to $\boldsymbol{a}$ point in V. The inclusions $\boldsymbol{c} \in \operatorname{cl} K \subset V$ imply that $\boldsymbol{c}$ is the metric projection of $\boldsymbol{a}$ on K. □

8.2.2 Tangent and Normal Cones at a Point

We recall that $C_{\boldsymbol{a}}(X)$ denotes the cone with apex $\boldsymbol{a} \in \mathbb{R}^n$ generated by a set $X \subset \mathbb{R}^n$ (see Definition 5.29).

Definition 8.23. Let $K \subset \mathbb{R}^n$ be a nonempty convex set and $\boldsymbol{z}$ be a point in $\operatorname{cl} K$. The cones

$$D_{\boldsymbol{z}}(K) = \operatorname{cl} C_{\boldsymbol{z}}(K) \quad \text{and} \quad T_{\boldsymbol{z}}(K) = D_{\boldsymbol{z}}(K) - \boldsymbol{z}$$

are called, respectively, the *support cone* and the *tangent cone* of K at $\boldsymbol{z}$. The sets

$$Q_{\boldsymbol{z}}(K) = \{\boldsymbol{x} \in \mathbb{R}^n : p_K(\boldsymbol{x}) = \boldsymbol{z}\} \quad \text{and} \quad N_{\boldsymbol{z}}(K) = Q_{\boldsymbol{z}}(K) - \boldsymbol{z}$$

are called, respectively, the *normal set* and the *normal cone* of K at $\boldsymbol{z}$.

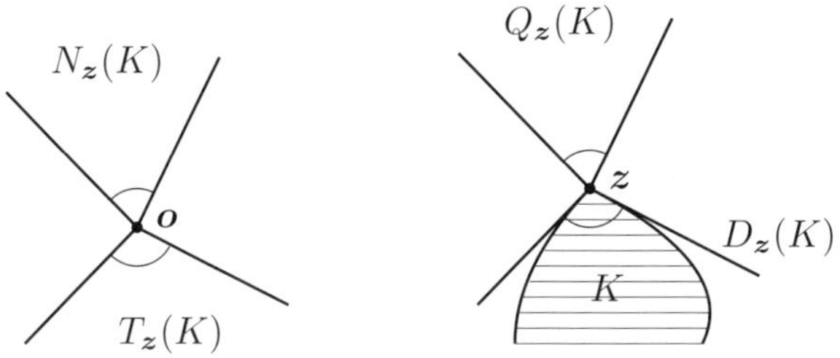

Fig. 8.6 Support, tangent, and normal cones of K at $\boldsymbol{z}$.

Theorem 8.24. *For a nonempty convex set $K \subset \mathbb{R}^n$ and a point $\boldsymbol{z} \in \operatorname{cl} K$, one has*

(*a*) $T_{\boldsymbol{z}}(K) = D_{\boldsymbol{o}}(K - \boldsymbol{z})$,
(*b*) $N_{\boldsymbol{z}}(K) = Q_{\boldsymbol{o}}(K - \boldsymbol{z})$,
(*c*) $N_{\boldsymbol{z}}(K) = (T_{\boldsymbol{z}}(K))^\circ = (D_{\boldsymbol{o}}(K - \boldsymbol{z}))^\circ = (C_{\boldsymbol{o}}(K - \boldsymbol{z}))^\circ = (K - \boldsymbol{z})^\circ$,
(*d*) $T_{\boldsymbol{z}}(K) = (N_{\boldsymbol{z}}(K))^\circ$.

Proof. Assertion (*a*) follows from Theorem 5.31:

$$T_{\boldsymbol{z}}(K) = \operatorname{cl} C_{\boldsymbol{z}}(K) - \boldsymbol{z} = \operatorname{cl}(C_{\boldsymbol{z}}(K) - \boldsymbol{z}) = \operatorname{cl} C_{\boldsymbol{o}}(K - \boldsymbol{z}) = D_{\boldsymbol{o}}(K - \boldsymbol{z}).$$

(*b*) This part follows from the obvious argument that $\boldsymbol{z}$ is the nearest to $\boldsymbol{u}$ point of $\operatorname{cl} K$ if and only if $\boldsymbol{o}$ is the nearest to $\boldsymbol{u} - \boldsymbol{z}$ point in $\operatorname{cl}(K - \boldsymbol{z})$. Equivalently, $\boldsymbol{u} \in Q_{\boldsymbol{z}}(K)$ if and only if $\boldsymbol{u} - \boldsymbol{z} \in Q_{\boldsymbol{o}}(K - \boldsymbol{z})$. Consequently,

$$N_{\boldsymbol{z}}(K) = Q_{\boldsymbol{z}}(K) - \boldsymbol{z} = Q_{\boldsymbol{o}}(K - \boldsymbol{z}).$$

(*c*) Due to assertions (*a*) and (*b*), it suffices to show that

$$Q_{\boldsymbol{o}}(K - \boldsymbol{z}) = (D_{\boldsymbol{o}}(K - \boldsymbol{z}))^\circ = (C_{\boldsymbol{o}}(K - \boldsymbol{z}))^\circ = (K - \boldsymbol{z})^\circ.$$

By Theorem 8.2, we have

$$(D_{\boldsymbol{o}}(K - \boldsymbol{z}))^\circ = (\operatorname{cl} C_{\boldsymbol{o}}(K - \boldsymbol{z}))^\circ = (K - \boldsymbol{z})^\circ.$$

So, it remains to prove the equality $Q_{\boldsymbol{o}}(K - \boldsymbol{z}) = (K - \boldsymbol{z})^\circ$. Since both sets $Q_{\boldsymbol{o}}(K - \boldsymbol{z})$ and $(K - \boldsymbol{z})^\circ$ contain $\boldsymbol{o}$, it suffices to verify that

$$Q_{\boldsymbol{o}}(K - \boldsymbol{z}) \setminus \{\boldsymbol{o}\} = (K - \boldsymbol{z})^\circ \setminus \{\boldsymbol{o}\}.$$

Indeed, Theorem 8.22 implies that a nonzero vector $\boldsymbol{u} \in \mathbb{R}^n$ belongs to $Q_{\boldsymbol{o}}(K - \boldsymbol{z})$ if and only if $K - \boldsymbol{z}$ lies in the closed halfspace $V = \{\boldsymbol{x} \in \mathbb{R}^n : \boldsymbol{x} \cdot \boldsymbol{u} \leqslant 0\}$, while Definition 8.1 shows that the latter inclusion holds if and only if $\boldsymbol{u} \in (K - \boldsymbol{z})^\circ$.

(*d*) Since $T_{\boldsymbol{z}}(K) = \operatorname{cl} C_{\boldsymbol{o}}(K - \boldsymbol{z})$ is a closed convex cone with apex $\boldsymbol{o}$, Theorem 8.2 and assertion (*c*) give

$$T_{\boldsymbol{z}}(K) = ((T_{\boldsymbol{z}}(K))^\circ)^\circ = (N_{\boldsymbol{z}}(K))^\circ. \qquad \square$$

A combination of Theorems 5.48, 8.4, and 8.24 implies the following corollary. We recall that $\operatorname{ort} X$ denotes the orthospace of the plane $\operatorname{aff} X$ (see Definition 2.47).

Corollary 8.25. *For a nonempty convex set* $K \subset \mathbb{R}^n$ *and a point* $\boldsymbol{z} \in \operatorname{cl} K$, *the following assertions hold.*

(*a*) $N_{\boldsymbol{z}}(K)$ *is a closed convex cone whose apex set is* $\operatorname{ort} K$.
(*b*) $N_{\boldsymbol{z}}(K) = \operatorname{ort} K$ *if and only if* $\boldsymbol{z} \in \operatorname{rint} K$.
(*c*) $N_{\boldsymbol{z}}(K) \neq \operatorname{ort} K$ *if and only if* $\boldsymbol{z} \in \operatorname{rbd} K$.

Theorem 8.26. *Let* $K_1, \ldots, K_r \subset \mathbb{R}^n$ *be nonempty convex sets,* $\mu_1, \ldots, \mu_r$ *be nonzero scalars, and* $\boldsymbol{z}_1, \ldots, \boldsymbol{z}_r$ *be points in* $\operatorname{cl} K_1, \ldots, \operatorname{cl} K_r$, *respectively. Then*

$$N_{\mu_1 \boldsymbol{z}_1 + \cdots + \mu_r \boldsymbol{z}_r}(\mu_1 K_1 + \cdots + \mu_r K_r) = \mu_1 N_{\boldsymbol{z}_1}(K_1) \cap \cdots \cap \mu_r N_{\boldsymbol{z}_r}(K_r),$$
$$T_{\mu_1 \boldsymbol{z}_1 + \cdots + \mu_r \boldsymbol{z}_r}(\mu_1 K_1 + \cdots + \mu_r K_r) = \operatorname{cl}(\mu_1 T_{\boldsymbol{z}_1}(K_1) + \cdots + \mu_r T_{\boldsymbol{z}_r}(K_r)).$$

Proof. An induction argument shows that the proof can be reduced to the case $r = 2$ (let $\mu_1 K_1 = \mu_1 K_1 + \mu_2\{\boldsymbol{o}\}$ if $r = 1$). By Corollary 3.57,

$$\mu_1 \boldsymbol{z}_1 + \mu_2 \boldsymbol{z}_2 \in \mu_1 \operatorname{cl} K_1 + \mu_2 \operatorname{cl} K_2 \subset \operatorname{cl}(\mu_1 K_1 + \mu_2 K_2).$$

Now, the first equality follows from a combination of Theorems 8.2 and 8.24:

$$N_{\mu_1 \boldsymbol{z}_1+\mu_2\boldsymbol{z}_2}(\mu_1 K_1 + \mu_2 K_2) = ((\mu_1 K_1 + \mu_2 K_2) - (\mu_1 \boldsymbol{z}_1 + \mu_2 \boldsymbol{z}_2))^\circ$$
$$= (\mu_1(K_1 - \boldsymbol{z}_1) + \mu_2(K_2 - \boldsymbol{z}_2))^\circ = (\mu_1(K_1 - \boldsymbol{z}_1))^\circ \cap (\mu_2(K_2 - \boldsymbol{z}_2))^\circ$$
$$= \mu_1(K_1 - \boldsymbol{z}_1)^\circ \cap \mu_2(K_2 - \boldsymbol{z}_2)^\circ = \mu_1 N_{\boldsymbol{z}_1}(K_1) \cap \mu_2 N_{\boldsymbol{z}_2}(K_2).$$

The second equality follows from the definitions of $S_{\boldsymbol{z}}(K)$ and $T_{\boldsymbol{z}}(K)$, Problem 1.13, and Corollary 5.60:

$$\begin{aligned}
&T_{\mu_1 \boldsymbol{z}_1+\mu_2\boldsymbol{z}_2}(\mu_1 K_1 + \mu_2 K_2)\\
&\quad = D_{\mu_1 \boldsymbol{z}_1+\mu_2\boldsymbol{z}_2}(\mu_1 K_1 + \mu_2 K_2) - (\mu_1 \boldsymbol{z}_1 + \mu_2 \boldsymbol{z}_2)\\
&\quad = \operatorname{cl}(C_{\mu_1 \boldsymbol{z}_1+\mu_2\boldsymbol{z}_2}(\mu_1 K_1 + \mu_2 K_2) - (\mu_1 \boldsymbol{z}_1 + \mu_2 \boldsymbol{z}_2))\\
&\quad = \operatorname{cl}(\mu_1 C_{\boldsymbol{z}_1}(K_1) + \mu_2 C_{\boldsymbol{z}_2}(K_2) - (\mu_1 \boldsymbol{z}_1 + \mu_2 \boldsymbol{z}_2))\\
&\quad = \operatorname{cl}(\mu_1(C_{\boldsymbol{z}_1}(K_1) - \boldsymbol{z}_1) + \mu_2(C_{\boldsymbol{z}_2}(K_2) - \boldsymbol{z}_2))\\
&\quad = \operatorname{cl}(\mu_1(\operatorname{cl} C_{\boldsymbol{z}_1}(K_1) - \boldsymbol{z}_1) + \mu_2(\operatorname{cl} C_{\boldsymbol{z}_2}(K_2) - \boldsymbol{z}_2))\\
&\quad = \operatorname{cl}(\mu_1(D_{\boldsymbol{z}_1}(K_1) - \boldsymbol{z}_1) + \mu_2(D_{\boldsymbol{z}_2}(K_2) - \boldsymbol{z}_2))\\
&\quad = \operatorname{cl}(\mu_1 T_{\boldsymbol{z}_1}(K_1) + \mu_2 T_{\boldsymbol{z}_2}(K_2)).
\end{aligned}$$

□

Theorem 8.27. *Let $K_1, \dots, K_r \subset \mathbb{R}^n$ be convex sets whose closures have nonempty intersection and let $\boldsymbol{z}$ be a point in $\operatorname{cl} K_1 \cap \cdots \cap \operatorname{cl} K_r$. Then*

$$N_{\boldsymbol{z}}(K_1) + \cdots + N_{\boldsymbol{z}}(K_r) \subset N_{\boldsymbol{z}}(K_1 \cap \cdots \cap K_r), \tag{8.13}$$
$$\operatorname{cl} C_{\boldsymbol{z}}(K_1 \cap \cdots \cap K_r) \subset \operatorname{cl} C_{\boldsymbol{z}}(K_1) \cap \cdots \cap \operatorname{cl} C_{\boldsymbol{z}}(K_r), \tag{8.14}$$
$$T_{\boldsymbol{z}}(K_1 \cap \cdots \cap K_r) \subset T_{\boldsymbol{z}}(K_1) \cap \cdots \cap T_{\boldsymbol{z}}(K_r). \tag{8.15}$$

Furthermore, equality in (8.13) *occurs if and only if it does in any of equalities* (8.14) *and* (8.15).

Proof. Choose any vectors $\boldsymbol{e}_i \in N_{\boldsymbol{z}}(K_i)$, $1 \leqslant i \leqslant r$. By Theorem 8.24, $\boldsymbol{e}_i \in (K_i - \boldsymbol{z})^\circ$. In other words, $(\boldsymbol{x} - \boldsymbol{z}) \cdot \boldsymbol{e}_i \leqslant 0$ for all $\boldsymbol{x} \in K_i$. So, if $\boldsymbol{x} \in K_1 \cap \cdots \cap K_r$, then

$$(\boldsymbol{x} - \boldsymbol{z}) \cdot (\boldsymbol{e}_1 + \cdots + \boldsymbol{e}_r) = (\boldsymbol{x} - \boldsymbol{z}) \cdot \boldsymbol{e}_1 + \cdots + (\boldsymbol{x} - \boldsymbol{z}) \cdot \boldsymbol{e}_r \leqslant 0.$$

The same Theorem 8.24 gives $\boldsymbol{e}_1 + \cdots + \boldsymbol{e}_r \in N_{\boldsymbol{z}}(K_1 \cap \cdots \cap K_r)$. Hence the inclusion (8.13) holds.

By Corollary 5.55, $C_{\boldsymbol{z}}(K_i) = \operatorname{cone}_{\boldsymbol{z}}(K_i)$. Hence Theorem 5.43 and the properties of closure (see page 10) give

$$\begin{aligned}
\operatorname{cl} C_{\boldsymbol{z}}(K_1 \cap \cdots \cap K_r) &= \operatorname{cl}(C_{\boldsymbol{z}}(K_1) \cap \cdots \cap C_{\boldsymbol{z}}(K_r))\\
&\subset \operatorname{cl} C_{\boldsymbol{z}}(K_1) \cap \cdots \cap \operatorname{cl} C_{\boldsymbol{z}}(K_r).
\end{aligned}$$

Finally,

$$\begin{aligned}T_{\boldsymbol{z}}(K_1\cap\cdots\cap K_r) &= \operatorname{cl} C_{\boldsymbol{z}}(K_1\cap\cdots\cap K_r)-\boldsymbol{z}\\ &\subset \operatorname{cl} C_{\boldsymbol{z}}(K_1)\cap\cdots\cap\operatorname{cl} C_{\boldsymbol{z}}(K_r)-\boldsymbol{z}\\ &= (\operatorname{cl} C_{\boldsymbol{z}}(K_1)-\boldsymbol{z})\cap\cdots\cap(\operatorname{cl} C_{\boldsymbol{z}}(K_r)-\boldsymbol{z})\\ &= T_{\boldsymbol{z}}(K_1)\cap\cdots\cap T_{\boldsymbol{z}}(K_r).\end{aligned}$$

The second assertion immediately follows from the identities below (see Theorems 8.2, 8.3, and 8.24):

$$\begin{aligned}\operatorname{cl} C_{\boldsymbol{z}}(K_1\cap\cdots\cap K_r) &= \boldsymbol{z}+(N_{\boldsymbol{z}}(K_1\cap\cdots\cap K_r))^\circ,\\ \operatorname{cl} C_{\boldsymbol{z}}(K_1)\cap\cdots\cap\operatorname{cl} C_{\boldsymbol{z}}(K_r) &= \boldsymbol{z}+(N_{\boldsymbol{z}}(K_1))^\circ\cap\cdots\cap(N_{\boldsymbol{z}}(K_r))^\circ\\ &= \boldsymbol{z}+(N_{\boldsymbol{z}}(K_1)+\cdots+N_{\boldsymbol{z}}(K_r))^\circ.\end{aligned}$$

and the fact that two closed convex cones with apex $\boldsymbol{o}$ coincide if and only if their polar cones coincide. □

The next theorem provides some sufficient conditions for equality in (8.14).

Theorem 8.28. *Let $K_1,\dots,K_r\subset\mathbb{R}^n$ be convex sets whose closures have nonempty intersection and let $\boldsymbol{z}$ be a point in $\operatorname{cl} K_1\cap\cdots\cap\operatorname{cl} K_r$. The following assertions hold.*

(*a*) *If $\operatorname{rint} K_1\cap\cdots\cap\operatorname{rint} K_r\neq\varnothing$, then*

$$\operatorname{cl} C_{\boldsymbol{z}}(K_1\cap\cdots\cap K_r)=\operatorname{cl} C_{\boldsymbol{z}}(K_1)\cap\cdots\cap\operatorname{cl} C_{\boldsymbol{z}}(K_r).$$

(*b*) *If the sets $K_1,\dots,K_r$ are closed and locally conic at $\boldsymbol{z}$, then all cones $C_{\boldsymbol{z}}(K_i)$, $1\leqslant i\leqslant r$, are closed and*

$$C_{\boldsymbol{z}}(K_1\cap\cdots\cap K_r)=C_{\boldsymbol{z}}(K_1)\cap\cdots\cap C_{\boldsymbol{z}}(K_r).$$

(*c*) *If the sets $K_1,\dots,K_p$, $1\leqslant p\leqslant r-1$ are closed and locally conic at $\boldsymbol{z}$ and $K_1\cap\cdots\cap K_p\cap\operatorname{rint} K_{p+1}\cap\cdots\cap\operatorname{rint} K_r\neq\varnothing$, then all cones $C_{\boldsymbol{z}}(K_i)$, $1\leqslant i\leqslant p$, are closed and*

$$\begin{aligned}&\operatorname{cl} C_{\boldsymbol{z}}(K_1\cap\cdots\cap K_r)\\ &\qquad= C_{\boldsymbol{z}}(K_1)\cap\cdots\cap C_{\boldsymbol{z}}(K_p)\cap\operatorname{cl} C_{\boldsymbol{z}}(K_{p+1})\cap\cdots\cap\operatorname{cl} C_{\boldsymbol{z}}(K_r).\end{aligned}$$

Proof. (*a*) By Theorem 3.48, $\operatorname{rint} K_i=\operatorname{rint}(\operatorname{cl} K_i)$, $1\leqslant i\leqslant r$. If a point $\boldsymbol{u}$ belongs to $\operatorname{rint} K_1\cap\cdots\cap\operatorname{rint} K_r$, which is $\operatorname{rint}(\operatorname{cl} K_1)\cap\cdots\cap\operatorname{rint}(\operatorname{cl} K_r)$, then

Theorem 5.49 shows that $\boldsymbol{u}$ belongs to $\operatorname{rint} C_{\boldsymbol{z}}(\operatorname{cl} K_1) \cap \cdots \cap \operatorname{rint} C_{\boldsymbol{z}}(\operatorname{cl} K_r)$. Using Theorem 3.54, we obtain

$$\operatorname{cl}(K_1 \cap \cdots \cap K_r) = \operatorname{cl} K_1 \cap \cdots \cap \operatorname{cl} K_r,$$
$$\operatorname{cl}(C_{\boldsymbol{z}}(\operatorname{cl} K_1) \cap \cdots \cap C_{\boldsymbol{z}}(\operatorname{cl} K_r)) = \operatorname{cl} C_{\boldsymbol{z}}(\operatorname{cl} K_1) \cap \cdots \cap \operatorname{cl} C_{\boldsymbol{z}}(\operatorname{cl} K_r).$$

Furthermore, the assumption $\boldsymbol{z} \in \operatorname{cl} K_1 \cap \cdots \cap \operatorname{cl} K_r$ and Corollary 5.61 give

$$C_{\boldsymbol{z}}(\operatorname{cl} K_1 \cap \cdots \cap \operatorname{cl} K_r) = C_{\boldsymbol{z}}(\operatorname{cl} K_1) \cap \cdots \cap C_{\boldsymbol{z}}(\operatorname{cl} K_r).$$

Because $\operatorname{cl} C_{\boldsymbol{z}}(X) = \operatorname{cl} C_{\boldsymbol{z}}(\operatorname{cl} X)$ for any set $X \subset \mathbb{R}^n$ (see Theorem 5.45), we conclude that

$$\begin{aligned}
&\operatorname{cl} C_{\boldsymbol{z}}(K_1 \cap \cdots \cap K_r) = \operatorname{cl} C_{\boldsymbol{z}}(\operatorname{cl}(K_1 \cap \cdots \cap K_r)) \\
&= \operatorname{cl} C_{\boldsymbol{z}}(\operatorname{cl} K_1 \cap \cdots \cap \operatorname{cl} K_r) = \operatorname{cl}(C_{\boldsymbol{z}}(\operatorname{cl} K_1) \cap \cdots \cap C_{\boldsymbol{z}}(\operatorname{cl} K_r)) \\
&= \operatorname{cl} C_{\boldsymbol{z}}(\operatorname{cl} K_1) \cap \cdots \cap \operatorname{cl} C_{\boldsymbol{z}}(\operatorname{cl} K_r) = \operatorname{cl} C_{\boldsymbol{z}}(K_1) \cap \cdots \cap \operatorname{cl} C_{\boldsymbol{z}}(K_r).
\end{aligned}$$

(*b*) Since the set $K_1 \cap \cdots \cap K_p$ is closed and locally conic at $\boldsymbol{z}$ (see Problem 7.2), Theorem 7.6 shows that all cones $C_{\boldsymbol{z}}(K_i)$, $1 \leqslant i \leqslant r$, are closed. By Corollary 5.61,

$$C_{\boldsymbol{z}}(K_1 \cap \cdots \cap K_r) = C_{\boldsymbol{z}}(K_1) \cap \cdots \cap C_{\boldsymbol{z}}(K_r).$$

(*c*) Let $M_1 = K_1 \cap \cdots \cap K_p$ and $M_2 = K_{p+1} \cap \cdots \cap K_r$. As above, the set M_1 is closed and locally conic at $\boldsymbol{z}$. Theorem 3.29 and the assumption of (*c*) give

$$M_1 \cap \operatorname{rint} M_2 = (K_1 \cap \cdots \cap K_p) \cap (\operatorname{rint} K_{p+1} \cap \cdots \cap \operatorname{rint} K_r) \neq \varnothing.$$

Consequently, $\operatorname{cl}(M_1 \cap M_2) = M_1 \cap \operatorname{cl} M_2$ (see Problem 3.15). By part (*b*), the cone $C_{\boldsymbol{z}}(M_1) = C_{\boldsymbol{z}}(K_1) \cap \cdots \cap C_{\boldsymbol{z}}(K_p)$ is closed and convex. Similarly to the proof of part (*a*), $C_{\boldsymbol{z}}(M_1)$ meets the set

$$\operatorname{rint} C_{\boldsymbol{z}}(\operatorname{cl} M_2) = \operatorname{rint} C_{\boldsymbol{z}}(\operatorname{cl} K_{p+1}) \cap \cdots \cap \operatorname{rint} C_{\boldsymbol{z}}(\operatorname{cl} K_r).$$

By the same Problem 3.15,

$$\operatorname{cl}(C_{\boldsymbol{z}}(M_1) \cap C_{\boldsymbol{z}}(\operatorname{cl} M_2)) = C_{\boldsymbol{z}}(K_1) \cap \operatorname{cl} C_{\boldsymbol{z}}(\operatorname{cl} M_2).$$

Finally, using parts (*a*) and (*b*), we obtain

$$\begin{aligned}
&\operatorname{cl} C_{\boldsymbol{z}}(K_1 \cap \cdots \cap K_r) = \operatorname{cl} C_{\boldsymbol{z}}(M_1 \cap M_2) = \operatorname{cl} C_{\boldsymbol{z}}(\operatorname{cl}(M_1 \cap M_2)) \\
&= \operatorname{cl} C_{\boldsymbol{z}}(M_1 \cap \operatorname{cl} M_2) = \operatorname{cl}(C_{\boldsymbol{z}}(M_1) \cap C_{\boldsymbol{z}}(\operatorname{cl} M_2)) \\
&= C_{\boldsymbol{z}}(M_1) \cap \operatorname{cl} C_{\boldsymbol{z}}(\operatorname{cl} M_2) = C_{\boldsymbol{z}}(M_1) \cap \operatorname{cl} C_{\boldsymbol{z}}(M_2) \\
&= C_{\boldsymbol{z}}(K_1) \cap \cdots \cap C_{\boldsymbol{z}}(K_p) \cap \operatorname{cl} C_{\boldsymbol{z}}(K_{p+1}) \cap \cdots \cap \operatorname{cl} C_{\boldsymbol{z}}(K_r).
\end{aligned}$$

□

8.2.3 Normal and Barrier Cones

Definition 8.29. For a nonempty convex set $K \subset \mathbb{R}^n$, the union

$$\operatorname{nor} K = \cup\, (N_{\boldsymbol{z}}(K) : \boldsymbol{z} \in \operatorname{cl} K)$$

is called the *normal cone* of K, and every nonzero vector $\boldsymbol{e} \in \operatorname{nor} K$ is called a *normal vector* of K. We let $\operatorname{nor} K = \mathbb{R}^n$ if $K = \varnothing$.

Examples. 1. The normal cone of a nonempty plane $L \subset \mathbb{R}^n$ is the orthospace $\operatorname{ort} L$ (see Theorem 8.37).

2. If $V = \{\boldsymbol{x} \in \mathbb{R}^n : \boldsymbol{x}\cdot\boldsymbol{e} \leqslant \gamma\}$, $\boldsymbol{e} \neq \boldsymbol{o}$, is a closed halfspace of $\mathbb{R}^n$, then $\operatorname{nor} V = \{\lambda\boldsymbol{e} : \lambda \geqslant 0\}$ (compare with Definition 2.29). A similar assertion holds for the open halfspace $W = \{\boldsymbol{x} \in \mathbb{R}^n : \boldsymbol{x}\cdot\boldsymbol{e} < \gamma\}$.

Theorem 8.30. *For a nonempty convex set $K \subset \mathbb{R}^n$, the following assertions hold.*

(a) $\operatorname{nor} K = \{\boldsymbol{o}\}$ *if and only if* $K = \mathbb{R}^n$.
(b) $\operatorname{nor} K = \mathbb{R}^n$ *if and only if* K *is bounded.*
(c) $\operatorname{nor} K$ *is a cone whose apex set is proper and contains* $\operatorname{ort} K$.
(d) $\operatorname{nor} K = \operatorname{ort} K$ *if and only if* K *is a plane.*
(e) *If* K *is not a plane, then* $\operatorname{nor} K = \cup\, (N_{\boldsymbol{z}}(K) : \boldsymbol{z} \in \operatorname{rbd} K)$.
(f) $\operatorname{nor} K = \operatorname{nor}(\operatorname{cl} K) = \operatorname{nor}(\operatorname{rint} K)$.

Proof. (a) If $K = \mathbb{R}^n$, then the equality $\operatorname{nor} K = \{\boldsymbol{o}\}$ is obvious. On the other hand, if $K \neq \mathbb{R}^n$, then $\operatorname{cl} K \neq \mathbb{R}^n$ according to Theorem 3.50. Hence there is a point $\boldsymbol{a} \in \mathbb{R}^n \setminus \operatorname{cl} K$. By Theorem 8.19, $\operatorname{cl} K$ contains the nearest to $\boldsymbol{a}$ point $\boldsymbol{c}$. Therefore, $\boldsymbol{a} \in Q_{\boldsymbol{c}}(K)$ and $\boldsymbol{o} \neq \boldsymbol{a} - \boldsymbol{c} \in N_{\boldsymbol{c}}(K) \subset \operatorname{nor} K$.

(b) Assume first that K is bounded and choose any nonzero vector $\boldsymbol{e} \in \mathbb{R}^n$. Since the linear functional $\varphi(\boldsymbol{x}) = \boldsymbol{x}\cdot\boldsymbol{e}$ is continuous, it attains a maximum value on the compact set $\operatorname{cl} K$ at a point $\boldsymbol{z}$. Let $\gamma = \boldsymbol{z}\cdot\boldsymbol{e}$. Then $\operatorname{cl} K$ lies in the closed halfspace $V = \{\boldsymbol{x} \in \mathbb{R}^n : \boldsymbol{x}\boldsymbol{e} \leqslant \gamma\}$ and $\boldsymbol{z}$ is the nearest to $\boldsymbol{e} + \boldsymbol{z}$ point in $\operatorname{cl} K$ (see Theorem 8.22). In other words, $\boldsymbol{e} + \boldsymbol{z} \in Q_{\boldsymbol{z}}(K)$, which gives $\boldsymbol{e} \in N_{\boldsymbol{z}}(K) \subset \operatorname{nor} K$. Hence $\operatorname{nor} K = \mathbb{R}^n$.

Conversely, suppose that K is not bounded. By Theorem 6.12, K contains a closed halfline h. According to Definition 2.23, we can write

$$h = \{(1-\lambda)\boldsymbol{u} + \lambda\boldsymbol{v} : \lambda \geqslant 0\} = \boldsymbol{u} + \{\lambda\boldsymbol{c} : \lambda \geqslant 0\},$$

where $\boldsymbol{u}$ and $\boldsymbol{v}$ are distinct points in K and $\boldsymbol{c} = \boldsymbol{v} - \boldsymbol{u} \neq \boldsymbol{o}$. Since the case $K = \mathbb{R}^n$ is obvious ($\operatorname{nor} K = \{\boldsymbol{o}\} \neq \mathbb{R}^n$), we may assume that $K \neq \mathbb{R}^n$. By assertion (a), $\operatorname{nor} K \neq \{\boldsymbol{o}\}$. Choose any nonzero vector $\boldsymbol{e} \in \operatorname{nor} K$. Then (see Definition 8.29) there is a point $\boldsymbol{z} \in \operatorname{cl} K$ such that $\boldsymbol{z}$ is the nearest to $\boldsymbol{e}+\boldsymbol{z}$ point in $\operatorname{cl} K$. Furthermore, Theorem 8.22 shows that $\operatorname{cl} K$ lies in the closed halfspace

$$V = \{\boldsymbol{x} \in \mathbb{R}^n : (\boldsymbol{x} - \boldsymbol{z})\cdot\boldsymbol{e} \leqslant 0\}.$$

In particular, $h \subset V$, which results in $(\boldsymbol{u} + \lambda\boldsymbol{c} - \boldsymbol{z})\cdot\boldsymbol{e} \leqslant 0$ for all $\lambda \geqslant 0$. Obviously, this condition is satisfied only if $\boldsymbol{c}\cdot\boldsymbol{e} \leqslant 0$, which shows that every vector $\boldsymbol{e} \in \operatorname{nor} K$ lies in the closed halfspace $\{\boldsymbol{x} \in \mathbb{R}^n : \boldsymbol{c}\cdot\boldsymbol{x} \leqslant 0\}$. Consequently, $\operatorname{nor} K \neq \mathbb{R}^n$.

(*c*) Given a point $\boldsymbol{z} \in \operatorname{cl} K$, Corollary 8.25 shows that the normal cone $N_{\boldsymbol{z}}(K)$ is a closed convex cone with proper apex set $\boldsymbol{z} + \operatorname{ort} K$. Hence $\operatorname{nor} K$, as the union of such cones, contains $\operatorname{ort} K$ in its proper apex set (see Corollary 5.9).

(*d*) The equality $\operatorname{nor} K = \operatorname{ort} K$ holds if and only if $N_{\boldsymbol{z}}(K) = \boldsymbol{z} + \operatorname{ort} K$ for all points $\boldsymbol{z} \in \operatorname{cl} K$. In this case, Corollary 8.25 shows that every point $\boldsymbol{z} \in \operatorname{cl} K$ belongs to $\operatorname{rint} K$. So, $\operatorname{nor} K = \operatorname{ort} K$ if and only if $\operatorname{rbd} K = \varnothing$. Finally, $\operatorname{rbd} K = \varnothing$ if and only if K is a plane (see Theorem 3.62).

(*e*) According to Theorem 3.62, $\operatorname{rbd} K \neq \varnothing$ provided K is not a plane. Corollary 8.25 shows that every normal cone $N_{\boldsymbol{z}}(K)$, where $\boldsymbol{z} \in \operatorname{cl} K$, contains $\operatorname{ort} K$, and $N_{\boldsymbol{z}}(K) = \operatorname{ort} K$ provided $\boldsymbol{z} \in \operatorname{rint} K$. So, we may exclude all points $\boldsymbol{z} \in \operatorname{rint} K$ in Definition 8.29.

(*f*) By Theorem 3.48, $\operatorname{cl}(\operatorname{cl} K) = \operatorname{cl} K = \operatorname{cl}(\operatorname{rint} K)$. So, the normal cones $N_{\boldsymbol{z}}(K)$, $N_{\boldsymbol{z}}(\operatorname{rint} K)$, and $N_{\boldsymbol{z}}(\operatorname{cl} K)$ coincide for any point $\boldsymbol{z} \in \operatorname{cl} K$. Now, the assertion immediately follows from Definition 8.29. □

Remarks. 1. Unlike apices of cones $N_{\boldsymbol{z}}(K)$, the apex set of $\operatorname{nor} K$ can be larger than a translate of $\operatorname{ort} K$. For instance, if K is the unit disk of the xy-plane in $\mathbb{R}^3$, then $\operatorname{ort} K$ is the z-axis, while $\operatorname{nor} K = \mathbb{R}^3$.

2. The cone $\operatorname{nor} K$ may be *neither closed nor convex*, even if K is closed (although Corollary 8.36 shows that both sets $\operatorname{cl}(\operatorname{nor} K)$ and $\operatorname{rint}(\operatorname{nor} K)$ are convex). Indeed, consider the set $K \subset \mathbb{R}^3$ be given by

$$K = \big\{(x, y, z) : |x| + |y| \leqslant 1,\ z \geqslant \max\{\tfrac{1}{1+x}, \tfrac{1}{1-x}, \tfrac{1}{1+y}, \tfrac{1}{1-y}\}\big\}. \qquad (8.16)$$

This set is closed and convex as the intersection of five closed convex set, given, respectively, by the following conditions: (*a*) $|x| + |y| \leqslant 1$;

$$(b)\ \ z \geqslant \tfrac{1}{1+x},\ x > -1; \quad (c)\ \ z \geqslant \tfrac{1}{1-x},\ x < 1;$$
$$(d)\ \ z \geqslant \tfrac{1}{1+y},\ y > -1; \quad (e)\ \ z \geqslant \tfrac{1}{1-y},\ y < 1.$$

Then $\operatorname{nor} K$ is the union of the open halfspace $\{(x,y,z) : z < 0\}$ and the lines $l = \{(x,y,0) : x = y\}$ and $l' = \{(x,y,0) : x = -y\}$. Clearly, $\operatorname{nor} K$ is neither closed nor convex.

3. It is easy to see that the normal cone of a convex set K in the plane is always convex.

Definition 8.31. The *barrier cone* of a nonempty set $X \subset \mathbb{R}^n$, denoted $\operatorname{bar} X$, is defined by

$$\operatorname{bar} X = \{\boldsymbol{e} \in \mathbb{R}^n : \exists\, \gamma = \gamma(\boldsymbol{e}) \in \mathbb{R} \text{ such that } \boldsymbol{x}\cdot\boldsymbol{e} \leqslant \gamma \text{ for all } \boldsymbol{x} \in X\}.$$

We let $\operatorname{bar} X = \mathbb{R}^n$ if $X = \varnothing$.

Remarks. 1. Clearly, $\boldsymbol{o} \in \operatorname{bar} X$ for every nonempty set $X \subset \mathbb{R}^n$. Indeed, $\boldsymbol{x}\cdot\boldsymbol{o} \leqslant 0$ for all $\boldsymbol{x} \in X$.

2. Various results on barrier cones of convex sets, given in this book, hold for the case of arbitrary sets. The obvious connection here is the equality $\operatorname{bar} X = \operatorname{bar}(\operatorname{conv} X)$.

Examples. 1. The barrier cone of a nonempty plane $L \subset \mathbb{R}^n$ is the subspace $\operatorname{ort} L$ (see Theorem 8.37). Consequently, $\operatorname{bar} L = \operatorname{nor} L$ (compare with an example on page 296).

2. The barrier cone of a closed halfspace $V = \{\boldsymbol{x} \in \mathbb{R}^n : \boldsymbol{x}\cdot\boldsymbol{e} \leqslant \gamma\}$, $\boldsymbol{e} \neq \boldsymbol{o}$, is the closed halfline $\{\lambda\boldsymbol{e} : \lambda \geqslant 0\}$ (see Theorem 2.93).

Theorem 8.32. *For a nonempty convex set $K \subset \mathbb{R}^n$, the following assertions hold.*

(*a*) $\operatorname{nor} K \subset \operatorname{bar} K$.
(*b*) $\operatorname{bar} K = \{\boldsymbol{o}\}$ *if and only if* $K = \mathbb{R}^n$.
(*c*) $\operatorname{bar} K = \mathbb{R}^n$ *if and only if* K *is bounded.*
(*d*) $\operatorname{bar} K$ *is a convex cone whose apex set is proper and contains* $\operatorname{ort} K$.
(*e*) $\operatorname{bar} K = \operatorname{ort} K$ *if and only if* K *is a plane.*
(*f*) $\operatorname{bar} K = \operatorname{bar}(\operatorname{cl} K) = \operatorname{bar}(\operatorname{rint} K)$.

Proof. (*a*) Let $\boldsymbol{e} \in \operatorname{nor} K$. Since the case $\boldsymbol{e} = \boldsymbol{o}$ is obvious, we assume that $\boldsymbol{e} \neq \boldsymbol{o}$. Then $\boldsymbol{e} \in N_{\boldsymbol{z}}(K) = Q_{\boldsymbol{z}}(K) - \boldsymbol{z}$ for a suitable point $\boldsymbol{z} \in \operatorname{cl} K$. Hence $p_K(\boldsymbol{e} + \boldsymbol{z}) = \boldsymbol{z}$. By Theorem 8.22, K lies in the closed halfspace $V = \{\boldsymbol{x} \in \mathbb{R}^n : \boldsymbol{x}\cdot\boldsymbol{e} \leqslant \gamma\}$, where $\gamma = \boldsymbol{z}\cdot\boldsymbol{e}$. Hence $\boldsymbol{e} \in \operatorname{bar} K$.

(*b*) If $K = \mathbb{R}^n$, then $\operatorname{bar} K = \{\boldsymbol{o}\}$. On the other hand, if $K \neq \mathbb{R}^n$, then a combination of Theorem 8.30 and assertion (*a*) gives $\{\boldsymbol{o}\} \neq \operatorname{nor} K \subset \operatorname{bar} K$.

(*c*) If K is bounded, then $\operatorname{nor} K = \mathbb{R}^n$ (see Theorem 8.30), and assertion (*a*) shows that $\operatorname{bar} K = \mathbb{R}^n$. Conversely, let K be unbounded. By Theorem 6.12, K contains a closed halfline h. According to Definition 2.23, we can write

$$h = \{(1-\lambda)\boldsymbol{u} + \lambda\boldsymbol{v} : \lambda \geqslant 0\} = \boldsymbol{u} + \{\lambda\boldsymbol{c} : \lambda \geqslant 0\},$$

where $\boldsymbol{u}$ and $\boldsymbol{v}$ are distinct points in K and $\boldsymbol{c} = \boldsymbol{v} - \boldsymbol{u} \neq \boldsymbol{o}$. Choose any scalar γ and a positive scalar $\lambda > (\gamma - \boldsymbol{u}\cdot\boldsymbol{c})/\boldsymbol{c}\cdot\boldsymbol{c}$. Then $(\boldsymbol{u} + \lambda\boldsymbol{c})\cdot\boldsymbol{c} > \gamma$. Since $\boldsymbol{u} + \lambda\boldsymbol{c} \in h \subset K$, Definition 8.31 show that $\boldsymbol{c} \notin \operatorname{bar} K$. Consequently, $\operatorname{bar} K \neq \mathbb{R}^n$.

(*d*) Let a vector $\boldsymbol{e} \in \operatorname{bar} K$ and a scalar γ satisfy the inequality $\boldsymbol{x}\cdot\boldsymbol{e} \leqslant \gamma$ whenever $\boldsymbol{x} \in K$. If $\mu \geqslant 0$, then $\boldsymbol{x}\cdot(\mu\boldsymbol{e}) \leqslant \mu\gamma$ for all $\boldsymbol{x} \in K$, which shows that $\mu\boldsymbol{e} \in \operatorname{bar} K$. Hence $\operatorname{bar} K$ is a cone with apex $\boldsymbol{o}$.

For the convexity of $\operatorname{bar} K$, choose vectors $\boldsymbol{e}_1, \boldsymbol{e}_2 \in \operatorname{bar} K$ and a scalar $\lambda \in [0,1]$. If γ_i satisfies the inequality $\boldsymbol{x}\cdot\boldsymbol{e}_i \leqslant \gamma_i$ for all $\boldsymbol{x} \in K$, $i = 1, 2$, then

$$\boldsymbol{x}\cdot((1-\lambda)\boldsymbol{e}_1 + \lambda\boldsymbol{e}_2) = (1-\lambda)\boldsymbol{x}\cdot\boldsymbol{e}_1 + \lambda\boldsymbol{x}\cdot\boldsymbol{e}_2 \leqslant (1-\lambda)\gamma_1 + \lambda\gamma_2$$

whenever $\boldsymbol{x} \in K$, implying the inclusion $(1-\lambda)\boldsymbol{e}_1 + \lambda\boldsymbol{e}_2 \in \operatorname{bar} K$. Hence $\operatorname{bar} K$ is a convex set.

Next, we observe that $\operatorname{ort} K \subset \operatorname{bar} K$. Indeed, this inclusion is obvious if $\operatorname{ort} K = \{\boldsymbol{o}\}$. Suppose that $\operatorname{ort} K \neq \{\boldsymbol{o}\}$ and choose a nonzero vector $\boldsymbol{e} \in \operatorname{ort} K$ and a point $\boldsymbol{u} \in K$. Then $\operatorname{dir} K$ lies in the $(n-1)$-dimensional subspace $S = \{\boldsymbol{x} \in \mathbb{R}^n : \boldsymbol{x}\boldsymbol{e} = 0\}$, implying that $\operatorname{aff} K$ lies in the hyperplane $H = \boldsymbol{u} + S$ (see Theorem 2.48). With $\gamma = \boldsymbol{u}\cdot\boldsymbol{e}$, we have

$$K \subset \operatorname{aff} K \subset \boldsymbol{u} + S = \{\boldsymbol{x} \in \mathbb{R}^n : \boldsymbol{c}\cdot\boldsymbol{e} = \gamma\}.$$

In particular, K lies in the closed halfspace $V = \{\boldsymbol{x} \in \mathbb{R}^n : \boldsymbol{x}\cdot\boldsymbol{e} \leqslant \gamma\}$. Hence $\boldsymbol{e} \in \operatorname{bar} K$, and $\operatorname{ort} K \subset \operatorname{bar} K$. Finally, since $\operatorname{ort} K$ is a subspace, Theorem 5.17 implies the inclusion $\operatorname{ort} K \subset \operatorname{ap}(\operatorname{bar} K)$.

(*e*) If K is a plane, then $\operatorname{bar} K = \operatorname{nor} K = \operatorname{ort} K$ (see Theorem 8.30 and an example on page 298). Conversely, if $\operatorname{bar} K = \operatorname{ort} K$, then the inclusions $\operatorname{ort} K \subset \operatorname{nor} K \subset \operatorname{bar} K = \operatorname{ort} K$ give $\operatorname{ort} K = \operatorname{nor} K$, implying that K is a plane (see Theorem 8.30).

(*f*) Definition 8.31 shows that a nonzero vector $\boldsymbol{e} \in \mathbb{R}^n$ belongs to $\operatorname{bar} K$ if and only if a suitable closed halfspace $V = \{\boldsymbol{x} \in \mathbb{R}^n : \boldsymbol{x}\cdot\boldsymbol{e} \leqslant \gamma\}$ contains K. Since V is a closed set and $\operatorname{cl} K = \operatorname{cl}(\operatorname{rint} K)$ (see Theorems 2.33 and 3.48), K lies in V if and only if each of the sets $\operatorname{cl} K$ and $\operatorname{rint} K$ lies in V. Consequently, $\boldsymbol{e} \in \operatorname{bar} K$ if and only if each of the inclusions $\boldsymbol{e} \in \operatorname{bar}(\operatorname{cl} K)$ and $\boldsymbol{e} \in \operatorname{bar}(\operatorname{rint} K)$ holds. □

Remarks. 1. A barrier cone of a closed convex set may be nonclosed. For instance, if $K = \{(x, y) : y \geqslant x^2\}$, then

$$\operatorname{nor} K = \operatorname{bar} K = \{\boldsymbol{o}\} \cup \{(x, y) : y < 0\}.$$

2. The cones $\operatorname{bar} K$ and $\operatorname{nor} K$ may be distinct. For instance, if $K \subset \mathbb{R}^2$ is a closed convex set bounded by a pair of asymptotic horizontal lines l_1 and l_2, as depicted in Figure 8.7, then

$$\operatorname{bar} K = \{(x, y) : x \leqslant 0\} \neq \operatorname{nor} K = \{\boldsymbol{o}\} \cup \{(x, y) : x < 0\}.$$

Nevertheless, $\operatorname{cl}(\operatorname{bar} K) = \operatorname{cl}(\operatorname{nor} K)$, according to Corollary 8.36.

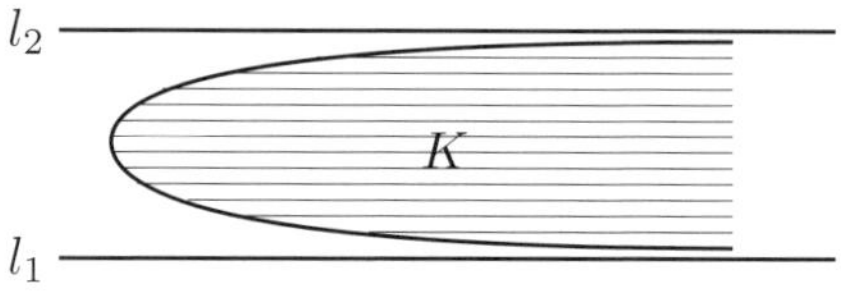

Fig. 8.7 Cones $\operatorname{nor} K$ and $\operatorname{bar} K$ may be distinct.

8.2.4 *Algebra of Barrier and Normal Cones*

Theorem 8.33. *If $K \subset \mathbb{R}^n$ is a convex set, $\boldsymbol{a}$ is a point in $\mathbb{R}^n$, and μ is a scalar, then*

$$\operatorname{bar}(\boldsymbol{a} + \mu K) = \mu \operatorname{bar} K, \quad \operatorname{nor}(\boldsymbol{a} + \mu K) = \mu \operatorname{nor} K.$$

Proof. Since the cases $\mu = 0$, $K = \varnothing$, and $K = \mathbb{R}^n$ are obvious (all the cones involved are either $\mathbb{R}^n$ or $\boldsymbol{o}$), we may assume that $\mu \neq 0$ and $K \neq \mathbb{R}^n$. Consequently, $\boldsymbol{a} + \mu K \neq \mathbb{R}^n$. By Theorems 8.30 and 8.32, none of the cones $\operatorname{nor} K$, $\operatorname{nor}(\boldsymbol{a} + \mu K)$, $\operatorname{bar} K$, and $\operatorname{bar}(\boldsymbol{a} + \mu K)$ is $\{\boldsymbol{o}\}$.

1. For the first equality, assume initially that $\mu > 0$. Let $\boldsymbol{e}$ be a nonzero vector in $\operatorname{bar}(\boldsymbol{a} + \mu K)$. Then there is a scalar γ such that $(\boldsymbol{a} + \mu \boldsymbol{x}) \cdot \boldsymbol{e} \leqslant \gamma$ for all $\boldsymbol{x} \in K$. Therefore, $\boldsymbol{x} \cdot \boldsymbol{e} \leqslant (\gamma - \boldsymbol{a} \cdot \boldsymbol{e})/\mu$ whenever $\boldsymbol{x} \in K$, which gives the inclusion $\boldsymbol{e} \in \operatorname{bar} K$. If $\mu < 0$, then the above argument gives the inequality, $\boldsymbol{x} \cdot (-\boldsymbol{e}) \leqslant (\gamma - \boldsymbol{a} \cdot \boldsymbol{e})/(-\mu)$, which results in the inclusion $-\boldsymbol{e} \in \operatorname{bar} K$. Thus $\operatorname{bar}(\boldsymbol{a} + \mu K) \subset \mu \operatorname{bar} K$. The opposite inequality is obtained in a similar way.

2. For the second equality, we observe that any point

$$\boldsymbol{z} \in \operatorname{cl}(\boldsymbol{a} + \mu K) = \boldsymbol{a} + \mu \operatorname{cl} K$$

can be written as $\boldsymbol{z} = \boldsymbol{a} + \mu\boldsymbol{u}$, where $\boldsymbol{u} \in \operatorname{cl} K$. Now, a combination of Theorems 8.2 and 8.24 gives

$$\begin{aligned}
&\operatorname{nor}(\boldsymbol{a} + \mu K) = \cup\,(N_{\boldsymbol{z}}(\boldsymbol{a} + \mu K) : \boldsymbol{z} \in \operatorname{cl}(\boldsymbol{a} + \mu K))\\
&= \cup\,((\boldsymbol{a} + \mu K - \boldsymbol{z})^\circ : \boldsymbol{z} \in \boldsymbol{a} + \mu \operatorname{cl} K)\\
&= \cup\,((\mu(K - \boldsymbol{u}))^\circ : \mu\boldsymbol{u} \in \mu \operatorname{cl} K) = \cup\,(\mu(K - \boldsymbol{u})^\circ : \boldsymbol{u} \in \operatorname{cl} K)\\
&= \mu \cup (N_{\boldsymbol{z}}(K) : \boldsymbol{u} \in \operatorname{cl} K) = \mu \operatorname{nor} K.
\end{aligned}$$

□

Theorem 8.34. *For convex sets $K_1, \ldots, K_r \subset \mathbb{R}^n$ and nonzero scalars $\mu_1, \ldots, \mu_r$, one has*

$$\mu_1 \operatorname{bar} K_1 \cap \cdots \cap \mu_r \operatorname{bar} K_r = \operatorname{bar}(\mu_1 K_1 + \cdots + \mu_r K_r), \tag{8.17}$$

$$\mu_1 \operatorname{nor} K_1 \cap \cdots \cap \mu_r \operatorname{nor} K_r \subset \operatorname{nor}(\mu_1 K_1 + \cdots + \mu_r K_r). \tag{8.18}$$

Proof. First, we are going to prove (8.17). If at least one of the set K_i, $1 \leqslant i \leqslant r$, is empty, then both parts of (8.17) are empty. So, we may assume that all $K_1, \ldots, K_r$ are nonempty. Since $\mu_i \operatorname{bar} K_i = \operatorname{bar}(\mu_i K_i)$, $1 \leqslant i \leqslant r$ (see Theorem 8.33), equality (8.17) can be rewritten as

$$\operatorname{bar}(\mu_1 K_1) \cap \cdots \cap \operatorname{bar}(\mu_r K_r) = \operatorname{bar}(\mu_1 K_1 + \cdots + \mu_r K_r).$$

First, we are going to show the inclusion

$$\operatorname{bar}(\mu_1 K_1) \cap \cdots \cap \operatorname{bar}(\mu_r K_r) \subset \operatorname{bar}(\mu_1 K_1 + \cdots + \mu_r K_r).$$

For this, choose a vector $\boldsymbol{e} \in \operatorname{bar}(\mu_1 K_1) \cap \cdots \cap \operatorname{bar}(\mu_r K_r)$. Since the case $\boldsymbol{e} = \boldsymbol{o}$ is obvious, we assume that $\boldsymbol{e} \neq \boldsymbol{o}$. Choose scalars $\gamma_1, \ldots, \gamma_r$ such that $\mu_i \boldsymbol{x}\cdot\boldsymbol{e} \leqslant \gamma_i$ for all $\boldsymbol{x} \in \mu_i K_i$, $1 \leqslant i \leqslant r$. Since a point $\boldsymbol{u} \in \mu_1 K_1 + \cdots + \mu_r K_r$ can be written as $\boldsymbol{u} = \mu_1 \boldsymbol{u}_1 + \cdots + \mu_r \boldsymbol{u}_r$, where $\boldsymbol{u}_i \in K_i$ for all $1 \leqslant i \leqslant r$, we have

$$\boldsymbol{u}\cdot\boldsymbol{e} = (\mu_1 \boldsymbol{u}_1 + \cdots + \mu_r \boldsymbol{u}_r)\cdot\boldsymbol{e} \leqslant \gamma_1 + \cdots + \gamma_r.$$

This inequality implies the inclusion $\boldsymbol{e} \in \operatorname{bar}(\mu_1 K_1 + \cdots + \mu_r K_r)$.

For the opposite inclusion,

$$\operatorname{bar}(\mu_1 K_1 + \cdots + \mu_r K_r) \subset \operatorname{bar}(\mu_1 K_1) \cap \cdots \cap \operatorname{bar}(\mu_r K_r),$$

choose a vector $\boldsymbol{e} \in \operatorname{bar}(\mu_1 K_1 + \cdots + \mu_r K_r)$. Since the case $\boldsymbol{e} = \boldsymbol{o}$ is obvious, we assume that $\boldsymbol{e} \neq \boldsymbol{o}$. Let a scalar γ satisfies the inequality $\boldsymbol{x}\cdot\boldsymbol{e} \leqslant \gamma$ for all $\boldsymbol{x} \in \mu_1 K_1 + \cdots + \mu_r K_r$. Fix points $\boldsymbol{z}_i \in K_i$, $2 \leqslant i \leqslant r$. For any point $\boldsymbol{x}_1 \in K_1$, the inequality

$$(\mu_1 \boldsymbol{x}_1 + \mu_2 \boldsymbol{z}_2 + \cdots + \mu_r \boldsymbol{z}_r)\cdot\boldsymbol{e} \leqslant \gamma$$

results in

$$\mu_1 \boldsymbol{x}_1 \cdot \boldsymbol{e} \leqslant \gamma - (\mu_2 \boldsymbol{z}_2 + \cdots + \mu_r \boldsymbol{z}_r) \cdot \boldsymbol{e}$$

Therefore, $\boldsymbol{e} \in \operatorname{bar}(\mu_1 K_1)$. In a similar way, $\boldsymbol{e} \in \operatorname{bar}(\mu_i K_i)$, $2 \leqslant i \leqslant r$. Thus $\boldsymbol{e} \in \operatorname{bar}(\mu_1 K_1) \cap \cdots \cap \operatorname{bar}(\mu_r K_r)$.

Next, we will prove (8.18). If at least one of the set K_i, $1 \leqslant i \leqslant r$, is empty, then both parts of (8.18) are empty. So, we may assume that all $K_1, \dots, K_r$ are nonempty. Since $\mu_i \operatorname{nor} K_i = \operatorname{nor}(\mu_i K_i)$, $1 \leqslant i \leqslant r$ (see Theorem 8.33), inclusion (8.18) can be rewritten as

$$\operatorname{nor}(\mu_1 K_1) \cap \cdots \cap \operatorname{nor}(\mu_r K_r) \subset \operatorname{nor}(\mu_1 K_1 + \cdots + \mu_r K_r).$$

Choose any vector $\boldsymbol{e} \in \operatorname{nor}(\mu_1 K_1) \cap \cdots \cap \operatorname{nor}(\mu_r K_r)$. For every $i = 1, \dots, r$, there is a point $\boldsymbol{z}_i \in \operatorname{cl}(\mu_i K_i)$ such that $\boldsymbol{e} \in N_{\boldsymbol{z}_i}(\mu_i K_i)$. Put $\boldsymbol{z} = \boldsymbol{z}_1 + \cdots + \boldsymbol{z}_r$ Corollary 3.57 gives the inclusion

$$\boldsymbol{z} \in \operatorname{cl}(\mu_1 K_1) + \cdots + \operatorname{cl}(\mu_i K_r) \subset \operatorname{cl}(\mu_1 K_1 + \cdots + \mu_r K_r).$$

By Theorem 8.26,

$$\begin{aligned} \boldsymbol{e} \in N_{\boldsymbol{z}_1}(\mu_1 K_1) \cap \cdots \cap N_{\boldsymbol{z}_r}(\mu_r K_r) &= N_{\boldsymbol{z}_1 + \cdots + \boldsymbol{z}_r}(\mu_1 K_1 + \cdots + \mu_r K_r) \\ &\subset \operatorname{nor}(\mu_1 K_1 + \cdots + \mu_r K_r). \end{aligned}$$ □

Remark. Inclusion (8.18) may be proper. Indeed, let

$$K_1 = \{(x, y) : x > 0, xy \geqslant 1\} \quad \text{and} \quad K_2 = \{(x, y) : x < 0, xy \geqslant -1\}.$$

Then $\operatorname{nor} K_1 \cap \operatorname{nor} K_2 = \{\boldsymbol{o}\}$ due to

$$\begin{aligned} \operatorname{nor} K_1 &= \{\boldsymbol{o}\} \cup \{(x, y) : x < 0, y < 0\}, \\ \operatorname{nor} K_2 &= \{\boldsymbol{o}\} \cup \{(x, y) : x > 0, y < 0\}. \end{aligned}$$

On the other hand, $K_1 + K_2 = \{(x, y) : y > 0\}$, which gives

$$\operatorname{nor}(K_1 + K_2) = \{(0, y) : y \leqslant 0\}.$$

8.2.5 *Polarity of Barrier, Normal, and Recession Cones*

Theorem 8.35. *For a nonempty convex set $K \subset \mathbb{R}^n$, one has*

$$\operatorname{rint}(\operatorname{rec}(\operatorname{cl} K))^\circ \subset \operatorname{nor} K \subset \operatorname{bar} K \subset (\operatorname{rec}(\operatorname{cl} K))^\circ. \tag{8.19}$$

Consequently,

$$\operatorname{rec}(\operatorname{cl} K) = (\operatorname{nor} K)^\circ = (\operatorname{bar} K)^\circ.$$

Proof. According to Theorems 8.30 and 8.32, $\operatorname{nor} K = \operatorname{nor}(\operatorname{cl} K)$ and $\operatorname{bar} K = \operatorname{bar}(\operatorname{cl} K)$. So, we may assume that K is closed. Then (8.19) becomes as

$$\operatorname{rint}(\operatorname{rec} K)^\circ \subset \operatorname{nor} K \subset \operatorname{bar} K \subset (\operatorname{rec} K)^\circ.$$

With $S = (\operatorname{lin} K)^\perp$, we have $K = (K \cap S) + \operatorname{lin} K$ (see Theorem 6.25). Furthermore, $S = (\operatorname{lin} K)^\circ$ because $\operatorname{lin} K$ is a subspace (see an example on page 269). Translating K by a suitable vector, we assume that $\boldsymbol{o} \in \operatorname{rint} K$.

First, we are going to prove that $\operatorname{rint}(\operatorname{rec} K)^\circ \subset \operatorname{nor} K$. Since this inclusion is obvious when $(\operatorname{rec} K)^\circ = \{\boldsymbol{o}\}$, we may suppose that $(\operatorname{rec} K)^\circ \neq \{\boldsymbol{o}\}$. Choose a nonzero vector $\boldsymbol{e} \in \operatorname{rint}(\operatorname{rec} K)^\circ$. Our goal is to find a point $\boldsymbol{z} \in K$ such that $p_K(\boldsymbol{e} + \boldsymbol{z}) = \boldsymbol{z}$. This argument will give $\boldsymbol{e} + \boldsymbol{z} \in Q_{\boldsymbol{z}}(K)$, or, equivalently, $\boldsymbol{e} \in Q_{\boldsymbol{z}}(K) - \boldsymbol{z} = N_{\boldsymbol{z}}(K) \subset \operatorname{nor} K$.

For the latter, we assert the existence of a closed halfspace

$$V_\gamma = \{\boldsymbol{x} \in \mathbb{R}^n : \boldsymbol{x} \cdot \boldsymbol{e} \leqslant \gamma\}, \quad \gamma \in \mathbb{R},$$

which contains $K \cap S$. Indeed, assume for a moment that no such a halfspace contains $K \cap S$. Then, for every integer $i \geqslant 1$, there is a point $\boldsymbol{x}_i \in K \cap S$ such that $\boldsymbol{x}_i \cdot \boldsymbol{e} > i$. The inequalities $\|\boldsymbol{x}_i\| \|\boldsymbol{e}\| \geqslant \boldsymbol{x}_i \cdot \boldsymbol{e} > i$ show that the sequence $\boldsymbol{x}_1, \boldsymbol{x}_2, \ldots$ is unbounded. By Theorem 6.13, there is a subsequence $\boldsymbol{x}'_1, \boldsymbol{x}'_2, \ldots$ of $\boldsymbol{x}_1, \boldsymbol{x}_2, \ldots$ with the following property: the unit vectors $\boldsymbol{c}_i = \boldsymbol{x}'_i / \|\boldsymbol{x}'_i\|$ tend to a unit vector $\boldsymbol{c} \in K \cap S$ such that the closed halfline $h = [\boldsymbol{o}, \boldsymbol{c}\rangle$ lies in $K \cap S$. A combination of Theorems 6.10 and 6.2 gives the inclusion

$$\boldsymbol{c} \in h \subset \operatorname{rec}(K \cap S) = \operatorname{rec} K \cap S.$$

If $M = \operatorname{span}(\operatorname{rec} K)$, then the subspace $L = M \cap S$ is the orthogonal complement of $\operatorname{lin} K$ within M. Clearly, $\operatorname{rec} K \cap S = \operatorname{rec} K \cap L$, and Theorem 8.6 shows that $\boldsymbol{c}\boldsymbol{e} < 0$. On the other hand, since the vectors $\boldsymbol{c}_1, \boldsymbol{c}_2, \ldots$ belong to $\mathbb{R}^n \setminus V_0$, the limit point $\boldsymbol{c}$ belongs to the closed halfspace $\{\boldsymbol{x} \in \mathbb{R}^n : \boldsymbol{x} \cdot \boldsymbol{e} \geqslant 0\}$, contrary to $\boldsymbol{c} \cdot \boldsymbol{e} < 0$. The obtained contradiction shows that $K \cap S \subset V_\gamma$ for a suitable scalar γ.

Next, we assert that the intersection of $K \cap S$ with every hyperplane

$$H_\gamma = \{\boldsymbol{x} \in \mathbb{R}^n : \boldsymbol{x} \cdot \boldsymbol{e} = \gamma\}, \quad \gamma \in \mathbb{R},$$

is compact. Indeed, this is obvious if $K \cap S$ is bounded. Suppose that $K \cap S$ is unbounded. By Theorem 8.6, $\boldsymbol{c} \cdot \boldsymbol{e} < 0$ for every nonzero point $\boldsymbol{c} \in \operatorname{rec} K \cap S$, implying that $H_0 \cap \operatorname{rec}(K \cap S) = \{\boldsymbol{o}\}$. Theorem 6.40 shows that the intersection of $K \cap S$ with every H_γ is compact.

Choose a pair of parallel hyperplanes H_μ and H_γ, $\mu < \gamma$, both meeting $K \cap S$, and denote by F the intersection of $K \cap S$ with the closed slab between H_μ and H_γ. We assert that F is compact. Indeed, assume for a moment that F is unbounded. By Theorem 6.12, F contains a closed halfline h, and Problem 2.12 shows that h lies in a hyperplane H_β, $\mu \leqslant \beta \leqslant \gamma$. Hence the set $H_\beta \cap (K \cap S)$ is unbounded, contrary to the above argument. Therefore, F is compact.

Since the linear functional $\varphi(\boldsymbol{x}) = \boldsymbol{x} \cdot \boldsymbol{e}$ is continuous (see Problem 1.18), it attains on F a maximum value α at a point $\boldsymbol{z} \in F$. Clearly, the halfspace V_α contains $K \cap S$ such that $\boldsymbol{z} \in H_\alpha$. We also observe that $K \subset V_\alpha$. Indeed, let $\boldsymbol{x}$ be a point in K. Since $K = (K \cap S) + \operatorname{lin} K$, we can write $\boldsymbol{x} = \boldsymbol{y} + \boldsymbol{c}$, where $\boldsymbol{y} \in K \cap S$ and $\boldsymbol{c} \in \operatorname{lin} K$. Because $\boldsymbol{ce} = 0$ (S and $\operatorname{lin} K$ are orthogonal subspaces), we have $\boldsymbol{x} \cdot \boldsymbol{e} = \boldsymbol{y} \cdot \boldsymbol{e} \leqslant \alpha$. Hence $\boldsymbol{x} \in V_\alpha$, and $K \subset V_\alpha$.

Since $\boldsymbol{z}$ is the nearest to $\boldsymbol{z} + \boldsymbol{e}$ point in V_α ($\boldsymbol{e}$ is a normal vector of V_α), it is the nearest to $\boldsymbol{z} + \boldsymbol{e}$ point in K. Therefore, $p_K(\boldsymbol{e} + \boldsymbol{z}) = \boldsymbol{z}$, as desired.

The inclusion $\operatorname{nor} K \subset \operatorname{bar} K$ is proved in Theorem 8.32. So, it suffices to show that $\operatorname{bar} K \subset (\operatorname{rec} K)^\circ$. Equivalently, that $\boldsymbol{ce} \leqslant 0$ whenever $\boldsymbol{c} \in \operatorname{bar} K$ and $\boldsymbol{e} \in \operatorname{rec} K$. Choose a scalar γ such that $\boldsymbol{x} \cdot \boldsymbol{c} \leqslant \gamma$ for all $\boldsymbol{x} \in K$. Let $\boldsymbol{u} \in K$. Then $\lambda \boldsymbol{e} + \boldsymbol{u} \in K$ for all $\lambda \geqslant 0$. Therefore, $(\lambda \boldsymbol{e} + \boldsymbol{u}) \cdot \boldsymbol{c} \leqslant \gamma$ for all $\lambda \geqslant 0$, which is possible only if $\boldsymbol{c} \cdot \boldsymbol{e} \leqslant 0$. □

Since the set $(\operatorname{rec}(\operatorname{cl} K))^\circ$ is closed (see Theorem 8.2), we obtain from Theorem 8.35 and Corollary 3.49 the following assertion.

Corollary 8.36. *For a nonempty convex set $K \subset \mathbb{R}^n$, one has*

$$\operatorname{cl}(\operatorname{bar} K) = \operatorname{cl}(\operatorname{nor} K) = (\operatorname{rec}(\operatorname{cl} K))^\circ,$$
$$\operatorname{rint}(\operatorname{bar} K) = \operatorname{rint}(\operatorname{nor} K) = \operatorname{rint}(\operatorname{rec}(\operatorname{cl} K))^\circ.$$

Consecutively, both sets $\operatorname{cl}(\operatorname{nor} K)$ *and* $\operatorname{rint}(\operatorname{nor} K)$ *are convex cones.* □

Theorem 8.37. *If C is a nonempty convex cone with apex $\boldsymbol{a}$, then*

$$\operatorname{bar} C = \operatorname{nor} C = (C - \boldsymbol{a})^\circ.$$

Consequently, both cones $\operatorname{bar} C$ *and* $\operatorname{nor} C$ *are closed. In particular,*

$$\operatorname{bar} L = \operatorname{nor} L = \operatorname{ort} L$$

for any nonempty plane L in $\mathbb{R}^n$, and $\operatorname{bar} F = \operatorname{nor} F$ *for any halfplane F of L.*

Proof. Since $\mathrm{cl}\, C$ is a closed convex cone with proper apex $\boldsymbol{a}$, a combination of Theorems 6.3 and 8.2 gives

$$\mathrm{rec}\,(\mathrm{cl}\, C) = \mathrm{cl}\, C - \boldsymbol{a} \quad \text{and} \quad (\mathrm{cl}\, C - \boldsymbol{a})^\circ = (C - \boldsymbol{a})^\circ.$$

Consequently, Corollary 8.36 implies that

$$\mathrm{cl}\,(\mathrm{bar}\, C) = \mathrm{cl}\,(\mathrm{nor}\, C) = (\mathrm{rec}\,(\mathrm{cl}\, C))^\circ = (\mathrm{cl}\, C - \boldsymbol{a})^\circ = (C - \boldsymbol{a})^\circ.$$

Hence it remains to show that $\mathrm{bar}\, C = \mathrm{nor}\, C = (\mathrm{cl}\, C - \boldsymbol{a})^\circ$.

Due to the inclusion $\mathrm{nor}\, C \subset \mathrm{bar}\, C$ (see Theorem 8.32), it suffices to prove that $\mathrm{nor}\, C$ contains $(\mathrm{cl}\, C - \boldsymbol{a})^\circ$. Indeed, let $\boldsymbol{e} \in (\mathrm{cl}\, C - \boldsymbol{a})^\circ$. Since the case $\boldsymbol{e} = \boldsymbol{o}$ is obvious, we assume that $\boldsymbol{e} \neq \boldsymbol{o}$. Then $\boldsymbol{x} \cdot \boldsymbol{e} \leqslant 0$ for all $\boldsymbol{x} \in \mathrm{cl}\, C - \boldsymbol{a}$, which means that $\mathrm{cl}\, C - \boldsymbol{a}$ lies in the closed halfspace $V = \{\boldsymbol{x} \in \mathbb{R}^n : \boldsymbol{x} \cdot \boldsymbol{e} \leqslant 0\}$. Because $\boldsymbol{o}$ is a nearest to $\boldsymbol{e}$ point in V and $\boldsymbol{o} \in \mathrm{cl}\, C - \boldsymbol{a}$, we see that $\boldsymbol{o}$ is a nearest to $\boldsymbol{e}$ point in $\mathrm{cl}\, C - \boldsymbol{a}$. Summing up, $\boldsymbol{e} \in \mathrm{nor}\,(\mathrm{cl}\, C - \boldsymbol{a})$.

The second assertion follows from the facts that any point $\boldsymbol{a} \in L$ is an apex of L and $(L - \boldsymbol{a})^\circ = (\mathrm{dir}\, L)^\circ = \mathrm{ort}\, L$. □

Problems and Notes for Chapter 8

Problems for Chapter 8

Problem 8.1. Let $C \subset \mathbb{R}^n$ be a nonempty convex cone with apex $\boldsymbol{o}$. Prove that $\mathrm{cl}\, C$ is a closed halfplane if and only if C° is a closed halfplane. In particular, $\mathrm{cl}\, C$ is a closed halfspace if and only if C° is a closed halfline.

Problem 8.2. Let $C \subset \mathbb{R}^n$, $n \geqslant 2$, be a nonempty convex cone with apex $\boldsymbol{o}$, which is not a plane. Prove that $\mathrm{cl}\, C \cap (-\mathrm{rint}\, C) = \varnothing$. Furthermore, show that $\mathrm{cl}\, C \cup (-\mathrm{rint}\, C) = \mathrm{span}\, C$ if and only if $\mathrm{cl}\, C$ is a closed halfplane of $\mathrm{span}\, C$ such that $\boldsymbol{o} \in \mathrm{rbd}\, C$.

Problem 8.3. Let $C \subset \mathbb{R}^n$, $n \geqslant 2$, be a nonempty convex cone with apex $\boldsymbol{o}$ such that $\{\boldsymbol{o}\} \neq C \neq \mathbb{R}^n$. Prove that the set $U = \mathbb{R}^n \setminus (\mathrm{cl}\, C \cup C^\circ)$ is a nonempty open cone with improper apex $\boldsymbol{o}$.

Problem 8.4. Let $C \subset \mathbb{R}^n$ be a nonempty convex cone with apex $\boldsymbol{o}$. Prove that

$$\mathrm{rint}\, C + \mathrm{rint}\, C^\circ = C + C^\circ = \mathbb{R}^n.$$

Problem 8.5. Given a closed convex cone $C \subset \mathbb{R}^n$ of positive dimension, prove the equivalence of the following conditions.

(*a*) C is a simplicial cone.

(*b*) C has a convex base which is a simplex.

(c) Every convex base of C is a simplex.

Problem 8.6. Let $C \subset \mathbb{R}^n$ be convex cone with apex $\boldsymbol{a}$ of dimension m, $1 \leqslant m \leqslant n$. Prove that C is line-free if and only if there is an $(m-1)$-simplicial cone with apex $\boldsymbol{a}$ containing C.

Problem 8.7. (Kirsch [165]) Prove that a nonempty proper convex set $K \subset \mathbb{R}^n$ is line-free if and only if is satisfies the following condition: for any point $\boldsymbol{a} \in \mathbb{R}^n \setminus \operatorname{cl} K$, there is a simplicial cone $\Gamma_{\boldsymbol{a}} \subset \mathbb{R}^n$ with apex $\boldsymbol{a}$ such that $K \subset \Gamma_{\boldsymbol{a}}$.

Problem 8.8. (Moreau [217]) Let $C \subset \mathbb{R}^n$ be a nonempty convex cone with apex $\boldsymbol{o}$, and let $\boldsymbol{x} \in \mathbb{R}^n$. Vectors $\boldsymbol{y} \in \operatorname{cl} C$ and $\boldsymbol{z} \in C^\circ$ are the metric projections of $\boldsymbol{x}$ on C and C°, respectively, if and only if $\boldsymbol{y}$ and $\boldsymbol{z}$ are orthogonal and satisfy the equality $\boldsymbol{x} = \boldsymbol{y} + \boldsymbol{z}$.

Problem 8.9. (Soltan [263]) Let E and F be closed convex sets in $\mathbb{R}^n$. Prove the equivalence of the following conditions.

(a) Every vector $\boldsymbol{u} \in \mathbb{R}^n$ is expressible as $\boldsymbol{u} = p_E(\boldsymbol{u}) + p_F(\boldsymbol{u})$.

(b) E is a closed convex cone with apex $\boldsymbol{o}$ and $F = E^\circ$.

Problem 8.10. Let $L \subset \mathbb{R}^n$ be a plane of positive dimension, D be a closed halfplane of L, and $\boldsymbol{a}$ be a point in $\operatorname{rbd} D$. Prove that D is expressible as $D = \{\boldsymbol{x} \in L : \boldsymbol{x} \cdot \boldsymbol{e} \leqslant \gamma\}$ if and only if $\boldsymbol{e}$ is a vector in $\operatorname{rint}(D - \boldsymbol{a})^\circ$ and $\gamma = \boldsymbol{a} \cdot \boldsymbol{e}$.

Problem 8.11. Let $L \subset \mathbb{R}^n$ be a plane of positive dimension, and D be a closed halfplane of L. Prove that for any point $\boldsymbol{a} \in \mathbb{R}^n \setminus D$, there is a closed halfspace V of $\mathbb{R}^n$ such that $L \cap V = D$ and $\boldsymbol{a} \notin V$.

Problem 8.12. Let $K \subset \mathbb{R}^n$ be a convex set which is not a plane, $\boldsymbol{z}$ be a point in $\operatorname{rint} K$, and $\boldsymbol{u}$ be a nearest to $\boldsymbol{z}$ point in $\operatorname{rbd} K$. Prove that K lies in the closed halfspace $V = \{\boldsymbol{x} \in \mathbb{R}^n : (\boldsymbol{x} - \boldsymbol{u}) \cdot (\boldsymbol{u} - \boldsymbol{z}) \leqslant 0\}$.

Problem 8.13. Given a convex set $K \subset \mathbb{R}^n$, prove the equivalence of the following conditions:

(a) $\dim(\operatorname{bar} K) = 1$, (b) $\dim(\operatorname{nor} K) = 1$,

(c) $\operatorname{cl} K$ is a hyperplane, a closed slab, or a closed halfspace.

Problem 8.14. Let $X \subset \mathbb{R}^n$ be a nonempty set, $\boldsymbol{c}$ be a point in X, and let $P_X(\boldsymbol{c})$ denote the set of all points $\boldsymbol{u} \in \mathbb{R}^n$ such that $\boldsymbol{c}$ is a nearest to $\boldsymbol{u}$ point in X (compare with Definition 8.23). Prove that the set $P_X(\boldsymbol{c})$ is closed and convex. Similarly, the set $Q_X(\boldsymbol{c})$ of all points $\boldsymbol{u} \in \mathbb{R}^n$ such that $\boldsymbol{c}$ is the only nearest to $\boldsymbol{u}$ point in X is convex (but not necessarily closed).

Problem 8.15. (Bunt [63, Theorem 50]) Let $X \subset \mathbb{R}^n$ be a nonempty closed set so that any point $\boldsymbol{a} \in \mathbb{R}^n$ has a unique nearest point in X. Prove that X is convex.

Problem 8.16. Let $K \subset \mathbb{R}^n$ be a line-free convex set and $H \subset \mathbb{R}^n$ be a hyperplane satisfying the condition $\operatorname{dir} H \cap \operatorname{rec}(\operatorname{cl} K) = \{o\}$. Prove the existence of a translate of H which is disjoint from $\operatorname{cl} K$.

Notes for Chapter 8

Cone polarity. The concept of cone polarity can be traced in the works of Minkowski [215, pp. 131–229] and Steinitz [266, § 19]. Theorem 8.3 (for the case $r = 2$) can be traced in Fenchel [105, p. 11].

Assertions (a), (c), and (e) of Theorem 8.4 can be found, respectively, in Gerstenhaber [118], Fenchel [105, p. 12] and Dragomirescu [96]. Theorem 8.5 is due to Soltan [260, Theorem 5.48]. Theorem 8.6 is proved in an equivalent form by Fenchel [105, Theorem 12]. Theorem 8.11 is given in the paper of Soltan [261] (see Stoker [270] for a weaker result regarding closed convex cones in $\mathbb{R}^3$), and Corollary 8.12 is due to Gaddum [114].

Given a closed convex cone $C \subset \mathbb{R}^n$ with apex $\boldsymbol{o}$, some properties of the difference $C - C^\circ$ are discussed by Levinson and Sherman [199].

Cones with base. Convex cones with base are used in the theory of ordered linear spaces (see, e. g. the books of Aliprantis and Tourky [2], Luc [205], and Peressini [229]).

Nearest points. Theorem 8.22 describes a well known property of convex sets (see, for instance, Carathéodory [65, p. 100]). The first assertion of Problem 8.14 is proved by Motzkin [219] for the case $n = 2$, by Pauc [227] for $n \geqslant 2$, and generalized by Phelps [230] for the case of any inner product space. The assertion of Problem 8.15 is due to Bunt [63, Theorem 50]; it is often attributed to Motzkin [218], who proved it a year later for the case $n = 2$ (see another proof of Jessen [160] for all $n \geqslant 2$).

Several generalizations of Moreau's assertion from Problem 8.8 are obtained by replacing the dot product with suitable functions. For instance, see (a) Han and Mangasarian [143, 144] for inner products of the form $\boldsymbol{x}^T(A + A^T)\boldsymbol{y}$, where A an $n \times n$ matrix; also (b) Ferreira and Németh [107] for some class of convex functions on $\mathbb{R}^n$. Soltan [263] proved the following characteristic properties of polar cones of each other.

(a) Nonempty closed convex sets E and F in $\mathbb{R}^n$ are polar cones of each other if and only if $E + F = \mathbb{R}^n$ and the metric projections $p_E(\boldsymbol{u})$ and $p_F(\boldsymbol{u})$ are orthogonal for every choice of $\boldsymbol{u} \in \mathbb{R}^n$.

(b) Nonempty convex sets E and F in $\mathbb{R}^n$ are polar cones of each other if and only every vector $\boldsymbol{u} \in \mathbb{R}^n$ is uniquely expressible as $\boldsymbol{u} = \boldsymbol{y} + \boldsymbol{z}$, where $\boldsymbol{y}$ and $\boldsymbol{z}$ are orthogonal vectors from E and F, respectively.

Without the requirement of orthogonality in assertion (b) above, one obtains a characterization of complementary planes (see Soltan [263]).

Normal and tangent cones at a point. The fact that every normal set $Q_{\boldsymbol{z}}(K)$ of a convex set $K \subset \mathbb{R}^n$ is a cone is explicitly stated by Klee [169]. The same author showed that a closed set $X \subset \mathbb{R}^n$ is convex provided it satisfies the following condition: for any point $\boldsymbol{a} \in \mathbb{R}^n \setminus X$ and any nearest to $\boldsymbol{a}$ point $\boldsymbol{c} \in X$, the normal set $N_{\boldsymbol{c}}(X)$ contains the closed halfline $[\boldsymbol{c}, \boldsymbol{a}\rangle$. Independently, Phelps [230] proved that a closed set $X \subset \mathbb{R}^n$ is convex if and only if every normal set $N_{\boldsymbol{c}}(X)$, $\boldsymbol{c} \in X$, is a cone (possibly, reduced to $\boldsymbol{c}$).

Clarke [75] (also see Rockafellar [242]) defined an important generalization of the tangent cone to the case of nonconvex sets. Namely, the tangent cone $T_{\boldsymbol{z}}(X)$ of a nonempty closed set $X \subset \mathbb{R}^n$ at a point $\boldsymbol{z} \in X$ consists of all vectors $\boldsymbol{e} \in \mathbb{R}^n$ satisfying the following condition: $\boldsymbol{e} = \lim_{k\to\infty} \boldsymbol{e}_k$ whenever $t_k \searrow 0$ and $\boldsymbol{z}_k \to \boldsymbol{z}$, where $\boldsymbol{z}_k \in X$ and $\boldsymbol{z}_k + t_k\boldsymbol{e}_k \in X$, $k \geqslant 1$. As shown by Clarke [75], the tangent cone $T_{\boldsymbol{z}}(X)$ is always convex, which provides the standard polarity argument in defining the normal cone at $\boldsymbol{z}$ as $N_{\boldsymbol{z}}(X) = (T_{\boldsymbol{z}}(X))^\circ$. See Treiman [280] for characterizations of the cones $T_{\boldsymbol{z}}(X)$ and $N_{\boldsymbol{z}}(X)$ and for a list of related literature.

Normal and barrier cones. The assertion $\operatorname{rec}(\operatorname{cl} K) = (\operatorname{bar} K)^\circ$ from Theorem 8.35 can be traced in Steinitz [267, § 26] for the case of closed convex sets. For barrier cones, see Rockafellar [241, Section 14].

Inner aperture. Larman [191] introduced the concept of *inner aperture* of a convex set $K \subset \mathbb{R}^n$, which was equivalently redefined by Brøndsted [53, 54] as

$$I(K) = \{\boldsymbol{e} \in \mathbb{R}^n : \ \forall \boldsymbol{x} \in \operatorname{aff} K \ \exists \lambda = \lambda(\boldsymbol{x}) \geqslant 0 \text{ such that } \boldsymbol{x} + \lambda\boldsymbol{e} \in K\}.$$

The inner aperture $I(K)$ is a convex cone with apex $\boldsymbol{o}$, contained in the subspace $\operatorname{dir} K$, and $\boldsymbol{o}$ is an improper apex of $I(K)$ provided K is not a plane. In analogy with the relation $\operatorname{rec}(\operatorname{cl} K) = (\operatorname{bar} K)^\circ$, Brøndsted [54] proved that $I(K) = (\operatorname{bar} K)^\Delta$, where the "polarity" operation X^Δ on a set $X \subset \mathbb{R}^n$ is defined by

$$X^\Delta = \{\boldsymbol{e} \in \mathbb{R}^n : \boldsymbol{x} \cdot \boldsymbol{e} < 0 \text{ for all } \boldsymbol{x} \in X \setminus \{\boldsymbol{o}\}\}.$$

Bair [15] showed that $I(K) \cup \{\boldsymbol{o}\} = \operatorname{rec}(\operatorname{cl} K) \setminus U$, where U is the union of all closed halflines with apex $\boldsymbol{o}$ which are translates of the boundary or asymptotic halflines of K.

Chapter 9

Bounds, Supports, and Asymptotes

9.1 Bounds, Supports, and Asymptotic Hyperplanes

9.1.1 *Bounding Hyperplanes*

Definition 9.1. Let $X \subset \mathbb{R}^n$ be a nonempty set and $H \subset \mathbb{R}^n$ be a hyperplane. We say that

(*a*) *H bounds X* if X lies in a closed halfspace determined by H.
(*b*) *H properly bounds X* if H bounds X such that $X \not\subset H$.
(*c*) *H strictly bounds X* if H bounds X such that $H \cap X = \varnothing$.
(*d*) *H strongly bounds X* if there is an open ρ-neighborhood $U_\rho(X)$ of X such that H bounds $U_\rho(X)$.

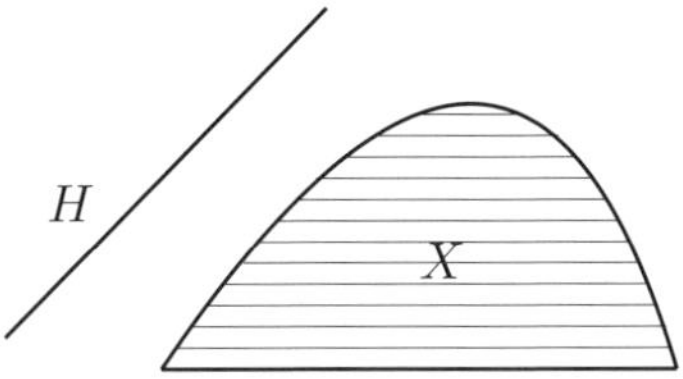

Fig. 9.1 A hyperplane H strongly bounds the set X.

Example. A hyperplane $H = \{\boldsymbol{x} \in \mathbb{R}^n : \boldsymbol{x}\cdot\boldsymbol{e} = \gamma\}$ bounds a closed ball $B_\rho(\boldsymbol{a}) \subset \mathbb{R}^n$ if and only if $\rho\|\boldsymbol{e}\| \leqslant |\gamma - \boldsymbol{a}\cdot\boldsymbol{e}|$ (see Problem 9.1). The same hyperplane strictly (equivalently, strongly) bounds $B_\rho(\boldsymbol{a})$ if and only if $\rho\|\boldsymbol{e}\| < |\gamma - \boldsymbol{a}\cdot\boldsymbol{e}|$.

A combination of Definition 2.29 and Corollary 2.92 implies the technical corollary below.

Corollary 9.2. *For a hyperplane* $H = \{\boldsymbol{x} \in \mathbb{R}^n : \boldsymbol{x}\cdot\boldsymbol{e} = \gamma\}$ *and a nonempty set* $X \subset \mathbb{R}^n$*, the following assertions hold.*

(*a*) *H bounds X if and only if*

$$\textit{either}\ \ \gamma \leqslant \inf\{\boldsymbol{x}\cdot\boldsymbol{e} : \boldsymbol{x} \in X\}\ \ \textit{or}\ \ \sup\{\boldsymbol{x}\cdot\boldsymbol{e} : \boldsymbol{x} \in X\} \leqslant \gamma.$$

(*b*) *H properly bounds X if and only if*

$$\textit{either}\ \ \gamma \leqslant \inf\{\boldsymbol{x}\cdot\boldsymbol{e} : \boldsymbol{x} \in X\} < \sup\{\boldsymbol{x}\cdot\boldsymbol{e} : \boldsymbol{x} \in X\}$$
$$\textit{or}\ \ \inf\{\boldsymbol{x}\cdot\boldsymbol{e} : \boldsymbol{x} \in X\} < \sup\{\boldsymbol{x}\cdot\boldsymbol{e} : \boldsymbol{x} \in X\} \leqslant \gamma.$$

(*c*) *H strictly bound X if and only if*

$$\textit{either}\ \gamma < \boldsymbol{x}\cdot\boldsymbol{e}\ \textit{for all}\ \boldsymbol{x} \in X\ \ \textit{or}\ \ \boldsymbol{x}\cdot\boldsymbol{e} < \gamma\ \textit{for all}\ \boldsymbol{x} \in X.$$

(*d*) *H strongly bounds X if and only if*

$$\textit{either}\ \ \gamma < \inf\{\boldsymbol{x}\cdot\boldsymbol{e} : \boldsymbol{x} \in X\}\ \ \textit{or}\ \ \sup\{\boldsymbol{x}\cdot\boldsymbol{e} : \boldsymbol{x} \in X\} < \gamma.$$ □

Theorem 9.3. *For a nonempty convex set* $K \subset \mathbb{R}^n$ *and a hyperplane* $H \subset \mathbb{R}^n$*, the following assertions hold.*

(*a*) *H properly bounds K if and only if $H \cap \operatorname{rint} K = \varnothing$.*
(*b*) *H strictly bounds K if and only if $H \cap K = \varnothing$.*
(*c*) *H strongly bounds K if and only if there is an open neighborhood $U_\rho(K)$ of K such that $H \cap U_\rho(K) = \varnothing$.*

Proof. (*a*) Suppose that H properly bounds K. Then K lies in a closed halfspace V determined by H such that $K \not\subset H$. Therefore, Theorem 3.67 gives the inclusion $\operatorname{rint} K \subset \operatorname{rint} V = V \setminus H$. Hence $H \cap \operatorname{rint} K = \varnothing$.

Conversely, let $H \cap \operatorname{rint} K = \varnothing$. Choose a point $\boldsymbol{u} \in \operatorname{rint} K$ and denote by W the open halfspace of $\mathbb{R}^n$ determined by H and containing $\boldsymbol{u}$. We assert that $\operatorname{rint} K \subset W$. Indeed, assume for a moment that the complementary to W closed halfspace V' contains a point $\boldsymbol{v} \in \operatorname{rint} K$. By Theorem 2.35, the semi-open segment $(\boldsymbol{u}, \boldsymbol{v}]$ meets H at a single point $\boldsymbol{w}$ (possibly, $\boldsymbol{v} = \boldsymbol{w}$). Theorem 3.24 implies that $\boldsymbol{w} \in \operatorname{rint} K$, in contradiction with the assumption $H \cap \operatorname{rint} K = \varnothing$. Therefore, $\operatorname{rint} K \subset W$. The closure, V, of W is a closed halfspace bounded by H. The inclusions $K \subset \operatorname{cl}(\operatorname{rint} K) \subset \operatorname{cl} W = V$ show that H properly bounds K.

(*b*) If H strictly bounds K, then $H \cap K = \varnothing$. Conversely, let $H \cap K = \varnothing$. Then $H \cap \operatorname{rint} K = \varnothing$, and assertion (*a*) shows that K lies in a closed halfspace determined by H. Thus H strictly bounds K.

(*c*) If H strongly bounds K, then there is an open neighborhood $U_\rho(K)$ of K such that $H \cap U_\rho(K) = \varnothing$. Conversely, suppose that $H \cap U_\rho(K) = \varnothing$ for a suitable open neighborhood $U_\rho(K)$ of K. Then $H \cap K = \varnothing$, and the above argument shows that K lies in a closed halfspace determined by H. Thus H strongly bounds K. □

Theorem 9.4. *If $K \subset \mathbb{R}^n$ is a nonempty convex set and $L \subset \mathbb{R}^n$ is a nonempty plane disjoint from* $\operatorname{rint} K$, *then there is a hyperplane $H \subset \mathbb{R}^n$ which contains L and properly bounds K.*

Proof. Since the cases $\dim L = n-1$ is obvious (see Theorem 9.3), we may assume that $\dim L \leqslant n-2$. Our goal is to enlarge L to a plane $L' \subset \mathbb{R}^n$ such that

$$\dim L' = \dim L + 1 \quad \text{and} \quad L' \cap \operatorname{rint} K = \varnothing.$$

A consecutive repetition of this enlargement procedure will give a desired hyperplane.

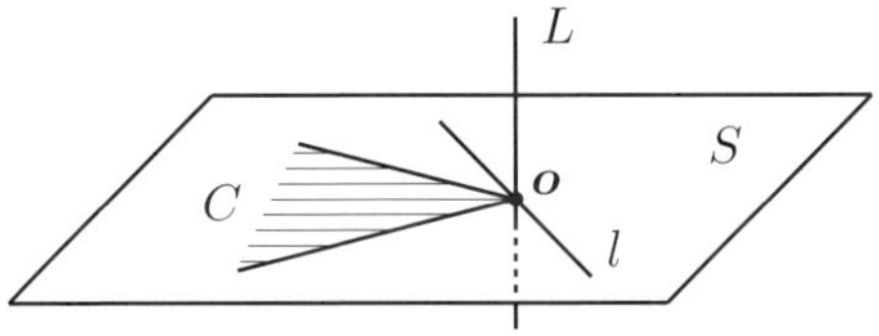

Fig. 9.2 Illustration to Theorem 9.4.

Translating both K and L on a suitable vector, we may suppose that L is a subspace. Denote by S the orthogonal complement of L (put $S = \mathbb{R}^n$ if $L = \{\boldsymbol{o}\}$). Then $\dim S = n - \dim L \geqslant 2$. Let f_S denote the orthogonal projection of $\mathbb{R}^n$ on S. By Theorems 3.15 and 3.43, the set $M = f_S(K)$ is convex and $\operatorname{rint} M = f_S(\operatorname{rint} K)$. Furthermore, $\boldsymbol{o} \notin \operatorname{rint} M$, since otherwise there would be a point

$$\boldsymbol{u} \in f_S^{-1}(\boldsymbol{o}) \cap \operatorname{rint} K = L \cap \operatorname{rint} K,$$

contrary to the assumption $L \cap \operatorname{rint} K = \varnothing$.

Consider the generated convex cone $C = C_{\boldsymbol{o}}(M)$. Theorem 5.48 shows that C is not a plane, and Corollary 5.27 gives the inclusion $\boldsymbol{o} \in \operatorname{rbd} C$. We assert the existence of a one-dimensional subspace $l \subset S$ satisfying the condition $l \cap \operatorname{rint} C = \varnothing$ (as depicted in Figure 9.2). Indeed, if C is a closed halfline, then a one-dimensional subspace $l \subset S$ orthogonal to C has the property $l \cap \operatorname{rint} C = \varnothing$. Suppose that $\dim C \geqslant 2$. By Theorem 3.68, $\dim(\operatorname{rbd} C) \geqslant \dim C - 1 \geqslant 1$. So, we can choose a nonzero point $\boldsymbol{v} \in \operatorname{rbd} C$, and Theorem 5.20 implies that the line $l = \langle \boldsymbol{o}, \boldsymbol{v} \rangle$ does not meet $\operatorname{rint} C$.

The inclusion $\boldsymbol{o} \in \operatorname{rbd} C$ and Theorem 5.49 show that

$$\operatorname{rint} C = C_{\boldsymbol{o}}(\operatorname{rint} M) \setminus \{\boldsymbol{o}\}.$$

Consequently, $\operatorname{rint} M \subset \operatorname{rint} C$, which gives $l \cap \operatorname{rint} M = \varnothing$ by the above argument. Finally, let $L' = L + l$. Clearly, L' is a subspace of dimension $\dim L + 1$ which contains L. We assert that $L' \cap \operatorname{rint} K = \varnothing$. Indeed, if $L' \cap \operatorname{rint} K \neq \varnothing$, then we would have

$$\varnothing \neq f_s(L' \cap \operatorname{rint} K) \subset f_s(L') \cap f_s(\operatorname{rint} K) = l \cap \operatorname{rint} M,$$

contrary to the above argument. □

Remark. If the plane L in Theorem 9.4 is disjoint from K, then the hyperplane H still can meet K. Indeed, let

$$K = \{(x, y) : y > 0\} \cup \{(x, 0) : x > 0\} \quad \text{and} \quad L = \{\boldsymbol{o}\}.$$

Then $L \cap K = \varnothing$, while every line through L meets K. Another example:

$$K = \{(x, y, z) : x \geqslant 0,\ y \geqslant 0,\ z \geqslant 0,\ xy \geqslant z^2\}, \quad L = \{(0, y, 1) : y \in \mathbb{R}\}.$$

It is easy to see that K is a closed convex cone with apex $\boldsymbol{o}$, and L is a line disjoint from K. Moreover, every line $L_a = \{(0, y, a) : y \in \mathbb{R}\}$, $a > 0$, is asymptotic to the planar convex set $K_a = \{(x, y, a) \in K\}$. These arguments easily imply that every 2-dimensional plane through L meets K.

Theorem 9.5. *Let $K \subset \mathbb{R}^n$ be a nonempty convex set whose relative boundary does not contain a closed halfline. If a plane $L \subset \mathbb{R}^n$ is disjoint from $\operatorname{cl} K$, then there is a hyperplane which contains L and strictly bounds $\operatorname{cl} K$.*

Proof. Since $L \cap \operatorname{rint} K \subset L \cap \operatorname{cl} K = \varnothing$, Theorem 9.4 shows the existence of a hyperplane $H \subset \mathbb{R}^n$ containing L and properly bounding $\operatorname{cl} K$.

If $H \cap \operatorname{cl} K = \varnothing$, then H is a desired hyperplane. Assume that H meets $\operatorname{cl} K$. Then K is not a plane (otherwise $K \subset H$ according to Theorem 2.34). Consequently, $\operatorname{rbd} K \neq \varnothing$ (see Theorem 3.62), implying that $H \cap \operatorname{cl} K \subset \operatorname{rbd} K$ according to Theorem 9.3. By the assumption, the set $H \cap \operatorname{rbd} K$ contains no closed halfline. Therefore, Theorem 6.12 implies that $H \cap \operatorname{cl} K$ is a nonempty compact set. According to Theorem 6.40, $\operatorname{dir} H \cap \operatorname{rec}(\operatorname{cl} K) = \{\boldsymbol{o}\}$.

By Theorem 2.17, we can write $H = \{\boldsymbol{x} \in \mathbb{R}^n : \boldsymbol{x} \cdot \boldsymbol{e} = \gamma\}$. Denote by V the closed halfspace of $\mathbb{R}^n$ determined by H and containing K. Without loss of generality, we may assume that $V = \{\boldsymbol{x} \in \mathbb{R}^n : \boldsymbol{x} \cdot \boldsymbol{e} \leqslant \gamma\}$ (see Definition 2.29). Choose a scalar $\gamma' < \gamma$ and consider the hyperplane $H' = \{\boldsymbol{x} \in \mathbb{R}^n : \boldsymbol{x} \cdot \boldsymbol{e} = \gamma'\}$. If P is the closed slab between H and H', then the set $\operatorname{cl} K \cap P$ is compact (see Problem 6.11). Since $\operatorname{cl} K \cap P$ is disjoint from L, there is a scalar $\rho > 0$ such that the open ρ-neighborhood $U_\rho(\operatorname{cl} K \cap P)$ does not meet L (see Problems 1.14 and 1.15).

If $\operatorname{cl} K \subset P$, then $U_\rho(\operatorname{cl} K) = U_\rho(\operatorname{cl} K \cap P)$, and Theorem 9.4 implies the existence of a hyperplane G which contains L and properly bounds $U_\rho(\operatorname{cl} K)$. Consequently,

$$G \cap \operatorname{cl} K \subset G \cap U_\rho(\operatorname{cl} K) = G \cap \operatorname{rint} U_\rho(\operatorname{cl} K) = \varnothing.$$

Suppose now that $\operatorname{cl} K \not\subset P$. Then $\operatorname{cl} K$ meets the closed halfspace $V' = \{\boldsymbol{x} \in \mathbb{R}^n : \boldsymbol{x} \cdot \boldsymbol{e} \leqslant \gamma'\}$. Furthermore,

$$\delta(H, \operatorname{cl} K \cap V') \geqslant \delta(H, H') = |\gamma - \gamma'| / \|\boldsymbol{e}\|,$$

as follows from Corollary 2.92. Choose a positive scalar r satisfying the condition $r < |\gamma - \gamma'| / \|\boldsymbol{e}\|$. Then

$$L \cap U_r(\operatorname{cl} K \cap V') \subset H \cap U_r(\operatorname{cl} K \cap V') = \varnothing.$$

This argument and the equalities

$$\begin{aligned} U_r(\operatorname{cl} K) &= U_r((\operatorname{cl} K \cap P) \cup (\operatorname{cl} K \cap V')) \\ &= U_r(\operatorname{cl} K \cap P) \cup U_r(\operatorname{cl} K \cap V') \end{aligned}$$

show that $U_r(\operatorname{cl} K) \cap L = \varnothing$. As above, there is a hyperplane G which contains L and misses $\operatorname{cl} K$. □

For the next result, we recall that nonempty sets X and Y in $\mathbb{R}^n$ are called strongly disjoint provided $\delta(X, Y) > 0$ (see page 11).

Theorem 9.6. *If a convex set $K \subset \mathbb{R}^n$ and a plane $L \subset \mathbb{R}^n$ are strongly disjoint, then there is a hyperplane $H \subset \mathbb{R}^n$ containing L and strongly bounding K.*

Proof. Clearly, both K and L are nonempty proper subsets of $\mathbb{R}^n$. Excluding the obvious case when L a hyperplane, we assume that $\dim L \leqslant n - 2$. Furthermore, translating both K and L on a suitable vector, we may suppose that L is a subspace. Let S be the orthogonal complement of L. Denote by f_S the orthogonal projection on S and put $M = f_S(K)$. Theorem 3.15 shows that M is a convex set. Choose a scalar $\rho > 0$ such that $L \cap U_\rho(K) = \varnothing$. Since $L = \operatorname{null} f_S$, we obtain $\boldsymbol{o} \notin f_S(U_\rho(K))$.

For any open ball $U_\rho(\boldsymbol{x}) \subset \mathbb{R}^n$, one has

$$f_S(U_\rho(\boldsymbol{x})) = U_\rho(f_S(\boldsymbol{x})) \cap S$$

(see Problem 9.12). This argument gives

$$\begin{aligned} f_S(U_\rho(K)) &= f_S(\cup (U_\rho(\boldsymbol{x}) : \boldsymbol{x} \in K)) = \cup (f_S(U_\rho(\boldsymbol{x})) : \boldsymbol{x} \in K) \\ &= \cup (U_\rho(f_S(\boldsymbol{x})) \cap S : f_S(\boldsymbol{x}) \in f_S(K)) \\ &= \cup (U_\rho(\boldsymbol{y}) \cap S : \boldsymbol{y} \in M) = U_\rho(M) \cap S. \end{aligned}$$

Consequently, $\boldsymbol{o} \notin U_\rho(M)$, implying that $\boldsymbol{o} \notin \operatorname{cl} M$. According to Theorem 8.22, $\operatorname{cl} M$ has a unique nearest to $\boldsymbol{o}$ point, say $\boldsymbol{c}$, such that $\operatorname{cl} M$ lies in the closed halfspace

$$V = \{\boldsymbol{x} \in \mathbb{R}^n : (\boldsymbol{x} - \boldsymbol{c}) \cdot (\boldsymbol{o} - \boldsymbol{c}) \leqslant 0\}.$$

The hyperplane $H = \{\boldsymbol{x} \in \mathbb{R}^n : \boldsymbol{x} \cdot \boldsymbol{c} = 0\}$ lies in the open halfspace $\mathbb{R}^n \setminus V$, and

$$\delta(M, H) = \delta(\boldsymbol{c}, H) = \|\boldsymbol{c}\| > 0$$

(see Theorem 2.91). Since $\boldsymbol{c} \in S = L^\perp$, one has $L \subset \{\boldsymbol{c}\}^\perp = H$.

We assert that $\delta(K, H) = \delta(M, H)$. Indeed, any point $\boldsymbol{x} \in K$ can be expressed as the sum of orthogonal vectors: $\boldsymbol{x} = f_S(\boldsymbol{x}) + (\boldsymbol{x} - f_S(\boldsymbol{x}))$, where $f_S(\boldsymbol{x}) \in S$ and $\boldsymbol{x} - f_S(\boldsymbol{x}) \in L$. If $\boldsymbol{z}$ is the nearest to $\boldsymbol{x}$ point in H, then $f_S(\boldsymbol{z}) = f_S(\boldsymbol{x}) - \boldsymbol{x} + \boldsymbol{z}$ is the nearest to $f_S(\boldsymbol{x})$ point in H. Consequently,

$$\begin{aligned} \delta(K, H) &= \inf \{\delta(\boldsymbol{x}, H) : \boldsymbol{x} \in K\} = \inf \{\|\boldsymbol{x} - \boldsymbol{z}\| : \boldsymbol{x} \in K\} \\ &= \inf \{\|f_S(\boldsymbol{x}) - f_S(\boldsymbol{z})\| : f_S(\boldsymbol{x}) \in M\} \\ &= \inf \{\delta(f_S(\boldsymbol{x}), H) : f_S(\boldsymbol{x}) \in M\} = \delta(M, H). \end{aligned}$$

Thus $\delta(K, H) > 0$, and H is a desired hyperplane. □

A combination of Theorems 9.4 and 9.6 implies the corollary below.

Corollary 9.7. *If K is a nonempty convex set in $\mathbb{R}^n$ and $\boldsymbol{a}$ is a point in $\mathbb{R}^n \setminus \operatorname{rint} K$ (respectively, in $\mathbb{R}^n \setminus \operatorname{cl} K$), then there is a hyperplane $H \subset \mathbb{R}^n$ through $\boldsymbol{a}$ properly (respectively, strongly) bounding K.* □

9.1.2 *Support and Asymptotic Hyperplanes*

Definition 9.8. We say that a hyperplane $H \subset \mathbb{R}^n$ *supports* (respectively, *properly supports*) a nonempty set $X \subset \mathbb{R}^n$ if H meets $\operatorname{cl} X$ and bounds (respectively, properly bounds) X.

Example. A hyperplane $H = \{\boldsymbol{x} \in \mathbb{R}^n : \boldsymbol{x}\cdot\boldsymbol{e} = \gamma\}$ supports a closed ball $B_\rho(\boldsymbol{a}) \subset \mathbb{R}^n$ (respectively, an open ball $U_\rho(\boldsymbol{a}) \subset \mathbb{R}^n$) if and only if $\rho\|\boldsymbol{e}\| = |\gamma - \boldsymbol{a}\cdot\boldsymbol{e}|$ (see Problem 9.1).

A combination of Corollaries 2.92 and 9.2 implies the technical corollary below.

Corollary 9.9. *For a hyperplane $H = \{\boldsymbol{x} \in \mathbb{R}^n : \boldsymbol{x}\cdot\boldsymbol{e} = \gamma\}$ and a nonempty set $X \subset \mathbb{R}^n$, the following assertions hold.*

(a) H supports X if and only if

$$\text{either } \gamma = \min\{\boldsymbol{x}\cdot\boldsymbol{e} : \boldsymbol{x} \in \operatorname{cl} X\} \text{ or } \max\{\boldsymbol{x}\cdot\boldsymbol{e} : \boldsymbol{x} \in \operatorname{cl} X\} = \gamma.$$

(b) H properly supports X if and only if

$$\text{either } \gamma = \min\{\boldsymbol{x}\cdot\boldsymbol{e} : \boldsymbol{x} \in \operatorname{cl} X\} < \sup\{\boldsymbol{x}\cdot\boldsymbol{e} : \boldsymbol{x} \in \operatorname{cl} X\}$$
$$\text{or } \inf\{\boldsymbol{x}\cdot\boldsymbol{e} : \boldsymbol{x} \in \operatorname{cl} X\} < \max\{\boldsymbol{x}\cdot\boldsymbol{e} : \boldsymbol{x} \in \operatorname{cl} X\} = \gamma. \qquad \square$$

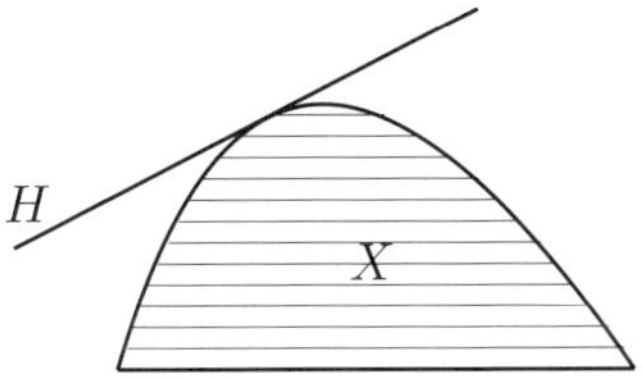

Fig. 9.3 A hyperplane H properly supports the set X.

The next corollary follows from Theorem 9.3.

Corollary 9.10. *A hyperplane $H \subset \mathbb{R}^n$ properly supports a nonempty convex set $K \subset \mathbb{R}^n$ if and only if H meets $\operatorname{cl} K$ such that any of the two equivalent conditions holds:*

$$(a)\ \ H \cap \operatorname{rint} K = \varnothing, \qquad (b)\ \ H \cap \operatorname{cl} K \subset \operatorname{rbd} K. \qquad \square$$

The corollary below is similar to Theorem 9.4.

Corollary 9.11. *If $K \subset \mathbb{R}^n$ is a nonempty convex set and $L \subset \mathbb{R}^n$ is a plane meeting $\operatorname{cl} K$ such that $L \cap \operatorname{rint} K = \varnothing$, then there is a hyperplane which contains L and properly supports K.*

Proof. Theorem 9.4 implies the existence of a hyperplane H containing L and properly bounding K. By Theorem 9.3, $H \cap \operatorname{rint} K = \varnothing$. This argument and Corollary 9.10 imply that H properly supports K. $\square$

Corollary 9.12. *If a convex set $K \subset \mathbb{R}^n$ is not a plane and F is a nonempty convex subset of* $\operatorname{rbd} K$, *then there is a hyperplane $H \subset \mathbb{R}^n$ properly supporting K such that $F \subset H$.*

Proof. Choose a point $\boldsymbol{a} \in \operatorname{rint} F$ (see Corollary 3.21). By Corollary 9.11, there is a hyperplane $H \subset \mathbb{R}^n$ properly supporting K at $\boldsymbol{a}$. Denote by V a closed halfspace determined by H and containing K. Since $\boldsymbol{a} \in H \cap F$ and $F \subset K \subset V$, the hyperplane H supports F. Consequently, Corollary 9.10 implies the inclusion $F \subset H$. □

We recall (see Definition 7.10) that a nonempty plane $L \subset \mathbb{R}^n$ is *asymptotic* to a nonempty set $X \subset \mathbb{R}^n$ (or is an *asymptote* of X) provided

$$L \cap \operatorname{cl} X = \varnothing \quad \text{and} \quad \delta(L, X) = 0,$$

where $\delta(L, X)$ denotes the *inf*-distance between L and X. The above definitions imply the following corollary.

Corollary 9.13. *If a hyperplane $H \subset \mathbb{R}^n$ bounds a nonempty set $X \subset \mathbb{R}^n$, then H either supports X, or strongly bounds X, or is asymptotic to X.* □

Theorem 9.3 implies the next corollary.

Corollary 9.14. *A hyperplane $H \subset \mathbb{R}^n$ is asymptotic to a nonempty convex set $K \subset \mathbb{R}^n$ if and only if $\delta(H, K) = 0$ and* $\operatorname{cl} K$ *lies in an open halfspace determined by H.* □

The next result complements Theorem 9.4.

Theorem 9.15. *If $K \subset \mathbb{R}^n$ is a nonempty convex set and $L \subset \mathbb{R}^n$ is a plane asymptotic to K, then there is a hyperplane $H \subset \mathbb{R}^n$ containing L such that H either properly supports K or is asymptotic to K.*

Proof. The definitions imply that L is disjoint from K. By Theorem 9.4, there is a hyperplane $H \subset \mathbb{R}^n$ which contains L and properly bounds K. If $H \cap \operatorname{cl} K \neq \varnothing$, then H properly supports K. If $H \cap \operatorname{cl} K = \varnothing$, then the inequality $\delta(H, K) \leqslant \delta(L, K) = 0$ implies that H is asymptotic to K. □

The example below show that Theorem 9.15 cannot be sharpened by asserting that the hyperplane H should be asymptotic to K.

Example. Let $K \subset \mathbb{R}^3$ be the sum of the closed unit ball $B_1(\boldsymbol{a})$ centered at $\boldsymbol{a} = (0, 0, 1)$ and the planar convex set $N = \{(x, y, 0) : x > 0,\ xy \geqslant 1\}$. It is easy to see that the x- and y-axes of $\mathbb{R}^3$ are asymptotic lines of K, while the xy-plane (which supports K) is the only 2-dimensional plane containing any of these lines and bounding K.

9.2 Containing and Supporting Halfspaces

9.2.1 *Containing Halfspaces*

Definition 9.16. Let $X \subset \mathbb{R}^n$ be a nonempty set and $V \subset \mathbb{R}^n$ be a closed halfspace containing X. If H is the boundary hyperplane of V, then we say that

(*a*) V *properly contains* X if V contains X such that $X \not\subset H$.
(*b*) V *strictly contains* X if V contains X such that $H \cap X = \varnothing$.
(*c*) V *strongly contains* K if there is an ρ-neighborhood $U_\rho(X)$ of X such that $U_\rho(X) \subset V$.

Example. A closed halfspace $V = \{\boldsymbol{x} \in \mathbb{R}^n : \boldsymbol{x}\cdot\boldsymbol{e} \leqslant \gamma\}$ contains a closed ball $B_\rho(\boldsymbol{a}) \subset \mathbb{R}^n$ (equivalently, strictly contains an open ball $U_\rho(\boldsymbol{a}) \subset \mathbb{R}^n$) if and only if $\rho\|\boldsymbol{e}\| \leqslant |\gamma - \boldsymbol{a}\cdot\boldsymbol{e}|$ (see Problem 9.1). The same halfspace strictly (equivalently, strongly) contains $B_\rho(\boldsymbol{a})$ if and only if $\rho\|\boldsymbol{e}\| < |\gamma - \boldsymbol{a}\cdot\boldsymbol{e}|$.

A combination of Definition 2.29 and Corollary 2.92 implies the technical corollary below.

Corollary 9.17. *For a closed halfspace* $V = \{\boldsymbol{x} \in \mathbb{R}^n : \boldsymbol{x}\cdot\boldsymbol{e} \leqslant \gamma\}$ *and a nonempty set* $X \subset \mathbb{R}^n$, *the following assertions hold.*

(*a*) *V contains X if and only if* $\sup\{\boldsymbol{x}\cdot\boldsymbol{e} : \boldsymbol{x} \in X\} \leqslant \gamma$.
(*b*) *V properly contains X if and only if*

$$\inf\{\boldsymbol{x}\cdot\boldsymbol{e} : \boldsymbol{x} \in X\} < \sup\{\boldsymbol{x}\cdot\boldsymbol{e} : \boldsymbol{x} \in X\} \leqslant \gamma.$$

(*c*) *V strictly contains X if and only if $\boldsymbol{x}\cdot\boldsymbol{e} < \gamma$ for all $\boldsymbol{x} \in X$.*
(*d*) *H strongly contains X if and only if* $\sup\{\boldsymbol{x}\cdot\boldsymbol{e} : \boldsymbol{x} \in X\} < \gamma$. □

Theorem 9.3 implies the following corollary.

Corollary 9.18. *Let a closed halfspace $V \subset \mathbb{R}^n$ with boundary hyperplane H contain a nonempty convex set $K \subset \mathbb{R}^n$. Then the following assertions hold.*

(*a*) *V properly contains K if and only if* $\operatorname{rint} K \subset \operatorname{int} V$.
(*b*) *V strictly contains K if and only if $K \subset \operatorname{int} V$.*
(*c*) *V strongly contains K if and only if there is an open neighborhood $U_\rho(K)$ of K such that $U_\rho(K) \subset \operatorname{int} V$.* □

Corollary 9.7 implies the following result.

Corollary 9.19. *If K is a nonempty convex set in $\mathbb{R}^n$ and $\boldsymbol{a}$ is a point in $\mathbb{R}^n \setminus \operatorname{rint} K$ (respectively, in $\mathbb{R}^n \setminus \operatorname{cl} K$), then there is a closed halfspace $V \subset \mathbb{R}^n$ properly containing (respectively, strongly containing) K such that $\boldsymbol{a} \in \operatorname{bd} V$.* □

Theorem 8.22 shows that for a nonempty convex set $K \subset \mathbb{R}^n$ and a point $\boldsymbol{a} \in \mathbb{R}^n \setminus \operatorname{cl} K$, there is a closed halfspace of $\mathbb{R}^n$ containing K and missing $\boldsymbol{a}$. The next result gives a similar assertion for the case of halfspaces strongly containing K.

Theorem 9.20. *Let $K \subset \mathbb{R}^n$ is a nonempty convex set. For any point $\boldsymbol{a} \in \mathbb{R}^n \setminus \operatorname{cl} K$, there is a closed halfspace of $\mathbb{R}^n$ which misses $\boldsymbol{a}$ and strongly contains K.*

Proof. Given a point $\boldsymbol{a} \in \mathbb{R}^n \setminus \operatorname{cl} K$, there is a unique nearest to $\boldsymbol{a}$ point $\boldsymbol{c} \in \operatorname{cl} K$ such that the closed halfspace

$$V = \{\boldsymbol{x} \in \mathbb{R}^n : (\boldsymbol{x} - \boldsymbol{c}) \cdot (\boldsymbol{a} - \boldsymbol{c}) \leqslant 0\}$$

contains K (see Theorem 8.22). We may express V as

$$V = \{\boldsymbol{x} \in \mathbb{R}^n : \boldsymbol{x} \cdot \boldsymbol{e} \leqslant \gamma\}, \quad \text{where} \quad \boldsymbol{e} = \boldsymbol{a} - \boldsymbol{c}, \quad \gamma = \boldsymbol{c} \cdot (\boldsymbol{a} - \boldsymbol{c}).$$

Then $\sup\{\boldsymbol{x} \cdot \boldsymbol{e} : \boldsymbol{x} \in K\} \leqslant \gamma$. Furthermore, $\gamma < \boldsymbol{a} \cdot \boldsymbol{c}$ because $\boldsymbol{a} \notin V$. If a scalar γ' satisfies the inequalities $\gamma < \gamma' < \boldsymbol{a} \cdot \boldsymbol{c}$, then Corollary 9.17 shows that the closed halfspace $V' = \{\boldsymbol{x} \in \mathbb{R}^n : \boldsymbol{x} \cdot \boldsymbol{e} \leqslant \gamma'\}$ misses $\boldsymbol{a}$ and strongly contains K. □

Theorem 9.21. *Let $K \subset \mathbb{R}^n$ be a nonempty convex set and $V \subset \mathbb{R}^n$ be a closed halfspace. There is a translate of V which contains K if and only if*

$$\operatorname{nor} V \subset \operatorname{bar} K. \tag{9.1}$$

Proof. Let $V = \{\boldsymbol{x} \in \mathbb{R}^n : \boldsymbol{x} \cdot \boldsymbol{e} \leqslant \gamma\}$, where $\boldsymbol{e}$ is a suitable nonzero vector of $\mathbb{R}^n$ and γ is a scalar (see Definition 2.29). Suppose that a translate V' of V contains K. Corollary 2.30 shows that V' can be expressed as $V' = \{\boldsymbol{x} \in \mathbb{R}^n : \boldsymbol{x} \cdot \boldsymbol{e} \leqslant \gamma'\}$ for a suitable scalar γ'. By Corollary 9.17, the inequality $\sup\{\boldsymbol{x} \cdot \boldsymbol{e} : \boldsymbol{x} \in K\} \leqslant \gamma'$ holds. Consecutively, the normal vector $\boldsymbol{e}$ of V belongs to $\operatorname{bar} K$ (see Definition 8.31). Hence (9.1) is satisfied.

Conversely, suppose that (9.1) is satisfied. We assert the existence of a translate of V which contains K. Indeed, choose a nonzero vector $\boldsymbol{e} \in \operatorname{nor} V$. Then $\boldsymbol{e} \in \operatorname{bar} K$, and Definition 8.31 implies the existence of a scalar γ' such that K lies in the closed halfspace $V' = \{\boldsymbol{x} \in \mathbb{R}^n : \boldsymbol{x} \cdot \boldsymbol{e} \leqslant \gamma'\}$. Obviously, V' is a translate of V. □

Theorem 9.22. *Let $K_1, \ldots, K_r$ be nonempty convex sets in $\mathbb{R}^n$ and $\mu_1, \ldots, \mu_r$ be nonzero scalars. A closed halfspace $V \subset \mathbb{R}^n$ contains the convex set $K = \mu_1 K_1 + \cdots + \mu_r K_r$ if and only if V can be expressed in the form $V = \mu_1 V_1 + \cdots + \mu_r V_r$, where $V_1, \ldots, V_r$ are closed halfspaces, satisfying the conditions:*

(*a*) *$K_i \subset V_i$ for all $1 \leqslant i \leqslant r$,*
(*b*) *$\mu_i V_i$ is a translate of V for all $1 \leqslant i \leqslant r$.*

Proof. If $V = \mu_1 V_1 + \cdots + \mu_r V_r$, where $V_1, \ldots, V_r$ are closed convex halfspaces, satisfying conditions (*a*) and (*b*), then

$$K = \mu_1 K_1 + \cdots + \mu_r K_r \subset \mu_1 V_1 + \cdots + \mu_r V_r = V.$$

Conversely, let V contain K. Then $V = \{\boldsymbol{x} \in \mathbb{R}^n : \boldsymbol{x}\!\cdot\!\boldsymbol{e} \leqslant \gamma\}$, where $\boldsymbol{e}$ is a nonzero vector in bar K. By Theorem 8.34, $\boldsymbol{e} \in \text{bar}\,(\mu_1 K_1) \cap \cdots \cap \text{bar}\,(\mu_r K_r)$ and $\gamma \in \mathbb{R}$. Put $\alpha = \sup\{\boldsymbol{x}\cdot\boldsymbol{e} : \boldsymbol{x} \in K\}$. Clearly, $\alpha \leqslant \gamma$. Let

$$\alpha_i = \begin{cases} \sup\{\boldsymbol{x}\cdot\boldsymbol{e} : \boldsymbol{x} \in K_i\} & \text{if } \mu_i > 0, \\ \inf\{\boldsymbol{x}\cdot\boldsymbol{e} : \boldsymbol{x} \in K_i\} & \text{if } \mu_i < 0. \end{cases}$$

Clearly,

$$\mu_i \alpha_i = \sup\{\boldsymbol{x}\cdot\boldsymbol{e} : \boldsymbol{x} \in \mu_i K_i\}, \quad 1 \leqslant i \leqslant r.$$

Furthermore, $\alpha = \mu_1 \alpha_1 + \cdots + \mu_r \alpha_r$. Put

$$\gamma_i = \alpha_i + \frac{\gamma - \alpha}{r \mu_i}, \quad 1 \leqslant i \leqslant r,$$

$$V_i = \begin{cases} \{\boldsymbol{x} \in \mathbb{R}^n : \boldsymbol{x}\cdot\boldsymbol{e} \leqslant \gamma_i\} & \text{if } \mu_i > 0, \\ \{\boldsymbol{x} \in \mathbb{R}^n : \boldsymbol{x}\cdot\boldsymbol{e} \geqslant \gamma_i\} & \text{if } \mu_i < 0. \end{cases}$$

Obviously, $K_i \subset V_i$ and

$$\mu_i V_i = \{\boldsymbol{x} \in \mathbb{R}^n : \boldsymbol{x}\cdot\boldsymbol{e} \leqslant \mu_i \gamma_i\}, \quad 1 \leqslant i \leqslant r,$$

which shows that all halfspaces $\mu_1 V_1, \ldots, \mu_r V_r$ are translates of V. Because

$$\begin{aligned} \mu_1 \gamma_1 + \cdots + \mu_r \gamma_r &= \mu_1 \big(\alpha_1 + \frac{\gamma - \alpha}{r \mu_1}\big) + \cdots + \mu_r \big(\alpha_r + \frac{\gamma - \alpha}{r \mu_r}\big) \\ &= \mu_1 \alpha_1 + \cdots + \mu_r \alpha_r + (\gamma - \alpha) = \gamma \end{aligned}$$

one has $\mu_1 V_1 + \cdots + \mu_r V_r = V$ (see Theorem 2.32). □

9.2.2 Intersections of Containing Halfspaces

Theorem 9.23. *If $\mathcal{V}$ is the family of all closed halfspaces of $\mathbb{R}^n$ containing (respectively, properly containing, strictly containing, or strongly containing) a nonempty convex set $K \subset \mathbb{R}^n$, then*

$$\operatorname{cl} K = \cap\,(V : V \in \mathcal{V}). \tag{9.2}$$

Proof. Since every closed halfspace $V \subset \mathcal{V}$ is a closed set (see Theorem 2.33), their intersection also is a closed set (see page 10). Consequently, the obvious inclusion $K \subset \cap\,(V : V \in \mathcal{V})$ gives $\operatorname{cl} K \subset \cap\,(V : V \in \mathcal{V})$. Hence it suffices to prove the opposite inclusion

$$\cap\,(V : V \in \mathcal{V}) \subset \operatorname{cl} K. \tag{9.3}$$

For this, choose a point $\boldsymbol{a} \in \mathbb{R}^n \setminus \operatorname{cl} K$. If $\boldsymbol{c}$ is the nearest to $\boldsymbol{a}$ point in $\operatorname{cl} K$, then Theorem 8.22 shows that $\operatorname{cl} K$ lies the closed halfspace

$$V = \{\boldsymbol{x} \in \mathbb{R}^n : (\boldsymbol{x} - \boldsymbol{c})\cdot(\boldsymbol{a} - \boldsymbol{c}) \leqslant 0\}. \tag{9.4}$$

Because $\boldsymbol{a} \notin V$ and $V \in \mathcal{V}$, inclusion (9.3) holds.

The cases of closed halfspaces properly containing, strictly containing, or strongly containing K are similar and use Theorem 9.20. □

Theorem 9.24. *For a nonempty convex set $K \subset \mathbb{R}^n$, the following assertions hold.*

(*a*) *Any family $\mathcal{F}$ of closed halfspaces of $\mathbb{R}^n$ satisfying the condition*

$$\operatorname{cl} K = \cap\,(V : V \in \mathcal{F}) \tag{9.5}$$

has a countable subfamily satisfying the same condition.

(*b*) *If a family $\mathcal{F}$ of closed halfspaces of $\mathbb{R}^n$ properly containing K satisfies condition* (9.5) *and* $\dim K \leqslant n - 1$, *then the family $\mathcal{F}$ is infinite.*

(*c*) *If a family $\mathcal{F}$ of closed halfspaces of $\mathbb{R}^n$ strongly containing K satisfies condition* (9.5), *then the family $\mathcal{F}$ is infinite and*

$$\operatorname{cl} K = \cap\,(\operatorname{int} V : V \in \mathcal{F}). \tag{9.6}$$

Proof. (*a*) De Morgan's Laws (see page 2) and (9.5) give

$$\mathbb{R}^n \setminus \operatorname{cl} K = \mathbb{R}^n \setminus (\cap\,(V : V \in \mathcal{F})) = \cup\,(\mathbb{R}^n \setminus V : V \in \mathcal{F}).$$

Because every set $\mathbb{R}^n \setminus V$ is open, Lindelöf's Theorem (see Problem 1.12) implies the existence of a countable subfamily $\mathcal{G} \subset \mathcal{F}$ with the property

$$\mathbb{R}^n \setminus \operatorname{cl} K = \cup\,(\mathbb{R}^n \setminus V : V \in \mathcal{G}).$$

Again using De Morgan's Laws, we obtain

$$\begin{aligned}\operatorname{cl} K &= \mathbb{R}^n \setminus (\mathbb{R}^n \setminus \operatorname{cl} K) = \mathbb{R}^n \setminus (\cup (\mathbb{R}^n \setminus V : V \in \mathcal{G}))\\ &= \cap (\mathbb{R}^n \setminus (\mathbb{R}^n \setminus V) : V \in \mathcal{G}) = \cap (V : V \in \mathcal{G}).\end{aligned}$$

(*b*) Choose a point $\boldsymbol{u} \in \operatorname{rint} K$ and let $V_1, \dots, V_r$ be any halfspaces from $\mathcal{F}$. Since $\boldsymbol{u} \in \operatorname{int} V_i$, there is a scalar $\rho_i > 0$ such that $U_{\rho_i}(\boldsymbol{u}) \subset \operatorname{int} V_i$, $1 \leqslant i \leqslant r$. Put $\rho = \min\{\rho_1, \dots, \rho_r\}$. Then $\rho > 0$. The assumption $\dim K \leqslant n - 1$ implies that $U_\rho(\boldsymbol{u}) \not\subset \operatorname{aff} K$. Hence there is a point $\boldsymbol{z} \in U_\rho(\boldsymbol{u}) \setminus \operatorname{cl} K$. Because

$$\boldsymbol{z} \in U_\rho(\boldsymbol{u}) \subset U_{\rho_1}(\boldsymbol{u}) \cap \dots \cap U_{\rho_r}(\boldsymbol{u}) \subset \operatorname{int} V_1 \cap \dots \cap \operatorname{int} V_r,$$

one has $\boldsymbol{z} \in V_1 \cap \dots \cap V_r \setminus \operatorname{cl} K$. Thus no finite subfamily of $\mathcal{F}$ satisfies (9.5). Consequently, $\mathcal{F}$ cannot be finite.

(*c*) Let $V_1, \dots, V_r$, $r \geqslant 1$, be any halfspaces from $\mathcal{F}$. Since V_i strongly contains K, there is a scalar $\rho_i > 0$ such that the open neighborhood $U_{\rho_i}(K)$ lies in $\operatorname{int} V_i$, $1 \leqslant i \leqslant r$ (see Problem 1.15). Consequently, the intersection

$$\operatorname{int} V_1 \cap \dots \cap \operatorname{int} V_r = \operatorname{int}(V_1 \cap \dots \cap V_r)$$

is a nonempty open set in $\mathbb{R}^n$. With $\rho = \min\{\rho_1, \dots, \rho_r\}$, we have $\rho > 0$ and

$$\begin{aligned}\operatorname{cl} K &\subsetneq U_\rho(K) \subset \operatorname{int} V_1 \cap \dots \cap \operatorname{int} V_r\\ &= \operatorname{int}(V_1 \cap \dots \cap V_r) \subsetneq V_1 \cap \dots \cap V_r.\end{aligned}$$

This argument and (9.5) show that $\mathcal{F} \neq \{V_1, \dots, V_r\}$. Thus the family $\mathcal{F}$ cannot be finite.

For (9.6), choose a halfspace $V \in \mathcal{F}$ and express it as

$$V = \{\boldsymbol{x} \in \mathbb{R}^n : \boldsymbol{x} \cdot \boldsymbol{e} \leqslant \gamma_{\boldsymbol{e}}\}$$

(see Definition 2.29). Since V strongly contains K, one has

$$\mu_{\boldsymbol{e}} = \sup\{\boldsymbol{x} \cdot \boldsymbol{e} : \boldsymbol{x} \in K\} < \gamma_{\boldsymbol{e}}$$

(see Corollary 9.2). Choose a scalar $\gamma'_{\boldsymbol{e}} \in (\mu_{\boldsymbol{e}}, \gamma_{\boldsymbol{e}})$. Then the closed halfspace

$$V' = \{\boldsymbol{x} \in \mathbb{R}^n : \boldsymbol{x} \cdot \boldsymbol{e} \leqslant \gamma'_{\boldsymbol{e}}\}$$

contains K such that $V' \subset \operatorname{int} V$. Therefore, $\operatorname{cl} K \subset V' \subset \operatorname{int} V$. Keeping this notation for every member of $\mathcal{F}$, we obtain

$$\operatorname{cl} K \subset \cap (V' : V \in \mathcal{F}) \subset \cap (\operatorname{int} V : V \in \mathcal{F}) \subset \cap (V : V \in \mathcal{F}) = \operatorname{cl} K,$$

which gives (9.6). □

Theorem 9.25. *If $\mathcal{V}$ is the family of all closed halfspaces of $\mathbb{R}^n$ each properly containing a nonempty convex set $K \subset \mathbb{R}^n$, then*

$$\operatorname{rint} K = \cap (\operatorname{int} V : V \in \mathcal{V}). \tag{9.7}$$

Proof. If $V \in \mathcal{V}$, then $\operatorname{rint} K \subset \operatorname{int} V$ according to Corollary 9.18. Hence

$$\operatorname{rint} K \subset \cap (\operatorname{int} V : V \in \mathcal{V}).$$

For the opposite inclusion, assume for a moment the existence of a point

$$\boldsymbol{u} \in (\cap (\operatorname{int} V : V \in \mathcal{V})) \setminus \operatorname{rint} K.$$

Since $\boldsymbol{u} \in \cap (V : V \in \mathcal{V}) = \operatorname{cl} K$ according to (9.5), we obtain that $\boldsymbol{u} \in \operatorname{rbd} K$. Now, Corollary 9.19 gives the existence of a closed halfspace $V_0 \subset \mathbb{R}^n$ which properly contains K such that $\boldsymbol{u} \in \operatorname{bd} V_0$. Because $V_0 \in \mathcal{V}$, we obtain a contradiction with the choice of $\boldsymbol{u}$. Summing up,

$$\cap (\operatorname{int} V : V \in \mathcal{V}) \subset \operatorname{rint} K. \qquad \square$$

Remark. Equality (9.7) does not hold for the case of any family $\mathcal{F}$ of closed halfspaces satisfying (9.5). Indeed, the closed halfplane

$$K = \{(x, y, 0) : x \geqslant 0\} \subset \mathbb{R}^3$$

is the intersection of three closed halfspaces

$$V_1 = \{(x, y, z) : x \geqslant 0\},\ V_2 = \{(x, y, z) : z \geqslant 0\},\ V_3 = \{(x, y, z) : z \leqslant 0\},$$

while $\operatorname{int} V_1 \cap \operatorname{int} V_2 \cap \operatorname{int} V_3 = \varnothing$.

Theorem 9.26. *Let $K \subset \mathbb{R}^n$ be a nonempty convex set, X be a dense subset of $\mathbb{S} \cap \operatorname{nor} K$ (where $\mathbb{S}$ denotes the unit sphere of $\mathbb{R}^n$), and Φ be a dense subset of $\mathbb{R}$. With*

$$\mu_{\boldsymbol{e}} = \sup \{\boldsymbol{x} \cdot \boldsymbol{e} : \boldsymbol{x} \in K\}, \quad \textit{where} \quad \boldsymbol{e} \in X,$$

$\operatorname{cl} K$ is expressible as the intersection of a countable family $\mathcal{F}$ of closed halfspaces of the form

$$V_\gamma(\boldsymbol{e}) = \{\boldsymbol{x} \in \mathbb{R}^n : \boldsymbol{x} \cdot \boldsymbol{e} \leqslant \gamma\},\ \boldsymbol{e} \in X,\ \gamma \in \Phi \cap [\mu_{\boldsymbol{e}}, \infty). \tag{9.8}$$

Furthermore, the family $\mathcal{F}$ can be chosen such that each member properly (respectively, strictly, or strongly) contains K.

Proof. Clearly, it suffices to consider the case when every member of $\mathcal{F}$ strongly contains K. Corollary 9.2 shows that for any scalar $\gamma \in (\mu_{\boldsymbol{e}}, \infty)$, a halfspace $V_\gamma(\boldsymbol{e})$ of the form (9.8) strongly contains K.

According to Theorem 9.23, it suffices to show that $\mathcal{F}$ satisfies (9.2). Equivalently, that for any given point $\boldsymbol{u} \in \mathbb{R}^n \setminus \operatorname{cl} K$, there is a halfspace $V_\gamma(\boldsymbol{e}) \in \mathcal{F}$ which misses $\boldsymbol{u}$ and strongly contains K.

Denote by $\boldsymbol{z}$ the nearest to $\boldsymbol{u}$ point in $\operatorname{cl} K$. By Theorem 8.22, the closed halfspace

$$V_1 = \{\boldsymbol{x} \in \mathbb{R}^n : (\boldsymbol{x} - \boldsymbol{z})\cdot(\boldsymbol{u} - \boldsymbol{z}) \leqslant 0\}$$

contains K. One can express V_1 as

$$V_1 = \{\boldsymbol{x} \in \mathbb{R}^n : \boldsymbol{x}\cdot\boldsymbol{e}_1 \leqslant \mu_1\}, \quad \text{where } \boldsymbol{e}_1 = \frac{\boldsymbol{u} - \boldsymbol{z}}{\|\boldsymbol{u} - \boldsymbol{z}\|} \quad \text{and} \quad \mu_1 = \boldsymbol{z}\cdot\boldsymbol{e}_1.$$

Clearly, $\boldsymbol{e}_1 \in \mathbb{S} \cap \operatorname{nor} K$ and $\boldsymbol{u}\cdot\boldsymbol{e}_1 > \mu_1$ due to $\boldsymbol{u} \in \mathbb{R}^n \setminus V$. Put $\beta_1 = \boldsymbol{u}\cdot\boldsymbol{e}_1$.

Our goal is to replace V_1 with a suitable halfspace $V_\gamma(\boldsymbol{e}) \in \mathcal{F}$ which misses $\boldsymbol{u}$ and strongly contains K. For this, we assume first that the cone $\operatorname{nor} K$ is one-dimensional. Then $\operatorname{nor} K$ is either a closed halfline or a line. If $\operatorname{nor} K$ is a closed halfline, then $\operatorname{nor} K = [\boldsymbol{o}, \boldsymbol{e}_1\rangle$, and Problem 8.13 shows that $\operatorname{cl} K$ is a closed halfspace of the form $\operatorname{cl} K = V_{\mu_1}(\boldsymbol{e}_1)$. Thus $X = \{\boldsymbol{e}_1\}$, and any halfspace $V_\gamma(\boldsymbol{e}_1)$, $\gamma \in \Phi \cap (\mu_1, \beta_1)$, is a desired one. Similarly, if $\operatorname{nor} K$ is the line $\langle \boldsymbol{o}, \boldsymbol{e}_1\rangle$, then the same problem shows that $\operatorname{cl} K$ is a hyperplane of the form

$$H = \{\boldsymbol{x} \in \mathbb{R}^n : \boldsymbol{x}\cdot\boldsymbol{e}_1 = \mu_1\}$$

or a closed slab of the form

$$P = \{\boldsymbol{x} \in \mathbb{R}^n : \nu_1 \leqslant \boldsymbol{x}\cdot\boldsymbol{e}_1 \leqslant \mu_1\}, \quad \nu_1 < \mu_1.$$

Consequently, $X = \{\boldsymbol{e}_1, -\boldsymbol{e}_1\}$, implying that $V_\gamma(\boldsymbol{e}_1)$ is a desired halfspace for any choice of $\gamma \in \Phi \cap (\mu_1, \beta_1)$.

Suppose now that $\dim(\operatorname{nor} K) = m \geqslant 2$. Corollary 8.36 shows that $\operatorname{rint}(\operatorname{nor} K)$ is an m-dimensional convex cone with an improper apex $\boldsymbol{o}$. The affine span of $\operatorname{rint}(\operatorname{nor} K)$ is an m-dimensional subspace of $\mathbb{R}^n$. Therefore, we can choose in $\mathbb{S} \cap \operatorname{rint}(\operatorname{nor} K)$ a linearly independent set $\{\boldsymbol{e}_1, \ldots, \boldsymbol{e}_m\}$. Then the set $\{\boldsymbol{o}, \boldsymbol{e}_1, \ldots, \boldsymbol{e}_m\}$ is affinely independent (see Theorem 2.52), and Corollary 5.8 implies that $\{\boldsymbol{o}\} \cup \operatorname{rint}(\operatorname{nor} K)$ contains the simplicial cone $\Gamma = \Gamma_{\boldsymbol{o}}(\boldsymbol{e}_1, \ldots, \boldsymbol{e}_m)$. According to Definition 8.29, there are points $\boldsymbol{z}_2, \ldots, \boldsymbol{z}_m \in \operatorname{cl} K$ such that $\boldsymbol{z}_i$ is the nearest to $\boldsymbol{e}_i + \boldsymbol{z}_i$ point in $\operatorname{cl} K$, $2 \leqslant i \leqslant m$. By Theorem 8.22, K lies in each of the closed halfspaces

$$V_i = \{\boldsymbol{x} \in \mathbb{R}^n : \boldsymbol{x}\cdot\boldsymbol{e}_i \leqslant \mu_i\}, \quad \text{where} \quad \mu_i = \boldsymbol{z}_i\cdot\boldsymbol{e}_i, \quad 2 \leqslant i \leqslant m.$$

Choose a vector $\boldsymbol{e} \in \mathbb{S} \cap \operatorname{rint} \Gamma$. Then $\boldsymbol{e} = \lambda_1 \boldsymbol{e}_1 + \cdots + \lambda_m \boldsymbol{e}_m$ for suitable positive scalars $\lambda_1, \ldots, \lambda_m$ (see Theorem 5.23). Let $\mu = \lambda_1 \mu_1 + \cdots + \lambda_m \mu_m$. We assert that the closed halfspace $V_\mu(\boldsymbol{e}) = \{\boldsymbol{x} \in \mathbb{R}^n : \boldsymbol{x} \cdot \boldsymbol{e} \leqslant \mu\}$ contains the set $M = V_1 \cap \cdots \cap V_m$. Indeed, if $\boldsymbol{x} \in M$, then $\boldsymbol{x} \cdot \boldsymbol{e}_i \leqslant \mu_i$ for all $1 \leqslant i \leqslant m$, which gives

$$\boldsymbol{x} \cdot \boldsymbol{e} = \boldsymbol{x} \cdot (\lambda_1 \boldsymbol{e}_1 + \cdots + \lambda_m \boldsymbol{e}_m) \leqslant \lambda_1 \mu_1 + \cdots + \lambda_m \mu_m = \mu,$$

implying the inclusion $K \subset M \subset V_\mu(\boldsymbol{e})$.

Theorem 2.55 shows the existence of a scalar $\delta > 0$ with the following property: for any point $\boldsymbol{e} \in B_\delta(\boldsymbol{e}_1) \cap \mathbb{S} \cap \operatorname{rint} \Gamma$, the positive scalars $\lambda_1, \lambda_2, \ldots, \lambda_m$ in the expression $\boldsymbol{e} = \lambda_1 \boldsymbol{e}_1 + \cdots + \lambda_m \boldsymbol{e}_m$ are so close to $1, 0, \ldots, 0$, respectively, that they satisfy the inequality

$$\lambda_1 (\boldsymbol{u} \cdot \boldsymbol{e}_1 - \mu_1) > \lambda_2 (\mu_2 - \boldsymbol{u} \cdot \boldsymbol{e}_2) + \cdots + \lambda_m (\mu_m - \boldsymbol{u} \cdot \boldsymbol{e}_m).$$

Equivalently,

$$\boldsymbol{u} \cdot \boldsymbol{e} = \boldsymbol{u} \cdot (\lambda_1 \boldsymbol{e}_1 + \cdots + \lambda_m \boldsymbol{e}_m) > \lambda_1 \mu_1 + \cdots + \lambda_m \mu_m = \mu,$$

which shows that $\boldsymbol{u} \notin V_\mu(\boldsymbol{e})$. Because X is dense in $\mathbb{S} \cap \operatorname{rint} \Gamma$ (see Problem 5.8), we may let $\boldsymbol{e} \in X$. By Corollary 8.36,

$$\boldsymbol{e} \in \operatorname{rint} \Gamma \subset \operatorname{rint} (\operatorname{nor} K) \subset \operatorname{nor} K.$$

Finally, put $\beta = \boldsymbol{u} \cdot \boldsymbol{e}$ and choose a scalar $\gamma \in \Phi \cap (\mu, \beta)$. By the above argument, the halfspace $V_\gamma(\boldsymbol{e}) \in \mathcal{F}$ misses $\boldsymbol{u}$ and strongly contains K. $\square$

Remarks. 1. The proof of Theorem 9.26 shows that the set $X \cap \operatorname{nor} K$ from (9.8) can be replaced with a slightly smaller set $X \cap \operatorname{rint} (\operatorname{nor} K)$.

2. The assumption that the set X is dense in $\mathbb{S} \cap \operatorname{nor} K$ cannot be dropped. For instance, if K is the unit disk of the plane $\mathbb{R}^2$ and $\mathcal{F}$ is a family of closed halfplanes whose intersection is K, then $\operatorname{nor} K = \mathbb{R}^2$ and the set of unit normal vectors of the halfplanes from $\mathcal{F}$ should be dense in the unit circle $\mathbb{S}$ of $\mathbb{R}^2$.

3. The example below shows that the closed halflines $[\mu_{\boldsymbol{e}}, \infty)$ cannot be replaced with bounded intervals. Indeed, let $K = \{(x, y) : y \geqslant x^2\}$ be the solid parabola in the plane. Then $\operatorname{nor} K = \{\boldsymbol{o}\} \cup \{(x, y) : y < 0\}$. It is easy to see that K is the intersection of the family $\mathcal{F}$ of closed halfplanes

$$V_\alpha = \{(x, y) : 2\alpha x - y \leqslant \alpha^2\}, \quad \alpha \in \mathbb{R},$$

whose boundary lines support K. With

$$\boldsymbol{e}_\alpha = \Big(\frac{2\alpha}{\sqrt{4\alpha^2 + 1}}, \frac{-1}{\sqrt{4\alpha^2 + 1}}\Big) \quad \text{and} \quad \gamma_\alpha = \frac{\alpha^2}{\sqrt{4\alpha^2 + 1}},$$

we can rewrite the inequality $2\alpha x - y \leqslant \alpha^2$ as $\boldsymbol{x}\cdot\boldsymbol{e}_\alpha \leqslant \gamma_\alpha$, where $\boldsymbol{e}_\alpha$ is a unit vector in $\operatorname{nor} K$ and $\boldsymbol{x} = (x, y)$, implying that every halfplane V_α can be expressed as $V_\alpha = \{\boldsymbol{x} \in \mathbb{R}^2 : \boldsymbol{x}\cdot\boldsymbol{e}_\alpha \leqslant \gamma_\alpha\}$. One has $\gamma_\alpha \geqslant 0$ because $\boldsymbol{o} \in K$. Furthermore, $\gamma_\alpha \to \infty$ as $\alpha \to \infty$; so there is no upper bound for the values of γ_α in $[0, \infty)$. It is easy to see that the normal vectors $\boldsymbol{e}_\alpha$ of any subfamily of $\mathcal{F}$ whose intersection is K must be dense in $\operatorname{nor} K$; consequently, the respective values γ_α must be dense in $[0, \infty)$.

The next result shows that normal directions of convex sets provide a suitable tool for describing their recession cones and lineality spaces (see Definitions 6.1 and 6.19).

Theorem 9.27. *Let $K \subset \mathbb{R}^n$ be a nonempty convex set and $X \subset \mathbb{R}^n \setminus \{\boldsymbol{o}\}$ be a nonempty set such that $\operatorname{cl} K$ is the intersection of a family $\mathcal{F} = \{V_{\gamma_{\boldsymbol{e}}}(\boldsymbol{e}) : \boldsymbol{e} \in X\}$ of closed halfspaces of the form*

$$V_{\gamma_{\boldsymbol{e}}}(\boldsymbol{e}) = \{\boldsymbol{x} \in \mathbb{R}^n : \boldsymbol{x}\cdot\boldsymbol{e} \leqslant \gamma_{\boldsymbol{e}}\}, \quad \boldsymbol{e} \in X. \tag{9.9}$$

Then $\operatorname{rec}(\operatorname{cl} K)$ *and* $\operatorname{lin}(\operatorname{cl} K)$ *are, respectively, the intersections of the families of closed halfspaces and $(n-1)$-dimensional subspaces of the form*

$$V_0(\boldsymbol{e}) = \{\boldsymbol{x} \in \mathbb{R}^n : \boldsymbol{x}\cdot\boldsymbol{e} \leqslant 0\}, \quad \boldsymbol{e} \in X,$$
$$S(\boldsymbol{e}) = \{\boldsymbol{x} \in \mathbb{R}^n : \boldsymbol{x}\cdot\boldsymbol{e} = 0\}, \quad \boldsymbol{e} \in X.$$

Proof. Let $C = \cap(V_0(\boldsymbol{e}) : \boldsymbol{e} \in X)$. First, we are going to prove the inclusion $\operatorname{rec}(\operatorname{cl} K) \subset C$. Since the case $\operatorname{rec}(\operatorname{cl} K) = \{\boldsymbol{o}\}$ is obvious, we may suppose that $\operatorname{rec}(\operatorname{cl} K) \neq \{\boldsymbol{o}\}$. Choose a nonzero vector $\boldsymbol{c} \in \operatorname{rec}(\operatorname{cl} K)$ and a point $\boldsymbol{u} \in K$. Then the closed halfline $[\boldsymbol{u}, \boldsymbol{c} + \boldsymbol{u}\rangle$ lies in $\operatorname{cl} K$ (see Definition 6.1). Therefore, $[\boldsymbol{u}, \boldsymbol{c}+\boldsymbol{u}\rangle \subset V_{\gamma_{\boldsymbol{e}}}(\boldsymbol{e})$ for every halfspace $V_{\gamma_{\boldsymbol{e}}}(\boldsymbol{e}) \in \mathcal{F}$. Equivalently, $(\lambda\boldsymbol{c} + \boldsymbol{u})\cdot\boldsymbol{e} \leqslant \gamma_{\boldsymbol{e}}$ for all $\lambda \geqslant 0$. As easy to see, these inequalities hold only if $\boldsymbol{c}\cdot\boldsymbol{e} \leqslant 0$. Hence $\boldsymbol{c} \in V_0(\boldsymbol{e})$ for all $\boldsymbol{e} \in X$, which gives $\boldsymbol{c} \in C$. Summing up, $\operatorname{rec}(\operatorname{cl} K) \subset C$.

Conversely, let $\boldsymbol{c} \in C$. Choose a vector $\boldsymbol{e} \in X$ and a point $\boldsymbol{u} \in \operatorname{cl} K$. Then $\boldsymbol{c}\cdot\boldsymbol{e} \leqslant 0$ and $\boldsymbol{u} \in V_{\gamma_{\boldsymbol{e}}}(\boldsymbol{e})$. Clearly,

$$(\boldsymbol{u} + \lambda\boldsymbol{c})\cdot\boldsymbol{e} \leqslant \boldsymbol{u}\cdot\boldsymbol{e} \leqslant \gamma_{\boldsymbol{e}} \quad \text{for all} \quad \lambda \geqslant 0.$$

Hence $\boldsymbol{u} + \lambda\boldsymbol{c} \in V_{\gamma_{\boldsymbol{e}}}(\boldsymbol{e})$ whenever $\lambda \geqslant 0$, which gives

$$\boldsymbol{u} + \lambda\boldsymbol{c} \in \cap(V_{\gamma_{\boldsymbol{e}}}(\boldsymbol{e}) : V_{\gamma_{\boldsymbol{e}}}(\boldsymbol{e}) \in \mathcal{F}) = \operatorname{cl} K \quad \text{for all} \quad \lambda \geqslant 0.$$

According to Definition 6.1, $\boldsymbol{c} \in \operatorname{rec}(\operatorname{cl} K)$. Thus $C \subset \operatorname{rec}(\operatorname{cl} K)$.

The second assertion of the theorem immediately follows from the equalities

$$\operatorname{lin}(\operatorname{cl} K) = \operatorname{rec}(\operatorname{cl} K) \cap (-\operatorname{rec}(\operatorname{cl} K)),$$
$$S(\boldsymbol{e}) = V_0(\boldsymbol{e}) \cap (-V_0(\boldsymbol{e})) \quad \text{for all} \quad \boldsymbol{e} \in X. \qquad \square$$

9.2.3 Support and Asymptotic Halfspaces

Definition 9.28. We say that a closed halfspace $V \subset \mathbb{R}^n$ *supports* a nonempty set $X \subset \mathbb{R}^n$ provided V contains X such that the boundary hyperplane H of V supports X. If $\boldsymbol{u}$ is a point in $H \cap \operatorname{cl} X$, then we will say that V supports X at $\boldsymbol{u}$. Furthermore, V *properly supports* X if H properly supports X. Similarly, V is *asymptotic* to X provided V contains X such that the boundary hyperplane H of V is asymptotic to X.

Examples. 1. A closed halfspace $V = \{\boldsymbol{x} \in \mathbb{R}^n : \boldsymbol{x}\cdot\boldsymbol{e} \leqslant \gamma\}$ properly supports a closed ball $B_\rho(\boldsymbol{a}) \subset \mathbb{R}^n$ (respectively, properly supports an open ball $U_\rho(\boldsymbol{a}) \subset \mathbb{R}^n$) if and only if $\rho\|\boldsymbol{e}\| = |\gamma - \boldsymbol{a}\cdot\boldsymbol{e}|$ (see Problem 9.1).

2. The closed halfspace $V = \{(x_1, \ldots, x_n) : x_1 \leqslant x_n\}$ is asymptotic to the solid elliptic hyperboloid

$$\{(x_1, \ldots, x_n) : x_1^2 + \cdots + x_{n-1}^2 + 1 \leqslant x_n^2,\ x_n > 0\}.$$

A combination of Definition 2.29 and Corollaries 2.92 and 9.9 implies the technical corollary below.

Corollary 9.29. *For a closed halfspace $V = \{\boldsymbol{x} \in \mathbb{R}^n : \boldsymbol{x}\cdot\boldsymbol{e} \leqslant \gamma\}$ and a nonempty set $X \subset \mathbb{R}^n$, the following assertions hold.*

(*a*) *V either supports X or is asymptotic to X if and only if*

$$\sup\{\boldsymbol{x}\cdot\boldsymbol{e} : \boldsymbol{x} \in X\} = \gamma.$$

(*b*) *V supports X if and only if* $\max\{\boldsymbol{x}\cdot\boldsymbol{e} : \boldsymbol{x} \in \operatorname{cl} X\} = \gamma$.

(*c*) *V properly supports X if and only if*

$$\inf\{\boldsymbol{x}\cdot\boldsymbol{e} : \boldsymbol{x} \in \operatorname{cl} X\} < \max\{\boldsymbol{x}\cdot\boldsymbol{e} : \boldsymbol{x} \in \operatorname{cl} X\} = \gamma. \qquad \square$$

Corollary 9.18 implies the following result.

Corollary 9.30. *Let a closed halfspace $V \subset \mathbb{R}^n$ with boundary hyperplane H contain a nonempty convex set $K \subset \mathbb{R}^n$. Then V properly supports K if and only if H meets* $\operatorname{cl} K$ *and* $\operatorname{rint} K \subset \operatorname{int} V$. $\square$

Theorem 8.22 shows that for a nonempty convex set $K \subset \mathbb{R}^n$ and a point $\boldsymbol{a} \in \mathbb{R}^n \setminus \operatorname{cl} K$ there is a closed halfspace of $\mathbb{R}^n$ containing K and missing $\boldsymbol{a}$. The next theorem establishes a similar result on properly supporting halfspaces. We recall that the relative boundary of a nonempty convex set $K \subset \mathbb{R}^n$ is nonempty if and only if K is not a plane (see Theorem 3.62).

Theorem 9.31. *If a convex set $K \subset \mathbb{R}^n$ is not a plane, then the following assertions hold.*

(*a*) *For any point* $\boldsymbol{a} \in \operatorname{rbd} K$*, there is a closed halfspace* $V \subset \mathbb{R}^n$ *properly supporting* K *at* $\boldsymbol{a}$.

(*b*) *For any point* $\boldsymbol{a} \in \mathbb{R}^n \setminus \operatorname{cl} K$*, there is a closed halfspace* $V \subset \mathbb{R}^n$ *properly supporting* K *and missing* $\boldsymbol{a}$.

(*c*) *If* $\operatorname{aff} K \neq \mathbb{R}^n$*, then for any points* $\boldsymbol{a} \in \mathbb{R}^n \setminus \operatorname{aff} K$ *and* $\boldsymbol{u} \in \operatorname{rbd} K$*, there is a closed halfspace* $V \subset \mathbb{R}^n$ *properly supporting* K *at* $\boldsymbol{u}$ *and missing* $\boldsymbol{a}$.

Proof. (*a*) Since $\{\boldsymbol{a}\}$ is a 0-dimensional plane disjoint with $\operatorname{rint} K$, Corollary 9.11 shows the existence of a hyperplane $H \subset \mathbb{R}^n$ containing $\boldsymbol{a}$ and properly supporting K. Denote by V the closed halfspace determined by H and containing K (see Theorem 9.3). Clearly, V supports K and $\boldsymbol{a} \in H = \operatorname{bd} V$. Since $K \not\subset H$, the halfspace V properly supports K at $\boldsymbol{a}$ (see Corollary 9.18).

(*b*) Given a point $\boldsymbol{a} \in \mathbb{R}^n \setminus \operatorname{cl} K$, there is a unique nearest to $\boldsymbol{a}$ point $\boldsymbol{c} \in \operatorname{cl} K$ such that the closed halfspace

$$V = \{\boldsymbol{x} \in \mathbb{R}^n : (\boldsymbol{x} - \boldsymbol{c})\cdot(\boldsymbol{a} - \boldsymbol{c}) \leqslant 0\}$$

contains K (see Theorem 8.22). If $\operatorname{aff} K = \mathbb{R}^n$, then $\dim K = n$, and Corollary 8.25 shows that $\boldsymbol{c} \in \operatorname{bd} K$. Then the boundary hyperplane H of V properly supports K. Hence V properly supports K. If $\operatorname{aff} K \neq \mathbb{R}^n$, then the conclusion follows from the argument below.

(*c*) Let $\operatorname{aff} K \neq \mathbb{R}^n$, $\boldsymbol{a} \in \mathbb{R}^n \setminus \operatorname{aff} K$, and $\boldsymbol{u} \in \operatorname{rbd} K$. By Corollary 9.12, there is a hyperplane $H \subset \mathbb{R}^n$ properly supporting K at $\boldsymbol{u}$ and one of the closed halfspaces determined by H properly supports K at $\boldsymbol{u}$. Denote this halfspace by V. Then the set $D = V \cap \operatorname{aff} K$ is a closed halfplane of $\operatorname{aff} K$ properly containing K. Problem 8.11 shows the existence of a closed halfspace $V' \subset \mathbb{R}^n$ such that $V' \cap \operatorname{aff} K = D$ and $\boldsymbol{a} \notin V'$. Since

$$K \not\subset H \cap \operatorname{aff} K = \operatorname{rbd} D = \operatorname{bd} V' \cap \operatorname{aff} K,$$

the halfspace V' properly supports K at $\boldsymbol{u}$. □

Corollary 9.32. *If a convex set* $K \subset \mathbb{R}^n$ *is not a plane and* F *is a nonempty convex subset of* $\operatorname{rbd} K$*, then there is a closed halfspace* $V \subset \mathbb{R}^n$ *properly supporting* K *such that* $F \subset \operatorname{bd} V$*. Furthermore, if* $\boldsymbol{a}$ *is a point in* $\operatorname{rint} F$ *and* $V \subset \mathbb{R}^n$ *is a closed halfspace supporting* K *at* $\boldsymbol{a}$*, then* $F \subset \operatorname{bd} V$.

Proof. Choose a point $\boldsymbol{a} \in \operatorname{rint} F$ (see Corollary 3.21). By Theorem 9.31, there is a closed halfspace $V \subset \mathbb{R}^n$ properly supporting K at $\boldsymbol{a}$. Since V supports F and $\boldsymbol{a} \in \operatorname{rint} F \cap \operatorname{bd} V$, Corollary 9.30 implies the inclusion $F \subset \operatorname{bd} V$. □

Theorem 9.33. *If a translate of a closed halfspace $V \subset \mathbb{R}^n$ contains a nonempty convex set $K \subset \mathbb{R}^n$, then there is a translate of V which either supports K or is asymptotic to K.*

Proof. Let $V = \{\boldsymbol{x} \in \mathbb{R}^n : \boldsymbol{x}\cdot\boldsymbol{e} \leqslant \gamma\}$ for a suitable nonzero vector $\boldsymbol{e} \in \mathbb{R}^n$ and a scalar γ. Suppose a translate V' of V contains K. By Corollary 2.30, $V' = \{\boldsymbol{x} \in \mathbb{R}^n : \boldsymbol{x}\cdot\boldsymbol{e} \leqslant \gamma'\}$ for a suitable scalar γ', and Corollary 9.2 gives $\gamma'' = \sup\{\boldsymbol{x}\cdot\boldsymbol{e} : \boldsymbol{x} \in K\} \leqslant \gamma'$. The above argument and Corollary 9.29 imply that the closed halfspace $V'' = \{\boldsymbol{x} \in \mathbb{R}^n : \boldsymbol{x}\cdot\boldsymbol{e} \leqslant \gamma''\}$ is a translate of V which either supports K or is asymptotic to K. □

Theorem 9.34. *Let $K \subset \mathbb{R}^n$ be a nonempty convex set and $V \subset \mathbb{R}^n$ be a closed halfspace. The following assertions hold.*

(*a*) *There is a translate of V which supports K if and only if*

$$\operatorname{nor} V \subset \operatorname{nor} K. \tag{9.10}$$

(*b*) *There is a translate of V which properly supports K if and only if*

$$\operatorname{nor} V \setminus \{\boldsymbol{o}\} \subset \operatorname{nor} K \setminus \operatorname{ort} K. \tag{9.11}$$

(*c*) *There is a translate of V which either supports K or is asymptotic to K if and only if*

$$\operatorname{nor} V \subset \operatorname{bar} K. \tag{9.12}$$

(*d*) *There is a translate of V which is asymptotic to K if and only if*

$$\operatorname{nor} V \subset (\operatorname{bar} K \setminus \operatorname{nor} K).$$

Proof. Let $V = \{\boldsymbol{x} \in \mathbb{R}^n : \boldsymbol{x}\cdot\boldsymbol{e} \leqslant \gamma\}$, where $\boldsymbol{e} \in \mathbb{R}^n$ is a suitable nonzero vector and γ is a scalar (the case of the opposite inequality $\boldsymbol{x}\cdot\boldsymbol{e} \leqslant \gamma$ is similar).

(*a*) Suppose that a translate V' of V supports K at a point $\boldsymbol{u} \in \operatorname{cl} K$. Clearly, $\boldsymbol{u} \in \operatorname{bd} V'$. Since $\boldsymbol{e} \in \operatorname{nor} V = \operatorname{nor} V'$ (see an example on page 296), $\boldsymbol{u}$ is the nearest to $\boldsymbol{e} + \boldsymbol{u}$ point in V'. Consequently, $\boldsymbol{u}$ is the nearest to $\boldsymbol{e} + \boldsymbol{u}$ point in $\operatorname{cl} K$, implying the inclusion $\boldsymbol{e} \in \operatorname{nor} K$ (see Definition 8.29). Hence (9.10) is satisfied.

Conversely, assume that (9.10) is satisfied. Choose a nonzero vector $\boldsymbol{e} \in \operatorname{nor} V$. According to Definition 8.29, there is a point $\boldsymbol{u} \in \operatorname{cl} K$ such that $\boldsymbol{u}$ is the nearest to $\boldsymbol{e} + \boldsymbol{u}$ point in $\operatorname{cl} K$. Theorem 8.22 shows that K lies in the closed halfspace $V = \{\boldsymbol{x} \in \mathbb{R}^n : (\boldsymbol{x} - \boldsymbol{u})\cdot\boldsymbol{e} \leqslant 0\}$. Expressing V as $V = \{\boldsymbol{x} \in \mathbb{R}^n : \boldsymbol{x}\cdot\boldsymbol{e} \leqslant \gamma\}$, where $\gamma = \boldsymbol{u}\cdot\boldsymbol{e}$, we conclude that $\boldsymbol{e} \in \operatorname{nor} V$.

(*b*) Suppose that a translate V' of V properly supports K at a point $\boldsymbol{u} \in \mathrm{cl}\, K$. Denote by H the boundary hyperplane of V. Clearly, $\boldsymbol{e} \in \mathrm{nor}\, V \subset \mathrm{ort}\, H$. By the above proved, $\boldsymbol{e} \in \mathrm{nor}\, K$. If $\boldsymbol{e}$ belonged to $\mathrm{ort}\, K$, then $\mathrm{dir}\, K \subset \{\boldsymbol{e}\}^{\perp} = \mathrm{dir}\, H$, implying the inclusion

$$\mathrm{aff}\, K = \boldsymbol{u} + \mathrm{dir}\, K \subset \boldsymbol{u} + \mathrm{dir}\, H = H,$$

which is impossible due to $K \not\subset H$. Hence (9.11) is satisfied.

Conversely, assume that (9.11) is satisfied. Choose a nonzero vector $\boldsymbol{e} \in \mathrm{nor}\, V$. Then $\boldsymbol{e} \in \mathrm{nor}\, K$, and, by the above proved, there is a translate V' of V supporting K. If the hyperplane $H' = \mathrm{bd}\, V'$ contained K, then $\boldsymbol{e} \in \mathrm{ort}\, H' \subset \mathrm{ort}\, K$, which is impossible. Hence $K \not\subset H'$, implying that V' properly supports K.

(*c*) Assume first the existence of a translate V' of V which either supports K or is asymptotic to K. A combination of Corollaries 2.30 and 9.29 shows that

$$V' = \{\boldsymbol{x} \in \mathbb{R}^n : \boldsymbol{x}\cdot\boldsymbol{e} \leqslant \gamma'\}, \quad \text{where} \quad \gamma' = \sup\{\boldsymbol{x}\cdot\boldsymbol{e} : \boldsymbol{x} \in K\}.$$

Then $\boldsymbol{e} \in \mathrm{bar}\, K$ according to Definition 8.31. Since $\boldsymbol{e} \in \mathrm{nor}\, V$, (9.12) is satisfied.

Conversely, assume that (9.12) holds. By Theorem 9.21, there is a translate of V which contains K, and Theorem 9.33 implies the existence of a translate of V which either supports K or is asymptotic to K.

(*d*) This part follows from assertions (*a*) and (*c*). □

Theorem 9.35. *Let $K_1, \dots, K_r$ be nonempty convex sets in $\mathbb{R}^n$, and $\mu_1, \dots, \mu_r$ be nonzero scalars. A closed halfspace $V \subset \mathbb{R}^n$ either supports or is asymptotic to the convex set $K = \mu_1 K_1 + \cdots + \mu_r K_r$ if and only if V can be expressed as $V = \mu_1 V_1 + \cdots + \mu_r V_r$, where $V_1, \dots, V_r$ are closed halfspaces satisfying the conditions:*

(*a*) *every V_i either supports or is asymptotic to K_i, $1 \leqslant i \leqslant r$,*
(*b*) *$\mu_i V_i$ is a translate of V for all $1 \leqslant i \leqslant r$.*

Proof. Assume first that V either supports or is asymptotic to K. Then V contains K, and Theorem 9.21 implies that V can be expressed as $V = \{\boldsymbol{x} \in \mathbb{R}^n : \boldsymbol{x}\cdot\boldsymbol{e} \leqslant \gamma\}$, where $\boldsymbol{e}$ is a nonzero vector in $\mathrm{bar}\, K$. According to Corollary 9.29, $\sup\{\boldsymbol{x}\cdot\boldsymbol{e} : \boldsymbol{x} \in K\} = \gamma$. By Theorem 8.34, $\boldsymbol{e} \in \mu_1 \mathrm{bar}\, K_1 \cap \cdots \cap \mu_r \mathrm{bar}\, K_r$. Consequently, the scalars $\gamma_1, \dots, \gamma_r$, defined by

$$\gamma_i = \begin{cases} \sup\{\boldsymbol{x}\cdot\boldsymbol{e} : \boldsymbol{x} \in K_i\} & \text{if } \mu_i > 0, \\ \inf\{\boldsymbol{x}\cdot\boldsymbol{e} : \boldsymbol{x} \in K_i\} & \text{if } \mu_i < 0, \end{cases}$$

are finite. Clearly,

$$\mu_i\gamma_i = \sup\{\boldsymbol{x}\cdot\boldsymbol{e} : \boldsymbol{x} \in \mu_i K_i\}, \quad 1 \leqslant i \leqslant r.$$

Furthermore, $\mu_1\gamma_1 + \cdots + \mu_r\gamma_r = \gamma$. Put

$$V_i = \begin{cases} \{\boldsymbol{x} \in \mathbb{R}^n : \boldsymbol{x}\cdot\boldsymbol{e} \leqslant \gamma_i\} & \text{if } \mu_i > 0, \\ \{\boldsymbol{x} \in \mathbb{R}^n : \boldsymbol{x}\cdot\boldsymbol{e} \geqslant \gamma_i\} & \text{if } \mu_i < 0. \end{cases}$$

Obviously, $K_i \subset V_i$ and

$$\mu_i V_i = \{\boldsymbol{x} \in \mathbb{R}^n : \boldsymbol{x}\cdot\boldsymbol{e} \leqslant \mu_i\gamma_i\}, \quad 1 \leqslant i \leqslant r,$$

which shows that all halfspaces $\mu_1 V_1, \ldots, \mu_r V_r$ are translates of V (see Corollary 2.30). Furthermore, $V = \mu_1 V_1 + \cdots + \mu_r V_r$ (see Theorem 2.32). Corollary 9.29 implies that every halfspace V_i either supports or is asymptotic to K_i, $1 \leqslant i \leqslant r$.

Conversely, if $V = \mu_1 V_1 + \cdots + \mu_r V_r$, where the closed halfspaces $V_1, \ldots, V_r$ satisfy conditions (a) and (b), then a similar to the above argument shows that V either supports or is asymptotic to K. □

The next result complements Theorem 9.35 and is used in the study of extreme and exposed faces of convex sets (see Theorem 12.13).

Theorem 9.36. *In terms of Theorem* 9.35*, let* $H, H_1, \ldots, H_r$ *denote, respectively, the boundary hyperplanes of* $V, V_1, \ldots, V_r$. *The following assertions hold.*

(a) *If every* H_i *supports* K_i, $1 \leqslant i \leqslant r$, *then* H *also supports* K *and*

$$\mu_1(H_1 \cap \operatorname{cl} K_1) + \cdots + \mu_r(H_r \cap \operatorname{cl} K_r) \subset H \cap \operatorname{cl} K. \tag{9.13}$$

(b) *If* H *meets* K, *then every* H_i *meets* K_i, $1 \leqslant i \leqslant r$, *and*

$$\mu_1(H_1 \cap K_1) + \cdots + \mu_r(H_r \cap K_r) = H \cap K. \tag{9.14}$$

Proof. Let $H = \{\boldsymbol{x} \in \mathbb{R}^n : \boldsymbol{x}\cdot\boldsymbol{e} = \gamma\}$ and

$$H_i = \{\boldsymbol{x} \in \mathbb{R}^n : \boldsymbol{x}\cdot\boldsymbol{e} = \gamma_i\}, \quad 1 \leqslant i \leqslant r,$$

where $\boldsymbol{e}$ is a nonzero vector in bar K and the scalars $\gamma, \gamma_1, \ldots, \gamma_r$ are defined in the proof of Theorem 9.35. Then $\mu_1\gamma_1 + \cdots + \mu_r\gamma_r = \gamma$, and Theorem 2.18 gives the equality $\mu_1 H_1 + \cdots + \mu_r H_r = H$.

(a) Choose any points $\boldsymbol{u}_i \in H_i \cap \operatorname{cl} K_i$, $1 \leqslant i \leqslant r$, and let $\boldsymbol{u} = \mu_1\boldsymbol{u}_1 + \cdots + \mu_r\boldsymbol{u}_r$. By the above argument,

$$\boldsymbol{u} \in \mu_1 H_1 + \cdots + \mu_r H_r = H,$$

and Problem 1.13 (and induction on r) implies the inclusion

$$\boldsymbol{u} \in \mu_1 \mathrm{cl}\, K_1 + \cdots + \mu_r \mathrm{cl}\, K_r \subset \mathrm{cl}\,(\mu_1 K_1 + \cdots + \mu_r K_r) = \mathrm{cl}\, K.$$

Hence H supports K and (9.13) holds.

(b) Choose any point $\boldsymbol{u} \in H \cap K$. Then $\boldsymbol{u} = \mu_1 \boldsymbol{u}_1 + \cdots + \mu_r \boldsymbol{u}_r$, where $\boldsymbol{u}_i \in K_i$, $1 \leqslant i \leqslant r$. The inequalities

$$\gamma = \boldsymbol{u} \cdot \boldsymbol{e} = (\mu_1 \boldsymbol{u}_1 + \cdots + \mu_r \boldsymbol{u}_r) \cdot \boldsymbol{e} \leqslant \mu_1 \gamma_1 + \cdots + \mu_r \gamma_r = \gamma$$

imply that $\mu_i \boldsymbol{u}_i \cdot \boldsymbol{e} = \mu_i \gamma_i$, or $\boldsymbol{u}_i \cdot \boldsymbol{e} = \gamma_i$ for all $1 \leqslant i \leqslant r$. Hence $\boldsymbol{u}_i \in H_i$, $1 \leqslant i \leqslant r$, giving the inclusion

$$H \cap K \subset \mu_1 (H_1 \cap K_1) + \cdots + \mu_r (H_r \cap K_r).$$

For the opposite inclusion, choose points $\boldsymbol{x}_i \in H_i \cap K_i$, $1 \leqslant i \leqslant r$, and let $\boldsymbol{x} = \mu_1 \boldsymbol{x}_1 + \cdots + \mu_r \boldsymbol{x}_r$. As above,

$$\boldsymbol{x} \in \mu_1 H_1 + \cdots + \mu_r H_r = H \quad \text{and} \quad \boldsymbol{x} \in \mu_1 K_1 + \cdots + \mu_r K_r = K.$$

Hence $\boldsymbol{x} \in H \cap K$, implying the inclusion

$$\mu_1 (H_1 \cap K_1) + \cdots + \mu_r (H_r \cap K_r) \subset H \cap K. \qquad \square$$

Remark. Inclusion (9.13) may be proper. Indeed, consider in $\mathbb{R}^3$ the closed convex sets

$$K_1 = \{(x, y, z) : x > 0,\ xy \geqslant 1,\ z \geqslant 0\},$$
$$K_2 = \{(x, y, z) : x < 0,\ xy \leqslant -1,\ z \geqslant 0\}.$$

Clearly, $K_1 + K_2$ is a nonclosed convex set, given by $\{(x, y, z) : y > 0,\ z \geqslant 0\}$. The closed halfspace $V = \{(x, y, z) : z \geqslant 0\}$ supports all three sets K_1, K_2, and $K_1 + K_2$. Therefore,

$$H = H_1 = H_2 = \{(x, y, z) : z = 0\}.$$

Consequently,

$$H_1 \cap K_1 + H_2 \cap K_2 = H \cap (K_1 + K_2) \neq H \cap \mathrm{cl}\,(K_1 + K_2).$$

9.2.4 *Intersections of Support Halfspaces*

Theorem 9.37. *For a nonempty convex set $K \subset \mathbb{R}^n$, the following assertions hold.*

(a) *If $\mathcal{F}$ is the family of all closed halfspaces of $\mathbb{R}^n$ supporting K, then*

$$\mathrm{cl}\, K = \cap (V : V \in \mathcal{F}). \tag{9.15}$$

(*b*) *If* K *is not a plane and* $\mathcal{F}$ *is the family of all closed halfspaces of* $\mathbb{R}^n$ *properly supporting* K, *then* (9.15) *holds. Furthermore,*

$$\operatorname{rbd} K = \cup\,(\operatorname{rbd} K \cap \operatorname{bd} V : V \in \mathcal{F}). \tag{9.16}$$

Proof. (*a*) Since every halfspace $V \subset \mathcal{F}$ is a closed set (see Theorem 2.33), their intersection also is a closed set (see page 10). Consequently, the obvious inclusion $K \subset \cap\,(V : V \in \mathcal{F})$ implies that $\operatorname{cl} K \subset \cap\,(V : V \in \mathcal{F})$. The opposite inclusion follows from Theorem 8.22.

(*b*) The case of closed halfspaces properly supporting K is similar to the above argument and uses Theorem 9.20. For (9.16), we first observe that the inclusion

$$\cup\,(\operatorname{rbd} K \cap \operatorname{bd} V : V \in \mathcal{F}) \subset \operatorname{rbd} K$$

is obvious. On the other hand, let $\boldsymbol{x} \in \operatorname{rbd} K$. Since $\{\boldsymbol{x}\}$ is a 0-dimensional plane disjoint with $\operatorname{rint} K$, Theorem 9.31 shows the existence of a closed halfspace $V \subset \mathbb{R}^n$ properly containing K such that $\boldsymbol{x} \in \operatorname{bd} V$. Hence

$$\operatorname{rbd} K \subset \cup\,(\operatorname{rbd} K \cap \operatorname{bd} V : V \in \mathcal{F}).$$ □

A combination of Theorems 9.24 and 9.37 implies the corollary below.

Corollary 9.38. *Let* $K \subset \mathbb{R}^n$ *be a convex set which is not a plane;* $\mathcal{F}$ *be a family of closed halfspaces of* $\mathbb{R}^n$ *properly supporting* K *and satisfying* (9.15). *The following assertions hold.*

(*a*) $\mathcal{F}$ *has a countable subfamily satisfying* (9.15).
(*b*) $\operatorname{rint} K = \cap\,(\operatorname{int} V : V \in \mathcal{F})$.
(*c*) *If* $\dim K \leqslant n-1$, *then* $\mathcal{F}$ *is infinite.* □

Theorem 9.39. *Let* $K \subset \mathbb{R}^n$ *be a nonempty convex set and* X *be a dense subset of* $\mathbb{S} \cap \operatorname{nor} K$ (*where* $\mathbb{S}$ *is the unit sphere of* $\mathbb{R}^n$). *With*

$$\mu_{\boldsymbol{e}} = \sup\{\boldsymbol{x}\cdot\boldsymbol{e} : \boldsymbol{x} \in K\} \quad \textit{for every} \quad \boldsymbol{e} \in X,$$

the following assertions hold.

(*a*) $\operatorname{cl} K$ *is the intersection of countably many closed halfspaces* (*each supporting* K) *of the form*

$$V_{\mu_{\boldsymbol{e}}}(\boldsymbol{e}) = \{\boldsymbol{x} \in \mathbb{R}^n : \boldsymbol{x}\cdot\boldsymbol{e} \leqslant \mu_{\boldsymbol{e}}\}, \quad \boldsymbol{e} \in X. \tag{9.17}$$

(*b*) *If, additionally,* K *is not a plane, then* $\operatorname{cl} K$ *is the intersection of countably many closed halfspaces properly supporting* K *of the form*

$$V_{\mu_{\boldsymbol{e}}}(\boldsymbol{e}) = \{\boldsymbol{x} \in \mathbb{R}^n : \boldsymbol{x}\cdot\boldsymbol{e} \leqslant \mu_{\boldsymbol{e}}\}, \quad \boldsymbol{e} \in X \setminus \operatorname{ort} K. \tag{9.18}$$

Proof. (*a*) According to Theorem 9.26, $\mathrm{cl}\,K$ is the intersection of a countable family $\mathcal{F}$ of closed halfspaces of the form (9.8). Given a halfspace $V_\gamma(\boldsymbol{e}) \in \mathcal{F}$, $\boldsymbol{e} \in X$, a combination of Corollary 9.29 and Theorem 9.34 shows that the halfspace $V_{\mu_e}(\boldsymbol{e})$ of the form (9.17) supports K such that

$$\mathrm{cl}\,K \subset V_{\mu_e}(\boldsymbol{e}) \subset V_\gamma(\boldsymbol{e}) \quad \text{for all} \quad V_\gamma(\boldsymbol{e}) \in \mathcal{F}.$$

Therefore,

$$\mathrm{cl}\,K \subset \cap\,(V_{\mu_e}(\boldsymbol{e}) : V_\gamma(\boldsymbol{e}) \in \mathcal{F}) \subset \cap\,(V_\gamma(\boldsymbol{e}) : V_\gamma(\boldsymbol{e}) \in \mathcal{F}) = \mathrm{cl}\,K,$$

which implies the desired equality

$$\mathrm{cl}\,K = \cap\,(V_{\mu_e}(\boldsymbol{e}) : V_\gamma(\boldsymbol{e}) \in \mathcal{F}).$$

(*b*) Corollary 8.25 and Definition 8.29 show that the subspace $\mathrm{ort}\,K$ lies in the apex set of the cone $\mathrm{nor}\,K$. Because K is not a plane, Theorem 5.25 shows that $\mathrm{ort}\,K \subset \mathrm{cl}\,(\mathrm{nor}\,K \setminus \mathrm{ort}\,K)$. According to Problem 5.8, the set $X \cap (\mathrm{nor}\,K \setminus \mathrm{ort}\,K)$ is dense in $\mathbb{S} \cap \mathrm{nor}\,K$. Repeating the argument of part (*a*) for the set $X \cap (\mathrm{nor}\,K \setminus \mathrm{ort}\,K)$, we obtain a countable family of closed halfspaces of the form (9.18) whose intersection is $\mathrm{cl}\,K$. Finally, Theorem 9.34 shows that every member of this family properly supports K. $\square$

Remark. Theorem 13.7 complements the above result by asserting that any family $\mathcal{F}$ of closed halfspaces supporting a plane $L \subset \mathbb{R}^n$ of dimension m, $0 \leqslant m \leqslant n-1$, contains a subfamily of r members, where $n - m + 1 \leqslant r \leqslant 2(n-m)$, whose intersection is L.

Sometimes it is convenient to express the closure of a convex set $K \subset \mathbb{R}^n$ of positive dimension m as the intersection of closed halfplanes of the plane $\mathrm{aff}\,K$. This can be done in two equivalent ways: either identifying $\mathrm{aff}\,K$ with $\mathbb{R}^m$ (first, translating K on a suitable vector $-\boldsymbol{u}$, where $\boldsymbol{u} \in K$), or considering subplanes of dimension $m-1$ (see Theorem 2.19) and closed halfplanes of $\mathrm{aff}\,K$ as intersections of $\mathrm{aff}\,K$ with suitable hyperplanes $H \subset \mathbb{R}^n$ and closed halfspaces $V \subset \mathbb{R}^n$ satisfying the conditions

$$\varnothing \neq H \cap \mathrm{aff}\,K \neq \mathrm{aff}\,K \quad \text{and} \quad \varnothing \neq V \cap \mathrm{aff}\,K \neq \mathrm{aff}\,K.$$

In either way, all results of this chapter can be reformulated in terms of $(m-1)$-dimensional subplanes and halfplanes of $\mathrm{aff}\,K$.

Additional results on this topic are given in Theorems 9.40 and 9.41.

Theorem 9.40. *Let $K \subset \mathbb{R}^n$ be a convex set which is not a plane and X be a dense subset of* $\mathrm{rbd}\,K$. *For every point $\boldsymbol{x} \in X$, choose a closed halfplane $D_{\boldsymbol{x}}$ of* $\mathrm{aff}\,K$ *properly supporting K at $\boldsymbol{x}$ and let $\mathcal{D} = \{D_{\boldsymbol{x}} : \boldsymbol{x} \in X\}$. Then* $\mathrm{cl}\,K = \cap\,(D_{\boldsymbol{x}} : D_{\boldsymbol{x}} \in \mathcal{D})$.

Proof. Theorem 9.31 shows that for any point $\boldsymbol{x} \in X$ there is a closed halfspaces $V_{\boldsymbol{x}} \subset \mathbb{R}^n$ properly supporting K at $\boldsymbol{x}$. Consequently, the closed halfplane $D_{\boldsymbol{x}} = V_{\boldsymbol{x}} \cap \operatorname{aff} K$ properly supports K at $\boldsymbol{x}$.

Because the inclusion $\operatorname{cl} K \subset \cap (D_{\boldsymbol{x}} : D_{\boldsymbol{x}} \in \mathcal{D})$ is obvious, it suffices to prove the opposite one. For this, choose any point $\boldsymbol{u} \in \operatorname{aff} K \setminus \operatorname{cl} K$ and a point $\boldsymbol{v} \in \operatorname{rint} K$. Then the open segment $(\boldsymbol{u}, \boldsymbol{v})$ meets $\operatorname{rbd} K$ at a unique point $\boldsymbol{z}$ (see Theorem 3.65). Expressing $\boldsymbol{z}$ as $\boldsymbol{z} = (1 - \lambda)\boldsymbol{u} + \lambda \boldsymbol{v}$ for a suitable scalar $0 < \lambda < 1$, we can write $\boldsymbol{v} = (1 - \lambda^{-1})\boldsymbol{u} + \lambda^{-1}\boldsymbol{z}$.

According to Theorem 3.17, there is a scalar $\rho > 0$ such that $\operatorname{aff} K \cap U_\rho(\boldsymbol{v}) \subset \operatorname{rint} K$. Because X is dense in $\operatorname{rbd} K$, one can choose a point $\boldsymbol{x} \in X$ satisfying the inequality $\|\boldsymbol{x} - \boldsymbol{z}\| < \lambda\rho$. Put $\boldsymbol{w} = (1 - \lambda^{-1})\boldsymbol{u} + \lambda^{-1}\boldsymbol{x}$. Then $\boldsymbol{w} \in \operatorname{aff} K$ because $\boldsymbol{w}$ is an affine combination of $\boldsymbol{u}$ and $\boldsymbol{x}$ (see Theorem 2.38). Furthermore, the inequalities

$$\|\boldsymbol{v} - \boldsymbol{w}\| = \|\lambda^{-1}(\boldsymbol{x} - \boldsymbol{z})\| < \lambda^{-1}(\lambda\rho) = \rho$$

show that $\boldsymbol{w} \in \operatorname{aff} K \cap U_\rho(\boldsymbol{v}) \subset \operatorname{rint} K$.

We assert that $\boldsymbol{u} \notin D_{\boldsymbol{x}}$. Indeed, assume for a moment that $\boldsymbol{u} \in D_{\boldsymbol{x}}$. Since $D_{\boldsymbol{x}}$ properly supports K, one has $\boldsymbol{w} \in \operatorname{rint} K \subset \operatorname{rint} D_{\boldsymbol{x}}$, and Theorem 2.35 implies the inclusions

$$\boldsymbol{x} = (1 - \lambda)\boldsymbol{u} + \lambda \boldsymbol{w} \in (\boldsymbol{u}, \boldsymbol{w}) \subset \operatorname{rint} D_{\boldsymbol{x}},$$

in contradiction with the assumption $\boldsymbol{x} \in \operatorname{rbd} D_{\boldsymbol{x}}$. Summing up,

$$\cap (D_{\boldsymbol{x}} : D_{\boldsymbol{x}} \in \mathcal{D}) \subset \operatorname{cl} K. \qquad \square$$

Let $K \subset \mathbb{R}^n$ be a convex set which is not a plane, and $\boldsymbol{u}$ be a point in $\operatorname{rbd} K$. We will say that $\boldsymbol{u}$ is a *regular point* of K provided there is a unique plane $L \subset \operatorname{aff} K$ of dimension $m - 1$ which supports K at $\boldsymbol{u}$.

Example. Given a closed ball $B_\rho(\boldsymbol{a}) \subset \mathbb{R}^n$ and a plane $L \subset \mathbb{R}^n$ through $\boldsymbol{a}$, of positive dimension m, consider the m-dimensional closed ball $B'_\rho(\boldsymbol{a}) = L \cap B_\rho(\boldsymbol{a})$. Then $\operatorname{aff} B'_\rho(\boldsymbol{a}) = L$ (see Theorem 2.68). By Problem 9.1 (formulated for the space L), every relative boundary point of $B'_\rho(\boldsymbol{a})$ is regular.

Theorem 9.41. *Let $K \subset \mathbb{R}^n$ be a convex set which is not a plane. The set of all regular points of K is dense in* $\operatorname{rbd} K$.

Proof. Choose any point $\boldsymbol{a} \in \operatorname{rbd} K$ (we know that $\operatorname{rbd} K \neq \varnothing$ according to Theorem 3.62) and a scalar $\varepsilon > 0$. Select a point $\boldsymbol{c} \in \operatorname{rint} K$. Theorem 3.47 shows that the open segment $(\boldsymbol{a}, \boldsymbol{c})$ lies in $\operatorname{rint} K$. Consider a point

$$\boldsymbol{z} = (1 - \lambda)\boldsymbol{a} + \lambda \boldsymbol{c}, \quad \text{where} \quad 0 < \lambda < \varepsilon / \|\boldsymbol{a} - \boldsymbol{c}\|.$$

Clearly, $0 < \|\boldsymbol{a} - \boldsymbol{z}\| < \varepsilon$. By Theorem 3.62, the set $\operatorname{rbd} K$ is closed. Hence there is a point $\boldsymbol{u} \in \operatorname{rbd} K$ such that $\|\boldsymbol{z} - \boldsymbol{u}\| = \delta(\boldsymbol{z}, \operatorname{rbd} K)$ (see Problem 1.14). Let $\mu = \|\boldsymbol{z} - \boldsymbol{u}\|$. Obviously, $0 < \mu \leqslant \varepsilon$ and the relative open set $U'_\mu(\boldsymbol{z}) = \operatorname{aff} K \cap U_\mu(\boldsymbol{z})$ lies in $\operatorname{rint} K$. By Theorem 3.48,

$$B'_\mu(\boldsymbol{z}) = \operatorname{cl} U'_\mu(\boldsymbol{z}) \subset \operatorname{cl}(\operatorname{rint} K) = \operatorname{cl} K.$$

Let $m = \dim(\operatorname{aff} K)$. According to Corollary 9.12, there is a hyperplane $H \subset \mathbb{R}^n$ through $\boldsymbol{u}$ which properly supports $\operatorname{cl} K$. By Theorem 2.19, the plane $G = H \cap \operatorname{aff} K$ is $(m-1)$-dimensional. Clearly, G properly supports $\operatorname{cl} K$ at $\boldsymbol{u}$. Therefore, G properly supports $B'_\mu(\boldsymbol{z})$. As discussed in the above example, G is the unique $(m-1)$-dimensional plane in $\operatorname{aff} K$ supporting $B'_\mu(\boldsymbol{z})$ at $\boldsymbol{u}$. Therefore, G is the unique $(m-1)$-dimensional plane in $\operatorname{aff} K$ supporting $\operatorname{cl} K$. Hence $\boldsymbol{u}$ is a regular point in $\operatorname{rbd} K$.

Since the scalar $\varepsilon > 0$ was chosen arbitrarily, the set of all regular points of K is dense in $\operatorname{rbd} K$. □

Since a dense subset of all regular points of a convex set $K \subset \mathbb{R}^n$ is dense in $\operatorname{rbd} K$, we can refine Theorem 9.40 as follows.

Corollary 9.42. *Let $K \subset \mathbb{R}^n$ be a convex set which is not a plane and X be a dense subset of the set of regular points of K. For every point $\boldsymbol{x} \in X$, choose a (unique) closed halfplane $D_{\boldsymbol{x}}$ of* $\operatorname{aff} K$ *properly supporting K at $\boldsymbol{x}$ and let $\mathcal{D} = \{D_{\boldsymbol{x}} : \boldsymbol{x} \in X\}$. Then* $\operatorname{cl} K = \cap(D_{\boldsymbol{x}} : D_{\boldsymbol{x}} \in \mathcal{D})$. □

9.2.5 *Support Properties of Convex Cones*

Theorem 9.43. *If a hyperplane $H \subset \mathbb{R}^n$ supports a convex cone $C \subset \mathbb{R}^n$, then* $\operatorname{ap}(\operatorname{cl} C) \subset H$.

Proof. Choose a point $\boldsymbol{u} \in H \cap \operatorname{cl} C$ and denote by V a closed halfspace determined by H and containing C. Clearly, $\operatorname{cl} C \subset V$.

Suppose first that $\boldsymbol{u} \in \operatorname{ap}(\operatorname{cl} C)$. By Theorem 5.14, $\operatorname{ap}(\operatorname{cl} C)$ is a plane. Since $\operatorname{ap}(\operatorname{cl} C)$ lies in V and $\boldsymbol{u}$ belongs to $H \cap \operatorname{ap}(\operatorname{cl} C)$, Theorem 2.34 gives the inclusion $\operatorname{ap}(\operatorname{cl} C) \subset H$.

Suppose now that $\boldsymbol{u} \notin \operatorname{ap}(\operatorname{cl} C)$ and choose a point $\boldsymbol{v} \in \operatorname{ap}(\operatorname{cl} C)$. Then the closed halfline $h = [\boldsymbol{v}, \boldsymbol{u}\rangle$ lies in $\operatorname{cl} C$. Choose a point $\boldsymbol{w} \in h \setminus [\boldsymbol{v}, \boldsymbol{u}]$. Then

$$\boldsymbol{u} \in (\boldsymbol{v}, \boldsymbol{w}) \subset h \subset \operatorname{cl} C \subset V,$$

and Corollary 2.36 implies that $\boldsymbol{v} \in H$. By the above argument (with $\boldsymbol{v}$ instead of $\boldsymbol{u}$), we obtain the inclusion $\operatorname{ap}(\operatorname{cl} C) \subset H$. □

Theorem 9.44. *If $C \subset \mathbb{R}^n$ is a convex cone, then no hyperplane $H \subset \mathbb{R}^n$ is asymptotic to C.*

Proof. Assume, for contradiction, the existence of a hyperplane $H \subset \mathbb{R}^n$ which is asymptotic to C. Since H bounds C, there is a closed halfspace $V \subset \mathbb{R}^n$ determined by H and containing C. Obviously, $\operatorname{cl} C \subset V$.

Let $H = \{\boldsymbol{x} \in \mathbb{R}^n : \boldsymbol{x} \cdot \boldsymbol{e} = \gamma\}$ (see Theorem 2.17). Without loss of generality, we may suppose that $V = \{\boldsymbol{x} \in \mathbb{R}^n : \boldsymbol{x} \cdot \boldsymbol{e} \leqslant \gamma\}$. According to Definition 7.10, $H \cap \operatorname{cl} C = \varnothing$ and $\delta(H, C) = 0$. Hence $\operatorname{cl} C$ lies in the open halfspace $W = V \setminus H = \{\boldsymbol{x} \in \mathbb{R}^n : \boldsymbol{x} \cdot \boldsymbol{e} < \gamma\}$ (see Definition 2.29). Choose an apex $\boldsymbol{a}$ of $\operatorname{cl} C$ and let $\mu = \boldsymbol{a} \cdot \boldsymbol{e}$. Clearly, $\mu < \gamma$ due to $\boldsymbol{a} \in W$. By Theorem 2.91,

$$\delta(\boldsymbol{a}, H) = \frac{\gamma - \boldsymbol{a} \cdot \boldsymbol{e}}{\|\boldsymbol{e}\|} = \frac{\gamma - \mu}{\|\boldsymbol{e}\|}.$$

The equality $\delta(H, C) = 0$ implies the existence of a point $\boldsymbol{u} \in C \subset W$ such that

$$0 < \delta(\boldsymbol{u}, H) < \frac{\gamma - \mu}{2\|\boldsymbol{e}\|}.$$

Let $\nu = \boldsymbol{u} \cdot \boldsymbol{e}$. Then $\nu < \gamma$ and, by the same Theorem 2.91,

$$\frac{\gamma - \nu}{\|\boldsymbol{e}\|} = \delta(\boldsymbol{u}, H) < \frac{\gamma - \mu}{2\|\boldsymbol{e}\|}.$$

Consequently, $\mu < 2\nu - \gamma < \nu$. Now, let

$$\boldsymbol{v} = (1 - \lambda)\boldsymbol{a} + \lambda \boldsymbol{u}, \quad \text{where} \quad \lambda = \frac{\gamma - \mu}{\nu - \mu}.$$

Since $\lambda > 0$, we have $\boldsymbol{v} \in [\boldsymbol{a}, \boldsymbol{u}\rangle \subset \operatorname{cl} C \subset W$. On the other hand,

$$\boldsymbol{v} \cdot \boldsymbol{e} = ((1 - \lambda)\boldsymbol{a} + \lambda \boldsymbol{u}) \cdot \boldsymbol{e} = \frac{\nu - \gamma}{\nu - \mu}\mu + \frac{\gamma - \mu}{\nu - \mu}\nu = \gamma.$$

Therefore, $\boldsymbol{v} \in H$, contrary to the choice $\boldsymbol{v} \in W$. The obtained contradiction shows that H cannot be asymptotic to C. □

Remark. The assertions of Theorems 9.43 and 9.44 do not hold for the case of planes whose dimension is less than $n - 1$. Indeed, consider the closed convex cone

$$C = \{(0, 0, z) : z \geqslant 0\} \cup \{t(x, 1, z) : t > 0, z \geqslant x^2\},$$

which is the closure of the cone generated by the solid parabola $\{(x, 1, z) : z \geqslant x^2\}$. The line $l = \{(x, 1, 0) : x \in \mathbb{R}\}$ supports C at $(0, 1, 0)$ but does not meet $\operatorname{ap} C = \{\boldsymbol{o}\}$. The line $l' = \{(1, y, 0) : y \in \mathbb{R}\}$ is asymptotic to C.

Theorem 9.45. *Let $C \subset \mathbb{R}^n$ be a convex cone which is not a plane, and let $L \subset \mathbb{R}^n$ be a plane satisfying the condition*

$$\varnothing \neq L \cap \operatorname{cl} C \subset \operatorname{ap}(\operatorname{cl} C).$$

There is a hyperplane $H \subset \mathbb{R}^n$ which contains L and supports C such that

$$H \cap \operatorname{cl} C = \operatorname{ap}(\operatorname{cl} C).$$

Proof. Choosing a point $\boldsymbol{c} \in L \cap \operatorname{cl} C$ and translating both planes L and $\operatorname{ap} C$ on $-\boldsymbol{c}$, we may suppose that L and $\operatorname{ap}(\operatorname{cl} C)$ are subspaces. First, we assert that the subspace $M = L + \operatorname{ap}(\operatorname{cl} C)$ satisfies the inclusion

$$M \cap \operatorname{cl} C \subset \operatorname{ap}(\operatorname{cl} C). \tag{9.19}$$

Indeed, assume the existence of a point

$$\boldsymbol{x} \in (M \cap \operatorname{cl} C) \setminus \operatorname{ap}(\operatorname{cl} C).$$

Since $\boldsymbol{x} \notin L$ (otherwise $\boldsymbol{x} \in L \cap \operatorname{cl} C \subset \operatorname{ap}(\operatorname{cl} C)$), the inclusion $\boldsymbol{x} \in M$ shows that it can be expressed as $\boldsymbol{x} = \boldsymbol{y} + \boldsymbol{z}$ such that both $\boldsymbol{y}$ and $\boldsymbol{z}$ are nonzero vectors in L and $\operatorname{ap}(\operatorname{cl} C)$, respectively (see Figure 9.4).

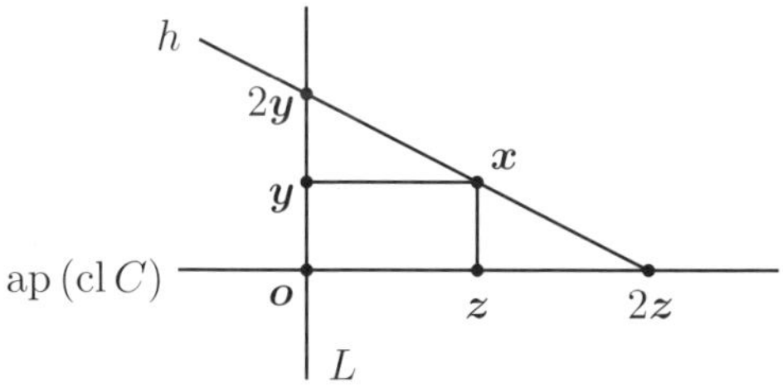

Fig. 9.4 Illustration to Theorem 9.45.

Because $2\boldsymbol{z} \in \operatorname{ap}(\operatorname{cl} C)$, the open halfline $h = (2\boldsymbol{z}, \boldsymbol{x}\rangle$ lies in $\operatorname{cl} C$. At the same time, $h \cap \operatorname{ap}(\operatorname{cl} C) = \varnothing$. For if h met $\operatorname{ap}(\operatorname{cl} C)$ at a point $\boldsymbol{u}$, then

$$\boldsymbol{x} \in h \subset \langle 2\boldsymbol{z}, \boldsymbol{u}\rangle \subset \operatorname{ap}(\operatorname{cl} C)$$

due to Theorem 5.14, which is impossible by the choice of $\boldsymbol{x}$. Finally,

$$2\boldsymbol{y} = (1-2)(2\boldsymbol{z}) + 2\boldsymbol{x} \in h \subset \operatorname{cl} C \setminus \operatorname{ap}(\operatorname{cl} C),$$

contrary to the assumption $2\boldsymbol{y} \in L \cap \operatorname{cl} C \subset \operatorname{ap}(\operatorname{cl} C)$. Summing up, (9.19) holds.

Because C is not a subspace, we have $\mathrm{ap}\,(\mathrm{cl}\, C) \subset \mathrm{rbd}\, C$ (see Corollary 5.27), which gives, combined with (9.19):

$$M \cap \mathrm{rint}\, C = M \cap (\mathrm{cl}\, C \setminus \mathrm{rbd}\, C) = (M \cap \mathrm{cl}\, C) \setminus \mathrm{rbd}\, C$$
$$\subset \mathrm{ap}\,(\mathrm{cl}\, C) \setminus \mathrm{rbd}\, C = \varnothing.$$

Therefore, $M \neq \mathbb{R}^n$, implying that $\dim M \leqslant n-1$. Denote by S the orthogonal complement of M and let $D = f_S(C)$ be the orthogonal projection of C on S. The set D is a convex cone with apex $\boldsymbol{o}$, as follows from Theorem 5.12. Furthermore, Theorem 6.25 implies that $\mathrm{cl}\, D$ is a line-free convex set, which shows that $\mathrm{ap}\,(\mathrm{cl}\, D) = \{\boldsymbol{o}\}$. By Theorem 3.43, $\mathrm{rint}\, D = f_S(\mathrm{rint}\, C) \neq \varnothing$. Clearly,

$$f_S(M) = \{\boldsymbol{o}\} \not\subset f_S(\mathrm{rint}\, C) = \mathrm{rint}\, D,$$

since otherwise there would be a point $\boldsymbol{u} \in f_S^{-1}(\boldsymbol{o}) \cap \mathrm{rint}\, C = M \cap \mathrm{rint}\, K$, contrary to $M \cap \mathrm{rint}\, C = \varnothing$.

Choose a nonzero vector $\boldsymbol{e} \in \mathrm{rint}\, D^\circ$. Then $\boldsymbol{e} \in S$ and $\boldsymbol{x} \cdot \boldsymbol{e} < 0$ for all $\mathrm{cl}\, D \setminus \{\boldsymbol{o}\}$ (see Theorem 8.6). Consequently, the subspace $T = \{\boldsymbol{x} \in \mathbb{R}^n : \boldsymbol{x} \cdot \boldsymbol{e} = 0\}$ supports D such that $T \cap \mathrm{cl}\, D = \{\boldsymbol{o}\}$. Because $\boldsymbol{e} \notin S^\perp$, Theorem 2.19 shows that the subspace $P = S \cap T$ has dimension $\dim S - 1$. Finally, let $H = M + S \cap T$. Then H is a hyperplane supporting $\mathrm{cl}\, C$ such that $H \cap \mathrm{cl}\, C = \mathrm{ap}\,(\mathrm{cl}\, C)$, as desired. □

Remarks. 1. The assertion of Theorem 9.45 cannot be modified by excluding the closure of C. Indeed, let

$$C = \{(x, y) : y > 0\} \cup \{(x, 0) : x \geqslant 0\} \quad \text{and} \quad L = \{\boldsymbol{o}\}.$$

Then $L \cap C = \mathrm{ap}\, C = \{\boldsymbol{o}\}$, and no line $H \subset \mathbb{R}^2$ satisfies the condition $H \cap C = \{\boldsymbol{o}\}$.

2. The assumption that C is not a plane is essential in Theorem 9.45. Indeed, if C is the x-axis of $\mathbb{R}^3$ and L is the yz-coordinate plane, then $L \cap C = \{\boldsymbol{o}\} \in \mathrm{ap}\,(\mathrm{cl}\, C)$, while $H = L$ is the only plane containing L (and not supporting C).

Theorem 9.46. *For a nonempty convex cone $C \subset \mathbb{R}^n$, the following assertions hold.*

(*a*) *If $\mathcal{H}$ is the family of hyperplanes each of them supporting C, then*

$$\mathrm{ap}\,(\mathrm{cl}\, C) = \cap\,(H : H \in \mathcal{H}).$$

(*b*) *If C is not a plane and $\mathcal{H}'$ is the family of hyperplanes each of them properly supporting C, then*

$$\mathrm{ap}\,(\mathrm{cl}\,C) = \cap\,(H : H \in \mathcal{H}').$$

(*c*) *If C is not a plane and $\mathcal{H}''$ is the family of hyperplanes each of them satisfying the condition $H \cap \mathrm{cl}\,C = \mathrm{ap}\,(\mathrm{cl}\,C)$, then*

$$\mathrm{ap}\,(\mathrm{cl}\,C) = \cap\,(H : H \in \mathcal{H}'').$$

Proof. (*a*) Choosing an apex $\boldsymbol{a}$ of C and translating C on $-\boldsymbol{a}$, we may assume that C is a convex cone with apex $\boldsymbol{o}$. Consequently, $\mathrm{cl}\,C$ is a convex cone with apex $\boldsymbol{o}$ (see Corollary 5.21), and Theorem 6.3 shows that $\mathrm{ap}\,(\mathrm{cl}\,C) = \mathrm{lin}\,(\mathrm{cl}\,C)$. By Theorem 9.45, every hyperplane $H \in \mathcal{H}$ is a subspace containing $\mathrm{ap}\,(\mathrm{cl}\,C)$.

Let $L = \cap\,(H : H \in \mathcal{H})$ and consider the family $\mathcal{F}$ of all closed halfspaces which contain C and are determined by the hyperplanes from $\mathcal{H}$. By Theorem 9.37, $\mathrm{cl}\,C = \cap\,(V : V \in \mathcal{F})$. This argument and Theorem 9.27 give $\mathrm{ap}\,(\mathrm{cl}\,C) = \mathrm{lin}\,(\mathrm{cl}\,C) = L$.

(*b*) Proof of this part is similar to that of (*a*).

(*c*) Since the inclusion $\mathrm{ap}\,(\mathrm{cl}\,C) \subset \cap\,(H : H \in \mathcal{H}'')$ is obvious, it remains to prove the opposite one. As above, we may suppose that $\mathrm{cl}\,C$ is a cone with apex $\boldsymbol{o}$ and $\mathrm{ap}\,(\mathrm{cl}\,C) = \mathrm{lin}\,(\mathrm{cl}\,C)$. Theorem 8.6 shows that an $(n-1)$-dimensional subspace $H = \{\boldsymbol{x} \in \mathbb{R}^n : \boldsymbol{x}\cdot\boldsymbol{e} = 0\}$ satisfies the condition $H \cap \mathrm{cl}\,C = \mathrm{ap}\,(\mathrm{cl}\,C)$ if and only if $\boldsymbol{e} \in \mathrm{rint}\,C^\circ$. Put $m = \dim\,(\mathrm{ap}\,(\mathrm{cl}\,C))$. A combination of Corollary 3.21 and Theorem 8.4 gives

$$\dim\,(\mathrm{rint}\,C^\circ) = \dim C^\circ = n - \dim\,(\mathrm{ap}\,(\mathrm{cl}\,C)) = n - m.$$

Choose in $\mathrm{rint}\,C^\circ$ a linearly independent set $\{\boldsymbol{e}_1, \ldots, \boldsymbol{e}_{n-m}\}$ and consider the $(n-1)$-dimensional subspaces

$$H_i = \{\boldsymbol{x} \in \mathbb{R}^n : \boldsymbol{x}\cdot\boldsymbol{e}_i = 0\}, \quad 1 \leqslant i \leqslant n-m.$$

By the above argument, $\mathrm{ap}\,(\mathrm{cl}\,C) \subset H_1 \cap \cdots \cap H_{n-m}$. Since the plane $H_1 \cap \cdots \cap H_{n-m}$ has dimension m (see Problem 2.7), Theorem 2.6 implies the equality $\mathrm{ap}\,(\mathrm{cl}\,C) = H_1 \cap \cdots \cap H_{n-m}$. Finally,

$$\cap\,(H : H \in \mathcal{H}'') \subset H_1 \cap \cdots \cap H_{n-m} = \mathrm{ap}\,(\mathrm{cl}\,C),$$

as desired. □

Theorem 9.47. *For a nonempty convex set $K \subset \mathbb{R}^n$, the following assertions hold.*

(*a*) $\mathrm{cl}\, K$ *is a convex cone if and only if* $\cap (H : H \in \mathcal{H}) \neq \varnothing$, *where* $\mathcal{H}$ *denotes the family of hyperplanes each supporting* K.

(*b*) *If, additionally,* K *is not a plane, then* $\mathrm{cl}\, K$ *is a convex cone if and only if* $\cap (H : H \in \mathcal{H}') \neq \varnothing$, *where* $\mathcal{H}'$ *denotes the family of hyperplanes each properly supporting* K.

Proof. (*a*) The "only if" part of the assertion immediately follows from Theorem 9.46. Conversely, suppose that $\cap (H : H \in \mathcal{H}) \neq \varnothing$ and choose a point $\boldsymbol{a} \in \cap (H : H \in \mathcal{H})$. We assert first that $\boldsymbol{a} \in \mathrm{cl}\, K$. Indeed, assume for a moment that $\boldsymbol{a} \notin \mathrm{cl}\, K$. By Theorem 9.31, there is a closed halfspace $V \subset \mathbb{R}^n$ supporting K and disjoint from $\boldsymbol{a}$. The boundary hyperplane H of V supports K and does not contain $\boldsymbol{a}$, contrary to the choice of $\boldsymbol{a}$. The obtained contradiction shows that $\boldsymbol{a}$ should be in $\mathrm{cl}\, K$.

Next, we assert that the set $M = \{\boldsymbol{a}\} \cup \mathrm{rint}\, K$ is a cone with apex $\boldsymbol{a}$. Since the case $K = \{\boldsymbol{a}\}$ is obvious, we may suppose that $K \neq \{\boldsymbol{a}\}$. Then $\mathrm{rint}\, K \neq \{\boldsymbol{a}\}$. Assume, for contradiction, the existence of a point $\boldsymbol{u} \in \mathrm{rint}\, K \setminus \{\boldsymbol{a}\}$ such that the closed halfline $h = [\boldsymbol{a}, \boldsymbol{u}\rangle$ does not lie in M. Because $[\boldsymbol{a}, \boldsymbol{u}] \subset M$ due to Theorem 3.47, the remaining open halfline $h' = h \setminus [\boldsymbol{a}, \boldsymbol{u}]$ does not lie in M. By Theorem 3.65, there is a unique point $\boldsymbol{z} \in h' \cap \mathrm{rbd}\, K$, and Corollary 9.12 shows the existence of a hyperplane H_0 properly supporting K at $\boldsymbol{z}$. We observe that H_0 does not contain $\boldsymbol{a}$ since otherwise $\boldsymbol{u} \in (\boldsymbol{a}, \boldsymbol{z}) \subset H_0$, contrary to the condition $H_0 \cap \mathrm{rint}\, K = \varnothing$. Thus $\boldsymbol{a}$ cannot be in $\cap (H : H \in \mathcal{H})$, a contradiction. Hence M is a cone with apex $\boldsymbol{a}$, which implies that $\mathrm{cl}\, K = \mathrm{cl}\, M$ is a convex cone with apex $\boldsymbol{a}$ (see Corollary 5.21).

(*b*) The proof of this part is similar to the proof of assertion (*a*), with reference on Theorem 9.31 instead of Theorem 9.31. □

Theorem 9.48. *Let* $K \subset \mathbb{R}^n$ *be a convex set which is not a plane, and let* $\boldsymbol{a}$ *be a point in* $\mathrm{rbd}\, K$. *Then the intersection of all closed halfspaces supporting* K *at* $\boldsymbol{a}$ *equals the closure of the generated cone* $C_{\boldsymbol{a}}(K)$.

Proof. Let $V \subset \mathbb{R}^n$ be a closed halfspace supporting K at $\boldsymbol{a}$ (see Theorem 9.31). If $\boldsymbol{x} \in K \setminus \{\boldsymbol{a}\}$, then, according to Theorem 2.35, the closed halfline $[\boldsymbol{a}, \boldsymbol{x}\rangle$ lies in V. Hence the generated cone $C_{\boldsymbol{a}}(K)$, as the union of such halflines, lies in V. Consequently, $\mathrm{cl}\, C_{\boldsymbol{a}}(K) \subset V$ due to the closedness of V (see Theorem 2.33). Hence $\mathrm{cl}\, C_{\boldsymbol{a}}(K)$ is included into the intersection, say D, of all closed halfspaces supporting K at $\boldsymbol{a}$.

For the opposite inclusion, choose any point $\boldsymbol{u} \in \mathbb{R}^n \setminus \mathrm{cl}\, C_{\boldsymbol{a}}(K)$. By Theorem 8.22, there is a unique point $\boldsymbol{v} \in \mathrm{cl}\, C_{\boldsymbol{a}}(K)$ such that the closed halfspace

$$V = \{\boldsymbol{x} \in \mathbb{R}^n : (\boldsymbol{x} - \boldsymbol{v})\cdot(\boldsymbol{u} - \boldsymbol{v}) \leqslant 0\}$$

contains $\operatorname{cl} C_{\boldsymbol{a}}(K)$. Since the boundary hyperplane H of V supports $\operatorname{cl} C_{\boldsymbol{a}}(K)$, and since $\boldsymbol{a}$ is an apex of $\operatorname{cl} C_{\boldsymbol{a}}(K)$ (see Corollary 5.21), Theorem 9.45 gives the inclusion $\boldsymbol{a} \in H$. This argument and the inclusions $\boldsymbol{a} \in K \subset \operatorname{cl} C_{\boldsymbol{a}}(K)$ show that V supports K at $\boldsymbol{a}$. Hence the set $D \subset V$ does not contain $\boldsymbol{u}$. Summing up, $D \subset \operatorname{cl} C_{\boldsymbol{a}}(K)$. □

Problems and Notes for Chapter 9

Problems for Chapter 9

Problem 9.1. Prove that a hyperplane $H = \{\boldsymbol{x} \in \mathbb{R}^n : \boldsymbol{x} \cdot \boldsymbol{e} = \gamma\}$ bounds (respectively, strongly bounds) a closed ball $B_\rho(\boldsymbol{a})$ if and only if $\rho\|\boldsymbol{e}\| \leqslant |\gamma - \boldsymbol{a}\cdot\boldsymbol{e}|$ (respectively, $\rho\|\boldsymbol{e}\| < |\gamma - \boldsymbol{a}\cdot\boldsymbol{e}|$). Consequently, H supports $B_\rho(\boldsymbol{a})$ if and only if $\rho\|\boldsymbol{e}\| = |\gamma - \boldsymbol{a}\cdot\boldsymbol{e}|$.

Problem 9.2. (Klee [173]) Let $K \subset \mathbb{R}^n$ be a nonempty proper convex set whose relative boundary does not contain a closed halfline, and $L \subset \mathbb{R}^n$ be a plane disjoint from $\operatorname{cl} K$. Prove the existence of a hyperplane which contains L and strictly bounds $\operatorname{cl} K$.

Problem 9.3. Let $K \subset \mathbb{R}^n$ be a nonempty convex set, $\boldsymbol{a}$ be a point in $\operatorname{cl} K$, and $U_\rho(\boldsymbol{a}) \subset \mathbb{R}^n$ be an open ball. Prove that a hyperplane $H \subset \mathbb{R}^n$ through $\boldsymbol{a}$ supports (respectively, properly supports) K if and only if H supports (respectively, properly supports) the set $U_\rho(\boldsymbol{a}) \cap K$.

Problem 9.4. (Dragomirescu [96]) Let $K \subset \mathbb{R}^n$ be a nonempty convex set, $\boldsymbol{a}$ be a point in $\mathbb{R}^n \setminus \operatorname{cl} K$, and $\boldsymbol{e}$ be a nonzero vector in $\operatorname{rint}(K - \boldsymbol{a})^\circ$. Prove that the hyperplane $H = \{\boldsymbol{x} \in \mathbb{R}^n : \boldsymbol{x}\cdot\boldsymbol{e} = \boldsymbol{a}\cdot\boldsymbol{e}\}$ strongly bounds K.

Problem 9.5. (Dragomirescu [96]) Let $K \subset \mathbb{R}^n$ be a nonempty convex set and $\boldsymbol{a}$ be a point in $\mathbb{R}^n \setminus \operatorname{cl} K$. Prove the equivalence of the following conditions.

(*a*) K is line-free.

(*b*) For every point $\boldsymbol{b} \in \mathbb{R}^n \setminus \{\boldsymbol{a}\}$, there is a hyperplane through $\boldsymbol{a}$ which misses $\boldsymbol{b}$ and strongly bounds K.

Problem 9.6. Let $K \subset \mathbb{R}^n$ be a convex set which is not a plane; F be a nonempty convex subset of $\operatorname{rbd} K$. Prove that $\operatorname{aff} F \cap \operatorname{rint} K = \varnothing$.

Problem 9.7. Let $C \subset \mathbb{R}^n$ be a nonempty proper convex cone with apex $\boldsymbol{o}$ and $L \subset \mathbb{R}^n$ be a proper subspace. Prove the equivalence of the following conditions.

(*a*) $L \cap \operatorname{cl} C \subset \operatorname{ap}(\operatorname{cl} C)$.

(*b*) $(L^\perp \setminus \{\boldsymbol{o}\}) \cap \operatorname{rint} C^\circ \neq \varnothing$.

Problem 9.8. (Steinitz [265, § 11]) Let $L \subset \mathbb{R}^n$ be a plane of positive dimension and X be a subset of L such that no halfplane of L contains X. Prove that $\operatorname{conv} X = L$.

Problem 9.9. Let X be a dense subset of the unit sphere $\mathbb{S}$ of $\mathbb{R}^n$ and Φ be a dense subset of $\mathbb{R}$. Given a nonempty line-free convex set $K \subset \mathbb{R}^n$, put $\mu_{\boldsymbol{e}} = \sup\{\boldsymbol{x}\cdot\boldsymbol{e} : \boldsymbol{x} \in K\}$, $\boldsymbol{e} \in X \cap \operatorname{nor} K$. Prove that $\operatorname{cl} K$ can be expressed as the intersection of a countable family $\mathcal{F}$ of closed halfspaces of the form

$$V_\gamma(\boldsymbol{e}) = \{\boldsymbol{x} \in \mathbb{R}^n : \boldsymbol{x}\cdot\boldsymbol{e} \leqslant \gamma\}, \ \boldsymbol{e} \in X \cap \operatorname{nor} K, \ \gamma \in \Phi \cap [\mu_{\boldsymbol{e}}, \infty). \tag{9.20}$$

Problem 9.10. Let $K \subset \mathbb{R}^n$ be a convex set which is not a plane, and $\mathcal{F}$ be a family of closed halfspaces whose intersection is $\operatorname{cl} K$. Prove that $\mathcal{F}$ contains a countable subfamily $\mathcal{G}$ such that $V \cap \operatorname{aff} K$ is a closed halfplane of $\operatorname{aff} K$ for all $V \in \mathcal{G}$ and $\operatorname{cl} K = \cap (V \cap \operatorname{aff} K : V \in \mathcal{G})$.

Problem 9.11. Let $X \subset \mathbb{R}^n$ be a nonempty set and

$$V_i = \{\boldsymbol{x} \in \mathbb{R}^n : \boldsymbol{x}\cdot\boldsymbol{e}_i \leqslant \mu_i\}, \quad 1 \leqslant i \leqslant r,$$

be closed halfspaces each containing X. Choose positive scalars $\lambda_1, \dots, \lambda_r$ such that

$$\boldsymbol{e} = \lambda_1\boldsymbol{e}_1 + \cdots + \lambda_r\boldsymbol{e}_r \neq \boldsymbol{o} \ \text{ and put } \ \mu = \lambda_1\mu_1 + \cdots + \lambda_r\mu_r.$$

Prove that the closed halfspace $V = \{\boldsymbol{x} \in \mathbb{R}^n : \boldsymbol{x}\boldsymbol{e} \leqslant \mu\}$ contains X. Furthermore, show that V properly (respectively, strictly, or strongly) contains X provided at least one of the halfspaces $V_1, \dots, V_r$ properly (respectively, strictly, or strongly) contains X.

Problem 9.12. Let f_L denote the orthogonal projection on a nonempty plane $L \subset \mathbb{R}^n$. Given a point $\boldsymbol{a} \in \mathbb{R}^n$ and a scalar $\rho > 0$, prove that

$$f_L(B_\rho(\boldsymbol{a})) = B_\rho(f_L(\boldsymbol{a})) \cap L \quad \text{and} \quad f_L(U_\rho(\boldsymbol{a})) = U_\rho(f_S(\boldsymbol{a})) \cap L.$$

Problem 9.13. Given an open halfspace $W = \{\boldsymbol{x} \in \mathbb{R}^n : \boldsymbol{x}\cdot\boldsymbol{e} < \gamma\}$, where $\boldsymbol{e}$ is a unit vector, and a point $\boldsymbol{a} \in \operatorname{bd} W$, prove the following assertions.

(a) The family of open balls $U_k(\boldsymbol{a} - k\boldsymbol{e})$, $k = 1, 2, \dots$, is nested, and W is their union.

(b) For any compact set $X \subset W$, there is an integer $m \geqslant 1$ such that $X \subset U_m(\boldsymbol{a} - m\boldsymbol{e})$.

Problem 9.14. Let $K \subset \mathbb{R}^n$ be a bounded convex set. Prove that $\operatorname{cl} K$ is the intersection of countably many closed balls (respectively, open balls) containing K.

Notes for Chapter 9

Bounding hyperplanes and containing halfspaces. Theorem 9.4 belongs to Rockafellar [241, Theorem 11.2] (see the previous articles of Mazur [210] and Bourbaki [45] for the case when K is an open convex set of a normed or topological vector space). The assertion of Corollary 9.7 is due to Carathéodory [65].

Following Kirsch [165], a family $\mathcal{F}$ of closed halfspaces of $\mathbb{R}^n$ is called *dense* provided for every affinely independent set $\{\boldsymbol{x}_0, \boldsymbol{x}_1, \ldots, \boldsymbol{x}_n\} \subset \mathbb{R}^n$ there is a halfspace $V \in \mathcal{F}$ such that $\boldsymbol{x}_1, \ldots, \boldsymbol{x}_n \in V$ and $\boldsymbol{x}_0 \notin V$. As proved in [165], given a dense family $\mathcal{F}$ of closed halfspaces in $\mathbb{R}^n$, every line-free closed convex set $K \subset \mathbb{R}^n$ is the intersection of closed halfspaces from $\mathcal{F}$ which contain K. Since every closed halfspace can be expressed as the intersection of denumerably many open halfspaces, Kirsch's result also holds for dense families of open halfspaces. Querió and Sà [234] proved the following results. Let S be a family of closed convex sets in $\mathbb{R}^n$, such that every closed convex set in the space is the intersection of a subfamily of S. Then every closed convex set in the space is the intersection of halfspaces belonging to S. They also obtained a sharper version of Problem 9.9 for the case of compact convex sets in $\mathbb{R}^n$ by showing that the condition on X to be dense in $\mathbb{S}$ is also necessary.

Clark [74] proved that a convex set $K \subset \mathbb{R}^n$ lies in a slab if and only if there is a finite upper bound for the radii of closed balls which are contained in K.

Support hyperplanes and halfspaces. Minkowski [214, §16] introduced the notion of support hyperplane of a set in $\mathbb{R}^n$ and showed that every point of a bounded convex hypersurface $S \subset \mathbb{R}^n$, containing $\boldsymbol{o}$ in its interior, belongs to a hyperplane supporting S. His proof uses a description of S as the "unit sphere" with respect to a real-valued function on $\mathbb{R}^n$, called the Minkowski distance nowadays (see [39] for further references). An extension of Minkowski's support theorem to the case of n-dimensional convex sets, possibly unbounded, was given by Carathéodory [66] and later by Steinitz [267, pp. 6–8].

If a closed set with interior points has the property that a support hyperplane passes through each of its boundary points, then the set is convex (see Bonnesen and Fenchel [39, pp. 6–7], Burago and Zalgaller [64], Mani-Levitska [206], and Valentine [285, Part IV] for various references on this result).

Theorem 9.43 can be found in the paper of Botts [44]. Problem 9.6 refines a result of Rockafellar [241, Theorem 11.6].

Asymptotic planes. Given an n-dimensional closed convex set $K \subset \mathbb{R}^n$, let $\alpha(K)$ denote the set of integers j between 1 and $n-1$ such that K admits an j-dimensional asymptotic plane. When K admits no boundary halfline, then $\alpha(K) = \varnothing$ or $\alpha(K) = \{1, \ldots, n-1\}$; when K is a convex cone, then $\alpha(K) = \varnothing$ or $\alpha(K) = \{1, \ldots, n-2\}$. Klee [177] posed the question to describe the sets $\alpha(K)$. The answer to this question was given by Klee [181] and Goossens [125]: For every set $J \subset \{1, \ldots, n-1\}$, there is an n-dimensional closed convex set $K \subset \mathbb{R}^n$ such that $\alpha(K) = J$.

Evenly convex set. According to Fenchel [104], a subset of $\mathbb{R}^n$ is called evenly convex if it is an intersection of open halfspaces. As follows from Theorem 9.24, every open or closed convex set in $\mathbb{R}^n$ is evenly convex. Various properties of evenly convex sets in $\mathbb{R}^n$ are studied by Goberna, Jornet, Rodríguez [121] and Klee, Maluta, Zanco [185]. The intersection of all evenly convex sets containing a given set $X \subset \mathbb{R}^n$ is called the evenly convex hull of X (see Goberna and Rodíguez [124] for some details).

Chapter 10

Separation Properties

10.1 Separation by Hyperplanes

10.1.1 *Separation of Arbitrary Sets*

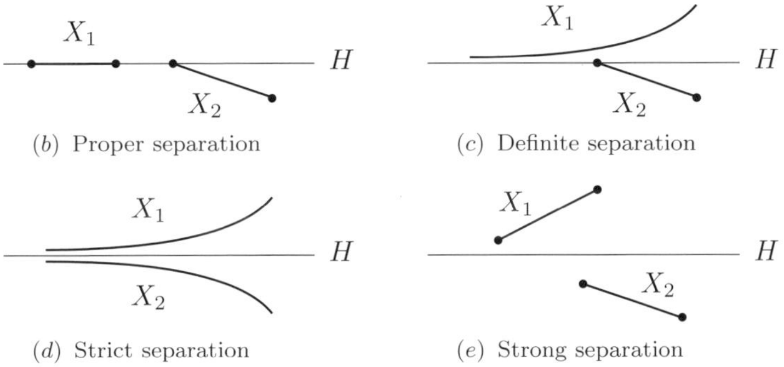

(b) Proper separation

(c) Definite separation

(d) Strict separation

(e) Strong separation

Fig. 10.1 Types of separation of sets.

Definition 10.1. Let X_1 and X_2 be nonempty sets in $\mathbb{R}^n$ and $H \subset \mathbb{R}^n$ be a hyperplane. We say that

(a) H *separates* X_1 and X_2 if X_1 and X_2 lie in the opposite closed half-spaces determined by H,
(b) H *properly separates* X_1 and X_2 if H separates X_1 and X_2 such that $X_1 \cup X_2 \not\subset H$,
(c) H *definitely separates* X_1 and X_2 if H separates X_1 and X_2 such that $X_1 \not\subset H$ and $X_2 \not\subset H$,
(d) H *strictly separates* X_1 and X_2 if H separates X_1 and X_2 such that both sets are disjoint from H (equivalently, X_1 and X_2 lie, respectively,

in the opposite open halfspaces determined by H),

(*e*) H *strongly separates* X_1 and X_2 if there is a scalar $\rho > 0$ such that the open ρ-neighborhoods $U_\rho(X_1)$ and $U_\rho(X_2)$ of these sets are separated by H.

Theorem 10.2. *For nonempty sets X_1 and X_2 in $\mathbb{R}^n$, the following assertions hold.*

(*a*) *X_1 and X_2 are separated by a hyperplane if and only if there is a nonzero vector $\boldsymbol{e} \in \mathbb{R}^n$ such that*

$$\sup\{\boldsymbol{x}\cdot\boldsymbol{e} : \boldsymbol{x} \in X_1\} \leqslant \inf\{\boldsymbol{x}\cdot\boldsymbol{e} : \boldsymbol{x} \in X_2\}. \tag{10.1}$$

(*b*) *X_1 and X_2 are properly separated by a hyperplane if and only if there is a nonzero vector $\boldsymbol{e} \in \mathbb{R}^n$ satisfying both* (10.1) *and*

$$\inf\{\boldsymbol{x}\cdot\boldsymbol{e} : \boldsymbol{x} \in X_1\} < \sup\{\boldsymbol{x}\cdot\boldsymbol{e} : \boldsymbol{x} \in X_2\}. \tag{10.2}$$

(*c*) *X_1 and X_2 are definitely separated by a hyperplane if and only if there is a nonzero vector $\boldsymbol{e} \in \mathbb{R}^n$ such that*

$$\begin{aligned} \inf\{\boldsymbol{x}\cdot\boldsymbol{e} : \boldsymbol{x} \in X_1\} &< \sup\{\boldsymbol{x}\cdot\boldsymbol{e} : \boldsymbol{x} \in X_1\} \\ &\leqslant \inf\{\boldsymbol{x}\cdot\boldsymbol{e} : \boldsymbol{x} \in X_2\} < \sup\{\boldsymbol{x}\cdot\boldsymbol{e} : \boldsymbol{x} \in X_2\}. \end{aligned} \tag{10.3}$$

(*d*) *X_1 and X_2 are strictly separated by a hyperplane if and only if there is a nonzero vector $\boldsymbol{e} \in \mathbb{R}^n$ such that both conditions below are satisfied:*

$$\begin{aligned} &\boldsymbol{u}\cdot\boldsymbol{e} < \inf\{\boldsymbol{x}\cdot\boldsymbol{e} : \boldsymbol{x} \in X_2\} \quad \textit{for every} \quad \boldsymbol{u} \in X_1, \\ &\sup\{\boldsymbol{x}\cdot\boldsymbol{e} : \boldsymbol{x} \in X_1\} < \boldsymbol{u}\cdot\boldsymbol{e} \quad \textit{for every} \quad \boldsymbol{u} \in X_2. \end{aligned} \tag{10.4}$$

(*e*) *X_1 and X_2 are strongly separated by a hyperplane if and only if there is a nonzero vector $\boldsymbol{e} \in \mathbb{R}^n$ such that*

$$\sup\{\boldsymbol{x}\cdot\boldsymbol{e} : \boldsymbol{x} \in X_1\} < \inf\{\boldsymbol{x}\cdot\boldsymbol{e} : \boldsymbol{x} \in X_2\}. \tag{10.5}$$

Proof. (*a*) Assume that X_1 and X_2 are separated by a hyperplane $H \subset \mathbb{R}^n$. According to Theorem 2.17, H can be expressed as

$$H = \{\boldsymbol{x} \in \mathbb{R}^n : \boldsymbol{x}\cdot\boldsymbol{e} = \gamma\}. \tag{10.6}$$

Denote by V_1 and V_2 the closed halfspaces determined by H (see Definition 2.29). We can write

$$V_1 = \{\boldsymbol{x} \in \mathbb{R}^n : \boldsymbol{x}\cdot\boldsymbol{e} \leqslant \gamma\} \quad \text{and} \quad V_2 = \{\boldsymbol{x} \in \mathbb{R}^n : \boldsymbol{x}\cdot\boldsymbol{e} \geqslant \gamma\}. \tag{10.7}$$

If $X_1 \subset V_1$ and $X_2 \subset V_2$, then

$$\sup\{\boldsymbol{x}\cdot\boldsymbol{e} : \boldsymbol{x} \in X_1\} \leqslant \gamma \leqslant \inf\{\boldsymbol{x}\cdot\boldsymbol{e} : \boldsymbol{x} \in X_2\}, \tag{10.8}$$

and if $X_1 \subset V_2$ and $X_2 \subset V_1$, then replacing $\boldsymbol{e}$ and γ with $-\boldsymbol{e}$ and $-\gamma$, respectively, we obtain (10.8). In either case, (10.1) holds.

Conversely, let X_1 and X_2 satisfy (10.1). Choose a scalar γ such that (10.8) holds. Then the closed halfspaces V_1 and V_2 from (10.7) contain X_1 and X_2, respectively. Hence X_1 and X_2 are separated by H.

(*b*) Let X_1 and X_2 be properly separated by a hyperplane H of the form (10.6). As above, we may assume that the inequalities (10.8) hold. Since $X_1 \cup X_2 \not\subset H$, at least one of the inequalities

$$\inf\{\boldsymbol{x}\cdot\boldsymbol{e} : \boldsymbol{x} \in X_1\} \leqslant \gamma \leqslant \sup\{\boldsymbol{x}\cdot\boldsymbol{e} : \boldsymbol{x} \in X_2\} \tag{10.9}$$

is strict, implying (10.2).

Conversely, if both (10.1) and (10.2) hold, then X_1 and X_2 are separated by a hyperplane H of the form (10.6), with γ satisfying (10.8). Clearly, (10.2) implies that $X_1 \cup X_2 \not\subset H$.

(*c*) If X_1 and X_2 are definitely separated by a hyperplane H, given in (10.6), then (10.3) is obvious. Conversely, if (10.3) holds and γ satisfies (10.8), then the hyperplane H from (10.6) definitely separates X_1 and X_2.

(*d*) Suppose that X_1 and X_2 are strictly separated by a hyperplane H, given by (10.6). Denote by W_1 and W_2 the open halfspaces determined by H (see Definition 2.29). We may assume that W_1 and W_2 are given by

$$W_1 = \{\boldsymbol{x} \in \mathbb{R}^n : \boldsymbol{x}\cdot\boldsymbol{e} < \gamma\} \quad \text{and} \quad W_2 = \{\boldsymbol{x} \in \mathbb{R}^n : \boldsymbol{x}\cdot\boldsymbol{e} > \gamma\}. \tag{10.10}$$

If $X_1 \subset W_1$ and $X_2 \subset W_2$, then

$$\boldsymbol{u}\cdot\boldsymbol{e} < \gamma \leqslant \inf\{\boldsymbol{x}\cdot\boldsymbol{e} : \boldsymbol{x} \in X_2\} \quad \text{for every} \quad \boldsymbol{u} \in X_1,$$
$$\sup\{\boldsymbol{x}\cdot\boldsymbol{e} : \boldsymbol{x} \in X_1\} \leqslant \gamma < \boldsymbol{u}\cdot\boldsymbol{e} \quad \text{for every} \quad \boldsymbol{u} \in X_2;$$

so, (10.4) is satisfied. If $X_1 \subset W_2$ and $X_2 \subset W_1$, then we replace $\boldsymbol{e}$ and γ with $-\boldsymbol{e}$ and $-\gamma$, respectively, to obtain (10.4).

Conversely, let (10.4) hold. With a scalar γ satisfying (10.8), the open halfspaces W_1 and W_2 from (10.10) contain X_1 and X_2, respectively. Hence X_1 and X_2 are strictly separated by H.

(*e*) Let X_1 and X_2 be strongly separated by a hyperplane H of the form (10.6). Then there is a scalar $\rho > 0$ such that the open ρ-neighborhoods $U_\rho(X_1)$ and $U_\rho(X_2)$ lie, respectively, in the closed halfspaces V_1 and V_2. Without loss of generality, we may suppose that V_1 and V_2 are given by (10.7) and $U_\rho(X_i) \subset V_i$, $i = 1, 2$. Choose a point $\boldsymbol{x} \in X_1$. Then

$$\boldsymbol{x} + \frac{\rho}{2\|\boldsymbol{e}\|}\boldsymbol{e} \in U_\rho(\boldsymbol{x}) \subset W_1,$$

which gives the following: $(\boldsymbol{x}+\frac{\rho}{2\|\mathbf{e}\|}\mathbf{e})\cdot\boldsymbol{e}<\gamma$, or $\boldsymbol{x}\cdot\boldsymbol{e}<\gamma-\frac{\rho\|\mathbf{e}\|}{2}$. Similarly, $\boldsymbol{x}\cdot\boldsymbol{e}>\gamma+\rho\|\mathbf{e}\|/2$ whenever $\boldsymbol{x}\in X_2$. Hence (10.5) holds.

Conversely, assume the existence of a nonzero vector $\boldsymbol{e}\in\mathbb{R}^n$ such that (10.5) holds. Let γ be a scalar satisfying

$$\sup\{\boldsymbol{x}\cdot\boldsymbol{e}:\boldsymbol{x}\in X_1\}<\gamma<\inf\{\boldsymbol{x}\cdot\boldsymbol{e}:\boldsymbol{x}\in X_2\}. \tag{10.11}$$

Since the linear functional $\varphi(\boldsymbol{x})=\boldsymbol{x}\cdot\boldsymbol{e}$ is continuous on $\mathbb{R}^n$ (see Problem 1.18), there is a scalar $\rho>0$ such that $\boldsymbol{x}\cdot\boldsymbol{e}<\gamma$ for all $\boldsymbol{x}\in U_\rho(X_1)$ and $\boldsymbol{x}\cdot\boldsymbol{e}>\gamma$ for all $\boldsymbol{x}\in U_\rho(X_2)$. This argument implies that the hyperplane (10.6) separates $U_\rho(X_1)$ and $U_\rho(X_2)$. □

Remark. In assertion (*e*) of Theorem 10.2, conditions (10.4) cannot be weakened to $\boldsymbol{x}_1\cdot\boldsymbol{e}<\boldsymbol{x}_2\cdot\boldsymbol{e}$ for all $\boldsymbol{x}_1\in X_1$ and $\boldsymbol{x}_2\in X_2$, even if both sets X_1 and X_2 are closed and convex. Indeed, let

$$X_1=\{(x,y):y\leqslant 0\}\quad\text{and}\quad X_2=\{(x,y):y\geqslant 2^x\}.$$

Clearly, the only vectors $\boldsymbol{e}\in\mathbb{R}^2$ to satisfy the inequalities $\boldsymbol{x}_1\cdot\boldsymbol{e}<\boldsymbol{x}_2\cdot\boldsymbol{e}$ for all $\boldsymbol{x}_1\in X_1$ and $\boldsymbol{x}_2\in X_2$ is of the form $\boldsymbol{e}=(a,0)$, where $a<0$, and the only line separating X_1 and X_2 is $l=\{(x,0):x\in\mathbb{R}\}$, which does not separates X_1 and X_2 strictly.

Corollary 10.3. *Let X_1 and X_2 be nonempty sets in $\mathbb{R}^n$ and $\mathbf{e}\in\mathbb{R}^n$ be a nonzero vector such that* (10.1) *is satisfied. The assertions below hold.*

(*a*) *A hyperplane of the form* (10.6) *separates X_1 and X_2 if and only if*

$$\sup\{\boldsymbol{x}\cdot\boldsymbol{e}:\boldsymbol{x}\in X_1\}\leqslant\gamma\leqslant\inf\{\boldsymbol{x}\cdot\boldsymbol{e}:\boldsymbol{x}\in X_2\}. \tag{10.12}$$

(*b*) *A hyperplane of the form* (10.6) *properly separates X_1 and X_2 if and only if both conditions* (10.12) *and*

$$\inf\{\boldsymbol{x}\cdot\boldsymbol{e}:\boldsymbol{x}\in X_1\}<\sup\{\boldsymbol{x}\cdot\boldsymbol{e}:\boldsymbol{x}\in X_2\}$$

are satisfied.

(*c*) *A hyperplane of the form* (10.6) *definitely separates X_1 and X_2 if and only if*

$$\inf\{\boldsymbol{x}\cdot\boldsymbol{e}:\boldsymbol{x}\in X_1\}<\sup\{\boldsymbol{x}\cdot\boldsymbol{e}:\boldsymbol{x}\in X_1\}\leqslant\gamma$$
$$\leqslant\inf\{\boldsymbol{x}\cdot\boldsymbol{e}:\boldsymbol{x}\in X_2\}<\sup\{\boldsymbol{x}\cdot\boldsymbol{e}:\boldsymbol{x}\in X_2\}.$$

(*d*) *A hyperplane of the form* (10.6) *strictly separates X_1 and X_2 if and only if $\boldsymbol{x}_1\cdot\boldsymbol{e}<\gamma<\boldsymbol{x}_2\cdot\boldsymbol{e}$ for all $\boldsymbol{x}_1\in X_1$ and $\boldsymbol{x}_2\in X_2$.*

(e) *A hyperplane of the form* (10.6) *strongly separates* X_1 *and* X_2 *if and only if* $\sup\{\boldsymbol{x}\cdot\boldsymbol{e} : \boldsymbol{x} \in X_1\} < \gamma < \inf\{\boldsymbol{x}\cdot\boldsymbol{e} : \boldsymbol{x} \in X_2\}$. $\square$

Sometimes it is convenient to deal with asymmetric types of set separation, as defined below.

Definition 10.4. Let nonempty sets X_1 and X_2 in $\mathbb{R}^n$ be separated by a hyperplane $H \subset \mathbb{R}^n$. We say that

(a) H *properly separates* X_1 *from* X_2 if $X_1 \not\subset H$,
(b) H *strictly separates* X_1 *from* X_2 if X_1 is disjoint from H,
(c) H *strongly separates* X_1 *from* X_2 if X_1 is strongly disjoint from H (see page 11).

The following examples illustrate Definition 10.4.

Examples. 1. The coordinate xy-plane of $\mathbb{R}^3$ is the only plane separating the planar disks $X_1 = \{(0, y, z) : y^2 + (z-1)^2 \leqslant 1\}$ and $X_2 = \{(x, y, 0) : x^2 + (y-1)^2 \leqslant 1\}$. This plane contains X_2 but not X_1.
2. The convex set $X_1 = \{(x, y) : y \geqslant 2^x\}$ is strictly separated from the closed halfplane $X_2 = \{(x, y) : y \leqslant 0\}$, while X_1 and X_2 are not strictly separated.

Similarly to Corollary 10.3, we can formulate the following criteria for the asymmetric types of set separation.

Corollary 10.5. *Let* X_1 *and* X_2 *be nonempty sets in* $\mathbb{R}^n$, *and* $\mathbf{e} \in \mathbb{R}^n$ *be a nonzero vector such that* (10.1) *is satisfied. The assertions below take place.*

(a) *A hyperplane of the form* (10.6) *properly separates* X_1 *from* X_2 *if and only if*
$$\inf\{\boldsymbol{x}\cdot\boldsymbol{e} : \boldsymbol{x} \in X_1\} < \sup\{\boldsymbol{x}\cdot\boldsymbol{e} : \boldsymbol{x} \in X_1\} \leqslant \gamma \leqslant \inf\{\boldsymbol{x}\cdot\boldsymbol{e} : \boldsymbol{x} \in X_2\}.$$
(b) *A hyperplane of the form* (10.6) *strictly separates* X_1 *from* X_2 *if and only if* $\boldsymbol{x}_1\cdot\boldsymbol{e} < \gamma \leqslant \boldsymbol{x}_2\cdot\boldsymbol{e}$ *for all* $\boldsymbol{x}_1 \in X_1$ *and* $\boldsymbol{x}_2 \in X_2$.
(c) *A hyperplane of the form* (10.6) *strongly separates* X_1 *from* X_2 *if and only if* $\sup\{\boldsymbol{x}\cdot\boldsymbol{e} : \boldsymbol{x} \in X_1\} < \gamma \leqslant \inf\{\boldsymbol{x}\cdot\boldsymbol{e} : \boldsymbol{x} \in X_2\}$. $\square$

Theorem 10.6. *Let* X_1 *and* X_2 *be nonempty sets and* H_1 *and* H_2 *be hyperplanes in* $\mathbb{R}^n$ *such that* H_i *separates (respectively, properly, strictly, or strongly)* X_i *from* X_{3-i}, $i = 1, 2$. *Then there is a hyperplane containing the set* $H_1 \cap H_2$ *and separating (respectively, properly, strictly, or strongly)* X_1 *and* X_2.

Proof. Let $H_i = \{\boldsymbol{x} \in \mathbb{R}^n : \boldsymbol{x}\cdot\boldsymbol{e}_i = \mu_i\}$, where $\boldsymbol{e}_i$ is a nonzero vector and μ_i is a scalar, $i = 1, 2$. Denote by V_i the closed halfspace determined by H_i and containing X_i. Without loss of generality, we may assume that

$$V_1 = \{\boldsymbol{x} \in \mathbb{R}^n : \boldsymbol{x}\cdot\boldsymbol{e}_1 \leqslant \mu_1\} \quad \text{and} \quad V_2 = \{\boldsymbol{x} \in \mathbb{R}^n : \boldsymbol{x}\cdot\boldsymbol{e}_2 \leqslant \mu_2\}.$$

Choose positive scalars λ_1 and λ_2 such that the vector $\boldsymbol{e} = \lambda_1\boldsymbol{e}_1 + \lambda_2(-\boldsymbol{e}_2)$ is not $\boldsymbol{o}$ and put

$$\mu = \lambda_1\mu_1 + \lambda_2(-\mu_2) \quad \text{and} \quad V = \{\boldsymbol{x} \in \mathbb{R}^n : \boldsymbol{x}\cdot\boldsymbol{e} \leqslant \gamma\}.$$

First, we assert that the set $H_1 \cap H_2$ lies in the hyperplane $H = \{\boldsymbol{x} \in \mathbb{R}^n : \boldsymbol{x}\cdot\boldsymbol{e} = \mu\}$. Indeed, this inclusion is obvious if $H_1 \cap H_2 = \varnothing$. Suppose that $H_1 \cap H_2 \neq \varnothing$ and choose any point $\boldsymbol{x} \in H_1 \cap H_2$. Then $\boldsymbol{x}\cdot\boldsymbol{e}_i = \mu_i$, $i = 1, 2$. Consequently,

$$\boldsymbol{x}\cdot\boldsymbol{e} = \boldsymbol{x}\cdot(\lambda_1\boldsymbol{e}_1 - \lambda_2\boldsymbol{e}_2) = \lambda_1\mu_1 - \lambda_2\mu_2 = \mu,$$

implying the inclusion $\boldsymbol{x} \in H$. Hence $H_1 \cap H_2 \subset H$.

Since the closed halfspace $V_2' = \{\boldsymbol{x} \in \mathbb{R}^n : \boldsymbol{x}\cdot(-\boldsymbol{e}_2) \leqslant -\mu_2\}$ contains X_1, Problem 9.11 shows that X_1 lies in V; furthermore, V properly contains (respectively, strictly, or strongly) X_1 provided V_1 properly contains (respectively, strictly, or strongly) X_1. We assert that X_2 lies in the opposite closed halfspace $V' = \{\boldsymbol{x} \in \mathbb{R}^n : \boldsymbol{x}\cdot\boldsymbol{e} \geqslant \mu\}$. Indeed, since

$$X_2 \subset V_1' = \{\boldsymbol{x} \in \mathbb{R}^n : \boldsymbol{x}\cdot\boldsymbol{e}_1 \geqslant \mu_1\},$$

any point $\boldsymbol{x} \in X_2$ satisfies the inequality

$$\boldsymbol{x}\cdot\boldsymbol{e} = \boldsymbol{x}\cdot(\lambda_1\boldsymbol{e}_1 + \lambda_2(-\boldsymbol{e}_2)) \geqslant \lambda_1\mu_1 + \lambda_2(-\mu_2) = \mu.$$

Summing up, H separates (respectively, properly, strictly, or strongly) X_1 from X_2 provided H_1 separates (respectively, properly, strictly, or strongly) X_1 from X_2.

A similar argument shows that the closed halfspace

$$V' = \{\boldsymbol{x} \in \mathbb{R}^n : \boldsymbol{x}\cdot(\lambda_1(-\boldsymbol{e}_1) + \lambda_2\boldsymbol{e}_2) \leqslant \lambda_1(-\mu_1) + \lambda_2\mu_2\}$$

properly contains (respectively, strictly, or strongly) X_2 provided H_2 properly (respectively, strictly, or strongly) separates X_2 from X_1. Consequently, H separates (respectively, properly, strictly, or strongly) X_2 from X_1 if H_2 separates (respectively, properly, strictly, or strongly) X_2 from X_1. □

The next theorem shows that a separation of two sets by a hyperplane can be reduced to the case of separation of a set from zero vector by a suitable parallel $(n-1)$-dimensional subspace.

Theorem 10.7. *For nonempty sets X_1 and X_2 in $\mathbb{R}^n$ and a hyperplane $H \subset \mathbb{R}^n$, the following assertions hold.*

(*a*) *A translate of* H *separates* (*respectively, strongly separates*) X_1 *and* X_2 *if and only if the* $(n-1)$*-dimensional subspace* $S = \operatorname{dir} H$ *separates* (*respectively, strongly separates*) $X_1 - X_2$ *and* $\{\boldsymbol{o}\}$.

(*b*) *A translate of* H *properly* (*respectively, strictly*) *separates one of the sets* X_1 *and* X_2 *from the other if and only if the* $(n-1)$*-dimensional subspace* $S = \operatorname{dir} H$ *properly* (*respectively, strictly*) *separates* $X_1 - X_2$ *from* $\{\boldsymbol{o}\}$.

Proof. According to Theorem 2.17, H can be expressed as

$$H = \{\boldsymbol{x} \in \mathbb{R}^n : \boldsymbol{x}\cdot\boldsymbol{e} = \gamma\}. \tag{10.13}$$

Theorem 2.18 shows that a translate H' of H and the $(n-1)$-dimensional subspace $S = \operatorname{dir} H$ are given by

$$H' = \{\boldsymbol{x} \in \mathbb{R}^n : \boldsymbol{x}\cdot\boldsymbol{e} = \gamma'\} \quad \text{and} \quad S = \{\boldsymbol{x} \in \mathbb{R}^n : \boldsymbol{x}\cdot\boldsymbol{e} = 0\}.$$

(a_1) Suppose that X_1 and X_2 are separated by a translate H' of H. As in Corollary 10.3, we may assume that

$$\sup\{\boldsymbol{x}\cdot\boldsymbol{e} : \boldsymbol{x} \in X_1\} \leqslant \gamma' \leqslant \inf\{\boldsymbol{x}\cdot\boldsymbol{e} : \boldsymbol{x} \in X_2\} \tag{10.14}$$

(in the case of opposite inequalities, we replace $\boldsymbol{e}$ and γ' with $-\boldsymbol{e}$ and $-\gamma'$, respectively). Then

$$\begin{aligned}\sup\{\boldsymbol{x}\cdot\boldsymbol{e} : \boldsymbol{x} \in X_1 - X_2\} &= \sup\{\boldsymbol{x}\cdot\boldsymbol{e} : \boldsymbol{x} \in X_1\} \\ &\quad - \inf\{\boldsymbol{x}\cdot\boldsymbol{e} : \boldsymbol{x} \in X_2\} \leqslant \gamma' - \gamma' = 0 = \boldsymbol{o}\cdot\boldsymbol{e}.\end{aligned} \tag{10.15}$$

The same Corollary 10.3 implies that S separates $X_1 - X_2$ and $\{\boldsymbol{o}\}$.

Conversely, let S separate $X_1 - X_2$ and $\{\boldsymbol{o}\}$. We may assume that $X_1 - X_2$ lies in the closed halfspace $V = \{\boldsymbol{x} \in \mathbb{R}^n : \boldsymbol{x}\cdot\boldsymbol{e} \leqslant 0\}$ (the case of the opposite closed halfspace is similar). Then

$$\begin{aligned}\sup\{\boldsymbol{x}\cdot\boldsymbol{e} : \boldsymbol{x} \in X_1\} &- \inf\{\boldsymbol{x}\cdot\boldsymbol{e} : \boldsymbol{x} \in X_2\} \\ &= \sup\{\boldsymbol{x}\cdot\boldsymbol{e} : \boldsymbol{x} \in X_1 - X_2\} \leqslant 0.\end{aligned}$$

Therefore, for any choice of γ' satisfying (10.14), the hyperplane H' separates X_1 and X_2.

(a_2) Suppose that X_1 and X_2 are strongly separated by a translate H' of H. As in Corollary 10.3, we may assume that

$$\sup\{\boldsymbol{x}\cdot\boldsymbol{e} : \boldsymbol{x} \in X_1\} < \gamma' < \inf\{\boldsymbol{x}\cdot\boldsymbol{e} : \boldsymbol{x} \in X_2\}$$

(in the case of opposite inequalities, we replace $\boldsymbol{e}$ and γ' with $-\boldsymbol{e}$ and $-\gamma'$, respectively). Choose a scalar $\delta > 0$ satisfying the conditions

$$\sup\{\boldsymbol{x}\cdot\boldsymbol{e} : \boldsymbol{x} \in X_1\} \leqslant \gamma' - \delta < \gamma' + \delta \leqslant \inf\{\boldsymbol{x}\cdot\boldsymbol{e} : \boldsymbol{x} \in X_2\}. \tag{10.16}$$

Then

$$\begin{aligned} \sup\{\boldsymbol{x}\cdot\boldsymbol{e} : \boldsymbol{x} \in X_1 - X_2\} &= \sup\{\boldsymbol{x}\cdot\boldsymbol{e} : \boldsymbol{x} \in X_1\} \\ &\quad - \inf\{\boldsymbol{x}\cdot\boldsymbol{e} : \boldsymbol{x} \in X_2\} \leqslant -2\delta < 0 = \boldsymbol{o}\cdot\boldsymbol{e}. \end{aligned} \tag{10.17}$$

The same Corollary 10.3 implies that S strongly separates $X_1 - X_2$ and $\{\boldsymbol{o}\}$.

Conversely, let S strongly separate $X_1 - X_2$ and $\{\boldsymbol{o}\}$. We may assume that $X_1 - X_2$ lies in the closed halfspace $V = \{\boldsymbol{x} \in \mathbb{R}^n : \boldsymbol{x}\cdot\boldsymbol{e} \leqslant \mu\}$ which misses $\boldsymbol{o}$. Then $\mu < 0$ (otherwise $\boldsymbol{o} \in V$) and

$$\begin{aligned} &\sup\{\boldsymbol{x}\cdot\boldsymbol{e} : \boldsymbol{x} \in X_1\} - \inf\{\boldsymbol{x}\cdot\boldsymbol{e} : \boldsymbol{x} \in X_2\} \\ &\qquad = \sup\{\boldsymbol{x}\cdot\boldsymbol{e} : \boldsymbol{x} \in X_1 - X_2\} \leqslant \mu < 0. \end{aligned}$$

Therefore, for any choice of γ' satisfying (10.16), the hyperplane H' strongly separates X_1 and X_2.

(b_1) Suppose that a translate H' of H properly separates X_1 from X_2 (the case of nontrivial separation of X_2 from X_1 is similar). We may assume that (10.14) holds, with $\inf\{\boldsymbol{x}\cdot\boldsymbol{e} : \boldsymbol{x} \in X_1\} < \gamma'$ (see Corollary 10.5). As in part (a), the set $X_1 - X_2$ lies in the closed halfspace $\{\boldsymbol{x} \in \mathbb{R}^n : \boldsymbol{x}\cdot\boldsymbol{e} \leqslant 0\}$ determined by S. Finally, the inequality

$$\begin{aligned} \inf\{\boldsymbol{x}\cdot\boldsymbol{e} : \boldsymbol{x} \in X_1 - X_2\} &= \inf\{\boldsymbol{x}\cdot\boldsymbol{e} : \boldsymbol{x} \in X_1\} \\ &\quad - \sup\{\boldsymbol{x}\cdot\boldsymbol{e} : \boldsymbol{x} \in X_2\} < \gamma' - \gamma' = 0 = \boldsymbol{o}\cdot\boldsymbol{e} \end{aligned}$$

shows that S properly separates $X_1 - X_2$ from $\{\boldsymbol{o}\}$.

Conversely, let S properly separate $X_1 - X_2$ from $\{\boldsymbol{o}\}$. Assume that $X_1 - X_2$ lies in the closed halfspace $V = \{\boldsymbol{x} \in \mathbb{R}^n : \boldsymbol{x}\cdot\boldsymbol{e} \leqslant 0\}$ (the case of the opposite closed halfspace is similar), with $\inf\{\boldsymbol{x}\cdot\boldsymbol{e} : \boldsymbol{x} \in X_1 - X_2\} < 0$. As above, for any choice of γ' satisfying (10.14), the hyperplane H' separates X_1 and X_2. The inequality

$$\begin{aligned} &\inf\{\boldsymbol{x}\cdot\boldsymbol{e} : \boldsymbol{x} \in X_1\} - \sup\{\boldsymbol{x}\cdot\boldsymbol{e} : \boldsymbol{x} \in X_2\} \\ &\qquad = \inf\{\boldsymbol{x}\cdot\boldsymbol{e} : \boldsymbol{x} \in X_1 - X_2\} < \boldsymbol{o}\cdot\boldsymbol{e} = 0 \end{aligned}$$

shows that at least one of the sets X_1 and X_2 does not lie in H'. Hence H' properly separates one of the sets X_1 and X_2 from the other.

(b_2) Suppose that a translate H' of H strictly separates X_1 from X_2 (the case of strict separation of X_2 from X_1 is similar). We may assume that (10.14) holds, with $\boldsymbol{x}_1\cdot\boldsymbol{e} < \gamma \leqslant \boldsymbol{x}_2\cdot\boldsymbol{e}$ whenever $\boldsymbol{x}_1 \in X_1$ and $\boldsymbol{x}_2 \in X_2$. Consequently,

$$(\boldsymbol{x}_1 - \boldsymbol{x}_2)\cdot\boldsymbol{e} < 0 = \boldsymbol{o}\cdot\boldsymbol{e} \quad \text{for all} \quad \boldsymbol{x}_1 - \boldsymbol{x}_2 \in X_1 - X_2,$$

and Corollary 10.5 shows that S strictly separates $X_1 - X_2$ from $\{\boldsymbol{o}\}$.

Conversely, let S strictly separate $X_1 - X_2$ from $\{\boldsymbol{o}\}$. Assume that $X_1 - X_2$ lies in the open halfspace $W = \{\boldsymbol{x} \in \mathbb{R}^n : \boldsymbol{x}\cdot\boldsymbol{e} < 0\}$ (the case of the opposite open halfspace is similar). So, $(\boldsymbol{x}_1 - \boldsymbol{x}_2)\cdot\boldsymbol{e} < 0$ whenever $\boldsymbol{x}_1 - \boldsymbol{x}_2 \in X_1 - X_2$. Equivalently, $\boldsymbol{x}_1\cdot\boldsymbol{e} < \boldsymbol{x}_2\cdot\boldsymbol{e}$ for all $\boldsymbol{x}_1 \in X_1$ and $\boldsymbol{x}_2 \in X_2$. Let

$$\gamma_1 = \sup\{\boldsymbol{x}_1\cdot\boldsymbol{e} : \boldsymbol{x}_1 = X_1\} \quad \text{and} \quad \gamma_2 = \sup\{\boldsymbol{x}_2\cdot\boldsymbol{e} : \boldsymbol{x}_2 = X_2\}.$$

Then $\gamma_1 \leqslant \gamma_2$ due to the inclusion $X_1 - X_2 \subset W$ and the inequality

$$\begin{aligned}\gamma_1 - \gamma_2 &= \sup\{\boldsymbol{x}\cdot\boldsymbol{e} : \boldsymbol{x} \in X_1\} - \inf\{\boldsymbol{x}\cdot\boldsymbol{e} : \boldsymbol{x} \in X_2\} \\ &= \sup\{\boldsymbol{x}\cdot\boldsymbol{e} : \boldsymbol{x} \in X_1 - X_2\} \leqslant 0.\end{aligned}$$

If $\gamma_1 < \gamma_2$, then, by assertion (a), the hyperplane H' strongly separates X_1 and X_2 for any choice of $\gamma' \in (\gamma_1, \gamma_2)$. Let $\gamma_1 = \gamma_2$ and put $\gamma' = \gamma_1 = \gamma_2$. If there is a point $\boldsymbol{x}_2 \in X_2$ such that $\boldsymbol{x}_2\cdot\boldsymbol{e} = \gamma'$, then $\boldsymbol{x}_1\cdot\boldsymbol{e} < \gamma'$ for all $\boldsymbol{x}_1 \in X_1$, implying that H' strictly separates X_1 from X_2. Similarly, if there is a point $\boldsymbol{x}_1 \in X_1$ such that $\boldsymbol{x}_1\cdot\boldsymbol{e} = \gamma'$, then $\gamma' < \boldsymbol{x}_2\cdot\boldsymbol{e}$ for all $\boldsymbol{x}_2 \in X_2$, implying that H' strictly separates X_2 from X_1. □

10.1.2 *Separation of Convex Sets*

Theorem 10.8. *For nonempty convex sets K_1 and K_2 in $\mathbb{R}^n$, the following conditions are equivalent.*

(a) *K_1 and K_2 are properly separated by a hyperplane.*
(b) $\operatorname{rint} K_1 \cap \operatorname{rint} K_2 = \varnothing$.

Proof. $(a) \Rightarrow (b)$ Assume, for instance, that a hyperplane $H \subset \mathbb{R}^n$ properly separates K_1 from K_2. Denote by V_1 and V_2 the closed halfspaces determined by H. Without loss of generality, we may assume that $K_1 \subset V_1$ and $K_2 \subset V_2$. Since $K_1 \not\subset H$, one has $H \cap \operatorname{rint} K_1 = \varnothing$ (see Theorem 3.67). Hence $\operatorname{rint} K_1 \subset V_1 \setminus H = \operatorname{int} V_1$, implying that

$$\operatorname{rint} K_1 \cap \operatorname{rint} K_2 \subset \operatorname{int} V_1 \cap V_2 = \varnothing.$$

$(b) \Rightarrow (a)$ By Theorem 3.10, the set $K = K_1 - K_2$ is convex. Since $\operatorname{rint} K = \operatorname{rint} K_1 - \operatorname{rint} K_2$ (see Theorem 3.34), one has $\boldsymbol{o} \notin \operatorname{rint} K$. Corollary 9.7 shows the existence of a hyperplane $H \subset \mathbb{R}^n$ through $\boldsymbol{o}$ properly bounding K. Equivalently, H properly separates K from $\{\boldsymbol{o}\}$, Theorem 10.7 implies the existence of a translate of H properly separating K_1 and K_2. □

Corollary 10.9. *Nonempty convex sets K_1 and K_2 in $\mathbb{R}^n$ are separated by a hyperplane if and only if any of the following conditions is satisfied.*

(*a*) $\dim(K_1 \cup K_2) \leqslant n-1$.
(*b*) $\operatorname{rint} K_1 \cap \operatorname{rint} K_2 = \varnothing$. □

Proof. Let K_1 and K_2 be separated by a hyperplane $H \subset \mathbb{R}^n$. If $K_1 \cup K_2 \subset H$, then $\dim(K_1 \cup K_2) \leqslant \dim H = n-1$. Suppose that $K_1 \cup K_2 \not\subset H$. Then H properly separates K_1 and K_2, and Theorem 10.8 implies condition (*b*).

Conversely, if condition (*a*) holds, then a hyperplane containing $K_1 \cup K_2$ trivially separates them. If condition (*b*) holds, then Theorem 10.8 implies the existence of a separating hyperplane. □

Theorem 10.10. *Let* $K_1 \subset \mathbb{R}^n$ *be a convex set which is not a plane and whose relative boundary does not contain a closed halfline. If a nonempty convex set* $K_2 \subset \mathbb{R}^n$ *satisfies the condition* $\operatorname{cl} K_1 \cap \operatorname{cl} K_2 = \varnothing$, *then there is a hyperplane strictly separating* $\operatorname{cl} K_1$ *from* $\operatorname{cl} K_2$.

Proof. Since $\operatorname{rint} K_1 \cap \operatorname{rint} K_2 \subset \operatorname{cl} K_1 \cap \operatorname{cl} K_2 = \varnothing$, Theorem 10.8 shows the existence of a hyperplane $H \subset \mathbb{R}^n$ properly separating $\operatorname{cl} K_1$ from $\operatorname{cl} K_2$.

If $H \cap \operatorname{cl} K_1 = \varnothing$, then H is a desired hyperplane. Assume that H meets $\operatorname{cl} K_1$. Then K_1 is not a plane (otherwise $K_1 \subset H$ according to Theorem 2.34). Consequently, $\operatorname{rbd} K_1 \neq \varnothing$ (see Theorem 3.62). According to Theorem 9.3, $H \cap \operatorname{cl} K_1 \subset \operatorname{rbd} K_1$. By the assumption, the set $H \cap \operatorname{rbd} K_1$ contains no closed halfline. Therefore, Theorem 6.12 implies that $H \cap \operatorname{cl} K_1$ is a nonempty compact set. According to Theorem 6.40, $\operatorname{dir} H \cap \operatorname{rec}(\operatorname{cl} K_1) = \{\boldsymbol{o}\}$.

By Theorem 2.17, we can write $H = \{\boldsymbol{x} \in \mathbb{R}^n : \boldsymbol{x} \cdot \boldsymbol{e} = \gamma\}$. Denote by V_1 and V_2 the closed halfspaces of $\mathbb{R}^n$ determined by H and containing K_1 and K_2, respectively. Without loss of generality, we may assume that $V_1 = \{\boldsymbol{x} \in \mathbb{R}^n : \boldsymbol{x} \cdot \boldsymbol{e} \leqslant \gamma\}$ (see Definition 2.29). Choose a scalar $\gamma' < \gamma$ and consider the hyperplane $H' = \{\boldsymbol{x} \in \mathbb{R}^n : \boldsymbol{x} \cdot \boldsymbol{e} = \gamma'\}$. Denote by P the closed slab between H and H'. The set $\operatorname{cl} K_1 \cap P$ is compact (see Problem 6.11). Since $\operatorname{cl} K_1 \cap P$ is disjoint from $\operatorname{cl} K_2$, there is a scalar $\rho > 0$ such that the open ρ-neighborhoods $U_\rho(\operatorname{cl} K_1 \cap P)$ and $U_\rho(K_2)$ do not meet (see Problems 1.14 and 1.15).

If $\operatorname{cl} K_1 \subset P$, then $U_\rho(\operatorname{cl} K_1) = U_\rho(\operatorname{cl} K_1 \cap P)$, and Theorem 10.8 implies the existence of a hyperplane G which properly separates $U_\rho(\operatorname{cl} K_1)$ and $U_\rho(\operatorname{cl} K_2)$. Denote by Q_1 and Q_2 the closed halfspaces determined by G and containing $U_\rho(\operatorname{cl} K_1)$ and $U_\rho(\operatorname{cl} K_2)$, respectively. Since both sets $U_\rho(\operatorname{cl} K_1)$ and $U_\rho(\operatorname{cl} K_2)$ are n-dimensional, none of them lies in G. Furthermore, the set $U_\rho(\operatorname{cl} K_1)$ is open, which gives $U_\rho(\operatorname{cl} K_1) \subset Q_1 \setminus G$ due to Theorem 3.67.

Consequently,

$$G \cap \operatorname{cl} K_1 \subset G \cap U_\rho(\operatorname{cl} K_1) = \varnothing,$$

and G is a desired hyperplane.

Suppose now that $\operatorname{cl} K_1 \not\subset P$. Then $\operatorname{cl} K_1$ meets the closed halfspace $V_1' = \{\boldsymbol{x} \in \mathbb{R}^n : \boldsymbol{x} \cdot \boldsymbol{e} \leqslant \gamma'\}$. Furthermore,

$$\delta(\operatorname{cl} K_1 \cap V_1', \operatorname{cl} K_2) \geqslant \delta(\operatorname{cl} K_1 \cap V_1', V_2) \geqslant \delta(H, H') = |\gamma - \gamma'|/\|\boldsymbol{e}\|,$$

as follows from Corollary 2.92. Choose a positive scalar r satisfying the condition $r < |\gamma - \gamma'|/(2\|\boldsymbol{e}\|)$. Then

$$U_r(\operatorname{cl} K_1 \cap V_1') \cap U_r(K_2) \subset U_r(V_1') \cap U_r(V_2) = \varnothing.$$

This argument and the equalities

$$\begin{aligned} U_r(\operatorname{cl} K_1) &= U_r((\operatorname{cl} K_1 \cap P) \cup (\operatorname{cl} K_1 \cap V_1')) \\ &= U_r(\operatorname{cl} K_1 \cap P) \cup U_r(\operatorname{cl} K_1 \cap V_1') \end{aligned}$$

show that $U_r(\operatorname{cl} K_1) \cap U_r(K_2) = \varnothing$. As above, there is a hyperplane G strictly separating $\operatorname{cl} K_1$ from $\operatorname{cl} K_2$. □

A combination of Theorems 10.6 and 10.10 implies the following corollary.

Corollary 10.11. *Let K_1 and K be convex sets in $\mathbb{R}^n$ both distinct from planes and such that $\operatorname{cl} K_1 \cap \operatorname{cl} K_2 = \varnothing$. If none of the sets $\operatorname{rbd} K_1$ and $\operatorname{rbd} K_2$ contains a closed halfline, then there is a hyperplane strictly separating $\operatorname{cl} K_1$ and $\operatorname{cl} K_2$.* □

Theorem 10.12. *If K_1 and K_2 are nonempty convex sets in $\mathbb{R}^n$, then the following conditions are equivalent.*

(*a*) *K_1 and K_2 are strongly separated by a hyperplane.*
(*b*) *$\delta(K_1, K_2) > 0$.*
(*c*) *$\boldsymbol{o} \notin \operatorname{cl}(K_1 - K_2)$.*

Proof. Because the equivalence of conditions (*b*) and (*c*) is shown in Problem 1.15, it suffices to prove that (*a*) ⇔ (*b*).

Assume first that K_1 and K_2 are strongly separated by a hyperplane $H \subset \mathbb{R}^n$. Equivalently, there is a scalar $\rho > 0$ such that the open ρ-neighborhoods $U_\rho(K_1)$ and $U_\rho(K_2)$ of these sets lie in the opposite halfspaces determined by H. Consequently, $U_\rho(K_1) \cap U_\rho(K_2) = \varnothing$, and Problem 1.15 gives the inequality $\delta(K_1, K_2) \geqslant 2\rho > 0$.

Conversely, suppose that $\delta(K_1, K_2) > 0$. By Problem 1.15, there is a scalar $\rho > 0$ such that $U_\rho(K_1) \cap U_\rho(K_2) = \varnothing$. Theorem 10.8 shows that $U_\rho(K_1)$ and $U_\rho(K_2)$ are separated by a hyperplane $H \subset \mathbb{R}^n$. Denote by V_1 and V_2 the closed halfspaces determined by H and containing $U_\rho(K_1)$ and $U_\rho(K_2)$, respectively. Since both sets $U_\rho(K_1)$ and $U_\rho(K_2)$ are n-dimensional, none of them lies in H. Furthermore, the sets $U_\rho(K_1)$ and $U_\rho(K_2)$ are open, which gives $U_\rho(K_i) \subset V_i \setminus H$, $i = 1, 2$, due to Theorem 3.67. Hence K_1 and K_2 are strongly separated by H. □

A combination of Problem 1.14 and Theorem 10.12 implies the following corollary.

Corollary 10.13. *Let K_1 and K_2 be nonempty convex sets in $\mathbb{R}^n$. If K_1 is bounded and $\operatorname{cl} K_1 \cap \operatorname{cl} K_2 = \varnothing$, then K_1 and K_2 are strongly separated by a hyperplane.* □

10.1.3 *Separation of Convex Cones*

If nonempty convex cones C_1 and C_2 in $\mathbb{R}^n$, with a common apex $\boldsymbol{a}$, are separated by a hyperplane $H \subset \mathbb{R}^n$, then H supports both C_1 and C_2. Consequently, Theorem 9.43 gives the inclusion

$$\operatorname{ap}(\operatorname{cl} C_1) \cup \operatorname{ap}(\operatorname{cl} C_2) \subset H. \tag{10.18}$$

In this regard, we introduce the following definition.

Definition 10.14. Let C_1 and C_2 be nonempty convex cones in $\mathbb{R}^n$, with a common apex $\boldsymbol{a}$. We say that a hyperplane $H \subset \mathbb{R}^n$ *sharply separates* C_1 and C_2 provided H separates C_1 and C_2 such that

$$H \cap \operatorname{cl} C_1 = \operatorname{ap}(\operatorname{cl} C_1) \quad \text{and} \quad H \cap \operatorname{cl} C_2 = \operatorname{ap}(\operatorname{cl} C_2). \tag{10.19}$$

Remarks. 1. Conditions (10.19) are equivalent to the inclusions

$$H \cap \operatorname{cl} C_1 \subset \operatorname{ap}(\operatorname{cl} C_1) \quad \text{and} \quad H \cap \operatorname{cl} C_2 \subset \operatorname{ap}(\operatorname{cl} C_2). \tag{10.20}$$

2. Conditions (10.19) are more restrictive then (10.18). Indeed, consider the closed convex cones

$$C_1 = \{(x, y, z) : z \leqslant 0\} \quad \text{and} \quad C_2 = \{(x, y, z) : x \geqslant 0, z \geqslant 0\}.$$

Then $\operatorname{ap} C_1$ coincides with the plane $H = \{(x, y, z) : z = 0\}$, while $\operatorname{ap} C_2$ is the y-axis of $\mathbb{R}^3$. Clearly, H separates C_1 and C_2 such that (10.18) holds, while $H \cap C_2 \not\subset \operatorname{ap} C_2$.

3. Corollary 5.27 shows that sharp separation of convex cones C_1 and C_2 is equivalent to their proper separation if and only if neither C_1 nor C_2 is a plane.

Theorem 10.15. *Let C_1 and C_2 be nonempty convex cones in $\mathbb{R}^n$, with a common apex $\boldsymbol{a}$. The following conditions are equivalent.*

(*a*) *C_1 and C_2 are sharply separated by a hyperplane.*
(*b*) $\mathrm{cl}\, C_1 \cap \mathrm{cl}\, C_2 = \mathrm{ap}\,(\mathrm{cl}\, C_1) \cap \mathrm{ap}\,(\mathrm{cl}\, C_2)$ *and at least one of the conditions below holds.*

(b_1) $\dim (C_1 \cup C_2) \leqslant n - 1$.
(b_2) *At least one of the cones C_1 and C_2 is not a plane.*

Proof. Translating both cones C_1 and C_2 on $-\boldsymbol{a}$, we may suppose that $\boldsymbol{a} = \boldsymbol{o}$ is their common apex. By Corollary 5.21, both sets $\mathrm{cl}\, C_1$ and $\mathrm{cl}\, C_2$ are closed convex cones with apex $\boldsymbol{o}$. A combination of Theorems 6.3 and 6.21 shows that

$$\mathrm{rec}\,(\mathrm{cl}\, C_i) = \mathrm{cl}\, C_i \quad \text{and} \quad \mathrm{lin}\,(\mathrm{cl}\, C_i) = \mathrm{ap}\,(\mathrm{cl}\, C_i), \quad i = 1, 2.$$

(a) $\Rightarrow$ (b) Denote by V_1 and V_2 the closed halfspaces determined by H and containing C_1 and C_2, respectively. Then

$$\begin{aligned} \mathrm{cl}\, C_1 \cap \mathrm{cl}\, C_2 &= (\mathrm{cl}\, C_1 \cap V_1) \cap (\mathrm{cl}\, C_2 \cap V_2) = (\mathrm{cl}\, C_1 \cap \mathrm{cl}\, C_2) \cap (V_1 \cap V_2) \\ &= (\mathrm{cl}\, C_1 \cap \mathrm{cl}\, C_2) \cap H = (\mathrm{cl}\, C_1 \cap H) \cap (\mathrm{cl}\, C_2 \cap H) \\ &= \mathrm{ap}\,(\mathrm{cl}\, C_1) \cap \mathrm{ap}\,(\mathrm{cl}\, C_2). \end{aligned}$$

If both cones C_1 and C_2 are subspaces, then Problem 10.2 shows that condition (b_1) holds.

(b) $\Rightarrow$ (a) If both cones C_1 and C_2 are subspaces, then $\dim (C_1 \cup C_2) \leqslant n - 1$ according to a combination of conditions (b_1) and (b_2). Choose a hyperplane $H \subset \mathbb{R}^n$ containing $C_1 \cup C_2$. Clearly, H sharply separates C_1 and C_2.

Let C_1 be a subspace and C_2 be not. Then $C_1 \neq \mathbb{R}^n$ and Theorem 9.45 shows the existence of a hyperplane $H \subset \mathbb{R}^n$ which contains C_1 and satisfies the condition $H \cap \mathrm{cl}\, C_2 = \mathrm{ap}\,(\mathrm{cl}\, C_2)$. Since $\mathrm{ap}\,(\mathrm{cl}\, C_1) = \mathrm{cl}\, C_1$, the hyperplane H sharply separates C_1 and C_2.

Finally, assume that both cones C_1 and C_2 are not subspaces. In particular, $\mathrm{cl}\, C_1 \neq \mathrm{ap}\,(\mathrm{cl}\, C_1)$. Denote by E_1 the orthogonal complement of $\mathrm{ap}\,(\mathrm{cl}\, C_1)$ within $\mathrm{span}\,(\mathrm{cl}\, C_1)$. A combination of Theorems 6.29 and 6.37 shows that $\mathrm{cl}\, C_1$ is expressible as $\mathrm{cl}\, C_1 = (\mathrm{cl}\, C_1 \cap E_1) + \mathrm{ap}\,(\mathrm{cl}\, C_1)$, where the convex cone $F_1 = \mathrm{cl}\, C_1 \cap E_1$ with apex $\boldsymbol{o}$ is line-free. Theorem 8.17 gives the existence of a compact convex set $B_1 \subset F_1 \setminus \{\boldsymbol{o}\} = F_1 \setminus \mathrm{ap}\,(\mathrm{cl}\, C_1)$ such that $F_1 = C_{\boldsymbol{o}}(B_1)$.

Consider the convex cone $C = \mathrm{ap}\,(\mathrm{cl}\, C_1) + \mathrm{cl}\, C_2$. Condition (b) gives

$$\begin{aligned}\mathrm{lin}\,(\mathrm{cl}\, C_1) \cap \mathrm{cl}\, C_2 &\subset \mathrm{cl}\, C_1 \cap \mathrm{cl}\, C_2 \\ &= \mathrm{lin}\,(\mathrm{cl}\, C_1) \cap \mathrm{lin}\,(\mathrm{cl}\, C_2) \subset \mathrm{lin}\,(\mathrm{cl}\, C_1) \cap \mathrm{cl}\, C_2.\end{aligned}$$

Hence $\mathrm{lin}\,(\mathrm{cl}\, C_1) \cap \mathrm{cl}\, C_2 = \mathrm{lin}\,(\mathrm{cl}\, C_1) \cap \mathrm{lin}\,(\mathrm{cl}\, C_2)$, and Theorem 7.18 shows that the cone C is closed and $\mathrm{lin}\, C$ coincides with the subspace $L = \mathrm{lin}\,(\mathrm{cl}\, C_1) + \mathrm{lin}\,(\mathrm{cl}\, C_2)$. We observe that $B_1 \cap C = \varnothing$. Indeed, suppose the existence of a point $\boldsymbol{x} \in B_1 \cap C$. Then $\boldsymbol{x} \in E_1 \setminus \{\boldsymbol{o}\}$ and $\boldsymbol{x} = \boldsymbol{y} + \boldsymbol{z}$ for suitable $\boldsymbol{x} \in \mathrm{lin}\,(\mathrm{cl}\, C_1)$ and $\boldsymbol{z} \in \mathrm{cl}\, C_2$. Consequently,

$$\boldsymbol{z} = \boldsymbol{x} - \boldsymbol{y} \in (F_1 \setminus \{\boldsymbol{o}\}) + \mathrm{lin}\,(\mathrm{cl}\, C_1) \subset \mathrm{cl}\, C_1 \setminus \mathrm{lin}\,(\mathrm{cl}\, C_1).$$

This argument and condition (b) give

$$\begin{aligned}\boldsymbol{z} &\in (\mathrm{cl}\, C_1 \setminus \mathrm{lin}\,(\mathrm{cl}\, C_1)) \cap \mathrm{cl}\, C_2 = (\mathrm{cl}\, C_1 \cap \mathrm{cl}\, C_2) \setminus \mathrm{lin}\,(\mathrm{cl}\, C_1) \\ &= (\mathrm{lin}\,(\mathrm{cl}\, C_1) \cap \mathrm{lin}\,(\mathrm{cl}\, C_2)) \setminus \mathrm{lin}\,(\mathrm{cl}\, C_1) = \varnothing.\end{aligned}$$

The obtained contradiction results in $B_1 \cap C = \varnothing$.

By Corollary 10.13, there is a hyperplane $H_1 \subset \mathbb{R}^n$ strongly separating B_1 and C. According to Theorem 2.17, H_1 can be expressed as $H_1 = \{\boldsymbol{x} \in \mathbb{R}^n : \boldsymbol{x}\cdot\boldsymbol{e}_1 = \gamma_1\}$. Without loss of generality, we may assume that B_1 lies in the open halfspaces $W_1 = \{\boldsymbol{x} \in \mathbb{R}^n : \boldsymbol{x}\cdot\boldsymbol{e}_1 < \gamma_1\}$. Consider the hyperplane

$$H_1' = \{\boldsymbol{x} \in \mathbb{R}^n : \boldsymbol{x}\cdot\boldsymbol{e}_1 = \gamma_1'\}, \quad \text{where} \quad \gamma_1' = \inf\{\boldsymbol{x}\cdot\boldsymbol{e}_1 : \boldsymbol{x} \in C\}.$$

Corollary 9.9 implies that H_1' either supports C or is asymptotic to C. Since C cannot have asymptotic hyperplanes (see Theorem 9.44), H_1' supports C. Therefore, $\mathrm{ap}\, C \subset H_1'$, as follows from Theorem 9.43. Because $\mathrm{ap}\, C$ is a subspace, H_1' is also a subspace, which gives $\gamma_1' = 0$.

By Corollary 10.5, H_1' strongly separates B_1 from C. In particular, B_1 lies in the open halfspace $W_1' = \{\boldsymbol{x} \in \mathbb{R}^n : \boldsymbol{x}\cdot\boldsymbol{e}_1 < 0\}$. Since every point $\boldsymbol{u} \in F_1 \setminus \{\boldsymbol{o}\}$ belongs to a suitable open halfline of the form $(\boldsymbol{o}, \boldsymbol{b}\rangle$, $\boldsymbol{b} \in B_1$, Theorem 2.35 shows that

$$F_1 \setminus \{\boldsymbol{o}\} = \cup\{(\boldsymbol{o}, \boldsymbol{x}\rangle : \boldsymbol{x} \in B_1\} \subset W_1'.$$

By Theorem 5.18, the set $G_1 = \mathrm{cl}\, C_1 \setminus \mathrm{ap}\,(\mathrm{cl}\, C_1)$ is convex. We assert that $G_1 \subset W_1'$. Indeed, Theorem 6.26 implies the equality

$$G_1 = (F_1 \setminus \{\boldsymbol{o}\}) + \mathrm{lin}\,(\mathrm{cl}\, C_1).$$

Thus any point $\boldsymbol{x} \in G_1$ can be written as $\boldsymbol{x} = \boldsymbol{u} + \boldsymbol{z}$, where $\boldsymbol{u} \in F_1 \setminus \{\boldsymbol{o}\}$ and $\boldsymbol{z} \in \mathrm{lin}\,(\mathrm{cl}\, C_1)$. Since $\boldsymbol{z}\cdot\boldsymbol{e}_1 = 0$ due to the inclusion $\boldsymbol{z} \in \mathrm{lin}\,(\mathrm{cl}\, C_1) \subset \mathrm{lin}\, C \subset H_1'$, we have $\boldsymbol{x}\cdot\boldsymbol{e}_1 = (\boldsymbol{u} + \boldsymbol{z})\cdot\boldsymbol{e}_1 < 0$, which gives the inclusion $\boldsymbol{x} \in W_1'$. Summing up, $G_1 \subset W_1'$. Equivalently, the closed halfspace $V_1' = \{\boldsymbol{x} \in$

$\mathbb{R}^n : \boldsymbol{x}\cdot\boldsymbol{e}_1 \leqslant 0\}$ strictly separates G_1 from C, such that $L \subset H_1' = \operatorname{bd} V_1'$. In particular, V_1' strictly separates G_1 from the convex set $G_2 = \operatorname{cl} C_2 \setminus \operatorname{lin}(\operatorname{cl} C_2)$.

In a similar way, there is a closed halfspace $V_2' = \{\boldsymbol{x} \in \mathbb{R}^n : \boldsymbol{x}\cdot\boldsymbol{e}_2 \leqslant 0\}$ which strictly separates F_2 from G_1 and contains L. By Theorem 10.6, there is a hyperplane $H \subset \mathbb{R}^n$ which strictly separates G_1 and G_2 and contains the subspace $L = \operatorname{lin}(\operatorname{cl} C_1) + \operatorname{lin}(\operatorname{cl} C_2)$. Thus $\operatorname{lin}(\operatorname{cl} C_i) = H \cap \operatorname{cl} C_i$, $i = 1, 2$, as desired. □

The next theorem gives a criterion for sharp separation of cones with a common apex in terms of their polar cones.

Theorem 10.16. *If C_1 and C_2 are convex cones in $\mathbb{R}^n$ with common apex $\boldsymbol{a}$, then the following conditions are equivalent.*

(a) C_1 and C_2 are sharply separated by a hyperplane $H \subset \mathbb{R}^n$.
(b) The set $E = \operatorname{rint}(C_1 - \boldsymbol{a})^\circ \cap (-\operatorname{rint}(C_2 - \boldsymbol{a})^\circ)$ has positive dimension.

Proof. Put $F_1 = C_1 - \boldsymbol{a}$ and $F_2 = C_2 - \boldsymbol{a}$. Clearly, both F_1 and F_2 are convex cones with common apex $\boldsymbol{o}$.

(a) ⇒ *(b)* Let C_1 and C_2 are sharply separated by a hyperplane of the form $H = \{\boldsymbol{x} \in \mathbb{R}^n : \boldsymbol{x}\cdot\boldsymbol{e} = \gamma\}$. Then F_1 and F_2 are sharply separated by the $(n-1)$-dimensional subspace

$$G = H - \boldsymbol{a} = \{\boldsymbol{x} \in \mathbb{R}^n : \boldsymbol{x}\cdot\boldsymbol{e} = \boldsymbol{o}\}.$$

Without loss of generality, we may assume that

$$F_1 \subset V_1 = \{\boldsymbol{x} \in \mathbb{R}^n : \boldsymbol{x}\cdot\boldsymbol{e} \leqslant 0\} \quad \text{and} \quad F_2 \subset V_2 = \{\boldsymbol{x} \in \mathbb{R}^n : \boldsymbol{x}\cdot\boldsymbol{e} \geqslant 0\}.$$

Assume first that both F_1 and F_2 are subspaces. By Theorem 10.15, $F_1 \cup F_2 \subset G$, which implies the inclusion $F_1 + F_2 \subset G$. Since the orthogonal complement $F_i^\perp$ is a subspace, one has $F_i^\circ = F_i^\perp$, $i = 1, 2$. Consequently,

$$\operatorname{rint} F_i^\circ = \operatorname{rint} F_i^\perp = F_i^\perp, \quad i = 1, 2.$$

and Theorem 8.3 implies

$$\boldsymbol{e} \in G^\perp \subset (F_1 + F_2)^\perp = F_1^\perp \cap F_2^\perp = F_1^\perp \cap (-F_2^\perp) = \operatorname{rint} F_1^\circ \cap (-\operatorname{rint} F_2^\circ).$$

Suppose next that F_1 is a subspace and F_2 is not (the case when F_2 is a subspace and F_1 is not is similar). Then $F_1 \subset G$ and, as above,

$$\operatorname{rint} F_1^\circ = F_1^\circ = F_1^\perp, \quad \text{while} \quad \operatorname{cl} F_2 \setminus \operatorname{lin}(\operatorname{cl} F_2) \subset \operatorname{int} V_2.$$

Choose a nonzero outward normal vector $\boldsymbol{e}$ of V_2. By Theorem 8.6,

$$\boldsymbol{e} \in G^\perp \cap \operatorname{rint} F_2^\circ \subset (-F_1^\perp) \cap \operatorname{rint} F_2^\circ = (-\operatorname{rint} F_1^\circ) \cap \operatorname{rint} F_2^\circ.$$

So, $-\boldsymbol{e} \in E$, and $\dim E > 0$.

Finally, suppose that neither F_1 nor F_2 is a subspace. Sharp separability of F_1 and F_2 by G gives the inclusions

$$\mathrm{cl}\, F_1 \setminus \mathrm{ap}\,(\mathrm{cl}\, F_1) \subset \mathrm{int}\, V_1 \quad \text{and} \quad \mathrm{cl}\, F_2 \setminus \mathrm{ap}\,(\mathrm{cl}\, F_2) \subset \mathrm{int}\, V_2.$$

By Theorem 8.6, $\boldsymbol{e} \in \mathrm{rint}\, F_1^\circ$ and $-\boldsymbol{e} \in \mathrm{rint}\, F_2^\circ$. Thus the set E has positive dimension.

(b) $\Rightarrow$ *(a)* Let $\dim E > 0$ and $\boldsymbol{e}$ be a nonzero vector in $E \cup (-E)$, then the above argument implies that the hyperplane G sharply separates F_1 and F_2. Consequently, H sharply separates C_1 and C_2. □

Corollary 10.17. *If $C \subset \mathbb{R}^n$ is a nonempty convex cone with apex $\boldsymbol{o}$, which is not a subspace, then C and its polar cone C° are sharply separated by an $(n-1)$-dimensional subspace. Furthermore, an $(n-1)$-dimensional subspace*

$$S = \{\boldsymbol{x} \in \mathbb{R}^n : \boldsymbol{x} \cdot \boldsymbol{e} = 0\}, \quad \textit{where} \quad \boldsymbol{e} \neq \boldsymbol{o},$$

sharply separates C and C° if and only if

$$\boldsymbol{e} \in E \cup (-E), \quad \textit{where} \quad E = \mathrm{rint}\, C \cap (-\mathrm{rint}\, C^\circ).$$

Proof. Since C is not a subspace, the orthogonal complement of $\mathrm{ap}\,(\mathrm{cl}\, C)$ within $\mathrm{span}\, C$ is not $\{\boldsymbol{o}\}$. Also, Theorem 8.4 shows that C° is not a subspace. Based on these arguments, the corollary can be proved in two distinct ways: since $\mathrm{cl}\, C \cap C^\circ = \{\boldsymbol{o}\}$ (see Theorem 8.2), sharp separation of C and C° follows from Theorem 10.15. Alternatively, since $\mathrm{rint}\, C \cap (-\mathrm{rint}\, C^\circ) \neq \{\boldsymbol{o}\}$ (see Theorem 8.11), sharp separation of C and C° follows from Theorem 10.16.

The last assertion of the corollary follows from Theorem 8.7. □

10.2 Various Types of Separation

10.2.1 *Separation of Convex Sets by Slabs*

We recall (see page 72) that a closed slab in $\mathbb{R}^n$ is a set of the form

$$M = \{\boldsymbol{x} \in \mathbb{R}^n : \gamma \leqslant \boldsymbol{x} \cdot \boldsymbol{e} \leqslant \gamma'\}, \tag{10.21}$$

where $\boldsymbol{e} \in \mathbb{R}^n$ is a suitable nonzero vector and $\gamma < \gamma'$. The opposite closed halfspaces determined by M are defined as

$$V = \{\boldsymbol{x} \in \mathbb{R}^n : \boldsymbol{x} \cdot \boldsymbol{e} \leqslant \gamma\} \quad \text{and} \quad V' = \{\boldsymbol{x} \in \mathbb{R}^n : \boldsymbol{x} \cdot \boldsymbol{e} \geqslant \gamma'\}. \tag{10.22}$$

Definition 10.18. We say that a closed slab $M \subset \mathbb{R}^n$ *separates* nonempty subsets X_1 and X_2 of $\mathbb{R}^n$ provided X_1 and X_2 lie in the opposite closed halfspaces determined by M.

Lemma 10.19. *For nonempty sets X_1 and X_2 in $\mathbb{R}^n$, the following assertions hold.*

(*a*) *X_1 and X_2 are strongly separated by a translate of a hyperplane $H \subset \mathbb{R}^n$ if and only if they are separated by a closed slab whose boundary hyperplanes are parallel to H.*
(*b*) *X_1 and X_2 are separated by a closed slab $M \subset \mathbb{R}^n$ if and only if $\boldsymbol{o}$ and $X_2 - X_1$ are separated by a suitable translate of M.*

Proof. (*a*) Suppose first that X_1 and X_2 are strongly separated by a translate of a hyperplane $H \subset \mathbb{R}^n$. Expressing H as $H = \{\boldsymbol{x} \in \mathbb{R}^n : \boldsymbol{x}\cdot\boldsymbol{e} = \mu\}$, where $\boldsymbol{e}$ is a nonzero vector and μ is a scalar, we may assume that (10.5) holds. With

$$\gamma = \sup\{\boldsymbol{x}\cdot\boldsymbol{e} : \boldsymbol{x} \in X_1\} \quad \text{and} \quad \gamma' = \inf\{\boldsymbol{x}\cdot\boldsymbol{e} : \boldsymbol{x} \in X_2\},$$

the slab (10.21) separates X_1 and X_2.

Conversely, if a closed slab of the form (10.21) separates X_1 and X_2, then, according to Corollary 10.3, for any scalar $\delta \in (\gamma, \gamma')$, the hyperplane $H' = \{\boldsymbol{x} \in \mathbb{R}^n : \boldsymbol{x}\cdot\boldsymbol{e} = \delta\}$ strongly separates X_1 and X_2. By Theorem 2.18, H' is a translate of H.

(*b*) Suppose first that X_1 and X_2 are separated by a closed slab M of the form (10.21). Without loss of generality, we may assume that $X_1 \subset V$ and $X_2 \subset V'$, where V and V' are closed halfspaces from (10.22). Theorem 10.7 implies that $\boldsymbol{o}$ and $X_2 - X_1$ lie, respectively, in the opposite closed halfspaces

$$W = \{\boldsymbol{x} \in \mathbb{R}^n : \boldsymbol{x}\cdot\boldsymbol{e} \leqslant 0\} \quad \text{and} \quad W' = \{\boldsymbol{x} \in \mathbb{R}^n : \gamma' - \gamma \leqslant \boldsymbol{x}\cdot\boldsymbol{e}\}.$$

Since W and W' are translates of V and V' on the same vector, we conclude that the slab $N = \{\boldsymbol{x} \in \mathbb{R}^n : 0 \leqslant \boldsymbol{x}\cdot\boldsymbol{e} \leqslant \gamma' - \gamma\}$ is a translate of M and separates $\boldsymbol{o}$ and $X_2 - X_1$. The converse assertion is proved in a similar way. □

In what follows, by the *width* of a slab we will mean the distance between its boundary hyperplanes. So, if a slab $M \subset \mathbb{R}^n$ is given by (10.21), then its width equals $(\gamma' - \gamma)/\|\boldsymbol{e}\|$ (see Corollary 2.92).

Theorem 10.20. *Let nonempty convex sets K_1 and K_2 in $\mathbb{R}^n$ be strongly disjoint. Then there is a unique slab $M \subset \mathbb{R}^n$ of width $\delta(K_1, K_2)$ separating K_1 and K_2, and the width of any other slab separating these sets is less then $\delta(K_1, K_2)$. Furthermore, if $\boldsymbol{e}$ is the metric projection of $\boldsymbol{o}$ on* $\operatorname{cl}(K_1 - K_2)$, *then* $\operatorname{nor} M = \operatorname{span}\{\boldsymbol{e}\}$.

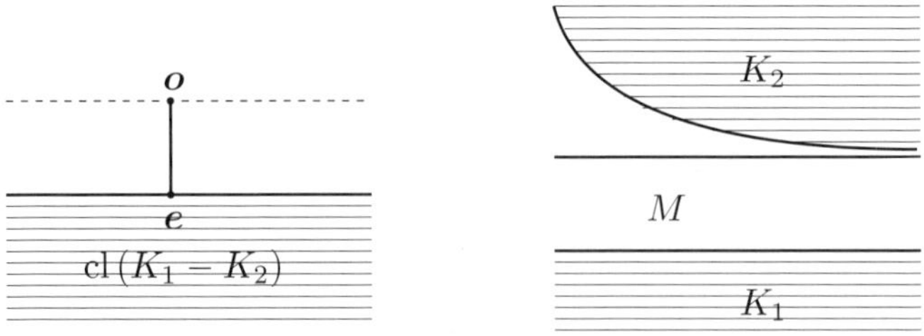

Fig. 10.2 A slab separating convex sets K_1 and K_2.

Proof. Let $\delta = \delta(K_1, K_2)$. The equality

$$\delta = \inf\{\|\boldsymbol{x}_1 - \boldsymbol{x}_2\| : \boldsymbol{x}_1 \in K_1, \boldsymbol{x}_2 \in K_2\}$$

can be rewritten as $\delta = \inf\{\|\boldsymbol{o} - \boldsymbol{x}\| : \boldsymbol{x} \in K_2 - K_1\}$, which shows that $\delta = \delta(K_1 - K_2, \boldsymbol{o})$. Since the set $K_1 - K_2$ is convex (see Theorem 3.10), Theorem 8.22 implies the existence of a unique point $\boldsymbol{e} \in \mathrm{cl}\,(K_1 - K_2)$ nearest to $\boldsymbol{o}$ and such that the closed halfspace

$$V = \{\boldsymbol{x} \in \mathbb{R}^n : (\boldsymbol{x} - \boldsymbol{e})\cdot\boldsymbol{e} \geqslant 0\} \tag{10.23}$$

contains $K_1 - K_2$. Let $\beta = \boldsymbol{e}\cdot\boldsymbol{e}$. The above argument shows that the closed slab

$$N = \{\boldsymbol{x} \in \mathbb{R}^n : 0 \leqslant \boldsymbol{x}\cdot\boldsymbol{e} \leqslant \beta\} \tag{10.24}$$

separates $K_1 - K_2$ and $\boldsymbol{o}$. Furthermore, since $\boldsymbol{e}$ is a normal vector of both hyperplanes

$$H_0 = \{\boldsymbol{x} \in \mathbb{R}^n : \boldsymbol{x}\cdot\boldsymbol{e} = 0\} \quad \text{and} \quad H_1 = \{\boldsymbol{x} \in \mathbb{R}^n : \boldsymbol{x}\cdot\boldsymbol{e} = \beta\},$$

and since $\boldsymbol{o} \in H_0$ and $\boldsymbol{e} \in H_1$, the width of N equals $\delta(H_0, H_1) = \|\boldsymbol{e}\| = \delta$ and $\boldsymbol{e}$ is a normal vector of N. Finally, Lemma 10.19 shows that a suitable translate of N separates K_1 and K_2.

Next, choose a closed slab $M \subset \mathbb{R}^n$ separating K_1 and K_2. Denote by ε the width of M. According to Lemma 10.19, a suitable translate M' of M separates $K_1 - K_2$ and $\boldsymbol{o}$. Clearly, the width of M' equals ε. Put

$$M' = \{\boldsymbol{x} \in \mathbb{R}^n : \mu \leqslant \boldsymbol{x}\cdot\boldsymbol{b} \leqslant \mu'\},$$

where $\boldsymbol{b} \in \mathbb{R}^n$ is a nonzero vector and $\mu < \mu'$. Since $K_1 - K_2$ and $\boldsymbol{o}$ lie in the opposite closed halfspaces determined by M', the closed segment $[\boldsymbol{o}, \boldsymbol{e}]$ meets the boundary hyperplanes

$$H = \{\boldsymbol{x} \in \mathbb{R}^n : \boldsymbol{x}\cdot\boldsymbol{b} = \mu\} \quad \text{and} \quad H' = \{\boldsymbol{x} \in \mathbb{R}^n : \boldsymbol{x}\cdot\boldsymbol{b} = \mu'\}$$

of M' at some points $\boldsymbol{u} \in H$ and $\boldsymbol{u}' \in H'$, as depicted in Figure 10.3. If $\boldsymbol{v}$ denotes the orthogonal projection of $\boldsymbol{u}$ on H', then

$$\varepsilon = \|\boldsymbol{u} - \boldsymbol{v}\| \leqslant \|\boldsymbol{u} - \boldsymbol{u}'\| \leqslant \|\boldsymbol{e}\| = \delta,$$

where the equality $\varepsilon = \delta$ holds if and only if $[\boldsymbol{u}, \boldsymbol{u}'] = [\boldsymbol{o}, \boldsymbol{e}]$, which occurs if and only if M' coincides with the closed slab N defined by (10.24). Hence M has width δ only if M is a translate of N; otherwise, the width of M is less than δ.

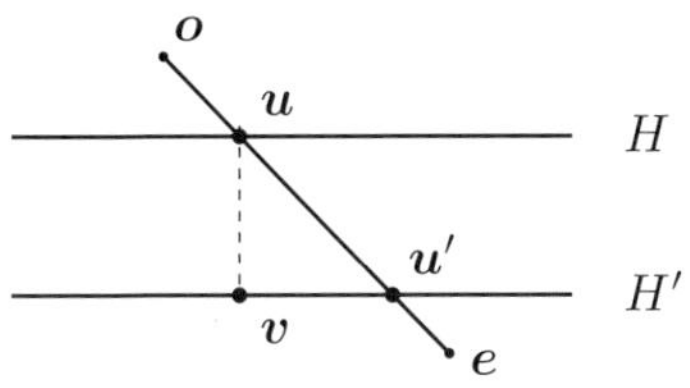

Fig. 10.3 Illustration to Theorem 10.20.

Finally, assume for a moment the existence of distinct closed slabs M_1 and M_2 of width δ, both separating K_1 and K_2. By the above argument, these slabs are translates of N. So, they can be expressed as

$$M_1 = \{\boldsymbol{x} \in \mathbb{R}^n : \gamma_1 \leqslant \boldsymbol{x}\cdot\boldsymbol{e} \leqslant \gamma_1'\}, \quad M_2 = \{\boldsymbol{x} \in \mathbb{R}^n : \gamma_2 \leqslant \boldsymbol{x}\cdot\boldsymbol{e} \leqslant \gamma_2'\}.$$

Put $\gamma_0 = \min\{\gamma_1, \gamma_2\}$ and $\gamma_0' = \max\{\gamma_1', \gamma_2'\}$. Then the slab

$$M_0 = \{\boldsymbol{x} \in \mathbb{R}^n : \gamma_0 \leqslant \boldsymbol{x}\cdot\boldsymbol{e} \leqslant \gamma_0'\}$$

separates K_1 and K_2, and the width δ_0 of M_0 equals $(\gamma_0' - \gamma_0)/\|\boldsymbol{e}\|$, which is greater than $\delta = (\gamma_1' - \gamma_1)/\|\boldsymbol{e}\| = (\gamma_2' - \gamma_2)/\|\boldsymbol{e}\|$. By the above argument, the slab

$$N_0 = \{\boldsymbol{x} \in \mathbb{R}^n : 0 \leqslant \boldsymbol{x}\cdot\boldsymbol{e} \leqslant \gamma_0' - \gamma_0\}$$

separates $K_1 - K_2$ and $\boldsymbol{o}$. Consequently, $\|\boldsymbol{e}\| = \|\boldsymbol{e} - \boldsymbol{o}\| \geqslant \delta_0$, contrary to $\|\boldsymbol{e}\| = \delta$. The obtained contradiction shows the uniqueness of a closed slab of width δ separating K_1 and K_2. □

The next theorem provides a sufficient condition for a separation of convex sets by slabs. We will need the following auxiliary lemma.

Lemma 10.21. *Let nonempty convex sets* K_1 *and* K_2 *in* $\mathbb{R}^n$ *satisfy the condition* $\mathrm{cl}\,(K_1) \cap \mathrm{cl}\,(K_2) = \varnothing$. *Then*

$$\mathrm{rec}\,(\mathrm{cl}\,K_1) \cap \mathrm{rec}\,(\mathrm{cl}\,K_2) \neq \{\boldsymbol{o}\}$$

provided any of the following two conditions is satisfied.

(*a*) *There are unbounded sequences* $\boldsymbol{x}_1, \boldsymbol{x}_2, \ldots$ *and* $\boldsymbol{y}_1, \boldsymbol{y}_2, \ldots$ *in* K_1 *and* K_2*, respectively, such that*

$$\lim_{i\to\infty} \inf \{\|\boldsymbol{x}_j - \boldsymbol{y}_j\| : j \geqslant i\} < \infty.$$

(*b*) $\delta(K_1, K_2) = 0$.

Proof. (*a*) Let unbounded sequences $\boldsymbol{x}_1, \boldsymbol{x}_2, \ldots \in K_1$ and $\boldsymbol{y}_1, \boldsymbol{y}_2, \ldots \in K_2$ satisfy condition (*a*) of the lemma. Then

$$\lim_{i\to\infty} \inf \{\|\boldsymbol{x}_j - \boldsymbol{y}_j\| : j \geqslant i\} \leqslant \alpha$$

for a suitable scalar $\alpha \geqslant 0$. Consider the unit vectors $\boldsymbol{u}_i = \boldsymbol{x}_i/\|\boldsymbol{x}_i\|$, $i \geqslant 1$. Choosing, if necessary, a suitable subsequence, we assume that the sequence $\boldsymbol{u}_1, \boldsymbol{u}_2, \ldots$ converges to a unit vector $\boldsymbol{u}$. Then $\boldsymbol{u} \in \operatorname{rec}(\operatorname{cl} K_1)$ according to Theorem 6.17. The same theorem and the equalities

$$\lim_{i\to\infty} \frac{\boldsymbol{y}_i}{\|\boldsymbol{x}_i\|} = \lim_{i\to\infty} \Big(\frac{\boldsymbol{y}_i - \boldsymbol{x}_i}{\|\boldsymbol{x}_i\|} + \frac{\boldsymbol{x}_i}{\|\boldsymbol{x}_i\|}\Big) = \boldsymbol{o} + \boldsymbol{u} = \boldsymbol{u}$$

show that $\boldsymbol{u} \in \operatorname{rec}(\operatorname{cl} K_2)$.

(*b*) Suppose that $\delta(K_1, K_2) = 0$ and choose sequences $\boldsymbol{x}_1, \boldsymbol{x}_2, \ldots \in K_1$ and $\boldsymbol{y}_1, \boldsymbol{y}_2, \ldots \in K_2$ satisfying the condition $\lim_{i\to\infty} \|\boldsymbol{x}_i - \boldsymbol{y}_i\| = 0$. We observe first that both set $X = \{\boldsymbol{x}_1, \boldsymbol{x}_2, \ldots\}$ and $Y = \{\boldsymbol{y}_1, \boldsymbol{y}_2, \ldots\}$ must be unbounded. Indeed, assume for a moment that X is bounded. Then X contains a subsequence $\boldsymbol{x}'_1, \boldsymbol{x}'_2, \ldots$ converging to a point $\boldsymbol{x} \in \operatorname{cl} X$. If $\boldsymbol{y}'_1, \boldsymbol{y}'_2, \ldots$ is the respective subsequence from Y, then the inequalities

$$\|\boldsymbol{x} - \boldsymbol{y}'_i\| \leqslant \|\boldsymbol{x} - \boldsymbol{x}'_i\| + \|\boldsymbol{x}'_i - \boldsymbol{y}'_i\|, \quad i \geqslant 1,$$

show that the set $Y' = \{\boldsymbol{y}'_1, \boldsymbol{y}'_2, \ldots\}$ is bounded. Choose in Y' a subsequence $\boldsymbol{y}''_1, \boldsymbol{y}''_2, \ldots$ converging to a point $\boldsymbol{y} \in \operatorname{cl} Y$. If $\boldsymbol{x}''_1, \boldsymbol{x}''_2, \ldots$ is the respective subsequence from $\boldsymbol{x}'_1, \boldsymbol{x}'_2, \ldots$, then

$$\|\boldsymbol{x} - \boldsymbol{y}\| = \lim_{i\to\infty} \|\boldsymbol{x}''_i - \boldsymbol{y}''_i\| = \lim_{i\to\infty} \|\boldsymbol{x}_i - \boldsymbol{y}_i\| = 0,$$

implying that $\boldsymbol{x} = \boldsymbol{y}$. This equality is in contradiction with the assumption $\operatorname{cl} K_1 \cap \operatorname{cl} K_2 = \varnothing$.

Hence both sequences $\boldsymbol{x}_1, \boldsymbol{x}_2, \ldots$ and $\boldsymbol{y}_1, \boldsymbol{y}_2, \ldots$ are unbounded, and $\operatorname{rec}(\operatorname{cl} K_1) \cap \operatorname{rec}(\operatorname{cl} K_2) \neq \{\boldsymbol{o}\}$ by the above argument. □

Theorem 10.22. *If* K_1 *and* K_2 *are nonempty convex sets in* $\mathbb{R}^n$ *such that* $\operatorname{cl} K_1 \cap \operatorname{cl} K_2 = \varnothing$ *and neither* K_1 *nor* K_2 *admits an asymptotic plane, then* $\delta(K_1, K_2) > 0$.

Proof. We will prove this assertion by induction on n. The case $n = 1$ is obvious. Assume that the assertion holds for all $n \leqslant m-1$, $m \geqslant 2$, and consider convex sets K_1 and K_2 in $\mathbb{R}^m$ which satisfy the hypothesis.

Suppose for a moment that $\delta(K_1, K_2) = 0$. By Lemma 10.21, the cone $\operatorname{rec}(\operatorname{cl} K_1) \cap \operatorname{rec}(\operatorname{cl} K_2)$ is not $\{\boldsymbol{o}\}$; so, it contains a closed halfline h with apex $\boldsymbol{o}$. We assert that no line parallel to h meets both $\operatorname{cl} K_1$ and $\operatorname{cl} K_2$. Indeed, assume for a moment the existence of a line l which is parallel to h and meets $\operatorname{cl} K_1$ and $\operatorname{cl} K_2$ at points $\boldsymbol{x}_1$ and $\boldsymbol{x}_2$, respectively. Clearly, one of the closed halflines $h_1 = \boldsymbol{x}_1 + h$ and $h_2 = \boldsymbol{x}_2 + h$ contains the other. Since $h_1 \subset \operatorname{cl} K_1$ and $h_2 \subset \operatorname{cl} K_2$, we obtain that $\operatorname{cl} K_1 \cap \operatorname{cl} K_2 \neq \varnothing$, contrary to the assumption.

Choose a new orthogonal basis $\{\boldsymbol{c}_1, \ldots, \boldsymbol{c}_m\}$ for $\mathbb{R}^m$ such that $h = [\boldsymbol{o}, \boldsymbol{c}_m\rangle$. Let

$$p : \mathbb{R}^m \to \mathbb{R}^{m-1} = \operatorname{span}\{\boldsymbol{c}_1, \ldots, \boldsymbol{c}_{m-1}\}$$

be the orthogonal projection of $\mathbb{R}^m$ on $\mathbb{R}^{m-1}$. Put $K_1' = p(K_1)$ and $K_2' = p(K_2)$. Clearly, $\operatorname{null} p = \langle \boldsymbol{o}, \boldsymbol{c}_m\rangle$. Because neither K_1 nor K_2 admits an asymptotic line parallel to h, Theorem 7.14 implies that $\operatorname{cl} K_1' = p(\operatorname{cl} K_1)$ and $\operatorname{cl} K_2' = p(\operatorname{cl} K_2)$. Since no line parallel to h meets both $\operatorname{cl} K_1$ and $\operatorname{cl} K_2$, one has

$$\operatorname{cl} K_1' \cap \operatorname{cl} K_2' = p(\operatorname{cl} K_1) \cap p(\operatorname{cl} K_2) = \varnothing.$$

We assert that neither K_1' nor K_2' admits an asymptotic plane. Indeed, if $L \subset \mathbb{R}^{m-1}$ were an asymptotic plane for K_i', then the plane $L + \langle \boldsymbol{o}, \boldsymbol{c}_m\rangle$ would be an asymptotic plane for K_i, which is impossible. By the induction hypothesis, $\delta(K_1', K_2') > 0$. Obviously, $\delta(K_1, K_2) \geqslant \delta(K_1', K_2') > 0$. □

Remark. The condition that neither K_1 nor K_2 admits an asymptotic plane is essential in Theorem 10.22. Indeed, let

$$K_1 = \{(x, y) : x > 0,\ xy \geqslant 1\} \quad \text{and} \quad K_2 = \{(x, y) : y \leqslant 0\}.$$

Then K_1 and K_2 are disjoint closed convex sets, while $\delta(K_1, K_2) = 0$.

10.2.2 *Pairs of Nearest Points*

Given nonempty sets X_1 and X_2 in $\mathbb{R}^n$, we will say that points $\boldsymbol{x}_1 \in X_1$ and $\boldsymbol{x}_2 \in X_2$ form a *nearest pair* provided $\|\boldsymbol{x}_1 - \boldsymbol{x}_2\| = \delta(X_1, X_2)$.

Theorem 10.23. *Let nonempty convex sets K_1 and K_2 in $\mathbb{R}^n$ be such that*

$$\operatorname{cl} K_1 \cap \operatorname{cl} K_2 = \varnothing \quad \textit{and} \quad \delta(K_1, K_2) = \|\boldsymbol{c}_1 - \boldsymbol{c}_2\|$$

for suitable points $\mathbf{c}_1 \in \operatorname{cl} K_1$ *and* $\mathbf{c}_2 \in \operatorname{cl} K_2$. *If* H_1 *and* H_2 *are hyperplanes through* $\mathbf{c}_1$ *and* $\mathbf{c}_2$, *respectively, both orthogonal to the closed segment* $[\mathbf{c}_1, \mathbf{c}_2]$, *then the closed slab between* H_1 *and* H_2 *separates* K_1 *and* K_2.

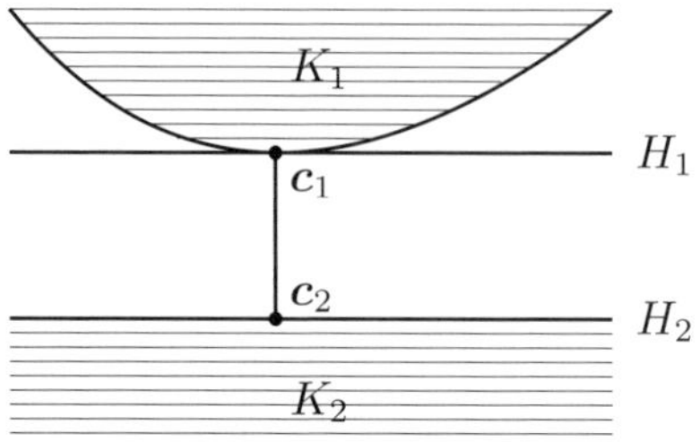

Fig. 10.4 Illustration to Theorem 10.23.

Proof. Clearly, $\boldsymbol{x}_1 \neq \boldsymbol{x}_2$. According to Theorem 8.22, the halfspaces

$$V_1 = \{\boldsymbol{x} \in \mathbb{R}^n : (\boldsymbol{x} - \boldsymbol{c}_1)\cdot(\boldsymbol{c}_1 - \boldsymbol{c}_2) \geqslant 0\},$$
$$V_2 = \{\boldsymbol{x} \in \mathbb{R}^n : (\boldsymbol{x} - \boldsymbol{c}_2)\cdot(\boldsymbol{c}_1 - \boldsymbol{c}_2) \leqslant 0\}$$

contain K_1 and K_2, respectively. So, the hyperplanes

$$H_1 = \{\boldsymbol{x} \in \mathbb{R}^n : (\boldsymbol{x} - \boldsymbol{c}_1)\cdot(\boldsymbol{c}_1 - \boldsymbol{c}_2) = 0\},$$
$$H_2 = \{\boldsymbol{x} \in \mathbb{R}^n : (\boldsymbol{x} - \boldsymbol{c}_2)\cdot(\boldsymbol{c}_1 - \boldsymbol{c}_2) = 0\}$$

support $\operatorname{cl} K_1$ and $\operatorname{cl} K_2$ at $\boldsymbol{c}_1$ and $\boldsymbol{c}_2$, respectively (see Figure 10.4). Consequently, the closed slab between H_1 and H_2 separates K_1 and K_2. □

If at least one of nonempty sets X_1 and X_2 in $\mathbb{R}^n$ is bounded, then $\delta(X_1, X_2) = \|\boldsymbol{x}_1 - \boldsymbol{x}_2\|$ for suitable points $\boldsymbol{x}_1 \in \operatorname{cl} X_1$ and $\boldsymbol{x}_2 \in \operatorname{cl} X_2$ (see Problem 1.14). The theorem below refines this assertion for the case of convex sets.

Theorem 10.24. *If nonempty convex sets* K_1 *and* K_2 *in* $\mathbb{R}^n$ *satisfy the condition* $\operatorname{rec}(\operatorname{cl} K_1) \cap \operatorname{rec}(\operatorname{cl} K_2) = \{\boldsymbol{o}\}$, *then* $\delta(K_1, K_2) = \|\boldsymbol{c}_1 - \boldsymbol{c}_2\|$ *for suitable points* $\boldsymbol{c}_1 \in \operatorname{cl} K_1$ *and* $\boldsymbol{c}_2 \in \operatorname{cl} K_2$.

Proof. Since the case $\operatorname{cl} K_1 \cap \operatorname{cl} K_2 \neq \varnothing$ is obvious, we may assume that K_1 and K_2 are disjoint. Let sequences $\boldsymbol{u}_1, \boldsymbol{u}_2, \ldots \in K_1$ and $\boldsymbol{v}_1, \boldsymbol{v}_2, \ldots \in K_2$ satisfy the condition $\lim_{i\to\infty} \|\boldsymbol{u}_i - \boldsymbol{v}_i\| = \delta(K_1, K_2)$. We assert that at least one of these sequences is bounded. Indeed, if both $\boldsymbol{u}_1, \boldsymbol{u}_2, \ldots$ and

$\boldsymbol{v}_1, \boldsymbol{v}_2, \ldots$ were unbounded, then, according to Lemma 10.21, we would have $\operatorname{rec}(\operatorname{cl} K_1) \cap \operatorname{rec}(\operatorname{cl} K_2) \neq \{\boldsymbol{o}\}$. Assume, for instance, that $\boldsymbol{u}_1, \boldsymbol{u}_2, \ldots$ is bounded. Then there is a subsequence $\boldsymbol{u}_{p_1}, \boldsymbol{u}_{p_2}, \ldots$ of $\boldsymbol{u}_1, \boldsymbol{u}_2, \ldots$ which converges to a point $\boldsymbol{c}_1 \in \operatorname{cl} K_1$. Choose an index r such that

$$\|\boldsymbol{u}_i - \boldsymbol{v}_i\| \leqslant \delta(K_1, K_2) + 1 \text{ if } i \geqslant r, \text{ and } \|\boldsymbol{c}_1 - \boldsymbol{u}_{p_i}\| \leqslant 1 \text{ if } p_i \geqslant r.$$

Then

$$\|\boldsymbol{c}_1 - \boldsymbol{v}_{p_i}\| \leqslant \|\boldsymbol{c}_1 - \boldsymbol{v}_{p_i}\| + \|\boldsymbol{v}_{p_i} - \boldsymbol{u}_{p_i}\| \leqslant \delta(K_1, K_2) + 2$$

for all $p_i \geqslant r$, implying that the sequence $\boldsymbol{v}_{p_1}, \boldsymbol{v}_{p_2}, \ldots$ is bounded. Therefore, there is a subsequence $\boldsymbol{v}_{q_1}, \boldsymbol{v}_{q_2}, \ldots$ of $\boldsymbol{v}_{p_1}, \boldsymbol{v}_{p_2}, \ldots$ which converges to a point $\boldsymbol{c}_2 \in \operatorname{cl} K_2$. Finally,

$$\|\boldsymbol{c}_1 - \boldsymbol{c}_2\| = \lim_{i \to \infty} \|\boldsymbol{u}_{q_i} - \boldsymbol{v}_{q_i}\| = \delta(K_1, K_2). \qquad \square$$

10.2.3 *Convex Halfspaces and Semispaces*

Theorem 10.25. *Any pair of disjoint convex subsets K_1 and K_2 of a convex set $K \subset \mathbb{R}^n$ can be separated by complementary convex subsets Q_1 and Q_2 of K:*

$$K_1 \subset Q_1, \quad K_2 \subset Q_2, \quad Q_1 \cup Q_2 = K, \quad Q_1 \cap Q_2 = \varnothing.$$

Proof. The case when at least one of the sets K_1 and K_2 is empty is obvious (if, for instance, $K_2 = \varnothing$, then let $Q_1 = K$ and $Q_2 = \varnothing$). So, we may assume that both sets K_1 and K_2 are nonempty.

Corollary 3.9 implies the existence of a maximal convex subset Q_1 of $K \setminus K_2$ which contains K_1. Similarly, there is a maximal convex subset Q_2 of $K \setminus Q_1$ containing K_2. To prove that $\{Q_1, Q_2\}$ is a desired pair of sets, it remains to prove the equality $Q_1 \cup Q_2 = K$.

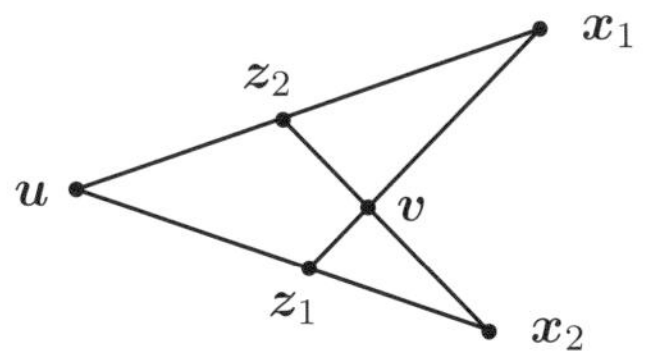

Fig. 10.5 Illustration to Theorem 10.25.

Assume, for contradiction, that $Q_1 \cup Q_2 \neq K$ and choose a point $\boldsymbol{u} \in K \setminus (Q_1 \cup Q_2)$. Put $Q_1' = \operatorname{conv}(\{\boldsymbol{u}\} \cup Q_1)$ and $Q_2' = \operatorname{conv}(\{\boldsymbol{u}\} \cup Q_2)$.

Obviously, $Q_1' \cup Q_2' \subset K$, Q_1 and Q_2 are nonempty proper subsets of Q_1' and Q_2', respectively. By the choice of Q_1 and Q_2, we have $Q_1 \cap Q_2' \neq \varnothing$ and $Q_1' \cap Q_2 \neq \varnothing$. Hence there are points $\boldsymbol{z}_1 \in Q_1 \cap Q_2'$ and $\boldsymbol{z}_2 \in Q_1' \cap Q_2$.

Corollary 4.14 implies the equalities

$$Q_i' = \operatorname{conv}(\{\boldsymbol{u}\} \cup Q_i) = \cup([\boldsymbol{u}, \boldsymbol{x}] : \boldsymbol{x} \in Q_i), \quad i = 1, 2.$$

Therefore, $\boldsymbol{z}_1 \in [\boldsymbol{u}, \boldsymbol{x}_2]$ and $\boldsymbol{z}_2 \in [\boldsymbol{u}, \boldsymbol{x}_1]$ for some points $\boldsymbol{x}_2 \in Q_2$ and $\boldsymbol{x}_1 \in Q_1$. By Lemma 2.26, the closed segments $[\boldsymbol{x}_1, \boldsymbol{z}_1]$ and $[\boldsymbol{x}_2, \boldsymbol{z}_2]$ have a common point $\boldsymbol{v}$, as depicted in Figure 10.5. The convexity of Q_1 and Q_2 gives

$$\boldsymbol{v} \in [\boldsymbol{x}_1, \boldsymbol{z}_1] \cap [\boldsymbol{x}_2, \boldsymbol{z}_2] \subset Q_1 \cap Q_2,$$

contrary to $Q_1 \cap Q_2 = \varnothing$. Summing up, $Q_1 \cup Q_2 = K$. □

Corollary 10.26. *Any pair of disjoint convex subsets K_1 and K_2 of a plane $L \subset \mathbb{R}^n$ can be separated by complementary convex sets Q_1 and Q_2:*

$$K_1 \subset Q_1, \quad K_2 \subset Q_2, \quad Q_1 \cup Q_2 = L, \quad Q_1 \cap Q_2 = \varnothing. \qquad \square$$

Remark. Corollary 10.26 cannot be extended to the case of more than two convex sets. For instance, the convex cones C_1, C_2, and C_3 in the plane, depicted in Figure 10.6, cannot be enlarged into pairwise disjoint (even pairwise non-overlapping) convex sets whose union is the entire plane.

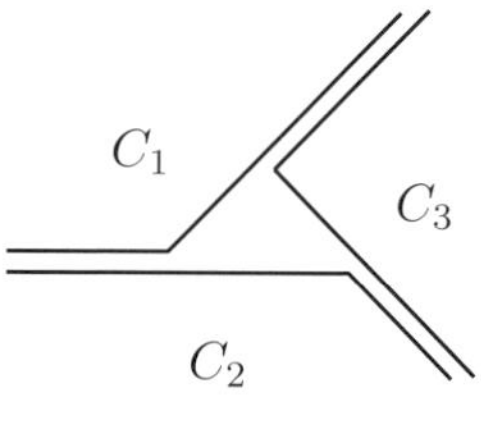

Fig. 10.6 Convex extensions of cones C_1, C_2 and C_3 do not cover the plane.

Corollary 10.26 gives a base to the following definition.

Definition 10.27. Let L be a nonempty plane in $\mathbb{R}^n$. A convex subset Q of L (possibly, $Q = \varnothing$ or $Q = L$) is called a *convex halfplane* of L provided its complement $L \setminus Q$ is a convex set. Convex halfplanes of $\mathbb{R}^n$ are called *convex halfspaces*.

In these terms, Corollary 10.26 states that any pair of disjoint convex subsets of a plane $L \subset \mathbb{R}^n$ (respectively, of $\mathbb{R}^n$) can be separated by complementary convex halfplanes of L (respectively, by convex halfspaces).

The next result gives a description of nonempty proper convex halfplanes of a given plane. Let $L \subset \mathbb{R}^n$ be a plane of positive dimension m, and let

$$L_0 \subset L_1 \subset \dots L_{m-1} \subset L_m = L, \quad \dim L_i = i, \quad 0 \leqslant i \leqslant m, \qquad (10.25)$$

be a sequence of subplanes of L. Denote by E_i an open halfplane of L_i determined by L_{i-1}, $1 \leqslant i \leqslant m$ (see Theorem 2.73).

Theorem 10.28. *Given a plane $L \subset \mathbb{R}^n$ of positive dimension m and nonempty complementary convex halfplanes Q and Q' of L, the following assertions hold.*

(*a*) *Then there is a sequence of planes of the form* (10.25) *and an integer $1 \leqslant r \leqslant m$ such that of Q and Q' are described as*

$$F_r = E_r \cup \dots \cup E_m, \quad F'_r = L_{r-1} \cup E'_r \cup \dots \cup E'_m, \quad 1 \leqslant r \leqslant m, \quad (10.26)$$

where E_i, E'_i are complementary open halfplanes of L_i determined by L_{i-1}.

(*b*) *Any pair of sets F_r and F'_r of the form* (10.26) *are complementary convex halfplanes of L.*

Proof. We will prove assertion (*a*) by induction on m.

Let $m = 1$. Then Q and Q' are complementary closed halflines of the line L. Denote by $\mathbf{c}$ the common endpoint of Q and Q', and suppose that $\mathbf{c} \in Q'$. Then the sets $E_1 = Q$ and $E'_1 = L \setminus Q$ are disjoint open halflines of L. Clearly, $Q = F_1 = E_1$ and $Q' = F'_1 = L_0 \cup E'_1$, where $L_0 = \{\mathbf{c}\}$.

Assume that the assertion holds for all $m \leqslant k - 1$, $k \geqslant 2$, and let Q and Q' be nonempty complementary convex halfplanes of a k-dimensional plane L. By Theorem 10.8, there is a hyperplane $H \subset \mathbb{R}^n$ separating Q and Q'. Since $L \not\subset H$, the plane $M = H \cap L$ is $(k-1)$-dimensional (see Theorem 2.19) and separates Q and Q' in L. Hence Q and Q' lie in the opposite closed halfplanes, say D and D', of L determined by M. We may assume that $Q \subset D$ and $Q' \subset D'$. Due to $Q \cup Q' = L$, the open halfplanes $E = \operatorname{rint} D$ and $E' = \operatorname{rint} D$ lie in Q and Q', respectively.

The equality

$$M = M \cap L = M \cap (Q \cup Q') = (M \cap Q) \cup (M \cap Q')$$

shows that M is the union of complementary convex subsets $P = M \cap Q$ and $P' = M \cap Q'$. If one of the sets P and P', say P, is empty, then $Q = E$ and $Q' = E' \cup M$. In this case, we can write $Q = E_k$, where $E_k = E$.

Suppose that both sets P and P' are nonempty. By the inductive assumption, one of these sets, say P, can be written as $P = E_r \cup \cdots \cup E_{k-1}$, where E_{k-1} is a suitable open halfplane of M. Letting $E_k = E$, we have $Q = E_r \cup \cdots \cup E_k$, as desired. Clearly, $Q' = L_{r-1} \cup E_r' \cup \cdots \cup E_k'$.

(*b*) Let F_r be a set of the form (10.26). First, we observe that F_r is a convex set. Indeed, if $\boldsymbol{u}, \boldsymbol{v} \in F_r$, then $\boldsymbol{u} \in E_i$ and $\boldsymbol{v} \in E_j$, where $r \leqslant i, j \leqslant m$. Suppose that $i \leqslant j$. If $i = j$, then $[\boldsymbol{u}, \boldsymbol{v}] \subset E_i \subset F_r$ by the convexity of E_i. If $i < j$, then $E_i \subset L_j = \operatorname{rbd} E_j$ and $(\boldsymbol{u}, \boldsymbol{v}) \subset E_j$ according to Theorem 3.47. Consequently, $[\boldsymbol{u}, \boldsymbol{v}] \subset E_j \cup E_j \subset F_r$.

The complementary set $L \setminus F_r$ can be written as

$$F_r' = L_{r-1} \cup E_r' \cup \cdots \cup E_m', \quad \text{where} \quad E_i' = L_i \setminus (L_{i-1} \cup E_i)$$

for all $r \leqslant i \leqslant m$. So, $L \setminus F_r = F_r'$. Similarly to the above argument, the set F_r' is convex. So, F_r a nonempty proper convex halfplane of L. $\square$

Corollary 10.29. *If Q and Q' are nonempty complementary convex halfplanes of an m-dimensional plane $L \subset \mathbb{R}^n$, then, in terms of Theorem* 10.28, *both Q and Q' are convex cones with common apex set L_{r-1}.*

Proof. Without loss of generality, we assume that $Q = F_r$ and $Q' = F_r'$ for a suitable integer $1 \leqslant r \leqslant m$.

Let $\boldsymbol{a}$ and $\boldsymbol{x}$ be points in L_{r-1} and Q, respectively. Then $\boldsymbol{x} \in E_i$ for a suitable $r \leqslant i \leqslant m$. Since the plane $L_{i-1} = \operatorname{rbd} E_i$, $r \leqslant i \leqslant m$, is the improper apex set of E_i (see an example on page 150), the inclusion $L_{r-1} \subset L_{i-1}$ gives $(\boldsymbol{a}, \boldsymbol{x}\rangle \subset E_i \subset Q$. Therefore, $L_{r-1} \subset \operatorname{ap} Q$.

Conversely, let $\boldsymbol{a}$ be any point in $\operatorname{ap} Q$. Then $\boldsymbol{a}$ is an apex of every open halfplane E_i, $r \leqslant i \leqslant m$. Hence

$$\boldsymbol{a} \in \operatorname{ap} E_r \cap \cdots \cap \operatorname{ap} E_m = L_{r-1} \cap \cdots \cap L_{m-1} = L_{r-1}.$$

Thus $\operatorname{ap} C \subset L_{r-1}$. Summing up, Q is a convex cone with the improper apex set L_{r-1}.

The argument for the case of Q' is similar (L_{r-1} is the proper apex set of Q' due to the inclusion $L_{r-1} \subset Q'$). $\square$

A special type of convex halfplanes, defined below, is of particular interest.

Definition 10.30. Let $L \subset \mathbb{R}^n$ be a plane positive dimension and $\boldsymbol{c}$ be a point in L. Any maximal convex subset of $L \setminus \{\boldsymbol{c}\}$ is called a *semiplane* of L at $\boldsymbol{c}$ and is denoted $Q_{\boldsymbol{c}}$. Convex semiplanes of $\mathbb{R}^n$ are called *semispaces.*

Theorem 10.31. *Let $L \subset \mathbb{R}^n$ be a plane of positive dimension. The family of all semiplanes of L is the smallest among all families $\mathcal{F}$ of convex subsets of L satisfying the following condition: every proper convex subset K of L is the intersection of some elements from $\mathcal{F}$.*

Proof. First, we observe that any proper convex subset K of L is the intersection of all semiplanes of L containing K. Indeed, for any point $\boldsymbol{c} \in L \setminus K$, there is a maximal convex subset of $L \setminus \{\boldsymbol{c}\}$ (that is, a semiplane of L) which contains K and misses $\boldsymbol{c}$ (see Corollary 3.9).

Now, suppose that $\mathcal{F}$ is a family of convex subsets of L such that every proper convex subset K of L is the intersection of some elements from $\mathcal{F}$. We assert that every semiplane $Q_{\boldsymbol{c}} \subset L$ belongs to $\mathcal{F}$. Indeed, choose a convex set $K \in \mathcal{F}$ such that $Q_{\boldsymbol{c}} \subset K$ and $\boldsymbol{c} \notin K$. By the maximality of $Q_{\boldsymbol{c}}$ in the family of all convex subsets of $L \setminus \{\boldsymbol{c}\}$ we conclude that $Q_{\boldsymbol{c}} = K$. $\square$

The structure of semiplanes is described in the following theorem.

Theorem 10.32. *Let $L \subset \mathbb{R}^n$ be a plane of positive dimension m, and $\boldsymbol{c}$ be a point in L. Every semiplane $Q_{\boldsymbol{c}}$ of L is a convex halfplane; it can be expressed as $Q_{\boldsymbol{c}} = E_1 \cup \cdots \cup E_m$, where all E_i are open halfplanes defined by suitable planes of the form* (10.25) *such that $L_0 = \{\boldsymbol{c}\}$.*

Proof. Let $Q_{\boldsymbol{c}}$ be a semiplane of L. By Theorem 10.28, there are complementary convex halfplanes Q and Q' of L containing, respectively, the disjoint convex sets $Q_{\boldsymbol{c}}$ and $\{\boldsymbol{c}\}$. Then $Q_{\boldsymbol{c}} = Q$ due to the maximality of $Q_{\boldsymbol{c}}$ in $L \setminus \{\boldsymbol{c}\}$. Theorem 10.28 shows that $Q_{\boldsymbol{c}} = F_r = E_r \cup \cdots \cup E_m$ for a suitable integer $1 \leqslant r \leqslant m$, where the open halfplanes E_i are determined by suitable planes of the form (10.25) such that $L_0 = \{\boldsymbol{c}\}$. Since the convex set F_1 contains $Q_{\boldsymbol{c}}$ and lies in $L \setminus \{\boldsymbol{c}\}$, the maximality of $Q_{\boldsymbol{c}}$ gives $Q_{\boldsymbol{c}} = F_1$. $\square$

Problems and Notes for Chapter 10

Problems for Chapter 10

Problem 10.1. Let $L \subset \mathbb{R}^n$ be a plane of dimension m, $0 \leqslant m \leqslant n-2$, and $\boldsymbol{u}, \boldsymbol{v} \in \mathbb{R}^n \setminus L$ be distinct points such that $\dim(L \cup \{\boldsymbol{u}, \boldsymbol{v}\}) = m+2$. Prove the existence of a hyperplane $H \subset \mathbb{R}^n$ which contains L and strictly separates $\boldsymbol{u}$ and $\boldsymbol{v}$.

Problem 10.2. Let L_1 and L_2 be nonempty planes in $\mathbb{R}^n$ such that $L_1 \cap L_2 \neq \varnothing$. Prove the equivalence of the following conditions:

(*a*) L_1 and L_2 are separated by a hyperplane,

(*b*) $\dim(L_1 \cup L_2) \leqslant n-1$, (*c*) $L_1 + L_2 \neq \mathbb{R}^n$.

Problem 10.3. Let K_1 and K_2 be nonempty convex sets in $\mathbb{R}^n$. Prove the equivalence of the following conditions.

(*a*) There is a translate $\boldsymbol{u} + K_1$ of K_1 such that $\boldsymbol{u} + K_1$ and K_2 are separated (respectively, strongly separated) by a hyperplane.

(*b*) $\operatorname{bar} K_1 \cap (-\operatorname{bar} K_2) \neq \{\boldsymbol{o}\}$.

Problem 10.4. (Nieuwenhuis [223]) Let K_1 and K_2 be nonempty line-free convex sets in $\mathbb{R}^n$ satisfying the conditions

$$\operatorname{cl} K_1 \cap \operatorname{cl} K_2 = \varnothing \quad \text{and} \quad \operatorname{rec}(\operatorname{cl} K_1) \cap \operatorname{rec}(\operatorname{cl} K_2) = \{\boldsymbol{o}\}.$$

Prove the existence of simplicial cones $\Gamma, \Gamma' \subset \mathbb{R}^n$ with a common apex $\boldsymbol{o}$ and of points $\boldsymbol{a}, \boldsymbol{b} \in \mathbb{R}^n$ such that

(*a*) $K_1 \subset \boldsymbol{a} + \Gamma$ and $K_2 \subset \boldsymbol{a} - \Gamma$,

(*b*) $K_1 \subset \boldsymbol{a} + \Gamma'$, $K_2 \subset \boldsymbol{b} - \Gamma'$, and $(\boldsymbol{a} + \Gamma') \cap (\boldsymbol{b} - \Gamma') = \varnothing$.

Problem 10.5. Let $C \subset \mathbb{R}^n$ be a convex cone with a proper apex $\boldsymbol{a}$ and $K \subset \mathbb{R}^n$ be a nonempty convex set which is separated from C by a hyperplane $H \subset \mathbb{R}^n$. Prove that the hyperplane H' through $\boldsymbol{a}$ parallel to H separates C and K.

Problem 10.6. Let K_1 and K_2 be nonempty bounded sets, with $\operatorname{cl} K_1 \cap \operatorname{cl} K_2 = \varnothing$. Prove the existence of disjoint closed balls B_1 and B_2 such that $K_1 \subset B_1$ and $K_2 \subset B_2$. Show that a similar assertion holds for the case of open balls.

Problem 10.7. (Grzybowski, Pallaschke, and Urbański [135]) We will say that a convex set $K \subset \mathbb{R}^n$ separates nonempty sets X_1 and X_2 provided K meets every closed segment $[\boldsymbol{x}_1, \boldsymbol{x}_2]$, where $\boldsymbol{x}_1 \in X_1$ and $\boldsymbol{x}_2 \in X_2$. Prove that a closed convex set K separates bounded sets X_1 and X_2 if and only if

$$X_1 + X_2 \subset K + \operatorname{conv}(X_1 \cup X_2).$$

Problem 10.8. Let C_1 and C_2 be nonempty convex cones in $\mathbb{R}^n$ satisfying the condition $\operatorname{cl} C_1 \cap \operatorname{cl} C_2 \neq \varnothing$ be separated by a hyperplane $H \subset \mathbb{R}^n$. Prove that $\operatorname{ap}(\operatorname{cl} C_1) \cup \operatorname{ap}(\operatorname{cl} C_2) \subset H$.

Problem 10.9. (Soltan [261]) Let C_1 and C_2 be convex cones in $\mathbb{R}^n$ with common apex $\boldsymbol{a}$. Prove that the following conditions are equivalent.

(*a*) There is a unique hyperplane $H \subset \mathbb{R}^n$ sharply separating C_1 and C_2.

(*b*) $H = \operatorname{aff}(\operatorname{ap}(\operatorname{cl} C_1) \cup \operatorname{ap}(\operatorname{cl} C_2))$.

(c) The set $E = \operatorname{rint}(C_1 - \boldsymbol{a})^\circ \cap (-\operatorname{rint}(C_2 - \boldsymbol{a})^\circ)$ is one-dimensional.

Problem 10.10. (Soltan [261]) Let $C \subset \mathbb{R}^n$ is a nonempty convex cone with apex $\boldsymbol{o}$, which is not a subspace. Prove that the following conditions are equivalent.

(a) There is a unique hyperplane separating (respectively, sharply separating) C and C°.

(b) The subspace $\operatorname{ap}(\operatorname{cl} C) + \operatorname{ap} C^\circ$ is a hyperplane.

(c) $\operatorname{cl} C$ is a closed halfplane of the form $\operatorname{cl} C = L + h$, where L is a subspace, with $0 \leqslant \dim L \leqslant n-1$, and h is a closed halfline with endpoint $\boldsymbol{o}$, orthogonal to L.

Notes for Chapter 10

Separation by hyperplanes. Separation properties of convex sets go up to Minkowski (see [215], p. 141), who showed that a pair of convex bodies in $\mathbb{R}^3$ is separated by a plane provided their interiors do not meet. Various generalizations of this result, predominantly for the case of infinite-dimensional vector spaces, are summarized by Klee in [183]. For separation of finitely many convex sets by hyperplanes and halfspaces see Gale and Klee [116] and the survey by Deumlich, Elster, and Nehse [92].

Theorem 10.8 is attributed to Rockafellar [241, Theorem 11.3]. A weaker version if this theorem was obtained earlier by Fenchel (see [105, page 48]), who proved that nonempty convex sets K_1 and K_2 in $\mathbb{R}^n$ satisfying the condition $\operatorname{rint} K_1 \cap \operatorname{rint} K_2 = \varnothing$ are separated (not necessarily properly) by a hyperplane. Under the additional assumption $\operatorname{aff}(K_1 \cup K_2) = \mathbb{R}^n$, Theorem 10.8 was independently proved by Klee [172]. Theorem 10.10 and Corollary 10.11 are due to Klee [173]. Also, Klee [183] provides various references related to Theorems 10.12 and 10.22. Dieudonné [93] showed that disjoint closed convex sets K_1 and K_2 in $\mathbb{R}^n$ are strongly separated by a hyperplane provided $\operatorname{rec} K_1 \cap \operatorname{rec} K_2 = \{\boldsymbol{o}\}$.

Bair and Jongmans [23] obtained various conditions for nontrivial separation of convex sets in linear space. Certain types of separation of a convex cone and a set in $\mathbb{R}^n$, related to generalized Lagrange multipliers are considered by Bigi and Pappalardo [35] and Castellani, Mastroeni, and Pappalardo [67].

Maximal separation theorems. Klee [182] (see also Gritzmann and Klee [128]) provides an elaborated classification of various types of separation of convex sets by hyperplanes in $\mathbb{R}^n$ and gives a series of *maximal separation* results regarding these types of separation. Following [182], given a pair $\{\mathcal{B}, \mathcal{C}\}$ of nonempty families of proper convex sets in $\mathbb{R}^n$, we say that $\mathcal{B}$ is maximal with respect to a type S of separation provided it satisfies the following conditions.

(a) The sets B and C are S-separated whenever $B \in \mathcal{B}$ and $C \in \mathcal{C}$ with $B \cap C = \varnothing$.

(b) For every $B \in \mathcal{B}$ there is $C \in \mathcal{C}$ such that $B \cap C = \varnothing$.

(c) For every proper convex set $Y \notin \mathcal{B}$ there is $C \in \mathcal{C}$ such that $Y \cap C = \varnothing$ but Y and C are not S-separated.

For instance, the following two families are mutually maximal with respect to strong separation (see also Gale and Klee [116]):

1. $\mathcal{B}$ is the family of all continuous convex sets (i. e., n-dimensional closed convex sets without boundary or asymptotic halflines), and $\mathcal{C}$ is the family of all closed convex sets.

2. $\mathcal{B}$ and $\mathcal{C}$ are the families of closed convex sets without asymptotic halflines.

Sharp separation of convex cones. Theorems 10.15 and 10.16 can be found in Soltan [261]. In terms of continuous linear functionals on a linear topological space, Theorem 10.15 was proved earlier by Klee [172] under the assumption $\mathrm{ap}\,(\mathrm{cl}\, C_1) \cap \mathrm{ap}\,(\mathrm{cl}\, C_2) = \{\boldsymbol{o}\}$, and by Bair and Gwinner [22] under the condition that $\mathrm{ap}\,(\mathrm{cl}\, C_1) \cap \mathrm{ap}\,(\mathrm{cl}\, C_2)$ is a subspace.

Separation by slabs. Theorem 10.20 is attributed to Dax [87]. De Wilde [89] proved that disjoint line-free closed convex sets K_1 and K_2 satisfying the condition $\mathrm{rec}\, K_1 \cap \mathrm{rec}\, K_2 = \{\boldsymbol{o}\}$ are separated by a closed slab $P \subset \mathbb{R}^n$ (not necessarily of maximal width) such that both sets $K_1 \cap P$ and $K_2 \cap P$ are singletons.

Convex halfspaces and semispaces. Corollary 10.26 is proved by Tukey [281] for the case of convex sets in any vector space E (the condition that E should be normed is superfluous); see also a later publication of Stone [271, Chapter 3, Theorem 7], and a related paper of Páles [225]. Theorem 10.28 is given (in equivalent terms) by Lassak [192] (see also Martínez-Legaz and Singer [208]). The notion of semispace and Theorems 10.31 and 10.32 (for the case $L = \mathbb{R}^n$) are due to Hammer [141, 142].

Chapter 11

Extreme Structure of Convex Sets

11.1 Extreme Faces

11.1.1 *Basic Properties of Extreme Faces*

Definition 11.1. Let $K \subset \mathbb{R}^n$ be a convex set. A nonempty convex subset F of K is called an *extreme face* of K if for any choice of points $\boldsymbol{x}, \boldsymbol{y} \in K$ and a scalar $0 < \lambda < 1$ the inclusion $(1-\lambda)\boldsymbol{x} + \lambda\boldsymbol{y} \in F$ implies that $\boldsymbol{x}, \boldsymbol{y} \in F$. The empty set $\varnothing$ is assumed to be an extreme face of K.

For instance, the planar closed convex set $K \subset \mathbb{R}^2$ depicted in Figure 11.1 has the following extreme faces: $\varnothing$, every singleton from the closed arc $\operatorname{bd} K \setminus (\boldsymbol{a}, \boldsymbol{c})$, the closed segment $[\boldsymbol{a}, \boldsymbol{c}]$, and K.

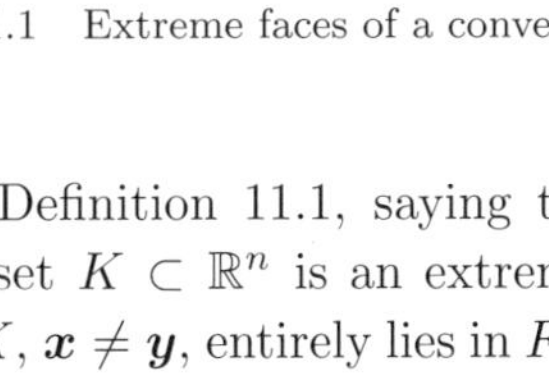

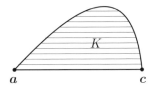

Fig. 11.1 Extreme faces of a convex sets.

We can reformulate Definition 11.1, saying that a nonempty convex subset F of the convex set $K \subset \mathbb{R}^n$ is an extreme face of K provided a closed segment $[\boldsymbol{x}, \boldsymbol{y}] \subset K$, $\boldsymbol{x} \neq \boldsymbol{y}$, entirely lies in F whenever the respective open segment $(\boldsymbol{x}, \boldsymbol{y})$ meets F.

Examples. 1. The extreme faces of a closed ball $B_\rho(\boldsymbol{a}) \subset \mathbb{R}^n$ are $\varnothing$, all singletons from its boundary sphere $S_\rho(\boldsymbol{a})$, and $B_\rho(\boldsymbol{a})$ (see Problem 11.1).

2. The extreme faces of an r-simplex $\Delta = \Delta(\boldsymbol{x}_1, \dots, \boldsymbol{x}_{r+1}) \subset \mathbb{R}^n$ are $\varnothing$ and all simplices $\Delta(\boldsymbol{z}_1, \dots, \boldsymbol{z}_t)$, where $\{\boldsymbol{z}_1, \dots, \boldsymbol{z}_t\}$ is a nonempty subset of $\{\boldsymbol{x}_1, \dots, \boldsymbol{x}_{r+1}\}$ (see Problem 11.2).

3. The extreme faces of a simplicial cone $\Gamma_{\boldsymbol{a}}(\boldsymbol{x}_1, \dots, \boldsymbol{x}_r) \subset \mathbb{R}^n$ are $\varnothing$, $\{\boldsymbol{a}\}$, and all simplicial cones $\Gamma_{\boldsymbol{a}}(\boldsymbol{z}_1, \dots, \boldsymbol{z}_t)$, where $\{\boldsymbol{z}_1, \dots, \boldsymbol{z}_t\}$ is a nonempty subset of $\{\boldsymbol{x}_1, \dots, \boldsymbol{x}_r\}$ (see Problem 11.3).

4. The extreme faces of a plane $L \subset \mathbb{R}^n$ are $\varnothing$ and L. If $\dim L = m > 0$ and D is a closed halfplane of L, then $\varnothing$, $\operatorname{rbd} D$, and D are the extreme faces of D (see Problem 11.4). If G is a closed slab of L, then the extreme faces of G are $\varnothing$, two $(m-1)$-dimensional planes determining G, and G (see Problem 11.5).

Theorem 11.2. *If $K \subset \mathbb{R}^n$ is a convex set and F is a convex subset of K, then F is an extreme face of K if and only if the following two conditions are satisfied: $K \cap \operatorname{aff} F = F$ and the set $K \setminus F$ is convex.*

Proof. Since both cases $F = \varnothing$ and $F = K$ are obvious, we may assume that F is a nonempty proper subset of K.

Let F be an extreme face of K. To show that $K \cap \operatorname{aff} F = F$, it suffices to prove the inclusion $K \cap \operatorname{aff} F \subset F$ (the opposite one is obvious). Choose points $\boldsymbol{x} \in K \cap \operatorname{aff} F$ and $\boldsymbol{y} \in \operatorname{rint} F$ ($\operatorname{rint} F \neq \varnothing$ according to Corollary 3.21). By Theorem 3.26, there is a scalar $0 < \lambda < 1$ such that $(1-\lambda)\boldsymbol{x} + \lambda \boldsymbol{y} \in F$. One has $\boldsymbol{x}, \boldsymbol{y} \in F$ because F is an extreme face of K. In particular, $\boldsymbol{x} \in F$, implying the inclusion $K \cap \operatorname{aff} F \subset F$.

Next, Definition 11.1 shows that $F \cap [\boldsymbol{x}, \boldsymbol{z}] = \varnothing$ whenever $\boldsymbol{x}, \boldsymbol{z} \in K \setminus F$. Because $[\boldsymbol{x}, \boldsymbol{z}] \subset K$ by the convexity of K, we obtain that $[\boldsymbol{x}, \boldsymbol{z}] \subset K \setminus F$. Hence the set $K \setminus F$ is convex.

Conversely, suppose that the set $K \setminus F$ is convex and $K \cap \operatorname{aff} F = F$. Choose points $\boldsymbol{x}, \boldsymbol{y} \in K$ and a scalar $0 < \lambda < 1$ with $\boldsymbol{z} = (1-\lambda)\boldsymbol{x} + \lambda \boldsymbol{y} \in F$. Then at least one of $\boldsymbol{x}, \boldsymbol{y}$ should belong to F (otherwise $[\boldsymbol{x}, \boldsymbol{y}] \subset K \setminus F$ by the convexity of $K \setminus F$). Let, for instance, $\boldsymbol{x} \in F$. Consequently,

$$\boldsymbol{y} = (1 - \lambda^{-1})\boldsymbol{x} + \lambda^{-1}\boldsymbol{z} \subset \operatorname{aff} F$$

(see Theorem 2.38), implying the inclusion $\boldsymbol{y} \in K \cap \operatorname{aff} F = F$. Summing up, F is an extreme face of K. $\square$

Theorem 11.3. *For a convex set $K \subset \mathbb{R}^n$, the following assertions hold.*

(*a*) *The intersection of any family of extreme faces of K is an extreme face of K.*

(*b*) *If* F *is an extreme face of* K *and* G *is an extreme face of* F, *then* G *is an extreme face of* K.

(*c*) *If* F *is an extreme face of* K *and* M *is a convex subset of* K, *then* $F \cap M$ *is an extreme face of* M. *In particular,* F *is an extreme face of* M *provided* $F \subset M$.

Proof. Since all three assertions are obvious for the case of empty sets, we may assume that all sets involved are nonempty.

(*a*) Let $\mathcal{F} = \{F_\alpha\}$ be a family of extreme faces of K. Put $F = \bigcap_\alpha F_\alpha$. Since the case $F = \varnothing$ is obvious, we may assume that F is nonempty. Choose points $\boldsymbol{x}, \boldsymbol{y} \in K$ and a scalar $0 < \lambda < 1$ such that $(1-\lambda)\boldsymbol{x}+\lambda\boldsymbol{y} \in F$. Then $(1-\lambda)\boldsymbol{x} + \lambda\boldsymbol{y} \in F_\alpha$ for all $F_\alpha \in \mathcal{F}$. Because every set $F_\alpha \in \mathcal{F}$ is an extreme face of K, we have $\boldsymbol{x}, \boldsymbol{y} \in F_\alpha$. Hence both $\boldsymbol{x}$ and $\boldsymbol{y}$ belong to $F = \bigcap_\alpha F_\alpha$. Therefore, F is an extreme face of K.

(*b*) The case $G = \varnothing$ is obvious; so, we may assume that G is nonempty. Choose points $\boldsymbol{x}, \boldsymbol{y} \in K$ and a scalar $0 < \lambda < 1$ such that $(1-\lambda)\boldsymbol{x}+\lambda\boldsymbol{y} \in G$. Then $(1-\lambda)\boldsymbol{x} + \lambda\boldsymbol{y} \in F$, and hence $\boldsymbol{x}, \boldsymbol{y} \in F$ because F is an extreme face of K. Since G is an extreme face of F, we obtain that $\boldsymbol{x}, \boldsymbol{y} \in G$. Thus G is an extreme face of K.

(*c*) Since the case $F \cap M = \varnothing$ is obvious, we may assume that $F \cap M$ is nonempty. Choose points $\boldsymbol{x}, \boldsymbol{y} \in M$ and a scalar $0 < \lambda < 1$ such that $(1-\lambda)\boldsymbol{x} + \lambda\boldsymbol{y} \in F \cap M$. Because $\boldsymbol{x}, \boldsymbol{y} \in K$ and F is an extreme face of K, we have $\boldsymbol{x}, \boldsymbol{y} \in F$. Thus $\boldsymbol{x}, \boldsymbol{y} \in F \cap M$, implying that $F \cap M$ is an extreme face of M. □

Theorem 11.4. *For a convex set* $K \subset \mathbb{R}^n$, *the following assertions hold.*

(*a*) *If* F *is an extreme face of* K, *then* $F = K \cap \operatorname{cl} F$; *in particular,* F *is closed provided* K *is closed.*

(*b*) *If* F *is an extreme face of* K *and* M *is a convex subset of* K *such that* $F \cap \operatorname{rint} M \neq \varnothing$, *then* $M \subset F$.

(*c*) *If* F *is an extreme face of* K *and* M *is a convex subset of* F, *then the set* $(K \setminus F) \cup M$ *is convex.*

(*d*) *If an extreme face* F *of* K *meets* $\operatorname{rint} K$, *then* $F = K$.

(*e*) *If* F *is a nonempty proper extreme face of* K, *then* $F \subset \operatorname{rbd} K$ *and* $\dim F \leqslant \dim K - 1$.

(*f*) *Distinct extreme faces of* K *have disjoint relative interiors.*

(*g*) *If* F *is an extreme face of* K *and* X *is a subset of* K, *then*

$$\operatorname{conv}(F \cap X) = F \cap \operatorname{conv} X.$$

Proof. Assertions (a), (b), and (g) are obvious when any of the sets F, K, M and X involved is empty (if $M = \varnothing$, then assertion (c) follows from Theorem 11.2). So, we may assume that all sets involved are nonempty.

(a) Because the inclusion $F \subset K \cap \operatorname{cl} F$ is obvious, it remains to prove that $K \cap \operatorname{cl} F \subset F$. For this, choose points $\boldsymbol{x} \in K \cap \operatorname{cl} F$ and $\boldsymbol{y} \in \operatorname{rint} F$ ($\operatorname{rint} F \neq \varnothing$ according to Corollary 3.21). Since the plane $\operatorname{aff} F$ is closed (see Corollary 2.21), we have $\operatorname{cl} F \subset \operatorname{aff} F$. Now, Theorem 3.26 implies the existence of a scalar $0 < \lambda < 1$ such that $(1 - \lambda)\boldsymbol{x} + \lambda \boldsymbol{y} \in F$. Therefore, $\boldsymbol{x} \in F$ because F is an extreme face of K. Summing up, $K \cap \operatorname{cl} F \subset F$.

(b) Let $\boldsymbol{x} \in M$. Choose a point $\boldsymbol{y} \in F \cap \operatorname{rint} M$. By Theorem 3.26, there is a scalar $0 < \lambda < 1$ such that $(1 - \lambda)\boldsymbol{x} + \lambda \boldsymbol{y} \in M$. Since $\boldsymbol{x}, \boldsymbol{y} \in K$ and F is a face of K, one has $\boldsymbol{x} \in F$. Hence $M \subset F$.

(c) Choose points $\boldsymbol{x}$ and $\boldsymbol{y}$ in $(K \setminus F) \cup M$. If both $\boldsymbol{x}$ and $\boldsymbol{y}$ belong to one of the sets $K \setminus F$ or M, then $[\boldsymbol{x}, \boldsymbol{y}]$ lies in this set by a convexity argument ($K \setminus F$ is convex according to Theorem 11.2). Let, for instance, $\boldsymbol{x} \in K \setminus F$ and $\boldsymbol{y} \in M$. We observe that $(\boldsymbol{x}, \boldsymbol{y}) \subset K \setminus F$. Indeed, assuming the existence of a point $\boldsymbol{z} \in F \cap (\boldsymbol{x}, \boldsymbol{y})$, we would obtain that $\boldsymbol{y} \in \langle \boldsymbol{x}, \boldsymbol{z} \rangle \subset \operatorname{aff} F$, contrary to $K \cap \operatorname{aff} F = \varnothing$ by the above theorem. Hence

$$[\boldsymbol{x}, \boldsymbol{y}] = [\boldsymbol{x}, \boldsymbol{y}) \cup \{\boldsymbol{y}\} \subset (K \setminus F) \cup M,$$

implying the convexity of $(K \setminus F) \cup M$.

(d) This part immediately follows from assertion (b), where $M = K$.

(e) By assertion (d), F should be disjoint from $\operatorname{rint} K$. Therefore, $F \subset \operatorname{rbd} K$. Theorem 3.67 implies that $\dim F \leqslant \dim K - 1$.

(f) If F and G are extreme faces of K such that $\operatorname{rint} F \cap \operatorname{rint} G \neq \varnothing$, then assertion (b) gives the equality $F = G$.

(g) Since the set $F \cap \operatorname{conv} X$ is convex, the inclusion $F \cap X \subset F \cap \operatorname{conv} X$ implies that $\operatorname{conv}(F \cap X) \subset F \cap \operatorname{conv} X$.

For the opposite inclusion, choose a point $\boldsymbol{x} \in F \cap \operatorname{conv} X$. By Theorem 4.3, $\boldsymbol{x}$ can be written as a convex combination $\boldsymbol{x} = \lambda_1 \boldsymbol{x}_1 + \cdots + \lambda_k \boldsymbol{x}_k$ of suitable points $\boldsymbol{x}_1, \ldots, \boldsymbol{x}_k$ from X. We may assume that all scalars $\lambda_1, \ldots, \lambda_k$ are positive (indeed, the exclusion of zero terms does not affect the equality $\boldsymbol{x} = \lambda_1 \boldsymbol{x}_1 + \cdots + \lambda_k \boldsymbol{x}_k$). According to Corollary 4.19, $\boldsymbol{x}$ belongs to the relative interior of the convex set $M = \operatorname{conv}\{\boldsymbol{x}_1, \ldots, \boldsymbol{x}_k\} \subset K$. Hence $\boldsymbol{x} \in F \cap \operatorname{rint} M$, and assertion (b) implies that $M \subset F$. Consequently, $\boldsymbol{x}_1, \ldots, \boldsymbol{x}_k \in F$. Thus $\boldsymbol{x}_1, \ldots, \boldsymbol{x}_k \in F \cap X$, and Theorem 4.3 shows that $\boldsymbol{x} \in \operatorname{conv}(F \cap X)$, confirming the inclusion $F \cap \operatorname{conv} X \subset \operatorname{conv}(F \cap X)$. $\square$

Remarks. 1. If F is an extreme face of a convex set $K \subset \mathbb{R}^n$, then $\operatorname{cl} F$ is not necessarily an extreme face of $\operatorname{cl} K$. For instance, if $K = \{\boldsymbol{o}\} \cup \{(x, y) :$

$y > 0\}$, then $\{\boldsymbol{o}\}$ is an extreme face of K, while the smallest extreme face of $\operatorname{cl} K = \{(x, y) : y \geqslant 0\}$ containing $\boldsymbol{o}$ is the x-axis of $\mathbb{R}^2$.

2. A combination of Corollary 3.9 and Theorems 12.2 and 12.3 implies that any convex subset F of $\operatorname{rbd} K$ lies in a maximal under inclusion extreme face G of K, which satisfies the inclusion $G \subset \operatorname{rbd} K$.

The next theorem shows that we can reduce the study of extreme faces of arbitrary convex sets to those with zero lineality.

Theorem 11.5. *Let a nonempty convex set $K \subset \mathbb{R}^n$ be expressed as $K = (K \cap S) + \operatorname{lin} K$, where $S \subset \mathbb{R}^n$ is a subspace complementary to* $\operatorname{lin} K$. *A nonempty set $F \subset \mathbb{R}^n$ is an extreme face of K if and only if $F = E + \operatorname{lin} K$, where E is an extreme face of $K \cap S$.*

Proof. Denote by f the linear projection on S along $\operatorname{lin} K$ (see page 7).

Choose an extreme face F of K and put $E = f(F)$. Clearly, $F \subset E + \operatorname{lin} K$. We assert that $F = E + \operatorname{lin} K$. Indeed, assume for a moment the existence of a point $\boldsymbol{x} \in (E + \operatorname{lin} K) \setminus F$. Since $f(\boldsymbol{x}) \in E$, there is a point $\boldsymbol{y} \in F$ such that $f(\boldsymbol{x}) = f(\boldsymbol{y})$. Consequently, $f(\boldsymbol{x} - \boldsymbol{y}) = \boldsymbol{o}$ and $\boldsymbol{x} - \boldsymbol{y} \in \operatorname{null} f = \operatorname{lin} K$. This argument shows that both $\boldsymbol{x}$ and $\boldsymbol{y}$ belong to the plane $L = \boldsymbol{y} + \operatorname{lin} K \subset K$. Put $\boldsymbol{z} = 2\boldsymbol{y} - \boldsymbol{x}$. Then $\boldsymbol{z} \in L$ (see Theorem 2.38) and $\boldsymbol{y} = (\boldsymbol{x} + \boldsymbol{z})/2 \in (\boldsymbol{x}, \boldsymbol{z})$. Consequently, $\boldsymbol{x}, \boldsymbol{z} \in F$ because F is an extreme face of K. The last is in contradiction with the choice of $\boldsymbol{x}$. Summing up, $F = E + \operatorname{lin} K$.

Next, we assert that E is an extreme face of $K \cap S$. Indeed, let points $\boldsymbol{x}, \boldsymbol{y} \in K \cap S$ and a scalar $0 < \lambda < 1$ be such that $\boldsymbol{z} = (1 - \lambda)\boldsymbol{x} + \lambda\boldsymbol{y} \in E$. Then $\boldsymbol{z} \in F$ because $E \subset F$. Since F is an extreme face of K, both $\boldsymbol{x}$ and $\boldsymbol{y}$ belong to F, and whence $\boldsymbol{x}, \boldsymbol{y} \in E$ due to $\boldsymbol{x} = f(\boldsymbol{x})$ and $\boldsymbol{y} = f(\boldsymbol{y})$. Summing up, E is an extreme face of $K \cap S$.

Conversely, let E be an extreme face of $K \cap S$. Put $F = E + \operatorname{lin} K$. Clearly, F is a convex subset of $K = (K \cap S) + \operatorname{lin} K$. Choose points $\boldsymbol{x}, \boldsymbol{y} \in K$ and a scalar $0 < \lambda < 1$ such that $\boldsymbol{z} = (1 - \lambda)\boldsymbol{x} + \lambda\boldsymbol{y} \in F$. Then $f(\boldsymbol{z}) = (1 - \lambda)f(\boldsymbol{x}) + \lambda f(\boldsymbol{y}) \in E$. Since E is an extreme face of $K \cap S$, we have $f(\boldsymbol{x}), f(\boldsymbol{y}) \in E$. Therefore, $\boldsymbol{x}, \boldsymbol{y} \in f^{-1}(E) = F$, implying that F is an extreme face of K. $\square$

Theorem 11.6. *If $C \subset \mathbb{R}^n$ is a convex cone, then every extreme face F of C is a convex cone, with* $\operatorname{ap} C \subset \operatorname{ap} F$. *If, additionally,* $\operatorname{ap} C \subset C$, *then* $\operatorname{ap} C$ *is the smallest nonempty extreme face of C.*

Proof. Since the case $F = \varnothing$ is obvious, we may assume that $F \neq \varnothing$. If $F \subset \operatorname{ap} C$, then $F = \operatorname{ap} C$ (see Problem 11.4). Suppose that $F \not\subset \operatorname{ap} C$ and choose any point $\boldsymbol{x} \in F \setminus \operatorname{ap} C$. Given a point $\boldsymbol{a} \in \operatorname{ap} C$, the open halfline $(\boldsymbol{a}, \boldsymbol{x}\rangle$ lies in C. Since $\boldsymbol{x} \in (\boldsymbol{z}, \boldsymbol{y})$ for any points $\boldsymbol{z} \in (\boldsymbol{a}, \boldsymbol{x})$ and $\boldsymbol{y}$ from the open halfline $h = (\boldsymbol{a}, \boldsymbol{x}\rangle \setminus (\boldsymbol{a}, \boldsymbol{x}]$, we conclude that $(\boldsymbol{a}, \boldsymbol{x}\rangle \subset F$. Hence F is a cone with apex $\boldsymbol{a}$. Furthermore, $\operatorname{ap} C \subset \operatorname{ap} F$.

Suppose that $\operatorname{ap} C \subset C$. To show that $\operatorname{ap} C$ is an extreme face of C, choose any points $\boldsymbol{x}, \boldsymbol{y} \in C$ and a scalar $0 < \lambda < 1$ such that the point $\boldsymbol{z} = (1-\lambda)\boldsymbol{x} + \lambda\boldsymbol{y}$ belongs to $\operatorname{ap} C$. Assume, for contradiction, that at least one of the points $\boldsymbol{x}$ and $\boldsymbol{y}$, say $\boldsymbol{x}$, does not belong to $\operatorname{ap} C$. Then $\boldsymbol{y} \notin \operatorname{ap} C$, since otherwise $\boldsymbol{x}$, as an affine combination of $\boldsymbol{z}$ and $\boldsymbol{y}$, should be in the plane $\operatorname{ap} C$:

$$\boldsymbol{x} = \frac{1}{1-\lambda}\boldsymbol{z} - \frac{\lambda}{1-\lambda}\boldsymbol{y} \in \langle \boldsymbol{z}, \boldsymbol{y} \rangle \subset \operatorname{ap} C$$

(see Theorems 2.38 and 5.14). The inclusion $\boldsymbol{z} \in \operatorname{ap} C$ implies that both closed halflines $[\boldsymbol{z}, \boldsymbol{x}\rangle$ and $[\boldsymbol{z}, \boldsymbol{y}\rangle$ lies in C. Consequently, the whole line $\langle \boldsymbol{z}, \boldsymbol{y} \rangle$ lies in C. According to Theorem 5.17, $\operatorname{ap} C$ is the largest plane among all planes containing $\boldsymbol{z}$ and lying in C. Thus $\boldsymbol{y}, \boldsymbol{z} \in \langle \boldsymbol{z}, \boldsymbol{y} \rangle \subset \operatorname{ap} C$, contrary to the assumption $\boldsymbol{x} \notin \operatorname{ap} C$. Summing up, $\operatorname{ap} C$ is an extreme face of C.

To prove that $\operatorname{ap} C$ is the smallest nonempty extreme face of C, choose any nonempty face F of C. By the above proved, $\operatorname{ap} C \subset \operatorname{ap} F \subset F$. □

11.1.2 *Generated Extreme Faces*

Theorem 11.3 gives a base to the definition below.

Definition 11.7. Let $K \subset \mathbb{R}^n$ be a convex set and X be a subset of K. The intersection of all extreme faces of K containing X is called *the extreme face of K generated by X* and denoted $F_X(K)$.

The corollary below follows from definitions and Theorem 11.3.

Corollary 11.8. *Given a convex set $K \subset \mathbb{R}^n$ and subsets X and Y of K, the following assertions hold.*

(*a*) *$X \subset F_X(K)$, with $X = F_X(K)$ if and only if X is an extreme face of K. In particular, $F_\varnothing(K) = \varnothing$ and $F_K(K) = K$.*
(*b*) *$F_X(K)$ is the smallest extreme face of K which contains X.*
(*c*) *$F_X(F_X(K)) = F_X(K)$.*
(*d*) *$F_X(K) = F_Y(K)$, where $Y = \operatorname{conv} X$.*

(*e*) $F_X(K) \subset F_Z(K)$ *if* $X \subset Z$. □

The next result deals with extreme faces generated by points.

Theorem 11.9. *For a nonempty convex set* $K \subset \mathbb{R}^n$ *and a point* $\boldsymbol{x} \in K$, *the following assertions hold.*

(*a*) $F_{\boldsymbol{x}}(K)$ *consists of* $\boldsymbol{x}$ *and all points* $\boldsymbol{y} \in K \setminus \{\boldsymbol{x}\}$ *such that* $\boldsymbol{x} \in (\boldsymbol{y}, \boldsymbol{z})$ *for a suitable* $\boldsymbol{z} \in K$.

(*b*) $F_{\boldsymbol{x}}(K)$ *is the largest among all convex subsets* C *of* K *satisfying the condition* $\boldsymbol{x} \in \operatorname{rint} C$.

(*c*) *If* G *is a convex subset of* K *and* $\boldsymbol{x} \in \operatorname{rint} G$, *then* $G \subset F_{\boldsymbol{x}}(K)$ *and* $\operatorname{rint} G \subset \operatorname{rint} F_{\boldsymbol{x}}(K)$.

(*d*) *If* F *is an extreme face of* K, *then* $F = F_{\boldsymbol{x}}(K)$ *if and only if* $\boldsymbol{x} \in \operatorname{rint} F$.

(*e*) $F_{\boldsymbol{x}}(K)$ *is the unique extreme face of* K *containing* $\boldsymbol{x}$ *in its relative interior.*

(*f*) *For a point* $\boldsymbol{y} \in K$, *the equality* $F_{\boldsymbol{x}}(K) = F_{\boldsymbol{y}}(K)$ *holds if and only if* $\boldsymbol{y} \in \operatorname{rint} F_{\boldsymbol{x}}(K)$.

(*g*) $F_{\boldsymbol{x}}(K) = K$ *if and only if* $\boldsymbol{x} \in \operatorname{rint} K$.

(*h*) $F_{\boldsymbol{x}}(K) \subset \operatorname{rbd} K$ *if and only if* $\boldsymbol{x} \in \operatorname{rbd} K$.

(*i*) *The apex set of the generated cone* $C_{\boldsymbol{x}}(K)$ *equals* $\operatorname{aff} F_{\boldsymbol{x}}(K)$.

Proof. (*a*) Denote by G the set consisting of $\boldsymbol{x}$ and all points $\boldsymbol{y} \in K \setminus \{\boldsymbol{x}\}$ such that $\boldsymbol{x} \in (\boldsymbol{y}, \boldsymbol{z})$ for a suitable $\boldsymbol{z} \in K$. To prove the equality $G = F_{\boldsymbol{x}}(K)$, it suffices to verify that G is the smallest extreme face of K containing $\boldsymbol{x}$.

First, we will show that G is a convex set. Indeed, let $\boldsymbol{v}_1, \boldsymbol{v}_2 \in G$. Since the case $\boldsymbol{v}_1 = \boldsymbol{v}_2 = \boldsymbol{x}$ is obvious, we may assume, for instance, that $\boldsymbol{v}_1 \neq \boldsymbol{x}$. Then $\boldsymbol{x} \in (\boldsymbol{v}_1, \boldsymbol{w}_1)$ for a suitable point $\boldsymbol{w}_1 \in K$. Clearly, $K \cap \langle \boldsymbol{v}_1, \boldsymbol{w}_1 \rangle \subset G$ by the convexity of K. If $\boldsymbol{v}_2 \in K \cap \langle \boldsymbol{v}_1, \boldsymbol{w}_1 \rangle$, then $[\boldsymbol{v}_1, \boldsymbol{v}_2] \subset K \cap \langle \boldsymbol{v}_1, \boldsymbol{w}_1 \rangle \subset G$. Hence it remains to consider the case when $\boldsymbol{v}_2 \notin K \cap \langle \boldsymbol{v}_1, \boldsymbol{w}_1 \rangle$. As above, $\boldsymbol{x} \in (\boldsymbol{v}_2, \boldsymbol{w}_2)$ for a suitable point $\boldsymbol{w}_2 \in K$. Given a point $\boldsymbol{v} \in [\boldsymbol{v}_1, \boldsymbol{v}_2]$, there is a point $\boldsymbol{w} \in [\boldsymbol{w}_1, \boldsymbol{w}_2]$ such that $\boldsymbol{x} \in (\boldsymbol{v}, \boldsymbol{w})$ (see Lemma 2.27 and Figure 11.2). Since both points $\boldsymbol{v}$ and $\boldsymbol{w}$ belong to K, we obtain the inclusion $\boldsymbol{v} \in G$. Hence $[\boldsymbol{v}_1, \boldsymbol{v}_2] \subset G$, implying the convexity of G.

Next, we are going to show that G is an extreme face of K. Indeed, let $\boldsymbol{y}, \boldsymbol{z}$ be points in K such that the open segment $(\boldsymbol{y}, \boldsymbol{z})$ meets G at a point $\boldsymbol{u}$. If $\boldsymbol{u} = \boldsymbol{x}$, then the inclusions $\boldsymbol{y}, \boldsymbol{z} \in G$ follow from the definition of G. Let $\boldsymbol{u} \neq \boldsymbol{x}$. Then $\boldsymbol{x} \in (\boldsymbol{u}, \boldsymbol{w})$ for a suitable point $\boldsymbol{w} \in G$. According to Lemma 2.26, there are points $\boldsymbol{y}' \in (\boldsymbol{z}, \boldsymbol{w})$ and $\boldsymbol{z}' \in (\boldsymbol{y}, \boldsymbol{w})$ such that $\boldsymbol{x} \in (\boldsymbol{y}, \boldsymbol{y}') \cap (\boldsymbol{z}, \boldsymbol{z}')$. Since $\boldsymbol{y}', \boldsymbol{z}' \in K$ by the convexity of K, we have

$\boldsymbol{y}, \boldsymbol{z} \in G$ according to the definition of G. Summing up, G is an extreme face of K.

Finally, let F be any extreme face of K containing $\boldsymbol{x}$. If $\boldsymbol{y} \in G \setminus \{\boldsymbol{x}\}$ and $\boldsymbol{z}$ is a point in K satisfying the inclusion $\boldsymbol{x} \in (\boldsymbol{y}, \boldsymbol{z})$, then Definition 11.1 shows that $\boldsymbol{x}, \boldsymbol{y} \in F$. Hence $G \subset F$, and G is the smallest extreme face of K containing $\boldsymbol{x}$. Consequently, $G = F_{\boldsymbol{x}}(K)$.

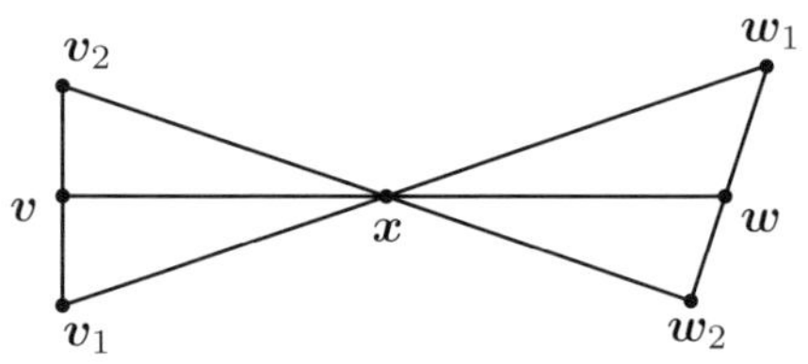

Fig. 11.2 Illustration to Theorem 11.9.

(*b*) The inclusion $\boldsymbol{x} \in \operatorname{rint} F_{\boldsymbol{x}}(K)$ follows from assertion (*a*) and Corollary 3.27. Let C be a convex subset of K such that $\boldsymbol{x} \in \operatorname{rint} C$. If $\boldsymbol{y} \in C \setminus \{\boldsymbol{x}\}$, then, by Corollary 3.27, there is a point $\boldsymbol{z} \in C \subset K$ such that $\boldsymbol{x} \in (\boldsymbol{y}, \boldsymbol{z})$. Hence $\boldsymbol{y} \in F_{\boldsymbol{x}}(K)$, and $C \subset F_{\boldsymbol{x}}(K)$.

(*c*) This part follows from Theorem 11.4 and assertion (*b*) above.

(*d*) If $F = F_{\boldsymbol{x}}(K)$, then $\boldsymbol{x} \in \operatorname{rint} F_{\boldsymbol{x}}(K) = \operatorname{rint} F$. Conversely, let $\boldsymbol{x} \in \operatorname{rint} F$. Then $F_{\boldsymbol{x}}(K) \subset F$ as the smallest extreme face of K that contains $\boldsymbol{x}$. On the other hand, $F \subset F_{\boldsymbol{x}}(K)$, since $F_{\boldsymbol{x}}(K)$ is the largest convex subset of K that contains $\boldsymbol{x}$ in its relative interior.

(*e*) The last assertion follows from Theorem 11.4 and assertion (*b*) above.

(*f*) If $F_{\boldsymbol{x}}(K) = F_{\boldsymbol{y}}(K)$, then $\boldsymbol{y} \in \operatorname{rint} F_{\boldsymbol{y}}(K) = \operatorname{rint} F_{\boldsymbol{x}}(K)$ by assertion (*b*) above. Conversely, let $\boldsymbol{y} \in \operatorname{rint} F_{\boldsymbol{x}}(K)$. Then $F_{\boldsymbol{x}}(K)$ and $F_{\boldsymbol{y}}(K)$ share a common relative interior point, $\boldsymbol{y}$. By Theorem 11.4, $F_{\boldsymbol{x}}(K) = F_{\boldsymbol{y}}(K)$.

(*g*) This part immediately follows from assertion (*d*) of Theorem 11.4.

(*h*) If $F_{\boldsymbol{x}}(K) \subset \operatorname{rbd} K$, then the inclusion $\boldsymbol{x} \in \operatorname{rbd} K$ is obvious. Conversely, let $\boldsymbol{x} \in \operatorname{rbd} K$. Assume for a moment the existence of a point $\boldsymbol{y} \in F_{\boldsymbol{x}}(K) \cap \operatorname{rint} K$. By assertion (*b*) above, there is a point $\boldsymbol{z} \in K$ such that $\boldsymbol{x} \in (\boldsymbol{y}, \boldsymbol{z})$, and Theorem 3.24 implies the inclusion $\boldsymbol{x} \in \operatorname{rint} K$, in contradiction with the choice of $\boldsymbol{x}$. Hence $F_{\boldsymbol{x}}(K) \subset \operatorname{rbd} K$.

(*i*) By assertion (*b*), $\boldsymbol{x} \in \operatorname{rint} F_{\boldsymbol{x}}(K) \subset K$. Consequently, a combination of Theorems 5.38 and 5.48 gives

$$\operatorname{aff} F_{\boldsymbol{x}}(K) = C_{\boldsymbol{x}}(F_{\boldsymbol{x}}(K)) \subset C_{\boldsymbol{x}}(K).$$

As follows from Theorem 5.17, $\operatorname{ap} C_{\boldsymbol{x}}(K)$ is the largest plane through $\boldsymbol{x}$ which lies in $C_{\boldsymbol{x}}(K)$. Therefore, $\operatorname{aff} F_{\boldsymbol{x}}(K) \subset \operatorname{ap} C_{\boldsymbol{x}}(K)$.

For the opposite inclusion, choose any point $\boldsymbol{u} \in \operatorname{ap} C_{\boldsymbol{x}}(K)$. If $\boldsymbol{u} = \boldsymbol{x}$, then the inclusion $\boldsymbol{u} \in \operatorname{aff} F_{\boldsymbol{x}}(K)$ is obvious. Let $\boldsymbol{u} \neq \boldsymbol{x}$. Then the line $l = \langle \boldsymbol{x}, \boldsymbol{u} \rangle$ lies in $C_{\boldsymbol{x}}(K)$. Hence there are points $\boldsymbol{y}, \boldsymbol{z} \in K \setminus \{\boldsymbol{x}\}$ such that $l = \langle \boldsymbol{y}, \boldsymbol{x}] \cup [\boldsymbol{x}, \boldsymbol{z} \rangle$. Since $F_{\boldsymbol{x}}(K)$ is an extreme face, we obtain that $\boldsymbol{y}, \boldsymbol{z} \in F_{\boldsymbol{x}}(K)$. Finally, $\boldsymbol{u} \in l = \operatorname{aff}\{\boldsymbol{y}, \boldsymbol{z}\} \subset \operatorname{aff} C_{\boldsymbol{x}}(K)$.

Summing up, $\operatorname{ap} C_{\boldsymbol{x}}(K) \subset \operatorname{aff} F_{\boldsymbol{x}}(K)$. □

The next theorem shows that the study of generated extreme faces $F_X(K)$ can be reduced to those generated by points.

Theorem 11.10. *If X is a nonempty subset of a convex set $K \subset \mathbb{R}^n$, then* $\operatorname{rint}(\operatorname{conv} X) \subset \operatorname{rint} F_X(K)$. *Consequently, $F_{\boldsymbol{u}}(K) = F_X(K)$ for every point* $\boldsymbol{u} \in \operatorname{rint}(\operatorname{conv} X)$.

Proof. Choose a point $\boldsymbol{u} \in \operatorname{rint}(\operatorname{conv} X)$. First, we are going to show that $F_{\boldsymbol{u}}(K)$ equals the set $G = \cup(F_{\boldsymbol{x}}(K) : \boldsymbol{x} \in \operatorname{conv} X)$. Since $F_{\boldsymbol{u}}(K) \subset G$, it suffices to prove the opposite inclusion. For this, choose any point $\boldsymbol{y} \in G$. If $\boldsymbol{y} \in \operatorname{conv} X$, then $\boldsymbol{u} \in (\boldsymbol{y}, \boldsymbol{z})$ for a suitable point $\boldsymbol{z} \in \operatorname{conv} X$ (see Theorem 3.24), which gives $\boldsymbol{y} \in F_{\boldsymbol{u}}(K)$. Assume that $\boldsymbol{y} \in G \setminus \operatorname{conv} X$. Then $\boldsymbol{y} \in F_{\boldsymbol{z}}(K)$ for a suitable point $\boldsymbol{z} \in \operatorname{conv} X$. Equivalently, there is a point $\boldsymbol{p} \in K$ such that $\boldsymbol{z} \in (\boldsymbol{p}, \boldsymbol{y})$. By Theorem 3.24 again, there is a point $\boldsymbol{q} \in \operatorname{conv} X$ with the property $\boldsymbol{u} \in (\boldsymbol{q}, \boldsymbol{z})$. According to Lemma 2.26, there is a point $\boldsymbol{v} \in (\boldsymbol{p}, \boldsymbol{q})$ satisfying the condition $\boldsymbol{u} \in (\boldsymbol{v}, \boldsymbol{y})$. Therefore, $\boldsymbol{y} \in F_{\boldsymbol{u}}(K)$. Summing up, $G \subset F_{\boldsymbol{u}}(K)$.

Next, the inclusion $X \subset G = F_{\boldsymbol{u}}(K)$ and Theorem 11.9 imply that $F_{\boldsymbol{u}}(K)$ is the smallest extreme face of K which contains X. Hence $F_{\boldsymbol{u}}(K) = F_X(K)$, and Theorem 11.9 gives $\boldsymbol{u} \in \operatorname{rint} F_{\boldsymbol{u}}(K) = \operatorname{rint} F_X(K)$. Summing up, $\operatorname{rint}(\operatorname{conv} X) \subset \operatorname{rint} F_X(K)$. □

Corollary 11.11. *Given a nonempty convex set $K \subset \mathbb{R}^n$, the following assertions hold.*

(*a*) *K is a disjoint union of* $\operatorname{rint} K$ *and the relative interiors of all proper extreme faces of K.*

(*b*) *The set $K \cap \operatorname{rbd} K$ is a disjoint union of the relative interiors of all proper extreme faces of K.*

Proof. (*a*) By Theorem 11.9, $\boldsymbol{x} \in \operatorname{rint} F(\boldsymbol{x})$ for all $\boldsymbol{x} \in K$. Consequently, the sets $\operatorname{rint} F_{\boldsymbol{x}}(K)$, $\boldsymbol{x} \in K$, cover the whole K. The same theorem states that $\operatorname{rint} F_{\boldsymbol{x}}(K) = \operatorname{rint} K$ if $\boldsymbol{x} \in \operatorname{rint} K$, and $F_{\boldsymbol{x}}(K) \subset K \cap \operatorname{rbd} K$ if $\boldsymbol{x} \in K \cap \operatorname{rbd} K$. Since the relative interiors of distinct extreme faces of K are disjoint (see Theorem 11.4), the extreme faces $F_{\boldsymbol{x}}(K)$, $\boldsymbol{x} \in K$, provide a desired partition.

(*b*) This part immediately follows from the argument of (*a*). □

11.1.3 *Algebra of Extreme Faces*

Theorem 11.12. *Let $\mathcal{F} = \{K_\alpha\}$ be a family of convex sets in $\mathbb{R}^n$, and let $K = \bigcap_\alpha K_\alpha$. The following assertions hold.*

(*a*) *If F_α is an extreme face of K_α, $K_\alpha \in \mathcal{F}$, then the set $F = \bigcap_\alpha F_\alpha$ is an extreme face of K.*

(*b*) *If the family $\mathcal{F}$ is finite, say $\mathcal{F} = \{K_1, \ldots, K_r\}$, then every extreme face G of K can be expressed as*

$$G = F_{K_1}(G) \cap \cdots \cap F_{K_r}(G),$$

where $F_{K_i}(G)$ is the extreme face of K_i generated by G, $1 \leqslant i \leqslant r$.

Proof. (*a*) Since the case $F = \varnothing$ is obvious, we may assume that F is nonempty. Choose points $\boldsymbol{x}, \boldsymbol{y} \in K$ such that the open segment $(\boldsymbol{x}, \boldsymbol{y})$ meets F. Then $\boldsymbol{x}, \boldsymbol{y} \in K_\alpha$ and $(\boldsymbol{x}, \boldsymbol{y}) \cap F_\alpha \neq \varnothing$ for every $K_\alpha \in \mathcal{F}$. Because F_α is an extreme face of K_α, one has $\boldsymbol{x}, \boldsymbol{y} \in F_\alpha$. Hence $\boldsymbol{x}, \boldsymbol{y} \in \bigcap_\alpha F_\alpha = F$, and F is an extreme face of K.

(*b*) The inclusion $G \subset F_{K_1}(G) \cap \cdots \cap F_{K_r}(G)$ trivially holds. So, it suffices to prove the opposite inclusion. Choose a point $\boldsymbol{z} \in \operatorname{rint} G$. By Theorem 11.10,

$$G = F_{\boldsymbol{z}}(K) \quad \text{and} \quad F_G(K_i) = F_{\boldsymbol{z}}(K_i) \quad \text{for all} \quad 1 \leqslant i \leqslant r.$$

So, it remains to show that $F_{\boldsymbol{z}}(K_1) \cap \cdots \cap F_{\boldsymbol{z}}(K_r) \subset F_{\boldsymbol{z}}(K)$. Indeed, let $\boldsymbol{x}$ be a point in $F_{\boldsymbol{z}}(K_1) \cap \cdots \cap F_{\boldsymbol{z}}(K_r)$. Since the case $\boldsymbol{x} = \boldsymbol{z}$ is obvious, we assume that $\boldsymbol{x} \neq \boldsymbol{z}$. Choose points $\boldsymbol{y}_i \in F_{\boldsymbol{z}}(K_i)$ such that $\boldsymbol{z} \in (\boldsymbol{x}, \boldsymbol{y}_i)$, $1 \leqslant i \leqslant r$. Because the open segment $(\boldsymbol{x}, \boldsymbol{y}) = (\boldsymbol{x}, \boldsymbol{y}_1) \cap \cdots \cap (\boldsymbol{x}, \boldsymbol{y}_r)$ lies in $K_1 \cap \cdots \cap K_r = K$ and contains $\boldsymbol{z}$, we obtain $\boldsymbol{x} \in F_{\boldsymbol{z}}(K)$. □

Remark. The second assertion of Theorem 11.12 does not hold if the family $\mathcal{F}$ is infinite. Indeed, consider the closed intervals $K_i = [0, 1 + \frac{1}{i}]$, $i \geqslant 1$. Then $K = K_1 \cap K_2 \cap \cdots = [0, 1]$, implying that $F_1(K) = \{1\}$. On the other hand, $F_1(K_i) = [0, 1 + \frac{1}{i}]$, which gives

$$F_1(K_1) \cap F_1(K_2) \cap \cdots = [0, 1] \neq F_1(K).$$

Theorem 11.13. *For convex sets $K_1, \dots, K_r$ in $\mathbb{R}^n$ and nonzero scalars $\mu_1, \dots, \mu_r$, let $K = \mu_1 K_1 + \cdots + \mu_r K_r$. The following assertions hold.*

(a) *Any extreme face F of K can be expressed as $F = \mu_1 F_1 + \cdots + \mu_r F_r$, where every F_i is an extreme face of K_i, $1 \leqslant i \leqslant r$.*
(b) *If, additionally, the extreme face F of K is nonempty and bounded, then the above faces $F_1, \dots, F_r$ are nonempty and bounded; also, they are uniquely determined by F.*

Proof. (a) Without loss of generality, we may assume that $F \neq \varnothing$. An induction argument shows that the proof can be reduced to the case $r = 2$ and $\mu_1\mu_2 \neq 0$ (let $\mu_1 K_1 = \mu_1 K_1 + \mu_2\{\boldsymbol{o}\}$ if $r = 1$).

We assert that the sets

$$F_1 = \{\boldsymbol{x}_1 \in K_1 : \exists\ \boldsymbol{x}_2 \in K_2 \ \text{ such that } \ \mu_1\boldsymbol{x}_1 + \mu_2\boldsymbol{x}_2 \in F\},$$
$$F_2 = \{\boldsymbol{x}_2 \in K_2 : \exists\ \boldsymbol{x}_1 \in K_1 \ \text{ such that } \ \mu_1\boldsymbol{x}_1 + \mu_2\boldsymbol{x}_2 \in F\}$$

are desired extreme faces. Clearly, both F_1 and F_2 are nonempty due to $F \neq \varnothing$. By a symmetry argument, it suffices to prove that F_1 is an extreme face of K_1. To show the convexity of F_1, choose points $\boldsymbol{x}_1, \boldsymbol{x}_2 \in F_1$ and a scalar $0 \leqslant \lambda \leqslant 1$. Then there are points $\boldsymbol{y}_1, \boldsymbol{y}_2 \in K_2$ such that both $\boldsymbol{z}_1 = \mu_1\boldsymbol{x}_1 + \mu_2\boldsymbol{y}_1$ and $\boldsymbol{z}_2 = \mu_1\boldsymbol{x}_2 + \mu_2\boldsymbol{y}_2$ belong to F. By the convexity of K_1, K_2, and F, we have

$$(1-\lambda)\boldsymbol{x}_1 + \lambda\boldsymbol{x}_2 \in K_1, \quad (1-\lambda)\boldsymbol{y}_1 + \lambda\boldsymbol{y}_2 \in K_2, \quad (1-\lambda)\boldsymbol{z}_1 + \lambda\boldsymbol{z}_2 \in F.$$

Since

$$\mu_1((1-\lambda)\boldsymbol{x}_1 + \lambda\boldsymbol{x}_2) + \mu_2((1-\lambda)\boldsymbol{y}_1 + \lambda\boldsymbol{y}_2) = (1-\lambda)\boldsymbol{z}_1 + \lambda\boldsymbol{z}_2,$$

we conclude that $(1-\lambda)\boldsymbol{x}_1 + \lambda\boldsymbol{x}_2 \in F_1$, and hence F_1 is convex.

To prove that F_1 is an extreme face of K_1, choose points $\boldsymbol{u}, \boldsymbol{v} \in K_1$ and a scalar $0 < \lambda < 1$ such that $(1-\lambda)\boldsymbol{u} + \lambda\boldsymbol{v} \in F_1$. Then there is a point $\boldsymbol{y} \in K_2$ with the property $\mu_1((1-\lambda)\boldsymbol{u} + \lambda\boldsymbol{v}) + \mu_2\boldsymbol{y} \in F$. Equivalently,

$$(1-\lambda)(\mu_1\boldsymbol{u} + \mu_2\boldsymbol{y}) + \lambda(\mu_1\boldsymbol{v} + \mu_2\boldsymbol{y}) \in F.$$

Since both points $\mu_1\boldsymbol{u} + \mu_2\boldsymbol{y}$ and $\mu_1\boldsymbol{v} + \mu_2\boldsymbol{y}$ belong to K, and since F is an extreme face of K, the points $\mu_1\boldsymbol{u} + \mu_2\boldsymbol{y}$ and $\mu_1\boldsymbol{v} + \mu_2\boldsymbol{y}$ belong to F. Hence $\boldsymbol{u}, \boldsymbol{v} \in F_1$, and F_1 is an extreme face of K_1.

It remains to prove the equality $F = \mu_1 F_1 + \mu_2 F_2$. First, we observe that the inclusion $F \subset \mu_1 F_1 + \mu_2 F_2$ follows from the equality $K = \mu_1 K_1 + \mu_2 K_2$ and the definition of F_1 and F_2. For the opposite inclusion, choose any point $\boldsymbol{u} \in \mu_1 F_1 + \mu_2 F_2$. Then $\boldsymbol{u} = \mu_1\boldsymbol{x}_1 + \mu_2\boldsymbol{y}_2$, where $\boldsymbol{x}_1 \in F_1$ and

$\boldsymbol{y}_2 \in F_2$. By the definitions of F_1 and F_2, there are points $\boldsymbol{x}_2 \in K_1$ and $\boldsymbol{y}_1 \in K_2$ such that both $\mu_1\boldsymbol{x}_1 + \mu_2\boldsymbol{y}_1$ and $\mu_1\boldsymbol{x}_2 + \mu_2\boldsymbol{y}_2$ are in F. By the convexity of F, the point

$$\boldsymbol{z} = \tfrac{1}{2}(\mu_1\boldsymbol{x}_1 + \mu_2\boldsymbol{y}_1) + \tfrac{1}{2}(\mu_1\boldsymbol{x}_2 + \mu_2\boldsymbol{y}_2)$$

belongs to F. Put $\boldsymbol{v} = \mu_1\boldsymbol{x}_2 + \mu_2\boldsymbol{y}_1$. Clearly, $\boldsymbol{v} \in K$ and $\boldsymbol{z} = \frac{1}{2}(\boldsymbol{u} + \boldsymbol{v})$. Therefore, $\boldsymbol{u}, \boldsymbol{v} \in F$ because F is an extreme face of K. Summing up, $\mu_1F_1 + \mu_2F_2 \subset F$.

(*b*) Let F be nonempty and bounded. Suppose that F has another representation as $F = \mu_1F_1' + \mu_2F_2'$, where F_1' and F_2' are extreme faces of K_1 and K_2, respectively. The above definition of F_1 and F_2 implies the inclusions $F_1' \subset F_1$ and $F_2' \subset F_2$. Therefore,

$$F = \mu_1F_1' + \mu_2F_2' \subset \mu_1F_1 + \mu_2F_2' \subset F,$$

which gives $\mu_1F_1' + \mu_2F_2' = \mu_1F_1 + \mu_2F_2'$. Since all sets μ_1F_1, μ_1F_1', and μ_2F_2' are bounded (see page 11), Theorem 3.60 shows that $\operatorname{cl}(\mu_1F_1') = \operatorname{cl}(\mu_2F_1)$. Thus $\operatorname{cl}F_1' = \operatorname{cl}F_1$. As a result,

$$F_1 = K_1 \cap \operatorname{cl}F_1 = K_1 \cap \operatorname{cl}F_1' = F_1'$$

(see Theorem 11.4). Similarly, $F_2 = F_2'$. □

Remarks. 1. The assumption on boundedness of the extreme face F in assertion (*b*) of Theorem 11.13 is essential. Indeed, let $K_1 = K_2 = h$, where $h \subset \mathbb{R}^n$ is a closed halfline originated at $\boldsymbol{o}$. Then the sum $K = K_1 + K_2 = h$, being its own trivial extreme face, can be expressed as the sum of extreme faces of K_1 and K_2 in two distinct ways: $K = K_1 + \{\boldsymbol{o}\} = K_1 + K_2$.

2. It might be that the sum $F_1 + F_2$ of extreme faces F_1 and F_2 of convex sets K_1 and K_2, respectively, is not an extreme face of $K_1 + K_2$. Indeed, if $K_1 = K_2 = \mathbb{B}$, where $\mathbb{B}$ is the closed unit ball of $\mathbb{R}^n$, then for a boundary point $\boldsymbol{x}$ of $\mathbb{B}$, the sets $\{\boldsymbol{x}\}$ and $\{-\boldsymbol{x}\}$ are 0-dimensional extreme faces of $\mathbb{B}$, while $\{\boldsymbol{x}\} + \{-\boldsymbol{x}\} = \{\boldsymbol{o}\}$ is not an extreme face of $K_1 + K_2 = 2\mathbb{B}$.

Theorem 11.14. *For an affine transformation $f : \mathbb{R}^n \to \mathbb{R}^m$ and convex sets $K \subset \mathbb{R}^n$ and $M \subset \mathbb{R}^m$, the following assertions hold.*

(*a*) *A subset G of $f(K)$ is an extreme face of $f(K)$ if and only if $f^{-1}(G) \cap K$ is an extreme face of K.*

(*b*) *If f is one-to-one, then a set F is an extreme face of K if and only if $f(F)$ is an extreme face of $f(K)$.*

(*c*) *If G is an extreme face of M, then $f^{-1}(G)$ is an extreme face of $f^{-1}(M)$.*

(*d*) *If* G *is a convex subset of* $M \cap \operatorname{rng} f$ *such that* $f^{-1}(G)$ *is an extreme face of* $f^{-1}(M)$, *then* G *is an extreme face of* $M \cap \operatorname{rng} f$.

Proof. Since the assertion is obvious if the respective sets are empty, we may assume that all involved sets are nonempty. Furthermore, the convexity of these sets follows from Theorem 3.15.

(*a*) Let G be an extreme face of $f(K)$. Choose points $\boldsymbol{x}, \boldsymbol{y} \in K$, a scalar $0 < \lambda < 1$ such that $\boldsymbol{z} = (1-\lambda)\boldsymbol{x} + \lambda\boldsymbol{y} \in f^{-1}(G) \cap K$. By Theorem 2.81, $f(\boldsymbol{z}) = (1-\lambda)f(\boldsymbol{x}) + \lambda f(\boldsymbol{y}) \in G$. Consequently, $f(\boldsymbol{x}), f(\boldsymbol{y}) \in G$. Hence $\boldsymbol{x}, \boldsymbol{y} \in f^{-1}(G) \cap K$, implying that $f^{-1}(G) \cap K$ is an extreme face of K.

Conversely, let $f^{-1}(G) \cap K$ be an extreme face of K. Choose points $\boldsymbol{x}, \boldsymbol{y} \in f(K)$ and a scalar $0 < \lambda < 1$ such that the point $\boldsymbol{z} = (1-\lambda)\boldsymbol{x} + \lambda\boldsymbol{y}$ belongs to G. Let $\boldsymbol{x}_0, \boldsymbol{y}_0$ be points in K with $f(\boldsymbol{x}_0) = \boldsymbol{x}$ and $f(\boldsymbol{y}_0) = \boldsymbol{y}$. Put $\boldsymbol{z}_0 = (1-\lambda)\boldsymbol{x}_0 + \lambda\boldsymbol{y}_0$. Then $\boldsymbol{z}_0 \in K$ by the convexity of K. From

$$f(\boldsymbol{z}_0) = (1-\lambda)f(\boldsymbol{x}_0) + \lambda f(\boldsymbol{y}_0) = (1-\lambda)\boldsymbol{x} + \lambda\boldsymbol{y} \in G$$

we conclude that $\boldsymbol{z}_0 \in f^{-1}(G) \cap K$. Since $f^{-1}(G) \cap K$ is an extreme face of K, one has $\boldsymbol{x}_0, \boldsymbol{y}_0 \in f^{-1}(G) \cap K$. Thus both $\boldsymbol{x} = f(\boldsymbol{x}_0)$ and $\boldsymbol{y} = f(\boldsymbol{y}_0)$ belong to $f(f^{-1}(G)) \subset G$, implying that G is an extreme face of $f(K)$.

(*b*) Let F be an extreme face of K. Choose points $\boldsymbol{x}, \boldsymbol{y} \in f(K)$ and a scalar $0 < \lambda < 1$ such that the point $\boldsymbol{z} = (1-\lambda)\boldsymbol{x} + \lambda\boldsymbol{y}$ belongs to $f(F)$. Denote by $\boldsymbol{x}_0$ and $\boldsymbol{y}_0$ the points in K satisfying the conditions $f(\boldsymbol{x}_0) = \boldsymbol{x}$ and $f(\boldsymbol{y}_0) = \boldsymbol{y}$. Put $\boldsymbol{z}_0 = (1-\lambda)\boldsymbol{x}_0 + \lambda\boldsymbol{y}_0$. Since

$$f(\boldsymbol{z}_0) = (1-\lambda)f(\boldsymbol{x}_0) + \lambda f(\boldsymbol{y}_0) = (1-\lambda)\boldsymbol{x} + \lambda\boldsymbol{y} = \boldsymbol{z} \in f(F)$$

we obtain that $\boldsymbol{z}_0 = f^{-1}(\boldsymbol{z}) \in F$ (f is one-to-one). Then $\boldsymbol{x}_0, \boldsymbol{y}_0 \in F$ because F is an extreme face of K. Consequently, $\boldsymbol{x}, \boldsymbol{y} \in f(F)$. Summing up, $f(F)$ is an extreme face of $f(K)$. The converse assertion follows from assertion (*a*) above.

(*c*) Choose points $\boldsymbol{x}, \boldsymbol{y} \in f^{-1}(M)$ and a scalar $0 < \lambda < 1$ such that $\boldsymbol{z} = (1-\lambda)\boldsymbol{x} + \lambda\boldsymbol{y} \in f^{-1}(G)$. Then $f(\boldsymbol{z}) \in f(f^{-1}(G)) \subset G$. Furthermore,

$$f(\boldsymbol{z}) = (1-\lambda)f(\boldsymbol{x}) + \lambda f(\boldsymbol{y}) \quad \text{and} \quad f(\boldsymbol{x}), f(\boldsymbol{y}) \in f(f^{-1}(M)) \subset M.$$

Since G is an extreme face of M, one has $f(\boldsymbol{x}), f(\boldsymbol{y}) \in G$. Hence $\boldsymbol{x}, \boldsymbol{y} \in f^{-1}(G)$, implying that $f^{-1}(G)$ is an extreme face of $f^{-1}(M)$.

(*d*) Choose points $\boldsymbol{x}, \boldsymbol{y} \in M \cap \operatorname{rng} f$ and a scalar $0 < \lambda < 1$ such that $\boldsymbol{z} = (1-\lambda)\boldsymbol{x} + \lambda\boldsymbol{y} \in G$. Let $\boldsymbol{x}_0, \boldsymbol{y}_0$ be points in $f^{-1}(M)$ such that $f(\boldsymbol{x}_0) = \boldsymbol{x}$ and $f(\boldsymbol{y}_0) = \boldsymbol{y}$. Put $\boldsymbol{z}_0 = (1-\lambda)\boldsymbol{x}_0 + \lambda\boldsymbol{y}_0$. Then

$$f(\boldsymbol{z}_0) = (1-\lambda)f(\boldsymbol{x}_0) + \lambda f(\boldsymbol{y}_0) = (1-\lambda)\boldsymbol{x} + \lambda\boldsymbol{y} = \boldsymbol{z} \in G,$$

implying that $\boldsymbol{z}_0 \in f^{-1}(G)$. Hence $\boldsymbol{x}_0, \boldsymbol{y}_0 \in f^{-1}(G)$, which gives the inclusions $\boldsymbol{x}, \boldsymbol{y} \in G$. Summing up, G is an extreme face of $M \cap \operatorname{rng} f$. $\square$

Remark. The assumption that f is one-to-one is essential in assertion (b) of Theorem 11.14. Indeed, if $K = \{(x, y) : x^2 + y^2 \leqslant 1\}$ is the unit disk of $\mathbb{R}^2$ and f is the orthogonal projection on the x-axis, then every singleton from $\operatorname{bd} K$ is a 0-dimensional extreme face of K, while the only 0-dimensional extreme faces of the closed segment $f(K) = \{(x, 0) : -1 \leqslant x \leqslant 1\}$ are its endpoints.

11.2 Extreme Points and Halflines

11.2.1 *Algebra of Extreme Points*

Reformulating Definition 11.1 for the case of singletons, we introduce the following notion.

Definition 11.15. Let $K \subset \mathbb{R}^n$ be a nonempty convex set. A point $\boldsymbol{x} \in K$ is called an *extreme point* of K if for any choice of points $\boldsymbol{y}, \boldsymbol{z} \in K$ and a scalar $0 < \lambda < 1$ the equality $\boldsymbol{x} = (1 - \lambda)\boldsymbol{y} + \lambda \boldsymbol{z}$ implies that $\boldsymbol{x} = \boldsymbol{y} = \boldsymbol{z}$. In what follows, $\operatorname{ext} K$ denotes the set of all extreme points of K. We put $\operatorname{ext} \varnothing = \varnothing$.

Examples. 1. The extreme points of a closed ball $B_\rho(\boldsymbol{a}) \subset \mathbb{R}^n$ are precisely the points of the boundary sphere $S_\rho(\boldsymbol{a})$ (see Problem 11.1).

2. The extreme points of an r-simplex $\Delta(\boldsymbol{x}_1, \ldots, \boldsymbol{x}_{r+1}) \subset \mathbb{R}^n$ are exactly its vertices, $\boldsymbol{x}_1, \ldots, \boldsymbol{x}_{r+1}$ (see Problem 11.2).

3. The only extreme point of an r-simplicial cone $\Gamma_{\boldsymbol{a}}(\boldsymbol{x}_1, \ldots, \boldsymbol{x}_r) \subset \mathbb{R}^n$ is its apex $\boldsymbol{a}$ (see Problem 11.3).

The above results on extreme faces derive the following corollary, which describes some basic properties of extreme points.

Corollary 11.16. *For a nonempty convex set $K \subset \mathbb{R}^n$ and a point $\boldsymbol{u} \in K$, the following assertions hold.*

(a) *The inclusion $\boldsymbol{u} \in \operatorname{ext} K$ holds if and only if the set $K \setminus \{\boldsymbol{u}\}$ is convex.*
(b) *If $\boldsymbol{u} \in \operatorname{ext} K$ and M is a convex subset of K containing $\boldsymbol{u}$, then $\boldsymbol{u} \in \operatorname{ext} M$.*
(c) *If $\boldsymbol{u} \in \operatorname{ext} K$, then $\boldsymbol{u} \in \operatorname{rbd} K$.*
(d) *If K has at least one extreme point, then $\operatorname{lin} K = \{\boldsymbol{o}\}$ (consequently, K is line-free provided it is closed).*
(e) *Given a scalar $\rho > 0$, one has $\boldsymbol{u} \in \operatorname{ext} K$ if and only if $\boldsymbol{u} \in \operatorname{ext} B_\rho(K)$.*

(*f*) *The inclusion* $\boldsymbol{u} \in \operatorname{ext} K$ *holds if and only if* $\boldsymbol{u}$ *is the only apex of the generated cone* $C_{\boldsymbol{u}}(K)$.

(*g*) *Let* $f : \mathbb{R}^n \to \mathbb{R}^m$ *be a one-to-one affine transformation. Then* $\boldsymbol{x} \in \operatorname{ext} K$ *if and only if* $f(\boldsymbol{x}) \in \operatorname{ext} f(K)$.

Proof. Due to the previous results of this chapter, it remains to prove assertion (*f*). We will proceed by contraposition. Suppose first that $\boldsymbol{u} \notin \operatorname{ext} K$. Then there are points $\boldsymbol{v}, \boldsymbol{w} \in K \setminus \{\boldsymbol{u}\}$ such that $\boldsymbol{u} \in (\boldsymbol{v}, \boldsymbol{w})$. Since $\boldsymbol{u}$ is an apex of $C_{\boldsymbol{u}}(K)$, the opposite closed halflines $[\boldsymbol{u}, \boldsymbol{v}\rangle$ and $[\boldsymbol{u}, \boldsymbol{w}\rangle$ lie in $C_{\boldsymbol{u}}(K)$. Consequently, the whole line $l = \langle \boldsymbol{v}, \boldsymbol{w}\rangle$ lies in $C_{\boldsymbol{u}}(K)$. By Theorem 5.17, the apex set of $C_{\boldsymbol{u}}(K)$ is the largest plane through $\boldsymbol{u}$ contained in $C_{\boldsymbol{u}}(K)$. Therefore, l lies in the apex set of $C_{\boldsymbol{u}}(K)$.

Conversely, suppose that the apex set of the cone $C_{\boldsymbol{u}}(K)$ contains another point $\boldsymbol{v}$. By Theorem 5.17, the line $l = \langle \boldsymbol{u}, \boldsymbol{v}\rangle$ lies in $C_{\boldsymbol{u}}(K)$. Put $\boldsymbol{w} = 2\boldsymbol{u} - \boldsymbol{v}$. Clearly, $\boldsymbol{w} \in l$ and $\boldsymbol{u} \in (\boldsymbol{v}, \boldsymbol{w})$. By the definition of $C_{\boldsymbol{u}}(K)$, there are points $\boldsymbol{v}', \boldsymbol{w}' \in K$ such that $\boldsymbol{v} \in (\boldsymbol{u}, \boldsymbol{v}'\rangle$ and $\boldsymbol{w} \in (\boldsymbol{u}, \boldsymbol{w}'\rangle$. Clearly, $\boldsymbol{u} \in (\boldsymbol{v}', \boldsymbol{w}')$, which show that $\boldsymbol{u} \notin \operatorname{ext} K$. □

Theorem 11.17. *Given nonempty convex sets* $K_1, \ldots, K_r$ *in* $\mathbb{R}^n$ *and nonzero scalars* $\mu_1, \ldots, \mu_r$, *let* $K = \mu_1 K_1 + \cdots + \mu_r K_r$. *The following assertions hold.*

(*a*) *If a point* $\boldsymbol{u} \in K$ *is uniquely expressible as* $\boldsymbol{u} = \mu_1 \boldsymbol{u}_1 + \cdots + \mu_r \boldsymbol{u}_r$, *where* $\boldsymbol{u}_i \in K_i$ *for all* $1 \leqslant i \leqslant r$, *then* $\boldsymbol{u} \in \operatorname{ext} K$.

(*b*) *Every point* $\boldsymbol{u} \in \operatorname{ext} K$ *is uniquely expressible as* $\boldsymbol{u} = \mu_1 \boldsymbol{u}_1 + \cdots + \mu_r \boldsymbol{u}_r$, *where* $\boldsymbol{u}_i \in \operatorname{ext} K_i$ *for all* $1 \leqslant i \leqslant r$.

Proof. (*a*) Let a point $\boldsymbol{u} \in K$ be uniquely expressible as $\boldsymbol{u} = \mu_1 \boldsymbol{u}_1 + \cdots + \mu_r \boldsymbol{u}_r$, where $\boldsymbol{u}_i \in K_i$ for all $1 \leqslant i \leqslant r$. Assume for a moment that $\boldsymbol{u}$ is not an extreme point of K. Then there are distinct points $\boldsymbol{v}, \boldsymbol{w} \in K$ and a scalar $0 < \lambda < 1$ such that $\boldsymbol{u} = (1 - \lambda)\boldsymbol{v} + \lambda \boldsymbol{w}$. We can write

$$\boldsymbol{v} = \mu_1 \boldsymbol{v}_1 + \cdots + \mu_r \boldsymbol{v}_r \quad \text{and} \quad \boldsymbol{w} = \mu_1 \boldsymbol{w}_1 + \cdots + \mu_r \boldsymbol{w}_r$$

for suitable points $\boldsymbol{v}_i, \boldsymbol{w}_i \in K_i$, $1 \leqslant i \leqslant r$. Clearly, there is an index $i \in \{1, \ldots, r\}$ such that $\boldsymbol{v}_i \neq \boldsymbol{w}_i$ (otherwise $\boldsymbol{v} = \boldsymbol{w}$). Let, for instance, $\boldsymbol{v}_1 \neq \boldsymbol{w}_1$. Put $\boldsymbol{z}_i = (1 - \lambda)\boldsymbol{v}_i + \lambda \boldsymbol{w}_i$, $1 \leqslant i \leqslant r$. Clearly, $\boldsymbol{v}_1 \neq \boldsymbol{z}_1 \neq \boldsymbol{w}_1$ and $\boldsymbol{z}_i \in K_i$ by the convexity of K_i, $1 \leqslant i \leqslant r$. Consequently,

$$\begin{aligned}
\boldsymbol{u} &= (1 - \lambda)\boldsymbol{v} + \lambda \boldsymbol{w} \\
&= (1 - \lambda)(\mu_1 \boldsymbol{v}_1 + \cdots + \mu_r \boldsymbol{v}_r) + \lambda(\mu_1 \boldsymbol{w}_1 + \cdots + \mu_r \boldsymbol{w}_r) \\
&= \mu_1((1 - \lambda)\boldsymbol{v}_1 + \lambda \boldsymbol{w}_1) + \cdots + \mu_r((1 - \lambda)\boldsymbol{v}_r + \lambda \boldsymbol{w}_r) \\
&= \mu_1 \boldsymbol{z}_1 + \cdots + \mu_r \boldsymbol{z}_r
\end{aligned}$$

is a new representation of $\boldsymbol{u}$, a contradiction. Hence $\boldsymbol{u} \in \operatorname{ext} K$.

(*b*) Let $\boldsymbol{u}$ be an extreme point of K. By Theorem 11.13, the 0-dimensional face $F = \{\boldsymbol{u}\}$ is uniquely expressible as $F = \mu_1 F_1 + \cdots + \mu_r F_r$, where every F_i is an extreme face of K_i, $1 \leqslant i \leqslant r$. Clearly, all sets $F_1, \ldots, F_r$ are singletons: $F_i = \{\boldsymbol{u}_i\}$, $1 \leqslant i \leqslant r$. Equivalently, $\boldsymbol{u}_i \in \operatorname{ext} K_i$, $1 \leqslant i \leqslant r$. □

11.2.2 *Topological Properties of Extreme Points*

The remaining part of this subsection deals with extreme points of closed convex sets. We recall (see Definition 6.36) that a convex set $K \subset \mathbb{R}^n$ is line-free if it contains no line.

Theorem 11.18. *For a nonempty closed convex set $K \subset \mathbb{R}^n$, the following assertions hold.*

(*a*) $\operatorname{ext} K \neq \varnothing$ *if and only if K is line-free.*
(*b*) *K has a unique extreme point, say $\boldsymbol{u}$, if and only if K is a line-free convex cone with apex $\boldsymbol{u}$.*

Proof. (*a*) Suppose that $\operatorname{ext} K \neq \varnothing$ and choose a point $\boldsymbol{u} \in \operatorname{ext} K$. Assume for a moment that K contains a line l. Then the line l' which is parallel to l and contains $\boldsymbol{u}$ entirely lies in K (see Corollary 6.11). Choosing points $\boldsymbol{y}, \boldsymbol{z} \in l' \setminus \{\boldsymbol{u}\}$ such that $\boldsymbol{u} \in (\boldsymbol{y}, \boldsymbol{z})$, we conclude that $\boldsymbol{u} \notin \operatorname{ext} K$. The obtained contradiction shows that K must be line-free.

Conversely, let K be line-free. We will show the existence of an extreme point of K by induction on $m = \dim K$. If $m = 0$, then K is a singleton and $\operatorname{ext} K = K$. If $\dim K = 1$, then K is either a closed segment $[\boldsymbol{u}, \boldsymbol{z}]$ or a closed halfline $[\boldsymbol{u}, \boldsymbol{z}\rangle$; in either case $\boldsymbol{u}$ is an extreme point of K.

Assume that the assertion holds for all line-free closed convex sets of dimension $r \leqslant m-1$, where $m \geqslant 2$, and let $K \subset \mathbb{R}^n$ be a line-free closed convex set of dimension m. According to Theorem 3.62, $\operatorname{rbd} K \neq \varnothing$. Choose a point $\boldsymbol{z} \in \operatorname{rbd} K$. By Theorem 11.9, the generated extreme face $F_{\boldsymbol{z}}(K)$ lies in $\operatorname{rbd} K$, and Theorem 11.4 implies that $F_{\boldsymbol{z}}(K)$ is a closed convex set of dimension $m-1$ or less. Since $F_{\boldsymbol{z}}(K)$ is line-free (as a subset of the line-free set K), it has, by the induction hypothesis, an extreme point $\boldsymbol{u}$. Finally, Theorem 11.3 shows that $\boldsymbol{u}$ is an extreme point of K.

(*b*) Suppose that K has exactly one extreme point, $\boldsymbol{u}$. As shown above, K is line-free. By induction on $m = \dim K$, we are going to prove that K is a cone with apex $\boldsymbol{u}$. The case $m = 0$ is obvious: $K = \{\boldsymbol{u}\}$. Let $\dim K = 1$. Then K is either a closed segment of the form $[\boldsymbol{u}, \boldsymbol{z}]$ or a closed halfline of the form $[\boldsymbol{u}, \boldsymbol{z}\rangle$. Since $[\boldsymbol{u}, \boldsymbol{z}]$ has two extreme points, K must be $[\boldsymbol{u}, \boldsymbol{z}\rangle$, which is a cone with apex $\boldsymbol{u}$.

Assume that the assertion holds for all closed convex sets of dimension $r \leqslant m - 1$, where $m \geqslant 2$, and let $K \subset \mathbb{R}^n$ be a closed convex set of dimension m with a unique extreme point, $\boldsymbol{u}$. Choose a point $\boldsymbol{y} \in K \setminus \{\boldsymbol{u}\}$ and consider the generated extreme face $F_{\boldsymbol{y}}(K)$. As above, $F_{\boldsymbol{y}}(K)$ is a line-free closed convex set, so it has an extreme point $\boldsymbol{z}$, which is also an extreme point of K (see Theorem 11.3). Hence $\boldsymbol{u} = \boldsymbol{z}$, which shows that $\boldsymbol{u}$ is the only extreme point of $F_{\boldsymbol{y}}(K)$. By the induction hypothesis, $F_{\boldsymbol{y}}(K)$ is a cone with apex $\boldsymbol{u}$. Therefore, $[\boldsymbol{u}, \boldsymbol{y}\rangle \subset F_{\boldsymbol{y}}(K) \subset K$. Because $\boldsymbol{y}$ was chosen arbitrarily in $K \setminus \{\boldsymbol{u}\}$, we conclude that K is a cone with apex $\boldsymbol{u}$.

Conversely, let K be a line-free convex cone with apex $\boldsymbol{u}$. If $K = \{\boldsymbol{u}\}$, then $\boldsymbol{u}$ is the only extreme point of K. Suppose that $K \neq \{\boldsymbol{u}\}$. Then no point $\boldsymbol{y} \in K \setminus \{\boldsymbol{u}\}$ is extreme because it belongs to an open halfline $(\boldsymbol{u}, \boldsymbol{y}\rangle \subset K$. It remains to show that $\boldsymbol{u}$ is an extreme point of K. Indeed, if $\boldsymbol{u}$ belonged to an open segment $(\boldsymbol{v}, \boldsymbol{w})$, where $\boldsymbol{v}, \boldsymbol{w} \in K \setminus \{\boldsymbol{u}\}$, then $\langle \boldsymbol{v}, \boldsymbol{w}\rangle = \langle \boldsymbol{v}, \boldsymbol{u}] \cup [\boldsymbol{u}, \boldsymbol{w}\rangle \subset K$, contrary to the assumption that K is line-free. □

The next result shows that every extreme point of a line-free closed convex set can be included into a "cap" of K of an arbitrary small diameter.

Theorem 11.19. *Let $K \subset \mathbb{R}^n$ be a nonempty line-free closed convex set. For a point $\boldsymbol{u} \in \operatorname{ext} K$ and a scalar $\delta > 0$, there is an open halfspace $W \subset \mathbb{R}^n$ such that $\boldsymbol{u} \in K \cap W \subset U_\delta(\boldsymbol{u})$.*

Proof. Put $M = (B_{2\delta}(\boldsymbol{u}) \cap K) \setminus U_\delta(\boldsymbol{u})$. Clearly, M is a compact set which lies in the convex set $K \setminus \{\boldsymbol{u}\}$ (see Theorem 11.2). According to Theorem 4.16, $\operatorname{conv} M$ is a compact convex set. Furthermore, $\boldsymbol{u} \notin \operatorname{conv} M$ due to the inclusion $\operatorname{conv} M \subset K \setminus \{\boldsymbol{u}\}$. By Theorem 10.12, there is a hyperplane $H \subset \mathbb{R}^n$ strongly separating $\boldsymbol{u}$ and $\operatorname{conv} M$. Denote by W the open halfspace determined by H and containing $\boldsymbol{u}$.

We assert that $K \cap W \subset U_\delta(\boldsymbol{u})$. Indeed, assume for a moment that $K \cap W$ contains a point $\boldsymbol{v} \notin U_\delta(\boldsymbol{u})$. Then $\|\boldsymbol{u} - \boldsymbol{v}\| \geqslant \delta$. The closed segment $[\boldsymbol{u}, \boldsymbol{v}]$ lies in $K \cap W$ by the convexity of $K \cap W$. On the other hand, one can choose a point $\boldsymbol{w} \in [\boldsymbol{u}, \boldsymbol{v}]$ such that $\delta \leqslant \|\boldsymbol{u} - \boldsymbol{w}\| \leqslant 2\delta$. Then $\boldsymbol{w} \in M \subset \operatorname{conv} M$, contrary to the condition $\operatorname{conv} M \cap W = \varnothing$. □

Remark. Closedness of K is an essential requirement in both Theorems 11.18 and 11.19. Indeed, consider the convex set $K = \{\boldsymbol{o}\} \cup \{(x, y) :$

$0 < y < 1\}$. Then $\operatorname{ext} K = \{\boldsymbol{o}\}$, while K contains lines and is not a cone. Furthermore, for any open halfplane W of $\mathbb{R}^2$ which contains $\boldsymbol{o}$, the set $K \cap W$ is unbounded.

The next result describes an important topological property of extreme points (see also page 411 for various references). We recall that an F_σ set in $\mathbb{R}^n$ is a countable union of closed sets (see page 11).

Theorem 11.20. *If $K \subset \mathbb{R}^n$ is a closed convex set, then* $\operatorname{ext} K$ *is a G_δ-set (equivalently, $K \setminus \operatorname{ext} K$ is an F_σ-set).*

Proof. The assertion is obvious if $K = \varnothing$. So, we may assume that $K \neq \varnothing$. For every positive integer $r \geqslant 1$, let X_r denote the set of midpoints of all closed segments of length $1/r$ contained in K.

We observe that $K \setminus \operatorname{ext} K = X_1 \cup X_2 \cup \dots$. Indeed, since the inclusion $X_1 \cup X_2 \cup \dots \subset K \setminus \operatorname{ext} K$ is obvious, it remain to verify the opposite one. For this, choose any point $\boldsymbol{u} \in K \setminus \operatorname{ext} K$. Then $\boldsymbol{u} \in [\boldsymbol{v}, \boldsymbol{w}]$ for suitable points $\boldsymbol{v}, \boldsymbol{w} \in K \setminus \{\boldsymbol{u}\}$. Choose a positive integer p satisfying the condition

$$\tfrac{1}{2p} \leqslant \min\{\|\boldsymbol{u} - \boldsymbol{v}\|, \|\boldsymbol{u} - \boldsymbol{w}\|\}$$

and then chose points $\boldsymbol{v}' \in [\boldsymbol{u}, \boldsymbol{v}]$ and $\boldsymbol{w}' \in [\boldsymbol{u}, \boldsymbol{w}]$ such that

$$\|\boldsymbol{u} - \boldsymbol{v}'\| = \|\boldsymbol{u} - \boldsymbol{w}'\| = \tfrac{1}{2p}.$$

Since $\boldsymbol{u}$ is the midpoint of $[\boldsymbol{v}', \boldsymbol{w}']$ and $\|\boldsymbol{v}' - \boldsymbol{w}'\| = \frac{1}{p}$, we conclude that $\boldsymbol{u} \in X_p$.

Next, we assert that every set X_r, $r \geqslant 1$, is closed. Indeed, let $\boldsymbol{u}_1, \boldsymbol{u}_2, \dots$ be a sequence of points in X_r which converges to a point $\boldsymbol{u}_0$. Then one can find an index j such that $\|\boldsymbol{u}_i - \boldsymbol{u}_0\| \leqslant 1$ for all $i \geqslant j$. Every point $\boldsymbol{u}_i$ is the midpoint of a suitable closed segment $[\boldsymbol{v}_i, \boldsymbol{w}_i] \subset K$ of length $1/r$. The inequality

$$\|\boldsymbol{v}_i - \boldsymbol{u}_0\| \leqslant \|\boldsymbol{v}_i - \boldsymbol{u}_i\| + \|\boldsymbol{u}_i - \boldsymbol{u}_0\| \leqslant \tfrac{1}{2r} + 1, \quad i \geqslant j,$$

show that the sequence $\boldsymbol{v}_j, \boldsymbol{v}_{j+1}, \dots$ is bounded. Hence it contains a convergent subsequence. Without loss of generality, we may suppose that $\boldsymbol{v}_j, \boldsymbol{v}_{j+1}, \dots$ converges to a point $\boldsymbol{v}_0$. Consequently,

$$\lim_{i \to \infty} \boldsymbol{w}_i = \lim_{i \to \infty} (2\boldsymbol{u}_i - \boldsymbol{v}_i) = 2\boldsymbol{u}_0 - \boldsymbol{v}_0.$$

This argument shows that $\boldsymbol{u}_0$ is the midpoint of the closed segment $[\boldsymbol{v}_0, \boldsymbol{w}_0]$. Since $\boldsymbol{v}_0, \boldsymbol{w}_0 \in K$ due to the closedness of K, one has $[\boldsymbol{v}_0, \boldsymbol{w}_0] \subset K$ by the convexity of K. Furthermore

$$\|\boldsymbol{v}_0 - \boldsymbol{w}_0\| = \lim_{i \to \infty} \|\boldsymbol{v}_i - \boldsymbol{w}_i\| = \tfrac{1}{r}.$$

Hence $\boldsymbol{u}_0 \in X_r$. Summing up, the set X_r is closed.

Finally, the set $K \setminus \operatorname{ext} K$ is an F_σ-set as a countable union of closed sets X_i, $i \geqslant 1$. Consequently, $\operatorname{ext} K$ is a G_δ-set. □

Remark. The set $\operatorname{ext} K$ of all extreme points of a closed (even compact) convex set $K \subset \mathbb{R}^n$, $n \geqslant 3$, may be nonclosed (see Problem 11.7 for the case $n = 2$). Indeed, let $K = \operatorname{conv}(\{\boldsymbol{u}, \boldsymbol{v}\} \cup C)$, where $\boldsymbol{u} = (0,0,1)$, $\boldsymbol{v} = (0,0,-1)$, and $C = \{(x,y,0) : (x-1)^2 + y^2 = 1\}$ is the circle in the xy-plane of $\mathbb{R}^3$. By Theorem 4.16, the set K is compact. It is easy to see that $\operatorname{ext} K = ((\{\boldsymbol{u}, \boldsymbol{v}\} \cup C) \setminus \{\boldsymbol{o}\})$, which is a nonclosed set: $\boldsymbol{o} \notin \operatorname{ext} K$ because $\boldsymbol{o} \in (\boldsymbol{u}, \boldsymbol{v})$.

11.2.3 *Convex Hulls of Extreme Points and Halflines*

Theorem 11.21. *Every compact convex set $K \subset \mathbb{R}^n$ is the convex hull of the set of its extreme points: $K = \operatorname{conv}(\operatorname{ext} K)$.*

Proof. The case $K = \varnothing$ is obvious: $\operatorname{ext} K = \varnothing$ and $\varnothing = \operatorname{conv} \varnothing$. Hence we may assume that K is nonempty. Another obvious case is when K is a singleton, say $K = \{\boldsymbol{u}\}$: then $\operatorname{ext} K = \{\boldsymbol{u}\}$ and $\{\boldsymbol{u}\} = \operatorname{conv}\{\boldsymbol{u}\}$.

So, we will suppose that $\dim K > 0$. Clearly, $\operatorname{conv}(\operatorname{ext} K) \subset K$ due to the inclusion $\operatorname{ext} K \subset K$ and the convexity of K. By induction on $m = \dim K$, we are going to prove the opposite inclusion $K \subset \operatorname{conv}(\operatorname{ext} K)$. If $\dim K = 1$, then K is a closed segment, say $[\boldsymbol{x}, \boldsymbol{z}]$; in this case, $\operatorname{ext} K = \{\boldsymbol{x}, \boldsymbol{z}\}$ and

$$K = [\boldsymbol{x}, \boldsymbol{z}] = \operatorname{conv}\{\boldsymbol{x}, \boldsymbol{z}\} = \operatorname{conv}(\operatorname{ext} K).$$

Assume that the assertion holds for all compact convex sets of dimension $r \leqslant m-1$, where $m \geqslant 2$, and let $K \subset \mathbb{R}^n$ be a compact convex set of dimension m. Choose a point $\boldsymbol{x} \in K$ and a line l through $\boldsymbol{x}$. Since K is compact, the set $K \cap l$ is a closed segment, say $[\boldsymbol{u}, \boldsymbol{v}]$ (possibly, $\boldsymbol{u} = \boldsymbol{v}$). Corollary 3.27 implies that both points $\boldsymbol{v}$ and $\boldsymbol{w}$ belong to $\operatorname{rbd} K$. Theorem 11.9 shows that the generated extreme faces $F_{\boldsymbol{u}}(K)$ and $F_{\boldsymbol{v}}(K)$ lie in $\operatorname{rbd} K$, and Theorem 11.4 implies that these faces are compact convex sets of dimension less than m. By the induction hypothesis,

$$F_{\boldsymbol{u}}(K) \subset \operatorname{conv}(\operatorname{ext} F_{\boldsymbol{u}}(K)) \quad \text{and} \quad F_{\boldsymbol{v}}(K) \subset \operatorname{conv}(\operatorname{ext} F_{\boldsymbol{v}}(K)).$$

Furthermore, Theorem 11.3 shows that

$$\operatorname{ext} F_{\boldsymbol{u}}(K) \cup \operatorname{ext} F_{\boldsymbol{v}}(K) \subset \operatorname{ext} K.$$

Finally, using Theorem 4.2, we conclude:

$$\begin{aligned} \boldsymbol{x} \in [\boldsymbol{u}, \boldsymbol{v}] &\subset \operatorname{conv}(F_{\boldsymbol{u}}(K) \cup F_{\boldsymbol{v}}(K)) \\ &\subset \operatorname{conv}(\operatorname{conv}(\operatorname{ext} F_{\boldsymbol{u}}(K)) \cup \operatorname{conv}(\operatorname{ext} F_{\boldsymbol{v}}(K))) \\ &= \operatorname{conv}(\operatorname{ext} F_{\boldsymbol{u}}(K) \cup \operatorname{ext} F_{\boldsymbol{v}}(K)) \\ &\subset \operatorname{conv}(\operatorname{ext} K). \end{aligned}$$

Hence $K \subset \operatorname{conv}(\operatorname{ext} K)$. □

The next result describes the "minimality" property of $\operatorname{ext} K$ (see also Problem 11.13 for a characteristic property of $\operatorname{ext} K$).

Theorem 11.22. *If $K \subset \mathbb{R}^n$ is a compact convex set and X is a subset of K, then $K = \operatorname{conv} X$ if and only if $\operatorname{ext} K \subset X$.*

Proof. Let $K = \operatorname{conv} X$. Assume for a moment the existence of a point $\boldsymbol{x} \in \operatorname{ext} K \setminus X$. Then $X \subset \operatorname{ext} K \setminus \{\boldsymbol{x}\} \subset K \setminus \{\boldsymbol{x}\}$. Since $K \setminus \{\boldsymbol{x}\}$ is a convex set (see Corollary 11.16), one has $K = \operatorname{conv} X \subset K \setminus \{\boldsymbol{x}\}$, a contradiction.

Conversely, if $\operatorname{ext} K \subset X$, then a combination of Theorems 4.2 and 11.21 gives $K = \operatorname{conv}(\operatorname{ext} K) \subset \operatorname{conv} X \subset K$. Therefore, $K = \operatorname{conv} X$. □

Remark. 1. The equality $K = \operatorname{conv}(\operatorname{ext} K)$ does not imply that the convex set $K \subset \mathbb{R}^n$ is bounded or closed. For instance, if $K \subset \mathbb{R}^2$ is the unbounded and nonclosed convex set given by $K = \{(x, y) : x > 0, \ y \geqslant x^2\}$, then $\operatorname{ext} K = \{(x, y) : x > 0, \ y = x^2\}$ and $K = \operatorname{conv}(\operatorname{ext} K)$.
2. A description of closed convex sets $K \subset \mathbb{R}^n$ satisfying the condition $K = \operatorname{conv}(\operatorname{ext} K)$ is given in Corollary 11.30.

Corollary 11.23. *For a compact convex set $K \subset \mathbb{R}^n$ of dimension m, the following assertions hold.*

(*a*) $\dim(\operatorname{ext} K) = m$.
(*b*) $\operatorname{ext} K$ *contains an affinely independent subset of* $m + 1$ *points.*
(*c*) *If* $K \neq \varnothing$, *then* $\operatorname{card}(\operatorname{ext} K) = m + 1$ *if and only if* K *is an* m*-simplex.*

Proof. (*a*) This part follows from a combination of Theorems 4.2 and 11.21.

(*b*) Corollary 2.66 shows that $\operatorname{ext} K$ contains an affinely independent subset of $m + 1$ points, with $\operatorname{card}(\operatorname{ext} K) = m + 1$ if and only if the set $\operatorname{ext} K$ is affinely independent.

(*c*) Suppose that $K \neq \varnothing$. If $\operatorname{ext} K$ is affinely independent, then a combination of Definition 3.5 and Theorem 11.21 shows that the set $K =$

$\mathrm{conv}\,(\mathrm{ext}\, K)$ is a simplex. Conversely, if K is a simplex, then $\mathrm{card}\,(\mathrm{ext}\, K) = m + 1$ (see Problem 11.2). □

A combination of Theorems 4.7 and 11.21 implies the following corollary.

Corollary 11.24. *If $K \subset \mathbb{R}^n$ is a nonempty compact convex set of dimension m, then every point $\boldsymbol{u} \in K$ is a positive convex combination of an affinely independent set of $m + 1$ or fewer extreme points of K.* □

Definition 11.25. An *extreme halfline* (also called an *extreme ray*) of a convex set $K \subset \mathbb{R}^n$ is a halfline $h \subset K$ which is an extreme face of K. In what follows, $\mathrm{extr}\, K$ denotes the union of extreme halflines of K (put $\mathrm{extr}\, K = \varnothing$ if K has no extreme halflines).

Example. The extreme halflines of an r-simplicial cone $\Gamma_{\boldsymbol{a}}(\boldsymbol{x}_1, \dots, \boldsymbol{x}_r)$ in $\mathbb{R}^n$ are exactly the closed halflines $[\boldsymbol{a}, \boldsymbol{x}_i\rangle$, $1 \leqslant i \leqslant r$ (see Problem 11.3).

Theorem 11.26. *If $K \subset \mathbb{R}^n$ is a closed convex set and h is an extreme halfline of K, then $h = \boldsymbol{u} + g$, where $\boldsymbol{u} \in \mathrm{ext}\, K$ and g is an extreme halfline of $\mathrm{rec}\, K$.*

Proof. Let $\boldsymbol{u}$ be the endpoint of h. Then $\boldsymbol{u}$ is an extreme point of h and whence is an extreme point of K (see Theorem 11.3). If h is expressed as $h = \{\boldsymbol{u} + t\boldsymbol{v} : t \geqslant 0\}$, $\boldsymbol{v} \neq \boldsymbol{o}$, then the inclusion $\boldsymbol{v} \in \mathrm{rec}\, K$ follows from Corollary 6.15. Consequently, the closed halfline $g = \{t\boldsymbol{v} : t \geqslant 0\}$ lies in $\mathrm{rec}\, K$.

Assume, for contradiction, that g is not an extreme halfline of $\mathrm{rec}\, K$. Choose a nonzero point $\boldsymbol{z} \in g$. Then the generated face $F = F_{\boldsymbol{z}}(\mathrm{rec}\, K)$ is larger than g. Hence there is a point $\boldsymbol{x} \in F \setminus g$. According to Theorem 11.9, $\boldsymbol{z} \in \mathrm{rint}\, F$. Consequently, Theorem 3.26 shows the existence of a point $\boldsymbol{y} \in \mathrm{rec}\, K$ such that $\boldsymbol{u} \in (\boldsymbol{x}, \boldsymbol{y})$. Because the closed halflines $g_1 = \{t\boldsymbol{x} : t \geqslant 0\}$ and $g_2 = \{t\boldsymbol{y} : t \geqslant 0\}$ lie in $\mathrm{rec}\, K$, their translates $h_1 = \boldsymbol{u} + g_1$ and $h_2 = \boldsymbol{u} + g_2$ lie in K (see Theorem 6.8). In particular, the points $\boldsymbol{u} + \boldsymbol{x}$ and $\boldsymbol{u} + \boldsymbol{y}$ belong to K. Clearly, the point $\boldsymbol{u} + \boldsymbol{z} \in h$ belongs to the open segment $(\boldsymbol{u} + \boldsymbol{x}, \boldsymbol{u} + \boldsymbol{y}) \subset K$, and this segment does not lie in h. Thus h cannot be an extreme halfline of K, a contradiction. □

The next lemma will be used in the proof of Theorem 11.28.

Lemma 11.27. *For a nonempty closed convex set $K \subset \mathbb{R}^n$, the following conditions are equivalent.*

(*a*) *There is a line l such that K is the union of parallel to l closed segments $[\boldsymbol{u}, \boldsymbol{v}]$, where $\boldsymbol{u}, \boldsymbol{v} \in \operatorname{rbd} K$.*
(*b*) *K is the convex hull of $\operatorname{rbd} K$.*
(*c*) *K is neither a plane nor a closed halfplane of $\operatorname{aff} K$.*

Proof. The assertion $(a) \Rightarrow (b)$ is obvious.

$(b) \Rightarrow (c)$ The set K cannot be a plane, since otherwise $\operatorname{rbd} K = \varnothing$ (see Theorem 3.62). Similarly, K cannot be a closed halfplane of $\operatorname{aff} K$, since otherwise $\operatorname{rbd} K$ would be a proper subset of $\operatorname{aff} K$ (see Theorem 3.68), which would give $K \neq \operatorname{rbd} K = \operatorname{conv}(\operatorname{rbd} K)$.

$(c) \Rightarrow (a)$ By Theorems 3.62 and 3.68, the set $\operatorname{rbd} K$ is nonempty and nonconvex. Hence there are distinct points $\boldsymbol{x}, \boldsymbol{y} \in \operatorname{rbd} K$ such that $\operatorname{rint} K \cap (\boldsymbol{x}, \boldsymbol{y}) \neq \varnothing$. Choose a point $\boldsymbol{u} \in \operatorname{rint} K \cap (\boldsymbol{x}, \boldsymbol{y})$ and consider the line $l = \langle \boldsymbol{x}, \boldsymbol{y} \rangle$. We assert that $K \cap l = [\boldsymbol{x}, \boldsymbol{y}]$. Indeed, because the inclusion $[\boldsymbol{x}, \boldsymbol{y}] \subset K \cap l$ is obvious, it suffices to prove the opposite inclusion. Assume for a moment the existence of a point $\boldsymbol{v} \in K \cap l \setminus [\boldsymbol{x}, \boldsymbol{y}]$. Then one of the points $\boldsymbol{x}, \boldsymbol{y}$, say $\boldsymbol{x}$, belongs to $(\boldsymbol{u}, \boldsymbol{v})$. In this case, $\boldsymbol{x} \in \operatorname{rint} K$ according to Theorem 3.24, which contradicts the assumption $\boldsymbol{x} \in \operatorname{rbd} K$. Hence $K \cap l \subset [\boldsymbol{x}, \boldsymbol{y}]$.

Next, we assert that for a given point $\boldsymbol{z} \in K$, the line $l' = (\boldsymbol{z} - \boldsymbol{x}) + l$ through $\boldsymbol{z}$ also meets K along a closed segment. Indeed, if $K \cap l'$ were unbounded, then a closed halfline h with endpoint $\boldsymbol{z}$ would lie in $K \cap l'$, implying that the closed halfline $h' = (\boldsymbol{x} - \boldsymbol{z}) + h$ would lie in $K \cap l$ (see Theorem 6.10), which is impossible due to $K \cap l = [\boldsymbol{x}, \boldsymbol{y}]$. Hence $K \cap l'$ is a closed segment, say $[\boldsymbol{u}, \boldsymbol{v}]$. Corollary 3.27 shows that both $\boldsymbol{u}$ and $\boldsymbol{v}$ belong to $\operatorname{rbd} K$. Since $\boldsymbol{z} \in [\boldsymbol{u}, \boldsymbol{v}]$, condition (a) is satisfied. □

Remark. It is easy to see (using Theorem 6.40) that a line $l \subset \mathbb{R}^n$ satisfies condition (a) of Lemma 11.27 if and only if l is a translate of an one-dimensional subspace $l' \subset \mathbb{R}^n$ such that $l' \subset \operatorname{dir} K$ and $l' \cap \operatorname{rec} K = \{\boldsymbol{o}\}$.

Theorem 11.28. *For a closed convex set $K \subset \mathbb{R}^n$, the following conditions are equivalent.*

(*a*) *K is line-free.*
(*b*) *$K = \operatorname{conv}(\operatorname{ext} K \cup \operatorname{extr} K)$.*
(*c*) *$K = \operatorname{conv}(\operatorname{ext} K) + \operatorname{rec} K$.*

Proof. Since the case $K = \varnothing$ is obvious, we may assume that K is nonempty.

$(a) \Rightarrow (b)$ The inclusion $\operatorname{conv}(\operatorname{ext} K \cup \operatorname{extr} K) \subset K$ is obvious. So, we are going to prove the opposite inclusion $K \subset \operatorname{conv}(\operatorname{ext} K \cup \operatorname{extr} K)$. This will be done by induction on $m = \dim K$. The cases $m = 0$ and $m = 1$ are simple: if $m = 0$, then K is a singleton and $K = \operatorname{ext} K$; if $m = 1$, then K is either a closed segment (which is the convex hull of its endpoints), or a closed halfline.

Assume that the assertion holds for all $r \leqslant m - 1$, $m \geqslant 2$, and let $K \subset \mathbb{R}^n$ be a line-free closed convex set of dimension m. Then K is neither a plane nor a closed halfplane. Choose a point $\boldsymbol{x} \in K$. By Lemma 11.27, there is a line l through $\boldsymbol{x}$ such that $K \cap l$ is a closed segment, $[\boldsymbol{y}, \boldsymbol{z}]$, where both $\boldsymbol{y}$ and $\boldsymbol{z}$ belong to $\operatorname{rbd} K$.

Theorem 11.9 shows that both generated extreme faces $F_{\boldsymbol{u}}(K)$ and $F_{\boldsymbol{v}}(K)$ lie in $\operatorname{rbd} K$, and Theorem 11.4 implies that these faces are closed convex sets of dimension less than m. Clearly, $F_{\boldsymbol{u}}(K)$ and $F_{\boldsymbol{v}}(K)$ are line-free. By the induction hypothesis,

$$F_{\boldsymbol{y}}(K) \subset \operatorname{conv}(\operatorname{ext} F_{\boldsymbol{y}}(K) \cup \operatorname{extr} F_{\boldsymbol{y}}(K)),$$
$$F_{\boldsymbol{z}}(K) \subset \operatorname{conv}(\operatorname{ext} F_{\boldsymbol{z}}(K) \cup \operatorname{extr} F_{\boldsymbol{z}}(K)).$$

According to Theorem 11.3, the extreme points and the extreme halflines of both sets $F_{\boldsymbol{y}}(K)$ and $F_{\boldsymbol{z}}(K)$ are also extreme points and extreme halflines of K. Hence

$$F_{\boldsymbol{y}}(K) \cup F_{\boldsymbol{z}}(K) \subset \operatorname{conv}(\operatorname{ext} K \cup \operatorname{extr} K),$$

which gives

$$\boldsymbol{x} \in \operatorname{conv}\{\boldsymbol{y}, \boldsymbol{z}\} \subset \operatorname{conv}(F_{\boldsymbol{y}}(K) \cup F_{\boldsymbol{z}}(K)) \subset \operatorname{conv}(\operatorname{ext} K \cup \operatorname{extr} K).$$

Hence $K \subset \operatorname{conv}(\operatorname{ext} K \cup \operatorname{extr} K)$.

$(b) \Rightarrow (c)$ Since $\operatorname{conv}(\operatorname{ext} K) \subset K$ due to the convexity of K, the inclusion $\operatorname{conv}(\operatorname{ext} K) + \operatorname{rec} K \subset K$ follows from Theorem 6.2. Hence it remains to prove the opposite inclusion. For this, choose any point $\boldsymbol{x} \in K$. Condition (b) and Theorem 4.3 show that $\boldsymbol{x}$ can be written as a positive convex combination

$$\boldsymbol{x} = \lambda_1 \boldsymbol{x}_1 + \cdots + \lambda_p \boldsymbol{x}_p + \lambda_{p+1} \boldsymbol{z}_{p+1} + \cdots + \lambda_q \boldsymbol{z}_q,$$

where $\boldsymbol{x}_1, \ldots, \boldsymbol{x}_p \in \operatorname{ext} K$ and $\boldsymbol{z}_{p+1}, \ldots, \boldsymbol{z}_q \in \operatorname{extr} K \setminus \operatorname{ext} K$ (possibly, $p = 0$ or $p = q$). Denote by $[\boldsymbol{u}_i, \boldsymbol{v}_i\rangle$ the extreme halfline of K containing $\boldsymbol{z}_i$, $p + 1 \leqslant i \leqslant q$. By Theorem 11.26, $\boldsymbol{u}_{p+1}, \ldots, \boldsymbol{u}_q \in \operatorname{ext} K$. Theorem 6.2 implies that

$$\boldsymbol{z}_i = \boldsymbol{u}_i + \gamma_i \boldsymbol{w}_i, \text{ where } \gamma_i > 0 \text{ and } \boldsymbol{w}_i = \boldsymbol{v}_i - \boldsymbol{u}_i \in \operatorname{rec} K, \ p + 1 \leqslant i \leqslant q.$$

Hence $\boldsymbol{x} = \boldsymbol{z} + \boldsymbol{w}$, where

$$\boldsymbol{z} = \lambda_1 \boldsymbol{x}_1 + \cdots + \lambda_p \boldsymbol{x}_p + \lambda_{p+1} \boldsymbol{u}_{p+1} + \cdots + \lambda_q \boldsymbol{u}_q$$

is a convex combination of points from $\operatorname{ext} K$, and

$$\boldsymbol{w} = \lambda_{p+1} \gamma_{p+1} \boldsymbol{w}_{p+1} + \cdots + \lambda_q \gamma_q \boldsymbol{w}_q$$

is a positive combination of points from $\operatorname{rec} K$. Therefore, $\boldsymbol{z} \in \operatorname{conv}(\operatorname{ext} K)$ and $\boldsymbol{w} \in \operatorname{rec} K$, as follows from Theorems 4.3 and 5.32. Summing up,

$$\boldsymbol{x} = \boldsymbol{z} + \boldsymbol{w} \in \operatorname{conv}(\operatorname{ext} K) + \operatorname{rec} K.$$

$(c) \Rightarrow (a)$ If K satisfies condition (c), then the set $\operatorname{ext} K$ should be nonempty. Indeed, if $\operatorname{ext} K = \varnothing$, then $K \neq \varnothing = \operatorname{conv}(\operatorname{ext} K) + \operatorname{rec} K$, a contradiction. By Theorem 11.18, K is line-free. $\square$

The next result extends Theorem 11.22 to the case of line-free closed convex sets. We will say that a set $X \subset \mathbb{R}^n$ is *coterminal* with a closed halfline $h \subset \mathbb{R}^n$, provided X meets every closed halfline contained in h.

Remark. An easy argument shows that a convex set $K \subset \mathbb{R}^n$ is conterminal with a closed halfline $h \subset \mathbb{R}^n$ if and only if $K \cap h$ is a closed halfline.

Theorem 11.29. *For a line-free closed convex set $K \subset \mathbb{R}^n$ and a subset X of K, the following assertions hold.*

(a) $K = \operatorname{conv} X$ *if and only if* $\operatorname{ext} K \subset X$ *and* X *is coterminal with every extreme halfline of* K.
(b) $K = \operatorname{conv} X$ *if and only if* $\operatorname{ext} K \subset X$ *and* $h = \operatorname{conv}(X \cap h)$ *for every extreme halfline of* K.
(c) $K = \operatorname{conv} X + \operatorname{rec} K$ *if and only if* $\operatorname{ext} K \subset X$.

Proof. (a) Let $K = \operatorname{conv} X$. Assume for a moment the existence of a point $\boldsymbol{x} \in \operatorname{ext} K \setminus X$. Then $X \subset \operatorname{ext} K \setminus \{\boldsymbol{x}\} \subset K \setminus \{\boldsymbol{x}\}$. Since $K \setminus \{\boldsymbol{x}\}$ is a convex set (see Corollary 11.16), one has $K = \operatorname{conv} X \subset K \setminus \{\boldsymbol{x}\}$, a contradiction. Thus $\operatorname{ext} K \subset K$.

Next, assume that X is not coterminal with an extreme halfline h of K. Then there is a point $\boldsymbol{v} \in h$ such that the open halfline $h' \subset h$ with endpoint $\boldsymbol{v}$ is disjoint from X. Consider the set $M = (K \setminus h) \cup [\boldsymbol{u}, \boldsymbol{v}]$, where $\boldsymbol{u}$ is the endpoint of h (possibly, $\boldsymbol{u} = \boldsymbol{v}$). It is easy to see that M contains X and is a proper subset of K. Theorem 11.4 shows that M is a convex set. Hence $\operatorname{conv} X \subset M \neq K$, a contradiction.

Conversely, let a subset X of K contain $\operatorname{ext} K$ and be coterminal with every extreme halfline of K. Then $\operatorname{conv} X$ contains $\operatorname{ext} K$ and is coterminal with every extreme halfline of K, which gives the inclusion $\operatorname{extr} K \subset \operatorname{conv} X$ (see a remark on page 398). Hence $\operatorname{ext} K \cup \operatorname{extr} K \subset \operatorname{conv} X$. A combination of Theorems 4.2 and 11.28 gives

$$K = \operatorname{conv}(\operatorname{ext} K \cup \operatorname{extr} K) \subset \operatorname{conv}(\operatorname{conv} X) = \operatorname{conv} X \subset K.$$

Summing up, $K = \operatorname{conv} X$.

(b) Let $K = \operatorname{conv} X$. By assertion ($a$), $\operatorname{ext} K \subset X$ and X is coterminal with every extreme halfline of K. If $h = [\boldsymbol{u}, \boldsymbol{v}\rangle$ is an extreme halfline of K, then $\boldsymbol{u} \in \operatorname{ext} K$. Consequently, $h = \operatorname{conv}(h \cap X)$.

Conversely, if $\operatorname{ext} K \subset X$ and $h = \operatorname{conv}(X \cap h)$ for every extreme halfline of K, then X is coterminal with every extreme halfline of K, and assertion (a) implies that $K = \operatorname{conv} X$.

(c) Let $K = \operatorname{conv} X + \operatorname{rec} K$. Assume for a moment the existence of a point $\boldsymbol{x} \in \operatorname{ext} K \setminus X$. Since $\boldsymbol{x} \in K$, one can write $\boldsymbol{x} = \boldsymbol{u} + \boldsymbol{v}$, where $\boldsymbol{u} \in \operatorname{conv} X$ and $\boldsymbol{v} \in \operatorname{rec} K$. We assert that $\boldsymbol{v} \neq \boldsymbol{o}$. Indeed, suppose that $\boldsymbol{v} = \boldsymbol{o}$. Then $\boldsymbol{x} = \boldsymbol{u} \in \operatorname{conv} X$. On the other hand, $X \subset \operatorname{ext} K \setminus \{\boldsymbol{x}\} \subset K \setminus \{\boldsymbol{x}\}$. Since the set $K \setminus \{\boldsymbol{x}\}$ is convex, one has $\boldsymbol{x} \in \operatorname{conv} X \subset K \setminus \{\boldsymbol{x}\}$, a contradiction. Hence $\boldsymbol{v} \neq \boldsymbol{o}$. The closed halfline $[\boldsymbol{o}, \boldsymbol{v}\rangle$ lies in $\operatorname{rec} K$, and Theorem 6.2 gives the inclusion

$$[\boldsymbol{u}, \boldsymbol{x}\rangle = [\boldsymbol{u}, \boldsymbol{u} + \boldsymbol{v}\rangle = \boldsymbol{u} + [\boldsymbol{o}, \boldsymbol{v}\rangle \subset K.$$

Because $\boldsymbol{x}$ is not the endpoint of the closed halfline $[\boldsymbol{u}, \boldsymbol{x}\rangle$, it cannot be extreme in K, contrary to the assumption $\boldsymbol{x} \in \operatorname{ext} K \setminus X$. Hence $\operatorname{ext} K \subset X$.

Conversely, let $\operatorname{ext} K \subset X$. Then $\operatorname{conv}(\operatorname{ext} K) \subset \operatorname{conv} X$. A combination of Theorems 6.2 and 11.28 gives

$$K = \operatorname{conv}(\operatorname{ext} K) + \operatorname{rec} K \subset \operatorname{conv} X + \operatorname{rec} K \subset K + \operatorname{rec} K = K.$$

Hence $K = \operatorname{conv} X + \operatorname{rec} K$. □

The corollary below, which refines Theorem 11.21, follows from Theorems 11.28 and 11.29.

Corollary 11.30. *A closed convex set $K \subset \mathbb{R}^n$ is the convex hull of* $\operatorname{ext} K$ *if and only if K is line-free and has no extreme halflines. If K is the convex hull of* $\operatorname{ext} K$, *then, given a set $X \subset K$, one has $K = \operatorname{conv} X$ if and only if* $\operatorname{ext} K \subset X$. □

A combination of Theorems 4.7 and 11.28 implies that any point of a nonempty line-free closed convex set $K \subset \mathbb{R}^n$ of dimension m can be expressed as a positive convex combination of $m + 1$ or fewer points from $\operatorname{ext} K \cup \operatorname{extr} K$. The following the theorem refines this assertion.

Theorem 11.31. *If $K \subset \mathbb{R}^n$ is a nonempty line-free closed convex set of dimension m, then every point $\boldsymbol{u} \in K$ is a positive convex combination of an affinely independent set of*

(*a*) *$m+1$ or fewer points from* $\operatorname{ext} K$, *or*
(*b*) *m or fewer points from* $\operatorname{ext} K \cup \operatorname{extr} K$.

Proof. Let $\boldsymbol{u} \in K$. Translating K on $-\boldsymbol{u}$, we may suppose that $\boldsymbol{u} = \boldsymbol{o}$. Indeed, a desired representation $\boldsymbol{u} = \gamma_1 \boldsymbol{z}_1 + \cdots + \gamma_t \boldsymbol{z}_t$ is equivalent to

$$\boldsymbol{o} = \gamma_1(\boldsymbol{z}_1 - \boldsymbol{u}) + \cdots + \gamma_t(\boldsymbol{z}_t - \boldsymbol{u}),$$

where $\boldsymbol{z}_i - \boldsymbol{u} = \operatorname{ext}(K - \boldsymbol{u}) \cup \operatorname{extr}(K - \boldsymbol{u})$ for all $1 \leqslant i \leqslant t$.

A combination of Theorems 4.6 and 11.28 implies that $\boldsymbol{o}$ is a positive convex combination of the form

$$\boldsymbol{o} = \lambda_1 \boldsymbol{x}_1 + \cdots + \lambda_p \boldsymbol{x}_p, \quad p \leqslant m+1, \tag{11.1}$$

where $X = \{\boldsymbol{x}_1, \ldots, \boldsymbol{x}_p\}$ is an affinely independent subset of $\operatorname{ext} K \cup \operatorname{extr} K$. If $X \subset \operatorname{ext} K$ or $p \leqslant m$, then the assertion holds. To obtaining a contradiction, we will suppose that no set $X \subset \operatorname{ext} K \cup \operatorname{extr} K$ satisfying (11.1) lies in $\operatorname{ext} K$ and that always $\operatorname{card} X = m+1$.

For a set $X = \{\boldsymbol{x}_1, \ldots, \boldsymbol{x}_{m+1}\}$ satisfying (11.1), we renumber its elements to have $X \setminus \operatorname{ext} K = \{\boldsymbol{x}_1, \ldots, \boldsymbol{x}_q\}$, where $1 \leqslant q \leqslant m+1$. Among all such sets X, we choose one with the smallest possible value of q. The equality

$$\boldsymbol{o} = \lambda_1 \boldsymbol{x}_1 + \cdots + \lambda_{m+1} \boldsymbol{x}_{m+1}, \tag{11.2}$$

shows that $\boldsymbol{o} \in \operatorname{aff} X$. Hence the plane $S = \operatorname{aff} X$ is an m-dimensional subspace and $S = \operatorname{span} X$. This argument and the equality

$$\lambda_1 \boldsymbol{x}_1 = -\lambda_2 \boldsymbol{x}_2 - \cdots - \lambda_{m+1} \boldsymbol{x}_{m+1} \tag{11.3}$$

imply that $\{-\boldsymbol{x}_2, \ldots, -\boldsymbol{x}_{m+1}\}$ is a basis for S. Consequently, the convex cone $C = \operatorname{cone}_{\boldsymbol{o}}\{-\boldsymbol{x}_2, \ldots, -\boldsymbol{x}_{m+1}\}$ is an m-simplicial cone with apex $\boldsymbol{o}$. Furthermore, (11.3) gives the inclusion $\lambda_1 \boldsymbol{x}_1 \in \operatorname{rint} C$ (see Theorem 5.23). Consequently, $\boldsymbol{x}_1 = \lambda_1^{-1}(\lambda_1 \boldsymbol{x}_1) \in \operatorname{rint} C$, as follows from Theorem 5.20.

Denote by h_1 the extreme halfline of K which contains $\boldsymbol{x}_1$. Let $\boldsymbol{u}_1$ be the endpoint of h_1. Then $\boldsymbol{u}_1 \in \operatorname{ext} K$ (see Theorem 11.26). We assert that $\boldsymbol{u}_1 \notin C$. Indeed, assume for a moment that $\boldsymbol{u}_1 \in C$. According to Definition 5.6, $\boldsymbol{u}_1$ can be written as a nonnegative combination of the form

$$\boldsymbol{u}_1 = \mu_2(-\boldsymbol{x}_2) + \cdots + \mu_{m+1}(-\boldsymbol{x}_{m+1}).$$

Rewriting this equality (with $\mu = 1 + \mu_2 + \cdots + \mu_{m+1}$) as

$$\boldsymbol{o} = \frac{1}{\mu}\boldsymbol{u}_1 + \frac{\mu_2}{\mu}\boldsymbol{x}_2 + \cdots + \frac{\mu_{m+1}}{\mu}\boldsymbol{x}_{m+1},$$

we obtain another representation for $\boldsymbol{o}$ as a convex combination of points from $\operatorname{ext} K \cup \operatorname{extr} K$, with $q-1$ points from $X \setminus \operatorname{ext} K$. The last is in contradiction with the choice of q. So, $\boldsymbol{u}_1 \notin C$.

The inclusion $\boldsymbol{o} \in K$ shows that $\operatorname{aff} K$ is an m-dimensional subspace. So, $\operatorname{aff} K = \operatorname{span} K$. Clearly, $\{-\boldsymbol{x}_2, \ldots, -\boldsymbol{x}_{m+1}\}$ is a basis for $\operatorname{span} K$, and a combination of Theorems 2.6 and 5.7 implies that $\operatorname{span} C = \operatorname{span} K$. Consequently, $\boldsymbol{x}_1 \in \operatorname{span} C$. According to Theorem 3.65, the open segment $(\boldsymbol{u}_1, \boldsymbol{x}_1)$ meets $\operatorname{rbd} C$ at a unique point, say $\boldsymbol{w}_1$. Then $\boldsymbol{w}_1 \in h_1 \setminus \{\boldsymbol{u}_1\} \subset \operatorname{extr} K$, and the inclusion $\boldsymbol{w}_1 \in \operatorname{rbd} C$ shows that $\boldsymbol{w}_1$ can be written as $\boldsymbol{w}_1 = \gamma_2(-\boldsymbol{x}_2) + \cdots + \gamma_{m+1}(-\boldsymbol{x}_{m+1})$, where $\gamma_2, \ldots, \gamma_{m+1} \geqslant 0$ and at least one of the scalars $\gamma_2, \ldots, \gamma_{m+1}$ is zero. With $\gamma = 1 + \gamma_2 + \cdots + \gamma_{m+1}$,

$$\boldsymbol{o} = \frac{1}{\gamma}\boldsymbol{w}_1 + \frac{\gamma_2}{\gamma}\boldsymbol{x}_2 + \cdots + \frac{\gamma_{m+1}}{\gamma}\boldsymbol{x}_{m+1}$$

is a new representation of $\boldsymbol{o}$ as a convex combination of points from $\operatorname{ext} K \cup \operatorname{extr} K$, with fewer than $m+1$ terms. The obtained contradiction proves assertion (b). □

11.2.4 *Special Extreme Faces and r-Extreme Sets*

A combination of Theorems 6.6, 6.25, and 11.28 implies the following result.

Corollary 11.32. *Let $K \subset \mathbb{R}^n$ be a nonempty closed convex set. If $L \subset \mathbb{R}^n$ is a plane complementary to* $\operatorname{lin} K$, *then*

$$K = \operatorname{conv}(\operatorname{ext}(K \cap L) \cup \operatorname{extr}(K \cap L)) + \operatorname{lin} K,$$
$$K = \operatorname{conv}(\operatorname{ext}(K \cap L)) + \operatorname{rec}(K \cap L) + \operatorname{lin} K,$$
$$K = \operatorname{conv}(\operatorname{ext}(K \cap L)) + \operatorname{rec} K.$$
□

Definition 11.33. Let $K \subset \mathbb{R}^n$ be a nonempty convex set. A nonempty extreme face F of $K \subset \mathbb{R}^n$ is called *planar* (respectively, *halfplanar*) if F is a plane (respectively, F is a halfplane).

Example. If $K \subset \mathbb{R}^3$ is a convex cone, given by $K = \{(x, y, z) : y \geqslant |x|\}$, then $\operatorname{ap} K = \operatorname{lin} K = \{(0, 0, z) : z \in \mathbb{R}\}$ is the only planar extreme face of K, and the halfplanes

$$D_1 = \{(x, x, z) : x \geqslant 0\} \quad \text{and} \quad D_2 = \{(x, -x, z) : x \leqslant 0\}$$

are the halfplanar extreme faces of K.

Clearly, the planar extreme faces of a line-free closed convex set $K \subset \mathbb{R}^n$ are its extreme points, and the halfplanar extreme faces of K are its extreme halflines.

Theorem 11.34. *Let $K \subset \mathbb{R}^n$ be a nonempty closed convex set expressed as $K = (K \cap S) + \operatorname{lin} K$, where S is a subspace of $\mathbb{R}^n$ complementary to $\operatorname{lin} K$. The following assertions hold.*

(*a*) *An extreme face F of K is planar if and only if $F = \boldsymbol{x} + \operatorname{lin} K$, where $\boldsymbol{x} \in \operatorname{ext}(K \cap S)$.*
(*b*) *An extreme face F of K is halfplanar if and only if $F = h + \operatorname{lin} K$, where h is an extreme halfline of $K \cap S$.*

Proof. Let F be an extreme face of K. According to Theorem 11.5, F can be expressed as $F = G + \operatorname{lin} K$, where G is an extreme face of $K \cap S$. Because the set $K \cap S$ is line-free (see Theorems 6.25 and 6.37), G also is line-free.

From the obvious relation $G = F \cap S$ it follows that F is a plane (respectively, a closed halfplane) if and only if G is a plane (respectively, a closed halfplane). Since $K \cap S$ is line-free, a plane in $K \cap S$ is a singleton, and a halfplane in $K \subset S$ is a closed halfline. Hence G is either an extreme point or an extreme halfline of $K \cap S$. □

The corollary below follows from Theorems 11.18 and 11.34.

Corollary 11.35. *Let $K \subset \mathbb{R}^n$ be a nonempty closed convex set. An extreme face F of K is planar if and only if F is a translate of $\operatorname{lin} K$. Furthermore, K has exactly one planar extreme face, F, if and only if K is a convex cone with $\operatorname{ap} K = F$.* □

The next result extends condition (*b*) of Theorem 11.28.

Theorem 11.36. *A nonempty closed convex set $K \subset \mathbb{R}^n$ is the convex hull of the union of its planar and halfplanar extreme faces. Similarly, K is the sum of $\operatorname{rec} K$ and the convex hull of the union of its planar extreme faces.*

Proof. Express K as a direct sum $K = (K \cap S) + \operatorname{lin} K$, where S is a subspace complementary to $\operatorname{lin} K$. By Corollary 11.32,

$$\begin{aligned} K &= \operatorname{conv}(\operatorname{ext}(K \cap S) \cup \operatorname{extr}(K \cap S)) + \operatorname{lin} K \\ &= \operatorname{conv}(\cup(\boldsymbol{x} + \operatorname{lin} K : \boldsymbol{x} \in \operatorname{ext}(K \cap S)) \\ &\qquad \cup (h + \operatorname{lin} K : h \text{ is an extreme halfline of } K \cap S)). \end{aligned}$$

This argument and Theorem 11.34 give

$$K = \mathrm{conv}\,(\cup\,(F : F \text{ is a planar or a halfplanar extreme face of } K)).$$

Similarly, Theorems 6.25 and 11.28 give

$$\mathrm{rec}\,K = \mathrm{rec}\,(K \cap S) + \mathrm{lin}\,K,$$
$$K \cap S = \mathrm{conv}\,(\mathrm{ext}\,(K \cap S)) + \mathrm{rec}\,(K \cap S),$$

which implies

$$\begin{aligned} K &= (\mathrm{conv}\,(\mathrm{ext}\,(K \cap S)) + \mathrm{rec}\,(K \cap S)) + \mathrm{lin}\,K \\ &= \mathrm{conv}\,(\cup\,(\boldsymbol{x} + \mathrm{lin}\,K : \boldsymbol{x} \in \mathrm{ext}\,(K \cap S))) + \mathrm{rec}\,(K \cap S) + \mathrm{lin}\,K \\ &= \mathrm{conv}\,(\cup\,(F : F \text{ is a planar extreme face of } K)) + \mathrm{rec}\,K. \end{aligned}$$ □

The next result extends Theorem 11.29 to the case of arbitrary closed convex sets.

Theorem 11.37. *For a nonempty closed convex set $K \subset \mathbb{R}^n$ and a subset X of K, the following assertions hold.*

(*a*) $K = \mathrm{conv}\,X$ *if and only if* $F = \mathrm{conv}\,(F \cap X)$ *for every planar or halfplanar extreme face* F *of* K.
(*b*) $K = \mathrm{conv}\,X + \mathrm{rec}\,K$ *if and only if* $F = \mathrm{conv}\,(F \cap X)$ *for every planar extreme face* F *of* K.

Proof. (*a*) Suppose that $K = \mathrm{conv}\,X$, and choose an extreme face F of K. By Theorem 11.4, one has

$$F = F \cap K = F \cap \mathrm{conv}\,X = \mathrm{conv}\,(F \cap X).$$

Conversely, suppose that X satisfies the hypothesis of assertion (*a*). By Theorem 11.36,

$$\begin{aligned} K &= \mathrm{conv}\,(\cup\,(F : F \text{ is a planar or halfplanar extreme face of } K)) \\ &= \mathrm{conv}\,(\cup\,(\mathrm{conv}\,(F \cap X) : F \text{ is a planar or halfplanar} \\ &\qquad\qquad \text{extreme face of } K)) \\ &\subset \mathrm{conv}\,(\mathrm{conv}\,X) = \mathrm{conv}\,X \subset K, \end{aligned}$$

implying the equality $K = \mathrm{conv}\,X$.

The proof of assertion (*b*) of the theorem is similar. □

Definition 11.38. Let K be a nonempty convex set in $\mathbb{R}^n$ and r be an integer satisfying the inequalities $0 \leqslant r \leqslant \dim K$. The union of all nonempty extreme faces of K of dimension r or less, is called the *r-extreme set* of K and denoted $\mathrm{ext}_r K$. The elements of $\mathrm{ext}_r K$ are called *r-extreme points* of K.

Another terminology used for the r-extreme set is the *r-skeleton.* Clearly, $\mathrm{ext}_0 K = \mathrm{ext}\, K$.

Theorem 11.39. *If $K \subset \mathbb{R}^n$ is a nonempty convex set of dimension m and r is an integer r satisfying the inequalities $0 \leqslant r \leqslant m$, then the following assertions hold.*

(a) $\mathrm{ext}_0 K \subset \mathrm{ext}_1 K \subset \cdots \subset \mathrm{ext}_{m-1} K = K \cap \mathrm{rbd}\, K \subset \mathrm{ext}_m K = K$.
(b) *The set $K \setminus \mathrm{ext}_r K$ is convex. Furthermore, $K \setminus \mathrm{ext}_r K$ is the union of relative interiors of extreme faces of K, each of dimension $r+1$ or more.*

Proof. (a) The inclusions

$$\mathrm{ext}_0 K \subset \mathrm{ext}_1 K \subset \cdots \subset \mathrm{ext}_{m-1} K \subset K \cap \mathrm{rbd}\, K$$

and the equality $\mathrm{ext}_m K = K$ follow from definitions and Theorem 11.4. The opposite inclusion, $K \cap \mathrm{rbd}\, K \subset \mathrm{ext}_{m-1} K$, is a consequence of Theorem 11.9.

(b) Let $\mathcal{F}_r$ denote the family of all extreme faces of K, each of dimension r or less. By the definition, $\mathrm{ext}_r K = \cup\, (F : F \in \mathcal{F}_r)$. The equality

$$K \setminus \mathrm{ext}_r K = K \setminus (\cup\, (F : F \in \mathcal{F}_r)) = \cap\, (K \setminus F : F \in \mathcal{F}_r)$$

shows that $K \setminus \mathrm{ext}_r K$ is convex as the intersection of convex sets $K \setminus F$, $F \in \mathcal{F}_r$ (see Theorem 11.2). The second part of assertion (b) follows from Corollary 11.11. □

Theorem 11.40. *Let $K \subset \mathbb{R}^n$ be a nonempty convex set of dimension m and r be an integer r satisfying the inequalities $0 \leqslant r \leqslant m$. For a point $\boldsymbol{x} \in K$, the following conditions are equivalent.*

(a) $\boldsymbol{x} \in \mathrm{ext}_r K$.
(b) $\dim F_{\boldsymbol{x}}(K) \leqslant r$.
(c) *No $(r+1)$-simplex $\Delta \subset K$ satisfies the inclusion $\boldsymbol{x} \in \mathrm{rint}\, \Delta$.*
(d) $\boldsymbol{x} \in \mathrm{ext}_r(K \cap B_\rho(\boldsymbol{x}))$ *for every closed ball $B_\rho(\boldsymbol{x}) \subset \mathbb{R}^n$, $\rho > 0$.*
(e) *The apex set of the generated cone $C_{\boldsymbol{x}}(K)$ has dimension at most r.*

Proof. (a) ⇒ (b) Let $\boldsymbol{x} \in \mathrm{ext}_r K$. According to Definition 11.38, there is an extreme face F of K of dimension r or less such that $\boldsymbol{x} \in F$. By Corollary 11.8, $F_{\boldsymbol{x}}(K)$ is the smallest extreme face of K containing $\boldsymbol{x}$. Hence $F_{\boldsymbol{x}}(K) \subset F$, which gives $\dim F_{\boldsymbol{x}}(K) \leqslant \dim F \leqslant r$.

(b) ⇒ (c) Let $\dim F_{\boldsymbol{x}}(K) \leqslant r$. Choose a simplex $\Delta \subset K$ such that $\boldsymbol{x} \in \mathrm{rint}\, \Delta$. Since $F_{\boldsymbol{x}}(K)$ is an extreme face of K, Theorem 11.4 implies the inclusion $\Delta \subset F_{\boldsymbol{x}}(K)$. Hence $\dim \Delta \leqslant \dim F_{\boldsymbol{x}}(K) \leqslant r$.

$(c) \Rightarrow (a)$ Let $\boldsymbol{x}$ be a point in K such that no $(r+1)$-simplex from K contains $\boldsymbol{x}$ in its relative interior. According to Theorem 11.9, $\boldsymbol{x} \in \operatorname{rint} F_{\boldsymbol{x}}(K)$. Therefore, there is a simplex $\Delta \subset F_{\boldsymbol{x}}(K)$ such that $\boldsymbol{x} \in \operatorname{rint} \Delta$ and $\dim \Delta = \dim F_{\boldsymbol{x}}(K)$ (see Theorem 3.22). By the assumption, $\dim \Delta \leqslant r$. Consequently, $\dim F_{\boldsymbol{x}}(K) \leqslant r$, and Definition 11.38 implies the inclusion $\boldsymbol{x} \in \operatorname{ext}_r K$.

$(a) \Leftrightarrow (d)$ Assume first that $\boldsymbol{x} \in \operatorname{ext}_r K$. By condition (c) above, no $(r+1)$-simplex $\Delta \subset K \cap B_\rho(\boldsymbol{x})$ contains $\boldsymbol{x}$ in its relative interior. Consequently, $\boldsymbol{x} \in \operatorname{ext}_r (K \cap B_\rho(\boldsymbol{x}))$.

Conversely, let $\boldsymbol{x} \in \operatorname{ext}_r (K \cap B_\rho(\boldsymbol{x}))$. Assume for a moment the existence of an $(r+1)$-simplex $\Delta = \Delta(\boldsymbol{x}_1, \dots, \boldsymbol{x}_{r+2}) \subset K$ satisfying the inclusion $\boldsymbol{x} \in \operatorname{rint} \Delta$. By Theorem 3.20, $\boldsymbol{x}$ can be written as a positive convex combination $\boldsymbol{x} = \lambda_1 \boldsymbol{x}_1 + \dots + \lambda_{r+2} \boldsymbol{x}_{r+2}$. Let

$$\delta = \max\{\|\boldsymbol{x} - \boldsymbol{x}_i\| : 1 \leqslant i \leqslant r+2\}$$

and

$$\boldsymbol{x}'_i = \boldsymbol{x} + \frac{\rho}{\delta}(\boldsymbol{x}_i - \boldsymbol{x}), \quad 1 \leqslant i \leqslant r+2.$$

A combination of Theorems 2.86 and 2.78 shows that the set $\{\boldsymbol{x}'_1, \dots, \boldsymbol{x}'_{r+2}\}$ is affinely independent. Furthermore,

$$\sum_{i=1}^{r+2} \lambda_i \boldsymbol{x}'_i = \sum_{i=1}^{r+2} \lambda_i \boldsymbol{x} + \frac{\rho}{\delta}\Big(\sum_{i=1}^{r+2} \lambda_i \boldsymbol{x}_i - \sum_{i=1}^{r+2} \lambda_i \boldsymbol{x}\Big) = \boldsymbol{x} + \frac{\rho}{\delta}(\boldsymbol{x} - \boldsymbol{x}) = \boldsymbol{x},$$

and Theorem 3.20 gives $\boldsymbol{x} \in \operatorname{rint} \Delta'$, where $\Delta' = \Delta(\boldsymbol{x}'_1, \dots, \boldsymbol{x}'_{r+2})$.

Since $\{\boldsymbol{x}'_1, \dots, \boldsymbol{x}'_{r+2}\} \subset \Delta$, one has

$$\Delta' = \operatorname{conv}\{\boldsymbol{x}'_1, \dots, \boldsymbol{x}'_{r+2}\} \subset \Delta \subset K.$$

Similarly, the inequalities

$$\|\boldsymbol{x} - \boldsymbol{x}'_i\| \leqslant \frac{\rho}{\delta}\|\boldsymbol{x} - \boldsymbol{x}_i\| \leqslant \rho, \quad 1 \leqslant i \leqslant r+2,$$

imply that $\Delta' \subset B_\rho(\boldsymbol{x})$. Summing up, $\Delta' \subset K \cap B_\rho(\boldsymbol{x})$. Consequently, $\boldsymbol{x} \notin \operatorname{ext}_r (K \cap B_\rho(\boldsymbol{x}))$, contrary to the assumption on $\boldsymbol{x}$.

(e) This part follows from assertion (b), Theorem 11.9, and the equalities

$$\dim(\operatorname{aff} F_{\boldsymbol{x}}(K)) = \dim F_{\boldsymbol{x}}(K) \quad \text{and} \quad \operatorname{aff} F_{\boldsymbol{x}}(K) = \operatorname{ap} C_{\boldsymbol{x}}(K). \qquad \square$$

The following corollary is a consequence of Theorem 11.5.

Corollary 11.41. *Let a nonempty convex set $K \subset \mathbb{R}^n$ of dimension m be expressed as a direct sum $K = (K \cap S) + \operatorname{lin} K$, where S is a subspace complementary to $\operatorname{lin} K$. If $\dim(\operatorname{lin} K) = p$, then $\operatorname{ext}_r K = \varnothing$ for all $0 \leqslant r \leqslant p-1$, and*

$$\operatorname{ext}_{p+q} K = \operatorname{ext}_q (K \cap S) + \operatorname{lin} K, \quad 0 \leqslant q \leqslant m-p. \qquad \square$$

The following lemma is used in the proof of Theorem 11.43.

Lemma 11.42. *Let $K \subset \mathbb{R}^n$ be a convex set of dimension $m \geqslant 2$ and F be an $(m-1)$-dimensional convex subset of* $\operatorname{rbd} K$. *For any point $\boldsymbol{z} \in \operatorname{rint} F$, there is an open ball $U_\rho(\boldsymbol{z}) \subset \mathbb{R}^n$ such that $\operatorname{rbd} K \cap U_\rho(\boldsymbol{z}) \subset \operatorname{rint} F$. Consequently, the set $\operatorname{rbd} K \setminus \operatorname{rint} F$ is closed.*

Proof. By Theorem 3.67, $\operatorname{aff} F \cap \operatorname{cl} K \subset \operatorname{rbd} K$; so, $\operatorname{aff} F \cap \operatorname{rint} K = \varnothing$. Choose a point $\boldsymbol{a} \in \operatorname{rint} K$. Then $\boldsymbol{a} \notin \operatorname{aff} F$, and the set $\{\boldsymbol{a}\} \cup F$ is m-dimensional (see Corollary 2.67). The inclusion $\{\boldsymbol{a}\} \cup F \subset K$ gives $\operatorname{aff}(\{\boldsymbol{a}\} \cup F) \subset \operatorname{aff} K$, and Theorem 2.6 shows that $\operatorname{aff}(\{\boldsymbol{a}\} \cup F) = \operatorname{aff} K$. Furthermore, Theorem 5.38 implies the equality $\operatorname{aff}(\operatorname{cone}_{\boldsymbol{a}} F) = \operatorname{aff} K$.

Let $\boldsymbol{z} \in \operatorname{rint} F$. By Theorem 5.49, $\boldsymbol{z} \in \operatorname{rint}(\operatorname{cone}_{\boldsymbol{a}} F)$, and Theorem 3.17 shows the existence of a scalar $\rho > 0$ satisfying the condition

$$\operatorname{aff} K \cap U_\rho(\boldsymbol{z}) = \operatorname{aff}(\operatorname{cone}_{\boldsymbol{a}} F) \cap U_\rho(\boldsymbol{z}) \subset \operatorname{rint}(\operatorname{cone}_{\boldsymbol{a}} F). \tag{11.4}$$

Finally, let $\boldsymbol{x} \in \operatorname{rbd} K \cap U_\rho(\boldsymbol{z})$. By the above argument,

$$\boldsymbol{x} \in \operatorname{aff} K \cap U_\rho(\boldsymbol{z}) \subset \operatorname{rint}(\operatorname{cone}_{\boldsymbol{a}} F) = \operatorname{cone}_{\boldsymbol{a}}(\operatorname{rint} F) \setminus \{\boldsymbol{a}\}.$$

Hence there is a point $\boldsymbol{u} \in \operatorname{rint} F$ such that $\boldsymbol{x} \in (\boldsymbol{a}, \boldsymbol{u}\rangle$ (see Theorem 5.38). According to Theorem 3.65, the halfline $(\boldsymbol{a}, \boldsymbol{u}\rangle$ meets $\operatorname{rbd} K$ at a unique point. Since both $\boldsymbol{x}$ and $\boldsymbol{u}$ belong to $(\boldsymbol{a}, \boldsymbol{u}\rangle \cap \operatorname{rbd} K$, we conclude that $\boldsymbol{x} = \boldsymbol{u} \in \operatorname{rint} F$. ☐

Theorem 11.43. *For a convex set $K \subset \mathbb{R}^n$ of dimension $m \geqslant 2$, one has*

$$\begin{aligned} \operatorname{ext}_{m-1} K &= K \cap \operatorname{cl}(\operatorname{ext}_{m-1} K), \\ \operatorname{ext}_{m-2} K &= K \cap \operatorname{cl}(\operatorname{ext}_{m-2} K). \end{aligned}$$

Consequently, both sets $\operatorname{ext}_{m-1} K$ and $\operatorname{ext}_{m-2} K$ are closed provided K is closed.

Proof. By Theorem 11.39, one has $\operatorname{ext}_{m-1} K = K \cap \operatorname{rbd} K \subset \operatorname{rbd} K$. Because the set $\operatorname{rbd} K$ is closed (see Theorem 3.62), we have

$$\begin{aligned} \operatorname{ext}_{m-1} K \subset K \cap \operatorname{cl}(\operatorname{ext}_{m-1} K) &\subset K \cap \operatorname{cl}(\operatorname{rbd} K) \\ &= K \cap \operatorname{rbd} K = \operatorname{ext}_{m-1} K. \end{aligned}$$

Hence $\operatorname{ext}_{m-1} K = K \cap \operatorname{cl}(\operatorname{ext}_{m-1} K)$.

Because the inclusion $\operatorname{ext}_{m-2} K \subset K \cap \operatorname{cl}(\operatorname{ext}_{m-2} K)$ is obvious, it suffices to prove the opposite one. Assume for a moment the existence of a point

$$\boldsymbol{x} \in K \cap \operatorname{cl}(\operatorname{ext}_{m-2} K) \setminus \operatorname{ext}_{m-2} K.$$

Then $\dim F_{\boldsymbol{x}}(K) \geqslant m-1$ according to Theorem 11.39. On the other hand, the inclusion $\boldsymbol{x} \in \mathrm{cl}\,(\mathrm{ext}_{m-2}K) \subset \mathrm{rbd}\,K$ and a combination of Theorems 11.4 and 11.9 shows that $F_{\boldsymbol{x}}(K)$ is a convex subsets of $\mathrm{rbd}\,K$ and $\dim F_{\boldsymbol{x}}(K) \leqslant m-1$. So $\dim F_{\boldsymbol{x}}(K) = m-1$. By Lemma 11.42, there is an open ball $U_\rho(\boldsymbol{x})$ such that $\mathrm{rbd}\,K \cap U_\rho(\boldsymbol{x}) \subset \mathrm{rint}\,F_{\boldsymbol{x}}(K)$. From here we obtain that $\boldsymbol{x}$ cannot belong to $\mathrm{cl}\,(\mathrm{ext}_{m-2}K)$, a contradiction. □

Remark. Theorem 11.43 does not hold for $\mathrm{ext}_r K$ if $r \leqslant m-3$. For instance, an example on page 393 describes a compact convex set $K \subset \mathbb{R}^3$ with nonclosed set $\mathrm{ext}_0 K$.

The next result is related to Theorems 11.28 and 11.29 and Corollary 11.30.

Theorem 11.44. *For a nonempty closed convex set $K \subset \mathbb{R}^n$ of dimension m and an integer $0 \leqslant r \leqslant m-1$, the following assertions hold.*

(*a*) *If* $\dim(\mathrm{lin}\,K) = r$, *then* $K = \mathrm{conv}\,(\mathrm{ext}_{r+1}K)$.
(*b*) *If* $\dim(\mathrm{lin}\,K) = r$ *and K has no extreme halfplane of dimension greater than r, then* $K = \mathrm{conv}\,(\mathrm{ext}_r K)$.

Proof. Let $K = (K \cap S) + \mathrm{lin}\,K$, where S is a subspace complementary to $\mathrm{lin}\,K$ and $K \cap S$ is a line-free closed convex set (see Theorem 6.25).

(*a*) By Theorem 11.28,

$$K \cap S = \mathrm{conv}\,(\mathrm{ext}\,(K \cap S) \cup \mathrm{extr}\,(K \cap S)).$$

By the definitions,

$$\mathrm{ext}\,(K \cap S) \cup \mathrm{extr}\,(K \cap S) \subset \mathrm{ext}_1(K \cap S).$$

Therefore, Corollary 11.41 shows that

$$K = (K \cap S) + \mathrm{lin}\,K \subset \mathrm{conv}\,(\mathrm{ext}_1(K \cap S)) + \mathrm{lin}\,K = \mathrm{conv}\,(\mathrm{ext}_{r+1}K).$$

(*b*) In the above notation, $K \cap S$ has no extreme halflines, and Corollary 11.30 gives

$$\begin{aligned} K &= (K \cap S) + \mathrm{lin}\,K \subset \mathrm{conv}\,(\mathrm{ext}_0(K \cap S)) + \mathrm{lin}\,K \\ &= \mathrm{conv}\,(\mathrm{ext}_r K). \end{aligned}$$ □

The next result is related to Theorem 11.31.

Theorem 11.45. *Let $K \subset \mathbb{R}^n$ be a closed convex set of dimension m, with* $\dim(\mathrm{lin}\,K) = r$. *Then every point $\boldsymbol{u} \in K$ is a positive convex combination of an affinely independent subset of*

(*a*) *$m - r + 1$ or fewer points from* $\operatorname{ext}_r K$*, or*
(*b*) *$m - r$ or fewer points from* $\operatorname{ext}_{r+1} K$.

Proof. Let $K = (K \cap S) + \operatorname{lin} K$, where S is a subspace complementary to $\operatorname{lin} K$ and $K \cap S$ is a line-free closed convex set of dimension $m - r$ (see Theorem 6.25). Denote by $\boldsymbol{z}$ the projection of $\boldsymbol{u}$ on S along $\operatorname{lin} K$. Then $\boldsymbol{z} \in K \cap S$, and Theorem 11.31 shows that $\boldsymbol{z}$ is a positive convex combination of an affinely independent subset X of

(a') $m - r + 1$ or fewer points from $\operatorname{ext}_0(K \cap S)$, or
(b') $m - r$ or fewer points from $\operatorname{ext}_1(K \cap S)$.

Let $Y = (\boldsymbol{u} - \boldsymbol{z}) + X$. Clearly, the set Y is affinely independent and $\operatorname{card} X = \operatorname{card} Y$. Since $\boldsymbol{u} - \boldsymbol{z} \in \operatorname{lin} K$, one has

$$Y = (\boldsymbol{u} - \boldsymbol{z}) + X \subset \operatorname{lin} K + (K \cap S) = K.$$

Furthermore,

$$\boldsymbol{u} = (\boldsymbol{u} - \boldsymbol{z}) + \boldsymbol{z} \in (\boldsymbol{u} - \boldsymbol{z}) + \operatorname{conv} X = \operatorname{conv}\left((\boldsymbol{u} - \boldsymbol{z}) + X\right) = \operatorname{conv} Y.$$

Finally, Corollary 11.41 implies that

$$Y \subset \begin{cases} \operatorname{ext}_r K & \text{if } X \subset \operatorname{ext}_0(K \cap S), \\ \operatorname{ext}_{r+1} K & \text{if } X \subset \operatorname{ext}_1(K \cap S). \end{cases}$$ □

Problems and Notes for Chapter 11

Problems for Chapter 11

Problem 11.1. Prove that the extreme faces of a closed ball $B_\rho(\boldsymbol{a}) \subset \mathbb{R}^n$ are $\varnothing$, all singletons from the boundary sphere $S_\rho(\boldsymbol{a})$, and $B_\rho(\boldsymbol{a})$.

Problem 11.2. Let $\Delta = \Delta(\boldsymbol{x}_1, \ldots, \boldsymbol{x}_{r+1})$ be an r-simplex in $\mathbb{R}^n$. Prove that the extreme faces of Δ are $\varnothing$ and all simplices $\Delta(\boldsymbol{z}_1, \ldots, \boldsymbol{z}_t)$, where $\{\boldsymbol{z}_1, \ldots, \boldsymbol{z}_t\}$ is a nonempty subset of $\{\boldsymbol{x}_1, \ldots, \boldsymbol{x}_{r+1}\}$.

Problem 11.3. Let $\Gamma = \Gamma_{\boldsymbol{a}}(\boldsymbol{x}_1, \ldots, \boldsymbol{x}_r)$ be an r-simplicial cone in $\mathbb{R}^n$. Prove that the extreme faces of Γ are $\varnothing$, $\{\boldsymbol{a}\}$, and all simplicial cones of the form $\Gamma_{\boldsymbol{a}}(\boldsymbol{z}_1, \ldots, \boldsymbol{z}_t)$, where $\{\boldsymbol{z}_1, \ldots, \boldsymbol{z}_t\}$ is a nonempty subset of $\{\boldsymbol{x}_1, \ldots, \boldsymbol{x}_r\}$.

Problem 11.4. Given a plane $L \subset \mathbb{R}^n$ of positive dimension and a closed half-plane D of L, prove the following assertions.

(*a*) The extreme faces of L are $\varnothing$ and L.

(*b*) The extreme faces of D are $\varnothing$, $\operatorname{rbd} D$, and D.

Problem 11.5. Let $K \subset \mathbb{R}^n$ be a closed convex set. Prove the following assertions.

(*a*) K has exactly two extreme faces if and only if K is a plane.

(*b*) K has exactly three extreme faces if and only if K is a closed halfplane.

(*c*) K has exactly four extreme faces if and only if K is a plane slab (see Definition 6.2.5).

Problem 11.6. (Roshchina, Sang, Yost [244]) Let $0 = d_0 < d_1 < \cdots < d_p = m$, where $m \leqslant n$, be an increasing sequence of integers. Prove the existence of a compact convex set $K \subset \mathbb{R}^n$ of dimension m such that the dimensions of extreme faces of K are given precisely by the numbers $d_0, d_1, \ldots, d_p$.

Problem 11.7. Prove that the set $\mathrm{ext}\, K$ of all extreme points of a closed convex set $K \subset \mathbb{R}^2$ is closed.

Problem 11.8. Let $K \subset \mathbb{R}^n$ be a nonempty closed convex set and $\boldsymbol{u}$ be a point in K. Prove the equivalence of the following conditions: (*a*) $\boldsymbol{u} \in \mathrm{ext}\, K$, (*b*) if midpoints of closed segments $[\boldsymbol{y}_i, \boldsymbol{z}_i] \subset K$, $i \geqslant 1$, converge to $\boldsymbol{u}$, then at least one of the sequences $\boldsymbol{y}_1, \boldsymbol{y}_2, \ldots$ and $\boldsymbol{z}_1, \boldsymbol{z}_2, \ldots$ converges to $\boldsymbol{u}$.

Problem 11.9. Let $K \subset \mathbb{R}^n$ be a nonempty convex set. Prove that a convex subset F of K is an extreme face of K if and only if F is an extreme face of $K \cap B_\rho(F)$, where $B_\rho(F)$ is the closed ρ-neighborhood of F.

Problem 11.10. Let $K \subset \mathbb{R}^n$ be a nonempty line-free closed convex set which is locally conic at a point $\boldsymbol{a} \in K$ (see Definition 7.5). Prove the existence of a scalar $\delta > 0$ such that $B_\delta(\boldsymbol{a}) \cap (\mathrm{ext}\, K \setminus \{\boldsymbol{a}\}) = \varnothing$.

Problem 11.11. Given a nonempty set $X \subset \mathbb{R}^n$ and a point $\boldsymbol{x} \in \mathrm{conv}\, X$, prove the equivalence of the following conditions.

(*a*) $\boldsymbol{x}$ is an extreme points of $\mathrm{conv}\, X$.

(*b*) $\boldsymbol{x} \in X$ and $\boldsymbol{x} \notin \mathrm{conv}\,(X \setminus \{\boldsymbol{x}\})$.

Problem 11.12. (Martínez-Legaz and Pintea [207]). Let a closed convex set $K \subset \mathbb{R}^n$ have a minimal subset X of K with the property $K = \mathrm{conv}\, X$. Prove that $X = \mathrm{ext}\, K$.

Problem 11.13. (Björck [36], Klee [171]) Let $X \subset \mathbb{R}^n$ be a nonempty set. Prove that $X = \mathrm{ext}\,(\mathrm{conv}\, X)$ if and only if the following property holds:

(*a*) $X \cap \mathrm{conv}\, Y \subset Y$ for every subset Y of X.

Suppose, additionally, that $\mathrm{conv}\, X$ is compact. Then $X = \mathrm{ext}\,(\mathrm{conv}\, X)$ if and only if conditions (*a*), (*b*), and (*c*) hold, where

(*b*) $\mathrm{cl}\, X \subset \mathrm{conv}\, X$,

(*c*) $\mathrm{cl}\, X$ is compact.

Problem 11.14. Let $K \subset \mathbb{R}^n$ be a nonempty line-free closed convex set. Prove that

$$K = \cap\,(C_{\boldsymbol{a}}(K) : \boldsymbol{a} \in \operatorname{ext} K), \quad \operatorname{rec} K = \cap\,(C_{\boldsymbol{a}}(K) - \boldsymbol{a} : \boldsymbol{a} \in \operatorname{ext} K).$$

Problem 11.15. Let $K \subset \mathbb{R}^n$ be a nonempty convex set. Prove the equivalence of the following conditions.

(*a*) For points $\boldsymbol{x} \in K$ and $\boldsymbol{y} \in \operatorname{cl} K$, the open segment $(\boldsymbol{x}, \boldsymbol{y})$ lies in K.

(*b*) For any extreme face F of $\operatorname{cl} K$, either $F \cap K = \varnothing$ or $\operatorname{rint} F \subset K$.

Notes for Chapter 11

Extreme points and faces. The concept of extreme point of a convex set is due to Minkowski [215, p. 159]. Extreme faces of convex sets were defined (and called *poonems*) in an equivalent form by Grünbaum [134, Section 2.4]. Comprehensive surveys on various properties of extreme points and extreme faces can be found in the works of Schneider [248, 249]. A general study of extreme faces of convex sets is made in Bair and Fourneau [20, Chapter 2].

Various parts of Theorems 11.2, 11.3, and 11.4 can be found in Grünbaum [134, Section 2.4]. Theorem 11.5 implies the assertion of Lommatzsch [204]: the minimum dimension of an extreme face of a closed convex set $K \subset \mathbb{R}^n$ equals the dimension of $\operatorname{lin} K$. Theorems 11.9, 11.10, 11.12 and Corollary 11.11 belong to Dubins [97] (see also Bourbaki [45, p. TVS II.87]). Theorem 11.19 can be found in Schneider [249, Lemma 1.4.6]. The second part of Problem 11.14 is proved by Bair [16].

Batson [29] proved that the set $\operatorname{ext} K$ of extreme points of an unbounded line-free closed convex set $K \subset \mathbb{R}^n$ is bounded provided the following two conditions are satisfied: (*a*) the set $\operatorname{text} K$ of all endpoints of extreme rays of K is bounded, (*b*) $\operatorname{conv}(\operatorname{ext} K) \subset \operatorname{cl}(\operatorname{conv}(\operatorname{text} K)) + (h - h)$ for every extreme halfline h of K.

Lattice properties of the family of extreme faces of a line-free closed convex cone in $\mathbb{R}^n$ are studied by Barker [27] and Lowey and Tam [203].

Extreme structure of the sum of convex sets. Theorem 11.13 is proved by Roy [245] for the case of compact convex sets. Theorem 11.17 goes back to Minkowski [215, p. 181], where it is proved for the case of compact convex sets in $\mathbb{R}^3$, with a simplified proof and a generalization to $\mathbb{R}^n$ by Fujiwara [113] (see also Bair, Fourneau, and Jongmans [21], Dalla [81], Husain and Tweddle [156]). Given convex sets K_1 and K_2 in $\mathbb{R}^n$, Bair [13, 14] formulates in terms of generated cones some conditions for the sum of points $\boldsymbol{x}_1 \in \operatorname{ext} K_1$ and $\boldsymbol{x}_2 \in \operatorname{ext} K_2$ be an extreme point of $K_1 + K_2$.

Husain and Tweddle [156] (see also Dalla [81]) proved that if K_1 and K_2 are nonempty compact convex sets in $\mathbb{R}^n$ then, for any point $\boldsymbol{x} \in \operatorname{ext} K_1$, there is a point $\boldsymbol{y} \in \operatorname{ext} K_2$ such that $\boldsymbol{x}+\boldsymbol{y} \in \operatorname{ext}(K_1+K_2)$. Roy [245] showed that a similar result holds for the case of extreme faces of K_1 and K_2 (see also Jongmans [161]).

The following result is proved by Soltan [257]. Let a line-free closed convex set $A \subset \mathbb{R}^n$ be the sum of closed convex sets B and C. Then both sets $B_0 = B + \operatorname{rec} A$ and $C_0 = C + \operatorname{rec} A$ are closed and satisfy the following conditions:

(a) for every point $\boldsymbol{a} \in \operatorname{ext} A$ there are unique points $\boldsymbol{b} \in \operatorname{ext} B_0$ and $\boldsymbol{c} \in \operatorname{ext} C_0$ such that $\boldsymbol{a} = \boldsymbol{b} + \boldsymbol{c}$,

(b) the sets

$$\operatorname{ext}_C B = \{\boldsymbol{x} \in \operatorname{ext} B : \exists\, \boldsymbol{y} \in \operatorname{ext} C \text{ such that } \boldsymbol{x} + \boldsymbol{y} \in \operatorname{ext} A\},$$
$$\operatorname{ext}_B C = \{\boldsymbol{x} \in \operatorname{ext} C : \exists\, \boldsymbol{y} \in \operatorname{ext} B \text{ such that } \boldsymbol{x} + \boldsymbol{y} \in \operatorname{ext} A\}$$

are dense in $\operatorname{ext} B_0$ and $\operatorname{ext} C_0$, respectively,

(c) $\operatorname{ext}_C B = \operatorname{ext} B$ and $\operatorname{ext}_B C = \operatorname{ext} C$ if both B and C are compact.

Topological properties of extreme points and extreme faces. Theorem 11.20 is proved by Klee [175] (see also Phelps [231, p. 5]). A generalization of this assertion to the case of $\operatorname{ext}_r K$ belongs to Schneider [249, p. 66].

To illustrate the complexity of the structure of $\operatorname{ext} K$, Klee [175, Example 6.11] shows the existence of a convex body $K \subset \mathbb{R}^3$ for which both sets $\operatorname{ext} K$ and $\operatorname{bd} K \setminus \operatorname{ext} K$ are dense in $\operatorname{bd} K$. An example of a 4-dimensional closed convex set K, with both sets $\operatorname{ext} K$ and $\operatorname{extr} K$ nonclosed, is given in Klee [174].

Klee [175] posed (for $n = 3$) the problem to "find a useful and simple characterization of the class $\mathcal{G}$ of subsets X of the unit sphere $\mathbb{S} \subset \mathbb{R}^n$ such that there is a homeomorphism of X onto the boundary of a convex body $K \subset \mathbb{R}^n$ which maps X onto $\operatorname{ext} K$." Some results in this direction were obtained by Bronsteĭn [49–52] and Collier [76]. An interesting assertion of Collier [77] states that given a convex body $K \subset \mathbb{R}^3$, every component of $\operatorname{cl}(\operatorname{ext} K) \setminus \operatorname{ext} K$ is a subset of a 1-dimensional extreme face of K.

Convex hulls of extreme points and faces. Theorem 11.21 goes back to Minkowski (see [215], pp. 157–161), who proved, without using the concept of convex hull, that a convex body $K \subset \mathbb{R}^3$ is the smallest among all convex bodies containing $\operatorname{ext} K$. In its present form Theorem 11.21 is proved by Steinitz [267, p. 16] for all $n \geqslant 3$. The equivalence of conditions (a) and (b) in Lemma 11.27 is due to Steinitz [267, § 26], and the equivalence of conditions (b) and (c) can be found in Fenchel [105, Chapter II, § 7] (also see Klee [174]). The equivalence of assertions (a) and (b) in Theorem 11.28 and assertion (a) of Theorem 11.29 are proved by Klee [174]. The equivalence of assertions (a) and (c) in Theorem 11.28 is explicitly shown by Grünbaum [134, p. 25]. Theorem 11.31 and the first assertion of Theorem 11.36 are proved by Klee in [179] and [176, Proposition 2.5], respectively.

Goberna, Martínez-Legaz, and Todorov [122] (also see He and Sun [149]) established the following variation of part (c) from Theorem 11.29: A line-free closed convex set $K \subset \mathbb{R}^n$ contains the smallest subset X (not necessarily convex) such that $K = X + \operatorname{rec} K$; this set X is the Pareto optimum of K with respect to $-\text{rec}\, K$, described by the condition $X = \{\boldsymbol{x} \in K : K \cap (\boldsymbol{x} - \text{rec}\, K) = \{\boldsymbol{x}\}\}$. Problem 11.11 complements Corollary 18.3.1 from the book of Rockafellar [241].

Given a nonempty closed convex set $K \subset \mathbb{R}^n$, denote by $\ell(K)$ the maximum number of elements in a nested sequence of nonempty extreme faces of K. Clearly, $\ell(K) \leqslant \dim K + 1$. Ito and Lourenço [157] proved the following results.

(*a*) If $K \subset \mathbb{R}^n$ is a nonempty compact convex set, then any point $\boldsymbol{u} \in K$ can be written as a convex combination of $\ell(K)$ or fewer points from $\text{ext}\, K$.

(*b*) If $C \subset \mathbb{R}^n$ is a line-free closed convex cone of positive dimension, then any point $\boldsymbol{u} \in C$ can be written as a convex combination of $\ell(C) - 1$ or fewer points from $\text{extr}\, C$.

Zamfirescu [297] extended Theorems 11.21 and 11.28 to the case of nonclosed convex sets, as follows.

(*a*) Let $K \subset \mathbb{R}^n$ be a bounded convex set. Then $K = \text{conv}\,(\text{ext}\, K)$ if and only if for every face F of K the set $\text{ext}\, F$ is dense in $\text{ext}\,(\text{cl}\, F)$.

(*b*) Let $K \subset \mathbb{R}^n$ be a line-free convex set. Then $K = \text{conv}\,(\text{ext}\, K \cup \text{extr}\, K)$ provided for every face F of K, the set $\text{ext}\, F \cup \text{extr}\, F$ is dense in $\text{ext}\,(\text{cl}\, F) \cup \text{extr}\,(\text{cl}\, F)$.

Another analog of Theorems 11.21 is due to Phu [232], who showed that any bounded convex set $K \subset \mathbb{R}^n$ is the convex hull of its γ-extreme points (a point $\boldsymbol{x} \in K$ is called γ-extreme, where γ is a given positive scalar, if the condition $\boldsymbol{x} \in [\boldsymbol{y}, \boldsymbol{z}] \subset K$ implies $\|\boldsymbol{x} - \boldsymbol{y}\| < \gamma$ or $\|\boldsymbol{x} - \boldsymbol{z}\| < \gamma$).

Motzkin [221] announced without proof that given a convex polytope $P \subset \mathbb{R}^n$, a point $\boldsymbol{z} \in P$, and positive integers $n_1, \ldots, n_s$, with $n_1 + \cdots + n_s = n + 1$, there are nonempty extreme faces $F_1, \ldots, F_s$ of P such that

$$\boldsymbol{z} \in \text{conv}\,(F_1 \cup \cdots \cup F_s) \quad \text{and} \quad \dim F_i \leqslant n_i - 1 \quad \text{for all} \quad i = 1, \ldots, s.$$

This assertion was confirmed by Lawrence and Soltan [195]. (Danielyan, Movsisyan, and Tatalyan [82] independently attempted to prove this assertion for the case of a compact convex set $K \subset \mathbb{R}^n$.) The paper [195] also contains the following Carathéodory-type results concerning a family $\{K_1, \ldots, K_r\}$ of nonempty line-free closed convex sets in $\mathbb{R}^n$:

(*a*) For every point $\boldsymbol{z} \in K_1 + \cdots + K_r$, there are nonempty extreme faces F_i of K_i, $1 \leqslant i \leqslant r$, such that $\boldsymbol{z} \in F_1 + \cdots + F_r$ and $\dim F_1 + \cdots + \dim F_r \leqslant n$.

(*b*) For every point $\boldsymbol{z} \in \text{conv}\,(K_1 \cup \cdots \cup K_r)$, there is an index set $I \subset \{1, \ldots, r\}$, with $|I| \leqslant n + 1$, and nonempty extreme faces F_i of K_i, $i \in I$, such that $\boldsymbol{z} \in \text{conv}\,(\cup F_i : i \in I)$ and $\sum (\dim F_i : i \in I) \leqslant n$. If, additionally, all $K_1, \ldots, K_r$ are compact, then $\sum (\dim F_i : i \in I) \leqslant n + 1 - |I|$.

Chapter 12

Exposed Structure of Convex Sets

12.1 Exposed Faces

12.1.1 *Basic Properties of Exposed Faces*

Definition 12.1. Let $K \subset \mathbb{R}^n$ be a convex set. An *exposed face* of K is a subset $G \subset K$ satisfying any of the following conditions:

(a) $G = H \cap K$ for a suitable hyperplane $H \subset \mathbb{R}^n$ supporting K,
(b) $G = \varnothing$ or $G = K$.

For instance, the plane closed convex set $K \subset \mathbb{R}^2$ depicted in Figure 12.1 has the following exposed faces: $\varnothing$, every singleton from the union of open circular arcs γ_1 and γ_2, the closed segments $[\boldsymbol{a}_1, \boldsymbol{a}_2]$ and $[\boldsymbol{c}_1, \boldsymbol{c}_2]$, and K. We observe that none of the singletons $\{\boldsymbol{a}_1\}, \{\boldsymbol{a}_2\}, \{\boldsymbol{c}_1\}$, and $\{\boldsymbol{c}_2\}$ is an exposed face of K (for instance, the only line supporting K at $\boldsymbol{a}_1$ is the horizontal line, which meets K at $[\boldsymbol{a}_1, \boldsymbol{a}_2]$).

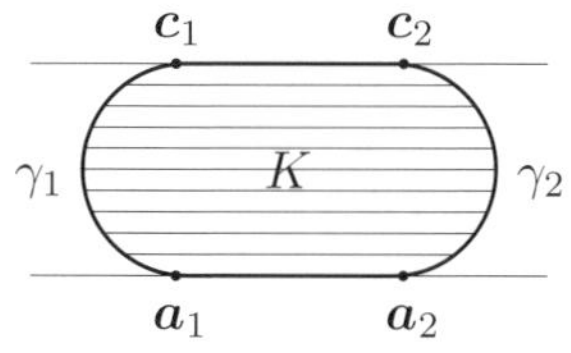

Fig. 12.1 Exposed faces of a plane closed convex set K.

Examples. 1. The exposed faces of a closed ball $B_\rho(\boldsymbol{a}) \subset \mathbb{R}^n$ are $\varnothing$, all singletons from its boundary sphere $S_\rho(\boldsymbol{a})$, and $B_\rho(\boldsymbol{a})$ (see Problem 12.1).

2. The exposed faces of an r-simplex $\Delta = \Delta(\boldsymbol{x}_1, \ldots, \boldsymbol{x}_{r+1}) \subset \mathbb{R}^n$ are $\varnothing$ and all simplices $\Delta(\boldsymbol{z}_1, \ldots, \boldsymbol{z}_t)$, where $\{\boldsymbol{z}_1, \ldots, \boldsymbol{z}_t\}$ is a nonempty subset of $\{\boldsymbol{x}_1, \ldots, \boldsymbol{x}_{r+1}\}$ (see Problem 12.2).

3. The exposed faces of a simplicial cone $\Gamma_{\boldsymbol{a}}(\boldsymbol{x}_1, \ldots, \boldsymbol{x}_r) \subset \mathbb{R}^n$ are $\varnothing$, $\{\boldsymbol{a}\}$, and all simplicial cones $\Gamma_{\boldsymbol{a}}(\boldsymbol{z}_1, \ldots, \boldsymbol{z}_t)$, where $\{\boldsymbol{z}_1, \ldots, \boldsymbol{z}_t\}$ is a nonempty subset of $\{\boldsymbol{x}_1, \ldots, \boldsymbol{x}_r\}$ (see Problem 12.3).

4. The exposed faces of a plane $L \subset \mathbb{R}^n$ are $\varnothing$ and L. If $\dim L = m > 0$ and D is a closed halfplane of L, then $\varnothing$, $\operatorname{rbd} D$, and D are the exposed faces of D (see Problem 12.5). If G is a closed slab of L, then the exposed faces of G are $\varnothing$, two $(m-1)$-dimensional planes determining G, and G (see Problem 12.6).

Remark. Every exposed face G of a convex set $K \subset \mathbb{R}^n$ is a convex set. Indeed, this is obvious if $G = \varnothing$ or $G = K$. If G is a nonempty proper subset of K, then, according to Definition 12.1, $G = H \cap K$ for a suitable hyperplane $H \subset \mathbb{R}^n$ supporting K. Consequently, the set G is convex as the intersection of convex sets H and K.

The next two theorems describe relations between extreme and exposed faces of a convex set.

Theorem 12.2. *Every exposed face of a convex set $K \subset \mathbb{R}^n$ is an extreme face of K.*

Proof. Let G be an exposed face of K. Excluding the obvious cases $G = \varnothing$ and $G = K$, we assume that G is a nonempty proper subset of K. Then there is a hyperplane $H \subset \mathbb{R}^n$ supporting K such that $G = H \cap K$. Denote by V a closed halfspace determined by H and containing K. Without loss of generality, we can write $V = \{\boldsymbol{x} \in \mathbb{R}^n : \boldsymbol{x} \cdot \boldsymbol{e} \leqslant \gamma\}$ for a nonzero vector $\boldsymbol{e} \in \mathbb{R}^n$ and a scalar $\gamma \in \mathbb{R}$ (see Definition 2.29).

Suppose that points $\boldsymbol{x}, \boldsymbol{y} \in K$ and a scalar $0 < \lambda < 1$ satisfy the condition $\boldsymbol{z} = (1-\lambda)\boldsymbol{x} + \lambda\boldsymbol{y} \in G$. Then $\boldsymbol{x}, \boldsymbol{y} \in V$ and $\boldsymbol{z} \in H$, giving

$$\gamma = \boldsymbol{z} \cdot \boldsymbol{e} = (1-\lambda)\boldsymbol{x} \cdot \boldsymbol{e} + \lambda\, \boldsymbol{y} \cdot \boldsymbol{e} \leqslant (1-\lambda)\gamma + \lambda\gamma = \gamma.$$

Therefore, $\boldsymbol{x} \cdot \boldsymbol{e} = \boldsymbol{y} \cdot \boldsymbol{c} = \gamma$, which gives the inclusion $\boldsymbol{x}, \boldsymbol{y} \in H \cap K = G$. Hence G is an extreme face of K. □

Figure 12.1 gives an example of a closed convex set $K \subset \mathbb{R}^n$ which has extreme faces which are not exposed. Theorem 12.2 can be partially converted, as shown below.

Theorem 12.3. *Let K be a convex set in $\mathbb{R}^n$. Every convex subset of K, which is maximal under inclusion in* $\operatorname{rbd} K$, *is an exposed face of K.*

Proof. Excluding the obvious case $\operatorname{rbd} K = \varnothing$, we may assume that $\operatorname{rbd} K \neq \varnothing$. Then K is not a plane (see Theorem 3.62). Let F be a convex subset of K which is maximal under inclusion in $\operatorname{rbd} K$. By Corollary 9.12, there is a hyperplane $H \subset \mathbb{R}^n$ properly supporting K and containing F, and Corollary 9.10 shows that $H \cap K \subset \operatorname{rbd} K$. Since the sets $H \cap K$ is convex, the inclusions $F \subset H \cap K \subset \operatorname{rbd} K$ and the assumption on F give the equality $F = H \cap K$. Consequently, F is an exposed face of K. □

Theorem 12.4. *If $K \subset \mathbb{R}^n$ is a convex set of positive dimension m, then every $(m-1)$-dimensional extreme face of K (if any) is its exposed face.*

Proof. Let F be an extreme face of K of dimension $m - 1$. Then $F \subset \operatorname{rbd} K$ according to Theorem 11.4. By Corollary 3.9, there is a maximal under inclusion convex set G satisfying the conditions $F \subset G \subset \operatorname{rbd} K$. Theorem 3.67 shows that $\dim G \leqslant m - 1$. On the other hand, $\dim G \geqslant \dim F = m - 1$ due to the inclusion $F \subset G$. So, $\dim G = m - 1$.

We assert that $F \cap \operatorname{rint} G \neq \varnothing$. Indeed, if $F \cap \operatorname{rint} G = \varnothing$, then $F \subset \operatorname{rbd} G$, and Theorem 3.67 would give $\dim F \leqslant \dim G - 1 = m - 2$.

Hence $F \cap \operatorname{rint} G \neq \varnothing$, implying the inclusion $G \subset F$ (see Theorem 11.4). Thus $F = G$, and Theorem 12.3 shows that F is an exposed face of K. □

The theorem below deals with set-theoretic properties of exposed faces of a given convex set.

Theorem 12.5. *For a convex set $K \subset \mathbb{R}^n$, the following assertions hold.*

(*a*) *The intersection of any family of exposed faces of K is an exposed face of K.*
(*b*) *If G is an exposed face of K and M is a convex subset of K, then $G \cap M$ is an exposed face of M. In particular, G is an exposed face of M provided $F \subset M$.*

Proof. (*a*) This part is a particular case of a more general assertion from Theorem 12.12.

(*b*) Since the cases $G = \varnothing$ and $G = K$ are obvious, we may assume that G is a nonempty proper exposed face of K. Similarly, we may suppose that $G \cap M \neq \varnothing$. Choose a hyperplane $H \subset \mathbb{R}^n$ which supports K such that $G = H \cap K$. Denote by V a closed halfspace determined by H and containing K. Then $\varnothing \neq G \cap M \subset H$ and $M \subset V$, which shows that H supports M. Furthermore,

$$H \cap M = H \cap (K \cap M) = (H \cap K) \cap M = G \cap M.$$

Hence $G \cap M$ is an exposed face of M. □

Remark. The property of being an exposed face is not hereditary (compare with assertion (*b*) of Theorem 11.3). Indeed, if K is a closed convex set depicted in Figure 12.1, then $\{\boldsymbol{a}_1\}$ is an exposed face of $[\boldsymbol{a}_1, \boldsymbol{a}_2]$, and $[\boldsymbol{a}_1, \boldsymbol{a}_2]$ is an exposed face of K, while $\{\boldsymbol{a}_1\}$ is not an exposed face of K.

Theorem 12.2 allows us to derive some properties of exposed faces from Theorem 11.4, as shown in the following corollary.

Corollary 12.6. *For a convex set $K \subset \mathbb{R}^n$, the following assertions hold.*

(*a*) *If G is an exposed faces of K and M is a convex subset of K such that $G \subset M$, then G is an exposed face of M.*
(*b*) *If G is an exposed face of K, then $G = K \cap \operatorname{cl} G$; in particular, G is closed provided K is closed.*
(*c*) *If G is an exposed face of K and M is a convex subset of K such that $G \cap \operatorname{rint} M \neq \varnothing$, then $M \subset G$.*
(*d*) *If G is an exposed face of K and M is a convex subset of G, then the set $(K \setminus G) \cup M$ is convex.*
(*e*) *If an exposed face G of K meets* $\operatorname{rint} K$, *then $G = K$.*
(*f*) *Every nonempty proper exposed face of K lies in* $\operatorname{rbd} K$ *and has dimension less than* $\dim K$.
(*g*) *Distinct exposed faces of K have disjoint relative interiors.*
(*h*) *If G is an exposed face of K and X is a subset of K, then*

$$\operatorname{conv}(G \cap X) = G \cap \operatorname{conv} X.$$ □

The following theorem gives a method to describe extreme faces of a convex sets in terms of exposed faces of its subsets.

Theorem 12.7. *Let $K \subset \mathbb{R}^n$ be a convex set and F be a nonempty proper extreme face of K. There is a sequence of convex sets*

$$F = G_r \subset G_{r-1} \subset \cdots \subset G_1 \subset G_0 = K$$

such that every G_i is a nonempty proper exposed face of G_{i-1}, $1 \leqslant i \leqslant r$.

Proof. Let $G_0 = K$. Theorem 11.4 shows that $F \subset G_0 \cap \operatorname{rbd} G_0$, and Corollary 9.12 implies the existence of a hyperplane H_1 which contains F and properly supports G_0. Put $G_1 = G_0 \cap H_1$. Then G_1 is a nonempty proper exposed face of G_0. Furthermore, $G_1 \subset \operatorname{rbd} G_0$ due to $H_1 \cap \operatorname{rint} K = \varnothing$. By Theorem 3.67, $\dim G_1 \leqslant \dim G_0 - 1$. Since $F \subset G_1$, Theorem 11.3 implies that F is an extreme face of G_1.

If $F = G_1$, the proof is done. If $F \neq G_1$, then $F \subset \operatorname{rbd} G_1$ by Theorem 11.4, and we can apply to G_1 and F the above argument. Repeating this procedure finitely many (no more than $\dim K - \dim F$) times, we obtain the desired conclusion. □

The next theorem shows that we can reduce the study of exposed faces of arbitrary convex sets to those with zero lineality.

Theorem 12.8. *Let a nonempty convex set $K \subset \mathbb{R}^n$ be expressed as $K = (K \cap S) + \operatorname{lin} K$, where S is a subspace complementary to* lin K. *A nonempty set $G \subset \mathbb{R}^n$ is an exposed face of K if and only if $G = E + \operatorname{lin} K$, where E is an exposed face of $K \cap S$.*

Proof. Since the case $\operatorname{lin} K = \{\boldsymbol{o}\}$ is obvious, we may assume that $\operatorname{lin} K$ is not $\{\boldsymbol{o}\}$.

Let G be an exposed face of K. Since the case $G = K$ is obvious, we may assume that $G \neq K$. Let $H \subset \mathbb{R}^n$ be a hyperplane such that $H \cap K = G$. By Theorem 12.2, G is an extreme face of K, and Theorem 11.5 implies that $G = E + \operatorname{lin} K$, where E is an extreme face of $K \cap S$. For every point $\boldsymbol{x} \in E$, one has $\boldsymbol{x} + \operatorname{lin} K \subset G \subset H$. Hence $H = (H \cap S) + \operatorname{lin} K$, and

$$\begin{aligned} H \cap K &= ((H \cap S) + \operatorname{lin} K) \cap ((K \cap S) + \operatorname{lin} K) \\ &= (H \cap K \cap S) + \operatorname{lin} K. \end{aligned}$$

Comparing this expression with that of $H \cap K = G = E + \operatorname{lin} K$, we obtain $H \cap (K \cap S) = E$. Furthermore, if V is a closed halfspace determined by H and containing K, then, V also contains $K \cap S$. Hence H supports $K \cap S$, implying that E is an exposed face of $K \cap S$.

Conversely, let E be an exposed face of $K \cap S$. If $E = K \cap S$, then $G = E + \operatorname{lin} K = K$. Suppose that $E \neq K \cap S$ and choose a hyperplane $H \subset \mathbb{R}^n$ supporting $K \cap S$ such that $E = H \cap (K \cap S)$. Since S does not lie in H (otherwise $E = K \cap S$), the plane $H \cap S$ has dimension $\dim S - 1$ (see Theorem 2.19). If V is a closed halfspace determined by H and containing K, then $V \cap S$ is a closed halfplane of S determined by $H \cap S$ and containing $K \cap S$. Put

$$G = E + \operatorname{lin} K \quad \text{and} \quad H' = (H \cap S) + \operatorname{lin} K.$$

Clearly, H' is a hyperplane with the property $H' \cap G = K$ such that the closed halfspace $V' = (V \cap S) + \operatorname{lin} K$ contains K. Hence H' properly supports K, and G is a nonempty proper exposed face of K. $\square$

12.1.2 *Generated Exposed Faces*

Theorem 12.5 gives a base to the following definition.

Definition 12.9. Let $K \subset \mathbb{R}^n$ be a convex set and X be a subset of K. The intersection of all exposed faces of K containing X is called *the exposed face of K generated by X* and denoted $G_X(K)$.

The corollary below follows from the definitions. We recall that $F_X(K)$ denotes the extreme face of K generated by a set $X \subset K$ (see Definition 11.7).

Corollary 12.10. *For a convex set $K \subset \mathbb{R}^n$ and subsets X and Y of K, the following assertions hold.*

(*a*) $X \subset F_X(K) \subset G_X(K)$, *with* $X = G_X(K)$ *if and only if X is an exposed face of K. In particular,* $G_\varnothing(K) = \varnothing$ *and* $G_K(K) = K$.
(*b*) $G_X(K)$ *is the smallest exposed face of K which contains X.*
(*c*) $G_X(G_X(K)) = G_X(K)$.
(*d*) $G_Y(K) = G_X(K)$ *if* $Y = \operatorname{conv} X$.
(*e*) $G_X(K) \subset G_Z(K)$ *if* $X \subset Z$. $\square$

The next result deals with exposed faces generated by points.

Theorem 12.11. *For a nonempty convex set $K \subset \mathbb{R}^n$ and a point $\boldsymbol{u} \in K$, the following assertions hold.*

(*a*) *If M is a convex subset of K and* $\boldsymbol{u} \in \operatorname{rint} M$, *then* $M \subset G_{\boldsymbol{u}}(K)$.
(*b*) *If G is an exposed face of K and* $\boldsymbol{u} \in \operatorname{rint} G$, *then* $G = F_{\boldsymbol{u}}(K) = G_{\boldsymbol{u}}(K)$.
(*c*) $G_{\boldsymbol{u}}(K) = K$ *if and only if* $\boldsymbol{u} \in \operatorname{rint} K$.
(*d*) $G_{\boldsymbol{u}}(K) \subset \operatorname{rbd} K$ *if and only if* $\boldsymbol{u} \in \operatorname{rbd} K$.

Proof. (*a*) A combination of Theorem 11.9 and Corollary 12.10 gives the inclusions

$$M \subset F_{\boldsymbol{u}}(K) \subset G_{\boldsymbol{u}}(K).$$

(*b*) Theorem 11.4 gives $G \subset F_{\boldsymbol{u}}(K)$. One has $G = F_{\boldsymbol{u}}(K) = G_{\boldsymbol{u}}(K)$ due to the inclusions

$$F_{\boldsymbol{u}}(K) \subset G_{\boldsymbol{u}}(K) \subset G.$$

(c) If $\boldsymbol{u} \in \operatorname{rint} K$, then $G_{\boldsymbol{u}}(K) = K$ by assertion (b). Conversely, let $G_{\boldsymbol{u}}(K) = K$. Assume, for contradiction, that $\boldsymbol{u} \in \operatorname{rbd} K$. Corollary 9.11 shows the existence of a hyperplane $H \subset \mathbb{R}^n$ which contains $\boldsymbol{u}$ and properly supports K. Then $H \cap \operatorname{rint} K = \varnothing$, implying that the exposed face $H \cap K$ of K lies in $\operatorname{rbd} K$. Consequently,

$$G_{\boldsymbol{u}}(K) \subset H \cap K \subset \operatorname{rbd} K \neq K,$$

contrary to the assumption. Hence $\boldsymbol{u} \in \operatorname{rint} K$.

(d) This part immediately follows from assertion (c). □

12.1.3 *Algebra of Exposed Faces*

The following theorem is similar to assertion (a) of Theorem 11.3.

Theorem 12.12. *Let $\mathcal{F} = \{K_\alpha\}$ be a family of convex sets in $\mathbb{R}^n$, with an exposed face G_α chosen in every $K_\alpha \in \mathcal{F}$. Then the set $G = \bigcap_\alpha G_\alpha$ is an exposed face of the convex set $K = \bigcap_\alpha K_\alpha$.*

Proof. Excluding the obvious cases, we may suppose that $\varnothing \neq G \neq K$ and $G_\alpha \neq K_\alpha$ for all $K_\alpha \in \mathcal{F}$. Denote by $\mathcal{H} = \{H_\alpha\}$ the family of hyperplanes in $\mathbb{R}^n$ such that $G_\alpha = H_\alpha \cap K_\alpha$ for all $K_\alpha \in \mathcal{F}$. Clearly, every hyperplane H_α properly supports K_α. Denote by V_α the closed halfspace determined by H_α and containing K_α. According to Theorem 2.4, the set $L = \bigcap_\alpha H_\alpha$ is a plane. Furthermore,

$$G = \bigcap_\alpha G_\alpha = \bigcap_\alpha (H_\alpha \cap K_\alpha) = (\bigcap_\alpha H_\alpha) \cap (\bigcap_\alpha K_\alpha) = L \cap K.$$

Put $m = \dim L$. Theorem 2.20 shows that $\mathcal{H}$ contains a subfamily $\{H_1, \dots, H_{n-m}\}$ of hyperplanes whose intersection is L. By Theorem 2.17, we can write

$$H_i = \{\boldsymbol{x} \in \mathbb{R}^n : \boldsymbol{x}\cdot\boldsymbol{e}_i = \gamma_i\}, \quad 1 \leqslant i \leqslant n - m.$$

Suppose that

$$V_i = \{\boldsymbol{x} \in \mathbb{R}^n : \boldsymbol{x}\cdot\boldsymbol{e}_i \leqslant \gamma_i\} \quad \text{for all} \quad 1 \leqslant i \leqslant n - m$$

(otherwise, replace $\boldsymbol{e}_i$ and γ_i with $-\boldsymbol{e}_i$ and $-\gamma_i$, respectively). Clearly, the closed convex set $P = V_1 \cap \dots \cap V_{n-m}$ contains K. Put

$$\boldsymbol{e} = \frac{1}{n-m}(\boldsymbol{e}_1 + \dots + \boldsymbol{e}_{n-m}) \quad \text{and} \quad \gamma = \frac{1}{n-m}(\gamma_1 + \dots + \gamma_{n-m}).$$

One has $\boldsymbol{e} \neq \boldsymbol{o}$ because the set $\{\boldsymbol{e}_1, \dots, \boldsymbol{e}_{n-m}\}$ is linearly independent (see Problem 2.7). Let

$$H = \{\boldsymbol{x} \in \mathbb{R}^n : \boldsymbol{x}\cdot\boldsymbol{e} = \gamma\} \quad \text{and} \quad V = \{\boldsymbol{x} \in \mathbb{R}^n : \boldsymbol{x}\cdot\boldsymbol{e} \leqslant \gamma\}.$$

We observe that $P \subset V$. Indeed, if $\boldsymbol{x} \in P$, then $\boldsymbol{x} \cdot \boldsymbol{e}_i \leqslant \gamma_i$ for all $1 \leqslant i \leqslant n-m$, which gives

$$\boldsymbol{x} \cdot \boldsymbol{e} = \frac{1}{n-m} \boldsymbol{x} \cdot (\boldsymbol{e}_1 + \cdots + \boldsymbol{e}_{n-m}) \leqslant \frac{1}{n-m} (\gamma_1 + \cdots + \gamma_{n-m}) = \gamma.$$

Since $G \subset L \subset H$ and $K \subset P \subset V$, the hyperplane H supports K.

Next, we assert that $H \cap P = L$. By the above argument, it suffices to show the inclusion $H \cap P \subset L$. Indeed, if $\boldsymbol{x} \in H \cap P$, then $\boldsymbol{x} \cdot \boldsymbol{e}_i \leqslant \gamma_i$ for all $1 \leqslant i \leqslant n-m$, and

$$\gamma = \boldsymbol{x} \cdot \boldsymbol{e} = \frac{1}{n-m} \boldsymbol{x} \cdot (\boldsymbol{c}_1 + \cdots + \boldsymbol{c}_{n-m}) \leqslant \frac{1}{n-m} (\gamma_1 + \cdots + \gamma_{n-m}) = \gamma.$$

Hence $\boldsymbol{x}_i \cdot \boldsymbol{e} = \gamma_i$ for all $1 \leqslant i \leqslant n-m$, which gives

$$\boldsymbol{x} \in H_1 \cap \cdots \cap H_{n-m} = L.$$

Finally, the equalities

$$G = L \cap K = (H \cap P) \cap K = H \cap (P \cap K) = H \cap K$$

imply that G is an exposed face of K. □

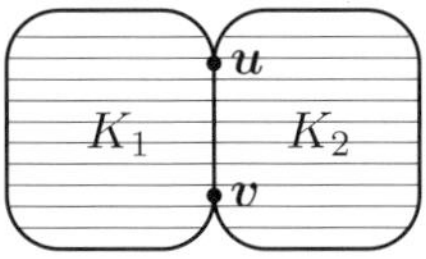

Fig. 12.2 Exposed faces of the intersection of convex sets.

Remark. Unlike extreme faces (see Theorem 11.12), an exposed face of the intersection of convex sets K_1 and K_2 in $\mathbb{R}^n$ may not be expressible as the intersection of exposed faces $G_1 \subset K_1$ and $G_2 \subset K_2$. For instance, let K_1 and K_2 be planar convex sets depicted in Figure 12.2. Then $\{\boldsymbol{u}\}$ is a 0-dimensional exposed face of $K_1 \cap K_2 = [\boldsymbol{u}, \boldsymbol{v}]$, while $[\boldsymbol{u}, \boldsymbol{v}]$ is the only exposed face of K_1 (and K_2) containing $\boldsymbol{u}$.

Theorem 12.13. *For convex sets $K_1, \ldots, K_r$ in $\mathbb{R}^n$ and nonzero scalars $\mu_1, \ldots, \mu_r$, let $K = \mu_1 K_1 + \cdots + \mu_r K_r$. The following assertions hold.*

(*a*) *Any exposed face G of K can be expressed as $G = \mu_1 G_1 + \cdots + \mu_r G_r$, where every G_i is an exposed face of K_i, $1 \leqslant i \leqslant r$.*

(*b*) *If, additionally, the exposed face G is nonempty and bounded, then the faces $G_1, \dots, G_r$ above are nonempty and bounded; also, they are uniquely determined by G.*

Proof. (*a*) Since the cases $K = \mathbb{R}^n$ and $G = \varnothing$ are obvious, we assume that $K \neq \mathbb{R}^n$ and $G \neq \varnothing$. Denote by H a hyperplane in $\mathbb{R}^n$ supporting K such that $H \cap K = G$. By Theorem 9.36, there are translates $H_1, \dots, H_r$ of H such that every H_i supports and meets K_i, and

$$G = H \cap K = \mu_1(H_1 \cap K_1) + \cdots + \mu_r(H_r \cap K_r).$$

Since every set $G_i = H_i \cap K_i$ is an exposed face of K_i, $1 \leqslant i \leqslant r$, we obtain a desired representation for G.

(*b*) Suppose that G is nonempty and bounded. According to Theorem 12.2, all the above exposed faces $G_1, \dots, G_r$ are extreme faces of $K, K_1, \dots, K_r$, respectively. Now, Theorem 11.13 implies that $G_1, \dots, G_r$ are nonempty bounded sets, uniquely determined by G. □

Remarks. 1. It might be that the sum $G_1 + G_2$ of exposed faces G_1 and G_2 of convex sets K_1 and K_2, respectively, is not an exposed face of $K_1 + K_2$. Indeed, if $K_1 = K_2 = B$, where $\mathbb{B}$ is the closed unit ball of $\mathbb{R}^n$, then for a boundary point $\boldsymbol{x}$ of $\mathbb{B}$, the sets $\{\boldsymbol{x}\}$ and $\{-\boldsymbol{x}\}$ are 0-dimensional exposed faces of $\mathbb{B}$ (see Problem 12.1), while $\{\boldsymbol{x}\} + \{-\boldsymbol{x}\} = \{\boldsymbol{o}\}$ is not an exposed face of $K_1 + K_2 = 2\mathbb{B}$.

2. The assumption on boundedness of the exposed face G in assertion (*b*) of Theorem 12.13 is essential. Indeed, let $K_1 = K_2 = h$, where $h \subset \mathbb{R}^n$ is a closed halfline originated at $\boldsymbol{o}$. Then the sum $K = K_1 + K_2 = h$, being its own trivial exposed face, can be expressed as the sum of exposed faces of K_1 and K_2 in two distinct ways: $K = K_1 + \{\boldsymbol{o}\} = K_1 + K_2$.

Theorem 12.14. *For an affine transformation $f : \mathbb{R}^n \to \mathbb{R}^m$ and convex sets $K \subset \mathbb{R}^n$ and $M \subset \mathbb{R}^m$, the following assertions hold.*

(*a*) *If G is an exposed face of M, then $f^{-1}(G)$ is an exposed face of $f^{-1}(M)$.*
(*b*) *If G is a convex subset of $M \cap \operatorname{rng} f$ such that $f^{-1}(G)$ is an exposed face of $f^{-1}(M)$, then G is an exposed face of $M \cap \operatorname{rng} f$.*
(*c*) *If G is an exposed face of $f(K)$, then $f^{-1}(G) \cap K$ is an exposed face of K.*
(*d*) *If f is one-to-one, then a set F is an exposed face of K if and only if $f(F)$ is an exposed face of $f(K)$.*

Proof. Without loss of generality, we will suppose that in all assertions (a)–(d) the transformation f is a linear transformation. Indeed, if $f(\boldsymbol{x}) = \boldsymbol{a} + g(\boldsymbol{x})$, where $\boldsymbol{a} \in \mathbb{R}^m$ and $g : \mathbb{R}^n \to \mathbb{R}^m$ is a linear transformation, then Theorem 2.77 shows that for any sets $X \subset \mathbb{R}^n$ and $Y \subset \mathbb{R}^m$ and a point $\boldsymbol{c} \in g^{-1}(\boldsymbol{a})$, one has

$$\operatorname{rng} f = \boldsymbol{a} + \operatorname{rng} g, \quad f(X) = \boldsymbol{a} + g(X), \quad f^{-1}(Y) = -\boldsymbol{c} + g^{-1}(Y).$$

(a) Excluding the obvious cases $G = \varnothing$ and $G = M$, we may assume that G is a nonempty proper subset of M. Then $G = H \cap M$ for a suitable hyperplane $H \subset \mathbb{R}^m$ properly supporting M. Denote by V the closed halfspace of $\mathbb{R}^m$ determined by H and containing M. Let

$$G' = G \cap \operatorname{rng} f, \quad H' = H \cap \operatorname{rng} f, \quad M' = M \cap \operatorname{rng} f, \quad V' = V \cap \operatorname{rng} f.$$

Since $\operatorname{rng} f$ is a plane, the set H' also is a plane. Clearly, $G' = H' \cap M'$. Excluding the obvious cases $G' = \varnothing$ and $G' = M'$ (which correspond to $f^{-1}(G) = \varnothing$ and $F^{-1}(G) = f^{-1}(M)$), we may suppose that G' is a nonempty proper subset of M'. Then $\operatorname{rng} f \not\subset H'$, since otherwise

$$\begin{aligned} G' &= G \cap \operatorname{rng} f = (H \cap M) \cap \operatorname{rng} f \\ &= M \cap (H \cap \operatorname{rng} f) = M \cap \operatorname{rng} f = M'. \end{aligned}$$

Therefore, V' is a closed halfplane of $\operatorname{rng} f$ determined by H' and containing M' (see Definition 2.69). Hence H' properly supports M'. According to Theorem 3.67, this argument gives $H' \cap \operatorname{rint} M' = \varnothing$. Due to

$$\begin{aligned} f^{-1}(G) &= f^{-1}(G') = f^{-1}(H' \cap M') \\ &= f^{-1}(H') \cap f^{-1}(M') = f^{-1}(H) \cap f^{-1}(M) \end{aligned}$$

it suffices to show that $f^{-1}(H')$ is a hyperplane in $\mathbb{R}^n$ properly supporting $f^{-1}(M')$. By Corollary 2.84, $f^{-1}(H')$ is a plane, and the obvious equality

$$f^{-1}(H') \cap \operatorname{rint} f^{-1}(M') = f^{-1}(H' \cap \operatorname{rint} M') = \varnothing.$$

shows that $f^{-1}(H')$ properly supports $f^{-1}(M')$.

Next, we assert that $f^{-1}(H')$ is a hyperplane of $\mathbb{R}^n$. Indeed, Theorem 2.19 shows that $\dim H' = \dim(\operatorname{rng} f) - 1$. Furthermore, the subspace $S = H' - H'$ lies in $\operatorname{rng} f$ and is a translate of H' (see Theorem 2.2). Hence $f^{-1}(S)$ is a translate of $f^{-1}(H')$. Therefore,

$$\begin{aligned} \dim f^{-1}(H') &= \dim f^{-1}(S) = \dim S + \dim(\operatorname{null} f) \\ &= \dim(\operatorname{rng} f) - 1 + \dim(\operatorname{null} f) = n - 1, \end{aligned}$$

which shows that $f^{-1}(H')$ is a hyperplane of $\mathbb{R}^n$.

(*b*) Excluding obvious cases, we may suppose that G and $f^{-1}(G)$ are nonempty proper subsets of $M \cap \operatorname{rng} f$ and $f^{-1}(M)$, respectively. Let $H \subset \mathbb{R}^n$ be a hyperplane properly supporting $f^{-1}(M)$ such that $f^{-1}(G) = H \cap f^{-1}(M)$. Because translates of $\operatorname{null} f$ lie in $f^{-1}(G)$ and $f^{-1}(M)$, respectively, the equality $f^{-1}(G) = H \cap f^{-1}(M)$ implies that H contains a translate of $\operatorname{null} f$. Choose a subspace $S \subset \mathbb{R}^n$ which is complementary to $\operatorname{null} f$. Then

$$f^{-1}(M) = M' + \operatorname{null} f, \quad f^{-1}(G) = G' + \operatorname{null} f, \quad H = H' + \operatorname{null} f,$$

where G' and M' are convex subset of S, and H' is a plane in S. Clearly,

$$G' = H' \cap M', \quad M = f(M'), \quad G = f(G'), \quad f(H) = f(H').$$

Since the linear transformation $h : S \to \operatorname{rng} f$ defined by $h(\boldsymbol{x}) = f(\boldsymbol{x})$ is invertible on S (see Problem 1.1), we have

$$G = f(G') = f(H' \cap M') = f(H') \cap f(M') = f(H') \cap M.$$

Furthermore,

$$\dim f(H') = \dim H' = \dim S - 1 = \dim(\operatorname{rng} f) - 1.$$

By Lemma 2.72, there is a hyperplane L of $\mathbb{R}^m$ with the property $L \cap \operatorname{rng} f = f(H')$. Clearly, L supports $M \cap \operatorname{rng} f$ such that $G = L \cap (M \cap \operatorname{rng} f)$. Summing up, G is a nonempty proper exposed face of $M \cap \operatorname{rng} f$.

(*c*) Excluding the obvious cases $G = \varnothing$ and $G = f(K)$, we may suppose that G is a nonempty proper subset of $f(K)$. Then there is a hyperplane L of $\mathbb{R}^m$ which properly supports $f(K)$ such that $G = L \cap f(K)$. Consequently, $L \cap \operatorname{rint} f(K) = \varnothing$. Since $K \subset f^{-1}(f(K))$, the equality

$$f^{-1}(G) = f^{-1}(L \cap f(K)) = f^{-1}(L) \cap f^{-1}(f(K))$$

gives

$$f^{-1}(G) \cap K = f^{-1}(L) \cap f^{-1}(f(K)) \cap K = f^{-1}(L) \cap K.$$

As shown in the proof of assertion (*a*) above, the hyperplane $f^{-1}(L) = f^{-1}(L \cap \operatorname{rng})$ supports $f^{-1}(K)$. Hence $f^{-1}(L)$ also supports $f^{-1}(G) \cap K$, implying that $f^{-1}(G) \cap K$ is an exposed face of K.

(*d*) Let $M = f(K)$ and $G = f(F)$. Then $K = f^{-1}(M)$ and $F = f^{-1}(G)$ since f is one-to-one. According to assertion (*a*) and (*c*), F is an exposed face of K if and only if G is an exposed face of M. □

Remark. An example on page 388 shows that the assumption on f to be one-to-one is essential in Theorem 12.14.

12.1.4 Cancelation Law

The next result is analogous to Theorem 3.60.

Theorem 12.15. *If K_1 and K_2 are convex sets and X is a nonempty compact set in $\mathbb{R}^n$ such that $K_1 + X = K_2 + X$, then $K_1 = K_2$.*

Proof. Since the case $K_1 = K_2 = \varnothing$ is obvious, we assume that both sets K_1 and K_2 are nonempty. By Theorem 4.16, the set $Y = \operatorname{conv} X$ is compact. Furthermore, Theorem 4.12 gives

$$\begin{aligned} K_1 + Y &= K_1 + \operatorname{conv} X = \operatorname{conv}(K_1 + X) \\ &= \operatorname{conv}(K_2 + X) = K_2 + \operatorname{conv} X = K_2 + Y. \end{aligned}$$

Consequently, Theorem 3.60 shows that $\operatorname{cl} K_1 = \operatorname{cl} K_2$.

We will prove the equality $K_1 = K_2$ by induction on $m = \dim(K_1 + Y)$. The case $m = 0$ is obvious: all three sets K_1, K_2 and Y are singletons.

Let $m = 1$. The 1-dimensional convex set $K_1 + Y$ may be a line, a closed halfline, or a closed segment. We consider each of these possibilities separately.

1. If $K_1 + Y$ is a line, l, then K_1 is a line l' parallel to l (because Y is bounded). In this case, the equality $\operatorname{cl} K_2 = \operatorname{cl} K_1 = l'$ gives $K_2 = l'$. So, $K_1 = K_2$.

2. Suppose that $K_1 + Y$ is a closed halfline h with endpoint $\boldsymbol{x}$. Then $\operatorname{cl} K_1 = \operatorname{cl} K_2$ is a closed halfline h', which is a translate of $\operatorname{cl} h$. Denote by $\boldsymbol{u}$ the endpoint of h' and express $\boldsymbol{x}$ as $\boldsymbol{x} = \boldsymbol{u} + \boldsymbol{v}$, where $\boldsymbol{v}$ is a point in $\operatorname{rbd} Y$. Clearly, the inclusion $\boldsymbol{x} \in h$ holds if and only if $\boldsymbol{u} \in K_1 \cap K_2$. Hence either $K_1 = h' = K_2$ or $K_1 = h' \setminus \{\boldsymbol{u}\} = K_2$.

3. Suppose that $K_1 + Y$ is a closed segment. Then $\operatorname{cl} K_1 = \operatorname{cl} K_2 = [\boldsymbol{u}, \boldsymbol{v}]$ (possibly, $\boldsymbol{u} = \boldsymbol{v}$). As above, each of the points $\boldsymbol{u}, \boldsymbol{v}$ either belongs to both K_1 and K_2, or misses both of them. This argument shows that $K_1 = K_2$.

Assume that the inductive assumption holds for all $m \leqslant r - 1$, where $r \geqslant 2$, and let $\dim(K_1 + Y) = r$. Since $\operatorname{cl} K_1 = \operatorname{cl} K_2$, Corollary 3.49 implies that

$$\operatorname{rint} K_1 = \operatorname{rint} K_2 \quad \text{and} \quad \operatorname{rbd} K_1 = \operatorname{rbd} K_2.$$

The representations

$$K_1 = \operatorname{rint} K_1 + (K_1 \cap \operatorname{rbd} K_1) \quad \text{and} \quad K_2 = \operatorname{rint} K_2 + (K_2 \cap \operatorname{rbd} K_2)$$

show that it suffices to prove the equality

$$K_1 \cap \operatorname{rbd} K_1 = K_2 \cap \operatorname{rbd} K_2.$$

Since the case

$$K_1 \cap \operatorname{rbd} K_1 = K_2 \cap \operatorname{rbd} K_2 = \varnothing$$

is obvious, we may assume that at least one of these sets, say $K_1 \cap \operatorname{rbd} K_1$, is nonempty. By Theorem 3.62, K_1 is not a plane, and the same theorem implies that K_2 also is not a plane.

For the inclusion

$$K_1 \cap \operatorname{rbd} K_1 \subset K_2 \cap \operatorname{rbd} K_2,$$

choose any point $\boldsymbol{u} \in K_1 \cap \operatorname{rbd} K_1$. By Corollary 9.32, there is a closed halfspace $V_1 \subset \mathbb{R}^n$ which properly supports K_1 at $\boldsymbol{u}$. Then

$$V_1 = \{\boldsymbol{x} \in \mathbb{R}^n : \boldsymbol{x}\cdot\boldsymbol{e} \leqslant \gamma_1\},$$

where $\boldsymbol{e}$ is a nonzero vector in $\operatorname{nor} V_1$ and $\gamma_1 = \sup\{\boldsymbol{x}\cdot\boldsymbol{e} : \boldsymbol{x} \in K_1\}$. Denote by V_2 the translate of V_1 which supports Y (V_2 exists due to the compactness of Y). Clearly,

$$V_2 = \{\boldsymbol{x} \in \mathbb{R}^n : \boldsymbol{x}\cdot\boldsymbol{e} \leqslant \gamma_2\},$$

where $\gamma_2 = \sup\{\boldsymbol{x}\cdot\boldsymbol{e} : \boldsymbol{x} \in Y\}$. By the compactness of Y, there is a point $\boldsymbol{v} \in Y$ such that $\boldsymbol{v}\cdot\boldsymbol{e} = \gamma_2$.

A combination of Theorems 8.32 and 8.34 gives

$$\boldsymbol{e} \in \operatorname{bar} K_1 = \operatorname{bar} K_1 \cap \mathbb{R}^n = \operatorname{bar} K_1 \cap \operatorname{bar} Y = \operatorname{bar}(K_1 + Y).$$

Therefore, Theorem 9.34 implies the existence of a translate V of V_1 which either supports $K_1 + Y$ or is asymptotic to $K_1 + Y$. As above,

$$V = \{\boldsymbol{x} \in \mathbb{R}^n : \boldsymbol{x}\cdot\boldsymbol{e} \leqslant \gamma\}, \quad \text{where} \quad \gamma = \sup\{\boldsymbol{x}\cdot\boldsymbol{e} : \boldsymbol{x} \in K_1 + Y\}.$$

Denote by H, H_1, and H_2 the boundary hyperplanes of the halfspaces V, V_1, and V_2, respectively. Because $\boldsymbol{u} \in H_1 \cap K_1$ and $\boldsymbol{v} \in H_2 \cap Y$, one has

$$\boldsymbol{u} + \boldsymbol{v} \in H_1 \cap K_1 + H_2 \cap Y \subset (H_1 + H_2) \cap (K_1 + Y) = H \cap (K_1 + Y).$$

Hence H meets $K_1 + Y$. Theorem 9.35 implies that V properly supports $K_1 + Y$. Since all three sets

$$H_1 \cap K, \quad H_2 \cap Y, \quad H \cap (K_1 + Y)$$

are nonempty proper exposed faces of K_1, Y, and $K_1 + Y$, respectively, Theorem 12.13 gives

$$\begin{aligned} H_1 \cap K_1 + H_2 \cap Y &= H \cap (K_1 + Y) = H \cap (K_2 + Y) \\ &= H_1 \cap K_2 + H_2 \cap Y. \end{aligned}$$

Both sets $H_1 \cap K_1$ and $H \cap (K_1 + Y)$ are convex, while $H_2 \cap Y$ is nonempty and compact. According to Corollary 12.6,

$$\dim (H \cap (K_1 + Y)) \leqslant \dim (K_1 + Y) - 1 = r - 1.$$

Therefore, the inductive assumption gives $H_1 \cap K_1 = H_1 \cap K_2$. So,

$$\boldsymbol{u} \in H_1 \cap K_1 = H_1 \cap K_2 \subset K_2 \cap \operatorname{rbd} K_2.$$

Summing up, $K_1 \cap \operatorname{rbd} K_1 \subset K_2 \cap \operatorname{rbd} K_2$. The proof of the opposite inclusion is similar. □

Corollary 12.16. *If a nonempty compact convex set $K \subset \mathbb{R}^n$ is the sum of convex sets K_1 and K_2, then both sets K_1 and K_2 are compact.*

Proof. According to Problem 1.13,

$$K = K_1 + K_2 \subset K_1 + \operatorname{cl} K_2 \subset \operatorname{cl} K_1 + \operatorname{cl} K_2 \subset \operatorname{cl} (K_1 + K_2) = \operatorname{cl} K = K.$$

So, $K_1 + \operatorname{cl} K_2 = \operatorname{cl} K_1 + \operatorname{cl} K_2$. The set K_2, as a summand of K, must be bounded. Therefore, $\operatorname{cl} K_2$ is compact. Theorem 12.15 gives $K_1 = \operatorname{cl} K_1$. Similarly, K_1 is bounded and $K_2 = \operatorname{cl} K_2$. Hence both K_1 and K_2 are compact. □

12.2 Exposed Points and Halflines

12.2.1 *Exposed Points*

Reformulating Definition 12.1 for the case of singletons, we introduce the following notion.

Definition 12.17. Let $K \subset \mathbb{R}^n$ be a nonempty convex set. A point $\boldsymbol{x} \in K$ is called an *exposed point* of K if there is a hyperplane supporting K such that $H \cap K = \{\boldsymbol{x}\}$. In what follows, $\exp K$ denotes the set of all exposed points of K. We put $\exp \varnothing = \varnothing$.

Examples. 1. The exposed points of a closed ball $B_\rho(\boldsymbol{a}) \subset \mathbb{R}^n$ are exactly the points of the boundary sphere $S_\rho(\boldsymbol{a})$ (see Problem 12.1).

2. The exposed points of an r-simplex $\Delta = \Delta(\boldsymbol{x}_1, \dots, \boldsymbol{x}_{r+1}) \subset \mathbb{R}^n$ are exactly its vertices, $\boldsymbol{x}_1, \dots, \boldsymbol{x}_{r+1}$ (see Problem 12.2).

3. The only exposed point of an r-simplicial cone $\Gamma_{\boldsymbol{a}}(\boldsymbol{x}_1, \dots, \boldsymbol{x}_r)$ in $\mathbb{R}^n$ is its apex $\boldsymbol{a}$ (see Problem 12.3).

Theorem 12.18. *If $K \subset \mathbb{R}^n$ is a closed convex set and $W \subset \mathbb{R}^n$ is an open halfspace such that the intersection $K \cap W$ is nonempty and bounded, then W contains an exposed point of K.*

Proof. Suppose that W is expressed as $W = \{\boldsymbol{x} \in \mathbb{R}^n : \boldsymbol{x}\cdot\boldsymbol{e} < \gamma\}$ for a suitable nonzero vector $\boldsymbol{e}$ and a scalar γ (see Definition 2.29). Choose a point $\boldsymbol{u} \in K \cap W$ and let $V = \{\boldsymbol{x} \in \mathbb{R}^n : \boldsymbol{x}\cdot\boldsymbol{e} \leqslant \gamma\}$. Then $\mu = \boldsymbol{u}\cdot\boldsymbol{e} < \gamma$. A combination of Corollary 3.52 and Theorem 3.54 implies that $K \cap V = \mathrm{cl}\,(K \cap W)$. Hence the set $K \cap V$ is compact as the closure of the bounded set $K \cap W$. Put $d = \mathrm{diam}\,(K \cap V)$, and let a scalar $\lambda > 0$ satisfy the inequality $d^2 < 2\lambda(\gamma - \mu)$. Put $\boldsymbol{v} = \boldsymbol{u} + \lambda\boldsymbol{e}$. The continuous function $\delta_{\boldsymbol{v}}(\boldsymbol{x}) = \|\boldsymbol{x} - \boldsymbol{v}\|$ (see Problem 1.9) attains its maximum value, ρ, on $K \cap V$ at a suitable point $\boldsymbol{z} \in K \cap V$. Equivalently, $K \cap V \subset B_\rho(\boldsymbol{v})$ and $\boldsymbol{z} \in S_\rho(\boldsymbol{v})$.

We assert that $\boldsymbol{z} \in W$. This is obvious if $K \subset W$. So, we assume that $K \not\subset W$. By the convexity of K, the boundary hyperplane $H = \{\boldsymbol{x} \in \mathbb{R}^n : \boldsymbol{x}\cdot\boldsymbol{e} = \gamma\}$ of V meets K (see Theorem 2.35). For any point $\boldsymbol{y} \in H \cap K$, one has $\boldsymbol{y}\cdot\boldsymbol{e} = \gamma > \mu$, which gives

$$\begin{aligned}\|\boldsymbol{y} - \boldsymbol{v}\|^2 &= \|\boldsymbol{y} - \boldsymbol{u} - \lambda\boldsymbol{e}\|^2 = \|\boldsymbol{y} - \boldsymbol{u}\|^2 - 2\lambda(\boldsymbol{y} - \boldsymbol{u})\cdot\boldsymbol{e} + \lambda^2\|\boldsymbol{e}\|^2\\ &\leqslant d^2 - 2\lambda(\gamma - \mu) + \lambda^2\|\boldsymbol{e}\|^2 < \lambda^2\|\boldsymbol{e}\|^2 = \|\boldsymbol{u} - \boldsymbol{v}\|^2\\ &\leqslant \|\boldsymbol{z} - \boldsymbol{v}\|^2 = \rho^2.\end{aligned}$$

Hence $\|\boldsymbol{y} - \boldsymbol{v}\| < \|\boldsymbol{z} - \boldsymbol{v}\|$, which shows that $\boldsymbol{y} \neq \boldsymbol{z}$. Therefore, $\boldsymbol{z}$ cannot be in $H \cap K$. Consequently, $\boldsymbol{z} \in K \cap W$.

Since $\boldsymbol{z}$ belongs to the sphere $S_\rho(\boldsymbol{v})$, the hyperplane

$$H = \{\boldsymbol{x} \in \mathbb{R}^n : (\boldsymbol{x} - \boldsymbol{v})\cdot(\boldsymbol{z} - \boldsymbol{v}) = 0\}$$

supports $B_\rho(\boldsymbol{v})$ such that $H \cap B_\rho(\boldsymbol{v}) = \{\boldsymbol{z}\}$ (see Theorem 8.22). Hence

$$\{\boldsymbol{z}\} \subset H \cap K \subset H \cap B_\rho(\boldsymbol{v}) = \{\boldsymbol{z}\}.$$

Consequently, $H \cap K = \{\boldsymbol{z}\}$, and $\boldsymbol{z}$ is an exposed point of K. □

Theorem 12.19. *If $K \subset \mathbb{R}^n$ is a closed convex set, then*

$$\exp K \subset \mathrm{ext}\, K \subset \mathrm{cl}\,(\exp K).$$

Proof. The inclusion $\exp K \subset \mathrm{ext}\, K$ follows from Definitions 11.15, 12.17 and Theorem 12.2. The second inclusion, $\mathrm{ext}\, K \subset \mathrm{cl}\,(\exp K)$, is obvious if $\mathrm{ext}\, K = \varnothing$. Suppose that $\mathrm{ext}\, K \neq \varnothing$ and choose a point $\boldsymbol{u} \in \mathrm{ext}\, K$. Given a scalar $\delta > 0$, Theorem 11.19 shows the existence of an open halfspace $W \subset \mathbb{R}^n$ such that $\boldsymbol{u} \in K \cap W \subset U_\delta(\boldsymbol{u})$. Next, Theorem 12.18 implies the existence of a point $\boldsymbol{v} \in \exp K$ located in $K \cap W$. Thus $\|\boldsymbol{u} - \boldsymbol{v}\| < \delta$. Since δ can be chosen arbitrarily small, we conclude that $\boldsymbol{u} \in \mathrm{cl}\,(\exp K)$. Summing up, $\mathrm{ext}\, K \subset \mathrm{cl}\,(\exp K)$. □

Remarks. 1. Both inclusions in Theorem 12.19 can be proper. For instance, if K is a plane convex body depicted in Figure 12.1, then

$$\mathrm{ext}\, K \setminus \exp K = \{\boldsymbol{a}_1, \boldsymbol{a}_2, \boldsymbol{c}_1, \boldsymbol{c}_2\}.$$

Similarly, if K is the 3-dimensional convex body described in a remark on page 393, then

$$\exp K = \mathrm{ext}\, K = ((\{\boldsymbol{u}, \boldsymbol{v}\} \cup C) \setminus \{\boldsymbol{o}\}).$$

Consequently, $\boldsymbol{o} \in \mathrm{cl}\,(\exp K) \setminus \mathrm{ext}\, K$.

2. Unlike extreme points (see Theorem 11.20), the set of exposed points of a closed convex set $K \subset \mathbb{R}^n$ may be a non-G_δ-set (see page 447).

Corollary 12.20. *If $K \subset \mathbb{R}^n$ is a compact convex set of dimension m, then the following assertions hold.*

(*a*) $\dim(\exp K) = m$.
(*b*) $\exp K$ *contains an affinely independent subset of $m + 1$ points.*
(*c*) *If $K \neq \varnothing$, then* $\mathrm{card}\,(\exp K) = m+1$ *if and only if K is an m-simplex.*

Proof. (*a*) This part follows from a combination of Theorems 2.64 and 12.19.

(*b*) Corollary 2.66 shows that $\exp K$ contains an affinely independent subset of $m+1$ points, with $\mathrm{card}\,(\exp K) = m+1$ if and only if the set $\exp K$ is affinely independent. Theorems 4.2 and 11.21 one has $\dim(\mathrm{ext}\, K) = m$. Consequently, Corollary 2.66 shows that $\mathrm{ext}\, K$ contains an affinely independent subset of $m + 1$ points. Thus $\mathrm{card}\,(\mathrm{ext}\, K) \geqslant m + 1$.

(*c*) Suppose that $K \neq \varnothing$. If $\exp K$ is affinely independent, then $\mathrm{ext}\, K = \exp K$, and a combination of Definition 3.5 and Theorem 11.21 shows that the set $K = \mathrm{conv}\,(\exp K)$ is a simplex. Conversely, if K is a simplex, then $\mathrm{card}\,(\exp K) = m + 1$ (see Problem 12.2). □

Theorem 12.21. *For a nonempty closed convex set $K \subset \mathbb{R}^n$, the following assertions hold.*

(*a*) $\exp K \neq \varnothing$ *if and only if K is line-free.*
(*b*) *K has a unique exposed point if and only if K is a line-free convex cone.*

Proof. (*a*) If $\exp K \neq \varnothing$, then Theorem 12.19 shows that $\mathrm{ext}\, K \neq \varnothing$. Consequently, K is line-free according to Theorem 11.18. Similarly, if K is line-free, then a combination of Theorems 11.18 and 12.19 implies that $\exp K \neq \varnothing$.

(*b*) Suppose first that K has a unique exposed point. Theorem 12.19 shows that $\operatorname{ext} K$ is a singleton. Therefore, K is a line-free convex cone according to Theorems 11.18.

Conversely, suppose that K is a line-free convex cone. By Theorem 11.18, K has a unique extreme point $\boldsymbol{u}$, which is the apex of K. Since the case $K = \{\boldsymbol{u}\}$ is obvious (then $\boldsymbol{u}$ is a unique exposed point of K), we may assume that $K \neq \{\boldsymbol{u}\}$. Consider the closed convex cone $C = K - \boldsymbol{u}$. Then $\operatorname{ap} C = \{\boldsymbol{o}\}$, and $\dim C^\circ = n$ (see Theorem 8.4). Choose a nonzero vector $\boldsymbol{e} \in \operatorname{rint} C^\circ$. Theorem 8.7 shows that the $(n-1)$-dimensional subspace $S = \{\boldsymbol{x} \in \mathbb{R}^n : \boldsymbol{x} \cdot \boldsymbol{e} = 0\}$ supports C such that $S \cap C = \{\boldsymbol{o}\}$. Consequently, the hyperplane $H = \boldsymbol{u} + S$ supports K such that $H \cap K = \{\boldsymbol{u}\}$, implying that $\boldsymbol{u} \in \exp K$. Finally, a combination of Theorems 11.18 and 12.19 shows that $\boldsymbol{u}$ is the unique exposed point of K. □

The following theorem shows that the family of hyperplanes supporting a given line-free closed convex set at its exposed points is sufficiently large.

Theorem 12.22. *Let $K \subset \mathbb{R}^n$ be a nonempty line-free closed convex set and $\mathcal{H}$ be the family of hyperplanes in $\mathbb{R}^n$ all supporting K such that $H \cap K$ is a singleton for any choice of $H \in \mathcal{H}$. The following assertions hold.*

(*a*) *The set of normal vectors of the hyperplanes from $\mathcal{H}$ is dense in the cone* $\operatorname{nor} K$.
(*b*) *K is the intersection of countably many closed halfspaces containing K and determined by the hyperplanes from $\mathcal{H}$.*

Proof. (*a*) It suffices to show that the set of unit normal vectors of the hyperplanes from $\mathcal{H}$ is dense in $\mathbb{S} \cap \operatorname{rint}(\operatorname{nor} K)$, where $\mathbb{S}$ is the unit sphere of $\mathbb{R}^n$. Choose a unit vector $\boldsymbol{e} \in \operatorname{rint}(\operatorname{nor} K)$. Let the scalars μ, γ, λ, the closed halfspace V, and the point $\boldsymbol{u} \in K \cap V$ be as in the proof of Theorem 12.18. Also, let

$$\boldsymbol{v} = \boldsymbol{u} + \lambda \boldsymbol{e} \quad \text{and} \quad d = \operatorname{diam}(K \cap V).$$

Denote by $\boldsymbol{z}$ a furthest from $\boldsymbol{v}$ point in $K \cap V$. Put

$$\rho = \|\boldsymbol{z} - \boldsymbol{v}\|, \quad \boldsymbol{c} = (\boldsymbol{z} - \boldsymbol{v})/\|\boldsymbol{z} - \boldsymbol{v}\|, \quad H = \{\boldsymbol{x} \in \mathbb{R}^n : \boldsymbol{x} \cdot \boldsymbol{c} = \rho\}.$$

Then $\boldsymbol{c}$ is a unit normal vector of the hyperplane H, and H supports K such that $\{\boldsymbol{z}\} = H \cap K$ (see the proof of Theorem 12.18).

We assert that $\boldsymbol{c} \to \boldsymbol{e}$ as $\lambda \to \infty$. First, we observe that

$$|\lambda - \rho| = \big|\|\boldsymbol{u} - \boldsymbol{v}\| - \|\boldsymbol{z} - \boldsymbol{v}\|\big| \leqslant \|\boldsymbol{u} - \boldsymbol{z}\| \leqslant d,$$

which gives $\rho - d \leqslant \lambda \leqslant \rho + d$ and $\lambda - d \leqslant \rho \leqslant \lambda + d$. Thus

$$\lim_{\lambda\to\infty} \frac{\lambda}{\rho} = \lim_{\lambda\to\infty} \frac{\rho}{\lambda} = 1.$$

Next,

$$\|\boldsymbol{u} - \boldsymbol{z}\|^2 = \|(\boldsymbol{u} - \boldsymbol{v}) - (\boldsymbol{z} - \boldsymbol{v})\|^2 = \|\lambda \boldsymbol{e} - \rho \boldsymbol{c}\|^2 = \lambda^2 + \rho^2 - 2\lambda\rho\, \boldsymbol{c}\cdot\boldsymbol{e}.$$

Therefore,

$$\begin{aligned}\lim_{\lambda\to\infty} \|\boldsymbol{c} - \boldsymbol{e}\|^2 &= \lim_{\lambda\to\infty} (\|\boldsymbol{c}\|^2 + \|\boldsymbol{e}\|^2 - 2\,\boldsymbol{c}\cdot\boldsymbol{e}) \\ &= 2 + \lim_{\lambda\to\infty} \frac{\|\boldsymbol{u} - \boldsymbol{z}\|^2 - \lambda^2 - \rho^2}{\lambda\rho} \\ &= 2 + \lim_{\lambda\to\infty} \frac{\|\boldsymbol{u} - \boldsymbol{z}\|^2}{\lambda\rho} - \lim_{\lambda\to\infty} \frac{\lambda}{\rho} - \lim_{\lambda\to\infty} \frac{\rho}{\lambda} = 0.\end{aligned}$$

Hence the set of unit normal vectors of the hyperplanes from $\mathcal{H}$ is dense in nor K.

(*b*) This part follows from (*a*) and Theorem 9.39. □

12.2.2 *Convex Hulls of Exposed Points and Halflines*

Theorem 12.23. *If $K \subset \mathbb{R}^n$ is compact convex set, then*

$$K = \operatorname{conv}(\operatorname{cl}(\exp K)) = \operatorname{cl}(\operatorname{conv}(\exp K)).$$

Proof. A combination of Theorems 11.22 and 12.19 implies the equality $K = \operatorname{conv}(\operatorname{cl}(\exp K))$. Since the set $\exp K$ is bounded, the equality

$$\operatorname{conv}(\operatorname{cl}(\exp K)) = \operatorname{cl}(\operatorname{conv}(\exp K))$$

follows from Theorem 4.16. □

Corollary 12.24. *For a compact convex set $K \subset \mathbb{R}^n$ and a subset X of K, any of the equalities*

$$K = \operatorname{cl}(\operatorname{conv} X) \quad \text{and} \quad K = \operatorname{conv}(\operatorname{cl} X) \tag{12.1}$$

holds if and only if $\exp K \subset \operatorname{cl} X$. Consequently, both equalities (12.1) take place provided X is a dense subset of $\exp K$.

Proof. Theorem 11.22 shows that $K = \operatorname{conv}(\operatorname{cl} X)$ if and only if $\operatorname{ext} K \subset \operatorname{cl} X$, which is equivalent to the inclusion

$$\exp K \subset \operatorname{ext} K \subset \operatorname{cl}(\operatorname{cl} X) = \operatorname{cl} X.$$

Since any subset X of K is bounded, the equality $\operatorname{cl}(\operatorname{conv} X) = \operatorname{conv}(\operatorname{cl} X)$ follows from Theorem 4.16. □

Corollary 12.25. *If $K \subset \mathbb{R}^n$ is a nonempty compact convex set of dimension m, then every point $\boldsymbol{x} \in \operatorname{rint} K$ is a positive convex combination of an affinely independent set of $m+1$ or fewer points from $\exp K$.*

Proof. According to Theorem 12.23, $K = \operatorname{cl}(\operatorname{conv}(\exp K))$. Therefore, Corollary 3.49 gives

$$\operatorname{rint} K = \operatorname{rint}(\operatorname{cl}(\operatorname{conv}(\exp K))) = \operatorname{rint}(\operatorname{conv}(\exp K)).$$

Finally, Theorem 4.7 shows that every point $\boldsymbol{u} \in \operatorname{rint} K$ is a positive convex combination of an affinely independent set of $m+1$ or fewer points from $\exp K$. □

Remark. The assertion of Corollary 12.25 fails if $\boldsymbol{u} \in \operatorname{rbd} K$. For instance, if $K \subset \mathbb{R}^2$ is the compact convex set depicted in Figure 12.1, then

$$\operatorname{conv}(\exp K) = K \setminus ([\boldsymbol{a}_1, \boldsymbol{a}_2] \cup [\boldsymbol{c}_1, \boldsymbol{c}_2]).$$

Consequently, no point $\boldsymbol{u} \in [\boldsymbol{a}_1, \boldsymbol{a}_2] \cup [\boldsymbol{c}_1, \boldsymbol{c}_2]$ can be written as a convex combination of exposed points of K.

Definition 12.26. An *exposed halfline* (also called an *exposed ray*) of a convex set $K \subset \mathbb{R}^n$ is a halfline $h \subset K$ which is an exposed face of K. In what follows, $\operatorname{expr} K$ denotes the union of exposed halflines of K (we put $\operatorname{expr} K = \varnothing$ if K has no exposed halflines).

Remarks. 1. Theorem 12.2 shows that every exposed halfline of a convex set $K \subset \mathbb{R}^n$ is its extreme halfline. Consequently, the endpoint of an exposed halfline of K is an extreme point of K, but not necessarily an exposed point of K. Indeed, let $K = \mathbb{B} + h \subset \mathbb{R}^2$, where $\mathbb{B}$ is the closed unit disk of $\mathbb{R}^2$ and h is the positive closed halfline of the x-axis. Then both closed halflines

$$h_1 = \{(x, -1) : x \geqslant 0\} \quad \text{and} \quad h_2 = \{(x, 1) : x \geqslant 0\}$$

are exposed halflines of K, while their endpoints, $(0, -1)$ and $(0, 1)$, are extreme but not exposed points of K.

2. For a closed convex set $K \subset \mathbb{R}^n$, $n \geqslant 3$, the inclusion $\operatorname{extr} K \subset \operatorname{cl}(\operatorname{expr} K)$ does not generally hold (compare with Theorem 12.19). For instance, let

$$M = \{(x, y, z) : x^2 + y^2 \leqslant 2x\sqrt{z},\ z \geqslant 0\}.$$

Every plane $z = a > 0$ meets M along the planar disk

$$\{(x, y, a) : (x - a)^2 + y^2 \leqslant a^2\}.$$

M is a line-free closed convex set and $h = \{(0,0,z) : z \geqslant 0\}$ is the only exposed halfline of M. Let

$$\boldsymbol{e} = (0,1,0), \quad M' = \boldsymbol{e} + M, \quad K = \operatorname{conv}(M \cup M').$$

Clearly, the closed halflines h and $h' = \boldsymbol{e} + h$ are the only extreme halflines of the line-free closed convex set K. Neither h nor h' is an exposed halfline of K because the coordinate x-plane is the only plane supporting K at h or h'. Consequently, $\operatorname{extr} K = h \cup h'$ and $\operatorname{expr} K = \varnothing$.

The following lemma will be used in the proof of Theorem 12.28.

Lemma 12.27. *Let $K \subset \mathbb{R}^n$ be a closed convex set and $H \subset \mathbb{R}^n$ be a hyperplane which meets K. If a point $\boldsymbol{u}$ belongs to $\exp(H \cap K) \setminus \operatorname{ext} K$, then the generated exposed face $G_{\boldsymbol{u}}(K)$ is one-dimensional and $\boldsymbol{u} \in \operatorname{rint} G_{\boldsymbol{u}}(K)$.*

Proof. Consider the closed convex set $M = H \cap K$. By the assumption, $\boldsymbol{u} \in \exp M \setminus \operatorname{ext} K$. Hence $K \neq \{\boldsymbol{u}\}$ and there is a hyperplane $L \subset \mathbb{R}^n$ supporting M such that $L \cap M = \{\boldsymbol{u}\}$. This argument shows that L properly supports M. We observe that $H \neq L$. Indeed, if $H = L$, then

$$H \cap K = (L \cap H) \cap K = L \cap (H \cap K) = L \cap M = \{\boldsymbol{u}\},$$

implying the inclusion $\boldsymbol{u} \in \exp K \subset \operatorname{ext} K$, which contradicts the assumption $\boldsymbol{u} \notin \operatorname{ext} K$. By Theorem 2.19, the plane $H \cap L$ has dimension $n-2$. Clearly, $H \cap L$ properly supports M.

Since $\boldsymbol{u} \notin \operatorname{ext} K$, there are distinct points $\boldsymbol{x}, \boldsymbol{z} \in K$ such that $\boldsymbol{u} \in (\boldsymbol{x}, \boldsymbol{z})$. We observe that $H \cap \{\boldsymbol{x}, \boldsymbol{z}\} = \varnothing$. Indeed, assume for a moment that at least one of these points, say $\boldsymbol{x}$, belongs to H. Then the inclusion $\boldsymbol{u} \in H \cap (\boldsymbol{x}, \boldsymbol{z})$ implies that $\boldsymbol{z} \in H$ (see Theorem 2.35). Consequently, $\boldsymbol{u} \in (\boldsymbol{x}, \boldsymbol{z}) \subset H \cap K$, in contradiction with the assumption $\boldsymbol{u} \in \exp(H \cap K)$. The same Theorem 2.35 shows that $\boldsymbol{x}$ and $\boldsymbol{z}$ belong to the opposite open halfspaces determined by H.

According to Corollary 9.11, there is a hyperplane $P \subset \mathbb{R}^n$ which contains $H \cap L$ and properly supports K. If $V \subset \mathbb{R}^n$ is the closed halfspace determined by H and containing K, the inclusion $[\boldsymbol{x}, \boldsymbol{z}] \subset V$ implies that $[\boldsymbol{x}, \boldsymbol{z}] \subset P$ (see Corollary 2.36). Thus $P \neq H$, implying that $H \cap P$ is an $(n-2)$-dimensional plane. Since $H \cap L \subset H \cap P$ and the plane have the same dimension $n-2$, they should coincide: $H \cap P = H \cap L$ (see Theorem 2.6).

We assert that the exposed face $P \cap K$ is a subset of the line $l = \langle \boldsymbol{x}, \boldsymbol{z} \rangle$. Indeed, assume for a moment the existence of a point $\boldsymbol{v} \in (P \cap K) \setminus l$.

Clearly, one of the closed segments $[\boldsymbol{x}, \boldsymbol{v}]$ and $[\boldsymbol{z}, \boldsymbol{v}]$, say $[\boldsymbol{x}, \boldsymbol{v}]$, meets H at a point $\boldsymbol{w} \notin l$. Therefore,

$$\begin{aligned} \boldsymbol{w} \in H \cap [\boldsymbol{x}, \boldsymbol{v}] \subset H \cap (K \cap P) &= (H \cap P) \cap (H \cap P) \\ &= (H \cap L) \cap M = \{\boldsymbol{u}\}, \end{aligned}$$

contrary to $\boldsymbol{u} \in l$ and $\boldsymbol{w} \notin l$.

Finally, since the generated exposed face $G_{\boldsymbol{u}}(K)$ lies in $P \cap K$, one has $G_{\boldsymbol{u}}(K) \subset l$. Because $G_{\boldsymbol{u}}(K) \neq \{\boldsymbol{u}\}$, the dimension of $G_{\boldsymbol{u}}(K)$ equals one. Since the endpoints of $G_{\boldsymbol{u}}(K)$, if any, should belong to $\operatorname{ext} K$ (see remark 1 on page 431), the point $\boldsymbol{u}$ belongs to $\operatorname{rint} G_{\boldsymbol{u}}(K)$. □

The next result extends Theorem 12.23.

Theorem 12.28. *For a closed convex set $K \subset \mathbb{R}^n$, the following conditions are equivalent.*

(*a*) *K is line-free.*
(*b*) *$K = \operatorname{cl}(\operatorname{conv}(\exp K \cup \operatorname{expr} K))$.*
(*c*) *$K = \operatorname{conv}(\operatorname{cl}(\exp K)) + \operatorname{rec} K = \operatorname{cl}(\operatorname{conv}(\exp K)) + \operatorname{rec} K$.*

Proof. Since the case $K = \varnothing$ is obvious, we may assume that K is nonempty.

(*a*) ⇒ (*b*) Due to the inclusion $\exp K \cup \operatorname{expr} K \subset K$, the convexity and closedness of K give

$$\operatorname{cl}(\operatorname{conv}(\exp K \cup \operatorname{expr} K)) \subset \operatorname{cl}(\operatorname{conv} K) = K.$$

For the opposite inclusion, let $K' = \operatorname{cl}(\operatorname{conv}(\exp K \cup \operatorname{expr} K))$. Assume for a moment that $K \not\subset K'$ and choose a point $\boldsymbol{u} \in K \setminus K'$. By Theorem 9.20, there is a hyperplane $H \subset \mathbb{R}^n$ strictly separating $\boldsymbol{u}$ and K'. Denote by W the open halfspace bounded by H and containing $\boldsymbol{u}$. Consider the closed convex set $M = H \cap K$, which is line-free as a subset of the line-free set K. According to Theorem 12.21, M has an exposed point $\boldsymbol{u}$. By Theorem 12.19, $\operatorname{ext} K \subset \operatorname{cl}(\exp K) \subset K'$. Therefore, $\boldsymbol{u} \notin \operatorname{ext} K$ due to $K' \cap \operatorname{cl} W = \varnothing$. By Lemma 12.27, the generated exposed face $G_{\boldsymbol{u}}(K)$ is one-dimensional and $\boldsymbol{u} \in \operatorname{rint} G_{\boldsymbol{u}}(K)$. Because $G_{\boldsymbol{u}}(K)$ does not lie in H, it meets the opposite to W open halfspace W' determined by H. Since the endpoints of $G_{\boldsymbol{u}}(K)$, if any, should belong to $\operatorname{ext} K$ (see remark 1 on page 431), $G_{\boldsymbol{u}}(K)$ does not have an endpoint in W'. This argument implies that $G_{\boldsymbol{u}}(K)$ is an exposed halfline of K, contrary to the assumption $\operatorname{expr} K \subset K' \subset W$. The obtained contradiction shows that $K \subset K'$.

$(a) \Rightarrow (c)$ According to Theorems 4.16 and 12.19, one has

$$K = \operatorname{conv}(\operatorname{ext} K) + \operatorname{rec} K \subset \operatorname{conv}(\operatorname{cl}(\exp K)) + \operatorname{rec} K$$
$$\subset \operatorname{cl}(\operatorname{conv}(\exp K)) + \operatorname{rec} K \subset K + \operatorname{rec} K = K,$$

implying both equalities of (c).

$(b) \Rightarrow (a)$ and $(c) \Rightarrow (a)$. If K contained a line, then $\operatorname{ext} K = \varnothing$ (see Theorems 11.18 and 12.21), implying that $\operatorname{expr} K = \varnothing$, which is impossible. Hence K must be line-free. □

Remarks. 1. Assertion (b) of Theorem 12.28 cannot be sharpened by stating that $K = \operatorname{conv}(\operatorname{cl}(\exp K \cup \operatorname{expr} K))$. Indeed, consider in $\mathbb{R}^3$ the line-free convex set $M = \operatorname{conv}(\Gamma \cup \{\boldsymbol{a}_1, \boldsymbol{a}_2\})$, where $\Gamma = \{(x, 0, z) : z = x^2\}$ is the parabola in the xz-plane of $\mathbb{R}^3$ and $\boldsymbol{a}_1 = (1, 1, 2)$ and $\boldsymbol{a}_2 = (-1, 1, 2)$. The set M is not closed, and its closure K equals $F \cup M$, where F is the half-slab of the plane $y = 1$, given by

$$F = \{(x, 1, z) : -1 \leqslant x \leqslant 1, z > 2\}.$$

It is easy to see that

$$\operatorname{ext} K = \exp K = \Gamma \cup \{\boldsymbol{a}_1, \boldsymbol{a}_2\},$$

while $\operatorname{extr} K$ is the union of two closed halflines

$$h_1 = \{(1, 1, z) : z \geqslant 2\} \quad \text{and} \quad h_2 = \{(-1, 1, z) : z \geqslant 2\}.$$

There is a unique plane through h_1 (or h_2) supporting K, and this plane contains F. Hence $\operatorname{expr} K = \varnothing$. Since the set $\exp K$ is closed, one has

$$K \neq M = \operatorname{conv}(\exp K) = \operatorname{conv}(\operatorname{cl}(\exp K \cup \operatorname{expr} K)).$$

2. The above example also shows that the inclusion

$$\operatorname{ext} K \cup \operatorname{extr} K \subset \operatorname{cl}(\exp K \cup \operatorname{expr} K)$$

does not, generally, hold. Indeed, in the above notation,

$$\operatorname{ext} K \cup \operatorname{extr} K = \Gamma \cup h_1 \cup h_2 \not\subset \Gamma \cup \{\boldsymbol{a}_1, \boldsymbol{a}_2\} = \operatorname{cl}(\exp K \cup \operatorname{expr} K).$$

3. If $\exp_1 K$ denotes the union of exposed points, segments, and halflines of a line-free closed convex set $K \subset \mathbb{R}^n$ (see Definitions 11.38 and 12.38), then

$$\operatorname{ext}_1 K \subset \exp_1 K \quad \text{and} \quad K = \operatorname{conv}(\operatorname{cl} \exp_1 K),$$

as follows from Theorems 12.41 and 12.42.

Corollary 12.29. *If a line-free closed convex set $K \subset \mathbb{R}^n$ has no exposed halflines, then $K = \text{cl}(\text{conv}(\exp K))$.* □

Remark. If a closed convex set $K \subset \mathbb{R}^n$, $n \geqslant 3$, satisfies the condition $K = \text{cl}(\text{conv}(\exp K))$, then it still can have exposed halflines. For instance, let K be the line-free closed convex set given by

$$K = \{(x, y, z) : (x - \sqrt{z})^2 + y^2 \leqslant z, \ z \geqslant 0\}.$$

It is easy to see that $h = \{(0, 0, z) : z \geqslant 0\}$ is the unique exposed halfline of K and $\exp K = \{\boldsymbol{o}\} \cup (\text{bd}\, K \setminus h)$. At the same time, $K = \text{conv}(\text{cl}(\exp K))$.

Corollary 12.30. *If $K \subset \mathbb{R}^n$ is a nonempty line-free closed convex set of dimension m, then every point $\boldsymbol{x} \in \text{rint}\, K$ is a positive convex combination of an affinely independent set of $m+1$ or fewer points from $\exp K \cup \text{expr}\, K$.*

Proof. A combination of Corollary 3.49 and Theorem 12.28 gives

$$\text{rint}\, K = \text{rint}\,(\text{cl}\,(\text{conv}\,(\exp K \cup \text{expr}\, K)) \subset \text{conv}\,(\exp K \cup \text{expr}\, K).$$

Consequently, Theorem 4.7 shows that every point $\boldsymbol{x} \in \text{rint}\, K$ is a positive convex combination of an affinely independent set of $m+1$ or fewer points from $\exp K \cup \text{expr}\, K$. □

The following result, which is an analog of Theorem 11.31, refines Corollary 12.30

Theorem 12.31. *If $K \subset \mathbb{R}^n$ is a nonempty line-free closed convex set of dimension m, then every point $\boldsymbol{u} \in \text{rint}\, K$ is a positive convex combination of an affinely independent set of*

(a) $m+1$ or fewer points from $\exp K$, or
(b) m or fewer points from $\exp K \cup \text{expr}\, K$.

Proof. Let $\boldsymbol{u} \in K$. Translating K on $-\boldsymbol{u}$, we may suppose that $\boldsymbol{u} = \boldsymbol{o}$. Indeed, a desired representation $\boldsymbol{u} = \gamma_1 \boldsymbol{z}_1 + \cdots + \gamma_t \boldsymbol{z}_t$ is equivalent to

$$\boldsymbol{o} = \gamma_1(\boldsymbol{z}_1 - \boldsymbol{u}) + \cdots + \gamma_t(\boldsymbol{z}_t - \boldsymbol{u}),$$

where $\boldsymbol{z}_i - \boldsymbol{u} = \exp(K - \boldsymbol{u}) \cup \text{expr}\,(K - \boldsymbol{u})$ for all $1 \leqslant i \leqslant t$.

Corollary 12.30 shows that $\boldsymbol{o}$ is a positive convex combination of the form

$$\boldsymbol{o} = \lambda_1 \boldsymbol{x}_1 + \cdots + \lambda_p \boldsymbol{x}_p, \quad p \leqslant m+1, \tag{12.2}$$

where $X = \{\boldsymbol{x}_1, \dots, \boldsymbol{x}_p\}$ is an affinely independent subset of $\exp K \cup \operatorname{expr} K$. If $X \subset \exp K$ or $p \leqslant m$, then the assertion holds. To obtaining a contradiction, we will suppose that no set $X \subset \exp K \cup \operatorname{expr} K$ satisfying (12.2) lies in $\exp K$ and that always $\operatorname{card} X = m + 1$.

For a set $X = \{\boldsymbol{x}_1, \dots, \boldsymbol{x}_{m+1}\}$ satisfying (12.2), we renumber its elements to have $X \setminus \exp K = \{\boldsymbol{x}_1, \dots, \boldsymbol{x}_q\}$, where $1 \leqslant q \leqslant m + 1$. Among all such sets X, we choose one with the smallest possible value of q. The equality

$$\boldsymbol{o} = \lambda_1 \boldsymbol{x}_1 + \cdots + \lambda_{m+1} \boldsymbol{x}_{m+1}, \tag{12.3}$$

shows that $\boldsymbol{o} \in \operatorname{aff} X$. Hence the plane $S = \operatorname{aff} X$ is an m-dimensional subspace and $S = \operatorname{span} X$. This argument and the equality

$$\lambda_1 \boldsymbol{x}_1 = -\lambda_2 \boldsymbol{x}_2 - \cdots - \lambda_{m+1} \boldsymbol{x}_{m+1} \tag{12.4}$$

imply that $\{-\boldsymbol{x}_2, \dots, -\boldsymbol{x}_{m+1}\}$ is a basis for S. Consequently, the convex cone $C = \operatorname{cone}_{\boldsymbol{o}}\{-\boldsymbol{x}_2, \dots, -\boldsymbol{x}_{m+1}\}$ is an m-simplicial cone with apex $\boldsymbol{o}$. Furthermore, (12.4) gives the inclusion $\lambda_1 \boldsymbol{x}_1 \in \operatorname{rint} C$ (see Theorem 5.23). Thus $\boldsymbol{x}_1 = \lambda_1^{-1}(\lambda_1 \boldsymbol{x}_1) \in \operatorname{rint} C$, as follows from Theorem 5.20.

The inclusion $\boldsymbol{o} \in K$ shows that $\operatorname{aff} K$ is an m-dimensional subspace. So, $\operatorname{aff} K = \operatorname{span} K$. Clearly, $\{-\boldsymbol{x}_2, \dots, -\boldsymbol{x}_{m+1}\}$ is a basis for $\operatorname{span} K$, and a combination of Theorems 2.6 and 5.7 implies that $\operatorname{span} C = \operatorname{span} K$.

Denote by h_1 the exposed halfline of K which contains $\boldsymbol{x}_1$. Let $\boldsymbol{u}_1$ be the endpoint of h_1. We assert that $\boldsymbol{u}_1 \notin \operatorname{rint} C$. Indeed, assume for a moment that $\boldsymbol{u}_1 \in \operatorname{rint} C$. According to Theorem 3.17, there is a scalar $\rho > 0$ such that

$$\operatorname{aff} K \cap U_\rho(\boldsymbol{u}_1) = \operatorname{aff} C \cap U_\rho(\boldsymbol{u}_1) \subset \operatorname{rint} C.$$

Since $\boldsymbol{u}_1 \in \operatorname{ext} K$ (see Remark 1 on page 431), Theorem 12.19 implies the existence of a point $\boldsymbol{v}_1 \in \exp K$ which belongs to $\operatorname{aff} K \cap U_\rho(\boldsymbol{u}_1)$. Hence $\boldsymbol{v}_1 \in \operatorname{rint} C$, and Theorem 5.23 shows that $\boldsymbol{u}_1$ can be written as a positive combination of the form

$$\boldsymbol{v}_1 = \mu_2(-\boldsymbol{x}_2) + \cdots + \mu_{m+1}(-\boldsymbol{x}_{m+1}).$$

Rewriting this equality (with $\mu = 1 + \mu_2 + \cdots + \mu_{m+1}$) as

$$\boldsymbol{o} = \frac{1}{\mu} \boldsymbol{v}_1 + \frac{\mu_2}{\mu} \boldsymbol{x}_2 + \cdots + \frac{\mu_{m+1}}{\mu} \boldsymbol{x}_{m+1},$$

we obtain another representation for $\boldsymbol{o}$ as a convex combination of points from $\exp K \cup \operatorname{expr} K$, with $q - 1$ points from $X \setminus \exp K$. The last is in contradiction with the choice of q. So, $\boldsymbol{u}_1 \notin \operatorname{rint} C$.

By the above argument, $\boldsymbol{u}_1 \in \operatorname{span} C$. According to Theorem 3.65, the closed segment $[\boldsymbol{u}_1, \boldsymbol{x}_1]$ meets $\operatorname{rbd} C$ at a unique point, say $\boldsymbol{w}_1$ (possibly, $\boldsymbol{w}_1 = \boldsymbol{u}_1$). Then $\boldsymbol{w} \in h \subset \operatorname{expr} K$, and the inclusion $\boldsymbol{w}_1 \in \operatorname{rbd} C$ shows that $\boldsymbol{w}$ can be written as $\boldsymbol{w}_1 = \gamma_2(-\boldsymbol{x}_2) + \cdots + \gamma_{m+1}(-\boldsymbol{x}_{m+1})$, where $\gamma_2, \ldots, \gamma_{m+1} \geqslant 0$ and at least one of the scalars $\gamma_2, \ldots, \gamma_{m+1}$ is zero. With $\gamma = 1 + \gamma_2 + \cdots + \gamma_{m+1}$,

$$\boldsymbol{o} = \frac{1}{\gamma}\boldsymbol{w}_1 + \frac{\gamma_2}{\gamma}\boldsymbol{x}_2 + \cdots + \frac{\gamma_{m+1}}{\gamma}\boldsymbol{x}_{m+1}$$

is a new representation for $\boldsymbol{o}$ as a convex combination of points from $\exp K \cup \operatorname{expr} K$, with fewer than $m+1$ terms. The obtained contradiction proves assertion (b). $\square$

Theorem 12.32. *For a line-free closed convex set $K \subset \mathbb{R}^n$ a subset X of K, any of the equalities*

$$K = \operatorname{conv}(\operatorname{cl} X) + \operatorname{rec} K \quad \text{and} \quad K = \operatorname{cl}(\operatorname{conv} X) + \operatorname{rec} K$$

holds if and only if $\exp K \subset \operatorname{cl} X$.

Proof. Since the case $K = \varnothing$ is obvious, we may suppose that $K \neq \varnothing$. If

$$K = \operatorname{conv}(\operatorname{cl} X) + \operatorname{rec} K, \tag{12.5}$$

then Theorem 11.29 gives $\exp K \subset \operatorname{ext} K \subset \operatorname{cl} X$. Conversely, suppose that $\exp K \subset \operatorname{cl} X$. In this case $\operatorname{ext} K \subset \operatorname{cl}(\exp K) \subset \operatorname{cl} X$ according to Theorem 12.19. Consequently, Theorem 11.29 implies (12.5).

Now, suppose that

$$K = \operatorname{cl}(\operatorname{conv} X) + \operatorname{rec} K. \tag{12.6}$$

To show the inclusion $\exp K \subset \operatorname{cl} X$, choose a point $\boldsymbol{u} \in \exp K$ and a scalar $\delta > 0$. Since $\boldsymbol{u} \in \operatorname{ext} K$, Theorem 11.19 implies the existence of an open halfspace $W \subset \mathbb{R}^n$ such that $\boldsymbol{u} \in K \cap W \subset U_\delta(\boldsymbol{u})$. We assert that X meets $K \cap W$. Indeed, assume for a moment that X entirely lies in the closed halfspace $V = \mathbb{R}^n \setminus W$. Then $\operatorname{cl}(\operatorname{conv} X) \subset K \cap V$. Since the set $K \setminus V = K \cap W$ is bounded, any closed halfline with apex $\boldsymbol{x} \in K \cap V$ should lie in $K \cap V$. This argument shows that

$$\boldsymbol{x} + \operatorname{rec} K \subset K \cap V$$

for any choice of $\boldsymbol{x} \in K \cap V$ (see Theorem 6.2). Therefore,

$$\begin{aligned} K &= \operatorname{cl}(\operatorname{conv} X) + \operatorname{rec} K = \cup(\boldsymbol{x} + \operatorname{rec} K : \boldsymbol{x} \in \operatorname{cl}(\operatorname{conv} X)) \\ &\subset \cup(\boldsymbol{x} + \operatorname{rec} K : \boldsymbol{x} \in K \cap V) \subset K \cap V \subset V, \end{aligned}$$

contrary to $\boldsymbol{u} \in K \setminus V$. Hence X meets $K \cap W$, and thus $X \cap U_\delta(\boldsymbol{u}) \neq \varnothing$. So, $\boldsymbol{u} \in \operatorname{cl} X$ because δ was chosen arbitrarily. Summing up, $\exp K \subset \operatorname{cl} X$.

Conversely, if $\exp K \subset \operatorname{cl} X$, then, by Theorem 4.16 and the above argument,

$$K = \operatorname{conv}(\operatorname{cl} X) + \operatorname{rec} K \subset \operatorname{cl}(\operatorname{conv} X) + \operatorname{rec} K \subset K,$$

which gives (12.6). □

12.2.3 *Special Exposed Faces and r-Exposed Sets*

A combination of Theorems 6.6, 6.25, and 12.28 implies the following result.

Corollary 12.33. *Let $K \subset \mathbb{R}^n$ be a nonempty proper closed convex set. If $L \subset \mathbb{R}^n$ is a plane complementary to* $\operatorname{lin} K$, *then*

$$\begin{aligned}
K &= \operatorname{cl}(\operatorname{conv}(\exp(K \cap L) \cup \operatorname{expr}(K \cap L))) + \operatorname{lin} K,\\
K &= \operatorname{conv}(\operatorname{cl}(\exp(K \cap L))) + \operatorname{rec}(K \cap L) + \operatorname{lin} K,\\
K &= \operatorname{cl}(\operatorname{conv}(\exp(K \cap L))) + \operatorname{rec}(K \cap L) + \operatorname{lin} K,\\
K &= \operatorname{conv}(\operatorname{cl}(\exp(K \cap L))) + \operatorname{rec} K,\\
K &= \operatorname{cl}(\operatorname{conv}(\exp(K \cap L))) + \operatorname{rec} K.
\end{aligned}$$ □

Definition 12.34. Let $K \subset \mathbb{R}^n$ be a nonempty convex set. A nonempty exposed face F of $K \subset \mathbb{R}^n$ is called *planar* (respectively, *halfplanar*) if F is a plane (respectively, F is a halfplane).

Example. If $K \subset \mathbb{R}^3$ is a convex cone, given by $K = \{(x, y, z) : y \geqslant |x|\}$, then $\operatorname{ap} K = \operatorname{lin} K = \{(0, 0, z) : z \in \mathbb{R}\}$ is the only planar exposed face of K, and the halfplanes

$$D_1 = \{(x, x, z) : x \geqslant 0\} \quad \text{and} \quad D_2 = \{(x, -x, z) : x \leqslant 0\}$$

are the halfplanar exposed faces of K.

Clearly, the planar exposed faces of a line-free closed convex set $K \subset \mathbb{R}^n$ are its exposed points, and the halfplanar exposed faces of K are its exposed halflines.

Theorem 12.35. *Let $K \subset \mathbb{R}^n$ be a nonempty closed convex set expressed as $K = (K \cap S) + \operatorname{lin} K$, where S is a subspace of $\mathbb{R}^n$ complementary to* $\operatorname{lin} K$. *The following assertions hold.*

(*a*) *An exposed face F of K is planar if and only if $F = \boldsymbol{x} + \operatorname{lin} K$, where $\boldsymbol{x} \in \exp(K \cap S)$.*

(*b*) *An exposed face F of K is halfplanar if and only if $F = h + \operatorname{lin} K$, where h is an exposed halfline of $K \cap S$.*

Proof. According to Theorem 12.8, an exposed face F of K can be expressed as $F = G + \operatorname{lin} K$, where G is an exposed face of $K \cap S$.

(*a*) If $F = \boldsymbol{x} + \operatorname{lin} K$, where $\boldsymbol{x} \in \exp(K \cap S)$, then F is a plane, which is an exposed face of K by the above argument. Conversely, let F a planar exposed face of K. Choose a point $\boldsymbol{x} \in G$. Then

$$\boldsymbol{x} + \operatorname{lin} K \subset G + \operatorname{lin} K = F.$$

On the other hand, Corollary 6.34 implies the opposite inclusion $F \subset \boldsymbol{x} + \operatorname{lin} K$.

(*b*) If $F = h + \operatorname{lin} K$, where h is an exposed halfline of $K \cap S$, then F is a halfplane (see Theorem 2.74), which is an exposed face of K by the above argument. Conversely, let F be a halfplanar exposed face of K. Theorem 2.74 shows that $F = g + L$, where g is a closed halfline with endpoint $\boldsymbol{o}$ and L is a plane. Denote by h the image of g upon the linear projection of g on S along $\operatorname{lin} K$. Since $F = G + \operatorname{lin} K$, it is easy to see that G is a closed halfline, h, with endpoint $\boldsymbol{o}$ (otherwise $G = \{\boldsymbol{o}\}$ and F is a translate of $\operatorname{lin} K$). Then, by the above argument, $F = h + \operatorname{lin} K$. □

The corollary below follows from Theorem 12.21 and the argument in the proof of Theorem 12.35.

Corollary 12.36. *Let $K \subset \mathbb{R}^n$ be a nonempty closed convex set. An exposed face F of K is planar if and only if F is a translate of* $\operatorname{lin} K$. *Furthermore, K has exactly one such exposed face, F, if and only if K is a convex cone with* $\operatorname{ap} K = F$. □

Remark. Every nonempty exposed face of a closed convex set $K \subset \mathbb{R}^n$ contains a translate of $\operatorname{lin} K$ (see Theorem 12.8). Therefore, planar exposed faces of K are minimal among all nonempty exposed faces of K. Figure 12.1 on page 413 shows that the converse assertion is not true: the exposed face $[\boldsymbol{a}_1, \boldsymbol{a}_2]$ is minimal, but is not a singleton.

The following result extends Theorem 12.28 to the case of closed convex sets in $\mathbb{R}^n$.

Theorem 12.37. *Every nonempty closed convex set $K \subset \mathbb{R}^n$ is the closure of the convex hull of the union of its planar and halfplanar exposed faces. Similarly, K is the sum of* $\operatorname{rec} K$ *and the closure of the convex hull (or the convex hull of the closure) of the union of its planar exposed faces.*

Proof. Express K as a direct sum $K = (K \cap S) + \operatorname{lin} K$, where S is a subspace complementary to $\operatorname{lin} K$. By Corollary 12.33,

$$\begin{aligned} K &= \operatorname{cl}(\operatorname{conv}(\exp(K \cap S) \cup \operatorname{expr}(K \cap S))) + \operatorname{lin} K \\ &= \operatorname{cl}(\operatorname{conv}(\cup(\boldsymbol{x} + \operatorname{lin} K : \boldsymbol{x} \in \exp(K \cap S)) \\ &\qquad \cup (h + \operatorname{lin} K : h \text{ is an exposed halfline of } K \cap S))). \end{aligned}$$

By Theorem 12.35, every planar exposed face of K is expressible as $\boldsymbol{x} + \operatorname{lin} K$, where $\boldsymbol{x} \in \exp K$, and every exposed halfplane of K is expressible as $h + \operatorname{lin} K$, where h is an exposed halfline of $K \cap S$. Hence

$$K = \operatorname{cl}(\operatorname{conv}(\cup(G : G \text{ is a planar or a halfplanar exposed face of } K))).$$

For the second assertion, observe first that the equalities

$$\begin{aligned} K \cap S &= \operatorname{conv}(\operatorname{cl}(\exp(K \cap S))) + \operatorname{rec}(K \cap S), \\ \operatorname{rec} K &= \operatorname{rec}(K \cap S) + \operatorname{lin} K \end{aligned}$$

give

$$\begin{aligned} K &= \operatorname{conv}(\operatorname{cl}(\exp(K \cap S)) + \operatorname{rec}(K \cap S)) + \operatorname{lin} K \\ &= \operatorname{conv}(\operatorname{cl}(\cup(\boldsymbol{x} + \operatorname{lin} K : \boldsymbol{x} \in \exp(K \cap S)))) + \operatorname{rec}(K \cap S) + \operatorname{lin} K \\ &= \operatorname{conv}(\operatorname{cl}(\cup(G : G \text{ is a planar exposed face of } K))) + \operatorname{rec} K. \end{aligned}$$

Similarly,

$$K = \operatorname{cl}(\operatorname{conv}(\cup(G : G \text{ is a planar exposed face of } K))) + \operatorname{rec} K. \qquad \square$$

Definition 12.38. Let K be a nonempty convex set in $\mathbb{R}^n$ and r be an integer satisfying the inequalities $0 \leqslant r \leqslant \dim K$. The union of all nonempty exposed faces of K of dimension r or less is called the *r-exposed set* of K and denoted $\exp_r K$. The elements of $\exp_r K$ are called *r-exposed points* of K.

Another terminology used for the r-exposed set is the *exposed r-skeleton*. Clearly, $\exp_0 K = \exp K$.

Theorem 12.39. *If $K \subset \mathbb{R}^n$ is a nonempty convex set of dimension m and r is an integer satisfying the inequalities $0 \leqslant r \leqslant m$, then the following assertions hold.*

(*a*) $\exp_r K \subset \operatorname{ext}_r K$.
(*b*) $\exp_0 K \subset \exp_1 K \subset \cdots \subset \exp_{m-1} K = K \cap \operatorname{rbd} K \subset \exp_m K = K$.
(*c*) *The set $K \setminus \exp_r K$ is convex and contains the union of relative interiors of exposed faces of K each of dimension $r + 1$ or more.*

(*d*) *For a point* $\boldsymbol{x} \in K$, *the following conditions are equivalent.*

(d_1) $\boldsymbol{x} \in \exp_r K$.

(d_2) $\dim G_{\boldsymbol{x}}(K) \leqslant r$.

(d_3) $\boldsymbol{x} \in \exp_r(K \cap B_\rho(\boldsymbol{x}))$ *for every closed ball* $B_\rho(\boldsymbol{x}) \subset \mathbb{R}^n$, $\rho > 0$.

Proof. (*a*) Let $\boldsymbol{x}$ be a point in $\exp_r K$. Then $\boldsymbol{x}$ belongs to an exposed face G of K such that $\dim G \leqslant r$. Since every exposed face of K also is an extreme face of K (see Theorem 12.2), we conclude that $\boldsymbol{x} \in \operatorname{ext}_r K$. Therefore $\exp_r K \subset \operatorname{ext}_r K$.

(*b*) This part follows from the definition.

(*c*) Let $\mathcal{G}_r$ denote the family of all exposed faces of K, each of dimension r or less. By the definition, $\exp_r K = \cup(G : G \in \mathcal{G}_r)$. The equality

$$K \setminus \exp_r K = K \setminus (\cup(G : G \in \mathcal{G}_r)) = \cap(K \setminus G : G \in \mathcal{G}_r)$$

shows that $K \setminus \exp_r K$ is convex as the intersection of convex sets $K \setminus G$, $G \in \mathcal{G}_r$ (see Theorem 11.2).

For the second part of (*c*), let a point $\boldsymbol{u} \in K$ belong to the relative interior of an exposed face G of K. Then $G = G_{\boldsymbol{u}}(K)$ according to Theorem 12.11. Hence any exposed face of K which contains $\boldsymbol{u}$ also contains G. So, no exposed face of dimension r or less may contain $\boldsymbol{u}$. In other words, $\boldsymbol{u} \in K \setminus \exp_r K$.

(*d*) First, we assert that (d_1) $\Leftrightarrow$ (d_2). Let $\boldsymbol{x} \in \exp_r K$. Then there is an exposed face G of K of dimension r or less such that $\boldsymbol{x} \in G$. By Theorem 12.12, $G_{\boldsymbol{x}}(K)$ is the smallest exposed face of K containing $\boldsymbol{x}$. Hence $G_{\boldsymbol{x}}(K) \subset G$, implying that

$$\dim G_{\boldsymbol{x}}(K) \leqslant \dim G \leqslant r.$$

Conversely, if $\dim G_{\boldsymbol{x}}(K) \leqslant r$, then $G_{\boldsymbol{x}}(K)$ is an exposed face of K which contains $\boldsymbol{x}$ and has dimension r or less. Therefore $\boldsymbol{x} \in \exp_r K$.

(d_1) $\Leftrightarrow$ (d_3). Let $\boldsymbol{x} \in \exp_r K$ and G be an exposed face of K which contains $\boldsymbol{x}$ and has dimension r or less. By Theorem 12.12, $G \cap B_\rho(\boldsymbol{x})$ is an exposed face of $K \cap B_\rho(\boldsymbol{x})$. Clearly,

$$\dim(G \cap B_\rho(\boldsymbol{x})) \leqslant \dim G \leqslant r,$$

which shows the inclusion $\boldsymbol{x} \in \exp_r(K \cap B_\rho(\boldsymbol{x}))$.

Conversely, let $\boldsymbol{x} \in \exp_r(K \cap B_\rho(\boldsymbol{x}))$ for a suitable closed ball $B_\rho(\boldsymbol{x}) \subset \mathbb{R}^n$. Choose an exposed face G of $K \cap B_\rho(\boldsymbol{x})$ which contains $\boldsymbol{x}$ and has dimension r or less. Let H be a hyperplane supporting $K \cap B_\rho(\boldsymbol{x})$ such that $G = H \cap (K \cap B_\rho(\boldsymbol{x}))$. Then H supports K (see Problem 9.3). Hence

$H \cap K$ is an exposed face of K containing $\boldsymbol{x}$. Furthermore, Theorem 3.51 gives

$$\dim (H \cap K) = \dim ((H \cap K) \cap B_\rho(\boldsymbol{x})) = \dim G \leqslant r.$$

Hence $\boldsymbol{x} \in \exp_r K$. □

The corollary below is a consequence of Theorem 12.8.

Corollary 12.40. *Let a convex set $K \subset \mathbb{R}^n$ of dimension m be expressed as a direct sum $K = (K \cap S) + \operatorname{lin} K$, where S is a subspace complementary to $\operatorname{lin} K$. If $\dim (\operatorname{lin} K) = p$, then $\exp_r K = \varnothing$ for all $0 \leqslant r \leqslant p-1$, and*

$$\exp_{p+q} K = \exp_q (K \cap S) + \operatorname{lin} K, \quad 0 \leqslant q \leqslant m-p. \qquad \square$$

A generalization of Theorem 12.19 to the case of r-extreme and r-exposed points is given in the next theorem.

Theorem 12.41. *Let $K \subset \mathbb{R}^n$ be a nonempty closed convex set of dimension m. Then*

$$\operatorname{ext}_r K \subset \operatorname{cl} (\exp_r K) \quad \textit{for all} \quad 0 \leqslant r \leqslant m.$$

Proof. The assertion is obvious for $r = m-1$ and $r = m$ due to

$$\operatorname{ext}_{m-1} K = \exp_{m-1} K = \operatorname{rbd} K \quad \text{and} \quad \operatorname{ext}_m K = \exp_m K = K$$

(see Theorems 11.39 and 12.39). So, one may assume that $r \leqslant m-2$.

We will prove the assertion by induction on r. Let $r = 0$. If K contains a line, then $\operatorname{ext}_0 K = \exp_0 K = \varnothing$ (see Corollaries 11.41 and 12.40). If K is line-free, then $\operatorname{ext}_0 K \subset \operatorname{cl} (\exp_0 K)$, as shown in Theorem 12.19.

Suppose that the assertion holds for all $s \leqslant r-1$, where $r \geqslant 1$, and choose a point $\boldsymbol{x} \in \operatorname{ext}_r K$ (the case $\operatorname{ext}_r K = \varnothing$ is obvious). Assume for a moment that $\boldsymbol{x} \notin \operatorname{cl} (\exp_r K)$. Then there is a closed ball $B_\delta(\boldsymbol{x}) \subset \mathbb{R}^n$ which is disjoint from $\operatorname{cl} (\exp_r K)$. Consequently, $B_\delta(\boldsymbol{x}) \cap \operatorname{ext}_{r-1} K = \varnothing$ due to

$$\operatorname{ext}_{r-1} K \subset \operatorname{cl} (\exp_{r-1} K) \subset \operatorname{cl} (\exp_r K).$$

The inclusion $\boldsymbol{x} \in \operatorname{ext}_r K \setminus \operatorname{ext}_{r-1} K$ means that the generated extreme face $F_{\boldsymbol{x}}(K)$ is r-dimensional. If $K = F_{\boldsymbol{x}}(K)$, then

$$\operatorname{extr}_{r-1} K = \operatorname{expr}_{r-1} K = K.$$

So, we may assume that $K \neq F_{\boldsymbol{x}}(K)$ and $\dim K \geqslant r+1$. As a nonempty proper subset of K, the extreme face $F_{\boldsymbol{x}}(K)$ lies in $\operatorname{rbd} K$ (see Theorem 11.4). Let $L = \operatorname{aff} F_{\boldsymbol{u}}(K)$. Then

$$\dim L = r \leqslant m-2 \quad \text{and} \quad L \cap \operatorname{rint} K = \varnothing,$$

as follows from Theorem 11.2. Choose a point $\boldsymbol{u} \in \operatorname{rint} K$ and a plane $N \subset \mathbb{R}^n$ through $\{\boldsymbol{x}, \boldsymbol{u}\}$ which is complementary to L. Then $\dim N = n-r$, as shown in Theorem 2.10.

We observe that $\boldsymbol{x}$ is an extreme point of the closed convex set $M = K \cap N$. Indeed, assume for a moment that $\boldsymbol{x}$ belongs to an open segment $(\boldsymbol{s}, \boldsymbol{a}) \subset M$. Since $\boldsymbol{x} \in \operatorname{rint} F_{\boldsymbol{x}}(K)$ (see Theorem 11.9), Problem 4.16 gives

$$\boldsymbol{x} \in \operatorname{rint} F_{\boldsymbol{x}}(K) \cup (\boldsymbol{s}, \boldsymbol{a}) \subset \operatorname{rint} (\operatorname{conv} (F_{\boldsymbol{x}}(K) \cup [\boldsymbol{s}, \boldsymbol{a}])).$$

Consequently, Theorem 11.4 shows that

$$\operatorname{conv} (F_{\boldsymbol{x}}(K) \cup [\boldsymbol{s}, \boldsymbol{a}]) \subset F_{\boldsymbol{x}}(K),$$

contrary to the condition

$$F_{\boldsymbol{x}}(K) \cap [\boldsymbol{s}, \boldsymbol{a}] \subset L \cap N = \{\boldsymbol{x}\}.$$

Due to the inclusion $\operatorname{ext}_0 M \subset \operatorname{cl} (\exp_0 M)$, there is a point $\boldsymbol{v} \in \exp_0 M$ such that $\|\boldsymbol{x} - \boldsymbol{v}\| \leqslant \delta$. Choose a plane $E \subset N$ of dimension $n - r - 1$ satisfying the condition $E \cap M = \{\boldsymbol{v}\}$. Since $E \cap \operatorname{rint} K = \varnothing$, Theorem 9.4 shows the existence of a hyperplane $H \subset \mathbb{R}^n$ which contains E and properly supports K. Clearly, H properly supports M. Hence $N \not\subset H$, and the plane $H \cap N$ is $(n - r - 1)$-dimensional (see Theorem 2.19). According to Theorem 2.6, the equality $\dim E = \dim (H \cap N)$ implies that $E = H \cap N$. Consequently,

$$H \cap M = H \cap (N \cap K) = (H \cap N) \cap K = E \cap K = \{\boldsymbol{v}\}.$$

Because $H \cap K$ is an exposed face of K containing $\boldsymbol{v}$, one has $G_{\boldsymbol{v}}(K) \subset H \cap K$. Furthermore,

$$\dim (H \cap K) \geqslant \dim G_{\boldsymbol{v}}(K) \geqslant r + 1$$

due to the assumption $\boldsymbol{v} \notin \exp_r K$. Put $Q = \operatorname{aff} (H \cap K)$. Then

$$\dim Q = \dim (H \cap K) \geqslant r + 1.$$

Both planes $E = H \cap N$ and Q lie in H. Hence their sum lies in the hyperplane $2H$. A combination of Theorems 2.6 and 2.7 gives

$$\begin{aligned}
\dim (N \cap Q) &= \dim (H \cap N) + \dim Q - \dim (H \cap N + Q) \\
&\geqslant \dim (H \cap N) + \dim Q - \dim (2H) \\
&= n - r - 1 + \dim Q - (n - 1) \\
&= \dim Q - r.
\end{aligned}$$

Because $(N \cap Q) \cap K \subset (H \cap N) \cap K = \{\boldsymbol{v}\}$, there is a plane $T \subset Q$ of dimension $\dim Q - 1$ which contains $N \cap Q$ and supports $H \cap K = K \cap Q$. By the above,

$$\dim T - \dim (N \cap Q) \leqslant (\dim Q - 1) - (\dim Q - r) = r - 1.$$

Finally, since T contains the exposed face $H \cap K$ of K, one has $F_{\boldsymbol{v}}(K) \subset G_{\boldsymbol{v}}(K) \subset T$. Because $\boldsymbol{v} \in \operatorname{rint} F_{\boldsymbol{v}}(K)$ and

$$(N \cap Q) \cap F_{\boldsymbol{v}}(K) \subset (N \cap Q) \cap K = \{\boldsymbol{v}\},$$

Theorem 3.39 gives

$$\begin{aligned}\dim F_{\boldsymbol{v}}(K) &= \dim (F_{\boldsymbol{v}}(K) \cup (N \cap Q)) - \dim (N \cap Q) \\ &\quad + \dim (F_{\boldsymbol{v}}(K) \cap (N \cap Q)) \\ &\leqslant \dim T - \dim (N \cap Q) + \dim \{\boldsymbol{v}\} \leqslant r - 1.\end{aligned}$$

Hence $\boldsymbol{v} \in \operatorname{ext}_{r-1} K \cap B_\delta(\boldsymbol{x})$, contrary to the assumption.

Summing up, $\boldsymbol{x} \in \operatorname{cl}(\exp_r K)$, and the inclusion $\operatorname{ext}_r K \subset \operatorname{cl}(\exp_r K)$ holds. □

The next result generalizes Theorem 12.28 and Corollary 12.29 (compare with Theorem 11.44).

Theorem 12.42. *For a nonempty closed convex set $K \subset \mathbb{R}^n$ of dimension m and an integer $0 \leqslant r \leqslant m$, the following assertions hold.*

(*a*) *If $\dim (\operatorname{lin} K) \leqslant r$, then*

$$K = \operatorname{conv}(\operatorname{cl}(\exp_{r+1} K)) = \operatorname{cl}(\operatorname{conv}(\exp_{r+1} K)).$$

(*b*) *If $\dim (\operatorname{lin} K) \leqslant r$ and K has no exposed halfplane of dimension greater than r, then*

$$K = \operatorname{conv}(\operatorname{cl}(\exp_r K)) = \operatorname{cl}(\operatorname{conv}(\exp_r K)).$$

Proof. Let $K = (K \cap S) + \operatorname{lin} K$, where S is a subspace complementary to $\operatorname{lin} K$ (see Theorem 6.25).

(*a*) Let $p = \dim (\operatorname{lin} K)$, where $p \leqslant r$. Since $K \cap S$ is a line-free closed convex set, Theorem 12.28 gives

$$K \cap S = \operatorname{conv}(\operatorname{cl}(\exp (K \cap S) \cup \operatorname{expr}(K \cap S)))$$

By the definitions,

$$\exp (K \cap S) \cup \operatorname{expr}(K \cap S) \subset \exp_1 (K \cap S),$$

Therefore, Corollary 12.40 shows that

$$K = (K \cap S) + \operatorname{lin} K \subset \operatorname{conv}(\operatorname{cl}(\exp_1(K \cap S))) + \operatorname{lin} K$$
$$= \operatorname{conv}(\operatorname{cl}(\exp_{p+1} K)) \subset \operatorname{conv}(\operatorname{cl}(\exp_{r+1} K)) \subset K.$$

Hence, $K = \operatorname{conv}(\operatorname{cl}(\exp_{r+1} K))$. Similarly, $K = \operatorname{cl}(\operatorname{conv}(\exp_{r+1} K))$.

(*b*) If K has no exposed halfplane of dimension greater than r, then Theorem 12.8 shows that the set $K \cap S$ has no exposed faces of positive dimension. By Corollary 12.29,

$$K \cap S = \operatorname{conv}(\operatorname{cl}(\exp_0(K \cap S))) = \operatorname{cl}(\operatorname{conv}(\exp_0(K \cap S))).$$

Therefore, Corollary 12.40 shows that

$$K = (K \cap S) + \operatorname{lin} K \subset \operatorname{conv}(\operatorname{cl}(\exp_0(K \cap S))) + \operatorname{lin} K$$
$$= \operatorname{conv}(\operatorname{cl}(\exp_p K)) \subset \operatorname{conv}(\operatorname{cl}(\exp_r K)) \subset K.$$

Hence, $K = \operatorname{conv}(\operatorname{cl}(\exp_r K))$. Similarly, $K = \operatorname{cl}(\operatorname{conv}(\exp_r K))$. □

The next result provides a generalization of Corollary 12.30.

Theorem 12.43. *Let $K \subset \mathbb{R}^n$ be a nonempty closed convex set of dimension m, with* $\dim(\operatorname{lin} K) = r$*. Then every point $\boldsymbol{x} \in \operatorname{rint} K$ is a positive convex combination of an affinely independent set of $m - r + 1$ or fewer points from* $\exp_{r+1} K$.

Proof. Let $K = (K \cap S) + \operatorname{lin} K$, where S is a subspace complementary to $\operatorname{lin} K$ (see Theorem 6.25). Then $K \cap S$ is a line-free closed convex set of dimension $m - r$.

Choose a point $\boldsymbol{x} \in \operatorname{rint} K$ and express it as $\boldsymbol{x} = \boldsymbol{y} + \boldsymbol{z}$, where $\boldsymbol{y} \in \operatorname{rint}(K \cap S)$ and $\boldsymbol{z} \in \operatorname{lin} K$ (see Corollary 6.32). By Corollary 12.30, $\boldsymbol{y}$ can be written as a positive convex combination $\boldsymbol{y} = \lambda_1 \boldsymbol{y}_1 + \cdots + \lambda_p \boldsymbol{y}_p$, where $p \leqslant m - r + 1$ and $\{\boldsymbol{y}_1, \ldots, \boldsymbol{y}_p\}$ is an affinely independent subset of $\exp(K \cap S) \cup \operatorname{expr}(K \cap S)$. Let $\boldsymbol{x}_i = \boldsymbol{y}_i + \boldsymbol{z}$, $1 \leqslant i \leqslant p$. Clearly, the set $\{\boldsymbol{x}_1, \ldots, \boldsymbol{x}_p\}$ is affinely independent and $\boldsymbol{x}_i \in \boldsymbol{y}_i + \operatorname{lin} K$ for all $1 \leqslant i \leqslant p$. Since

$$\exp(K \cap S) \cup \operatorname{expr}(K \cap S) \subset \exp_1(K \cap S),$$

Corollary 12.40 implies that

$$\boldsymbol{x}_1, \ldots, \boldsymbol{x}_p \in \exp_1 K + \operatorname{lin} K = \exp_{r+1} K,$$

and the equality

$$\boldsymbol{x} = \boldsymbol{y} + \boldsymbol{z} = \lambda_1 \boldsymbol{y}_1 + \cdots + \lambda_p \boldsymbol{y}_p + \boldsymbol{z} = \lambda_1(\boldsymbol{y}_1 + \boldsymbol{z}) + \cdots + \lambda_p(\boldsymbol{y}_p + \boldsymbol{z})$$
$$= \lambda_1 \boldsymbol{x}_1 + \cdots + \lambda_p \boldsymbol{x}_p$$

shows that $\boldsymbol{x}$ is a positive convex combination of an affinely independent set of $p\,(\leqslant m - r + 1)$ points from $\exp_{r+1} K$. □

Problems and Notes for Chapter 12

Problems for Chapter 12

Problem 12.1. Prove that the exposed faces of a closed ball $B_\rho(\boldsymbol{a}) \subset \mathbb{R}^n$ are $\varnothing$, all singletons from the boundary sphere $S_\rho(\boldsymbol{a})$, and $B_\rho(\boldsymbol{a})$.

Problem 12.2. Let $\Delta = \Delta(\boldsymbol{x}_1, \ldots, \boldsymbol{x}_{r+1})$ be an r-simplex in $\mathbb{R}^n$. Prove that the exposed faces of Δ are $\varnothing$ and all simplices $\Delta(\boldsymbol{z}_1, \ldots, \boldsymbol{z}_t)$, where $\{\boldsymbol{z}_1, \ldots, \boldsymbol{z}_t\}$ is a nonempty subset of $\{\boldsymbol{x}_1, \ldots, \boldsymbol{x}_{r+1}\}$.

Problem 12.3. Let $\Gamma = \Gamma_{\boldsymbol{a}}(\boldsymbol{x}_1, \ldots, \boldsymbol{x}_r)$ be an r-simplicial cone in $\mathbb{R}^n$. Prove that the exposed faces of Γ are the empty set, $\{\boldsymbol{a}\}$, and all simplicial cones $\Gamma_{\boldsymbol{a}}(\boldsymbol{z}_1, \ldots, \boldsymbol{z}_t)$, where $\{\boldsymbol{z}_1, \ldots, \boldsymbol{z}_t\}$ is a nonempty subset of $\{\boldsymbol{x}_1, \ldots, \boldsymbol{x}_r\}$.

Problem 12.4. Let $K \subset \mathbb{R}^n$ be a convex set. Prove that a subset G of K is an exposed face of K if and only if G is an exposed face of $K \cap B_\rho(G)$, where $B_\rho(G)$ is the ρ-neighborhood of G.

Problem 12.5. Given a plane $L \subset \mathbb{R}^n$ of positive dimension and a closed halfplane D of L, prove the following assertions.

(*a*) The exposed faces of L are $\varnothing$ and L.

(*b*) The exposed faces of D are $\varnothing$, $\operatorname{rbd} D$, and D.

Problem 12.6. Let $K \subset \mathbb{R}^n$ be a closed convex set. Prove the following assertions.

(*a*) K has exactly two exposed faces if and only if K is a plane.

(*b*) K has exactly three exposed faces if and only if K is a closed halfplane.

(*c*) K has exactly four exposed faces if and only if K is a plane closed slab.

Problem 12.7. Let $0 = d_0 < d_1 < \cdots < d_p = m$, where $m \geqslant n$, be an increasing sequence of integers. Prove the existence of a compact convex set $K \subset \mathbb{R}^n$ of dimension m such that the dimensions of exposed faces of K are given precisely by the numbers $d_0, d_1, \ldots, d_p$.

Problem 12.8. Let $K \subset \mathbb{R}^n$ be a nonempty line-free closed convex set. Prove that $K = \cap\, (C_{\boldsymbol{a}}(K) : \boldsymbol{a} \in \exp K)$ and $\operatorname{rec} K = \cap\, (C_{\boldsymbol{a}}(K) - \boldsymbol{a} : \boldsymbol{a} \in \exp K)$.

Problem 12.9. Let $K \subset \mathbb{R}^n$ be a compact convex set with more than one point and $\boldsymbol{e}$ be a unit vector in $\mathbb{R}^n$. Prove that for any scalar $\varepsilon > 0$, there is a unit vector $\boldsymbol{c} \in \mathbb{R}^n$ such that $\|\boldsymbol{e} - \boldsymbol{c}\| < \varepsilon$ and distinct hyperplanes H and H' both orthogonal to $\boldsymbol{c}$ and supporting K satisfy the condition: the sets $K \cap H$ and $K \cap H'$ are singletons.

Notes for Chapter 12

Extreme and exposed faces. Theorem 12.7 derives from an argument of Grünbaum [134, p. 20], who uses the terms "faces" and "poonems" for exposed and extreme faces, respectively. The same approach is used by Fedotov [102] to describe $\text{ext}_r K$ is terms of $\exp_r K$.

Theorem 12.15 is obtained by Iusem, Martínez-Legaz, and Todorov [158], and independently by Soltan [260, Theorem 8.11].

Generated exposed faces. Brown [57] showed that for every convex body $K \subset \mathbb{R}^n$, $n \leqslant 5$, and an exposed face G of K, either there is an exposed point $\boldsymbol{x}$ of K that belongs to $\text{rbd}\, G$, or there is a point $\boldsymbol{x} \in \text{rbd}\, G$ such that $G_{\boldsymbol{x}}(K) = G$. It is an open question whether this assertion holds for all $n \geqslant 5$.

Exposed structure of the sum of convex sets. The following result is proved in Soltan [257]. Let a line-free closed convex set $A \subset \mathbb{R}^n$ be the sum of closed convex sets B and C. Then both sets $B_0 = B + \text{rec}\, A$ and $C_0 = C + \text{rec}\, A$ are closed and satisfy the following conditions:

(a) for every point $\boldsymbol{a} \in \exp A$ there are unique points $\boldsymbol{b} \in \exp B_0$ and $\boldsymbol{c} \in \exp C_0$ such that $\boldsymbol{a} = \boldsymbol{b} + \boldsymbol{c}$,

(b) the sets

$$\exp_C B = \{\boldsymbol{x} \in \exp B : \exists\, \boldsymbol{y} \in \exp C \text{ such that } \boldsymbol{x} + \boldsymbol{y} \in \exp A\},$$
$$\exp_B C = \{\boldsymbol{x} \in \exp C : \exists\, \boldsymbol{y} \in \exp B \text{ such that } \boldsymbol{x} + \boldsymbol{y} \in \exp A\}$$

are dense in $\exp B_0$ and $\exp C_0$, respectively.

De Wilde [89] obtained some separation results that involve exposed points of line-free closed convex sets.

Topological properties of exposed faces. Theorem 12.19 is due to Klee [175]. In the same paper Klee gave an example of convex body $K \subset \mathbb{R}^3$ such that $\exp K$ is not a G_δ-set, and of a convex body $M \subset \mathbb{R}^3$ such that all sets $\text{bd}\, M \setminus \text{ext}\, M$, $\text{ext}\, M \setminus \exp M$, and $\exp M$ are dense in $\text{bd}\, M$. Corson [80], negatively answering a question from [175], gave an example of a convex body $K \subset \mathbb{R}^3$ such that $\exp K$ is of the first category.

Choquet, Corson, and Klee [72] showed that for every closed convex set $K \subset \mathbb{R}^n$, the set $\exp K$ is the union of an F_σ-set, a G_δ-set, and $n-2$ sets each of which is the intersection of an F_σ-set and a G_δ-set. Holický and Laczkovich [152], negatively answering a question from [72], gave an example of a convex body $K \subset \mathbb{R}^3$ for which $\exp K$ is not the intersection of an F_σ-set and a G_δ-set.

Theorem 12.41 is proved by Asplund [6] for the case of convex bodies (compact convex sets with nonempty interior). Reiter and Stavrakas [239] proved that the metric space of all exposed faces of a convex body $K \subset \mathbb{R}^n$ is compact (in the Hausdorff metric) if and only if all sets $\exp_r K$ are compact, $0 \leqslant r \leqslant \dim K - 1$.

Exposed faces of polar cones. Given a closed convex cone $C \subset \mathbb{R}^n$ with apex

$\boldsymbol{o}$ and an exposed face B of C, the set

$$B^* = C^\circ \cap F^\perp = \{\boldsymbol{x} \in C^\circ : \boldsymbol{x} \cdot \boldsymbol{z} = 0 \text{ for all } \boldsymbol{z} \in B\}$$

is called the *conjugate to B face* of C°. Obviously, B^* is an exposed face of C° and $(B^*)^* = B$. Problem 8.8 shows that for any vector $\boldsymbol{x} \in \mathbb{R}^n$ there is an exposed face B of C such that $\boldsymbol{x}$ is uniquely expressible as the sum $\boldsymbol{x} = \boldsymbol{y} + \boldsymbol{z}$, where $\boldsymbol{y} \in B$ and $\boldsymbol{z} \in B^*$. Soltan [263] proved that given an exposed face B of C, there is an $(n-1)$-dimensional subspace $S \subset \mathbb{R}^n$ containing B and separating C and B^* such that $\operatorname{rint} C$ and $\operatorname{rint} B^*$ lie in the opposite open halfspaces of $\mathbb{R}^n$ determined by S. If the cone C is polyhedral, then $\dim B + \dim B^* = n$ for any exposed face B of C (see Tam [276]).

Exposed representations. Theorem 12.23 is proved by Straszewicz [273], and Theorem 12.28 is due to Klee [175].

A point $\boldsymbol{x}$ of a nonempty compact convex set $K \subset \mathbb{R}^n$ is called *bare* provided there is a closed ball $B_\rho(\boldsymbol{a}) \subset \mathbb{R}^n$ containing K such that $\boldsymbol{x}$ belongs to the boundary sphere of $B_\rho(\boldsymbol{a})$. Clearly, every bare point of K is its exposed point, but not converse. Orland [224] (for $n = 2$) and Berberian [32] (for all $n \geqslant 2$) proved that a nonempty compact convex set $K \subset \mathbb{R}^n$ is the closed convex hull of its bare points. The second part of Problem 12.8 is proved by De Wilde [89].

Affine and exposed diameters. A chord $[\boldsymbol{a}, \boldsymbol{c}]$ of a convex body $K \subset \mathbb{R}^n$ is called an *affine diameter* of K provided there are distinct parallel hyperplanes $H_{\boldsymbol{a}}$ and $H_{\boldsymbol{c}}$ supporting K at $\boldsymbol{a}$ and $\boldsymbol{c}$, respectively. The following properties of affine diameters are widely used (see, e. g., the survey Soltan [256]):

(a) an affine diameter of a convex body $K \subset \mathbb{R}^n$ is a longest chord of K in its direction,

(b) any point $\boldsymbol{u}$ of a convex body $K \subset \mathbb{R}^n$ belongs to an affine diameter of K.

We will say that a chord $[\boldsymbol{a}, \boldsymbol{c}]$ of a convex body $K \subset \mathbb{R}^n$ is an *exposed affine diameter* of K if there are distinct parallel hyperplanes $H_{\boldsymbol{a}}$ and $H_{\boldsymbol{c}}$ supporting K such that $K \cap H_{\boldsymbol{a}} = \{\boldsymbol{a}\}$ and $K \cap H_{\boldsymbol{c}} = \{\boldsymbol{c}\}$. Soltan [253] showed that a convex body $K \subset \mathbb{R}^n$ has at least n exposed affine diameters (with precisely n such diameters if and only if K is an octahedron). Refining Theorem 12.23, the same author proved that every compact convex set $K \subset \mathbb{R}^n$ is the closure of the convex hull of its exposed diameters (see [258]).

Chapter 13

Polyhedra

13.1 Polyhedra as Intersections of Halfspaces

13.1.1 *General Systems of Halfspaces*

Definition 13.1. A set $P \subset \mathbb{R}^n$ is called a *polyhedron* if either $P = \mathbb{R}^n$ or P is the intersection of finitely many closed halfspaces of $\mathbb{R}^n$:

$$P = V_1 \cap \cdots \cap V_r. \tag{13.1}$$

In the latter case, we say that the family $\mathcal{F} = \{V_1, \ldots, V_r\}$ *represents* P. A bounded convex polyhedron is called a *polytope*. The empty set $\varnothing$ is assumed to be a polytope.

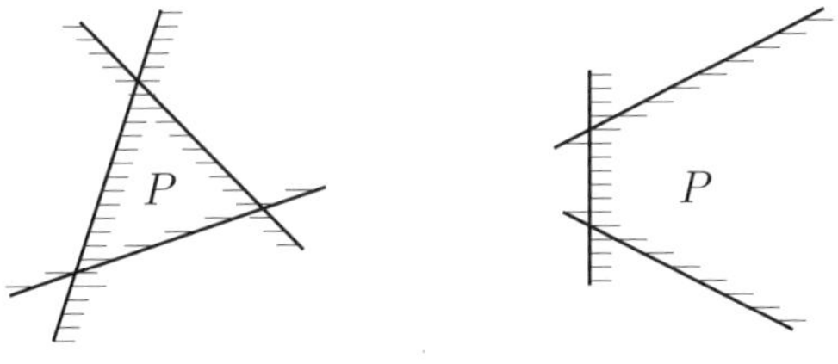

Fig. 13.1 A polytope and a polyhedron in the plane.

Corollary 13.2. *Every polyhedron $P \subset \mathbb{R}^n$ is a closed convex set.*

Proof. Excluding the obvious cases $P = \varnothing$ and $P = \mathbb{R}^n$, we assume that P is a nonempty proper subset of $\mathbb{R}^n$. Then P is the intersection of finitely many closed halfspaces. Because every closed halfspace is a closed convex set (see Theorem 2.33 and an example on page 75), and since the intersection of every family of closed convex sets is also a closed convex set (see, respectively, page 10 and Theorem 3.8), we obtain that P is closed and convex. □

Examples. 1. Every plane $L \subset \mathbb{R}^n$ is a polyhedron. Indeed, excluding the obvious cases $L = \varnothing$ and $L = \mathbb{R}^n$, suppose that L is a nonempty proper plane. Then L is the intersection of suitable $n - \dim L + 1$ closed halfspaces (see Problem 13.1).

2. Every closed halfplane $D \subset \mathbb{R}^n$ is a polyhedron. Since the case when D is a halfspace is obvious, we may assume that $\dim D = m \leqslant n - 1$. According to Definition 2.69, D is the intersection of the plane aff D and a suitable closed halfspace. Because aff D is an m-dimensional plane, it can be expressed as the intersection of $n - m + 1$ closed halfspaces (see the above example). Hence D is the intersection of $n - m + 2$ closed halfspaces.

3. Every simplex $\Delta = \Delta(\boldsymbol{x}_1, \dots, \boldsymbol{x}_{r+1}) \subset \mathbb{R}^n$ is a polytope. Indeed, if $r = 0$, then $\Delta = \{\boldsymbol{x}_1\}$ is the intersection of $n + 1$ closed halfspaces (see the first example above). Suppose that $r \geqslant 1$. Then a combination of Definition 3.5 and Theorem 2.75 shows that Δ is the intersection of $r + 1$ closed halfplanes

$$D_i = \{\lambda_1 \boldsymbol{x}_1 + \cdots + \lambda_{r+1} \boldsymbol{x}_{r+1} : \lambda_1 + \cdots + \lambda_{r+1} = 1, \ \lambda_i \geqslant 0\}, \ 1 \leqslant i \leqslant r + 1.$$

If $r = n$, then every D_i is a halfspace, and Δ is the intersection of $n + 1$ closed halfspaces. Let $1 \leqslant r \leqslant n - 1$. Then every D_i is the intersection of the r-dimensional plane aff Δ and a closed halfspace V_i, $1 \leqslant i \leqslant r + 1$. So, we can write $\Delta = \text{aff}\, \Delta \cap V_1 \cap \cdots \cap V_r$. Since aff Δ is the intersection of $n - r + 1$ closed halfspaces (see the first example above), Δ is the intersection of $(n - r + 1) + (r + 1) = n + 2$ closed halfspaces.

4. Every r-simplicial cone $\Gamma = \Gamma_{\boldsymbol{a}}(\boldsymbol{x}_1, \dots, \boldsymbol{x}_r) \subset \mathbb{R}^n$ is a polyhedron. Indeed, Definition 5.6 and Theorem 2.75 show that Γ is the intersection of r closed halfplanes

$$D_i = \{(1 - \lambda_1 - \cdots - \lambda_r)\boldsymbol{a} + \lambda_1 \boldsymbol{x}_1 + \cdots + \lambda_r \boldsymbol{x}_r : \lambda_i \geqslant 0\}, \ 1 \leqslant i \leqslant r.$$

If $r = n$, then every D_i is a halfspace, and C is the intersection of n closed halfspaces. Let $1 \leqslant r \leqslant n - 1$. Then every D_i is the intersection of the r-dimensional plane aff Γ and a suitable closed halfspace V_i, $1 \leqslant i \leqslant r$. So, $\Gamma = \text{aff}\, \Gamma \cap V_1 \cap \cdots \cap V_r$. Since the r-dimensional plane aff C is the intersection of $n - r + 1$ closed halfspaces (see the first example above), Γ is the intersection of $(n - r + 1) + r = n + 1$ closed halfspaces.

Theorem 13.3. *Let a nonempty proper polyhedron $P \subset \mathbb{R}^n$ be represented by a family $\mathcal{F} = \{V_1, \ldots, V_r\}$ of closed halfspaces. Then $\mathcal{F}$ can be partitioned into two subfamilies, say $\mathcal{G} = \{V_1, \ldots, V_s\}$ and $\mathcal{H} = \{V_{s+1}, \ldots, V_r\}$, such that the following assertions hold (let $\mathcal{G} = \varnothing$ if P is a plane and $\mathcal{H} = \varnothing$ if $\dim P = n$).*

(*a*) *Every set $D_i = V_i \cap \operatorname{aff} P$, $1 \leqslant i \leqslant s$, is a closed halfplane of* aff P, *and*

$$P = D_1 \cap \cdots \cap D_s = V_1 \cap \cdots \cap V_s \cap \operatorname{aff} P. \tag{13.2}$$

(*b*) *Furthermore,*

$$\begin{aligned} \operatorname{rint} P &= \operatorname{rint} D_1 \cap \cdots \cap \operatorname{rint} D_s \\ &= \operatorname{int} V_1 \cap \cdots \cap \operatorname{int} V_s \cap \operatorname{aff} P, \end{aligned} \tag{13.3}$$

$$\begin{aligned} \operatorname{rbd} P &= (P \cap \operatorname{rbd} D_1) \cup \cdots \cup (P \cap \operatorname{rbd} D_s) \\ &= (P \cap \operatorname{bd} V_1) \cup \cdots \cup (P \cap \operatorname{bd} V_s). \end{aligned} \tag{13.4}$$

(*c*) *Every halfspace V_i, $s + 1 \leqslant i \leqslant r$, contains* aff P, *and*

$$\operatorname{aff} P = V_{s+1} \cap \cdots \cap V_r. \tag{13.5}$$

Proof. First, we consider the special cases when P is a plane or $\dim P = n$.

If P is a plane, then equalities (13.2), (13.3), and (13.4) are excluded due to $\mathcal{G} = \varnothing$, and (13.5) becomes (13.1).

Similarly, if $\dim P = n$, then (13.5) is excluded due to $\mathcal{H} = \varnothing$. Furthermore, $\operatorname{aff} P = \mathbb{R}^n$, $D_i = V_i$, $1 \leqslant i \leqslant r$, and (13.2) becomes (13.1). Equalities (13.3) and (13.4) hold due to the equalities

$$\begin{aligned} \operatorname{int} P &= \operatorname{int} (V_1 \cap \cdots \cap V_r) = \operatorname{int} V_1 \cap \cdots \cap \operatorname{int} V_r, \\ \operatorname{bd} P &= \operatorname{bd} (V_1 \cap \ldots V_r) = (P \cap \operatorname{bd} V_1) \cup \cdots \cup (P \cap \operatorname{bd} V_s), \end{aligned}$$

and (13.4) is excluded because $\mathcal{H} = \varnothing$.

So, we may assume that P is not a plane and $\dim P \leqslant n - 1$.

(*a*) Denote by $\mathcal{G}$ the family of all halfspaces from $\mathcal{F}$ which do not contain aff P. We observe that $\mathcal{G} \neq \varnothing$, since otherwise

$$P \subset \operatorname{aff} P \subset V_1 \cap \cdots \cap V_r = P,$$

contrary to the assumption $P \neq \operatorname{aff} P$. Without loss of generality, we may suppose that $\mathcal{G} = \{V_1, \ldots, V_s\}$, $1 \leqslant s \leqslant r$. Then every set $D_i = V_i \cap \operatorname{aff} P$, $1 \leqslant i \leqslant s$, is a closed halfplane of aff P (see Definition 2.69). Clearly,

$$P \subset D_1 \cap \cdots \cap D_s = V_1 \cap \cdots \cap V_s \cap \operatorname{aff} P.$$

For the opposite inclusion, choose a point $\boldsymbol{z} \in \operatorname{aff} P \setminus P$. Then $\boldsymbol{z} \in \operatorname{aff} P \setminus V_i$ for a suitable halfspace $V_i \in \mathcal{F}$. Therefore, $V_i \in \mathcal{G}$, which gives

$$V_1 \cap \cdots \cap V_s \cap \operatorname{aff} P \subset P.$$

Summing up, equalities (13.2) hold.

(*b*) The inclusions $P \subset D_i \subset \operatorname{aff} P$, $1 \leqslant i \leqslant r$, give $\operatorname{aff} P = \operatorname{aff} D_i$, as follows from Theorem 2.42. Therefore, $\operatorname{rint} P \subset \operatorname{rint} D_i$ according to Theorem 3.18. A combination of Theorems 3.29 and 2.70 gives

$$\begin{aligned}\operatorname{rint} P &= \operatorname{rint}(D_1 \cap \cdots \cap D_s) = \operatorname{rint} D_1 \cap \cdots \cap \operatorname{rint} D_s \\ &= (\operatorname{int} V_1 \cap \operatorname{aff} P) \cap \cdots \cap (\operatorname{int} V_s \cap \operatorname{aff} P) \\ &= \operatorname{int} V_1 \cap \cdots \cap \operatorname{int} V_s \cap \operatorname{aff} P.\end{aligned}$$

Equalities (13.4) follow from (13.3) (see Corollary 3.64 and Problem 3.17):

$$\begin{aligned}\operatorname{rbd} P &= (P \cap \operatorname{rbd} D_1) \cup \cdots \cup (P \cap \operatorname{rbd} D_s) \\ &= (P \cap (\operatorname{bd} V_1 \cap \operatorname{aff} P)) \cup \cdots \cup (P \cap (\operatorname{bd} V_s \cap \operatorname{aff} P)) \\ &= (P \cap \operatorname{bd} V_1) \cup \cdots \cup (P \cap \operatorname{bd} V_s).\end{aligned}$$

(*c*) First, we assert that the family $\mathcal{H} = \mathcal{F} \setminus \mathcal{G}$ is nonempty. For this, choose a point $\boldsymbol{u} \in \operatorname{rint} P$. There should be at least on halfspace $V_i \in \mathcal{F}$ with the property $\boldsymbol{u} \in \operatorname{bd} V_i$. Indeed, otherwise

$$\boldsymbol{u} \in \operatorname{int} V_1 \cap \cdots \cap \operatorname{int} V_r = \operatorname{int}(V_1 \cap \cdots \cap V_r) = \operatorname{int} P,$$

contrary to the assumption $\dim P \leqslant n-1$ (see Corollary 3.21). If a halfspace $V_i \in \mathcal{F}$ satisfies the condition $\boldsymbol{u} \in \operatorname{bd} V_i$, then Theorem 3.67 shows that $P \subset \operatorname{bd} V_i$. Because $\operatorname{bd} V_i$ is a hyperplane, one has $\operatorname{aff} P \subset \operatorname{bd} V_i \subset V_i$. Summing up, $\mathcal{H} \neq \varnothing$. Equivalently, $s < r$.

Denote by $\mathcal{H}'$ the family of all halfspaces V_i from $\mathcal{H}$ with the property $\boldsymbol{u} \in \operatorname{bd} V_i$. As shown above, $\mathcal{H}' \neq \varnothing$, and $\operatorname{aff} P \subset \operatorname{bd} V_i$ for every $V_i \in \mathcal{H}'$. Renumbering the elements of $\mathcal{H}$, we suppose that $\mathcal{H}' = \{V_{s+1}, \dots, V_t\}$, where $s+1 \leqslant t \leqslant r$. Then

$$\operatorname{aff} P \subset \operatorname{bd} V_{s+1} \cap \cdots \cap \operatorname{bd} V_t \subset V_{s+1} \cap \cdots \cap V_t.$$

We assert that

$$\operatorname{aff} P = \operatorname{bd} V_{s+1} \cap \cdots \cap \operatorname{bd} V_t = V_{s+1} \cap \cdots \cap V_t. \tag{13.6}$$

Assume for a moment the existence of a point

$$\boldsymbol{v} \in (V_{s+1} \cap \cdots \cap V_t) \setminus \operatorname{aff} P.$$

Then $[\boldsymbol{u}, \boldsymbol{v}] \subset V_{s+1} \cap \cdots \cap V_t$ by a convexity argument, while $(\boldsymbol{u}, \boldsymbol{v}) \cap \operatorname{aff} P = \varnothing$ (see Theorem 2.35). Furthermore, $\boldsymbol{u} \in \operatorname{int} V_i$ for every halfspace

$V_i \in \mathcal{F} \backslash \mathcal{H}'$. Indeed, $\boldsymbol{u} \in \operatorname{int} V_i$ for all $1 \leqslant i \leqslant s$ due to (13.3), and $\boldsymbol{u} \in \operatorname{int} V_i$ for all $i \geqslant t+1$ (if $t \leqslant r-1$) by the choice of $\mathcal{H}'$. Therefore, there is a scalar $\rho > 0$ such that $B_\rho(\boldsymbol{u}) \subset V_i$ for all $V_i \in \mathcal{F} \setminus \mathcal{H}'$. Choose a point $\boldsymbol{w} \in (\boldsymbol{u}, \boldsymbol{v})$ such that $\|\boldsymbol{u} - \boldsymbol{w}\| \leqslant \rho$. By the above argument,

$$\begin{aligned} \boldsymbol{w} &\in \cap (V_i : V_i \in \mathcal{H}') \cap B_\rho(\boldsymbol{u}) \subset \cap (V_i : V_i \in \mathcal{H}') \cap (V_i : V_i \in \mathcal{F} \setminus \mathcal{H}') \\ &= V_1 \cap \cdots \cap V_r = P, \end{aligned}$$

in contradiction with $(\boldsymbol{u}, \boldsymbol{v}) \cap \operatorname{aff} P = \varnothing$. Summing up, equalities (13.6) hold. Finally, if $t \leqslant r - 1$, then every halfspace V_i, $t + 1 \leqslant i \leqslant r$, contains $\operatorname{aff} P$, which gives

$$\operatorname{aff} P = V_{s+1} \cap \cdots \cap V_t = V_{s+1} \cap \cdots \cap V_r. \qquad \square$$

Remark. It might happen that a polyhedron $P \subset \mathbb{R}^n$ is the intersection of an infinite family of closed halfspaces; furthermore, this family may not contain a finite subfamily whose intersection is P. For instance, the solid triangle

$$P = \{(x, y) : x \geqslant 0,\ y \geqslant 0,\ x + y \leqslant 1\} \subset \mathbb{R}^2$$

is the intersection of the infinite family $\mathcal{F}$ of closed halfplanes

$$\begin{aligned} V = \{(x, y) : x \geqslant 0\}, \quad V' = \{(x, y) : y \geqslant 0\}, \\ V_k = \{(x, y) : y \leqslant \tfrac{k+1}{k}(1 - x)\}, \quad k \geqslant 1. \end{aligned}$$

It is easy to see that no finite subfamily of $\mathcal{F}$ represents P.

13.1.2 *Irredundant Systems of Halfspaces*

Definition 13.4. A family $\mathcal{F} = \{V_1, \dots, V_r\}$ of closed halfspaces representing a nonempty proper polyhedron $P \subset \mathbb{R}^n$ is called *irredundant* (also called *irreducible*) provided either $r = 1$, or $r \geqslant 2$ and

$$P \neq V_1 \cap \cdots \cap V_{i-1} \cap V_{i+1} \cap \cdots \cap V_r \quad \text{for all} \quad 1 \leqslant i \leqslant r.$$

Similarly, if P is not a plane, then a family $\mathcal{D} = \{D_1, \dots, D_s\}$ of closed halfplanes of the plane $\operatorname{aff} P$ which represent P is called *irredundant* if either $s = 1$, or $s \geqslant 2$ and

$$P \neq D_1 \cap \cdots \cap D_{i-1} \cap D_{i+1} \cap \cdots \cap D_s \quad \text{for all} \quad 1 \leqslant i \leqslant s.$$

Remark. If a family $\mathcal{F} = \{V_1, \dots, V_r\}$ of closed halfspaces represents a nonempty proper polyhedron $P \subset \mathbb{R}^n$, then, by a finiteness argument, $\mathcal{F}$ contains an irredundant subfamily.

Theorem 13.5. *Let a nonempty proper polyhedron $P \subset \mathbb{R}^n$ be represented by a family $\mathcal{F} = \{V_1, \dots, V_r\}$ of closed halfspaces. If $\mathcal{F}$ is partitioned into subfamilies, say $\mathcal{G} = \{V_1, \dots, V_s\}$ and $\mathcal{H} = \{V_{s+1}, \dots, V_r\}$, satisfying conditions (a)–(c) of Theorem 13.3, then the following assertions hold (let $s = 0$ if P is a plane and $s = r$ if $\dim P = n$).*

(*a*) *$\mathcal{F}$ is irredundant if and only if both families $\mathcal{G}$ and $\mathcal{H}$ are irredundant with respect to conditions (13.2) and (13.5).*

(*b*) *If P is not a plane, then $\mathcal{G}$ is irredundant if and only if either $s = 1$, or $s \geqslant 2$ and*

$$\operatorname{bd} V_i \cap \operatorname{rint}(V_1 \cap \dots \cap V_{i-1} \cap V_{i+1} \cap \dots \cap V_r) \neq \varnothing, \quad 1 \leqslant i \leqslant s. \tag{13.7}$$

(*c*) *If P is not a plane, then the family $\mathcal{D} = \{D_1, \dots, D_s\}$ of closed halfplanes $D_i = V_i \cap \operatorname{aff} P$ of $\operatorname{aff} P$, $1 \leqslant i \leqslant s$, is irredundant if and only if either $s = 1$, or $s \geqslant 2$ and*

$$\operatorname{rbd} D_i \cap \operatorname{rint}(D_1 \cap \dots \cap D_{i-1} \cap D_{i+1} \cap \dots \cap D_s) \neq \varnothing, \quad 1 \leqslant i \leqslant s. \tag{13.8}$$

(*d*) *If $\dim P \leqslant n - 1$, then $\mathcal{H}$ is irredundant if and only if*

$$\operatorname{aff} P = \operatorname{bd} V_{s+1} \cap \dots \cap \operatorname{bd} V_r = V_{s+1} \cap \dots \cap V_r \tag{13.9}$$

and none of the sets

$$C_i = V_{s+1} \cap \dots \cap V_{i-1} \cap V_{i+1} \cap \dots \cap V_r, \quad s + 1 \leqslant i \leqslant r,$$

is a plane.

Proof. (*a*) This part of the proof follows from Theorem 13.3 and the method the subfamilies $\mathcal{G}$ and $\mathcal{H}$ were chosen.

(*b*) Let $\mathcal{G}$ be irredundant. Suppose for a moment that $s \geqslant 2$ and at least one of the halfspaces from $\mathcal{G}$, say V_1, does not satisfy (13.7). Then $\operatorname{rint}(V_2 \cap \dots \cap V_s) \subset \operatorname{int} V_1$, and Theorem 3.48 gives

$$V_2 \cap \dots \cap V_s = \operatorname{cl}(\operatorname{rint}(V_2 \cap \dots \cap V_s)) \subset \operatorname{cl}(\operatorname{int} V_1) = V_1.$$

Consequently, $P = V_2 \cap \dots \cap V_s \cap \operatorname{aff} P$, contrary to irredundancy of $\mathcal{G}$.

Conversely, if $s \geqslant 2$ and $\mathcal{G}$ satisfies (13.7), then no halfspace V_i, $1 \leqslant i \leqslant s$, can be omitted in the representation $P = V_1 \cap \dots \cap V_r$, implying that $\mathcal{G}$ is irredundant.

(*c*) An equivalent assertion for the halfplanes D_i follows from the relations $\operatorname{rbd} D_i = \operatorname{bd} V_i \cap \operatorname{aff} P$, $1 \leqslant i \leqslant s$ (see Corollary 3.64).

(*d*) Suppose first that $\mathcal{H}$ is irredundant. Then (13.9) follows from the proof of assertion (*c*) of Theorem 13.3. From (13.7) one has $C_i \not\subset V_i$ for all $s + 1 \leqslant i \leqslant r$. Assume for a moment that a set C_i, $s + 1 \leqslant i \leqslant r$, is a plane. Then C_i contains a point $\boldsymbol{u} \in \operatorname{int} V_i$ (see Theorem 2.70). On the other hand, (13.9) implies that

$$\boldsymbol{u} \in C_i \cap \operatorname{int} V_i \subset C_i \cap V_i = \operatorname{aff} P = \operatorname{bd} V_{s+1} \cap \cdots \cap \operatorname{bd} V_r \subset \operatorname{bd} V_i,$$

a contradiction.

Conversely, suppose that (13.9) holds, and that no set C_i, $s+1 \leqslant i \leqslant r$, is a plane. Since $V_{s+1} \cap \cdots \cap V_r$ is a plane, one has

$$C_i \neq V_{s+1} \cap \cdots \cap V_r = \operatorname{aff} P, \quad s+1 \leqslant i \leqslant r.$$

Therefore, the family $\mathcal{H}$ is irredundant. □

Remark. Since $\operatorname{rbd} D_i = \operatorname{bd} V_i \cap \operatorname{aff} P$, $1 \leqslant i \leqslant s$, conditions (13.8) can be written as

$$\operatorname{bd} V_i \cap \operatorname{rint}(D_1 \cap \cdots \cap D_{i-1} \cap D_{i+1} \cap \cdots \cap D_s) \neq \varnothing, \quad 1 \leqslant i \leqslant s. \quad (13.10)$$

Theorem 13.6. *If a polyhedron $P \subset \mathbb{R}^n$ of positive dimension is not a plane, then there is a unique irredundant family of closed halfplanes of* aff P *representing* P.

Proof. Let $\mathcal{G} = \{D_1, \ldots, D_s\}$ and $\mathcal{G}' = \{D'_1, \ldots, D'_t\}$ be irredundant families of closed halfplanes of aff P both representing P:

$$P = D_1 \cap \cdots \cap D_s = D'_1 \cap \cdots \cap D'_t.$$

By a symmetry argument, it suffices to show that $\mathcal{G} \subset \mathcal{G}'$. Choose a halfplane $D_i \in \mathcal{G}$. Clearly, $\operatorname{rbd} D_i$ is a plane of dimension $m-1$, where $m = \dim P$ (see an example on page 113). Consider the convex set $K_i = \operatorname{rbd} D_i \cap P$. We assert that

$$K_i \subset \operatorname{rbd} P \quad \text{and} \quad \operatorname{aff} K_i = \operatorname{rbd} D_i.$$

Indeed, the inclusion $K_i \subset \operatorname{rbd} P$ follows from (13.4). Let

$$C_i = D_1 \cap \cdots \cap D_{i-1} \cap D_{i+1} \cap \cdots \cap D_s.$$

The inclusion $P \subset C_i \subset \operatorname{aff} P$ gives $\operatorname{aff} C_i = \operatorname{aff} P_i$ (see Theorem 2.42). Furthermore,

$$K_i = \operatorname{rbd} D_i \cap P = (\operatorname{rbd} D_i \cap D_i) \cap C_i = \operatorname{rbd} D_i \cap C_i.$$

Since (13.8) can be rewritten as $\operatorname{rbd} D_i \cap \operatorname{rint} C_i \neq \varnothing$, Theorem 3.38 gives

$$\operatorname{aff} K_i = \operatorname{rbd} D_i \cap \operatorname{aff} C_i = \operatorname{rbd} D_i \cap \operatorname{aff} P = \operatorname{rbd} D_i.$$

Consequently,

$$\dim K_i = \dim(\operatorname{aff} K_i) = \dim(\operatorname{rbd} D_i) = m-1.$$

Now, choose a point $\boldsymbol{u} \in \operatorname{rint} K_i$. Since $\boldsymbol{u} \in K_i \subset \operatorname{rbd} P$, the equality

$$\operatorname{rbd} P = (P \cap \operatorname{rbd} D'_1) \cup \cdots \cup (P \cap \operatorname{rbd} D'_t)$$

implies the existence of a halfplane $D'_j \in \mathcal{G}'$ such that $\boldsymbol{u} \in P \cap \operatorname{rbd} D'_j$. By Theorem 3.67, $K_i \subset \operatorname{rbd} D'_j$. Because $\dim K_i = \dim(\operatorname{rbd} D'_j) = m - 1$, Corollary 2.65 shows that $\operatorname{rbd} D_i = \operatorname{aff} K_i = \operatorname{rbd} D'_j$.

Finally, since both closed halfplanes D_i and D'_j contain P and are bounded by the same $(m-1)$-dimensional subplane $\operatorname{rbd} D_i = \operatorname{rbd} D'_j$ of $\operatorname{aff} P$, Theorem 2.70 gives $D_i = D'_j$. So, $D_i \in \mathcal{G}'$, implying the inclusion $\mathcal{G} \subset \mathcal{G}'$. □

Remark. A similar to Theorem 13.6 assertion does not hold in the case of irredundant representations by halfspaces. For instance, the halfplane $P = \{(x, y, 0) : x \geqslant 0\} \subset \mathbb{R}^3$ can be irredundantly expressed by closed halfspaces in distinct ways. For instance,

$$P = V_1 \cap V_2 \cap V_3 \quad \text{and} \quad P = V_1 \cap V_2 \cap V'_3,$$

where V_1 and V_2 are given by the inequalities $z \geqslant 0$ and $z \leqslant 0$, respectively;

$$V_3 = \{(x, y, z) : z \leqslant x\} \quad \text{and} \quad V'_3 = \{(x, y, z) : z \leqslant -x\}.$$

Theorem 13.7. *If a nonempty proper plane $L \subset \mathbb{R}^n$ of dimension m is represented by an irredundant family $\mathcal{H} = \{V_1, \dots, V_p\}$ of closed halfspaces, then*

$$n - m + 1 \leqslant p \leqslant 2(n - m).$$

Proof. Since the family $\mathcal{H}$ is irredundant, one has $L = \operatorname{bd} V_1 \cap \cdots \cap \operatorname{bd} V_p$ according to Theorem 13.5. Choose a vector $\boldsymbol{u} \in L$ and translate all $V_1, \dots, V_p$ on the same vector $-\boldsymbol{u}$. Then L becomes an m-dimensional subspace and every V_i can be written as

$$V_i = \{\boldsymbol{x} \in \mathbb{R}^n : \boldsymbol{x} \cdot \boldsymbol{c}_i \leqslant 0\} = \{\boldsymbol{c}_i\}^\circ, \quad 1 \leqslant i \leqslant p.$$

By Theorem 8.3,

$$L = V_1 \cap \cdots \cap V_p = \{\boldsymbol{c}_1\}^\circ \cap \cdots \cap \{\boldsymbol{c}_p\}^\circ = \{\boldsymbol{c}_1, \dots, \boldsymbol{c}_p\}^\circ,$$

and a combination of Theorems 8.2 and 5.46 gives

$$\begin{aligned} L^\perp &= L^\circ = (\{\boldsymbol{c}_1, \dots, \boldsymbol{c}_p\}^\circ)^\circ \\ &= \operatorname{cl}(\operatorname{cone}_{\boldsymbol{o}}\{\boldsymbol{c}_1, \dots, \boldsymbol{c}_p\}) = \operatorname{cone}_{\boldsymbol{o}}\{\boldsymbol{c}_1, \dots, \boldsymbol{c}_p\}. \end{aligned}$$

Hence $\operatorname{cone}_{\boldsymbol{o}}\{\boldsymbol{c}_1, \dots, \boldsymbol{c}_p\}$ is an $(n-m)$-dimensional subspace. Corollary 5.54 shows the existence of a set $Y \subset \{\boldsymbol{c}_1, \dots, \boldsymbol{c}_p\}$ such that

$$\operatorname{cone}_{\boldsymbol{o}} Y = \operatorname{cone}_{\boldsymbol{o}}\{\boldsymbol{c}_1, \dots, \boldsymbol{c}_p\} \quad \text{and} \quad n - m + 1 \leqslant \operatorname{card} Y \leqslant 2(n - m).$$

Let $Y = \{\boldsymbol{c}_{i_1}, \ldots, \boldsymbol{c}_{i_r}\}$, where $r = \operatorname{card} Y$ and $1 \leqslant i_1 < \cdots < i_r \leqslant p$. As above,

$$L = (L^\circ)^\circ = (\operatorname{cone}_{\boldsymbol{o}} Y)^\circ = (\operatorname{cone}_{\boldsymbol{o}}\{\boldsymbol{c}_{i_1}, \ldots, \boldsymbol{c}_{i_r}\})^\circ$$
$$= \{\boldsymbol{c}_{i_1}, \ldots, \boldsymbol{c}_{i_r}\}^\circ = \{\boldsymbol{c}_{i_1}\}^\circ \cap \cdots \cap \{\boldsymbol{c}_{i_r}\}^\circ = V_{i_1} \cap \cdots \cap V_{i_r}.$$

Since the representation $L = V_1 \cap \cdots \cap V_p$ is irredundant, one has

$$\{V_1, \ldots, V_p\} = \{V_{i_1}, \ldots, V_{i_r}\},$$

which gives the equality $Y = \{\boldsymbol{c}_1, \ldots, \boldsymbol{c}_p\}$. Hence $p = r$. □

A combination of Theorems 13.5, 13.6, and 13.7 implies the following corollary.

Corollary 13.8. *For a nonempty proper polyhedron $P \subset \mathbb{R}^n$ of positive dimension m, there is a nonnegative integer s such that every irredundant family $\mathcal{F} = \{V_1, \ldots, V_r\}$ of closed halfspaces representing P is uniquely partitioned into subfamilies $\mathcal{G}$ and $\mathcal{H}$, say $\mathcal{G} = \{V_1, \ldots, V_s\}$ and $\mathcal{H} = \{V_{s+1}, \ldots, V_r\}$ (let $s = 0$ if P is a plane and $s = r$ if $\dim P = n$), satisfying conditions (a)–(c) of Theorem 13.5. Furthermore, $n - m + 1 \leqslant r - s \leqslant 2(n - m)$.* □

Remark. There is no upper bound on each of the numbers s and r in Corollary 13.8. For instance, if $P \subset \mathbb{R}^2$ is a convex polygon with s sides, then $s = r$ and every irredundant family of closed halfplanes representing P consists of exactly s members.

13.2 Faces and Representations of Polyhedra

13.2.1 *Faces of Polyhedra*

Definition 13.9. Let $P \subset \mathbb{R}^n$ be a polyhedron which is not a plane, and $\mathcal{F} = \{V_1, \ldots, V_r\}$ be an irredundant family of closed halfspaces representing P. If the subfamily $\mathcal{G} = \{V_1, \ldots, V_s\}$ of $\mathcal{F}$ satisfies conditions (a) and (b) of Theorem 13.5, then the sets

$$Q_i = \operatorname{bd} V_i \cap P, \quad 1 \leqslant i \leqslant s,$$

are called *facets* of P.

The following examples of facets immediately follow from examples on page 450.

Examples. **1.** A closed halfplane $D \subset \mathbb{R}^n$ has a unique facet, $\operatorname{rbd} D$.

2. The facets of an r-simplex $\Delta(\boldsymbol{x}_1, \ldots, \boldsymbol{x}_{r+1}) \subset \mathbb{R}^n$, $r \geqslant 1$, are exactly the $(r-1)$-simplices

$$\Delta_i = \Delta(\boldsymbol{x}_1, \ldots, \boldsymbol{x}_{i-1}, \boldsymbol{x}_{i+1}, \ldots, \boldsymbol{x}_{r+1}), \quad 1 \leqslant i \leqslant r+1.$$

3. The facets of an r-simplicial cone $\Gamma_{\boldsymbol{a}}(\boldsymbol{x}_1, \ldots, \boldsymbol{x}_r) \subset \mathbb{R}^n$, $r \geqslant 2$, are exactly the $(r-1)$-simplicial cones

$$\Gamma_{\boldsymbol{a}}(\boldsymbol{x}_1, \ldots, \boldsymbol{x}_{i-1}, \boldsymbol{x}_{i+1}, \ldots, \boldsymbol{x}_r), \quad 1 \leqslant i \leqslant r.$$

Remark. In terms of Definition 13.9, every facet Q_i of P can be expressed as

$$Q_i = \operatorname{rbd} D_i \cap P, \quad \text{where} \quad D_i = V_i \cap \operatorname{aff} P, \quad 1 \leqslant i \leqslant s. \tag{13.11}$$

A combination of Theorems 13.5 and 13.6 shows that the family of facets of a polyhedron $P \subset \mathbb{R}^n$ is uniquely determined and does not depend on the choice of an irredundant family $\mathcal{F}$ of closed halfspaces representing P.

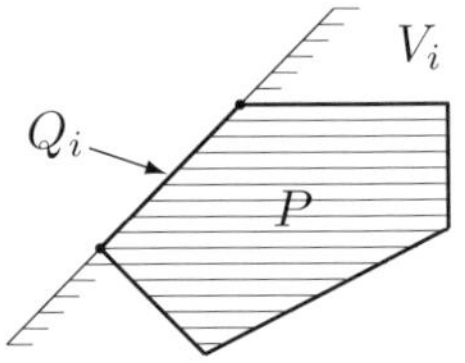

Fig. 13.2 A facet Q_i of the polyhedron P.

Theorem 13.10. *Let $P \subset \mathbb{R}^n$ be a polyhedron of positive dimension m, which is not a plane. Then every facet of P is a polyhedron. Furthermore, a set F is a facet of P if and only if F is an $(m-1)$-dimensional exposed face of P.*

Proof. Let $\mathcal{F} = \{V_1, \ldots, V_r\}$ be an irredundant family of closed halfspaces representing P, and let $Q_i = \operatorname{bd} V_i \cap P$, $1 \leqslant i \leqslant s$, be the facets of P.

Denote by V_i' the opposite to V_i closed halfspace determined by the hyperplane $\operatorname{bd} V_i$, $1 \leqslant i \leqslant s$. Clearly, $\operatorname{bd} V_i = V_i \cap V_i'$, which gives

$$Q_i = \operatorname{bd} V_i \cap P = (V_i \cap V_i') \cap (V_1 \cap \cdots \cap V_r).$$

Therefore, Q_i is a polyhedron as a finite intersection of closed halfspaces.

For the second assertion, suppose first that F is a facet of P, say $F = Q_i$. Because the hyperplane $\operatorname{bd} V_i$ properly supports P along Q_i, the set Q_i is a nonempty proper exposed face of P (see Definition 12.1). By Corollary 12.6, $\dim Q_i \leqslant \dim P - 1 = m - 1$. For the opposite inequality, since $P \subset M$, where

$$M = D_1 \cap \cdots \cap D_{i-1} \cap D_{i+1} \cap \cdots \cap D_s$$

and $D_j = V_j \cap \operatorname{aff} P$, $1 \leqslant j \leqslant s$, we have $\dim P \leqslant \dim M$. Combining (13.10) and Corollary 3.42, one has

$$\dim Q_i = \dim (\operatorname{bd} V_i \cap M) = \dim M - 1 \geqslant \dim P - 1 = m - 1.$$

Summing up, $\dim Q_i = m - 1$.

Conversely, suppose that F is an $(m-1)$-dimensional exposed face of P. Choose a hyperplane $H \subset \mathbb{R}^n$ satisfying the condition $H \cap P = F$. One of the closed halfspaces, say V, determined by H properly supports P along F. Clearly, V does not contain $\operatorname{aff} P$; so, $V \cap \operatorname{aff} P$ is a closed halfplane of $\operatorname{aff} P$. As shown in the proof of Theorem 13.6, there is an index $1 \leqslant i \leqslant s$ such that $V \cap \operatorname{aff} P = V_i \cap \operatorname{aff} P$. Therefore,

$$\begin{aligned} F &= \operatorname{bd} V \cap P = (\operatorname{bd} V \cap \operatorname{aff} P) \cap P = \operatorname{rbd} (V \cap \operatorname{aff} P) \cap P \\ &= \operatorname{rbd} (V_i \cap \operatorname{aff} P) \cap P = (\operatorname{bd} V_i \cap \operatorname{aff} P) \cap P = \operatorname{bd} V_i \cap P = Q_i. \end{aligned}$$

Hence F is a facet of P. □

Theorem 13.11. *Let $P \subset \mathbb{R}^n$ be a polyhedron of positive dimension m, which is not a plane. Then the following assertions hold.*

(*a*) *Any nonempty proper extreme face of P lies within a facet of P.*
(*b*) *The relative boundary of P is the union of all facets of P.*
(*c*) *Every extreme face of P is a polyhedron.*
(*d*) *Given nonempty proper extreme faces F_j and F_k of P such that $F_j \subset F_k$ and $\dim F_j = j < \dim F_k = k$, there is a sequence of extreme faces F_i of P satisfying the conditions*

$$F_j \subset F_{j+1} \subset \cdots \subset F_k \quad \textit{and} \quad \dim F_i = i, \quad j \leqslant i \leqslant k.$$

(*e*) *Every nonempty proper extreme face F of P is the intersection of all facets of P containing F.*

Proof. Let $\mathcal{F} = \{V_1, \dots, V_r\}$ be an irredundant family of closed halfspaces representing P such that $Q_i = \operatorname{bd} V_i \cap P$, $1 \leqslant i \leqslant s$, are the facets of P.

(*a*) Choose a nonempty proper extreme face F of P. Then $F \subset \operatorname{rbd} P$ according to Theorem 11.4. Let $\boldsymbol{x}$ be a point in $\operatorname{rint} F$. Then $\boldsymbol{x} \in \operatorname{rbd} P$, and Theorem 13.3 shows the existence of a halfspace V_i, $1 \leqslant i \leqslant s$, such that $\boldsymbol{x} \in \operatorname{bd} V_i \cap P$. Then $F \subset \operatorname{bd} V_i$ (see Theorem 3.67), implying the inclusion $F \subset \operatorname{bd} V_i \cap P = Q_i$.

(b) According to Theorem 13.10, every facet F of P is a $(m-1)$-dimensional exposed face of P, while Theorem 11.4 asserts that $F \subset \operatorname{rbd} P$. Hence the union of all facets of P lies in $\operatorname{rbd} P$. Conversely, any point $\boldsymbol{x} \in \operatorname{rbd} P$ lies within the generated exposed face $G_{\boldsymbol{x}}(P) \subset \operatorname{rbd} P$ (see Theorem 12.11). Since $G_{\boldsymbol{x}}(P)$ is a nonempty proper exposed face of P, Theorem 13.10 implies the existence of a facet G of P which contains $G_{\boldsymbol{x}}(P)$. In particular, $\boldsymbol{x} \in G$. This argument shows that $\operatorname{rbd} P$ lies in the union of all facets of P.

The proofs of assertions (c)–(e) use induction on $m = \dim P$. Let $m = 1$. Then P is either a closed segment, $[\boldsymbol{a}, \boldsymbol{b}]$, or a closed halfline with an endpoint $\boldsymbol{a}$. Then $\{\boldsymbol{a}\}$ and $\{\boldsymbol{b}\}$ (respectively, $\{\boldsymbol{a}\}$) are all nonempty proper extreme faces of P. In this case, all three assertions (c)–(e) hold.

Assume that assertions (c)–(e) hold for all $m \leqslant r-1$, where $2 \leqslant r \leqslant n$, and let $P \subset \mathbb{R}^n$ be a polyhedron of dimension r which is not a plane.

(c) Let F be an extreme face of P. If $F = \varnothing$ or $F = P$, then P is a polyhedron. Assume that F is a nonempty proper extreme face of P. By assertion (a) above, there is a facet Q_i of P containing F. According to Theorem 13.10, Q_i is a polyhedron of dimension $r-1$. Since F is an extreme face of Q_i (see Theorem 11.3), the induction hypothesis implies that F is a polyhedron.

(d) The case $k - j = 1$ is obvious. So, we assume that $k - j \geqslant 2$. Since F_k is a polyhedron, and since F_j is an extreme face of F_k, assertion (a) shows the existence of a facet, say F_{k-1} of F_k containing F_j. By the induction hypothesis, there is a sequence $F_j \subset F_{j+1} \subset \cdots \subset F_{k-1}$ of extreme faces of F_k such that $\dim F_i = i$ for all $j \leqslant i \leqslant k-1$. According to Theorem 11.3, all sets $F_j, \ldots, F_{k-1}$ are extreme faces of P. So, the sequence $F_j \subset F_{j+1} \subset \cdots \subset F_k$ is a desired one.

(e) Let F be a nonempty proper extreme face of P. By assertion (a) above, F lies within a facet Q_i of P. By the induction hypothesis, F is the intersection of all facets of Q_i containing F. We contend that every facet of Q_i is the intersection of two facets of P. Indeed, consider the closed halfplanes $E_j = \operatorname{bd} V_i \cap V_j, \quad j \neq i,\ 1 \leqslant i \leqslant r$. Form $Q_i = \operatorname{bd} V_i \cap P$ we obtain

$$\begin{aligned} Q_i &= \operatorname{bd} V_i \cap (V_1 \cap \cdots \cap V_{i-1} \cap V_{i+1} \cap \cdots \cap V_r) \\ &= E_1 \cap \cdots \cap E_{i-1} \cap E_{i+1} \cap \cdots \cap E_r. \end{aligned}$$

Choose in $\{E_1, \dots, E_{i-1}, E_{i+1}, \dots, E_r\}$ an irredundant subfamily, say $\{E'_1, \dots, E'_p\}$, of closed halfplanes of $\operatorname{bd} V_i$ representing Q_i. Then $\operatorname{rbd} E'_k \cap Q_i$, $1 \leqslant k \leqslant p$, are the facets of Q_i. Thus Q_i can be expressed as the intersection of two facets of P:

$$\begin{aligned}\operatorname{rbd} E'_k \cap Q_i &= \operatorname{rbd}(\operatorname{bd} V_i \cap V'_k) \cap (\operatorname{bd} V_i \cap P)\\ &= (\operatorname{bd} V_i \cap \operatorname{bd} V'_k) \cap (\operatorname{bd} V_i \cap P)\\ &= (\operatorname{bd} V_i \cap P) \cap (\operatorname{bd} V'_k \cap P).\end{aligned}$$

Consequently, F is the intersection of all facets of P containing F. □

Corollary 13.12. *For any polyhedron $P \subset \mathbb{R}^n$, the families of extreme faces and exposed faces of P coincide.*

Proof. Because every exposed face of P is its extreme face (see Theorem 12.2), it suffices to prove the converse assertion. Let F be an extreme face of P. Excluding the obvious cases $F = \varnothing$ and $F = P$, we assume that F is a nonempty proper subset of P. By Theorem 13.11, F is the intersection of facets of P containing F. Since every facet of P is an exposed face of P, Theorem 12.12 implies that F is an exposed face of P. □

Theorem 13.13. *For a nonempty closed convex set $K \subset \mathbb{R}^n$, the following conditions are equivalent.*

(*a*) *K is a polyhedron.*
(*b*) *K has at most finitely many extreme faces.*
(*c*) *K has at most finitely many exposed faces.*

Proof. Excluding the obvious cases, we assume that K is a proper subset of $\mathbb{R}^n$ of positive dimension. If K is a plane, then K is a polyhedron (see an example on page 450) without nonempty proper extreme or exposed faces. So, we suppose that K is not a plane.

(*a*) ⇒ (*b*) If K is a polyhedron, then K has at most finitely many facets. By Theorem 13.11, every nonempty proper extreme face of K is an intersection of its facets. Consequently, K has at most finitely many extreme faces.

(*b*) ⇒ (*c*) This part immediately follows from Theorem 12.2.

(*c*) ⇒ (*a*) Let $G_1, \dots, G_r$, $r \geqslant 1$, be all nonempty proper exposed faces of K. Denote by V_i a closed halfspace containing K such that its boundary hyperplane H_i satisfies the condition $G_i = K \cap H_i$, $1 \leqslant i \leqslant r$. Since every hyperplane H_i properly supports K, no halfspace V_i contains $\operatorname{aff} K$.

We are going to show that K coincides with the polyhedron

$$P = V_1 \cap \cdots \cap V_r \cap \operatorname{aff} K.$$

Due to the obvious inclusion $K \subset P$, it suffices to show that $P \subset K$. Assume for a moment the existence of a point $\boldsymbol{z} \in P \setminus K$. Choose a point $\boldsymbol{x} \in \operatorname{rint} K$. Since $\boldsymbol{z} \in P \subset \operatorname{aff} K$, the open segment $(\boldsymbol{x}, \boldsymbol{z})$ contains a point $\boldsymbol{u} \in \operatorname{rbd} K$ (see Theorem 3.65). By Corollary 12.10, there is a nonempty proper exposed face G of K which contains $\boldsymbol{u}$. By the assumption, $G = G_j$ for a suitable index $j \in \{1, \ldots, r\}$. Since $\boldsymbol{x} \in P \subset V_j$, we have $[\boldsymbol{x}, \boldsymbol{z}] \subset V_j$ due to the convexity of V_j. Furthermore, $\boldsymbol{u} \in H_j = \operatorname{bd} V_j$. Consequently, $[\boldsymbol{x}, \boldsymbol{z}] \subset \langle \boldsymbol{x}, \boldsymbol{z} \rangle \subset H_j$ (see Corollary 2.36). Hence $\boldsymbol{x} \in K \cap H_j = G_j \subset \operatorname{rbd} K$, contrary to the choice of $\boldsymbol{x}$. Summing up, $P \subset K$. □

The following result complements Theorem 13.13.

Theorem 13.14. *For a closed convex set $K \subset \mathbb{R}^n$ distinct from a plane, the following conditions are equivalent.*

- (a) *K is a polyhedron.*
- (b) *K has at most finitely extreme faces which are maximal under inclusion in* $\operatorname{rbd} K$.
- (c) *K has at most finitely exposed faces which are maximal under inclusion in* $\operatorname{rbd} K$.

Proof. The equivalence of conditions (b) and (c) follows from Corollary 13.12, while Theorem 13.13 shows that $(a) \Rightarrow (c)$.

Hence it remains to prove that $(c) \Rightarrow (a)$. The case $\operatorname{rbd} K = \varnothing$ is obvious (in this case, K is a plane according to Theorem 3.62, which is a polyhedron). So, we may assume that $\operatorname{rbd} K \neq \varnothing$. Let $G_1, \ldots, G_r$ be all exposed faces of K which are maximal under inclusion in $\operatorname{rbd} K$. We assert that $G_1 \cup \cdots \cup G_r = \operatorname{rbd} K$. Indeed, the inclusion $G_1 \cup \cdots \cup G_r \subset \operatorname{rbd} K$ is obvious. If $\boldsymbol{x}$ is any point in $\operatorname{rbd} K$, then, by Corollary 3.9, the convex set $\{\boldsymbol{x}\}$ lies in a maximal under inclusion convex subset G of $\operatorname{rbd} K$, and Theorem 12.3 shows that G is an exposed face of K, which is maximal under inclusion in $\operatorname{rbd} K$. Hence $G = G_i$ for a suitable $1 \leqslant i \leqslant r$. Consequently, $\operatorname{rbd} K \subset G_1 \cup \cdots \cup G_r$.

Denote by H_i a hyperplane satisfying the condition $G_i = H_i \cap K$, and let V_i be the closed halfspace determined by H_i and containing K, $1 \leqslant i \leqslant r$. Theorem 9.40 shows that $K = V_1 \cap \cdots \cap V_r \cap \operatorname{aff} K$. Hence K is a polyhedron. □

One more result generalizes Theorem 13.13 by dealing with planar and halfplanar faces (compare with Theorems 11.36 and 12.37).

Theorem 13.15. *For a nonempty closed convex set $K \subset \mathbb{R}^n$, the following conditions are equivalent.*

(a) *K is a polyhedron.*
(b) *K has at most finitely many planar and halfplanar extreme faces.*
(c) *If $S \subset \mathbb{R}^n$ is a subspace complementary to $\operatorname{lin} K$, then $K \cap S$ has at most finitely many extreme points and extreme halflines.*
(d) *K has at most finitely many planar and halfplanar exposed faces.*
(e) *If $S \subset \mathbb{R}^n$ is a subspace complementary to $\operatorname{lin} K$, then $K \cap S$ has at most finitely many exposed points and exposed halflines.*

Proof. Since the equivalence of conditions (a)–(e) is obvious if K is a plane, we may assume that K is not a plane.

(a) $\Rightarrow$ (b) This part follows from Theorem 13.13.

(b) $\Rightarrow$ (c) If $S \subset \mathbb{R}^n$ is a subspace complementary to $\operatorname{lin} K$, then the convex set $K \cap S$ is closed and line-free (see Theorems 6.25 and 6.37). By Theorem 11.5, a subset F of K is an extreme face of K if and only if it can be expressed as $F = E + \operatorname{lin} K$, where E is an extreme face of $K \cap S$. Clearly, F is a planar or halfplanar extreme face of K if and only if so is E for $K \cap S$. Hence condition (b) implies that $K \cap S$ has at most finitely many planar and halfplanar extreme faces. Finally, since $K \cap S$ is line-free, its planar extreme faces are points and its halfplanar extreme faces are halflines.

(c) $\Rightarrow$ (a) First, we assert that the set $K \cap S$ is polyhedral. By Theorem 13.13, it suffices to show that $K \cap S$ has at most finitely many extreme faces. For this, choose an extreme face E of $K \cap S$. Then E is a line-free closed convex set (see Theorem 11.4), and, by Theorem 11.28, E is the convex hull of the set $\operatorname{ext} E \cup \operatorname{extr} E$. Furthermore, $\operatorname{ext} E \subset \operatorname{ext}(K \cap S)$ and $\operatorname{extr} E \subset \operatorname{extr}(K \cap S)$ (see Theorem 11.3). Since the sets $\operatorname{ext}(K \cap S)$ and $\operatorname{extr}(K \cap S)$ are finite, there are at most finitely many sets of the form $\operatorname{conv}(X \cup Y)$, where $X \subset \operatorname{ext}(K \cap S)$ and Y is a family of halflines from $\operatorname{extr}(K \cap S)$. In conclusion, $K \cap S$ has at most finitely many extreme faces.

Because every extreme face of K can be expressed in the form $F = E + \operatorname{lin} K$, where E is an extreme face of $K \cap S$ (see Theorem 6.25), the family of extreme faces of K is finite.

(b) $\Rightarrow$ (d) and (c) $\Rightarrow$ (e). These assertions follow from the fact that every planar exposed face (respectively, every halfplanar exposed face) of

K is its planar extreme face (respectively, every halfplanar extreme face). The last assertion derives from Theorems 11.34 and 12.35.

$(d) \Rightarrow (e)$ This part is analogous to that of $(b) \Rightarrow (c)$, with a reference on Theorem 12.8 instead of Theorem 11.5.

$(e) \Rightarrow (c)$ Since $K \cap S$ has at most finitely many exposed points, Theorem 12.19 shows that $K \cap S$ has at most finitely many extreme points. It remains to show that $K \cap S$ has at most finitely many extreme halflines. For this, we observe first that $K \cap S$ has at most finitely many exposed segments. Indeed, since the endpoints of every exposed segment of $K \cap S$ are extreme points of $K \cap S$, their finiteness implies the finiteness of the family of exposed segments of $K \cap S$. According to Theorem 12.41, the set $\mathrm{ext}_1(K \cap S)$ lies in the closure of $\mathrm{exp}_1(K \cap S)$, which is a finite union of extreme segments and extreme halflines of $K \cap S$. Hence $\mathrm{ext}_1(K \cap S) = \mathrm{exp}_1(K \cap S)$, implying that $K \cap S$ has at most finitely many extreme halflines. □

13.2.2 *Polyhedra as Convex Hulls and Sums*

Theorem 13.16. *A closed convex set $K \subset \mathbb{R}^n$ is a polyhedron if and only if it is the convex hull of finitely many points and closed halflines.*

Proof. Because the case $K = \varnothing$ is obvious, we may suppose that K is nonempty.

1. Assume first that K is a polyhedron. By Theorem 13.15, K has at most finitely many extreme faces.

1a. If K is line-free, then K has at most finitely many extreme points and extreme halflines (which are closed halflines). According to Theorem 11.28, K is the convex hull of these points and halflines.

1b. Suppose that the polyhedron K is not line-free and let $r = \dim(\mathrm{lin}\, K)$. We assert that K is the convex hull of finitely many closed halflines. Choose a basis $\{\boldsymbol{c}_1, \dots, \boldsymbol{c}_r\}$ for $\mathrm{lin}\, K$ and let $\boldsymbol{c}_{r+1} = -(\boldsymbol{c}_1 + \dots + \boldsymbol{c}_r)$. Then $\{\boldsymbol{c}_1, \dots, \boldsymbol{c}_{r+1}\}$ is an affine basis for $\mathrm{lin}\, K$ (see Theorem 2.54). Since $\boldsymbol{o}$ can be written as a positive convex combination

$$\boldsymbol{o} = \frac{1}{r+1}\boldsymbol{o} = \frac{1}{r+1}(\boldsymbol{c}_1 + \dots + \boldsymbol{c}_{r+1}),$$

Consider the closed halflines $h_i = \{\lambda \boldsymbol{c}_i : \lambda \geqslant 0\}$, $1 \leqslant i \leqslant r+1$. Theorem 3.20 gives

$$\boldsymbol{o} \in \mathrm{rint}\, \Delta(\boldsymbol{c}_1, \dots, \boldsymbol{c}_{r+1}) = \mathrm{rint}\,(\mathrm{conv}\, \{\boldsymbol{c}_1, \dots, \boldsymbol{c}_{r+1}\}).$$

Now, combination Theorems 5.38 and 5.48, we obtain

$$\begin{aligned} \mathrm{lin}\, K &= \mathrm{cone}_{\boldsymbol{o}}\{\boldsymbol{c}_1, \dots, \boldsymbol{c}_{r+1}\} = \mathrm{conv}\, C_{\boldsymbol{o}}\{\boldsymbol{c}_1, \dots, \boldsymbol{c}_{r+1}\} \\ &= \mathrm{conv}\,(h_1 \cup \dots \cup h_{r+1}). \end{aligned}$$

Choose a subspace $S \subset \mathbb{R}^n$ complementary to $\operatorname{lin} K$. By Theorem 13.15, the set $K \cap S$ is a polyhedron. Theorem 13.15 shows that $K \cap S$ has at most finitely many extreme points, say $\boldsymbol{x}_1, \ldots, \boldsymbol{x}_p$, and at most finitely many extreme halflines, say $g_1, \ldots, g_q$. Let $\boldsymbol{z}_j$ denote the endpoint of g_j, $1 \leqslant j \leqslant q$. Theorem 11.5 gives

$$F_j = \boldsymbol{x}_j + \operatorname{lin} K, \quad 1 \leqslant j \leqslant p,$$

are the planar extreme faces of K, and

$$G_j = \operatorname{conv}((\boldsymbol{z}_j + \operatorname{lin} K) \cup g_j), \quad 1 \leqslant j \leqslant q,$$

are the extreme halfplanes of K. By the above argument,

$$\begin{aligned} F_j &= \boldsymbol{x}_j + \operatorname{lin} K = \boldsymbol{x}_j + \operatorname{conv}(h_1 \cup \cdots \cup h_{r+1}) \\ &= \operatorname{conv}(h_1(\boldsymbol{x}_j) \cup \cdots \cup h_{r+1}(\boldsymbol{x}_j)), \end{aligned}$$

where $h_i(\boldsymbol{x}_j) = \boldsymbol{x}_j + h_i$, $1 \leqslant i \leqslant r+1$, are closed halflines with common endpoint $\boldsymbol{x}_j$. Similarly (see Problem 4.4),

$$\begin{aligned} G_j &= \operatorname{conv}((\boldsymbol{z}_j + \operatorname{conv}(h_1 \cup \cdots \cup h_{r+1}) \cup g_j)) \\ &= \operatorname{conv}(h_1(\boldsymbol{z}_j) \cup \cdots \cup h_{r+1}(\boldsymbol{z}_j) \cup g_j), \quad 1 \leqslant j \leqslant q, \end{aligned}$$

where $h_i(\boldsymbol{z}_j) = \boldsymbol{z}_j + h_i$, $1 \leqslant i \leqslant r+1$, are closed halflines with common endpoint $\boldsymbol{z}_j$. Finally, a combination of Theorems 11.36 and 4.2 gives

$$\begin{aligned} K &= \operatorname{conv}(F_1 \cup \cdots \cup F_p \cup G_1 \cup \cdots \cup G_q) \\ &= \operatorname{conv}\Big(\bigcup_{j=1}^{p} \operatorname{conv}(h_1(\boldsymbol{x}_j) \cup \cdots \cup h_{r+1}(\boldsymbol{x}_j)) \\ &\qquad \cup \bigcup_{j=1}^{q} \operatorname{conv}(g_j \cup h_1(\boldsymbol{z}_j) \cup \cdots \cup h_{r+1}(\boldsymbol{z}_j))\Big) \\ &= \operatorname{conv}\Big(\bigcup_{j=1}^{p} h_1(\boldsymbol{x}_j) \cup \cdots \cup h_{r+1}(\boldsymbol{x}_j) \\ &\qquad \cup \bigcup_{j=1}^{q} g_j \cup h_1(\boldsymbol{z}_j) \cup \cdots \cup h_{r+1}(\boldsymbol{z}_j)\Big), \end{aligned}$$

implying that K is the convex hull of finitely many closed halflines.

2. Conversely, suppose that the closed convex set K is the convex hull of suitable points $\boldsymbol{x}_1, \ldots, \boldsymbol{x}_p$ and closed halflines $h_1, \ldots, h_q$. Denote by $\boldsymbol{z}_i$ the endpoint of h_i, $1 \leqslant i \leqslant q$, and let

$$X = \{\boldsymbol{x}_1, \ldots, \boldsymbol{x}_p\} \cup h_1 \cup \cdots \cup h_q.$$

If F is a halfplanar extreme face of K, then, by Theorem 11.37, $F = \operatorname{conv}(F \cap X)$. It is easy to see that if a closed halfline h_j meets F, then either $h_j \subset F$ or $h_j \cap F = \{\boldsymbol{z}_j\}$. Consequently, the set $F \cap X$ is the union of a subset of $\{\boldsymbol{x}_1, \ldots, \boldsymbol{x}_p\}$ and some of the closed halflines $h_1, \ldots, h_q$. Because there are at most finitely many sets of the form $F \cap X$, the family of planar and halfplanar extreme faces of K is finite. According to Theorem 13.15, K is a polyhedron. $\square$

Remark. The closedness of K in Theorem 13.16 is essential. Thus, the convex hull of the point $\boldsymbol{u} = (1, 0)$ and the closed halfline $h = \{(x, 0) : x \geqslant 0\}$ is not closed; whence this convex hull is not a polyhedron.

Theorem 13.17. *A convex set $K \subset \mathbb{R}^n$ is a polyhedron if and only if there are finite sets X and Y in $\mathbb{R}^n$ such that $K = \operatorname{conv} X + \operatorname{cone}_{\boldsymbol{o}} Y$.*

Proof. Because the case $K = \varnothing$ is obvious, we may suppose that K is nonempty.

Assume first that K is a polyhedron. By Theorem 13.16, K can be expressed as the convex hull of suitable points, say $\boldsymbol{x}_1, \dots, \boldsymbol{x}_p$, and closed halflines, say $h_1, \dots, h_q$. Denote by $\boldsymbol{z}_i$ the endpoint of h_i and let $h_i' = h_i - \boldsymbol{z}_i$, $1 \leqslant i \leqslant q$. Clearly, $h_1', \dots, h_q'$ are closed halflines with common endpoint $\boldsymbol{o}$. Put

$$X = \{\boldsymbol{x}_1, \dots, \boldsymbol{x}_p, \boldsymbol{z}_1, \dots, \boldsymbol{z}_q\} \quad \text{and} \quad C = \operatorname{conv}(h_1' \cup \dots \cup h_q').$$

We assert that $K = \operatorname{conv} X + C$. Indeed, observe first that $\operatorname{conv} X \subset K$ by the convexity of K. Next, Theorem 6.10 implies the inclusion $\boldsymbol{x} + h_i' \subset K$ for all $\boldsymbol{x} \in \operatorname{conv} X$ and all $1 \leqslant i \leqslant q$. Hence $\boldsymbol{x} + C \subset K$ by a convexity argument. Therefore,

$$\operatorname{conv} X + C = \cup\, (\boldsymbol{x} + C : \boldsymbol{x} \in \operatorname{conv} X) \subset K.$$

For the opposite inclusion, choose a point $\boldsymbol{u} \in K$. By the assumption on K, one can express $\boldsymbol{u}$ as a convex combination

$$\boldsymbol{u} = \lambda_1 \boldsymbol{v}_1 + \dots + \lambda_r \boldsymbol{v}_r + \mu_1 \boldsymbol{w}_1 + \dots + \mu_s \boldsymbol{w}_s,$$

where $\boldsymbol{v}_1, \dots, \boldsymbol{v}_r \in X$ and $\boldsymbol{w}_1, \dots, \boldsymbol{w}_s \in h_1 \cup \dots \cup h_q$. If $\boldsymbol{w}_i \in h_{m(i)}$, where $1 \leqslant m(i) \leqslant q$, then put $\boldsymbol{w}_i' = \boldsymbol{w}_i - \boldsymbol{z}_{m(i)}$. In these terms, $\boldsymbol{w}_s' \in h_{m(i)}'$ and

$$\begin{aligned} \boldsymbol{u} &= (\lambda_1 \boldsymbol{v}_1 + \dots + \lambda_r \boldsymbol{v}_r + \mu_1 \boldsymbol{z}_{m(1)} + \dots + \mu_s \boldsymbol{z}_{m(s)}) \\ &\quad + (\mu_1 \boldsymbol{w}_1' + \dots + \mu_s \boldsymbol{w}_s') \\ &\in \operatorname{conv} X + \operatorname{cone}_{\boldsymbol{o}}\{\boldsymbol{w}_1', \dots, \boldsymbol{w}_q'\} \\ &\subset \operatorname{conv} X + \operatorname{conv}(h_1' \cup \dots \cup h_q') \\ &= \operatorname{conv} X + C. \end{aligned}$$

Summing up, $K \subset \operatorname{conv} X + C$.

Finally, choosing nonzero points $\boldsymbol{y}_i \in h_i$, $1 \leqslant i \leqslant q$, and letting $Y = \{\boldsymbol{y}_1, \dots, \boldsymbol{y}_q\}$, we obtain

$$\begin{aligned} K &= \operatorname{conv} X + \operatorname{conv}(h_1' \cup \dots \cup h_q') \\ &= \operatorname{conv} X + \operatorname{cone}_{\boldsymbol{o}}\{\boldsymbol{y}_1, \dots, \boldsymbol{y}_q\} \\ &= \operatorname{conv} X + \operatorname{cone}_{\boldsymbol{o}} Y. \end{aligned}$$

Conversely, suppose that $K = \operatorname{conv} X + \operatorname{cone}_{\boldsymbol{o}} Y$ for finite sets X and Y in $\mathbb{R}^n$. First, we assert that K is closed. Indeed, $\operatorname{cone}_{\boldsymbol{o}} Y$ is closed by Theorem 5.46. Therefore, K is closed as the sum of the compact set $\operatorname{conv} X$ and the closed set $\operatorname{cone}_{\boldsymbol{o}} Y$ (see Problem 1.13 and Theorem 4.16).

Let $X = \{\boldsymbol{x}_1, \dots, \boldsymbol{x}_p\}$ and $Y = \{\boldsymbol{y}_1, \dots, \boldsymbol{y}_q\}$. Excluding the obvious cases, we suppose that none of the points $\boldsymbol{y}_1, \dots, \boldsymbol{y}_q$ is $\boldsymbol{o}$. Denote by g_j the closed halfline with endpoint $\boldsymbol{o}$ containing $\boldsymbol{y}_j$ and let $h_{ij} = \boldsymbol{x}_i + g_j$ for all $1 \leqslant i \leqslant p$ and $1 \leqslant j \leqslant q$. Since $g_j \subset \operatorname{cone}_{\boldsymbol{o}} Y$, one has

$$h_{ij} \subset \operatorname{conv} X + \operatorname{cone}_{\boldsymbol{o}} Y = K.$$

We assert that K equals the set

$$M = \operatorname{conv}\left(\cup\left(h_{ij} : 1 \leqslant i \leqslant p,\ 1 \leqslant j \leqslant q\right)\right).$$

The inclusion $M \subset K$ holds due to the convexity of K and the above argument. Hence it suffices to prove that $K \subset M$. For this, choose a point $\boldsymbol{u} \in K$. By the assumption on K, we can write $\boldsymbol{u} = \boldsymbol{x} + \boldsymbol{y}$, where $\boldsymbol{x} \in \operatorname{conv} X$ and $\boldsymbol{y} \in \operatorname{cone}_{\boldsymbol{o}} Y$. The points $\boldsymbol{x}$ and $\boldsymbol{y}$ are expressible, respectively, as convex combinations of points $\boldsymbol{x}_1, \dots, \boldsymbol{x}_p \in X$ and a nonnegative combination of points $\boldsymbol{y}_1, \dots, \boldsymbol{y}_q \in \operatorname{cone}_{\boldsymbol{o}} Y$:

$$\boldsymbol{x} = \lambda \boldsymbol{x}_1 + \cdots + \lambda_r \boldsymbol{x}_r \quad \text{and} \quad \boldsymbol{y} = \mu_1 \boldsymbol{y}_1 + \cdots + \mu_q \boldsymbol{y}_q.$$

Without loss of generality, we may assume that all scalars $\lambda_1, \dots, \lambda_p$ are positive and $\mu_1 + \cdots + \mu_q > 0$. Then $\boldsymbol{u}$ can be written as

$$\boldsymbol{u} = \sum_{i=1}^{p} \sum_{j=1}^{q} \frac{\lambda_i \mu_j}{\mu_1 + \cdots + \mu_q} \left(\boldsymbol{x}_i + \frac{\mu_1 + \cdots + \mu_q}{\lambda_i p} \boldsymbol{y}_j\right). \tag{13.12}$$

Clearly, every point

$$\boldsymbol{z}_{ij} = \boldsymbol{x}_i + \frac{\mu_1 + \cdots + \mu_q}{\lambda_i p} \boldsymbol{y}_j, \quad 1 \leqslant i \leqslant p, \quad 1 \leqslant j \leqslant q,$$

belongs to the closed halfline h_{ij}, and

$$\boldsymbol{u} = \sum_{i=1}^{p} \sum_{j=1}^{q} \frac{\lambda_i \mu_j}{\mu_1 + \cdots + \mu_q} \boldsymbol{z}_{ij}$$

is their convex combination. Hence $\boldsymbol{u} \in M$, and $K \subset M$.

Finally, Theorem 13.16 implies that K is a polyhedron. □

Theorem 13.18. *The following assertions hold.*

(*a*) *A bounded convex set in* $\mathbb{R}^n$ *is a polytope if and only if it is the convex hull of finitely many points.*

(*b*) *A convex cone in* $\mathbb{R}^n$ *is polyhedral if and only if it is the convex hull of finitely many closed halflines with common endpoint.*
(*c*) *A convex set in* $\mathbb{R}^n$ *is a polyhedron if and only if it is the sum of a polytope and a polyhedral cone with apex* $\boldsymbol{o}$.
(*d*) *A convex set in* $\mathbb{R}^n$ *is a polyhedron if and only if it is the closure of the convex hull of finitely many points and closed halflines.*
(*e*) *A convex set in* $\mathbb{R}^n$ *is a polyhedron if and only if it is the closure of the convex hull of a polytope and a polyhedral cone.*

Proof. Without loss of generality, we may assume that convex sets $K \subset \mathbb{R}^n$ involved in assertions (*a*)–(*e*) are nonempty. Furthermore, we assume that K is expressed as $K = (K \cap S) + \operatorname{lin} K$, where $S \subset \mathbb{R}^n$ is a subspace complementary to $\operatorname{lin} K$ (see Theorem 6.25).

Assertion (*a*) is a particular cases of Theorem 13.17 (corresponding to the cases $Y = \{\boldsymbol{o}\}$).

(*b*) Let $K \subset \mathbb{R}^n$ be a convex cone. Assume first that K is polyhedral. Then $K \cap S$ is a line-free polyhedral cone, with a unique apex, say $\boldsymbol{a}$. Then $\boldsymbol{a}$ is the only extreme point of K (see Theorem 11.18), and Theorem 13.15 shows that $K \cap S$ has at most finitely many extreme halflines, say $h_1, \dots, h_p$. Furthermore, each of the closed halflines $g_1, \dots, g_q$ has $\boldsymbol{a}$ as the endpoint. According to Theorem 11.34, every set $F_j = g_j + \operatorname{lin} K$, $1 \leqslant j \leqslant q$, is an extreme halfplane of K, and Theorem 11.36 shows that $K = \operatorname{conv}(F_1 \cup \dots \cup F_p)$. It was shown in the proof of Theorem 13.16 that $\operatorname{lin} K$ can be expressed as the convex hull of closed halfline, say $h_1, \dots, h_p$ with common endpoint $\boldsymbol{o}$. Finally, Theorem 4.2 gives

$$\begin{aligned} K &= \operatorname{conv}(F_1 \cup \dots \cup F_q) = \operatorname{conv}((g_1 + \operatorname{lin} K) \cup \dots \cup (g_q + \operatorname{lin} K)) \\ &= \operatorname{conv}((g_1 \cup (\boldsymbol{a} + \operatorname{lin} K)) \cup \dots \cup (g_q \cup (\boldsymbol{a} + \operatorname{lin} K))) \\ &= \operatorname{conv}\Big(\bigcup_{j=1}^{q} (g_j \cup \operatorname{conv}((\boldsymbol{a} + h_1) \cup \dots \cup (\boldsymbol{a} + h_p)))\Big) \\ &= \operatorname{conv}\Big(\bigcup_{j=1}^{q} (g_j \cup (\boldsymbol{a} + h_1) \cup \dots \cup (\boldsymbol{a} + h_p))\Big). \end{aligned}$$

(*c*) By Theorem 13.17, a convex set $K \subset \mathbb{R}^n$ is a polyhedron if and only if there are finite sets X and Y in $\mathbb{R}^n$ such that $K = \operatorname{conv} X + \operatorname{cone}_{\boldsymbol{o}} Y$. The above argument shows that $\operatorname{conv} X$ is a polytope, and $\operatorname{cone}_{\boldsymbol{o}} Y$ is a polyhedral cone.

(*d*) If a convex set $K \subset \mathbb{R}^n$ is a polyhedron, then Theorem 13.16 shows that K can be expressed as the convex hull of finitely many points and closed halflines. Conversely, suppose that K is the closure of the convex hull of a finite set $X = \{\boldsymbol{x}_1, \dots, \boldsymbol{x}_p\}$ of points and a finite family of closed halflines $h_1, \dots, h_q$:

$$K = \mathrm{cl}\,(\mathrm{conv}\,(X \cup h_1 \cup \dots \cup h_q)).$$

We assert that K is a polyhedron. Indeed, let $\boldsymbol{z}_i$ denote the endpoint of h_i, and put $h'_i = h_i - \boldsymbol{z}_i$, $1 \leqslant i \leqslant q$. Clearly, the closed halflines $h'_1, \dots, h'_q$ have $\boldsymbol{o}$ as a common endpoint, and

$$\mathrm{conv}\,(h'_1 \cup \dots \cup h'_q) = \mathrm{cone}_{\boldsymbol{o}}\{\boldsymbol{y}_1, \dots, \boldsymbol{y}_q\},$$

where $\boldsymbol{y}_i$ is a nonzero point in h'_i, $1 \leqslant i \leqslant q$. Since $h_i = \boldsymbol{z}_i + h'_i \subset K$, Theorem 6.10 shows that $\boldsymbol{x}_i + h'_j \subset K$ for all $1 \leqslant i \leqslant p$ and $1 \leqslant j \leqslant q$. Consequently, with $Z = \{\boldsymbol{x}_1, \dots, \boldsymbol{x}_p, \boldsymbol{z}_1, \dots, \boldsymbol{z}_q\}$, one has

$$X \cup h_1 \cup \dots \cup h_q \subset Z + (h'_1 \cup \dots \cup h'_q) \subset K.$$

By a convexity argument,

$$\begin{aligned}\mathrm{conv}\,(X \cup h_1 \cup \dots \cup h_q) &\subset \mathrm{conv}\,(Z + (h'_1 \cup \dots \cup h'_q))\\ &= \mathrm{conv}\,Z + \mathrm{conv}\,(h'_1 \cup \dots \cup h'_q)\\ &= \mathrm{conv}\,Z + \mathrm{cone}_{\boldsymbol{o}}\{\boldsymbol{y}_1, \dots, \boldsymbol{y}_q\} \subset K.\end{aligned}$$

According to Theorem 13.17, the set $M = \mathrm{conv}\,Z + \mathrm{cone}_{\boldsymbol{o}}\{\boldsymbol{y}_1, \dots, \boldsymbol{y}_q\}$ is a polyhedron. Therefore, M is closed, which gives

$$K = \mathrm{cl}\,(\mathrm{conv}\,(X \cup h_1 \cup \dots \cup h_q)) \subset M \subset K.$$

Hence $K = M$, implying that K is a polyhedron.

(e) By assertion (c) above, K is a polyhedron if and only if it is the sum of a polytope Q and a polyhedral cone C with apex $\boldsymbol{o}$. If $\boldsymbol{u}$ is a point in Q, then, as shown in Problem 7.7,

$$K = Q + C = \mathrm{cl}\,(\mathrm{conv}\,(Q \cup (\boldsymbol{u} + C))). \qquad \square$$

13.3 General Properties of Polyhedra

13.3.1 *Algebra of Polyhedra*

Theorem 13.19. *If $P_1, \dots, P_r \subset \mathbb{R}^n$ are polyhedra, then both sets*

$$P = P_1 \cap \dots \cap P_r \quad \text{and} \quad Q = \mathrm{cl}\,(\mathrm{conv}\,(P_1 \cup \dots \cup P_r))$$

also are polyhedra.

Proof. By the definition, every P_i is the intersection of a finite family $\mathcal{F}_i$ of closed halfspaces, $1 \leqslant i \leqslant r$. Therefore, $P_1 \cap \cdots \cap P_r$ is a polyhedron as the intersection of the finite family $\mathcal{F}_1 \cup \cdots \cup \mathcal{F}_r$ of closed halfspaces.

Expressing every polyhedron P_i as the convex hull of a finite set X_i and finitely many closed halflines $h_{i1}, \ldots, h_{iq_i}$ (see Theorem 13.16), we obtain, by Theorem 4.2, that

$$\begin{aligned} Q &= \mathrm{cl}\,(\mathrm{conv}\,(P_1 \cup \cdots \cup P_r)) \\ &= \mathrm{cl}\,(\mathrm{conv}\,(\bigcup_{i=1}^{r} \mathrm{conv}\,(X_i \cup h_{i1} \cup \cdots \cup h_{iq_i}))) \\ &= \mathrm{cl}\,(\mathrm{conv}\,(\bigcup_{i=1}^{r} X_i \cup h_{i1} \cup \cdots \cup h_{iq_i})). \end{aligned}$$

According to Theorem 13.18, Q is a polyhedron. □

Theorem 13.20. *For polyhedra $P_1, \ldots, P_r \subset \mathbb{R}^n$ and scalars $\mu_1, \ldots, \mu_r$, the set $P = \mu_1 P_1 + \cdots + \mu_r P_r$ is a polyhedron.*

Proof. Without loss of generality, we may assume that all sets $P_1, \ldots, P_r$ are nonempty. An induction argument shows that the proof can be reduced to the case $r = 2$ and $\mu_1\mu_2 \neq 0$ (consider $\mu_1 P_1 + \mu_2\{\boldsymbol{o}\}$ if, for instance, $r = 1$ or $\mu_2 = 0$). By Theorem 13.16, $P_i = \mathrm{conv}\, X_i + \mathrm{cone}_{\boldsymbol{o}} Y_i$, for suitable finite sets X_i and Y_i in $\mathbb{R}^n$, $i = 1, 2$. With $Y_i' = Y_i \cup \{\boldsymbol{o}\}$, one has $\mathrm{cone}_{\boldsymbol{o}} Y_i' = \mathrm{cone}_{\boldsymbol{o}} Y_i$, $i = 1, 2$. By Theorems 4.12 and 5.41

$$\begin{aligned} \mu_1 P_1 + \mu_2 P_2 &= (\mathrm{conv}\,(\mu_1 X_1) + \mathrm{cone}_{\boldsymbol{o}}(\mu_1 Y_1)) \\ &\qquad + (\mathrm{conv}\,(\mu_2 X_2) + \mathrm{cone}_{\boldsymbol{o}}(\mu_2 Y_2)) \\ &= (\mathrm{conv}\,(\mu_1 X_1) + \mathrm{conv}\,(\mu_2 X_2)) \\ &\qquad + (\mathrm{cone}_{\boldsymbol{o}}(\mu_1 Y_1') + \mathrm{cone}_{\boldsymbol{o}}(\mu_2 Y_2')) \\ &= \mathrm{conv}\,(\mu_1 X_1 + \mu_2 X_2) + \mathrm{cone}_{\boldsymbol{o}}(\mu_1 Y_1' + \mu_2 Y_2'). \end{aligned}$$

Since both sets $\mu_1 X_1 + \mu_2 X_2$ and $\mu_1 Y_1' + \mu_2 Y_2'$ are finite, Theorem 13.16 shows that the set $\mu_1 P_1 + \mu_2 P_2$ is a polyhedron. □

Theorem 13.21. *If $f : \mathbb{R}^n \to \mathbb{R}^m$ is an affine transformation and $P \subset \mathbb{R}^n$ and $Q \subset \mathbb{R}^m$ are polyhedra, then both sets $f(P)$ and $f^{-1}(Q)$ are polyhedra.*

Proof. Let $f(\boldsymbol{x}) = \boldsymbol{z} + g(\boldsymbol{x})$, where $\boldsymbol{z} \in \mathbb{R}^m$ and $g : \mathbb{R}^n \to \mathbb{R}^m$ is a linear transformation. By Theorem 13.16, $P = \mathrm{conv}\, X + \mathrm{cone}_{\boldsymbol{o}} Y$ for finite sets X and Y in $\mathbb{R}^n$. A combination of Theorems 4.15 and 5.44 gives

$$\begin{aligned} f(P) &= \boldsymbol{z} + g(P) = \boldsymbol{z} + g(\mathrm{conv}\, X + \mathrm{cone}_{\boldsymbol{o}} Y) \\ &= \boldsymbol{z} + \mathrm{conv}\, g(X) + \mathrm{cone}_{\boldsymbol{o}} g(Y). \end{aligned}$$

By Theorem 13.16, the set $T = \operatorname{conv} g(X) + \operatorname{cone}_o g(Y)$ is a polyhedron; so $f(P)$ also is a polyhedron as a translate of T.

Since the subspace $\operatorname{rng} g$ is a polyhedron (see an example on page 450), the set $T = Q \cap \operatorname{rng} g$ also is a polyhedron. Clearly, $g^{-1}(Q) = g^{-1}(T)$. By Theorem 13.17, $T = \operatorname{conv} X + \operatorname{cone}_o Y$ for finite sets $X = \{\boldsymbol{x}_1, \dots, \boldsymbol{x}_p\}$ and $Y = \{\boldsymbol{y}_1, \dots, \boldsymbol{y}_q\}$ in $\operatorname{rng} g$. Choose in $\mathbb{R}^n$ sets $X' = \{\boldsymbol{x}'_1, \dots, \boldsymbol{x}'_p\}$ and $Y' = \{\boldsymbol{y}'_1, \dots, \boldsymbol{y}'_q\}$ such that $g(\boldsymbol{x}'_i) = \boldsymbol{x}_i$ and $g(\boldsymbol{y}'_j) = \boldsymbol{y}_j$ for all $1 \leqslant i \leqslant p$ and $1 \leqslant j \leqslant q$. By Theorem 13.17 again, the set $T' = \operatorname{conv} X' + \operatorname{cone}_o Y'$ is polyhedral. A combination of Theorems 4.15 and 5.44 gives

$$\begin{aligned} g(T') &= g(\operatorname{conv} X' + \operatorname{cone}_o Y') = \operatorname{conv} g(X') + \operatorname{cone}_o g(Y') \\ &= \operatorname{conv} X + \operatorname{cone}_o Y = T. \end{aligned}$$

Finally, according to Theorem 13.20, the set

$$g^{-1}(Q) = g^{-1}(T) = T' + \operatorname{null} g$$

is a polyhedron. □

13.3.2 *Support and Separation Properties of Polyhedra*

Theorem 13.22. *No polyhedron $P \subset \mathbb{R}^n$ has asymptotic planes.*

Proof. Let $L \subset \mathbb{R}^n$ be a plane of positive dimension. Since L is a polyhedron (see an example on page 450), Theorem 13.20 shows that the set $P + L$ also is a polyhedron. In particular, $P + L$ is closed, and Theorem 7.13 shows that L cannot be asymptotic for P. □

Theorem 13.23. *Given a nonempty polyhedron $P \subset \mathbb{R}^n$, its recession, barrier, and normal cones are polyhedral. Furthermore, if P is represented by a finite family of closed halfspaces*

$$V_i = \{\boldsymbol{x} \in \mathbb{R}^n : \boldsymbol{x} \cdot \boldsymbol{e}_i \leqslant \gamma_i\}, \quad 1 \leqslant i \leqslant p,$$

then

$$\operatorname{rec} K = \bigcap_{i=1}^{p} \{\boldsymbol{x} \in \mathbb{R}^n : \boldsymbol{x} \cdot \boldsymbol{e}_i \leqslant 0\}, \tag{13.13}$$

$$\operatorname{bar} K = \operatorname{nor} K = \operatorname{cone}_o\{\boldsymbol{e}_1, \dots, \boldsymbol{e}_p\}. \tag{13.14}$$

Proof. Equality (13.13) follows from Theorem 9.27. Hence the recession cone $\operatorname{rec} K$ is polyhedral as the intersection of finitely many closed halfspaces.

For the second assertion, we first observe that all halfspaces

$$W_i = \{\boldsymbol{x} \in \mathbb{R}^n : \boldsymbol{x} \cdot \boldsymbol{e}_i \leqslant 0\}, \quad 1 \leqslant i \leqslant p,$$

are closed convex cones with apex $\boldsymbol{o}$. Since $W_i^\circ = h_i = [\boldsymbol{o}, \boldsymbol{e}_i\rangle$, Corollary 8.36 and Theorem 8.3 give

$$\begin{aligned}\mathrm{cl}\,(\mathrm{bar}\,K) &= (\mathrm{rec}\,K)^\circ = (W_1 \cap \cdots \cap W_p)^\circ = \mathrm{cl}\,(\mathrm{conv}\,(h_1 \cup \cdots \cup h_p)) \\ &= \mathrm{cl}\,(\mathrm{cone}_{\boldsymbol{o}}\{\boldsymbol{e}_1, \dots, \boldsymbol{e}_p\}) = \mathrm{cone}_{\boldsymbol{o}}\{\boldsymbol{e}_1, \dots, \boldsymbol{e}_p\}\end{aligned}$$

because $\mathrm{cone}_{\boldsymbol{o}}\{\boldsymbol{e}_1, \dots, \boldsymbol{e}_p\}$ is a closed set (see Theorem 5.46). The obvious inclusion $\{\boldsymbol{e}_1, \dots, \boldsymbol{e}_p\} \subset \mathrm{bar}\,K$ and the convexity of $\mathrm{bar}\,K$ (see Theorem 8.32), imply that $\mathrm{cone}_{\boldsymbol{o}}\{\boldsymbol{e}_1, \dots, \boldsymbol{e}_p\} \subset \mathrm{bar}\,K$. Hence $\mathrm{bar}\,K = \mathrm{cone}_{\boldsymbol{o}}\{\boldsymbol{e}_1, \dots, \boldsymbol{e}_p\}$, and $\mathrm{bar}\,K$ is a polyhedron according to Theorem 13.18.

Since K has no asymptotic planes (see Theorem 13.22), $\mathrm{bar}\,K \setminus \mathrm{nor}\,K = \varnothing$. This argument and the inclusion $\mathrm{nor}\,K \subset \mathrm{bar}\,K$ shows that $\mathrm{nor}\,K = \mathrm{bar}\,K$. □

Theorem 13.24. *If P is a nonempty proper polyhedron and $\boldsymbol{a}$ is a point in P, then the generated cone $C_{\boldsymbol{a}}(P)$ is polyhedral. If, additionally, P is represented by a finite family $\mathcal{F}$ of closed halfspaces and $\mathcal{G}$ is the family of all halfspaces $V \in \mathcal{F}$ satisfying the condition $\boldsymbol{a} \in \mathrm{bd}\,V$, then $C_{\boldsymbol{a}}(P) = \cap\,(V : V \in \mathcal{G})$. Furthermore, if $\mathcal{G} = \{V_1, \dots, V_r\}$, where*

$$V_i = \{\boldsymbol{x} \in \mathbb{R}^n : \boldsymbol{x} \cdot \boldsymbol{e}_i \leqslant \gamma_i\}, \quad 1 \leqslant i \leqslant r,$$

then the normal set $Q_{\boldsymbol{a}}(P)$ equals $\boldsymbol{a} + \mathrm{cone}_{\boldsymbol{o}}\{\boldsymbol{e}_1, \dots, \boldsymbol{e}_r\}$.

Proof. Since the assertion is obvious when P is a singleton, we may assume that $P \neq \{\boldsymbol{a}\}$. If $\boldsymbol{a} \in \mathrm{rint}\,P$, then $C_{\boldsymbol{a}}(P) = \mathrm{aff}\,P$ (see Theorem 5.48), implying that $C_{\boldsymbol{a}}(P)$ is a polyhedron.

Suppose that $\boldsymbol{a} \in \mathrm{rbd}\,P$. Then $\mathcal{G} \neq \varnothing$, as follows from Theorem 13.3. Let $Q = \cap\,(P : P \in \mathcal{G})$. Given a halfspace $V \in \mathcal{G}$, the obvious inclusion $P \subset V$ and Theorem 2.35 show that $[\boldsymbol{a}, \boldsymbol{x}\rangle \subset V$ for every point $\boldsymbol{x} \in P$. Therefore, Theorem 5.38 gives

$$C_{\boldsymbol{a}}(P) = \cup\,([\boldsymbol{a}, \boldsymbol{x}\rangle : \boldsymbol{x} \in P \setminus \boldsymbol{a}) \subset V, \quad V \in \mathcal{G}.$$

Consequently, $C_{\boldsymbol{a}}(P) \subset Q$.

For the opposite inclusion, $Q \subset C_{\boldsymbol{a}}(P)$, choose a point $\boldsymbol{x} \in Q$. Then $[\boldsymbol{a}, \boldsymbol{x}] \subset Q$ by a convexity argument. Since the inclusion $\boldsymbol{x} \in P$ is obvious when $\boldsymbol{x} = \boldsymbol{a}$, we assume that $\boldsymbol{x} \neq \boldsymbol{a}$. If $\mathcal{G} = \mathcal{F}$, then $Q = P \subset C_{\boldsymbol{a}}(P)$. Suppose that $\mathcal{G} \neq \mathcal{F}$. Then $\boldsymbol{a} \in \mathrm{int}\,V$ for every halfspace $V \in \mathcal{F} \setminus \mathcal{G}$. By the finiteness of $\mathcal{F} \setminus \mathcal{G}$, there is a point $\boldsymbol{z} \in (\boldsymbol{a}, \boldsymbol{x})$ so close to $\boldsymbol{x}$ that $\boldsymbol{z} \in V$ for every $V \in \mathcal{F} \setminus \mathcal{G}$. Hence $\boldsymbol{z} \in Q \cap (V : V \in \mathcal{F} \setminus \mathcal{G}) = P$, and $\boldsymbol{x} \in [\boldsymbol{a}, \boldsymbol{z}\rangle \subset C_{\boldsymbol{a}}(P)$. Summing up, $Q \subset C_{\boldsymbol{a}}(P)$.

Finally, combining Theorems 5.31 and 8.24, we obtain

$$Q_{\boldsymbol{a}}(P) = \boldsymbol{a} + (C_{\boldsymbol{a}}(P) - \boldsymbol{a})^{\circ} = \boldsymbol{a} + (C_{\boldsymbol{o}}(P - \boldsymbol{a}))^{\circ}.$$

Since

$$C_{\boldsymbol{o}}(P - \boldsymbol{a}) = \bigcap_{i=1}^{r} \{\boldsymbol{x} \in \mathbb{R}^n : \boldsymbol{x}\cdot\boldsymbol{e}_i \leqslant 0\},$$

Theorems 8.37 and 13.23 give

$$(C_{\boldsymbol{o}}(P - \boldsymbol{a}))^{\circ} = \operatorname{nor} C_{\boldsymbol{o}}(P - \boldsymbol{a}) = \operatorname{cone}_{\boldsymbol{o}}\{\boldsymbol{e}_1, \dots, \boldsymbol{e}_r\}.$$

Summing up, $Q_{\boldsymbol{a}}(P) = \boldsymbol{a} + \operatorname{cone}_{\boldsymbol{o}}\{\boldsymbol{e}_1, \dots, \boldsymbol{e}_r\}$. □

Corollary 13.25. *Let $P \subset \mathbb{R}^n$ be a polyhedron which is not a plane, and let $\boldsymbol{a}$ be a point in $\operatorname{rbd} P$. If G is an exposed face of P generated by $\boldsymbol{a}$, then the normal set $Q_{\boldsymbol{a}}(P)$ is a polyhedral cone of dimension $n - \dim G$.*

Proof. Translating P on $-\boldsymbol{a}$, we may suppose that $\boldsymbol{a} = \boldsymbol{o}$. By Theorem 13.24, the generated cone $C_{\boldsymbol{o}}(P)$ is polyhedral, and Theorem 11.9 shows that $\operatorname{lin} C_{\boldsymbol{o}}(P) = \operatorname{span} G$. Since $\boldsymbol{o} \in \operatorname{span} G$, we have

$$\dim(\operatorname{lin} C_{\boldsymbol{o}}(P)) = \dim(\operatorname{span} G) = \dim G.$$

Theorems 8.24 implies that $Q_{\boldsymbol{o}}(P) = (C_{\boldsymbol{o}}(P))^{\circ}$. Finally, Theorem 8.4 shows that

$$\dim Q_{\boldsymbol{o}}(P) = n - \dim(\operatorname{lin} C_{\boldsymbol{o}}(P)) = n - \dim G. \qquad \square$$

Theorem 13.26. *If P is a nonempty proper polyhedron and $\boldsymbol{a}$ is a point in $\mathbb{R}^n \setminus P$, then the set $\operatorname{cl} C_{\boldsymbol{a}}(P)$ is a polyhedral cone.*

Proof. By Theorem 13.16, P can be expressed as the convex hull of finitely many points, say $\boldsymbol{x}_1, \dots, \boldsymbol{x}_p$, and finitely many closed halflines, say $h_1, \dots, h_q$. Let $\boldsymbol{z}_j$ be the endpoint of h_j, $1 \leqslant j \leqslant q$. Denote by h'_j the translate of h_j on $\boldsymbol{a} - \boldsymbol{z}_j$ (so that $\boldsymbol{a}$ is the endpoint of h'_j), and put $h''_j = [\boldsymbol{a}, \boldsymbol{z}_i\rangle$, $1 \leqslant j \leqslant q$. It is easy to see that $h_j \subset \operatorname{cone}_{\boldsymbol{a}}(h'_j \cup h''_j)$. Theorem 6.10 gives

$$h'_j \cup h''_j \subset \operatorname{cl} C_{\boldsymbol{a}}(P), \quad 1 \leqslant j \leqslant q.$$

Clearly, every closed halfline $g_i = [\boldsymbol{a}, \boldsymbol{x}_i\rangle$ lines in $C_{\boldsymbol{a}}(P)$, $1 \leqslant i \leqslant p$. Using Theorems 5.30 and 5.38, we obtain

$$\begin{aligned} C_{\boldsymbol{a}}(P) &= C_{\boldsymbol{a}}(\operatorname{conv}(\{\boldsymbol{x}_1, \dots, \boldsymbol{x}_p\} \cup h_1 \cup \cdots \cup h_p)) \\ &= \operatorname{cone}_{\boldsymbol{a}}(\{\boldsymbol{x}_1, \dots, \boldsymbol{x}_p\} \cup h_1 \cup \cdots \cup h_p) \\ &\subset \operatorname{cone}_{\boldsymbol{a}}((\bigcup_{i=1}^{p} g_i) \cup (\bigcup_{j=1}^{q} \operatorname{cone}_{\boldsymbol{a}}(h'_j \cup h''_j))) \\ &= \operatorname{cone}_{\boldsymbol{a}}((\bigcup_{i=1}^{p} g_i) \cup (\bigcup_{j=1}^{q} h'_j \cup h''_j)) \\ &= \operatorname{conv}((\bigcup_{i=1}^{p} g_i) \cup (\bigcup_{j=1}^{q} h'_j \cup h''_j)) \\ &\subset \operatorname{cl}(\operatorname{cone}_{\boldsymbol{a}} P). \end{aligned}$$

By Theorem 13.18, the set

$$Q = \operatorname{conv}\,((\bigcup_{i=1}^{p} g_i) \cup (\bigcup_{j=1}^{q} h'_j \cup h''_j))$$

is a polyhedral cone with apex $\boldsymbol{a}$. Therefore, $\operatorname{cl} C_{\boldsymbol{a}}(P) = Q$, which shows that $\operatorname{cl} C_{\boldsymbol{a}}(P)$ is a polyhedral cone. □

Remark. The generated cone $C_{\boldsymbol{a}}(P)$ in Theorem 13.26 can be non-closed and thus non-polyhedral (see an example in Figure 13.3).

Theorems 13.22 and 10.22 imply the following result.

Corollary 13.27. *If P_1 and P_2 are nonempty disjoint polyhedra in $\mathbb{R}^n$, then there is a hyperplane $H \subset \mathbb{R}^n$ strongly separating P_1 and P_2.* □

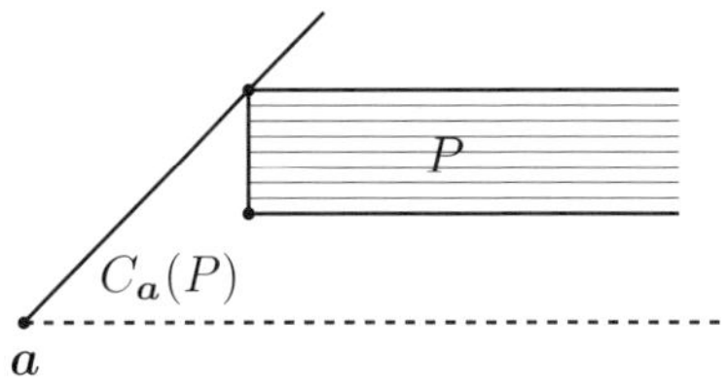

Fig. 13.3 Illustration to Theorem 13.26.

Theorem 13.28. *Given nonempty proper polyhedra P_1 and P_2 in $\mathbb{R}^n$, the following assertions hold.*

(*a*) *P_1 is properly separated from P_2 by a hyperplane if and only if $\operatorname{rint} P_1 \cap P_2 = \varnothing$.*
(*b*) *P_1 and P_2 are definitely separated by a hyperplane if and only if*

$$\operatorname{rint} P_1 \cap P_2 = P_1 \cap \operatorname{rint} P_2 = \varnothing.$$

Proof. (*a*) Assume first that P_1 is properly separated from P_2 by a hyperplane $H \subset \mathbb{R}^n$ (see Definition 10.4). If V_1 and V_2 are the opposite closed halfspaces determined by H, then, without loss of generality, may assume that $P_1 \subset V_1$ and $P_2 \subset V_2$. Since V_1 properly contains P_1 (see Definition 9.16), Corollary 9.18 gives the inclusion $\operatorname{rint} P_1 \subset \operatorname{int} V_1$. Hence

$$\operatorname{rint} P_1 \cap P_2 \subset \operatorname{int} V_1 \cap V_2 = \varnothing.$$

The converse assertion will be proved by induction on n. The case $n = 1$ is obvious. Suppose that the inductive assumption holds for all $n \leqslant m - 1$, where $m \geqslant 2$, and let P_1 and P_2 be nonempty proper polyhedra in $\mathbb{R}^m$ satisfying the condition $\operatorname{rint} P_1 \cap P_2 = \varnothing$.

If $P_1 \cap P_2 = \varnothing$, then P_1 and P_2 are strongly separated by a hyperplane (see Corollary 13.27). So, we may assume that $P_1 \cap P_2 \neq \varnothing$. Since $\operatorname{rint} P_1 \cap \operatorname{rint} P_2 = \varnothing$, Theorem 10.8 implies the existence of a hyperplane $H \subset \mathbb{R}^m$ separating P_1 and P_2 such that at least one of these polyhedra does not lie in H. If $P_1 \not\subset H$, then H is a desired hyperplane. Hence we may suppose that $P_1 \subset H$ and $P_2 \not\subset H$.

Put $Q = H \cap P_2$. Clearly, Q is an exposed face of P_2. Furthermore, $\operatorname{rint} P_1 \cap Q = \varnothing$ and $P_1 \cap Q \neq \varnothing$, because otherwise

$$P_1 \cap P_2 = (P_1 \cap P_2) \cap H = P_1 \cap (H \cap P_2) = P_1 \cap Q = \varnothing,$$

contrary to the assumption $P_1 \cap P_2 \neq \varnothing$. By the inductive assumption, there is an $(n-2)$-dimensional subplane $G \subset H$ properly separating P_1 from Q. Obviously, the set $E = G \cap Q$ is an exposed face of Q. Hence E is an exposed face of P_2. A combination of Theorems 11.3 and 12.2 shows that E is an extreme face of P_2. Consequently, Corollary 13.12 implies that E is an exposed face of P_2.

Denote by S the orthogonal complement of the $(n-2)$-dimensional subspace $\operatorname{dir} G$, and let f be the orthogonal projection of $\mathbb{R}^n$ on S. Clearly, $f(G)$ is a singleton, say $\mathbf{c}$, which belongs to the line $l = H \cap S$. By Theorem 13.21, $D_2 = f(P_2)$ is a convex polygonal set. Furthermore, l supports D_2. By the assumption $P_1 \not\subset G$, the closed convex set $D_1 = f(P_1)$, which lies in l, is distinct from $\{\mathbf{c}\}$. Since $D_1 \cap D_2 = \{\mathbf{c}\}$, there is a line $l' \subset S$ through $\mathbf{c}$ which separates D_1 and D_2 such that $D_1 \not\subset l'$. Consequently, the hyperplane $H' = l' + \operatorname{dir} G$ separates P_1 and P_2 such that $P_1 \not\subset P_2$, as desired.

(*b*) This part immediately follows from (*a*) and Theorem 10.6. □

Remark. None of the assertions from Theorem 13.28 for the case of general closed convex sets. For instance, let K_1 and K_2 be planar circular disks in $\mathbb{R}^3$, given by

$$K_1 = \{(x, y, 0) : x^2 + (y-1)^2 \leqslant 1\},$$
$$K_2 = \{(0, y, z) : y^2 + (z-1)^2 \leqslant 1\}.$$

Clearly, $\operatorname{rint} K_1 \cap K_2 = K_1 \cap \operatorname{rint} K_2 = \varnothing$. On the other hand, the coordinate xy-plane (which contains K_1) is the only plane separating K_1 and K_2.

13.3.3 *Polyhedral Approximations of Convex Sets*

Theorem 13.29. *Let $K \subset \mathbb{R}^n$ be a convex set and $P_1, \dots, P_r \subset \mathbb{R}^n$ be polyhedra such that $K \subset P_1 \cup \dots \cup P_r$. Then there is a polyhedron $P \subset \mathbb{R}^n$ satisfying the inclusions $K \subset P \subset P_1 \cup \dots \cup P_r$.*

Proof. Since the cases $K = \varnothing$ and $P_1 \cup \dots \cup P_r = \mathbb{R}^n$ are obvious, we may assume that both K and $P_1 \cup \dots \cup P_r$ are nonempty proper subsets of $\mathbb{R}^n$. Consequently, all polyhedra $P_1, \dots, P_r$ are nonempty proper subsets of $\mathbb{R}^n$. Then every P_i can be expressed as the intersection of respective closed halfspaces $V_j^{(i)}$:

$$P_i = V_1^{(i)} \cap \dots \cap V_{p_i}^{(i)}, \quad 1 \leqslant i \leqslant r.$$

If $W_j^{(i)}$ denotes the complementary to $V_j^{(i)}$ open halfspace of $\mathbb{R}^n$, then

$$\begin{aligned}\mathbb{R}^n \setminus P_i &= \mathbb{R}^n \setminus (V_1^{(i)} \cap \dots \cap V_{p_i}^{(i)}) = (\mathbb{R}^n \setminus V_1^{(i)}) \cup \dots \cup (\mathbb{R}^n \setminus V_{p_i}^{(i)}) \\ &= W_1^{(i)} \cup \dots \cup W_{p_i}^{(i)}, \quad 1 \leqslant i \leqslant r.\end{aligned}$$

Consequently,

$$\begin{aligned}\mathbb{R}^n \setminus (P_1 \cup \dots \cup P_r) &= (\mathbb{R}^n \setminus P_1) \cap \dots \cap (\mathbb{R}^n \setminus P_r) \\ &= (W_1^{(1)} \cup \dots \cup W_{p_1}^{(1)}) \cap \dots \cap (W_1^{(r)} \cup \dots \cup W_{p_r}^{(r)}) \\ &= M_1 \cup \dots \cup M_q,\end{aligned}$$

where each of the sets $M_1, \dots, M_q$ has the form

$$W_{j_1}^{(1)} \cap \dots \cap W_{j_r}^{(r)}, \quad 1 \leqslant j_i \leqslant p_i.$$

Every set M_j is open and convex (possibly, empty) as the intersection of open halfspaces. Furthermore, $K \cap (M_1 \cup \dots \cup M_q) = \varnothing$ due to the inclusion $K \subset P_1 \cup \dots \cup P_r$.

According to Theorem 10.8, there is a hyperplane $H_j \subset \mathbb{R}^n$ properly separating K from M_j (if M_j is empty, then choose any halfplane $H_j \subset \mathbb{R}^n$ bounding K). Let E_j denote a closed halfspace determined by H_j and containing K. Clearly, $E_j \cap M_j = \varnothing$ for all $1 \leqslant j \leqslant q$.

Finally, consider the polyhedron $P = E_1 \cap \dots \cap E_j$. Then $K \subset P$ and

$$\begin{aligned}P \cap (M_1 \cup \dots \cup M_q) &= (P \cap M_1) \cup \dots \cup (P \cap M_q) \\ &\subset (E_1 \cap M_1) \cup \dots \cup (E_q \cap M_q) = \varnothing.\end{aligned}$$

So, $P \subset \mathbb{R}^n \setminus (M_1 \cup \dots \cup M_q) = P_1 \cup \dots \cup P_r$, as desired. □

Theorem 13.30. *Let $P \subset \mathbb{R}^n$ be a nonempty polyhedron and $K_1, \ldots, K_r \subset \operatorname{aff} P$ be closed convex sets whose intersection is P. Then there are polyhedra $P_1, \ldots, P_r \subset \operatorname{aff} P$ such that*

$$P_1 \cap \cdots \cap P_r = P \quad \text{and} \quad K_i \subset P_i, \quad 1 \leqslant i \leqslant r. \tag{13.15}$$

Proof. Let $F_1, \ldots, F_q$ be the facets of P. Choose any facet F_i of P and a point $\boldsymbol{u} \in \operatorname{rint} F_i$. Problem 3.17 shows that $\boldsymbol{u} \in \operatorname{rbd} K_j$ for a suitable set K_j, $1 \leqslant j \leqslant r$. By Corollary 9.32, there is a closed halfspace, say V_{ij}, properly supporting K_j such that $\boldsymbol{u} \in \operatorname{bd} V_{ij}$. Consequently, Corollary 9.32 implies that $F_i \subset \operatorname{bd} V_{ij}$.

Let $D_{ij} = V_{ij} \cap \operatorname{aff} P$. Then D_{ij} is a closed halfplane of $\operatorname{aff} K$, and Corollary 3.64 implies the inclusion $\boldsymbol{u} \in \operatorname{bd} V_{ij} \cap \operatorname{aff} P = \operatorname{rbd} D_{ij}$.

For every index j, $1 \leqslant j \leqslant r$, denote by $\mathcal{D}_j$ the family of all closed halfplane D_{ij} defined above (possibly, $\mathcal{D}_j = \varnothing$). Since $F_1 \cup \cdots \cup F_q = \operatorname{rbd} P$ (see Theorem 13.11), Theorem 9.40 shows that

$$P = \cap\,(D_{ij} : D_{ij} \in \mathcal{D}_1 \cup \cdots \cup \mathcal{D}_r).$$

The above argument implies that the polyhedra $P_1, \ldots, P_r$ defined by

$$P_j = \begin{cases} \cap\,(D_{ij} : D_{ij} \in \mathcal{D}_j) & \text{if } \mathcal{D}_j \neq \varnothing, \\ \operatorname{aff} P & \text{if } \mathcal{D}_j = \varnothing \end{cases} \tag{13.16}$$

satisfy (13.15). □

Remark. The assumption $K_1, \ldots, K_r \subset \operatorname{aff} P$ is essential in Theorem 13.30. Indeed, the polyhedron $\{\boldsymbol{o}\}$ is the intersection of circular disks $K_1 = \{(x, y) : (x-1)^2 + y^2 \leqslant 1\}$ and $K_2 = \{(x, y) : (x+1)^2 + y^2 \leqslant 1\}$. It is easy to see that no two polyhedra P_1 and P_2 in $\mathbb{R}^2$ satisfy the conditions $P_1 \cap P_2 = \{\boldsymbol{o}\}$ and $K_i \subset P_i$, $i = 1, 2$.

Theorem 13.31. *Given a nonempty bounded convex set $K \subset \mathbb{R}^n$, a point $\boldsymbol{a} \in \operatorname{rint} K$, and a scalar $\varepsilon > 0$, there is a polytope $P \subset \mathbb{R}^n$ satisfying the conditions*

$$\boldsymbol{a} \in \operatorname{rint} P \quad \text{and} \quad P \subset K \subset (1+\varepsilon)P - \varepsilon\boldsymbol{a}.$$

If, additionally, K is compact, then the vertices of P can be chosen in the set $\operatorname{ext} K$.

Proof. Let $m = \dim K$. According to Theorem 3.22, there is an m-simplex $\Delta = \Delta(\boldsymbol{z}_1, \ldots, \boldsymbol{z}_{m+1}) \subset K$ such that $\boldsymbol{a} \in \operatorname{rint} \Delta$. On the other hand, Definition 3.16 shows the existence of a scalar $\rho > 0$ with the property

$\operatorname{aff} K \cap U_\rho(\boldsymbol{a}) \subset K$. Clearly, the family of open balls $U_{\varepsilon\rho}(\boldsymbol{x})$, $\boldsymbol{x} \in K$, covers the bounded set K. So, one can find (see page 11) a finite set $Y \subset K$ such that $K \subset \cup (U_{\varepsilon\rho}(\boldsymbol{x}) : \boldsymbol{x} \in Y)$.

Put $X = Y \cup \{\boldsymbol{z}_1, \dots, \boldsymbol{z}_{m+1}\}$. Clearly, $X \subset K$. Combining Theorem 2.42 and Corollary 2.66, we obtain

$$\operatorname{aff} K = \operatorname{aff} \{\boldsymbol{z}_1, \dots, \boldsymbol{z}_{m+1}\} \subset \operatorname{aff} X \subset \operatorname{aff} K.$$

So, $\operatorname{aff} K = \operatorname{aff} \{\boldsymbol{z}_1, \dots, \boldsymbol{z}_{m+1}\} = \operatorname{aff} X$. These equalities, combined with Theorem 4.2, give

$$\operatorname{aff} K = \operatorname{aff} \{\boldsymbol{z}_1, \dots, \boldsymbol{z}_{m+1}\} = \operatorname{aff} \varDelta = \operatorname{aff} (\operatorname{conv} X).$$

Let $P = \operatorname{conv} X$. Theorem 13.18 shows that P is a polytope. By the above argument, $\varDelta \subset P \subset K$ and $\operatorname{aff} K = \operatorname{aff} \varDelta = \operatorname{aff} P$. Theorem 3.18 gives the inclusions $\boldsymbol{a} \in \operatorname{rint} \varDelta \subset \operatorname{rint} P$. On the other hand,

$$\begin{aligned} K &\subset \cup (U_{\varepsilon\rho}(\boldsymbol{x}) : \boldsymbol{x} \in Y) \subset \cup (U_{\varepsilon\rho}(\boldsymbol{x}) : \boldsymbol{x} \in X) \\ &= U_{\varepsilon\rho}(X) = X + U_{\varepsilon\rho}(\boldsymbol{o}). \end{aligned} \tag{13.17}$$

Consequently, the inclusions $X \subset K$ and (13.17), combined with Theorem 4.12, give

$$\begin{aligned} P &= \operatorname{conv} X \subset K \subset \operatorname{conv} (X + U_{\varepsilon\rho}(\boldsymbol{o})) \\ &= \operatorname{conv} X + U_{\varepsilon\rho}(\boldsymbol{o}) = P + U_{\varepsilon\rho}(\boldsymbol{o}) = U_{\varepsilon\rho}(P). \end{aligned}$$

With $\mu = 1 + \varepsilon$, Theorem 3.44 shows that the set $\boldsymbol{a} + (1 + \varepsilon)(P - \boldsymbol{a})$ contains $\operatorname{aff} P \cap U_{\varepsilon\rho}(P)$. Summing up,

$$P \subset K = \operatorname{aff} P \cap K \subset \operatorname{aff} P \cap U_{\varepsilon\rho}(P) \subset (1 + \varepsilon)P - \varepsilon\boldsymbol{a}. \tag{13.18}$$

Suppose, additionally, that K is compact. Then $K = \operatorname{conv} (\operatorname{ext} K)$ according to Theorem 11.21. Consequently, $\operatorname{rint} K = \operatorname{rint} (\operatorname{conv} (\operatorname{ext} K))$. Since $\boldsymbol{a} \in \operatorname{rint} K$, a combination of Theorem 4.18 and Corollary 4.19 shows the existence of a finite set $Z \subset \operatorname{ext} K$ such that $\operatorname{aff} Z = \operatorname{aff} K$ and $\boldsymbol{a} \in \operatorname{rint} (\operatorname{conv} Z)$. Because the set $\operatorname{ext} K$ is bounded, one can find (see page 11) a finite set $Y \subset \operatorname{ext} K$ such that $\operatorname{ext} K \subset \cup (U_{\varepsilon\rho}(\boldsymbol{x}) : \boldsymbol{x} \in Y)$.

Put $X = Y \cup Z$. Similarly to the above argument, $X \subset \operatorname{ext} K \subset U_{\varepsilon\rho}(X)$. Furthermore, let $P = \operatorname{conv} X$. One has $\boldsymbol{a} \in \operatorname{rint} P$ and

$$\begin{aligned} P &= \operatorname{conv} X \subset \operatorname{conv} (\operatorname{ext} K) = K \subset \operatorname{conv} (X + U_{\varepsilon\rho}(\boldsymbol{o})) \\ &= \operatorname{conv} X + U_{\varepsilon\rho}(\boldsymbol{o}) = P + U_{\varepsilon\rho}(\boldsymbol{o}) = U_{\varepsilon\rho}(P). \end{aligned}$$

Finally, with $\mu = 1 + \varepsilon$, Theorem 3.44 implies inclusion (13.18). □

Remarks. 1. An obvious modification of the proof of Theorem 13.31 allows us to choose the set X within $\exp K$ provided K is compact.

2. A similar to Theorem 13.31 assertion does not hold for the case of unbounded convex sets. For instance, let $K = \{(x, y) : y \geqslant x^2\}$ be a solid parabola, $\boldsymbol{a}$ be a point in $\operatorname{int} P$, and ε be a positive scalar. It is easy to see that no polyhedron $P \subset \mathbb{R}^2$ satisfies the inclusions $P \subset K \subset (1+\varepsilon)P - \varepsilon\boldsymbol{a}$.

Problems and Notes for Chapter 13

Problems for Chapter 13

Problem 13.1. Prove that every nonempty proper plane $L \subset \mathbb{R}^n$ of dimension m is an intersection of $n - m + 1$ closed halfspaces. Furthermore, each such representation of L is irredundant.

Problem 13.2. Let a nonempty proper plane $L \subset \mathbb{R}^n$ be the intersection of a family $\mathcal{F}$ of closed halfspaces, each containing L in its boundary hyperplane. Prove that $L = \cap\,(\operatorname{bd} V : V \in \mathcal{F})$.

Problem 13.3. (Bair [11]) Let a convex polytope $P \subset \mathbb{R}^n$ be the sum of convex sets A and B. Prove that both A and B are convex polytopes.

Problem 13.4. (Bair [12]) Let a convex polyhedron $P \subset \mathbb{R}^n$ be the sum of convex sets A and B, where B is bounded. Prove that A is a polyhedron.

Problem 13.5. (Klee [184]) Prove that a convex set $K \subset \mathbb{R}^n$ is a polytope if and only if it is not the union of an infinite increasing sequence of pairwise distinct convex sets.

Problem 13.6. Let $P \subset \mathbb{R}^n$ be a nonempty polyhedron of dimension m and $K_1, \dots K_r$ be m-dimensional convex subsets of P with the properties:

$$K_1 \cup \cdots \cup K_r = P, \quad \operatorname{rint} K_i \cap \operatorname{rint} K_j = \varnothing \quad \text{for all} \quad 1 \leqslant i \neq j \leqslant r.$$

Prove that the sets $\operatorname{cl} K_1, \dots, \operatorname{cl} K_r$ are polyhedra.

Notes for Chapter 13

Polyhedra and polytopes. The concept of polyhedron appeared as a geometric interpretation of the solution set of a system of linear inequalities on the turn of 20th century. It was related to questions of compatibility and dependence between the inequalities and geometric structure of the solution (its dimension, boundedness and unboundedness, distribution of vertices, edges, and faces of any dimension) in terms of ranks and determinants of the system. An extensive historic literature on this topic can be found, for instance, in the collection of articles edited by Kuhn and Tucker [189] and in the monograph of Černikov [69].

Minkowski [215, p. 135–138] defined a convex polytope in $\mathbb{R}^3$ as the set of all convex combinations of finitely many given points. Alternatively, Steinitz [268, Part IV]) introduced convex polyhedra as bounded intersections of finitely many closed halfspaces. Motzkin [220, § 4,5] already distinguishes *convex polytopes* (convex hulls of finite sets) and *polyhedral sets* (intersections of finitely many closed halfspaces).

Polyhedra as planar sections of special sets. Every n-dimensional polytope with f facets is a plane section of an $(f-1)$-dimensional simplex (see Davis [86] and Grünbaum [134, page 71]), while every n-dimensional polytope with v vertices is the image of an affine projection of an $(v-1)$-dimensional simplex (see Grünbaum [134, page 72]). Naumann [222] showed that every n-dimensional polytope with f facets is a section of a $2^n(n+1)f$-dimensional cube. Epifanov [101] proved that for every n-dimensional polyhedron P with m facets in $\mathbb{R}^n$, there is an $(n+m-1)$-dimensional vector space E containing $\mathbb{R}^n$ as a subspace and a solid polyhedral cone $C \subset E$ with m facets and pairwise orthogonal $(n+m-2)$ faces such that $P = E \cap C$. Davis [86] showed that every line-free polyhedral cone is affine-equivalent to the intersection of the positive hyperoctant with a linear subspace, and is affine-equivalent to the image of an affine projection of the positive hyperoctant.

Faces and extreme representations of polyhedra. Various results on facial structure of polyhedra can be found in the books of Brøndsted [55, § 8] and Grünbaum [134, Section 2.6].

For the case of compact convex sets, Theorem 13.16 goes back to Minkowski [215, p. 135–138], who proved that the convex hull, P, of finitely many points in $\mathbb{R}^3$ can be expressed as the intersection of finitely many closed halfspaces each supporting P. In its present form, Theorem 13.16 is proved by Weyl [293], who also showed that a polyhedral cone in $\mathbb{R}^n$ is the positive hull of finitely many points. Part (*c*) of Theorem 13.18 is due to Motzkin [220, p. 47] (see also Rubinštein [247]).

Support and separation properties. Corollary 13.27 is due to Motzkin [220, Theorem G5] (see also Černikov [68]). Part (*a*) of Theorem 13.28 is proved by Rockafellar [241, Theorem 20.2] in a more general setting: P_1 is any convex set and P_2 is a polyhedron. Special types of separation of disjoint polyhedra by hyperplanes are considered by Bourgin [46] and Write [295].

Approximations by polyhedra. Theorem 13.29 can be found in Webster [291, pp. 114-115]. Theorem 13.30 is a improved version of an assertion of Eggleston [99], given there under the implicit assumption $\dim P = n$. Hoffman [151] showed that if a polyhedron $P \subset \mathbb{R}^n$ is the union of closed convex sets $K_1, \dots, K_r$, there are polyhedra $P_i \subset K_i$, $1 \leqslant i \leqslant r$, such that $P = P_1 \cup \cdots \cup P_r$. If $K_1, \dots, K_r \subset \mathbb{R}^n$ are pairwise disjoint convex sets, then there are convex polyhedra $P_1, \dots, P_r \subset \mathbb{R}^n$ of which no two have common interior points and such that $K_i \subset P_i$ for all $1 \leqslant i \leqslant r$ (Stoelinga [269, Theorem 31]). The first part of Theorem 13.31 belongs to Minkowski [215, p. 139].

Various results on polyhedra. Lawrence and Soltan [194] proved for any family $\mathcal{F} = \{X_1, \ldots, X_k\}$ of nonempty sets in $\mathbb{R}^n$, the intersection

$$T(\mathcal{F}) = \cap\,(\mathrm{conv}\,\{\boldsymbol{x}_1, \ldots, \boldsymbol{x}_k\} : \boldsymbol{x}_i \in X_i,\ 1 \leqslant i \leqslant k)$$

is a convex polytope. Moreover, if $k = n + 1$ and $\dim T(\mathcal{F}) = n$, then $T(\mathcal{F})$ is a simplex. If, in addition, all sets $X_1, \ldots, X_{n+1}$ are bounded, then the simplex $T(\mathcal{F})$ is the intersection of $n + 1$ closed halfspaces $V_1, \ldots, V_{n+1}$, such that every V_i supports the subfamily $\mathcal{F} \setminus \{X_i\}$, $1 \leqslant i \leqslant n + 1$.

Characteristic properties of polyhedra. Bol [38] showed that a compact convex set $K \subset \mathbb{R}^3$ is a polytope if and only if all orthogonal projections of K on 2-dimensional planes are polygons. Klee [176] (completing an argument of Mirkil [216]) proved that a convex cone $C \subset \mathbb{R}^n$ is polyhedral if and only if all orthogonal projections of C on 2-dimensional planes are closed. Bastiani [28] showed that a closed convex cone $D \subset \mathbb{R}^n$ distinct from a plane is polyhedral if and only if all generated cones $C_{\boldsymbol{p}}(D)$, $\boldsymbol{p} \in \mathrm{rbd}\,D$, are closed.

Klee [176] proved that a convex set $K \subset \mathbb{R}^n$ is polyhedral if and only if any of the following assertions holds:

(a) all sections of K by 2-dimensional planes through a given point $\boldsymbol{u} \in \mathrm{rint}\,K$ are polygonal,

(b) all sections of K by 3-dimensional planes through a given point $\boldsymbol{u} \in \mathbb{R}^n$ are polyhedral,

(c) all orthogonal projections on 3-dimensional planes of $\mathbb{R}^n$ are polyhedral.

If, additionally, K is bounded, then K is a polytope is and only if all its orthogonal projections on 2-dimensional planes are polyhedral.

Boundedly polyhedral sets. According to Klee [176], a sets $X \subset \mathbb{R}^n$ is called *boundedly polyhedral* provided its intersection with every polytope is polyhedral. The following properties of boundedly polyhedral sets are proved in [176]. For any closed convex sets $K \subset \mathbb{R}^n$ and any $\varepsilon > 0$ there are boundedly polyhedral sets P and Q such that $P \subset K \subset B_\varepsilon(P)$ and $K \subset Q \subset B_\varepsilon(K)$ (compare with Theorem 13.31).

A convex set $K \subset \mathbb{R}^n$ is boundedly polyhedral if and only if any of the following assertions holds:

(a) all sections of K by 2-dimensional planes through a given point $\boldsymbol{u} \in \mathrm{rint}\,K$ are boundedly polygonal,

(b) all sections of K by 3-dimensional planes through a given point $\boldsymbol{u} \in \mathbb{R}^n$ are boundedly polyhedral,

(c) with $2 \leqslant i \leqslant n$, all closed orthogonal projections of K on i-dimensional planes of $\mathbb{R}^n$ are boundedly polyhedral.

(d) K is closed and all generated cones $C_{\boldsymbol{z}}(K)$, $\boldsymbol{z} \in K$, are closed.

Choquet simplices. In 1956 Choquet [70] defined a simplex (afterwards called a *Choquet simplex*) as a convex set S in a linear space $\mathbb{E}$ of any dimension such that the intersection of any two directly homothetic copies of S, if nonempty, is again a homothetic copy of S, possibly degenerated into a point:

$$(\boldsymbol{u} + \lambda S) \cap (\boldsymbol{v} + \mu S) = \boldsymbol{w} + \nu S, \quad \boldsymbol{u}, \boldsymbol{v}, \boldsymbol{w} \in \mathbb{E}, \quad \lambda, \mu, \nu \geqslant 0.$$

By using the technique of measure representation (see Phelps [231]), it was shown later that finite-dimensional compact Choquet simplices are precisely the simplices in the usual sense, i.e., they are the convex hulls of finite affinely independent sets of points. Independently, Rogers and Shephard [243] proved that a convex body K in $\mathbb{R}^n$ is a simplex if and only if any nonempty intersection of K and a translate of K is a direct homothetic copy of K, possibly degenerated into a point:

$$K \cap (\boldsymbol{u} + K) = \boldsymbol{v} + \lambda K, \quad \boldsymbol{u}, \boldsymbol{v} \in \mathbb{R}^n, \quad \lambda \geqslant 0.$$

Generalizing this statement, Soltan [254] proved that convex bodies K_1 and K_2 in $\mathbb{R}^n$ are directly homothetic simplices if and only if all n-dimensional intersections $K_1 \cap (\boldsymbol{u} + K_2)$, $\boldsymbol{u} \in \mathbb{R}^d$, belong to at most countably many homothety classes of convex bodies. See the survey [255] for further references and results on Choquet simplices.

Closed unbounded Choquet simplices in $\mathbb{R}^n$ are simplicial cones and direct sums of subspaces and simplicial cones (see Bair and Fourneau [18, 19, 109]). Following Rockafellar [241, p. 154], a generalized n-simplex $S \subset \mathbb{R}^n$ is defined as the direct sum of an m-simplex and a simplicial $(n-m)$-cone, $0 \leqslant m \leqslant n$:

$$S = \operatorname{conv}\{\boldsymbol{x}_0, \boldsymbol{x}_1, \ldots, \boldsymbol{x}_m\} + \sum_{i=m+1}^{n} [\boldsymbol{x}_0, x_i\rangle, \quad 0 \leqslant m \leqslant n,$$

where $\boldsymbol{x}_0, \boldsymbol{x}_1, \ldots, \boldsymbol{x}_n$ are affinely independent points and $[\boldsymbol{x}_0, \boldsymbol{x}_i\rangle$ means the halfline through $\boldsymbol{x}_i$ originated at $\boldsymbol{x}_0$. Fourneau [110] showed that a line-free n-dimensional closed convex set $K \subset \mathbb{R}^n$ is a generalized n-simplex if and only if all n-dimensional intersections $K \cap (\boldsymbol{v} + K)$, $\boldsymbol{v} \in \mathbb{R}^n$, are directly homothetic to K. Generalizing this statement, Soltan [259] proved that n-dimensional line-free closed convex sets K_1 and K_2 in $\mathbb{R}^n$ are generalized simplices if and only if all n-dimensional intersections $K_1 \cap (\boldsymbol{v} + K_2)$, $\boldsymbol{v} \in \mathbb{R}^n$, belong to a unique direct homothety class of convex sets.

M-decomposable sets. Based on Theorem 13.18, Goberna, González, A. Iusem, Martínez-Legaz, and Todorov (see [119], [120], [122]) introduced and studied various properties of M-decomposable convex sets, which are defined as sums of compact convex sets and convex cones.

Chapter 14

Solutions to Problems

Chapter 1

Problem 1.1. The linearity of h is obvious. We state that h is onto. Indeed, let $\boldsymbol{u} \in \operatorname{rng} f$ and $\boldsymbol{x} \in \mathbb{R}^n$ be such that $f(\boldsymbol{x}) = \boldsymbol{u}$. Since $\mathbb{R}^n = S + \operatorname{null} f$, one can write $\boldsymbol{x} = \boldsymbol{x}_0 + \boldsymbol{y}$, where $\boldsymbol{x}_0 \in S$ and $\boldsymbol{y} \in \operatorname{null} f$. Then

$$h(\boldsymbol{x}_0) = f(\boldsymbol{x}_0) = f(\boldsymbol{x} - \boldsymbol{y}) = f(\boldsymbol{x}) = \boldsymbol{u},$$

implying the equality $\operatorname{rng} h = \operatorname{rng} f$. Furthermore, h is invertible because

$$\dim S = n - \dim(\operatorname{null} f) = \dim(\operatorname{rng} f).$$

For the equality $f^{-1}(\boldsymbol{x}) = h^{-1}(\boldsymbol{x}) + \operatorname{null} f$, choose a point $\boldsymbol{u} \in f^{-1}(\boldsymbol{x})$. Then $\boldsymbol{u} = \boldsymbol{u}_0 + \boldsymbol{y}$, where $\boldsymbol{u}_0 \in S$ and $\boldsymbol{y} \in \operatorname{null} f$. The equalities $\boldsymbol{x} = f(\boldsymbol{u}) = f(\boldsymbol{u}_0) = h(\boldsymbol{u}_0)$ show that $\boldsymbol{u}_0 = h^{-1}(\boldsymbol{x})$ and $\boldsymbol{u} = \boldsymbol{u}_0 + \boldsymbol{y} \in h^{-1}(\boldsymbol{x}) + \operatorname{null} f$. Conversely, if $\boldsymbol{u} \in h^{-1}(\boldsymbol{x}) + \operatorname{null} f$ and $\boldsymbol{u} = \boldsymbol{u}_0 + \boldsymbol{y}$, where $\boldsymbol{u}_0 = h^{-1}(\boldsymbol{x}) \in S$ and $\boldsymbol{y} \in \operatorname{null} f$, then $\boldsymbol{u} \in f^{-1}(\boldsymbol{x})$ due to

$$f(\boldsymbol{u}) = f(\boldsymbol{u}_0) = f(h^{-1}(\boldsymbol{x})) = h(h^{-1}(\boldsymbol{x})) = \boldsymbol{x}.$$

Problem 1.2. Let f be a projection in $\mathbb{R}^n$. Put $S = \operatorname{rng} f$ and $T = \operatorname{null} f$. So, every vector $\boldsymbol{x} \in \mathbb{R}^n$ is uniquely expressible as $\boldsymbol{x} = \boldsymbol{y} + \boldsymbol{z}$ such that $\boldsymbol{y} \in S$, $\boldsymbol{z} \in T$, and $f(\boldsymbol{x}) = \boldsymbol{y}$. Then $f(\boldsymbol{y}) = \boldsymbol{y}$, which gives

$$(f \circ f)(\boldsymbol{x}) = f(f(\boldsymbol{x})) = f(\boldsymbol{y}) = \boldsymbol{y} = f(\boldsymbol{x}).$$

Conversely, assume that a linear transformation $f : \mathbb{R}^n \to \mathbb{R}^n$ satisfies the condition $f \circ f = f$. Let $S = \operatorname{rng} f$. We state that $S = \{\boldsymbol{y} \in \mathbb{R}^n : f(\boldsymbol{y}) = \boldsymbol{y}\}$. Indeed, if $f(\boldsymbol{y}) = \boldsymbol{y}$, then $\boldsymbol{y} \in \operatorname{rng} f$. Conversely, if $\boldsymbol{y} \in \operatorname{rng} f$, then $\boldsymbol{y} = f(\boldsymbol{x})$ for a suitable vector $\boldsymbol{x} \in \mathbb{R}^n$. Therefore, $f(\boldsymbol{y}) = f(f(\boldsymbol{x})) = f(\boldsymbol{x}) = \boldsymbol{y}$.

Put $T = \{\boldsymbol{x} - f(\boldsymbol{x}) : \boldsymbol{x} \in \mathbb{R}^n\}$. We claim that $T = \operatorname{null} f$. Indeed, from

$$f(\boldsymbol{x} - f(\boldsymbol{x})) = f(\boldsymbol{x}) - f(f(\boldsymbol{x})) = \boldsymbol{o}$$

if follows that $T \subset \operatorname{null} f$. Conversely, if $\boldsymbol{x} \in \operatorname{null} f$, then $\boldsymbol{x} = \boldsymbol{x} - f(\boldsymbol{x}) \in T$.

It remains to show that the subspaces S and T are complementary. First, we observe that $S + T = \mathbb{R}^n$ because every vector $\boldsymbol{x} \in \mathbb{R}^n$ can be written as

$$\boldsymbol{x} = f(\boldsymbol{x}) + (\boldsymbol{x} - f(\boldsymbol{x})) \in S + T.$$

Next, choose a vector $\boldsymbol{y} \in S \cap T$. Then $f(\boldsymbol{y}) = \boldsymbol{y}$ and $\boldsymbol{y} = \boldsymbol{z} - f(\boldsymbol{z})$ for a suitable vector $\boldsymbol{z} \in \mathbb{R}^n$. This gives

$$\boldsymbol{y} = f(\boldsymbol{y}) = f(\boldsymbol{z} - f(\boldsymbol{z})) = f(\boldsymbol{z}) - f(f(\boldsymbol{z})) = \boldsymbol{o},$$

implying the equality $S \cap T = \{\boldsymbol{o}\}$. Hence the sum $S + T = \mathbb{R}^n$ is direct, and f is a projection on S along T.

Problem 1.3. The assertion is obvious when both vectors $\boldsymbol{x}$ and $\boldsymbol{y}$ are $\boldsymbol{o}$. Suppose that $\boldsymbol{y} \neq \boldsymbol{o}$. For any scalar $t \in \mathbb{R}$, we have

$$0 \leqslant \|\boldsymbol{x} - t\boldsymbol{y}\|^2 = (\boldsymbol{x} - t\boldsymbol{y})\cdot(\boldsymbol{x} - t\boldsymbol{y}) = \boldsymbol{x}\cdot\boldsymbol{x} - 2t\boldsymbol{x}\cdot\boldsymbol{y} + t^2\boldsymbol{y}\cdot\boldsymbol{y}.$$

With $t = \dfrac{\boldsymbol{x}\cdot\boldsymbol{y}}{\boldsymbol{y}\cdot\boldsymbol{y}}$, the above inequality becomes

$$\begin{aligned} 0 \leqslant \left\|\boldsymbol{x} - \frac{\boldsymbol{x}\cdot\boldsymbol{y}}{\boldsymbol{y}\cdot\boldsymbol{y}}\boldsymbol{y}\right\|^2 &= \boldsymbol{x}\cdot\boldsymbol{x} - 2\frac{(\boldsymbol{x}\cdot\boldsymbol{y})^2}{\boldsymbol{y}\cdot\boldsymbol{y}} + \frac{(\boldsymbol{x}\cdot\boldsymbol{y})^2}{\boldsymbol{y}\cdot\boldsymbol{y}} \\ = \boldsymbol{x}\cdot\boldsymbol{x} - \frac{(\boldsymbol{x}\cdot\boldsymbol{y})^2}{\boldsymbol{y}\cdot\boldsymbol{y}} &= \|\boldsymbol{x}\|^2 - \frac{(\boldsymbol{x}\cdot\boldsymbol{y})^2}{\|\boldsymbol{y}\|^2}. \end{aligned} \tag{14.1}$$

So, $|\boldsymbol{x}\cdot\boldsymbol{y}| \leqslant \|\boldsymbol{x}\|\,\|\boldsymbol{y}\|$, and equality occurs if and only if

$$\left\|\boldsymbol{x} - \frac{\boldsymbol{x}\cdot\boldsymbol{y}}{\boldsymbol{y}\cdot\boldsymbol{y}}\boldsymbol{y}\right\|^2 = 0, \quad \text{or} \quad \boldsymbol{x} = \frac{\boldsymbol{x}\cdot\boldsymbol{y}}{\boldsymbol{y}\cdot\boldsymbol{y}}\boldsymbol{y}.$$

Consequently, $\boldsymbol{x}\cdot\boldsymbol{y} = \|\boldsymbol{x}\|\,\|\boldsymbol{y}\|$ if and only if the scalar $t = \dfrac{\boldsymbol{x}\cdot\boldsymbol{y}}{\boldsymbol{y}\cdot\boldsymbol{y}}$ is nonnegative.

Problem 1.4. From Problem 1.3, we obtain

$$\begin{aligned} \|\boldsymbol{x} + \boldsymbol{y}\|^2 &= (\boldsymbol{x} + \boldsymbol{y})\cdot(\boldsymbol{x} + \boldsymbol{y}) = \boldsymbol{x}\cdot\boldsymbol{x} + 2\boldsymbol{x}\cdot\boldsymbol{y} + \boldsymbol{y}\cdot\boldsymbol{y} \\ &= \|\boldsymbol{x}\|^2 + 2\boldsymbol{x}\cdot\boldsymbol{y} + \|\boldsymbol{y}\|^2 \leqslant \|\boldsymbol{x}\|^2 + 2\|\boldsymbol{x}\|\,\|\boldsymbol{y}\| + \|\boldsymbol{y}\|^2 = (\|\boldsymbol{x}\| + \|\boldsymbol{y}\|)^2. \end{aligned}$$

Thus $\|\boldsymbol{x} + \boldsymbol{y}\| \leqslant \|\boldsymbol{x}\| + \|\boldsymbol{y}\|$, and equality holds if and only if $\boldsymbol{x}\cdot\boldsymbol{y} = \|\boldsymbol{x}\|\,\|\boldsymbol{y}\|$. According to Problem 1.3, the last condition occurs if and only if one of the vectors $\boldsymbol{x}$ and $\boldsymbol{y}$ is a nonnegative scalar multiple of the other.

Problem 1.5. According to Problem 1.4,

$$\|\boldsymbol{x} - \boldsymbol{y}\| = \|(\boldsymbol{x} - \boldsymbol{z}) + (\boldsymbol{z} - \boldsymbol{y})\| \leqslant \|\boldsymbol{x} - \boldsymbol{z}\| + \|\boldsymbol{z} - \boldsymbol{y}\|,$$

and the equality $\|\boldsymbol{x}-\boldsymbol{y}\| = \|\boldsymbol{x}-\boldsymbol{z}\|+\|\boldsymbol{z}-\boldsymbol{y}\|$ gives $\boldsymbol{x}-\boldsymbol{z} = \lambda(\boldsymbol{x}-\boldsymbol{y})$ for a suitable nonnegative scalar λ. If $\|\boldsymbol{x}-\boldsymbol{y}\| = \|\boldsymbol{x}-\boldsymbol{z}\| + \|\boldsymbol{z}-\boldsymbol{y}\|$, then $\|\boldsymbol{x}-\boldsymbol{z}\| \leqslant \|\boldsymbol{x}-\boldsymbol{y}\|$, which implies $\lambda = \|\boldsymbol{x}-\boldsymbol{z}\|/\|\boldsymbol{x}-\boldsymbol{y}\| \leqslant 1$.

Conversely, suppose that $\boldsymbol{x}-\boldsymbol{z} = \lambda(\boldsymbol{x}-\boldsymbol{y})$, where $0 \leqslant \lambda \leqslant 1$. Then $\boldsymbol{z}-\boldsymbol{y} = (1-\lambda)(\boldsymbol{x}-\boldsymbol{y})$ and

$$\|\boldsymbol{x}-\boldsymbol{z}\| + \|\boldsymbol{z}-\boldsymbol{y}\| = \lambda\|\boldsymbol{x}-\boldsymbol{y}\| + (1-\lambda)\|\boldsymbol{x}-\boldsymbol{y}\| = \|\boldsymbol{x}-\boldsymbol{y}\|.$$

Problem 1.6. The triangle inequality $\|\boldsymbol{x}-\boldsymbol{y}\| \leqslant \|\boldsymbol{x}-\boldsymbol{z}\| + \|\boldsymbol{z}-\boldsymbol{y}\|$ gives

$$\|\boldsymbol{x}-\boldsymbol{y}\| - \|\boldsymbol{z}-\boldsymbol{y}\| \leqslant \|\boldsymbol{x}-\boldsymbol{z}\|. \tag{14.2}$$

Similarly, the inequality $\|\boldsymbol{y}-\boldsymbol{z}\| \leqslant \|\boldsymbol{y}-\boldsymbol{x}\| + \|\boldsymbol{x}-\boldsymbol{z}\|$ implies

$$-\|\boldsymbol{x}-\boldsymbol{z}\| \leqslant \|\boldsymbol{y}-\boldsymbol{x}\| - \|\boldsymbol{y}-\boldsymbol{z}\| = \|\boldsymbol{x}-\boldsymbol{y}\| - \|\boldsymbol{z}-\boldsymbol{y}\|. \tag{14.3}$$

Combining (14.2) and (14.3), we obtain $\big|\|\boldsymbol{x}-\boldsymbol{y}\| - \|\boldsymbol{z}-\boldsymbol{y}\|\big| \leqslant \|\boldsymbol{x}-\boldsymbol{z}\|$.

The equality $\big|\|\boldsymbol{x}-\boldsymbol{y}\| - \|\boldsymbol{z}-\boldsymbol{y}\|\big| = \|\boldsymbol{x}-\boldsymbol{z}\|$ holds if and only if one of the equalities

$$\|\boldsymbol{x}-\boldsymbol{y}\| = \|\boldsymbol{x}-\boldsymbol{z}\| + \|\boldsymbol{z}-\boldsymbol{y}\|, \quad \|\boldsymbol{y}-\boldsymbol{z}\| = \|\boldsymbol{y}-\boldsymbol{x}\| + \|\boldsymbol{x}-\boldsymbol{z}\| \tag{14.4}$$

takes place. By Problem 1.5, the first equality from (14.4) occurs if and only if $\boldsymbol{x}-\boldsymbol{z} = \lambda(\boldsymbol{x}-\boldsymbol{y})$ for a suitable scalar $0 \leqslant \lambda \leqslant 1$; equivalently, it occurs if and only if $\boldsymbol{z}-\boldsymbol{y} = (1-\lambda)(\boldsymbol{x}-\boldsymbol{y})$. By the same problem, the second equality from (14.4) occurs if and only if $\boldsymbol{y}-\boldsymbol{x} = \lambda(\boldsymbol{y}-\boldsymbol{z})$ for a suitable scalar $0 \leqslant \lambda \leqslant 1$; equivalently, it occurs if and only if $\boldsymbol{x}-\boldsymbol{y} = \lambda(\boldsymbol{z}-\boldsymbol{y})$.

Problem 1.7. The equalities

$$\begin{aligned}\frac{\|\boldsymbol{x}+t\boldsymbol{z}\| - \|\boldsymbol{x}\|}{t} &= \frac{\|\boldsymbol{x}+t\boldsymbol{z}\| - \|\boldsymbol{x}\|}{t} \cdot \frac{\|\boldsymbol{x}+t\boldsymbol{z}\| + \|\boldsymbol{x}\|}{\|\boldsymbol{x}+t\boldsymbol{z}\| + \|\boldsymbol{x}\|} \\ &= \frac{\|\boldsymbol{x}+t\boldsymbol{z}\|^2 - \|\boldsymbol{x}\|^2}{t(\|\boldsymbol{x}+t\boldsymbol{z}\| + \|\boldsymbol{x}\|)} = \frac{\boldsymbol{x}\cdot\boldsymbol{x} + 2t\boldsymbol{x}\cdot\boldsymbol{z} + t^2\boldsymbol{z}\cdot\boldsymbol{z} - \boldsymbol{x}\cdot\boldsymbol{x}}{t(\|\boldsymbol{x}+t\boldsymbol{z}\| + \|\boldsymbol{x}\|)} \\ &= \frac{2\boldsymbol{x}\cdot\boldsymbol{z} + t\boldsymbol{z}\cdot\boldsymbol{z}}{\|\boldsymbol{x}+t\boldsymbol{z}\| + \|\boldsymbol{x}\|}\end{aligned}$$

imply the existence of the desired limit:

$$\begin{aligned}\lim_{t\to0}\frac{\|\boldsymbol{x}+t\boldsymbol{z}\| - \|\boldsymbol{x}\|}{t} &= \lim_{t\to0}\frac{2\boldsymbol{x}\cdot\boldsymbol{z} + t\boldsymbol{z}\cdot\boldsymbol{z}}{\|\boldsymbol{x}+t\boldsymbol{z}\| + \|\boldsymbol{x}\|} \\ &= \frac{\lim_{t\to0}(2\boldsymbol{x}\cdot\boldsymbol{z} + t\boldsymbol{z}\cdot\boldsymbol{z})}{\lim_{t\to0}(\|\boldsymbol{x}+t\boldsymbol{z}\| + \|\boldsymbol{x}\|)} = \frac{2\boldsymbol{x}\cdot\boldsymbol{z}}{2\|\boldsymbol{x}\|} = \frac{\boldsymbol{x}\cdot\boldsymbol{z}}{\|\boldsymbol{x}\|}.\end{aligned}$$

Problem 1.8. (*a*) Suppose that $B_\rho(\boldsymbol{c}) \subset B_\mu(\boldsymbol{e})$. Since the case $\boldsymbol{c} = \boldsymbol{e}$ is obvious, we may assume that $\boldsymbol{c} \neq \boldsymbol{e}$. The point $\boldsymbol{a} = \boldsymbol{c} + \rho\dfrac{\boldsymbol{c}-\boldsymbol{e}}{\|\boldsymbol{c}-\boldsymbol{e}\|}$ belongs to $B_\rho(\boldsymbol{c})$ because

$\|\boldsymbol{a}-\boldsymbol{c}\| = \rho$. Therefore, $\boldsymbol{a} \in B_\rho(\boldsymbol{c}) \subset B_\mu(\boldsymbol{e})$. The equality

$$\boldsymbol{a}-\boldsymbol{e} = (\boldsymbol{c}-\boldsymbol{e}) + \rho\frac{\boldsymbol{c}-\boldsymbol{e}}{\|\boldsymbol{c}-\boldsymbol{e}\|} = \big(1+\frac{\rho}{\|\boldsymbol{c}-\boldsymbol{e}\|}\big)(\boldsymbol{c}-\boldsymbol{e})$$

can be rewritten as $\boldsymbol{c}-\boldsymbol{e} = \lambda(\boldsymbol{a}-\boldsymbol{e})$, where $0 < \lambda < 1$. Consequently, Problem 1.5 and the inclusion $\boldsymbol{a} \in B_\mu(\boldsymbol{e})$ imply the inequality

$$\|\boldsymbol{c}-\boldsymbol{e}\| + \rho = \|\boldsymbol{e}-\boldsymbol{c}\| + \|\boldsymbol{c}-\boldsymbol{a}\| = \|\boldsymbol{e}-\boldsymbol{a}\| \leqslant \mu.$$

Conversely, let $\|\boldsymbol{c}-\boldsymbol{e}\| \leqslant \mu - \rho$. For any point $\boldsymbol{x} \in B_\rho(\boldsymbol{c})$, one has

$$\|\boldsymbol{x}-\boldsymbol{e}\| \leqslant \|\boldsymbol{x}-\boldsymbol{c}\| + \|\boldsymbol{c}-\boldsymbol{e}\| \leqslant \rho + \|\boldsymbol{c}-\boldsymbol{e}\| \leqslant \mu.$$

Hence $\boldsymbol{x} \in B_\mu(\boldsymbol{e})$, which gives the inclusion $B_\rho(\boldsymbol{c}) \subset B_\mu(\boldsymbol{e})$.

Assume now that $U_\rho(\boldsymbol{c}) \subset U_\mu(\boldsymbol{e})$. Then

$$B_\rho(\boldsymbol{c}) = \operatorname{cl} U_\rho(\boldsymbol{c}) \subset \operatorname{cl} U_\mu(\boldsymbol{e}) = B_\mu(\boldsymbol{e}),$$

and the above argument gives the inequality $\|\boldsymbol{c}-\boldsymbol{e}\| + \rho \leqslant \mu$.

Conversely, let $\|\boldsymbol{c}-\boldsymbol{e}\| + \rho \leqslant \mu$. For any point $\boldsymbol{x} \in U_\rho(\boldsymbol{c})$, one has

$$\|\boldsymbol{x}-\boldsymbol{e}\| \leqslant \|\boldsymbol{x}-\boldsymbol{c}\| + \|\boldsymbol{c}-\boldsymbol{e}\| < \rho + \|\boldsymbol{c}-\boldsymbol{e}\| \leqslant \mu.$$

Hence $\boldsymbol{x} \in U_\mu(\boldsymbol{e})$, which gives the inclusion $U_\rho(\boldsymbol{c}) \subset U_\mu(\boldsymbol{e})$.

(*b*) Since the "if" part is obvious, it remains to prove the "only if" part. So, let $B_\rho(\boldsymbol{c}) = B_\mu(\boldsymbol{e})$. Applying assertion (*a*) twice, we obtain $\|\boldsymbol{c}-\boldsymbol{e}\| \leqslant \mu - \rho$ and $\|\boldsymbol{e}-\boldsymbol{c}\| \leqslant \rho - \mu$. Thus $\rho = \mu$ and $\boldsymbol{c} = \boldsymbol{e}$. The case of open balls is similar.

Problem 1.9. Choose a vector $\boldsymbol{x}_0 \in \mathbb{R}^n$ and an infinite sequence of vectors $\boldsymbol{x}_1, \boldsymbol{x}_2, \dots$ from $\mathbb{R}^n$ converging to $\boldsymbol{x}_0$. By Problem 1.6,

$$|\delta_{\boldsymbol{c}}(\boldsymbol{x}_i) - \delta_{\boldsymbol{c}}(\boldsymbol{x}_0)| = \big|\|\boldsymbol{x}_i-\boldsymbol{c}\| - \|\boldsymbol{x}_0-\boldsymbol{c}\|\big| \leqslant \|\boldsymbol{x}_i-\boldsymbol{x}_0\|,$$

which gives $\lim_{i\to\infty} \delta_{\boldsymbol{c}}(\boldsymbol{x}_i) = \delta_{\boldsymbol{c}}(\boldsymbol{x}_0)$. Hence $\delta_{\boldsymbol{c}}(\boldsymbol{x})$ is continuous on $\mathbb{R}^n$.

Problem 1.10. Assume first that $\lim_{i\to\infty} \lambda_j^{(i)} = \lambda_j$ for all $1 \leqslant j \leqslant r$. Put $\gamma = \max\{\|\boldsymbol{b}_1\|, \dots, \|\boldsymbol{b}_r\|\}$. Given an $\varepsilon > 0$, there is an index i_0 such that $|\lambda_j^{(i)} - \lambda_j| < \varepsilon/(\gamma r)$ for all $j = 1, \dots, r$ and $i \geqslant i_0$. Hence

$$\|\boldsymbol{x}_i - \boldsymbol{x}\| = \|\sum_{j=1}^{r}(\lambda_j^{(i)} - \lambda_j)\boldsymbol{b}_j\| \leqslant \sum_{j=1}^{r}|\lambda_j^{(i)} - \lambda_j|\,\|\boldsymbol{b}_j\| \leqslant r\frac{\varepsilon}{\gamma r}\gamma = \varepsilon$$

for all $i \geqslant i_0$, which results in $\lim_{i\to\infty} \boldsymbol{x}_i = \boldsymbol{x}$.

Conversely, suppose that $\lim_{i\to\infty} \boldsymbol{x}_i = \boldsymbol{x}$. Put $\boldsymbol{u}_i = \boldsymbol{x}_i - \boldsymbol{x}$ and choose an orthonormal basis $\{\boldsymbol{e}_1, \dots, \boldsymbol{e}_r\}$ for span $\{\boldsymbol{b}_1, \dots, \boldsymbol{b}_r\}$. We can write

$$\boldsymbol{u}_i = \sum_{k=1}^{r} \mu_k^{(i)} \boldsymbol{e}_k, \quad i \geqslant 1, \quad \text{and} \quad \boldsymbol{b}_j = \sum_{k=1}^{r} a_{jk} \boldsymbol{e}_k, \quad 1 \leqslant j \leqslant r.$$

The $r \times r$ matrix $A = (a_{jk})$ is invertible because the set $\{\boldsymbol{b}_1, \dots, \boldsymbol{b}_r\}$ is linearly independent. From $\lim_{i\to\infty} \boldsymbol{u}_i = \boldsymbol{o}$ and the equalities

$$\|\boldsymbol{u}_i\|^2 = \big(\sum_{k=1}^{r} \mu_k^{(i)} \boldsymbol{e}_k\big)\cdot\big(\sum_{k=1}^{r} \mu_k^{(i)} \boldsymbol{e}_k\big) = \big(\mu_1^{(i)}\big)^2 + \cdots + \big(\mu_r^{(i)}\big)^2, \quad i \geqslant 1,$$

we obtain that $\lim_{i\to\infty} \mu_k^{(i)} = 0$ for all $1 \leqslant k \leqslant r$. Rewriting $\boldsymbol{u}_i$ as

$$\begin{aligned}\boldsymbol{u}_i &= \sum_{j=1}^{r} (\lambda_j^{(i)} - \lambda_j)\boldsymbol{b}_j = \sum_{j=1}^{r} (\lambda_j^{(i)} - \lambda_j) \sum_{k=1}^{r} a_{jk} \boldsymbol{e}_k \\ &= \sum_{k=1}^{r} \big(\sum_{j=1}^{r} (\lambda_j^{(i)} - \lambda_j) a_{jk}\big) \boldsymbol{e}_k\end{aligned}$$

one has

$$\mu_k^{(i)} = \sum_{j=1}^{r} (\lambda_j^{(i)} - \lambda_j) a_{jk}, \quad 1 \leqslant k \leqslant r, \quad i \geqslant 1. \tag{14.5}$$

Since A is invertible, (14.5) is solvable for $\lambda_j^{(i)} - \lambda_j$:

$$\lambda_j^{(i)} - \lambda_j = \sum_{k=1}^{r} b_{jk} \mu_k^{(i)}, \quad 1 \leqslant j \leqslant r, \quad i \geqslant 1.$$

Therefore,

$$\lim_{i\to\infty} (\lambda_j^{(i)} - \lambda_j) = \lim_{i\to\infty} \big(\sum_{k=1}^{r} b_{jk} \mu_k^{(i)}\big) = \sum_{k=1}^{r} b_{jk} \lim_{i\to\infty} \mu_k^{(i)} = 0, \quad 1 \leqslant j \leqslant r.$$

Hence $\lim_{i\to\infty} \lambda_j^{(i)} = \lambda_j$ for all $1 \leqslant j \leqslant r$.

Problem 1.11. Complete the set $\{\boldsymbol{b}_1, \dots, \boldsymbol{b}_r\}$ into a basis $\{\boldsymbol{b}_1, \dots, \boldsymbol{b}_n\}$ for $\mathbb{R}^n$. Let $\boldsymbol{b}_i = (b_{i1}, \dots, b_{in})$, $1 \leqslant i \leqslant n$. From linear algebra we know that the determinant of the $n \times n$ matrix $B = (b_{ij})$ is a nonzero scalar, say α. Since $\det B$ is a polynomial of degree n in the n^2 variables $b_{11}, b_{12}, \dots, b_{nn}$, it is a continuous function on the Cartesian product $\mathbb{R}^{n^2}$, equipped with the max-norm. Hence there is a $\rho > 0$ such that the determinant of an $n \times n$ matrix $B' = (b'_{ij})$ satisfies the condition $|\det B' - \alpha| < |\alpha|/2$ provided $|b_{ij} - b'_{ij}| < \rho$ for all $1 \leqslant i, j \leqslant n$. Interpreting the rows of B' as vectors $\boldsymbol{b}'_i = (b'_{i1}, \dots, b'_{in})$, $1 \leqslant i \leqslant n$, we obtain that the inclusions $\boldsymbol{b}'_i \in U_\rho(\boldsymbol{b}_i)$, $1 \leqslant i \leqslant n$, imply the inequalities

$$|b_{ij} - b'_{ij}| \leqslant \big(\sum_{j=1}^{n} (b_{ij} - b'_{ij})^2\big)^{1/2} = \|\boldsymbol{b}_i - \boldsymbol{b}'_i\| < \rho, \quad 1 \leqslant i, j \leqslant n.$$

Consequently, $|\det B'| \geqslant |\alpha|/2 > 0$, which shows that the set $\{\boldsymbol{b}'_1, \ldots, \boldsymbol{b}'_n\}$ is linearly independent.

Problem 1.12. Because the case $X = \varnothing$ is obvious, we may assume that $X \neq \varnothing$. Choose any point $\boldsymbol{x} \in X$ and an open set $F_{\boldsymbol{x}} \in \mathcal{F}$ which contains $\boldsymbol{x}$. Let $U_{\rho_{\boldsymbol{x}}}(\boldsymbol{x})$ be an open ball which lies in $F_{\boldsymbol{x}}$. Choose a rational scalar $r_{\boldsymbol{x}}$ satisfying the condition $0 < r_{\boldsymbol{x}} \leqslant \rho_{\boldsymbol{x}}$ and a point $\boldsymbol{c}_{\boldsymbol{x}} \in U_{r_{\boldsymbol{x}}}(\boldsymbol{x})$ with all rational coordinates. By Problem 1.8, $\boldsymbol{x} \in U_{r_{\boldsymbol{x}}}(\boldsymbol{e}_{\boldsymbol{x}}) \subset U_{\rho_{\boldsymbol{x}}}(\boldsymbol{x}) \subset F_{\boldsymbol{x}}$.

Since the set of points of $\mathbb{R}^n$ with all rational coordinates is denumerable, the above argument shows the existence of a countable family $\mathcal{U}$ of open balls of the form $U_{r_{\boldsymbol{x}}}(\boldsymbol{e}_{\boldsymbol{x}})$ whose union covers X. Choose for each open ball $U_{r_{\boldsymbol{x}}}(\boldsymbol{e}_{\boldsymbol{x}}) \in \mathcal{U}$ an open set from $\mathcal{F}$ containing the closed ball and denote this set by $F_{r_{\boldsymbol{x}}}(\boldsymbol{e}_{\boldsymbol{x}})$. The above argument shows that the countable subfamily $\{F_{r_{\boldsymbol{x}}}(\boldsymbol{e}_{\boldsymbol{x}})\}$ of $\mathcal{F}$ covers X.

Problem 1.13. Omitting the obvious case when at least one of the sets X and Y is empty, we suppose that both X and Y are nonempty. For the inclusion $\lambda \operatorname{cl} X + \mu \operatorname{cl} Y \subset \operatorname{cl}(\lambda X + \mu Y)$, choose a point $\boldsymbol{z} \in \lambda \operatorname{cl} X + \mu \operatorname{cl} Y$. Then $\boldsymbol{z} = \lambda \boldsymbol{x} + \mu \boldsymbol{y}$, with $\boldsymbol{x} \in \operatorname{cl} X$ and $\boldsymbol{y} \in \operatorname{cl} Y$. Let $\boldsymbol{x} = \lim_{i\to\infty} \boldsymbol{x}_i$ and $\boldsymbol{y} = \lim_{i\to\infty} \boldsymbol{y}_i$ for suitable sequences $\boldsymbol{x}_1, \boldsymbol{x}_2, \dots \in X$ and $\boldsymbol{y}_1, \boldsymbol{y}_2, \dots \in Y$. Then

$$\boldsymbol{z} = \lambda \boldsymbol{x} + \mu \boldsymbol{y} = \lim_{i\to\infty} \lambda \boldsymbol{x}_i + \lim_{i\to\infty} \mu \boldsymbol{y}_i = \lim_{i\to\infty} (\lambda \boldsymbol{x}_i + \mu \boldsymbol{y}_i) \in \operatorname{cl}(\lambda X + \mu Y).$$

The inclusion $\lambda \operatorname{cl} X + \mu \operatorname{cl} Y \subset \operatorname{cl}(\lambda X + \mu Y)$ gives

$$\operatorname{cl}(\lambda \operatorname{cl} X + \mu \operatorname{cl} Y) \subset \operatorname{cl}(\operatorname{cl}(\lambda X + \mu Y)) = \operatorname{cl}(\lambda X + \mu Y).$$

Conversely, the inclusion $\lambda X + \mu Y \subset \lambda \operatorname{cl} X + \mu \operatorname{cl} Y$ implies

$$\operatorname{cl}(\lambda X + \mu Y) \subset \operatorname{cl}(\lambda \operatorname{cl} X + \mu \operatorname{cl} Y).$$

(a) Assume that at least one of the sets X and Y, say X, is bounded. By the above proved, it suffices to verify the inclusion $\operatorname{cl}(\lambda X + \mu Y) \subset \lambda \operatorname{cl} X + \mu \operatorname{cl} Y$. Choose a point $\boldsymbol{z} \in \operatorname{cl}(\lambda X + \mu Y)$. Then $\boldsymbol{z} = \lim_{i\to\infty} \boldsymbol{z}_i$ for a suitable sequence $\boldsymbol{z}_1, \boldsymbol{z}_2, \dots$ from $\lambda X + \mu Y$. Every $\boldsymbol{z}_i$ can be written as $\boldsymbol{z}_i = \lambda \boldsymbol{x}_i + \mu \boldsymbol{y}_i$, where $\boldsymbol{x}_i \in X$ and $\boldsymbol{y}_i \in Y$, $i \geqslant 1$. Since X is bounded, the sequence $\boldsymbol{x}_1, \boldsymbol{x}_2, \dots$ contains a subsequence $\boldsymbol{x}_{i_1}, \boldsymbol{x}_{i_2}, \dots$ converging to a point $\boldsymbol{x} \in \operatorname{cl} X$. Consequently,

$$\boldsymbol{z} - \lambda \boldsymbol{x} = \lim_{k\to\infty} \boldsymbol{z}_{i_k} - \lim_{k\to\infty} \lambda \boldsymbol{x}_{i_k} = \lim_{k\to\infty} (\boldsymbol{z}_{i_k} - \lambda \boldsymbol{x}_{i_k}) = \lim_{k\to\infty} \mu \boldsymbol{y}_{i_k} \in \mu \operatorname{cl} Y,$$

and $\boldsymbol{z} = \lambda \boldsymbol{x} + (\boldsymbol{z} - \lambda \boldsymbol{x}) \in \lambda \operatorname{cl} X + \mu \operatorname{cl} Y$.

(b) Assume now that there are independent subspaces S and T such that $X \subset S$ and $Y \subset T$. Since $\lambda X + \mu Y \subset S + T$, and since the subspace $S + T$ is closed, one has $\operatorname{cl}(\lambda X + \mu Y) \subset S + T$. As above, it suffices to show that $\operatorname{cl}(\lambda X + \mu Y) \subset \lambda \operatorname{cl} X + \mu \operatorname{cl} Y$. Let $\boldsymbol{z} \in \operatorname{cl}(\lambda X + \mu Y)$. Then $\boldsymbol{z} = \lim_{i\to\infty} \boldsymbol{z}_i$, where all $\boldsymbol{z}_1, \boldsymbol{z}_2, \dots$ are in $\lambda X + \mu Y$. Furthermore, $\boldsymbol{z}_i = \lambda \boldsymbol{x}_i + \mu \boldsymbol{y}_i$, $i \geqslant 1$, with

$\boldsymbol{x}_1, \boldsymbol{x}_2, \dots$ in X and $\boldsymbol{y}_1, \boldsymbol{y}_2, \dots$ in Y. Problem 1.10 implies that $\boldsymbol{z}_i \to \boldsymbol{z}$ if and only if $\boldsymbol{x}_i \to \boldsymbol{x}$ and $\boldsymbol{y}_i \to \boldsymbol{y}$, where $\boldsymbol{z} = \lambda\boldsymbol{x} + \mu\boldsymbol{y}$. So,

$$\boldsymbol{z} = \lim_{i\to\infty} (\lambda\boldsymbol{x}_i + \mu\boldsymbol{y}_i) = \lambda \lim_{i\to\infty} \boldsymbol{x}_i + \mu \lim_{i\to\infty} \boldsymbol{y}_i = \lambda\boldsymbol{x} + \mu\boldsymbol{y} \in \lambda \operatorname{cl} X + \mu \operatorname{cl} Y.$$

Problem 1.14. Since the case $X \cap Y \neq \varnothing$ is obvious, we may assume that X and Y are disjoint. Choose sequences of points $\boldsymbol{x}_1, \boldsymbol{x}_2, \dots \in X$ and $\boldsymbol{y}_1, \boldsymbol{y}_2, \dots \in X$ satisfying the condition $\lim_{i\to\infty} \|\boldsymbol{x}_i - \boldsymbol{y}_i\| = \delta(X, Y)$. Because X is bounded, there is a subsequence $\boldsymbol{x}_{i_1}, \boldsymbol{x}_{i_2}, \dots$ of $\boldsymbol{x}_1, \boldsymbol{x}_2, \dots$ which converges to a point $\boldsymbol{x} \in \operatorname{cl} X$. Let an index p be such that $\|\boldsymbol{x}_i - \boldsymbol{y}_i\| \leqslant \delta(X, Y) + 1$ for all $i \geqslant p$ and $\|\boldsymbol{x}_{i_r} - \boldsymbol{x}\| \leqslant 1$ for all $i_r \geqslant p$. Then

$$\|\boldsymbol{x} - \boldsymbol{y}_{i_r}\| \leqslant \|\boldsymbol{x} - \boldsymbol{x}_{i_r}\| + \|\boldsymbol{x}_{i_r} - \boldsymbol{y}_{i_r}\| \leqslant \delta(X, Y) + 2, \quad i_r \geqslant p,$$

implying that the sequence $\boldsymbol{y}_{i_1}, \boldsymbol{y}_{i_2}, \dots$ is bounded. Therefore, there is a subsequence $\boldsymbol{y}_{j_1}, \boldsymbol{y}_{j_2}, \dots$ of $\boldsymbol{y}_{i_1}, \boldsymbol{y}_{i_2}, \dots$ which converges to a point $\boldsymbol{y} \in \operatorname{cl} Y$. Finally,

$$\|\boldsymbol{x} - \boldsymbol{y}\| = \lim_{s\to\infty} \|\boldsymbol{x}_{j_s} - \boldsymbol{y}_{j_s}\| = \delta(X, Y).$$

Problem 1.15. We will prove contrapositive assertions.

(*a*) Let $\delta(X, Y) = 0$. Then $\lim_{i\to\infty} \|\boldsymbol{x}_i - \boldsymbol{y}_i\| = 0$ for suitable sequences $\boldsymbol{x}_1, \boldsymbol{x}_2, \dots \in X$ and $\boldsymbol{y}_1, \boldsymbol{y}_2, \dots \in Y$. Therefore, $\boldsymbol{o} = \lim_{i\to\infty} (\boldsymbol{x}_i - \boldsymbol{y}_i) \in \operatorname{cl}(X - Y)$.

Conversely, let $\boldsymbol{o} \in \operatorname{cl}(X - Y)$. Then $\lim_{i\to\infty} \boldsymbol{z}_i = \boldsymbol{o}$ for a suitable sequence of points $\boldsymbol{z}_1, \boldsymbol{z}_2, \dots \in X - Y$ Writing $\boldsymbol{z}_i = \boldsymbol{x}_i - \boldsymbol{y}_i$, where $\boldsymbol{x}_i \in X$ and $\boldsymbol{y}_i \in Y$, $i \geqslant 1$, one has

$$\delta(X, Y) \leqslant \lim_{i\to\infty} \|\boldsymbol{x}_i - \boldsymbol{y}_i\| = 0.$$

(*b*) Let $\delta(X, Y) < \rho$. Choose points $\boldsymbol{x} \in X$ and $\boldsymbol{y} \in Y$ satisfying the condition $\|\boldsymbol{x} - \boldsymbol{y}\| < \rho$. If $\boldsymbol{z} = (\boldsymbol{x} + \boldsymbol{y})/2$, then

$$\|\boldsymbol{x} - \boldsymbol{z}\| = \|\boldsymbol{y} - \boldsymbol{z}\| = \|\boldsymbol{x} - \boldsymbol{y}\|/2 < \rho/2,$$

which gives the inclusion $\boldsymbol{z} \in U_{\rho/2}(\boldsymbol{x}) \cap U_{\rho/2}(\boldsymbol{y}) \subset U_{\rho/2}(X) \cap U_{\rho/2}(Y)$.

Conversely, let $U_{\rho/2}(X) \cap U_{\rho/2}(Y) \neq \varnothing$. Choose a point $\boldsymbol{z} \in U_{\rho/2}(X) \cap U_{\rho/2}(Y)$ and points $\boldsymbol{x} \in X$ and $\boldsymbol{y} \in Y$ such that $\|\boldsymbol{x} - \boldsymbol{z}\| < \rho/2$ and $\|\boldsymbol{x} - \boldsymbol{z}\| < \rho/2$. This argument and Problem 1.5 give

$$\delta(X, Y) \leqslant \|\boldsymbol{x} - \boldsymbol{y}\| \leqslant \|\boldsymbol{x} - \boldsymbol{z}\| + \|\boldsymbol{z} - \boldsymbol{y}\| < \rho/2 + \rho/2 = \rho.$$

(*c*) This part is similar to (*b*).

Problem 1.16. (*a*) Choose any point $\boldsymbol{x} \in \operatorname{cl} X$. Then $\boldsymbol{x} = \lim_{i\to\infty} \boldsymbol{x}_i$ for a suitable sequence $\boldsymbol{x}_1, \boldsymbol{x}_2, \dots$ of points from X. By the continuity of f,

$$f(\boldsymbol{x}) = f(\lim_{i\to\infty} \boldsymbol{x}_i) = \lim_{i\to\infty} f(\boldsymbol{x}_i) \in \operatorname{cl} f(X).$$

Hence $f(\operatorname{cl} X) \subset \operatorname{cl} f(X)$, which gives $\operatorname{cl} f(\operatorname{cl} X) \subset \operatorname{cl} f(X)$.

The opposite inclusion derives from the obvious implications

$$X \subset \operatorname{cl} X \;\Rightarrow\; f(X) \subset f(\operatorname{cl} X) \;\Rightarrow\; \operatorname{cl} f(X) \subset \operatorname{cl} f(\operatorname{cl} X).$$

(*b*) Assume that X is bounded. Due to assertion (*a*), it suffices to prove the inclusion $\operatorname{cl} f(X) \subset f(\operatorname{cl} X)$. For this, choose any point $\boldsymbol{z} \in \operatorname{cl} f(X)$. Then $\boldsymbol{z} = \lim_{i\to\infty} \boldsymbol{z}_i$ for a suitable sequence $\boldsymbol{z}_1, \boldsymbol{z}_2, \dots$ of points from $f(X)$. Choose points $\boldsymbol{x}_1, \boldsymbol{x}_2, \dots \in X$ such that $f(\boldsymbol{x}_i) = \boldsymbol{z}_i$ for all $i \geqslant 1$. Because the set X is bounded, there is a subsequence $\boldsymbol{x}_{i_1}, \boldsymbol{x}_{i_2}, \dots$ of $\boldsymbol{x}_1, \boldsymbol{x}_2, \dots$ which converges to a point $\boldsymbol{x} \in \operatorname{cl} X$. By the continuity of f,

$$\boldsymbol{z} = \lim_{k\to\infty} \boldsymbol{z}_{i_k} = \lim_{k\to\infty} f(\boldsymbol{x}_{i_k}) = f(\lim_{k\to\infty} \boldsymbol{x}_{i_k}) = f(\boldsymbol{x}) \in f(\operatorname{cl} X).$$

(*c*) The obvious equality $f^{-1}(Y) = f^{-1}(Y \cap \operatorname{rng} f)$ immediately gives

$$\operatorname{cl} f^{-1}(Y) = \operatorname{cl} f^{-1}(Y \cap \operatorname{rng} f).$$

For the first inclusion, $\operatorname{cl} f^{-1}(Y \cap \operatorname{rng} f) \subset f^{-1}(\operatorname{cl}(Y \cap \operatorname{rng} f))$, choose a point $\boldsymbol{x} \in \operatorname{cl} f^{-1}(Y \cap \operatorname{rng} f)$ and a sequence $\boldsymbol{x}_1, \boldsymbol{x}_2, \dots \in f^{-1}(Y \cap \operatorname{rng} f)$ converging to $\boldsymbol{x}$. Then $f(\boldsymbol{x}_i) \in Y \cap \operatorname{rng} f$ for all $i \geqslant 1$, which gives

$$f(\boldsymbol{x}) = f(\lim_{i\to\infty} \boldsymbol{x}_i) = \lim_{i\to\infty} f(\boldsymbol{x}_i) \in \operatorname{cl}(Y \cap \operatorname{rng} f).$$

So, $\boldsymbol{x} \in f^{-1}(\operatorname{cl}(Y \cap \operatorname{rng} f))$. The second inclusion, $f^{-1}(\operatorname{cl}(Y \cap \operatorname{rng} f)) \subset f^{-1}(\operatorname{cl} Y)$, immediately follows from $\operatorname{cl}(Y \cap \operatorname{rng} f)) \subset \operatorname{cl} Y$.

Problem 1.17. (*a*) If $Y \subset \mathbb{R}^m$ is a closed set, then, by Problem 1.16,

$$f^{-1}(Y) \subset \operatorname{cl} f^{-1}(Y) \subset f^{-1}(\operatorname{cl}(Y \cap \operatorname{rng} f)) \subset f^{-1}(\operatorname{cl} Y) = f^{-1}(Y).$$

Consequently, $f^{-1}(Y) = \operatorname{cl} f^{-1}(Y)$, which shows the closedness of $f^{-1}(Y)$.

(*b*) Let Y be an open set in $\mathbb{R}^m$. Since the case $f^{-1}(Y) = \varnothing$ is obvious, we may assume that $f^{-1}(Y) \neq \varnothing$. Choose a point $\boldsymbol{u} \in f^{-1}(Y)$ and let $\boldsymbol{z} = f(\boldsymbol{u})$. Then $\boldsymbol{z} \in Y$. Choose in Y an open ball $U_\rho(\boldsymbol{z})$. By the continuity of f, there is a scalar $\delta > 0$ such that $f(\boldsymbol{x}) \in U_\rho(\boldsymbol{z})$ whenever $\boldsymbol{x} \in U_\delta(\boldsymbol{u})$. So, $f(U_\delta(\boldsymbol{u})) \subset U_\rho(\boldsymbol{z}) \subset Y$. Hence $U_\delta(\boldsymbol{u}) \subset f^{-1}(f(U_\delta(\boldsymbol{u}))) \subset f^{-1}(Y)$, implying that $f^{-1}(Y)$ is an open set.

Problem 1.18. For any vector $\boldsymbol{x} = (x_1, \dots, x_n) \in \mathbb{R}^n$,

$$f(\boldsymbol{x}) = f(x_1\boldsymbol{e}_1 + \cdots + x_n\boldsymbol{e}_n) = x_1 f(\boldsymbol{e}_1) + \cdots + x_n f(\boldsymbol{e}_n) = (y_1, \dots, y_m),$$

where $y_i = \sum_{j=1}^{n} a_{ji} x_j$, $1 \leqslant i \leqslant m$. Using the Cauchy-Schwarz inequality

$$(\beta_1\gamma_1 + \cdots + \beta_n\gamma_n)^2 \leqslant (\beta_1^2 + \cdots + \beta_n^2)(\gamma_1^2 + \cdots + \gamma_n^2),$$

we obtain

$$y_i^2 = \Big(\sum_{j=1}^{n} a_{ji}x_j\Big)^2 \leqslant \sum_{j=1}^{n} a_{ji}^2 \sum_{j=1}^{n} x_j^2 = \sum_{j=1}^{n} a_{ji}^2 \|\boldsymbol{x}\|^2.$$

Hence

$$\|f(\boldsymbol{x})\|^2 = \sum_{i=1}^{m} y_i^2 \leqslant \sum_{i=1}^{m}\sum_{j=1}^{n} a_{ji}^2 \|\boldsymbol{x}\|^2 = L^2\|\boldsymbol{x}\|^2.$$

For the continuity of f, choose a vector $\boldsymbol{x}_0 \in \mathbb{R}^n$ and a sequence of vectors $\boldsymbol{x}_1, \boldsymbol{x}_2, \ldots$ in $\mathbb{R}^n$ converging to $\boldsymbol{x}_0$. The inequality

$$\|f(\boldsymbol{x}_i) - f(\boldsymbol{x}_0)\| = \|f(\boldsymbol{x}_i - \boldsymbol{x}_0)\| \leqslant L\|\boldsymbol{x}_i - \boldsymbol{x}_0\|$$

implies that $\lim_{i\to\infty} f(\boldsymbol{x}_i) = f(\boldsymbol{x}_0)$. So, f is continuous on $\mathbb{R}^n$.

Problem 1.19. Let $X \subset \mathbb{R}^n$ be a set in $\mathbb{R}^n$. According to Problem 1.16, $f(\operatorname{cl} X) \subset \operatorname{cl} f(X)$. So, it remains to prove the opposite inclusion. The linear transformation $h : \mathbb{R}^n \to \operatorname{rng} f$ defined by $h(\boldsymbol{x}) = f(\boldsymbol{x})$ is invertible, and Problem 1.18 shows that the inverse linear transformation $h^{-1} : \operatorname{rng} f \to \mathbb{R}^n$ is continuous. Again by Problem 1.16,

$$h^{-1}(\operatorname{cl} f(X)) \subset \operatorname{cl} h^{-1}(f(X)) = \operatorname{cl}(h^{-1}\circ f)(X) = \operatorname{cl} X.$$

Consequently, $\operatorname{cl} f(X) = (f\circ h^{-1})(\operatorname{cl} f(X)) = f(h^{-1}(\operatorname{cl} f(X))) \subset f(\operatorname{cl} X)$.

Problem 1.20. Due to Problem 1.16, it suffices to prove the inclusion

$$f^{-1}(\operatorname{cl}(Y \cap \operatorname{rng} f)) \subset \operatorname{cl} f^{-1}(Y \cap \operatorname{rng} f).$$

Choose a subspace $S \subset \mathbb{R}^n$ complementary to $\operatorname{null} f$ and consider the mapping $h : S \to \operatorname{rng} f$ defined by $h(\boldsymbol{x}) = f(\boldsymbol{x})$. By Problem 1.1, h is an invertible linear transformation. Problem 1.18 shows that the inverse mapping $h^{-1} : \operatorname{rng} f \to \mathbb{R}^n$ is continuous. Consequently, Problem 1.16 (with h^{-1} instead of f) implies

$$h^{-1}(\operatorname{cl}(Y \cap \operatorname{rng}\,)) \subset \operatorname{cl} h^{-1}(Y \cap \operatorname{rng} f).$$

Because the subspace $\operatorname{null} f$ is a closed set, a combination of Problems 1.1 and 1.13 gives

$$\begin{aligned} f^{-1}(\operatorname{cl}(Y \cap \operatorname{rng}\,)) &= h^{-1}(\operatorname{cl}(Y \cap \operatorname{rng}\,)) + \operatorname{null} f \subset \operatorname{cl} h^{-1}(Y \cap \operatorname{rng} f) + \operatorname{null} f \\ &\subset \operatorname{cl}(h^{-1}(Y \cap \operatorname{rng} f) + \operatorname{null} f) = \operatorname{cl} f^{-1}(Y \cap \operatorname{rng} f). \end{aligned}$$

Chapter 2

Problem 2.1. By Theorem 2.4, $\operatorname{dir} L'_\alpha = \operatorname{dir} L_\alpha$ for every plane $L_\alpha \in \mathcal{F}$. Choose points $\boldsymbol{c} \in \bigcap_\alpha L_\alpha$ and $\boldsymbol{c}' \in \bigcap_\alpha L'_\alpha$. Then $L_\alpha = \boldsymbol{c} + \operatorname{dir} L_\alpha$ and $L'_\alpha = \boldsymbol{c}' + \operatorname{dir} L_\alpha$ (see

Theorem 2.2). So,

$$\bigcap_\alpha L'_\alpha = \bigcap_\alpha (\mathbf{c}' + \operatorname{dir} L_\alpha) = (\mathbf{c}' - \mathbf{c}) + \bigcap_\alpha (\mathbf{c} + \operatorname{dir} L_\alpha) = (\mathbf{c}' - \mathbf{c}) + \bigcap_\alpha L_\alpha.$$

Problem 2.2. If one of the planes, say $L_\gamma \in \mathcal{F}$, contains all the others, then $M = L_\gamma$. Conversely, assume that M is a plane. Since the case $M = \varnothing$ is obvious, one may assume that $M \neq \varnothing$. Also, we may suppose that all planes $L_\alpha \in \mathcal{F}$ are nonempty. We will prove the inclusion $M \in \mathcal{F}$ by induction on $m = \dim M$. The case $m = 0$ is obvious. Let $m \geqslant 1$. If $m = 1$, then at least one $L_\alpha \in \mathcal{F}$ coincides with M (otherwise, each L_α would be a singleton and $\bigcup_\alpha L_\alpha$ would be a countable subset of M).

Suppose that the induction hypothesis holds for all $m \leqslant k-1$, where $k \geqslant 2$, and let $\dim M = k$. Assume, for contradiction, that $M \notin \mathcal{F}$. Choose in M a plane N of dimension $k-1$. Then $N = \bigcup_\alpha (N \cap L_\alpha)$ is a countable union of planes $N \cap L_\alpha$ (see Theorem 2.4). By the induction hypothesis, there is a plane $L_\beta \in \mathcal{F}$ such that $N \cap L_\beta = N$, which gives $N = L_\beta \in \mathcal{F}$ (indeed, if $N \neq L_\beta$ then $M = L_\beta \in \mathcal{F}$, contrary to the assumption).

Let $l \subset M$ be a line which meets N at a single point, $\mathbf{z}$. For every point $\mathbf{x} \in l$, consider the plane $N_\mathbf{x} = (\mathbf{x} - \mathbf{z}) + N$. Clearly, the family $\mathcal{G} = \{N_\mathbf{x} : \mathbf{x} \in l\}$ is uncountable and consists of pairwise disjoint translates of N all lying in M. On the other hand, $\mathcal{G} \subset \mathcal{F}$ because every plane $N_\mathbf{x}$ is a member of $\mathcal{F}$ by the above argument. Since $\mathcal{F}$ is countable, the family $\mathcal{G}$ should also be countable, a contradiction. Hence $M \in \mathcal{F}$.

Problem 2.3. Let $L_i = \mathbf{a}_i + S_i$, where $S_i = \operatorname{dir} L_i$, $i = 1, 2$. Consider the subspace

$$S = \operatorname{span}\{\mathbf{a}_1 - \mathbf{a}_2\} + S_1 + S_2 = \operatorname{span}\{\{\mathbf{a}_1 - \mathbf{a}_2\} \cup S_1 \cup S_2\}$$

and the plane $L = \mathbf{a}_1 + S$. Due to $L_1 + L_2 = (\mathbf{a}_1 + \mathbf{a}_2) + S_1 + S_2$, we obtain

$$\dim L = \dim S \leqslant \dim (S_1 + S_2) + 1 = \dim (L_1 + L_2) + 1.$$

Clearly, $L_1 = \mathbf{a}_1 + S_1 \subset \mathbf{a}_1 + S = L$. The equality $S = (\mathbf{a}_1 - \mathbf{a}_2) + S$ gives

$$L_2 = \mathbf{a}_2 + S_2 \subset \mathbf{a}_2 + S = \mathbf{a}_2 + (\mathbf{a}_1 - \mathbf{a}_2) + S = \mathbf{a}_1 + S = L.$$

Problem 2.4. By Theorem 2.15, L_1 and L_2 lie, respectively, in the disjoint parallel planes L'_1 and L'_2, which are translates of the subspace $S = \operatorname{dir} L_1 + \operatorname{dir} L_2$. Hence $S \neq \mathbb{R}^n$. Consequently, $L_1 + L_2$, as a translate of S, is a proper plane.

Problem 2.5. Assume first that the planes L_1 and L_2 are complementary. By Theorem 2.11, there are complementary subspaces S_1 and S_2 and a point $\mathbf{c} \in \mathbb{R}^n$ such that $L_1 = \mathbf{c} + S_1$ and $L_2 = \mathbf{c} + S_2$. Denote by $\mathbf{c}_i$ the projection of $\mathbf{c}$ on

S_i along S_{3-i}, $i = 1, 2$. Clearly, $\boldsymbol{c} = \boldsymbol{c}_1 + \boldsymbol{c}_2$. Let $\boldsymbol{a} = \boldsymbol{c}_2 - \boldsymbol{c}_1$. According to Theorem 2.2,

$$\boldsymbol{a} = \boldsymbol{c} - 2\boldsymbol{c}_1 \in \boldsymbol{c} + S_1 = L_1, \quad -\boldsymbol{a} = \boldsymbol{c} - 2\boldsymbol{c}_2 \in \boldsymbol{c} + S_2 = L_2.$$

Choose any vector $\boldsymbol{u} \in \mathbb{R}^n$. Then $\boldsymbol{u}$ is uniquely expressible as $\boldsymbol{u} = \boldsymbol{z}_1 + \boldsymbol{z}_2$, where $\boldsymbol{z}_1 \in S_1$ and $\boldsymbol{z}_2 \in S_2$. Therefore, $\boldsymbol{u} = \boldsymbol{u}_1 + \boldsymbol{u}_2$, where $\boldsymbol{u}_1 = \boldsymbol{z}_1 + \boldsymbol{a} \in L_1$ and $\boldsymbol{u}_2 = \boldsymbol{z}_2 - \boldsymbol{a} \in L_2$ by the above theorem. If $\boldsymbol{u} = \boldsymbol{u}'_1 + \boldsymbol{u}'_2$ for suitable vectors $\boldsymbol{u}'_1 \in L_1$ and $\boldsymbol{u}'_2 \in L_2$, then $\boldsymbol{u} = (\boldsymbol{u}'_1 - \boldsymbol{a}) + (\boldsymbol{u}'_2 + \boldsymbol{a})$, where $\boldsymbol{u}'_1 - \boldsymbol{a} \in S_1$ and $\boldsymbol{u}'_2 + \boldsymbol{a} \in S_2$. Due to the uniqueness of the representation $\boldsymbol{u} = \boldsymbol{z}_1 + \boldsymbol{z}_2$, one has $\boldsymbol{z}_1 = \boldsymbol{u}'_1 - \boldsymbol{a}$ and $\boldsymbol{z}_2 = \boldsymbol{u}'_2 + \boldsymbol{a}$. Consequently, $\boldsymbol{u}'_1 = \boldsymbol{z}_1 + \boldsymbol{a} = \boldsymbol{u}_1$ and $\boldsymbol{u}'_2 = \boldsymbol{z}_2 - \boldsymbol{a} = \boldsymbol{u}_2$.

Conversely, given nonempty planes L_1 and L_2, suppose that every vector $\boldsymbol{u} \in \mathbb{R}^n$ is uniquely expressible as $\boldsymbol{u} = \boldsymbol{u}_1 + \boldsymbol{u}_2$, where $\boldsymbol{u}_1 \in L_1$ and $\boldsymbol{u}_2 \in L_2$. This argument shows that $L_1 + L_2 = \mathbb{R}^n$. Expressing the planes L_1 and L_2 as $L_i = \boldsymbol{c}_i + S_i$, where $\boldsymbol{c}_i \in \mathbb{R}^n$ and S_i is a subspace, $i = 1, 2$, we have to show that S_1 and S_2 are complementary. Clearly,

$$S_1 + S_2 = L_1 + L_2 - (\boldsymbol{c}_1 + \boldsymbol{c}_2) = \mathbb{R}^n - (\boldsymbol{c}_1 + \boldsymbol{c}_2) = \mathbb{R}^n.$$

By Theorem 2.8, $L_1 \cap L_2 \neq \varnothing$. Choose a point $\boldsymbol{c} \in L_1 \cap L_2$. Theorem 2.2 shows that $L_i = \boldsymbol{c} + S_i$, $i = 1, 2$. Similarly to the above argument, we obtain that every vector $\boldsymbol{x} \in \mathbb{R}^n$ is uniquely expressible as $\boldsymbol{x} = \boldsymbol{x}_1 + \boldsymbol{x}_2$, where $\boldsymbol{x}_i \in S_i$, $i = 1, 2$. Thus S_1 and S_2 are complementary subspaces, implying that the planes L_1 and L_2 are complementary.

Problem 2.6. Let $L_i = \boldsymbol{a}_i + S_i$, where $\boldsymbol{a}_i \in \mathbb{R}^n$ and S_i is a subspace, $i = 1, 2$.

1. Suppose first that L_1 and L_2 are parallel. Theorem 2.6 and the assumption $\dim L_1 \leqslant \dim L_2$ show that L_2 contains a translate of L_1. Therefore, $S_1 \subset S_2$ by Theorem 2.13. For condition (a), suppose that $L_1 \cap L_2 \neq \varnothing$. If $\boldsymbol{a} \in L_1 \cap L_2$, then Theorem 2.2 gives $L_1 = \boldsymbol{a} + S_1 \subset \boldsymbol{a} + S_2 = L_2$. For condition (b), choose a point $\boldsymbol{c} \in \mathbb{R}^n$ such that $\boldsymbol{c} + L_1 \subset L_2$ and put $\boldsymbol{b} = \boldsymbol{c} + \boldsymbol{a}_1$. Then

$$\boldsymbol{b} = \boldsymbol{c} + \boldsymbol{a}_1 \in \boldsymbol{c} + \boldsymbol{a}_1 + S_1 = \boldsymbol{c} + L_1 \subset L_2.$$

So, $L_2 = \boldsymbol{b} + S_2$ according to Theorem 2.2. Let $S = \operatorname{span}\{\{\boldsymbol{c}\} \cup S_2\}$ and $L = \boldsymbol{b} + S$. We will show that L is a desired plane. Indeed, $S_1 - \boldsymbol{c} \subset S_2 - \boldsymbol{c} \subset S$ and

$$\dim L = \dim S \leqslant \dim S_2 + 1 = \dim L_2 + 1.$$

This argument yields the inclusions

$$L_1 = \boldsymbol{a}_1 + S_1 = \boldsymbol{b} + (S_1 - \boldsymbol{c}) \subset \boldsymbol{b} + S = L, \;\; L_2 = \boldsymbol{b} + S_2 \subset \boldsymbol{b} + S = L.$$

2. Conversely, let L_1 and L_2 satisfy conditions (a) and (b). Since the case $L_1 \subset L_2$ implies the parallelism of L_1 and L_2, we may assume that $L_1 \cap L_2 = \varnothing$.

If $S = \operatorname{dir} L$, then $S_1 \cup S_2 \subset S$ according to Theorem 2.13. Furthermore,

$$\dim S = \dim L \leqslant \dim L_2 + 1 = \dim S_2 + 1. \tag{14.6}$$

Assume, for contradiction, that L_1 and L_2 are not parallel. Then $S_1 \not\subset S_2$ by the same Theorem 2.13. Thus $S \neq S_2$, which gives $\dim S = \dim S_2 + 1$. So, $S = \operatorname{span}(S_1 \cup S_2) = S_1 + S_2$.

Put $m = \dim S_2$. Choose a vector $\boldsymbol{b}_1$ in $S_1 \setminus S_2$ and a basis $\{\boldsymbol{b}_2, \ldots, \boldsymbol{b}_{m+1}\}$ for S_2. Then $\{\boldsymbol{b}_1, \boldsymbol{b}_2, \ldots, \boldsymbol{b}_{m+1}\}$ is a basis for S. Since both $\boldsymbol{a}_1$ and $\boldsymbol{a}_2$ belong to L, their difference $\boldsymbol{a}_1 - \boldsymbol{a}_2$ is in $L - L = S$. Therefore, $\boldsymbol{a}_1 - \boldsymbol{a}_2$ can be expressed as a linear combination $\boldsymbol{a}_1 - \boldsymbol{a}_2 = \lambda_1 \boldsymbol{b}_1 + \lambda_2 \boldsymbol{b}_2 + \cdots + \lambda_{m+1} \boldsymbol{b}_{m+1}$.

Put $\boldsymbol{c} = \lambda_2 \boldsymbol{b}_2 + \cdots + \lambda_{m+1} \boldsymbol{b}_{m+1}$. Clearly, $\boldsymbol{c} \in S_2$ and $\boldsymbol{a}_1 - \lambda_1 \boldsymbol{b}_1 = \boldsymbol{a}_2 + \boldsymbol{c}$. Furthermore, $\boldsymbol{a}_1 - \lambda_1 \boldsymbol{b}_1 \in \boldsymbol{a}_1 + S_1 = L_1$ and $\boldsymbol{a}_2 + \boldsymbol{c} \in \boldsymbol{a}_2 + S_2 = L_2$, contrary to the assumption $L_1 \cap L_2 = \varnothing$. Hence L_1 and L_2 are parallel.

Problem 2.7. With $\boldsymbol{x} = (x_1, \ldots, x_n)$ and $\boldsymbol{e}_i = (e_{1i}, \ldots, e_{ni})$, the intersection $H_1 \cap \cdots \cap H_r$ is the solution set of the system of r linear equations in $x_1, \ldots, x_n$:

$$\boldsymbol{x} \cdot \boldsymbol{e}_i = e_{1i} x_1 + \cdots + e_{ni} x_n = \gamma_i, \quad 1 \leqslant i \leqslant r.$$

From linear algebra we know that this solution set is a translate of an $(n-r)$-dimensional subspace if and only if the $r \times n$ matrix (e_{ij}) has rank r. This condition occurs if and only if the set $\{\boldsymbol{e}_1, \ldots, \boldsymbol{e}_r\}$ is linearly independent.

Problem 2.8. If X is an m-dimensional plane, then, according to Theorem 2.68, the set $B_\rho(\boldsymbol{c}) \cap X$ is an m-dimensional ball. Conversely, assume the existence of a scalar $\rho > 0$ such that the set $B_\rho(\boldsymbol{c}) \cap X$ is an m-dimensional ball for every choice of a point $\boldsymbol{c} \subset X$. Suppose for a moment that X is not a plane. Then $L = \operatorname{aff} X \neq X$. Choose points $\boldsymbol{u} \in X$ and $\boldsymbol{v} \in L \setminus X$. By Theorem 2.38, the line $\langle \boldsymbol{u}, \boldsymbol{v} \rangle$ lies in L. Denote by $(\boldsymbol{w}, \boldsymbol{v})$ the largest open segment in $[\boldsymbol{u}, \boldsymbol{v}]$ which is disjoint with X (possibly, $\boldsymbol{w} = \boldsymbol{v}$), and choose a point $\boldsymbol{s} \in [\boldsymbol{u}, \boldsymbol{w}] \cap X$ such that $\|\boldsymbol{s} - \boldsymbol{w}\| \leqslant \rho/2$.

By the assumption, the set $B_\rho(\boldsymbol{s}) \cap X$ is an m-dimensional ball. Hence there is an m-dimensional plane M through $\boldsymbol{s}$ such that $B_\rho(\boldsymbol{s}) \cap X = B_\rho(\boldsymbol{s}) \cap M$. On the other hand, $L = \operatorname{aff}(B_\rho(\boldsymbol{s}) \cap X)$ because $B_\rho(\boldsymbol{s}) \cap X$ is an m-dimensional subset of L (see Corollary 2.65). This argument and Theorem 2.68 give

$$L = \operatorname{aff}(B_\rho(\boldsymbol{s}) \cap X) = \operatorname{aff}(B_\rho(\boldsymbol{s}) \cap M) = M.$$

Consequently, $B_\rho(\boldsymbol{s}) \cap X = B_\rho(\boldsymbol{s}) \cap M = B_\rho(\boldsymbol{s}) \cap L$. Furthermore, $\boldsymbol{v} \neq \boldsymbol{w}$, since otherwise $\boldsymbol{v} \in B_\rho(\boldsymbol{s}) \cap L = B_\rho(\boldsymbol{s}) \cap X \subset X$.

Now, choose a point $\boldsymbol{r} \in (\boldsymbol{v}, \boldsymbol{w})$ such that $\|\boldsymbol{r} - \boldsymbol{w}\| \leqslant \rho/2$. Then

$$\|\boldsymbol{r} - \boldsymbol{s}\| \leqslant \|\boldsymbol{s} - \boldsymbol{w}\| + \|\boldsymbol{w} - \boldsymbol{r}\| \leqslant \rho \quad \text{and} \quad \boldsymbol{r} \in \langle \boldsymbol{u}, \boldsymbol{v} \rangle \subset L,$$

implying that $\boldsymbol{r} \in B_\rho(\boldsymbol{s}) \cap L = B_\rho(\boldsymbol{s}) \cap X \subset X$, contrary to $(\boldsymbol{w}, \boldsymbol{v}) \cap X = \varnothing$. Summing up, X is a plane.

If, additionally, the set X is closed, then $\boldsymbol{v} \neq \boldsymbol{w}$ in the above argument. Hence every scalar $\rho = \rho(\boldsymbol{w}) > 0$ satisfies the desired conclusion.

Problem 2.9. *Lemma* 2.25. (*a*) Since $\boldsymbol{u}, \boldsymbol{v} \in \langle \boldsymbol{x}, \boldsymbol{y} \rangle$, we can write them in the form (2.8) for suitable scalars λ and μ. Clearly, $\lambda \neq \mu$ because $\boldsymbol{u} \neq \boldsymbol{v}$.

To prove the inclusion $\langle \boldsymbol{u}, \boldsymbol{v} \rangle \subset \langle \boldsymbol{x}, \boldsymbol{y} \rangle$, choose a point $\boldsymbol{w} \in \langle \boldsymbol{u}, \boldsymbol{v} \rangle$. Then $\boldsymbol{w} = (1 - \xi)\boldsymbol{u} + \xi \boldsymbol{v}$ for a suitable scalar ξ. With $\nu = \lambda - \lambda\xi + \mu\xi$, one has

$$\begin{aligned} \boldsymbol{w} &= (1 - \xi)((1 - \lambda)\boldsymbol{x} + \lambda \boldsymbol{y}) + \xi((1 - \mu)\boldsymbol{x} + \mu \boldsymbol{y}) \\ &= (1 - \lambda + \lambda\xi - \mu\xi)\boldsymbol{x} + (\lambda - \lambda\xi + \mu\xi)\boldsymbol{y} = (1 - \nu)\boldsymbol{x} + \nu \boldsymbol{y} \in \langle \boldsymbol{x}, \boldsymbol{y} \rangle. \end{aligned}$$

Conversely, let $\boldsymbol{z} \in \langle \boldsymbol{x}, \boldsymbol{y} \rangle$. Then $\boldsymbol{z} = (1 - \eta)\boldsymbol{x} + \eta \boldsymbol{y}$ for a suitable scalar η. Solving (2.8) for $\boldsymbol{x}$ and $\boldsymbol{y}$, we obtain

$$\boldsymbol{x} = \frac{\mu}{\mu - \lambda}\boldsymbol{u} - \frac{\lambda}{\mu - \lambda}\boldsymbol{v} \quad \text{and} \quad \boldsymbol{y} = \frac{1 - \lambda}{\mu - \lambda}\boldsymbol{v} - \frac{1 - \mu}{\mu - \lambda}\boldsymbol{u}.$$

Put $\alpha = (\eta - \lambda)/(\mu - \lambda)$. Then

$$\begin{aligned} \boldsymbol{z} &= (1 - \eta)\Big(\frac{\mu}{\mu - \lambda}\boldsymbol{u} - \frac{\lambda}{\mu - \lambda}\boldsymbol{v}\Big) + \eta\Big(\frac{1 - \lambda}{\mu - \lambda}\boldsymbol{v} - \frac{1 - \mu}{\mu - \lambda}\boldsymbol{u}\Big) \\ &= \Big(1 - \frac{\eta - \lambda}{\mu - \lambda}\Big)\boldsymbol{u} + \frac{\eta - \lambda}{\mu - \lambda}\boldsymbol{v} = (1 - \alpha)\boldsymbol{u} + \alpha \boldsymbol{v} \in \langle \boldsymbol{u}, \boldsymbol{v} \rangle. \end{aligned}$$

(*b*) Let $\boldsymbol{u} = (1 - \lambda)\boldsymbol{x} + \lambda \boldsymbol{y}$, where $\lambda > 0$. For the inclusion $[\boldsymbol{x}, \boldsymbol{u}\rangle \subset [\boldsymbol{x}, \boldsymbol{y}\rangle$, choose a point $\boldsymbol{v} \in [\boldsymbol{x}, \boldsymbol{u}\rangle$. Then $\boldsymbol{v} = (1 - \eta)\boldsymbol{x} + \eta \boldsymbol{u}$ for a suitable scalar $\eta \geqslant 0$. Due to $\lambda\eta \geqslant 0$, we obtain

$$\boldsymbol{v} = (1 - \eta)\boldsymbol{x} + \eta((1 - \lambda)\boldsymbol{x} + \lambda \boldsymbol{y}) = (1 - \lambda\eta)\boldsymbol{x} + \lambda\eta \boldsymbol{y} \in [\boldsymbol{x}, \boldsymbol{y}\rangle.$$

Conversely, let $\boldsymbol{z} \in [\boldsymbol{x}, \boldsymbol{y}\rangle$. Then $\boldsymbol{z} = (1 - \mu)\boldsymbol{x} + \mu \boldsymbol{y}$ for a suitable scalar $\mu \geqslant 0$. Since $\boldsymbol{y} = (1 - \lambda^{-1})\boldsymbol{x} + \lambda^{-1}\boldsymbol{u}$ and $\mu\lambda^{-1} \geqslant 0$, one has

$$\boldsymbol{z} = (1 - \mu)\boldsymbol{x} + \mu((1 - \lambda^{-1})\boldsymbol{x} + \lambda^{-1}\boldsymbol{u}) = (1 - \mu\lambda^{-1})\boldsymbol{x} + \mu\lambda^{-1}\boldsymbol{u} \in [\boldsymbol{x}, \boldsymbol{u}\rangle.$$

Furthermore, $(\boldsymbol{x}, \boldsymbol{u}\rangle = [\boldsymbol{x}, \boldsymbol{u}\rangle \setminus \{\boldsymbol{x}\} = [\boldsymbol{x}, \boldsymbol{y}\rangle \setminus \{\boldsymbol{x}\} = (\boldsymbol{x}, \boldsymbol{y}\rangle$.

(*c*) Let $\boldsymbol{w} = (1 - \mu)\boldsymbol{x} + \mu \boldsymbol{y}$, where $\mu \in \mathbb{R}$. Choose a point $\boldsymbol{v} \in [\boldsymbol{x}, \boldsymbol{y}]$. Then $\boldsymbol{v} = (1 - \lambda)\boldsymbol{x} + \lambda \boldsymbol{y}$ for a suitable $\lambda \in [0, 1]$. Assume first that $\lambda \leqslant \mu$. Leaving aside the case $\lambda = \mu = 0$ (which corresponds to $\boldsymbol{v} = \boldsymbol{w} = \boldsymbol{x}$), we suppose that $\mu > 0$. Since $0 \leqslant \lambda/\mu \leqslant 1$, we obtain

$$\boldsymbol{v} = \Big(1 - \frac{\lambda}{\mu}\Big)\boldsymbol{x} + \frac{\lambda}{\mu}\Big((1 - \mu)\boldsymbol{x} + \mu \boldsymbol{y}\Big) = \Big(1 - \frac{\lambda}{\mu}\Big)\boldsymbol{x} + \frac{\lambda}{\mu}\boldsymbol{w} \in [\boldsymbol{x}, \boldsymbol{w}].$$

Suppose now that $\lambda \geqslant \mu$. Excluding the case $\lambda = \mu = 1$ (which corresponds to $\boldsymbol{v} = \boldsymbol{w} = \boldsymbol{y}$), we suppose that $\mu < 1$. Then $0 \leqslant (\lambda - \mu)/(1 - \mu) \leqslant 1$, and the equality

$$\boldsymbol{v} = \frac{1-\lambda}{1-\mu}\Big((1-\mu)\boldsymbol{x} + \mu\boldsymbol{y}\Big) + \frac{\lambda-\mu}{1-\mu}\boldsymbol{y} = \Big(1 - \frac{\lambda-\mu}{1-\mu}\Big)\boldsymbol{w} + \frac{\lambda-\mu}{1-\mu}\boldsymbol{y}$$

yields the inclusion $\boldsymbol{v} \in [\boldsymbol{w}, \boldsymbol{y}]$.

(*d*) We will assume that $\mu > 0$ (the case $\mu < 0$ is similar). If $\boldsymbol{u} \in (\boldsymbol{x}, \boldsymbol{v})$, then $\boldsymbol{u} = (1-\gamma)\boldsymbol{x} + \gamma\boldsymbol{v}$ for a suitable $\gamma \in (0,1)$. Consequently,

$$\boldsymbol{u} = (1-\gamma)\boldsymbol{x} + \gamma\boldsymbol{v} = (1-\gamma)\boldsymbol{x} + \gamma((1-\mu)\boldsymbol{x} + \mu\,\boldsymbol{y}) = (1-\gamma\mu)\boldsymbol{x} + \gamma\mu\boldsymbol{y}.$$

Comparing these expressions with $\boldsymbol{u} = (1-\lambda)\boldsymbol{x} + \lambda\boldsymbol{y}$, we conclude that $\lambda = \gamma\mu < \mu$.

Conversely, suppose that $0 < \lambda < \mu$. By assertion (*a*) above, $\langle \boldsymbol{x}, \boldsymbol{v}\rangle = \langle \boldsymbol{x}, \boldsymbol{y}\rangle$. Therefore, $\boldsymbol{u} \in \langle \boldsymbol{x}, \boldsymbol{v}\rangle$, and we can write $\boldsymbol{u} = (1-\alpha)\boldsymbol{x} + \alpha\boldsymbol{v}$ for a suitable scalar $\alpha \in \mathbb{R}$. Then

$$\boldsymbol{u} = (1-\alpha)\boldsymbol{x} + \alpha\boldsymbol{v} = (1-\alpha)\boldsymbol{x} + \alpha((1-\mu)\boldsymbol{x} + \alpha\boldsymbol{y}) = (1-\alpha\mu)\boldsymbol{x} + \alpha\mu\boldsymbol{y}.$$

As above, $\lambda = \alpha\mu$. So, $\alpha = \lambda/\mu \in (0,1)$, and $\boldsymbol{u} \in (\boldsymbol{x}, \boldsymbol{v})$.

Lemma 2.26. (*a*) Let $\boldsymbol{u} = (1-\lambda)\boldsymbol{x} + \lambda\boldsymbol{y}$ and $\boldsymbol{v} = (1-\mu)\boldsymbol{x} + \mu\boldsymbol{z}$, where $\lambda, \mu \in [0,1]$. Excluding the case $\lambda = \mu = 1$ (which corresponds to $\boldsymbol{u} = \boldsymbol{y}$ and $\boldsymbol{v} = \boldsymbol{z}$), we may assume that $\lambda\mu < 1$. Put

$$\boldsymbol{p} = \frac{(1-\lambda)(1-\mu)}{1-\lambda\mu}\boldsymbol{x} + \frac{\lambda(1-\mu)}{1-\lambda\mu}\boldsymbol{y} + \frac{(1-\lambda)\mu}{1-\lambda\mu}\boldsymbol{z}, \ \xi = \frac{1-\mu}{1-\lambda\mu}, \ \eta = \frac{1-\lambda}{1-\lambda\mu}.$$

Obviously, $\xi, \eta \in [0,1]$, and the equalities

$$\boldsymbol{p} = \Big(1 - \frac{1-\mu}{1-\lambda\mu}\Big)\boldsymbol{z} + \frac{1-\mu}{1-\lambda\mu}\Big((1-\lambda)\boldsymbol{x} + \lambda\boldsymbol{y}\Big) = (1-\xi)\boldsymbol{z} + \xi\boldsymbol{u},$$
$$\boldsymbol{p} = \Big(1 - \frac{1-\lambda}{1-\lambda\mu}\Big)\boldsymbol{y} + \frac{1-\lambda}{1-\lambda\mu}\Big((1-\mu)\boldsymbol{x} + \mu\boldsymbol{z}\Big) = (1-\eta)\boldsymbol{y} + \eta\boldsymbol{v}$$

imply the inclusion $\boldsymbol{p} \in [\boldsymbol{u}, \boldsymbol{z}] \cap [\boldsymbol{v}, \boldsymbol{y}]$.

(*b*) Given a point $\boldsymbol{q} \in [\boldsymbol{u}, \boldsymbol{z}]$, we can write $\boldsymbol{q} = (1-\gamma)\boldsymbol{z} + \gamma\boldsymbol{u}$ for a suitable scalar $\gamma \in [0,1]$. Leaving aside the case $\lambda = \gamma = 1$ (corresponding to $\boldsymbol{q} = \boldsymbol{u} = \boldsymbol{y}$), we may assume that $\lambda\gamma < 1$. Let

$$\zeta = 1 - \lambda\gamma, \quad \nu = \frac{1-\gamma}{1-\lambda\gamma}, \quad \boldsymbol{w} = (1-\nu)\boldsymbol{x} + \nu\boldsymbol{z}.$$

Then $\zeta, \nu \in [0,1]$ and $\boldsymbol{w} \in [\boldsymbol{x}, \boldsymbol{z}]$. Finally, the equalities

$$\begin{aligned}\boldsymbol{q} &= (1-\gamma)\boldsymbol{z} + \gamma\,\boldsymbol{u} = (1-\gamma)\boldsymbol{z} + \gamma((1-\lambda)\boldsymbol{x} + \lambda\,\boldsymbol{y}) \\ &= \lambda\gamma\,\boldsymbol{y} + (1-\lambda\gamma)\Big(\frac{(1-\lambda)\gamma}{1-\lambda\gamma}\boldsymbol{x} + \frac{1-\gamma}{1-\lambda\gamma}\,\boldsymbol{z}\Big) \\ &= (1-\zeta)\boldsymbol{y} + \zeta((1-\nu)\boldsymbol{x} + \nu\,\boldsymbol{z}) = (1-\zeta)\boldsymbol{y} + \zeta\boldsymbol{w}\end{aligned}$$

give the inclusion $\boldsymbol{q} \in [\boldsymbol{w}, \boldsymbol{y}]$.

Lemma 2.27. Let $\boldsymbol{u} = (1-\lambda)\boldsymbol{u}_1 + \lambda\boldsymbol{u}_2$ for a suitable $0 \leqslant \lambda \leqslant 1$. Excluding the obvious cases $\boldsymbol{u} = \boldsymbol{u}_1$ and $\boldsymbol{u} = \boldsymbol{u}_2$, we may assume that $0 < \lambda < 1$. Since $\boldsymbol{v}_i$ belongs to the closed halfline $[\boldsymbol{u}_i, \boldsymbol{z}\rangle \setminus [\boldsymbol{u}_i, \boldsymbol{z})$, we can write $\boldsymbol{v}_i = \gamma_i\boldsymbol{z} + (1-\gamma_i)\boldsymbol{u}_i$, where $\gamma_i \geqslant 1$, $i = 1, 2$. Let

$$\mu = \frac{\lambda\gamma_1}{\lambda\gamma_1 + (1-\lambda)\gamma_2} \quad \text{and} \quad \gamma = \frac{\gamma_1\gamma_2}{\lambda\gamma_1 + (1-\lambda)\gamma_2}.$$

Clearly, $0 < \mu < 1$ and $\gamma \geqslant 1$. Put $\boldsymbol{v} = (1-\mu)\boldsymbol{v}_1 + \mu\boldsymbol{v}_2$. Then $\boldsymbol{v} \in [\boldsymbol{v}_1, \boldsymbol{v}_2]$ and

$$\begin{aligned}\boldsymbol{v} = (1-\mu)\boldsymbol{v}_1 + \mu\boldsymbol{v}_2 &= \frac{(1-\lambda)\gamma_2}{\lambda\gamma_1 + (1-\lambda)\gamma_2}(\gamma_1\boldsymbol{z} + (1-\gamma_1)\boldsymbol{u}_1) \\ &\quad + \frac{\lambda\gamma_1}{\lambda\gamma_1 + (1-\lambda)\gamma_2}(\gamma_2\boldsymbol{z} + (1-\gamma_2)\boldsymbol{u}_2) \\ = \frac{\gamma_1\gamma_2}{\lambda\gamma_1 + (1-\lambda)\gamma_2}\boldsymbol{z} &+ \Big(1 - \frac{\gamma_1\gamma_2}{\lambda\gamma_1 + (1-\lambda)\gamma_2}\Big)((1-\lambda)\boldsymbol{u}_1 + \lambda\boldsymbol{u}_2) \\ = \gamma\boldsymbol{z} + (1-\gamma)\boldsymbol{u}.&\end{aligned}$$

Consequently, $\boldsymbol{z} = (1-\gamma^{-1})\boldsymbol{u} + \gamma^{-1}\boldsymbol{v} \in [\boldsymbol{u}, \boldsymbol{v}]$ due to $0 < \gamma^{-1} \leqslant 1$.

Lemma 2.28. Clearly, $\boldsymbol{u} = (1-\lambda)\boldsymbol{y} + \lambda\boldsymbol{z}$, where $0 < \lambda < 1$, and

$$\boldsymbol{y}' = (1-\mu)\boldsymbol{x} + \mu\boldsymbol{y}, \quad \boldsymbol{z}' = (1-\gamma)\boldsymbol{x} + \gamma\boldsymbol{z}, \quad \text{where} \;\; \mu > 0, \; \gamma > 0.$$

Put

$$\alpha = \frac{\lambda\mu}{(1-\lambda)\gamma + \lambda\mu} \quad \text{and} \quad \beta = \frac{\gamma\mu}{(1-\lambda)\gamma + \lambda\mu}.$$

Furthermore, let $\boldsymbol{u}' = (1-\alpha)\boldsymbol{y}' + \alpha\boldsymbol{z}'$. Then $\boldsymbol{u}' \in (\boldsymbol{y}', \boldsymbol{z}')$ because $0 < \alpha < 1$.

Also,

$$\begin{aligned}
\boldsymbol{u}' &= (1-\alpha)\boldsymbol{y}' + \alpha\boldsymbol{z}' = (1-\alpha)((1-\mu)\boldsymbol{x} + \mu\boldsymbol{y}) + \alpha((1-\gamma)\boldsymbol{x} + \gamma\boldsymbol{z}) \\
&= ((1-\alpha)(1-\mu) + \alpha(1-\gamma))\boldsymbol{x} + (1-\alpha)\mu\boldsymbol{y} + \alpha\gamma\boldsymbol{z} \\
&= \Big(\frac{(1-\lambda)\gamma}{(1-\lambda)\gamma + \lambda\mu}(1-\mu) + \frac{\lambda\mu}{(1-\lambda)\gamma + \lambda\mu}(1-\gamma)\Big)\boldsymbol{x} \\
&\qquad + \frac{(1-\lambda)\gamma}{(1-\lambda)\gamma + \lambda\mu}\mu\boldsymbol{y} + \frac{\lambda\mu}{(1-\lambda)\gamma + \lambda\mu}\gamma\boldsymbol{z} \\
&= \frac{(1-\lambda)\gamma + \lambda\mu - \gamma\mu}{(1-\lambda)\gamma + \lambda\mu}\boldsymbol{x} + \frac{\gamma\mu}{(1-\lambda)\gamma + \lambda\mu}((1-\lambda)\boldsymbol{y} + \lambda\boldsymbol{z}) \\
&= (1-\beta)\boldsymbol{x} + \beta\boldsymbol{u} \in h.
\end{aligned}$$

Problem 2.10. Let $H = \{\boldsymbol{x} \in \mathbb{R}^n : \boldsymbol{x}\boldsymbol{e} = \gamma\}$ (see Theorem 2.17). Without loss of generality, assume that $\boldsymbol{a}$ belongs to the open halfspace $W = \{\boldsymbol{x} \in \mathbb{R}^n : \boldsymbol{x}\cdot\boldsymbol{e} < \gamma\}$. Denote by $\boldsymbol{u}$ the point at which h meets H. Then $\boldsymbol{a}\cdot\boldsymbol{e} < \gamma = \boldsymbol{u}\cdot\boldsymbol{e}$, and Lemma 2.25 shows that $h = (\boldsymbol{a}, \boldsymbol{u}\rangle$. Consequently, $h' = (\boldsymbol{b} - \boldsymbol{a}) + (\boldsymbol{a}, \boldsymbol{u}\rangle = (\boldsymbol{b}, \boldsymbol{u}'\rangle$, where $\boldsymbol{u}' = \boldsymbol{b} - \boldsymbol{a} + \boldsymbol{u}$. Clearly,

$$\boldsymbol{b}\cdot\boldsymbol{e} = (\boldsymbol{a} - \boldsymbol{u} + \boldsymbol{u}')\cdot\boldsymbol{e} = (\boldsymbol{a} - \boldsymbol{u})\cdot\boldsymbol{e} + \boldsymbol{u}'\cdot\boldsymbol{e} < \boldsymbol{u}'\cdot\boldsymbol{e}. \tag{14.7}$$

Every point $\boldsymbol{x} \in (\boldsymbol{b}, \boldsymbol{u}'\rangle$ can be written as $\boldsymbol{x} = (1-\lambda)\boldsymbol{b} + \lambda\boldsymbol{u}'$, $\lambda > 0$, which gives

$$\boldsymbol{x}\cdot\boldsymbol{e} = (1-\lambda)\boldsymbol{b}\cdot\boldsymbol{e} + \lambda\boldsymbol{u}'\cdot\boldsymbol{e}. \tag{14.8}$$

Assume first that $\boldsymbol{b} \in W$. Then $\boldsymbol{b}\cdot\boldsymbol{e} < \gamma$, and h' meets H at the point

$$\boldsymbol{x}_0 = (1-\lambda_0)\boldsymbol{b} + \lambda_0\boldsymbol{u}' \in [\boldsymbol{b}, \boldsymbol{u}'\rangle, \quad \text{where} \quad \lambda_0 = \frac{\gamma - \boldsymbol{b}\cdot\boldsymbol{e}}{(\boldsymbol{u}' - \boldsymbol{b})\cdot\boldsymbol{e}} > 0.$$

Indeed,

$$\boldsymbol{x}_0\cdot\boldsymbol{e} = \frac{\boldsymbol{u}'\cdot\boldsymbol{e} - \gamma}{(\boldsymbol{u}' - \boldsymbol{b})\cdot\boldsymbol{e}}\boldsymbol{b}\cdot\boldsymbol{e} + \frac{\gamma - \boldsymbol{b}\cdot\boldsymbol{e}}{(\boldsymbol{u}' - \boldsymbol{b})\cdot\boldsymbol{e}}\boldsymbol{u}'\cdot\boldsymbol{e} = \gamma.$$

If $\boldsymbol{b} \notin W$, then $\boldsymbol{b}\cdot\boldsymbol{e} > \gamma$ (since $\boldsymbol{b} \notin H$), and a combination of (14.7) and (14.8) shows that

$$\boldsymbol{x}\cdot\boldsymbol{e} = ((1-\lambda)\boldsymbol{b} + \lambda\boldsymbol{u}')\cdot\boldsymbol{e} = \boldsymbol{b}\cdot\boldsymbol{e} + \lambda(\boldsymbol{u}'\cdot\boldsymbol{e} - \boldsymbol{b}\cdot\boldsymbol{e}) > \boldsymbol{b}\cdot\boldsymbol{e} > \gamma.$$

Hence $h' \cap H = \varnothing$.

Problem 2.11. Let $m = \dim L$. Denote by M_α the $(m-1)$-dimensional plane which determines a halfplane $F_\alpha \in \mathcal{F}$ (see Corollary 2.71). Theorem 2.70 shows that F_α can be described in one of the following ways:

$$F_\alpha = \{\boldsymbol{x} \in L : \boldsymbol{x}\cdot\boldsymbol{c}_\alpha \leqslant \gamma_\alpha\}, \quad F_\alpha = \{\boldsymbol{x} \in L : \boldsymbol{x}\cdot\boldsymbol{c}_\alpha < \gamma_\alpha\},$$

where $\boldsymbol{c}_\alpha$ is a nonzero vector in $N_\alpha = \operatorname{dir} L \cap \operatorname{ort} M_\alpha$. It is easy to see that N_α is a 1-dimensional subspace.

(a) Assume first that U is a halfplane and denote by the $(m-1)$-dimensional plane which determines U. From Theorem 2.34 we conclude that each plane M_α is a translate of M. In this case, all subspaces N_α coincide. Hence all $\boldsymbol{c}_\alpha$ are scalar multiples of the same nonzero vector, say $\boldsymbol{c}$. This argument shows that the family $\mathcal{F}$ is nested.

Conversely, suppose that the family $\mathcal{F}$ is nested. By the above argument, the planes $\{M_\alpha\}$ are pairwise parallel. Hence they have a common normal vector $\boldsymbol{c} \in \operatorname{dir} L$. Without loss of generality, we may assume that each F_α has one of the forms

$$F_\alpha = \{\boldsymbol{x} \in L : \boldsymbol{x}\cdot\boldsymbol{c} \leqslant \mu_\alpha\}, \quad F_\alpha = \{\boldsymbol{x} \in L : \boldsymbol{x}\cdot\boldsymbol{c} < \mu_\alpha\}.$$

Then, with $\mu = \sup \mu_\alpha$, the set U has one of the forms

$$U = \{\boldsymbol{x} \in \mathbb{R}^n : \boldsymbol{x}\cdot\boldsymbol{c} \leqslant \mu\}, \quad U = \{\boldsymbol{x} \in \mathbb{R}^n : \boldsymbol{x}\cdot\boldsymbol{c} < \mu\}.$$

Hence U is a halfplane of L. The proof of part (b) of the problem is similar.

Problem 2.12. Choose a point $\boldsymbol{u} \in [\boldsymbol{y}, \boldsymbol{z}\rangle$. Then $\boldsymbol{u} = (1-\lambda)\boldsymbol{y}+\lambda\boldsymbol{z}$ for a suitable scalar $\lambda \geqslant 0$. The inclusion $\boldsymbol{u} \in P$ gives

$$\gamma \leqslant \boldsymbol{u}\cdot\boldsymbol{e} = (1-\lambda)\boldsymbol{y}\cdot\boldsymbol{e} + \lambda\,\boldsymbol{z}\cdot\boldsymbol{e} = \boldsymbol{y}\cdot\boldsymbol{e} + \lambda(\boldsymbol{z}\cdot\boldsymbol{e} - \boldsymbol{y}\cdot\boldsymbol{e}) \leqslant \gamma'. \tag{14.9}$$

It is easy to see that the inequalities (14.9) hold for all $\lambda \geqslant 0$ only if $\boldsymbol{y}\cdot\boldsymbol{e} = \boldsymbol{z}\cdot\boldsymbol{e}$. Let $\beta = \boldsymbol{y}\cdot\boldsymbol{e}$. By the above, $\gamma \leqslant \beta \leqslant \gamma'$. If $\boldsymbol{v}$ is a point in $\langle\boldsymbol{y}, \boldsymbol{z}\rangle$, then $\boldsymbol{v} = (1-\mu)\boldsymbol{y}+\mu\boldsymbol{z}$ for a suitable scalar μ. Consequently, $\boldsymbol{v}\cdot\boldsymbol{e} = (1-\mu)\boldsymbol{y}\cdot\boldsymbol{e} + \mu\,\boldsymbol{z}\cdot\boldsymbol{e} = \beta$, implying that the line $\langle\boldsymbol{y}, \boldsymbol{z}\rangle$ lies in P.

Problem 2.13. Suppose that a translate $Z = \boldsymbol{c} + Y$ of Y lies in $\operatorname{aff} X$. Then $\operatorname{aff} Z \subset \operatorname{aff} X$ according to Theorem 2.42. Choose a point $\boldsymbol{y} \in Y$ and let $\boldsymbol{z} = \boldsymbol{c}+\boldsymbol{y}$. Then $\boldsymbol{z} \in Z$. By Theorem 2.46, $Z - \boldsymbol{z} \subset \operatorname{aff} X - \boldsymbol{z} = \operatorname{aff}(X - \boldsymbol{z})$. Since $\boldsymbol{o} = \boldsymbol{z}-\boldsymbol{z} \in Z-\boldsymbol{z} \subset \operatorname{aff}(X-\boldsymbol{z})$, the plane $\operatorname{aff}(X-\boldsymbol{z})$ is a subspaces. Theorem 2.2 gives

$$\operatorname{aff}(X - \boldsymbol{z}) + \operatorname{aff} Z = \operatorname{aff}(X - \boldsymbol{z}) + Z = \operatorname{aff} X,$$

and a combination of Theorems 2.2 and 2.46 implies

$$\begin{aligned} \operatorname{aff}(X + Y) &= \operatorname{aff}(X - \boldsymbol{z} + Z + \boldsymbol{y}) = \operatorname{aff}(X - \boldsymbol{z}) + \operatorname{aff} Z + \boldsymbol{y} = \operatorname{aff} X + \boldsymbol{y}, \\ \operatorname{aff} X + Y &= \operatorname{aff} X + (\boldsymbol{y} - \boldsymbol{z}) + Z = \operatorname{aff}(X - \boldsymbol{z}) + Z + \boldsymbol{y} = \operatorname{aff} X + \boldsymbol{y}. \end{aligned}$$

Problem 2.14. Let $L_i = \operatorname{aff} X_i$, $1 \leqslant i \leqslant r$. By Theorem 2.46,

$$\operatorname{aff}(\mu_1 X_1 + \cdots + \mu_r X_r) = \mu_1 L_1 + \cdots + \mu_r L_r.$$

Theorem 2.2 gives $L_i = \boldsymbol{a}_i + S_i$, where S_i is a subspace, $1 \leqslant i \leqslant r$. Since

$$\mu_1 L_1 + \cdots + \mu_r L_r = (\mu_1\boldsymbol{a}_1 + \cdots + \mu_r\boldsymbol{a}_r) + (\mu_1 S_1 + \cdots + \mu_r S_r),$$

one has $\dim(\mu_1 S_1 + \cdots + \mu_r S_r) = m$. Let $\mathcal{F} = \{\mu_1 S_1, \ldots, \mu_r S_r\}$. Choose recursively a maximal subset $\mathcal{H} = \{\mu_{i_1} S_{i_1}, \ldots, \mu_{i_k} S_{i_k}\} \subset \mathcal{F}$ such that

$$\mu_{i_{j+1}} S_{i_{j+1}} \not\subset \mu_{i_1} S_{i_1} + \cdots + \mu_{i_j} S_{i_j}, \quad 1 \leqslant j \leqslant k-1.$$

Clearly, $k \leqslant m \leqslant r$ and $\mu_{i_1} S_{i_1} + \cdots + \mu_{i_k} S_{i_k} = \mu_1 S_1 + \cdots + \mu_r S_r$, since otherwise we could further enlarge $\mathcal{H}$. Let $I = \{i_1, \ldots, i_k\}$ and $J = \{1, \ldots, r\} \setminus I$. Then $\operatorname{card} I \leqslant m$ and

$$\begin{aligned} \operatorname{aff}(\mu_1 X_1 + \cdots + \mu_r X_r) &= \mu_1 L_1 + \cdots + \mu_r L_r \\ &= (\mu_1 \boldsymbol{a}_1 + \cdots + \mu_r \boldsymbol{a}_r) + \sum_{i \in I} \mu_i S_i = \sum_{i \in J} \mu_i \boldsymbol{a}_i + \sum_{i \in I} \mu_i (\boldsymbol{a}_i + S_i) \\ &= \sum_{i \in J} \mu_i \boldsymbol{a}_i + \sum_{i \in I} \mu_i L_i = \sum_{i \in J} \mu_i \boldsymbol{a}_i + \sum_{i \in I} \mu_i \operatorname{aff} X_i. \end{aligned}$$

Problem 2.15. $(a) \Rightarrow (b)$ The case $r = 1$ is obvious because the singleton $\{\boldsymbol{a}_1\}$ is affinely independent and the nonzero vector $\boldsymbol{a}_1' = (\boldsymbol{a}_1, 1)$ is linearly independent. Suppose that $r \geqslant 2$. We consider the contrapositive assertion. Let the set $\{\boldsymbol{a}_1', \ldots, \boldsymbol{a}_r'\}$ be linearly dependent. Then there are scalars $\nu_1, \ldots, \nu_r$, not all zero, such that $\nu_1 \boldsymbol{a}_1' + \cdots + \nu_r \boldsymbol{a}_r' = \boldsymbol{o}'$, where $\boldsymbol{o}' = (\boldsymbol{o}, 0)$. Writing down this equality for the $(n+1)$-th coordinates of the vectors, we obtain that $\nu_1 + \cdots + \nu_r = 0$. This argument and Definition 2.50 show that the set $\{\boldsymbol{a}_1, \ldots, \boldsymbol{a}_r\}$ is affinely dependent.

The proof of $(b) \Rightarrow (a)$ is similar.

Problem 2.16. Since the case $r = 1$ is obvious (any point $\boldsymbol{a}_1'$ is affinely independent), we may assume that $r \geqslant 2$. By Theorem 2.52, the set of vectors $\boldsymbol{c}_i = \boldsymbol{a}_i - \boldsymbol{a}_1$, $2 \leqslant i \leqslant r$, is linearly independent. There is a scalar $\delta > 0$ (see Problem 1.11) such that for any choice of vectors $\boldsymbol{c}_i' \in U_\delta(\boldsymbol{c}_i)$, $2 \leqslant i \leqslant r$, the set $\{\boldsymbol{c}_2', \ldots, \boldsymbol{c}_r'\}$ is linearly independent.

Now, let $\rho = \delta/2$ and choose points $\boldsymbol{a}_i' \in U_\rho(\boldsymbol{a}_i)$, $1 \leqslant i \leqslant r$. Put $\boldsymbol{c}_i' = \boldsymbol{a}_i' - \boldsymbol{a}_1'$, $2 \leqslant i \leqslant r$. The inequalities

$$\begin{aligned} \|\boldsymbol{c}_i - \boldsymbol{c}_i'\| &= \|(\boldsymbol{a}_i - \boldsymbol{a}_1) - (\boldsymbol{a}_i' - \boldsymbol{a}_1')\| = \|(\boldsymbol{a}_i - \boldsymbol{a}_i') + (\boldsymbol{a}_1 - \boldsymbol{a}_1')\| \\ &\leqslant \|\boldsymbol{a}_i - \boldsymbol{a}_i'\| + \|\boldsymbol{a}_1 - \boldsymbol{a}_1'\| < \rho + \rho = \delta, \quad 2 \leqslant i \leqslant r, \end{aligned}$$

and the choice of δ show that the set $\{\boldsymbol{c}_2', \ldots \boldsymbol{c}_r'\}$ is linearly independent. Therefore, by Theorem 2.52, $\{\boldsymbol{a}_1', \ldots, \boldsymbol{a}_r'\}$ is affinely independent.

Problem 2.17. Let f be an affine projection; that is $f(\boldsymbol{x}) = \boldsymbol{c} + g(\boldsymbol{x} - \boldsymbol{c})$, where $g : \mathbb{R}^n \to \mathbb{R}^n$ is a linear projection and $\boldsymbol{c} \in \mathbb{R}^n$. Since $g \circ g = g$ (see Problem 1.2), one has

$$\begin{aligned} (f \circ f)(\boldsymbol{x}) &= f(f(\boldsymbol{x})) = f(\boldsymbol{c} + g(\boldsymbol{x} - \boldsymbol{c})) = \boldsymbol{c} + g(\boldsymbol{c} + g(\boldsymbol{x} - \boldsymbol{c}) - \boldsymbol{c}) \\ &= \boldsymbol{c} + g(g(\boldsymbol{x} - \boldsymbol{c})) = \boldsymbol{c} + g(\boldsymbol{x} - \boldsymbol{c}) = f(\boldsymbol{x}). \end{aligned}$$

Conversely, let $f : \mathbb{R}^n \to \mathbb{R}^n$ be an affine transformation with the property $f \circ f = f$. Put $\boldsymbol{e} = f(\boldsymbol{o})$ and $g(\boldsymbol{x}) = f(\boldsymbol{x}) - \boldsymbol{e}$. Then $g(\boldsymbol{x})$ is a linear transformation such that

$$g(\boldsymbol{e}) = f(\boldsymbol{e}) - \boldsymbol{e} = f(f(\boldsymbol{o})) - f(\boldsymbol{o}) = f(\boldsymbol{o}) - f(\boldsymbol{o}) = \boldsymbol{o}.$$

Furthermore,

$$\begin{aligned}(g\circ g)(\boldsymbol{x}) &= g(g(\boldsymbol{x})) = g(f(\boldsymbol{x}) - \boldsymbol{e}) = g(f(\boldsymbol{x})) - g(\boldsymbol{e}) = g(f(\boldsymbol{x})) - \boldsymbol{o}\\ &= g(f(\boldsymbol{x})) = f(f(\boldsymbol{x})) - \boldsymbol{e} = f(\boldsymbol{x}) - \boldsymbol{e} = g(\boldsymbol{x}).\end{aligned}$$

Hence g is a linear projection, and $f(\boldsymbol{x}) = \boldsymbol{e} + g(\boldsymbol{x}) = \boldsymbol{e} + g(\boldsymbol{x} - \boldsymbol{e})$ is an affine projection.

Chapter 3

Problem 3.1. If K is closed, then for every line $l \subset \mathbb{R}^n$, the set $K \cap l$ is closed. Conversely, let $K \subset \mathbb{R}^n$ be a convex set such that for every line $l \subset \mathbb{R}^n$, the set $K \cap l$ is closed. Choose a point $\boldsymbol{z} \in \operatorname{rint} K$. For a given point $\boldsymbol{x} \in \operatorname{cl} K$ consider the line $l = \langle \boldsymbol{x}, \boldsymbol{z}\rangle$. By Theorem 3.47, $K \cap l$ contains the semi-open segment $[\boldsymbol{z}, \boldsymbol{x})$. Furthermore, $[\boldsymbol{z}, \boldsymbol{x}] \subset K \cap l \subset K$ because $K \cap l$ is closed. Hence $\boldsymbol{x} \in K$, and K is closed.

If $K \subset \mathbb{R}^n$ a relatively open convex set and l is a line meeting K, then Corollary 3.31 implies that $K \cap l = \operatorname{rint} K \cap l = \operatorname{rint}(K \cap l)$. Hence $K \cap l$ is relatively open. Conversely, let a convex set $K \subset \mathbb{R}^n$ be such that for every line l the intersection $K \cap l$ is relatively open. To prove that K is relatively open, it suffices to show that every point $\boldsymbol{x} \in K$ belongs to $\operatorname{rint} K$. Since the case $K = \{\boldsymbol{x}\}$ is obvious, we may assume that $\dim K \geqslant 1$. Let $\boldsymbol{y} \in K$ be a point distinct from $\boldsymbol{x}$, and l be the line through $\boldsymbol{x}$ and $\boldsymbol{y}$. Since the relatively open set $K \cap l$ contains both $\boldsymbol{x}$ and $\boldsymbol{y}$, there is a scalar $\gamma > 1$ such that $\gamma\boldsymbol{x} + (1-\gamma)\boldsymbol{y} \in K \cap l \subset K$. By Theorem 3.26, $\boldsymbol{x} \in \operatorname{rint} K$.

Problem 3.2. Because the "if" part in both assertions is obvious, it suffices to prove the "only if" part. The assertions are obvious when $\operatorname{card} X \leqslant 1$; so, we may suppose that $\operatorname{card} X \geqslant 2$.

(a) Denote by Φ the set of scalars $0 < \eta < 1$ such that $(1-\eta)\boldsymbol{x} + \eta\boldsymbol{y} \in X$ whenever $\boldsymbol{x}, \boldsymbol{y} \in X$. We assert that $(1-\lambda)\eta_1 + \lambda\eta_2 \in \Phi$ for any choice of $\eta_1, \eta_2 \in \Phi$. Indeed, if

$$\boldsymbol{z}_1 = (1-\eta_1)\boldsymbol{x} + \eta_1\boldsymbol{y} \quad \text{and} \quad \boldsymbol{z}_2 = (1-\eta_2)\boldsymbol{x} + \eta_2\boldsymbol{y}, \tag{14.10}$$

then $\boldsymbol{z}_1, \boldsymbol{z}_2 \in X$, which gives $(1-\lambda)\boldsymbol{z}_1 + \lambda\boldsymbol{z}_2 \in X$. Consequently, the inclusion

$$(1 - (1-\lambda)\eta_1 - \lambda\eta_2)\boldsymbol{x} + ((1-\lambda)\eta_1 + \lambda\eta_2)\boldsymbol{y} = (1-\lambda)\boldsymbol{z}_1 + \lambda\boldsymbol{z}_2 \in X$$

shows that $(1-\lambda)\eta_1 + \lambda\eta_2 \in \Phi$.

The above property of Φ, combined with the inclusion $\{0, \lambda, 1\} \subset \Phi$, obviously implies that Φ is dense in $[0, 1]$. Furthermore, for points $\boldsymbol{z}_1$ and $\boldsymbol{z}_2$ of the form

(14.10), one has $\|z_1 - z_2\| = |\eta_1 - \eta_2| \, \|x - y\|$, implying that the set $X \cap [x, y]$ is dense in $[x, y]$ for any choice of points $x, y \in X$.

Assume for a moment that the set X is not convex. Then there are distinct points $x, y \in X$ and a scalar $0 < \mu < 1$ such that the point $z = (1 - \mu)x + \mu y$ does not belong to X. Then $\mu \neq \lambda$. Without loss of generality, we may suppose that $\mu < \lambda$ (the case $\mu > \lambda$ is similar). Because $X \cap [x, y]$ is relatively open in the line $\langle x, y \rangle$, it contains a semi-open segment $[x, v)$. Choosing v in (x, z), we may assume that $[x, v] \subset [x, z)$. For any point $c \in [x, v)$, let $c' = (1 - \lambda^{-1})c + \lambda^{-1}z$. Then $c' \notin X$, since otherwise $z = (1 - \lambda)c + \lambda c' \in X$, contrary to the choice of z. Clearly, the points c' fulfil the semi-open segment $(v', x'] \subset (z, y)$, where

$$x' = (1 - \lambda^{-1})x + \lambda^{-1}z \quad \text{and} \quad v' = (1 - \lambda^{-1})v + \lambda^{-1}z.$$

By the above argument, $X \cap (v', x'] = \varnothing$. On the other hand, $X \cap [x, y]$ being dense in $[x, y]$, should contain points from $(v', x']$. The obtained contradiction implies the inclusion $[x, y] \subset X$. Summing up, the set X is convex.

(*b*) Choose a pair of distinct points $x, y \in X$ and let $u = (1-\lambda)x + \lambda y$. Next, put $v = (1 - \lambda)u + \lambda y$. Then $v \in X$ and

$$v = (1 - \lambda)((1 - \lambda)x + \lambda y) + \lambda y = (1 - (2\lambda - \lambda^2))x + (2\lambda - \lambda^2)y \in X.$$

Since $0 < 2\lambda - \lambda^2 < 1$, assertion (*a*) shows that the set X is convex.

Furthermore, with $u_1 = u$ and $u_{k+1} = (1 - \lambda)x + \lambda u_k$, $k \geqslant 1$, we obtain by induction on k that $u_k = (1 - \lambda^k)x + \lambda^k u \in X \cap \langle x, u \rangle$ or all $k \geqslant 1$.

Because $\lambda^k \to \infty$ as $k \to \infty$, the convexity of X gives $[x, u\rangle \subset X$. Similarly, $[u, x\rangle \subset X$, which gives the inclusion $\langle x, u \rangle \subset X$. Theorem 2.38 shows that X is a plane.

Problem 3.3. Because the "if" part in both assertions is obvious, it suffices to prove the "only if" part. The assertions are obvious when $\operatorname{card} X \leqslant 1$; so, we assume that $\operatorname{card} X \geqslant 2$.

(*a*) Choose points $x, y \in X$, and assume for a moment that $[x, y] \not\subset X$. Since the set $X \cap [x, y]$ is closed, its complement $[x, y] \setminus X$ is a relatively open subset of $[x, y]$; so, it can be expressed as a disjoint union of countably many open subsegments. Let (u, v) be one of these subsegments. Then $u, v \in X$ because X is closed. By the assumption, X contains a point $w = (1 - \lambda)u + \lambda v$, with $0 < \lambda < 1$. Then $w \in (u, v)$, contradicting the choice of (u, v). Hence X is a convex set.

(*b*) Assume for a moment that X is not a plane. Then there are distinct points $x, y \in X$ such that $\langle x, y \rangle \not\subset X$. As above, $\langle x, y \rangle \setminus X$ can be expressed as a disjoint union of countably many open segments (possibly including open halflines) of $\langle x, y \rangle$.

We first assert that $X \cap \langle x, y \rangle$ is convex. Indeed, suppose that this is not the case. Then $X \cap \langle x, y \rangle$ contains a pair of distinct points u and v such that $X \cap (u, v) = \varnothing$. By the assumption, there is a scalar $1 < \lambda < 2$ such that the point $z = (1 - \lambda)u + \lambda v$ belongs to $X \cap \langle x, y \rangle$. Similarly, there is a scalar $1 < \mu < 2$

for which the point $\boldsymbol{w} = (1-\mu)\boldsymbol{z} + \mu\boldsymbol{v}$ belongs to $X \cap \langle \boldsymbol{x}, \boldsymbol{y}\rangle$. On the other hand,

$$\begin{aligned}\boldsymbol{w} &= (1-\mu)((1-\lambda)\boldsymbol{u} + \lambda\boldsymbol{v}) + \mu\boldsymbol{v}\\ &= (1-(1-\lambda)(1-\mu))\boldsymbol{v} + (1-\lambda)(1-\mu)\boldsymbol{u} \in (\boldsymbol{u}, \boldsymbol{v})\end{aligned}$$

because $0 < (1-\lambda)(1-\mu) < 1$. Since this inclusion contradicts the assumption $X \cap (\boldsymbol{u}, \boldsymbol{v}) = \varnothing$, the set $X \cap \langle \boldsymbol{x}, \boldsymbol{y}\rangle$ should be convex.

Next, the 1-dimensional closed convex set $X \cap \langle \boldsymbol{x}, \boldsymbol{y}\rangle$ distinct from the whole line $\langle \boldsymbol{x}, \boldsymbol{y}\rangle$ is either a closed segment $[\boldsymbol{a}, \boldsymbol{c}]$ or a closed halfline $[\boldsymbol{a}, \boldsymbol{c}\rangle$. In either case, the point $\boldsymbol{u} = (1-\lambda)\boldsymbol{c} + \lambda\boldsymbol{a}$ belongs to $\langle \boldsymbol{x}, \boldsymbol{y}\rangle \setminus X$, contrary to the assumption. The obtained contradiction shows that $\langle \boldsymbol{x}, \boldsymbol{y}\rangle \subset X$, and Theorem 2.38 implies that X is a plane.

Problem 3.4. Because the case $K = \varnothing$ is obvious, we may assume that $K \neq \varnothing$. The proof is organized by induction on $m = \dim K$. The case $m = 0$ is trivial. Suppose that the assertion is true for all $m \leqslant r-1$, where $r \geqslant 1$, and let K be a convex set of dimension r contained in the union $M = \cup(L_\alpha : L_\alpha \in \mathcal{F})$.

By Corollary 3.7, there is an r-simplex $\Delta = \Delta(\boldsymbol{x}_1, \dots, \boldsymbol{x}_{r+1}) \subset K$ such that $\operatorname{aff} \Delta = \operatorname{aff} K$. We are going to prove that Δ is contained in one of the planes $L_\alpha \in \mathcal{F}$. Indeed, assume, for contradiction, that this is not the case and consider the $(r-1)$-simplex $\Delta' = \Delta(\boldsymbol{x}_1, \dots, \boldsymbol{x}_r)$. Since $\Delta' \subset K \subset M$ and $\dim \Delta' = r-1$, the inductive assumption implies the existence of a plane $L' \in \mathcal{F}$ which contains Δ'. For any scalar $0 < \lambda < 1$, let $\boldsymbol{x}_i(\lambda) = (1-\lambda)\boldsymbol{x}_{r+1} + \lambda\boldsymbol{x}_i$, $1 \leqslant i \leqslant r$. The homothety $f(\boldsymbol{x}) = \boldsymbol{x}_{r+1} + \lambda(\boldsymbol{x} - \boldsymbol{x}_{r+1})$ maps the set $\{\boldsymbol{x}_1, \dots, \boldsymbol{x}_r\}$ onto $\{\boldsymbol{x}_1(\lambda), \dots, \boldsymbol{x}_r(\lambda)\}$. Therefore, a combination of Theorems 2.78 and 2.86 implies that the set $X(\lambda) = \{\boldsymbol{x}_1(\lambda), \dots, \boldsymbol{x}_r(\lambda)\}$ is affinely independent. Consequently, its convex hull is an $(r-1)$-simplex of the form $\Delta(\lambda) = \Delta(\boldsymbol{x}_1(\lambda), \dots, \boldsymbol{x}_r(\lambda))$. By a convexity argument, $\Delta(\lambda) \subset \Delta \subset K \subset M$, and the inductive assumption implies that $\Delta(\lambda)$ is included into a plane, say $L(\lambda) \in \mathcal{F}$.

We observe that $\Delta \cap L(\lambda) = \Delta(\lambda)$. Indeed, assume for a moment that $\Delta \cap L(\lambda)$ contains a point $\boldsymbol{x}_0 \in \Delta \setminus \Delta(\lambda)$. Then the set $\{\boldsymbol{x}_0\} \cup X(\lambda)$ is affinely independent (see Theorem 2.54). In this case, $\dim(\{\boldsymbol{x}_0\} \cup X(\lambda)) = r$. Since the plane $\operatorname{aff}(\{\boldsymbol{x}_0\} \cup X(\lambda))$ lies in $\operatorname{aff}\Delta$ and both planes have dimension r, we obtain that they coincide. Consequently, $\Delta \subset \operatorname{aff}\Delta = \operatorname{aff}(\{\boldsymbol{x}_0\} \cup X(\lambda)) \subset L(\lambda)$, contrary to the assumption on Δ.

Clearly, the simplices $\Delta(\lambda)$, $0 < \lambda < 1$, are pairwise disjoint. Hence the respective planes $L(\lambda)$ are pairwise distinct. Let $\mathcal{G} = \{L(\lambda) : 0 < \lambda < 1\}$. The above argument shows that $\mathcal{G}$ is an uncountable subfamily of $\mathcal{F}$, which is impossible by the assumption on $\mathcal{F}$. The obtained contradiction shows the existence of a plane $L_\alpha \in \mathcal{F}$ which contains Δ. Consequently, $K \subset \operatorname{aff} K = \operatorname{aff}\Delta \subset L_\alpha$, as desired.

Problem 3.5. Put $M = (\lambda+\mu)K + \frac{|\lambda|+|\mu|-|\lambda+\mu|}{2}(K - K)$. If $\lambda + \mu \geqslant 0$, then, by Theorem 3.4,

$$\begin{aligned}
M &= \lambda K + \mu K + 0(K - K) = \lambda K + \mu K \ \text{ if } \ \lambda \geqslant 0 \ \text{ and } \ \mu \geqslant 0, \\
M &= (\lambda + \mu)K - \mu(K - K) = \lambda K + \mu K \ \text{ if } \ \lambda \geqslant -\mu \geqslant 0, \\
M &= (\lambda + \mu)K - \lambda(K - K) = \lambda K + \mu K \ \text{ if } \ \mu \geqslant -\lambda \geqslant 0.
\end{aligned}$$

If $\lambda + \mu \leqslant 0$, then replace K with $-K$, μ with $-\lambda$, and λ with $-\mu$ in the above argument.

Problem 3.6. We consider the case $\lambda > \gamma$ (all other cases $\lambda < \gamma$, $\mu > \delta$, or $\mu < \delta$ are similar). By Theorem 3.4, $\lambda K = \gamma K + (\lambda - \gamma)K$. Hence

$$\gamma K + (\lambda - \gamma)K + \mu M = \gamma K + \delta M,$$

and Theorem 3.60 implies that $(\lambda - \gamma)K + \mu M = \delta M$. Similarly,

1) $(\lambda - \gamma)K = (\delta - \mu)M$ if $\delta > \mu$,
2) $(\lambda - \gamma)K = \{\boldsymbol{o}\}$ and $K = \{\boldsymbol{o}\} = \frac{\delta - \mu}{\lambda - \gamma}M$ if $\delta = \mu$,
3) $(\lambda - \gamma)K + (\mu - \delta)M = \{\boldsymbol{o}\}$ and $K = M = \{\boldsymbol{o}\}$ if $\delta < \mu$.

Summing up, $K = \frac{\delta - \mu}{\lambda - \gamma}M$.

Problem 3.7. For the first assertion, choose points $\boldsymbol{x}, \boldsymbol{y} \in A$ and a scalar $0 \leqslant \lambda \leqslant 1$. Using Theorem 3.4 twice, we obtain

$$\begin{aligned}
(1 - \lambda)\boldsymbol{x} + \lambda \boldsymbol{y} + K_1 &= (1 - \lambda)\boldsymbol{x} + \lambda \boldsymbol{y} + (1 - \lambda)K_1 + \lambda K_1 \\
&= (1 - \lambda)(\boldsymbol{x} + K_1) + \lambda(\boldsymbol{y} + K) \subset (1 - \lambda)K_2 + \lambda K_2 = K_2.
\end{aligned}$$

Consequently, $(1 - \lambda)\boldsymbol{x} + \lambda \boldsymbol{y} \in A$, which shows the convexity of A.

The second assertion follows from an obvious relation $B = K_2 - K_1$ and Theorem 3.4.

Problem 3.8. Choose any distinct points $\boldsymbol{x}, \boldsymbol{y} \in \operatorname{cl} K \cap B_\rho(\boldsymbol{c})$, and denote by $\boldsymbol{z}$ the midpoint of $[\boldsymbol{x}, \boldsymbol{y}]$: $\boldsymbol{z} = (\boldsymbol{x} + \boldsymbol{y})/2$. Clearly, $\boldsymbol{z} \in [\boldsymbol{x}, \boldsymbol{y}] \subset \operatorname{cl} K$ by the convexity of $\operatorname{cl} K$ (see Theorem 3.46). By the parallelogram identity (see page 8),

$$\begin{aligned}
\|\boldsymbol{z} - \boldsymbol{c}\|^2 &= \|\tfrac{1}{2}(\boldsymbol{x} - \boldsymbol{c}) + \tfrac{1}{2}(\boldsymbol{y} - \boldsymbol{c})\|^2 = \tfrac{1}{4}\|(\boldsymbol{x} - \boldsymbol{c}) + (\boldsymbol{y} - \boldsymbol{c})\|^2 \\
&= \tfrac{1}{4}(2\|\boldsymbol{x} - \boldsymbol{c}\|^2 + 2\|\boldsymbol{y} - \boldsymbol{c}\|^{-} \|\boldsymbol{x} - \boldsymbol{y}\|^2) \\
&\leqslant \tfrac{1}{4}(2\rho^2 + 2\rho^2 - \|\boldsymbol{x} - \boldsymbol{y}\|^2) < \rho^2.
\end{aligned}$$

Hence $\|\boldsymbol{z} - \boldsymbol{c}\| < \rho$. Let $\delta = \rho - \|\boldsymbol{z} - \boldsymbol{c}\|$. Then $U_\delta(\boldsymbol{z}) \subset U_\rho(\boldsymbol{c})$, as shown in Problem 1.8. By Theorem 3.51, there is a point $\boldsymbol{u} \in \operatorname{rint} K \cap U_\delta(\boldsymbol{z}) \subset \operatorname{rint} K \cap U_\rho(\boldsymbol{c})$.

(*a*) Since $\operatorname{cl}(X \cap B_\rho(\boldsymbol{c})) \subset \operatorname{cl} X \cap B_\rho(\boldsymbol{c})$ for any set $X \subset \mathbb{R}^n$, it suffices to show that $\operatorname{cl} K \cap B_\rho(\boldsymbol{c}) \subset \operatorname{cl}(K \cap B_\rho(\boldsymbol{c}))$. If $\boldsymbol{x} \in \operatorname{cl} K \cap B_\rho(\boldsymbol{c})$, then, in the above terms, $[\boldsymbol{u}, \boldsymbol{x}) \subset \operatorname{rint} K \subset K$, as follows from Theorem 3.47. Hence

$$\boldsymbol{x} \in \operatorname{cl}[\boldsymbol{u}, \boldsymbol{x}) \subset \operatorname{cl}(\operatorname{rint} K \cap B_\rho(\boldsymbol{c})) \subset \operatorname{cl}(K \cap B_\rho(\boldsymbol{c})).$$

Summing up, $\operatorname{cl} K \cap B_\rho(\boldsymbol{c}) \subset \operatorname{cl}(K \cap B_\rho(\boldsymbol{c}))$.

(*b*) Since $\operatorname{rint} K \cap U_\rho(\boldsymbol{c}) \neq \varnothing$, Theorem 3.29 gives

$$\operatorname{rint}(K \cap B_\rho(\boldsymbol{c})) = \operatorname{rint} K \cap \operatorname{rint} B_\rho(\boldsymbol{c}) = \operatorname{rint} K \cap U_\rho(\boldsymbol{c}).$$

(*c*) From the above proved we obtain that

$$\begin{aligned}
&\operatorname{rbd} K \cap B_\rho(\boldsymbol{c}) = (\operatorname{cl} K \setminus \operatorname{rint} K) \cap B_\rho(\boldsymbol{c}) \\
&= (\operatorname{cl} K \cap B_\rho(\boldsymbol{c})) \setminus (\operatorname{rint} K \cap B_\rho(\boldsymbol{c})) \\
&\subset (\operatorname{cl} K \cap B_\rho(\boldsymbol{c})) \setminus (\operatorname{rint} K \cap U_\rho(\boldsymbol{c})) \\
&= \operatorname{cl}(K \cap B_\rho(\boldsymbol{c})) \setminus \operatorname{rint}(K \cap B_\rho(\boldsymbol{c})) = \operatorname{rbd}(K \cap B_\rho(\boldsymbol{c})).
\end{aligned}$$

Problem 3.9. Choose any point $\boldsymbol{x} \in \operatorname{rint} K \setminus \{\boldsymbol{u}\}$. By Corollary 3.27, there is a point $\boldsymbol{v} \in K \setminus \{\boldsymbol{x}\}$ such that $\boldsymbol{x} \in [\boldsymbol{u}, \boldsymbol{v})$. Hence $\operatorname{rint} K \subset \cup([\boldsymbol{u}, \boldsymbol{v}) : \boldsymbol{v} \in K)$.

If $\boldsymbol{u} \in \operatorname{rint} K$, then, by Theorem 3.24, every semi-open segment $[\boldsymbol{u}, \boldsymbol{v})$, where $\boldsymbol{v} \in K$, lies in $\operatorname{rint} K$. Consequently, $\cup([\boldsymbol{u}, \boldsymbol{v}) : \boldsymbol{v} \in K) \subset \operatorname{rint} K$, and equality holds if and only if $\boldsymbol{u} \in \operatorname{rint} K$.

Problem 3.10. Choose a point $\boldsymbol{z} \in \operatorname{rint} X + \operatorname{rint} Y$, and express it as $\boldsymbol{z} = \boldsymbol{x} + \boldsymbol{y}$, where $\boldsymbol{x} \in \operatorname{rint} X$ and $\boldsymbol{y} \in \operatorname{rint} Y$. Choose a scalar $\rho > 0$ such that $K = \operatorname{aff} X \cap U_\rho(\boldsymbol{x}) \subset X$ and $M = \operatorname{aff} Y \cap U_\rho(\boldsymbol{y}) \subset X$. Clearly, K and M are convex sets, and Theorem 2.68 shows that K and M satisfy the conditions:

$$\boldsymbol{x} \in K \subset X, \quad \dim K = \dim X, \quad \boldsymbol{y} \in M \subset Y, \quad \dim M = \dim Y.$$

Furthermore, Theorem 3.29 implies that both sets K and M are relatively open as intersections of open balls and planes. A combination of Theorem 2.64 and 3.34 shows that $K+M$ is a relatively open convex set, with $\dim(K+M) = \dim(X+Y)$. Since $\boldsymbol{z} \in K + M \subset X + Y$, Theorem 2.68 implies that $\boldsymbol{z} \in \operatorname{rint}(X+Y)$.

Let the planes $\operatorname{aff} X$ and $\operatorname{aff} Y$ are independent. By the above argument, it suffices to show that $\operatorname{rint}(X+Y) \subset \operatorname{rint} X + \operatorname{rint} Y$. Choose a point $\boldsymbol{z} \in \operatorname{rint}(X+Y)$. By Theorem 2.68, there is a relatively open convex set $N = \operatorname{aff}(X+Y) \cap U_\rho(\boldsymbol{z})$ of dimension $\dim(X+Y)$ such that $\boldsymbol{z} \in N \subset X + Y$. Express $\boldsymbol{z}$ as $\boldsymbol{z} = \boldsymbol{x} + \boldsymbol{y}$, where $\boldsymbol{x} \in X$ and $\boldsymbol{y} \in Y$. Let $X' = X - \boldsymbol{x}$, $Y' = Y - \boldsymbol{y}$, and $N' = N - \boldsymbol{z}$. Then the subspaces $S = \operatorname{span} X' = \operatorname{aff} X - \boldsymbol{x}$ and $T = \operatorname{span} Y' = \operatorname{aff} Y - \boldsymbol{y}$ are independent, N' is a relatively open convex set of dimension $\dim(X' + Y')$, satisfying the conditions $\boldsymbol{o} \in N' \subset X' + Y'$, and

$$X' \subset S, \quad Y' \subset T, \quad \operatorname{span}(X' + Y') = S + T.$$

Corollary 3.31 and Theorem 3.34 show that the sets

$$K = S \cap (N + T) \quad \text{and} \quad M = (N + S) \cap T$$

are convex and relatively open. Furthermore, $\dim K = \dim S = \dim X$ and $\dim M = \dim T = \dim Y$. Since $\boldsymbol{x} \in K \subset X$ and $\boldsymbol{y} \in M \subset Y$, Theorem 2.68 shows that $\boldsymbol{x} \in \operatorname{rint} K$ and $\boldsymbol{y} \in \operatorname{rint} Y$. Summing up, $\operatorname{rint}(X+Y) \subset \operatorname{rint} X + \operatorname{rint} Y$.

Problem 3.11. Let $\boldsymbol{u} = \frac{1}{m+1}(\boldsymbol{u}_1 + \cdots + \boldsymbol{u}_{m+1})$, $\delta = \max\{\|\boldsymbol{u} - \boldsymbol{u}_1\|, \dots, \|\boldsymbol{u} - \boldsymbol{u}_{m+1}\|\}$. Assume first that $\boldsymbol{z} \in \operatorname{rint} K$. Choose a scalar $\rho > 0$ such that $\operatorname{aff} K \cap U_\rho(\boldsymbol{z}) \subset K$ and consider the direct homothety

$$f(\boldsymbol{x}) = \begin{cases} \boldsymbol{z} + \frac{1}{2}(\boldsymbol{x} - \boldsymbol{u}) & \text{if } \delta \leqslant \rho, \\ \boldsymbol{z} + \frac{\rho}{2\delta}(\boldsymbol{x} - \boldsymbol{u}), & \text{if } \delta > \rho. \end{cases}$$

Put $\boldsymbol{z}_i = f(\boldsymbol{u}_i)$, $1 \leqslant i \leqslant m+1$. As easily seen,

$$\boldsymbol{z}_1, \dots, \boldsymbol{z}_{m+1} \in \operatorname{aff} K \quad \text{and} \quad \boldsymbol{z} = \tfrac{1}{m+1}(\boldsymbol{z}_1 + \cdots + \boldsymbol{z}_{m+1}).$$

By Theorem 2.78, the set $\{\boldsymbol{z}_1, \dots, \boldsymbol{z}_{m+1}\}$ is affinely independent. Clearly, the simplices $\Delta(\boldsymbol{u}_1, \dots, \boldsymbol{u}_{m+1})$ and $\Delta(\boldsymbol{z}_1, \dots, \boldsymbol{z}_{m+1})$ are homothetic. Furthermore,

$$\|\boldsymbol{z} - \boldsymbol{z}_i\| = \begin{cases} \frac{1}{2}\|\boldsymbol{u} - \boldsymbol{u}_i\| \leqslant \frac{\rho}{2}, & \text{if } \delta \leqslant \rho, \\ \frac{\rho}{2\delta}\|\boldsymbol{u} - \boldsymbol{u}_i\| < \frac{\rho}{2}, & \text{if } \delta > \rho. \end{cases}$$

In either case, $\boldsymbol{z}_1, \dots, \boldsymbol{z}_{m+1} \in U_\rho(\boldsymbol{z})$. Hence

$$\Delta(\boldsymbol{z}_1, \dots, \boldsymbol{z}_{m+1}) = \operatorname{conv}\{\boldsymbol{z}_1, \dots, \boldsymbol{z}_{m+1}\} \subset \operatorname{aff} K \cap U_\rho(\boldsymbol{z}) \subset K.$$

Conversely, let $\Delta = \Delta(\boldsymbol{z}_1, \dots, \boldsymbol{z}_{m+1}) \subset K$ be a simplex such that $\boldsymbol{z} = \frac{1}{m+1}(\boldsymbol{z}_1 + \cdots + \boldsymbol{z}_{m+1})$. By Theorem 3.20, $\boldsymbol{z} \in \operatorname{rint} \Delta$. Since $\operatorname{aff} \Delta = \operatorname{aff} K$, Theorem 3.18 gives $\boldsymbol{z} \in \operatorname{rint} \Delta \subset \operatorname{rint} K$.

Problem 3.12. If X lies in an m-simplex $\Delta \subset \operatorname{aff} X$, then X is bounded as a subset of the compact set Δ (see Theorem 3.6). Conversely, suppose that X is bounded. Since the case $\dim X = 0$ is obvious, we may assume that $\dim X > 0$. Choose any m-simplex $\Delta' = \Delta(\boldsymbol{u}_1, \dots, \boldsymbol{u}_{m+1}) \subset \operatorname{aff} X$. Translating Δ' on $\boldsymbol{a} - \frac{1}{m+1}(\boldsymbol{u}_1 + \cdots + \boldsymbol{u}_{m+1})$, we may assume that $\boldsymbol{a}$ is the center of Δ'. Since $\boldsymbol{a} \in \operatorname{rint} \Delta'$ (see Theorem 3.20), there is an open ball $U_\delta(\boldsymbol{a}) \subset \mathbb{R}^n$ such that

$$\operatorname{aff} X \cap U_\delta(\boldsymbol{a}) = \operatorname{aff} \Delta' \cap U_\delta(\boldsymbol{a}) \subset \Delta'.$$

Choose a scalar $\rho > 0$ such that $X \subset U_\rho(\boldsymbol{a})$ (see page 11). Let $\gamma = \rho/\delta$. Clearly,

$$U_\rho(\boldsymbol{a}) = U_{\gamma\delta}(\boldsymbol{a}) = (1 - \gamma)\boldsymbol{a} + \gamma U_\delta(\boldsymbol{a}).$$

Put $z_i = a + \gamma(u_i - a)$, $1 \leqslant i \leqslant m+1$. As in Problem 3.11, the m-simples $\Delta = \Delta(u_1, \ldots, u_{m+1})$ is homothetic to Δ': $\Delta = a + \gamma(\Delta' - a)$. Therefore,

$$\begin{aligned} X &\subset \operatorname{aff} X \cap U_\rho(a) = \operatorname{aff} X \cap ((1-\gamma)a + \gamma U_\delta(a)) \\ &= (1-\gamma)a + \gamma(\operatorname{aff} X \cap U_\delta(a)) \subset (1-\gamma)a + \gamma\Delta' = \Delta. \end{aligned}$$

Problem 3.13. Excluding the obvious case $F = \varnothing$, we will assume that $F \neq \varnothing$. For the convexity of M, choose points $x, y \in M$. If $x, y \in F$, then $[x, y] \subset F \subset M$ by the convexity of F. Similarly, $[x, y] \subset \operatorname{rint} K \subset M$ if $x, y \in \operatorname{rint} K$ (see Corollary 3.25). Finally, if $x \in F$ and $y \in \operatorname{rint} K$, then $(x, y] \subset \operatorname{rint} K$ according to Theorem 3.47. Thus $[x, y] = \{x\} \cup (x, y] \subset F \cup \operatorname{rint} K = M$.

The inclusions $\operatorname{rint} K \subset M \subset \operatorname{cl} K$ and Theorem 3.48 give $\operatorname{cl} K = \operatorname{cl}(\operatorname{rint} K) \subset \operatorname{cl} M \subset \operatorname{cl} K$. Hence $\operatorname{cl} M = \operatorname{cl} K$. The equality $\operatorname{rint} M = \operatorname{rint} K$ follows from Corollary 3.49.

Problem 3.14. Choose points $z_i \in \mu_i K_i$ and put $K_i' = \mu_i K_i - z_i$, $1 \leqslant i \leqslant r$. Then $o \in K_i'$, $1 \leqslant i \leqslant r$. By Theorem 2.42 (see also page 5),

$$\operatorname{span}(K_1' + \cdots + K_r') = \sum_{i=1}^{r} \operatorname{span} K_i',$$

By a dimension argument, there is an index set $I \subset \{1, \ldots, r\}$ such that $\operatorname{card} I \leqslant m$ and

$$\operatorname{span}(K_1' + \cdots + K_r') = \sum_{i \in I} \operatorname{span} K_i'.$$

A combination of Theorems 3.18 and 3.34 gives

$$\begin{aligned} \operatorname{rint}(K_1' + \cdots + K_r') &= \operatorname{rint} K_1' + \cdots + \operatorname{rint} K_r' \\ &\subset \sum_{i \in I} \operatorname{rint} K_i' + \sum_{i \notin I} K_i' \subset \operatorname{rint}(K_1' + \cdots + K_r'). \end{aligned}$$

Hence, $\operatorname{rint}(K_1' + \cdots + K_r') = \sum_{i \in I} \operatorname{rint} K_i' + \sum_{i \notin I} K_i'$. Finally,

$$\begin{aligned} \operatorname{rint}(\mu_1 K_1 + \cdots + \mu_r K_r) &= (z_1 + \cdots + z_r) + \operatorname{rint}(K_1' + \cdots + K_r') \\ &= (z_1 + \cdots + z_r) + \sum_{i \in I} \operatorname{rint} K_i' + \sum_{i \notin I} K_i' \\ &= \sum_{i \in I} \mu_i \operatorname{rint} K_i + \sum_{i \notin I} \mu_i K_i. \end{aligned}$$

Problem 3.15. Due to the obvious inclusion

$$\operatorname{cl}(K_1 \cap K_2) \subset \operatorname{cl} K_1 \cap \operatorname{cl} K_2 = K_1 \cap \operatorname{cl} K_2,$$

it remains to prove the opposite one. Choose any point $x \in K_1 \cap \operatorname{cl} K_2$, and let $y \in K_1 \cap \operatorname{rint} K_2$. Then $(x, y) \subset K_1$ by the convexity of K_1 and $(x, y) \subset \operatorname{rint} K_2$

according to Theorem 3.47. Consecutively, $(1-\lambda)\boldsymbol{x}+\lambda\boldsymbol{y} \in K_1 \cap \operatorname{rint} K$ for all $0 \leqslant \lambda < 1$. Hence

$$\boldsymbol{x} = \lim_{\lambda \to 0} (1-\lambda)\boldsymbol{x}+\lambda\boldsymbol{y} \in \operatorname{cl}(K_1 \cap \operatorname{rint} K_2) \subset \operatorname{cl}(K_1 \cap K_2).$$

Summing up, $K_1 \cap \operatorname{cl} K_2 \subset \operatorname{cl}(K_1 \cap K_2)$.

Problem 3.16. (*a*) It suffices to show that for any point $\boldsymbol{u} \in K$ and a scalar $\rho > 0$ there is a point $\boldsymbol{v} \in X \cap \operatorname{rint} K$ with the property $\boldsymbol{v} \in U_\rho(\boldsymbol{u})$.

By Theorem 3.51, there is a point $\boldsymbol{w} \in \operatorname{rint} K \cap U_\rho(\boldsymbol{u})$. Let $\delta = \rho - \|\boldsymbol{u}-\boldsymbol{w}\|$. Clearly, $\delta > 0$ and $U_\delta(\boldsymbol{w}) \subset U_\rho(\boldsymbol{u})$ according to Problem 1.8. Theorem 3.17 gives the existence of a scalar $0 < \varepsilon \leqslant \delta$ with the property $\operatorname{aff} K \cap U_\varepsilon(\boldsymbol{w}) \subset \operatorname{rint} K$. Since $\operatorname{aff} K = \operatorname{aff} M$ (see Corollary 2.65), one has $\operatorname{rint} K \subset \operatorname{rint} M$, as follows from Theorem 3.18. Therefore,

$$\operatorname{aff} M \cap U_\varepsilon(\boldsymbol{w}) = \operatorname{aff} K \cap U_\varepsilon(\boldsymbol{w}) \subset \operatorname{rint} K \subset \operatorname{rint} M.$$

Because the set X is dense in M and $\boldsymbol{w} \in \operatorname{rint} M \subset M$, there is a point

$$\boldsymbol{v} \in X \cap U_\varepsilon(\boldsymbol{w}) \subset X \cap U_\rho(\boldsymbol{u}) \subset U_\rho(\boldsymbol{u}).$$

Finally, due to the inclusions $X \subset M \subset \operatorname{aff} M$, one has

$$\begin{aligned} \boldsymbol{v} \in X \cap U_\varepsilon(\boldsymbol{w}) &= (X \cap \operatorname{aff} M) \cap U_\varepsilon(\boldsymbol{w}) = (X \cap \operatorname{aff} K) \cap U_\varepsilon(\boldsymbol{w}) \\ &= X \cap (U_\varepsilon(\boldsymbol{w}) \cap \operatorname{aff} K) \subset X \cap \operatorname{rint} K. \end{aligned}$$

(*b*) Choose any point $\boldsymbol{u} \in M$ and a scalar $\rho > 0$. It suffices to show the existence of a point $\boldsymbol{v} \in (X \cap \operatorname{rint} M) \setminus \operatorname{cl} K$ which belongs to $U_\rho(\boldsymbol{u})$.

Theorem 3.51 shows that the set $\operatorname{rint} M \cap U_\rho(\boldsymbol{u})$ is nonempty and has dimension $\dim M$. Therefore, $\operatorname{rint} M \cap U_\rho(\boldsymbol{u}) \not\subset \operatorname{cl} K$, since otherwise

$$\operatorname{aff} M = \operatorname{aff}(\operatorname{rint} M \cap U_\rho(\boldsymbol{u})) \subset \operatorname{aff}(\operatorname{cl} K) = \operatorname{aff} K,$$

contrary to the assumption $\dim K < \dim M$.

Choose a point $\boldsymbol{w} \in (\operatorname{rint} M \cap U_\rho(\boldsymbol{u})) \setminus \operatorname{cl} K$ and let $\delta = \rho - \|\boldsymbol{u}-\boldsymbol{w}\|$. Clearly, $\delta > 0$ and $U_\delta(\boldsymbol{w}) \subset U_\rho(\boldsymbol{u})$ according to Problem 1.8. Because the set $\operatorname{cl} K$ is closed, there is a scalar $0 < \varepsilon' \leqslant \delta$ such that $\operatorname{cl} K \cap U_{\varepsilon'}(\boldsymbol{w}) = \varnothing$. Theorem 3.17 gives the existence of a scalar $0 < \varepsilon \leqslant \varepsilon'$ with the property $\operatorname{aff} M \cap U_\varepsilon(\boldsymbol{w}) \subset \operatorname{rint} M$. Since X is dense in M, there is a point $\boldsymbol{v} \in X$ such that $\boldsymbol{v} \in U_\varepsilon(\boldsymbol{w}) \subset U_\rho(\boldsymbol{u})$. By the argument from part (*a*), $\boldsymbol{v} \in X \cap \operatorname{rint} M$.

Problem 3.17. By Theorem 3.29, $\operatorname{rint} K \subset \bigcap_\alpha \operatorname{rint} K_\alpha$. Hence $\operatorname{rint} K \cap \operatorname{rbd} K_\alpha = \varnothing$ for all $K_\alpha \in \mathcal{F}$, implying that

$$\operatorname{rbd} K \cap \operatorname{rbd} K_\alpha = (\operatorname{cl} K \setminus \operatorname{rint} K) \cap \operatorname{rbd} K_\alpha = \operatorname{cl} K \cap \operatorname{rbd} K_\alpha.$$

Therefore, $\bigcup_\alpha (\mathrm{rbd}\, K \cap \mathrm{rbd}\, K_\alpha) = \bigcup_\alpha (\mathrm{cl}\, K \cap \mathrm{rbd}\, K_\alpha)$. Furthermore, because $\mathrm{cl}\, K \subset \mathrm{cl}\, K_\alpha$ for all $K_\alpha \in \mathcal{F}$, one has

$$\mathrm{cl}\, K \cap \mathrm{rbd}\, K_\alpha = \mathrm{cl}\, K \cap (\mathrm{cl}\, K_\alpha \setminus \mathrm{rint}\, K_\alpha) = \mathrm{cl}\, K \setminus \mathrm{rint}\, K_\alpha.$$

Consequently, Theorem 3.29 gives

$$\begin{aligned}\bigcup_\alpha (\mathrm{cl}\, K \cap \mathrm{rbd}\, K_\alpha) &= \bigcup_\alpha (\mathrm{cl}\, K \setminus \mathrm{rint}\, K_\alpha) = \mathrm{cl}\, K \setminus (\bigcap_\alpha \mathrm{rint}\, K_\alpha) \\ &\subset \mathrm{cl}\, K \setminus \mathrm{rint}\, (\bigcap_\alpha K_\alpha) = \mathrm{cl}\, K \setminus \mathrm{rint}\, K = \mathrm{rbd}\, K.\end{aligned}$$

Since $\mathcal{F}$ is finite, one has $\mathrm{rint}\, (\bigcap_\alpha K_\alpha) = \bigcap_\alpha \mathrm{rint}\, K_\alpha$ (see Theorem 3.29), implying the desired equality.

Problem 3.18. If the set X is convex and $U_\rho(\boldsymbol{x})$ is an open ball centered at a point $\boldsymbol{x} \in X$, then $U_\rho(\boldsymbol{x}) \cap X$ is convex as the intersection of convex sets $U_\rho(\boldsymbol{x})$ and X; whence X is locally convex.

1. Conversely, suppose that X is locally convex. First, we observe that any two points of X can be joined by a polygonal line in X. Indeed, let $\boldsymbol{u} \in X$ and denote by $C_{\boldsymbol{u}}$ the part of X accessible from $\boldsymbol{u}$ by polygonal lines. By the assumption, for any point $\boldsymbol{v} \in C_{\boldsymbol{u}}$, there is an open ball $U_\rho(\boldsymbol{v})$ such that $X \cap U_\rho(\boldsymbol{v})$ is convex. So, if $\mathit{\Gamma}$ is a polygonal line in X joining $\boldsymbol{u}$ and $\boldsymbol{v}$, then $\mathit{\Gamma} \cup [\boldsymbol{v}, \boldsymbol{x}]$ is a polygonal line joining $\boldsymbol{u}$ and a point $\boldsymbol{x} \in X \cap U_\rho(\boldsymbol{v})$. Consequently, $X \cap U_\rho(\boldsymbol{v}) \subset C_{\boldsymbol{v}}$. Similarly, if $\boldsymbol{w} \in \mathrm{cl}\, C_{\boldsymbol{u}}$ and $X \cap U_\delta(\boldsymbol{w})$ is convex, then choosing a point $\boldsymbol{z} \in C_{\boldsymbol{u}} \cap X \cap U_\delta(\boldsymbol{w})$ and a polygonal line $\mathit{\Lambda} \subset X$ joining $\boldsymbol{u}$ and $\boldsymbol{z}$, we see that $\mathit{\Lambda} \cup [\boldsymbol{z}, \boldsymbol{w}]$ is a polygonal line joining $\boldsymbol{u}$ and $\boldsymbol{w}$ in X. Therefore, $\boldsymbol{w} \in C_{\boldsymbol{u}}$. Summing up, $C_{\boldsymbol{u}}$ is both closed and open in X, implying that $C_{\boldsymbol{u}} = X$. Hence any two points in X can be joined by a polygonal line through $\boldsymbol{u}$.

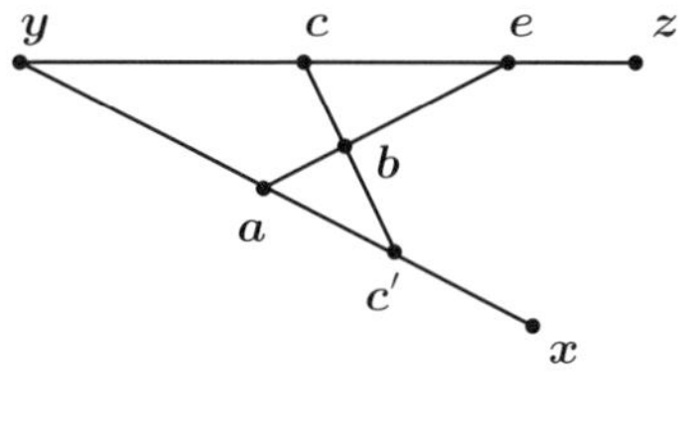

2. Next, we assert that $[\boldsymbol{x}, \boldsymbol{z}] \subset X$ provided $[\boldsymbol{x}, \boldsymbol{y}] \cup [\boldsymbol{y}, \boldsymbol{z}] \subset X$ for a suitable point $\boldsymbol{y} \in X$. For this, consider on $[\boldsymbol{y}, \boldsymbol{z}]$ the set E of those points $\boldsymbol{c}$ for which there exists a point $\boldsymbol{c}' \in (\boldsymbol{x}, \boldsymbol{y})$ such that the triangle $\mathit{\Delta}(\boldsymbol{c}, \boldsymbol{c}', \boldsymbol{y})$ lies in X (see the above picture). The set E is not empty because $\boldsymbol{y} \in E$. Obviously, $[\boldsymbol{c}, \boldsymbol{y}] \subset E$ for every point $\boldsymbol{c} \in E$.

We are going to prove that E is both closed and open in $[\boldsymbol{y}, \boldsymbol{z}]$, thus implying the equality $E = [\boldsymbol{y}, \boldsymbol{z}]$. Choose a point $\boldsymbol{c} \in E \setminus \{\boldsymbol{z}\}$ and let $\rho > 0$ be such that the set $X \cap U_\rho(\boldsymbol{c})$ is convex. We observe that $[\boldsymbol{y}, \boldsymbol{z}] \cap U_\rho(\boldsymbol{c}) \subset E$. Indeed, if $\boldsymbol{e} \in (\boldsymbol{y}, \boldsymbol{c}) \cap U_\rho(\boldsymbol{c})$, then $\mathit{\Delta}(\boldsymbol{e}, \boldsymbol{c}', \boldsymbol{y}) \subset \mathit{\Delta}(\boldsymbol{c}, \boldsymbol{c}', \boldsymbol{y}) \subset X$.

Suppose that $\boldsymbol{e} \in (\boldsymbol{c}, \boldsymbol{z}) \cap U_\rho(\boldsymbol{c})$. Select a point $\boldsymbol{b} \in (\boldsymbol{c}, \boldsymbol{c}') \cap U_\rho(\boldsymbol{c})$. If $\boldsymbol{a}$ is the intersection of $(\boldsymbol{x}, \boldsymbol{y})$ with the line $\langle \boldsymbol{e}, \boldsymbol{b} \rangle$, then, as easy to see, the triangle $\Delta(\boldsymbol{e}, \boldsymbol{a}, \boldsymbol{y})$ lies in X, implying that one can put $\boldsymbol{e}' = \boldsymbol{a}$. Summing up, E is open in $[\boldsymbol{y}, \boldsymbol{z}]$. Assume now that $\boldsymbol{e} \in \operatorname{cl} E$. We select an open ball $U_\rho(\boldsymbol{e})$ such that the set $X \cap U_\rho(\boldsymbol{e})$ is convex. Within this set, choose on $(\boldsymbol{y}, \boldsymbol{e})$ a point $\boldsymbol{c} \in E$ and a point $\boldsymbol{b} \in (\boldsymbol{c}, \boldsymbol{c}') \cap X \cap U_\rho(\boldsymbol{e})$. As above, the line $\langle \boldsymbol{e}, \boldsymbol{b} \rangle$ meets $(\boldsymbol{x}, \boldsymbol{y})$ at a point $\boldsymbol{a}$ such that $\Delta(\boldsymbol{e}, \boldsymbol{a}, \boldsymbol{y}) \subset X$. Hence $\boldsymbol{e} \in E$, and E is closed in $[\boldsymbol{y}, \boldsymbol{z}]$.

Thus $E = [\boldsymbol{y}, \boldsymbol{z}]$ and, therefore, there is a point $\boldsymbol{z}' \in (\boldsymbol{x}, \boldsymbol{y})$ such that $\Delta(\boldsymbol{z}, \boldsymbol{z}', \boldsymbol{y}) \subset X$. Consider the set F of those points $\boldsymbol{a} \in [\boldsymbol{x}, \boldsymbol{y}]$ for which $\Delta(\boldsymbol{z}, \boldsymbol{a}, \boldsymbol{y}) \subset X$. Clearly, $F \neq \varnothing$ since $\boldsymbol{z}' \in F$. Similarly, to the above argument, F is both open and closed in $[\boldsymbol{x}, \boldsymbol{y}]$, implying that $F = [\boldsymbol{x}, \boldsymbol{y}]$. Hence $[\boldsymbol{x}, \boldsymbol{z}] \subset X$.

3. Finally, choose points $\boldsymbol{u}, \boldsymbol{v} \in X$ and a polygonal line $\Gamma \subset X$ joining $\boldsymbol{u}$ and $\boldsymbol{v}$. Consecutively replacing any two adjacent closed segments $[\boldsymbol{x}, \boldsymbol{y}]$ and $[\boldsymbol{y}, \boldsymbol{z}]$ from Γ by a closed segment $[\boldsymbol{x}, \boldsymbol{z}] \subset X$, we conclude that $[\boldsymbol{u}, \boldsymbol{v}] \subset X$. Hence X is convex.

Problem 3.19. Clearly, both mappings $f(\boldsymbol{x}) = \boldsymbol{a} + g(\boldsymbol{x})$ and $\varphi(\boldsymbol{x}) = \alpha + \boldsymbol{x}\cdot\boldsymbol{c}$ are affine transformations. It suffices to show that the h-image of every closed segment $[\boldsymbol{x}, \boldsymbol{z}] \subset \mathbb{R}^n$ is the closed segment $[h(\boldsymbol{x}), h(\boldsymbol{z})]$. Indeed, choose a point

$$\boldsymbol{y} = (1-\lambda)\boldsymbol{x} + \lambda\boldsymbol{z} \in [\boldsymbol{x}, \boldsymbol{z}], \quad 0 \leqslant \lambda \leqslant 1,$$

and let $\mu = \lambda\varphi(\boldsymbol{z})/((1-\lambda)\varphi(\boldsymbol{x}) + \lambda\varphi(\boldsymbol{z}))$. Then $0 < \mu < 1$ because both $\varphi(\boldsymbol{x})$ and $\varphi(\boldsymbol{z})$ are either positive or negative. Therefore,

$$\begin{aligned} h(\boldsymbol{y}) &= \frac{(1-\lambda)f(\boldsymbol{x}) + \lambda f(\boldsymbol{z})}{(1-\lambda)\varphi(\boldsymbol{x}) + \lambda\varphi(\boldsymbol{z})} \\ &= \frac{(1-\lambda)\varphi(\boldsymbol{x})}{(1-\lambda)\varphi(\boldsymbol{x}) + \lambda\varphi(\boldsymbol{z})}\,\frac{f(\boldsymbol{x})}{\varphi(\boldsymbol{x})} + \frac{\lambda\varphi(\boldsymbol{z})}{(1-\lambda)\varphi(\boldsymbol{x}) + \lambda\varphi(\boldsymbol{z})}\,\frac{f(\boldsymbol{z})}{\varphi(\boldsymbol{z})} \\ &= (1-\mu)\,h(\boldsymbol{x}) + \mu\,h(\boldsymbol{z}) \in [h(\boldsymbol{x}), h(\boldsymbol{z})]. \end{aligned}$$

Hence $h([\boldsymbol{x}, \boldsymbol{z}]) \subset [h(\boldsymbol{x}), h(\boldsymbol{z})]$.

Conversely, let $\boldsymbol{u} = (1-\delta)\,h(\boldsymbol{x}) + \delta\,h(\boldsymbol{z})$, where $0 < \delta < 1$. Put

$$\gamma = \delta\varphi(\boldsymbol{x})/(\delta\varphi(\boldsymbol{x}) + (1-\delta)\varphi(\boldsymbol{z})).$$

Then $0 < \gamma < 1$ and $\boldsymbol{v} = (1-\gamma)\boldsymbol{x} + \gamma\boldsymbol{z} \in (\boldsymbol{x}, \boldsymbol{z})$. As above, $\boldsymbol{u} = h(\boldsymbol{v}) \in h([\boldsymbol{x}, \boldsymbol{z}])$, which gives $[h(\boldsymbol{x}), h(\boldsymbol{z})] \subset h([\boldsymbol{x}, \boldsymbol{z}])$.

Problem 3.20. For any point $\boldsymbol{x} \in \mathbb{R}^n$, one has

$$\begin{aligned} \|\boldsymbol{x} - \boldsymbol{a}\| &= \|(\lambda_1 + \cdots + \lambda_r)\boldsymbol{x} - (\lambda\boldsymbol{a}_1 + \cdots + \lambda_r\boldsymbol{a}_r)\| \\ &= \|\lambda_1(\boldsymbol{x} - \boldsymbol{a}_1) + \cdots + \lambda_r(\boldsymbol{x} - \boldsymbol{a}_r)\| \leqslant \lambda_1\|\boldsymbol{x} - \boldsymbol{a}_1\| + \cdots + \lambda_r\|\boldsymbol{x} - \boldsymbol{a}_r\|. \end{aligned}$$

Conversely, suppose that a point $\boldsymbol{a} \in \mathbb{R}^n$ satisfies the above inequality for any choice of $\boldsymbol{x} \in \mathbb{R}^n$. Translating $\{\boldsymbol{a}_1, \dots, \boldsymbol{a}_r, \boldsymbol{a}\}$ on a suitable vector, we may assume that $\lambda_1 \boldsymbol{a}_1 + \cdots + \lambda_r \boldsymbol{a}_r = \boldsymbol{o}$. So, we intend to prove that $\boldsymbol{a} = \boldsymbol{o}$.

Assume, for contradiction, that $\boldsymbol{a} \neq \boldsymbol{o}$. Let $\boldsymbol{z} = -\boldsymbol{a}$. Using the differentiability property of the norm function (see Problem 1.7), one has

$$\lim_{\gamma \to \infty} \sum_{i=1}^{r} \lambda_i (\|\gamma \boldsymbol{z} - \boldsymbol{a}_i\| - \|\gamma \boldsymbol{z}\|) = \lim_{\gamma \to \infty} \sum_{i=1}^{r} \lambda_i \frac{\|\boldsymbol{z} - \gamma^{-1} \boldsymbol{a}_i\| - \|\boldsymbol{z}\|}{\gamma^{-1}}$$
$$= \lim_{t \to 0} \sum_{i=1}^{r} \lambda_i \frac{\|\boldsymbol{z} - t\boldsymbol{a}_i\| - \|\boldsymbol{z}\|}{t} = \sum_{i=1}^{r} \lambda_i \lim_{t \to 0} \frac{\|\boldsymbol{z} - t\boldsymbol{a}_i\| - \|\boldsymbol{z}\|}{t}$$
$$= \sum_{i=1}^{r} \lambda_i \Big(-\frac{\boldsymbol{z} \cdot \boldsymbol{a}_i}{\|\boldsymbol{z}\|} \Big) = -\frac{\boldsymbol{z} \cdot (\lambda_1 \boldsymbol{a}_1 + \cdots + \lambda_r \boldsymbol{a}_r)}{\|\boldsymbol{z}\|} = -\frac{\boldsymbol{z} \cdot \boldsymbol{o}}{\|\boldsymbol{z}\|} = 0.$$

Hence there is a sufficiently large scalar γ_0 such that

$$\sum_{i=1}^{r} \lambda_i (\|\gamma_0 \boldsymbol{z} - \boldsymbol{a}_i\| - \|\gamma_0 \boldsymbol{z}\|) < \|\boldsymbol{a}\|.$$

Finally, let $\boldsymbol{x} = \gamma_0 \boldsymbol{z}$. Then $\boldsymbol{o} \in (\boldsymbol{x}, \boldsymbol{a})$, which gives

$$\sum_{i=1}^{r} \lambda_i \|\boldsymbol{x} - \boldsymbol{a}_i\| < \sum_{i=1}^{r} \lambda_i \|\boldsymbol{x}\| + \|\boldsymbol{a}\| = \|\boldsymbol{x}\| + \|\boldsymbol{a}\| = \|\boldsymbol{x} - \boldsymbol{a}\|,$$

contrary to the hypothesis. Hence $\boldsymbol{a}$ must be $\boldsymbol{o}$.

Problem 3.21. First, we will show the convexity of the solid quadrics.

1. *Solid ellipsoid.* If $\boldsymbol{y}, \boldsymbol{z} \in A$, then $\boldsymbol{y} = (y_1, \dots, y_n)$ and $\boldsymbol{z} = (z_1, \dots, z_n)$ such that

$$a_1 y_1^2 + \cdots + a_n y_n^2 \leqslant 1, \quad a_1 z_1^2 + \cdots + a_n z_n^2 \leqslant 1.$$

For every scalar $\lambda \in [0, 1]$, one has

$$(1 - \lambda)\boldsymbol{y} + \lambda \boldsymbol{z} = ((1 - \lambda)y_1 + \lambda z_1, \dots, (1 - \lambda)y_n + \lambda z_n).$$

Using the obvious inequalities

$$((1 - \lambda)y_i + \lambda z_i)^2 \leqslant (1 - \lambda)y_i^2 + \lambda z_i^2, \quad 1 \leqslant i \leqslant n, \tag{14.11}$$

we obtain

$$\begin{aligned} &a_1((1 - \lambda)y_1 + \lambda z_1)^2 + \cdots + a_n((1 - \lambda)y_n + \lambda z_n)^2 \\ &\quad \leqslant (1 - \lambda)(a_1 y_1^2 + \cdots + a_n y_n^2) + \lambda(a_1 z_1^2 + \cdots + a_n z_n^2) \\ &\quad \leqslant (1 - \lambda) + \lambda = 1. \end{aligned}$$

Hence $(1 - \lambda)\boldsymbol{y} + \lambda \boldsymbol{z} \in A$, which shows that A is convex.

2. *Solid elliptic paraboloid.* If $\boldsymbol{y}, \boldsymbol{z} \in B$, then $\boldsymbol{y} = (y_1, \ldots, y_n)$ and $\boldsymbol{z} = (z_1, \ldots, z_n)$ such that

$$a_1 y_1^2 + \cdots + a_{n-1} y_{n-1}^2 \leqslant a_n y_n, \quad a_1 z_1^2 + \cdots + a_{n-1} z_{n-1}^2 \leqslant a_n z_n.$$

For every scalar $\lambda \in [0, 1]$, one has

$$(1-\lambda)\boldsymbol{y} + \lambda \boldsymbol{z} = ((1-\lambda)y_1 + \lambda z_1, \ldots, (1-\lambda)y_n + \lambda z_n).$$

From (14.11) we obtain

$$\begin{aligned} &a_1((1-\lambda)y_1 + \lambda z_1)^2 + \cdots + a_{n-1}((1-\lambda)y_{n-1} + \lambda z_{n-1})^2 \\ &\quad \leqslant (1-\lambda)(a_1 y_1^2 + \cdots + a_{n-1} y_{n-1}^2) + \lambda(a_1 z_1^2 + \cdots + a_{n-1} z_{n-1}^2) \\ &\quad \leqslant a_n((1-\lambda)y_n + \lambda z_n). \end{aligned}$$

Hence $(1-\lambda)\boldsymbol{y} + \lambda \boldsymbol{z} \in B$, which shows that B is convex.

3. *Solid elliptic cone.* If $\boldsymbol{y}, \boldsymbol{z} \in C$, then $\boldsymbol{y} = (y_1, \ldots, y_n)$ and $\boldsymbol{z} = (z_1, \ldots, z_n)$ such that $y_n, z_n \geqslant 0$ and

$$a_1 y_1^2 + \cdots + a_{n-1} y_{n-1}^2 \leqslant a_n y_n^2, \quad a_1 z_1^2 + \cdots + a_{n-1} z_{n-1}^2 \leqslant a_n z_n^2.$$

For every scalar $\lambda \in [0, 1]$, one has

$$(1-\lambda)\boldsymbol{y} + \lambda \boldsymbol{z} = ((1-\lambda)y_1 + \lambda z_1, \ldots, (1-\lambda)y_n + \lambda z_n).$$

Clearly, $(1-\lambda)y_n + \lambda z_n \geqslant 0$. The Cauchy-Schwarz inequality

$$(\alpha_1 \beta_1 + \cdots + \alpha_{n-1} \beta_{n-1})^2 \leqslant (\alpha_1^2 + \cdots + \alpha_{n-1}^2)(\beta_1^2 + \cdots + \beta_{n-1}^2)$$

gives

$$\begin{aligned} &a_1 y_1 z_1 + \cdots + a_{n-1} y_{n-1} z_{n-1} \\ &= (\sqrt{a_1} y_1)(\sqrt{a_1} z_1) + \cdots + (\sqrt{a_{n-1}} y_{n-1})(\sqrt{a_{n-1}} z_{n-1}) \\ &\leqslant [(a_1 y_1^2 + \cdots + a_{n-1} y_{n-1}^2)(a_1 z_1^2 + \cdots + a_{n-1} z_{n-1}^2)]^{1/2} \\ &\leqslant \left[(a_n y_n^2)(a_n z_n^2)\right]^{1/2} = a_n y_n z_n. \end{aligned}$$

This inequality gives

$$\begin{aligned}
&a_1((1-\lambda)y_1+\lambda z_1)^2+\cdots+a_{n-1}((1-\lambda)y_1+\lambda z_1)^2\\
&=(1-\lambda)^2(a_1y_1^2+\cdots+a_{n-1}y_{n-1}^2)\\
&\quad+2(1-\lambda)\lambda(a_1y_1z_1+\cdots+a_{n-1}y_{n-1}z_{n-1})\\
&\quad+\lambda^2(a_1z_1^2+\cdots+a_{n-1}z_{n-1}^2)\\
&\leqslant(1-\lambda)^2a_ny_n^2+2(1-\lambda)\lambda\,a_ny_nz_n+\lambda^2a_nz_n^2\\
&=a_n((1-\lambda)y_n+\lambda z_n)^2.
\end{aligned}$$

Hence $(1-\lambda)\boldsymbol{y}+\lambda\boldsymbol{z}\in C$, which shows that C is convex.

4. *Solid elliptic hyperboloid.* If $\boldsymbol{y},\boldsymbol{z}\in D$, then $\boldsymbol{y}=(y_1,\ldots,y_n)$ and $\boldsymbol{z}=(z_1,\ldots,z_n)$ such that $y_n,z_n>0$ and

$$a_1y_1^2+\cdots+a_{n-1}y_{n-1}^2+1\leqslant a_ny_n^2,\quad a_1z_1^2+\cdots+a_{n-1}z_{n-1}^2+1\leqslant a_nz_n^2.$$

For every scalar $\lambda\in[0,1]$, one has

$$(1-\lambda)\boldsymbol{y}+\lambda\boldsymbol{z}=((1-\lambda)y_1+\lambda z_1,\ldots,(1-\lambda)y_n+\lambda z_n).$$

Clearly, $(1-\lambda)y_n+\lambda z_n>0$. The Cauchy-Schwarz inequality

$$(\alpha_1\beta_1+\cdots+\alpha_n\beta_n)^2\leqslant(\alpha_1^2+\cdots+\alpha_n^2)(\beta_1^2+\cdots+\beta_n^2)$$

gives

$$\begin{aligned}
&a_1y_1z_1+\cdots+a_{n-1}y_{n-1}z_{n-1}+1\\
&=(\sqrt{a_1}y_1)(\sqrt{a_1}z_1)+\cdots+(\sqrt{a_{n-1}}y_{n-1})(\sqrt{a_{n-1}}z_{n-1})+1\\
&\leqslant[(a_1y_1^2+\cdots+a_{n-1}y_{n-1}^2+1)(a_1z_1^2+\cdots+a_{n-1}z_{n-1}^2+1)]^{1/2}\\
&\leqslant\left[(a_ny_n^2)(a_nz_n^2)\right]^{1/2}=a_ny_nz_n.
\end{aligned}$$

This inequality gives

$$\begin{aligned}
&a_1((1-\lambda)y_1+\lambda z_1)^2+\cdots+a_{n-1}((1-\lambda)y_1+\lambda z_1)^2+1\\
&=(1-\lambda)^2(a_1y_1^2+\cdots+a_{n-1}y_{n-1}^2+1)\\
&\quad+2(1-\lambda)\lambda(a_1y_1z_1+\cdots+a_{n-1}y_{n-1}z_{n-1}+1)\\
&\quad+\lambda^2(a_1z_1^2+\cdots+a_{n-1}z_{n-1}^2+1)\\
&\leqslant(1-\lambda)^2a_ny_n^2+2(1-\lambda)\lambda\,a_ny_nz_n+\lambda^2a_nz_n^2\\
&=a_n((1-\lambda)y_n+\lambda z_n)^2.
\end{aligned}$$

Hence $(1-\lambda)\boldsymbol{y}+\lambda\boldsymbol{z}\in D$, which shows that D is convex.

Chapter 4

Problem 4.1. As shown in an example on page 75, the closed ball $B_\rho(\boldsymbol{a})$ is a convex set, which gives the inclusion $\mathrm{conv}_2 S_\rho(\boldsymbol{a}) \subset B_\rho(\boldsymbol{a})$ (see page 129). Conversely, let $\boldsymbol{u} \in B_\rho(\boldsymbol{a})$. If $\boldsymbol{u} = \boldsymbol{a}$, then choose any point $\boldsymbol{v} \in S_\rho(\boldsymbol{a})$ and let $\boldsymbol{w} = 2\boldsymbol{a} - \boldsymbol{v}$. From $\|\boldsymbol{w} - \boldsymbol{a}\| = \|\boldsymbol{u} - \boldsymbol{a}\| = \rho$ we conclude that $\boldsymbol{w} \in S_\rho(\boldsymbol{a})$. Furthermore, $\boldsymbol{a} = \frac{1}{2}(\boldsymbol{u} + \boldsymbol{w}) \in \mathrm{conv}_2 S_\rho(\boldsymbol{a})$. Suppose that $\boldsymbol{u} \neq \boldsymbol{a}$ and put

$$\boldsymbol{u}_1 = \boldsymbol{a} - \frac{\rho}{\|\boldsymbol{u} - \boldsymbol{a}\|}(\boldsymbol{u} - \boldsymbol{a}), \quad \boldsymbol{u}_2 = \boldsymbol{a} + \frac{\rho}{\|\boldsymbol{u} - \boldsymbol{a}\|}(\boldsymbol{u} - \boldsymbol{a}).$$

Then $\|\boldsymbol{u}_1 - \boldsymbol{a}\| = \|\boldsymbol{u}_2 - \boldsymbol{a}\| = \rho$, implying the inclusions $\boldsymbol{u}_1, \boldsymbol{u}_2 \in S_\rho(\boldsymbol{a})$. With $\lambda = (\rho + \|\boldsymbol{u} - \boldsymbol{a}\|)/(2\rho)$, one has $0 \leqslant \lambda \leqslant 1$ and

$$\begin{aligned}\boldsymbol{u} &= \frac{\rho - \|\boldsymbol{u} - \boldsymbol{a}\|}{2\rho}\Big(\boldsymbol{a} - \frac{\rho}{\|\boldsymbol{u} - \boldsymbol{a}\|}(\boldsymbol{u} - \boldsymbol{a})\Big) \\ &\quad + \frac{\rho + \|\boldsymbol{u} - \boldsymbol{a}\|}{2\rho}\Big(\boldsymbol{a} + \frac{\rho}{\|\boldsymbol{u} - \boldsymbol{a}\|}(\boldsymbol{u} - \boldsymbol{a})\Big) \\ &= \frac{\rho - \|\boldsymbol{u} - \boldsymbol{a}\|}{2\rho}\boldsymbol{u}_1 + \frac{\rho + \|\boldsymbol{u} - \boldsymbol{a}\|}{2\rho}\boldsymbol{u}_2 = (1 - \lambda)\boldsymbol{u}_1 + \lambda \boldsymbol{u}_2 \in \mathrm{conv}_2 S_\rho(\boldsymbol{a}).\end{aligned}$$

Problem 4.2. Without loss of generality, we may put $X = \{\boldsymbol{x}_1, \dots, \boldsymbol{x}_t\}$, where $t \leqslant r$. Since Δ is a convex set (see Theorem 3.6), one has $\mathrm{conv}_2(\Delta_X \cup \Delta_Y) \subset \Delta$. For the opposite inclusion, choose any point $\boldsymbol{u} \in \Delta$. Then $\boldsymbol{u}$ can be written as a convex combination $\boldsymbol{u} = \lambda_1 \boldsymbol{x}_1 + \cdots + \lambda_{r+1}\boldsymbol{x}_{r+1}$. Put $\lambda = \lambda_1 + \cdots + \lambda_t$. Clearly, $\boldsymbol{u} \in \Delta_Y \subset \Delta$ if $\lambda = 0$, and $\boldsymbol{u} \in \Delta$ if $\lambda = 1$. Suppose that $0 < \lambda < 1$ and consider the convex combinations

$$\boldsymbol{x} = \frac{\lambda_1}{\lambda}\boldsymbol{x}_1 + \cdots + \frac{\lambda_t}{\lambda}\boldsymbol{x}_t \quad \text{and} \quad \boldsymbol{y} = \frac{\lambda_{t+1}}{1 - \lambda}\boldsymbol{x}_{t+1} + \cdots + \frac{\lambda_{r+1}}{1 - \lambda}\boldsymbol{x}_{r+1}.$$

Then $\boldsymbol{x} \in \Delta_X$, $\boldsymbol{y} \in \Delta_Y$, and $\boldsymbol{u} = \lambda\boldsymbol{x} + (1 - \lambda)\boldsymbol{y} \in [\boldsymbol{x}, \boldsymbol{y}] \subset \mathrm{conv}_2(\Delta_X \cup \Delta_Y)$.

Problem 4.3. The set $\mathrm{int}\, Q$ is convex as the intersection of open halfspaces

$$W_i = \{(x_1, \dots, x_n) : x_i > 0\}, \quad W_i' = \{(x_1, \dots, x_n) : x_i < 1\}, \quad 1 \leqslant i \leqslant n,$$

(see an example on page 75 and Theorem 3.8). So, $\mathrm{conv}_2 X \subset \mathrm{conv}\, X \subset \mathrm{int}\, Q$. For the opposite inclusions, choose any point $\boldsymbol{u} = (u_1, \dots, u_n) \in \mathrm{int}\, Q$. Reordering the coordinates in $\mathbb{R}^n$, we may assume that all $u_1, \dots, u_k$ are rational and all $u_{k+1}, \dots, u_n$ are irrational (possibly, $k = 0$ or $k = n$). If $k = 0$, then $\boldsymbol{u} \in X \subset \mathrm{conv}_2 X$. Suppose that $k \geqslant 1$. Choose an irrational number $\xi > 0$ so small that $0 < u_i - \xi$ and $u_i + \xi < 1$ for all $1 \leqslant i \leqslant k$. Put

$$\boldsymbol{x} = (u_1', \dots, u_k', u_{k+1}, \dots, u_n) \quad \text{and} \quad \boldsymbol{z} = (u_1'', \dots, u_k'', u_{k+1}, \dots, u_n),$$

where $u_i' = u_i + \xi$ and $u_i'' = u_i - \xi$, $1 \leqslant i \leqslant k$. Clearly, $\boldsymbol{x}, \boldsymbol{z} \in X$ and $\boldsymbol{u} = (\boldsymbol{x} + \boldsymbol{z})/2$. Hence $\boldsymbol{u} \in \mathrm{conv}_2 X$.

Problem 4.4. Translating all sets M, D, and h on $-\boldsymbol{a}$, we may assume that $\boldsymbol{a} = \boldsymbol{o}$. The inclusions $\operatorname{conv}_2(h \cup M) \subset \operatorname{conv}(h \cup M) \subset D$ follow from the convexity of D (see an example on page 75). For the opposite inclusions, choose any point $\boldsymbol{x} \in D$. By Theorem 2.74, $D = h + M$. Consequently, $\boldsymbol{x} = \boldsymbol{u} + \boldsymbol{v}$, where $\boldsymbol{u} \in h$ and $\boldsymbol{v} \in M$. Since $2\boldsymbol{u} \in h$ and $2\boldsymbol{v} \in M$, one has $\boldsymbol{x} = \frac{1}{2}(2\boldsymbol{u} + 2\boldsymbol{v}) \in [\boldsymbol{u}, \boldsymbol{v}] \subset \operatorname{conv}_2(h \cup M)$.

Problem 4.5. (*a*) Let $\boldsymbol{c} \in L_1 \cap L_2$. A combination of Theorem 2.49 and Corollary 4.14 shows that it suffices to prove the equality

$$\operatorname{conv}_2(L_1 \cup L_2) = L_1 + L_2 - \boldsymbol{c}.$$

Theorem 4.12 shows that this equality can be rewritten as

$$\operatorname{conv}_2(S_1 \cup S_2) = S_1 + S_2,$$

where $S_1 = L_1 - \boldsymbol{c}$ and $S_2 = L_2 - \boldsymbol{c}$. Clearly, both S_1 and S_2 are subspaces. Because the subspace $S_1 + S_2$ is a convex set which contains $S_1 \cup S_2$, one has $\operatorname{conv}_2(S_1 \cup S_2) \subset S_1 + S_2$. For the opposite inclusion, choose a point $\boldsymbol{u} \in S_1 + S_2$. Then $\boldsymbol{u} = \boldsymbol{u}_1 + \boldsymbol{u}_2$, where $\boldsymbol{u}_1 \in S_1$ and $\boldsymbol{u}_2 \in S_2$. Consequently, $2\boldsymbol{u}_1 \in S_1$ and $2\boldsymbol{u}_2 \in S_2$, which gives

$$\boldsymbol{u} = \tfrac{1}{2}(2\boldsymbol{u}_1 + 2\boldsymbol{u}_2) \in [2\boldsymbol{u}_1, 2\boldsymbol{u}_2] \subset \operatorname{conv}_2(S_1 \cup S_2).$$

(*b*) As above, Corollary 4.14 shows that it suffices to prove the equality

$$\operatorname{conv}_2(L_1 \cup L_2) = L_1 \cup G \cup L_2.$$

Since all the sets L_1, G, L_2 are convex, and since L_1 and L_2 belong to the opposite components of $\operatorname{rbd} G$, we easily conclude that the set $L_1 \cup G \cup L_2$ is convex. Consequently, $\operatorname{conv}_2(L_1 \cup L_2) \subset L_1 \cup G \cup L_2$.

Conversely, since $L_1 \cup L_2 \subset \operatorname{conv}_2(L_1 \cup L_2)$, it suffices to prove the inclusion $G \subset \operatorname{conv}_2(L_1 \cup L_2)$. For this, choose a point $\boldsymbol{x} \in G$. Assertion (*a*) of Theorem 2.49 implies that $\boldsymbol{x} \in \langle \boldsymbol{c}_1, \boldsymbol{c}_2 \rangle$ for suitable points $\boldsymbol{c}_1 \in L_1$ and $\boldsymbol{c}_2 \in L_2$. Because G lies strictly between L_1' and L_2', we conclude that, in fact, $\boldsymbol{x} \in (\boldsymbol{c}_1, \boldsymbol{c}_2) \subset \operatorname{conv}_2(L_1 \cup L_2)$.

Problem 4.6. If $L_1 \cap L_2 \neq \varnothing$, then, by Problem 4.5, $\operatorname{conv}(L_1 \cup L_2)$ coincides with the plane $\operatorname{aff}(L_1 \cup L_2)$, which is a closed set (see Corollary 2.21). If L_1 and a translate of L_2, then, by the same theorem, $\operatorname{conv}(L_1 \cup L_2)$ is the closed slab of $\operatorname{aff}(L_1 \cup L_2)$ bounded by L_1 and L_2, which also is a closed set.

Conversely, let $\operatorname{conv}(L_1 \cup L_2)$ be a closed set. If $L_1 \cap L_2 = \varnothing$, then, by Problem 4.5, $\operatorname{conv}(L_1 \cup L_2) = L_1 \cup G \cup L_2$, where G is the open slab of $\operatorname{aff}(L_1 \cup L_2)$ bounded by the parallel planes $L_1' = L_1 + \operatorname{dir} L_2$ and $L_2' = L_2 + \operatorname{dir} L_1$ (see Theorem 2.15). Since the closure of $\operatorname{conv}(L_1 \cup L_2)$ equals $L_1' \cup G \cup L_2'$, one has $L_1 = L_1'$ and $L_2 = L_2$. The latter is possible if and only if $\operatorname{dir} L_1 = \operatorname{dir} L_2$, that is, when L_1 is a translate of L_2.

Problem 4.7. Since $X \subset \operatorname{conv} X$, we have $\operatorname{diam} X \leqslant \operatorname{diam}(\operatorname{conv} X)$. For the opposite inequality, choose points $\boldsymbol{u}$ and $\boldsymbol{v}$ in $\operatorname{conv} X$. By Theorem 4.3, both $\boldsymbol{u}$ and $\boldsymbol{v}$ are convex combinations of points from X: $\boldsymbol{u} = \lambda_1 \boldsymbol{y}_1 + \cdots + \lambda_k \boldsymbol{y}_k$ and $\boldsymbol{v} = \mu_1 \boldsymbol{z}_1 + \cdots + \mu_m \boldsymbol{z}_m$. Then

$$\begin{aligned}
\|\boldsymbol{u} - \boldsymbol{v}\| &= \Big\|\boldsymbol{u} - \sum_{i=1}^{m} \mu_i \boldsymbol{z}_i\Big\| = \Big\|\sum_{i=1}^{m} \mu_i (\boldsymbol{u} - \boldsymbol{z}_i)\Big\| \leqslant \sum_{i=1}^{m} \mu_i \big\|\boldsymbol{u} - \boldsymbol{z}_i\big\| \\
&= \sum_{i=1}^{m} \mu_i \Big\|\sum_{j=1}^{k} \lambda_j \boldsymbol{y}_j - \boldsymbol{z}_i\Big\| = \sum_{i=1}^{m} \mu_i \Big\|\sum_{j=1}^{k} \lambda_j (\boldsymbol{y}_j - \boldsymbol{z}_i)\Big\| \\
&\leqslant \sum_{i=1}^{m} \mu_i \sum_{j=1}^{k} \lambda_j \big\|\boldsymbol{y}_j - \boldsymbol{z}_i\big\| \leqslant \sum_{i=1}^{m} \mu_i \sum_{j=1}^{k} \lambda_j \operatorname{diam} X = \operatorname{diam} X.
\end{aligned}$$

Hence $\operatorname{diam}(\operatorname{conv} X) = \sup\{\|\boldsymbol{u} - \boldsymbol{v}\| : \boldsymbol{u}, \boldsymbol{v} \in \operatorname{conv} X\} \leqslant \operatorname{diam} X$.

Problem 4.8. Since the closed ball $B_\rho(\boldsymbol{o})$ is convex (see examples on page 75), Theorem 4.12 gives

$$B_\rho(\operatorname{conv} X) = B_\rho(\boldsymbol{o}) + \operatorname{conv} X = \operatorname{conv}(B_\rho(\boldsymbol{o}) + X) = \operatorname{conv} B_\rho(X).$$

The proof of the second equality, $\operatorname{conv} U_\rho(X) = U_\rho(\operatorname{conv} X)$, is similar.

Problem 4.9. Excluding the obvious case when at least one of the sets K, X, Y is empty, we may assume that all three are nonempty. A combination of Theorems 3.4 and 4.12 and of Corollary 4.14 gives

$$\begin{aligned}
&\operatorname{conv}((K + X) \cup (K + Y)) \\
&\quad = \cup(\lambda_1 \operatorname{conv}(K + X) + \lambda_2 \operatorname{conv}(K + Y) : \lambda_1, \lambda_2 \geqslant 0, \lambda_1 + \lambda_2 = 1) \\
&\quad = \cup(\lambda_1 (K + \operatorname{conv} X) + \lambda_2 (K + \operatorname{conv} Y) : \lambda_1, \lambda_2 \geqslant 0, \lambda_1 + \lambda_2 = 1) \\
&\quad = \cup(K + \lambda_1 \operatorname{conv} X + \lambda_2 \operatorname{conv} Y : \lambda_1, \lambda_2 \geqslant 0, \lambda_1 + \lambda_2 = 1) \\
&\quad = K + \operatorname{conv}(X \cup Y).
\end{aligned}$$

Problem 4.10. According to Corollary 4.14, the inclusion $\boldsymbol{x} \in \operatorname{conv}(\{\boldsymbol{y}\} \cup K)$ gives $\boldsymbol{x} = (1 - \lambda)\boldsymbol{y} + \lambda \boldsymbol{u}$ for a suitable point $\boldsymbol{u} \in K$ and a scalar $\lambda \in [0, 1]$. Similarly, $\boldsymbol{y} \in \operatorname{conv}(\{\boldsymbol{x}\} \cup K)$ gives $\boldsymbol{y} = (1 - \mu)\boldsymbol{x} + \mu \boldsymbol{z}$ for a point $\boldsymbol{z} \in K$ and a scalar $\mu \in [0, 1]$. Assume for a moment that $\boldsymbol{x} \neq \boldsymbol{y}$. Then $\lambda\mu \neq 0$, which gives $0 < \lambda(1 - \mu) + \mu \leqslant 1$. From

$$\begin{aligned}
\boldsymbol{x} &= (1 - \lambda)\boldsymbol{y} + \lambda \boldsymbol{u} = (1 - \lambda)((1 - \mu)\boldsymbol{x} + \mu \boldsymbol{z}) + \lambda \boldsymbol{u} \\
&= (1 - \lambda)(1 - \mu)\boldsymbol{x} + (1 - \lambda)\mu \boldsymbol{z} + \lambda \boldsymbol{u},
\end{aligned}$$

we obtain that $\boldsymbol{x} = \dfrac{\mu - \lambda\mu}{\lambda + \mu - \lambda\mu} \boldsymbol{z} + \dfrac{\lambda}{\lambda + \mu - \lambda\mu} \boldsymbol{u} \in [\boldsymbol{z}, \boldsymbol{u}] \subset K$ by the convexity of K. The last contradicts the assumption $\boldsymbol{x} \notin K$. Consequently, $\boldsymbol{x} = \boldsymbol{y}$.

Problem 4.11. We will prove the contrapositive assertion. Let a set $X = \{\boldsymbol{x}_1, \dots, \boldsymbol{x}_r\}$ be affinely dependent. Then there are scalars $\nu_1, \dots, \nu_r$, not all zero, such that $\nu_1 \boldsymbol{x}_1 + \cdots + \nu_r \boldsymbol{x}_r = \boldsymbol{o}$ and $\nu_1 + \cdots + \nu_r = 0$. Put $I = \{i : \nu_i \geqslant 0\}$ and $J = \{i : \nu_i < 0\}$. Then both sets I and J are nonempty and partition the set $\{1, \dots, r\}$. Furthermore,

$$\sum_{i \in I} \nu_i \boldsymbol{x}_i = \sum_{i \in J} (-\nu_j)\, \boldsymbol{x}_j. \tag{14.12}$$

Put $\nu = \sum_{i \in I} \nu_i$. Clearly, $\nu > 0$ and $\sum_{i \in J} (-\nu_j) = \nu$. Dividing both parts of (14.12) by ν, we obtain

$$\sum_{i \in I} \frac{\nu_i}{\nu} \boldsymbol{x}_i = \sum_{i \in J} \frac{(-\nu_j)}{\nu} \boldsymbol{x}_j. \tag{14.13}$$

The left-hand side of (14.13) is a convex combination of points from the set $Y = \{\boldsymbol{x}_i : i \in I\}$; while the right-hand side of (14.13) is a convex combination of points from the set $Z = \{\boldsymbol{x}_j : j \in J\}$. Hence $Y \cap Z = \varnothing$ and $\operatorname{conv} Y \cap \operatorname{conv} Z \neq \varnothing$.

Since $\varnothing \neq \operatorname{conv} Y \cap \operatorname{conv} Z \subset \operatorname{aff} Y \cap \operatorname{aff} Z$, it remains to show that $(c) \Rightarrow (a)$. So, let $X = Y \cup Z$ be a partition of X such that $\operatorname{aff} Y \cap \operatorname{aff} Z \neq \varnothing$. Choose a point $\boldsymbol{x} \in \operatorname{aff} Y \cap \operatorname{aff} Z$. By Theorem 2.44, $\boldsymbol{x}$ can be expressed as affine combinations, $\boldsymbol{x} = \lambda_1 \boldsymbol{y}_1 + \cdots + \lambda_k \boldsymbol{y}_k = \mu_1 \boldsymbol{z}_1 + \cdots + \mu_m \boldsymbol{z}_m$, of suitable points $\boldsymbol{y}_1, \dots, \boldsymbol{y}_k \in Y$ and $\boldsymbol{z}_1, \dots, \boldsymbol{z}_m \in Z$. Definition 2.50 and the equalities

$$\begin{gathered} \lambda_1 \boldsymbol{y}_1 + \cdots + \lambda_k \boldsymbol{y}_k - \mu_1 \boldsymbol{z}_1 - \cdots - \mu_m \boldsymbol{z}_m = \boldsymbol{o}, \\ \lambda_1 + \cdots + \lambda_k - \mu_1 - \cdots - \mu_m = 0 \end{gathered}$$

show that the set $\{\boldsymbol{y}_1, \dots, \boldsymbol{y}_k, \boldsymbol{z}_1, \dots, \boldsymbol{z}_m\}$ is affinely dependent. Hence X is affinely dependent.

Problem 4.12. By Theorem 4.17, $\operatorname{conv}(\operatorname{rint} X) \subset \operatorname{rint}(\operatorname{conv} X)$. Hence it suffices to prove the opposite inclusion, $\operatorname{rint}(\operatorname{conv} X) \subset \operatorname{conv}(\operatorname{rint} X)$.

We observe that $\operatorname{aff} X = \operatorname{aff}(\operatorname{rint} X) = \operatorname{aff}(\operatorname{cl}(\operatorname{rint} X))$. Indeed, $\operatorname{aff}(\operatorname{rint} X) \subset \operatorname{aff} X \subset \operatorname{aff}(\operatorname{cl}(\operatorname{rint} X))$ due to the assumption $X \subset \operatorname{cl}(\operatorname{rint} X)$. The opposite inclusion follows from Corollary 2.43: $\operatorname{aff}(\operatorname{cl}(\operatorname{rint} X)) = \operatorname{aff}(\operatorname{rint} X) \subset \operatorname{aff} X$. This argument and Theorem 4.2 imply that

$$\operatorname{aff}(\operatorname{conv}(\operatorname{rint} X)) = \operatorname{aff}(\operatorname{conv} X) = \operatorname{aff}(\operatorname{conv}(\operatorname{cl}(\operatorname{rint} X))).$$

Now, a combination of Theorems 3.18, 3.48, and 4.17 gives

$$\begin{aligned} \operatorname{rint}(\operatorname{conv} X) &\subset \operatorname{rint}(\operatorname{conv}(\operatorname{cl}(\operatorname{rint} X))) \subset \operatorname{rint}(\operatorname{cl}(\operatorname{conv}(\operatorname{rint} X))) \\ &= \operatorname{rint}(\operatorname{conv}(\operatorname{rint} X)) \subset \operatorname{conv}(\operatorname{rint} X). \end{aligned}$$

Problem 4.13. Since $X \subset Y \subset \operatorname{cl} X$, a combination of Theorems 4.2 and 4.16 gives

$$\operatorname{conv} X \subset \operatorname{conv} Y \subset \operatorname{conv}(\operatorname{cl} X) \subset \operatorname{cl}(\operatorname{conv} X).$$

So, $\mathrm{cl}\,(\mathrm{conv}\, X) = \mathrm{cl}\,(\mathrm{conv}\, Y)$. Next, the equality $\mathrm{rint}\,(\mathrm{conv}\, X) = \mathrm{rint}\,(\mathrm{conv}\, Y)$ follows from Corollary 3.49.

Problem 4.14. By Theorem 4.12, $\boldsymbol{x} = \mu_1\boldsymbol{x}_1 + \cdots + \mu_r\boldsymbol{x}_r$, where $\boldsymbol{x}_i \in \mathrm{conv}\, X_i$, $1 \leqslant i \leqslant r$. Without loss of generality, we may assume that none of the scalars $\mu_1, \ldots, \mu_r$ is zero. Theorem 4.7 shows that every $\boldsymbol{x}_i$ is a convex combination, $\boldsymbol{x}_i = \lambda_1^{(i)}\boldsymbol{z}_1^{(i)} + \cdots + \lambda_{p_i}^{(i)}\boldsymbol{z}_{p_i}^{(i)}$, where

$$Z_i = \{\boldsymbol{z}_1^{(i)}, \ldots, \boldsymbol{z}_{p_i}^{(i)}\} \subset X_i,\ p_i^{(i)} \leqslant m+1,\ 1 \leqslant i \leqslant r.$$

Furthermore, choose $Z_1, \ldots, Z_r$ such that the sum $p_1 + \cdots + p_r$ is minimal possible. Then none of the coefficients $\lambda_1^{(i)}, \ldots, \lambda_{p_i}^{(i)}$, $1 \leqslant i \leqslant r$, is zero. Renumbering $X_1, \ldots, X_r$, we suppose that $p_1 \geqslant \cdots \geqslant p_r$.

We assert that Z_{m+1} is a singleton. Suppose for a moment that this is not the case. Then each of $Z_1, \ldots, Z_{m+1}$ contains two or more points, and $\boldsymbol{x}_i \neq \boldsymbol{z}_1^{(i)}$, $1 \leqslant i \leqslant m+1$. Since the set $\{\mu_1(\boldsymbol{x}_1 - \boldsymbol{z}_1^{(1)}), \ldots, \mu_{m+1}(\boldsymbol{x}_{m+1} - \boldsymbol{z}_1^{(m+1)})\}$ belongs to the m-dimensional subspace $\mathrm{dir}\,(\mu_1 X_1 + \cdots + \mu_r X_r)$, it is linearly dependent. Hence there are scalars $\gamma_1, \ldots, \gamma_{m+1}$, not all zero, such that

$$\gamma_1\mu_1(\boldsymbol{x}_1 - \boldsymbol{z}_1^{(1)}) + \cdots + \gamma_{m+1}\mu_{m+1}(\boldsymbol{x}_{m+1} - \boldsymbol{z}_1^{(m+1)}) = \boldsymbol{o}.$$

For a scalar $t \in \mathbb{R}$, put $\boldsymbol{y}_i(t) = \boldsymbol{x}_i + \gamma_i t(\boldsymbol{x}_i - \boldsymbol{z}_1^{(i)})$, $1 \leqslant i \leqslant m+1$. Then

$$\begin{aligned}\boldsymbol{y}_i(t) &= (1+\gamma_i t)\boldsymbol{x}_i - \gamma_i t\boldsymbol{z}_1^{(i)} = (1+\gamma_i t)(\lambda_1^{(i)}\boldsymbol{z}_1^{(i)} + \cdots + \lambda_{p_i}^{(i)}\boldsymbol{z}_{p_i}^{(i)}) - \gamma_i t\boldsymbol{z}_1^{(i)}\\ &= ((1+\gamma_i t)\lambda_1^{(i)} - \gamma_i t)\boldsymbol{z}_1^{(i)} + (1+\gamma_i t)\lambda_2^{(i)}\boldsymbol{z}_2^{(i)} + \cdots + (1+\gamma_i t)\lambda_{p_i}^{(i)}\boldsymbol{z}_{p_i}^{(i)},\end{aligned}$$

where $(1+\gamma_i t)\lambda_1^{(i)} - \gamma_i t + (1+\gamma_i t)\lambda_2^{(i)} + \cdots + (1+\gamma_i t)\lambda_{p_i}^{(i)} = 1$.

Since $\boldsymbol{y}_i(0) = \boldsymbol{x}_i$, there is a scalar $t_0 \neq 0$ such that all scalars

$$(1+\gamma_i t_0)\lambda_1^{(i)} - \gamma_i t_0, (1+\gamma_i t_0)\lambda_2^{(i)}, \ldots, (1+\gamma_i t_0)\lambda_{p_i}^{(i)},\ 1 \leqslant i \leqslant m+1,$$

are nonnegative and at least one of them is zero. In other words, $\boldsymbol{y}_i(t_0) \in \mathrm{conv}\, Z_i$, $1 \leqslant i \leqslant m+1$, and there is an index $i \in \{1, \ldots, m+1\}$ such that $\boldsymbol{y}_i(t_0)$ belongs to the convex hull of a proper subset of Z_i. Because

$$\boldsymbol{x} = \mu_1\boldsymbol{y}_1(t_0) + \cdots + \mu_{m+1}\boldsymbol{y}_{m+1}(t_0) + \mu_{m+2}\boldsymbol{x}_{m+2} + \cdots + \mu_r\boldsymbol{x}_r,$$

we obtained a contradiction with the minimality of $p_1 + \cdots + p_r$. Hence Z_{m+1} is a singleton. Consequently, all $Z_{m+2}, \ldots, Z_r$ are singletons by the assumption $p_{m+1} \geqslant \cdots \geqslant p_r$.

Problem 4.15. Denote by M the set of all points of the form

$$\lambda_1\boldsymbol{x}_1 + \cdots + \lambda_r\boldsymbol{x}_r, \text{ where } \boldsymbol{x}_i \in \mathrm{rint}\,(\mathrm{conv}\, X_i) \text{ for all } 1 \leqslant i \leqslant r. \qquad (14.14)$$

Then the statement is $M = \operatorname{rint}(\operatorname{conv}(X_1 \cup \cdots \cup X_r))$. First, we establish the convexity of M. Indeed, let

$$\boldsymbol{x} = \gamma \boldsymbol{x}_1 + \cdots + \gamma \boldsymbol{x}_r \quad \text{and} \quad \boldsymbol{y} = \mu_1 \boldsymbol{y}_1 + \cdots + \mu_r \boldsymbol{y}_r$$

be points from M, expressed in the form (14.14). For any $0 < \lambda < 1$, the point $\boldsymbol{z} = (1-\lambda)\boldsymbol{x} + \lambda \boldsymbol{y}$ can be written as a positive convex combination $\boldsymbol{z} = \eta_1 \boldsymbol{z}_1 + \cdots + \eta_r \boldsymbol{z}_r$, where

$$\eta_i = (1-\lambda)\gamma_i + \lambda \mu_i \quad \text{and} \quad \boldsymbol{z}_i = \frac{(1-\lambda)\gamma_i}{\eta_i}\boldsymbol{x}_i + \frac{\lambda \mu_i}{\eta_i}\boldsymbol{y}_i, \quad 1 \leqslant i \leqslant r.$$

By Theorem 3.24, $\boldsymbol{z}_i \in (\boldsymbol{x}_i, \boldsymbol{y}_i) \subset \operatorname{rint}(\operatorname{conv} X_i)$, $1 \leqslant i \leqslant r$, which implies the inclusion $\boldsymbol{z} \in M$. Summing up, M is a convex set.

Next, we will prove the inclusion $M \subset \operatorname{rint}(\operatorname{conv}(X_1 \cup \cdots \cup X_r))$. For this, choose a point $\boldsymbol{x} \in M$, expressed in the form (14.14). By Theorem 4.18, every $\boldsymbol{x}_i$ is a positive convex combination $\boldsymbol{x}_i = \mu_1^{(i)} \boldsymbol{y}_1^{(i)} + \cdots + \mu_{p_i}^{(i)} \boldsymbol{y}_{p_i}^{(i)}$, where

$$\boldsymbol{y}_1^{(i)}, \ldots, \boldsymbol{y}_{p_i}^{(i)} \in X_i \quad \text{and} \quad \operatorname{aff}\{\boldsymbol{y}_1^{(i)}, \ldots, \boldsymbol{y}_{p_i}^{(i)}\} = \operatorname{aff} X_i, \quad 1 \leqslant i \leqslant r.$$

Clearly,

$$\boldsymbol{x} = \lambda_1 \mu_1^{(1)} \boldsymbol{y}_1^{(1)} + \cdots + \lambda_1 \mu_{p_1}^{(1)} \boldsymbol{y}_{p_1}^{(1)} + \cdots + \lambda_r \mu_1^{(r)} \boldsymbol{y}_1^{(r)} + \cdots + \lambda_r \mu_{p_r}^{(r)} \boldsymbol{y}_{p_r}^{(r)}$$

is a positive convex combination of all points from the set

$$Y = \{\boldsymbol{y}_1^{(1)}, \ldots, \boldsymbol{y}_{p_1}^{(1)}, \ldots, \boldsymbol{y}_1^{(r)}, \ldots, \boldsymbol{y}_{p_r}^{(r)}\}.$$

Furthermore, Theorems 2.42 and 4.2 give

$$\begin{aligned} \operatorname{aff} Y &= \operatorname{aff}(\operatorname{aff}\{\boldsymbol{y}_1^{(1)}, \ldots, \boldsymbol{y}_{p_1}^{(1)}\} \cup \cdots \cup \operatorname{aff}\{\boldsymbol{y}_1^{(r)}, \ldots, \boldsymbol{y}_{p_r}^{(r)}\}) \\ &= \operatorname{aff}(\operatorname{aff} X_1 \cup \cdots \cup \operatorname{aff} X_r) = \operatorname{aff}(X_1 \cup \cdots \cup X_r) \\ &= \operatorname{aff}(\operatorname{conv}(X_1 \cup \cdots \cup X_r)). \end{aligned}$$

Therefore, $\boldsymbol{x} \in \operatorname{rint}(\operatorname{conv}(X_1 \cup \cdots \cup X_r))$ according to Theorem 4.18.

Next, we assert that $\operatorname{conv}(X_1 \cup \cdots \cup X_r) \subset \operatorname{cl} M$. For this, let $\boldsymbol{x} \in \operatorname{conv}(X_1 \cup \cdots \cup X_r)$. To prove the inclusion $\boldsymbol{x} \in \operatorname{cl} M$, it suffices to show that for every $\rho > 0$ there is a point $\boldsymbol{z} \in M$ such that $\|\boldsymbol{x} - \boldsymbol{z}\| < \rho$. By Theorem 4.13, $\boldsymbol{x}$ can be expressed as a convex combination

$$\boldsymbol{x} = \gamma_1 \boldsymbol{x}_1 + \cdots + \gamma_r \boldsymbol{x}_r, \quad \text{where} \quad \boldsymbol{x}_1 \in \operatorname{conv} X_1, \ldots, \boldsymbol{x}_r \in \operatorname{conv} X_r.$$

Renumbering the sets $X_1, \ldots, X_r$, we may assume that all positive scalars γ_i are placed before all zero scalars:

$$\gamma_1, \ldots, \gamma_s > 0 \quad \text{and} \quad \gamma_{s+1} = \cdots = \gamma_r = 0, \quad 1 \leqslant s \leqslant r.$$

Consider first the case $s = r$. Since $\operatorname{conv} X_i \subset \operatorname{cl}(\operatorname{rint}(\operatorname{conv} X_i))$ due to Theorem 3.48, there are points $\boldsymbol{z}_i \in \operatorname{rint}(\operatorname{conv} X_i)$ satisfying the conditions $\|\boldsymbol{x}_i - \boldsymbol{z}_i\| < \rho$ for all $1 \leqslant i \leqslant r$. Let $\boldsymbol{z} = \gamma_1 \boldsymbol{z}_1 + \cdots + \gamma_r \boldsymbol{z}_r$. Then $\boldsymbol{z} \in M$ and

$$\|\boldsymbol{x} - \boldsymbol{z}\| \leqslant \gamma_1 \|\boldsymbol{x}_1 - \boldsymbol{z}_1\| + \cdots + \gamma_r \|\boldsymbol{x}_r - \boldsymbol{z}_r\| < \rho.$$

Now, suppose that $s < r$ and put

$$\mu = \frac{\rho}{2}, \quad \delta = \max\{\|\boldsymbol{x}_1\|, \ldots, \|\boldsymbol{x}_r\|\}, \quad 0 < \varepsilon < \min\Big\{\gamma_1, \ldots, \gamma_s, \frac{\rho}{4\delta s}\Big\}.$$

As above, choose points $\boldsymbol{z}_i \in \operatorname{rint}(\operatorname{conv} X_i)$ such that $\|\boldsymbol{x}_i - \boldsymbol{z}_i\| < \mu$ for all $1 \leqslant i \leqslant r$. Then

$$\|\boldsymbol{z}_i\| \leqslant \|\boldsymbol{z}_i - \boldsymbol{x}_i\| + \|\boldsymbol{x}_i\| < \mu + \delta, \quad 1 \leqslant i \leqslant r.$$

Consider the positive convex combination

$$\boldsymbol{z} = (\gamma_1 - \varepsilon)\boldsymbol{z}_1 + \cdots + (\gamma_s - \varepsilon)\boldsymbol{z}_s + \frac{\varepsilon s}{r - s}\boldsymbol{z}_{s+1} + \cdots + \frac{\varepsilon s}{r - s}\boldsymbol{z}_r.$$

Clearly, $\boldsymbol{z} \in M$ and

$$\begin{aligned}
\|\boldsymbol{x} - \boldsymbol{z}\| &\leqslant \|\gamma_1 \boldsymbol{x}_1 - (\gamma_1 - \varepsilon)\boldsymbol{z}_1\| + \cdots + \|\gamma_s \boldsymbol{x}_s - (\gamma_s - \varepsilon)\boldsymbol{z}_s\| \\
&\quad + \frac{\varepsilon s}{r - s}\|\boldsymbol{z}_{s+1}\| + \cdots + \frac{\varepsilon s}{r - s}\|\boldsymbol{z}_r\| \\
&= \|(\gamma_1 - \varepsilon)(\boldsymbol{x}_1 - \boldsymbol{z}_1) + \varepsilon \boldsymbol{x}_1\| + \cdots + \|(\gamma_s - \varepsilon)(\boldsymbol{x}_s - \boldsymbol{z}_s) + \varepsilon \boldsymbol{x}_s\| \\
&\quad + \frac{\varepsilon s}{r - s}\big(\|\boldsymbol{z}_{s+1}\| + \cdots + \|\boldsymbol{z}_r\|\big) \\
&\leqslant (\gamma_1 - \varepsilon)\|\boldsymbol{x}_1 - \boldsymbol{z}_1\| + \cdots + (\gamma_s - \varepsilon)\|\boldsymbol{x}_s - \boldsymbol{z}_s\| \\
&\quad + \varepsilon\big(\|\boldsymbol{x}_1\| + \cdots + \|\boldsymbol{x}_s\|\big) + \frac{\varepsilon s}{r - s}\big(\|\boldsymbol{z}_{s+1}\| + \cdots + \|\boldsymbol{z}_r\|\big) \\
&< (\gamma_1 + \cdots + \gamma_s - \varepsilon s)\mu + \varepsilon s \delta + \frac{\varepsilon s}{r - s}(r - s)(\delta + \mu) \\
&= \mu + 2\varepsilon \delta s < \frac{\rho}{2} + 2\frac{\rho}{4\delta s}\delta s = \rho.
\end{aligned}$$

Summing up, $\boldsymbol{x} \in \operatorname{cl} M$, and $\operatorname{conv}(X_1 \cup \cdots \cup X_r) \subset \operatorname{cl} M$.

Finally, Theorem 3.48 gives

$$M \subset \operatorname{rint}(\operatorname{conv}(X_1 \cup \cdots \cup X_r)) \subset \operatorname{rint}(\operatorname{cl} M) = \operatorname{rint} M \subset M,$$

implying the equality $M = \operatorname{rint}(\operatorname{conv}(X_1 \cup \cdots \cup X_r))$.

Problem 4.16. The inclusion

$$\operatorname{rint}(\operatorname{conv}(K_1 \cup \cdots \cup K_r)) \subset \operatorname{conv}(\operatorname{rint} K_1 \cup \cdots \cup \operatorname{rint} K_r). \tag{14.15}$$

immediately follows from Problem 4.15. Assume that the relative interiors of $K_1, \ldots, K_r$ have a point $\boldsymbol{c}$ in common. Due to (14.15), it suffices to prove that

$$\text{conv}\,(\text{rint}\,K_1 \cup \cdots \cup \text{rint}\,K_r) \subset \text{rint}\,(\text{conv}\,(K_1 \cup \cdots \cup K_r)).$$

First, we assert that $\boldsymbol{c} \in \text{rint}\,(\text{conv}\,(K_1 \cup \cdots \cup K_r))$. Indeed, let $\boldsymbol{x}$ be any point in $\text{conv}\,(K_1 \cup \cdots \cup K_r)$. By Theorem 4.13, $\boldsymbol{x}$ is a convex combination, $\boldsymbol{x} = \lambda_1 \boldsymbol{x}_1 + \cdots + \lambda_r \boldsymbol{x}_r$, of suitable points $\boldsymbol{x}_i \in K_i$, $1 \leqslant i \leqslant r$. Theorem 3.26 shows the existence of scalars $\gamma_1, \ldots, \gamma_r > 1$ such that the point $\boldsymbol{z}_i = \gamma_i \boldsymbol{c} + (1 - \gamma_i)\boldsymbol{x}_i$ belongs to K_i for all $1 \leqslant i \leqslant r$. Replacing every γ_i with $\gamma = \min\{\gamma_1, \ldots, \gamma_r\}$, Lemma 2.25 and convexity of K_i imply the inclusions

$$\boldsymbol{y}_i = \gamma \boldsymbol{c} + (1 - \gamma)\boldsymbol{x}_i \in [\boldsymbol{c}, \boldsymbol{z}_i] \subset [\boldsymbol{c}, \boldsymbol{x}_i] \subset K_i \quad \text{for all} \quad 1 \leqslant i \leqslant r.$$

Finally, with $\boldsymbol{y} = \lambda_1 \boldsymbol{y}_1 + \cdots + \lambda_r \boldsymbol{y}_r$, we obtain

$$\boldsymbol{y} = \lambda_1(\gamma \boldsymbol{c} + (1 - \gamma)\boldsymbol{x}_i) + \cdots + \lambda_r(\gamma \boldsymbol{c} + (1 - \gamma)\boldsymbol{x}_r) = \gamma \boldsymbol{c} + (1 - \gamma)\boldsymbol{x}.$$

Since $0 < \gamma^{-1} < 1$, Theorem 3.26 gives

$$\boldsymbol{c} = (1 - \gamma^{-1})\boldsymbol{x} + \gamma^{-1}\boldsymbol{y} \in \text{rint}\,(\text{conv}\,(K_1 \cup \cdots \cup K_r)).$$

Next, choose any point $\boldsymbol{z} \in \text{conv}\,(\text{rint}\,K_1 \cup \cdots \cup \text{rint}\,K_r)$. As above, $\boldsymbol{z}$ is a convex combination, $\boldsymbol{z} = \mu_1 \boldsymbol{z}_1 + \cdots + \mu_r \boldsymbol{z}_r$, of some points $\boldsymbol{z}_i \in \text{rint}\,K_i$, $1 \leqslant i \leqslant r$. By Theorem 3.26, there are scalars $\eta_1, \ldots, \eta_r > 1$ such that the point $\boldsymbol{u}_i = \eta_i \boldsymbol{z}_i + (1 - \eta_i)\boldsymbol{c}$ belongs to K_i for all $1 \leqslant i \leqslant r$. As above, we may suppose that $\eta_1 = \cdots = \eta_r = \eta$. Let $\boldsymbol{u} = \mu_1 \boldsymbol{u}_1 + \cdots + \mu_r \boldsymbol{u}_r$. Then $\boldsymbol{u} \in \text{conv}\,(K_1 \cup \cdots \cup K_r)$, and Theorem 3.26 gives

$$\boldsymbol{z} = (1 - \eta^{-1})\boldsymbol{c} + \eta^{-1}\boldsymbol{u} \in \text{rint}\,(\text{conv}\,(K_1 \cup \cdots \cup K_r)).$$

Problem 4.17. Let $Y = \{\boldsymbol{y}_1, \ldots, \boldsymbol{y}_r\}$. We proceed by induction on $r \geqslant 2$. Let $r = 2$. By Theorem 4.8, $\boldsymbol{y}_1$ can be expressed as a convex combination $\boldsymbol{y}_1 = \lambda_1 \boldsymbol{x}_1 + \cdots + \lambda_p \boldsymbol{x}_p + \lambda_{p+1} \boldsymbol{y}_2$, there $\boldsymbol{x}_1, \ldots, \boldsymbol{x}_p \in X$, $p \leqslant m$. Similarly, there are q ($\leqslant m$) points $\boldsymbol{z}_2, \ldots, \boldsymbol{z}_{q+1} \in X$ such that $\boldsymbol{y}_2$ can be written as a convex combination $\boldsymbol{y}_2 = \mu_1 \boldsymbol{y}_1 + \mu_2 \boldsymbol{z}_2 + \cdots + \mu_{q+1} \boldsymbol{z}_{q+1}$. Both scalars λ_{p+1} and μ_1 are distinct from 1 due to $\boldsymbol{y}_1 \neq \boldsymbol{y}_2$. Solving the equality

$$\begin{aligned} \boldsymbol{y}_1 &= \lambda_1 \boldsymbol{x}_1 + \cdots + \lambda_p \boldsymbol{x}_p + \lambda_{p+1} \boldsymbol{y}_2 \\ &= \lambda_1 \boldsymbol{x}_1 + \cdots + \lambda_p \boldsymbol{x}_p + \lambda_{p+1}(\mu_1 \boldsymbol{y}_1 + \mu_2 \boldsymbol{z}_2 + \cdots + \mu_{q+1} \boldsymbol{z}_{q+1}) \end{aligned}$$

for $\boldsymbol{y}_1$, we obtain the convex combination

$$\begin{aligned} \boldsymbol{y}_1 = {} & \frac{\lambda_1}{1 - \lambda_{p+1}\mu_1}\boldsymbol{x}_1 + \cdots + \frac{\lambda_p}{1 - \lambda_{p+1}\mu_1}\boldsymbol{x}_p \\ & + \frac{\lambda_{p+1}\mu_2}{1 - \lambda_{p+1}\mu_1}\boldsymbol{z}_2 + \cdots + \frac{\lambda_{p+1}\mu_{q+1}}{1 - \lambda_{p+1}\mu_1}\boldsymbol{z}_{q+1}. \end{aligned}$$

Similarly, $\boldsymbol{y}_2$ can be expressed as a convex combination

$$\boldsymbol{y}_2 = \frac{\lambda_1\mu_1}{1-\lambda_{p+1}\mu_1}\boldsymbol{x}_1 + \cdots + \frac{\lambda_p\mu_1}{1-\lambda_{p+1}\mu_1}\boldsymbol{x}_p + \frac{\mu_2}{1-\lambda_{p+1}\mu_1}\boldsymbol{z}_2 + \cdots + \frac{\mu_{q+1}}{1-\lambda_{p+1}\mu_1}\boldsymbol{z}_{q+1}.$$

Hence $\{\boldsymbol{y}_1, \boldsymbol{y}_2\}$ lies in the convex hull of $2m$ or fewer points from X.

Assume that the assertion holds for all $r \leqslant s$ and let $Y = \{\boldsymbol{y}_1, \ldots, \boldsymbol{y}_{s+1}\}$ be a set of $s+1$ points in $\operatorname{conv} X$, $s \geqslant 2$. By the induction hypothesis, there is a set $Z_0 \subset X$ of cardinality sm such that $\boldsymbol{y}_1, \ldots, \boldsymbol{y}_s \in \operatorname{conv} Z_0$. Choose a point $\boldsymbol{z}_0 \in Z_0$. By Theorem 4.8, there are t ($\leqslant m$) points $\boldsymbol{u}_1, \ldots, \boldsymbol{u}_t \in X$ such that $\boldsymbol{y}_{s+1}$ can be written as a convex combination $\boldsymbol{y}_{s+1} = \alpha_1\boldsymbol{u}_1 + \cdots + \alpha_t\boldsymbol{u}_t + \alpha_{t+1}\boldsymbol{z}_0$. Therefore, $\boldsymbol{y}_{s+1} \in \operatorname{conv}\{\boldsymbol{u}_1, \ldots, \boldsymbol{u}_t, \boldsymbol{z}_0\}$. Put $Z = Z_0 \cup \{\boldsymbol{u}_1, \ldots, \boldsymbol{u}_t\}$. Clearly, $Z \subset X$, $\operatorname{card} Z \leqslant (s+1)m$, and $Y \subset \operatorname{conv} Z$.

Problem 4.18. Let $Y = \{\boldsymbol{y}_1, \ldots, \boldsymbol{y}_r\}$. Without loss of generality, we may suppose that $\boldsymbol{y}_1, \ldots, \boldsymbol{y}_r$ are pairwise distinct. A combination of Theorem 4.18 and Corollary 4.19 shows the existence of a finite subset Z_i of X such that $\boldsymbol{y}_i \in \operatorname{rint}(\operatorname{conv} Z_i)$ and $\operatorname{aff} X = \operatorname{aff} Z_i$, $1 \leqslant i \leqslant r$. Put $Z = Z_1 \cup \cdots \cup Z_r$. Then $\operatorname{aff} X = \operatorname{aff} Z$, and Theorem 3.18 gives

$$Y \subset \bigcup_{i=1}^{r} \operatorname{rint}(\operatorname{conv} Z_i) \subset \operatorname{rint}\Big(\bigcup_{i=1}^{r} \operatorname{conv} Z_i\Big) \subset \operatorname{rint}\Big(\operatorname{conv}\Big(\bigcup_{i=1}^{r} Z_i\Big)\Big) = \operatorname{rint}(\operatorname{conv} Z).$$

Put $m = \dim(\operatorname{conv} Z)$. Denote by l the line through $\boldsymbol{y}_1$ and $\boldsymbol{y}_2$ and let $[\boldsymbol{v}_0, \boldsymbol{w}_0]$ be the closed segment at which l meets the compact convex set $\operatorname{conv} Z$. Since any finite union of planes of dimension $m-1$ is nowhere dense in $\operatorname{aff} Z$, we can choose, similar to the proof of Theorem 4.20, a point $\boldsymbol{v} \in \operatorname{rbd}(\operatorname{conv} Z)$ sufficiently close to $\boldsymbol{v}_0$ and written as a positive convex combination $\boldsymbol{v} = \lambda_1\boldsymbol{x}_1 + \cdots + \lambda_m\boldsymbol{x}_m$ of an affinely independent set $\{\boldsymbol{x}_1, \ldots, \boldsymbol{x}_m\} \subset Z$.

For every $\boldsymbol{y}_i$, denote by $\boldsymbol{w}_i$ the point of intersection of the open halfline $(\boldsymbol{v}, \boldsymbol{y}_i\rangle$ with $\operatorname{rbd}(\operatorname{conv} Z)$. By Theorem 4.7, $\boldsymbol{w}_i$ can be written as a convex combination

$$\boldsymbol{w}_i = \mu_1^{(i)}\boldsymbol{z}_1^{(i)} + \cdots + \mu_{k_i}^{(i)}\boldsymbol{z}_{k_i}^{(i)}, \quad 1 \leqslant i \leqslant r,$$

of an affinely independent subset $U_i = \{\boldsymbol{z}_1^{(i)}, \ldots, \boldsymbol{z}_{k_i}^{(i)}\}$ of Z, where $k_i \leqslant m+1$. Excluding all zero multiples $0\boldsymbol{z}_j^{(i)}$, we assume that $\mu_1^{(i)}, \ldots, \mu_{k_i}^{(i)}$ are positive. By Theorem 4.18, $\boldsymbol{w}_i \in \operatorname{rint}(\operatorname{conv} U_i)$. We assert that $k_i \leqslant m$. Indeed, if $k_i = m+1$, then $\operatorname{aff} U_i = \operatorname{aff} Z$ and, according Theorem 3.18,

$$\boldsymbol{w}_i \in \operatorname{rint}(\operatorname{conv} U_i) \subset \operatorname{rint}(\operatorname{conv} Z),$$

in contradiction with the choice of $\boldsymbol{w}_i$ in $\operatorname{rbd}(\operatorname{conv} Z)$. Hence $k_i \leqslant m$.

Since $\boldsymbol{y}_i \in (\boldsymbol{v}, \boldsymbol{w}_i)$, one can write $\boldsymbol{y}_i = (1-\eta_i)\boldsymbol{v} + \eta_i \boldsymbol{w}_i$, where $0 < \eta_i < 1$. Consequently, $\boldsymbol{y}_i$ is a positive convex combination,

$$\boldsymbol{y}_i = (1-\eta_i)\lambda_1 \boldsymbol{x}_1 + \cdots + (1-\eta_i)\lambda_m \boldsymbol{x}_m + \eta_i \mu_1^{(i)} \boldsymbol{z}_1^{(i)} + \cdots + \eta_i \mu_{k_i}^{(i)} \boldsymbol{z}_{k_i}^{(i)}.$$

With $V = \{\boldsymbol{x}_1, \ldots, \boldsymbol{x}_m\}$, Theorem 4.18 gives

$$\boldsymbol{y}_i \in \operatorname{rint}(\operatorname{conv}(V \cup U_i)), \quad V \cup U_i \subset Z, \quad \operatorname{card}(V \cup U_i) \leqslant 2m, \quad 1 \leqslant i \leqslant r.$$

By a continuity argument, the point $\boldsymbol{v}$ can be chosen so close to $\boldsymbol{v}_0$ that both points $\boldsymbol{w}_1$ and $\boldsymbol{w}_2$ belongs to one of the sets $\operatorname{rint}(\operatorname{conv} U_1)$ and $\operatorname{rint}(\operatorname{conv} U_2)$. For convenience of notation, we suppose that $\boldsymbol{w}_1, \boldsymbol{w}_2 \in \operatorname{rint}(\operatorname{conv} U_2)$. Finally, a combination of Theorems 3.18 and 4.2 and Problem 4.16 results in

$$\begin{aligned} Y &\subset \bigcup_{i=2}^{r} \operatorname{rint}(\operatorname{conv}(V \cup U_i)) \subset \operatorname{rint}\big(\bigcup_{i=2}^{r} \operatorname{conv}(V \cup U_i)\big) \\ &\subset \operatorname{rint}\big(\operatorname{conv}\big(V \cup \big(\bigcup_{i=2}^{r} U_i\big)\big)\big). \end{aligned}$$

Clearly, $\operatorname{card}(V \cup (\bigcup_{i=2}^{r} U_i)) \leqslant rm$.

Problem 4.19. Since the case $Y = \varnothing$ is obvious (the existence of a finite set $Z \subset X$ with $\operatorname{aff} Z = \operatorname{aff} X$ is proved in Theorem 2.61), we may assume that $Y \neq \varnothing$. Let $m = \dim X$. Theorem 4.18 and Corollary 4.19 show that for every point $\boldsymbol{x} \in Y$ there is a finite set $Z_{\boldsymbol{x}} \subset X$ such that $\boldsymbol{x} \in \operatorname{rint}(\operatorname{conv} Z_{\boldsymbol{x}})$ and $\operatorname{aff} Z_{\boldsymbol{x}} = \operatorname{aff} X$. Since the set Y is compact, the relatively open cover $\{\operatorname{rint}(\operatorname{conv} Z_{\boldsymbol{x}}) : \boldsymbol{x} \in Y\}$ of Y contains a finite subcover $\{\operatorname{rint}(\operatorname{conv} Z_{\boldsymbol{x}_i}) : 1 \leqslant i \leqslant r\}$. Let $Z = Z_{\boldsymbol{x}_1} \cup \cdots \cup Z_{\boldsymbol{x}_r}$. Then $\operatorname{aff} Z = \operatorname{aff} X$, and a combination of Theorems 3.18 and 4.2 gives

$$\begin{aligned} Y &\subset \bigcup_{i=1}^{r} \operatorname{rint}(\operatorname{conv} Z_{\boldsymbol{x}_i}) \subset \operatorname{rint}\big(\bigcup_{i=1}^{r} \operatorname{conv} Z_{\boldsymbol{x}_i}\big) \\ &\subset \operatorname{rint}\big(\operatorname{conv}\big(\bigcup_{i=1}^{r} Z_{\boldsymbol{x}_i}\big)\big) = \operatorname{rint}(\operatorname{conv} Z). \end{aligned}$$

Chapter 5

Problem 5.1. Without loss of generality, we may put $X = \{\boldsymbol{x}_1, \ldots, \boldsymbol{x}_t\}$, where $1 \leqslant t \leqslant r-1$. Assume for a moment that the set $\Gamma_X \cap \Gamma_Y$ contains a point $\boldsymbol{u} \neq \boldsymbol{a}$. Then $\boldsymbol{u}$ can be expressed as nonnegative combinations

$$\begin{aligned} \boldsymbol{u} &= \boldsymbol{a} + \lambda_1(\boldsymbol{x}_1 - \boldsymbol{a}) + \cdots + \lambda_t(\boldsymbol{x}_t - \boldsymbol{a}) \\ &= \boldsymbol{a} + \lambda_{t+1}(\boldsymbol{x}_{t+1} - \boldsymbol{a}) + \cdots + \lambda_r(\boldsymbol{x}_r - \boldsymbol{a}), \end{aligned}$$

where not all scalars $\lambda_1, \ldots, \lambda_r$ are zero. Then the equality

$$\begin{aligned} \lambda_1 \boldsymbol{x}_1 + &\cdots + \lambda_t \boldsymbol{x}_t - \lambda_{t+1}\boldsymbol{x}_{t+1} - \cdots - \lambda_r \boldsymbol{x}_r \\ &- (\lambda_1 + \cdots + \lambda_t - \lambda_{t+1} - \cdots - \lambda_r)\boldsymbol{a} = \boldsymbol{o}, \end{aligned}$$

shows that the set $\{\boldsymbol{a}, \boldsymbol{x}_1, \dots, \boldsymbol{x}_r\}$ is affinely dependent, which is impossible by the definition of $\Gamma_{\boldsymbol{a}}(\boldsymbol{x}_1, \dots, \boldsymbol{x}_r)$. Hence $\Gamma_X \cap \Gamma_Y = \{\boldsymbol{a}\}$.

For the second assertion, we first observe $\operatorname{conv}_2(\Gamma_X \cup \Gamma_Y) \subset \Gamma$ due to the convexity of Γ (see Corollary 4.14). Conversely, let $\boldsymbol{u} \in \Gamma$. Then $\boldsymbol{u}$ can be written as a nonnegative combination $\boldsymbol{u} = \boldsymbol{a} + \lambda_1(\boldsymbol{x}_1 - \boldsymbol{a}) + \cdots + \lambda_r(\boldsymbol{x}_r - \boldsymbol{a})$. Let

$$\begin{aligned} \boldsymbol{x} &= \boldsymbol{a} + 2\lambda_1(\boldsymbol{x}_1 - \boldsymbol{a}) + \cdots + 2\lambda_t(\boldsymbol{x}_t - \boldsymbol{a}), \\ \boldsymbol{y} &= \boldsymbol{a} + 2\lambda_{t+1}(\boldsymbol{x}_{t+1} - \boldsymbol{a}) + \cdots + 2\lambda_r(\boldsymbol{x}_r - \boldsymbol{a}). \end{aligned}$$

Clearly, $\boldsymbol{x} \in \Gamma_X$, $\boldsymbol{y} \in \Gamma_Y$, and $\boldsymbol{u} = (\boldsymbol{x} + \boldsymbol{y})/2$. Hence $\Gamma \subset \operatorname{conv}_2(\Gamma_X \cup \Gamma_Y)$.

Problem 5.2. Let $X = \{\boldsymbol{x}_1, \dots, \boldsymbol{x}_{r+1}\}$ and $X_i = X \setminus \{\boldsymbol{x}_i\}$, $1 \leqslant i \leqslant r+1$. Combining Theorem 4.2 and Corollary 4.5, we observe that

$$\boldsymbol{a} \in \operatorname{rint} \Delta \subset \Delta = \operatorname{conv} X \subset \operatorname{aff} X = \operatorname{aff} \Delta.$$

Therefore, Theorems 5.38 and 5.48 imply that

$$\Gamma_i = C_{\boldsymbol{a}}(\Delta_i) = C_{\boldsymbol{a}}(\operatorname{conv} X_i) = \operatorname{cone}_{\boldsymbol{a}} X_i \subset \operatorname{cone}_{\boldsymbol{a}} X = \operatorname{aff} X = \operatorname{aff} \Delta.$$

Hence $\Gamma_1 \cup \cdots \cup \Gamma_{r+1} \subset \operatorname{aff} \Delta$. For the opposite inclusion we observe that $\boldsymbol{a}$ is a positive convex combination of X: $\boldsymbol{a} = \mu_1 \boldsymbol{x}_1 + \cdots + \mu_{r+1}\boldsymbol{x}_{r+1}$ (see Theorem 3.20). By Theorem 2.52, this representation of $\boldsymbol{a}$ is unique. Hence $\boldsymbol{a}$ cannot be expressed as an affine combination of points from X_i, $1 \leqslant i \leqslant r+1$. The same theorem shows that every set $\{\boldsymbol{a}\} \cup X_i$, $1 \leqslant i \leqslant r+1$, is affinely independent. Because $\operatorname{cone}_{\boldsymbol{a}} X = \operatorname{aff} X$, Theorem 5.37 shows that any given point $\boldsymbol{x} \in \operatorname{aff} X$ belongs to $\operatorname{cone}_{\boldsymbol{a}} X_i$ for a suitable index $i \in \{1, \dots, r+1\}$. Thus $\boldsymbol{x} \in C_{\boldsymbol{a}}(\Delta_i) = \Gamma_i$, which gives $\operatorname{aff} X \subset \Gamma_1 \cup \cdots \cup \Gamma_{r+1}$.

For the second assertion, choose any distinct indices i and j from $\{1, \dots, r+1\}$. If a point $\boldsymbol{u}$ belongs to $\operatorname{rint} \Gamma_j$, then, by Corollary 5.52, it can be written as

$$\boldsymbol{u} = \boldsymbol{a} + \sum_{i=1}^{r+1} \lambda_i(\boldsymbol{x}_i - \boldsymbol{a}), \ i \neq j,$$

where all scalars $\lambda_1, \dots, \lambda_{j-1}, \lambda_{j+1}, \dots, \lambda_{r+1}$ are positive. This argument, combined with Definition 5.6, shows that $\boldsymbol{u} \notin \Gamma_i$ for all $i \in \{1, \dots, r+1\} \setminus \{j\}$. Therefore, $\Gamma_i \cap \operatorname{rint} \Gamma_j = \varnothing$ if $i \neq j$.

Problem 5.3. Suppose first that $\Gamma = \Gamma'$. Since a simplicial cone has a unique apex (see Theorem 5.7), one has $\boldsymbol{a} = \boldsymbol{c}$. By the same theorem, $r = \dim \Gamma = \dim \Gamma' = s$. For the last condition, we observe first that the case $r = 1$ is obvious: $(\boldsymbol{a}, \boldsymbol{x}_1\rangle = (\boldsymbol{a}, \boldsymbol{z}_1\rangle$ if and only if $\boldsymbol{z}_1 \in (\boldsymbol{a}, \boldsymbol{x}_1\rangle$ (see Lemma 2.25). Suppose that $r \geqslant 2$. According to Definition 5.6, the points $\boldsymbol{z}_1, \dots, \boldsymbol{z}_r$ can be expressed as nonnegative combinations

$$\boldsymbol{z}_j = \boldsymbol{a} + \lambda_1^{(j)}(\boldsymbol{x}_1 - \boldsymbol{a}) + \cdots + \lambda_r^{(j)}(\boldsymbol{x}_r - \boldsymbol{a}), \quad 1 \leqslant j \leqslant r. \tag{14.16}$$

Assume for, contradiction, that the open halfline $h_1 = (\boldsymbol{a}, \boldsymbol{x}_1\rangle$ is disjoint from $\{\boldsymbol{z}_1, \dots, \boldsymbol{z}_r\}$. Then $\{\boldsymbol{z}_1, \dots, \boldsymbol{z}_r\} \subset \Gamma \setminus h_1$. Since any point $\boldsymbol{u} \in h_1$ is expressible as $\boldsymbol{u} = \boldsymbol{a} + \mu(\boldsymbol{a} - \boldsymbol{x}_1)$, $\mu \geqslant 0$, a remark on page 155 implies that

$$\lambda_2^{(j)} + \cdots + \lambda_r^{(j)} > 0 \quad \text{for all} \quad 2 \leqslant j \leqslant r.$$

Furthermore, the inclusion $\boldsymbol{x}_1 \in \Gamma'$ shows that $\boldsymbol{x}_1$ is a nonnegative combination of the form

$$\boldsymbol{x}_1 = \boldsymbol{a} + \gamma_1(\boldsymbol{z}_1 - \boldsymbol{a}) + \cdots + \gamma_r(\boldsymbol{z}_r - \boldsymbol{a}), \tag{14.17}$$

where $\gamma_2 + \cdots + \gamma_r > 0$. Indeed, otherwise $\boldsymbol{x}_1 = \boldsymbol{a} + \gamma_1(\boldsymbol{z}_1 - \boldsymbol{a})$, where $\gamma_1 \neq 0$, implying the inclusion $\boldsymbol{z}_1 = \boldsymbol{a} + \gamma_1^{-1}(\boldsymbol{x}_1 - \boldsymbol{a}) \in (\boldsymbol{a}, \boldsymbol{x}_1\rangle$, which contradicts the assumption $\boldsymbol{z}_1 \notin (\boldsymbol{a}, \boldsymbol{x}_1\rangle$. Combining (14.16) and (14.17), we obtain

$$\boldsymbol{x}_1 = \boldsymbol{a} + \sum_{j=1}^{r} \gamma_j \big(\sum_{i=1}^{r} \lambda_i^{(j)}(\boldsymbol{x}_i - \boldsymbol{a})\big), \quad \sum_{j=2}^{r} \gamma_j \sum_{i=2}^{r} \lambda_i^{(j)} > 0. \tag{14.18}$$

On the other hand, a remark on page 155 shows that

$$\boldsymbol{x}_1 = \boldsymbol{a} + 1(\boldsymbol{x}_1 - \boldsymbol{a}) + 0(\boldsymbol{x}_2 - \boldsymbol{a}) + \cdots + 0(\boldsymbol{x}_{r+1} - \boldsymbol{a})$$

is the only way to express $\boldsymbol{x}_1$ as a nonnegative combination of the form (5.3), in contradiction with (14.18). Summing up, every open halfline $h_i = (\boldsymbol{a}, \boldsymbol{x}_i\rangle$ should contain a point from $\{\boldsymbol{z}_1, \dots, \boldsymbol{z}_r\}$. Since the halflines $h_1, \dots, h_r$ are pairwise disjoint, each of them contains a unique point from $\{\boldsymbol{z}_1, \dots, \boldsymbol{z}_r\}$.

Conversely, suppose that all conditions of the theorem are satisfied. The inclusions $\boldsymbol{z}_i \in (\boldsymbol{a}, \boldsymbol{x}_i\rangle$, $1 \leqslant i \leqslant r$, show that $\boldsymbol{z}_i = \boldsymbol{a} + \gamma_i(\boldsymbol{x}_i - \boldsymbol{a})$, where $\gamma_i > 0$ for all $1 \leqslant i \leqslant r$. To prove the equality $\Gamma = \Gamma'$, choose any point $\boldsymbol{u} \in \Gamma$. Then $\boldsymbol{u}$ can be written as a nonnegative combination

$$\boldsymbol{u} = \boldsymbol{a} + \lambda_1(\boldsymbol{x}_1 - \boldsymbol{a}) + \cdots + \lambda_r(\boldsymbol{x}_r - \boldsymbol{a}).$$

Therefore, $\boldsymbol{u} = \boldsymbol{a} + \lambda_1\gamma_1^{-1}(\boldsymbol{z}_1 - \boldsymbol{a}) + \cdots + \lambda_r\gamma_r^{-1}(\boldsymbol{z}_r - \boldsymbol{a}) \in \Gamma'$. Similarly, if a point $\boldsymbol{u} \in \Gamma'$ is written as a nonnegative combination

$$\boldsymbol{u} = \boldsymbol{a} + \mu_1(\boldsymbol{z}_1 - \boldsymbol{a}) + \cdots + \mu_r(\boldsymbol{z}_r - \boldsymbol{a}),$$

then $\boldsymbol{u} = \boldsymbol{a} + \gamma_1\mu_1(\boldsymbol{x}_1 - \boldsymbol{a}) + \cdots + \gamma_r\mu_r(\boldsymbol{x}_r - \boldsymbol{a}) \in \Gamma$.

Problem 5.4. By Theorem 5.38, $\operatorname{conv} C = \operatorname{cone}_{\boldsymbol{a}} C$. If

$$\boldsymbol{x} = (1 - k)\boldsymbol{a} + \boldsymbol{x}_1 + \cdots + \boldsymbol{x}_k = \boldsymbol{a} + 1(\boldsymbol{x}_1 - \boldsymbol{a}) + \cdots + 1(\boldsymbol{x}_k - \boldsymbol{a}),$$

where $\boldsymbol{x}_1, \dots, \boldsymbol{x}_k \in C$ and $\{\boldsymbol{a}, \boldsymbol{x}_1, \dots, \boldsymbol{x}_k\}$ is affinely independent, then, by Theorem 5.35, $\boldsymbol{x}$ belongs to $\operatorname{cone}_{\boldsymbol{a}} C \setminus \{\boldsymbol{a}\} = \operatorname{conv} C \setminus \{\boldsymbol{a}\}$.

Conversely, suppose that $\boldsymbol{x} \in \operatorname{cone}_{\boldsymbol{a}} C \setminus \{\boldsymbol{a}\}$. By the same theorem, $\boldsymbol{x}$ can be written as a positive combination

$$\boldsymbol{x} = \boldsymbol{a} + \lambda_1(\boldsymbol{x}_1 - \boldsymbol{a}) + \cdots + \lambda_k(\boldsymbol{x}_k - \boldsymbol{a}), \quad \boldsymbol{x}_1, \ldots, \boldsymbol{x}_k \in C, \quad k \geqslant 1,$$

where $\{\boldsymbol{a}, \boldsymbol{x}_1, \ldots, \boldsymbol{x}_k\}$ is affinely independent. Because $C - \boldsymbol{a}$ is a cone with apex $\boldsymbol{o}$ (see Corollary 5.11) and $\boldsymbol{x}_i - \boldsymbol{a} \in C - \boldsymbol{a}$, the vector $\boldsymbol{z}_i = \lambda_i(\boldsymbol{x}_i - \boldsymbol{a})$ belongs to $C - \boldsymbol{a}$ for all $1 \leqslant i \leqslant k$. The set $\{\boldsymbol{z}_1, \ldots, \boldsymbol{z}_k\}$ is linearly independent because the set $\{\boldsymbol{x}_1 - \boldsymbol{a}, \ldots, \boldsymbol{x}_k - \boldsymbol{a}\}$ is linearly independent (see Theorem 2.52). By the same theorem, the set $\{\boldsymbol{a}, \boldsymbol{u}_1, \ldots, \boldsymbol{u}_k\}$, where $\boldsymbol{u}_i = \boldsymbol{a} + \boldsymbol{z}_i \in C$, $1 \leqslant i \leqslant k$, is affinely independent. Finally,

$$\boldsymbol{x} = \boldsymbol{a} + \boldsymbol{z}_1 + \cdots + \boldsymbol{z}_k = (1 - k)\boldsymbol{a} + \boldsymbol{u}_1 + \cdots + \boldsymbol{u}_k.$$

Problem 5.5. By Theorem 5.4, the set $C' = \{\boldsymbol{a}\} \cup C$ is a cone with proper apex $\boldsymbol{a}$. We assert that C can be replaced with C'. Since this fact is obvious if $C = \{\boldsymbol{a}\}$, we may assume the existence of a point $\boldsymbol{b} \in C \setminus \{\boldsymbol{a}\}$.

First, $\operatorname{aff} C' = \operatorname{aff} C$, as shown in Theorem 5.4. Next, we observe that $C' - C' = C - C$. Indeed, because the inclusion $C - C \subset C' - C'$ is obvious, it suffices to show that $C' - C' \subset C - C$. Let $\boldsymbol{x} \in C' - C'$. Then $\boldsymbol{x} = \boldsymbol{u} - \boldsymbol{v}$, where both $\boldsymbol{u}, \boldsymbol{v} \in C'$. If $\boldsymbol{u}, \boldsymbol{v} \in C$, then $\boldsymbol{x} = \boldsymbol{u} - \boldsymbol{v} \in C - C$. Similarly, if $\boldsymbol{u} = \boldsymbol{v} = \boldsymbol{a}$, then $\boldsymbol{x} = \boldsymbol{a} - \boldsymbol{a} = \boldsymbol{o} = \boldsymbol{b} - \boldsymbol{b} \in C - C$. Suppose that $\boldsymbol{u} \neq \boldsymbol{a}$ and $\boldsymbol{v} = \boldsymbol{a}$. Then

$$\boldsymbol{x} = \boldsymbol{u} - \boldsymbol{a} = (\boldsymbol{a} + 2(\boldsymbol{u} - \boldsymbol{a})) - (\boldsymbol{a} + 1(\boldsymbol{u} - \boldsymbol{a})) \in C - C.$$

Similarly, if $\boldsymbol{u} = \boldsymbol{a}$ and $\boldsymbol{v} \neq \boldsymbol{a}$, then

$$\boldsymbol{x} = \boldsymbol{a} - \boldsymbol{v} = (\boldsymbol{a} + 1(\boldsymbol{v} - \boldsymbol{a})) - (\boldsymbol{a} + 2(\boldsymbol{v} - \boldsymbol{a})) \in C - C.$$

Summing up, $C' - C' \subset C - C$. For the last equality,

$$\operatorname{conv}(C' \cup (2\boldsymbol{a} - C')) = \operatorname{conv}(C \cup (2\boldsymbol{a} - C)), \tag{14.19}$$

we observe that

$$\boldsymbol{a} = \tfrac{1}{2}(\boldsymbol{b} + (2\boldsymbol{a} - \boldsymbol{b})) \in [\boldsymbol{b}, 2\boldsymbol{a} - \boldsymbol{b}] \subset \operatorname{conv}(C \cup (2\boldsymbol{a} - C)).$$

Therefore,

$$\begin{aligned} C \cup (2\boldsymbol{a} - C) \subset C' \cup (2\boldsymbol{a} - C') &= \{\boldsymbol{a}\} \cup C \cup (2\boldsymbol{a} - C) \\ &\subset \operatorname{conv}(C \cup (2\boldsymbol{a} - C)), \end{aligned}$$

and Theorem 4.2 gives (14.19)

By the above argument and Theorem 4.12, the desired equalities are equivalent to

$$\operatorname{aff} C' = \boldsymbol{a} + \operatorname{conv} C' - \operatorname{conv} C' = \operatorname{conv}(C' \cup (2\boldsymbol{a} - C')), \tag{14.20}$$

which we intend to prove. For the first equality in (14.20), let $F = \operatorname{conv} C' - \boldsymbol{a}$. Then F is a convex cone with apex $\boldsymbol{o}$, and a combination of Theorems 2.42 and 4.12 gives

$$\operatorname{span} F = \operatorname{aff} C' - \boldsymbol{a}, \quad F - F = \operatorname{conv} C' - \operatorname{conv} C'.$$

Consequently, the desired equality $\operatorname{aff} C' = \boldsymbol{a} + \operatorname{conv} C' - \operatorname{conv} C'$ can be rewritten as $\operatorname{span} F = F - F$. Choose any point $\boldsymbol{x} \in \operatorname{span} F$. Theorem 3.14 shows the existence of points $\boldsymbol{u}, \boldsymbol{v} \in F$ such that $\boldsymbol{x} = (1-\lambda)\boldsymbol{u} + \lambda\boldsymbol{v}$ for a suitable scalar $\lambda \in \mathbb{R}$. Using Theorem 5.5, we obtain

$$\boldsymbol{x} = \begin{cases} (\boldsymbol{u} + \lambda\boldsymbol{v}) - (\lambda\boldsymbol{u}) \in F - F & \text{if } \lambda > 0, \\ 2\boldsymbol{u} - \boldsymbol{v} \in F - F & \text{if } \lambda = 0, \\ (1-\lambda)\boldsymbol{u} - (-\lambda\boldsymbol{v}) \in F - F & \text{if } \lambda < 0, \end{cases}$$

which gives the inclusion $\operatorname{span} F \subset F - F$. Conversely, if $\boldsymbol{x} \in F - F$, then $\boldsymbol{x} = \boldsymbol{y} - \boldsymbol{z}$ for suitable points $\boldsymbol{y}, \boldsymbol{z} \in F$. Expressing $\boldsymbol{x}$ as the affine combination $\boldsymbol{x} = 2(\frac{1}{2}\boldsymbol{y}) - \boldsymbol{z}$ of points $\frac{1}{2}\boldsymbol{y}, \boldsymbol{z} \in F$, we obtain that $\boldsymbol{x} \in \operatorname{span} F$. The second equality in (14.20) follows from Theorem 5.10 (with $r = 2$, $\boldsymbol{a}_1 = \boldsymbol{a}_2 = \boldsymbol{a}$, and $\mu_1 = 1$, $\mu_2 = -1$).

Problem 5.6. Let a convex set $K \subset \mathbb{R}^n$ satisfy the equality $K = \boldsymbol{a} + \mu K$, with $\boldsymbol{a} \neq \boldsymbol{o}$ and $\mu > 0$. If $\mu \neq 1$, then

$$K = \frac{1}{1-\mu}\boldsymbol{a} + \mu(K - \frac{1}{1-\mu}\boldsymbol{a}) = \boldsymbol{c} + \mu(K - \boldsymbol{c}),$$

where $\boldsymbol{c} = (1-\mu)^{-1}\boldsymbol{a}$, which shows that K is a convex cone with apex $\boldsymbol{c}$ (see Theorem 5.3). If $\mu = 1$, then, by induction on r, we obtain that $K = \pm r\boldsymbol{a} + K$ for all $r \geqslant 1$, and a convexity argument shows that K contains a translate of the line $\langle \boldsymbol{o}, \boldsymbol{a} \rangle$. The proof of the converse assertion is similar.

Problem 5.7. Put $D = \operatorname{conv}(C_{\boldsymbol{u}}(K) \cup C_{\boldsymbol{v}}(K))$ and let $\boldsymbol{w} = (1-\lambda)\boldsymbol{u} + \lambda\boldsymbol{v}$ for a suitable scalar $0 < \lambda < 1$. For the inclusion $C_{\boldsymbol{w}}(K) \subset D$, choose any point $\boldsymbol{x} \in C_{\boldsymbol{w}}(K)$. If $\boldsymbol{x} \in \langle \boldsymbol{u}, \boldsymbol{v} \rangle$, then $\boldsymbol{x}$ belongs to one of the closed halflines $[\boldsymbol{u}, \boldsymbol{w}\rangle$ and $[\boldsymbol{v}, \boldsymbol{w}\rangle$, which implies that $\boldsymbol{x}$ belongs to one of the cones $C_{\boldsymbol{u}}(K)$ and $C_{\boldsymbol{v}}(K)$. Let $\boldsymbol{x} \notin \langle \boldsymbol{u}, \boldsymbol{v} \rangle$. The closed halfline $[\boldsymbol{w}, \boldsymbol{x}\rangle$ meets K at a point $\boldsymbol{x}'$. Since the case $\boldsymbol{x} \in [\boldsymbol{w}, \boldsymbol{x}'] \subset K \subset D$ is obvious, we may assume that $\boldsymbol{x}' \in (\boldsymbol{w}, \boldsymbol{x})$. Clearly, $[\boldsymbol{w}, \boldsymbol{x}\rangle = [\boldsymbol{w}, \boldsymbol{x}'\rangle$ (see Lemma 2.25). The inclusion $\boldsymbol{x} \in (\boldsymbol{w}, \boldsymbol{x}'\rangle \setminus (\boldsymbol{w}, \boldsymbol{x}')$ implies that $\boldsymbol{x} = (1-\mu)\boldsymbol{w} + \mu\boldsymbol{x}'$ for a suitable scalar $\mu > 1$. Put $\boldsymbol{u}' = (1-\mu)\boldsymbol{u} + \mu\boldsymbol{x}'$ and $\boldsymbol{v}' = (1-\mu)\boldsymbol{v} + \mu\boldsymbol{x}'$. Then

$$\begin{aligned} \boldsymbol{x} &= (1-\mu)\boldsymbol{w} + \mu\boldsymbol{x}' = (1-\lambda)((1-\mu)\boldsymbol{u} + \mu\boldsymbol{x}') + \lambda((1-\mu)\boldsymbol{v} + \mu\boldsymbol{x}') \\ &\in (1-\lambda)[\boldsymbol{u}, \boldsymbol{x}'\rangle + \lambda[\boldsymbol{v}, \boldsymbol{x}'\rangle \subset \operatorname{conv}([\boldsymbol{u}, \boldsymbol{x}'\rangle \cup [\boldsymbol{v}, \boldsymbol{x}'\rangle) \subset D. \end{aligned}$$

For the opposite inclusion, $D \subset C_{\boldsymbol{w}}(K)$, consider first any point $\boldsymbol{r} \in (\boldsymbol{u}, \boldsymbol{w})$. With the above notation, Lemma 2.26 shows that the open segment $(\boldsymbol{r}, \boldsymbol{x})$ meets

$[\boldsymbol{u}, \boldsymbol{x}']$ at a point $\boldsymbol{r}' \in K$. Hence $\boldsymbol{x} \in C_{\boldsymbol{r}}(K)$, which implies the inclusion $C_{\boldsymbol{w}}(K) \subset C_{\boldsymbol{r}}(K)$. In a similar way, $C_{\boldsymbol{r}}(K) \subset C_{\boldsymbol{w}}(K)$. Summing up, $C_{\boldsymbol{r}}(K) = C_{\boldsymbol{w}}(K)$ whenever $\boldsymbol{r} \in (\boldsymbol{u}, \boldsymbol{w})$.

By Corollary 4.14, D consists of all convex combinations $\boldsymbol{t} = (1-\gamma)\boldsymbol{p} + \gamma\boldsymbol{q}$, where $\boldsymbol{p} \in C_{\boldsymbol{u}}(K)$, $\boldsymbol{q} \in C_{\boldsymbol{w}}(K)$, and $0 \leqslant \gamma \leqslant 1$. Since the cases $\gamma = 0$ and $\gamma = 1$ correspond to the descriptions of $C_{\boldsymbol{u}}(K)$ and $C_{\boldsymbol{v}}(K)$, respectively, we may suppose that $0 < \gamma < 1$. As above, the convexity of K implies the existence of a scalar $\delta > 0$ such that $\boldsymbol{p} = (1-\delta)\boldsymbol{u} + \delta\boldsymbol{p}'$ and $\boldsymbol{q} = (1-\delta)\boldsymbol{v} + \delta\boldsymbol{q}'$, where both points $\boldsymbol{p}'$ and $\boldsymbol{p}'$ belong to K. Let $\boldsymbol{r} = (1-\gamma)\boldsymbol{u} + \gamma\boldsymbol{v}$ and $\boldsymbol{t}' = (1-\gamma)\boldsymbol{p}' + \gamma\boldsymbol{q}'$. Then $\boldsymbol{r} \in (\boldsymbol{u}, \boldsymbol{v})$, $\boldsymbol{t}' \in (\boldsymbol{p}', \boldsymbol{q}') \subset K$, and

$$\begin{aligned} \boldsymbol{t} &= (1-\gamma)\boldsymbol{p} + \gamma\boldsymbol{q} = (1-\gamma)[(1-\delta)\boldsymbol{u} + \delta\boldsymbol{p}'] + \gamma[(1-\delta)\boldsymbol{v} + \delta\boldsymbol{q}'] \\ &= (1-\delta)\boldsymbol{r} + \delta\boldsymbol{t}' \in [\boldsymbol{r}, \boldsymbol{t}') \subset C_{\boldsymbol{r}}(K) = C_{\boldsymbol{w}}(K). \end{aligned}$$

Problem 5.8. (*a*) It suffices to show that for any point $\boldsymbol{u} \in C \cap S_\rho(\boldsymbol{a})$ and a scalar $\delta > 0$ there is a point $\boldsymbol{v} \in X \cap \operatorname{rint} C$ with the property $\boldsymbol{v} \in U_\delta(\boldsymbol{u})$. Without loss of generality, we suppose that $\delta < \rho$. Translating all sets C, D, and X on the vector $-\boldsymbol{a}$, we may assume that $\boldsymbol{a} = \boldsymbol{o}$.

By Theorem 3.51, there is a point $\boldsymbol{w} \in \operatorname{rint} C \cap U_{\delta/2}(\boldsymbol{u})$. Clearly,

$$\big|\rho - \|\boldsymbol{w}\|\big| = \big|\|\boldsymbol{u}\| - \|\boldsymbol{w}\|\big| \leqslant \|\boldsymbol{u} - \boldsymbol{w}\| < \delta/2.$$

Let $\boldsymbol{w}' = (\rho/\|\boldsymbol{w}\|)\boldsymbol{w}$. Then $\boldsymbol{w}' \in S_\rho(\boldsymbol{o})$, and Theorem 5.20 shows that $\boldsymbol{w}' \in \operatorname{rint} C$. Furthermore,

$$\begin{aligned} \|\boldsymbol{u} - \boldsymbol{w}'\| &= \|\boldsymbol{u} - (\rho/\|\boldsymbol{w}\|)\boldsymbol{w}\| \leqslant \|\boldsymbol{u} - \boldsymbol{w}\| + \|\boldsymbol{w} - (\rho/\|\boldsymbol{w}\|)\boldsymbol{w}\| \\ &< \delta/2 + \big|\|\boldsymbol{w}\| - \rho\big| < \delta/2 + \delta/2 = \delta. \end{aligned}$$

Consequently, $\boldsymbol{w}' \in U_\delta(\boldsymbol{u})$. Let $\mu = \delta - \|\boldsymbol{u} - \boldsymbol{w}'\|$. Clearly, $\mu > 0$ and $U_\mu(\boldsymbol{w}') \subset U_\delta(\boldsymbol{u})$ (see Problem 1.8). Theorem 3.17 gives the existence of a scalar $0 < \varepsilon \leqslant \mu$ with the property $\operatorname{aff} C \cap U_\varepsilon(\boldsymbol{w}') \subset \operatorname{rint} C$. Since $\operatorname{aff} C = \operatorname{aff} D$ (see Corollary 2.65), one has $\operatorname{rint} C \subset \operatorname{rint} D$, as follows from Theorem 3.18. Therefore,

$$\operatorname{aff} D \cap U_\varepsilon(\boldsymbol{w}') = \operatorname{aff} C \cap U_\varepsilon(\boldsymbol{w}') \subset \operatorname{rint} C \subset \operatorname{rint} D.$$

Because the set X is dense in $D \cap S_\rho(\boldsymbol{o})$, there is a point $\boldsymbol{v} \in X \cap U_\varepsilon(\boldsymbol{w}') \subset X \cap U_\delta(\boldsymbol{u})$. Finally, due to the inclusions $X \subset D \subset \operatorname{aff} D$, one has

$$\begin{aligned} \boldsymbol{v} \in X \cap U_\varepsilon(\boldsymbol{w}') &= X \cap (\operatorname{aff} D \cap U_\varepsilon(\boldsymbol{w}')) \\ &= X \cap (\operatorname{aff} C \cap U_\varepsilon(\boldsymbol{w}')) \subset X \cap \operatorname{rint} C. \end{aligned}$$

(*b*) Choose any point $\boldsymbol{u} \in D \cap S_\rho(\boldsymbol{a})$ and a scalar $\delta > 0$. It suffices to show the existence of a point from $X \cap (\operatorname{rint} D \setminus \operatorname{cl} C)$ which belongs to $U_\delta(\boldsymbol{u})$. Without loss of generality, we suppose that $\delta < \rho$. Translating all sets C, D, and X on the vector $-\boldsymbol{a}$, we may assume that $\boldsymbol{a} = \boldsymbol{o}$.

Theorem 3.51 shows that the set rint $D \cap U_\delta(\boldsymbol{u})$ is nonempty and has dimension $\dim D$. Therefore, rint $D \cap U_\delta(\boldsymbol{u}) \not\subset \operatorname{cl} C$, since otherwise

$$\operatorname{aff}(\operatorname{rint} D \cap U_\delta(\boldsymbol{u})) \subset \operatorname{aff}(\operatorname{cl} C) = \operatorname{aff} C,$$

contrary to the assumption $\dim C < \dim D$. Choose a point

$$\boldsymbol{w} \in \operatorname{rint} D \cap U_\delta(\boldsymbol{u}) \cap S_\rho(\boldsymbol{o}) \setminus \operatorname{cl} C$$

and let $\mu = \delta - \|\boldsymbol{u} - \boldsymbol{w}\|$. Clearly, $\mu > 0$ and $U_\mu(\boldsymbol{w}) \subset U_\delta(\boldsymbol{u})$ (see Problem 1.8). Because the set $\operatorname{cl} C$ is closed, there is a scalar $0 < \varepsilon' \leqslant \mu$ such that $\operatorname{cl} C \cap U_{\varepsilon'}(\boldsymbol{w}) = \varnothing$. Theorem 3.17 gives the existence of a scalar $0 < \varepsilon \leqslant \varepsilon'$ with the property $\operatorname{aff} D \cap U_\varepsilon(\boldsymbol{w}) \subset \operatorname{rint} D$. Since X is dense in $D \cap S_\rho(\boldsymbol{o})$, there is a point $\boldsymbol{v} \in X$ such that $\boldsymbol{v} \in U_\varepsilon(\boldsymbol{w}) \subset U_\delta(\boldsymbol{u})$. By the above argument, $\boldsymbol{v} \in \operatorname{rint} D \setminus \operatorname{cl} C$.

Problem 5.9. Let $m = \dim L$. By Corollary 2.60, the plane M has dimension $m + 1$. Clearly, $L' \subset M$ and $L \cap L' = \varnothing$. A combination of Theorems 2.34 and 2.70 implies that L lies in an open halfplane, say E, of M determined by L'. By a similar argument, for any point $\boldsymbol{x} \in L$, the open halfline $(\boldsymbol{a}, \boldsymbol{x}\rangle$ lies in E. Hence $C_{\boldsymbol{a}}(L) = \cup([\boldsymbol{a}, \boldsymbol{x}\rangle : \boldsymbol{x} \in L) \subset \{\boldsymbol{a}\} \cup E$.

For the opposite inclusion, we observe first that by Lemma 2.72 and Theorem 2.70, there is a vector $\boldsymbol{c} \in \mathbb{R}^n \setminus \operatorname{ort} L$ and a scalar γ' satisfying the conditions

$$L' = \{\boldsymbol{x} \in M : \boldsymbol{x} \cdot \boldsymbol{c}\} = \gamma' \quad \text{and} \quad E = \{\boldsymbol{x} \in M : \boldsymbol{x} \cdot \boldsymbol{c} < \gamma'\}.$$

Clearly, $L = \{\boldsymbol{x} \in M : \boldsymbol{x} \cdot \boldsymbol{c}\} = \gamma$, where $\gamma < \gamma'$. Now, let

$$\boldsymbol{u} \in E, \quad \lambda = (\gamma - \gamma')/(\boldsymbol{u} \cdot \boldsymbol{c} - \gamma'), \quad \text{and} \quad \boldsymbol{v} = \boldsymbol{a} + \lambda(\boldsymbol{u} - \boldsymbol{a}).$$

Then $\lambda > 0$, and the equality

$$\boldsymbol{v} \cdot \boldsymbol{c} = (\boldsymbol{a} + \lambda(\boldsymbol{u} - \boldsymbol{a})) \cdot \boldsymbol{c} = \gamma' + \frac{\gamma - \gamma'}{\boldsymbol{u} \cdot \boldsymbol{c} - \gamma'}(\boldsymbol{u} \cdot \boldsymbol{c} - \gamma') = \gamma$$

shows that $\boldsymbol{v} \in L$. Hence $\boldsymbol{v} \in C_{\boldsymbol{a}}(L)$, and the inclusion $E \subset C_{\boldsymbol{a}}(L)$ holds. Since $\boldsymbol{a} \in C_{\boldsymbol{a}}(L)$, one has $\{\boldsymbol{a}\} \cup E \subset C_{\boldsymbol{a}}(L)$.

Problem 5.10. Since the cases $\Lambda = \varnothing$ and $K = \varnothing$ and obvious, we may assume that both sets Λ and K are nonempty. Choose any points $\boldsymbol{x}_1, \boldsymbol{x}_2 \in \Lambda K$ and a scalar $\mu \in [0, 1]$. We have to show that the point $\boldsymbol{x} = (1 - \mu)\boldsymbol{x}_1 + \mu \boldsymbol{x}_2$ belongs to ΛK. Clearly, $\boldsymbol{x}_i = \lambda_i \boldsymbol{z}_i$ for suitable scalars $\lambda_i \geqslant 0$ and points $\boldsymbol{z}_i \in K$, $i = 1, 2$. Since both cases $\lambda_1 = \mu = 0$ (corresponds to $\boldsymbol{x} = \boldsymbol{x}_1 = \boldsymbol{o} \in 0K \subset \Lambda K$) and $\lambda_2 = 1 - \mu = 0$ (corresponding to $\boldsymbol{x} = \boldsymbol{x}_2 = \boldsymbol{o} \in 0K \subset \Lambda K$) are obvious, we may suppose that the scalar $\xi = (1 - \mu)\lambda_1 + \mu\lambda_2$ is positive. Furthermore, $\xi \in \Lambda$ by the convexity of Λ. With $\gamma = \mu\lambda_2/\xi$, both scalars γ and $1 - \gamma = (1 - \mu)\lambda_1/\xi$

belong to $[0,1]$. Then the point $\boldsymbol{z} = (1-\gamma)\boldsymbol{z}_1 + \gamma\boldsymbol{z}_2$ belongs to K. Finally,

$$\begin{aligned}\boldsymbol{x} &= (1-\mu)\boldsymbol{x}_1 + \mu\boldsymbol{x}_2 = (1-\mu)\lambda_1\boldsymbol{z}_1 + \mu\lambda_2\boldsymbol{z}_2 \\ &= \xi\Big(\frac{(1-\mu)\lambda_1}{\xi}\boldsymbol{z}_1 + \frac{\mu\lambda_2}{\xi}\boldsymbol{z}_2\Big) = \xi((1-\gamma)\boldsymbol{z}_1 + \gamma\boldsymbol{z}_2) = \xi\boldsymbol{z} \in \Lambda K.\end{aligned}$$

Problem 5.11. Theorem 5.38 shows that $P(K_1, \boldsymbol{x}) = \mathrm{cone}_{\boldsymbol{x}} K_1$ for any $\boldsymbol{x} \in K_2$. Hence $U(K_1, K_2)$ is a convex set, as the intersection of convex sets $P(K_1, \boldsymbol{x})$, $\boldsymbol{x} \in K_2$ (see Theorem 3.8).

For the convexity of $P(K_1, K_2)$, choose points $\boldsymbol{x}, \boldsymbol{y} \in P(K_1, K_2)$ and a scalar $\lambda \in [0,1]$. Then $\boldsymbol{x} = (1-\eta)\boldsymbol{x}_2 + \eta\boldsymbol{x}_1$ and $\boldsymbol{y} = (1-\theta)\boldsymbol{y}_2 + \theta\boldsymbol{y}_1$ for some points $\boldsymbol{x}_1, \boldsymbol{y}_1 \in K_1$, $\boldsymbol{x}_2, \boldsymbol{y}_2 \in K_2$ and scalars $\eta, \theta \geqslant 1$. Put $\mu = (1-\lambda)\eta + \lambda\theta$. Clearly, $\mu \geqslant 1$ and

$$\begin{aligned}(1-\lambda)\boldsymbol{x} + \lambda\boldsymbol{y} &= (1-\lambda)((1-\eta)\boldsymbol{x}_2 + \eta\boldsymbol{x}_1) + \lambda((1-\theta)\boldsymbol{y}_2 + \theta\boldsymbol{y}_1) \\ &\in (1-\lambda)((1-\eta)K_2 + \eta K_1) + \lambda((1-\theta)K_2 + \theta K_1) \\ &= ((1-\lambda)(1-\eta)K_2 + \lambda(1-\theta)K_2) \\ &\qquad + ((1-\lambda)\eta K_1 + \lambda\theta K_1).\end{aligned}$$

By Theorem 3.4, $(1-\lambda)\eta K_1 + \lambda\theta K_1 = ((1-\lambda)\eta + \lambda\theta)K_1 = \mu K_1$. Similarly,

$$\begin{aligned}&(1-\lambda)(1-\eta)K_2 + \lambda(1-\theta)K_2 = -((1-\lambda)(\eta-1)K_2 + \lambda(\theta-1)K_2) \\ &= -((1-\lambda)(\eta-1) + \lambda(\theta-1))K_2 = (1-\mu)K_2.\end{aligned}$$

Therefore,

$$\begin{aligned}(1-\lambda)\boldsymbol{x} + \lambda\boldsymbol{y} &\in (1-\mu)K_2 + \mu K_1 = \cup((1-\mu)\boldsymbol{x} + \mu K_1 : \boldsymbol{x} \in K_2) \\ &\subset \cup((1-\gamma)\boldsymbol{x} + \gamma K_1 : \gamma \geqslant 1 \text{ and } \boldsymbol{x} \in K_2) = P(K_1, K_2),\end{aligned}$$

which proves the convexity of $P(K_1, K_2)$.

Chapter 6

Problem 6.1. Denote by C the set of all vectors $\boldsymbol{e} \in \mathbb{R}^n$ satisfying the problem condition. The inclusion $\mathrm{rec}\, X \subset C$ is obvious: for any $\boldsymbol{e} \in \mathrm{rec}\, X$ and $\boldsymbol{x} \in X$, let $\varepsilon(\boldsymbol{e}, \boldsymbol{x}) = 1$. Conversely, choose any point $\boldsymbol{e} \in C$. Assume for a moment that $\boldsymbol{e} \notin \mathrm{rec}\, X$. Then $\boldsymbol{e} \neq \boldsymbol{o}$ and there is a point $\boldsymbol{u} \in X$ such that the closed halfline $h = [\boldsymbol{u}, \boldsymbol{u} + \boldsymbol{e}\rangle$ does not lie in X. The complement of $h \setminus X$ is an open set in h, whence it can be expressed as a countable disjoint union of open segments (one of them may be an open halfline). Denote by $(\boldsymbol{v}, \boldsymbol{w})$ one of such segments. Furthermore, we suppose that $\boldsymbol{v} \in (\boldsymbol{u}, \boldsymbol{w})$ if $(\boldsymbol{v}, \boldsymbol{w})$ is bounded. Since $\boldsymbol{v} \in X$, there should be a positive scalar $\varepsilon(\boldsymbol{e}, \boldsymbol{v})$ such that $\lambda\boldsymbol{e} + \boldsymbol{v} \in X$ whenever $0 < \lambda < \varepsilon(\boldsymbol{e}, \boldsymbol{u})$. On the other hand, we can choose λ so small that $\lambda\boldsymbol{e} + \boldsymbol{v} \in (\boldsymbol{v}, \boldsymbol{w})$, contrary to the condition $X \cap (\boldsymbol{v}, \boldsymbol{w}) = \varnothing$. Summing up, $C \subset \mathrm{rec}\, X$.

Problem 6.2. If $\boldsymbol{e} \in \operatorname{rec} K$, then $\boldsymbol{e} + \boldsymbol{x} \in K$ for all $\boldsymbol{x} \in K$, which implies the inclusion $\boldsymbol{e} + K \subset K$. Conversely, let $\boldsymbol{e}$ be a point in $\mathbb{R}^n$ satisfying the condition $\boldsymbol{e} + K \subset K$. By induction on $r \geqslant 1$, we easily conclude that $r\boldsymbol{e} + K \subset K$ for all $r \geqslant 1$. Choose a point $\boldsymbol{x} \in K$, a scalar $\lambda \geqslant 0$, and a positive integer r such that $\lambda \leqslant r$. Then $\lambda \boldsymbol{e} \in [\boldsymbol{o}, r\boldsymbol{e}]$, and

$$\lambda \boldsymbol{e} + \boldsymbol{x} \in [\boldsymbol{o}, r\boldsymbol{e}] + \boldsymbol{x} = [\boldsymbol{x}, r\boldsymbol{e} + \boldsymbol{x}] \subset K$$

by the convexity of K. Hence $\boldsymbol{e} \in \operatorname{rec} K$ according to Definition 6.1.

For the second equality, let $\boldsymbol{e} \in \operatorname{rec} K$. Then $\boldsymbol{e} + \boldsymbol{x} \in K$, or, equivalently, $\boldsymbol{e} \in K - \boldsymbol{x}$ for every point $\boldsymbol{x} \in K$. Thus $\boldsymbol{e} \in \cap (K - \boldsymbol{x} : \boldsymbol{x} \in K)$. Conversely, suppose that $\boldsymbol{e} \in \cap (K - \boldsymbol{x} : \boldsymbol{x} \in K)$. Then $\boldsymbol{e} + \boldsymbol{x} \in K$ for all $\boldsymbol{x}$ in K. Equivalently, $\boldsymbol{e} + K \subset K$. Consequently, $\boldsymbol{e} \in \operatorname{rec} K$ by the above argument.

Problem 6.3. First, we are going to prove that the set

$$L = \{\boldsymbol{e} \in \mathbb{R}^n : \pm \boldsymbol{e} + K \subset K\}$$

equals $\operatorname{lin} K$. Indeed, if $\boldsymbol{e} \in \operatorname{lin} K$, then $\pm \boldsymbol{e} + \boldsymbol{x} \in K$ for all $\boldsymbol{x} \in K$, which implies the inclusion $\pm \boldsymbol{e} + K \subset K$. Hence $\operatorname{lin} K \subset L$.

Conversely, let $\boldsymbol{e} \in L$. By induction on $r \geqslant 1$, we easily conclude that $\pm r\boldsymbol{e} + K \subset K$ for all $r \geqslant 1$. Choose a point $\boldsymbol{x} \in K$, a scalar $\lambda \in \mathbb{R}$, and a positive integer r such that $|\lambda| \leqslant r$. Then $\lambda \boldsymbol{e} \in [-r\boldsymbol{e}, r\boldsymbol{e}]$, and

$$\lambda \boldsymbol{e} + \boldsymbol{x} \in [-r\boldsymbol{e}, r\boldsymbol{e}] + \boldsymbol{x} = [\boldsymbol{x} - r\boldsymbol{e}, \boldsymbol{x} + r\boldsymbol{e}] \subset K$$

by the convexity of K. Hence $\boldsymbol{e} \in \operatorname{lin} K$. Summing up, $L \subset \operatorname{lin} K$.

Next, we assert that L coincides with the set $L' = \{\boldsymbol{e} \in \mathbb{R}^n : \boldsymbol{e} + K = K\}$. Indeed, for every vector $\boldsymbol{e} \in L$, one has

$$K = \boldsymbol{o} + K = \boldsymbol{e} + (-\boldsymbol{e} + K) \subset \boldsymbol{e} + K \subset K,$$

implying the equality $\boldsymbol{e} + K = K$. Hence $\boldsymbol{e} \in L'$.

Conversely, let $\boldsymbol{e} \in L'$. Then $\boldsymbol{e} + K = K$. Furthermore,

$$-\boldsymbol{e} + K = -\boldsymbol{e} + (\boldsymbol{e} + K) = \boldsymbol{o} + K = K,$$

which gives the inclusion $\boldsymbol{e} \in L$. Summing up, $L = L'$.

Problem 6.4. If a vector $\boldsymbol{e} \in \mathbb{R}^n$ satisfies the problem condition, then $\boldsymbol{e} \in \operatorname{rec}(\operatorname{cl} K)$, as follows from Theorem 6.17. Conversely, let $\boldsymbol{e} \in \operatorname{rec}(\operatorname{cl} K)$. By Theorem 6.17, there is a sequence of points $\boldsymbol{x}_1, \boldsymbol{x}_2, \dots$ in K and a sequence of positive scalars $\mu_1, \mu_2, \dots$ tending to 0 such that $\boldsymbol{e} = \lim_{i \to \infty} \mu_i \boldsymbol{x}_i$. Choose any point $\boldsymbol{x} \in K$ and let $\boldsymbol{c}_i = \mu_i(\boldsymbol{x}_i - \boldsymbol{x})$, $i \geqslant 1$. Then

$$\lim_{i \to \infty} \boldsymbol{c}_i = \lim_{i \to \infty} \mu_i(\boldsymbol{x}_i - \boldsymbol{x}) = \lim_{i \to \infty} \mu_i \boldsymbol{x}_i - \lim_{i \to \infty} \mu_i \boldsymbol{x} = \boldsymbol{e} - \boldsymbol{o} = \boldsymbol{e}.$$

Put $\boldsymbol{x}_i = \boldsymbol{x} + \mu_i^{-1}\boldsymbol{c}_i$, $i \geqslant 1$. Clearly,

$$\boldsymbol{x}_i = \boldsymbol{x} + \mu_i^{-1}(\mu_i(\boldsymbol{x}_i - \boldsymbol{x})) = \boldsymbol{x} + \boldsymbol{x}_i - \boldsymbol{x} = \boldsymbol{x}_i \in K, \quad i \geqslant 1.$$

Choose any sequence of positive scalars $\lambda_1, \lambda_2, \ldots$ tending to 0. For any integer $j \geqslant 1$, there is an integer i_j such that $\mu_{i_j} < \lambda_j$. Clearly, $i_j \to \infty$ as $i \to \infty$ because $\lambda_j \to 0$. Let $\boldsymbol{x}'_j = \boldsymbol{x} + \lambda_j^{-1}\boldsymbol{c}_{i_j}$, $j \geqslant 1$. Since $0 < \lambda_j^{-1} < \mu_{i_j}^{-1}$ for all $j \geqslant 1$, the convexity of K gives

$$\boldsymbol{x}'_j = \boldsymbol{x} + \lambda_j^{-1}\boldsymbol{c}_{i_j} \in (\boldsymbol{x}, \boldsymbol{x} + \mu_{i_j}^{-1}\boldsymbol{c}_{i_j}) = (\boldsymbol{x}, \boldsymbol{x}_{i_j}) \subset K, \quad j \geqslant 1.$$

Furthermore,

$$\lim_{j\to\infty} \lambda_j \boldsymbol{x}'_j = \lim_{j\to\infty} (\lambda_j \boldsymbol{x} + \lambda_j \lambda_j^{-1}\boldsymbol{c}_{i_j}) = \boldsymbol{o} + \lim_{j\to\infty} \boldsymbol{c}_{i_j} = \boldsymbol{e}.$$

Problem 6.5. For the first equality, Theorem 6.4 implies that it is sufficient to prove the inclusion $\operatorname{rec}(K_1 \cap K_2) \subset \operatorname{rec} K_1 \cap \operatorname{rec} K_2$. Since the case $\operatorname{rec}(K_1 \cap K_2) \subset \{\boldsymbol{o}\}$ is obvious, we may assume that $\operatorname{rec}(K_1 \cap K_2) \not\subset \{\boldsymbol{o}\}$. Choose any nonzero vector $\boldsymbol{e} \in \operatorname{rec}(K_1 \cap K_2)$. Given a point $\boldsymbol{u} \in K_1 \cap K_2$, the halfline $h = [\boldsymbol{u}, \boldsymbol{e} + \boldsymbol{u}\rangle$ lies in $K_1 \cap K_2$. Consequently, Corollary 6.16 shows that $\boldsymbol{e} \in \operatorname{rec} K_1 \cap \operatorname{rec} K_2$.

Problem 6.6. (a) If K is a plane, then $\operatorname{lin} K = \operatorname{dir} K$ and $\dim(\operatorname{lin} K) = m$. Conversely, assume that $\dim(\operatorname{lin} K) = m$. Choose a point $\boldsymbol{x} \in K$. Then $\boldsymbol{x} + \operatorname{lin} K \subset K$ according to Theorem 6.20. Consequently, $\boldsymbol{x} + \operatorname{lin} K \subset \operatorname{aff} K$. Since $\dim(\operatorname{aff} K) = m$, Theorem 2.6 implies the equality $\boldsymbol{x} + \operatorname{lin} K = \operatorname{aff} K$. Hence $K = \operatorname{aff} K$, and K is a plane.

(b) If K is a halfplane of a slab of a plane $L \subset \mathbb{R}^n$, then $\dim(\operatorname{lin} K) = m - 1$. Conversely, assume that $\dim(\operatorname{lin} K) = m - 1$. Since the case $m = 1$ is obvious (K is either a closed halfline or a closed segment), we may suppose that $m \geqslant 2$. By Theorem 6.25, $K = (K \cap S) + \operatorname{lin} K$, where S is a complementary to $\operatorname{lin} K$ subspace of $\mathbb{R}^n$. Then

$$\dim(K \cap S) = \dim K - \dim(\operatorname{lin} K) = 1,$$

which shows that $K \cap S$ is either a closed halfline or a closed segment. Therefore, K is either a halfplane or a slab (compare with Theorem 2.74).

(c) Assume first that K contains an $(m-1)$-dimensional plane L. Choose points $\boldsymbol{a} \in L$ and $\boldsymbol{u} \in \operatorname{cl} K$. By Corollary 6.11, the plane $L' = (\boldsymbol{u} - \boldsymbol{a}) + L$ lies in $\operatorname{cl} K$. Corollary 6.34 shows that $L' \subset \boldsymbol{u} + \operatorname{lin}(\operatorname{cl} K) \subset \operatorname{cl} K$. Hence

$$m - 1 = \dim L' \leqslant \dim(\operatorname{lin}(\operatorname{cl} K)) \leqslant \dim(\operatorname{cl} K) = \dim K = m.$$

Assertions (a) and (b) imply that K is a plane, a halfplane, or a plane slab. The converse assertion is obvious.

Problem 6.7. Because the implications $(c) \Rightarrow (b) \Rightarrow (a)$ are obvious, it remains to show that $(a) \Rightarrow (c)$. Since $\operatorname{rbd} K = \operatorname{rbd}(\operatorname{cl} K)$, we may assume that the set

K is closed. Under this assumption, (c) becomes equivalent to the assertion that K is a closed plane slab. Choose a subspace $S \subset \mathbb{R}^n$ complementary to $\operatorname{lin} K$ and express K as $K = (K \cap S) + \operatorname{lin} K$, where $\operatorname{lin}(K \cap S) = \{\boldsymbol{o}\}$ (see Theorem 6.25). Then $\operatorname{rbd} K = \operatorname{rbd}(K \cap S) + \operatorname{lin} K$ according to Theorem 6.31. Furthermore, $K \cap S$ is line-free (see Theorem 6.37).

First, we observe that $\dim(K \cap S) \geqslant 1$. Indeed, if $\dim(K \cap S) = 0$, then $K \cap S$ is a singleton and K is translate of $\operatorname{lin} K$, implying that $\operatorname{rbd} K = \varnothing$. Next, suppose that $\dim(K \cap S) = 1$. Then $K \cap S$ is either a closed segment, a closed halfline, or a line. Under this assumption, $\operatorname{rbd}(K \cap S)$ is disconnected (namely, consists of two points) if and only if $K \cap S$ is a closed segment. Therefore, $\operatorname{rbd} K$ is disconnected (namely, consists of two translates of $\operatorname{lin} K$) if and only if K is a closed slab.

Finally, we are going to show that the set $\operatorname{rbd} K$ is connected provided $\dim(K \cap S) \geqslant 2$. For this, we will show that any pair of points $\boldsymbol{u}, \boldsymbol{v} \in \operatorname{rbd}(K \cap S)$ can be connected by a continuous curve which lies in $\operatorname{rbd}(K \cap S)$. Indeed, this is obvious if $[\boldsymbol{u}, \boldsymbol{v}] \subset \operatorname{rbd}(K \cap S)$. Alternatively, let $(\boldsymbol{u}, \boldsymbol{v}) \subset \operatorname{rint}(K \cap S)$ (see Theorem 3.67). Choose in S a 2-dimensional plane L which contains $(\boldsymbol{u}, \boldsymbol{v})$. From $K \cap L = (K \cap S) \cap L$ we conclude that the 2-dimensional closed convex set $K \cap L$ is line-free. It is easy to see that $\boldsymbol{u}$ and $\boldsymbol{v}$ are endpoints of a continuous arc $\Gamma(\boldsymbol{u}, \boldsymbol{v}) \subset \operatorname{rbd}(K \cap L)$. Since $\operatorname{rbd}(K \cap L) = \operatorname{rbd}(K \cap S) \cap L$ (see Corollary 3.64), we conclude that $\Gamma(\boldsymbol{u}, \boldsymbol{v}) \subset \operatorname{rbd}(K \cap S)$.

Now, let $\boldsymbol{x}$ and $\boldsymbol{z}$ be points in $\operatorname{rbd} K$. Denote by $\boldsymbol{x}_0$ and $\boldsymbol{z}_0$, respectively, the images of $\boldsymbol{x}$ and $\boldsymbol{z}$ upon the linear projection on S along $\operatorname{lin} K$. Clearly, $[\boldsymbol{x}, \boldsymbol{x}_0] \subset \operatorname{rbd} K$ and $[\boldsymbol{z}, \boldsymbol{z}_0] \subset \operatorname{rbd} K$. By the above proved, $\boldsymbol{x}_0$ and $\boldsymbol{z}_0$ are connected by a continuous arc $\Gamma(\boldsymbol{x}_0, \boldsymbol{z}_0) \subset \operatorname{rbd}(K \cap S) = \operatorname{rbd} K \cap S$. Summing up, the continuous path $[\boldsymbol{x}, \boldsymbol{x}_0] \cup \Gamma(\boldsymbol{x}_0, \boldsymbol{z}_0) \cup [\boldsymbol{z}_0, \boldsymbol{z}]$ joins $\boldsymbol{x}$ and $\boldsymbol{z}$ in $\operatorname{rbd} K$.

Problem 6.8. If $\operatorname{rbd} K = \varnothing$, then K is a plane (see Theorem 3.62). Assume that $\operatorname{rbd} K \neq \varnothing$. If $\operatorname{rbd} K$ is connected, then, being closed (see Theorem 3.62), it is convex, as shown in Problem 3.18. Therefore, K is a closed halfplane, as follows from Theorem 3.68. Finally, if $\operatorname{rbd} K$ is disconnected, then K is a plane slab, as shown in Problem 6.7.

Problem 6.9. Clearly, the assertion holds when C is a closed halfline. Therefore, we may assume that $\dim C \geqslant 2$. Since the case $\lambda_1 = \cdots = \lambda_r = 0$ is obvious, we let $\lambda_1 + \cdots + \lambda_r > 0$.

$(a) \Rightarrow (b)$ Assume, for contradiction, that a sequence of nonnegative combinations

$$\boldsymbol{z}_i = \lambda_1^{(i)} \boldsymbol{x}_1^{(i)} + \cdots + \lambda_r^{(i)} \boldsymbol{x}_r^{(i)}, \quad \lambda_1^{(i)} + \cdots + \lambda_r^{(i)} > 0, \quad i \geqslant 1,$$

of linearly independent sets $\{\boldsymbol{x}_1^{(i)}, \dots, \boldsymbol{x}_r^{(i)}\}$ of unit vectors from C satisfies the conditions

$$\|\boldsymbol{z}_i\| < \frac{1}{i}(\lambda_1^{(i)} + \cdots + \lambda_r^{(i)}), \quad i \geqslant 1.$$

Put $\lambda_i = \lambda_1^{(i)} + \cdots + \lambda_r^{(i)}$ and $\mu_j^{(i)} = \lambda_j^{(i)}/\lambda_i$ for all $i \geqslant 1$ and $1 \leqslant j \leqslant r$. Every point $\boldsymbol{u}_i = \boldsymbol{z}_i/\lambda_i$ is a convex combination of the form

$$\boldsymbol{u}_i = \mu_1^{(i)}\boldsymbol{x}_1^{(i)} + \cdots + \mu_r^{(i)}\boldsymbol{x}_r^{(i)}, \quad i \geqslant 1,$$

and satisfies the inequality $\|\boldsymbol{u}_i\| < \frac{1}{i}$. Therefore, $\lim_{i\to\infty} \boldsymbol{u}_i = \boldsymbol{o}$. Using Cantor's diagonalization argument, we may assume the existence of limits

$$\lim_{i\to\infty} \mu_j^{(i)} = \mu_j, \quad \lim_{i\to\infty} \boldsymbol{x}_j^{(i)} = \boldsymbol{x}_j, \quad 1 \leqslant j \leqslant r.$$

Clearly,

$$\boldsymbol{o} = \lim_{i\to\infty} \boldsymbol{u}_i = \lim_{i\to\infty} (\mu_1^{(i)}\boldsymbol{x}_1^{(i)} + \cdots + \mu_r^{(i)}\boldsymbol{x}_r^{(i)}) = \mu_1\boldsymbol{x}_1 + \cdots + \mu_r\boldsymbol{x}_r$$

is a convex combination of suitable unit vectors $\boldsymbol{x}_1, \ldots, \boldsymbol{x}_r$ from C (possibly, the set $\{\boldsymbol{x}_1, \ldots, \boldsymbol{x}_r\}$ is linearly dependent). Excluding all terms of the form $0\boldsymbol{x}_j$, we obtain a representation of $\boldsymbol{o}$ as a positive convex combination of suitable unit vectors $\boldsymbol{x}_{j_1}, \ldots, \boldsymbol{x}_{j_p} \in C$, $2 \leqslant p \leqslant r$. Corollary 4.19 shows that $\boldsymbol{o} \in \operatorname{rint}(\operatorname{conv}\{\boldsymbol{x}_{j_1}, \ldots, \boldsymbol{x}_{j_p}\})$. Consequently, Theorem 5.48 implies that

$$\operatorname{span}\{\boldsymbol{x}_{j_1}, \ldots, \boldsymbol{x}_{j_p}\} = \operatorname{cone}_{\boldsymbol{o}}\{\boldsymbol{x}_{j_1}, \ldots, \boldsymbol{x}_{j_p}\} \subset \operatorname{conv} C.$$

Finally, by Theorem 5.17, the subspace $S = \operatorname{span}\{\boldsymbol{x}_{j_1}, \ldots, \boldsymbol{x}_{j_p}\}$ lies in $\operatorname{ap}(\operatorname{conv} C)$. Since $\dim S \geqslant 1$, we obtain a contradiction with the condition $\operatorname{ap}(\operatorname{conv} C) = \{\boldsymbol{o}\}$.

$(b) \Rightarrow (a)$ Assume for a moment that the subspace $\operatorname{ap}(\operatorname{conv} C)$ is larger than $\{\boldsymbol{o}\}$. Choose a unit vector $\boldsymbol{x} \in \operatorname{ap}(\operatorname{conv} C)$. Since $\dim C \geqslant 2$, the cone C is larger that $\operatorname{span}\{\boldsymbol{x}\}$. So, there is a unit vector $\boldsymbol{u} \in C \setminus \operatorname{span}\{\boldsymbol{x}\}$. Let

$$\boldsymbol{x}_\lambda = \frac{\lambda\boldsymbol{u} - \boldsymbol{x}}{\|\lambda\boldsymbol{u} - \boldsymbol{x}\|}, \quad \lambda \geqslant 0.$$

Obviously, $\boldsymbol{x}_\lambda$ is a unit vector in C and the set $\{\boldsymbol{x}, \boldsymbol{x}_\lambda\}$ is linearly independent for all $\lambda > 0$. Furthermore, since $\boldsymbol{x}_\lambda \to -\boldsymbol{x}$ while $\lambda \to 0$, one has $\frac{1}{2}\boldsymbol{x} + \frac{1}{2}\boldsymbol{x}_\lambda \to \boldsymbol{o}$ as $\lambda \to 0$, contrary to condition (a).

Problem 6.10. Without loss of generality, we may assume that $\boldsymbol{a} = \boldsymbol{o}$. By Theorem 5.4, $\operatorname{conv} C$ is a convex cone with apex $\boldsymbol{o}$. Choose a point $\boldsymbol{x} \in \operatorname{cl}(\operatorname{conv} C)$ and a sequence of points $\boldsymbol{x}_1, \boldsymbol{x}_2, \ldots$ from $\operatorname{conv} C$ converging to $\boldsymbol{x}$. Since the case $\boldsymbol{x} = \boldsymbol{o}$ is obvious ($\boldsymbol{o} \in C \subset \operatorname{conv} C$), we may suppose that $\boldsymbol{x} \neq \boldsymbol{o}$, and whence that none of the points $\boldsymbol{x}_1, \boldsymbol{x}_2, \ldots$ is $\boldsymbol{o}$. According to Theorem 5.35, every $\boldsymbol{x}_i$ can be written as a positive combination

$$\boldsymbol{x}_i = \lambda_1^{(i)}\boldsymbol{x}_1^{(i)} + \cdots + \lambda_{k_i}^{(i)}\boldsymbol{x}_{k_i}^{(i)}, \quad i \geqslant 1,$$

where the set $\{\boldsymbol{o}, \boldsymbol{x}_1^{(i)}, \dots, \boldsymbol{x}_{k_i}^{(i)}\}$ is affinely independent (equivalently, the set $\{\boldsymbol{x}_1^{(i)}, \dots, \boldsymbol{x}_{k_i}^{(i)}\}$ is linearly independent). With

$$\boldsymbol{u}_j^{(i)} = \boldsymbol{x}_j^{(i)}/\|\boldsymbol{x}_j^{(i)}\| \quad \text{and} \quad \mu_j^{(i)} = \lambda_j^{(i)}\|\boldsymbol{x}_j^{(i)}\|, \quad i \geqslant 1, \;\; 1 \leqslant j \leqslant k_i,$$

every point $\boldsymbol{x}_i$ is expressed as the positive combination of the linearly independent set $\{\boldsymbol{u}_1^{(i)}, \dots, \boldsymbol{u}_{k_i}^{(i)}\}$ of unit vectors from C. Since all integers $k_1, k_2, \dots$ are bounded above by n, we can choose a subsequence $k_{i_1}, k_{i_2}, \dots$ of $k_1, k_2, \dots$ such that $k_{i_1} = k_{i_2} = \cdots = k$ for a suitable k. Furthermore, because all vectors $\boldsymbol{u}_j^{(i)}$ belong to the compact set $C \cap \mathbb{S}$, every sequence $\boldsymbol{u}_j^{(i_1)}, \boldsymbol{u}_j^{(i_2)}, \dots, 1 \leqslant j \leqslant k$, contains a converging subsequence. Finally, using Cantor's diagonalization argument, we may assume that every point $\boldsymbol{x}_i$ is written as a positive combination

$$\boldsymbol{x}_i = \mu_1^{(i)}\boldsymbol{u}_1^{(i)} + \cdots + \mu_k^{(i)}\boldsymbol{u}_k^{(i)}, \quad i \geqslant 1,$$

where the set $\{\boldsymbol{u}_1^{(i)}, \dots, \boldsymbol{u}_k^{(i)}\}$ is linearly independent and every sequence $\boldsymbol{u}_j^{(1)}, \boldsymbol{u}_j^{(2)}, \dots$ converges to a suitable point $\boldsymbol{u}_j \in C$, $1 \leqslant j \leqslant k$.

Since the case $\dim C = 1$ is obvious (C is a closed halfline with endpoint $\boldsymbol{o}$), we may assume that $\dim C \geqslant 2$. By Problem 6.8, there is a scalar $\gamma > 0$ such that

$$\|\boldsymbol{x}_i\| = \|\mu_1^{(i)}\boldsymbol{u}_1^{(i)} + \cdots + \mu_k^{(i)}\boldsymbol{u}_k^{(i)}\| \geqslant \gamma(\mu_1^{(i)} + \cdots + \mu_r^{(i)}), \quad i \geqslant 1.$$

Because $\lim_{i\to\infty}\|\boldsymbol{x}_i\| = \|\boldsymbol{x}\|$, every sequence $\mu_j^{(1)}, \mu_j^{(2)}, \dots, 1 \leqslant j \leqslant k$, is bounded. Again using a diagonalization argument, we assume that every sequence $\mu_j^{(1)}, \mu_j^{(2)}, \dots$ converges to a suitable scalar $\mu_j \geqslant 0$, $1 \leqslant j \leqslant k$. Finally,

$$\begin{aligned}\boldsymbol{x} &= \lim_{i\to\infty} \boldsymbol{x}_i = \lim_{i\to\infty} (\mu_1^{(i)}\boldsymbol{u}_1^{(i)} + \cdots + \mu_k^{(i)}\boldsymbol{u}_k^{(i)}) \\ &= \mu_1\boldsymbol{u}_1 + \cdots + \mu_k\boldsymbol{u}_k \in \operatorname{conv} C.\end{aligned}$$

Problem 6.11. $(a) \Rightarrow (b)$ Suppose that the set $\operatorname{cl} K \cap P'$ is bounded, where P' is a translate of P. Choose a hyperplane H which lies between the boundary hyperplane of P' and meets $\operatorname{cl} K$. Clearly, H is a translate of S. Then $\operatorname{cl} K \cap H$ is bounded, and Theorem 6.40 implies (b).

$(b) \Rightarrow (a)$ Assume, for contradiction, the existence of a translate P' of P that the set $\operatorname{cl} K \cap P'$ is unbounded. Choose a point $\boldsymbol{u} \in \operatorname{cl} K \cap P'$. By Theorem 6.12, $\operatorname{cl} K \cap P'$ contains a closed halfline h with endpoint $\boldsymbol{u}$. Therefore, the closed halfline $h' = h - \boldsymbol{u}$ lies in $\operatorname{rec}(\operatorname{cl} K)$, as follows from Corollary 6.15. Problem 2.12 shows that the line l containing h lies in P'. Consequently, l is parallel to the boundary hyperplanes of P' and whence is parallel to S (see Theorem 2.34). Finally, Theorem 2.13 implies the inclusion $h' \subset l - \boldsymbol{u} \subset S$, contrary to the assumption $\operatorname{rec}(\operatorname{cl} K) \cap S = \{\boldsymbol{o}\}$.

Chapter 7

Problem 7.1. Theorem 7.7 shows that the equality $\mathrm{cl}\, C_{\boldsymbol{a}}(K) = \boldsymbol{a} + \mathrm{rec}\,(\mathrm{cl}\, K)$ holds if and only if $C_{\boldsymbol{a}}(K) \subset \boldsymbol{a} + \mathrm{rec}\,(\mathrm{cl}\, K)$. Since $\boldsymbol{a} + \mathrm{rec}\,(\mathrm{cl}\, K)$ is a cone with apex $\boldsymbol{a}$, the last inclusion is satisfied if and only if $K \subset \boldsymbol{a} + \mathrm{rec}\,(\mathrm{cl}\, K)$.

Problem 7.2. An induction argument shows that the proof can be reduced to the case $r = 2$ (let $\mu_1 X_1 + \mu_2 X_2 = \mu_1 X_1 + \mu_2\{o\}$ if $r = 1$).

(*a*) By Definition 7.5, there is a cone $C_i \subset \mathbb{R}^n$ with apex $\boldsymbol{a}_i$ and a closed ball $B_{\rho_i}(\boldsymbol{a}_i) \subset \mathbb{R}^n$ such that $X_i \cap B_{\rho_i}(\boldsymbol{a}_i) = C_i \cap B_{\rho_i}(\boldsymbol{a}_i)$, $i = 1, 2$. Without loss of generality we may assume that both scalars μ_1, μ_2 are not zero. Let $\rho = \min\{|\mu_1|\rho_1, |\mu_2|\rho_2\}$. Then $\rho > 0$ and (see page 9)

$$\begin{aligned}(\mu_1 X_1 + \mu_2 X_2) \cap B_\rho(\mu_1\boldsymbol{a}_1 + \mu_2\boldsymbol{a}_2) &\\ = \mu_1(X_1 \cap B_{\rho/|\mu_1|}(\boldsymbol{a}_1)) + \mu_2(X_2 \cap B_{\rho/|\mu_2|}(\boldsymbol{a}_2)) &\\ = \mu_1(C_1 \cap B_{\rho/|\mu_1|}(\boldsymbol{a}_1)) + \mu_2(C_2 \cap B_{\rho/|\mu_2|}(\boldsymbol{a}_2)) &\\ = (\mu_1 C_1 + \mu_2 C_2) \cap B_\rho(\mu_1\boldsymbol{a}_1 + \mu_2\boldsymbol{a}_2). &\end{aligned}$$

Since $\mu_1 C_1 + \mu_2 C_2$ is a cone with apex $\mu_1\boldsymbol{a}_1 + \mu_2\boldsymbol{a}_2$ (see Theorem 5.10), we obtain that the set $\mu_1 X_1 + \mu_2 X_2$ is locally conic at $\mu_1\boldsymbol{a}_1 + \mu_2\boldsymbol{a}_2$.

(*b*) By Definition 7.5, there is a cone $C_i \subset \mathbb{R}^n$ with apex $\boldsymbol{a}_i$ and a closed ball $B_{\rho_i}(\boldsymbol{a}_i) \subset \mathbb{R}^n$ such that $X_i \cap B_{\rho_i}(\boldsymbol{a}_i) = C_i \cap B_{\rho_i}(\boldsymbol{a}_i)$, $i = 1, 2$. Let $\rho = \min\{\rho_1, \rho_2\}$. Then $\rho > 0$ and

$$\begin{aligned}(X_1 \cap X_2) \cap B_\rho(\boldsymbol{a}) &= (X_1 \cap B_\rho(\boldsymbol{a})) \cap (X_2 \cap B_\rho(\boldsymbol{a}))\\ &= (C_1 \cap B_\rho(\boldsymbol{a})) \cap (C_2 \cap B_\rho(\boldsymbol{a})) = (C_1 \cap C_2) \cap B_\rho(\boldsymbol{a}).\end{aligned}$$

Since $C_1 \cap C_2$ is a cone with apex $\boldsymbol{a}$ (see Corollary 5.9), we obtain that the set $X_1 \cap X_2$ is locally conic at $\boldsymbol{a}$.

Problem 7.3. Assume first that L is asymptotic to X. Since $p_T(L) = L \cap T = \{\boldsymbol{c}\}$, one has $\boldsymbol{c} \notin p_T(\mathrm{cl}\, X)$. The equality $\delta(L, X) = 0$ shows the existence of sequences $\boldsymbol{u}_1, \boldsymbol{u}_2, \ldots \in L$ and $\boldsymbol{v}_1, \boldsymbol{v}_2, \ldots \in X$ such that $\lim_{i\to\infty} \|\boldsymbol{u}_i - \boldsymbol{v}_i\| = 0$. Then $p_T(\boldsymbol{u}_i) = \boldsymbol{c}$ and $p_T(\boldsymbol{v}_i) \in p_T(X)$ for all $i \geq 1$. Because

$$\lim_{i\to\infty} \|\boldsymbol{c} - p_T(\boldsymbol{v}_i)\| = \lim_{i\to\infty} \|p_T(\boldsymbol{u}_i) - p_T(\boldsymbol{v}_i)\| \leqslant \lim_{i\to\infty} \|\boldsymbol{u}_i - \boldsymbol{v}_i\| = 0,$$

we obtain the inclusion $\boldsymbol{c} \in \mathrm{cl}\, p_T(X)$.

Conversely, let $\boldsymbol{c} \in \mathrm{cl}\, p_T(X) \setminus p_T(\mathrm{cl}\, X)$. As above, the condition $p_T(L) = \{\boldsymbol{c}\} \not\subset p_T(\mathrm{cl}\, X)$ show that $L \cap \mathrm{cl}\, X = \varnothing$. Next, let $\boldsymbol{w}_1, \boldsymbol{w}_2, \ldots$ be a sequence of points in $p_T(X)$ such that $\boldsymbol{c} = \lim_{i\to\infty} \boldsymbol{w}_i$. Choose in X points $\boldsymbol{v}_1, \boldsymbol{v}_2, \ldots$ satisfying the conditions $p_T(\boldsymbol{v}_i) = \boldsymbol{w}_i$ for all $i \geq 1$. Now, let $\boldsymbol{u}_i = \boldsymbol{c} + (\boldsymbol{v}_i - \boldsymbol{w}_i)$, $i \geq 1$. Then $\boldsymbol{u}_i \in \boldsymbol{c} + S = L$ because $\boldsymbol{v}_i - \boldsymbol{w}_i \in T^\perp = S$. Furthermore,

$$\delta(L, X) \leqslant \lim_{i\to\infty} \|\boldsymbol{u}_i - \boldsymbol{v}_i\| = \lim_{i\to\infty} \|\boldsymbol{c} + (\boldsymbol{v}_i - \boldsymbol{w}_i) - \boldsymbol{v}_i\| = \lim_{i\to\infty} \|\boldsymbol{c} - \boldsymbol{w}_i\| = 0.$$

Summing up, L is an asymptote of X.

Problem 7.4. By Theorem 7.13, $S_X \neq \varnothing$. First, we assert that

$$(\mathrm{cl}\, X + S) \cap S_X = \varnothing.$$

Indeed, assume for a moment the existence of a point $\boldsymbol{x} \in (\mathrm{cl}\, X + S) \cap S_X$. Then $\boldsymbol{x} = \boldsymbol{y} + \boldsymbol{z}$ for suitable points $\boldsymbol{y} \in \mathrm{cl}\, X$ and $\boldsymbol{z} \in S$. By the definition of S_X, there is a plane $L \subset S_X$ which is a translate of S, contains $\boldsymbol{x}$, and is asymptotic to X. The first two conditions give $L = \boldsymbol{x} + S$ (see Theorem 2.2), while the third one implies $\mathrm{cl}\, X \cap L = \varnothing$. On the other hand, $\boldsymbol{y} = \boldsymbol{x} - \boldsymbol{z} \in \boldsymbol{x} + S = L$, a contradiction.

Next, we are going to prove that

$$\mathrm{cl}\,(X + S) \subset (\mathrm{cl}\, X + S) \cup S_X.$$

Clearly, this inclusion is equivalent to the following:

$$\mathrm{cl}\,(X + S) \setminus (\mathrm{cl}\, X + S) \subset S_X.$$

By Lemma 7.11, $\mathrm{cl}\,(X + S) \setminus (\mathrm{cl}\, X + S) \neq \varnothing$. Choose any point

$$\boldsymbol{x} \in \mathrm{cl}\,(X + S) \setminus (\mathrm{cl}\, X + S)$$

and consider the plane $L = \boldsymbol{x} + S$. We state that L is an asymptote of X. For this, we need to verify that $\mathrm{cl}\, X \cap L = \varnothing$ and $\delta(X, L) = 0$. Indeed, assume for a moment that $\mathrm{cl}\, X \cap L \neq \varnothing$ and choose a point $\boldsymbol{z} \in \mathrm{cl}\, X \cap L$. Then $\boldsymbol{x} - \boldsymbol{z} \in L - L = S$, which gives the inclusion $\boldsymbol{x} = \boldsymbol{z} + (\boldsymbol{x} - \boldsymbol{z}) \in \mathrm{cl}\, X + S$, contrary to the choice of $\boldsymbol{x}$. For the equality $\delta(X, L) = 0$, choose a sequence $\boldsymbol{x}_1, \boldsymbol{x}_2, \ldots$ of points from $X + S$ which converge to $\boldsymbol{x}$. Let $\boldsymbol{x}_i = \boldsymbol{y}_i + \boldsymbol{u}_i$, where $\boldsymbol{y}_i \in X$ and $\boldsymbol{u}_i \in S$, $i \geqslant 1$. Then $\boldsymbol{x} - \boldsymbol{u}_i \in \boldsymbol{x} + S = L$ and

$$\begin{aligned}\delta(X, L) &\leqslant \lim_{i\to\infty} \|\boldsymbol{y}_i - (\boldsymbol{x} - \boldsymbol{u}_i)\| \\ &= \lim_{i\to\infty} \|(\boldsymbol{x}_i - \boldsymbol{u}_i) - (\boldsymbol{x} - \boldsymbol{u}_i)\| = \lim_{i\to\infty} \|\boldsymbol{x}_i - \boldsymbol{x}\| = 0.\end{aligned}$$

Summing up, L is an asymptotic plane of X. Thus $\boldsymbol{x} \in L \subset S_X$.

Finally, it remains to show that

$$(\mathrm{cl}\, X + S) \cup S_X \subset \mathrm{cl}\,(X + S).$$

Due to Lemma 7.11, it suffices to prove the inclusion $S_X \subset \mathrm{cl}\,(X + S)$. Choose any point $\boldsymbol{x} \in S_X$. Then there is a plane $L \subset \mathbb{R}^n$ which is a translate of S, contains $\boldsymbol{x}$, and is an asymptote of X. Clearly, $L = \boldsymbol{x} + S$ (see Theorem 2.2). Since L is an asymptote of X, there is a sequence $\boldsymbol{z}_1, \boldsymbol{z}_2, \ldots$ of points from X such that $\lim_{i\to\infty} \delta(\boldsymbol{z}_i, L) = 0$. Denote by $\boldsymbol{u}_i$ the orthogonal projection of $\boldsymbol{z}_i$ on L, $i \geqslant 1$.

By Theorem 2.90, $\|\boldsymbol{z}_i - \boldsymbol{u}_i\| = \delta(\boldsymbol{z}_i, L)$. Put $\boldsymbol{x}_i = \boldsymbol{z}_i + (\boldsymbol{x} - \boldsymbol{u}_i)$, $i \geqslant 1$. Since $\boldsymbol{x} - \boldsymbol{u}_i \in L - L = S$, one has $\boldsymbol{x}_i \in \boldsymbol{z}_i + S \subset X + S$. Furthermore,

$$\lim_{i\to\infty} \|\boldsymbol{x} - \boldsymbol{x}_i\| = \lim_{i\to\infty} \|\boldsymbol{z}_i - \boldsymbol{u}_i\| = \lim_{i\to\infty} \delta(\boldsymbol{z}_i, L) = 0.$$

Consequently, $\boldsymbol{x} \in \mathrm{cl}\,(X + S)$. Summing up, $S_X \subset \mathrm{cl}\,(X + S)$.

Problem 7.5. As in the proof of Theorem 7.14, we can equivalently reformulate our assertions replacing f with g. By Theorem 7.14, $N_X \neq \varnothing$. First, we assert that $g(\mathrm{cl}\, X) \cap g(N_X) = \varnothing$.

Indeed, assume for a moment the existence of a point $\boldsymbol{x} \in g(\mathrm{cl}\, X) \cap g(N_X)$. Then $\boldsymbol{x} = g(\boldsymbol{y}) = g(\boldsymbol{z})$ for suitable points $\boldsymbol{y} \in \mathrm{cl}\, X$ and $\boldsymbol{z} \in N_X$. By the definition of N_X, the plane $L = \boldsymbol{z} + \mathrm{null}\, g$ lies in N_X and is asymptotic to X. In particular, $\mathrm{cl}\, X \cap L = \varnothing$. On the other hand, the equality $g(\boldsymbol{y} - \boldsymbol{z}) = g(\boldsymbol{y}) - g(\boldsymbol{z}) = \boldsymbol{o}$ implies the inclusion $\boldsymbol{y} - \boldsymbol{z} \in \mathrm{null}\, g$. Consequently,

$$\boldsymbol{y} = \boldsymbol{z} - (\boldsymbol{z} - \boldsymbol{y}) \in \boldsymbol{z} + \mathrm{null}\, g = L,$$

a contradiction with $\mathrm{cl}\, X \cap L = \varnothing$.

Next, we are going to prove that

$$\mathrm{cl}\, g(X) \subset g(\mathrm{cl}\, X) \cup g(N_X). \tag{14.21}$$

Clearly, this inclusion is equivalent to the following:

$$\mathrm{cl}\, g(X) \setminus g(\mathrm{cl}\, X) \subset g(N_X).$$

To prove this inclusion, choose any point $\boldsymbol{u} \in \mathrm{cl}\, g(X) \setminus g(\mathrm{cl}\, X)$. Let T be a subspace of $\mathbb{R}^n$ complementary to $\mathrm{null}\, g$. Denote by Y the projection of $\mathrm{cl}\, X$ on T along $\mathrm{null}\, g$. Then $\mathrm{cl}\, X + \mathrm{null}\, g = Y + \mathrm{null}\, g$, which gives

$$g(\mathrm{cl}\, X) = g(\mathrm{cl}\, X + \mathrm{null}\, g) = g(Y + \mathrm{null}\, g) = g(Y).$$

From linear algebra we know that the mapping $h : T \to \mathrm{rng}\, g$ defined by $h(\boldsymbol{x}) = g(\boldsymbol{x})$ is an invertible linear transformation. Clearly, $g(Y) = h(Y)$.

Because $g(X)$ lies in the subspace $\mathrm{rng}\, g$, which is a closed set, we have $\mathrm{cl}\, g(X) \subset \mathrm{rng}\, g$. Hence there is a point $\boldsymbol{z} \in T$ such that $g(\boldsymbol{z}) = h(\boldsymbol{z}) = \boldsymbol{u}$. Consider the plane $L = \boldsymbol{z} + \mathrm{null}\, g$. Obviously, $\boldsymbol{u} = f(\boldsymbol{z}) \in f(L)$.

We assert that $L \subset N_X$. For this, we first observe that $\mathrm{cl}\, X \cap L = \varnothing$. Indeed, otherwise choosing a point $\boldsymbol{v} \in \mathrm{cl}\, X \cap L$, we will have $L = \boldsymbol{v} + \mathrm{null}\, g$. Then

$$\boldsymbol{u} = g(\boldsymbol{z} + \mathrm{null}\, g) = g(L) = g(\boldsymbol{v} + \mathrm{null}\, g) = g(\boldsymbol{v}) \in g(\mathrm{cl}\, X),$$

contrary to the choice of $\boldsymbol{u}$ in $\mathrm{cl}\, f(X) \setminus f(\mathrm{cl}\, X)$.

Now, we intend to show that the plane L is asymptotic to X. Indeed, choose any scalar $\varepsilon > 0$. Because $\boldsymbol{u} \in \mathrm{cl}\, g(X)$, there is a point $\boldsymbol{u}' \in g(X)$ such that $\|\boldsymbol{u} - \boldsymbol{u}'\| < \varepsilon/\beta$, where β is a Lipschitz constant of h^{-1}. Choose a point $\boldsymbol{y}' \in X$

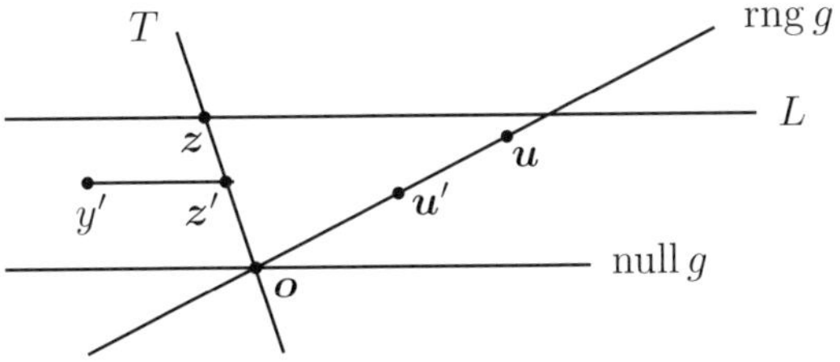

Fig. 14.1 Illustration to Problem 7.5.

with the property $g(\boldsymbol{y}') = \boldsymbol{u}'$ and denote by $\boldsymbol{z}'$ the projection of $\boldsymbol{y}'$ on T along $\operatorname{null} g$ (see Figure 14.1). Clearly, $h(\boldsymbol{z}') = g(\boldsymbol{z}') = g(\boldsymbol{y}') = \boldsymbol{u}'$. Consequently,

$$\delta(\boldsymbol{y}', L) = \delta(\boldsymbol{z}', L) \leqslant \|\boldsymbol{z}' - \boldsymbol{z}\| = \|h^{-1}(\boldsymbol{u}' - \boldsymbol{u})\| \leqslant \beta\|\boldsymbol{u}' - \boldsymbol{u}\| < \varepsilon.$$

Therefore,

$$\delta(X, L) = \inf\{\delta(\boldsymbol{x}, L) : \boldsymbol{x} \in X\} = 0,$$

which implies that L is asymptotic to X. Finally, (14.21) follows from

$$\boldsymbol{u} = g(\boldsymbol{z}) \in g(L) \subset g(N_X).$$

It remains to prove that $g(\operatorname{cl} X) \cup g(N_X) \subset \operatorname{cl} g(X)$. Because $g(\operatorname{cl} X) \subset \operatorname{cl} g(X)$, it suffices to show that $g(N_X) \subset \operatorname{cl} g(X)$. For this, choose any point $\boldsymbol{x} \in N_X$. Denote by L a translate of $\operatorname{null} g$ which contains $\boldsymbol{x}$. Clearly, $L = \boldsymbol{x} + \operatorname{null} g \subset N_X$. Thus L is an asymptotic plane of X. According to Lemma 7.12, $L \subset \operatorname{cl}(X + \operatorname{null} g)$. Hence, for a given $\varepsilon > 0$, there is a point $\boldsymbol{y} \in X + \operatorname{null} g$ such that $\|\boldsymbol{x} - \boldsymbol{y}\| < \varepsilon/\alpha$, where α is a Lipschitz constants of g. Let $\boldsymbol{y} = \boldsymbol{u} + \boldsymbol{v}$, where $\boldsymbol{u} \in X$ and $\boldsymbol{v} \in \operatorname{null} g$. Then $g(\boldsymbol{y}) = g(\boldsymbol{u}) \in g(X)$ and

$$\|g(\boldsymbol{x}) - g(\boldsymbol{u})\| = \|g(\boldsymbol{x}) - g(\boldsymbol{y})\| \leqslant \alpha\|\boldsymbol{x} - \boldsymbol{y}\| < \varepsilon.$$

Summing up, $g(\boldsymbol{x}) \in \operatorname{cl} g(X)$, which gives $g(N_X) \subset \operatorname{cl} g(X)$.

Problem 7.6. Let $S = \operatorname{dir} L$. By Theorem 7.15, there is a unit vector $\boldsymbol{e} \in S \cap \operatorname{rec}(\operatorname{cl} K)$. Denote by l the span of $\boldsymbol{e}$. Choose a 2-dimensional subspace T which contains l and meets $\operatorname{rint} K$. It is easy to see that a translate $l' \subset T$ of l either supports K or is asymptotic to K. If l' supports K at a point $\boldsymbol{a} \in \operatorname{cl} K$, then the closed halfline $h = \boldsymbol{a} + \{\lambda\boldsymbol{e} : \lambda \geqslant 0\}$ lies in $\operatorname{cl} K$ due to $\boldsymbol{e} \in \operatorname{rec}(\operatorname{cl} K)$. If l' is asymptotic to K and $\boldsymbol{a}$ is any point on l', then, as easy to see, one of the closed halflines of l' with endpoint $\boldsymbol{a}$ is asymptotic to K.

Problem 7.7. Let $\boldsymbol{x}$ be a point in $\operatorname{conv}(K \cup (\boldsymbol{u} + C)$. Since both sets K and $\boldsymbol{u}+C$ are convex, Corollary 4.14 shows that $\boldsymbol{x}$ can be written as $\boldsymbol{x} = (1-\lambda)\boldsymbol{y}+\lambda\boldsymbol{z}$, where $\boldsymbol{y} \in K$, $\boldsymbol{z} \in \boldsymbol{u} + C$, and $0 \leqslant \lambda \leqslant 1$. Furthermore, $\boldsymbol{z} = \boldsymbol{u} + \boldsymbol{v}$, where $\boldsymbol{v} \in C$.

By the convexity of K, the point $(1-\lambda)\boldsymbol{y}+\lambda\boldsymbol{u}$ lies in K, and $\lambda\boldsymbol{v}\in C$ since C is a cone with apex $\boldsymbol{o}$. Consequently,

$$\boldsymbol{x}=((1-\lambda)\boldsymbol{y}+\lambda\boldsymbol{u})+\lambda\boldsymbol{v}\in K+C.$$

Hence $\operatorname{conv}(K\cup(\boldsymbol{u}+C))\subset K+C$, and

$$\operatorname{cl}(\operatorname{conv}(K\cup(\boldsymbol{u}+C)))\subset\operatorname{cl}(K+C).$$

For the opposite inclusion, choose a point $\boldsymbol{x}\in K+C$. Then $\boldsymbol{x}$ can be written as $\boldsymbol{x}=\boldsymbol{y}+\boldsymbol{z}$, where $\boldsymbol{y}\in K$ and $\boldsymbol{z}\in C$. Since the case $\boldsymbol{z}=\boldsymbol{o}$ is obvious, we may assume that $\boldsymbol{z}\neq\boldsymbol{o}$. Put $h=[\boldsymbol{o},\boldsymbol{z}\rangle$. Because the closed halfline $\boldsymbol{u}+h$ lies in $\boldsymbol{u}+C$, it also lies in $\operatorname{conv}(K\cup(\boldsymbol{u}+C))$. Theorem 6.10 gives

$$\boldsymbol{x}=\boldsymbol{y}+\boldsymbol{z}\in\boldsymbol{y}+h\subset\operatorname{cl}(\operatorname{conv}(K\cup(\boldsymbol{u}+C))).$$

Summing up, $K+C\subset\operatorname{cl}(\operatorname{conv}(K\cup(\boldsymbol{u}+C)))$, and

$$\operatorname{cl}(K+C)\subset\operatorname{cl}(\operatorname{conv}(K\cup(\boldsymbol{u}+C))).$$

Problem 7.8. (*a*) Let $X\subset\mathbb{R}^n$ be an F_σ-set. By the definition, X can be expressed as a countable union of closed sets $\{Y_i\}$. Renumbering these sets, we may assume that they are given as a sequence $Y_1,Y_2,\ldots$. Choose any point $\boldsymbol{c}\in X$ and consider the set

$$X_i=B_i(\boldsymbol{c})\cap(Y_1\cup\cdots\cup Y_i),\quad i\geqslant 1.$$

Since every set $Y_1\cup\cdots\cup Y_i$ is closed as a finite union of closed sets, all sets $X_1,X_2,\ldots$ are compact. Obviously,

$$X_1\subset X_2\subset\ldots\quad\text{and}\quad X_1\cup X_2\cup\cdots=X.$$

Furthermore, let X be convex. The inclusions $X_i\subset\operatorname{conv}X_i$, $i\geqslant 1$, give

$$X=X_1\cup X_2\cup\cdots\subset\operatorname{conv}X_1\cup\operatorname{conv}X_2\cup\cdots\subset X.$$

Hence $X=\operatorname{conv}X_1\cup\operatorname{conv}X_2\cup\ldots$. Since all sets $\operatorname{conv}X_1,\operatorname{conv}X_2,\ldots$ are compact (see Theorem 4.16), we obtain a desired representation of X.

(*b*) Let $f:\mathbb{R}^n\to\mathbb{R}^m$ be an affine transformation, and X be an F_σ-set in $\mathbb{R}^n$. By the proved in part (*a*) above, X can be expressed as the union of a sequence of compact sets, say $X_1,X_2,\ldots$. Then

$$f(X)=f(X_1\cup X_2\cup\ldots)=f(X_1)\cup f(X_2)\cup\ldots.$$

Since all sets $f(X_1),f(X_2),\ldots$ are compact (see Problem 1.16), we obtain a desired representation of X.

(c) Let $X \subset \mathbb{R}^n$ be an F_σ-set. By the proved in part (a), X can be expressed as the union of an increasing sequence of compact sets $X_1, X_2, \ldots$. By Theorem 4.2,

$$\operatorname{conv} X = \operatorname{conv}(X_1 \cup X_2 \cup \ldots) = \operatorname{conv} X_1 \cup \operatorname{conv} X_2 \cup \cdots \subset \operatorname{conv} X,$$

which gives the equality $\operatorname{conv} X = \operatorname{conv} X_1 \cup \operatorname{conv} X_2 \cup \ldots$. Since all sets $\operatorname{conv} X_1, \operatorname{conv} X_2, \ldots$ are compact (see Theorem 4.16), we obtain a desired representation of $\operatorname{conv} X$.

For the second part of (c), we express X as the union of closed sets $X_1, X_2, \ldots$. Let

$$Y_i = (X_1 \cup \cdots \cup X_i) \cap \{\boldsymbol{x} \in \mathbb{R}^n : \tfrac{1}{i} \leqslant \|\boldsymbol{x} - \boldsymbol{a}\| \leqslant i\}, \quad i \geqslant 1.$$

Clearly, $X = Y_1 \cup Y_2 \cup \ldots$, every set Y_i is compact, and $\boldsymbol{o} \notin Y_i$, $i \geqslant 1$. By Theorem 5.45, all generated cones $C_{\boldsymbol{a}}(Y_1), C_{\boldsymbol{a}}(Y_2), \ldots$ are closed. Therefore, the generated cone

$$C_{\boldsymbol{a}}(X) = C_{\boldsymbol{a}}(Y_1 \cup Y_2 \cup \ldots) = C_{\boldsymbol{a}}(Y_1) \cup C_{\boldsymbol{a}}(Y_2) \cup \ldots$$

is an F_σ-set. Finally, the equality $\operatorname{cone}_{\boldsymbol{a}} X = \operatorname{conv} C_{\boldsymbol{a}}(X)$ and the first part of (c) imply that $\operatorname{cone}_{\boldsymbol{a}} X$ is an F_σ-set.

(d) According to part (a), X and Y can be expressed as unions of increasing sequences of compact sets: $X = X_1 \cup X_2 \cup \ldots$ and $Y = Y_1 \cup Y_2 \cup \ldots$. Clearly, the sum $X + Y$ is the union of countably many sets of the form $X_i + Y_j$, where $i, j \geqslant 1$. Since every set $X_i + Y_j$ is compact, the sum $X + Y$ is an F_σ-set. Finally, if both X and Y are convex sets, then $X + Y$ also is convex (see Theorem 3.4).

Chapter 8

Problem 8.1. Let L denote the span of C. Assume first that $\operatorname{cl} C$ is a closed halfplane. According to Theorem 2.74, $\operatorname{cl} C$ can be expressed as the sum of a suitable subspace $M \subset \mathbb{R}^n$ of dimension $\dim L - 1$ and any closed halfline h with endpoint $\boldsymbol{o}$ which lies in L but not in M. Clearly, $\operatorname{ap}(\operatorname{cl} C) = M$ (see an example on page 160). Choose h be orthogonal to M. Then the line $l = \operatorname{span} h$ is the orthogonal complement of M within L, and Theorem 8.5 shows that C° is the sum of the subspace $M' = \operatorname{ap} C^\circ$ and the closed halfline $h' = C^\circ \cap l$ which does not lie in M'. By Theorem 2.74, C° is a closed halfplane. The proof of converse assertion is similar.

Problem 8.2. First, we observe that $\boldsymbol{o} \in \operatorname{rbd} C$, as follows from Corollary 5.27. Assume for a moment that $\operatorname{cl} C \cap (-\operatorname{rint} C) \neq \varnothing$ and chose a vector $\boldsymbol{e} \in \operatorname{cl} C \cap (-\operatorname{rint} C)$. Then $\boldsymbol{e} \neq \boldsymbol{o}$ due to $\boldsymbol{o} \notin \operatorname{rint} C$. Because $-\boldsymbol{e} \in \operatorname{rint} C$, Theorem 3.47 gives $\boldsymbol{o} \in (\boldsymbol{e}, -\boldsymbol{e}) \subset \operatorname{rint} C$, a contradiction. Summing up, $\operatorname{cl} C \cap (-\operatorname{rint} C) = \varnothing$.

If $\operatorname{cl} C$ is a closed halfplane of $\operatorname{span} C$ such that $\boldsymbol{o} \in \operatorname{rbd} C$, then $-\operatorname{rint} C$ is the open halfplane of $\operatorname{span} C$ complementary to $\operatorname{cl} C$ (see Theorem 2.73 and an example on page 84). Conversely, suppose that $\operatorname{cl} C \cup (-\operatorname{rint} C) = \operatorname{span} C$. If E denotes the orthogonal complement of $\operatorname{ap}(\operatorname{cl} C)$ within $\operatorname{span} C$, then Theorem 8.5

shows that the cone $C^\circ \cap E$ is not a subspace and has positive dimension. Choose a nonzero vector $\boldsymbol{e} \in \operatorname{rint}(C^\circ \cap E) = \operatorname{rint} C^\circ \cap E$. By Theorem 8.6, $\boldsymbol{x} \cdot \boldsymbol{e} < 0$ for all vectors $\boldsymbol{x} \in \operatorname{cl} C \setminus \operatorname{ap}(\operatorname{cl} C)$. This argument and Theorem 8.7 show that the hyperplane $H = \{\boldsymbol{x} \in \mathbb{R}^n : \boldsymbol{x} \cdot \boldsymbol{e} = 0\}$ meets $\operatorname{cl} C$ along $\operatorname{ap}(\operatorname{cl} C)$ such that $\operatorname{cl} C \setminus \operatorname{ap}(\operatorname{cl} C)$ lies in the open halfspace $W = \{\boldsymbol{x} \in \mathbb{R}^n : \boldsymbol{x} \cdot \boldsymbol{e} < 0\}$. Consequently, $\operatorname{cl} C$ lies in the closed halfplane $D = \operatorname{span} C \cap \{\boldsymbol{x} \in \mathbb{R}^n : \boldsymbol{x} \cdot \boldsymbol{e} \leqslant 0\}$, implying that $-\operatorname{rint} C$ contains the complementary open halfplane $F = \operatorname{span} C \setminus D$.

We assert that $-\operatorname{rint} C = F$. Indeed, assume for a moment the existence of a nonzero vector $\boldsymbol{c} \in (-\operatorname{rint} C) \setminus F$ and denote by l the 1-dimensional subspace spanned by $\boldsymbol{c}$. Clearly, $l \setminus \{\boldsymbol{o}\}$ is the union of open halflines $h \subset \operatorname{cl} C$ and $h' \subset (-\operatorname{rint} C)$ such that $h \cap F = \varnothing$. Furthermore, $h + F = \operatorname{span} C$. Since the cone $-\operatorname{rint} C$ is convex and $h \cup F \subset -\operatorname{rint} C$ (see Theorem 5.20), we have $\operatorname{span} C = h + F \subset -\operatorname{rint} C$. Consequently, $-\operatorname{rint} C = \operatorname{span} C$, which gives $C = \operatorname{span} C$ (see Theorem 3.50), contrary to the assumption $C \neq \operatorname{span} C$. Thus $-\operatorname{rint} C = F$ and $C = D$.

Problem 8.3. By Corollary 5.9, the closed set $\operatorname{cl} C \cup C^\circ$ is a cone with proper apex $\boldsymbol{o}$. Therefore, Theorem 5.2 shows that U is an open cone with improper apex $\boldsymbol{o}$. Assume, for contradiction, that $U = \varnothing$. Then $\operatorname{cl} C \cup C^\circ = \mathbb{R}^n$. Choose a nonzero vector $\boldsymbol{e} \in \operatorname{rint} C^\circ$. By Theorems 8.6 and 8.7, the hyperplane $S = \{\boldsymbol{x} \in \mathbb{R}^n : \boldsymbol{x} \cdot \boldsymbol{e} = 0\}$ meets $\operatorname{cl} C$ along $\operatorname{ap}(\operatorname{cl} C)$ such that $\operatorname{cl} C \setminus \operatorname{ap}(\operatorname{cl} C)$ lies in the open halfspace $W = \{\boldsymbol{x} \in \mathbb{R}^n : \boldsymbol{x} \cdot \boldsymbol{e} < 0\}$. Consequently, $\operatorname{cl} C$ lies in the closed halfspace $V = \{\boldsymbol{x} \in \mathbb{R}^n : \boldsymbol{x} \cdot \boldsymbol{e} \leqslant 0\}$. Then C° contains the complementary closed halfspace $V' = \{\boldsymbol{x} \in \mathbb{R}^n : \boldsymbol{x} \cdot \boldsymbol{e} \geqslant 0\}$.

We state that $C^\circ = V'$. Indeed, assume for a moment that W contains a nonzero vector $\boldsymbol{c} \in C^\circ$. Then the whole closed halfline $h = \{\lambda \boldsymbol{e} : \lambda \geqslant 0\}$ belongs to C°, and Problem 4.4 shows that $V = \operatorname{conv}(S \cup h)$ lies in C°. Consequently, $\mathbb{R}^n = V \cup V' \subset C^\circ$, which is impossible due to $C \neq \{\boldsymbol{o}\}$ (see Theorem 8.2). Consequently, $C^\circ = V'$ and $\operatorname{cl} C = V$, which gives $\operatorname{cl} C \cap C^\circ = S \neq \{\boldsymbol{o}\}$, contrary to $\operatorname{cl} C \cap C^\circ = \{\boldsymbol{o}\}$, as follows from Theorem 8.2. The obtained contradiction gives $U \neq \varnothing$.

Problem 8.4. The obvious inclusions $\operatorname{rint} C + \operatorname{rint} C^\circ \subset C + C^\circ \subset \mathbb{R}^n$ show that it suffices to prove the equality $\operatorname{rint} C + \operatorname{rint} C^\circ = \mathbb{R}^n$. Since the latter is obvious when $C = \{\boldsymbol{o}\}$ or $C = \mathbb{R}^n$, we may assume that $\{\boldsymbol{o}\} \neq C \neq \mathbb{R}^n$.

Next, suppose that C is a subspace, say S. Then $C^\circ = S^\perp$ is a subspace too (see an example on page 269). Any given vector $\boldsymbol{x} \in \mathbb{R}^n$ is uniquely expressible as a sum $\boldsymbol{x} = \boldsymbol{y} + \boldsymbol{z}$, where $\boldsymbol{y} \in S$ and $\boldsymbol{z} \in S^\perp$. Since $\operatorname{rint} S = S$ and $\operatorname{rint} S^\perp = S^\perp$ (see an example on page 84), we obtain the inclusion

$$\boldsymbol{x} \in \operatorname{rint} S + \operatorname{rint} S^\perp = \operatorname{rint} C + \operatorname{rint} C^\circ.$$

Finally, it remains to consider the case when C is not a subspace. Then the orthogonal complement S of $\operatorname{ap}(\operatorname{cl} C)$ within $\operatorname{span} C$ has positive dimension (see Lemma 8.10). Choose any vector $\boldsymbol{x} \in \mathbb{R}^n$. By Problem 8.8, $\boldsymbol{x} = \boldsymbol{y} + \boldsymbol{z}$, where $\boldsymbol{y}$ and $\boldsymbol{z}$ are metric projections of $\boldsymbol{x}$ on $\operatorname{cl} C$ and C°, respectively. Lemma 8.10

shows the existence of a nonzero vector $\boldsymbol{e} \in \operatorname{rint} C \cap S$ such that $-\boldsymbol{e} \in \operatorname{rint} C^\circ \cap S$. Now, Corollary 5.21 gives

$$\boldsymbol{y} + \boldsymbol{e} \in \operatorname{cl} C + \operatorname{rint} C = \operatorname{rint} C \quad \text{and} \quad \boldsymbol{z} - \boldsymbol{e} \in C^\circ + \operatorname{rint} C^\circ = \operatorname{rint} C^\circ.$$

Therefore, $\boldsymbol{x} = (\boldsymbol{y} + \boldsymbol{e}) + (\boldsymbol{z} - \boldsymbol{e})$ is a desired representation.

Problem 8.5. Choose an apex $\boldsymbol{a}$ of C, and put $m = \dim C$.

$(a) \Rightarrow (c)$ If C is a simplicial cone, then it can be expressed as $C = \Gamma_{\boldsymbol{a}}(\boldsymbol{x}_1, \dots, \boldsymbol{x}_m)$, where $\{\boldsymbol{a}, \boldsymbol{x}_1, \dots, \boldsymbol{x}_m\}$ is an affinely independent set in $\mathbb{R}^n$ (see Definition 5.6 and Theorem 5.7). By Theorem 5.7, C is line-free. Hence Theorem 8.17 implies the existence of a convex base, say B, of C. Theorem 8.14 shows that $B = C \cap \operatorname{aff} B$. Furthermore, every open halfline $(\boldsymbol{a}, \boldsymbol{x}_i\rangle$ meets H at a point $\boldsymbol{z}_i$, $1 \leqslant i \leqslant m$. By Problem 5.3, the set $\{\boldsymbol{a}, \boldsymbol{z}_1, \dots, \boldsymbol{z}_m\}$ is affinely independent and $C = \Gamma_{\boldsymbol{a}}(\boldsymbol{z}_1, \dots, \boldsymbol{z}_m)$. Consider the simplex $\Delta = \Delta(\boldsymbol{z}_1, \dots, \boldsymbol{z}_m)$. Corollary 8.14 shows that $\operatorname{aff} B = \operatorname{aff} \Delta$, and Corollary 5.56 gives $C = C_{\boldsymbol{a}}(\Delta)$ This argument results in

$$B = C \cap \operatorname{aff} B = C \cap \operatorname{aff} \Delta = \Delta.$$

$(c) \Rightarrow (b)$ This part is obvious.

$(b) \Rightarrow (a)$ If C has a convex base which is a simplex, say Δ, then a combination of Corollary 5.56 and Theorem 8.14 shows that the cone $C = C_{\boldsymbol{a}}(\Delta)$ is simplicial.

Problem 8.6. Assume first that C is line-free. Then $(C - \boldsymbol{a})^\circ \neq \{\boldsymbol{o}\}$ because $C - \boldsymbol{a} \neq \mathbb{R}^n$ (see Theorem 8.2). Choose a nonzero vector $\boldsymbol{e} \in (\operatorname{rint}(C - \boldsymbol{a}))^\circ$. By Theorem 8.7, the subspace $T = \{\boldsymbol{x} \in \mathbb{R}^n : \boldsymbol{x} \cdot \boldsymbol{e} = 0\}$ satisfies the condition $\operatorname{ap}(\operatorname{cl} C - \boldsymbol{a}) \cap T = \{\boldsymbol{o}\}$. Theorem 6.40 implies that a hyperplane $H \subset \mathbb{R}^n$ of the form $H = \{\boldsymbol{x} \in \mathbb{R}^n : \boldsymbol{x} \cdot \boldsymbol{e} = \gamma\}$, where $\gamma < \boldsymbol{a} \cdot \boldsymbol{e}$, meets C along a bounded set. Moreover,

$$C = C_{\boldsymbol{a}}(H \cap C) = \cup\,([\boldsymbol{a}, \boldsymbol{x}\rangle : \boldsymbol{x} \in H \cap C)$$

according to Theorem 8.16.

Let $M = \operatorname{aff}(H \cap C)$, and let $\boldsymbol{b}$ be a point in $\operatorname{rint} C \cap H$. Problem 3.12 shows the existence of an $(m-1)$-simplex $\Delta = \Delta(\boldsymbol{x}_1, \dots, \boldsymbol{x}_m) \subset M$ which contains $C \cap H$. By Corollary 5.56, $\Gamma_{\boldsymbol{a}} = C_{\boldsymbol{a}}(\Delta)$ is an $(m-1)$-simplicial cone. Clearly, $C \subset \Gamma_{\boldsymbol{a}}$.

Conversely, suppose the existence of an $(m-1)$-simplicial cone $\Gamma_{\boldsymbol{a}} = C_{\boldsymbol{a}}(\Delta)$ containing C. Since the set Δ is compact, Theorem 8.17 shows that $\Gamma_{\boldsymbol{a}}$ is line-free. Consequently, C is line-free as a subset of $\Gamma_{\boldsymbol{a}}$.

Problem 8.7. By Corollary 7.21, K is line-free if and only if for any given point $\boldsymbol{a} \in \mathbb{R}^n \setminus \operatorname{cl} K$ the generated cone $C_{\boldsymbol{a}}(K)$ is line-free. Now the assertion follows from Problem 8.6.

Problem 8.8. Assume first that vectors $\boldsymbol{y} \in \operatorname{cl} C$ and $\boldsymbol{z} \in C^\circ$ are orthogonal and satisfy the equality $\boldsymbol{x} = \boldsymbol{y} + \boldsymbol{z}$. Then

$$(\boldsymbol{x} - \boldsymbol{y})\cdot\boldsymbol{y} = (\boldsymbol{x} - \boldsymbol{z})\cdot\boldsymbol{z} = \boldsymbol{y}\cdot\boldsymbol{z} = 0.$$

Choose a point $\boldsymbol{u} \in \operatorname{cl} C$. Since $(\operatorname{cl} C)^\circ = C^\circ$, one has $\boldsymbol{u}\cdot(\boldsymbol{x} - \boldsymbol{y}) = \boldsymbol{u}\cdot\boldsymbol{z} \leqslant 0$. Therefore,

$$\begin{aligned}\|\boldsymbol{x} - \boldsymbol{u}\|^2 - \|\boldsymbol{x} - \boldsymbol{y}\|^2 &= (\boldsymbol{x} - \boldsymbol{u})\cdot(\boldsymbol{x} - \boldsymbol{u}) - (\boldsymbol{x} - \boldsymbol{y})\cdot(\boldsymbol{x} - \boldsymbol{y}) \\ &= (\boldsymbol{u}\cdot\boldsymbol{u} - 2\boldsymbol{u}\cdot\boldsymbol{y} + \boldsymbol{y}\cdot\boldsymbol{y}) + (2\boldsymbol{x}\cdot\boldsymbol{y} - 2\boldsymbol{y}\cdot\boldsymbol{y}) + (2\boldsymbol{u}\cdot\boldsymbol{y} - 2\boldsymbol{u}\cdot\boldsymbol{x}) \\ &= \|\boldsymbol{u} - \boldsymbol{y}\|^2 + 2(\boldsymbol{x} - \boldsymbol{y})\cdot\boldsymbol{y} - 2\boldsymbol{u}\cdot(\boldsymbol{x} - \boldsymbol{y}) \geqslant \|\boldsymbol{u} - \boldsymbol{y}\|^2 \geqslant 0.\end{aligned}$$

Hence $\|\boldsymbol{x} - \boldsymbol{u}\| \geqslant \|\boldsymbol{x} - \boldsymbol{y}\|$, implying that $\boldsymbol{y}$ is the metric projection of $\boldsymbol{x}$ on $\operatorname{cl} C$. Similarly, if $\boldsymbol{v} \in C^\circ$, then $(\boldsymbol{x} - \boldsymbol{z})\cdot\boldsymbol{v} = \boldsymbol{y}\cdot\boldsymbol{v} \leqslant 0$, which gives

$$\begin{aligned}\|\boldsymbol{x} - \boldsymbol{v}\|^2 - \|\boldsymbol{x} - \boldsymbol{z}\|^2 &= (\boldsymbol{x} - \boldsymbol{v})\cdot(\boldsymbol{x} - \boldsymbol{v}) - (\boldsymbol{x} - \boldsymbol{z})\cdot(\boldsymbol{x} - \boldsymbol{z}) \\ &= (\boldsymbol{v}\cdot\boldsymbol{v} - 2\boldsymbol{v}\cdot\boldsymbol{z} + \boldsymbol{z}\cdot\boldsymbol{z}) + (2\boldsymbol{x}\cdot\boldsymbol{z} - 2\boldsymbol{z}\cdot\boldsymbol{z}) + (2\boldsymbol{z}\cdot\boldsymbol{v} - 2\boldsymbol{x}\cdot\boldsymbol{v}) \\ &= \|\boldsymbol{v} - \boldsymbol{z}\|^2 + 2(\boldsymbol{x} - \boldsymbol{z})\cdot\boldsymbol{z} - 2(\boldsymbol{x} - \boldsymbol{z})\cdot\boldsymbol{v} \geqslant \|\boldsymbol{v} - \boldsymbol{z}\|^2 \geqslant 0.\end{aligned}$$

Therefore, $\boldsymbol{z}$ is the metric projection of $\boldsymbol{x}$ on C°.

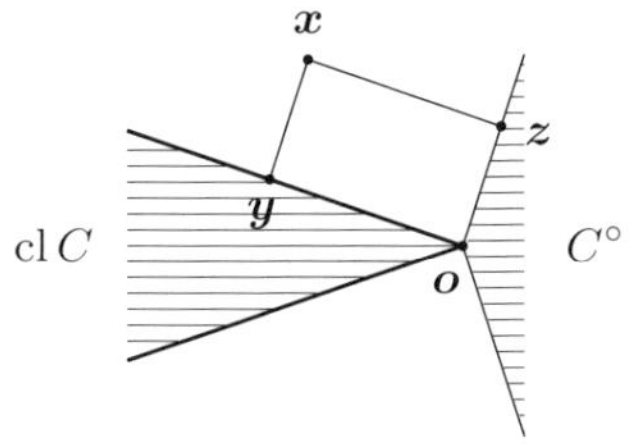

Conversely, let $\boldsymbol{y}$ and $\boldsymbol{z}$ be the metric projections of $\boldsymbol{x}$ on C and C°, respectively. We first observe that $\boldsymbol{o}$ is the nearest to $\boldsymbol{x} - \boldsymbol{y}$ point in $\operatorname{cl} C$. Indeed, let $\boldsymbol{u}$ be any point in $\operatorname{cl} C$. Then $\boldsymbol{u} + \boldsymbol{y} \in \operatorname{cl} C$ because $\operatorname{cl} C$ is a convex cone with apex $\boldsymbol{o}$ (see Theorem 5.28). This argument gives

$$\|(\boldsymbol{x} - \boldsymbol{y}) - \boldsymbol{u}\| = \|\boldsymbol{x} - (\boldsymbol{u} + \boldsymbol{y})\| \geqslant \|\boldsymbol{x} - \boldsymbol{y}\| = \|(\boldsymbol{x} - \boldsymbol{y}) - \boldsymbol{o}\|,$$

where the equality holds if and only if $\boldsymbol{u} = \boldsymbol{o}$. Therefore, Theorem 8.22 implies that

$$(\boldsymbol{x} - \boldsymbol{y})\cdot\boldsymbol{v} = ((\boldsymbol{x} - \boldsymbol{y}) - \boldsymbol{o})\cdot(\boldsymbol{v} - \boldsymbol{o}) \leqslant 0 \quad \text{for all} \quad \boldsymbol{v} \in \operatorname{cl} C.$$

Equivalently, $\boldsymbol{x} - \boldsymbol{y} \in (\operatorname{cl} C)^\circ = C^\circ$.

Next, we assert that $\boldsymbol{y}$ and $\boldsymbol{x} - \boldsymbol{y}$ are orthogonal. Since this is true when $\boldsymbol{y} = \boldsymbol{o}$ or $\boldsymbol{x} = \boldsymbol{y}$, we may suppose that both vectors $\boldsymbol{y}$ and $\boldsymbol{x} - \boldsymbol{y}$ are not $\boldsymbol{o}$. Because $\boldsymbol{y}$ is the metric projection of $\boldsymbol{x}$ on $\operatorname{cl} C$, and since $\lambda\boldsymbol{y} \in \operatorname{cl} C$ for all $\lambda \geqslant 0$, we have

$\|\boldsymbol{x}-\lambda\boldsymbol{y}\| \geqslant \|\boldsymbol{x}-\boldsymbol{y}\|$, with equality if and only if $\lambda = 1$. Hence $\lambda = 1$ is the double zero of the quadratic function

$$\|\boldsymbol{x}-\lambda\boldsymbol{y}\|^2 - \|\boldsymbol{x}-\boldsymbol{y}\|^2 = \lambda^2\boldsymbol{y}\cdot\boldsymbol{y} - 2\lambda\,\boldsymbol{x}\cdot\boldsymbol{y} + 2\boldsymbol{x}\cdot\boldsymbol{y} - \boldsymbol{y}\cdot\boldsymbol{y},$$

which happens if and only if $\boldsymbol{x}\cdot\boldsymbol{y} = \boldsymbol{y}\cdot\boldsymbol{y}$. Hence $(\boldsymbol{x}-\boldsymbol{y})\cdot\boldsymbol{y} = 0$.

Finally, let $\boldsymbol{w} = \boldsymbol{x}-\boldsymbol{y}$. By the above proved, $\boldsymbol{w} \in C^\circ$ and $\boldsymbol{w}\cdot\boldsymbol{y} = 0$. As shown in the proof of the "if" part, the conditions $\boldsymbol{x} = \boldsymbol{y}+\boldsymbol{w}$ and $\boldsymbol{w}\cdot\boldsymbol{y} = 0$ imply that $\boldsymbol{w}$ is the metric projection of $\boldsymbol{x}$ on C°. The uniqueness of this projection and the choice of $\boldsymbol{z}$ give $\boldsymbol{w} = \boldsymbol{z}$. Summing up,

$$\boldsymbol{y}\cdot\boldsymbol{z} = \boldsymbol{w}\cdot\boldsymbol{z} = 0 \quad \text{and} \quad \boldsymbol{x} = \boldsymbol{y}+\boldsymbol{w} = \boldsymbol{y}+\boldsymbol{z}.$$

Problem 8.9. $(a) \Rightarrow (b)$ Clearly, both sets E and F are nonempty. First, we consider the obvious case when at least one of the sets, say E, equals $\{\boldsymbol{o}\}$. Choose any vector $\boldsymbol{u} \in \mathbb{R}^n$. By the assumption,

$$\boldsymbol{u} = p_E(\boldsymbol{u}) + p_F(\boldsymbol{u}) = \boldsymbol{o} + p_F(\boldsymbol{u}) \in F.$$

Consequently, $F = \mathbb{R}^n = \{\boldsymbol{o}\}^\circ = E^\circ$, as desired.

Hence we may assume that none of the sets E and F is $\{\boldsymbol{o}\}$. Choose a nonzero vector $\boldsymbol{u} \in E$. Then $\boldsymbol{u} = p_E(\boldsymbol{u})$, which gives $p_F(\boldsymbol{u}) = \boldsymbol{u} - p_E(\boldsymbol{u}) = \boldsymbol{o}$. Hence $\boldsymbol{o}$ is the nearest to $\boldsymbol{u}$ point if F. By Theorem 8.22, F lies in the closed halfspace $V = \{\boldsymbol{x} \in \mathbb{R}^n : \boldsymbol{x}\cdot\boldsymbol{u} \leqslant 0\}$. Thus,

$$F \subset \{\boldsymbol{x} \in \mathbb{R}^n : \boldsymbol{x}\cdot\boldsymbol{u} \leqslant 0 \text{ for all } \boldsymbol{u} \in E\} = E^\circ.$$

Conversely, choose any point $\boldsymbol{u} \in E^\circ$. Then $p_E(\boldsymbol{u}) = \boldsymbol{o}$, which gives

$$\boldsymbol{u} = p_E(\boldsymbol{u}) + p_F(\boldsymbol{u}) = \boldsymbol{o} + p_F(\boldsymbol{u}) = p_F(\boldsymbol{u}).$$

Hence $\boldsymbol{u} \in F$, implying the inclusion $E^\circ \subset F$. Similarly, $E = F^\circ$, which shows that E is a closed convex cone with apex $\boldsymbol{o}$.

$(b) \Rightarrow (a)$ This part follows from Problem 8.8.

Problem 8.10. Let $F = D - \boldsymbol{a}$. Then F is a closed halfplane of the subspace $S = L - \boldsymbol{a}$ such that $\boldsymbol{o} \in \operatorname{ap} F = \operatorname{rbd} F$. If $\boldsymbol{e} \in \operatorname{rint} F^\circ$, then the hyperplane $H = \{\boldsymbol{x} \in \mathbb{R}^n : \boldsymbol{x}\cdot\boldsymbol{e} = 0\}$ meets F along $\operatorname{ap} F$ (see Theorem 8.7). Consequently, $\boldsymbol{e} \in \operatorname{ort}(\operatorname{ap} F) \setminus \operatorname{ort} L$, and Theorem 2.70 shows that F equals one of the closed halfplanes

$$D_1 = \{\boldsymbol{x} \in S : \boldsymbol{x}\cdot\boldsymbol{e} \leqslant 0\} \quad \text{and} \quad D_2 = \{\boldsymbol{x} \in S : \boldsymbol{x}\cdot\boldsymbol{e} \geqslant 0\}.$$

Furthermore, the inclusion $\boldsymbol{e} \in \operatorname{rint} F^\circ$ implies that F lies in the closed halfspace $V = \{\boldsymbol{x} \in \mathbb{R}^n : \boldsymbol{x}\cdot\boldsymbol{e} \leqslant 0\}$. Thus $F = D_1 = V \cap S$. With $\gamma = \boldsymbol{a}\cdot\boldsymbol{e}$, we have

$$D = \boldsymbol{a} + F = \boldsymbol{a} + \{\boldsymbol{x} \in S : \boldsymbol{x}\cdot\boldsymbol{e} \leqslant 0\} = \{\boldsymbol{x} \in L : \boldsymbol{x}\cdot\boldsymbol{e} \leqslant \gamma\}.$$

In a similar way, the equality $D = \{\boldsymbol{x} \in L : \boldsymbol{x}{\cdot}\boldsymbol{e} \leqslant \gamma\}$ implies that $\boldsymbol{e} \in \operatorname{rint}(D - \boldsymbol{a})^\circ$ and $\gamma = \boldsymbol{a}{\cdot}\boldsymbol{e}$.

Problem 8.11. Let $m = \dim L$. Since the case $L = \mathbb{R}^n$ is obvious (let $V = D$), we may assume that $m \leqslant n - 1$. Denote by M the subplane of L of dimension $m - 1$ which determines D (see Definition 2.69). Choosing a point $\boldsymbol{s} \in M$ and translating all sets involved on the same vector $-\boldsymbol{s}$, we may assume that both planes L and M are subspaces. Denote by N the orthogonal complement of L. Let $\boldsymbol{b} \in D$ be the unit vector orthogonal to N. A combination of Theorems 2.74 and 8.5 shows that the halfplane D and its polar halfplane D° (see Problem 8.1) can be expressed as $D = M + h_1$ and $D^\circ = N + h_2$, where $h_1 = \{\lambda \boldsymbol{b} : \lambda \geqslant 0\}$ and $h_2 = \{\lambda \boldsymbol{b} : \lambda \leqslant 0\}$ are the opposite closed halflines with endpoint $\boldsymbol{o}$.

Because $\mathbb{R}^n$ is the orthogonal sum of the subspaces $l = \operatorname{span}\{\boldsymbol{b}\}$, M, and N, one can write $\boldsymbol{a} = \mu\boldsymbol{b} + \boldsymbol{d} + \boldsymbol{c}$, where $\mu \in \mathbb{R}$, $\boldsymbol{d} \in M$ and $\boldsymbol{c} \in N$. We observe that $\boldsymbol{c} \neq \boldsymbol{o}$ if $\mu \geqslant 0$ (otherwise $\boldsymbol{a} = \mu\boldsymbol{b} + \boldsymbol{d} \in D$, contrary to the assumption $\boldsymbol{a} \notin D$). Let

$$\boldsymbol{e} = \begin{cases} -\boldsymbol{b} & \text{if } \mu < 0, \\ -\boldsymbol{b} + \boldsymbol{c} & \text{if } \mu = 0, \\ -\|\boldsymbol{c}\|^2\boldsymbol{b} + 2\mu\boldsymbol{c} & \text{if } \mu > 0 \end{cases}$$

and put $V = \{\boldsymbol{x} \in \mathbb{R}^n : \boldsymbol{x} \cdot \boldsymbol{e} \leqslant 0\}$. Theorem 2.74 shows that $\boldsymbol{e} \in \operatorname{rint} D^\circ$. Consequently, $D = V \cap L = \{\boldsymbol{x} \in L : \boldsymbol{x}{\cdot}\boldsymbol{e} \leqslant 0\}$ (see Problem 8.10). On the other hand,

$$\begin{aligned} &\boldsymbol{a}{\cdot}\boldsymbol{e} = (\mu\boldsymbol{b} + \boldsymbol{d} + \boldsymbol{c}){\cdot}(-\boldsymbol{b}) = -\mu > 0 && \text{if } \mu < 0, \\ &\boldsymbol{a}{\cdot}\boldsymbol{e} = (\boldsymbol{d} + \boldsymbol{c}){\cdot}(-\boldsymbol{b} + \boldsymbol{c}) = \|\boldsymbol{c}\|^2 > 0 && \text{if } \mu = 0, \\ &\boldsymbol{a}{\cdot}\boldsymbol{e} = (\mu\boldsymbol{b} + \boldsymbol{d} + \boldsymbol{c}){\cdot}(-\|\boldsymbol{c}\|^2\boldsymbol{b} + 2\mu\boldsymbol{c}) = \mu\|\boldsymbol{c}\|^2 > 0 && \text{if } \mu > 0. \end{aligned}$$

So, $\boldsymbol{a} \notin V$, as desired.

Problem 8.12. One has $\operatorname{rbd} K \neq \varnothing$ because K is not a plane (see Theorem 3.62). The existence of a nearest to $\boldsymbol{z}$ point $\boldsymbol{u} \in \operatorname{rbd} K$ follows from the continuity of the function $\delta_{\boldsymbol{z}}(\boldsymbol{x}) = \|\boldsymbol{x} - \boldsymbol{z}\|$ on $\mathbb{R}^n$ and the closedness of $\operatorname{rbd} K$ (see Theorem 3.62). Assume for a moment that $K \not\subset V$ and choose a point $\boldsymbol{v} \in K \setminus V$. Denote by L the 2-dimensional plane spanned by $\{\boldsymbol{u}, \boldsymbol{v}, \boldsymbol{z}\}$. Let F (respectively, G) be the open halfplane of L determined by the line $l = \langle \boldsymbol{u}, \boldsymbol{v}\rangle$ and missing $\boldsymbol{z}$ (respectively, determined by the line $m = \langle \boldsymbol{u}, \boldsymbol{z}\rangle$ and missing $\boldsymbol{v}$). Put $Q = F \cap G$.

We assert that $K \cap Q = \varnothing$. Indeed, assume for a moment the existence of a point $\boldsymbol{x} \in K \cap Q$. According to Theorem 2.35, the closed segment $[\boldsymbol{x}, \boldsymbol{v}]$ meets the line m at a point $\boldsymbol{y} \in F$. Furthermore, $\boldsymbol{y} \in K$ by the convexity of K. On the other hand, the inclusion $\boldsymbol{y} \in K$ is impossible because, according to Theorem 3.65, the open halfline $[\boldsymbol{z}, \boldsymbol{u}\rangle \setminus [\boldsymbol{z}, \boldsymbol{u}]$ must be disjoint from K. Hence $K \cap Q = \varnothing$.

Denote by h the closed halfline with endpoint $\boldsymbol{z}$ which meets the line l and is perpendicular to it. Since h meets Q, we have $h \not\subset K$. By Theorem 3.67, h meets $\operatorname{rbd} K$ at a point $\boldsymbol{c} \in L \setminus F$. Denote by $\boldsymbol{w}$ the point at which h meets l. Then

$\boldsymbol{c} \in [\boldsymbol{w}, \boldsymbol{z}]$, which gives $\|\boldsymbol{c} - \boldsymbol{z}\| \leqslant \|\boldsymbol{w} - \boldsymbol{z}\| < \|\boldsymbol{u} - \boldsymbol{z}\|$, in contradiction with the choice of $\boldsymbol{u}$. The obtained contradiction proves the inclusion $K \subset P$.

Problem 8.13. The equivalence of conditions (a) and (b) follows from the equalities below (which result from Theorem 3.46 and Corollary 8.36):

$$\dim(\operatorname{bar} K) = \dim(\operatorname{cl}(\operatorname{bar} K)) = \dim(\operatorname{cl}(\operatorname{nor} K)) = \dim(\operatorname{nor} K).$$

$(a) \Rightarrow (c)$ Since $\operatorname{bar} K$ is a convex cone with apex $\boldsymbol{o}$ (see Theorem 8.32), the condition $\dim(\operatorname{bar} K) = 1$ implies that $\operatorname{bar} K$ is either a closed halfline with endpoint $\boldsymbol{o}$ or a one-dimensional subspace of $\mathbb{R}^n$. If $\operatorname{bar} K$ is a closed halfline, then an example on page 269 and Theorem 8.35 show that $\operatorname{rec}(\operatorname{cl} K) = (\operatorname{bar} K)^\circ$ is a closed halfspace. If $\operatorname{bar} K$ is a line, then $\operatorname{rec}(\operatorname{cl} K) = (\operatorname{bar} K)^\perp$ is an $(n-1)$-dimensional subspace. In either case, Corollary 6.15 shows that $\operatorname{cl} K$ contains a hyperplane. Consequently, $n - 1 \leqslant \dim K \leqslant n$, and Problem 6.6 implies that $\operatorname{cl} K$ is a hyperplane, a closed slab, or a closed halfspace. The converse assertion is obvious.

Problem 8.14. For the closedness of $P_X(\boldsymbol{c})$, choose any sequence of points $\boldsymbol{x}_1, \boldsymbol{x}_2, \dots$ in $P_X(\boldsymbol{c})$ which converges to a point $\boldsymbol{x} \in \mathbb{R}^n$. For any point $\boldsymbol{v} \in X$, one has

$$\|\boldsymbol{x} - \boldsymbol{c}\| = \lim_{i\to\infty} \|\boldsymbol{x}_i - \boldsymbol{c}\| \leqslant \lim_{i\to\infty} \|\boldsymbol{x}_i - \boldsymbol{v}\| = \|\boldsymbol{x} - \boldsymbol{v}\|.$$

Hence $\boldsymbol{x} \in P_X(\boldsymbol{c})$, which shows the closedness of $P_X(\boldsymbol{c})$.

For the convexity of $P_X(\boldsymbol{c})$, we observe first that for any points $\boldsymbol{x}, \boldsymbol{y}$ and $\boldsymbol{z}$ in $\mathbb{R}^n$ and any scalar $\lambda \in \mathbb{R}$, the following identity holds:

$$\|(\lambda\boldsymbol{x} + (1-\lambda)\boldsymbol{y}) - \boldsymbol{z}\|^2 = \lambda\|\boldsymbol{x} - \boldsymbol{z}\|^2 - \lambda(1-\lambda)\|\boldsymbol{x} - \boldsymbol{y}\|^2 + (1-\lambda)\|\boldsymbol{y} - \boldsymbol{z}\|^2.$$

Next, translating X on $-\boldsymbol{c}$, we may assume that $\boldsymbol{c} = \boldsymbol{o}$. Choose any points $\boldsymbol{x}, \boldsymbol{y} \in P_X(\boldsymbol{o})$ and let $\boldsymbol{w} = \lambda\boldsymbol{x} + (1-\lambda)\boldsymbol{y}$. Given a point $\boldsymbol{v} \in X$, we have $\|\boldsymbol{x}\| \leqslant \|\boldsymbol{x} - \boldsymbol{v}\|$ and $\|\boldsymbol{y}\| \leqslant \|\boldsymbol{y} - \boldsymbol{v}\|$. Using twice the above identity (with $\boldsymbol{z} = \boldsymbol{o}$ and $\boldsymbol{z} = \boldsymbol{v}$, respectively), one has

$$\begin{aligned}\|\boldsymbol{w}\|^2 &= \|(\lambda\boldsymbol{x} + (1-\lambda)\boldsymbol{y})\|^2 = \lambda\|\boldsymbol{x}\|^2 - \lambda(1-\lambda)\|\boldsymbol{x} - \boldsymbol{y}\|^2 + (1-\lambda)\|\boldsymbol{y}\|^2 \\ &\leqslant \lambda\|\boldsymbol{x} - \boldsymbol{v}\|^2 - \lambda(1-\lambda)\|\boldsymbol{x} - \boldsymbol{y}\|^2 + (1-\lambda)\|\boldsymbol{y} - \boldsymbol{v}\|^2 = \|\boldsymbol{w} - \boldsymbol{v}\|^2.\end{aligned}$$

So, $\|\boldsymbol{w}\| \leqslant \|\boldsymbol{w} - \boldsymbol{v}\|$, implying the inclusion $\boldsymbol{w} \in P_X(\boldsymbol{o})$. Replacing the above inequalities with strict ones, we obtain the proof of the convexity of $Q_X(\boldsymbol{c})$.

Problem 8.15. Suppose that X satisfies the assumption but is not convex. Then one can choose points $\boldsymbol{x}, \boldsymbol{y} \in X$ such that $[\boldsymbol{x}, \boldsymbol{y}] \not\subset X$. Since $[\boldsymbol{x}, \boldsymbol{y}] \setminus X$ is an open subset of the line $\langle \boldsymbol{x}, \boldsymbol{y} \rangle$, it can be expressed as the union of open segments. Let $(\boldsymbol{u}, \boldsymbol{v})$ be one of these segments. Then $\boldsymbol{u}, \boldsymbol{v} \in X$ and $(\boldsymbol{u}, \boldsymbol{v}) \cap X = \varnothing$. Denote by $\boldsymbol{w}$ the middle point of $(\boldsymbol{u}, \boldsymbol{v})$ and choose a closed ball $B_\delta(\boldsymbol{w}) \subset \mathbb{R}^n$ disjoint from X. An elementary geometry argument shows that $\|\boldsymbol{v} - \boldsymbol{w}\|^2/(2\delta)$ is an upper bound for the radius of any closed ball $B_r(\boldsymbol{x})$ containing $B_\delta(\boldsymbol{w})$ such that

$U_r(\boldsymbol{x}) \cap X = \varnothing$. Clearly, the set of centers of all such closed balls is a compact set. Then a compactness argument shows the existence of a closed ball $B_\rho(\boldsymbol{a})$ of maximal radius ρ satisfying the conditions $B_\delta(\boldsymbol{w}) \subset B_\rho(\boldsymbol{a})$ and $U_r(\boldsymbol{a}) \cap X = \varnothing$. By the maximality of ρ, the closed ball $B_\rho(\boldsymbol{a})$ meets X at a point $\boldsymbol{c}$, and the assumption on X implies the uniqueness of $\boldsymbol{c}$. If the spheres $S_\delta(\boldsymbol{w})$ and $S_\rho(\boldsymbol{a})$ have a point in common, denote this (unique) point by $\boldsymbol{e}$; otherwise let $\boldsymbol{e} = \boldsymbol{w}$. Clearly, $\boldsymbol{e} \neq \boldsymbol{c}$ and there is a scalar $\varepsilon_0 > 0$ such that the translate $\varepsilon(\boldsymbol{e}-\boldsymbol{c})+U_\rho(\boldsymbol{a})$ contains $B_\delta(\boldsymbol{w})$ for all $0 < \varepsilon < \varepsilon_0$.

Let H be the hyperplane through $\boldsymbol{a}$ and normal to $\boldsymbol{e}-\boldsymbol{c}$. Denote by V_1 and V_2 the opposite closed halfspaces determined by H such that $\boldsymbol{c} \in V_1$. Since $B_\rho(\boldsymbol{a}) \cap X = \{\boldsymbol{c}\} \in \operatorname{int} V_1$, there is an $\varepsilon_1 \in (0, \varepsilon_0)$ such that the set $(\varepsilon(\boldsymbol{e}-\boldsymbol{c})+B_\rho(\boldsymbol{a})) \cap V_2$ is disjoint from X for all $0 < \varepsilon < \varepsilon_1$. Obviously, $(\varepsilon(\boldsymbol{e}-\boldsymbol{c})+B_\rho(\boldsymbol{a})) \cap V_1 \subset U_\rho(\boldsymbol{a})$ for all $\varepsilon > 0$. Summing up,

$$B_\delta(\boldsymbol{w}) \subset \varepsilon(\boldsymbol{e}-\boldsymbol{c}) + U_\rho(\boldsymbol{a}) \quad \text{and} \quad (\varepsilon(\boldsymbol{e}-\boldsymbol{c}) + B_\rho(\boldsymbol{a})) \cap X = \varnothing$$

for all $0 < \varepsilon < \varepsilon_1$. Therefore, choosing a particular value $\varepsilon \in (0, \varepsilon_1)$, we can find a scalar $\gamma > \rho$ such that the larger closed ball $B' = \varepsilon(\boldsymbol{e}-\boldsymbol{c}) + B_\gamma(\boldsymbol{a})$ still satisfies the conditions $B_\delta \subset B'$ and $B' \cap X = \varnothing$. Since the latter contradicts the choice of ρ, the set X must be convex.

Problem 8.16. The assertion trivially holds if K is empty or is a singleton. Suppose that $\dim K > 0$. By Theorem 6.40, all sections of $\operatorname{cl} K$ by translates of H are bounded. Corollary 8.9 implies the existence of a nonzero vector $\boldsymbol{e} \in \operatorname{ort} H \cap \operatorname{rint}(\operatorname{rec} K)^\circ$. In particular, $\boldsymbol{e} \in (\operatorname{rec} K)^\circ = \operatorname{rint}(\operatorname{bar} K)$ according to Corollary 8.36. By Definition 8.31, there is a scalar $\gamma = \gamma(\boldsymbol{e})$ such that $\boldsymbol{x}\cdot\boldsymbol{e} \leqslant \gamma$ for all $\boldsymbol{x} \in \operatorname{cl} K$. Choose any scalar $\gamma' > \gamma$. Then $\boldsymbol{x}\cdot\boldsymbol{e} < \gamma'$ for all $\boldsymbol{x} \in \operatorname{cl} K$. Consequently, the hyperplane $H' = \{\boldsymbol{x} \in \mathbb{R}^n : \boldsymbol{x}\cdot\boldsymbol{e} = \gamma'\}$ is disjoint from $\operatorname{cl} K$. Finally, Theorem 2.18 shows that H' is a translate of H.

Chapter 9

Problem 9.1. Clearly, H bounds (respectively, strongly bounds) $B_\rho(\boldsymbol{a})$ if and only if $\rho \leqslant \delta(\boldsymbol{a}, H)$ (respectively, $\rho < \delta(\boldsymbol{a}, H)$). Because $\delta(\boldsymbol{a}, H) = |\gamma - \boldsymbol{a}\cdot\boldsymbol{e}|/\|\boldsymbol{e}\|$ (see Theorem 2.91), H bounds (respectively, strongly bounds) $B_\rho(\boldsymbol{a})$ if and only if $\rho\|\boldsymbol{e}\| \leqslant |\gamma - \boldsymbol{a}\cdot\boldsymbol{e}|$ (respectively, $\rho\|\boldsymbol{e}\| < |\gamma - \boldsymbol{a}\cdot\boldsymbol{e}|$).

Problem 9.2. Theorem 9.4 shows the existence of a hyperplane $H \subset \mathbb{R}^n$ containing L and properly bounding $\operatorname{cl} K$. If $H \cap \operatorname{cl} K = \varnothing$, then H is a desired hyperplane. Assume that H meets $\operatorname{cl} K$. Then, according to Theorem 9.3, $H \cap \operatorname{cl} K \subset \operatorname{rbd} K$. By the hypothesis, the set $H \cap \operatorname{rbd} K$ contains no closed halfline. Therefore, Theorem 6.12 implies that $H \cap \operatorname{cl} K$ is a nonempty compact set. According to Theorem 6.40, $\operatorname{dir} H \cap \operatorname{rec}(\operatorname{cl} K) = \{\boldsymbol{o}\}$.

By Theorem 2.17, we can write $H = \{\boldsymbol{x} \in \mathbb{R}^n : \boldsymbol{x}\cdot\boldsymbol{e} = \gamma\}$. Denote by V the closed halfspace of $\mathbb{R}^n$ determined by H and containing K. Without loss of generality, we may assume that $V = \{\boldsymbol{x} \in \mathbb{R}^n : \boldsymbol{x}\cdot\boldsymbol{e} \leqslant \gamma\}$ (see Definition 2.29).

Choose a scalar $\gamma' < \gamma$ and consider the hyperplane $H' = \{\boldsymbol{x} \in \mathbb{R}^n : \boldsymbol{x}\cdot\boldsymbol{e} = \gamma'\}$. If P is the closed slab between H and H', then the set $\operatorname{cl} K \cap P$ is compact (see Problem 6.11). Since $\operatorname{cl} K \cap P$ is disjoint from L, there is a scalar $\rho > 0$ such that the open ρ-neighborhood $U_\rho(\operatorname{cl} K \cap P)$ does not meet L (see Problems 1.14 and 1.15).

If $\operatorname{cl} K \subset P$, then $U_\rho(\operatorname{cl} K) = U_\rho(\operatorname{cl} K \cap P)$, and Theorem 9.4 implies the existence of a hyperplane G which contains L and properly bounds $U_\rho(\operatorname{cl} K)$. Consequently,

$$G \cap \operatorname{cl} K \subset G \cap U_\rho(\operatorname{cl} K) = G \cap \operatorname{rint} U_\rho(\operatorname{cl} K) = \varnothing.$$

Suppose now that $\operatorname{cl} K \not\subset P$. Then $\operatorname{cl} K$ meets the closed halfspace $V' = \{\boldsymbol{x} \in \mathbb{R}^n : \boldsymbol{x}\cdot\boldsymbol{e} \leqslant \gamma'\}$. Furthermore,

$$\delta(H, \operatorname{cl} K \cap V') \geqslant \delta(H, H') = |\gamma - \gamma'|/\|\boldsymbol{e}\|,$$

as follows from Corollary 2.92. Choose a positive scalar r satisfying the condition $r < |\gamma - \gamma'|/\|\boldsymbol{e}\|$. Then

$$L \cap U_r(\operatorname{cl} K \cap V') \subset H \cap U_r(\operatorname{cl} K \cap V') = \varnothing.$$

This argument and the equalities

$$\begin{aligned} U_r(\operatorname{cl} K) &= U_r((\operatorname{cl} K \cap P) \cup (\operatorname{cl} K \cap V')) \\ &= U_r(\operatorname{cl} K \cap P) \cup U_r(\operatorname{cl} K \cap V') \end{aligned}$$

show that $L \cap U_r(\operatorname{cl} K) = \varnothing$. As above, there is a hyperplane G which contains L and misses $\operatorname{cl} K$.

Problem 9.3. Suppose first that a hyperplane H through $\boldsymbol{a}$ supports K. Since the case $K \subset H$ is obvious, we may assume that H properly supports K. Then $H \cap \operatorname{rint} K = \varnothing$ according to Corollary 9.10. By Theorem 3.51, $\operatorname{rint} K \cap U_\rho(\boldsymbol{a}) \neq \varnothing$. This argument and Theorems 3.29 give the inclusion

$$\operatorname{rint}(K \cap U_\rho(\boldsymbol{a})) = \operatorname{rint} K \cap U_\rho(\boldsymbol{a}) \subset \operatorname{rint} K.$$

So,

$$H \cap \operatorname{rint}(K \cap U_\rho(\boldsymbol{a})) \subset H \cap \operatorname{rint} K = \varnothing,$$

implying that H properly supports $U_\rho(\boldsymbol{a}) \cap K$.

Conversely, suppose that H supports $K \cap U_\rho(\boldsymbol{a})$. If $K \cap U_\rho(\boldsymbol{a}) \subset H$, then

$$K \subset \operatorname{aff} K = \operatorname{aff}(K \cap U_\rho(\boldsymbol{a})) \subset H$$

according to Theorems 2.42 and 3.51. Let H properly support $K \cap U_\rho(\boldsymbol{a})$. By Corollary 9.10 and Theorem 3.29,

$$H \cap (\operatorname{rint} K \cap U_\rho(\boldsymbol{a})) = H \cap \operatorname{rint}(K \cap U_\rho(\boldsymbol{a})) = \varnothing.$$

Assume, for contradiction, that H does not support K properly. The above argument shows the existence of a point $\boldsymbol{u} \in (H \cap \operatorname{rint} K) \setminus U_\rho(\boldsymbol{a})$. By Theorem 3.47, $(\boldsymbol{a}, \boldsymbol{u}) \subset \operatorname{rint} K$. Let a point $\boldsymbol{v} \in (\boldsymbol{a}, \boldsymbol{u})$ be so close to $\boldsymbol{a}$ that $\|\boldsymbol{a} - \boldsymbol{v}\| < \rho$. Then $\boldsymbol{v} \in (\boldsymbol{a}, \boldsymbol{u}) \subset \langle \boldsymbol{a}, \boldsymbol{u} \rangle \subset H$ (see Theorem 2.38) and

$$\boldsymbol{v} \in \operatorname{rint} K \cap U_\rho(\boldsymbol{a}) = \operatorname{rint}(U_\rho(\boldsymbol{a}) \cap K),$$

in contradiction with $H \cap \operatorname{rint}(K \cap U_\rho(\boldsymbol{a})) = \varnothing$. Hence H properly supports K.

Problem 9.4. Replacing $K - \boldsymbol{a}$ with K, we may assume that $\boldsymbol{a} = \boldsymbol{o}$. Choose any nonzero vector $\boldsymbol{e} \in \operatorname{rint} K^\circ$. By Corollary 9.19, there is a closed halfspace of the form $V = \{\boldsymbol{x} \in \mathbb{R}^n : \boldsymbol{x} \cdot \boldsymbol{c} \leqslant 0\}$ strongly bounding K. Consequently, the scalar $\varepsilon = \sup\{\boldsymbol{x} \cdot \boldsymbol{c} : \boldsymbol{x} \in K\}$ is negative, as follows from Corollary 9.17. By Theorem 3.26, there is a scalar $\gamma > 1$ such that $\gamma \boldsymbol{e} + (1 - \gamma)\boldsymbol{c} \in K^\circ$. Theorem 5.20 gives the inclusion $\boldsymbol{e} + t\boldsymbol{c} \in K^\circ$, where $t = (1 - \gamma)/\gamma < 0$.

Since $\boldsymbol{e} \in K^\circ$, the scalar $\delta = \sup\{\boldsymbol{x} \cdot \boldsymbol{e} : \boldsymbol{x} \in K\}$ is nonnegative. We assert that $\delta < 0$. Indeed, assume for a moment that $\delta = 0$. Then there is a point $\boldsymbol{u} \in K$ such that $\boldsymbol{u} \cdot \boldsymbol{e} > (-t\varepsilon)/2$. Consequently,

$$\boldsymbol{u} \cdot (\boldsymbol{e} + t\boldsymbol{c}) > -\frac{t\varepsilon}{2} + t\varepsilon = \frac{t\varepsilon}{2} > 0,$$

contrary to the inclusion $\boldsymbol{e} + t\boldsymbol{c} \in K^\circ$. Hence $\delta < 0$, and Corollary 9.2 shows that the hyperplane $H = \{\boldsymbol{x} \in \mathbb{R}^n : \boldsymbol{x} \cdot \boldsymbol{e} = 0\}$ strongly bounds K.

Problem 9.5. Replacing $K - \boldsymbol{a}$ with K, we may assume that $\boldsymbol{a} = \boldsymbol{o}$.

$(a) \Rightarrow (b)$ Assume first that K is line-free. Then the generated convex cone $C = C_{\boldsymbol{o}}(K)$ is line-free (see Corollary 7.21). By Theorem 8.4, the polar cone $K^\circ = C^\circ$ has nonempty interior. Choose a nonzero vector $\boldsymbol{e} \in \operatorname{int} K^\circ$ such that the $(n-1)$-dimensional subspace $H = \{\boldsymbol{x} \in \mathbb{R}^n : \boldsymbol{x} \cdot \boldsymbol{e} = 0\}$ misses $\boldsymbol{b}$ (if all such $(n-1)$-dimensional subspaces G contained $\boldsymbol{b}$, then their normal vectors would lie in a $\{\boldsymbol{b}\}^\perp$, contrary to the n-dimensionality of K°). Problem 9.4 shows that H strongly bounds K.

$(b) \Rightarrow (a)$ Suppose, for contradiction, that K contains a line $l = \{\boldsymbol{u} + t\boldsymbol{v} : t \in \mathbb{R}\}$, where $\boldsymbol{u} \in K$ and $\boldsymbol{v} \neq \boldsymbol{o}$. By the assumption, there is a hyperplane $H = \{\boldsymbol{x} \in \mathbb{R}^n : \boldsymbol{x} \cdot \boldsymbol{e} = 0\}$ strongly bounding K and missing $\boldsymbol{v}$. Without loss of generality, we may assume that K lies in the open halfspace $W = \{\boldsymbol{x} \in \mathbb{R}^n : \boldsymbol{x} \cdot \boldsymbol{c} < 0\}$ determined by H. Consequently, $l \subset K \subset W$, which gives $(\boldsymbol{u} + t\boldsymbol{v}) \cdot \boldsymbol{e} < 0$ for all $t \in \mathbb{R}$. Obviously, the last is possible only if $\boldsymbol{v} \cdot \boldsymbol{e} = 0$. Thus, $\boldsymbol{v} \in H$, contrary to the assumption.

Problem 9.6. According to Theorem 3.67, the plane $L = \operatorname{aff} F$ satisfies the condition $L \cap \operatorname{cl} K \subset \operatorname{rbd} K$. Hence $L \cap \operatorname{rint} K = \varnothing$.

Problem 9.7. $(a) \Rightarrow (b)$ By Theorem 9.45, there is an $(n-1)$-dimensional subspace $S \subset \mathbb{R}^n$ which contains L and supports C such that $S \cap \operatorname{cl} C = \operatorname{ap}(\operatorname{cl} C)$.

Then $S^\perp \subset L^\perp$. Theorem 8.7 gives

$$\varnothing \neq (S^\perp \setminus \{\boldsymbol{o}\}) \cap \operatorname{rint} C^\circ \subset (L^\perp \setminus \{\boldsymbol{o}\}) \cap \operatorname{rint} C^\circ.$$

(*b*) $\Rightarrow$ (*a*) Choose a vector $\boldsymbol{e} \in (L^\perp \setminus \{\boldsymbol{o}\}) \cap \operatorname{rint} C^\circ$. By Theorem 8.7, the $(n-1)$-dimensional subspace $S = \{\boldsymbol{x} \in \mathbb{R}^n : \boldsymbol{x} \cdot \boldsymbol{e} = 0\}$ satisfies the condition $S \cap \operatorname{cl} C = \operatorname{ap}(\operatorname{cl} C)$. Since $L \subset \{\boldsymbol{e}\}^\perp = S$, one has $L \cap \operatorname{cl} C \subset S \cap \operatorname{cl} C = \operatorname{ap}(\operatorname{cl} C)$.

Problem 9.8. Assume, for contradiction, that $\operatorname{conv} X \neq L$. Then $\operatorname{cl}(\operatorname{conv} X) \neq L$ due to Theorem 3.50. Choose a point $\boldsymbol{a} \in L \setminus \operatorname{cl}(\operatorname{conv} X)$. Theorem 8.22 shows the existence of a nearest to $\boldsymbol{a}$ point $\boldsymbol{c} \in \operatorname{cl}(\operatorname{conv} X)$ such that the closed halfspace $V = \{\boldsymbol{x} \in \mathbb{R}^n : (\boldsymbol{x} - \boldsymbol{c}) \cdot (\boldsymbol{a} - \boldsymbol{c}) \leqslant 0\}$ contains $\operatorname{cl}(\operatorname{conv} X)$. Since $\boldsymbol{a} \in L \setminus V$, the set $L \cap V$ is a closed halfplane of L (see Definition 2.69) containing X. The last contradicts the assumption on X.

Problem 9.9. Since the cone $\operatorname{rec}(\operatorname{cl} K)$ is line-free (see Theorem 6.37), a combination of Theorems 8.4 and 8.35 shows that the set $\operatorname{cl}(\operatorname{nor} K) = (\operatorname{rec}(\operatorname{cl} K))^\circ$ is n-dimensional. By Theorem 3.46, the cone $\operatorname{nor} K$ also is n-dimensional. Hence $X \cap \operatorname{nor} K$ is dense in $\mathbb{S} \cap \operatorname{nor} K$ (see Problem 5.8), and the assertion follows from Theorem 9.26.

Problem 9.10. By Theorem 9.24, $\mathcal{F}$ contains a countable subfamily $\mathcal{F}'$ whose intersection is $\operatorname{cl} K$. Let $\mathcal{G}$ be the family of all halfspaces from $\mathcal{F}'$ which do not contain $\operatorname{aff} K$. The family $\mathcal{G}$ is nonempty, since otherwise

$$\operatorname{cl} K \subset \operatorname{aff} K \subset \cap (V : V \in \mathcal{F}') = \operatorname{cl} K,$$

which gives $K = \operatorname{cl} K = \operatorname{aff} K$, contrary to the assumption that K is not a plane (see Theorem 3.50). Clearly, $V \cap \operatorname{aff} K$ is a closed halfplane of $\operatorname{aff} K$ for all $V \in \mathcal{G}$. If $\mathcal{G} = \mathcal{F}'$, then

$$\begin{aligned}\operatorname{cl} K &= \operatorname{cl} K \cap \operatorname{aff} K = (\cap (V : V \in \mathcal{F}')) \cap \operatorname{aff} K \\ &= \cap (V \cap \operatorname{aff} K : V \in \mathcal{F}') = \cap (V \cap \operatorname{aff} K : V \in \mathcal{G}).\end{aligned}$$

Similarly, if $\mathcal{G} \neq \mathcal{F}'$, then

$$\begin{aligned}\operatorname{cl} K &= \operatorname{cl} K \cap \operatorname{aff} K = (\cap (V : V \in \mathcal{F}')) \cap \operatorname{aff} K \\ &= (\cap(V \cap \operatorname{aff} K : V \in \mathcal{G})) \cap (\cap(V \cap \operatorname{aff} K : V \in \mathcal{F}' \setminus \mathcal{G})) \\ &= \cap (V \cap \operatorname{aff} K : V \in \mathcal{G}) \cap \operatorname{aff} K = \cap (V \cap \operatorname{aff} K : V \in \mathcal{G}).\end{aligned}$$

Problem 9.11. For the first part, we observe that the halfspace V contains the set $M = V_1 \cap \cdots \cap V_r$. Indeed, if $\boldsymbol{x} \in M$, then $\boldsymbol{x} \cdot \boldsymbol{e}_i \leqslant \mu_i$ for all $1 \leqslant i \leqslant r$, which gives

$$\boldsymbol{x} \cdot \boldsymbol{e} = \boldsymbol{x} \cdot (\lambda_1 \boldsymbol{e}_1 + \cdots + \lambda_r \boldsymbol{e}_r) \leqslant \lambda_1 \mu_1 + \cdots + \lambda_r \mu_r = \mu. \tag{14.22}$$

Therefore, $X \subset M \subset V$. If a halfspace V_i properly bounds X, then there is a point $\boldsymbol{x} \in X$ for which $\boldsymbol{x} \cdot \boldsymbol{e}_i < \mu_i$. Then (14.22) gives $\boldsymbol{x} \cdot \boldsymbol{e} < \mu$. Consequently,

V properly bounds X. The case when at least one of the halfspaces $V_1, \dots, V_r$ strictly (respectively, strongly) bounds X is similar.

Problem 9.12. Translating both sets L and $B_\rho(\boldsymbol{a})$ on the same suitable vector, we may suppose that L is a subspace. Furthermore, since the case $L = \{\boldsymbol{o}\}$ is obvious, we let $\dim L > 0$. Put $m = \dim L$, choose an orthonormal basis $\{\boldsymbol{b}_1, \dots, \boldsymbol{b}_m\}$ for L, and complete it into an orthonormal basis $\{\boldsymbol{b}_1, \dots, \boldsymbol{b}_n\}$ for $\mathbb{R}^n$. Let $\boldsymbol{x} \in B_\rho(\boldsymbol{a})$. With $\boldsymbol{a} = a_1\boldsymbol{b}_1 + \cdots + a_n\boldsymbol{b}_n$ and $\boldsymbol{x} = x_1\boldsymbol{b}_1 + \cdots + x_n\boldsymbol{b}_n$, we have

$$\begin{aligned} \|f_L(\boldsymbol{a}) - f_L(\boldsymbol{a})\| &= \sqrt{(a_1 - x_1)^2 + \cdots + (a_m - x_m)^2} \\ &\leqslant \sqrt{(a_1 - x_1)^2 + \cdots + (a_n - x_n)^2} = \|\boldsymbol{a} - \boldsymbol{x}\| \leqslant \rho. \end{aligned}$$

Hence $f_L(B_\rho(\boldsymbol{a})) \subset B_\rho(f_L(\boldsymbol{a})) \cap L$. For the opposite inclusion, let $\boldsymbol{x} \in B_\rho(f_L(\boldsymbol{a})) \cap L$. Then $\boldsymbol{x} = x_1\boldsymbol{b}_1 + \cdots + x_m\boldsymbol{b}_m$ and $\|f_L(\boldsymbol{a}) - \boldsymbol{x}\| \leqslant \rho$. With

$$\boldsymbol{y} = x_1\boldsymbol{b}_1 + \cdots + x_m\boldsymbol{b}_m + a_{m+1}\boldsymbol{b}_{m+1} + \cdots + a_n\boldsymbol{b}_n,$$

we have $f_L(\boldsymbol{y}) = \boldsymbol{x}$ and $\|\boldsymbol{a} - \boldsymbol{y}\| = \|f_L(\boldsymbol{a}) - \boldsymbol{x}\| \leqslant \rho$. Hence $B_\rho(f_L(\boldsymbol{a})) \cap L \subset f_L(B_\rho(\boldsymbol{a}))$. The case of open ball $U_\rho(\boldsymbol{a})$ is similar.

Problem 9.13. (*a*) First, we assert that $U_k(\boldsymbol{a} - k\boldsymbol{e}) \subset U_{k+1}(\boldsymbol{a} - (k+1)\boldsymbol{e})$ for all $k \geqslant 1$. Indeed, if $\boldsymbol{x} \in U_k(\boldsymbol{a} - k\boldsymbol{e})$, then the inequalities

$$\|\boldsymbol{x} - (\boldsymbol{a} - (k+1)\boldsymbol{e})\| = \|(\boldsymbol{x} - (\boldsymbol{a} - k\boldsymbol{e})) + \boldsymbol{e}\| \leqslant \|\boldsymbol{x} - (\boldsymbol{a} - k\boldsymbol{e})\| + \|\boldsymbol{e}\| \leqslant k + 1$$

imply the inclusion $\boldsymbol{x} \in U_{k+1}(\boldsymbol{a} - (k+1)\boldsymbol{e})$. Hence $U_k(\boldsymbol{a} - k\boldsymbol{e}) \subset U_{k+1}(\boldsymbol{a} - (k+1)\boldsymbol{e})$.

Next, we assert that $U_k(\boldsymbol{a} - k\boldsymbol{e}) \subset W$ for all $k \geqslant 1$. Indeed, if $\boldsymbol{x} \in U_k(\boldsymbol{a} - k\boldsymbol{e})$, then $\|\boldsymbol{x} - (\boldsymbol{a} - k\boldsymbol{e})\| < k$. Squaring this inequality, we obtain

$$(\boldsymbol{x} - \boldsymbol{a} + k\boldsymbol{e})\cdot(\boldsymbol{x} - \boldsymbol{a} + k\boldsymbol{e}) < k^2.$$

Equivalently,

$$\|\boldsymbol{x} - \boldsymbol{a}\|^2 + 2k(\boldsymbol{x} - \boldsymbol{a})\cdot\boldsymbol{e} + k^2\|\boldsymbol{e}\|^2 < k^2.$$

Since $\|\boldsymbol{e}\| = 1$, one has $(\boldsymbol{x} - \boldsymbol{a})\cdot\boldsymbol{e} < 0$, or $\boldsymbol{x}\cdot\boldsymbol{e} < \boldsymbol{a}\cdot\boldsymbol{e} = \gamma$. Thus $\boldsymbol{x} \in W$.

It remains to show that $W \subset U_1(\boldsymbol{a} - \boldsymbol{e}) \cup U_2(\boldsymbol{a} - 2\boldsymbol{e}) \cup \cdots$. For this, choose any point $\boldsymbol{u} \in W$. Put $\mu = \boldsymbol{u}\cdot\boldsymbol{e}$, $\boldsymbol{b} = \boldsymbol{a} - (\gamma - \mu)\boldsymbol{e}$, and $\delta = \|\boldsymbol{u} - \boldsymbol{b}\|$. Let an integer m satisfy the inequality

$$m > \frac{(\gamma - \mu)^2 + \delta^2}{2(\gamma - \mu)}.$$

Put $\boldsymbol{a}' = \boldsymbol{a} - m\boldsymbol{e}$. Because $\|\boldsymbol{a}' - \boldsymbol{b}\| = \|(\gamma - \mu - m)\boldsymbol{e}\| = |\gamma - \mu - m|$ and

$$\begin{aligned} (\boldsymbol{u} - \boldsymbol{b})\cdot(\boldsymbol{a}' - \boldsymbol{b}) &= (\boldsymbol{u} - \boldsymbol{a} + (\gamma - \mu)\boldsymbol{e})\cdot((\boldsymbol{a} - m\boldsymbol{e}) - (\boldsymbol{a} - (\gamma - \mu)\boldsymbol{e})) \\ &= (\gamma - \mu - m)(\boldsymbol{u} - \boldsymbol{a} + (\gamma - \mu)\boldsymbol{e})\cdot\boldsymbol{e} = (\gamma - \mu - m)(\mu - \gamma + \gamma - \mu) = 0, \end{aligned}$$

we obtain by the choice of m:

$$\begin{aligned}\|\boldsymbol{u}-\boldsymbol{a}'\|^2 &= \|(\boldsymbol{u}-\boldsymbol{b})-(\boldsymbol{a}'-\boldsymbol{b})\|^2 = \|(\boldsymbol{u}-\boldsymbol{b})\|^2+\|(\boldsymbol{a}'-\boldsymbol{b})\|^2 \\ &= \delta^2+(\gamma-\mu-m)^2 = ((\gamma-\mu)^2-2(\gamma-\mu)m+\delta^2)+m^2 < m^2.\end{aligned}$$

Consequently, $\|\boldsymbol{u}-\boldsymbol{a}'\| < m$. Hence $\boldsymbol{u} \in U_m(\boldsymbol{a}') = U_m(\boldsymbol{a}-m\boldsymbol{e})$, as desired.

(*b*) Since the union of open balls $U_k(\boldsymbol{a}-k\boldsymbol{e})$, $k \geqslant 1$, equals W, these balls form an open cover of X. By a compactness argument, there are finitely many such balls whose union contains X. Because this family of balls is nested, the largest one contains X.

Problem 9.14. Theorem 9.26 and a remark on page 322 imply that $\operatorname{cl} K$ is the intersection of a countably many open halfspaces $W_1, W_2, \ldots \subset \mathbb{R}^n$. Problem 9.13 shows the existence of open balls U_i satisfying the inclusions $\operatorname{cl} K \subset U_i \subset W_i$, $i \geqslant 1$. Because the set $\operatorname{cl} K$ is compact, every open ball U_i contains a closed ball B_i such that $\operatorname{cl} K \subset B_i$, $i \geqslant 1$. Thus

$$\operatorname{cl} K \subset B_1 \cap B_2 \cap \cdots \subset U_1 \cap U_2 \cap \cdots \subset W_1 \cap W_2 \cap \cdots = \operatorname{cl} K,$$

implying the desired representations $\operatorname{cl} K = B_1 \cap B_2 \cap \cdots = U_1 \cap U_2 \cap \cdots$.

Chapter 10

Problem 10.1. First, we observe that $L \cap (\boldsymbol{u}, \boldsymbol{v}) = \varnothing$. Indeed, assume, for contradiction, the existence of a point $\boldsymbol{x} \in L \cap (\boldsymbol{u}, \boldsymbol{v})$. Then $\boldsymbol{x} = (1-\lambda)\boldsymbol{u} + \lambda\boldsymbol{v}$ for a suitable scalar $0 < \lambda < 1$. Then $\boldsymbol{u} \in \langle \boldsymbol{x}, \boldsymbol{v}\rangle = \operatorname{aff}\{\boldsymbol{x}, \boldsymbol{v}\} \subset \operatorname{aff}\{L \cup \{\boldsymbol{v}\}\}$. Since $\dim(\{\boldsymbol{v}\} \cup L) = m+1$ (see Corollary 2.67), we obtain a contradiction with the assumption $\dim(L \cup \{\boldsymbol{u}, \boldsymbol{v}\}) = m+1$. So, $L \cap (\boldsymbol{u}, \boldsymbol{v}) = \varnothing$.

Denote by $\boldsymbol{w}$ the midpoint of the closed segment $[\boldsymbol{u}, \boldsymbol{v}]$. Since $\boldsymbol{w} \notin L$, the plane $M = \operatorname{aff}(\{\boldsymbol{w}\} \cup L)$ has dimension $m+1$ due to the same corollary. We observe that $M \cap \{\boldsymbol{u}, \boldsymbol{v}\} = \varnothing$. Indeed, assume for a moment that $\boldsymbol{u} \in M$. Then $\boldsymbol{v} = 2\boldsymbol{w} - \boldsymbol{u} \in M$ according to Theorem 2.38. Consequently, $\{\boldsymbol{u}, \boldsymbol{v}\} \cup L \subset M$, contrary to $\dim(L \cup \{\boldsymbol{u}, \boldsymbol{v}\}) = m+2$.

By Theorem 9.4, there is a hyperplane $H \subset \mathbb{R}^n$ which contains M and misses $\boldsymbol{u}$. Clearly, $\boldsymbol{v} \notin H$, since otherwise $\boldsymbol{u} = 2\boldsymbol{w} - \boldsymbol{v} \in H$ due to Theorem 2.38 and the inclusion $\boldsymbol{w} \in M \subset H$. If both $\boldsymbol{u}$ and $\boldsymbol{v}$ belonged to an open halfspace W determined by H, then $\boldsymbol{w} \in (\boldsymbol{u}, \boldsymbol{v}) \subset W$ by the convexity of W (see an example on page 75), contrary to the inclusion $\boldsymbol{w} \in M \subset H$. Hence H strictly separates $\boldsymbol{u}$ and $\boldsymbol{v}$.

Problem 10.2. Choose a point $\boldsymbol{u} \in L_1 \cap L_2$.

(*a*) $\Rightarrow$ (*b*) and (*a*) $\Rightarrow$ (*c*). If L_1 and L_2 are separated by a hyperplane, then L_1 and L_2 lie in the opposite closed halfspaces, say V_1 and V_2, determined by H. Then $\boldsymbol{u} \in V_1 \cap V_2 = H$, and Theorem 2.34 implies the inclusion $L_1 \cup L_2 \subset H$.

Consequently, aff $(L_1 \cup L_2) \subset H$, and

$$\dim (L_1 \cup L_2) = \dim (\text{aff } (L_1 \cup L_2)) \leqslant \dim H = n - 1.$$

Furthermore, the inclusion $L_1 \cup L_2 \subset H$ implies that $L_1 + L_2 \subset H + H = \boldsymbol{v} + H$ for a suitable point $\boldsymbol{v} \in \mathbb{R}^n$. Consequently, $L_1 + L_2 \neq \mathbb{R}^n$.

$(b) \Rightarrow (a)$ If $\dim (L_1 \cup L_2) \leqslant n - 1$, then, by the above, argument, the dimension of the plane $L = \text{aff } (L_1 \cup L_2)$ is $n-1$ or less. Hence L lies in a suitable hyperplane $H \subset \mathbb{R}^n$. Consequently, H (trivially) separates L_1 and L_2.

$(c) \Rightarrow (a)$ Suppose $L_1 + L_2 \neq \mathbb{R}^n$. Then $L_1 + L_2$ is a plane of dimension $\leqslant n - 1$. Choose a hyperplane $H \subset \mathbb{R}^n$ which contains $L_1 + L_2$. Since $L_1 - \boldsymbol{u}$ and $L_2 - \boldsymbol{u}$ are subspaces, we have

$$(L_1 - \boldsymbol{u}) \cup (L_2 - \boldsymbol{u}) \subset (L_1 - \boldsymbol{u}) + (L_2 - \boldsymbol{u}) \subset H - 2\boldsymbol{u}.$$

Thus $L_1 \cup L_2$ lies in the hyperplane $H - \boldsymbol{u}$, which (trivially) separates L_1 and L_2.

Problem 10.3. $(a) \Rightarrow (b)$ Let $\boldsymbol{u} \in \mathbb{R}^n$ be a vector such that $\boldsymbol{u} + K_1$ and K_2 are separated by a hyperplane $H \subset \mathbb{R}^n$. Denote by V_1 and V_2 closed halfspaces determined by H and containing $\boldsymbol{u} + K_1$ and K_2, respectively. Without loss of generality, we may write

$$V_1 = \{\boldsymbol{x} \in \mathbb{R}^n : \boldsymbol{x}\cdot\boldsymbol{e} \leqslant \gamma\} \quad \text{and} \quad V_2 = \{\boldsymbol{x} \in \mathbb{R}^n : \boldsymbol{x}\cdot\boldsymbol{e} \geqslant \gamma\}.$$

A combination of Definition 8.31 and Theorem 8.33 shows that $\boldsymbol{e} \in \text{bar } (\boldsymbol{u}+K_1) = \text{bar } K_1$ and $-\boldsymbol{e} \in \text{bar } K_2$. Consequently, $\boldsymbol{e} \in \text{bar } K_1 \cap (-\text{bar } K_2)$.

$(b) \Rightarrow (a)$ Choose a nonzero vector $\boldsymbol{e} \in \text{bar } K_1 \cap (-\text{bar } K_2)$. By Definition 8.31, there are scalars γ_1 and γ_2 such that $\boldsymbol{x}_1\cdot\boldsymbol{e} \leqslant \gamma_1$ and $\boldsymbol{x}_2\cdot\boldsymbol{e} \leqslant \gamma_2$ whenever $\boldsymbol{x}_1 \in K_1$ and $\boldsymbol{x}_2 \in -K_2$. Let $\mu = (\gamma_1 + \gamma_2 + 1)/\|\boldsymbol{e}\|^2$ and $\boldsymbol{u} = \mu\boldsymbol{e}$. For any point $\boldsymbol{x} \in K_2$, we have

$$(\boldsymbol{u} + \boldsymbol{x})\cdot\boldsymbol{e} = (\mu\boldsymbol{e} + \boldsymbol{x})\cdot\boldsymbol{e} \geqslant (\gamma_1 + \gamma_2 + 1) - \gamma_2 = \gamma_1 + 1.$$

Hence

$$\sup \{\boldsymbol{x}\cdot\boldsymbol{e} : \boldsymbol{x} \in K_1\} \leqslant \gamma_1 < \gamma_1 + 1 \leqslant \inf \{\boldsymbol{x}\cdot\boldsymbol{e} : \boldsymbol{x} \in K_2\},$$

and Theorem 10.2 shows that K_1 and K_2 are strongly separated by a hyperplane.

Problem 10.4. (a) By Theorem 10.24, there is a pair of points $\boldsymbol{c}_1 \in \text{cl } K_1$ and $\boldsymbol{c}_2 \in \text{cl } K_2$ such that $\delta(K_1, K_2) = \|\boldsymbol{c}_1 - \boldsymbol{c}_2\|$. Theorem 10.23 asserts that if H_1 and H_2 are the hyperplanes through $\boldsymbol{c}_1$ and $\boldsymbol{c}_2$, respectively, both orthogonal to the closed segment $[\boldsymbol{c}_1, \boldsymbol{c}_2]$, then the closed slab between H_1 and H_2 separates K_1 and K_2. Let V_1 and V_2 be the opposite closed halfspaces bounded by H_1 and H_2, such that $K_1 \subset V_1$ and $K_2 \subset V_2$.

Denote by $\boldsymbol{a}$ the midpoint of $[\boldsymbol{c}_1, \boldsymbol{c}_2]$ and consider the convex set $K_2' = 2\boldsymbol{a} - K_2$. Since rec (cl K_2') $= -$rec (cl K_2), Theorem 7.22 shows that the set $K = \text{conv } (K_1 \cup K_2')$ is line-free. Furthermore, K lies in V_1. We have $\boldsymbol{a} \notin \text{cl } K$ because $\boldsymbol{a} \notin V_1$. By Corollary 7.21, the generated cone $C = C_{\boldsymbol{a}}(K)$ is line-free. Problem 8.6 gives the existence of a simplicial cone $\Lambda \subset \mathbb{R}^n$ with apex $\boldsymbol{a}$ satisfying the condition

$C \subset \Lambda$. Clearly, $K_1 \subset \Lambda$ and $K_2 \subset 2\boldsymbol{a} - \Lambda$, which shows that the simplicial cone $\Gamma = \Lambda - \boldsymbol{a}$ is a required one.

(*b*) Similarly to the argument of part (*a*), consider some points $\boldsymbol{a}_1, \boldsymbol{a}_2 \in (\boldsymbol{c}_1, \boldsymbol{c}_2)$ such that $\boldsymbol{a}_1 \in (\boldsymbol{c}_1, \boldsymbol{a}_2)$. Choose a simplicial cone $\Lambda \subset \mathbb{R}^n$ with apex $\boldsymbol{a}_1$ which contains the generated cone $C_{\boldsymbol{a}_1}(K_1 \cup (2\boldsymbol{a} - K_2))$. Then the simplicial cone $\Gamma' = \Lambda - \boldsymbol{a}$ is a desired one.

Problem 10.5. By Theorem 2.17, H can be written as $H = \{\boldsymbol{x} \in \mathbb{R}^n : \boldsymbol{x} \cdot \boldsymbol{e} = \gamma\}$ for a nonzero vector $\boldsymbol{e} \in \mathbb{R}^n$ and a scalar γ. Without loss of generality, we may suppose that C and K belong, respectively, to the opposite closed halfspaces

$$V_1 = \{\boldsymbol{x} \in \mathbb{R}^n : \boldsymbol{x} \cdot \boldsymbol{e} \leqslant \gamma\} \quad \text{and} \quad V_2 = \{\boldsymbol{x} \in \mathbb{R}^n : \boldsymbol{x} \cdot \boldsymbol{e} \geqslant \gamma\},$$

Put

$$\gamma' = \sup \{\boldsymbol{x} \cdot \boldsymbol{e} : \boldsymbol{x} \in C\} \quad \text{and} \quad H' = \{\boldsymbol{x} \in \mathbb{R}^n : \boldsymbol{x} \cdot \boldsymbol{e} = \gamma'\}.$$

By Theorem 2.18, the hyperplane H' is parallel to H, and Theorem 10.2 shows that H' separates C and K. Since $\delta(C, H) = 0$ (see Corollary 2.92), H' either supports C or is asymptotic to C. Since C has no asymptotic hyperplane (see Theorem 9.44), H' must support C. Finally, Theorem 9.43 gives the inclusion $\boldsymbol{a} \in \operatorname{ap} C \subset H'$.

Problem 10.6. Choose points $\boldsymbol{u}_1 \in \operatorname{cl} K_1$ and $\boldsymbol{u}_2 \in \operatorname{cl} K_2$ satisfying the condition $\|\boldsymbol{u}_1 - \boldsymbol{u}_2\| = \delta(K_1, K_2)$. Put $\boldsymbol{e} = \boldsymbol{u}_1 - \boldsymbol{u}_2$, and let $\gamma_i = \boldsymbol{u}_i \cdot \boldsymbol{e}$, $i = 1, 2$. Without loss of generality, we may assume that $\gamma_1 < \gamma_2$. Choose scalars μ_1 and μ_2 such that $\gamma_1 < \mu_1 < \mu_2 < \gamma_2$. Then the open halfspaces

$$W_1 = \{\boldsymbol{x} \in \mathbb{R}^n : \boldsymbol{x} \cdot \boldsymbol{e} < \mu_1\} \quad \text{and} \quad W_2 = \{\boldsymbol{x} \in \mathbb{R}^n : \boldsymbol{x} \cdot \boldsymbol{e} > \mu_2\}$$

contain $\operatorname{cl} K_1$ and $\operatorname{cl} K_2$, respectively. Problem 9.13 shows the existence of open balls U_1 and U_2 satisfying the inclusions $\operatorname{cl} K_i \subset U_i \subset W_i$, $i = 1, 2$. Obviously, $U_1 \cap U_2 \subset W_1 \cap W_2 = \varnothing$. Because the sets $\operatorname{cl} K_1$ and $\operatorname{cl} K_2$ are compact, there are closed balls B_1 and B_2 such that $\operatorname{cl} K \subset B_i \subset U_i$, $i \geqslant 1$. Clearly, $B_1 \cap B_2 = \varnothing$.

Problem 10.7. Assume first that K separates X_1 and X_2. Choose any points $\boldsymbol{x}_1 \in X_1$ and $\boldsymbol{x}_2 \in X_2$. The assumption $[\boldsymbol{x}_1, \boldsymbol{x}_2] \cap K \neq \varnothing$ shows the existence of a scalar $0 \leqslant \lambda \leqslant 1$ such that $(1 - \lambda)\boldsymbol{x}_1 + \lambda \boldsymbol{x}_2 \in K$. Using Theorem 4.13 and the convexity of K, we obtain

$$\boldsymbol{x}_1 + \boldsymbol{x}_2 = [(1 - \lambda)\boldsymbol{x}_1 + \lambda \boldsymbol{x}_2] + [\lambda \boldsymbol{x}_1 + (1 - \lambda)\boldsymbol{x}_2] \in K + \operatorname{conv}(X_1 \cup X_2).$$

Consequently, $X_1 + X_2 \subset K + \operatorname{conv}(X_1 \cup X_2)$.

Conversely, let $X_1 + X_2 \subset K + \operatorname{conv}(X_1 \cup X_2)$. Choose any points $\boldsymbol{x}_1 \in X_1$ and $\boldsymbol{x}_2 \in X_2$. Then

$$X_1 = (X_1 + \boldsymbol{x}_2) - \boldsymbol{x}_2 \subset (X_1 + X_2) - \boldsymbol{x}_2 \subset K + \operatorname{conv}(X_1 \cup X_2) - \boldsymbol{x}_2.$$

Similarly, $X_2 \subset K + \operatorname{conv}(X_1 \cup X_2) - \boldsymbol{x}_1$. Consequently, Problem 4.9 gives

$$\begin{aligned}
\boldsymbol{o} + \operatorname{conv}(X_1 \cup X_2) &= \operatorname{conv}(X_1 \cup X_2) \\
&\subset \operatorname{conv}(((K - \boldsymbol{x}_2) + \operatorname{conv}(X_1 \cup X_2)) \cup ((K - \boldsymbol{x}_1) + \operatorname{conv}(X_1 \cup X_2))) \\
&= \operatorname{conv}((K - \boldsymbol{x}_2) \cup (K - \boldsymbol{x}_1)) + \operatorname{conv}(X_1 \cup X_2) \\
&= K + \operatorname{conv}\{-\boldsymbol{x}_1, -\boldsymbol{x}_2\} + \operatorname{conv}(X_1 \cup X_2) \\
&= (K + [-\boldsymbol{x}_1, -\boldsymbol{x}_2]) + \operatorname{conv}(X_1 \cup X_2).
\end{aligned}$$

Theorem 4.16 implies that the set $\operatorname{conv}(X_1 \cup X_2)$ is bounded. Consequently, the above argument and Theorem 3.60 give the inclusion $\boldsymbol{o} \in K + [-\boldsymbol{x}_1, -\boldsymbol{x}_2]$. Hence there is a point $\boldsymbol{u} \in K \cap [\boldsymbol{x}_1, \boldsymbol{x}_2]$.

Problem 10.8. Denote by V_1 and V_2 the opposite closed halfspaces determined by H such that $C_1 \subset V_1$ and $C_2 \subset V_2$. Choose a point $\boldsymbol{c} \in \operatorname{cl} C_1 \cap \operatorname{cl} C_2$. Then $\boldsymbol{c} \in V_1 \cap V_2 = H$, which shows that H supports both C_1 and C_2. Consequently, Theorem 9.43 gives the inclusion $\operatorname{ap}(\operatorname{cl} C_1) \cup \operatorname{ap}(\operatorname{cl} C_2) \subset H$.

Problem 10.9. $(a) \Leftrightarrow (c)$ Analysis of the proofs of Theorems 10.15 and 10.16 shows that a hyperplane $H = \{\boldsymbol{x} \in \mathbb{R}^n : \boldsymbol{x}\cdot\boldsymbol{e} = \boldsymbol{a}\cdot\boldsymbol{e}\}$, $\boldsymbol{e} \neq \boldsymbol{o}$, sharply separates C_1 and C_2 if and only if $\boldsymbol{e} \in E \cup (-E)$. On the other hand, Theorem 2.17 shows that the uniqueness of H occurs if and only if the normal vectors of hyperplanes through $\boldsymbol{a}$ sharply separating C_1 and C_2 are multiples of $\boldsymbol{e}$. Consequently, H is unique if and only if the set E is one-dimensional.

$(b) \Leftrightarrow (c)$ Problem 10.8 shows that any hyperplane separating C_1 and C_2 should contain the set $\operatorname{ap}(\operatorname{cl} C_1) \cup \operatorname{ap}(\operatorname{cl} C_2)$. Hence condition (b) implies the uniqueness of H. Conversely, suppose that condition (c) holds. According to Theorem 8.11, the set E is one-dimensional if and only if its closure,

$$D = (C_1 - \boldsymbol{a})^\circ \cap (-(C_2 - \boldsymbol{a})^\circ),$$

is one-dimensional. Hence D is a closed halfline with endpoint $\boldsymbol{o}$. A combination of Theorem 8.3 and Problem 8.1 shows that the set

$$D^\circ = ((C_1 - \boldsymbol{a})^\circ \cap (-(C_2 - \boldsymbol{a})^\circ))^\circ = \operatorname{cl}((C_1 - \boldsymbol{a}) - (C_2 - \boldsymbol{a}))$$

is a closed halfspace, whose apex set is a subspace $G \subset \mathbb{R}^n$ of dimension $n-1$. Furthermore, the assumption on C_1 and C_2, combined with Theorem 7.18 gives

$$\begin{aligned}
G &= \operatorname{ap}(\operatorname{cl}(C_1 - \boldsymbol{a})) - \operatorname{ap}(\operatorname{cl}(C_2 - \boldsymbol{a})) \\
&= \operatorname{span}(\operatorname{ap}(\operatorname{cl} C_1 - \boldsymbol{a}) \cup \operatorname{ap}(\operatorname{cl} C_2 - \boldsymbol{a})) \\
&= \operatorname{aff}(\operatorname{ap}(\operatorname{cl} C_1) \cup \operatorname{ap}(\operatorname{cl} C_2)) - \boldsymbol{a}.
\end{aligned}$$

Summing up, $H = G + \boldsymbol{a} = \operatorname{aff}(\operatorname{ap}(\operatorname{cl} C_1) \cup \operatorname{ap}(\operatorname{cl} C_2))$.

Problem 10.10. $(a) \Leftrightarrow (b)$ A combination of Corollary 2.43 and Theorem 8.11 show that $\dim D = \dim E$, where

$$D = \operatorname{cl} C \cap (-C^\circ) \quad \text{and} \quad E = \operatorname{rint} C \cap (-\operatorname{rint} C^\circ).$$

Consequently, combining Corollary 10.17 and Problem 10.9 and with the equality

$$\operatorname{span}(\operatorname{ap}(\operatorname{cl} C) \cup \operatorname{ap} C^\circ) = \operatorname{ap}(\operatorname{cl} C) + \operatorname{ap} C^\circ,$$

we obtain the equivalence of conditions (a) and (b).

$(b) \Leftrightarrow (c)$ Assume first that $\operatorname{ap}(\operatorname{cl} C) + \operatorname{ap} C^\circ$ is a hyperplane and let $L = \operatorname{ap}(\operatorname{cl} C)$. Denote by E the orthogonal complement of $\operatorname{ap}(\operatorname{cl} C)$ within $\operatorname{span} C$. Then Theorem 8.5 implies the equality $\dim E = 1$. Consequently, the set $h = \operatorname{cl} C \cap E$ is a closed halfline with endpoint $\boldsymbol{o}$, orthogonal to L. This argument and a combination of Theorems 2.74 and 6.25 immediately imply that $\operatorname{cl} C$ is a closed halfplane of the form

$$\operatorname{cl} C = \operatorname{ap}(\operatorname{cl} C) + (\operatorname{cl} C \cap S) = L + h.$$

Conversely, suppose that $\operatorname{cl} C = L + h$, where L is a subspace L and h is a closed halfline h with endpoint $\boldsymbol{o}$, orthogonal to L. Then

$$\operatorname{lin} C = L \quad \text{and} \quad \operatorname{span} C = L + (h \cup -h).$$

If E is the one-dimensional subspace containing h, then E is the orthogonal complement of $\operatorname{lin} C$ within $\operatorname{span} C$. A combination of Theorems 6.25 and 8.5 shows that

$$\begin{aligned}\dim(\operatorname{ap} C^\circ) &= \dim(\operatorname{span}(\operatorname{cl} C))^\perp = n - \dim(\operatorname{span} C)\\ &= n - \dim(\operatorname{ap}(\operatorname{cl} C)) - 1.\end{aligned}$$

Since the subspaces $\operatorname{ap}(\operatorname{cl} C)$ and $\operatorname{ap} C^\circ$ are orthogonal, their sum $\operatorname{ap}(\operatorname{cl} C) + \operatorname{ap} C^\circ$ is a hyperplane.

Chapter 11

Problem 11.1. Theorem 11.4 shows that every extreme singleton of $B_\rho(\boldsymbol{a})$ belongs to $\operatorname{bd} B_\rho(\boldsymbol{a}) = S_\rho(\boldsymbol{a})$. Conversely, let $\{\boldsymbol{x}\} \subset S_\rho(\boldsymbol{a})$. We assert that $\boldsymbol{x}$ is an extreme point of $B_\rho(\boldsymbol{a})$. Indeed, assume for a moment that $\boldsymbol{x} = (1-\lambda)\boldsymbol{y} + \lambda\boldsymbol{z}$ for distinct points $\boldsymbol{y}, \boldsymbol{z} \in B_\rho(\boldsymbol{a})$ and a scalar $0 < \lambda < 1$. The inequalities

$$\rho = \|\boldsymbol{x} - \boldsymbol{a}\| = \|(1-\lambda)\boldsymbol{y} + \lambda\boldsymbol{z} - \boldsymbol{a}\| \leqslant (1-\lambda)\|\boldsymbol{y} - \boldsymbol{a}\| + \lambda\|\boldsymbol{z} - \boldsymbol{a}\| \leqslant \rho$$

give $\|\boldsymbol{y}-\boldsymbol{a}\| = \|\boldsymbol{z}-\boldsymbol{a}\| = \rho$. Denote by $\boldsymbol{c}$ the midpoint of $[\boldsymbol{x},\boldsymbol{y}]$. According to the parallelogram law (see page 8),

$$\begin{aligned}\|\boldsymbol{a}-\boldsymbol{c}\|^2 &= \|\boldsymbol{a}-\tfrac{1}{2}(\boldsymbol{x}+\boldsymbol{y})\|^2 = \tfrac{1}{4}\|(\boldsymbol{a}-\boldsymbol{x})+(\boldsymbol{a}-\boldsymbol{y})\|^2\\ &= \tfrac{1}{4}(2\|\boldsymbol{a}-\boldsymbol{x}\|^2 + 2\|\boldsymbol{a}-\boldsymbol{y}\|^2 - \|\boldsymbol{x}-\boldsymbol{y}\|^2)\\ &= \tfrac{1}{4}(4\rho^2 - \|\boldsymbol{x}-\boldsymbol{y}\|^2) < \rho^2.\end{aligned}$$

Hence $\|\boldsymbol{a}-\boldsymbol{c}\| < \rho$, which gives the inclusion $\boldsymbol{c} \in U_\rho(\boldsymbol{a}) = \operatorname{int} B_\rho(\boldsymbol{a})$. By Theorem 3.24,

$$\boldsymbol{x} \in (\boldsymbol{y},\boldsymbol{z}) = (\boldsymbol{y},\boldsymbol{c}] \cup [\boldsymbol{c},\boldsymbol{z}) \subset U_\rho(\boldsymbol{a}),$$

contrary to the assumption $\boldsymbol{x} \in S_\rho(\boldsymbol{a})$. The obtained contradiction shows that $\boldsymbol{x} \in \operatorname{ext} B_\rho(\boldsymbol{a})$ and $S_\rho(\boldsymbol{a}) \subset \operatorname{ext} B_\rho(\boldsymbol{a})$.

It remains to show that every extreme face F of $B_\rho(\boldsymbol{a})$ containing two or more points coincides with $B_\rho(\boldsymbol{a})$. Let $\boldsymbol{y}$ and $\boldsymbol{z}$ be distinct points in F. We first assert that $(\boldsymbol{y},\boldsymbol{z}) \subset U_\rho(\boldsymbol{a})$. Indeed, choose a point $\boldsymbol{x} \in (\boldsymbol{y},\boldsymbol{z})$ and express it as $\boldsymbol{x} = (1-\lambda)\boldsymbol{y} + \lambda\boldsymbol{z}$, where $0 < \lambda < 1$. If at least one of the points $\boldsymbol{y},\boldsymbol{z}$, say $\boldsymbol{y}$, satisfies the inequality $\|\boldsymbol{y}-\boldsymbol{a}\| < \rho$, then

$$\begin{aligned}\|\boldsymbol{x}-\boldsymbol{a}\| &= \|(1-\lambda)(\boldsymbol{y}-\boldsymbol{a}) + \lambda(\boldsymbol{z}-\boldsymbol{a})\|\\ &\leqslant (1-\lambda)\|\boldsymbol{y}-\boldsymbol{a}\| + \lambda\|\boldsymbol{z}-\boldsymbol{a}\| < \rho,\end{aligned}$$

implying the inclusion $\boldsymbol{x} \in U_\rho(\boldsymbol{a})$. Assume now that both points $\boldsymbol{y},\boldsymbol{z}$ belong to the sphere $S_\rho(\boldsymbol{a})$. By the above proved, $(\boldsymbol{y},\boldsymbol{z}) \subset U_\rho(\boldsymbol{a})$. Summing up, $(\boldsymbol{y},\boldsymbol{z}) \subset U_\rho(\boldsymbol{a})$. Finally, $F \cap U_\rho(\boldsymbol{a}) \neq \varnothing$ due to $(\boldsymbol{y},\boldsymbol{z}) \subset F$, and Theorem 11.4 gives $F = B_\rho(\boldsymbol{a})$

Problem 11.2. First, we are going to show that every simplex $\Delta' = \Delta(\boldsymbol{z}_1,\dots,\boldsymbol{z}_t)$ whose vertices $\boldsymbol{z}_1,\dots,\boldsymbol{z}_t$ are chosen in $\{\boldsymbol{x}_1,\dots,\boldsymbol{x}_{r+1}\}$ is an extreme face of Δ. Renumbering, if necessary, we may assume that $\boldsymbol{z}_i = \boldsymbol{x}_i$ for all $1 \leqslant i \leqslant t$. Let points $\boldsymbol{x},\boldsymbol{y} \in \Delta$ and a scalar $0 < \lambda < 1$ be such that $\boldsymbol{z} = (1-\lambda)\boldsymbol{x} + \lambda\boldsymbol{y} \in \Delta'$. The points $\boldsymbol{x}$ and $\boldsymbol{y}$ can be written as convex combinations of $\{\boldsymbol{x}_1,\dots,\boldsymbol{x}_{r+1}\}$, and $\boldsymbol{z}$ can be written as a convex combination of $\{\boldsymbol{x}_1,\dots,\boldsymbol{x}_t\}$:

$$\begin{aligned}\boldsymbol{x} = \alpha_1\boldsymbol{x}_1 + \dots + \alpha_{r+1}\boldsymbol{x}_{r+1},\ \ \boldsymbol{y} = \beta_1\boldsymbol{x}_1 + \dots + \beta_{r+1}\boldsymbol{x}_{r+1},\\ \boldsymbol{z} = \gamma\boldsymbol{x}_1 + \dots + \gamma_t\boldsymbol{x}_t.\end{aligned}$$

On the other hand,

$$\boldsymbol{z} = ((1-\lambda)\alpha_1 + \lambda\beta_1)\boldsymbol{x}_1 + \dots + ((1-\lambda)\alpha_{r+1} + \lambda\beta_{r+1})\boldsymbol{x}_{r+1}.$$

Since the set $\{\boldsymbol{x}_1,\dots,\boldsymbol{x}_{r+1}\}$ is affinely independent, the representation of $\boldsymbol{z}$ as a convex combination of $\{\boldsymbol{x}_1,\dots,\boldsymbol{x}_{r+1}\}$ is unique (see Theorem 2.52). So,

$$\begin{aligned}(1-\lambda)\alpha_i + \lambda\beta_i &= \gamma_i,\ \ 1 \leqslant i \leqslant t,\\ (1-\lambda)\alpha_i + \lambda\beta_i &= 0,\ \ t+1 \leqslant i \leqslant r+1.\end{aligned}$$

Because the scalars $\alpha_1, \dots, \alpha_{r+1}, \beta_1, \dots, \beta_{r+1}$ are nonnegative, the above equalities imply that $\alpha_i = \beta_i = 0$ for all $t+1 \leqslant i \leqslant r+1$. Hence both $\boldsymbol{x}$ and $\boldsymbol{y}$ lie in Δ', i. e., Δ' is an extreme face of Δ.

Conversely, let F be an extreme face of Δ. Choose a point $\boldsymbol{x} \in F$. Since $\boldsymbol{x} \in \Delta$, it can be written as a convex combination $\boldsymbol{x} = \lambda_1 \boldsymbol{x}_1 + \cdots + \lambda_{r+1}\boldsymbol{x}_{r+1}$. We assert that $\boldsymbol{x}_i \in F$ for every positive scalar λ_i, $1 \leqslant i \leqslant r+1$. Indeed, this is obvious if $\lambda_i = 1$: then $\lambda_j = 0$ for all $j \neq i$ and $1\boldsymbol{x}_i = \boldsymbol{x} \in F$. Suppose $0 < \lambda_i < 1$ and consider the convex combination

$$\boldsymbol{z} = \frac{\lambda_1}{1-\lambda_i}\boldsymbol{x}_1 + \cdots + \frac{\lambda_{i-1}}{1-\lambda_i}\boldsymbol{x}_{i-1} + \frac{\lambda_{i+1}}{1-\lambda_i}\boldsymbol{x}_{i+1} + \cdots + \frac{\lambda_{r+1}}{1-\lambda_i}\boldsymbol{x}_{r+1}$$

Since $\boldsymbol{z} \in \Delta$, the equality $\boldsymbol{x} = \lambda_i \boldsymbol{x}_i + (1-\lambda_i)\boldsymbol{z}$ and Definition 11.1 show that $\boldsymbol{x}_i, \boldsymbol{z} \in F$. Finally, denote by $\boldsymbol{z}_1, \dots, \boldsymbol{z}_t$ the vertices of Δ that belong to F. By the above argument, $F \subset \Delta(\boldsymbol{z}_1, \dots, \boldsymbol{z}_t)$. On the other hand, since F is convex, we have $\Delta(\boldsymbol{z}_1, \dots, \boldsymbol{z}_t) \subset F$. Summing up, $F = \Delta(\boldsymbol{z}_1, \dots, \boldsymbol{z}_t)$.

Problem 11.3. First, we are going to show that $\boldsymbol{a}$ is an extreme point of Γ. Indeed, assume for a moment that $\boldsymbol{a} = (1-\lambda)\boldsymbol{u} + \lambda\boldsymbol{v}$ for points $\boldsymbol{u}, \boldsymbol{v} \in \Gamma$ and a scalar $0 < \lambda < 1$. According to Definition 5.6, we can write

$$\begin{aligned} \boldsymbol{u} &= \boldsymbol{a} + \alpha_1(\boldsymbol{x}_1 - \boldsymbol{a}) + \cdots + \alpha_r(\boldsymbol{x}_r - \boldsymbol{a}), \quad \alpha_1, \dots, \alpha_r \geqslant 0, \\ \boldsymbol{v} &= \boldsymbol{a} + \beta_1(\boldsymbol{x}_1 - \boldsymbol{v}) + \cdots + \beta_r(\boldsymbol{x}_r - \boldsymbol{a}), \quad \beta_1, \dots, \beta_r \geqslant 0. \end{aligned}$$

Then

$$\boldsymbol{a} = \boldsymbol{a} + ((1-\lambda)\alpha_1 + \lambda\beta_1)(\boldsymbol{x}_1 - \boldsymbol{a}) + \cdots + ((1-\lambda)\alpha_r + \lambda\beta_r)(\boldsymbol{x}_r - \boldsymbol{a}),$$

or

$$((1-\lambda)\alpha_1 + \lambda\beta_1)(\boldsymbol{x}_1 - \boldsymbol{a}) + \cdots + ((1-\lambda)\alpha_r + \lambda\beta_r)(\boldsymbol{x}_r - \boldsymbol{a}) = \boldsymbol{o}. \tag{14.23}$$

By Theorem 2.52, the set $\{\boldsymbol{x}_1 - \boldsymbol{a}, \dots, \boldsymbol{x}_r - \boldsymbol{a}\}$ is linearly independent. So, (14.23) gives

$$(1-\lambda)\alpha_1 + \lambda\beta_1 = \cdots = (1-\lambda)\alpha_r + \lambda\beta_r = 0.$$

Since all scalars α_i, β_i, $1 \leqslant i \leqslant r$, are nonnegative and $0 < \lambda < 1$, we have

$$\alpha_1 = \cdots = \alpha_r = \beta_1 = \cdots = \beta_r = 0,$$

which shows that $\boldsymbol{a} = \boldsymbol{u} = \boldsymbol{v}$; i. e., $\boldsymbol{a}$ is an extreme point of Γ.

Next, we are going to prove that every simplicial cone $\Gamma' = \Gamma_{\boldsymbol{a}}(\boldsymbol{z}_1, \dots, \boldsymbol{z}_t)$ whose vertices $\boldsymbol{z}_1, \dots, \boldsymbol{z}_t$ are chosen in $\{\boldsymbol{x}_1, \dots, \boldsymbol{x}_r\}$ is an extreme face of Γ. Renumbering, if necessary, we may assume that $\boldsymbol{z}_i = \boldsymbol{x}_i$ for all $1 \leqslant i \leqslant t$. Let $\boldsymbol{x}, \boldsymbol{y} \in \Gamma$ and $0 < \lambda < 1$ be such that $\boldsymbol{z} = (1-\lambda)\boldsymbol{x} + \lambda\boldsymbol{y} \in \Gamma'$. Then $\boldsymbol{x}, \boldsymbol{y}$, and $\boldsymbol{z}$

can be written as

$$\begin{aligned} \boldsymbol{x} &= \boldsymbol{a} + \alpha_1(\boldsymbol{x}_1 - \boldsymbol{a}) + \cdots + \alpha_r(\boldsymbol{x}_r - \boldsymbol{a}), \ \alpha_1, \ldots, \alpha_r \geqslant 0, \\ \boldsymbol{y} &= \boldsymbol{a} + \beta_1(\boldsymbol{x}_1 - \boldsymbol{a}) + \cdots + \beta_r(\boldsymbol{x}_r - \boldsymbol{a}), \ \beta_1, \ldots, \beta_r \geqslant 0, \\ \boldsymbol{z} &= \boldsymbol{a} + \gamma_1(\boldsymbol{x}_1 - \boldsymbol{a}) + \cdots + \gamma_t(\boldsymbol{x}_t - \boldsymbol{a}), \ \gamma_1, \ldots, \gamma_t \geqslant 0. \end{aligned}$$

On the other hand, the equality $\boldsymbol{z} = (1-\lambda)\boldsymbol{x} + \lambda\boldsymbol{y}$ gives

$$\boldsymbol{z} = \boldsymbol{a} + ((1-\lambda)\alpha_1 + \lambda\beta_1)(\boldsymbol{x}_1 - \boldsymbol{a}) + \cdots + ((1-\lambda)\alpha_r + \lambda\beta_r)(\boldsymbol{x}_r - \boldsymbol{a}).$$

Since the set $\{\boldsymbol{a}, \boldsymbol{x}_1, \ldots, \boldsymbol{x}_r\}$ is affinely independent, the above representations of $\boldsymbol{z}$ coincide (see Theorem 2.52):

$$\begin{aligned} (1-\lambda)\alpha_i + \lambda\beta_i &= \gamma_i, \ \ 1 \leqslant i \leqslant t, \\ (1-\lambda)\alpha_i + \lambda\beta_i &= 0, \ \ t+1 \leqslant i \leqslant r. \end{aligned}$$

Because the scalars $\alpha_1, \ldots, \alpha_r, \beta_1, \ldots, \beta_r$ are nonnegative, the last $r-t-1$ equalities imply that $\alpha_i = \beta_i = 0$ for all $t+1 \leqslant i \leqslant r$. Hence both $\boldsymbol{x}$ and $\boldsymbol{y}$ lie in Γ', i. e., Γ' is an extreme face of Γ.

Conversely, let F be an extreme face of Γ. By Theorem 11.6, F is a convex cone with apex $\boldsymbol{a}$. Choose a point $\boldsymbol{x} \in F$. Since $\boldsymbol{x} \in \Gamma$, it can be written as

$$\boldsymbol{x} = \boldsymbol{a} + \lambda_1(\boldsymbol{x}_1 - \boldsymbol{a}) + \cdots + \lambda_r(\boldsymbol{x}_r - \boldsymbol{a}), \ \lambda_1, \ldots, \lambda_r \geqslant 0.$$

We assert that $\boldsymbol{x}_i \in F$ for all positive scalars λ_i, $1 \leqslant i \leqslant r$. Indeed, consider the points

$$\begin{aligned} \boldsymbol{y} &= \boldsymbol{a} + 2\lambda_i(\boldsymbol{x}_i - \boldsymbol{a}), \\ \boldsymbol{z} &= \boldsymbol{a} + 2\lambda_1(\boldsymbol{x}_1 - \boldsymbol{a}) + \cdots + 2\lambda_r(\boldsymbol{x}_r - \boldsymbol{a}) - 2\lambda_i(\boldsymbol{x}_i - \boldsymbol{a}). \end{aligned}$$

Since $\boldsymbol{y}, \boldsymbol{z} \in \Gamma$ and $\boldsymbol{x} = (\boldsymbol{y}+\boldsymbol{z})/2 \in (\boldsymbol{y}, \boldsymbol{z})$, we conclude that both $\boldsymbol{y}$ and $\boldsymbol{z}$ belong to F. By the above, $\boldsymbol{x}_i = \boldsymbol{a} + \lambda_i^{-1}\lambda_i(\boldsymbol{x}_i - \boldsymbol{a}) \in F$.

Let $\boldsymbol{z}_1, \ldots, \boldsymbol{z}_t$ be all points from $\{\boldsymbol{x}_1, \ldots, \boldsymbol{x}_r\}$ that lie in F. By the above argument, we have $F \subset \Gamma_{\boldsymbol{a}}(\boldsymbol{z}_1, \ldots, \boldsymbol{z}_t)$. On the other hand, $\Gamma_{\boldsymbol{a}}(\boldsymbol{z}_1, \ldots, \boldsymbol{z}_t) \subset F$ because F is a convex cone with apex $\boldsymbol{a}$. Summing up, $F = \Gamma_{\boldsymbol{a}}(\boldsymbol{z}_1, \ldots, \boldsymbol{z}_t)$.

Problem 11.4. (*a*) It suffices to show that any nonempty proper subset X of L is not an extreme face of L. Indeed, choose points $\boldsymbol{x} \in X$ and $\boldsymbol{y} \in L \setminus X$. Then the point $\boldsymbol{z} = 2\boldsymbol{x} - \boldsymbol{y}$ belongs to L as an affine combination of $\{\boldsymbol{x}, \boldsymbol{y}\}$ (see Theorem 2.38). Now, Definition 11.1 implies that X is not an extreme face of L.

(*b*) To prove that $\operatorname{rbd} D$ is an extreme face of D, chose distinct points $\boldsymbol{x}, \boldsymbol{y} \in D$ and a scalar $0 < \lambda < 1$ such that the point $\boldsymbol{z} = (1-\lambda)\boldsymbol{x} + \lambda\boldsymbol{y}$ lies in $\operatorname{rbd} D$. By Theorem 2.38, $\boldsymbol{x}, \boldsymbol{y} \in \langle \boldsymbol{x}, \boldsymbol{y} \rangle \subset \operatorname{rbd} D$. Thus $\operatorname{rbd} D$ is an extreme face of D.

Let F be an extreme face of D. If F lies in $\operatorname{rbd} D$, then Theorem 11.3 shows that F is an extreme face of $\operatorname{rbd} D$. Since $\operatorname{rbd} D$ is a plane (see an example of page 113), assertion (*a*) implies that F is either $\varnothing$ or $\operatorname{rbd} D$.

Finally, if F meets $\operatorname{rint} D$, then $F = D$, as follows from Theorem 11.4.

Problem 11.5. First, we observe that every nonempty convex set $K \subset \mathbb{R}^n$ has two (trivial) extreme faces: $\varnothing$ and K.

(a) According to Theorem 11.9, K has exactly two extreme faces if and only if $\operatorname{rbd} K = \varnothing$. When K is closed, this happens if and only if K is a plane (see Theorem 3.62).

(b) If K is a closed halfplane, then it has exactly three extreme faces: $\varnothing$, $\operatorname{rbd} K$, and K (see Problem 11.4). Conversely, assume that K has exactly three extreme faces. By the above argument, K is not a plane and $\operatorname{rbd} K \neq \varnothing$. According to Theorem 11.9, every point $\boldsymbol{x} \in \operatorname{rbd} K$ belongs to the generated extreme face $F_{\boldsymbol{x}}(K) \subset \operatorname{rbd} K$. Thus assumption ($b$) shows that $F_{\boldsymbol{x}}(K) = F_{\boldsymbol{y}}(K)$ for any choice of points $\boldsymbol{x}, \boldsymbol{y} \in \operatorname{rbd} K$. This argument proves that $\operatorname{rbd} K$ is a convex set, implying that K is a halfplane (see Theorem 3.68).

(c) If K is a plane slab, then, as easy to see, K has exactly four extreme faces: $\varnothing$, two boundary planes, and K. Conversely, assume that K has exactly four extreme faces. Assertion (b) that K is not a closed halfplane and $\operatorname{rbd} K$ is not convex. Hence there are distinct points $\boldsymbol{u}, \boldsymbol{v} \in \operatorname{rbd} K$ such that $[\boldsymbol{u}, \boldsymbol{v}] \not\subset \operatorname{rbd} K$. Consider the generated extreme faces $F_{\boldsymbol{u}}(K)$ and $F_{\boldsymbol{v}}(K)$. By Theorem 11.9, $F_{\boldsymbol{u}}(K) \cup F_{\boldsymbol{v}}(K) \subset \operatorname{rbd} K$. If the set $\operatorname{rbd} K \setminus (F_{\boldsymbol{u}}(K) \cup F_{\boldsymbol{v}}(K))$ contained a point $\boldsymbol{w}$, then $F_{\boldsymbol{w}}(K)$ would be a new extreme face of K. Hence $\operatorname{rbd} K = F_{\boldsymbol{u}}(K) \cup F_{\boldsymbol{v}}(K)$. Similarly, if $F_{\boldsymbol{u}}(K) \cap F_{\boldsymbol{v}}(K) \neq \varnothing$, then, according to Theorem 11.12, $F_{\boldsymbol{u}}(K) \cap F_{\boldsymbol{v}}(K)$ would be a new extreme face of K. Summing up, $\operatorname{rbd} K$ is a disjoint union of two sets, which implies that K is a closed slab (see Problem 6.7).

Problem 11.6. The proof is organized by induction on $m = 1$. If $m = 1$, then $d_0 = 0$ and $d_1 = 1$. Let K be a closed segment, say $K = [\boldsymbol{u}, \boldsymbol{v}]$. Clearly, the dimensions of extreme faces of K are given precisely by the numbers 0 and 1.

Assume that the inductive assumption holds for $m \leqslant r - 1$, where $2 \leqslant r \leqslant n$, and let $0 = d_0 < d_1 < \cdots < d_p = r$ be an increasing sequence of integers. By the assumption, there is a compact convex set $M \subset \mathbb{R}^n$ of dimension d_{p-1} such that the dimensions of extreme faces of K are given precisely by the numbers $d_0, d_1, \ldots, d_{p-1}$. Translating K on a suitable vector, we may assume that $\operatorname{aff} K$ is a subspace. Choose an r-dimensional subspace $S \subset \mathbb{R}^n$ containing $\operatorname{aff} K$ and consider the closed unit ball $B = \mathbb{B} \cap S$ of S. Since the boundary points of B and the whole B are the only nonempty extreme faces of B (see Problem 11.1), Theorem 11.13 implies that the nonempty extreme faces of the set $K = B + M$ are K and translates of the extreme faces of M. Consequently, the dimensions of extreme faces of K are given precisely by the numbers $d_0, d_1, \ldots, d_p$.

Problem 11.7. Since the case $K = \varnothing$ is obvious, we may assume that $K \neq \varnothing$. By Theorem 11.18 is line-free. If $\dim K = 0$, then K is a singleton, $\{\boldsymbol{x}\}$, and $\operatorname{ext} K = \{\boldsymbol{x}\}$ is a closed set. Let $\dim K \geqslant 1$. A combination of Theorems 11.4 and 11.9 shows that $\operatorname{ext} K \subset \operatorname{rbd} K$ and $F_{\boldsymbol{x}}(K) \subset \operatorname{rbd} K$ for any point $\boldsymbol{x} \in \operatorname{rbd} K$.

Suppose that a sequence of points $\boldsymbol{x}_1, \boldsymbol{x}_2, \ldots$ from $\operatorname{ext} K$ converges to a point $\boldsymbol{x}_0 \in K$. Clearly, $\boldsymbol{x}_0 \in \operatorname{rbd} K$ due to the closedness of $\operatorname{rbd} K$ (see Theorem 3.62).

Assume for a moment that $\boldsymbol{x}_0 \notin \operatorname{ext} K$. Then the extreme face $F_{\boldsymbol{x}_0}(K)$ is a one-dimensional convex set which lies in $\operatorname{rbd} K$ and contains $\boldsymbol{x}_0$ in its relative interior. Hence there are distinct points $\boldsymbol{u}, \boldsymbol{v} \in F_{\boldsymbol{x}_0}(K)$ such that $\boldsymbol{x}_0 \in (\boldsymbol{u}, \boldsymbol{v}) \subset \operatorname{rbd} K$. Consequently, there is an integer $r \geqslant 1$ such that $\boldsymbol{x}_i \in (\boldsymbol{u}, \boldsymbol{v})$ for all $i \geqslant r$. Therefore, $\boldsymbol{x}_i \notin \operatorname{ext} K$, $i \geqslant r$, a contradiction. Summing up, $\boldsymbol{x}_0 \in \operatorname{ext} K$, and $\operatorname{ext} K$ is a closed set.

Problem 11.8. $(a) \Rightarrow (b)$. Let $\boldsymbol{u} \in \operatorname{ext} K$. Suppose that the midpoints $\boldsymbol{u}_i$ of closed segments $[\boldsymbol{y}_i, \boldsymbol{z}_i] \subset K$, $i \geqslant 1$, converge to $\boldsymbol{u}$. Replacing the sequence $[\boldsymbol{y}_1, \boldsymbol{z}_1], [\boldsymbol{y}_2, \boldsymbol{z}_2], \dots$ with a suitable subsequence, we may suppose that $\|\boldsymbol{u} - \boldsymbol{u}_i\| \leqslant \frac{1}{i}$ for all $i \geqslant 1$.

Assume, for contradiction, that neither $\boldsymbol{y}_1, \boldsymbol{y}_2, \dots$ nor $\boldsymbol{z}_1, \boldsymbol{z}_2, \dots$ converges to $\boldsymbol{u}$. Hence there is an integer $r \geqslant 1$ and a scalar $\rho > 0$ such that

$$\|\boldsymbol{u} - \boldsymbol{y}_i\| \geqslant \rho \quad \text{and} \quad \|\boldsymbol{u} - \boldsymbol{z}_i\| \geqslant \rho, \quad i \geqslant r.$$

Consequently, the inequalities

$$\|\boldsymbol{u}_i - \boldsymbol{y}_i\| \geqslant \|\boldsymbol{u} - \boldsymbol{y}_i\| - \|\boldsymbol{u} - \boldsymbol{u}_i\| \geqslant \rho - \tfrac{1}{i}, \quad i \geqslant 1,$$
$$\|\boldsymbol{u}_i - \boldsymbol{z}_i\| \geqslant \|\boldsymbol{u} - \boldsymbol{z}_i\| - \|\boldsymbol{u} - \boldsymbol{u}_i\| \geqslant \rho - \tfrac{1}{i}, \quad i \geqslant 1,$$

show the existence of an integer $s \geqslant r$ with the property

$$\|\boldsymbol{u} - \boldsymbol{y}_i\| \geqslant \tfrac{\rho}{2} \quad \text{and} \quad \|\boldsymbol{u} - \boldsymbol{z}_i\| \geqslant \tfrac{\rho}{2}, \quad i \geqslant r.$$

Choose points $\boldsymbol{y}_i' \in [\boldsymbol{u}_i, \boldsymbol{y}_i]$ and $\boldsymbol{z}_i' \in [\boldsymbol{u}_i, \boldsymbol{z}_i]$ satisfying the conditions

$$\|\boldsymbol{u}_i - \boldsymbol{y}_i'\| = \|\boldsymbol{u}_i - \boldsymbol{z}_i'\| = \tfrac{\rho}{2}, \quad i \geqslant s.$$

Clearly, $\boldsymbol{u}_i$ is the midpoint of the closed segment $[\boldsymbol{y}_i', \boldsymbol{z}_i']$. Consequently, $\boldsymbol{z}_i' - \boldsymbol{u}_i = -(\boldsymbol{y}_i' - \boldsymbol{u}_i)$. Put $\boldsymbol{e}_i = \boldsymbol{y}_i' - \boldsymbol{u}_i$, $i \geqslant 1$. Then $\boldsymbol{z}_i' - \boldsymbol{u}_i = -\boldsymbol{e}_i$, $i \geqslant 1$. Since the sequence $\boldsymbol{e}_1, \boldsymbol{e}_1, \dots$ lies in the sphere $S_{\rho/2}(\boldsymbol{o})$, which is a compact set, there is a subsequence $\boldsymbol{e}_{i_1}, \boldsymbol{e}_{i_2}, \dots$ which converges to a vector $\boldsymbol{e} \in S_{\rho/2}(\boldsymbol{o})$. Then

$$\lim_{k\to\infty} \boldsymbol{y}_{i_k}' = \lim_{k\to\infty} (\boldsymbol{u}_{i_k} + \boldsymbol{e}_{i_k}) = \boldsymbol{u} + \boldsymbol{e}, \quad \lim_{k\to\infty} \boldsymbol{z}_{i_k}' = \lim_{k\to\infty} (\boldsymbol{u}_{i_k} - \boldsymbol{e}_{i_k}) = \boldsymbol{u} - \boldsymbol{e}.$$

Summing up, $\boldsymbol{u}$ is the midpoint of the closed segment $[\boldsymbol{u}+\boldsymbol{e}, \boldsymbol{u}-\boldsymbol{e}] \subset K$, contrary to the assumption $\boldsymbol{u} \in \operatorname{ext} K$. The obtained contradiction shows that at least one of the sequences $\boldsymbol{y}_1, \boldsymbol{y}_2, \dots$ and $\boldsymbol{z}_1, \boldsymbol{z}_2, \dots$ converges to $\boldsymbol{u}$.

$(b) \Rightarrow (a)$. Suppose that $\boldsymbol{u} \in K \setminus \operatorname{ext} K$. Then $\boldsymbol{u} = (\boldsymbol{y}, \boldsymbol{z})$ for suitable points $\boldsymbol{y}, \boldsymbol{z} \in K \setminus \{\boldsymbol{u}\}$. Obviously, one can replace $(\boldsymbol{y}, \boldsymbol{z})$ with a smaller open segment $(\boldsymbol{y}', \boldsymbol{z}')$ such that $\boldsymbol{u}$ is the midpoint of $(\boldsymbol{y}', \boldsymbol{z}')$. Put $\boldsymbol{y}_i = \boldsymbol{y}'$ and $\boldsymbol{z}_i = \boldsymbol{z}'$, $i \geqslant 1$. Then $\boldsymbol{u}$ is the midpoint of every closed segment $[\boldsymbol{y}_i, \boldsymbol{z}_i]$, $i \geqslant 1$, while neither $\boldsymbol{y}_1, \boldsymbol{y}_2, \dots$ nor $\boldsymbol{z}_1, \boldsymbol{z}_2, \dots$ converges to $\boldsymbol{u}$.

Problem 11.9. Since the cases $F = \varnothing$ and $F = K$ are obvious, we may assume that $\varnothing \neq F \neq K$. If F is an extreme face of K, then, according to Theorem 11.3, F is an extreme face of the convex subset $K \cap B_\rho(F)$ of K (see Corollary 3.11).

Conversely, let F be an extreme face of the set $M = K \cap B_\rho(F)$. Choose points $\boldsymbol{x}, \boldsymbol{y} \in K$ and a scalar $0 < \lambda < 1$ such that the point $\boldsymbol{z} = (1-\lambda)\boldsymbol{x} + \lambda\boldsymbol{y}$ belongs to F. Assume for a moment that at least one of the points $\boldsymbol{x}, \boldsymbol{y}$, say $\boldsymbol{x}$, is not in F. Then $\boldsymbol{x} \notin \operatorname{cl} F$ according to Theorem 11.4. So, we can choose a point $\boldsymbol{x}' \in (\boldsymbol{x}, \boldsymbol{z}) \setminus \operatorname{cl} F$ such that $\boldsymbol{x}' \in B_\rho(F)$. Let $\boldsymbol{y}'$ be a point in $(\boldsymbol{z}, \boldsymbol{y})$ such that $\boldsymbol{y}' \in B_\rho(F)$. Clearly, $\boldsymbol{z} \in (\boldsymbol{x}', \boldsymbol{y}') \subset B_\rho(F)$ by the convexity of $B_\rho(F)$, as follows from Corollary 3.11. Furthermore, $(\boldsymbol{x}', \boldsymbol{y}') \subset K$ by the convexity of K. Hence $(\boldsymbol{x}', \boldsymbol{y}') \subset M$. Since F is an extreme face of M, one should have $\boldsymbol{x}' \in F$, contrary to the choice of $\boldsymbol{x}'$. The obtained contradiction shows that F is an extreme face of K.

Problem 11.10. Since the case $K = \{\boldsymbol{a}\}$ is obvious (then K is locally conic at $\boldsymbol{a}$ and $\operatorname{ext} K \setminus \{\boldsymbol{a}\} = \varnothing$), we may suppose that $K \neq \{\boldsymbol{a}\}$. According to Definition 7.5, there is a cone $C \subset \mathbb{R}^n$ with apex $\boldsymbol{a}$ and a scalar $\rho > 0$ such that $K \cap B_\rho(\boldsymbol{a}) = C \cap B_\rho(\boldsymbol{a})$. Let $\delta = \rho/2$. We assert that $B_\delta(\boldsymbol{a}) \cap (\operatorname{ext} K \setminus \{\boldsymbol{a}\}) = \varnothing$. Indeed, assume for a moment the existence of a point $\boldsymbol{u} \in B_\delta(\boldsymbol{a}) \cap (\operatorname{ext} K \setminus \{\boldsymbol{a}\})$. Put $\boldsymbol{v} = 2\boldsymbol{u} - \boldsymbol{a}$. Then $\boldsymbol{v} = \boldsymbol{a} + 2(\boldsymbol{u} - \boldsymbol{a}) \in C$ due to Definition 5.1. Furthermore,

$$\|\boldsymbol{v} - \boldsymbol{a}\| = 2\|\boldsymbol{u} - \boldsymbol{a}\| \leqslant 2\delta = \rho.$$

Hence $\boldsymbol{v} \in C \cap B_\rho(\boldsymbol{a}) = K \cap B_\rho(\boldsymbol{a})$. By the convexity of K, one has $\boldsymbol{u} = \frac{1}{2}(\boldsymbol{a} + \boldsymbol{v}) \in (\boldsymbol{a}, \boldsymbol{v}) \subset K$. Consequently, $\boldsymbol{u} \notin \operatorname{ext} K$, in contradiction with the choice of $\boldsymbol{u}$. Thus $B_\delta(\boldsymbol{a}) \cap (\operatorname{ext} K \setminus \{\boldsymbol{a}\}) = \varnothing$.

Problem 11.11. $(a) \Rightarrow (b)$ Let $\boldsymbol{x}$ be an extreme point of $\operatorname{conv} X$. By Theorem 11.2, the set $\operatorname{conv} X \setminus \{\boldsymbol{x}\}$ is convex. Suppose that $\boldsymbol{x} \notin X$. The inclusion $\boldsymbol{x} \in \operatorname{conv} X = \operatorname{conv}(X \setminus \{\boldsymbol{x}\})$ and Theorem 4.3 show that $\boldsymbol{x}$ is expressible as a convex combination $\boldsymbol{x} = \lambda_1\boldsymbol{x}_1 + \cdots + \lambda_k\boldsymbol{x}_k$ of suitable points $\boldsymbol{x}_1, \ldots, \boldsymbol{x}_k \in X \setminus \{\boldsymbol{x}\}$. The same theorem gives

$$\begin{aligned}\boldsymbol{x} \in \operatorname{conv}\{\boldsymbol{x}_1, \ldots, \boldsymbol{x}_k\} &\subset \operatorname{conv}(X \setminus \{\boldsymbol{x}\}) \\ &\subset \operatorname{conv}(\operatorname{conv} X \setminus \{\boldsymbol{x}\}) = \operatorname{conv} X \setminus \{\boldsymbol{x}\},\end{aligned}$$

which is impossible. Hence $\boldsymbol{x} \in X$. Similarly, the inclusion $X \setminus \{\boldsymbol{x}\} \subset \operatorname{conv} X \setminus \{\boldsymbol{x}\}$ implies that $\operatorname{conv}(X \setminus \{\boldsymbol{x}\}) \subset \operatorname{conv} X \setminus \{\boldsymbol{x}\}$. Hence $\boldsymbol{x} \notin \operatorname{conv}(X \setminus \{\boldsymbol{x}\})$.

$(b) \Rightarrow (a)$ Let a point $\boldsymbol{x} \in X$ satisfy the condition $\boldsymbol{x} \notin \operatorname{conv}(X \setminus \{\boldsymbol{x}\})$. Assume for a moment that $\boldsymbol{x}$ is not an extreme point of $\operatorname{conv} X$. Then $\boldsymbol{x} = (1-\lambda)\boldsymbol{u} + \lambda\boldsymbol{v}$ for suitable points $\boldsymbol{u}$ and $\boldsymbol{v}$ from $\operatorname{conv} X \setminus \{\boldsymbol{x}\}$ and a scalar $0 < \lambda < 1$. By Theorem 4.3, $\boldsymbol{u}$ and $\boldsymbol{v}$ can be written as convex combinations

$$\boldsymbol{u} = \mu_1\boldsymbol{u}_1 + \cdots + \mu_p\boldsymbol{u}_p \quad \text{and} \quad \boldsymbol{v} = \gamma_1\boldsymbol{v}_1 + \cdots + \gamma_q\boldsymbol{v}_q,$$

where $\boldsymbol{u}_1, \ldots, \boldsymbol{u}_p, \boldsymbol{v}_1, \ldots, \boldsymbol{v}_q \in X \setminus \{\boldsymbol{x}\}$. The same theorem gives the inclusion

$$\boldsymbol{x} \in \operatorname{conv}\{\boldsymbol{u}, \boldsymbol{v}\} \subset \operatorname{conv}(\{\boldsymbol{u}_1, \ldots, \boldsymbol{u}_p, \boldsymbol{v}_1, \ldots, \boldsymbol{v}_q\} \setminus \{\boldsymbol{x}\}) \subset \operatorname{conv}(X \setminus \{\boldsymbol{x}\}),$$

a contradiction. Hence $\boldsymbol{x} \in \operatorname{ext}(\operatorname{conv} X)$.

Problem 11.12. First, we state that $\operatorname{ext} K \subset X$. Indeed, this inclusion is obvious when $\operatorname{ext} K = \varnothing$. If $\operatorname{ext} K \neq \varnothing$, then Theorem 11.18 shows that K is line-free, and the inclusion $\operatorname{ext} K \subset X$ follows from Corollary 11.30.

Assume, for contradiction, the existence of a point $\boldsymbol{x} \in X \setminus \operatorname{ext} K$. By Problem 11.11, $\boldsymbol{x} \in \operatorname{conv}(X \setminus \{\boldsymbol{x}\})$. Consequently,

$$\begin{aligned} K = \operatorname{conv} X &= \operatorname{conv}(\{\boldsymbol{x}\} \cup (X \setminus \{\boldsymbol{x}\})) \subset \operatorname{conv}(\{\boldsymbol{x}\} \cup \operatorname{conv}(X \setminus \{\boldsymbol{x}\})) \\ &= \operatorname{conv}(\operatorname{conv}(X \setminus \{\boldsymbol{x}\})) = \operatorname{conv}(X \setminus \{\boldsymbol{x}\}) \subset \operatorname{conv} X. \end{aligned}$$

Hence $K = \operatorname{conv}(X \setminus \{\boldsymbol{x}\})$, contrary to the minimality of X. So, $X = \operatorname{ext} K$.

Problem 11.13. 1. Let $X = \operatorname{ext}(\operatorname{conv} X)$ and choose any subset Y of X. Assume for a moment that $X \cap \operatorname{conv} Y \not\subset Y$ and let $\boldsymbol{x} \in (X \cap \operatorname{conv} Y) \setminus Y$. Then $Y \subset X \setminus \{\boldsymbol{x}\}$, implying the inclusion $\boldsymbol{x} \in \operatorname{conv} Y \subset \operatorname{conv}(X \setminus \{\boldsymbol{x}\})$. Problem 11.11 shows that $\boldsymbol{x} \notin \operatorname{ext} X$, contrary to the choice of $\boldsymbol{x}$. Hence $X \cap \operatorname{conv} Y \subset Y$

Conversely, let X satisfy condition (a). Choose any point $\boldsymbol{x} \in X$ and let $Y = X \setminus \{\boldsymbol{y}\}$. By the assumption, $X \cap \operatorname{conv}(X \setminus \{\boldsymbol{x}\}) \subset X \setminus \{\boldsymbol{x}\}$. Therefore, $\boldsymbol{x} \notin \operatorname{conv}(X \setminus \{\boldsymbol{x}\})$, and Problem 11.11 shows that $\boldsymbol{x} \in \operatorname{ext} X$.

2. Assume, additionally, that $\operatorname{conv} X$ is compact. Then $\operatorname{cl} X \subset \operatorname{conv} X$, implying that $\operatorname{cl} X$ is compact and $\operatorname{cl} X \subset \operatorname{conv} X$.

Conversely, let a set $X \subset \mathbb{R}^n$ satisfy conditions (a)–(c). By condition (a), one has $X = \operatorname{ext}(\operatorname{conv} X)$. Condition (c) shows that X is bounded. Then $\operatorname{conv} X$ also is bounded (see Theorem 4.16). Theorem 4.16 and condition (b) give

$$\operatorname{cl}(\operatorname{conv} X) = \operatorname{conv}(\operatorname{cl} X) \subset \operatorname{conv}(\operatorname{conv} X) = \operatorname{conv} X.$$

Hence $\operatorname{conv} X$ is compact.

Problem 11.14. 1. Since the inclusion $K \subset \cap(C_{\boldsymbol{a}}(K) : \boldsymbol{a} \in \operatorname{ext} K)$ is obvious, it remains to prove the opposite one. Dropping the obvious case when K is a singleton, we may assume that $\dim K \geqslant 1$. Choose any point $\boldsymbol{u} \in \mathbb{R}^n \setminus K$. By Theorem 8.22, K contains a unique nearest to $\boldsymbol{u}$ point $\boldsymbol{c}$, and K lies in the closed halfspace

$$V = \{\boldsymbol{x} \in \mathbb{R}^n : (\boldsymbol{x} - \boldsymbol{c}) \cdot (\boldsymbol{u} - \boldsymbol{c}) \leqslant 0\}.$$

Clearly, $\boldsymbol{u}$ belongs to the boundary hyperplane $H = \{\boldsymbol{x} \in \mathbb{R}^n : (\boldsymbol{x} - \boldsymbol{c}) \cdot (\boldsymbol{u} - \boldsymbol{c}) = 0\}$ of V. A combination of Theorems 11.4 and 11.9 shows that the generated extreme face $F_{\boldsymbol{c}}(K)$ is a closed convex set and $\boldsymbol{c} \in \operatorname{rint} F_{\boldsymbol{c}}(K)$. Then $F_{\boldsymbol{c}}(K) \subset H$ according to Theorem 9.3. The set $F_{\boldsymbol{c}}(K)$ is line-free as a subset of the line-free set K. By Theorem 11.18, $F_{\boldsymbol{c}}(K)$ has at least one extreme point, say $\boldsymbol{a}$, and Theorems 11.4 implies the inclusion $\boldsymbol{a} \in \operatorname{ext} K$. Since $K \neq \{\boldsymbol{a}\}$, the generated cone $C_{\boldsymbol{a}}(K)$

is the union of closed halflines $[\boldsymbol{a}, \boldsymbol{x}\rangle$, where $\boldsymbol{x} \in K \setminus \{\boldsymbol{a}\}$ (see Definition 5.29). This argument and Theorem 2.35 imply the inclusion $C_{\boldsymbol{a}}(K) \subset V$. Consequently, $\boldsymbol{u} \notin C_{\boldsymbol{a}}(K)$, due to $\boldsymbol{u} \notin V$.

2. For the equality $\operatorname{rec} K = \cap (C_{\boldsymbol{a}}(K) - \boldsymbol{a} : \boldsymbol{a} \in \operatorname{ext} K)$, we observe first that $\boldsymbol{a} + \operatorname{rec} K \subset K \subset C_{\boldsymbol{a}}(K)$ for every point $\boldsymbol{a} \in \operatorname{ext} K$ (see Theorem 6.2). Hence

$$\operatorname{rec} K \subset \cap (C_{\boldsymbol{a}}(K) - \boldsymbol{a} : \boldsymbol{a} \in \operatorname{ext} K).$$

Conversely, suppose that a closed halfline h with endpoint $\boldsymbol{o}$ does not lie in $\operatorname{rec} K$. Choose a point $\boldsymbol{z} \in \operatorname{rint} K$. Corollary 6.15 shows that the closed halfline $h_{\boldsymbol{z}} = \boldsymbol{z} + h$ does not lie in K, and Corollary 3.66 implies that $h_{\boldsymbol{z}}$ meets $\operatorname{rbd} K$ at a unique point $\boldsymbol{u} \in \operatorname{rbd} K$ such that the open halfline $h_{\boldsymbol{u}} = \boldsymbol{u} + h$ is disjoint from K. According to Corollary 9.12, there is a hyperplane $H \subset \mathbb{R}^n$ containing $\boldsymbol{u}$ and properly supporting K. We observe that $h_{\boldsymbol{u}} \cap H = \varnothing$. Indeed, if a point $\boldsymbol{v}$ belonged to $h_{\boldsymbol{u}} \cap H$, then $\boldsymbol{z} \in \langle \boldsymbol{u}, \boldsymbol{v} \rangle \subset H$, contrary to the condition $H \cap \operatorname{rint} K = \varnothing$. Denote by V the closed halfspace determined by H and containing K. Clearly, $h_{\boldsymbol{u}} \cap V = \varnothing$. Similarly to the argument from part 1, we find an extreme point $\boldsymbol{a}$ of the set $F_{\boldsymbol{u}}(K) \subset H$. Since $C_{\boldsymbol{a}}(K) \subset V$ and the open halfline $h_{\boldsymbol{a}} = (\boldsymbol{a} - \boldsymbol{u}) + h_{\boldsymbol{u}}$ is disjoint from V (see Problem 2.10), one has $h_{\boldsymbol{a}} \cap C_{\boldsymbol{a}}(K) = \varnothing$. Consequently, $h = h_{\boldsymbol{a}} - \boldsymbol{a} \not\subset C_{\boldsymbol{a}}(K) - \boldsymbol{a}$. Summing up,

$$\cap (C_{\boldsymbol{a}}(K) - \boldsymbol{a} : \boldsymbol{a} \in \operatorname{ext} K) \subset \operatorname{rec} K.$$

Problem 11.15. $(a) \Rightarrow (b)$ Let F be an extreme face of $\operatorname{cl} K$ such that $F \cap K \neq \varnothing$. Choose a point $\boldsymbol{u} \in F \cap K$. Since the case $F = \{\boldsymbol{u}\}$ is obvious ($\operatorname{rint} F = \{\boldsymbol{u}\} \subset K$), we may assume that $\dim F > 0$. By Problem 3.9 and assertion (a),

$$\operatorname{rint} F \subset \cup ([\boldsymbol{u}, \boldsymbol{v}) : \boldsymbol{v} \in F) \subset \{\boldsymbol{u}\} \cup K = K.$$

$(b) \Rightarrow (a)$ Choose any pair of points $\boldsymbol{x} \in K$ and $\boldsymbol{y} \in \operatorname{cl} K$. Denote by $\boldsymbol{z}$ the midpoint of $(\boldsymbol{x}, \boldsymbol{y})$ and by F the generated face of $\boldsymbol{z}$ in $\operatorname{cl} K$. By Theorem 11.9, $\boldsymbol{z} \in \operatorname{rint} F$. Therefore, $(\boldsymbol{x}, \boldsymbol{y}) = (\boldsymbol{x}, \boldsymbol{z}] \cup [\boldsymbol{z}, \boldsymbol{y}) \subset \operatorname{rint} F$ according to Theorem 3.24. Furthermore, $[\boldsymbol{x}, \boldsymbol{y}] \subset F$ because F is a closed set (see Theorem 11.4). Since $\boldsymbol{x} \in F \cap K$, assertion (b) gives $(\boldsymbol{x}, \boldsymbol{y}) \subset \operatorname{rint} F \subset K$.

Chapter 12

Problem 12.1. Corollary 12.6 implies that every exposed singleton of $B_\rho(\boldsymbol{a})$ belongs to $S_\rho(\boldsymbol{a})$. Conversely, let $\boldsymbol{z} \in S_\rho(\boldsymbol{a})$. Problem 9.1 shows that

$$H = \{\boldsymbol{x} \in \mathbb{R}^n : (\boldsymbol{x} - \boldsymbol{z}) \cdot (\boldsymbol{z} - \boldsymbol{a}) = 0\}$$

is the only hyperplane through $\boldsymbol{z}$ which supports $B_\rho(\boldsymbol{a})$ such that $H \cap B_\rho(\boldsymbol{a}) = \{\boldsymbol{z}\}$. Hence $\boldsymbol{z} \in \exp B_\rho(\boldsymbol{a})$. Summing up, $S_\rho(\boldsymbol{a}) = \exp B_\rho(\boldsymbol{c})$.

It remains to prove that every exposed face G of $B_\rho(\boldsymbol{a})$ containing two or more points coincides with $B_\rho(\boldsymbol{a})$. Indeed, according to Theorem 12.2, G is an extreme face of $B_\rho(\boldsymbol{a})$, and Problem 11.1 shows that $G = B_\rho(\boldsymbol{a})$.

Problem 12.2. If G is an exposed face of the simplex $\Delta = \Delta(\boldsymbol{x}_1, \ldots, \boldsymbol{x}_{r+1})$, then, by Theorem 12.2, G is an extreme face of Δ, and Problem 11.2 shows that either $G = \varnothing$ or $G = \Delta(\boldsymbol{z}_1, \ldots, \boldsymbol{z}_t)$, where $\{\boldsymbol{z}_1, \ldots, \boldsymbol{z}_t\}$ is a nonempty subset of $\{\boldsymbol{x}_1, \ldots, \boldsymbol{x}_{r+1}\}$.

Conversely, choose a simplex $\Delta' = \Delta(\boldsymbol{z}_1, \ldots, \boldsymbol{z}_t)$, where $\{\boldsymbol{z}_1, \ldots, \boldsymbol{z}_t\}$ is a nonempty subset of $\{\boldsymbol{x}_1, \ldots, \boldsymbol{x}_{r+1}\}$. We are going to prove that Δ' is an exposed face of Δ. Since the case $\Delta' = \Delta$ is obvious, we assume that Δ' is a nonempty proper subset of Δ. Let $L = \operatorname{aff}\{\boldsymbol{z}_1, \ldots, \boldsymbol{z}_t\}$. Because Δ' is an extreme face of Δ, Theorem 11.2 gives $\Delta \cap L = \Delta'$. This argument and Problem 4.2 show that L is disjoint from $\Delta'' = \Delta(\boldsymbol{z}_{t+1}, \ldots, \boldsymbol{z}_{r+1})$. Furthermore, L and Δ'' are strongly disjoint because L is closed (see Problem 1.15 and Corollary 2.21) and Δ'' is compact (see Theorem 4.16). According to Theorem 9.6, there is a hyperplane $H \subset \mathbb{R}^n$ containing L and strongly disjoint from Δ''.

Finally, we assert that $H \cap \Delta = \Delta'$ (which implies that Δ' is an exposed face of Δ). Indeed, assume for a moment the existence of a point $\boldsymbol{u} \in (H \cap \Delta) \setminus \Delta'$. By the above argument, $\boldsymbol{u} \notin \Delta''$. Therefore, $\boldsymbol{u}$ can be written as a convex combination $\boldsymbol{u} = (1-\lambda)\boldsymbol{u}' + \lambda\boldsymbol{u}''$ of suitable points $\boldsymbol{u}' \in \Delta'$ and $\boldsymbol{u}'' \in \Delta''$, where $0 < \lambda < 1$ (see Problem 4.2). In this case,

$$\boldsymbol{u}'' = \lambda^{-1}\boldsymbol{u} + (1 - \lambda^{-1})\boldsymbol{u}' \in \langle \boldsymbol{u}, \boldsymbol{u}' \rangle \subset H,$$

contrary to $H \cap \Delta'' = \varnothing$. Hence $H \cap \Delta = \Delta'$.

Problem 12.3. If G is an exposed face of the simplicial cone $\Gamma = \Gamma_{\boldsymbol{a}}(\boldsymbol{x}_1, \ldots, \boldsymbol{x}_r)$, then, by Theorem 12.2, G is an extreme face of Γ, and Problem 11.3 shows that either $G = \varnothing$, or $G = \{\boldsymbol{a}\}$, or $G = \Delta(\boldsymbol{z}_1, \ldots, \boldsymbol{z}_t)$, where $\{\boldsymbol{z}_1, \ldots, \boldsymbol{z}_t\}$ is a nonempty subset of $\{\boldsymbol{x}_1, \ldots, \boldsymbol{x}_r\}$.

Conversely, choose a simplicial cone $\Gamma' = \Gamma_{\boldsymbol{a}}(\boldsymbol{z}_1, \ldots, \boldsymbol{z}_t)$, where $\{\boldsymbol{z}_1, \ldots, \boldsymbol{z}_t\}$ is a nonempty subset of $\{\boldsymbol{x}_1, \ldots, \boldsymbol{x}_r\}$. We are going to prove that Γ' is an exposed face of Γ. Since the case $\Gamma' = \Gamma$ is obvious, we assume that Γ' is a proper subset of Γ. Let $L = \operatorname{aff}\{\boldsymbol{z}_1, \ldots, \boldsymbol{z}_t\}$. Because Γ' is an extreme face of Γ, Theorem 11.2 gives $\Gamma \cap L = \Gamma'$. This argument and Problem 5.1 show that L is disjoint from the simplex $\Delta = \Delta(\boldsymbol{z}_{t+1}, \ldots, \boldsymbol{z}_r)$. Furthermore, L and Δ are strongly disjoint because L is closed (see Corollary 2.21) and Δ is compact (see Problem 1.15 and Theorem 4.16). According to Theorem 9.6, there is a hyperplane $H \subset \mathbb{R}^n$ containing L and strongly disjoint from Δ.

Finally, we assert that $H \cap \Gamma = \Gamma'$ (which implies that Γ' is an exposed face of Γ). Indeed, assume for a moment the existence of a point $\boldsymbol{u} \in (H \cap \Gamma) \setminus \Gamma'$. By the above argument, $\boldsymbol{u} \notin \Delta$. Therefore, $\boldsymbol{u}$ can be written as a convex combination $\boldsymbol{u} = (1-\lambda)\boldsymbol{u}' + \lambda\boldsymbol{u}''$ of points $\boldsymbol{u}' \in \Gamma'$ and $\boldsymbol{u}'' \in \Gamma'' = \Gamma_{\boldsymbol{a}}(\boldsymbol{z}_{t+1}, \ldots, \boldsymbol{z}_r)$, where

$0 < \lambda < 1$ (see Problem 5.1). In this case,

$$\boldsymbol{u}'' = \lambda^{-1}\boldsymbol{u} + (1 - \lambda^{-1})\boldsymbol{u}' \in \langle \boldsymbol{u}, \boldsymbol{u}' \rangle \subset H,$$

contrary to $H \cap \Delta = \varnothing$. Hence $H \cap \Gamma = \Gamma'$.

Problem 12.4. Since the cases $G = \varnothing$ and $G = K$ are obvious, we may assume that $\varnothing \neq G \neq K$. Put $M = K \cap B_\rho(G)$.

Let G be a nonempty proper exposed face of K. Choose a hyperplane $H \subset \mathbb{R}^n$ supporting K such that $G = H \cap K$. Then

$$G = G \cap B_\rho(G) = H \cap K \cap B_\rho(G) = H \cap M.$$

Because $H \cap \operatorname{rint} K = \varnothing$ and $B_\rho(G)$ meets $\operatorname{rint} K$ (see Theorem 3.51), H properly supports M. Therefore, G is a nonempty proper exposed face of M.

Conversely, suppose that G is a nonempty proper exposed face of M, and let H be a hyperplane supporting M such that $G = H \cap M$. We assert that $G = H \cap K$. Since $G = H \cap M \subset H \cap K$, it suffices to verify the opposite inclusion $H \cap K \subset G$. Assume for a moment the existence of a point $\boldsymbol{x} \in (H \cap K) \setminus G$. Choose a point $\boldsymbol{z} \in G$. Then $[\boldsymbol{x}, \boldsymbol{z}] \subset H \cap K$ by a convexity argument. The closed segment $[\boldsymbol{x}, \boldsymbol{z}]$ meets $\operatorname{cl} G$ along a closed segment $[\boldsymbol{u}, \boldsymbol{z}]$. Clearly, $[\boldsymbol{x}, \boldsymbol{u}] \subset G$ and $\boldsymbol{u} \in (\boldsymbol{x}, \boldsymbol{z}]$ due to $G = K \cap \operatorname{cl} G$ (see Corollary 12.6). If $\boldsymbol{v}$ is a point in $(\boldsymbol{u}, \boldsymbol{x})$ satisfying the inequality $\|\boldsymbol{u} - \boldsymbol{v}\| \leqslant \rho$, then

$$\boldsymbol{v} \in ((H \cap K) \cap B_\rho(G)) \setminus G = (H \cap M) \setminus G = G \setminus G = \varnothing,$$

a contradiction. Hence $G = H \cap K$. So, G is an exposed face of K.

Problem 12.5. Since every exposed face of a convex set is its extreme face (see Theorem 12.2) both assertions follow from Problem 11.4.

Problem 12.6. Since every exposed face of a convex set is its extreme face (see Theorem 12.2) all three assertions follow from Problem 11.5.

Problem 12.7. The proof is organized by induction on $m = 1$. If $m = 1$, then $d_0 = 0$ and $d_1 = 1$. Let K be a closed segment, say $K = [\boldsymbol{u}, \boldsymbol{v}]$. Clearly, the dimensions of exposed faces of K are given precisely by the numbers 0 and 1.

Assume that the inductive assumption holds for $m \leqslant r - 1$, where $2 \leqslant r \leqslant n$, and let $0 = d_0 < d_1 < \cdots < d_p = r$ be an increasing sequence of integers. By the assumption, there is a compact convex set $M \subset \mathbb{R}^n$ of dimension d_{p-1} such that the dimensions of exposed faces of K are given precisely by the numbers $d_0, d_1, \ldots, d_{p-1}$. Translating K on a suitable vector, we may assume that $\operatorname{aff} K$ is a subspace. Choose an r-dimensional subspace $S \subset \mathbb{R}^n$ containing $\operatorname{aff} K$ and consider the closed unit ball $B = \mathbb{B} \cap S$ of S. Since the boundary points of B and the whole B are the only nonempty extreme faces of B (see Problem 12.1), Theorem 12.13 implies that the nonempty exposed faces of the set $K = B + M$

are K and translates of the exposed faces of M. Consequently, the dimensions of exposed faces of K are given precisely by the numbers $d_0, d_1, \dots, d_p$.

Problem 12.8. Let $M = \cap (C_{\boldsymbol{a}}(K) - \boldsymbol{a} : \boldsymbol{a} \in \exp K)$. By Theorem 6.2, we have $\boldsymbol{a} + \operatorname{rec} K \subset K \subset C_{\boldsymbol{a}}(K)$ for every point $\boldsymbol{a} \in \exp K$. Hence $\operatorname{rec} K \subset M$. Since every cone $C_{\boldsymbol{a}}(K)$, with $\boldsymbol{a} \in \exp K$, lies in aff K, the set M lies in dir K.

The opposite inclusion is obvious if $M = \{\boldsymbol{o}\}$. So let $M \neq \{\boldsymbol{o}\}$. Assume, for contradiction, that a closed halfline h with endpoint $\boldsymbol{o}$ lies in dir K but not in rec K. Choose a point $\boldsymbol{u} \in \operatorname{rint} K$. Corollary 6.15 implies that the closed halfline $\boldsymbol{u} + h$ does not lie in K, and Corollary 3.66 shows that this halfline meets rbd K at a single point $\boldsymbol{v}$ such that $[\boldsymbol{u}, \boldsymbol{v}\rangle \setminus [\boldsymbol{u}, \boldsymbol{v}] \subset \operatorname{aff} K \setminus K$. Choose any point $\boldsymbol{w} \in [\boldsymbol{u}, \boldsymbol{v}\rangle \setminus [\boldsymbol{u}, \boldsymbol{v}]$. By Theorem 12.22, there is a point $\boldsymbol{a} \in \exp K$ and a closed halfspace V supporting K such that $\operatorname{bd} V \cap K = \{\boldsymbol{a}\}$ and $\boldsymbol{w} \notin V$. Theorem 2.35 implies the inclusions $K \subset C_{\boldsymbol{a}}(K) \subset V$. Since $\boldsymbol{a} + h \not\subset V$ (otherwise $\boldsymbol{u} + h \subset V$), one has $\boldsymbol{a} + h \not\subset C_{\boldsymbol{a}}(K)$. Consequently, $h \not\subset C_{\boldsymbol{a}}(K) - \boldsymbol{a}$, which shows that $h \not\subset M$.

Problem 12.9. Consider the compact convex set $K^* = K - K$. By Theorem 12.22, there is a hyperplane $G \subset \mathbb{R}^n$ such that $K^* \cap G$ is a singleton and one of the unit normals of G, say $\mathbf{c}$, satisfies the inequality $\|\mathbf{e} - \mathbf{c}\| < \varepsilon$. Clearly, $K^* \cap G \neq K^*$ since K is not a singleton. Denote by H and H' the hyperplanes which are translates of G and support, respectively, the sets K and $-K$ from the same side. By Theorem 9.36,

$$K^* \cap G = (K + (-K)) \cap G = K \cap H + (-K) \cap H'.$$

Hence that both sets $K \cap H$ and $(-K) \cap H'$ should be singletons. Finally, the singletons $K \cap H$ and $K \cap (-H')$ are distinct due to $K^* \cap G \neq K^*$.

Chapter 13

Problem 13.1. Translating L on a suitable vector, we may suppose that L is a subspace. Choose a basis $\{\boldsymbol{b}_1, \dots, \boldsymbol{b}_m\}$ for L and a basis $\{\boldsymbol{b}_{m+1}, \dots, \boldsymbol{b}_n\}$ for $L^\perp$. Let $\boldsymbol{b}_{n+1} = -(\boldsymbol{b}_{m+1} + \cdots + \boldsymbol{b}_n)$. We assert that L is the intersection of $n - m + 1$ closed halfspaces

$$V_i = \{\boldsymbol{x} \in \mathbb{R}^n : \boldsymbol{x} \cdot \boldsymbol{b}_i \leqslant 0\}, \quad m + 1 \leqslant i \leqslant n + 1.$$

Since the inclusion $L \subset V_{m+1} \cap \cdots \cap V_{n+1}$ is obvious, it remains to prove the opposite inclusion. For this, choose a point $\boldsymbol{x} \in V_{m+1} \cap \cdots \cap V_{n+1} \subset L$. Let $\boldsymbol{x} = \xi_1 \boldsymbol{b}_1 + \cdots + \xi_n \boldsymbol{b}_n$. The inequalities $\boldsymbol{x} \cdot \boldsymbol{b}_i \leqslant 0$, $m + 1 \leqslant i \leqslant n + 1$ imply that

$$\xi_{m+1} \leqslant 0, \dots, \xi_{n+1} \leqslant 0, \ \xi_{m+1} + \cdots + \xi_{n+1} \geqslant 0.$$

Hence $\xi_{m+1} = \cdots = \xi_{n+1} = 0$, and $\boldsymbol{x} \in L$. Consequently, $V_{m+1} \cap \cdots \cap V_{n+1} \subset L$. Theorem 13.7 shows that every family of $n - m + 1$ closed halfspaces whose intersection is L should be irredundant.

Problem 13.2. The equality $L = \cap\,(\mathrm{bd}\,V : V \in \mathcal{F})$ follows from

$$L \subset \cap\,(\mathrm{bd}\,V : V \in \mathcal{F}) \subset \cap\,(V : V \in \mathcal{F}) = L.$$

Problem 13.3. Since the polytope P is compact, Corollary 12.16 shows that both sets A and B are compact. Suppose that P is represented by an irredundant family $\mathcal{F} = \{V_1, \ldots, V_r\}$ of closed halfspaces. Let

$$V_i = \{\boldsymbol{x} \in \mathbb{R}^n : \boldsymbol{x}\cdot\boldsymbol{c}_i \leqslant \gamma_i\}, \quad 1 \leqslant i \leqslant r.$$

Because the hyperplane $\mathrm{bd}\,V_i$ supports P, there is a point $\boldsymbol{u}_i \in \mathrm{bd}\,V_i \cap P$. From $P = A + B$ we deduce the existence of points $\boldsymbol{a}_i \in A$ and $\boldsymbol{b}_i \in B$ such that $\boldsymbol{u}_i = \boldsymbol{a}_i + \boldsymbol{b}_i$. Put $\alpha_i = \boldsymbol{a}_i\cdot\boldsymbol{c}_i$ and $\beta_i = \boldsymbol{b}_i\cdot\boldsymbol{c}_i$. Clearly, $\alpha_i + \beta_i = (\boldsymbol{a}_i + \boldsymbol{b}_i)\cdot\boldsymbol{c}_i = \gamma_i$. Put

$$V_i' = \{\boldsymbol{x} \in \mathbb{R}^n : \boldsymbol{x}\cdot\boldsymbol{c}_i \leqslant \alpha_i\} \quad \text{and} \quad V_i'' = \{\boldsymbol{x} \in \mathbb{R}^n : \boldsymbol{x}\cdot\boldsymbol{c}_i \leqslant \beta_i\}.$$

Then $V_i' + V_i'' = V_i$.

We assert that $A \subset V_i'$ and $B \subset V_i''$. Indeed, assuming the existence of a point $\boldsymbol{a} \in A \setminus V_i'$, we would obtain that $\boldsymbol{a} + \boldsymbol{b}_i \in (A+B) \setminus V_i$ due to $(\boldsymbol{a}+\boldsymbol{b}_i)\cdot\boldsymbol{c}_i > \alpha_i + \beta_i$, contrary to the inclusion $A + B = P \subset V_i$.

Let $A_0 = V_1' \cap \cdots \cap V_r'$ and $B_0 = V_1'' \cap \cdots \cap V_r''$. From

$$P = A + B \subset A_0 + B \subset A_0 + B_0 \subset P$$

we obtain that $A + B = A_0 + B$. Theorem 3.60 gives $A = A_0$. Similarly, $B = B_0$. Summing up, both A and B are polytopes.

Problem 13.4. The obvious inclusions

$$P = A + B \subset A + \mathrm{cl}\,B \subset \mathrm{cl}\,(A + B) = P$$

show that $P = A + \mathrm{cl}\,B$. Suppose that P is represented by an irredundant family $\mathcal{F} = \{V_1, \ldots, V_r\}$ of closed halfspaces. Let

$$V_i = \{\boldsymbol{x} \in \mathbb{R}^n : \boldsymbol{x}\cdot\boldsymbol{c}_i \leqslant \alpha_i\}, \quad 1 \leqslant i \leqslant r.$$

Since $\mathrm{cl}\,B$ is compact, the continuous function $\varphi_i(\boldsymbol{x}) = \boldsymbol{x}\cdot\boldsymbol{c}_i$ attains a maximum value, β_i, on $\mathrm{cl}\,B$.

Because P is closed, there is a point $\boldsymbol{x}_i \in P$ such that $\varphi_i(\boldsymbol{x}_i) = \alpha_i$. Let $\boldsymbol{x}_i = \boldsymbol{a}_i + \boldsymbol{b}_i$, where $\boldsymbol{a}_i \in A$ and $\boldsymbol{b}_i \in \mathrm{cl}\,B$. Then $\varphi(\boldsymbol{b}_i) \leqslant \beta_i$, and

$$\begin{aligned}\gamma_i &= \sup\{\varphi(\boldsymbol{x}) : \boldsymbol{x} \in A\} \geqslant \varphi(\boldsymbol{a}_i)\\ &= \varphi(\boldsymbol{x}_i - \boldsymbol{b}_i) = \varphi(\boldsymbol{x}_i) - \varphi(\boldsymbol{b}_i) \geqslant \alpha_i - \beta_i.\end{aligned}$$

On the other hand, if $\boldsymbol{y}_i$ is a point in $\mathrm{cl}\,B$ with $\varphi_i(\boldsymbol{y}_i) = \beta_i$, then $\boldsymbol{x} + \boldsymbol{y}_i \in P$ for every point $\boldsymbol{x} \in A$, implying that

$$\varphi_i(\boldsymbol{x}) + \beta_i = \varphi_i(\boldsymbol{x}) + \varphi_i(\boldsymbol{y}_i) = \varphi(\boldsymbol{x} + \boldsymbol{y}_i) \leqslant \alpha_i.$$

Therefore, $\gamma_i = \sup\{\varphi(\boldsymbol{x}) : \boldsymbol{x} \in A\} \leqslant \alpha_i - \beta_i$, which gives $\alpha_i + \beta_i = \gamma_i$. Put

$$V_i' = \{\boldsymbol{x} \in \mathbb{R}^n : \boldsymbol{x}\cdot\boldsymbol{c}_i \leqslant \beta_i\} \quad \text{and} \quad V_i'' = \{\boldsymbol{x} \in \mathbb{R}^n : \boldsymbol{x}\cdot\boldsymbol{c}_i \leqslant \gamma_i\}.$$

Then $V_i' + V_i'' = V_i$. By the above argument, $A \subset V_i''$ and $B \subset V_i'$.

Let $A_0 = V_1'' \cap \cdots \cap V_r''$ and $B_0 = V_1' \cap \cdots \cap V_r'$. From

$$P = A + B \subset A_0 + B \subset A_0 + B_0 \subset P$$

we conclude that $A + B = A_0 + B$. Theorem 3.60 gives $A = A_0$. Hence A is a polyhedron.

Problem 13.5. Suppose that K is a polytope. According to Theorem 13.18, K is the convex hull of a finite set, say $X = \{\boldsymbol{v}_1, \ldots, \boldsymbol{v}_p\}$. Assume that K is the union of an infinite increasing sequence of convex sets $K_1 \subset K_2 \subset \cdots$. Therefore, for any point $\boldsymbol{v}_i \in X$ there is an index r_i such that $\boldsymbol{v}_i \in K_{r_i}$. Let $r = \max\{r_1, \ldots, r_p\}$. Then $K = \operatorname{conv}\{\boldsymbol{v}_1, \ldots, \boldsymbol{v}_p\} \subset K_r = K$. Consequently, $K_j = K$ for all $j \geqslant r$.

Conversely, suppose that a closed convex set $K \subset \mathbb{R}^n$ is not a polytope. Assume first that K is bounded. Then $\operatorname{ext} K$ contains a denumerable set, say $X = \{\boldsymbol{v}_1, \boldsymbol{v}_2, \ldots\}$. It is easy to see that every set $K_i = K \setminus X \cup \{\boldsymbol{v}_1, \ldots, \boldsymbol{v}_i\}$, $i \geqslant 1$, is convex, the sequence $K_1, K_2, \ldots$ consists of infinitely many pairwise distinct sets, and $K = K_1 \cup K_2 \cup \cdots$. Suppose now that K is unbounded. Choose a point $\boldsymbol{a} \in K$ and consider the sets $K_i = K \cap B_i(\boldsymbol{a})$, $i \geqslant 1$. Clearly, the sequence $K_1, K_2, \ldots$ consists of infinitely many pairwise distinct convex sets, and $K = K_1 \cup K_2 \cup \cdots$.

Problem 13.6. Given distinct sets K_i and K_j, Theorem 10.8 shows the existence of a hyperplane, say H_{ij}, properly separating K_i from K_j. In particular, H_{ij} does not contain $\operatorname{aff} P$. Denote by $V_{ij}^{(i)}$ and $V_{ij}^{(j)}$ the closed halfspaces determined by H_{ij} and containing, respectively, K_i and K_j. Let

$$P_i = P \cap V_{i1}^{(i)} \cap \cdots \cap V_{ii-1}^{(i)} \cap \cap V_{ii+1}^{(i)} \cap \cdots \cap V_{ir}^{(i)}, \quad 1 \leqslant i \leqslant r.$$

Clearly, $P_1, \ldots, P_r$ are polyhedra, and $\operatorname{cl} K_i \subset P_i \subset P$, $1 \leqslant i \leqslant r$. Since $\operatorname{aff} K_i = \operatorname{aff} P_i = \operatorname{aff} P$, one has $\operatorname{rint} K_i \subset \operatorname{rint} P_i$ (see Theorem 3.18). Conversely, if $\boldsymbol{x} \in \operatorname{rint} P_i$, then

$$\begin{aligned} \boldsymbol{x} &\in P \setminus (P_1 \cup \cdots \cup P_{i-1} \cup P_{i+1} \cup \cdots \cup P_r) \\ &\subset P \setminus (K_1 \cup \cdots \cup K_{i-1} \cup K_{i+1} \cup \cdots \cup K_r) = K_i. \end{aligned}$$

Thus $\operatorname{rint} P_i \subset K_i$, which gives $\operatorname{rint} P_i \subset \operatorname{rint} K_i$. Finally, the equality $\operatorname{rint} K_i = \operatorname{rint} P_i$ gives $\operatorname{cl} K_i = P_i$, $1 \leqslant i \leqslant r$.

Bibliography

[1] M. J. Adler, C. Fadiman, P. W. Goetz (eds), Great books of the western world. No. 11 (Euclid, Archimedes, Nicomachus), Encyclopaedia Britannica, Chicago, 1990.

[2] C. D. Aliprantis, R. Tourky, Cones and duality. Amer. Math. Soc., Providence, RI, 2007.

[3] K. A. Ariyawansa, W. C. Davidon, K. D. McKennon, A characterization of convexity-preserving maps from a subset of a vector space into another vector space. J. London Math. Soc. 64 (2001), 179–190.

[4] Z. Artstein, Discrete and continuous bang-bang and facial spaces or: look for the extreme points. SIAM Rev. 22 (1980), 172–185.

[5] S. Artstein-Avidan, B. A. Slomka, The fundamental theorems of affine and projective geometry revisited. Commun. Contemp. Math. 19 (2017), no. 5, 1650059.

[6] E. Asplund, A k-extreme point is the limit of k-exposed points. Israel J. Math. 1 (1963), 161–162.

[7] C. Audet, A short proof on the cardinality of maximal positive bases. Optim. Lett. 5 (2011), 191–194.

[8] A. Auslender, Closedness criteria for the image of a closed set by a linear operator. Numer. Funct. Anal. Optim. 17 (1996), 503–515.

[9] A. Auslender, M. Teboulle, Asymptotic cones and functions in optimization and variational inequalities. Springer, New York, 2003.

[10] J. Bair, Cones asymptotes et cones caractéristiques. Bull. Soc. Roy. Sci. Liège 40 (1971), 428–437.

[11] J. Bair, Une mise au point sur la décomposition des convexes. Bull. Soc. Roy. Sci. Liège 44 (1975), 698–705.

[12] J. Bair, Une étude des sommans d'un polyedre convexe. Bull. Soc. Roy. Sci. Liège 45 (1976), 307–311.

[13] J. Bair, À propos des points extrêmes de la somme de deux ensembles convexes. Bull. Soc. Roy. Sci. Liège 48 (1979), 262–264.

[14] J. Bair, Sur la structure extrémale de la somme de deux convexes. Canad. Math. Bull. 22 (1979), 1–7.

[15] J. Bair, A geometric description of the inner aperture of a convex set. Acta

Math. Acad. Sci. Hungar. 38 (1981), no. 1-4, 237–240.
[16] J. Bair, Some characterizations of the asymptotic cone and the lineality space of a convex set. Math. Operationsforsch. Statist. Ser. Optim. 12 (1981), 173–176.
[17] J. Bair, Critères de fermeture pour l'enveloppe convexe d'une réunion. Bull. Soc. Roy. Sci. Liège 52 (1983), 72–79.
[18] J. Bair, R. Fourneau, Étude géométrique des espaces vectoriels. Une introduction. Lecture Notes in Math. Vol. 489, Springer-Verlag, Berlin, 1975.
[19] J. Bair, R. Fourneau, Une démonstration géométrique du theoreme de Choquet-Kendall. Comment. Math. Univ. Carol. 16 (1975), 683–691.
[20] J. Bair, R. Fourneau, Étude géometrique des espaces vectoriels. II. Polyèders et polytopes convexes. Lecture Notes in Math. Vol. 802, Springer-Verlag, Berlin, 1980.
[21] J. Bair, R. Fourneau, F. Jongmans, Vers la domestication de l'extrémisme. Bull. Soc. Roy. Sci. Liège 46 (1977), 126–132.
[22] J. Bair, J. Gwinner, Sur la séparation vraie de cônes convexes. Arkiv Mat. 16 (1978), 207–212.
[23] J. Bair, F. Jongmans, La séparation vraie dans un espace vectoriel. Bull. Soc. Roy. Sci. Liège 41 (1972), 163–170.
[24] J. Bair, F. Jongmans, Relations entre l'enveloppe conique, le cône d'infinitude et la gaine d'un ensemble convexe. Bull. Soc. Roy. Sci. Liège 52 (1983), 17–21.
[25] J. Bair, F. Jongmans, Sur l'énigme de l'enveloppe conique fermée. Bull. Soc. Roy. Sci. Liège 52 (1983), 285–294.
[26] I. Bárány, R. Karasev, Notes about the Carathéodory number. Discrete Comput. Geom. 48 (2012) 783–792.
[27] G. P. Barker, The lattice of faces of a finite dimensional cone. Linear Algebra Appl. 7 (1973), 71–82.
[28] A. Bastiani, Polyèdres convexes de dimension quelconque. C. R. Acad. Sci. Paris 247 (1958) 1943–1946.
[29] R. G. Batson, Necessary and sufficient condition for boundedness of extreme points of unbounded convex sets. J. Math. Anal. Appl. 130 (1988), 365–374.
[30] G. Beer, Recession cones of nonconvex sets and increasing functions. Proc. Amer. Math. Soc. 73 (1979), 228–232.
[31] M. K. Bennett, Lattices of convex sets. Trans. Amer. Math. Soc. 234 (1977), 279–288.
[32] S. K. Berberian, Compact convex sets in innner product spaces. Amer. Math. Monthly 74 (1967), 702–705.
[33] G. M. Bergman, On lattices of convex sets in $\mathbb{R}^n$. Algebra Universalis 53 (2005), 357–395.
[34] C. Bessaga, A note on universal Banach spaces of a finite dimension. Bull. Acad. Polon. Sci. Sér. Sci. Math. Astronom. Phys. 6 (1958), 97–101.
[35] G. Bigi, M. Pappalardo, Regularity conditions for the linear separation of sets. J. Optim. Theory Appl. 99 (1998), 533–540.
[36] G. Björck, The set of extreme points of a compact convex set. Arkiv Math. 3 (1958), 463–468.

[37] D. Blackwell, M. A. Girshick, Theory of games and statistical decisions. Willey & Sons, New York, 1954.

[38] G. Bol, Über Eikörper mit Vieleckschatten. Math. Z. 48 (1942), 227–246.

[39] T. Bonnesen, W. Fenchel, Theorie der konvexen Körper. Springer, Berlin, 1934. English translation: Theory of convex bodies. BCS Associates, Moscow, ID, 1987.

[40] W. E. Bonnice, V. L. Klee, The generation of convex hulls. Math. Ann. 152 (1963), 1–29.

[41] V. A. Borovikov, On the intersection of a sequence of simplices. Uspehi Matem. Nauk 7 (1952), 179–180.

[42] J. M. Borwein, W. B. Moors, Stability of closedness of convex cones under linear mappings. J. Convex Anal. 16 (2009), 699–705.

[43] J. M. Borwein, W. B. Moors, Stability of closedness of convex cones under linear mappings. II. J. Nonlinear Anal. Optim. 1 (2010), 1–7.

[44] T. Botts, On convex sets in linear normed spaces. Bull. Amer. Math. Soc. 48 (1942), 150–152.

[45] N. Bourbaki, Topological vector spaces. Chapters 1–5. Elements of mathematics. Springer, Berlin, 1987.

[46] D. G. Bourgin, Restricted separation of polyhedra. Portugaliae Math. 11 (1952), 133–136.

[47] M. Breen, Admissible kernels for starshaped sets. Proc. Amer. Math. Soc. 82 (1981), 622–628.

[48] T. Bromek, J. Kaniewski, Linear images and sums of closed cones in Euclidean spaces. Bull. Acad. Polon. Sci. Sér. Sci. Math. Astronom. Phys. 24 (1976), 231–238.

[49] E. M. Bronshteĭn, Extremal boundaries of finite-dimensional convex compacta. Optimizatsiya No. 26 (1981), 119–128.

[50] E. M. Bronshteĭn, One-dimensional extremal boundaries of three-dimensional compact convex sets. Optimizatsiya No. 51 (1992), 34–46.

[51] E. M. Bronshteĭn, On convex compacta with given extreme points. Russian Acad. Sci. Dokl. Math. 49 (1994), 497–500.

[52] E. M. Bronshteĭn, On compact convex sets with given extreme points. Siberian Math. J. 36 (1995), 17–23

[53] A. Brøndsted, Intersections of translates of convex sets. Mathematika 24 (1977), 122–129.

[54] A. Brøndsted, The inner aperture of a convex set. Pacific J. Math. 72 (1977), 335–340.

[55] A. Brøndsted, An introduction to convex polytopes. Springer, New York, 1983.

[56] A. Brøndsted, Continuous barycenter functions on convex polytopes. Exposition. Math. 4 (1986), 179–187.

[57] A. L. Brown, Chebyshev sets and facial systems of convex sets in finite-dimensional spaces. Proc. London Math. Soc. 41 (1980), 297–339.

[58] H. Brunn, Über Ovale und Eiflächen. Inaugural Dissertation, Straub, München, 1887.

[59] H. Brunn, Über Kurven ohne Wendepunkte. Habilitationsschrift, Acker-

mann, München, 1889.
[60] H. Brunn, Über das durch eine belibige endliche Figur bestimmte Eigebilde. In: St. Meyer (ed), Ludwig Boltzmann Festschrift, pp. 94–104, Barth, Leipzig, 1904.
[61] H. Brunn, Zur Theorie der Eigebiete. Arch. Math. Phys. 17 (1911), 289–300.
[62] H. Brunn, Über Kerneigebiete. Math. Ann. 73 (1913), 436–440.
[63] L. N. H. Bunt, Bijdrage tot de theorie der convexe puntverzamelingen. Proefschrifft Groningen, Noord-Hollandsche Uitgevers Maatschappij, Amsterdam, 1934.
[64] Yu. D. Burago, V. A. Zalgaller, Sufficient conditions for convexity. J. Soviet Math. 16 (1978), 395–434.
[65] C. Carathéodory, Über den Variabilitätsbereich der Koefficienten von Potenzreihen, die gegebene Werte nicht annehmen. Math. Ann. 64 (1907), 95–115.
[66] C. Carathéodory, Über den Variabilitätsbereich der Fourierschen Konstanten von positiven harmonischen Funktionen. Rend. Circ. Mat. Palermo 32 (1911), 193–217.
[67] M. Castellani, G. Mastroeni, M. Pappalardo, Separation of sets, Lagrange multipliers, and totally regular extremum problems. J. Optim. Theory Appl. 92 (1997), 249–261.
[68] S. N. Černikov, Theorems on separation of convex polyhedral sets. Soviet Math. Dokl. 2 (1961), 838–840.
[69] S. N. Černikov, Linear inequalities. Izdat. Nauka, Moscow, 1968.
[70] G. Choquet, Unicité des représentations intégrales au moyen de points extrémaux dans les cônes convexes réticulés. C. R. Acad. Sci. Paris 243 (1956), 555–557.
[71] G. Choquet, Ensembles et cônes convexes faiblement complets. C. R. Acad. Sci. Paris 254 (1962), 1908–1910 and 2123–2125.
[72] G. Choquet, H. H. Corson, V. L. Klee, Exposed points of convex sets. Pacif. J. Math. 17 (1966), 33–43.
[73] A. Chubarev, I. Pinelis, Fundamental theorem of geometry without the 1-to-1 assumption. Proc. Amer. Math. Soc. 127 (1999), 2735–2744.
[74] C. Clark, On convex sets of finite width. J. London Math. Soc. 43 (1968), 513–516.
[75] F. H. Clarke, Generalized gradients and applications. Trans. Amer. Math. Soc. 205 (1975), 247–262.
[76] J. B. Collier, On the set of extreme points of a convex body. Proc. Amer. Math. Soc. 47 (1975), 184–186.
[77] J. B. Collier, On the facial structure of a convex body. Proc. Amer. Math. Soc. 61 (1976), 367–370.
[78] W. D. Cook, Carathéodory's theorem with linear constraints. Canad. Math. Bull. 17 (1974), 189–191.
[79] W. D. Cook, R. J. Webster, Carathéodory's theorem. Canad. Math. Bull. 15 (1972), 293.
[80] H. H. Corson, A compact convex set in E^3 whose exposed points are of the first category. Proc. Amer. Math. Soc. 16 (1965), 1015–1021.

[81] L. Dalla, On the measure of the one-skeleton of the sum of convex compact sets. J. Austral. Math. Soc. Ser. A. 42 (1987), 385–389.
[82] E. A. Danielyan, G. S. Movsisyan, K. R. Tatalyan, Generalization of the Carathéodory theorem. Akad. Nauk Armenii Dokl. 92 (1991), 69–75.
[83] L. Danzer, B. Grünbaum, V. L. Klee, Helly's theorem and its relatives. In: V. L. Klee (ed), Convexity. Proc. Sympos. Pure Math. Vol. VII, pp. 101–179, Amer. Math. Soc., Providence, RI, 1963.
[84] G. Darboux, Sur le théorème fondamental de la géométrie projective. Math. Ann. 17 (1880), 55–61.
[85] C. Davis, Theory of positive linear dependence. Amer. J. Math. 76 (1954), 733–746.
[86] C. Davis, Remarks on a previous paper. Michigan Math. J. 2 (1954), 23–25.
[87] A. Dax, The distance between two convex sets. Linear Algebra Appl. 416 (2006), 184–213.
[88] J. A. De Loera, J. Rambau, F. Santos, Triangulations. Structures for algorithms and applications. Springer, Berlin, 2010.
[89] M. De Wilde, Some properties of the exposed points of finite dimensional convex sets. J. Math. Anal. Appl. 99 (1984), 257–264.
[90] J.-P. Dedieu, Cône asymptote d'un ensemble non convexe. Application à l'optimisation. C. R. Acad. Sci. Paris 285 (1977), A501–A503.
[91] D. Derry, Convex hulls of simple space curves. Canad. J. Math. 8 (1956), 383–388.
[92] R. Deumlich, K.-H. Elster, R. Nehse, Recent results on separation of convex sets. Math. Operationsforsch. Statist. Ser. Optim. 9 (1978), 273–296.
[93] J. Dieudonné, Sur la séparation des ensembles convexes. Math. Ann. 163 (1966), 1–3.
[94] L. L. Dines, Convex extensions and linear inequalities. Bull. Amer. Math. Soc. 42 (1936), 353–365.
[95] L. L. Dines, N. H. McCoy, On linear inequalities. Trans. Roy. Soc. Canada. Sect. III. 27 (1933), 37–70.
[96] M. Dragomirescu, A strong separation theorem with a side-condition. Rev. Roumaine Math. Pures Appl. 39 (1994), 27–35.
[97] L. E. Dubins, On extreme points of convex sets. J. Math. Anal. Appl. 5 (1962), 237–244.
[98] E. Egerváry, On the smallest convex cover of a simple arc of space-curve. Publ. Math. Debrecen 1 (1949), 65–70.
[99] H. G. Eggleston, Intersection of convex sets. J. London Math. Soc. 5 (1972), 753–754.
[100] L. Eifler, Compactly generated cones. Tamkang J. Math. 18 (1987), 41–43.
[101] G. V. Epifanov, Universality of sections of cubes. Math. Notes 2 (1967), 540–541.
[102] V. P. Fedotov, The notions of a face of a convex compactum. Ukrain. Geom. Sb. No. 21 (1978), 131–141.
[103] W. Fenchel, Über Krümmung und Windung geschlossener Raumkurven. Math. Ann. 101 (1929), 238–252.
[104] W. Fenchel, A remark on convex sets and polarity. Comm. Sém. Math.

Univ. Lund. Tome Suppl. M. Riesz (1952), 82–89.
[105] W. Fenchel, Convex cones, sets, and functions. Mimeographed lecture notes. Spring Term 1951, Princeton University, 1953.
[106] W. Fenchel, Convexity through the ages. In: P. M. Gruber, J. M. Wills (eds), Convexity and its applications, pp. 120–130, Birkhäuser, Basel, 1983.
[107] O. P. Ferreira, S. Z. Németh, Generalized projections onto convex sets. J. Global Optim. 52 (2012), 831–842.
[108] T. Fischer, Strong unicity and alternation for linear optimization. J. Optim. Theory Appl. 69 (1991), 251–267.
[109] R. Fourneau, Some results on the geometry of Choquet simplices. J. Geometry 9 (1977), 143–147.
[110] R. Fourneau, Nonclosed simplices and quasi-simplices. Mathematika 24 (1977), 71–85.
[111] J. Frenkel, Géométrie pour l'élève-professur. Hermann, Paris, 1973.
[112] B. Fuglede, Continuous selection in a convexity theorem of Minkowski. Exposition. Math. 4 (1986), 163–178.
[113] M. Fujiwara, Über den Mittelkörper zweier konvexer Körper. Sci. Rep. Tôhoku Univ. 5 (1916), 275–283.
[114] J. W. Gaddum, A theorem on convex cones with applications to linear inequalities. Proc. Amer. Math. Soc. 3 (1952), 957–960.
[115] D. Gale, Linear combination of vectors with non-negative coefficients (solution of problem 4395). Amer. Math. Monthly 59 (1952), 46–47.
[116] D. Gale, V. Klee, Continuous convex sets. Math. Scand. 7 (1959), 379–391.
[117] R. J. Gardner, Geometric tomography. Cambridge University Press, New York, 1995. Second edition: 2006.
[118] M. Gerstenhaber, Theory of convex polyhedral cones. In: T. C. Koopmans (ed), Activity analysis of production and allocation, Chapter XVIII, pp. 298–316, Wiley, New York, NY, 1951.
[119] M. A. Goberna, E. González, J. E. Martínez-Legaz, M. I. Todorov, Motzkin decomposition of closed convex sets. J. Math. Anal. Appl. 364 (2010), 209–221.
[120] M. A. Goberna, A. Iusem, J. E. Martínez-Legaz, M. I. Todorov, Motzkin decomposition of closed convex sets via truncation. J. Math. Anal. Appl. 400 (2013), 35–47.
[121] M. A. Goberna, V. Jornet, M. M. L. Rodríguez, On linear systems containing strict inequalities. Linear Algebra Appl. 360 (2003), 151–171.
[122] M. A. Goberna, J. E. Martínez-Legaz, M. I. Todorov, On Motzkin decomposable sets and functions. J. Math. Anal. Appl. 372 (2010), 525–537.
[123] M. A. Goberna, V. Jornet, M. M. L. Rodríguez, On the characterization of some families of closed convex sets. Beitr. Algebra Geom. 43 (2002), 153–169.
[124] M. A. Goberna, M. M. L. Rodríguez, Analyzing linear systems containing strict inequalities via evenly convex hulls. European J. Oper. Res. 169 (2006), 1079–1095.
[125] P. Goossens, Hyperbolic sets and asymptotes. J. Math. Anal. Appl. 116 (1986), 604–618.

[126] J. Green, W. P. Heller, Mathematical analysis and convexity with applications to economics. In: K. J. Arrow, M. D. Intriligator (eds), Handbook of mathematical economics. Vol. I, pp. 15–52. Amsterdam, North-Holland, 1981.

[127] P. Gritzmann, V. Klee, External tangents and closedness of cone+subspace. J. Math. Anal. Appl. 188 (1994), 441–457.

[128] P. Gritzmann, V. Klee, Separation by hyperplanes in finite-dimensional vector spaces over Archimedean ordered fields. J. Convex Anal. 5 (1998), 279–301

[129] P. M. Gruber, Über den Durchscnitt einer abnehmenden Folge von Parallelepipeden. Elem. Math. 32 (1977), 13–15.

[130] P. M. Gruber, Durchsnitte und Vereinigungen monotoner Folgen spezieller konvexer Körper. Abh. Math. Sem. Univ. Hamburg 49 (1979), 189–197.

[131] P. M. Gruber, Zur Geschichte der Konvexgeometrie und der Geometrie der Zahlen. In: G. Fischer, et al. (eds), Ein Jahrhundert Mathematik 1890–1990, pp. 421–455, Deutsche Mathematiker Vereinigung, Freiburg, 1990.

[132] P. M. Gruber, History of convexity. In: P. M. Gruber, J. M. Wills (eds), Handbook of convex geometry. Vol. A, pp. 1–15, North-Holland, Amsterdam, 1993.

[133] B. Grünbaum, On a problem of S. Mazur. Bull. Res. Council Israel. Sect. F 7 (1957/1958), 133–135.

[134] B. Grünbaum, Convex polytopes. Interscience, New York, 1967. Second edition: Springer, New York, 2003.

[135] J. Ġrzybowski, D. Pallaschke, R. Urbański, Data pre-classification and the separation law for closed bounded convex sets. Optim. Methods Softw. 20 (2005), 219–229.

[136] R. Grząślewicz, A universal convex set in Euclidean space. Colloq. Math. 45 (1981), 41–44.

[137] J. Grzybowski, R. Urbański, Affine straight lines in family of bounded closed convex sets. Rend. Circ. Mat. Palermo 53 (2004), 225–230.

[138] W. Gustin, On the interior of the convex hull of a Euclidean set. Bull. Amer. Math. Soc. 53 (1947), 299–301.

[139] H. Hadwiger, Minkowskische Addition und Subtraktion beliebiger Punktmengen und die Theoreme von Erhard Schmidt. Math. Z. 53 (1950), 210–218.

[140] H. Hadwiger, Vorlesungen über Inhalt, Oberfläche und Isoperimetrie. Springer, Berlin, 1957.

[141] P. C. Hammer, Maximal convex sets. Duke Math. J. 22 (1955), 103–106.

[142] P. C. Hammer, Semispaces and the topology of convexity. In: V. L. Klee (ed), Convexity. Proc. Sympos. Pure Math. Vol. VII, pp. 305–316, Amer. Math. Soc., Providence, RI, 1963.

[143] S. P. Han, O. L. Mangasarian, A conjugate decomposition of the Euclidean space. Proc. Nat. Acad. Sci. U.S.A. 80 (1983), 5156–5157.

[144] S. P. Han, O. L. Mangasarian, Conjugate cone characterization of positive definite and semidefinite matrices. Linear Algebra Appl. 56 (1984), 89–103.

[145] H. Hancock, Development of the Minkowski geometry of numbers. Macmil-

lan, New York, 1939.

[146] O. Hanner, H. Rådström, A generalization of a theorem of Fenchel. Proc. Amer. Math. Soc. 2 (1951), 589–593.

[147] W. Hansen, V. L. Klee, Interesection theorems for positive sets. Proc. Amer. Math. Soc. 22 (1969), 450–457.

[148] W. Hare, H. Song, On the cardinality of positively linearly independent sets. Optim. Lett. 10 (2016), 649–654.

[149] Y. He, J. Sun, Minimum recession-compatible subsets of closed convex sets. J. Global Optim. 52 (2012), 253–263.

[150] J. Hjelmslev, Contribution à la géométrie infinitesimale de la courbe reelle. Overs. Danske Vidensk. Selk. Föhr. (1911), 433–494.

[151] A. J. Hoffman, On the covering of polyhedra by polyhedra. Proc. Amer. Math. Soc. 23 (1969), 123–126.

[152] P. Holický, M. Laczkovich, Descriptive properties of the set of exposed points of compact convex sets in $\mathbb{R}^3$. Proc. Amer. Math. Soc. 132 (2004), 3345–3347.

[153] R. Howe, On the tendency toward convexity of the vector sum of sets. Cowles Foundation Discussion Papers, No. 538, pp. 1–17. Yale University, CN, 1979.

[154] H. Hu, O. Wang, Closedness of a convex cone and application by means of the end set of a convex set. J. Optim. Theory Appl. 150 (2011), 52–64.

[155] A. P. Huhn, On nonmodular n-distributive lattices. I. Lattices of convex sets. Acta Sci. Math. (Szeged) 52 (1988), 35–45.

[156] T. Husain, I. Tweddle, On the extreme points of the sum of two convex sets. Math. Ann. 188 (1970), 113–122.

[157] M. Ito, B. F. Lourenço, A bound on the Carathéodory number. Linear Algebra Appl. 532 (2017), 347–363.

[158] A. Iusem, J. E. Martínez-Legaz, M. I. Todorov, Motzkin predecomposable sets. J. Global Optim. 60 (2014), 635–647.

[159] R. E. Jamison, Contractions of convex sets. Proc. Amer. Math. Soc. 62 (1976), 129–130.

[160] B. Jessen, Zwei Zätze über konvexe Punktmengen. (Danish) Mat. Didsskr. B (1940), 66–70.

[161] F. Jongmans, Réflexions sur l'art de sauver la face. Bull. Soc. Roy. Sci. Liège 45 (1976), 294–306.

[162] S.-M. Jung, A characterization of injective linear transformations. J. Convex Anal. 17 (2010), 293–299.

[163] J. A. Kalman, Continuity and convexity of projections and barycentric coordinates in convex polyhedra. Pacific J. Math. 11 (1961), 1017–1022.

[164] S. Karlin, L. S. Shapley, Geometry of reduced moment spaces. Proc. Nat. Acad. Sci. U.S.A. 35 (1949), 673–677.

[165] A. Kirsch, Konvexe Figuren als Durchschnitte abzählbar vieler Halbräume. Arch. Math. (Basel) 18 (1967), 313–319.

[166] T. H. Kjeldsen, From measuring tool to geometrical object: Minkowski's development of the concept of convex bodies. Arch. Hist. Exact Sci. 62 (2008), 59–89.

[167] T. H. Kjeldsen, Egg-forms and measure-bodies: different mathematical practices in the early history of the modern theory of convexity. Sci. Context 22 (2009), 85–113.
[168] T. H. Kjeldsen, History of convexity and mathematical programming: connections and relationships in two episodes of research in pure and applied mathematics of the 20th century. In: R. Bhatia, et al. (eds), Proc. Internat. Congress Math. Vol. IV, pp. 3233–3257, Hindustan Book Agency, New Delhi, 2010.
[169] V. L. Klee, A characterization of convex sets. Amer. Math. Monthly 56 (1949), 247–249.
[170] V. L. Klee, Convex sets in linear spaces. II. Duke Math. J. 18 (1951), 875–883.
[171] V. L. Klee, A note on extreme points. Amer. Math. Monthly 62 (1955), 30–32.
[172] V. L. Klee, Separation properties of convex cones. Proc. Amer. Math. Soc. 6 (1955), 313–318.
[173] V. L. Klee, Strict separation of convex sets. Proc. Amer. Math. Soc. 7 (1956), 735–737.
[174] V. L. Klee, Extremal structure of convex sets. Arch. Math. 8 (1957), 234–240.
[175] V. L. Klee, Extremal structure of convex sets. II. Math. Z. 69 (1958), 90–104.
[176] V. L. Klee, Some characterizations of convex polyhedra. Acta Math. 102 (1959), 79–107.
[177] V. L. Klee, Asymptotes and projections of convex sets. Math. Scand. 8 (1960), 356–362.
[178] V. L. Klee, The generation of affine hulls. Acta Sci. Math. (Szeged) 24 (1963), 60–81.
[179] V. L. Klee, On a theorem of Dubins. J. Math. Anal. Appl. 7 (1963), 425–427.
[180] V. L. Klee, A theorem on convex kernels. Mathematika 12 (1965), 89–93.
[181] V. L. Klee, Asymptotes of convex bodies. Math. Scand. 20 (1967), 89–90.
[182] V. L. Klee, Maximal separation theorems for convex sets. Trans. Amer. Math. Soc. 134 (1968), 133–147.
[183] V. L. Klee, Separation and support properties of convex sets–a survey. In: A. V. Balakrishnan (ed), Control theory and the calculus of variations, pp. 235–303, Academic Press, New York, 1969.
[184] V. L. Klee, Unions of increasing and intersections of decreasing sequences of convex sets. Israel J. Math. 12 (1972), 70–78.
[185] V. L. Klee, E. Maluta, C. Zanco, Basic properties of evenly convex sets. J. Convex Analysis 14 (2007), 137–148.
[186] W. Koenen, The Kuratowski closure problem in the topology of convexity. Amer. Math. Monthly 73 (1966), 704–708.
[187] H. Komiya, Closedness of cones. J. Math. Anal. Appl. 81 (1981), 320–326.
[188] H. Kramer, Über einen Satz von Fenchel. Mathematica (Cluj) 9 (1967), 281–283.
[189] H. W. Kuhn, A. W. Tucker (eds), Linear inequalities and related systems.

Princeton, N. J., 1956.
[190] A. V. Kuz'minyh, Affineness of convex-invariant mappings. Siberian Math. J. 16 (1975), 918–922.
[191] D. G. Larman, On the inner aperture and intersections of convex sets. Pacific J. Math. 55 (1974), 219–232.
[192] M. Lassak, Convex half-spaces. Fund. Math. 120 (1984), 7–13.
[193] J. Lawrence, Intersections of descending sequences of affinely equivalent convex bodies. Beitr. Algebra Geom. 54 (2013), 669–676.
[194] J. Lawrence, V. Soltan, The intersection of convex transversals is a convex polytope. Beitr. Algebra Geom. 50 (2009), 283–294.
[195] J. Lawrence, V. Soltan, Carathéodory-type results for the sums and unions of convex sets. Rocky Mountain J. Math. 43 (2013), 1675–1688.
[196] J. Lawrence, V. Soltan, On unions and intersections of nested families of cones. Beitr. Algebra Geom. 57 (2016), 655–665.
[197] H. Lenz, Einige Anwendungen der projektiven Geometrie auf Fragen der Flächentheorie. Math. Nachr. 18 (1958), 346–359.
[198] A. A. Lepin, On a method of forming the convex hull. Učen. Zap. Belorussk. Univ. Ser. Mat. (1959), No. 1, 27–30.
[199] N. Levinson, T. O. Sherman, The sum of the intersections of a cone with a linear subspace and of dual cone with orthogonal complementary subspace. J. Combin. Theory 1 (1966), 338–349.
[200] B. Li, Y. Wang, A new characterization for isometries by triangles. New York J. Math. 15 (2009), 423–429.
[201] B. Li, Y. Wang, A new characterization of line-to-line maps in the upper plane. Filomat 27 (2013), 127–133.
[202] B. Li, Y. Wang, Fundamental theorem of geometry without the surjective assumption. Trans. Amer. Math. Soc. 368 (2016), 6819–6834.
[203] R. Loewy, B. S. Tam, Complementation in the face lattice of a proper cone. Linear Algebra Appl. 79 (1986), 195–207.
[204] K. Lommatzsch, Über Extremalmengen und Stützhyperebenen konvexer Mengen. Wiss. Z. Humboldt-Univ. Berlin. Math.-naturwiss. R. 30 (1981), 411–413.
[205] D. T Luc, Theory of vector optimization. Springer, Berlin, 1989.
[206] P. Mani-Levitska, Characterizations of convex sets. In: P. M. Gruber, J. M. Wills (eds), Handbook of convex geometry. Vol. A, pp. 19–41, North-Holland, Amsterdam, 1993.
[207] J. E. Martínez-Legaz, C. Pintea, Closed convex sets of Minkowski type. J. Math. Anal. Appl. 444 (2016), 1195–1202
[208] J. E. Martínez-Legaz, I. Singer, The structure of hemispaces in R^n. Linear Algebra Appl. 110 (1988), 117–179.
[209] R. D. Mauldin (ed), The Scottish book. Birkäuser, Boston, 1981.
[210] S. Mazur, Über konvexe Mengen in linearen normierten Räumen. Studia Math. 4 (1933) 70–84.
[211] R. L. McKinney, Positive bases for linear spaces. Trans. Amer. Math. Soc. 103 (1962), 131–148.
[212] K. Meng, V. Roshchina, X. Yang, On local coincidence of a convex set and

its tangent cone. J. Optim. Theory Appl. 164 (2015), 123–137.
[213] W. Meyer, D. C. Kay, A convexity structure admits but one real linearization of dimension greater than one. J. London Math. Soc. 7 (1973), 124–130.
[214] H. Minkowski, Geometrie der Zahlen. I. Teubner, Leipzig, 1896; II. Teubner, Leipzig, 1910.
[215] H. Minkowski, Gesammelte Abhandlungen. Bd 2. Teubner, Leipzig, 1911.
[216] H. Mirkil, New characterizations of polyhedral cones. Canad. J. Math. 9 (1957), 1–4.
[217] J.-J. Moreau, Décomposition orthogonale d'un espace hilbertien selon deux cônes mutuellement polaires. C. R. Acad. Sci. Paris 255 (1962), 238–240.
[218] T. Motzkin, Sur quelques propriétés caractéristiques des ensembles convexes. Rend. Reale Accad. Lincei, Classe Sci. Fis., Mat. Nat. 21 (1935), 562–567.
[219] T. Motzkin, Sur quelques propriétés caractéristiques des ensembles bornés non convexes. Rend. Reale Accad. Lincei, Classe Sci. Fis., Mat. Nat. 21 (1935), 773–779.
[220] T. Motzkin, Beiträge zur Theorie der linearen Ungleichungen (Inaug. Diss., Univ. Basel). Azriel, Jerusalem, 1936. English translation: Contributions to the theory of linear inequalities. In: D. Cantor, et al. (eds), Theodore S. Motzkin: Selected Papers, pp. 1–80, Birkhäuser, Boston, 1983.
[221] T. Motzkin, Extensions of the Minkowski-Carathéodory theorem on convex hulls. Abstract 65T-385. Notices Amer. Math. Soc. 12 (1965), 705.
[222] H. Naumann, Beliebige konvexe Polytope als Schnitte und Projektionen höherdimensionaler Würfel, Simplizes und Masspolytope. Math. Z. 65 (1956), 91–103.
[223] J. W. Nieuwenhuis, About separation by cones. J. Optim. Theory Appl. 41 (1983), 473–479.
[224] G. H. Orland, On a class of operators. Proc. Amer. Math. Soc. 15 (1964), 75–79.
[225] Z. Páles, Separation theorems for convex sets and convex functions with invariance properties. Lecture Notes in Econom. and Math. Systems, Vol. 502, pp. 279–293, Springer, Berlin, 2001.
[226] G. Pataki, On the closedness of the linear image of a closed convex cone. Math. Oper. Res. 32 (2007), 395–412.
[227] C. Pauc, Sur la relation entre un point et une de ses projections sur un ensemble. Rev. Sci. Paris 77 (1939), 657–658.
[228] J.-P. Penot, A metric approach to asymptotic analysis. Bull. Sci. Math. 127 (2003), 815–833.
[229] A. L. Peressini, Ordered topological vector spaces. Harper & Row, New York, 1967.
[230] R. R. Phelps, Convex sets and nearest points. Proc. Amer. Math. Soc. 8 (1957), 790–797.
[231] R. R. Phelps, Lectures on Choquet's theorem. Van Nostrand, Princeton, 1966.
[232] H. X. Phu, Representation of bounded convex sets by rational convex hull of its gamma-extreme points. Numer. Funct. Anal. Optim. 15 (1994), 915–

920.
[233] K. A. Post, Star extension of plane convex sets. Indag. Math. 26 (1964), 330–338.
[234] J. F. Queiró, E. M. Sà, On separation properties of finite-dimensional compact convex sets. Proc. Amer. Math. Soc. 124 (1996), 259–264.
[235] J. Radon, Mengen konvexen Körper, die einen gemeinamen punkt enhalten. Math. Ann. 83 (1921), 113–115.
[236] H. Rådström, An embedding theorem for spaces of convex sets. Proc. Amer. Math. Soc. 3 (1952), 165–169.
[237] J. R. Reay, Generalizations of a theorem of Carathéodory. Mem. Amer. Math. Soc. No. 54 (1965).
[238] J. R. Reay, Unique minimal representations with positive bases. Amer. Math. Monthly 73 (1966), 253–261.
[239] H. B. Reiter, N. M. Stavrakas, On the compactness of the hyperspace of faces. Pacific J. Math. 73 (1977), 193–196.
[240] C. V. Robinson, Spherical theorems of Helly type and congruence indices of spherical caps. Amer. J. Math. 64 (1942), 260–272.
[241] R. T. Rockafellar, Convex analysis. Princeton Universty Press, Princeton, NJ, 1970.
[242] R. T. Rockafellar, Clarke's tangent cones and the boundaries of closed sets in $\mathbb{R}^n$. Nonlinear Anal. 3 (1978), 145–154.
[243] C. A. Rogers, G. C. Shephard, The difference body of a convex body. Arch. Math. (Basel) 8 (1957), 220–233.
[244] V. Roshchina, T. Sang, D. Yost, Compact convex sets with prescribed facial dimensions. In: D. R. Wood et al. (eds), 2016 MATRIX Annals, pp. 167–175, Springer, Cham, 2018.
[245] A. K. Roy, Facial structure of the sum of two compact convex sets. Math. Ann. 197 (1972), 189–196.
[246] H. Rubin, O. Wesler, A note on convexity in euclidean n-space. Proc. Amer. Math. Soc. 9 (1958), 522–523.
[247] G. Š. Rubinštein, The general solution of a finite system of linear inequalities. Uspehi Matem. Nauk 9 (1954), 171–177.
[248] R. Schneider, Boundary structure and curvature of convex bodies. In: P. M. Gruber, J. M. Wills (eds), Contributions to geometry, pp. 13–59, Birkhäuser, Basel, 1979.
[249] R. Schneider, Convex bodies: the Brunn-Minkowski theory. Cambridge University Press, Cambridge, 1993. Second edition: 2014.
[250] A. V. Shaĭdenko-Künzi, On mappings preserving convexity, Amer. Math. Soc. Transl. Ser. 2, Vol. 163, pp. 155–163, Amer. Math. Soc., Providence, RI, 1995.
[251] C. R. Smith, A characterization of star-shaped sets. Amer. Math. Monthly 75 (1968), 368.
[252] V. Soltan, Introduction to the axiomatic theory of convexity. Ştiinţa, Chişinău, 1984.
[253] V. Soltan, Lower bounds for the number of extremal and exposed diameters of a convex body. Studia Sci. Math. Hungar. 28 (1993), 99–104.

[254] V. Soltan, A characterization of homothetic simplices. Discrete Comput. Geom. 22 (1999), 193–200.
[255] V. Soltan, Choquet simplexes in finite dimension–a survey. Expo. Math. 22 (2004), 301–315.
[256] V. Soltan, Affine diameters of convex-bodies–a survey. Expo. Math. 23 (2005), 47–63.
[257] V. Soltan, Addition and subtraction of homothety classes of convex sets. Beitr. Algebra Geom. 47 (2006), 351–361.
[258] V. Soltan, Convex sets with homothetic projections. Beitr. Algebra Geom. 51 (2010), 237–249.
[259] V. Soltan, A characteristic intersection property of generalized simplices. J. Convex Anal. 18 (2011), 529–543.
[260] V. Soltan, Lectures on convex sets. World Scientific, Hackensack, NJ, 2015.
[261] V. Soltan, Polarity and separation of cones. Linear Algebra Appl. 538 (2018), 212–224.
[262] V. Soltan, Asymptotic planes and closedness conditions for linear images and vector sums of sets. J. Convex Anal. 25 (2018), 1183–1196.
[263] V. Soltan, Moreau-type characterizations of polar cones. Linear Algebra Appl. 567 (2019), 45–62.
[264] R. M. Starr, Quasi-equilibria in markets with non-convex preferences. Econometrica 37 (1969), 25–38.
[265] E. Steinitz, Bedingt konvergente Reihen und konvexe Systeme. I–IV. J. Reine Angew. Math. 143 (1913), 128–175.
[266] E. Steinitz, Bedingt konvergente Reihen und konvexe Systeme. V. J. Reine Angew. Math. 144 (1914), 1–40.
[267] E. Steinitz, Bedingt konvergente Reihen und konvexe Systeme. VI, VII. J. Reine Angew. Math. 146 (1916), 1–52.
[268] E. Steinitz, Polyeder und Raumeinteilungen. In: W. Fr. Meyer, H. Mohrmann (eds), Encyklopädie der Mathematischen Wissenschaften, Band III-1-2, pp. 1–139, Teubner, Leipzig, 1916.
[269] T. G. D. Stoelinga, Convexe Puntverzemelingen. (Dutch) Proefschrifft Groningen, Amsterdam, Noordhollandsche Uitgeversmaatschappij, 1932.
[270] J. J. Stoker, Unbounded convex sets. Amer. J. Math. 62 (1940), 165–179.
[271] M. H. Stone, Convexity. Mimeographed Lectures. University of Chicago, Fall Quarter, 1946. Chicago, 1946.
[272] S. Straszewicz, Beiträge zur Theorie der konvexen Punktmengen. Inaugural Dissertation, Meier, Zürich, 1914.
[273] S. Straszewicz, Über exponierte Punkte abgeschlossener Punktmengen. Fund. Math. 24 (1935), 139–143.
[274] C. H. Sung, B. S. Tam, On the cone of a finite dimensional compact convex set at a point. Linear Algebra Appl. 90 (1987), 47–55.
[275] E. Swift, On the condition that a point transformation of the plane be a projective transformation. Bull. Amer. Math. Soc. 10 (1904), 247–254.
[276] B. S. Tam, A note on polyhedral cones. J. Austral. Math. Soc. Ser. A 22 (1976), 456–461.
[277] T. Tamura, Remarks on the convexity of connected sets. J. Gakugei Toku-

shima Univ. 3 (1953), 24–27.
[278] H. Tietze, Über Konvexheit im kleinen und im grossen und über gewisse den Punkten einer Menge zugeordnete Dimensionszahlen. Math. Z. 28 (1928), 697–707.
[279] F. A. Toranzos, Radial functions of convex and star-shaped bodies. Amer. Math. Monthly 74 (1967), 278–280.
[280] J. S. Treiman, Characterization of Clarke's tangent and normal cones in finite and infinite dimensions. Nonlinear Anal. 7 (1983), 771–783.
[281] J. W. Tukey, Some notes on the separation of convex sets. Portugal. Math. 3 (1942), 95–102.
[282] H. Tverberg, A separation property of plane convex sets. Math. Scand. 45 (1979), 255–260.
[283] M. Ujvári, On a closedness theorem. Pure Math. Appl. 15 (2004), 469–486.
[284] M. Ujvári, On closedness conditions, strong separation, and convex duality. Acta Cybernet. 21 (2013), 273–285.
[285] F. A. Valentine, Convex sets. McGraw-Hill, New York, 1964.
[286] O. Veblen, J. H. C. Whitehead, The foundations of differential geometry. Cambridge University Press, Cambridge, 1932.
[287] Z. Waksman, M. Epelman, On point classification in convex sets. Math. Scand. 38 (1976), 83–96.
[288] J. L. Walsh, On the transformation of convex point sets. Ann. Math. 22 (1921), 262–266.
[289] J. Warren, Barycentric coordinates for convex polytopes. Adv. Comput. Math. 6 (1996), 97–108.
[290] D. Watson, A refinement of theorems of Kirchberger and Carathéodory. J. Austral. Math. Soc. 15 (1973), 190–192.
[291] R. J. Webster, Convexity. Oxford University Press, Oxford, 1994.
[292] F. Wehrung, M. V. Semenova, Sublattices of lattices of convex subsets of vector spaces. Algebra Logic 43 (2004), 145–161.
[293] H. Weyl, Elementare Theorie der konvexen Polyeder. Comment. Math. Helv. 7 (1935), 290–306.
[294] D. Wolfe, Metric inequalities and convexity. Proc. Amer. Math. Soc. 40 (1973), 559–562.
[295] S. E. Write, On the dimension of a face exposed by proper separation of convex polyhedra. Discrete Comp. Geom. 43 (2010), 467–476.
[296] K. D. Yaksubaev, Construction of a linear closed prototype of convex sets of type F_σ. Dokl. Akad. Nauk UzSSR No. 12 (1983), 4–6.
[297] T. Zamfirescu, Minkowski's theorem for arbitrary convex sets. European J. Combin. 29 (2008), 1956–1958.
[298] L. Zhou, A simple proof of the Shapley-Folkman theorem. Econom. Theory 3 (1993), 371–372.

Author Index

Subject Index

About the Technical Reviewer

Phil Nickinson spent 11 years at a daily newspaper as a sports clerk, page designer, news copy editor and news editor before becoming a full-time online editor with Smartphone Experts in December 2009. He cut his teeth on Windows Mobile and currently is editor of AndroidCentral.com, host of its weekly podcast and has more Android devices than he knows what to do with. Phil lives in Florida with his wife and two daughters.

Acknowledgments

A book like this takes many people to successfully complete. We would like to thank Apress for believing in us and our unique style of writing.

We would like to thank our Editors, Steve, Jim and Laurin, and the entire editorial team at Apress.

We would like to thank our families for their patience and support in allowing us to pursue projects such as this one.

A special thank you to the good folks at Verizon Wireless for lending us DROIDS to complete this book.

Portions of this book contain pictures that are modifications based on work created and shared by Google, and used according to terms described in the Creative Commons 3.0 Attribution License.

Part I

Quick Start Guide

In your hands is one of the most exciting devices to hit the market in quite some time: a new DROID smartphone. This Quick Start Guide will help get you and your new DROID up and running in a hurry. You'll learn all about the buttons, switches, and ports, and how to use the innovative and responsive touch screen and multitask. Our App Reference Tables introduce you to both the built-in apps and some valuable additions from the Android Market—and serve as a quick way to find out how to accomplish a task.

Getting Around Quickly

This Quick Start Guide is meant to be just that – a section that can help you jump right in and find information in this book, as well as learn the basics of how to get around and enjoy your DROID right away.

We'll start with the nuts and bolts in our "Learning Your Way Around" section, which covers what all the keys, buttons, switches, and symbols mean and do on your DROID. In this section, you'll see some handy features such as multitasking and adding and removing widgets. You'll also learn how to interact with the menus, submenus, and set switches – tasks that are required in almost every application on your DROID. You'll also find out how to read your connectivity status and what to do when you travel on an airplane.

> **TIP:** Check out Chapter 2: "Typing, Voice, Copy, and Search" for great typing tips and other helpful things.

In the "Touch Screen Basics" section, we will help you learn how to touch, swipe, flick, zoom, and more.

Later, in the "App Reference Tables," section, we've organized the app icons into general categories, so you can quickly browse through the icons and jump to a section in the book to learn more about the app a particular icon represents. This guide also includes several handy tables designed to help you get up and running with your DROID quickly:

- Getting Started (Table 2)
- Stay Organized (Table 3)
- Be Entertained (Table 4)
- Stay Informed (Table 5)
- Network Socially (Table 6)
- Be Productive (Table 7)

So let's get started!

Learning Your Way Around

To help you get comfortable with your DROID, we start with the basics – what the buttons, keys, and switches do – and then move into how you start apps and navigate the menus. Probably the most important status indicator on your DROID, besides the battery, is the one that shows network status in the upper-right corner. Understanding what these status icons mean is crucial to getting the most out of your DROID.

Keys, Buttons, and Switches

Figures 1 and 2 show all the things you can do with the buttons, keys, switches, and ports on your DROID, DROID 2, DROID 2 Global, and DROID X. Go ahead and try out a few things to see what happens. Tap the **Search** button (the **Magnifying Glass** icon) on the bottom of your phone, then tap the **Microphone** icon to try out **Voice Actions**. Next, swipe left or right to check out more **Home** screens, and then double-click your **Home** button (the **House** icon) to bring up **Voice Commands**. Long press (press and hold) the **Home** button to multitask. Have some fun getting acquainted with your device.

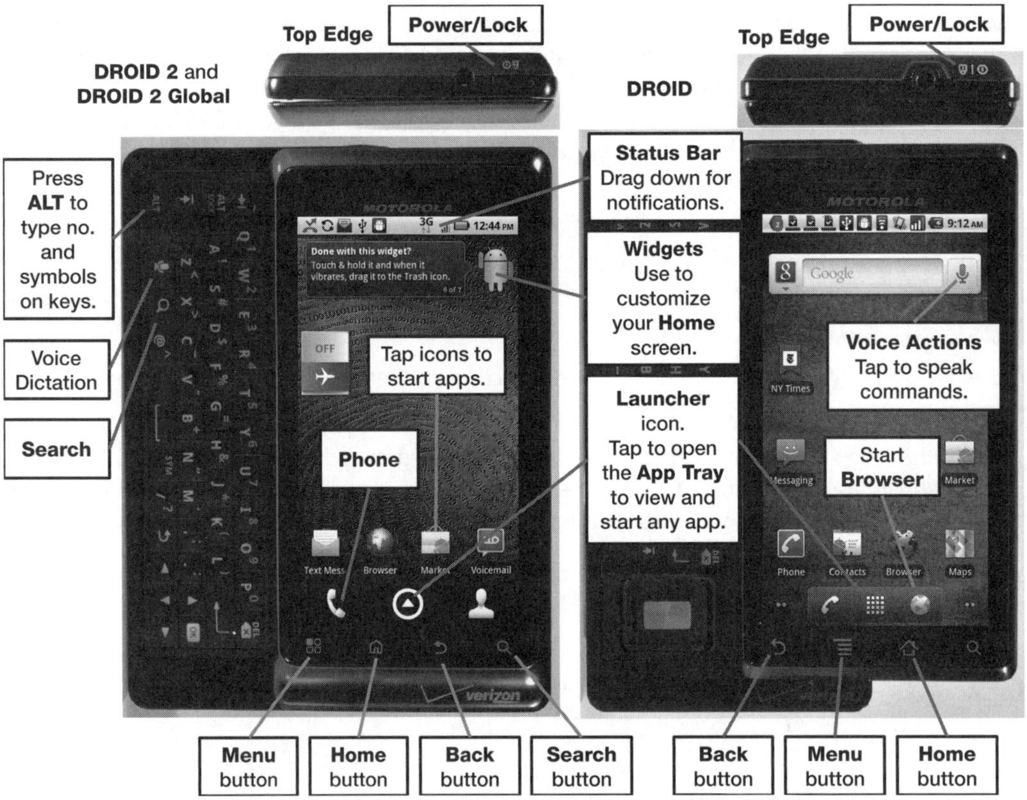

Figure 1. *The buttons, ports, and keys on the DROID, DROID 2, and DROID 2 Global.*

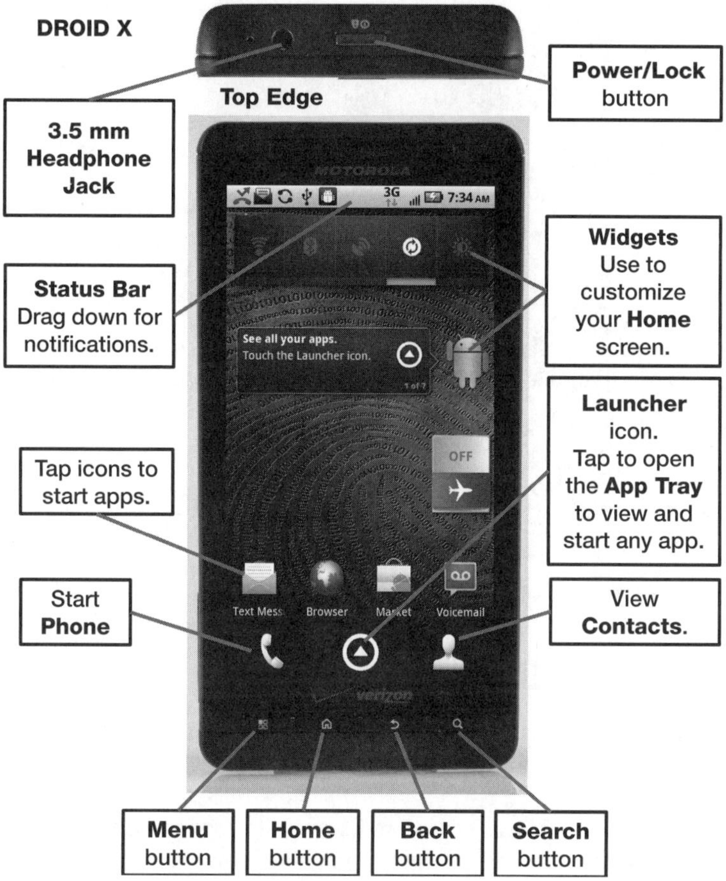

Figure 2. *The buttons, ports, and keys on the DROID X.*

The Launcher Icon

You may wonder where all your icons are kept. Swiping left or right won't show you them. To see them all, you need to tap the **Launcher** icon at the bottom of your main **Home** screen. Tap the **Launcher** icon to see all your icons in the **App Tray**. There are usually more icons than are visible on a single screen; you need to slide your finger up or down to see all the icons (see Figure 3).

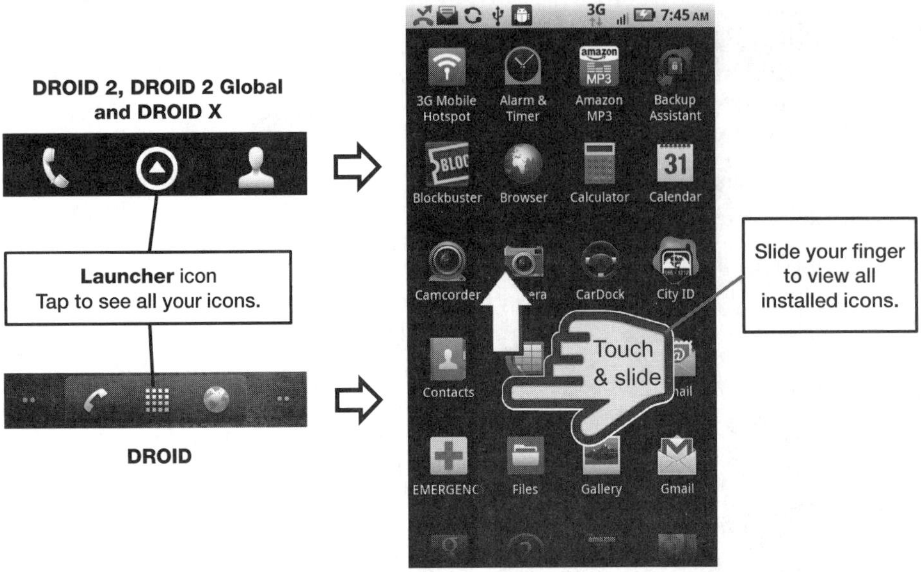

Figure 3. *Use the **Launcher** icon to see all your application icons in the **App Tray**.*

The Four Buttons Along the Bottom

In addition to tapping or the touching the screen, you can use the **Menu**, **Home**, **Back** and **Search** buttons to help you navigate around your DROID.

Menu button

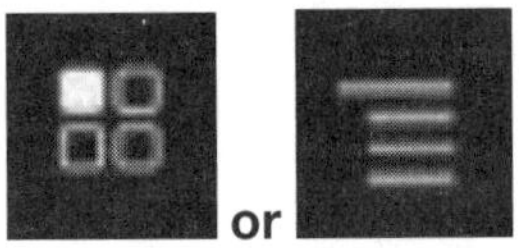

Press the **Menu** button to see a menu for the currently open app or the **Home** screen. Press the button again to hide the menu. Inside any app, long press the **Menu** button to see the virtual keyboard.

TIP: Press the **Menu** button, then press and hold it to see shortcuts to the various menu commands appear.

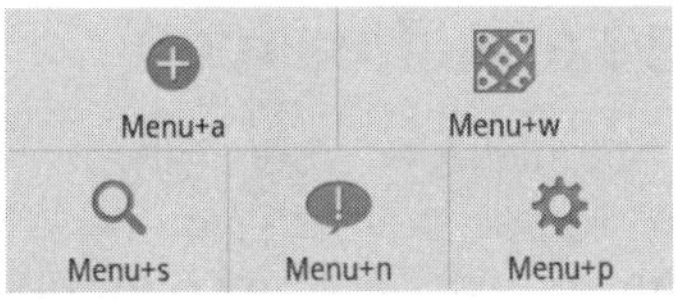

Home button

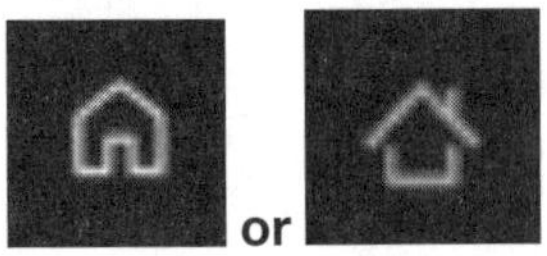

Press the **Home** button once to jump right to your main (center) **Home** screen.

Double-press the **Home** button to start **Voice Command**. This command allows you to speak commands to your DROID. (You can learn more about **Voice Command** in Chapter 2)

Long press to multitask and see the eight most recent apps you have opened.

Back button

Press the **Back** button to back out of any screen to the previous screen. Continue pressing to exit an app back to the **Home** screen.

Search button

Tap the **Search** button to bring up the Google **Search** screen.

Long press to bring up the **Voice Actions** screen, where you can speak commands and perform voice Google searches. (You can learn more about **Voice Actions** in Chapter 2.)

Short Tap and Long Press

As you just saw, the buttons do different things when you press them quickly or press and hold them. The same thing works on the touch screen.

Short Tap

Icons: A short tap of any icon will start the app.

Items inside apps: Tapping items inside apps such as calendar events, contacts, or picture thumbnails will usually expand the selected item to view more details or a larger version of the item.

Switches: A short tap of a switch will set it to **Off** or **On**.

Widgets: A short tap of a widget will do an action in the widget such as advance a screen, open the widget, or open an associated app. For example, touching a **Tips and Tricks** widget will open the tips so you can view them in more detail.

Long Press	**Icons or Widgets**: Long press an icon or widget to move it around the screen, between **Home** screens, or delete it from the **Home** screen. (Don't worry: you can get the icon back on your **Home** screen after you press the **Launcher** icon.) **Items inside apps**: Pressing and holding items inside apps such as calendar events, contacts, or pictures will usually give you a context-sensitive menu. For example, long pressing a contact will give the option to view, call, send a text message to, share an email with, or delete the contact.

Adding, Removing, and Moving Widgets and Icons

You will quickly find that your DROID is highly customizable, starting with your **Home** screen. The DROID 2 and DROID X have seven **Home** screens, and the DROID has five.

> **TIP:** We explain more details about how to use widgets and move icons in Chapter 6: "Organize Your Home Screen: Icons and Widgets."

To add a widget, shortcut, folder, or change the wallpaper, you long press anywhere you see a blank spot on a **Home** screen. This brings up a menu that lets you add any number of items including Motorola widgets, Android widgets, shortcuts, folders or even change your wallpaper (see the image to the right).

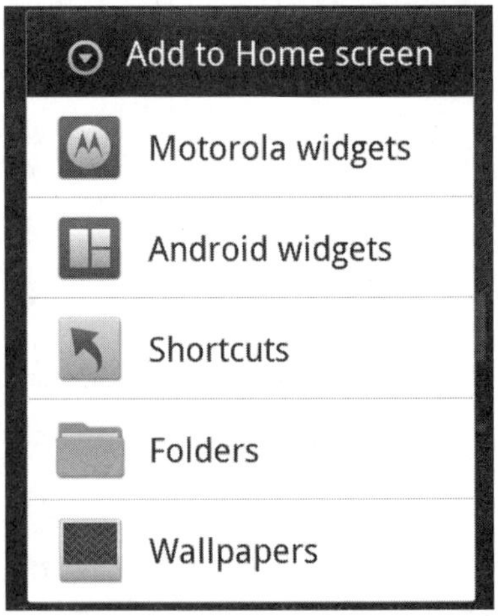

To remove a widget or other item, you long press until it gets highlighted and drag it down to the **Trashcan** icon at the bottom of the screen.

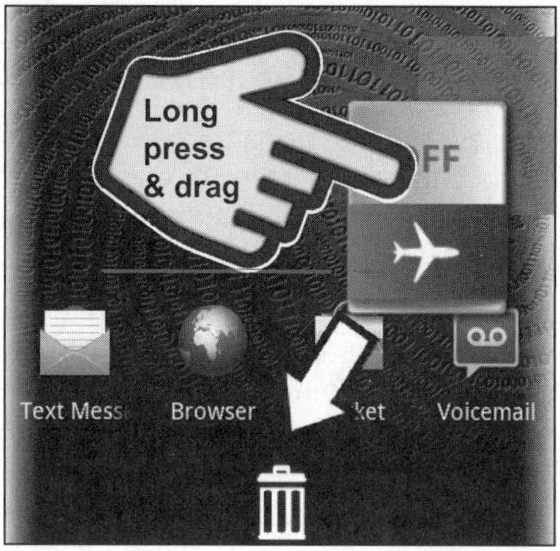

To move a widget, icon, or other **Home** screen item, long press and drag that item around the screen. To move it to a different **Home** screen, drag your finger to the very edge of the screen.

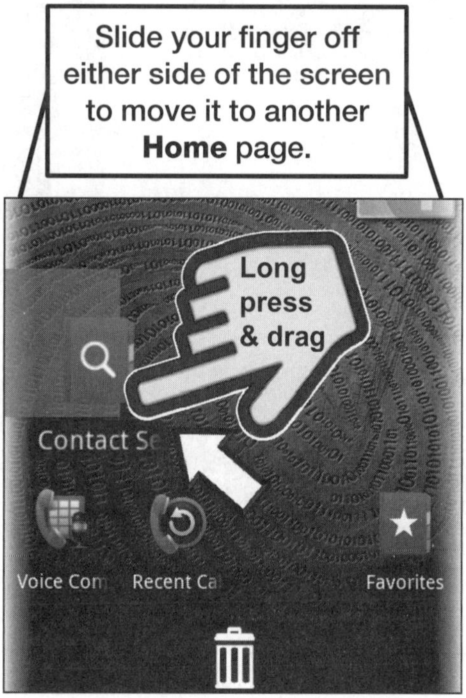

Switching Apps (aka Multitasking)

If you are like most smartphone users these days, you definitely want to be doing more than one thing at a time. For example, you might like listening to free **Pandora** Internet radio while reading and replying to your email or text messages. The DROID is built to multitask (see Figure 4).

Long press the **Home** button to bring up the recent apps in the middle of the screen, and then tap the icon of any app you want to start. If you don't see the icon you want, then press the **Home** button again to see the entire **Home** screen. Repeat these steps to jump back to the app you just left. The nice thing is that the app you just left is always shown as the first app in the top-left position in the list of recent apps.

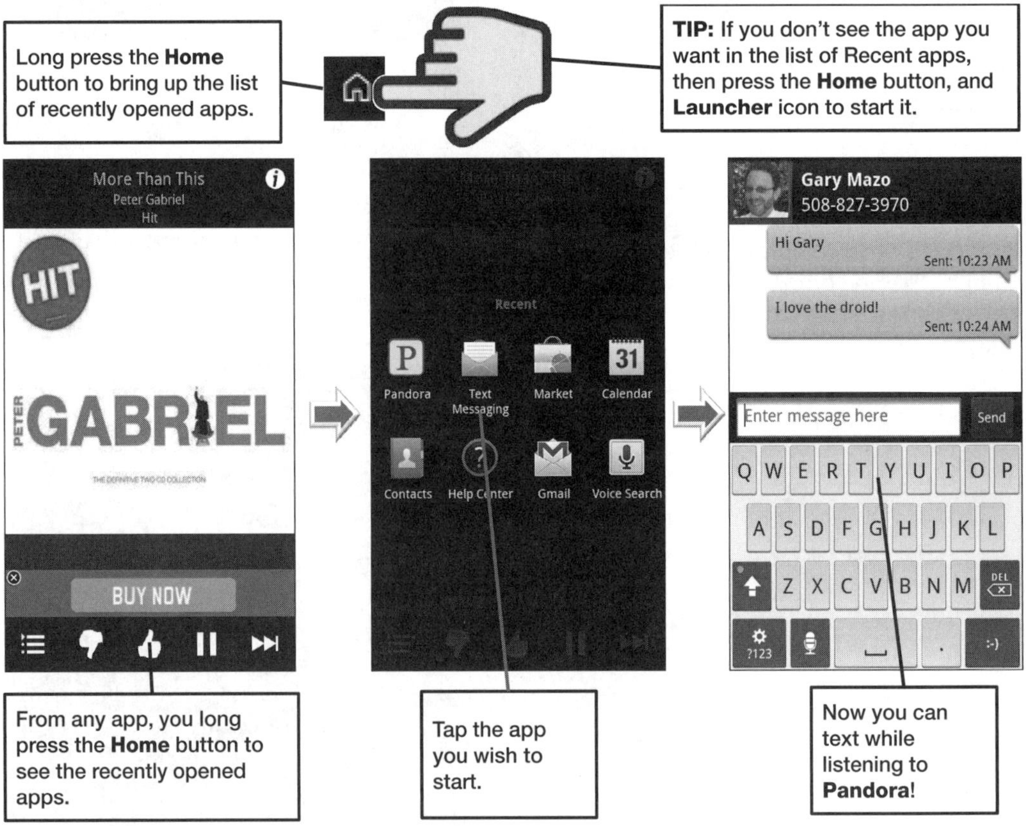
Long press the Home button to bring up the list of recently opened apps.
TIP: If you don't see the app you want in the list of Recent apps, then press the Home button, and Launcher icon to start it.
More Than This
Peter Gabriel
Hit
HIT
PETER GABRIEL
BUY NOW
Recent
Pandora
Text Messaging
Market
Calendar
Contacts
Help Center
Gmail
Voice Search
Gary Mazo
508-827-3970
Hi Gary
Sent: 10:23 AM
I love the droid!
Sent: 10:24 AM
Enter message here
Send
From any app, you long press the Home button to see the recently opened apps.
Tap the app you wish to start.
Now you can text while listening to Pandora!

Figure 4. *Multitasking or switching apps by long pressing the **Home** button.*

Starting and Exiting an App

To start any app, you simply tap the app's icon with your finger.

To close the app and exit back to the **Home** screen, press the **Back** button.

If you press the **Home** button, you can leave the app running in the background and start another app.

Menus, Submenus, and Checkboxes

Once you are in an app, you can select any menu item by simply touching it. Using the **Settings** app as an example, tap **Wireless & networks**, and then tap **Airplane mode** to set the checkbox (with a green check mark) and turn on **Airplane mode** (see Figure 5).

Submenus are any menus below the main menu.

NOTE: Switches set to **On** are green, while switches set to **Off** are gray. You can back up to the previous screen or menu by pressing the **Back** button on the bottom of your DROID. For example, if you're in the **Wireless & networks** menu, you can press the **Back** button to return to the main the **Settings** menu.

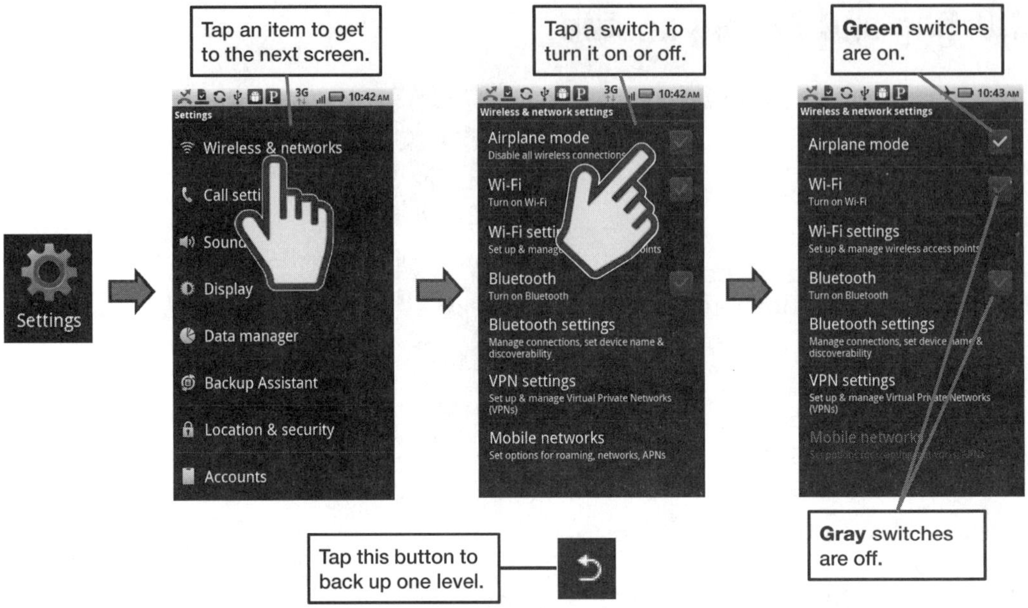

Figure 5. *Selecting menu items, navigating submenus, and setting switches.*

Reading the Top Status Bar

Most of the functions on your DROID work only when you are connected to the Internet (e.g., email, your browser, the wireless sync to **Google Contacts** and **Google Calendar**, Android Market, and so on), so you need to know when you're connected. The top status bar also has many other status icons that are helpful to know. Understanding how to read the status bar can save you time and frustration.

There are a wide variety of status icons on the top bar, and you can see various status examples in Figure 6.

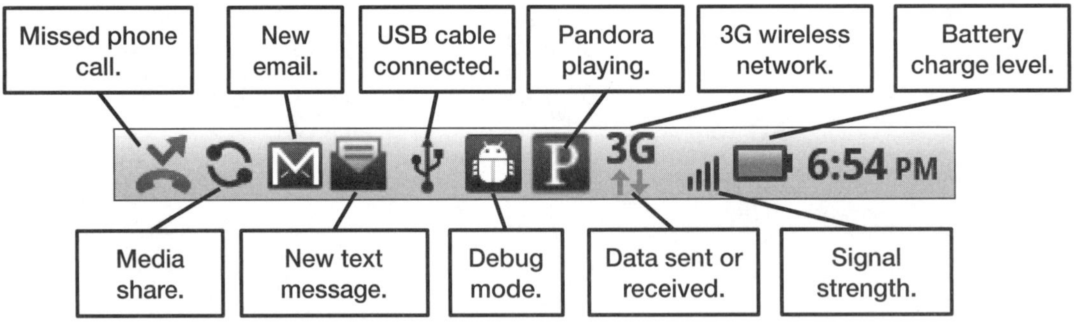

Figure 6. *Reading your top status bar icons.*

TIP: If you touch and hold the status bar, then you can see today's date.

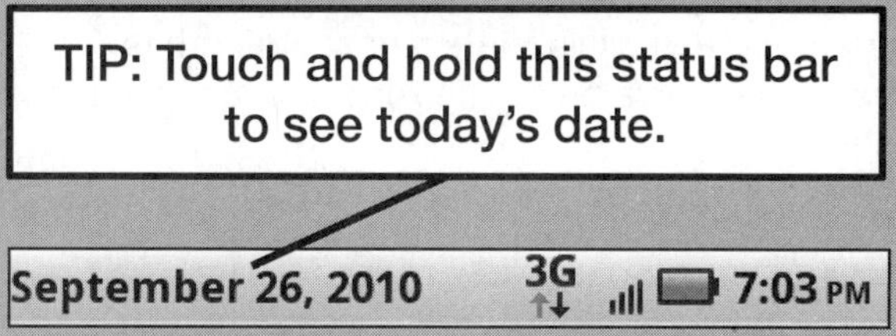

Seeing More Detailed Status Messages

Touch the very top of the device (above the screen), then swipe your finger down onto the screen to see detailed status messages. Press the **Back** button to hide the detailed status messages.

Notice that there are two areas of status messages: **Ongoing** and **Notifications**.

Ongoing Status Items

In the example to the right, the **Ongoing** status section shows that we are playing Pandora internet radio, we have USB debugging connected, and we have a USB connection to our computer.

Notification Items

The **Notification** section in the image to the right shows that have 152 new email messages. We have a possible connection for **Media Share**, and we've missed three calls.

TIP: Tap any item to learn more about it. For example, tapping the three missed calls will show you each caller, so you can immediately call them back if you wish.

Understanding the Data Connectivity Symbols

You can read the strength of your data connection and see when data is being transferred by looking at the top status bar. The cellular data signal strength is represented by an icon that shows from one to five bars. The **Airplane** icon indicates that you are in **Airplane mode**.

Strong   Weak   Radio Off – Airplane Mode

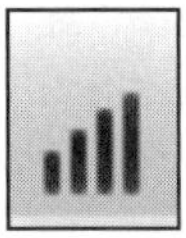

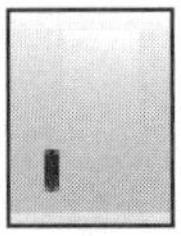

The two arrows under the 3G show when data is being sent and received.

| Data sent. | Data received. |

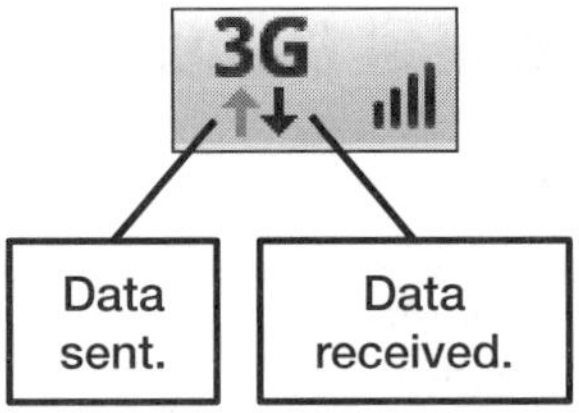

You can tell whether you are connected to a network, as well as the general speed of the connection, by looking at the right side of your DROID's top status bar. Table 1 shows typical examples of what you might see on this status bar.

Table 1. *How to Tell When You Are Connected.*

In the upper-right corner, if you see letters and symbols...	Cellular Network (Phone, Web, Text, Email, Sync, and Skype)	Wi-Fi Network (Web, Email, Sync, and Skype)
3G ↑↓ ⨪ or 🛜	✓ (Connected)	✓ (Connected)
🛜 ✈	✗ (Not connected)	✓ (Connected)
✈	✗ (Not connected)	✗ (Not connected)

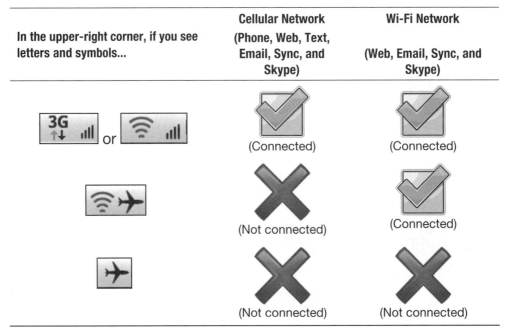

Chapter 5: "Wi-Fi and 3G Connectivity" shows you how to connect your DROID to a Wi-Fi or 3G Cellular Data Network.

Flying on an Airplane – Airplane Mode

When you fly on an airplane, the flight crew will ask you to turn off all portable electronic devices for takeoff and landing. Then, when you get to altitude, they will say "all approved electronic devices" can be set back to **On**.

TIP: Check out the "International Travel" section of Chapter 5 for many money saving tips you can take advantage of when you travel overseas with your DROID.

If you need to turn off your DROID, long press the **Power** button on the top-right edge, and then tap **Power off**.

Follow these steps to enable **Airplane Mode**:

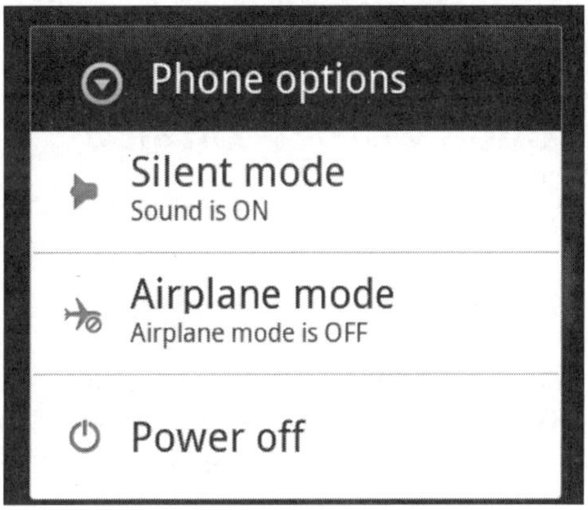

1. Long press the **Power** button on the top edge of your DROID.
2. Tap **Airplane mode** from the pop-up menu.
3. Notice that the **Airplane mode** icon in the top status bar replaces the 3G and signal bars. Your phone, web, wireless sync, and any applications that require an Internet connect will not work in **Airplane mode**.

TIP: Some airlines do have in-flight Wi-Fi networks. On those flights, you may want to turn your Wi-Fi back **On** at the appropriate time.

You can turn your Wi-Fi connection **Off** or **On** by following these steps:

1. Tap the **Settings** icon.
2. Tap **Wireless & networks**.

3. To enable the Wi-Fi connections, set the switch next to **Wi-Fi** to **On** (Green = **On**, Gray = **Off**).
4. To disable the Wi-Fi, set the same switch to **Off** (gray).
5. Tap **Wi-Fi settings** to select the Wi-Fi network and follow the steps the flight attendant provides to connect to the in-flight Wi-Fi.

NOTE: On the DROID 2, DROID 2 Global, and DROID X, you have an **Airplane mode** widget you can add to your **Home** screen.

Touch Screen Basics

In this section, we will describe how to interact with the DROID's touch screen.

Touch Screen Gestures

The DROID has an amazingly sensitive and intuitive touch screen. If you own a DROID, DROID 2, or DROID 2 Global, you also can slide out the physical keyboard to type. We show you more typing tips in Chapter 2.

Typing on the touch-screen keyboard will take a little effort to master. With a little practice, though, you'll soon become comfortable interacting with your DROID.

You can do almost anything on your DROID by using a combination of the following:

- Touch screen "gestures"
- Touching icons, widgets, or soft keys on the screen
- Tapping the **Menu, Home, Back,** or **Search** buttons at the bottom of your device (earlier in the chapter, we explained what each button does)

The following sections describe the various gestures you can use on an DROID.

Tapping and Swiping or Flicking

To start an app, confirm a selection, select a menu item, or select an answer, simply tap the screen. To move quickly through **Home** screens, contacts, lists, and the music library in **List** mode, flick from side-to-side or up and down. Figure 7 shows both of these gestures.

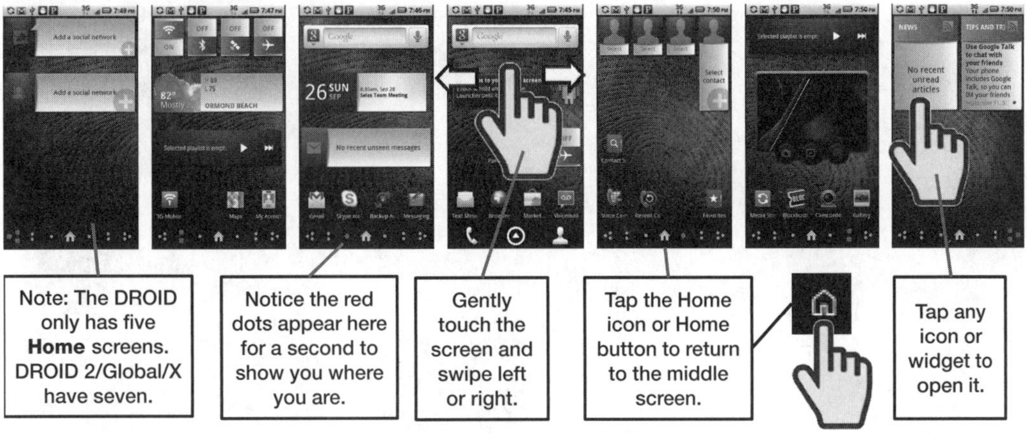

Note: The DROID only has five Home screens. DROID 2/Global/X have seven.
Notice the red dots appear here for a second to show you where you are.
Gently touch the screen and swipe left or right.
Tap the Home icon or Home button to return to the middle screen.
Tap any icon or widget to open it.

Figure 7. *Swipe left or right to see all your* ***Home*** *screens.*

Swiping

To swipe, gently touch and move your finger as shown in Figure 8 to move between pictures. Swiping up and down also works in lists, such as the **Contacts** list.

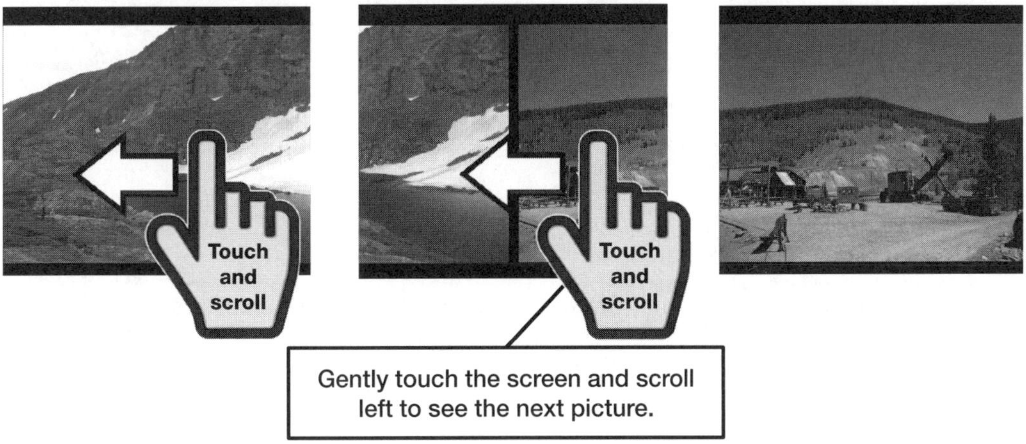

Touch and scroll
Touch and scroll
Gently touch the screen and scroll left to see the next picture.

Figure 8. *Touch and swipe to move between pictures and up and down lists.*

Scrolling

Scrolling is as simple as touching the screen and sliding your finger in the direction you want to scroll (see Figure 9). You can use this technique in messages (email), the **Browser** app, menus, and more.

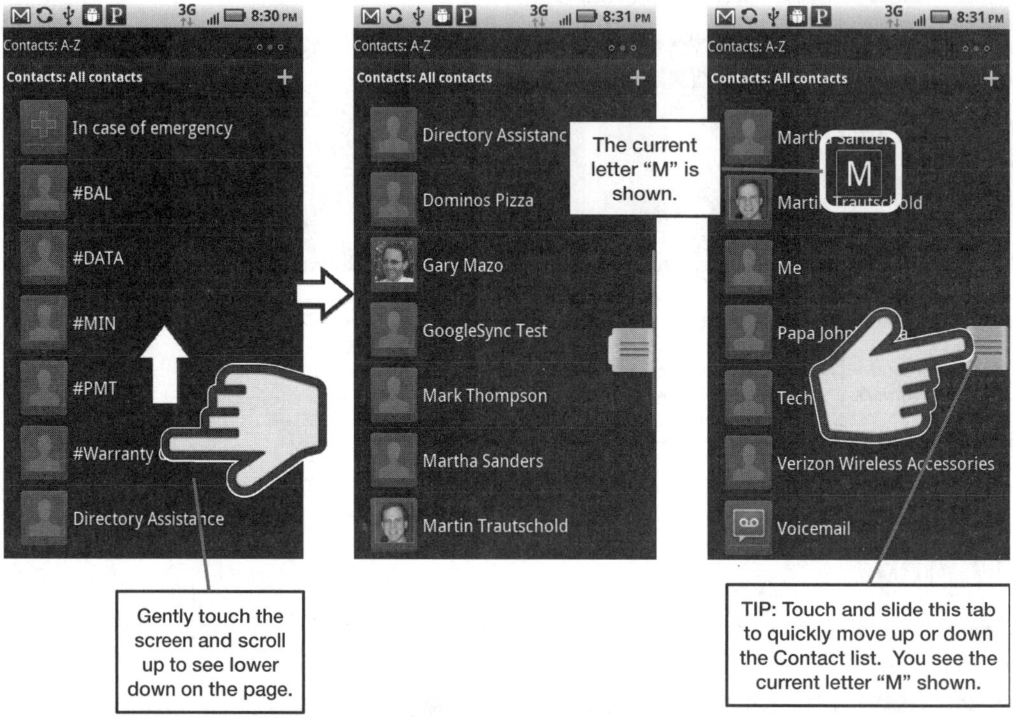

Figure 9. *Touch and slide your finger to scroll around contacts, a web page, a zoomed picture, and more.*

Double-Tapping

You can double-tap the screen to zoom in and then double-tap again to zoom back out. This works in many places, such as web pages, mail messages, and pictures (see Figure 10).

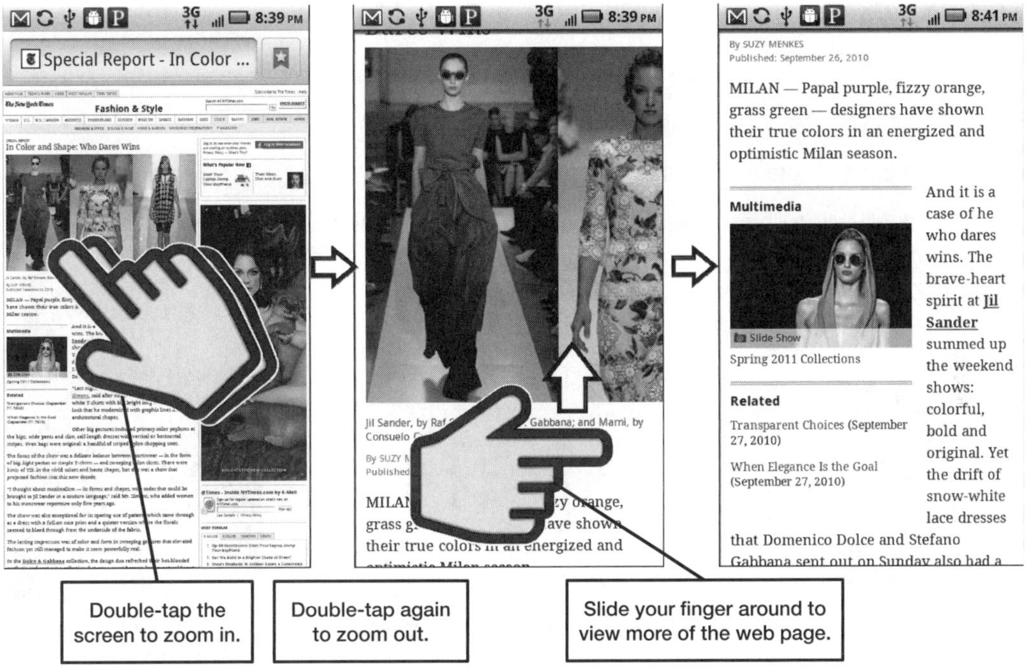

Figure 10. *Double-tapping to zoom in or out.*

Pinching

You can also pinch open or closed to zoom in or out. This works in many places, including web pages, mail messages, and pictures (see Figure 11). Follow these steps to zoom in using the *pinching* feature:

1. To zoom in, place two fingers that touch each other on the screen.
2. Gradually slide your fingers open. The screen zooms in.

Follow these steps to zoom out using the pinching feature.

1. To zoom out, place two fingers with space between them on the screen.
2. Gradually slide your fingers closed, so they touch. The screen zooms out.

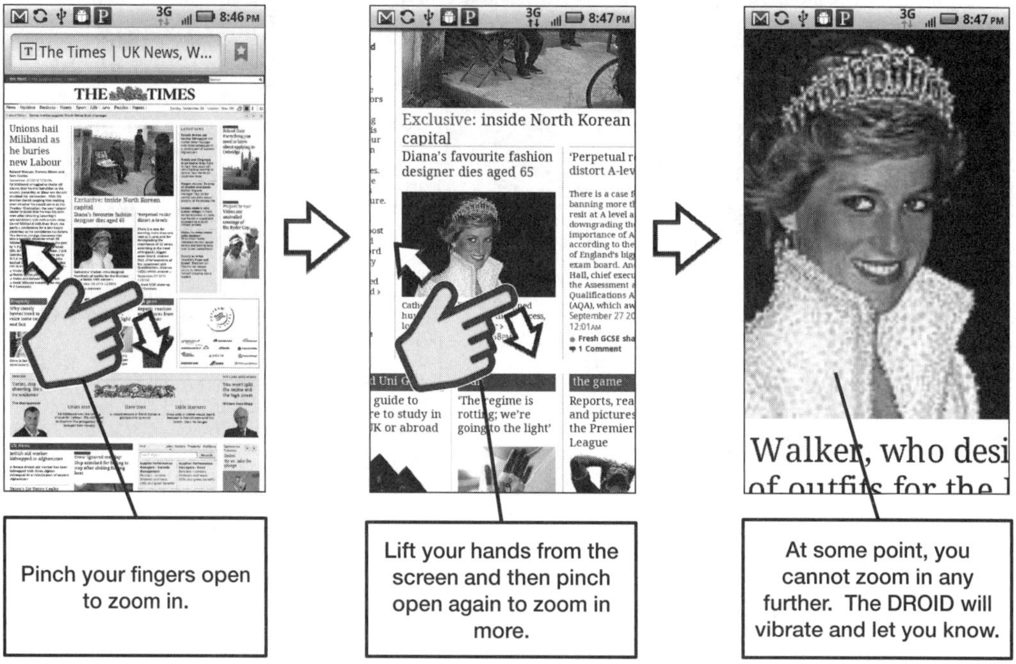

Pinch your fingers open to zoom in.
Lift your hands from the screen and then pinch open again to zoom in more.
At some point, you cannot zoom in any further. The DROID will vibrate and let you know.

App Reference Tables

This section gives you a number of handy reference tables that group together the various apps that are pre-installed on your DROID by their functionality. Also included in the tables are other useful apps you can download from the Android Marketplace. Each table gives you a brief description of the app and tells you where you can find more information about it in this book.

Getting Started

Table 2 provides some quick links to help you connect your DROID to the Web (using Wi-Fi or 3G); buy and enjoy songs or videos; lock your DROID or power it off; unlock your DROID; and more.

Table 2. *Getting Started.*

To Do This...	Use This...	Where to Learn More
Turn the DROID on or off.	 The **Power/Lock** button: Press and hold this key located on the DROID's top edge.	Getting Started – Ch. 1
See all icons on your DROID in the **App Tray**.	  **Launcher**	Getting Started – Ch. 1
Adjust settings and connect to the Internet (via Wi-Fi or 3G).	 **Settings > Wi-Fi** or **Settings > General > Network**	Wi-Fi and 3G – Ch. 5
Return to the **Home** screen.	 or  The **Home** button	Getting Started – Ch. 1
Unlock the DROID.	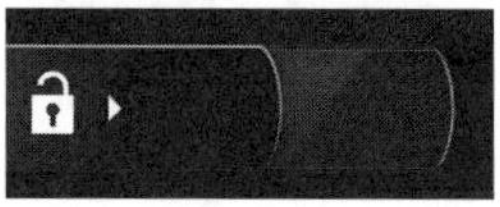 Slide your finger to unlock your DROID.	Getting Started – Ch. 1
Completely power down your DROID.	Press and hold the **Power/Lock** button. Tap this button to power the device off. 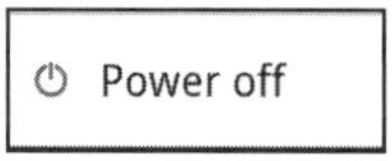	Getting Started – Ch. 1
Sync addresses, calendar, email, and notes with your main account.	**Google Sync**	Google Sync – Ch. 3 Other Sync Methods – Ch. 4
Share and sync Share and sync music, videos, and pictures.	**Picasa**, **Facebook**, **MySpace**, **Photobucket**  **DoubleTwist**	Photos – Ch. 18 Media Sync – Ch. 25

Stay Connected and Organized

Table 3 provides links for everything from organizing and finding your contacts to managing your calendar, working with email, sending messages, getting driving directions, calling people, and more.

Table 3. *Staying Connected and Organized.*

To Do This…	Use This…	Where to Learn More
Manage your contact names and numbers.	**Contacts**	Contacts – Ch. 12
Manage your calendar.	**Calendar**	Calendar – Ch. 13
Surf the Web	**Browser**	Browser – Ch. 11
Call your friends.	**Phone**	Phone – Ch. 7
Call and chat with friends.	**Skype mobile**	Social Networking and Skype – Ch. 20
Control your iPod and DROID with your voice. (Press and hold the **Home** button.)	**Voice Command**	Typing, Voice, Copy, and Search – Ch. 2
Search Google and control your phone by simply speaking.	**Voice Search Voice Action**	Typing, Voice, Copy, and Search – Ch. 2

To Do This...	Use This...	Where to Learn More
Send text, picture, and video messages.	**Text Messaging** **Messages**	SMS and MMS – Ch. 10
View and send email.	**Gmail** **Email**	Email – Ch. 9
Find just about anything, get directions, avoid traffic, and more.	**Maps**	Maps – Ch. 19

Entertaining Yourself

You can have lots of fun with your DROID; Table 4 shows you a few ways to get started. For example, you can use your DROID to view movies and TV shows, as well as to check out free Internet radio with **Pandora.** If you already use a Kindle, you can sync all your Kindle books to your DROID and enjoy them right away. You can also choose from thousands of free and paid apps from the **Marketplace** to make your DROID even more amazing, fun, and useful. Finally, y ou can buy new music on your DROID from the **Amazon MP3** app.

Table 4. *Being Entertained.*

To Do This...	Use This...	Where to Learn More
Buy music.	**Amazon MP3**	Music – Ch. 14
Rent a video.	**Blockbuster** **Netflix Mobile Queue**	Videos, TV, and More – Ch. 15
Use your computer to sync, buy music apps, and listen to music and other content.	**DoubleTwist**	Media Sync – Ch. 25

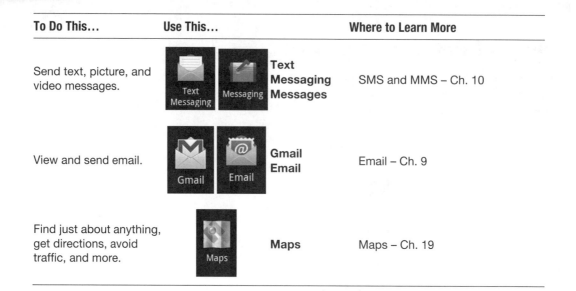

To Do This...	Use This...		Where to Learn More
Browse and download apps right to your DROID.		**Market**	DROID App Guide – Appendix A
See playlists, artists,songs, albums, audiobooks, and more.		**Music**	Music – Ch. 14
Listen to free Internet radio.		**Pandora**	Music – Ch. 14
Read your Kindle books.		**Kindle**	New Media and E-Books – Ch. 16
Look at, zoom in on, and organize your pictures.		**Gallery**	Photos and Videos – Ch. 18
Take and shere pictures.		**Camera**	Photos and Videos – Ch. 18
Take and share videos.		**Camcorder**	Photos and Videos – Ch. 18
Watch a video from YouTube.		**YouTube**	Videos, TV, and More – Ch. 15
Play a game.		**Games** **Need For Speed**	Fun and Games – Ch. 22

Staying Informed

You can also use your DROID to read your favorite magazine or newspaper with up-to-the-minute vibrant pictures and videos (see Table 5). Or, you can use it to check out the latest weather reports.

Table 5. *Staying Informed.*

To Do This...	Use This...	Where to Learn More
Check your favorite radio news program.	 **NPR News**	Market – Ch. 17
Read the newspaper.	 **New York Times**	Market – Ch. 17
Check the weather.	 **The Weather Channel**	Market – Ch. 17
Check out the latest headlines.	 **AP Mobile**	Market – Ch. 17

Networking Socially

You can also use your DROID to connect to and stay up-to-date with friends, colleagues, and professional networks using the social networking tools on your DROID (see Table 6).

Table 6. *Networking Socially.*

To Do This...	Use This...	Where to Learn More
Skype.	**Skype**	Social Networking and Skype – Ch. 20
Network on **LinkedIn**.	**LinkedIn**	Social Networking – Ch. 20
Stay connected with friends on **Facebook**.	**Facebook**	Social Networking – Ch. 20
Follow your favorites on **Twitter**.	**Twitter**	Social Networking – Ch. 20

Being Productive

A DROID can also help you be more productive. You can use it to access and read just about any PDF file or other document with the **GoodReader** app. You can also take notes with the basic **Notes** app or step up to the advanced **Evernote** app, which has amazing capabilities for integrating audio, pictures, and text notes, as well as the ability to sync everything to a web site. You can also use your DROID to set an alarm, calculate a tip, see what direction you are walking in, and record a voice memo (see Table 7).

Table 7. *Being Productive.*

To Do This...	Use This...	Where to Learn More
Take and organize your notes in a whole new way.	**Evernote**	Notes and Documents– Ch. 21
Move files between your DROID and your computer	**Dropbox**	Notes and Documents – Ch. 21
Use folders to organize your icons.	**Folders**	Icons and Widgets – Ch. 6
Set an alarm, countdoun timer, and more.	**Alarm & Timer**	Utilities – Ch. 23
Calculate a tip or find the cosine of 30 degrees.	**Calculator**	Utilities – Ch. 23
Open and edit Microsoft Office documents.	**Quickoffice**	Notes and Documents – Ch. 21

Part II

Introduction

Welcome to your new DROID—and to the book that tells you what you need to know to get the most out of it. In this part we show you how the book is organized and where to go to find what you need. We even show you how to get some great tips and tricks sent right to your DROID via short email messages.

Introduction

Congratulations on Your New DROID!

You hold in your hands perhaps the most powerful smartphones available today, a phone that is also a media player, e-book reader, gaming machine, life organizer, and just about everything else available today: the DROID.

The DROID can do just about any other smartphone on the market. In a beautiful package, the DROID will have you placing phone calls, listening to music, playing games, surfing the web, checking email, and organizing your busy life in no time.

NOTE: Take a look at Chapter 17: "Exploring the Android Market,"and Appendix A: "Droid App Guide" where we show you how to get all the greatest apps to boost the performance and fun of your new DROID!

With your DROID, you can view your photos and interact with them using intuitive touch-screen gestures. You can pinch, zoom, rotate, and email your photos—all by using simple gestures.

Interact with your content like never before. News sites and web sites look amazing due to the incredibly clear and crisp large touch screen display. Flip through stories, videos, and pictures, and interact with your news.

Manage your media library with ease. The **Music** app features an intuitive interface, letting you choose music, watch videos, organize playlists, and more—all in an effortless and fun way on the DROID's high definition–quality screen.

Update your **Facebook** status and receive push alerts—all on your DROID.

Stay connected to the web and your email with the 3G wireless and built-in Wi-Fi connection of the DROID. All the latest high-speed protocols are supported, so you can always be in touch and get the latest content. You can even turn your DROID into a Wi-Fi Hotspot to connect your laptop and up to four other devices to the internet just about anywhere you get a signal. The DROID and DROID 2 also include a slide-out keyboard to type out emails and notes when you use the device in Landscape mode.

Getting the Most out of *Droids Made Simple*

Read this book cover-to-cover if you choose, but you can also peruse it in a modular fashion, by chapter or topic. Maybe you just want to check out the Android **Market** app, try the **Kindle** app, set up your email or contacts, or just load up your phone with music using **doubleTwist** on your computer (see chapter 25: "DROID Media Sync"). You can do all this and much more with our book.

Be sure to check out our DROID App Guide in Appendix A at the end of the book to explore more than 80 apps that the authors have reviewed and tested to help you get the most out of your DROID.

You will soon realize that your DROID is a very powerful device. There are, however, many secrets "locked" inside, which we help you "unlock" throughout this book.

Take your time—this book can help you understand how to best use and have fun with your new DROID. Think back to when you tried to use your first Windows or Mac computer. It took a little while to get familiar with how to do things. It's the same with the DROID. This book will help you get up to speed and learn all the best tips and tricks more quickly.

Also remember that devices this powerful are not always easy to grasp—at first.

You will get the most out of your DROID if you can read a section and then try out what you just read. We all know that reading and then doing an activity gives us a much higher retention rate than simply reading alone.

So, in order to learn and remember what you learn, we recommend the following:

Read a little, try a little on your DROID, and repeat!

Referring to your DROID

In this book, we generally use the word DROID to mean DROID, DROID 2, DROID 2 Global and DROID X. Occasionally, we will specifically call out DROID 2/X, when we say that we mean this feature works on the DROID 2, DROID 2 Global and DROID X, but not the original DROID. Similarly, if we say only on the DROID or original DROID, we mean that the specified feature works on the original DROID, not the DROID 2/X.

How This Book Is Organized

Knowing how this book is organized will help you quickly locate things that are important to you. Here we show you the main organization of this book. Remember to take advantage of the abridged table of contents, detailed table of contents, and comprehensive index. All of these elements can help you quickly pinpoint items of interest to you.

Day in the Life of a DROID User

Located inside the front and back covers, the "Day in the Life of an DROID User" reference is an excellent guide to your phone's features, providing ideas on how to use your DROID and lots of easy-to-access, cross-referenced chapter numbers. So, if you see something you want to learn, simply thumb to that page and learn it—all in just a few minutes.

Part I: Quick Start Guide

Learning Your Way Around: Learn about the buttons and switches on the DROID, how to read your status bar, how to start and exit the apps, multitask, turn on **Airplane** mode and more.

Touch Screen Basics: This book's many practical and informative screen shots will help you quickly learn how to touch, swipe, flick, zoom, and more with your DROID's touch screen.

App Reference Tables: Quickly skim the icons or apps grouped by category. Get a thumbnail of what all the apps do on your DROID, including a pointer to the relevant chapter numbers so you can jump right to the details of how to get the most out of each app in this book.

Part II: Introduction

You are here now . . .

Part III: You and Your DROID

This is the meat of the book, organized in 24 easy-to-understand chapters, all of them packed with loads of pictures to guide you every step of the way.

Part IV: DROID Media Sync

In Chapter 25, learn how to use **doubleTwist** to sync your music, playlists, videos, podcasts and more to your DROID from your Windows or Mac computer. Also, buy music using the Amazon MP3 store, check out cool apps in the Android Market and locate and subscribe to podcasts all in **doubleTwist** on your computer. We also show you how to use Mass Storage mode to transfer media and documents using the USB cable. The more of your media you can load on your DROID, the more fun you will have with it.

Appendix A: DROID App Guide

Learn about over 80 apps in the following categories Microsoft Office document editing, printing, file management, virus protection, backup, security, presentation software, web conferencing, note taking and mind-mapping, to-do and task lists, expenses and finance, travel, health and medicine, law and legal, real estate, sales force automation, retail, project management, education and training, social media, information technology, and other apps.

Quickly Locating Tips, Cautions, and Notes

If you flip through this book, you can instantly see specially formatted **TIPS**, **CAUTIONS**, and **NOTES** that highlight important facts about using the DROID. For example, if you want to find all the special tips relevant to using the **Calendar**, you can flip to the Calendar chapter and search for these highlighted nuggets of information.

TIPS, **CAUTIONS**, and **NOTES** are all formatted like this, with a gray background, to help you see them more quickly.

Free DROID Email Tips

Finally, check out the author's web site at `www.madesimplelearning.com` for a series of very useful "bite-sized" chunks of DROID tips and tricks. We have taken a selection of the great tips out of this book and even added a few new ones. Click the "Free Tips" link and register for your tips in order to receive a tip right in your DROID inbox about once a week. Learning in small chunks is a great way to master your DROID!

Part III

You and Your DROID. . .

This is the heart of *Droids Made Simple*. In this section, you'll find clearly labeled chapters—each explaining the key features of your DROID. You'll see that most chapters focus on an individual app or a specific type of application. Many of the chapters discuss applications that come with your DROID, but we also include some fun and useful apps you can download from the Android Market. Sure, the DROID can help you get work done, but it's for a whole lot more, too. We finish with some handy troubleshooting tips that can help if your DROID isn't working quite right.

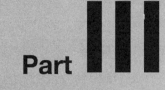

Chapter 1

Getting Started

In this chapter, we will tell you everything you need to know to get up and running on your DROID. This chapter will go into a little more detail than you saw in the Quick Start Guide. First, we will go around the outside of the DROID and describe what each of the buttons and keys on your device does.

Second, we will go inside the device and show you how to remove the battery and install a memory card. You will also learn how to charge your DROID and get the most out of your battery life with our battery life tips.

Third, we will show you how to connect your DROID to your email, social networking, and other accounts, so you can have your life in the palm of your hand and stay up-to-date with anyone at anytime and anywhere!

Fourth, we will show you how to clean your DROID and protect it, and even how to give it some "bling" with stylish cases.

Fifth, we will show you how to use a password to protect your personal and/or confidential information on your DROID.

Finally, we will end this chapter with a discussion and some tips on saving money on your DROID phone plans. There are various options you can add or remove from your phone plan that can save you a good sum of money.

TIP: Be sure to check out the Quick Start Guide if you haven't already done so. This guide explains what all the buttons on your DROID do, how to navigate around the touch screen, and where to find other important information in this book.

Getting to Know Your DROID

In this section, we will show you how to use everything you get in the box with your DROID. We will also give you some DROID battery and charging tips.

What Is Included in the Box

The DROID box may seem small, but it contains everything you need to get started (your DROID, a USB cable, and wall plug adapter) except for a good manual – which is why we wrote this book!

The USB Cable

This is the cable that connects your DROID to your computer. This cable also doubles as your power cable when you plug it into the wall plug adapter.

The Wall Plug Adapter

You will also see your wall plug adapter in the box. This adapter allows you to charge your DROID directly from a wall outlet without having your computer around. All you do is plug the USB cable into this wall adapter and the other end into your DROID.

Now let's explore some of DROID's basic features . . .

The DROID wall plug adapter.
Plug your USB cable in here.

Power and Lock Button

To power on your DROID, press and hold the **Power/Lock** button on the top-right edge of the DROID for a few seconds (see the figure to the right). Simply tapping this button quickly won't power on the DROID if it is completely off – you really need to hold it until you see the DROID power on.

Tap this button once to put your DROID in **Sleep** mode when the screen goes off; this saves your battery life.

Tap the screen again quickly to wake the device and bring the screen back on.

Power/Lock
Press and hold to power on/off.
Press quickly to sleep or wake.

Top edge of DROID 2

NOTE: The **Power/Lock** button may be at a slightly different location, depending on your DROID model.

Volume Buttons

The **Volume** buttons located on the upper-right side of your DROID perform multiple functions, depending on the context in which you press them.

When you press these buttons, you will see the type of volume you are changing with an on-screen pop-up window, as shown below.

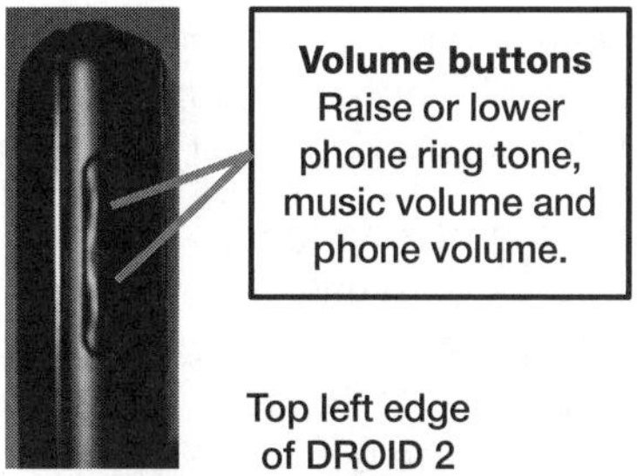

Top left edge of DROID 2

If you are not playing any media (e.g., a song, video, or other content), then these **Volume** buttons will change your phone's ringer volume.

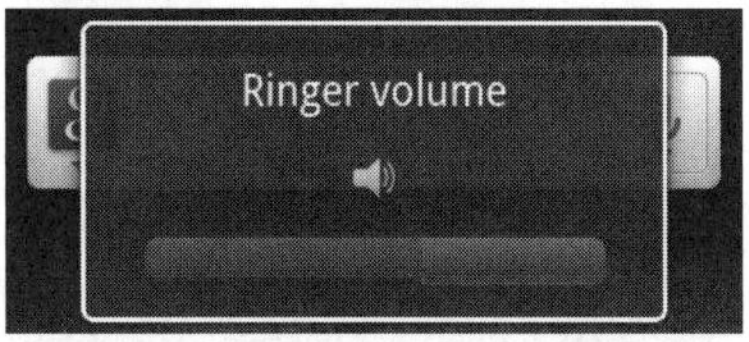

When you are playing music, watching a video, or listening to other media, then pressing the **Volume** buttons will adjust the playback (speaker) volume.

When you're on a phone call, these **Volume** buttons will change the volume of the caller.

Finally, when you are in the **Camera** or **Camcorder** apps, the **Volume** buttons will zoom in or out.

Slide to Unlock

When you first power on your DROID or wake it from **Sleep** mode, you will see two sliders at the bottom of the screen.

Touch and drag the **Lock** icon slider from left to right to unlock your phone. If you have assigned a password to your DROID, then you will need to enter your password to unlock the device.

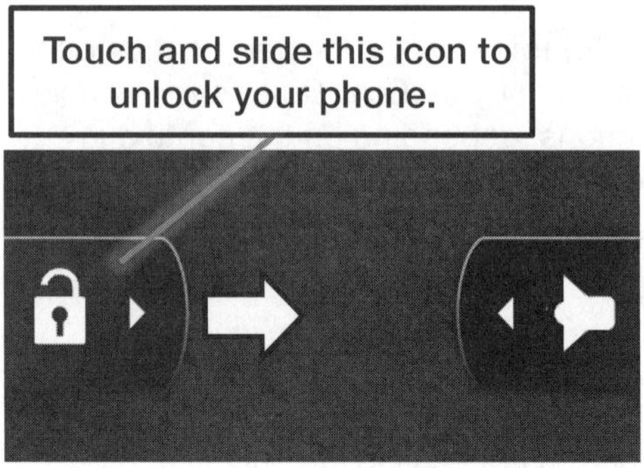

Slide to Mute

The other slider that appears next to the **Lock** icon when the phone wakes up is the **Slide to Mute** icon.

Touch and drag the **Slide to Mute** icon slider from right to left to **Mute** or **Unmute** your phone ringer, as shown in Figure 1–1.

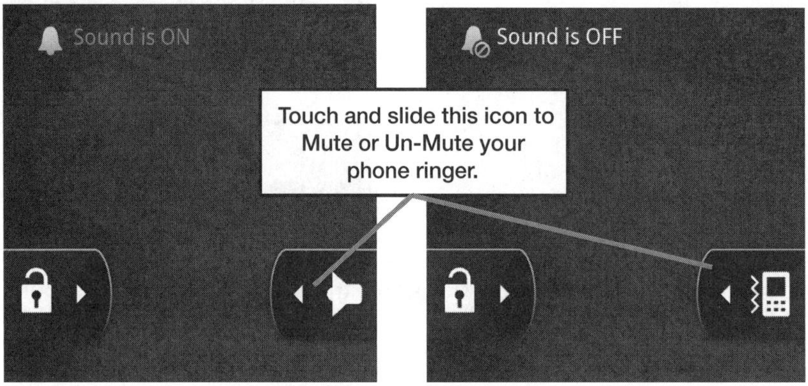

Figure 1–1. *Slide to **Mute** or **Unmute** your phone ringer.*

Using the Four Buttons Along the Bottom

Along the bottom of your DROID, you will find four buttons: **Menu**, **Home**, **Back**, and **Search**. Note that these buttons do not appear in the same order on all DROID models. (Be sure to check out the Quick Start Guide earlier in this book to learn what each of the buttons does.)

The DROID 2's four buttons appear in this order: **Menu**, **Home**, **Back**, and **Search**.

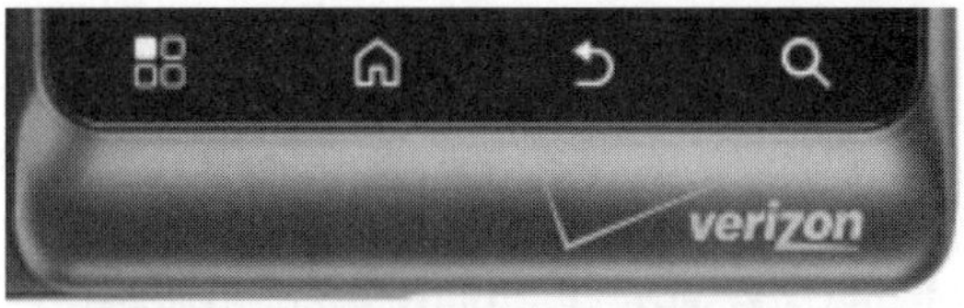

The DROID X's four buttons appear in this order: **Menu**, **Home**, **Back**, and **Search**.

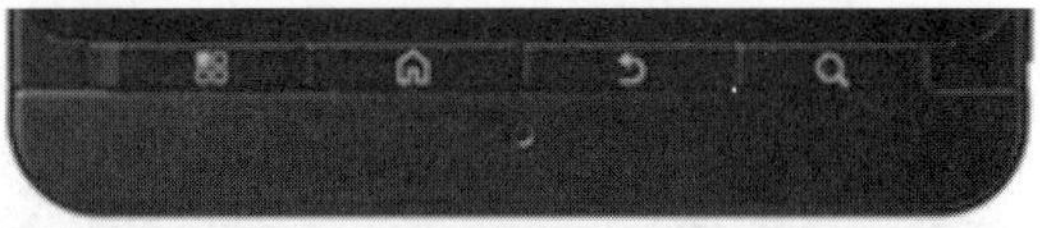

The DROID's four buttons appear in this order: **Back**, **Menu**, **Home**, and **Search**.

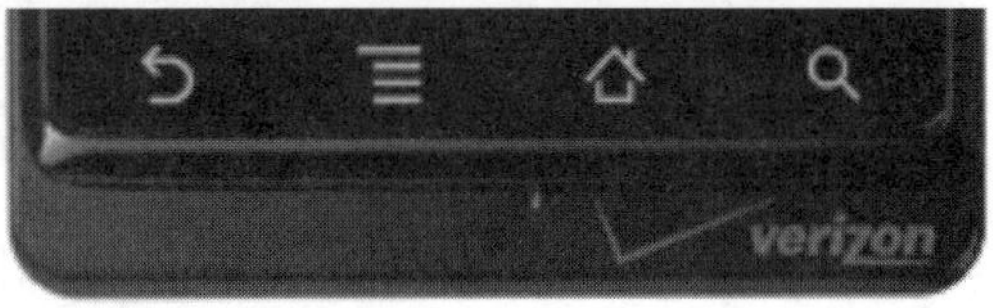

Slide-out Keyboard (for DROID and DROID 2)

If you own the DROID, DROID 2, or DROID 2 Global, then you have a physical slide-out keyboard. Slide it out from the left side of the phone and turn the phone to **Landscape** mode to type using the keyboard (see Figure 1–2). We recommend typing with two thumbs on the keyboard – it will help you type a little faster (see Chapter 2: "Typing, Voice, Copy, and Search" to learn more about these keyboards).

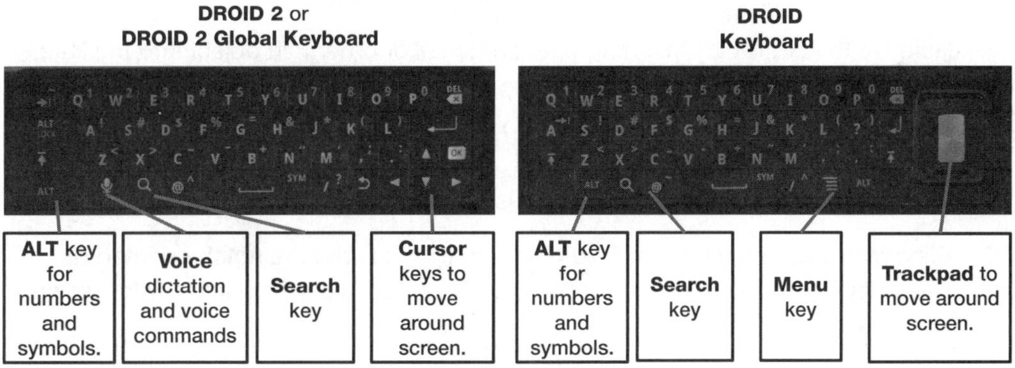

Figure 1–2. *The DROID and DROID 2 slide-out keyboards.*

Dedicated Camera Button

On the lower-right side of your DROID, you will see a small button. This is your dedicated camera button. Press and hold it for about a second to start the **Camera** app.

You can take a picture by pressing this button once the **Camera** app has started. You can learn more about taking pictures in Chapter 18: "Taking Photos and Video."

Camera button. Press to start the **Camera** app or take a picture.
verizon

Inserting a Memory Card and Removing the Battery

You need to open the back of your phone to remove and replace the battery, as well as to remove or insert a MicroSD memory card. You also need to open the back of your DROID 2 Global to remove or insert a SIM card.

In order to get at the battery and memory card slots, you need to do the following.

1. Power off your phone.
2. Remove the back cover by sliding it down and lifting it up.
3. Remove the battery by inserting your fingernail or other thin object into the little space by the white or silver tab on the battery that says **BATTERY REMOVAL HERE.** The DROID X has a plastic tab you can pull to remove the battery.
4. Figure 1–3 shows where the memory card slots are located on your phone. You find these slots in different places on the DROID X and DROID 2 models.
5. Slide the media card (MicroSD format) into the slot with the notch oriented as shown in Figure 1–3; the metal contacts of the media card must be facing down as you slide it in.

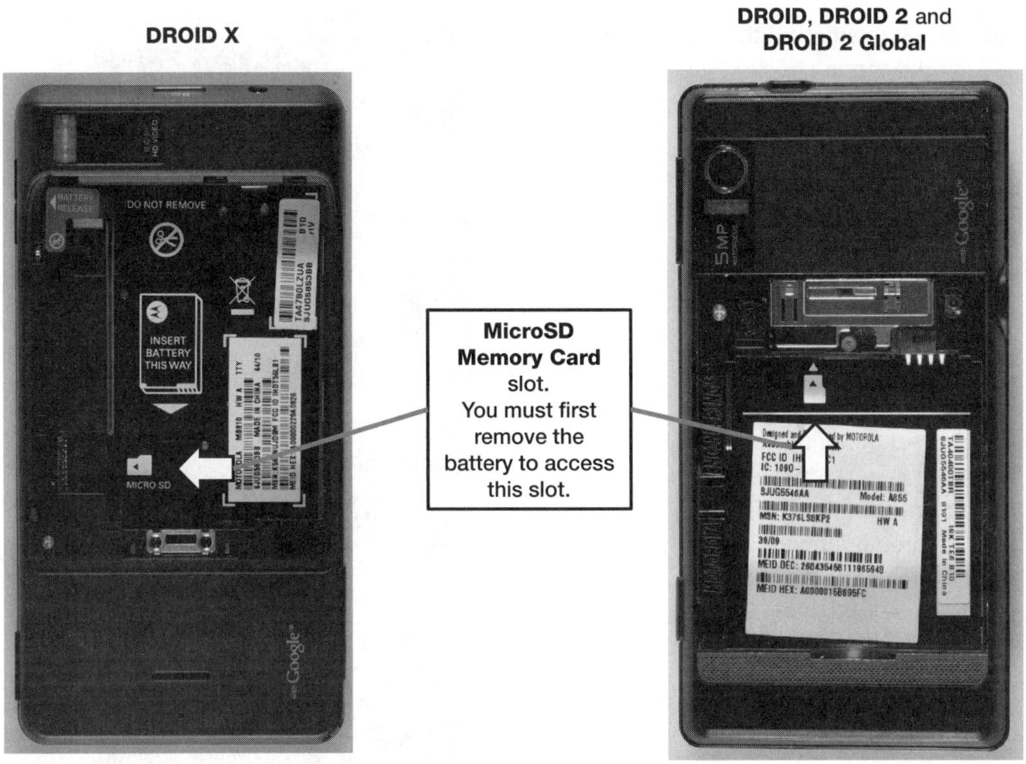

Figure 1–3. *How to insert a memory card into your DROID.*

Adjusting the Date, Time, Time Zone, and 24-Hour Format

Usually, the date and time is set and adjusted automatically on your DROID using the wireless network. However, there may be a few adjustments you might want to make, such as your time and date formats. The following sections will show you how to adjust everything related to your device's date, time, and time zones.

When you travel with your DROID, your time zone will usually adjust automatically. Keep an eye on the time and see if it has adjusted to local time when you land in a new time zone. If not, then you can change the time zone by following these steps:

1. Touch the **Settings** icon.
2. Swipe up to see the items at the bottom of the list and touch **Date & time**.

3. You can see that the **Automatic** box at the top is checked (set to **On**) in the figure to the right. If you want to manually adjust the date, time zone, and/or time, then you have to uncheck the **Automatic** setting (i.e., set it to **Off**).

4. To set the date, tap the **Set date** option and make adjustments by tapping either the **+** and **-** icons or the items themselves. If you tap an item such as the number 29 (as shown in the image to the right), you can type numbers to set the day.
5. When done, tap the **Set** button.
6. Tap **Select time zone** to adjust the time zone. Next, swipe up or down on the next screen and tap the correct time zone.
7. Tap **Set time** to adjust the time. The **Set time** screen is very much like the controls you use to set the date. Use the **+** or **-** icons or tap inside the numbers to make adjustments with the keyboard. Tap **Set** when you are done.

8. To use the 24-hour time format (e.g., 16:00 instead of 4:00 PM), you tap **Use 24-hour time format**.
9. To adjust the date format, tap **Select date format**.

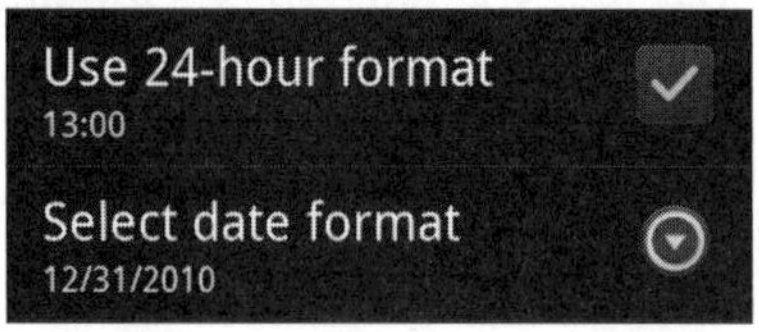

10. You will now see a pop-up window showing various date options. Tap your selection to choose it.

11. This brings you back to the main **Date & time** settings screen. Press the **Home** button to return to your **Home** screen.

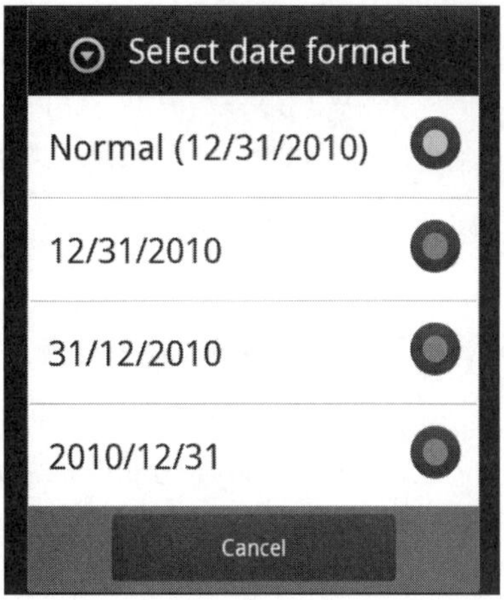

Adjusting the DROID's Brightness

The wonderful screen on your DROID is also one of the largest consumers of your battery life. The default is automatic brightness, which uses the built-in light sensor to adjust the brightness of the screen. When it is darker in your immediate environment, the auto-brightness control will dim the screen. When it is bright or sunny, the screen will be automatically brightened, so it is easier to read.

You may want to play with this setting and see how it affects your battery life. First, try the **Automatic brightness** setting (the default) and see how it works.

If you find that the automatic brightness screen seems a little too bright, then you may want to play with this setting and dim the screen. A dimmer screen will help you conserve battery power.

If you want to adjust the brightness manually, follow these steps:

1. Touch the **Launcher** icon to see all your apps.
2. Tap the **Settings** icon.
3. Tap **Display**.
4. Tap **Brightness**.
5. Tap **Automatic brightness** to set this option to **Off** (i.e., to make the **Checkmark** icon go gray).
6. Move the slider control to adjust the brightness.
7. Tap **OK** to save your settings.

TIP: You can also use the Android **Power Control** widget to adjust brightness.

TIP: Setting the brightness lower will help you save battery life. A little less than halfway across seems to work fine.

Charging Your DROID and Battery Life Tips

Your DROID may already have some battery life, but you might want to charge it completely, so you can enjoy uninterrupted hours of use after you get it set up.

TIP: We recommend charging your DROID every night, especially if you use it a lot during the day for phone calls or web browsing.

Charging From the Wall Outlet

The fastest way to charge your DROID is to use the adapter and plug it directly into the wall outlet.

TIP: Some newer cars have built-in power outlets (just like your home). These outlets let you plug in your DROID power cord. Note that these outlets are sometimes buried in the middle console behind the front seat.

Charging from Your Computer

You can also charge your DROID when you plug it into your computer.

TIP: Try powering your DROID with different USB ports on your computer. Some USB ports share a bus and have less power, while others have their own bus and more power.

For best charging, you should have your computer plugged into the wall outlet. If your computer is not connected to the wall power outlet, your DROID will charge, but at a slower rate.

Keep in mind that if your laptop computer goes to sleep or you close the screen, your DROID will stop charging.

Stopping the Verizon VCAST Video Auto-play (Windows PC)

If you plug your DROID into a Windows PC, you might see your web browser pop up and start automatically playing a Verizon VCAST promotional video. If your computer speaker volume is turned up relatively high, you might be really surprised by this video. It can get quite loud!

Fortunately, there is a way to turn off the auto-play feature of this video:

1. Right-click the **Motorola** icon in your **Windows** tray, as shown in Figure 1–4.
2. Select **When phone connects, launch.**
3. Select **Nothing**.

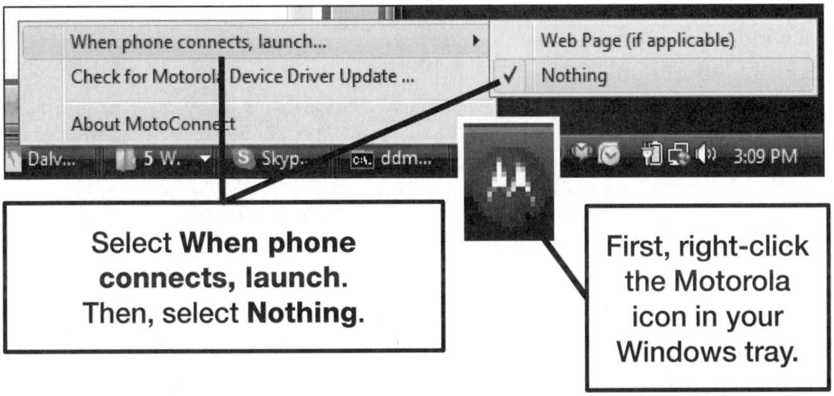

Figure 1–4. *Stopping the VCAST video from auto-playing.*

What Drains the Battery the Fastest?

The main drains on your DROID battery are the large touch-screen display and the wireless radios (cellular, Wi-Fi, and Bluetooth). As long as you know this, you can increase your battery life – we will show you some cool battery extending tips in the next section. But the short version is this: you want to keep the screen dimmer and turn it off (**Sleep** mode) as soon as you don't need it. You also want to keep your radio usage down or limit it to those times when you need it. For example, if you don't need your GPS location enabled, then turn it off. If you don't need to use your device for 15 minutes, then you can have the data updates turned off automatically. We will show you how to do this with the **Battery Manager** feature in the "Using the Battery Manager" section.

The Power Control Widget

A very convenient tool to help prolong your battery life without having to dig through the **Settings** app is the **Power Control** widget. Follow these steps to add this widget to your **Home** screen:

1. Press the **Menu** button.
2. Tap **Add**.
3. Tap **Android widgets**.
4. Tap **Power Control**.

The Power Control widget will look like the image to the right. From left to right, the buttons will:

- Turn on/off the Wi-Fi radio.
- Turn on/off the Bluetooth radio.
- Turn on/off the GPS receiver.
- Turn on/off the wireless sync to Google.
- Adjust your screen brightness. Tap once to brighten, tap again to go to auto brightness, and tap yet again to dim the screen.

Getting More Out of Each Charge

Following these tips will help you extend your DROID's battery life:

1. **Put the DROID into Sleep mode whenever possible**: Tap the **Power/Lock** button on the upper-right edge of the device to put it into **Sleep** mode whenever you are not using it. We use the term **Sleep** mode loosely; this feature really just turns off the screen to save the battery.

2. **Turn off Wi-Fi when not needed**: The Wi-Fi antenna uses power even if you are not connected to a Wi-Fi network, so turn it off when you don't need it. Turn off Wi-Fi by going to **Settings** > **Wireless & Networks** > Set **Wi-Fi** to **Off**.

TIP: DROID 2 and X users can swipe left to the second of the seven **Home** screens, and then tap the Wi-Fi switch that looks like this to turn Wi-Fi off or on. 

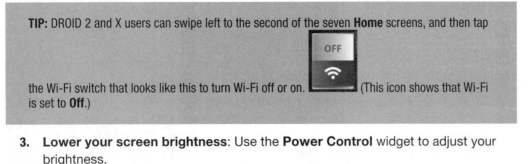

(This icon shows that Wi-Fi is set to **Off**.)

3. **Lower your screen brightness**: Use the **Power Control** widget to adjust your brightness.
4. **Set a Shorter Screen timeout**: Shortening the time your DROID takes to turn off the screen when it's not being used can help you extend your effective battery life. To do this, tap **Settings**, then **Display**, and **Screen timeout**. Set **Screen timeout** as short as possible – you can set it as short as 15 seconds, if you can stand it. Sometimes it get annoying when the screen keeps turning off,so play with this setting a bit to find the best interval for how you use the device.
5. **Disable GPS when not needed**: Use the **Power Control** widget to turn off GPS.
6. **Turn off Bluetooth when not needed**: Use the **Power Control** widget to turn off Bluetooth.
7. **Adjust your Battery Manager settings to a Saver mode**: See the "Using the Battery Manager" section later in this chapter to learn how to adjust these settings.

Long-Term Battery Life

The DROID uses a rechargeable battery that will lose its ability to maintain a charge over time and has only a limited number of cycles during its useful life. You can extend the life of your DROID battery by making sure you run it down completely at least once a month. The rechargeable battery will last longer if you do this complete draining once a month.

Using the Battery Manager

To see the **Battery Manager** (DROID 2 and DROID X) shown to the right, tap your **Settings** icon, then swipe up and tap **Battery Manager**. The **Checkmark** in the large **Battery** icon shows that the DROID is currently being charged.

The image to the right shows a full battery charge with 100% battery power remaining.

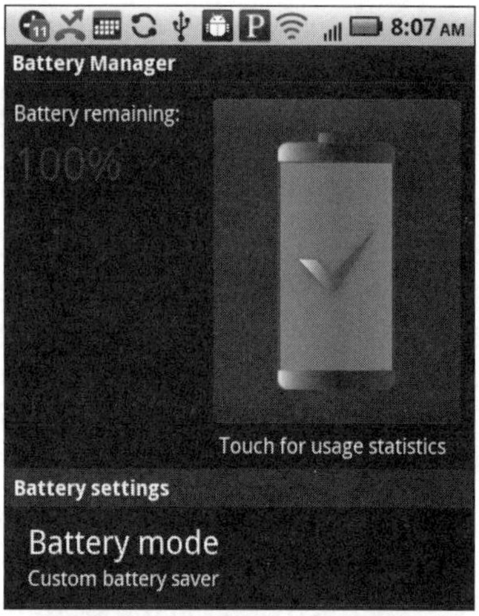

NOTE: On the original DROID, go to Settings/About phone and you can view the battery status and usage.

Tap the **Battery** icon to see how much of your battery is being consumed by various processes.

In most cases, you will see that the display takes the majority of your power.

The display and the data updates – which mean your radios such as Wi-Fi and Cell – consume the majority of your power.

Press the **Back** button to return to the **Battery Manager** screen.

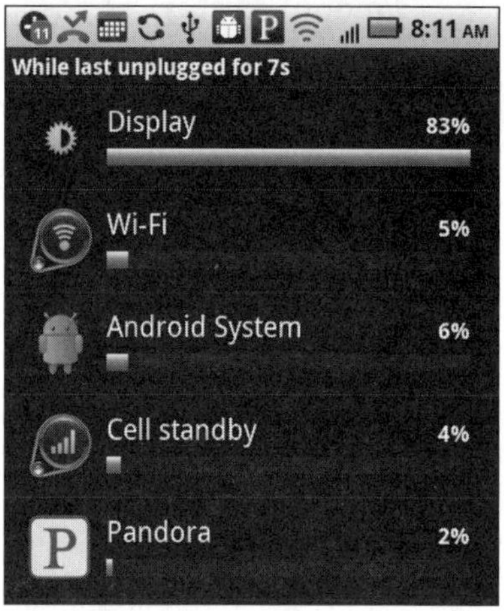

Tap **Battery mode** at the bottom of the screen to see available modes, as shown to the right.

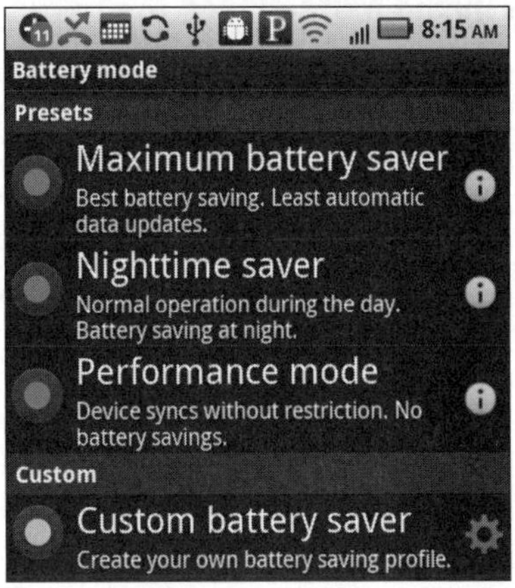

To save battery life, you need to reduce the frequency of automatic data updates and make the screen a little dimmer. The top two settings will allow you to get more battery life.

The **Performance mode** setting gives you the highest performance, but will consume the battery the fastest.

With **Custom battery saver**, you can fine-tune the settings to fit your needs. To learn more details about each preset, tap the **Information** icon. 

To adjust the custom battery saver settings, tap the **Gear** icon. 

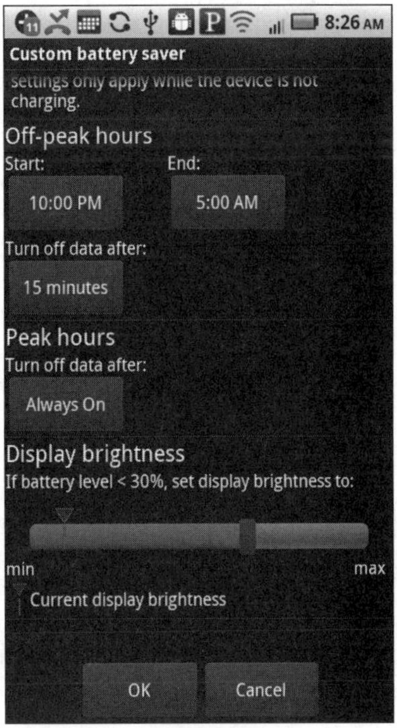

Now you can fine-tune your **Off-peak hours**, **Peak hours**, how soon to turn off the data (your radios), and the **Display brightness**.

When you're done, tap the **OK** button at the bottom.

Press the **Home** button to return to your **Home** screen.

As you can see, selecting the correct **Battery mode** is a tradeoff between your performance needs and the battery life.

Finding More Places to Charge Your Droid

No matter what you do, you will want to find more places and more ways to charge your DROID if you use it a lot. Table 1–1 shows you some other options for charging your DROID besides using your power cord or connecting it to your computer.

Table 1–1. *Other Places and Ways to Charge Your DROID.*

Option	Description
Airport Charging Station	Most airports have wall sockets available today where you can top off your DROID's battery while you are waiting for your flight. Some airports have labeled charging stations, and others simply have wall sockets that may even be hidden behind chairs or other objects. You may have to do a bit of hunting to beat out all those other power-hungry travelers!
Car Charger Accessory	If you are using your DROID heavily for phone calls during the day, you may want to invest in a car charger or some other way to give your DROID a little more juice in the middle of a long day. These chargers plug directly into the cigarette lighter socket in your car. These run about US $15–25.
Car Power Inverter	If you are taking a long car trip, you can buy a power inverter to convert your 12V car power outlets into a power outlet that lets you plug in your DROID charger. Do a web search for "power inverter for cars" to find many options for under US $50. This is a small price to pay for hours of enjoyment on your DROID!

Setting up Accounts on Your DROID

You need to link to at least one Google account on your DROID. You can also connect to many other types of accounts (e.g., email, social networking, corporate, and photo). In this section, we explain the basics of setting up various types of accounts.

Setting up Your Google Account

When you power on your DROID for the first time and slide to unlock it, certain apps such as the **Calendar**, **Contacts**, and **Email** require that you connect your DROID to a Google account. This is because your DROID is running the Android operating system created by Google. It was designed from the ground up to be wirelessly connected to a Google account.

If you don't have a Google account already, you can create a free one with a Gmail address by registering at `http://mail.google.com/mail/signup`.

To connect your DROID to Google, you need to enter your Gmail account login and password and follow the steps outlined to log in. We show you the detailed steps of how to do this in Chapter 3: "Sync Your DROID with Your Google Account." Chapter 3 also shows you how to transfer your contacts from your computer to your Google account, so you can sync them to your phone.

TIP: If your workplace uses **Google Apps for Enterprise**, then you could use this ID as your Google account. However, unless your phone is part of an enterprise deployment, the wiser course of action is to use a personal Google account and add the **Google Apps** information as an additional email account. That way, you don't lose your phone data if you switch jobs.

Using Other Google Services

We'll cover how to use other Google services in greater detail in Chapter 3, but virtually everything on Android phones is handled through your Google account. You should set up and explore these tools on the Web for a better understanding of how they work on your phone.

If you purchase apps in the Android Market, you'll use your Google account and the **Google Checkout** app to complete the transaction. The default email account is **Gmail**, and the default calendar is **Google Calendar**.

Here are a few of the default Google services you'll get to know as you use your DROID: **Gmail** (Google's email program), **Google Calendar**, **Google Maps**, **Google Checkout** (a payment system like the one from PayPal), **Picasa** (a photo upload and sharing site like **Flickr**), and **YouTube** (a video upload and sharing service).

Setting up More Types of Accounts

Except for the first Google account, which the phone automatically guides you through at the login screen, every other account requires that you access your **Accounts** page. Follow these steps to do so:

1. Tap the **Launcher** icon in the middle of the bottom of the **Home** screen.
2. Tap the **My Accounts** icon.
3. Tap the **Add account** button at the bottom.
4. Select your type of account from the screen, as shown to the right. Use **Corporate Sync** icon for a **Microsoft Exchange** account. You may see fewer or more account type options depending on the number of apps you have downloaded from the Android Market.
5. Follow the on-screen instructions to enter your login information. You may need to accept a license agreement if this is the first time you are accessing an app you just downloaded.

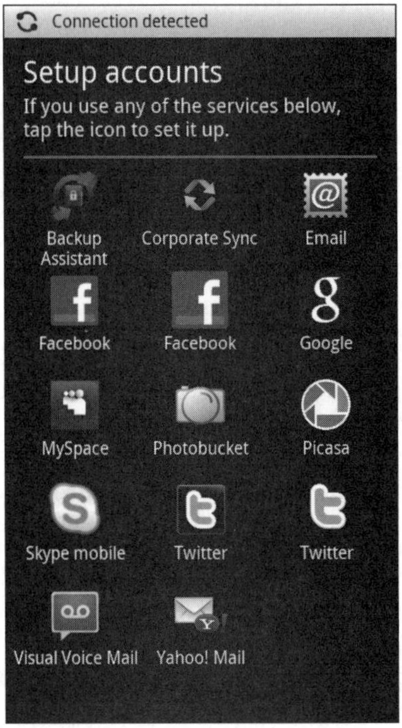

Finding Your Email

After you get your accounts set up, you may be wondering where you can find your email. There are two icons for email on your DROID: **Email** and **Gmail**. The one called **Email** handles all email accounts except Gmail. And, as you might imagine, you use the **Gmail** app for all your Gmail accounts.

This is the app you use for all your non-Gmail email accounts.

This is the app you use for your Gmail account(s).

TIP: Sometimes the **Corporate Setup** option might not work for your **Exchange** or other type of server.

In this case, you can turn to third-party apps for a solution. **TouchDown** is a $20 app from NitroDesk that allows **Exchange ActiveSync** with Push email. This company also offers a free trial, so do take advantage of it to make sure the app works with your **Exchange** service. For example, it is not supported with some **Exchange Server 2003** configurations. The newest version of **TouchDown** also supports other types of servers, such as **Zimbra**, **Kerio**, **Novell GroupWise**, **Sun Java Communication Suite**, **Oracle's Oracle Collaboration Suite** (OCS) and **Beehive Suite**, and other servers. Check out NitroDesk's website at `www.nitrodesk.com` for more information.

If you use **TouchDown**, you'll have a separate email, calendar, and task list. It will all look familiar to **Microsoft Outlook** users; however, it doesn't sync this data with your **Google Calendar**. You can download **TouchDown** from the Android Market or by visiting `http://www.nitrodesk.com`.

Securing Your DROID

Your DROID can hold a great deal of valuable information. This is especially true if you save information such as the Social Security numbers and birthdates of your family members in your **Contacts** list. It's a good idea to make sure that anyone who picks up your DROID can't access all that information!

Setting a Screen Lock

To set a screen lock to protect your DROID, tap your **Settings** icon, then **Location & Security settings**. Next, scroll down and tap **Set up screen lock** to see the list of lock options, as shown to the right:

- **None**: Disable security lock.
- **Pattern**: Draw a pattern of dots.
- **PIN**: Set a numeric PIN code.
- **Password**: Enter a password.

Tap **Pattern** to draw a pattern with your finger across nine dots as shown to the right.

You will need to draw the pattern twice to confirm it.

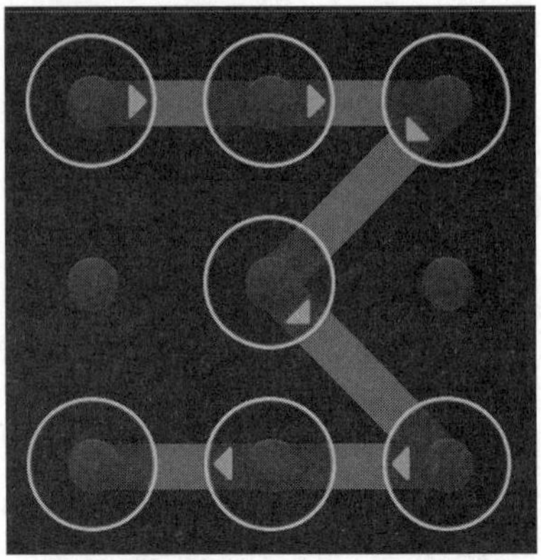

The defaults are **no visible pattern** or **tactile feedback** – you may want to enable these. If someone is watching you do this, however, then that person may be able to see your pattern code.

By default the DROID will not lock for **20 minutes**, so we recommend tapping the **Security lock timer** and setting it to a smaller interval. The most secure setting is **When display is off**, which means to lock the device every time the display times out or turns off with the **Power/Lock** button.

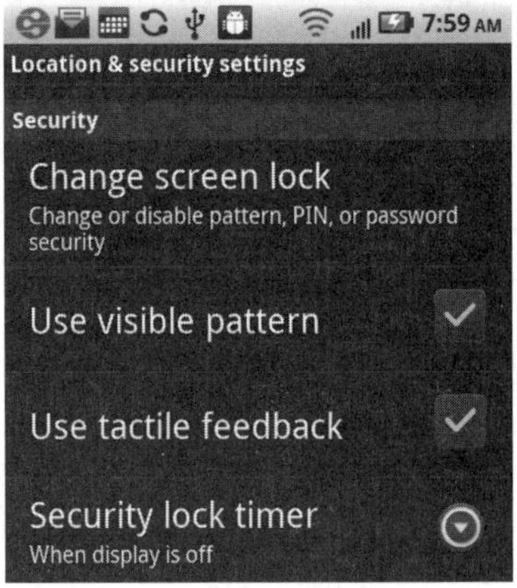

If you tapped **PIN** for the security method, then you would need to type a numeric PIN of at least four digits. You can definitely use more, if you like. In this screen shot to the right, we used nine digits.

Make sure that you use a PIN that you will remember.

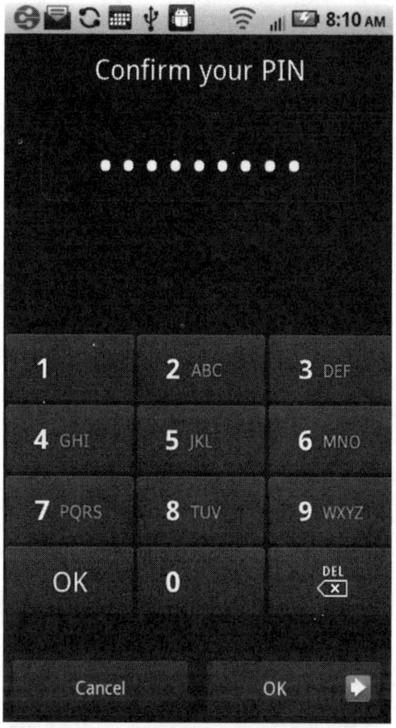

Tap **password** to use a complex password that can include letters, numbers, and symbols.

Again, your password must be at least four characters long.

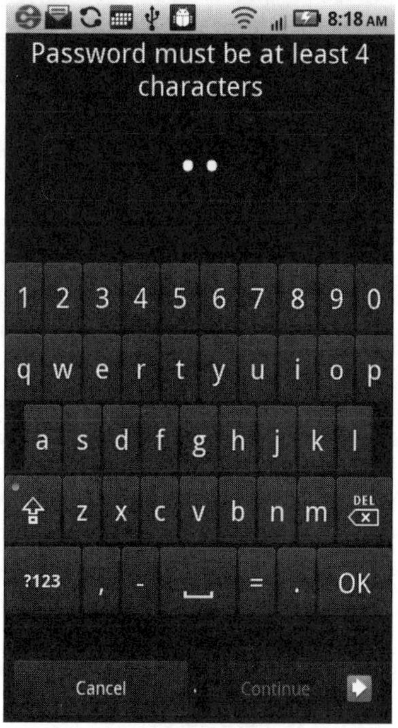

Locking Your DROID

Your DROID will lock when one of the following occurs:

- You press the **Power/Lock** button on the top edge, and the security lock timer is set to **When display is off**.
- The security lock timer expires.

Unlocking Your DROID

To unlock your DROID, you first slide the lock slider to see the lock code entry screen. Next, enter your pattern, PIN, or password to unlock the phone.

You get five attempts to try your unlock code. After the fifth try, the DROID will lock up for 30 seconds before you can try again.

If you enter 20 incorrect unlock codes, then your DROID will ask you to enter your Google account information to unlock it.

CAUTION: We have heard that sometimes even entering the correct Google information will not allow you to unlock your DROID. If this occurs, then you will need to perform a **Hard Reset**. See Chapter 24: "Troubleshooting" for detailed information on performing a **Hard Reset**.

Maintaining Your DROID

Now that you have your DROID fully charged, you will want to know how to safely clean the screen and keep it protected with various cases.

Cleaning Your DROID Screen

After using your DROID a little while, you will see that your fingers (or other fingers besides yours) have left smudges and oil on the formerly pristine screen. You will want to know how to safely clean the screen. One way to keep the screen cleaner throughout the day is to place a protective screen cover on the DROID, which may also have the added benefit of cutting down on glare (as discussed in the next section).

We also recommend the following:

1. Turn off your DROID by pressing and holding the **Sleep/Power** key on the top edge, and tap **Power off** from the **Phone Options** pop-up menu.
2. Remove any cables, such as the USB sync cable.
3. Rub the screen with a dry, soft, lint-free cloth (such as a cloth supplied to clean eyeglasses or something similar).

4. If the dry cloth does not work, then try adding a very tiny bit of water to dampen the cloth. If you use a damp cloth, try not to get any water in the openings.

CAUTION: Never use household cleaners, abrasive cleaners such as SoftScrub, or ammonia-based cleaners such as Windex, alcohol, aerosol sprays, or solvents on your device.

Cases and Protective Covers for Your DROID

Once you have your DROID in your hands, you will notice how beautifully it is constructed. You will also notice that it can be fairly slippery. At some point, it could even slip out of your hands, rock around a bit, or have the back get scratched when you are typing on it.

We recommend buying a protective case for your DROID. Average cases run about US $10 – 40, and fancy leather cases can cost US $100 or more. Spending a little to protect your DROID, which costs $200 or more, makes good sense.

Where to Buy Your Covers

You can purchase your DROID protective cover at any of the following locations:

- Amazon.com (`www.amazon.com`)
- Android Central (`www.androidcentral.com`)
- Handheld Items (`www.handhelditems.com`)

You could also do a web search for "DROID cases" or"DROID protective covers."

TIP: You *may* be able to use a case designed for another type of smartphone for your DROID. If you go this route to save some money, just make sure your DROID fits securely in the case or cover.

What to Buy

The following sections provide information on the available types of cases to choose from and their price ranges.

Rubber / Silicone / Decorative Cases ($10–30)

What these do: Provide a cushioned grip, absorb DROID bumps and bruises, and isolate the edges of the phone (antennas) from your fingers.

Pros: Deliver inexpensive, colorful, and comfortable DROID protection.

Cons: Provide a less professional look than a leather case.

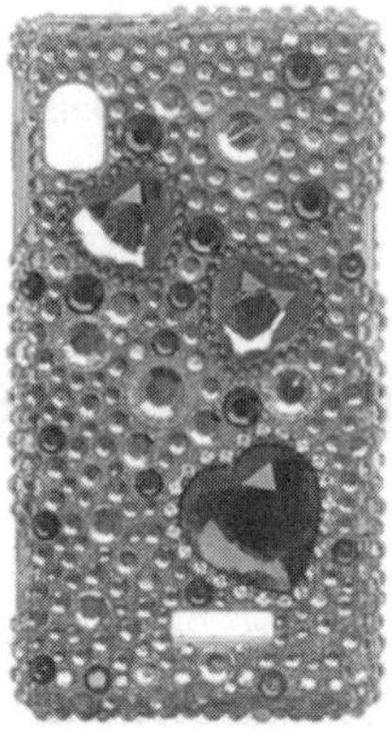

Motorola A955 Droid 2 Full Diamond Graphic Case – Purple Heart Image courtesy of www.handhelditems.com.

Waterproof Cases ($10–40)

What these do: Provide waterproof protection for your DROID and allow you to safely use the DROID near water (in the rain, at the pool, at the beach, on a boat, and so on).

Pros: Provide good water protection.

Cons: May make the touch screen harder to use; these covers typically do not protect your DROID from drops or bumps.

Hard Plastic / Metal Case ($20–40)

What these do: Provide hard, solid protection against scratches and bumps and short drops.

Pros: Provide good protection.

Cons: Add some bulk and weight. You may need to remove such covers when charging because the DROID might overheat in that circumstance.

Leather or Special Cases ($50–100+)

What these do: Provide more of a luxury feel and protect the DROID.

Pros: Provide the more upscale look-and-feel of leather; such cases also protect the front and the back of the device.

Cons: Cost more than other cases; add bulk and weight.

Front Screen Glass Protectors ($5–40)

What these do: Protect the DROID's screen from scratches.

Pros: Help prolong life of your DROID and protect against scratches; most covers like this also decrease screen glare.

Cons: Some may increase glare or may affect the touch sensitivity of the screen.

Saving Money on Phone Plans

You know you have to purchase a voice and data plan to use your DROID; however, fully understanding the available optional bells and whistles can help you save some money.

> **TIP:** Check with your workplace to see if it offers special deals on Verizon DROID and rate plans. You might be able to save some money through such a plan.

Data Packages

At the time of publishing, Verizon offers two data plans that give you access to the mobile web, email, and contacts or calendar sync capabilities of your device. Both plans offered by Verizon feature unlimited calls. Option 1 is an unlimited corporate plan; Option 2 is a regular or "personal" plan. The corporate plan is about $45 / month, and the personal plan is about $30 / month. The only real difference between the plans is that the corporate plan allows you to connect your DROID to a corporate email service, such as **Microsoft Exchange Active Sync**, **IBM Lotus Notes Traveler**, or **Good Mobile Messaging**.

> **TIP:** Your DROID comes with built-in wireless sync to your **Google Contacts**, **Email**, and **Calendar** programs. This means that, unless you need a special corporate email connection, you should be able to go for the personal data plan and save some money.

Text, Picture, and Video Messaging Plans

Usually, the texting add-on service plans can be added or removed from your plan at any time. Texting rates are all for both inbound and outbound messages. At publishing time, Verizon bills $0.20 for text messages and $0.25 for picture messaging. As soon as you exceed 25 text messages in a month, you should opt for the basic 250 message plan; it is cheaper. Text, picture, and video messaging plans range from 250 to 5000 messages.

TIP: Save Money on Text, Video, and Picture Messaging

You need to watch your monthly bill closely. If you find out that you have gone over $20 in total messaging charges, you may be able to cut the bill down to $20 by calling Verizon and asking to be switched over to an unlimited messaging plan. If you sweet talk them, they will usually institute the plan retroactively, so it reduces your current phone bill.

Equipment Insurance Plans

You may be offered a total equipment protection plan that will allow you to get a new phone if yours is damaged or lost. Please note that many of these plans have deductibles of $90 or more and caps on how many phones per year (usually two) that you can recover. New DROIDs with a service plan cost about $200; without a service plan, the phones usually cost $500 – $600. Because this is what you might have to pay to buy a phone to replace your lost or damaged DROID, it seems to make financial sense to get the insurance. At about $8 / month, you are paying $96 per year. If you add on the $90 deductible, you are still at just $186 in the first year and $276 for the second year. This is well below the $500 - $600 price to buy a new DROID without a service plan.

NOTE: You can typically add insurance only during the first 30 days after you sign up for the original service plan.

Mobile Wi-Fi Hot Spot

A very cool and useful feature on the DROID is the **Mobile Wi-Fi Hot Spot** feature (also known as **Wireless Internet Access**). We describe this feature in more In Chapter 5: “Wi-Fi and 3G Connectivity.” This feature allows you to turn your DROID into a secure, private, mobile Wi-Fi hot spot for up to five devices. A good example of this is when you are traveling in a car or sitting in a location waiting for something, and you need an Internet connection for your laptop. In this case, you can use the DROID to connect your laptop, iPod touch, iPad, or any other device that needs a Wi-Fi Internet connection. Usually, the Wi-Fi connection speeds are quite respectable!

TIP: Controlling Costs With Reminders

You can set a DROID **Calendar Reminder** to help you remember to turn off temporary service plan features.

For example, Verizon (like most phone companies) bills in advance for additional add-on services such as **Mobile Hotspot** or **Visual Voicemail**. Our recommendation is that you put the service on a recurring monthly reminder on your DROID calendar. This reminder should be set to a day or two before your monthly billing cycle ends. When the reminder comes up, you should check whether there are any features that you want to turn off. For example, if you are going on a summer trip and need the **Mobile Hotspot** feature for just a month, the reminder would help you save $20 or more by making sure you turn off the service when you are done with it.

Enhanced Voicemail Plans

Verizon also sells **Visual Voice Mail** or **Premium Voice Mail** services on an ad-hoc basis. This means you can turn the features on or off at will, with no penalties. The **Visual Voice Mail** service gives you the ability to see a list of all your messages on your DROID and tap to listen to them in any order. It's a nice feature. The **Premium Voice Mail** service gives you twice the storage for your voice mail messages. At the time of publishing, the Premium plan lets you store 40 messages instead of the standard 20. If you are careful about cleaning out your inbox, then the standard 20 messages included for free should be adequate.

TIP: If you use the free **Google Voice** service, you can get free **Visual Voice Mail** service. You can learn more about **Google Voice** in Chapter 7: "Making Phone Calls."

International Plans

If you travel to Canada or Mexico, Verizon offers special plans that allow you to fully use your DROID for voice and data while traveling. Again, these plans can be turned on or off at will. So, if you are only going to be in Mexico a week or so, just remember to turn off the plan when you return home. We cover international travel in more detail in Chapter 5: "Wi-Fi and 3G Connectivity."

Also, if you call from the US to other countries, be sure to inquire about international access plans that give you reduced calling rates to overseas phone numbers.

If you travel to other countries, check with Verizon and get the company to enable your phone for international travel. You should also have someone explain all the various plans, as well as any limitations to your coverage when you travel.

TIP: Plans for Cruise Ships!

There are even special plans and rates that will allow you to make calls and stay connected to your data plan when you take a cruise. Rates and plans vary by cruise line, so give Verizon a call to see what is available.

Now that you know all about the various service plans, you are ready to start using your new DROID!

Chapter 2

Typing, Voice, Copy, and Search

In this chapter, we will show you the ins-and-outs of using the DROID keyboards. We will cover both the virtual on-screen keyboard and the slide out keyboard (for DROID and DROID 2).

The on-screen keyboard comes in two flavors: **Portrait** mode (vertical/smaller) and **Landscape** mode (horizontal/larger). You will also learn how to use the innovative *Swype* typing, where you drag your finger across the virtual keyboard to Swype words. You can also fine-tune your keyboard to vibrate, click, or be silent when you press keys.

Your DROID also comes with accessibility features such as **Voice ReadOuts** (where the DROID reads the screen out loud) and **Zoom Mode** (where you zoom in on a section of the screen for easier reading).

You may also choose to skip typing altogether and use the DROID's voice recognition software to speak your text – it works amazingly well. Your DROID also comes with a **Voice Command** app to help you control the DROID itself with your voice.

Finally, we will show you how to use the DROID's copy-and-paste features, as well as the highly flexible **Google Search** and **Voice Search** functions.

Typing on Your DROID

When you first use your DROID, you will quickly find two on-screen keyboard orientations on the device. The first (and smaller of the two) is visible when you hold your DROID in a vertical orientation (**Portrait** mode); the second orientation is the larger **Landscape** mode that you see when you hold the DROID in a horizontal orientation. The nice thing is that you can use whichever orientation works best for you.

Two Ways to Type: Multi-touch and Swype

Your DROID also gives you two ways to type text on your virtual keyboards: Multi-touch and Swype. We recommend trying both methods and seeing which one works best for you.

Quickly Switch Between Multi-touch and Swype

While you can change typing modes in the **Settings** app, one of the fastest ways to change typing modes is to long-press on any word. This will bring up a pop-up window from which you can select **Input method** at the bottom of the list to change your typing method.

Multi-touch Typing

With the standard Multi-touch typing method, you tap each key only once, just as you do on a standard keyboard.

TIP: You know are using the Multi-touch keyboard if you see the **?123** key or the **ABC** key (which is visible after you quickly tap the **?123** key) in the lower-left corner of your virtual keyboard.

Various Multi-Touch Keyboards

You will see the entire keyboard show uppercase letters when you press the **Shift** key or you are typing at the beginning of a sentence (DROID automatically capitalizes the first letter). You can tap the **?123** key to access numbers and basics symbols, and then tap the **ALT** key to access the advanced symbols keyboard. Tap the **ABC** key to return to the letter keyboard (see Figure 2–1).

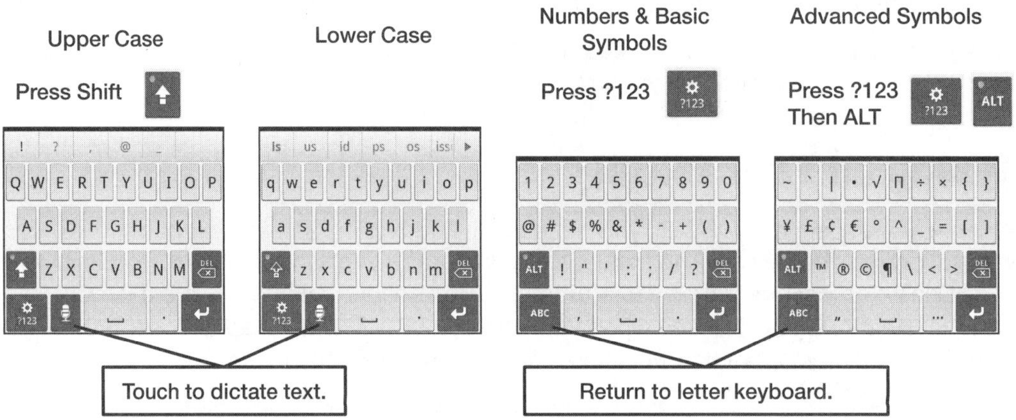

Figure 2–1. *Accessing number and symbol keyboards in Multi-touch mode.*

Accessing Settings and Switching Typing Modes from Multi-touch

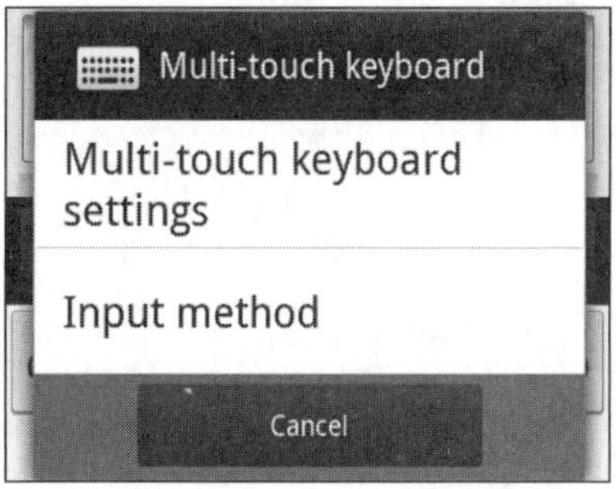

Assume you want to quickly access the **Settings** program or switch typing modes while using the Multi-touch keyboard. From the Multi-touch keyboard, press and hold the **Settings** / **?123** key  in the lower-left corner of the keyboard to bring up the Multi-touch keyboard dialog box. You can also get to this by pressing and holding any input box.

Tap **Multi-touch keyboard settings** or **Input method** to access either function.

> **NOTE:** On the original DROID, you have the option of Android keyboard settings instead of Multi-touch keyboard settings.

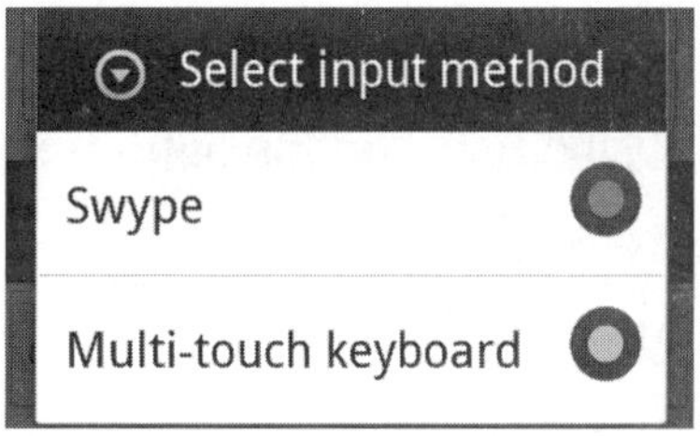

From the **Input method** dialog, you can switch between the **Swype** and **Multi-touch** keyboards.

Swype Typing (DROID 2 and DROID X only)

When using the Swype typing method, you touch-and-drag your finger around the screen to cross each of the letters in the word you are typing. With Swype, you only lift your finger after completely touching all the letters.

> **TIP:** You know you are in the Swype mode of typing if you see the **Swype** (stylized S) key or the **OPT** key (which is visible after pressing the SYM key) in the lower-left corner on your virtual keyboard.

Follow these steps to type the word "hope" with Swype:

1. Touch the keyboard on first letter (h).
2. Drag your finger across the next letters (o) and (p).
3. Lift your finger on the last letter (e).

Using Swype to type the word "hope."

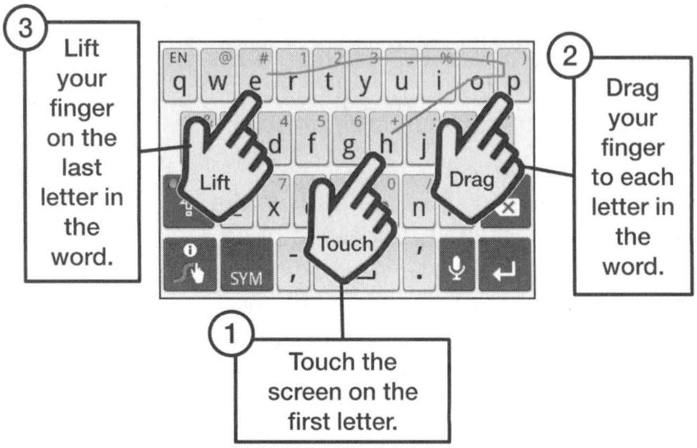

Swype Help and the Built-in Swype Tutorial

If you want to learn more about Swype, the on-board help is pretty good.

To access **Swype Help**, long-press the **Swype** key in the lower-left corner.

To access **Swype Tutorial**, tap the **Tutorial** button at the bottom of the **Help** screen. You will see an **Options** button; this button provides quick access to the settings screens. Follow these steps to navigate through the available options:

- Tap the **Next** button to move forward (the **Next** button is the green, right-facing arrow)
- Tap the **Previous** button to move back a step (the **Previous** button is the green, left-facing arrow).
- Tap the **Close** button (signified by a red "X") or the **Back** key to close the tutorial.

Word Choice Window

If Swype cannot figure out what you are trying to type, it shows you a **Word Choice** window. Follow these tips to navigate the **Word Choice** window:

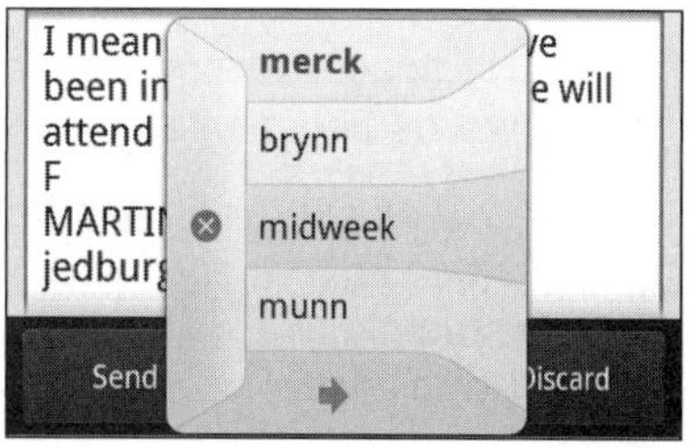

- Tap any word to select it or press the **Space** key to select the **bold** word (in this image, **merck** is bold.)
- Tap the arrow on the bottom of the window to see more choices.
- Tap the **X** button on the left side to close the window.

Various Swype Keyboard Layouts

As with the Multi-touch keyboard, you can use Swype to access various number and symbol keyboards using the buttons in the lower-left corner (see Figure 2–2).

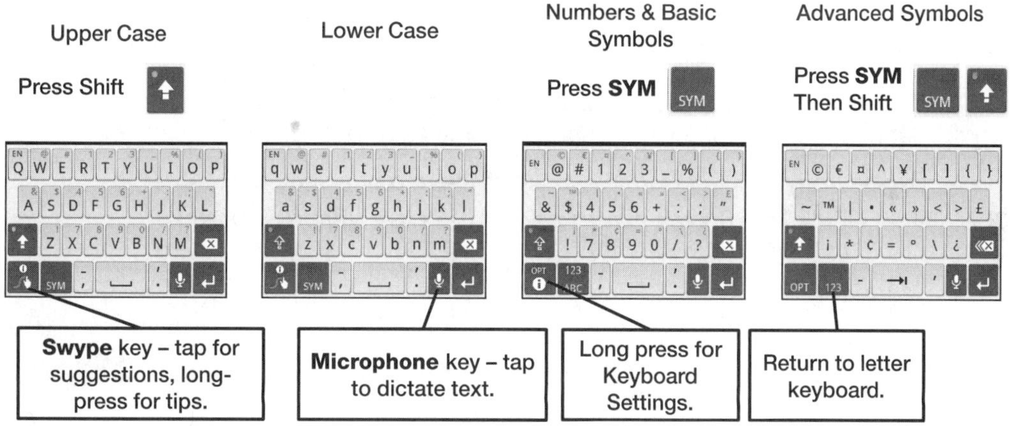

Figure 2–2. *Accessing numbers and symbol keyboards in Swype typing mode.*

Getting to Keyboard Options from the Swype Keyboard

To quickly access options to change keyboard settings or to change from Swype typing back to Multi-touch typing, you press the **SYM** key, and then press and hold (long press) the **OPT** key in the lower-left corner of the keyboard.

Swype Tips and Tricks

Here are some things to keep in mind while using Swype:

- Touch and hold the **Swype** key for tips and tricks about how to use Swype.
- No spaces are necessary. Skip pressing the **Space** key between words because spaces are automatically inserted.
- You can enter double letters by "scribbling" on the letter or making a little circle on the letter you want to double as you drag your finger. For example, when typing "hello," scribble a little on the letter "l" to get the double letter.
- If you want to change the last word typed, tap the **Swype** key to see a list of alternate words appear, and then tap the correct word.
- If you get tired of Swype typing, try using the **Microphone** key next to the **Space** key to dictate your text (see the "Dictating Your Text" section in later in this chapter for more information).
- Sometimes you may want to type in ALL CAPS. Start Swyping as you do normally, then slide your finger until it is above the keyboard. Next, make a circle, and then continue Swyping the rest of the word. It will appear in ALL CAPS.
- If you just can't type a word using the Swyping motion, you can go back to tapping each letter; that still works as a back-up plan.
- Swype from the period to the space at the end of a sentence, rather than tapping each separately – this will save you time and make the next letter you type automatically uppercase.

The Device Keyboard (DROID 2 and Similar)

If you have a DROID 2 or DROID with a slide out keyboard – also known as the Device keyboard – you can slide it out to type (see Figure 2–3).

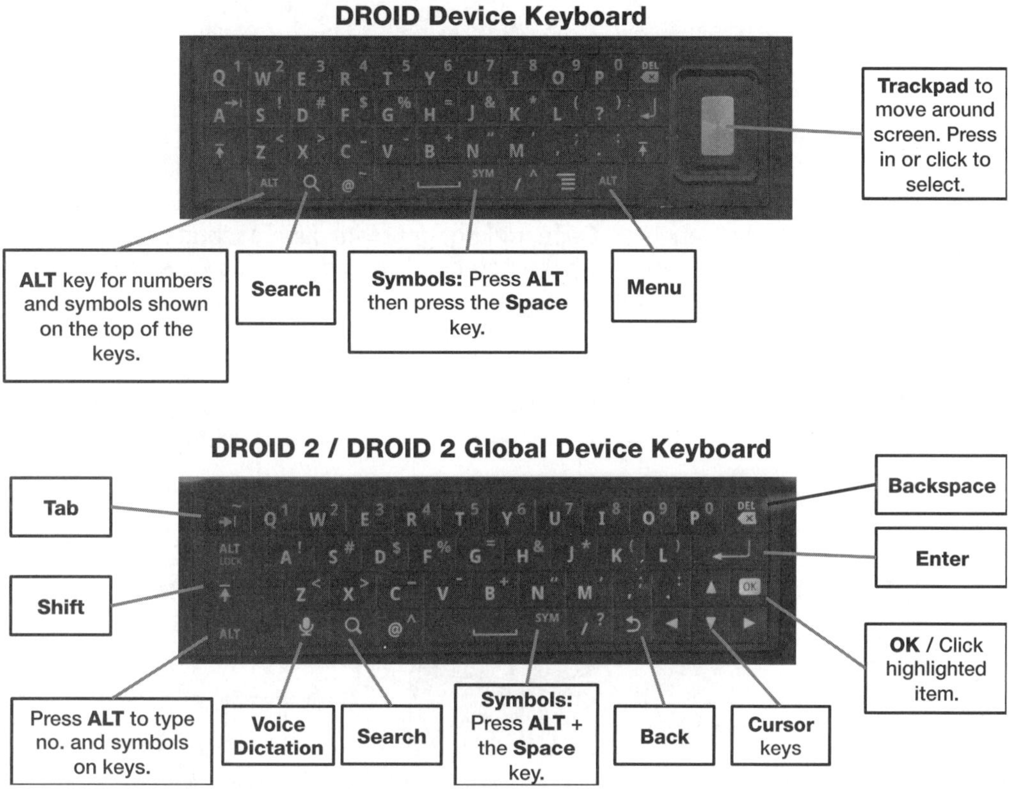

Figure 2–3. *The DROID and DROID 2 slide-out physical keyboard.*

Device Keyboard Options

You can set a few options on your Device keyboard.

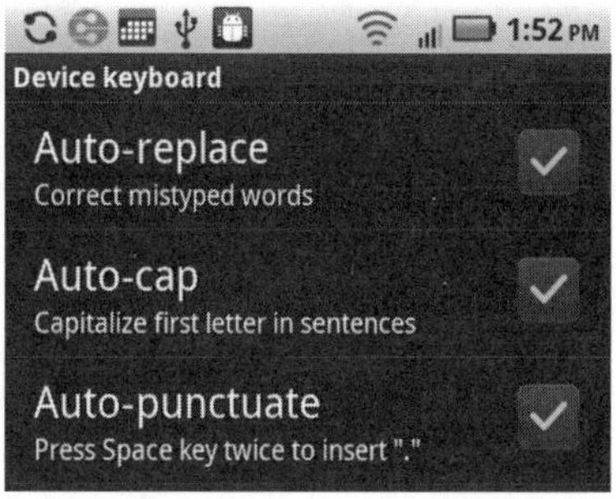

From your **Settings** icon, tap **Language & Keyboard** settings, and then tap **Device keyboard**. This brings up a menu with three options: **Auto-replace**, **Auto-cap**, and **Auto-punctuate**.

Auto-replace automatically corrects misspelled words.

Auto-cap capitalizes the first letter in sentences.

Auto-punctuate allows you to press the **Space** key twice to insert a period at the end of a sentence.

Dictating Your Text

If you get tired of typing, tap the little **Microphone** key to the left or right of the **Space** key on the virtual keyboard. This allows you to dictate your text (see Figure 2–4). In informal testing, the authors found that this dictation worked amazingly well! However, one author's 10-year-old daughter, Cece, did not get quite such accurate results. "My name is Cece," she said. But the DROID typed out, "I'm a meanie cc." OK, so it wasn't quite right, but it gave us a good laugh.

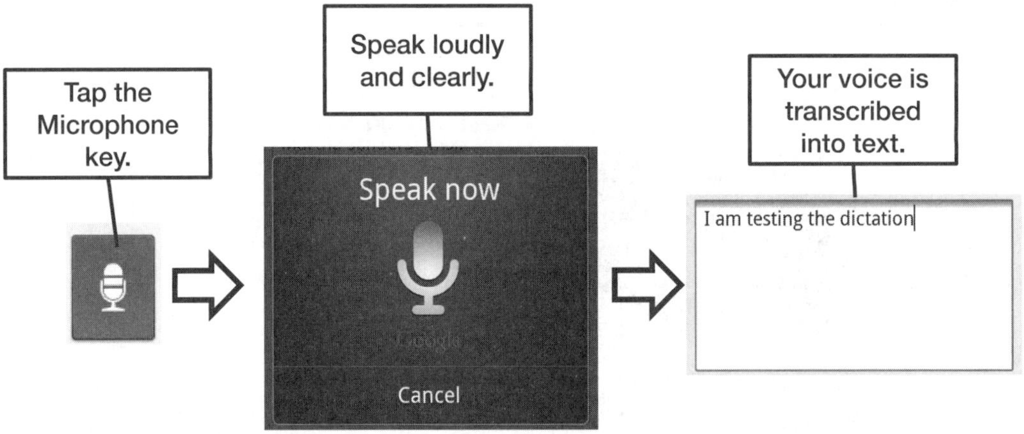

Figure 2–4. *Using the **Microphone** key to dictate text for emails,text messages, notes, and more.*

TIP: To enhance the translation of your dictated words, try these tips:

- Dictate in a quiet place.
- Try to speak slowly and enunciate your words clearly.
- Speak loudly.

Quickly Delete a Word at a Time

To save your finger when you need to delete an entire word or series of words, simply press and hold the **Backspace** key (see the image to the right) to delete a word at a time.

This brings up a little window pop-up saying **Delete Word Back** (again, see the image to the right).

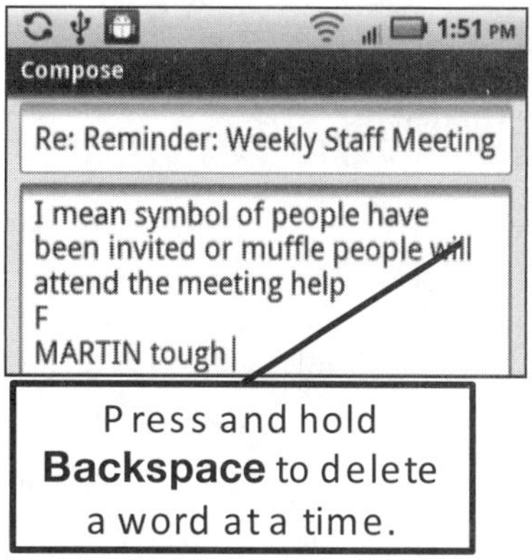

Typing on the Screen With Two Thumbs

You will find when you first start out with your DROID that you can type most easily with one finger – usually your index finger – while holding the DROID with the other hand.

After a little while, you should be able to experiment with thumb typing (as you see so many people doing with other phones). Once you practice a little, typing with two thumbs instead of a single finger will really boost your speed. Just be patient; it takes practice to become proficient typing quickly with two thumbs.

TIP: If you have large hands and fingers, try flipping your DROID on its side to get the larger landscape keyboard!

You will eventually notice that the on-screen keyboard touch sensitivity assumes you are typing with two thumbs. This means that the letters on the left side of your keyboard are meant to be pressed on their left side, and the keys on the right are meant to be touched on their right side (see Figure 2–5).

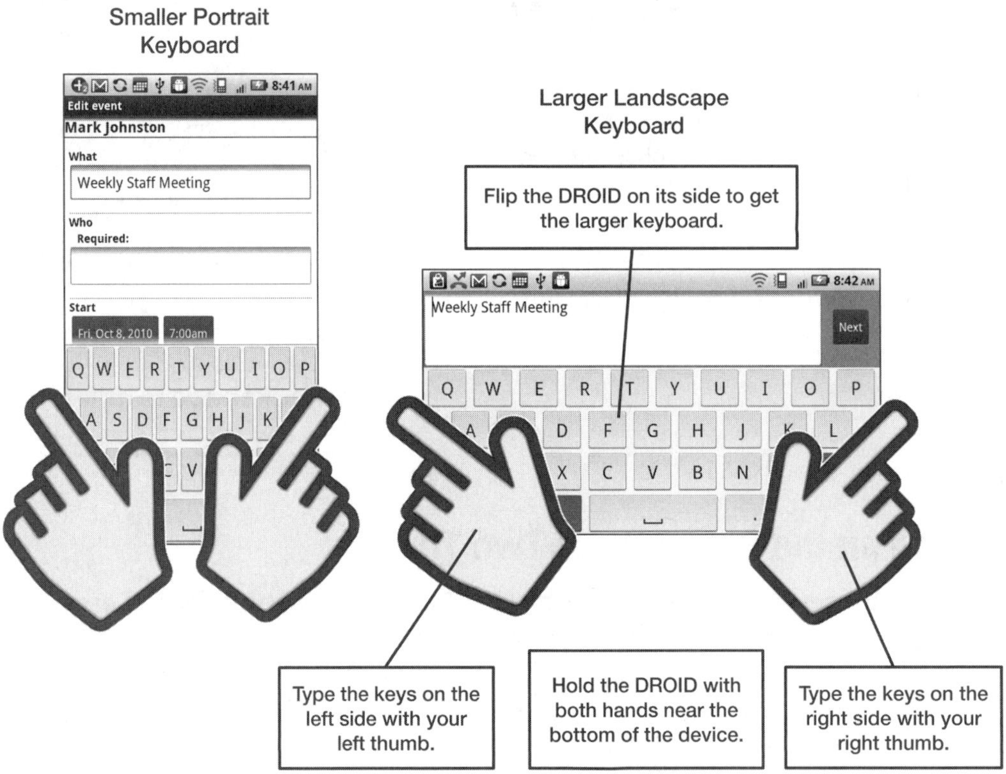
Smaller Portrait Keyboard
Larger Landscape Keyboard
Flip the DROID on its side to get the larger keyboard.
Edit event
Mark Johnston
What
Weekly Staff Meeting
Who
Required:
Start
Fri, Oct 8, 2010
7:00am
Weekly Staff Meeting
Next
Type the keys on the left side with your left thumb.
Hold the DROID with both hands near the bottom of the device.
Type the keys on the right side with your right thumb.

Figure 2–5. *Typing with two thumbs can be much faster than using a single finger.*

Moving the Cursor Around the Screen

As you type, you will want to precisely position the cursor in the text, so you can edit it or switch between fields on a form, such as a calendar. The following sections explain how to do this.

Tapping to Move the Cursor

You can tap anywhere inside a text entry area on the screen to move the cursor. You might do this to correct a sentence you typed in an email message or to jump between fields in a calendar event, as shown in Figure 2–6.

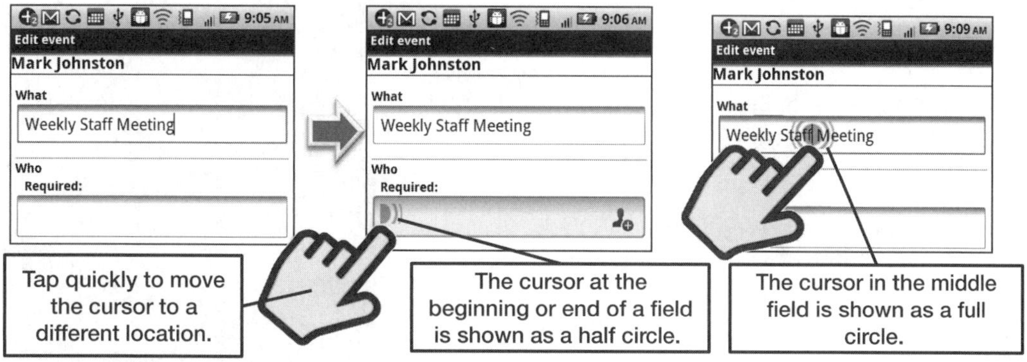

Figure 2–6. *Tap the screen to move the cursor around.*

> **NOTE:** The full and half circles do not appear on the original DROID.

Sliding to Exactly Position the Cursor

If you need to position the cursor on the screen more precisely, then you need to touch and slide your finger around the screen. You will see a little window appear above your finger; this tells you where the cursor is located, as shown in the figure to the right.

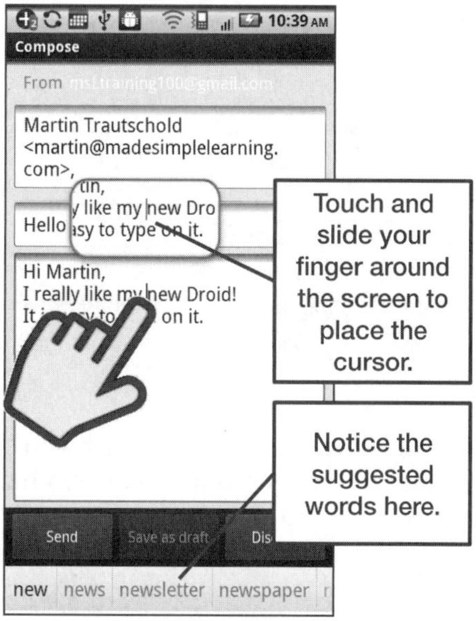

> **TIP:** You will see suggested replacement words based on the word closest to or under the cursor. In the figure to the right, notice that the word "new" is touched. Consequently, the related suggestions are "news," "newsletter," and "newspaper."

Keyboard Vibration and Sounds

When using the on-screen keyboard, you might find it useful to have some audible or sensory feedback as you press the virtual keys. Follow these steps to toggle a click sound or a short vibration on or off:

1. Tap the **Settings** icon.
2. Tap **Language & keyboard**.
3. Tap the type of keyboard you are using (**Multi-touch keyboard** or **Swype**).
4. For the **Multi-touch keyboard**, you can turn **Vibrate** on or off on keypress and **Sound** on or off on keypress by tapping either option. If you are using the Swype keyboard, you can only turn **Vibrate** on or off on keypress.

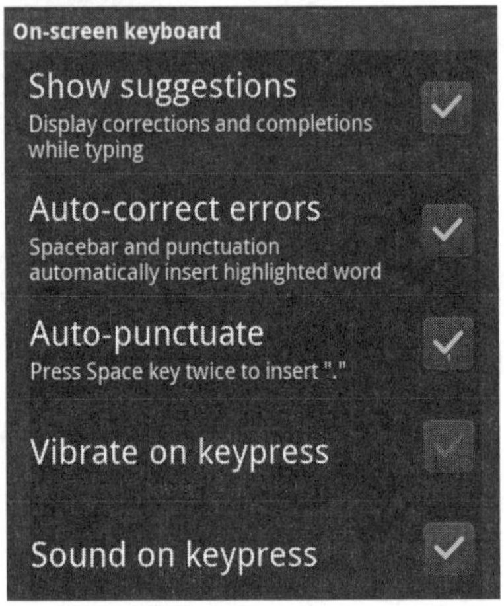

Saving Time With Suggested Words

On the Multi-touch keyboard, you will see a line of suggested words appear in a gray bar directly above the keyboard. These suggestions appear whether you have an on-screen keyboard or physical keyboard, as long as you have the option enabled in your settings.

> **TIP:** If you never see the suggested words appear above the keyboard, then you will have to enable suggestions by selecting your **Settings** app > **Language & keyboard** > **Multi-touch keyboard** or **Swype keyboard**, and then setting **Show suggestions** to on (a **Green check mark** button next to the option means it is selected**)**.

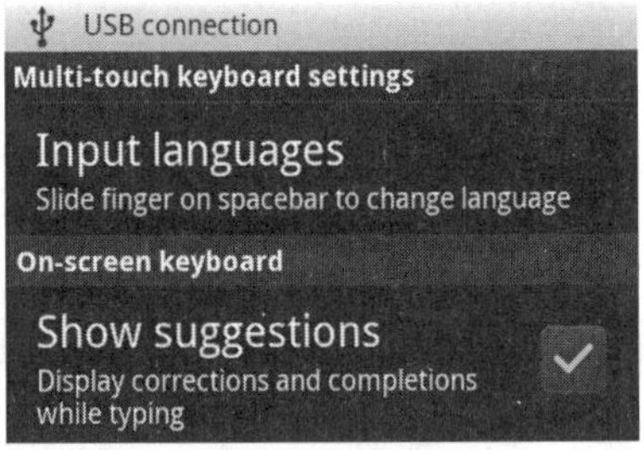

You can save yourself time when you see the correct word guessed (shown in **bold** font) by just pressing the **Space** key at the bottom of the keyboard to select that word (see Figure 2–7).

In this example, we wanted to type the word "Martin" but misspelled it as "marton." The suggested words box displayed the correct spelling of his name with the highlighted word. Since the correct suggestion is highlighted, we can simply tap the **Space** key to select it. If the correct word were not highlighted, then we would need to tap the correct word to select it from the list.

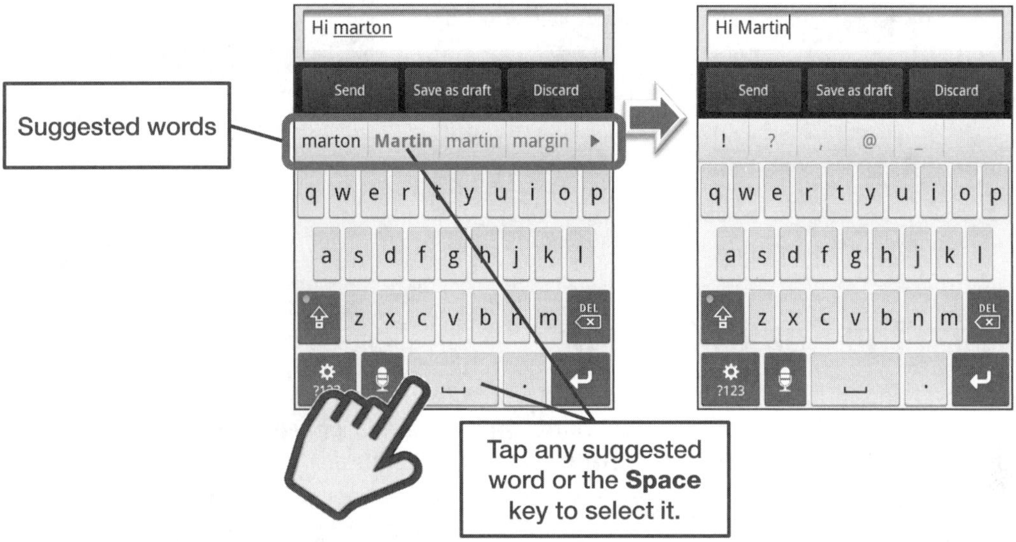

Figure 2–7. *Tap any suggested word to select it or press the **Space** key to select the highlighted word.*

> **TIP:** The **Auto-Replace** feature also looks through your **Contacts** list to make suggestions. For example, if Martin Trautschold was in your **Contacts** list, you would see "Trautschold" come up as a suggested word after typing "Trau" – the figure to the right shows an example of this.

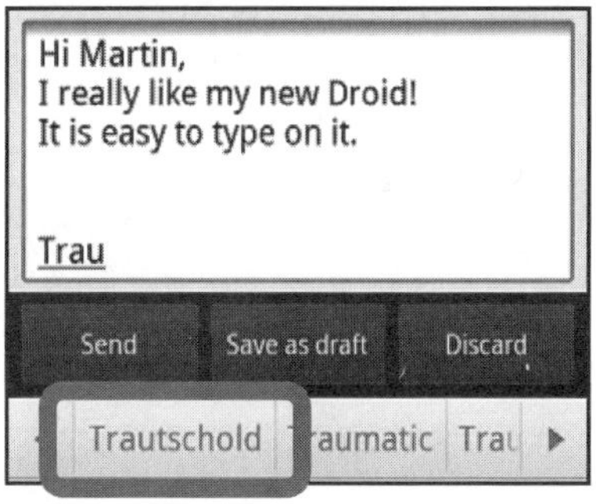

After you get used to tapping the suggested words and using the **Space** key to select highlighted suggestions, you will come to appreciate just how much time this feature saves you.

Sometimes when you hit the **Space** key, the wrong word is selected. In this case, you simply need to press the **Backspace** key  to see the originally typed word. At this point, you can select one of the other suggestions or keep typing.

TIP: With the **Auto-Replace** feature, you can save time by not typing the apostrophe in many common contractions, such as "wont" and "cant." The suggested words will show you the contraction spelled correctly. If the correction is highlighted, press the **Space** key to select it. If not, then tap the word.

Using the Spell Checker

Working hand-in-hand with the suggested words feature is the DROID's built-in spell checker and its user dictionary. Most of the time, your misspelled words will be caught and corrected automatically by the **Auto-correction** feature.

CAUTION: At the time of publishing, if you ignore the suggested correction or the misspelled word is not in the dictionary, then it will not be corrected before you send your email message. The spell checker on the DROID does not check words after you finish typing them; it only checks words as you type them.

Adding Words to the Custom User Dictionary

You can add words to the built-in user dictionary, so they are not auto-corrected by the DROID. Follow these steps to do so:

1. Tap **Settings**.
2. Tap **Language & keyboard**.
3. Tap **User dictionary**.
4. Press the **Menu** key and select **Add**.
5. Type in your new word for the user dictionary and click **OK**.
6. Repeat for as many words as you want to add.

Editing or Deleting Words From the User Dictionary

Navigate to the **User dictionary** screen, as shown in the above image.

Next, long-press (press and hold) any word and then select **Edit** or **Delete**.

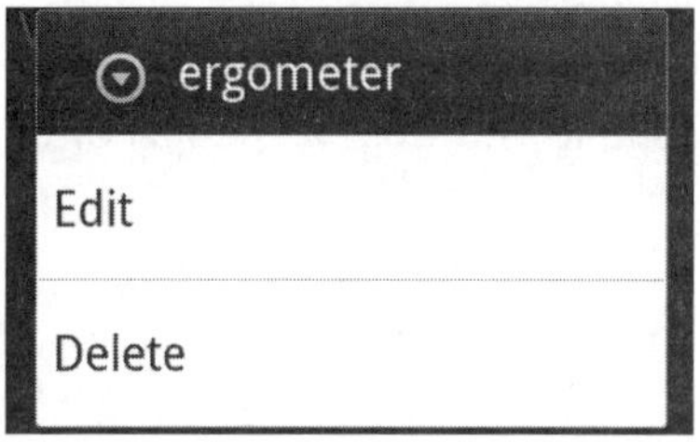

Clearing Out the User Dictionary

It's possible that you will end up adding misspelled words to your user dictionary. If at some point you find that your user dictionary has too many misspelled words, then you can give it a fresh start by clearing out all the custom words. Follow these steps to do so:

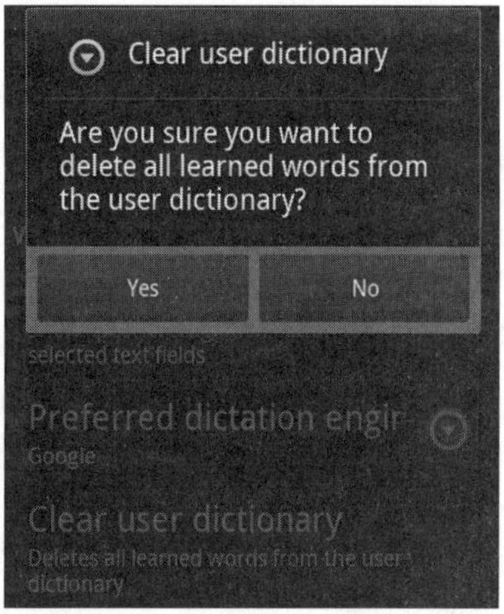

1. Tap **Settings**.
2. Tap **Language & keyboard**.
3. Tap **Multi-touch keyboard**.
4. Tap **Clear user dictionary**.
5. Tap **Yes** to confirm that you want to clear the user dictionary.

The preceding process will clear out all custom words added to your DROID dictionary.

Accessibility Options

There are a number of useful features on the DROID to help with accessibility. For example, the **Voice Readouts** feature will read text on the screen to you. It will tell you what you tap on, what choices are selected, and even read email. If you like to see things in a larger size, you can turn on the **Zoom Mode** feature; this chapter's "Using Zoom Mode to Magnify a Portion of the Screen" section explains how to do so.

Accessibility – Voice Readouts (Reads the Screen)

One cool feature of the DROID is that it has a **Voice Readouts** feature you can turn on to make the DROID will speak anything you tap on the screen. You can even get it to read to you from any email or other text-based document.

TIP: Use a set of headphones when listening to **Voice Readouts** to better hear what is being said and to avoid disturbing others.

Follow these steps to enable **Voice Readouts**:

1. Tap the **Settings** icon.
2. Tap **Accessibility** near the bottom of the page.
3. Tap **Accessibility** again to check it.
4. Tap **Voice Readouts** to check it.

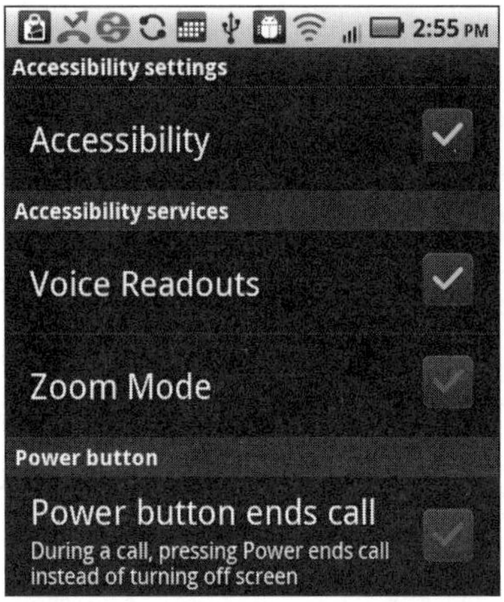

TIP: The **Voice Readouts** gestures are a little different from the normal gestures. A single tap will result in the item being read to you. Once you enable this feature, you have to double-tap an item to select it or set a switch.

NOTE: The original DROID has the options of TalkBack, SoundBack and KickBack. TalkBack recites menu options, app titles, contacts, etc when scrolling through and opening. SoundBack plays a tone when scrolling menus and a different one everytime a menu is selected. KickBack is like SoundBack but with vibration.

Using Zoom Mode to Magnify a Portion of the Screen

You may want to turn on the **Zoom Mode** feature if you find that the text, icons, buttons, or anything else on the screen is a little too hard to see. Zoom mode is not available on the original DROID.

Follow these steps to enable **Zoom Mode**:

1. Tap the **Settings** icon.
2. Tap **Accessibility** near the bottom of the page.
3. Make sure **Accessibility** at the top is checked.
4. Tap **Zoom Mode** to check it.
5. You will see a warning that the **Zoom Mode** will collect everything you type. Tap **OK** to continue.

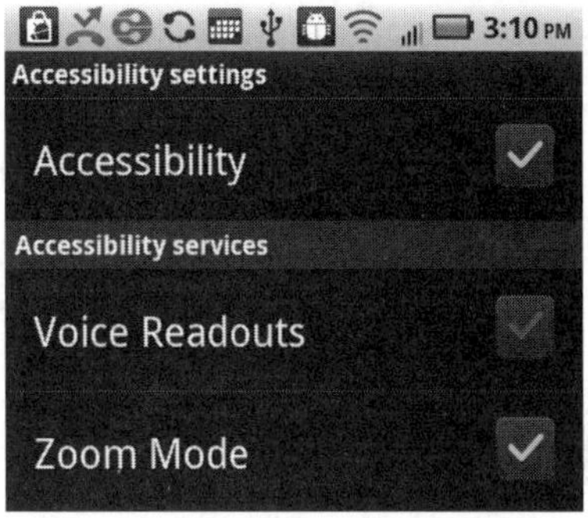
3:10 PM
Accessibility settings
Accessibility
Accessibility services
Voice Readouts
Zoom Mode

With the **Zoom Mode** turned on, you will see a box with gray bars on the top and bottom. In the upper- and lower-left corners, you will also see + (plus) and – (minus) symbols. Tap the + symbol to zoom in and the – symbol to zoom out. Drag either gray bar to move the screen around and zoom in on something else (see Figure 2–8).

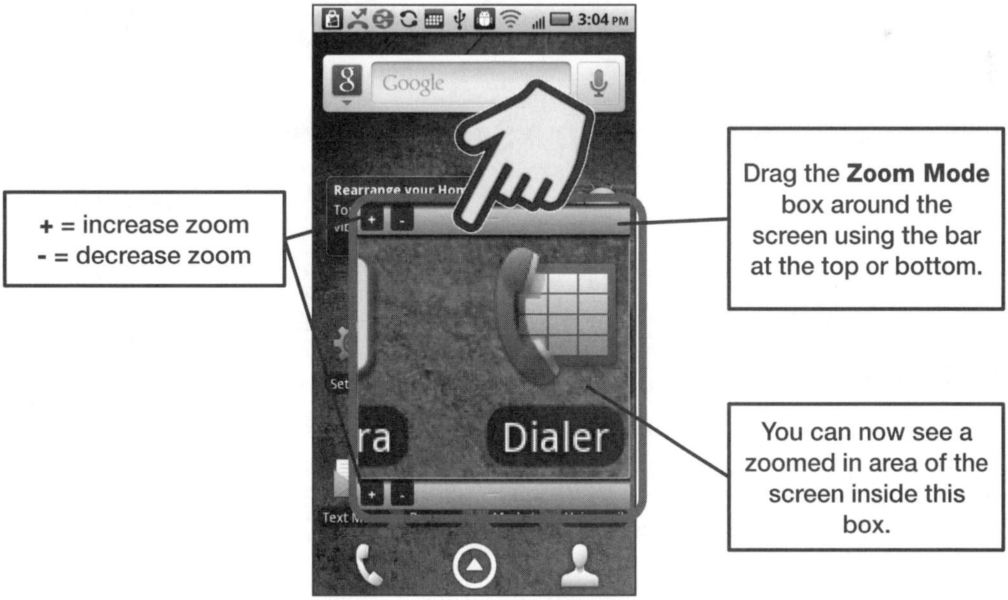
+ = increase zoom
- = decrease zoom
Drag the **Zoom Mode** box around the screen using the bar at the top or bottom.
You can now see a zoomed in area of the screen inside this box.
3:04 PM
Google
Dialer

Figure 2–8. *Using **Zoom Mode** to expand a portion of your DROID screen.*

Double Tap the Home Button Options

The default setting for when you double-click the **Home** button is to start up **Voice Command**, but you can adjust this behavior in your settings. Follow these steps to do so:

1. Tap your **Settings** icon.
2. Tap **Applications**.
3. Tap **Double tap home launch**.
4. Select from any of the available options.

Note that this list is longer than the screen, so scroll up and down to see all the options available.

You can choose from **Browser**, **Camera**, **Contacts**, **Dialer**, **Gmail**, **Maps**, **Messaging**, **Music**, **News**, **Social Networking**, **Text Messaging**, **Voice Command**, and **Voice Search**.

> **NOTE:** Double tap home launch is not available on the original DROID.

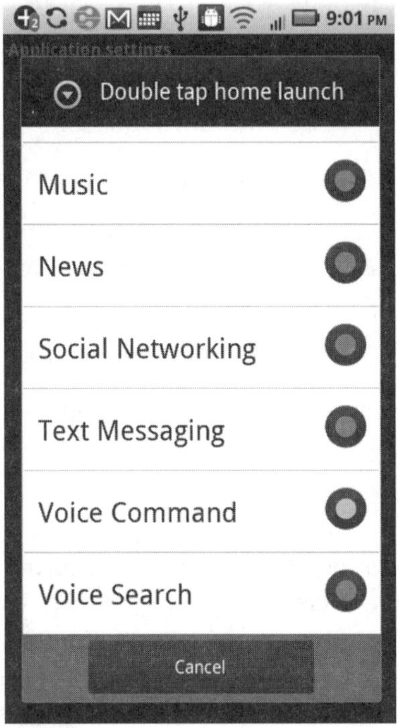

Pressing and Holding to Type Symbols (and More)

You might wonder how you type symbols that aren't shown on the standard keyboard.

> **TIP:** You can type more symbols than are shown on the screen.

To do so, simply press and hold a letter, number, or symbol that is related to the symbol you want. If you see three dots (...) in the lower-left corner of a letter, that's your tip that pressing and holding that letter will show you more character options (see the figure showing the letter "E" to the right).

Several characters have special, additional characters associated with them, including all the vowels; the letters "C" and "N"; and the **.** (period) and **$** (dollar) symbols.

For example, follow these steps if you want to type the **¥** (yen) symbol:

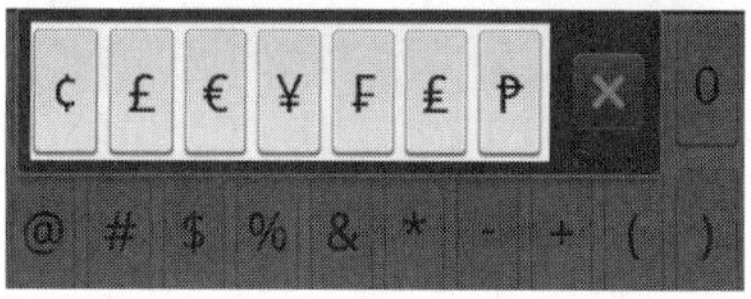

- Press and hold the **$** key until you see the other options.
- Slide up your finger to highlight the **¥** symbol and then let go with your finger over that symbol.

Toggling Caps Lock

You double-tap the **Shift** key to turn on **Caps Lock**. You know **Caps Lock** is turned on when the little dot in the upper-left corner of the key turns green. An easy way to see whether **Caps Lock** is on: all the letters on the keyboard will be shown in UPPER CASE.

To turn off **Caps Lock**, simply press the **Shift** key again.

Quickly Changing a Word

If you want to quickly change a single word, tap it once to put the cursor in it, as shown in the figure to the right. The cursor in this image is in the middle of the word "quickly." This causes the suggested words shown at the bottom of the image to be related to the word "quickly." If you see the suggestion you want, tap it. If not, tap the gray triangle in the lower-right corner to see more suggestions.

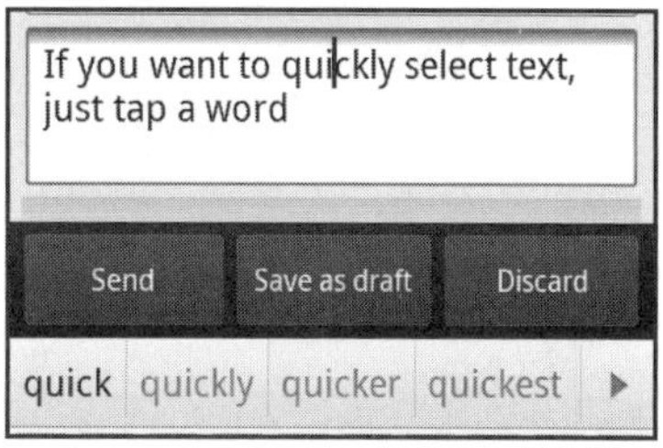

Quickly Change Text

Sometimes you need to quickly change or delete some text you are typing. Follow these steps to do so:

1. Begin by selecting the desired text by double-tapping it. Notice that the word you tapped is now selected.

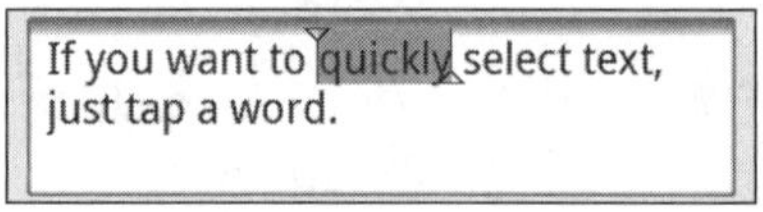

If you want to quickly select text,
just tap a word.

2. Adjust the selection by dragging the blue handles.

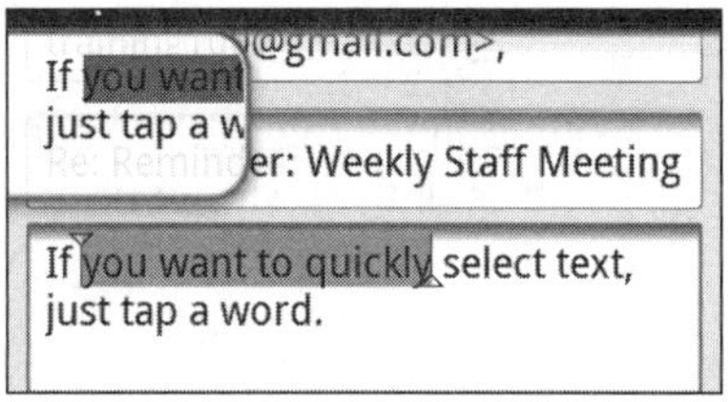

If you want
just tap a w
er: Weekly Staff Meeting
If you want to quickly select text,
just tap a word.

3. To erase the selected text, press the **Backspace** key. 

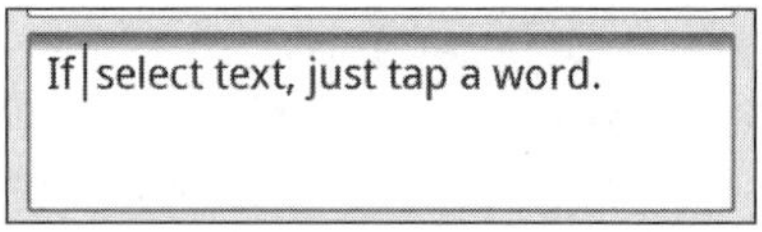

If select text, just tap a word.

4. To replace the text, simply start typing. The text will be instantly replaced by the letters you type.

Changing the Language and Keyboard Options

The DROID includes a few keyboard options to make typing on your DROID easier. The keyboard options are located in the **Settings** app. Follow these steps to change these settings:

1. Tap the **Settings** icon.
2. Tap **Language & keyboard**.
3. To change your language, tap **Select Locale** and select another language.
4. To change your typing method, tap **Input Method** and select between **Multi-touch keyboard** and **Swype**.
5. To adjust the settings particular to the various keyboards, tap each type of keyboard: **Swype**; M**ulti-touch keyboard**; and if you happen to have a device keyboard (DROID 2 models), **Device keyboard**.
6. Tap **User dictionary** to make adjustments to your user dictionary. For example, you might add, edit, or delete words to your custom dictionary.

If you changed your local language to **Español**, then you would see all the labels and menus on the DROID change to Spanish.

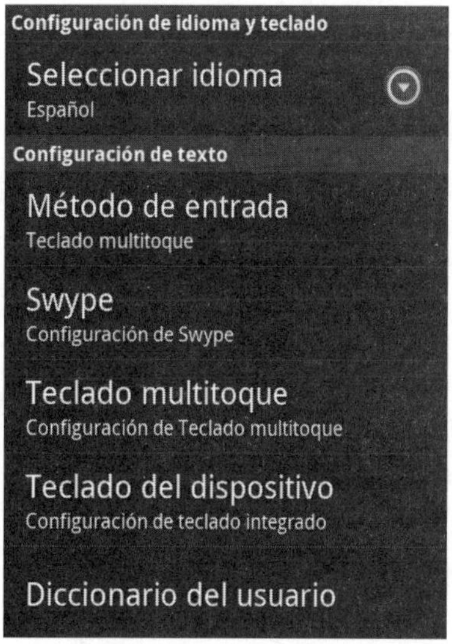

Voice Command

The DROID 2 and DROID X comes with a nice feature called **Voice Command** that allows you to control many aspects of your phone using your voice.

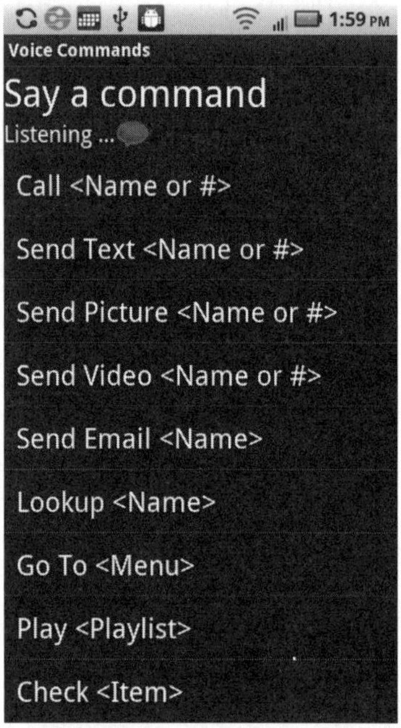

Double-pressing the **Home** button will usually bring up **Voice Command**. You can also tap the icon to start it.

If the default behavior of double-clicking the **Home** button has been changed (DROID 2 and DROID X only), you can change it back to start **Voice Command** by following the steps we showed you in this chapter's earlier section, "Double-Clicking the Home Button Options."

The **Voice Command** feature lets you do the following:

- Call a person by name or number.
- Send a text message.
- Send a picture.
- Send a video.
- Send an email.
- Lookup someone's contact information.
- Go to a menu.
- Play a playlist.

When you say (or tap) **Check,** you can check any number of status items, as shown in the image to the right.

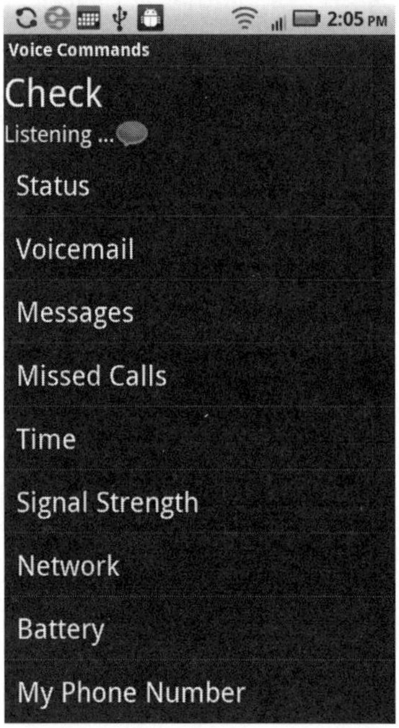

Texting by Voice

The **Voice Command** feature lets you do a lot of cool things. For example, you can use this feature to dictate and send an email to someone. Follow these steps to do so:

- Say “Send text” and then the person’s name.
- Select the desired phone number or email address when prompted.
- You will see a screen with the addressee. Tap the **Enter message here** window to write the message.
- You can continue dictating the message by tapping the **Microphone** icon next to the **Space** key on your keyboard. In the image to the right, we just finished dictating the exact message shown as underlined. If the dictation was incorrect, one tap of the **Backspace** key would erase the entire underlined text.
- Tap the **Send** button to send the message.

NOTE: On the original DROID, just start the Messaging app and then touch the Microphone icon to speak your text.

Using Copy-and-Paste

The ability to copy-and-paste is very useful for saving time and increasing your accuracy. You can use this feature for taking text from your email (such as meeting details) and pasting it into your Calendar. Or, you may want to simply copy an email address from one place in a form to another to save yourself the time and trouble of retyping it. There are lots of places to use copy-and-paste; the more comfortable you are with it, the more you will use it.

Selecting Text

If you are reading or typing text, you can select text by double-tapping a word and then dragging the handles as we showed you in the "Quickly Change Text" section earlier in this chapter. You can also select text by long-pressing it and choosing **Select all** or **Select text**.

TIP: In the **Browser**, you can save an image to your DROID by pressing and holding it until you see a pop-up menu. Tap **Save image** to save the image or **Set as wallpaper** to set the image as your DROID wallpaper.

Cutting, Copying, and Pasting Text

Once you have the text that you wish to copy or cut highlighted, long-press again and tap **Copy** or **Cut**.

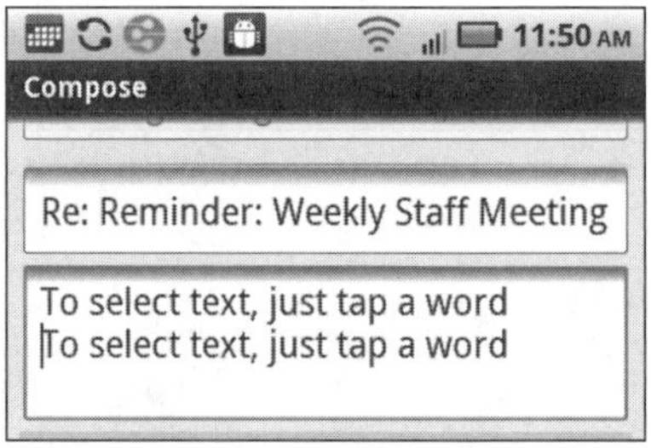

If you want to paste the text in the same screen, reposition the cursor by tapping the screen, and then long-press and select **Paste**.

If you want to paste the text into another app, use the multitasking steps shown in the next section.

Jumping Between Apps or Multitasking

After you copy text, you may want to paste it into another app. The easiest way to jump between apps is to use the **Recent** list. Follow these steps to paste text you copy into another app:

1. Copy or cut your text.
2. Long-press the **Home** button to bring up the **Recent** list of the eight most recent apps you have opened on your DROID.
3. If you just left an app running in the background, you will be able to find it in the top left position of the **Recent** list. Tap any of the eight apps to jump to it.

TIP: If you don't see the app you want to start in the **Recent** list of eight apps, then tap the **Home** button, and then the **Launcher** icon to see all your apps. From here, you can fire up the app you need.

4. Paste the text by pressing and holding the screen and selecting **Paste** from the pop-up.
5. Long-press the **Home** button again to see **Recent** list and tap the app you just left to jump back to it.

Finding Things with Google Search

As you might expect from the creator of the most popular search engine on the planet, Google's Android operating system has a very nice search feature. You even have a dedicated **Search** button on the bottom of your DROID, as shown to the left. Follow these steps to start a search for something on your DROID:

1. Press the **Search** button on the bottom of the DROID and start typing your search word or phrase.
2. Instantly, you will see matching items and search suggestions appear. Tap the item or search suggestion you desire to view or use.

Modifying Your Search

It's also a simple matter to change the parameters of your search. Tap the icon (usually the Google "g" ) just to the left of the search box to adjust what you're searching for. The default setting for search is **All**, but you can also search only the **Web** or **Apps** from the Android Market.

If you want to see more items that you can search for, tap the **Settings** icon in the upper-right corner of the pop-up window with the search icons.

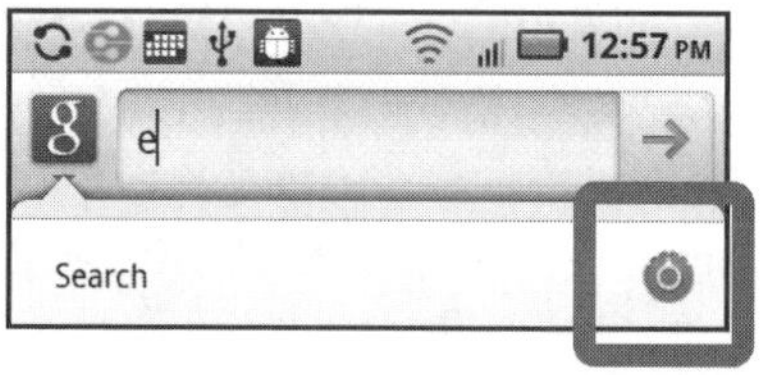

From this window, you can check or uncheck various items to include in the search. By default, only the **Web** and **Apps** options are checked.

For example, you can check the **Kindle** app to search titles and authors of your Kindle books.

You can also check **Contacts** to search your DROID contact list.

Or, you can check **Music** to quickly search for artists, album names, and song names.

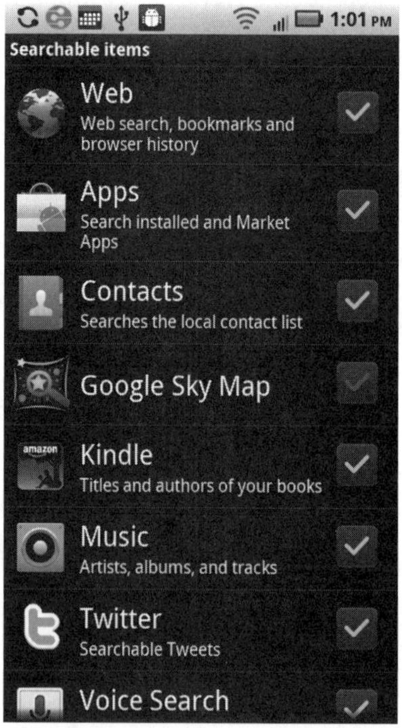

Performing the Search

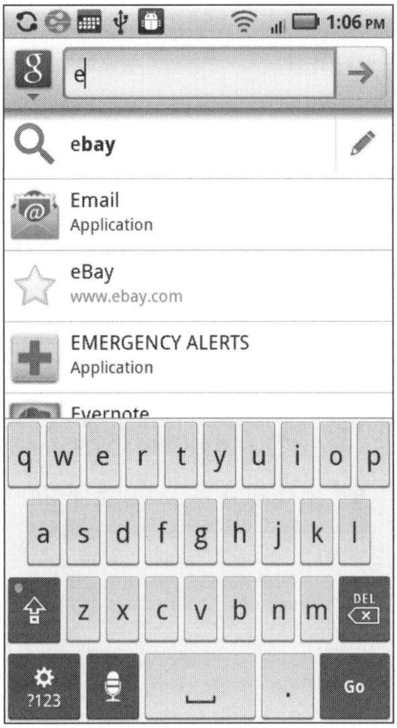

The figure to the right shows the results of typing the letter "e" into the **Search** window. Doing so displays a number of apps and suggested search terms.

You have several options at this point. You can tap a given app icon to jump into that app or tap the search term to execute a Google search with your search term.

You can also tap the **Pencil** icon to the right of the search term to select the chosen search term and continue typing more in the **Search** box.

> **TIP**: To hide the virtual keyboard, swipe your finger down from the top of the keyboard toward the bottom.

Performing a Voice Search

After you bring up the **Search** window, tap the **Microphone** icon to the right of the **Search** window to use your voice to dictate the search.

Speak loudly and clearly when you see the **Speak now** window as shown to the right.

Once you stop speaking, the DROID will take a second and do its best to type what you just spoke and search for it.

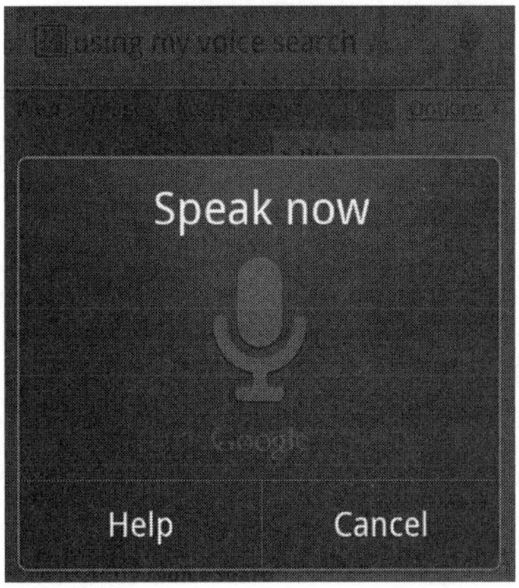

In the image shown to the right, we searched for "pizza 32174." This search quickly found all pizza restaurants in the ZIP code 32174.

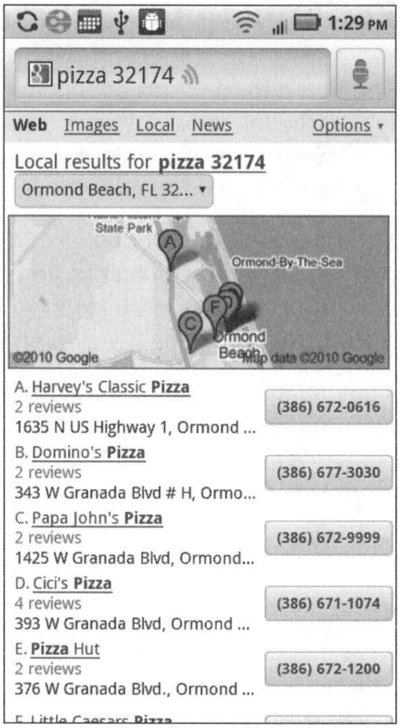

You could search for anything, including the following:

- Golf courses
- Bars
- Plumbers
- Libraries
- Grocery stores
- Panera Bread stores
- Your favorite gas station and more!

Adding the Google Search Widget to Your Home Screen

If you use the search feature often, you will want to add a **Google Search** widget to your **Home** screen. Follow these steps to do so:

1. The **Google Search** widget requires four empty spaces across to be placed. Locate such an empty space on your **Home** screen and press and hold the screen.
2. Select **Android widgets**.
3. Select Google Search.

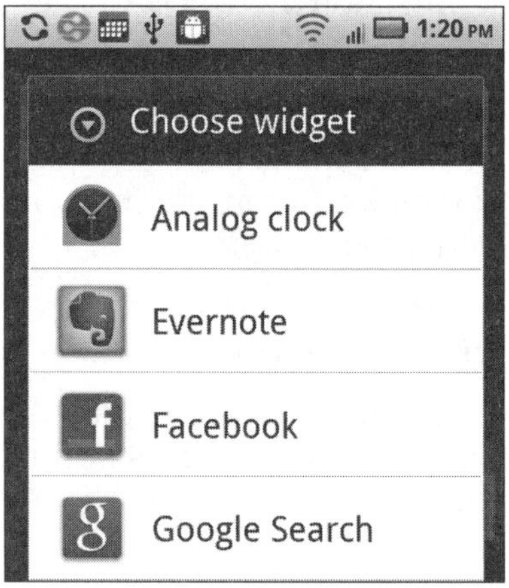

Now you have instant access to **Google Search** right from your **Home** screen.

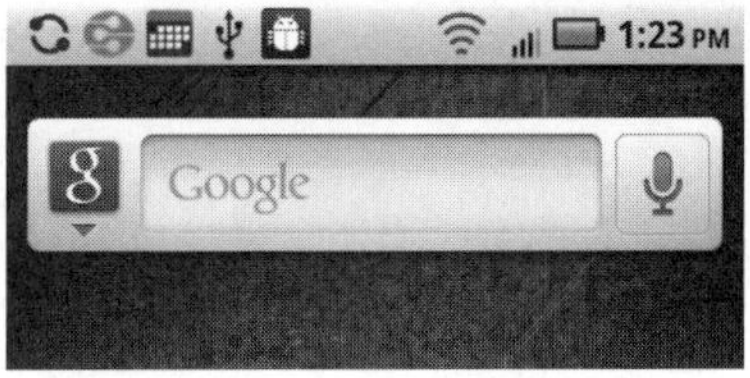

Chapter 3

Sync Your DROID With Your Google Account

In order to use a DROID, you must have a Google Account. You can still check your work email and use other accounts, but you must have at least one Google Account. When you activate a new DROID, setting up a Google Account is the second thing you'll do, right after activating the phone number itself with Verizon Wireless.

Your Google Account serves as your default contact list, email, and calendar on your DROID. You will be able to view, manage and update your Google Contacts and Calendar using the **Contacts** and **Calendar** apps on your DROID. You can add other services such as Twitter, Facebook, and **Microsoft Exchange** email (we'll get to that in Chapters 9 and 20); however, by default all new contacts you create directly on your phone go into your Google Account.

Activating Your Phone

If you already have a DROID in your hand, chances are you've already activated it. If you're still considering a purchase or an upgrade, here are the steps for activation:

1. Tap on the screen. You'll see an Android with a hand.
2. Follow the instructions on the screen.
3. Your phone will dial a special number for activation.
4. You'll see a message that your phone has been activated.
5. It may take up to 15 minutes for service. Meanwhile, you can press the **Next** key to begin a tutorial on activating and using your DROID. This tutorial will also guide you through setting up your Google Account.

Activating Your Google Account

You have two basic choices: you can use a Google Account you've already created, or you can create one. Once you've completed this step, you can also go back and add multiple Google Accounts; thus, if you have one email for home and another for work, it's no problem.

Figure 3–1 shows the activation steps for creating a new Google Account, and Figure 3–2 shows the steps for signing in with your existing account.

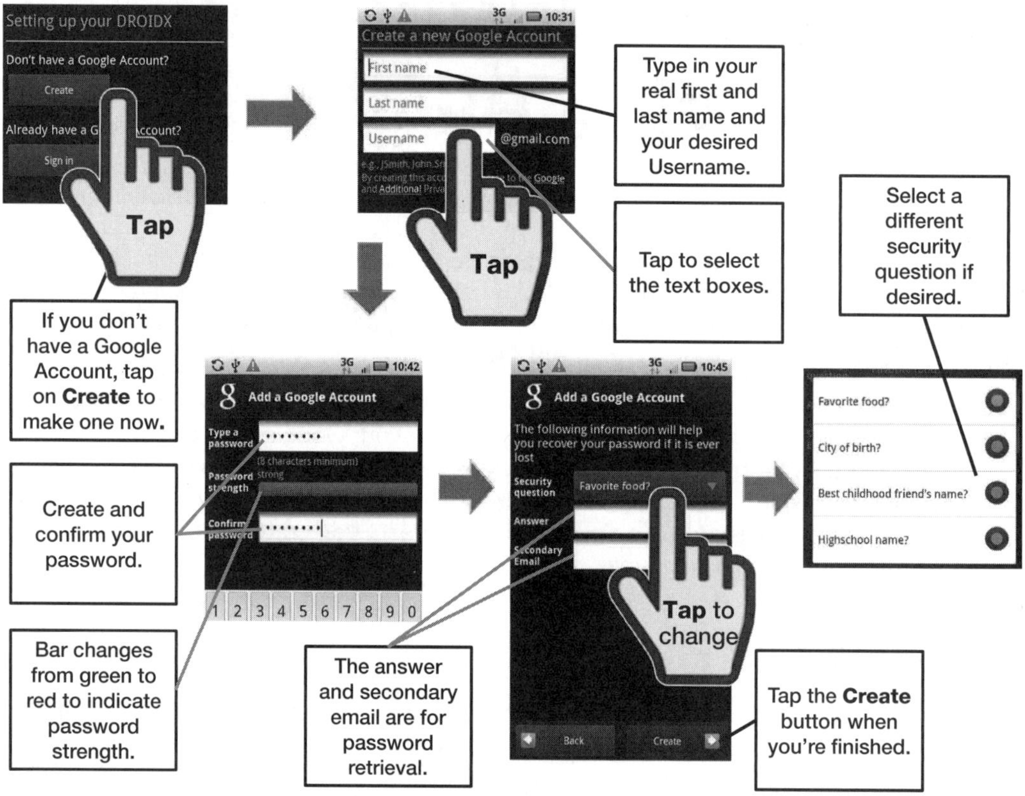

Figure 3–1. *Setting up a new Google account.*

Follow these steps to create a new Google Account:

1. If this is your *first* Google Account, press the **Create** button.
2. Enter your first and last name and your desired **Gmail** username in the appropriate boxes.
3. Google will check to see if your username is available. If not, it will suggest alternatives.
4. Once you've selected an available username, you need to create a password.

5. Google will indicate if this is a "strong" password. Our experience is that Google overrates the strength of passwords, so try to create a password that has at least one number, punctuation character, and at least one capital letter.
6. Next you'll need to add security questions in case you ever forget your password. You can click the security question to select a different question.

CAUTION: Keep in mind that identity thiefs can use Google to find obvious answers about you, so avoid questions such as your city of birth or high school. You could also make up an answer to these questions that you'll remember but strangers can't research.

7. Now you should enter a secondary email address. This is where your password reset instructions will be emailed.
8. Agree to the Google Terms of Service by pressing **I agree, Next**.
9. Congratulations! You're the proud owner of a new Google Account.

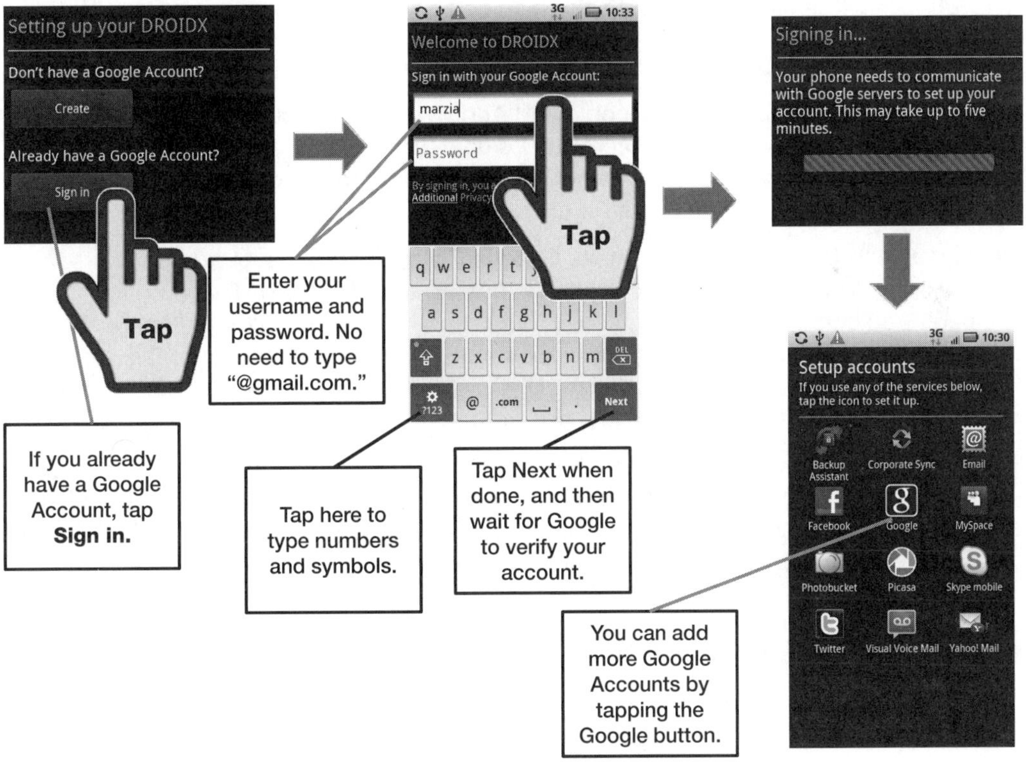

Figure 3–2. *Adding an existing Google Account.*

Follow these steps to sign into an existing Google account:

1. If you already have a Google Account, tap the **Sign in** button.
2. Enter your username and password in the appropriate fields.
3. Google will verify your username and password.
4. Congratulations, your phone is set up to sync with your Google Account.

TIP: You don't need to type your full **Gmail** address in the **Username** field. Just type the part that goes before the @ sign, and your DROID will autocomplete the rest.

Once you're done adding the first account, the DROID will ask if you'd like to add more Google Accounts. To add multiple accounts, just press the **Google** button and repeat the same steps. When you add new Google Accounts, you'll need to decide if you want to sync your contacts, **Gmail**, or both (you'll learn more about this in the "Fine Tuning Your DROID Sync" section later in this chapter).

The Wonders of the Cloud

One of the driving forces behind Google and Android is the concept of *cloud computing*. Rather than rely on a single computer with a single processor, Google relies on a bank of servers on the Internet for nearly every product or service it produces—in other words, these apps and services are "in the cloud."

When you send **Gmail** messages, make **Google Calendar** entries, or add contacts to your **Google Contact** list, that information is available on your DROID and on any computer you connect to the Internet with when you log into your Google Account. You do not need to connect your DROID to your computer to get the benefit of this syncing. And if you should accidentally break your DROID, you'll probably be very sad, but you won't be without your Google contacts or calendar.

Have you ever seen someone update their Facebook status or send out a frustrated email where they beg everyone to send them their contact information again to replace all the numbers they lost with their broken phone? As long as you stick with DROIDs and other Android-powered phones, that will never be you.

More Reasons You Need a Google Account

If you purchase apps in the Android Market, you'll use your Google account and **Google Checkout** to complete the transaction. The default email account on your DROID is **Gmail**, and the default calendar is **Google Calendar**.

Table 3–1 lists a few of the Google services you'll get to know as you use your phone.

Table 3–1. *Drilling Down on Google Account Services.*

	Gmail:	**Gmail** is a free web-based email service, but it's good enough to replace those email accounts your Internet service provider gives you. We'll talk about email in greater detail later. Make sure you register for an account. Some Android phones will not let you activate them without one.
	Google Voice:	**Google Voice** is a VoIP (Voice over IP) service that allows you to use a single phone number to forward your calls, create a visual voicemail message with text transcription, and make low cost international long-distance calls. **Google Voice** isn't a pure VoIP service at this time. You still need a phone in order to use it, although you can initiate calls from your computer using **Gmail**.
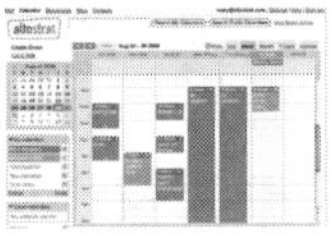	**Google Calendar:**	**Google Calendar** works a bit differently from the calendar in **Microsoft Outlook**. It includes standard features like events and invitations, but **Google Calendar** is meant to be even more collaborative. You manage **Google Calendar** by adding multiple calendars and sharing them with others. For instance, you can have a calendar you allow colleagues to see but not edit, a calendar team members can all edit, and another calendar of fully public events.
	Google Maps:	You're probably already familiar with this map application. **Google Maps** is the engine behind most of your phone's geographically sensitive apps. **Google Maps** does more than provide you driving directions; it can also give you walking and public transport directions. This is invaluable when you're on the road.
	Google Checkout:	**Google Checkout** is a tool for buyers and merchants to complete credit card transactions without revealing the credit card info to the merchant. It's a competitor to PayPal. You'll need to set up an account with this service that contains your credit card information if you want to purchase apps from the Android Market.
	Picasa:	The web-based albums in **Picasa Web Albums** are Google's answer to Flickr. If you want to upload pictures from your phone to the Web, this is the default location for sharing such pictures on the DROID. You may want to set up your account with albums and public or private sharing permissions if you need to share photos as part of your job. It's more efficient to upload photos to **Picasa**

than it is to send them as email attachments, though you can do both. **Picasa** also has a desktop program you can use for syncing and editing photos.

YouTube: If you have any reason to take quick videos with your DROID, set up a YouTube account with your preferred username beforehand. You can upload videos directly instead of offloading them to your desktop computer first. You can also use a YouTube account to comment, rate, and add videos to playlists.

TIP: Most DROIDs do not come with a SIM card. The exception is the DROID 2 Global. This is because they run on Verizon Wireless' CDMA network and not on a GSM network. This means you won't be able to transfer contacts by popping a SIM card between phones, but your Google contacts will still sync with new Android phones.

What to Do When your Calendar and Contacts are Not Already in Google

If you already use Google for your Contacts and Calendar, by setting up your Google Account on your DROID, these contacts and calendar events flow automatically and wirelessly to your DROID. You're done.

But what happens if your contacts and calendar are stored in Microsoft Outlook, the Apple Address Book, another desktop application or even just your old phone?

TIP: See Chapter 4: "Other Sync Methods" to learn about ways to sync or share information between Microsoft Outlook, Apple Address Book, Microsoft Entourage, iCal, ACT!, GoldMine, Lotus Notes and other applications and your DROID.

Getting Information from Your Old Phone

In the case that you only have your contacts and calendar stored on your old phone and not in any other application, you should try to transfer that information into Google directly or into another desktop application or file.

How you get this accomplished varies widely based on the type of phone you have, so we are not able to provide step-by-step instructions here. Instead, you have a couple of options: Option one, ask your Verizon representative for assistance. Option two, do a web search that specifically identifies your phone and says "transfer contacts from (my phone name) to DROID or Google." If you can figure out how to get your contacts from your phone into a desktop application, or file of the format vCard or .csv (comma separated variable) format, then you will be able to follow the steps below to import them into Google.

> **NOTE**: If you can get the data to **Google Contacts** at `http://www.google.com/contacts`, it will appear on your phone after you setup your Google Account on your DROID. The backup plan, if none of the sync or import options work is that you can add your contacts manually. See Chapter 12: "Using your Contacts."

One-time Import of Contact Information from Outlook or the Apple Address Book

Follow these steps to perform a one-time import of your contacts from **Outlook** or the **Apple Address Book** to your DROID phone:

1. Export your contacts as a `CSV` or `vCard` file.
2. Use the import link on the upper-right corner of **Google Contacts** as shown in Figure 3–3.

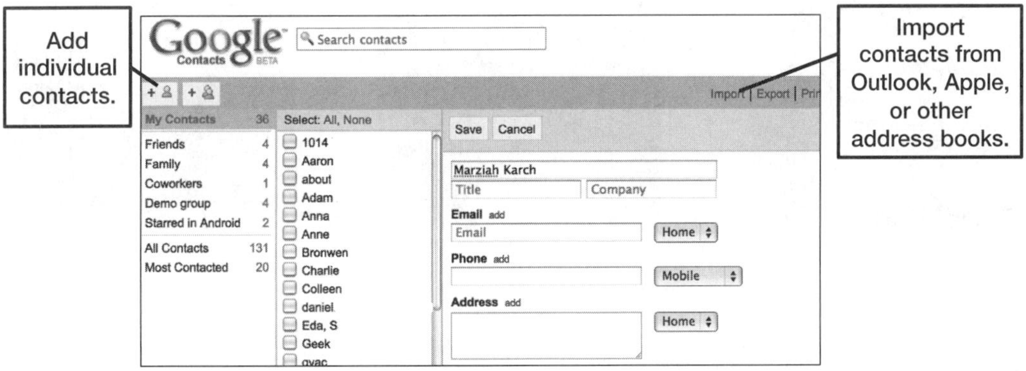

Figure 3–3. *Importing contacts from* ***Google Contacts****.*

Fine Tuning Your DROID Sync

When you add Google Accounts, you choose how much information you want to sync. You may not want to check your work email on your phone, or you may want to read your email but ignore the contact list. It's your choice. The three basic Google Account items you can sync on a DROID are your **Google Calendar**, your **Gmail**, and your **Google Contacts**.

Figure 3–4 guides you through the steps for enabling and disabling Google syncing; follow these steps to do so:

1. From the **Home** screen, press the **Menu** button.
2. Tap **Settings**, then tap **Accounts**.
3. Tap the Google account you wish to modify.
4. Green checks indicate syncing, and gray checks indicate syncing is disabled. Tap on the checkbox to switch syncing on or off for a service.
5. Press the **Back** button until you return to the **Home** screen.

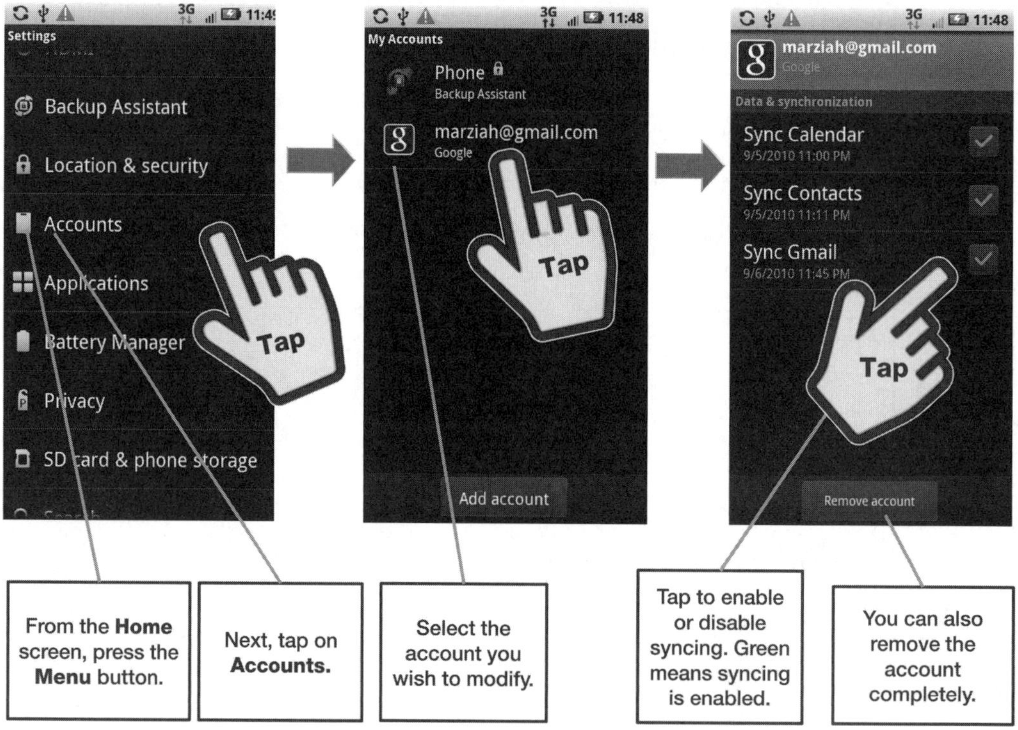

Figure 3–4. *Syncing Google accounts.*

Chapter 4

Other Sync Methods

In Chapter 3: "Sync Your DROID with Your Google Account," you learned how to connect your DROID to your Google account to wirelessly sync your Google contacts, calendar, and Gmail. That works well if you already use Google for your personal information, but it will not work for you if you store your contacts, calendar, and other items in your computer using software such as Microsoft Outlook, iCal, or Microsoft Entourage.

In this chapter, we show you alternative sync methods to sync some or all of your personal information to your DROID. Some of these options are free, others cost about US$40. We hope that this chapter will give you a good overview of the options and help you implement the one that best suits your needs. While we don't intend to replace the software manufacturer's level of documentation and troubleshooting materials, we do hope to give you enough to get up and running.

NOTE: We cover how to sync media and other files such as documents using free software options in Chapter 25: "DROID Media Sync." We discuss ways to sync your email in Chapter 9: "Email on Your DROID."

While there are other options out there to sync to your DROID, we decided to cover a few of the more popular software products. You may find other options at different prices or even other free options, but beware that we have heard of people losing all their calendar entries by trying some free sync options not listed in this book.

CAUTION: For any sync solution, including the options listed in this book, we do recommend saving a backup file of the personal information on your computer just as a precaution in case something goes wrong with the sync process. It's always safer to have that backup handy.

Table 4–1 provides a brief overview of the sync options covered in this chapter.

Table 4–1. *Alternative Sync Methods for Your DROID.*

Type of Information	Google Calendar Sync	Missing Sync for Android	Companion Link
Outlook Contacts (Windows)	-	Yes	Yes
Outlook Calendar (Windows)	Yes	Yes	Yes
Outlook Tasks (Windows)	-	Coming	Yes (to **DejaOffice** app)
Outlook Notes (Windows)	-	Yes (to **Fliq Notes** app)	Yes (to **DejaOffice** app)
Outlook 2011 for Mac	-	-	-
iCal (Mac)	-	Yes	-
Entourage Calendar (Mac)	-	Yes (To **Fliq Calendar** app)	-
Entourage Notes (Mac)	-	Yes (To **Fliq Notes** app)	-
Address Book (Mac)	-	Yes	-
Entourage Address Book (Mac)	-	Yes	-
ACT! (Windows)	-	-	Yes
Novell GroupWise (Windows)	-	-	Yes
Lotus Notes (Windows)	-	-	Yes
GoldMine (Windows)	-	-	Yes

Backing up Your Outlook Data

Before you start syncing your Outlook data, we recommend taking a backup copy of your Outlook data. This may be a single file or several files, depending on how your system is setup.

1. Determine the file name and folder containing your Outlook data. From the Outlook menu, select **File** > **Data File Management**.

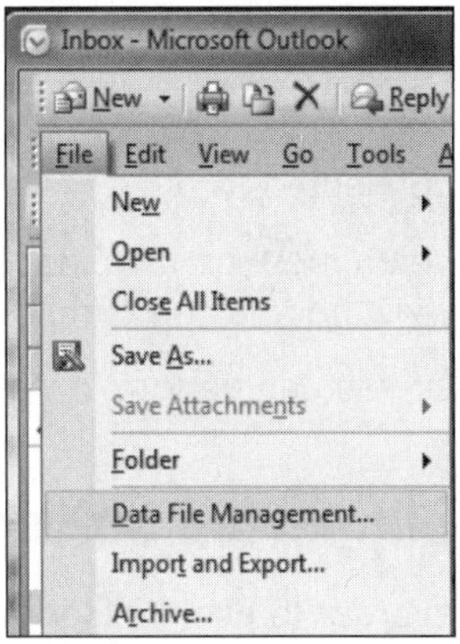

2. Look under the Filename column for both the data file name (usually Outlook.pst or Archive.pst) and the folder name which is listed right after the file name.
3. Close Outlook to close all the data files.

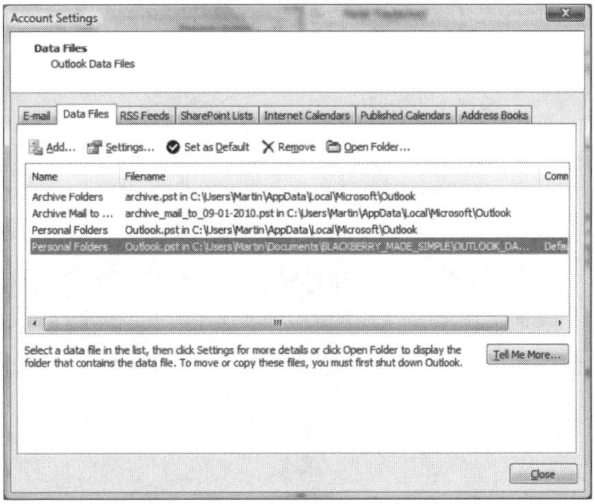

4. After Outlook is closed, use your **Windows Explorer** (shortcut **Windows Key + E**) to locate all the data files and copy them to a backup location. We recommend something not on your computer, e.g. an external hard disk, CD/DVD, cloud storage, or a USB thumb drive.

Google Calendar Sync for Outlook (Windows PC)

Google provides a free application to sync your Microsoft Outlook Calendar to your Google Calendar. From the Google Calendar, you use the standard DROID sync software described in Chapter 3 to sync Google to your DROID. It is a two-step sync: Outlook to Google Calendar to DROID Calendar. For this sync to stay updated, your computer needs to be turned on, Microsoft Outlook needs to be running, along with the Google Calendar sync, and you need an Internet connection. If any of these are not working, the sync will stop. It should catch up next time everything is on and connected.

Learn more on the Google Calendar Sync here:
www.google.com/support/calendar/bin/answer.py?answer=89955

Or you can perform a web search for "Google Outlook Calendar Sync."

Supported Software (What You Need)

In order to use the Google Calendar Sync, you need the following:

- A Windows PC running XP, Vista, or Windows 7
- Microsoft Outlook 2003, 2007, 2010 (32-bit).
- An active Internet connection
- No special software on your DROID—you just use the standard **Calendar** app and sync it to the Google Calendar.

> **NOTE:** No Mac support was available to Sync your Google calendar to Outlook 2011 or Entourage on the Mac. You can sync your Google calendar to iCal by setting up your Google calendar as a CalDAV account under iCal preferences. Also, Outlook 2010 (64–bit) for Windows was not yet supported, however we understand Google is working on this issue and may have it resolved soon.

Before You Install

We highly recommend doing the following:

- Turning off any other sync applications that are syncing to your Outlook Calendar.
- Making a backup copy of your Outlook Data File. See the "How to Backup Your Outlook Data" section in this chapter.

Downloading and Installing

Download the software from Google. To find the software, go to www.google.com/sync/index.html. You will most likely see a web page similar to the one shown in Figure 4–1.

Click on **PC** to go to the Google Outlook Calendar Sync for Windows page. Follow the instructions on the website to install the software.

Figure 4–1. *Google Sync Services Website.*

Setting Up the Software

1. Start up the software. After you have installed Google Sync, start it up by typing Google Calendar Sync in your Windows search box. Click on the software which appears at the very top of the list under Programs.

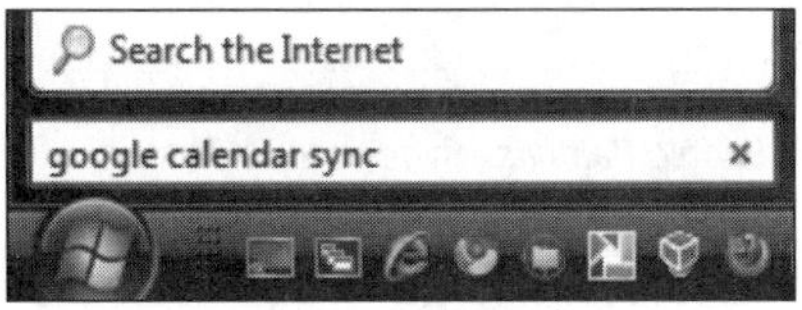

2. The program may not appear on the screen but will start up in a Windows Tray icon. You need to right-click the tray icon and select **Options** to see the setup window.

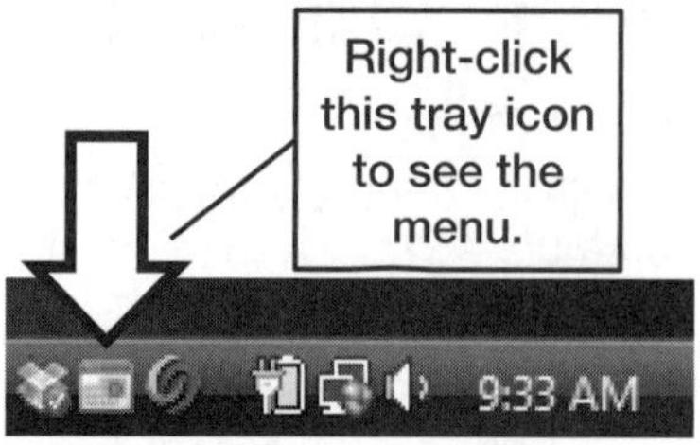

Figure 4–2. *Google Calendar Sync Setup Window.*

3. Enter your Google Calendar username and password in the window shown in Figure 4–2.

4. Then choose your **Sync Options**. You probably want the **2-way** sync unless you have a strong reason for 1-way sync. With the **2-way** sync, any updates you make on your DROID Calendar will by synced to Google and back to Outlook and vice-versa. Everything is kept up-to-date.

- **2-way**
  Gives you full two-way synchronization between your Google Calendar (and DROID Calendar) and your Microsoft Outlook Calendar.
- **1-way: Google to Outlook**
  Syncs only from your Google Calendar (DROID Calendar) into your Outlook Calendar.
- **1-way: Outlook to Google**
  Syncs only from your Outlook Calendar to your Google Calendar (DROID Calendar).

5. You can then adjust the sync frequency. The default is the shortest duration of 10 minutes; however, you can make it longer interval if you desire.
6. When you are done, click **Save** to start the calendar sync from Outlook to Google.
7. Remember to follow the steps shown in Chapter 3 to get your DROID Calendar to sync wirelessly with your Google Calendar.

Now all appointments from your Google calendar should flow into both your Outlook Calendar and your DROID Calendar app. Pretty soon, all three calendars should be identical.

Troubleshooting Google Calendar Sync

Common issues and potential resolutions with Google Calendar Sync are shown in Table 4–2.

Table 4–2. *Common Google Calendar Sync Issues and Resolutions*

Common Issue	Resolution
Error connecting to Outlook.	Make sure Outlook is running on your computer.
Error connecting to Google.	Make sure you have an Internet connection from your computer.
Calendar events are not updated on your DROID.	Make sure your account is setup correctly on your DROID. Make sure your computer is powered on, Outlook is running and it has an internet connection.

Google also provides a wealth of troubleshooting resources. The fastest way to find them is to do a web search for “Trouble Syncing Google Calendar with Outlook.” Then look for a link in the upper right corner of the page that says "Google Calendar Sync Troubleshooting."

Some of the help topics covered online include:

- Events not syncing at all.
- Events display the wrong time after syncing.
- Calendar alarms aren't syncing correctly.
- Trouble connecting to Outlook.
- Privacy settings don't match.
- Description or error codes 0, 1008, 1011, 2008, 2013, 2016, and more.

Hopefully, with the online resources and the description in this book, you should be armed with enough information to get your Google Calendar Sync set up and running smoothly to your DROID.

Missing Sync for Android

The Missing Sync for Android from Mark/Space, Inc. provides you the ability to sync from your Windows or Mac PC to your DROID. Items that you can sync are:

- Outlook or Entourage Calendar, Contacts, and Notes
- Music from iTunes or Windows Media Player
- Photos, Videos, Ringtones, Documents and Files, Podcasts, and Call History.

Learn more about the Missing Sync for Android here:
www.markspace.com/products/android/missing-sync-android.html

> **NOTE:** The software vendor does not provide a free trial of the software, so we encourage you to read the online reviews and comments about the software before purchasing it.

Reviews of the Missing Sync for Android

We found a review of the Missing Sync for Android at publishing time. You should do a quick web search to see if there are any new or updated reviews at the time you are reading this. Try searching for "Missing Sync Android Review."

PC Magazine review for the Missing Sync for Android v1.4 (August 2, 2010) at

`www.pcmag.com/article2/0,2817,2367260,00.asp`

The review gave the application 2 out of 5 stars and said it worked as advertized; however, it was challenging to use and needed more features to justify the $40 price. This review also included comments from people who have purchased the software. Keep in mind that the *PC Magazine* review and comments are based on version 1.4 of the software, so that if the vendor has released an update, some or all of the concerns of the reviewers may have already been addressed.

Supported Software (What You Need)

In order to use the Missing Sync for Android, you need the following:

- A Windows PC running XP, Vista (32-bit only), or Windows 7 (32-bit or 64–bit) and Microsoft Outlook 2003, 2007, 2010 (32-bit)
- A Mac OS 10.5.6 Leopard or later, or Snow Leopard running Entourage 2004, 2008, or Apple iCal
- iTunes or Windows Media Player to sync media
- An active Internet connection
- Three free apps for your DROID to fully take advantage of the synced data. All three are available in the Android **Market** app.
    - **The Missing Sync for Android**
    - **Fliq Calendar** (Required for a 2-way calendar sync)
    - **Fliq Notes** (Required to sync notes)

NOTE: At the time of publishing, the Missing Sync for Android did not support Office 2011 for Mac.

Before You Install

We highly recommend doing the following:

- Turning off any other sync applications that are syncing to Outlook.
- Making a backup copy of your Outlook Data File. See the "How to Backup Your Outlook Data" section in this chapter.

Downloading and Installing PC or Mac Software

There is no free trial available of the Missing Sync, so you need to purchase it in order to download it.

1. Buy and download the software from the www.markspace.com site.
2. Double-click the downloaded file to get the installation started.

3. Once you have installed the software, locate and double-click on the application icon on your PC or Mac. You should see a settings screen similar to the one shown in Figure 4–3.

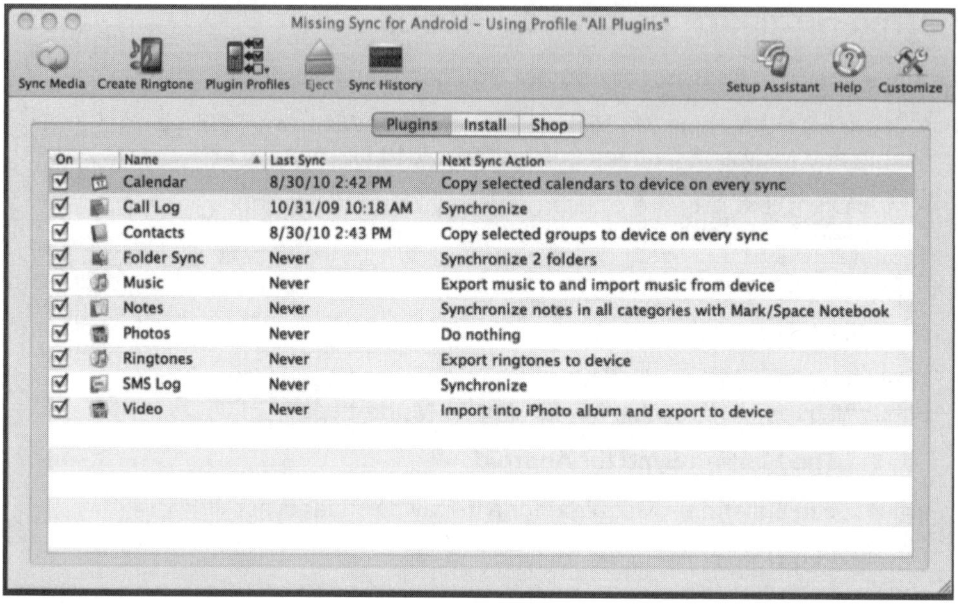

Figure 4–3. *Missing Sync for Android Sync Setup Window.*

4. Check or un-check items to turn the sync on or off.

5. Double-click a particular item to see more details about that item and to be able to customize the synchronization settings. For example, clicking the Calendar row at the top will show you the **Calendar Settings** dialog box shown to the right.

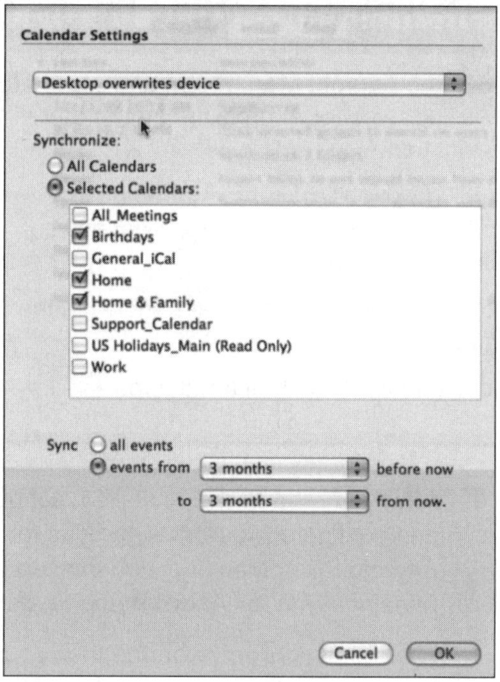

6. Repeat this procedure to set up the sync for all the various data types: **Call Logs**, **Contacts**, **Folders** (easily share documents and files), **Music**, **Notes** (sync to Fliq Notes from), **Photos**, **Ringtones** (create and sync), **SMS Log** (keep track of SMS messages received and sent), **Video** (copy and re-format videos to play on your DROID or copy videos recorded on your DROID to your computer).

7. When you are done configuring all the data types, you are ready to connect or pair your DROID with the desktop Missing Sync software using the Setup Assistant by clicking the **Setup Assistant** icon. At the end of this process, you will receive a 5-digit passkey to enter on the Missing Sync app on your DROID to complete the pairing. Follow the steps below to finish the setup on your DROID.

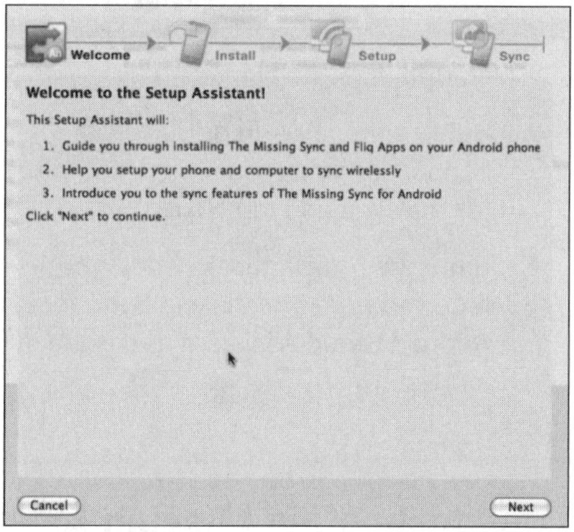

Installing and Setting Up the DROID Software

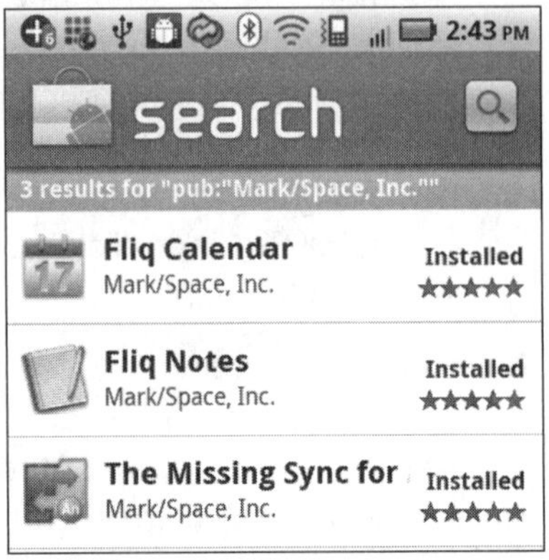

Go to the Android **Market** app and search for "Missing Sync for Android," then tap that application.

Scroll down a little and tap **View more applications** to see all the applications by Mark/Space, the developer.

You should see three apps as shown in the figure to the right.

Go ahead and install all three apps.

1. After you have already installed, set up, and configured your Missing Sync for Android PC or Mac app, you can tap the **Missing Sync for Android** app to start it.
2. Tap the **Synchronize** button then answer the question "Do you have the Missing Sync for Android installed on your Mac or Windows System?" as a **Yes** to get started.
3. Setup either a Wi-Fi or Bluetooth connection from your computer to use for the sync and tap **Next**.
4. Enter the 5-digit passkey that you received from the Missing Sync for Android Setup Assistant to get the sync started.

Troubleshooting Missing Sync

The Missing Sync website provides a knowledgebase of common issues and suggested solutions. From your web browser, go to support.markspace.com and click on **Technical Support: Missing Sync for Android** to see all the support topics.

Some of the help topics covered online include:

- Configuring Entourage 2008.
- Unable to locate Missing Sync in the Android Market.

- Some or all contacts are missing after a sync.
- Error: the location of Outlook is unknown.
- Seeing an error message during the sync for either Windows or Mac.

Hopefully, with the online resources and the description in this book, you should be armed with enough information to get the Missing Sync for Android set up and running smoothly to your DROID.

CompanionLink for Android

CompanionLink (CL) software provides various software products to sync to your DROID. With CompanionLink, you have three sync options:

Option 1: **Direct USB Sync Android direct via USB**	This will sync from your PC through the USB cable directly to your DROID. If you choose this option, you need to install the DejaOffice app on your DROID.	You need to install a new app on your DROID called **DejaOffice** to use this option.
Option 2: **Sync to DROID via Google**	This will sync your data from your PC to your Google Account. Then you use the setup process in Chapter 3 to setup the sync from Google to your DROID.	No additional software needs to be installed on your DROID; you use the standard **Contacts** and **Calendar** apps.
Option 3: **Sync to DROID via the CL Hosted Servers**	This is like the Google sync option, except that your data goes to the CompanionLink servers instead of Google.	You need to install a new app on your DROID called **DejaOffice** to use this option.

The **CompanionLink for Outlook** software provides two-way sync from Outlook to Google (then you sync your DROID to Google) or Outlook direct via a USB cable to your DROID.

In this book, we have chosen to focus on the **CompanionLink for Outlook** product with the setup that it syncs to your Google Account. We do cover how to get the **DejaOffice** app installed on your DROID should you desire to use the USB or CL Hosted Server sync options. The steps shown here should be able to help you get a feel for how all the various CompanionLink sync options work.

CompanionLink supports a number of PC software applications to sync to your DROID:

- Microsoft Outlook
- ACT! (Contact Management Software)
- Palm Desktop and Pimlical
- Lotus Notes
- GoldMine
- Novell GroupWise

NOTE: CompanionLink does not support any Mac software as of publishing.

Get all the latest information about CompanionLink's products by visiting their website:

`www.companionlink.com/`

Reviews of CompanionLink and DejaOffice App

We found a couple of reviews of CompanionLink at publishing time. Please note that these two reviews are already a bit dated (six months old) as of publishing time, so you keep in mind that any negative comments could already have been resolved in newer versions of CompanionLink.

Droid Forums.net Review of CompanionLink (March 7, 2010):

Disclaimer: CompanionLink is a sponsor of this particular forum.

`www.droidforums.net/forum/droid-news/27927-video-review-companionlink-software.html`

Andronica's Blog Review of CompanionLink USB (February 16, 2010):

`androinica.com/2010/02/16/sync-outlook-and-android-with-companionlink-software-review/`

DejaOffice (Android App) created by CompanionLink to store Outlook data on your DROID. Read the customer reviews in the Android **Market** for this app.

Keep in mind that you only need to install the **DejaOffice** app if you want to use the USB cable to sync directly from Outlook to your DROID and skip the sync to Google. If you go with the USB sync option, you have the benefit of not having your data pass through Google; however, you have the added challenge of installing and using a new contacts, calendar, notes, and task app on your DROID.

Supported Software (What You Need)

In order to use the Missing Sync for Android, you need the following:

- A Windows PC running XP, Vista, or Windows 7 (32-bit or 64–bit) and Microsoft Outlook 2000, 2003, 2007, 2010 (32-bit and 64–bit) or Microsoft Outlook Business Contact Manager 2007 and 2010
- An active Internet connection
- A Google, Gmail, or Google Apps account
- **DejaOffice** app (free) on your DROID only if you choose the USB or CL Hosted Server Sync options.

NOTE: At the time of publishing, the CompanionLink did not support any Mac software applications includingOffice 2011 for Mac.

Before You Install

We highly recommend doing the following:

- Turning off any other sync applications that are syncing to Outlook.
- Making a backup copy of your Outlook Data File. See the "How to Backup Your Outlook Data" section in this chapter.

Google Calendar First Time User

If you are setting up a new Google account and have not yet added any events to your Google Calendar, you need to go into Google and add a new event in order for CompanionLink to work.

The first time you go into the Google Calendar, you will see a screen similar to the one shown to the right. Click **Continue** to set up your calendar.

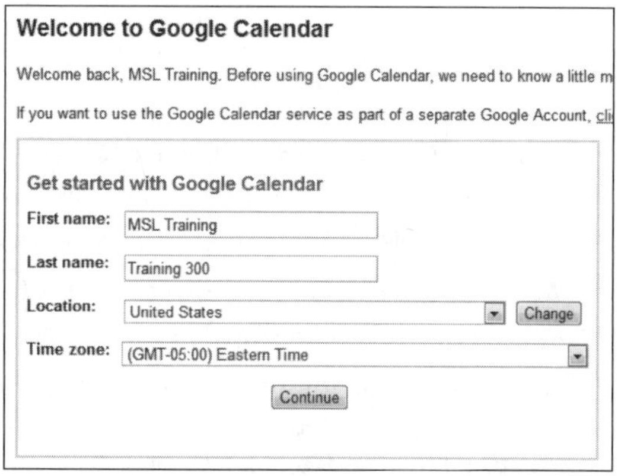

Downloading and Installing the PC Software

We highly recommend downloading the 14–day free trial version of the CompanionLink software by going to www.companionlink.com/downloads/.

This will allow you to test out the software before you buy it. At publishing time, pricing for CompanionLink for Outlook was $39.95.

The Free Evaluation Software page asks for your email, the product, and which PC software you want to sync with as well as what type of phone you use.

> **TIP:** Video Tutorials on CompanionLink's Website
>
> Visit www.companionlink.com/support for some good technical support information as well as some great video tutorials explaining how to setup their software.

1. Double-click the downloaded file to get the installation started.
2. Once you have installed the software, click your **Windows logo** in the lower left corner and select **All Programs**. Scroll down to Companion Link and click on **Companion Link Setup** as shown to the right.

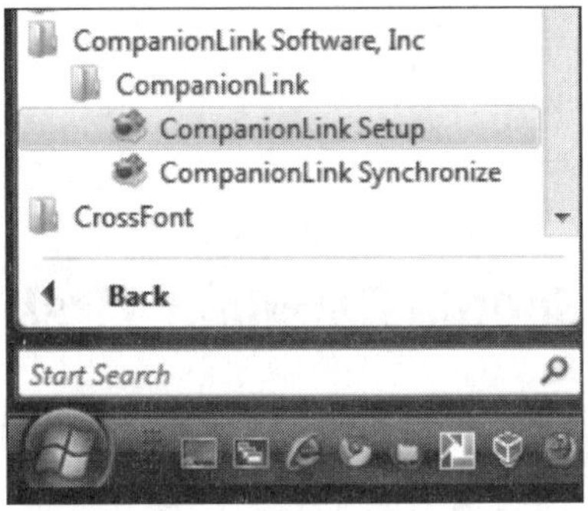

3. You should see a settings screen similar to the one shown in Figure 4–4. Click the drop-down menu that says **(Select your sync target)** to choose one of the Android options.
   - **Android (CL Secure Hosted Sync)** — This will sync your data via the Companion Link secure server then to your DROID.
   - **Android direct via USB** — This will sync from your PC through the USB cable directly to your DROID. If you choose this option, you need to install the **DejaOffice** app on your DROID.
   - **Android via Google** — This will sync your data from your PC to your Google Account. Then you use the setup process in Chapter 3 to setup the sync from Google to your DROID.

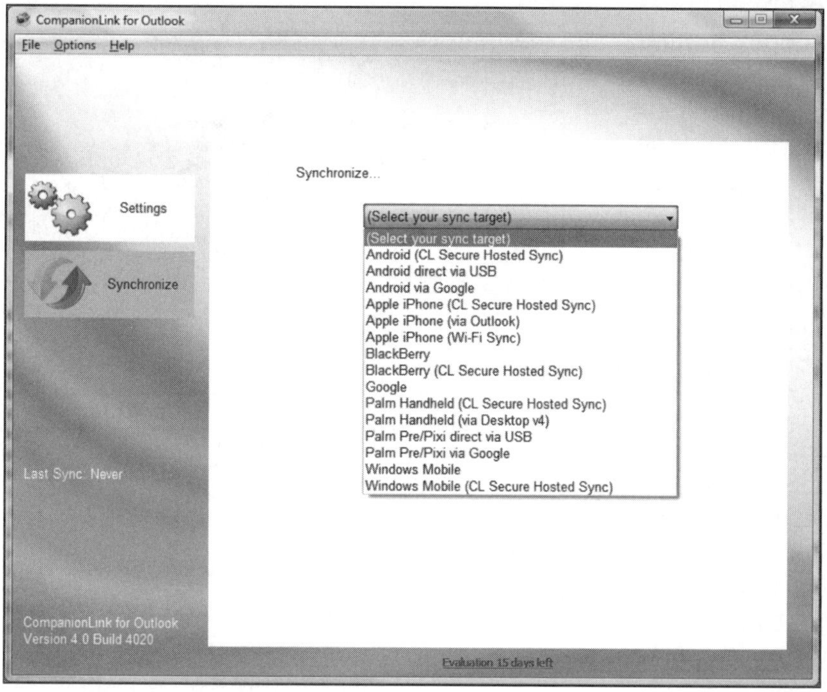

Figure 4–4. *CompanionLink for Outlook Sync Setup Window.*

4. Next, click **(Select your Contact Manager)** to select your software. In this case, we selected Microsoft Outlook 2000 - 2010.

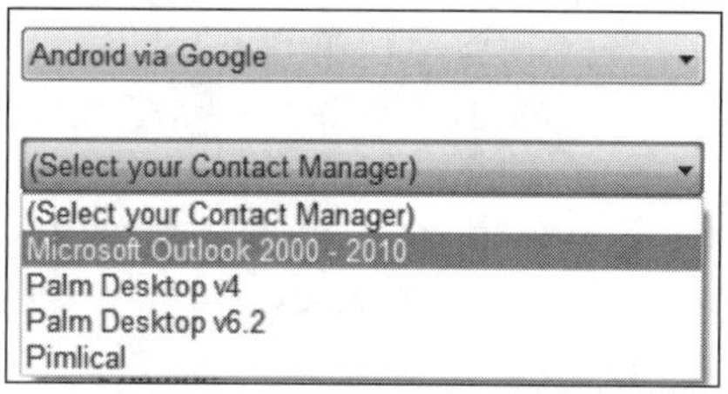

5. Click **Google Settings** from the main settings screen to see the **Google Settings** pop-up window shown to the right. Enter your Google Account **email** and **password**. Then, if you want to adjust which Google Calendars to sync (if you have more than one), click the **Google Calendars** tab and make adjustments. When you are done, click the **OK** button to save your settings.

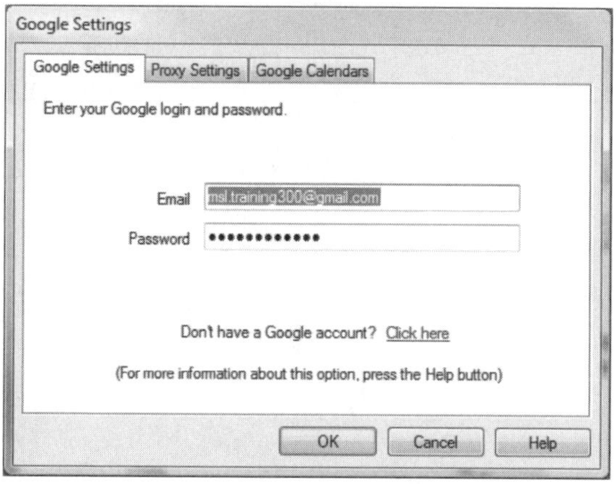

6. Next, click **Outlook Settings** from the main setup screen to see the screen to the right. Make adjustments to Selected Categories, if you choose.

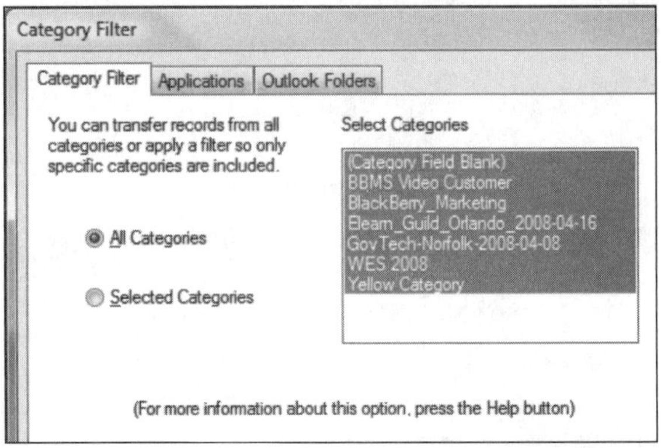

7. Click the **Applications** tab to adjust how to sync your data. Each check shows you want to sync Contacts, Calendar, or Tasks. You can **Sync Both Ways**, (1-way) **Outlook to Google** or (1-way) **Google to Outlook**.

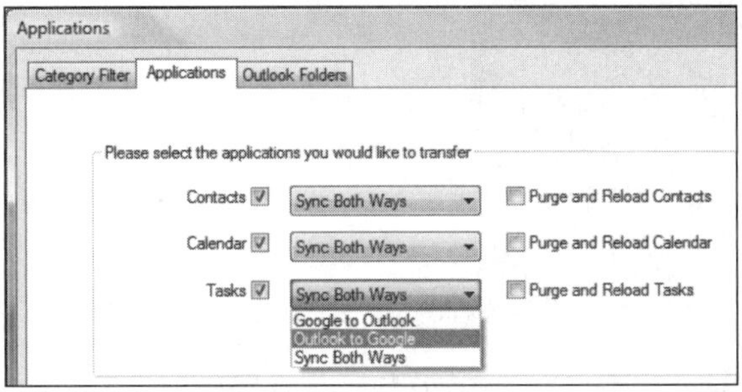

8. Click the **Outlook Folders** tab to adjust which folders are synced from Outlook. Check the box to **Include SubFolders** if you desire. Finally, click the **OK** button to save your settings.

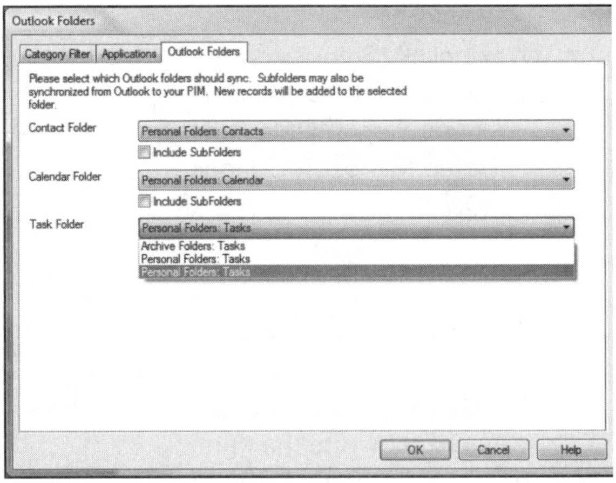

9. Next, click **Advanced Settings** from the main window to make further adjustments. Most people will not need to use this area. But it might be good to browse the screens and see if anything might need adjusting. The things you can change are syncing personal category from Outlook, how to handle sync conflicts (Outlook wins, Google wins, create duplicate record, keep latest change), date range for the calendar sync, and name sorting order (First, Last or Last, First). You can also adjust field mapping, phone number formats, and the sync settings as shown in the Applications tab of the Outlook Settings window shown above in step 7.

10. Finally, click the **Auto-Sync** button to setup how often CompanionLink will sync between Outlook and Google. The default is set to Synchronize Manually, but you may want to sync every **15 minutes** or some other setting. Also, you can set the Sync Time to only operate between certain times— usually the work day is fine because who needs to sync changes in the middle of the night? Click **OK** to save your settings.

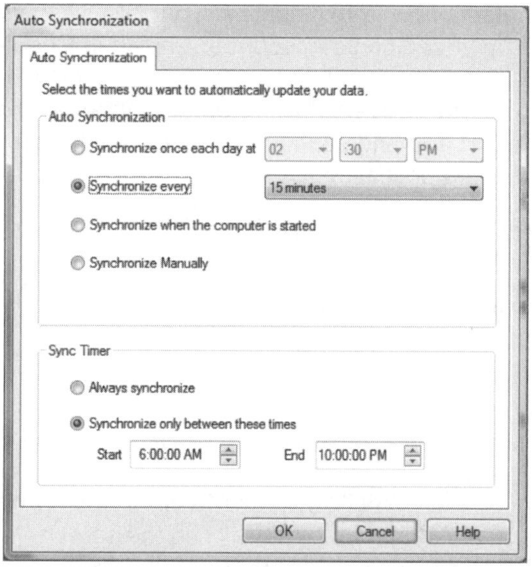

11. Now, to get started, click the **Synchronize** button in the left column of the main window.

12. Finally, if you have not already done so, you need to follow the steps in Chapter 3 to set up access from your DROID to the Google Account to which you just synced all your Outlook Contact and Calendar information.

NOTE: Working with a Google Calendar Sync Error.

After you click the **Synchronize** button, if you see an error telling you that CompanionLink was unable to read your Google Calendar, then you need to complete setting it up. See the "Google Calendar First-Time User" section above.

Installing and Setting Up the DROID Software (Only Required for USB Sync or CL Hosted Server options)

> **NOTE:** This step is only required if you are using the USB direct sync option or the Companion Link (CL) Hosted Server sync option from Outlook to Android.

If you are using the sync method just described above from Outlook to Google, you don't need to install any additional software on your DROID. Instead, your calendar and contacts flow from Google to your DROID native **Contacts** and **Calendar** apps after you setup your Google Account as shown in Chapter 3.

To locate the DejaOffice app, go to the Android **Market** app and search for "DejaOffice," then tap that application.

You should see the app as shown in the figure to the right.

Go ahead and install the **DejaOffice - Outlook** app.

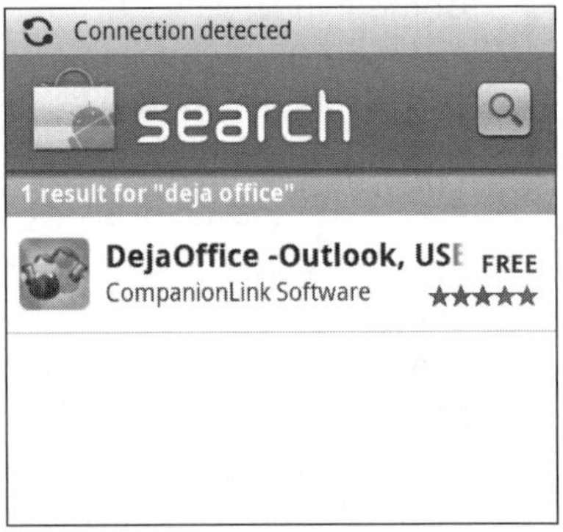

1. After installing **DejaOffice**, tap the **icon** to start it up. You will see the screen shown to the right. You will probably want to check both boxes for **Sync with the Android Contacts App** and **Sync with Android Calendar App**. This allows Caller ID and e-mail to work well on your DROID. Tap the **Configure** button to adjust these sync settings. Tap **OK** when done to start the sync with the **Contacts** and **Calendar** apps.

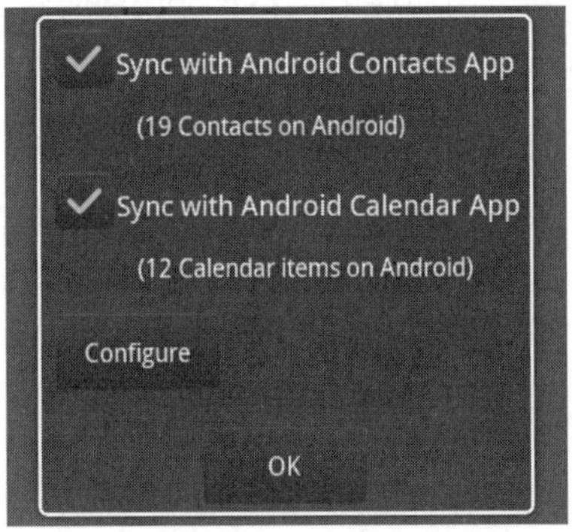

2. Then, you can tap any of the icons within DejaOffice to access your **DejaContacts** (Outlook Contacts), **DejaCalendar** (Outlook Calendar), **DejaTasks** (Outlook Tasks), **DejaNotes** (Outlook Notes), **DejaToday** (Today view), **Sync** (force a sync), **Settings** (all settings including sync settings), **Categories** (Outlook Categories), **Read Android Data** (sync with your DROID Contacts and Calendar), and **Release Notes** (notes from the developer about this release of the DejaOffice app).

3. Think of **DejaOffice** as your mobile version of Outlook on your DROID.

Troubleshooting CompanionLink

The CompanionLink website provides a FAQ (frequently asked questions), video tutorials, Set Up Guides, and the ability to contact tech support via e-mail or purchase premium support for US$99 for 12 months (as of publishing time).

Check out all these support options by visiting www.companionlink.com/support/

You may also be able to find additional support and troubleshooting tips by doing a web search for "CompanionLink Google (your issue description)."

Hopefully, with the online resources and the description in this book, you should be armed with enough information to get the CompanionLink set up and running smoothly.

Chapter 5

Wi-Fi and 3G Connectivity

Let's face it, you wouldn't buy a DROID if you didn't want to go online. Going online doesn't just mean using the built-in web browser in Android. Whether you're using apps or checking your contact list, talking on the phone is just about the only activity that doesn't involve using your data plan. That's one of the reasons you can't buy a DROID phone without also purchasing a data plan. Trust us: you wouldn't want to do so, anyway.

In this chapter, we'll talk about getting online with your DROID. Android works best when it works online. Google purchased and developed the Android platform around the idea of storing data online or "in the cloud," rather than just on the device. We'll go over the different ways your phone can access data and how to get the fastest connection with the least amount of battery drain.

Understanding Your Connection

The status or notification bar on the upper part of your DROID shows you what type of connection you have available and the relative strength of the connection. Figure 5–1 shows various types of connection information you might see in your DROID's status bar.

Back in the days when a car phone meant the phone was built into your car, cell towers carried an analog signal that was much like a radio signal, but at a different frequency. This was the first generation of wireless, and it's no longer in use. The second generation of technology, or **2G**, is what Verizon Wireless calls **1X**. Instead of using an analog signal, 2G networks are digital. Many phone carriers use the GSM (Global System for Mobile Communications) standard to deliver 2G data, but Verizon and your DROID use CDMA (Coe-Division Multiple Access).

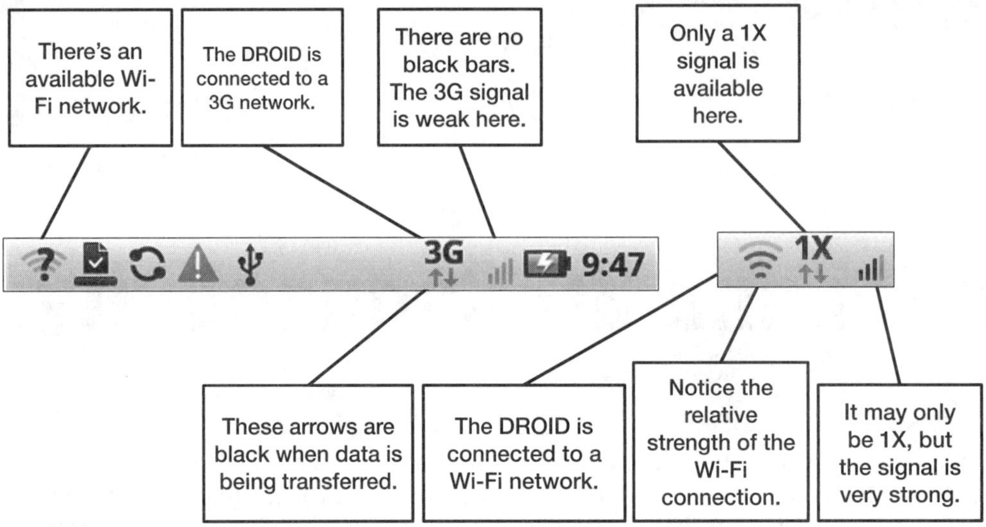

Figure 5–1. *Sample status bar notifications.*

Using 1X is like using an old dial-up modem. It works, but it isn't fast, and you'll have to spend a lot of time waiting for email to load, apps to sync, and pictures to upload. However, 1X signals are available everywhere your phone can connect to the Verizon network, and it doesn't take much battery power to use 1X.

3G and 4G

CDMA was upgraded to **3G** with yet another acronym, **EV-DO**, for Evolution-Data Optimized. 3G networks have huge speed advantages over 1X networks, but these are still slower than the average coffee shop Wi-Fi hotspot.

The fourth generation of wireless technology in the US, **4G**, has begun a rollout in select US cities at the time of writing. Sprint has begun its 4G rollout, and Verizon expects to offer full 4G coverage by 2013, including all those areas that only get a 1X signal now. Verizon purchased some of the spectrum previously used by analog television broadcasts in order to build its next-generation network.

Verizon is using **LTE** (Long Term Evolution) technology, which Verizon promises will work much better in rural areas and will be more internationally compatible than the current CDMA network. It will also give phones a large speed boost, so we can do things such as reliably video conference from our mobile phones. LTE is a high-speed, long distance Internet signal that can be used for home and phone networks, and it will likely be used the same way Wi-Fi is used on smartphones today. You might think of it as a form of Wi-Fi that can be broadcast for miles instead of several dozen feet.

The bad news is that you will *not* be able to connect to 4G networks on your DROID, DROID X or DROID 2. Your phone must be specifically made to take advantage of 4G networks. And although we expect an announcement of a 4G version of DROID soon, it did not happen by the time this book went to press.

The good news is that you don't need 4G in order to have a fast connection. You can still use Wi-Fi; and as Verizon rolls out 4G networks, it'll undoubtedly sell portable devices you can use to receive 4G signals and create your own portable Wi-Fi hotspot.

Wi-Fi

Wi-Fi signals are generally the fastest way to connect your DROID to the Internet. This is the same technology that connects laptops and other wireless devices to networks. It's fast but short-range, and it's not the same signal that telecommunications companies (telcos) send over cell towers. In order to connect to a Wi-Fi network, you have to be within range of the signal, and you have to be authorized to use the network.

Some book stores, fast food chains, and restaurants offer free Wi-Fi access networks to anyone within range of the signal. Connecting is easy; Figure 5–2 shows you how.

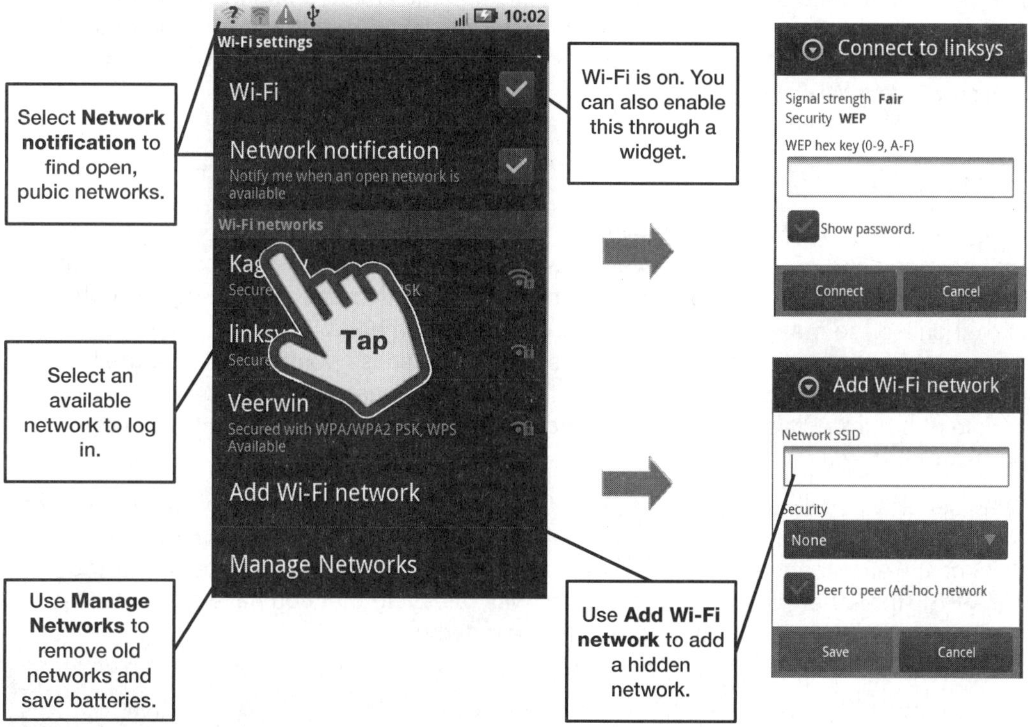

Figure 5–2. *Connecting to a Wi-Fi network.*

Follow these steps to connect to a Wi-Fi network:

1. Go to your **Home** screen.
2. Tap the **Settings** icon.
3. Select **Wireless & Networks**. Then tap **Wi-Fi settings**.

4. Check the boxes to turn on Wi-Fi and receive a notification when an open Wi-Fi network is within range.
5. Tap an available network to log into it. If the network requires a password, you'll only have to type it here once. Your DROID will remember your settings next time you use this service.

If you want to add a private network that doesn't broadcast its SSID, tap **Add Wi-Fi network**. If you want to remove networks you no longer use, such as hotels you visited only once, tap **Manage Networks**.

> **TIP:** You can also use a Wi-Fi widget from your **Home** screen to toggle Wi-Fi on and off. See Chapter 6, "Organize Your Home Screens" for specifics on adding widgets.

The clear advantage of Wi-Fi is speed. Wi-Fi is superior when it comes to uploading or watching videos. Its chief disadvantage is distance. If you're not within range of a Wi-Fi network all day or even most of the day, you can extend your DROID's battery life by turning off your Wi-Fi signal.

Wi-Fi Security

A big consideration with Wi-Fi is security. If you're using an encrypted connection, this isn't as much of a problem. However, those convenient, open Wi-Fi access points at the coffee shop may in theory expose your phone to unwanted eavesdropping through an exploit called the **man-in-the-middle** attack where someone intercepts your unencrypted information before it reaches the Wi-Fi access point. It's also sometimes called a **bucket brigade** attack, so think of the attacker as a stranger standing in the middle of a bucket brigade with the chance to see the contents of each bucket that passes by.

Wi-Fi security usually involves some sort of password protection to access the network. An older, less secure method of connecting to Wi-Fi is WEP. A more secure method is WPA or WPA2. Most personal networks, like your router at home, can be set to use WPA-PSK (pre-shared key.) This is a fancy way of saying that you have to type in a password or passphrase to get access to the network.

> **CAUTION:** If you have a choice in the matter, avoid relying on WEP for your Wi-Fi security. It's an old standard and very easy to crack.

Businesses that want to sell or restrict access to their network use a form of WPA-enterprise. This type of connection usually requires you to log in when you open your first web page, and it compares your username with a list of authorized users. In some cases, you don't actually have to log in, but you do have to click something to agree to the location's terms of service. This is still part of WPA security.

If you aren't required to log into anything, you don't need to click **OK** to agree to the access rules, and you don't need a password to get onto the network – then chances are that you're using an open Wi-Fi access point. A skilled hacker may be able to intercept your signal.

CAUTION: Unless you've installed security software, avoid entering passwords or sending sensitive information on open Wi-Fi networks.

Bluetooth

Bluetooth is a super short-range technology meant as more of a wire replacement than a way to get onto the Internet. Bluetooth can be used to communicate with a wireless headset or your laptop, and some apps can use Bluetooth to transfer files between your phone and your computer.

For more on Bluetooth, read Chapter 8: "Bluetooth on Your DROID."

GPS

GPS stands for global positioning system. It's one of the few acronyms in this chapter worth spelling out, because the long name explains what it does. GPS triangulates your position through satellite signals. This isn't the only way your phone can tell where you are, but it's the most commonly used method.

Devices that use maps or tag your photos by location typically rely on a GPS signal to do this. Android can also supplement this with the location of nearby cell towers and the location of any Wi-Fi networks you're using. However, plenty of apps require a GPS signal to tell you what movies are showing nearby or the location of the nearest Thai restaurant.

GPS activity is represented on the top of your screen as a satellite.  If you have the GPS feature activated, you may notice it activating when you open your Web browser, even if you aren't doing anything directly map related. This is usually to sense your location for local search results and ads. You can turn GPS off when you're not using it to save your phone's battery.

Creating a Wi-Fi Hotspot with Your DROID

If you pay for mobile hotspot access, which is an add-on service for DROID phones, you can use Verizon's **3G Mobile Hotspot** app to turn your DROID into a wireless modem for your laptop or other device, as shown in Figure 5–3. It's still sharing a 3G connection to a laptop, so it is not super fast; however, it will let you use your laptop's larger screen and keyboard while you're out of Wi-Fi range. You can share your connection with up to five devices, so you could also share a connection with a friend.

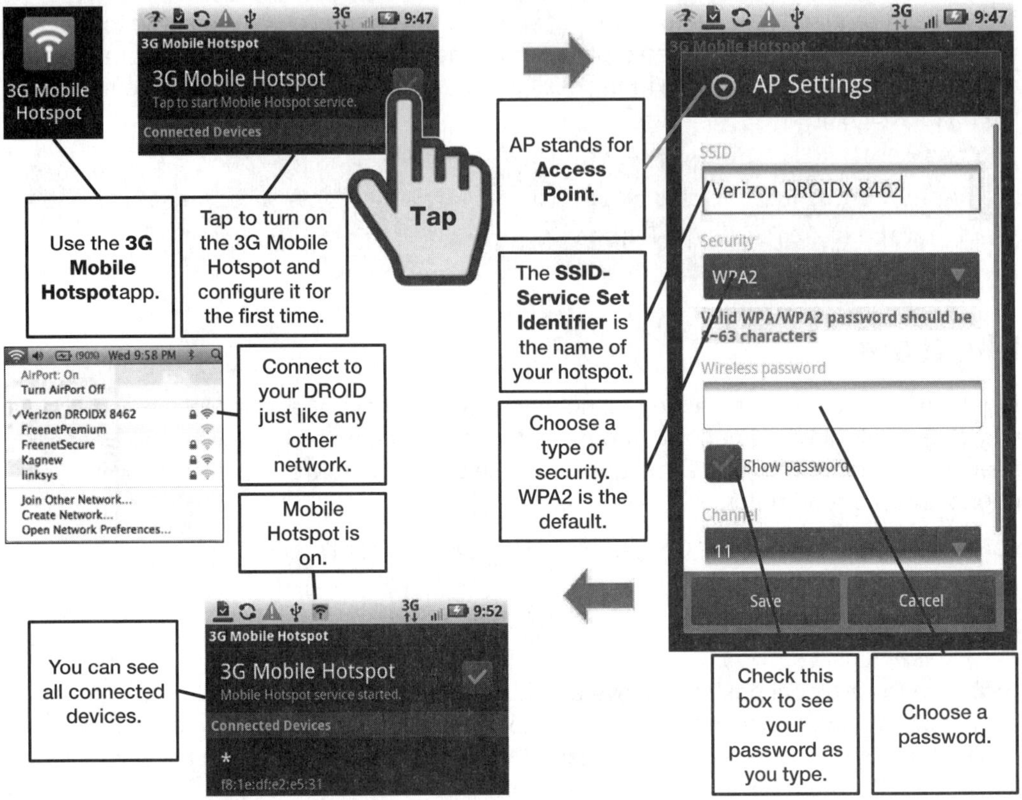

Figure 5–3. *Creating a Wi-Fi hotspot.*

Follow these steps to connect to a mobile hotspot:

1. Launch the **3G Mobile Hotspot** app.
2. Tap the checkbox to start **Mobile Hotspot** service.
3. The first time you launch the app, you'll be prompted to set up your **AP Settings**. AP stands for access point.
4. Give your mobile network a name (SSID) or leave the default setting.
5. Choose a security method; WPA2 is preferred.
6. Choose a password.
7. Save your settings.
8. Connect to your network from your laptop, iPad, or other Wi-Fi device just as if it were another Wi-Fi access point.

> **TIP:** Creating a Wi-Fi hotspot uses a lot of energy and drains a battery quickly. Connect your DROID to your laptop with a USB cable to keep the phone charged while surfing.

Users on Verizon's support forum have also reported that June Fabric's **PDAnet** program works on their DROID and allows them to tether their 3G connection through their USB port. It's a $30 app instead of a recurring fee. However, using **PDAnet** may violate Verizon's terms of service, so proceed at your own risk.

Troubleshooting Connections

The top of your phone will indicate which types of signals you're using and the relative strengths of those signals, as shown in Figure 5–1. You'll also see a pair of side-by-side **Up** and **Down** arrows; these arrows indicate an active data transfer.

Wi-Fi is indicated by a dot with curved lines above it to indicate a point giving out signal. GPS is depicted as a satellite, and it only shows up when it is actively being used. Bluetooth uses the trademarked Bluetooth symbol.

If you're having trouble with your signal, first check the top of your screen to make sure that you have an adequate signal and that you are using the network you expected to use. If you're using Wi-Fi, make sure you are correctly signed into the network.

Roaming

When you wander outside the range of cell towers that belong to your carrier (Verizon), you start **roaming**. You may be billed for roaming fees if roaming is not covered by your service plan. You may also need to add I-Dial or an international plan to your phone in order to check email or use data on your DROID in Mexico or Canada. See `http://b2b.vzw.com/international/naroaming.htm` for more information on this topic.

Follow these steps to turn off data roaming on your DROID and avoid roaming charges:

1. Go to your app tray and tap **Settings**.
2. Tap **Data manager**.
3. Tap **Data Delivery**.
4. Make sure the checkbox next to **Data Roaming** is grayed out. If you do select it, you'll see a warning box that roaming may cost you money.

> **NOTE:** Data roaming is different from voice roaming. You disable voice roaming through the **Wireless and networks** settings instead of the **Data manager**.

Managing Power

All of this connectivity comes at a cost. In order to save battery power, you should disable services you aren't using. This is especially true when you are traveling and can't charge your phone immediately. Keep your GPS off unless you're using a map or other app that hooks into your location. Turn off Wi-Fi and Bluetooth when you're not actively using them.

In order to make all this management easier, your phone has a **Power Control** widget that lets you toggle your signal on and off with a touch. 

The DROID even comes with two styles of widgets for toggling your connections. Simply tap the connection on or off from your **Home** screen. For more information on adding and rearranging widgets, look at Chapter 6: "Organize Your Home Screens."

Managing Syncing

In addition to turning connections on and off, you can control how often your connections sync. Figure 5–4 shows how the **Data manager** settings can help you save battery time, as well as how to disable data roaming. On the DROID 2 and DROID X, you can also turn off data completely for those times you really need to preserve your battery.

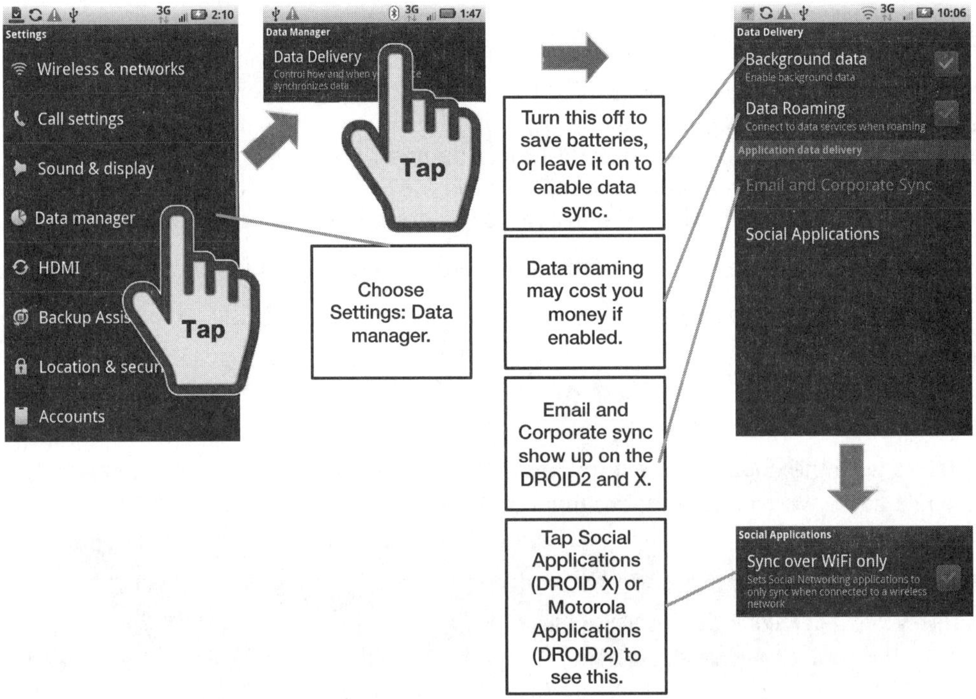

Figure 5–4. *The DROID's data management settings.*

Airplane Mode

If you're on an airplane, on a cruise ship, or in an area you know qualifies as roaming, you may want to just put your phone in **Airplane mode**. Toggling this mode on means you will not get any connection at all, whether it's by data, cell, or GPS. The following steps explain how to toggle **Airplane mode** on and off:

1. To turn on **Airplane mode**, hold down the **Power** key until you see an alert window.
2. Tap **Airplane mode**.
3. Use the same steps to exit **Airplane mode**.

Your DROID also comes with a **Motorola** widget to toggle **Airplane mode** on and off.

> **TIP:** In addition to using **Airplane mode** to avoid roaming charges, you can use it to avoid calls, emails, and other distractions during important meetings.

VPN

A **VPN**, or Virtual Private Network, allows you to log into your workplace intranet and enjoy the security of your corporate firewall without having to be hardwired into the network. Some places require this in order to access **Exchange** email or view sensitive corporate files.

This technology is natively supported on your DROID, but the implementation on your DROID won't necessarily work with every VPN setup. We're still waiting on an Android version of **AnyConnect**, but Cisco assures us it will be done around the time this book goes to press.

Follow these steps to log into a VPN from your DROID:

1. Go to the **Home** screen and then press the **Menu** button.
2. Tap **Settings**. Then tap **Wireless & network settings**.
3. Next, select **VPN settings**. If you've already configured a VPN, it will be available here.
4. Otherwise, you'll need to select **add a VPN**.

You'll need to obtain the specific format and settings from your workplace. These settings include PPTP, L2TP, L2TP/IPSec with pre-shared key (PSK), and L2TP/IPSec CRT (certificate based). If your workplace doesn't support one of these protocols, you'll need to work with your company's IT department to see if there's any other way to log in securely.

VNC

VNC, or Virtual Network Computing, provides a way to share screens remotely and control one device from another, even if that device runs on a different platform. If you leave your office or home computer on at all times and your office allows it, then you can use VNC to check documents, email, or execute work tasks from wherever you are. VNC can be used with Macs, Windows, and Linux computers.

In order to use VNC securely, you should pair it with VPN. There are several VNC clients available for Android, including **Android VNC Viewer**, **Remote VNC**, and **PhoneMyPC**.

Android's Web Browser

Android has a full featured web browser based on **Webkit**, as shown in the figure on the right. It uses the same codebase that is used in the **Chrome** and **Safari** web browsers. Generally, this browser behaves just like other web browsers. You can also download alternative browsers like the **Dolphin** browser or **Opera**.

One thing the Android **Browser app** is *not* is **Internet Explorer**. You may encounter sites that absolutely will not work unless you use **IE**. You may also encounter websites that won't work without plug-ins and extensions that aren't available on Android. If you find this is the case, you might be able to get around this limitation by using VNC and launching **IE** from your remote computer.

For more information, turn to Chapter 11: "Surfing the Web."

Chapter 6

Organize Your Home Screen: Icons and Widgets

In this chapter, we'll look at ways to customize your **Home** screen. This way you're not stuck with a boring collection of apps, and you're not stuck using the screen arrangement that your DROID came with. You can customize your phone to match your personality and the way you use the device by adding and rearranging icons and widgets.

Home Screen, Sweet Home Screen

Icons on your **Home** screen are just shortcuts to the actual apps, so don't worry about deleting them or moving them around. You're not actually deleting the apps. Think of your **Home** screen as one long screen that can't be displayed all at once. Alternatively, you can think of it as a series of several pages of screens. The **Home** screen, shown to the right, is where you can store your favorite apps or display your favorite wallpaper. With your phone in **Portrait** mode, swipe your finger sideways to flip between the pages of your **Home** screen.

How you organize your **Home** screens is entirely up to you. A method we've found useful is to create a theme for every page. One page might be dedicated to social networking apps, while another page might be dedicated to email and office productivity. A third page might be dedicated to games, restaurants, and entertainment tools.

The App Tray

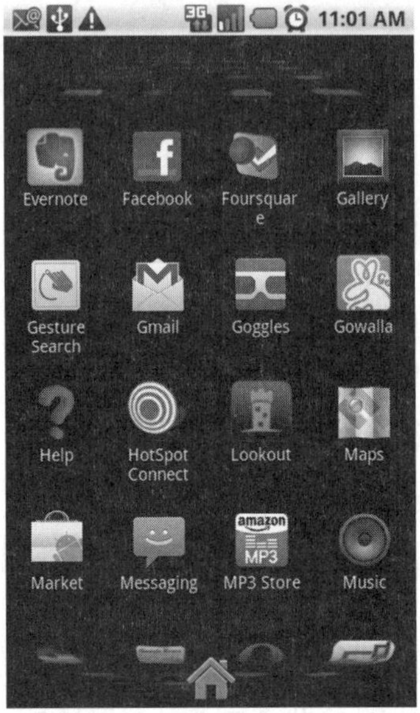

The **App** tray, which is shown to the right, holds all of your apps, including apps you're already using as icons on your **Home** screen.

Sometimes you'll also see apps that don't do anything by themselves, but add features to an existing app, such as pro feature upgrades.

You open the **App** tray by tapping your **Launcher** icon at the bottom center of the screen. You can also scroll through apps by swiping your finger up and down on the screen. See Figure 3 in the Quick Start Guide for images of the Launcher icons on the DROID, DROID 2, DROID 2 Global and DROID X. Once the **App** tray is open, you can launch apps by tapping them. Click the **Back** or **Home** button to return to the **Home** screen.

Adding App Icons to Your Home Screen

Let's review the **long press**, because you'll need it here. If you press down on an item and keep pressing for a few seconds, this is what Android calls a **long click.** However, you don't physically click the screen, so we'll call it a long "press" for clarity. You'll generally feel some haptic feedback when you use the long press – the DROID will vibrate slightly to let you know that you've done something different than a regular tap.

A long press can be programmed into apps. On the **Home** screen, a long press is used to add and remove items. Figure 6–1 illustrates how to add an icon to your **Home** screen.

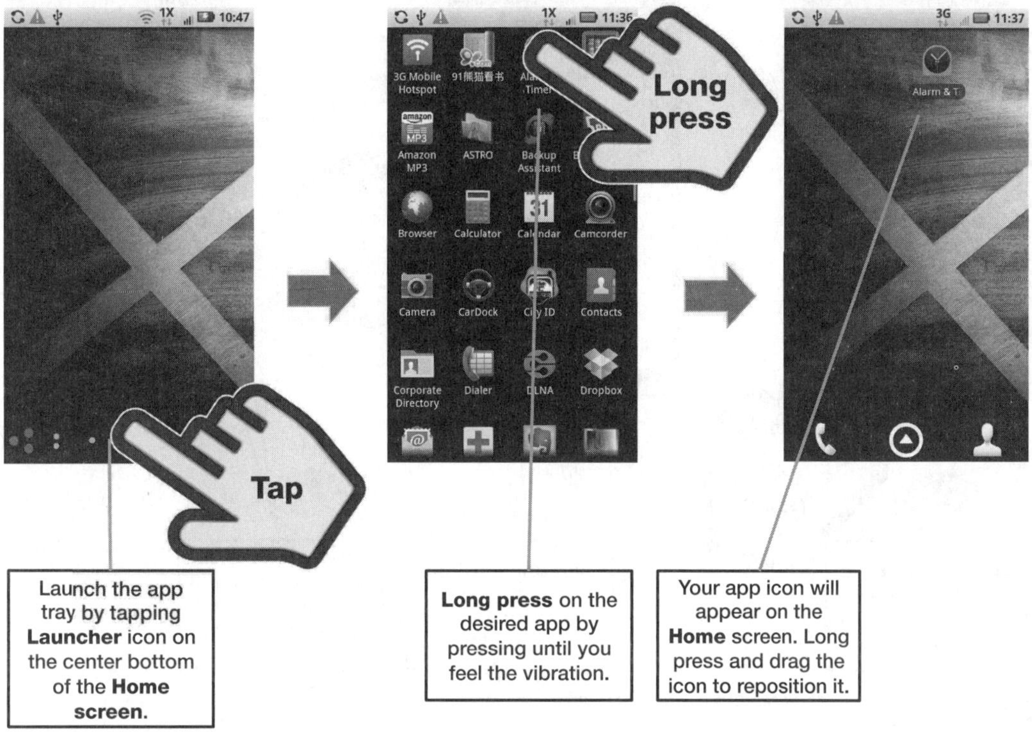
Tap
Long press
Launch the app tray by tapping **Launcher** icon on the center bottom of the **Home** screen.
Long press on the desired app by pressing until you feel the vibration.
Your app icon will appear on the **Home** screen. Long press and drag the icon to reposition it.

Figure 6–1. *Adding icons to the* ***Home*** *screen.*

Follow these steps to add an app to your **Home** screen:

1. Navigate to page of the **Home** screen you want to modify.
2. Open the **App** tray and find the icon for the app you want to launch.
3. Long press the app and keep pressing.
4. The **App** tray will vanish after a few seconds, and you'll see the **Home** screen.
5. Continue to press down, and then drag your app to the desired position.
6. Release your finger.

Adding Bookmarks to Your Home Screen

You're not limited to just **App** icons. You can also long press a bookmark to add it to your **Home** screen as a shortcut. It will be added to whichever **Home** screen page you viewed immediately before launching the **Browser** app. You can read more about using the **Browser** app in Chapter 11: "Surfing the Web." However, if you're already familiar with using the **Browser** app and adding bookmarks, Figure 6–2 illustrates how to add a **Browser** app bookmark to your **Home** screen.

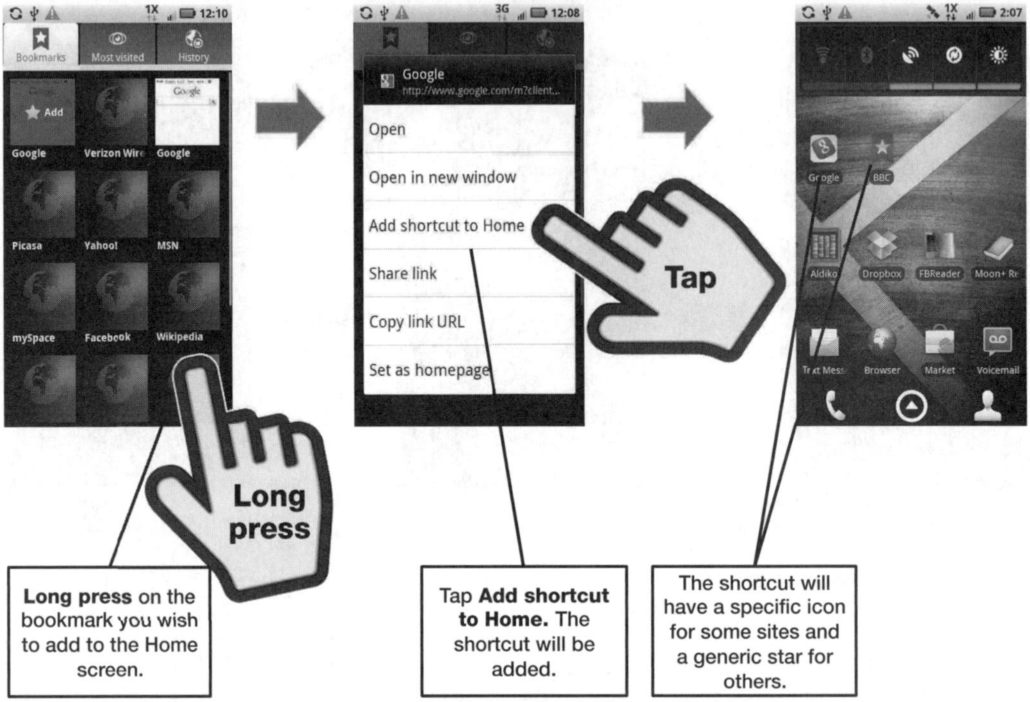
Long press
Tap
Long press on the bookmark you wish to add to the Home screen.
Tap Add shortcut to Home. The shortcut will be added.
The shortcut will have a specific icon for some sites and a generic star for others.
Bookmarks
Most visited
History
Open
Open in new window
Add shortcut to Home
Share link
Copy link URL
Set as homepage

Figure 6–2. *Adding bookmarks to the **Home** screen.*

Follow these steps to add a bookmark from the **Browser** app to your **Home** screen:

1. Open the **Browser** app.
2. Press the **Menu** button.
3. Tap **Bookmarks**.
4. Long press on the desired bookmark.
5. Tap **Add shortcut to home**.

If you're already at the **Home** screen, you can add a bookmark by adding a shortcut after the initial long press.

Adding Shortcuts

When you long press on the **Home** screen, you'll see the menu shown to the right. One of your menu choices is **Shortcuts**.

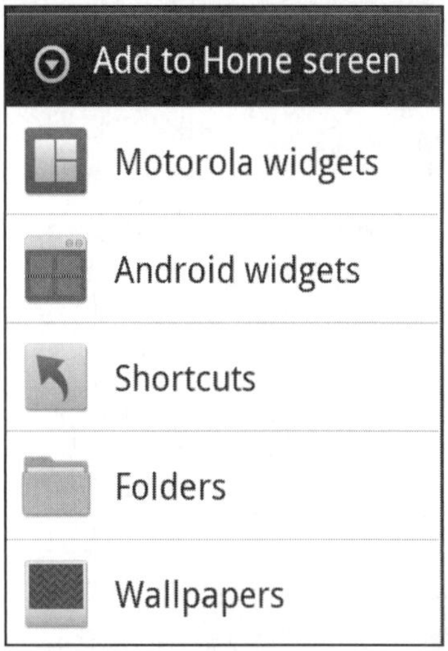

Shortcuts aren't reserved for bookmarks. You can add shortcuts for individual contacts, Gmail labels, items in the **Settings** menu, and more. Some apps may also have shortcut options. For example, **Foursquare**, a social location app, allows you to save locations as shortcuts for easy check-ins. And **Aldiko**, an eBook reader, allows you to save shortcuts to individual books in your library.

All of your **Home** screen app icons are actually shortcuts, and they can be added through the **Shortcuts** menu, too. We still prefer adding them with the method illustrated in Figure 6–1 because it's easy to accidentally click the wrong app when you add them through the **Shortcuts** menu.

Adding Folders

You can add as many contacts to your **Home** page as will fit. But why stop there? Instead of adding a single contact, you can add a folder, as shown to the right. Follow these steps to create a folder on your **Home** screen:

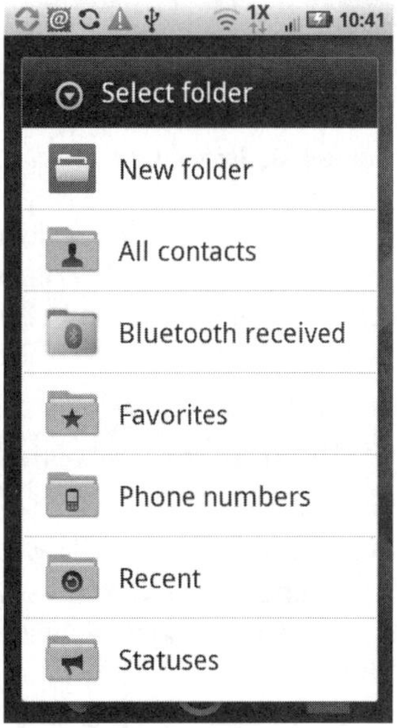

1. Long press on the **Home** screen.
2. Select **Folders.**
3. Tap your desired folder.

You can either add an empty **New folder** or choose from the many "smart" folders that have already been predefined and filled with content, such as **Statuses** or **All contacts**.

Move items into blank folders by long clicking the item and dragging it on top of the folder, just as you would on a computer desktop.

CAUTION: Removing a folder from your **Home** screen also removes any shortcuts you've moved into that folder.

Widgets

Widgets are like miniature applications that run on your **Home** screen. They can be interactive or passive, and they can be used for everything from instant access to the time and weather, to access to your personal finances at a glance. Figure 6–3 shows some example widgets.

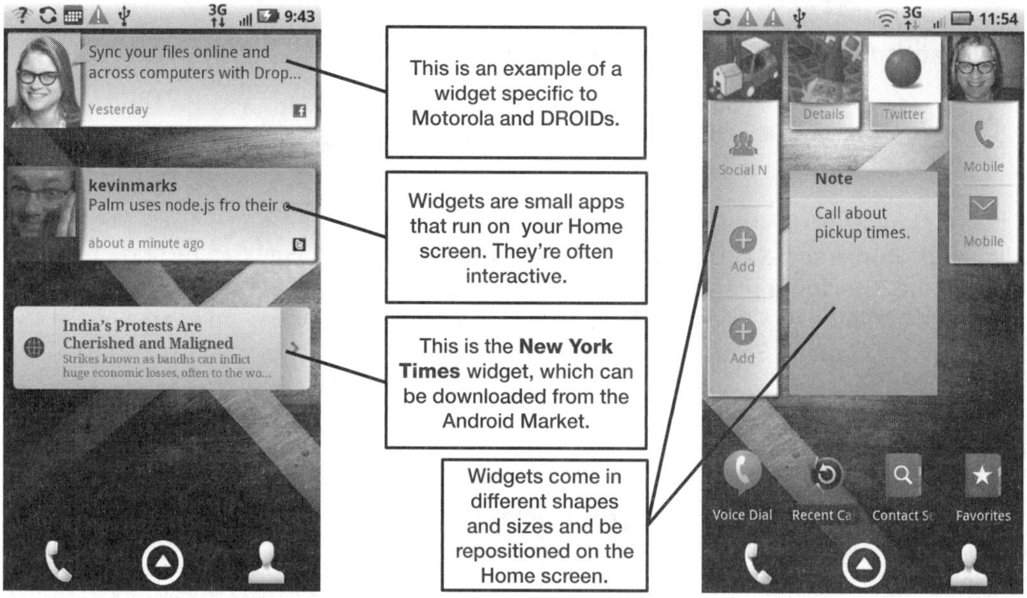

Figure 6–3. *Some Android widgets.*

Your DROID comes with two basic styles of widgets: the standard Android widget and the Motorola widget. Motorola widgets were created as part of the Motorola DROID modifications, and they behave a bit differently than standard widgets by letting you resize them. We'll cover both Android and Motorola widgets.

NOTE: The Motorola widgets are not available on the original DROID.

You use the same basic process to add widgets to your **Home** screen that you use to add icons and bookmarks. Figure 6–4 illustrates the process with the Android **Home** screen **Tips** widget. This widget was designed to give you tips on how to use your phone.

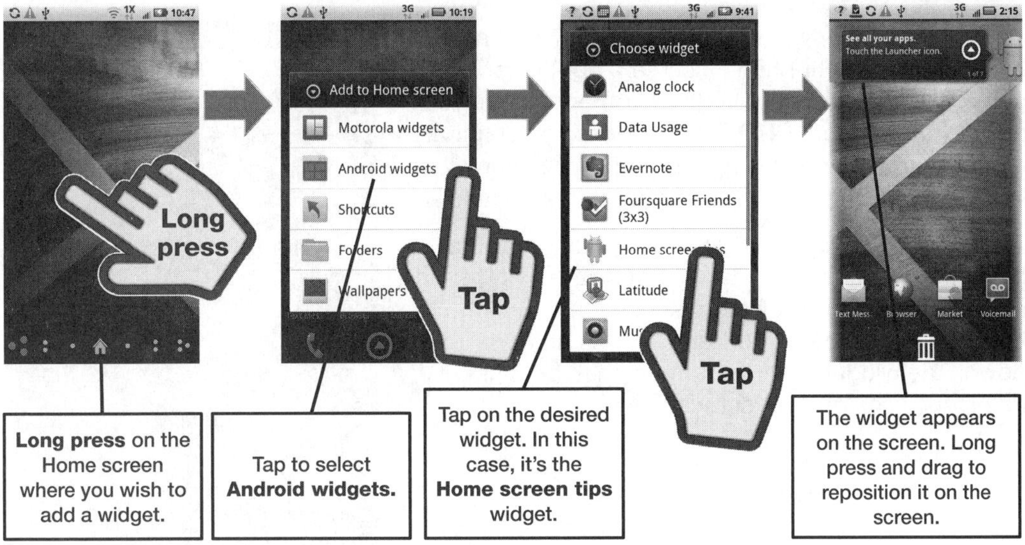

Figure 6–4. *Adding widgets.*

Follow these steps to add a widget to your **Home** screen:

1. **Long press** on the **Home** screen.
2. Tap **Android widgets**.
3. Tap your desired widget.
4. Your widget will be added. Now you can long click and drag the widget to a new location on the screen, if desired. You can even drag it to another **Home** screen page.

Android Widget Size

When you add widgets to the **Home** screen, most widgets are already a predetermined size and shape. Either they fit in the space given – or they do not. If you long-click a widget, you can move it within a screen. However, you cannot make it larger or smaller. Many app developers make multiple versions of the same widget and label them with size-specific information such as "3×2" or "4×1."

You might wonder what these measurements mean. If you measure your phone's screen, not by its physical screen size, but by the number of icons you can store on it, your DROID has a **Home** screen that measures four app icons by four app icons. Each icon is rectangular in shape to allow space for text labels.

The widget measurement generally goes horizontally by vertically. Thus, a widget that measures 4×1, such as the **Power Control** widget shown on the right, takes up the space of four apps across and one down. The widget has five buttons on it, but it is only four icons wide on your screen.

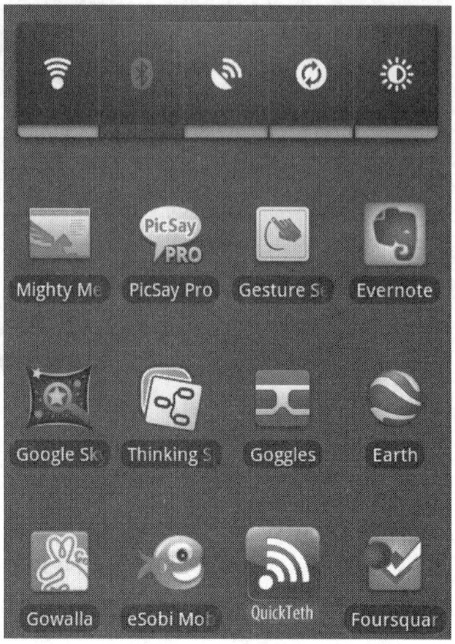

Motorola Widgets

Your DROID ships with a few Android widgets, and you can download a nearly endless supply from the Android Market. Your DROID 2, DROID 2 Global, and DROID X also ships with Motorola widgets that aren't available for download. Table 6–1 has a rundown of the different Motorola widgets and what they do.

Table 6–1. *Motorola widgets.*

	**Airplane mode toggle**: This widget allows you to toggle your phone in and out of **Airplane mode**. It's useful for flying or for times when you absolutely don't want to be interrupted.
	**Bluetooth toggle:** Like the **Airplane mode** toggle, this is a simple on and off switch for enabling and disabling Bluetooth.
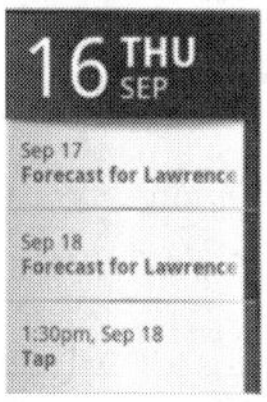	**Calendar widget:** This is a simple calendar that ties into both the **Google Calendar** and the **Corporate Sync Calendar**.

Contact quick tasks: This widget is useful for your frequent contacts like friends and relatives. You can add shortcuts to call, email, text, or even send them a Twitter or Facebook message.

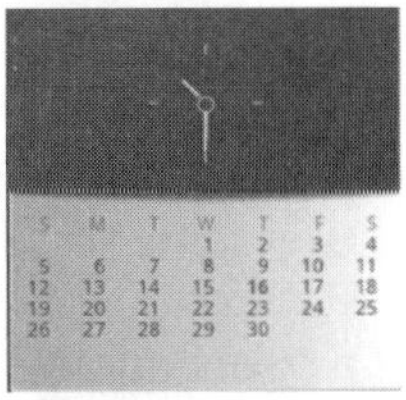

Date and Time: Unlike the **Calendar** widget, the **Date and Time** widget isn't about your appointments. This is strictly a clock for displaying the current date and time. You can select the color and choose between a digital and analog time display.

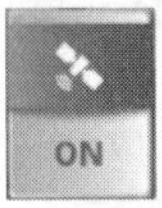

GPS toggle: Like the other toggles, this is a simple on/off switch. This one controls your DROID GPS.

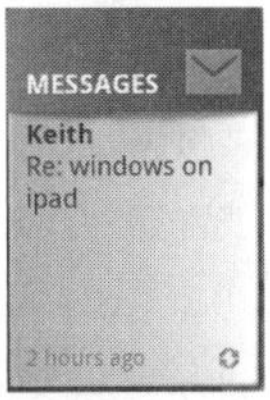

Messages: This widget displays email messages from any account you have linked to your DROID. It also displays SMS/MMS text messages and status updates from social networking accounts.

News: The **News** widget displays news items or blog entries. You can add feeds by URL or subscribe to preset bundles of news feeds.

Photo slideshow: This is a gallery widget that displays an interactive slideshow of photos you've taken from your DROID. You can also use it as a quick shortcut to your DROID's gallery and camera apps. You can only have one Photo slideshow widget at a time.

Photo Widget: This widget lets you display a single photo as if it were an app icon or other shortcut. You can add as many of these as you wish, but each one takes space on your **Home** screen, and each widget can only display a single photo.

Social Networking: This widget displays a feed of updates from your linked social networking services such as Twitter, Facebook, and MySpace.

Social Status: This widget is the other end of the social networking widget. It allows you to update your linked social networking services with status updates. You can update individual services, such as just Facebook or just Twitter; or you can update all of them at once with the same message.

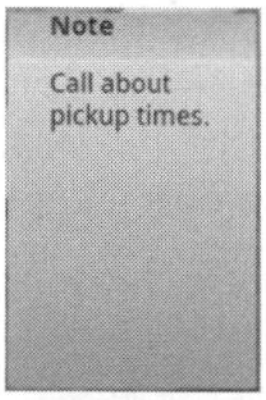

Sticky Note: This is the DROID version of a paper sticky note. You can write new notes to yourself and display them until you change the note. It's not for long-term storage, and it doesn't pull the information from other sites.

Weather: This widget displays weather information with a nice graphic to suggest current conditions. The information displayed in this widget comes from `AccuWeather.com`.

WiFi toggle: This is a simple on/off toggle for using Wi-Fi for your DROID. Toggle it off when you're out of Wi-Fi range to save battery time.

Motorola widgets have a lot in common with Android widgets, but they also give you an extra bonus in flexibility. Motorola widgets are resizable, so they'll take up as much or as little space as you allow them. Figure 6–5 shows you how this works.

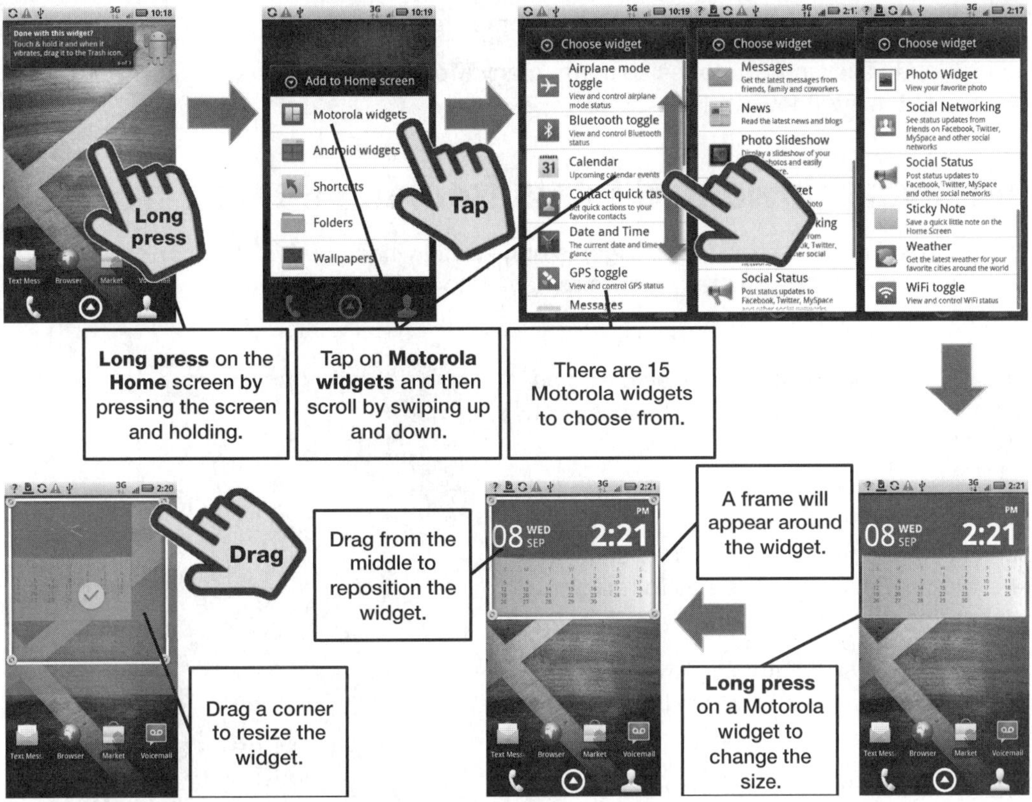

Figure 6–5. *Working with Motorola widgets.*

Adding and Resizing Motorola Widgets

Follow these steps to add a Motorola widget to your **Home** screen; note that you use the same long press procedure you would to add an Android widget:

1. Long press on the **Home** screen.
2. Tap **Motorola widgets.**
3. Slide your finger along the screen to scroll through the widget list.
4. Tap to select the desired widget.

Once you've added a widget, you can follow these steps to resize that widget to fit on your screen:

1. Long press on a **Motorola widget** until your DROID vibrates and you see a green halo around the widget.
2. A frame appears around the widget. Drag the corners of the frame to resize the widget.

3. Tap anywhere else on the screen to complete the action.

In addition to allowing you to resize them, many **Motorola widgets** also allow you to customize the info they display.

Removing Unwanted Items from the Home Screen

The Android **Home** screen **Tips** widget is only useful when you first get started with your DROID. After a while, you may want to remove it to make space for other app icons, widgets, or shortcuts.

Removing widgets is as simple as dragging them to the trash, as shown in the image on the right.

In fact, this is the same process you use to remove any unwanted item from the **Home** screen, whether it's an app icon, a shortcut, or a widget. Follow these steps to remove a widget from the **Home** screen:

1. Long press the item you want to remove until you feel the haptic feedback.
2. The item will get a green halo, and you'll see a trashcan on the bottom of the screen. Drag the item to the trash.

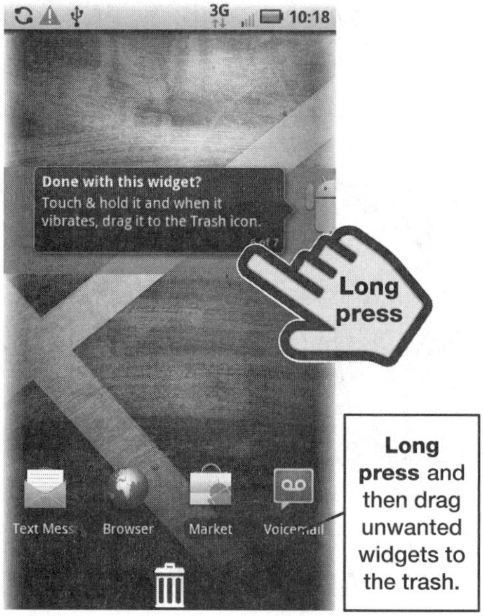

Don't worry. You haven't permanently deleted the app or widget. If you change your mind, you can add it back to your **Home** screen.

Wallpaper

So far we've covered widgets, shortcuts, and folders. The last option you see when you long press on the screen is **wallpapers**. Wallpapers provide an easy way to personalize your phone, whether it's by downloading artwork or using a photo you shot from your DROID.

There are three basic types of wallpapers on your DROID: **Live wallpapers**, **Media gallery**, and **Wallpapers**. Figure 6–6 illustrates their basic differences.

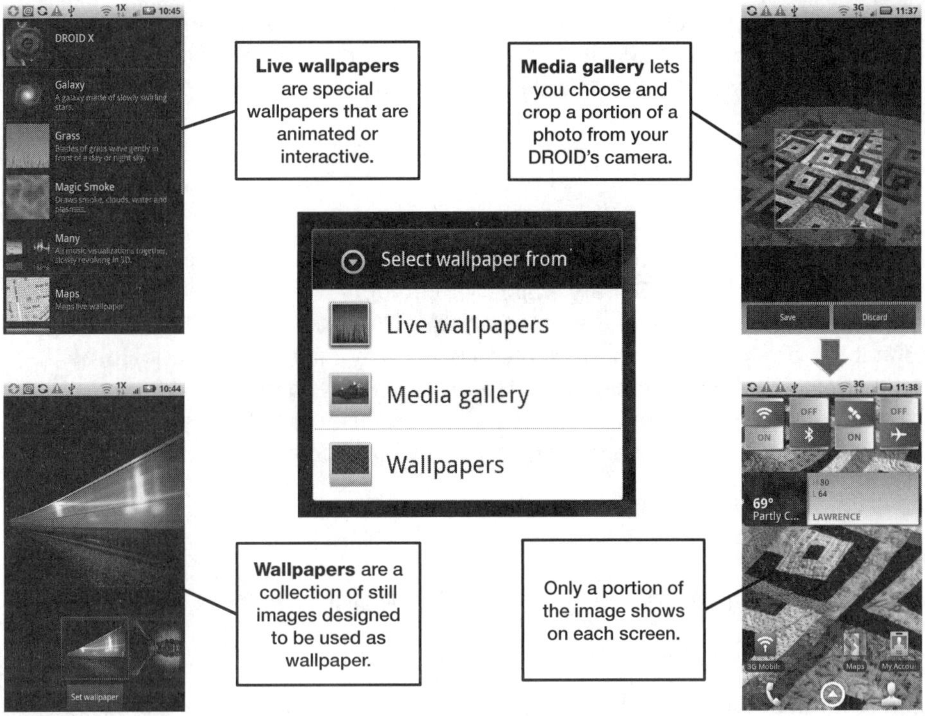

Figure 6–6. *Wallpaper types.*

Live Wallpapers

Android 2.1 introduced **Live wallpapers**. These are background wallpapers that can be animated and made to react to events on your phone. For example, they might respond to you touching the phone, the time of day, or the sounds playing on your phone.

Several **Live wallpapers** come with your DROID, including the red, lit robotic eye that you see in the DROID commercials. You can also download new wallpapers both as free and paid apps.

The screen on the right shows the **Shake Them All!** wallpaper by Yougli; one of the **Live wallpapers** included in this app shows the Android logo falling in slow-motion on your screen. It reacts when you shake the screen or touch the androids as they fall.

The disadvantage of **Live wallpapers** is that they take more power to display, especially if they have interactive features.

Live wallpapers are essentially small apps running in the background, which means they can occasionally crash or cause issues with other running apps. If you find your phone crashing frequently or losing battery power rapidly, try switching to a different wallpaper.

Media Gallery

You might want to use a photo you took on your DROID as your wallpaper; this is easy to do, and it falls into the **Media Gallery** wallpaper category. Keep in mind that you can only use a portion of the photo, and the entire photo will not display at once. You might also notice that the portion of the image you see is actually thinner than your **Home** screens. As you scroll between screens, the **Home** screen wallpaper will appear to move more slowly than the foreground images. Follow these steps to display a photo you took with your DROID as the wallpaper on your **Home** screen:

1. Long press the **Home** screen.
2. Tap **Wallpapers**.
3. Tap **Media gallery** (DROID 2/X) or tap Gallery (DROID).
4. Choose a photo you want to use.
5. Select an area of the photo to use. By default, the photo starts with a small, central crop. You can drag the red frame edges until you've selected the portion of the photo you want to use.
6. Tap **Save**.

Chapter 7

Making Phone Calls

The DROID is capable of so many cool things that it's easy to forget that it's also a very powerful phone. In this chapter, we will cover the many DROID features you would expect from a high-end smartphone. You can dial by name, save time by using your recent call logs, dial by voice, use speed dial numbers, and use **Basic** or **Visual Voicemail** features. You can also use your DROID to initiate a conference call among several people.

You can also customize your phone, message, and other ringtones. You can even set a custom ringtone for individual contacts in your address book – this is a great way to know who is calling without looking at your phone. It is easy to use your own music as ringtones on a DROID, but a little harder to use your own music as custom ringtones for contacts or text messages. We will show you how to do all these things, and a few more nice tricks, as well. We will also show you how to purchase ringtones from the Amazon MP3 and Android Market online stores.

Getting Started with the Phone (Dialer)

Start your phone by using the **Phone** icon located at the bottom-left corner of every **Home** screen. You can also start the phone by using the **Dialer** icon or Phone icon on the original DROID.

Finding Your DROID Phone Number

Maybe you just received your new DROID and don't yet know the phone number. Don't worry; you can find your number in the **Settings** app by following these steps:

1. Tap the **Settings** icon.
2. Scroll down and tap **About Phone**.
3. Finally, tap **Status**. Your number is listed under **My phone number**, as shown in the image to the right.

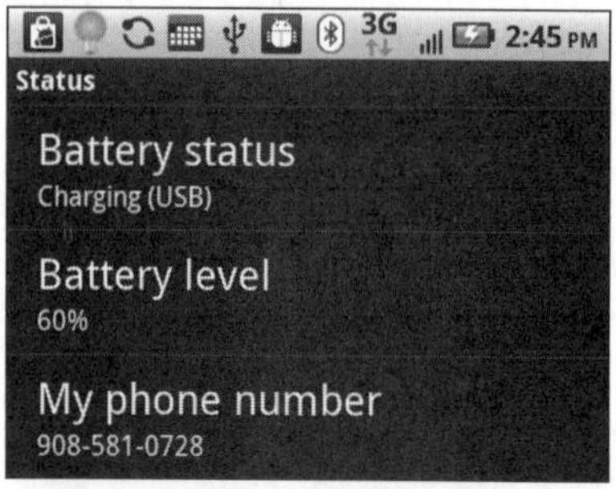

> **TIP:** You can also check your phone number using **Voice Command** on the DROID 2 and DROID X. From the **Phone dialer** screen, tap the **Voice Command** icon just to the right of the **Green phone** icon and say, "Check my phone number."

Muting the Phone Ringer - Slide for Mute

Occasionally, you want to silence your phone ringer. Follow these simple steps silence your DROID's ringer:

1. Tap the **Power/Lock** key on the top of your phone to turn off the screen (see Figure 7–1).
2. Tap the same **Power/Lock** key to bring up the **Lock** screen.
3. Slide the **Speaker** icon to the right to turn it to **Vibrate** mode. Notice that the **Speaker** icon changes to show a **Vibrating phone** icon.
4. To turn off **Vibrate** mode, repeat the procedure. Notice that the icon returns to the **Speaker** icon.

> **TIP:** You can also press and hold the **Volume Down** button to mute your ringer.

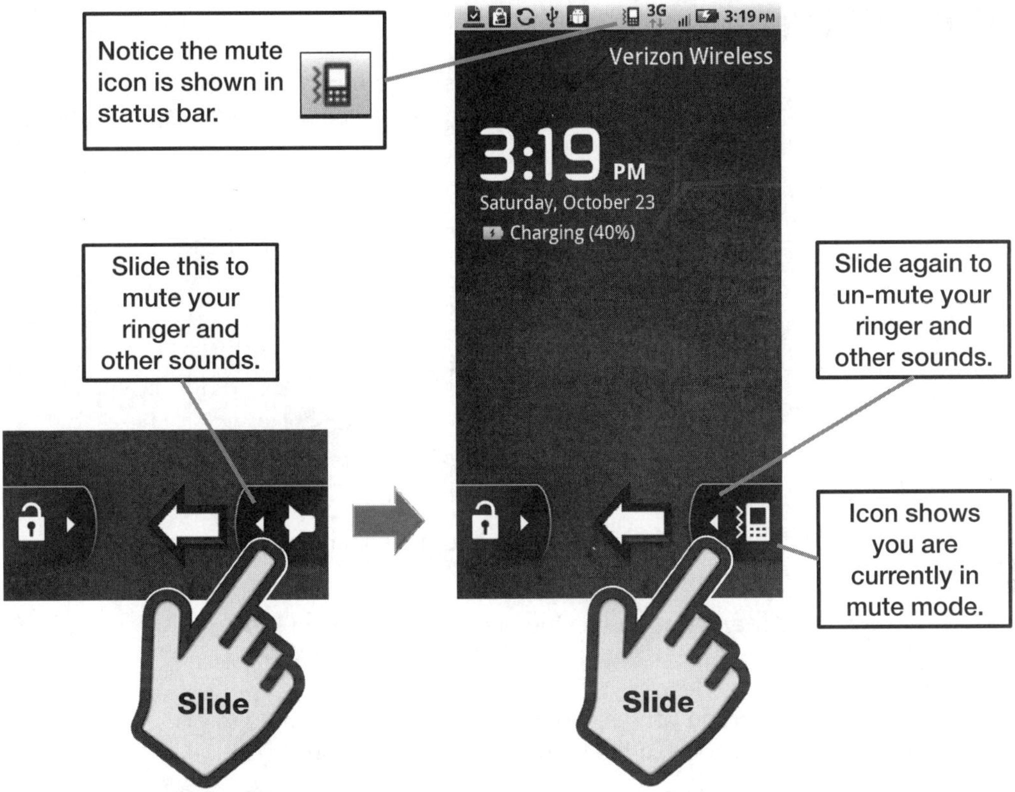
Notice the mute icon is shown in status bar.
Verizon Wireless
3:19 PM
Saturday, October 23
Charging (40%)
Slide this to mute your ringer and other sounds.
Slide again to un-mute your ringer and other sounds.
Icon shows you are currently in mute mode.
Slide
Slide

Figure 7–1. *Slide to mute or unmute your phone.*

Examining Different Phone Views

Your phone keypad can be shown by tapping the **Dialer** soft key. There are three other soft keys along the top that give you different ways to use your phone: **Recent**, which shows recent calls placed, missed, or received; **Contacts**, which shows your contact list; and **Favorites**, which shows your favorite phone numbers (see Figure 7–2).

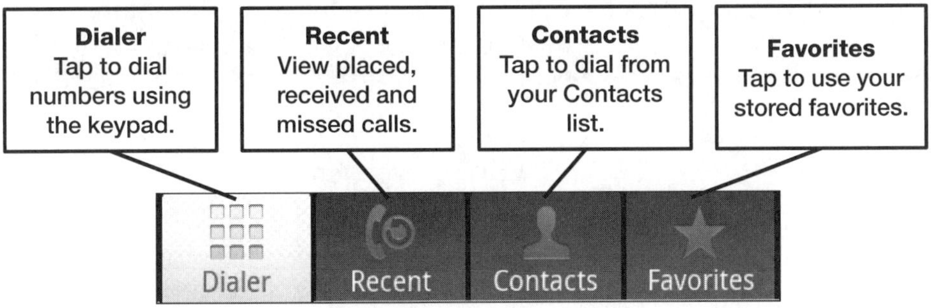
Dialer
Tap to dial numbers using the keypad.
Recent
View placed, received and missed calls.
Contacts
Tap to dial from your Contacts list.
Favorites
Tap to use your stored favorites.
Dialer
Recent
Contacts
Favorites

Figure 7–2. *These soft keys show you different views of your phone.*

Using the Dialer Keypad

The simplest way to make a call is to use the **Dialer**. The numbers on the screen are large, so it's easy to dial a number. Follow these steps to use the **Dialer** app or Phone app on the original DROID to make a call:

1. Tap the **Dialer** icon (see Figure 7–3).
2. If you do not see the keypad to dial, tap the **Dialer** soft key 

at the top-left portion of the screen.
3. Now you can simply start dialing by tapping number keys.
4. If you make a mistake, press the **Backspace** key  next to the window where the numbers appear.
5. If you need to type a **Plus** (+) sign for an international number, press and hold the **Zero** (**0** +) key.
6. When you are done dialing, press the green **Phone** key at the bottom . If the person you are dialing has a picture attached to his contact record, you will see the picture appear (see Figure 7–3).
7. When you are done with the call, press the **End call** key (the **Red phone** icon) 

.

> **TIP:** Sometimes you have to enter a pause in a phone number for a couple of seconds, then dial another number such as an extension or a password. You can dial a pause by pressing the **Menu** button and selecting **Add Pause**. You see a comma appear next to the phone number, which indicates a two-second pause. If you need a wait to be inserted, then press the **Menu** button and select **Add wait**. A wait will pause in dialing and wait for you to tap a button to continue dialing.

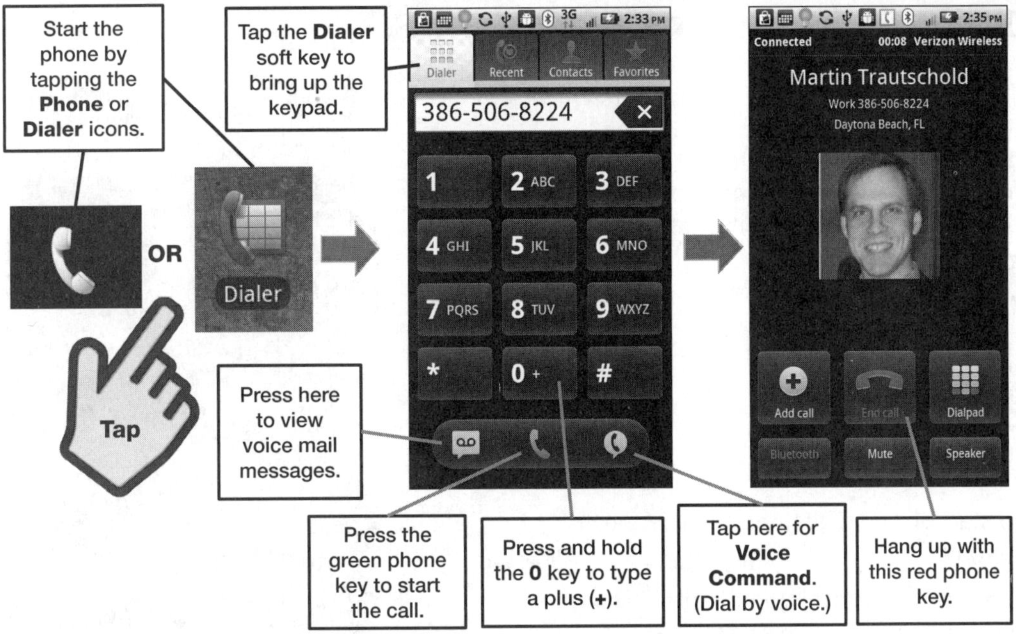

Figure 7–3. *Dialing phone numbers with your DROID keypad.*

> **TIP:** You can redial the number you just called by tapping the **Green phone** key on the **Dialer** pad. You will see the most recent number you dialed appear; tap the **Green phone** key again to place the call.

Muting Yourself on a Call

If you want some privacy during the call, tap the **Mute** button in the bottom-middle of the screen. You know the call is muted when you see the green bar at the bottom of the button lit up as shown in the figure to the right. Tap it again to unmute yourself.

Dialing Digits While on a Call

Sometimes you need to dial numbers while already on a call. Common examples include dialing an extension or dialing the first few letters of someone's last name to look up her name in automated directory.

To start dialing digits, tap the **Dialpad** button shown to the right.

Tap the same button to hide the dialpad, which has changed to say **Hide**.

Using the Speaker Phone

Tap the **Speaker** key to use the speakerphone on your DROID. Like the **Mute** key, the speakerphone is on when the green bar is lit on the bottom of the key.

Talking Hands-free with a Bluetooth Headset or Bluetooth Car Stereo

If you are in one of the many states or provinces where you cannot legally hold your DROID while driving a car, you will want to purchase a Bluetooth headset or use a Bluetooth car stereo connection to talk hands-free (please see Chapter 8: "Bluetooth on Your DROID" for more information on this topic).

Tap the **Bluetooth** key in the lower-left corner of the screen to switch audio over to your Bluetooth device. As with the **Mute** feature, you know **Bluetooth** is on when the green bar is lit on the bottom of the key.

Check out the "Voice Dialing" section later in this chapter for more information.

Opening Other Apps While on a Call

It's easy to multitask on your DROID. This is especially useful when you are on a phone call. For example, you can press the **Home** button to jump back to the **Home** screen and start another app, or long-press the **Home** button to jump to a recently used app.

Some common examples of useful multitasking while on a call include the following:

- Checking and scheduling a new **Calendar** event.
- Looking up a name, address, or phone number in the **Contacts** app.
- Finding a message in your **Email** app to use during the call.

Getting Back to the Phone From Any App

Returning to your phone call from any app is easy. Follow these steps to get back to your call in progress from the **Home** screen or any other app on your DROID (see Figure 7–4):

1. Start by swiping your finger down from the top status bar.
2. Tap the **Phone** call in progress to return to the current call or tap the **Red phone** icon to hang up.

You can also return to a call by long-pressing the **Home** button to bring up your **Recent** window of apps, and then selecting the **Dialer** icon.

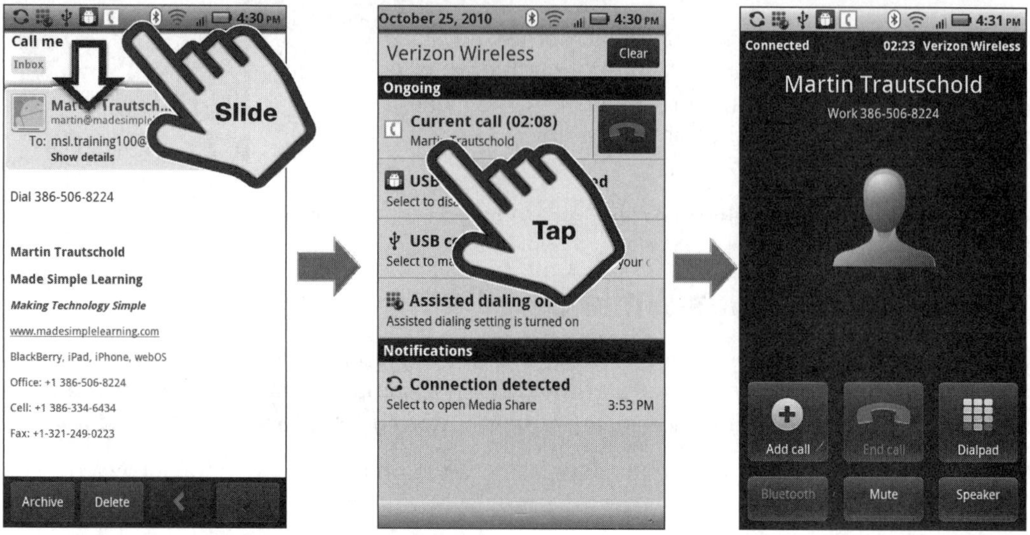

Figure 7–4. *Getting back to the call in progress from any app.*

Using Speed Dial on Your DROID (DROID 2 and DROID X only)

The **Speed Dial** feature lets you use the number keys on the phone's dialpad to quickly dial phone numbers. You can press and hold a number to call the associated speed dial number.

> **TIP:** The only speed dial number that is pre-assigned is the number one (**1**), which dials your voicemail by default. You speed dial your voicemail by pressing and holding the **1** key.

Adding Speed Dial Numbers

You have a few options for adding new numbers to speed dial, including the following:

- Press and hold any unassigned number on the dialpad to set it as a new speed dial number. Next, select a contact to assign to that number to.
- From any **Phone** screen, press the **Menu** key and select **Speed dial setup**. Next, tap any number that has not yet been assigned to select a contact to assign to it.
- From the **Recent** list, you can add speed dial numbers by following the steps shown in the "Assigning a Recent Caller a Speed Dial Number" section later in this chapter.

- From the **Contacts** view in the phone, press and hold any contact name and select **Edit speed dial** to assign this contact a speed dial number.

Using Speed Dial

It's easy to use the DROID's speed dial feature to make a call. Follow these steps to do so:

1. Bring up the keypad on the phone by tapping the **Phone** icon and then tapping the **Dialer** soft key.
2. Press and hold the correct speed dial number on the keypad to dial its associated number.

Using the Recent View (Call Logs)

The **Recent** view serves as your call log of outgoing, incoming, and missed calls.

Touching the **Recent** soft key displays a list of all your recent calls.

Tap the **View** bar just under the soft keys to filter your view by **All calls**, **Missed calls**, **Received calls**, or **Outgoing calls** (see Figure 7–5).

NOTE: The original DROID uses a Call log soft key instead of one that says Recent.

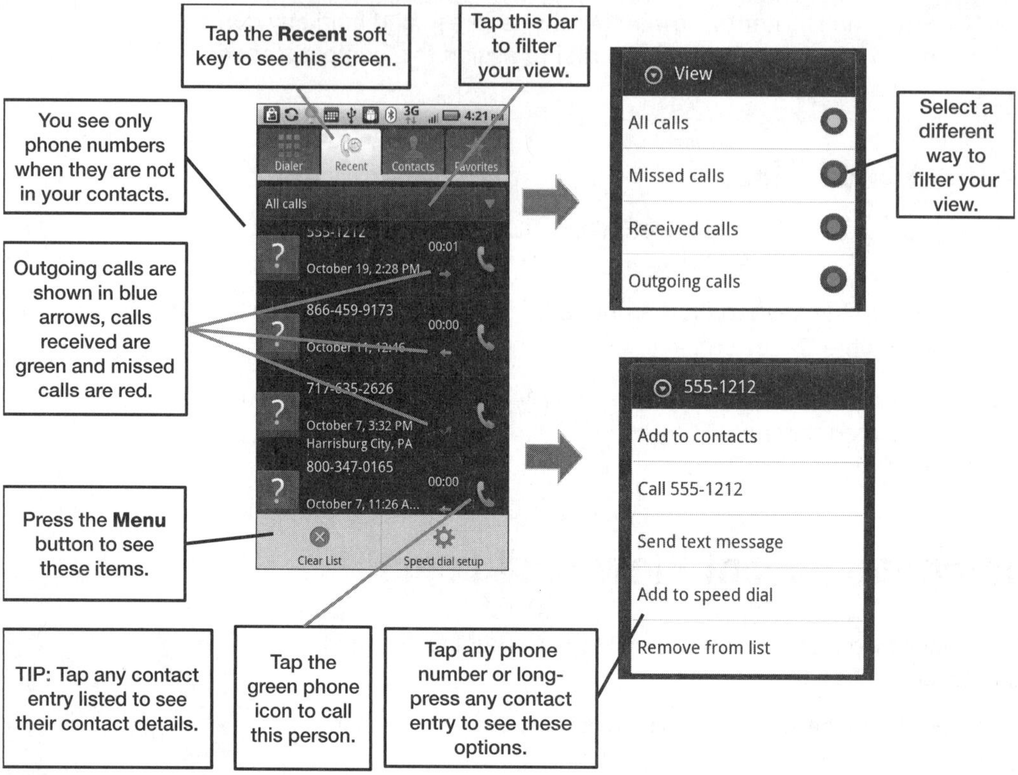

Figure 7–5. *Working with your **Recent** screen.*

Placing a Call from the Recent View

It's a simple matter to place a call when looking at the **Recent** view. All you need to do is tap the **Green phone** icon next to a recent caller to call that contact or number.

Viewing Contact Details from the Recent View

It's also easy to view the contact information for someone listed in the **Recent** view. Simply tap the contact's name to view their contact details.

Adding a Number to Your Contacts List

Sometimes you will want to add the phone number of a recent caller to your **Contacts** list. Follow these steps to do so:

1. Tap the phone number listed in the **Recent** or Call log list to see a pop-up asking whether you want to **Add to contacts**.
2. You will be asked whether you want to add the number to an **Existing** contact or to create a **New** contact with this number. If you select **Existing**, then you will need to choose a contact from your list.

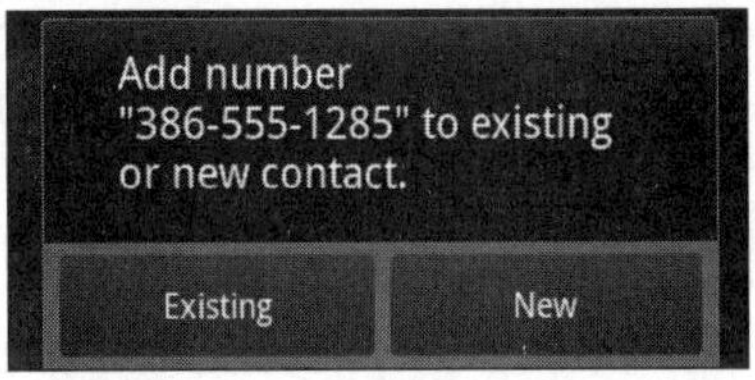

3. If you select **New**, then you need to create a new contact with the number listed in the **Recent** view.

Sending a Text Message to a Recent Caller

Another cool DROID ability: You can send a text message to a recent caller. To do so, long-press the name or phone number in the **Recent** list and select **Send text message**.

Assigning a Recent Caller a Speed Dial Number (DROID 2 and DROID X only)

It's also a simple matter to assign a speed dial number to a recent caller. Follow these steps to do so:

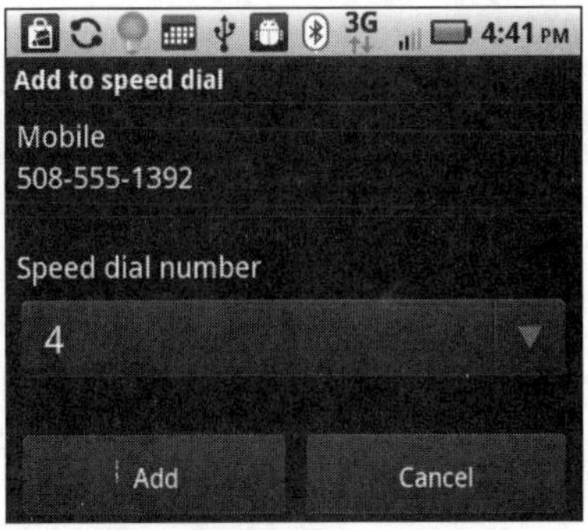

1. Long-press the name or phone number in the **Recent** list and select **Add to speed dial**.
2. The next screen (shown to the right) displays a default number (4, in this case) as the next available speed dial number. Tap that number to change it to another number, if you desire.

3. When done, tap **Add** at the bottom to assign this speed dial number to a recent caller.

Tapping the Contact Icon to Access Other Features

When you tap the icon shown to the left of the contact name in your **Recent** list, a little **Quick Access** pop-up appears above the icon. This pop-up allows you to view the contact's details; or to call, text, email, or map the contact (see Figure 7–6). If the contact has a picture – such as the one for Gary Mazo shown to the right or just a blank shoulder/head – then you know the person is in your **Contacts** list. If you see **Question mark** icon, then you know the phone number is not connected to any contact.

or

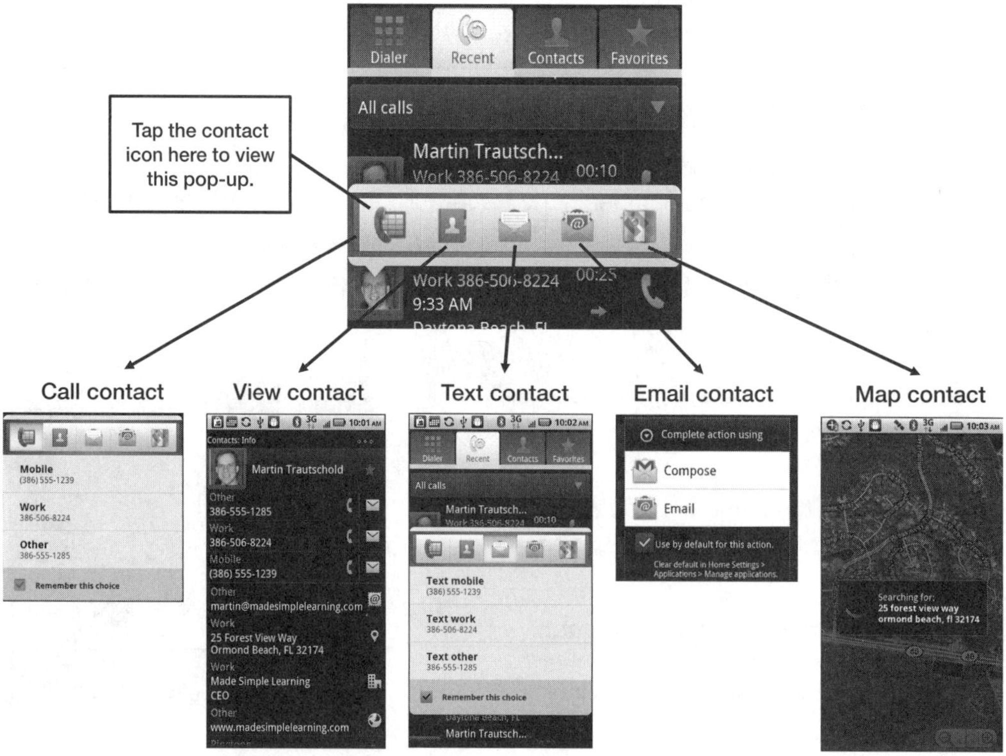

Figure 7–6. *Using the* ***Quick Access*** *contact pop-up from the* ***Recent*** *screen.*

NOTE: On the original DROID, you only see the contact icons in the Contacts and Favorites sections of the Phone app.

NOTE: Certain icons will not appear in the **Quick Access** window if you do not have the related information stored in your **Contacts** app. For example, if you do not have an address for a contact, then the **Map** icon will not appear. Similarly, if you do not have an email address, then the **Email** icon will not show up. This **Quick Access** window also appears in your **Contacts** app when you tap the icon to the left of the contact name.

Tapping the Question Mark Icon to Add, Call, or Text a Recent Caller

You have two clues that a given phone number is not in your **Contacts** list. First, you will see a **Question mark** icon (**?**) instead of a face or blank head/shoulders in the icon. Second, you will see a phone number instead of a contact name.

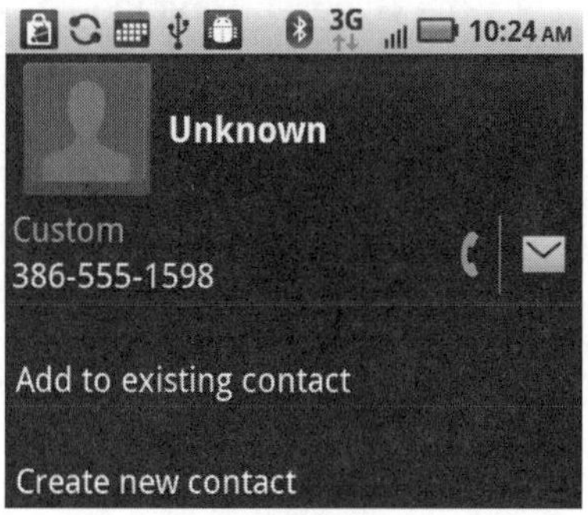

To quickly add this number to an existing or new contact, tap the **Question mark** icon to see this screen.

From this screen, you can choose **Add to existing contact** if this is a phone number for someone already in your address book.

Tap **Create new contact** to associate this number with a new contact entry.

Clearing Out or Erasing Your Recent List

Sometimes you may want to delete all the numbers in your recent list. Follow these steps to do so:

1. Press the **Menu** button.
2. Select **Clear list** to erase all the entries in the **Recent** list.

Placing Calls From Contacts

One of the great things about having all your contact information in your phone is that it becomes very easy to place calls from your **Contacts** list on the DROID. Follow these steps to call someone in your **Contacts** list:

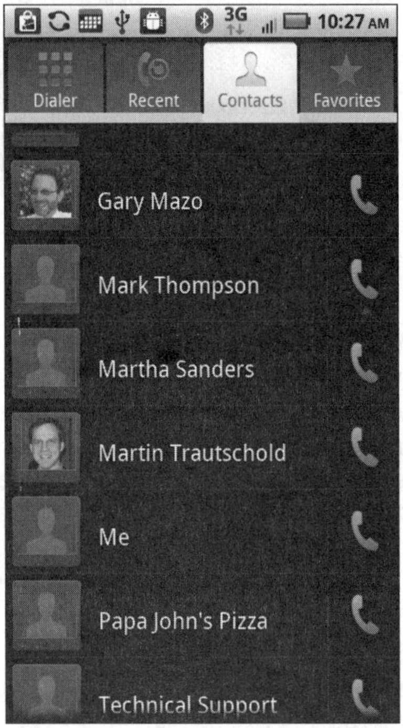

1. If you are not in your **Dialer** app, tap the **Dialer** icon to start it up.
2. Touch the **Contacts** soft key at the top.
3. Locate a contact to call using one of the following methods:
    a. Swipe up or down through the list.
    b. Press the **Menu** key and select **Search**.
    c. If you have a physical keyboard, slide it out and start typing a name.
4. When you find the contact entry you want, tap the **Green phone** icon next to his name.
5. If the contact has more than one phone number, you need to select one. If you want the DROID to remember this is your default choice, check the box next to **Remember this choice.** See the "Calling Favorites" section of this chapter for more information on how to do this.

Using Favorites

Your **Favorites** view shows both contacts you have specifically assigned as **Favorites** and frequently called contacts. This is a nice view because it can save you time by putting your contacts just a tap away.

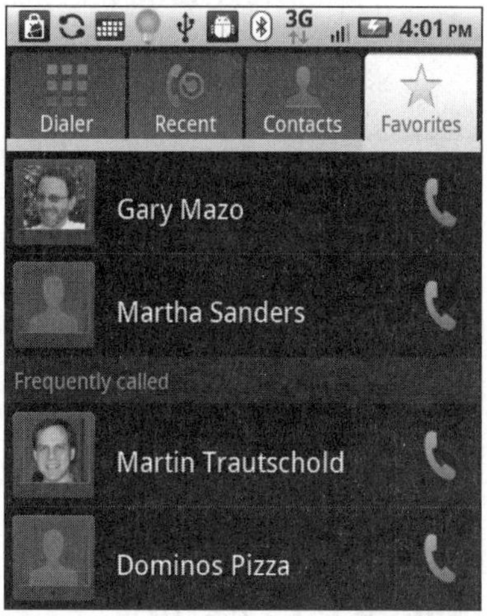

All your favorites are shown at the top, and the **Frequently called** contacts are displayed at the bottom.

> **TIP:** You can clear out the **Frequently called** contacts by pressing the **Menu** key and selecting **Clear frequent list.**

Adding New Favorites

It is easy to add new favorites to your list from the **Contacts** view in the **Dialer** app. Follow these steps to do so:

1. If you are not in the **Dialer** app, tap the **Dialer** icon to start it up.
2. Touch the **Contacts** soft key at the top to view your **Contacts** list.

3. To search for a contact, press the **Menu** button and select **Search.**
4. Start typing a few letters of the contact's first, last, or company name to find the contact quickly.

> **TIP:** If you have a physical keyboard, then you can also slide out the keyboard and start typing to find a particular contact.

5. You can simply swipe up or down to find a contact.

6. Long-press the contact you want to add as a **Favorite** and select **Add to favorites.**

> **TIP**: You can also make any contact a **Favorite** by just touching the star next to his name in your **Contacts** app.

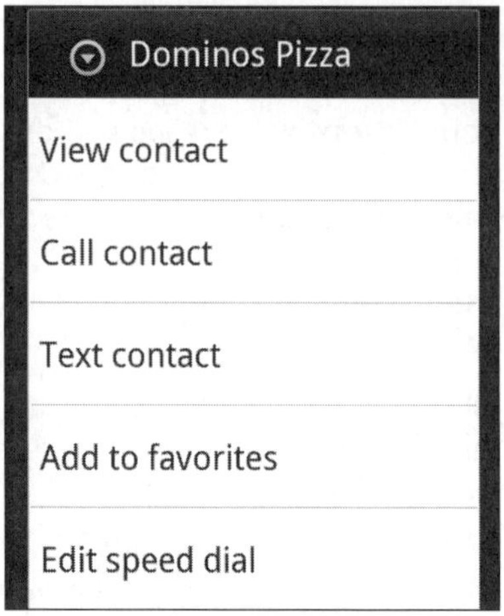

Calling a Favorite

To call any **Favorite,** touch the **Green phone** icon next to the name of the contact in your **Favorites** screen.

If the contact has several phone numbers, then you will be asked to select the phone number to call.

> **TIP:** If you almost always use the same number for that contact, then you can press the **Remember this choice** option in the dialog box that pops up, as shown in the image to the right.

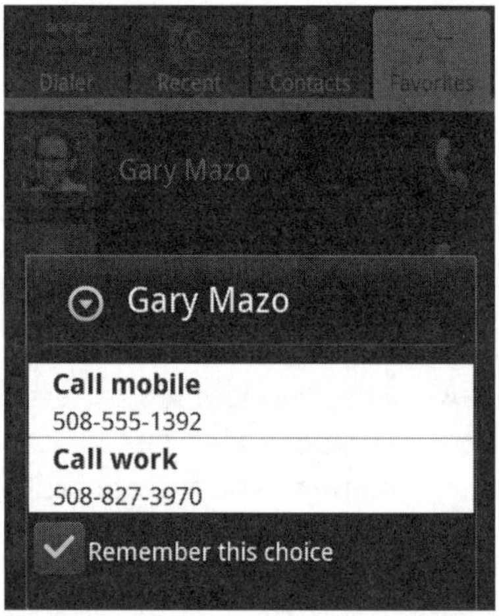

Voice Dialing

You can use your voice to dial names and phone numbers from a variety of places on your DROID. Here are three options you have for dialing a call with your voice:

- Tap the **Voice dialing** icon in the lower-left corner of the **Dialer** app's **Dialer** screen. 
- Press and hold the **Search** button  on the bottom of your DROID until you see the **Speak now** box appear in the middle of the screen. If you have a slide-out keyboard, you can do the same thing with the **Search** button on the keyboard. You can also double-press the **Home** button on both the DROID X and DROID 2 to accomplish this task.
- Press the button on the side of your Bluetooth headset or the call button on your Bluetooth car stereo to start **Voice Command**.

Voice Command

The **Voice Command** feature on your DROID enables you to place calls using your voice.

Note that you can use **Voice Command** to dial a contact by name or number; send a text message; dictate an email; check for missed calls; look up a contact entry; play a playlist; check messages; check signal strength, network availability, or battery charge; and much more.

We cover **Voice Command** fully in Chapter 2: "Typing, Voice, Copy, and Search."

Quickly Checking Missed Calls

If you have a Bluetooth headset, you can quickly check for all missed calls by pressing the button on the headset and saying, "Check missed calls."

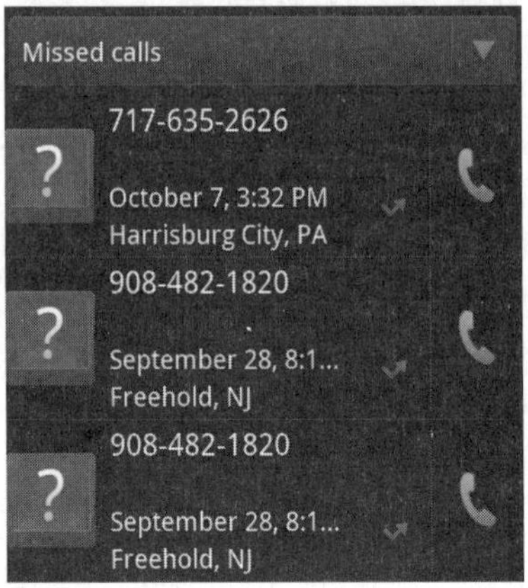

> **TIP:** You don't need a Bluetooth headset for this trick. You can also do this by tapping the **Voice dialing** icon in the lower-right corner of the **Dialer** app's **Dialpad** screen.

Conference Calling

You can get a conference call going fairly easily on your DROID. Follow these steps to do so:

1. Call the first person and press the **Add call** button.

2. Call the second person using any of the methods described in this chapter: **Dialer**, **Recent**, **Contacts**, **Favorites**, or **Speed Dial**.

> **TIP:** You know someone is on the phone because the top status bar is now green, as shown to the right. On the original DROID, you won't see a green status bar but you will see a Green Phone icon.

3. Once the second person is connected, you see a large **Green phone** icon in the center of the screen *and* the green **Merge calls** button. Tap **Merge calls** to put the callers together in a conference call.

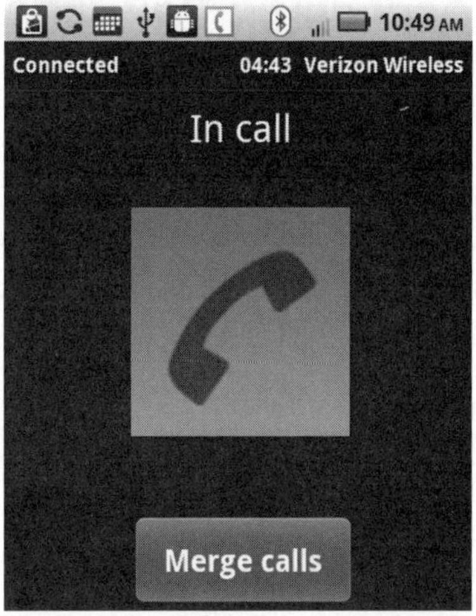

4. Once you have connected the callers, you have the following options:
   - Tap **End last call** to hang up with the second caller.
   - Tap the **Add call** key to add another person to the conference call.
   - Tap **End call** to hang up on all the callers.

Accessing Voicemail

Your DROID comes with the basic **Voicemail** features and an optional enhanced voicemail system called **Visual Voicemail** that costs an extra $2.99 per month at publishing time. **Visual Voicemail** is a nice feature because it allows you to quickly see all voicemails and play them in any order. You don't have to listen to each message to see who called; instead, you can just tap the message you want to hear.

Setting up Basic Voicemail

Setting up basic voicemail features on your DROID requires only a handful of steps:

1. Tap the **Dialer icon** or Phone icon on the original DROID.
2. Tap the **Voicemail** icon  in the bottom row of soft keys next to the **Green phone** key.
3. Tap **Call Voicemail.** If this is the first time you are using voicemail with this phone number, then follow the prompts to set up your voicemail and select a pass code. Otherwise, enter your passcode if requested and follow the prompts to retrieve your voicemail messages.

> **TIP:** Your default voicemail password might be the last four digits of your phone number. If you are unsure, give that a try.

Getting Your DROID to Enter Your Voicemail Password

Normally, you have to type in your voicemail password or PIN every time you call voicemail. This tip will allow you to have it automatically entered for you by your DROID:

1. Start your **Contacts** app.

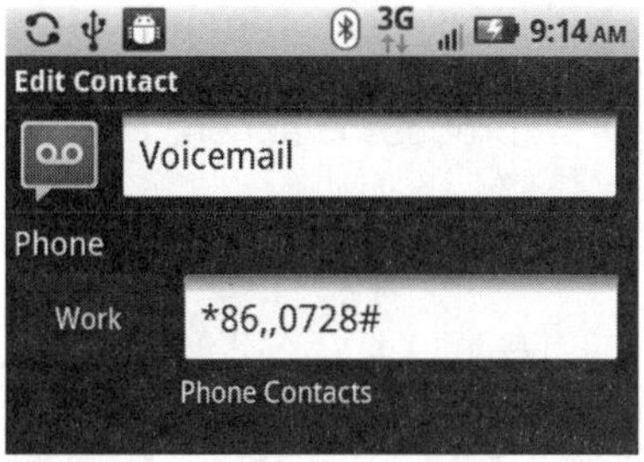

2. If you see a contact called **Voicemail**, press the **Menu** button and select **Edit**. Otherwise, create a new contact and call it **Voicemail**.
3. In the phone number box, type ***86,,,nnnn#**, where **nnnn** is your four-digit voicemail password or PIN. You will need to tap the ***#(** key to see the **Pause** key. Each comma (**,**) is really a two-second pause. Remember to put the pound sign (**#**) at the end of your entry.
4. Now you can call the **Voicemail** contact and have it automatically dial your password. You can make this process even easier by associating your **Voicemail** contact with your speed dial for the number **1**. This will enable you to access your voicemail with a single long-press of the **1** key on your phone's dialpad.
5. Start your **Dialer** app, press the **Menu** key, and select **Speed dial setup**.

6. Tap the **Minus** key  to the right of the top speed dial to remove it.
7. Now tap the top item (**1**) to set it as the **Voicemail** contact that you just edited or created. It should look similar to the screen shown to the right when you are done.

8. Next, press and hold the **1** key on the dialpad of your **Phone** app to automatically call the **Voicemail** contact and have your password entered automatically. You will see a short pause, after which your password will be sent in a little pop-up window that says, "Sending Tones." The figure to the right shows this process in action.

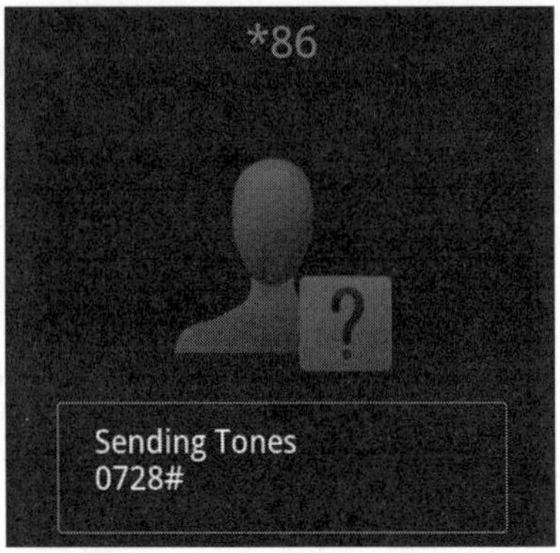

TIP: You can use this same trick to add pauses and waits to other numbers when dialing extensions, dialing phone and other passwords, or even when accessing numbers. You need to set up new contacts for each number you want to call, after which you can quickly dial these numbers and their access codes or passwords without entering them manually – and without dialing them each time. We definitely recommend you secure your DROID with a password or other method if you choose to enter sensitive information such as bank access passwords in your DROID contacts.

Visual Voicemail

As we mentioned earlier in this chapter, the **Visual Voicemail** feature allows you to instantly see who has left you voicemail messages. It also lets you listen to them or delete them in any order. At publishing time, this was a paid add-on service from Verizon that costs US $2.99 per month.

TIP: You can get free **Visual Voicemail**-like features from other places. For example, you can choose one of the free apps from the Amazon Market such as **YouMail Visual Voicemail** or **Visual VoiceMail**. The other method is to use the free **Google Voice** app; you'll learn more about this app in the "Using Google Voice" section later in this chapter.

Follow these steps to access the **Visual Voicemail** service from your DROID:

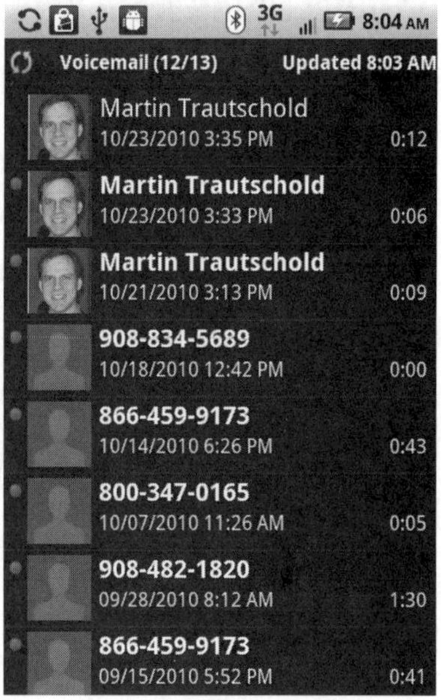

1. Tap the **Dialer** icon.
2. Tap the **Voicemail** icon  in the bottom row of soft keys next to the **Green phone** key.
3. If this is the first time you are using **Visual Voicemail**, then you need to tap the **Subscribe to Visual Voicemail** button at the bottom of the screen and **Accept** the license agreement.
4. Enter your voicemail password and click **Login**.
5. At this point, you should see your **Visual Voicemail** mailbox, as shown in the image to the right. The blue dot 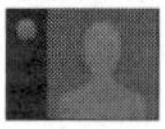 in the left column indicates that you have new messages you haven't listened to yet.
6. Tap any entry to listen to it, delete it, or call the person back. From this detail screen, you can do the following:
   - **Rewind**
   - **Play / Pause**
   - **Fast forward**
   - **Call back** the person who left the message
   - Toggle the **Speaker** on or off
   - **Delete** a message

Unsubscribing from Visual Voicemail

If you decide you want to turn off and stop paying for the **Visual Voicemail** service, you can do so, but the wireless carrier wants to make absolutely certain you want to stop using the service; you need to confirm your choice four times!

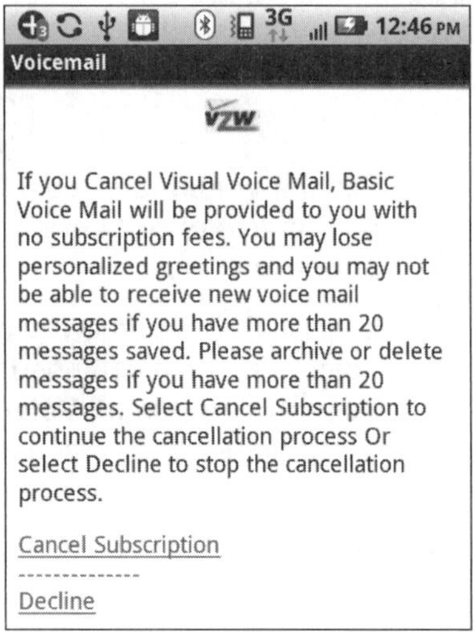

Follow these steps to cancel the **Visual Voicemail** service:

1. Bring up your **Visual Voicemail** inbox, as just explained.
2. Press the **Menu** key and select **More**.
3. Tap **Unsubscribe** from the menu.
4. Tap **Unsubscribe** from the pop-up window.
5. You may see a warning similar to the one shown to the right. If you're sure you want to cancel your subscription, tap **Cancel Subscription**.
6. You will also have another screen that asks you again to **Confirm Cancel Subscription**.
7. You may also be presented with a survey that you can choose to ignore. From this point, it takes about five minutes to cancel.

Deleting All Visual Voicemails at Once

The **Visual Voicemail** service provides the option of deleting all your voicemails at once. Follow these steps to do so:

1. Press the **Menu** button and tap **More**.
2. Select **Delete All Voicemails**.

Archiving or Deleting Multiple Messages

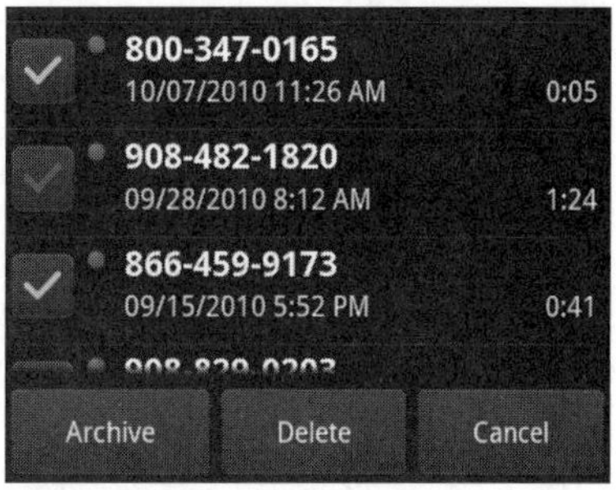

Sometimes you may want to archive or delete multiple voicemail messages at once. Follow these steps to do so:

1. From your **Visual Voicemail** inbox, press the **Menu** button and tap **Select Multiple**.
2. Now each message you tap will be selected with a **Green checkmark** icon.
3. Tap **Archive** or Mark as Heard to save all the selected messages.
4. Tap **Delete** or Eraseto delete all the selected voicemails.

Changing Your Voicemail Ringtone and Vibration

You can also change your voicemail ringtones and vibration settings from your **Visual Voicemail** inbox. Follow these steps to do so:

1. From your **Visual Voicemail** inbox, press the **Menu** button and tap **Settings**.
2. To change the ringtone you hear when you receive a new voicemail message, tap **Select ringtone** and change it.
3. Tap **Vibrate** to turn the vibration feature on or off when you receive new messages.

Using Google Voice

If you already use the **Google Voice** app, you will want to install it on your DROID. If you are new to **Google Voice**, then you should know that this is a great service that gives you the following features for free:

- *Ring all your phones simultaneously* – This feature allows people to reach you whether you are at home, at your office, or out and about with your DROID by calling a new **Google Voice** number that in turn rings all your phones, wherever you might be.
- *Receive a free **Visual Voicemail**-type service* – This feature lets you see all your voicemails listed on your DROID screen. You can listen to these voicemails in any order.
- *Receive voicemails transcribed as email or SMS text messages* – This feature provides a very convenient way to read and respond to voicemail messages.
- *Display your Google Voice number as your caller ID* – This feature shows your **Google Voice** number when people call your DROID.
- *Send free text messages* – This feature lets you use **Google Voice** to send free text messages from your DROID.

You can learn more about **Google Voice** by viewing these video tutorials produced by Google at this URL: `http://www.google.com/googlevoice/about.html`.

NOTE: At publishing time, you could not use **Google Voice** outside the United States.

Installing the Google Voice App

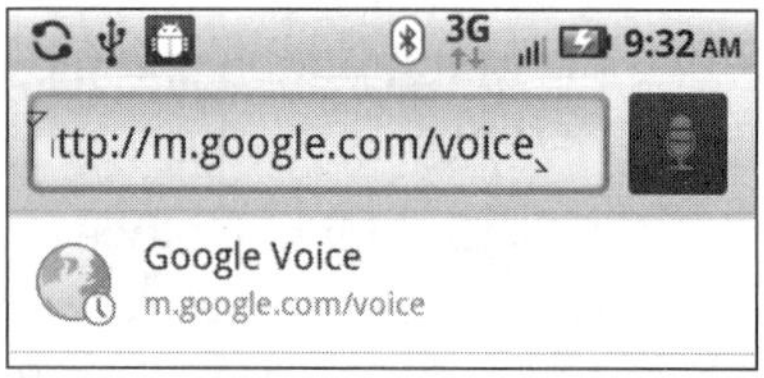

Obviously, you'll need to install the **Google Voice** app before you can use it. Fortunately, you can acquire this app for free from the Android Market. We initially tried to find it by searching for "Google Voice"; however, we found it hard to locate the app with that search criteria. Instead, we recommend starting your **Browser** app and going to `http://m.google.com/voice`. Next, click the **Download from Market** button. Once you're in the **Market** app, follow the steps described to install the app.

Setting Up Google Voice

To start **Google Voice**, tap the **Voice** icon. You may think you should look for a **Google Voice** icon; however, the icon is simply called **Voice**.

The first time you start the app, you need to log in. You need to use the Google account you used to set up your **Google Voice** account. This account could be the same or different from your default Google account on your DROID.

After logging in, you have a few things to set up. Tap the **Next** button on the bottom of the screen to continue, and then follow the instructions to select and verify your DROID phone number.

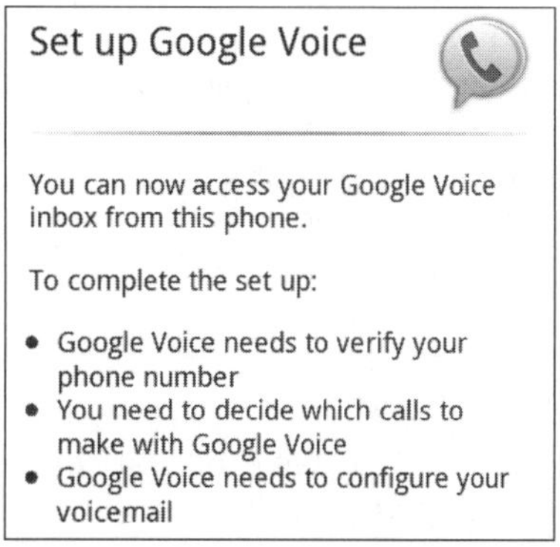

Next, you can choose whether you want to use **Google Voice** to make all calls, no calls, international calls, or be prompted each time you make a call. Choose the option that works best for you. The example shown in this chapter uses **all calls**, as shown in the figure to the right.

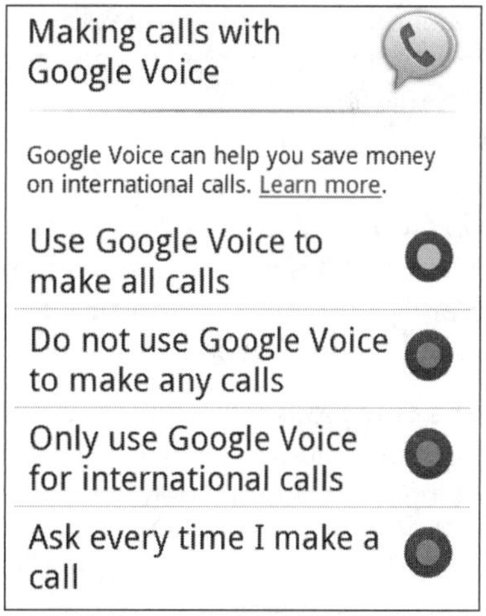

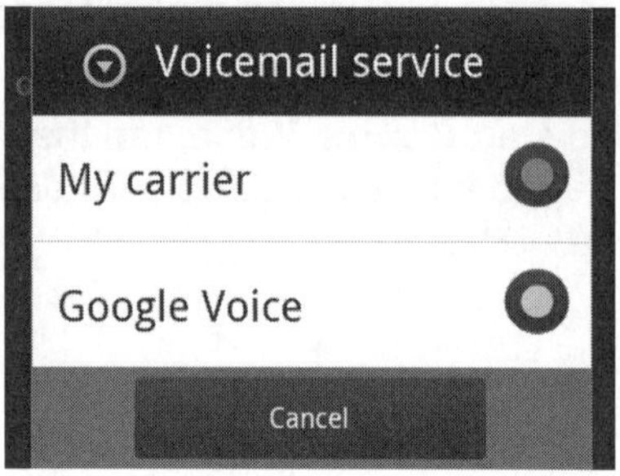

Now select whether you want to use the **Google Voice** app's **Voicemail** service on your DROID. We recommend doing this because it gives you free visual voicemail and transcription of all your voicemail messages.

Make sure that you tap **Google Voice** when shown the screen to the right. You can also get to this screen from the **Settings** app > Call **settings** > **Voicemail service**.

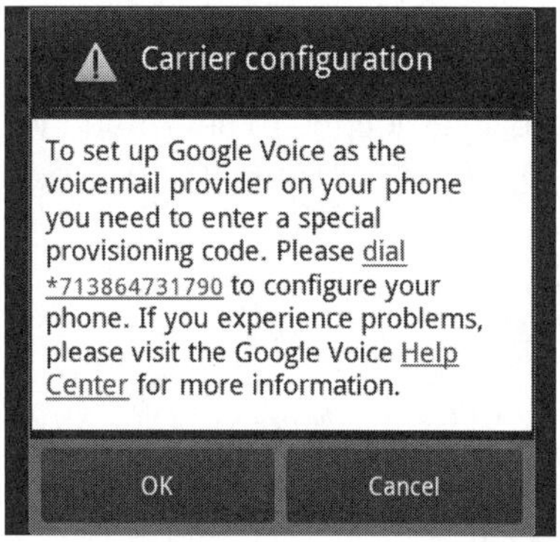

At this point, you will see a screen that asks you to dial a number – the example shown to the right asks us to "dial *713864731790." This number consists of *71, plus our **Google Voice** number.

Tap the underlined number to dial it from your DROID.

You will then see a pop-up window called **Call Settings**. This window displays a status message that shows your **Reading** settings. When this process is complete, your voicemail should be correctly set to **Google Voice**.

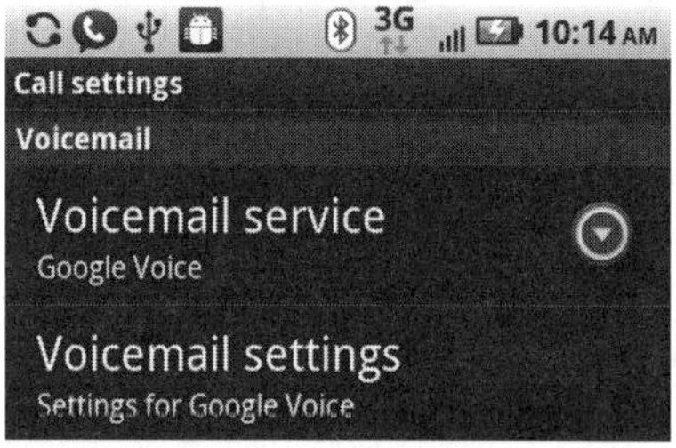

You will know if you have successfully set up **Google Voice** as your voicemail service if you see it shown under **Voicemail Service** on the **Call settings** screen, as shown in the image to the right.

The **Google Voice** service will now route all your callers to the **Google Voice** voicemail system when they call you.

TIP: You can adjust many of the **Google Voice** app's initial settings by starting the **Voice** app, pressing the **Menu** button, tapping **More**, and then tapping **Settings**.

Placing Calls With Google Voice

The nice thing about using **Google Voice** is that it is seamlessly integrated into your DROID phone and voicemail systems.

To place a call, all you need to do is launch the **Phone** app; and then call a **Recent**, **Favorite**, or other contact as you would normally. If you selected for all calls to be placed with Google Voice as shown above, then the call will automatically be placed using **Google Voice**.

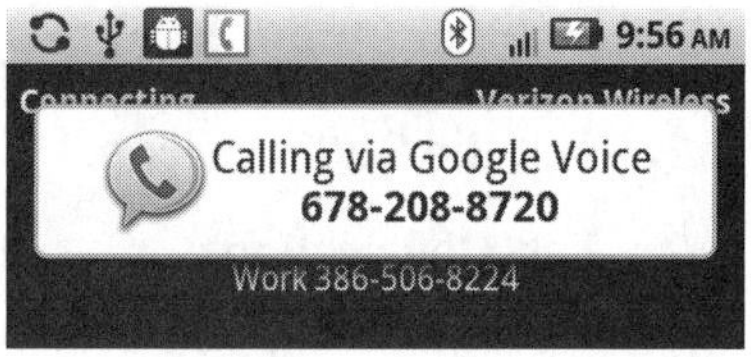

The **caller ID** displayed on the phone of the person you are calling will show your **Google Voice** number, not the number of your DROID. That way, the person will call you back on your **Google Voice** number. This also means that you can use the enhanced voicemail services of **Google Voice**.

Retrieving Google Voice Voicemails

You might think you should be able to tap the regular **Voicemail** icon next to the **Green phone** icon in your dialpad to retrieve your voicemail messages. However, this does not work. Instead, you need to tap the **Voice** app to see all your **Google Voice** voicemail messages.

> **NOTE**: You can also retrieve **Google Voice** voicemails from any desktop computer browser.

You can tap any message in the list to view the entire transcript and play the voice message.

When viewing an individual message, you can press the **Menu** button to accomplish the following tasks:

- **Call** the person back.
- **Text** that person.
- **View contact** details (if that person is in your address book).
- **Add a Star** or press **More** to **Delete** the message.

Customizing Phone Options and Settings

You can customize your DROID phone by going into the **Settings** app. Follow these steps to do so:

1. Tap the **Settings** icon.
2. Tap **Call settings**.

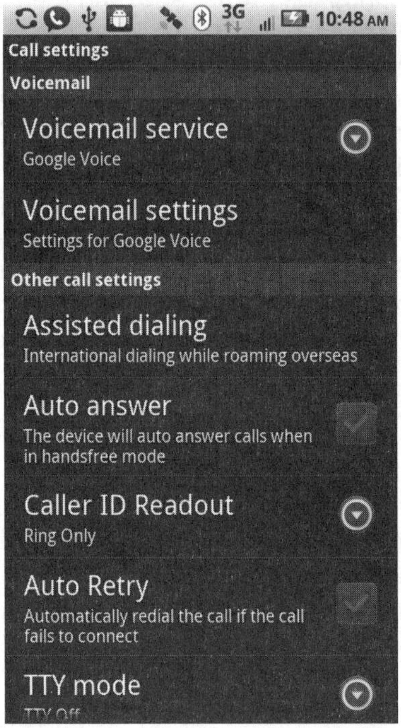

From this screen, you can configure the following options:

- **Voicemail service** and **settings** – Choose whether to use your carrier or Google Voice to handle voicemail.
- **Assisted dialing** – This option helps you place calls when you are roaming overseas.
- **Auto answer** – Check this box to have the DROID automatically answer calls when the device is in **Hands-free** mode (e.g., when the device is connected to your Bluetooth car stereo or headset).
- **Caller ID Readout** – This option reads out the contact name or phone number via the DROID speaker. You can set this option to ring only (the default value); speak the **Caller ID** and then ring; or **Caller ID repeat**. The last option keeps repeating the name or number, which can be a little annoying!
- **Auto Retry** – If a call fails, this option prompts the DROID to automatically try the same number again.

- **TTY Mode** – This option specifies the deaf accessibility settings for the DROID. Possible values for this option are **Off**, **Full**, **HCO**, or **VCO**.
- **DTMF Tones** – This option sets the length of the tones you hear when you press a key on the dialpad between **Normal** (default) and **Long**. This can be useful if you are calling automated systems with a bad connection; making the tones longer can help the system recognize the digits you input.
- **HAC Mode Settings** – These settings govern hearing-aid compatibility.
- **Voice Privacy** – This option adds an extra layer of encryption to your voice calls; it is set to **Checked** (on) by default.
- **Show Dialpad –** This option enables you to set the phone to automatically show the dialpad after you are connected to a phone number. The default shows the dialpad only for **Voicemail**, **toll-free**, and **900 numbers**; however, you can change this setting to show the dialpad for **Custom phone numbers** or even **All calls**.

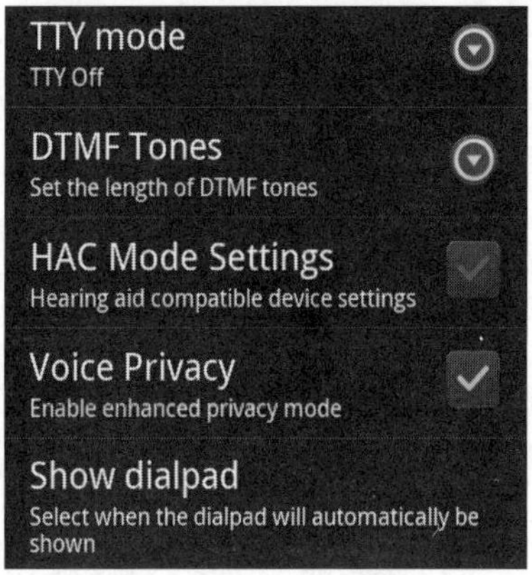

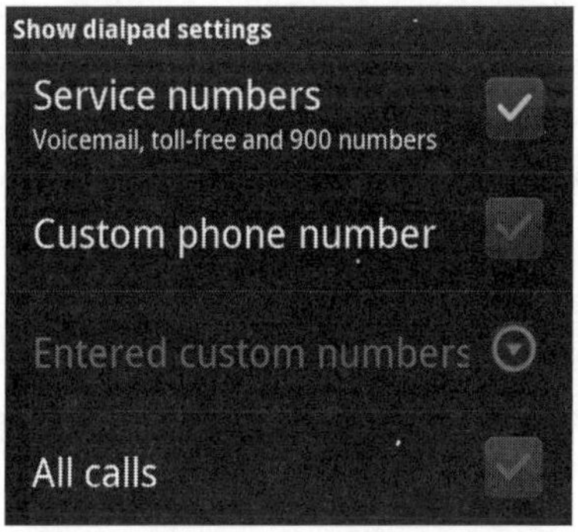

Sounds, Ring Tones, and Vibration

The DROID will alert you to incoming calls, voicemails received, and other features with unique sounds or vibrations. These can easily be adjusted using your **Music** and **Settings** apps. You can also assign unique ringtones to specific people from your **Contacts** apps.

Using Your Own Music As Your Ringtone

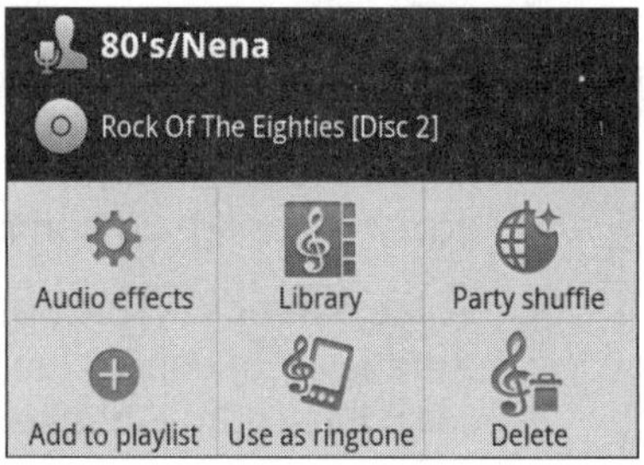

One cool DROID feature: You're able to make your own music serve as your ringtone. For this to work, you need to sync the desired music to your DROID.

If you want to use a custom ringtone, you need to follow the steps shown in this chapter's "Copying Custom Ringtones and Alerts to your DROID" section.

Begin by starting the **Music** app and playing the song you want to use as the main ringtone on your DROID.

In the example shown to the right, we have started an old favorite: *99 Red Balloons* by Nena.

Press the **Menu** button and select **Use as ringtone**.

TIP: Check out Chapter 25: "DROID Media Sync" to learn how to get your music onto your DROID.

Using Custom Ringtones and Alerts

You can customize the way your DROID sounds by using the preloaded DROID ringtones and notifications. Or you can use your own custom ringtones and notifications. There are a few extra steps involved before you can use custom ringtones; however, the extra effort can be worth it if you want to hear a particular sound or song when a friend calls or sends you a new text message. The extra effort can also be worth it if you want a particular sound or song to serve as an alarm.

Selecting New Phone Ringtones and Notification Ringtones

Your DROID comes with a number of fun ringtones and alerts already preloaded. In this section, we show you how to select these preloaded items.

> **TIP:** The steps described in this section will also work for custom ringtones and alarms – once you have them synced to your device. You will learn how to sync these to your DROID later in this chapter.

Follow these steps to select from the DROID's preloaded (or previously synced) ringtones and alerts:

1. Tap your **Settings** icon.
2. Tap **Sound**.
3. Tap **Phone ringtone** to listen to and select a new ringtone. Tap **OK** when done. All the standard ringtones and any new ringtones you have added to the `media/audio/ringtones` folder on your media card will appear in this list.
4. Tap **Notification ringtone** to set a new ringtone for your notifications. Tap **OK** when done. All the standard notification ringtones and any new ringtones you have added to the `media/audio/notifications` folder on your media card will appear in this list.

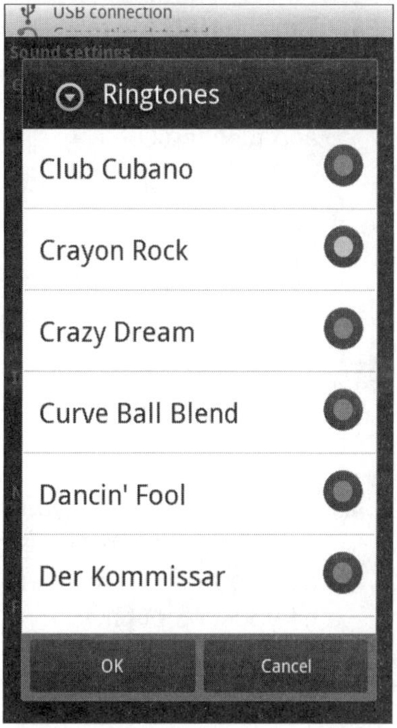

Copying Custom Ringtones and Alerts to Your DROID

You cannot select and use your own music or custom ringtones until you copy them into the correct folder on your DROID. In this section, we will show you how to copy your own music and ringtones to your DROID. Once you do this, you can enjoy custom ringtones for your phone, contacts, alerts, notifications, and text messages. Note that you need a MicroSD format media card to perform this step. Compatible file formats for ringtones are files with these extensions: MP3, MID, AAC, and WMA. Follow these steps to copy your own music or ringtones to your DROID:

1. Connect your DROID to your computer with the USB cable.
2. Drag your finger down from the top of the status bar to open your **Notifications** screen, and then tap the **USB connection** setting.

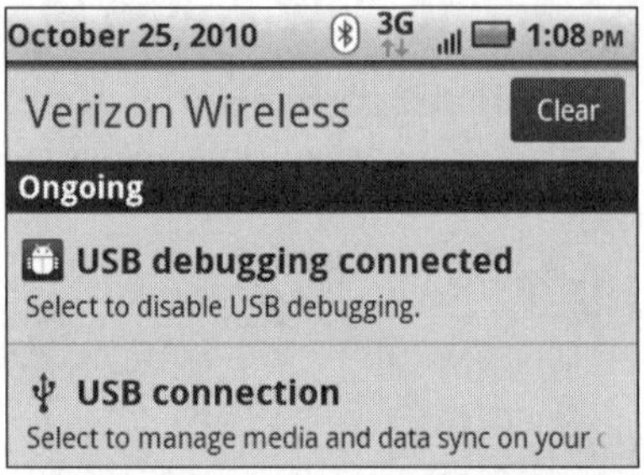

3. Tap the **USB Mass Storage** option (as shown in the figure to the right), and then tap OK.

4. You should now see your DROID media card appear as a new disk drive letter on your computer.

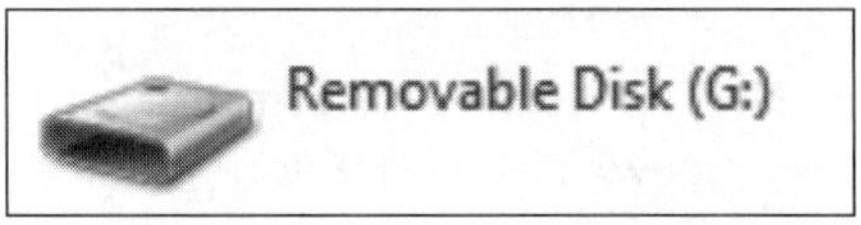

5. If you don't see the **media** folder, then you need to create one. Next, you need to create the folders as shown in the image to the right. Be sure to place the **audio** folder inside **media**, and then place **ringtones** and **notifications** folders inside **audio**.

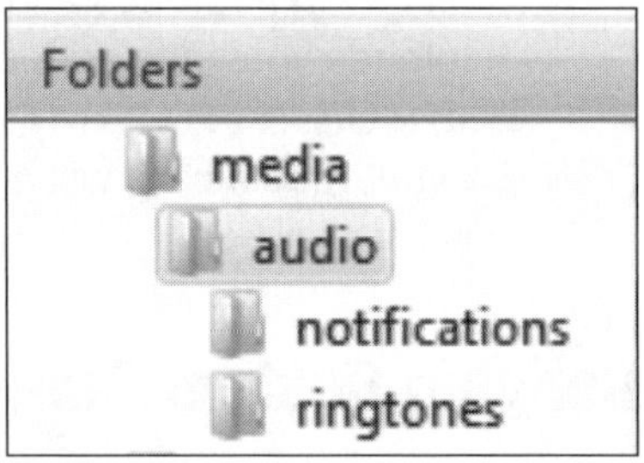

6. Now copy (or drag-and-drop) the ringtone or notification alert into the correct folder, as described previously:

   - The **Ringtones** folder stores ringtones for the phone and contacts.
   - The **Notifications** folder stores ringtones for alerts and text messages.

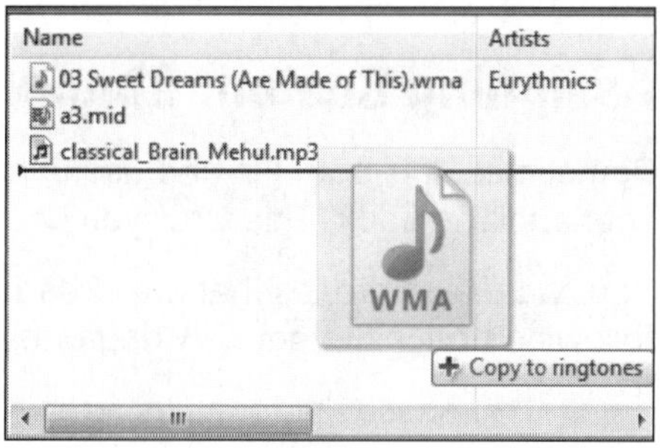

7. Next, disconnect your DROID by unplugging the USB cable from your computer. This step is important; otherwise, your DROID will not be able to see the new ringtones and notifications you have added.

Selecting a New Ringtone for Text Messaging

You will be able to use these same steps for both the preloaded ringtones and any customized ringtones you have added. Follow these steps to select a new ringtone for a text message:

1. Tap the **Messaging** icon.
2. Press the **Menu** button and select **Messaging settings.**
3. Tap **Select Ringtone** in the **Text messaging settings** area.
4. Swipe up and down to see all the ringtones.
5. Tap a ringtone to listen to it and select it.
6. Tap **OK** when you are done.

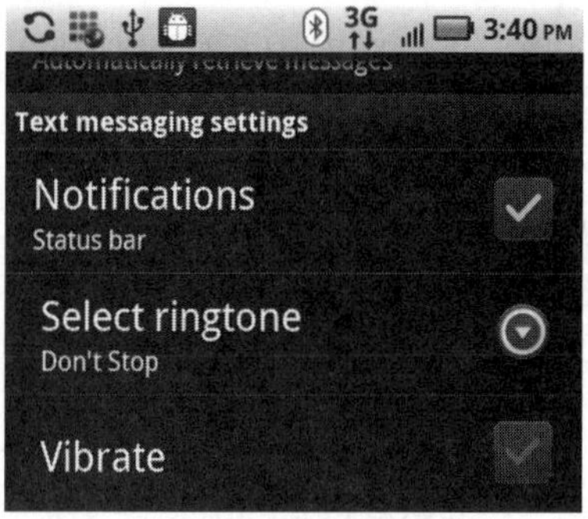

Assigning Custom Ringtones to Contacts

Sometimes, it is both fun and useful to give a unique ringtone to a certain contacts in your address book. This way, you know who is calling without looking at your phone.

You can use ringtones that are already on your DROID, or you can use one of the following options to get new ringtones:

- Purchase ringtones using the **Amazon MP3** or **DROID Market** apps on your DROID.
- Create or download ringtones to your computer, and then copy them to your DROID.
- Use your own songs synced to your DROID as ringtones.

For example, one of the authors (Gary) sets the ringtone for his son Daniel to the ring tone of Elton John's "Daniel." You need to edit a person's information in **Contacts** to change his ringtone. Follow these steps to do so:

1. Tap the **Contacts** icon.
2. Tap the contact you wish to change (in this case, **Gary Mazo**).
3. Press the **Menu** button and select **Edit**.
4. Swipe to the very bottom and tap the **Additional info** gray bar.

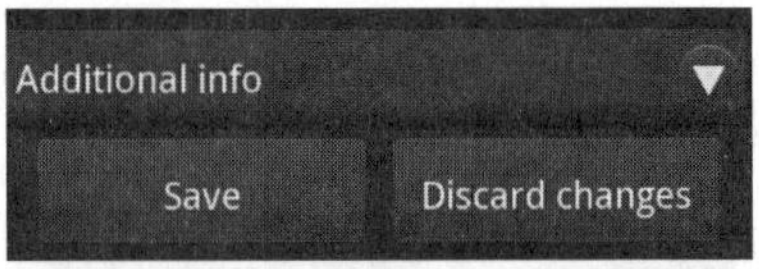

5. Swipe again to the very bottom of the screen.
6. Tap the **Call Handling Ringtone** option. It will say **No ringtone specified** when you first change it.

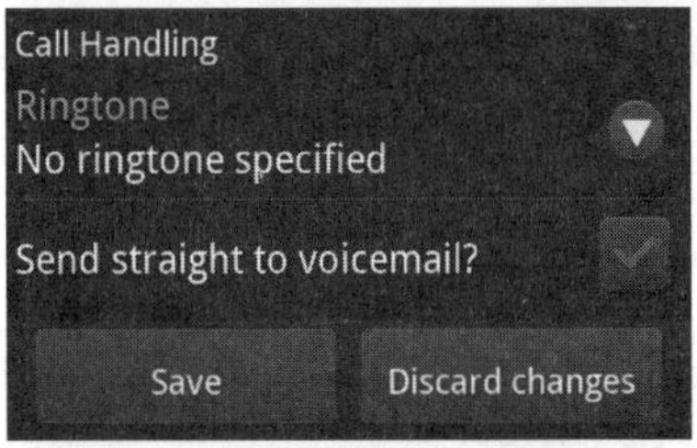

7. Tap any ringtone to select it and tap **OK** to save it. Now you will see the name of the selected ringtone on the same screen.
8. Tap **Save** to save your changes.

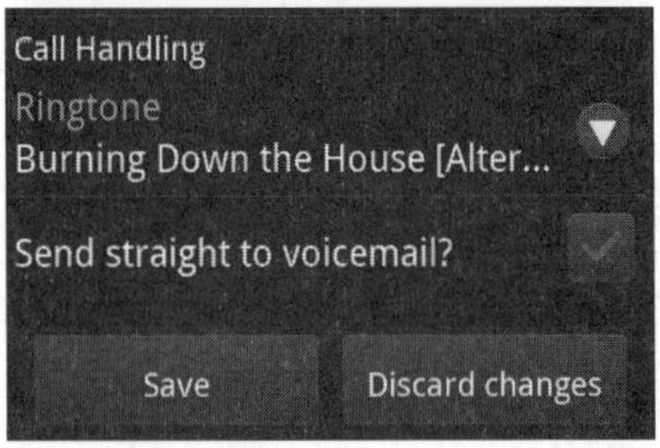

The next time this contact calls you, you will hear the newly selected custom ringtone.

NOTE: On the original DROID, press the Menu button, then press Options and then select Ringtone and choose a unique ringtone.

Purchasing a Ringtone from the Amazon MP3

The **Amazon MP3** app connects you to a site where you can purchase files that can serve as ringtones. Follow these steps to purchase ringtones from this site:

1. Tap the **Amazon MP3** icon.
2. Type **ringtone** in the **Amazon** search window and tap the search button (where the enter key is usually located).
3. You will then see all items that match. The image to the right shows **Albums** or groups of ringtones. Tap **Songs** at the top of the list to view individual ringtones. Most individual ringtones are US $0.89 or US $0.99.
4. Tap the price or **FREE** button to purchase or download the ringtone to your DROID.

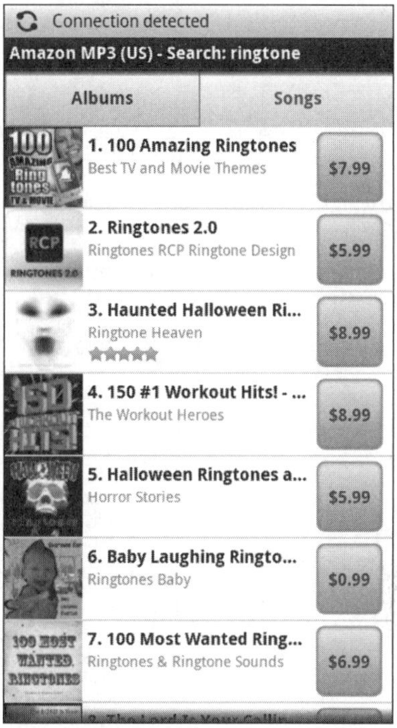

After you purchase your ringtone, follow the steps described in the preceding sections to assign your new ringtone to your phone or a contact.

Buy Ringtones from the Android Market

You can also buy ringtones using the Android **Market** app. Start the **Market** app and type ringtone into the **Search** window. You will find both ringtones and apps that help you create ringtones. Once you find a ringtone or app that interests you, follow the instructions provided to acquire the file(s) in question. At this point, you can follow the steps described in the preceding sections to assign your new ringtone to your phone or a contact.

Chapter 8

Bluetooth on Your DROID

In this chapter, we will show you how to pair your DROID with any Bluetooth device, whether it is another computer, stereo speakers, or a wireless headset.

Thanks to the technology known as A2DP, you can also stream your music to a capable Bluetooth stereo.

> **NOTE:** You must have a capable third-party Bluetooth adapter or Bluetooth stereo to stream your music via Bluetooth technology. Also, there is AVRCP profile support, so many music controls on a Bluetooth device (like Play, Pause, or Skip) can be operated from your DROID.

Think of Bluetooth as a short-range, wireless technology that allows your DROID to connect to various peripheral devices without wires.

Bluetooth is believed to be named after a Danish Viking and king, Harald Blåtand, whose name has been translated as *Bluetooth*. King Blåtand lived in the tenth century and is famous for uniting Denmark and Norway. Similarly, Bluetooth technology unites computers and telecom. His name, according to legend, is from his very dark hair, which was unusual for Vikings. Blåtand means dark complexion. There's also a popular story that the king loved to eat blueberries, so much so that his teeth became stained with the color blue.

Sources:

- `http://cp.literature.agilent.com/litweb/pdf/5980-3032EN.pdf`
- `www.cs.utk.edu/~dasgupta/bluetooth/history.htm`
- `www.britannica.com/eb/topic-254809/Harald-I`

Understanding Bluetooth

Bluetooth allows your DROID to communicate with things wirelessly. Bluetooth is a small radio that transmits from each device. Before you can use a peripheral with the DROID, you have to "pair" it with that device to connect it to the peripheral. Many Bluetooth devices can be used up to 30 feet away from the DROID.

Among other things, the DROID works with Bluetooth headphones, Bluetooth stereo systems and adapters, Bluetooth keyboards, Bluetooth car stereo systems, Bluetooth headsets, and hands-free devices. The DROID supports A2DP, which is known as Stereo Bluetooth.

Turning On Bluetooth

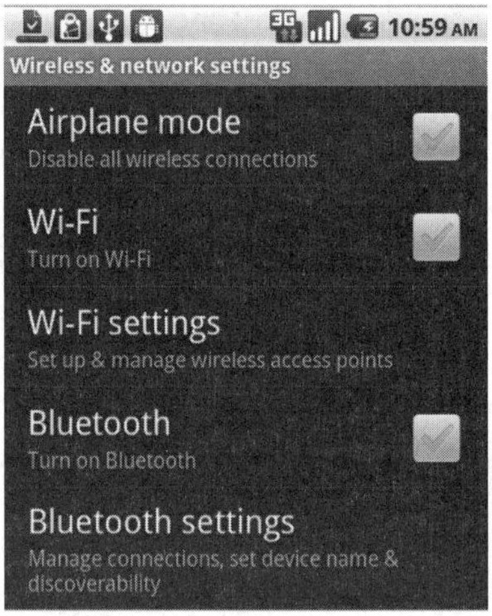

The first step to using Bluetooth is to turn the Bluetooth radio **On**.

1. From your Home screen, tap your **Menu** button.
2. Then, touch **Settings.**
3. Touch **Wireless & Networks** at the top of the list.
4. You will see **Bluetooth** in the list.
5. By default, Bluetooth is initially **Off** on the DROID. Touch the box to turn it to the **On** position. You will see a green check mark appear in the box.

TIP: Bluetooth is an added drain on your battery. If you don't plan on using Bluetooth for a period of time, think about turning the switch back to **Off**.

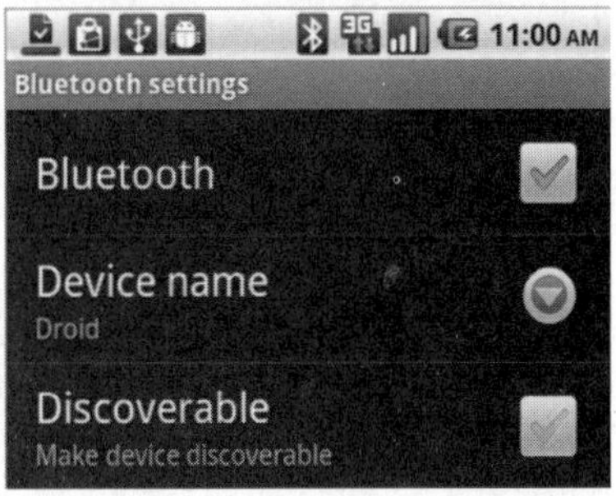

Pairing with a Bluetooth Device

Your primary uses for Bluetooth might be with Bluetooth headphones, Bluetooth stereo adapters, or a Bluetooth headset. Any Bluetooth headphones should work well with your DROID. To start using any Bluetooth device, you need to first pair (connect) it with your DROID.

Pairing with a Headset or Any Bluetooth Device

As soon as you turn Bluetooth **On**, the DROID will begin to search for any nearby Bluetooth device—like a Bluetooth headset or stereo adapter (see Figure 8–1). For the DROID to find your Bluetooth device, you need to put that device into "pairing mode." Read the instructions that came with your headset carefully—usually there are a combination of buttons to push to achieve this.

TIP: Some headsets require you to press and hold a button for five seconds until you see a series of flashing blue or red/blue lights. Some accessories automatically start up in pairing mode.

Once the DROID detects the Bluetooth device, it will attempt to automatically pair with it. If pairing takes place automatically, there is nothing more for you to do.

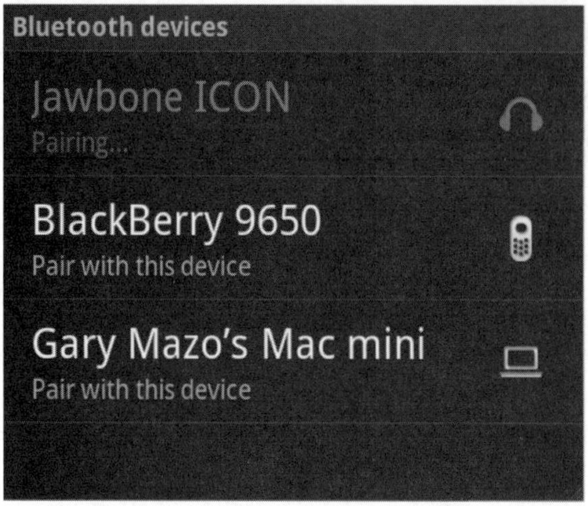

Figure 8–1. *Bluetooth device discovered and in process of pairing.*

NOTE: In the case of a Bluetooth device, such as a computer, you may be asked to enter a series of numbers (passkey) on the keyboard itself or confirm that a passkey is being shown. See Figure 8–2.

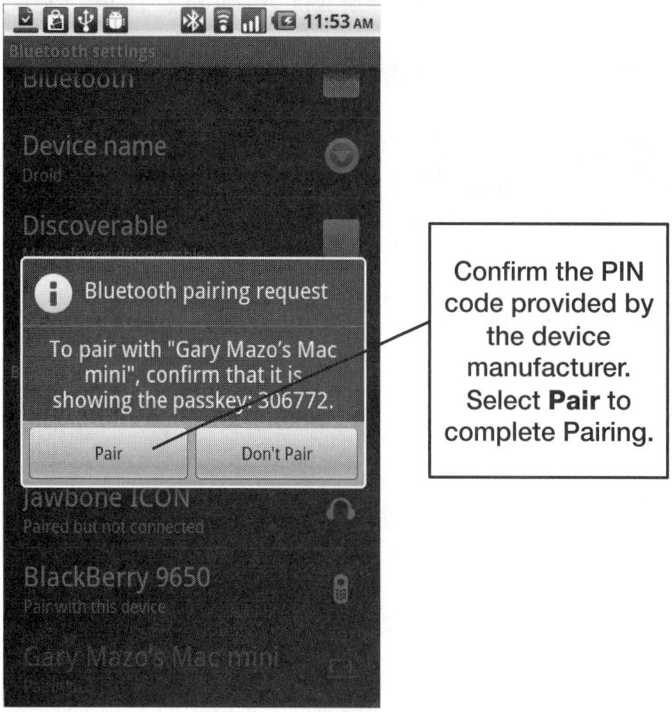

Figure 8–2. *Select Pair to connect to another Bluetooth Device that requires a passkey.*

Newer headsets like the Aliph Jawbone ICON, used here, will automatically pair with your DROID. Simply put the headset into pairing mode and turn on Bluetooth on the DROID—that's all you have to do!

Pairing will be automatic, and you should never have to re-pair the headset again.

Using the Bluetooth Headset

If your headset is properly paired and on, all incoming calls should be routed to your headset. Usually you can just press the main button on the headset to answer the call or answer it on the DROID.

Move the phone away from your face (while the DROID is dialing), and you should see the indicator showing you that the Bluetooth headset is in use. In the image you see that the **Bluetooth** icon is activated.

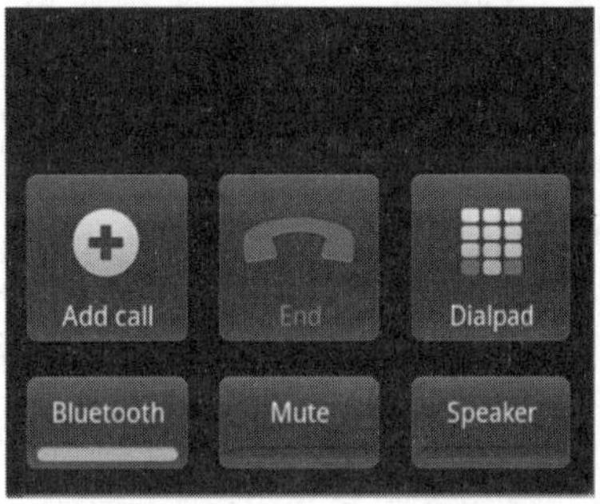

You will also see the options to send the call to the **Speaker** or **Mute** the call. You can change this at any point while you are on the call.

Just choose to send the call to any of the options shown, and you will see the small **Speaker** icon move to the current source being used for the call (Figure 8–3).

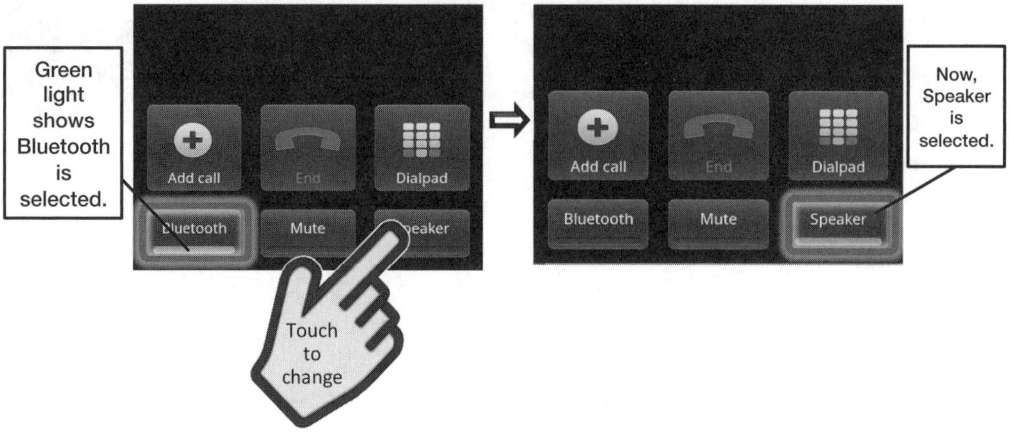

Figure 8–3. *Changing from Bluetooth headset back to the DROID while on a call.*

Bluetooth Stereo (A2DP)

One of the great features of today's advanced Bluetooth technology is the ability to stream your music without wires via Bluetooth. The fancy name for this technology is A2DP, but it is simply known as Stereo Bluetooth.

Connecting to a Stereo Bluetooth Device

The first step to using Stereo Bluetooth is to connect to a capable Stereo Bluetooth device. This can be a car stereo with this technology built in, a pair of Bluetooth headphones or speakers, or even newer headsets like the Jawbone ICON.

Put the Bluetooth device into pairing mode as per the manufacturer's instructions, and then go to the Bluetooth setting page from the **Settings** icon, as we showed you earlier in the chapter.

Once connected, you will see the new Stereo Bluetooth device listed under your Bluetooth devices. Sometimes it will simply be listed as "Headset." Just touch the device, and you will see the name of the actual device next to the **Bluetooth** tab in the next screen, as shown here.

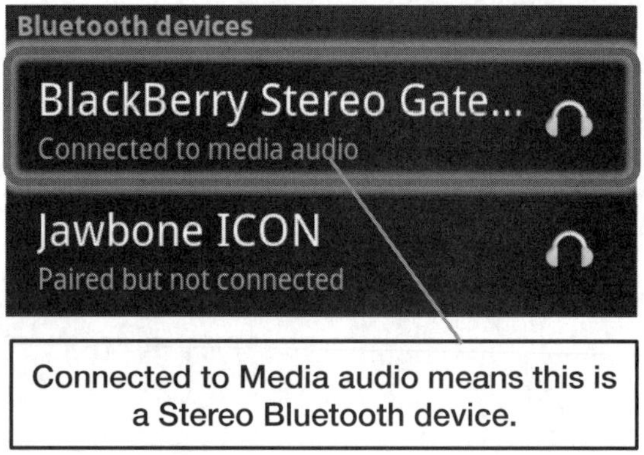

Next, tap your **Music** icon and start up any song, playlist, podcast, or video music library. You will now notice a small **Bluetooth** icon in the middle of top status bar. Touch the **Volume** buttons on the side of your DROID to adjust the volume of the paired Bluetooth device streaming your music (see Figure 8–4).

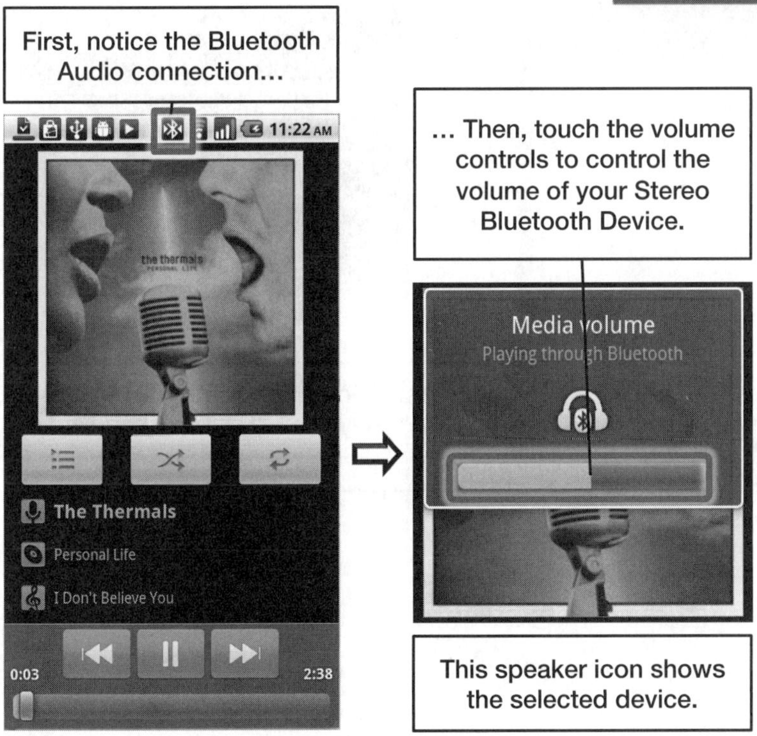

Figure 8–4. *Adjusting the volume of a Bluetooth device.*

Disconnecting a Bluetooth Device

Sometimes, you might want to disconnect a Bluetooth device from your DROID.

It is easy to get this done. Get into the Bluetooth settings as you did earlier in this chapter. Touch and hold the device you want to disconnect in order to bring up the next screen, then tap the **Disconnect & Unpair** button, and confirm your choice.

> **NOTE:** Bluetooth has a range of only about 30 feet, so if you are not nearby or not using a Bluetooth device, turn off **Bluetooth**. You can always turn it back on when you are actually going to be using it.

This will delete the Bluetooth profile from the DROID. (See Figure 8–5.)

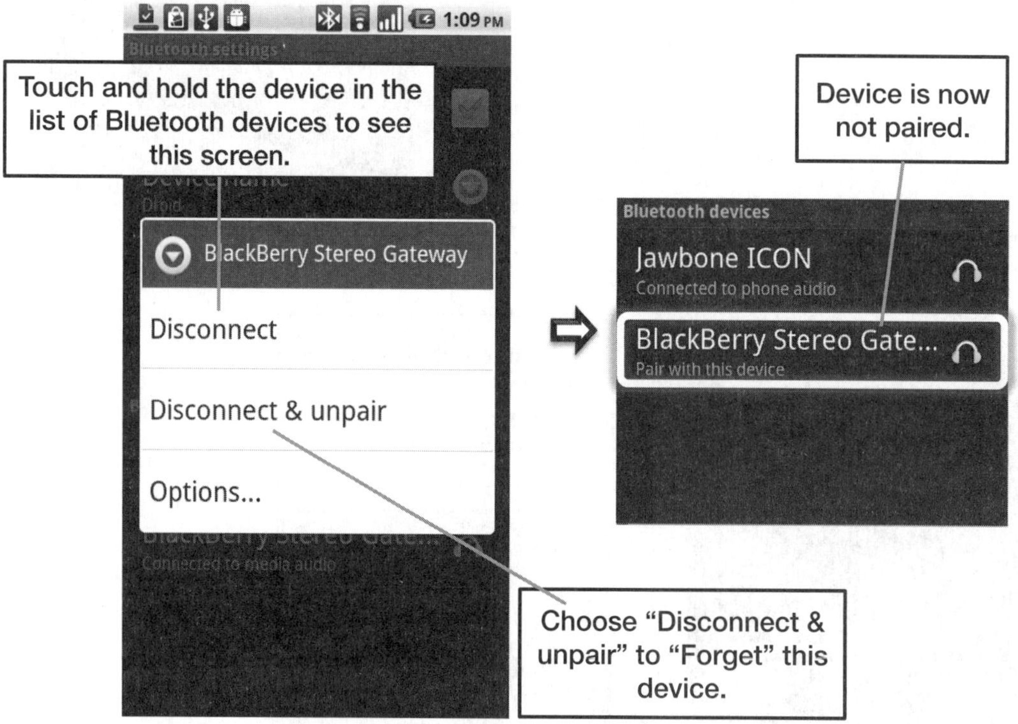

Figure 8–5. *Disconnecting and unpairing a Bluetooth device.*

Chapter 9

Email on Your DROID

One of the big reasons to have a smartphone is to keep on top of your email. Your DROID allows you to keep track of both personal and corporate email. It works with Microsoft Exchange accounts, Yahoo! Email, and, of course, your Gmail account. You can sync your phone with all your accounts, and you can create signatures to let your recipients know you're responding from your phone, if you choose.

DROIDs offers many options for email:

- The Gmail app
- The Email app (includingYahoo accounts)
- Outlook Web Access
- Corporate Sync Accounts
- The Web Browser
- Third-party apps

Your DROID also comes with a Motorola widget that creates a universal messaging inbox on your Home screen, as shown to the right. Messages sent to all your accounts can be read without having to launch any email apps.

You can read more about widgets in Chapter 6: "Organize Your Home Screens: Icons and Widgets."

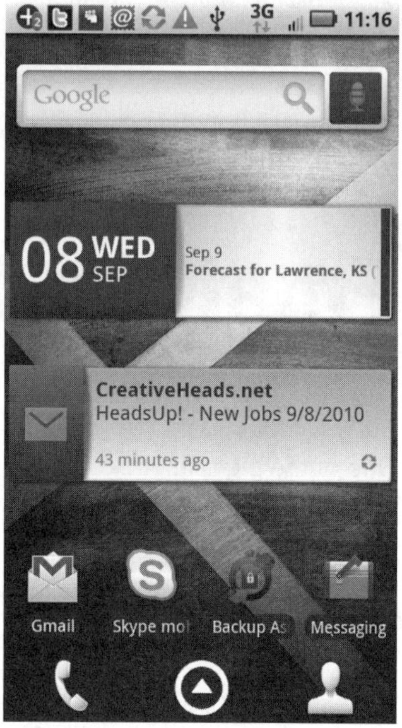

Getting to Know Gmail

Gmail is the default email app for Android phones, and your DROID ships with the Gmail app icon displaying on the **Home** screen. Go to `http://mail.google.com` to get started with a free Gmail account.

To understand the **Gmail** app, it's helpful to first understand Gmail on the Web. You'll also need to visit Gmail on the Web in order to get the best use of Gmail by setting up filters and experimenting with new features.

Gmail is arguably the best free email service available. There's no automatic tagline on the end of your messages advertising that you're using a free email. You don't have to pay extra in order to use a desktop or mobile app to access your email. The spam filtering is above average, and you get plenty of storage space. In fact, Gmail works so well that many business users have come to rely on the service through the enterprise **Google Apps** suite.

TIP: Although it's not a faux pas to use a Gmail address for professional correspondence, you can use Google Apps to send and receive Gmail through custom business domains. If you own a small- to medium-sized business, you can take advantage of their services from either the limited free "standard" account or the $50 per-user per-year "premium" account. If you qualify for the free standard version, you can set up Google Apps as a free email service for a domain name you already own. For more information, visit Google Apps at www.google.com/a.

Understanding Gmail Mobile

There are two basic ways to access Gmail from your phone. You can use either the **Gmail** app, or the phone's web browser to access Gmail from the Web. When you use your web browser, by default you'll see a mobile version of Gmail that is trimmed down and simpler to use on phones.

The Gmail app on Android uses **push email**. That means that you don't need to keep checking a web site. Your email is always on, always ready to receive new messages. This is just like keeping your desktop email client on in the background when you use your laptop. It's the big advantage of the Android **Gmail** app over your phone's web browser, though the browser does have a few features missing in the **Gmail** app.

When new messages arrive, by default you'll see a notice in the status bar. You can drag down the status bar and click the notification to launch the **Gmail** app.

We will go over quite a few web-based features in this chapter, but there is an important reason to get to know them first. Many of these features cannot be changed from the **Gmail** app in Android or Gmail's mobile browser version.

Inbox and Archive

Gmail doesn't have folders. Rather, Gmail uses labels. We'll get to that next, so let's just say that for most purposes, there are only two places for email you want to keep: the inbox and the archive.

There are two places for email you don't want to keep: trash and spam. Generally you'll want to mark spammy messages appropriately before deleting them, because this helps train the spam filters to recognize unwanted messages.

If you don't ever want a message again, by all means delete it. Email sent to the trash is permanently deleted after thirty days. However, messages you might need later should be archived. To archive a message from the Web, select the check box next to the message, and then press the Archive button. It's on the left side of the buttons above the inbox, as shown in Figure 9–1.

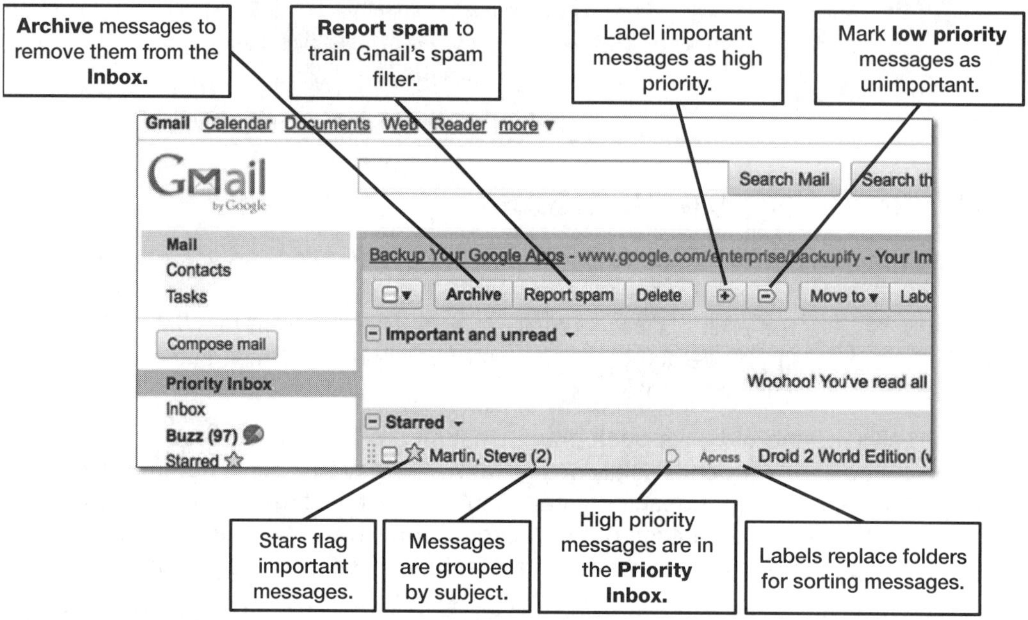

Figure 9–1. *Gmail Web.*

When you archive a message, you move it out of the inbox. You can still find the message by using the "all mail" label or by using the Gmail search box. For messages of low importance, you may even want to set up a filter that archives the messages immediately so they never clog your inbox. We'll explain how to do that later in this chapter.

NOTE: Gmail messages are grouped into conversations. Rather than showing each message in the order they arrived, conversations are clusters of messages to and from a person or group. The messages are stacked together, so you can view the conversation in context, and they appear chronologically in your inbox according to the last message received. Any actions you apply to one message in the conversation will apply to all of them. If you want to view or act on the messages individually, just click the "Expand all" link to the right of the message.

If you keep seeing a super long conversation that you'd rather ignore, use the "mute" option to archive the current and future messages in that conversation. The messages will still be available and marked as unread. They just won't be in your inbox.

Priority Inbox

Google introduced a new feature called the **Priority Inbox.** What this is meant to do is filter your important email from what some call **bacn**. Bacn (pronounced "bacon") messages are those newsletters, alerts, and coupons that you did at some point sign up to receive and you probably do want to read, just not right now. They're not really spam, but they're not really important. They're bacn.

The Priority Inbox flags important, unread messages and displays them at the top of your Gmail inbox on the Web. You can train Gmail to better recognize which messages are important and which are not by flagging them,  and you can customize the Priority Inbox to also separate items with specific labels, as shown earlier in Figure 9–1.

Behind the scenes, this is really just a new way to display a label or star.

Labels

Many email accounts work by allowing you to place email messages in folders. Gmail would prefer you use labels. What is the difference? A single piece of email can exist only in a single folder. You'd have to copy an email message for it to be both in the "work" and "tax related" folders, but it can have multiple labels.

Use labels to organize your messages by topic. You can click one of the labels on the left side of the screen (as shown in Figure 9–1) in order to view only messages with that particular label, including messages that have been archived.

Gmail automatically creates the following labels:

- Inbox
- Buzz
- Starred
- Chats
- Sent Mail
- Drafts
- All Mail
- Spam
- Trash

You can create other labels as needed. We sometimes set up temporary labels for upcoming conferences or events and then remove or hide the labels after the conference.

You'll notice that **Starred** is also a label. Click the empty star ☆ to the left of a message in order to "star" it or give it the star label. That highlights the message with a yellow star. ☆ Since you can apply more than one label to an item, adding both a star and a different label could emphasize urgent messages or highlight items that needed a response or required action.

Creating and Deleting Labels

You can create labels many ways on the Web.

1. Click the **Labels** button at the top of your inbox.

2. Click **Manage labels** from the drop-down menu, as shown on the right.

3. You now are in the **Label settings** area. You can enter new labels by typing into the box labeled **Create a new label**.

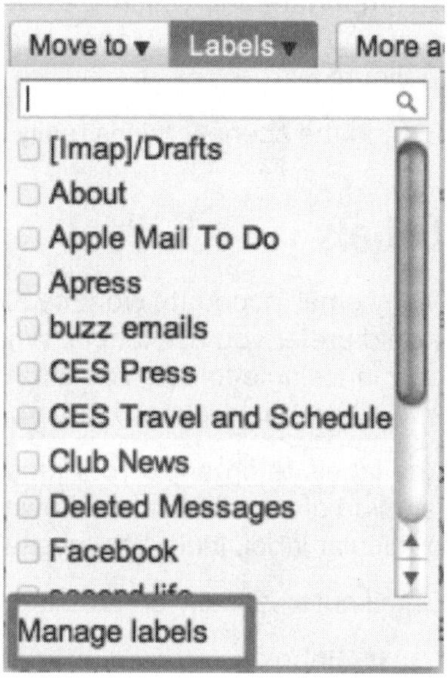

You can also get to this menu by clicking **settings** on the upper right-hand side of the screen and then clicking **Labels**. Another method is to click **More** on the bottom of your label list on the right side of the screen and then **Manage labels**.

> **TIP:** You're limited to 40 characters in a label, but it would be wise to keep it even shorter. Long labels mean less space for your message previews.

You can edit a label by clicking the label name in the Label settings and then typing the new name in. You can show or hide labels by clicking the link to the right of the labels, and you can delete them as well. Deleting a label does not delete the messages.

Automatic Filters

Automatic filters are probably one of the most powerful tools in any email program. When combined with labels, Gmail lets you do quite a lot with filters.

1. Click the check box next to one or more messages.
2. Click the **More actions** button.
3. Select **Filter messages like these**.
4. Gmail will try to guess the criteria you're using, such as messages from a certain sender or messages containing a particular subject line. If the guess is wrong, you can change the criteria. Once you've got the correct criteria, click the **Next Step** button.
5. Now you choose an action. Your choices include archive (the Skip the Inbox option), star, mark as read, apply a label, forward it, delete it, or never flag it as spam. If you're using Priority Inbox, this is a choice as well. You can select more than one action for an item, such as starring and never marking it as spam.
6. Next, create the filter. You can also select the check box to apply that filter to any previous messages that matched your criteria.

We use filters to automatically prioritize messages from business contacts with stars and subject labels. We de-clutter by archiving distracting mailing lists and other bacon items we may want to read later or notifications from Facebook and Twitter. We also make sure important senders never have their messages marked as spam.

> **NOTE**: You can create an easy filter for a group or project by creating a custom email address. Your Gmail address supports adding words to your address by adding them on with a plus sign. For instance, you can have everyone involved with a project send messages to `YourUserName+YourProject@gmail.com`. Add a filter for messages sent to that specific address, and then apply the desired label.

The Settings Menu

We've already explored filters and labels. There are many other options on the settings menu worth mentioning. To adjust settings, click the **Settings** link on the upper right of the Gmail Web screen, just next to your email address.

General Settings

The first tab offers some general settings. Make sure your browser connection is set to **always use https://**. That makes sure you use a more secure connection to check your email when using a web browser. It's also the default setting, so if you don't have either one selected, it's still using the secure connection.

The other important thing to note is that you can create text signatures and set automatic vacation replies through the appropriate boxes here. Be aware that any signatures you create here will *not* translate to your phone. You have to set those up separately.

If you change anything, be sure to press the **Save Changes** button before moving on.

Accounts

You can add additional email accounts through Gmail and check and respond to them from the same inbox as your Gmail account. They have to be a standard POP3 account, but that includes most web-based email and email accounts offered through Internet service providers. That generally does not include Exchange accounts.

The settings are shown in Figure 9–2. To add an email account to your Web Gmail account, do the following:

1. Log into Gmail on the Web.
2. Click **Settings**. It's on the upper right of your browser window.
3. Click **Accounts** and **Import**.
4. Click **Add POP3 email account**.
5. Enter your email address, password, and any specific settings provided by your ISP.
6. Decide how you handle messages on the old account. Do you delete them as they're imported or are they left on the server?
7. Decide if you'd like to add a custom label (by default your email address will be the label) and if you'd like to automatically archive new messages.

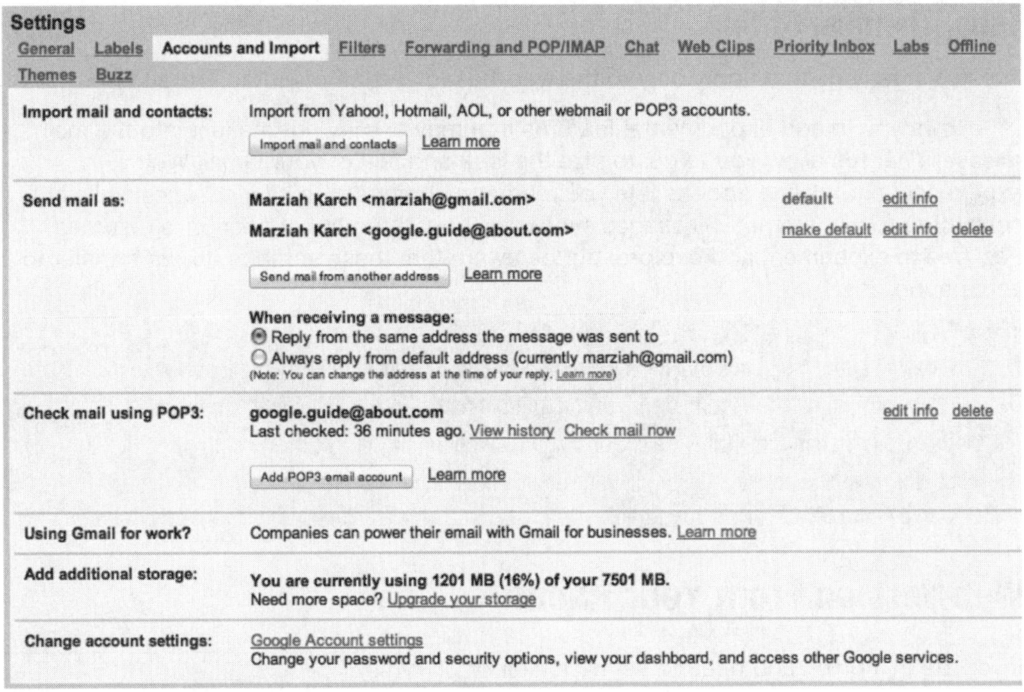

Figure 9–2. *Adding email accounts to a single Gmail.*

If you add accounts, you'll want to decide if you should respond from the address that received the email or always use your default email address. We find it less confusing to respond with the same account that originated the email. Your default address is the address you'll use to compose new messages.

You can also add more accounts from your DROID without joining them in your Gmail account. It's just a matter of deciding how you'd rather manage your email accounts.

Forwarding and POP/IMAP

You can automatically forward a copy of each mail message to a different account and either keep, archive, or delete the original message. This applies to *all* messages to that account, but you can forward selectively by creating a filter.

For accessing email on your Android phone, you'll want to enable **IMAP** (Internet Message Access Protocol). This is the mail protocol that allows your account to sync with your phone. You can also enable **POP** (Post Office Protocol) if you wish, but this isn't necessary for Android access.

Labs, Themes, Offline

These are settings that apply only to the web-based version of Gmail. **Gmail Labs**  allows you to add experimental features that may or may not make it into the main release. **Themes** allow you to customize the look and feel of your Gmail Web experience, and **offline access** lets you read and compose Gmail messages while not connected to the Internet. Messages sync once your Internet connection is resumed. Feel free to experiment and explore, but be aware that these settings do *not* transfer to your phone.

> **NOTE**: One interesting Labs tool is called **Green Robot**. This add-on turns the icons of chat buddies into robots if they're currently using Android for their chat session. It works only for Android, so you can't tell if they're chatting from an iPhone or Blackberry. As with other Gmail Labs, this doesn't change anything in your Gmail phone app.

Web Version From Your Phone

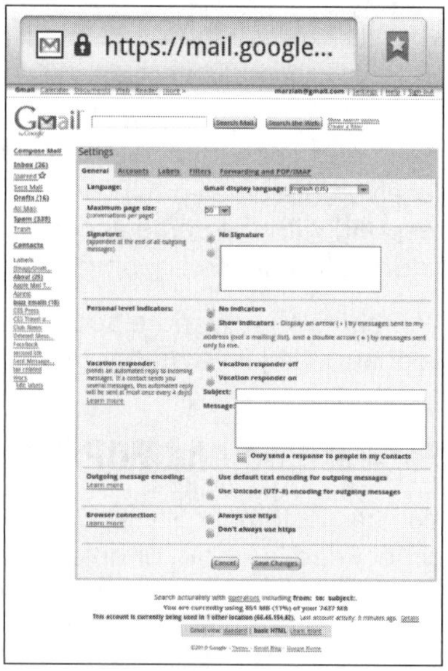

If you are in a pinch and need to set up a filter or create a label, you can still do this from your phone. It just involves a bit of wrangling.

1. Point your DROID browser to `http://mail.google.com`.
2. When you are logged in, scroll to the very bottom of the screen. You'll see that you're viewing Gmail in: Mobile.
3. Click the link next to that that says **Desktop**.

What you see should be similar to the figure on the right. It's tiny, so you'll need to magnify your view and scroll around to navigate. You will still have fewer options than you would on your laptop's web browser. However, you still have all the options you need for effective phone use.

Sending and Replying to Email

Let's return to the **Gmail** app on your DROID.  Sending email with Gmail is illustrated in Figure 9–3.

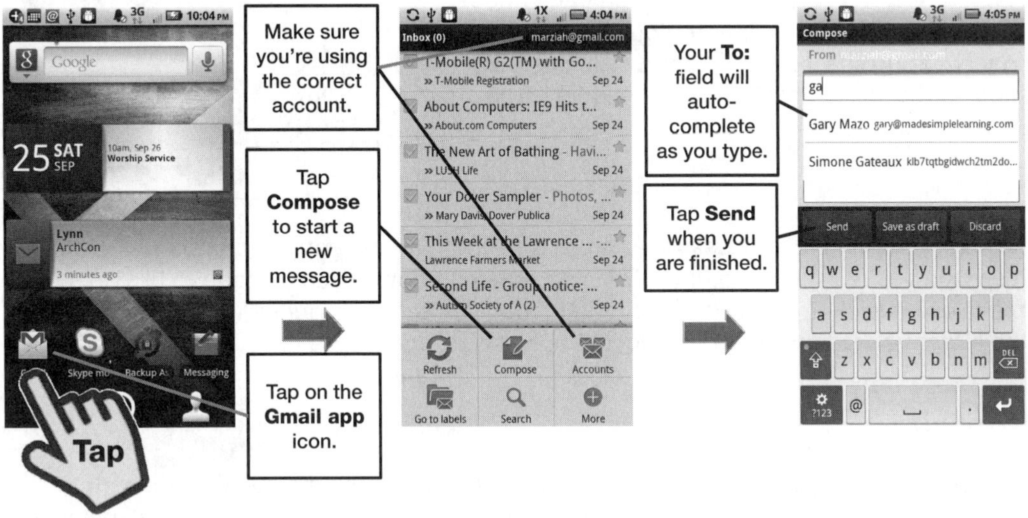

Figure 9–3. *Composing Gmail messages.*

1. Navigate to the correct account, click the menu button, and select **Compose**. If you're using a DROID or DROID 2 with a keyboard, you can slide it out or just use the virtual keyboards.

2. Start entering an address in the **To:** field, and Google will attempt to auto-complete the email address from your contact list. If this is to a new contact, you'll just have to type the whole thing out. Use your finger to navigate to the next fields.

3. If you want to add a picture attachment or more recipients, press the menu button again. You'll have the option to add BCC and CC recipients and attachments from your phone's camera gallery. You can also choose to take a new picture to attach. Picture files are the only type of attachment supported on the default Android Gmail app, but you can still forward messages that contain other types of attachments.

4. When you're done with your message, tap **Send**.

To reply to a message, open that message, and tap the **Reply** button. This menu remains sticky at the top of the screen, even if you scroll through a long message.

As with desktop email programs, you can choose the **Reply** or **Reply All** option. You can also choose to forward messages. Expand your options by tapping the left-facing triangle, as shown in the image on the right.

If you are replying to a message, Android will automatically copy and append the entire message you're replying to. If you're used to paring down this message to highlight only the relevant section or insert something in the middle, you're out of luck. You can't edit the attached previous message, so just note the relevant parts in text.

Search

It's easy to get trapped into navigating through messages by the subject line and preview, but sometimes there's a faster way to find what you need. Google is known for search, so it's unsurprising to find a well-supported search tool within Gmail. Whenever you're in the Gmail app, press the physical **Search** button on your DROID, and you can search through your messages. The search tool will auto-suggest as you type.

Custom Signatures

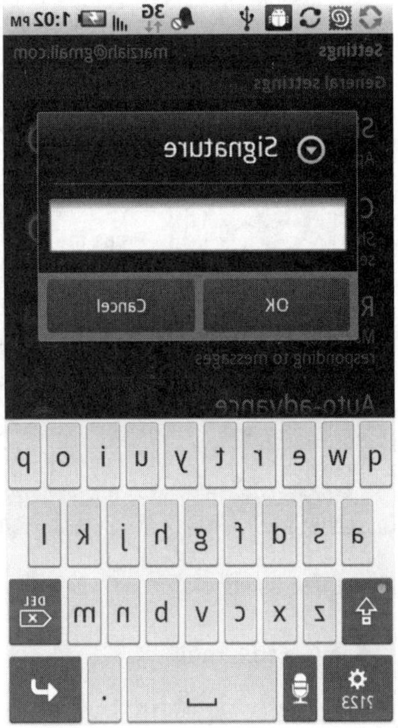

If you set a signature on Gmail on the Web, that signature *doesn't* get included on email you send from your phone. This gives you the chance to make a custom signature from your phone—perhaps something indicating that you're using a phone, so your recipient is more willing to forgive short messages and the occasional typo.

To set your custom signature, do the following:

1. Go to your Gmail inbox.
2. Press the **Menu** button.
3. Next, select **Settings**. On the original DROID, you need to touch More and then Settings.You'll see the **Signature** setting, and you can use this to create a text-only signature.
4. When you're done, hit **Save**. That signature will apply only to messages sent from your phone for that account.

Notifications

While you're editing settings, it's a good time to think about notifications. Do you want a ringtone every time you get a message? Do you want the phone to vibrate? Do you want an update in your status bar? Or, you may want your DROID to do nothing, so you can review new email when you choose? These are options listed under "Notification settings."

By default, your DROID will use the robotic "Droid" ringer, which may be startling the first time you get a message.

We get a lot of messages, so we silence the ringers and keep the option Email notifications checked, so we can glance at the Notifications bar to find new messages. Figure 9–4 illustrates how to silence the email notifications.

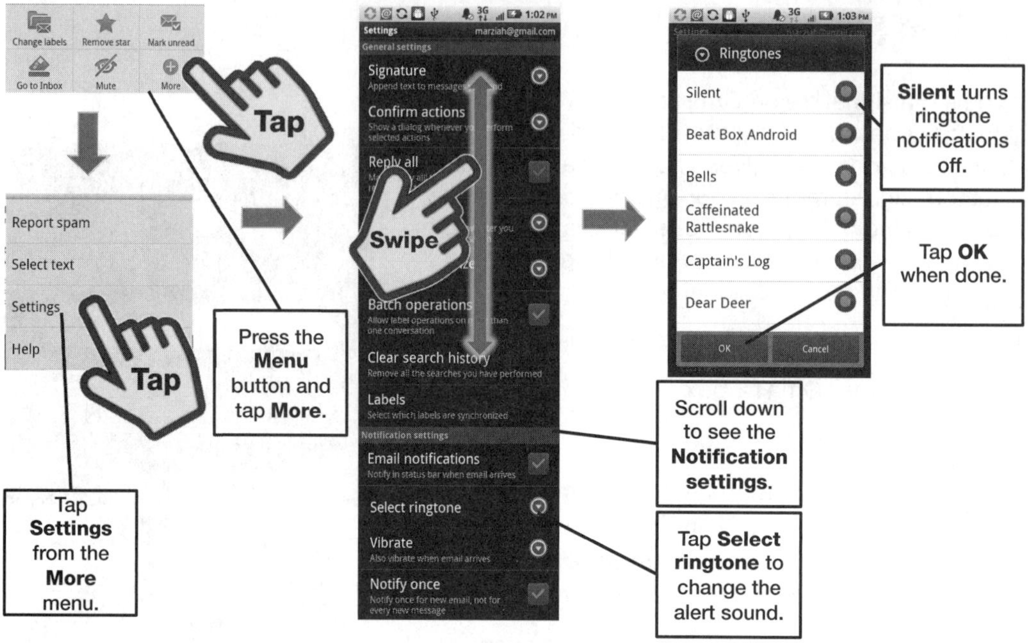

Figure 9–4. *Ringtones.*

1. Go to Gmail.
2. Press the **Menu** button.
3. Tap **More.**
4. Tap **Settings.**
5. Scroll down the menu to the **Notifications settings** and tap **Select ringtone.**
6. Choose **Silent.**
7. Tap the **OK** button.

Labels

Another way you can cut down on your inbox clutter is to sync only certain Gmail labels. Choose **Labels** in from the **Settings** menu of the account you want to change, and you'll see the **Synchronization** menu as shown on the right.

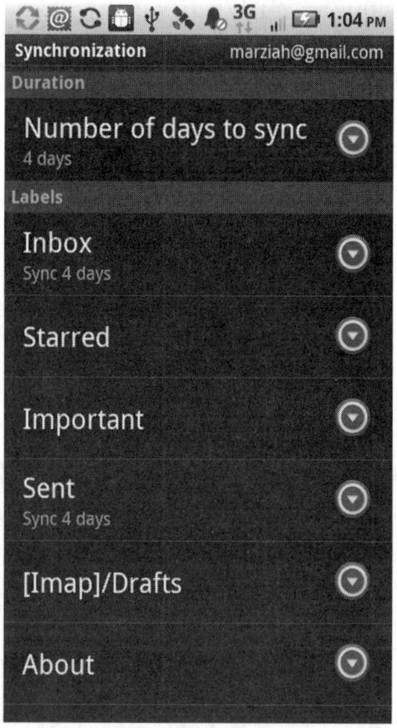

You can choose how far back you want to sync messages in your inbox and choose which labels to sync on a case-by-case basis. You could put an automatic label on some of your bacn newsletters, and then choose not to sync them to your phone, or you could choose to sync work-related emails only on weekdays.

Not syncing labels doesn't mean you can't still find the information, just like archiving a message doesn't mean it is inaccessible. Searching your inbox will still retrieve old messages. It just saves some phone memory and syncing time for things you don't need instantly available every time you launch the **Gmail** app.

Confirm Delete

If you check this item, you'll get an extra dialog every time you try to delete a message. If you're pretty sure with your fingers, leave it unchecked. If you're nervous that you'll have a butter finger moment and accidentally delete an important email, leave it checked.

Talk and Other Missing Gmail Features

Gmail on the Web has a chat window with Google Talk integration. Rather than accessing Chat through the Android Gmail app, you'll use the separate Google **Talk** app on your phone. 

We cover Google Talk in Chapter 10: "SMS, MMS, and Instant Messaging.

Google Task List

Another feature you may notice missing from the **Gmail** app is a task list. It's a very handy to-do list tool. It's not included in the Gmail app. However, you can still use the task list. Simply navigate your web browser to `http://mail.google.com/tasks`.

You can also make a shortcut for your Home screen. 

1. Create a bookmark of this address by pressing the star in your browser bar.
2. Go to your Home screen and **long press**.
3. Select **Shortcut**.
4. Select **Bookmark**, and then find the task list.

Buzz

Google Buzz is a social networking component of Gmail. We'll talk about social networking tools in more detail in Chapter 20, "Social Media and Skype." Buzz isn't supported in the Gmail app on Android, but you can download a widget from Google that allows you to post updates, your location, and photos from your phone.

Multiple Gmail Accounts

You can set up multiple Gmail accounts on your DROID and manage them all from the Gmail app.

The figure on the right shows multiple Gmail accounts. Unread messages are shown to the right of each account. If you click an account, you'll see only the inbox of that account. You can always get back to the view shown to the right by clicking the menu button and selecting **Accounts.**

To add another Gmail account to your DROID, follow the same steps you'd use to add any other type of email account, as shown in Figure 9–5.

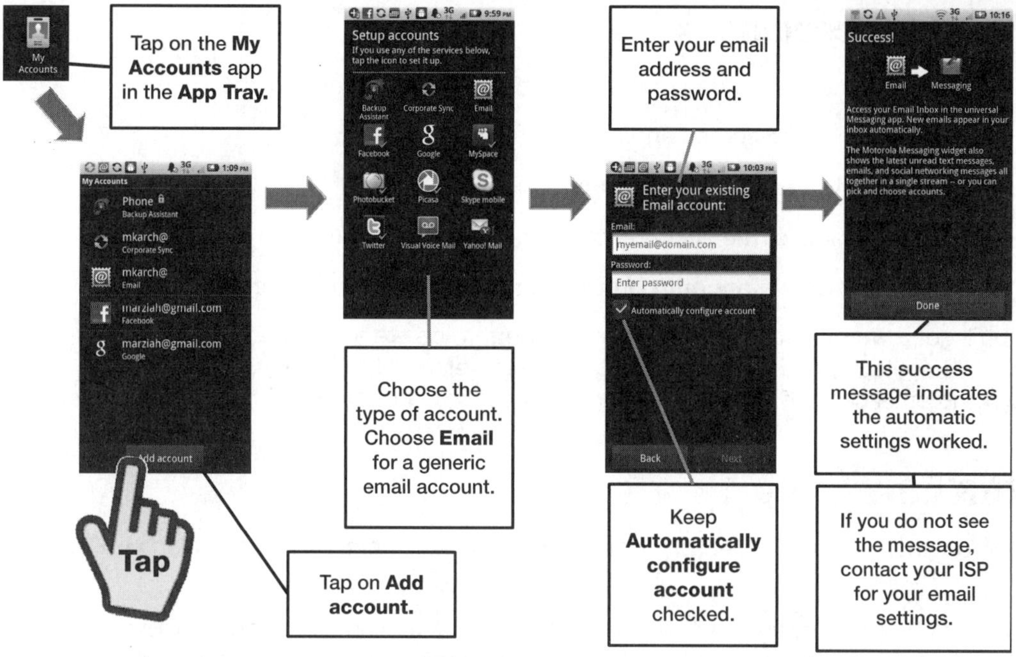

Figure 9–5. *Adding accounts.*

You can also add accounts from directly within the Gmail app by pressing the **Menu** button, going to **Accounts**, and then tapping the **Add account** button.

When you add another Google account, you'll be prompted to specify which parts of that account you want to sync. Your choices depend on what services you've used, but for email accounts, you'll have the choice to sync Gmail and Contacts.

Switching Between Accounts

It's important to keep track of which account you're using when you read or send messages. To switch between accounts:

1. Press the **Menu** button.
2. Tap **Accounts.**
3. Tap the email inbox you wish to switch to.

Or

1. When composing an email, tap the **From:** field.
2. A pop-up window will show your Gmail accounts.
3. Tap your choice.

Deleting Accounts

Deleting accounts is a reverse of the same process that created them.

1. Go to the **Home** screen and press the **Menu** button.
2. Select **Settings** then tap **Accounts.**
3. Tap the name of the account you want to delete.
4. Tap the **Remove account** button on the bottom of the screen.

You'll get a warning message that you're about to delete an account, the email, and the synced contacts, and you'll need to confirm to delete. Alternatively, you could just stop syncing an account if you wanted to retain your contacts without checking the email.

The Email App

Android includes the **Gmail** app for adding Gmail accounts, but there's also an **Email** app for checking mail with non-Gmail accounts.  Depending on your Exchange server's settings, this account can sometimes be used to check Exchange accounts as well as standard email accounts that use POP or IMAP protocols.

Just like the Gmail app, you can add more than one account to the **Email** app. The **Email** app also syncs with your DROID **Messaging** widget and **Universal Inbox.**

In addition to the Email app, your DROID also has a **Yahoo! email** app for adding your Yahoo! account.

Exchange Accounts on Android

Android 2.2 supports Exchange email through the Email app. As we mentioned earlier, Verizon charges for corporate email sync, so you may have limited success using Exchange email through the standard **Email** app, you may have problems syncing your calendar, and you may not see your Global Address List (GAL).

Turning Off Ringtones in the Email App

Turning off the "Droid" ringtone in the Email app is similar to the way you do it in Gmail.

1. Launch the **Email** app.
2. Press the **Menu** button.
3. Tap **Email settings**.
4. Tap **Notifications**.

5. Tap **Select ringtone**.

6. Select **Silent**.

Of course, if you do want an audio notification of new emails, you can use this method to change your ringtone to any ringtone sound you'd like.

Outlook Web Access

If you can't add an Exchange account through the **Email** app, you might be able to use an Outlook Web Access or OWA account instead. You have to actively check for email yourself instead of getting notification that you've got a new message. If you use OWA to access mail, just set up a bookmark on your **Home** screen for quick access.

Corporate Sync Accounts

If you have no luck adding an Exchange email to the **Email** app, you may need to use a Corporate Sync account.  Corporate Sync accounts are designed to be compliant with security standards mandated by corporate email systems. Corporate email accounts can also be **remote wiped**, or erased remotely if you lose your phone.

The disadvantage to Corporate Sync accounts is that as of the time of this publication, Verizon charges more for these accounts. That's why it's important to remove any accounts you don't need.

Universal Inbox

Your DROID organizes all your non-Gmail email accounts, your text messages, and social networking status updates into a **Universal Inbox** in the **Messaging** app, so you can read all your messages from one central location. The Universal Inbox is shown in Figure 9–6. Please see Chapter 10 for more on using Instant Messaging and this Universal Inbox.

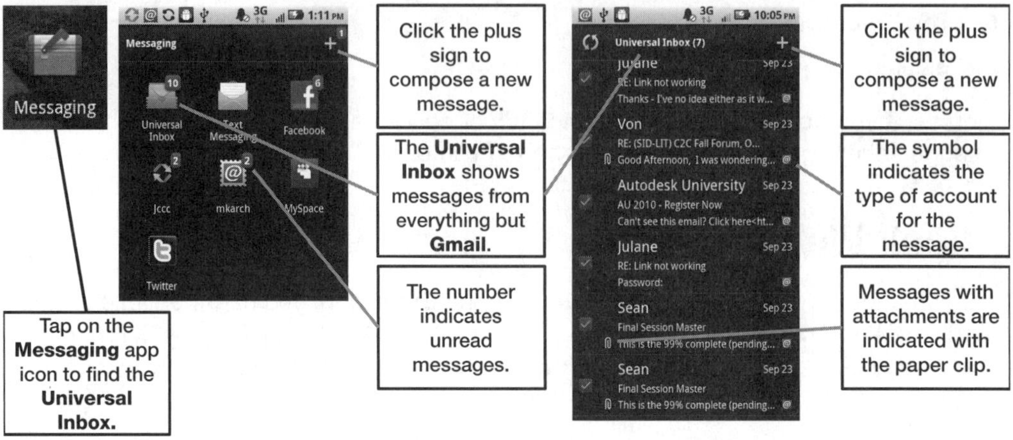

Figure 9–6. *The Universal Inbox.*

You can navigate to your email messages in many ways:

- Launch the **Email** app in the App Tray.
- Use the **Messaging** app and the Universal Inbox.
- Drag down the **Notification bar** whenever you see a notification that you've got a new message.
- Use the **Messaging widget** on the **Home** screen.

Your **Universal Inbox** is available through the **Messaging** app, but it's also available through the **Messaging widget**. By using the widget, you can read previews of your email messages right from your Home screen, and even compose replies. Figure 9–7 shows quick ways to check email.

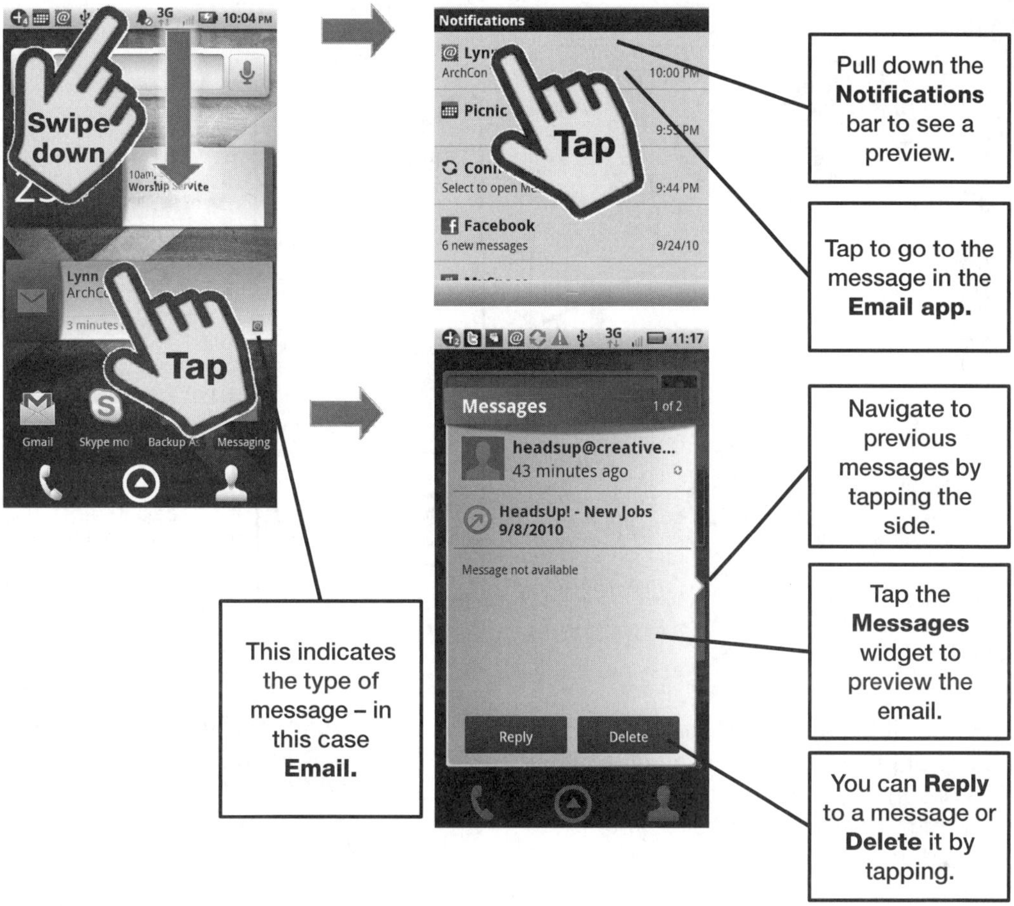

Figure 9–7. *Two ways to check email.*

Reading Attachments

Your ability to read attachments is going to depend on the type of attachment, the app you're using, and the software you have installed on your phone. If an email has embedded pictures, just as with many desktop email programs, you'll generally see them. Gmail doesn't automatically download pictures, but you can tap **Show Pictures** to download and see them.

Other attachments require you to click to download and view, as shown in Figure 9–8.

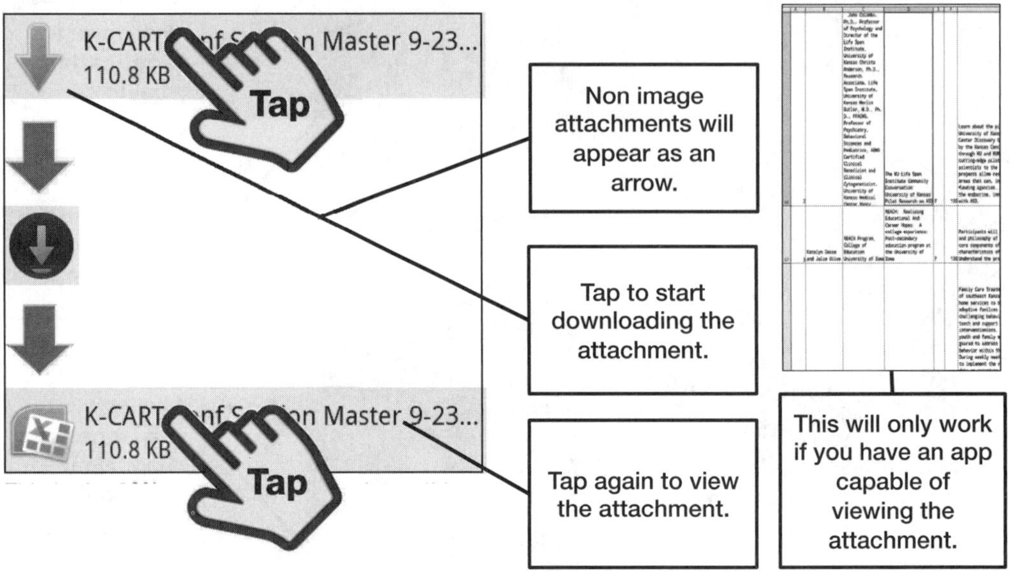
110.8 KB
Tap
Non image attachments will appear as an arrow.
Tap to start downloading the attachment.
110.8 KB
Tap
Tap again to view the attachment.
This will only work if you have an app capable of viewing the attachment.

Figure 9–8. *Emailing attachments.*

If your DROID has more than one app capable of handling an attachment, you'll see a dialog box that lets you choose how you want to handle the file. Click your preferred app, and you can view your attachment.

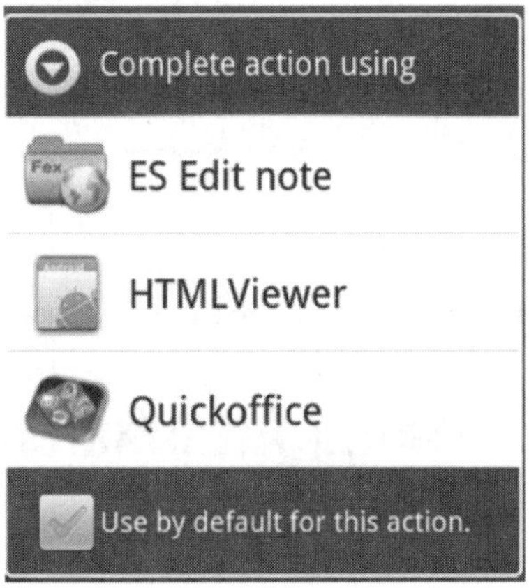
Complete action using
ES Edit note
HTMLViewer
Quickoffice
Use by default for this action.

If your DROID doesn't have any app capable of viewing an attachment, you can still download it in case you get an app later.

Third-Party Apps

If you want an alternative to Corporate Sync, you may be able to use third-party apps. One popular app is **TouchDown** by NitroDesk. You can get more information here: www.nitrodesk.com/dk_touchdownFeatures.aspx.

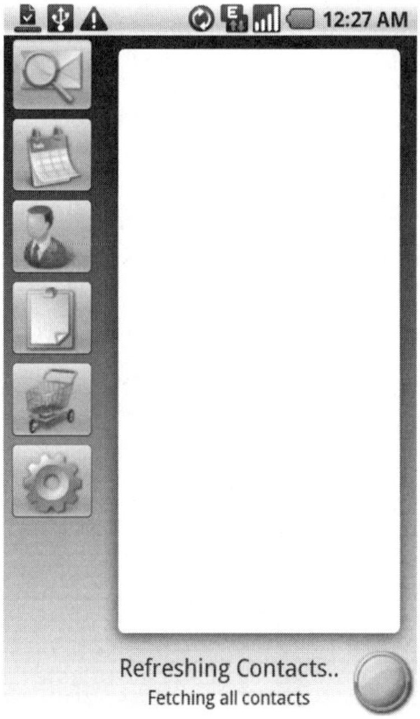

TouchDown comes in both a free and paid version. The free version lets you check email and get the day's calendar, while the $19.99 paid version, shown to the right, allows you to sync your Exchange email, accept and send task and event requests, use the Global Address List, etc. TouchDown also supports security policy enforcement, so it should pass muster with most IT departments. You can also use it through OWA if you can't get it to work with ActiveSync. There's a fully functional free trial, so you can test to make sure everything works before you purchase it.

TouchDown is a solid app. Not only does it provide most of the features of Outlook in your pocket, it also includes several widget options to keep your Exchange info handy whenever you use your phone. However, it doesn't mesh your email, calendar, and task information with your other Android calendar contact, or task lists. On one hand, it's handy to have business separate from personal life, and on the other hand, it would be nice to have access to your Global Address List when using your main contact list.

Chapter **10**

SMS, MMS, and Instant Messaging

SMS stands for *Short Messaging Service*, and it is commonly referred to as *text messaging* or simply *texting*. Text messages are usually limited to 160 characters, and they are a great way to quickly touch base with someone without interrupting them with a voice call. Sometimes you can text someone and receive a text reply when it would be impossible or difficult to make a voice call.

Icon on DROID

A related technology is *Multimedia Messaging Service* (MMS), which lets you send a message with pictures, audio, or video.

DROID 2/2 Global/X

In this chapter, we will cover how to send and receive SMS text and MMS picture/video messages on your DROID.

As shown to the right, your DROID has only one **Messaging** icon, but on your DROID 2/X, you will see two icons **Messaging** and **Text Messaging**. The **Text Messaging** icon is the faster way to get into messaging when you have a DROID 2/X.

You will also learn how to send a text message from your **Contacts** app and how to send a picture as an MMS message from your **Gallery** app.

SMS Text Messaging on your DROID

Text messaging has become one of the most popular services on cell phones today. While it is still used more extensively in Europe and Asia, it is growing in popularity in North America.

The concept is very simple; instead of placing a phone call, you send a short message to someone's handset. It is much less disruptive than a phone call; and you may have friends, colleagues, or co-workers who do not own a DROID – so email may not be an option.

One of the this book's authors uses text messaging with his children all the time – this is how the generation his kids are part of communicates. "R u coming home 4 dinner?" "Yup." There you have it: meaningful dialog with an 18-year-old – short, instant and easy.

Composing SMS Text Messages

Composing an SMS message is much like sending an email. The beauty of an SMS message is that it arrives on virtually any handset and is quite simple to reply to.

Composing an SMS Message from the Messaging App

There are a couple of ways to send text messages on your DROID. The easiest way is to touch the **Messaging** icon (DROID) and **Text Messaging** icon (DROID 2/X) on the **Home** screen.

When you first start the app, you most likely won't have any messages, so the screen will be blank. Once you get started with SMS messaging, you will have a list of messages and current "open" discussions with your contacts.

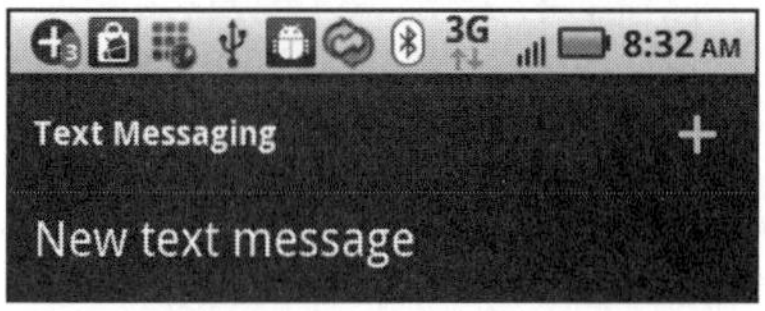

Follow these steps to send a new SMS message:

1. Tap the **Messaging** icon (DROID) or **Text Messaging** icon (DROID 2/X).
2. Touch **New message** or **New text message** at the top of the screen.
3. The cursor will immediately go to the **To:** line. Touch the **To:** field and start typing in the name of your contact. Or, you can tap the person with the plus icon as shown in the figure to the right to select a contact from your Contacts list.
4. If you want to just type someone's mobile phone number, then press the **?123** button and dial the number.

5. When you find the contact you wish to use, touch the name and it will appear in the **To:** line (see Figure 10–1).

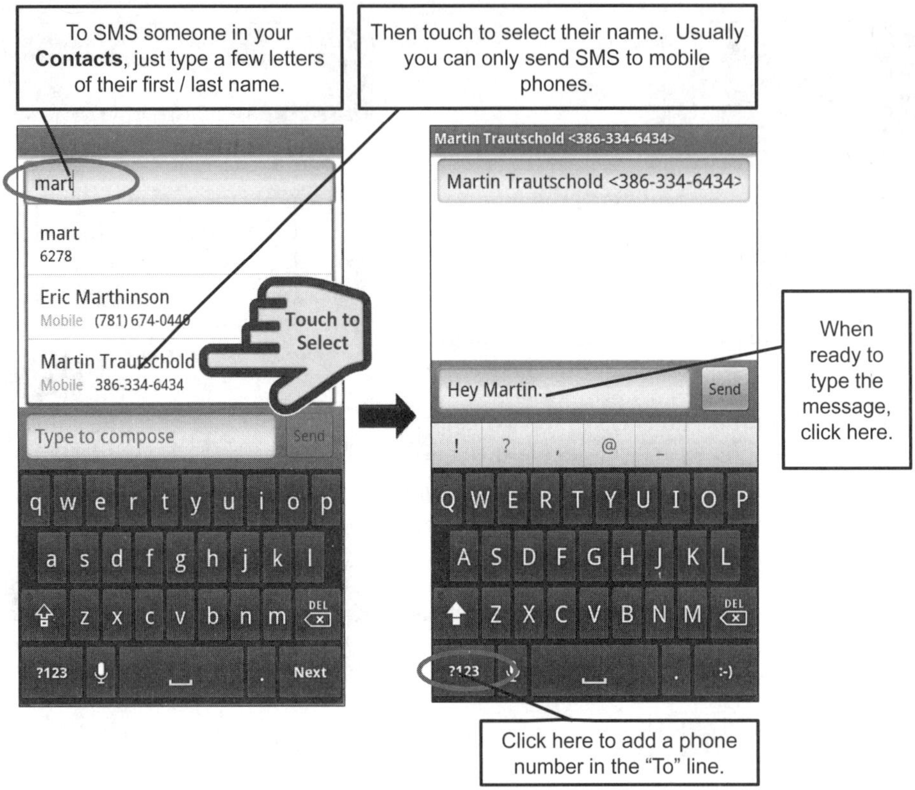
To SMS someone in your Contacts, just type a few letters of their first / last name.
Then touch to select their name. Usually you can only send SMS to mobile phones.
mart
mart
6278
Eric Marthinson
Mobile (781) 674-0440
Martin Trautschold
Mobile 386-334-6434
Touch to Select
Type to compose
Send
Martin Trautschold <386-334-6434>
Martin Trautschold <386-334-6434>
Hey Martin.
Send
When ready to type the message, click here.
Click here to add a phone number in the "To" line.

Figure 10–1. *Choosing a recipient for an SMS message.*

6. When you are ready to type the SMS message, touch anywhere in the box in the middle of the screen (next to the **Send** button).
7. The keyboard will be displayed. Just type in your message and then touch **Send** when you are done.

Martin Trautschold <386-334-6434>
Martin Trautschold <386-334-6434>
Hey Martin. This is a text from the Droid.
Send

> **NOTE**: There is no character counter in the Android OS, so if your messages go over 160 characters, they will be divided into two messages.

TIP: If you prefer, you can use the larger **Landscape** mode keyboard for sending text messages. It can be easier to type with the larger keys on the DROID X or the keyboard on the DROID and DROID 2, especially when your fingers are a little larger, or it is hard to see the smaller keys.

Options After Sending a Text

Once the text has been sent, the window changes to a *threaded* discussion window between you and the contact. The text that you sent displays with a white background. When your contact replies, his message will appear in a blue background. If you have a contact picture for the recipient, that will show up in the display, as well.

To leave the SMS screen, touch the **Back** key a couple of times; or, you can just touch the **Home** key to go back to your **Home** screen.

NOTE: If the message fails to send, it is usually because of a low wireless signal. When you get to a stronger signal area, the message should send.

Messaging on DROID

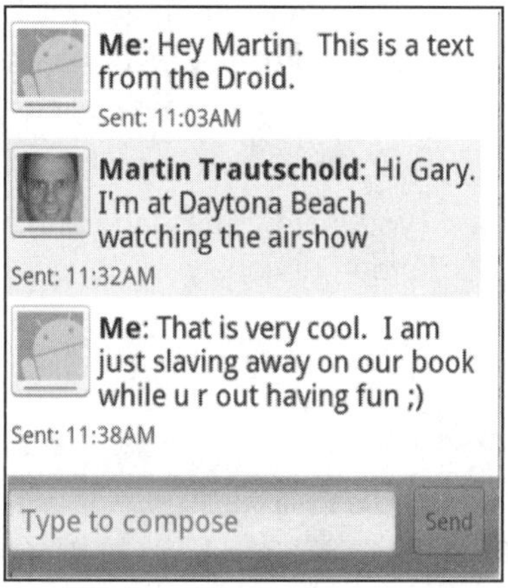

DROID 2/2 Global/X

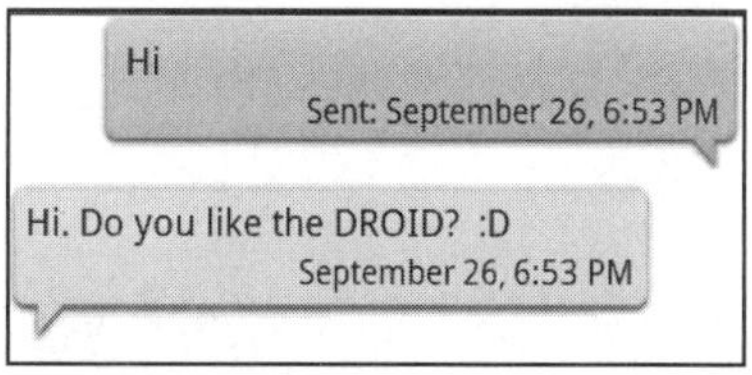

At this point, you can send another text message following the steps just outlined. You can also call the contact or view his contact info.

To initiate a call to the contact you are texting with, touch the image of the contact in the threaded message to bring up a pop-up window. In the example on the right, you could call Martin by touching the **Call** button. To look at his contact info, you can touch the **Contact Info** button.

Composing an SMS Message from Contacts

You also have the ability to send a text message right from your Contacts list on your DROID. Follow these steps to do so:

1. Tap your **Contacts** icon.
2. Find the contact you wish to send a text to by searching or scrolling through **Contacts**.
3. At the right-hand side of the phone numbers in the contact info, there will be a **Messaging** icon (see Figure 10–2). Touch that icon and you will be taken to the messaging screen (as shown previously).
4. Type in your message and follow the steps listed previously.

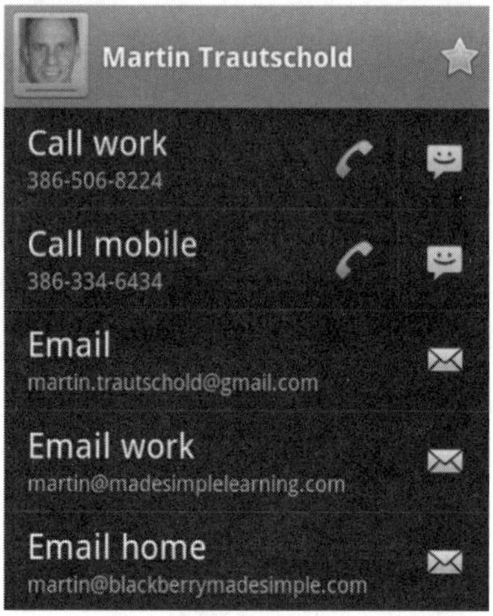

NOTE: Remember that you can only send SMS messages to a mobile number.

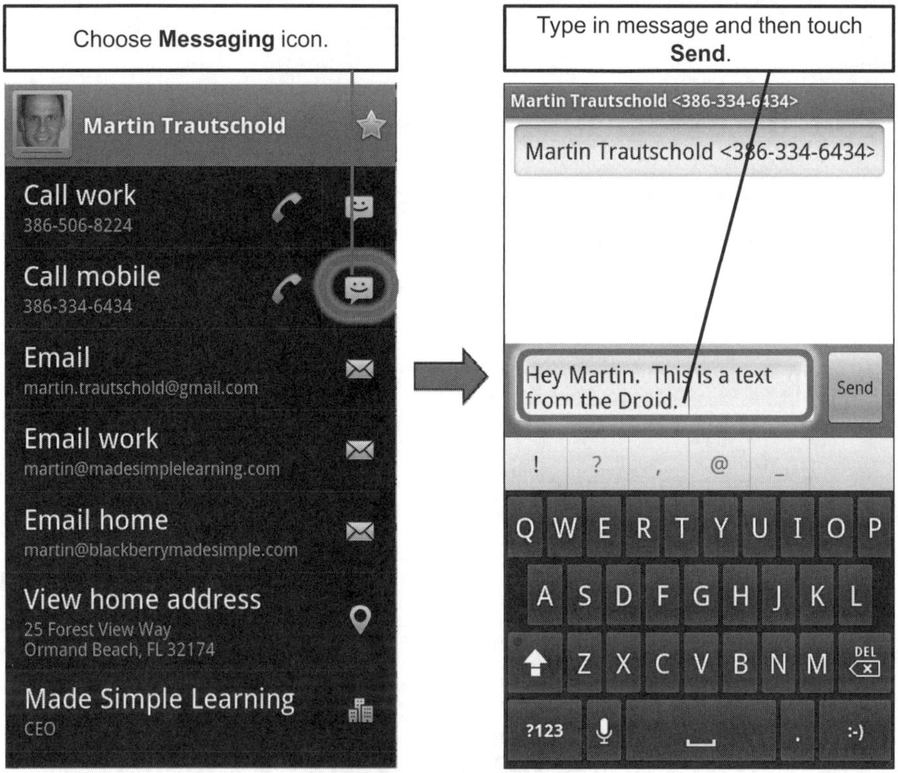

Figure 10–2. *Sending an SMS message from your* ***Contacts*** *app.*

Replying to a Text Message

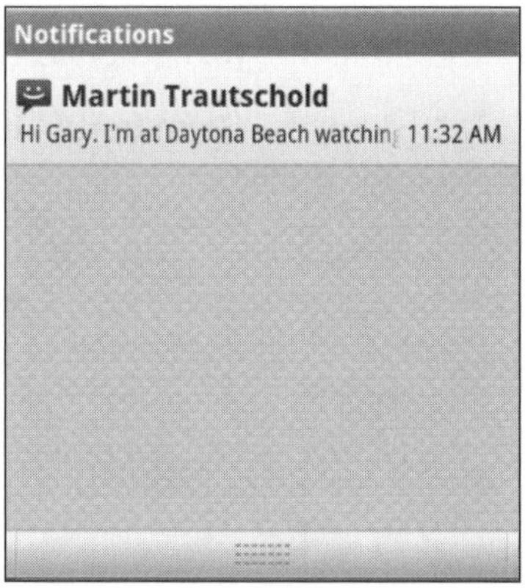

When a text message is received, your DROID will play an indicator tone or vibrate – or both – depending on your settings. A notification will appear on the screen in the notification bar at the top.

Responding to a text is easy. Simply pull down the indicator bar, touch the message response, and then touch **Type to compose** to enter a response.

> **NOTE**: If your screen is locked, you will not see the message. Just slide the **Lock** tab and you will be able to pull down the indicator bar to see the message.

Viewing Stored Messages

Once you begin a few threaded messages, they will be stored in the **Messaging** (DROID) or Text Messaging (DROID 2/X) app. Touch the **Messaging** icon to scroll through your message threads.

Sometimes you will want to continue an earlier conversation with someone. Follow these steps to do so:

1. Touch the thread you want to continue. The conversation will open up, showing you a threaded view of the previous messages.
2. Touch the text box and type your message.
3. Touch the **Send** button to continue the conversation.

Messaging Notification Options

There are a couple of options available to you with respect to how your DROID reacts when an SMS message arrives. Follow these steps to customize those options:

1. Start your **Text Messaging** app and press the **Menu** button.
2. Touch **Settings (**DROID) or **Messaging Settings** (DROID 2/X).

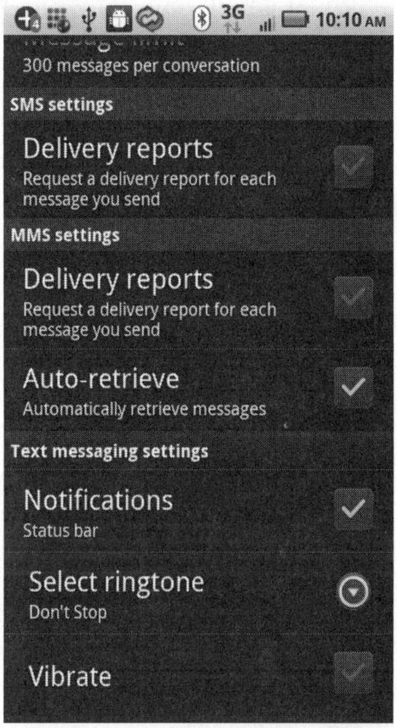

3. Scroll down a bit further and you will see a drop down that says **Select Ring Tone**. Touch this and you can choose the tone for the SMS message. You are limited to the choices offered (usually 12 or more); you can also choose **None**.
4. Choose your preferred sound for incoming SMS message notifications and then touch the **OK** button in the lower-left corner to finalize your selection.
5. You can also set your DROID to vibrate for every text message received by checking the box next to **Vibrate**.

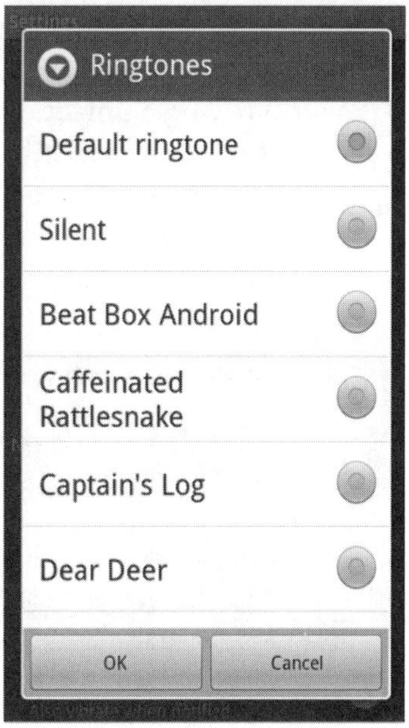

Multimedia Messaging

The **Messaging** app provides the necessary tools to send and receive multimedia messages in MMS format, including picture and video messages. MMS messages appear right in the messaging window, just like your SMS text messages.

> **NOTE**: You can send multimedia messages from your DROID that includeimages, videos, locations (from maps), audio (from **Voice Memo**), and vCard files (from **Contacts**) in MMS format.

The Messaging App

Follow these steps to send a picture to someone else in MMS format:

1. Touch the **Messaging** icon to start messaging, just as you did to initiate an SMS message.

2. Press the **Menu** button and then select **Attach** (DROID) or Insert (DROID 2/X) or **Add Subject.** This will put you into **MMS** mode; a notification on the screen will let you know you've entered this mode.

> **NOTE:** The screen shown to the right is from a DROID. On your DROID 2/X you will see a slightly different screen that is titled with the word **Insert** but it has almost all the same options.

DROID

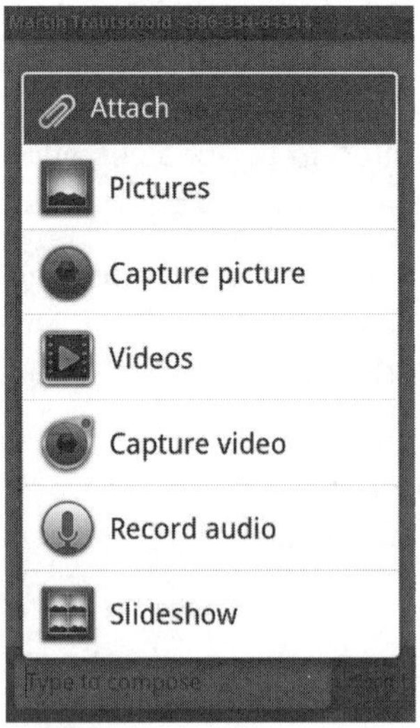

DROID 2/2 Global/X

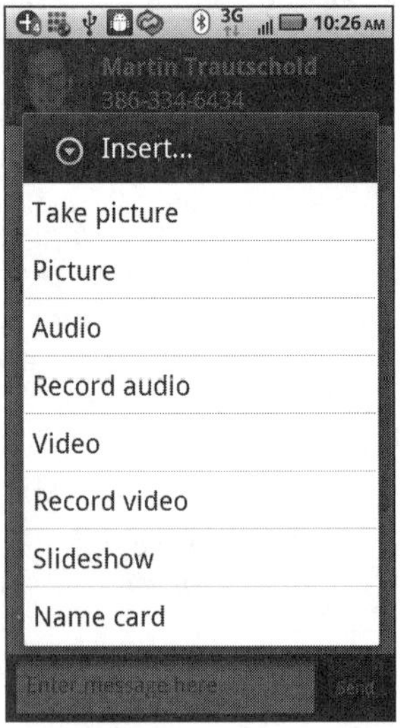

3. To take a photo, follow the instructions in Chapter 18: "Take Photos and Videos." If you touch **Pictures** or **Videos**, just navigate through your pictures/videos and find the item you would like to add to your message (see Figure 10–3).

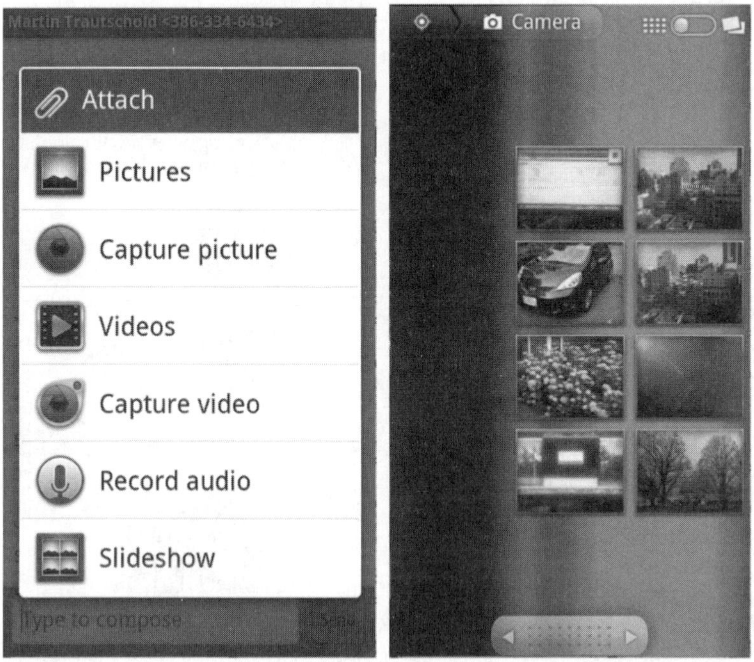

Figure 10–3. *Choosing an existing photo to send in MMS format.*

4. Touch the picture you wish to send as an MMS message and you will see the picture load into the small window. You may see a warning that the picture is too large to send, tap **Resize** to shrink the image so it can be sent.

5. Select a recipient and type in a short note if you like.
6. Touch the **Send** button.

If you already have a threaded discussion with that particular contact, then the picture will show up in that threaded discussion.

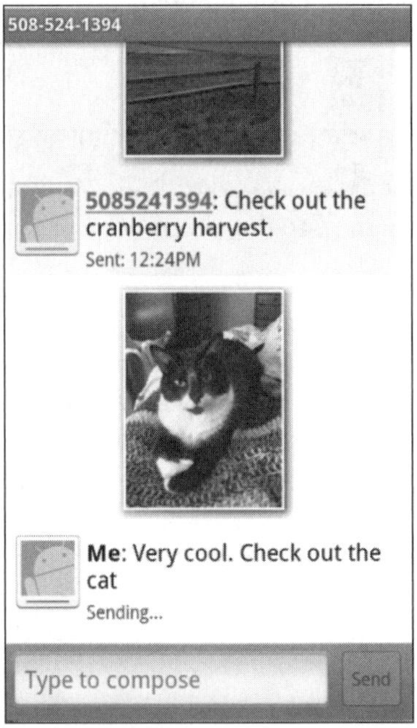

NOTE: You can continue to exchange images and text in the middle of a threaded discussion. You can always scroll through to see the entire discussion – pictures and all!

Choosing a Picture from Your Gallery to Send via MMS

A second way to send an MMS message is to go straight to your **Gallery** app and choose a picture. Follow these steps to do so:

DROID

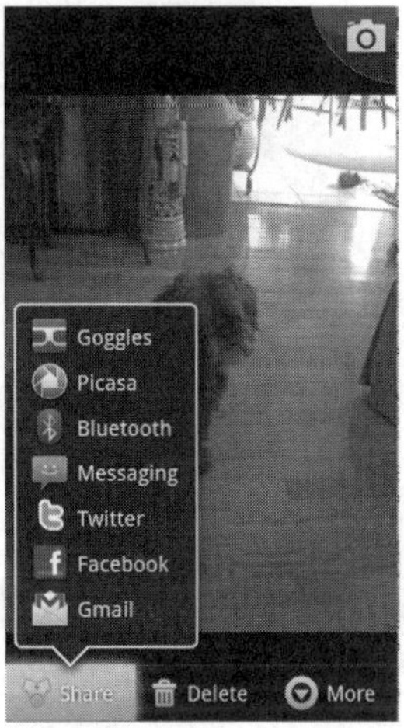

DROID 2/2 Global/X

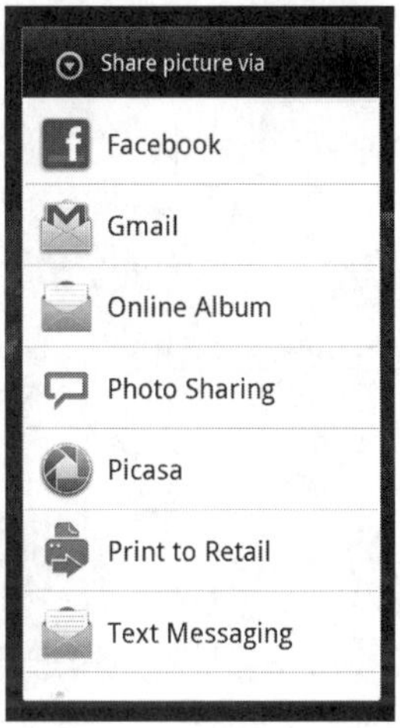

1. Start your **Gallery** app and navigate through your pictures (see Chapter 18: "Take Photos and Videos" for more information on how to do this).
2. To send only one picture, touch the picture you wish to send.
3. Press the **Menu** button and tap **Share**.You will now see **Messaging** or Text Messaging in the list of options and other services that are installed on your DROID.
4. Choose **Messaging** and the photo will load into the message, just as it did previously.

Sending Multiple Pictures

You can also send multiple pictures in an MMS message. Start your **Gallery** app as you did in the previous section. On the DROID, touch and hold one picture until you see a check mark. On the DROID 2/X, press the **Menu** button and choose **Select items**. Now tap as many pictures as you want to add. This will highlight your pictures and display them with a green check mark in the corner of the picture's box (see Figure 10–4).

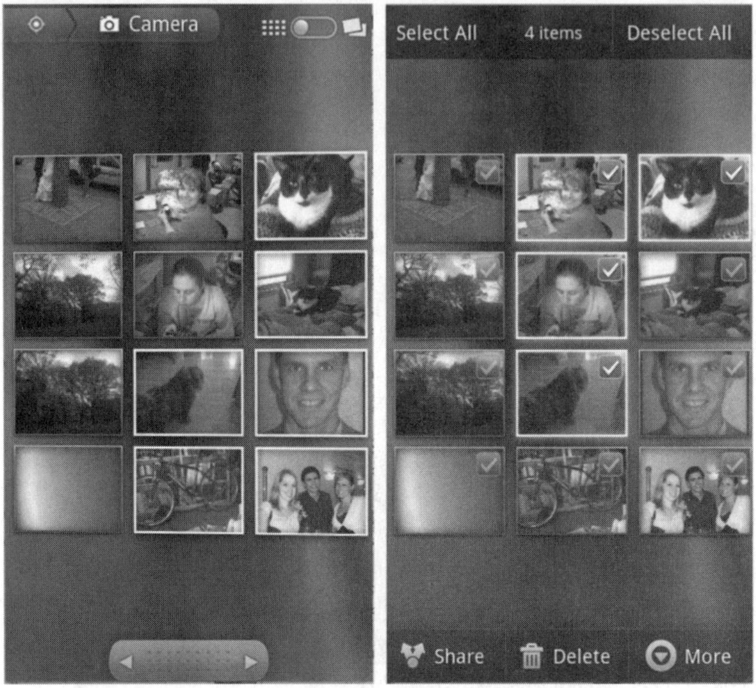

Figure 10–4. *Selecting multiple photos to send in an MMS message.*

Once you have chosen all the pictures you want to send, on the DROID, touch the **Share** button and choose **Messaging**. On the DROID 2/X, press the **Menu** button and select **Share** then tap **Text Messaging** and the pictures will appear in the message, along with an arrow that allows the recipient to play a slideshow of the images.

Instant Messaging on Your DROID

So far we have covered SMS and MMS messaging in this chapter. However, there are many other ways to stay in touch with your friends and use your DROID as a messaging device. One option for staying in touch with friends and family members is to use one of the many instant messaging apps available.

The advantages of instant messaging, as opposed to SMS or MMS messaging, are as follows:

- Instant messaging is usually free from additonal charges (assuming you have a data plan).
- Instant messages reach your intended audience immediately.
- Instant messages let you have quick, ongoing chats in real time.

Google Talk

Your DROID is made to operate in the Google world, so it makes sense to start with the instant messaging app designed by Google: **Google Talk**.

Google Talk is an instant messaging client available to anyone with a Gmail or Google account. You can also invite contacts to become **Google Talk** "chat buddies" by sending them a **Google Talk** invite.

Follow these steps to invite someone to be part of your **Google Talk** contacts:

1. Start **Talk** from the **Home** screen.
2. Press the **Menu** button and select **Add friend**.
3. Type in the **Send chat invitation** window until you see a match with one of your contacts (see Figure 10–5).
4. Tap **Send Invitation** to send the invitation.

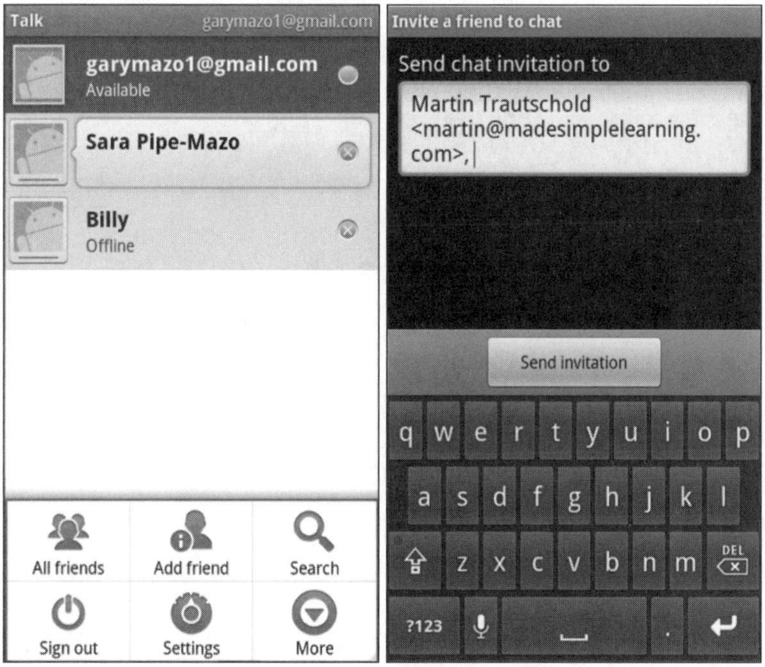

Figure 10–5. *Sending a **Googe Chat** invitation to a contact.*

Using Google Talk

Using Google Talk is just like using your **Messaging** app. Start the **Google Talk** app and choose the contact you wish to chat with. Type your message in the **Type to compose** box and then touch **Send**.

You can keep your chat window open to have a running chat; alternatively, you can rely on the **Notification** icon to tell you when you have a new chat message.

NOTE: Even if your contact is not online, you can still post your chat message. In this case, your contact will see it as soon as he or she logs on.

AIM and Other Instant Messaging Apps

The Andoid Market is filled with instant messaging apps. **AIM** is particularly popular chat program, and it can also be tied to your **Facebook** app for chat.

Follow these steps to download **AIM** from the Android Market.

1. Start up the Android Market (see Chapter 17: "Exploring the Android Market" for more information on how to do this).
2. Search for **AIM** and download the app.

Using AIM

You start the **AIM** app by touching the icon on your **Home** screen. Choose whether to log in using you standard AIM account or by using your Facebook account, as shown in Figure 10–6. Input your login information and then chat as you would using the **Google Talk** app or your **Messaging** app.

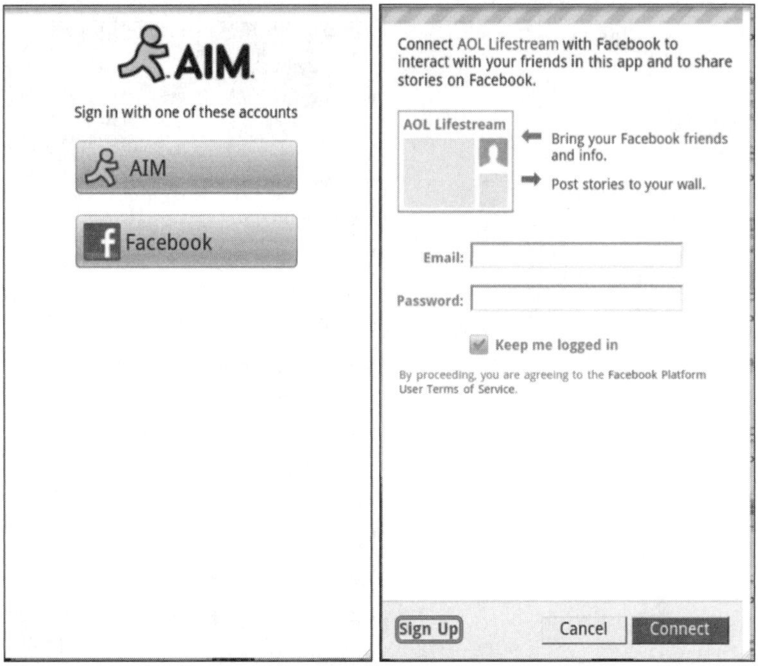

Figure 10–6. *Logging in to your AIM account using AIM or Facebook.*

Chapter 11

Surfing the Web

Now, we'll take you through one of the most fun things to do on your DROID: surfing the Web. You may have heard web surfing on the DROID is a better experience than ever before—we agree! We'll show you how to touch, zoom around, and interact with the Web like never before with the web browser on your DROID. You'll learn how to set and use bookmarks, quickly find things with the search engine, open and switch between multiple browser windows, and even easily copy text and graphics from web pages.

Web Browsing on the DROID

You can browse the web to your heart's content via your DROID's Wi-Fi or 3G connection. Like other smartphones using a Webkit browser, your DROID has one of the most capable mobile browsing experiences available today. Web pages look very much like web pages on your computer. With the DROID's ability to zoom in, you don't have to worry about the smaller screen size inhibiting your web browsing experience. In short, web browsing is a much more satisfying experience on the DROID.

Choose to browse in portrait or landscape mode, whichever you prefer. Quickly zoom into a video by double-tapping it or pinching open on it, which is natural to you because those are the motions you use to zoom in text and graphics.

Why Do Some Videos and Sites Not Appear? (Flash Player Required)

Some web sites are designed with Adobe Flash Player. At the time of this writing, the DROID fully support Adobe Flash through the Flash "lite" app that is making its way onto select Android phones.

Go to the Android Market, look for Flash Player, and install in. Learn how to download apps from the Android Market in Chapter 17: "Exploring the Android Market."

If you tap a video and the video does not play, or you see something like "Flash Plugin Required," "Download the Latest Flash Plugin to view this video," or "Adobe Flash Required to view this site," you will need to update your operating system to view the video or web page.

An Internet Connection Is Required

You do need an Internet connection on your DROID via Wi-Fi or 3G to browse the web. Check out the Chapter 5: "Wi-Fi and 3G Connectivity" to learn more.

Launching the Web Browser

You should find the web browser icon on your initial **Home** screen. Usually the **Browser** icon is in the bottom row of icons on your Home screen.

Touch the **Browser** icon, and you will be taken to the browser's home page. Most likely, this will be the Google start page.

Just turn your DROID on its side to see the same page in wider landscape mode. As you find web sites you like, you can set bookmarks to easily jump to these sites. We will show you how to do that later in this chapter.

View web pages in portrait mode.

Or, turn your DROID on its side to see them in landscape mode.

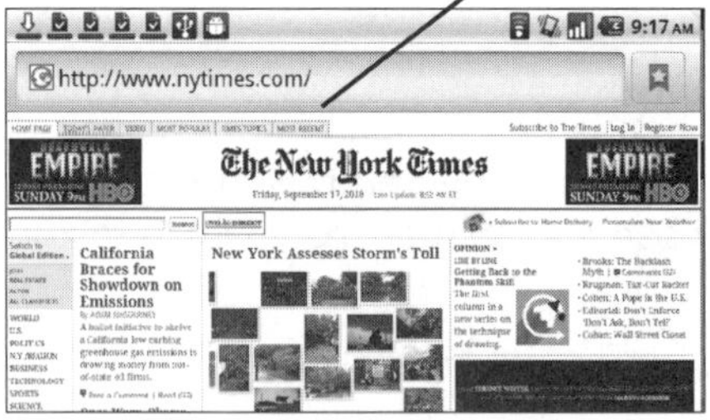

Layout of Web Browser Screen

Figure 11–1 shows how a web page looks in the browser and the different actions you can take in the browser.

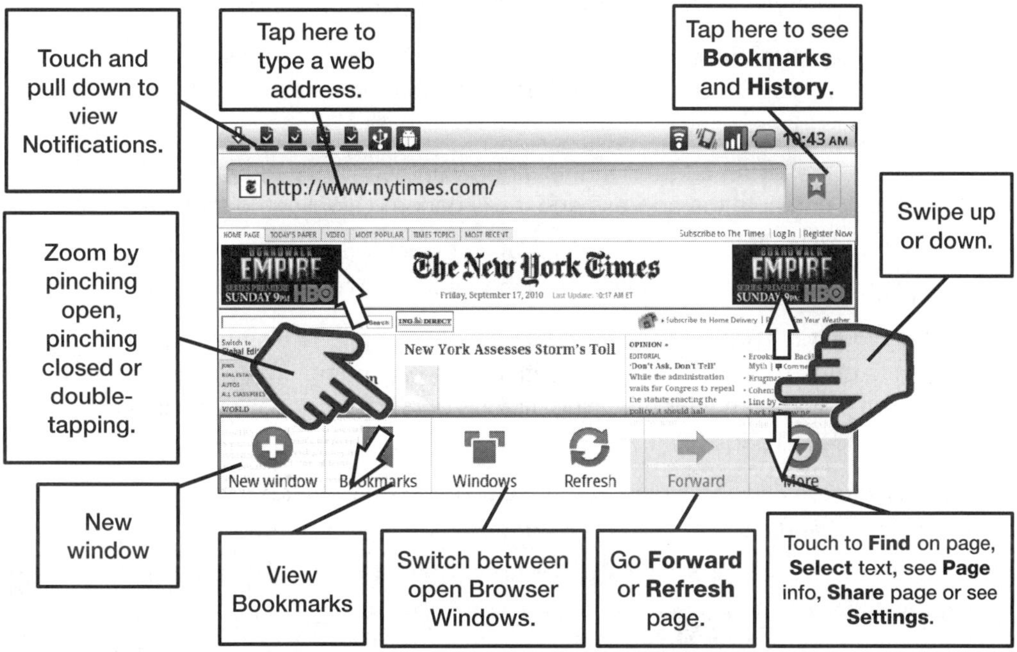

Figure 11–1. *Web browser page layout.*

NOTE: On your DROID 2/X the icons shown on the menu items may appear slightly different than shown in figure 11-1, but the names and functions they perform will be the same.

As you look at your screen, notice the **Address Bar** in the upper left side of the screen. This displays the current web address. You can also type search words right in the address bar. If you type words and click Go or the Enter key, then the DROID assumes you want to do a web search. By default, this is set to Google search, but you can change that if you want.

At the bottom of the screen are five icons: **Back**, **Forward**, **Add Bookmark, Bookmarks**, and **Pages View**.

Typing a Web Address

The first thing you'll want to learn is how to get to your favorite web pages. Just like on your computer, you type in the web address (URL) into the browser.

1. To start, tap the **Address Bar** at the top of the browser as shown in Figure 11–2. You'll then see the keyboard appear and the window for the address bar expand.
2. If there is already an address in the window and you want to erase it, just touch and hold the address and it will become highlighted and the keyboard will pop up.
3. Start typing your web address (you don't need the **www.**).
4. When you start typing, you may see suggestions appear. Just tap any of them to go to that page. The suggestions are very complete because they are pulled from your browsing history, bookmarks, the web address (URL), and web page titles.
5. When you are finished typing, tap the **Go** key to go to that page.

TIP: Don't type the **www.** because it's not necessary. Remember to use the **colon**, **forward slash**, **underscore**, **dot**, and **.com** keys at the bottom to save time.

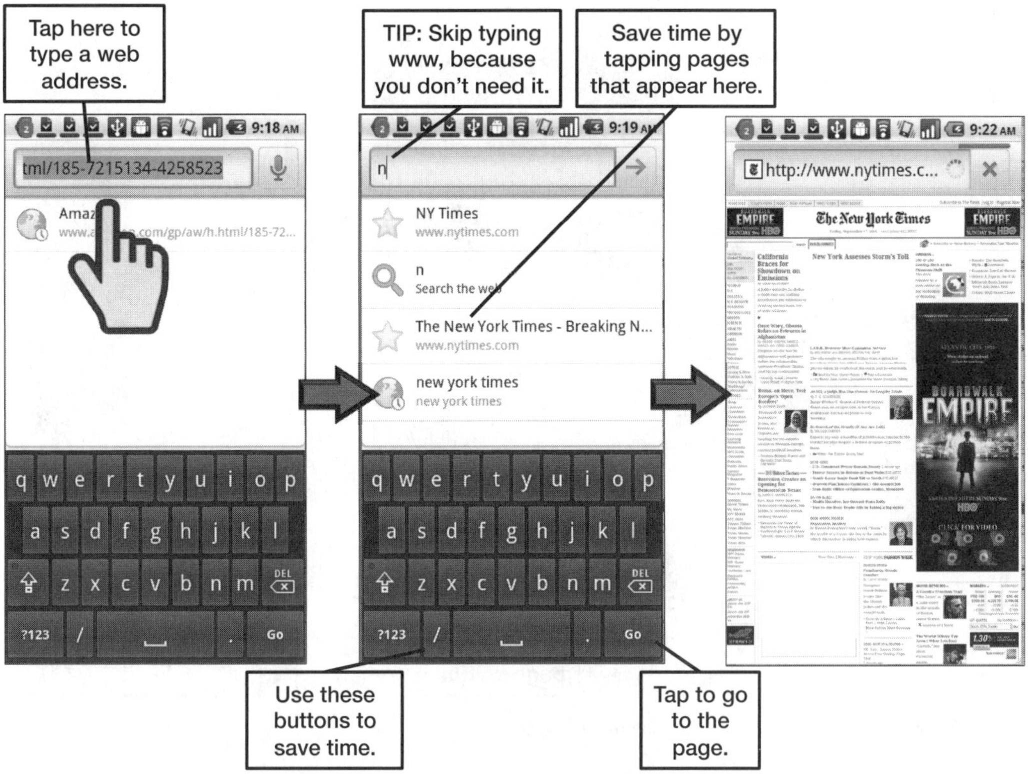
Tap here to type a web address.
TIP: Skip typing www, because you don't need it.
Save time by tapping pages that appear here.
Use these buttons to save time.
Tap to go to the page.
NY Times
www.nytimes.com
Search the web
The New York Times - Breaking N...
new york times
http://www.nytimes.c...

Figure 11–2. *Typing a web address.*

> **TIP:** If you turn the DROID sideways, you get a landscape keyboard which might be easier for those with "big thumbs" to use.

Moving Backward or Forward Through Open Web Pages

Now that you know how to enter web addresses, you'll probably be jumping to various web sites. The **Forward** and **Back** buttons make it very easy to go to recently visited pages in either direction, as Figure 11–3 shows. Touch the **Back** button (the soft key at the bottom of the DROID) to go backwards to the last page visited.

To move forwards, touch the **Menu** button, then select the **Forward** soft key.

Let's say you were looking at the news on *The New York Times* web site, and you jumped to ESPN to check sports scores. To go back to *The New York Times* page, just touch the **Back** button. To return to the ESPN site again, touch the **Menu** key and then the **Forward** arrow.

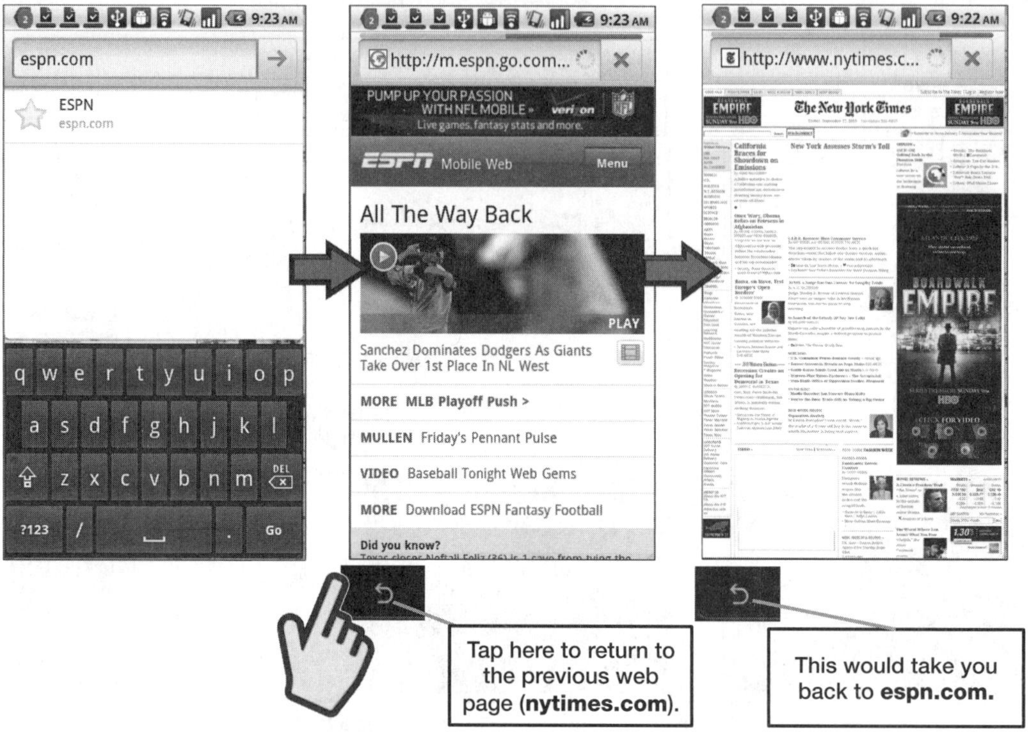

Figure 11–3. *Returning to a previously viewed web page.*

Using the Windows Menu Command

When you press the **Menu** button and then select the **New window** button, the DROID keeps track of all the open windows in the browser. Just press the **Menu** button again and choose the **Windows** button to see all the open Browser windows.

The URL for each window is now listed. Just touch the desired URL to jump right to that open browser window.

In the example shown in Figure 11–4, we touched a link that opened a new browser window. The only way to get back to the old one was to press the Menu button and tap the **Windows** button and select the desired page.

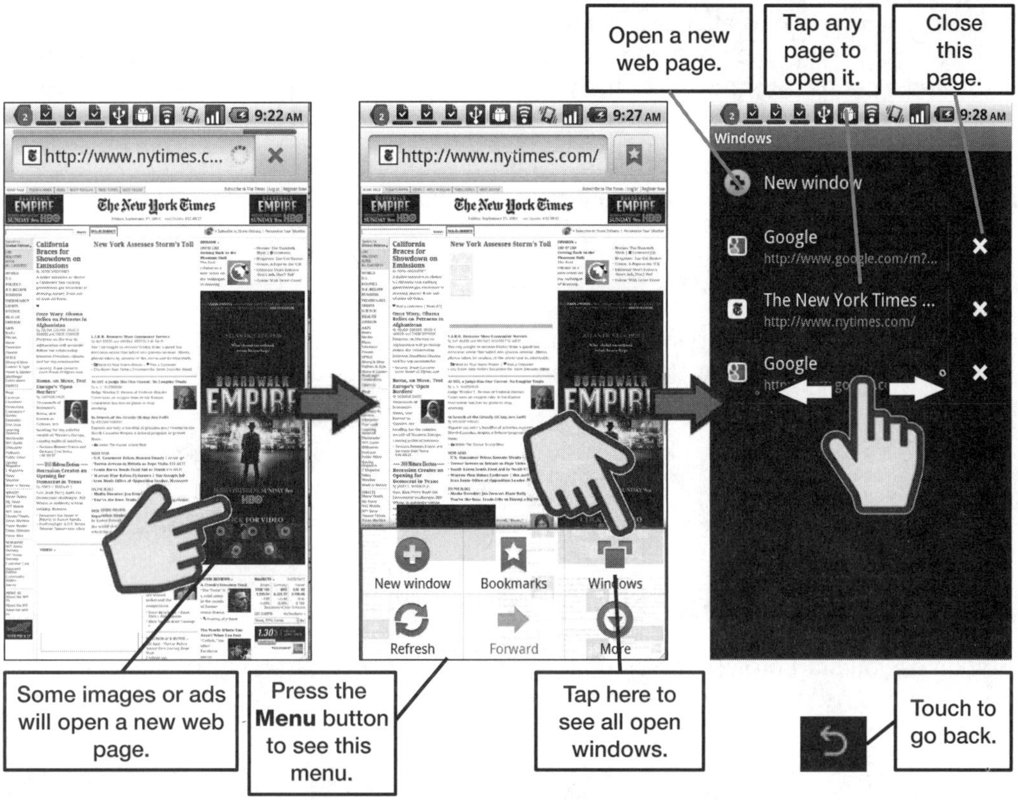

Figure 11–4. *Jumping between open web pages on the DROID.*

Zooming In and Out in Web Pages

Zooming in and out of web pages is very easy on the DROID. There are two primary ways of zooming—double tapping, and pinching.

Double-tapping

If you tap twice on a web page, the page will zoom in on that particular column. This lets you hone in on exactly the right place on the web page, which is very helpful for pages that aren't formatted for a mobile screen.

To zoom out, just double-tap once more. See how this looks graphically in the "Quick Start Guide" earlier in this book.

Pinching

This technique lets you zoom in on a particular section of a page. It takes a little bit of practice but will soon become second nature. Take a look in the "Quick Start Guide" to see graphically how it looks.

Place your thumb and forefinger close together at the section of the web page you wish to zoom into. Slowly pinch out, separating your fingers. You will see the web page zoom in. It takes a couple of seconds for the web page to focus, but it will zoom in and be very clear in a short while.

To zoom out to where you were before, just start with your fingers apart and move them slowly together; the page will zoom out to its original size.

Activating Links from Web Pages

When you're surfing the Web, often you'll come across a link that will take you to another web site. Simply touch the link and you will jump to a new page.

NOTE: Once you jump to a new page from a link, the old page can still be found using the technique shown in Figure 11–4 above.

Working with Browser Bookmarks

As soon as you start browsing a bit on your DROID, you will want to quickly access your favorite web sites. One good way to do this is to add bookmarks for one-tap access to web sites.

TIP: You can sync your Bookmarks from your computer's web browser. Check out Chapter 3: "Sync to your Google account" for more details.

Adding a New Bookmark

Adding new bookmarks on your DROID is just a few taps away.

1. To add a new bookmark for the web page you are currently viewing, tap the **Bookmark** icon to the right of the web address window.
2. The first box says **Add**, tap that box.

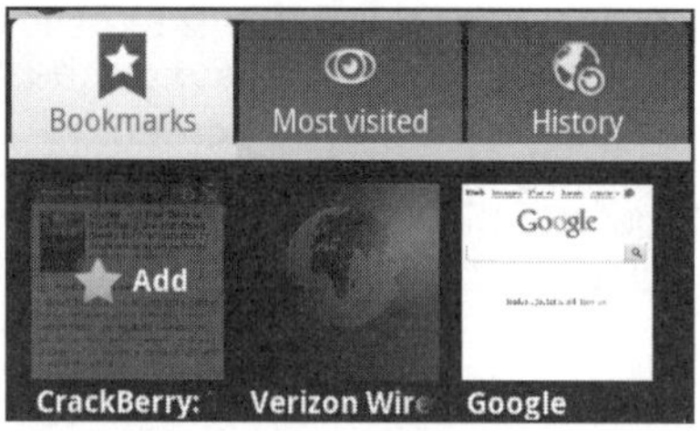

3. We recommend that you edit the bookmark name to something short and recognizable.
4. Make sure the location is correct.
5. When you're finished, tap the **OK** button.

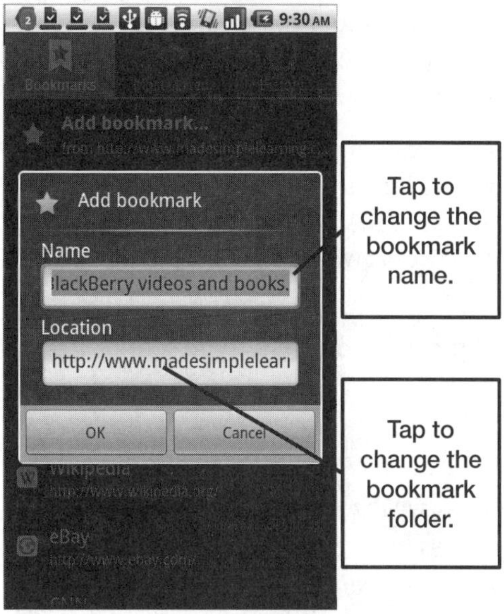

Using Bookmarks and History

Once you have set a few bookmarks, it is easy to view and work with them. In the same area, you can also see and use your web browsing history. A very useful tool on your DROID is the ability to browse the web from your **History**, just as you would on a computer.

1. Tap the **Bookmarks** icon  at the top of the page.
2. Swipe up or down to view all your bookmarks.
3. Tap any bookmark to jump to that web page.

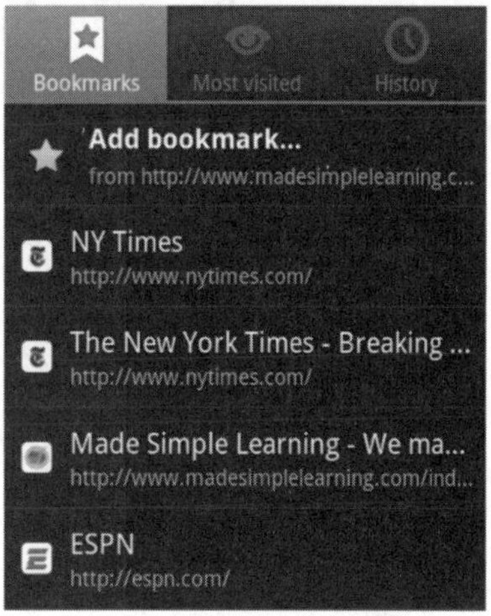

4. Tap the **History** button to view your recent history of visited web pages.

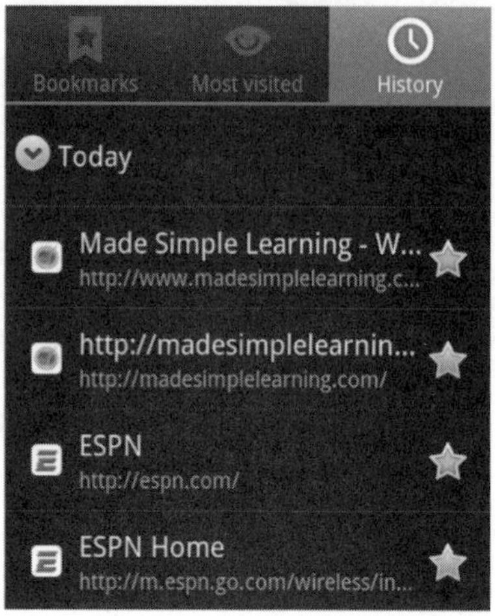

5. At the top of the list, you see **Today**, which shows the sites you visited today; at the bottom of the list, you see **Older**, which shows you previous days.
6. Tap any history item to go to that web page.

> **TIP:** To clear your history, tap the **Menu** button and then select **Clear history**. You can also clear your history, cookies, and cache in the **Settings** app. Tap **Menu** from an open Browser page and tap **More** then scroll down to **Settings**, scroll to the bottom and tap **Clear Cache, Clear History, Clear all cookie data, Clear form data, Clear location access** or **Clear passwords**.

Managing Your Bookmarks

It is very easy to accumulate quite a collection of bookmarks, since setting them up is so easy. You may find you no longer need a particular bookmark, or you may want to organize them by adding new folders.

Like other lists on your DROID, you can reorder your bookmarks' list and remove entries.

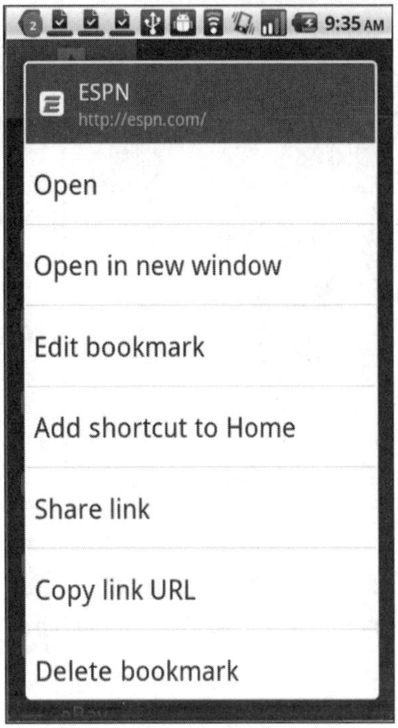

1. View your **Bookmarks** list as you did previously.
2. Touch and hold, then choose the **Edit bookmark** button to edit the name or location.
3. To delete a bookmark, follow the same procedure as above and simply scroll to **Delete bookmark**.
4. To copy the link URL, touch and hold the bookmark and scroll to **Copy link URL**.
5. To set the bookmark as your Home Page, follow the same procedure and scroll down to **Set as homepage**.
6. To toggle between thumbnail and list view, press the **Menu** button and choose the appropriate view.
7. When you are finished, tap the **Back** button and you will return to the previous web page.

Browser Tips and Tricks

Now that you know the basics of how to get around, we will cover a few useful tips and tricks to make web browsing more enjoyable and quicker on your DROID.

Finding Something on a Web Page

Sometimes, you need to find something specific on a particular web page. Fortunately, it is easy to get a **Find** on page search box. Just touch the **Menu** button and then touch **More** or touch and hold **Menu + f**, as shown in Figure 11–5. Then just type in the word or phrase you are searching for. This only works on DROIDs with physical keyboards.

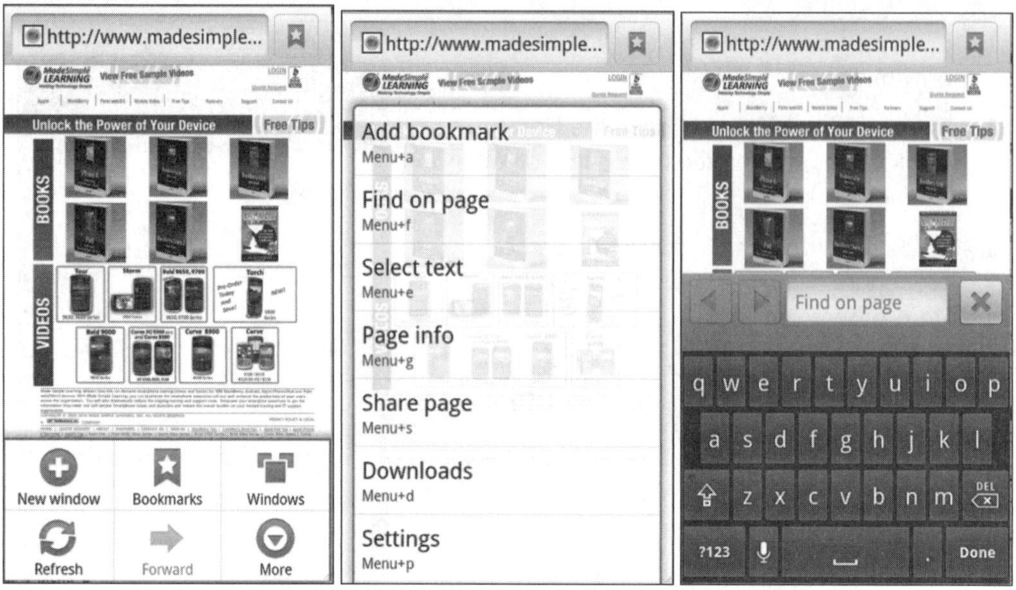

Figure 11–5. *Use the Find command on a page the browser.*

Emailing a Web Page

Sometimes while browsing, you find a page so compelling you just have to send it to a friend or colleague. Touch the **Menu** key and then touch **More** select **Share Page** (see Figure 11–6). You have the option to share the page via Bluetooth, Facebook, Gmail, Messaging, SiteShot or Twitter.

NOTE: You can also hold the Menu button and the **S** key to bring up the share menu.

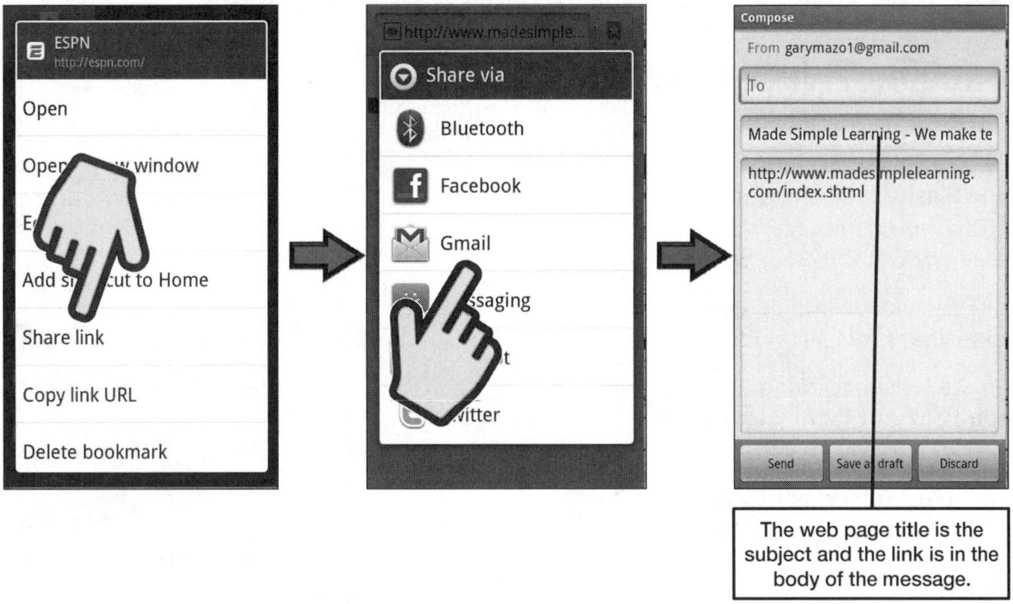

Figure 11–6. *Share a link to a web page.*

Printing a Web Page

The DROID (at the time of this writing) does not have a built-in **Print** command. You have a couple of options, but neither is very simple.

- *Option 1:* Email yourself or a colleague the web page link and print it from a computer. If you are traveling and staying at a hotel with a business center, you may be able to send it to someone at the business center or front desk to print the page.
- *Option 2:* Buy a network printing app from the **Android Market** that allows you to print to a networked printer. Of course, this only works if you have access to a networked printer. It's usually best if you do this from your home or office network and can get help setting up, as doing so can be challenging.

Watching Videos in Browser

You will find videos in many web sites. You will be able to play most but not all videos.

If the Flash Player installs properly on your DROID, you can play Flash videos right inside the web browser window.

YouTube videos should also play right inside your browser window.

Tap the screen to bring up the player controls if they have disappeared.

Some videos you click on may actually need to download onto the device. For these videos, pull down the notification window to monitor the download and touch on the file once it is downloaded. Your video player should launch and play the clip.

> **TIP:** Check out all the video player tips and tricks in Chapter 15: "Viewing Videos, TV Shows, and More."

Saving or Copying Text and Graphics

From time to time, you may see text or a graphic you want to copy from a web site. We tell you briefly how to do this in this section, but to see how to get it done graphically, including using the **Cut** and **Paste** functions, please see the "Copy and Paste" section in Chapter 2: "Typing, Voice, Copy and Search." Here's a quick look:

To copy text, touch the **Menu** button and then tap **More**. Choose **Select text**. Now, drag your finger across the screen to highlight text to copy. As soon as you let go, you will see the words **Text copied to clipboard** appear on the screen (see Figure 11-7).

> **CAUTION:** As soon as you release your finger from the screen, you will see the **Copy to clipboard** icon. Make sure you have all the text you want highlighted before you release your finger!

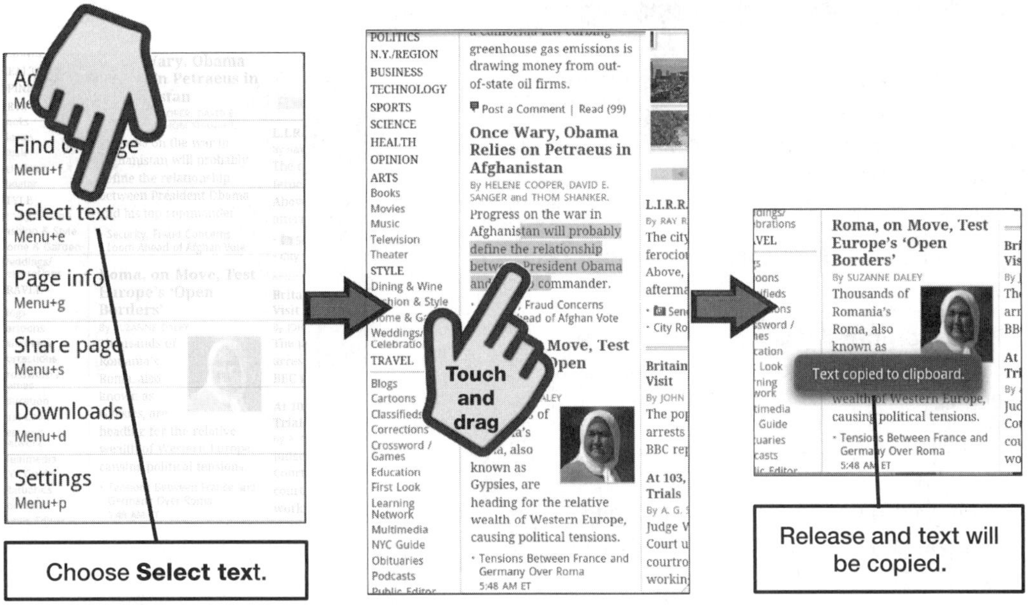

Figure 11–7. *Select and copy text from a web page to the clipboard.*

To **Save** or **Copy** a graphic, touch and hold the picture or image until you see the pop-up asking if you would like to **Save** or **View** the image or **Set as wallpaper**.

Remember Form Data and Passwords

Remembering form data and passwords is a great way to save time typing your personal information including usernames and passwords on web sites. The **Remember Form Data** tool can remember and fill in information required in web forms.

Once this is enabled, just go to any web page that has a field to fill out. if you have your DROID set to remember your login information, you should see it displayed already in the proper fields. Just touch the **Login** button to enter the site.

CAUTION: Having your name and password entered automatically means that anyone who picks up your DROID will be able to access your personal sites and information.

For Usernames and Passwords

The first time you go to a web site where you have to enter a username and password, you type them and press **Submit** or **Enter**. At that time, AutoFill will ask if you want to remember them.

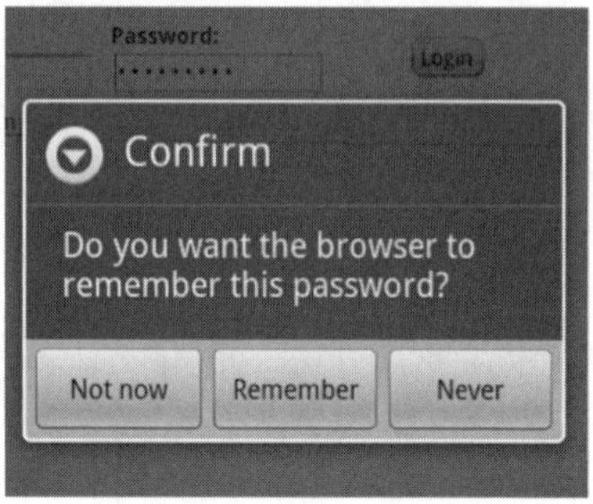

Your options are **Not now**, **Remember** or **Never**.

Tap **Remember** if you want them to be remembered and next time automatically entered.

The next time you visit this login page, your username and password will be automatically filled in.

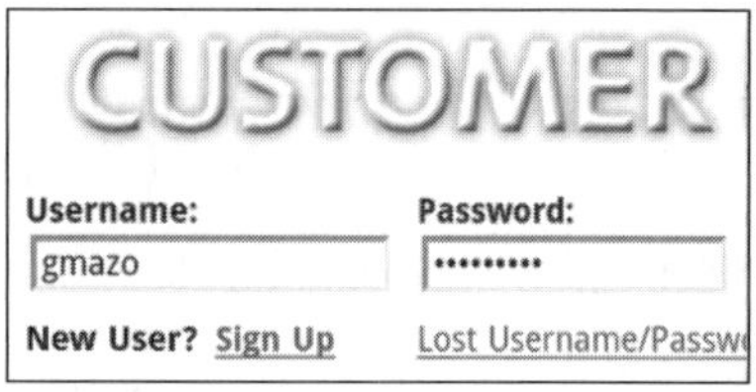

Adding a Web Page Icon to Your Home Screen

If you love a web site or page, it's very easy to add it as an icon to your **Home Screen**. That way, you can instantly access the web page without going through the **Browser ➤ Bookmarks** bookmark selection process. You'll save lots of steps by putting the icon on your **Home Screen.** This is especially good for quickly launching web apps, like Gmail or Buzz from Google, or web app games.

Here's how to add the icon:

1. Touch the **Bookmark** icon next to the web address.
2. Touch and hold any bookmark to see the menu list of options.
3. Choose **Add shortcut to home**.
4. Touch the **Home** button and the icon will be on the home screen.

Adjusting the Browser Settings

There are settings you can adjust in your Browser app.

1. Press the **Menu** button from any browser page and select **More**.
2. Select **Settings** at the bottom.
3. Choose to adjust **Text size**, **Default zoom**, **Text encoding**, **Enable plugins**, **Set home page**, and more from choosing the Drop down arrow button.
4. Choose to **Open pages in overview**, **Block pop-up windows**, **Load images**, **Auto-fit pages**, **Enable java script**, **Open in background** and more by placing a green "check" in the radio boxes

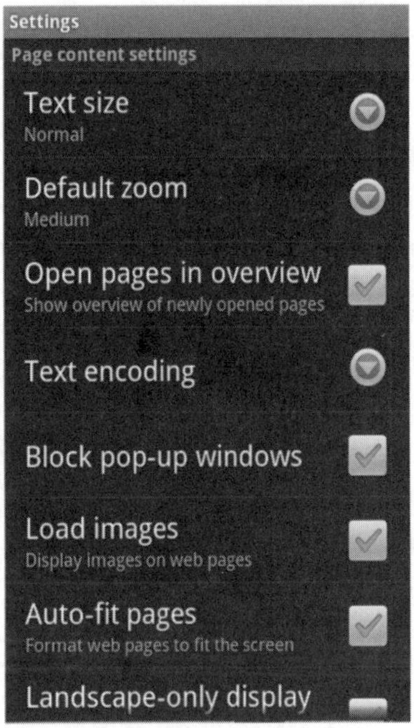

Adjusting Security Settings

Under the **Security settings** heading, **Remember passwords, Clear passwords** and **Show security warnings** can all be adjusted. You can modify any of these by either opening the drop down menu or putting a check in the appropriate radio box.

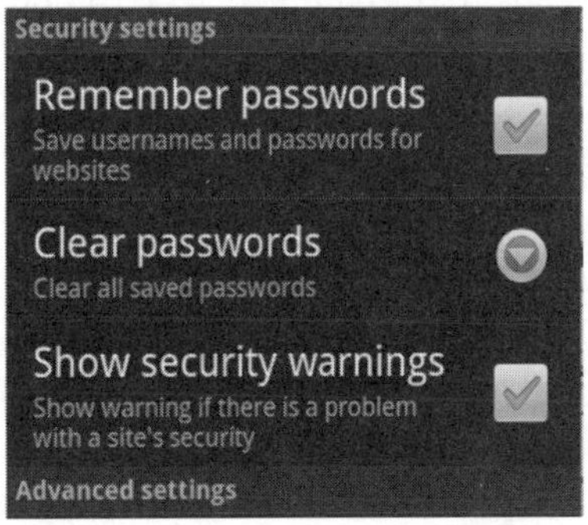

NOTE: Many popular sites like Facebook require JavaScript to be ON. Turn it on in the Page content settings.

Speeding Up Your Browser by Clearing History and Cookies

In the middle of the Browser settings screen, you can see a heading marked **Privacy settings**. The **Clear History**, **Clear all Cookie data,** and **Clear Cache** drop down arrows are all next to the appropriate item.

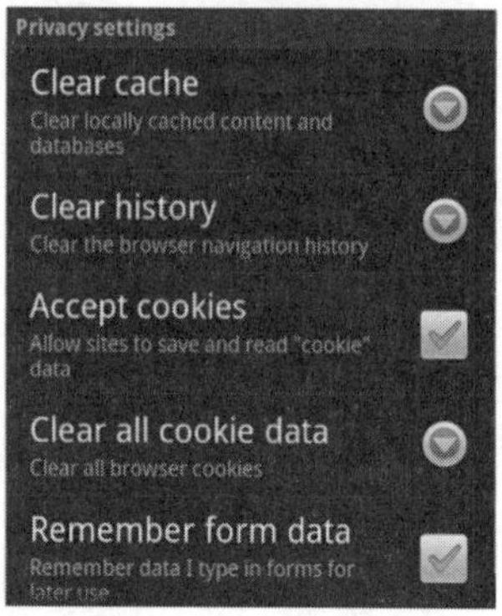

If you notice your web browsing getting sluggish, it's probably a good time to clear out all three of these by tapping them and confirming your choices.

TIP: Clearing the history, cookies, and cache is also a good privacy measure, as it prevents others from seeing where you've been browsing.

Remembering Form Data

As we showed you earlier in this chapter, remembering form data is a convenient way to have your browser automatically fill out web page forms that ask for your name, address, phone number, and even username and password. It can save you a tremendous amount of time typing and retyping your name and other information.

To enable **Remember form data**, follow these steps:

1. Go to the Browser settings as shown previously.

2. Scroll down to **Privacy settings**.
3. Put a check mark in the **Remember form data** box.

4. To remember all usernames and passwords, scroll down to **Security settings**.
5. Put a check mark in the **Remember passwords** box.

Chapter 12

Working with Contacts

Your DROID gives you immediate access to all your important information. Just like your computer, your DROID can store thousands of contacts for easy retrieval. In this chapter we'll show you how to add new contacts (including from an email message), customize your contacts by adding notes and nicknames, organize your contacts with groups, quickly search or scroll through contacts, and even show a contact's location with the DROID **Maps** app. We will also show you how to customize the **Contacts** view so it is sorted and displayed just the way you like it. Finally, you will learn a few troubleshooting tips that will save you some time when you run into difficulties.

The beauty of the DROID is how it integrates all of the apps so you can email and map your contacts right from the contact entry.

Loading Your Contacts onto the DROID

In Chapter 3: "Sync Your DROID to Your Google Account," we show you how to load your contacts onto the DROID using your Mac or Windows computer. You can also use your Google account information to seamlessly and wirelessly sync at all times. Various snyc methods are described in the Chapter 3: "Sync Your DRIOD to your Google Account" and in Chapter 4: "Other Sync Methods."

TIP: You can add new contact entries from email messages you receive. Learn how in the "Adding Contacts from Email Messages" section later in this chapter.

When Is Your Contact List Most Useful?

The **Contacts** app is most useful when three things are true:

1. You have many names and addresses in it.
2. You continually add new information as it becomes available.
3. You can easily find contacts.

Two Simple Rules to Improve Your Contact List

Here are a couple of basic rules to help make your contact list on your DROID more useful.

Rule 1: Add anything and everything to your contacts.

You never know when you might need that obscure restaurant name, or that plumber's number, etc.

Rule 2: As you add entries, make sure you think about how to find them in the future (First name, Last name, Company).

We have many tips and tricks in this chapter to help you enter names so that they can be instantly located when you need them.

> **TIP:** Here's a good way to find restaurants. Whenever you enter a restaurant into your contacts list, put the entire restaurant name in the First name field and type the word "restaurant" into the Last name field. Then when you type the letters "rest," you should instantly find all your restaurants!

Adding a New Contact Right on Your DROID

You can always add your contacts right on your DROID. This is handy when you're away from your computer—but have your DROID—and need to add someone to your contacts. It's very easy to do. Here's how.

Start the Contacts App

1. From your Home screen, touch the **Contacts** icon and then the **Menu** button. Tap the **New contact** button to add a new contact, as shown in the figure to the right.

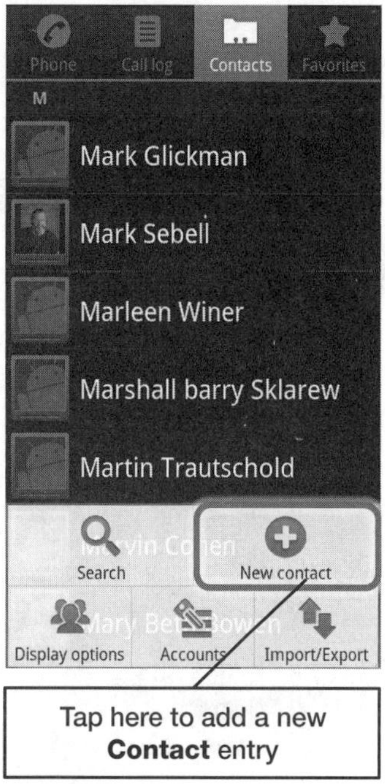

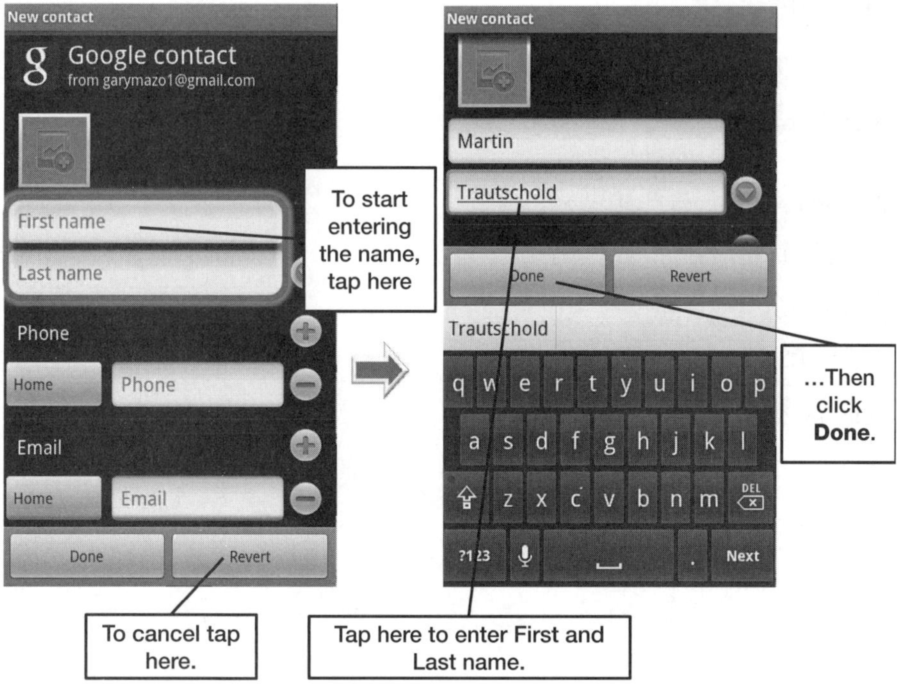

Figure 12–1. *Entering a new contact name.*

2. Touch the **First name** and **Last name** buttons to enter the new contact's first and last names. You can also add a company name by touching the green + sign next to **Organization**.

> **TIP**: Keep in mind that the contacts search feature uses first, last, and company names. When you add or edit contacts, adding a special word to the company name can help you find a particular contact later. For example, adding the words "Cece friend" to the **Company** field can help you find all of Cece's friends quickly using the search feature.

3. Under the **First Last** button are more buttons, as shown in Figure 12–2. Each is activated by touching either the drop-down arrow or green + to the right of the category name. Touch the green + again to add another line—for example, a second phone number or email.

Figure 12–2. *Available contact fields.*

Adding a New Phone Number

Touch the **Phone** field and use the number keyboard to input the phone number.

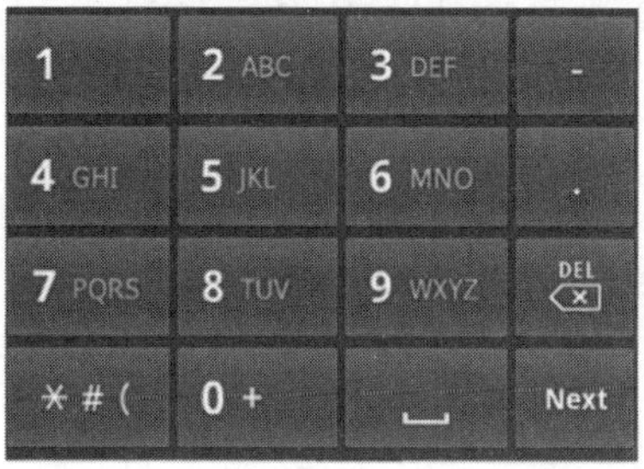

TIP: Don't worry about parentheses, dashes, or dots—the DROID will put the number into the correct format. Just type the digits of the area code and number. If you know the country code, it's a good idea to put that in as well.

Next, choose which type of phone number it is. There are nine fields you can choose from, including an **Other** field if you find that none of the built-in fields apply.

TIP: Sometimes you need to add a pause to a phone number—for example, when the phone number is for someone at an organization that requires you to dial the main number and then an extension. This is easy to do on the DROID. You just add a Pause which shows up as a comma between the main number and the extension like this: 386-555-7687, 19323. To add a pause, tap the * # (key in the lower left corner of the keypad and tap Pause. When you dial this number from your DROID, the phone would dial the main number, pause for two seconds, and then dial the extension. If you need a longer pause, simply add more commas.

Adding an Email Address and Web Site

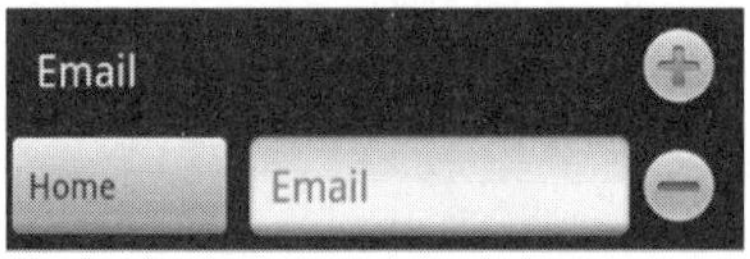

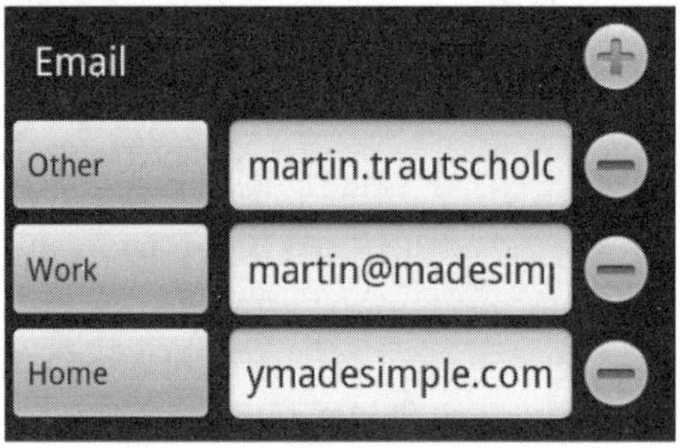

Touch the **Email** field and enter the email address for your contact. You can also touch the tab to the left of the email address and select whether this is a home, work, or other email address.

Under the **Organization** field you'll also find a **More** button. Touch the **More** button and scroll to the bottom. Touch the green + sign and a new field will pop up for the address of your contact's web site.

NOTE: If you use Facebook on the DROID, it will automatically look for a Facebook homepage to integrate into the contact info.

TIP: Suppose you met someone at the bus stop—someone you wanted to remember. Of course, you should enter your new friend's first and last names (if you know it), but also enter the words "bus stop" in the **Company name** field. Then when you type the words "bus" or "stop," you should instantly find everyone you met at the bus stop, even if you can't remember their names!

Adding the Address

Below the **Email** field are the fields for adding the **Postal** address. Input the **Street**, **City**, **State** and **Zip Code**. You can also specify the **Country** and whether this is a home or work address.

When you are done, just touch the **Done** button right below the **Address** field.

Adding a Photo to Contacts

From the New Contact screen we've been working in, just touch the **Add Photo** icon at the top of the Contact editing window.

After you touch the **Add Photo** icon, you'll see that you can

- Take a Photo
- Select a Photo from Gallery

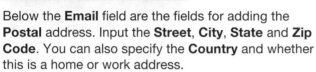

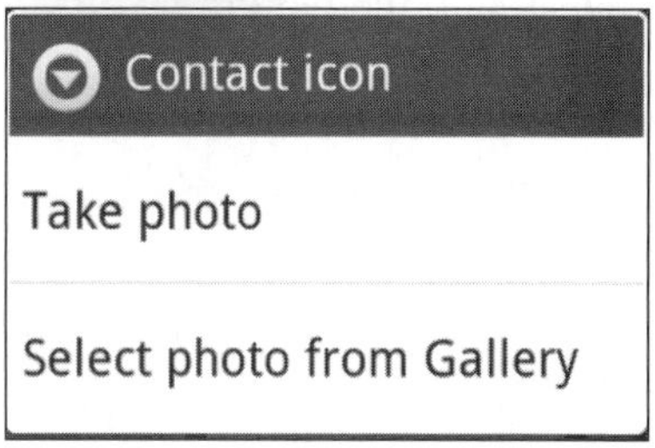

If there's a photo already in place, you can

- Remove a Photo icon
- Change a Photo icon

To choose an existing photo, select the photo album where the picture is located and touch the corresponding tab. When you see the picture you want to use, just touch it.

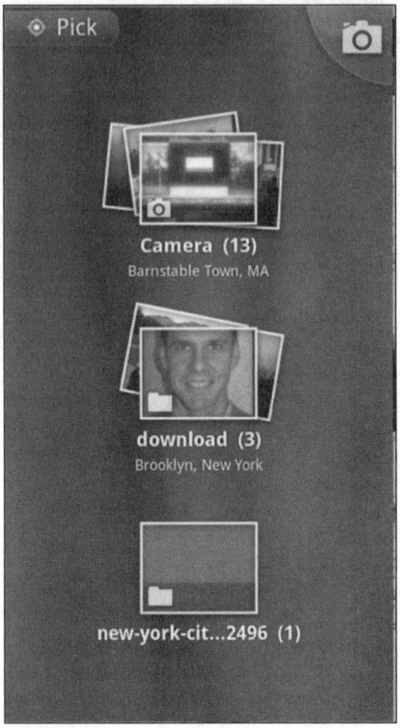

You'll notice that the top and bottom of the photo are now grayed out and that you can manipulate the picture by moving it and then arranging it in the picture window.

Once the picture is sitting where you want it, touch the **Save** button in the lower left corner and that picture will be set for the contact.

TIP: If you just moved into a new neighborhood, it can be quite daunting to remember everyone's name. A good practice to follow is to add the word "neighbor" into the **Company Name** field for every neighbor you meet. Then, to instantly call up all your neighbors, simply type the letters "neigh" to find everyone you've met!

Searching Your Contacts

Let's say you need to find a specific phone number or email address. Just touch your **Contacts** icon as you did previously and then touch the **Search** button on the DROID and you'll see a search box at the top of your **Contacts** list, as in Figure 12–3.

Figure 12–3. *The contacts search box.*

NOTE: On your DROID you may not see the Android contact icon to the left of your search window, but it still works the same way!

Enter the first few letters of any of these three searchable fields:

- First Name
- Last Name
- Company Name

The DROID begins to filter immediately and displays only those contacts that match the letters typed.

> **TIP:** To further narrow the search, hit the space key and type a few more letters.

When you see the correct name, just touch it and that individual's contact information will appear.

Quickly Jump to a Letter by Scrolling and Sliding on the Contact Card

If you start scrolling through your contacts, you will see a small Contact Card icon on the right edge of the screen. Drag it up or down, and you can quickly advance by letter through the alphabet.

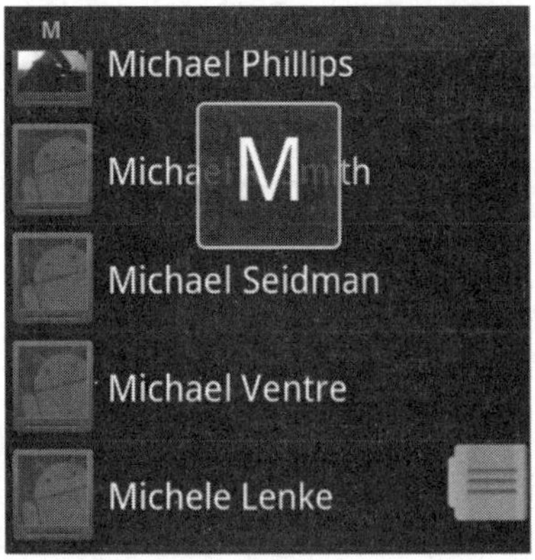

Search by Flicking

If you don't want to manually input letters, you can just move your finger and flick from the bottom up, and you'll see your contacts move quickly on the screen. Just continue to flick or scroll until you see the name you want. Tap the name and the contact information will appear.

Adding Contacts from Email Messages

Often you'll receive an email message and realize that the contact is not in your address book. Adding a new contact from an email message is easy.

Open the email message from the contact you'd like to add to your contacts list. Then, in the email message's **From** field, just touch the **Android** icon next to the name of the sender next to the **From:** tag.

If the sender is not in your address book, you'll be taken to a screen that lets you choose whether to add that email address to an existing contact or to create a new one.

If you select **Add contact**, you'll be taken to the same New Contact screen you saw earlier (Figure 12–1). Just make sure you select **Create new contact** at the top.

Once you select **Create new contact**, you will see the contact editing screen. Add any other pertinent information for this contact (the email and name will automatically be put in) and then select **Done**.

Linking Contacts to Another App

You might have contact information for the sender of the email message in another app on the phone. With the DROID it is easy to link these contacts together.

In this example, Martin, the sender of the email message, is one of my Facebook contacts, and I want to link his picture and birthday to my DROID contact information. Here is how I can link his contact information in my DROID to the information I have in Facebook.

1. I add him to my contacts, as shown previously.
2. I start up my **Facebook** app. See Chapter 20: "Social Networking and Skype" for more information on the topic.
3. I find my contact information for Martin to verify that he is in my **Facebook** app.
4. I touch the **Menu** button.
5. I choose **Settings** in the lower left-hand corner.
6. I then scroll down to **Sync Contacts** (see Figure 12–4).
7. I choose one of the following options: **Sync all, Sync with existing contacts.** or **Remove Facebook data**. In this case, I will choose **Sync with existing contacts**.
8. Martin's picture and updated information are then brought into his contact information on my DROID.

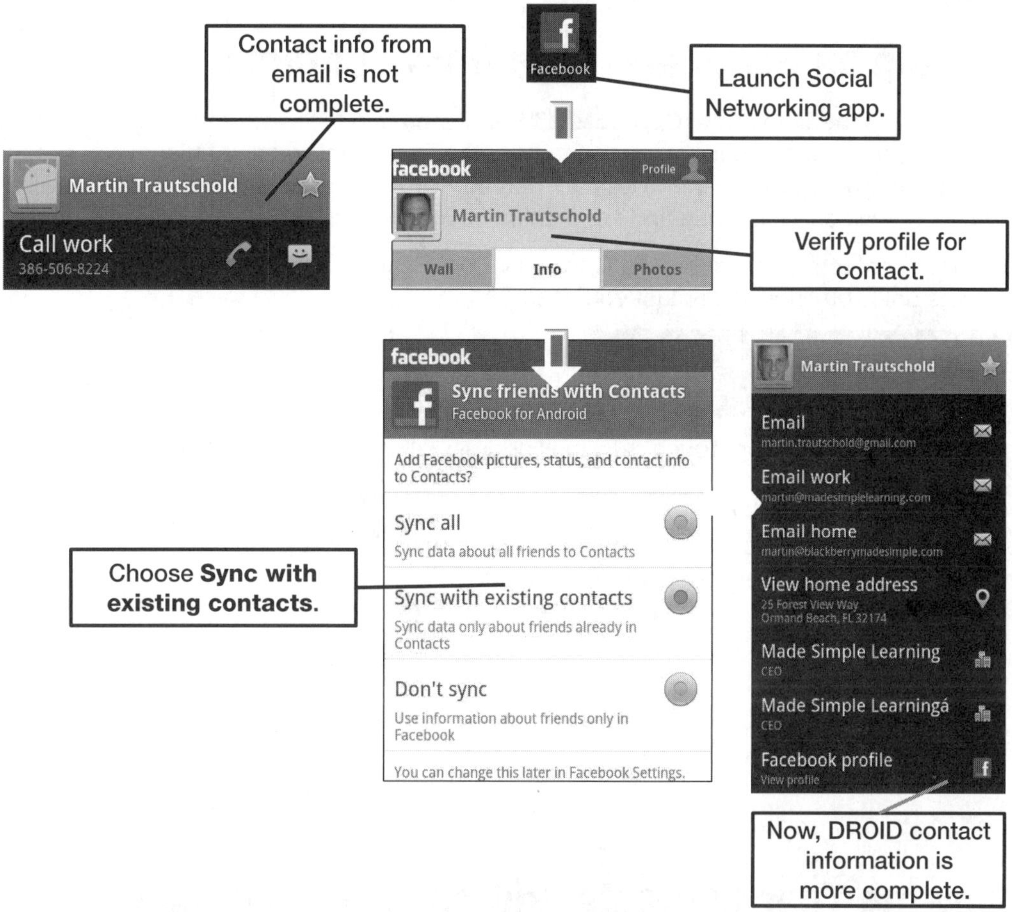

Figure 12–4. *Linking social networking contact to an existing contact profile.*

TIP: Learning the names of parents of your school-age children's friends can be fairly challenging. In the **First** field, however, you can add not just your child's friend's name but the parents' names as well (e.g., **First: Samantha (Mom: Susan, Dad: Ron)**). Then in the **Company** field, add in the name of your child and "school friend" (e.g., **Cece school friend**). Just typing your child's name in your **All Contacts** list's search box brings up every person you ever met at your child's school. Now you can say, "Hello, Susan, great to see you again!" without missing a beat. *Try your best to covertly look up the name.*

Sending a Picture to a Contact

If you want to send a picture to a contact, you can do that from the **Photos** app. (See Chapter 18: "Taking Photos and Videos.")

Sending an Email Message from Contacts

Since many of the core apps (**Contacts**, **EMail**, **Gmail** and **Messages**) are fully integrated, one app can easily trigger another. So, if you want to send an email message to one of your contacts, open the contact and tap the email address. The **Mail** app will launch, and you can compose and send an email message to this person.

Start your contacts by touching the **Contacts** icon. Either search or flick through your contacts until you find the contact you need.

In the contact information, touch the email address of the contact you'd like to use.

You'll see that the **Email** program launches automatically with the contact's name in the **To:** field of the email message. Type and send the message.

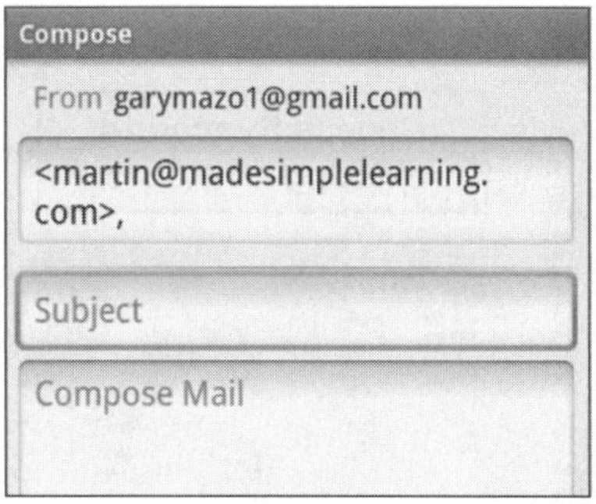

Showing Your Contacts Addresses on the Map

One of the great things about the DROID is its integration with Google Maps. This is very evident in the **Contacts** app. Let's say you want to map the home or work address of any contact in your address book. In the old days (pre-DROID), you'd have to use Google, MapQuest, or some other program and laboriously retype or copy and paste the address information. This is very time-consuming—but you don't have to do this on the DROID.

Simply open the contact as you did earlier. This time, touch the address at the bottom of the contact information.

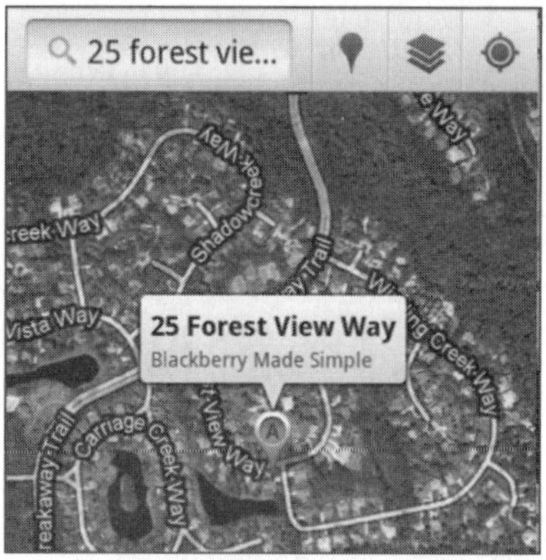

Your **Maps** app (which is powered by Google Maps) immediately loads and drops a marker at the exact location of the contact. The contact name will appear above the marker. (The satellite imagery is ©2010 Google.)

> **NOTE:** The old company name of **Made Simple Learning** was **BlackBerry Made Simple**, that's why it is showing up on Google Maps.

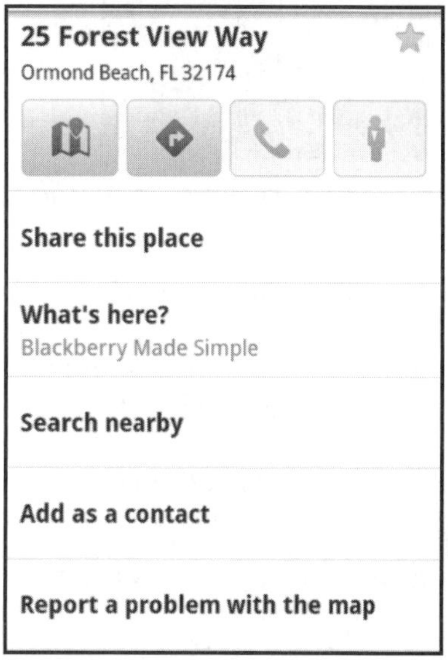

Touch the tab on the top of the marker to get to the info screen.

Now you can select **Share this place** or **Search nearby**.

Touch the **Directions** icon and then touch the **Driving Navigation**, **Walking Navigation**, or **Get directions**.

What if you had just typed the address into your **Maps** app instead of clicking from your contact list? In that case, you might want to touch **Add as a Contact** to add this address.

> **TIP:** To return to your contact information, tap the **Map** button and then touch the **Back** button.

Contact History and Social Networking Screens

On your DROID 2 and DROID X, you can see Contact History and Social Networking feeds for your contacts by swiping left and right from the contact detail screen. (See Figure 12-5.) You can also see history and social networking information for all your contacts together if you swipe left or right from the Contact List view.

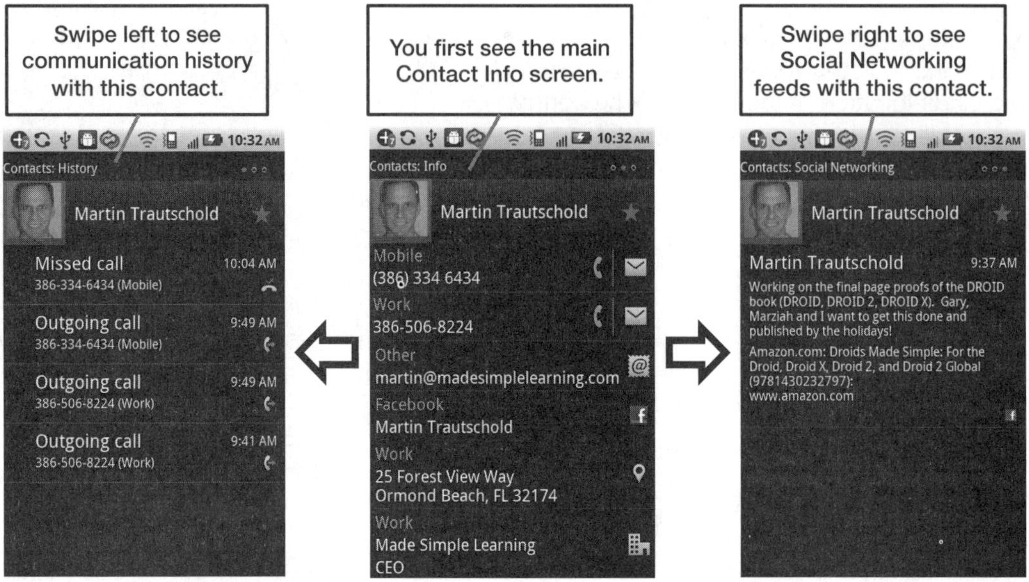

Figure 12–5. *Swipe left or right to see Contact History and Social Networking screens.*

Contacts Troubleshooting

Sometimes, your **Contacts** app might not work the way you expect. If you don't see all your contacts, review the steps in the Chapter 3: "Sync Your DRIOD to your Google Account" or Chapter 4: "Other Sync Methods" on how to sync with your address book application. Make sure you have selected **All Groups** in the settings in your contact management app.

TIP: If you are syncing with another contact application, such as **Contacts** in Gmail, make sure you select the option closest to **All Contacts** rather than a subset like a particular group.

Chapter 13

Managing Your Calendar

The DROID makes the old calendar that used to hang on the fridge obsolete. In this chapter, we will show you how to utilize the **Calendar** app of the DROID to its full potential. We will show you how to schedule appointments, how to manage multiple calendars, how to change views on your calendar, and even how to deal with meeting invitations.

NOTE: For most of this chapter, we will talk about syncing your DROID calendar with another calendar because it is nice to have your calendar accessible on your DROID and other places. If you choose, you can also use your DROID in a *standalone* mode, where you do not sync to any other calendar. In the latter case, all the steps we describe for events, viewing, and managing events still apply equally to you.

Managing Your Busy Life on Your DROID

The **Calendar** app is a powerful and easy-to-use application that helps you manage your appointments, keep track of what you have to do, set reminder alarms, and even create and respond to meeting invitations for Microsoft **Exchange** users.

Accessing Your Calendar

The **Calendar** icon is usually visible once you touch your **Launcher** icon. Once all applications are visible on the **Home** screen, scroll to the **Calendar** icon.

TIP: If you use your DROID's **Calendar** app often, you can also create a **Calendar** widget (see Chapter 6: "Organize your Home Screens: Icons and Widgets"); this **Calendar** widget will show today's date and any upcoming appointments for the day.

If you maintain a calendar on your computer or on a website such as **Google Calendar**, you can synchronize or share that calendar with your DROID (see Chapter 3: "Sync Your DROID with Your Google Account" and Chapter 4: "Other Sync Methods" for more information on syncing).

After you set up the calendar sync, all of your PC, Mac or online calendar appointments will be synced with your DROID calendar automatically, based on your sync settings (see Figure 13–1).

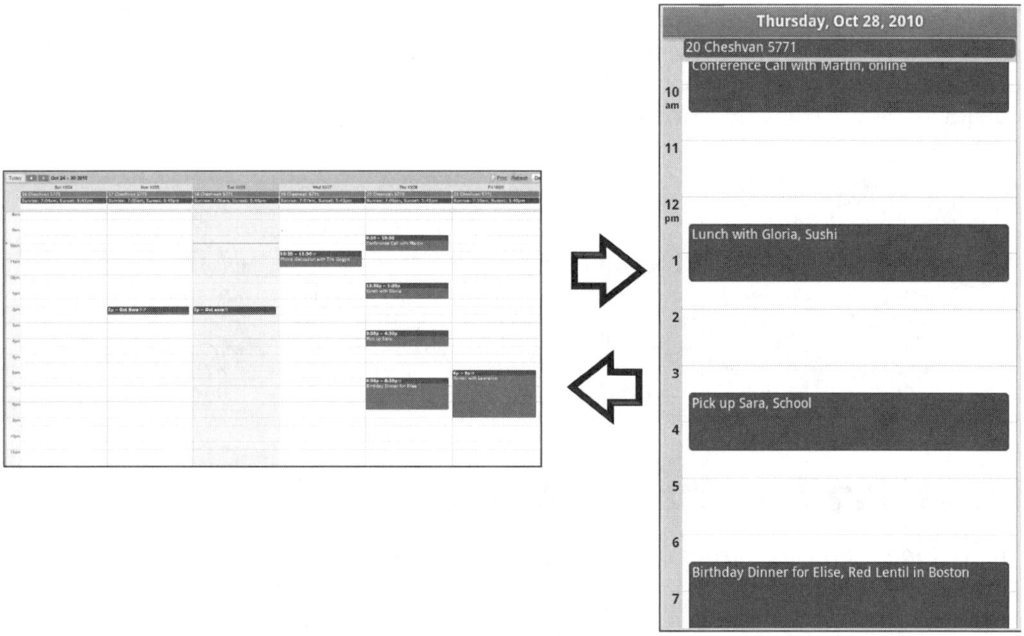

Figure 13–1. *Syncing a PC, Mac, or online calendar to a DROID.*

Viewing Your Schedule and Getting Around

The default view for the **Calendar** app shows your **Day** view. This view shows you at a glance any upcoming appointments for your day. Appointments are shown in your calendar (see Figure 13–2). If you happen to have multiple calendars set up on your computer, such as **Work** and **Home**, then appointments from the different calendars will display as different colors on your DROID calendar.

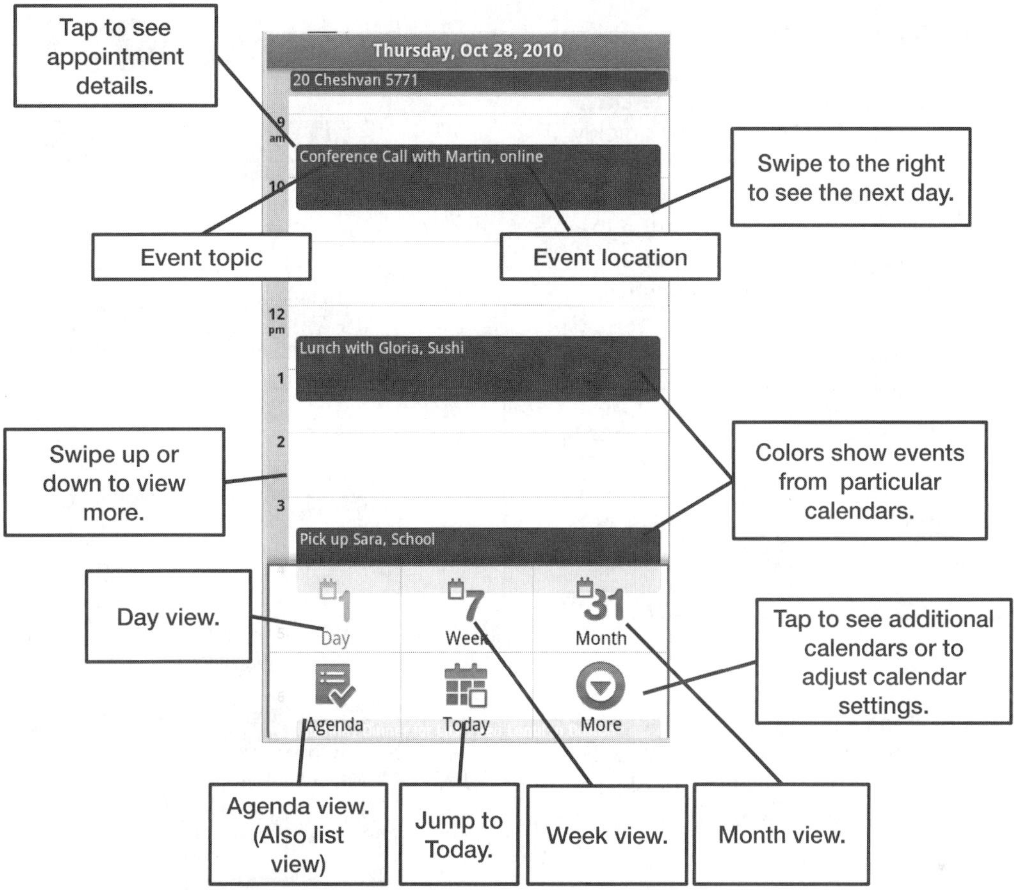

Figure 13–2. *The **Calendar** app's **Day** view layout.*

You can manipulate the calendar in various ways:

- **Move a day at a time**: If you swipe left or right, you move forward or backward a day.
- **Change views**: Press the **Menu** key and then use the **Agenda**, **Day**, **Week**, and **Month** buttons at the bottom to change the view.
- **Jump to today**: Press the **Menu** key and then touch the **Today** button located in the bottom-center of the screen.

Switching Between the Four Calendar Views

Your **Calendar** app comes with four views: **Day**, **Week**, **List** (**Agenda**), and **Month**. You can switch views by pressing the **Menu** button and selecting the view.

- **Day view**: When you start the DROID's **Calendar** app, the default view is usually the **Day** view. This allows you to quickly see everything you have scheduled for the day. You can bring up buttons to change the view by pressing the **Menu** button; your view options will appear at the bottom of the **Calendar** app.

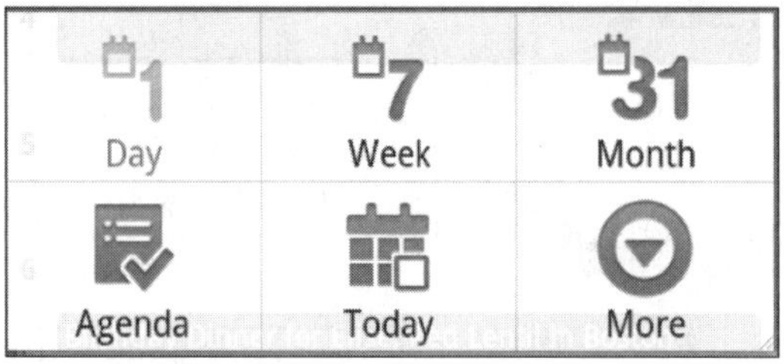

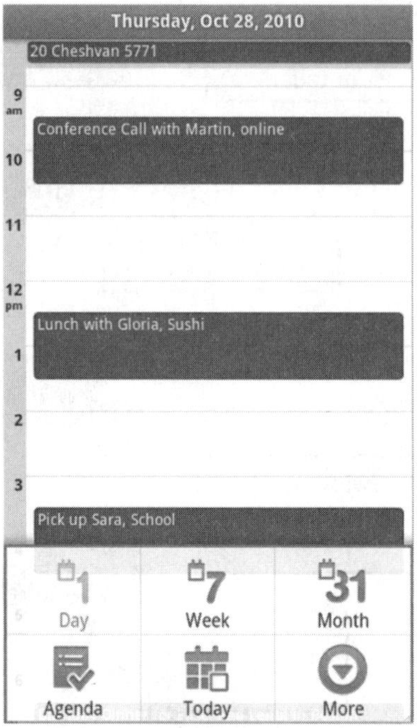

- **Week View:** Touch the **Week** button to see an overview of the current week. Different colored boxes denote appointments.
- **Agenda view**: Touch the **Agenda** button at the bottom to see a list of your appointments.

  Depending on how much you have scheduled, you could see the next day's or even the next week's worth of scheduled events.

  Swipe up or down to see more events.

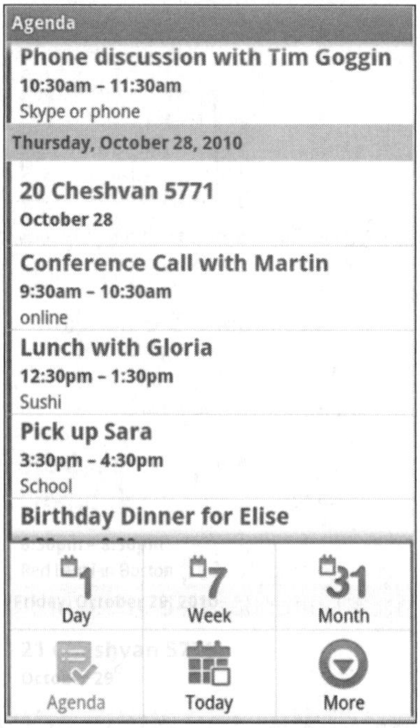

- **Month view**: Touch the **Month** button at the bottom to see a layout of the full month. Days with appointments have a small dot in them.

TIP: To return to the **Today** view, press the **Menu** button and touch the **Today** button at the bottom of the screen.

You can go to the next month by swiping up; you can go to the previous month by swiping down to return to the previous month.

Adding New Calendar Events

You can easily add new events or appointments right on your DROID. These new events and appointments will be synced (i.e., shared with) your computer the next time the sync takes place.

Adding a New Appointment

As you might expect, you simply long-press or double tap the screen at a particular time to set an appointment.

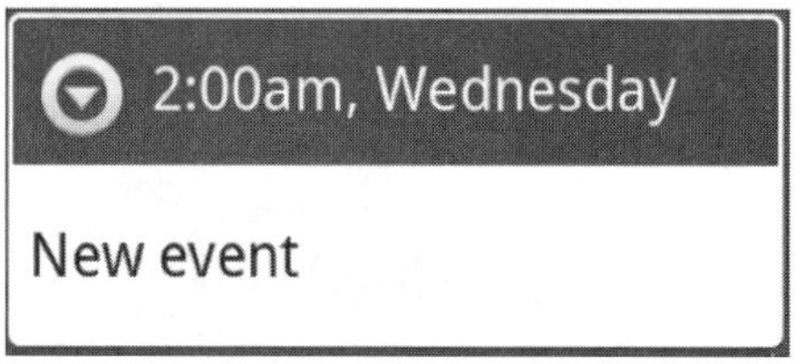

To add a new calendar event from any **Calendar** view, follow these steps:

1. Touch the day or time block (while in **Day** view) for which you want to schedule an appointment and then touch **New event**. The **Event details** screen will be shown.

2. Next, touch the box marked **What.**

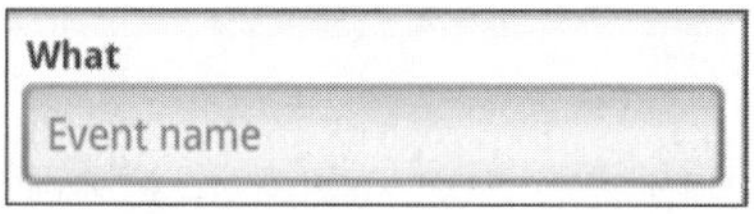

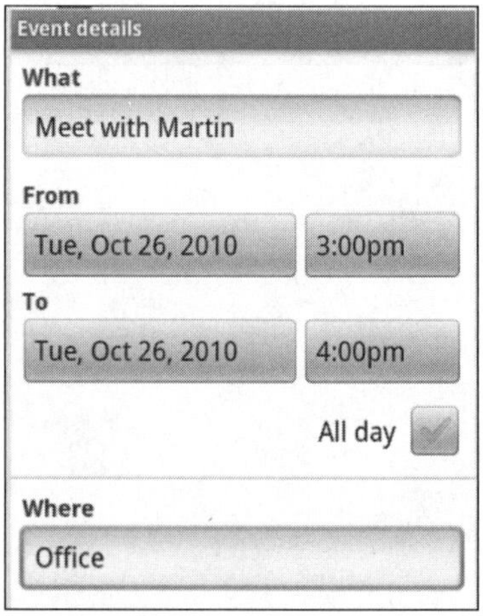

   Type in a title for the event, then scroll down to the **Where** box and type in a location. For example, you might type "Meet with Martin" as the title and input the location as "Office." Or, you might choose to type "Lunch with Martin" and then choose a very expensive restaurant in New York City.

3. Touch the **From** or **To** tab to adjust the timing of the event. To change the date, touch the day and date field to bring up the month, date, and year adjustments. Next, touch the **+** or **-** buttons above and below to adjust the date and start time of the appointment. Touch the **Time** field and set the starting and ending times. When done, touch the **Set** key.

4. The **Where** box includes a **Description** box; enter the event's location in the latter box.

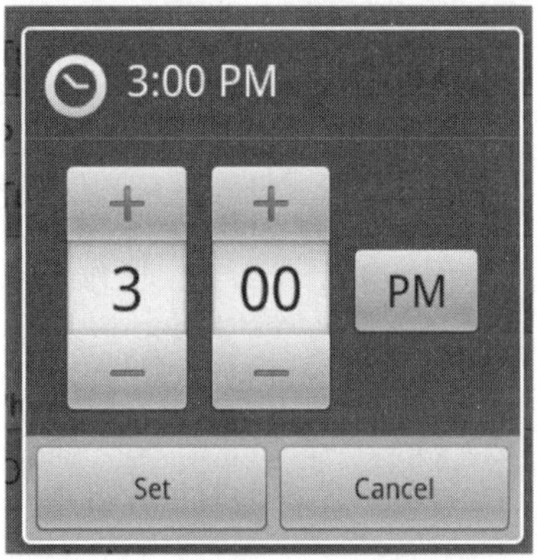

5. Alternatively, you can set an all-day event by touching the box next to **All-day**; this sets the switch to **ON**.

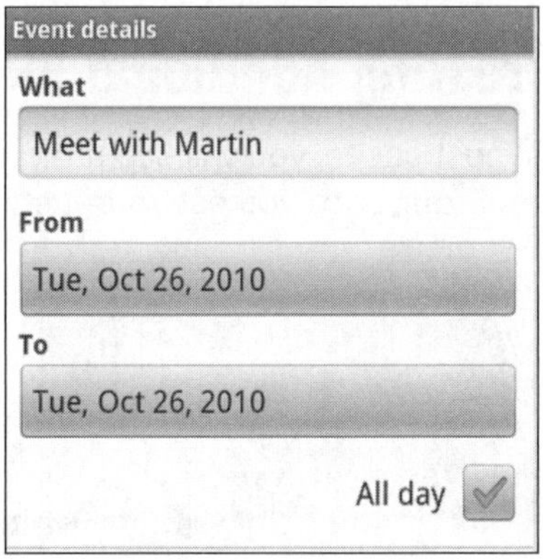

Setting Calendar Reminders

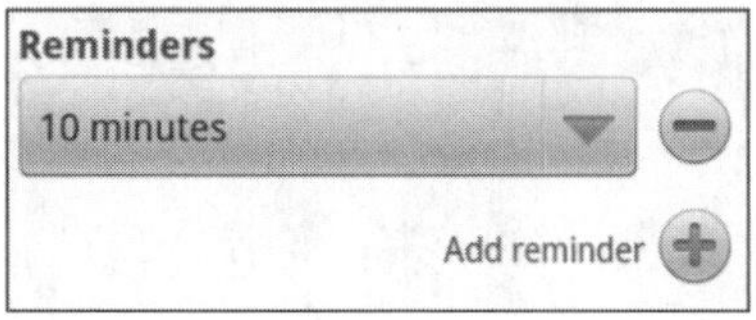

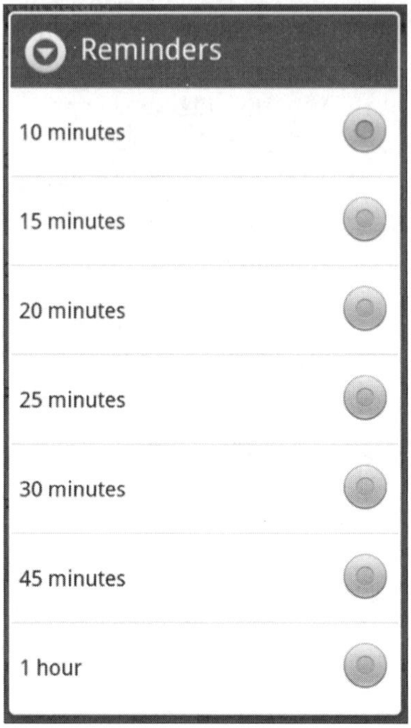

You can have your DROID give you an audible reminder, or *alert*, about an upcoming appointment. Alerts can help you keep from forgetting an important event. Follow these steps to create an alert:

1. Touch the **Reminders** tab and then select the option for a reminder alarm. You can have no alarm at all or set a reminder anytime from one minute before the event all the way to one week before it, depending on what works best for you.
2. Once you make your selection, you will automatically return to the **Event** screen.

Creating Additional Alerts

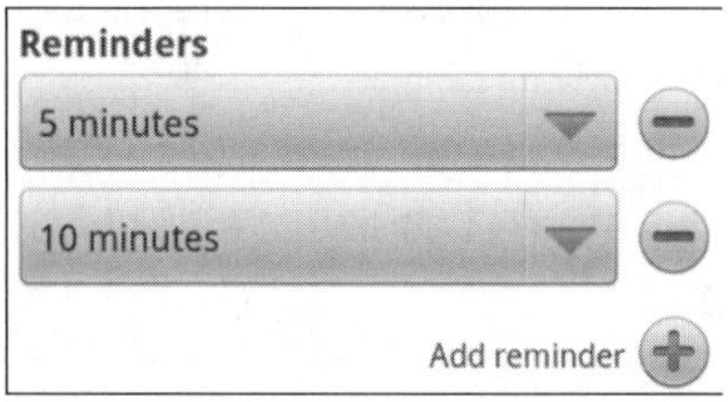

In most cases, you will see a tab that says **Add reminder** once you set your first reminder.

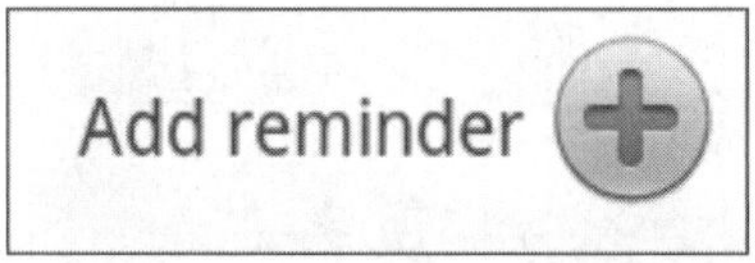

NOTE: You can set up to five reminders for any appointment.

You can set an additional reminder to another time before or after the first reminder. Some people find a second alert very helpful for remembering critical events or appointments.

TIP: Here's a practical example that illustrates when you might want to set up two calendar reminders.

If your child has a doctor or dentist appointment, then you might want to set the first reminder to go off the night before. This will remind you to write a note to the school and give it to your child.

You can then set the second reminder for 45 minutes prior to the appointment time. This will leave you enough time to pick up your child from school and get to the appointment.

Adding Recurring Events

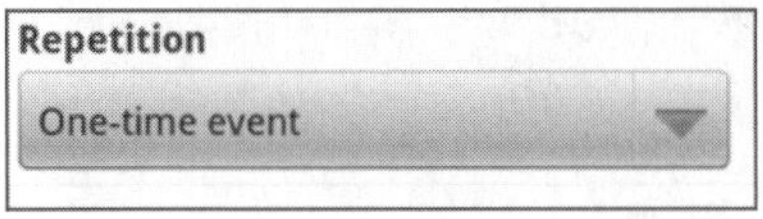

Some of your appointments happen every day, week, or month at the same time. Follow these steps if you are scheduling a repeating or recurring appointment:

1. Touch the **Repetition** tab and then select the correct time interval from the list.
2. Once you select the repetition, you will return to the main **Event** screen.

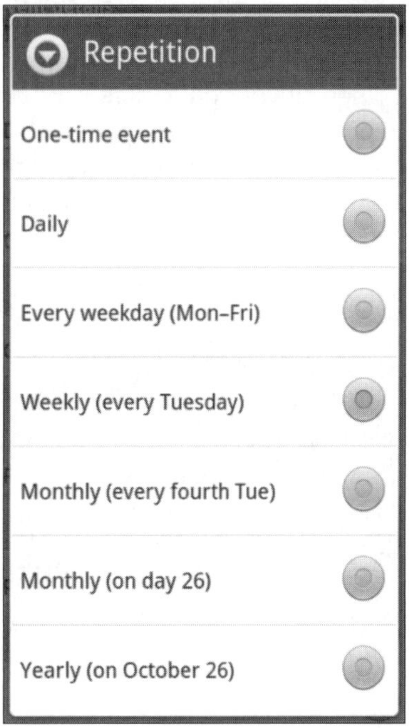

Choosing Which Calendar to Sync

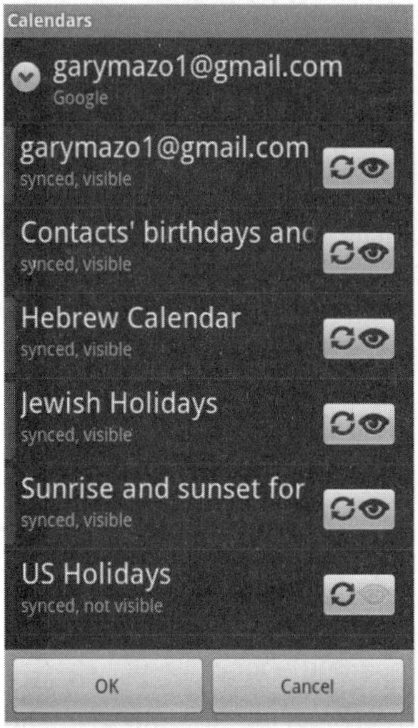

If you use more than one calendar in **Outlook**, **Google**, **iCal**, or some other program, then you will have various calendars available to you when you sync your DROID with that program.

> **NOTE:** If you create an event and choose an **Exchange** or **Google** calendar, then you'll see an option to invite other users to the event.

To see all your calendars, touch the **Menu** button, touch **More**, and then touch **Calendars**. Tap the calendar you want to use to highlight the **Sync** and **Visible** icons; this ensures that events are synced and visible in your DROID calendar.

> **NOTE:** To schedule an appointment in a calendar other than the default, start a new event and touch the **Calendar** dropdown. You should see your available calendars for scheduling the new appointment.
>
> Touch the specific calendar you want to place the new appointment under.

Editing Appointments

Sometimes, the details of an appointment may change and need to be adjusted (see Figure 13–3). Fortunately, it's easy to revise an appointment on your DROID:

1. Tap the appointment that you want to change.
2. Press the **Menu** key and then tap the **Edit** button to see the **Edit** screen showing the appointment details.

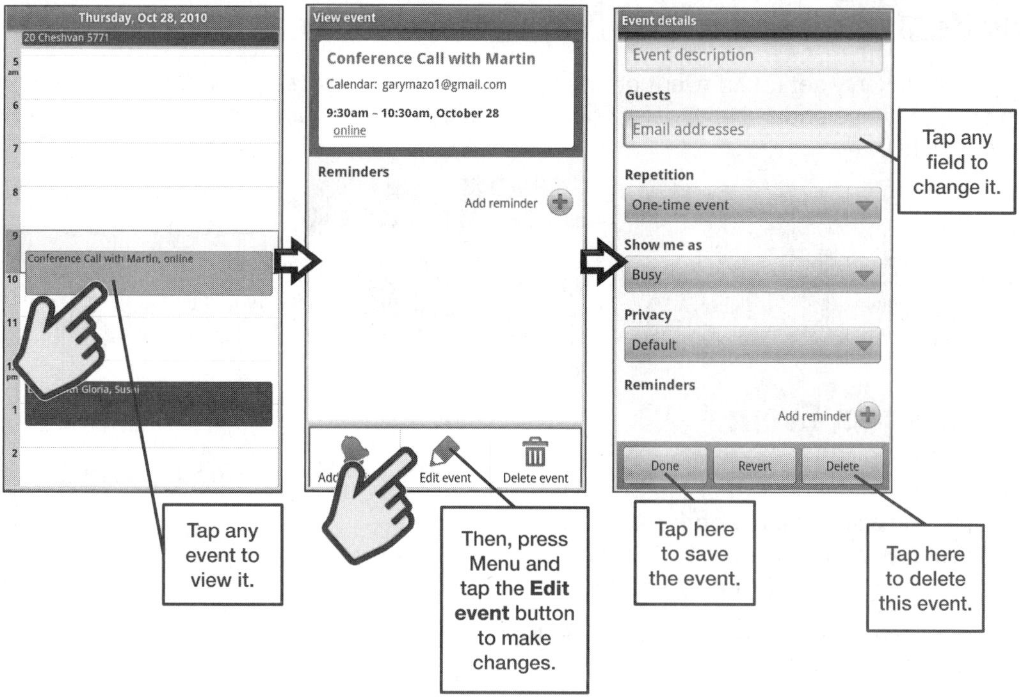

Figure 13–3. *Editing an appointment.*

3. Now just touch the tab in the field you need to adjust. For example, you can change the time of this appointment by touching the **From** or **To** tab, and then adjusting the time for the event's starting or ending time. Any field can be adjusted or changed.

4. When you are done, touch the **Done** button in the lower-left corner.

Editing a Repeating Event

You edit a recurring or repeating event in exactly the same manner as any other event. The only difference is that you will be asked a question before you edit the event.

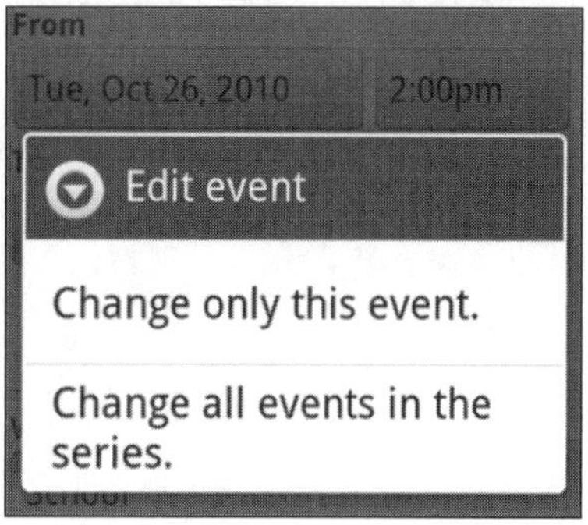

Tap **Change only this event** if you want to make changes to only this instance of the repeating event.

Tap **Change all events in the series** if you want to make changes to all instances of this repeating event.

Switching an Event to a Different Calendar

If you mistakenly set up an event on the wrong calendar, then tap the **Calendar** button to change the calendar. Next, select one of the different calendars you have synced to your DROID.

> **NOTE**: Remember that only those calendars configured to sync with your computer will update when selected. If you switch to a calendar that doesn't sync with your computer, the change will only be visible on the DROID, and the event will no longer display on your computer.

Deleting an Event

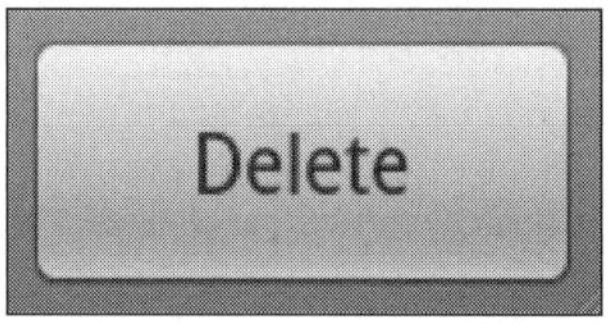

Notice that, at the bottom of the **Edit** screen, you also have the option to delete this event. Simply touch the **Delete** button at the bottom of the screen to do so.

You can also delete events from the Event Details screen by pressing the **Menu** button and selecting **Delete event**.

Accepting Meeting Invitations

For those who use **Microsoft Exchange**, **Microsoft Outlook**, or **Entourage** regularly, meeting invitations become a way of life. If you receive a meeting invitation in your email, accepting the invitation automatically places the appointment in your calendar.

On your DROID, you will see the invitations you accept placed into your calendar immediately.

> **NOTE:** If you use an **Exchange** calendar or a **Google** calendar, you can invite people and reply to meeting invitations on your DROID (see Chapter 4: "Other Sync Methods" to learn more about this subject).

When a meeting invitation goes to your Gmail or Exchange account, you will receive an email with the invitation; however, the meeting will automatically get placed into your calendar, as shown to the right.

Just pull down the **Attending** drop-down window and respond with either **Yes**, **Maybe**, or **No**.

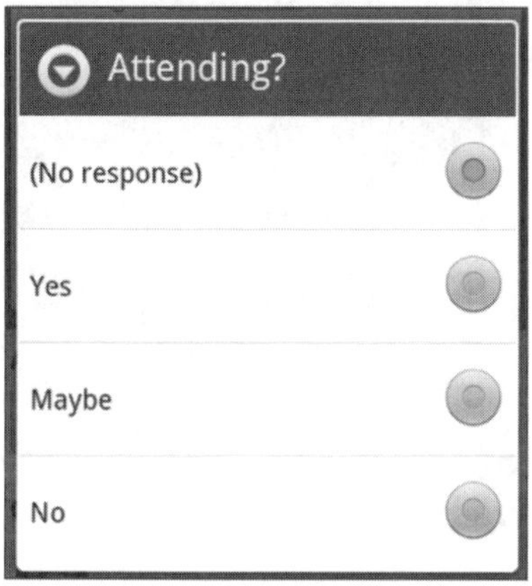

NOTE: Responding to a meeting invitation requires that you open your **Calendar** app, touch the meeting invite, and then send your response from there.

Calendar Settings

You are able to adjust a few settings in your **Calendar** app; you can find these by touching the **Menu** key from inside the **Calendar** app and then choosing **More** > **Settings**. Follow these steps to adjust these options:

1. Tap the **Settings** icon.

2. To **Hide declined events**, just put a check in the box.
3. Scroll down to **Set alerts & notifications** and touch the drop down arrow to select either an **Alert** or a **Status bar notification**.

NOTE: On the DROID 2/X, you can customize **Week view** to show either the **Work week view** (5 days) or the **Full week view** (all 7 days).

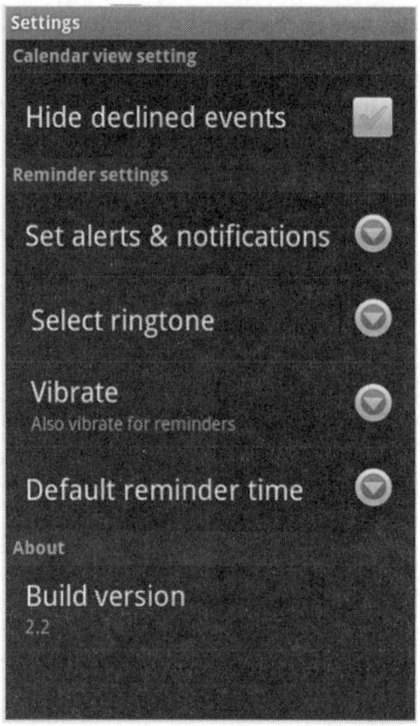

4. Scroll down to **Select ringtone** and touch the drop-down arrow to choose from one of the built-in ringtones for your alert.
5. Touch the drop-down box next to **Vibrate** and choose when you want your DROID to vibrate for alerts: **Always**, **Only when silent**, or **Never**.
6. Touch the **Default reminder** tab and select a default reminder time prior to calendar events. You can choose to have no reminder, or you can select any time interval from one minute to one week prior to the appointment.

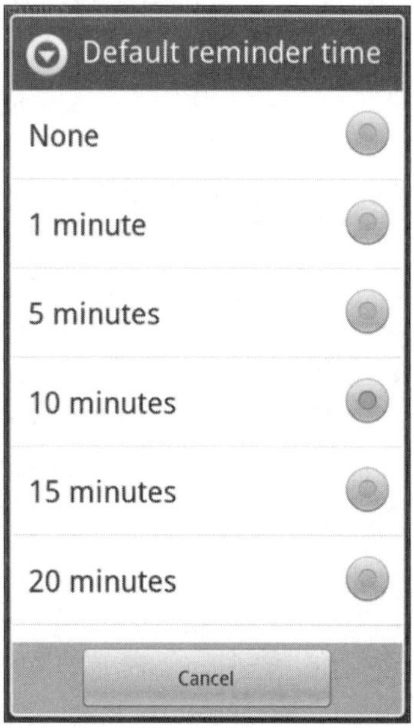

Chapter 14

Enjoying Your Music

This chapter shows you how to turn your DROID into a terrific music player. We'll show you how to play and organize the music you buy from the Amazon MP3 store or sync from your computer, how to view playlists in a variety of ways, and how to quickly find songs.

> **TIP:** Learn how to load your iTunes or Windows Media Player music and playlists into **doubleTwist** in Chapter 25: "DROID Media Sync"; this will enable you to sync them with your DROID.

And you'll learn how to stream music using the **Pandora** applications. With these applications, you can select from a number of Internet radio stations or create your own station by typing in your favorite artist's name – and it's all free.

Your DROID as a Music Player

Your DROID is probably one of the best music players on the market today. The touch screen makes it easy to interact with and manage your music, playlists, cover art, and the organization of your music library. You can even connect your DROID to your home or car stereo via Bluetooth, so you can listen to beautiful stereo sound from your DROID!

> **TIP:** Check out Chapter 8: "Bluetooth on your DROID" to learn how to hook up your DROID to your Bluetooth stereo speakers or car stereo.

Whether you use the built-in DROID **Music** app or an Internet radio app like **Pandora**, you'll find you have unprecedented control over your music on the DROID.

Buying Music from the Amazon MP3 App

You can purchase music and ringtones right on your DROID from the Amazon MP3 app.

1. Tap the **Amazon MP3** app to get started.

2. Now you will see the main screen of the app with the **Search** window at the top, links for **Bestselling Albums**, **Bestselling Songs** and **Browse by Genre** as well as the featured **Free Song of the Day** and the **Album Daily Deal**.

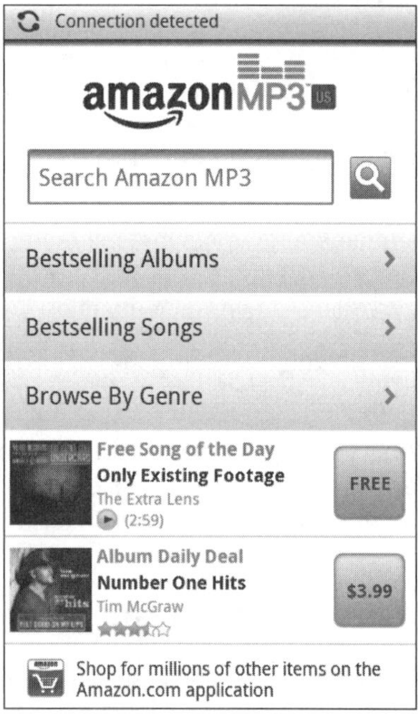

3. Tap **Bestselling Albums** to see a list of albums for sale. You will see similar screens when you tap **Bestselling Songs.**
4. To purchase a song or album, simply tap the **price**, enter your amazon.com account information and confirm your purchase.

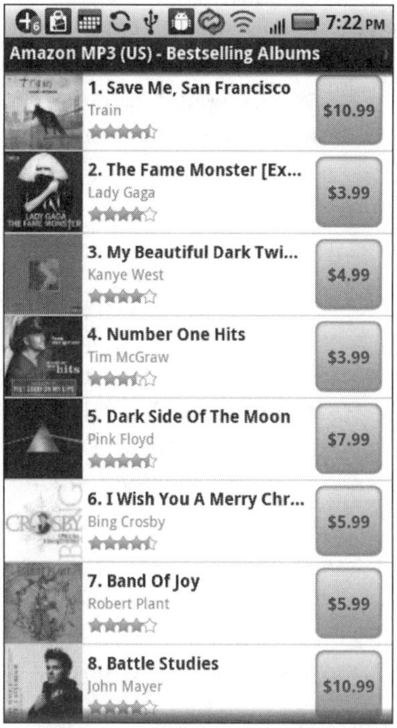

5. If you prefer to browse for a song, tap Browse by Genre and then select any sub-genre until you get to the list of songs or albums you desire.
6. To view only songs, tap the **Songs** tab at the top of the screen.
7. To view only full albums, tap the **Albums** tab at the top.

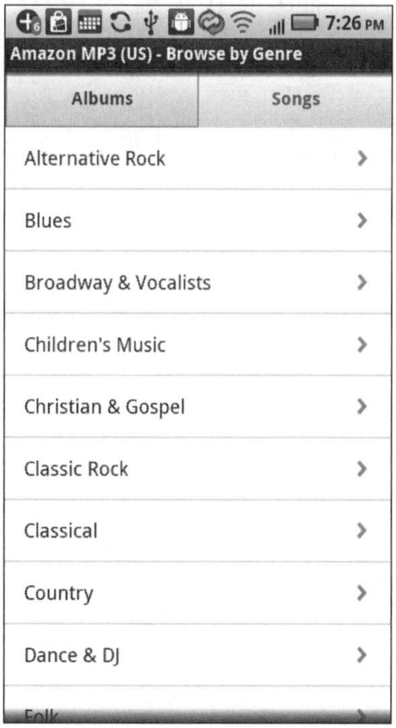

8. You can also search for a particular song, album or artist by using the search field in the first screen you see when you enter the app.

Viewing and Playing Downloaded Songs and Albums

1. From the main screen in the **Amazon MP3** app, press the **Menu** button and select **Downloads** to see all content you have downloaded.

2. Now you will see all the songs and albums downloaded. You will know the item has been successfully downloaded when you see the checkmark next to it and the word Downloaded under it.

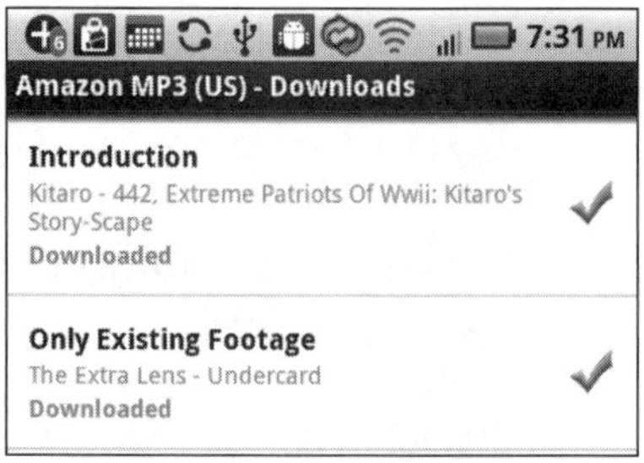

3. Tap any item to start playing it in your DROID Music player.

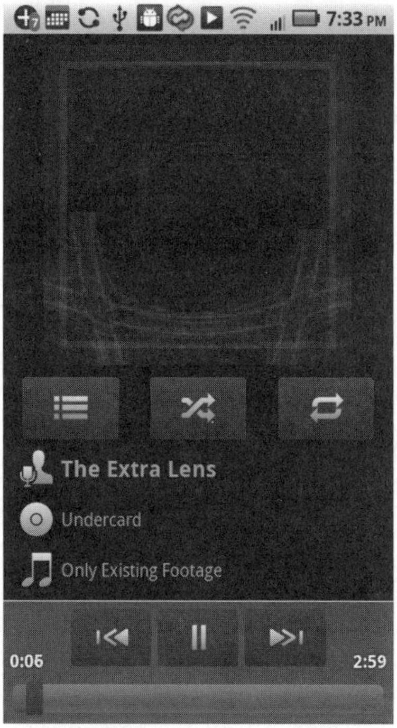

Enter an Amazon Gift Code, Log Out and Adjust Settings

You can enter an Amazon Give Card code, log out or adjust settings from the Setting screen.

1. From the main screen in the **Amazon MP3** app, press the **Menu** button and select **Settings**.

2. From the **Settings** screen you can:

- Tap **Sign out** to log out.
- Choose to turn on or off the **Auto-resume downloads** setting.
- **Clear cache** – this can sometimes speed up the app as well as free up memory on your DROID.
- **Enter a claim code** – enter an Amazon.com Gift card or code.

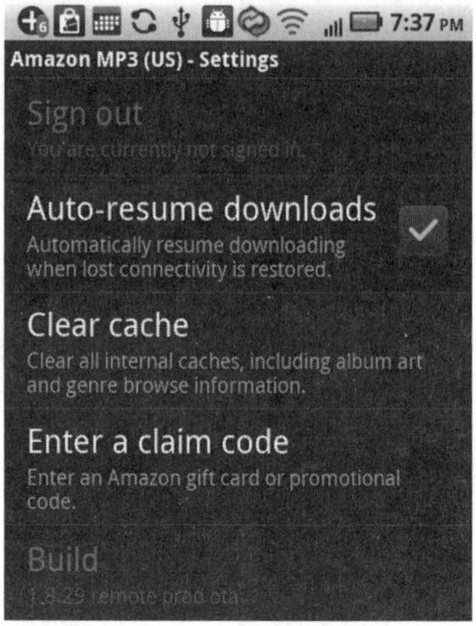

The Music App

Most music is handled through the **Music** app – you find the icon for this app on the **Home** screen of the DROID.

Touch the **Music** icon and, as Figure 14–1 shows, you'll see four soft keys across the top:

- **Artists:** Lets you see an alphabetical list of artists that is searchable like your address book.
- **Albums:** Lets you see your music organized by album title with cover art (also searchable).
- **Songs:** Lets you see an alphabetical list of songs (also searchable).
- **Playlists:** Lets you see synced playlists from your computer plus playlists created on the DROID.

Figure 14–1. *The **Music** app's soft keys.*

Changing the View in the Music App

The **Music** app is very flexible in how it lets you display and categorize your music. Sometimes you want to look at your songs listed by the artist. Other times you might prefer seeing your library organized by album name. The DROID lets you easily change the view to help manage and play just the music you want at a given moment.

The Artists View

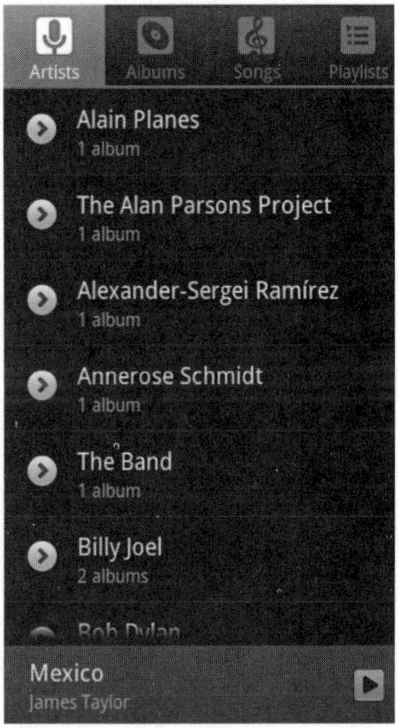

The **Artists** view lists all the artists on your DROID; or, if you are in a playlist, it lists the artists in that playlist.

Flick through the list to move to the first letter of the artist's name.

When you find the artist's name, touch it and all the songs and albums by that artist will be listed, with a picture of the album art to the left.

TIP: You can use the same navigation and search features in the **Music** app that you do in the **Contacts** app (the address book).

The Albums View

The music on your DROID is also organized by albums, which you'll see when you touch the **Albums** icon.

Again, you can scroll through the album covers to find the album you are looking for.

Once you start scrolling, you will see the **Search** tab on the right that you can "pull" down to quickly advance through the alphabetical list of albums.

When you choose an album, all the songs on that album will be listed.

To go back, just touch the **Back** button.

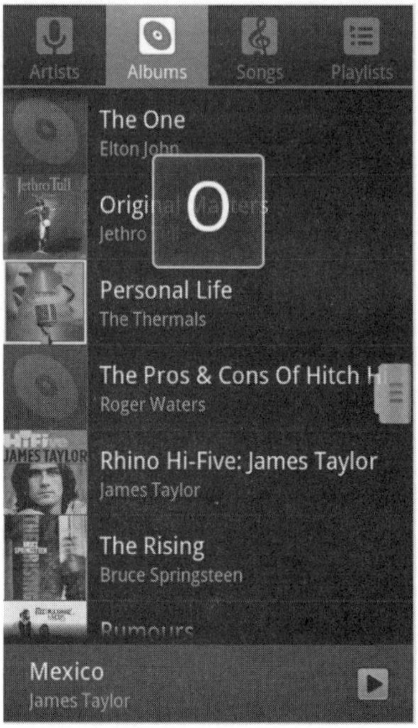

The Songs View

Touching the **Songs** button displays a list of every song on your DROID.

If you know the name of the song, flick through the list or touch the first letter of the song in the alphabetical list to the right.

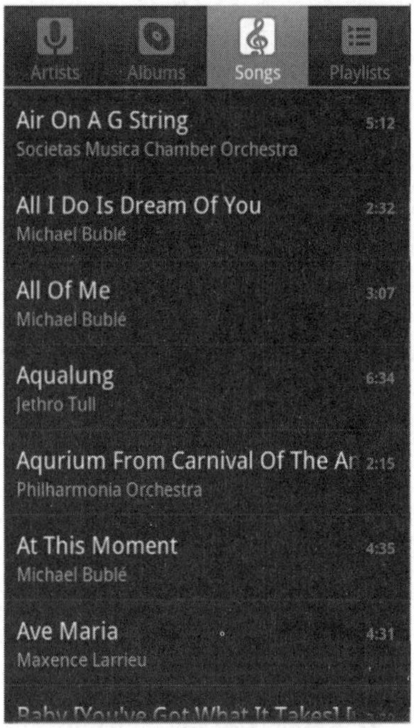

Creating Playlists on the DROID

The DROID lets you create a playlist from recently added music. You can change the playlist whenever you want, removing old songs and adding new ones – it couldn't be easier!

To create a new playlist on the DROID, touch the **Recently added** tab under **Playlists**.

Touch the **Menu** button and select **Save as playlist.** Give your playlist a unique name (we'll call this one "New playlist 1"), and then touch **Save**.

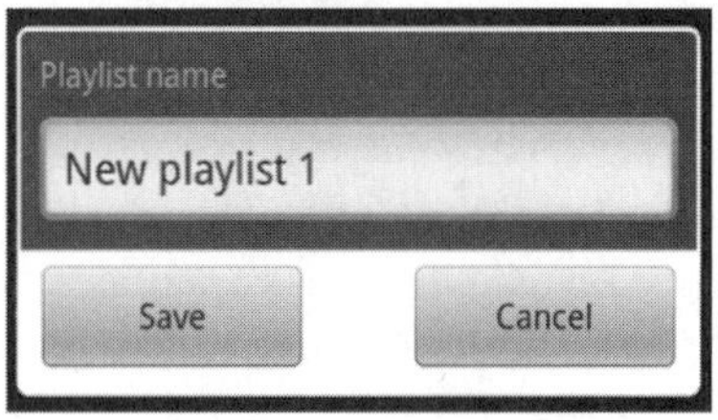

Follow these steps to delete a song:

1. Touch and hold a song in the playlist to bring up a menu list with six options.
2. Tap the **Delete** button towards the bottom of the list. The song will be removed from your music library on the DROID.

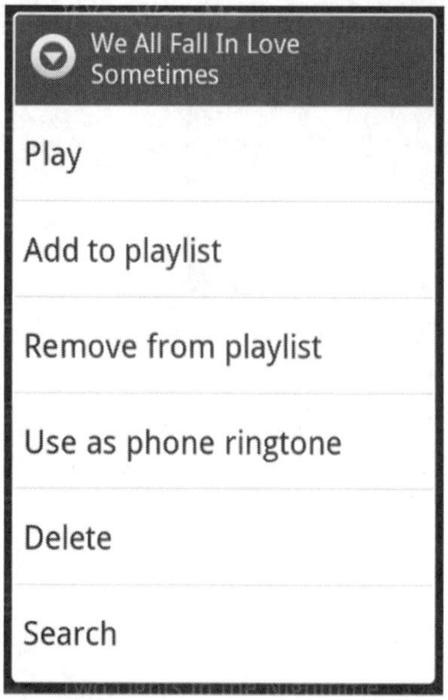

Follow these steps to move a song up or down in a playlist:

1. Touch a playlist to display the songs.
2. Touch and hold the three gray bars to the left of the song.
3. Drag the song up or down and then let go.
4. Touch the **Back** button to exit the **Playlist** view.

To move a song, touch here and drag up or down.

Searching for Music

Every view from your **Music** app (e.g., **Playlists**, **Artists**, **Songs**) can have a search window at the top of the screen, as shown in Figure 14–2. From any of the music views, just touch the **Search** button on the DROID. Tap once in the **Search** window and type a few letters of the name of an artist, album, playlist, video, or song to instantly see a list of all matching items. This is the best way to quickly find something to listen to on your DROID.

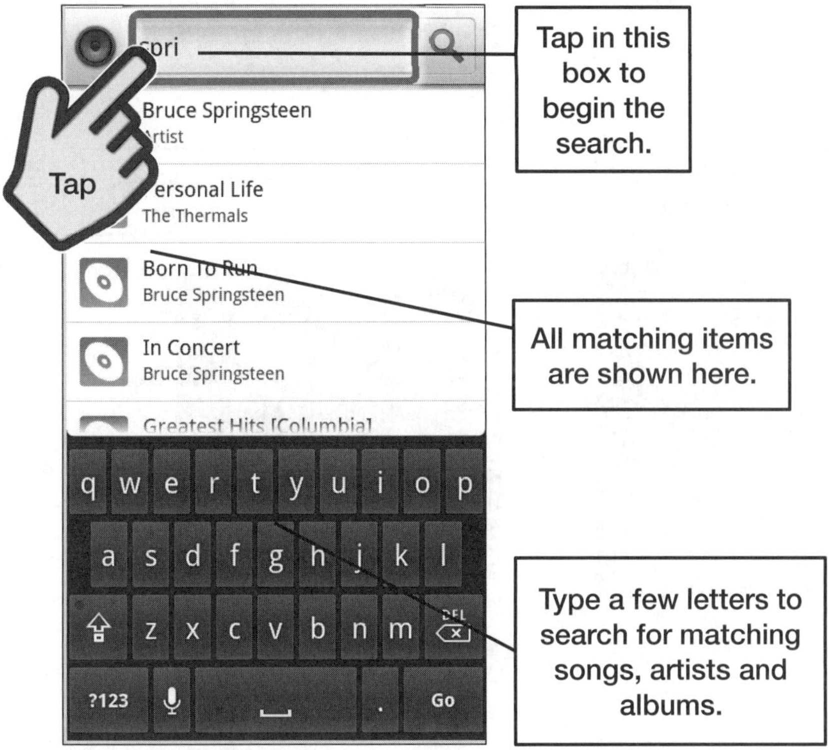

Figure 14–2. *Finding music.*

Viewing Songs in an Album

When you're in **Albums** view, just touch an album cover or name and the screen will slide, showing you the songs on that album (see Figure 14–3).

To see the songs on an album that is playing, tap the **List** button. The album cover will turn over, revealing all the songs on that album. The song that is playing will have a small blue arrow next to it.

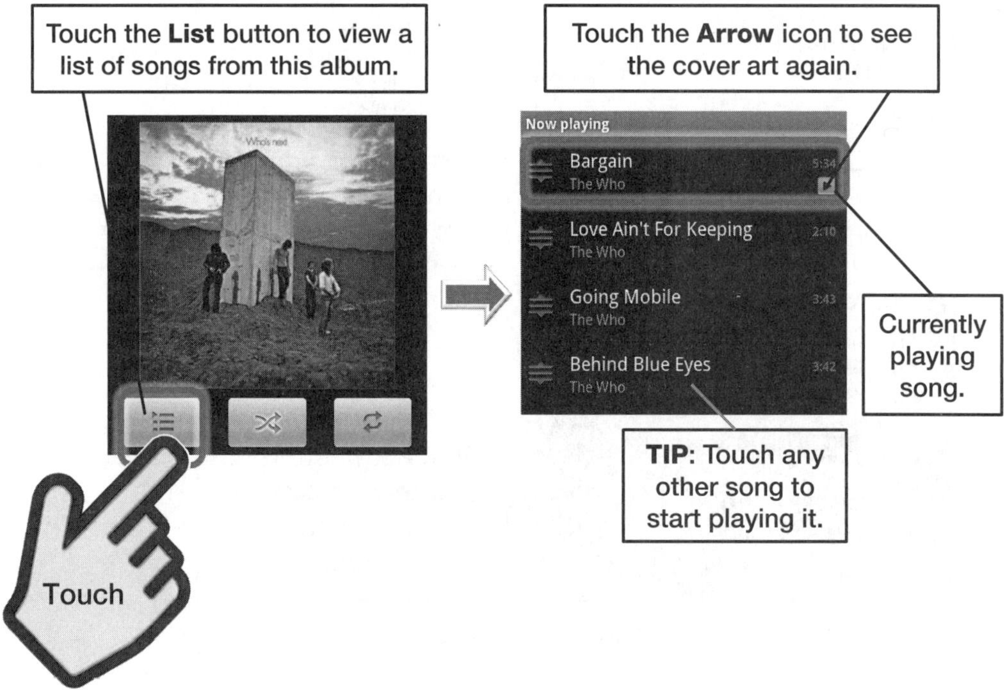
Touch the List button to view a list of songs from this album.
Touch the Arrow icon to see the cover art again.
Now playing
Bargain
The Who
5:34
Love Ain't For Keeping
The Who
2:10
Going Mobile
The Who
3:43
Behind Blue Eyes
The Who
3:42
Currently playing song.
TIP: Touch any other song to start playing it.
Touch

Figure 14–3. *Touch the **List** button to see the songs on a particular album.*

Playing Your Music

Now that you know how to find your music, it's time to play it! Find a song or browse to a playlist using any of the methods mentioned above. Simply tap the song name and it will begin to play.

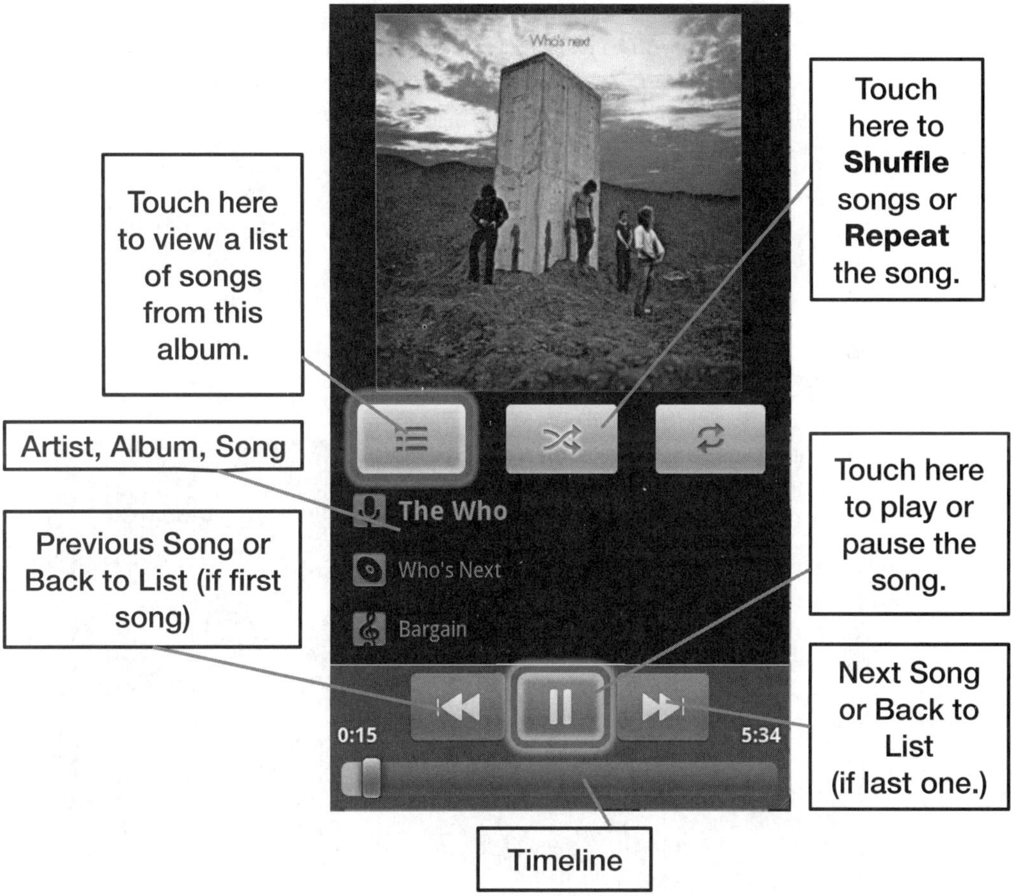
Who's next
Touch here to Shuffle songs or Repeat the song.
Touch here to view a list of songs from this album.
Artist, Album, Song
The Who
Who's Next
Bargain
Previous Song or Back to List (if first song)
Touch here to play or pause the song.
Next Song or Back to List (if last one.)
0:15
5:34
Timeline

This screen shows the name of the artist, album, and song underneath the picture of the album cover.

Along the bottom of the screen, you'll find the **Timeline** slider bar and the **Previous Song**, **Play/Pause**, and **Next Song** buttons.

To see other songs on the album, just tap the **List** button.

You can also touch the **Shuffle** or **Repeat** buttons under the album artwork.

Pausing and Playing

Tap the **Pause** symbol (if a song is playing) or the play arrow (if the music is paused) to stop or resume the song.

Playing the Previous or Next Song

If you are in a playlist, touching the **Next Song** arrow (to the right of the **Play/Pause** button) advances you to the next song in the list. If you are searching through your music by album, touching **Next Song** moves you to the next song on the album. Touching the **Previous Song** button does the reverse.

NOTE: If you're at the beginning of a song, **Previous Song** takes you to the preceding song. If the song is already playing, **Previous Song** goes to the beginning of the current song (and a second tap would take you to the previous song).

Adjusting the Volume

You can adjust the volume on your DROID by using the external **Volume** buttons on the side of the phone.

The external **Volume** buttons are on the upper-right side of the device. Press the **Volume Up** key (the top button) or the **Volume Down** key to raise or lower the volume. You'll see the **Media volume** control move as you adjust the volume.

TIP: To quickly mute the sound, press and hold the **Volume Down** key and the volume will eventually reduce to zero.

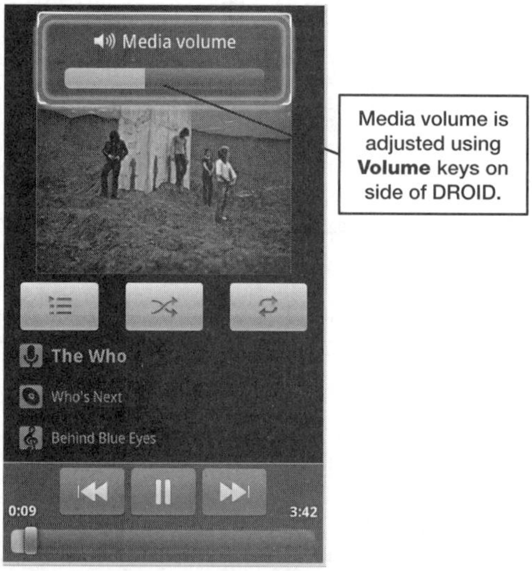

TIP: If you hold down the **Previous Song** control, the song will rewind; if you hold down the **Next Song** control, it will fast forward.

Repeating, Shuffling, and Moving Around in a Song

In play mode, under the album art, you will see the controls for **Repeat** and **Shuffle**.

At the bottom of the **Now playing** screen, you will see a timeline below the **Play/Pause** indicator that shows you where you are in the song.

Moving to Another Part of a Song

Slide the scrubber bar to the right and you'll see the elapsed time of the song (displayed to the far right) change accordingly. If you are looking for a specific section of the song, drag the slider, then let go and listen to see if you're in the right place.

Repeating One Song or All Songs

To repeat all the songs you're listening to, touch the **Repeat** symbol at the left of the top controls twice until you see it turn green and display **Repeating all songs**.

To repeat the current song in the playlist, song list, or album, touch the **Repeat** icon again until it displays **Repeating current song**.

To turn off the **Repeat** feature, press the icon until it turns gray again.

Shuffling Your Playlist

If you are listening to a playlist or album or any other category or list of music, you might decide you don't want to listen to the songs in order. You can touch the **Shuffle** symbol so the music will play in random order. You know **Shuffle** is turned on when the icon is green; when it's gray, the **Shuffle** feature is off.

Now Playing

Sometimes you're having so much fun exploring your options for playlists or albums that you get deeply buried in a menu – and then find yourself just wanting to get back to the song you're listening to. Fortunately, this is always easy to do – you can just touch the **Now Playing** icon at the bottom of most of the music screens.

Viewing Other Songs on the Album

You may decide you want to listen to another song from the same album rather than going to the next song in the playlist or genre list.

In the upper-right corner of the **Now Playing** screen, you'll see a small button with three lines on it.

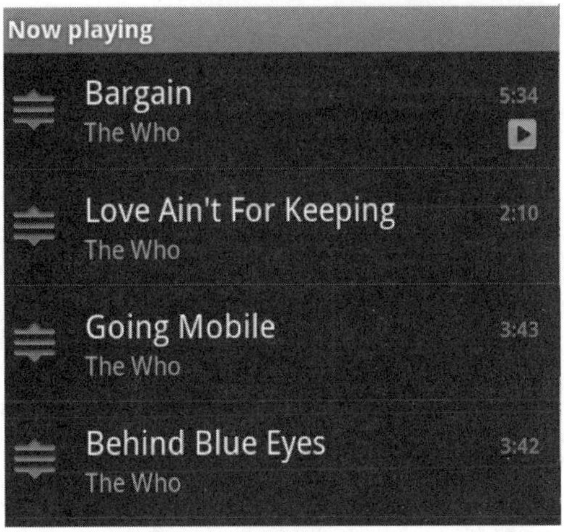

Tap that button and the view switches to a list of all the songs on that album on your DROID.

Touch another song on the list and that song will begin to play.

NOTE: If you were in the middle of a playlist and you jump to another song from an album, you won't be taken back to that playlist. To return to that playlist, you'll need to go back to your playlist library.

Exploring Your Music Options

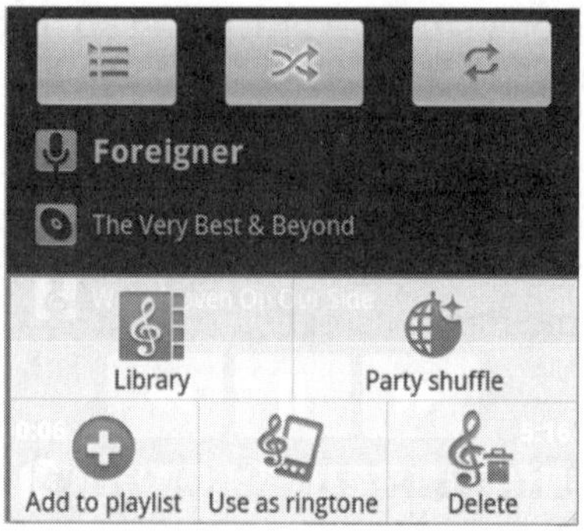

There are a few options available to you when you are in the **Now playing** screen. From this screen, touch the **Menu** key and you should see five soft keys: **Library**, **Party shuffle**, **Add to playlist**, **Use as ringtone**, and **Delete**.

Library will take you back to your music library.

Party shuffle will take you out of your playlist and arrange a random shuffling of music.

Add to playlist allows you to add the current song to any playlist on your DROID.

Use as Ringtone allows you to use the current song as the general ringtone for the device.

Delete will delete the song from your DROID.

NOTE: On the DROID 2/X, you will see another menu item called **Audio Effects**. Use this to adjust things like which speakers are connected and adjust the sound profile or use an equalizer.

Listening to Free Internet Radio (Pandora)

While your DROID gives you unprecedented control over your personal music library, there may be times when you want to listen to some other music.

TIP: A basic **Pandora** account is free. It can save you considerable money compared with buying many new songs from Amazon MP3.

Pandora grew out of the Music Genome Project, a huge undertaking in which a large team of musical analysts looked at just about every song ever recorded and then developed a complex algorithm of attributes to associate with each song.

NOTE: Pandora may have some competition by the time you read this book. Right now there's one other competitor called **Slacker Personal Radio**, but there will probably be more. If you want to find more options, try searching the Android Market for "Internet Radio." Also, please note that **Pandora** is a US-only application and **Slacker** is available only in the U.S. and Canada. **Spotfly** is a similar app for Europe. More options should begin to pop up for international users.

Getting Started with Pandora

Pandora lets you design your own unique radio stations built around artists you enjoy. Best of all, it is completely free!

Start by downloading the **Pandora** app from the Android Market. Just go to the Market and search for "Pandora."

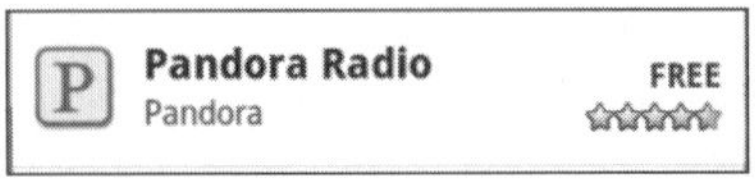

Now just touch the **Pandora** icon to start.

NOTE: Some users have reported sound issues with Android 2.2 and Pandora, but they are being addressed.

The first time you start **Pandora**, you'll be asked to either create an account or to sign in if you already have an account. Just fill in the appropriate information – an email address and a password are required – and you can start designing your own music listening experience.

Pandora is also available for your Windows or Mac computer, as well as for most smartphone platforms. If you already have a Pandora account, all you have to do is sign in.

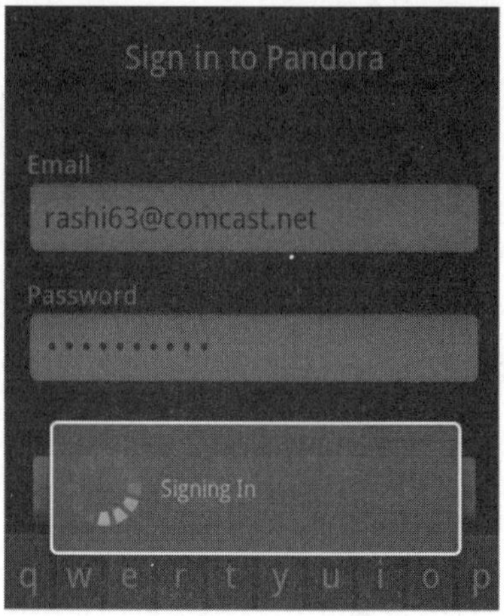

Pandora's Main Screen

Your stations are listed on the screen. Just touch one and it will begin to play. Usually, the first song will be from the actual artist chosen, and the next songs will be from similar artists.

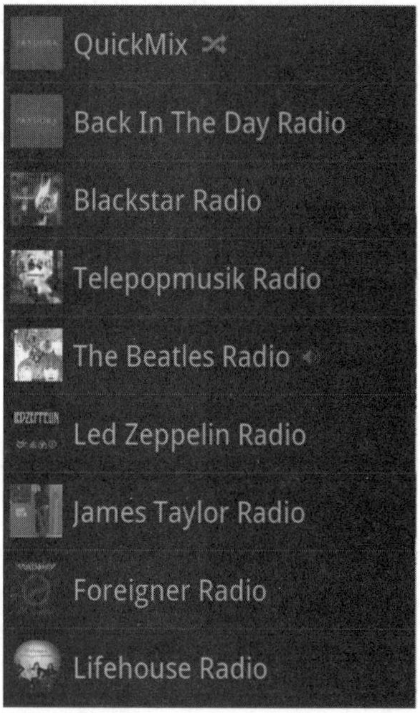

Once you select a station, the music begins to play. You'll see the current song displayed, along with album art – very much like when you play a song using the **Music** app.

Touch the **Information** icon in the upper-right corner, and you'll see a bio of the artist, which changes with each new song.

Thumbs Up or Thumbs Down in Pandora

If you like a particular song, touch the **Thumbs-up** icon and you'll hear more from that artist.

Alternatively, if you don't like an artist on this station, touch the **Thumbs-down** icon and you won't hear that artist again.

If you like, you can pause a song and come back to it later. Or, you can skip to the next selection in your station.

NOTE: With a free Pandora account, you are limited to a certain amount of skips per hour. Also, you'll occasionally hear advertising. To get rid of these annoyances, you can upgrade to a paid "Pandora One" account.

Pandora's Menu

From the **Now playing** screen, press the **Menu** button on your DROID. Touch this and you can **Bookmark** the artist or song, go to **Amazon MP3** to buy music from this artist, or **Share** the station with someone in your **Contacts** list.

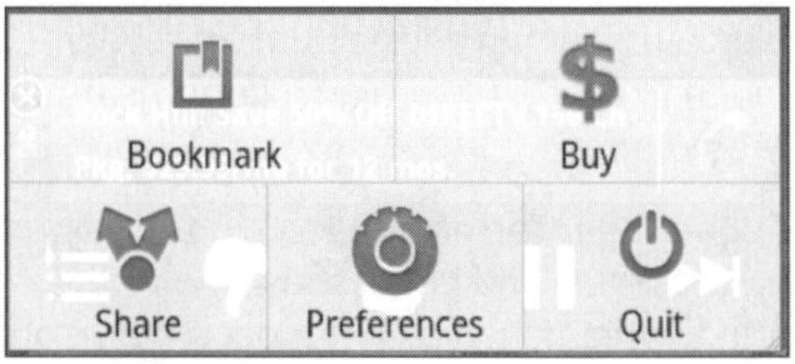

You can also adjust your Pandora **Preferences**.

Creating a New Station in Pandora

Creating a new station couldn't be easier. Start by pressing the **Menu** key when at **Station list** screen of **Pandora**.

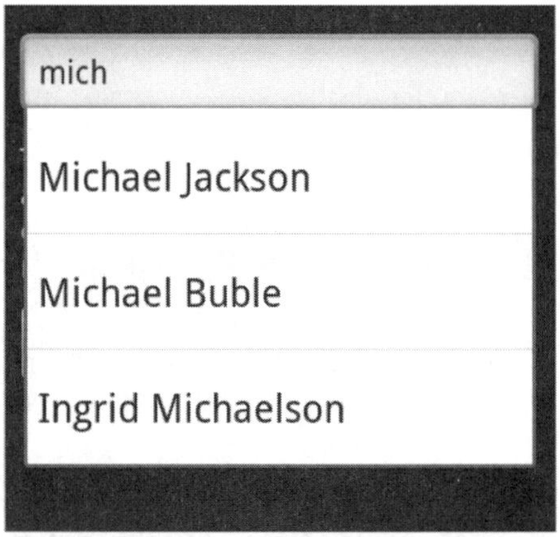

Just touch the **Create Station** button along the bottom row. Type in the name of an artist, song, or composer.

When you find what you are looking for, touch the selection and **Pandora** will immediately start to build a station around your choice.

You can also touch **Genre** and build a station around a particular genre of music.

You'll then see the new station listed with your other stations.

You can build up to 100 stations in Pandora.

> **TIP:** You can organize your stations by pressing the **By Date** or **ABC** buttons at the top of the screen.

Adjusting Pandora's Settings – Your Account, Upgrading, and More

You can sign out of your Pandora account, adjust the audio quality, and even upgrade to Pandora One (which removes advertising) by tapping the **Preferences** icon after pressing the **Menu** key from the **Now playing** screen (see Figure 14–4.)

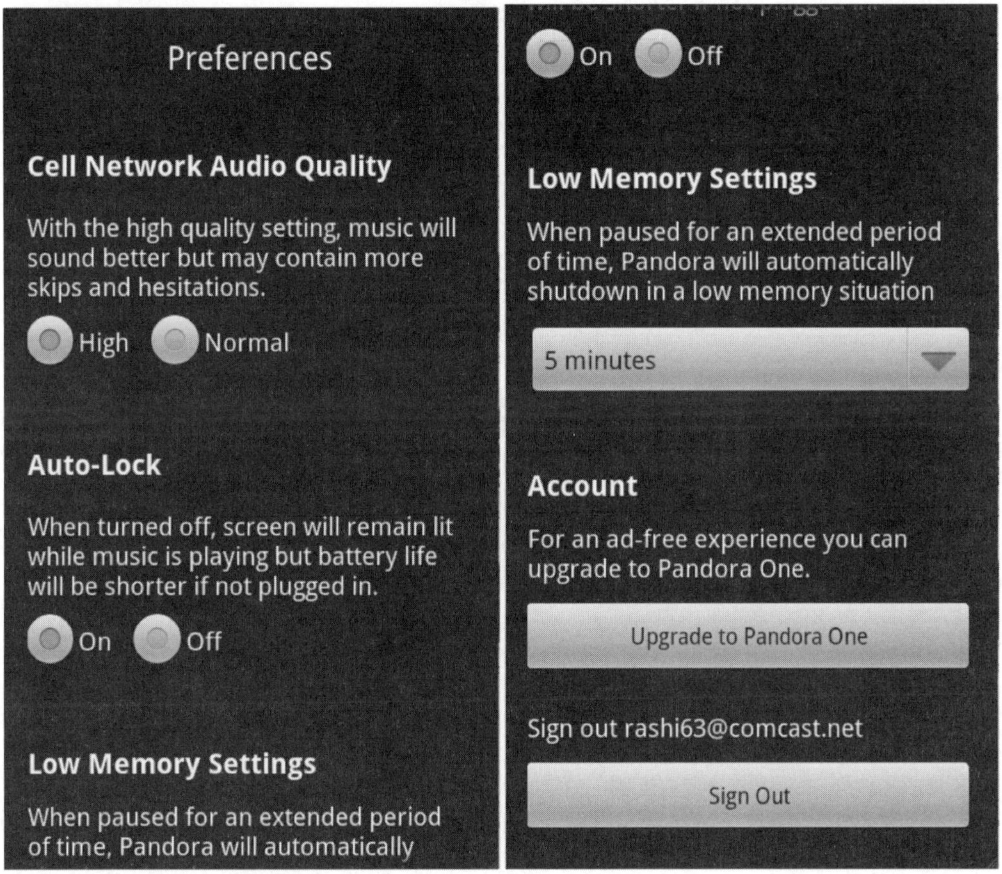

Figure 14–4. *The **Preferences** options in Pandora.*

To sign out, tap your account name.

To adjust the sound quality, move the switch under **Cell Network Audio Quality** to either **High** or **Normal**. When you are on a cellular network, setting this to **Off** is probably better; otherwise, you may hear more skips and pauses in the playback.

When you are on a strong Wi-Fi connection, you can set this to **High** for better quality. See Chapter 5: "Wi-Fi and 3G Connectivity" to learn more about the various connections.

To save your battery life, you should set the **Auto-Lock** to **On**, which is the default. If you want the force the screen to stay lit, then switch this to **Off**.

To remove all advertising, tap the **Upgrade to Pandora One** button. A web browser window will open, and you'll be taken to the Pandora web site to enter your credit card information. At the time of publishing, the annual account cost is $36.00, but that may be different by the time you read this book.

Chapter 15

Viewing Videos, TV Shows, and More

The DROID is an amazing "media consumption" device. Nowhere is this more apparent than in the various video-viewing applications available for it.

This chapter shows you how to watch movies, TV shows, podcasts, and music videos on your DROID. You can buy or download many videos for free from the Android Market or through doubleTwist (see Chapter 25: "DROID Media Sync" for more information). According to some sources, you will be able to link your DROID to your Netflix account by early 2011 (other video rental services will likely follow soon), allowing you to watch streaming TV shows and movies.

You can also use your DROID to watch YouTube videos, as well as to view videos from the Web in your **Browser** app and in various other apps available from the Android Market. DROID X and DROID 2 phone ship with the Blockbuster app for renting and watching movies from Blockbuster on your phone.

> **NOTE:** These apps change quickly. We expect that services such as **Hulu plus** will also make their way to the DROID. We suggest that you go to the Android Market frequently and type in "videos" to see the changing options.

Your DROID as a Video Player

The DROID is not only a capable music player; it is a fantastic portable video player. The wide screen, fast processor, good pixel density, and great operating system make watching anything from music videos to TV shows and full-length motion pictures a real joy. The size of the DROID is perfect for watching clips or shows while commuting or traveling on an airplane. It is also great for the kids in the back seat of long car trips. The decent battery life means you can even go on a short flight and not run out of power! However, if you need more power, you can buy a power inverter or DROID charger for your car to keep the DROID charged even longer (see Chapter 1: "Getting Started" for battery tips).

Loading Videos onto Your DROID

You can load videos onto your DROID just as you do with your music – through doubleTwist (see Chapter 25). You can also use the **USB Load** feature on your DROID or **Media Share.**

NOTE: Videos (e.g., DVDs) can be *ripped* (i.e., copied) to your computer and then synced to your DROID. Make sure you don't violate any copyright laws in the process of doing so! Video conversion software is widely available on the Web. The optimal output settings for ripping videos to play on the DROID are to use MP4 with the following video/audio codecs and settings:

Video
Resolution: Up to WVGA (854x480)
Codec: H.264, Baseline profile
Bitrate: 2.5 Mbps
Framerate: 24fps

Audio
Codec: AAC stereo au

Watching Videos on the DROID

To watch videos, touch the **Gallery** icon, which is usually on **Home** screen of the DROID.

NOTE: You can also watch videos from the **YouTube** icon, the **Browser** icon, and other video-related apps you download from the Android Market.

Video Categories

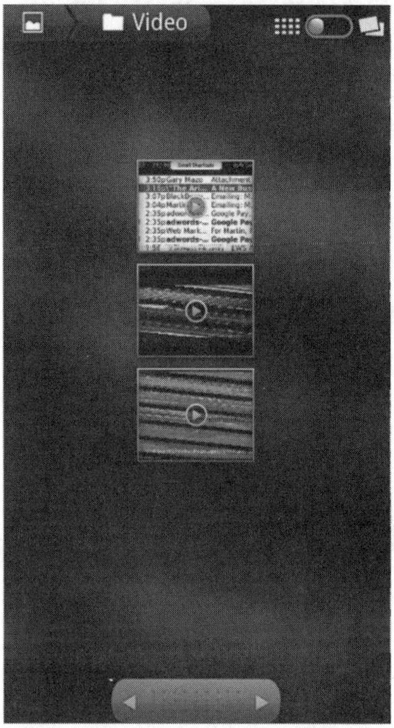

In the **Gallery** app, you will see separate folders labeled **Camera**, **Downloads**, and **Videos**. Touch the **Videos** folder to go to the **Videos** section; if you have movies loaded on the DROID, they will be listed in this folder.

You won't see your videos broken down by any particular category such as music videos, podcasts, or movies. Instead, all your videos will simply be shown in a single, continuous list.

On the DROID, to get more information on a specific video, touch and hold that video to highlight it, and then press the **Menu** key and select **More**. Next, touch **Get info** to display the details of the video.

On the DROID 2/X, tap the video thumbnail to view it in full screen, then tap the information (I with circle) to see the details screen.

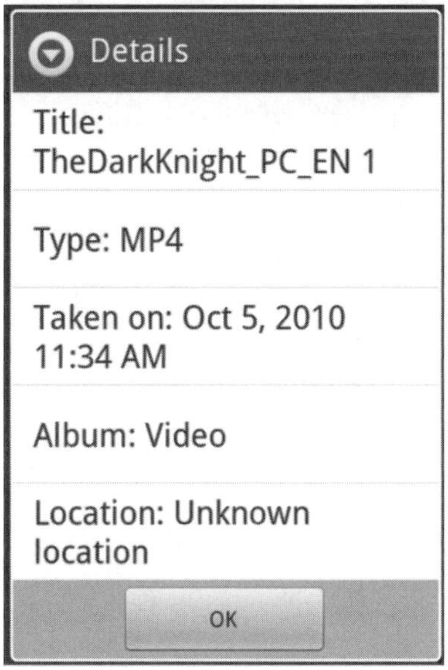

Playing a Movie

On the DROID touch the movie you wish to watch, and it will begin to play (see Figure 15–1). On the DROID 2/X, tap the video thumbnail, then tap the video again to start playing it. Most videos take advantage of the relatively large screen real estate of the DROID to play in widescreen or landscape mode; just turn your DROID sideways to watch them.

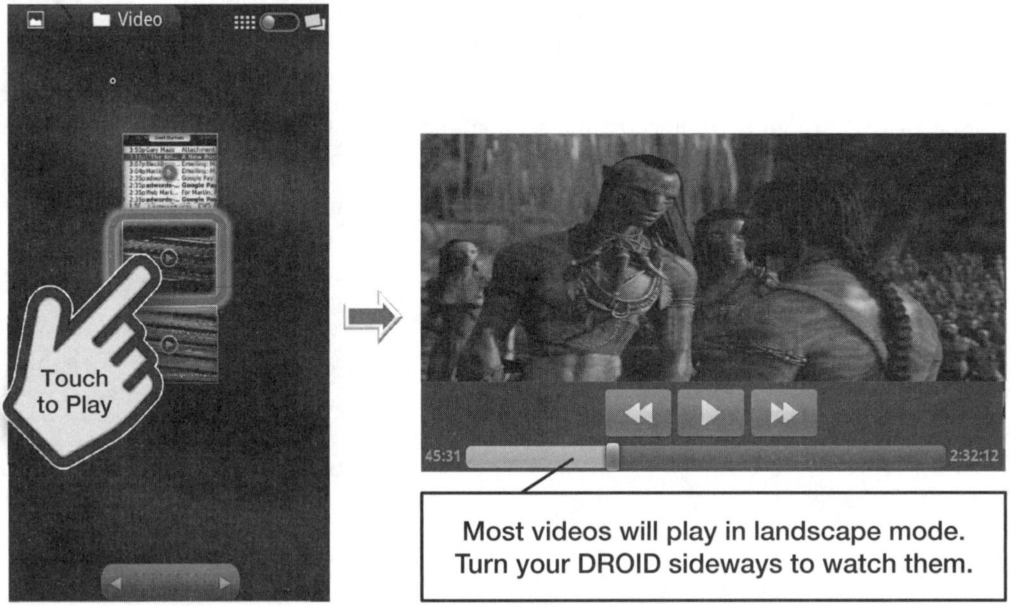

Figure 15–1. *Playing a video.*

When the video first starts to play, there are no menus and no controls. Indeed, there is nothing on the screen except for the video.

To Pause or Access Controls

You can touch anywhere on the screen to make the control bars and options in the **Gallery** media app visible (see Figure 15–2). Most controls and options are very similar to those in the **Music** player. Tap the **Pause** button and the video will pause.

> **NOTE:** On the DROID 2/X, you will not have Rewind or Fast Forward buttons, instead you can drag the slider bar back or forth to rewind or fast forward.

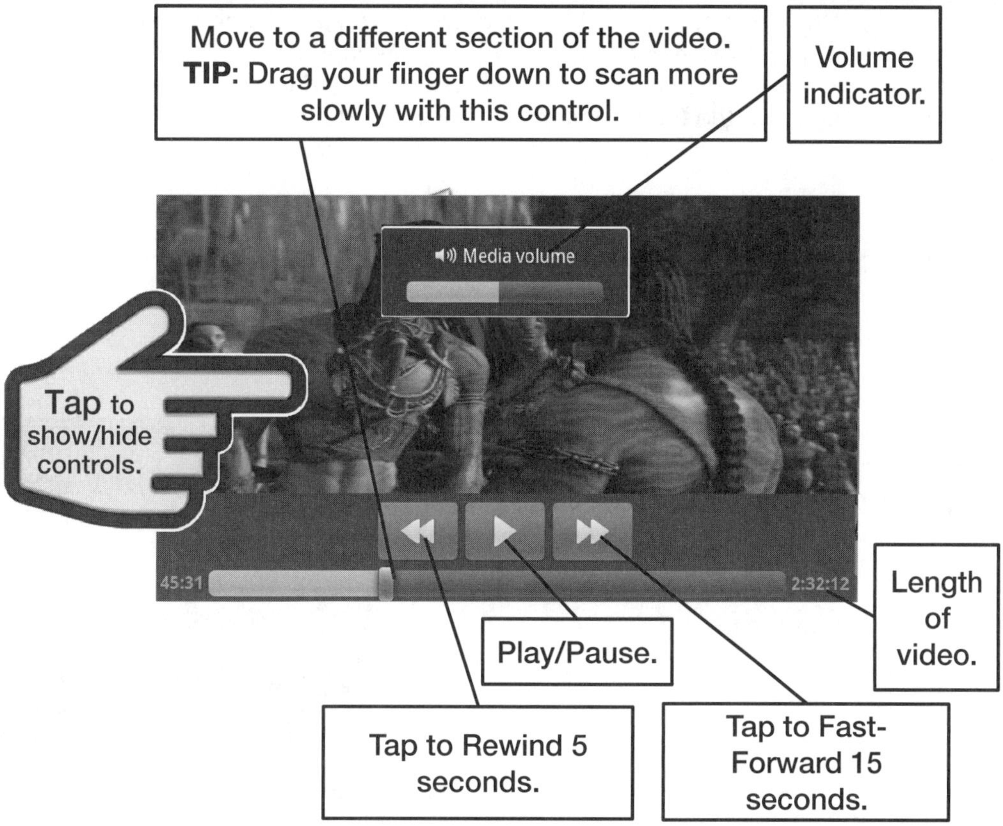
Move to a different section of the video.
TIP: Drag your finger down to scan more slowly with this control.
Volume indicator.
Media volume
Tap to show/hide controls.
45:31
2:32:12
Length of video.
Play/Pause.
Tap to Rewind 5 seconds.
Tap to Fast-Forward 15 seconds.

Figure 15–2. *The video controls in the **Gallery** media app.*

Fast-Forward or Rewind the Video

On the DROID, on either side of the **Play/Pause** button, you can see the typical **Fast-Forward** and **Rewind** buttons. To advance 15 seconds in the video, touch the **Fast-Forward** button (to the right of **Play/Pause**). When you get to the desired spot, release the button, and the video will begin playing normally.

To rewind in five-second intervals, tap the **Rewind** button. To rewind to a specific part or location, move the slider bar to the desired part of the video. On the DROID 2/X, use the slider bar at the bottom to move around the video.

NOTE: There is no way to adjust the volume using on-screen controls. The volume rocker switch on the side of the DROID controls media volume.

Using the Time Slider Bar

At the bottom of the video screen is a slider that shows you the elapsed time of the video. If you know exactly (or approximately) which point in the video you wish to watch, just hold and drag the slider to that location. Some people find this to be a little more exact than holding down the **Fast-Forward** or **Rewind** Buttons.

Other Video Players

The **Gallery** app is a very limited video player. It functions fine for basic viewing and it is built in to the DROID, which is convenient.

However, there are many other media players in the Android Market that you can download and use in place of the **Gallery** app.

Once you have more than one video player installed, you will get a pop-up window asking which video player you wish to use to open the selected video.

One video player that consistently gets great reviews is the **Act One Video Player**. You can download this app from the Android Market (see Chapter 17: "Exploring the Android Market" for more information on using this marketplace) and search for "video players" or "Act One."

Download and install the app, and it will now appear on your **Home** screen.

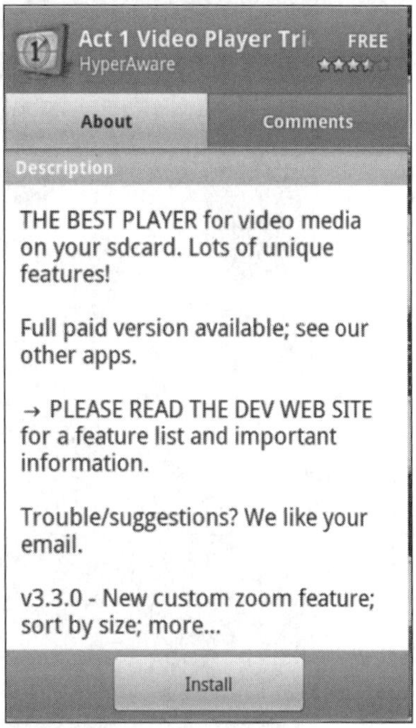

Deleting Videos

Sometimes you might want to delete a video (to save space on your DROID). To do so, touch and hold a video from the video list. You will see three soft keys appear at the bottom of the screen; the middle key is **Delete**. Touch the **Delete** key (see Figure 15–3) and confirm that you want to delete the file.

NOTE: Deleting a video deletes the video only from your DROID – a copy will still remain in your video library, assuming that you have synced your DROID with your computer after purchasing or ripping the video. This means you can reload it onto your DROID at a later date.

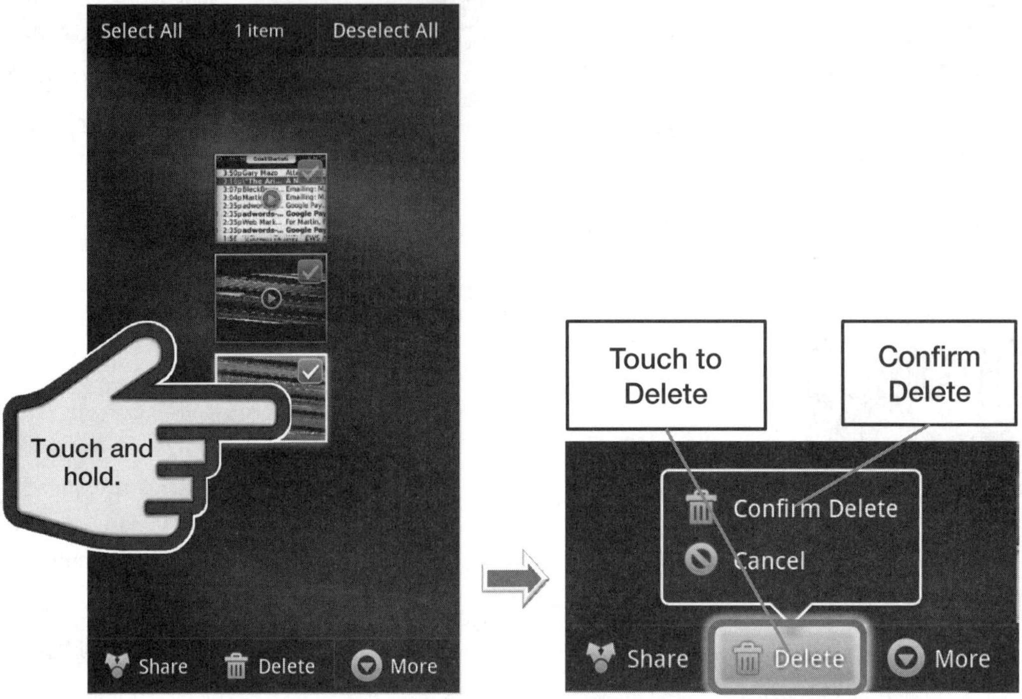

Figure 15–3. *Deleting a video.*

> **NOTE:** On your DROID 2/X, the Play, Delete, Share and Edit buttons will be in a drop down menu in the middle of the screen instead of along the bottom as shown.

Using YouTube on your DROID

Watching YouTube videos is certainly one of the most popular things for people to do on their computers these days. YouTube is as close to you as your DROID.

Your DROID's **Home** screen includes a **YouTube** icon; touch the **YouTube** icon and you will be taken to the **YouTube** app.

Searching for Videos

When you first start the **YouTube** app, you usually see the **Featured** videos on YouTube that day.

Just scroll through the video choices as you do in other apps.

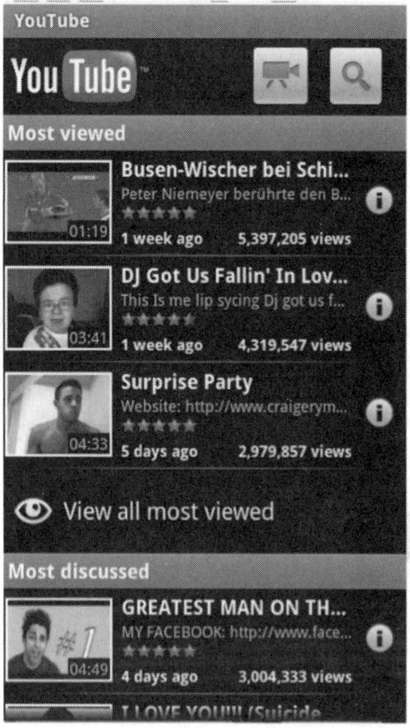

Using the Bottom Icons

Touch the **Menu** key and the **YouTube** app will display five icons along the bottom: **Search**, **Upload**, **My account**, **Categories**, and **Settings**. Each is fairly self-explanatory.

To see the videos that YouTube is featuring that day, scroll down to the **Featured** list. To see the list of most-viewed videos, scroll down to the **Most Viewed** icon.

After you watch a particular video, you will have the option to set it as a favorite on **YouTube** for easy retrieval later on. If you have set bookmarks, they will appear when you touch the **Favorite** icon.

You can also **Rate**, **Comment** on, **Share**, or **Flag** a video after you watch it.

You can also search the huge library of YouTube videos from your DROID. Touch the **Search** box as in previous apps, and the keyboard will pop up. Type in a phrase, topic, or even the name of a video.

The example to the right shows a search for the newest Made Simple Learning video tutorial, with the user entering "Made Simple Learning" to see a list of such videos.

When the user finds a video she wants to watch, she can touch it to see more information. She can even rate the video by touching it during playback and selecting a rating.

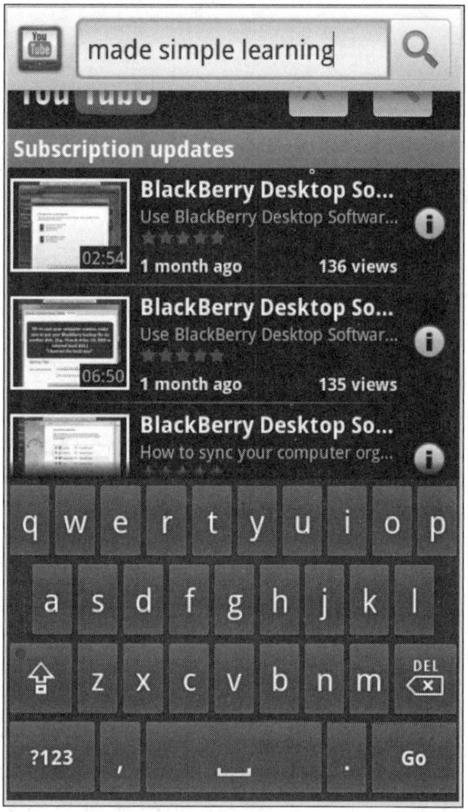

Playing Videos

Once you make your choice, touch the video you want to watch. Your DROID will begin playing the YouTube video in **Landscape** mode. There is no way to force the playback into **Portrait** mode, so you will need to turn the DROID sideways to watch the video (see Figure 15–4).

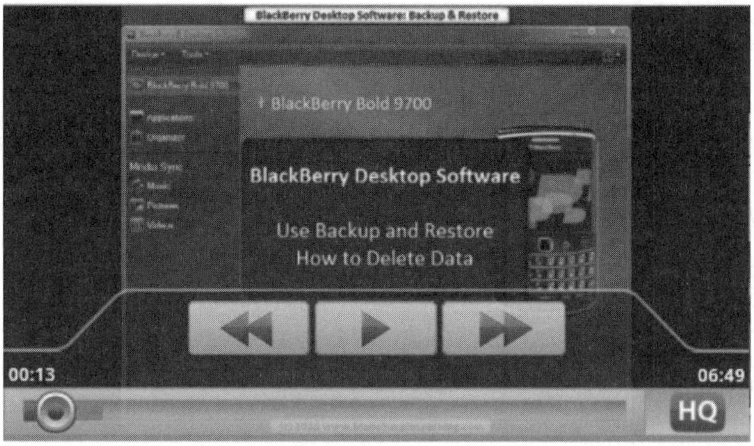

Figure 15–4. *Playing a video in **Landscape** mode.*

Adjusting the DROID's Video Controls

Once the video begins to play, the on-screen controls disappear, so you see only the video. To stop, pause, or activate any other options while the video is playing, just tap the screen (see Figure 15–5).

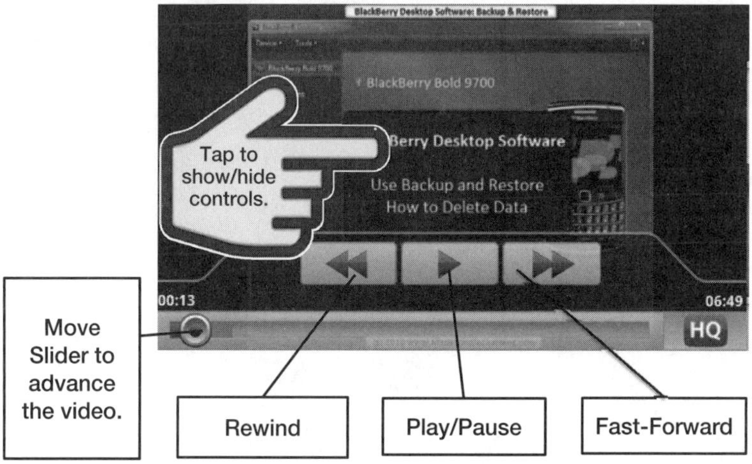

Figure 15–5. *Selecting playback and other video options in YouTube.*

The on-screen options in the **YouTube** app are very similar to the options you see when watching other videos. Along the bottom is a slider that shows your place in the video. To move to another place in the video, just drag the slider.

To fast-forward through the video, tap the **Fast-Forward** arrow. To move in reverse, tap the **Rewind** arrow.

Press the **Menu** key and six more soft keys appear: **Captions**, **Details**, **Rate**, **Favorite**, **Share**, and **More**.

To set a favorite, touch the **Favorite** icon at the far left.

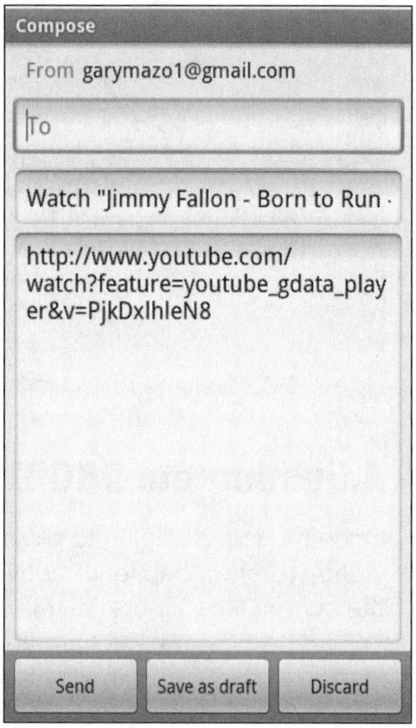

To email the video, Follow these steps:

1. Touch the **Share** icon. Your email will start with the link to the video in the body of your email.
2. Type the recipient's name (see Chapter 9: "Email on your DROID" for more information on how to send content via email).
3. Write a short note to provide some context for the link to the recipient, if you so desire.
4. Press the **Send** button.

Clearing Your History

For a variety of reasons, you may want to clear your browsing history on your DROID.

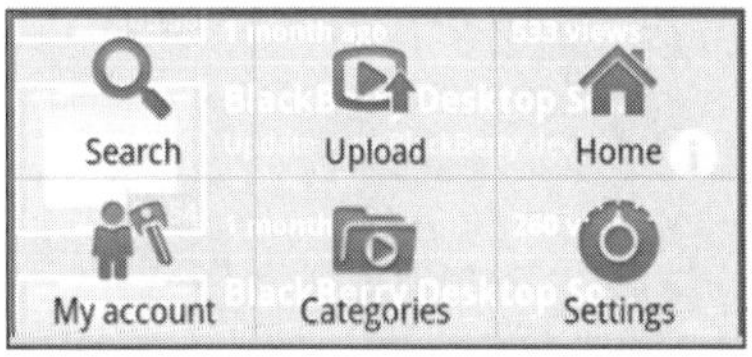

Follow these steps to clear the history log in your **Browser** app:

1. Touch the **Menu** key and then touch **Settings.**
2. Scroll down to **Clear Search History** and touch it.
3. Click the **OK** button in the center to complete the process of clearing your history.

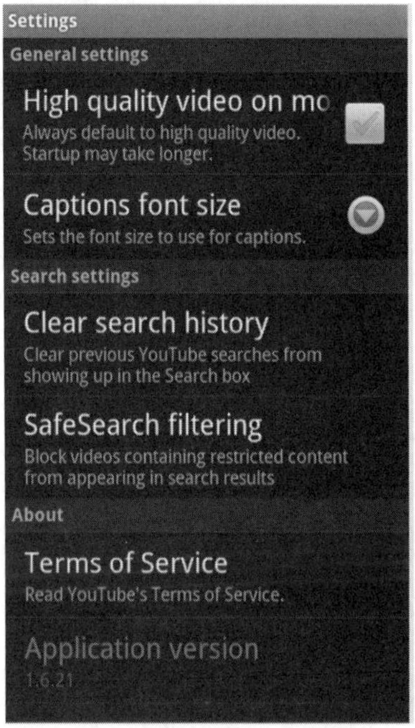

Using Netflix and Hulu on the DROID

In recent years, Netflix has become a leading source of video rentals for consumers. Relatively recently, Netflix has also added on-demand video streaming. This streaming can even be delivered wirelessly to computers and set-top boxes for your TV.

At the time of writing, Netflix is developing a new app for streaming movies on the DROID; it should be available by late 2010 or early 2011.

Hulu is a **Flash**-based video service for watching recent TV episodes. While you can get to the Hulu website on your DROID, the **Flash lite** player available for the DROID will not play the episodes available on Hulu. There is a **Hulu plus** subscription service available for other smartphone platforms; this same service is due to make its appearance on Android in the not-too-distant future.

Chapter 16

New Media: Reading Newspapers, Magazines, and E-books

Your DROID has the potential to replace your newspaper, favorite magazines, and even your book library. We're not saying it's time to give up paper books completely. We like them, too. However, you may find that your phone is a surprisingly good reading device. In this chapter, we'll explore ways to get your news and reading done without going to the bookstore or newspaper stand.

Newspapers on the DROID

Remember the days when newspapers were delivered to the house? Invariably, if there was one puddle in the sidewalk, that was where the newspaper landed! You took it out of that plastic bag, shook it off, and tried to make out what was in section two—the section that got soaked.

Well, those days may be gone forever. You now have the opportunity to interact with the news and even get your paper delivered every day—but to your DROID instead of your driveway.

Many newspapers and news sites are developing apps for the DROID, with new apps seeming to appear every day. Figure 16–1 shows two apps for reading popular newspapers.

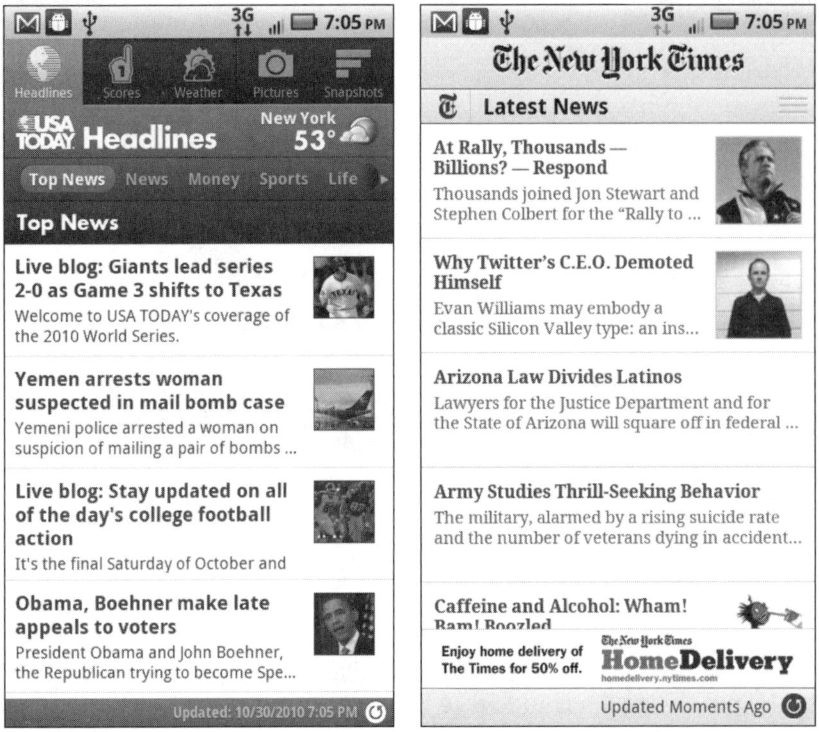

Figure 16–1. *The front pages of two newspaper apps.*

Popular Choices: The New York Times and USA Today

The New York Times and *USA Today* both have large circulations, but each paper has taken a different approach to bringing you the news on the DROID.

NOTE: You can always go and visit the dedicated web site for any news source. Some are optimized for the DROID, while others offer you a full web experience. Some require registration or a paid subscription to view the paper's full content.

You must first find, download, and install a news app on your DROID to use these. Here are the steps:

1. Locate your desired news app in the Android Market. You may find one or more news apps in the **Featured** section, and there's also a direct link to **News & Weather** under **apps** on the top of the Android Market home page.

2. Browse or search for your desired news app, just as you would for any other app.

3. Once you locate the desired news app, download it as you would any other app.

4. Once the app is downloaded, tap its icon to launch it.

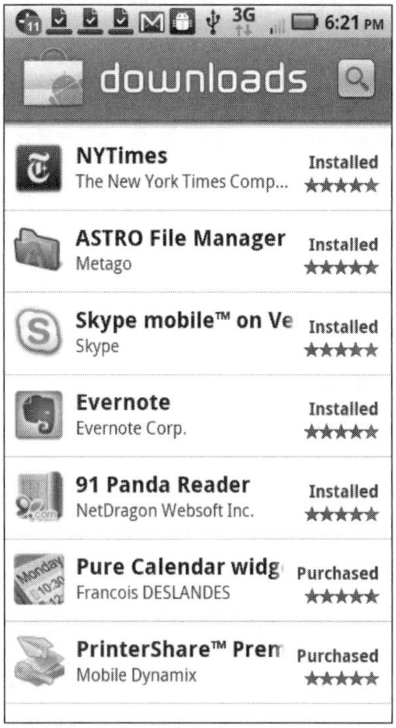

NOTE: Many news apps are free. Some are free to try, but require you to buy them to continue receiving them. Others offer limited free content, but you need to subscribe to gain access to their full content. See Chapter 17: "The Android Market" for more information.

The New York Times App

The New York Times offers a slimmed-down version of the paper in its free Android app. By default, you'll see the latest headlines. Figure 16–2 shows basic navigation from one section to the next. Simply pull down the section menu and tap.

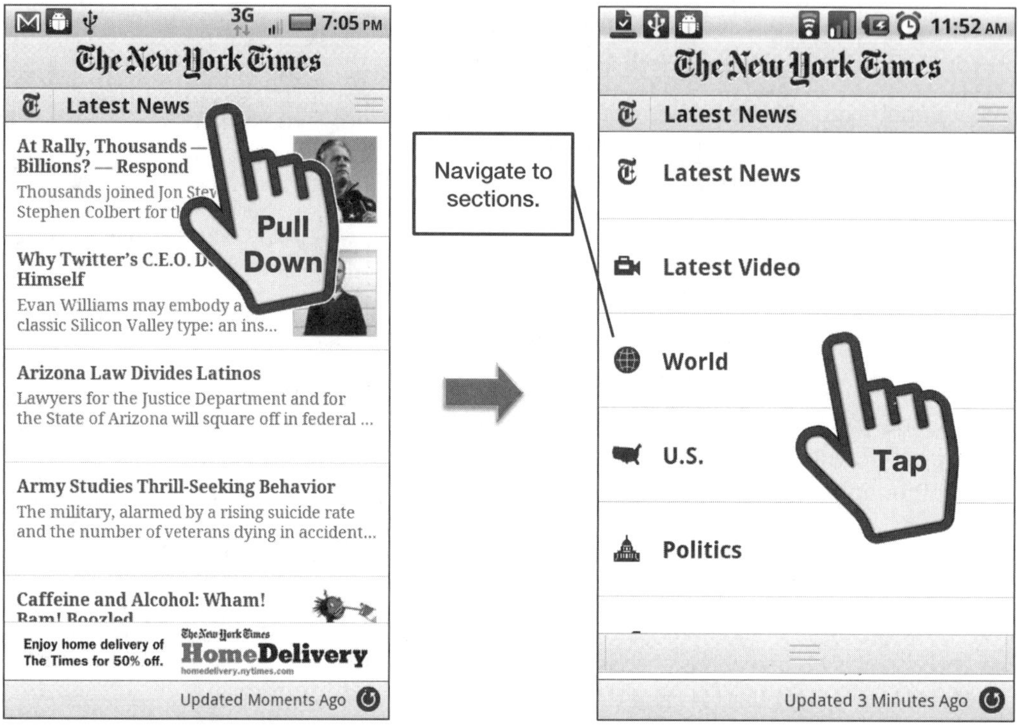

Figure 16–2. *Navigating to another section.*

Navigating **The New York Times** app is as simple as touching an article and scrolling through. Drag your finger up and down to scroll through an article, and drag your finger sideways to navigate to the next or previous article in the section.

To go back to the **Home** page, use the back button.You can also share an article with friends using e-mail, Twitter, Facebook, text messaging or any other app designed to handle Android share requests.

The **New York Times** app comes with a handy widget. Rather than launching the full app and navigating to your favorite section, you can add the widget to your **Home** screen and display headlines from your favorite section.

Tap on the small arrow on the side to navigate to the next headline, and tap on a heading to view the full article. To read more about widgets, see Chapter 6: "Organize Your Home Screens: Icons and Widgets."

USA Today App

While not as big and colorful as the physical paper version, you can still get USA Today on your phone. The app is available in the **News & Weather** category of the Android Market. Note that as of press time, it doesn't ship with a widget.

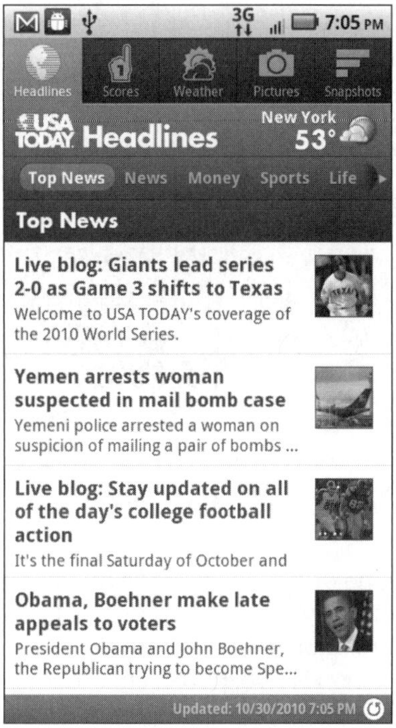

Download the app as you did the other news apps.

The app will detect your current location if you've enabled the GPS on your phone. That means the weather section can give you weather for your location as well as other regions.

The sections of the paper are near the top of the home screen. Just slide from right to left and then touch the section of the paper you want to read.

In addition, you'll find sections for **Headlines, Scores, Weather, Pictures**, and **Snapshots** along the top of the screen.

Moving Through and Enjoying Content

After you play for a while with all these news sites, you'll begin to realize that there is no real standard for moving around. This means you'll need to become familiar with each app's own way of navigating articles and returning to the main screen. Here's a short guide for generally navigating these types of apps; these features are common to **The New York Times** and **USA Today** apps:

- **Showing or Hiding Control Buttons:** Pressing the **Menu** button will generally show all your options for that article, picture, or video.
- **Getting to the Details of an Article:** Usually, you just scroll through the articles, just as you would on a web page.
- **Viewing a Video:** Tap a video to start playing it. Usually, this plays the video in the same manner as any other video. See Chapter 15: "Viewing Videos, TV Shows, and More" to learn how to navigate videos on your DROID.

- **Expanding a Video or Image Size:** You can try pinching open in the video or image and then double-tapping it. You can also try rotating to landscape mode.
- **Reducing a Video or Image Size:** You can try pinching closed inside the video or image. You can also try rotating back to portrait mode.

Adjusting Font Sizes and Sharing, E-mailing, or Saving an Article

The various apps for reading newspapers and other content usually include a button or icon for changing the font size. That same button or another one near it may also allow you to share, save, or e-mail an article to a friend. Some apps allow you to share the article with a social networking site, such as Facebook or Twitter.

> **TIP:** Almost all newspaper or magazine apps let you change font sizes and e-mail or otherwise share an article. Look for a button or icon that says **Tools**, **Options**, **Settings**, or something similar. In some apps, the font-size adjustment option shows as small **A** and large **A** icons.

News Widgets

One of the advantages of using your phone for news is that you're not restricted to a single source. There are many news aggregators, such as Google or Yahoo News, and there are lots of ways to view news headlines on topics that interest you.

On your DROID X, DROID 2, and DROID Global, you can use the Motorola **News** widget to view custom bundles of news feeds on subject and categories that interest you. On all DROIDs, you can also use the Android **News and Weather** widget. To learn more about using widgets, read Chapter 6.

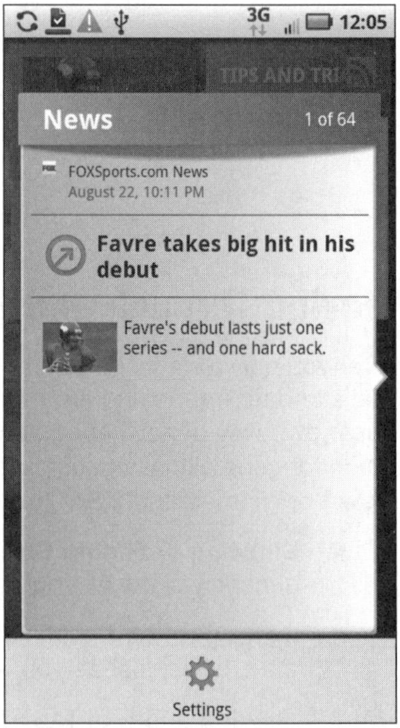

Magazines on Android

It is no secret that both newspapers and magazines have suffered declines in readership over the last few years. Android offers a totally new way of reading magazines that might just give the media industry the boost it needs.

Pictures are incredibly clear and brilliant in magazines on your DROID. Navigation is usually easy, and stories seem to come to life, much more so than in their print counterparts. Add video and sound integration right into the magazine, and you can see how the DROID truly enhances the magazine reading experience.

Some magazines, such as *TIME Magazine*, include links to live or frequently updated content. These might be called **Newsfeeds**, **Live Edition**, or **Updates**. Check for them in any magazine you purchase—they will give you the most up-to-date information.

TIP: Make sure to check the user ratings for a magazine or other app before you purchase it. Doing so may save you some money and some grief!

The Android Market offers some individual magazines for purchase (however, these magazines sometimes include limited content for free). You can also check out the magazine readers that provide samples of many magazines that allow you to subscribe to weekly or monthly delivery of a given magazine.

Unlike newspapers, only a few magazines are available for free.

TIME Mobile is one of the few available free magazines for Android. The navigation is similar to *The New York Times* and other newspaper apps. On the right, you can see how navigating to different sections is accomplished by dragging down the sections menu from the top of the screen.

Other magazines are beginning to be offered throgh E-book stores, such as the Nook, Kindle, and Kobo markets.

Comic Books on Your DROID

One genre of "new media" poised for a comeback with the advent of mobile is the comic book. Reading a comic book on your DROID really makes the pages come alive. Readers can also appreciate comics that would otherwise be harder to find, such as lesser known works or imported Japanese manga.

As this goes to press, there's still no official Marvel or DC Comic app for Android. However, DC Comics announced at the 2010 Comicon that they were working on one. There was no word from Marvel, but they do offer an iPhone/iPad app.

If you'd like to view free comics from lesser-known artists, the Android Comic Reader app allows you to browse and download content directly to your phone. Navigation is similar to other magazine and e-book readers. The example to the right is a story by Cory Doctorow called "iRobot."

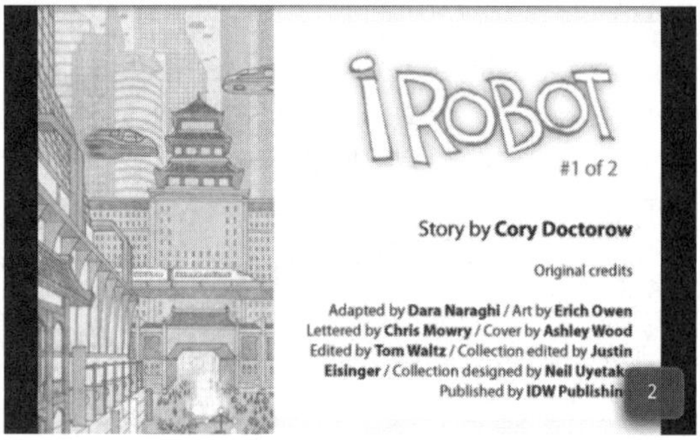

The DROID as a PDF Reader

Another way to get e-books and other content to phones is through PDF files. There are several apps capable of viewing PDF content, including Adobe's Acrobat Reader.

You can find the **Acrobat Reader** in the Android Market. See Chapter 17: "Exploring the Android Market.

Files can be sent as e-mail attachments, downloaded from websites, or transferred directly to your SD card. Clicking on a PDF file should open them in an appropriate reader.

We also like to use **Dropbox**, an Internet-based file sharing and storage app to transfer files to our DROIDs. You can find a quick link to download Dropbox in **Appendix A**. See Chapter 21: "Working with Notes and Documents" for a description of how **Dropbox** works.

E-Books

Unlike the iPhone, Android doesn't come with a standard e-book reader. That doesn't leave you out of options, however. You can choose from a large selection of readers, both for open and proprietary book formats. Keep in mind that there are several formats for reading books, and some companies put Digital Rights Management (DRM) on their e-books to prevent them from being illegally distributed or read in unauthorized readers.

Figure 16–3 shows just a sampling of available e-book readers for Android.

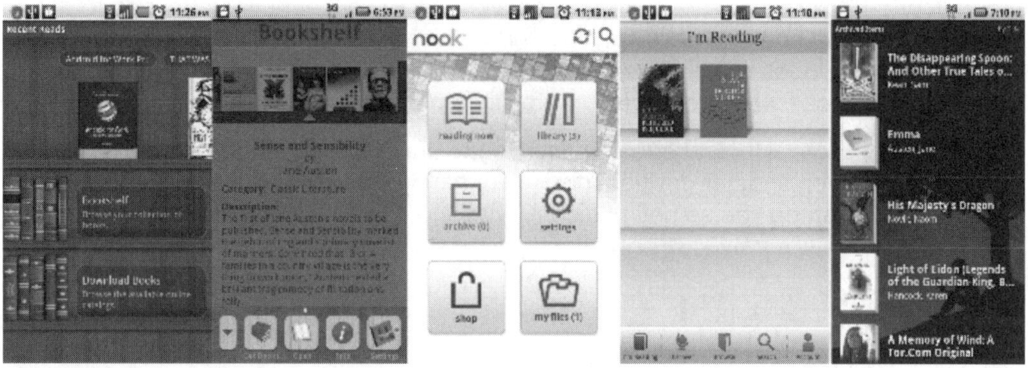

Figure 16–3. *Android e-book readers.*

If you have already begun creating a digital library, you'll want to stick with an app that is compatible with your content. If you're just getting started, there are a few things to keep in mind as you choose a reading app:

- **Proprietary formats**: Epub is one of the most widely accepted book formats, but even it comes in DRM and non-DRM varieties. Check to see if the book you purchase is protected by DRM or an open offering.
- **Storage**: Where are your books stored? Can you download a new copy if your phone crashes? Can you pick up your book on one device and finish on another?
- **Availability**: How many books are available from your online bookstore? If you can only buy a few books, and those books can only be read on a single reader, it's quickly a case of diminishing returns.
- **Price**: What's the average price for a book? Publishers are moving from a model where e-book stores offer variable pricing, so look at several titles.
- **Stability**: The digital world is a very rapidly changing place. There are no guarantees that your favorite store will stay in business.

NOTE: Many titles have a sample download. This is a great idea if you are not sure that you want to purchase the book. Just download a sample, and you can always purchase the full book from within the sample.

Proprietary Readers

Proprietary e-book readers are extremely convenient, because they're attached to a bookstore for easy book purchasing. You don't have to install anything extra or think about where your book will be stored. Just register for an account, click to purchase, and pay for the book when your credit card bill comes due.

The reading apps we'll look at all also allow you to start reading on one device and continue reading on another device. This is very handy if you've got a dedicated e-book reader, iPad, or laptop you sometimes carry with you.

Kindle Reader

The Amazon Kindle app is available through the Android Market. Amazon has made a Kindle app for several different mobile devices, so you can use it on Android phones, iPods, iPhones, laptops, and, of course, Amazon's dedicated Kindle e-book reading device.

If you've registered for an Amazon account, you can purchase and download books for the Kindle app directly from your DROID and read them from your DROID or any other device that runs a Kindle app.

You can find free Kindle books as well as paid books, although you must be registered in order to purchase these free downloads.

Kindle books do not transfer to non-Kindle readers, and they do not use standard ePub format.

TIP: If you use a Kindle device, don't worry about signing in from your DROID. You can have several devices tied to your single account. You will be able to enjoy all the books you purchased for your Kindle right on the Kindle app on your DROID. In some books, the publisher might limit this capability, but we have never run into this limit.

Just touch the Kindle app and either sign in to your Kindle account or press the **Menu** button and select create a new account with a user name and password.

Once you sign in, you will see your Kindle books on the home page. You can touch either a **book cover** to start reading or **Get Books** to start shopping in the Kindle store.

> **NOTE:** Touching **Get Books** from the Kndle menu will start up your **Browser**. From there you can purchase Kindle books. Once you are done, you will need to exit **Browser** and start up the Kindle app once again. If you just purchased a title, then you may see a button that says something like Go to Kindle for Android. Tap that button to jump right back to Kindle and start enjoying your new book.

To read a Kindle book, touch the **book cover**. The book will open.

To see the options for reading, just press the **Menu** button, and they will be along the bottom row of icons.

You can add a bookmark by touching the plus (+) button. Once the bookmark is set, the plus (+**)** turns to a minus (-).

You can go to the cover, Table of Contents, or beginning of the book (or specify any location in the book) by touching the **Go To** from the menu.

The font, as well as the color of the page, can be adjusted. One very interesting feature is the ability to change the page to **Black**, which is great when reading at night.

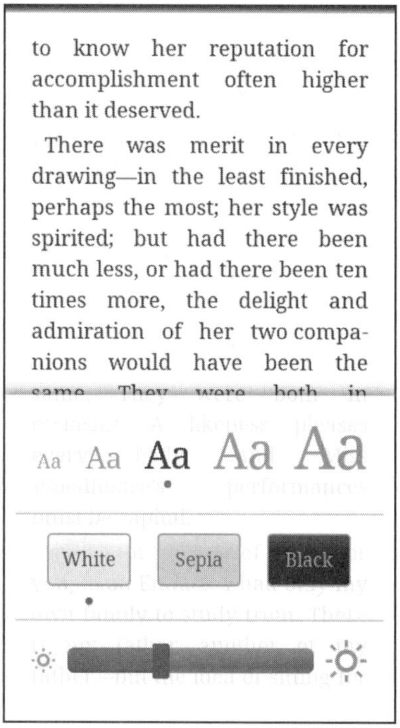

To advance pages, either swipe from right to left, or touch the right-hand side of the page. To go back a page, just swipe from left to right or touch the left-hand side of the page.

Tap the screen and a slider appears at the bottom, which you can move to advance to any page in the book.

To return to your list of books, just press the **Back** button or the Home option from the Kindle menu.

Nook

The Nook is Barnes & Noble Bookstore's dedicated e-reader. The Nook reader is actually a modified Android tablet, so it's not surprising that Barnes & Noble has released a **Nook** app for Android.

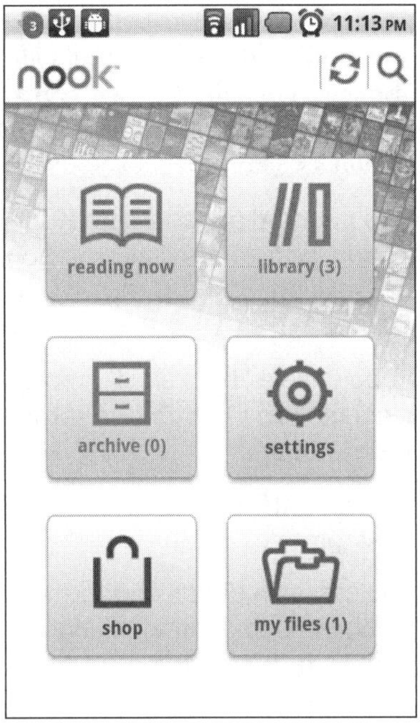

Register with an account either through the Nook app on your DROID or on the Web.

One of the selling points for the Nook app is that it allows you to lend books to other Nook users for up to two weeks. (Shortly before this book went to press, Amazon announced they were going to do the same for Kindle books.)

Open E-Readers

In addition to commercial e-book readers like the Kindle and Nook apps, you can use open, alternative readers that support common formats. The disadvantage is that many of these apps are not attached to a bookstore, so purchasing books is not as convenient.

We'll start with **Kobo**, which looks like a commercial e-reader with a brick-and-mortar bookstore backing it, but it is also an open reader. That means it has the polish of a proprietary reader with fewer content restrictions.

Kobo

Kobo is an e-book reading device owned by Borders and other investors. Just as with the Kindle and Nook apps, you can register with an account and download books directly to the **Kobo** app. It was formerly known as Shortcovers.

Like the Kindle reader, the Kobo reader asks you first to sign in to your existing Kobo Books account. All of your existing Kobo Books will then be available for reading.

Kobo uses a "bookshelf" approach similar to many readers. Tap the **book cover** for whichever book you wish to open.

Or, touch the **List** tab to see your books organized in a list format.

You can also directly go to the Kobo store to purchase books by touching the **Discover** or **Browse** buttons at the bottom

However, Kobo also supports and sells standard Epub and Adobe Digital Editions book formats in their store. That means you may be able to read e-books from public libraries and independent booksellers on your Kobo, too. You can also take books you buy for your Kobo and transfer them to another reader if you prefer.

Touch the **Settings** button brings up other buttons for seeing information about the book and adjusting the page transition style and font. Touch any of the buttons to make adjustments to your viewing.

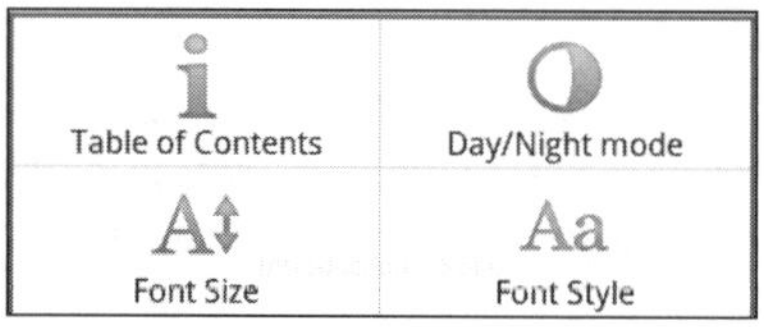

Aldiko

Aldiko is a free e-book reading app that reads DRM-free Epub formatted books. You can download some books through the Android Market for use with Aldiko, or you can download any Epub book and place it in the **eBooks** folder on your DROID's SD card.

O'Reilly Media sells DRM-free e-books through the Android Market specifically for use with **Aldiko.**

Libris

Libris comes in both a free version, Libris Lite, and a paid $2.99 app. The paid app offers support for more book formats, including Mobipocket, Epub, Libris, PalmDoc, and plain text, but it can't read any format with DRM.

You can use Libris to read books on your computer desktop (if your computer has Java installed) or on your Android, but the books are not stored online like they are on some e-readers, so you'll still need to physically sync them between devices.

That said, if you're reading free books or buying them through the Libris-supported Fictionwise online library, Libris has a nice user interface and supplies a free tool for converting text to Libris format at www.hillbillyinteractive.com.

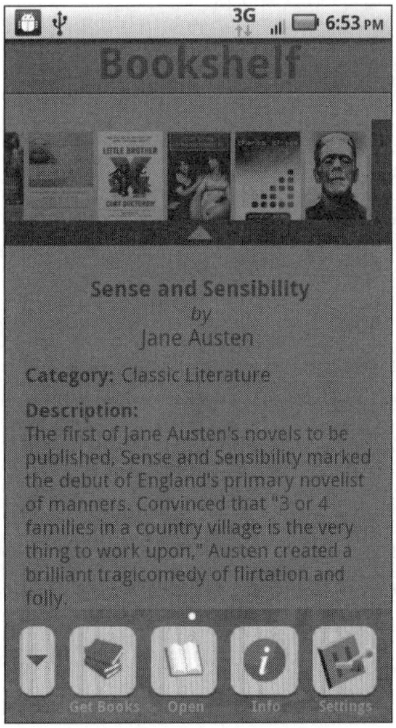

Reading E-Books

Just as with newspaper apps, there are no standards, but there are a series of things that are generally true of most e-reading apps for your DROID. When you first launch an e-reading app, nearly all of them offer a tips menu for navigation.

- **Horizontal and Vertical:** You can switch between horizontal and vertical views for easier reading. Tilt your screen, and the text will reflow to follow. Some readers also let you lock the orientation, so accidental screen movements don't reflow your page.
- **Bookshelves:** Most apps are organized around a bookshelf analogy. You'll see a series of book covers showing your available library of choices.

Touch any title in your library to open it for reading. The book will open to the very first page, which is often the title page or other front matter in the book.

- **Pages**: It's just text. You could just see a very long column of text that scrolls, but nearly every reader has kept the analogy of pages of text to make it much easier to navigate. Swipe your finger sideways to flip between pages. Some apps even show page-curl animations as you do this.
- **Chapters:** You're not stuck just navigating page by page. You can navigate from chapter to chapter or between bookmarks, or in the case of the Kindle app, to a navigational number point in the text. Press the **Menu** button, and you'll usually see a way to navigate directly to a chapter, section, or bookmark.

- **Contextual Menus:** In most readers, you can look up words, make notes, or create bookmarks.

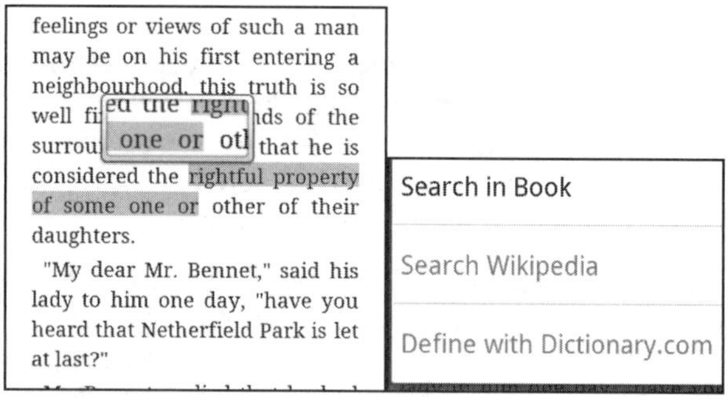

feelings or views of such a man may be on his first entering a neighbourhood, this truth is so well fi ... nds of the surrou ... that he is considered the rightful property of some one or other of their daughters.
"My dear Mr. Bennet," said his lady to him one day, "have you heard that Netherfield Park is let at last?"
Search in Book
Search Wikipedia
Define with Dictionary.com

Sometimes pressing the **Menu** button does that, and sometimes it's done by long-pressing on a word in the book. Generally, you look up word definitions by launching a Web browser to either an online dictionary or Wikipedia.

- **Search:** You can also search to find specific text within a book. Just look for the magnifying glass button. You can usually either press the **Search** button or the **Menu** button to find it.
- **Customize Your Experience:** There's almost always a **Settings** menu, and you can usually find this by pressing the **Menu** button. Some readers have more choices than others, but most allow you to increase or decrease font size, and switch from day or night view (which usually just switches from black text on a white screen to white text on a black screen). In some cases, you can also make the background a tan color for less eye strain.

Chapter 17

The Android Market

The Android Market is the primary place for downloading apps for your DROID.

Right now, there are several versions of Android shipping on a wide variety of phones, and the Android Market sells apps for all of them. So how do you know if your DROID will run the latest Twitter app? The general rule is that, if you can see an app in the Android Market on your phone, you can run it. Developers can exclude incompatible devices from seeing their apps in the Android Market. However, you should always read the app description just to be sure.

In this chapter, we'll go into more detail about how you find and download apps, how you can leave feedback, and how you can try before you buy. You'll also learn about must-have apps to download right now.

Using QR Codes

Before we go further, let's introduce the concept of QR (quick response) codes. You may have seen these square barcodes on objects or websites. The QR code was patented by the Japanese company, Denso Wave. Rather than restrict the use of such codes with licensing fees, Denso Wave chose to allow anyone to generate or use QR codes without having to pay a fee, and their use has been growing as smartphone use has grown.

The example to the right goes to `http://zxing.appspot.com/generator`, which is a free QR code generator.

QR codes can contain all sorts of information, including map locations, URLs, notes, names, phone numbers, and product identification. You don't have to worry about scanning them right-side up; upside down and sideways will work, too. They're easily read by phone cameras, so they make an ideal way to offer information to phone users without requiring a lot of typing. In fact, you may want to print a QR code on the back of your next business card, so smartphone users can scan in your contact information immediately. You can generate your own codes at `http://zxing.appspot.com/generator`.

Your DROID likely did not ship with a barcode reader, but it's easy to get one. There are countless apps in the Android Market that allow you to scan QR codes, including **Google Goggles** and ZXing's **Barcode Scanner**.

In this chapter, we'll use QR codes whenever possible. If you're reading this book with your DROID and (and not reading this book *on* your DROID), just use the QR code to get to a given app faster.

Browsing the Android Market

You can visit the Android Market Showcase on the Web at www.android.com/market; however, this site will only show you a fraction of the available apps. Google has plans to change this in the future, but for now you'll need to use your phone to see the apps available for DROID. Launch the **Android Market** app from your phone's application tray or desktop. The initial page will look similar to the figure on the right, with buttons for **Apps**, **Games**, and **Verizon**; a splash banner; and a list of featured apps.

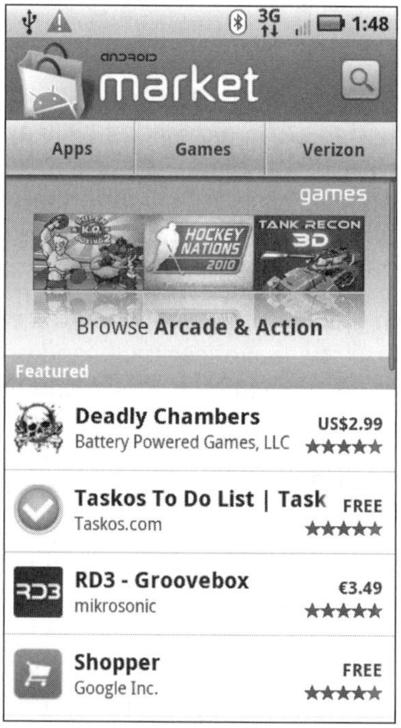

NOTE: You can also browse available apps on the Web at `www.androlib.com`, `www.appbrain.com`, and `www.androidzoom.com`. These are ad-sponsored sites that pull data from the Android Market; however, as third-party sites, they don't always produce identical search results.

Notice that each app lists a rating out of five stars, as well as the app's price. You can tap an app to read more details about it, including user reviews. Sometimes you may want to browse through the featured apps to see what is new; sometimes you know exactly what you want; and sometimes you want to browse, but only within a category, such as productivity apps or shopping. Any of these approaches is valid.

If you know exactly what you want or how to frame what you want precisely, use the **Search** button at the top of the screen. You can search for a name or keywords. For instance, searching for "Twitter" would show you both the official **Twitter** app and apps that use the Twitter service in their description, such as **HootSuite** or **Touiter**.

Navigating by Category

Follow these steps to navigate the Android Market by category:

1. Tap the Apps button.
2. Tap a category to select it.
3. You'll see three new buttons at the top: Top paid, Top free, and Just in. By default, the Top paid category is selected, but you can switch to free or recent apps by tapping the appropriate button.

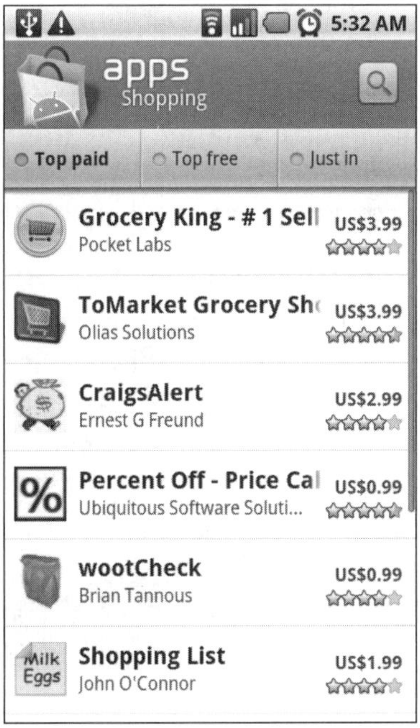

NOTE: Apps in the Android Market are weighted by popularity and not strictly listed by rating. This is because it's easy for an app to get a five-star rating if only one person has rated the app.

4. Tap the name of an app to see the details page associated with that app. You'll see something that resembles Figure 17–1; and as you scroll down, you'll see the app's name, rating, two screen captures, the price, and a description of the app submitted by the developer. You'll also see any website and contact information the developer or publisher has provided, such as an email address and phone number.

   You'll also see links to any other apps that the developer or publisher may have created. If the app is deceptive or malicious, the very bottom of the page gives you the option to flag it.

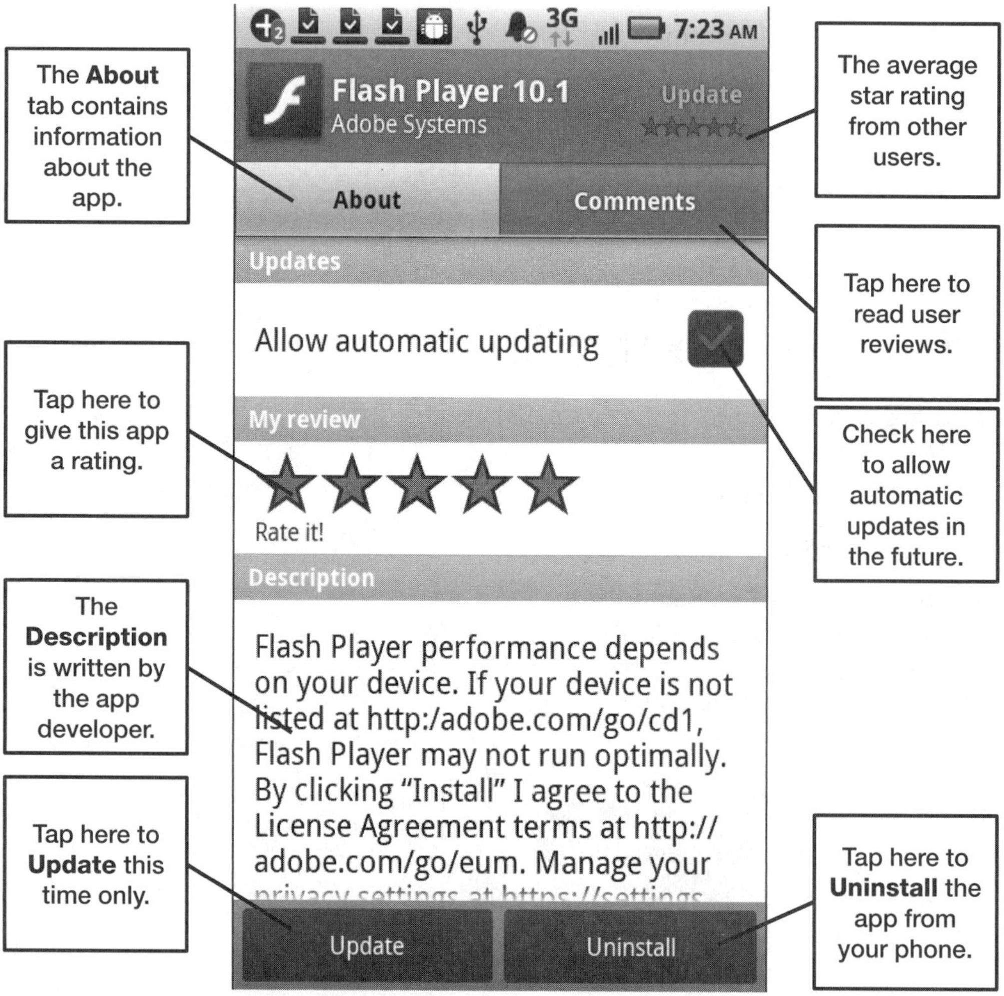

***Figure 17–1** App details.*

5. Tap on the **Comments** button to see user ratings and comments.

6. Comments aren't located at the bottom of the page; rather, they have their own section, as shown to the right. There are **Up** and **Down** buttons under each rating. This gives site users a quick-and-easy way to give a thumbs-up or thumbs-down to a given comment. You can rate comments as **helpful**, **unhelpful**, or **spam**. This is similar to the way users can rate Amazon.com reviews.

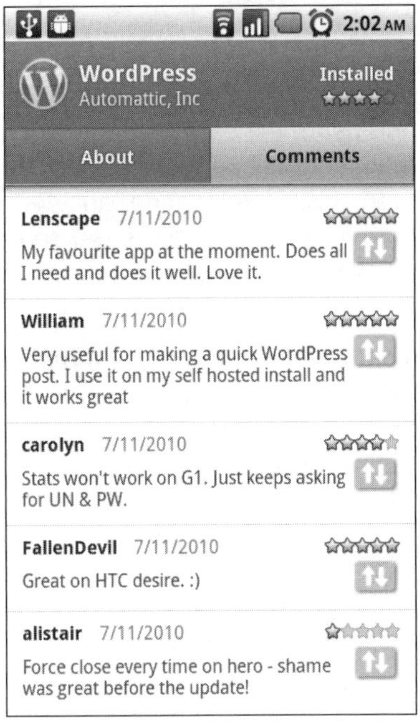

Paying for Apps

The Android Market has a huge selection of free apps, but there are times when it's worth it to pick up a paid app. You use **Google Checkout** to buy your app, so if you have not already signed up for a Google Checkout account, you should do so.

Currently, T-Mobile is the only supported carrier for direct billing (it supports US-dollar transactions only); however, Verizon is rumored to be planning a separate Verizon app market that would support direct billing.

Google Checkout is an online payment-processing system. Register for a Google Checkout account by going to `https://checkout.google.com/buyerSignup` and using the same Google account you use as the main account for your phone. You can enter your credit card information, and Google will store it. Be sure to register in advance, so you can avoid the hassle of entering credit card info on a phone keyboard.

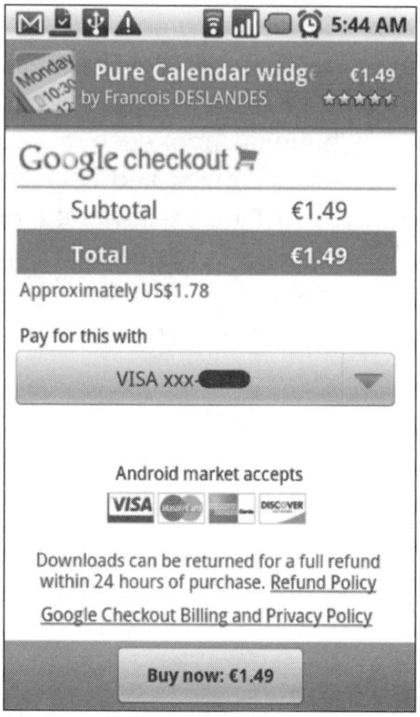

You can purchase apps in most foreign currencies using Google Checkout and your credit card, as long as your credit card allows foreign currency transactions. Google will give you an estimate of what the price is in US dollars, as shown in the screen capture to the right. However, your credit card may charge you a fee for currency conversion or use a different exchange rate, and that won't be reflected in the estimate.

NOTE: You have 24 hours after purchasing an app to "return" it for a full refund. Your refund price does not include any transaction fees for foreign currency.

Downloading Apps

Other than payment processing, the basic steps to downloading an app are the same:

1. Go to the Android Market by tapping the **Android Market** app icon.
2. Navigate to the details page.
3. Tap the Install button at the bottom of the screen.

   Android will confirm that you want to download the app, and it will also show you specific information about application permissions. In most cases, the uses are quite legitimate, but you should read them carefully to make sure a word puzzle game doesn't have access to dial your phone, for example.

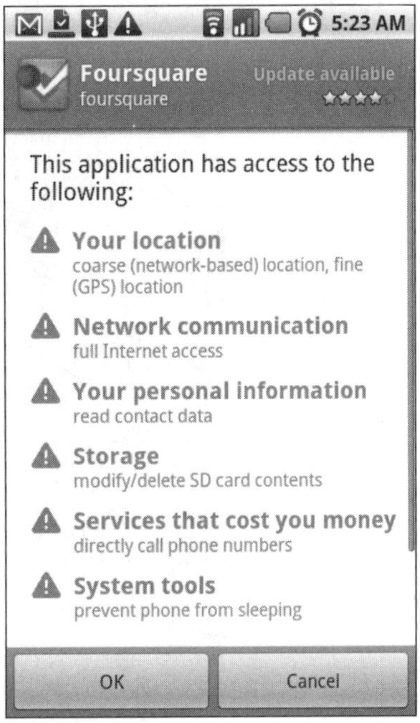

4. If everything looks good, tap the **OK** button.

   Your download will start. You'll see a progress bar in the Android Market, as well as a notice in your **Notification** bar that a download is in progress. The notice will change once the download is complete. There's no need to keep using the **Android Market** app while you're downloading. The download will continue in the background.

Installing Apps

In most cases, downloading the app installs it automatically. If there are any widgets included with the app, you'll need to long-tap the **Home** screen to install them. If there are additional steps, the developer should provide instructions about these. In the case of some paid apps, you may have two downloads. The first is a trial version, and the second is a separate key that doesn't actually do anything other than unlock the full features from the first download.

NOTE: Android 2.2 (Froyo) allows developers to give you the option to install apps on your SD card instead of the phone's hard disk. This can potentially save some space. However, it is up to the developer to allow the option in his app (see Figure 17–2).

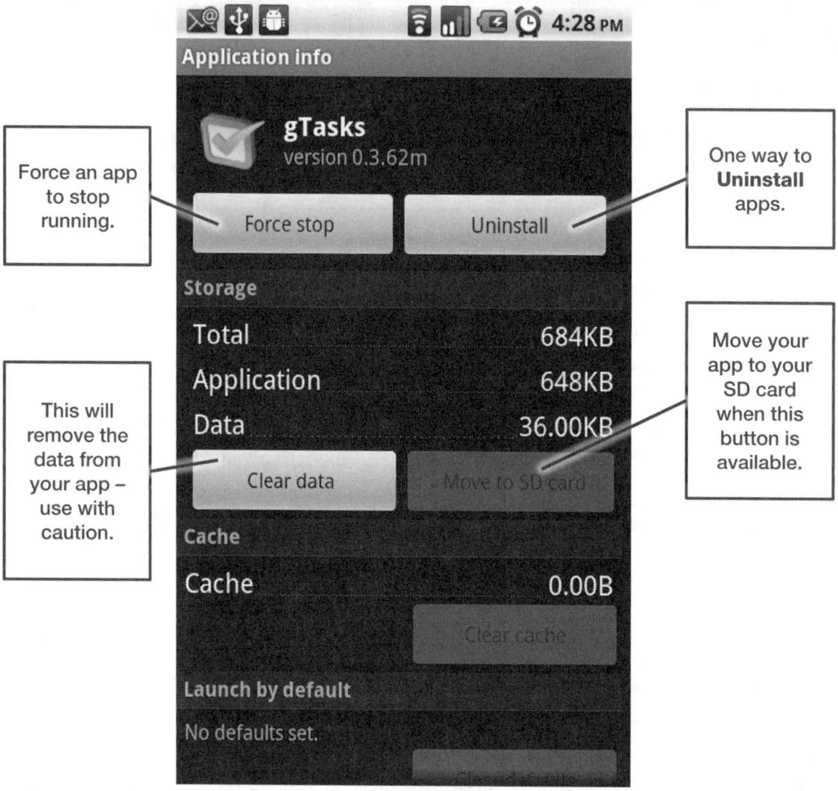

Figure 17–2. *Installing an app to the SD card.*

NOTE: to get to the screen above, go to Settings -> Applications -> Manage Applications and then tap on the application for which you wish to see the Application info.

Uninstalling Apps

There are two basic ways to uninstall an app. The more complicated method is to go to the **Home screen**, press the **Menu** button, and then tap **Applications** > "**Manage applications**." Tap the app you want to remove, and then tap **Uninstall.**

The second, much easier method is to find your app in the Android Market, navigate to the details page for that app, and tap the **Uninstall** button. If you paid for the app, you'll see an **Uninstall & refund** button. You have 24 hours after a purchase to return it for a refund. However, you must do this through the Android Market.

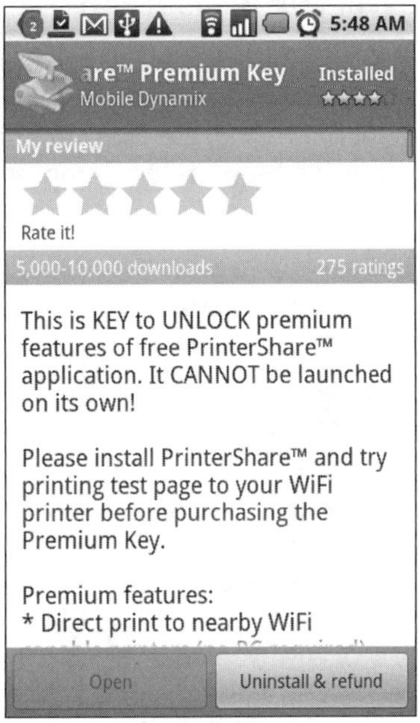

Follow these steps to get a refund on a purchase:

1. Go to the Android Market by pressing the store's app icon within 24 hours of a purchase.
2. Press the **Menu** button.
3. Tap **Downloads**.
4. You'll see the name of all apps you've recently downloaded. Tap the name of the app you wish to uninstall.
5. Tap the **Uninstall & refund** button.

   You'll see a warning window telling you that you're about to uninstall an app. You'll also be told whether you can install the app again at no charge. In the case of paid apps, not only must you pay for it again (since you're being refunded), but you cannot return it a second time for a refund.

6. Android also collects data on why you chose to uninstall an app. Select a reason or "I'd rather not say," and then click **OK**.

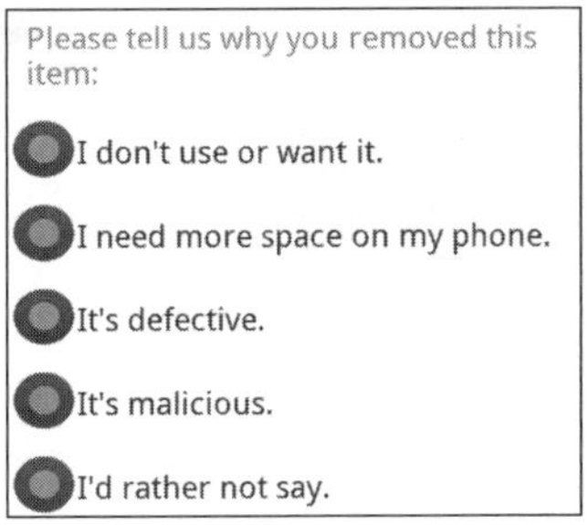

Updating Apps

Occasionally developers will add features or bug fixes. When there's an update available, you'll see an alert in your **Notification** bar at the top of your **Home** screen.

You can tap the alert to go directly to the Android Market, or you can launch the **Android Market** app, press the **Menu** button, and then tap **Downloads.**

Tap an app to view the details, and then tap the **Update** button at the bottom of the screen.

The process for updating an app is very similar to the process for downloading it. You can update everything at once by tapping **Update all**.

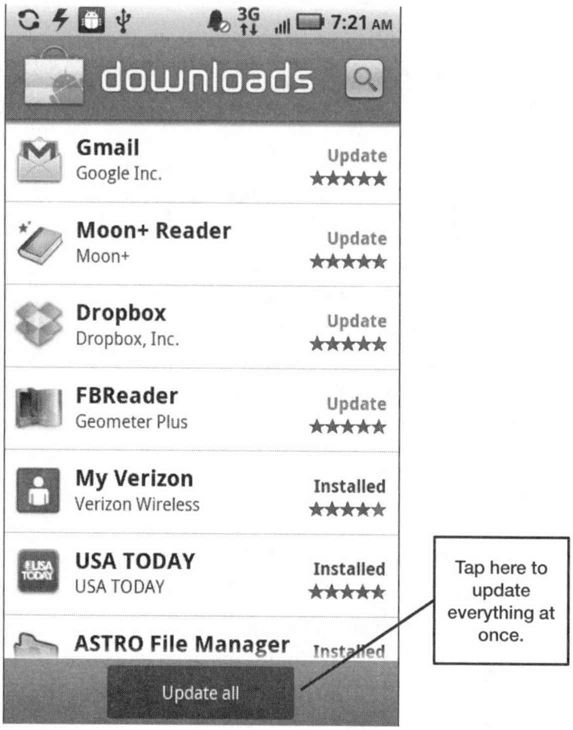

For an app that you'll unhesitatingly update, you can also specify that you'll allow automatic updates on that app's **About** page in the Android Market, as seen in Figure 17–1.

Checking the **Allow automatic updating** box means available updates will automatically install as soon as they're available, unless the app changes permissions, in which case you'll need to update the app manually.

CAUTION: Read the comments before updating an app or allowing automatic updates. Sometimes an update will break something or not work with particular phone models.

Rating and Commenting

If you've installed an app, you may want to go back and tell other users how well it works. Even a comment verifying that it works on your model of phone is helpful. The first step is to rate the app.

The ratings area is at the top of the app's detail page under the **My review** section. Tap the **Stars** icon, and you'll see a rating screen resembling the figure on the right. Indicate the number of stars the app deserves by dragging your finger from left to right. Once you are finished, tap **OK**.

Once you've rated an app, you'll see a link to post a comment right under your rating. You can only comment on apps you've rated.

Installing Apps Outside the Android Market

In most cases, the Android Market is all you need to find apps for your phone. However, developers aren't required to offer their apps through the market. App developers in countries that don't yet support paid apps through the Android Market may want to sell paid versions of their apps outside the market, for example. This also allows developers to create alternative app markets, like SlideME (`http://slideme.org`). You can download apps from other locations, but you need to enable downloads from unknown sources to authorize this ability.

Follow these steps to enable apps from unknown sources:

1. Go to the **Home** screen
2. Press the **Menu** button.
3. Tap **Applications**. On the original DROID, touch Settings and then applications.
4. Check the box next to **Unknown sources**.

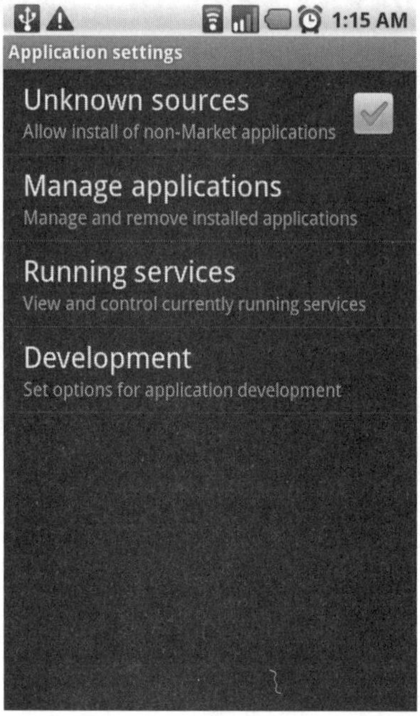

Now you can download apps that aren't in the Android Market.

If you know the location of an app, you can navigate there through your phone's **Browser** app and install it. You can also install apps delivered by email.

Android apps have the `.apk` extension. Download the APK file, and then tap the message that the download is complete in your **Notification** bar. You'll see a screen asking you if you'd like to install the app, similar to the figure on the right.

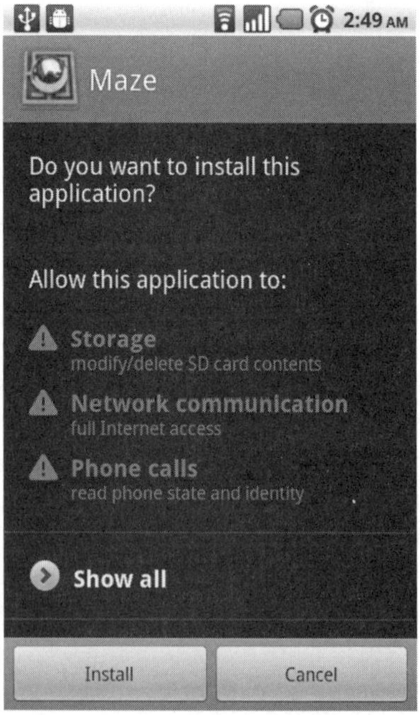

Keep in mind that this isn't a move without risks. Apps could have been removed from the Android Market for malicious activity, and those apps can be remotely removed from your phone when you install them using the Android Market. You can still see the permissions required by the app, so be mindful of these permissions before you install it.

Ten Apps to Install Right Now

There are a lot of quality apps available, so here are a few of our favorite general-use apps. We've provided the QR code, so if you have a barcode-scanning app already installed, such as **Barcode Scanner** (which might have made the list if we had more room), you can scan in the code and navigate directly to the app on the Android Market. We've also listed additional apps in Appendix A at the end of this book.

ShopSavvy

ShopSavvy is a longtime favorite free app. It debuted on the Android and really showed off the potential of the phone. It uses your camera to scan in barcodes and comparison shop with both local and online items. Local shopping results are sometimes limited because the app is only as accurate as the data available online.

You can set price alerts, tweet about your scans (although this gets obnoxious), view your history, create a wish list, and more. If you're searching for an item that has neither a barcode nor a store sticker over the barcode, you can also enter your search terms by hand.

Best of all, **ShopSavvy** supports QR codes; so if you download this app first, you can scan any of the QR codes you see in this book.

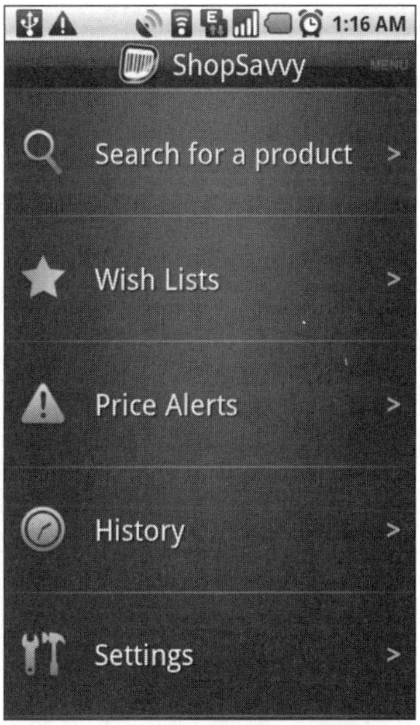

Lookout Mobile

Lookout Mobile is a free app that provides three valuable services: phone location, virus protection, and backup. You can enable or disable the services as you see fit. For remote location and backup, you need to register with the company's website at `www.mylookoutmobile.com`. You can also schedule backups for times when you know your phone will be in its charger.

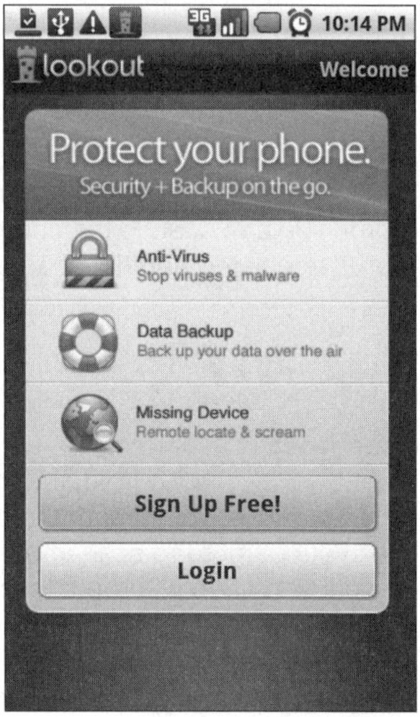

Virus protection may not seem like a huge deal right now; however, it could become a more significant problem as smartphones and the Android platform in particular gain in popularity. **Lookout Mobile** scans apps as you download them, which means downloads do take slightly longer to install.

If you ever lose your phone, the **Lookout Mobile** app's **Remote locate & scream** feature might help you find it.

Mint.com

Mint.com is Intuit's free, online personal finance software. The official **Mint.com** app is also free, and it gives you access to your personal finances. You can also download a widget that gives you access to this software on your DROID. You can choose whether the phone remembers your data or requires a password each time you log in. This is as simple as making the appropriate choices in the app's **Settings** menu, which you access by pressing the **Menu** button.

> **CAUTION:** If you choose to install the **Mint.com** widget portion, you're exposing your financial data to anyone within eyesight.

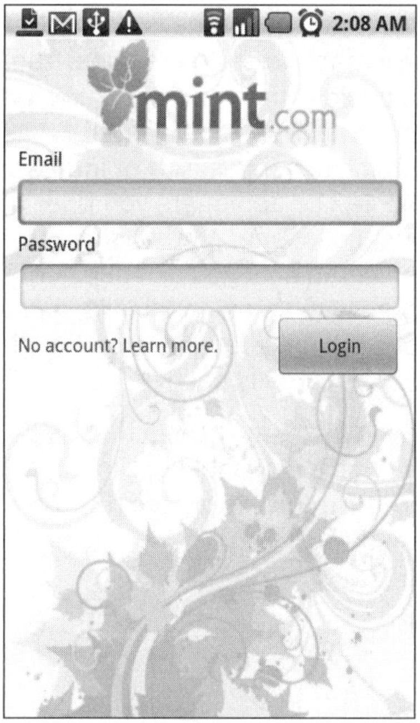

Cooking Capsules Taster

Cooking Capsules is an innovative approach to selling recipes. The **Taster** module is free, but the **Brunch** app is a premium app. However, at $0.99, even the premium version isn't going to break the bank. The app offers you brief video instructions on preparation; a shopping list of ingredients you can check off as you purchase them; and finally, a checklist of cooking instructions based on the video you've already seen.

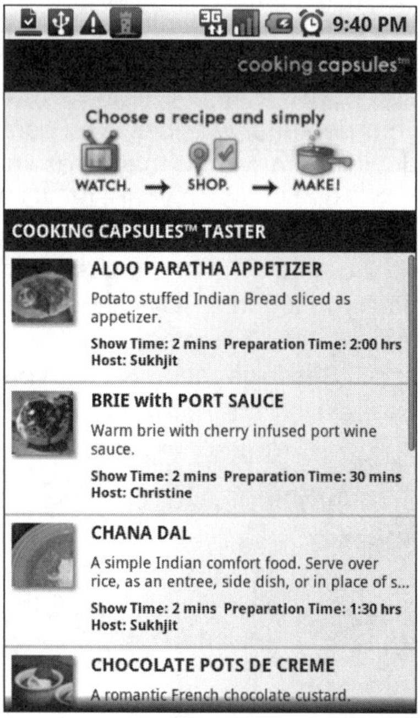

Even if you're not much of a chef, this app will have you cooking like a pro. Try the free **Taster** app before moving on to other modules.

Evernote

Evernote is a web service available at **www.evernote.com** that allows you to keep virtual scrapbooks of notes, pictures, web clippings, and audio files. You can add tags and search terms to your notes and access them anywhere on the Web. This means you can make a grocery list and have it available on your phone. Or, you might take a picture of something on the road and examine it from your desktop computer. You might even take a quick audio note or picture in the parking garage, so you can remember where you parked. The app also includes a widget.

The **Evernote** app is free, as is the basic Evernote service. There is also a premium Evernote subscription service with higher storage limits.

Pandora

Pandora is an Internet radio service that lets you create custom "radio stations" based around a song or artist. The playlist won't consist entirely of songs from that act, but it will share common style features as analyzed by the Music Genome Project. You can refine the choices by clicking a **Like** or **Dislike** button. It's a great way to find new music you didn't realize you liked; it's also a great way to listen to music on your phone without having to download it.

Pandora offers free (ad-sponsored) listening for 40 hours per month. You can upgrade to a premium account for $36 a year; the premium service features unlimited listening and no ads. The Android app is free either way, but it's subject to the same limits as your user account.

New York Times

The official **New York Times** app is a free app that lets you see the newspaper's content in a format optimized for mobile viewing. It's also easier to carry around than a full paper. There's no registration required, and at this point there appears to be no advertising for anything other than home delivery of *The Times*.

There is also a widget version of this app.

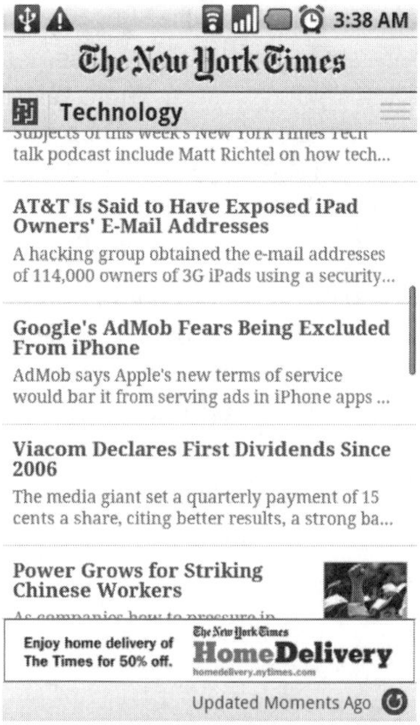

Google Voice

Google Voice is Google's free phone-call forwarding and visual-voicemail service. It allows you to use a single telephone number as your point of contact, even if you're not always at the same phone. You can also use it for free SMS text messages and discount international calls.

The picture on the right is courtesy of Google, and it shows how **Google Voice** can also give you a text transcript of your voicemail messages.

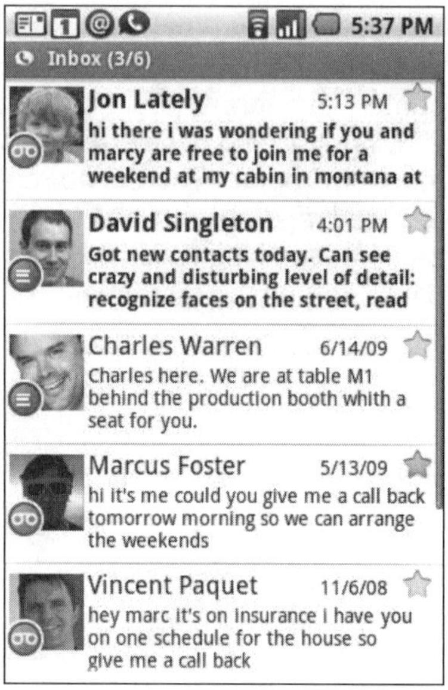

WikiMobile Encyclopedia (Bonfire)

WikiMobile Encyclopedia may not be the most accurate encyclopedia, but it's certainly a great first start. If you find yourself looking things up in Wikipedia all the time, it makes sense to have an app that can do that for you. This encyclopedia uses less bandwidth than your web browser, so you get the results faster. The app lets you swipe through pages one at a time; it also includes a **Back** or **Forward** option for navigating densely linked pages.

Yelp

There are countless restaurant-finding apps for phones these days, but **Yelp** has a vibrant user community and ratings for just about every location. You can use this app to find a restaurant, bank, gas station, or drugstore near you, at home, or on the road. The app's straightforward interface and copious reviews make this a must-have for anyone who travels or just likes to eat out.

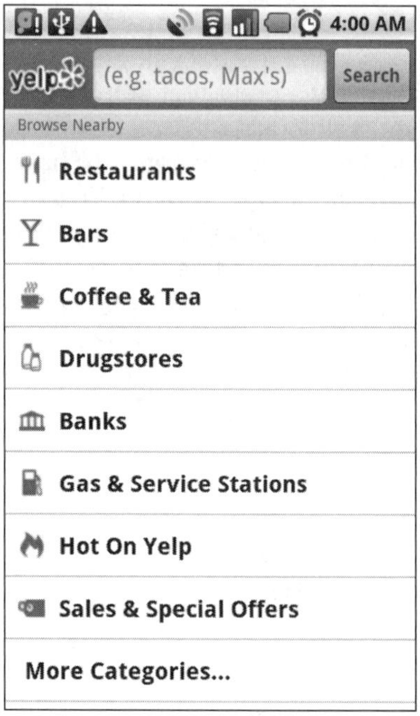

Chapter 18

Taking Photos and Videos

Camera phones are not a new phenomenon, but only recently have those cameras improved to the point that you can leave your point-and-shoot behind and still end up with decent photos. Not every phone is created equal when it comes to cameras, and the same is true for DROIDs. The DROID X sports a higher resolution camera than the DROID 2 or the DROID, but all three are capable of replacing your point-and-shoot in a pinch.

Whether you need to document work sites or scan bar codes, this chapter will get you up and running with Android photography. We'll discuss a few digital camera basics, like megapixels and image size. We'll also talk about how to adjust your camera for different lighting conditions, and how to turn the flash on and off.

You'll also learn how to enhance your photos and share them by email, Internet, and MMS, (picture texting). Finally, you'll learn about using your phone to take video footage, and some of the apps that will make your photography and video session shine.

NOTE: If you're using an original DROID, your camera screens and settings are going to look somewhat different from what is pictured in this chapter.

Understanding Your Camera

Before going into the nitty gritty of shooting photos, let's discuss the camera on your phone. Since there is no standard, one-size-fits-all Android phone, there's no standard Android camera. However, there are a few things most phone cameras have in common.

So far, no DROID will match the quality you'll find in a DSLR (digital single-lens reflex) camera. This is the type of professional camera with a separate lens and body that lets you adjust just about everything and change lenses for specific purposes, such as long distances or wide angles. DSLRs are expensive, large, and heavy, and we don't have the technology to fit them in a phone . . . yet. Likewise, you won't find the video quality in a phone that you will in a dedicated digital video camera, and phones just don't have the memory to store large, uncompressed video files.

However, if your job does not require professional high-end photography or video, it is entirely possible to use your phone for these purposes, and avoid having to carry around of two or three separate devices.

Megapixels and Image Size

Each square on a monitor or phone display is a pixel. A megapixel is a million pixels, or 1000×1000 pixels. Webcams are generally either a low-quality .3 megapixels (close to the size of old standard-definition television broadcasts) or 1.3 megapixels, the size of an SXGA (1280×1024) monitor. Neither of those is large enough to yield satisfying print results, because of yet another dimension, pixels per inch (ppi) (also called dots per inch, or dpi).

When you display images on a monitor, 72 dpi looks fine. However, if you print that same image, it will look horrible at that resolution. You'll be able to see every pixel. If you're printing, you want an image somewhere around the 250 to 300 dpi range for good print results; most professionals use 300 dpi as the standard. That means to get a quality 8×10-inch photo, you need a camera with at least 5 megapixels for a 250 dpi print and 7.2 megapixels for a 300 dpi print.

The Motorola DROID and DROID 2 cameras both have 5-megapixel cameras with flash. The DROID X has an 8-megapixel camera, so both will handle 5x7 prints, and the DROID X will do better for 8x10 prints. Some other smartphones have has an 8-megapixel camera on the back and a 1.3-megapixel camera on the front for video conferencing, and this may be something we'll see from Verizon Wireless in future DROIDs.

Video resolution is lower than print resolution. High-definition (HD) video is at maximum just slightly bigger than 2 megapixels. However, video struggles against the amount of space it takes up, so most phones do not support HD video capture, and those that do generally make some sacrifices. Full resolution HD video can be as large as 1080x1920 pixels at 60 frames per second. Needless to say, that's not a resolution you'll see on phones at this point. Many entry-level video cameras don't even support that rate.

The DROID X will shoot 720p at 24fps. That means it will shoot a video of 1280x720 pixels at 24 **frames per second**. It also uses three microphones to capture better audio. The DROID 2 will capture **standard-definition** video at 720x480 resolution and 30 frames per second. The DROID will capture the same **standard-definition** video at 720x480 resolution at up to 24 frames per second.

Focus

Point-and-shoot cameras come with either **fixed focus** or **autofocus**. Fixed-focus cameras are optimized to take a photo with the same focus—usually from a couple of feet to infinity. They use the same aperture opening and shutter speed for every single picture. That means anything too close will be out of focus. This is the type of focus you get with disposable cameras, because it's cheap and doesn't require any sort of adjustment on the user's end.

Autofocus cameras change the focus by using software and hardware adjustments. The biggest difference you'll notice as a user is that you can focus on things very near the camera lens, such as bar codes. It also means you'll have more out-of-focus pictures, since the autofocus might not always work quite as well as you'd hoped, but the overall picture quality will be better. Your DROID camera comes with autofocus, but some settings can use fixed focus benefits. Some settings, such as **portrait mode**, can also take advantage of tap-to-focus. Just tap the screen to focus on that portion of the image.

Zoom

Zoom is another popular feature. There are two types of zoom: **optical** and **digital.** Optical zooms use the camera's lens (the camera optics) to magnify part of the photo frame. You can still get a high-resolution photo from an optical zoom. Digital zoom is just a software solution in which the camera makes part of the picture look bigger. It's the illusion of zoom without adding any detail to the picture, and this is the type of zoom you find on phones. When possible, it's best to ignore digital zoom and just stand closer to the subject of your photo or video. However, that's not always practical or possible, and that's where digital zoom is handy.

Taking a Picture

A lot of this chapter is going to depend on which phone you use. Not only are there differences in physical hardware, but the interface is different between the DROID and DROID 2/X camera and galleries.

To take a picture, press the physical **Camera** button on your phone as shown here. You can also launch the **Camera** app from the app tray or an icon on your **Home** screen, but the button is the easiest shortcut. Other software may also allow you to use your phone's camera, and we'll cover some of it in this chapter.

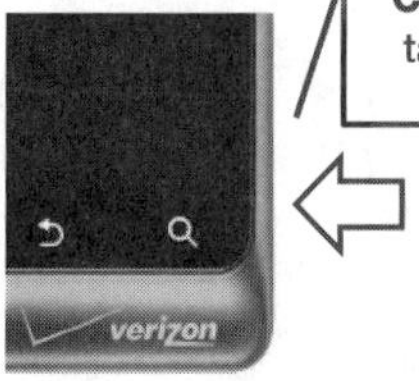

Camera button. Press to start the **Camera** app or take a picture.

When taking pictures, you can hold your phone horizontally or vertically, but the interface tends to work best when the phone is held on its side for a horizontally framed photo. Unlike with some apps, the position of the buttons on the screen will not change as you rotate the phone orientation. (This isn't true if you're using an original DROID.)

Figure 18–1 shows a typical screen.

To take a photo, tap the shutter button on the screen or press the physical **Camera** button on the side of your phone again. Using the Camera button may prove to be less shaky than trying to hold your phone steady as you poke at the screen, but either method will shoot your picture.

Figure 18–1. *The Camera screen (for DROID 2 and DROID X only).*

You have more options than simple pictures. Your DROID will let you choose the mode best for your images by using **Scenes**, and even apply some **Effects** before shooting your picture. Figure 18–2 shows some of the options available by dragging the Settings tray to the left.

You can also adjust focus by pressing on the desired focal point in some modes. Tap the focal point of your image on the screen *gently* and wait. You'll see brackets around the new focal point.

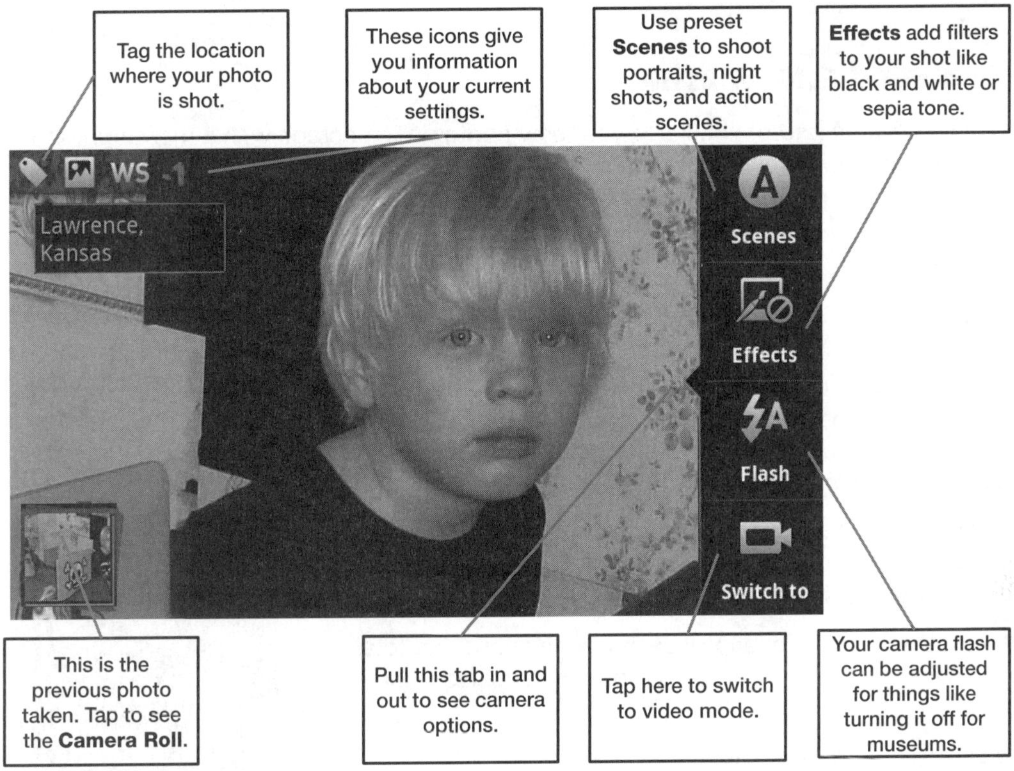
Tag the location where your photo is shot.
These icons give you information about your current settings.
Use preset **Scenes** to shoot portraits, night shots, and action scenes.
Effects add filters to your shot like black and white or sepia tone.
WS -1
Lawrence, Kansas
Scenes
Effects
Flash
Switch to
This is the previous photo taken. Tap to see the **Camera Roll**.
Pull this tab in and out to see camera options.
Tap here to switch to video mode.
Your camera flash can be adjusted for things like turning it off for museums.

Figure 18–2. *Photo settings.*

Flash Modes

One of the most common things you may want to control when taking photos is whether you use a flash. In most cases, the DROID camera will automatically detect the lighting conditions and make the right choice, but there are cases like shots lit from behind where you want the flash on, or museums and performances where you want the flash off, no matter what the camera seems to think is appropriate.

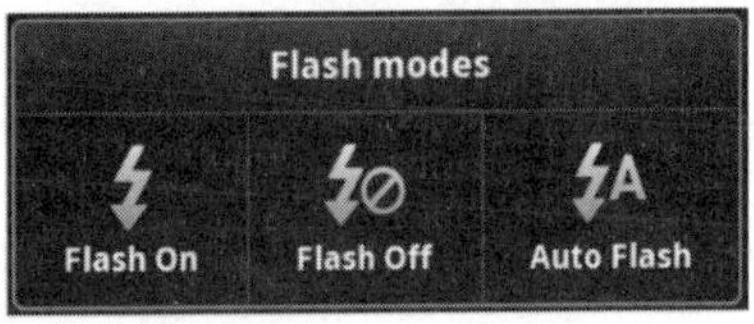
Flash modes
Flash On
Flash Off
Auto Flash

Turn the flash on and off by sliding out the camera options tray and then tapping the Flash button. Your choices are **Flash On**, **Flash Off**, and **Auto Flash**, as shown in the image here.

Reviewing and Sharing

Once you snap a photo, you'll see your new photo on the screen with a message to tap the screen for more options. If you tap the screen, you'll see buttons for **Share**, **Set as**, **Quick Upload**, and **Delete.** Delete is pretty self explanatory, so let's talk about the other three options.

Share

Sharing a photo lets you send it to someone by email, upload it to Facebook, or even send your photos to a retail printer. Android apps can take advantage of photo sharing, so your sharing choices can include all sorts of apps, as shown here.

Your DROID ships with the ability to share with **Bluetooth**, **Email**, **Facebook**, **Picasa**, **Photobucket**, **Text Messaging**, and **Print to Retail**.

The **Print to Retail** option will search for a supported retail outlet where you can pick up your photos, such as Costco and CVS.

To share a photo by email:

1. Snap a photo.
2. Tap the screen.
3. Tap the **Share** icon.
4. Choose **Email** or **Gmail**, depending on the account you wish to use.
5. Compose your email message and press **Send**.

To share a photo on **Facebook** or **Photobucket**:

1. Snap a photo.
2. Tap the screen.
3. Tap the **Share** icon.
4. Choose **Photo Sharing**.
5. Choose the appropriate account. Your choices depend on which accounts are set up to sync with your phone, but they may include **Facebook**, **Photobucket**, **Picasa**, and **MySpace**.
6. Choose a **Caption**. Some accounts also let you choose a photo title if desired.
7. Press **Send**.

To send your photo to a retail store for printing:

1. Snap a photo.
2. Tap the screen.
3. Tap the **Share** icon.
4. Tap **Send to Retail.**
5. If your GPS is enabled, your DROID will find a retail location near you. Alternatively, you can tap the **Search by City or Zip Code** button instead. Use this when you are on vacation or sending photos to another city for someone else to pick up.
6. You'll see a listing of stores, the price of prints, and their distance from your location.
7. Tap a store name, and you'll also see the store hours.
8. Tap **Choose This Store**.
9. Enter your contact information and specify the number of prints. You'll be given the total including tax, and you can drive to the store and pick up your photos, usually within an hour.

Set As

You use the **Set as** button to use a picture to personalize part of your phone. You can set a picture as a **Contact**, your profile picture, or your Home screen **Wallpaper.**

To create a Contact icon from a photo, do the following:

1. Snap a picture.
2. Tap the screen.
3. Tap **Set as**.
4. Tap **Contacts**.
5. Choose a contact from your contact list.
6. You'll see your photo with a red outline marking the cropped area for your icon. Drag the corners to expand or contract the selection, and drag from the middle of the square to move the center of focus.
7. When you've completed your choice, tap **Save**.

Quick Upload

Sometimes you don't care about adding descriptions or fancy settings. You just want to post pictures at the touch of a button and sort out the rest later.

You can set your camera to quickly upload your photos to a favorite web album, such as **Facebook**, **Picasa**, or **Photobucket**, through the **Quick Upload** feature.

When you tap the **Quick Upload** button for the first time, you'll be prompted to specify where you want to upload your photos. After that, tapping the **Quick Upload** button will go directly to the web album you've chosen. Use the **Share** button instead of Quick Upload if you want to add comments and titles to your pictures before you upload them.

The Camera Roll

When you take a photo, you see an image for review briefly on the screen. After that, the review image goes to the bottom corner of the screen, as shown in Figure 18–2. If you tap the previous image, you'll see the **Camera roll,** as shown in Figure 18–3.

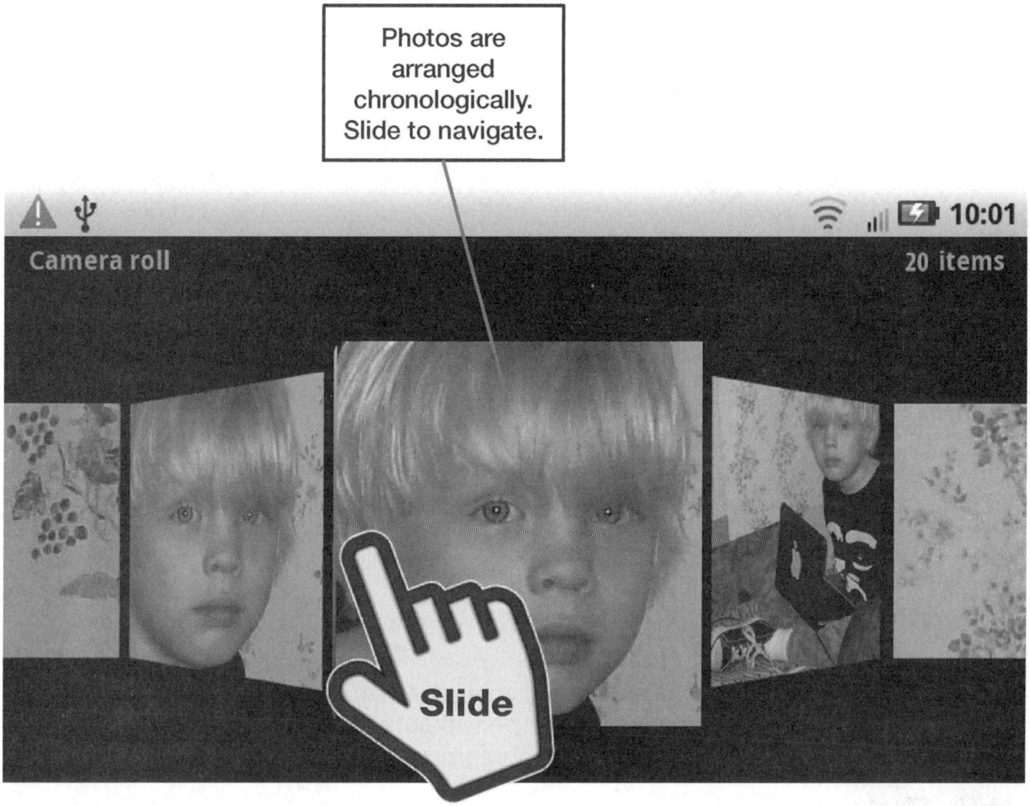

Figure 18–3. *Camera roll.*

Drag your finger left and right to progress through the photos. They're organized chronologically, and you'll periodically see text indicating when the photos in your camera roll were taken, such as **Today**, **Last Week**, or **Last Month**.

> **NOTE:** You still have all the same options for sharing, editing, and deleting photos as you do when you tap the review image shortly after snapping a photo. All you need to do is press the **Menu** button to see them.

Scenes

Scenes are a quick way to adjust your camera for specific situations. You can adjust for lighting conditions, such as indoor shots, or focus situations like portraits. To change the Scene settings, slide the tray out, as shown in Figure 18–1 and 18–2, and then tap the **Scenes** button. 

You can swipe side to side to navigate through the different scenes. The camera will show you a preview of your scene as you go. Your scene choices for still pictures are shown in Table 18–1.

> **NOTE:** On the original DROID, touch the settings icon and then scroll down to Scene Mode. The icons pictured here are not available on the original DROID. The icons may look slightly different on your DROID 2 depending on your software version.

Table 18-1. *Scenes*

Scene Icon	Description
	This is the default setting. Auto allows the camera to decide how to handle images and is the best general choice.
	Landscape mode sets your camera to fixed focus, so you can capture a landscape without the camera attempting to focus on objects in the foreground.
	Night portraits will attempt to adjust for very low lighting conditions.

Scene Icon	Description
	Macros are close-ups and detail shots. Use this to focus on something small and near the camera.
	Use this for better sunset pictures. Rather than focusing on the low lighting in the foreground, the camera will capture the pretty colors in the background and silhouette anything in the foreground.
	Use this scene setting if you're having problems with your hands shaking when you try to take pictures.
	Portraits are for taking pictures of people. Remember to gently tap to select an area for focus. In this case, you'll want to focus on the face.
	The Sport setting is for taking pictures of fast-moving objects, like runners, horses, or people jumping in the air.

Creating a Panorama

You can create a panorama using your DROID by using the Panorama Assist feature. It's not completely intuitive, but once you figure it out, you can have some fantastic panoramas. Rather than one continuous shot, a panorama is a long, skinny shot made from stitching photos together. Figure 18–4 shows how this works.

NOTE: Panorama assist is only available on the DROID 2 and DROID X.

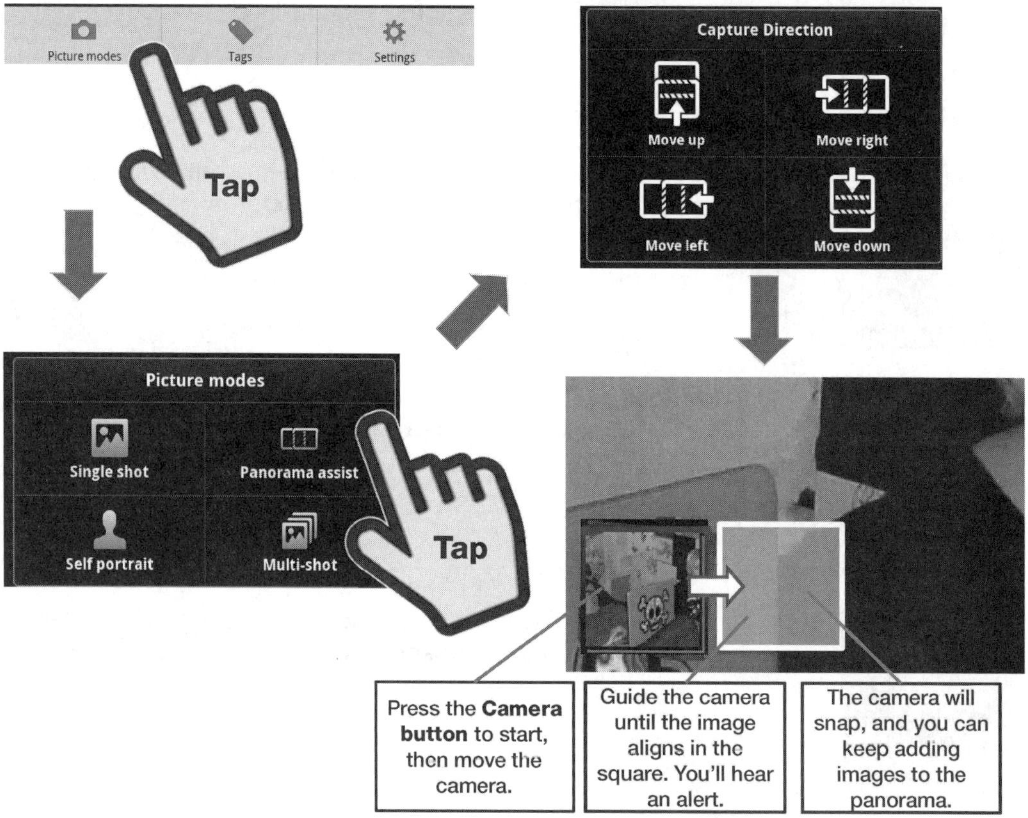

Figure 18–4. Creating a panorama.

1. Press the **Menu** button.
2. Tap **Picture modes**.
3. Tap **Panorama assist**.
4. Select a camera direction. You can go left to right, right to left, up to down, or down to up.

5. Start by taking your first photo. If you're working left to right, start on the leftmost side of your panorama and snap the photo.

6. Move your camera slowly in the direction you've selected. For instance, if you're working left to right, move the camera slowly to the right.

7. Pay attention to the preview window on the bottom left of the screen. You'll see an arrow and two squares that will show you where your phone is positioned in relation to the last photo.

8. When you've lined up your next photo, the camera will beep and shoot the picture by itself.

9. Proceed to the next photo. If you do not want to take the full six photos, you can stop the series by tapping the square symbol on the upper right side of the screen. Otherwise, continue.

10. When you've taken six photos or stopped the panorama, your DROID will stitch the photos together to form your panoramic image.

Camera Settings

The Camera Settings menu lets you change advanced features, such as the Quick Upload album, the length of time you preview images, or even ISO settings on your phone's camera. To get to the Camera Settings menu, press the **Menu** button and then tap **Settings.** On the original DROID, touch the Settings icon (farthest to the right) and scroll down for settings to adjust.

Your choices are shown in Figure 18–5.

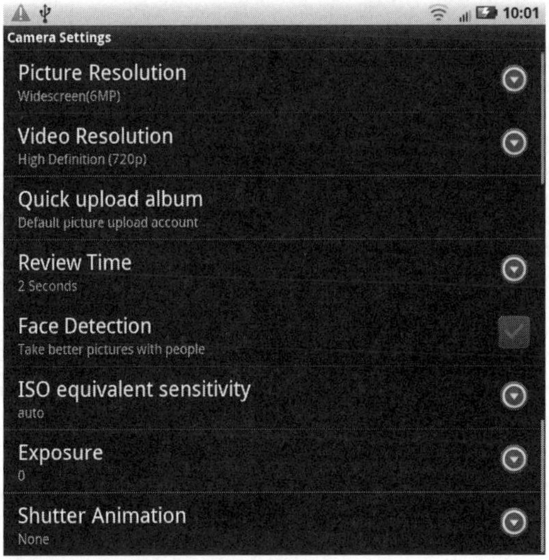

Figure 18–5. *Camera Settings*

Let's explore these settings.

Picture Resolution

This option allows you to set the size of your photos. Larger pictures will print better, but they take more storage space. If you're using your DROID only to take photos for the Web, you probably don't need the highest setting, but if you'd like to print some of your photos and have room on your phone's memory card, increase this.

The highest resolution on a DROID or DROID 2 is 5 megapixels, and the highest on a DROID X is 8 megapixels.

Video Resolution

This is the size of video you shoot with your camera. Although your DROID X may be capable of shooting HD video, you'll probably get better overall balance of quality and storage space by using the VGA (640 x 480) setting. That's the resolution of analog TVs. It can be uploaded to YouTube and edited in programs like Windows Movie Maker and iMovie.

If you're sending video text messages or sending videos by email, you'll want to make the video even smaller. Try QCIF (176x144) or QVGA (320 x 240). Those are tiny videos to watch, but they take much up less storage space as an attachment to a message.

Quick Upload Album

We've already discussed setting up a **Quick upload** album. Here is where you'd change the location. You may want to change locations prior to a trip in order to make it easier to sort the photos once you get back.

Review Time

Review Time is the amount of time a photo shows on your screen after you take it. Increase the time for a longer review or decrease it to take the next picture faster.

Face Detection

Check this box if you take lots of pictures of people and want the camera to focus on their faces rather than focusing on the object nearest the center of the frame. Leave it unchecked if you primarily take landscape photos or detail shots of items other than people.

ISO Equivalent Sensitivity

The **ISO Equivalent sensitivity** has to do with the light sensitivity of the camera. In film, the ISO setting is used to determine shutter speed, and this setting is used to simulate that. Faster speeds mean grainier/poorer quality pictures. Slower speeds mean more chance of shaky hands and motion blur. In most cases, your best bet is to leave this setting as **Auto** and use **Scenes** to compensate for different lighting situations.

Exposure

This is a simulation of film and exposure or how light-sensitive the sensors on the camera are. Just as with ISO settings, your best bet is to leave the setting as is and adjust for specific situations by using **Scenes**.

Shutter Animation

The shutter animation setting just controls whether you see an animation when you snap a picture. It's entirely a matter of personal preference.

Effects

Not only can you optimize your camera for certain lighting conditions and subjects, you can add effects to photos as you shoot them.

1. Launch the **Camera** app as shown here by pressing your **Camera** button or tapping the **Camera** app icon.
2. Open the **Photo Capture Settings** tray tab by sliding your finger on the right side of the screen.
3. Tap the **Effects** button.
4. Slide and drag your finger to the left or right to navigate through the Effects options. You'll see an on-screen preview of the effect.

Your effect choices are **Normal, Black and White, Negative, Sepia, Solarize, Red Tint, Green Tint,** and **Blue Tint.**

Tags

Tags are optional metadata that you store with your picture to let you know the location where a photo was shot or add an additional label to classify a photo. Use this to keep track of exactly where you took photos on vacations or during field work or add custom tags for events. Keep in mind that others will also be able to see your metadata if you upload these tagged photos to the Web.

In order for location tags to work, your phone GPS must be enabled. You can also disable location or **geotags** separately from custom tags. Figure 18–6 shows how this works.

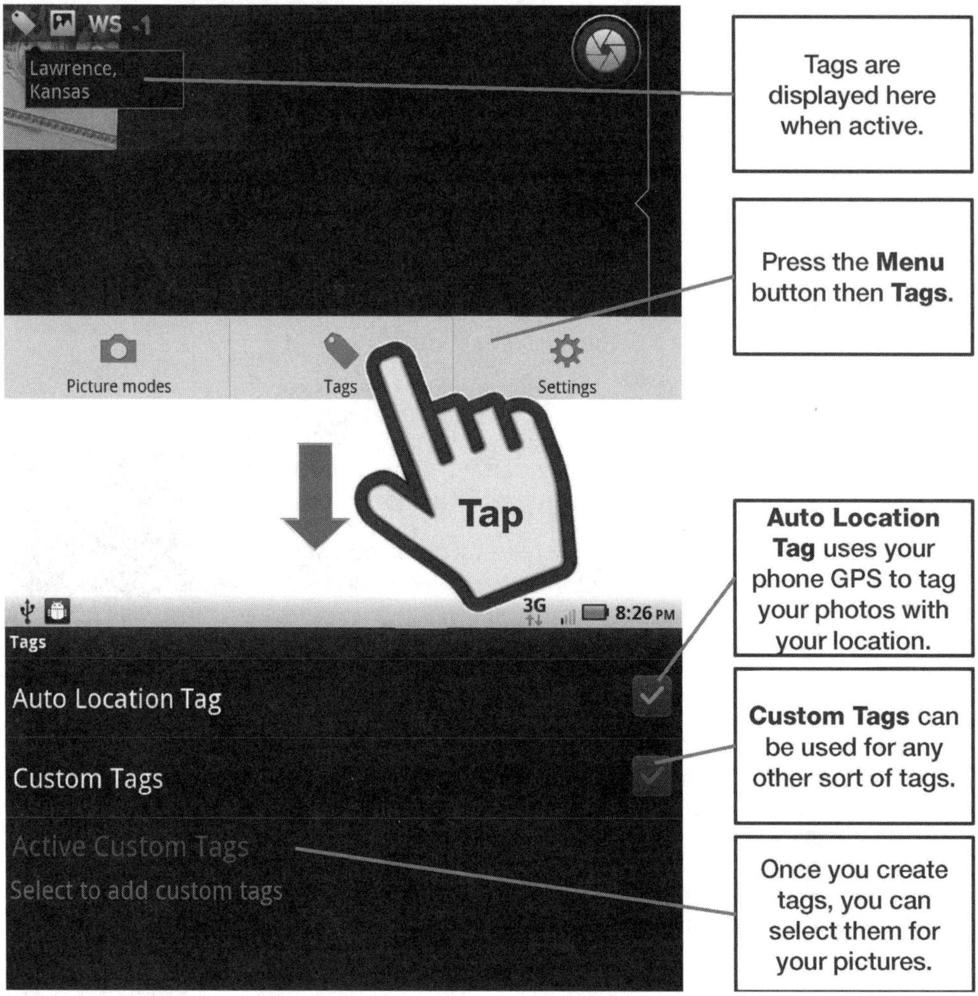
WS
Lawrence, Kansas
Tags are displayed here when active.
Press the Menu button then Tags.
Picture modes
Tags
Settings
Tap
3G
8:26 PM
Tags
Auto Location Tag
Custom Tags
Active Custom Tags
Select to add custom tags
Auto Location Tag uses your phone GPS to tag your photos with your location.
Custom Tags can be used for any other sort of tags.
Once you create tags, you can select them for your pictures.

Figure 18–6. *Adding tags.*

1. From the Camera app, press the **Menu** button.
2. Tap **Tags.**
3. Check the box next **Auto Location Tag** if you want the camera to automatically geotag your images. Your GPS must be enabled for this to work.
4. Check the box next to **Custom Tags** to create a custom tag.
5. If you've already created a tag, it will be listed under **Active Custom Tags**. Otherwise, you should tap **Select to add custom tags**.

NOTE: On the original DROID, touch the Store Location icon (second from right) to turn on geotagging.

Shooting Video

You can shoot video using the same Camera app and slightly adjusted settings. Drag out the **Photo Capture Settings** tray, as shown to the right, and tap the bottom button: **Switch to.** On the original DROID, just move the slider from the **Photo** to the **Video** icon.

Notice that the button shows a video camera. This will switch your camera to video mode. You can switch back the same way by tapping the **Switch to** button, only the button will now show a still camera icon.

You'll notice a few other differences in video mode, as shown in Figure 18–6. The icons along the top left still show tags, and rather than showing your resolution in megapixels, it will show your video resolution.

Rather than a shutter, you'll see a red record button on the upper right. Tap it once to start recording. Tap it again to stop or pause your recording.

The button will change to show a square DVD-style stop symbol in the middle to indicate that you're recording, and you'll see an overlay of the time elapsed on the screen.

Figure 18–7. *Video recording.*

Scenes and Effects

Just as with still photos, you can slide out the ***Video* Capture Settings** tray, as shown here. Just as with still photos, you can choose **Scenes** and **Effects.**

Just as with still photos, you just need to tap the **Scenes** button and slide through the available choices. Table 18-2 shows the available Scene settings for your DROID.

Scenes for video adjusts audio quality as well as video.

Table 18-2. *Video Scenes*

Scene Setting Icon	Description
	The **Everyday** setting is a good overall choice. It records in stereo and is suitable for most situations.
	Use the **Concert** settings for recording concerts or other loud, stereo sources.
	**Narrative** is used when you're voicing over the video as you film it, such as film of the Grand Canyon while talking about your impressions of the area. The sound picks up behind the camera.
	The **Outdoors** setting tries to reduce the amount of wind noise.
	**Subject** video is good for standard interviews and other situations where you're filming someone else in front of the camera. The audio is recorded from the front to avoid accidentally picking up random noises from your end.

Effects for video offer exactly the same choices as effects for still pictures: **Normal, Black and White, Negative, Sepia, Solarize, Red Tint, Green Tint,** and **Blue Tint.**

Flash and Video

What if you want to light up a dark scene? You can use the same light you'd use for the camera flash by using the Light button on the Video Capture Settings tray.

Your choices are rather simple: on or off, as shown in the image here. Although turning on the light will help in very dark situations, it's not a substitute for adequate lighting and it won't work miracles.

Video Modes

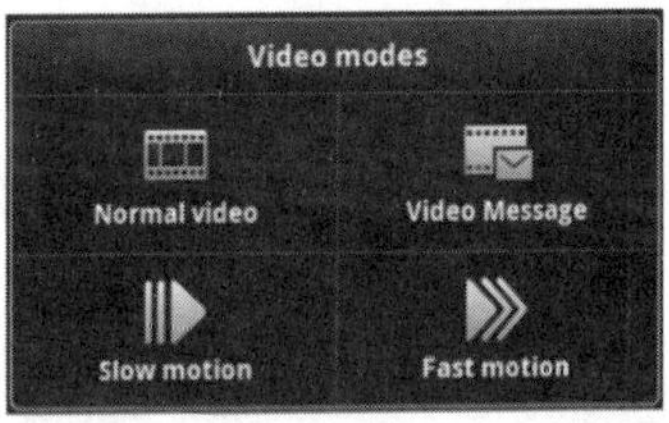

Just as you can switch to different picture modes for cool features like panoramas, you can switch video modes on your DROID for sophisticated options.

1. Launch the **Camera** app.
2. Switch to **Video mode**.
3. Press the **Menu** button.
4. Tap **Video Modes**.

You'll see choices like those on the right except for the original DROID. **Normal video** is self-explanatory, and this is the setting you'll use to return to your regular video settings. **Slow motion** video captures extra frames, so when you play the video back, everything appears to be slowed down. Fast motion does the opposite and skips frames. **Video Message** is used for sending MMS video messages.

When you select either **Slow motion** or **Video Message**, your video resolution will drop. Your DROID X doesn't support shooting slow-motion video at HD resolution, and the DROID and DROID 2 don't support shooting HD video at all.

NOTE: You can't make a **Video Message** out of a video you've already created in a higher resolution, so be sure to change your settings *before* you shoot your video.

Sharing Videos

Once you create video, you can share that video using the **Share** button. Just as with still pictures, you can tap the screen right after the video is shot, or you can go to the **Camera roll,** tap the video, and then press the **Menu** button.

When you create videos, the primary way to upload and share them is through YouTube. This works really well for personal videos, but it's a bit trickier for corporate videos, because each YouTube "channel" is limited to one user account and one password. Eventually, there may be an easy way to upload to a collaborative channel. Google is currently experimenting with methods to allow users to upload submissions to another channel.

YouTube offers unlimited storage for videos and two basic privacy settings, private and public. Private videos can be shared with small groups of other users. You can also upload videos privately at first and make them public later. Public videos are searchable and are automatically closed-captioned using speech-to-text technology.

Videos on YouTube are limited to 15 minutes unless you are part of YouTube's Partner Program for commercial content creators. More information on YouTube's Partner Program is available at www.youtube.com/t/partnerships_faq.

In order to upload videos to YouTube, you must have a YouTube account, and it must be linked to a Google account. However, large files require a Wi-Fi connection.

NOTE: When you give public photos and videos tags and descriptions, keep search in mind. Use terms that you think people searching for that video would use, and do not skip fields.

The **Share** button will also work for sending your video as an email attachment or text message. However, you can't make a video smaller from your phone, so if you've shot the video in VGA or higher resolution, you'll need to upload it to YouTube or transfer it to your computer before trying to send it anywhere else.

Even though Photobucket supports video, you can't upload videos directly to your Photobucket account as of the time of publication.

The Gallery

Photos taken from your camera are stored on your phone's memory card, and you can review these using your **Camera** app. However, you might also have photos you've moved or downloaded from other sources. You can use the **Gallery** app to return to your **Camera roll** and browse through your pictures.

Tap an album to view the pictures within it. You can enlarge photos using the pinch-to-zoom motion. You can view pictures either as a Camera roll or as a grid of preview icons depending on how you hold your phone. Tilt your phone horizontally to view Camera rolls and vertically to view an image grid. Tap an individual picture to view it.

TIP: You can upload or delete multiple photos at once through the gallery by viewing an album and pressing the **Menu** button twice. Gray check marks will appear on each photo or video. Select multiple items by pressing each one. You can then mass-delete them with the **Delete** button, or upload them to **Picasa** or **YouTube** using the **Share** button.

Using Picasa

Picasa is Google's online and desktop photo software, and it is the default photo service for unmodified Android phones. It is one of your choices when using the **Share** button, and it provides an easy way to upload a photo for embedding into a web page.

Technically, Picasa and Picasa Web Albums are two separate products, but the distinction is fuzzy, since you can sync online and desktop photos. Picasa Web Albums is located on the Web at http://picasaweb.google.com.

Think of Picasa Web Albums as Google's answer to Flickr. You can upload photos and organize them into albums. Each album can have privacy settings, captions, tags, and location information (geotags). When using the service in a standard web browser, it also has facial recognition capabilities. So far this hasn't been translated into the phone version, but it's only a matter of time.

Picasa doesn't provide unlimited storage. At the time of publication, it provides 1GB of free storage, and anything beyond that must be purchased from Google on an annual basis.

There are three basic privacy levels for Picasa:

- *Public*: This is just as it sounds. Your album is visible to anyone and can be found in search.
- *Unlisted*: Google will give you an obscure URL, which you can distribute as you see fit. This is not actually private; it's just hard to guess. However, anyone with the URL can see your album and pass that URL on to other people, so it is a poor security setting for anything you really need to remain private.
- *Sign-in Required*: You specify who can see the album. You enter the name of specific users' Google accounts, and only those people can see your album, and only when they are logged in. You can add and remove anyone from your "shared with" list.

To upload a photo to Picasa, click the **Share** button when viewing a picture in the Gallery app or reviewing a picture in the Camera app.

1. Select **Picasa**.
2. Choose a **Google account**.
3. Enter a caption.
4. Select an album.
5. Tap **Upload.**

Using Photos As Wallpaper and Widgets

You can use a photo as your **Home** screen wallpaper. When you've shot a photo, you can set this immediately, but you can also use photos stored on your phone's card. This is explained in further detail in Chapter 6.

To make a wallpaper from a photo, you can use the **Set As** button, but you can also:

1. Long-press the **Home** screen.
2. Select **Wallpapers**.
3. Select **Gallery**.
4. Navigate to the photo you want to use. You'll see an outline around part of the photo indicating where it will be cropped to fit as your wallpaper.
5. Drag your finger to expand, shrink, or move the cropping area.
6. Click **Save** when you are done.

Using Photos for Contact Icons

You can also add contact photos by browsing to the photo in the Gallery and clicking **Menu ➤ More ➤ Set as**, and then clicking the **Contact** icon. You'll be given the option to crop the image, just as you are for wallpaper. Browse to the contact you want to replace (this is easiest if you use your trackball to scroll through your contacts, so you can avoid accidentally selecting the wrong one). Select the correct contact, and then click **Save**.

This option is also available immediately after shooting a photo through the **Set as** button, so when you enter a new friend or business partner's contact information into your phone, take a quick picture of them and add their picture to the contact info.

Copying Photos to Your Computer

You can get photos from your phone to your computer in many different ways. The method you use depends on the bandwidth you have available, your privacy concerns with the photos, and your personal style. You can email photos to yourself or upload them to Picasa and download them to your desktop from the Web. This may present privacy issues if the photos are sensitive, and it may just take too long if you've taken a lot of photos.

Syncing media is explained in more detail in Chapter 25: "DROID Media Sync."

Editing Photos

You can apply effects and choose Scenes, but what if you want to do something more advanced? What if you want to add an effect after you've already taken the photo or edit out red eye? Fear not, there are tons of apps that allow you to edit photos directly from your phone, and your DROID even ships with advanced photo editing tools, which were used on the image shown here.

Nothing offers the same quality you'd get from a desktop photo-editing program, but you're using this with a phone camera, not the latest SLR.

To edit a photo, do the following:

1. Go to the **Gallery app** and navigate to the photo you wish to edit.
2. Press the **Menu** button.
3. Tap **Edit**.
4. Tap **Advanced editing**.
5. If you don't have any other advanced editing software installed, this will launch the **Photo Workshop** app. Otherwise, you'll have a choice.

The advanced editing tools allow you to add frames and stamps, but you can also use it to change the color settings, resize, and crop photos.

Photoshop Mobile

Adobe Photoshop is probably the most trusted name in photo-editing software, and Adobe has expanded to also offer a mobile version of its product, as shown here. It's not nearly as full-featured as the desktop version, but it is considerably cheaper. The current price is free.

Photoshop Mobile doesn't let you take new photos from within the program, but it lets you work with the photos you already have. Think of it as a Gallery app alternative. It allows you to edit a variety of photos features, including soft focus, saturation, tint, cropping, and color effects. You can upload photos to a free Photoshop.com account (you'll be prompted to create an account if you don't have one already).

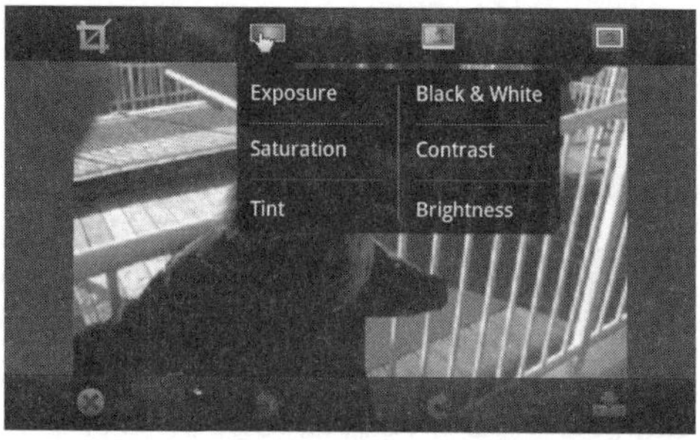

Once your photos have been uploaded, you can share and edit them from your Photoshop.com account.

PicSay

PicSay is probably the best known of several photo-editing apps that allow you to make artistic and novel changes to photos before uploading them, as shown here. PicSay comes in a free trial version and a paid app (about $4.00 as of publishing time). Google Checkout will convert the currency if you buy the app through the Android Market. The trial version of PicSay has an older version of the interface than the for-pay version, and limits the size of pictures. The pro version also offers more editing options.

PicSay allows you to apply an impressive amount of effects to photos. Not only does it allow you make whimsical edits like applying fake mustaches or novelty eyeballs, but it has an impressive list of very practical effects. You can use it to edit out red-eye, add captions, or add grain and other textures to photos.

Similar photo-editing apps include Pic Paint and Camera Illusion, along with Photo Workshop, which ships with your DROID. See Chapter 17, "The Android Market" for more information on purchasing and installing apps.

Other Photo Apps

If you prefer not to use Picasa for your online photo albums, you can use Photobucket or Facebook. You can also use any other service, so long as you have the proper app installed. You can also share photos using Bump, an app that allows you to exchange contact info by physically bumping the phone of another user. Bump was previously an iPhone-only app, but it is now cross-platform compatible with Android users.

Android allows developers to have access to camera controls, so many apps allow you to take photos or use the camera. Price comparison software like ShopSavvy and Compare Everywhere use the camera to scan bar codes. Evernote allows you to take and attach snapshots to notes. Camera Pro and Snap Photo Pro are paid apps that offer higher-end camera features like a timer and grid marks for easier photo composition. Camera Pro even offers to replace the Camera app as the default camera.

Google launched an experimental app for searching with pictures, called Goggles. Goggles is best when used to scan man-made objects like DVD covers, text, and famous buildings. Goggles analyzes any photo you take with it and attempts to identify the object and find it in search. If it can't identify the object, it looks for visually similar images.

Goggles is currently not much more than a novelty, but in the future it may end up being an easier way to search than typing search terms into your phone.

Printing

If you want to print photos, documents, or other files without downloading them to your PC or sending them to CVS, you can use the PrinterShare app from the Android Market. Download the free Mac or PC desktop component from http://printershare.com, and install the Android app on your phone. This lets you share that computer's printer access. You can use a trial version to make sure it's compatible with your network and do some limited printing. The $4.95 pro Android app allows you to print directly to Wi-Fi printers and doesn't have a page limit.

Remember that the resolution on your DROID or DROID 2 is still only enough for quality 5×7-inch prints or smaller.

Chapter 19

Finding Your Way With Maps

One of the big advantages of owning a smartphone is that, not only do you have a mobile computer with you at all times, but you also have a compass, map, and restaurant guide.

This chapter will discuss using **Google Maps** and other location-conscious apps on your phone for both business and pleasure. You'll learn how to use your phone for driving directions, deciding where to eat, and letting your friends know where to find you.

There are a lot of apps that use maps, but in order to do so, those apps have to know where your phone is. In general, phones know where they are by using the following:

- GPS (global positioning satellites)
- Cell phone towers
- WPS (Wireless Positioning System)

There are dozens global positioning satellites orbiting the Earth. Your DROID's GPS unit attempts to find the signal from at least three of them and triangulate your position. However, this requires your phone to have a chip that detects GPS signals and be in an area that can detect them. If you're indoors underground or around lots of tall buildings, your phone might not pick up a GPS signal.

Your location can also be estimated using relative positions to cell phone towers. This isn't as accurate as GPS because cell towers are positioned for better signal reception, not triangulation, so there are generally not three overlapping points for positioning.

The third method of locating your phone comes from using a map of known public Wi-Fi spots. It's a method that works well in urban areas and indoors – precisely the places where GPS does poorly. Because it only requires a Wi-Fi signal, it even works on laptops, netbooks, and tablets.

If you combine all three methods, you end up with a phone that usually knows where it is.

You can enable and disable your phone's ability to trace your location by using the **GPS Toggle** widget, which is a Motorola widget for phones that support them. 

You can also do the following:

1. Press the **Menu** button while on the **Home** screen.
2. Tap **Settings**.
3. Tap **Location & Security**.
4. Check the options under **My Location**.

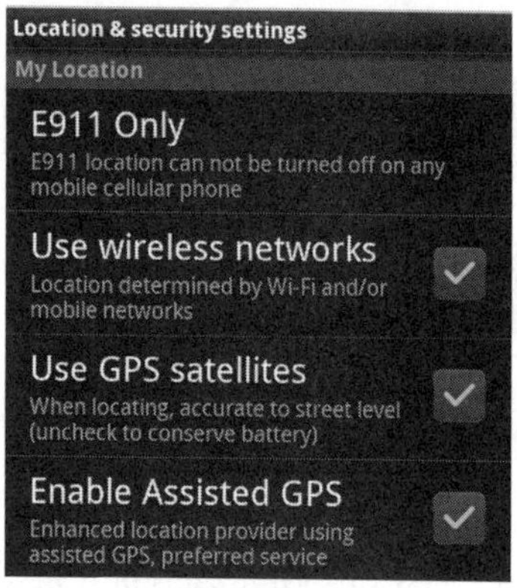

You can choose to enable Wi-Fi tracking, GPS satellite tracking, or assisted GPS, which uses a combination of location methods. The more services you enable, the better your phone will be at determining your location.

Understanding Google Maps

Your DROID ships with **Google Maps**, and it comes with several related apps, all of which will be covered in this chapter. **Google Maps**-related apps include **Google Maps**, **Places**, **Street View**, and **Navigator**. You can either launch these apps individually or launch them from within **Google Maps**.

Google Maps works with your phone's GPS; and if you have location sensing enabled on your DROID, **Google Maps** will determine your location when you launch the app. You can also use it to search for distant locations.

You can move your view of the map with your fingers, and you can also use pinch-to-zoom motions to enlarge and shrink the area you're displaying. You can also use the **+** and **–** buttons to shrink and zoom the image. Tap anywhere on the map, and **Google Maps** will attempt to tell you the address of that location.

Press the **Menu** button to see more options. Remember that if you're ever lost, you can use the **My Location** tool and **Google Maps** to try to determine your location.

You'll see yourself as a point on the map with a light-blue circle around it. Because there are a lot of variables that affect accuracy, the larger circle shows where you *could* be.

The smaller the circle, the more accurate the prediction. Your location may be indicated as a blue arrow if the DROID can determine which direction your phone is facing or moving. If you don't see any circle around your blue dot or arrow, then your location is as accurate as possible - usually about 3 meters (about 10 feet).

Driving Directions

If you just feel like exploring an area, use the **Search** button. You can either use the physical **Menu** button or the button that appears when you press the **Menu** button from the **Home** screen. This is useful for answering questions such as "What's near 131st street?" or "Where is Uganda?" It's not, however, for directions on how to get to places.

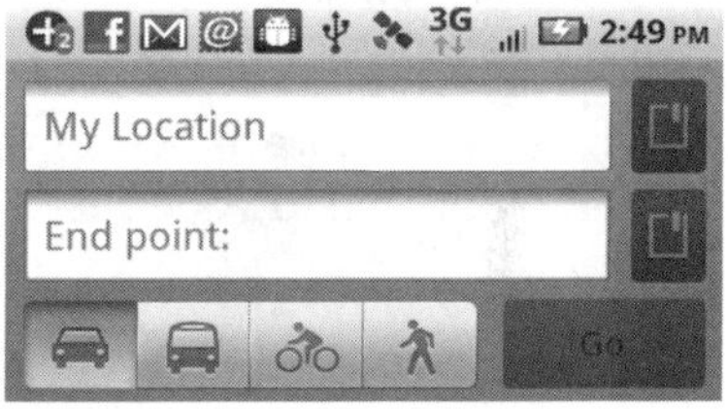

Follow these steps to get actual driving directions:

1. Press the Menu button.
2. Tap Directions.

You'll see fields for **My Location** and **End point**, as shown to the right. The **My Location** field assumes that you want directions from your current location. If you want to use a different address, you'll need to enter it here.

There's also a handy **Bookmark** button  next to the **My Location** and **End point** fields. This lets you choose from your current location, a place you point to on the map, the address of one of your contacts, or any location you've starred.

You'll also see a series of buttons below these fields for choosing what mode of transportation you need.  You can choose car, public transportation, bicycle, and walking directions. This is a lifesaver if you're trying to get anywhere in a big city without a car.

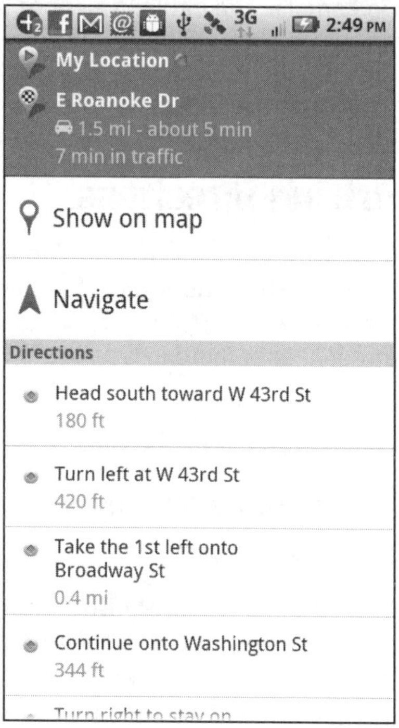

Once you've settled on a start point, endpoint, and means of transportation, tap **Go**. You'll see a list of text-based step-by-step directions.

Tap **Show on map** to see the route displayed on the map instead. If you leave your phone's GPS on, you can even see your progress as you go. We've used this to navigate in cities without annoying the locals by pausing too long to figure out directions.

Press the phone's **Back** button to get back to text directions. From here you can also press the phone's **Menu** button to get updated directions, reverse the directions for the trip back, or report a problem with the directions.

Press the **Navigate** button to hear spoken directions that guide you to your destination.

Map Layers

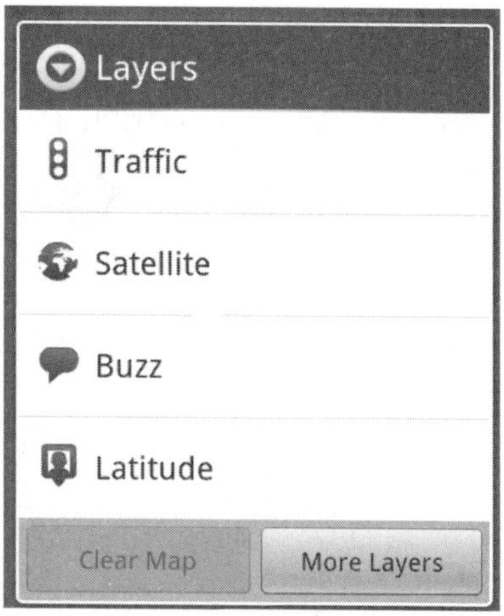

Google Maps for Android works by displaying information as a series of information-layer overlays. If you're familiar with **Google Earth**, it works the same way. These layers can be turned on and off individually. Follow these steps to use this app:

1. On the DROID 2/X press the **Menu** button and then **Layers** to see some of the available layers, as shown to the right. This list may also scroll. On the original DROID, Layers is activated by the Layers icon at the top right of the screen.
2. Tap the **More Layers** button to see even more layers.
3. Tap each layer to toggle it on or off. You can toggle more than one of these layers at a time.

Traffic

Much of the information you see in **Google Maps** comes from contracts with third parties, and the traffic information is no exception. Traffic information is only available for large cities, and it's shown by color-coding the roads. Green indicates smooth traffic, yellow indicates delays, and red indicates major snarls. Traffic information can change rapidly, so don't expect absolute accuracy. It's also difficult (and dangerous) to check while you're actually on the road.

Satellite

Satellite info comes from a variety of third-party imaging sources as well as Google, and those photos are stitched together and superimposed on the map information. An example is shown to the right.

The images are usually great, but sometimes the stitching process distorts the image. Consequently, there are times when an address appears to be in the wrong location, and you'll notice patches of ground with different image quality.

Satellite images are also not necessarily up to date. The photos for any given area could be several years old. Google will often buy new images when something major happens in an area, such as Hurricane Katrina, the BP oil spill, or the earthquake in Haiti, but don't be surprised if the satellite image of your house doesn't include your recently built garage.

Buzz

Google Buzz is a social networking service. You can make posts in Google Buzz that include your location information, and that adds your post to the **Buzz** layer on **Google Maps**. Using this layer, you can see Buzz posts that were made nearby. The **Google Buzz** layer shows those posts as little quote bubbles over the map. Tap a quote to see the post.

Google Labs Layers

The **Google Labs** layer is shown to the right. It is a collection of experimental features you can turn on and off. They're not always reliable, and they don't always last, but some Google Labs "graduates" have turned into solid and popular features, such as the public transportation directions in the **Google Maps** app.

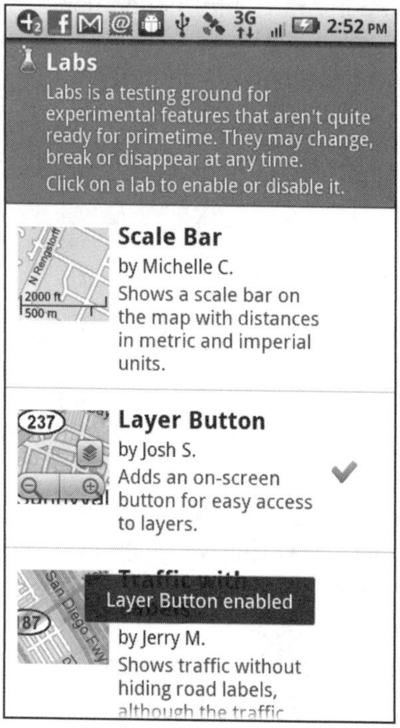

Most main Google products have their own set of Google Labs experiments, and quite often (as in **Gmail**) those features just won't work on your phone. **Google Maps** is an exception to this general rule. Follow these steps to toggle **Google Labs** layers on or off:

1. To get to **Google Labs**, press the **Menu** button while in **Google Maps**.
2. Tap **More.**
3. Tap **Labs.**
4. Tap on individual layers to toggle them on or off. You can enable or disable layers at will.

Google does use the relative popularity of Google Labs projects as one factor in determining what stays and what goes from an application or service.

Location Sharing With Latitude

Google's Latitude service is a way to let your social network know where you are. You can use it to make sure people know you made your flight, or let your contacts know you've got a trip in their city.

Currently, you can only share information with mutual friends, which means you must invite your friends to share Latitude information with you, and they must accept the invitation. You can also use the Latitude service from a laptop or desktop computer, so it doesn't depend on everyone owning a phone. You have four global choices for sharing Latitude location information:

- **Detect automatically:** You just let your phone report where you are to your friends.
- **Set your location:** You can manually update your location (and lie about where you are if you wish).
- **Hide your location:** Nobody sees your location, but you can still see your friends.
- **Turn Latitude off:** Your friends can't see you where you are, nor can you can see where they are.

Keep in mind that your friends are the only ones who can see any of this, and settings for individuals will override global choices. When Latitude was initially released, there was concern that someone could be stalked by having this feature turned on without his knowledge, so you may receive an email letting you know you've joined Latitude or that you have turned on location tracking.

To add friends to your Latitude account from the **Google Maps** app, take the following steps:

1. Press the **Menu** button.
2. Tap **Join Latitude**.
3. Press **Menu**.
4. Tap **Add friends**.
5. Choose to add a friend through your contact list or by the friend's email address.

The friend will receive an email inviting her to join Latitude or accept your request. When someone sends you an invitation, you'll receive an email asking if you'd like to ignore the request, share your location back, or accept their request and hide your location.

If all of this sounds a bit too personal, you can ratchet it down a notch for more casual business contacts.

For example, you can manage friends on an individual level by tapping **Latitude** and then tapping a contact's name.

You can see where someone is on a map, contact them (through email, Google Talk, and so on), get navigation directions to visit them, remove them as a friend, and set specific privacy settings.

Tap **Sharing options**, and you can choose to do the following:

- Share the best available location (most likely your exact location)
- Share city information only
- Hide your information

You can change these settings later or tweak them by relationship level. For example, you might let your spouse know your exact location and your business contacts know your city only when you travel. You can also globally shut down location information by hiding or manually entering just your city name when you don't want to broadcast your location.

Google Place Pages

Sometimes you want more information about a particular location. In the web version of **Google Maps**, Google has moved toward a system where each location has a *place page*. On Android, that means each location has a well-organized detail page with tabs. Double-tap a location or tap the location bubble, and you'll see the location details, as shown on the right.

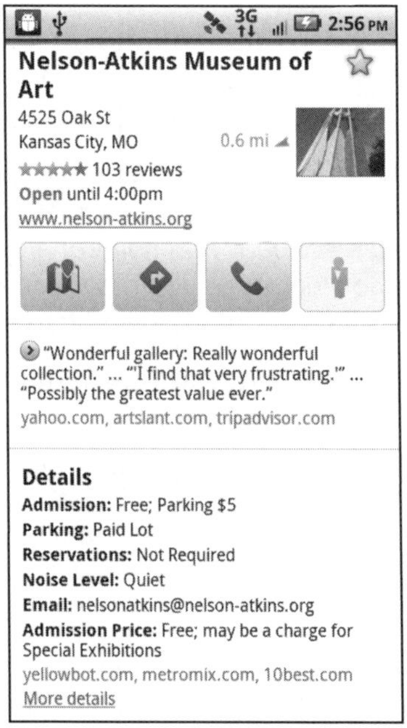

Not every location will have so many details, so what you see will depend on the information available for the location. Also note that this is a fairly new feature, so sometimes the details themselves are off.

Some details you can find include reviews, a location's website, driving directions, a location's distance from you, and the ability to share info about a location on social networks. Tap the **Places** icon to start the **Places** app.

If there's an available phone number, you can call it by tapping the **Phone** button.

> **NOTE:** If you need to give someone an address in a hurry, go to the location's place page, tap **Share this place**, and send it as an SMS text message. If that person has a smartphone, they can tap the link you sent her and use **Google Maps** to get directions.

Google Places

Rather than randomly finding spots on a map, you can search for nearby attractions by using the **Places** app. Although it's a separate app, it's really just a different interface for the same Google Maps database. You can use **Places** to browse for apps by category or search individually. Once you've landed on an item's Page Place, you can get navigation directions, phone the business, view the location in Street View, and more. Figure 19–1 illustrates the power of **Places.** Tap the **Places** icon to start the **Places** app.

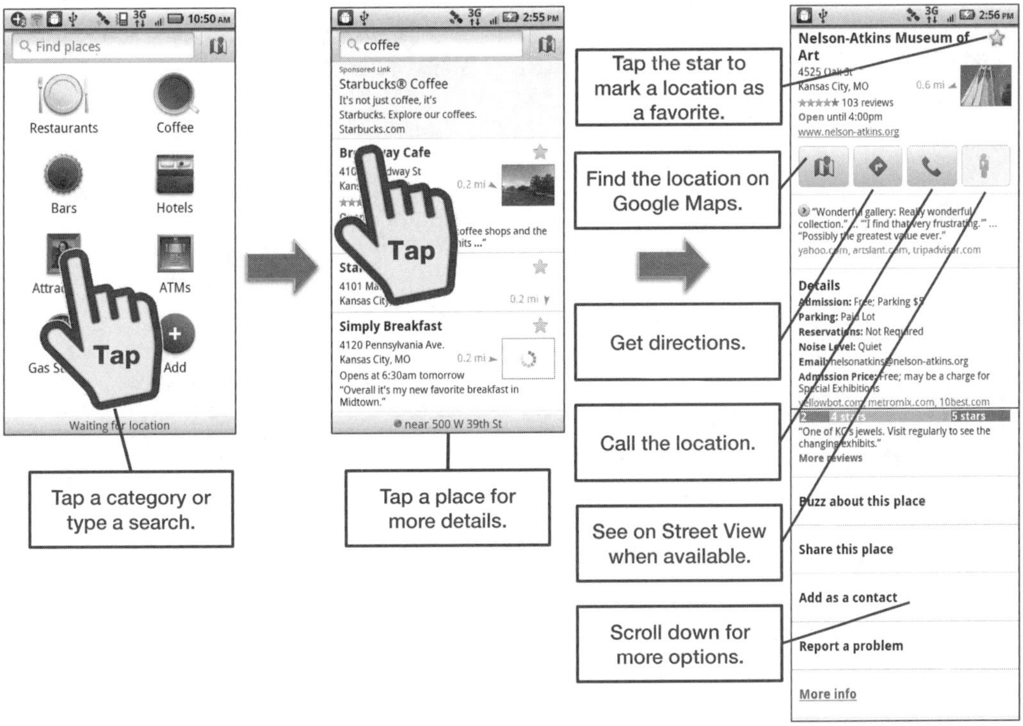

Figure 19–1. *Leveraging the **Google Places** app.*

Starring Locations

You can also use the location details to add a star to a location. Tap the **Star** icon in the upper-right corner of the screen. It will glow yellow when the location is starred. Simply tap it again to remove the star. Using stars is like saving bookmarks for web pages. This feature enables you to easily find locations you visit frequently or need to find quickly.

Follow these steps to access your list of starred items in **Google Maps**:

1. Press **Menu**.
2. Tap **More**. On the original DROID, tap **Starred Places.**
3. Tap **Starred items**.

NOTE: When you travel for business, put a star on the convention center and your hotel before you arrive; this will give you instant access to addresses, driving directions, relevant phone numbers, and nearby restaurants.

Google Street View

At first the idea of the Google Street View service seemed a bit creepy: Google used cars with mounted cameras and other equipment to take 360-degree photos of roads everywhere. It's still a bit creepy in light of the company's admission that it inadvertently snooped on users in open Wi-Fi hotspots (this is another reason to use caution when using unencrypted hotspots).

That aside, Google Street View is an amazing tool for figuring out where you need to be. We use it when we have an appointment at a new location to see what the building looks like before we arrive; we also use it to check whether there's any parking or tricky intersections along the way.

Follow these steps to get to Street View:

1. Go to the location details page.
2. Tap the **Street View** button, which looks like a person with a triangle on his chest. Not every location has a Street View; if this button is grayed out, that option is unavailable.
3. Navigate by dragging around the picture with your finger to pan around the scene.
4. Go further up or down the road by tapping the arrows. The yellow line shows you the path the Street View car took as it traveled.
5. Exit Street View with the **Back** button on your phone.

TIP: Street View uses large pictures and takes some bandwidth, so you should only attempt to use it if you've got a fairly strong signal or you are in a Wi-Fi hotspot. If you are not in a Wi-Fi hotspot, be aware that Street View uses a lot of data, so we recommend only using it if you have a large or unlimited wireless data plan.

Using Your Phone for GPS Navigation

If you have a phone with maps, wouldn't it be nice if you didn't also need a car GPS? As it happens, you can indeed use your phone as a GPS. **Google Maps** includes navigation instructions, but it also includes a separate app called **Navigation** that you can use to get directions in a hurry. **Navigation** is not just for driving directions; you can also use it for walking, biking, and public transport directions.

Like **Places**, **Navigation** isn't so much a separate app as it is an interface to get to the same Google Maps data. When you get driving directions, choose the **Navigate** option. You can select **Navigate** from within **Google Maps** or just launch the **Navigation** app. Directions will be spoken from your DROID's speaker or Bluetooth headset when attached.

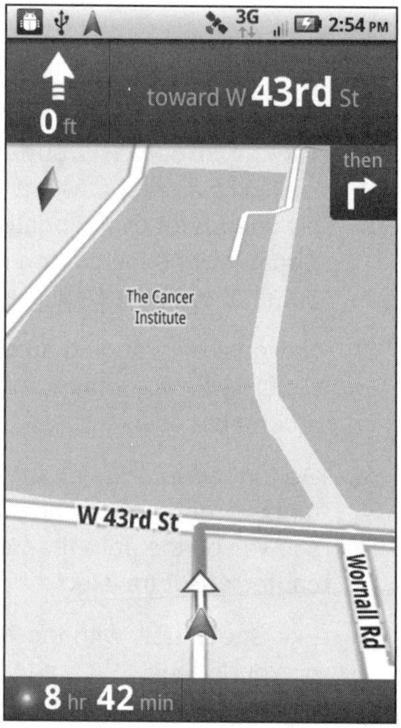

Just as with dedicated GPS navigation systems, **Navigation** will attempt to compensate if you take a wrong turn or choose to take an alternate road. It also has text-to-speech, which means you'll hear "Turn left on West Highway 50" and not just "Turn left in 1000 feet." As with any GPS navigation system, this works better for some roads than others. Sometimes the Google Maps data will reflect a less commonly used name for a road, and sometimes the pronunciation will be off.

As shown on the right, **Navigation** displays map information for driving, the length of time it will take to reach your destination, and the direction of your next turn.

As you near your destination, **Navigation** will show you Street View, so you can glance (or better yet, have a passenger glance) to see where you're headed.

Unlike some commercial navigation systems, you cannot choose new voices or change the appearance of your "car" on the map.

You can purchase special accessories for your DROID such as car mounts for charging your phone while using it to navigate. You can also use a cigarette lighter adapter to power your phone while navigating. It's a good idea to have some sort of dock or charger because navigation eats up a lot of battery juice.

Google isn't the only navigation app for DROIDs. There are a growing number of commercial navigation apps available for download, and DROIDs also come with **VZ Navigator**, which currently costs an extra $10 per month to use. Because Google's navigation tools are free and very useful, we feel they're the best bet. However, you may find a must-have feature, such as live rerouting to avoid traffic jams (available in **VZ Navigator**), that make a commercial app worth the extra money. It's still cheaper than buying a standalone GPS unit.

CarDock

In addition to **Google Maps**, **Google Navigation**, and **Google Places**, your DROID also has a Google app called **CarDock**, a tool specifically made for using your phone in the car. **CarDock** assumes that you will mainly be using your phone as a speaker phone to dial your contacts; using voice commands to find driving directions; and playing music.

The buttons on **CarDock** are intentionally big, so you can press them with a glance instead of a stare. Figure 19–2 illustrates the various **CarDock** options. When someone calls you, you still have to drag to answer the call, which is unfortunate.

Tap the large **Microphone** button to launch **Voice Search**. You can use **Voice Search** to find a location. Be sure to say, "Navigate to" and then your destination. This tells **Voice Search** that you want driving directions, not just to search for a location on the map.

CarDock will launch the **Navigation** app in order to find your destination. You may see a prompt to choose whether you want to use **VZ Navigator** or **Google Navigation** to complete your task. Choose **Navigation** unless you're paying for the VZ Navigator service.

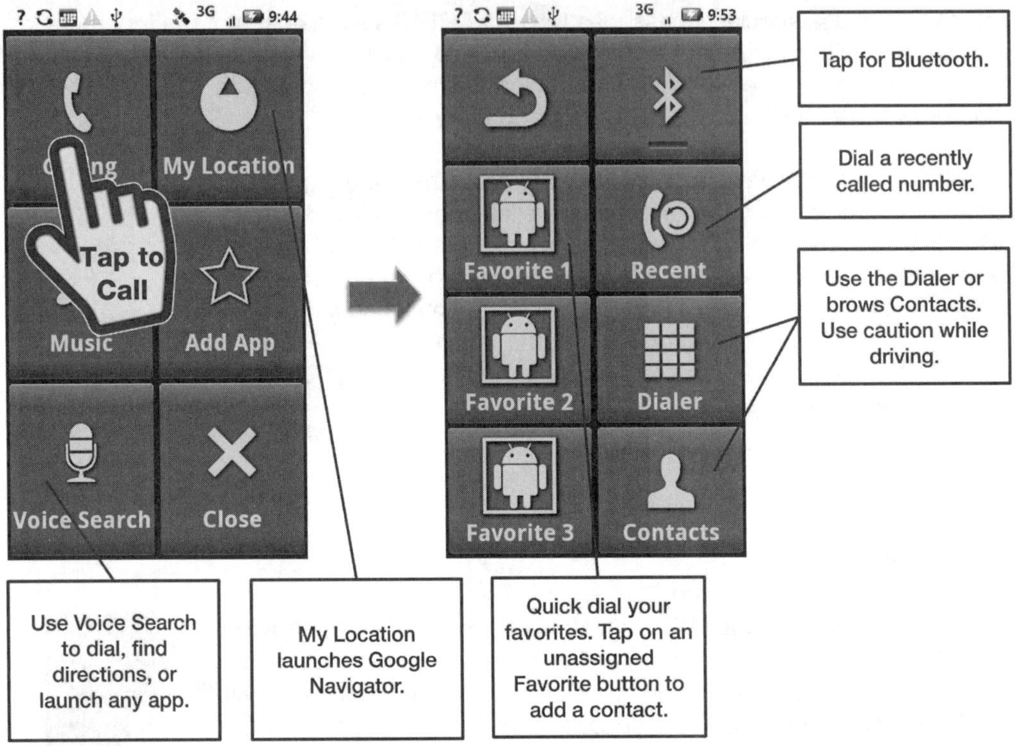

Figure 19–2. ***CarDock*** *as seen in* ***Night*** *mode.*

NOTE: On the original DROID, this is called Car Home and the icons are somewhat different.

Adding a Favorite App

You may notice that **CarDock** has six buttons, and one of them is labeled **Add App**. You can use this button to launch **Pandora**, the **FM Radio** app on your DROID, **Amazon's MP3 Player**, or any other app you frequently use while driving. Follow these steps to add an app to **CarDock**:

1. Press down and hold your finger on the **Add App** button (the long-press gesture)
2. You will see a prompt. Scroll through the list of available apps, and tap the one you wish to add.
3. The **Add App** button will be replaced by the name of the app you've chosen.

Alternatively, you can assign an app to this button with the following steps:

1. From the **CarDock** app, press the **Menu** button.

2. Tap **Preferences.**
3. Select **Custom.**
4. Choose an app.

You must use the **Preferences** method if you've already chosen a custom app for **CarDock** and want to change it to something else.

Email and Text Directions

Many apps link to **Google Maps**, and Google provides plenty of alternative ways to find directions and locations. If someone sends you a location in **Gmail**, Google will sense that the information is an address and attempt to automatically create a **Google Maps** link from it. Likewise, if you receive an SMS message with a **Google Maps** address link, you can use the link to launch **Google Maps**.

Making Your Own Maps

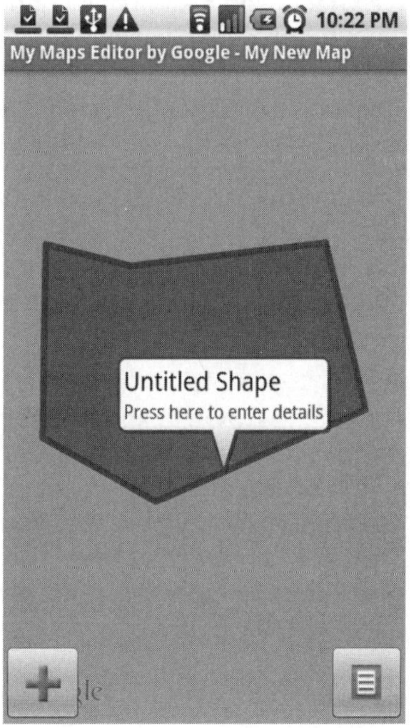

You may have noticed that one of your options in **Google Maps** layers is **My maps**. You're not limited to Google's layers in order to make a map. You can actually create your own map as a layer to **Google Maps**.

To create your own maps, you can either use **Google Maps** on a desktop computer, or you can download the **My Maps Editor** app from Google. This lets you add photos, lines, shapes, markers, and new addresses. Follow these steps to do so:

1. Tap a location.
2. Tap the + button.
3. Select the type of content you want to add.

   Photos can only be added if you shoot them from your camera at that moment.

4. Once you're done adding an element, you can add details to it. For instance, you could put a shape around the area of a convention where your company will locate its booth.

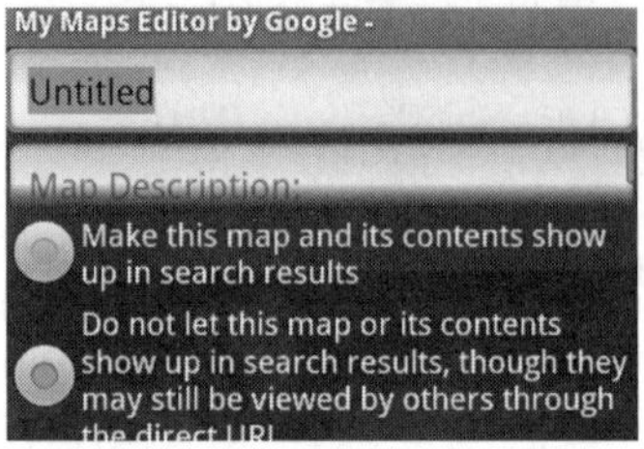

5. Once you've added the elements you need, tap the **Edit** button and choose whether this is a personal, private, or public map.

6. Give your map a name, and tap the **Save** button.

All your maps will be available in the **My maps** layer, but only maps you've made public will be visible to other people or available in search.

Location-Based Social Media and Games

We've focused a lot in this chapter on **Google Maps** specifically. However, one recent trend worth noting is location-sensitive social media. There are restaurant and service guides like **Yelp** and **Geodelic**, as well as apps that tag photos or messages with your location. Your DROID camera can also tag your photos with the location they were taken. See Chapter 18: "Photos and Video" for more information.

The **Latitude** and **Buzz services** use some of these geo-tagging features, and **Twitter** has likewise enabled the ability to give location information. However, two rising stars in this field are **Gowalla** and **Foursquare.** Both were created around the same time and offer very similar features. App developers are also free to take advantage of the Google Maps library and Android's location-sensing features.

In both the **Gowalla** and **Foursquare** services, the object is to check into locations. You need to use a phone or other location-sensing mobile device. You can't just manually type in a location. The whole point is to actually *be* there. You can share these check-ins with nobody, your friends, your Facebook page, or the whole world via Twitter.

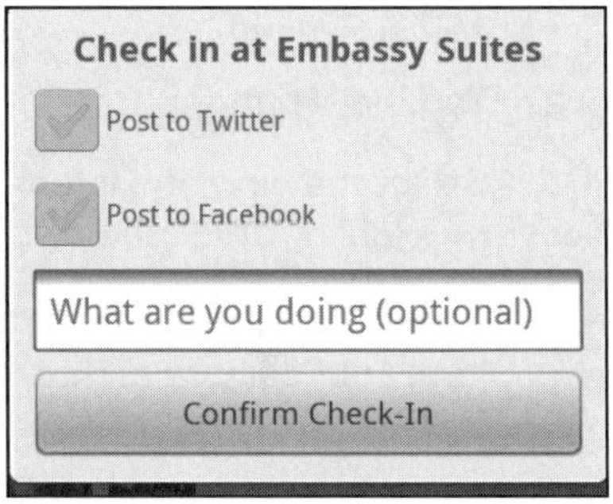

Both **Foursquare** and **Gowalla** are working with businesses and cities to offer features like coupons and specials for users who check in. This unique form of advertising may become very popular because it gives you an obvious way to measure the effectiveness of an advertising campaign.

Foursquare

Foursquare was cofounded by ex-Googler Dennis Crowley, who worked on a similar project, Dodgeball, which was purchased and abandoned by Google. However, **Foursquare** takes the social network to a new and different level by adding a gaming component. Dennis Crowley explained it as a method to combine exploring your city with gaming and a little bit of Boy Scout pride in earning merit badges and trophies.

Certain combinations of check-ins in Foursquare earn badges. Some examples include checking into the same location three times, finding five karaoke bars, finding three places in Chicago with photo booths, or checking in after 3 a.m. on a school night.

Checking into the same location regularly could also earn you a "mayorship." The picture to the right shows a check in that resulted in winning a new mayorship. The mayor of a location is the person who has checked in most frequently in the last two months, so you need to keep checking in to maintain the title.

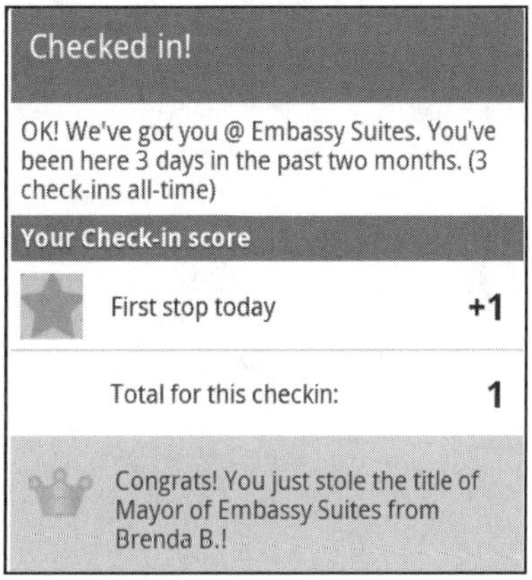

However, rapid checking in is disabled to prevent too many people from gaming the system.

Gowalla

While Foursquare is a bit of a competitive game, **Gowalla** is more of a personal exploration and virtual geocaching tool. You can earn pins, similar to the badges in Foursquare, and you maintain a passport of places you've visited. You can also create and travel on tours of different check-in locations.

Gowalla also gives users a few virtual items they can leave or exchange at locations. Examples would be blankets, avocados, and espresso machines. To the right, you can see the luggage tags left at a location. When you encounter a virtual item at a location, you can see the history of who owned the item, encouraging the frequent exchange of these virtual geocaches.

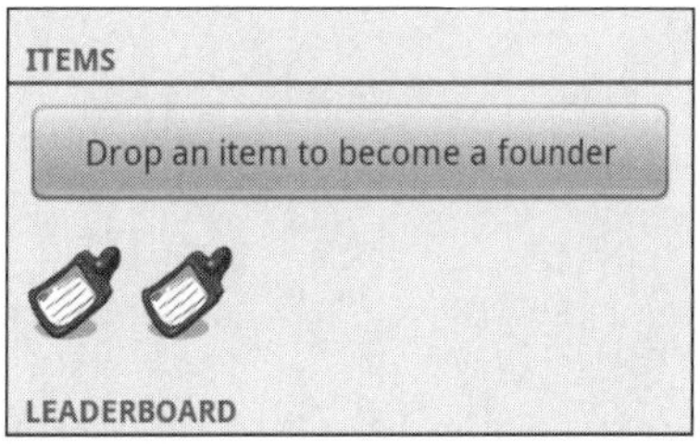

Rapid check-ins are fine with Gowalla, since there's no mayorship. It's also much more sensitive to proximity than Foursquare, so you need to be pretty close to a location to check in. This makes it ideal for walking tours and pub crawls, but not so great for check-ins within a building or anyplace where the GPS will not work well.

In the next chapter, we'll explore social media in more detail.

Chapter 20

Social Media and Skype

Social media can broadly be defined as Internet sites and apps designed around social interaction. The Internet, has always centered on communication, and now that form of communication entered the mainstream. As we write this, the *Facebook* site at `www.facebook.com` is the most popular website in the world.

These days, being social can also be good business, so much so that it's become a regular component of CRM (customer-relationship management). Social media keeps your customers updated with your latest projects, keeps them excited about your products and services, and lets them tell you where you should go next. Social media also helps you personally network with your colleagues or find your next opportunity.

Social media done badly can also be bad business for you or your company, so think twice about what you post and exercise some common sense. People have been fired for casual posts they've made on their blog, Facebook, Twitter, and other social media websites.

In this chapter, we'll explore some of the social media tools available for Android, as well as how you can make efficient use of your social time both on and off the clock.

Motorola Widgets

Your DROID comes ready to sync with several social networking sites, including Facebook, Twitter, and MySpace. You can learn more about syncing with your accounts in Chapter 3: "Sync to your Google Account." In addition, there are two Motorola widgets you can use with your social media. These are covered in more detail in Chapter 6: "Organize Your Home Screens."

Let's begin by reviewing the two social networking Motorola widgets: **Social Networking** and **Social Status** (see Figure 20–1).

> **NOTE:** The original DROID has a separate Facebook and Twitter widgets as opposed to the Social Networking widget.

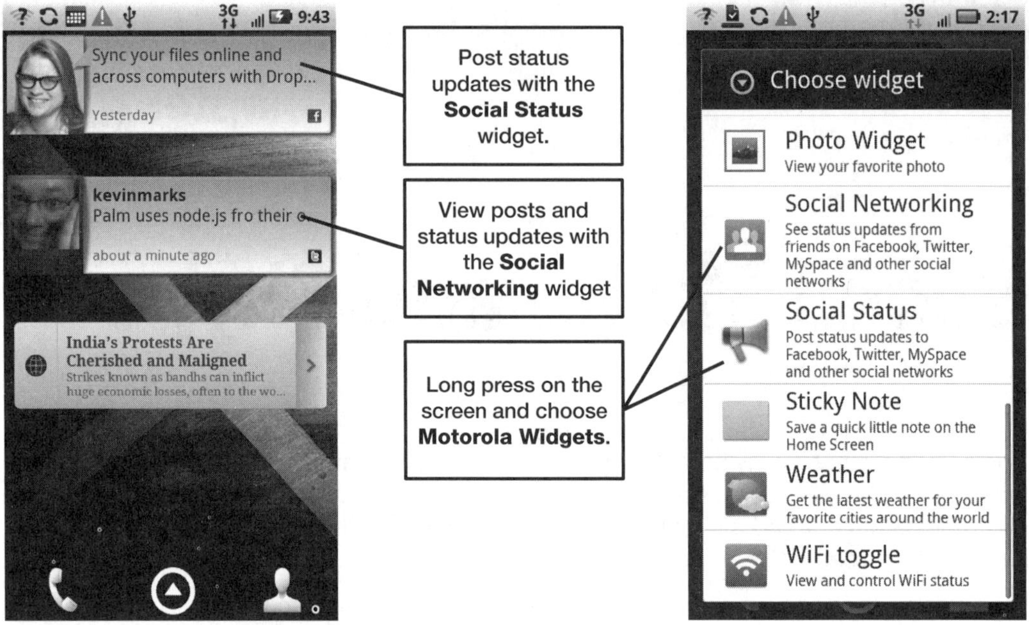

Figure 20–1. *The social networking Motorola widgets.*

Both of these widgets are interactive. Tap on a widget to expand it for more choices. The **Social Status** widget expands to allow you to make status updates to your Twitter and Facebook accounts. The **Social Networking** widget expands to let you read, comment, **Like**, **Retweet**, and otherwise interact with your social networks (see Figure 20–2).

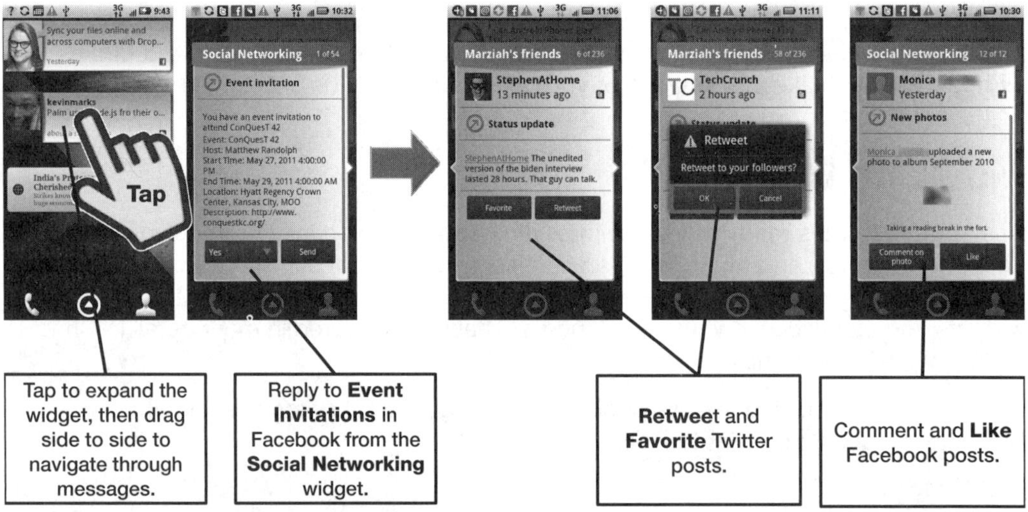

Figure 20–2. *The expanded **Social Networking** widget.*

It's possible you'll never need another social networking app for your DROID. It's also possible you'll find features or social networks missing from your collection if you rely only on the included Motorola widgets.

Twitter and Microblogs

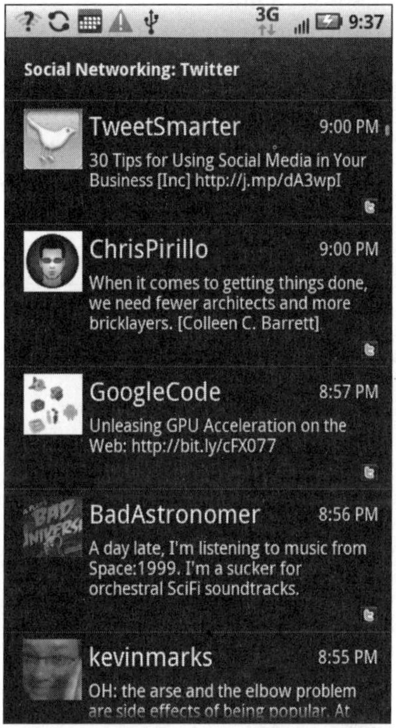

Twitter is part of a new generation of short, public blogging tools known as *microblogs*. Twitter is a rapidly evolving service that essentially started out as a blog-like public collection of SMS messages from a given user. Twitter posts, or *tweets*, are limited to 140 characters; this limitation reflects their start in SMS. However, Twitter is also available from the Web, so many users are not accessing it with their phone at all. Twitter gained popularity in part because it allowed a lot of open use from third-party tools. Some (but not all) of those tools have made their way to Android, and Twitter features are built into your **Social Networking** and **Social Status** widgets, as well as many other apps, including those shown on the right.

You might wonder why you would use Twitter. The short messages are great for pointing out items of interest, letting your friends know what you're feeling, or critiquing a conference as it happens.

Twitter is good for business, too. If your business is transportation, you can let the riders know about delays. If your business is weather dependent, you can let your customers know if you need to make cancellations. You can also use it to advertise specials, promote your latest accomplishments, or deepen your customer engagement by having a conversation about their needs. Or you can use Twitter to listen to what your customers or colleagues in your industry are saying. Chris Brogan has an excellent blog post on the subject of Twitter and business at this URL: `www.chrisbrogan.com/50-ideas-on-using-twitter-for-business`.

The Mechanics and Culture of Twitter

We tend to think of Twitter like a noisy party where everyone has to speak loudly. As you walk through the party, you can hear snippets of conversations, but you can't always be sure who is being addressed. We're not the only one to make that observation. In fact, there's an entire book that uses the party analogy: *Social Media Is a Cocktail Party* by Jim Tobin and Lisa Braziel (CreateSpace, 2008).

The basic mechanics of Twitter are this. Twitter is a free service available at `http://twitter.com`. Every user can make 140 character posts in her own Twitter "stream." You can follow other users, and they can follow you. You have a stream of all the posts from people you follow. You can choose to make your stream public or visible only to those you preapprove as followers, but you can't specify privacy on individual posts. The number of followers a person has can be taken as a measure of authority, though it's not an absolute measure. Ashton Kutcher's Twitter account is far more popular than Eric Schmidt's account, even though many would argue that the CEO of Google has more authority in his tweets.

Many of Twitter's conventions are ad hoc creations designed to work around some of Twitter's shortcomings. Twitter is an unthreaded conversation stream. You can make a post in a stream as a reply to someone else, but your response remains in your Twitter feed, not the other person's. In order to indicate replies, people began using the `@username` convention. For example, replies to Marziah would be `@marziah`. Eventually, Twitter worked this into the system and began making automatic links to users based on this, letting users know whenever they had a new `@reply` or `@mention`.

Retweets

If someone says something you agree with, like, or want to repeat, you don't repeat it, you retweet it. A *retweet* is a repeat that gives credit to the original author. The convention for retweets is `RT: @username`, followed by the repeated message. Twitter also picked up on this convention and now allows you to press a button and retweet messages with the other user's icon to indicate the source. However, some users still opt for the old `RT: @username` style retweets because this approach allows them to add commentary before the retweet.

Hashtags

Searching for relevant information on a given topic is difficult, so users began putting unlikely character combinations into their posts to tag them or help them sort related posts in searches.

I'm intentionally tweeting during the digital nation presentation #sidlit
8:29 AM Aug 5th via Twitter for Android

The # (hash) symbol became the common marker for these tags, so the *hashtag* was born. Sometimes conventions or advertising campaigns mention a particular hashtag, and sometimes the hashtag is spontaneous and viral.

URL Shorteners

You've only got 140 characters, so you don't have room for long URLs. A new class of service emerged that would simply forward links from a shorter URL to a longer one. URL shorteners themselves became shorter to save space, so services like *TinyURL.com* ended up competing with services like *bit.ly* and *ow.ly*. Many of these services also added value, such as metrics for the shortened URLs.

NOTE: You should avoid using shortened URLs when they're not necessary. You are relying on a third party to forward your link when you use a shortened URL. That third party could go out of business or have an outage at any time, making you look unprofessional for having a broken link. In fact, this is exactly what happened to the *tr.im* shortening service. In April of 2010, the company announced it was going out of the URL shortening business and that it was ceasing all forwards at the end of 2010. Some URLs are also "nicer" than others when it comes to forwarding your links in a way that search engines can easily crawl.

TwitPic

TwitPic was created by Noah Everett to solve the problem of photo sharing on Twitter. It's a separate service, but interconnected. You must register for a TwitPic account at `http://twitpic.com`, but your Twitter username is your TwitPic username, and each photo has its own comment thread on TwitPic. Many phone Twitter apps tie into TwitPic or use similar services, so uploading and linking a photo is a quick process that can be accomplished entirely from your phone.

Your DROID doesn't support directly uploading photos to Twitter from the **Social Networking** app at this time, so you'll need to use a service like TwitPic or a Twitter client that supports such a feature to attach pictures to your updates.

Direct Messages

Direct messages – or *DMs* – are short, private messages you can send to a follower that is also following you. They're almost like email message, but you're still constrained to 140 characters. You can send a direct message using a button for that specific purpose.

Finding Twitter Apps

There are many Twitter apps available for Android, including the official **Twitter** app from the Twitter team, which is shown to the right. That doesn't mean the official app is the best or the only possibility you should consider. You can also use the **Social Networking** app that comes with your DROID, and many third-party apps provide enhanced features not available in the official app.

It may be helpful to make a list of priority features before deciding on an app. Most apps also come with lite and premium options, so you can try before you buy. Some factors in your decision may include the following:

- Support for multiple accounts
- URL shortening
- Link metrics
- Speed
- The ability to upload and link pictures or video
- The ability to create and track custom keyword searches
- Support for both viewing and adding lists
- Ability to follow, unfollow, and block accounts
- Ability to easily send retweets and direct messages
- Available widgets
- Background sync

Hootsuite is our favorite Twitter app for a balance of features and price. You can use it to manage multiple accounts, and the interface makes it clear which account is posting.

If you buy the premium app, you can also track visitors through the URL-shortening service, ow.ly. The image on the right shows our favorite extra feature: scheduled tweets. You can use this feature to precompose press releases or turn 3:00 a.m. deep thoughts into mid-afternoon topics of conversation. Using the feature is as simple as pressing the **Calendar** icon when composing a message and choosing when to send it.

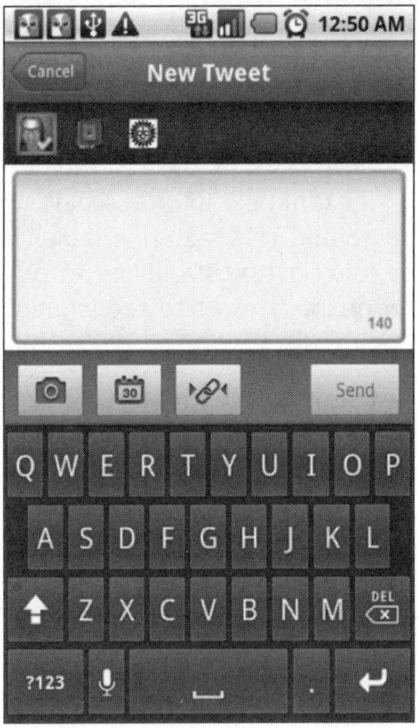

Twidroyd and **Touiteur** (pronounced *Twitter*) are a pair of popular apps that feature both free and paid versions.

> **Twidroyd** was formerly known as **Twidroid**, but its publisher, TweetUp, changed the name when it purchased the service to avoid any branding confusion with the LucasArts trademarked term, *droid*. Verizon Wireless actually licensed the term *droid* for your DROID.

Touiteur offers nice features like trend search, but you must upgrade to the €1.99 (about $2.60) version to upload pictures or manage multiple accounts.

Tweetcaster (shown on the right) is a $4.99 app with a pleasant user interface that supports saved searches. This feature enables you to keep checking certain hashtags and keywords. You can also search for nearby tweets to see what's trending in your local community.

Seesmic is a full-featured free app that supports multiple accounts and is also available for desktop computers.

Twitter's official app is easy to use and supports all the official Twitter features, such as lists, trends, and `@mentions`. Lists are shown to the right. The official **Twitter** app also comes with a nice widget, a splash screen with a bird that flaps its wings and lists trending topics. However, the official **Twitter** app doesn't support multiple accounts at this time.

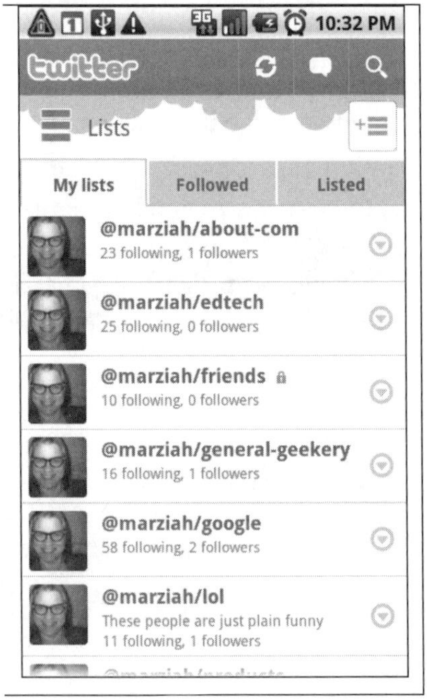

As you consider choosing a favorite Twitter app, you need to keep a few things in mind. Nearly all of the apps that support Twitter allow you to change how often they check for new tweets. Unless you absolutely, positively must be notified of new tweets, you should turn this down to a reasonable, battery-conserving setting of 30 minutes or longer.

Many apps allow you to use either your own URL-shortening service or use one supplied with the individual apps. You should pick a service that offers you tracking and metrics, even if you don't think you'll use them just yet. It's easier to have the data than it is to wish you had the data. `Bit.ly` and `Ow.ly` both allow data tracking and are popular choices.

Yammer

If you want the instant communication of Twitter, but you want to restrict access to only your company, you may want to try *Yammer* (`www.yammer.com`). Yammer creates a Twitter-like atmosphere that is only accessible to people with the same email domain. Yammer also adds threaded conversations to the tweets and organizational charts, so it has value beyond a simple Twitter imitation. Recently, it has added the ability to create communities across multiple email domains, so it's possible to have a partnership community with vendors and customers or a group of close friends with different email domains.

Yammer makes an official app. It's not robust, but it does allow you to post and see status updates. Fortunately, the intra-company nature of the Yammer service means that you don't need the same bells and whistles for communication that you would for Twitter.

Other Microblogs

Twitter stole most of the microblog spotlight, but you may prefer a different platform for your message. There's no shortage of microblogging services, so there's no chance of naming them all. Some, like *Jaiku*, seem to be short on dedicated apps. Others have a large selection of dedicated apps, but none of these other microblogs are included in the **Social Networking** widget on DROID.

Tumblr is worthy of mention. It allows short microblog posts, but it also supports multimedia better than Twitter by directly embedding it in threads. It also enables threaded responses to posts. Tumblr users create rich *Tumblogs* with pictures, videos, links, and text.

There are multiple Android apps that support Tumblr, both free and premium. Just as with Twitter and Facebook, there's also an official **Tumblr** app. The free **ttTumblr** app shown to the right offers a lot of features in conjunction with a simple user interface.

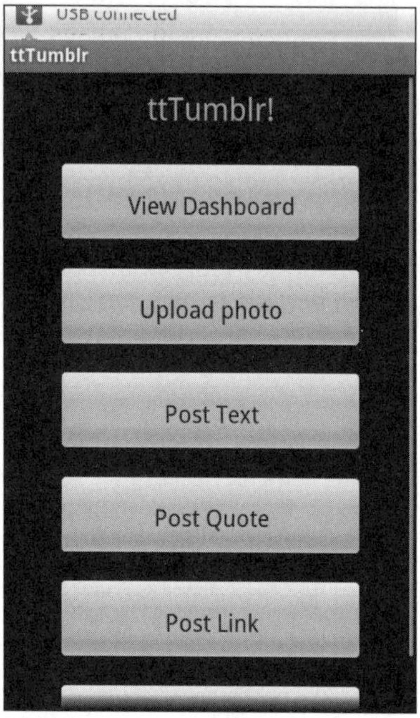

Plurk is another microblogging alternative to Twitter and Facebook. It organizes posts on a timeline and assigns "karma" for participation. Like Tumblr, it allows easier multimedia embedding, and it has built-in privacy settings to make it useful for both small and large groups. Posts are organized on a scrolling horizontal timeline, which allows threaded responses. The web timeline interface makes it challenging to translate to a phone app; however, the **PlurQ** app (shown on the right) does a good job of implementing this feature on a phone.

Social Bookmarking

Social bookmarking is a method of sharing sites you like. In the purest form, *Delicious*, one of the leading social bookmarking services (`www.delicious.com`), allows you to save a bookmark with a quick note and tags to organize the bookmark by category. You can also network with other users to see their bookmarks and measure the relative popularity of bookmarks by seeing how many other users have marked a particular site.

Instead of using a dedicated, standalone Delicious-based app, you typically use the service through add-ons to your browser. When an app such as **Bookmarking for Delicious** is installed, it adds an option from within your web browser to add bookmarks to your Delicious account. To add bookmarks, make sure **Bookmarking for Delicious** is installed and then follow these steps:

1. Tap **Menu**.
2. Tap **More.**
3. Tap **Share page**.
4. Choose **Delicious.**

For more information on the built-in **Browser** app, please refer to Chapter 11, "Surfing the Web."

Digg (`http://digg.com`) is another popular social bookmarking site. Unlike Delicious, Digg also adds a threaded comment discussion. Digg emphasizes quantity and focuses on showing the most popular links of the moment as a method of crowd-sourced news. Users can *digg* or *bury* items to see what stories float to the top. Similar services include *Reddit* and *Slashdot*. Sites like these can attract huge crowds of visitors to a business's site, so it makes good business sense for blogs that can handle high levels of traffic to leverage social bookmarking sites to entice people to visit.

Android apps are available for all three social bookmarking services. You just need to decide how involved you need to be from your phone. For example, do you need to comment and submit bookmarks, or do you just want to see what the currently trending articles are? Do you want a full app or just a widget?

Facebook

The Facebook social network started out as a simple virtual yearbook for college students, but it has since morphed into one of the most popular websites in the world. At the time of this book's publication, it's even more popular than Google search in terms of the sheer volume of page views on the site.

Facebook intends for people to use their actual names instead of pseudonyms. Users are also intended to share information with small to large circles of acquaintances. However, Facebook has been facing increasing scrutiny over its privacy policies and confusing security settings. When you use Facebook, the wisest course of action is to assume anything you say is completely visible to the world; indeed, you should make this same assumption for *any* website you post information to.

Facebook allows multiple types of posts, from quick status updates to photos, videos, and longer notes. You can also link to articles, videos, and pictures hosted outside of Facebook, as well as add apps that incorporate games, group reading lists, and more. Facebook is also moving toward a universal **Like** button that allows you to interact with pages and websites outside of Facebook.

You might wonder how you manage both personal and business contacts on Facebook. You can do so in a couple of ways. One approach I *don't* recommend is to create multiple accounts. If you create multiple accounts using your real name, it will only serve to confuse you and your contacts when they try to add you as a *friend*.

The two approaches you can take are either to friend everyone and assign them to friend groups through the privacy settings or to create a *fan page*. Fan pages (officially, Facebook just calls them "pages") got their name from the way people used to add them to their feed. Users would "become a fan" of a given page. Facebook has changed this mechanism to a simple **Like** button.

TIP: Currently, Facebook has a 5,000-friend limit on personal accounts. If you anticipate reaching that limit between clients, fans, and good friends, then you need a fan page. Even if you don't anticipate an overwhelming deluge of clients and business contacts friending you, it may still be disturbing to manage personal and work acquaintances in the same social space.

Creating Fan Pages

You set up fan pages through the **Ads and Pages** application. If you don't have any pages, search for "Ads and Pages" from within Facebook. We suggest using a desktop browser to set this feature up.

The advantage of using a fan page is that you can make a fan page an official company presence without needing to be friends with any of the fans of the page. The disadvantage is that you do not see the activity fans generate anywhere outside of that page. Whether or not you want your business contacts mingling with your classmates and relatives is a personal decision. However, you should decide how you want to handle the situation before you get your first friend request. It's a lot easier to have separate spaces established in advance than it is to move everyone over to new spaces later.

CAUTION: Whether or not you are Facebook friends with colleagues, business partners, or customers, it's just bad business to badmouth *any* of them. They may not be able to see what you've said, but it's not hard to copy-and-paste. The last thing you need is for casual gossip to get back to the person you badmouthed. People have been fired for less.

Facebook Apps

Your DROID already supports Facebook through the **Social Networking** app. Original DROID owners can use Facebook's official app, which is shown to the right. Facebook's official app is great for personal networking. You can also use it to sync status updates with your contacts, check into locations, and you can upload photos directly to Facebook.

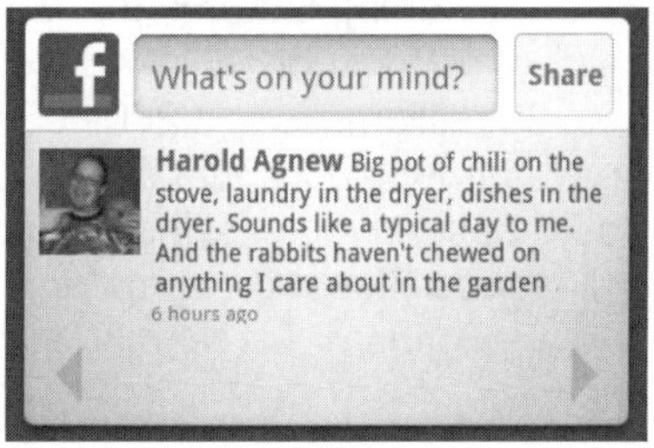

Personal Facebook management isn't a problem, but managing fan pages and groups from your phone requires going beyond the default **Social Networking** widget. The easiest method to post directly to fan pages we've found is through the *Ping.fm* service. Ping.fm (which is also the URL) allows you to cross-post to an impressive number of social media sites at once. We'll cover cross-posting in more detail at the end of this chapter.

LinkedIn

LinkedIn is a social networking site for professionals. It's designed primarily as a place to hang your resume, cross-network with business partners, give and receive recommendations, and offer status updates about your latest accomplishments. Although it's something you may think of as a tool for job seekers, it's a good idea to build and maintain your network even when you aren't actively looking for work.

You can establish yourself as a trusted source in your community by joining groups and recommending colleagues. Chances are you will need to look for work at some point, and it's better to have connections and trust already established than to try to build them when your situation is desperate.

LinkedIn has grown in popularity among business users by adding features for use beyond a simple chart of connections. LinkedIn claims to have more than 65 million users in 200 countries. As it has grown in popularity, LinkedIn has added features that go far beyond job seeking. You can network with colleagues in user groups, add your Twitter feed, and add applications like reading lists and document sharing.

Android doesn't sport an official LinkedIn app at the time of this book's publication, and there are only a couple of third-party choices. **Linked,** by JUPE, is shown to the right. **Linked** is an ad-sponsored app that offers basic status updates and reading, allows you to see your contacts, and allows you to search and send contact requests.

Blogging

Blogs – short for *weblogs* – started out as a series of manually maintained updates with no ability to comment. However, today blogs are a thriving, interactive format used worldwide. Many businesses use blogs to keep customers informed about their products, issue press releases, or just put a human face on their company. Freelancers often keep blogs as a way to promote themselves. In some cases, the blog itself has become the business, with advertising and market tie-ins generating enough revenue for the blogger to quit his day job.

Blogs are generally intended to be public and visible, so it's vital that you and your boss be clear on your intentions when it comes to corporate blogging. If you maintain a personal blog, it should go without saying that you need to be careful what you say about your boss or customers, even when blogging under a pseudonym.

The standard format for most blogs is that the newest entry goes at the top, with older entries following it. The blog page itself uses either the RSS or Atom format for blog aggregation to make it easier for viewers to read the blog or find new updates without having to visit the blog itself. Feeds can be full, partial, or headline only. While full feeds are certainly the most convenient for readers, they also make it easier for content thieves to steal blog entries and claim them as their own.

Phone Posts

Most blog platforms offer a method to email blog entries. Some also offer a way to post blog entries via SMS text message. Some, like *LiveJournal*, even offer a way to call and voice-record a message. In LiveJournal's case, users can then manually transcribe the voice recording, so you can call in with a quick update (e.g., "It's a girl!" or "Accident on the 435 bridge") without having to enter text.

As Android and smartphones become more popular, blogging platforms have also discovered the value of providing a native phone app for making and managing posts. **Blogaway** and other third-party apps support Google's own blogging service, called *Blogger*; however – and inexplicably – Google hasn't released an official app for its blogging service at the time of writing.

WordPress

WordPress deserves special mention because it is the most popular blogging platform in the world. It can be used for content management that goes beyond blogging; however, blogs remain the core functionality that drives the popularity of WordPress.

WordPress is open source and free. It can be templated and modified to run on corporate sites, and it can power personal blogs as well. There are a large variety of plug-ins and extensions from both free and premium developers available for this service.

WordPress is supported on Android through a native **WordPress** app, which is shown to the right. This app lets you post messages with formatting, tag posts, and geotag posts; the app also lets you manage comments. You can also add photos and video to your posts.

You aren't offered as many options for templating and administration as you'd see in a desktop browser. However, you probably don't *want* as many options when you're trying to type them in on a slide-out keyboard or touchscreen. If you need more access on the road, you can log into your account from your DROID's **Browser** app.

Bump

It's possible you may prefer to socially network the old-fashioned way. You still can. The **Bump** app, shown to the right, is available for both iPhones and Android. This app allows you to share your contact information by launching the application and then literally touching another Bump user's phone.

You're not limited to just DROID users. Android and iPhone Bump users can share contact info with each other this way. You do need a reasonable network connection because the **Bump** app transmits your information over cell or Wi-Fi networks, not Bluetooth. Android users can also use the **Bump** app to share free apps from phone to phone.

Buzz

Google has been trying to compete in the social media arena, but so far it hasn't made much progress. One of its latest endeavors is its *Google Buzz* (Buzz) service. This service is part of Gmail, but it behaves like a separate service. Google has been heavily promoting Buzz, so it may end up becoming more popular as a social and self-promotional tool.

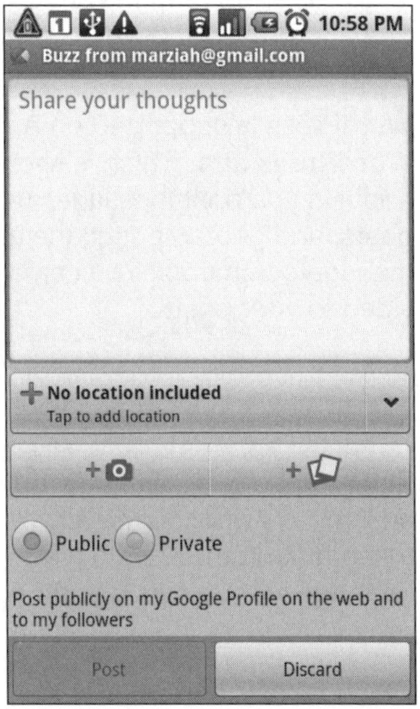

Buzz allows for long posts that can embed photos and videos, as shown to the right. You can create private or public posts and follow the posts of your contacts. Posts in your *Buzz stream* are often bumped to the top of the list based on who last replied. Thus, the more popular Buzz users tend to dominate the conversation. You can feed Twitter posts into Buzz to allow threaded comments on them, but Buzz posts do not feed back to Twitter.

The **Google Buzz** app for Android is a widget that enables quick posting of photos and location information. It also enables you to adjust your privacy settings. Reading Google Buzz is still handled through the mobile web browser interface; however, this is something that we hope will change with time.

Cross-Posting

Once you're up and running with all these social media services, many of which use similar posting formats, you might wonder how you manage your time posting to them. Fortunately, you can take advantage of cross-posting tools that let you focus on the tool or format that is easiest and/or most rewarding for you.

The Motorola widget, **Social Status**, does exactly this. Follow these steps to use the **Social Status** widget:

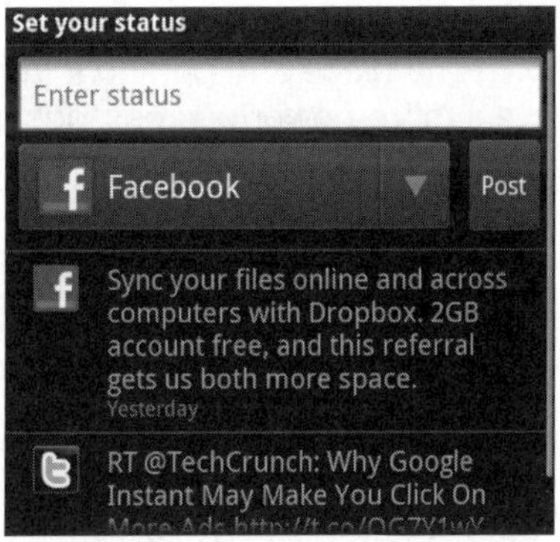

1. Tap the **Social Status** widget to enter a new status.
2. Tap the selection box to choose a service for updates.
3. Your choices are **Facebook**, **MySpace**, **Twitter**, and **All services**. Choose **All services** to cross-post your update.
4. Tap **Post** to finish posting the update.

Increasingly, apps are offering built-in cross-posting to and from Twitter and Facebook. While you can use the **Social Status** widget to make cross-posts individually, you may want to create other types of cross-posts.

If you're primarily a blogger, but want to add tweets to announce new blog entries, one way to do so is through *Twitterfeed* at http://twitterfeed.com. This is a free service that takes just about any blog feed and translates it into a shortened Twitter or Facebook post. You specify any prefix or suffix and how you want the post to be shortened, as shown to the right.

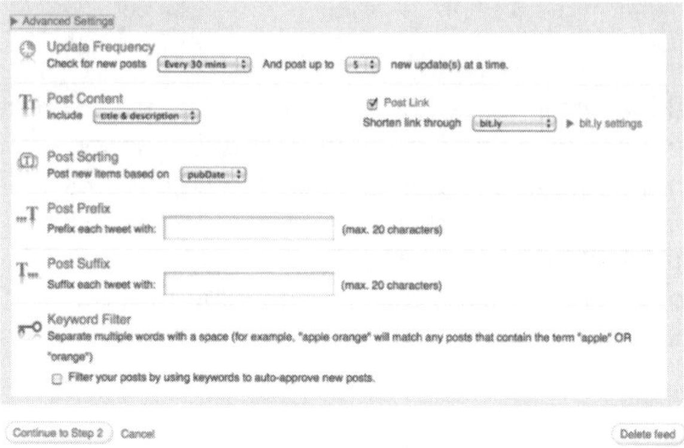

RSS Graffiti (www.facebook.com/RSS.Graffiti) is a service for porting blog posts into Facebook fan pages. There are many other cross-posting solutions as well, including free and paid apps.

If you want to go beyond simply scooping a feed from one place and putting it into another, you can use a more powerful cross-posting service. Ping.fm, as mentioned earlier, is a free service that can cross-post to Facebook fan pages.

Ping.fm can also cross-post to an impressive variety of social networks, blogs, and microblogs. From within Ping.fm, you can also make groups of media to post to; for instance, you could have a "press release" group that goes to your business Twitter account, Facebook fan page, WordPress blog, and Delicious bookmark. Any post you make to that group is automatically cross-posted.

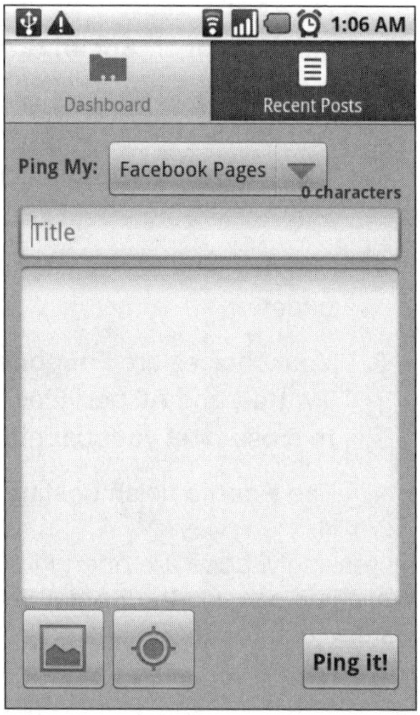

> **NOTE:** Ping.fm is the web service, not the DROID app that enables cross-posting. The DROID app for using the Ping.fm service is called **AnyPost**.

AnyPost, which is shown to the right, is a fantastic free Android client for Ping.fm. You can use it to post to services one at a time or as a group.

A similar service for Android is *Moby*. However, this service is more blog-oriented and does not let you post to as many services.

NOTE: As far as cross-posting is concerned, there's a fine line between posting the same message to multiple groups and simply spamming. The more places you cross-post, the more places you'll also have to monitor comments.

Aggregating Content With Readers

If you want to read all your content in one place rather than posting it, you'll want an *aggregator* (aka *feed reader*). Feed readers take feeds from other sources and pile them into one place for easy reading. Tweets, blog posts, news items, and even Google searches are delivered as feeds that you can add to a feed reader, and many blogs add handy links for adding feeds. The universal symbol for an RSS feed is this: . When using most Android browsers while logged into your Google account, you can simply click the **RSS** icon in a blog to launch the **Google Reader** app. This app allows you to add a site's feed to your Google Reader account.

Google Reader is a robust feed reader that lets you organize feeds by category, share likes, mark favorites, and leave comments. It also keeps track of the last item you read.

Unfortunately, official Google Reader support is currently only handled through the mobile web interface, which is shown to the right. The **Google Reader** app is capable, but it suffers from a few shortcomings. The most important interface issue when using the **Google Reader** app through the web interface is that you can't use the **Back** button on your phone. It's a hard habit to break when you're used to navigating apps, but the **Back** button will exit your browser instead of going back to the previous feed.

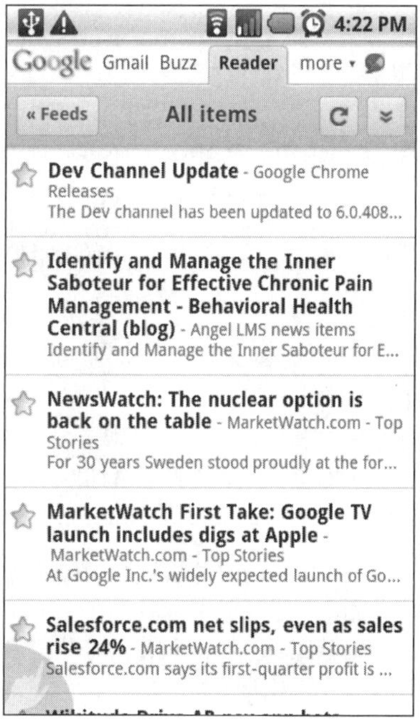

Some third-party apps support the Google Reader service, including free apps like **NetaShare** and paid apps like **eSobi**. You can specify whether you want feeds to sync in the background (do *not* choose this option if you're concerned about battery life). You can also use it to specify how many feeds should be fetched at a time.

Making Phone Calls and More with Skype

Social networking is all about keeping in touch with our friends, colleagues, and family. Passive communication through sites such as `www.facebook.com` and `www.myspace.com` is nice, but sometimes there is just no substitute for hearing someone's voice.

Amazingly, you can make phone calls using the **Skype** app from your DROID. Calls to other Skype users anywhere in the world are free. A nice thing about Skype is that it works on computers and many mobile devices, including iPhones, other Android phones, BlackBerry smartphones, and other mobile devices. You will be charged for calls to mobile phones and landlines, but the rates are reasonable.

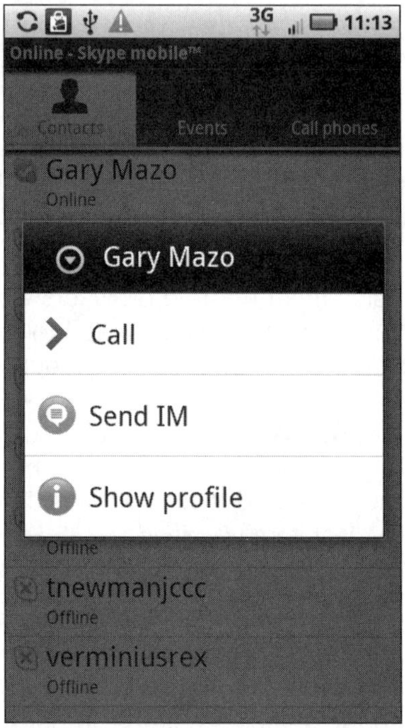

Your DROID ships with Skype already installed. You don't need to download anything to use it.

NOTE: As of October 2010, Verizon Wireless customers can only use Skype on 3G networks; however, Skype users in the US on other networks can only use Skype on Wi-Fi.

Creating Your Skype Account on Your DROID

If you need to set up your Skype account and have not already done so on your computer (see the "Using Skype on Your Computer" section later in this chapter), then follow these steps to set up **Skype** on your DROID:

1. Tap the **Skype** icon from your **Home** screen.
2. Tap the **Create Account** button.
3. Tap **Accept** if you accept the **No Emergency Calls** pop-up warning window.
4. Enter your **Full Name** and **Email**.
5. Enter your **Skype Name** and **Password**.
6. Choose whether to receive Skype news and offers.
7. Tap the **Create Account** button to create your account.

Log in to the Skype App

After you create your account, you're ready to log in to **Skype** on your DROID. To do so, follow these steps:

1. If you are not already in **Skype**, tap the **Skype** icon from your **Home** screen.
2. Type your **Skype Name** and **Password**.
3. Tap the **Sign In** button in the lower-left corner.
4. Check the box labeled **Sign in to Skype automatically**. You should not have to enter this log in information again; it is saved in **Skype**. The next time you tap **Skype**, it will automatically log you in.

Finding and Adding Skype Contacts

Once you log into the **Skype** app, you will want to start communicating with people. To do so, you will have to find them and add them to your Skype contacts list:

1. If you are not already in the **Skype** app, tap the **Skype** icon from your **Home** screen and log in, if asked.
2. Press the **Menu** button.

3. Tap **Add a Contact**.
4. Tap **Search Skype Directory** and then type someone's first and last name or **Skype Name**.
5. Tap the **Magnifying Glass** button to locate that person.
6. Once you see the person you want to add, tap his name.

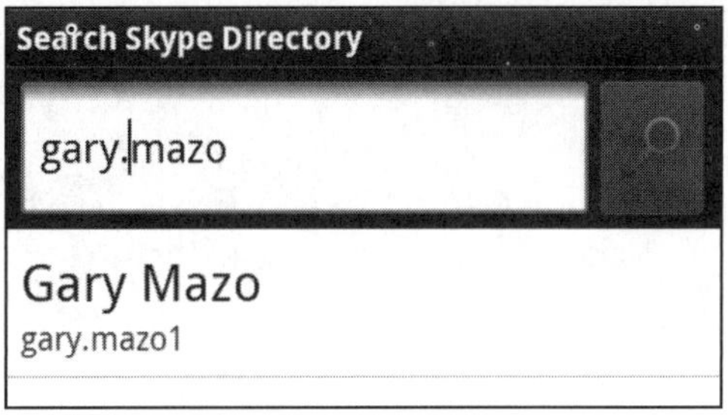

7. Tap **Add Contact**.
8. Adjust the invitation message appropriately.
9. Tap the **Send** button to send this person an invitation to become one of your **Skype** contacts.

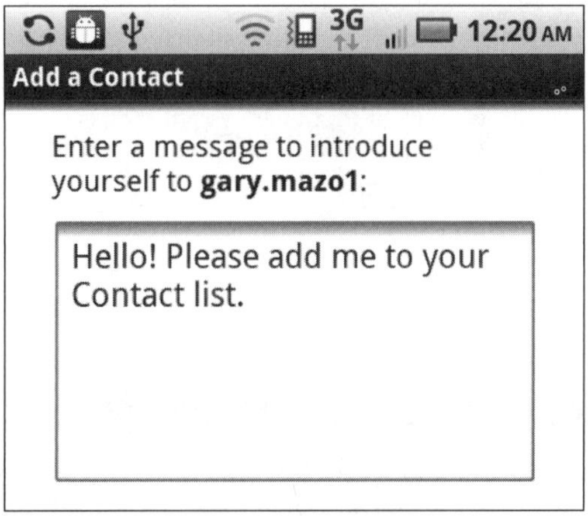

10. Repeat the procedure to add more contacts.
11. When you are done, tap the **Contacts** soft key at the bottom.
12. Tap **All Contacts** from the **Groups** screen to see all new contacts you have added.
13. Once this person accepts you as a contact, you will see him listed as a contact in your **All Contacts** screen.

TIP: Sometimes you want to get rid of a Skype contact. You can remove or block a contact by tapping her name from the contact list. Press the **Menu** button and select either **Remove** or **Block**.

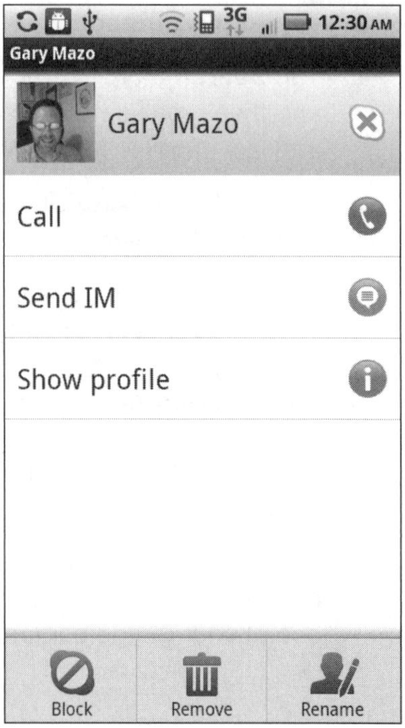

Making Calls With Skype on Your DROID

So far you have created your account and added your contacts. Now you are ready to finally make that first call with **Skype** on your DROID. Follow these steps to do so:

1. If you are not already in **Skype**, tap the **Skype** icon from your **Home** screen and log in, if asked.
2. Tap the **Contacts** soft key at the top.
3. Tap the contact name you wish to call (see Figure 20–3).
4. Tap the **Call** button.
5. You may see a **Skype** option and a **Mobile** or other phone option. Tap **Skype** to make the free call. Making any other call requires that you pay for it with *Skype Credits*.

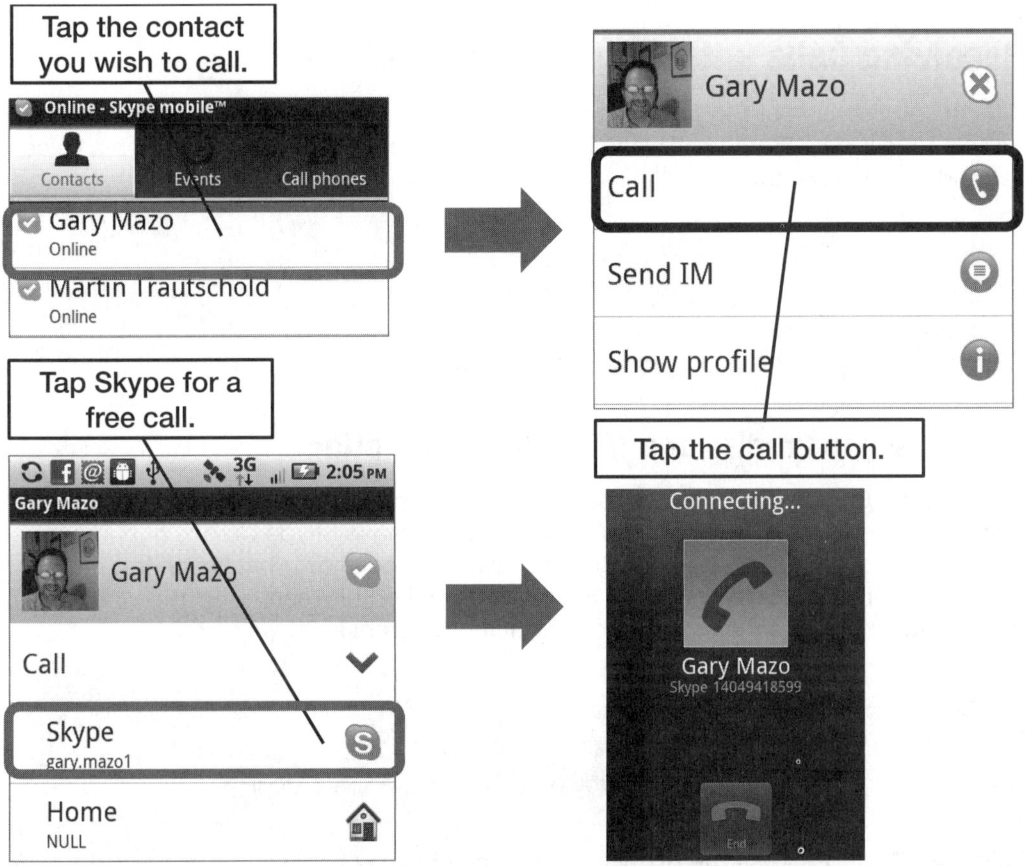

Figure 20–3. *Placing calls from Skype on your DROID.*

NOTE: You can call toll free numbers for free using **Skype Out** on your DROID. The following notice comes from the Skype website at `www.skype.com`:

"The following countries and number ranges are supported and are free of charge to all users. We're working on the rest of the world. France: +33 800, +33 805, +33 809 Poland: +48 800 UK: +44 500, +44 800, +44 808 USA: +1 800, +1 866, +1 877, +1 888 Taiwan: +886 80"

Receiving Calls with Skype on your DROID

With the new version of **Skype**, you can have **Skype** running in the background and still be able to receive a **Skype** call when it comes in. In theory, you can even be on a voice call and answer your **Skype** call!

> **TIP:** If you want to call someone whom you know uses Skype on her DROID, just send her a quick email or give her a quick call to alert her to the fact you would like to talk to her using the **Skype** app.

Buying Skype Credits or a Monthly Subscription

Skype-to-**Skype** calls are free. However, if you want to call people on their landlines or mobile phones from **Skype**, then you will need to purchase Skype Credits or purchase a monthly subscription plan. If you try to purchase the credits or subscription from within the **Skype** app, it will take you to the Skype website. For this reason, we recommend using a web browser on your phone or computer to purchase these credits.

> **TIP:** You may want to start with a limited amount of Skype Credits to try out the service before you sign up for a subscription plan. Subscription plans are the way to go if you plan on using Skype a lot for non-Skype callers (e.g., regular landlines and mobile phones).

Follow these steps to use the **Browser** app to buy Skype Credits:

1. Tap the **Browser** icon.
2. Type `www.skype.com` in the top address bar and tap **Go**.
3. Tap the **Sign In** link at the top of the page.
4. Enter your **Skype Name** and **Password**, and then tap **Sign me in**.
5. If you are not already on your **Account** screen, tap the **Account** tab at the right end of the top nav bar. At this point, you can choose to buy credits or a subscription.
6. Tap the **Buy pre-pay credit** button to purchase a fixed amount of credits.
7. Tap the **Get a subscription** button to buy a monthly subscription account.
8. Finally, complete the payment instructions for either type of purchase.

Chatting with Skype

In addition to making phone calls, you can also chat via text with other **Skype** users from your DROID. Starting a chat is very similar to starting a call; follow these steps to do so:

1. If you are not already in **Skype**, tap the **Skype** icon from your **Home** screen and log in, if asked.
2. Tap the **Contacts** soft key at the top.
3. Tap the name of the contact you wish to chat with (see Figure 20–4).
4. Tap **Send IM.**
5. Type your chat text and press the **Send** button. Your chat will appear at the top of the screen.

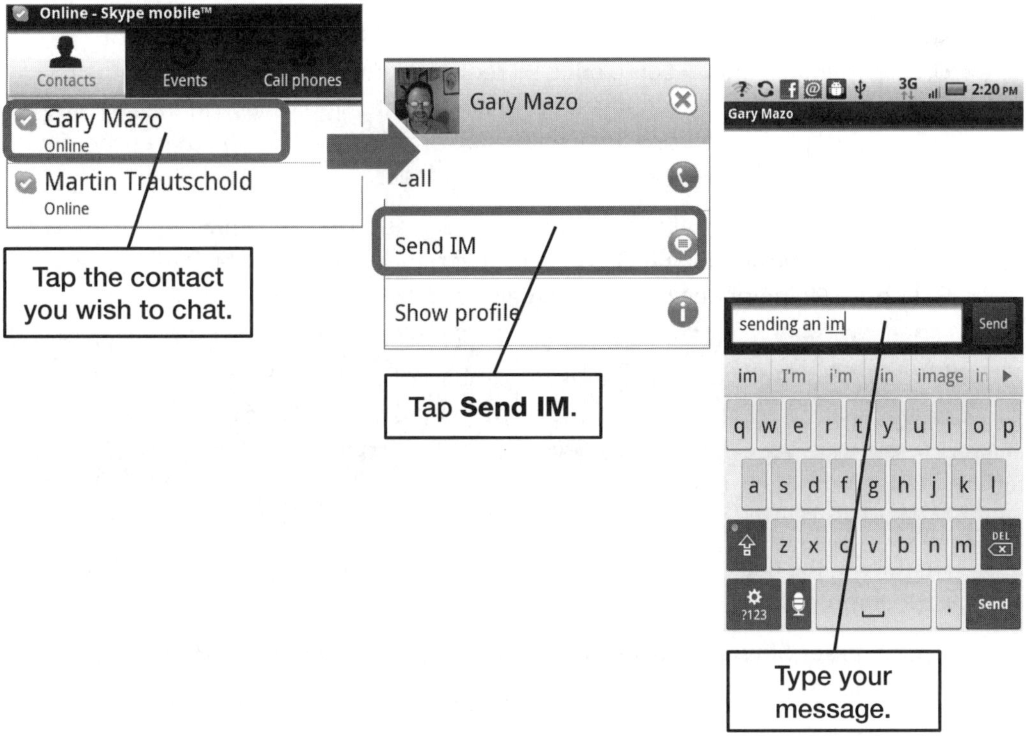

Figure 20–4. *Chatting with Skype on your DROID.*

Adding Skype to Your Computer

You can use the **Skype** app on your computer, as well. We will show you how this works next. You can also use **Skype** to make video calls on your computer if you also have a web cam hooked up.

> **NOTE:** When you call from your computer to a DROID, you will not be able to make a video call.

To create a Skype account and download the **Skype** app to your computer, follow these steps:

1. Open a web browser on your computer.
2. Go to: `www.skype.com`.
3. Click the **Join** link at the top of the page.
4. Create your account by completing all required information and clicking the **Continue** button. Notice that you only have to enter information in the required fields, which are denoted with an asterisk. For example, you do not need to enter your gender, birthdate, and mobile phone number.
5. You are now done with the account setup process. Next, you are presented with the option of buying Skype Credits; however, this is not required for the free **Skype**-to-**Skype** phone calls, video calls, or chats.

> **TIP:** You only need to pay for **Skype** if you want to call someone who is not using Skype. For example, calls to phones on landlines or mobile phones (not using the **Skype** app) will cost you. At publishing time, pay-as-you-go rates were about US 2.1 cents; monthly subscriptions ranged from about US $3 - $14 for various calling plans.

6. Next, click the `Get Skype` link in the top nav bar of the site to download **Skype** to your computer.
7. Click the **Get Skype for Windows** button or the **Get Skype for Mac** button.
8. Follow the instructions to install the software. For more information on downloading and installing software, see the "Getting iTunes Software" section in Chapter 30: "iTunes User Guide.'
9. Once the software is installed, launch it and log in using your Skype account.

You are ready to initiate (or receive) phone calls, video calls, and chats with anyone else using the Skype service, including all your friends with the **Skype** app installed on their iPhones.

Chapter 21

Working With Notes and Documents

In this chapter, we will give you an overview of two popular and free notepad apps (**AK Notepad** and **Evernote**) that you can install on your DROID because nothing is preloaded other than a simple sticky-note style Motorola widget. See Chapter 6, "Organizing Your Home Screens" for more information. Dozens of notepad choices exist, so it is not possible to cover everything available; therefore, we have chosen two apps that have a lot of popular support and some great features.

Next, we will look at an easy way to transfer your documents between your computer and your DROID using handy and free software called **DropBox**. This app allows you to drag-and-drop files to a folder on your computer. Next, those files are copied or synchronized to your web-based DropBox account. Finally, after installing the **DropBox** app, you can access those same files on your DROID.

Finally, so many people use **Microsoft Office** that we will also cover how to open and edit **Word** files, **Excel** spreadsheets, and **PowerPoint** presentation files on your DROID. We will show you two apps in this arena: **Quickoffice**, which may be pre-installed on the DROID 2 and DROID X models; and the full version of **Documents to Go**. Both of these apps allow you to open, view, create, and edit **Office** documents.

Finding and Installing These Apps

You can obtain all the apps we discuss in this chapter from the Android Market (see Chapter 17: "Exploring the Android Market" for more information). The easiest way to find apps is to tap the **Search** icon and type the exact name of the app in the **Search** window. Because there are dozens of alternatives in this case, you may be overwhelmed at the number of apps if you type in only the general category. Once you find the app you want, tap it and install it. If it is an app you want to use frequently, move it to one of your DROID **Home** screens by long-pressing it and dragging it.

Notes-Based Apps on Your DROID

We will cover two of the many notes-based apps that you can use on your DROID. The **AK Notepad** app is more of a standard, easy-to-use notepad app. **Evernote** is a more of a full-featured note-taking app that includes the ability to attach multimedia items to your notes, including pictures, files, and voice notes.

AK Notepad App

To use the **AK Notepad** app, begin by downloading and installing the **AK Notepad** app from the Android Market.

If you are used to other notepad apps from other devices such as an iPhone, iPad, or BlackBerry, this app will look familiar. The default **Note** view displays yellow-lined notepad paper.

Adding and Labeling (Tagging) New Notes

It's easy to add and label new notes. Simply tap the **Add note** line at the top of the main window to start composing a new note.

Now just start typing your note. Be aware that the first line or first few words become the title of your note.

To add a label or tag to organize your notes, simply precede the label with a number sign (#) like this: `#mytag`.

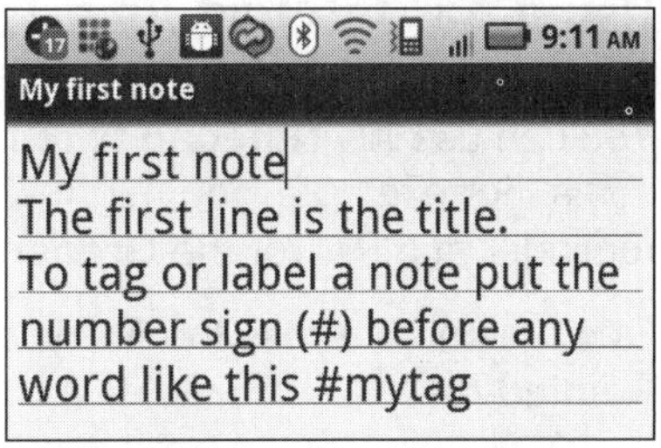

When done typing, press the **Back** button on your DROID to save your note and return to the list of notes.

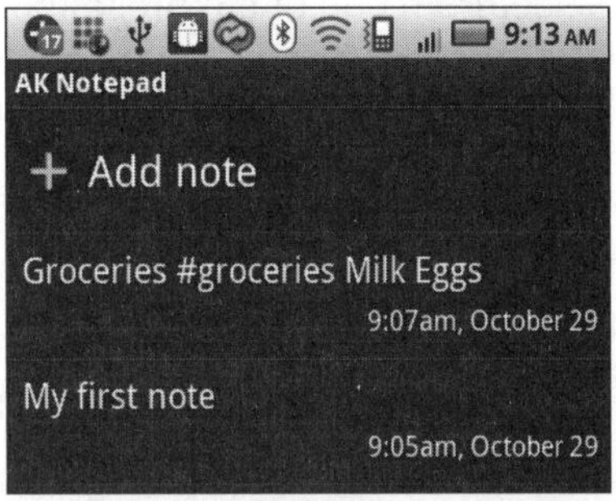

If you are not satisfied with the pre-assigned title of a note, like the **Groceries** list example in the previous screen shot, you can tap the note to open it. Press the **Menu** button and select **Edit title**. Adjust the title and tap **OK**.

Now you see the newly adjusted title in the list of notes.

To view a list of only those notes that contain certain labels or tags, press the **Menu** button and select **Labels**. Next, tap the label of the group of notes you want to view.

At this point, you will see only the notes with that selected label.

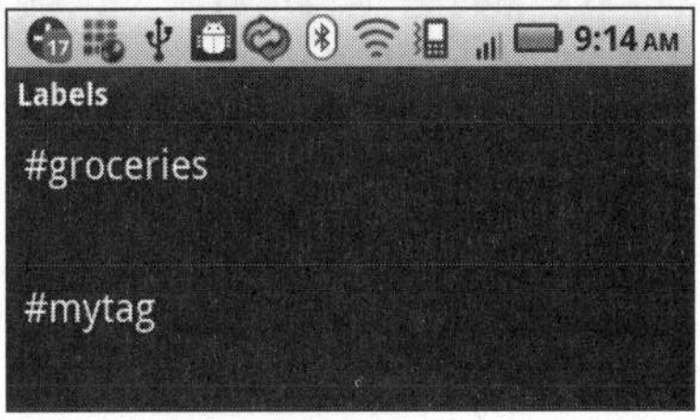

Using AK Notepad As a To-Do Alarm Reminder

You can use **AK Notepad** to remind you of to-do items. For example, you might want to set a reminder at 5 PM for the **Groceries** list item.

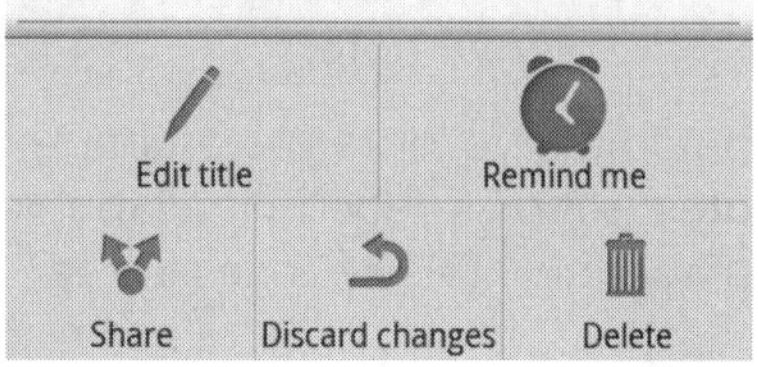

Open the note, press the **Menu** button, and select **Remind me**.

Select from one of the preset durations or select **Custom date/time** if you need another reminder time, and then select the time you want. In this case, we would select 5:00 PM today.

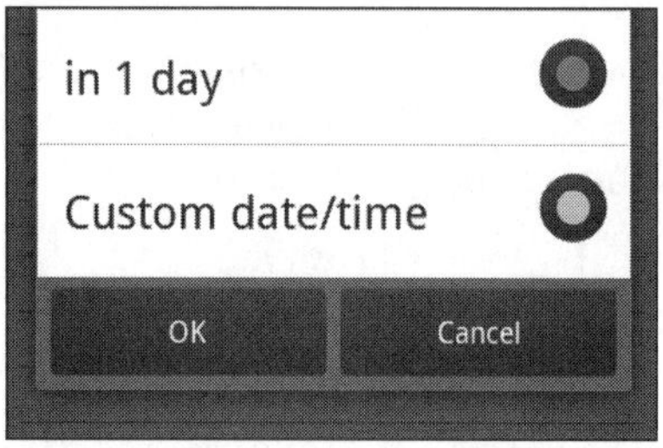

When the reminder rings on your DROID, it will give you a vibration and usually a ringtone. (You can change the way **AK Notepad** alerts you in the app by pressing **Menu** button > **Settings** > **Notification settings**.)

The alert will pop up in the very top status bar. Drag your finger down from the top of the DROID screen to see all the reminders. Now you can tap the **Groceries** list to view your list as you go through the store.

> **TIP: Keeping Track of What You Buy**
>
> As you walk through the store, you can edit your **Groceries** list and put a space before each item as you drop it into your shopping cart. This way, you can be sure not to miss anything on your list!

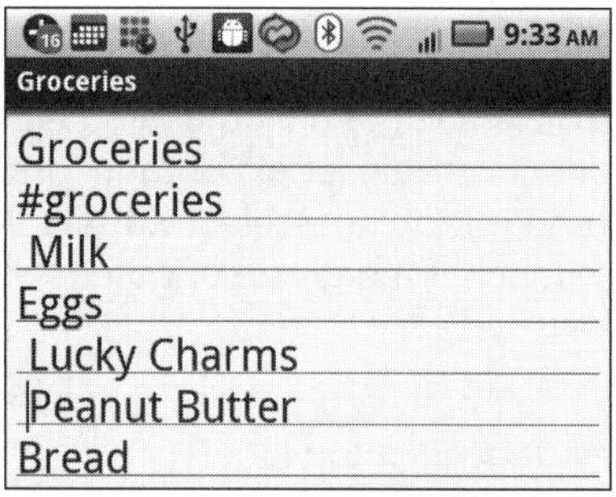

Pinning a Note to Your Home Screen

You can pin or place an icon for any of the notes in **AK Notepad** as icons on your **Home** screen for quick access. From the

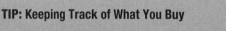

AK Notepad list view of your notes, long-press the note and select **Pin Note to Home Screen**.

Next, press the **Home** button to jump to your **Home** screen and swipe left or right a few times until you see the new **Note** icon.

Repeat this process for all notes you want to have quick access to on your **Home** screen.

Sync Your AK Notepad Notes to Catch.com

AK Notepad provides a free service that will sync all your notes to your own account on the Catch.com website. This can provide you with the following: an instant backup of all your notes, the ability to view your notes on your computer, and the ability to add new notes by typing them on your full computer keyboard instead of the small keyboard on the DROID.

Step 1 – Go to Catch.com and setup a new free account. You could also do this from the **AK Notepad** settings screen, if you wish. You will probably receive a confirmation email from Catch.com to verify it was you who signed up for the free account. You need to click the link in the email to verify your status.

Step 2 – Open up **AK Notepad** on your DROID, press the **Menu** button, and select **Settings**. Scroll down to the **Catch Sync** section and tap **Sign in**.

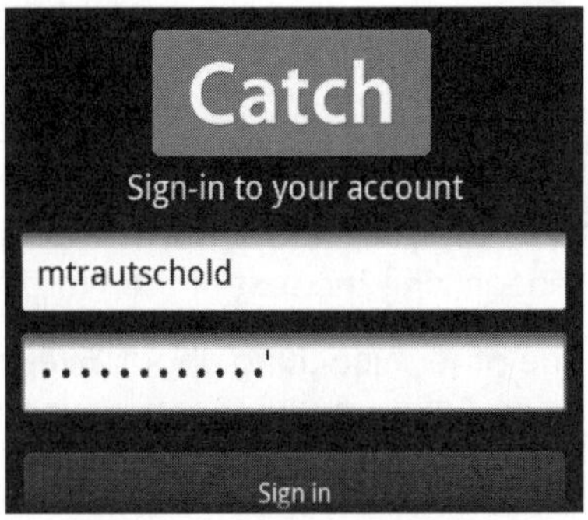

Enter the **username** and **password** you used to set up your Catch.com account and tap the **Sign In** button.

Now your list of notes on your DROID will instantly display any new notes you have added from Catch.com, as well as the **Welcome to Catch.com** note.

The great part is that everything is now kept synchronized between your DROID and the Catch.com website. This means you can now enter or edit notes both places, and the changes will be reflected in both places (see Figure 21–1).

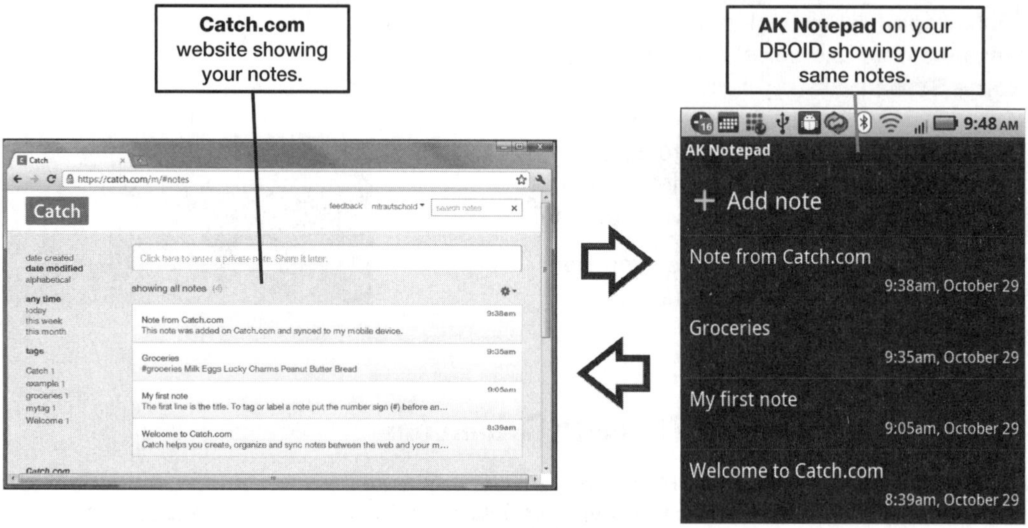

Figure 21–1. *Syncing notes between Catch.com and the DROID **AK Notes** app.*

Evernote App

Like **AK Notepad**, **Evernote** provides the ability to write and sync notes to a website. However, **Evernote** is designed to be a more comprehensive note-taking system that includes the ability to add multimedia notes. For example, you can add voice, pictures, videos, and even geotags (elements that indicate the GPS location) to all your notes.

With **Evernote**, you can even find text inside images. For example, if you take a picture of a receipt with the word "Starbucks" on it, **Evernote** can find that receipt later by the recognizing the text of the word "Starbucks" in the image.

The other nice thing about **Evernote** is that there are apps for multiple mobile devices, so you can view notes you synchronize from your DROID, PC, or Mac on an iPod touch, iPad, iPhone, or BlackBerry.

Getting Started with Evernote

To begin working with **Evernote** , download and install the app from the Android Market.

Once **Evernote** is downloaded and installed, you need to tap the **Evernote** icon.

The first time you use **Evernote**, you will be prompted either to sign in or create a free account. Tap **Create account** to set up your free account or enter your **Username** and **Password** and tap **Sign in**.

Adding and Tagging Notes

After logging in, you see the main screen.

Evernote's main screen gives you various options for adding, tagging and viewing your notes:

- **New note** (add a new note which can include attachments of pictures, audio, video or files)
- **Snapshot** (take a picture with your DROID camera)
- **All Notes** (view all your notes)
- **Tags** (view your notes organized by their tags)
- **Notebooks** (view your notebooks)
- **Search** (use the **Evernote** powerful search feature)

Viewing and Finding Notes

To view your notes, press **All Notes**, **Tags** or **Notebooks** from the main screen.

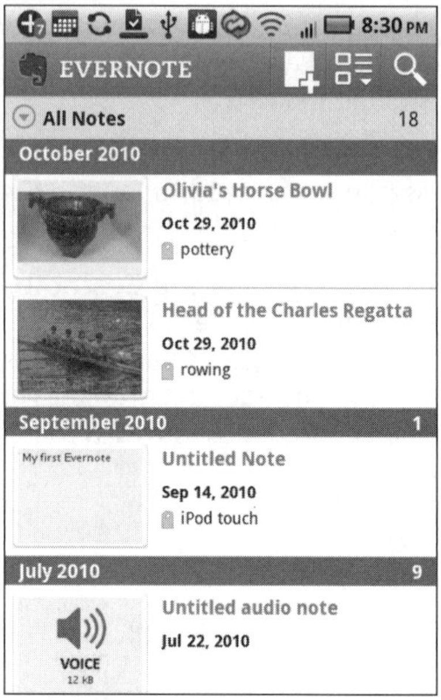

Follow these steps to find a note:

1. Press the **Search** icon from the main screen.
2. Type a word or few words to search for notes containing this text.

Keep in mind that the search engine will usually be able to locate images that contain text that match your search, as well.

Adding, Tagging and Organizing Notes

Follow these steps to add a new note:

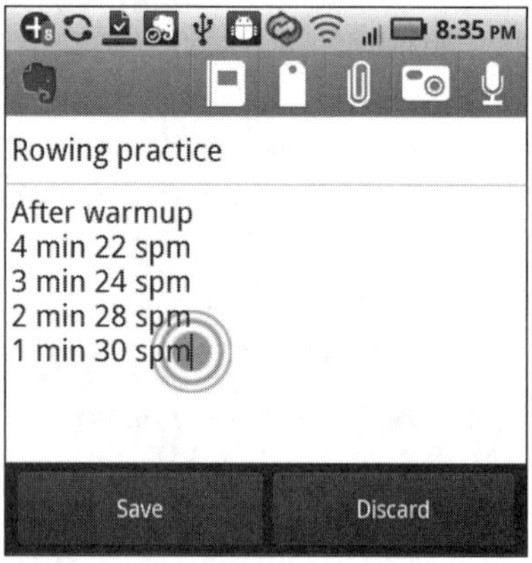

1. Tap the **New note** icon from the main screen.
2. Give your note a unique title and type any text in the textbox below the title for your note.
3. To assign this note to a specific **Notebook**, tap the **Notebook** icon  and select a notebook.

4. To assign one or more tags to this note, tap the **Tag** icon .

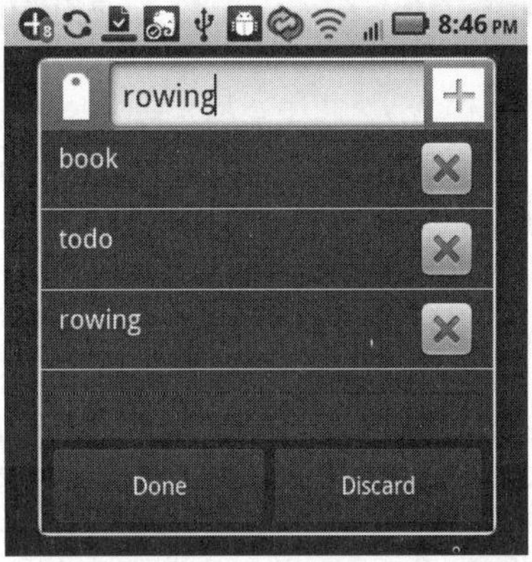

5. From this screen you may type a new tag and click the plus sign (+) in the upper right corner to add It to the list. Tap the X next to any tag to remove it.

6. Or you may press the **Tag** icon in the upper left corner to select from tags you have already created.

7. You may select as few or as many tags as you would like to assign to this note, tap the **OK** button at the bottom.

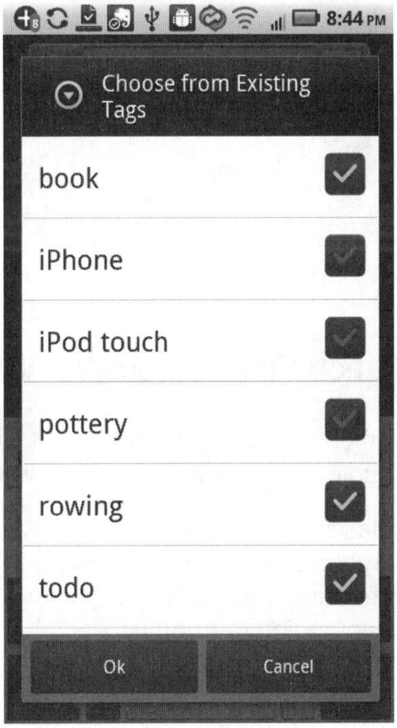

Adding Snapshots (Taking a Picture to Add to the Note)

You may want to add a picture or snapshot to a note. From either the main menu or the note detail view, click the **Snapshot** icon  to bring up your camera; this will enable you to grab a picture and save it as a note.

The picture is also geotagged with your current GPS location in **Evernote**; this enables you to track where you took it. You can even take a picture of a document and have **Evernote** find words in the image of the document.

Attach Pictures, Audio, Video or Files to Notes

Sometimes you will want to upload a file with a note. Tap the **Attach**  (paperclip) icon from the note detail view screen to select a picture or other file from your DROID to upload and attach to a note.

To locate pictures you have taken on your DROID, tap the `dcim` folder and then tap the `Camera` folder.

Select **Pictures** to browse and attach a picture to the note.

Select **Audio** to attach an audio file (music track or sound recorder file).

Select **Video** to attach a video (you must be a paid or Premium **Evernote** user to use this feature).

Select **File** to browse to a file on your DROID or SD Card to attach to the note.

Adding Text to, Emailing, Deleting, Creating a Shortcut, or Editing a Note

When you are viewing a note, you have several options that you can get to by pressing the **Menu** button. For example, you can **Edit**, **Email** (send the note as an attachment), **Delete**, and **Refresh** the note. You can also view **Note info**, including details about the note such as its **Title**, **Notebook**, **Tags**, **Date created** and **Last updated**, and **Location**. If you are viewing a multimedia note such as a picture or audio file, you can **Append text** (add text to the note), **Email**, view **Note info**, **Delete**, or **View in full size** (see the image in full size on the screen).

You can also create an icon or shortcut to this note on your DROID Home screen by selecting **Create shortcut** from the menu.

Viewing or Updating Evernote on Your Computer

As we mentioned earlier, all your notes get synced to the Evernote website wirelessly and automatically. You can then log in to your Evernote.com account from your PC, Mac, iPad, iPod touch, iPhone, or BlackBerry to check out or update your notes (see Figure 21-2). This is a great feature if you have multiple devices, and you would like to stay up-to-date or add notes from any of them.

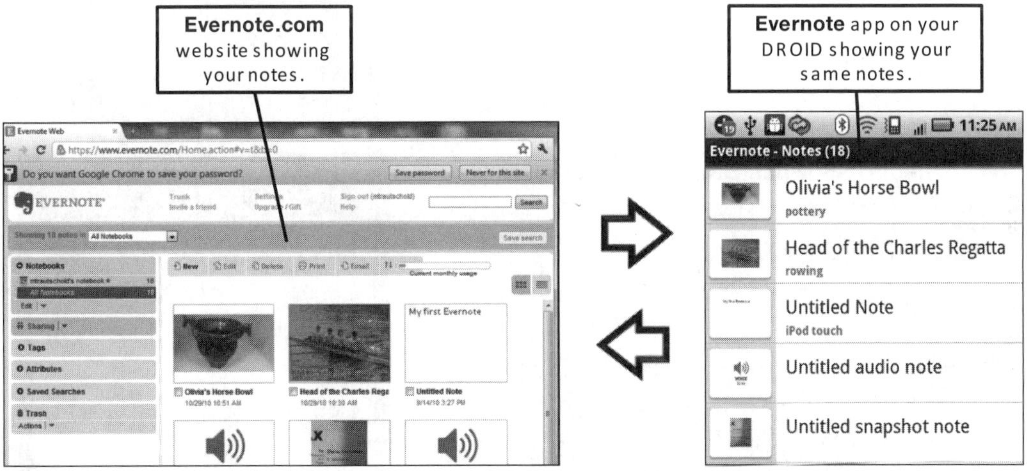

Figure 21–2. *Your notes are synced between the Evernote.com website and your DROID Evernote app.*

Sharing Files and Documents

Your DROID comes with the built-in ability to share files between your DROID's MicroSD format media card (also known as the **SD card**) and your computer using the USB cable. You can also use third-party apps to make the sharing process easier and more seamless. We cover one such app called **DropBox** in this section.

NOTE: You will need an SD card to transfer files to and from your DROID. This section assumes you have an SD card installed in your DROID (most DROID models come with an SD card pre-installed). Check out our "Quick Start Guide" at the beginning of this book for more information about how to open up your DROID and install a card if you need.

Sharing Files with USB Mass Storage Mode

When using the **USB Mass Storage** mode, the SD card in your DROID looks like another disk drive letter on your computer. This means you can drag-and-drop files between the SD card and your computer. Follow these steps to do so:

1. Connect your DROID to your computer using the USB cable.
2. Drag your finger down from the top to see your status messages.
3. Look in the **Ongoing** section for **USB connection** and tap it.

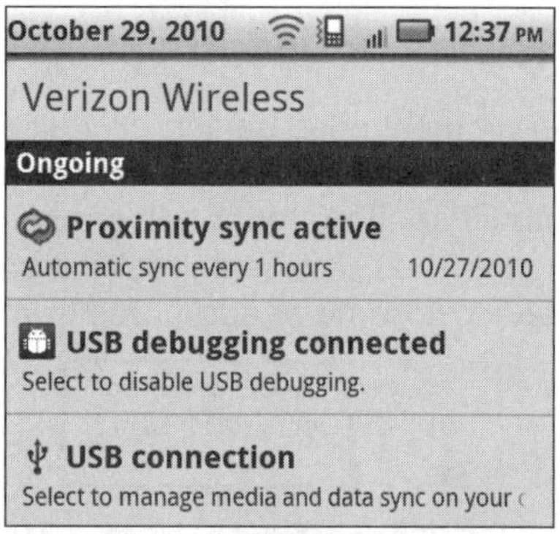

4. Select **USB Mass Storage** and tap **OK**.

NOTE: If you are in **USB Mass Storage** mode, your computer can see and access the files on your DROID SD card; however your DROID cannot. In order to view or access any files (e.g., pictures, music, and videos) on your DROID, you need to switch back to **Charge Only** mode on this screen. Or you can simply unplug your DROID from your computer.

5. Now you will see your DROID SD card appear as a removable disk on your computer. In the image to the right, the DROID has shown up as **Removable Disk (G:)**. Note that your pictures/videos and music are stored in the following locations on your DROID:
   - Your camera pictures and videos are stored in `dcim / Camera` folder.
   - Your music is stored in the `music` folder.

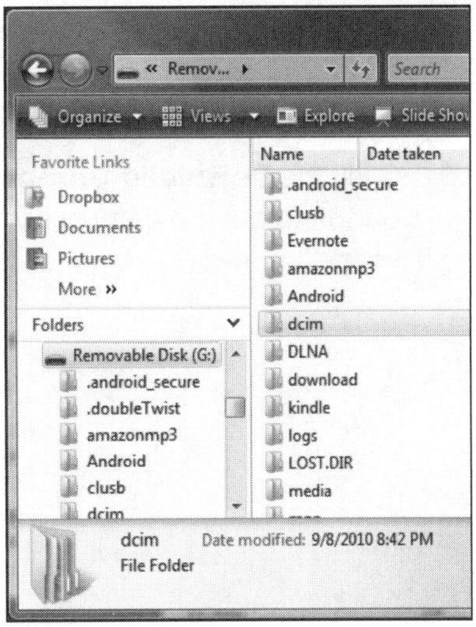

6. You can now open up folders on your DROID SD card and drag-and-drop files to and from your computer.

Dropbox File and Document Sharing

If you want more seamless and easier-to-use file sharing, try the **Dropbox** app. You will need to install the **Dropbox** software on both your computer (PC or Mac) and your DROID.

At the time of publishing, **Dropbox** was a free application and service for up to 2 GB (gigabytes) of storage. If you want more storage space, you will have to pay a monthly fee. 50 GB of storage costs $9.99 per month, and 100 GB of storage costs $19.99 per month.

> **TIP:** You can also use **Dropbox** as a backup service for your important files. If your DROID or computer crashes, you will still have a backup copy of your files in your Dropbox account on that company's servers.

Installing Dropbox on Your Computer (PC or Mac)

You need to install the **Dropbox** app on your computer before you can use it to drag-and-drop files into the `Dropbox` folder. Files dropped into this folder are synchronized with the **Dropbox** app folders on your DROID for easy retrieval. Follow these steps to acquire and set up the **Dropbox** app on your computer:

7. Open a web browser on your computer and go to `www.dropbox.com`.

8. Click the **Download Dropbox** button to get the software for your Windows PC, Mac, or Linux computer.

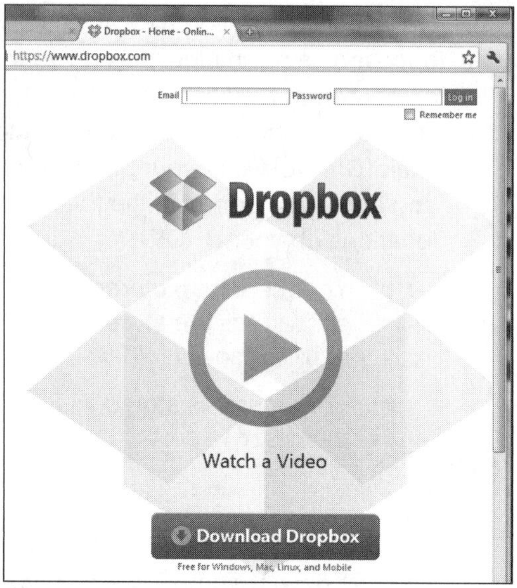

9. Double-click the file you downloaded to start the **Dropbox** installation. You will be prompted to create your **Drobpox** account. Enter your information and click **Next**.

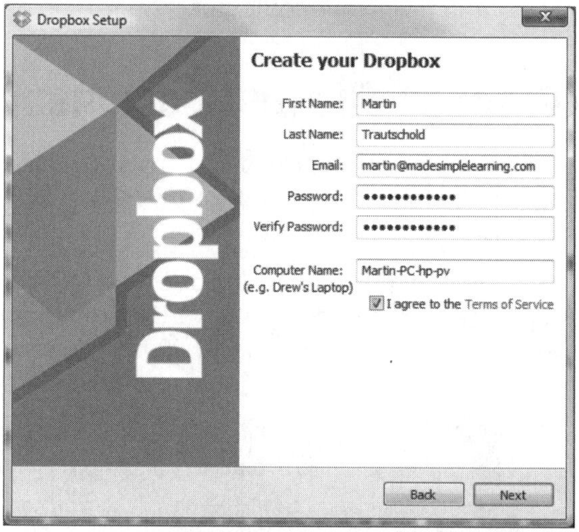

10. Select your **Dropbox** folder size on the next screen. The current rates are as follows:

- 2G = Free
- 50 GB = $9.99/month
- 100 GB = $19.99/month.

11. Next, you will see some tour screens explaining how **Dropbox** works, where to set up your **Dropbox** folder on your computer, and how to quickly access the folder using the **Dropbox** tray icon shown to the right.

If you click the **Dropbox** tray icon, you can see the following commands:

- **Open Dropbox Folder**
- **Launch Dropbox Website**
- **Recently Changed Files**
- Your usage status (showing 0.1% of 2.0 GB used)
- Your status of syncs (All files up to date)
- **Help**
- **Get More Space**
- **Preferences**
- **Exit**

12. Learn more about how to use the **Dropbox** application by viewing the online video tutorials for the app at `www.dropbox.com/tour`.

Installing Dropbox on Your DROID

Once you have set up **Dropbox** on your computer, you're ready to repeat the process on your DROID. Follow these steps to do so:

1. Install the **Dropbox** app from the Android Market and tap the **Dropbox** icon to start it.
2. You need an account to get going, so choose one of these options:

   - **I'm already a Dropbox user –** This allows you to enter your username and password.
   - **I'm new to Dropbox –** This allows you to set up a new account.

Go ahead and enter your login credentials or set up your account.

3. After you log in, the next time you open the **Dropbox** app, you will immediately be taken to your **Dropbox** shared folders.

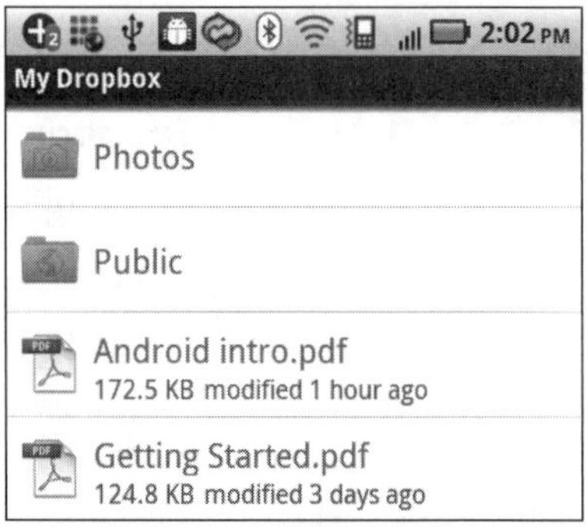

4. Tap any folder to open it or tap any document to open and view it. First, the file is downloaded to your DROID; second, you are asked how you want to open and view the file.

5. Depending on the type of file you tap, you will either see the file open immediately (like a picture) or be asked which app you would like to use to open the file. In the image to the right, we tapped a `.pdf` file, so we were asked which application to use. If you always want to use the same app, then tap the **Checkbox** icon at the bottom next to the text that says, **Use by default for this action**.

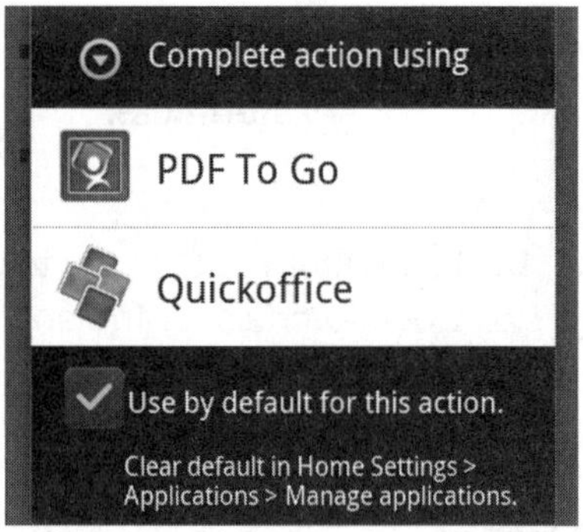

Moving Files From Your Computer to Your DROID

Once the software is set up on your computer and your DROID, you can drag-and-drop or copy-and-paste files into your `Dropbox` folder on your PC or Mac. Within a few minutes of doing this, that document will appear in the **Dropbox** app on your DROID. The same thing works in reverse: if you place a new file in the `Dropbox` folder on your DROID, it will appear in minutes in the `Dropbox` folder on your computer (see Figure 21–3).

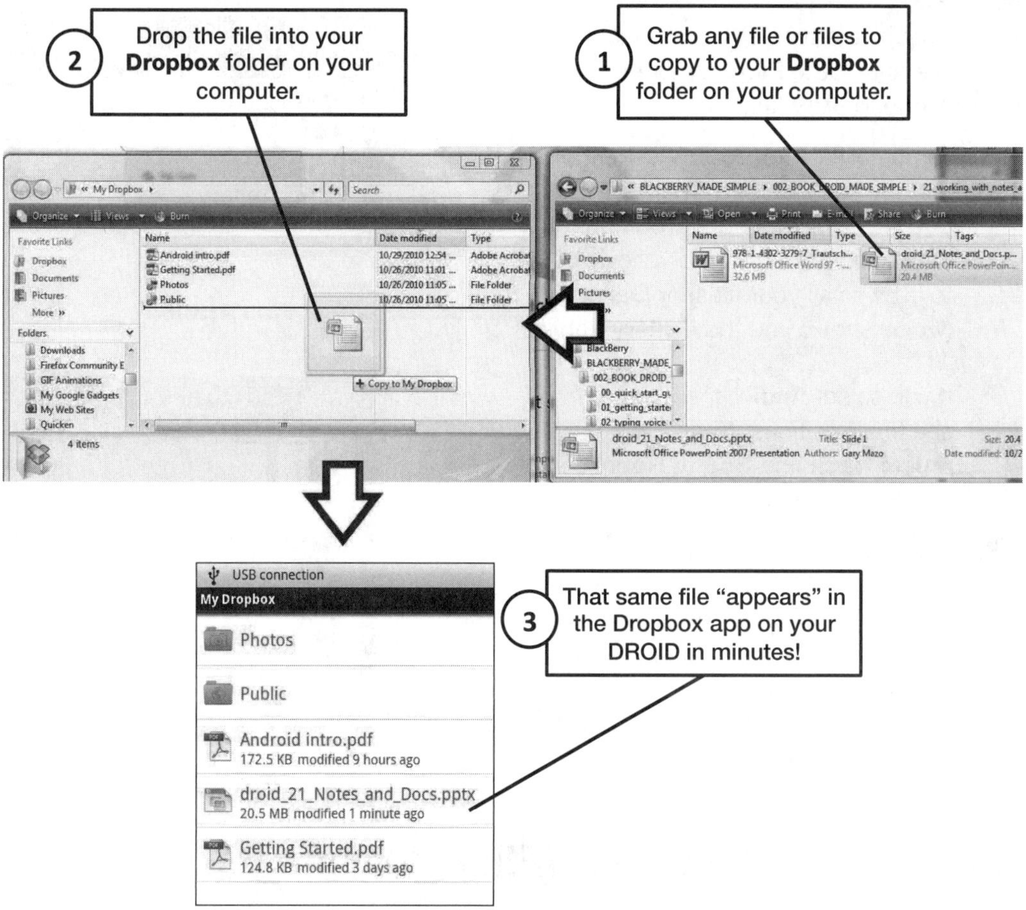

Figure 21–3. *Copy files into your* `Dropbox` *folder on your computer, and they appear in the* `Dropbox` *folder on your DROID.*

Moving Files from Your DROID

Follow these steps to move files from your DROID to your computer and Dropbox account:

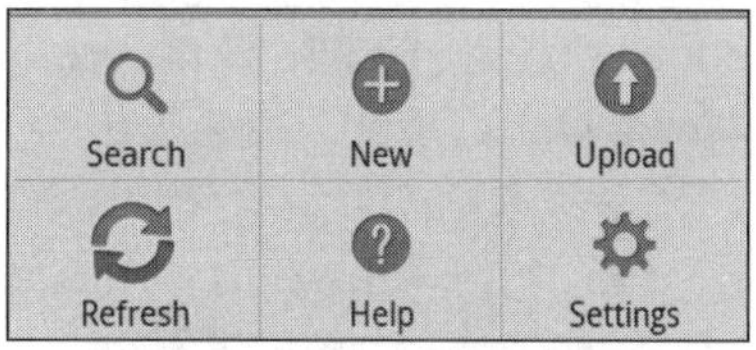

1. Tap the **Drobox** icon to start it.
2. Press the **Menu** button and select **Upload**.

3. Specify the type of file you would like to upload: **Picture**, **Video**, **Audio**, or **Any file**.

   If you select **Picture** or **Video**, then you will see a screen similar to the one shown to the right. You can then choose **Files** (which allows you to browse all your files) or **Gallery** (which shows you the **Gallery** app).

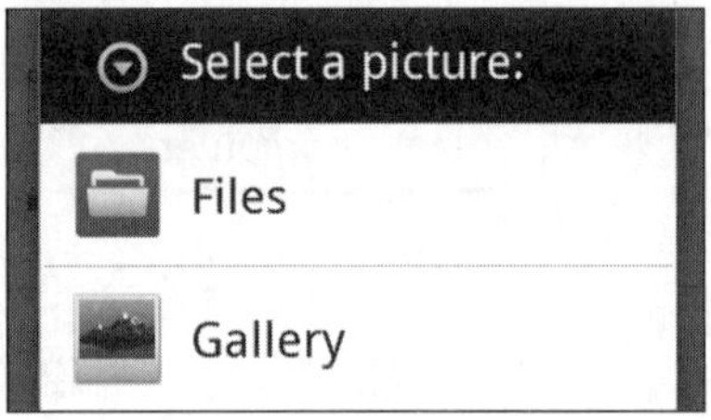

   If you select **Audio**, then you will see these options: **Files**, **Select music track** and **Sound Recorder** (record sound now).

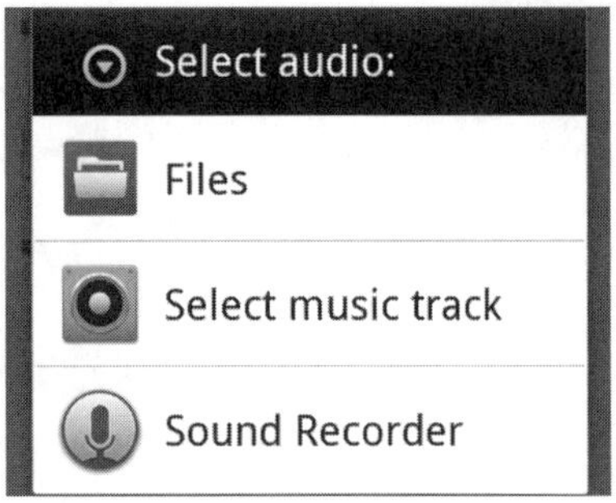

4. Once you select your file, it will be automatically uploaded and saved on your Dropbox account on the dropbox.com server. After a very short time, you will also see that same file in your `Dropbox` folder on your computer.

Working With Microsoft Office Documents

Microsoft Office documents are ubiquitous, and it's helpful to be able to view them on your DROID. Fortunately, you can open and view **Microsoft Office** documents on your DROID with the free version of **Documents to Go**. If you want to create and edit documents, then you have to use **Quickoffice Connect Mobile Suite**, which may already be loaded on your DROID, or the full version of **Documents to Go Full Version** for USD $14.99. If you need to purchase **Quickoffice Connect Mobile Suite**, the regular price Is USD $19.99. Both products have fairly strong customer reviews and are available on the Android Market.

> **TIP:** Check your list of applications; you may already have the full version of **Quickoffice** pre-installed on your DROID. Be sure to check whether you already have it before you buy any third-party software for creating or editing **Office** documents.

The ability to open and view documents on your DROID means you can stay productive on the road. When you add the ability to edit these documents and forward them to colleagues in email messages, then you can really boost your mobile productivity. Go ahead and edit that document while waiting for lunch, waiting at the airport, riding the train, or flying in an airplane.

NOTE: Imagine putting the core of **Microsoft Office** on your DROID and just how many many features and functions it includes. We could easily write 50 pages or more about either **Quickoffice** or **Documents to Go**; however, we will do our best to stick to just the basics to help you get started and become productive.

Finding Product Reviews

You will find product reviews both on the Web and in individual user reviews on the Android Market site. Do a web search to find the latest reviews and comments; to facilitate your research, we've gathered links to a few reviews at major sites:

- Review of **Quickoffice** with some comparison to **Documents to Go** (ZDnet - 6/9/2010):

  `www.zdnet.com/blog/cell-phones/quickoffice-brings-cloud-document-access-and-editing-to-google-android/3996`

- Review of **Documents to Go** (ZDnet - 5/28/2009):

  `www.zdnet.com/blog/cell-phones/review-documents-to-go-rocks-google-android-with-unique-office-functions/1307?tag=rbxccnbzd1`

- Review of **Documents to Go 2.0** (Brighthand – 11/11/2009):

  `www.brighthand.com/default.asp?newsID=15697`

NOTE: Keep in mind that the reviews and comments are based on specific versions of the apps; if a vendor has released an update, some or all of the concerns of the reviewers may have already been addressed.

Moving Documents to and From Your DROID

As we discussed in the "Dropbox" section of this chapter, you can use apps like Dropbox to easily transfer files between your computer and your DROID. Everything is shared and synchronized wirelessly.

Another popular way to move documents is to attach them to email messages. You can receive and send attachments to yourself and others. Learn more about working with email in Chapter 9: "Email on Your DROID." Follow these steps to transfer documents between your computer and DROID via email:

1. When you receive email attachments, they will be listed at the bottom of the message. Tap the **Preview** button to open the attachment.

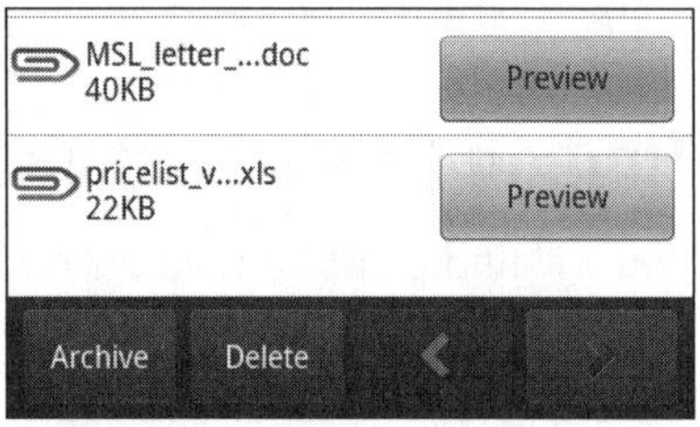

2. You may be asked to select a particular app to open this attachment type.
3. Tap the app you wish to use to open and view the document.

4. If you are composing a message, you can attach a file or document by pressing the **Menu** button and selecting **Attach**.
5. Browse to the file location and select one or more files to attach to that email message.

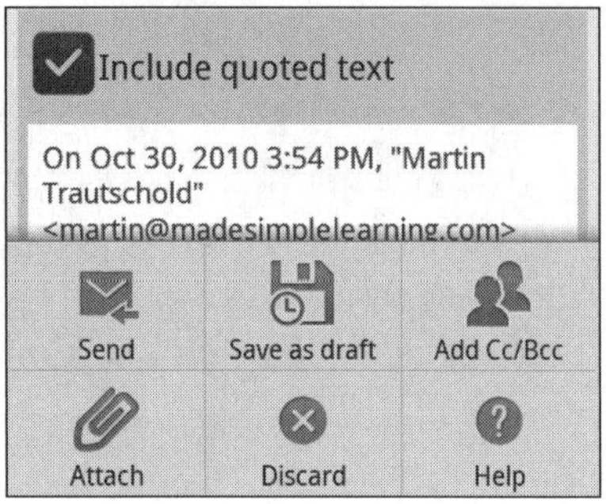

Quickoffice

You may already have **Quickoffice** pre-installed on your DROID. Take a look through your app icons; if you don't see it, then you can purchase it for about $10 from the Android Market.

As we mentioned previously, you can start **Quickoffice** by selecting it as the app to open an email attachment with. We'll start by drilling down on the app itself:

1. Tap the **Quickoffice** app to start it.
2. You will see the main screen of **Quickoffice**. From this screen, you can select between the various **Quickoffice** apps and **QuickPDF**, as well as browse for files on your DROID.

   Tap **Update** to check for updates to the software.

   Tap **Support** to load the **User Guide**, bring up **Frequently Asked Questions**, or **Submit a Support Ticket** to the software publisher about an issue you are having.

   Tap **Explore** to follow Quickoffice on social networks, view the Quickoffice blog, read news releases, and provide feedback.

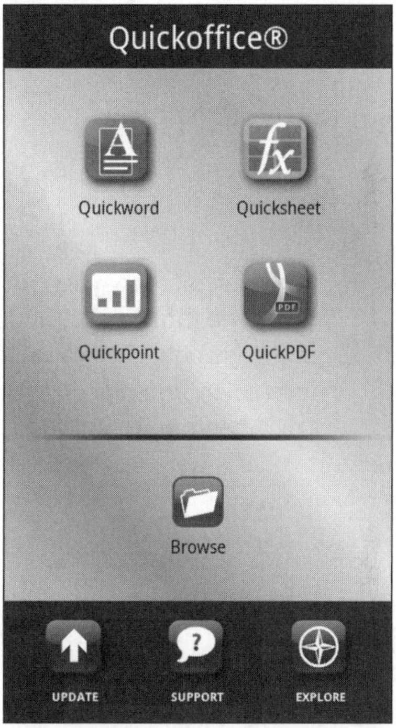

3. If you tap **Quickword**, **Quicksheet**, or **Quickpoint**, you will see a screen similar to the one shown to the right that asks whether you want to **Create New Document** (this option is not available in **Quickpoint**), browse the **SD Card** (browse the SD memory card), or see a list of **Recent Documents**.

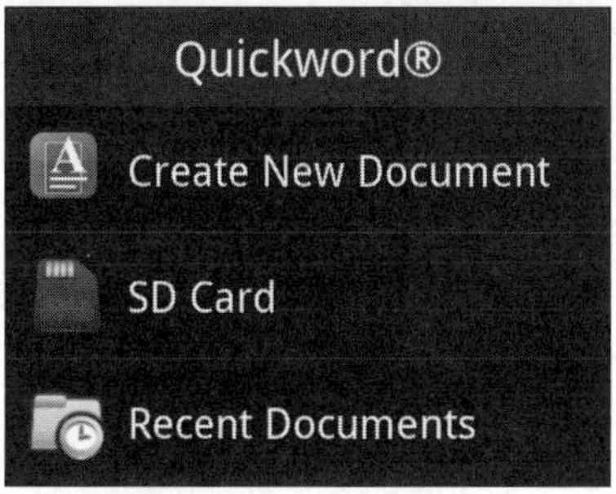

Formatting Text in Quickword

Once you have a **Quickword** document open, you can change its text formatting by following these steps:

1. Select text by double-tapping it, and then tap the screen above or below the selected text to expand the selection.
2. Press the **Menu** button and select **Format**.

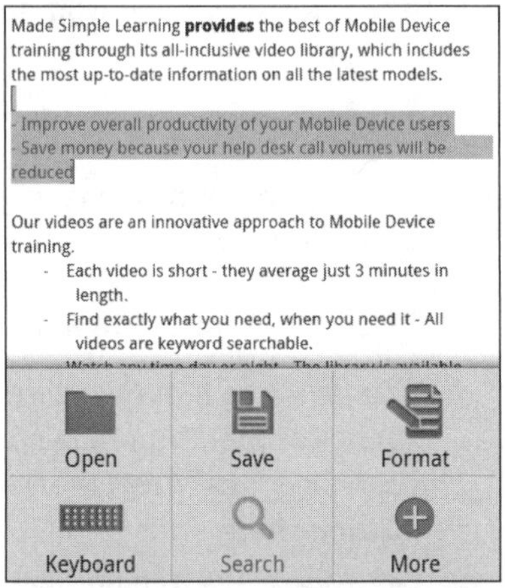

> **TIP:** To show the keyboard, select **Keyboard** from the menu.

3. From this screen, you can set the font style (bold, italic, underline, or strikethrough), font face, font size, font color, and highlight color. Tap **OK** when done.

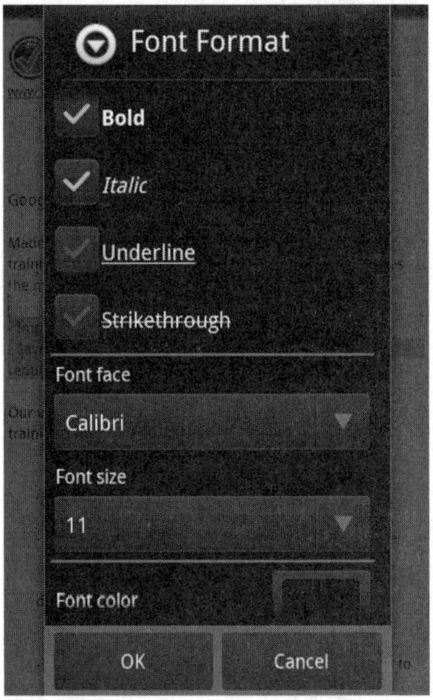

Getting Around in Quickoffice

Once you understand that you get to most of the commands by pressing the **Menu** button, you can access all the functionality in the **Quickoffice** apps. What follows is a list of menu commands accessible from the various apps:

- **Quickword menu commands** – Open, Save, Format, Keyboard (show/hide), Search (find text), More, Page View (shows entire page), New (new file), Save As, Properties (document properties), Updates (check for app updates), About, and Help.
- **Quicksheet menu commands** – Open, Save, Worksheet (jump to different worksheet), Keyboard (show/hide), Number Format (set as General, Number, Currency, Date, Time, and so on), More, New, Save As, Search (for text), Font Format (same as Format in **Quickword**), Go To Cell (type in a cell reference such as A10 to jump to it), Properties (document properties), Updates (check for app updates), About, and Help.
- **Quickpoint menu commands** – Open, Save, Save As, Go To Slide (jump to slide number), Start Slideshow, More, Properties (document properties), Updates (check for app updates), About, and Help.
- **Quickpdf menu commands** – Open, Reading View (reflows the text on the page so it is more easy to read and does not require scrolling left and right), Go To Page, Bookmarks (view bookmarks in the file), Find (search for text), More, Rotate (rotate the page left or right), Updates (check for app updates), Properties (document properties), About, and Help.

Editing Text in Quickpoint

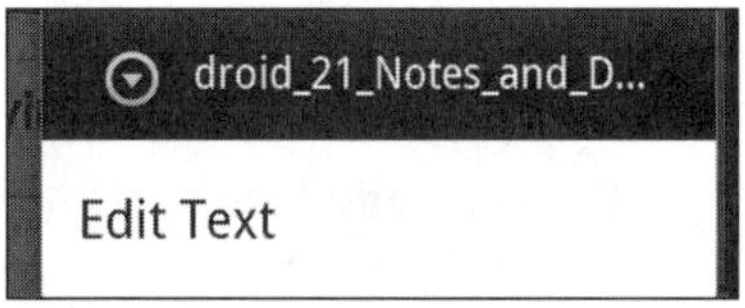

In order to edit text on slides in **Quickpoint**, you need to long-press the text you wish to edit and select **Edit Text** from the pop-up window. Next, you will see the text on a new screen. Tap anywhere to position the cursor and use the keyboard to change the text. Tap **OK** when done.

NOTE: Editing Text Inside Graphics – Quickpoint is the Clear Winner

At the time of publishing, **Quickpoint** lets you edit any text on a **PowerPoint** slide by long-pressing it. This works whether the text is in the main text area or inside a graphic such as a callout. However, **Slideshow To Go** from **Documents To Go** only allows you to edit text placed directly on the slide; it does not allow you to edit text inside boxes, callouts, or other graphics.

Zooming in Quickoffice

The best way to zoom in or out is to pinch open or pinch closed with your fingers on the screen. Double-tapping only works to zoom in or out in **Quickpoint**; otherwise, it will move the cursor and start a selection.

Documents to Go – the Full Version

The **Documents To Go - Full Version** (version 3.0) costs $14.99 at publishing time. The key difference between the full and the free version is that the full version of **Documents to Go** gives you the added ability to create, edit, and send **Microsoft Office** and **Adobe PDF** files. You can buy the full version of the app from the Android Market. **Documents To Go** consists of three main programs: **Word To Go** (for **Word** documents), **SlideShow To Go** (for **PowerPoint** documents), and **Sheet To Go** (for **Excel** documents).

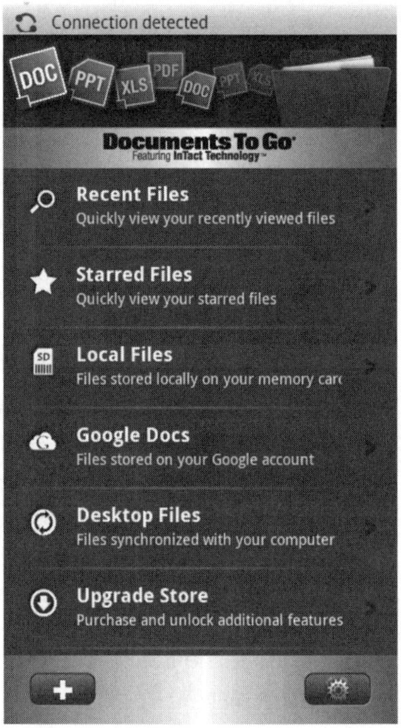

Tap the **Documents to Go** icon to start it. From the main screen, you can tap any of the following options:

- **Recent Files –** View recently opened files.
- **Starred Files –** View files you have starred as your favorites.
- **Local Files –** View files on your SD memory card.
- **Google Docs –** Access files stored on your Google account.
- **Desktop Files –** Access files you have synchronized from your computer.
- **Upgrade Store –** Buy additional features.
- **Plus sign –** (lower-left corner) Click to create a new **Word**, **Excel**, or **PowerPoint** document.
- **Settings Gear –** (lower-right corner) Click to access settings.

Word To Go Tips

You'll probably spend the most time using the **Word To Go** app. In the upcoming sections, we'll look at several tips for getting the most out of this app.

Zooming in or out

Tap the screen once to bring up the (-) and (+) **Magnifying glass** buttons at the bottom of the screen. Tap these buttons to zoom out or in. Note that pinching open/closed does not work.

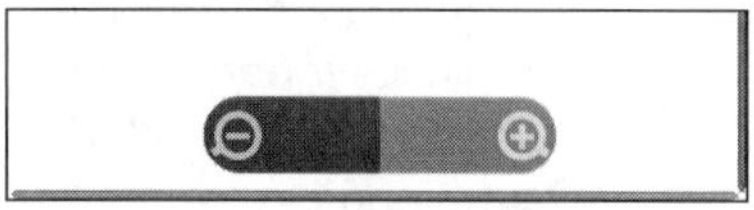

Selecting Text

Obviously, you'll want to select text to copy-and-paste or otherwise reformat it. Follow these steps to do so:

1. Long-press to bring up the **Edit** pop-up window. Try to position your finger exactly where you want to start selecting text because it does not allow you to change the starting point of your selection. It works best to zoom in as much as possible first to make the words larger.
2. Tap **Selection Mode**.
3. Drag your finger across the screen to adjust the selected text. If you see that your starting point for the selection is incorrect, press the **Back** button and start at Step 1 again.
4. Long-press again to choose from the menu. Options include **Cut**, **Copy**, **Font** (change font size, style, type), and **Bullets & Numbering**.

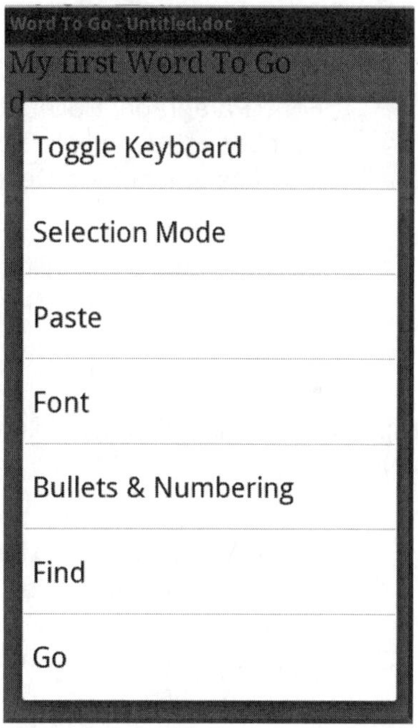

Menu Commands

You may be amazed at the number of features and functions available to you in **Word To Go**. Just press the **Menu** button to see the following commands:

- **File –** New, Open, Close, Save, Save As, and Send via Email
- **Edit** (this brings up the same options as long-pressing the text) **–** Toggle Keyboard, Start/Cancel Selection, Select All, Cut, Copy, Paste, and Undo
- **View –** Zoom, Find, Go, Table of Contents, Comments, Footnotes, and Endnotes
- **Format –** Bold, Italic, Underline, Font, Paragraph, Bullets & Numbering, Hyperlink, Bookmark, Increase Indent, and Decrease Indent
- **Insert –** Page Break, Bookmark, Hyperlink, Table, and Comment

Tapping **More** brings up the following menus and options:

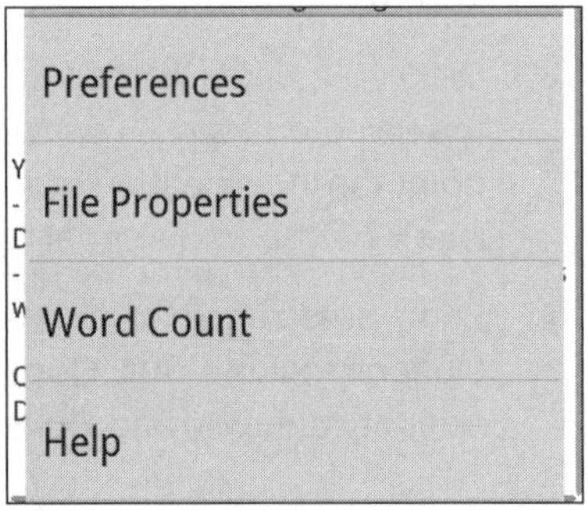

- **Preferences –** Format for new files (Word 97-2004 or Word 07-2008), Name, Initial, options for Track Changes (such as how insertions and deletions are shown and what colors are used).
- **File Properties –** Name, Type, Location, Size, and Last Modified
- **Word Count**
- **Help –** Check for Updates, About, and Help

Sending a File Via Email, Bluetooth, Dropbox, and More

The **Send a File** command works in all the **Documents To Go** applications. Follow these steps to send an email from **Documents To Go**:

1. Press the **Menu** button, select **File**, and then **Send via Email**.
2. You will see a screen similar to the one shown to the right. You will only see **Dropbox** as an option if you have it installed. You may see other applications you have installed as options, as well.

   Select the method you would like to use to send your file.

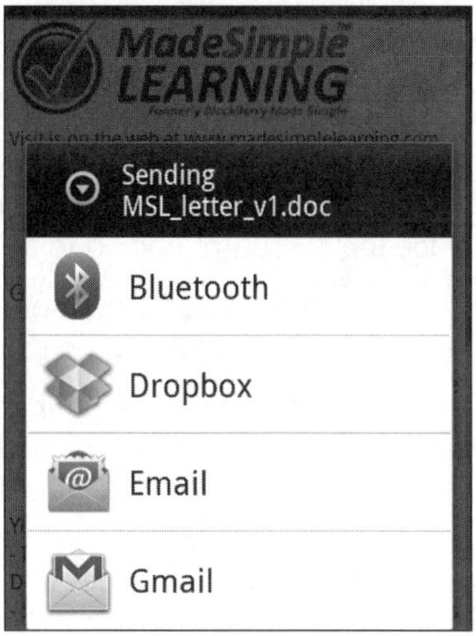

Slideshow To Go Tips

The **Slideshow To Go** app lets you edit **PowerPoint** documents. The next section looks at ways to get the most out of this app.

For example, once you get the hang of using the **Menu** button, you can use the long-press to bring up the **Slideshow To Go** menu. This menu lets you do almost anything possible from this app.

Remember that you can flip your DROID on its side to make a slide fit its screen better.

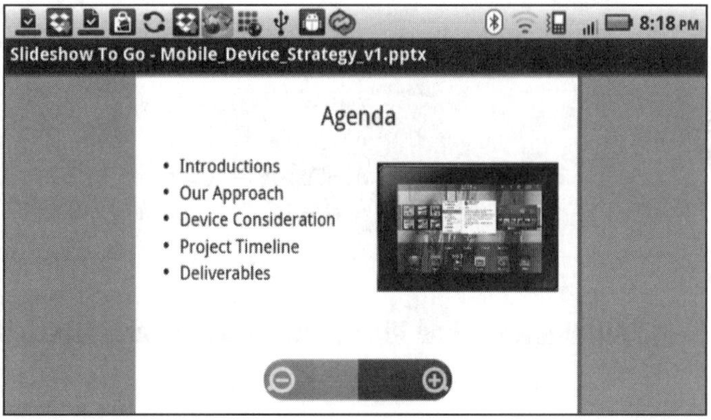

Editing Slide Text (Using Outline View)

Editing text on a slide is easy to do. Simply long-press anywhere on the slide and select **Edit Slide Text**. This switches you into **Outline** view.

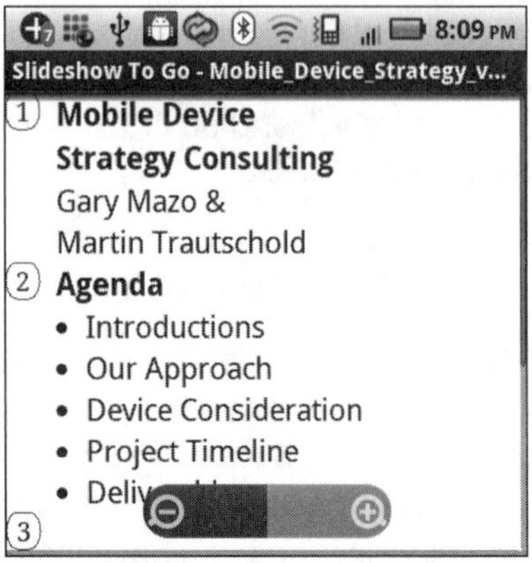

This will give you a screen similar to the one shown to the right. Tap your finger anywhere to place the cursor for editing text.

Tap the screen to zoom in or out using the **Zoom** buttons on the bottom of the screen.

Use the keyboard to type your changes. You can select, format, and copy / paste text by following the steps described in the "Word To Go Tips" section.

To insert a new bullet item, press the **Menu** button and select **Insert** > **New Bullet Item**.

From this **Insert** menu, you can also **Insert Slide** (insert a new blank slide) or insert a **Duplicate Slide**.

NOTE: At the time of publishing, you could not use **Slideshow To Go** to edit text inside a slide's objects. This holds true for callout boxes or any other kinds of graphics. It is good to know that **Quickoffice**'s **Quickpoint** app does allow you to edit text inside graphics in a **PowerPoint** file.

Switching Views (Slide View, Outline View, and Notes View)

Follow these steps to change your current view of a slide:

1. Press the **Menu** button and select **View**.
2. Select **Slide View** to view the slides, **Outline** to view the text on the slides, or **Notes** to view notes on the slide.

Moving Around

It's easy to navigate between slides in **Slide To Go**. Follow these steps to do so:

1. Swipe left or right to move between slides.
2. Long-press the screen and select **Go To Slide** to jump to a particular slide.

Sheet To Go Tips

The **Sheet to Go** app is **Document To Go**'s **Excel** viewing and editing app. As when using **Word To Go** and **Slideshow To Go**, you can use the **Menu** button, **Zoom** buttons, and drag your finger around the screen to do almost anything you need to in the app.

Moving Around the Spreadsheet

Tap your finger and drag it around the screen to move around the spreadsheet. Press the **Menu** button and select **View** to move to other **Worksheets** or **Go** to jump to the beginning (Home) or End of the current spreadsheet or to a specific cell.

Selecting and Editing a Cell

Here are a couple tips for selecting and editing a cell. Tap the cell to select it, and then tap your finger in the edit box at the top of the screen. Now you can edit the cell text. Be sure to start all formulas with an **Equals sign** character (=). You can input this sign by pressing the **?123** key and then the **ALT** key. Or, you can press the **Menu** button and select **Operators**.

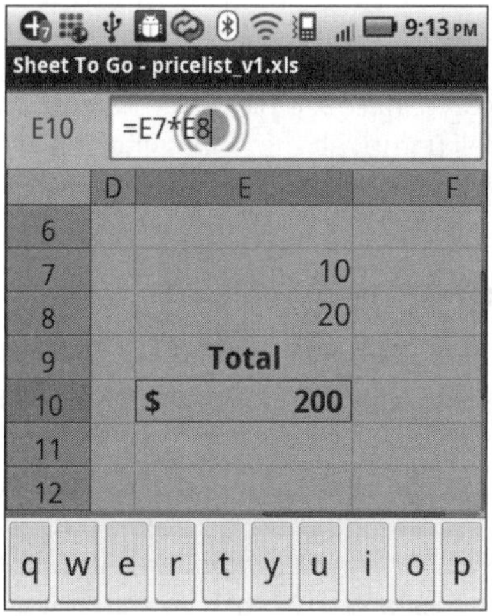

> **TIP:** While editing a formula on your computer in **Excel**, you can just click cells to reference them. However, to add a cell reference in a formula on your DROID, you need to either type out the cell reference (e.g. "E8"); or press the **Menu** button, select **Cell Reference**, and then tap the cell.

Adding Functions

While you are editing a cell, press the **Menu** button and select **Function**.

This displays a list of virtually every function available in **Excel**.

To narrow the list, tap the **All** dropdown list at the top of the window. Next, you can filter for categories such as **Financial**, **Date & Time**, **Math & Trig**, **Statistical**, and so on.

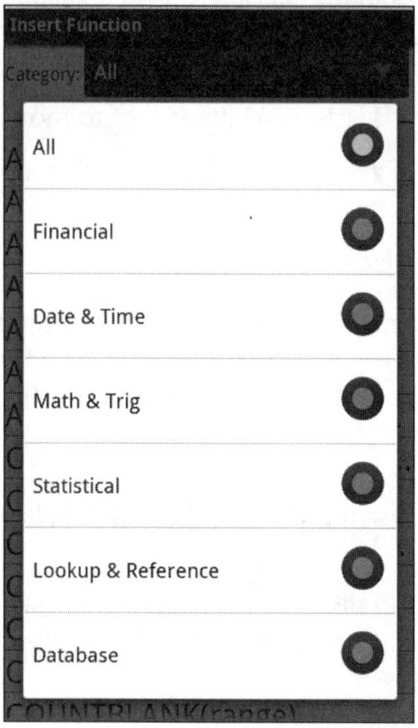

Freeze Panes

You can freeze panes or hold all the cells above and to the left of the currently selected cell unmovable by pressing the **Menu** button, selecting **View**, then choosing **Freeze Panes**. Repeat this procedure to **Unfreeze Panes**.

Switching Worksheets

To move between worksheets in a spreadsheet workbook, press the **Menu** button, select **View**, and then **Worksheets**.

Inserting Various Elements

You can insert a function, AutoSum, sheet, row, or column by pressing the **Menu** button and selecting **Insert**. At this point, you can choose your preferred option.

Selecting an Entire Row or Column

It's also easy to select an entire row or column. Tap the row header (number) on the left side of the screen to select the row. Similarly, you can tap the column header (letter) on the top of the screen to select that column.

Adjusting Row and Column Sizes (and Hiding or Unhiding Them)

Now let's look at some tips for adjusting the sizes of rows and columns. Press the **Menu** button, tap **More**, and then tap **Row** or **Column** to see a menu similar to the one shown to the right.

From this menu, you can accomplish the following tasks: select a row or column, adjust a row width or column height, autofit (for columns), or hide or unhide a row or column.

Many of the **Menu** commands are similar to what we described in this chapter's "Word To Go Tips" section, including **Send as Email, Save**, and **Save As**.

Chapter 22

Fun and Games

Your DROID excels at many things. It is a multimedia workhorse, and it can keep track of your busy life. Your DROID also serves as a nice gaming device. You can even find versions of popular games for the device that you might expect to find only on dedicated gaming consoles.

The DROID brings many advantages to portable gaming: the high-resolution screen delivers realistic visuals; the high-quality audio provides great sound effects; and the accelerometer allow you to interact with your games in a way that many PCs and dedicated gaming consoles (outside of the Wii) don't. For example, in racing games, the last feature lets you steer your car by turning the DROID as you hold it.

The DROID is also great for lots of other fun stuff such as following your local football team. You can even use the DROID as a musical instrument with great apps like **xPiano** (which we will show you later in the chapter.)

> **NOTE:** There is enough fun stuff to do with the DROID that we routinely discover that the DROID has disappeared from its charger and we have to yell out: "Where is my DROID? I need to finish this book!"

Using the DROID as a Gaming Device

The DROID includes a built-in accelerometer, which is essentially a device that detects movement (acceleration).

Combine the accelerometer with a fantastic screen, lots of memory, and a fast processor, and you have the makings of a great gaming platform. With literally thousands of gaming titles to choose from, you can play virtually any type of game you wish on your DROID.

With most games, you can even take a phone call and come back to the exact place you left off when the call ends. This means no more restarts!

> **NOTE:** Some games do require that you have an active network connection through Wi-Fi to engage in multiplayer games.

With the DROID, you can play a driving game and use the DROID itself to steer. You do this simply by turning the device. You can touch the DROID to brake or tilt it forward to accelerate.

The game on the right, **Raging Thunder**, is so fun and fast that it might make you car sick!

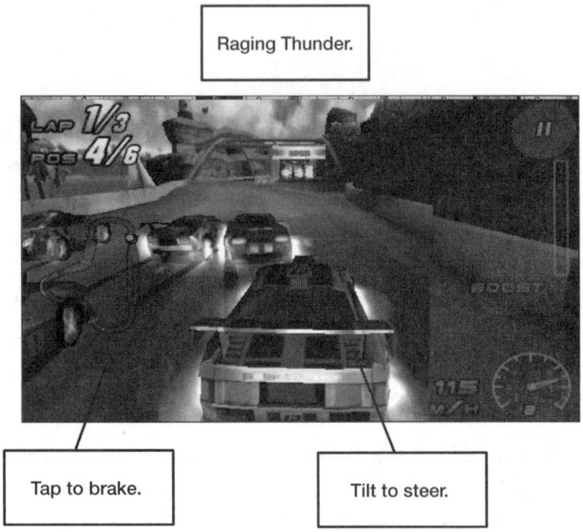
Raging Thunder.
Tap to brake.
Tilt to steer.

Or, you can try a fishing game, where you feel like you are fishing from a real boat! In the **Fishing 2 Go** app shown here, you flick your DROID to cast the line then rotate your finger on the screen to reel in your fish.

Fishin 2 Go
Wind with your finger to reel in your fish.

If music/rhythm games are your thing, then you will find many such programs in the Android Market. Popular console games such as **Guitar Hero** (and many others) have been ported to the DROID.

On some games, such as the new **Guitar Hero**, you really have to "strum" to keep pace and score points.

The DROID also has a very fast processor and a sophisticated graphics chip. Bundling these together with the accelerometer gives you a very capable gaming device.

Acquiring Games and Other Fun Apps

As is the case for all DROID apps, games can be found in the Android Market (see Figure 21-1). You can get them either through **doubleTwist** (See Chapter 25: "DROID Media Sync") on your computer or through the device's built-in **Android Market** app.

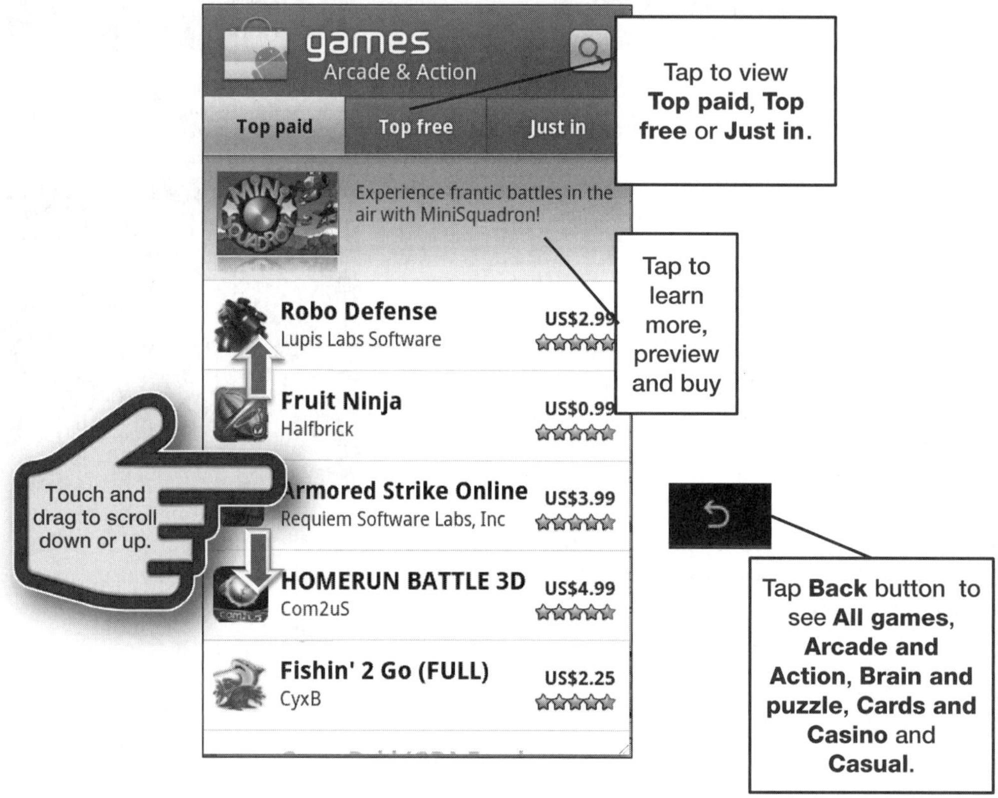

Figure 22–1. *The layout of the Android Market's **Games** section.*

To get a game, fire up the **Android Market**, as you did in Chapter 17. Next, go to the **Games** tab. You will also find many games in the **Featured** section of the Android Market. Figure 22–2 shows the **App Purchase** page for a game available for the DROID.

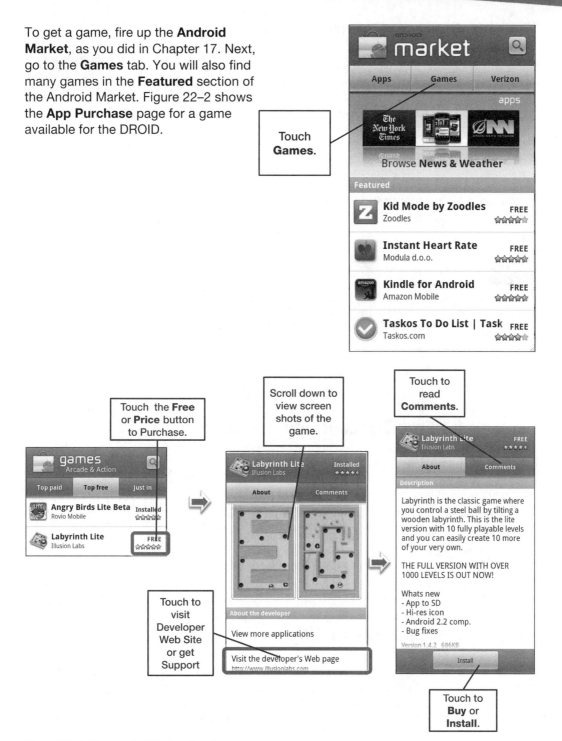

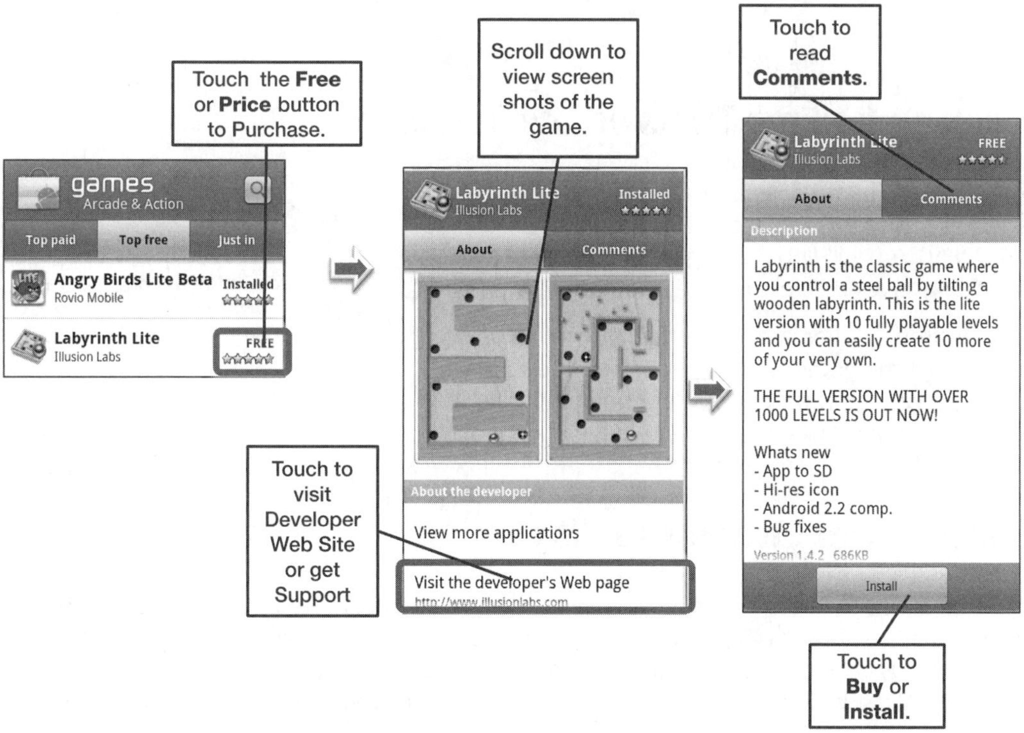

Figure 22–2. *The layout of the **App Purchase** page.*

Reading Reviews Before You Buy

Many of the games have user reviews that are worth perusing. Sometimes, you can get a good sense of the game before you buy it. If you find a game that looks interesting, don't be afraid to do a simple Google search to see whether any mainstream media outlets have performed a full review.

Just touch the **Comments** tab to see user reviews of any game.

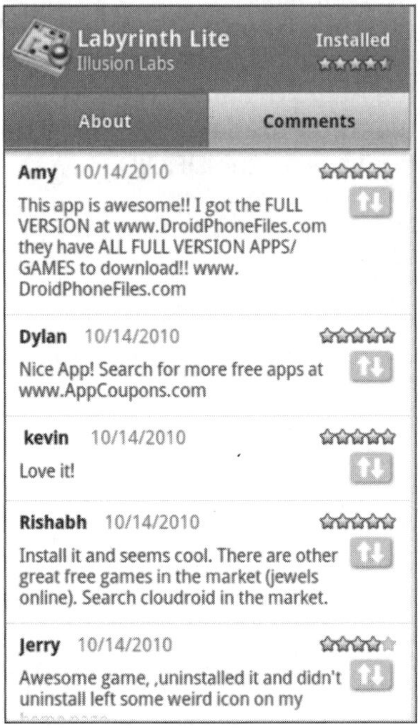

CAUTION: Be aware that some of the reviews may contain explicit or foul language.

Looking for Free Trials or Lite Versions

Increasingly, game developers are giving users free trials of their games to see if they like them before they buy. You will find many games have both a *lite* version and a *full* version in the Android Market.

Some "free" games are supported by the inclusion of ads within the game. Other games are free to start, but require in-app purchases for continued play or additional features.

Being Careful When You Play

You might use the DROID to cast your line in a fishing game, as you would in real life. You can also move around a bit in driving and first person shooter games. So be mindful of your surroundings as you play! For example, make sure you have a good grip on your device, so it doesn't slip out of your hand; we recommend a good silicone case to help with this.

CAUTION: Games such as **Angry Birds** can be quite addictive!

Two-Player Games

The DROID really opens up the possibility for two-player gaming. In this example, we are playing checkers against one another, using the DROID as a game board.

You can find similar two-person gaming apps for other board games, such as chess or checkers.

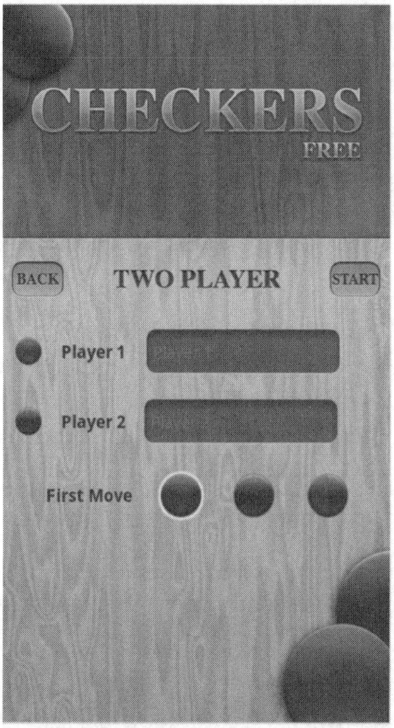

Online and Wireless Games

The DROID also allows online and wireless, peer-to-peer gaming (if the game supports it). Many new games are incorporating this technology. In **Raging Thunder**, for example, you can play against multiple players on their own devices.

The example to the right shows the screen presented when a user chooses the **Multi Player** option from the **Raging Thunder** menu. At this point, the user now chooses **Internet** to go online and join a race against opponents.

NOTE: If you just want to play against a friend who is nearby, select **Wi-Fi** mode for multiplayer games. If you just want to play against new people, try going online for a league race or game.

Playing Music Games with Your DROID

The DROID's relatively large screen means that you can even install a piano keyboard on your DROID and play music. There are a number of music-related games available; check out the **Arcade** subcategory of the **Games** category in the Android Market to see what's available.

One of the apps that was in the Top 5 of the **Free DROID** apps category when we were writing this book was **xPiano**, which turns your DROID into an on-screen piano.

Download the app and just have fun with it!

If you have children, they might enjoy it, as well.

NOTE: Some apps want to use your location and notify you of this when downloading. You can always say "no."

Piano	xPiano	FREE
	cyandroid	★ ★ ★ ★

About	Comments

Description

Really sorry, update stopped.
See my website for the reason.

A Piano App.

+4 octaves piano keyboard
+12 instruments
+Adjustable piano keyboard width
+Multi Touch(Android2.1~)
+Record and Play
+2 sample songs to play

xPiano+, which has 5 octaves, 128 instruments and more songs, is available. Check [View more applications]

Install

Other Fun Stuff: Football on the DROID

There are many great apps that can provide you with endless hours of entertainment on the DROID. Since Verizon is the official provider of the NFL now, they have an **NFL Mobile** app in the Verizon section of the Android Market.

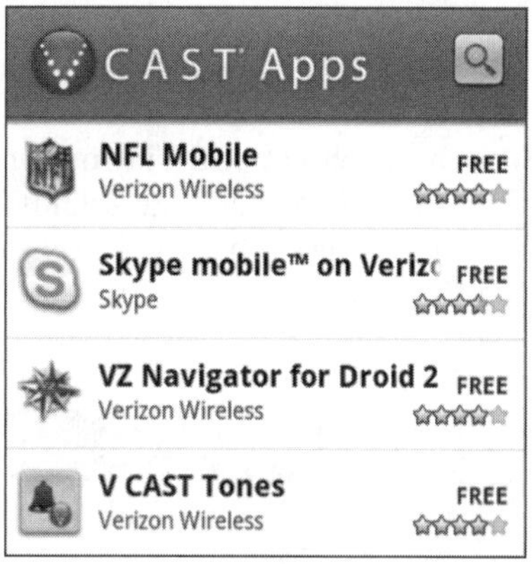

Start up the Android Market and then touch the **Verizon** tab.

NFL Mobile should be one of the first apps you see. Touch the app and then choose Install on the next screen.

In the **NFL Mobile** app, you press the **Menu** key to see the main menu of options. When you first register the app, you pick your favorite team. The favorite team on the DROID in this example is set to the Patriots.

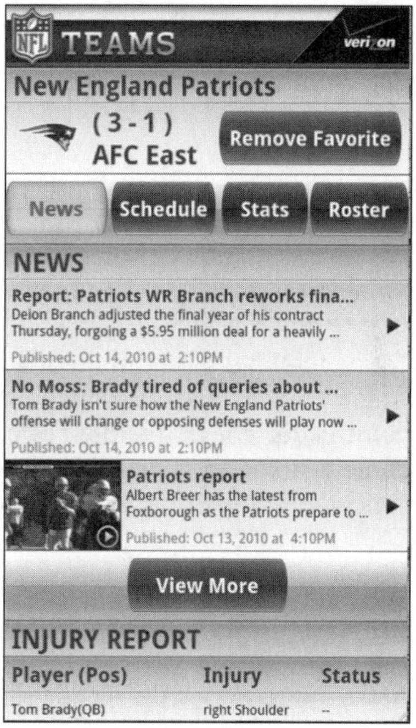

Touch the **My Team** tab to go to your team's page. So, if this team is playing, then the view automatically goes to that team's game first. If this team is not playing, then the app displays a recap of the team's previous game. Alternatively, it might list the details of the team's next game.

Touch the **News and Videos** tab and you can see the days NFL headlines. You can also watch video highlights by touching the **Videos** tab.

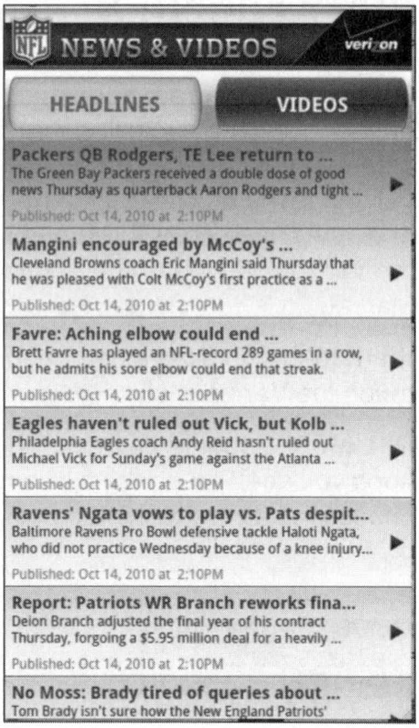

Touch the **Live** tab from the menu to watch TV broadcasts of live games. If no game is currently on and an upcoming game is going to be televised, that will be indicated at the bottom of the **Live** screen.

Chapter 23

Utilities: Clock, Calculator, and Weather

The DROID is useful for a great number of tasks, and it comes preloaded with several powerful and interesting utilities. One of the nice aspects about having a DROID is that many of its simplest and easiest-to-use apps and abilities are things that you will find yourself using quite frequently.

In this chapter, we will walk you through how to use several such apps and features, including your clock, the built-in **Calculator** app, and the **Weather** app. Specifically, we will show you how to set the clock's alarms, including how to use the snooze feature and dismiss alarms. We will also show you how to use both the **Basic** and **Advanced** modes of the built-in calculator. Finally, we will show you, not only how to configure the built-in **Weather** app, but also how to download other free weather apps you might want to add to your DROID. We will also show you how to add a **Weather and Clock** widget to your **Home** screen for easy viewing.

Your DROID can replace your wristwatch and even your alarm clock. You can use it to set multiple alarms – even a different alarm for every day of the week. Finally, you can set up a widget for the clock right on your **Home** screen. You can do all this in the **Clock** app on the DROID and in the **Alarm and Timer** app on the DROID 2 and DROID X.

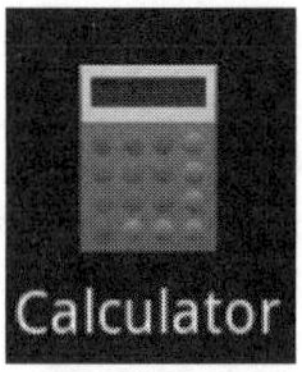

Another extremely useful application is your DROID's built-in **Calculator** app. You can use this app to determine the tip for your meal or to perform other simple calculations. For example, you might use it to determine how much 120 licenses of the Made Simple Learning video tutorials would cost a company.

Your DROID's **Weather** app is part of the **News and Weather** app, and you can use it to look up the weather for the next few days in your own city (or any other city in the world).

TIP: You can always find additional utilities in the Android Market. Check out Chapter 17: "Exploring the Android Market" for more information on the DROID's official marketplace.

The Clock App (for DROID)

Your DROID comes with a built-in clock that also provides various alarm options. Touch the **Clock** icon to start it.

On the DROID, touch the Clock icon to see the screen shown to the right. You will immediately see the current time and date.

Beneath this date and time information, you will see a snapshot of the local weather.

On the DROID, you will also see four soft keys at the bottom of the screen: **Alarm Clock**, **Photo Slideshow**, **Music**, and **Home**.

To launch a slideshow of all your photos from the **Clock** app, touch the **Slideshow** icon.

To jump to your **Music** app, touch the **Music** icon.

To return to the **Home** screen, touch the **Home** icon.

The Alarm Clock (for DROID)

The DROID's alarm clock feature is flexible and powerful. You can use it to easily set multiple alarms. For example, you might set one alarm to wake you up on weekdays and a separate alarm on weekends. You can even set an additional alarm to wake you up from your Tuesday and Sunday afternoon naps at 3 pm.

To get started, tap the **Alarm** icon in the lower row of soft keys of the clock on the DROID.

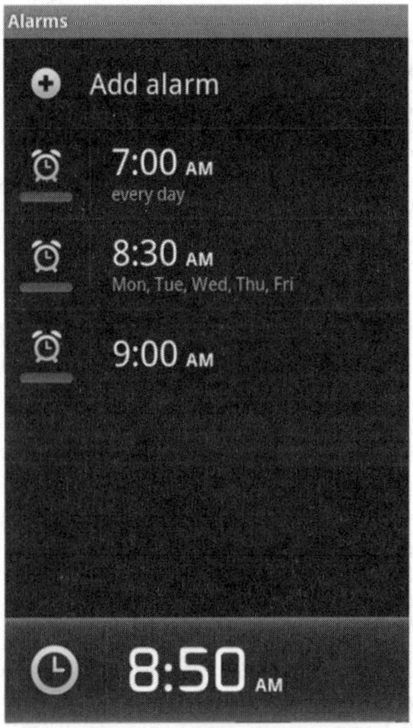

This will display any alarms you have set. If there are no alarms set, tap the **Add alarm** tab at the top to add a new one.

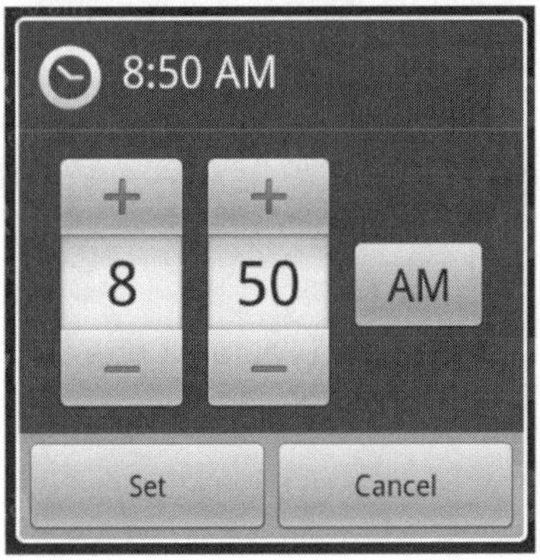

You can adjust the time of the alarm by using the + (plus) and – (minus) keys above and below the numbers to set the time.

Touch the **AM** icon and it will change to **PM**.

Touching **Set** takes you to the **Alarm Options** screen.

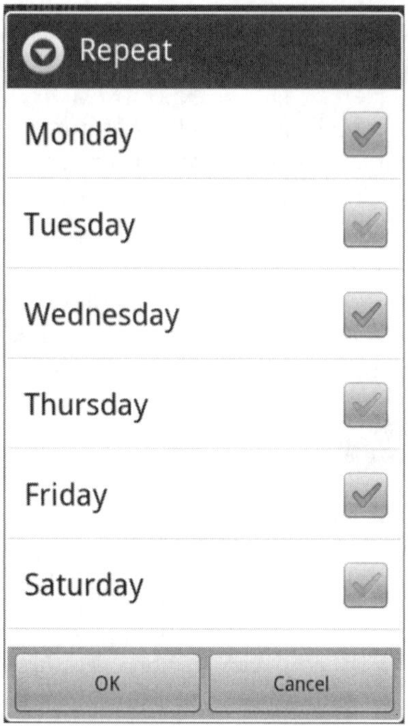

From this screen, you can adjust the **Repeat**, **Ringtone**, **Vibration**, and **Label** of the alarm.

If this is a one-time alarm, then leave the **Repeat** option set at **Never**. This setting will cause the alarm to automatically be set to **Off** after it rings.

If the alarm does repeat, then adjust the repeating function of the alarm by touching the **Repeat** tab. Touch the days of the week you would like the new alarm to be active.

> **TIP:** You may touch as many or as few days as you want.

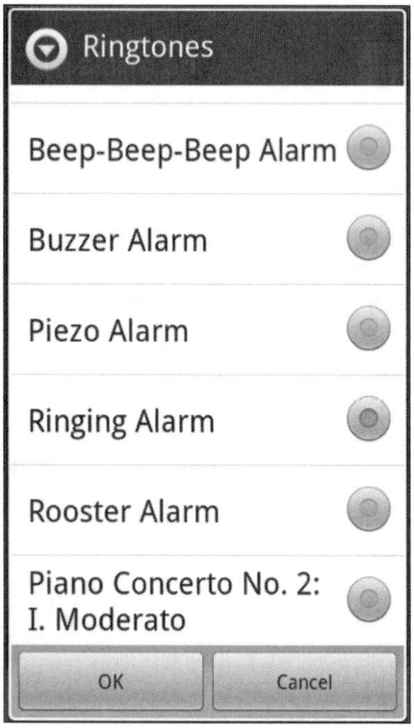

You can adjust the sound the alarm makes by touching the **Ringtone** tab and then choosing an alarm sound from the list.

For silent alarms, set the sound to **Silent** at the top of the list to have an onscreen silent alarm – no sound will be made.

Tap **OK** when you are done adjusting your **Ringtone** settings.

To adjust the **Snooze** feature, press the **Menu** key from the **Clock** app and then choose **Settings**. Snooze will be in the default **10 minutes** position.

NOTE: The pre-set **Snooze** duration is 10 minutes; however, you can adjust that value to anywhere from 5 to 30 minutes.

You can rename your alarm by touching the **Label** tab. The keyboard will launch, and you can type in a new name for that particular alarm.

Be sure to give your alarm a name that is easy to recognize.

NOTE: If you want to use the alarm feature to wake up in the morning at different times on different days, you will need to set an alarm for each day of the week by following the aforementioned procedure.

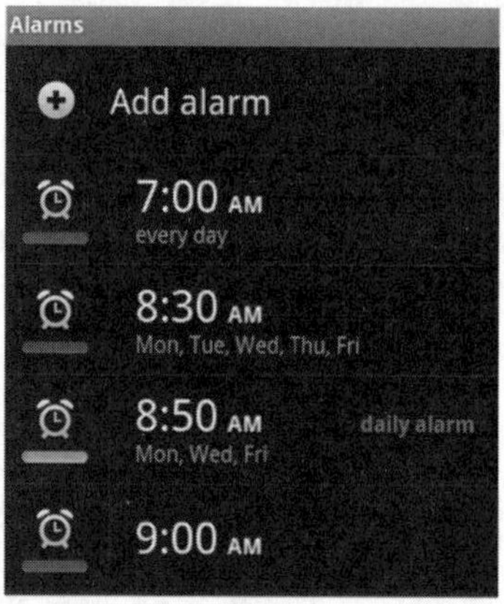

NOTE: An alarm will not turn your DROID on if it is completely powered off. However, if your DROID is in **Sleep** mode (see Chapter 1: "Getting Started"), then your alarms will ring just fine.

Using the Alarm (for DROID 2/X)

On the DROID 2/X, you get to the Alarm using the **Alarm & Timer** icon.

Settings are very similar to the DROID Alarm, with some minor variations. Here is how to add a new alarm on the DROID 2/X.

1. Tap **Alarm & Timer.**
2. Tap the **Alarm** tab at the top to make sure you are on the Alarm screen as shown to the right.
3. Press the **Menu** button and select **Add alarm**.

4. Now, you will see the **Set alarm** screen as shown to the right where you can adjust many aspects of the alarm such as:

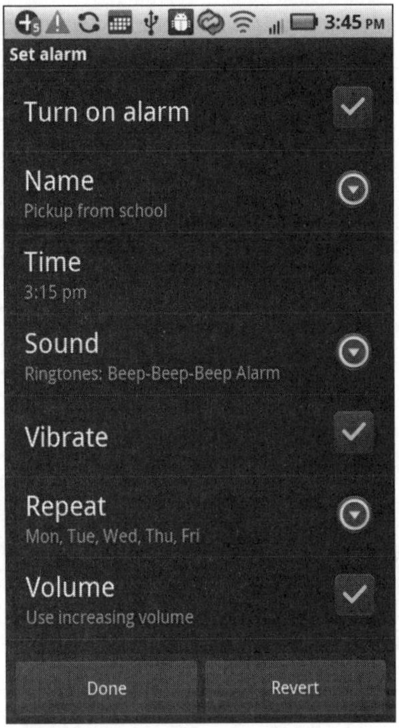

- **Turn on alarm** - place a checkmark here to activate this alarm.
- **Name -** Set a easily recognizable name such as "Pickup from school"
- **Time -** Adjust the time for the alarm.
- **Sound -** Adjust the ringtone, you can even select a video to play if you choose!
- **Vibrate -** Check this box to make the device to vibrate as well as ring at alarm time.
- **Repeat** - Choose which days of the week for this alarm to be active. In this case we just wanted week days.
- **Volume -** Check this box to have the sound increase in volume for the alarm unless you silence it.

Using the Timer (for DROID 2/X)

On the DROID 2/X the **Timer** app provides a count-down timer that can prove handy in a number of situations. For example, you might use it in lieu of a kitchen timer to remind you to take something out of the oven in 30 minutes or to ensure that you cook your pasta for exactly eight minutes.

Or, you might use it to remind yourself to turn off the sprinkler in one hour.

All of these situations are great reasons to use the **Timer** app.

Follow these steps to use the built-in **Timer** app:

1. Tap the **Timer** tab to enter **Timer** mode. You can add or subtract time in minute and hour increments by pressing the **+** and **–** buttons.
2. Tap **Start** to start the timer.
3. Tap **Cancel** if you need to stop the timer before it goes off; otherwise, it will go off with a ringing sound and vibration when the countdown expires.

NOTE: You can continue to use the phone for other tasks without stopping the timer.

The Calculator App

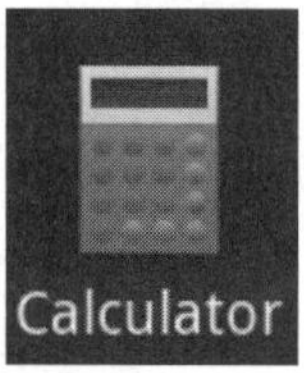

Another handy app included on your DROID is the **Calculator** app. The DROID's **Calculator** app can handle almost anything a typical family will throw its way, performing both basic and scientific calculations.

Viewing the Basic Calculator (Portrait Mode)

Click the **Calculator** icon to start the **Calculator** app.

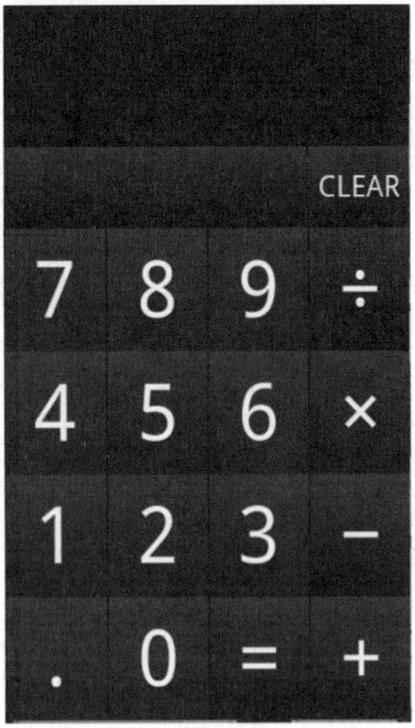

When first started, the **Calculator** application is a "basic" calculator. All functions are activated by simply touching the corresponding key to perform the desired action.

If you need more advanced functions, simply press the **Menu** key and then touch **Advanced panel** soft key.

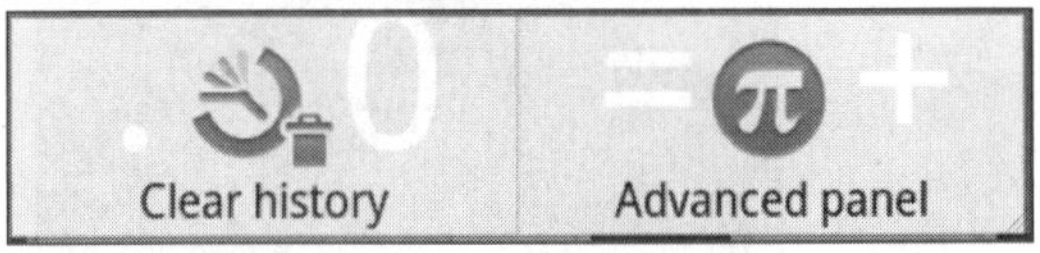

Viewing the Advanced Panel

Once your **Calculator** app is in **Advanced** mode, turn the DROID sideways to enter **Landscape** mode (horizontal). This gives you a bit more room to work with your calculations.

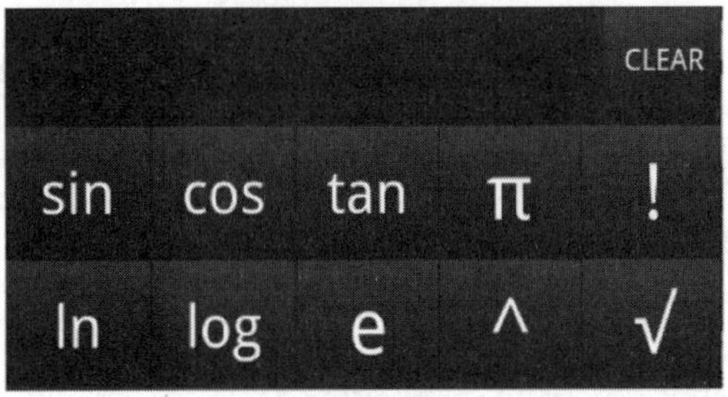

The Weather App

The DROID also comes with a very useful and easy-to-use **Weather** app built in.

The location of the **Weather** app is initially in the **News and Weather** app 

, the icon for which is on the **Home** screen.

The DROID 2 and DROID X devices also feature a separate **News** app, which is an RSS reader. For more information on the Motorola widgets that ship with your DROID, see Chapter 6, "Organizing Your Home Screens.

You can have the **Weather** app automatically set up your location, or you can manually set up another location to check the weather forecast.

Getting Started with the Weather App

You start the **Weather** app by tapping the **News and Weather** icon. By default, the DROID will use your GPS location to find the closest city or town. If the DROID is not able to do that, you will need to add your location manually.

NOTE: By default, GPS is *not* turned on at first boot on the DROID 2 and DROID X devices. Thus the **News and Weather** app may use your wireless connection or cell-tower triangulation, but it won't use GPS by default.

The original DROID, however, has GPS turned on by default.

Follow these steps to adjust the settings of the **Weather** app:

1. Touch the **Menu** button to reveal two soft keys at the bottom: **Refresh** and **Settings**.
2. Touch **Settings** and then choose the **Weather settings**.

3. The **Use my location** box should have a check in it. You can uncheck the box and then touch the **Set** location tab to input a new ZIP code.
4. To see a detailed hourly forecast, touch the **i** button.
5. Drag your finger along the chart to see the temperature and humidity throughout the day (see Figure 23-1).

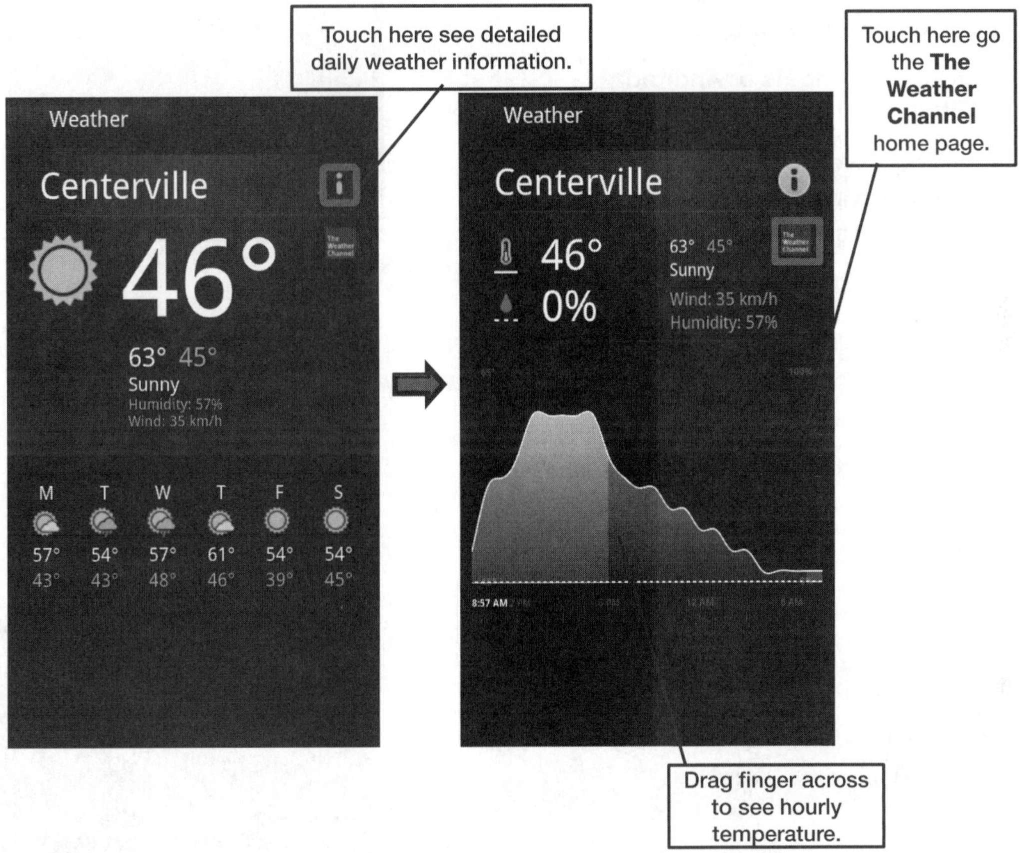

Figure 23–1. *The current and hourly views of the weather from the built-in **Weather** app.*

Adding a Weather Widget

One of the great things about the software on your DROID is that it is highly customizable. One way to customize it is to add widgets to your **Home** screen. A widget is essentially a live, updating shortcut to another app.

It's easy to add a **Weather**, **Clock**, or other widget to an available **Home** screen on the DROID. For more on widgets, see Chapter 6: "Organize your Home Screens: Icons and Widgets."

Follow these steps to add a **Weather** widget:

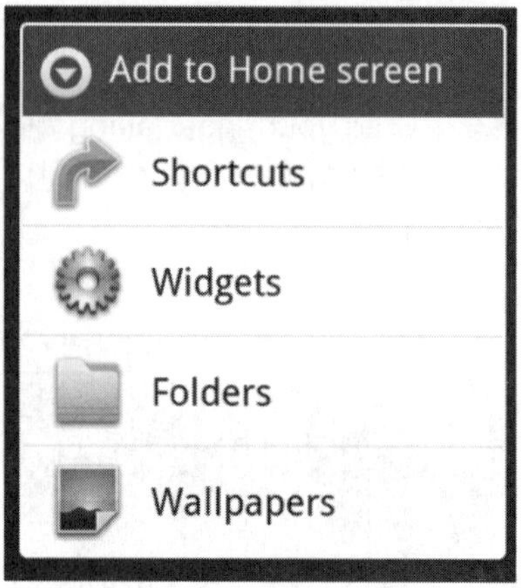

1. Slide to a blank **Home** screen – either to the left or right of the current **Home** screen.
2. Touch and hold anywhere on the screen until the **Add to Home screen** menu appears.
3. Choose **Widgets** or **Android Widgets** from the menu.
4. Choose **News and Weather** to make a widget for the **News and Weather** app.
5. On the DROID 2/X, you will then be asked to configure the widget to show **Weather**, **News**, or **News and Weather**.

NOTE: You can also choose an analog-style clock; doing so will display such a clock as a widget on your **Home** screen.

Other Weather Apps

The **Weather** app bundled with the DROID is certainly functional, but there are alternatives available. Most of the weather apps are free in the Android Market, but some also offer premium versions for a modest fee.

NOTE: Most of the free weather apps are supported by ads in the app. For the most part, these ads are not intrusive.

The easiest way to find alternative weather apps is to go to the Android Market and touch the **Apps** icon at the top of the screen. The store actually includes a distinct **News and Weather** category. In the **Weather** category, touch **Top Free** at the top and then search for apps. You can learn more about downloading apps in Chapter 17: "Exploring the Android Market."

The Weather Channel

The Weather Channel is one of the preeminent weather authorities today. The Weather Channel website can be accessed right from the DROID's **Weather** app.

Just touch the small **Weather Channel** icon on your **News and Weather** home page to visit the site.

When you first go to the site, you will input your ZIP code or address, so a custom home page with your weather can be created. This home page shows the current weather. You can scroll down the page to see **Hourly**, **36 Hour**, and **10 Day** forecasts.

Scroll further down the page to see local video, weekend forecasts, weather tools, and more.

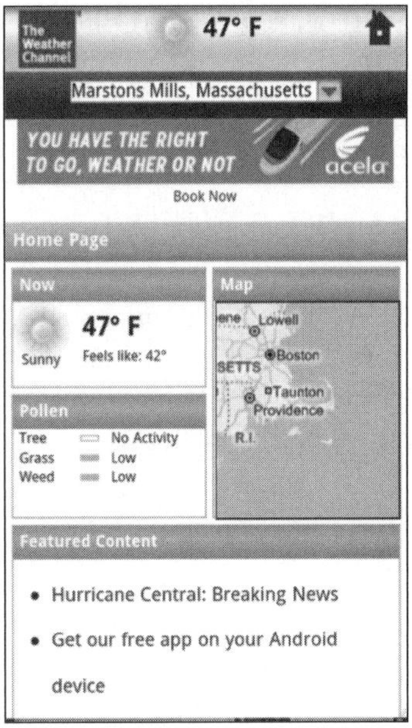

AccuWeather

Another weather authority, AccuWeather, has put together a very comprehensive weather app for the DROID.

You can download this app from the Android Market, as explained earlier.

When you fire up the **AccuWeather** app, you will be prompted to use your location for determining local weather – we recommend allowing AccuWeather to do this.

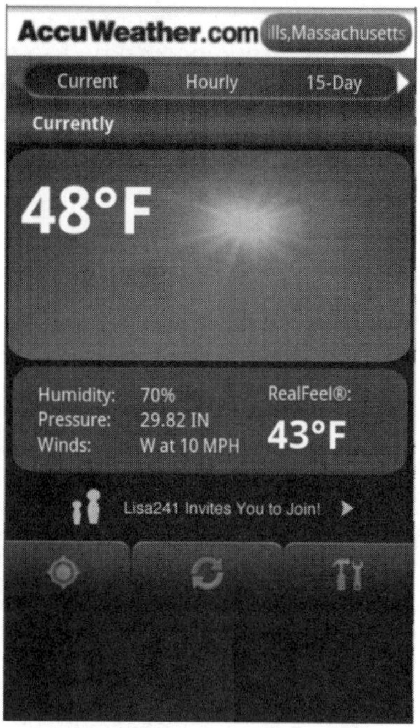

The home page of the app shows you the current temperature and conditions, along with a graphic of what the sky should look like where you are. There are soft keys to show different views.

The upper level of soft keys at the top of the screen shows buttons for **Current**, **Hourly**, and **15 Day** forecasts. There is also a soft key for **Map**, **Video Indices Alarms**, **Alerts**, and **Risk**.

The bottom of the app's screen includes the following function keys: **Location**, **Refresh**, and **Preferences**.

Chapter 24

Troubleshooting

The DROID is usually very reliable. Occasionally, as with your computer or any complicated electronic device, you might have to reset the device or troubleshoot a problem. In this chapter, we will give you some useful tools to help get your DROID back up and running as quickly as possible. We will start with some basic, quick troubleshooting and move into more in-depth problems and resolutions in the "Advanced Troubleshooting" section.

We will also cover some other odds-and-ends related to your DROID and give you a list of resources where you can find more help for your DROID.

Basic Troubleshooting

We will begin by covering a few basic tips and tricks to get your DROID back up and running.

What to Do If the DROID Stops Responding

Sometimes, your DROID won't respond to your touch because it freezes in the middle of a program. If this happens, try these steps to try to revive your DROID:

1. Press the **Home** button once to see whether the app you're in can be closed; if things go well, you will jump out to the **Home** screen.

2. Long-press the **Home** button to see whether you can bring up the list of **Recent** apps. Next, try touching one of the other apps to switch to that app and see whether you can get back to the **Home** screen by pressing the **Home** button again from the new app.

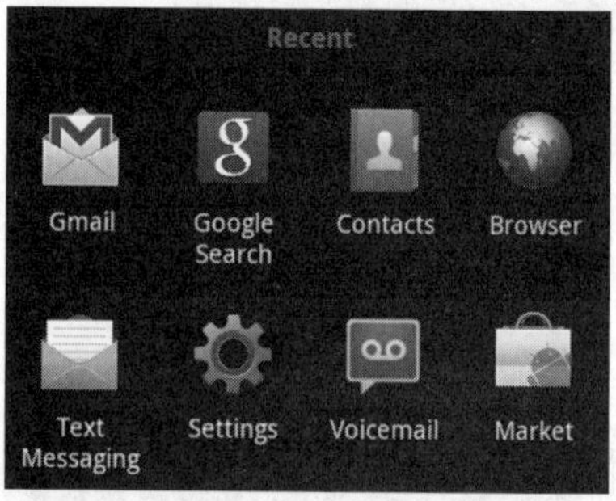

3. If your DROID continues to be unresponsive, try pressing and holding the **Power/Lock** button on the top of the phone until you bring up the **Phone** options. Next, tap **Power off**, and then turn your phone back on by pressing and holding the **Power/Lock** button.

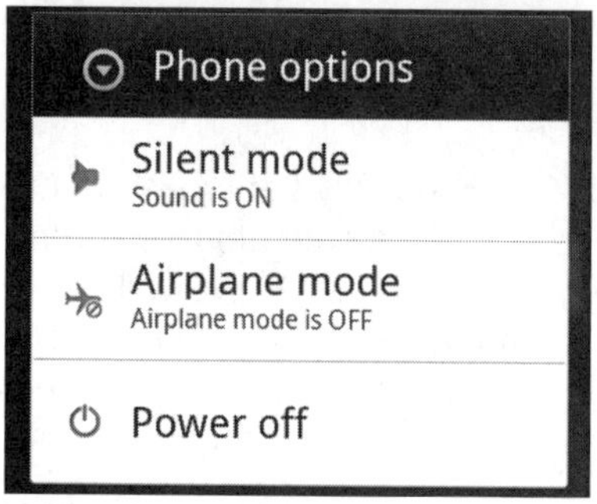

4. Make sure your DROID isn't running out of power. Tap the **Settings** icon, then **Battery Manager** on the DROID 2 and DROID X. If you have 15% power or less, you should recharge your DROID right away.
5. Try performing a *battery-pull*. Begin by removing the battery and replacing it. Power off the phone as described previously, if possible, and then open the battery cover door and remove the battery. Wait a few seconds and replace the battery and door. Next, power on the phone and see whether everything is working again.

If these steps don't work, or if your DROID seems to be getting stuck with particular apps, then read the "Managing Your Apps" section. If your DROID still won't work after trying all these steps, then you will need to perform a factory reset of your DROID or look at more advanced troubleshooting techniques or additional resources that we'll touch on later in this chapter.

Managing Your Apps

You can do a few things to manage and troubleshoot your applications on your DROID in your **Settings** app. We'll cover some of these in the upcoming sections.

Forcing an App to Stop

Occasionally, you will want to force one or more applications to quit or stop. This is called a *Force stop* on your DROID. Follow these steps to stop an app:

1. Tap your **Settings** app.
2. Tap **Applications**.
3. Tap **Manage applications**.
4. Tap the **Running** tab at the top of the screen to see all the apps currently running.

5. Locate the app that is causing you trouble.
6. Tap the **Force stop** button in the top-left portion of the screen to force the app to stop running.

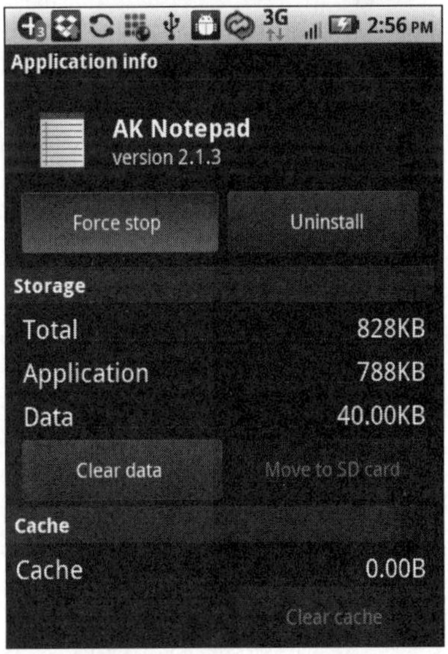

> **TIP:** You might want to tap the **Clear cache** button to reset the memory for this app, and then try restarting it.

Resolving Memory Problems

We all love to install cool new apps on our DROID. However, at some point, the love must come to an end when we receive an "Out of Memory" error. At that point, we can either remove unused apps or try to move some apps from our main internal memory to our SD card.

Deleting Apps

Follow the steps described in the "Forcing an App to Stop" section and tap the **Uninstall** button. If the app you want to uninstall is not running, then you need to tap the **Downloaded** tab at the top of the screen instead of the **Running** tab to find it.

Moving Apps to Your SD Card

Some apps, but not all, will allow you to move them from your main DROID internal memory to your SD Card. This can save you space on your DROID and allow you to install more apps.

Follow the same steps shown in the "Force an App to Stop" section and tap the **Move to SD card** button. If the button is not clickable or grayed out, then the app cannot be moved to the SD card.

If you see an error message that says something like "the application has failed to be moved because there is not enough storage left," then you may simply need to put your SD card into **Charge Only** mode or disconnect your DROID from your computer (see Chapter 25; "DROID Media Sync" for more information).

Once you have moved an app to your SD Card, you can move it back to your phone by tapping the **Move to phone** button; you find this button located in the same space.

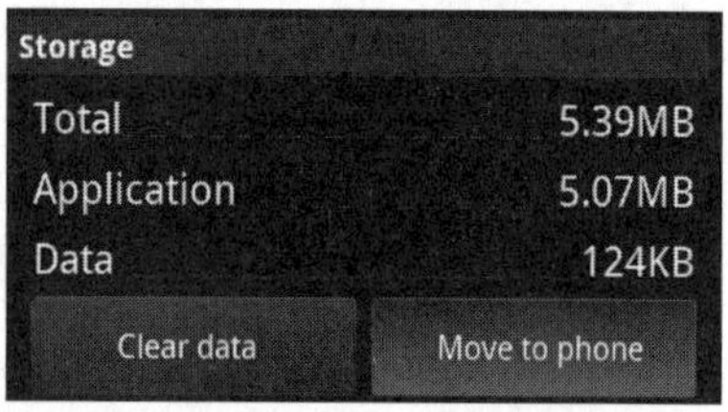

> **TIP**: If you do need to free up some space on your SD card, try moving some of your media files to your computer and deleting them from your SD card. This should free up SD card space, so you can move apps (see Chapter 25: "DROID Media Sync" for more information).

Changing the Launch by Default Setting

Sometimes you set the default open or launch setting for an app and want to change it later. For example, you might originally set the default to open **Microsoft Word** documents to the free version of **Documents To Go**. If you later purchase **Quickoffice**, then you might want to change this default to **Quickoffice**, which provides more complete editing tools. In this section, we will show you how to accomplish this.

When you open certain files, you may see a dialog box similar to the one shown at the right. If you check the **Use by default option for this action** box at the bottom of this dialog, then you have associated this type of file to the **Launch by default** for the selected app. In this image, **Word To Go** has been selected. You can change this default selection by following the same steps you used to set this option.

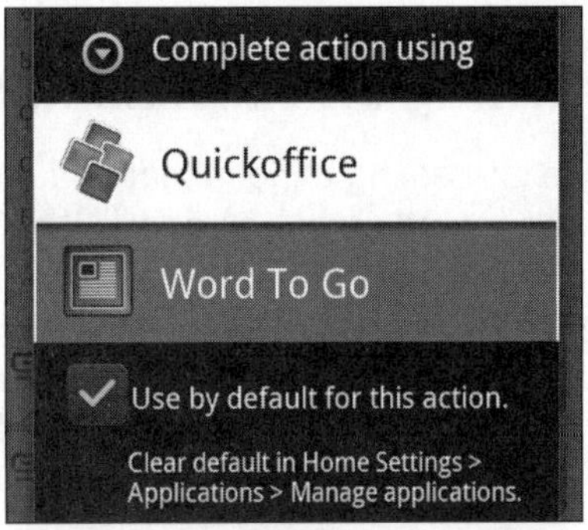

Follow the steps shown in the "Force an App to Stop" section and tap the **Clear defaults** button in the **Launch by default** section to deselect an app as the default app for opening a given file type.

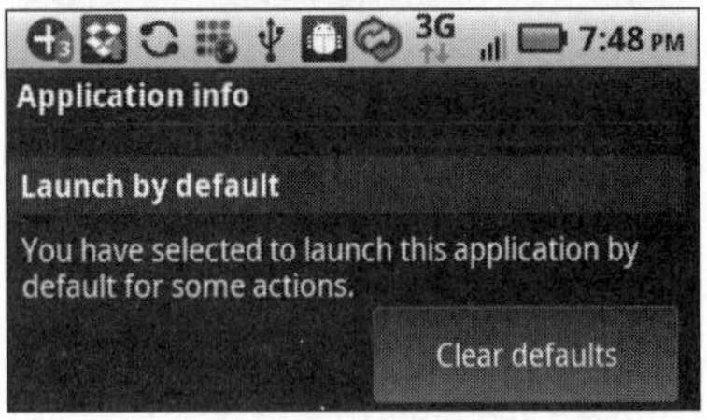

Resolving Issues With Placing a Phone Call, Syncing With Google, or Browsing the Web

There are several reasons you might not be able to place a call, sync with Google, or browse the web. One simple reason is that your DROID might be in **Airplane** mode. You can tell if you are in **Airplane** mode if you see the **Airplane** icon in the top status bar, as shown to the right.

Turning Off Airplane Mode

The interesting thing is that you may not have turned on **Airplane** mode; it could have been turned on by the phone itself. If you are out of an area with good wireless coverage for 15 minutes or more, your DROID will switch to **Airplane** mode to conserve battery life. You just need to turn off **Airplane** mode to fix this problem.

If you try to place a call when in **Airplane** mode, you will see the message shown to the right.

Tap **Yes** to turn off **Airplane** mode and make your call.

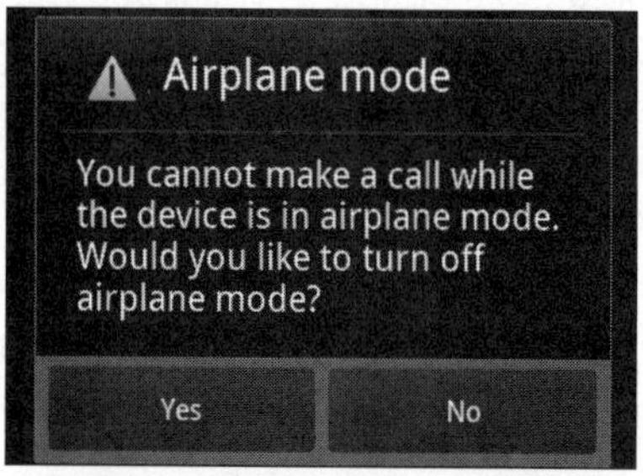

When you are trying to browse the web, the message is not quite as straightforward as the one shown to the right.

To quickly turn off **Airplane** mode, follow these steps:

1. Press and hold the **Power/Lock** button on the top of your DROID.
2. Tap the **Airplane mode** button.

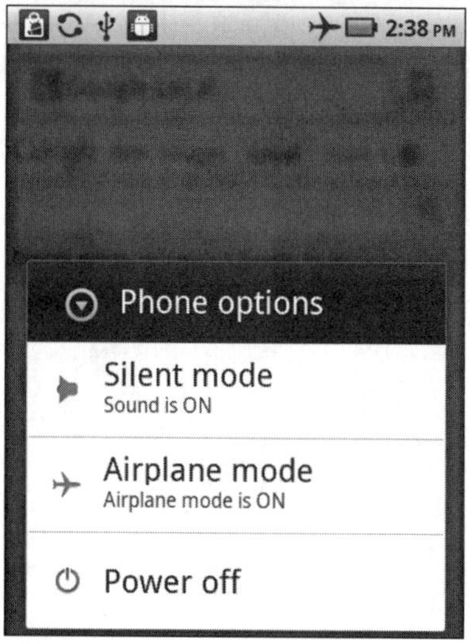

Turning **Airplane** mode off should allow you to browse the web and make phone calls, assuming you are in a place with good wireless cellular coverage.

Cycling Your Wi-Fi Connection

Another trick that can help you establish or re-establish connectivity is to *cycle* your Wi-Fi connection off and on. This might help with your Internet connection in locations where you are using Wi-Fi to connect. Follow these steps to cycle your connection:

1. Tap your **Settings** icon.
2. Tap **Wireless & networks**.
3. Tap **Wi-Fi** to turn it off (it is off when the checkmark next to Wi-Fi is gray).
4. Once the Wi-Fi connection is off, tap **Wi-Fi** again to turn it back on (it is on when the checkmark is green).

Resolving Sound Issues in Music or Video

Few things are more frustrating than hoping to listen to music or watch a video, only to hear no sound coming from your DROID. Usually, there is an easy fix for this problem:

1. Check the volume by using the **Volume Up** key in the upper-right edge of your DROID. You might have accidentally lowered the volume all the way or muted it.

2. If you are using wired headphones from the headphone jack, unplug your headphones, and then put them back in. Sometimes, the headset jack isn't connected well.
3. If you are using wireless Bluetooth headphones or a Bluetooth (car) stereo setup, then try these steps:
    a. Check the volume setting (if available on the headphones or stereo).
    b. Check to make sure that the Bluetooth device is connected. Follow these steps to do so:
        i. Tap the **Settings** icon.
        ii. Tap **Wireless & network settings**, and then make sure the box is checked next to **Bluetooth**.
        iii. Tap **Bluetooth settings** and make sure you see your device listed under Bluetooth devices at the bottom of the screen. Also, make sure that its status is **Connected to phone audio** or **Connected to media audio**.
        iv. If it is not connected, then see Chapter 8: "Bluetooth on Your DROID" to learn how to reconnect it.

NOTE: Sometimes you may actually be connected to a Bluetooth device and not know it. If you are connected to a Bluetooth stereo device or connected to your car stereo's Bluetooth (and the car stereo volume is turned down), no sound will come out of the DROID itself.

4. Make sure the song or video you want to play is not in **Pause** mode. If you see the **Play** button on the screen, then your song or video is currently paused.

If none of these steps helps, check out the "Additional Troubleshooting and Help Resources" section later in this chapter. Finally, if that does not help, then contact the store or business that sold you your DROID for assistance.

Resolving Problems When Making Purchases

So you have this cool new device, and now you want to buy some fun apps or music from the **Android Market** or the **Amazon MP3** store. Sometimes, you may receive an error message or a message that says you are not allowed to make a purchase. Follow these steps to resolve these issues:

1. Both stores require an active Internet connection. Make sure you have an active Wi-Fi or 3G connection. For assistance, check out Chapter 5: "Wi-Fi and 3G Connectivity."

2. Verify that you have an active Google Checkout account. We show you how to set up your Android Market account in Chapter 17: "Exploring the Android Market"; similarly, we show you how to set up your Amazon MP3 account in Chapter 14: "Enjoying Your Music."

Advanced Troubleshooting

If you've tried the tips and tricks in the preceding sections and you're still having issues, then you may need to resort to more advanced techniques. In the sections that follow, we will walk you through some more advanced troubleshooting steps.

Performing a Factory Data Reset

One technique that can help when others fail is to perform a **Factory Data Reset**. This procedure will work if you can still turn on your phone and get to the **Settings** app; however, you should use it only as a last resort. If you cannot get into the **Settings** app, then you need to do a battery-pull or hard reset, as described in this chapter's "What to do if the DROID Stops Responding" section.

CAUTION: Performing a **Factory Data Reset** will erase all the data you have on your DROID. If you are syncing to Google or another application, all the data synced automatically will be saved. Don't worry about pictures and videos stored on the memory card; those will not be lost when you do this reset.

Follow these steps to reset your DROID:

1. Tap the **Settings** icon.
2. Tap **Privacy**.
3. Tap **Factory Data Reset.**
4. Tap **Reset Phone** to start the **Factory Data Reset** process.
5. Once the process is complete, you will need to set up your phone as you did when you first got it out of the box (see Chapter 1: "Getting Started").

Increasing Your Text Message Limit

Sometimes, you may find you receive an error message while you are texting that says you have reached your message limit. The default is usually 200 messages per conversation. You can increase or decrease this limit inside the **Text Messaging** app by following these steps:

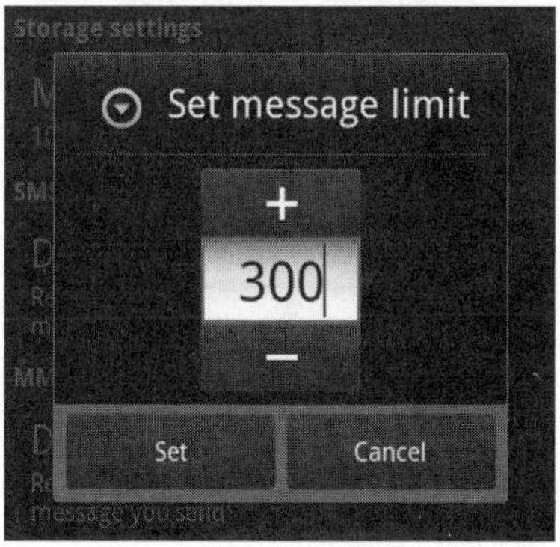

1. Tap the **Text Messaging** icon.
2. Press the **Menu** button and select **Messaging** settings.
3. Tap **Message limit** at the top.
4. You will see a screen similar to the one shown to the right. Use the + or - buttons or simply tap the number itself to type a new number. You can select between 10 and 999 messages for the limit.

Additional Troubleshooting and Help Resources

Sometimes you may encounter a particular issue or question that you cannot find an answer to in this book. In the following sections, we provide some good resources that you can access from the DROID and from your computer's web browser. The Motorola support site and knowledgebase are helpful if you are facing a troubleshooting problem that is proving especially difficult to resolve. The DROID 2/DROID X-related web blogs and forums are good places to locate answers and even ask questions about unique issues you might be facing.

The Motorola DROID Support Pages

To get to the Motorola DROID support pages, follow these steps.

1. On your DROID or computer's web browser, go to `www.motorola.com`.
2. Click **SUPPORT** in the top-right corner of the main navigation bar.
3. Click **Android support** in the right column.
4. You should now see a screen similar to the one shown in Figure 24–1. From this screen, you can click **VIEW SUPPORT DETAILS** under your DROID phone in the right column.

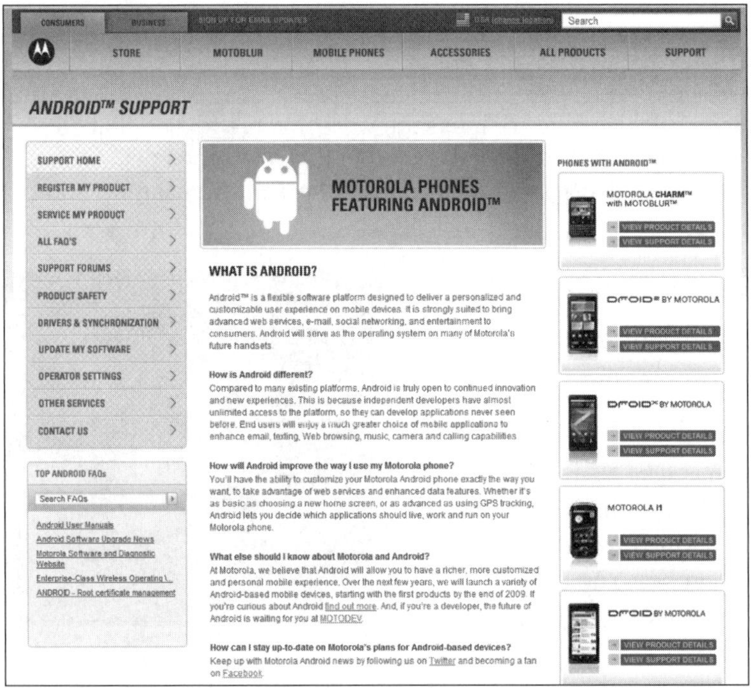

Figure 24–1. *The Motorola Android Support Page.*

DROID-Related Blogs

One of the great things about owning a DROID is that you immediately become part of the worldwide camaraderie of DROID owners.

Many DROID owners would be classified as "enthusiasts" and are part of any number of DROID user groups. These user groups, along with various forums and websites, serve as great resources for DROID users.

Many of these resources are available right from your DROID, and others are websites that you might want to visit on your computer.

Sometimes you might want to connect with other DROID enthusiasts, ask a technical question, or keep up with the latest and greatest rumors. The blogs are a great place to do that.

Here are a few popular DROID (DROID, DROID 2, DROID X, etc.) blogs:

`www.androidcentral.com`

`www.droid-life.com`

`www.theandroidblog.com`

`www.droidx.net`

`www.droidblog.net`

`www.technocrati.com`

`www.droid-forum.com`

`www.engadget.com/droid`

> **TIP:** Before you post a new question on any of these blogs, please do a search on the blog to make sure your question has not already been asked and answered. Also, make sure you are posting your question on the right section (e.g., DROID) of the blog. Otherwise, you may incur the wrath of the community for not doing your homework first!

Finally, you can do a web search for "DROID blogs" or "DROID news and reviews" to locate more blogs.

Part IV

Sync Media to your DROID

Your DROID is a great mobile device for enjoying all your media - pictures, videos, ringtones, and music. You can take great pictures and videos right on your DROID. You will want to know how to transfer all your pictures and videos on your computer for safekeeping.

You may also want to know how to easily transfer your music library, playlists, videos and pictures from your computer to your DROID. We cover two simple solutions including **doubleTwist** and using Mass Storage mode to transfer media and documents between your computer (PC or Mac) and your DROID. Soon, you will be able to fill your DROID will all sorts of media.

With **doubleTwist**, you can also browse the Android Market, search for and subscribe to podcasts, and purchase music from the Amazon MP3 market. Sometimes doing this on your larger computer screen and keyboard is a bit easier than on the DROID itself.

Chapter 25

DROID Media Sync

In the final chapter of this book, we will cover a couple good options for helping you sync or transfer media (e.g., music, pictures, and videos) between your computer and your DROID. These software products will help you sync your **iTunes** or **Windows Media Player** music and playlists, as well as your pictures and videos from your computer (PC and Mac). You can also use these tools to transfer the pictures and videos you take on your DROID back to your computer for safekeeping.

The media transfer and sync options we cover are listed in Table 25–1; this table also lists when you might want to choose each option.

Table 25–1. *Media and Sync Options for Your DROID.*

Solution	When to Use	Compatibility & Price
Drag-and-drop elements using your USB cable connection.	To quickly transfer one or more files between your computer and your DROID.	PC and Mac Free
Use **doubleTwist** – an "iTunes-like" media player and sync tool for your DROID.	To sync **iTunes** or **Windows Media Player** playlists and related media.	PC and Mac Free
Dropbox document and file sync (See Chapter 21 for details.)	To transfer files and documents wirelessly between your computer and DROID.	PC and Mac Free (up to 2GB)

> **NOTE:** You can find several alternative media sync options not covered in the preceding table. For example, your DROID comes with the **Media Sync** app, however, it only syncs media to PCs, so we do not cover it here. Verizon supplies an app called **VCAST Media Sync**; however, it also works only for PCs, so we do not cover it in this chapter, either. There are also many other solutions available on the market to sync media to your DROID. If none of the options described in our book serve your needs, please do a web search for "DROID media sync." During our research, we saw that Salling Software (`www.salling.com`) had both Mac and PC versions of its **Media Sync** program. Finally, **Winamp** (PC-only, `www.winamp.com`) lets you play your media on your computer (like iTunes) and also sync media to your DROID.

Where to Enjoy Your Media on Your DROID

You can view, play, or listen to media on your DROID in a few different programs. As you have seen, we have chapters devoted to each type of media supported on the DROID in this book (see Table 25–2).

Table 25–2: *Playing Media on Your DROID.*

App	Type of Media	Chapter
	Songs and Podcasts	Chapter 14: "Enjoying Your Music"
	Pictures and Videos	Chapter 18: "Taking Photos and Videos" Chapter 15: "View Videos, TV Shows, and More"

Moving Files With Your USB Connection

Your DROID's USB connection provides the most basic method for transferring files between your DROID and your computer. It works not just for media files, but for all types of files, including **Microsoft Office** documents, PDF files, or anything else. What this method does not provide is the ability to sync your playlists from **iTunes** or **Windows Media Player** or perform any sort of compression or optimization for the music, pictures, or videos you transfer to your DROID.

CAUTION: You may not be able to play video files that you have just dragged-and-dropped onto your DROID SD card using a USB connection. This is because most video files need to be encoded in a specific format to play on your DROID.

Selecting a USB Connection Mode

When you first connect your DROID to your computer with the USB cable, you may see a screen asking what connection mode you would like to use. We recommend the **Mass Storage** mode in order to sync media. You can change your mode while you are connected to your computer; see the "Changing USB Connection Modes" section in this chapter for more details. The following connection modes are available:

- **Mass Storage Mode** – This is the recommended mode for syncing media. It allows only your computer to view and update the DROID media card files (your DROID cannot access the SD memory card in this mode). The SD card in your DROID appears as a removable disk (see image to the right). This mode is required to use **doubleTwist**, and it allows you to copy files back-and-forth between your computer and your DROID. If you try to access any files from your DROID stored on the media card while in this mode, you will receive the following error: "The SD card cannot be found." You need to change to a **Charge Only** mode to see files from your DROID.

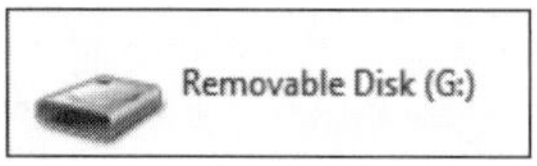

- **Windows Media Sync** – This mode allows both your Windows computer and the DROID to see and access files on the media card. Your DROID appears as a portable device (as shown in the image to the right) inside **Windows Explorer**. You would use this method to sync when using **Windows Media Player**.

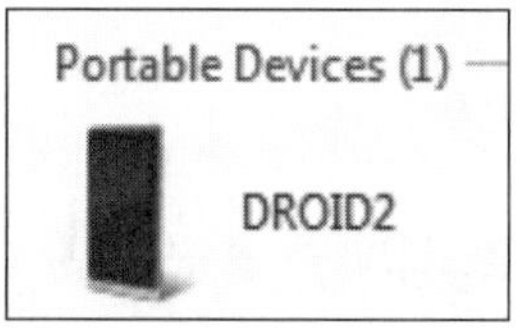

- **PC Mode** – This mode also allows both your computer and the DROID to see and access the media card; like the **Windows Media Sync** method, your DROID appears as a portable device.
- **Charge Only** – This mode is really for charging only, and it does not make the SD card in the DROID visible to the computer. You cannot use this mode to transfer media. However, since the SD card is visible to your DROID in this mode, you can play your media on your DROID when connected in this mode.

Changing USB Connection Modes

After you have plugged your DROID into your computer, you can change your connection mode from any of your **Home** screens. Follow these steps to do so:

1. Connect your DROID to your computer using the USB cable.
2. Drag your finger down from the top of any **Home** screen to see the notifications and then tap the **USB connection** item in the **Ongoing** section near the top.

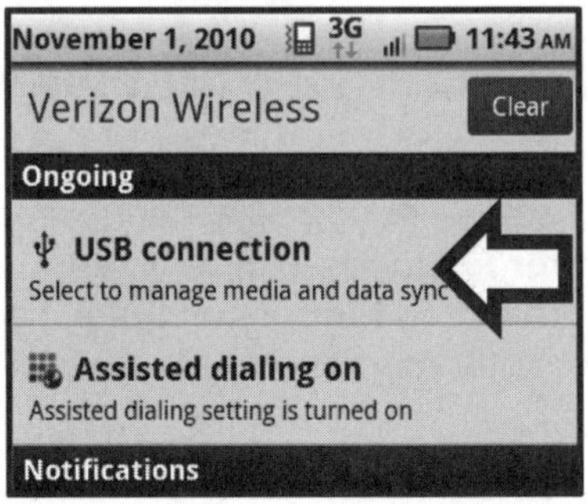

3. Tap any mode you wish to select. To sync media, tap the **USB Mass Storage** mode and tap **OK**.

   To access your SD media card and charge your DROID from the USB connection, select **Charge Only**.

Dragging-and-Dropping Files

Once you have connected your DROID to your computer in **Mass Storage** mode, you are ready to drag-and-drop files between your computer and your DROID SD card. Follow these steps to do so:

Step 1: Open your DROID SD card window on your computer

On your Windows computer, open **Windows Explorer** and look for the removable disk that is your DROID SD card.

On your Mac, use your **Finder** to locate the removable disk that represents the DROID SD card.

Step 2: Drag-and-drop files between your computer and the DROID SD card

Open a window on your computer with the type of media you want to drag-and-drop. This media could be your music files, pictures, or videos. Next, locate the correct folder on your DROID SD card for the same type of media. Table 25–3 lists typical locations for common media types on your DROID.

Table 25–3: *Typical Folder Locations for Media on Your DROID SD Card.*

Type of Media	Folder Location on the DROID SD Card
Pictures or videos taken with the DROID camera	Dcim / Camera
Ringtones for the phone	Media / Audio / Ringtones
Notifications (e.g., alarms)	Media / Audio / Notifications
Music	Music
Files from **Documents to Go**	Documents

Once you have the source and destination folders open, you can drag-and-drop files between the folders to transfer the files. For example, to copy a **Microsoft Office** document from your computer to the SD card on the DROID, you would drag-and-drop it into the documents folder, as shown in Figure 25–1.

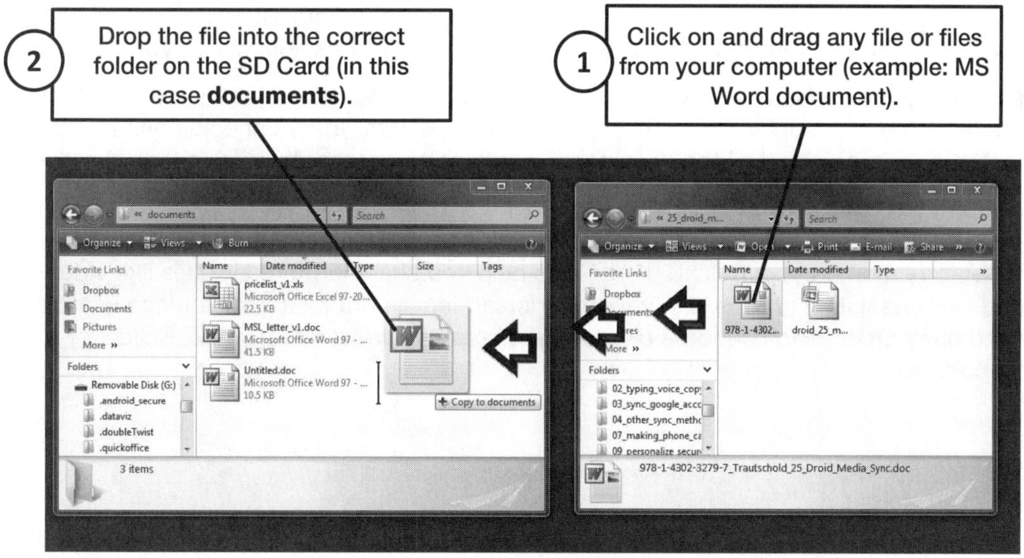

Figure 25–1. *Dragging-and-dropping files between your computer and the DROID SD card.*

Disconnecting Your DROID Safely (Don't Skip!)

Because of the way your DROID handles files, you will want to be sure to **Eject** the DROID from your computer prior to yanking out the USB cable.

On either a Windows or Mac computer, you can right-click the disk drive that represents your DROID SD card.

This image is from a Windows PC, but the view from your Mac's file explorer will be similar. The DROID in this image is **Removable Disk (G:)**. Select **Eject** from the menu and then unplug your DROID.

On a Mac, you can also drag the **Disk** icon on your desktop to the **Trash** to safely eject it.

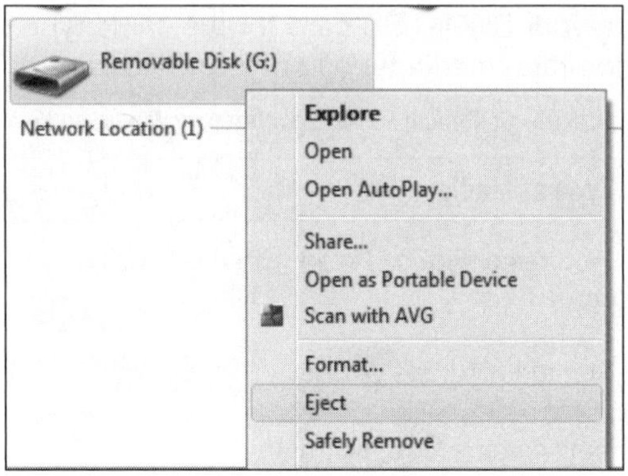

Using doubleTwist

doubleTwist is one of several software alternatives to **iTunes** and **Windows Media Player** that work well for syncing music, pictures, and videos to your DROID. In addition to the DROID, **doubleTwist** is available for many other smartphones and devices. Also, **doubleTwist**, like iTunes, lets you purchase apps from the Android Market, subscribe to podcasts, and buy music from the Amazon MP3 store (just as you can on your DROID).

According to its official website (`www.doubletwist.com`), **doubleTwist** was founded on the following philosophy: To be a "unifying media platform that connects consumers with all their media and all their devices, regardless of whether they are online or offline."

doubleTwist has several compelling features. For example, it can import all your **iTunes** or **Windows Media Player** information – including playlists – and then allow you to sync this information on your DROID. It can also help you share large media files like videos and high-resolution photos with your friends and family. For example, you can use it to send baby pictures to Grandma or lots of pictures you've taken on your DROID to your friends.

Downloading and Installing doubleTwist

The first step to getting your media onto your DROID using **doubleTwist** is to download and install the application on your PC or Mac. Follow these steps to do so:

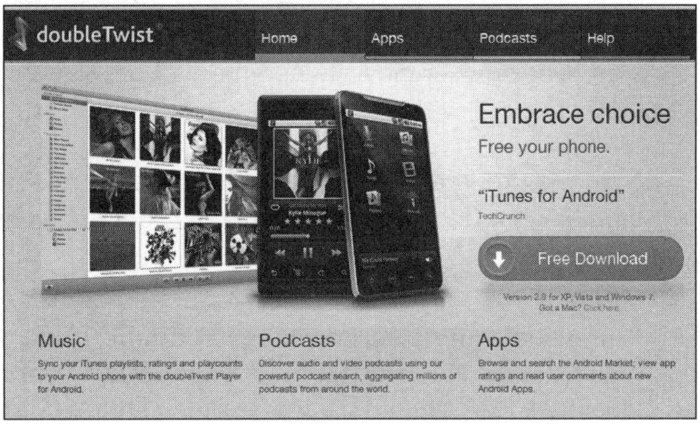

1. From your computer, visit `www.doubletwist.com` and click the **Free Download** button; the website will sense which version of the software is right for your computer.

2. Download the install file and follow the directions for setup.

3. On a Mac, simply drag the **doubleTwist** icon to your applications folder.

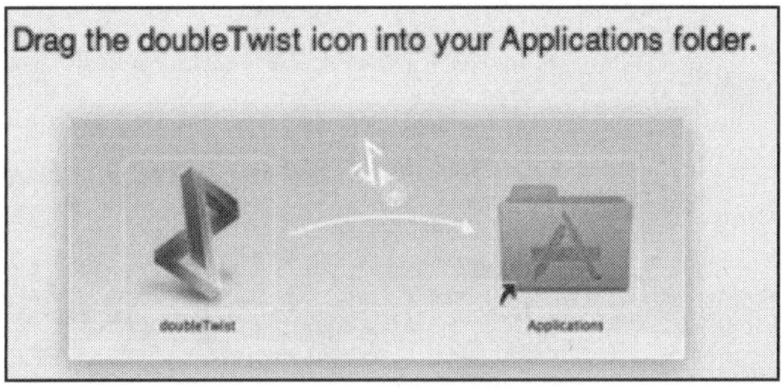

   On a PC, double-click the installation file in your `Downloads` directory if it did not start automatically.

4. You will be prompted to create a doubleTwist account unless you have one already. Enter your username, email, and password for a one-time setup of your doubleTwist account.

   A few moments after setting up your account, you will receive a confirmation email letting you know that your doubleTwist account is now active.

Getting Started With doubleTwist

Once you have the program installed on your computer, you are ready to set it up and start syncing your playlists and media from **iTunes** or **Windows Media Player** to your DROID.

Start the program by double-clicking the **doubleTwist** icon on your desktop or locating the program in your **Start** menu.

You'll want to get familiar with what **doubleTwist** offers. In the top of **doubleTwist**'s left column, click the **Android Market**, **Podcast Search**, and **Music Store** options to see what they look like. If you are familiar with the **iTunes** app, you will already know how to use these features (see Figure 25–2).

Figure 25–2. *Accessing the Android Market, Podcast Search, and the Music Store (Amazon MP3) from doubleTwist.*

Importing Your Playlists

You're now ready to import your playlists from **iTunes** or the **Windows Media Player**:

1. Click **Library** from the main menu at the top.
2. Select either **Import iTunes Playlists** or **Import WMP Playlists** (**Windows Media Player**) to import your playlists.
3. You will then see a warning message that any changes made in **doubleTwist** to the same playlists will be lost and replaced with the current playlists from **iTunes** or **WMP**. Click **Import** to continue.

4. When the process is complete, you will see all your playlists in the left column under **PLAYLISTS**. Click any playlist to see its contents in the main window.

Connecting Your DROID to doubleTwist

In order to sync media or just drag-and-drop items onto your DROID in **doubleTwist**, you have to connect it to your computer with the USB cable. Follow these steps to do so:

1. Follow the steps shown in the "Changing USB Connections Mode" section in this chapter to set your connection to **USB Mass Storage** mode.

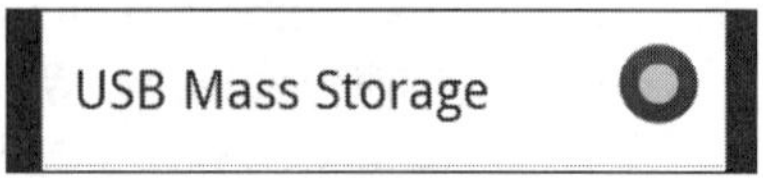

2. You should immediately see your device listed at the very bottom of the left column. It may show up as Motorola A955 (as shown in the image to the right) or as another Motorola model instead of DROID, DROID 2, or DROID X.

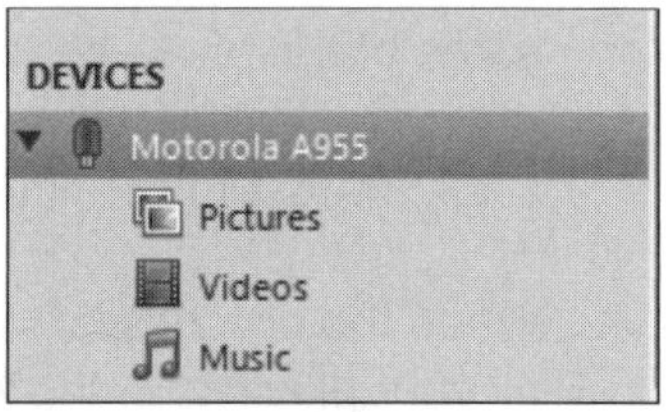

TIP: Troubleshooting USB Connection Problems

If you do not see your DROID in **doubleTwist**, then try using another USB port on your computer. Also, double-check that you are in **Mass Storage** mode and not some other USB connection mode.

Syncing Music and Podcast Subscriptions Automatically

After you have successfully connected your DROID and it is visible in **doubleTwist**, click your DROID in the left column under **DEVICES** to highlight it. (Remember: It may show up as Motorola A955 or another Motorola model.) At this point, you should see the **General** tab in the main window, as shown in Figure 25–3. Check the box next to **Music** to sync your music, and then check the box next to **Subscriptions** to sync your podcast subscriptions.

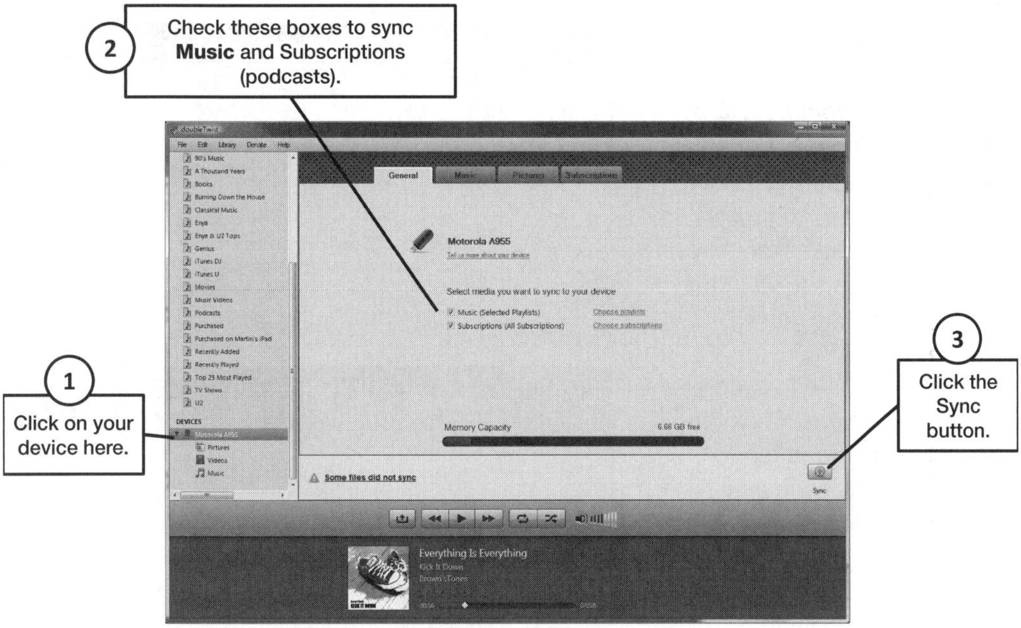

Figure 25–3. *Setting up automatic syncing of your music and podcasts with **doubleTwist**.*

Fine Tuning the Sync

If you have a large amount of music or podcasts, you will want to fine-tune the sync process by clicking the appropriate tabs at the top of the window. Follow these steps to use **doubleTwist** to fine-tune how your DROID syncs various types of media:

1. Begin by syncing your music. Click the **Music** tab to selectively sync only certain playlists.
2. Click **Only the selected**.
3. Place checkmarks only next to the playlists you want to sync to your DROID.

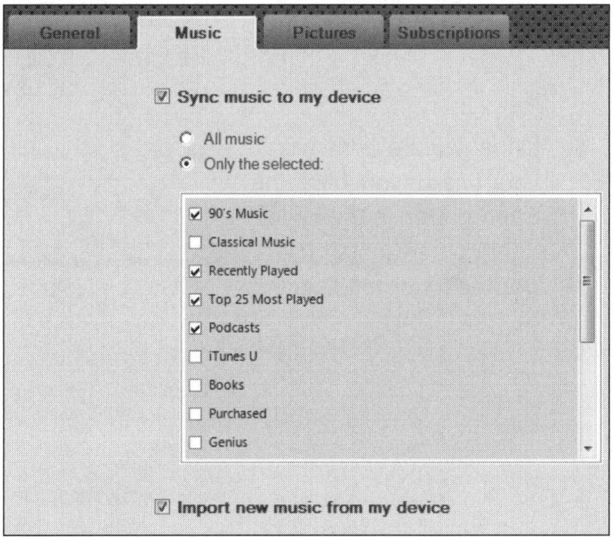

4. Next, import new pictures and videos that you have taken on your DROID to your computer. Click the **Pictures** tab and check the box next to **Import new pictures from my device**, as shown in the image to the right.

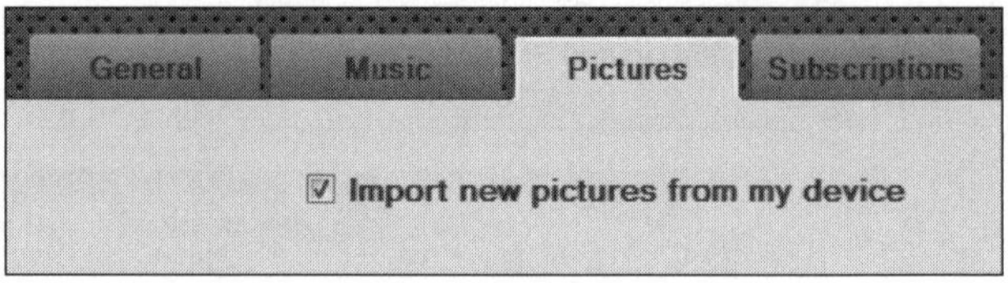

5. Now it's time to sync your subscriptions. Click the **Subscriptions** tab to selectively sync only certain podcasts.
6. Click **Only the selected**.
7. Check off only the podcasts you want to sync to your DROID.
8. You can also check **Only sync** at the bottom and select from only a few recent shows or only episodes you haven't yet played.

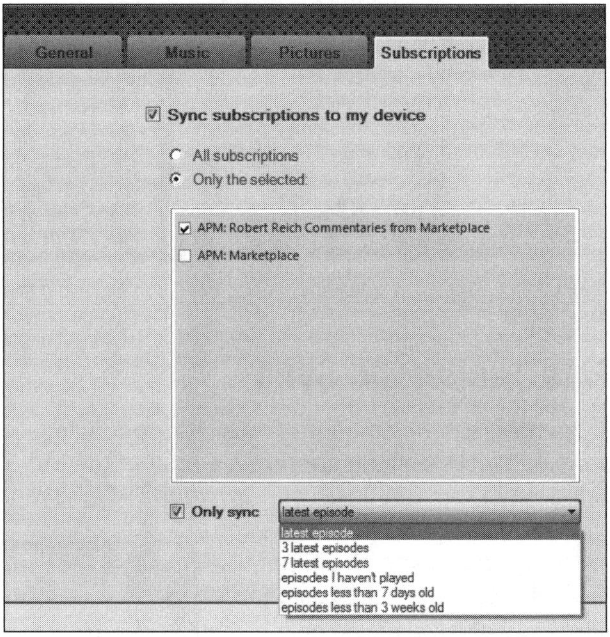

9. Once you are done with your fine-tuning, click the **Sync** button in the lower-right corner to start the **doubleTwist** media sync.

Dragging-and-Dropping Media Onto Your DROID

If you want a little more control over what is syncing, you can easily drag-and-drop individual items (e.g., songs and podcasts) onto your DROID or entire playlists or albums.

With your DROID connected and visible, grab a song, playlist, podcast, or other item in **doubleTwist** and drop it onto your DROID (it's listed under DEVICES in the bottom of the left column). Each item you drop will be synced to your DROID (see Figure 25–4).

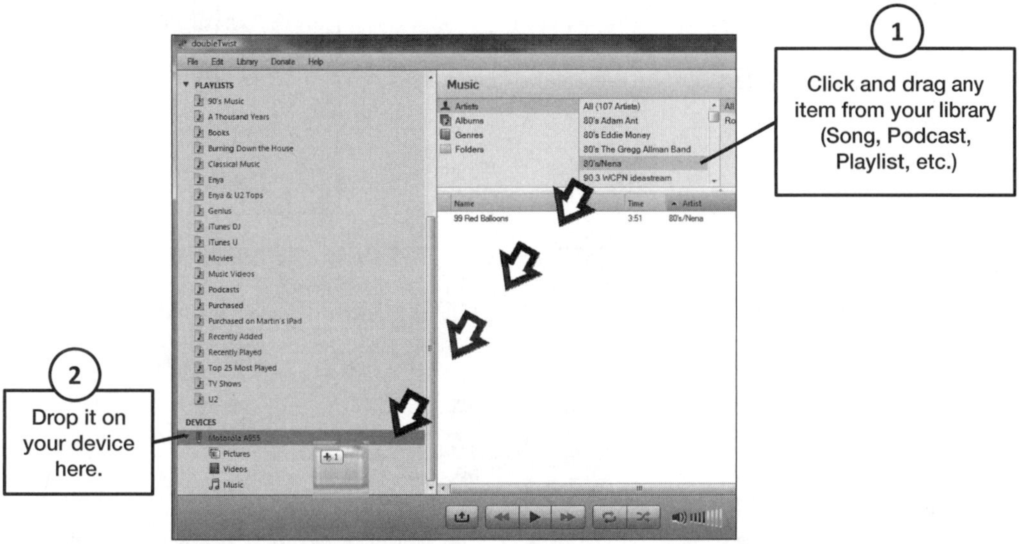

1
Click and drag any item from your library (Song, Podcast, Playlist, etc.)
2
Drop it on your device here.

Figure 25–4. *Dragging-and-dropping media (e.g., music, playlists, and podcasts) with **doubleTwist**.*

My Music App Won't Play My Music

Occasionally, you will see the following error after you sync or drag-and-drop music files onto your DROID: "Sorry the player does not support this type of audio file." If you see this, try powering your DROID off and then on again. If that does not correct the problem, open the back of the device, remove the battery for a few seconds, and then put the battery back in again. This usually fixes the problem.

Finding and Subscribing to Podcasts With doubleTwist

You can find just about any podcast you want using **doubleTwist**. Follow these steps to do so:

1. Click **Podcast Search** in the upper portion of the left column in **doubleTwist**.

2. Type in a word or a few words to help you find your podcast in the main **Search** window and click **Search**. In the example to the right, we searched for the word "marketplace." You can play the most recent episode of a podcast by clicking the **Play latest episode** button. To subscribe to a podcast, click the podcast title.

3. Click the **Subscribe with doubleTwist** button to subscribe to a podcast.

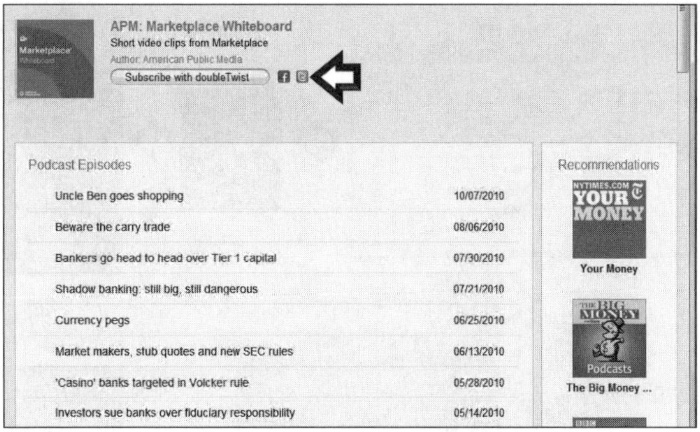

Finding Apps in the Android Market With doubleTwist

Many users find it more convenient to quickly browse and find great apps using a desktop computer and keyboard rather than their DROID devices. However, your DROID gives you yet an easier option for finding and downloading great apps. Specifically, you can combine a free **Barcode scanner** app, QR codes, and the **doubleTwist** app to simplify the process of locating and acquiring apps from the Android Market. Follow these steps to do so:

1. Click **Android Market** in the upper portion of **doubleTwist**'s left column (see Figure 25–5).
2. Click any app or perform a search to locate the app you wish to purchase or download.
3. Click the **Barcode** button, also known as the *QR code*, to enlarge it on the screen.
4. Use the **Barcode scanner** app on your DROID to scan the QR-code. (You can download this app for free from the Android Market.)
5. Tap the **Open browser** button once the barcode is recognized.
6. This places you in this app's detail page in the **Android Market** app. Click **Install** or **Buy** at the bottom of the screen to get the app on your DROID.

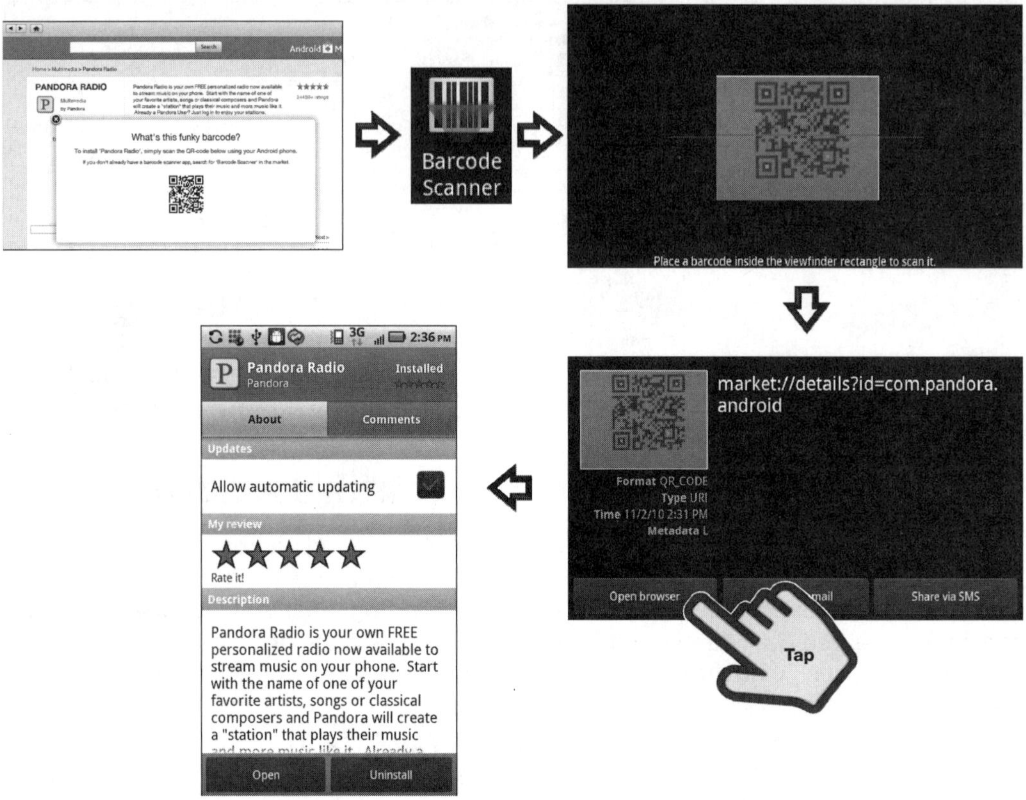

Figure 25–5. *Locating and downloading apps to your DROID with the* ***Barcode Scanner*** *app and* ***doubleTwist****.*

Buying Music from the Amazon MP3 Store in doubleTwist

It's possible to purchase music from the Amazon MP3 store using your DROID's built-in **Amazon MP3** app; however, you can also buy from this store directly from the **doubleTwist** app. Follow these steps to do so:

1. Click the **Music Store** link in the top portion of **doubleTwist**'s left column to see the Amazon MP3 music store.
2. Enter a search term or click any album you see.
3. Click the **Buy** button to purchase and download the desired song or album (see Figure 25–6).
4. You will need to sign into your Amazon.com account to complete the purchase.
5. The song or album will be downloaded to your **doubleTwist** library.
6. Use the techniques described earlier in this chapter to sync the song or album to your DROID.

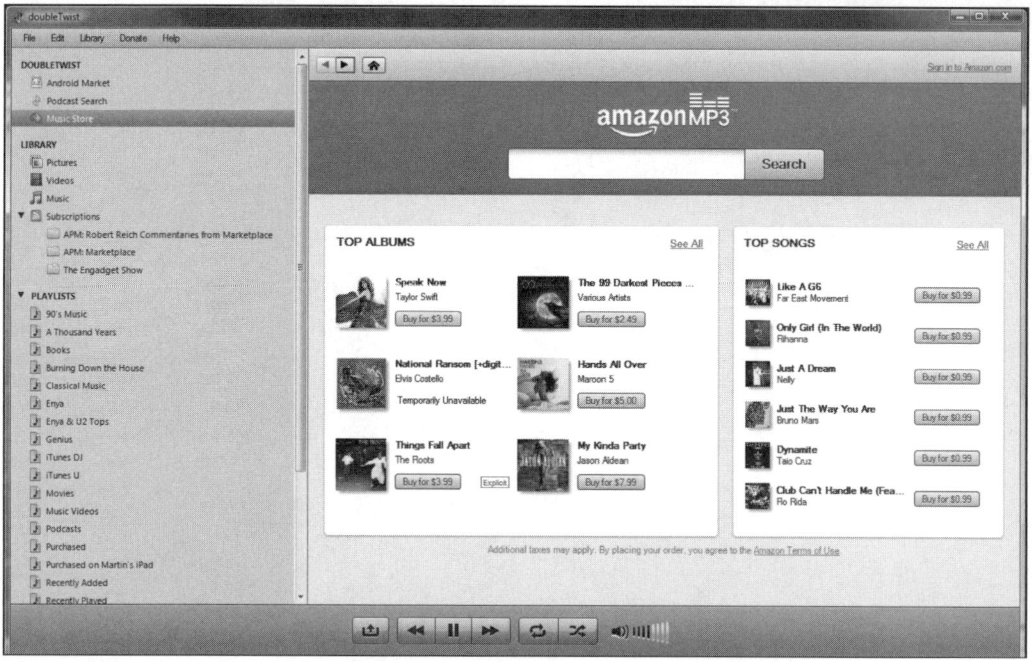

Figure 25–6. *Finding and purchasing music in the Amazon MP3 store from* ***doubleTwist****.*

Appendix

DRIOD App Guide

This appendix includes information on apps you can download and install for your DROID. We've included QR codes, so you can use a barcode scanner app on your DROID to navigate directly to a given app. Read Chapter 17, "Exploring the Android Market," for more information on QR codes.

CAUTION: We want to make you aware that apps may change frequently. Some of the apps listed in this guide will have been updated, others may be gone altogether, and new and improved ones added by the time you read this book. Please use the apps listed in this guide as a starting point to become familiar with what is possible and help you find some useful apps.

Document Tools

If you work in an office, you may need to deal with documents. At the time of publication, there are no solutions to reading **Microsoft OneNote** on Android. That may come in the future, but for now we recommend using **Evernote** instead.

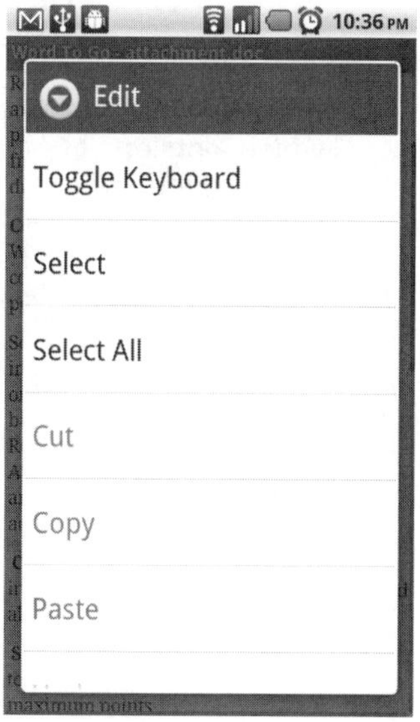

Of the document-management software we tested, DataViz's **Docs to Go** and **QuickOffice Mobile Suite** from **QuickOffice** were the most stable and offered the best interfaces for both viewing and editing documents. However, you may have different results with different types of documents, so the best approach is to use free trial versions of apps that interest you and compare how well these apps display the types of documents you typically use. **Docs to Go** allows some editing and formatting, and it will even handle **DOCX** files. It's not a replacement for a full-sized word processor, but it is a good complement to one.

> **TIP:** Check out Chapter 21: "Working with Notes and Documents" for descriptions of how to use **Docs to Go** and **QuickOffice**.

ThinkFree Mobile is another popular choice, although in our testing it suffered from more formatting errors when trying to display documents. Table A–1 shows these and other document tools.

Table A–1. *Document Tools.*

App Name	Price	Notes	QR Code
Docs to Go, by DataViz	Free trial/$14.99	The free trial allows you to view documents, and the paid version allows you to edit them. It supports **Word** and **Excel** formats; the paid version also supports **PowerPoint**.	
QuickOffice Mobile Suite from QuickOffice, Inc.	US$9.99	This is a complete software solution offering good capabilities for creating and editing Microsoft Office documents on your DROID. For some PowerPoint editing it has better capabilities than **Docs to Go**.	
OfficeSuite Viewer, by Mobile Systems	Free trial/$1.99	The free trial is for 30 days. The app supports **Word**, **Excel**, **PowerPoint**, and **PDF** files. This app had a high failure rate in our testing when trying to open files.	
ThinkFree Office Mobile, by ThinkFree Mobile	Free/$9.99	The free version is just a viewer. The paid version can be purchased either item by item (e.g., word processing, spreadsheets, and slideshow) or all at once. It will download and open **Google Docs**, and it has a built-in file browser. Overall, this app has a nice feature set; in my testing, however, it did not display documents well and often made them difficult to read with black backgrounds.	
Adobe Reader, by Adobe	Free	Allows you to view PDF attachments, but does not let you create or edit them. This is stable, free, and offered directly from Adobe. It also supports pinch-to-zoom and viewing from the Web.	

Printing

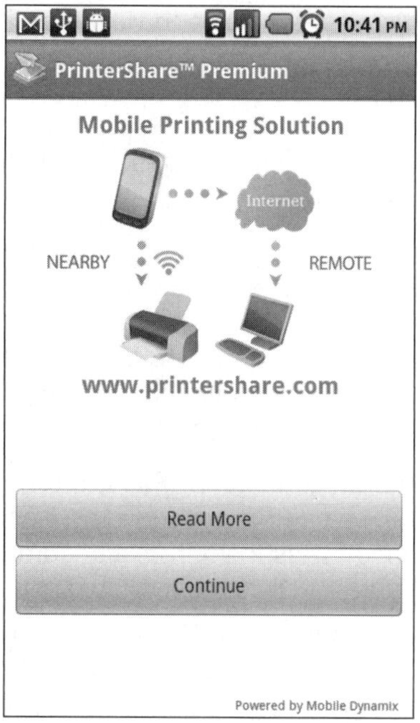

If you've got a document ready for printing, why not print it directly from your phone? **PrinterShare** lets you do exactly that. However, as of publishing time, it did not yet support printing **Word** or **Excel** files, so this app is most useful for printing photos, not **Office** documents. You also need to install desktop software in order to print to non-Wi-Fi printers. The free trial lets you print a test page to verify that it will work.

As Android gains popularity, you can expect to see more printing solutions; however, at the time of publication, it's a lonely field. Table A–2 shows **PrinterShare**, the only app in this category.

Table A–2. *Printing Apps.*

App Name	Price	Notes	QR Code
PrinterShare Droid Print, by Mobile Dynamix	Free trial/$4.95	Use the free trial's test-print feature before committing to a purchase. Be aware that **Office** files are not supported, but **PDF** files and photos are.	

File Management

It's nice to be able to view or edit your document attachments, but it's also nice to know where those files are stored on your SD card without having to connect your phone to a computer in order to do basic file-management tasks. Your DROID comes with the **Files** app for file management.

File-management software lets you see and move your files; it also enables you to break things if you don't know what you're doing or you aren't careful. Use caution when moving files and renaming folders. Of the apps we've evaluated, **EStrongs File Explorer** was our favorite, but there are many strong, free choices.

We like to combine **File Explorer** with **Dropbox**, which allows us to share and sync files through the Internet and access them from any computer or mobile device. Table A–3 lists the available file-management options.

> **TIP:** To learn more about Dropbox, check out Chapter 21. To learn about which folders store all your pictures, videos, and ringtones, check out Chapter 25: "DROID Media Sync.

Table A–3. *File-Management Apps.*

App Name	Price	Notes	QR Code
EStrongs File Explorer, by EStrongs	Free	This app allows file copying, file moving, multiple file selections, ZIP expansion, app management, and FTP (File Transfer Protocol) and Bluetooth file transfer.	
AndExplorer, by LYSESOFT	Free	This app allows file copying, file moving, GZIP, and other file-management functions. It features a straightforward user interface.	

App Name	Price	Notes	QR Code
File Manager, by Apollo Software	Free	This app has a cleaner interface than **EStrongs File Explorer**, but it does not include support for Bluetooth file transfer.	
Linda File Manager, by Nylinda.com	Free	**Linda File Manager** is also a solid choice, but it also lacks Bluetooth support.	
Dropbox, by Dropbox	Free	This app allows **Dropbox** users to access and share files between users and computers as if they share a common folder. Basic accounts are free, and premium accounts offer more storage. Visit the website at www.dropbox.com for more info. See Chapter 21 for more details about **Dropbox**.	

Virus Protection, Backups, and Security

Viruses may seem like a remote worry; however, as mobile devices become more popular, the bad guys will figure out better ways to distribute them. However, losing your phone and/or having it stolen are big potential problems right now, especially if you store sensitive data on your phone. It may be a job requirement that you be able to wipe business data from your phone if it is stolen.

Your DROID already ships with the ability to remote wipe and back up your phone data.

Lookout Mobile is our top pick in this category. It's free and provides three services within one app. If you feel the virus protection is unnecessary, you can disable that feature and use only the **Data Backup** and **Missing Device** features (see the figure to the right). We also appreciate that it sends you an email whenever you use remote location to make sure your online account hasn't been breached.

Table A–4 shows virus protection and other security apps.

Table A–4. *Virus Protection, Backups, and Security Apps.*

App Name	Price	Notes	QR Code
Lookout Mobile Security Free, by Lookout	Free	This app provides virus scans, remote location, and file backup. The remote location can either show you where your phone is on a map (if GPS is enabled) or emit a loud alarm.	
KeePassDroid, by Brian Pellin	Free	This is a password manager based on the open source **KeePass** project. You can combine this with **Dropbox** to make a cloud-based password safe.	
WaveSecure, by WaveSecure	$19 per year	**WaveSecure** offers theft protection with remote location, remote lock, and remote wiping of the data on the SD card.	
Norton Security Beta, by NortonMobile	Free trial	This is a beta release with an unknown pricing model for the full release. Its features include virus scanning, call screening, remote wipe, and remote lock.	
Super Private Conversation, by Superdroid.net	Free	This app blocks unwanted SMS and phone calls and filters specified SMS conversations for privacy.	

Presentation Software

Most of the document-management software listed earlier can handle viewing **PowerPoint** files. Chances are slim that you'll need to actually present *from* your Android phone; however, if you need to, you can use **Docs to Go** or the slideshow feature in your phone gallery in a pinch.

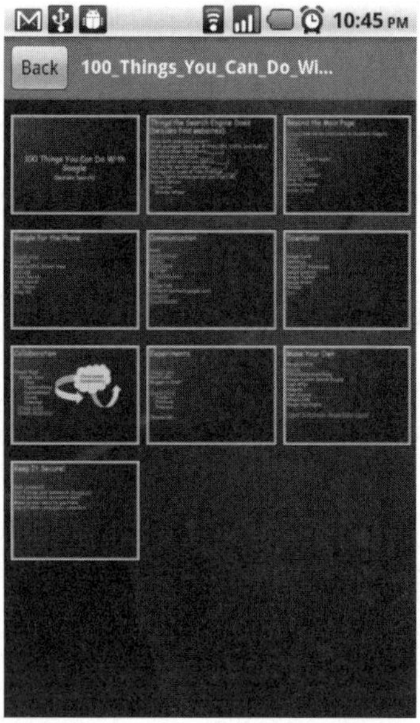

At this time, the DROID and DROID 2 do not support TV out, so you can't just plug the device into a TV or monitor and see video images. This is a hardware – not software – limitation, so most of us actually giving presentations using Android phones are stuck projecting them from an overhead projector. However, the DROID X *does* come with TV-out capabilities, which means you can present slideshows from your phone. This also means you can pair them with portable projectors that take standard TV connections, use the **MightyMeeting** app, and leave the PC behind. Table A–5 shows some presentation options.

Table A–5. *Presentation Software.*

App Name	Price	Notes	QR Code
MightyMeeting, by MightyMeeting	Free	This app enables you to give **PowerPoint** and **Keynote** presentations from your DROID. It can be used to either lead or attend live conferences with invited attendees. Files must first be uploaded to your MightyMeeting account before they become available as presentations. More information is available at www.mightymeeting.com.	
Oration Sensation, by EpiCache	Free	**Oration Sensation** is a presentation timer that offers vibrating timed alerts at preset intervals. You can save a list of presentation types, such as "short sales pitch" or "conference presentation," and keep your phone in your pocket. The alerts will let you know when it's time to switch slides or move to Q&A.	

App Name	Price	Notes	QR Code
PPT Remote for PowerPoint, by Johan Brodin	Free	This lets you use your phone as a Bluetooth remote for **PowerPoint** presentations. Test it beforehand (obviously) because not all computer and device combinations will be compatible.	
Gmote 2.0, by Marc Stogaltis and Mimi Sun	Free/$2.99	**Gmote** is a general Wi-Fi remote control for your PC that can be used for music files as well as **PowerPoint**. It's cross-platform compatible, so you can use this to control Mac media, as well. However, it requires server software to be installed on the computer you want to control, making it a no-no for some work environments. The $2.99 version is "donateware" for the developers.	

Web Conferencing

At this time, none of the DROID models covered in this book have a front-facing video camera, so they can't be used for video conferencing. However, Web conferencing from your phone is starting to emerge as a real possibility. WebEx and GoToMeeting don't have official apps at the time of publication, but that may change as their competitors eliminate the need for a PC. As mentioned earlier, **MightyMeeting** offers the ability to host and attend live conferences without requiring a laptop client. Table A–6 shows some handy apps for managing web conferences.

Table A–6. *Web Conferencing Apps.*

App Name	Price	Notes	QR Code
MightyMeeting, by MightyMeeting	Free	This app uses www.mightymeeting.com. If you own a phone with TV-out capability, you can present directly from your phone to the screen. You can also use this for live meetings with live chat feeds from users on a variety of phone platforms or using the Web. (Currently, mobile attendees must have US phone numbers.)	

App Name	Price	Notes	QR Code
Fuze Meeting, by FuzeBox	See note	Fuze Meeting is a subscription web-conferencing service. The dedicated Android app is free, but the presenter must have a Fuze Meeting account. More information is available at www.fuzemeeting.com.	
Vibrate During Meetings, by Sidetop Software	$2.99	This doesn't create web conferences; rather, it makes you more polite during conferences and other scheduled meetings. When your calendar says you have a meeting, your phone will automatically switch to **Vibrate** mode.	

Notetaking and Mind Mapping

You may not want to take traditional notes during a meeting from your phone unless you're a very fast thumb typist. However, you may want to access notes you've taken earlier or leave yourself quick sticky notes. Some apps are also tied into to-do lists.

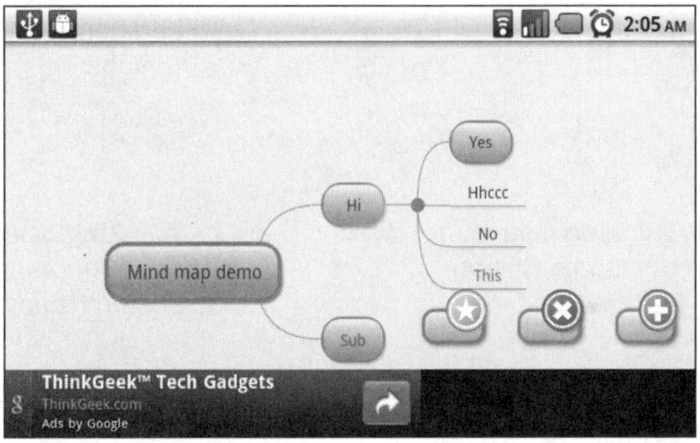

You may prefer mind mapping to traditional notetaking. Android can help with that, too. **Thinking Space**, shown in the image to the right, is one of several mind-mapping applications that allow you to diagram words and ideas visually, rather than by creating bullet-point lists.

Table A–7 shows **Thinking Space** and other notetaking options for Android.

Table A–7. *Notetaking Mind Mapping Apps.*

App Name	Price	Notes	QR Code
ColorNote Notepad Notes, by Notes	Free	This is a virtual sticky-note app for your phone. It lets you take quick, color-coded notes or make simple to-do lists.	
Notebook, by Darkgreener	Free/£.99 (about $1.55)	This notetaking app uses an old-fashioned book font for a more formal feel, but it also has some nice features. The full version lets you import email and provides password protection.	
Mind Map Memo, by Takahicorp	Free/¥180 (about $2.08)	This is a simple mind-mapping app. The paid version provides a few additional features like extra node options.	
Thinking Space, by Charlie Chilton	Free	**Thinking Space** is a full-featured mind-mapping tool that offers a lot of customization options.	

Email Management

We discussed email in Chapter 9: "Email on Your DROID." Android has very capable native options for email; however, in the corporate world, you may need more than what comes in the Android box, and you may not want to pay Verizon for monthly access to Corporate Sync email accounts.

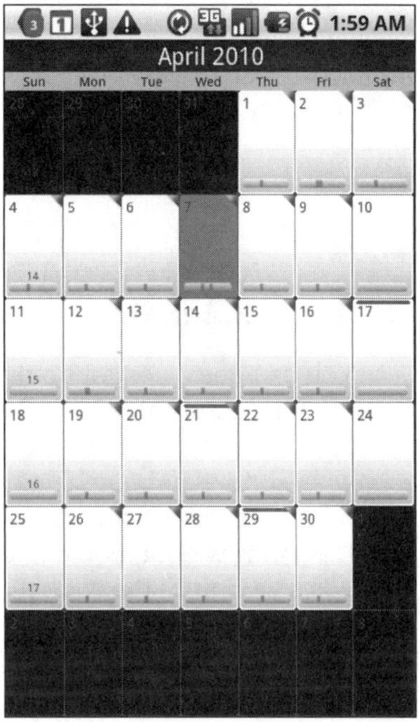

Table A–8 shows some email-management apps.

Table A–8. *Email-Management Software.*

App Name	Price	Notes	QR Code
Exchange by TouchDown, by NitroDesk	Free trial /$19.95	**TouchDown** supports security policies and **ActiveSync**. You can also specify times to turn off push notifications. **TouchDown** also comes with a variety of widgets. It does not merge your **Gmail** and **Exchange** calendars or tasks. After the trial expires, you can still use some, but not all of its features.	
RoadSync 2.0, by DataViz	Free trial/$9.95	This app offers features similar to **TouchDown**. After the trial expires, the product is disabled.	

App Name	Price	Notes	QR Code
SpamDrain, by _SpamDrain_	$30 per year	**SpamDrain** is a web-based spam-filtering service. All filtered messages are still available via the website, and messages not marked as spam are delivered to your inbox. The app comes with a 30-day trial.	

To-Do Lists

You can use a widget bookmark for **Google Tasks**, but it's easier and more efficient to use a dedicated app. Google didn't write it, but **gTasks ToDo** syncs with Google and combines Android-level app power with an intuitive user interface. If you use the **Pure Calendar** widget, you can display **gTasks To-Do** items as part of the widget.

Table A–9 shows **gTasks ToDo** and other to-do apps.

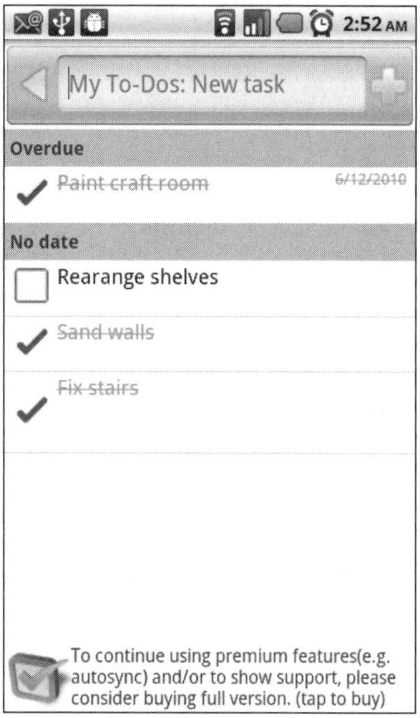

Table A–9. *To-Do List Software.*

App Name	Price	Notes	QR Code
gTasks ToDo for Android, by SSI	Free	Provides online and offline **Google Tasks** syncing. This feels like the missing app Google should have created. There is a paid version of the app that includes autosyncing, but it is not sold through the Android Market.	

App Name	Price	Notes	QR Code
Astrid Task/Todo List, by Todoroo	Free	**Astrid** is an open source task-management tool that syncs with **Remember the Milk**. It's a solid choice for anyone who doesn't need **Google Tasks** syncing.	
Got To Do Lite, by Slamjibe Software	Free/£2.00 (about $3.10)	This app is based on the "Getting Things Done" system (created by David Allen).	

Expenses and Finance

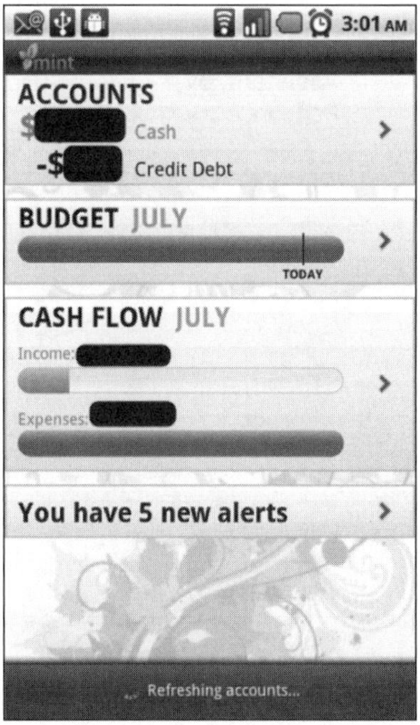

Tracking expenses is a necessary evil. We were very happy to learn about web-based services like **Expensify** and **Mint.com** (see the figure to the right). These services handle most of the data–entry process for you, letting you just see the results.

For personal finances, **Mint** is our favorite. You must have a `Mint.com` account to use the app. It makes it easy to keep track of your accounts, portfolio, and budget. It can also send you email or text alerts for events you specify, such as when you are charged a banking fee, when you make a large purchase, and when your bills are due.

Table A–10 shows some of our favorite finance apps.

Table A–10. *Expenses and Finance Software.*

App Name	Price	Notes	QR Code
Mint.com Personal Finance, by Intuit	Free	You must have a **Mint.com** account to use the **Mint.com Personal Finance** app, but both the service and the app are free. **Mint.com** makes its money through sponsored offers for credit cards and other financial services.	
Expensify Expense Reports, by Expensify	Free trial/$4.95	**Expensify** is a web-based service for creating expense reports "that don't suck." Most transactions come directly from your credit card as you charge them, but this app is used for entering cash transactions and taking photos of other receipts that are not automatically entered. More information is available at www.expensify.com.	
Personal Assistant, by Pageonce	Free/$7.00	**Personal Assistant** combines bank and credit-card management with travel itineraries, frequent flyer mileage, phone minutes, Netflix, and portfolio management.	
Finance, by Google	Free	This is Google's official app for **Google Finance**. It offers multiple-portfolio support and stock quotes.	

Travel

If you have to travel for work, you'll appreciate travel apps that track your mileage, list your schedule, help you find places to eat, and make sure you don't say anything embarrassing to the locals. **Kayak** is a great app for comparing airfares across all airlines, hotel prices and even learning information about your local airport.

Google Maps includes public transportation directions when available, but there are also third-party apps for specific cities available, so don't forget to search the market before you travel. **Google Maps for Android** also includes the **Places** app, which allows you to browse nearby locations by category.

Our favorite travel app, aside from the preinstalled **Google Maps**, is **TripIt**. It allows you to see your itinerary and share it with close contacts, while providing general information to professional contacts or the public in general. **TripIt** can also be tied in with LinkedIn and Facebook.

Table A–11 shows **TripIt** and some other travel app options.

Table A–11. *Travel Apps.*

App Name	Price	Notes	QR Code
TripIt Travel Organizer, by TripIt	Free	**TripIt** is a web-based service that tracks your travel itinerary and mileage. The phone app gives you your flight schedule and appointment information on your phone.	
Kayak Flight & Hotel Search by Kayak	Free	Allows quick searches of airfares across all airlines and web sites, hotels, cars, check flight status, and even airport information such as "Where is the gift shop?	

App Name	Price	Notes	QR Code
Gaia GPS, by Trailbehind	$4.99	This app provides offline road and topographic maps for times when you're traveling out of cell tower range. It's aimed primarily at outdoor sports, but it's still very useful for traveling to rural areas.	
Geodelic, by Geodelic Systems	Free	This is also known as **Sherpa** on some phones. **Geodelic** finds nearby restaurants, cafes, attractions, and so on. You can browse by category of food or distance.	
Google Translate, by Google	Free	This is Google's official app for its Google Translate service. Not only does this app translate text to and from more than 50 languages, but it also has an audio-pronunciation guide.	

Health and Medicine

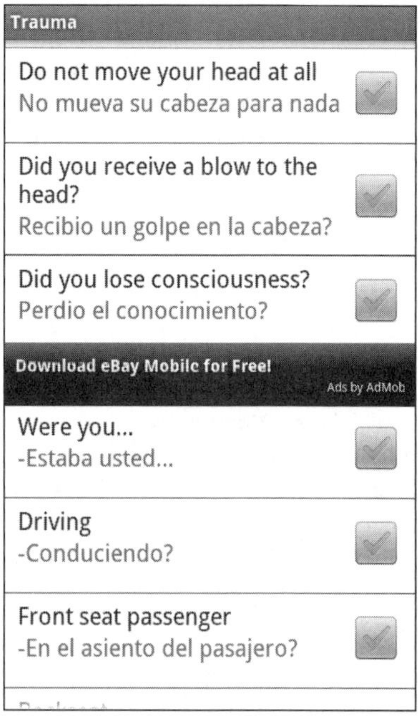

Health-care professionals were among the first to see the immediate usefulness of mobile technology, and app writers haven't ignored this. **Epocrates** has a long tradition of offering its reference materials on PDAs and other mobile devices, and pharmaceutical companies are making their own apps.

Medical Spanish Audio is our favorite app of this group. It lists Spanish phrases by category, such as trauma assessments.

If you check the box next to a question, you can bookmark it for reference; and if you click a question, the app will pronounce it aloud for you.

See Table A-12 for a list of health and medicine apps.

Table A–12. *Health and Medicine Apps.*

App Name	Price	Notes	QR Code
Epocrates Rx, by Epocrates	Free	**Epocrates Rx** is a free reference app for drug information. At the time of publication, the software is still in beta, so **Epocrates** may choose to charge for a premium service or full version in the future.	
Skyscape Medical Resources, by Skyscape	Free	This is a general medical reference that includes prescription and over-the-counter medicine.	
Medical Spanish Audio Lite, by Mavro	Free/$6.99	This is an app for non-Spanish-speaking care providers, enabling them to communicate with Spanish-speaking patients. The paid version removes the ads.	
Medical Mnemonics, by Regular Rate and Rhythm Software	$1.99	So you're not a medical professional... yet? This app helps students study with a library of mnemonics; it also lets them create and submit their own.	

Law and Legal

For those working in or around law offices, there are reference materials and news sources available for the DROID. **DroidLaw** is a good example of such an app. You can use it to sort through legal procedures, but you can also use it to track many popular legal blogs and news feeds.

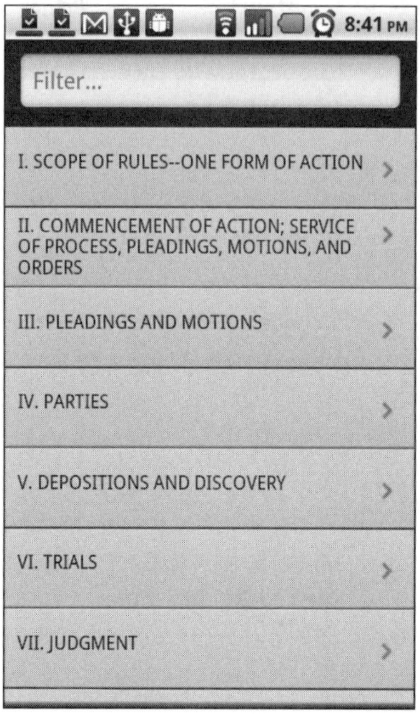

For legal students, there are lots of study guides and flashcard games. There are also a few specialized apps for calculating billing hours and target dates. Legal students may also consider apps like **Locale** that automatically turn the phone's ringer off at certain locations, such as the courthouse. See Table A–13 for a list of law and legal apps.

Table A–13. *Law and Legal Apps.*

App Name	Price	Notes	QR Code
DroidLaw, by BigTwit Software	Free/varies	**DroidLaw** is a legal reference app. The base app is free, but you can expand it through paid modules that contain the material you need, such as Supreme Court cases ($2.99) or the United States Code ($3.99).	
LangLearner Legal Dictionary, by LangLearner	Free	This app is a simple dictionary of legal terms for lawyers, legal assistants, and people studying law.	

Wathen Legal News, by Genwi	Free	This app provides international legal news stories and allows you to comment.
Lawyer's Calendar Calc, by Hawkmoon Software	Free/$1.99	This app calculates target dates and the number of workdays between two dates for legal billing purposes.

Real Estate

Real estate agents can benefit from many generalized apps, such as to-do lists, galleries, email apps, and notetaking apps. There are also a number of apps using the MLS database both for professionals and consumers. **Zillow**, for example, is a Google Maps-based app that makes estimates of house values and shows current listings. If you click a property listing, it will show a picture of the property and details. It provides useful general information for both consumers and agents.

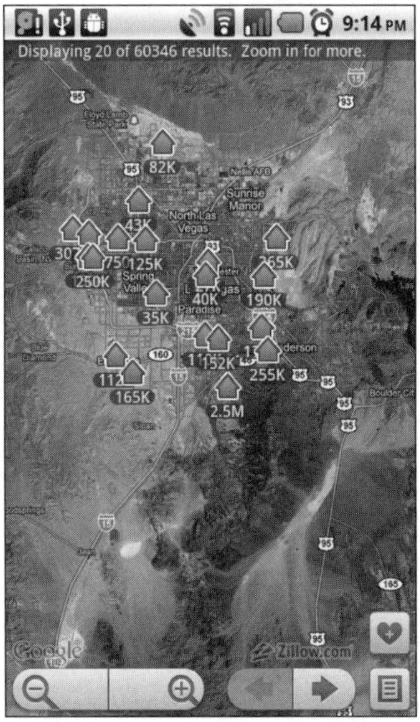

See Table A–14 for more real estate apps.

Table A–14. *Real Estate Apps.*

App Name	Price	Notes	QR Code
Mortgage Calculator, by Siva G	Free/$4.99	This is a fairly straightforward mortgage and autopayment estimator with PMI (private mortgage insurance). It relies on data from `Bankrate.com`. The pro version offers more options.	
Mortgage Pro, by Skynet Creations	Free	This is a mortgage calculator that helps buyers evaluate A– or 30-year mortgages, points, balloon payments, and other mortgage options.	
RE/Max University, by Mediafly	Free	This app was written specifically for RE/Max agents, and it includes corporate communications and training videos.	
Real Estate Vocabulary Quiz, by Upward Mobility	$2.99	Studying for a real estate broker exam? This app offers vocabulary quizzes. The company also offers many state-specific versions.	
Zillow Real Estate, by Zillow	Free	This app comes from the same makers as the Zillow website. This app gives estimates (“Zestimates”) of property value and shows listings on a map.	

Sales and CRM

CRM (customer relationship management) seems like something that was meant to be done from your phone. You can find tools that work with **Microsoft CRM**, and Salesforce.com is developing tools that are not yet on the market. There are also plenty of smaller CRM companies that are willing to work with Android, such as Simply Sales, and several developers have introduced standalone apps for the freelancer.

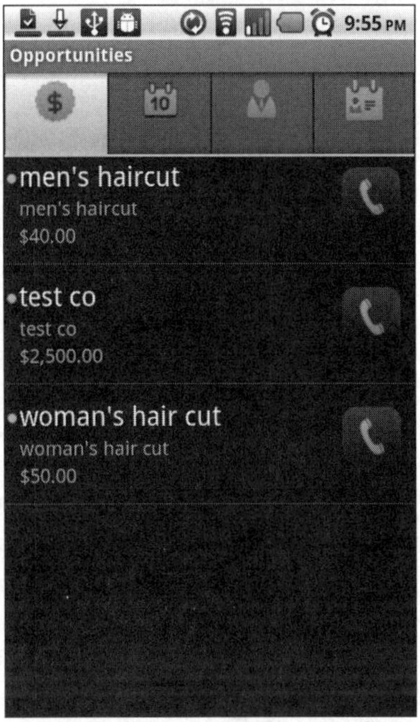

If you're interested in using CRM apps, try a few out to see what works best for you. See Table A–15 for a list of sales and CRM apps.

Table A–15. *Sales and CRM Apps.*

App Name	Price	Notes	QR Code
Locale, by two forty four a.m.	$9.99	**Locale** sets your phone's ring behavior based on the time, who is calling, what is on your schedule, and the phone's location. Set it to stop ringing during sales meetings or prioritize some calls over others. Many other apps also work with the **Locale** app's settings.	
Mobile CRM for MSCRM, by Softtrends Software Private Ltd	Free trial/$99.00	This app is a third-party solution for connecting to **Microsoft CRM**. The full version is expensive, so take advantage of the fully functional trial to see whether it will serve your needs.	

Simply Sales, by MyOlive.net Small Business Portal	Free	**Simply Sales** is a simple CRM app that integrates with **Gmail** and **Google Maps**. The app was written by MyOlive.net, which also offers a small business CRM portal.	

Retail

Small business retailers will really appreciate being able to accept and process credit cards from their phones. For the most part, these apps are free; however, you must have a merchant account, and the apps and services in this space charge membership and/or transaction fees for credit-card processing.

The **FaceCash** app promises an innovative way to let retailers and merchants handle transactions. Rather than carrying a credit card, a user relies on **FaceCash** to show merchants an ID that prominently showcases a picture of her face and transfers the funds from an account with **FaceCash** rather than a credit card. A user can choose to tie her bank account to the payment service or rely on transfers from other customers (such as her parents). You can think of this service as PayPal with an easier ID system. However, **FaceCash** isn't useful if nobody accepts it, and nobody asks to use it; therefore, most merchants will still need to process credit cards.

See Table A–16 for a selection of money-processing retail apps.

Table A–16. *Retail Sales Apps.*

App Name	Price	Notes	QR Code
Pocket Verifier, by MerchantAnywhere	Free app/$299 hardware	This app uses your phone's built-in Bluetooth to process credit cards from a device sold separately. It only runs on Android 2.1 and higher. More information is available at www.merchantanywhere.com.	

App Name	Price	Notes	QR Code
Mobile Credit Card Processing, by Merchant Swipe	Free	It requires you to manually input the credit card information into the phone, rather than giving you a swipe reader. More information is available at www.merchantswipe.com.	
Square, by Square	Free	Square is a relatively new company that's been making a splash with its payment system. Square officially opened for business in October 2010.	
FaceCash, by Think Computer Corporation	Free	**FaceCash** enables a purchaser to use her phone for payments instead of carrying a credit card. The purchaser uses the app for payment. It shows the merchant a picture of the true phone owner's face for security verification, and no paper or plastic needs to change hands. Both the purchaser and retailer must have FaceCash accounts.	

Finance

Android lends itself to personal-finance and portfolio-management software. One example is Google's own **Google Finance** app.This app allows you to track general stock direction, portfolios, and financial news.

There are also many apps written specifically by banks for their customers. You should always double-check the author and read reviews before downloading an app that claims to have been written by your bank.

You'll also find currency converters, MBA study guides, and basic expense-management software. See Table A–17 for a few apps that stand out either by offering more features or superior look-and-feel.

Table A–17. *Finance Apps.*

App Name	Price	Notes	QR Code
Google Finance, by Google	Free	This is the official **Google Finance** app; it includes stock quotes and financial news.	
Personal Assistant, by Pageonce	Free/ $7.00	**Personal Assistant** combines investment-portfolio and personal-bill management with flight itineraries, cell phone minutes, and the ability to monitor your Netflix queue.	
pFinance, by BiShiNews	Free	This app is a personal-finance manager that also tracks portfolios and financial news. What makes this app so nice are the simple calculators in the **Finance** column. For example, the app includes things like a tip calculator, an interest calculator, and a currency converter.	

Project Management

Project managers have to keep track of a lot of parts and people in order to do their jobs. It seems ideal for them to be able to do some of that tracking from a mobile device, rather than having to lug around a laptop or transfer handwritten notes.

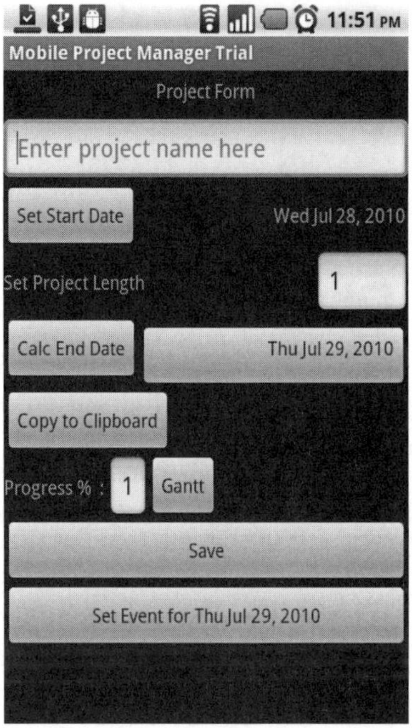

Mobile Project Manager is a very capable mobile app. It can import from **MS Project** format, create Gantt charts, and send reports via email. You can add both location and contact resources, as well as attach files from your phone's SD card.

Mobile Project Manager does have an overly exuberant **Help** screen that tries to pop up every time you enter a file; however, you can uncheck the box to display the **Help** screen at all times, removing this annoyance.

See Table A–18 for a list of project-management apps.

Table A–18. *Project-Management Apps.*

App Name	Price	Notes	QR Code
Upvise Pro, by Unyverse	Free trial	This app handles simple CRM and project management.	
Mobile Project Manager, by Hawkmoon Software	Free/$2.99	This is a simple project manager that allows you to import, export, and share **Excel** files.	

App Name	Price	Notes	QR Code
Time Tracker, by Sean E Russell	Free	This is a simple time-tracking app for figuring out the time spent on a project or task. It also generates reports on projects and tasks.	

Education and Training

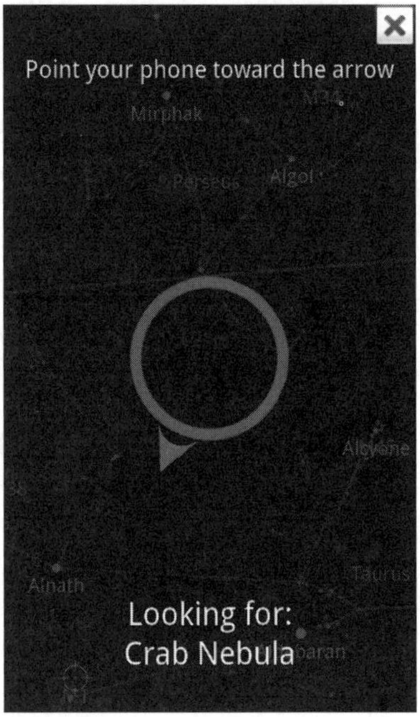

Higher education institutions have begun offering apps and services aimed at students. The apps and services include maps, enrollment information, and access to the campus learning-management system. The developers for both the Blackboard and Moodle services have recognized that students may want to access coursework with a mobile phone. For example, the developers of **Moodle** have worked on optimizing the mobile browser experience for general usability on all mobile platform – without a dedicated app. Meanwhile, Blackboard is partnering with Sprint to offer a free native app for Sprint customers. Schools that wish to support other phone networks pay an additional licensing fee.

Currently, most Android apps aimed at students focus on E-Books, flash cards, grammar, and study guides. There are some true gems among these apps. **Google Sky Map** shows a view of the stars that changes as you change the angle you hold your phone at (as if your phone has become a virtual telescope). Likewise, **Google Earth** provides a virtual globe with layers of customizable information about the planet. See Table A–19 for a list of education and training apps.

Table A–19. *Education and Training Apps.*

App Name	Price	Notes	QR Code
Grade Rubric, by Android for Academics	Free/$0.99	This is a simple grading app for teachers that use rubrics. The paid version can email final grades. It does not tie into a learning-management system at this time. The company is developing a version that will sync with **Google Docs** and create a grading spreadsheet.	

App Name	Price	Notes	QR Code
Google Sky Map, by Google	Free	This app shows a map of the stars, but that is a simple explanation. It's one of the apps we regularly pull out when people ask why they'd want a smartphone.	
Blackboard Mobile, by Blackboard	Free	This app only works with institutions that are running Blackboard's learning management software with the mobile plug-in. It allows students to participate in distance-learning courses from their phone.	
Formulas Lite/Formula Droid, by Abhishek Kumar	Free/€1.50 (about $1.95)	This is a scientific calculator and formula reference guide for students; it includes things like periodic tables and a web reference. The paid version removes ads.	

Social Media

Social media has become an important part of doing business and interacting with the world. Table A–20 lists a few of the more popular social media apps.

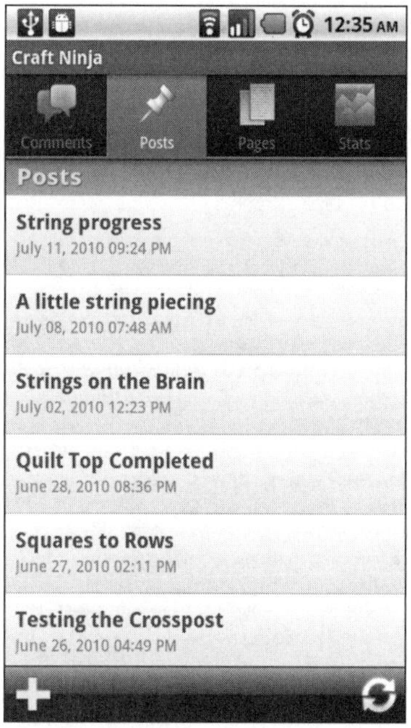

Table A–20. *Social Media Apps.*

App Name	Price	Notes	QR Code
HootSuite, by HootSuite	Free/$2.99	**HootSuite** is our top pick for Twitter on Android. Schedule tweets, manage multiple accounts, and see analytics. The paid version offers more features.	
Facebook, by Facebook	Free	The official **Facebook** app is different from the Motorola Widget on the DROID X and DROID 2 phones.	
WordPress, by Automattic	Free	If you have a WordPress blog, whether on `WordPress.com` or on your own server, this app will let you post to it (provided that your blog is WordPress 2.7 or higher and has the correct server settings enabled).	
Google Buzz, by Google	Free	The simple **GoogleBuzz** widget allows you to make quick posts with pictures and/or location information to the Google Buzz service.	

Information Technology

Your IT staff may have already installed, modified, and mastered the apps for Android covered in this section; however, if you're looking into using an Android phone, it's good to know you can manage your server or answer help-desk tickets from your phone. Not only are there a wide variety of IT tools for Android, but a large portion of them are free and written for the convenience of IT professionals, such as the difficult-to-read but very useful **ConnectBot**, which enables SSH (Secure Shell) connections from your phone.

Table A–21 lists a few apps that don't require special phone hacks to work.

Table A–21. *IT Apps.*

App Name	Price	Notes	QR Code
Zendesk for Android, by Zendesk	Free	This app is for existing Zendesk customers. It allows you to remotely track and manage help-desk tickets.	
android-vnc-viewer, by androidVNC team + antlersoft	Free	This is a simple, open source VNC (virtual native client) viewer for Android. It connects to **TightVNC**, **RealVNC**, and **Apple Remote Desktop**.	

ConnectBot, by Kenny Root and Jeffrey Sharkey	Free	**ConnectBot** is a simple, open source SSH client for Android. It's difficult to see the tiny text, but it makes up for this by giving you the ability to copy-and-paste.
IPConfig, by Mankind	Free	This is a simple utility that provides statistics about your current Wi-Fi connection, including your IP address and DHCP server.
AndFTP, by __LYSESOFT	Free	**AndFTP** is an FTP and SFTP (Secure FTP) app for Android. It also provides open, rename, cut, paste, delete, and other basic functions. Although we doubt you'd want to use this app to set up and configure websites regularly, it works well for fixing small problems quickly.

Other Apps

Finally, we'll review a small handful of apps that are worth mentioning, even if they don't quite fit into one of the broader categories. It may logically belong in the education app category, but the **Kindle** app likely won't replace textbooks until Amazon adds better audio navigation for the user interface. However, for the mobile professional, E-Books are starting to come of age. Amazon chose an E-Book format incompatible with industry-standard ePub books, but made up for it by developing apps for most mobile platforms, including Android.

The user interface for **Kindle for Android** is more intuitive than the standalone Kindle eReader because you can swipe your finger to turn pages rather than pressing a button. You also don't get a month-long battery life on your phone; however, the books are still easy to read, even on a small screen. For more information on E-book readers, see Chapter 16, "New Media.

Firehouse Scheduler is another great app. It was created for emergency responders, and it is designed to keep track of shift schedules.

Finally, if you or your business owns a car, **aCar** is a great app to use to make sure you take care of it. Table A–22 lists all three of these apps.

Table A–22. *Other Apps.*

App Name	Price	Notes	QR Code
Kindle for Android, by Amazon	Free	Purchase, download, and read Kindle books from your phone. The app keeps track of your downloads and reading progress between devices, so you can start a book on your kindle and finish reading on your Android phone.	
Firehouse Scheduler, by Leaky Nozzle	$5.99	This app is an emergency responder's scheduler. It's designed to track shifts, vacation time, paydays, sick time, and traded shifts. Leaky Nozzle also has a variation specific to the New York Fire Department.	
aCar, by Armond Avanes	Free	This app tracks car expenses, maintenance, and mileage. It has reminders for regular maintenance items like oil changes. This app is useful whether you're maintaining a business or personal car.	

Index

Special Characters and Numerics

A

C

D

E

F

G

H

I

J

K

L

M

O

P

Q

R

S

T

U

W

X

Y

Z